Term	Meaning	Example
cysto-	Bladder or sac	Cystocele (hernia of a bladder)
-cyte-	Cell	Erythrocyte (red blood cell)
cyto-	Cell	Cytoskeleton (supportive fibers inside a cell)
de-	Away from	Dehydrate (remove water)
derm-	Skin	Dermatology (study of the skin)
di-	Two	Diploid (two sets of chromosomes)
dia-	Through, apart, across	Diapedesis (ooze through)
dis-	Reversal, apart from	Dissect (cut apart)
-duct-	Draw	Abduct (lead away from)
-dynia	Pain	Mastodynia (breast pain)
dys-	Difficult, bad	Dysmentia (bad mind)
e-	Out, away from	Eviscerate (take out viscera)
ec-	Out from	Ectopic (out of place)
ecto-	On outer side	Ectoderm (outer skin)
-ectomy	Cut out	Appendectomy (cut out the appendix)
-edem-	Swell	Myoedema (swelling of a muscle)
em-	In	Empyema (pus in)
-emia	Blood	Anemia (deficiency of blood)
en-	In	Encephalon (in the brain)
endo-	Within	Endometrium (within the uterus)
entero-	Intestine	Enteritis (inflammation of the intestine)
epi-	Upon, on	Epidermis (on the skin)
erythro-	Red	Erythrocyte (red blood cell)
eu-	Well, good	Euphoria (well-being)
ex-	Out, away from	Exhalation (breathe out)
exo-	Outside, on outer side	Exogenous (originating outside)
extra-	Outside	Extracellular (outside the cell)
-ferent	Carry	Afferent (carrying to the central nervous system)
-form	Expressing resemblance	Fusiform (resembling a fusion)
gastro-	Stomach	Gastrodynia (stomach ache)
-genesis	Produce, origin	Pathogenesis (origin of disease)
gloss-	Tongue	Hypoglossal (under the tongue)
glyco-	Sugar, sweet	Glycolysis (breakdown of sugar)
-gram	A drawing	Myogram (drawing of a muscle contraction)
-graph	Instrument that records	Myograph (instrument for measuring muscle contraction)
hem-	Blood	Hemolysis (breakdown of blood)
hemi-	Half	Hemiplegia (paralysis of half of the body)
hepato-	Liver	Hepatitis (inflammation of the liver)
hetero-	Different, other	Heterozygous (different genes for a trait)
hist-	Tissue	Histology (study of tissues)
homeo-	Same	Homeostasis (state of staying the same)
hydro-	Wet, water	Hydrocephalus (fluid within the head)
hyper-	Over, above, excessive	Hypertrophy (overgrowth)
hypo-	Under, below, deficient	Hypotension (low blood pressure)
-ia	Expressing condition	Neuralgia (pain in nerve)
-iatr-	Treat, cure	Pediatrics (treatment of children)
-id	Expressing condition	Flaccid (state of being weak)
im-	Not	Impermeable (not permeable)
in-	In, into	Injection (forcing fluid into)
infra-	Below	Infraorbital (below the eye)
inter-	Between	Intercostal (between the ribs)

Continued on inside back cover.

Fifth Edition

Anatomy & Physiology

Rod R. Seeley

Idaho State University

Trent D. Stephens

Idaho State University

Philip Tate

Phoenix College

Boston   Burr Ridge, IL   Dubuque, IA   Madison, WI   New York   San Francisco   St. Louis
Bangkok   Bogotá   Caracas   Lisbon   London   Madrid   Mexico City   Milan
New Delhi   Seoul   Singapore   Sydney   Taipei   Toronto

McGraw-Hill Higher Education

A Division of The **McGraw-Hill** *Companies*

ANATOMY & PHYSIOLOGY, FIFTH EDITION

 This book is printed on recycled, acid-free paper containing 10% postconsumer waste.

1 2 3 4 5 6 7 8 9 0 VNH/VNH 0 9 8 7 6 5 4 3 2 1 0

ISBN 0-07-289917-4

Vice president and editorial director: *Kevin Kane*
Publisher: *Colin H. Wheatley*
Sponsoring editor: *Kristine Tibbetts*
Developmental editor: *Patricia Hesse*
Marketing manager: *Heather K. Wagner*
Project manager: *Jill R. Peter*
Production supervisor: *Laura Fuller*
Designer: *K. Wayne Harms*
Senior photo research coordinator: *Lori Hancock*
Senior supplement coordinator: *Audrey A. Reiter*
Compositor: *Carlisle Communications, Ltd.*
Typeface: *10/12 Garamond*
Printer: *Von Hoffmann Press, Inc.*

Cover design: *Jeff Storm*
Cover photo: *Mitchel Gray/SuperStock*
Photo research: *Feldman and Associates*

The credit section for this book begins on page 1065 and is considered an extension of the copyright page.

Library of Congress Cataloging-in-Publication Data
Seeley, Rod R.
  Anatomy & physiology/Rod R. Seeley, Trent D. Stephens, Philip
Tate.—5th ed.
    p.   cm.
  Includes index.
  ISBN 0-07-289917-4
  1. Human physiology.   2. Human anatomy.   I. Stephens, Trent D.
II. Tate, Philip.   III. Title.
  [DNLM: 1. Physiology.   2. Anatomy.   QT 104 S452a 2000]
QP34.5.S4   2000
612—dc21
DNLM/DLC
for Library of Congress   98-54873
                                                            CIP

http://www.mhhe.com/biosci/ap/seeleyap/

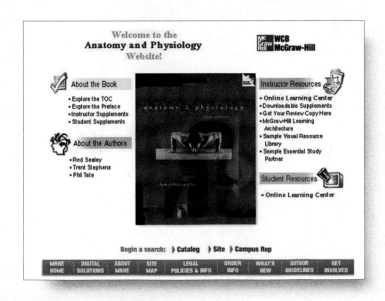

Essential Study Partner CD-ROM

A free study partner that engages, investigates, and reinforces what you are learning from your textbook. You'll find the **Essential Study Partner** for *Anatomy & Physiology* by Seeley, Stephens, and Tate to be a complete, interactive student study tool packed with hundreds of animations and learning activities. From quizzes to interactive diagrams, you'll find that there has never been a better study partner to ensure the mastery of core concepts. Best of all, it's **FREE** with your new textbook purchase.

The unit pop-up menu is accessible at anytime within the program. Clicking on the current unit will bring up a menu of other units available in the program.

The topic menu contains an interactive list of the available topics. Clicking on any of the listings within this menu will open your selection and will show the specific concepts presented within this topic. Clicking any of the concepts will move you to your selection. You can use the UP and DOWN arrow keys to move through the topics.

To the right of the arrows is a row of icons that represent the number of screens in a concept. There are three different icons, each representing different functions that a screen in that section will serve. The screen that is currently displayed will highlight yellow and visited ones will be checked.

Along the bottom of the screen you will find various navigational aids. At the left are arrows that allow you to page forward and backward through text screens or interactive exercise screens. You can also use the LEFT and RIGHT arrows on your keyboard to perform the same function.

The film icon represents an animation screen.

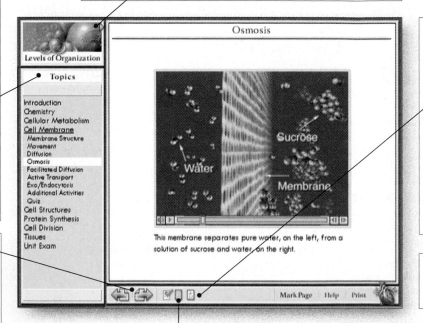

The activity icon represents an interactive learning activity.

The page icon represents a page of informational text.

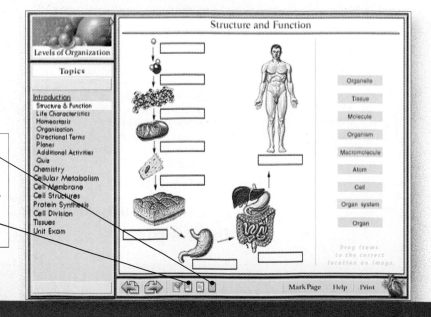

In addition

to the study activities and quizzes found in the *Essential Study Partner,* a film icon 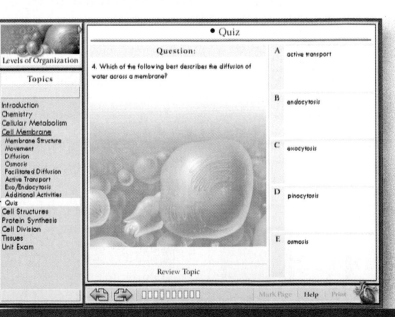 is placed in the text beside topics and concepts that are animated on the CD-ROM. These animations will ease you into a better understanding of the most difficult topics.

The heart icon allows you to access additional controls, navigation, and exit.

The "Audio" button opens up the controls for the audio. You may turn the sound on or off and adjust the volume.

The Bookmarks button lets you access the pages that you have marked with the "Mark Page" button.

"Search" will help you locate specific content in the CD.

"Text Guide" will find the correlation between areas in the textbook and screens in the CD.

Use the "Glossary" to find definitions of underlined words.

"Internet" will launch your web browser and take you to the text's website.

"Exit" will leave the program.

The "Mark Page" button allows you to tag a screen. The tagged screen then appears in the Bookmarks.

"Help" will bring you to the guided help walkthrough.

"Print" will print the current screen's content area.

Additional Activities

Levels of Organization

Topics

Introduction
Chemistry
Cellular Metabolism
Cell Membrane
  Membrane Structure
  Movement
  Diffusion
  Osmosis
  Facilitated Diffusion
  Active Transport
  Exo/Endocytosis
  Additional Activities
  Quiz
Cell Structures
Protein Synthesis
Cell Division
Tissues
Unit Exam

1. Name as many molecules as you can that are found in the cell membrane.

Reveal Answer

2. Ions and proteins are _____ to the cell membrane, whereas lipids are _____ to the cell membrane.

3. The membrane carrier, receptor, enzymatic, and channel are all composed of _____ molecules.

Audio
Bookmarks
Search
Text Guide
Glossary
Internet
Exit

Mark Page | Help | Print

Each topic ends with a quiz for review of the material. The quiz has a "Review Topic" feature for additional study.

Quiz

Levels of Organization

Topics

Introduction
Chemistry
Cellular Metabolism
Cell Membrane
  Membrane Structure
  Movement
  Diffusion
  Osmosis
  Facilitated Diffusion
  Active Transport
  Exo/Endocytosis
  Additional Activities
  Quiz
Cell Structures
Protein Synthesis
Cell Division
Tissues
Unit Exam

Question:

4. Which of the following best describes the diffusion of water across a membrane?

A active transport

B endocytosis

C exocytosis

D pinocytosis

E osmosis

Review Topic

Mark Page | Help | Print

New Version 2.0!

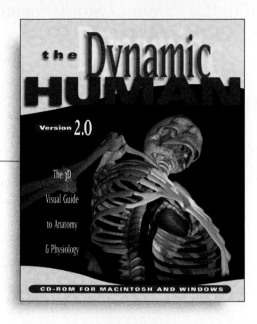

by WCB/McGraw-Hill and Engineering Animation, Inc.

1998 • CD-ROMs (two CD set) for Macintosh and Windows • ISBN 0-697-38935-9

This powerful version of *The Dynamic Human* CD-ROM builds upon the best in anatomy and physiology technology. **Version 2.0** features added detail and depth—expanding upon a variety of topics and incorporating the World Wide Web.

Features

- The World Wide Web provides the ideal opportunity to implement and greatly extend the powers of this updated CD-ROM. The all-new *Dynamic Human Version 2.0* web site features an immense amount of supplemental information, including:
  - approximately 20 to 30 readings per body system
  - extensive links to additional informational sites on the World Wide Web
  - supplemental visuals, including animations, video, and vivid illustrations
  - correlation guides for many WCB/McGraw-Hill texts
  - on-line self-testing
- An updated design and interface make navigation much simpler with easy-to-use pop-up menus.
- Each body system can be studied through four standard content areas: anatomy, explorations, clinical concepts, and histology.
- A self-quizzing section is available for each body system.
- Powerful 3D images and even more animations enhance the thorough content and draw your students into the wonder of anatomy and physiology.

Contents

Human Body • Skeletal System • Muscular System • Nervous System • Endocrine System • Cardiovascular System • Lymphatic System • Digestive System • Respiratory System • Urinary System • Reproductive System

System Requirements

IBM/PC or compatible	Macintosh
486/66 or better (Pentium recommended)	Power PC
Windows 95 or newer	System 7.1 or newer
16 MB RAM or better	16 MB RAM or better
640 x 480 x 256 color monitor	640 x 480 x 256 color monitor
CD-ROM drive	CD-ROM drive
(transfer rate of 300 kbs or better)	(transfer rate of 300 kbs or better)
SoundBlaster compatible audio card	Mouse
Mouse	

This distinctive icon appears in appropriate figure legends of the fifth edition of *Anatomy & Physiology* by Seeley, Stephens, and Tate and correlates information to specific modules found on the new *Dynamic Human CD-ROM, Version 2.0.*

Brief Contents

Contents

Part Two
Support and Movement

Chapter 5
Integumentary System

Chapter 6
Skeletal System: Bones and Bone Tissue

Chapter 7
Skeletal System: Gross Anatomy

Chapter 8
Articulations and Movement

Chapter 9
Receptor Responses and Membrane Potentials

Contents

Chapter 14
Peripheral Nervous System: Cranial Nerves and Spinal Nerves

Chapter 15
The Senses

Chapter 16
Autonomic Nervous System

Chapter 17
Functional Organization of the Endocrine System

Chapter 18
Endocrine Glands

Contents

Chapter 25
Nutrition, Metabolism, and Temperature Regulation

Chapter 26
Urinary System

Chapter 27
Water, Electrolytes, and Acid–Base Balance

Part Five
Reproduction and Development

Chapter 28
Reproductive System

Chapter 29
Development, Growth, Aging, and Genetics

Appendices

Glossary   1028

Credits   1065

Index   1067

About the Authors

Rod R. Seeley   Professor of Physiology, Idaho State University

With a B.S. in zoology from Idaho State University and an M.S. and Ph.D. in zoology from Utah State University, Rod Seeley has built a solid reputation as a widely published author of journal and feature articles, a popular public lecturer, and an award-winning instructor. Very much involved in the methods and mechanisms that help students learn, he contributes to this text his teaching expertise and proven ability to communicate effectively in any medium.

Trent D. Stephens   Professor of Anatomy and Embryology, Idaho State University

An award-winning educator, Trent Stephens teaches human anatomy, neuroanatomy, and embryology. His skill as a biological illustrator has greatly influenced every illustration in this text. With B.S. and M.S. degrees in zoology from Brigham Young University and a Ph.D. in anatomy from the University of Pennsylvania, Trent Stephens has also published numerous scientific papers and books. His students continually rate him highly on their evaluations—you will too!

Philip Tate   Instructor of Anatomy and Physiology, Phoenix College

From the community college to the private 4-year college, Phil Tate has taught anatomy and physiology to all levels of students: nursing and allied health, physical education, and biology majors. At San Diego State University, Phil earned B.S. degrees in both mathematics and zoology and a M.S. in ecology. He earned his doctorate in biological education from Idaho State University.

Preface

Human anatomy and physiology courses present tremendous challenges to both students and teachers. Acquisition of basic anatomical and physiological facts is essential to the study of anatomy and physiology, but it is also important for students to develop the ability to solve practical, real-life problems related to the knowledge they have acquired. It is impossible to memorize all of the body's responses to all possible situations. Students who have accumulated background knowledge and who are prepared to reason effectively can accurately anticipate responses to new situations and are better prepared to be effective citizens and health care professionals. In addition, it is not possible for students to learn all of the details of anatomy and physiology that are known. Selecting the most important information to provide a solid understanding of anatomy and physiology and to prepare students to solve problems effectively are major challenges for teachers and for authors.

We have written each edition of *Anatomy & Physiology* with the same major intention: to help students learn basic anatomy and physiology. We chose to present the major concepts that provide a current understanding of the subject. We presented the information in a readable form that seeks to **explain** rather than dictate so that concepts may be truly understood rather than simply memorized. When teaching beginning students, it is important not to obscure the "big picture" with an overwhelming deluge of detail. It is also important to provide enough pieces of information to allow the students to solve basic problems. It is our goal to present basic content at an appropriate level and in a way that supports the development of problem-solving skills that emphasize the practical application of concepts in anatomy and physiology to real-life situations.

Anatomy & Physiology is unique in its approach to the development of problem-solving skills. The fifth edition provides the teacher with a great deal of flexibility and assistance. It can be used very successfully to focus content and learning of the vocabulary of anatomy and physiology. The fifth edition can also provide an introduction to problem-solving techniques that can be emphasized to a greater extent as students progress through the course. It can also be used for courses that emphasize problem-solving and application of concepts to clinical situations.

Themes

We have chosen to emphasize the following two major themes throughout this text: **the relationship between structure and function** and **homeostasis.**

Just as the structure of a hammer makes it well-suited for the function of pounding nails, the **structures** of specific cells, tissues, and organs within the body allow them to perform specific **functions** effectively. For example, muscle cells contain proteins that make contraction possible, and bone cells surround themselves with a mineralized matrix that provides strength and support. Knowledge of structure and function relationships makes it easier to understand anatomy and physiology and greatly enhances one's appreciation for the subject.

Homeostasis, the maintenance of an internal environment within an acceptably narrow range of values, is necessary for the survival of the human body. The emphasis in this book is on how mechanisms operate to maintain homeostasis. Because failure of these mechanisms also illustrates how they work, pathological conditions that result in dysfunction, disease, and possibly death are also presented. Changes in response to increased physical exercise or aging also illustrate how these mechanisms work. Consideration of pathology, exercise, and aging adds relevance and interest, makes the material more meaningful, and enhances the background of the people who plan to pursue areas related to health. The two themes—the relationship between structure and function and homeostasis—combined with the book's strong problem-solving orientation and numerous clinical and other related examples, make this text unique among anatomy and physiology texts.

General Features

The following general features are combined to distinguish *Anatomy & Physiology* from other texts:

1. **Essential Study Partner.** The *Essential Study Partner* CD-ROM is an interactive student study tool packed with hundreds of animations and learning activities. From quizzes to interactive diagrams, students will find that

there has never been a more exciting way to study anatomy and physiology. A self-quizzing feature allows students to check their knowledge of a topic before moving on to a new module. Additional unit exams give students the opportunity to review coverage for a more complete understanding. This CD-ROM tutorial supports and enhances the material presented in *Anatomy & Physiology,* 5th edition and is packaged free when students purchase a new textbook. In addition to the study activities and quizzes, a film icon ▌ is placed in the text beside topics and concepts that are animated on the CD-ROM. These animations will ease students into a better understanding of the most difficult topics.

2. **Online Learning Center** http://www.mhhe.com/biosci/ap/seeleyap/ When you use *Anatomy & Physiology,* 5th edition, you are getting much more than a textbook. A multitude of learning opportunities are at your fingertips on our password protected web site. You'll gain access to our Online Learning Center where you'll find quizzes, links, case studies, clinical applications, and a world of ways of explore anatomy and physiology.

The Online Learning Center features both Student Resources and Instructor Resources. For students who want an edge in their course, they'll have immediate access to additional study questions, problem-solving exercises, applications, and much more—all in one place. The Online Learning Center gives students the password to success. A free password to the Online Learning Center Student Resources is available with the purchase of a new textbook.

Instructors will gain access to lecture outlines, illustrations from the textbook, templates to help build a web site course, suggested classroom activities, and a host of ways to enhance their anatomy and physiology course.

3. **Systematic Presentation.** The systematic presentation of content is designed to be consistent with the problem-solving approach of the text. Explanations are based on a conceptual framework that allows students to tie together individual pieces of information. Simple facts are presented first, and explanations are developed in a logical sequence. Special care has been taken to provide explanations that are thorough enough to support the problem-solving emphasis of the book, but without making the explanations ponderous.

4. **Balanced Coverage.** We offer balanced coverage of anatomy and physiology. Some texts emphasize anatomy at the expense of physiology coverage. As a result, when health professionals return to school for further training, it is invariably because they need a better understanding of physiology. Other texts do not adequately integrate anatomy with physiology and fail to help students understand how structures carry out functions. This text provides a solid foundation in anatomy as well as thorough coverage of physiology. Two chapters in this text are particularly illustrative of the emphasis we put on providing adequate coverage of physiology. These are Chapter 9, "Receptor Responses and Membrane

Potentials," and Chapter 17, "Functional Organization of the Endocrine System." These chapters present current information that makes it easier to understand the relationships between the molecular structure of membranes and the functions of organ systems.

5. **Problem-Solving Examples.** Problem-solving is encouraged by the addition of relevant contextual examples related to both anatomy and physiology. Clinical information should not be an end in itself. In some texts, mere clinical descriptions or medical terminology represent a significant portion of the material. Some texts also provide lists of pathologies with brief explanations. This text provides clinical examples to promote interest and demonstrate relevance, but clinical information is used primarily to illustrate the **application** of basic knowledge. The ability to apply information is a skill that will always be an asset for students, even after knowledge learned today is no longer current. We encourage students using *Anatomy & Physiology* to apply the knowledge they have gained through problem-solving to their professional and private lives.

6. **Problem-Solving Questions.** We present systematic inclusion of questions that require the solution of practical problems. At best, some anatomy and physiology texts include a few "thought" questions that, for the most part, involve a restatement or a summary of content. Yet once students understand the material well enough to state it in their own words, it only seems logical for them to proceed to the next step—that is, to apply the knowledge to hypothetical situations. This text features two sets of problem-solving questions in every chapter: Predict questions and Develop Your Reasoning Skills questions. These questions provide students with an opportunity and challenge because we believe that practice in solving problems greatly enhances problem-solving skills.

This text helps to develop problem-solving skills in several ways. First, all the information necessary to solve a problem is presented at a level that is sufficiently simple to avoid unnecessary confusion. Second, the opportunity to practice problem-solving is made available through Predict questions embedded within the chapter material and the Develop Your Reasoning Skills questions found at the end of each chapter. Third, answers and explanations for the Predict questions are included at the end of the book in Appendix F. Explanations for Develop Your Reasoning Skills questions are presented in the Instructor's Manual. The explanations illustrate the methods used to solve problems and provide a model for the development of problem-solving skills. When students are exposed to the reasoning used to correctly solve a problem, they are more likely to be able to successfully apply that reasoning to future problems. The acquisition of problem-solving skills is necessary for a complete understanding of anatomy and physiology. It is fun, and it makes it possible for the student to deal with the many problems that occur as part of professional and everyday life.

New to this Edition

It is impossible to list every change made in the fifth edition of *Anatomy & Physiology,* but the major highlights of the revision are:

1. We took great care to make the text even more inviting to read and easier for students to use. We have reviewed all of the figures in the text and changes have been made to many of the figures to improve their organization, color coordination, and style. Some figures have been replaced to make it easier for students to grasp anatomical concepts and physiological mechanism. Additional homeostasis figures have been added where it is appropriate to do so to make it easier for students to develop a working knowledge of homeostatic mechanisms. As always, we have examined illustrations, and used reviewer's comments, to avoid labelling errors and inconsistencies.

2. We have attempted to organize the text to optimize the relationship between the figures and the relevant text material by moving a number of figures within the chapters.

3. The chapters were reviewed with the objective of updating information and increasing the clarity of explanations.

4. Selected chapters were substantially revised to update the information, improve organization of concepts to make them easier for students to learn, and to update information. These include:

   Chapter 7: The description and the illustrations of the skull to present the complete skull before the individual bones are described. The chapter was reorganized to bring the figures and figure call-outs closer together.

   Chapter 9: The first part of this chapter has been rewritten to organize and categorize receptors into better organized categories which are accurate, simple, and clear. This is a major change in this chapter which should make receptors easier for students to under-

stand. Changes in the figures reflect the changes in the categories of receptors. Changes in figures have also been included to illustrate more clearly how the receptors produce a response. Clarification of membrane channels and the role they play in the action potential are more clearly illustrated.

Chapter 17: Explanations have been modified to be consistent with those included in chapter 9. Illustrations have been reorganized to conform with the changed classification of receptors.

Chapter 20: The Electrical Properties section has been updated and rewritten to better explain the role of cardiac muscle ion channels in establishing the resting membrane potential, producing the action potentials that stimulate cardiac muscle to contract, and generating the spontaneous action potentials responsible for the heart's autorhythmicity. A new figure (figure 20.15) showing the action potential in the SA node has been added. The Cardiac Cycle section has been rewritten. First, an overview of the main events of the cardiac cycle is provided, accompanied by a new figure (figure 20.17) showing valve positions and blood flow during each stage of the cardiac cycle. Then, the details of each stage

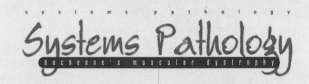

Systems Pathology
systems pathology
duchenne's muscular dystrophy

DUCHENNE'S MUSCULAR DYSTROPHY

A couple became concerned about their 3-year-old boy when they noticed that he was much weaker than other boys his age and the differences appeared to become more obvious as time passed. He had difficulty sitting, standing, and walking. He seemed clumsy and he fell often. He had difficulty climbing stairs, and he often got from a sitting position on the floor to a standing position by using his hands and arms to climb up his legs. His muscles appeared to be poorly developed. The couple took their son to a physician to have him examined. After several kinds of tests, they were informed that their son had Duchenne's muscular dystrophy.

BACKGROUND INFORMATION

Duchenne's muscular dystrophy (DMD) is usually identified in children at around 3 years of age when the parents notice slow motor development with progressive weakness and muscle wasting. Typically, muscular weakness begins in the pelvic girdle, causing a waddling gait. Temporary enlargement of the calf muscles is apparent in 80% of cases. Rising from the floor by "climbing up the legs" is characteristic and is caused by weakness of the lumbar and gluteal muscles. Within 3–5 years, muscles of the shoulder girdle become involved. Wasting of the muscles contribute to muscular atrophy and deformity of the skeleton. People with DMD are usually unable to walk by 10–12 years of age, and few live beyond age 20. There is no effective treatment to prevent the progressive deterioration of muscles in DMD.

Duchenne's muscular dystrophy results from an abnormal gene located on the X chromosome, at a position called Xp21, and is therefore a sex-linked (X-linked) condition. Although the gene is carried by females, DMD affects males almost exclusively. This position, or gene locus, is responsible for producing a protein called dystrophin, which plays a role in attaching myofibrils to and regulating the activity of other proteins in the plasma membrane. Dystrophin is thought to protect muscles cells against mechanical stress in the normal individual. In DMD, part of the gene at Xp21 is missing, and the protein it produces malfunctions, resulting in abnormal contractions and progressive muscular weakness.

10 PREDICT

A boy with Duchenne's muscular dystrophy developed pulmonary edema and then pneumonia. His physician diagnosed the condition in the following way: the pulmonary edema was the result of heart failure and the increased fluid in the lungs acted as site where bacteria invaded and grew. The fact that the boy could not breath deeply or cough effectively made the condition worse. Explain how a boy with DMD might develop heart failure and ineffective respiratory movements.

✓ Answer in Appendix F

298

The control center responds to information from the receptor.

The activity of the effector changes.

Increase in the variable is detected by the receptor.

Decrease in the variable is caused by the response of the effector.

Normal range — Value increases / Value decreases

Normal range — Homeostasis is maintained

Decrease in the variable is detected by the receptor.

Increase in the variable is caused by the response of the effector.

The control center responds to information from the receptor.

The activity of the effector changes.

are discussed and correlated with a more detailed figure (figure 20.18). Figure 20.18, which graphs the detailed events of the cardiac cycle, has been reorganized and the text has been correlated with each stage. Table 20.2, which summarizes the events of the cardiac cycle has been reorganized and rewritten and placed on facing pages so that all of the table can be viewed at once. The section on Mean Arterial Blood Pressure has been rewritten to better explain the relationships between heart rate, stroke volume, and blood pressure. A new flowchart figure (figure 20.20) clarifies these relationships. A new figure (figure 20.9) showing how the heart valves function has been added. A new flowchart figure (figure 20.10) showing the route of blood flow through the heart has been added. Other figures are being redrawn for sizing (e.g., figure 20.1), to correct errors (e.g., figure 20.2), or reorganized to better show the relationships between parts of the figure (e.g., figure 20.12).

Chapter 26: This chapter has been redone extensively. The physiology of the kidney is described more clearly and several figures have been added to illustrate kidney functions. The transport of solutes across the wall of the nephron and the mechanisms which function to regulate urine concentration are better illustrated and explained. The descriptions of the formation of concentrated and dilute urine have been rewritten. Several figures were altered to clarify the anatomy of the kidney and blood flow through the kidney. Six new figures have been added to illustrate the transport processes in the kidney with greater clarity and to illustrate the mechanisms that maintain a high concentration of solute in the medulla of the kidney and the mechanisms that regulate the concentration of urine.

Chapter 27: This chapter has been rewritten to clarify the regulation of water and electrolyte balance and pH balance. Figures have been modified to better illustrate the regulation of solutes such as potassium ions. The homeostasis figure format has been used for this purpose. A new figure to make the role of the kidney in the regulation of body fluid pH is included.

Learning Aids

As the amount of information in a textbook increases, it becomes more and more difficult for students to organize the material in their minds, determine the main points, and evaluate the progress of their learning. Above all, the text must be an effective teaching tool. Because each student may learn best in a different way, a variety of teaching and learning aids are provided.

1. **Chapter Outline.** Each chapter begins with a list of the section headings for the chapter. The purpose is to give an overview of the organization of the chapter so that students see how the chapter is organized.

2. **Chapter Objectives.** Each chapter begins with a series of learning objectives. The objectives are not a detailed cataloging of everything to be learned in the chapter; rather, they emphasize the important facts, topics, and concepts to be covered. The chapter objectives are a conceptual framework to which additional material will be added as the chapter is read in detail.

3. **Boldfaced Terms.** Key terms in the chapter are set in boldface for student identification. Most of these terms are included in the glossary at the end of the book.

4. **Vocabulary Aids.** Learning anatomy and physiology is, in many ways, like learning a new language. A basic terminology must be mastered to communicate effectively. In cases where it is instructionally valuable, the derivation or origin of key words is given. In their original language, words are often descriptive, and knowing the original meaning can enhance understanding and make it easier to remember the definition of the word. Common prefixes, suffixes, and combining forms of many biologic terms appear on the inside of the front and back covers of the text and provide additional information on the derivation of words. When the pronunciation of a word is complex, a pronunciation guide is included. Simply being able to pronounce a word correctly is often the key to remembering it. The glossary, which collects the most important terms into one location for easy reference, also has a pronunciation guide.

5. **Clinical Notes.** The Clinical Notes are designed to provide relevant and interesting examples to enhance the background of students who plan to pursue areas related to health. Other examples related to sports medicine or everyday experiences are included when they reinforce basic concepts. The Clinical Notes appear right after concepts are presented, and in so doing, the relevance of the concepts is immediately apparent, helping the student to better appreciate and understand them.

6. **Clinical Focus Boxes.** The boxed essays (see example on page xxii) are expanded versions of the Clinical Notes that permit a more detailed or complete coverage of a topic. Subjects covered include pathologies, current research, sports medicine, exercise physiology, pharmacology, and clinical applications. They are designed to not only illustrate the chapter content but also stimulate interest.

7. **Predict Questions.** Clinical Notes or Clinical Focus boxes can illustrate how a concept works, but a Predict question requires *application* of a concept. When reading a text, it is very easy to become a passive learner; everything seems very clear to passive learners until they attempt to use the information. The Predict questions convert the passive learner into an active learner who must use new information to solve a problem. The answer to this kind of question is not a mere restatement of fact, but rather a prediction and analysis of the data, the synthesis of an experiment, or the evaluation and weighing of the important variables of the problem. For example, "Given a stimulus, predict how a system will respond." Or, "Given a clinical condition, explain why the

Anatomy

Chapter Six

Skeletal System: Bones and Bone Tissue

Objectives

1. Name the components of the skeletal system, and list its major functions.
2. Define the different types of cartilage cells, and describe cartilage matrix.
3. Describe the perichondrium and the formation of cartilage by appositional and interstitial growth.
4. Name the major bone shapes, and describe their anatomy.
5. Describe the composition and organization of bone matrix.
6. Define the different types of bone cells, and describe their functions and their origins.
7. Describe the features that characterize woven, cancellous, and compact bone.
8. Name the two patterns of bone formation, and describe the features of each.
9. Describe bone growth, and explain how it differs from the growth of cartilage.
10. List the nutritional and hormonal requirements for bone growth.
11. Explain how bone remodeling occurs, and describe the conditions in which it occurs.
12. Describe the effects of mechanical strain and of inadequate stress on bone.
13. Describe the process of bone repair, the cells involved, and the types of tissue produced.
14. Explain the role of bone in calcium homeostasis.

Part Two

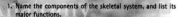

---

observed symptoms occurred." Develop Your Reasoning Skills questions, located at the end of chapters, are additional practice problems that help to develop the skills necessary to solve problems. Answers are given for the Predict questions at the end of the text in Appendix F. Not only are possible answers given for the questions, but explanations are provided that demonstrate the process of problem-solving.

8. **Tables.** The book contains many tables that have several uses. They provide more specific information than that included in the text discussion, allowing the text to concentrate on the general or main points of a topic. The tables also summarize some aspects of the chapter's content, providing a convenient way to find information quickly. Often, a table is designed to accompany an illustration, so a written description and a visual presentation are combined to communicate information more effectively.

9. **Homeostasis Figures.** Homeostasis is illustrated using these figures, which provide a summary of the functions of a system and the means by which that system regulates a parameter within a narrow range of values. Homeostasis is a major theme of this text, and the homeostasis figures reinforce that theme effectively.

10. **Systems Pathology.** These new boxes added to each systems chapter represent a modified case study. Their goal is to show how each body system is influenced by the condition described in the case study. A Predict question follows each Systems Pathology reading.

11. **Chapter Summary.** As the student reads the chapter, details may obscure the overall picture. The chapter summary is an outline that briefly states the important facts and concepts and provides a perspective of the "big picture."

12. **Content Review Questions.** The Content Review questions are another method used in this text to transform the passive learner into an active one. The questions systematically cover the content and require students to summarize and restate the content in their own words.

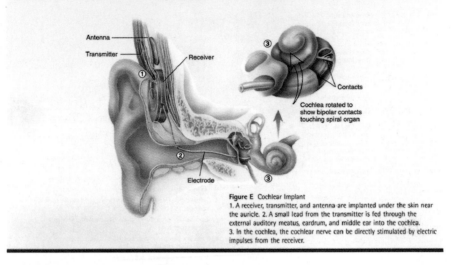

Clinical Focus Deafness and Functional Replacement of the Ear

Deafness can have many causes. In general, there are two categories of deafness: conduction and sensorineural (or nerve) deafness. **Conduction deafness** involves a mechanical deficiency in transmission of sound waves from the external ear to the spiral organ and may often be corrected surgically. Hearing aids help people with such hearing deficiencies by boosting the sound volume reaching the ear. **Sensorineural deafness** involves the spiral organ or nerve pathways and is more difficult to correct.

Research is currently being conducted on ways to replace the hearing pathways with electric circuits. One approach involves the direct stimulation of nerves by electric impulses. There has been considerable success in the area of cochlear nerve stimulation. Certain types of sensorineural deafness in which the hair cells of the spiral organ are impaired can now be partially corrected. Prostheses are available that consist of a microphone for picking up the initial sound waves; a microelectronic processor for converting the sound into

electric signals; a transmission system for relaying the signals to the inner ear; and a long, slender electrode that is threaded into the cochlea. This electrode delivers electric signals directly to the endings of the cochlear nerve (figure E). High frequency sounds are picked up by the microphone and transmitted through specific circuits to terminate near the oval window, whereas low-frequency sounds are transmitted farther up the cochlea to cochlear nerve endings near the helicotrema.

Figure E Cochlear Implant
1. A receiver, transmitter, and antenna are implanted under the skin near the auricle. 2. A small lead from the transmitter is fed through the external auditory meatus, eardrum, and middle ear into the cochlea. 3. In the cochlea, the cochlear nerve can be directly stimulated by electric impulses from the receiver.

13. **Develop Your Reasoning Skills Questions.** Following mastery of the Content questions and therefore chapter content, the Develop Your Reasoning Skills questions require the application of content to new situations. These are not essay questions that involve the restatement or summarization of chapter content. Instead, they provide additional practice in problem-solving and promote the development and acquisition of problem-solving skills.

14. **Multimedia Tie-ins.** *The Dynamic Human* CD-ROM is correlated to many figures. A Dynamic Human 🏃 icon appears in appropriate figure legends. The McGraw-Hill *Life Science Animations* Videotape Series is also correlated to many figures, and videotape icons appear in relevant figure legends. A complete listing of these correlations appears at the end of this preface.

15. **Appendices.** Appendix A is a table of measurements that helps the student relate the metric system to the more familiar English system when determining the size or weight of a structure. Appendix B helps the student understand the shorthand of scientific notation. Appendix C defines various methods for reporting the concentration of solutions and explains the rationale behind how various solutions are described. Appendix D explains the concept of pH and how it is measured. Appendix E contains tables of routine clinical test results along with normal values of clinical significance. Reference to this appendix provides students with the homeostatic values of many common substances in the blood and urine. Also, the importance of laboratory testing in the diagnosis and treatment of illnesses

becomes readily apparent to the students. Appendix F lists answers to the Predict questions.

Supplemental Materials

1. **Essential Study Partner.** (007-228407-2) This CD-ROM is an interactive student study tool packed with hundreds of animations and learning activities. From quizzes to interactive diagrams, your students will find that there has never been a more exciting way to study anatomy and physiology. A self-quizzing feature allows students to check their knowledge of a topic before moving on to a new module. Additional unit exams give students the opportunity to review coverage for a more complete understanding. The ESP is packaged free with textbooks.

2. **Online Learning Center.** **http://www.mhhe.com/biosci/ap/seeleyap/** Students and Instructors gain access to a world of opportunities through this password protected web site. Students will find quizzes, activities, links, and much more. Instructors will find all the enhancement tools needed for teaching online, or for incorporating technology in the traditional course.

3. **Course Solutions.** Designed specifically to help the instructor with their individual course needs, **Course Solutions** will assist the instructor in integrating their syllabus with the fifth edition of *Anatomy & Physiology* and the state-of-the-art new media tools that support them.

4. **Laboratory Manual.** (007-290753-3) Written by Eric Wise, this lab manual has been revised and closely tied to the 5th edition of the text.

5. **Student Study Guide.** (007-289918-2) Written by Philip Tate and James Kennedy, this Study Guide builds on the same teaching goals as the main text. Students are encouraged to use their recall and synthesis skills to complete the Study Guide objectives, which mirror and supplement the text objectives.

6. **Instructor's Manual and Test Item File.** (007-289921-2) The Instructor's Manual revised by Barbara Wiggins of Delaware Technical and Community College and includes an overview of changes in the new edition, suggested course outlines, suggestions for integrating the lecture and lab, learning strategies, organizing themes, answers to concept questions, and list of transparencies. The Test Item File, written by Dorothy Martin and Sandra Larson of Black Hawk College, includes a wide range of useful multiple-choice, matching, and essay questions.

7. **MicroTest III.** Available in Windows 3.5 (007-289919-0) and Macintosh 3.5 (007-289920-4). A computerized test generator for use with the text allows for quick creation of tests based on questions from the test item file and requires no programming experience.

8. **Transparencies.** (007-289922-0) A set of 600 full-color acetate transparencies. The figures that appear were chosen by the authors to be the most useful in lecture presentations.

9. **McGraw-Hill Visual Resource Library.** (007-228408-0) A CD-ROM containing all of the line art with an easy-to-use interface program enabling the user to quickly move among the images and create a multimedia presentation.

10. **A World Wide Web home page** exists for *Anatomy & Physiology*. The address is http://www.mhhe.com/biosci/ap/seeleyap/. The page contains links to relevant anatomy and physiology sites, as well as to appropriate McGraw-Hill multimedia products.

11. ***The Dynamic Human*** CD-ROM Version 2.0 (0-697-38935) illustrates the important relationships between anatomic structures and their functions in the human body. Realistic computer animation and three-dimensional visualizations are the premier features of this CD-ROM. Various figures throughout this text are correlated to modules of *The Dynamic Human*. See the end of the preface for a detailed listing of figures.

12. ***The Dynamic Human Videodisc*** (0-697-38937-5) contains all the animations (200+) from the CD-ROM. A barcode directory is also available.

13. ***McGraw-Hill Life Science Animations Videotape Series*** is a series of five videotapes containing 53 animations that cover many of the key physiologic processes. Another videotape containing similar animations is also available, entitled *Physiological Concepts of Life Science*. Various figures throughout this text are correlated to animations from the *Life Science Animations*. See the end of the preface for a detailed listing of figures.

Tape 1: Chemistry, The Cell, Energetics (0-697-25068-7)

Tape 2: Cell Division, Heredity, Genetics, Reproduction and Development (0-697-25069-5)

Tape 3: Animal Biology I (0-697-25070-9)

Tape 4: Animal Biology II (0-697-25071-7)

Tape 5: Plant Biology, Evolution, and Ecology (0-697-26600-1)

Tape 6: Physiological Concepts of Life Science (0-697-21512-1)

14. **Life Science Animations 3D CD-ROM.** More than 120 animations that illustrate key biological processes are available at your fingertips on this exciting CD-ROM. This CD contains all of the animations found on the *Essential Study Partner* and much more. The animations can be imported into presentation programs, such as PowerPoint. Imagine the benefit of showing the animations during lecture. (0-07-234296-X)

15. **Life Science Animations 3D** Videotape (0-07-29065-2). Featuring 42 animations of key biologic processes, this tape contains 3D animations and is fully narrated. Various figures throughout this text are correlated to video animations. See the end of the preface for a detailed listing of figures.

16. ***Explorations in Cell Biology and Genetics*** CD-ROM (0-697-37908-6 hybrid) contains interactive concepts related to key topics covered in an anatomy and physiology course. The CD-ROM can be used by an instructor in lecture or placed in a lab or resource center for students and is available for use with Macintosh and IBM Windows computers.

17. ***Life Science Living Lexicon*** CD-ROM (0-697-37993-0 hybrid) contains a comprehensive collection of life science terms, including definitions of their roots, prefixes, and suffixes as well as audio pronunciations and illustrations. The Lexicon is student-interactive, featuring quizzing and notetaking capabilities.

18. ***The Virtual Physiology Lab*** CD-ROM (0-697-37994-9 hybrid) contains 10 dry labs of the most common and important physiology experiments.

19. ***Anatomy and Physiology Videodisc*** (0-697-27716-X) is a four-sided videodisc containing more than 30 animations of physiologic processes, as well as line art and micrographs. A bar code directory is also available.

20. ***McGraw-Hill Anatomy and Physiology Video Series*** consists of the following:

    1. Internal Organs and the Circulatory System of the Cat (0-697-13922-0);
    2. Blood Cell Counting, Identification, and Grouping (0-697-11629-8);

3. Introduction to the Human Cadaver and Prosection (0-697-11177-6); and

4. Introduction to Cat Dissection: Musculature (0-697-11630-1).

21. **Study Cards for Anatomy and Physiology** (0-697-26447-5) by Van De Graaff and colleagues is a boxed set of (300) 3 × 5-inch cards. It serves as a well-organized and illustrated synopsis of the structure and function of the human body. The Study Cards offer a quick and effective way for students to review human anatomy and physiology.

22. **Coloring Review Guide to Anatomy and Physiology** (0-697-17109-4) by Robert and Judith Stone emphasizes learning through the process of color association. The Coloring Guide provides a thorough review of anatomic and physiologic concepts.

23. **Atlas of the Skeletal Muscles,** third edition (007-290332-5) by Robert and Judith Stone is a guide to the structure and function of human skeletal muscles. The illustrations help students locate muscles and understand their actions.

24. **Laboratory Atlas of Anatomy and Physiology,** second edition (0-697-39480-8) by Eder and colleagues is a full-color atlas containing histology, human skeletal anatomy, human muscular anatomy, dissections, and reference tables.

Acknowledgments

No modern textbook is solely the work of the authors. To adequately acknowledge the support of loved ones is not possible. They have had the patience and understanding to tolerate our frustrations and absence, and they have been willing to provide assistance and undying encouragement. We also wish to express our gratitude to the staff of McGraw-Hill for their help and encouragement. We sincerely appreciate Kris Tibbetts and Pat Hesse for their hours of work, suggestions, and tremendous patience and encouragement. We also thank our production, art, and photo editors, Jill Peter and Lori Hancock, who spent many hours turning manuscript into a book. The McGraw-Hill employees with whom we have worked are excellent professionals. They have been consistently helpful and their efforts are appreciated. Their commitment to this project has clearly been more than a job to them.

We also thank the many illustrators who worked on the development and execution of the illustration program for the fifth edition of *Anatomy & Physiology*. The art program for this text represents a monumental effort, and we appreciate their contribution to the overall appearance and pedagogical value of the illustrations.

Finally, we sincerely thank the reviewers and the teachers who have provided us with excellent constructive criticism. The remuneration they received represents only a token payment for their efforts. To conscientiously review a textbook requires a true commitment and dedication to excellence in teaching. Their helpful criticisms and suggestions for improvement were significant contributions that we greatly appreciate. We acknowledge them by name in the next section.

Rod Seeley
Trent Stephens
Phil Tate

Reviewers

Latifeh Amini-Kormi
Worcester State College

Linda M. Bacha
Camden County College

P. Bagavandoss
Kent State University

Frank Baker
Golden West College

Sarah M. Bales
Moraine Valley Community College

Tim Ballard
University of North Carolina-Wilmington

Robert Bauman, Jr.
Amarillo College

Moges Bizuneh
Ivy Tech State College

Leonard I. Borack
Rutgers University

J. B. Boren
Missouri Baptist College

Gary Brady
Spokane Falls Community College

Mary Teresa Brandon
Dona Ana Branch Community College

Stan Braude
Washington University

Sara Brenizer
Shelton State Community College

James Bridger
Prince George's Community College

Heather Brient-Johnson
Inver Hills Community College

Nishi Bryska
University of North Carolina-Charlotte

Thomas E. Byrne
Roane State Community College

John R. Capeheart
University of Houston-Downtown

Jennifer Carr Burtwistle
Northeast Community College

Christopher Chabot
Plymouth State College

Carolyn C. Clarke
Jefferson Community College

Wade L. Collier
Manatee Community College, South Campus

Mark S. Condon
Rockland Community College

W. Wade Cooper
Shelton State Community College

James J. Copi
Madonna University

Camille P. Cress
University of Arkansas at Little Rock

John R. Crooks
Iowa Wesleyan College

Leonard V. Crowley
Century College

Judith Anne D'Aleo
Plymouth State College

Bob Davis
Cabarrus College of Health Sciences

John W. Davis
Benedictine College

Judith K. Davis
Florida Community College at Jacksonville

Annette Dawson
Stephen F. Austin State University

Brent DeMars
Lakeland Community College

Michael A. Dorset
Cleveland State Community College

William E. Dunscombe
Union County College

John H. Dustman
Indiana University Northwest

Donna L. Ellis
Camden County College

Joan T. Fieldus
Laramie County Community College

Ann M. Findley
Northeast Louisiana University

Kathleen Anne Flickinger
Iowa State University

James E. Forbes
Hampton University

Dee Forrest
Western Wyoming Community College

Pamela B. Fouché
Walters State Community College

Kim T. Fredricks
Viterbo College

Frank Furbush
Henderson Community College

Patrick Galliart
North Iowa Area Community College

Louis A. Giacinti
Milwaukee Area Technical College

Chaya Gopalan
St. Louis Community College

Donald W. Green
San Juan College

Timothy A. Hacker
Concordia University, Wisconsin

Thomas R. Halcomb
Northwest Shoals Community College

Cecil M. Hampton
Jefferson College

Steve Hardin
Ozarks Technical Community College

Larry E. Hibbert
Ricks College

Angie Huxley
Pima Community College-West Campus

Ronald Jenkins
North Iowa Area Community College

Marsha Jones
Southwestern Community College

Lloyd M. Kahn
Hudson County Community College

Kamal I. Kamal
Valencia Community College-West Campus

Suzanne Kempke
Armstrong Atlantic State University

George S. Kendrick
Navarro College

Shelley A. Kirkpatrick
St. Francis College, PA

Steven G. Kish
Muskingum Area Technical College

William C. Kleinelp, Jr.
Middlesex County College

Karen M. LaFleur
Greenville Technical College

Marian G. Langer
St. Francis College, PA

William Langley
Butler County Community College

Carolyn J. Lebsack
Linn-Benton Community College

Jeffrey Lee
Essex County College

Joe Leffelman
Volunteer State Community College

George J. Leslie
Springfield Technical Community College

Jerri K. Lindsey
Tarrant County College-NE Campus

J. Mitchell Lockhart
Valdosta State University

Karen McCort
El Paso Community College

Margaret A. Maher
University of Wisconsin-La Crosse

Bonita L. Makin
Cambria-Rowe Business College

David L. Mapes
State Technical Institute at Memphis

Elden W. Martin
Bowling Green State University

William J. Mathena
Kaskaskia College

Craighton S. Mauk
Prestonsburg Community College

Donna Maus
Fort Scott Community College

Karen McCort
El Paso Community College

John A. Mecham
Catawba College

Roberta M. Meehan
University of Northern Colorado

Robert Moldenhauer
St. Clair County Community College

A. Kenneth Moore
Seattle Pacific University

Tina Moore
Midwestern State University

Alfredo Muńoz
University of Texas/Texas Southmost College

Joseph Murray
Blue Ridge Community College

John J. Natalini
Quincy University

S. R. Neubauer
Rogers State University

Claire R. Oakley
Rocky Mountain College

Tammy O'Brien
Augusta Technical Institute

Thomas Oeltmann
Vanderbilt University

Kerry L. Openshaw
Bemidji State University

Amy G. Ouchley
Northeast Louisiana University

Glenn Perrigo
Texas A&M University-Kingsville

Randa A. Pinkston
Blue Ridge Community College

Angela R. Porta
Kean University

Samirsubas Raychoudhury
Benedict College

David C. Reff
Middle Georgia College

Jean Revie
Northland Pioneer College

Jackie Reynolds
Richland College

Laura H. Ritt
Burlington County College

Kenneth E. Roth
Eastern Mennonite University

John Rousseau
Montgomery College

Michael W. Ruhl
Vernon Regional Junior College

Laura Jones Rutter
Lindenwood University

Jane Salisbury
Edison State Community College

May Linda Samuel
Benedict College

Melvin Schmidt
McNeese State University

Marilyn M. Shannon
Indiana University-Purdue University Fort Wayne

Aida Shehata
Prince George's Community College

Eileen Kennedy Shull
Scott Community College

Richard Sims
Jones County Junior College

Jeffery L. Smith
Delgado Community College

Michael E. Smith
Valdosta State University

Erich K. Stabenau
Bradley University

Janet Steele
University of Nebraska at Kearney

Jane E. Stephens
Delta State University

Ralph W. Stevens III
Old Dominion University

Barbara L. Stewart
J. Sargeant Reynolds Community College

David L. Swanson
University of South Dakota

Kathryn B. Sympson
Florida Keys Community College

Todd Templeton
Metropolitan Community College

Kenneth Thomas
Hillsborough Community College

Kimberly T. Turk
Mitchell Community College

Victoria L. Veigl
University of Louisville

Robert Vick
Elon College

F. R. Voorhees
Central Missouri State University

Burton J. Webb
Indiana Wesleyan University

Janice J. Weber
Huron University

Terry P. Wheeler
Halifax Community College

Ricky K. Wong
Los Angeles Trade-Technical College

Jeanne M. Workman
Duquesne University

Xiaobo Yu
Kean University

Dynamic Human Version 2.0 🏃 Correlation Guide

Life Science Animations (LSA) 📼 Correlation Guide

Figure 2.4 LSA 1	Formation of an Ionic Bond	Figure 15.33 LSA 26	Organ of Static Equilibrium	Figure 24.3 LSA 33	Peristalsis
Figure 2.27 LSA 11	ATP as an Energy Carrier	Figure 15.34 LSA 26	Organ of Static Equilibrium	Figure 24.25 LSA 34	Digestion of Carbohydrates
Figure 3.1 LSA 2	Journey into a Cell	Figure 15.35 Figure 15.36 LSA 26	Organ of Static Equilibrium	Figure 24.26 LSA 36	Digestion of Lipids
Figure 3.11 LSA 4	Cellular Secretion	Figure 19.12 LSA 40	A, B, O Blood Types	Figure 24.30 LSA 35	Digestion of Proteins
Figure 3.12 LSA 3	Endocytosis	Figure 19.13 LSA 40	A, B, O Blood Types	Figure 25.2 LSA 11	ATP as an Energy Carrier
Figure 3.27 LSA 6	Oxidative Respiration	Figure 20.1 LSA 37	Blood Circulation	Figure 25.3 LSA 5 LSA 6	Glycolysis Oxidative Respiration
Figure 3.28 LSA 16 LSA 17	Transcription of a Gene Protein Synthesis	Figure 20.9 LSA 32	The Cardiac Cycle and Production of Sounds	LSA 7	The Electron Transport Chain and the Production of ATP
Figure 3.29 LSA 16	Transcription of a Gene	Figure 20.13 LSA 38	Production of Electrocardiogram	Figure 25.4 LSA 5	Glycolysis
Figure 3.31 LSA 17	Protein Synthesis	Figure 20.14 LSA 38	Production of Electrocardiogram	Figure 25.6 LSA 6	Oxidative Respiration
Figure 3.32 LSA 12	Mitosis	Figure 20.16 LSA 38	Production of Electrocardiogram	Figure 25.7 LSA 7	The Electron Transport Chain and the Production of ATP
Figure 3.33 LSA 15	DNA Replication				
Figure 3.34 LSA 12	Mitosis	Figure 20.17 LSA 32	The Cardiac Cycle and Production of Sounds	Figure 25.8 LSA 36	Digestion of Lipids
Figure 3.35 LSA 13	Meiosis	Figure 20.18 LSA 32 LSA 38	The Cardiac Cycle and Production of Sounds Production of Electrocardiogram	Figure 28A LSA 13 Figure 28.4 LSA 19	Meiosis Spermatogenesis
Figure 3.36 LSA 14	Crossing over				
Figure 9.7 LSA 28	Peptide Hormone Action (cAMP)			Figure 28.11 LSA 20	Oogenesis
Figure 10.3 LSA 29	Levels of Muscle Structure	Figure 22.15 Figure 22.20 LSA 41	B-Cell Immune Response	Figure 28.12 LSA 20	Oogenesis
Figure 10.8 Figure 10.13 LSA 30	Sliding Filament Model of Muscle Contraction	Figure 22.16 LSA 42	Structure and Function of Antibodies	Figure 29.5 LSA 21	Human Embryonic Development
				Figure 29.6 LSA 21	Human Embryonic Development
Figure 10.12 LSA 23	Saltatory Nerve Conduction	Figure 22.17 Figure 22.18 LSA 42	Structure and Function of Antibodies	Figure 29.7 LSA 21	Human Embryonic Development
Figure 12.7 Figure 12.10 LSA 22	Formation of Myelin Sheath				
Figure 12.13 LSA 23	Saltatory Nerve Conduction	Figure 22.19		Figure 29.10 LSA 21	Human Embryonic Development
Figure 12.19 LSA 25	Reflex Arcs	Figure 22.20 LSA 43	Types of T-Cells		
Figure 13.2 LSA 21	Human Embryonic Development	Figure 22.20 LSA 44	Relationship of Helper T-Cells and Killer T-Cells		

Life Science Animations 3D-Videotape ▢ Correlation Guide

▤ Essential Study Partner CD-ROM Animations Correlation Guide

Chapter One

The Human Organism

Objectives

1. Explain the importance of understanding the relationship between structure and function.

2. Define the terms anatomy and physiology, and identify the different ways in which they can be studied.

3. Describe the chemical, organelle, cell, tissue, organ, organ system, and whole organism levels of organization.

4. List the 11 organ systems, and indicate the major functions of each.

5. List the characteristics of life.

6. Explain the importance of studying other animals to help understand human anatomy and physiology.

7. Define homeostasis. Give an example of a negative-feedback system and a positive-feedback system, and describe the relationship of each to homeostasis.

8. Describe the anatomic position. Use the directional terms in table 1.1 to describe specific body structures.

9. Name and describe the three major planes of the body or of an organ.

10. List the terms used to describe different regions or parts of the body.

11. Describe two ways to subdivide the abdominal region.

12. Define the terms thoracic cavity, abdominal cavity, pelvic cavity, and mediastinum.

13. Define serous membrane, and explain the relationship between parietal and visceral serous membranes.

14. Name the membranes that line the walls and cover the organs of each body cavity, and name the fluid found inside each cavity.

15. Define mesentery, and describe its function.

16. Define the term retroperitoneal, and list examples of retroperitoneal organs.

You are about to begin a wondrous adventure, the study of the human body. The knowledge you gain will be useful for many reasons, and soon you will be immersed in the details of the subject. Don't forget, however, to appreciate the beauty of what you are studying. The human body is an amazing and marvelous construction that is maintained by a complex system of checks and balances.

The study of anatomy and physiology is essential for those who plan a career in the health sciences, because a sound knowledge of structure and function is necessary for health professionals to perform their duties adequately. Knowledge of anatomy and physiology is also beneficial to the nonprofessional. This background improves your ability to understand your body in health and disease, evaluate recommended treatments, critically review advertisements and reports in the popular literature, and interact rationally with health professionals.

Human anatomy and physiology is the study of the structure and function of the human body. Knowledge of both structure and function allows one to understand how the body responds to a stimulus. For example, eating a candy bar results in an increase in blood sugar (the stimulus). Knowledge of the pancreas allows one to predict that the pancreas will secrete insulin (the response). Insulin moves into blood vessels and is transported to cells, where it increases the movement of sugar from the blood into cells, providing the cells with a source of energy. As glucose moves into cells, blood sugar levels decrease.

Knowledge of structure and function also provides the basis for understanding disease. In one type of diabetes mellitus, for example, the pancreas does not secrete adequate amounts of insulin. Without adequate insulin, not enough sugar moves into cells, which deprives them of a needed source of energy, and they malfunction.

anatomy the body is studied system by system, which is the approach taken in this and most other introductory textbooks. A system is a group of structures that have one or more common functions. Examples are the circulatory, nervous, respiratory, skeletal, and muscular systems. In **regional anatomy** the body is studied area by area, which is the approach taken in most graduate programs at medical and dental schools. Within each region, such as the head, abdomen, or arm, all systems are studied simultaneously.

Surface anatomy is the study of the external form of the body and its relation to deeper structures. For example, the sternum (breastbone) and parts of the ribs can be seen and palpated (felt) on the front of the chest. These structures can be used as landmarks to identify regions of the heart and points on the chest at which certain heart sounds can best be heard. **Anatomic imaging** involves the use of radiographs (x-rays), ultrasound, magnetic resonance imaging (MRI), and other technologies to create pictures of internal structures. Both surface anatomy and anatomic imaging provide important information in diagnosing disease.

Clinical Note

Humans are not structurally identical. For instance, one person may have longer fingers than another person. Despite this kind of variability, most humans exhibit a basic pattern. Normally, we each have 10 fingers. Sometimes, however, anatomic anomalies occur—structures that are unusual and different from the normal pattern. For example, some individuals have 12 fingers.

Anatomic anomalies can vary in severity from the relatively harmless to the life-threatening because they compromise normal function. For example, each kidney is normally supplied by one blood vessel, but in some individuals a kidney can be supplied by two blood vessels. Either way, the kidney receives adequate blood. On the other hand, in the condition called "blue baby" syndrome certain blood vessels arising from the heart of an infant are not attached in their correct locations; blood is not effectively pumped to the lungs, resulting in tissues not receiving adequate oxygen.

Anatomy

Anatomy is the scientific discipline that investigates the body's structure. For example, the shape and size of the bones can be described. In addition, anatomy examines the relationship between the structure of a body part and its function. Just as the structure of a hammer makes it well suited for pounding nails, the structure of a specific body part allows it to perform a particular function effectively. For example, bones can provide strength and support because bone cells surround themselves with a hard, mineralized substance. Understanding the relationship between structure and function makes it easier to understand and appreciate anatomy.

Anatomy can be considered at many different levels. Some structures are so small that they are best studied using a microscope. **Cytology** (sī-tol′ō-jē) examines the structural features of cells, and **histology** (his-tol′ō-jē) is the study of tissues, which are cells and the materials surrounding them.

Gross anatomy, the study of structures that can be examined without the aid of a microscope, can be approached from either a systemic or regional perspective. In **systemic**

Physiology

Physiology is the scientific investigation of the processes or functions of living things. In physiology it is important to recognize structures as dynamic rather than unchanging. The major goals of physiology are to understand and predict the responses of the body to stimuli and to understand how the body maintains conditions within a narrow range of values in the presence of a continually changing environment.

Like anatomy, physiology can be considered at many different levels. For example, **cell physiology** deals with the processes of cells, **neurophysiology** considers the nervous system, and **human physiology** studies the biologic processes of humans. Physiology often examines systems rather than regions because most physiologic processes involve the interactions of one or more systems.

▐ Structural and Functional Organization

The body can be considered conceptually at seven structural levels: the chemical, organelle, cell, tissue, organ, organ system, and complete organism (figures 1.1 and 1.2).

Chemical

The structural and functional characteristics of all organisms are determined by their chemical makeup. The **chemical** level of organization involves interactions between atoms and their combinations into molecules. The function of a molecule is related intimately to its structure. For example, collagen molecules are strong, ropelike fibers that give skin structural strength and flexibility. With old age, the structure of collagen changes, and the skin becomes fragile and is torn more easily. A brief overview of chemistry is presented in chapter 2.

Organelle

An **organelle** (or′gă-nel) is a small structure contained within a cell that performs one or more specific functions. For example, the nucleus is an organelle containing the cell's hereditary information. Organelles are discussed in chapter 3.

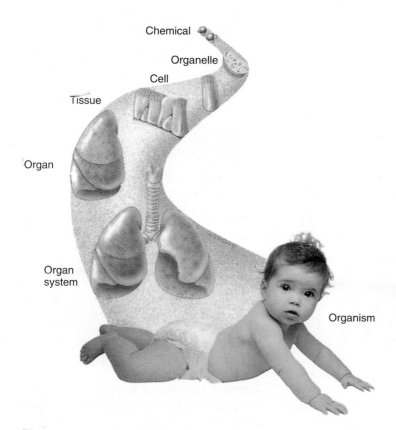

Figure 1.1   Levels of Organization

Seven levels of organization for the human body are the chemical, organelle, cell, tissue, organ, organ system, and organism levels.

Cell

Cells are the basic living units of all plants and animals. Although cell types differ in their structure and function, they have many characteristics in common. Knowledge of these characteristics and their variations is essential to a basic understanding of anatomy and physiology. The cell is discussed in chapter 3.

Tissue

A group of cells with similar structure and function, together with the extracellular substances located between them, form a **tissue.** The many tissues that make up the body are classified into four primary tissue types: epithelial, connective, muscle, and nervous. Tissues are discussed in chapter 4.

Organ

Organs are composed of two or more tissue types that perform one or more common functions. The skin, stomach, eye, and heart are examples of organs.

Organ System

An **organ system** is a group of organs classified as a unit because of a common function or set of functions. In this text the body is considered to have 11 major organ systems: the integumentary, skeletal, muscular, nervous, endocrine, cardiovascular, lymphatic, respiratory, digestive, urinary, and reproductive systems (see figure 1.2).

Organism

An **organism** is any living thing considered as a whole, whether composed of one cell such as a bacterium or of trillions of cells such as a human. The human organism is a complex of organ systems, all mutually dependent on one another.

1	P R E D I C T

One type of diabetes is a disorder in which the pancreas (an organ) fails to produce insulin, which is a chemical normally made by pancreatic cells and released into the circulation. List as many levels of organization as you can in which this disorder could be corrected.

✔ *Answer in Appendix F*

The Human Organism

Humans are organisms and have many characteristics in common with other organisms. The most important common feature of all organisms is life. Essential characteristics of life are organization, metabolism, responsiveness, growth and development, and reproduction.

Clinical Focus   Anatomic Imaging

Anatomic imaging has revolutionized medical science. It has been estimated that during the past 20 years as much progress has been made in clinical medicine as in all its previous history combined, and anatomic imaging has made a major contribution to that progress. Anatomic imaging allows medical personnel to look inside the body with amazing accuracy and without the trauma and risk of exploratory surgery. Although most of the technology of anatomic imaging is very new, the concept and earliest technology are quite old.

X-rays were first used in medicine by Wilhelm Roentgen (1845-1923) in 1895 to see inside the body. They were called x-rays because no one knew what they were. The rays, extremely shortwave electromagnetic radiation (see chapter 2), can pass through the body and expose a photographic plate to form a **radiograph** (rā′dē-ō-graf). Bones and radiopaque dyes absorb the rays and create underexposed areas that appear white on the photographic film (figure A). X-rays have been in common use for many years and have numerous applications. A major limitation of x-rays is that they give only a flat, two-dimensional (2-D) image of the body, which is a three-dimensional (3-D) structure. Almost everyone has had a radiograph taken, either to visualize a broken bone or to check for a cavity in a tooth.

Ultrasound is the second oldest imaging technique. When it was first developed in the early 1950s as an extension of World War II sonar technology, it used high-frequency sound waves. The sound waves are emitted from a transmitter–receiver placed on the skin over the area that is scanned. The sound waves strike internal organs and are reflected back to the receiver on the skin. Even though the basic technology is fairly old, the most important advances in the field occurred only after it became possible to analyze the reflected sound waves by computer. Once the computer analyzes the pattern of sound waves, the information is transferred to a monitor, where the result is visualized as an ultrasound image called a **sonogram** (son′ō-gram) (figure B). One of the more recent advances in ultrasound technology is the ability of the more advanced computers to analyze changes in position through time and to display those changes as "real time" movements. Among other medical uses, ultrasound commonly is used to evaluate the condition of the fetus during pregnancy.

Computer analysis is also the basis of another major medical breakthrough in imaging. **Computed tomographic** (tō′mō-graf′ik) **(CT) scans,** developed in 1972 and originally called **computerized axial tomographic (CAT) scans,** are computer-analyzed x-ray images. A low-intensity x-ray tube is rotated through a 360-degree arc around the patient, and the images are fed into a computer. The computer then constructs the image of a "slice" through the body at the point at which the x-ray beam was focused and rotated (figure C). It is also possible with some computers to take several scans short distances apart and stack the slices to produce a 3-D image of a part of the body (figure D).

Dynamic spatial reconstruction (DSR) takes CT one step further. Instead of using a

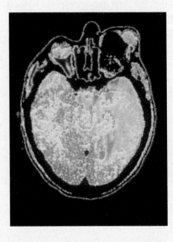

Figure C   Computed Tomography
Transverse section through the skull. Note the tumor causing the eye to bulge. ✗

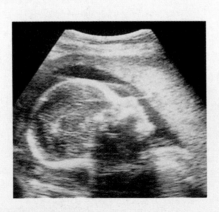

Figure A   X-ray   Radiograph produced by x-rays shows a lateral view of the head and neck. ✗

Figure B   Ultrasound   Sonogram produced with ultrasound shows a lateral view of the head and neck of a fetus within the uterus. ✗

Figure D   Computed Tomography (CT)
Stacking of images acquired using CT technology. ✗

single rotating x-ray machine to take single slices and add them together, DSR uses about 30 x-ray tubes. The images from all the tubes are compiled simultaneously, rapidly producing a 3-D image. Because of the speed of the process, multiple images can be compiled to show changes through time, giving the system a dynamic quality. This system allows us to move away from seeing only static structure and toward seeing dynamic structure and function.

Digital subtraction angiography (an'jē-og'rǎ-fē) **(DSA)** is also one step beyond CT scans. A 3-D radiographic image of an organ such as the heart is made and stored in a computer. A radiopaque dye is injected into the circulation, and a second radiographic computer image is made. The first image is subtracted from the second one, greatly enhancing the differences, with the primary difference being the presence of the injected dye (figure E). These computer images can be dynamic and can be used, for example, to guide a catheter into a coronary artery during angioplasty, which is the insertion of a tiny balloon into a coronary artery to compress material clogging the artery.

Magnetic resonance imaging (MRI), which directs radio waves at a person lying inside a large electromagnetic field, is also based on principles that have been known

for years but have been applied only recently to medicine. The magnetic field causes the protons of various atoms to align (see chapter 2). Because of the large amounts of water in the body, the alignment of hydrogen atom protons is at present most important in this imaging system. Radio waves of certain frequencies, which change the alignment of the hydrogen atoms, then are directed at the patient. When the radio waves are turned off, the hydrogen atoms realign in accordance with the magnetic field. The time it takes the hydrogen atoms to realign is different for various tissues of the body. These differences can be analyzed by computer to produce very clear sections through the body (figure F). The technique is also very sensitive in detecting some forms of cancer and can detect a tumor far more readily than can a CT scan.

Positron emission tomographic (PET) scans are able to identify the metabolic states of various tissues. This technique is particularly useful in analyzing the brain. When cells are active, they are using energy. The energy they need is supplied by the breakdown of glucose (blood sugar). If radioactively treated, or "labeled," glucose is given to a patient, the active cells take up the labeled glucose. As the radioactivity in the glucose decays, positively

charged subatomic particles called positrons are emitted. When the positrons collide with electrons, the two particles annihilate each other, and gamma rays are given off. The gamma rays can be detected, pinpointing the cells that are metabolically active (figure G).

Whenever the human body is exposed to x-rays, ultrasound, electromagnetic fields, or radioactively labeled substances, there is potential risk. In the medical application of anatomic imaging, the risk must be weighed against the benefit. Numerous studies have been conducted and are still being done to determine the outcomes of diagnostic and therapeutic exposures to x-rays.

The risk of anatomic imaging is minimized by using the lowest possible doses that provide the necessary information. For example, it is well known that x-rays can cause cell damage, particularly to the reproductive cells. As a result of this knowledge, the number of x-rays and the level of exposure are kept to a minimum, the x-ray beam is focused as closely as possible to avoid scattering of the rays, areas of the body not being x-rayed are shielded, and personnel administering x-rays are shielded. There are no known risks from ultrasound or electromagnetic fields at the levels used in diagnosis.

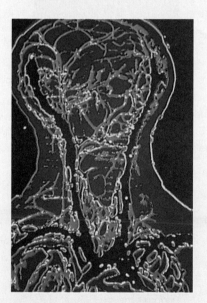

Figure E   Digital Subtraction Angiography (DSA)   Reveals the major blood vessels supplying the head and upper limbs.

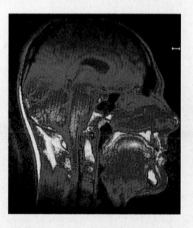

Figure F   Magnetic Resonance Imaging (MRI)   Shows a lateral view of the head and neck. ✗

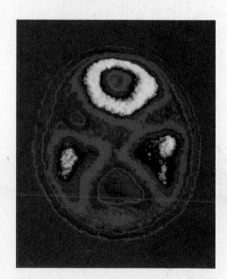

Figure G   Positron Emission Tomography (PET)   Shows a transverse section through the skull. ✗

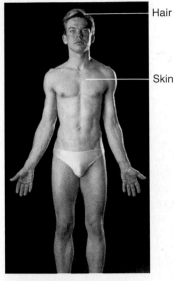

Hair

Skin

Integumentary System
Provides protection, regulates
temperature, prevents water loss,
and produces vitamin D precursors.
Consists of skin, hair, nails, and
sweat glands.

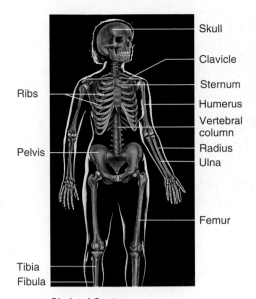

Skull

Clavicle

Sternum

Humerus

Vertebral
column

Ribs

Radius

Pelvis

Ulna

Femur

Tibia

Fibula

Skeletal System
Provides protection and support,
allows body movements, produces
blood cells, and stores minerals and
fat. Consists of bones, associated
cartilages, and joints.

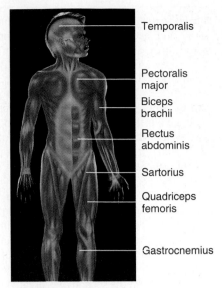

Temporalis

Pectoralis
major

Biceps
brachii

Rectus
abdominis

Sartorius

Quadriceps
femoris

Gastrocnemius

Muscular System
Produces body movements,
maintains posture, and produces body
heat. Consists of muscles attached
to the skeleton.

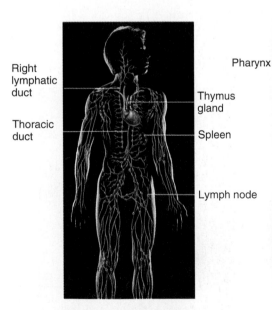

Right
lymphatic
duct

Thoracic
duct

Thymus
gland

Spleen

Lymph node

Lymphatic System
Removes foreign substances from
the blood and lymph, combats
disease, maintains tissue fluid
balance, and absorbs fats from the
digestive tract. Consists of the
lymph vessels, lymph nodes, and
other lymph organs.

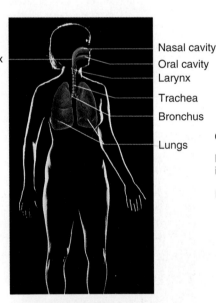

Pharynx

Nasal cavity

Oral cavity

Larynx

Trachea

Bronchus

Lungs

Respiratory System
Exchanges oxygen and carbon
dioxide between the blood and air and
regulates blood pH. Consists of the
lungs and respiratory passages.

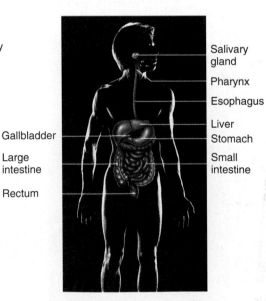

Salivary
gland

Pharynx

Esophagus

Gallbladder

Liver

Large
intestine

Stomach

Small
intestine

Rectum

Digestive System
Performs the mechanical and
chemical processes of digestion,
absorption of nutrients, and
elimination of wastes. Consists of
the mouth, esophagus, stomach,
intestines, and accessory organs.

Figure 1.2   Organ Systems of the Body

(continued)

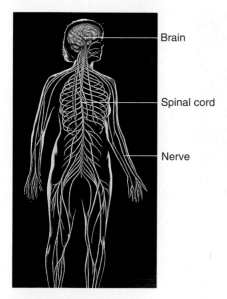

Nervous System
A major regulatory system that detects sensations and controls movements, physiologic processes, and intellectual functions. Consists of the brain, spinal cord, nerves, and sensory receptors.

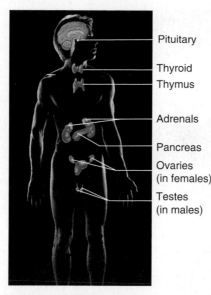

Endocrine System
A major regulatory system that influences metabolism, growth, reproduction, and many other functions. Consists of glands, such as the pituitary, that secrete hormones.

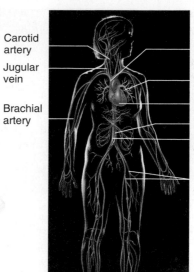

Cardiovascular System
Transports nutrients, waste products, gases, and hormones throughout the body; plays a role in the immune response and the regulation of body temperature. Consists of the heart, blood vessels, and blood.

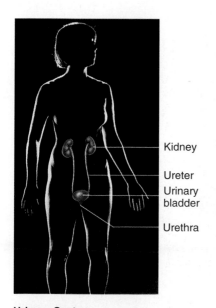

Urinary System
Removes waste products from the blood and regulates blood pH, ion balance, and water balance. Consists of the kidneys, urinary bladder, and ducts that carry urine.

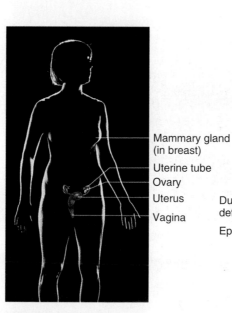

Female Reproductive System
Produces oocytes and is the site of fertilization and fetal development; produces milk for the newborn; produces hormones that influence sexual functions and behaviors. Consists of the ovaries, vagina, uterus, mammary glands, and associated structures.

Male Reproductive System
Produces and transfers sperm cells to the female and produces hormones that influence sexual functions and behaviors. Consists of the testes, accessory structures, ducts, and penis.

Figure 1.2   Organ Systems of the Body

(continued)

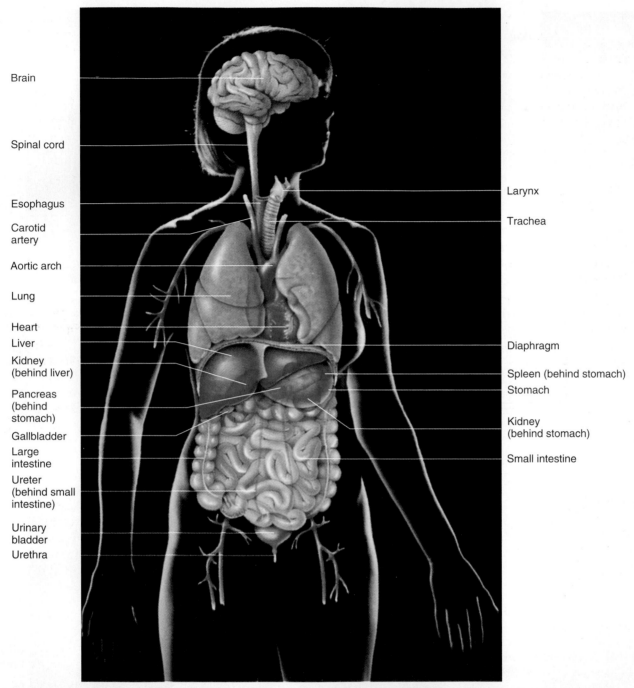

The relationship of different organ systems to one another.

Figure 1.2 Organ Systems of the Body

Characteristics of Life

Organization

Living things are highly organized. All living things are composed of cells or, in the case of viruses, must be contained within cells to perform living functions. Cells in turn are composed of highly organized organelles, which depend on the precise organization of large molecules. Dis-

ruption of this organized state can result in loss of functions and death.

Metabolism

Metabolism (mě-tab′ō-lizm) is the ability to use energy to perform vital functions, such as growth, movement, and reproduction. Plants can capture energy from sunlight, and humans obtain energy from food.

Responsiveness

An organism is responsive if it can sense changes in the environment and make adjustments that help maintain its life. Responses include movement toward food or water and away from danger or poor environmental conditions. Organisms can also make adjustments that maintain their internal environment. For example, if body temperature increases in a hot environment, sweat glands produce sweat, which can lower body temperature back toward normal levels.

Growth and Development

Growth results from the ability of cells to increase in size or number, producing an overall enlargement of all or part of the organism. **Development** includes the changes an organism undergoes through time; it begins with fertilization and ends at death. The greatest developmental changes occur before birth, but many changes continue after birth, and some continue throughout life. Development usually involves growth, but it also involves differentiation and morphogenesis. **Differentiation** is changes in cell structure and function from generalized to specialized, and **morphogenesis** (mōr-fō-jen′ĕ-sis) is changes in the shape of tissues, organs, and the entire organism. For example, following fertilization, generalized cells specialize to become specific cell types, such as skin, bone, muscle, or nerve cells. As the cells specialize, tissues and organs take shape.

Reproduction

Reproduction is the formation of new cells or new organisms. Without reproduction of cells, growth and development are not possible. Without reproduction of the organism, there is only extinction for the species.

Biomedical Research

Humans share many characteristics with other organisms, and much of our knowledge about humans has come from studying other organisms. For example, the study of single-celled bacteria has provided much information about human cells. Some biomedical research, however, cannot be accomplished using single-celled organisms or isolated cells. For example, great progress in open-heart surgery and kidney transplantation was made possible by perfecting techniques on other mammals before attempting them on humans. Strict laws govern the use of animals in biomedical research—laws designed to ensure minimum suffering on the part of the animal and to discourage unnecessary experimentation.

Although much can be learned from studying other organisms, the ultimate answers to questions about humans can be obtained only from humans, because other organisms are often different from humans in significant ways.

Homeostasis

Homeostasis (hō′mē-ō-stā′sis) is the existence and maintenance of a relatively constant environment within the body. Each cell of the body is surrounded by a small amount of fluid, and the normal functions of each cell depend on the maintenance of its fluid environment within a narrow range of conditions, including volume, temperature, and chemical content. These conditions are called **variables** because their values can change. For example, body temperature is a variable that can increase in a hot environment or decrease in a cold one.

Homeostatic mechanisms, such as sweating or shivering, normally maintain body temperature near an ideal normal value, or **set point** (figure 1.3). Note that these mechanisms are not able to maintain body temperature precisely at the set point. Instead, body temperature increases and decreases slightly around the set point, producing a normal range of values. As long as body temperature remains within this normal range, homeostasis is maintained.

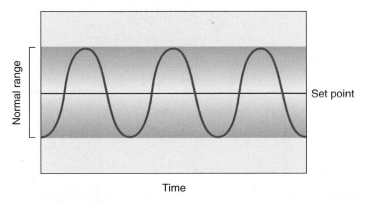

Figure 1.3   Homeostasis

Homeostasis is the maintenance of a variable around an ideal normal value, or set point. The value of the variable fluctuates around the set point, establishing a normal range of values.

The organ systems help to control the internal environment so that it remains relatively constant. For example, the digestive, respiratory, circulatory, and urinary systems function together so that each cell in the body receives adequate oxygen and nutrients and so that waste products do not accumulate to a toxic level. If the fluid surrounding cells deviates from homeostasis, the cells do not function normally and can even die. Disruption of homeostasis results in disease and sometimes death.

Negative Feedback

Most systems of the body are regulated by **negative-feedback** mechanisms that maintain homeostasis. *Negative* means that any deviation from the set point is made smaller or is resisted. Many negative-feedback mechanisms have three components: the **receptor,** which monitors the value of some variable such as blood pressure; the **control center,** which establishes the set point around which the variable is maintained; and the **effector,** which can change the value of the variable. A devia-

tion from the set point is called a **stimulus.** The receptor detects the stimulus and informs the control center, which analyzes the input from the receptor. The control center sends output to the effector, and the effector produces a **response,** which tends to return the variable back toward the set point (figure 1.4).

The maintenance of normal blood pressure is an example of a negative-feedback mechanism that maintains homeostasis. Normal blood pressure is important because it is responsible for moving blood from the heart to tissues. The blood supplies the tissues with oxygen and nutrients and removes waste products. Thus normal blood pressure is required to ensure that tissue homeostasis is maintained.

Receptors that monitor blood pressure are located within large blood vessels near the heart, the control center for blood pressure is in the brain, and the heart is the effector. Blood pressure depends in part on contraction (beating) of the heart: as heart rate increases, blood pressure increases; as heart rate decreases, blood pressure decreases.

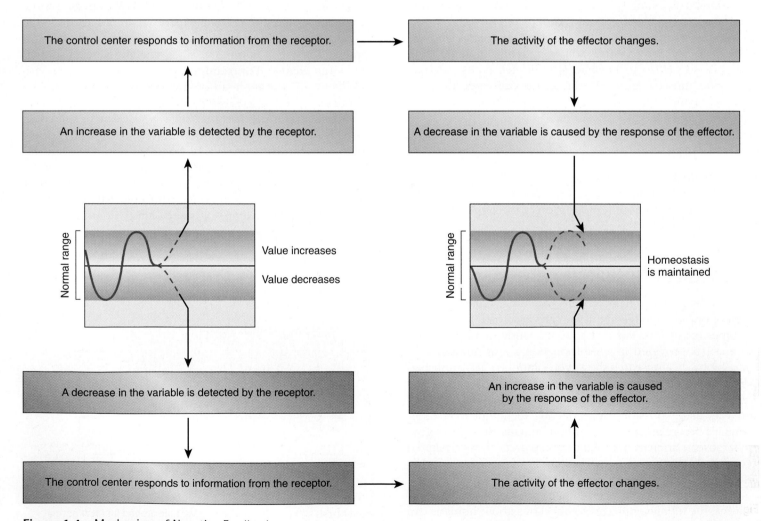

Figure 1.4   Mechanism of Negative Feedback

An increase or a decrease of a variable is detected by a receptor. That information is sent to a control center, which modifies the activity of an effector. The response of the effector maintains homeostasis by keeping the value of the variable within a normal range.

If blood pressure increases slightly, the receptors detect the increased blood pressure and send that information to the control center in the brain. The control center causes heart rate to decrease, resulting in a decrease in blood pressure. If blood pressure decreases slightly, the receptors inform the control center, which increases heart rate, producing an increase in blood pressure. As a result, blood pressure constantly rises and falls within a normal range of values (figure 1.5).

Although homeostasis is the maintenance of a normal range of values, this does not mean that all variables are maintained within the same narrow range of values at all times. Some situations occur during which a deviation from the usual range of values can be beneficial. For example, during exercise the normal range for blood pressure differs from the range under resting conditions, and the blood pressure is significantly elevated (figure 1.6). The elevated blood pressure is required to deliver blood to muscles so that muscle cells are supplied with the extra nutrients and oxygen they need to maintain their increased rate of activity.

2   P R E D I C T

Explain how negative-feedback mechanisms control respiratory rates when a person is at rest and when a person is exercising.

✔ *Answer in Appendix F*

Positive Feedback

Positive-feedback responses are not homeostatic and are rare in healthy individuals. *Positive* implies that, when a deviation from a normal value occurs, the response of the system is to make the deviation even greater (figure 1.7). Positive feedback therefore usually creates a "vicious cycle," leading away from homeostasis and, in some cases, resulting in death.

Inadequate delivery of blood to cardiac (heart) muscle is an example of positive feedback. Contraction of cardiac muscle generates blood pressure and moves blood through blood vessels to tissues. A system of blood vessels on the

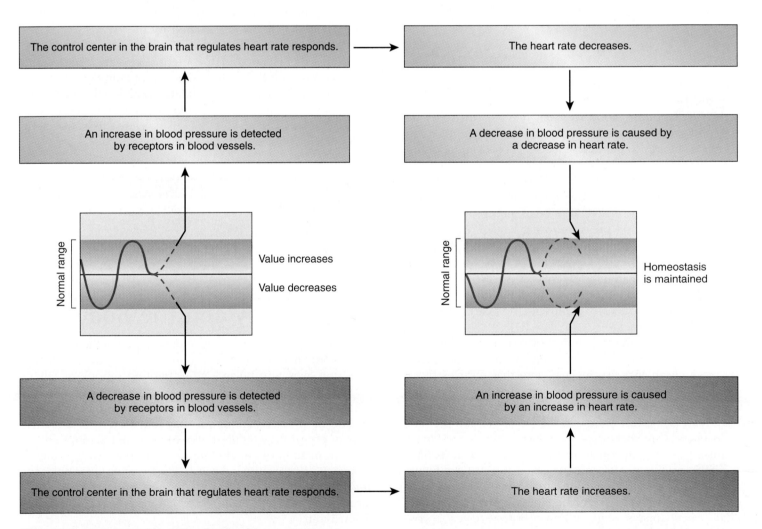

Figure 1.5 Example of Negative Feedback

Blood pressure is maintained within a normal range by negative-feedback mechanisms. An increase in blood pressure is detected by receptors within large blood vessels near the heart. Consequently, regulatory changes are initiated in the brain that cause the heart rate to decrease, resulting in a decrease in blood pressure. When a decrease in blood pressure is detected, an increase in heart rate results, and blood pressure increases.

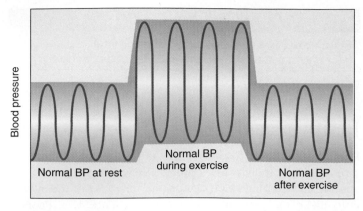

Figure 1.6   Changes in Blood Pressure During Exercise

During exercise the demand for oxygen by muscle tissue increases. To meet this demand, blood pressure (BP) increases, causing an increase in blood flow to the tissues. The increased blood pressure is not an abnormal or nonhomeostatic condition but is a resetting of the normal homeostatic range to meet the increased demand. The reset range is higher and broader than the resting range. After exercise ceases, the range returns to that of the resting condition.

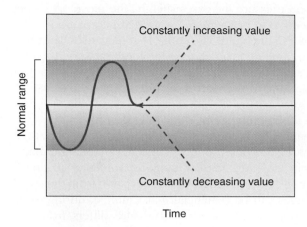

Figure 1.7   Positive Feedback

Deviations from the normal set point value cause an additional deviation away from that value in either a positive or negative direction.

outside of the heart provides cardiac muscle with a blood supply sufficient to allow normal contractions to occur. In effect, the heart pumps blood to itself. Just as with other tissues, blood pressure must be maintained to ensure adequate delivery of blood to cardiac muscle. Following extreme blood loss, blood pressure decreases to the point that delivery of blood to cardiac muscle is inadequate. As a result, cardiac muscle homeostasis is disrupted, and cardiac muscle does not function normally. The heart pumps less blood, which causes the blood pressure to drop even further. This additional decrease in blood pressure means that even less blood is delivered to cardiac muscle, and the heart pumps even less blood, which again decreases the blood pressure (figure 1.8). The process continues until the blood pressure is too low to sustain the cardiac muscle, the heart stops beating, and death results.

Following a moderate amount of blood loss (e.g., after a person donates a pint of blood), negative-feedback mechanisms produce an increase in heart rate that restores blood pressure. If blood loss is severe, however, negative-feedback mechanisms may not be able to maintain homeostasis, and the positive-feedback effect of an ever-decreasing blood pressure can develop. Circumstances in which negative-feedback mechanisms are not adequate to maintain homeostasis illustrate a basic principle. Many disease states result from failure of negative-feedback mechanisms to maintain homeostasis. Medical therapy seeks to overcome illness by aiding negative-feedback mechanisms (e.g., a transfusion reverses a constantly decreasing blood pressure and restores homeostasis).

A few positive-feedback mechanisms do operate in the body under normal conditions, but in all cases they are eventually limited in some way. Birth is an example of a normally

occurring positive-feedback mechanism. Near the end of pregnancy, the uterus is stretched by the baby's larger size. This stretching, especially around the opening of the uterus, stimulates contractions of the uterine muscles. The uterine contractions push the baby against the opening of the uterus, stretching it further. This stimulates additional contractions that result in additional stretching. This positive-feedback sequence ends only when the baby is delivered from the uterus and the stretching stimulus is eliminated.

3   **P R E D I C T**

Is the sensation of thirst associated with a negative- or a positive-feedback mechanism? Explain.

✔ *Answer in Appendix F*

Terminology and the Body Plan

When you first study anatomy and physiology, the number of new words may seem overwhelming. Nonetheless, you will need this new vocabulary. When writing reports or talking with colleagues, you must use correct terminology to avoid confusion and errors. Learning these new words can be easier and more interesting if you pay attention to their derivation, or **etymology** (et'uh-mol'o-jē). Most of the terms are derived from Latin or Greek and are descriptive in the original languages. For example, *foramen* is a Latin word for hole, and *magnum* means large. The foramen magnum is therefore a large hole in the skull through which the spinal cord attaches to the brain.

Words are often modified by adding a prefix or suffix. The suffix "-itis" means an inflammation, so appendicitis is an inflammation of the appendix. As new terms are introduced in this text, their meanings are often explained. The glossary

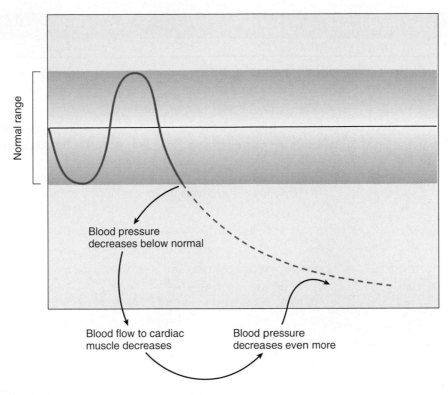

Figure 1.8   Example of Harmful Positive Feedback

A decrease in blood pressure below the normal range causes decreased blood flow to the heart. The heart is unable to pump enough blood to maintain blood pressure, and blood flow to the heart muscle decreases. Thus the ability of the heart to pump decreases further, and blood pressure decreases even more.

and the list of word roots, prefixes, and suffixes on the inside front and back covers of the textbook provide additional information about the new terms.

Directional Terms

When describing parts of the body, it is often important to refer to their relative positions, and directional terms have been developed to facilitate such references. A series of important directional terms is presented in table 1.1. It is important to become familiar with these terms as soon as possible because you will see them repeatedly throughout the text.

Directional terms always refer to the body in the anatomic position (figure 1.9), regardless of its actual position. The **anatomic position** refers to a person standing erect with the feet facing forward, arms hanging to the sides, and palms of the hands facing forward with the thumbs to the outside. Right and left are retained as directional terms in anatomic terminology. Up is replaced by **superior,** down by **inferior,** front by **anterior,** and back by **posterior.**

In humans, superior is synonymous with **cephalic** (se-fal'ik), which means toward the head, because, when we are in the anatomic position, the head is the highest point. In humans, the term inferior is synonymous with **caudal** (kaw'dăl), which means toward the tail, which would be located at the end of the vertebral column if humans had tails. The terms

cephalic and caudal can be used to describe directional movements on the trunk, but they are not used to describe directional movements on the limbs.

The word *anterior* means that which goes before, and **ventral** means belly. The anterior surface of the human body is therefore the ventral surface, or belly, because the belly "goes first" when we are walking. The word *posterior* means that which follows, and **dorsal** means back. The posterior surface of the body is the dorsal surface, or back, which follows as we are walking.

---

4   P R E D I C T

The anatomic position of a cat refers to the animal standing erect on all four limbs and facing forward. On the basis of the etymology of the directional terms, what two terms indicate movement toward the head? What two terms mean movement toward the back? Compare these terms with those referring to a human in the anatomic position.

✔ *Answer in Appendix F*

Proximal means nearest, whereas **distal** means to be distant. These terms are used to refer to linear structures, such as the limbs, in which one end is near some other structure and the other end is farther away. Each limb is attached at its

Table 1.1   Directional Terms for Humans

Term	Etymology*	Definition	Example
Right		Toward the right side of the body	The right ear
Left		Toward the left side of the body	The left eye
Superior	L., higher	A structure above another	The chin is superior to the navel.
Inferior	L., lower	A structure below another	The navel is inferior to the chin.
Cephalic	G. *kephale*, head	Closer to the head than another structure (usually synonymous with superior)	The chin is cephalic to the navel.
Caudal	L. *cauda*, a tail	Closer to the tail than another structure (usually synonymous with inferior)	The navel is caudal to the chin.
Anterior	L., before	The front of the body	The navel is anterior to the spine.
Posterior	L. *posterus*, following	The back of the body	The spine is posterior to the breastbone.
Ventral	L. *ventr-*, belly	Toward the belly (synonymous with anterior)	The navel is ventral to the spine.
Dorsal	L. *dorsum*, back	Toward the back (synonymous with posterior)	The spine is dorsal to the breastbone.
Proximal	L. *proximus*, nearest	Closer to the point of attachment to the body than another structure	The elbow is proximal to the wrist.
Distal	L. *di-* plus *sto*, to stand apart or be distant	Farther from the point of attachment to the body than another structure	The wrist is distal to the elbow.
Lateral	L. *latus*, side	Away from the midline of the body	The nipple is lateral to the breastbone.
Medial	L. *medialis*, middle	Toward the midline of the body	The bridge of the nose is medial to the eye.
Superficial	L. *superficialis*, toward the surface	Toward or on the surface (not shown in figure)	The skin is superficial to muscle.
Deep	O.E. *deop*, deep	Away from the surface, internal (not shown in figure)	The lungs are deep to the ribs.

*Origin and meaning of the word: L., Latin; G., Greek: O.E., Old English.

proximal end to the body, and the distal end, such as the hand, is farther away.

Medial means toward the midline, and **lateral** means away from the midline. The nose is located in a medial position in the face, and the eyes are lateral to the nose. The term **superficial** refers to a structure close to the surface of the body, and **deep** is toward the interior of the body. The skin is superficial to muscle and bone.

Two positional terms used in anatomy should also be mentioned. **Prone** means to lie or be placed with the anterior surface down, and **supine** means to lie or be placed with the anterior surface facing up.

---

5   P R E D I C T

Describe in as many directional terms as you can the relationship between your kneecap and your heel.

✔ *Answer in Appendix F*

▯ Planes

At times it is conceptually useful to describe the body as having imaginary flat surfaces called **planes** passing through it (figure 1.10*a*). A plane divides or sections the body, making it

possible to "look inside" and observe the body's structures. A **sagittal** (saj′i-tăl) plane runs vertically through the body and separates it into right and left portions. The word sagittal literally means "the flight of an arrow" and refers to the way the body would be split by an arrow passing anteriorly to posteriorly. A **midsagittal**, or a **median**, plane divides the body into equal right and left halves, and a **parasagittal** plane runs vertically through the body to one side of the midline. A **transverse**, or **horizontal**, plane runs parallel to the ground and divides the body into superior and inferior portions. A **frontal**, or **coronal** (kōr′ŏ-năl), plane runs vertically from right to left and divides the body into anterior and posterior parts.

Organs are often sectioned to reveal their internal structure (figure 1.11). A cut through the long axis of the organ is a **longitudinal** section, and a cut at right angles to the long axis is a **cross**, or **transverse**, section. If a cut is made across the long axis at other than a right angle, it is called an **oblique** section.

Body Regions

A number of terms are used when referring to different regions or parts of the body (figure 1.12). The upper limb is divided into the arm, forearm, wrist, and hand. The **arm** extends from the shoulder to the elbow, and the **forearm**

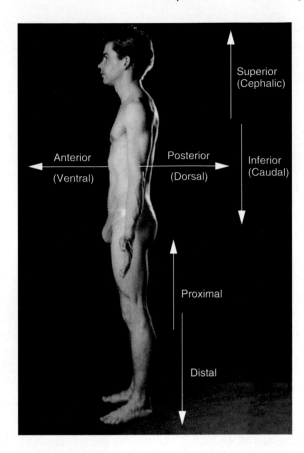

Figure 1.9   Directional Terms

All directional terms are in relation to a person in the anatomic position: a person standing with the feet and palms of the hands facing forward with the thumbs to the outside. ✗

extends from the elbow to the wrist. The lower limb is divided into the thigh, leg, ankle, and foot. The **thigh** extends from the hip to the knee, and the **leg** extends from the knee to the ankle. Note that, contrary to popular usage, the terms arm and leg refer to only a part of the respective limb.

The central region of the body consists of the **head, neck,** and **trunk.** The trunk can be divided into the **thorax** (chest), **abdomen** (region between the thorax and pelvis), and **pelvis** (the inferior end of the trunk associated with the hips).

The abdomen is often subdivided superficially into **quadrants** by two imaginary lines—one horizontal and one vertical—that intersect at the navel (figure 1.13*a*). The quadrants formed are the right-upper, left-upper, right-lower, and left-lower quadrants. In addition to these quadrants, the abdomen is sometimes subdivided into nine **regions** by four imaginary lines: two horizontal and two vertical. These four lines create an imaginary tic-tac-toe figure on the abdomen, resulting in nine regions: epigastric, right and left hypochondriac, umbilical, right and left lumbar, hypogastric, and right and left iliac (figure 1.13*b*). The quadrants or regions are used by clinicians as reference points for locating the underlying organs. For example, the appendix is located in the right-lower quadrant, and the pain of an acute appendicitis is usually felt there.

6   **P R E D I C T**

Using figures 1.2 (p. 6–8) and 1.13, determine in which quadrant each of the following organs is located: spleen, gallbladder, kidneys, most of the stomach, and most of the liver.

✔ *Answer in Appendix F*

Body Cavities

The body contains many cavities, among which are the nasal, cranial, and abdominal cavities. Some of these open to the outside of the body, and some do not. Introductory anatomy and physiology textbooks sometimes describe a dorsal cavity, in which the brain and spinal cord are found, and a ventral body cavity that contains all the trunk cavities. The concept of a dorsal cavity is not described in standard works on anatomy. There are no embryonic, anatomic, or histologic parallels between the fluid-filled space around the central nervous system and the trunk cavities. Discussion in this chapter is therefore limited to the major trunk cavities that do not open to the outside.

The trunk contains three large cavities: the thoracic, the abdominal, and the pelvic (figure 1.14). The **thoracic cavity** is surrounded by the rib cage and separated from the abdominal

Figure 1.10   Planes of Section of the Body

In the center of the illustration, planes of section through the whole body are indicated by "glass" sheets. Actual sections through the head, hip, and abdomen are also shown.

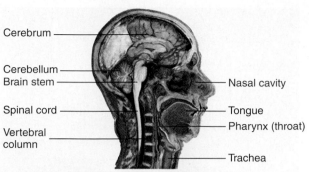

Cerebrum

Cerebellum
Brain stem
 — Nasal cavity

Spinal cord
 — Tongue
 — Pharynx (throat)

Vertebral
column

 — Trachea

Midsagittal section of the head

(b)

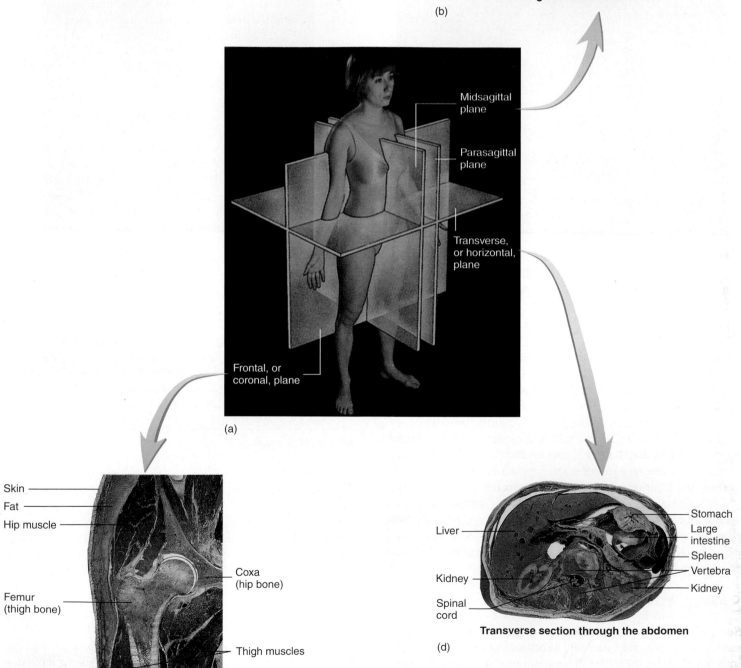

Midsagittal plane

Parasagittal plane

Transverse, or horizontal, plane

Frontal, or coronal, plane

(a)

Skin
Fat
Hip muscle

Femur
(thigh bone)

Coxa
(hip bone)

Thigh muscles

Frontal section through the right hip

(c)

Liver

Kidney

Spinal
cord

Stomach
Large
intestine
Spleen
Vertebra
Kidney

Transverse section through the abdomen

(d)

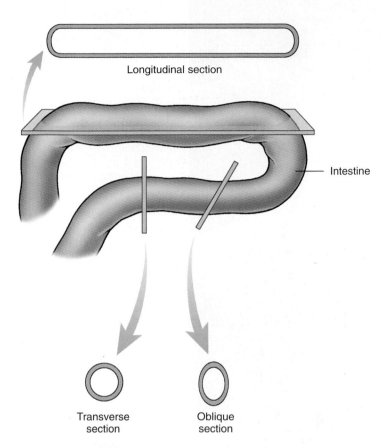

Longitudinal section

Intestine

Transverse
section

Oblique
section

Figure 1.11 Planes of Section Through an Organ

In the center of the illustration, planes of section through the small intestine are indicated by "glass" sheets. The views of the small intestine after sectioning are also shown. Although the small intestine is basically a tube, the sections appear quite different in shape.

cavity by the muscular diaphragm. It is divided into right and left parts by a median structure called the **mediastinum** (me′dē-as-tī′nŭm; middle wall). The mediastinum is a partition containing the heart, thymus gland, trachea, esophagus, and other structures such as blood vessels and nerves. The two lungs are located on either side of the mediastinum.

The **abdominal cavity** is bounded primarily by the abdominal muscles and contains the stomach, intestines, liver, spleen, pancreas, and kidneys. The **pelvic cavity** is a small space enclosed by the bones of the pelvis and contains the urinary bladder, part of the large intestine, and the internal reproductive organs. The abdominal and pelvic cavities are not physically separated and sometimes are called the **abdominopelvic cavity.**

Serous Membranes

Serous (sēr′ŭs) **membranes** cover the organs of the trunk cavities and line the trunk cavities. Imagine an inflated balloon into which a fist has been pushed (figure 1.15). The fist represents an organ, the inner balloon wall in contact with the fist represents the **visceral** (vis′er-ăl; organ) **serous membrane** covering the organ, and the outer part of the balloon wall represents the **parietal** (pă-rī′ĕ-tăl; wall) **serous membrane.**

The cavity or space between the visceral and parietal serous membranes is normally filled with a thin, lubricating film of serous fluid produced by the membranes. As organs rub against the body wall or against another organ, the combination of serous fluid and smooth serous membranes reduces friction. The thoracic cavity contains three serous membrane-lined cavities: a pericardial cavity and two pleural cavities.

The **pericardial** (per-i-kar′dē-ăl; around the heart) **cavity** surrounds the heart (figure 1.16a). The heart is covered by the visceral pericardium and is contained within a connective tissue sac that is lined with the parietal pericardium. The pericardial cavity, which contains pericardial fluid, is located between the visceral and parietal pericardia.

Each lung is surrounded by a **pleural** (plūr′ăl; associated with the ribs) **cavity** and covered by visceral pleura (figure 1.16b). The inner surface of the thoracic wall, the lateral surfaces of the mediastinum, and the superior surface of the diaphragm are lined by the parietal pleura. The pleural cavity is located between the visceral and parietal pleurae and contains pleural fluid.

The abdominopelvic cavity contains a serous membrane-lined cavity called the **peritoneal** (per′i-tō-nē′ăl; to stretch over) **cavity** (figure 1.16c). Many of the organs of the abdominopelvic cavity are covered by visceral peritoneum. The wall of the abdominopelvic cavity and the inferior surface of the diaphragm are lined with parietal peritoneum. The peritoneal cavity is located between the visceral and parietal peritonea and contains peritoneal fluid.

> ### Clinical Note
>
> The serous membranes can become inflamed, usually as a result of an infection. **Pericarditis** (per′i-kar-dī′tis) is inflammation of the pericardium, **pleurisy** (plūr′i-sē) is inflammation of the pleura, and **peritonitis** (per′i-tō-nī′tis) is inflammation of the peritoneum.

The visceral peritoneum of some abdominopelvic organs is connected to the parietal peritoneum on the body wall or to the visceral peritoneum of other abdominopelvic organs by **mesenteries** (mes′en-ter-ēz), which consist of two layers of peritoneum fused together (see figure 1.16c). The mesenteries anchor the organs to the body wall and provide a pathway for nerves and blood vessels to reach the organs. Other abdominopelvic organs are more closely attached to the body wall and do not have mesenteries. They are covered by the parietal peritoneum and are said to be **retroperitoneal** (re′trō-per′i-tō-nē′ăl; behind the peritoneum). The retroperitoneal organs include the kidneys, the adrenal glands, the pancreas, parts of the intestines, and the urinary bladder (see figure 1.16c).

7 **P R E D I C T**

Explain how an organ can be located within the abdominopelvic cavity but not be within the peritoneal cavity.

✔ *Answer in Appendix F*

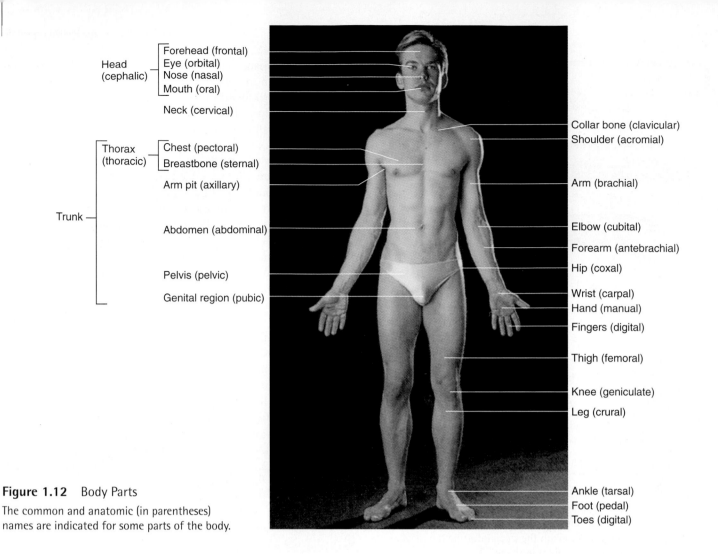

Figure 1.12 Body Parts

The common and anatomic (in parentheses) names are indicated for some parts of the body.

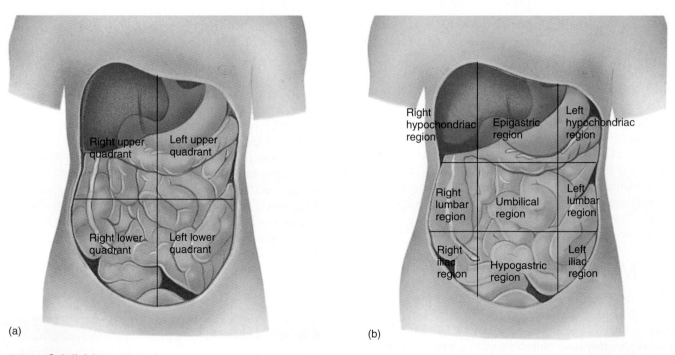

(a)

(b)

Figure 1.13 Subdivisions of the Abdomen

Lines are superimposed over internal organs to demonstrate the relationship of the organs to the subdivisions. (*a*) Abdominal quadrants consist of four subdivisions. (*b*) Abdominal regions consist of nine subdivisions.

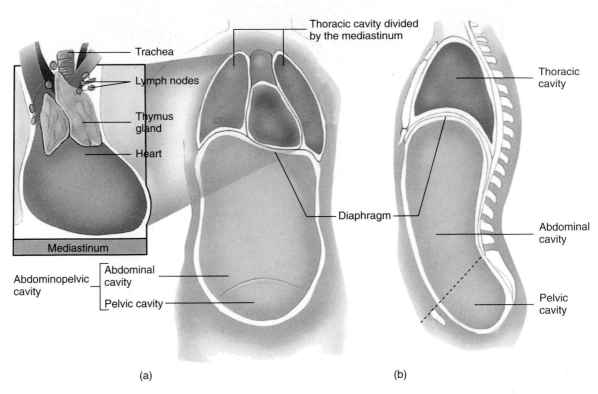

Figure 1.14   Trunk Cavities

(*a*) Anterior view showing the major trunk cavities. The diaphragm separates the thoracic cavity from the abdominal cavity. The mediastinum, which includes the heart, divides the thoracic cavity. (*b*) Sagittal view of trunk cavities. The dashed line shows the division between the abdominal and pelvic cavities. The mediastinum has been removed to show the thoracic cavity.

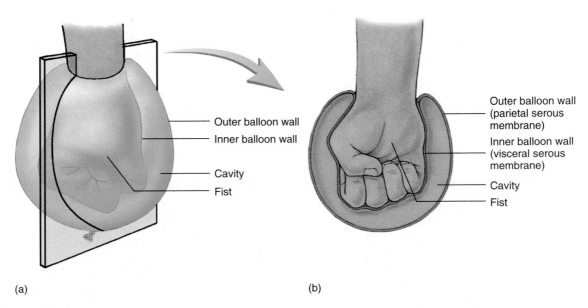

Figure 1.15   Serous Membranes

(*a*) Fist pushing into a balloon. A "glass" sheet indicates the location of a cross section through the balloon. (*b*) Interior view produced by the cross section in (*a*). The fist represents an organ, and the walls of the balloon the serous membranes. The inner wall of the balloon represents a visceral serous membrane in contact with the fist (organ). The outer wall of the balloon represents a parietal serous membrane.

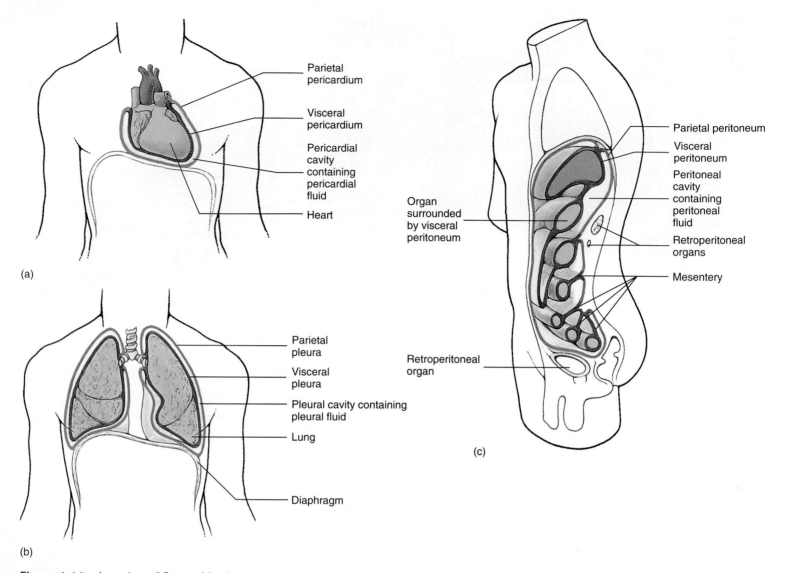

Figure 1.16   Location of Serous Membranes

(*a*) Frontal section showing the pericardial membranes and pericardial cavity. (*b*) Frontal section showing the pleural membranes and pleural cavities. (*c*) Sagittal section through the abdominopelvic cavity showing the peritoneum, peritoneal cavity, mesenteries, and retroperitoneal organs.

Summary

A functional knowledge of anatomy and physiology can be used to solve problems concerning the body when healthy or diseased.

Anatomy

1. Anatomy is the study of the body's structures.
   - Cytology examines cells, and histology examines tissues.
   - Gross anatomy emphasizes organs from a systemic or regional perspective.
2. Surface anatomy uses superficial structures to locate deeper structures, and anatomic imaging is a noninvasive technique for identifying deep structures.

Physiology

1. Physiology is the study of the body's functions.
2. The discipline can be approached from a cellular, an organismal (e.g., human), or systems point of view.

Structural and Functional Organization

1. Basic chemical characteristics are responsible for the structure and functions of life.
2. Organelles are small structures within cells that perform specific functions.
3. Cells are the basic living units of plants and animals and have many common characteristics.

4. Tissues are groups of cells of similar structure and function and their associated extracellular substances. The four primary tissue types are epithelial, connective, muscle, and nervous tissues.
5. Most organs are structures composed of two or more tissues that perform specific functions.
6. Organs are arranged into the 11 organ systems of the human body (see figure 1.2).
7. Organ systems interact to form a whole, functioning organism.

The Human Organism
Characteristics of Life

Humans have many characteristics such as organization, metabolism, responsiveness, growth and development, and reproduction in common with other organisms.

Biomedical Research

Much of what is known about humans is derived from research on other organisms.

Homeostasis

Homeostasis is the condition in which body functions, fluids, and other factors of the internal environment are maintained at levels suitable to support life.

Negative Feedback

1. Negative-feedback mechanisms operate to maintain homeostasis.
2. Many negative-feedback mechanisms consist of a receptor, control center, and effector.

Positive Feedback

1. Positive-feedback mechanisms usually increase deviations from normal.
2. Although a few positive-feedback mechanisms normally exist in the body, most positive-feedback mechanisms are harmful.

Terminology and the Body Plan
Directional Terms

1. A human standing erect with feet facing forward, arms hanging to the side, and palms facing forward is in the anatomic position.
2. Directional terms always refer to the anatomic position, no matter what the actual position of the body (see table 1.1).

Planes

1. Planes of the body
   - A midsagittal, or median, section divides the body into equal left and right parts. A parasagittal section produces unequal left and right parts.
   - A transverse (horizontal) plane divides the body into superior and inferior parts.
   - A frontal (coronal) plane divides the body into anterior and posterior parts.
2. Sections of an organ
   - A longitudinal section of an organ divides it along the long axis.
   - A cross (transverse) section cuts at a right angle to the long axis of an organ.
   - An oblique section cuts across the long axis of an organ at an angle other than a right angle.

Body Regions

1. The body can be divided into the limbs, upper and lower, and a central region consisting of the head, neck, and trunk regions.
2. Superficially the abdomen can be divided into quadrants or nine regions. These divisions are useful for locating internal organs or describing the location of a pain or tumor.

Body Cavities

1. The thoracic cavity is subdivided by the mediastinum.
2. The diaphragm separates the thoracic and abdominal cavities.
3. The pelvic cavity is surrounded by the pelvic bones.

Serous Membranes

1. The trunk cavities are lined by serous membranes. The parietal portion of a serous membrane lines the wall of the cavity, and the visceral portion is in contact with the internal organs.
   - The serous membranes secrete fluid that fills the space between the visceral and parietal membranes. The serous membranes protect organs from friction.
   - The pleural membranes surround the lungs, the pericardial membranes surround the heart, and the peritoneal membranes line the abdominal and pelvic cavities and surround their organs.
2. Mesenteries are parts of the peritoneum that hold the abdominal organs in place and provide a passageway for blood vessels and nerves to the organs.
3. Retroperitoneal organs are located "behind" the parietal peritoneum.

Content Review

1. Why is it important to understand the relationship between structure and function?
2. Define anatomy and physiology.
3. List seven structural levels at which the body can be considered conceptually.
4. Define a tissue. What are the four primary tissue types?
5. Define organ and organ system. List the 11 organ systems of the body and their functions.
6. Describe five characteristics of life.
7. Why is it important to realize that humans share many characteristics with other animals?
8. What is meant by the term homeostasis? If a deviation from homeostasis occurs, what mechanism restores it?
9. Define positive feedback. Why are positive-feedback mechanisms often harmful?
10. Why is knowledge of the etymology of anatomic and physiologic terms useful?
11. What is the anatomic position?

12. List two terms that in humans indicate toward the head. Name two terms that mean the opposite.

13. List two terms that indicate the back in humans. What two terms mean the front?

14. Define the following terms, and give the word that means the opposite: proximal, lateral, and superficial.

15. Define the three planes of the body. What is the difference between a parasagittal section and a midsagittal section?

16. What is the difference between the arm and the upper limb and the difference between the leg and the lower limb?

17. Describe the quadrant and the nine-region methods of subdividing the abdominal region. What is the purpose of these divisions?

18. Define the thoracic, abdominal, and pelvic cavities. What is the mediastinum?

19. Differentiate between parietal and visceral serous membranes. What is the function of the serous membranes?

20. Name the serous membranes lining each of the body cavities.

21. What are mesenteries? Explain their function.

22. What are retroperitoneal organs? List four examples.

Develop Your Reasoning Skills

1. Exposure to a hot environment causes the body to sweat. The hotter the environment, the greater the sweating. Two anatomy and physiology students are arguing about the mechanisms involved: Student A claims that they are positive feedback, and student B claims they are negative feedback. Do you agree with student A or student B and why?

2. The following observations were made on a patient who had suffered a bullet wound: Heart rate elevated and rising. Blood pressure very low and dropping. After bleeding was stopped and a blood transfusion was given, blood pressure increased. Which of the following statements is (are) consistent with these observations?
   a. Negative-feedback mechanisms are occasionally inadequate without medical intervention.
   b. The transfusion interrupted a positive-feedback mechanism.
   c. The transfusion interrupted a negative-feedback mechanism.
   d. The transfusion was not necessary.
   e. a and b

3. Provide the correct directional term for the following statement: When a boy is standing on his head, his nose is _____ to his mouth.

4. Complete the following statements, using the correct directional terms for a human being. Note that more than one term can apply.
   a. The navel is _____ to the nose.
   b. The nipple is _____ to the lung.
   c. The arm is _____ to the forearm.
   d. The little finger is _____ to the index finger.

5. The esophagus is a muscular tube that connects the pharynx (throat) to the stomach. In which quadrant and region is the esophagus located? In which quadrant and region is the urinary bladder located?

6. Given the following procedures:
   1. Make an opening into the mediastinum.
   2. Lay the patient on his back.
   3. Lay the patient face down.
   4. Make an incision through the pericardial serous membranes.
   5. Make an opening into the abdomen.
   Which of the procedures should be accomplished to expose the anterior surface of a patient's heart?
   a. 2, 1, 4
   b. 2, 5, 4
   c. 3, 1, 4
   d. 3, 5, 4

7. During pregnancy, which of the mother's body cavities increases most in size?

8. A bullet enters the left side of a man, passes through the left lung, and lodges in the heart. Name in order the serous membranes and their cavities through which the bullet passes.

9. A woman falls while skiing and accidentally is impaled by her ski pole. The pole passes through the abdominal body wall and into and through the stomach, pierces the diaphragm, and finally stops in the left lung. List in order the serous membranes the pole pierces.

Web Site Link

For a listing of the most current web sites related to this chapter, please visit the Seeley home page at:
http://www.mhhe.com/biosci/ap/seeleyap/

The Chemical Basis of Life

Objectives

1. Define the terms matter, mass, weight, element, and atom.

2. Describe the subatomic particles of an atom using the terms atomic number, mass number, and electron cloud.

3. Define isotope, atomic mass, mole, and molar mass.

4. Explain ionic bonding, covalent bonding, and metallic bonding. Describe polar covalent bonds and hydrogen bonds.

5. Distinguish between a molecule and a compound. Define dissociate and electrolyte.

6. Describe and give an example of synthesis, decomposition, dehydration, hydrolysis, oxidation–reduction, and reversible reactions.

7. List the factors that affect the rate of chemical reactions.

8. Define potential and kinetic energy. Describe electric, electromagnetic, chemical, mechanical, and heat energy.

9. Describe the chemical potential energy changes that take place in metabolism and photosynthesis. Use adenosine triphosphate (ATP) as an example.

10. List the properties of water that make it important for living organisms.

11. Define solution, solvent, and solute. Describe osmolality and milliosmole.

12. Define acid and base, and differentiate between a strong acid or base and a weak acid or base.

13. Describe the pH scale, and define salt and buffer.

14. Explain the importance of oxygen and carbon dioxide to living organisms.

15. Describe the chemical structure of carbohydrates, and state the role of carbohydrates in the body.

16. List and describe the importance of the major types of lipids.

17. Describe the different structural levels of proteins. Define enzymes and state their functions.

18. Contrast the structure and function of deoxyribonucleic acid (DNA) and ribonucleic acid (RNA).

19. Explain the function of ATP.

A basic knowledge of chemistry—the scientific discipline that deals with the composition and structure of substances and with the reactions they undergo—is essential for understanding anatomy and physiology, because all of the structures of the body are composed of chemicals and all of the functions of the body result from chemical reactions. For example, the physiologic processes of digestion, muscle contraction, and metabolism and the generation of nerve impulses can all be described in chemical terms. Many abnormal conditions and their treatments can also be explained in chemical terms, even though their symptoms appear as malfunctions in specific organ systems. For example, Parkinson's disease, which has the symptom of uncontrolled shaking movements, results from a shortage of a chemical called dopamine in certain nerve cells of the brain. It is treated by giving patients another chemical that is converted to dopamine by brain cells.

This chapter outlines some basic chemical principles and emphasizes the relationship of these principles to living organisms. It is not a comprehensive review of chemistry, but it does review some of the basic chemical principles that make anatomy and physiology more understandable. You should refer to this chapter when chemical phenomena are discussed later in the text.

Basic Chemistry
Matter, Mass, and Weight

All living and nonliving things are composed of **matter,** which is anything that occupies space and has mass. **Mass** is the amount of matter in an object, and **weight** is the gravitational force acting on an object of a given mass. For example, the weight of an apple results from the force of gravity "pulling" on the apple's mass.

1 ▌ P R E D I C T

The difference between mass and weight can be illustrated by considering an astronaut. How does an astronaut's mass and weight in outer space compare with his mass and weight on the earth's surface?

✔ *Answer in Appendix F*

The international unit for mass is the **kilogram (kg),** which is the mass of a platinum–iridium cylinder kept at the International Bureau of Weights and Measurements in France. The mass of all other objects is compared with this cylinder. For example, the mass of your textbook is approximately 2.3 kg. An object with 1/1000 the mass of a kilogram is defined to have a mass of 1 **gram (g).**

Chemists use a balance to determine the mass of objects. Although we commonly refer to weighing an object on a balance, we actually are "massing" the object. For example, a mechanical balance compares an unknown's mass with objects of known mass. A balance produces the same results on a mountaintop as at sea level.

Elements and Atoms

An **element** is the simplest type of matter with unique chemical properties. To date, 112 elements are known. A list of the elements commonly found in the human body is given in table 2.1. About 96% of the weight of the body results from the elements oxygen, carbon, hydrogen, and nitrogen.

An **atom** is the smallest particle of an element that has the chemical characteristics of that element. An element is composed of atoms of only one kind. For example, the element carbon is composed of only carbon atoms, and the element oxygen is composed of only oxygen atoms.

An element, or an atom of that element, often is represented by a symbol. Usually the first letter or letters of the element's name are used—for example, C for carbon, H for hydrogen, Ca for calcium, and Cl for chlorine. Occasionally the symbol is taken from the Latin, Greek, or Arabic name for the element—for example, Na from the Latin word *natrium* is the symbol for sodium.

▊ Atomic Structure

The characteristics of living and nonliving matter result from the structure, organization, and behavior of atoms. Neutrons, protons, and electrons are the three major types of subatomic particles that form atoms. **Neutrons** have no electric charge, **protons** have positive charges, and **electrons** have negative charges. The positive charge of a proton is equal in magnitude to the negative charge of an electron. Because equal numbers of protons and electrons occur in an atom, the individual charges cancel each other, and the atom is electrically neutral.

Protons and neutrons form the **nucleus,** and electrons are located around the nucleus (figure 2.1*a*). The nucleus accounts for 99.97% of an atom's mass, but only 1 ten-trillionth of its volume. Most of the volume of an atom is occupied by the electrons. Although it is impossible to know precisely where any given electron is located at any particular moment, the region where it is most likely to be found can be represented by an **electron-density diagram** (figure 2.1*b*). The likelihood of locating an electron at a specific point in a region correlates with the darkness of that region in the diagram. The greater the number of dots, the greater the likelihood of finding the electron there at any given moment. Electron-density diagrams are sometimes called **electron "clouds."**

Atomic Number and Mass Number

The **atomic number** of an element is equal to the number of protons in each atom, and because the number of electrons and protons is equal, the atomic number also indicates the

Table 2.1   Some Common Elements

Element	Symbol	Atomic Number	Mass Number	Atomic Mass	Percent in Human Body by Weight (%)	Percent in Human Body by Number of Atoms (%)
Hydrogen	H	1	1	1.008	9.5	63.0
Carbon	C	6	12	12.01	18.5	9.5
Nitrogen	N	7	14	14.01	3.3	1.4
Oxygen	O	8	16	16.00	65.0	25.5
Fluorine	F	9	19	19.00	Trace	Trace
Sodium	Na	11	23	22.99	0.2	0.3
Magnesium	Mg	12	24	24.31	0.1	0.1
Phosphorus	P	15	31	30.97	1.0	0.22
Sulfur	S	16	32	32.07	0.3	0.05
Chlorine	Cl	17	35	35.45	0.2	0.03
Potassium	K	19	39	39.10	0.4	0.06
Calcium	Ca	20	40	40.08	1.5	0.31
Chromium	Cr	24	52	51.00	Trace	Trace
Manganese	Mn	25	55	54.94	Trace	Trace
Iron	Fe	26	56	55.85	Trace	Trace
Cobalt	Co	27	59	58.93	Trace	Trace
Copper	Cu	29	63	63.55	Trace	Trace
Zinc	Zn	30	64	65.39	Trace	Trace
Selenium	Se	34	80	78.96	Trace	Trace
Molybdenum	Mo	42	98	95.94	Trace	Trace
Iodine	I	53	127	126.9	Trace	Trace

number of electrons. Each element is uniquely defined by the number of protons in the atoms of that element. For example, only hydrogen atoms have one proton, only carbon atoms have six protons, and only oxygen atoms have eight protons (figure 2.2; see table 2.1).

Scientists have been able to create new elements by changing the number of protons in the nuclei of existing elements. Protons, neutrons, or electrons from one atom are accelerated to very high speeds and then smashed into the nucleus of another atom. The resulting changes in the nucleus produces a new element with a new atomic number. To date, 20 elements with an atomic number greater than 92 have been synthesized in this fashion. These artificially produced elements are usually unstable, and they quickly convert back to more stable elements.

Protons and neutrons have about the same mass, and they are responsible for most of the mass of atoms. Electrons, on the other hand, have very little mass. The **mass number** of an element is the number of protons plus the number of neutrons in each atom. For example, the mass number for carbon is 12 because it has six protons and six neutrons.

2   P R E D I C T

The atomic number of potassium is 19, and the mass number is 39. What is the number of protons, neutrons, and electrons in an atom of potassium?

✔ *Answer in Appendix F*

Isotopes and Atomic Mass

Isotopes (ī′sō-tōpz) are two or more forms of the same element that have the same number of protons and electrons but a different number of neutrons. Thus isotopes have the same atomic number but different mass numbers. For example, there are three isotopes of hydrogen: hydrogen, deuterium, and tritium. All three isotopes have one proton and one electron, but hydrogen has no neutrons in its nucleus, deuterium has one neutron, and tritium has two neutrons (figure 2.3). Isotopes can be denoted using the symbol of the element preceded by the mass number (number of protons and neutrons) of the isotope. Thus hydrogen is ^{1}H, deuterium is ^{2}H, and tritium is ^{3}H.

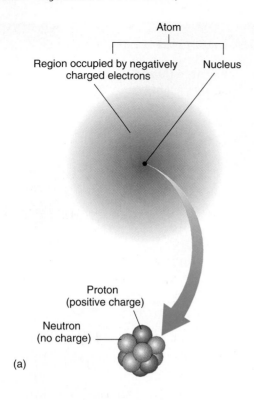

Atom

Region occupied by negatively   Nucleus
charged electrons

Proton
(positive charge)

Neutron
(no charge)

(a)

Nucleus

Electron
density
diagram

(b)

Figure 2.1   Features of an Atom

(a) Most of the volume of an atom is occupied by rapidly moving, negatively charged electrons. The tiny, dense nucleus consists of positively charged protons and uncharged neutrons. (b) Electron-density diagram, or electron cloud, showing probable location of an electron relative to the nucleus.

Individual atoms have very little mass. For example, a hydrogen atom has a mass of 1.67×10^{-24} g (see appendix B for an explanation of the scientific notation of numbers). To avoid using such small numbers, a system of relative atomic mass is used. In this system, a **unified atomic mass unit (u),** or **dalton (D),** is 1/12 the mass of ^{12}C, a carbon atom with six protons and six neutrons. Thus ^{12}C has an atomic mass of exactly 12 u. A naturally occurring sample of carbon, however, contains mostly ^{12}C but also a small quantity of other carbon isotopes such as ^{13}C, which has six protons and seven neutrons. The **atomic mass** of an element is the *average* mass of its naturally occurring isotopes, taking into account the relative abundance of each isotope. For example, the atomic mass of the element carbon is 12.01 u (see table 2.1), which is slightly more than 12 u because of the additional mass of the small amount of other carbon isotopes. Because the atomic mass is an average, a sample of carbon can be treated as if all the carbon atoms have an atomic mass of 12.01 u.

The Mole and Molar Mass

Avogadro's number is the number of atoms in exactly 12 g of ^{12}C. This enormous number is 6.022×10^{23}. A **mole** of a substance contains Avogadro's number of entities, such as atoms. The **molar mass** of a substance is the mass of one mole of the substance expressed in grams. Because 12 g of ^{12}C is used as the standard, the atomic mass of an entity expressed in unified atomic mass units is the same as the molar mass expressed in grams. Thus, carbon atoms have an atomic mass of 12.01 u, and 12.01 g of carbon have Avogadro's number (1 mol) of carbon atoms.

Just as a grocer sells eggs in lots of 12 (a dozen), a chemist groups atoms in lots of 6.022×10^{23} (Avogadro's number, 1 mol). The molar mass is used as a convenient way to determine the number of atoms in a sample of an element. For example, 1.008 g of hydrogen (1 mol) has the same number of atoms as 12.01 g of carbon (1 mol).

3	P R E D I C T

Are there more or fewer atoms in 12.01 g of carbon than in 12.01 g of magnesium? (*Hint:* Use table 2.1.)

✔ *Answer in Appendix F*

▌Electrons and Chemical Bonding

The chemical behavior of an atom is determined largely by its outermost electrons. **Chemical bonding** occurs when the outermost electrons are transferred or shared between atoms. Chemical bonding can be grouped into three categories: ionic, covalent, and metallic bonding.

Ionic Bonding

An atom is electrically neutral because it has an equal number of protons and electrons. If an atom loses or gains electrons, the number of protons and electrons are no longer equal, and a charged particle called an **ion** (ī′on) is formed. After an atom loses an electron, it has one more proton than it has electrons and is positively charged. For example, a sodium atom (Na) can lose an electron to become a positively charged sodium ion (Na^+) (figure 2.4a). After an atom gains an electron, it has one more electron than it has protons and is negatively charged. For example, a chlorine atom (Cl) can accept an electron to become a negatively charged chloride ion (Cl^-).

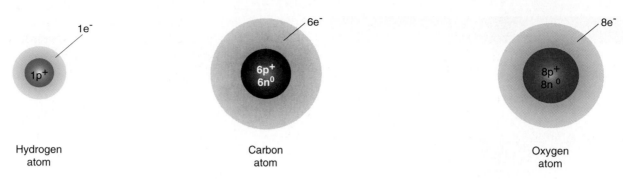

Figure 2.2   Hydrogen, Carbon, and Oxygen Atoms

Within the nucleus, the number of positively charged protons (p^+) and uncharged neutrons (n^0) is indicated. The negatively charged electrons (e^-) are around the nucleus. Atoms are electrically neutral because the number of protons and electrons within an atom are equal.

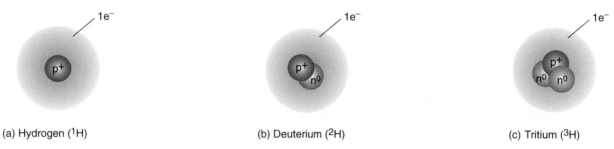

Figure 2.3   Isotopes of Hydrogen

(a) Hydrogen has one proton and no neutrons in its nucleus. (b) Deuterium has one proton and one neutron in its nucleus. (c) Tritium has one proton and two neutrons in its nucleus.

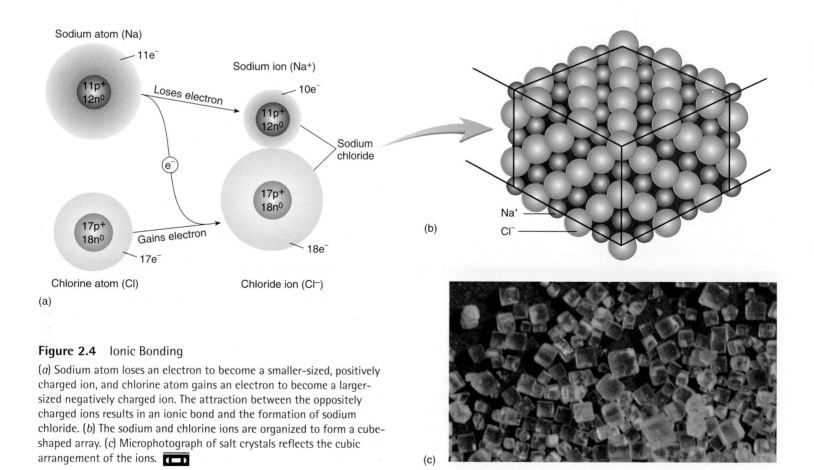

Figure 2.4   Ionic Bonding

(a) Sodium atom loses an electron to become a smaller-sized, positively charged ion, and chlorine atom gains an electron to become a larger-sized negatively charged ion. The attraction between the oppositely charged ions results in an ionic bond and the formation of sodium chloride. (b) The sodium and chlorine ions are organized to form a cube-shaped array. (c) Microphotograph of salt crystals reflects the cubic arrangement of the ions.

Table 2.2 Important Ions		
Common Ions	Symbols	Functions
Calcium	Ca^{2+}	Bones, teeth, blood clotting, muscle contraction, release of neurotransmitters
Sodium	Na^+	Membrane potentials, water balance
Potassium	K^+	Membrane potentials
Hydrogen	H^+	Acid–base balance
Hydroxide	OH^-	Acid–base balance
Chloride	Cl^-	Water balance
Bicarbonate	HCO_3^-	Acid–base balance
Ammonium	NH_4^+	Acid–base balance
Phosphate	PO_4^{3-}	Bone, teeth, energy exchange, acid–base balance
Iron	Fe^{2+}	Red blood cell formation
Magnesium	Mg^{2+}	Necessary for enzymes
Iodide	I^-	Present in thyroid hormones

No interaction between the two hydrogen atoms because they are too far apart.

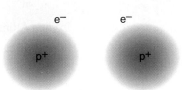

The positively charged nucleus of each hydrogen atom begins to attract the electron of the other.

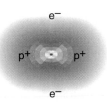

A covalent bond is formed when the electrons are shared between the nuclei because the electrons are equally attracted to each nucleus.

Figure 2.5   Covalent Bonding

Positively charged ions are called **cations** (kat′ī-onz), and negatively charged ions are called **anions** (an′ī-onz). Because oppositely charged ions are attracted to each other, cations and anions tend to remain close together, which is called **ionic** (ī-on′ik) **bonding.** For example, sodium and chloride ions are held together by ionic bonding to form an array of ions called sodium chloride, or table salt (see figures 2.4b and c). Some ions commonly found in the body are listed in table 2.2.

Covalent Bonding

Covalent bonding results when atoms share one or more pairs of electrons. The resulting combination of atoms is called a molecule. An example is the covalent bond between two hydrogen atoms to form a hydrogen molecule (figure 2.5). Each hydrogen atom has one electron. As the atoms get closer together, the positively charged nucleus of each atom begins to attract the electron of the other atom. At an optimal distance, the two nuclei mutually attract the two electrons, and each electron is shared by both nuclei. The two hydrogen atoms are now held together by a covalent bond.

When an electron pair is shared between two atoms, a **single covalent bond** results. A single covalent bond can be represented by a single line between the symbols of the atoms involved (for example, H—H). A **double covalent bond** results when two atoms share four electrons, two from each atom. When a carbon atom combines with two oxygen atoms to form carbon dioxide, two double covalent bonds are formed. Double covalent bonds are indicated by a double line between the atoms (O═C═O).

When electrons are shared equally between atoms, as in a hydrogen molecule, the bonds are called **nonpolar covalent bonds.** Atoms bound to one another by a covalent bond do not always share their electrons equally, however, because the nucleus of one atom attracts the electrons more strongly than does the nucleus of the other atom. Bonds of this type are called **polar covalent bonds** and are common in both living and nonliving matter.

Polar covalent bonds can result in polar molecules, which are electrically asymmetric. For example, oxygen atoms attract electrons more strongly than do hydrogen atoms. When an oxygen atom and two hydrogen atoms are bound by covalent bonds to form a water molecule, the electrons are located in the vicinity of the oxygen nucleus more than in the vicinity of the hydrogen nuclei. Because electrons have a negative charge, the oxygen side of the molecule is slightly more negative than the hydrogen side (figure 2.6).

Metallic Bonding

In covalent and ionic bonding, electrons are shared, or transferred, between two atoms. In **metallic bonding,** the outermost electrons of atoms are shared equally among all the atoms in the sample. The electrons form a "sea" of electrons, in which the electrons move freely between the nuclei of all

Figure 2.6   Polar Covalent Bonds

(*a*) Water molecule formed when two hydrogen atoms form covalent bonds with an oxygen atom. (*b*) Electron pairs (indicated by dots) are shared between the hydrogen atoms and oxygen. The electrons are shared unequally, as shown by the electron cloud (*yellow*) not coinciding with the dashed outline. Consequently, the oxygen side of the molecule has a slight negative charge (indicated by δ⁻) and the hydrogen side of the molecule has a slight positive charge (indicated by δ⁺).

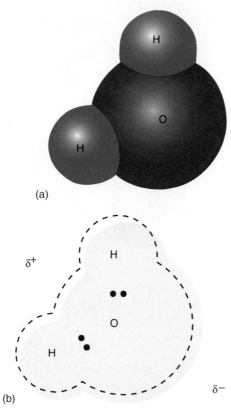

the atoms. The positively charged nuclei are surrounded and held in place by the negatively charged electrons. Metals are excellent conductors of electricity because their electrons move freely between nuclei.

Molecules and Compounds

A **molecule** is formed when two or more atoms chemically combine to form a structure that behaves as an independent unit. The resulting combination can be represented by a **molecular formula,** which consists of the symbols of the atoms forming the molecule plus subscripts denoting the number of each type of atom. For example, the chemical formula for glucose (a sugar) is $C_6H_{12}O_6$, indicating that glucose has 6 carbon, 12 hydrogen, and 6 oxygen atoms (table 2.3).

Most covalent substances consist of molecules because their atoms form distinct units as a result of the joining of the atoms to each other by a pair of shared electrons. The atoms that combine to form a molecule can be of the same type, such as two hydrogen atoms combining to form a hydrogen molecule. More typically, a molecule consists of two or more different types of atoms, such as two hydrogen atoms and an oxygen atom forming water. Thus, a glass of water consists of a collection of individual water molecules positioned next to one another.

Table 2.3   Picturing Molecules

Representation	Hydrogen	Carbon Dioxide	Glucose
Chemical Formula Shows the kind and number of atoms present	H_2	CO_2	$C_6H_{12}O_6$
Electron-dot Formula The bonding electrons are shown as dots between the symbols of the atoms	H:H Single covalent bond	O::C::O Double covalent bond	Not used for complex molecules
Bond-line Formula The bonding electrons are shown as lines between the symbols of the atoms	H—H Single covalent bond	O=C=O Double covalent bond	CH₂OH, O, OH, HO, OH, OH
Models Atoms are shown as different sized and colored spheres	Hydrogen atom	Oxygen atom Carbon atom	

A **compound** is a substance composed of two or more *different* types of atoms that are chemically combined. A hydrogen molecule is not a compound because it does not consist of different types of atoms. Many molecules are compounds, however. For example, a water molecule is a covalent compound.

On the other hand, ionic compounds are not molecules because the ions are held together by the force of attraction between opposite charges. A piece of sodium chloride does not consist of individual sodium chloride molecules positioned next to each other. Instead, table salt is an organized array of individual sodium and individual chloride ions in which each charged ion is surrounded by several ions of the opposite charge (see figure 2.4b). Sodium chloride is an example of a substance that is a compound but is not a molecule.

The **formula unit** is used to describe the relative number of cations and anions in an ionic compound. For example, sodium chloride is represented as NaCl because it is an array of ions in which one Na^+ ion occurs for each Cl^- ion.

The molecular formula and formula unit can be used to determine the **molecular mass** of a molecule or compound. The molecular mass is the sum of the atomic masses of the atoms or ions in the molecular formula or formula unit. The term molecular mass is used for convenience for ionic compounds, even though ions are not molecules. For example, the atomic mass of sodium is 22.99 and chloride is 35.45. The molecular mass of NaCl is therefore 58.44 (22.99 + 35.45).

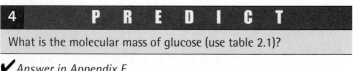

4   P R E D I C T

What is the molecular mass of glucose (use table 2.1)?

✔ *Answer in Appendix F*

Intermolecular Forces

Intermolecular forces result from the weak electrostatic attractions between the oppositely charged parts of different molecules, or between ions and molecules. Intermolecular forces are much weaker than the forces that produce chemical bonding.

Hydrogen Bonds

Molecules with polar covalent bonds have positive and negative "ends." Intermolecular force results from the attraction of the positive end of one polar molecule to the negative end of another polar molecule. When hydrogen forms a covalent bond with oxygen, nitrogen, or fluorine, the resulting molecule becomes very polarized. If the positively charged hydrogen of one molecule is attracted to the negatively charged oxygen, nitrogen, or fluorine of another molecule, a **hydrogen bond** is formed. For example, the positively charged hy-

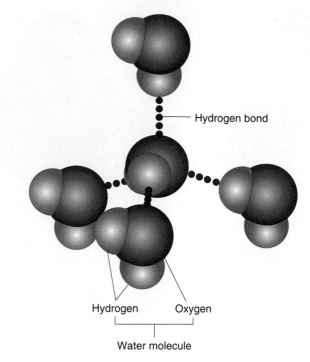

Figure 2.7   Hydrogen Bonds

The positive hydrogen part of one water molecule forms a hydrogen bond (*red dotted line*) with the negative oxygen part of another water molecule. As a result, hydrogen bonds hold the water molecules together.

drogen atoms of a water molecule form hydrogen bonds with the negatively charged oxygen atoms of other water molecules (figure 2.7).

Hydrogen bonds play an important role in determining the shape of complex molecules because the hydrogen bonds between different polar parts of the molecule hold the molecule in its normal three-dimensional shape (see the sections Proteins and Nucleic Acids: DNA and RNA later in this chapter).

Table 2.4 summarizes the important characteristics of chemical bonding (ionic, covalent, and metallic) and intermolecular forces (hydrogen bonds).

Solubility and Dissociation

The ability of one substance to dissolve in another is called **solubility.** Charged and polar substances, such as sodium chloride and glucose, dissolve in water readily, whereas nonpolar substances such as oils do not. Substances dissolve in water when they become surrounded by water molecules. If the positive and negative ends of the water molecules are attracted more to the charged ends of other molecules than they are to each other, the hydrogen bonds between the ends of the water molecules are broken, and the water molecules surround the other molecules, which become dissolved in the water.

Table 2.4   Comparison of Bonds

Definition	Charge Distribution	Example
Ionic Bond		
Complete transfer of electrons between two atoms	Separate positively charged and negatively charged ions	$Na^+ \; Cl^-$ Sodium chloride
Polar Covalent Bond		
Unequal sharing of electrons between two atoms	Slight positive charge (δ^+) on one side of the molecule and slight negative charge (δ^-) on the other side of the molecule	H δ^+ Oδ^- H Water
Nonpolar Covalent Bond		
Equal sharing of electrons between two atoms	Charge evenly distributed among the atoms of the molecule	H—C—H (with H above and below) Methane
Metallic Bond		
Equal sharing of electrons among all the atoms in a sample	Charge evenly distributed among all the atoms of a sample	Array of atoms in a sample of gold
Hydrogen Bond		
Attraction of oppositely charged ends of one polar molecule to another polar molecule	Charge distribution within the polar molecules results from polar covalent bonds	O----H—O (with H atoms) Water molecules

When ionic compounds dissolve in water, their ions **dissociate,** or separate, from one another because the cations are attracted to the negative ends of the water molecules, and the anions are attracted to the positive ends of the water molecules. For example, when sodium chloride dissociates in water, the sodium and chloride ions separate, and water molecules surround and isolate the ions, keeping them in solution (figure 2.8). Note that there are no sodium chloride molecules in a salt solution.

When molecules (covalent compounds) dissolve in water, the molecules usually remain intact even though they are surrounded by water molecules. Thus, in a glucose solution, glucose molecules are surrounded by water molecules. Parts of some molecules, however, can dissociate in water. These molecules are called acids (see section on Acids and Bases).

Cations and anions that dissociate in water are sometimes called **electrolytes** (ē-lek′trō-lītz) because they have the capacity to conduct an electric current, which is the flow of charged particles. An electrocardiogram (ECG) is a recording of electric currents produced by the heart. These currents can be detected by electrodes on the surface of the body be-

cause the ions in the body fluids conduct electric currents. Molecules that do not dissociate form solutions that do not conduct electricity and are called **nonelectrolytes.**

Chemical Reactions

In a **chemical reaction,** atoms, ions, molecules, or compounds interact either to form or to break chemical bonds. The substances that enter into a chemical reaction are called the **reactants,** and the substances that result from the chemical reaction are called the **products.**

For our purposes, three important points can be made about chemical reactions. First, in some reactions, less complex reactants are combined to form a larger, more complex product. An example is the synthesis of the complex molecules of the human body from basic "building blocks" obtained in food (figure 2.9*a*). Second, in other reactions, a reactant can be broken down, or decomposed, into simpler, less complex products. An example is the breakdown of food molecules into basic building blocks. (figure 2.9*b*). Third, atoms are generally associated with other atoms through chemical

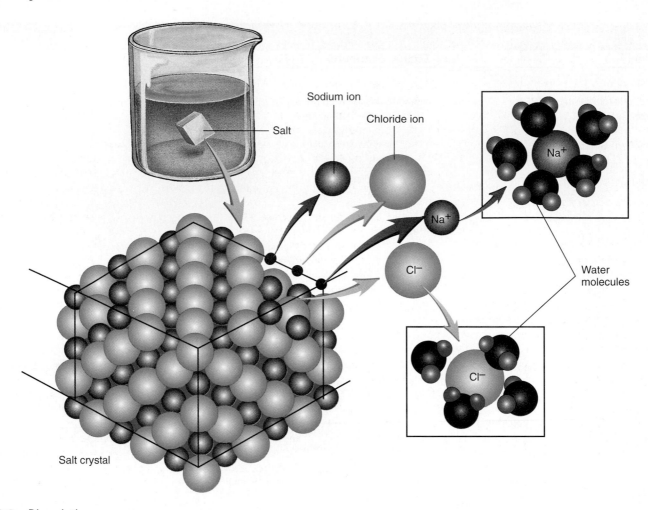

Figure 2.8 Dissociation

Sodium chloride (table salt) dissociating in water. The positively charged sodium ions (Na^+) are attracted to the negative oxygen (*red*) end of the water molecule, and the negatively charged chlorine ions (Cl^-) are attracted to the positively charged hydrogen (*blue*) end of the water molecule.

bonding or intermolecular forces; therefore, to synthesize new products or break down reactants it is necessary to change the relationship between atoms.

Synthesis Reactions

When two or more reactants chemically combine to form a new and larger product, the process is called a **synthesis reaction.** An example of a synthesis reaction is the combination of two amino acids to form a dipeptide (figure 2.10*a*). In this particular synthesis reaction, water is removed from the amino acids as they are bound together. Synthesis reactions in which water is a product are called **dehydration** (water out) **reactions.** Note that old chemical bonds are broken and new chemical bonds are formed as the atoms rearrange as a result of the synthesis reaction.

Another example of a synthesis reaction in the body is the formation of adenosine triphosphate (ATP). In ATP, A stands for adenosine, T stands for tri- or three, and P stands for phosphate group (PO_4^{3-}). Thus, ATP consists of adenosine and three phos-

phate groups (see the end of this chapter for the details of the structure of ATP). ATP is synthesized from adenosine diphosphate (ADP), which has two phosphate groups, and an inorganic phosphate ($H_2PO_4^-$), which is often symbolized as P_i.

$$
\begin{array}{ccc}
\text{A-P-P} \;+\; & P_i & \rightarrow \;\; \text{A-P-P-P} \\
\text{(ADP)} & \text{(Inorganic} & \text{(ATP)} \\
 & \text{phosphate)} &
\end{array}
$$

The molecules characteristic of life, such as ATP, proteins, carbohydrates, lipids, and nucleic acids, are produced by synthesis reactions. All of the synthesis reactions that occur within the body are referred to collectively as **anabolism** (ă-nab′ō-lizm). The growth, maintenance, and repair of the body could not take place without anabolic reactions.

Decomposition Reactions

The term *decompose* means to break down into smaller parts. A **decomposition reaction** is the reverse of a synthesis reaction—a larger reactant is chemically broken down into two

Clinical Focus   Radioactive Isotopes and X-rays

Protons, neutrons, and electrons are responsible for the chemical properties of atoms. They also have other properties that can be useful in a clinical setting. For example, they have been used to develop methods for examining the inside of the body.

Radioactive isotopes have unstable nuclei that spontaneously change to form more stable nuclei. As a result, either new isotopes or new elements are produced. In this process of nuclear change, alpha particles, beta particles, and gamma rays are emitted from the nuclei of radioactive isotopes. Alpha (α) particles are positively charged helium ions (He^{2+}), which consist of two protons and two neutrons. Beta (β) particles are electrons formed as neutrons change into protons. The electrons are ejected from the nuclei, and the protons remain in the nuclei. Gamma (γ) rays are a form of electromagnetic radiation (high-energy photons) released from nuclei as they lose energy.

All isotopes of an element have the same atomic number, and their chemical behavior is very similar. For example, ^3H (tritium) can substitute for ^1H (hydrogen), and either ^{125}iodine or ^{131}iodine can substitute for ^{126}iodine in chemical reactions.

Radioactive isotopes are commonly used by clinicians and researchers because sensitive measuring devices can detect their radioactivity, even when they are present in very small amounts. Several procedures that are used to determine the concentration of substances such as hormones depend on the incorporation of small amounts of radioactive isotopes, such as ^{125}iodine, into the substances being measured. Disorders of the thyroid gland, the adrenal gland, and the reproductive organs can be more accurately diagnosed using these procedures.

Radioactive isotopes are also used to treat cancer. Some of the particles released from isotopes have a very high energy content and can penetrate and destroy tissues. Thus radioactive isotopes can be used to destroy tumors because rapidly growing tissues such as tumors are more sensitive to radiation than healthy cells. Radiation can also be used to sterilize materials that cannot be exposed to high temperatures (e.g., some fabric and plastic items used during surgical procedures). In addition, radioactive emissions can be used to sterilize food and other items.

X-rays are electromagnetic radiations with a much shorter wavelength than visible light. When electric current is used to heat a filament to very high temperatures, energy of the electrons becomes so great that some electrons are emitted from the hot filament. When these electrons strike a positive electrode at high speeds, they release some of their energy in the form of x-rays.

X-rays do not penetrate dense material as readily as they penetrate less dense material, and x-rays can expose photographic film. Consequently, an x-ray beam can pass through a person and onto photographic film. Dense tissues of the body absorb the x-rays, and in these areas the film is underexposed, appearing white or light in color on the developed film. On the other hand, the x-rays readily pass through less dense tissue, and the film in these areas is overexposed and appears black or dark in color. For example, in an x-ray film of the skeletal system the dense bones are white, and the less dense soft tissues are dark, often so dark that no details can be seen. Because the dense bone material is clearly visible, x-rays can be used to determine if bones are broken or have other abnormalities.

Soft tissues can be photographed by using low-energy x-rays. For example, mammograms are low-energy x-rays of the breast that can be used to detect tumors, because tumors are slightly denser than normal tissue.

Radiopaque substances are dense materials that absorb x-rays. If a radiopaque liquid is given to a patient, the liquid assumes the shape of the organ into which it is placed. For example, if a barium solution is swallowed, the outline of the upper digestive tract can be photographed using x-rays to detect such abnormalities as ulcers.

---

or more smaller products. The breakdown of a disaccharide (a type of carbohydrate) into glucose molecules (figure 2.10*b*) is an example. Note that this particular reaction requires that water be split into two parts and that each part be contributed to one of the new glucose molecules. Reactions that use water in this manner are called **hydrolysis** (hī-drol′i-sis; water dissolution) **reactions.**

The breakdown of ATP to ADP and an inorganic phosphate is another example of a decomposition reaction.

$$\begin{array}{cccc}\text{A-P-P-P} & \rightarrow & \text{A-P-P} & + & P_i \\ \text{(ATP)} & & \text{(ADP)} & & \text{(Inorganic} \\ & & & & \text{phosphate)}\end{array}$$

The decomposition reactions that occur in the body are collectively called **catabolism** (kă-tab′-ō-lizm). They include the digestion of food molecules in the intestine and within cells, the breakdown of fat stores, and the breakdown of foreign matter and microorganisms in certain blood cells that function to protect the body. All of the anabolic and catabolic reactions in the body are collectively defined as **metabolism.**

Oxidation–Reduction Reactions

Chemical reactions that result from the exchange of electrons between the reactants are called oxidation–reduction reactions. For example, when sodium and chlorine react to form sodium chloride, the sodium atom loses an electron, and the chlorine atom gains an electron. The loss of an electron by an atom is called **oxidation,** and the gain of an electron is called **reduction.** The transfer of the electron can be complete, resulting in an ionic bond, or it can be a partial transfer, resulting in a covalent bond. Because the complete or partial loss of an electron by one atom is accompanied by the gain of that electron by another atom, these reactions are called **oxidation–reduction reactions.** Synthesis and decomposition reactions can be oxidation–reduction reactions. Thus, it is possible for a chemical reaction to be described in more than one way.

Reversible Reactions

A **reversible reaction** is a chemical reaction in which the reaction can proceed from reactants to products or from products to reactants. When the rate of product formation is equal to the rate of the reverse reaction, the reaction system is said to be at **equilibrium.** At equilibrium the amount of reactants relative to the amount of products remains constant.

The following analogy may help to clarify the concept of reversible reactions and equilibrium. Imagine a trough containing water. The trough is divided into two compartments by a partition, but the partition contains holes that allow water to move freely between the compartments. Because water can move in either direction, this is like a reversible reaction. Let the water in the left compartment be the reactant and the water in the right compartment be the product. At equilibrium, the amount of reactant relative to the amount of product in each compartment is always the same because the partition allows water to pass between the two compartments until the level of water is the same in both compartments. If additional water is added to the reactant, water flows from the left compartment through the partition to the right compartment until

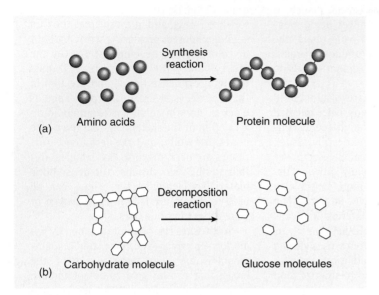

(a) Amino acids → Protein molecule (Synthesis reaction)

(b) Carbohydrate molecule → Glucose molecules (Decomposition reaction)

Figure 2.9   Synthesis and Decomposition Reactions

(*a*) Synthesis reaction in which amino acids, the basic "building blocks" of proteins, combine to form a protein molecule. (*b*) Decomposition reaction in which a complex carbohydrate breaks down into smaller glucose molecules, which are the "building blocks" of carbohydrates.

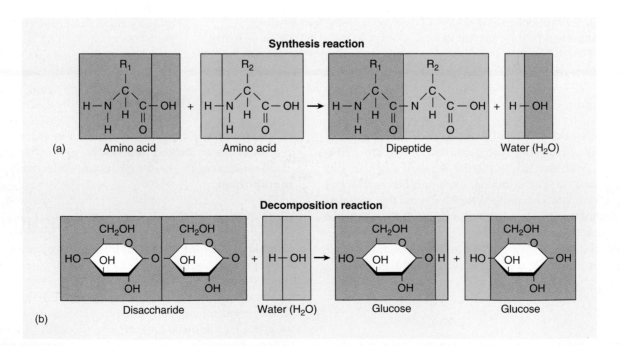

Figure 2.10   Dehydration and Hydrolysis Reactions

(*a*) Synthesis reaction in which two amino acids combine to form a dipeptide. This reaction is also a dehydration reaction because it results in the removal of a water molecule from the amino acids. (*b*) Decomposition reaction in which a disaccharide breaks apart to form glucose molecules. This reaction is also a hydrolysis reaction because it involves the splitting of a water molecule.

the level of water is the same in both compartments. Likewise, if additional reactants are added to a reaction mixture, some will form product until equilibrium is reestablished. Unlike this analogy, however, the amount of the reactants compared with the amount of products of most reversible reactions is not one to one. Depending on the specific reversible reaction, there can be one part reactant to two parts product, two parts products to one part reactant, or many other possibilities.

An important reversible reaction in the human body involves carbon dioxide and hydrogen ions. The reaction between carbon dioxide (CO_2) and water (H_2O) to form carbonic acid (H_2CO_3) is reversible. Carbonic acid then separates by a reversible reaction to form hydrogen ions (H^+) and bicarbonate ions (HCO_3^-):

$$CO_2 + H_2O \rightleftarrows H_2CO_3 \rightleftarrows H^+ + HCO_3^-$$

If carbon dioxide is added to water, additional carbonic acid forms, which, in turn, causes more hydrogen ions and bicarbonate ions to form. The amount of hydrogen and bicarbonate ions relative to carbon dioxide therefore remains constant. Maintaining a constant level of hydrogen ions is necessary for proper functioning of the nervous system. This can be achieved, in part, by regulating blood carbon dioxide levels. For example, slowing down the respiration rate causes blood carbon dioxide levels to increase.

6	P R E D I C T

If the respiration rate increases, carbon dioxide is eliminated from the blood. What effect does this change have on blood hydrogen ion levels?

✔ *Answer in Appendix F*

Rate of Chemical Reactions

The rate at which a chemical reaction proceeds is influenced by several factors, including how easily the substances react with one another, their concentrations, the temperature, and the presence of a catalyst.

Reactants

Reactants differ from one another in their ability to undergo chemical reactions. For example, iron corrodes much more rapidly than does stainless steel. For this reason, the iron bars forming the skeleton of the Statue of Liberty were replaced in 1986 with stainless steel bars.

Concentration

Within limits, the greater the concentration of the reactants, the greater the rate at which a given chemical reaction proceeds. This occurs because, as the concentration of reactants increases, they are more likely to come into contact with one another. For example, the normal concentration of oxygen inside cells enables oxygen to come into contact with other molecules, producing the chemical reactions necessary for life. If the oxygen concentration decreases, the rate of chemical reactions decreases. This decrease can impair cell function and even result in death.

Temperature

The rate of chemical reactions also increases when the temperature is increased. As temperature increases, reactants move at faster speeds, and they collide with one another more frequently and with greater force, increasing the likelihood of a chemical reaction. When a person has a fever of only a few degrees, reactions occur throughout the body at an accelerated rate, resulting in increased activity in the organ systems such as increased heart and respiratory rates. When body temperature drops, various metabolic processes slow. The clumsy movement of very cold fingers results largely from the reduced rate of chemical reactions in cold muscle tissue.

Catalyst

At normal body temperatures most chemical reactions would proceed very slowly if it were not for the action of the body's catalysts. A **catalyst** (kat′ă-list) is a substance that increases the rate at which a chemical reaction proceeds without itself being permanently changed or depleted. **Enzymes** (en′zīmz) are protein molecules in the body that act as catalysts. Many of the chemical reactions that occur in the body require enzymes, and chemical events in cells are regulated primarily by mechanisms that control either the concentration or the activity of enzymes. Enzymes are considered in greater detail later in this chapter.

Energy

Energy, unlike matter, does not occupy space and has no mass. **Energy** is defined as the capacity to do **work,** that is, to move matter. Energy can be subdivided into potential energy and kinetic energy. **Potential energy** is stored energy that could do work but is not doing so. For example, a coiled spring stores potential energy. It could push against an object and move the object, but as long as the spring does not uncoil, no work is accomplished. **Kinetic** (ki-net′ik) **energy** is energy caused by the movement of an object and is the form of energy that actually does work. An uncoiling spring pushing an object causing it to move is an example. When potential energy is released, it becomes kinetic energy, thus doing work.

According to the conservation of energy principle, energy can be neither created nor destroyed. Potential energy, however, can be converted into kinetic energy, and kinetic energy can be converted into potential energy. Potential and kinetic energy can be found in many different forms. Of particular interest to the study of human organisms are electric, electromagnetic, chemical, mechanical, and heat energy.

Electric Energy

Electric energy involves the movement of ions or electrons. Examples are nerve impulses and the electric current supplying a lightbulb. A nerve impulse, which carries messages from one part of the body to another, results from the movement of ions across cell membranes. Nerve impulses are discussed in chapter 9.

Electromagnetic Energy

Electromagnetic energy is energy that moves in waves analogous to the waves produced by dropping a pebble into a pond of water. Electromagnetic waves, however, are disturbances of electric and magnetic fields rather than disturbances of a material substance such as water or air. Energy waves of different wavelengths make up the electromagnetic spectrum. The parts of the electromagnetic spectrum, listed from the shortest to the longest wavelengths, and some of their important uses include the following: γ (gamma) rays (radiation therapy), x-rays (medical examination), ultraviolet light (stimulation of vitamin D production, also responsible for sunburn), visible light (activation of chemicals in the eye, resulting in vision), infrared radiation (loss or gain of heat), microwaves (electric appliances, radar), and radio waves (radio, television, magnetic resonance imaging for medical examination).

Chemical Energy

Consider two balls attached by a relaxed spring. To pull the balls apart and stretch the spring, energy must be put into this system. As the spring is stretched, potential energy increases. When the stretched spring recoils and pulls the balls closer together, potential energy decreases. Although there are no springs in atoms, oppositely charged particles, such as electrons and protons, attract one another. When work is done to move an electron away from the nucleus, it is like stretching the spring, and potential energy increases. When an electron moves closer to the nucleus, potential energy decreases.

Again consider the two balls connected by a relaxed spring. In order to push the balls together and compress the spring, energy must be put into this system. As the spring is compressed, potential energy increases. When the compressed spring expands, potential energy decreases. Similarly charged particles, such as two negatively charged electrons or two positively charged nuclei, repel each other. As similarly charged particles move closer together, their potential energy increases, much like compression of a spring, and as they move further apart, their potential energy decreases.

The **chemical energy** of a substance results from the relative positions and interactions among its charged subatomic particles. Some substances, such as food molecules, contain more potential energy than other substances, such as waste products. The difference in potential energy between food and waste products is used by living systems for many activities such as growth, repair, movement, and heat production.

Metabolism

In metabolism, chemical bonds are formed as molecules and compounds are synthesized, and chemical bonds are broken as molecules and compounds are broken apart. As these covalent and ionic bonds are formed and broken, the relative positions of electrons and nuclei change, producing changes in chemical energy. The formation of a chemical bond results in a decrease of potential energy and the release of energy, whereas breaking a chemical bond results in an increase of potential energy and requires an input of energy.

In any given chemical reaction, the potential energy contained in the chemical bonds of the reactants can be compared to the potential energy in the chemical bonds of the products. If the potential energy in the chemical bonds of the reactants is greater than that of the products, then energy is released by the reaction. For example, the hydrolysis of ATP to ADP results in the release of energy.

$$ATP + H_2O \quad \rightarrow \quad ADP + H_2PO_4^- \quad + \quad Energy$$
(More potential energy in reactants)   (Less potential energy in products)

For simplicity, the H_2O is often not shown in this reaction, and P_i is used to represent inorganic phosphate ($H_2PO_4^-$). For this reaction to occur, the bonds of ATP and H_2O are broken, which requires the input of energy, and the bonds of $H_2PO_4^-$ are formed, which results in the release of energy. As a result of breaking the existing bonds and forming new bonds, these products have less potential energy than the reactants, and energy is released (figure 2.11a). Note that the energy released does not come from breaking the phosphate bond of ATP, because breaking a chemical bond requires the input of energy. It is commonly stated, however, that the breakdown of ATP results in the release of energy, which is true when the overall reaction is considered. The energy released when ATP is broken down can be used in the synthesis of other molecules, to do work, such as muscle contraction, or to produce heat.

If the potential energy in the chemical bonds of the reactants is less than that of the products, then energy must be supplied for the reaction to occur. For example, the synthesis of ATP from ADP.

$$ADP + H_2PO_4^- \quad + \quad Energy \rightarrow \quad ATP + H_2O$$
(Less potential energy in reactants)   (More potential energy in products)

For this reaction to occur, the bonds of $H_2PO_4^-$ are broken, which requires the input of energy, and the bonds of ATP and H_2O are formed, which results in the release of energy. As a result of the breaking of existing bonds, the formation of new bonds, and the input of energy, these products have more potential energy than the reactants (figure 2.11b). Some of this increased potential energy can be used later when ATP is broken down (see preceding discussion).

When ATP is synthesized, where does the energy come from? The answer is that the energy comes from the chemical

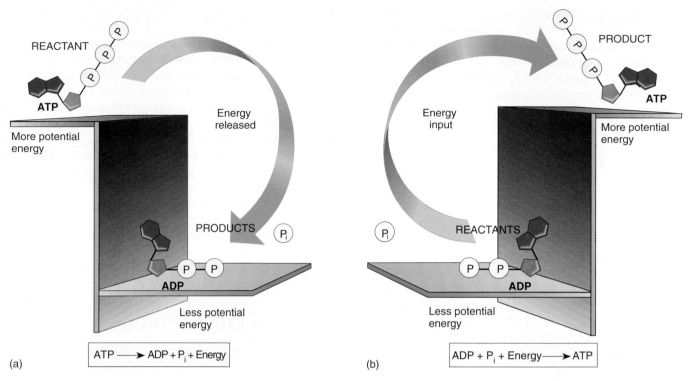

Figure 2.11   Energy and Chemical Reactions

In each figure the upper shelf represents a higher energy level, and the lower shelf represents a lower energy level. (*a*) Reaction in which energy is released as a result of the breakdown of ATP. (*b*) Reaction in which the input of energy is required for the synthesis of ATP.

bonds in food molecules. Just as energy is released when ATP is broken down, energy is released when food molecules are broken down. But then, where does the energy in food molecules come from? The answer is photosynthesis.

Photosynthesis

The energy that makes almost all life on earth possible ultimately comes from the sun. In the process of **photosynthesis,** plants capture the energy in sunlight and incorporate it into chemical bonds. The captured energy "excites" electrons, raising them to a higher potential energy level. As the electrons return to a lower energy state, they release energy, some of which is incorporated into the chemical bonds of glucose ($C_6H_{12}O_6$), which is a primary food molecule used in metabolism by both plants and animals. The overall reaction for photosynthesis is:

$6\ CO_2 + 12\ H_2O + \text{Light energy} \rightarrow C_6H_{12}O_6 + 6\ H_2O + 6\ O_2$
  (Less potential   (More potential
energy in reactants)   energy in products)

Mechanical Energy

Mechanical energy is energy resulting from the position or movement of an object. The potential energy in a spring that is converted to kinetic energy when the compressed spring uncoils is an example of mechanical energy. In the human body, chemical energy is converted into mechanical energy

that results in body movements such as walking or the beating of the heart.

Heat Energy

Heat is the energy that flows between objects that are at different temperatures. Temperature is a measure of how hot or cold a substance is relative to another substance. Heat is always transferred from a hotter object to a cooler object, such as from a hot stove top to a finger.

All other forms of energy can be converted into heat energy. For example, when a moving object comes to rest, its kinetic energy is converted into heat energy by friction. The potential energy in chemical bonds can also be released as heat energy during chemical reactions. The body temperature of humans is maintained by heat produced in this fashion.

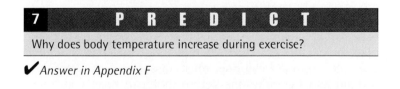

7	P R E D I C T

Why does body temperature increase during exercise?

✔ *Answer in Appendix F*

Inorganic Chemistry

Originally it was believed that inorganic substances were those that came from nonliving sources and organic substances were those extracted from living organisms. As the science of chemistry developed, however, it became apparent that organic substances could be manufactured in the laboratory. As defined currently, **inorganic chemistry** deals with those substances that do not contain carbon, whereas **organic chemistry** is the study of carbon-containing substances. These definitions have a few exceptions. For example, carbon dioxide and carbon monoxide are classified as inorganic molecules.

Water

Approximately two-thirds of the human body is water; and plasma, the liquid portion of blood, is 92% water. A molecule of **water** is composed of one atom of oxygen joined to two atoms of hydrogen by covalent bonds. Water molecules are polar, with a partial positive charge associated with the hydrogen atoms and a partial negative charge associated with the oxygen atom. Hydrogen bonds form between the positively charged hydrogen atoms of one water molecule and the negatively charged oxygen atoms of another water molecule. These hydrogen bonds organize the water molecules into a lattice that holds the water molecules together (see figures 2.6 and 2.7).

Water has physical and chemical properties well suited for its many functions in living organisms. These properties are outlined in the following sections.

Stabilizing Body Temperature

Water has a high **specific heat,** meaning that a relatively large amount of heat is required to raise its temperature; therefore it tends to resist large temperature fluctuations. When water evaporates, it changes from a liquid to a gas, and because heat is required for that process, the evaporation of water from the surface of the body rids the body of excess heat.

Protection

Water is an effective lubricant that provides protection against damage resulting from friction. For example, tears protect the surface of the eye from the rubbing of the eyelids. Water also forms a fluid cushion around organs that helps to protect them from trauma. The cerebrospinal fluid that surrounds the brain is an example.

Chemical Reactions

Many of the chemical reactions necessary for life do not take place unless the reacting molecules are dissolved in water. For example, sodium chloride must dissociate in water into sodium and chloride ions before they can react with other ions. Water also directly participates in many chemical reactions. As previously mentioned, a dehydration reaction is a synthesis reaction in which water is produced, and a hydrolysis reaction is a decomposition reaction that requires a water molecule (see figure 2.10).

Mixing Medium

Water can mix with other substances to form solutions, suspensions, and colloids. A **solution** is any liquid that contains dissolved substances. For example, sweat is a solution in which sodium chloride and other substances are dissolved in water. A **suspension** is a liquid that contains nondissolved materials that settle out of the liquid unless it is continually shaken. Red blood cells in plasma are a suspension. A **colloid** (kol'oyd) is a liquid that contains nondissolved materials that do not settle out of the liquid. Water and proteins inside cells form a colloid.

In living organisms the complex fluids inside and outside cells consist of solutions, suspensions, and colloids. The ability of water to mix with other substances enables it to act as a medium for transport. Body fluids such as plasma transport nutrients, gases, and waste products from one part of the body to another.

Solution Concentrations

The liquid portion of a solution is the **solvent,** and the substances dissolved in the solvent are **solutes** (sol'yūtz). For example, water is the solvent, and sodium chloride is the solute in a sodium chloride solution. The concentration of solute particles dissolved in solvents can be expressed in several ways. One common way is to indicate the percent of solute by weight per volume of solution. For example, a 10% solution of sodium chloride can be made by dissolving 10 g of sodium chloride into enough water to make 100 mL of solution.

Physiologists often determine concentrations in **osmoles** (os'mōlz), which express the number of particles in a solution. A particle can be an atom, ion, or molecule. An osmole (osm) is 1 mole (Avogadro's number) of particles in 1 kilogram (kg) of water. The **osmolality** (os-mō-lal'i-tē) of a solution is a reflection of the number, not the type, of particles in a solution. Thus, a 1 osmolal solution contains 1 osmole of particles per kilogram of solution, but the particles can be all one type or a complex mixture of different types of particles.

Because the concentration of particles in body fluids is so low, the measurement **milliosmole** (mOsm), 1/1000 of an osmole, is used. Most body fluids have a concentration of about 300 mOsm and contain many different ions and molecules. The concentration of body fluids is important because it influences the movement of water into or out of cells (see chapter 3). Appendix C contains more information on calculating concentrations.

Acids and Bases

Many molecules and compounds are classified as acids or bases. For most purposes an **acid** is defined as a proton donor. Because a hydrogen atom without its electron is a proton (H^+), any substance that releases hydrogen ions is an acid.

For example, hydrochloric acid (HCl) forms hydrogen ions (H^+) and chloride ions (Cl^-) in solution and therefore is an acid.

$$HCl \rightarrow H^+ + Cl^-$$

A **base** is defined as a proton acceptor, and any substance that binds to (accepts) hydrogen ions is a base. Many bases function as proton acceptors by releasing hydroxide ions (OH^-) when they dissociate. For example, the base sodium hydroxide (NaOH) dissociates to form sodium and hydroxide ions:

$$NaOH \rightarrow Na^+ + OH^-$$

The hydroxide ions are proton acceptors that combine with hydrogen ions to form water:

$$OH^- + H^+ \rightarrow H_2O$$

Acids and bases are classified as strong or weak. Strong acids or bases dissociate almost completely when dissolved in water. Consequently, they release almost all of their hydrogen or hydroxide ions. The more completely the acid or base dissociates, the stronger it is. For example, hydrochloric acid is a strong acid because it completely dissociates in water.

$$HCl \rightarrow H^+ + Cl^-$$
Not freely reversible

Weak acids or bases only partially dissociate in water. Consequently, they release only some of their hydrogen or hydroxide ions. For example, when acetic acid (CH_3COOH) is dissolved in water, some of it dissociates, but some of it remains in the undissociated form. An equilibrium is established between the ions and the undissociated weak acid.

$$CH_3COOH \rightleftharpoons CH_3COO^- + H^+$$
Freely reversible

For a given weak acid or base, the amount of the dissociated ions relative to the weak acid or base is a constant.

The pH Scale

The pH scale is a means of referring to the hydrogen ion concentration in a solution (figure 2.12). Pure water is defined as a **neutral solution** and has a pH of 7. A neutral solution has equal concentrations of hydrogen and hydroxide ions. Solutions with a pH less than 7 are **acidic** and have a greater concentration of hydrogen ions than hydroxide ions. **Alkaline** (al′kă-līn), or **basic,** solutions have a pH greater than 7 and have fewer hydrogen ions than hydroxide ions.

The symbol pH stands for power (p) of hydrogen ion (H^+) concentration. The power is a factor of 10, which means that a change in the pH of a solution by 1 pH unit represents a 10-fold change in the hydrogen ion concentration. For example, a solution of pH 6 has a hydrogen ion concentration 10 times greater than a solution of pH 7 and 100 times greater than a solution of pH 8. As the pH value becomes smaller, the solution has more hydrogen ions and is more acidic, and as the pH value becomes larger, the solution has fewer hydrogen ions and is more basic. Appendix D considers pH in greater detail.

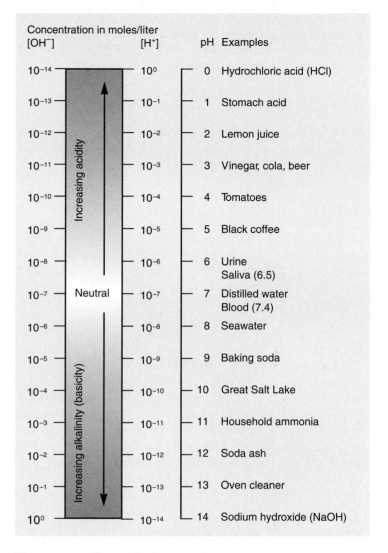

Figure 2.12   The pH Scale

A pH of 7 is considered neutral. Values less than 7 are acidic (the lower the number, the more acidic). Values greater than 7 are basic (the higher the number, the more basic). Representative fluids and their approximate pH values are listed.

Clinical Note

The normal pH range for human blood is 7.35–7.45. **Acidosis** results if blood pH drops below 7.35, in which case the nervous system becomes depressed, and the individual can become disoriented and possibly comatose. **Alkalosis** results if blood pH rises above 7.45. Then the nervous system becomes overexcitable, and the individual can be extremely nervous or have convulsions. Both acidosis and alkalosis can be fatal.

Salts

A **salt** is a compound consisting of a cation other than a hydrogen ion and an anion other than a hydroxide ion. Salts are formed by the interaction of an acid and a base in which the hydrogen ions of the acid are replaced by the positive ions of the base. For example, in a solution when hydrochloric acid (HCl) reacts with the base sodium hydroxide (NaOH), the salt sodium chloride (NaCl) is formed.

$$HCl + NaOH \rightarrow NaCl + H_2O$$
$$\text{(Acid)}\quad \text{(Base)}\quad \text{(Salt)}\quad \text{(Water)}$$

Typically, when salts such as sodium chloride dissociate in water, they form positively and negatively charged ions (see figure 2.8).

Buffers

The chemical behavior of many molecules changes as the pH of the solution in which they are dissolved changes. For example, many enzymes work best within narrow ranges of pH. The survival of an organism depends on its ability to regulate body fluid pH within a narrow range. Deviations from the normal pH range for human blood are life-threatening.

One way body fluid pH is regulated involves the action of buffers, which resist changes in solution pH when either acids or bases are added. A **buffer** is a solution of a conjugate acid–base pair in which the acid component and the base component occur in similar concentrations. A conjugate base is everything that remains of an acid after the hydrogen ion (proton) is lost. A conjugate acid is formed when a hydrogen ion is transferred to the conjugate base. Two substances related in this way are a **conjugate acid–base pair.** For example, carbonic acid (H_2CO_3) and bicarbonate ion (HCO_3^-), formed by the dissociation of carbonic acid, are a conjugate acid–base pair.

$$H_2CO_3 \rightleftarrows H^+ + HCO_3^-$$

In the forward reaction, carbonic acid loses a hydrogen ion to produce bicarbonate ion, which is a conjugate base. In the reverse reaction, a hydrogen ion is transferred to the bicarbonate ion (conjugate base) to produce carbonic acid, which is a conjugate acid.

For a given condition, this reversible reaction results in an equilibrium, in which the amounts of carbonic acid relative to the amounts of hydrogen ion and bicarbonate ions remains constant. The conjugate acid–base pair can resist changes in pH because of this equilibrium. If an acid is added to a buffer, the hydrogen ions from the added acid can combine with the base component of the conjugate acid–base pair. As a result, the concentration of hydrogen ions does not increase as much as it would without this reaction. For example, if hydrogen ions are added to a carbonic acid solution, many of the hydrogen ions combine with bicarbonate ions to form carbonic acid.

On the other hand, if a base is added to a buffered solution, the conjugate acid can release hydrogen ions to counteract the effects of the added base. For example, if hydroxide ions are added to a carbonic acid solution, the hydroxide ions combine with hydrogen ions to form water. As

the hydrogen ions are incorporated into water, carbonic acid dissociates to form hydrogen and bicarbonate ions, maintaining the hydrogen ion concentration (pH) within a normal range.

The greater the buffer concentration, the more effective it is in resisting a change in pH, but buffers cannot entirely prevent some change in the pH of a solution. For example, when an acid is added to a buffered solution, the pH decreases but not to the extent it would have without the buffer. Several very important buffers are found in living systems and include bicarbonate, phosphates, amino acids, and proteins as components.

---

8 ■ **P R E D I C T**

Dihydrogen phosphate ion ($H_2PO_4^-$) and monohydrogen phosphate ion (HPO_4^{2-}) form the phosphate buffer system.

$$H_2PO_4^- \rightleftarrows H^+ + HPO_4^{2-}$$

Identify the conjugate acid and conjugate base in the phosphate buffer system. Explain how they function as a buffer when either hydrogen or hydroxide ions are added to the solution.

✔ *Answer in Appendix F*

Oxygen

Oxygen (O_2) is an inorganic molecule consisting of two oxygen atoms bound together by a double covalent bond. About 21% of the gas in the atmosphere is oxygen, and it is essential for most animals. Oxygen is required by humans in the final step of a series of reactions in which energy is extracted from food molecules (see chapters 3 and 25).

Carbon Dioxide

Carbon dioxide (CO_2) consists of one carbon atom bound by double covalent bonds to two oxygen atoms. Carbon dioxide is produced when organic molecules such as glucose are metabolized within the cells of the body (see chapters 3 and 25). Much of the energy stored in the covalent bonds of glucose is transferred to other organic molecules when glucose is broken down, and carbon dioxide is released. Once carbon dioxide is produced, it is eliminated from the cell as a metabolic by-product, transferred to the lungs by blood, and exhaled during respiration. If carbon dioxide is allowed to accumulate within cells, it becomes toxic.

Organic Chemistry

The ability of carbon to form covalent bonds with other atoms makes possible the formation of the large, diverse, complicated molecules necessary for life. A series of carbon atoms bound together by covalent bonds constitutes the "backbone" of many large molecules. Variation in the length of the carbon chains and the combination of atoms bound to the carbon backbone allows for the formation of a wide variety of molecules. For example, some protein molecules have thousands of carbon atoms bound by covalent bonds to one another or to other atoms, such as nitrogen, sulfur, hydrogen, and oxygen.

The four major groups of organic molecules essential to living organisms are carbohydrates, lipids, proteins, and nucleic acids. Each of these groups has specific structural and functional characteristics.

Carbohydrates

Carbohydrates are composed primarily of carbon, hydrogen, and oxygen atoms and range in size from small to very large. In most carbohydrates, for each carbon atom there are two hydrogen atoms and one oxygen atom. Note that the ratio of hydrogen atoms to oxygen atoms is two to one, the same as in water. They are called carbohydrates because each carbon (carbo) is "watered," or hydrated. The large number of oxygen atoms in carbohydrates makes them relatively polar molecules. Consequently, they are soluble in polar solvents such as water.

Monosaccharides

Large carbohydrates are composed of numerous, relatively simple building blocks called **monosaccharides** (mon-ō-sak'ă-rīdz; the prefix mono- means one; the term saccharide means sugar), or simple sugars. Monosaccharides commonly contain three carbons (trioses), four carbons (tetroses), five carbons (pentoses), or six carbons (hexoses).

The monosaccharides most important to humans include both five- and six-carbon sugars. Common six-carbon sugars, such as glucose, fructose, and galactose, are **isomers** (ī'sō-merz), which are molecules that have the same number and types of atoms but differ in their three-dimensional arrangement (figure 2.13). Glucose, or blood sugar, is the major carbohydrate found in the blood and is a major nutrient for most cells of the body. Fructose and galactose are also important dietary nutrients. Important five-carbon sugars include ribose and deoxyribose (see figure 2.24), which are components of ribonucleic acid (RNA) and deoxyribonucleic acid (DNA), respectively.

Disaccharides

Disaccharides (dī-sak'ă-rīdz; *di-* means two) are composed of two simple sugars bound together through a dehydration

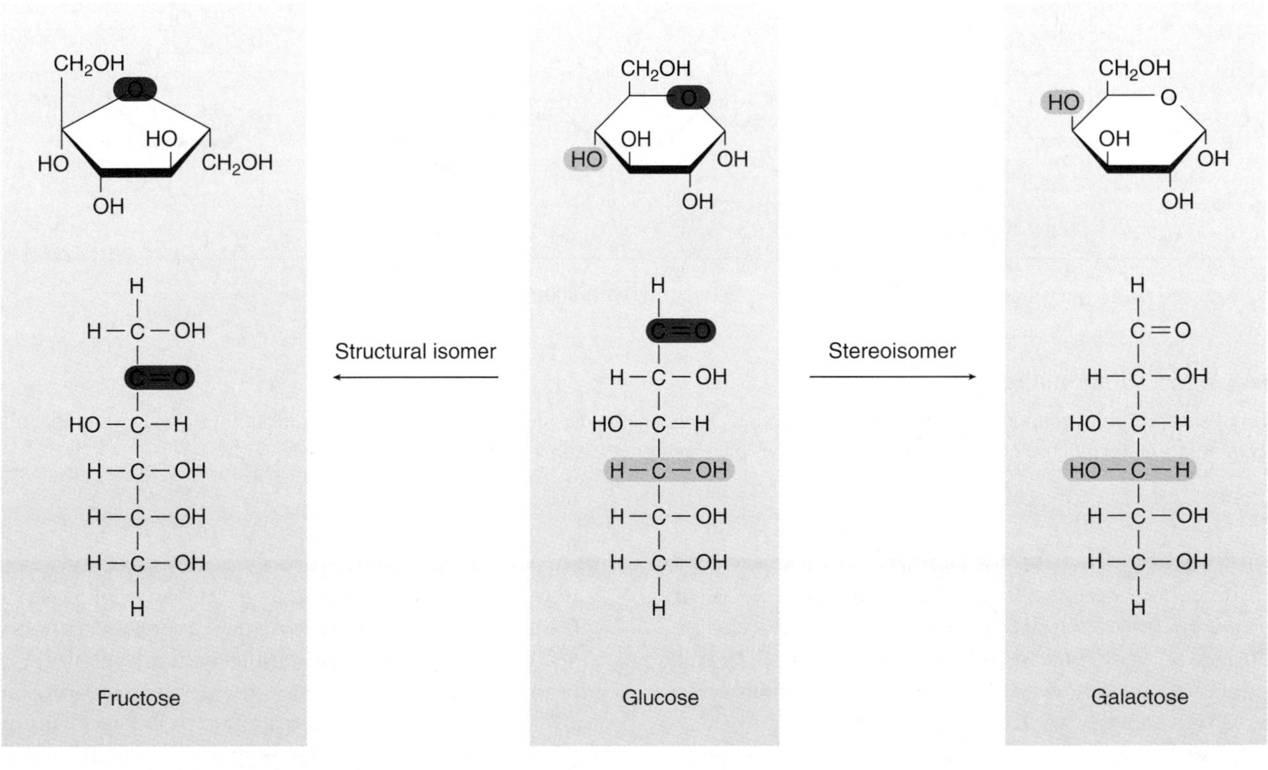

Figure 2.13   Monosaccharides

These monosaccharides almost always form a ring-shaped molecule. They are represented as linear models to more readily illustrate the relationships between the atoms of the molecules. Fructose is a structural isomer of glucose because it has identical chemical groups bonded in a different arrangement in the molecule (*indicated by red shading*). Galactose is a stereoisomer of glucose because it has exactly the same groups bonded to each carbon atom but located in a different three-dimensional orientation (*indicated by yellow shading*).

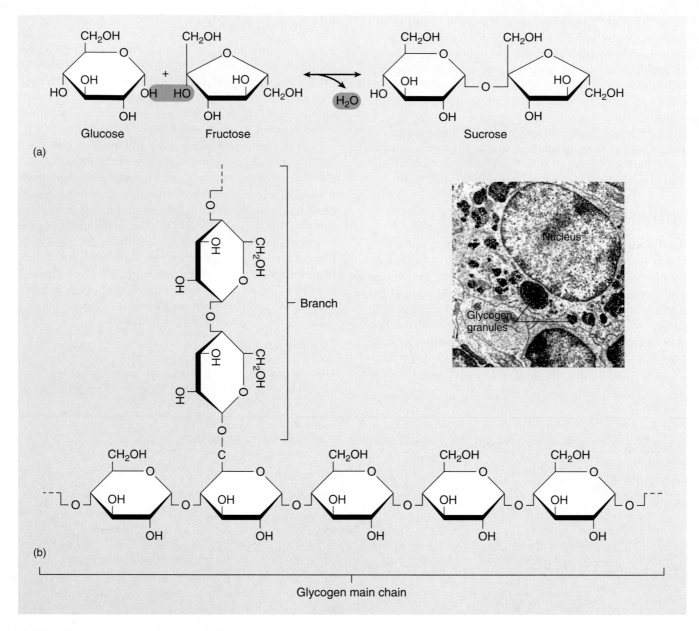

Figure 2.14 Disaccharide and Polysaccharide

(*a*) Formation of sucrose, a disaccharide, by a dehydration reaction involving glucose and fructose (monosaccharides). (*b*) Glycogen is a polysaccharide formed by combining many glucose molecules. The photo shows glycogen granules in a liver cell.

reaction. Glucose and fructose, for example, combine to form a disaccharide called **sucrose** (table sugar) plus a molecule of water (figure 2.14*a*). Several disaccharides are important to humans, including sucrose, lactose, and maltose. Lactose, or milk sugar, is glucose combined with galactose; and maltose, or malt sugar, is two glucose molecules joined together.

Polysaccharides

Polysaccharides (pol-ē-sak′ă-rīdz; the prefix *poly-* means many) consist of many monosaccharides bound together to form long chains that are either straight or branched. **Glyco-**

gen, or animal starch, is a polysaccharide composed of many glucose molecules (figure 2.14*b*). Because glucose can be metabolized rapidly and the resulting energy can be used by cells, glycogen is an important storage molecule. A substantial amount of the glucose that is metabolized to produce energy for muscle contraction during exercise is stored in the form of glycogen in the cells of the liver and skeletal muscles.

Starch and **cellulose** are two important polysaccharides found in plants, and both are composed of long chains of glucose. Plants use starch as a storage molecule in the same way that animals use glycogen, and cellulose is an important structural component of plant cell walls. When humans ingest

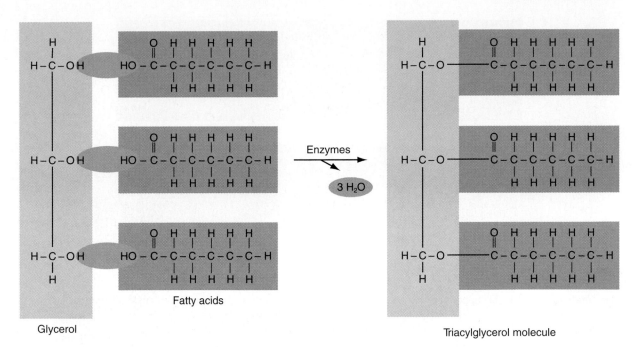

Figure 2.15   Triacylglycerols
Production of a triacylglycerol from one glycerol molecule and three fatty acids.

plants, the starch can be broken down and used as an energy source. Humans, however, do not have the digestive enzymes necessary to break down cellulose. The cellulose is eliminated in the feces, where it provides bulk. Table 2.5 summarizes the role of carbohydrates in the body.

Table 2.5	Role of Carbohydrates in the Body
Role	**Example**
Structure	Ribose forms part of RNA and ATP molecules, and deoxyribose forms part of DNA.
Energy	Monosaccharides (glucose, fructose, galactose) can be used as energy sources. Disaccharides (sucrose, lactose, maltose) and polysaccharides (starch, glycogen) must be broken down to monosaccharides before they can be used for energy. Glycogen is an important energy-storage molecule in muscles and in the liver.
Bulk	Cellulose forms bulk in the feces.

Lipids

Lipids are a second major group of organic molecules common to living systems. Like carbohydrates, they are composed principally of carbon, hydrogen, and oxygen; but other elements, such as phosphorus and nitrogen, are minor components of some lipids. Lipids contain a lower ratio of oxygen to carbon than do carbohydrates, which makes them less polar. Consequently, lipids can be dissolved in nonpolar organic solvents, such as alcohol or acetone, but they are relatively insoluble in water. The definition of lipids is so general that several different kinds of molecules, such as fats, phospholipids, steroids, and prostaglandins, fit into this category.

Fats are a major type of lipid. Like carbohydrates, fats are ingested and broken down by hydrolysis reactions in cells to release energy for use by those cells. Conversely, if intake exceeds need, excess chemical energy from any source can be stored in the body as fat for later use as energy is needed. Fats also provide protection by surrounding and padding organs, and under-the-skin fats act as an insulator to prevent heat loss.

Triacylglycerols (trī-as′il-glis′er-olz) constitute 95% of the fats in the human body. Triacylglycerols, which are sometimes called triglycerides (trī-glis′er-īdz), consist of two different types of building blocks: glycerol and fatty acids. **Glycerol** is a three-carbon molecule with a hydroxyl group attached to each carbon atom, and **fatty acids** consist of a straight chain of carbon atoms with a carboxyl group attached at one end (figure 2.15). A **carboxyl** (kar-bok′sil) **group** (—COOH) consists of both an oxygen atom and a hydroxyl group attached to a carbon atom. The carboxyl group is responsible for the acidic nature of the molecule because it releases hydrogen ions into solution. Glycerols can be described according to the number and kinds of fatty acids that combine with glycerol through dehydration reactions. Monoacylglycerols have one fatty acid, diacylglycerols have two fatty acids, and triacylglycerols have three fatty acids bound to glycerol.

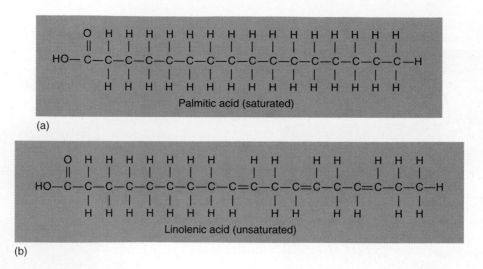

Figure 2.16   Fatty Acids

(*a*) Palmitic acid (saturated with no double bonds between the carbons). (*b*) Linolenic acid (unsaturated with three double bonds between the carbons).

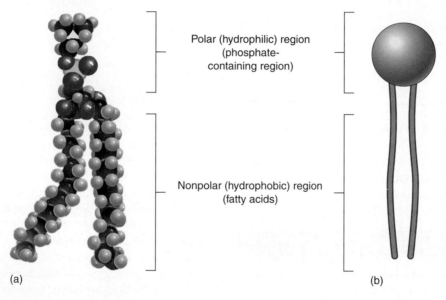

Figure 2.17   Phospholipids

(*a*) Molecular model of a phospholipid. (*b*) Simplified way in which phospholipids are often depicted.

Fatty acids differ from one another according to the length and the degree of saturation of their carbon chains. Most naturally occurring fatty acids contain an even number of carbon atoms, with 14- to 18-carbon chains being the most common. A fatty acid is **saturated** (figure 2.16) if it contains only single covalent bonds between the carbon atoms. Sources of saturated fats include beef, pork, whole milk, cheese, butter, eggs, coconut oil, and palm oil. The carbon chain is **unsaturated** if it has one or more double covalent bonds between carbon atoms. Because the double covalent bonds can occur anywhere along the carbon chain, many types of unsaturated fatty acids with an equal degree of unsaturation are possible. **Monounsaturated** fats, such as olive and peanut oils, have one double covalent bond between carbon atoms. **Polyunsaturated fats,** such as safflower, sunflower, corn, or fish oils, have two or more double covalent bonds between carbon atoms. Unsaturated fats are the best type of fats in the diet because unlike saturated fats they do not contribute to the development of cardiovascular disease.

Phospholipids are similar to triacylglycerols, except that one of the fatty acids bound to the glycerol is replaced by a molecule containing phosphate and, usually, nitrogen (figure 2.17). They are polar at the end of the molecule to which the phosphate is bound and nonpolar at the other end. The polar end of the molecule is attracted to water, and the nonpolar end is repelled by water. Phospholipids are important components of cell membranes (see chapter 3).

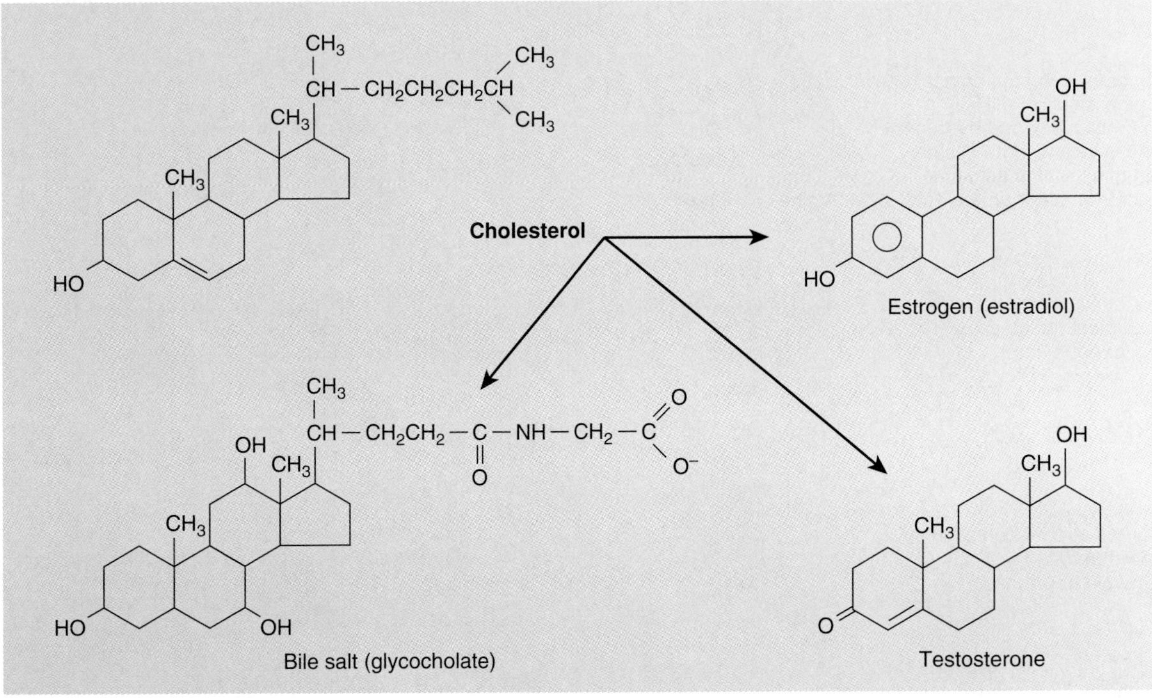

Figure 2.18 Steroids

Steroids are four-ringed molecules that differ from one another according to the groups attached to the rings. Cholesterol, the most common steroid, can be modified to produce other steroids.

Prostaglandins (pros′tă-glan′dinz), **thromboxanes** (thromb′box-zānz), and **leukotrienes** (lū-kō-trī′ēnz) are lipids derived from fatty acids. They are made in most cells and are important regulatory molecules. Among their numerous effects is their role in the response of tissues to injuries. Prostaglandins have been implicated in regulating the secretion of some hormones, blood clotting, some reproductive functions, and many other processes. Many of the therapeutic effects of aspirin and other anti-inflammatory drugs result from their ability to inhibit prostaglandin synthesis.

Steroids differ in chemical structure from other lipid molecules, but their solubility characteristics are similar. All steroid molecules are composed of carbon atoms bound together into four ringlike structures (figure 2.18). Important steroid molecules include cholesterol, bile salts, estrogen, progesterone, and testosterone. Cholesterol is an important steroid because other molecules are synthesized from it. For example, bile salts, which increase fat absorption in the intestines, are derived from cholesterol, as are the reproductive hormones estrogen, progesterone, and testosterone. In addition, cholesterol is an important component of cell membranes. Although high levels of cholesterol in the blood increase the risk of cardiovascular disease, a certain amount of cholesterol is vital for normal function.

Another class of lipids is the **fat-soluble vitamins.** Their structures are not closely related to one another, but they are nonpolar molecules essential for many normal functions of the body. Table 2.6 lists the functions of lipids in the body.

Table 2.6   Role of Lipids in the Body

Role	Example
Protection	Fat surrounds and pads organs.
Insulation	Fat under the skin prevents heat loss. Myelin surrounds nerve cells and electrically insulates the cells from one another.
Regulation	Steroid hormones regulate many physiologic processes. For example, estrogen and testosterone are sex hormones responsible for many of the differences between males and females. Prostaglandins help regulate tissue inflammation and repair.
Vitamins	Fat-soluble vitamins perform a variety of functions. Vitamin A forms retinol, which is necessary for seeing in the dark; active vitamin D promotes calcium uptake by the small intestine; vitamin E promotes wound healing; and vitamin K is necessary for the synthesis of proteins responsible for blood clotting.
Structure	Phospholipids and cholesterol are important components of cell membranes.
Energy	Lipids can be stored and broken down later for energy; per unit of weight, they yield more energy than carbohydrates or proteins.

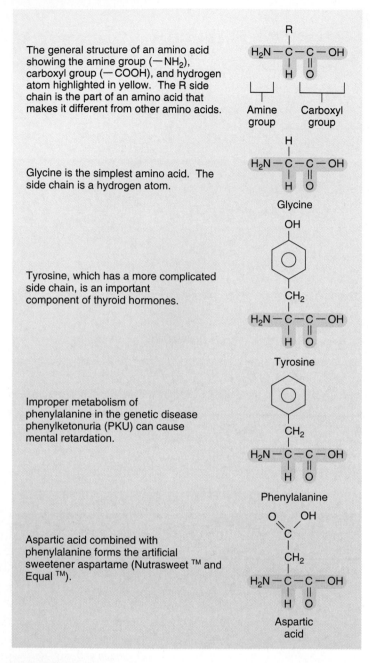

The general structure of an amino acid showing the amine group (—NH₂), carboxyl group (—COOH), and hydrogen atom highlighted in yellow. The R side chain is the part of an amino acid that makes it different from other amino acids.

Glycine is the simplest amino acid. The side chain is a hydrogen atom.

Tyrosine, which has a more complicated side chain, is an important component of thyroid hormones.

Improper metabolism of phenylalanine in the genetic disease phenylketonuria (PKU) can cause mental retardation.

Aspartic acid combined with phenylalanine forms the artificial sweetener aspartame (Nutrasweet ™ and Equal ™).

Figure 2.19   Amino Acids

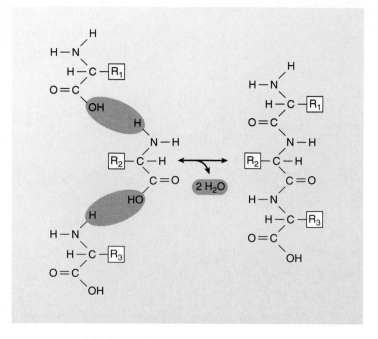

Figure 2.20   Peptide Bonds

(*left*) Dehydration reaction between three amino acids to form (*right*) a tripeptide. One water molecule (H_2O) is given off for each peptide bond formed.

Protein Structure

The basic building blocks for proteins are the 20 **amino** (ă-mē′no) **acid** molecules. Each amino acid has an amine (ă-mēn′) group (—NH₂), a carboxyl group (—COOH), a hydrogen atom, and a side chain designated by the symbol R attached to the same carbon atom. The side chain can be a variety of chemical structures, and the differences in the side chains make the amino acids different from one another (figure 2.19).

Covalent bonds formed between amino acid molecules during protein synthesis are called **peptide bonds** (figure 2.20). A dipeptide is two amino acids bound together by a peptide bond, a tripeptide is three amino acids bound together by peptide bonds, and a polypeptide is many amino acids bound together by peptide bonds. Proteins are polypeptides composed of hundreds of amino acids. Because there are 20 different amino acids and because each amino acid can be located at any position along a polypeptide chain, the potential number of different protein molecules is enormous.

The **primary structure** (figure 2.21*a*) of a protein is determined by the sequence of the amino acids bound by peptide bonds. The **secondary structure** (figure 2.21*b*) results from the folding or bending of the polypeptide chain caused by the hydrogen bonds between amino acids. Two common shapes that result are helices or pleated sheets. The ability of proteins to function depends on their shape. If the hydrogen bonds that maintain the shape of the protein are broken, the protein becomes nonfunctional. This change in shape is called

Proteins

All **proteins** contain carbon, hydrogen, oxygen, and nitrogen bound together by covalent bonds, and most proteins contain some sulfur. In addition, some proteins contain small amounts of phosphorus, iron, and iodine. The molecular mass of proteins can be very large. For the purpose of comparison, the molecular mass of water is approximately 18, sodium chloride 58, and glucose 180; but the molecular mass of proteins ranges from approximately 1000 to several million.

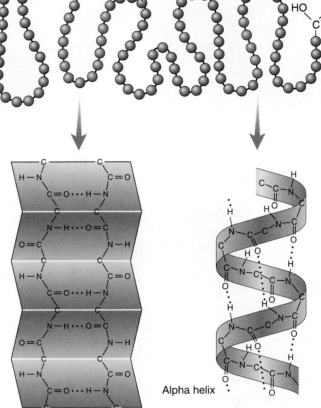

(a) Primary structure—the amino acid sequence

Amino acids

Peptide bond

(b) Secondary structure with folding as a result of hydrogen bonding (dotted red lines)

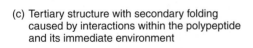

Pleated sheet

Alpha helix

(c) Tertiary structure with secondary folding caused by interactions within the polypeptide and its immediate environment

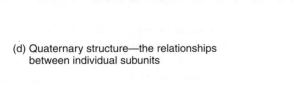

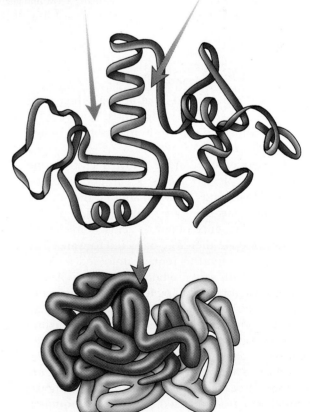

(d) Quaternary structure—the relationships between individual subunits

Figure 2.21   Protein Structure

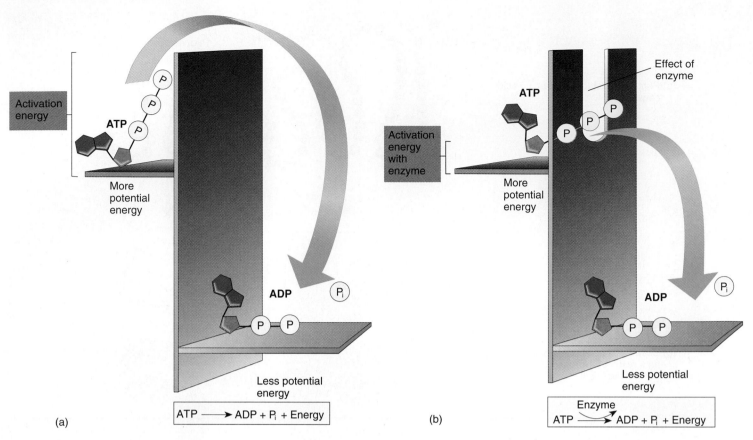

Figure 2.22   Activation Energy and Enzymes

(*a*) Activation energy is needed to change ATP to ADP. The upper shelf represents a higher energy level, and the lower shelf represents a lower energy level. The "wall" extending above the upper shelf represents the activation energy. Even though energy is given up moving from the upper to the lower shelf, the activation energy "wall" must be overcome before the reaction can proceed. (*b*) The enzyme lowers the activation energy, making it easier for the reaction to proceed.

denaturation, and it can be caused by abnormally high temperatures or changes in the pH of body fluids. An everyday example of denaturation is the change in the proteins of egg whites when they are cooked.

The **tertiary structure** (figure 2.21*c*) results from the folding of the helices or pleated sheets. It can be caused by the formation of covalent bonds between sulfur atoms of one amino acid and sulfur atoms in another amino acid located at a different place in the sequence of amino acids. Some amino acids are quite polar and are therefore attracted to water. The polar portions of proteins tend to remain unfolded so as to maximize their contact with water, whereas the less polar regions tend to fold into a globular shape to minimize their contact with water. Changes in only a few amino acids in the sequence can markedly change the protein's tertiary structure. Because the function of a protein is influenced by its three-dimensional structure, changes that influence the tertiary structure of the protein can also affect the function of the protein.

If two or more proteins associate to form a functional unit, the individual proteins are called subunits. The **quaternary structure** refers to the spatial relationships between the individual subunits (figure 2.21*d*).

Enzymes

Proteins perform many roles in the body, including acting as enzymes. An **enzyme** is a protein catalyst that increases the rate at which a chemical reaction proceeds without the enzyme being permanently changed. For a chemical reaction to begin, energy must be provided. For example, heat in the form of a spark is required to start the reaction between oxygen and gasoline vapor. Once some oxygen molecules react with gasoline, the energy released can start additional reactions.

The **activation energy** is the minimum energy that the reactants must have to start a chemical reaction. For most of the chemical reactions in the body, the activation energy is so high that few chemical reactions can occur naturally without help. Enzymes provide that help by lowering the activation energy (figure 2.22). With an enzyme, the rate of a chemical reaction can take place more than a million times faster than without the enzyme.

The three-dimensional shape of enzymes is critical for their normal function because it determines the structure of the enzyme's **active site.** According to the **lock-and-key model** of

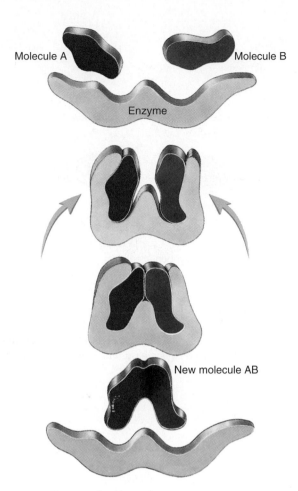

Molecule A   Molecule B

Enzyme

New molecule AB

Figure 2.23   Enzyme Action

The enzyme brings the two reacting molecules together. This is possible because the reacting molecules "fit" the shape of the enzyme (lock-and-key model). After the reaction, the unaltered enzyme can be used again.

Table 2.7	Role of Proteins in the Body
Role	**Example**
Regulation	Enzymes control chemical reactions. Hormones regulate many physiologic processes; for example, insulin affects glucose transport into cells.
Transport	Hemoglobin transports oxygen and carbon dioxide in the blood. Plasma proteins transport many substances in the blood. Proteins in cell membranes control the movement of materials into and out of the cell.
Protection	Antibodies and complement protect against microorganisms and other foreign substances.
Contraction	Actin and myosin in muscle are responsible for muscle contraction.
Structure	Collagen fibers form a structural framework in many parts of the body. Keratin adds strength to skin, hair, and nails.
Energy	Proteins can be broken down for energy; per unit of weight, they yield as much energy as carbohydrates.

enzyme action, reactants must bind to a specific active site on the enzyme. That is, the shape of the reactants and the active site must fit together as a key fits into a lock. This view of enzymes and reactants as rigid structures fitting together has been modified by the **induced fit model,** in which the enzyme is able to slightly change shape and better fit the reactants.

At the active site, reactants are brought into close proximity (figure 2.23), reducing the activation energy of the reaction. After the reactants combine, they are released from the active site, and the enzyme is capable of catalyzing additional reactions. Slight changes in the structure of an enzyme can destroy the ability of the active site to function. Enzymes are very sensitive to changes in temperature or pH, which can influence their structure.

9   P R E D I C T

Describe how changing one amino acid in an enzyme could affect the function of that enzyme.

✔ *Answer in Appendix F*

To be functional, some enzymes require additional, nonprotein substances called **cofactors.** The cofactor can be an ion, such as magnesium or zinc, or an organic molecule. Cofactors that are organic molecules, such as certain vitamins, may be referred to as **coenzymes.** Cofactors normally form part of the enzyme's active site and are required to make the enzyme functional.

Enzymes are highly specific because their active site can bind only to certain reactants. Each enzyme catalyzes a specific chemical reaction and no others. Many different enzymes are therefore needed to catalyze the many chemical reactions of the body. Enzymes often are named by adding the suffix- *ase* to the name of the molecules on which they act. For example, an enzyme that catalyzes the breakdown of lipids is a **lipase** (lip′ās), and an enzyme that breaks down proteins is called a **protease** (prō′tē-ās).

Enzymes control the rate at which most chemical reactions proceed in living systems. Consequently, they control essentially all cellular activities. At the same time, the activity of enzymes themselves is regulated by several mechanisms that exist within the cells. Some mechanisms control the enzyme concentration by influencing the rate at which the enzymes are synthesized, and others alter the activity of existing enzymes. Much of what is known about the regulation of cellular activity involves knowledge of how enzyme activity is controlled. The role of enzymes and other proteins is summarized in table 2.7.

Nucleic Acids: DNA and RNA

The nucleic acids are another group of very important organic molecules. **Deoxyribonucleic** (dē-oks′ē-rī′bō-nū-klē′ik) **acid (DNA)** is the genetic material of cells. The information directing the chemical processes that occur in organisms and therefore determine their characteristics is contained in DNA. **Ribonucleic** (rī′bō-nū-klē′ik) **acid (RNA)** is structurally related to DNA, and there are three types of RNA that play important roles in protein synthesis. In chapter 3 the means by which DNA and RNA direct the functions of the cell are described.

The **nucleic** (nū-klē′ik, nū-klā′ik) **acids** are large molecules composed of carbon, hydrogen, oxygen, nitrogen, and phosphorus. Both DNA and RNA consist of basic building blocks called **nucleotides** (nū′klē-ō-tīdz). Each nucleotide is composed of a monosaccharide to which a nitrogenous organic base and a phosphate group are attached (figure 2.24). The monosaccharide is deoxyribose for DNA, and ribose for RNA. The organic bases are thymine (thī′mēn, thī′min), cytosine (sī′tō-sēn), and uracil (yūr′ă-sil), which are single-ringed pyrimidines (pī-rim′i-dēnz), and adenine (ad′ĕ-nēn) and guanine (gwahn′ēn), which are double-ringed purines (pyūr′ēnz) (figure 2.25). The nucleotides are joined together in a chain by covalent bonds to form the nucleic acids.

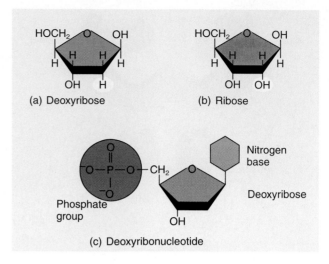

Figure 2.24   Components of Nucleotides

(*a*) Deoxyribose sugar that forms nucleotides used in DNA production. (*b*) Ribose sugar that forms nucleotides used in RNA production. Note that deoxyribose is ribose minus an oxygen atom. (*c*) Deoxyribonucleotide consisting of deoxyribose, a nitrogen base, and a phosphate group.

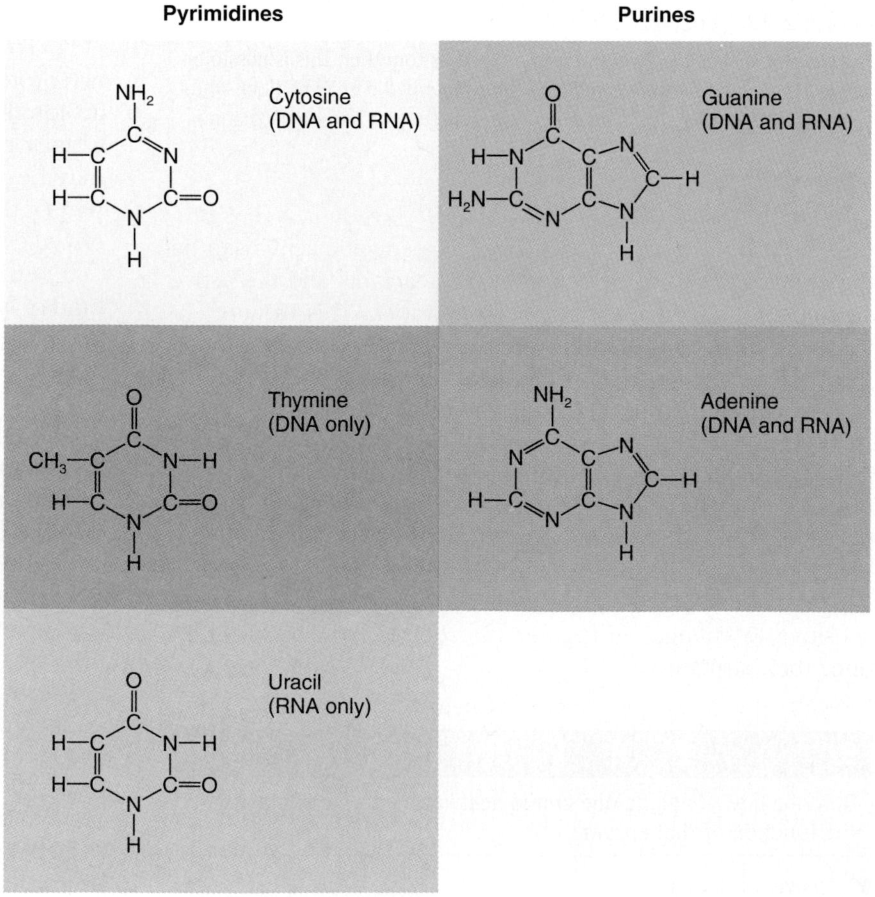

Figure 2.25   Nitrogenous Organic Bases

The organic bases found in nucleic acids are separated into two groups. Purines are double-ringed molecules, and pyrimidines are single-ringed molecules.

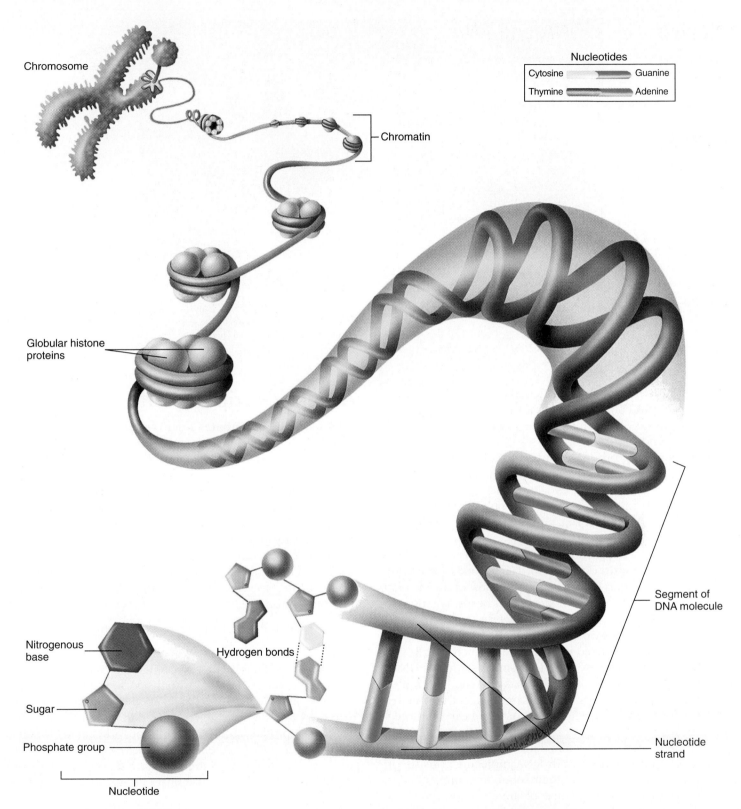

Nucleotides

| Cytosine | | Guanine |
| Thymine | | Adenine |

Chromosome

Chromatin

Globular histone
proteins

Segment of
DNA molecule

Nitrogenous
base

Hydrogen bonds

Sugar

Nucleotide
strand

Phosphate group

Nucleotide

Figure 2.26 Structure of DNA

Nucleotides join to form two strands. The nucleotides of one strand are joined by hydrogen bonds to the nucleotides of the other strand to form a DNA molecule. Associated with the DNA molecule are globular histone proteins. Usually the DNA molecule is stretched out, resembling a string of beads, and is called chromatin. During cell division, however, the chromatin condenses to form bodies called chromosomes.

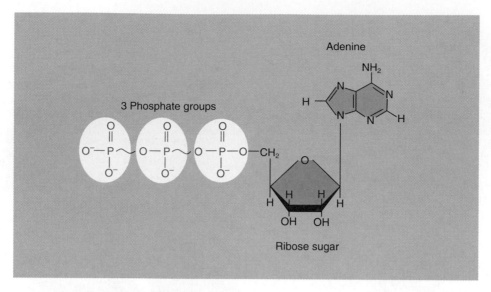

Figure 2.27   Adenosine Triphosphate (ATP) Molecule

DNA has two strands of nucleotides (figure 2.26). Each nucleotide of DNA contains one of the organic bases: adenine, thymine, cytosine, or guanine. The organic bases of one strand are bound to the organic bases of the other strand by hydrogen bonds to produce a twisted ladderlike structure called a helix. Adenine binds only to thymine because the structure of these organic bases allows two hydrogen bonds to form between them. Cytosine binds only to guanine because the structure of these organic bases allows three hydrogen bonds to form between them.

DNA molecules are associated with globular histone proteins to form **chromatin** (krō′ma-tin). The histone proteins are involved with regulating DNA function. For most of the life of a cell, chromatin is organized as a string with beads. During cell division, however, the chromatin condenses into structures called **chromosomes** (krō′mō-sōmz). DNA, chromatin, and chromosomes are considered in greater detail in chapter 3.

RNA has a structure similar to a single strand of DNA. Like DNA, four different nucleotides make up the RNA molecule, and the organic bases are the same, except that uracil substitutes for thymine (see figure 2.25). Uracil can bind only to adenine.

The sequence of organic bases in DNA molecules stores genetic information. Each DNA molecule consists of millions of organic bases, and their sequence ultimately determines the type and sequence of amino acids found in protein molecules. Because enzymes are proteins, DNA structure determines the rate and type of chemical reactions that occur in cells by controlling enzyme structure. The information contained in DNA therefore ultimately defines all cellular activities. Other proteins, such as collagen, that are coded by DNA determine many of the structural features of humans.

Adenosine Triphosphate

Adenosine triphosphate (ă-den′ō-sēn trī-fos′fāt) **(ATP)** is an especially important organic molecule found in all living organisms. It consists of adenosine and three phosphate groups (figure 2.27). Adenosine is the sugar ribose with the organic base adenine. The potential energy stored in the covalent bond between the second and third phosphate groups is important to living organisms because it provides the energy used in nearly all of the chemical reactions within cells.

The catabolism of glucose and other nutrient molecules results in chemical reactions that release energy. Some of that energy is used to synthesize ATP from ADP and an inorganic phosphate group (P_i):

$$\text{ADP} + P_i + \text{Energy (from catabolism)} \rightarrow \text{ATP}$$

The transfer of energy from nutrient molecules to ATP involves a series of oxidation–reduction reactions in which a high-energy electron is transferred from one molecule to the next molecule in the series. In chapter 25 the oxidation–reduction reactions of metabolism are considered in greater detail.

Once produced, ATP is used to provide energy for other chemical reactions (anabolism) or to drive cell processes such as muscle contraction. In the process ATP is converted back to ADP and an inorganic phosphate group.

$$\text{ATP} \rightarrow \text{ADP} + P_i + \text{Energy}$$
$$\text{(for anabolism and other cell processes)}$$

ATP is often called the energy currency of cells because ATP is capable of both storing and providing energy. The concentration of ATP is maintained within a narrow range of values, and essentially all energy-requiring chemical reactions stop when there is inadequate ATP.

Summary

Chemistry is the study of the composition, structure, and properties of substances and the reactions they undergo. Much of the structure and function of healthy or diseased organisms can be understood at the chemical level.

Basic Chemistry
Matter, Mass, and Weight

1. Matter is anything that occupies space.
2. Mass is the amount of matter in an object, and weight results from the force exerted by earth's gravity on matter.

Elements and Atoms

1. An element is the simplest type of matter with unique chemical and physical properties.
2. An atom is the smallest particle of an element that has the chemical characteristics of that element. An element is composed of only one kind of atom.
3. Atoms consist of protons, neutrons, and electrons.
   - Protons are positively charged, electrons are negatively charged, and neutrons have no charge.
   - Protons and neutrons are found in the nucleus, and electrons are located around the nucleus, which can be represented by an electron cloud.
4. The atomic number is the number of protons in an atom. The mass number is the sum of the protons and the neutrons.
5. Isotopes are atoms that have the same atomic number but different mass numbers.
6. The atomic mass of an element is the average mass of its naturally occurring isotopes weighted according to their abundances.
7. The molar mass is the weight in grams of one mole (Avogadro's number) of a substance.

Electrons and Chemical Bonding

1. The chemical behavior of atoms is determined mainly by their outermost electrons. A chemical bond occurs when atoms share or transfer electrons.
2. Ions are atoms that have gained or lost electrons.
   - An atom that loses an electron becomes positively charged and is called a cation. An anion is an atom that becomes negatively charged after accepting an electron.
   - Ionic bonding is the attraction of the oppositely charged cation and anion to each other.
3. A covalent bond is the sharing of electron pairs between atoms. A polar covalent bond results when the sharing of electrons is unequal and can produce a polar molecule that is electrically asymmetric.
4. In metallic bonding the outermost electrons of atoms are shared equally among all the atoms in a sample.

Molecules and Compounds

1. A molecule is two or more atoms chemically combined to form a structure that behaves as an independent unit. A compound is two or more *different* types of atoms chemically combined.
2. The molecular formula represents the number and kinds of atoms in a molecule. The formula unit represents the relative number of cations and ions in an ionic compound.
3. The molecular mass is the sum of the atomic masses of the atoms or ions in the molecular formula or formula unit.

Intermolecular Forces

1. A hydrogen bond is the weak attraction that occurs between the oppositely charged regions of polar molecules. Hydrogen bonds are important in determining the three-dimensional structure of large molecules.
2. Solubility is the ability of one substance to dissolve in another. Ionic substances that dissolve in water by dissociation are electrolytes. Molecules that do not dissociate are nonelectrolytes.

Chemical Reactions
Synthesis Reactions

1. Synthesis reactions are the chemical combination of two or more substances to form a new or larger substance.
2. Dehydration reactions are synthesis reactions in which water is produced.
3. Anabolism is the sum of all the synthesis reactions in the body.

Decomposition Reactions

1. Decomposition reactions are the chemical breakdown of a larger substance to two or more different smaller substances.
2. Hydrolysis reactions are decomposition reactions in which water is depleted.
3. All of the decomposition reactions in the body are called catabolism.

Oxidation–Reduction Reactions

Oxidation–reduction reactions involve the complete or partial transfer of electrons between atoms.

Reversible Reactions

Reversible reactions produce an equilibrium condition in which the amount of reactants relative to the amount of products remains constant.

Rate of Chemical Reactions

The rate of chemical reactions can be affected by the nature of the reactants, the concentration of the reactants, the temperature, and catalyst (enzymes).

Energy

Energy is the ability to do work. Potential energy is stored energy, and kinetic energy is energy resulting from movement of an object.

Electric Energy

Electric energy involves the movements of ions or electrons and is responsible for nerve impulses.

Electromagnetic Energy

Electromagnetic energy moves in waves.

Chemical Energy

1. Chemical bonds are a form of potential energy.
2. Chemical reactions in which the products have less potential energy than the reactants release energy. The energy can be lost as heat, used to synthesize molecules, or do work.
3. Chemical reactions in which the products contain more potential energy than the reactants require the input of energy.
4. Photosynthesis incorporates energy from the sun into chemical bonds, which can be broken, providing energy for the formation of ATP, which provides energy for many cellular processes.

Mechanical Energy

Mechanical energy is energy resulting from the position or movement of an object.

Heat Energy

1. Heat energy is energy that flows between objects that are at different temperatures.
2. Heat energy is released in chemical reactions and is responsible for body temperature.

Inorganic Chemistry

Inorganic chemistry is mostly concerned with noncarbon-containing substances but does include some carbon-containing substances, such as carbon dioxide and carbon monoxide.

Water

1. Water is a polar molecule composed of one atom of oxygen and two atoms of hydrogen.
2. Water stabilizes body temperature, protects against friction and trauma, makes chemical reactions possible, directly participates in chemical reactions (e.g., dehydration and hydrolysis reactions), and is a mixing medium (e.g., solutions, suspensions, and colloids).

Solution Concentrations

1. A liquid (solvent) containing a dissolved substance (solute) is a solution.
2. An osmole contains 1 mole (Avogadro's number) of particles (i.e., atoms, ions, or molecules) in 1 kilogram water. A milliosmole is 1/1000 of an osmole.

Acids and Bases

1. Acids are proton (i.e., hydrogen ion) donors, and bases (e.g., hydroxide ion) are proton acceptors.
2. A strong acid or base almost completely dissociates in water. A weak acid or base partially dissociates.

The pH Scale

1. A neutral solution has an equal number of hydrogen ions and hydroxide ions and is assigned a pH of 7.

2. Acid solutions, in which the number of hydrogen ions is greater than the number of hydroxide ions, have pH values less than 7.
3. Basic, or alkaline, solutions have more hydroxide ions than hydrogen ions and a pH greater than 7.

Salts

A salt is a molecule consisting of a cation other than hydrogen and an anion other than hydroxide. Salts are formed when acids react with bases.

Buffers

A buffer is a solution of a conjugate acid–base pair that resists changes in pH when acids or bases are added to the solution.

Oxygen

Oxygen is necessary in the reactions that extract energy from food molecules in living organisms.

Carbon Dioxide

During metabolism when the organic molecules are broken down, carbon dioxide and energy are released.

Organic Chemistry

Organic molecules contain carbon atoms bound together by covalent bonds.

Carbohydrates

1. Monosaccharides are the basic building blocks of other carbohydrates. They, especially glucose, are important sources of energy. Examples are ribose, deoxyribose, glucose, fructose, and galactose.
2. Disaccharide molecules are formed by dehydration reactions between two monosaccharides. They are broken apart into monosaccharides by hydrolysis reactions. Examples of disaccharides are sucrose, lactose, and maltose.
3. Polysaccharides are many monosaccharides bound together to form long chains. Examples include cellulose, starch, and glycogen.

Lipids

1. Triacylglycerols are composed of glycerol and fatty acids. One, two, or three fatty acids can attach to the glycerol molecule.
   - Fatty acids are straight chains of carbon molecules with a carboxyl group. Fatty acids can be saturated (only single covalent bonds between carbon atoms) or unsaturated (one or more double covalent bonds between carbon atoms).
   - Energy is stored in fats.
2. Phospholipids are lipids in which a fatty acid is replaced by a phosphate-containing molecule. Phospholipids are a major structural component of cell membranes.

3. Steroids are lipids composed of four interconnected ring molecules. Examples include cholesterol, bile salts, and sex hormones.
4. Other lipids include fat-soluble vitamins, prostaglandins, thromboxanes, and leukotrienes.

Proteins

1. The building blocks of protein are amino acids, which are joined by peptide bonds.
2. The number, kinds, and arrangement of amino acids determine the primary structure of a protein. Hydrogen bonds between amino acids determine secondary structure, and hydrogen bonds between amino acids and water determine tertiary structure. Interactions between different protein subunits determine quaternary structure.
3. Enzymes are specialized protein catalysts that lower the activation energy for chemical reactions. Enzymes speed up chemical reactions but are not consumed or altered in the process.
4. Activation energy is the minimum energy that the reactants must have to start a chemical reaction.
5. The active sites of enzymes bind only to specific reactants.
6. Cofactors are ions or organic molecules such as vitamins that are required for some enzymes to function.

Nucleic Acids: DNA and RNA

1. The basic unit of nucleic acids is the nucleotide, which is a monosaccharide with an attached phosphate and organic base.
2. DNA nucleotides contain the monosaccharide deoxyribose and the organic bases adenine, thymine, guanine, or cytosine. DNA occurs as a double strand of joined nucleotides and is the hereditary material of cells.
3. RNA nucleotides are composed of the monosaccharide ribose. The organic bases are the same as for DNA, except that thymine is replaced with uracil.

Adenosine Triphosphate

ATP stores energy derived from catabolism. The energy is released from ATP and is used in anabolism and other cell processes.

Content Review

1. Define chemistry. Why is an understanding of chemistry important for studying human anatomy and physiology?
2. Define matter, mass, weight, element, and atom.
3. Describe the structure of a single atom. Contrast the charge and the weight of the subatomic particles.
4. Define atomic number, mass number, isotope, atomic mass, mole, and molar mass.
5. Describe ionic, covalent, and metallic bonding. Define cation and anion.
6. Distinguish between a molecule and a compound. Define molecular formula and formula unit.
7. Define solubility and dissociate. Distinguish between an electrolyte and a nonelectrolyte.
8. How do polar covalent bonds result in hydrogen bonds and dissociation?
9. Define a chemical reaction. Contrast what occurs in synthesis and decomposition reactions. How do anabolism, catabolism, and metabolism relate to synthesis and decomposition reactions?
10. Describe a dehydration and a hydrolysis reaction.
11. Define oxidation–reduction reaction.
12. Describe reversible reactions. What is meant by the equilibrium condition in freely reversible reactions?
13. List four factors that affect the rate of chemical reactions. How must each factor change to increase the rate of reaction?
14. Define energy. How are potential and kinetic energy different from each other?
15. Describe electric energy, electromagnetic energy, chemical energy, mechanical, and heat energy.
16. Describe the relationship between the potential energy in reactants and products and the release or input of energy in chemical reactions. Use ATP as an example.
17. Define inorganic and organic chemistry.
18. List four functions that water performs in living systems.
19. Define solution, solute, and solvent. What is the osmolality of a solution?
20. Define acid and base. Describe the pH scale. What is the difference between a strong acid or base and a weak acid or base?
21. What is a salt? What is a buffer, and why are buffers important to organisms?
22. What are the functions of oxygen and carbon dioxide in living systems?
23. Name the four major types of organic molecules important to life.
24. Name the basic building blocks of carbohydrates, fats, proteins, and nucleic acids.
25. Distinguish between fats, phospholipids, and steroids. Name an example of each.
26. What makes proteins different from one another? Define peptide bond.
27. Describe the primary, secondary, tertiary, and quaternary structures of proteins.
28. Chemically, what type of organic molecule is an enzyme? What do enzymes do, and how do they work? Define cofactor and coenzyme.
29. What are the structural and functional differences between DNA and RNA?
30. Describe the structure of ATP. What role does this molecule play in energy exchange?

Develop Your Reasoning Skills

1. Iron has an atomic number of 26 and a mass number of 56. How many protons, neutrons, and electrons are in an atom of iron? If an atom of iron lost three electrons, what would the charge of the resulting ion be? Write the correct symbol for this ion.

2. Which of each of the following pairs of terms applies to the reaction that results in the formation of fatty acids and glycerol from a triacylglycerol molecule?
   a. Decomposition or synthesis reaction
   b. Anabolism or catabolism
   c. Dehydration or hydrolysis reaction

3. A mixture of chemicals is warmed slightly. As a consequence, although no more heat is added, the solution becomes very hot. Explain what occurred to make the solution so hot.

4. Two solutions, when mixed together at room temperature, produce a chemical reaction. When the solutions are boiled and allowed to cool to room temperature before mixing, however, no chemical reaction takes place. Explain.

5. In terms of the potential energy in the food, explain why eating food is necessary for increasing muscle mass.

6. Solution A has a pH of 2, and solution B has a pH of 8. If equal amounts of solutions A and B are mixed, is the resulting solution acidic or basic?

7. Given a buffered solution that is based on the following equilibrium:

$$CO_2 + H_2O \rightleftarrows H_2CO_3 \rightleftarrows H^+ + HCO_3^-$$

what happens to the pH of the solution if $NaHCO_3$ is added to the solution?

8. An enzyme E catalyzes the following reaction:

$$A + B \xrightarrow{E} C$$

The product C, however, binds to the active site of the enzyme in a reversible fashion and keeps the enzyme from functioning. What happens if A and B are continually added to a solution that contains a fixed amount of the enzyme?

9. Given the materials commonly found in a kitchen, explain how one could distinguish between a protein and a lipid.

10. A student is given two unlabeled substances: one a typical phospholipid and one a typical protein. She is asked to determine which substance is the protein and which is the phospholipid. The available techniques allow her to determine the elements in each sample. How can she identify each substance?

Web Site Link

For a listing of the most current web sites related to this chapter, please visit the Seeley home page at:
http://www.mhhe.com/biosci/ap/seeleyap/

Chapter Three

Structure and Function of the Cell

Objectives

1. Describe the structure of the plasma membrane. Explain why the plasma membrane is more permeable to lipid-soluble substances and small molecules than to large water-soluble substances.

2. Describe the structure and function of the nucleus and nucleoli.

3. Define cytoplasm, cytosol, and organelle.

4. Contrast microtubules, microfilaments, and intermediate filaments.

5. Compare the structure and function of rough and smooth endoplasmic reticulum.

6. Explain the role in secretion of the Golgi apparatus and secretory vesicles.

7. Distinguish between lysosomes and peroxisomes.

8. Describe the structure and function of mitochondria.

9. Describe centrioles, spindle fibers, cilia, flagella, and microvilli.

10. Describe the factors that affect the rate and the direction of diffusion of a solute in a solvent.

11. Explain the role of osmosis in controlling the movement of water across the plasma membrane. Compare isotonic, hypertonic, and hypotonic solutions with isosmotic, hyperosmotic, and hyposmotic solutions.

12. Describe mediated transport, and explain the characteristics of specificity, competition, and saturation.

13. Describe the processes of facilitated diffusion, active transport, secondary active transport, phagocytosis, pinocytosis, and exocytosis.

14. Define cell metabolism, and contrast aerobic and anaerobic respiration.

15. Describe the process of protein synthesis.

16. Explain what is accomplished during mitosis and cytokinesis.

17. Describe the events of meiosis, and explain how they result in the production of genetically unique individuals.

The human body is made up of trillions of cells. If each of these cells was about the size of a standard brick, we could build a colossal structure in the shape of a human over $5\frac{1}{2}$ miles (10 km) high! Obviously, there are many differences between a cell and a brick. Not only is a cell much smaller than a brick, but an average-sized cell is one-fifth the size of the smallest dot you can make on a sheet of paper with a sharp pencil! Also in marked contrast to bricks, that tiny cell is very much alive.

The cell is the structural and functional unit of all living organisms. All human cells originate from a single fertilized cell. During development, cell division and specialization give rise to trillions of cells with a wide variety of cell types, such as nerve, muscle, bone, fat, and blood cells. Each cell type has important characteristics, which are critical to the normal function of the body as a whole. One of the important reasons for maintaining homeostasis is to keep the trillions of cells that form the body functioning normally.

Although cells may have quite different structures and functions, all cells share some common characteristics (figure 3.1). The **plasma** (plaz′mă), or **cell, membrane** forms the outer boundary of the cell, through which the cell interacts with its external environment. The **nucleus** (nū′klē-ŭs) is usually located centrally and functions to direct cell activities, most of which take place in the **cytoplasm** (sī′tō-plazm), located between the plasma membrane and the nucleus.

This chapter presents the structure and function of the components of a cell. It is a brief overview of cell biology and offers adequate background information for the remainder of this text.

How We Learn About Cells

Because most cells are too small to be seen with the unaided eye, it is necessary to use microscopes to study cells. **Light microscopes** allow us to visualize some general features of cells. **Electron microscopes,** however, must be used to study the fine structure of cells. A **scanning electron microscope (SEM)** allows us to see features of the cell surface and the surfaces of internal structures. A **transmission electron microscope (TEM)** allows us to see "through" parts of the cell and thus to discover other aspects of cell structure. If you are not somewhat familiar with these types of microscopes, you should turn to the discussion on microscopic imaging in chapter 4.

Usually, cells must be killed before they can be studied under the microscope. As a result, microscopic techniques only allow us to study cell structure, and the function can only be implied. Other techniques, such as tissue culture, must be employed to study cell function directly; however, tissue culture also has limitations. Normal cells can only be grown for a short time in tissue culture, and some cell types cannot be grown at all this way. Therefore, as with most fields of science, what we know about cell structure and function must be derived from combinations of numerous observations and experiments.

One of the many things we can learn about cell function from tissue culture is that cells are very active. Many cell types actively move about in tissue culture. The plasma membrane and the internal scaffolding of the cell can reform rapidly, changing the shape of the cell and projecting membrane ruffles as the cell probes its microenvironment. Some muscle cells, whose normal function is to contract, continue to contract for quite some time when placed in tissue culture. For example, muscle cells taken from the heart and placed into tissue culture continue to contract rhythmically, approximating the rhythm of the beating heart.

Cells are constantly interacting with their immediate environment through the plasma membrane. Ions, such as sodium, potassium, and calcium are constantly being exchanged between the cytoplasm of the cell and the fluids surrounding the cell. Cells actively modify their microenvironment, and the nature of that environment can dramatically affect cell function. By concentrating certain ions on one side or the other of the plasma membrane, cells can develop charge differences across the plasma membrane, which allows cells to function like microscopic batteries. This charged condition is an important feature of a living cell's normal function.

Plasma Membrane

The **plasma membrane** is the outermost component of a cell. Substances outside the plasma membrane are **extracellular,** sometimes referred to as **intercellular** (between the cells), and substances inside it are **intracellular.** The functions of the plasma membrane are to enclose and support the cell contents and to determine what moves into and out of the cell. Other important functions of the plasma membrane are recognition of and communication with other cells.

The plasma membrane consists of 45%–50% lipids, 45%–50% proteins, and 4%–8% carbohydrates (figure 3.2). The predominant lipids are phospholipids and cholesterol. **Phospholipids** readily assemble to form a **lipid bilayer,** a double layer of lipid molecules, because they have a polar (charged) head and a nonpolar (uncharged) tail (see chapter 2). The polar **hydrophilic** (water-loving) heads are exposed to water inside and outside the cell, whereas the nonpolar **hydrophobic** (water-fearing) tails face one another in the interior of the plasma membrane. The other major lipid in the plasma membrane is **cholesterol** (see chapter 2), which is interspersed among the phospholipids and accounts for about a third of the total lipids in the plasma membrane. Cholesterol is too hydrophobic to extend to the hydrophilic surface of the membrane but lies within the hydrophobic region of the phospholipids. The amount of cholesterol in a given membrane is a major factor in determining the fluid nature of the membrane, which is critical to its function.

The modern concept of the plasma membrane, the **fluid-mosaic model,** suggests that the plasma membrane is neither rigid nor static in structure but is highly flexible and can change its shape and composition through time. The lipid

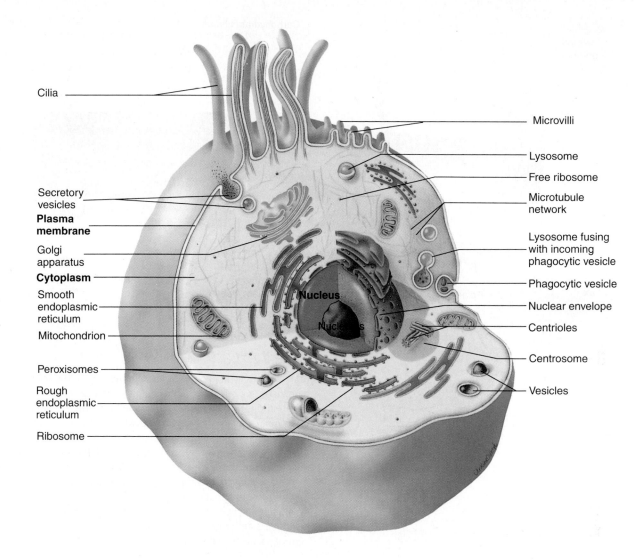

Cilia

Microvilli

Lysosome

Free ribosome

Secretory
vesicles

**Plasma
membrane**

Microtubule
network

Golgi
apparatus

Lysosome fusing
with incoming
phagocytic vesicle

Cytoplasm

Smooth
endoplasmic
reticulum

Phagocytic vesicle

Nucleus

Nuclear envelope

Nucleolus

Mitochondrion

Centrioles

Centrosome

Peroxisomes

Rough
endoplasmic
reticulum

Vesicles

Ribosome

Figure 3.1   The Cell

A generalized human cell showing the plasma membrane, nucleus, and cytoplasm with its organelles. Although no single cell contains all these organelles, many cells contain a large number of them. 🏃 📼

bilayer functions as a liquid in which other molecules such as proteins "float." The fluid nature of the lipid bilayer has several important consequences. It provides an important means of distributing molecules within the plasma membrane. In addition, slight damage to the membrane can be repaired because the phospholipids tend to reassemble around damaged sites and seal them closed. In addition, the fluid nature of the lipid bilayer enables membranes to fuse with one another.

Although the basic structure of the plasma membrane is determined mainly by its lipids, the functions of the plasma membrane are determined mainly by its proteins. Some protein molecules, called **integral,** or **intrinsic proteins,** penetrate the lipid bilayer from one surface to the other (figure 3.3), whereas other proteins, called **peripheral,** or **extrinsic proteins,** are attached to either the inner or outer surfaces of the lipid bilayer. Integral proteins consist of regions made up of amino acids with hydrophobic R

groups and other regions of amino acids with hydrophilic R groups. The hydrophobic regions are located within the hydrophobic part of the membrane, and the hydrophilic regions are located at the inner or outer surface of the membrane or line channels through the membrane. Peripheral proteins are usually bound to integral proteins. Some membrane proteins form channels through the membrane (figure 3.4) or act as carrier molecules. Other membrane proteins are receptors, markers, enzymes, or structural supports in the membrane. The ability of membrane proteins to function depends on their three-dimensional shape.

Channel proteins (figure 3.5a) are one or more integral proteins arranged so that they form a tiny channel through the plasma membrane. The hydrophobic regions of the proteins face outward toward the hydrophobic part of the cell membrane, and the hydrophilic regions of the protein line the channel. Small molecules or ions of the right shape, size,

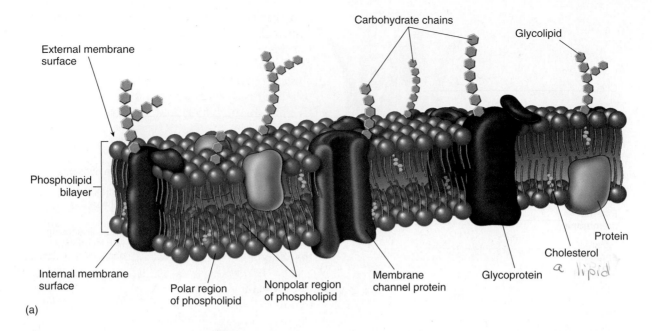

External membrane surface

Carbohydrate chains

Glycolipid

Phospholipid bilayer

Internal membrane surface

Polar region of phospholipid

Nonpolar region of phospholipid

Membrane channel protein

Glycoprotein   *a lipid*

Cholesterol

Protein

(a)

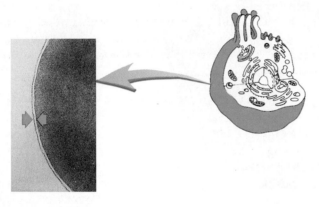

(b)

Figure 3.2   Cell Membrane

(*a*) Fluid-mosaic model of the plasma membrane. The membrane is composed of a bilayer of phospholipids and cholesterol with proteins "floating" in the membrane. The nonpolar hydrophobic region of each phospholipid molecule is directed toward the center of the membrane, and the polar hydrophilic region is directed toward the water environment either outside or inside the cell. (*b*) Transmission electron micrograph of the cell membrane of a human red blood cell, with the membrane indicated by the blue arrows. Proteins at either surface of the lipid bilayer stain more readily than the lipid bilayer does and give the membrane the appearance of consisting of three parts: the two dark outer parts are proteins and the phospholipid heads, and the lighter central part is the phospholipid tails.

and charge can pass through the channel. The charges in the hydrophilic part of the channel protein determine which types of ions can pass through the channel.

The function of a channel protein is determined by its shape. The channel can be open or closed, depending on the shape of the channel proteins. Some channel proteins change shape to open the channel when a ligand binds to a specific receptor site on the protein. This is called a **ligand-gated** channel. Other channel proteins change shape to open the channel when there is a change in charge across the cell membrane. This is called a **voltage-gated** channel.

Receptor molecules (figure 3.5*b*) are proteins in the cell membrane with an exposed **binding site** on the outer cell surface, which can attach to specific **ligand** (lī′gand, meaning, the molecule bound by the receptor) molecules. The receptors and the ligands they bind are part of an intercellular communication system that facilitates coordination of cell activities. For example, a nerve cell can release a chemical messenger that diffuses to a muscle cell and binds to its receptor. The binding acts as a signal that triggers a response, such as con-

traction in the muscle cell. The same chemical messenger would have no effect on another cell that lacks the receptor molecule. Some receptor molecules function by means of a **G protein** complex located on the inner surface of the cell membrane. G proteins may function in one of several ways. For example, when a ligand such as a hormone attaches to the receptor molecule, the G protein complex binds guanosine triphosphate (GTP) and is activated. The activated G protein, in turn, activates adenylate cyclase, which catalyzes the conversion of adenosine triphosphate (ATP) to cyclic adenosine monophosphate (cAMP). cAMP functions as a **second messenger** inside the cell, stimulating a variety of cell functions.

Marker molecules are cell surface molecules that allow cells to identify and attach to each other. They are mostly **glycoproteins** (proteins with attached carbohydrates) or **glycolipids** (lipids with attached carbohydrates) (figure 3.5*c*). Examples include recognition of the oocyte by the sperm cell and the ability of the immune system to distinguish between self-cells and foreign cells, such as bacteria or donor cells in an organ transplant. Intercellular com-

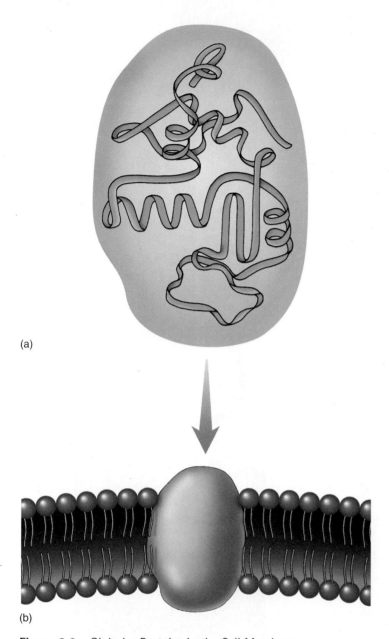

(a)

(b)

Figure 3.3   Globular Proteins in the Cell Membrane

(*a*) Proteins are commonly depicted as ribbons (see chapter 2). The domain occupied by the protein ribbon can be enclosed by a 3-D shaded region. (*b*) The shaded region can be depicted as a 3-D globular integral protein inserted into the cell membrane.

munication and recognition are important because cells are not isolated entities and they must work together to ensure normal body functions.

Nucleus

The **nucleus,** which contains most of the genetic information of the cell, is a large, membrane-bound structure usually located near the center of the cell. It may be spherical, elongated, or lobed, depending on the cell type. All cells of the

body have a nucleus at some point in their life cycle, although some cells, such as red blood cells (also called red blood corpuscles), lose their nuclei as they develop. Other cells, such as skeletal muscle cells and certain bone cells, called osteoclasts, contain more than one nucleus. The nucleus is surrounded by a **nuclear envelope** (figure 3.6) composed of two membranes separated by a space. At many points on the surface of the nuclear envelope, the inner and outer membranes fuse to form porelike structures, the **nuclear pores.** Molecules move between the nucleus and the cytoplasm through these nuclear pores.

Deoxyribonucleic acid (DNA) and associated proteins are dispersed throughout the nucleus as thin strands about 4–5 nanometers (nm) in diameter (see appendix A). The proteins include **histones** (his′tōnz) and other proteins that play a role in the regulation of DNA function. The DNA and protein strands can be stained with dyes and are called **chromatin** (krō′ma-tin; meaning colored material). Chromatin is distributed throughout the nucleus but is more condensed and more readily stained in some areas than in others. The more highly condensed chromatin apparently is less functional than the more evenly distributed chromatin, which stains lighter. During cell division the chromatin condenses to form the more solid bodies called **chromosomes** (colored bodies).

DNA ultimately determines the structure of proteins (protein synthesis is described later in this chapter). Many structural components of the cell and all the enzymes, which regulate most chemical reactions in the cell, are proteins. By determining protein structure, DNA therefore ultimately controls the structural and functional characteristics of the cell. DNA does not leave the nucleus, but functions by means of an intermediate, **ribonucleic acid (RNA),** which can leave the nucleus. DNA determines the structure of messenger RNA (mRNA), ribosomal RNA (rRNA), and transfer RNA (tRNA) (all described in more detail later). mRNA moves out of the nucleus through the nuclear pores into the cytoplasm, where it determines the structure of proteins.

Clinical Note

The **Human Genome Project** is an ambitious international project, which began in 1990, with the 15-year goal of mapping and sequencing the entire human genome by the year 2005. The **genome** is the total of all the genes contained within each cell. One goal of the Human Genome Project is to construct a map indicating where each of the approximately 70,000–100,000 genes is located on the human chromosomes. The other major goal of the project is to determine the sequence of the estimated 3 billion base pairs (bp) that make up the human DNA molecules. To date, many genes implicated in human genetic disorders, such as Huntington's disease, cystic fibrosis, neurofibromatosis, and colon cancer genes, have been mapped and sequenced. It is hoped that by knowing for what proteins the genes implicated in these and other genetic disorders are coded, and by determining the functions of those proteins, we will be able to more effectively treat these diseases.

Some regions of a protein are helical. Each helical region can be depicted as a cylinder.

In some membrane proteins, the helical regions form a circle with a channel in the center.

Protein

The ring of cylinders can be depicted as a 3-D globular structure with a channel in the center. This is called a channel protein.

The channel protein can be depicted cut in half to show the channel.

The cut channel protein is depicted in location within the cell membrane.

Figure 3.4   Channel Protein

Because mRNA synthesis occurs within the nucleus, cells without nuclei accomplish protein synthesis only as long as the mRNA produced before the nucleus degenerates remains functional. The nuclei of developing red blood cells are expelled from the cells before the red blood cells enter the blood, where they survive without a nucleus for about 120 days. In comparison, many cells with nuclei, such as nerve and skeletal muscle cells, survive as long as the individual person survives.

A **nucleolus** (nū-klē′ō-lŭs) is a somewhat rounded, dense region within the nucleus that lacks a surrounding membrane (see figure 3.6). There is usually one nucleolus per

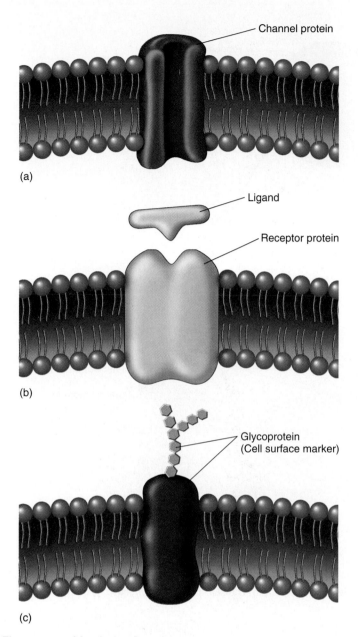

(a)

(b)

(c)

Figure 3.5   Membrane Proteins

(*a*) Channel protein. Integral membrane proteins forming a tiny channel through the plasma membrane. (*b*) Receptor protein. A protein in the cell membrane with an exposed binding site, which can attach to specific ligands. Once the ligand is bound to the receptor, other portions of the receptor can activate other molecules, such as enzymes or G proteins. (*c*) Cell surface marker. Glycoproteins on the cell surface allow cells to identify and attach to one another.

nucleus, but several smaller, accessory nucleoli may also be seen in some nuclei, especially during the latter phases of cell division. The nucleolus incorporates portions of 10 chromosomes (five pairs), called **nucleolar organizer regions.** These regions contain DNA from which rRNA is produced. Within the nucleolus, the subunits of ribosomes are manufactured (see section on Ribosomes).

Cytoplasm

Cytoplasm, the cellular material outside the nucleus but inside the plasma membrane, is about half cytosol and half organelles.

Cytosol

Cytosol (sī′tō-sol) consists of a fluid portion, a cytoskeleton, and cytoplasmic inclusions. The fluid portion of cytosol is a solution with dissolved ions and molecules and a colloid with suspended molecules, especially proteins. Many of these proteins are enzymes that catalyze the breakdown of molecules for energy or the synthesis of sugars, fatty acids, nucleotides, amino acids, and other molecules.

Cytoskeleton

The **cytoskeleton** supports the cell and holds the nucleus and organelles in place. It is also responsible for cell movements, such as changes in cell shape or movement of cell organelles. The cytoskeleton consists of three groups of proteins: microtubules, actin filaments, and intermediate filaments (figure 3.7).

   Microtubules are hollow tubules composed primarily of protein units called **tubulin.** The microtubules are about 25 nm in diameter, with walls that are about 5 nm thick. Microtubules vary in length but are normally several micrometers (μm) long. Microtubules play a variety of roles within cells. They help provide support and structure to the cytoplasm of the cell, much like an internal scaffolding. They are involved in the process of cell division and form essential components of certain cell organelles, such as centrioles, spindle fibers, cilia, and flagella.

   Actin filaments, or **microfilaments,** are small fibrils about 8 nm in diameter that form bundles, sheets, or networks in the cytoplasm of cells. Actin filaments provide structure to the cytoplasm and mechanical support for microvilli. Actin filaments support the plasma membrane and define the shape of the cell. Changes in cell shape involve the breakdown and reconstruction of actin filaments. Actin filaments are involved in cell movement. Cell motility in cells that can move about is accomplished by changes in cell shape mediated by the actin cytoskeleton. Muscle cells contain a large number of highly organized actin filaments responsible for the muscle's contractile capabilities (see chapter 10).

   Intermediate filaments are protein fibers about 10 nm in diameter. They provide mechanical strength to cells. For example, intermediate filaments support the extensions of nerve cells, which have a very small diameter but can be a meter in length.

Cytoplasmic Inclusions

The cytosol also contains **cytoplasmic inclusions,** which are aggregates of chemicals either produced by the cell or taken in by the cell. For example, lipid droplets or glycogen

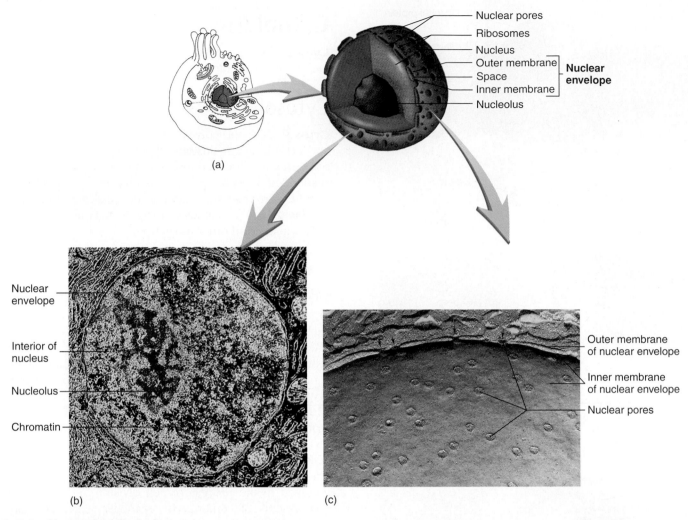

Figure 3.6   Nucleus and Nucleolus

(*a*) The nuclear envelope consists of inner and outer membranes that become fused at the nuclear pores. The nucleolus is a condensed region of the nucleus not bounded by a membrane and consisting mostly of RNA and protein. (*b*) Transmission electron micrograph of the nucleolus within the nucleus. (*c*) Scanning electron micrograph showing the inner nuclear membrane of the nuclear envelope and the nuclear pores (*arrowheads*).

granules store energy-rich molecules; hemoglobin in red blood cells transports oxygen; melanin is a pigment that colors the skin, hair, and eyes; and **lipochromes** (lip′ō-krōmz) are pigments that increase in amount with age. Dust, minerals, and dyes can also accumulate in the cytoplasm.

Organelles

Organelles are small structures within cells that are specialized for particular functions, such as manufacturing proteins or producing ATP. Most organelles have membranes that are similar to the plasma membrane. The membranes separate the organelles from the rest of the cytoplasm, creating a subcellular compartment with its own enzymes that is capable of carrying out its own unique chemical reactions. The nucleus is an example of an organelle.

The number and type of cytoplasmic organelles within each cell are related to the specific structure and function of

the cell. Cells secreting large amounts of protein contain well-developed organelles that synthesize and secrete protein, whereas cells actively transporting substances such as sodium ions across their plasma membrane contain highly developed organelles that produce ATP. The following sections describe the structure and main functions of the major cytoplasmic organelles found in cells.

Ribosomes

Ribosomes (rī′bō-sōms) are the sites of protein synthesis. Each ribosome is composed of a large subunit and a smaller one. The ribosomal subunits, which consist of **ribosomal RNA (rRNA)** and proteins, are assembled separately in the nucleolus of the nucleus. The ribosomal subunits then move through the nuclear pores into the cytoplasm, where they assemble to form the functional ribosome during protein synthesis (figure 3.8). Ribosomes can be found free in the cyto-

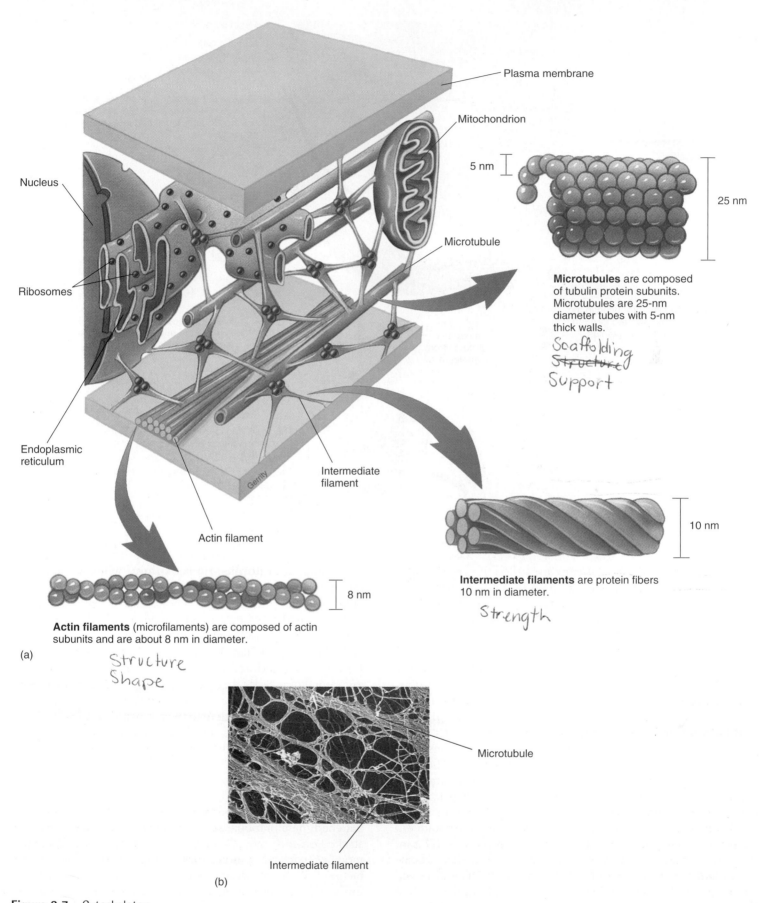

Plasma membrane

Mitochondrion

5 nm

25 nm

Nucleus

Ribosomes

Microtubule

Microtubules are composed of tubulin protein subunits. Microtubules are 25-nm diameter tubes with 5-nm thick walls.

Scaffolding
Structure
Support

Endoplasmic reticulum

Intermediate filament

Actin filament

10 nm

Intermediate filaments are protein fibers 10 nm in diameter.

Strength

8 nm

Actin filaments (microfilaments) are composed of actin subunits and are about 8 nm in diameter.

(a)

Structure
Shape

Microtubule

Intermediate filament

(b)

Figure 3.7 Cytoskeleton

(a) Diagram of the cytoskeleton. (b) Scanning electron micrograph of the cytoskeleton.

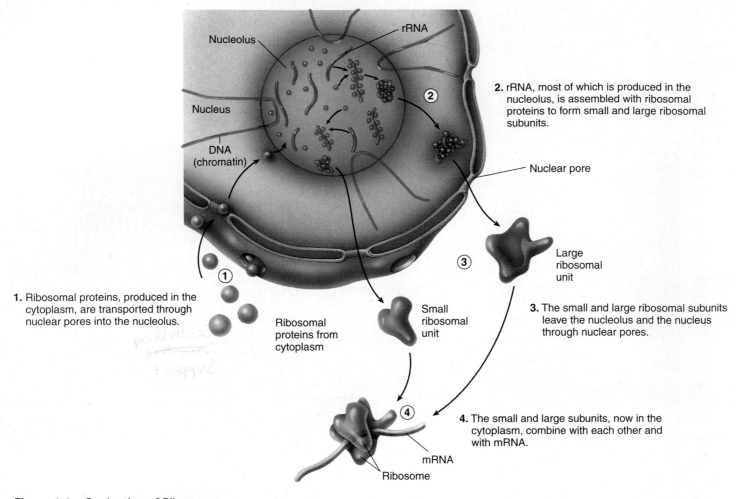

2. rRNA, most of which is produced in the nucleolus, is assembled with ribosomal proteins to form small and large ribosomal subunits.

1. Ribosomal proteins, produced in the cytoplasm, are transported through nuclear pores into the nucleolus.

3. The small and large ribosomal subunits leave the nucleolus and the nucleus through nuclear pores.

4. The small and large subunits, now in the cytoplasm, combine with each other and with mRNA.

Figure 3.8   Production of Ribosomes

plasm or associated with a membrane called the endoplasmic reticulum. **Free ribosomes** primarily synthesize proteins used inside the cell, whereas endoplasmic reticulum ribosomes can produce proteins that are secreted from the cell.

Endoplasmic Reticulum

The outer membrane of the nuclear envelope is continuous with a series of membranes distributed throughout the cytoplasm of the cell, collectively referred to as the **endoplasmic reticulum** (en′dō-plas′mik re-tik′yū-lŭm; network inside the cytoplasm) (figure 3.9). The endoplasmic reticulum consists of broad, flattened, interconnecting sacs and tubules. The interior spaces of those sacs and tubules are called **cisternae** (sister′nē) and are isolated from the rest of the cytoplasm.

Rough endoplasmic reticulum is endoplasmic reticulum with attached ribosomes. The ribosomes of the rough endoplasmic reticulum produce proteins for secretion and for internal use. The amount and configuration of the endoplasmic reticulum within the cytoplasm depend on the cell type and function. Cells with abundant rough endoplasmic reticulum synthesize large amounts of protein that are secreted for use outside the cell.

Smooth endoplasmic reticulum, which is endoplasmic reticulum without attached ribosomes, manufactures lipids, such as phospholipids, cholesterol, steroid hormones, and carbohydrates such as glycogen. Cells that synthesize large amounts of lipid contain dense accumulations of smooth endoplasmic reticulum. Enzymes required for lipid synthesis are associated with the membranes of the smooth endoplasmic reticulum. Smooth endoplasmic reticulum also participates in the detoxification processes by which enzymes act on chemicals and drugs to change their structure and reduce their toxicity. The smooth endoplasmic reticulum of skeletal muscle stores calcium ions that function in muscle contraction.

Golgi Apparatus

The **Golgi** (gōl′jē) **apparatus** (figure 3.10) is composed of flattened membranous sacs, containing cisternae, that are stacked on each other like dinner plates. The Golgi apparatus modifies, packages, and distributes proteins and lipids manufactured by the rough and smooth endoplasmic reticula (figure 3.11). Proteins produced at the ribosomes of the rough endoplasmic reticulum are surrounded by a **vesicle** (ves′i-kl), or little sac, that forms from the membrane of the endoplasmic

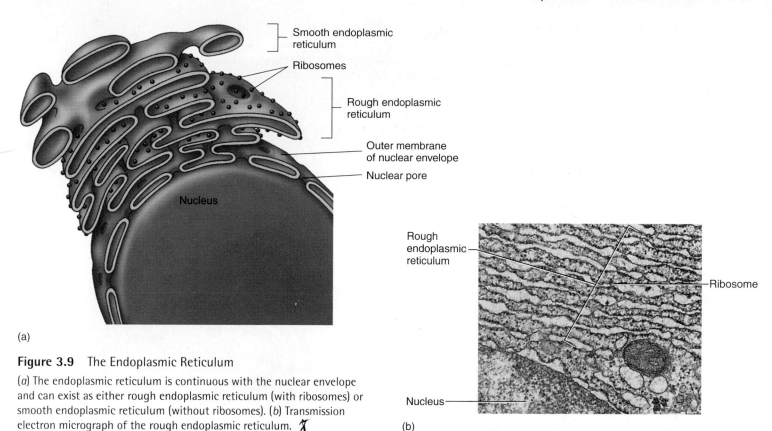

(a)

Figure 3.9   The Endoplasmic Reticulum

(*a*) The endoplasmic reticulum is continuous with the nuclear envelope and can exist as either rough endoplasmic reticulum (with ribosomes) or smooth endoplasmic reticulum (without ribosomes). (*b*) Transmission electron micrograph of the rough endoplasmic reticulum. ✗

(b)

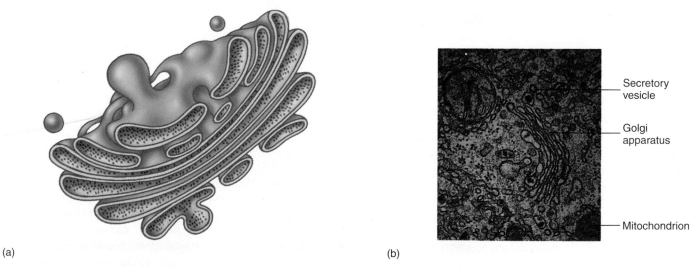

(a)   (b)

Figure 3.10   Golgi Apparatus

(*a*) The Golgi apparatus is composed of flattened membranous sacs, containing cisternae, and resembles a stack of dinner plates or pancakes. (*b*) Transmission electron micrograph of the Golgi apparatus. ✗

reticulum. The vesicle moves to the Golgi apparatus, fuses with the membrane of the Golgi apparatus, and releases the protein into the cisterna of the Golgi apparatus. The Golgi apparatus concentrates and, in some cases, chemically modifies the proteins by synthesizing and attaching carbohydrate molecules to the proteins to form glycoproteins or attaching lipids to proteins to form lipoproteins. The proteins are then pack-

aged into vesicles that pinch off from the margins of the Golgi apparatus and are distributed to various locations. Some vesicles carry proteins to the plasma membrane where the proteins are secreted from the cell by exocytosis; other vesicles contain proteins that become part of the plasma membrane; and still other vesicles contain enzymes that are used within the cell.

1. Proteins are produced at ribosomes on the surface of the rough endoplasmic reticulum and are transferred into the cisterna as they are produced.

2. The proteins are surrounded by vesicles that form from the membrane of the endoplasmic reticulum.

3. The vesicle moves from the endoplasmic reticulum to the Golgi apparatus, fuses with the membrane of the Golgi apparatus, and releases the protein into the cisterna of the Golgi apparatus.

4. The Golgi apparatus concentrates and, in some cases, modifies the proteins into glycoproteins or lipoproteins.

5. The proteins are packaged into vesicles that form from the membrane of the Golgi apparatus.

6. Some vesicles contain enzymes that are used within the cell.

7. Some vesicles carry proteins to the plasma membrane where the proteins are secreted from the cell by exocytosis.

8. Some vesicles contain proteins that become part of the plasma membrane.

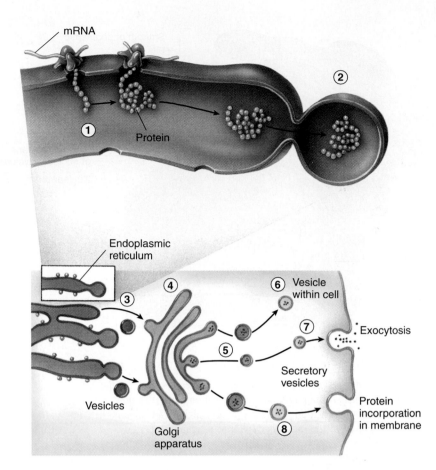

Figure 3.11   Function of the Golgi Apparatus

The Golgi apparatuses are most numerous and most highly developed in cells that secrete large amounts of protein or glycoproteins, such as cells in the salivary glands and the pancreas.

Secretory Vesicles

The membrane-bound **secretory vesicles** (see figure 3.11) that pinch off from the Golgi apparatus move to the surface of the cell, their membranes fuse with the plasma membrane, and the contents of the vesicle are released to the exterior by exocytosis. The membranes of the vesicles are then incorporated into the plasma membrane.

Secretory vesicles accumulate in many cells, but their contents frequently are not released to the exterior until a signal is received by the cell. For example, secretory vesicles that contain the hormone insulin do not release it until the concentration of glucose in the blood increases and acts as a signal for the secretion of insulin from the cells.

Lysosomes

Lysosomes (lī′sō-sōmz) are membrane-bound vesicles that pinch off from the Golgi apparatus (see figure 3.11). They contain a variety of hydrolytic enzymes that function as intracellu-

lar digestive systems. Vesicles taken into the cell fuse with the lysosomes to form one vesicle and to expose the phagocytized materials to hydrolytic enzymes (figure 3.12). Various enzymes within lysosomes digest nucleic acids, proteins, polysaccharides, and lipids. Certain white blood cells have large numbers of lysosomes that contain enzymes to digest phagocytized bacteria. Lysosomes also digest organelles of the cell that are no longer functional in a process called **autophagia** (aw′tō-fā′jē-ă, meaning self-eating). Furthermore, when tissues are damaged, ruptured lysosomes within the damaged cells release their enzymes, which digest both damaged and healthy cells. In other cells the lysosomes move to the plasma membrane, and the enzymes are secreted by exocytosis. For example, the normal process of bone remodeling involves the breakdown of bone tissue by specialized bone cells. Enzymes responsible for that degradation are released into the extracellular fluid from lysosomes produced by those cells.

Clinical Note

Some diseases result from nonfunctional lysosomal enzymes. For example, **Pompe's disease** results from the inability of lysosomal enzymes to break down glycogen. The glycogen accumulates in large amounts in the heart, liver, and skeletal muscles, an accumulation that often leads to heart failure. **Familial hyperlipoproteinemia** is a

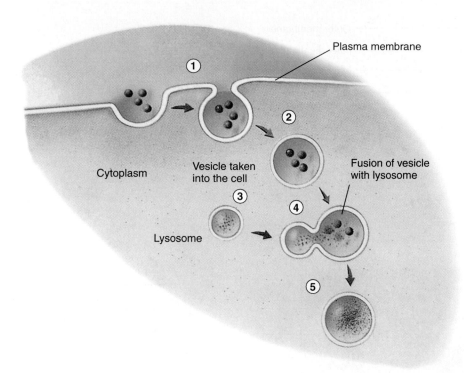

Plasma membrane

Cytoplasm

Vesicle taken
into the cell

Fusion of vesicle
with lysosome

Lysosome

1. Vesicles containing materials
   from outside the cell are
   taken into the cell.

2. The vesicle is pinched off from
   the plasma membrane and
   becomes a separate vesicle.

3. A lysosome approaches the vesicle.
4. The lysosome fuses with the vesicle.

5. The enzymes from the lysosome mix with
   the material in the vesicle, and the
   enzymes digest the material.

Figure 3.12   Action of Lysosomes

group of genetic disorders characterized by the accumulation of large amounts of lipids in phagocytic cells that lack the normal enzymes required to break down the lipid droplets. Symptoms include abdominal pain, enlargement of the spleen and liver, and eruption of yellow nodules in the skin filled with the affected phagocytic cells. **Mucopolysaccharidoses,** such as **Hurler's syndrome,** are diseases in which lysosomal enzymes are unable to break down mucopolysaccharides (glycosaminoglycans), so these molecules accumulate in the lysosomes of connective tissue cells and nerve cells. People affected by these diseases suffer mental retardation and skeletal deformities.

Peroxisomes

Peroxisomes (per-ok′si-sōmz) are membrane-bound vesicles that are smaller than lysosomes. Peroxisomes contain enzymes that break down fatty acids and amino acids. Hydrogen peroxide (H_2O_2), which can be toxic to the cell, is a by-product of that breakdown. Peroxisomes also contain the enzyme **catalase,** which breaks down hydrogen peroxide to water and oxygen. Cells that are active in detoxification, such as liver and kidney cells, have many peroxisomes.

Mitochondria

Mitochondria (mī-tō-kon′drē-ă) usually are illustrated as small, rod-shaped structures (figure 3.13). In living cells, time-lapse photomicrography reveals that mitochondria constantly change shape from spherical to rod-shaped or even to long, threadlike structures. Mitochondria are the major sites of ATP

production, which is the major energy source for most endergonic chemical reactions within the cell. Each mitochondrion has an inner and outer membrane separated by an intermembranous space. The outer membrane has a smooth contour, but the inner membrane has numerous infoldings called **cristae** (kris′tē) that project like shelves into the interior of the mitochondria.

A complex series of mitochondrial enzymes forms two major enzyme systems that are responsible for oxidative metabolism and most ATP synthesis (see chapter 25). The enzymes of the citric acid (or Krebs) cycle are found in the **matrix,** which is the substance located in the space formed by the inner membrane. The enzymes of the electron transport chain are embedded within the inner membrane. Cells with a greater energy requirement have more mitochondria with more cristae than cells with lower energy requirements. Within the cytoplasm of a given cell, the mitochondria are more numerous in areas in which ATP is used. For example, mitochondria are numerous in cells that perform active transport and are packed near the membrane where active transport occurs.

Increases in the number of mitochondria result from the division of preexisting mitochondria. When muscles enlarge as a result of exercise, the number of mitochondria within the muscle cells increases to provide the additional ATP required for muscle contraction.

The information for making some mitochondrial proteins is stored in DNA contained within the mitochondria themselves, and those proteins are synthesized on ribosomes within the mitochondria. The structure of many other mitochondrial

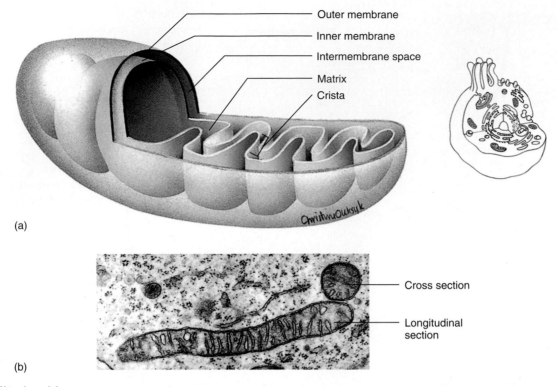

Outer membrane
Inner membrane
Intermembrane space
Matrix
Crista

(a)

Cross section

Longitudinal
section

(b)

Figure 3.13 Mitochondrion

(*a*) Typical mitochondrion structure. (*b*) Transmission electron micrograph of mitochondria in longitudinal and cross section.

proteins is determined by nuclear DNA, however, and these proteins are synthesized on ribosomes within the cytoplasm and then transported into the mitochondria. Both the mitochondrial DNA and mitochondrial ribosomes are very different from those within the nucleus and cytoplasm of the cell, respectively. Mitochondrial DNA is a closed circle of about 16,500 bp coding for 37 genes, compared with the open strands of nuclear DNA, which is composed of 3 billion bp coding for 100,000 genes. In addition, unlike nuclear DNA, mitochondrial DNA does not have associated proteins.

1   P R E D I C T

Describe the structural characteristics of cells that are highly specialized to do the following: (a) synthesize and secrete proteins; (b) actively transport substances into the cell; (c) synthesize lipids; and (d) phagocytize foreign substances.

✔ *Answer in Appendix F*

Clinical Note

Half of the nuclear DNA of an individual is derived from the mother, and half is derived from the father; but mitochondrial DNA comes only from the mother. The mitochondria of the sperm cell from the father are not incorporated into the oocyte at the time of fertilization. Because only the mother's mitochondrial DNA is passed down from generation to generation, maternal pedigrees are much easier to trace using mitochondrial DNA than with nuclear DNA. This unique quality

of mitochondria has been used in a number of studies, from reuniting mothers or grandmothers with lost children to searching for the origins of the human species. A number of degenerative disorders affecting the nervous system, heart, or kidneys have been linked to mutations in mitochondrial DNA. The study of these disorders is providing some valuable clues to the aging process.

Centrioles and Spindle Fibers

The **centrosome** (sen′trō-sōm) is a specialized zone of cytoplasm close to the nucleus that contains two **centrioles** (sen′trē-ōlz). Each centriole is a small, cylindrical organelle about 0.3–0.5 μm in length and 0.15 μm in diameter, and the two centrioles are normally oriented perpendicular to each other within the centrosome (see figure 3.1). The wall of the centriole is composed of nine evenly spaced, longitudinally oriented, parallel units, or triplets. Each unit consists of three parallel microtubules joined together (figure 3.14).

The centrosome is the center of microtubule formation. Microtubules, in turn, appear to influence the distribution of actin and intermediate filaments. Through its control of microtubule formation, the centrosome is therefore closely involved in determining cell shape and movement. The microtubules extending from the centrosomes are very dynamic—constantly growing and shrinking.

Before cell division, the two centrioles double in number, the centrosome divides into two, and one centrosome, containing two centrioles, moves to each end of the cell. Mi-

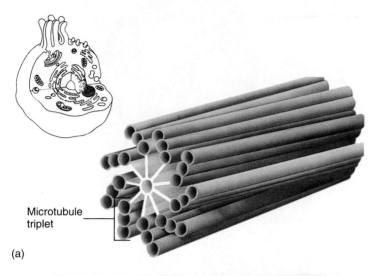

Microtubule
triplet

(a)

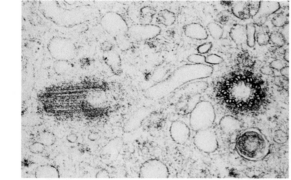

(b)

Figure 3.14   Centriole

(a) Structure of a centriole, which comprises nine triplets of microtubules. Each triplet contains one complete microtubule fused to two incomplete microtubules. (b) Transmission electron micrograph of a pair of centrioles, which are normally located together near the nucleus. One is shown in cross section, and one in longitudinal section. ✗

crotubules called **spindle fibers** extend out in all directions from the centrosome. These microtubules grow and shrink even more rapidly than those of nondividing cells. If the extended end of a spindle fiber comes in contact with a **kinetochore** (ki-nē′tō-kōr), a specialized region on each chromosome, the spindle fiber attaches to the kinetochore and stops growing or shrinking. Eventually spindle fibers from each centromere bind to the kinetochores of all the chromosomes. The chromosomes are pulled apart and moved by the microtubules toward the two centrosomes during cell division (see the section on Cell Division near the end of the chapter).

Cilia and Flagella

Cilia (sil′ē-ă) are appendages that project from the surface of cells and are capable of movement. They are usually limited to one surface of a given cell and vary in number from one to thousands per cell. Cilia are cylindrical in shape, about 10 μm in length and 0.2 μm in diameter, and the shaft of each cilium

is enclosed by the plasma membrane. Two centrally located microtubules and nine peripheral pairs of fused microtubules extend from the base to the tip of each cilium (figure 3.15a). Movement of the microtubules past each other, a process that requires energy from ATP, is responsible for movement of the cilia. A **basal body** (a modified centriole) is located in the cytoplasm at the base of the cilium. Cilia are numerous on surface cells that line the respiratory tract and the female reproductive tract. In these regions cilia move in a coordinated fashion, with a power stroke in one direction and a recovery stroke in the opposite direction (figure 3.15b). Their motion moves materials over the surface of the cells. For example, cilia in the trachea move mucus embedded with dust particles upward and away from the lungs. This action helps keep the lungs clear of debris.

Flagella (flă-jel′ă) have a structure similar to cilia but are longer (55 μm), and there is usually only one per cell. Furthermore, whereas cilia move small particles across the cell surface, flagella move the cell. For example, each sperm cell is propelled by a single flagellum. In contrast to cilia, which have a power stroke and a recovery stroke, flagella move in a whiplike fashion.

Microvilli

Microvilli (mī′krō-vil′ī) (figure 3.16) are cylindrically shaped extensions of the plasma membrane about 0.5–1 μm in length and 90 nm in diameter. Normally many microvilli are on each cell, and they function to increase the cell surface area. A student looking at photographs may confuse microvilli with cilia. Microvilli, however, are only one tenth to one twentieth the size of cilia. Individual microvilli can usually only be seen with an electron microscope, whereas cilia can be seen with a light microscope. Microvilli do not move, and they are supported with actin filaments, not microtubules. Microvilli are found in the intestine, kidney, and other areas in which absorption is an important function. In certain locations of the body, microvilli are highly modified to function as sensory receptors. For example, elongated microvilli in hair cells of the inner ear respond to sound.

Cell Functions

Although the structure and functions of individual cell components can be described (table 3.1), to understand how a cell functions, the interactions between the parts must be considered. For example, the active transport of many food molecules into the cell by the plasma membrane requires proteins, such as carrier molecules, and ATP manufactured within the cell. The ATP is produced in the cytosol and in mitochondria, and the proteins are produced on ribosomes. The production of proteins requires amino acids transported into the cell by the plasma membrane, ATP produced by the mitochondria, and assembly instructions provided by the nucleus. Many of the proteins produced by ribosomes are enzymes that control chemical reactions in the plasma membrane, mitochondria, ribosomes, nucleus,

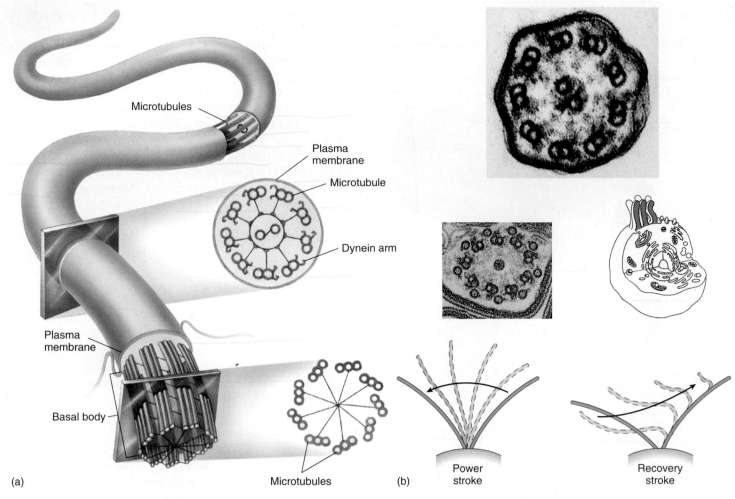

Figure 3.15 Cilia and Flagella

(*a*) Ciliary or flagellar structures. The shaft is composed of nine microtubule doublets around its periphery and two in the center. Dynein arms are proteins that connect one pair of microtubules to another pair. Dynein arm movement, which requires ATP, causes the microtubules to slide past each other, resulting in bending or movement of the cilium or flagellum. A basal body attaches the cilium or flagellum to the plasma membrane. (*b*) Ciliary movement, showing power and recovery strokes.

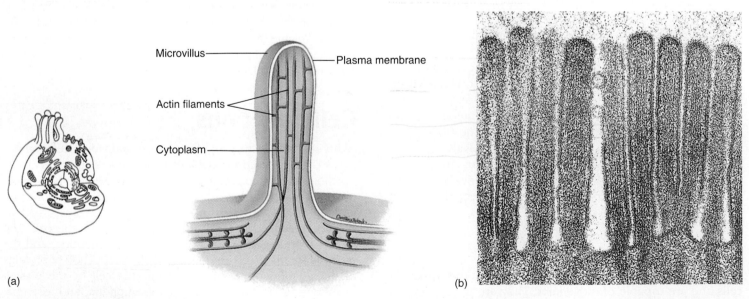

Figure 3.16 Microvillus

(*a*) A microvillus is a tiny tubular extension of the cell and contains cytoplasm and some actin filaments (microfilaments). (*b*) Transmission electron micrograph of microvilli.

Table 3.1 Summary of Cell Parts

Cell Parts	Structure	Function
Plasma Membrane	Lipid bilayer composed of phospholipids and cholesterol with proteins that extend across or are buried in either surface of the lipid bilayer	Outer boundary of cells that controls entry and exit of substances; receptor molecules function in intercellular communication; marker molecules enable cells to recognize one another
Nucleus		
Nuclear envelope	Double membrane enclosing the nucleus; the outer membrane is continuous with the endoplasmic reticulum; nuclear pores extend through the nuclear envelope	Separates nucleus from cytoplasm and regulates movement of materials into and out of the nucleus
Chromatin	Dispersed thin strands of DNA, histones, and other proteins; condenses to form chromosomes during cell division	DNA regulates protein (e.g., enzyme) synthesis and therefore the chemical reactions of the cell; DNA is the genetic or hereditary material
Nucleolus	One to four dense bodies consisting of ribosomal RNA and proteins	Assembly site of large and small ribosomal subunits
Cytoplasm: Cytosol		
Fluid part	Water with dissolved ions and molecules; colloid with suspended proteins	Contains enzymes that catalyze decomposition and synthesis reactions; ATP is produced in glycolysis reactions
Cytoskeleton		
Microtubules	Hollow cylinders composed of the protein tubulin; 25 nm in diameter	Support the cytoplasm and form centrioles, spindle fibers, cilia, and flagella; responsible for cell movements
Actin filaments	Small fibrils of the protein actin; 8 nm in diameter	Support the cytoplasm, form microvilli, responsible for cell movements
Intermediate filaments	Protein fibers; 10 nm in diameter	Support the cytoplasm
Cytoplasmic inclusions	Aggregates of molecules manufactured or ingested by the cell; may be membrane-bound	Function depends on the molecules: energy storage (lipids, glycogen), oxygen transport (hemoglobin), skin color (melanin), and others
Cytoplasm: Organelles		
Ribosome	Ribosomal RNA and proteins form large and small subunits; attached to endoplasmic reticulum or free	Site of protein synthesis
Rough endoplasmic reticulum	Membranous tubules and flattened sacs with attached ribosomes	Protein synthesis and transport to Golgi apparatus
Smooth endoplasmic reticulum	Membranous tubules and flattened sacs with no attached ribosomes	Manufactures lipids and carbohydrates; detoxifies harmful chemicals; stores calcium
Golgi apparatus	Flattened membrane sacs stacked on each other	Modification, packaging, and distribution of proteins and lipids for secretion or internal use
Secretory vesicle	Membrane-bound sac pinched off Golgi apparatus	Carries proteins and lipids to cell surface for secretion
Lysosome	Membrane-bound vesicle pinched off Golgi apparatus	Contains digestive enzymes
Peroxisome	Membrane-bound vesicle	One site of lipid and amino acid degradation and breaks down hydrogen peroxide
Mitochondria	Spherical, rod-shaped, or threadlike structures; enclosed by double membrane; inner membrane forms projections called cristae	Major site of ATP synthesis when oxygen is available
Centrioles	Pair of cylindrical organelles in the centrosome, consisting of triplets of parallel microtubules	Centers for microtubule formation; determine cell polarity during cell division; form the basal bodies of cilia and flagella
Spindle fibers	Microtubules extending from the centrosome to chromosomes and other parts of the cell (i.e., aster fibers)	Assist in the separation of chromosomes during cell division
Cilia	Extensions of the plasma membrane containing doublets of parallel microtubules; 10 μm in length	Move materials over the surface of cells
Flagellum	Extension of the plasma membrane containing doublets of parallel microtubules; 55 μm in length	In humans, responsible for movement of spermatozoa
Microvilli	Extension of the plasma membrane containing microfilaments	Increase surface area of the plasma membrane for absorption and secretion; modified to form sensory receptors

and other parts of the cell. Thus a picture of mutual interdependence of cell parts emerges when the whole cell is examined. Later topics, cell metabolism, protein synthesis, the cell life cycle, and meiosis, illustrate the interactions of cell parts that result in a functioning cell.

Movement Through the Plasma Membrane

The plasma membrane separates the extracellular material from the intracellular material and is **selectively permeable,** that is, it allows only certain substances to pass through it. The intracellular material has a different composition from the extracellular material, and the survival of the cell depends on the maintenance of these differences. Enzymes, other proteins, glycogen, and potassium ions are found in higher concentrations intracellularly; and sodium, calcium, and chloride ions are found in greater concentrations extracellularly. In addition, nutrients must continually enter the cell, and waste products must exit, but the volume of the cell remains unchanged. Because of the plasma membrane's permeability characteristics and its ability to transport molecules selectively, the cell is able to maintain homeostasis. Rupture of the membrane, alteration of its permeability characteristics, or inhibition of transport processes can disrupt the normal concentration differences across the plasma membrane and lead to cell death.

Substances move across the plasma membrane in four ways: (1) directly through the lipid bilayer, (2) through membrane channels, (3) with carrier molecules in the membrane, or (4) in vesicles. Molecules that are soluble in lipids, such as oxygen, carbon dioxide, and steroids, pass through the plasma membrane readily by dissolving in the lipid bilayer. The lipid bilayer acts as a barrier to most polar substances, which are not soluble in lipids.

Membrane channels are composed of large protein molecules that extend from one surface of the plasma membrane to the other (see figures 3.4 and 3.5). Several channel types exist, and each type allows molecules of only a certain size range to pass through it. In addition, most channels are positively charged, and because like charges repel one another, positive ions pass through the channels less readily than do neutral or negatively charged molecules of the same size. Chloride ions (Cl^-) pass through the membrane channels relatively easily, but sodium ions (Na^+) and potassium ions (K^+) pass through channels more slowly. Some water can pass directly through the phospholipid bilayer of the cell membrane, but rapid movement of water across many cell membranes apparently occurs through membrane channels. Water can move through channels with solutes such as Cl^-.

Large polar substances, such as glucose and amino acids, cannot pass through the plasma membrane in significant amounts because they are too large to move through membrane channels and their polar nature prevents their dissolving in the lipid bilayer. Protein molecules within the membrane function as carrier molecules, which combine with large

polar substances on one side of the membrane and transport them to the other side of the membrane.

Large polar molecules, small pieces of matter, and even whole cells can be transported across the plasma membrane in a **vesicle,** which is a membrane-bound sac. Because of the fluid nature of membranes, the vesicle and the plasma membrane fuse, allowing the contents of the vesicle to cross the plasma membrane.

▯ Diffusion

A solution consists of one or more substances called **solutes** dissolved in the predominant liquid or gas, which is called the **solvent. Diffusion** can be viewed as the tendency for solute molecules to move from an area of higher concentration to an area of lower concentration in solution (figure 3.17). Diffusion is a product of the constant random motion of all atoms, molecules, or ions in a solution. Because more solute particles exist in an area of higher concentration than in an area of lower concentration and because the particles move randomly, the chances are greater that solute particles will move from the higher to the lower concentration than in the opposite direction. Thus the overall, or net, movement is from the area of higher concentration to that of lower concentration. At equilibrium, the net movement of solutes stops, although the random molecular motion continues, and the movement of solutes in any one direction is balanced by an equal movement in the opposite direction (see figure 3.17). The movement and distribution of smoke or perfume throughout a room in which there are no air currents or of a dye throughout a beaker of still water are examples of diffusion.

A concentration difference exists when the concentration of a solute is greater at one point than at another point in a solvent. The concentration difference between two points divided by the distance between those two points is called the **concentration gradient.** Solutes diffuse with their concentration gradients (from a higher to a lower concentration) until an equilibrium is achieved. For a given concentration difference between two points in a solution, the concentration gradient is larger if the distance between the two points is small, and the concentration gradient is smaller if the distance between the two points is large.

The rate of diffusion is influenced by the magnitude of the concentration gradient, the temperature of the solution, the size of the diffusing molecules, and the viscosity of the solvent. The greater the concentration gradient, the greater is the number of solute particles moving from a higher to a lower concentration. As the temperature of a solution increases, the speed at which all molecules move increases, resulting in a greater diffusion rate. Small molecules diffuse through a solution more readily than do large ones. **Viscosity** is a measure of how easily a liquid flows; thick solutions, such as syrup, are more viscous than water. Diffusion occurs more slowly in viscous solvents than in thin, watery solvents.

Diffusion of molecules is an important means by which substances move between the extracellular and intracellular fluids in the body. Substances that can diffuse through either

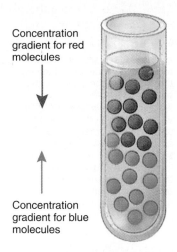
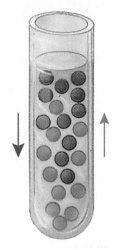

Concentration
gradient for red
molecules

Concentration
gradient for blue
molecules

(a) One solution (red balls representing
one type of molecule) is layered onto
a second solution (blue balls
representing a second type of
molecule). There is a concentration
gradient for the red molecules from
the red solution into the blue solution
because there are no red molecules
in the blue solution. There is also a
concentration gradient for the blue
molecules from the blue solution into
the red solution because there are
no blue molecules in the red
solution.

(b) Red molecules move down their
concentration gradient into the
blue solution (*red arrow*), and
the blue molecules move down
their concentration gradient into
the red solution (*blue arrow*).

(c) Red and blue molecules are
distributed evenly throughout the
solution. Although the red and
blue molecules continue to move
randomly, an equilibrium exists,
and no net movement occurs
because no concentration
gradient exists.

Figure 3.17   Diffusion

the lipid bilayer or the membrane channels can pass through the plasma membrane. Some nutrients enter and some waste products leave the cell by diffusion, and maintenance of the appropriate intracellular concentration of these substances depends to a large degree on diffusion. For example, if the extracellular concentration of oxygen is reduced, inadequate oxygen diffuses into the cell, and normal cell function cannot occur.

2	P R E D I C T

Urea is a toxic waste produced inside cells. It diffuses from the cells into the blood and is eliminated from the body by the kidneys. What would happen to the intracellular and extracellular concentration of urea if the kidneys stopped functioning?

✔ *Answer in Appendix F*

Osmosis

Osmosis (os-mō′sis) is the diffusion of water (solvent) across a selectively permeable membrane, such as a plasma membrane. A selectively permeable membrane is a membrane that allows water but not all the solutes dissolved in the water to diffuse through the membrane. Water diffuses from a solution with proportionately more water, across a selectively permeable membrane, and into a solution with proportionately less

water. Because solution concentrations are defined in terms of solute concentrations and not in terms of water content (see chapter 2), water diffuses from the less concentrated solution (fewer solutes, more water) into the more concentrated solution (more solutes, less water). Osmosis is important to cells because large volume changes caused by water movement disrupt normal cell function.

Osmotic pressure is the force required to prevent the movement of water by osmosis across a selectively permeable membrane. The osmotic pressure of a solution can be determined by placing the solution into a tube that is closed at one end by a selectively permeable membrane (figure 3.18). The tube is then immersed in distilled water. Water molecules move by osmosis through the membrane into the tube, forcing the solution to move up the tube. As the solution rises into the tube, its weight produces hydrostatic pressure that moves water out of the tube back into the distilled water surrounding the tube. At equilibrium, net movement of water stops, which means the movement of water into the tube by osmosis is equal to the movement of water out of the tube caused by hydrostatic pressure. The osmotic pressure of the solution in the tube is equal to the hydrostatic pressure that prevents net movement of water into the tube.

The osmotic pressure of a solution provides information about the tendency for water to move by osmosis across a selectively permeable membrane. Because water moves from less concentrated solutions (fewer solutes, more water) into

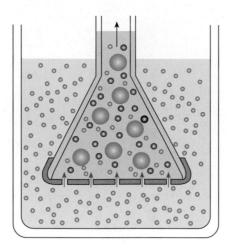

Because the tube contains salt ions *(green and red spheres)* as well as water molecules *(blue spheres)*, the tube has proportionately less water than is in the beaker, which contains only water. The water molecules diffuse with their concentration gradient into the tube *(blue arrows)*. Because the salt ions cannot leave the tube, the total fluid volume inside the tube increases, and fluid moves up the glass tube *(black arrow)* as a result of osmosis.

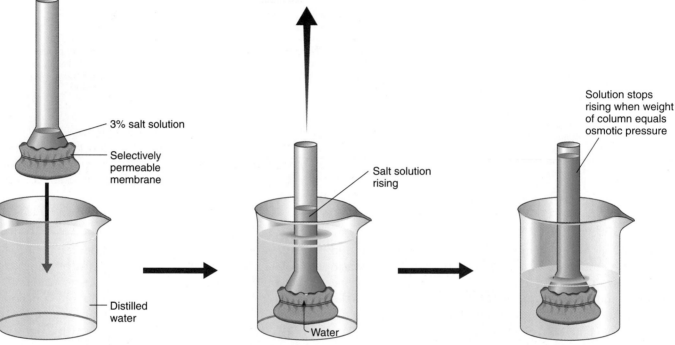

3% salt solution

Selectively permeable membrane

Distilled water

Salt solution rising

Water

Solution stops rising when weight of column equals osmotic pressure

(a) The end of a tube containing a 3% salt solution (green) is closed at one end with a selectively permeable membrane, which allows water molecules to pass through it but retains the salt ions within the tube.

(b) The tube is immersed in distilled water. Water moves into the tube by osmosis (see inset above).

(c) Water continues to move into the tube until the weight of the column of water in the tube (hydrostatic pressure) exerts a downward force equal to the osmotic force moving water molecules into the tube. The hydrostatic pressure that prevents net movement of water into the tube is equal to the osmotic pressure of the solution in the tube.

Figure 3.18   Osmosis

more concentrated solutions (more solutes, less water), the greater the concentration of a solution (the less water it has), the greater the tendency for water to move into the solution, and the greater the osmotic pressure to prevent that movement. Thus, the greater the concentration of a solution, the greater the osmotic pressure of the solution, and the greater the tendency for water to move into the solution.

3   P R E D I C T

Given the experiment in figure 3.18, what would happen to osmotic pressure if the membrane were not selectively permeable but instead allowed all solutes and water to pass through it?

✔ *Answer in Appendix F*

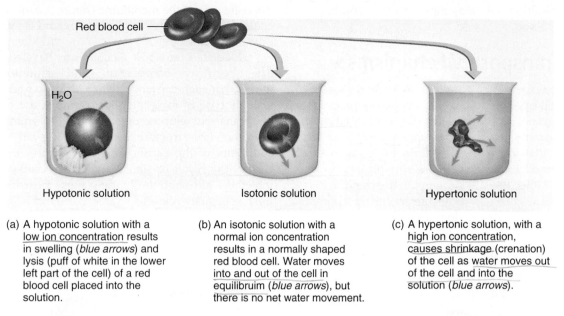

Red blood cell

H₂O

Hypotonic solution

Isotonic solution

Hypertonic solution

(a) A hypotonic solution with a low ion concentration results in swelling (*blue arrows*) and lysis (puff of white in the lower left part of the cell) of a red blood cell placed into the solution.

(b) An isotonic solution with a normal ion concentration results in a normally shaped red blood cell. Water moves into and out of the cell in equilibruim (*blue arrows*), but there is no net water movement.

(c) A hypertonic solution, with a high ion concentration, causes shrinkage (crenation) of the cell as water moves out of the cell and into the solution (*blue arrows*).

Figure 3.19   Effects of Hypotonic, Isotonic, and Hypertonic Solutions on Red Blood Cells

Three terms describe the osmotic pressure of solutions. Solutions with the same concentration of solute particles (see chapter 2) have the same osmotic pressure and are referred to as **isosmotic** (ī′sos-mot′ik). The solutions are still isosmotic even if the types of solute particles in the two solutions differ from each other. If one solution has a greater concentration of solute particles and therefore a greater osmotic pressure than another solution, the first solution is said to be **hyperosmotic** (hī′per-oz-mot′ik) compared with the more dilute solution. The more dilute solution, with the lower osmotic pressure, is **hyposmotic** (hī-pos-mot′ik) compared with the more concentrated solution.

Three additional terms describe the tendency of cells to shrink or swell when placed into a solution. If a cell is placed into a solution in which it neither shrinks nor swells, the solution is said to be **isotonic** (ī′sō-ton′ik). If a cell is placed into a solution and water moves out of the cell by osmosis, causing the cell to shrink, the solution is called **hypertonic** (hī-per-ton′ik). If a cell is placed into a solution and water moves into the cell by osmosis, causing the cell to swell, the solution is called **hypotonic** (hī-pō-ton′ik) (figure 3.19a).

An isotonic solution may be isosmotic. Because isosmotic solutions have the same concentration of solutes and water as the cytoplasm of the cell, there is no net movement of water, and the cell neither swells nor shrinks (figure 3.19b). Hypertonic solutions can be hyperosmotic and have a greater concentration of solute molecules and a lower concentration of water than the cytoplasm of the cell. Therefore water moves by osmosis from the cell into the hypertonic solution, causing the cell to shrink, a process called **crenation** (krē-nā′shŭn) (figure 3.19c). Hypotonic solutions can be hyposmotic and have a smaller concentration of solute molecules and a greater concentration of water than the cytoplasm of the cell. Therefore water moves by osmosis into the cell, causing

it to swell. If the cell swells enough, it can rupture, a process called **lysis** (lī′sis) (see figure 3.19a). Solutions injected into the circulatory system or the tissues must be isotonic because crenation or swelling of cells disrupts their normal function and can lead to cell death.

The *-osmotic* terms refer to the concentration of the solutions, and the *-tonic* terms refer to the tendency of cells to swell or shrink. These terms should not be used interchangeably. Not all isosmotic solutions are isotonic. For example, it is possible to prepare a solution of glycerol and a solution of mannitol that are isosmotic to the cytoplasm of the cell. Because the solutions are isosmotic, they have the same concentration of solutes and water as the cytoplasm. Glycerol, however, can diffuse across the plasma membrane, and mannitol cannot. When glycerol diffuses into the cell, the solute concentration of the cytoplasm increases, and its water concentration decreases. Therefore, water moves by osmosis into the cell, causing it to swell, and the glycerol solution is both isosmotic and hypotonic. In contrast, mannitol cannot enter the cell, and the isosmotic mannitol solution is also isotonic.

Filtration

Filtration results when a partition containing small holes is placed in a stream of moving liquid. Particles small enough to pass through the holes move through the partition with the liquid, but particles larger than the holes are prevented from moving beyond the partition. In contrast to diffusion, filtration depends on a pressure difference on either side of the partition. The liquid moves from the side of the partition with the greater pressure to the side with the lower pressure.

Filtration occurs in the kidneys as a step in urine formation. Blood pressure moves fluid from the blood through a partition, or filtration membrane. Ions and small molecules pass through the partition, whereas most proteins and blood cells remain in the blood.

Mediated Transport Mechanisms

Many essential molecules, such as amino acids and glucose, cannot enter the cell by simple diffusion, and many products, such as proteins, cannot exit the cell by diffusion. **Mediated transport mechanisms** involve carrier molecules within the plasma membrane that move large, water-soluble molecules or electrically charged molecules across the plasma membrane. The carrier molecules are proteins that extend across the plasma membrane. Once a molecule to be transported binds to the carrier molecule on one side of the membrane (figure 3.20*a*), the three-dimensional shape of the carrier molecule changes, and the transported molecule is moved to the opposite side of the membrane (figure 3.20*b*). The carrier molecule then resumes its original shape and is available to transport other molecules.

Mediated transport mechanisms have three characteristics: specificity, competition, and saturation. **Specificity** means that each carrier molecule binds to and transports only a single type of molecule. For example, the carrier molecule that transports glucose does not bind to amino acids or ions. The chemical structure of the binding site determines the specificity of the carrier molecule (figure 3.21*a*). **Competition** is the result of similar molecules binding to the carrier molecule. Although the binding sites of carrier molecules exhibit specificity, closely related substances may bind to the same binding site. The substance in the greater concentration or the substance that binds to the binding site more readily is transported across the plasma membrane at the greater rate

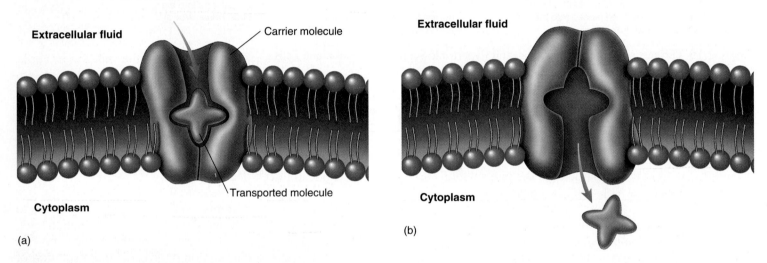

(a)

(b)

Figure 3.20   Mediated Transport by a Carrier Molecule

(*a*) The carrier molecule binds with a molecule from one side of the plasma membrane. (*b*) The carrier molecule changes shape and releases the molecule on the other side of the plasma membrane.

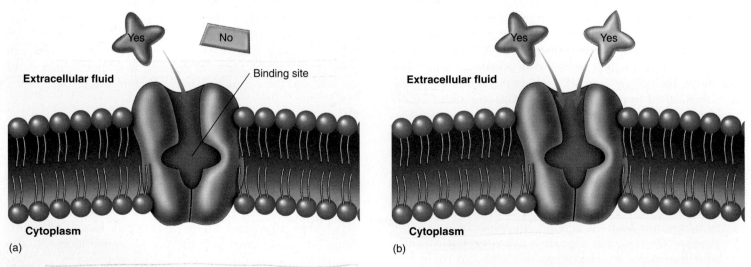

(a)

(b)

Figure 3.21   Mediated Transport: Specificity and Competition

(*a*) Specificity. Only molecules that are the right shape to bind to the binding site are transported. (*b*) Competition. Similarly shaped molecules can compete for the same binding site.

(figure 3.21*b*). **Saturation** means that the rate of transport of molecules across the membrane is limited by the number of available carrier molecules. As the concentration of a transported substance increases, more carrier molecules have their binding sites occupied. The rate at which the substance is transported increases; however, once the concentration of the substance is increased so that all the binding sites are occupied, the rate of transport remains constant, even though the concentration of the substance increases further (figure 3.22).

There are three kinds of mediated transport: facilitated diffusion, active transport, and secondary active transport.

Facilitated Diffusion

Facilitated diffusion is a carrier-mediated process that moves substances into or out of cells from a higher to a lower concentration. Facilitated diffusion does not require metabolic energy to transport substances across the plasma membrane. The rate at which molecules are transported is directly proportional to their concentration gradient up to the point of saturation, when all the carrier molecules are being used. Then the rate of transport remains constant at its maximum rate.

4	P R E D I C T

The transport of glucose into and out of most cells, such as muscle and fat cells, occurs by facilitated diffusion. Once glucose enters a cell, it is rapidly converted to other molecules, such as glucose-6-phosphate or glycogen. What effect does this conversion have on the ability of the cell to acquire glucose? Explain.

✔ *Answer in Appendix F*

▌Active Transport

Active transport is a carrier-mediated process that requires energy provided by ATP (figure 3.23). Movement of the transported substance to the opposite side of the membrane and its subsequent release from the carrier molecule are fueled by the breakdown of ATP. The maximum rate at which active transport proceeds depends on the number of carrier molecules in the plasma membrane and the availability of adequate ATP. Active-transport processes are important because they can move substances against their concentration gradient from a lower concentration to a higher concentration. Consequently, they have the ability to accumulate substances on one side of the plasma membrane at concentrations many times greater than those on the other side. Active transport can also move substances from higher to lower concentrations.

Clinical Note

Cystic fibrosis is a genetic disorder that affects chloride ion channels. There are three types of cystic fibrosis. In about 70% of cases, a defective channel protein is produced that fails to reach the cell membrane from its site of production inside the cell. In the remaining cases, the channel protein is incorporated into the cell membrane but does not function normally. In some cases, the channel protein fails to bind ATP. In others, ATP is bound to the channel protein, but the channel does not open.

Active-transport mechanisms may also exchange one substance for another. For example, the sodium–potassium exchange pump moves sodium out of cells and potassium into cells (see figure 3.23). The result is a higher concentration of sodium outside the cell and a higher concentration of potassium inside the cell (see chapter 9).

Secondary Active Transport

Secondary active transport involves the active transport of an ion such as sodium out of a cell, establishing a concentration gradient, with a higher concentration of the ions outside the cell. The diffusion of the ion back into the cell, down its concentration gradient, provides the energy necessary to transport a different ion or some other molecule into the cell. For example, glucose is transported from the lumen of the intestine into epithelial cells by secondary active transport (figure 3.24). This process requires two carrier molecules: (1) a sodium–potassium exchange pump actively transports Na^+ ions out of the cell, and (2) the other carrier molecule facilitates the diffusion of Na^+ ions and glucose into the cell. Both Na^+ ions and glucose are necessary for the carrier molecule to function.

The movement of Na^+ ions down their concentration gradient provides the energy to move glucose molecules into the cell against their concentration gradient. Thus glucose can accumulate at concentrations higher inside the cell than outside. Because the movement of glucose ions against their concentration gradient results from the formation of a concentration gradient of Na^+ ions by an active transport mechanism, the process is called secondary active transport.

The ions or molecules moved by secondary active transport can move in the same direction as or in a different direction across the membrane than the ion that enters the cell by diffusion down its concentration gradient. In **cotransport**, or **symport**, movement is in the same direction. For example, glucose, fructose, and amino acids move with Na^+ ions into cells of the intestine and kidneys. In **countertransport**, or **antiport**, ions or molecules move in opposite directions. For example, the internal pH of cells is maintained by countertransport, which moves H^+ ions out of the cell as Na^+ ions move into the cell.

5	P R E D I C T

In cardiac (heart) muscle cells, the concentration of intracellular Ca^{2+} ions affects the force of heart contraction. The higher the intracellular Ca^{2+} ion concentration, the greater the force of contraction. Na^+/Ca^{2+} countertransport helps to regulate intracellular Ca^{2+} ion levels by transporting Ca^{2+} ions out of cardiac muscle cells. Given that digitalis slows the transport of Na^+ ions, should the heart beat more or less forcefully? Explain.

✔ *Answer in Appendix F*

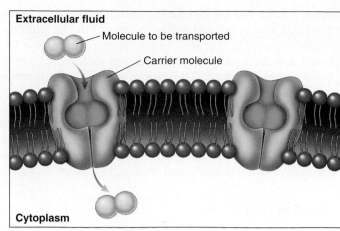

Extracellular fluid

Molecule to be transported

Carrier molecule

Cytoplasm

1. When only a few molecules are present outside the cell (the extracellular concentration is low), only a few molecules can be transported into the cell, and therefore the transport rate is low (the rate of transport is limited by the number of molecules available).

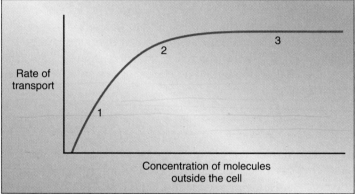

Rate of transport

Concentration of molecules outside the cell

The rate of transport of molecules into a cell is plotted against the concentration of those molecules outside the cell. At 1, there are more carrier molecules than molecules for transport. At 2, the concentration of molecules available for transport increases until all the carrier molecules are involved in transporting molecules (that is, the system is saturated). At 3, even though the concentration of molecules for transport has increased, the rate of transport is limited by the number of carrier molecules.

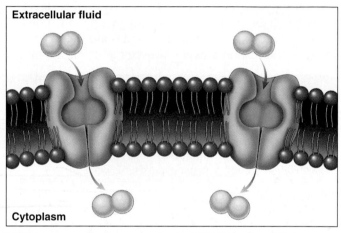

Extracellular fluid

Cytoplasm

2. When more molecules are present outside the cell, as long as enough carrier molecules are available, more molecules can be transported, and therefore the transport rate increases. When all the carrier molecules are involved in transporting molecules, the system is saturated.

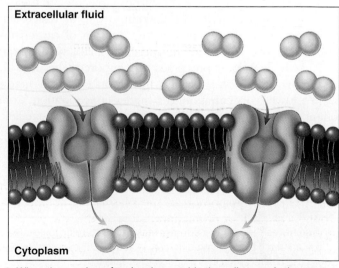

Extracellular fluid

Cytoplasm

3. When the number of molecules outside the cell exceeds the number of carrier molecules, the transport rate is limited by the number of carrier molecules.

Figure 3.22   Mediated Transport: Saturation

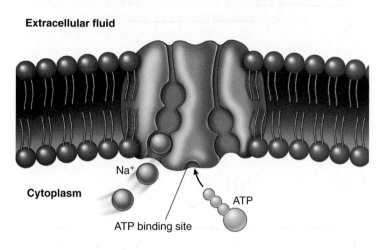

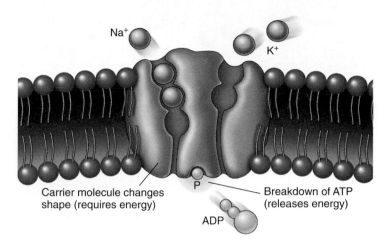

1. Three Na⁺ ions and ATP bind to the carrier molecule.

2. The ATP breaks down to adenosine diphosphate and releases energy. The carrier molecule changes shape, and the Na⁺ ions are transported across the membrane.

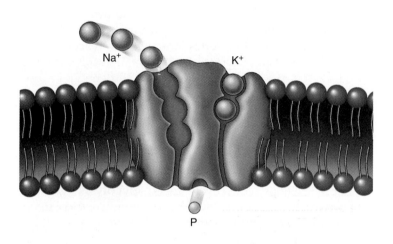

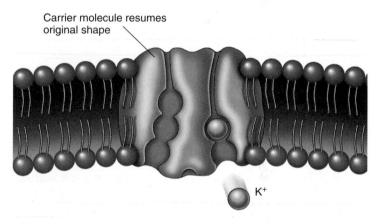

3. The Na⁺ ions diffuse away from the carrier molecule, two K⁺ ions bind to the carrier molecule, and the phosphate is released.

4. The carrier molecule changes shape, transporting K⁺ ions across the membrane, and the K⁺ ions diffuse away from the carrier molecule. The carrier molecule can again bind to Na⁺ ions and ATP.

Figure 3.23   Sodium–Potassium Exchange Pump

Endocytosis and Exocytosis

Endocytosis (en′dō-sī-tō′sis) includes both phagocytosis and pinocytosis and refers to the bulk uptake of material through the plasma membrane by the formation of a vesicle. A vesicle is a membrane-bound sac found within the cytoplasm of a cell. A portion of the plasma membrane wraps around a particle or droplet and fuses so that the particle or droplet is surrounded by a membrane. That portion of the membrane then "pinches off" so that the particle or droplet, surrounded by a membrane, is within the cytoplasm of the cell, and the plasma membrane is left intact.

Phagocytosis (fag′ō-sī-tō′sis) literally means cell-eating (figure 3.25a and b) and applies to endocytosis when solid particles are ingested and phagocytic vesicles are formed. White blood cells and some other cell types phagocytize bacteria, cell debris, and foreign particles. Phagocytosis is therefore important in the elimination of harmful substances from the body.

Pinocytosis (pin′ō-sī-tō′sis) means cell-drinking and is distinguished from phagocytosis in that smaller vesicles are formed and they contain molecules dissolved in liquid rather than particles (figure 3.25c and d). Pinocytosis often forms vesicles near the tips of deep invaginations of the plasma membrane. It is a common transport phenomenon in a variety of cell types and occurs in certain cells of the kidneys, epithelial cells of the intestines, cells of the liver, and cells that line capillaries.

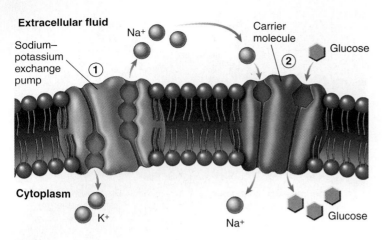

Extracellular fluid

Sodium–potassium exchange pump

Na⁺

Carrier molecule

Glucose

① ②

Cytoplasm

K⁺

Na⁺

Glucose

Figure 3.24   Secondary Active Transport

The example here is cotransport of Na⁺ ions and glucose.

1. A sodium–potassium exchange pump maintains a concentration of Na⁺ ions that is higher outside the cell than inside.
2. The Na⁺ ions diffuse back into the cell by facilitated diffusion, assisted by a carrier molecule that also facilitates the diffusion of glucose. The diffusing Na⁺ ions provide energy that can be used to move glucose against its concentration gradient.

Endocytosis can exhibit specificity. For example, cells that phagocytize bacteria and necrotic tissue do not phagocytize healthy cells. The plasma membrane may contain specific receptor molecules that recognize certain substances and allow them to be transported into the cell by phagocytosis or pinocytosis. This is called **receptor-mediated endocytosis,** and the receptor sites combine only with certain molecules (figure 3.25e). This mechanism increases the rate at which specific substances are taken up by the cells. Cholesterol and growth factors are examples of molecules that can be taken into a cell by receptor-mediated endocytosis. Both phagocytosis and pinocytosis require energy in the form of ATP and therefore are active processes. Because they involve the bulk movement of material into the cell, however, phagocytosis and pinocytosis do not exhibit either the degree of specificity or saturation that active transport exhibits.

In some cells, secretions accumulate within vesicles. These secretory vesicles then move to the plasma membrane, where the membrane of the vesicle fuses with the plasma membrane and the content of the vesicle is expelled from the cell. This process is called **exocytosis** (ek′sō-sī-tō′sis) (figure 3.26). Secretion of digestive enzymes by the pancreas, of mucus by the salivary glands, and of milk by the mammary glands are examples of exocytosis. In some respects the process is similar to phagocytosis and pinocytosis but occurs in the opposite direction.

Table 3.2 summarizes and compares the mechanisms by which different kinds of molecules are transported across the plasma membrane.

Cell Metabolism

Cell metabolism is the sum of all the catabolic (decomposition) and anabolic (synthesis) reactions in the cell. The breakdown of food molecules such as carbohydrates, lipids, and proteins releases energy that is used to synthesize ATP. Each ATP molecule contains a portion of the energy originally stored in the chemical bonds of the food molecules. The ATP molecules are smaller "packets" of energy that can be used to drive other chemical reactions or processes such as active transport.

The production of ATP takes place in the cytosol and in mitochondria through a series of chemical reactions (see chapter 25 for details). Energy from food molecules is transferred to ATP in a controlled fashion. If the energy in food molecules were released all at once, the cell literally would burn up.

The breakdown of the sugar glucose, such as from sugar found in a candy bar, is used to illustrate the production of ATP from food molecules. Once glucose is transported into a cell, a series of reactions takes place within the cytosol. These chemical reactions, collectively called **glycolysis** (glī-kol′i-sis), convert the glucose to pyruvic acid. Pyruvic acid can enter different biochemical pathways, depending on oxygen availability (figure 3.27).

Aerobic (ār-ō′bik) **respiration** occurs when oxygen is available. The pyruvic acid molecules enter mitochondria and, through another series of chemical reactions, collectively called the citric acid cycle and the electron transport chain, are converted to carbon dioxide and water. Aerobic respiration can produce 36–38 ATP molecules from the energy contained in each glucose molecule.

Several important points should be noted about aerobic respiration. First, the quantities of ATP produced through aerobic respiration are absolutely necessary to maintain the energy-requiring chemical reactions of life in human cells. Second, aerobic respiration requires oxygen because the last chemical reaction that takes place in aerobic respiration is the combination of oxygen with hydrogen to form water. If this reaction does not take place, the reactions immediately preceding it do not occur either. This explains why breathing oxygen is necessary for human life: without oxygen, aerobic respiration is inhibited, and the cells do not produce enough ATP to sustain life. Finally, during aerobic respiration the carbon atoms of food molecules are separated from one another to form carbon dioxide. Thus the carbon dioxide humans breathe out comes from the food they eat.

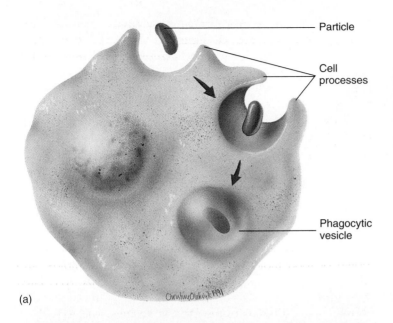

Particle

Cell processes

Phagocytic vesicle

(a)

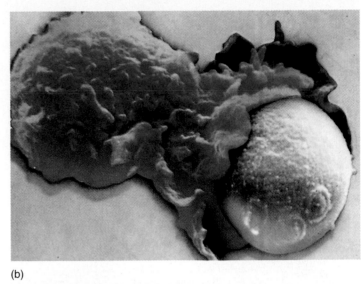

(b)

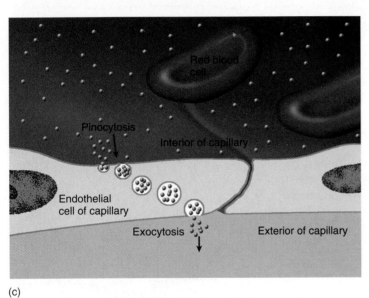

Red blood cell

Pinocytosis

Interior of capillary

Endothelial cell of capillary

Exocytosis

Exterior of capillary

(c)

Pinocytotic vesicles

Interior of capillary

Capillary wall

Exterior of capillary

(d)

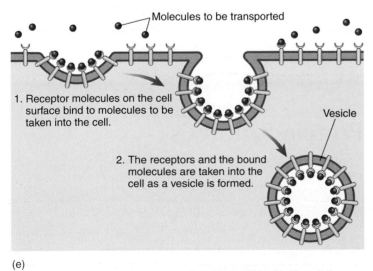

Molecules to be transported

1. Receptor molecules on the cell surface bind to molecules to be taken into the cell.

Vesicle

2. The receptors and the bound molecules are taken into the cell as a vesicle is formed.

(e)

Figure 3.25   Endocytosis

(*a*) Phagocytosis. (*b*) Transmission electron micrograph of phagocytosis. (*c*) Pinocytosis is much like phagocytosis, except the cell processes and therefore the vesicles formed are much smaller and the material inside the vesicle is liquid rather than particulate. Pinocytotic vesicles form on the internal side of a capillary, are transported across the cell, and open by exocytosis outside the capillary. (*d*) Transmission electron micrograph of pinocytosis. (*e*) Receptor-mediated endocytosis.

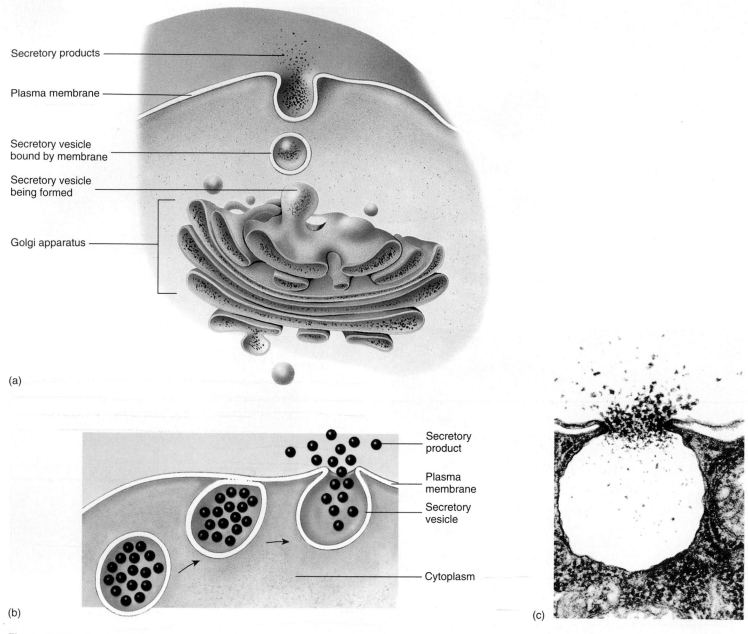

Secretory products

Plasma membrane

Secretory vesicle bound by membrane

Secretory vesicle being formed

Golgi apparatus

(a)

Secretory product

Plasma membrane

Secretory vesicle

Cytoplasm

(b)

(c)

Figure 3.26   Exocytosis

(*a*) A secretory vesicle is formed inside the cell from the Golgi apparatus and contains products manufactured by the cell. (*b*) The vesicle moves to the plasma membrane, fuses with the membrane, then opens to the outside, and dumps its contents into the extracellular space. (*c*) Transmission electron micrograph of exocytosis.

Anaerobic (an-ă-rō'bik) **respiration** occurs without oxygen and includes the conversion of pyruvic acid to lactic acid. There is a net production of two ATP molecules for each glucose molecule used. Anaerobic respiration does not produce as much ATP as aerobic respiration, but it does allow the cells to function for short periods when oxygen levels are too low for aerobic respiration to provide all the needed ATP. For example, during intense exercise, when aerobic respiration has depleted the oxygen supply, anaerobic respiration can provide additional ATP.

Protein Synthesis

Normal cell structure and function would not be possible without proteins (figure 3.28), which form the cytoskeleton and other structural components of cells and function as transport molecules, receptors, and enzymes. In addition, proteins secreted from cells perform vital functions: collagen is a structural protein that gives tissues flexibility and strength; enzymes control the chemical reactions of food digestion in the

Table 3.2   Comparison of Membrane Transport Mechanisms

Transport Mechanism	Description	Substances Transported	Example
Diffusion	Random movement of molecules results in net movement from areas of higher to lower concentration.	Lipid-soluble molecules dissolve in the lipid bilayer and diffuse through it; ions and small molecules diffuse through membrane channels.	Oxygen, carbon dioxide, and lipids such as steroid hormones dissolve in the lipid bilayer; Cl^- ions and urea move through membrane channels.
Osmosis	Water diffuses across a selectively permeable membrane.	Water diffuses through membrane channels.	Water moves from the stomach into the blood.
Filtration	Liquid moves through a partition that allows some, but not all, of the substances in the liquid to pass through it; movement is due to a pressure difference across the partition.	Liquid and substances pass through holes in the partition.	Filtration in the kidneys allows removal of everything from the blood except proteins and blood cells.
Facilitated diffusion	Carrier molecules combine with substances and move them across the plasma membrane; no ATP is used; substances are always moved from areas of higher to lower concentration; it exhibits the characteristics of specificity, saturation, and competition.	Substances too large to pass through membrane channels and too polar to dissolve in the lipid bilayer are transported.	Glucose moves by facilitated diffusion into muscle cells and fat cells.
Active transport	Carrier molecules combine with substances and move them across the plasma membrane; ATP is used; substances can be moved from areas of lower to higher concentration; it exhibits the characteristics of specificity, saturation, and competition.	Substances too large to pass through channels and too polar to dissolve in the lipid bilayer are transported; substances that are accumulated in concentrations higher on one side of the membrane than on the other are transported.	Ions such as Na^+, K^+, and Ca^{2+} are actively transported.
Secondary active transport	Ions are moved across the cell membrane by active transport, which establishes a concentration gradient; ATP is required; ions then move back down their concentration gradient by facilitated diffusion, and another ion or molecule moves with the diffusion ion (cotransport) or in the opposite direction (countertransport).	Some sugars, amino acids, and ions are transported.	The concentration gradient is established by Na^+; glucose is cotransported with Na^+ ions into intestinal epithelial cells; in many cells, H^+ ions are countertransported opposite of Na^+ ions.
Endocytosis	The plasma membrane forms a vesicle around the substances to be transported, and the vesicle is taken into the cell; this requires ATP; in receptor-mediated endocytosis specific substances are ingested.	Phagocytosis takes in cells and solid particles; pinocytosis takes in molecules dissolved in liquid.	Immune system cells called phagocytes ingest bacteria and cellular debris; most cells take in substances through pinocytosis.
Exocytosis	Materials manufactured by the cell are packaged in secretory vesicles that fuse with the plasma membrane and release their contents to the outside of the cell; this requires ATP.	Secreted proteins and lipids are transported.	Digestive enzymes, hormones, neurotransmitters, and glandular secretions are transported, and cell waste products are eliminated.

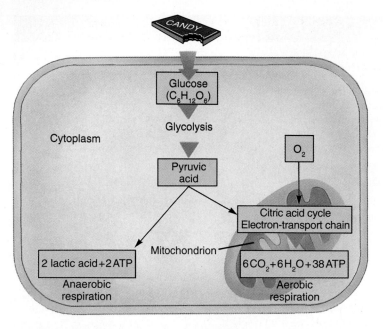

Figure 3.27   Overview of Cell Metabolism

Aerobic respiration requires oxygen and produces more ATP per glucose molecule than does anaerobic metabolism.

intestines; and protein hormones regulate the activities of many tissues.

Ultimately, the production of all the proteins in the body is under the control of DNA. Recall from chapter 2 that the building blocks of DNA are nucleotides containing adenine (A), thymine (T), cytosine (C), and guanine (G). The nucleotides form two antiparallel strands of nucleic acids. The term antiparallel means that the strands are parallel but extend in opposite directions. Each strand has a 5′ (phosphate) end and a 3′ (hydroxyl) end. The sequence of the nucleotides in the DNA is a method of storing information. Every three nucleotides, called a **triplet,** code for an amino acid, and amino acids are the building blocks of proteins. All of the triplets required to code for the synthesis of a specific protein are called a **gene.**

The production of proteins from the stored information in DNA involves two steps: transcription and translation, which can be illustrated with an analogy. Suppose a cook wants a recipe that is found only in a reference book in the library. Because the book cannot be checked out, the cook makes a handwritten copy, or **transcription,** of the recipe. Later, in the kitchen the information contained in the copied recipe is used to prepare the meal. The changing of something

1. DNA contains the information necessary to produce proteins.

2. Transcription of DNA results in mRNA, which is a copy of the information in DNA needed to make a protein.

3. The mRNA leaves the nucleus and goes to a ribosome.

4. Amino acids, the building blocks of proteins, are carried to the ribosome by tRNAs.

5. In the process of translation, the information contained in mRNA is used to determine the number, kinds, and arrangement of amino acids in the protein.

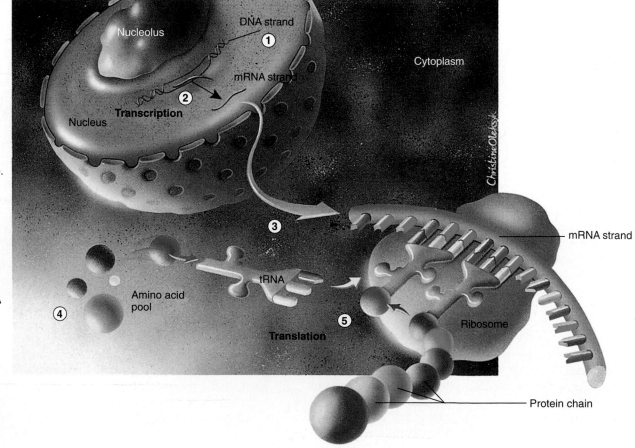

Figure 3.28   Overview of Protein Synthesis

from one form to another (from recipe to meal) is called **translation.** In this analogy, DNA is the reference book that contains many recipes for making different proteins. DNA, however, is too large a molecule to pass through the nuclear envelope to go to the ribosomes (the kitchen) where the proteins are prepared. Just as the reference book stays in the library, DNA remains in the nucleus. Therefore, through transcription, the cell makes a copy of the information in DNA necessary to make a particular protein (the recipe). The copy, which is called **messenger RNA (mRNA),** travels from the nucleus to ribosomes in the cytoplasm, where the information in the copy is used to construct a protein (i.e., translation). Of course, to turn a recipe into a meal, the actual ingredients are needed. The ingredients necessary to synthesize a protein are amino acids. Specialized transport molecules, called **transfer RNA (tRNA),** carry the amino acids to the ribosome (see figure 3.28).

In summary, the synthesis of proteins involves transcription, making a copy of part of the stored information in DNA, and translation, converting that copied information into a protein. The details of transcription and translation are considered next.

Transcription

Transcription is the synthesis of mRNA on the basis of the sequence of nucleotides in DNA. It occurs when the double strands of a DNA segment separate and RNA nucleotides pair with DNA nucleotides (figure 3.29). Nucleotides pair with each other according to the following rule: adenine pairs with thymine or uracil, and cytosine pairs with guanine. DNA contains thymine, but thymine is replaced by uracil in RNA. Adenine, thymine, cytosine, and guanine nucleotides of DNA therefore pair with uracil, adenine, guanine, and cytosine nucleotides of mRNA, respectively. This pairing relationship between nucleotides ensures that the information in DNA is transcribed correctly to mRNA. The RNA nucleotides combine through dehydration reactions catalyzed by RNA polymerase enzymes to form a long mRNA segment. The elongation of all nucleic acids, both DNA and RNA, occurs in the same chemical direction: from the 5′ to the 3′ end of the molecule. The mRNA molecule contains the information required to determine the sequence of amino acids in a protein. The information, called the **genetic code,** is carried in groups of three nucleotides called **codons.** The number and sequence of codons in the mRNA are determined by the number and sequence of sets of three nucleotides, called triplets, in the segments of DNA that were transcribed. For example, the triplet code of CTA in DNA results in the codon GAU in mRNA, which codes for aspartic acid. Each codon codes for a specific amino acid. There are 64 possible mRNA codons, but only 20 amino acids are in proteins. As a result, the genetic code is redundant because more than one codon codes for some amino acids. For example, CGA, CGG, CGT, and CGC all code for the amino acid alanine, and UUU and UAC both code for phenylalanine. Some codons do not code for amino acids but perform other functions. AUG and sometimes GUG act as signals

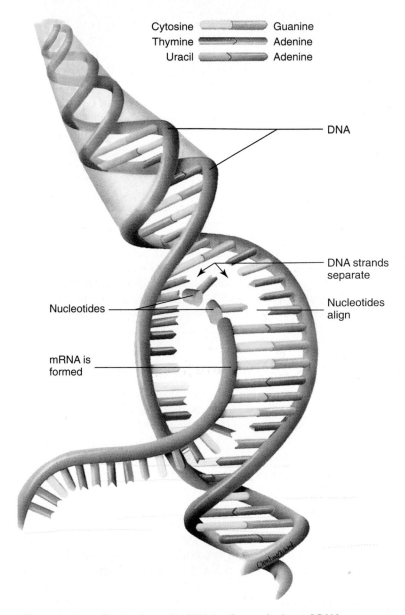

Cytosine ⟶ Guanine
Thymine ⟶ Adenine
Uracil ⟶ Adenine

DNA

DNA strands separate

Nucleotides

Nucleotides align

mRNA is formed

Figure 3.29 Formation of mRNA by Transcription of DNA

A segment of the DNA molecule is opened, and RNA polymerase (an enzyme that is not shown) assembles nucleotides into mRNA according to the base pair combinations shown in the inset. Thus the sequence of nucleotides in DNA determines the sequence of nucleotides in mRNA. As nucleotides are added, an mRNA molecule is formed.

for starting the transcription of a stretch of DNA to RNA. Three codons, UAA, UGA, and UAG, act as signals for stopping the transcription of DNA to RNA. The region of a DNA molecule between the codon starting transcription and the codon stopping transcription is transcribed into a stretch of RNA and is called a **transcription unit.** A transcription unit codes for a protein or part of a protein. A transcription unit is not necessarily a gene. A gene is a functional unit, and some regulatory genes don't code for proteins. A molecular definition of a **gene** is all of the nucleic acid sequences necessary to make a functional RNA or protein.

Not all of a continuous stretch of DNA may code for parts of a protein. Regions of the DNA that code for parts of the protein are called **exons,** whereas those regions of the DNA that do not code for portions of the protein are called **introns.** Both the exon and intron regions of the DNA may be transcribed into mRNA. An mRNA containing introns is called a **pre-mRNA.** After a stretch of pre-mRNA has been transcribed, the introns can be removed and the exons spliced together by enzymes to produce the functional mRNA (figure 3.30). These changes in the mRNA are called **posttranscriptional processing.**

Clinical Note

Hemoglobin is an oxygen-carrying protein molecule composed of four polypeptides. **Thalassemia** is a group of genetic disorders in which one or more of the polypeptides of hemoglobin is produced in decreased amounts as the result of defective posttranscriptional processing. The decreased amount of hemoglobin in the blood causes anemia, the reduction in the oxygen-carrying capacity of the blood.

▌Translation   transfer RNA

The synthesis of a protein at the ribosome in response to the codons of mRNA is called translation. In addition to mRNA, translation requires tRNA and ribosomes. Ribosomes consist of **ribosomal RNA (rRNA)** and proteins. Like mRNA, tRNA and rRNA are produced in the nucleus by transcription.

The function of tRNA is to match a specific amino acid to a specific codon of mRNA. To do this, one end of each kind of tRNA combines with a specific amino acid. Another part of the tRNA has an **anticodon,** which consists of three nucleotides. On the basis of the pairing relationships between nucleotides, the anticodon can combine only with its matched codon. For example, the tRNA that binds to aspartic acid has the anticodon CUA, which combines with the codon GAU of mRNA. Therefore the codon GAU codes for aspartic acid.

Ribosomes align the codons of the mRNA with the anticodons of tRNA and then join the amino acids of adjacent tRNA molecules. As the amino acids are joined together, a chain of amino acids, or a protein, is formed. The step-by-step process of protein synthesis at the ribosome is described in detail in figure 3.31.

Many proteins are longer when first made than they are in their final, functional state. These proteins are called **proproteins,** and the extra piece of the molecule is cleaved off by enzymes to make the proprotein into a functional protein. Many proteins are enzymes, and the proproteins of those enzymes are called **proenzymes.** If many proenzymes were made within cells as functional enzymes, they could digest the cell that made them. Instead, they are made as proenzymes and are not converted to active enzymes until they reach a protected region of the body, such as inside the small intestine, where they are functional. Many proteins have side chains, such as polysaccharides, added to them following translation. Some proteins are composed of two or more amino acid chains that are joined after each chain is produced

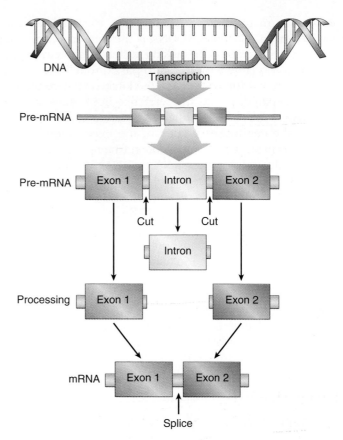

Figure 3.30   Posttranscriptional Change in mRNA

An intron is cleaved from between two exons and is discarded. The exons are spliced together to make the functional mRNA.

on separate ribosomes. These various modifications to proteins are referred to as **posttranslational processing.**

After the initial part of mRNA is used by a ribosome, another ribosome can attach to the mRNA and begin to make a protein. The resulting cluster of ribosomes attached to the mRNA is called a **polyribosome.** Each ribosome in a polyribosome produces an identical protein, and polyribosomes are an efficient way to use a single mRNA molecule to produce many copies of the same protein.

6   P R E D I C T

Explain how changing one nucleotide within a DNA molecule of a cell could change the structure of a protein produced by that cell. What effect would this change have on the protein's function?

✔ *Answer in Appendix F*

Regulation of Protein Synthesis

All of the cells in the body, except for sex cells, have the same DNA. The transcription of mRNA in cells is regulated, however, so that all portions of all DNA molecules are not continually transcribed. The proteins associated with DNA in the nucleus play a role in regulating the transcription. As cells

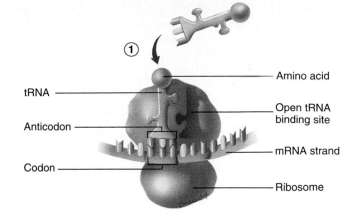

1. To start protein synthesis a ribosome binds to mRNA. The ribosome also has two binding sites for tRNA, one of which is occupied by a tRNA with its amino acid. Note that the codon of mRNA and the anticodon of tRNA are aligned and joined. The other tRNA binding site is open.

tRNA ——

Anticodon ——

Codon ——

—— Amino acid

—— Open tRNA binding site

—— mRNA strand

—— Ribosome

2. By occupying the open tRNA binding site the next tRNA is properly aligned with mRNA and with the other tRNA.

3. An enzyme within the ribosome catalyzes a synthesis reaction to form a peptide bond between the amino acids. Note that the amino acids are now associated with only one of the tRNAs.

4. The ribosome shifts position by three nucleotides. The tRNA without the amino acid is released from the ribosome, and the tRNA with the amino acids takes its position. A tRNA binding site is left open by the shift. Additional amino acids can be added by repeating steps **2** through **4**. Eventually a stop codon in the mRNA ends the production of the protein, which is released from the ribosome.

Ribosome moves to next codon of mRNA strand

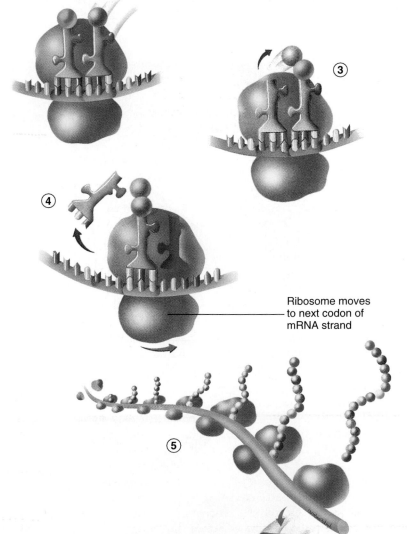

5. This is an overview of protein synthesis.

Figure 3.31   Translation of mRNA to Produce a Protein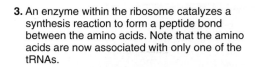

differentiate and become specialized for specific functions during development, part of the DNA becomes nonfunctional and is not transcribed, whereas other segments of DNA remain very active. For example, in most cells the DNA coding for hemoglobin is nonfunctional, and little if any hemoglobin is synthesized. In developing red blood cells, however, the DNA coding for hemoglobin is functional, and hemoglobin synthesis occurs rapidly.

Protein synthesis in a single cell is not normally constant, but it occurs more rapidly at some times than others. Regulatory molecules that interact with the nuclear proteins can either increase or decrease the transcription rate of specific DNA segments. For example, thyroxine, a hormone released by cells of the thyroid gland, enters cells such as skeletal muscle cells, interacts with specific nuclear proteins, and increases specific types of mRNA transcription. Consequently, the production of certain proteins increases. As a result, an increase in the number of mitochondria and an increase in metabolism occur in these cells.

Cell Life Cycle

The **cell life cycle** includes the changes a cell undergoes from the time it is formed until it divides to produce two new cells. The life cycle of a cell can be divided into interphase and cell division, or mitosis (figure 3.32).

Interphase

Interphase is the time between cell divisions. Ninety percent or more of the life cycle of a typical cell is spent in interphase. During this time the cell carries out the metabolic activities necessary for life and performs its specialized functions such as secreting digestive enzymes. In addition, the cell prepares for cell division. This preparation includes an increase in cell size, because many cell components double in quantity, and a replication of the cell's DNA. The centrioles within the centrosome are also duplicated. Consequently, when the cell di-

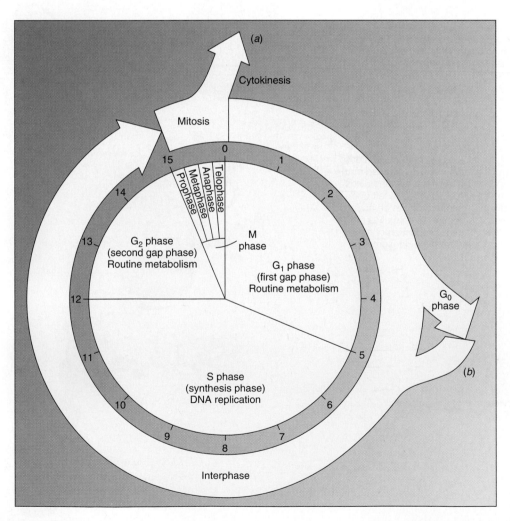

Figure 3.32   Cell Cycle

The cell cycle is divided into interphase and mitosis. Interphase is divided into G_1, S, and G_2 subphases. During G_1 and G_2, the cell carries out routine metabolic activities. During the S phase DNA is replicated. (*a*) Following mitosis, two cells are formed by the process of cytokinesis. Each new cell begins a new cell cycle. (*b*) Many cells exit the cell cycle and enter the G_0 phase, where they remain until stimulated to divide, at which point they reenter the cell cycle.

vides, each new cell receives the organelles and DNA necessary for continued functioning.

Interphase can be divided into three subphases, called G_1, S, and G_2. During G_1 (the first *gap* phase) and G_2 (the second *gap* phase), the cell carries out routine metabolic activities. During the S phase (the *synthesis* phase) the DNA is replicated (new DNA is synthesized). Many cells in the body do not divide for days, months, or even years. These "resting" cells exit the cell cycle and enter what is called the G_0 phase, in which they remain until stimulated to divide.

DNA Replication

DNA **replication** is the process by which two new strands of DNA are made, using the two existing strands as templates. During interphase, DNA and its associated proteins appear as dis-

persed chromatin threads within the nucleus. When DNA replication begins, the two strands of each DNA molecule separate from each other for some distance (figure 3.33). Each strand is then a template, or pattern, for the production of a new strand of DNA, which is formed as new nucleotides pair with the existing nucleotides of each strand of the separated DNA molecule. The production of the new nucleotide strands is catalyzed by DNA **polymerase,** which adds new nucleotides at the 3′ end of the growing strands. One strand, called the **leading strand,** is formed as a continuous strand, whereas the other strand, called the **lagging strand,** is formed in short segments going in the opposite direction. The short segments are then spliced by **DNA ligase.** As a result of DNA replication, two identical DNA molecules are produced (see figure 3.33). Each of the two new DNA molecules has one strand of nucleotides derived from the original DNA molecule and one newly synthesized strand.

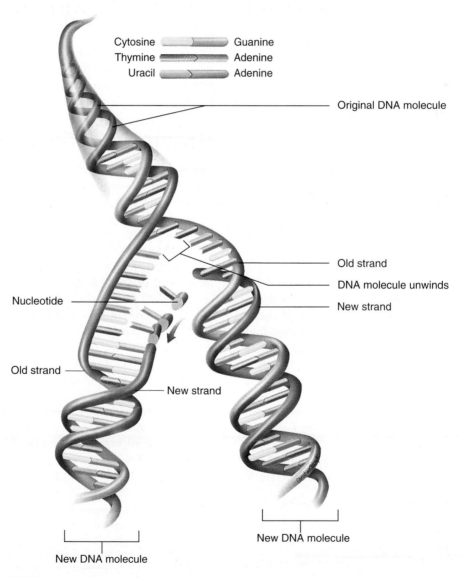

Cytosine ▭ Guanine
Thymine ▭ Adenine
Uracil ▭ Adenine

Original DNA molecule

Old strand

DNA molecule unwinds

Nucleotide

New strand

Old strand

New strand

New DNA molecule

New DNA molecule

Figure 3.33   Replication of DNA

Replication of DNA during interphase produces two identical molecules of DNA. The strands of the DNA molecule separate from each other, and each strand functions as a template on which another strand is formed. The base-pairing relationship between nucleotides determines the sequence of nucleotides in the newly formed strands.

Cell Division

The new cells necessary for growth and tissue repair are produced by cell division. A parent cell divides to form two daughter cells, each of which has the same amount and type of DNA as the parent cell. Because DNA determines cell structure and function, the daughter cells have the same structure and perform the same functions as the parent cell.

Cell division involves two major events: the division of the nucleus to form two new nuclei, and the division of the cytoplasm to form two new cells, each of which contains one of the newly formed nuclei. The division of the nucleus occurs by mitosis, and the division of the cytoplasm is called cytokinesis.

Stop for test

Mitosis

Mitosis (mī-tō′sis) is the division of the nucleus into two nuclei, each of which has the same amount and type of DNA as the original nucleus. The DNA, which was dispersed as chromatin in interphase, condenses in mitosis to form chromosomes. In each of the human **somatic** (sō-mat′ik) cells, which include all cells except the sex cells, there are 46 chromosomes, which are referred to as a **diploid** (dip′loyd) number of chromosomes. Sex cells have half the number of chromosomes as somatic cells (see following section on Meiosis). The 46 chromosomes in somatic cells are organized into 23 pairs of chromosomes. Twenty-two of these pairs are called **autosomes.** Each member of an autosomal pair of chromosomes looks structurally alike, and together they are called a **homologous** (hŏ-mol′ō-gŭs) pair of chromosomes. One member of each autosomal pair is derived from the person's father, and the other is derived from the mother. The remaining pair of chromosomes comprises the sex chromosomes. In females, the sex chromosomes look alike, and each is called an **X chromosome.** In males, the sex chromosomes do not look alike. One chromosome is an X chromosome, and the other is smaller and is called a **Y chromosome.**

For convenience, mitosis is divided into four stages: **prophase, metaphase, anaphase,** and **telophase** (tel′ō-fāz). Although each stage represents major events, mitosis is a continuous process, and no discrete jumps occur from one stage to another. Learning the characteristics associated with each stage is helpful, but a more important concept is how each daughter cell obtains the same number and type of chromosomes as the parent cell. The major events of mitosis are summarized in figure 3.34.

Cytokinesis

Cytokinesis (sī′tō-ki-nē′sis) is the division of the cytoplasm of the cell to produce two new cells. Cytokinesis begins in anaphase, continues through telophase, and ends in the following interphase (see figures 3.34*d–f*). The first sign of cytokinesis is the formation of a **cleavage furrow,** or puckering of the cell membrane, which forms midway between the centrioles. A contractile ring composed primarily of actin filaments pulls the plasma membrane inward, dividing the cell into two halves. Cytokinesis is complete when the two halves separate to form two new cells.

▮Meiosis

The formation of all the cells of the body except for sex cells occurs by mitosis. Sex cells are formed by **meiosis** (mī-ō′sis), a process in which the nucleus undergoes two divisions resulting in four nuclei, each containing half as many chromosomes as the parent cell. The daughter cells that are produced by cytokinesis differentiate into **gametes** (gam′ētz), or sex cells. The gametes are reproductive cells—**sperm cells** in males and **oocytes** (egg cells) in females. Each gamete not only has half the number of chromosomes found in a somatic cell but also has one chromosome from each homologous pair found in the parent cell. The complement of chromosomes in a gamete is referred to as a **haploid** number. Oocytes contain one autosomal chromosome from each of the 22 homologous pair and an X chromosome. Sperm cells have 22 autosomal chromosomes and either an X or Y chromosome. During fertilization, when a sperm cell fuses with an oocyte, the normal number of 46 chromosomes in 23 pairs is reestablished. The sex of the baby is determined by the sperm cell that fertilizes the oocyte. The sex is male if a Y chromosome is carried by the sperm cell that fertilizes the oocyte, and female if the sperm cell carries an X chromosome.

Clinical Focus   Genetic Engineering

We are living at the beginning of an exciting era, when the genetic bases of many human illnesses are rapidly being revealed. As we discover the defective genes associated with these diseases and learn the nature and function of the proteins they encode, we are greatly improving our ability to understand and therefore to treat many of these diseases. Once we have learned the basis of a given disease, a number of approaches are possible for treating it, such as genetic engineering or other molecular techniques. For example, the gene for insulin has been inserted into a bacterial genome, enabling the bacterium to produce large quantities of human insulin, increasing its availability and functional quality. Antibodies are being developed that will target specific cells or cell surface marker molecules associated with diseases such as arthritis or cancer. Within a few years, it may be possible to introduce a functional copy of a gene into the cells of a person who has a defective gene.

There is a negative side to this technology, however. Many people are concerned that the introduction of foreign genes into bacteria and human cells may have unexpected side effects. Many genes have multiple functions, and there is a danger that we may begin using gene therapy before we know all the ramifications. There is also a great concern about how far genetic engineering should be allowed to go. What range of "genetic defects" should humanity be allowed to change, or should there be no limit? For example, when we discover the genes involved in controlling human height, should parents be allowed to use gene therapy to increase a child's height so that he or she can be better at basketball? An even more immediate concern is to what extent should a person's genetic code be made public. For example, should a medical insurance company or employer be allowed to see a person's genetic profile in order to set insurance premiums or make employment judgments? If a person is shown to have a gene for muscular dystrophy, should the person's insurance company be given that information? Also of current concern is whether a person or company should be able to patent and thus to own a human gene.

The first division during meiosis is divided into four stages: prophase I, metaphase I, anaphase I, and telophase I (figure 3.35). As in prophase of mitosis, the nuclear envelope degenerates, spindle fibers form, and the already duplicated chromosomes become visible. Each chromosome consists of two chromatids joined by a centromere. In prophase I, however, the four chromatids of a homologous pair of chromosomes join together, or **synapse,** (si-naps) to form a **tetrad** (four). In metaphase I the tetrads align at the equatorial plane, and in anaphase I each pair of homologous chromosomes separates and moves toward opposite ends of the cell. For each pair of homologous chromosomes, one daughter cell receives one member of the pair, and the other daughter cell receives the other member. Thus each daughter cell has 23 chromosomes, each of which is composed of two chromatids. Telophase I, with cytokinesis, is similar to telophase of mitosis, and two daughter cells are produced.

Interkinesis (in-ter-ki-nē′sis) is the time between the formation of the daughter cells and the second meiotic division. There is no duplication of DNA during interkinesis. The second division of meiosis also has four stages: prophase II, metaphase II, anaphase II, and telophase II. These stages occur much as they do in mitosis, except that there are 23 chromosomes instead of 46. The chromosomes align at the equatorial plane in metaphase II, and their chromatids split apart in anaphase II. The chromatids then are called chromosomes, and each new cell receives 23 chromosomes. Table 3.3 compares mitosis and meiosis.

In addition to reducing the number of chromosomes in a cell from 46 to 23, meiosis is also responsible for genetic diversity for two reasons. First, there is a random distribution of the genetic material received from each parent. One member of each homologous pair of chromosomes was derived from the person's father and the other member from the person's mother. The homologous chromosomes align randomly during metaphase I; when they split apart, each daughter cell receives some of the father's and some of the mother's genetic material. The amount of each that the daughter cell receives is determined by chance, however.

Second, when tetrads are formed, some of the chromatids may break apart, and part of one chromatid from one homologous pair may be exchanged for part of another chromatid from the other homologous pair (figure 3.36). This exchange is called **crossing over;** as a result, chromatids with different DNA content are formed.

With random assortment of homologous chromosomes and crossing over, the possible number of gametes with different genetic makeup is practically unlimited. When the different gametes of two individuals unite, it is virtually certain that the resulting genetic makeup never has occurred before and never will occur again.

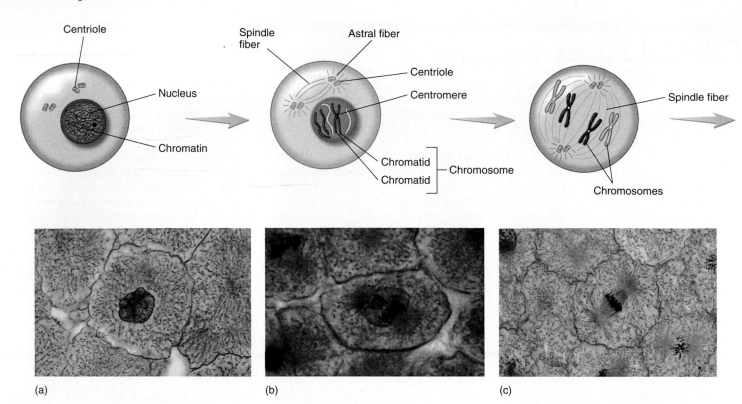

Figure 3.34   Mitosis

(*a*) **Interphase.** DNA, which is dispersed as chromatin, replicates. The two strands of each DNA molecule separate, and a copy of each strand is made. Consequently, two identical DNA molecules are produced. The pair of centrioles replicate to produce two pairs of centrioles.

(*b*) **Prophase.** Chromatin strands condense to form chromosomes. Each chromosome is composed of two identical strands called **chromatids,** which are joined together at one point by a specialized region called the **centromere.** Each chromatid contains one of the DNA molecules replicated during interphase. One pair of centrioles moves to each side, or pole, of the cell. Microtubules form near the centrioles and project in all directions. Some of the microtubules end blindly and are called **astral fibers.** Others, known as **spindle fibers** project toward an invisible line, called the **equator,** and either overlap with fibers from other centrioles or attach to the centromeres of the chromosomes. At the end of prophase the nuclear envelope degenerates, and the nucleoli disappear.

(*c*) **Metaphase.** The chromosomes align along the equator with spindle fibers from each pair of centrioles attached to their centromeres.

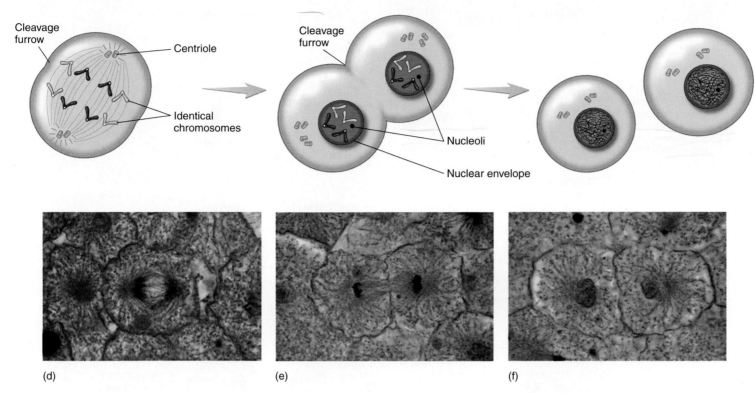

Cleavage furrow

Centriole

Identical chromosomes

Cleavage furrow

Nucleoli

Nuclear envelope

(d)

(e)

(f)

Figure 3.34   Mitosis

(*d*) **Anaphase.** The centromeres separate, and each chromatid is then referred to as a chromosome. Thus, when the centromeres divide, the chromosome number doubles, and there are two identical sets of chromosomes. The two sets of chromosomes are pulled by the spindle fibers toward the poles of the cell. Separation of the chromatids signals the beginning of anaphase, and by the time anaphase has ended, the chromosomes have reached the poles of the cell. The beginning of cytokinesis is evident during anaphase; along the equator of the cell the cytoplasm becomes narrower as the plasma membrane pinches inward.

(*e*) **Telophase.** The migration of each set of chromosomes is complete. A new nuclear envelope develops from the endoplasmic reticulum, and the nucleoli reappear. During the latter portion of telophase the spindle fibers disappear, and the chromosomes unravel to become less distinct chromatin threads. The nuclei of the two daughter cells assume the appearance of interphase nuclei, and the process of mitosis is complete.

(*f*) **Interphase.** Cytokinesis, which continued from anaphase through telophase becomes complete when the plasma membranes move close enough together at the equator of the cell to fuse, completely separating the two new daughter cells, each of which now has a complete set of chromosomes (a diploid number of chromosomes) identical to the parent cell.

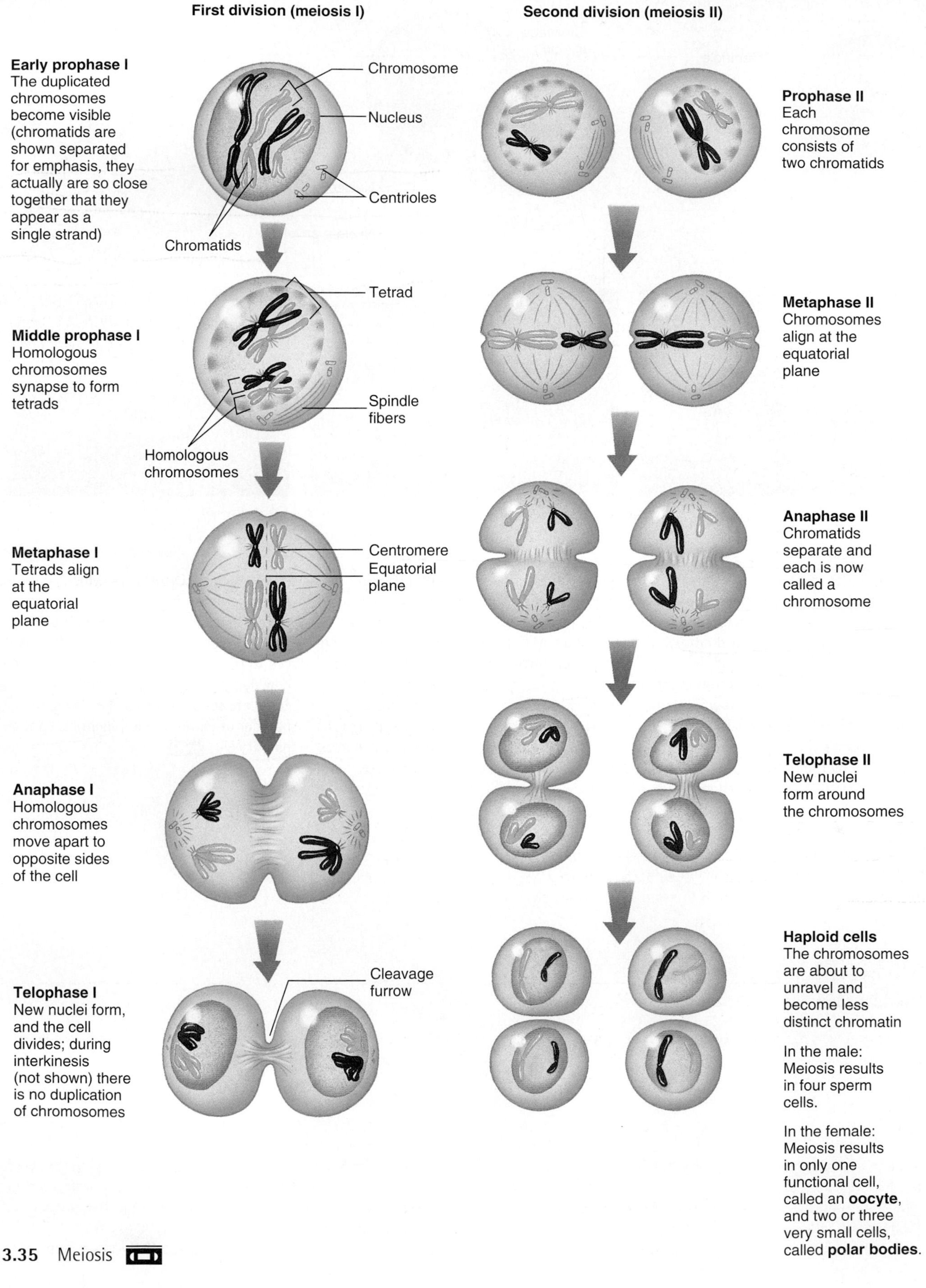

First division (meiosis I)

Second division (meiosis II)

Early prophase I
The duplicated chromosomes become visible (chromatids are shown separated for emphasis, they actually are so close together that they appear as a single strand)

Chromosome

Nucleus

Centrioles

Chromatids

Prophase II
Each chromosome consists of two chromatids

Middle prophase I
Homologous chromosomes synapse to form tetrads

Tetrad

Spindle fibers

Homologous chromosomes

Metaphase II
Chromosomes align at the equatorial plane

Metaphase I
Tetrads align at the equatorial plane

Centromere
Equatorial plane

Anaphase II
Chromatids separate and each is now called a chromosome

Anaphase I
Homologous chromosomes move apart to opposite sides of the cell

Telophase II
New nuclei form around the chromosomes

Telophase I
New nuclei form, and the cell divides; during interkinesis (not shown) there is no duplication of chromosomes

Cleavage furrow

Haploid cells
The chromosomes are about to unravel and become less distinct chromatin

In the male: Meiosis results in four sperm cells.

In the female: Meiosis results in only one functional cell, called an **oocyte**, and two or three very small cells, called **polar bodies**.

Figure 3.35   Meiosis

Table 3.3	Comparison of Mitosis and Meiosis	
Feature	**Mitosis**	**Meiosis**
Time of DNA replication	Interphase	Interphase
Number of cell divisions	One	Two; there is no replication of DNA in between the two meiotic divisions.
Cell produced	The two daughter cells are genetically identical to the parent cell; each daughter cell has a diploid number of chromosomes.	Gametes each different from the parent cell and each other; the gametes have a haploid number of chromosomes; in males, four gametes (sperm cells); in females, one gamete (oocyte) and two or three polar bodies.
Function	New cells are formed during growth or tissue repair; new cells have identical DNA and can perform the same functions as the parent cells.	Gametes are produced for reproduction; during fertilization the chromosomes from the haploid gametes unite to restore the diploid number typical of somatic cells; genetic variability is increased because of random distribution of chromosomes during meiosis and crossing over.

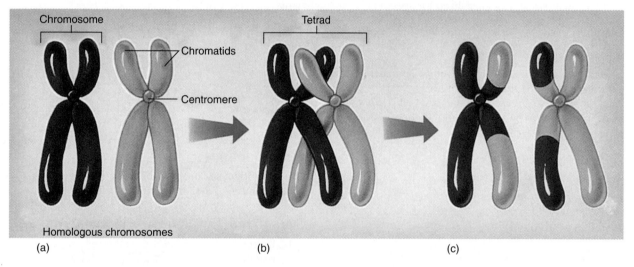

(a)   (b)   (c)

Figure 3.36   Crossing Over

Crossing over may occur during prophase I of meiosis. (*a*) A pair of replicated homologous chromosomes. (*b*) Chromatids of the homologous chromosomes form a tetrad. The chromatids are crossed in two places. The chromatids may break at the points of crossing and become fused to the opposite chromosome, resulting in crossing over. (*c*) Genetic material is exchanged following crossing over of the chromatids.

Summary

1. The plasma membrane forms the outer boundary of the cell.
2. The nucleus directs the activities of the cell.
3. Cytoplasm, between the nucleus and plasma membrane, is where most cell activities take place.

Plasma Membrane

1. The plasma membrane passively or actively regulates what enters or leaves the cell.
2. The plasma membrane is composed of a phospholipid bilayer in which proteins float (fluid-mosaic model). The proteins function as membrane channels, carrier molecules, receptor molecules, marker molecules, enzymes, and structural components of the membrane.

Nucleus

1. The nuclear envelope consists of two separate membranes with nuclear pores.
2. DNA and associated proteins are found inside the nucleus as chromatin. DNA is the hereditary material of the cell and controls the activities of the cell through the production of proteins through RNA.
3. Proteins play a role in the regulation of DNA activity.

4. Nucleoli consist of RNA and proteins and are the sites of ribosomal subunit assembly.

Cytoplasm

Cytosol

Cytosol consists of a fluid part (the site of chemical reactions), the cytoskeleton, and cytoplasmic inclusions.

Cytoskeleton

The cytoskeleton supports the cell and enables cell movements. It consists of protein fibers.

1. Microtubules are hollow tubes composed of the protein tubulin. They form spindle fibers and are components of centrioles, cilia, and flagella.
2. Actin filaments are small protein fibrils that provide structure to the cytoplasm or cause cell movements.
3. Intermediate filaments are protein fibers that provide structural strength to cells.

Organelles

Organelles are subcellular structures specialized for specific functions.

Ribosomes

1. Ribosomes consist of small and large subunits manufactured in the nucleolus and assembled in the cytoplasm.
2. Ribosomes are the sites of protein synthesis.
3. Ribosomes can be free or associated with the endoplasmic reticulum.

Endoplasmic Reticulum

1. The endoplasmic reticulum is an extension of the outer membrane of the nuclear envelope and forms tubules or sacs (cisternae) throughout the cell.
2. The rough endoplasmic reticulum has ribosomes and is a site of protein synthesis.
3. The smooth endoplasmic reticulum lacks ribosomes and is involved in lipid production, detoxification, and calcium storage.

Golgi Apparatus

The Golgi apparatus is a series of closely packed, modified cisternae that function to modify, package, and distribute lipids and proteins produced by the endoplasmic reticulum.

Secretory Vesicles

Secretory vesicles are membrane-bound sacs that carry substances from the Golgi apparatus to the plasma membrane, where the contents of the vesicle are released by exocytosis.

Lysosomes

1. Lysosomes are membrane-bound sacs containing hydrolytic enzymes. Within the cell the enzymes break down phagocytized material and nonfunctional organelles (autophagia).
2. Enzymes released from the cell by lysis or enzymes secreted from the cell can digest extracellular material.

Peroxisomes

Peroxisomes are membrane-bound sacs containing enzymes that digest fatty acids and amino acids and enzymes that catalyze the breakdown of hydrogen peroxide.

Mitochondria

1. Mitochondria are the major sites of the production of ATP, which is used as an energy source by cells.
2. The mitochondria have a smooth outer membrane and an inner membrane that is infolded to produce cristae.
3. Mitochondria contain their own DNA, can produce some of their own proteins, and can replicate independently of the cell.

Centrioles and Spindle Fibers

1. Centrioles are cylindrical organelles located in the centrosome, a specialized zone of the cytoplasm. The centrosome is the site of microtubule formation.
2. Spindle fibers are involved in the separation of chromosomes during cell division.

Cilia and Flagella

1. Movement of materials over the surface of the cell is facilitated by cilia.
2. Flagella, much longer than cilia, propel sperm cells.

Microvilli

Microvilli increase the surface area of the plasma membrane for absorption or secretion.

Cell Functions

Understanding the interactions between cell parts is necessary to understand the functions of whole cells.

Movement Through the Plasma Membrane

1. Lipid-soluble molecules pass through the plasma membrane readily by dissolving in the lipid bilayer.
2. Small molecules pass through membrane channels. Most channels are positively charged, allowing negatively charged ions and neutral molecules to pass through more readily than positively charged ions.
3. Large polar substances (e.g., glucose and amino acids) are transported through the membrane by carrier molecules.
4. Larger pieces of material enter cells in vesicles.

Diffusion

1. Diffusion is the movement of a substance from an area of higher concentration to one of lower concentration (with a concentration gradient).
2. The concentration gradient is the difference in solute concentration between two points divided by the distance separating the points.
3. The rate of diffusion increases with an increase in the concentration gradient, an increase in temperature, a decrease in molecular size, and a decrease in viscosity.
4. The end result of diffusion is a uniform distribution of molecules.
5. Diffusion requires no expenditure of energy.

Osmosis

1. Osmosis is the diffusion of water (solvent) across a selectively permeable membrane.
2. Osmotic pressure is the force required to prevent the movement of water across a selectively permeable membrane.
3. Isosmotic solutions have the same concentration of solute particles, hyperosmotic solutions have a greater concentration, and hyposmotic solutions have a lesser concentration of solute particles than a reference solution.
4. Cells placed in an isotonic solution neither swell nor shrink. In a hypertonic solution they shrink (crenate), and in a hypotonic solution they swell and may burst (lyse).

Filtration

1. Filtration is the movement of a liquid through a partition with holes that allow the liquid, but not everything in the liquid, to pass through them.
2. Liquid movement results from a pressure difference across the partition.

Mediated Transport Mechanisms

1. Mediated transport is the movement of a substance across a membrane by means of a carrier molecule. The substances transported tend to be large, water-soluble molecules.
   - The carrier molecules have binding sites that bind with either a single transport molecule or a group of similar transport molecules. This selectiveness is called specificity.
   - Similar molecules can compete for carrier molecules, with each reducing the rate of transport of the other.
   - Once all the carrier molecules are in use, the rate of transport cannot increase further (saturation).
2. There are three kinds of mediated transport.
   - Facilitated diffusion moves substances with their concentration gradient and does not require energy expenditure (ATP).
   - Active transport can move substances against their concentration gradient and requires ATP. An exchange pump is an active-transport mechanism that simultaneously moves two substances in opposite directions across the plasma membrane.
   - In secondary active transport, an ion is moved across the cell membrane by active transport, and the energy produced by the ion diffusing back down its concentration gradient can transport another molecule, such as glucose, against its concentration gradient.

Endocytosis and Exocytosis

1. Endocytosis is the bulk movement of materials into cells.
   - Phagocytosis is the bulk movement of solid material into cells by the formation of a vesicle.
   - Pinocytosis is similar to phagocytosis, except that the ingested material is much smaller or is in solution.
2. Exocytosis is the secretion of materials from cells by vesicle formation.
3. Endocytosis and exocytosis use vesicles, can be specific for the substance transported, and require energy.

Cell Metabolism

1. Aerobic respiration requires oxygen and produces carbon dioxide, water, and 36–38 ATP molecules from a molecule of glucose.

2. Anaerobic respiration does not require oxygen and produces lactic acid and two ATP molecules from a molecule of glucose.

Protein Synthesis

1. Information stored in DNA is copied to mRNA.
2. The mRNA goes to ribosomes where it directs the synthesis of proteins.

Transcription

1. DNA unwinds and, through nucleotide pairing, produces mRNA (transcription).
2. The genetic code, which codes for amino acids, consists of codons, which are a sequence of three nucleotides in mRNA.

Translation

1. The mRNA moves through the nuclear pores to ribosomes.
2. Transfer RNA (tRNA), which carries amino acids, interacts at the ribosome with mRNA. The anticodons of tRNA bind to the codons of mRNA, and the amino acids are joined to form a protein (translation).

Regulation of Protein Synthesis

1. Cells become specialized because of inactivation of certain parts of the DNA molecule and activation of other parts.
2. The level of DNA activity and thus protein production can be controlled internally or can be affected by regulatory substances secreted by other cells.

Cell Life Cycle
Interphase

1. Interphase is the period between cell divisions.
2. DNA unwinds, and each strand produces a new DNA molecule during interphase.

Cell Division

1. Mitosis is the replication of the nucleus of the cell, and cytokinesis is division of the cytoplasm of the cell.
2. Humans have 22 pairs of homologous chromosomes called autosomes. Females also have two X chromosomes, and males also have an X chromosome and a Y chromosome.
3. Mitosis is a continuous process divided into four stages.
   - *Prophase.* Chromatin condenses to become visible as chromosomes. Each chromosome consists of two chromatids joined at the centromere. Centrioles move to opposite poles of the cell, and astral fibers and spindle fibers form. Nucleoli disappear, and the nuclear envelope degenerates.
   - *Metaphase.* Chromosomes align at the equatorial plane.
   - *Anaphase.* The chromatids of each chromosome separate at the centromere. Each chromatid then is called a chromosome. The chromosomes migrate to opposite poles.
   - *Telophase.* Chromosomes unravel to become chromatin. The nuclear envelope and nucleoli reappear.
4. Cytokinesis begins with the formation of the cleavage furrow during anaphase. It is complete when the plasma membrane comes together at the equator, producing two new daughter cells.

Meiosis

1. Meiosis results in the production of gametes (oocytes or sperm cells).
2. All gametes receive one half of the homologous autosomes (one from each homologous pair). Oocytes also receive an X chromosome. Sperm cells have an X or a Y chromosome.
3. There are two cell divisions in meiosis. Each division has four stages (prophase, metaphase, anaphase, and telophase) similar to those in mitosis.

- In the first division tetrads form, crossing over occurs, and homologous chromosomes are distributed randomly. Two cells are formed, each with 23 chromosomes. Each chromosome has two chromatids.
- In the second division the chromatids of each chromosome separate, and each cell receives 23 chromatids, which then are called chromosomes.

4. Genetic variability is increased by crossing over and random assortment of chromosomes.

Content Review

1. Define the terms plasma membrane, nucleus, and cytoplasm.
2. What is the function of the plasma membrane? How do phospholipids, cholesterol, and proteins contribute to that function? Describe the fluid-mosaic model of the plasma membrane.
3. List and describe the parts of cytosol.
4. What is the function of the cytoskeleton? Name the three groups of proteins of which it is made.
5. Describe the structure of the nuclear envelope.
6. What is the difference between chromatin and chromosomes? Name the two components of chromatin and explain their functions.
7. Where are ribosomes assembled and what kinds of molecules are found in them? Give the function of ribosomes.
8. What is the endoplasmic reticulum? Contrast rough and smooth endoplasmic reticulum according to structure and function.
9. Describe the Golgi apparatus and state its function.
10. Where are secretory vesicles produced? What are their contents and how are they released?
11. What is a lysosome? What is a peroxisome? Explain the function of lysosomes and peroxisomes.
12. Describe the structure of mitochondria. Name the important molecule produced by mitochondria. For what is this molecule used?
13. Describe the structure and function of the centromere, centrioles, spindle fibers, cilia, flagella, and microvilli.
14. How do large lipid-soluble molecules move across the plasma membrane?
15. How do small molecules, water- or lipid-soluble, pass through the membrane? What effect does the electric charge of molecules have on the ease of passage?
16. Define the term diffusion. How do the concentration gradient, temperature, molecule size, and viscosity affect the rate of diffusion?
17. Define the terms osmosis and osmotic pressure.
18. What happens to a cell that is placed in an isotonic solution? In a hypertonic or hypotonic solution? What are crenation and lysis?
19. Define the term filtration.
20. What is mediated transport? Explain the basis for specificity, competition, and saturation of transport mechanisms.
21. Contrast active transport and facilitated diffusion in relation to energy expenditure and movement of substances with or against their concentration gradients.
22. What are secondary active transport, cotransport, and countertransport?
23. Name three ways in which phagocytosis, pinocytosis, and exocytosis are similar. How do they differ?
24. Define glycolysis. Where does it take place?
25. Contrast the production of ATP by aerobic and anaerobic respiration. What happens to the oxygen we breathe in? The carbon dioxide we breathe out comes from where?
26. Explain what happens during transcription. How does mRNA get to the ribosomes?
27. Describe translation. What kinds of RNA are involved? How are codons and anticodons involved in the synthesis of proteins?
28. What are posttranscriptional and posttranslational processing?
29. Distinguish between mitosis and cytokinesis.
30. List the events that occur during interphase, prophase, metaphase, anaphase, and telophase of mitosis.
31. Define the term meiosis, and describe the events that occur during meiosis. What happens to the number of chromosomes during meiosis?
32. How are the chromosomes of males and females different?
33. What are two ways in which genetic variability is increased?

Develop Your Reasoning Skills

1. Given the following data from electron micrographs of a cell, predict the major function of the cell:
   Moderate number of mitochondria
   Well-developed rough endoplasmic reticulum
   Moderate number of lysosomes
   Well-developed Golgi apparatus
   Dense nuclear chromatin
   Numerous vesicles

2. Why does a surgeon irrigate a surgical wound from which a tumor has been removed with sterile distilled water rather than with sterile physiologic saline?

3. Solution A is hyperosmotic to solution B. If solution A is separated from solution B by a selectively permeable membrane, does water move from solution A into solution B or vice versa? Explain.

4. A dialysis membrane is selectively permeable, and substances smaller than proteins are able to pass through it. If you wanted to use a dialysis machine to remove only urea (a small molecule) from blood, what could you use for the dialysis fluid?
   a. A solution that is isotonic and contains only large molecules, such as protein
   b. A solution that is isotonic and contains the same concentration of all substances except that it has no urea
   c. Distilled water, which contains no ions or dissolved molecules
   d. Blood, which is isotonic and contains the same concentration of all substances, including urea

5. A researcher wants to determine the nature of the transport mechanism that moved substance X into a cell. She could measure the concentration of substance X in the extracellular fluid and within the cell, as well as the rate of movement of substance X into the cell. She does a series of experiments and gathers the data shown in the graph. Choose the transport process that is consistent with the data.
   a. Diffusion
   b. Active transport
   c. Facilitated diffusion
   d. Not enough information to make a judgment

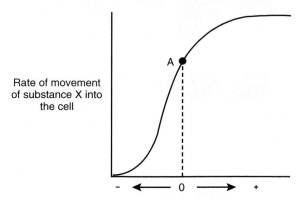

Graph depicting the rate of movement of substance X from a fluid into a cell (y axis) versus the concentration of substance X within the cell (x axis). At point "A" the extracellular concentration of substance X is equal to the intracellular concentration of substance X (designated 0 on the x axis).

6. If you had the ability to inhibit mRNA synthesis with a drug, explain how you could distinguish between proteins released from secretory vesicles in which they had been stored and proteins released from cells in which they have been newly synthesized.

Web Site Link

For a listing of the most current web sites related to this chapter, please visit the Seeley home page at:
http://www.mhhe.com/biosci/ap/seeleyap/

Chapter Four

Histology: The Study of Tissues

Objectives

1. List the characteristics used to classify tissues into one of the four major tissue types.

2. List the features that characterize epithelium.

3. Describe the characteristics that are used to classify the various epithelial types.

4. For each epithelial type, list its number of cell layers, cell shapes, major cellular organelles, and surface specializations and the functions to which each type is adapted.

5. Explain why junctional complexes between cells are important to the normal function of epithelium.

6. Define the term gland, and describe the two major categories of glands.

7. List the features that characterize connective tissue.

8. List the major large molecules of the connective tissue matrix, and explain their functions in the matrix.

9. List the major categories of connective tissue, and describe the characteristics of each.

10. List the general characteristics of muscle.

11. Name the main types of muscles, and list their major characteristics.

12. Describe the characteristics of nervous tissue.

13. Name the three embryonic germ layers, and describe the function of mesenchyme.

14. List the functional and structural characteristics of serous, mucous, and synovial membranes.

15. Describe the process of inflammation, and explain why inflammation is protective to the body.

16. Describe the major events involved in tissue repair.

In some ways, the human body can be compared to a complex machine such as a car. Both consist of many parts, each of which is made of materials consistent with its specialized function. For example, the windows of a car are made of transparent glass, the tires are made of synthetic rubber reinforced with a variety of fibers, the engine is made of a variety of metal parts, and the hoses that move water, air, and gasoline are made of synthetic rubber or plastic. All parts of an automobile cannot be made of a single type of material. Metal capable of withstanding the heat of the engine cannot be used for windows or tires. Similarly, the many parts of the human body are made of collections of specialized cells and the materials surrounding them. For example, muscle cells, which contract to produce movements of the body, look different and have different functions than those of epithelial cells, which protect, secrete, or absorb. Also, cells in the retina of the eye, specialized to detect light allowing us to see, do not contract like muscle cells or exhibit the functions of epithelial cells.

Collections of similar cells and the substances surrounding them are called **tissues.** The structure of the cells and the composition of the noncellular substances surrounding cells, called the **extracellular matrix,** are characteristics used to classify tissue types. The four **primary tissue types** from which all of the organs of the body are formed are:

1. Epithelial tissue
2. Connective tissue
3. Muscle tissue
4. Nervous tissue

Epithelial and connective tissues are the most diverse in form. For this reason the different types of epithelial and connective tissues are classified by structure, including cell shape, relationship of cells to one another, and the material making up the extracellular matrix. In contrast, muscle and nervous tissues are classified mainly by function.

The tissues of the body are interdependent. For example, muscle tissue cannot produce movement unless it receives oxygen carried by red blood cells, and new bone tissue cannot be formed unless epithelial tissue absorbs calcium and other nutrients from the digestive tract. When cancer or some other disease destroys the tissues of the liver, all other tissues in the body die.

The microscopic study of tissues is **histology** (his-tol′ō-jē). Much information about the health of a person can be gained by examining tissues. For example, some red blood cells are shaped differently in people suffering from sickle cell disease than they are in people with iron-deficiency anemia, white blood cells are different in people who have leukemia than in people who have infections, and epithelial cells from respiratory passages are different in people with chronic bronchitis than in people with lung cancer. Examining tissue samples from individuals with these disorders can distinguish the specific disease.

This chapter discusses the structure of the major tissue types and their functional characteristics. The structure and function of tissues are so closely related that a student should be able to predict the function of a tissue when given its structure and vice versa. Knowledge of tissue structure and function is important in understanding how cells are organized to form tissues, organs, organ systems, and the complete organism.

Epithelial Tissue

Characteristics common to all types of **epithelium** (pl., epithelia) are (figure 4.1):

1. Epithelium consists almost entirely of cells, with very little extracellular material between them.
2. Epithelium covers surfaces such as the outside of the body and the lining of the digestive tract, the vessels, and many body cavities; or it forms structures such as glands, which are derived developmentally from the body surfaces.
3. Most epithelial tissues have one **free surface** that is not associated with other cells and a **basal surface.** The basal surface of most epithelial tissues is attached to a **basement membrane.** The basement membrane is a specialized type of extracellular material that is secreted by the epithelial cells on the side opposite their free surface and by connective tissue cells; it helps attach the epithelial cells to the underlying tissues, and it plays an important role in supporting and guiding cell migration during tissue repair. Some epithelial tissues (e.g., cells in some endocrine glands) do not have a free surface or a basal surface with a basement membrane. Also, epithelium of some lymph vessels and liver sinusoids do have free and basal surfaces but do not have basement membranes.
4. Specialized cell contacts, such as tight junctions and desmosomes, bind adjacent epithelial cells together.
5. Blood vessels do not penetrate the basement membrane to reach the epithelium; thus all gases and nutrients carried in the blood must reach the epithelium by diffusing across the basement membrane from blood vessels in the underlying connective tissue. In epithelia with many layers of cells, the most metabolically active cells are close to the basement membrane, and cells die as they move farther away from the basement membrane.
6. Epithelial cells retain the ability to undergo mitosis and therefore are able to replace damaged cells with new epithelial cells.

Classification of Epithelium

The major types of epithelia and their distributions are illustrated in figure 4.2. Epithelium is classified primarily according to the number of cell layers and the shape of the cells.

There are three major types of epithelium based on the number of cell layers in each type:

1. **Simple epithelium** consists of a single layer of cells, with each cell extending from the basement membrane to the free surface.
2. **Stratified epithelium** consists of more than one layer of cells, only one of which is adjacent to the basement membrane.

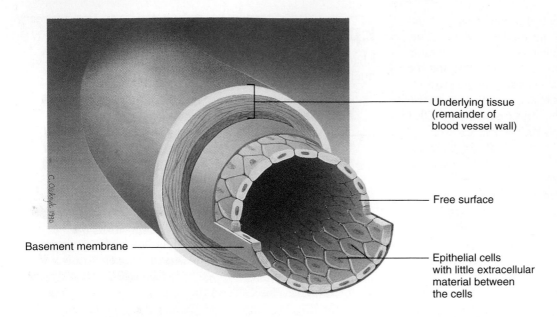

- Underlying tissue (remainder of blood vessel wall)

- Free surface

Basement membrane ———

- Epithelial cells with little extracellular material between the cells

Figure 4.1   Characteristics of Epithelium

Epithelium lining a blood vessel illustrates the following characteristics: little extracellular material between cells, a free surface, a basement membrane attaching epithelial cells to underlying tissues. No capillaries penetrate the basement membrane to provide a blood supply to epithelial cells from the underlying tissues. ⚡

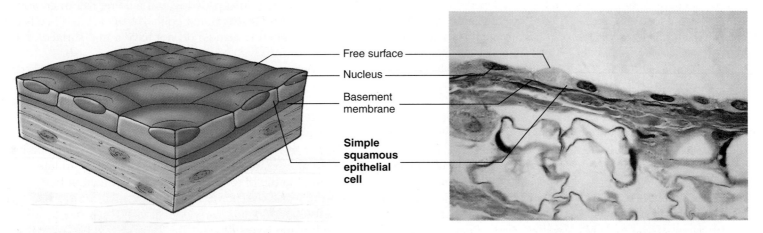

Free surface ——
Nucleus ——
Basement membrane ——
Simple squamous epithelial cell

Types of Epithelium

(a) **Simple squamous epithelium**

Location: Lining of blood and lymphatic vessels (endothelium) and small ducts, alveoli of the lungs, loop of Henle in kidney tubules, lining of serous membranes (mesothelium), and inner surface of the eardrum.

Structure: Single layer of flat, often hexagonal cells. The nuclei appear as bumps when viewed as a cross section because the cells are so flat.

Function: Diffusion, filtration, some protection against friction, secretion, and absorption.

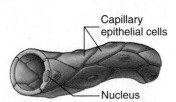

Capillary epithelial cells

Nucleus

Figure 4.2   Types of Epithelium ⚡

3. **Pseudostratified epithelium** consists of epithelial cells that are all attached to the basement membrane, but only some of the cells reach the free surface. This epithelium is called pseudostratified because, although it consists of a single cell layer, it appears multilayered. The arrangement of the nuclei gives a stratified appearance.

There are three types of epithelium based on the epithelial cell shapes:

1. **Squamous** (skwā′mŭs; flat) cells are flat or scalelike.
2. **Cuboidal** (cubelike) cells are cube-shaped; about as wide as they are tall.
3. **Columnar** (tall and thin, similar to a column) cells are taller than they are wide.

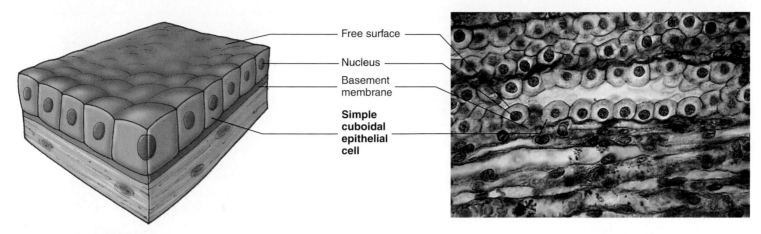

Free surface

Nucleus

Basement membrane

Simple cuboidal epithelial cell

Types of Epithelium

(b) **Simple cuboidal epithelium**

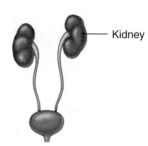

Kidney

Location: Kidney tubules, glands and their ducts, choroid plexus of the brain, lining of terminal bronchioles of the lungs, and surface of the ovaries.

Structure: Single layer of cube-shaped cells; some cells have microvilli (kidney tubules) or cilia (terminal bronchioles of the lungs).

Function: Active transport and facilitated diffusion result in secretion and absorption by cells of the kidney tubules; secretion by cells of glands and choroid plexus; movement of mucus-containing particles out of the terminal bronchioles by ciliated cells.

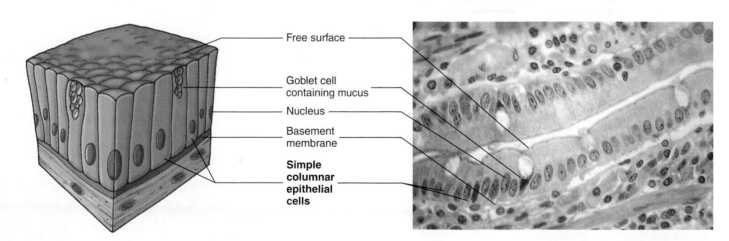

Free surface

Goblet cell containing mucus

Nucleus

Basement membrane

Simple columnar epithelial cells

Types of Epithelium

(c) **Simple columnar epithelium**

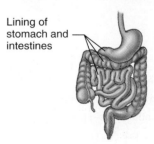

Lining of stomach and intestines

Location: Glands and some ducts, bronchioles of lungs, auditory tubes, uterus, uterine tubes, stomach, intestines, gallbladder, bile ducts, and ventricles of the brain.

Structure: Single layer of tall, narrow cells. Some cells have cilia (bronchioles of lungs, auditory tubes, uterine tubes, and uterus) or microvilli (intestines).

Function: Movement of particles out of the bronchioles of the lungs; partially responsible for the movement of the oocyte through the uterine tubes by ciliated cells. Secretion by cells of the glands, the stomach, and the intestine. Absorption by cells of the intestine.

Figure 4.2   *(continued)*

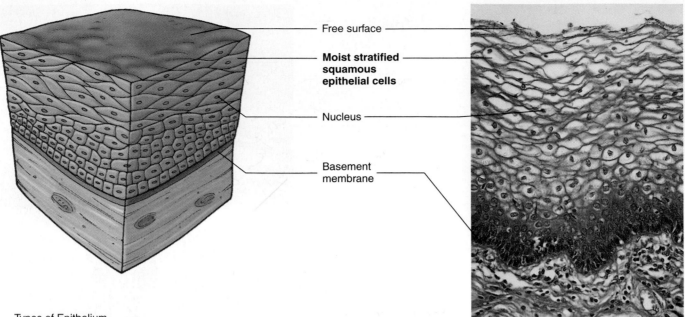

Free surface

Moist stratified squamous epithelial cells

Nucleus

Basement membrane

Types of Epithelium

(d) **Stratified squamous epithelium**

Location: Moist—mouth, throat, larynx, esophagus, anus, vagina, inferior urethra, and cornea. Keratinized—skin.

Structure: Multiple layers of cells that are cuboidal in the basal layer and progressively flattened toward the surface. The epithelium can be moist or keratinized. In moist stratified squamous epithelium the surface cells retain a nucleus and cytoplasm. In keratinized cells the cytoplasm is replaced by keratin, and the cells are dead.

Function: Protection against abrasion and infection.

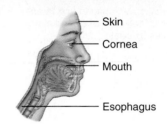

Skin

Cornea

Mouth

Esophagus

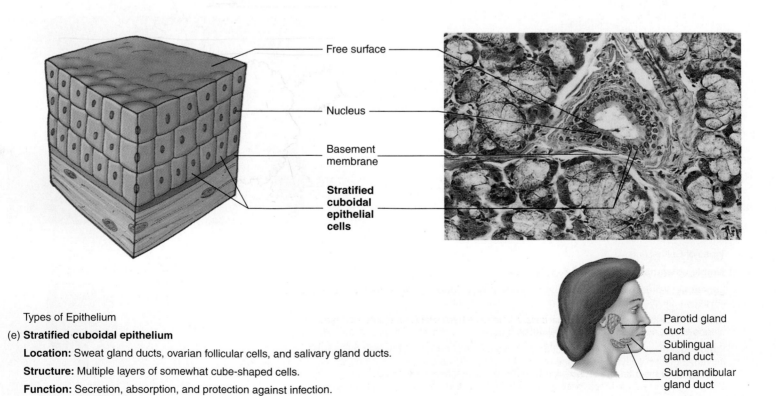

Free surface

Nucleus

Basement membrane

Stratified cuboidal epithelial cells

Types of Epithelium

(e) **Stratified cuboidal epithelium**

Location: Sweat gland ducts, ovarian follicular cells, and salivary gland ducts.

Structure: Multiple layers of somewhat cube-shaped cells.

Function: Secretion, absorption, and protection against infection.

Parotid gland duct
Sublingual gland duct
Submandibular gland duct

Figure 4.2 *(continued)*

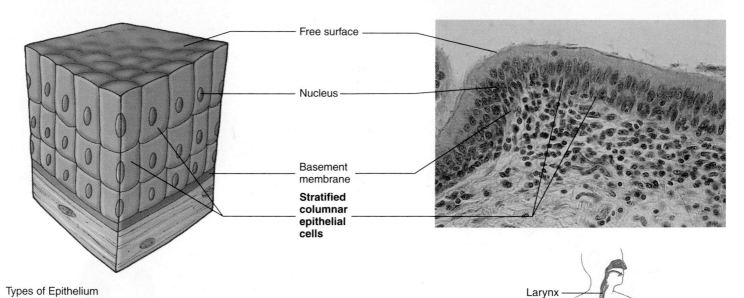

Free surface

Nucleus

Basement membrane

Stratified columnar epithelial cells

Larynx

Types of Epithelium

(f) Stratified columnar epithelium

Location: Mammary gland duct, larynx, and a portion of the male urethra.

Structure: Multiple layers of cells, with tall, thin cells resting on layers of more cuboidal cells. The cells are ciliated in the larynx.

Function: Protection and secretion.

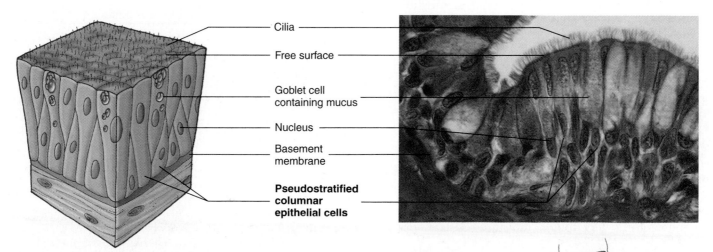

Cilia

Free surface

Goblet cell containing mucus

Nucleus

Basement membrane

Pseudostratified columnar epithelial cells

Trachea

Bronchus

Type of Epithelium

(g) Pseudostratified columnar epithelium

Location: Lining of nasal cavity, nasal sinuses, auditory tubes, pharynx, trachea, and bronchi of lungs.

Structure: Single layer of cells; some cells are tall and thin and reach the free surface, and others do not; the nuclei of these cells are at different levels and appear stratified; the cells are almost always ciliated and are associated with goblet cells that secrete mucus onto the free surface.

Function: Synthesize and secrete mucus onto the free surface and move mucus (or fluid) that contains foreign particles over the surface of the free surface and from passages.

Figure 4.2 *(continued)*

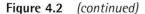

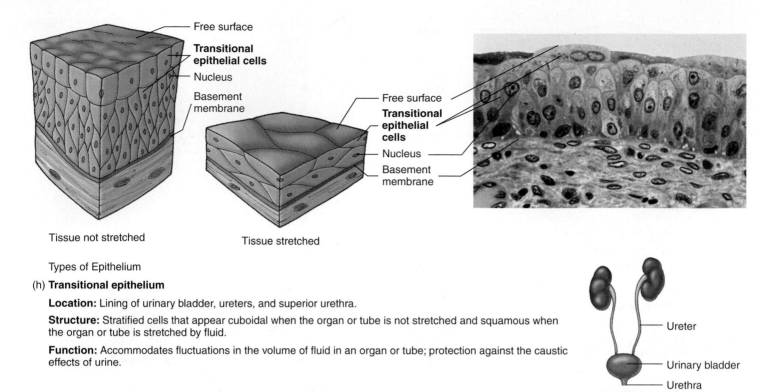

Free surface
Transitional epithelial cells
Nucleus
Basement membrane

Free surface
Transitional epithelial cells
Nucleus
Basement membrane

Tissue not stretched

Tissue stretched

Types of Epithelium

(h) **Transitional epithelium**

Location: Lining of urinary bladder, ureters, and superior urethra.

Structure: Stratified cells that appear cuboidal when the organ or tube is not stretched and squamous when the organ or tube is stretched by fluid.

Function: Accommodates fluctuations in the volume of fluid in an organ or tube; protection against the caustic effects of urine.

Ureter

Urinary bladder

Urethra

Figure 4.2   *(continued)*

In most cases an epithelium is given two names, such as simple squamous, stratified squamous, simple columnar, or pseudostratified columnar. The first name indicates the number of layers, and the second indicates the shape of the cells (table 4.1) at the free surface.

Stratified squamous epithelium can be classified further as either moist or keratinized, according to the condition of the outermost layer of cells. In both types the deepest layers are composed of living cells. In **moist stratified squamous epithelium,** found in areas such as the mouth, esophagus, rectum, and vagina, the outermost layers also consist of living cells. Because the outer layers of cells are living, a layer of fluid covers them, which makes them moist. In contrast, **keratinized stratified squamous epithelium,** found in the skin (see chapter 5), has outer layers composed of dead cells containing the hard protein keratin. The dead, keratinized cells give the tissue a very tough, moisture-resistant, dry character.

A unique type of stratified epithelium called **transitional epithelium** lines the urinary bladder and ureters, structures in which considerable expansion can occur. The shape of the cells and the number of cell layers vary, depending on whether it is stretched or not. The surface cells and the underlying cells are roughly cuboidal or columnar when the epithelium is not stretched, and become more flattened or squamouslike when the epithelium is stretched. Also, the number of layers of epithelial cells decreases in response to stretch. As the epithelium is stretched, the epithelial cells have the ability to shift on one another so that the number of layers decreases from five or six to two or three.

Table 4.1   Classification of Epithelium

Number of Layers or Category	Shape of Cells
Simple (single layer of cells)	Squamous Cuboidal Columnar

Stratified (more than one layer of cells)	Squamous
	    Moist
	    Keratinized
	Cuboidal (very rare)
	Columnar (very rare)

Pseudostratified (modification of simple epithelium)	Columnar
Transitional (modification of stratified epithelium)	Roughly cuboidal to columnar when not stretched and squamouslike when stretched

Functional Characteristics

Epithelial tissues have many functions (table 4.2), including forming a barrier between a free surface and the underlying tissues, and secreting, transporting, and absorbing selected molecules. The type and arrangement of organelles within each cell (see chapter 3), the shape of cells, and the organization of cells within each epithelial type reflect these functions. Accordingly, specialization of epithelial cells can be understood best in terms of the functions they perform.

Cell Layers and Cell Shapes

Simple epithelium with its single layer of cells is found in organs in which the principal functions are diffusion (lungs), filtration (kidneys), secretion (glands), or absorption (intestines). The selective movement of materials through epithelium would be hindered by a stratified epithelium, which is found in areas in which protection is a major function. The multiple layers of cells in stratified epithelium are well adapted for a protective role because, as the outer cells are damaged, they are replaced by cells from deeper layers and a continuous barrier of epithelial cells is maintained in the tissue. Stratified squamous epithelium is found in areas of the body in which abrasion can occur, such as the skin, mouth, throat, esophagus, anus, and vagina.

Differing functions are also reflected in cell shape. Cells that allow substances to diffuse through them and that filter are normally flat and thin. For example, simple squamous epithelium forms blood and lymph capillaries, the alveoli (air sacs) of the lungs, and parts of the kidney tubules. Cells that secrete or absorb are usually cuboidal or columnar. They have greater cytoplasmic volume compared with that of squamous epithelium; this cytoplasmic volume results from the presence of the organelles responsible for the tissues' functions. For example, pseudostratified columnar epithelium, which secretes large amounts of mucus, lines the respiratory tract (see chapter 23) and contains large **goblet cells,** which are specialized columnar epithelial cells. The goblet cells contain abundant organelles responsible for the synthesis and secretion of mucus, such as ribosomes, endoplasmic reticulum, Golgi apparatuses, and secretory vacuoles filled with mucus.

1   P R E D I C T

Explain the consequences of (a) having moist stratified epithelium rather than simple columnar epithelium lining the digestive tract, (b) having moist stratified squamous epithelium rather than keratinized stratified squamous epithelium in skin, and (c) having simple columnar epithelium rather than moist stratified squamous epithelium lining the mouth.

✔ *Answer in Appendix F*

Cell Surfaces

The surfaces of epithelial cells can be divided into three categories: (1) a free surface that faces away from underlying tissue, (2) a surface that faces other cells, such as the lateral surfaces in simple epithelia, and (3) a surface that faces the basement membrane, the basal surface. The free surfaces of epithelia can be smooth, contain microvilli, be ciliated, or be folded.

Smooth surfaces reduce friction. Simple squamous epithelium with a smooth surface forms the covering of serous membranes. The lining of blood vessels is a simple squamous epithelium that reduces friction as blood flows through the vessels (see chapter 21).

Microvilli and cilia were described in chapter 3. Microvilli greatly increase surface area and are found in cells involved in absorption or secretion, such as the lining of the

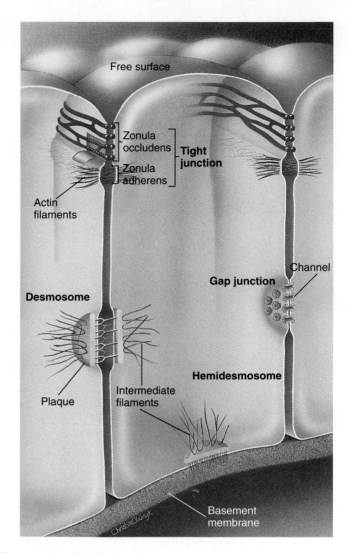

Figure 4.3   Cell Connections

Tight junctions consist of a zonula adherens and zonula occludens. Desmosomes, hemidesmosomes, and gap junctions are also shown, although few cells have all of these different connections.

small intestine (see chapter 24). Cilia propel materials across the surface of the cell. Simple ciliated cuboidal, simple ciliated columnar, and pseudostratified ciliated columnar epithelia are in the respiratory tract (see chapter 23), where mucus that contains foreign substances, such as dust particles, is removed from the respiratory passages by the movements of the cilia.

Transitional epithelium has a rather unusual cell membrane specialization: more rigid sections of membrane are separated by very flexible regions in which the cell membrane is folded. When transitional epithelium is stretched, the folded regions of the cell membrane can unfold. Transitional epithelium is specialized to expand. It is found in the urinary bladder, ureters, and the upper portion of the urethra.

Cell Connections

Lateral and basilar surfaces have structures that serve to hold cells to one another or to the basement membrane (figure 4.3). These structures do three things: (1) they mechanically bind

Table 4.2   Function and Location of Epithelial Tissue

Function	Simple Squamous Epithelium	Simple Cuboidal Epithelium	Simple Columnar Epithelium
Diffusion	Blood and lymph capillaries, alveoli of lungs, thin segment of loop of Henle		
Filtration	Bowman's capsule of kidney		
Secretion or absorption	Mesothelium (serous fluid)	Choroid plexus (cerebrospinal fluid), part of kidney tubule, many glands and their ducts	Stomach, small intestine, large intestine, uterus, many glands
Protection (against friction and abrasion)	Endothelium, mesothelium		
Movement of mucus (ciliated)		Terminal bronchioles of lungs	Bronchioles of lungs, auditory tubes, uterine tubes, uterus
Capable of great stretching			
Miscellaneous	Lines the inner part of the eardrum, smallest ducts of glands	Surface of ovary, inside lining of eye (pigmented epithelium of retina), ducts of glands	Bile duct, gallbladder, ependyma (lining of brain ventricles and central canal of spinal cord), ducts of glands

the cells together; (2) they help form a permeability barrier; and (3) they provide a mechanism for intercellular communication. Epithelial cells secrete glycoproteins that attach the cells to the basement membrane and to one another. This relatively weak binding between cells is reinforced by **desmosomes** (dez′mō-sōmz), disk-shaped structures with especially adhesive glycoproteins that bind cells to one another (see figure 4.3). Many desmosomes are found in epithelia that are subjected to stress, such as the stratified squamous epithelium of the skin. **Hemidesmosomes,** similar to one-half of a desmosome, attach epithelial cells to the basement membrane.

Tight junctions hold cells together and form a permeability barrier (see figure 4.3). They consist of a zonula adherens and a zonula occludens, which are found in close association with each other (see figure 4.3). The **zonula adherens** (zō′nū-lăh ad-hě′renz) is located between the cell membranes of adjacent cells and acts like a weak glue that holds cells together. The zonulae adherens are best developed in simple epithelial tissues; they form a girdle of adhesive glycoprotein around the lateral surface of each cell and bind adjacent cells together. These connections are not as strong as desmosomes.

The **zonula occludens** (ō-klood′enz) forms a permeability barrier. It is formed by cell membranes of adjacent cells that join one another in a jigsaw fashion to form a tight seal (see figure 4.3). Near the free surface of simple epithelial cells,

a ring formed by the zonulae occludens completely surrounds each cell and binds adjacent cells together. The zonulae occludens prevent the passage of materials between cells. Thus water and other substances must pass through the epithelial cells, which can actively regulate what is absorbed or secreted. Zonulae occludens are found in areas in which a layer of simple epithelium forms a permeability barrier. For example, water can diffuse through epithelial cells, and most nutrients are transported by active transport, cotransport, or facilitated diffusion through the epithelial cells of the intestine.

A **gap junction** is a small protein channel that provides a means of intercellular communication by allowing the passage of ions and small molecules between cells (see figure 4.3). The exact function of gap junctions in epithelium is not clear, but they are important in coordinating the function of cardiac and smooth muscle tissues. Because ions can pass through the gap junctions from one cell to the next, electric signals can pass from cell to cell to coordinate the contraction of cardiac and smooth muscle cells. Thus electric signals that originate in one cell of the heart can spread from cell to cell and cause the entire heart to contract. The gap junctions between cardiac muscle cells are found in specialized cell-to-cell connections called **intercalated disks.** Gap junctions between ciliated epithelial cells can function to coordinate the movements of the cilia.

Stratified Squamous Epithelium	Stratified Cuboidal Epithelium	Stratified Columnar Epithelium	Pseudostratified Columnar Epithelium	Transitional Epithelium
Skin (epidermis), cornea, mouth and throat, epiglottis, larynx, esophagus, anus, vagina				
			Larynx, nasal cavity, paranasal sinus, nasopharynx, auditory tube, trachea, bronchi of lungs	
				Urinary bladder, ureter, upper part of urethra
Lower part of urethra, sebaceous gland duct	Sweat gland ducts	Part of male urethra, epididymis, ductus deferens, mammary gland duct	Part of male urethra, salivary gland duct	

Glands

Glands are secretory organs. Most glands are composed primarily of epithelium, with a supporting network of connective tissue. They develop from an infolding or outfolding of epithelium in the embryo. If the gland maintains an open contact with the epithelium from which it developed, a duct is present. Glands with ducts are called **exocrine** (ek′sō-krin) **glands,** and their ducts are lined with epithelium. Alternatively, some glands become separated from the epithelium of their origin. Glands that have no ducts are called **endocrine** (en′dō-krin) **glands.** Their cellular products, which are called **hormones** (hōr′mōnz), are secreted into the bloodstream and carried throughout the body.

Most exocrine glands are composed of many cells and are called **multicellular glands,** but some exocrine glands are composed of a single cell and are called **unicellular glands** (figure 4.4a; see figure 4.2c). **Goblet** cells of the respiratory system are unicellular glands that secrete mucus. Multicellular glands can be classified further according to the structure of their ducts (figure 4.4b–g). Glands that have ducts with few branches are called **simple,** and glands with ducts that branch repeatedly are called **compound.** Further classification is based on whether the ducts end in **tubules** (small tubes) or saclike structures called

acini (as′ĭ-nī; meaning grapes and suggesting a cluster of grapes or small sacs) or **alveoli** (al-ve′ō-lī, meaning a hollow sac). Tubular glands can be classified as straight or coiled. Most tubular glands are simple and straight, simple and coiled, or compound and coiled. Acinar glands can be simple or compound.

Exocrine glands can also be classified according to how products leave the cell. **Merocrine** (mer′ō-krin) glands, such as water-producing sweat glands and the exocrine portion of the pancreas, secrete products with no loss of actual cellular material (figure 4.5a). Secretions are either actively transported or packaged in vesicles and then released by the process of exocytosis at the free surface of the cell. **Apocrine** (ap′ō-krin) glands, such as the milk-producing mammary glands, discharge fragments of the gland cells in the secretion (figure 4.5b). Products are retained within the cell, and large portions of the cell are pinched off to become part of the secretion. **Holocrine** (hol′ō-krin) glands, such as sebaceous (oil) glands of the skin, shed entire cells (figure 4.5c). Substances accumulate in the cytoplasm of each epithelial cell, the cell ruptures and dies, and the entire cell becomes part of the secretion.

Endocrine glands are so variable in their structure that they are not classified easily. They are described in chapters 17 and 18.

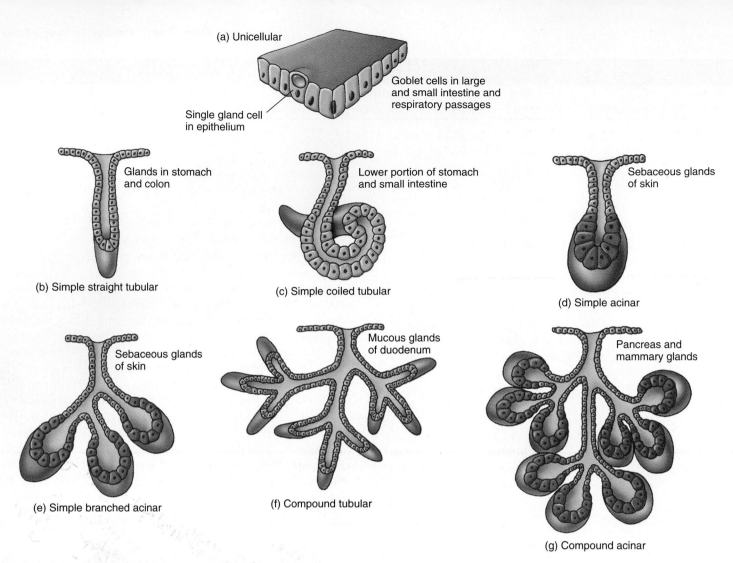

Figure 4.4 Structure of Exocrine Glands

(a) Unicellular

Single gland cell in epithelium

Goblet cells in large and small intestine and respiratory passages

Glands in stomach and colon

(b) Simple straight tubular

Lower portion of stomach and small intestine

(c) Simple coiled tubular

Sebaceous glands of skin

(d) Simple acinar

Sebaceous glands of skin

(e) Simple branched acinar

Mucous glands of duodenum

(f) Compound tubular

Pancreas and mammary glands

(g) Compound acinar

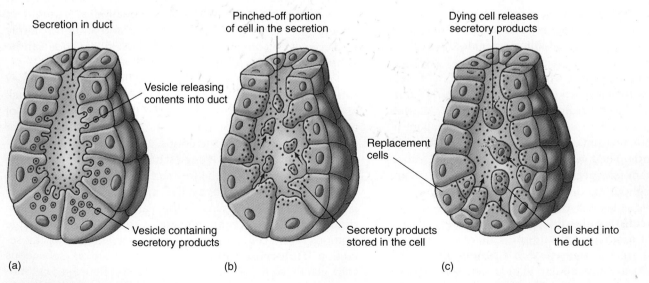

Secretion in duct

Pinched-off portion of cell in the secretion

Dying cell releases secretory products

Vesicle releasing contents into duct

Replacement cells

Vesicle containing secretory products

Secretory products stored in the cell

Cell shed into the duct

(a) (b) (c)

Figure 4.5 Exocrine Glands and Secretion Types

Exocrine glands classified according to the type of secretion. (*a*) **Merocrine gland.** Cells of the gland produce vesicles that contain secretory products, and the vesicles empty their contents into the duct through exocytosis. (*b*) **Apocrine gland.** Secretory products are stored in the cell near the lumen of the duct. A portion of the cell near the duct that contains the secretory products is actually pinched off the cell and joins the secretion. (*c*) **Holocrine gland.** Secretory products are stored in the cells of the gland. Entire cells are shed by the gland and become part of the secretion. The lost cells are replaced by other cells deeper in the gland.

Connective Tissue

The essential structural characteristic that distinguishes connective tissue from the other three tissue types is that it consists of cells separated from each other by abundant extracellular matrix. This nonliving extracellular matrix is the basis for the classification of connective tissue into subgroups.

Connective Tissue Cells

The specialized cells of the various connective tissues produce the extracellular matrix. The names of the cells end with suffixes that identify the cell functions as blasts, cytes, or clasts. **Blasts** create the matrix, **cytes** maintain it, and **clasts** break it down for remodeling. For example, fibroblasts are cells that form fibrous connective tissue, and chondrocytes are cells that maintain cartilage (*chondro-* refers to cartilage). Osteoblasts form bone (*osteo-* means bone), osteocytes maintain it, and osteoclasts break it down (see chapter 6).

The extracellular matrix has three major components: (1) protein fibers, (2) ground substance consisting of nonfibrous protein and other molecules, and (3) fluid. The structure of the matrix gives connective tissue types most of their functional characteristics, such as the ability of bones and cartilage to bear weight, of tendons and ligaments to withstand tension, and of dermis of the skin to withstand punctures, abrasions, and other abuses.

Protein Fibers of the Matrix

Three types of protein fibers—collagen, reticular, and elastic fibers—help form connective tissue.

Collagen (kol′lă-jen) is the most common protein in the body and accounts for one-fourth to one-third of the total body protein, which is approximately 6% of the total body weight. Each collagen molecule resembles a microscopic rope consisting of three polypeptide chains coiled together. Collagen is very strong and flexible but quite inelastic. There are at least 10 different types of collagen, many of which are specific to certain tissues.

Reticular (re-tik′yū-lăr, meaning netlike) fibers are actually very fine collagen fibers and therefore are not a chemically distinct category of fibers. They are very short, thin fibers that branch to form a network and appear different microscopically from other collagen fibers. Reticular fibers are not as strong as most collagen fibers, but networks of reticular fibers fill space between tissues and organs.

Elastic fibers contain a protein called **elastin** (ĕ-las′tin). As the name suggests, this protein is elastic with the ability to return to its original shape after being distended or compressed. Elastin gives the tissue in which it is found an elastic quality. Elastin molecules look like tiny coiled springs, and individual molecules are cross-linked to produce a large interwoven meshwork of springlike molecules that extend through the entire tissue.

Other Matrix Molecules

Two types of large, nonfibrous molecules called hyaluronic acid and proteoglycans are part of the extracellular matrix. These molecules constitute most of the **ground substance** of the matrix, the shapeless background against which the collagen fibers are seen through the microscope. The molecules themselves, however, are not shapeless but are highly structured. **Hyaluronic** (hī′ă-lū-ron′ik, meaning glassy appearance) **acid** is a long, unbranched polysaccharide chain composed of repeating disaccharide units. It gives a very slippery quality to the fluids that contain it; for that reason, it is a good lubricant for joint cavities (see chapter 8). Hyaluronic acid is also found in large quantities in connective tissue and is the major component of the vitreous humor of the eye (see chapter 15). A **proteoglycan** (prō′tē-ō-glī′kan, meaning formed from proteins and polysaccharides) is a large molecule that consists of numerous polysaccharides, each attached at one end to a common protein core. These **proteoglycan monomers** resemble minute pine tree branches. The protein core is the branch of the tree, and the proteoglycans are the needles. The protein cores of proteoglycan monomers can attach to a molecule of hyaluronic acid to form a **proteoglycan aggregate.** The aggregate resembles a complete pine tree, with hyaluronic acid represented by the tree trunk and the proteoglycan monomers forming the limbs. Proteoglycans trap large quantities of water, which gives them the capacity to return to their original shape when compressed or deformed.

Classification of Connective Tissue

Connective tissue types blend into one another, and the transition points cannot be defined precisely. As a result, the classification scheme for connective tissues is somewhat arbitrary. There are three major categories of connective tissues, however, based on the following prominent features: (1) primarily protein fibers in the extracellular matrix, (2) both protein fibers and ground substance in the extracellular matrix, and (3) a fluid extracellular matrix (table 4.3).

Matrix with Fibers as the Primary Feature

Two subtypes of connective tissue have protein fibers in the extracellular matrix as a prominent feature: fibrous connective tissue and special connective tissue.

Fibrous Connective Tissue

In **fibrous connective tissue,** protein fibers in the extracellular matrix predominate, but the types of fibrous connective tissue can be divided into two subtypes, depending on the amount of fibrous protein present. In **loose connective**

Table 4.3   Classification of Connective Tissue

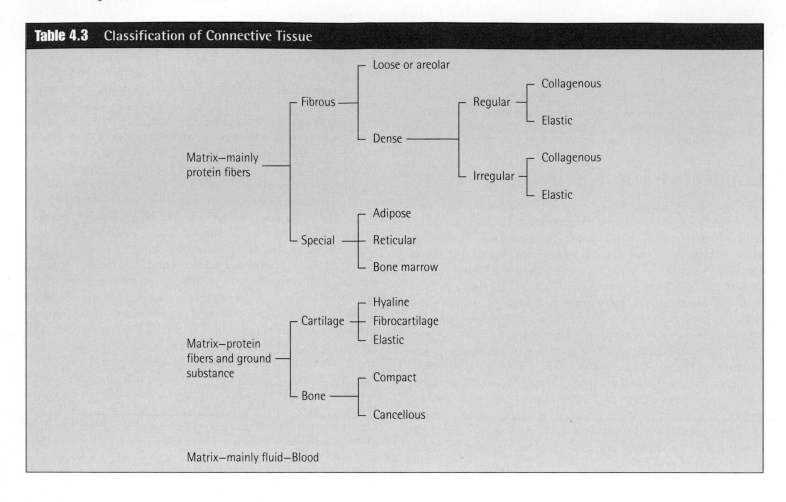

Matrix—mainly fluid—Blood

tissue (figure 4.6*a*), which is sometimes referred to as **areolar** (ă-rē′ō-lar, area) **tissue,** the protein fibers form a lacy network with numerous fluid-filled spaces. In **dense connective tissue,** protein fibers form thick bundles and fill nearly all of the extracellular space.

Areolar tissue is the "loose packing" material of most organs and other tissues and attaches the skin to underlying tissues. It contains collagen, reticular, and elastic fibers and a variety of cells. For example, fibroblasts produce the fibrous matrix, macrophages move through the tissue engulfing bacteria and cell debris, mast cells contain chemicals that help mediate inflammation, and lymphocytes are involved in immunity. The loose packing of areolar tissue is often associated with other connective tissue types such as reticular tissue and fat (adipose tissue).

Dense connective tissue can be subdivided into two major groups: regular and irregular. The cells of developing dense connective tissue are spindle-shaped **fibroblasts.** Once a fibroblast becomes completely surrounded by matrix, it is a **fibrocyte.** In **dense regular connective tissue** (figure 4.6*b* and *c*), the protein fibers of the extracellular matrix are oriented predominantly in one direction. Dense regular connective tissue forms structures such as tendons, which connect muscles to bones (see chapter 11) and ligaments, which connect bone to bone (see chapter 8). **Dense irregu-**

lar connective tissue (figure 4.6*d* and *e*) contains protein fibers arranged as a meshwork of randomly oriented fibers. Alternatively, the fibers within a given layer of irregular dense connective tissue can be oriented in one direction, whereas the fibers of adjacent layers are oriented at nearly right angles to that layer. Dense irregular connective tissue forms most of the dermis of the skin, which is the tough, inner portion of the skin.

The collagen fibers of dense connective tissue resist stretching and give the tissue considerable strength in the direction of the fiber orientation. Dense regular connective tissue forms tough, cablelike structures that have considerable strength in one direction. Dense irregular connective tissue forms sheets of connective tissue that have strength in many directions, but less strength in any single direction than does regular connective tissue.

2   P R E D I C T

Scars consist of dense irregular connective tissue in which the fibers are collagen. Vitamin C is required for collagen synthesis. Predict the effect of scurvy, which is a nutritional disease caused by vitamin C deficiency, on wound healing.

✔ *Answer in Appendix F*

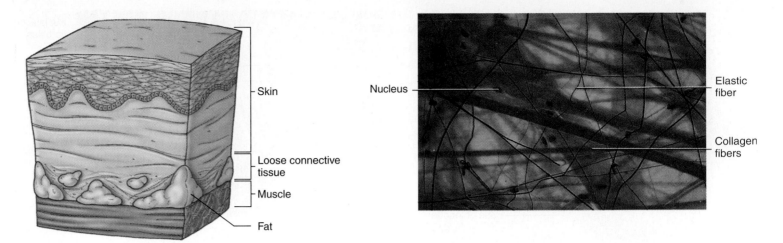

Skin

Loose connective tissue

Muscle

Fat

Nucleus

Elastic fiber

Collagen fibers

Connective Tissues

(a) **Areolar, or loose, connective tissue**

Location: Widely distributed throughout the body; substance on which epithelial basement membranes rest; packing between glands, muscles, and nerves. Attaches the skin to underlying tissues.

Structure: Cells (for example, fibroblasts, macrophages, and lymphocytes) within a fine network of mostly collagen fibers. Often merges with denser connective tissue.

Function: Loose packing, support, and nourishment for the structures with which it is associated.

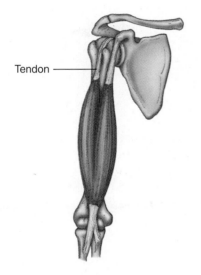

Tendon

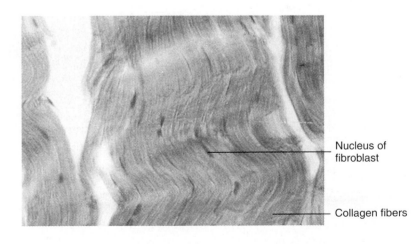

Nucleus of fibroblast

Collagen fibers

Connective Tissues

(b) **Dense regular collagenous connective tissue**

Location: Tendons (attach muscle to bone) and ligaments (attach bones to each other).

Structure: Matrix composed of collagen fibers running in somewhat the same direction.

Function: Ability to withstand great pulling forces exerted in the direction of fiber orientation, great tensile strength, and stretch resistance.

Figure 4.6 Connective Tissues 𝑋

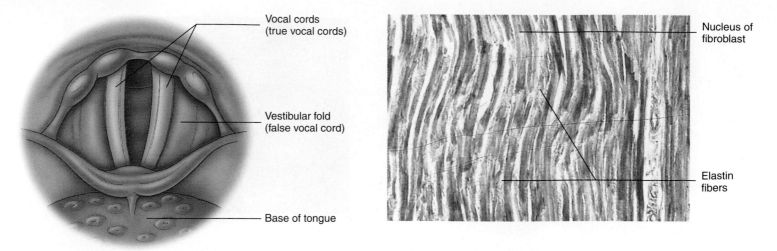

Vocal cords
(true vocal cords)

Vestibular fold
(false vocal cord)

Base of tongue

Nucleus of
fibroblast

Elastin
fibers

Connective Tissues

(c) **Dense regular elastic connective tissue**

Location: Ligaments between the vertebrae and along the dorsal aspect of the neck (nucha) and in the vocal cords.

Structure: Matrix composed of regularly arranged collagen fibers and elastin fibers.

Function: Capable of stretching and recoiling like a rubber band with strength in the direction of fiber orientation.

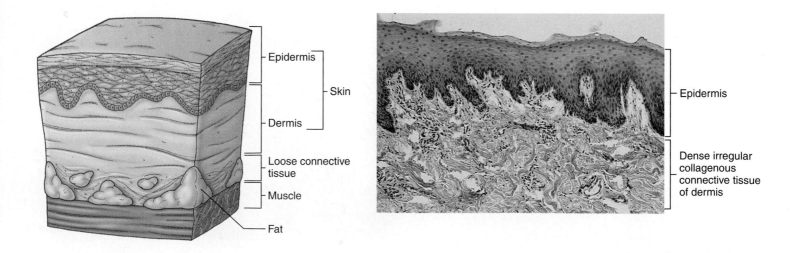

Epidermis

Skin

Dermis

Loose connective
tissue

Muscle

Fat

Epidermis

Dense irregular
collagenous
connective tissue
of dermis

Connective Tissues

(d) **Dense irregular collagenous connective tissue**

Location: Sheaths; most of the dermis of the skin; organ capsules and septa; outer covering of body tubes.

Structure: Matrix composed of collagen fibers that run in all directions or in alternating planes of fibers oriented in a somewhat single direction.

Function: Tensile strength capable of withstanding stretching in all directions.

Figure 4.6 *(continued)*

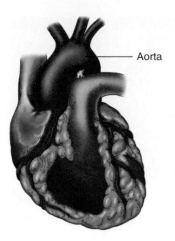

Aorta

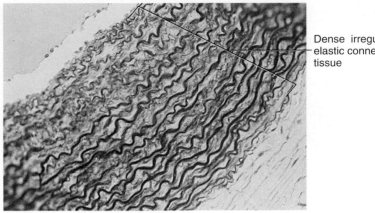

Dense irregular elastic connective tissue

Connective Tissues

(e) Dense irregular elastic connective tissue

Location: Elastic arteries.

Structure: Matrix composed of bundles and sheets of collagenous and elastin fibers oriented in multiple directions.

Function: Capable of strength with stretching and recoil in several directions.

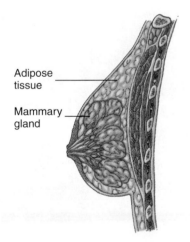

Adipose tissue

Mammary gland

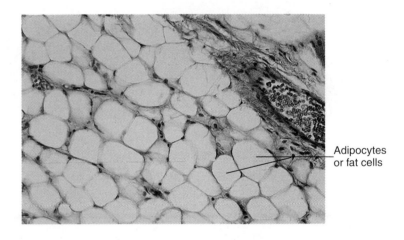

Adipocytes or fat cells

Connective Tissues

(f) Adipose tissue

Location: Predominantly in subcutaneous areas, mesenteries, renal pelvis, around kidneys, attached to the surface of the colon, mammary glands, and in loose connective tissue that penetrates into spaces and crevices.

Structure: Little extracellular material surrounding cells. The adipocytes, or fat cells, are so full of lipid that the cytoplasm is pushed to the periphery of the cell.

Function: Packing material, thermal insulator, energy storage, and protection of organs against injury from being bumped or jarred.

Figure 4.6   *(continued)*

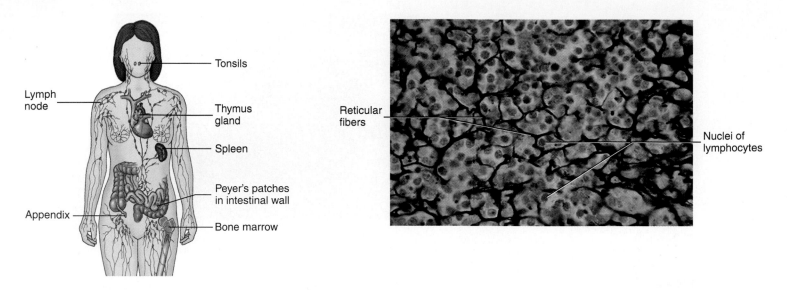

Tonsils

Lymph
node

Thymus
gland

Spleen

Peyer's patches
in intestinal wall

Appendix

Bone marrow

Reticular
fibers

Nuclei of
lymphocytes

Connective Tissues

(g) **Reticular tissue**

Location: Within the lymph nodes, spleen, and bone marrow.

Structure: Fine network of reticular fibers irregularly arranged.

Function: Provides a superstructure for the lymphatic and hemopoietic tissues.

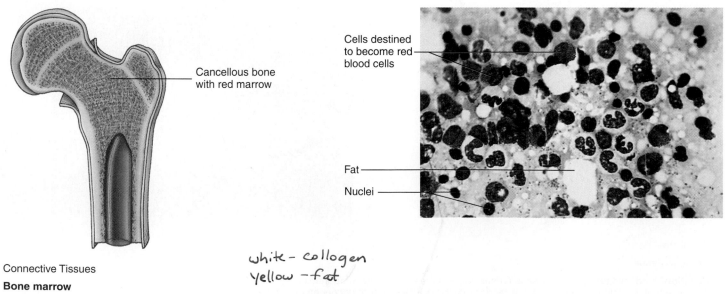

Cancellous bone
with red marrow

Cells destined
to become red
blood cells

Fat

Nuclei

white – collogen
yellow – fat

Connective Tissues

(h) **Bone marrow**

Location: Within marrow cavities of bone. Two types: yellow marrow (mostly adipose tissue) in the shafts of long bones; and red marrow (hemopoietic or blood-forming tissue) in the ends of long bones and in short, flat, and irregularly shaped bones.

Structure: Reticular framework with numerous blood-forming cells (red marrow).

Function: Production of new blood cells (red marrow).

Figure 4.6   *(continued)*

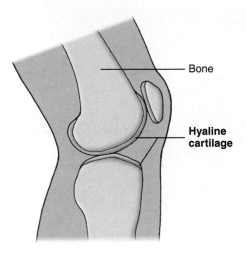

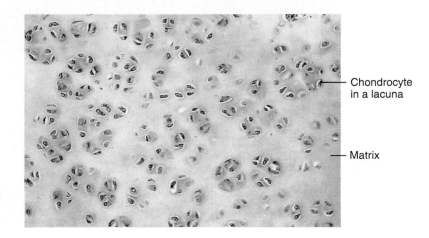

Bone

Hyaline cartilage

Chondrocyte in a lacuna

Matrix

Connective Tissues

(i) **Hyaline cartilage**

Location: Growing long bones, cartilage rings of the respiratory system, costal cartilage of ribs, nasal cartilage, articulating surface of bones, and the embryonic skeleton.

Structure: Collagen fibers are small and evenly dispersed in the matrix, making the matrix appear transparent. The cartilage cells, or chondrocytes, are found in spaces, or lacunae, within the firm, but flexible matrix.

Function: Allows growth of long bones. Provides rigidity with some flexibility in the trachea, bronchi, ribs, and nose. Forms rugged, smooth, yet somewhat flexible articulating surfaces. Forms the embryonic skeleton.

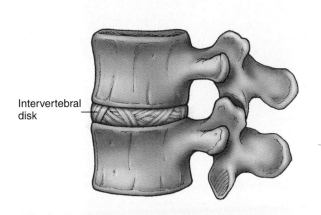

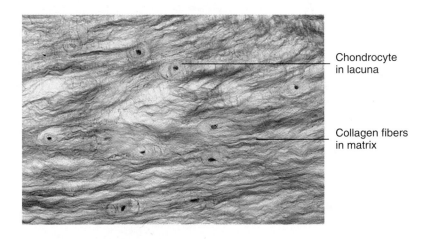

Intervertebral disk

Chondrocyte in lacuna

Collagen fibers in matrix

Connective Tissues

(j) **Fibrocartilage**

Location: Intervertebral disks, symphysis pubis, articular disks (e.g., knee and temporomandibular [jaw] joints).

Structure: Collagenous fibers similar to those in hyaline cartilage. The fibers are more numerous than in other cartilages and are arranged in thick bundles.

Function: Somewhat flexible and capable of withstanding considerable pressure. Connects structures subjected to great pressure.

Figure 4.6   *(continued)*

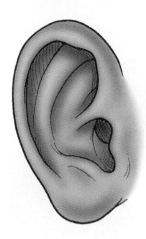

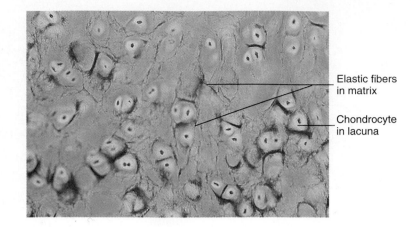

Elastic fibers in matrix

Chondrocyte in lacuna

Connective Tissues

(k) Elastic cartilage

Location: External ear, epiglottis, and auditory tubes.

Structure: Similar to hyaline cartilage, but matrix also contains elastin fibers.

Function: Provides rigidity with even more flexibility than hyaline cartilage because elastic fibers return to their original shape after being stretched.

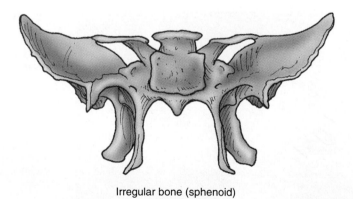

Irregular bone (sphenoid)

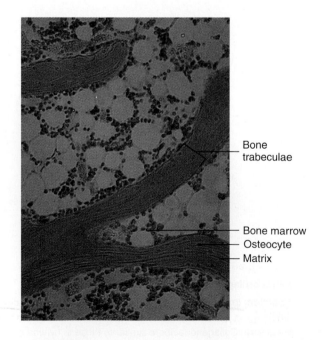

Bone trabeculae

Bone marrow
Osteocyte
Matrix

Connective Tissues

(l) Cancellous bone

Location: In the interior of the bones of the skull, vertebrae, sternum, and pelvis; also found in the ends of the long bones.

Structure: Latticelike network of scaffolding characterized by trabeculae with large spaces between them filled with hemopoietic tissue. The osteocytes, or bone cells, are located within lacunae in the trabeculae.

Function: Acts as a scaffolding to provide strength and support without the greater weight of compact bone.

Figure 4.6 *(continued)*

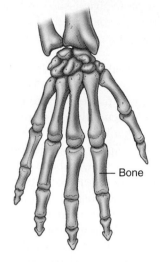

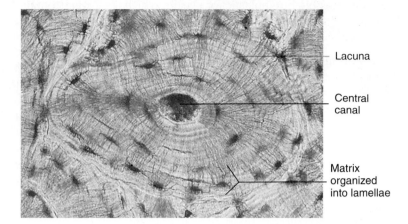

Lacuna

Central
canal

Matrix
organized
into lamellae

Bone

Connective Tissues

(m) Compact bone

Location: Outer portions of all bones and the shafts of long bones.

Structure: Hard, bony matrix predominates. Many osteocytes are located within lacunae that are distributed in a circular fashion around the central canals. Small passageways connect adjacent lacunae.

Function: Provides great strength and support. Forms a solid outer shell on bones that keeps them from being easily broken or punctured.

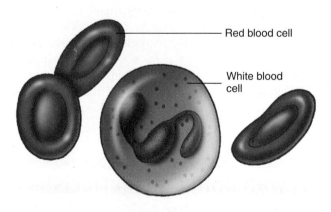

Red blood cell

White blood
cell

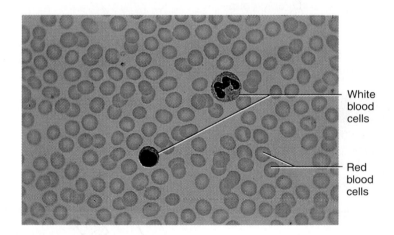

White
blood
cells

Red
blood
cells

Connective Tissues

(n) Blood

Location: Within the blood vessels. Produced by the hemopoietic tissues. White blood cells frequently leave the blood vessels and enter the interstitial spaces.

Structure: Blood cells and a fluid matrix.

Function: Transports oxygen, carbon dioxide, hormones, nutrients, waste products, and other substances. Protects the body from infections and is involved in temperature regulation.

Figure 4.6   *(continued)*

The predominance in dense connective tissue of either collagen, which is quite flexible but inelastic, or elastin, which is flexible and elastic, is the basis for further classification of dense connective tissue (see figure 4.6b–e):

1. **Dense regular collagenous** connective tissue forms tendons and most ligaments. Both tendons and most ligaments consist almost entirely of thick bundles of densely packed parallel collagen fibers. The orientation of the collagen fibers in one direction makes the tendons and ligaments very strong in that direction. Because collagen is a white protein, tendons and most ligaments appear white.

   Although their general structures are similar, the major histologic differences between tendons and ligaments include the following: (a) collagen fibers of ligaments are often less compact; (b) some fibers of many ligaments are not parallel; and (c) ligaments usually are more flattened than tendons and form sheets or bands of tissue.

2. **Dense regular elastic** connective tissue forms some elastic ligaments such as the **nuchal** (nū′kăl, meaning back of neck) ligament, which lies along the posterior of the neck and helps hold the head upright. Elastic ligaments consist of parallel bundles of collagen fibers and abundant elastic fibers. The elastin in elastic ligaments gives them a slightly yellow color.

3. **Dense irregular collagenous** connective tissue is characteristic of the dermis of the skin (see chapter 5) and of the connective tissue capsules that surround organs such as the kidney and spleen. Bundles of collagen fibers are oriented in many directions in dense irregular collagenic connective tissue.

4. **Dense irregular elastic** connective tissue helps form the walls of large arteries. Abundant elastic fibers oriented in many directions form layers in dense irregular elastic tissue.

---

3   P R E D I C T

Explain the advantages of having elastic ligaments that extend from vertebra to vertebra in the vertebral column and why it would be a disadvantage if tendons, which connect skeletal muscles to bone, were elastic.

✔ *Answer in Appendix F*

Special Connective Tissue

Adipose tissue, reticular tissue, and hemopoietic tissue are special types of connective tissue. **Adipose** (ad′ĭ pōs, meaning fat) **tissue** (figure 4.6f) consists of **adipocytes** (ad′ĭ-pō-sĭtz), or fat cells, which contain large amounts of lipid. The lipid pushes the rest of the cell contents to the periphery, so that each cell appears to contain a large, centrally located lipid droplet with a thin layer of cytoplasm around it. Unlike other connective tissue types, adipose tissue is composed of large cells and a small amount of reticular matrix. Adipose tissue

functions as an insulator, a protective tissue, and a site of energy storage. Lipids take up less space per calorie than either carbohydrates or proteins and therefore are well adapted for energy storage.

Adipose tissue exists in both yellow (white) and brown forms. Yellow adipose tissue is by far the most abundant. At birth, a human's yellow adipose tissue is white but turns yellow with age because of the accumulation of pigments such as carotene, a plant pigment that humans can metabolize as a source of vitamin A. Brown adipose tissue is found only in specific areas of the body such as the axillae (armpits), neck, and near the kidneys. The brown color results from the cytochrome pigments in its numerous mitochondria and its abundant blood supply. Although brown fat is much more prevalent in babies than in adults, it is difficult to distinguish brown fat from yellow fat because the color difference between them is not great. Brown fat is specialized to generate heat as a result of oxidative metabolism of lipid molecules in mitochondria and can play a significant role in body temperature regulation in newborn babies.

Reticular (rĕ-tik′yū-lăr) tissue forms the framework of lymphatic tissue, bone marrow, and liver (figure 4.6g). It is characterized by a network of reticular fibers and a number of cell types. The reticular fibers are produced by **reticular cells,** which remain closely attached to the fibers. The spaces between the reticular fibers can contain a wide variety of cells, such as dendritic cells, which look very much like reticular cells but are part of the immune system, (see chapter 22), lymphocytes, macrophages, and other blood cells.

Another type of special connective tissue is **hemopoietic** (hē′mō-poy-et′ik), or blood-forming, tissue. Most of the hemopoietic tissue is found in **bone marrow** (mar′ō) (figure 4.6h), which is the soft connective tissue in the cavities of bones. There are two types of bone marrow: **yellow marrow** and **red marrow** (see chapter 6). Yellow marrow consists of yellow adipose tissue, and red marrow consists of hemopoietic tissue surrounded by a framework of reticular fibers. Hemopoietic tissue produces red and white blood cells and is described in detail in chapter 19.

Matrix with Both Protein Fibers and Ground Substance

Cartilage

Cartilage (kar′ti-lij) is composed of cartilage cells, or **chondrocytes** (kon′drō-sītz), located in spaces called **lacunae** (lă-kū′nē) within an extensive and relatively rigid matrix. The matrix contains protein fibers, ground substance, and fluid. The protein fibers are collagen fibers or, in some cases, collagen and elastic fibers. The ground substance consists of proteoglycans and other organic molecules. Most of the proteoglycans in the matrix form aggregates with hyaluronic acid. Within the cartilage matrix, proteoglycan aggregates function as minute sponges capable of trapping large quantities of water. This trapped water allows cartilage to spring back after being compressed. The collagen fibers give cartilage considerable strength.

The surface of cartilage is surrounded by a layer of dense irregular connective tissue called the **perichondrium** (per-i-kon′drē-ŭm). Cartilage has no blood vessels or nerves except those of the perichondrium, it therefore heals very slowly after an injury because the cells and nutrients necessary for tissue repair cannot reach the damaged area easily.

There are three types of cartilage:

1. **Hyaline** (hī′ă-lin) **cartilage** has large amounts of both collagen fibers and proteoglycans (figure 4.6*i*). Fine collagen fibers are evenly dispersed throughout the ground substance, and in joints hyaline cartilage has a very smooth surface. Specimens appear to have a glassy, translucent matrix when viewed through the microscope. It is found in areas in which strong support and some flexibility are needed, such as in the rib cage and the cartilage within the trachea and bronchi (see chapter 23). Hyaline cartilage also covers the surfaces of bones that move smoothly against each other in joints. It forms most of the skeleton before it is replaced by bone in the embryo, and it is involved in growth that increases the length of bones (see chapter 6).
2. **Fibrocartilage** has more collagen fibers than proteoglycans (figure 4.6*j*). Compared with hyaline cartilage, fibrocartilage has much thicker bundles of collagen fibers dispersed through its matrix. Fibrocartilage is slightly compressible and very tough. It is found in areas of the body where a great deal of pressure is applied to joints, such as the knee, the jaw, and between vertebrae.
3. **Elastic cartilage** has elastic fibers in addition to collagen and proteoglycans (figure 4.6*k*). The numerous elastic fibers are dispersed throughout the matrix of elastic cartilage. It is found in areas, such as the external ears, which have rigid but elastic properties.

---

4   P R E D I C T

One of several changes caused by rheumatoid arthritis in joints is the replacement of hyaline cartilage with dense irregular collagenous connective tissue. Predict the effect of replacing hyaline cartilage with fibrous connective tissue.

✔ *Answer in Appendix F*

Bone

Bone is a hard connective tissue that consists of living cells and mineralized matrix. Bone matrix has an organic and an inorganic portion. The organic portion consists of protein fibers, primarily collagen, and other organic molecules. The mineral, or inorganic, portion consists of specialized crystals called **hydroxyapatite** (hī-drok′sē-ap-ă-tīt), which contain calcium and phosphate. The strength and rigidity of the mineralized matrix allow bones to support and protect other tissues and organs of the body. Bone cells, or **osteocytes** (os′tē-ō-sītz), are located within holes in the matrix, which are called lacunae and are similar to the lacunae of cartilage.

There are two types of bone:

1. **Cancellous** (kan-sel′ŭs), or **spongy, bone** has spaces between **trabeculae** (tră-bek′yū-lē, meaning beams), or plates, of bone and therefore resembles a sponge (figure 4.6*l*).
2. **Compact bone** is more solid with almost no space between many thin layers, or **lamellae** (lă-mel′ă, pl. lă-mel′ē) of bone (figure 4.6*m*).

Bone, unlike cartilage, has a rich blood supply. For this reason, bone can repair itself much more readily than can cartilage. Bone is described more fully in chapter 6.

Predominantly Fluid Matrix

Blood is unusual among the connective tissues because the matrix between the cells is liquid (figure 4.6*n*). The cells of most other connective tissues are more or less stationary within a relatively rigid matrix, but blood cells are free to move within a fluid matrix. Some blood cells leave the bloodstream and wander through other tissues. The liquid matrix of blood allows it to flow rapidly through the body, carrying food, oxygen, waste products, and other materials. The matrix of blood is also unusual in that most of it is produced by cells contained in other tissues rather than by blood cells. Blood is discussed more fully in chapter 19.

Muscle Tissue

The main characteristic of **muscle tissue** is that it is contractile and therefore responsible for movement. Muscle contraction is accomplished by the interaction of contractile proteins, which are described in chapter 10. Muscles contract to move the entire body, to pump blood through the heart and blood vessels, and to decrease the size of hollow organs, such as the stomach and urinary bladder.

The three types of muscle tissue are skeletal, cardiac, and smooth muscle. The types of muscle tissue are grouped according to both structure and function (table 4.4). Muscle tissue grouped according to structure is either **striated** (strī′āt-ĕd), in which microscopic bands or striations can be seen in muscle cells, or **nonstriated.** When classified according to function, a muscle is **voluntary,** meaning that it is consciously controlled, or **involuntary,** meaning that it is not normally consciously controlled. Thus the three muscle types are striated voluntary, or **skeletal muscle** (figure 4.7*a*); striated involuntary, or **cardiac muscle** (figure 4.7*b*); and nonstriated involuntary, or **smooth muscle** (figure 4.7*c*).

Table 4.4   Comparison of Muscle Types

Features	Skeletal Muscle	Cardiac Muscle	Smooth Muscle
Location	Attached to bones	Heart	Walls of hollow organs, blood vessels, eyes, glands, and skin.
Cell shape	Very long, cylindrical cells (1–40 mm in length and may extend the entire length of the muscle; 10–100 μm in diameter).	Cylindrical cells that branch (100–500 μm in length; 100–200 μm in diameter).	Spindle-shaped cells (15–200 μm in length; 5–10 μm in diameter).
Nucleus	Multinucleated, peripherally located.	Single, centrally located.	Single, centrally located.
Striations	Yes	Yes	No
Control	Voluntary	Involuntary	Involuntary
Ability to contract spontaneously	No	Yes	Yes
Function	Body movement	Contraction provides the major force for moving blood through the blood vessels.	Movement of food through the digestive tract, emptying of the urinary bladder, regulation of blood vessel diameter, change in pupil size, contraction of many gland ducts, movement of hair, and many more functions.
Special features		Branching fibers, intercalated disks join the cells to each other (gap junctions).	Gap junctions

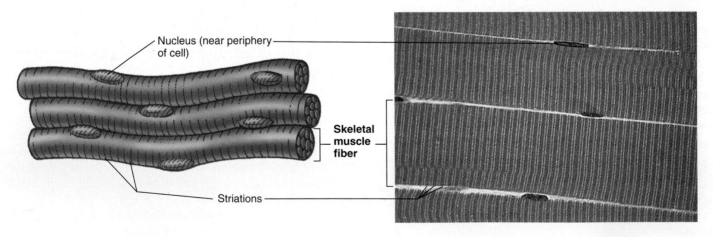

Nucleus (near periphery of cell)

Skeletal muscle fiber

Striations

Muscle Tissue

(a) **Skeletal muscle**

Location: Attaches to bone.

Structure: Skeletal muscle cells or fibers appear striated (banded). Cells are large, long, and cylindrical, with many nuclei located at the periphery.

Function: Movement of the body; under voluntary control.

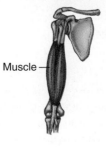

Muscle

Figure 4.7   Muscle Tissue

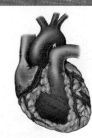

Nucleus (central)

Cardiac muscle cell

Intercalated disks (special junctions between cells)

Striations

Muscle Tissue

(b) **Cardiac muscle**

Location: Cardiac muscle is in the heart.

Structure: Cardiac muscle cells are cylindrical and striated and have a single, centrally located, nucleus. They are branched and connected to one another by intercalated disks.

Function: Pumps the blood; under involuntary control.

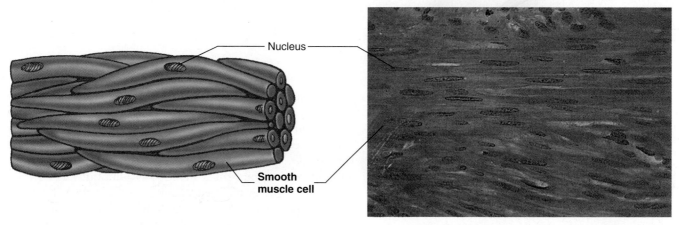

Nucleus

Smooth muscle cell

Muscle Tissue

(c) **Smooth muscle**

Location: Smooth muscle is in hollow organs such as the stomach and intestine.

Structure: Smooth muscle cells are tapered at each end, are not striated, and have a single nucleus.

Function: Regulates the size of organs, forces fluid through tubes, controls the amount of light entering the eye, and produces "goose flesh" in the skin; under involuntary control.

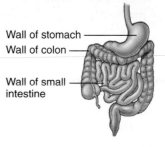

Wall of stomach

Wall of colon

Wall of small intestine

Figure 4.7   *(continued)*

For most people, the term muscle means skeletal muscle (see chapter 10). It constitutes the meat of animals and represents a large portion of the total weight of the human body. Skeletal muscle, as the name implies, attaches to the skeleton and, by contracting, causes the major body movements. Cardiac muscle is the muscle of the heart (see chapter 20), and contraction of cardiac muscle is responsible for pumping blood. Smooth muscle is widespread throughout the body and is responsible for a wide range of functions, such as movements in the digestive, urinary, and reproductive systems.

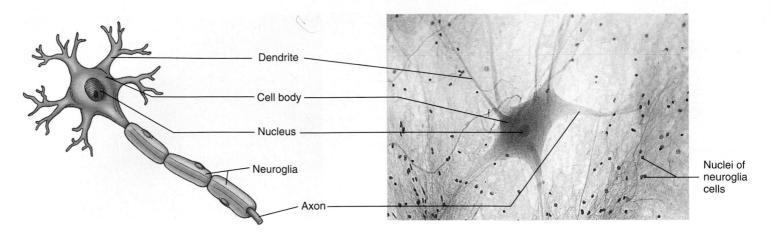

Neurons

(a) **Multipolar neuron**

Location: Neurons are located in the brain, spinal cord, and in ganglia.

Structure: The neuron consists of dendrites, a cell body, and a long axon. Neuroglia, or support cells, surround the neurons.

Function: Neurons transmit information in the form of action potentials, store "information," and in some way integrate and evaluate data. Neuroglia support, protect, and form specialized sheaths around axons.

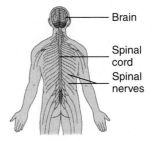

Figure 4.8   Neurons 𝑋

Nervous Tissue

The fourth and final class of tissue is **nervous tissue** (see chapter 12). It is found in the brain, spinal cord, and nerves and is characterized by the ability to conduct electric signals called **action potentials.** It consists of neurons, which are responsible for this conductive ability, and support cells called neuroglia.

Neurons, or **nerve cells** (figure 4.8), are the actual conducting cells of nervous tissue. They are composed of three major parts: cell body, dendrites, and axon. The **cell body** contains the nucleus and is the site of general cell functions. Dendrites and axons are two types of nerve cell processes, both consisting of projections of cytoplasm surrounded by membrane. **Dendrites** (den′drītz) usually receive electric signals and conduct them toward the cell body. They are much shorter than axons and usually taper to a fine tip. **Axons** (ak′sonz) usually conduct action potentials away from the cell body. They can be much longer than dendrites, and they have a constant diameter along their entire length.

Neurons that possess several dendrites and one axon are called **multipolar neurons** (figure 4.8*a*). Neurons that possess a single dendrite and an axon are called **bipolar neurons.** Some very specialized neurons, called **unipolar neurons** (figure 4.8*b*), have only one axon and no dendrites.

Within each subgroup are many shapes and sizes of neurons, especially in the brain and the spinal cord.

Neuroglia (nū-rog′lē-ăh, meaning nerve glue) are the support cells of the brain, spinal cord, and peripheral nerves (figure 4.9). The term neuroglia originally referred only to the support cells of the central nervous system, but it is now also applied to cells in the peripheral nervous system. Neuroglia nourish, protect, and insulate neurons. Neurons and neuroglial cells are described in greater detail in chapters 12 and 13.

Embryonic Tissue

Approximately 13 or 14 days after fertilization, the cells that give rise to a new individual form a slightly elongated disk consisting of two layers called ectoderm and endoderm. Cells of the ectoderm then migrate between the two layers to form a third layer called mesoderm. Ectoderm, mesoderm, and endoderm are called germ layers because the beginning of all the tissues of the adult can be traced back to one of them. All four tissue types are derived from the three germ layers during early development of the embryo. The germ layers and their derivatives are described more fully in chapter 29 (also see table 29.1).

The **endoderm** (en′dō-derm), the inner layer, forms the lining of the digestive tract and its derivatives. The **mesoderm** (mez′ō-derm), the middle layer, forms tissues such as

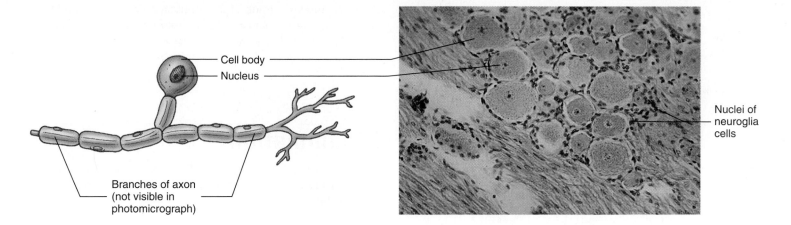

Nuclei of neuroglia cells

Neurons

(b) **Unipolar neuron**

Location: Cell bodies are located in ganglia outside of the brain and spinal cord.

Structure: The neuron consists of a cell body with one axon.

Function: Conducts action potentials from the surface of the body to the brain or spinal cord.

Figure 4.8   *(continued)*

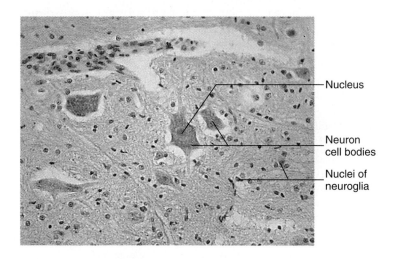

Figure 4.9   Neuroglia

muscle, bone, and blood vessels. The **ectoderm** (ek′tō-derm), the outer layer, forms the skin, and a portion of the ectoderm, called the **neuroectoderm** (nūr-ō-ek′tō-derm) becomes the nervous system (see chapter 13). Groups of cells that break away from the neuroectoderm during development, called **neural crest cells,** give rise to parts of the peripheral nerves (see chapters 14 and 16), skin pigment (see chapter 5), and many tissues of the face.

Mesenchyme (mez′en-kīm) is the embryonic tissue from which connective tissues arise. It forms early in embryonic development from mesoderm and neural crest cells (see

chapter 29). Mesenchyme consists of irregularly shaped cells, abundant ground substance, and delicate reticular fibers. It appears that some mesenchymal cells remain in the adult and give rise to new connective tissue cells as tissue repair occurs following injury.

Membranes

A membrane is a thin sheet or layer of tissue that covers a structure or lines a cavity. Most membranes are formed from epithelium and the connective tissue on which it rests. The three major categories of internal membranes are mucous membranes, serous membranes, and synovial membranes.

A **mucous** (myū′kŭs) **membrane** consists of epithelial cells, their basement membrane, a thick layer of loose connective tissue called the **lamina propria** (lam′i-nă prō′prē-ăh), and, sometimes, a layer of smooth muscle cells. Mucous membranes line cavities and canals that open to the outside of the body, such as the digestive, respiratory, excretory, and reproductive passages (figure 4.10). Many, but not all, mucous membranes contain goblet cells or multicellular mucous glands, which secrete a viscous substance called **mucus** (myū′kŭs). The functions of the mucous membranes vary, depending on their location, and include protection, absorption, and secretion.

A **serous** (ser′ŭs) **membrane** consists of three components: a layer of simple squamous epithelium called **mesothelium** (mez-ō-thē′lē-ŭm), its basement membrane, and a delicate layer of loose connective tissue. Serous membranes line cavities such as the pericardial, pleural, and

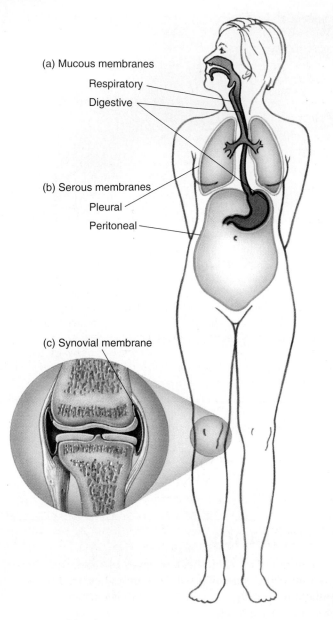

(a) Mucous membranes

Respiratory

Digestive

(b) Serous membranes

Pleural

Peritoneal

(c) Synovial membrane

Figure 4.10   Membranes

(*a*) Mucous membranes line cavities that open to the outside and often contain mucous glands, which secrete mucus. (*b*) Serous membranes line cavities that do not open to the exterior, do not contain glands, but do secrete serous fluid. (*c*) Synovial membranes line cavities that surround synovial joints.

peritoneal cavities that do not open to the exterior (see figure 4.10). Serous membranes do not contain glands but are moistened by a small amount of fluid, called **serous fluid,** produced by the serous membranes. The serous fluid lubricates the serous membranes and makes their surfaces slippery. Serous membranes protect the internal organs from friction, help hold them in place, and act as selectively permeable barriers that prevent the accumulation of large amounts of fluid within the serous cavities.

Synovial (si-nō′vē-ăl) **membranes** consist of modified connective tissue cells either intermixed with part of the

dense connective tissue of the joint capsule or separated from the capsule by areolar or adipose tissue. Synovial membranes line freely movable joints (see chapter 8) (see figure 4.10). They produce a fluid rich in hyaluronic acid, which makes the joint fluid very slippery, facilitating smooth movement within the joint.

Inflammation

The inflammatory response occurs when tissues are damaged (figure 4.11) or in association with an immune response. Although there are many possible agents of injury, such as microorganisms, cold, heat, radiant energy, chemicals, electricity, or mechanical trauma, the inflammatory response to all causes is similar. The inflammatory response mobilizes the body's defenses, isolates and destroys microorganisms and other injurious agents, and removes foreign materials and damaged cells so that tissue repair can proceed. The details of the inflammatory response are presented in chapter 22.

Inflammation produces five major signs: redness, heat, swelling, pain, and disturbance of function. Although unpleasant, these processes usually benefit recovery, and each of the symptoms can be understood in terms of events that occur during the inflammatory response.

After a person is injured, chemical substances called **mediators of inflammation** are released or activated in the tissues and the adjacent blood vessels. The mediators include histamine, kinins, prostaglandins, leukotrienes, and others. Some mediators induce dilation of blood vessels and produce the symptoms of redness and heat. Dilation of blood vessels is beneficial because it increases the speed with which white blood cells and other substances important for fighting infections and repairing the injury are brought to the site of injury.

Mediators of inflammation also stimulate pain receptors and increase the permeability of blood vessels, allowing the movement of materials such as clotting proteins and white blood cells out of the blood vessels and into the tissue, where they can deal directly with the injury. As proteins from the blood move into the tissue, they change the osmotic relationship between the blood and the tissue. Water follows the proteins by osmosis, and the tissue swells, producing **edema** (e-dē′mă). Edema increases the pressure in the tissue, which can also stimulate neurons and cause the sensation of pain.

Clotting proteins found in blood diffuse into the interstitial spaces and form a clot. Clotting of blood also occurs in the more severely injured blood vessels. The effect of clotting is to isolate the injurious agent and to separate it from the remainder of the body. Foreign particles and microorganisms that are present at the site of injury are "walled off" from tissues by the clotting process. Pain, limitation of movement resulting from edema, and tissue destruction all contribute to the disturbance of function. This disturbance can be valuable because it warns the person to protect the injured structure from further damage. Sometimes the inflammatory response lasts longer or is more intense than is desirable, and drugs are used to suppress the symptoms. Antihistamines block the effects of

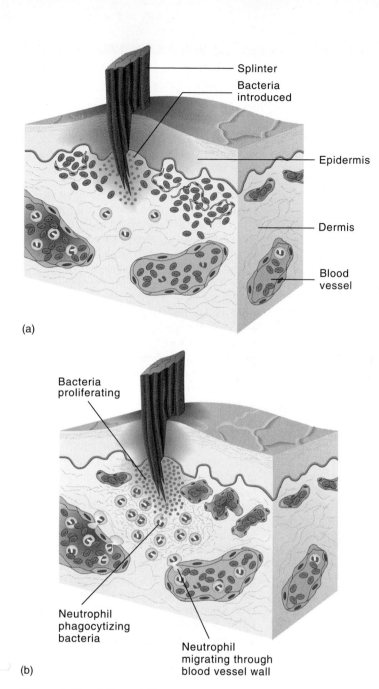

(a)

(b)

Figure 4.11   Inflammation

(a) A splinter in the skin causes tissue damage and releases bacteria. Dilated blood vessels (capillaries) cause skin to become red, and fluid from the tissues and vessels causes swelling. (b) White blood cells (e.g., neutrophils and macrophages) leave the blood vessels and arrive at the site of bacterial infection, where they begin to phagocytize bacteria. Additional connective tissue fibers are formed to contain the infection and keep it from spreading. ✗

histamine, aspirin prevents the synthesis of prostaglandins, and cortisone reduces the release of several mediators of inflammation. On the other hand, the inflammatory response by itself may not be enough to combat the effects of injury or fight off an infection. Medical intervention such as administering antibiotics may be required.

5   **P R E D I C T**

In some injuries, tissues are so severely damaged that there are areas in which cells are killed and blood vessels are destroyed. For injuries such as these, where do the signs of inflammation such as redness, heat, edema, and pain occur?

✔ *Answer in Appendix F*

Clinical Note

When the agent responsible for an injury is not removed or if there is some interference with the process of healing, the inflammatory response persists and is called chronic inflammation. For example, an infection of the lung can result in a brief period of inflammation followed by repair, but a prolonged infection results in chronic inflammation, which results in tissue destruction and permanent damage to the lung. Also, chronic inflammation of the stomach or small intestine may result in an ulcer. Prolonged infections, prolonged exposure to irritants such as silica in the lung, or abnormal immune responses can result in chronic inflammation. White blood cells invade areas of chronic inflammation, and ultimately healthy tissues are destroyed and replaced by a fibrous connective tissue, which is an important cause of the loss of organ function. Chronic inflammation of the lungs, the liver, the kidney, or other vital organs can lead to death.

Tissue Repair

Tissue repair is the substitution of viable cells for dead cells, and it can occur by regeneration or replacement. In **regeneration** (rē′jen-er-ā′shŭn) the new cells are the same type as those that were destroyed, and normal function is usually restored. In **replacement** a new type of tissue develops that eventually causes scar production and the loss of some tissue function. Most wounds heal through regeneration and replacement; which process dominates depends on the tissues involved and the nature and extent of the wound.

Cells can be classified into three groups called labile, stable, or permanent cells, according to their ability to regenerate. **Labile** cells, including cells of the skin, mucous membranes, and hemopoietic and lymphoid tissues, continue to divide throughout life. Damage to these cells can be repaired completely by regeneration. **Stable** cells, such as connective tissues and glands, including the liver, pancreas, and endocrine glands, do not divide after growth ceases; but they do retain the ability to divide and are capable of regeneration in response to injury. **Permanent** cells have very limited ability to replicate, and, if killed, they are usually replaced by a different type of cell. Neurons fit this category, although neurons are able to recover from damage. If the cell body of a neuron is not destroyed, the neuron can replace a damaged axon or dendrite; but if the neuron cell body is destroyed, the neuron cannot regenerate. Evidence indicates that some neurons of the central nervous system can undergo mitosis, although the degree to which mitosis occurs and its functional significance is not clear. Skeletal and cardiac muscle cells also have very limited ability to regenerate,

Clinical Focus   Microscopic Imaging

We see objects because light either passes through them or is reflected off them and enters our eyes (see chapter 15). We are limited, however, in what we can see with the unaided eye. Without the aid of magnifying lenses, the smallest object we can resolve is approximately 100 µm, or 0.1 mm, in diameter, which is approximately the size of a fine pencil dot. Resolution is a measure of the ability to distinguish detail in small objects. For example, we must use a microscope to resolve structures less than 100 µm in diameter.

There are two basic types of microscopes: light microscopes and electron microscopes. As their names imply, light microscopes use light to produce an image, and electron microscopes use beams of electrons. **Light microscopes** usually use transmitted light, which is light that passes through the object being examined, but some light microscopes are equipped to use reflected light. Glass lenses are used in light microscopes to magnify images, and images can either be observed directly by looking into the microscope, or the light from the images can be used to expose photographic film to make a photomicrograph of the images. Video cameras are sometimes used to record images. The resolution of light microscopes is limited by the wavelength of light, the lower limit of which is approximately 0.1 µm—about the size of a small bacterium.

A **biopsy** is the process of removing tissue from living patients for diagnostic examination. For example, changes in tissue structure allow pathologists to identify tumors and to distinguish between noncancerous (benign) and cancerous (malignant) tumors. Light microscopy is used on a regular basis to examine biopsy specimens. Light microscopy is used instead of electron microscopy because less time and effort are required to prepare materials for examination, and the resolution is adequate to diagnose most conditions that cause changes in tissue structure.

Because images are usually produced using transmitted light, tissues to be examined must be cut very thinly to allow the light to pass through them. Sections are routinely cut between 1 and 20 µm thick to make them thin enough for light microscopy. To cut such thin sections, the tissue must be fixed or frozen, which is a process that preserves the tissue and makes it more rigid. Fixed tissues are then embedded in some material, such as wax or plastic, that makes the tissue rigid enough for cutting into sections. The frozen sections, which can be prepared rapidly, are rigid enough for sectioning, but tissue embedded in wax or plastic can be cut much thinner, which makes the image seen through the microscope clearer. Because most tissues are colorless and transparent when thinly sectioned, the tissue must be stained with a colored dye so that the structural details can be seen. As a result, the colors seen in color photomicrographs are not the true colors of the tissue but instead are the colors of the stains used. The color of the stain can also provide specific information about the tissue, because special stains color only certain structures.

To see objects much smaller than a cell, such as cell organelles, an **electron microscope**, which has a limit of resolution of approximately 0.1 nm, must be used; 0.1 nm is about the size of some molecules. In objects viewed through an electron microscope, a beam of electrons either is passed through objects using a transmission electron microscope (TEM) or is reflected off the surface of objects using a scanning electron microscope (SEM). The electron beam is focused with electromagnets. For both processes the specimen must be fixed, and for TEM the specimen must be embedded in plastic and thinly sectioned (0.01–0.15 µm thick). Care must be taken when examining specimens in an electron microscope because a focused electron beam can cause most tissues to quickly disintegrate. Furthermore, the electron beam is not visible to the human eye; thus it must be directed onto a fluorescent or photographic plate on which the electron beam is converted into a visible image. Because the electron beam does not transmit color information, electron micrographs are black and white unless color enhancement has been added using computer technology.

The magnification ability of SEM is not as great as that of TEM; however, depth of focus of SEM is much greater, allowing for the production of a clearer three-dimensional image of the tissue structure.

---

although they can repair themselves. In contrast, smooth muscle readily regenerates following injury.

Skin repair is a good example of wound repair (figure 4.12). The basic pattern of the repair is the same for other tissues, especially ones covered by epithelium. If the edges of the wound are close together such as in a surgical incision, the wound heals by a process called primary union, or primary intention. If the edges are not close together, or if there has been extensive loss of tissue, the process is called secondary union, or secondary intention.

In **primary union,** the wound fills with blood, and a clot forms (see chapter 19). The clot contains a threadlike protein, **fibrin** (fī′brin), that binds the edges of the wound together. The surface of the clot dries to form a **scab,** which seals the wound and helps prevent infection. An inflammatory response induces vasodilation, bringing increased numbers of blood cells and other substances to the area. Blood vessel permeability increases, resulting in edema. Fibrin and blood cells move into the wounded tissues because of the increased vascular permeability. The fibrin acts to isolate and wall off microorganisms and other foreign matter. Some of the white blood cells that move into the tissue are phagocytic cells called **neutrophils** (nū′trō-filz; see figure 4.12b). They ingest bacteria, thus helping to fight infection, and they also ingest tissue debris, clearing the area for repair. Neutrophils are killed in this process and can accumulate as a mixture of dead cells and fluid called **pus** (pŭs).

Fibroblasts from surrounding connective tissue migrate into the clot and produce collagen and other extracellular matrix components. Capillaries grow from blood vessels at the edge of the wound and revascularize the area, and fibrin in the clot is broken down and removed. The result is the replacement of the clot by a delicate connective tissue, called **granulation tissue,** which consists of fibroblasts, collagen,

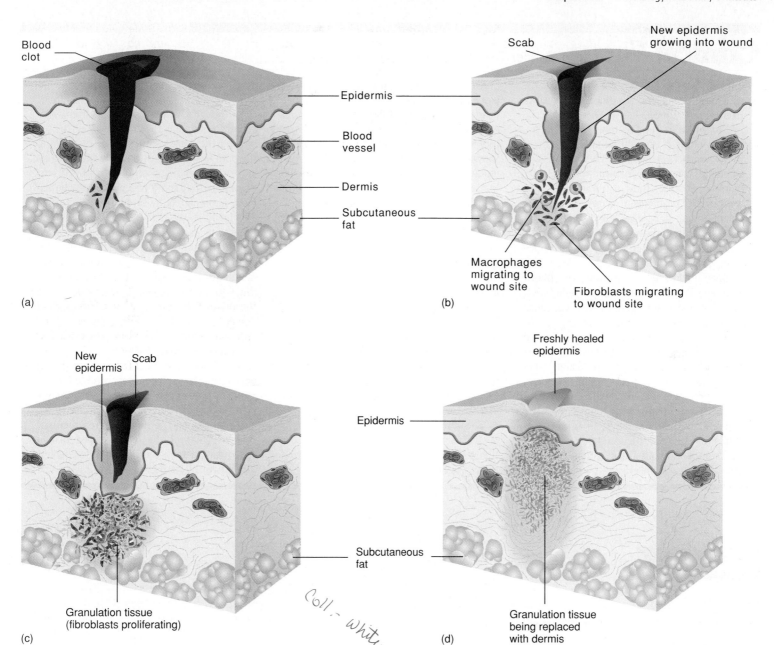

Figure 4.12 Tissue Repair

(*a*) Fresh wound cuts through the epithelium (epidermis) and underlying connective tissue (dermis), and a clot forms. (*b*) Approximately 1 week after the injury, a scab is present, and epithelium (new epidermis) is growing into the wound. (*c*) Approximately 2 weeks after the injury, the epithelium has grown completely into the wound, and granulation tissue has formed. (*d*) Approximately 1 month after the injury, the wound has completely closed, the scab has been sloughed, and the granulation tissue is being replaced with dermis.

and capillaries. A large amount of granulation tissue sometimes persists as a **scar** (skar), which at first is bright red because of vascularization of the tissue. Later, the scar blanches and becomes white, as collagen accumulates and the vascular channels are compressed.

Repair by **secondary union** proceeds in a fashion similar to healing by primary union, but there are some differences. Because the wound edges are far apart, the clot may not close the gap completely, and it takes the epithelial cells much longer to regenerate and cover the wound. With in-

creased tissue damage, the degree of the inflammatory response is greater, there is more cell debris for the phagocytes to remove, and the risk of infection is greater. Much more granulation tissue forms, and **wound contraction** occurs as a result of the contraction of fibroblasts in the granulation tissue. Wound contraction leads to disfiguring and debilitating scars. Thus it is advisable to suture a large wound so that it can heal by primary rather than secondary union. Healing is faster, the risk of infection is lowered, and the degree of scarring is reduced.

Clinical Focus   Cancer Tissue

Cancer (kan'ser) refers to a malignant, spreading tumor and the illness that results from such a tumor.

Tumor (tŭ'mōr) refers to any swelling, although modern usage has limited the term to swellings that involve neoplastic tissue. Oncology (ong-kol'ō-jē, meaning the study of tumors) is the study of tumors and their associated problems. Neoplasm (nē'ō-plazm) means new growth and refers to abnormal tissue growth resulting in unusually rapid cellular proliferation that continues after normal growth of the tissue has stopped or slowed considerably. A neoplasm can be either malignant (mă-lig'nănt, meaning with malice or intent to cause harm), able to spread and become worse, or benign (bē-nīn', meaning kind), not inclined to spread and not likely to become worse. Although benign tumors are usually less dangerous than malignant tumors, they can cause problems. As the benign tumor enlarges, it can compress surrounding tissues and impair their functions. In some cases (e.g., brain tumors), the result can be death.

Malignant tumors can spread by local growth and expansion or by metastasis (mĕ-tas'tă-sis, meaning moving to another place), which results from tumor cells separating from the main neoplasm and being carried by the lymphatic or circulatory system to a new site, where a second neoplasm forms. A carcinoma (kar-si-nō'mă) is a malignant neoplasm derived from epithelial tissue. A sarcoma (sar-kō'mă) is a malignant neoplasm derived from connective tissue.

Malignant neoplasms lack the normal growth control that is exhibited by most other adult tissues, and in many ways they resemble embryonic tissue. Rapid growth is one characteristic of embryonic tissue, but as the tissue begins to reach its adult size and function, it slows or stops growing completely. This cessation of growth is controlled at the individual cell level. Cancer results when a cell or group of cells, for some reason, breaks away from that control. This breaking loose involves the genetic machinery and can be induced by viruses, environmental toxins, and other causes.

The illness associated with cancer usually occurs as the tumor invades and destroys the healthy surrounding tissues, eliminating their functions.

Cancer therapy concentrates primarily on trying to confine and then kill the malignant cells. This goal is accomplished currently by killing the tissue with x-rays or lasers, by removing the tumor surgically, or by treating the patient with drugs that kill rapidly dividing cells. The major problem with current therapy is that some cancers cannot be removed completely by surgery or killed completely by x-rays and laser therapy. These treatments can also kill normal tissue adjacent to the tumor. Drugs used in cancer therapy kill not only cancer tissue but also other rapidly growing tissues, such as bone marrow, where new blood cells are produced, and the lining of the intestinal tract. Loss of these tissues can result in anemia, caused by the lack of red blood cells, and nausea, caused by the loss of the intestinal lining.

Promising anticancer therapies are being developed in which cells responsible for immune responses can be stimulated to recognize tumor cells and destroy them. A major advantage in such anticancer treatments is that the cells of the immune system can specifically attack the tumor cells and not other, healthy tissues.

Summary

Epithelial Tissue

1. Epithelium consists of cells with little extracellular matrix, covers surfaces, has a basement membrane, and does not have blood vessels.
2. The basement membrane is secreted by the epithelial cells and attaches the epithelium to the underlying tissues.

Classification of Epithelium

1. Simple epithelium has a single layer of cells, stratified epithelium has two or more layers, and pseudostratified epithelium has a single layer that appears stratified.
2. Cells can be squamous (flat), cuboidal, or columnar.
3. Stratified squamous epithelium can be moist or keratinized.
4. Transitional epithelium is stratified, with cells that can change shape from cuboidal to flattened.

Functional Characteristics

1. Simple epithelium is usually involved in diffusion, filtration, secretion, or absorption. Stratified epithelium serves a protective role. Squamous cells function in diffusion and filtration. Cuboidal or columnar cells, with a larger cell volume that contains many organelles, secrete or absorb.
2. A smooth free surface reduces friction (mesothelium and endothelium), microvilli increase absorption (intestines), and cilia move materials across the free surface (respiratory tract and uterine tubes). Transitional epithelium has a folded surface that allows the cell to change shape, and the number of cells making up the epithelial layers changes.
3. Cells are bound together mechanically by glycoproteins, desmosomes, and the zonulae adherens and to the basement membrane by hemidesmosomes. The zonulae occludens and zonulae adherens form a permeability barrier or tight junction, and gap junctions allow intercellular communication.

Glands

1. Glands are organs that secrete. Exocrine glands secrete through ducts, and endocrine glands release hormones that are absorbed directly into the blood.
2. Glands are classified as unicellular or multicellular. Goblet cells are unicellular glands. Multicellular exocrine glands have ducts, which are simple or compound (branched). The ducts can be tubular or end in small sacs (acini or alveoli). Tubular glands can be straight or coiled.
3. Glands are classified according to their mode of secretion. Merocrine glands (pancreas) secrete substances as they are

produced, apocrine glands (mammary glands) accumulate secretions that are released when a portion of the cell pinches off, and holocrine glands (sebaceous glands) accumulate secretions that are released when the cell ruptures and dies.

Connective Tissue

Connective tissue is distinguished by its extracellular matrix.

Connective Tissue Cells

The extracellular matrix results from the activity of specialized connective tissue cells; in general, blast cells form the matrix, cyte cells maintain it, and clast cells break it down. Fibroblasts form protein fibers of many connective tissues, osteoblasts form bone, and chondroblasts form cartilage.

Protein Fibers of the Matrix

1. Collagen fibers structurally resemble ropes. They are strong and flexible but resist stretching.
2. Reticular fibers are fine collagen fibers that form a branching network that supports other cells and tissues.
3. Elastin fibers have a structure similar to a spring. After being stretched they tend to return to their original shape.

Other Matrix Molecules

1. Hyaluronic acid makes fluids slippery.
2. Proteoglycan aggregates trap water, giving tissues resiliency.

Classification of Connective Tissue

Connective tissue is classified according to the type of protein and the proportions of protein, ground substance, and fluid in the matrix.

Matrix with Fibers as the Primary Feature

1. Loose (areolar) connective tissue has many different cell types and a random arrangement of protein fibers with space between the fibers. This tissue fills spaces around the organs and attaches the skin to underlying tissues.
2. Dense regular connective tissue is composed of fibers arranged in one direction, providing strength in a direction parallel to the fiber orientation. There are two types of dense regular connective tissue: collagenous (tendons and most ligaments) and elastic (ligaments of vertebrae).
3. Dense irregular connective tissue has fibers organized in many directions, producing strength in different directions. There are two types of dense irregular connective tissue: collagenous (capsules of organs and dermis of skin) and elastic (large arteries).
4. Adipose tissue has fat cells (adipocytes) filled with lipid and very little extracellular matrix (a few reticular fibers).
   • Adipose tissue functions as energy storage, insulation, and protection.
   • Adipose tissue can be yellow (white) or brown. Brown fat is specialized for generating heat.
5. Reticular tissue is a network of reticular fibers and forms the framework of lymphoid tissue, bone marrow, and the liver.
6. Yellow bone marrow is a site of fat storage, and red bone marrow is the site of blood cell formation.

Matrix with Both Protein Fibers and Ground Substance

1. Cartilage has a relatively rigid matrix composed of protein fibers and proteoglycan aggregates. The major cell type is the chondrocyte, which is located within lacunae.
   • Hyaline cartilage has evenly dispersed collagen fibers that provide rigidity with some flexibility. Examples include the costal cartilage, the covering over the ends of bones in joints, the growing portion of long bones, and the embryonic skeleton.
   • Fibrocartilage has collagen fibers arranged in thick bundles; can withstand great pressure; and is found between vertebrae, in the jaw, and in the knee.
   • Elastic cartilage is similar to hyaline cartilage, but it has elastin fibers. It is more flexible than hyaline cartilage. It is found in the external ear.
2. Bone cells, or osteocytes, are located in lacunae that are surrounded by a mineralized matrix (hydroxyapatite) that makes bone very hard. Cancellous bone has spaces between bony trabeculae, and compact bone is more solid.

Predominantly Fluid Matrix

Blood cells are suspended in a fluid matrix.

Muscle Tissue

1. Muscle tissue has the ability to contract.
2. Skeletal (striated voluntary) muscle attaches to bone and is responsible for body movement. Skeletal muscle cells are long, cylindrically shaped cells with many peripherally located nuclei.
3. Cardiac (striated involuntary) muscle cells are cylindrical, branching cells with a single, central nucleus. Cardiac muscle is found in the heart and is responsible for pumping blood through the circulatory system.
4. Smooth (nonstriated involuntary) muscle forms the walls of hollow organs, the iris of the eye, and other structures. Its cells are spindle-shaped with a single, central nucleus.

Nervous Tissue

1. Nervous tissue has the ability to conduct electric impulses and is composed of neurons (conductive cells) and neuroglia (support cells).
2. Neurons have cell processes called dendrites and axons. The dendrites can receive electric impulses, and the axons can conduct them. Neurons can be multipolar (several dendrites and an axon), bipolar (one dendrite and one axon), or unipolar (one axon).

Embryonic Tissue

1. The endoderm, mesoderm, and ectoderm are the primary germ layers from which all adult structures arise.
2. Mesenchyme, formed mainly from mesoderm, is the embryonic tissue from which connective tissues arise.

Membranes

1. Mucous membranes consist of epithelial cells, basement membrane, the lamina propria, and, sometimes, smooth muscle cells; they line cavities that open to the outside and often contain mucous glands, which secrete mucus.
2. Serous membranes line cavities that do not open to the exterior, do not contain glands, but do secrete serous fluid.
3. Synovial membranes are formed by connective tissue and line joint cavities.

Inflammation

1. The function of the inflammatory response is to isolate injurious agents from the rest of the body and to attack and destroy the injurious agent.
2. The inflammatory response produces five symptoms: redness, heat, swelling, pain, and disturbance of function.

Tissue Repair

1. Tissue repair is the substitution of viable cells for dead ones. Tissue repair occurs by regeneration or replacement.
   - Labile cells divide throughout life and can undergo regeneration.

- Stable cells do not ordinarily divide after growth is complete but can regenerate if necessary.
- Permanent cells cannot replicate. If killed, permanent tissue is repaired by replacement.

2. Tissue repair by primary union occurs when the edges of the wound are close together. Secondary union occurs when the edges are far apart.

Content Review

1. Define histology and tissues. What two things distinguish one type of tissue from another?
2. Name the four primary tissue types, and give the general basis for defining them.
3. List the characteristics of epithelium.
4. What is the basement membrane, and how is it produced?
5. Name two general characteristics that are used to classify epithelium. On the basis of this classification scheme, describe the different kinds of epithelium.
6. Why is pseudostratified epithelium not considered a stratified epithelium? Describe transitional epithelium.
7. What kind of functions would a single layer of epithelium be expected to perform? A stratified layer?
8. In locations in which diffusion or filtration is occurring, what shape would you expect the epithelial cells to be?
9. Why are cuboidal or columnar cells found where secretion or absorption is occurring?
10. What is the function of an epithelial free surface that is smooth, has cilia, has microvilli, or is folded? Give an example of epithelium in which each surface type is found.
11. Name the ways in which epithelial cells are bound to one another and to the basement membrane.
12. Define the term gland. Distinguish between exocrine and endocrine glands. Describe the classification scheme for exocrine glands on the basis of their duct systems.
13. Describe three different ways in which exocrine glands release their secretions. Give an example for each method.
14. What is the major characteristic that distinguishes connective tissue from other tissues?
15. Explain the difference between connective tissue cells that are termed blast, cyte, or clast cells.
16. Contrast the structure and characteristics of collagen fibers, reticular fibers, and elastin fibers.
17. What three components are found in the extracellular matrix of connective tissue? How are they used to classify connective tissue?
18. Describe the fiber arrangement in loose (areolar) connective tissue. What functions does this tissue accomplish?
19. What is the function of reticular tissue?

20. Structurally and functionally, what is the difference between dense regular connective tissue and dense irregular connective tissue?
21. Name the two kinds of dense regular connective tissue, and give an example of each. Do the same for dense irregular connective tissue.
22. What features of the extracellular matrix distinguish adipose tissue from other connective tissue? What is an adipocyte?
23. List the functions of adipose tissue. Name the two types of adipose tissue. Which one is important in generating heat?
24. What are red marrow and yellow marrow?
25. Describe the components of cartilage. How do hyaline cartilage, elastic cartilage, and fibrocartilage differ in structure and function? Give an example of each.
26. Describe the components of bone. Differentiate between cancellous and compact bone.
27. What characteristic separates blood from the other connective tissues?
28. Functionally, what is unique about muscle? Contrast the structure of skeletal, cardiac, and smooth muscle cells. Which of the muscle types is under voluntary control? What tasks does each type perform?
29. Functionally, what is unique about nervous tissue? What do neurons and neuroglia accomplish?
30. What is the difference between a dendrite and an axon? Describe the structure of multipolar, bipolar, and unipolar neurons.
31. What are the three embryonic germ layers, and what is mesenchyme?
32. Compare serous and mucous membranes according to the type of cavity they line and their secretions.
33. What is the function of the inflammatory response? Name the five symptoms of the inflammatory response, and explain how each is produced.
34. Define the term tissue repair. Differentiate between tissue repair that occurs by regeneration and by replacement.
35. Differentiate labile cells, stable cells, and permanent cells. Give examples of each type. What is the significance of these cell types to tissue repair?
36. Describe the process of tissue repair. Contrast healing by primary union and by secondary union.

Develop Your Reasoning Skills

1. Given the observation that a tissue has more than one layer of cells lining a free surface, (1) list the possible tissue types that exhibit those characteristics, and (2) explain what additional observations need to be made to identify the tissue as a specific tissue type.
2. A patient suffered from kidney failure a few days after he was exposed to a toxic chemical. A biopsy of his kidney indicated that many of the thousands of epithelium-lined tubules that make up the kidney had lost the layer of simple cuboidal epithelial cells that normally line them, although the basement membranes appeared to be mostly intact. Predict how likely this person is to fully recover.
3. Compare the cell shapes and surface specializations of an epithelium that functions to resist abrasion with those of an epithelium that functions to carry out absorption of materials.
4. Tell how to distinguish between a gland that produces a merocrine secretion and a gland that produces a holocrine secretion. Assume that you have the ability to chemically analyze the composition of secretions.
5. Indicate whether the following statement is appropriate or not: "If a tissue is capable of contracting, is under involuntary control, and has mononucleated cells, it is smooth muscle." Explain your answer.
6. Antihistamines block the effect of a chemical mediator of inflammation called histamine, which is released during the inflammatory response. What effect does administering antihistamines have on the inflammatory response, and is use of an antihistamine beneficial?

Web Site Link

For a listing of the most current web sites related to this chapter, please visit the Seeley home page at:
http://www.mhhe.com/biosci/ap/seeleyap/

Integumentary System

Objectives

1. Describe the structure and function of the hypodermis.

2. Name and describe the two layers of the dermis.

3. Explain the basis for dividing the epidermis into strata. List and describe each stratum.

4. Describe the events occurring during keratinization that produce a skin resistant to abrasion and water loss.

5. Contrast thick skin and thin skin.

6. Discuss melanocytes and the way they produce and transfer melanin.

7. Explain how melanin, carotene, and blood affect skin color.

8. Distinguish between lanugo, vellus, and terminal hair.

9. Describe the structure of a hair and the sheaths that surround the hair. Discuss the phases of hair growth.

10. Describe the production of "goose flesh."

11. Describe the glands of the skin and their secretions.

12. Describe the parts of a nail, and explain how nails are produced.

13. Discuss the functions of the skin, hair, nails, and glands.

14. Describe the production of vitamin D by the body and its functions.

15. Describe the changes that occur in the integumentary system with age.

issues combine to form organs, and organs combine to form organ systems. The **integumentary system** consists of the skin and accessory structures, such as hair, nails, and glands. The term integument means covering, and the integumentary system is familiar to most people because it covers the outside of the body. We apply lotion to our skin, color our hair, trim our nails, and try to prevent sweating with antiperspirants. Although we are often concerned with how the integumentary system looks, it has many important functions that go beyond appearance. The integumentary system protects internal structures, prevents the entry of infectious agents, reduces water loss, regulates body temperature, produces vitamin D, and detects stimuli such as touch, pain, and temperature.

Hypodermis

Just as a house rests on a foundation, the skin rests on the **hypodermis** (hī-pō-der′mis), which attaches it to underlying bone and muscle and supplies it with blood vessels and nerves (figure 5.1). The hypodermis consists of loose connective tissue with collagen and elastin fibers. The main types of cells within the hypodermis are fibroblasts, adipose cells, and macrophages. The hypodermis, which is not part of the skin, is sometimes called **subcutaneous** (sŭb-kyū-tā′nē-ŭs) **tissue,** or **superficial fascia** (fash′ē-ă).

Approximately half of the body's stored fat is in the hypodermis, although the amount and location vary with age, sex, and diet. For example, newborn infants have a large amount of fat, which accounts for their chubby appearance; and women have more fat than men, especially over the thighs, buttocks, and breasts. Fat in the hypodermis functions as padding and insulation and is responsible for some of the differences in body shape between men and women.

> **Clinical Note**
>
> The hypodermis can be used to estimate total body fat. The skin is pinched at selected locations, and the thickness of the fold of skin and underlying hypodermis is measured. The thicker the fold, the greater is the amount of total body fat. Clinically, the hypodermis is the site of subcutaneous injections.

Skin

The skin is made up of two major tissue layers. The **dermis** (der′mis, meaning skin) is a layer of connective tissue that is connected to the hypodermis. The **epidermis** (ep-i-der′mis, meaning on the dermis) is a layer of epithelial tissue that rests on the dermis (see figure 5.1). If the hypodermis is the foundation on which the house rests, the dermis forms most of the house, and the epidermis is its roof.

Dermis

The dermis is connective tissue with fibroblasts, a few adipose cells, and macrophages. Collagen is the main connective tissue fiber, but elastin and reticular fibers are also present. The dermis is responsible for most of the structural strength of the skin. Compared with the hypodermis, adipose cells and blood vessels are scarce in the dermis. Nerve endings, hair follicles, smooth muscles, glands, and lymphatics extend into the dermis (see figure 5.1).

> **Clinical Note**
>
> The dermis is that part of an animal hide from which leather is made. The epidermis of the skin is removed, and the fibrous dermis is treated with chemicals in a process called tanning. Clinically the dermis in humans is sometimes the site of such injections as the tuberculin skin test.

dense   irregular

The dermis is divided into two layers (see figure 5.1; figure 5.2): the deeper **reticular** (re-tik′yū-lăr) **layer** and the more superficial **papillary** (pap′i-lār-ē) **layer.** The reticular layer, which is dense, irregular connective tissue, is the main layer of the dermis. It is continuous with the hypodermis and forms a mat of irregularly arranged fibers that are resistant to stretching in many directions. The elastin and collagen fibers are oriented more in some directions than in others and produce **cleavage,** or **tension, lines** in the skin (figure 5.3). Knowledge of cleavage line directions is important to a surgeon because an incision made across the cleavage lines is more likely to gap, increasing the likelihood of infection and producing considerable scar tissue, than is an incision made parallel with the cleavage lines. In addition, if the skin is overstretched, the dermis may rupture, leaving lines that are visible through the epidermis. These lines, called **striae** (strī′ē), or **stretch marks,** may develop on the abdomen and breasts of a woman during pregnancy.

The papillary layer derives its name from projections called **papillae** (pă-pil′ē) that extend toward the epidermis (see figure 5.2). The papillary layer is less dense than the reticular layer and is sometimes called loose connective tissue because it has thin fibers that are somewhat loosely arranged. The papillary layer also contains a large number of blood vessels that supply the overlying avascular epidermis with nutrients, remove waste products, and aid in regulating body temperature.

Epidermis

The epidermis is stratified squamous epithelium separated from the dermis by a basement membrane. The epidermis is not as thick as the dermis, contains no blood vessels, and is nourished by diffusion from capillaries of the papillary layer (see figures 5.1 and 5.2). Most of the cells of the epidermis are **keratinocytes** (ke-rat′i-nō-sītz), which produce a protein mixture called **keratin** (ker′ă-tin). Keratinocytes are responsible for the structural strength and permeability characteristics

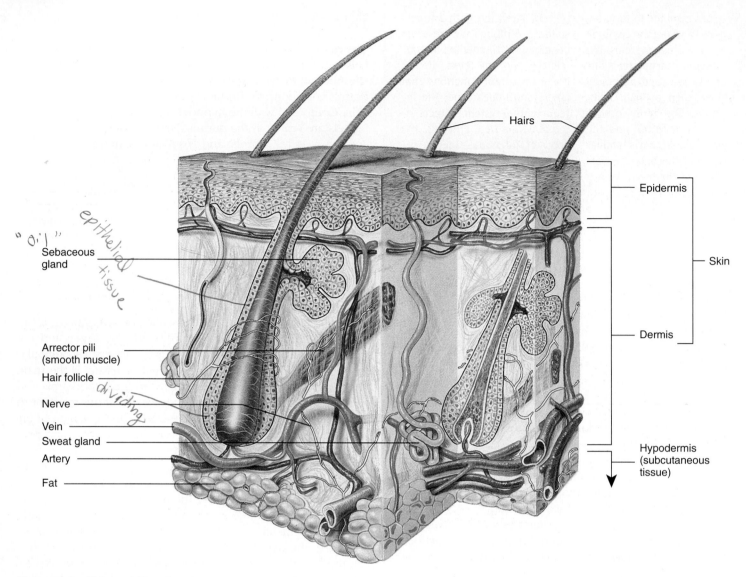

Figure 5.1   Skin and Hypodermis

The figure represents a block of skin (dermis and epidermis), hypodermis, and accessory structures (hairs and glands).

of the epidermis. Other cells of the epidermis include **melanocytes** (mel′ă-nō-sītz), which contribute to skin color, and **Langerhans' cells,** which are part of the immune system (see chapter 22).

Cells are produced in the deepest layers of the epidermis by mitosis. As new cells are formed, they push older cells to the surface where they slough off, or **desquamate** (des′kwă-māt). The outermost cells in this stratified arrangement protect the cells underneath, and the deeper replicating cells replace cells lost from the surface. As they move from the deeper epidermal layers to the surface, the cells change shape and chemical composition. This process is called **keratinization** (ker′ă-tin-i-zā′shŭn) because the cells become filled with keratin. During keratinization these cells eventually die and produce an outer layer of cells that resists abrasion and forms a permeability barrier.

Clinical Note

The study of keratinization is important because many skin diseases result from malfunctions in this process. For example, large scales of epidermal tissue are sloughed off in psoriasis (sō-rī′ă-sis; see "Clinical Focus: Clinical Disorders of the Integumentary System" at the end of the chapter). By comparing normal with abnormal keratinization, scientists may be able to develop effective therapies.

Although keratinization is a continual process, distinct transitional stages can be recognized as the cells change. On the basis of these stages, the many layers of cells in the epidermis are divided into regions, or **strata** (sing., stratum) (see figure 5.2 and figure 5.4). From the deepest to the most

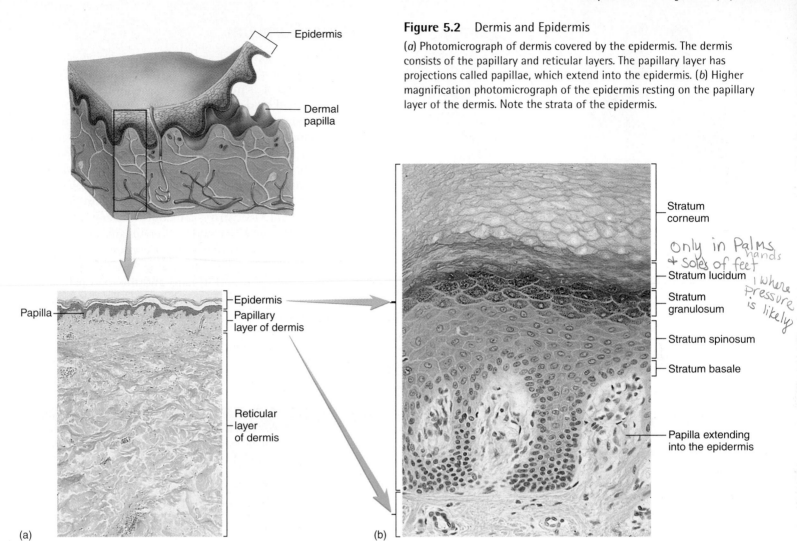

Epidermis

Dermal
papilla

Stratum
corneum

Stratum lucidum

Stratum
granulosum

Stratum spinosum

Stratum basale

Papilla extending
into the epidermis

only in palms
+ hands
Soles of feet
where
pressure
is likely

Figure 5.2   Dermis and Epidermis

(*a*) Photomicrograph of dermis covered by the epidermis. The dermis consists of the papillary and reticular layers. The papillary layer has projections called papillae, which extend into the epidermis. (*b*) Higher magnification photomicrograph of the epidermis resting on the papillary layer of the dermis. Note the strata of the epidermis.

Papilla

Epidermis
Papillary
layer of dermis

Reticular
layer
of dermis

(a)

(b)

superficial, these five strata are observed: stratum basale, stratum spinosum, stratum granulosum, stratum lucidum, and stratum corneum. The number of cell layers in each stratum and even the number of strata in the skin vary, depending on their location in the body.

Stratum Basale

The deepest portion of the epidermis is a single layer of cuboidal or columnar cells, the **stratum basale** (bā′să-lē) (see figures 5.2 and 5.4). Structural strength is provided by hemidesmosomes, which anchor the epidermis to the basement membrane, and by desmosomes, which hold the keratinocytes together (see chapter 4). Keratinocytes are strengthened internally by keratin fibers (intermediate filaments) that insert into the desmosomes. Keratinocytes undergo mitotic divisions approximately every 19 days. One daughter cell becomes a new stratum basale cell and divides again, but the other daughter cell is pushed toward the surface and becomes **keratinized** (ker′ă-ti-nīzd). It takes approximately 40–56 days for the cell to reach the epidermal surface and desquamate.

Stratum Spinosum

Superficial to the stratum basale is the **stratum spinosum** (spī-nō′sŭm), consisting of 8–10 layers of many-sided cells (see figures 5.2 and 5.4). As the cells in this stratum are pushed to the surface, they flatten; desmosomes are broken apart, and new desmosomes are formed. During preparation for microscopic observation, the cells usually shrink from one another, except where they are attached by desmosomes, causing the cells to appear spiny—hence the name stratum spinosum. Additional keratin fibers and lipid-filled, membrane-bound organelles called **lamellar** (lam′ĕ-lăr, lă-mel′ăr) **bodies** are formed inside the keratinocytes. A limited amount of cell division takes place in this stratum, and for this reason the stratum basale and stratum spinosum are sometimes considered a single stratum called the **stratum germinativum** (jer′mi-nă-tīv′ŭm). Mitosis does not occur in the more superficial strata.

Stratum Granulosum

The **stratum granulosum** (gran-yū-lō′sŭm) consists of two to five layers of somewhat flattened, diamond-shaped cells

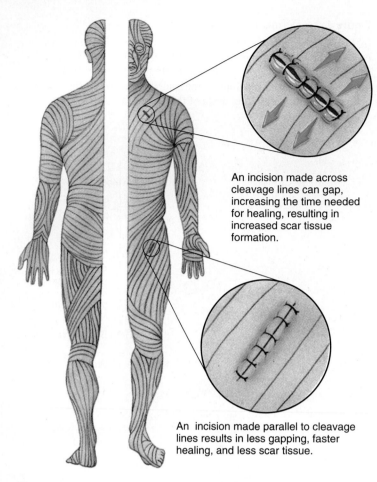

An incision made across cleavage lines can gap, increasing the time needed for healing, resulting in increased scar tissue formation.

An incision made parallel to cleavage lines results in less gapping, faster healing, and less scar tissue.

Figure 5.3   Cleavage Lines

The orientation of collagen fibers produces cleavage, or tension, lines in the skin.

with long axes that are oriented parallel to the surface of the skin (see figures 5.2 and 5.4). This stratum derives its name from the nonmembrane-bound protein granules of **kerato-hyalin** (ker′ă-tō-hī′ă-lin), which accumulate in the cyto-plasm of the cell. The lamellar bodies of these cells move to the cell membrane and release their lipid contents into the intercellular space. Inside the cell, a protein envelope forms beneath the cell membrane. In the most superficial layers of the stratum granulosum, the nucleus and other organelles degenerate, and the cell dies. Unlike the other organelles, however, the keratin fibers and keratohyalin granules do not degenerate.

Stratum Lucidum

The **stratum lucidum** (lū′si-dŭm) appears as a thin, clear zone above the stratum granulosum (see figures 5.2 and 5.4) and consists of several layers of dead cells with indistinct boundaries. Keratin fibers are present, but the keratohyalin, which was evident as granules in the stratum granulosum, has dispersed around the keratin fibers, and the cells appear somewhat transparent. The stratum lucidum is present in only a few areas of the body (see Thick and Thin Skin).

Stratum Corneum

The last and most superficial stratum of the epidermis is the **stratum corneum** (kōr′nē-ŭm) (see figures 5.2 and 5.4). This stratum is composed of as many as 25 or more layers of dead squamous cells joined by desmosomes. Eventually the desmosomes break apart, and the cells are desquamated from the surface of the skin. Dandruff is an example of desquamation of the stratum corneum of the scalp. Less noticeably, cells are continually shed as clothes rub against the body or as the skin is washed.

The stratum corneum consists of **cornified cells,** which are dead cells with a hard protein envelope that are filled with the protein keratin. **Keratin** is a mixture of keratin fibers and keratohyalin. The envelope and the keratin are responsible for the structural strength of the stratum corneum. The type of keratin found in the skin is soft keratin. Another type of keratin, hard keratin, is found in nails and the external parts of hair. Cells containing hard keratin are more durable than cells with soft keratin and do not desquamate.

Surrounding the cells are the lipids released from the lamellar bodies. The lipids are responsible for many of the permeability characteristics of the skin. Table 5.1 summarizes the structures and functions of the skin and hypodermis.

1	**P R E D I C T**

Some drugs are administered by applying the drug to the skin (e.g., a nicotine skin patch to help a person stop smoking). The drug diffuses through the epidermis to blood vessels in the dermis. What kind of substances can pass easily through the skin by diffusion? What kind have difficulty?

✔ *Answer in Appendix F*

Thick and Thin Skin

Skin is classified as thick or thin on the basis of the structure of the epidermis. **Thick skin** has all five epithelial strata, and the stratum corneum has many layers of cells. Thick skin is found in areas subject to pressure or friction, such as the palms of the hands, the soles of the feet, and the fingertips. The papillae of the dermis underlying thick skin are in parallel, curving ridges that shape the overlying epidermis into fingerprints and footprints. The ridges increase friction and improve the grip of the hands and feet.

Fingerprints were first used in criminal investigation in 1880 by Henry Faulds, a Scottish medical missionary. The identity of a thief who had been drinking purified alcohol from the dispensary was confirmed by a greasy fingerprint on a bottle.

Thin skin is found over the rest of the body and is more flexible than thick skin. Each stratum contains fewer layers of cells than are found in thick skin; the stratum granulosum frequently consists of only one or two layers of cells, and the stratum lucidum generally is absent. The dermis under thin skin projects upward as separate papillae and does not produce the ridges seen in thick skin. Hair is found only in thin skin.

Superficial

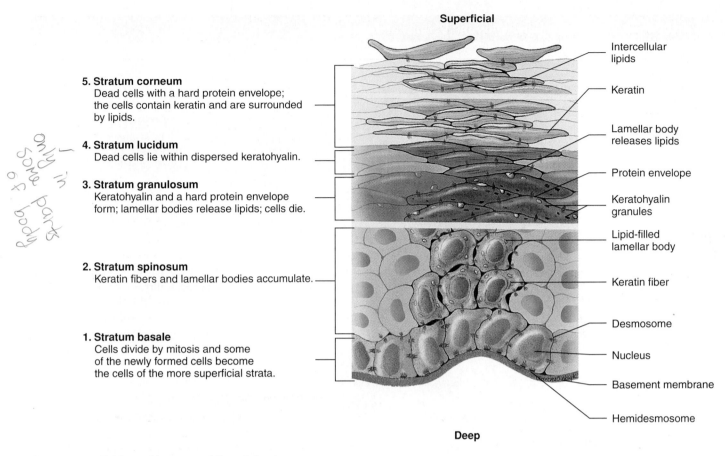

5. **Stratum corneum**
   Dead cells with a hard protein envelope;
   the cells contain keratin and are surrounded
   by lipids.

4. **Stratum lucidum**
   Dead cells lie within dispersed keratohyalin.

3. **Stratum granulosum**
   Keratohyalin and a hard protein envelope
   form; lamellar bodies release lipids; cells die.

2. **Stratum spinosum**
   Keratin fibers and lamellar bodies accumulate.

1. **Stratum basale**
   Cells divide by mitosis and some
   of the newly formed cells become
   the cells of the more superficial strata.

(margin note: only in some parts of body)

Intercellular lipids

Keratin

Lamellar body releases lipids

Protein envelope

Keratohyalin granules

Lipid-filled lamellar body

Keratin fiber

Desmosome

Nucleus

Basement membrane

Hemidesmosome

Deep

Figure 5.4 Epidermal Layers and Keratinization

The entire skin, including both the epidermis and the dermis, varies in thickness from 0.5 mm in the eyelids to 5 mm for the back and shoulders. The terms thin and thick, which refer to the epidermis only, should not be used when total skin thickness is considered. Most of the difference in total skin thickness results from variation in the thickness of the dermis. For example, the skin of the back is thin skin, whereas that of the palm is thick skin; however, the total skin thickness of the back is greater than that of the palm because there is more dermis in the skin of the back.

In skin subjected to friction or pressure, the number of layers in the stratum corneum greatly increases, producing a thickened area called a **callus** (kal′ŭs). The skin over bony prominences may develop a cone-shaped structure called a **corn**. The base of the cone is at the surface, but the apex extends deep into the epidermis, and pressure on the corn may be quite painful. Calluses and corns can develop in both thin and thick skin.

Skin Color

Skin color is determined by pigments in the skin, by blood circulating through the skin, and by the thickness of the stratum corneum. **Melanin** (mel′ă-nin) is the term used to describe a group of pigments responsible for skin, hair, and eye color. Melanin is believed to provide protection against ultraviolet light from the sun. Large amounts of melanin are found in certain regions of the skin, such as freckles, moles, nipples, are-olae of the breasts, the axillae, and the genitalia. Other areas of the body, such as the lips, the palms of the hands, and the soles of the feet, contain less melanin.

In the production of melanin, the amino acid tyrosine is converted to dopaquinone (dō′pă-kwī-nōn) by the enzyme **tyrosinase** (tī′rō-sī-nās). Dopaquinone can be converted to a variety of related molecules, most of which are brown to black pigments, but some of which are yellowish or reddish.

Melanin is produced by **melanocytes** (mel′ă-nō-sītz), irregularly shaped cells with many long processes that extend between the keratinocytes of the stratum basale and the stratum spinosum (figure 5.5). The Golgi apparatuses of the melanocytes package melanin into vesicles called **melanosomes** (mel′ă-nō-sōmz), which move into the cell processes of the melanocytes. Keratinocytes phagocytize (see chapter 3) the tips of the melanocyte cell processes, thereby acquiring melanosomes. Although all keratinocytes can contain melanin, only the melanocytes produce it.

Melanin production is determined by genetic factors, hormones, and exposure to light. Genetic factors are primarily responsible for the variations in skin color between different races and among people of the same race. The amount and types of melanin produced by the melanocytes, and the size, number, and distribution of the melanosomes, is genetically determined. Skin colors are not determined by the number of melanocytes because all races have essentially the same number. Although many genes are responsible for skin color,

Table 5.1   Comparison of the Skin (Epidermis and Dermis) and Hypodermis

Part	Structure	Function
Epidermis	Superficial part of skin; stratified squamous epithelium; composed of four or five strata	Barrier that prevents water loss and the entry of chemicals and microorganisms; protects against abrasion and ultraviolet light; produces vitamin D; gives rise to hair, nails, and glands
Stratum corneum	Most superficial strata of the epidermis; 25 or more layers of dead squamous cells	Provision of structural strength by keratin within cells; prevention of water loss by lipids surrounding cells; desquamation of most superficial cells resists abrasion
Stratum lucidum	Three to five layers of dead cells; appears transparent; present in thick skin, absent in most thin skin	Dispersion of keratohyalin around keratin fibers
Stratum granulosum	Two to five layers of flattened, diamond-shaped cells	Production of keratohyalin granules; lamellar bodies release lipids from cells; cells die
Stratum spinosum	A total of 8–10 layers of many-sided cells	Production of keratin fibers; formation of lamellar bodies
Stratum basale	Deepest strata of the epidermis; single layer of cuboidal or columnar cells; basement membrane of the epidermis attaches to the dermis	Production of cells of the most superficial strata; melanocytes produce and contribute melanin, which protects against ultraviolet light
Dermis	Deep part of skin; connective tissue composed of two layers	Responsible for the structural strength and flexibility of the skin; the epidermis exchanges gases, nutrients, and waste products with blood vessels in the dermis
Papillary layer	Papillae projects toward the epidermis; loose connective tissue	Brings blood vessels close to the epidermis; papillae form fingerprints and footprints
Reticular layer	Mat of collagen and elastin fibers; dense, irregular connective tissue	Main fibrous layer of the dermis; strong in many directions; forms cleavage lines
Hypodermis	Not part of the skin; loose connective tissue with abundant fat deposits	Attaches the dermis to underlying structures; fat tissue provides energy storage, insulation, and padding; blood vessels and nerves from the hypodermis supply the dermis

a single mutation (see chapter 3) can prevent the manufacture of melanin. **Albinism** (al'bi-nizm) usually is a recessive genetic trait causing an inability to produce tyrosinase. The result is a deficiency or absence of pigment in the skin, hair, and eyes.

During pregnancy, certain hormones cause an increase in melanin production in the mother, which in turn causes darkening of the nipples, areolae, and genitalia. The cheekbones, forehead, and chest also may darken, resulting in the "mask of pregnancy," and a dark line of pigmentation may appear on the midline of the abdomen. Diseases such as Addison's disease, which cause an increased secretion of certain hormones, also cause increased pigmentation.

Exposure to ultraviolet light darkens melanin already present and stimulates melanin production, resulting in tanning of the skin.

The location of pigments and other substances in the skin affects the color produced. If a dark pigment is located in the dermis or hypodermis, light reflected off the dark pigment can be scattered by collagen fibers of the dermis to produce a blue color. The same effect produces the blue color of the sky as light is reflected from dust particles in the air. The deeper within the dermis or hypodermis any dark pigment is located, the bluer the pigment appears because of the light-scattering effect of the overlying tissue. This effect causes the blue color of tattoos, bruises, and some superficial blood vessels.

Carotene (kar'ō-tēn) is a yellow pigment found in plants such as carrots and corn. Humans normally ingest carotene and use it as a source of vitamin A. Carotene is lipid-soluble, and, when large amounts of carotene are consumed, the excess accumulates in the stratum corneum and in the adipose cells of the dermis and hypodermis, causing the skin to develop a yellowish tint that slowly disappears once carotene intake is reduced.

Blood flowing through the skin imparts a reddish hue, and, when blood flow increases (e.g., during blushing, anger, and the inflammatory response), the red color intensifies. A decrease in blood flow such as occurs in shock can make the skin appear pale, and a decrease in the blood oxygen content produces **cyanosis** (sī-ă-nō'sis), a bluish skin color.

---

2   P R E D I C T

Explain the differences in skin color between (a) palms of the hands and the lips; (b) palms of the hands of a person who does heavy manual labor and one who does not; (c) anterior and posterior surfaces of the forearm; and (d) genitals and the soles of the feet.

✔ *Answer in Appendix Γ*

Clinical Focus   Burns

Burns are classified according to the depth of the burn and the extent of surface area involved. On the basis of depth, burns are either partial-thickness or full-thickness burns (figure A). **Partial-thickness burns** are divided into first- and second-degree burns. **First-degree burns** involve only the epidermis and are red and painful, and slight edema (swelling) may be present. They can be caused by sunburn or brief exposure to hot or cold objects, and they heal in a week or so without scarring.

Second-degree burns damage the epidermis and the dermis. If there is minimal dermal damage, symptoms include redness, pain, edema, and blisters. Healing takes approximately 2 weeks, and there is no scarring. If the burn goes deep into the dermis, however, the wound appears red, tan, or white, may take several months to heal, and might scar. In all second-degree burns the epidermis regenerates from epithelial tissue in hair follicles and sweat glands, as well as from the edges of the wound.

Full-thickness burns are also termed **third-degree burns.** The epidermis and the dermis are completely destroyed, and deeper tissue may also be involved. Third-degree burns are often surrounded by first- and second-degree burns. Although the areas that have first- and second-degree burns are painful, the region of third-degree burn is usually painless because of destruction of sensory receptors. Third-degree burns appear white, tan, brown, black, or deep cherry red in color. Skin can regenerate in a third-degree burn only from the edges, and skin grafts are often necessary.

Deep partial-thickness and full-thickness burns take a long time to heal and form scar tissue with disfiguring and debilitating wound contracture. Skin grafts are performed to prevent these complications and to speed healing. In a split skin graft, the epidermis and part of the dermis are removed from another part of the body and are placed over the burn. Interstitial fluid from the burned area nourishes the graft until it becomes vascularized. Meanwhile, the donor tissue produces new epidermis from epithelial tissue in the hair follicles and sweat glands such as occurs in superficial second-degree burns.

Other types of grafts are possible, and in cases in which a suitable donor site is not practical, artificial skin or grafts from human cadavers or from pigs are used. These techniques are often unsatisfactory because the body's immune system recognizes the graft as a foreign substance and rejects it. A solution to this problem is laboratory-grown skin. A piece of healthy skin from the burn victim is removed and placed in a flask with nutrients and hormones that stimulate rapid growth. The skin that is produced

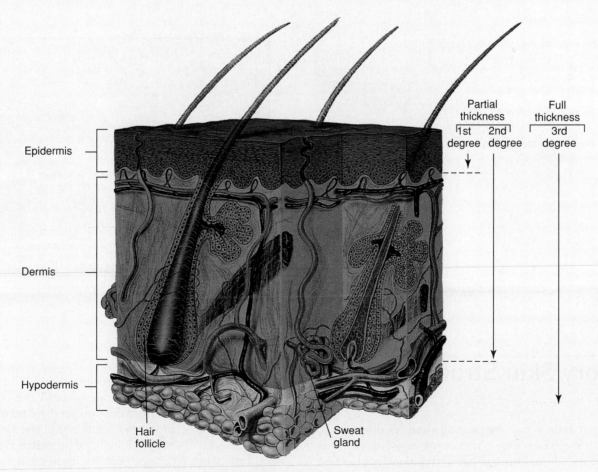

Figure A   Burns   Parts of the skin damaged by burns of different degrees.

(continued)

consists only of epidermis and does not contain glands or hair.

For an adult, the surface area that is burned can be conveniently estimated by "the rule of nines," in which the body is divided into areas that are approximately 9% or multiples of 9% of the total body surface (figure B). For younger patients, surface area relationships are different. For example, in an infant the head and neck are 21% of total surface area, whereas in an adult they are 9%. For burn victims younger than age 15, tables specifically developed for this age group should be consulted.

Figure B   The Rule of Nines   (*a*) In an adult, surface areas can be estimated using the rule of nines: each major area of the body is 9%, or a multiple of 9%, of the total body surface area. (*b*) In infants and children the head represents a larger proportion of surface area. The rule of nines is not as accurate for children, as can be seen in this 5-year-old child.

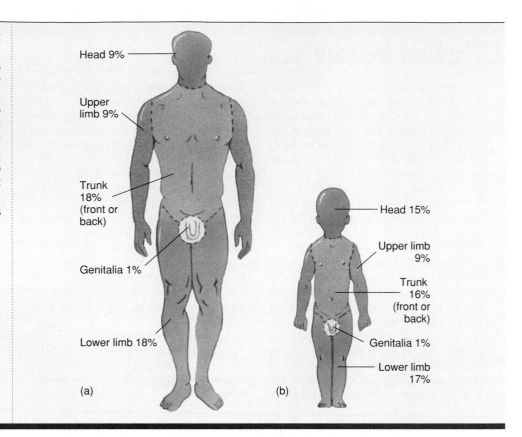

1. Melanosome is produced by Golgi apparatus of the melanocyte.

2. Melanosomes move into melanocyte cell processes.

3. Epithelial cells phagocytize the tips of the melanocyte cell processes.

4. These melanosomes are within epithelial cells.

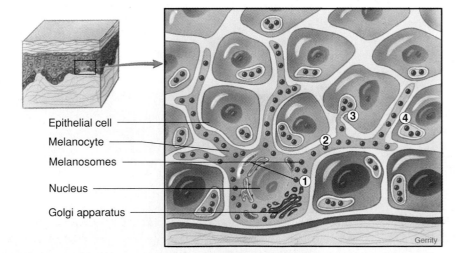

Epithelial cell
Melanocyte
Melanosomes
Nucleus
Golgi apparatus

Figure 5.5   Melanin Transfer from Melanocyte to Keratinocytes
Melanocytes make melanin, which is packaged into melanosomes and transferred to many keratinocytes.

Accessory Skin Structures
Hair

The presence of **hair** is one of the characteristics common to all mammals; if the hair is dense and covers most of the body surface, it is called fur. In humans, hair is found everywhere on the skin except the palms, soles, lips, nipples, parts of the external genitalia, and the distal segments of the fingers and toes.

By the fifth or sixth month of fetal development, delicate unpigmented hair called **lanugo** (lă-nū′gō), which covers the fetus, is produced. Near the time of birth the lanugo of the scalp, eyelids, and eyebrows is replaced by **terminal hairs,** which are long, coarse, and pigmented. The lanugo on the rest of the body is shed and replaced by **vellus** (vel′ŭs) **hairs,** which are short, fine, and usually unpigmented. At puberty much of the vellus hair is replaced by terminal hair, especially in the pubic and axillary regions. The hair of the chest, legs,

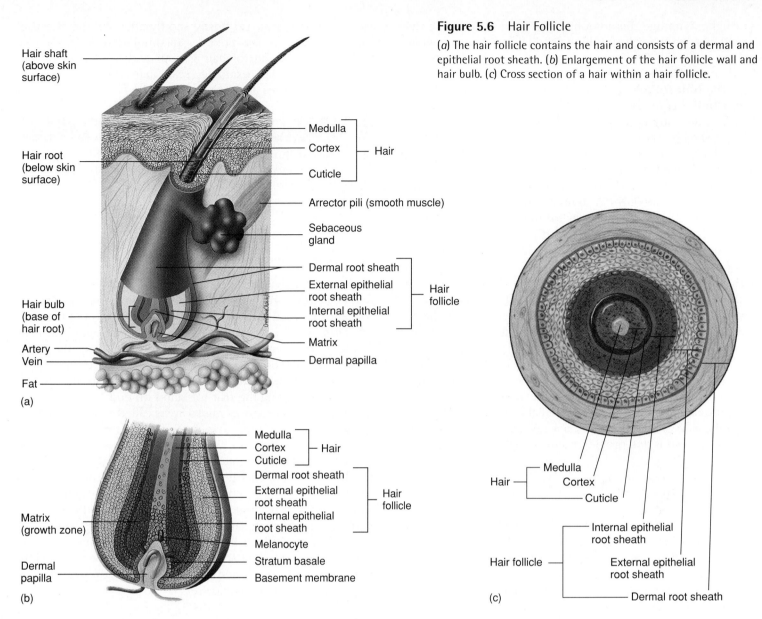

Figure 5.6   Hair Follicle

(*a*) The hair follicle contains the hair and consists of a dermal and epithelial root sheath. (*b*) Enlargement of the hair follicle wall and hair bulb. (*c*) Cross section of a hair within a hair follicle.

and arms is approximately 90% terminal hair in males compared with approximately 35% in females. In males the vellus hairs of the face are replaced by terminal hairs to form the beard. The beard, pubic, and axillary hair are signs of sexual maturity. In addition, pubic and axillary hair may function as wicks for dispersing odors produced by secretions from specialized glands in the pubic and axillary regions. It also has been suggested that pubic hair provides protection against abrasion during intercourse, and axillary hair reduces friction when the arms move.

Hair Structure

A hair is divided into the **shaft** and **root** (figure 5.6*a*). The shaft protrudes above the surface of the skin, and the root is located below the surface. The base of the root is expanded to form the **hair bulb** (figure 5.6*b*). Most of the root and the shaft of the hair are composed of columns of dead keratinized

epithelial cells arranged in three concentric layers: the medulla, the cortex, and the cuticle (figure 5.6*c*). The **medulla** (me-dūl′ă) is the central axis of the hair and consists of two or three layers of cells containing soft keratin. The **cortex** forms the bulk of the hair and consists of cells containing hard keratin. The **cuticle** (kyū′ti-kl) is a single layer of cells that forms the hair surface. The cuticle cells contain hard keratin, and the edges of the cuticle cells overlap like shingles on a roof.

Hard keratin contains more sulfur than does soft keratin. When hair burns, the sulfur combines with hydrogen to form hydrogen sulfide, which produces the unpleasant odor of rotten eggs. In some animals such as sheep, the cuticle edges of the hair are raised and during textile manufacture catch each other and hold together to form threads.

As with the skin, varying amounts and types of melanin cause different shades of hair color. Hair color is controlled by several genes, and dark hair color is not necessarily dominant over light. With age the amount of melanin in hair may decrease, causing the color of the hair to fade or become white (i.e., no melanin). Gray hair is usually a mixture of unfaded, faded, and white hairs.

The **hair follicle** consists of a **dermal root sheath** and an **epithelial root sheath** (see figure 5.6b). The dermal root sheath is the portion of the dermis that surrounds the epithelial root sheath. The epithelial root sheath is divided into an external and an internal part. At the opening of the follicle the external epithelial root sheath has all the strata found in thin skin. Deeper in the hair follicle the number of cells decreases until at the hair bulb only the stratum germinativum is present. This has important consequences for the repair of the skin. If the epidermis and the superficial part of the dermis are damaged, the undamaged part of the hair follicle that lies deep in the dermis can be a source of new epithelium. The internal epithelial root sheath has raised edges that mesh closely with the raised edges of the hair cuticle and hold the hair in place. When a hair is pulled out, the internal epithelial root sheath usually comes out as well and is plainly visible as whitish tissue around the root of the hair.

The hair bulb is an expanded knob at the base of the hair root (see figure 5.6a and b). Inside the hair bulb is a mass of undifferentiated epithelial cells, the **matrix,** which produces the hair and the internal epithelial root sheath. The dermis of the skin projects into the hair bulb as a papilla and contains blood vessels that provide nourishment to the cells of the matrix.

Hair Growth

Hair is produced in cycles that involve a **growth stage** and a **resting stage.** During the growth stage, hair is formed by cells of the matrix that differentiate, become keratinized, and die. The hair grows longer as cells are added at the base of the hair root. Eventually hair growth stops; the hair follicle shortens and holds the hair in place. After the resting period a new cycle begins, and a new hair replaces the old hair, which falls out of the hair follicle. Thus loss of hair normally means that the hair is being replaced. The length of each stage depends on the hair—eyelashes grow for approximately 30 days and rest for 105 days, whereas scalp hairs grow for a period of 3 years and rest for 1–2 years. At any given time an estimated 90% of the scalp hairs are in the growing stage, and there is a normal loss of approximately 100 scalp hairs per day.

The most common kind of permanent hair loss is "pattern baldness." Hair follicles are lost, and the remaining hair follicles revert to producing vellus hair, which is very short, transparent, and for practical purposes invisible. Although more common and more pronounced in certain men, baldness may also occur in women. Genetic factors and the hormone testosterone are involved in causing pattern baldness.

The average rate of hair growth is approximately 0.3 mm per day, although hairs grow at different rates even in the same approximate location. Cutting, shaving, or plucking hair does not alter the growth rate or the character of the hair, but hair can feel coarse and bristly shortly after shaving because the short hairs are less flexible. Maximum hair length is determined by the rate of hair growth and the length of the growing phase. For example, scalp hair can become very long, but eyelashes are short.

3   P R E D I C T

Marie Antoinette's hair supposedly turned white overnight after she heard she would be sent to the guillotine. Explain why you believe or disbelieve this story.

✔ *Answer in Appendix F*

Muscles

Associated with each hair follicle are smooth muscle cells, the **arrector pili** (ă-rek′tōr pī′lī), which extend from the dermal root sheath of the hair follicle to the papillary layer of the dermis (see figure 5.6a). Normally the hair follicle and the hair inside it are at an oblique angle to the surface of the skin. When the arrector pili muscles contract, however, they pull the follicle into a position more perpendicular to the surface of the skin, causing the hair to "stand on end." Movement of the hair follicles produces raised areas called "goose flesh," or "goose bumps."

Contraction of the arrector pili muscles occurs in response to cold or to frightening situations, and in animals with fur the response increases the thickness of the fur. When the response results from cold temperatures, it is beneficial because the fur traps more air and thus becomes a better insulator. In a frightening situation the animal appears larger and more ferocious, which might deter an attacker. It is unlikely that humans, with their sparse amount of hair, derive any important benefit from either response and probably retain this trait as an evolutionary holdover.

Glands

The major glands of the skin are the **sebaceous** (sē-bā′shŭs) **glands** and the **sweat glands** (figure 5.7). Sebaceous glands, located in the dermis, are simple or compound alveolar glands that produce **sebum** (sē′bŭm), an oily, white substance rich in lipids. Because sebum is released by the lysis and death of the secretory cells, sebaceous glands are classified as holocrine glands (see chapter 4). Most sebaceous glands are connected by a duct to the upper part of the hair follicles from which the sebum oils the hair and the skin surface. This prevents drying and provides protection against some bacteria. A few sebaceous glands located in the lips, in the eyelids (meibomian glands), and in the genitalia are not associated with hairs but open directly onto the skin surface.

There are two types of sweat glands, and at one time it was believed that one released its secretions in a merocrine fashion and the other in an apocrine fashion (see chapter 4). Accordingly, they were called merocrine and apocrine sweat

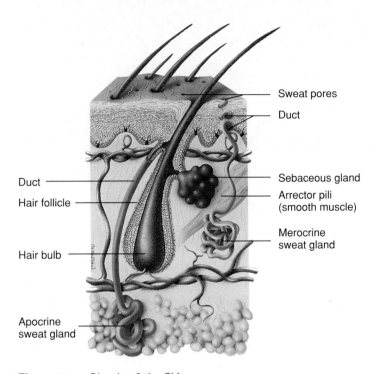

Sweat pores

Duct

Duct

Hair follicle

Hair bulb

Apocrine
sweat gland

Sebaceous gland

Arrector pili
(smooth muscle)

Merocrine
sweat gland

Figure 5.7   Glands of the Skin

Merocrine sweat glands open to the surface of the skin. Apocrine sweat glands and sebaceous glands open into hair follicles.

glands. It is now known that apocrine sweat glands release their secretions in a merocrine fashion. Traditionally, they are still referred to as apocrine sweat glands.

Merocrine (mer′ō-krin) **sweat glands,** the most common type, are simple coiled tubular glands that open directly onto the surface of the skin through sweat pores (see figure 5.7). Merocrine sweat glands can be divided into two parts: the deep coiled portion, which is located mostly in the dermis, and the duct, which passes to the surface of the skin. The coiled part of the gland produces an isotonic fluid that is mostly water but also contains some salts (mainly sodium chloride) and small amounts of ammonia, urea, uric acid, and lactic acid. As this fluid moves through the duct, sodium chloride moves by active transport from the duct back into the body, conserving salts. The resulting hyposmotic fluid that leaves the duct is called **sweat.** When the body temperature starts to rise above normal levels, the sweat glands produce sweat, which evaporates and cools the body. Sweat also can be released in the palms, soles, and axillae as a result of emotional stress.

Merocrine sweat glands are most numerous in the palms of the hands and the soles of the feet but are absent from the margin of the lips, the labia minora, and the tips of the penis and clitoris. Only a few mammals such as humans and horses have merocrine sweat glands in the hairy skin. Dogs, on the other hand, keep cool by water lost through panting instead of sweating.

Apocrine (ap′ō-krin) **sweat glands** are compound coiled tubular glands that usually open into hair follicles superficial to the opening of the sebaceous glands (see figure 5.7). These glands are found in the axillae and genitalia (scrotum and labia majora) and around the anus. They become active at puberty as a result of the influence of sex hormones. Their secretions contain organic substances such as 3-methyl-2-hexenoic acid that are essentially odorless when first released, but that are quickly metabolized by bacteria to cause what commonly is known as body odor. Many mammals use scent as a means of communication, and it has been suggested that the activity of apocrine sweat glands may be a sign of sexual maturity.

Other skin glands include the ceruminous glands of the external auditory meatus, which produce cerumen (earwax), and the mammary glands.

Nails

The distal joints of primate digits have nails, whereas most other mammals have claws or hooves. The nails protect the ends of the digits, aid in manipulation and grasping of small objects, and are used for scratching.

The **nail** consists of the proximal **nail root** and the distal **nail body** (figure 5.8*a*). The nail root is covered by skin, and the nail body is the visible portion of the nail. The lateral and proximal edges of the nail are covered by skin called the **nail fold,** and the edges are held in place by the **nail groove** (figure 5.8*b*). The stratum corneum of the nail fold grows onto the nail body as the **eponychium** (ep-ō-nik′ē-ŭm), or cuticle. Beneath the free edge of the nail body is the **hyponychium** (hī′pō-nik′ē-ŭm), a thickened region of the stratum corneum (figure 5.8*c*).

The nail root and the nail body attach to the **nail bed,** the proximal portion of which is the **nail matrix.** Only the stratum germinativum is present in the nail bed and nail matrix. The nail matrix is thicker than the nail bed and produces most of the nail, although the nail bed does contribute. The nail bed is visible through the clear nail and appears pink because of blood vessels in the dermis. A small part of the nail matrix, the **lunula** (lū′nū-lă), is seen through the nail body as a whitish, crescent-shaped area at the base of the nail. The lunula, seen best on the thumb, appears white because the blood vessels cannot be seen through the thicker nail matrix.

The nail is stratum corneum that contains hard keratin. The nail cells are produced in the nail matrix and pushed distally over the nail bed. Nails grow at an average rate of 0.5–1.2 mm per day, and fingernails grow more rapidly than toenails. Nails, like hair, grow from the base. Unlike hair, they grow continuously throughout life and do not have a resting phase.

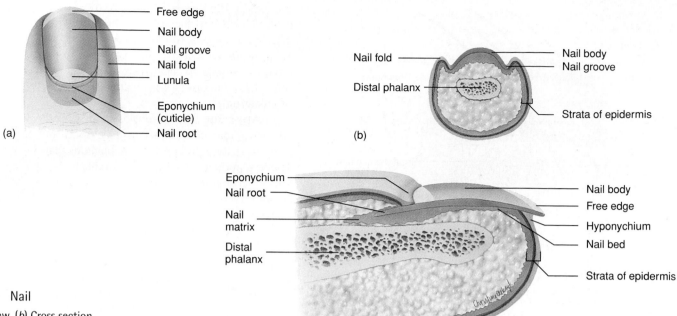

Figure 5.8   Nail

(*a*) Dorsal view. (*b*) Cross section.
(*c*) Longitudinal section.

Functions of the Integumentary System
Protection

The integumentary system performs many protective functions. The intact skin forms a physical barrier that prevents the entry of microorganisms and other foreign substances into the body. Secretions from skin glands are slightly acidic and produce an environment unsuitable for some microorganisms; the skin also contains components of the immune system that act against microorganisms (see chapter 22). The skin is a permeability barrier that determines what substances can diffuse into or out of the body surface, and it is especially important in preventing water loss.

> **Clinical Note**
>
> Some lipid-soluble substances readily pass through the epidermis. Lipid-soluble medications can be administered by applying them to the skin, after which the medication slowly diffuses through the skin into the blood. For example, a medicine used to prevent sea sickness is impregnated into a patch, and the patch is placed behind the ear. Nicotine patches are used to help reduce withdrawal symptoms in those attempting to quit smoking.

The integumentary system provides protection against abrasion. As the outer cells of the stratum corneum are desquamated, they are replaced by cells from the stratum basale. Calluses develop in areas subject to heavy friction or pressure. Hair on the head and in the pubic and axillary regions also protects against abrasion. Hair and melanin in the skin protect against ultraviolet radiation, which can damage the DNA of cells. When exposure to ultraviolet light increases, the amount of melanin in the skin increases, providing additional protection. The eyebrows prevent sweat from entering the eyes; the eyelashes protect the eyes from foreign objects; hair in the nose and ears prevents the entry of dust and other foreign materials; and nails protect the ends of the digits from damage and can be used in defense.

Temperature Regulation

Body temperature tends to increase as a result of exercise, fever, or an increase in environmental temperature. Homeostasis is maintained by the loss of excess heat. The blood vessels (arterioles) in the dermis dilate and allow more blood to flow through the skin, thus transferring heat from deeper tissues to the skin. To counteract environmental heat gain or to get rid of excess heat produced by the body, sweat is produced. The sweat spreads over the surface of the skin, and as it evaporates, heat is lost from the body.

If body temperature begins to drop below normal, heat can be conserved by a decrease in the diameter of dermal blood vessels, thus reducing blood flow to the skin. With less warm blood flowing through the skin, however, the skin temperature decreases. If the skin temperature drops below approximately 15°C, blood vessels dilate, which helps to prevent tissue damage from the cold.

Contraction of the arrector pili muscles causes hair to stand on end, but with the sparse amount of hair covering the body, this does not significantly reduce heat loss in humans. Hair on the head, however, is an effective insulator. General temperature regulation is considered in chapter 25.

4	P R E D I C T

You may have noticed that, on very cold winter days, people's ears and noses turn red. Can you explain why this happens?

✔ *Answer in Appendix F*

Vitamin D Production

Vitamin D synthesis begins in skin exposed to ultraviolet light, and humans can produce all the vitamin D they require by this process if enough ultraviolet light is available. Because humans live indoors and wear clothing, however, their exposure to ultraviolet light may not be adequate for the manufacture of sufficient vitamin D. For example, people living in cold climates may have inadequate exposure to ultraviolet light because they remain indoors or are covered by warm clothing when outdoors. Fortunately, vitamin D can also be ingested and absorbed in the intestine. Natural sources of vitamin D are liver (especially fish liver), egg yolks, and dairy products (e.g., butter, cheese, and milk). In addition, the diet can be supplemented with vitamin D in fortified milk or vitamin pills.

Vitamin D synthesis begins when the precursor molecule, 7-dehydrocholesterol (7-dē-hī′drō-kō-les′ter-ol) is exposed to ultraviolet light and is converted into cholecalciferol (kō′lē-kal-sif′er-ol). The cholecalciferol is released into the blood and modified by hydroxylation (hydroxide ion is added) in the liver and kidneys to form active vitamin D (calcitriol; kal-si-trī′ol). Vitamin D functions as a hormone to stimulate uptake of calcium and phosphate from the intestines, to promote their release from bones, and to reduce calcium loss from the kidneys, resulting in increased blood calcium and phosphate levels. Adequate levels of these minerals are necessary for normal bone metabolism (see chapter 6), and calcium is required for normal nerve and muscle function (see chapters 9 and 10).

Sensation

The integumentary system is well supplied with sensory receptors, including touch receptors in the epidermis and dermal papillae and pain, heat, cold, and pressure receptors in the dermis and deeper tissues. Hair follicles (but not the hair) are well innervated, and movement of the hair can be detected by sensory receptors surrounding the base of the hair follicle. Sensory receptors are discussed in more detail in chapter 15.

Excretion

Excretion is the removal of waste products from the body. In addition to water and salts, sweat contains a small amount of waste products, such as urea, uric acid, and ammonia. Large amounts of sweat can be lost from the body, especially during vigorous exercise in a hot environment, and lost water and salts must be replaced to restore fluid and electrolyte homeostasis. Compared with the excretory organs, however, the quantity of waste products eliminated in the sweat is insignificant.

Effects of Aging on the Integumentary System

As the body ages, the blood flow to the skin is reduced, and the skin becomes thinner, is more easily damaged, and repairs more slowly. Elastic fibers in the dermis decrease in number and diameter, and the skin tends to sag. A loss of subcutaneous tissue, especially in the face, also causes sagging, wrinkled skin.

A decrease in the activity of sebaceous and sweat glands results in dry skin and poor thermoregulatory ability. The decrease in ability to sweat can contribute to death from heat prostration in elderly individuals who do not take proper precautions.

The number of functioning melanocytes generally decreases, but in some localized areas, especially on the hands and the face, melanocytes increase in number to produce age spots. (Age spots are different from freckles, which are caused by an increase in melanin production and not an increase in melanocyte numbers.) White or gray hairs also occur because of a decrease in or lack of melanin production.

Skin that is exposed to sunlight appears to age more rapidly than nonexposed skin. This effect is observed on areas of the body such as the face and hands that receive sun exposure (figure 5.9). The effects of chronic sun exposure on the skin, however, are different from the effects of normal aging. In skin exposed to sunlight, normal elastic fibers are replaced by an interwoven mat of thick elasticlike material, the number of collagen fibers decreases, and the ability of keratinocytes to divide is impaired.

Clinical Note

Retin-A (tretinoin; tret′i-nō-in) is a vitamin A derivative that is being used to treat skin wrinkles. It appears to be effective in treating fine wrinkles on the face, such as those caused by long-term exposure to the sun, but is not effective in treating deep lines. One ironic side effect of Retin-A use is increased sensitivity to the sun's ultraviolet rays. Doctors prescribing this cream caution their patients to always use a sun block when they are going to be outdoors.

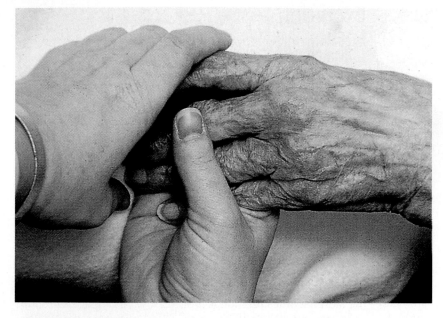

Figure 5.9   Skin Changes with Age

As we age, skin loses elasticity and wrinkles form.

Clinical Focus   Clinical Disorders of the Integumentary System

The Integumentary System as a Diagnostic Aid

The integumentary system is useful in diagnosis because it is easily observed and often reflects events occurring in other parts of the body. For example, **cyanosis** (sī-ă-nō′sis), a bluish color to the skin that results from decreased blood oxygen content, is an indication of impaired circulatory function or respiratory function. When red blood cells wear out, they are broken down, and part of their contents are excreted by the liver as bile pigments into the intestine. **Jaundice** (jawn′dis), a yellowish skin color, occurs when there are excess bile pigments in the blood. If the liver is damaged by a disease such as viral hepatitis, bile pigments are not excreted and accumulate in the blood.

Rashes and lesions in the skin can be symptomatic of problems elsewhere in the body. For example, scarlet fever results from a bacterial infection in the throat. The bacteria releases a toxin into the blood that causes the pink-red rash for which this disease was named. In allergic reactions (see chapter 22), a release of histamine into the tissues produces swelling and reddening. The development of a rash (hives) in the skin can indicate an allergy to ingested foods or drugs such as penicillin.

The condition of the skin, hair, and nails is affected by nutritional status. In vitamin A deficiency the skin produces excess keratin and assumes a characteristic sandpaper texture, whereas in iron-deficiency anemia the nails lose their normal contour and become flat or concave (spoon-shaped).

The hair concentrates many substances that can be detected by laboratory analysis, and comparison of a patient's hair to a "normal" hair can be useful in diagnosis. For example, lead poisoning results in high levels of lead in the hair. The use of hair analysis as a screening test to determine the health or nutritional status of an individual remains unreliable, however.

Bacterial Infections

Staphylococcus aureus is commonly found in pimples, boils, and carbuncles and causes **impetigo** (im-pe-tī′gō), a disease of the skin that usually affects children and that is characterized by small blisters containing pus that easily rupture and form a thick, yellowish crust. *Streptococcus pyogenes* causes **erysipelas** (er-i-sip′ĕ-las), swollen red patches in the skin. Burns are often infected by *Pseudomonas aeruginosa*, producing a characteristic blue-green pus caused by bacterial pigment.

Acne is a disorder of the hair follicles and sebaceous glands that affects almost everyone at some time or another. Although the exact cause of acne is unknown, four factors are believed to be involved: hormones, sebum, abnormal keratinization within the hair follicle, and the bacterium *Propionibacterium acnes.* The lesions apparently begin with a hy-perproliferation of the hair follicle epidermis, and many cells are desquamated. These cells are abnormally sticky and adhere to one another to form a mass of cells mixed with sebum that blocks the hair follicle. During puberty, hormones, especially testosterone, stimulate the sebaceous glands to increase sebum production. Because both the adrenal gland and the testes produce testosterone, the effect is seen in both males and females. An accumulation of sebum behind the blockage produces a whitehead, which may continue to develop into a blackhead or a pimple. A blackhead results if the opening of the hair follicle is pushed open by the accumulating cornified cells and sebum. Although there is general agreement that dirt is not responsible for the black color of blackheads, the exact cause of the black color is disputed. A pimple develops if the wall of the hair follicle ruptures. Once the wall of the follicle ruptures, *P. acnes* and other microorganisms stimulate an inflammatory response that results in the formation of a red pimple filled with pus. If tissue damage is extensive, scarring occurs.

Viral Infections

Some of the well-known viral infections of the skin include **chickenpox** (varicella-zoster), **measles**, **German measles** (rubella), and **cold sores** (herpes simplex). **Warts**, which are caused by a viral infection of the epidermis, are generally harmless and usually disappear without treatment.

(continued)

Fungal Infections

Ringworm is a fungal infection that affects the keratinized portion of the skin, hair, and nails and produces patchy scaling and an inflammatory response. The lesions are often circular with a raised edge and in ancient times were thought to be caused by worms. Several species of fungus cause ringworm in humans and are usually described by their location on the body; in the scalp the condition is ringworm, in the groin it is jock itch, and in the feet it is athlete's foot.

Decubitus Ulcers

Decubitus (dē-kyū′bi-tŭs) **ulcers,** also known as bedsores or pressure sores, develop in patients who are immobile (e.g., bedridden or confined to a wheelchair). The weight of the body, especially in areas over bony projections such as the hip bones and heels, compresses tissues and causes **ischemia** (is-kē′mē-ă), or reduced circulation. The consequence is destruction, or **necrosis** (nĕ-krō′sis), of the hypodermis and deeper tissues that is followed by death of the skin. Once the skin dies, microorganisms gain entry to produce an infected ulcer.

Bullae

Bullae (bul′ē) are fluid-filled areas in the skin that develop when tissues are damaged, and the resultant inflammatory response produces edema. Infections or physical injuries can cause bullae or lesions in different layers of the skin.

Psoriasis

The cause of **psoriasis** (sō-rī′ă-sis) is unknown, although there may be a genetic component. An increase in mitotic activity in the stratum basale, abnormal keratinization, and elongation of the dermal papillae toward the skin surface result in a thicker-than-normal stratum corneum that desquamates to produce large, silvery scales. If the scales are scraped away, bleeding occurs from the blood vessels at the top of the dermal papillae. Psoriasis is a chronic disease that can be controlled but as yet has no cure.

Eczema and Dermatitis

Eczema (ek′zĕ-mă) and **dermatitis** (der-mă-tī′tis) are inflammatory conditions of the skin. Cause of the inflammation can be allergy; infection; poor circulation; or exposure to physical factors, such as chemicals, heat, cold, or sunlight.

Birthmarks

Birthmarks are congenital (present at birth) disorders of the capillaries in the dermis of the skin. Usually they are only of concern for cosmetic reasons. A **strawberry birthmark** is a mass of soft, elevated tissue that appears bright red to deep purple in color. In 70% of patients, strawberry birthmarks disappear spontaneously by the age of 7. **Port-wine stains** appear as flat, dull red or bluish patches that persist throughout life.

Vitiligo

Vitiligo (vit-i-lī′gō) is the development of patches of white skin because the melanocytes in the affected area are destroyed, apparently by an autoimmune response (see chapter 22).

Moles

A **mole** is an elevation of the skin that is variable in size and is often pigmented and hairy. Histologically, a mole is an aggregation, or "nest," of melanocytes in the epidermis or dermis. They are a normal occurrence, and most people have 10–20 moles, which appear in childhood and enlarge until puberty.

Cancer

Skin cancer is the most common type of cancer (figure C). Although chemicals and radiation (x-rays) are known to induce cancer, the development of skin cancer is most often associated with exposure to ultraviolet radiation from the sun, and, consequently, most skin cancers develop on the face or neck. The group of people most likely to have skin cancer are fair-skinned (i.e., have less protection from the sun) or are older than 50 (have had long exposure to the sun).

Basal cell carcinoma (kar-si-nō′mă), the most frequent skin cancer, begins in the stratum basale and extends into the dermis to produce an open ulcer. Surgical removal or radiation therapy cures this type of cancer, and fortunately there is little danger that the cancer will spread, or metastasize (mĕ-tas′tă-sīz), to other areas of the body if treated in time. **Squamous cell carcinoma** develops from stratum spinosum keratinocytes that continue to divide as they produce keratin. Typically the result is a nodular, keratinized tumor confined to the epidermis, but it can invade the dermis, metastasize, and cause death. **Malignant melanoma** (mel′ă-nō′mă) is a less common form of skin cancer that arises from melanocytes, usually in a preexisting mole. The melanoma can appear as a large, flat, spreading lesion or as a deeply pigmented nodule. Metastasis is common, and, unless diagnosed and treated early in development, this cancer is often fatal. Other types of skin cancer are possible (e.g., metastasis from other parts of the body to the skin).

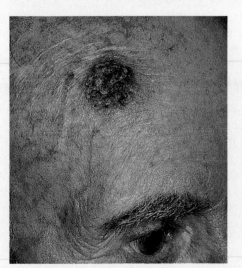

(a)

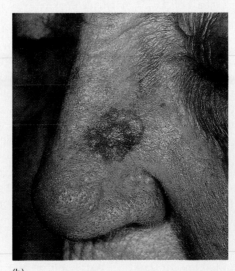

(b)

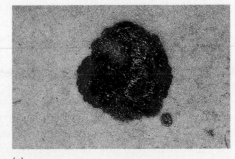

(c)

Figure C Cancer of the Skin

(*a*) Basal cell carcinoma.
(*b*) Squamous cell carcinoma.
(*c*) Malignant melanoma.

Systems Pathology

burns

BURNS

Mr. S is a 23-year-old man who had difficulty falling asleep at night. He often stayed up late watching TV or reading until he fell asleep. Mr. S is also a chain smoker. One night he took several sleeping pills. Unfortunately, he fell asleep before putting out his cigarette, which started a fire. As a result, Mr. S was severely burned, receiving full-thickness and partial-thickness burns (figure D*a*). He was rushed to the emergency room and was eventually transferred to a burn unit.

For the first day after his accident, his condition was critical because he went into shock. Administration of large volumes of intravenous fluid stabilized his condition. As part of his treatment, Mr. S was also given a high-protein, high-calorie diet.

A week later, dead tissue was removed from the most serious burns (figure D*b*), and a skin graft was performed. Despite the use of topical antimicrobial drugs and sterile bandages, some of the burns became infected. An additional complication was the development of a venous thrombosis in his leg.

Although the burns were painful and the treatment was prolonged, Mr. S made a full recovery. He no longer smokes.

BACKGROUND INFORMATION

When large areas of skin are severely burned, systemic effects are produced that can be life-threatening. Within minutes of a major burn injury, there is increased permeability in the capillaries, which are the small blood vessels in which fluid, gases, nutrients, and waste products are normally exchanged between the blood and tissues. This increased permeability occurs at the burn site and throughout the body. As a result, fluid and electrolytes (see chapter 2) are lost from the burn wound and into tissue spaces. The loss of fluid decreases blood volume, which decreases the ability of the heart to pump blood. The resulting decrease in blood delivery to tissues can cause tissue damage, shock, and even death. Treatment consists of administering intravenous fluid at a faster rate than it leaks out of the capillaries. Although this can reverse the shock and prevent death, fluid continues to leak into tissue spaces causing pronounced **edema,** a swelling of the tissues.

Typically, after 24 hours, capillary permeability returns to normal, and the amount of intravenous fluid administered can be greatly decreased. How burns result in capillary permeability changes is not well understood. It is clear that following a burn immunologic and metabolic changes occur that affect not only capillaries, but the rest of the body as well. For example, mediators of inflammation (see chapter 4), which are released in response to the tissue damage, contribute to changes in capillary permeability throughout the body.

Substances released from the burn may also play a role in causing cells to function abnormally. Burn injuries result in an almost immediate hypermetabolic state that persists until wound closure. Also contributing to the increased metabolism is a resetting of the temperature control center in the brain to a higher temperature and an increase in the hormones released by the endocrine system. For example, epinephrine and norepinephrine from the adrenal glands increase cell metabolism. Compared with a normal body temperature of approximately 37°C (98.6°F), a body temperature of 38.5°C (101.3°F) is typical in burn patients, despite the higher loss of water by evaporation from the burn.

In severe burns, the increased metabolic rate can result in weight loss as great as 30%–40% of the patient's preburn weight. To help compensate, caloric intake may double or even triple. In addition, the need for protein, which is necessary for tissue repair, is greater.

The skin normally maintains homeostasis by preventing the entry of microorganisms. Because burns damage and even completely destroy the skin, microorganisms can cause infections. For this reason, burn patients are maintained in an aseptic environment, which attempts to prevent the entry of microorganisms into the wound. They are also given antimicrobial drugs, which kill microorganisms or suppress their growth. **Debridement,** (dā-brēd-mon′), the removal of dead tissue from the burn, helps to prevent infections by cleaning the wound and removing tissue in which infections could develop. Skin grafts, performed within a week of the injury, also prevent infections by closing the wound and preventing the entry of microorganisms.

Despite these efforts, however, infections still are the major cause of death of burn victims. Depression of the immune system during the first or second week after the injury contributes to the high infection rate. The thermally altered tissue is recognized as a foreign substance that stimulates the immune system. As a result, the immune system is overwhelmed as immune system cells become less effective and production of the chemicals that normally provide resistance to infections decreases (see chapter 22). The greater the magnitude of the burn, the greater the depression of the immune system, and the greater the risk of infection.

Venous thrombosis, the development of a clot in a vein, is also a complication of burns. Blood normally forms a clot when exposed to damaged tissue, such as at a burn site, but the clot can block blood flow, resulting in tissue destruction. In addition, the concentration of chemicals in the blood that cause clotting increases for two reasons: loss of fluid from the burn and the increased release of clotting factors from the liver.

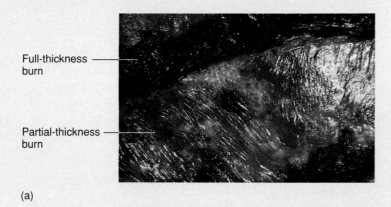

(a)

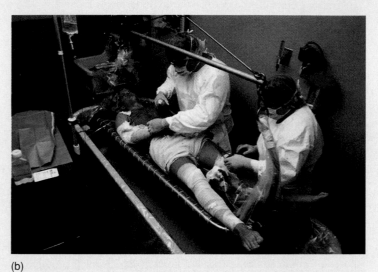

(b)

Full-thickness burn

Partial-thickness burn

Figure D   Burn Victim

(*a*) Partial and full-thickness burns. (*b*) Patient in a burn unit.

System Interactions	
System	**Interactions**
Skeletal	Red bone marrow replaces red blood cells (erythrocytes) destroyed in the burnt skin.
Muscular	Loss of muscle mass resulting from the hypermetabolic state caused by the burn.
Nervous	Pain is sensed in the partial-thickness burns. The temperature-regulatory center in the brain is set to a higher temperature, contributing to increased body temperature. Abnormal K^+ ion concentrations disturb normal nervous system activity: elevated levels are caused by release of K^+ ions from damaged tissues; low levels can be caused by rapid loss of these K^+ ions in fluid from the burn.
Endocrine	Increased secretion of epinephrine and norepinephrine from the adrenal gland in response to the injury contributes to increased body temperature by increasing cell metabolism.
Cardiovascular	Increased capillary permeability causes decreased blood volume, resulting in decreased blood delivery to tissues, edema, and shock. The pumping effectiveness of the heart is impaired by electrolyte imbalance and substances released from the burn. Increased blood clotting causes venous thrombosis. Preferential delivery of blood to the injury promotes healing.
Lymphatic and Immune	Inflammation increases in response to tissue damage. Later, depression of the immune system can result in infection.
Respiratory	Airway obstruction caused by edema. Increased respiration rate caused by increased metabolism and lactic acid buildup.
Digestive	Decreased blood delivery as a result of the burn causes degeneration of the intestinal lining and liver. Bacteria from the intestine can cause systemic infections. The liver releases blood-clotting factors in response to the injury. Increased nutrients necessary to support increased metabolism and for repair of the integumentary system are absorbed.
Urinary	The kidneys compensate for the increased fluid loss caused by the burn by greatly reducing or even stopping urine production. Decreased blood volume causes decreased blood flow to the kidneys, which reduces urine output, but can cause kidney tissue damage. Hemoglobin, released from red blood cells damaged in the burnt skin, can obstruct urine flow in the kidneys.

Summary

The integumentary system consists of the skin, hair, nails, and a variety of glands. The skin protects, helps regulate body temperature, produces vitamin D, and detects stimuli.

Hypodermis

1. Located beneath the dermis, the hypodermis is loose connective tissue that contains collagen and elastin fibers.
2. The hypodermis attaches the skin to underlying structures and is a site of fat storage.

Skin
Dermis

1. The dermis is connective tissue divided into two layers.
2. The reticular layer is the main layer. It is dense, irregular connective tissue consisting mostly of collagen.
3. The papillary layer has projections called papillae and is loose connective tissue that is well supplied with capillaries.

Epidermis

1. The epidermis is stratified squamous epithelium divided into five strata.
2. The stratum basale consists of keratinocytes, which produce the cells of the more superficial strata.
3. The stratum spinosum consists of several layers of cells held together by many desmosomes. The stratum basale and the stratum spinosum are sometimes called the stratum germinativum.
4. The stratum granulosum consists of cells filled with granules of keratohyalin. Cell death occurs in this stratum.
5. The stratum lucidum consists of a layer of dead transparent cells.
6. The stratum corneum consists of many layers of dead squamous cells. The most superficial cells are desquamated.
7. Keratinization is the transformation of the living cells of the stratum basale into the dead squamous cells of the stratum corneum.
   - Keratinized cells are filled with keratin and have a protein envelope, both of which contribute to structural strength. The cells are also held together by many desmosomes.
   - Intercellular spaces are filled with lipids that contribute to the impermeability of the epidermis to water.
8. Soft keratin is found in skin and the inside of hairs, whereas hard keratin occurs in nails and the outside of hairs. Hard keratin makes cells more durable, and these cells do not desquamate.

Thick and Thin Skin

1. Thick skin has all five epithelial strata. The dermis under thick skin produces fingerprints and footprints.
2. Thin skin contains fewer cell layers per stratum, and the stratum lucidum is usually absent. Hair is found only in thin skin.

Skin Color

1. Melanocytes produce melanin inside melanosomes and then transfer the melanin to keratinocytes. The size and distribution of melanosomes determine skin color. Melanin production is determined genetically but can be influenced by hormones and ultraviolet light (tanning).
2. Carotene, an ingested plant pigment, can cause the skin to appear yellowish.
3. Increased blood flow produces a red skin color, whereas a decreased blood flow causes a pale skin. Decreased oxygen content in the blood results in a bluish color called cyanosis.

Accessory Skin Structures
Hair

1. Lanugo, fetal hair, is replaced near the time of birth by terminal hairs (eyelids, eyebrows, and scalp) and vellus hairs. At puberty vellus hairs can be replaced with terminal hairs.
2. Hair is dead keratinized epithelial cells consisting of a central axis of cells with soft keratin, the medulla, which is surrounded by a cortex of cells with hard keratin. The cortex is covered by the cuticle, a single layer of cells filled with hard keratin.
3. Hair color is determined by the amount and kind of melanin present.
4. A hair has three parts: the shaft, the root, and the hair bulb.
5. The hair bulb produces the hair in cycles involving a growth stage and a resting stage.

Muscles

Contraction of the arrector pili muscles, which are smooth muscles, causes hair to "stand on end" and produces "goose flesh."

Glands

1. Sebaceous glands produce sebum, which oils the hair and the surface of the skin.
2. Merocrine sweat glands produce sweat that cools the body. Apocrine sweat glands produce an organic secretion that can be broken down by bacteria to cause body odor.
3. Other skin glands include ceruminous glands and the mammary glands.

Nails

1. The nail consists of a nail root and a nail body resting on the nail bed.
2. Part of the nail root, the nail matrix, produces the nail body, which is several layers of cells containing hard keratin.

Functions of the Integumentary System
Protection

The skin prevents the entry of microorganisms, acts as a permeability barrier, and provides protection against abrasion and ultraviolet light.

Temperature Regulation

1. Through dilation and constriction of blood vessels, the skin controls heat loss from the body.
2. Sweat glands produce sweat that evaporates and lowers body temperature.

Vitamin D Production

1. Skin exposed to ultraviolet light produces cholecalciferol that is modified in the liver and then in the kidneys to form active vitamin D.
2. Vitamin D increases blood calcium levels by promoting calcium uptake from the intestine, release of calcium from bone, and reduction of calcium loss from the kidneys.

Sensation

The skin contains sensory receptors for pain, touch, hot, cold, and pressure that allow proper response to the environment.

Excretion

Skin glands remove small amounts of waste products (e.g., urea, uric acid, and ammonia) but are not important in excretion.

Effects of Aging on the Integumentary System

1. As the body ages, blood flow to the skin is reduced, the skin becomes thinner, and elasticity is lost.
2. Sweat and sebaceous glands are less active, and the number of melanocytes decreases.

Content Review

1. Name the components of the integumentary system.
2. List the functions of the integumentary system.
3. Describe the structure and the function of the hypodermis.
4. What type of tissue is the dermis? How is the dermis different from the hypodermis?
5. Name the two layers of the dermis, and contrast them to each other.
6. What are cleavage lines in the skin, and why are they important?
7. What kind of tissue is the epidermis? Name and describe the five strata of the epidermis. In which strata are new cells formed by mitosis? Which strata have live cells, and which have dead cells?
8. Define keratinization. Describe the structural features that result from keratinization and make the epidermis structurally strong and resistant to water loss.
9. Distinguish between soft and hard keratin, and state where each type of keratin is found.
10. Compare thick and thin skin. Is hair found in thick skin or thin skin?
11. What is a callus? A corn?
12. Which cells of the epidermis produce melanin? What happens to the melanin once it is produced? What factors determine the amount of melanin produced in the skin?
13. How do melanin, carotene, and blood affect skin color?
14. When and where are lanugo, vellus, and terminal hairs found in the skin?
15. Define the root, shaft, and hair bulb of a hair. Describe the three parts of the root or shaft seen in cross section.
16. What determines hair color?
17. Describe the parts of a hair follicle. How is the epithelial root sheath important in the repair of the skin?
18. In what part of a hair does growth take place? What are the stages of hair growth?
19. What determines the length of hair? How does baldness occur?
20. What happens when the arrector pili muscles of the skin contract?
21. What secretion is produced by the sebaceous glands? What is the function of the secretion?
22. Which glands of the skin are responsible for cooling the body? Which glands are involved with the production of body odor?
23. Name the parts of a nail. Which part produces the nail? What is the lunula?
24. How does the skin provide protection?
25. How does the skin assist in the regulation of body temperature?
26. Where is cholecalciferol produced and then modified into vitamin D? What are the functions of the vitamin D?
27. What kind of sensory receptors are found in the skin, and why are they important?
28. What substances are excreted by skin glands? Is the skin an important site of excretion?
29. List the changes that occur in the skin with increasing age.

Develop Your Reasoning Skills

1. A woman has stretch marks on her abdomen, yet she states that she has never been pregnant. Is this possible?
2. The skin of infants is more easily penetrated and injured by abrasion than that of adults. Based on this fact, which stratum of the epidermis is probably much thinner in infants than that in adults?
3. Melanocytes are found primarily in the stratum basale of the epidermis. In reference to their function, why does this location make sense?
4. Harry Fastfeet, a white man, jogs on a cold day. What color would you expect his skin to be (a) just before starting to run, (b) during the run, and (c) 5 min after the run?
5. Why are your eyelashes not a foot long? Your fingernails?
6. Given what you know about the cause of acne, propose some ways to prevent or treat the disorder.
7. A patient has an ingrown toenail, a condition in which the nail grows into the nail fold. Would cutting the nail away from the nail fold permanently correct this condition? Why or why not?

Web Site Link

For a listing of the most current web sites related to this chapter, please visit the Seeley home page at:
http://www.mhhe.com/biosci/ap/seeleyap/

Chapter Six

Skeletal System: Bones and Bone Tissue

Objectives

1. Name the components of the skeletal system, and list its major functions.

2. Define the different types of cartilage cells, and describe cartilage matrix.

3. Describe the perichondrium and the formation of cartilage by appositional and interstitial growth.

4. Name the major bone shapes, and describe their anatomy.

5. Describe the composition and organization of bone matrix.

6. Define the different types of bone cells, and describe their functions and their origins.

7. Describe the features that characterize woven, cancellous, and compact bone.

8. Name the two patterns of bone formation, and describe the features of each.

9. Describe bone growth, and explain how it differs from the growth of cartilage.

10. List the nutritional and hormonal requirements for bone growth.

11. Explain how bone remodeling occurs, and describe the conditions in which it occurs.

12. Describe the effects of mechanical strain and of inadequate stress on bone.

13. Describe the process of bone repair, the cells involved, and the types of tissue produced.

14. Explain the role of bone in calcium homeostasis.

Part Two

The skeletal system includes bone, cartilage, tendons, and ligaments. The term skeleton is derived from a Greek word meaning dried, indicating that the skeleton is the dried, hard parts left after the softer parts are removed. Most of us are familiar with the skeleton as a collection of bones that supports our bodies. Even with the flesh and organs removed, the skeleton is easily recognized as human. Despite its association with death, however, the skeletal system actually consists of dynamic, living tissues that are capable of growth, adapt to stress, and undergo repair after injury.

This chapter considers the structure and functions of the skeletal system with an emphasis on bone. The anatomy and histology of bone is described, followed by an explanation of how bone is initially formed in the fetus. The ways in which bones increase in length and diameter are also described, as well as the nutritional factors and regulatory mechanisms that affect bone growth. The ability of bone to change through remodeling and to repair following injury is presented, followed by a consideration of the role of the bones in maintaining calcium homeostasis.

Functions of the Skeletal System

The skeletal system provides support and protection, allows body movements, stores minerals and fats, and is the site of blood cell production.

1. *Support*. Rigid, strong bone is well suited for bearing weight and is the major supporting tissue of the body. Cartilage provides a firm, yet flexible support within certain structures, such as the nose, external ear, rib cartilages, and trachea. Ligaments are strong bands of fibrous connective tissue that attach to bones and hold them together.
2. *Protection*. Bone is hard and protects the organs it surrounds. For example, the skull encloses and protects the brain, and the vertebrae surround the spinal cord. The rib cage protects the heart, lungs, and other organs of the thorax.
3. *Movement*. Skeletal muscles attach to bones by tendons, which are strong bands of connective tissue. Contraction of the skeletal muscles moves the bones, producing body movements. Joints, which are formed where two or more bones come together, permit and control the movement between bones. Smooth cartilage covers the ends of bones within some joints, allowing the bones to move freely. Ligaments allow some movement between bones but prevent excessive movements.
4. *Storage*. Some minerals in the blood are taken into bone and stored. Should blood levels of these minerals decrease, the minerals are released from bone into the blood. The principal minerals stored are calcium and phosphorus. Fat (adipose tissue) is also stored within bone cavities. If needed, the fats are released into the blood and used by other tissues as a source of energy.
5. *Blood Cell Production*. Many bones contain cavities filled with bone marrow that gives rise to blood cells and platelets (see chapter 19).

Cartilage

Cartilage comes in three types: hyaline cartilage, fibrocartilage, and elastic cartilage (see chapter 4). Although each type of cartilage can provide support, hyaline cartilage is most intimately associated with bone. Most of the bones in the body develop from hyaline cartilage, the growth of bones in length results from hyaline cartilage growth, and the repair of damaged bone often involves hyaline cartilage. For these reasons, an understanding of hyaline cartilage is important.

Hyaline cartilage consists of specialized cells that produce a matrix surrounding the cells. The cells that produce new cartilage matrix are **chondroblasts** (kon′drō-blastz; *chondro* is from the Greek word *chondrion* and means cartilage). When a chondroblast is surrounded by matrix, it becomes a **chondrocyte** (kon′drō-sīt), which is a rounded cell that occupies a space within the matrix called a **lacuna** (lă-kū′nă) (figure 6.1). The matrix contains collagen, which provides strength, and proteoglycans, which make cartilage resilient by trapping water (see chapter 4).

Most cartilage is covered by a double-layered connective tissue sheath, the **perichondrium** (per-i-kon′drē-ŭm) (see figure 6.1). The outer layer of the perichondrium is dense irregular connective tissue containing fibroblasts. The inner, more delicate layer has fewer fibers and contains chondroblasts. Blood vessels and nerves penetrate the outer layer of the perichondrium but do not enter the cartilage matrix, so that nutrients must diffuse through the cartilage matrix to reach the chondrocytes. **Articular** (ar-tik′yū-lăr) **cartilage,** which is the cartilage covering the ends of bones where they come together to form joints, has no perichondrium, blood vessels, or nerves.

Cartilage grows in two different ways. Through **appositional growth,** chondroblasts in the perichondrium lay down new matrix and add new chondrocytes to the outside of the tissue, and through **interstitial growth** chondrocytes within the tissue divide and add more matrix from the inside (see figure 6.1).

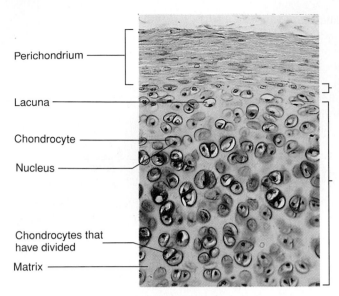

Perichondrium

Lacuna

Chondrocyte

Nucleus

Chondrocytes that
have divided

Matrix

Appositional growth
(New cartilage is added to the surface of the
older cartilage by chondroblasts from the inner
layer of the perichondrium)

Interstitial growth
(New cartilage is formed within the older cartilage
by chondrocytes that divide and produce
additional matrix)

Figure 6.1   Hyaline Cartilage

Photomicrograph of hyaline cartilage covered by perichondrium.
Chondrocytes within lacunae are surrounded by cartilage matrix.

Bone Anatomy

Bone Shapes

Individual bones can be classified according to their shape as long, short, flat, or irregular (figure 6.2). **Long bones** are longer than they are wide. Most of the bones of the upper and lower limbs are long bones. **Short bones** are about as broad as they are long. They are nearly cube-shaped or round and are exemplified by the bones of the wrist (carpals) and ankle (tarsals). **Flat bones** have a relatively thin, flattened shape and are usually curved. Examples of flat bones are certain skull bones, the ribs, the breastbone (sternum), and the shoulder blades (scapulae). **Irregular bones,** such as the vertebrae and facial bones, have shapes that do not fit readily into the other three categories.

Structure of a Long Bone

Each growing long bone has three major components: a diaphysis, an epiphysis, and an epiphyseal plate (figure 6.3a and table 6.1). The **diaphysis** (dī-af′i-sis), or shaft, is composed primarily of **compact bone,** which is mostly bone matrix surrounding a few small spaces. The **epiphysis** (e-pif′i-sis), or end of the bone, consists primarily of **cancellous** (kan′sĕ-lŭs), or **spongy, bone,** which has many small spaces or cavities surrounded by bone matrix. The outer surface of the epiphysis is a layer of compact bone, and within joints the epiphyses are covered by articular cartilage. The **epiphyseal** (ep-i-fiz′ē-ăl), or **growth, plate** is hyaline cartilage located between the epiphysis and diaphysis. Growth in bone length occurs at the epiphyseal plate, but, when bone stops growing in length, the epiphyseal plate becomes ossified and is called the **epiphyseal line** (figure 6.3b).

In addition to the small spaces within cancellous bone and compact bone, the diaphysis of a long bone can have a

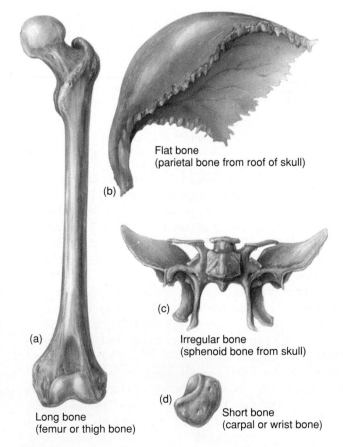

Flat bone
(parietal bone from roof of skull)

(b)

(c)

Irregular bone
(sphenoid bone from skull)

(a)

(d)

Short bone
(carpal or wrist bone)

Long bone
(femur or thigh bone)

Figure 6.2   Bone Shapes

large space called the **medullary cavity.** The cavities of cancellous bone and the medullary cavity are filled with marrow (mar′ō). The spaces within the cancellous bone of the proximal epiphyses of the larger adult long bones contain **red marrow,** which is the site of blood cell formation. The medullary cavity of the adult diaphysis normally is filled with

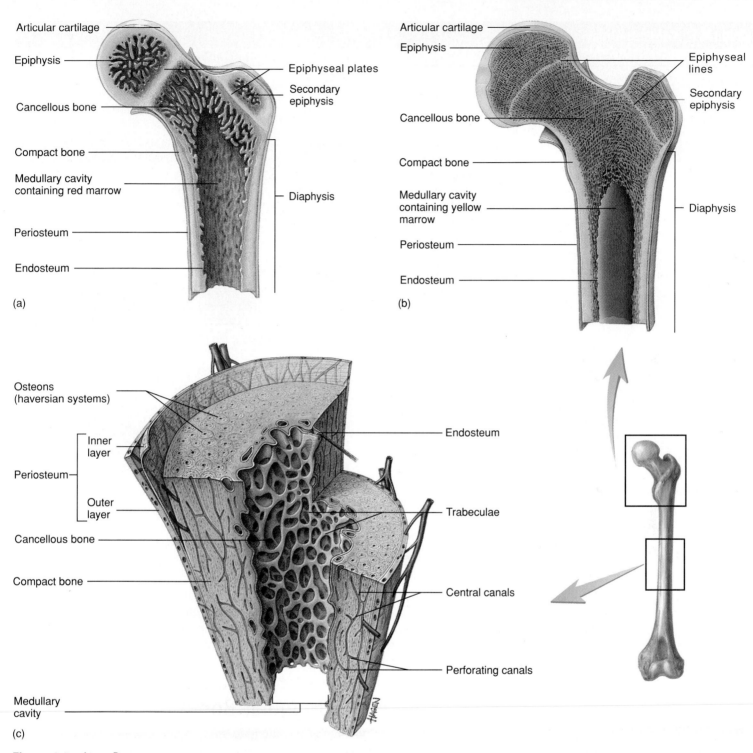

Figure 6.3   Long Bone

(a) Young long bone (the femur) showing epiphysis, epiphyseal plates, and diaphysis. (b) Adult long bone with epiphyseal lines. (c) Internal features of a portion of the diaphysis in (a). ✗

yellow marrow, which is mostly adipose tissue. In general, yellow marrow is associated with the long bones of the limbs, and red marrow is associated with the rest of the skeleton (figure 6.4). Children's bones have more red marrow than do adult bones. Children even have red marrow located in the diaphyses of long bones. As children mature, the red marrow in their limbs is replaced with yellow marrow.

Bone is covered by a double-layered connective tissue sheath called the **periosteum** (per-ē-os′tē-ŭm) (figure 6.3c). The outer fibrous layer is dense, irregular collagenous connective tissue that contains blood vessels and nerves. The inner layer is a single layer of bone cells, which includes osteoblasts, osteoclasts, and osteoprogenitor cells (see the section titled "Bone Cells" later in this chapter). Where tendons

Table 6.1   Gross Anatomy of Long Bone

Part	Description	Part	Description
Diaphysis	Shaft of the bone	Epiphyseal plate	Area of hyaline cartilage between the diaphysis and epiphysis; cartilage growth followed by endochondral ossification results in bone growth in length
Epiphyses	Ends of the bone		
Periosteum	Double-layered connective tissue membrane covering the outer surface of bone except where there is articular cartilage; ligaments and tendons attach to bone through the periosteum; blood vessels and nerves from the periosteum supply the bone; the periosteum is the site of bone growth in diameter	Cancellous (spongy) bone	Bone having many small spaces; found mainly in the epiphysis; arranged into trabeculae
		Compact bone	Dense bone with few internal spaces organized into osteons; forms the diaphysis and covers the cancellous bone of the epiphyses
Endosteum	Thin connective tissue membrane lining the inner cavities of bone	Medullary cavity	Large cavity within the diaphysis
Articular cartilage	Thin layer of hyaline cartilage covering a bone where it forms a joint (articulation) with another bone	Red marrow	Connective tissue in the spaces of cancellous bone; the site of blood cell production
		Yellow marrow	Fat stored within the medullary cavity

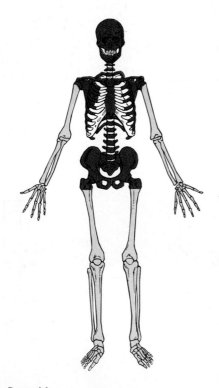

Figure 6.4   Bone Marrow
Distribution of red marrow and yellow marrow in an adult.

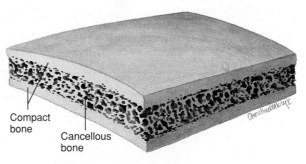

Compact bone   Cancellous bone

Figure 6.5   Structure of a Flat Bone
Outer layers of compact bone surround cancellous bone.

ities in cancellous and compact bone (see figure 6.3). The endosteum is a single layer of cells, which includes osteoblasts, osteoclasts, and osteoprogenitor cells (see the section titled "Bone Cells" later in this chapter).

Structure of Flat, Short, and Irregular Bones

Flat bones usually have no diaphyses or epiphyses, and they contain an interior framework of cancellous bone sandwiched between two layers of compact bone (figure 6.5). Short and irregular bones have a composition similar to the epiphyses of long bones. They have compact bone surfaces that surround a cancellous bone center with small spaces that usually are filled with marrow. Short and irregular bones are not elongated and have no diaphyses. Certain regions of these bones, however, such as the processes of irregular bones, possess epiphyseal growth plates and therefore have small epiphyses.

Some of the flat and irregular bones of the skull have air-filled spaces called **sinuses** (see chapter 7), which are lined by mucous membranes.

and ligaments attach to bone, the collagen fibers of the tendon or ligament become continuous with those of the periosteum. In addition, some of the collagen fibers of the tendons or ligaments penetrate the periosteum into the outer part of the bone. These bundles of collagen fibers are called **perforating,** or **Sharpey's, fibers,** and they further strengthen the attachment of the tendons or ligaments to the bone.

The **endosteum** (en-dos'tē-ŭm) is a membrane that lines the medullary cavity of the diaphysis and the smaller cav-

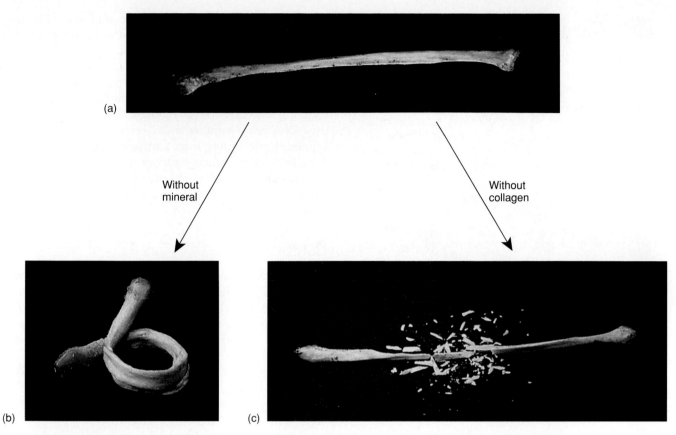

Figure 6.6   Effects of Changing the Bone Matrix

(*a*) Normal bone. (*b*) Demineralized bone, in which collagen is the primary remaining component, can be bent without breaking. (*c*) When collagen is removed, mineral is the primary remaining component, making the bone so brittle it is easily shattered.

Bone Histology

Bone consists of extracellular bone matrix and bone cells. The composition of the bone matrix is responsible for the characteristics of bone. The bone cells produce the bone matrix, become entrapped within the bone matrix, and break down the bone matrix so that new matrix can replace the old matrix.

Bone Matrix

By weight, mature bone matrix normally is approximately 35% organic and 65% inorganic material. The organic material primarily consists of collagen and proteoglycans. The inorganic material primarily consists of a calcium phosphate crystal called **hydroxyapatite** (hī-drok′sē-ap-ă-tīt), which has the molecular formula $Ca_{10}(PO_4)_6(OH)_2$.

The collagen and mineral components are responsible for the major functional characteristics of bone. Bone matrix might be said to resemble reinforced concrete. Collagen, like reinforcing steel bars, lends flexible strength to the matrix, whereas the mineral components, like concrete, give the matrix compression (weight-bearing) strength.

If all the mineral is removed from a long bone, collagen becomes the primary constituent, and the bone becomes overly flexible. On the other hand, if the collagen is removed from the bone, the mineral component becomes the primary constituent, and the bone is very brittle (figure 6.6).

2   P R E D I C T

In general, the bones of elderly people break more easily than the bones of younger people. Give as many possible explanations as you can for this observation.

✔ *Answer in Appendix F*

Bone Cells

Bone cells can be categorized as osteoblasts, osteocytes, and osteoclasts, each of which has different functions and origins.

Osteoblasts

Osteoblasts (os′tē-ō-blastz) have an extensive endoplasmic reticulum and numerous ribosomes. They produce collagen and proteoglycans, which are packaged into vesicles by the Golgi apparatus and released from the cell. Osteoblasts also form vesicles that accumulate Ca^{2+} ions, PO_4^{2-} ions, and various enzymes. The vesicles bud off of the osteoblast, and their contents are used to form hydroxyapatite crystals. As a result of these processes, mineralized bone matrix is formed.

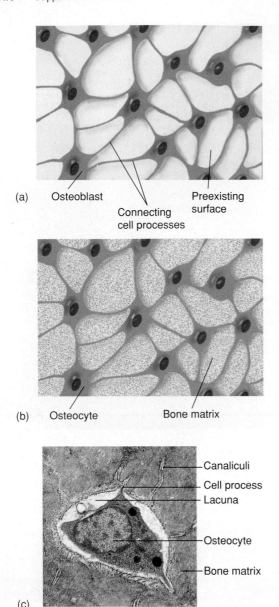

(a) Osteoblast
Connecting cell processes
Preexisting surface

(b) Osteocyte
Bone matrix

(c)
Canaliculi
Cell process
Lacuna
Osteocyte
Bone matrix

Figure 6.7 Ossification

(*a*) Osteoblasts on a preexisting surface, such as cartilage or bone. The cell processes of different osteoblasts join together. (*b*) Osteoblasts have produced bone matrix. The osteoblasts are now osteocytes. (*c*) Photomicrograph of an osteocyte in a lacuna with cell processes in the canaliculi.

Ossification (os′i-fi-kā-shŭn), or **osteogenesis** (os′tē-ō-jen′ĕ-sis), is the formation of bone by osteoblasts. Elongated cell processes from osteoblasts connect to cell processes of other osteoblasts. The osteoblasts then form an extracellular bony matrix that surrounds the cells and their processes (figure 6.7).

Osteocytes

Once an osteoblast becomes surrounded by bone matrix, it is a mature bone cell called an **osteocyte** (os′tē-ō-sīt). Osteo-

cytes become relatively inactive compared with most osteoblasts, but it is possible for them to produce components needed to maintain the bone matrix.

The spaces occupied by the osteocyte cell bodies are called **lacunae** (lă-kū′nē), and the spaces occupied by the osteocyte cell processes are called **canaliculi** (kan-ă-lik′yū-lī, meaning little canals) (see figure 6.7). In a sense, the cells and their processes form a "mold" around which the matrix is formed. Bone differs from cartilage in that the processes of bone cells are in contact with one another through the canaliculi. This allows nutrients to pass from cell to cell through the canaliculi rather than having to diffuse through the mineralized matrix.

Osteoclasts

Osteoclasts (os′tē-ō-klastz), which are large cells with several nuclei, are responsible for the **resorption,** or breakdown, of bone. The plasma membrane of osteoclasts forms many projections called a **ruffled border,** which is sealed against the bone matrix. Hydrogen ions are pumped across the ruffled border, producing an acid environment that causes decalcification of the bone matrix. The osteoclasts also release enzymes that digest the protein components of the matrix. Some of the breakdown products of bone resorption can be taken into the osteoclast by endocytosis.

The resorption of bone by osteoclasts is assisted by osteoblasts. Osteoclasts break down bone best when they are in direct contact with mineralized bone matrix; however, bone normally is covered by a thin layer of unmineralized organic matrix. Osteoblasts produce enzymes that break down the unmineralized organic matrix, which then enables the osteoclasts to come into contact with the mineralized bone matrix.

Origin of Bone Cells

Connective tissue develops embryologically from mesenchymal cells (see chapter 4). Some of the mesenchymal cells become **stem cells,** which have the ability to replicate and give rise to more specialized cell types. **Osteoprogenitor cells** are stem cells that have the ability to become osteoblasts or chondroblasts. Near capillaries where oxygen concentrations are higher, osteoprogenitor cells become osteoblasts, whereas farther from capillaries, where oxygen concentrations are lower, they become chondroblasts. Osteoprogenitor cells are located in the inner layer of the perichondrium, the inner layer of the periosteum, and in the endosteum. From these locations, they can be a potential source of new osteoblasts or chondroblasts.

Osteoblasts are derived from osteoprogenitor cells, and osteocytes are derived from osteoblasts. Whether or not osteocytes freed from their surrounding bone matrix by resorption can revert to active osteoblasts is a debated issue. Osteoclasts are not derived from osteoprogenitor cells but instead from stem cells in red bone marrow (see chapter 19). The bone marrow stem cells that give rise to a type of white blood cell, called a monocyte, apparently also are the source of osteoclasts. The multinucleated osteoclasts probably result from the fusion of many stem cell descendants.

Woven and Lamellar Bone

Bone tissue can be classified as woven or lamellar bone according to the organization of collagen fibers within the bone matrix. In **woven bone,** the collagen fibers are randomly oriented in many directions. Woven bone is first formed during fetal development or during the repair of a fracture. After it is formed, woven bone is broken down by osteoclasts, and new matrix is formed by osteoblasts. This process is called **remodeling** and is discussed later in this chapter. Woven bone is remodeled to form lamellar bone.

Lamellar bone is mature bone that is organized into thin sheets or layers approximately 3–7 micrometers (μm) thick called **lamellae** (lă-mel′ē). In general, the collagen fibers of one lamella lie parallel to one another but at an angle to the collagen fibers in the adjacent lamellae. Osteocytes, within their lacunae, are arranged in layers sandwiched between lamellae.

Cancellous and Compact Bone

Woven or lamellar bone can be further classified according to the amount of bone matrix relative to the amount of space within the bone. Cancellous bone is characterized by less bone matrix and more space than compact bone, which has more bone matrix and less space than cancellous bone.

Cancellous bone (figure 6.8a) consists of interconnecting rods or plates of bone called **trabeculae** (tră-bek′yū-lē, meaning beam). Between the trabeculae are spaces that in life are filled with bone marrow and blood vessels. Cancellous bone is sometimes called spongy bone because of its porous appearance.

Most trabeculae are thin (50–400 μm), consisting of several lamellae with osteocytes located between the lamellae (figure 6.8b). Each osteocyte is associated with other osteocytes through canaliculi. Usually no blood vessels penetrate the trabeculae, so osteocytes must obtain nutrients through their canaliculi. The surfaces of trabeculae are covered with a single layer of cells consisting mostly of osteoblasts with a few osteoclasts.

Trabeculae are oriented along the lines of stress within a bone (figure 6.9). If the direction of weight-bearing stress is changed slightly (e.g., because of a fracture that heals improperly), the trabecular pattern realigns with the new lines of stress.

Compact bone (figure 6.10) is denser and has fewer spaces than cancellous bone. Blood vessels enter the substance of the bone itself, and the lamellae of compact bone are primarily oriented around those blood vessels. Vessels that run parallel to the long axis of the bone are contained within **central,** or **haversian** (ha-ver′shan), **canals.** Central canals are lined with endosteum and contain blood vessels, nerves, and loose connective tissue. **Concentric lamellae** are circular layers of bone matrix that surround a common center, the

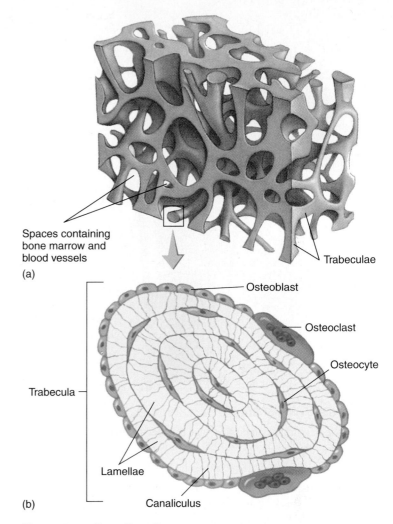

Spaces containing bone marrow and blood vessels

(a)

Trabeculae

Osteoblast

Osteoclast

Osteocyte

Trabecula

Lamellae

(b)

Canaliculus

Figure 6.8   Cancellous Bone

(a) Beams of bone, the trabeculae, surround spaces in the bone. In life, the spaces are filled with red or yellow bone marrow and with blood vessels. (b) Transverse section of a trabecula.

central canal. An **osteon** (os′tē-on), or **haversian system,** consists of a single central canal, its contents, and associated concentric lamellae and osteocytes. In cross section the osteon resembles a circular target; the "bull's-eye" of the target is the central canal, and 4–20 concentric lamellae form the rings. Osteocytes are located in lacunae between the lamellar rings, and canaliculi radiate between lacunae across the lamellae, producing the appearance of minute cracks across the rings of the target.

In between the osteons are **interstitial lamellae** (see figure 6.10), which are remnants of older osteons that were partially removed during bone remodeling. The outer surfaces of compact bone are covered by **circumferential lamellae,** which are flat plates that extend around the bone (see figure 6.10). In some bones, such as certain bones of the face, the layer of compact bone can be so thin that no osteons exist, and the compact bone is composed of only circumferential lamellae.

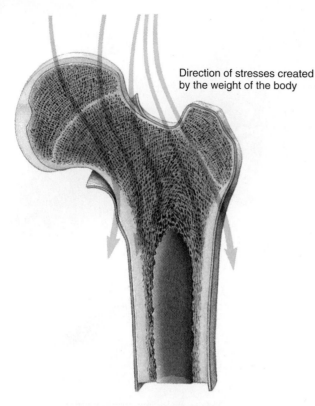

Direction of stresses created by the weight of the body

Figure 6.9   Lines of Stress

Long bone (femur) showing trabeculae oriented along lines of stress (*arrows*).

The osteocytes receive nutrients and eliminate waste products through the canal system within compact bone. Blood vessels from the periosteum or medullary cavity enter the bone through **perforating,** or **Volkmann's, canals,** which run perpendicular to the long axis of the bone (see figure 6.10). Perforating canals are not surrounded by concentric lamellae but pass through the concentric lamellae of osteons. The central canals receive blood vessels from perforating canals. Nutrients in the blood vessels enter the central canals, pass into the canaliculi, and are transported through the cytoplasm of the osteocytes that occupy the canaliculi and lacunae to the most peripheral cells within each osteon. Waste products are removed in the reverse direction.

---

3   P R E D I C T

Compact bone has a specialized canal system for the transport of nutrients and waste products. Why isn't such a system necessary in cancellous bone?

✔ *Answer in Appendix F*

Bone Development

During fetal development there are two patterns of bone formation called **intramembranous** and **endochondral ossification.** The terms describe the tissues in which bone for-

mation takes place: intramembranous ossification in connective tissue membranes and endochondral ossification in cartilage. Both methods of ossification initially produce woven bone that is then remodeled. After remodeling, bone formed by intramembranous ossification cannot be distinguished from bone formed by endochondral ossification. Table 6.2 compares intramembranous and endochondral ossification.

Intramembranous Ossification

Intramembranous ossification begins at approximately the fifth week of development with the formation of a mesenchymal membrane. Ossification of the membrane begins at approximately the eighth week of development and is completed by approximately 2 years of age. Many skull bones, part of the mandible (lower jaw), and the diaphyses of the clavicles (collarbones) develop by intramembranous ossification.

1. Embryonic mesenchyme condenses around the developing brain to form a membrane of connective tissue with randomly oriented, delicate collagen fibers. Some of the mesenchymal cells become osteoprogenitor cells, which become osteoblasts. The osteoblasts produce bone matrix that surrounds the collagen fibers of the connective tissue membrane, and the osteoblasts become osteocytes. As a result of this process, many tiny trabeculae of woven bone are formed (figure 6.11*a*). Additional osteoblasts gather on the surfaces of the trabeculae and produce more bone, causing the trabeculae to become larger and longer.

2. Cancellous bone is formed as the trabeculae join together to form an interconnected network of trabeculae separated by spaces (figure 6.11*b*). Cells within the spaces of the cancellous bone specialize to form red bone marrow. As cancellous bone is formed, cells surrounding the developing bone specialize and form the periosteum. Osteoblasts from the periosteum lay down bone matrix to form an outer surface of compact bone. Thus the end products of intramembranous bone formation are bones with outer compact bone surfaces and cancellous centers (see figure 6.5). Remodeling converts woven bone to mature bone and contributes to the final shape of the bone.

3. The locations where ossification begins in the membrane are called **centers of ossification.** As ossification proceeds, some of the trabeculae radiate out in many directions from each center of ossification (figure 6.11*c*). The production of bone by intramembranous ossification results in enlargement of the centers of ossification and the formation of many of the bones of the skull.

The larger membrane-covered spaces between the developing skull bones that have not yet been ossified are called **fontanels,** or soft spots (figure 6.12) (see chapter 8). The bones eventually grow together, and all the fontanels are usually closed by 2 years of age.

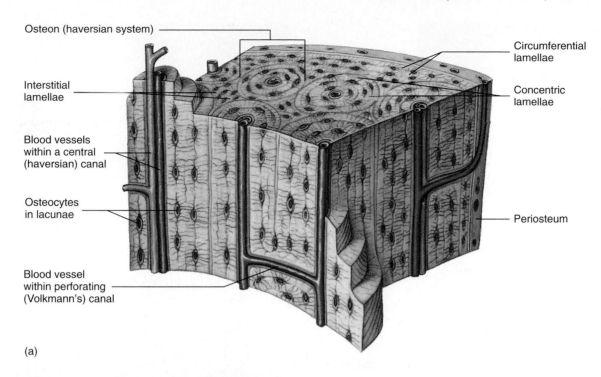

Osteon (haversian system)

Interstitial lamellae

Blood vessels within a central (haversian) canal

Osteocytes in lacunae

Blood vessel within perforating (Volkmann's) canal

Circumferential lamellae

Concentric lamellae

Periosteum

(a)

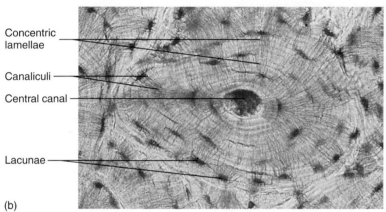

Concentric lamellae

Canaliculi

Central canal

Lacunae

(b)

Figure 6.10   Compact Bone

(*a*) Compact bone consists mainly of osteons, which are concentric lamellae surrounding blood vessels within central canals. The outer surface of the bone is formed by circumferential lamellae, and bone between the osteons consists of interstitial lamellae.
(*b*) Photomicrograph of an osteon. 🕺

Table 6.2 Comparison of Intramembranous and Endochondral Ossification	
Intramembranous Ossification	**Endochondral Ossification**
Embryonic mesenchyme forms a collagen membrane containing osteoprogenitor cells.	Embryonic mesenchyme becomes chondroblasts that produce a cartilage template, which is surrounded by the perichondrium.
No stage is comparable.	Chondrocytes hypertrophy, the cartilage matrix is calcified, and the chondrocytes die.
Embryonic mesenchyme forms the periosteum, which contains osteoblasts.	The perichondrium becomes the periosteum when osteoprogenitor cells within the periosteum become osteoblasts.
Osteoprogenitor cells become osteoblasts at centers of ossification; internally the osteoblasts produce bone matrix; externally the periosteal osteoblasts produce bone.	Blood vessels and osteoblasts from the periosteum invade the calcified cartilage template; internally these osteoblasts produce bone matrix at primary ossification centers (and later at secondary ossification centers); externally the periosteal osteoblasts produce bone.
Intramembranous bone is remodeled and is indistinguishable from endochondral bone.	Endochondral bone is remodeled and is indistinguishable from intramembranous bone

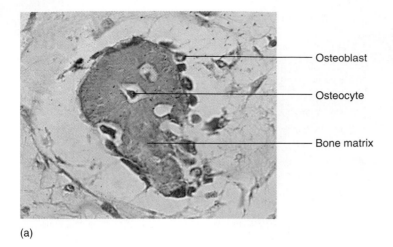

(a)

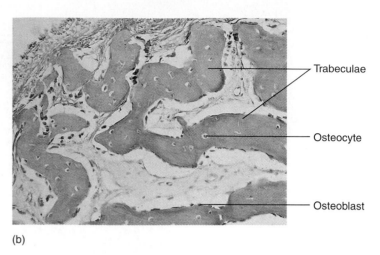

(b)

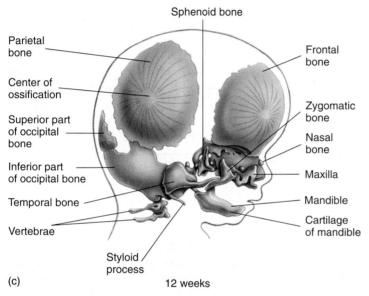

(c)   12 weeks

Figure 6.11   Intramembranous Ossification

(*a*) Photomicrograph of a cross section of a trabecula. Osteocytes are surrounded by newly formed bone matrix. Osteoblasts are forming a ring on the outer surface of the trabecula. As they lay down additional bone matrix, the trabecula increases in size. (*b*) Lower magnification photomicrograph than (*a*), showing cancellous bone, which has been formed as a result of the enlargement and interconnections of many trabeculae. (*c*) Twelve-week-old fetus showing skull bones that develop by intramembranous ossification (*yellow*). Also shown are bones formed by endochondral ossification (*blue*).

Endochondral Ossification

Endochondral ossification begins with the formation of cartilage at approximately the end of the fourth week of development. Some of this cartilage starts to ossify at approximately the eighth week of development, whereas endochondral ossification of some cartilage might not begin until as late as age 18–20 years. Bones of the base of the skull, part of the mandible, the epiphyses of the clavicles, and most of the remaining skeletal system develop through the process of endochondral ossification (see figures 6.11 and 6.12):

1. Endochondral ossification begins as mesenchymal cells aggregate in regions of future bone formation. The mesenchymal cells become chondroblasts, which produce a hyaline cartilage model having the approximate shape of the bone that will later be formed (figure 6.13*a*). As the chondroblasts become surrounded by cartilage matrix they become chondrocytes. The cartilage model is surrounded by the perichondrium, except where a joint will form connecting one bone to another bone. Not shown in figure 6.13, the

perichondrium is continuous with tissue that will become the joint capsule (see chapter 8).

2. When blood vessels invade the perichondrium surrounding the cartilage model (figure 6.13*b*), osteoprogenitor cells within the perichondrium become osteoblasts. The perichondrium becomes the periosteum when the osteoblasts begin to produce bone. The osteoblasts produce compact bone on the surface of the cartilage model, forming a **bone collar.** Two other events are occurring at the same time that the bone collar is forming. First, the cartilage model increases in size as a result of interstitial and appositional cartilage growth. Second, the chondrocytes in the center of the cartilage model **hypertrophy** (hī-per′trō-fē), or enlarge, and the matrix between the enlarged cells becomes mineralized with calcium carbonate. At this point the cartilage is referred to as **calcified cartilage** (see figure 6.13*b*). The chondrocytes in this calcified area eventually die, leaving enlarged lacunae with thin walls of calcified matrix.

3. Blood vessels grow into the enlarged lacunae of the calcified cartilage (figure 6.13*c*). The connective tissue

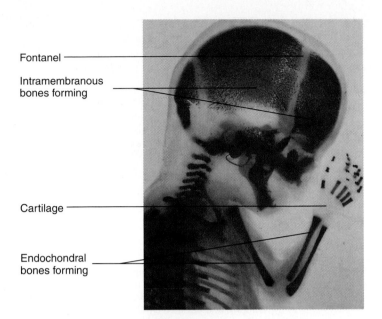

Fontanel

Intramembranous bones forming

Cartilage

Endochondral bones forming

Figure 6.12   Bone Formation

18-week-old fetus showing intramembranous and endochondral ossification. Intramembranous ossification occurs at centers of ossification in the flat bones of the skull. Endochondral ossification has formed bones in the diaphyses of long bones. The epiphyses are still cartilage at this stage of development.

surrounding the blood vessels brings in osteoblasts and osteoclasts from the periosteum. The osteoblasts produce bone on the surface of the calcified cartilage, forming bone trabeculae, which changes the calcified cartilage of the diaphysis into cancellous bone. This area of bone formation is called the **primary ossification center.**

4. As bone development proceeds, the cartilage model continues to grow, more perichondrium becomes periosteum, the bone collar thickens and extends further along the diaphysis, and additional cartilage within the diaphysis is calcified and transformed into cancellous bone (figure 6.13*d*). Remodeling converts woven bone to mature bone and contributes to the final shape of the bone. Osteoclasts remove bone from the center of the diaphysis to form the medullary cavity, and cells within the medullary cavity specialize to form red bone marrow.

5. In long bones the diaphysis is the primary ossification center, and additional sites of ossification, called **secondary ossification centers,** appear in the epiphyses (figure 6.13*e*). The events occurring at the secondary ossification centers are the same as those occurring at the primary ossification centers, except that the spaces in the epiphyses do not enlarge to form a medullary cavity as in the diaphysis. Primary ossification centers appear during early fetal development, whereas secondary ossification centers appear in the proximal epiphysis of the femur, humerus, and tibia about 1 month before birth. A baby is considered full term if

one of these three ossification centers can be seen on radiographs at the time of birth. At about 18–20 years of age the last secondary ossification center appears in the medial epiphysis of the clavicle.

6. Replacement of cartilage by bone continues in the cartilage model until all the cartilage, except that in the epiphyseal plate and on articular surfaces, has been replaced by bone (figure 6.13*f*). The epiphyseal plate, which exists throughout an individual's growth, and the articular cartilage, which is a permanent structure, are derived from the original embryonic cartilage model.

7. In mature bone, cancellous and compact bone are fully developed and the epiphyseal plate has become the epiphyseal line. The only cartilage present is the articular cartilage at the ends of the bone (figure 6.13*g*). All of the original perichondrium that surrounded the cartilage model has become periosteum.

4	P R E D I C T

During endochondral ossification, calcification of cartilage results in the death of chondrocytes. Later in the process, ossification of the bone matrix does not result in the death of osteocytes. Explain.

✔ *Answer in Appendix F*

Bone Growth

Unlike cartilage, bones cannot grow by interstitial growth. Bones increase in size only by **appositional growth,** the formation of new bone on the surface of cartilage or older bone.

5	P R E D I C T

Explain why bones cannot undergo interstitial growth, as does cartilage.

✔ *Answer in Appendix F*

Growth at the Epiphyseal Plate

In a long bone the epiphyseal plate separates the epiphysis from the diaphysis (figure 6.14*a*). Bones with long projections, such as the processes of vertebrae (see chapter 7), also have epiphyseal plates. Long bones and bony projections can increase in length because of growth at the epiphyscal plate. This elongation involves the formation of new cartilage by interstitial cartilage growth followed by appositional bone growth on the surface of the cartilage.

The epiphyseal plate is organized into four zones (figure 6.14*b*). The **zone of resting cartilage** is nearest to the epiphysis and contains randomly arranged chondrocytes that do not divide rapidly. The chondrocytes in the **zone of proliferation** produce new cartilage through interstitial cartilage growth. The chondrocytes divide and form columns resembling stacks of plates or coins. In the **zone of hypertrophy,**

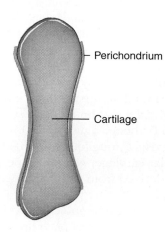

Perichondrium

Cartilage

(a) A cartilage model, surrounded by perichondrium, is produced by chondroblasts that become chondrocytes enclosed by cartilage matrix.

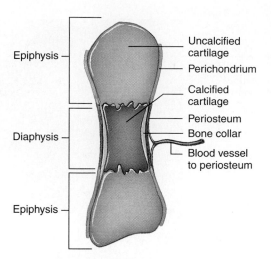

Epiphysis

Uncalcified cartilage

Perichondrium

Calcified cartilage

Periosteum

Diaphysis

Bone collar

Blood vessel to periosteum

Epiphysis

(b) The perichondrium of the diaphysis becomes the periosteum, and a bone collar is produced. Internally, the chondrocytes hypertrophy, and calcified cartilage is formed.

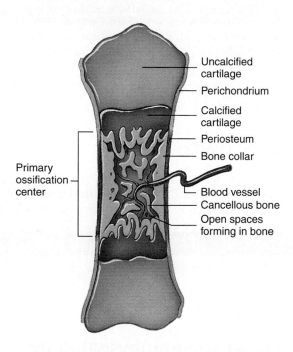

Uncalcified cartilage

Perichondrium

Calcified cartilage

Periosteum

Bone collar

Primary ossification center

Blood vessel

Cancellous bone

Open spaces forming in bone

(c) A primary ossification center forms as blood vessels and osteoblasts invade the calcified cartilage. The osteoblasts lay down bone matrix, forming cancellous bone.

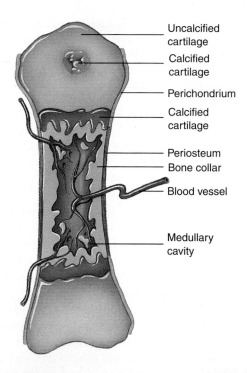

Uncalcified cartilage

Calcified cartilage

Perichondrium

Calcified cartilage

Periosteum

Bone collar

Blood vessel

Medullary cavity

(d) The process of bone collar formation, cartilage calcification, and cancellous bone production continues. Calcified cartilage begins to form in the epiphysis. A medullary cavity begins to form in the center of the diaphysis.

Figure 6.13 Endochondral Ossification

the chondrocytes produced in the zone of proliferation mature and enlarge. Thus a maturation gradient exists in each column: cells nearer to the epiphysis are younger and are actively proliferating, whereas cells progressively nearer the diaphysis are older and are undergoing hypertrophy. The **zone of calcification** is very thin, consisting of cartilage matrix mineralized with calcium carbonate. The hypertrophied chondrocytes die, and blood vessels from the diaphysis grow into the area. The

connective tissue surrounding the blood vessels contains osteoblasts from the endosteum. The osteoblasts line up on the surface of the calcified cartilage and through appositional bone growth deposit new bone matrix, which is later remodeled.

As new cartilage cells are formed in the zone of proliferation, and as these cells enlarge in the zone of hypertrophy, the overall length of the diaphysis increases (figure 6.15). The thickness of the epiphyseal plate does not increase, however,

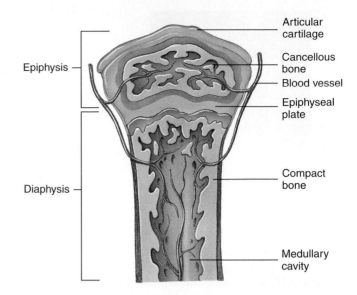

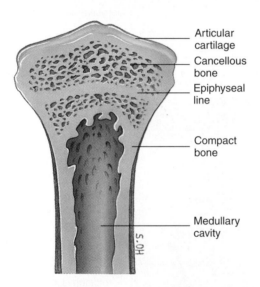

(e) Secondary ossification centers form in the epiphyses long bones.

(f) The original cartilage model is almost completely ossified. Unossified cartilage becomes the epiphyseal plate and the articular cartilage over the articular surface at the ends of the bone.

(g) Mature bone in which the epiphyseal plate has become the epiphyseal line and all the cartilage in the epiphysis, except the articular cartilage, has become bone.

Figure 6.13   Endochondral Ossification *(continued)*

because the rate of cartilage growth on the epiphyseal side of the plate is equal to the rate at which cartilage is replaced by bone on the diaphyseal side of the plate.

As the bones achieve normal adult size, growth in bone length ceases because the epiphyseal plate is ossified and becomes the epiphyseal line. This event, called closure of the epiphyseal plate, occurs between approximately 12 and 25 years of age, depending on the bone.

6   P R E D I C T

A 15-year-old football player is tackled during a game, and the epiphyseal plate of the left femur is damaged (figure 6.16). What are the results of such an injury, and why is recovery difficult?

✔ *Answer in Appendix F*

Growth at Articular Cartilage

Just as growth of cartilage in the epiphyseal plate results in an increase in bone length, growth of articular cartilage results in larger epiphyses. In addition, growth of articular cartilage can result in an increase in the size of bones that do not have an epiphysis, such as short bones. The process of growth in articular cartilage is similar to that occurring in the epiphyseal plate, except that the chondrocyte columns are not as pronounced. The chondrocytes near the surface of the articular cartilage are similar to those in the zone of resting cartilage of the epiphyseal plate. In the deepest part of the articular cartilage, nearer to bone tissue, the cartilage is calcified, dies, and is ossified to form new bone.

When the epiphyses reach their full size, the growth of cartilage and its replacement by bone ceases. The articular

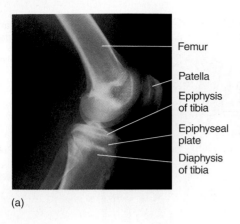

(a)

Epiphyseal side

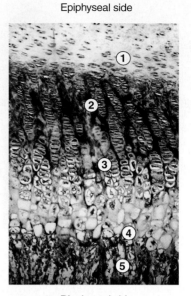

Diaphyseal side

1. **Zone of resting cartilage.** Cartilage attaches to the epiphysis.

2. **Zone of proliferation.** New cartilage is produced on the epiphyseal side of the plate as the chondrocytes divide and form stacks of cells.

3. **Zone of hypertrophy.** Chondrocytes mature and enlarge.

4. **Zone of calcification.** Matrix is calcified, and chondrocytes die.

5. **Ossified bone.** The calcified cartilage on the diaphyseal side of the plate is replaced by bone.

Figure 6.14   Epiphyseal Plate

(*a*) Radiograph of the knee, showing the epiphyseal plate of the tibia (shin bone). Because cartilage does not appear readily on x-ray film, the epiphyseal plate appears as a black area between the white diaphysis and the epiphyses. (*b*) Zones of the epiphyseal plate. ✗

(b)

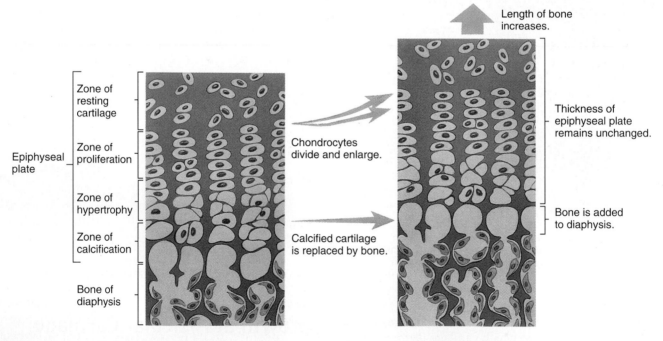

Figure 6.15   Bone Growth in Length at the Epiphyseal Plate

New cartilage is formed on the epiphyseal side of the plate at the same rate that new bone is formed on the diaphyseal side of the plate. Consequently, the epiphyseal plate remains the same thickness, but the length of the diaphysis increases.

cartilage, however, persists throughout life and does not become ossified as does the epiphyseal plate.

7   P R E D I C T

After bone growth ceases, why doesn't the articular cartilage become ossified?

✔ *Answer in Appendix F*

Other Bone Growth

Appositional growth is responsible for the increase in diameter of long bones and an increase in the size of other bones. Osteoprogenitor cells from the periosteum become osteoblasts and through appositional bone growth deposit new bone matrix, which is later remodeled.

In cancellous bone, appositional bone growth adds bone matrix to the outer surface of trabeculae. In compact

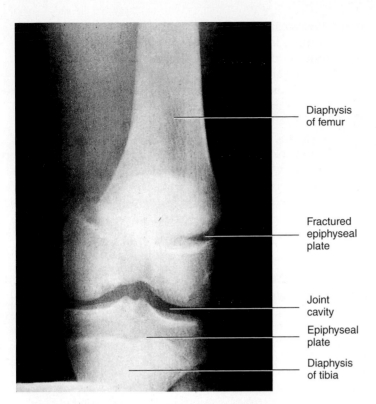

Diaphysis
of femur

Fractured
epiphyseal
plate

Joint
cavity

Epiphyseal
plate

Diaphysis
of tibia

Figure 6.16   Fracture of the Epiphyseal Plate

Radiograph of a child's knee. The femur (thigh bone) is separated from the tibia (leg bone) by a joint cavity. The epiphyseal plate of the femur is fractured, separating the diaphysis from the epiphysis.

bone, appositional bone growth is responsible for the formation of circumferential and concentric lamellae (see the section titled "Bone Remodeling" later in the chapter).

Factors Affecting Bone Growth

Bones of an individual's skeleton usually reach a certain length, thickness, and shape through the processes described in the previous sections. The potential shape and size of a bone and an individual's final adult height are determined genetically, but factors such as nutrition and hormones can greatly modify the expression of those genetic factors.

Nutrition

Because bone growth requires chondroblast and osteoblast proliferation, any metabolic disorder that affects the rate of cell proliferation or the production of collagen and other matrix components affects bone growth, as does the availability of calcium or other minerals needed in the mineralization process.

The long bones of a child sometimes exhibit lines of arrested growth, which are transverse regions of greater bone density crossing an otherwise normal bone. These lines are caused by greater calcification below the epiphyseal plate of a bone, where it has grown at a slower rate during an illness or severe nutritional deprivation. They demonstrate that illness or malnutrition during the time of bone growth can

cause a person to be shorter than he or she would have been otherwise.

Certain vitamins are important in very specific ways to bone growth. **Vitamin D** is necessary for the normal absorption of calcium from the intestines (see chapters 5 and 24). Vitamin D can be synthesized by the body or ingested. Its rate of synthesis is increased when the skin is exposed to sunlight.

Insufficient vitamin D in children causes **rickets,** a disease resulting from reduced mineralization of the organic matrix of bone. Children with rickets can have bowed bones and inflamed joints. During the winter in northern climates if children are not exposed to sufficient sunlight, vitamin D can be taken as a dietary supplement to prevent rickets. Vitamin D deficiency can also be caused by the body's inability to absorb fats in which vitamin D is soluble. This condition can occur in adults who suffer from digestive disorders and can be one cause of "adult rickets," or **osteomalacia** (os′tē-ō-mă-lā′shē-ă), which is a softening of the bones as a result of calcium depletion.

Vitamin C is necessary for collagen synthesis by osteoblasts. Normally, as old collagen is broken down, new collagen is synthesized to replace it. Vitamin C deficiency results in bones and cartilage that are deficient in collagen because collagen synthesis is impaired. In children, vitamin C deficiency can cause growth retardation. In children and adults, vitamin C deficiency can result in **scurvy,** which is marked by ulceration and hemorrhage in almost any area of the body because of the lack of normal collagen synthesis in connective tissues. Wound healing, which requires collagen synthesis, is hindered in patients with vitamin C deficiency. In extreme cases the teeth can fall out because the ligaments that hold them in place break down.

Hormones

Hormones are very important in bone growth. **Growth hormone** from the anterior pituitary increases general tissue growth (see chapters 17 and 18), including overall bone growth, by stimulating interstitial cartilage growth and appositional bone growth. **Thyroid hormone** is also required for normal growth of all tissues, including cartilage; therefore, a decrease in this hormone can result in decreased size of the individual. **Sex hormones** also influence bone growth. Estrogen (a class of female sex hormones) and testosterone (a male sex hormone) initially stimulate bone growth, which accounts for the burst of growth at the time of puberty when production of these hormones increases. Both hormones also stimulate ossification of epiphyseal plates, however, and thus the cessation of growth. Females usually stop growing earlier than males because estrogens cause a quicker closure of the epiphyseal plate than does testosterone. Because their entire growth period is somewhat shorter, females usually do not reach the same height as males. Decreased levels of testosterone or estrogen can prolong the growth phase of the epiphyseal plates, even though the bones grow more slowly. Growth is very complex, however, and is influenced by many factors in addition to sex hormones, such as other hormones, genetics, and nutrition.

8   P R E D I C T

A 12-year-old female has an adrenal tumor that produces large amounts of estrogen. If untreated, what effect will this condition have on her growth for the next 6 months? On her height when she is 18?

✔ *Answer in Appendix F*

Bone Remodeling

Bone remodeling is the removal of old bone by osteoclasts and the deposition of new bone by osteoblasts. It converts woven bone into lamellar bone, and it is involved in bone growth, changes in bone shape, the adjustment of the bone to stress, bone repair, and calcium ion regulation in the body. For example, as a long bone increases in length and diameter, the size of the medullary cavity also increases (figure 6.17). Otherwise, the bone would consist of nearly solid bone matrix and would be very heavy. A cylinder with the same height, weight, and composition as a solid rod but with a greater diameter can support much more weight than the rod without bending. Bone therefore has a mechanical advantage as a cylinder rather than as a rod. The relative thickness of compact bone is maintained by the removal of bone on the inside by osteoclasts and the addition of bone to the outside by osteoblasts.

Remodeling is also responsible for the formation of new osteons in compact bone. This process occurs in two ways (figure 6.18). First, a few osteoclasts in the periosteum remove bone, resulting in groove formation along the surface of the bone. Periosteal capillaries lie within these grooves and become surrounded as the osteoblasts of the periosteum form new bone. Additional lamellae then are added around the capillary until an osteon results. Second, within already existing osteons, osteoclasts enter a central canal through the blood vessels and begin to remove bone from the center of the osteon, resulting in an enlarged channel through the bone. New concentric lamellae are then formed around the vessels until the new osteon fills the area occupied by the old osteon.

Osteons with their concentric lamellae are constantly being removed by osteoclasts, and new osteons are being formed by osteoblasts. This process, however, leaves portions of older osteons, called interstitial lamellae, between the newly developed osteons (see figure 6.10).

Clinical Note

Remodeling, the formation of additional bone, alteration in trabecular alignment to reinforce the scaffolding, or other changes can modify the strength of the bone in response to the amount of stress applied to it. Mechanical stress applied to bone increases osteoblast activity in bone tissue, and removal of mechanical stress decreases osteoblast activity. Under conditions of reduced stress, such as when a person is bedridden or paralyzed, osteoclast activity continues at a nearly normal rate, but osteoblast activity is reduced, resulting in a decrease in bone density. In addition, pressure in bone causes an electrical change that decreases the bone reabsorption activity of osteoclasts. Applying weight (pressure) to a broken bone therefore speeds the healing process. Weak pulses of electric current applied to a broken bone sometimes are used clinically to speed the healing process.

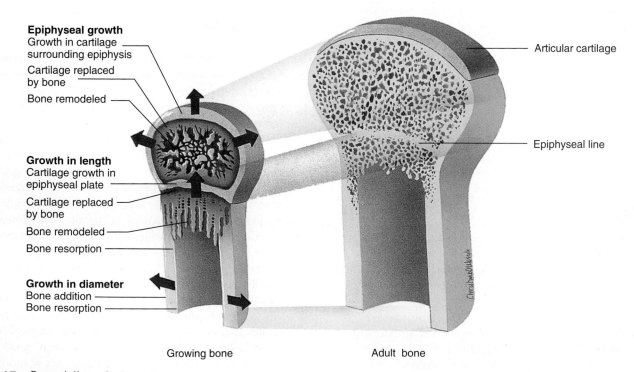

Epiphyseal growth
Growth in cartilage surrounding epiphysis
Cartilage replaced by bone
Bone remodeled

Growth in length
Cartilage growth in epiphyseal plate
Cartilage replaced by bone
Bone remodeled
Bone resorption

Growth in diameter
Bone addition
Bone resorption

Articular cartilage

Epiphyseal line

Growing bone   Adult bone

Figure 6.17   Remodeling of a Long Bone

The diameter of the bone increases as a result of bone growth on the outside of the bone, and the size of the medullary cavity increases because of bone resorption. The diaphysis increases in length and the epiphysis enlarges as new cartilage is formed and replaced by bone, which is remodeled.

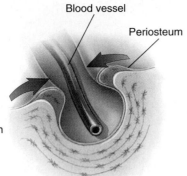

Blood vessel

Periosteum

The surface of a bone consists of grooves and ridges. Blood vessels in the periosteum tend to lie in the grooves. New bone is added to the ridges (arrows), building them up.

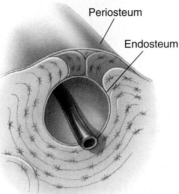

Periosteum

Endosteum

When the bone built on adjacent ridges meets, the groove is transformed into a tunnel. The periosteum of the groove becomes the endosteum of the tunnel.

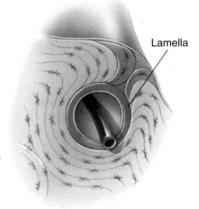

Lamella

Appositional growth by osteoblasts from the endosteum results in the formation of a new lamella.

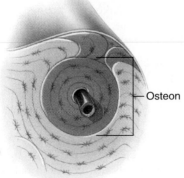

Osteon

Additional bone apposition fills in the tunnel, thus completing the osteon.

Figure 6.18   Formation of a New Osteon

Bone Repair

Bone is a living tissue that can undergo repair following damage to the bone. This process has five major steps.

1. *Hematoma formation* (figure 6.19*a*). When bone is fractured, the blood vessels in the bone and surrounding periosteum are damaged, and a hematoma forms. A **hematoma** (hē-mă-tō′mă) is a localized mass of blood released from blood vessels, but confined within an organ or space. Usually the blood in a hematoma forms a clot, which consists of fibrous proteins that stop the bleeding. Disruption of blood vessels in the central canals results in inadequate blood delivery to osteocytes, and bone tissue adjacent to the fracture site dies. Inflammation and swelling of tissues around the bone often occur following the injury.

2. *Internal callus formation* (figure 6.19*b*). A **callus** (kal′ŭs) is a mass of tissue that forms at a fracture site and connects the broken ends of the bone. An **internal callus** forms *between* the ends of the broken bone, and in the marrow cavity if the fracture occurs in the diaphysis of a long bone. Several days after the fracture, blood vessels grow into the clot. As the clot dissolves (see chapter 19), phagocytic cells clean up cell debris, osteoclasts break down dead bone tissue, and fibroblasts produce collagen and other extracellular materials to form a delicate granulation tissue (see chapter 4). As the fibroblasts continue to produce collagen fibers, a denser fibrous network, which helps to hold the bone together, is produced. Chondro-blasts derived from osteoprogenitor cells of the periosteum and endosteum begin to produce cartilage in the fibrous network, resulting in the formation of fibrocartilage. As these events are occurring, osteoprogenitor cells in the endosteum become osteoblasts, producing new bone that contributes to the internal callus.

3. *External callus formation* (figure 6.19*b*). The **external callus** forms a collar *around* the opposing ends of the bone fragments. Osteoprogenitor cells from the periosteum become osteoblasts, which produce bone, and chondroblasts, which produce cartilage. Cartilage production is more rapid than bone production, and cartilage from either side of the break eventually grows together. The external callus is a bone–cartilage collar that stabilizes the ends of the broken bone. In modern medical practice, stabilization of the bone is assisted by using a cast or surgical implantation of metal supports.

4. *Cartilage ossification* (figure 6.19*c*). Like the cartilage models formed during development, the cartilage in the external callus is replaced by woven, cancellous bone through endochondral ossification. The result is a stronger external callus. Even as the fibrocartilage of the internal callus is forming and replacing the hematoma, osteoblasts from the periosteum and endosteum enter the fibrocartilage and begin to produce bone. Eventually the fibrocartilage of the internal callus is replaced by woven, cancellous bone, which further stabilizes the broken bone.

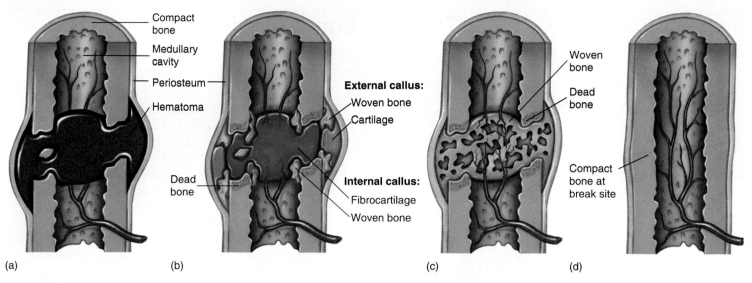

Figure 6.19   Bone Repair

(*a*) Hematoma formation following a fracture. (*b*) Callus formation. The internal callus replaces the hematoma. The external callus provides support. (*c*) Callus ossification. Woven, cancellous bone replaces the cartilage of the internal and external callus. (*d*) Remodeling of bone replaces the woven bone of the callus and the dead bone adjacent to the fracture site with compact bone. Healing is complete.

5. *Remodeling of bone* (figure 6.19*d*). Filling the gap between bone fragments with an internal callus of woven bone is not the end of the repair process because woven bone is not as structurally strong as the original lamellar bone. Repair is not complete until the woven bone of the internal callus and the dead bone adjacent to the fracture site are replaced by compact bone in which osteons from both sides of the break extend across the fracture line to "peg" the bone fragments together. This remodeling process takes time and may not be complete even after a year. As the internal callus is remodeled and becomes stronger, there is less need for the external callus, which is reduced in size by osteoclast activity. Eventually, repair may be so complete that no evidence of the break remains, however, the repaired zone usually remains slightly thicker than the adjacent bone. If the fracture occurred in the diaphysis of a long bone, remodeling also restores the medullary cavity.

Clinical Note

Before formation of compact bone between the broken ends of a bone can take place, the appropriate substrate must be present. Normally this is the woven, cancellous bone of the internal callus. If formation of the internal callus is prevented by infections, bone movements, or the nature of the injury, then nonunion of the bone occurs. This condition can be treated by surgically implanting an appropriate substrate such as living bone taken from another site in the body or dead bone from cadavers. Other substrates have also been used. For example, a specific marine coral calcium phosphate is converted into a predominantly hydroxyapatite biomatrix that is very much like cancellous bone.

Calcium Homeostasis

Bones play an important role in regulating blood calcium levels, which must be maintained within narrow limits for functions such as muscle contraction and membrane potentials to occur normally (see chapters 9, 10, and 12). Bone is the major storage site for calcium in the body, and movement of calcium into and out of bone helps to determine blood calcium levels. Calcium moves into bone as osteoblasts build new bone and out of bone as osteoclasts break down bone (figure 6.20). When osteoblast and osteoclast activity is balanced, the movement of calcium into and out of a bone is equal.

When blood calcium levels are too low, osteoclast activity increases. More calcium is released by osteoclasts from bone into the blood than is removed by osteoblasts from the blood to make new bone. Consequently, there is a net movement of calcium from bone into blood, and blood calcium levels increase. Conversely, if blood calcium levels are too high, osteoclast activity decreases. Less calcium is released by osteoclasts from bone into the blood than is taken from the blood by osteoblasts to produce new bone. As a result, there is a net movement of calcium from the blood to bone, and blood calcium levels decrease.

Parathyroid hormone (PTH) from the parathyroid glands (see figure 17.1) is the major regulator of blood calcium levels. If the blood calcium level decreases, the secretion of parathyroid hormone increases, resulting in increased osteoclast activity, which causes increased bone breakdown and increased blood calcium levels (see figure 6.20). The effect of PTH on osteoclasts is indirect, however, because they do not have receptors for PTH. Instead, receptors for PTH are on osteoblasts, which respond to PTH by releasing an

Clinical Focus   Bone Fractures

Bone fractures are classified in several ways. The most commonly used classification involves the severity of injury to the soft tissues surrounding the bone. An **open fracture** (formerly called compound) occurs when an open wound extends to the site of the fracture or when a fragment of bone protrudes through the skin. If the skin is not perforated, the fracture is called a closed fracture (formerly called simple). If the soft tissues around a closed fracture are damaged, the fracture is complicated.

Two other terms to designate fractures are **incomplete**, in which the fracture does not extend completely across the bone, and **complete,** in which the bone is broken into at least two fragments (figure A*a*). An incomplete fracture that occurs on the convex side of the curve of the bone is a **greenstick fracture. Hairline fractures** are incomplete fractures in which the two sections of bone do not separate; they are common in skull fractures.

Comminuted (kom′i-nū-ted) **fractures** are complete fractures in which the bone breaks into more than two pieces—-usually two major fragments and a smaller fragment (figure A*b*). **Impacted fractures** are those in which one fragment is driven into the cancellous portion of the other fragment (figure A*c*).

Fractures are also classified according to the direction of the fracture within a bone. **Linear** fractures run parallel to the long axis of the bone, and **transverse** fractures are at right angles to the long axis (see figure A*b*). **Spiral** fractures have a helical course around the bone, and **oblique** fractures run obliquely in relation to the long axis (figure A*d*). **Dentate** fractures have rough, toothed, broken ends, and **stellate** fractures have breakage lines radiating from a central point.

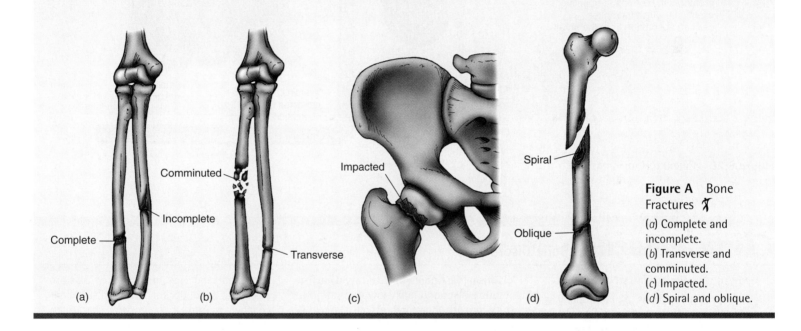

Figure A   Bone Fractures ✗
(*a*) Complete and incomplete.
(*b*) Transverse and comminuted.
(*c*) Impacted.
(*d*) Spiral and oblique.

osteoclast-stimulating factor. In addition, osteoblasts respond to PTH by releasing enzymes that result in the breakdown of the layer of unmineralized organic bone matrix covering bone, making the mineralized bone matrix available to osteocytes.

PTH also regulates blood calcium levels by increasing calcium uptake in the small intestine (see figure 6.20). Increased PTH promotes the formation of vitamin D in the kidneys, and vitamin D increases the absorption of calcium from the small intestine. PTH also increases the reabsorption of calcium from urine in the kidneys, which reduces calcium lost in the urine (see figure 6.20).

Tumors that secrete large amounts of parathyroid hormone can cause so much bone breakdown that bones are weakened and fracture easily. On the other hand, an increase in blood calcium levels results in less parathyroid hormone secretion, decreased osteoclast activity, reduced calcium release from bone, and decreased blood calcium levels.

Calcitonin (kal-si-tō′nin), secreted from the thyroid gland (see figure 17.1), decreases osteoclast activity (see figure 6.20). An increase in blood calcium levels stimulates the thyroid gland to secrete calcitonin, which inhibits osteoclast activity. PTH and calcitonin are described more fully in chapters 18 and 27.

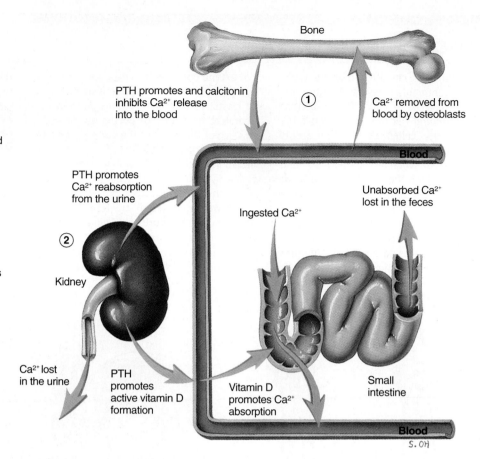

1. Osteoclasts break down bone and release calcium into the blood, and osteoblasts remove calcium from the blood to make bone. PTH regulates blood calcium levels by indirectly stimulating osteoclast activity, resulting in increased calcium release into the blood. Calcitonin plays a minor role in calcium maintenance by inhibiting osteoclast activity.

2. In the kidneys, PTH also increases calcium reabsorption and promotes the formation of active vitamin D, which increases calcium absorption from the small intestine.

Bone

PTH promotes and calcitonin inhibits Ca²⁺ release into the blood

①

Ca²⁺ removed from blood by osteoblasts

Blood

PTH promotes Ca²⁺ reabsorption from the urine

②

Kidney

Ingested Ca²⁺

Unabsorbed Ca²⁺ lost in the feces

Ca²⁺ lost in the urine

PTH promotes active vitamin D formation

Vitamin D promotes Ca²⁺ absorption

Small intestine

Blood
S. OH

Figure 6.20   Calcium Homeostasis 🏃

Clinical Focus   Bone Disorders

Growth and Development Disorders

Giantism is a condition of abnormally increased height that usually results from excessive cartilage and bone formation at the epiphyseal plates of long bones. The most common type of giantism, **pituitary giantism,** results from excess secretion of pituitary growth hormone. The large stature of some individuals, however, can result from genetic factors rather than from abnormal levels of growth hormone.

Acromegaly (ak-rō-meg′ă-lē) is also caused by excess pituitary growth hormone secretion; however, acromegaly involves growth of connective tissue, including bones, after the epiphyseal plates have ossified. The effect mainly involves increased diameter of all bones and is most strikingly apparent in the face and hands. Many pituitary giants also develop acromegaly later in life.

Dwarfism, the condition in which a person is abnormally short, is the opposite of

giantism (see figure B*a*). **Pituitary dwarfism** results when abnormally low levels of pituitary growth hormone affect the whole body, thus producing a small person who is normally proportioned. **Achondroplastic** (ă-kon-drō-plas′tik) **dwarfism,** results in a disproportionate shortening of the long bones. It is more common than proportionate dwarfing and produces a person with a nearly normal-sized trunk and head but shorter-than-normal limbs. Most cases of achondroplastic dwarfism are the result of genetic defects that cause deficient or improper growth of the cartilage model, especially the epiphyseal plate, and often involve deficient collagen synthesis. Often the cartilage matrix does not have its normal integrity, and the chondrocytes of the epiphysis cannot form their normal columns, even though rates of cell proliferation may be normal.

Osteogenesis imperfecta (os′tē-ō-jen′ĕ-sis im-per-fek′tă), a group of genetic disor-

ders producing very brittle bones that are easily fractured, occurs because insufficient collagen is formed to properly strengthen the bones. Intrauterine fractures of the extremities usually heal in poor alignment, causing the limbs to appear bent and short (figure B*b*). Several other hereditary disorders of bone mineralization involve the enzymes responsible for normal phosphate or calcium metabolism. They closely resemble rickets and result in weak bones.

Bacterial Infections

Osteomyelitis (os′tē-ō-mī-ĕ-lī′tis) is bone inflammation that often results from bacterial infection.

It can lead to complete destruction of the bone. *Staphylococcus aureus,* often introduced into the body through wounds, is a common cause of osteomyelitis (figure B*c*). Bone tuberculosis, a specific type of osteomyelitis, results from spread of the tubercular bacterium (*Mycobacterium tuber-*

culosis) from the initial site of infection such as the lungs to the bones through the circulatory system.

Tumors

There are many types of tumors, with a wide range of resultant bone defects and prognoses (figure B*d*). Tumors can be benign or malignant. Malignant bone tumors can metastasize to other parts of the body, or they can spread to bone from metastasizing tumors elsewhere in the body.

Decalcification

Osteomalacia (os′tē-ō-mă-lā′shē-ă), or the softening of bones, results from calcium depletion from bones. If the body has an unusual need for calcium, such as during pregnancy when growth of the fetus requires large amounts of calcium, it can be removed from the mother's bones, which consequently become soft and weakened.

Osteoporosis, which is a major disorder of decalcification, is discussed in the Systems Pathology later in this chapter.

Figure B   Bone Disorders
(*a*) Giant and dwarf.

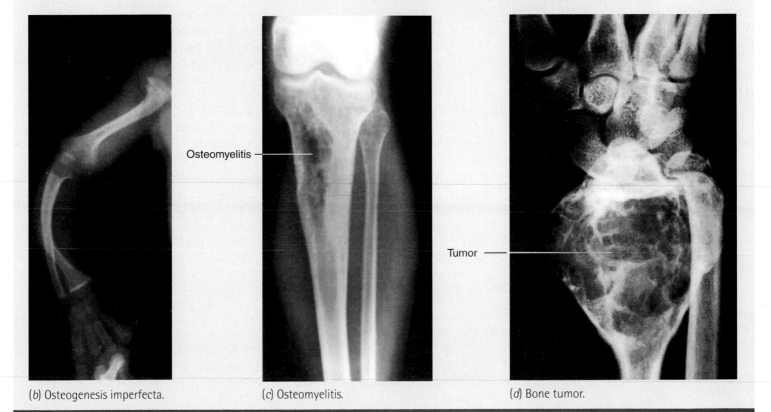

(*b*) Osteogenesis imperfecta.   (*c*) Osteomyelitis.   (*d*) Bone tumor.

Systems Pathology
Osteoporosis

OSTEOPOROSIS

Mrs. B is a 70-year-old grandmother. Since she was a teenager, she has been a heavy smoker. She is typically sedentary, seldom goes outside, did not have the best dietary habits, and was underweight. One of her favorite yearly events was the family picnic on the fourth of July. During the picnic, misfortune struck when Mrs. B tripped on a lawn sprinkler and fell. She was unable to stand because of severe hip pain, so she was rushed to the hospital where a radiograph revealed that the neck of her femur was fractured (figure C*a*) and that she had osteoporosis (figure C*b*).

It was decided that hip replacement surgery was indicated. Before the surgery could be performed, however, a fat embolism from the fracture site lodged in her lungs, making it difficult for her to breathe. The surgery was postponed and the fracture immobilized until she recovered from the fat embolism. Three weeks after the accident, Mrs. B had a successful hip transplant and began physical therapy. She appeared to be on the road to recovery, but 6 weeks after the surgery she developed persistent pain and edema in her hip. A bone biopsy confirmed a postoperative infection that was successfully treated with antibiotics.

BACKGROUND INFORMATION

Osteoporosis (os′tē-ō-pō-rō′sis), or porous bone, results from reduction in the overall quantity of bone tissue. It occurs when the rate of bone reabsorption exceeds the rate of bone formation. The loss of bone mass makes bones so porous and weakened that they become deformed and prone to fracture. The occurrence of osteoporosis increases with age. In both men and women, bone mass starts to decrease at about age 40 and continually decreases thereafter. Women can eventually lose approximately half, and men a quarter, of their cancellous bone. Osteoporosis is 2.5 times more common in women than in men.

In postmenopausal women, the decreased production of the female sex hormone, estrogen, can cause osteoporosis. Estrogen is secreted by the ovaries, and it normally contributes to the maintenance of normal bone mass by inhibiting the stimulatory effects of PTH on osteoclast activity. Following menopause, estrogen production decreases, resulting in degeneration of cancellous bone, especially in the vertebrae of the spine and the bones of the forearm. Collapse of the vertebrae can cause a decrease in height or, in more severe cases, produce kyphosis, or a "dowager's hump," in the upper back.

Conditions that result in decreased estrogen levels, other than menopause, can also cause osteoporosis. Examples include removal of the ovaries before menopause, extreme exercise to the point of amenorrhea (lack of menstrual flow), anorexia nervosa (self-starvation), and cigarette smoking.

In males, reduction in testosterone levels can cause loss of bone tissue. Decreasing testosterone levels are usually less of a problem for men than decreasing estrogen levels are for women for two reasons. First, because males have denser bones than females, loss of some bone tissue has less of an effect. Second, testosterone levels generally don't decrease significantly until after age 65, and even then the rate of decrease is often slow.

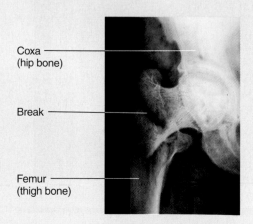

Coxa
(hip bone)

Break

Femur
(thigh bone)

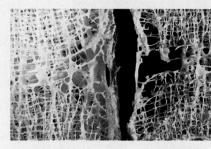

Normal bone   Osteoporotic bone

Figure C Osteoporosis

(*a*) Radiograph of a broken hip. A "broken hip" is actually a break of the femur (thigh bone) in the hip region. (*b*) Photomicrograph of normal bone and osteoporotic bone.

Overproduction of PTH, which results in overstimulation of osteoclast activity, can also cause osteoporosis.

Inadequate dietary intake or absorption of calcium can contribute to osteoporosis. Absorption of calcium from the small intestine decreases with age, and individuals with osteoporosis often have insufficient intake of calcium, vitamin D, and vitamin C. Drugs that interfere with calcium uptake or use can also increase the risk of osteoporosis.

Finally, osteoporosis can result from inadequate exercise or disuse caused by fractures or paralysis. Significant amounts of bone are lost after 8 weeks of immobilization.

Treatments for osteoporosis are designed to reduce bone loss or increase bone formation or both. Increased dietary calcium and vitamin D can increase calcium uptake and promote bone formation. Daily doses of 1000–1500 mg of calcium and 800 IU (20 μg) of vitamin D are recommended. Exercise, such as walking or using light weights, also appears to be effective not only in reducing bone loss but in increasing bone mass.

In postmenopausal women, estrogen replacement therapy decreases osteoclast activity, which reduces bone loss, but does not result in an increase in bone mass because osteoclast activity still exceeds osteoblast activity. Calcitonin (Mi-

acalcin), which inhibits osteoclast activity, is now available as a nasal spray. Calcitonin can be used to treat osteoporosis in men and women and has been shown to produce a slight increase in bone mass. Alendronate (Fosamax) belongs to a class of drugs called bisphosphonates. Bisphosphonates bind to hydroxyapatite and inhibit bone resorption by osteoclasts. Alendronate increases bone mass and reduces fracture rates even more effectively than calcitonin. Slow-releasing sodium fluoride (Slow Fluoride) in combination with calcium citrate (Citracal) also appears to increase bone mass.

Early diagnosis of osteoporosis may lead to the use of more preventative treatments. Instruments that measure the absorption of photons (particles of light) by bone are currently used, of which dual-energy x-ray absorptiometry (DEXA) is considered the best.

9 P R E D I C T

What advice should Mrs. B give to her granddaughter so that the granddaughter will be less likely to develop osteoporosis when she is Mrs. B's age?

✔ *Answer in Appendix F*

System Interactions

System	Interactions
Integumentary	Decreased exposure to sunlight because of an indoor life-style, reduces vitamin D production and decreases calcium absorption. Surgical wounds through the skin can allow the entry of bacteria, resulting in postoperative infections.
Muscular	A sedentary life-style and decreased body weight reduces stress on bone and contributes to osteoporosis. Muscle atrophy and weakness make it difficult to maintain balance, which increases the likelihood of falling and injury. Following hip replacement surgery, physical therapy places stress on the bones and improves muscular strength.
Nervous	Pain sensations following the injury and during rehabilitation help to prevent further injury.
Endocrine	Although not a factor in this case of osteoporosis, elevated PTH (usually from a benign parathyroid tumor) or elevated thyroid hormone (Grave's disease) can result in excessive osteoclast activity. Calcitonin is being used to treat osteoporosis.
Cardiovascular	Blood clotting following the injury starts the process of tissue repair. Blood cells are carried to the injury site to fight infections and remove cell debris. Blood vessels grow into the recovering tissue, providing nutrients and removing waste products.
Lymphatic and Immune	Immune cells resist infections and release chemicals that promote tissue repair. New immune cells are produced in bone marrow.
Respiratory	Excessive smoking lowers estrogen levels, which increases bone loss. A fat embolism from a fractured bone can impair respiration.
Digestive	Inadequate calcium and vitamin D in the diet or inadequate calcium absorption by the digestive system can contribute to osteoporosis.
Urinary	Calcium released from the bones is excreted through the urinary system.
Reproductive	Decreased estrogen levels following menopause contribute to osteoporosis.

Summary

Functions of the Skeletal System

1. The skeletal system consists of bones, cartilage, tendons, and ligaments.
2. The skeletal system supports the body, protects organs it surrounds, allows body movements, stores minerals and fats, and is the site of blood cell production.

Cartilage

1. Chondroblasts produce cartilage and become chondrocytes. Chondrocytes are located in lacunae surrounded by matrix.
2. The matrix of cartilage contains collagen fibers (for strength) and proteoglycans (trap water).
3. The perichondrium surrounds cartilage.
   - The outer layer contains fibroblasts.
   - The inner layer contains chondroblasts.
4. Cartilage grows by appositional and interstitial growth.

Bone Anatomy
Bone Shapes

Individual bones can be classified as long, short, flat, or irregular.

Structure of a Long Bone

1. The diaphysis is the shaft of a long bone, and the epiphyses are the ends.
2. The epiphyseal plate is the site of bone growth in length.
3. The medullary cavity is a space within the diaphysis.
4. Red marrow is the site of blood cell production, and yellow marrow consists of fat.
5. The periosteum covers the outer surface of bone.
   - The outer layer contains blood vessels and nerves.
   - The inner layer contains osteoblasts, osteoclasts and osteoprogenitor cells.
   - Perforating fibers hold the periosteum, ligaments, and tendons in place.
6. The endosteum lines cavities inside bone and contains osteoblasts, osteoclasts, and osteoprogenitor cells.

Structure of Flat, Short, and Irregular Bones

Flat, short, and irregular bones have an outer covering of compact bone surrounding cancellous bone.

Bone Histology
Bone Matrix

1. Collagen provides flexible strength.
2. Hydroxyapatite provides compressional strength.

Bone Cells

1. Osteoblasts produce bone matrix and become osteocytes.
   - Osteoblasts connect to one another through cell processes and surround themselves with bone matrix to become osteocytes.
   - Osteocytes are located in lacunae and are connected to one another through canaliculi.
2. Osteoclasts (with assistance from osteoblasts) break down bone.
3. Osteoblasts originate from osteoprogenitor cells, whereas osteoclasts originate from stem cells in red bone marrow.

Woven and Lamellar Bone

1. Woven bone has collagen fibers oriented in many different directions. It is remodeled to form lamellar bone.
2. Lamellar bone is arranged in thin layers, called lamellae, which have collagen fibers oriented parallel to one another.

Cancellous and Compact Bone

1. Cancellous bone has many spaces.
   - Lamellae combine to form trabeculae, beams of bone that interconnect to form a latticelike structure with spaces filled with bone marrow and blood vessels.
   - The trabeculae are oriented along lines of stress and provide structural strength.
2. Compact bone is dense with few spaces.
   - Compact bone consists of organized lamellae: circumferential lamellae cover the outer surface of compact bones; concentric lamellae surround central canals, forming osteons; interstitial lamellae are remnants of lamellae left after bone remodeling.
   - Canals within compact bone provide a means for the exchange of gases, nutrients, and waste products. From the periosteum or endosteum perforating canals carry blood vessels to central canals, and canaliculi connect central canals to osteocytes.

Bone Development
Intramembranous Ossification

1. Some skull bones, part of the mandible, and the diaphyses of the clavicles develop from membranes.
2. Within the membrane at centers of ossification, osteoblasts produce bone along the membrane fibers to form cancellous bone.
3. Beneath the periosteum osteoblasts lay down compact bone to form the outer surface of the bone.
4. Fontanels are areas of membrane that are not ossified at birth.

Endochondral Ossification

1. Most bones develop from a cartilage model.
2. The cartilage matrix is calcified, and chondrocytes dies. Osteoblasts form bone on the calcified cartilage matrix, producing cancellous bone.
3. An outer surface of compact bone is formed beneath the periosteum by osteoblasts.
4. Primary ossification centers form in the diaphysis during fetal development. Secondary ossification centers form in the epiphyses.
5. Articular cartilage on the ends of bones and the epiphyseal plate is cartilage that does not ossify.

Bone Growth

Growth at the Epiphyseal Plate

1. Epiphyseal plate growth involves the interstitial growth of cartilage followed by appositional bone growth on the cartilage.
2. Epiphyseal plate growth results in an increase in the length of the diaphysis and bony processes. Bone growth in length ceases when the epiphyseal plate becomes ossified and forms the epiphyseal line.

Growth at Articular Cartilage

1. Articular cartilage growth involves the interstitial growth of cartilage followed by appositional bone growth on the cartilage.
2. Articular cartilage growth results in larger epiphyses and an increase in the size of bones that do not have epiphyseal plates.

Other Bone Growth

1. Appositional bone growth beneath the periosteum increases the diameter of long bones and the size of other bones.
2. Appositional bone growth increases the size of trabeculae and is necessary for the formation of circumferential and concentric lamellae.

Factors Affecting Bone Growth

1. Genetic factors determine bone shape and size. The expression of genetic factors can be modified.
2. Factors that alter the mineralization process or production of organic matrix, such as deficiencies in vitamins D and C, can affect bone growth.
3. Growth hormone, thyroid hormone, estrogen, and testosterone stimulate bone growth.
4. Estrogen and testosterone cause increased bone growth and closure of the epiphyseal plate.

Bone Remodeling

1. Remodeling converts woven bone to lamellar bone and allows bone to change shape, adjust to stress, repair, and regulate body calcium levels.
2. Bone adjusts to stress by adding new bone and by realignment of bone through remodeling.

Bone Repair

1. Fracture repair begins with the formation of a hematoma.
2. The hematoma is replaced by the internal callus consisting of fibrocartilage.
3. The external callus is a bone–cartilage collar that stabilizes the ends of the broken bone.
4. The internal and external calluses are ossified to become woven bone.
5. Woven bone is remodeled.

Calcium Homeostasis

PTH increases bone breakdown and thus increases blood calcium levels. Calcitonin has the opposite effect.

Content Review

1. What are the components of the skeletal system?
2. List the functions of the skeletal system.
3. Describe the structure of cartilage. Where are chondrocytes and chondroblasts found?
4. What is the perichondrium? Describe its structure.
5. How does cartilage grow?
6. Describe the four basic shapes of individual bones.
7. Define diaphysis, epiphysis, epiphyseal plate, and epiphyseal line.
8. Define red and yellow marrow. Where are they located in an adult and in a child?
9. Where are the periosteum and the endosteum located?
10. Name the extracellular components of the bony matrix, and explain their contribution to the strength of bone.
11. What are osteoblasts, osteocytes, and osteoclasts?
12. Describe the structure of cancellous bone. What are trabeculae, and what is their function?
13. Describe the structure of compact bone. What is an osteon? How does compact bone differ from cancellous bone?
14. Name the three types of lamellae found in compact bone.
15. Describe the canal system that brings nutrients to osteocytes.
16. How are cancellous bone and compact bone formed during intramembranous and endochondral ossification?
17. When and where do primary and secondary ossification centers appear?
18. How does a bone increase in width? Describe the increase in length of bone at the epiphyseal plate and articular cartilage.
19. Name the two basic components of bone that are affected by nutritional status. How do vitamins D and C affect bone growth?
20. Describe the conditions that result from too little or too much growth hormone in children and in adults.
21. What effects do estrogen and testosterone have on bone growth? How do these effects account for the average height difference observed in men and women?
22. What cells are involved in bone remodeling? What is accomplished by remodeling bones?
23. How does bone adjust to stress? Describe the role of osteoblasts and osteoclasts in this process. What happens to bone that is not subjected to stress?
24. Describe the repair of a broken bone.
25. Name the hormones that regulate calcium levels in the body. Describe the effect of these hormones on the activity of bone cells.

Develop Your Reasoning Skills

1. When a person develops Paget's disease, for unknown reasons the collagen fibers in the bone matrix run randomly in all directions. In addition, the amount of trabecular bone decreases. What symptoms would you expect to observe?

2. When closure of the epiphyseal plate occurs, the cartilage of the plate is replaced by bone. Does this occur from the epiphyseal side of the plate or the diaphyseal side?

3. Assume that two patients have identical breaks in the femur (thigh bone). If one is bedridden and the other has a walking cast, which patient's fracture heals faster? Explain.

4. Explain why running helps prevent osteoporosis in the elderly. Does the benefit include all bones or mainly those of the legs and spine?

5. Astronauts can experience a dramatic decrease in bone density while in a weightless environment. Explain how this happens and suggest a way to slow the loss of bone tissue.

6. Would a patient suffering from kidney failure be more likely to develop osteomalacia or osteoporosis? Explain.

7. In some cultures eunuchs were responsible for guarding harems, which are the collective wives of one male. Eunuchs are males who, as boys, were castrated. Castration removes the testes, the major site of testosterone production in males. Because testosterone is responsible for the sex drive in males, the reason for castration is obvious. As a side effect of this procedure, the eunuchs grew to above-normal heights. Can you explain why?

8. When a long bone is broken, blood vessels at the fracture line are severed. The formation of blood clots stops the bleeding. Within a few days bone tissue on both sides of the fracture site dies. The bone only dies back a certain distance from the fracture line, however. Explain.

9. A patient has hyperparathyroidism because of a tumor in the parathyroid gland that produces excessive amounts of PTH. What effect does this hormone have on bone? Would administration of large doses of vitamin D help the situation? Explain.

Web Site Link

For a listing of the most current web sites related to this chapter, please visit the Seeley home page at:
http://www.mhhe.com/biosci/ap/seeleyap/

Chapter Seven

Skeletal System: Gross Anatomy

Objectives

1. Name the major bony landmarks, and explain the functional significance of each.

2. Describe the major features of the skull as seen from the superior, posterior, lateral, inferior, and frontal views.

3. Describe the major features of the floor of the cranial vault, including the foramina and what passes through them.

4. List the bones of the cranial vault and of the face.

5. Describe the four major curvatures of the vertebral column, explain what causes them, and indicate when these curvatures develop.

6. List the features that characterize the vertebrae of the cervical, thoracic, lumbar, and sacral regions.

7. Give the number of and explain the difference between true, false, and floating ribs.

8. Describe the shape of the three parts of the sternum and their relationship to the ribs.

9. Describe the bones of the pectoral girdle, and point out surface features that can be seen on a living human.

10. List the major features of the humerus, and give the function of each.

11. Describe the major features of the ulna and radius, and explain how these two bones interact when the radius is rotated around the ulna.

12. List the eight carpal bones, and describe the carpal tunnel.

13. Indicate the skeletal differences between the thumb and the fingers.

14. Describe the coxa in relation to the three fused bones composing it.

15. List and explain the differences between the male and the female pelves.

16. Describe the head and neck of the femur, and compare them to those of the humerus.

17. Describe the relationships among the tibia, fibula, and femur.

18. List the tarsal bones, and describe the relationships among the tibia, fibula, talus, and calcaneus.

Part Two

Without a skeletal system, we would have very little shape. We wouldn't be able to move much either, because most muscles act on bone to produce movement, often pulling on the bones with considerable force. Human bones are very strong, and can resist tremendous bending and compression forces without breaking. Nonetheless, each year nearly 2 million Americans manage to break one.

The skeletal system includes the bones, cartilages, ligaments, and tendons. The content of this chapter is confined almost entirely to the study of bones and some major cartilages. Ligaments are described in chapter 8 in relation to joints, and tendons are described in chapter 11 in relation to muscles.

Dried, prepared bones are used to study skeletal gross anatomy. This allows the major features of individual bones to be seen clearly without being obstructed by associated soft tissues, such as muscles, tendons, ligaments, cartilage, nerves, and blood vessels. As a consequence, however, it is easy to ignore the important relationships among bones and soft tissues and the fact that living bones have soft tissue, such as the periosteum (see chapter 6).

The named bones are divided into two categories: (1) the axial skeleton and (2) the appendicular skeleton. The axial skeleton consists of the skull, hyoid bone, vertebral column, and thoracic cage (rib cage). The appendicular skeleton consists of the limbs and their girdles.

General Considerations

It is traditional to list 206 bones in the average adult skeleton (table 7.1 and figure 7.1), although the actual number varies from person to person and decreases with age as some bones become fused.

Many of the anatomic features of bones are listed in table 7.2. Most of these features are based on the relationship between the bones and associated soft tissues. If a bone possesses a **tubercle** (lump) or **process** (projection), it is usually because a ligament or tendon was attached to that lump or projection during life. If a bone has a smooth, articular surface, that surface was part of a joint and was covered with articular cartilage. If the bone has a **foramen** (fō-rā′men, meaning, a hole) in it, that hole was occupied by something such as a nerve or blood vessel. Some bones contain mucous membrane-lined air spaces called **sinuses.** These bones are composed of paper-thin, translucent compact bone only and have little or no cancellous center (see chapter 6).

Axial Skeleton

The **axial skeleton** is divided into the skull, hyoid bone, vertebral column, and thoracic cage. The axial skeleton forms the upright axis of the body. It also protects the brain, the spinal cord, and the vital organs housed within the thorax.

Skull

The skull protects the brain; supports the organs of vision, hearing, smell, and taste; and provides a foundation for the structures that take air, food, and water into the body. When the skull is disassembled, the mandible is easily separated from the rest of the skull, which remains intact. Special effort is needed to separate the other bones. For this reason, it is convenient to think of the skull, except for the mandible, as a single unit. The top of the skull is usually cut off to reveal its interior. The skull has ridges, lines, processes, and plates—important for the attachment of muscles or for articulations between the bones of the skull. There are also a number of spaces and ridges inside the skull. Selected features of the intact skull are listed in table 7.3.

Superior View of the Skull

The skull appears quite simple when viewed from above. Only four bones are seen from this view: the frontal bone, two parietal bones, and a small part of the occipital bone. The paired **parietal bones** are joined at the midline by the **sagittal suture,** and the parietal bones are connected to the **frontal bone** by the **coronal suture** (figure 7.2).

Clinical Note

Sutural bones are usually small and bilateral and in many cases are apparently genetically determined. A large midline bone, called an Inca bone, may form at the junction of the lambdoid and sagittal sutures. The bone was common in the skulls of Incas and is still present in their Andean descendants.

Posterior View of the Skull

The parietal and occipital bones are the major structures seen from the posterior view (figure 7.3). The parietal bones are joined to the **occipital bone** by the **lambdoid** (lam′doyd; the shape resembles the Greek letter lambda) **suture.** Occasionally, extra small bones called **sutural** (sū′chūr-ăl) **bones** form along the lambdoid suture.

An **external occipital protuberance** is present on the posterior surface of the occipital bone (see figure 7.3). It can be felt through the scalp at the base of the head and varies considerably in size from person to person. The external occipital protuberance is the site of attachment of the **ligamentum nuchae** (nū′kē, nape of neck), an elastic ligament that extends down the neck and helps keep the head erect by pulling on the occipital region of the skull. **Nuchal lines** are a set of small ridges that extend laterally from the protuberance and are the points of attachment for several neck muscles.

Table 7.1   Number of Named Bones Listed by Category

Bones			Number	Bones			Number
Axial Skeleton				**Appendicular Skeleton**			
Skull				*Pectoral Girdle*			
Cranial vault				Scapula			2
Paired	Parietal		2	Clavicle			2
	Temporal		2	*Upper Limb*			
Unpaired	Frontal		1	Humerus			2
	Sphenoid		1	Ulna			2
	Occipital		1	Radius			2
	Ethmoid		1	Carpals			16
Face				Metacarpals			10
Paired	Maxilla		2	Phalanges			28
	Zygomatic		2		Total Upper Limb and Girdle		64
	Palatine		2	*Pelvic Girdle*			
	Lacrimal		2	Coxa			2
	Nasal		2	*Lower Limb*			
	Inferior nasal concha		2	Femur			2
Unpaired	Mandible		1	Tibia			2
	Vomer		1	Fibula			2
Auditory ossicles				Patella			2
Malleus			2	Tarsals			14
Incus			2	Metatarsals			10
Stapes			2	Phalanges			28
	Total Skull		28		Total Lower Limb and Girdle		62
					Total Appendicular Skeleton		126
Hyoid			1				
Vertebral Column					Total Axial Skeleton		80
Cervical vertebrae			7		Total Appendicular Skeleton		126
Thoracic vertebrae			12		Total Bones		206
Lumbar			5				
Sacrum			1				
Coccyx			1				
	Total Vertebral Column		26				
Thoracic Cage (Rib Cage)							
Ribs			24				
Sternum			1				
	Total Thoracic Cage		25				
	Total Axial Skeleton		80				

Clinical Note

The ligamentum nuchae and neck muscles in humans are not as strong as comparable structures in other animals; therefore, the human bony prominence and lines of the posterior skull are not as well developed as in those animals. The location of the human foramen magnum allows the skull to balance above the vertebral column and allows for an upright posture. Thus human skulls require less ligamental and muscular effort to balance the head on the spinal column than do the skulls of other animals, including other primates such as chimpanzees, whose skulls are not balanced over the vertebral column. The presence of small nuchal lines in hominids (i.e., animals with an upright stance like humans) reflects this decreased musculature and is one way by which anthropologists can establish probable upright posture in hominids.

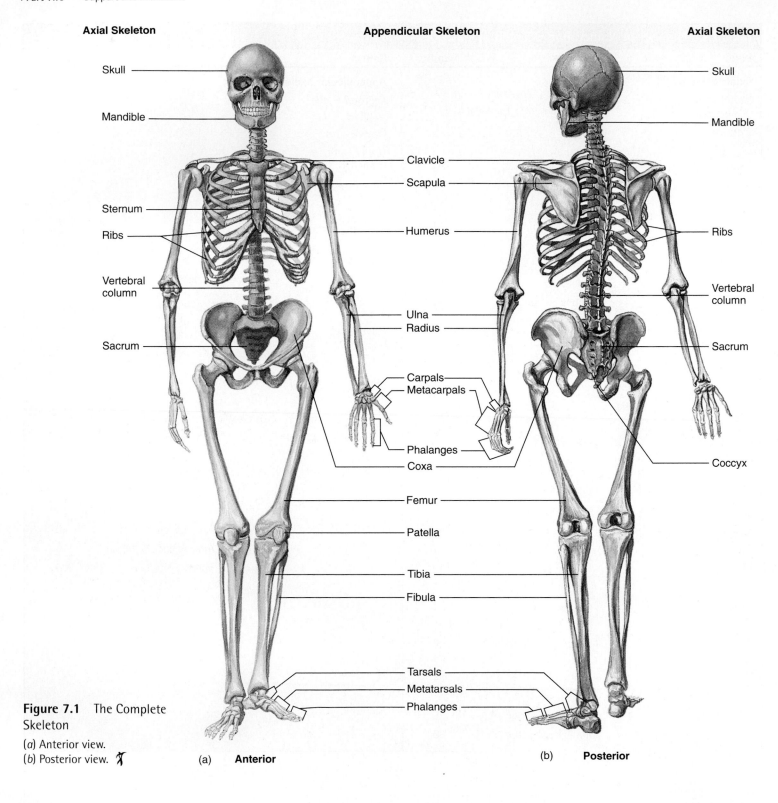

Axial Skeleton

Appendicular Skeleton

Axial Skeleton

Skull

Mandible

Sternum

Ribs

Vertebral column

Sacrum

Clavicle

Scapula

Humerus

Ulna

Radius

Carpals

Metacarpals

Phalanges

Coxa

Femur

Patella

Tibia

Fibula

Tarsals

Metatarsals

Phalanges

Skull

Mandible

Ribs

Vertebral column

Sacrum

Coccyx

Figure 7.1 The Complete Skeleton

(*a*) Anterior view.
(*b*) Posterior view.

(a)   **Anterior**

(b)   **Posterior**

Lateral View of the Skull

The parietal bone and the squamous part of the temporal bone form a large part of the side of the head (figure 7.4). The term temporal means related to time, and the temporal bone is so named because the hair of the temples is often the first to turn white, indicating the passage of time. The **squamous** **suture** joins these bones. A prominent feature of the temporal bone is a large hole, the **external auditory meatus** (mē-ā′tŭs; meaning, passageway or tunnel), which transmits sound waves toward the eardrum. The external ear, or auricle, surrounds the meatus. Just posterior and inferior to the external auditory meatus is a large inferior projection, the **mastoid**

Table 7.2   General Anatomic Terms for Various Features of Bones

Term	Description	Term	Description
Body	Main part	**Projections (continued)**	
Head	Enlarged (often rounded) end	Trochanter	Tuberosities on the proximal femur
Neck	Constriction between head and body	Epicondyle	Near or above a condyle
Margin or border	Edge	Lingula	Flat, tongue-shaped process
Angle	Bend	Hamulus	Hook-shaped process
Ramus	Branch off the body (beyond the angle)	Cornu	Horn-shaped process
Condyle	Smooth, rounded articular surface		
Facet	Small, flattened articular surface	**Openings**	
		Foramen	Hole
Ridges		Canal or meatus	Tunnel
Line or linea	Low ridge	Fissure	Cleft
Crest or crista	Prominent ridge	Sinus or labyrinth	Cavity
Spine	Very high ridge		
		Depressions	
Projections		Fossa	General term for a depression
Process	Prominent projection	Notch	Depression in the margin of a bone
Tubercle	Small, rounded bump	Fovea	Little pit
Tuberosity or tuber	Knob; larger than a tubercle	Groove or sulcus	Deeper, narrow depression

Table 7.3   Processes and Other Features of the Skull

Feature	Bone on Which Feature Is Found	Description
External Features		
Alveolar process	Mandible, Maxilla	Ridges on mandible and maxilla containing the teeth
Angle	Mandible	Posterior, inferior corner of mandible
Coronoid process	Mandible	Attachment point for the temporalis muscle
Genu	Mandible	Chin (resembles a bent knee)
Horizontal plate	Palatine	Posterior third of the hard palate
Mandibular condyle	Mandible	Region where the mandible articulates with the skull
Mandibular fossa	Temporal	Depression where the mandible articulates with the skull
Mastoid process	Temporal	Enlargement posterior to the ear; attachment site for several muscles that move the head
Nuchal lines	Occipital	Attachment points for several posterior neck muscles
Occipital condyle	Occipital	Point of articulation between the skull and the vertebral column
Palatine process	Maxilla	Anterior two-thirds of the hard palate
Pterygoid hamulus	Sphenoid	Hooked process on the inferior end of the medial pterygoid plate, around which the tendon of one palatine muscle passes; an important dental landmark
Pterygoid plates (medial and lateral)	Sphenoid	Bony plates on the inferior aspect of the sphenoid bone; the lateral pterygoid plate is the site of attachment for two muscles of mastication (chewing)
Ramus	Mandible	Portion of the mandible superior to the angle
Styloid process	Temporal	Attachment site for three muscles (to tongue, pharynx, and hyoid bone) and some ligaments
Temporal lines	Parietal	Where the temporalis muscle, which closes the jaw, attaches
Internal Features		
Crista galli	Ethmoid	Process in the anterior part of the cranial vault to which one of the connective tissue coverings of the brain (dura mater) connects
Petrous portion	Temporal	Thick, interior part of temporal bone; contains middle and inner ears and auditory ossicles
Sella turcica	Sphenoid	Bony structure resembling a saddle in which the pituitary gland is located

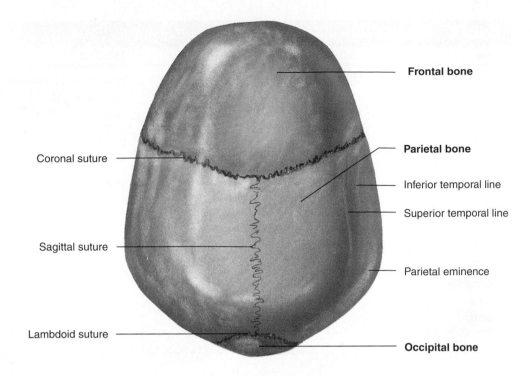

Figure 7.2 Skull as Seen from the Superior View

Frontal bone

Coronal suture

Parietal bone

Inferior temporal line

Superior temporal line

Sagittal suture

Parietal eminence

Lambdoid suture

Occipital bone

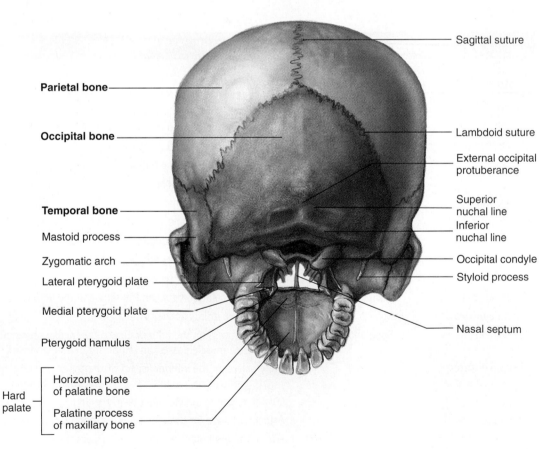

Figure 7.3 Skull as Seen from the Posterior View

Sagittal suture

Parietal bone

Occipital bone

Lambdoid suture

External occipital protuberance

Temporal bone

Superior nuchal line

Mastoid process

Inferior nuchal line

Zygomatic arch

Occipital condyle

Lateral pterygoid plate

Styloid process

Medial pterygoid plate

Pterygoid hamulus

Nasal septum

Horizontal plate of palatine bone

Hard palate

Palatine process of maxillary bone

(mas'toyd, meaning resembling a breast) **process.** The process can be seen and felt as a prominent lump just posterior to the ear (see figure 7.4). The process is not solid bone but is filled with cavities called the **mastoid air cells,** which are connected to the middle ear. Important neck muscles involved in rotation of the head attach to the mastoid process. The superior and inferior **temporal lines,** which are attachment points of the temporalis muscle, one of the major muscles of mastication, arch across the lateral surface of the parietal bone (see figure 7.4).

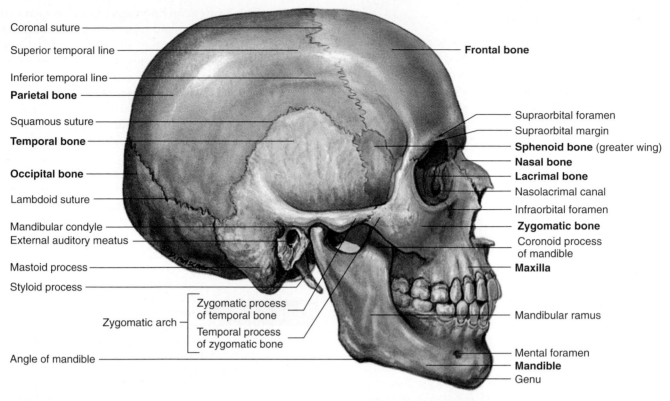

Coronal suture

Superior temporal line

Inferior temporal line

Parietal bone

Squamous suture

Temporal bone

Occipital bone

Lambdoid suture

Mandibular condyle
External auditory meatus

Mastoid process

Styloid process

Zygomatic arch — {
Zygomatic process
of temporal bone
Temporal process
of zygomatic bone
}

Angle of mandible

Frontal bone

Supraorbital foramen
Supraorbital margin
Sphenoid bone (greater wing)
Nasal bone
Lacrimal bone
Nasolacrimal canal
Infraorbital foramen
Zygomatic bone
Coronoid process
of mandible
Maxilla

Mandibular ramus

Mental foramen
Mandible
Genu

Figure 7.4   Lateral View of the Skull as Seen from the Right Side

The lateral surface of the **greater wing** of the **sphenoid** (sfē′noyd, meaning wedge-shaped) **bone** is immediately anterior to the temporal bone (see figure 7.4). Although appearing to be two bones, one on each side of the skull, the sphenoid bone is actually a single bone that extends completely across the skull. Anterior to the sphenoid bone is the **zygomatic** (zī′gō-mat′ik, meaning a bar or yoke) **bone,** or cheek bone (see figure 7.4), which can be easily seen and felt on the face (figure 7.5).

The **zygomatic arch,** which consists of joined processes from the temporal and zygomatic bones, forms a bridge across the side of the skull (see figure 7.4). The zygomatic arch is easily felt on the side of the face, and the muscles on either side of the arch can be felt as the jaws are opened and closed (see figure 7.5).

The **maxilla** is anterior and inferior to the zygomatic bone to which it is joined (see figure 7.4). The **mandible** is inferior to the maxilla and articulates posteriorly with the temporal bone (see figure 7.4). The mandible consists of two main portions: the **body,** which extends anteroposteriorly, and the **ramus** (branch), which extends superiorly from the

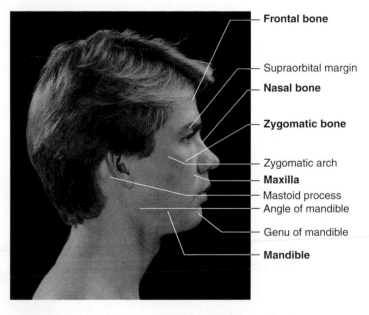

Frontal bone

Supraorbital margin
Nasal bone

Zygomatic bone

Zygomatic arch
Maxilla
Mastoid process
Angle of mandible

Genu of mandible

Mandible

Figure 7.5   Lateral View of Bony Landmarks on the Face

body toward the temporal bone. The superior end of the ramus has a **condylar process,** which articulates with the mandibular fossa of the temporal bone, and the **coronoid** (kōr′ŏ-noyd, meaning shaped like a crow's beak) **process** to which the powerful temporalis muscle, one of the chewing muscles, attaches. The maxilla contains the superior set of teeth, and the mandible contains the inferior teeth.

Frontal View of the Skull

The major structures seen from the frontal view are the frontal bone (forehead), the zygomatic bones (cheeks), the maxillae (upper jaw), and the mandible (lower jaw) (figure 7.6). The teeth, which are very prominent in this view, are discussed in chapter 24. Many bones of the face can be easily felt through the skin of the face (figure 7.7).

From this view the most prominent openings into the skull are the orbits and the nasal cavity. The **orbits** are cone-shaped fossae with their apices directed posteriorly (see figure 7.6; figure 7.8). They are called orbits because of the rotation of the eyes within the fossae. The bones of the orbits provide both protection for the eyes and attachment points for the muscles that move the eyes. The major portion of each eyeball is within the orbit, and the portion of the eye visible from the outside is relatively small. Each orbit contains blood vessels, nerves, and fat, as well as the eyeball and the muscles that move it. The bones forming the orbit are listed in table 7.4.

The orbit has several openings through which structures communicate between it and other cavities. The nasolacrimal duct passes from the orbit into the nasal cavity through the **nasolacrimal canal** and carries tears from the eyes to the nasal cavity. The optic nerve for the sense of vision passes from the eye through the **optic foramen** at the posterior apex of the orbit and enters the cranial vault. Two fissures in the posterior region of the orbit provide openings through which nerves and vessels communicate with structures in the orbit or pass to the face.

The **nasal cavity** (table 7.5 and figure 7.9; see figure 7.6) has a pear-shaped opening anteriorly and is divided into right and left halves by a **nasal septum** (sep'tŭm, meaning wall). The bony part of the nasal septum consists primarily of the vomer and the perpendicular plate of the ethmoid bone. The anterior part of the nasal septum is formed by hyaline cartilage.

> ### Clinical Note
>
> The superolateral corner of the orbit, where the zygomatic and frontal bones join, is a weak point in the skull that is easily fractured by a severe blow to that region of the head. The bone tends to collapse into the orbit, resulting in an injury that is difficult to repair.

> ### Clinical Note
>
> The nasal septum usually is located in the midsagittal plane until a person is 7 years old. Thereafter it tends to deviate, or bulge slightly to one side or the other. The septum can deviate abnormally at birth or, more commonly, as a result of injury. The deviation can be severe enough to block the nasal passage on one side and interfere with normal breathing. Severe deviation requires surgery to repair.

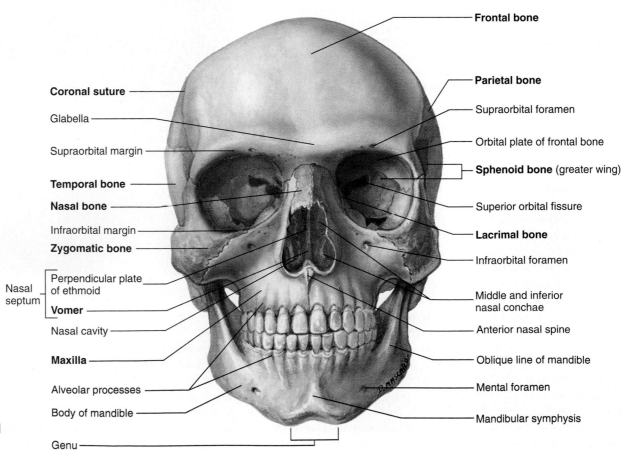

Figure 7.6   Skull as Seen from the Frontal View

Coronal suture
Glabella
Supraorbital margin
Temporal bone
Nasal bone
Infraorbital margin
Zygomatic bone
Nasal septum — Perpendicular plate of ethmoid
Vomer
Nasal cavity
Maxilla
Alveolar processes
Body of mandible
Genu

Frontal bone
Parietal bone
Supraorbital foramen
Orbital plate of frontal bone
Sphenoid bone (greater wing)
Superior orbital fissure
Lacrimal bone
Infraorbital foramen
Middle and inferior nasal conchae
Anterior nasal spine
Oblique line of mandible
Mental foramen
Mandibular symphysis

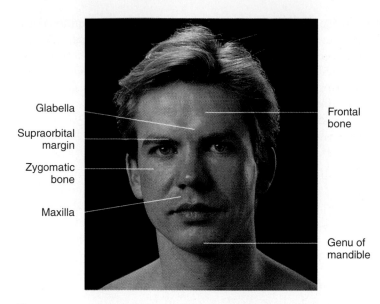

Glabella

Supraorbital
margin

Zygomatic
bone

Maxilla

Frontal
bone

Genu of
mandible

Figure 7.7   Anterior View of Bony Landmarks on the Face

Table 7.4	Bones Forming the Orbit (see figures 7.6 and 7.8)
Bone	**Part of Orbit**
Frontal	Roof
Sphenoid	Roof and lateral wall
Zygomatic	Lateral wall
Maxilla	Floor
Lacrimal	Medial wall
Ethmoid	Medial wall
Palatine	Medial wall

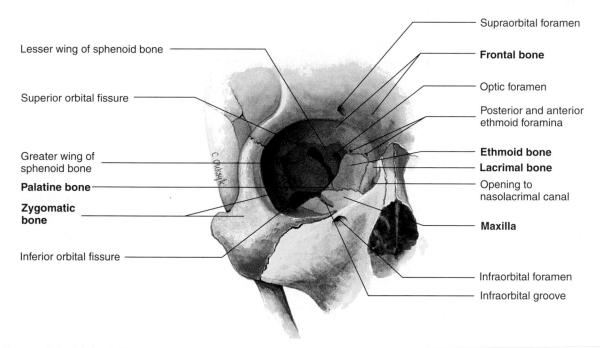

Lesser wing of sphenoid bone

Superior orbital fissure

Greater wing of
sphenoid bone

Palatine bone

**Zygomatic
bone**

Inferior orbital fissure

Supraorbital foramen

Frontal bone

Optic foramen

Posterior and anterior
ethmoid foramina

Ethmoid bone

Lacrimal bone

Opening to
nasolacrimal canal

Maxilla

Infraorbital foramen

Infraorbital groove

Figure 7.8   Bones of the Right Orbit

The external part of the nose, formed mostly of hyaline cartilage, is almost entirely absent in the dried skeleton and is represented mainly by the nasal bones and the frontal processes of the maxillary bones, which form the bridge of the nose.

2   P R E D I C T

A direct blow to the nose may result in a "broken nose." List at least three bones that may be broken.

✔ *Answer in Appendix F*

The lateral wall of the nasal cavity has three bony shelves, the **nasal conchae** (kon′kē, resembling a conch shell), which are directed inferiorly (see figure 7.9). The inferior nasal concha is a separate bone, and the middle and superior nasal conchae are projections from the ethmoid bone. The conchae function to increase the surface area in the nasal cavity, facilitating moistening, removal of particles, and warming of the air inhaled through the nose.

Several of the bones associated with the nasal cavity have large cavities within them called the **paranasal sinuses,** which open into the nasal cavity (figure 7.10). The sinuses

Table 7.5   Bones Forming the Nasal Cavity (see figures 7.6 and 7.9)

Bone	Part of Nasal Cavity
Frontal	Roof
Nasal	Roof
Sphenoid	Roof
Ethmoid	Roof, septum, and lateral wall
Inferior nasal concha	Lateral wall
Lacrimal	Lateral wall
Maxilla	Floor
Palatine	Floor and lateral wall
Vomer	Septum

decrease the weight of the skull and act as resonating chambers during voice production. Compare the normal voice to the voice of a person who has a cold and whose sinuses are "stopped up." The sinuses are named for the bones in which they are located and include the frontal, maxillary, ethmoidal, and sphenoidal sinuses.

Interior of the Cranial Vault

When the floor of the cranial vault is viewed from above with the roof cut away, it can be divided roughly into anterior, middle, and posterior fossae, which are formed as the developing skull conforms to the shape of the brain (figure 7.11).

A prominent ridge, the **crista galli** (kris′tă găl′ē, meaning rooster's comb), is located in the center of the an-

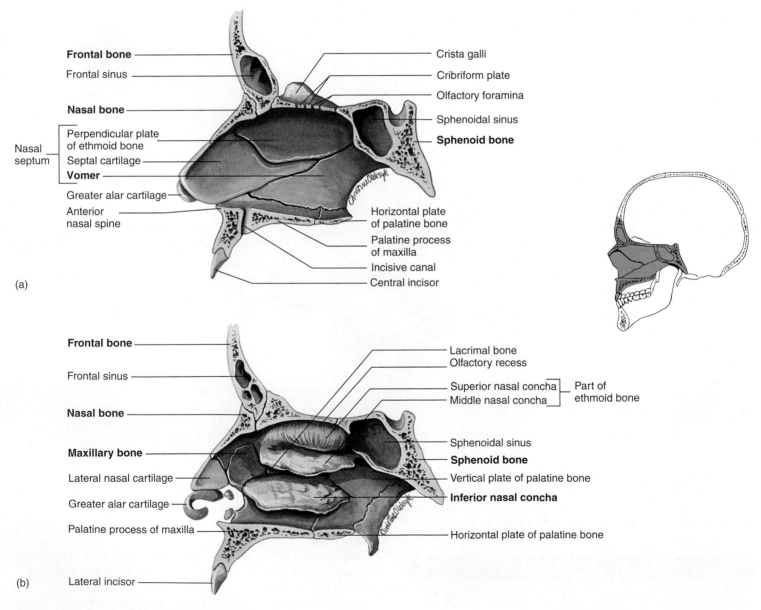

Figure 7.9   Bones of the Nasal Cavity

(a) Nasal septum as seen from the left nasal cavity. (b) Right lateral nasal wall as seen from inside the nasal cavity (nasal septum removed).

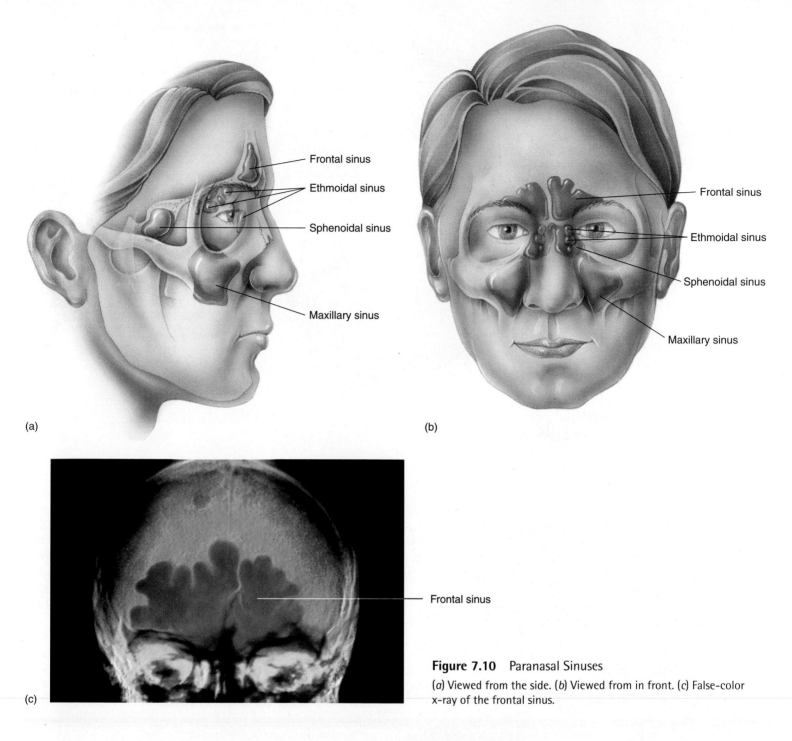

(a)

(b)

(c)

Figure 7.10   Paranasal Sinuses

(*a*) Viewed from the side. (*b*) Viewed from in front. (*c*) False-color x-ray of the frontal sinus.

terior fossa. The crista galli is a point of attachment for one of the **meninges** (mĕ-nin′jēz), a thick connective tissue membrane that supports and protects the brain (see chapter 13). On either side of the crista galli is a fossa for an olfactory bulb, which receives the olfactory nerves for the sense of smell. The floor of each olfactory fossa is formed by the **cribriform** (krib′ri-fōrm) **plate** of the ethmoid bone. The olfactory nerves extend from the cranial vault into the roof of the nasal cavity through sievelike perforations in the cribriform plate called **olfactory foramina** (see figure 7.9*a* and chapter 15).

Clinical Note

The cribriform plate may be fractured in an automobile accident if the driver's nose strikes the steering wheel, in which case cerebrospinal (sĕ-rē′brō-spī-năl) fluid from the cranial cavity may leak through the fracture into the nose. This leakage is a dangerous sign and requires immediate medical attention because risk of infection is very high.

One self-defense maneuver in martial arts involves an open-handed upward chop with the base of the hand to an assailant's nose, a movement that may break the cribriform plate.

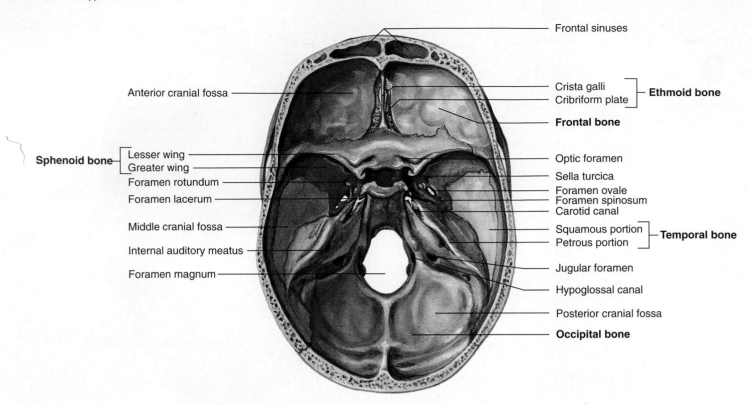

Figure 7.11   Floor of the Cranial Vault

The roof of the skull has been removed, and the floor is viewed from above.

A central prominence located within the floor of the cranial vault is formed by the body of the sphenoid bone. This prominence is modified into a structure resembling a saddle, the **sella turcica** (sel′ă tŭr′si-kă, meaning Turkish saddle), which is occupied by the pituitary gland during life. The petrous part of the temporal bone is on each side of and slightly posterior to the sella turcica. This thick bony ridge is hollow and contains the middle and inner ears.

The prominent **foramen magnum,** through which the spinal cord and brain are connected, is located in the posterior fossa. The other foramina of the skull and the structures passing through them are listed in table 7.6.

Inferior View of the Skull

Seen from below with the mandible removed, the base of the skull is complex, with a number of foramina and specialized surfaces (figure 7.12). The foramen magnum passes through the occipital bone just slightly posterior to the center of the skull base. **Occipital condyles,** the smooth points of articulation between the skull and the vertebral column, are located on the lateral and anterior margins of the foramen magnum.

The major entry and exit points for blood vessels that supply the brain can be seen from this view. Blood reaches the brain through the internal carotid arteries, which pass through the **carotid canals,** and the vertebral arteries, which pass through the foramen magnum. Immediately after the internal carotid artery enters the carotid canal, it turns medially

almost 90 degrees, continues through the carotid canal, again turns almost 90 degrees, and enters the cranial cavity through the superior part of the **foramen lacerum** (lă-ser′um). A thin plate of bone separates the carotid canal from the middle ear, therefore, making it possible for a person to hear his own heartbeat, for example, when he is frightened or after he has run. Most blood leaves the brain through the internal jugular veins, which exit through the **jugular foramina** located lateral to the occipital condyles.

Two long, pointed **styloid** (stī′loyd, meaning stylus- or pen-shaped) **processes** project from the floor of the temporal bone (see figures 7.4 and 7.12). Three muscles involved in movement of the tongue, hyoid bone, and pharynx attach to each process. The **mandibular fossa,** where the mandible articulates with the rest of the skull, is anterior to the mastoid process at the base of the zygomatic arch.

The posterior opening of the nasal cavity is bounded on each side by the vertical bony plates of the sphenoid bone: the **medial pterygoid** (ter′i-goyd, meaning wing-shaped) **plate** and the **lateral pterygoid plate.** The medial and lateral pterygoid muscles, which help move the mandible, attach to the lateral plate (see chapter 11). The **vomer** forms the posterior portion of the nasal septum and can be seen between the medial pterygoid plates in the center of the nasal cavity.

The **hard palate** forms the floor of the nasal cavity. Sutures join four bones to form the hard palate; the palatine processes of the two maxillary bones form the anterior two-

Table 7.6   Skull Foramina, Fissures, and Canals (see figures 7.11 and 7.12)

Opening	Bone Containing the Opening	Transmitted Structures
Carotid canal	Temporal	Carotid artery and carotid sympathetic nerve plexus
Ethmoid foramina, anterior and posterior (see figure 7.8)	Between frontal and ethmoid	Anterior and posterior ethmoid nerves
External auditory meatus	Temporal	Sound waves enroute to eardrum
Foramen lacerum	Between temporal, occipital, and sphenoid	The foramen is filled with cartilage during life; carotid canal and pterygoid canal cross its superior part but do not actually pass through it
Foramen magnum	Occipital	Spinal cord, accessory nerves, and vertebral arteries
Foramen ovale	Sphenoid	Mandibular division of trigeminal nerve
Foramen rotundum	Sphenoid	Maxillary division of trigeminal nerve
Foramen spinosum	Sphenoid	Middle meningeal artery
Hypoglossal canal	Occipital	Hypoglossal nerve
Incisive foramen (canal)	Between maxillae	Incisive nerve
Inferior orbital fissure	Between sphenoid and maxilla	Infraorbital nerve and vessels and zygomatic nerve
Infraorbital foramen	Maxilla	Infraorbital nerve
Internal auditory meatus	Temporal	Facial nerve and vestibulocochlear nerve
Jugular foramen	Between temporal and occipital	Internal jugular vein, glossopharyngeal nerve, vagus nerve, and accessory nerve
Mandibular foramen (see figure 7.13l)	Mandible	Inferior alveolar nerve to mandibular teeth
Mental foramen (see figure 7.4)	Mandible	Mental nerve
Nasolacrimal canal (see figure 7.4)	Between lacrimal and maxilla	Nasolacrimal (tear) duct
Olfactory foramina (see figure 7.9)	Ethmoid	Olfactory nerves
Optic foramen	Sphenoid	Optic nerve and ophthalmic artery
Palatine foramina, anterior and posterior	Palatine	Palatine nerves
Pterygoid canal (see figure 7.13d)	Sphenoid	Sympathetic and parasympathetic nerves to the face
Sphenopalatine foramen	Between palatine and sphenoid	Nasopalatine nerve and sphenopalatine vessels
Stylomastoid foramen	Temporal	Facial nerve
Superior orbital fissures (see figures 7.6 and 7.8)	Sphenoid	Oculomotor nerve, trochlear nerve, ophthalmic division of trigeminal nerve, abducens nerve, and ophthalmic veins
Supraorbital foramen or notch (see figures 7.6 and 7.8)	Frontal	Supraorbital nerve and vessels
Zygomaticofacial foramen (see figure 7.13h)	Zygomatic	Zygomaticofacial nerve
Zygomaticotemporal foramen (on posterior surface of zygomatic bone; not shown on any figure)	Zygomatic	Zygomaticotemporal nerve

thirds of the palate, and the horizontal plates of the two palatine bones form the posterior one-third of the palate. The tissues of the soft palate extend posteriorly from the hard, or bony, palate. The hard and soft palates separate the nasal cavity from the mouth, enabling humans to eat and breathe at the same time.

Clinical Note

During development, the facial bones sometimes fail to fuse with one another. A **cleft lip** results if the maxillae do not form normally, and a **cleft palate** occurs when the palatine processes of the maxillae do

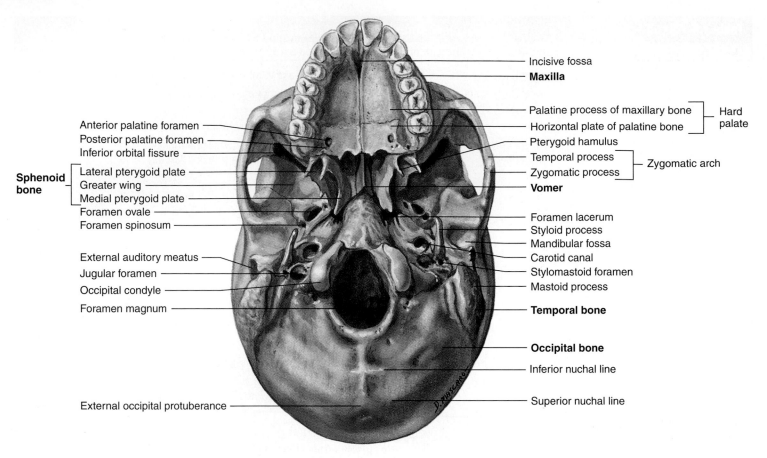

Figure 7.12   Inferior View of the Skull

not fuse with one another. A cleft palate produces an opening between the nasal and oral cavities, making it difficult to eat or drink or to speak distinctly. An artificial palate may be inserted into a newborn's mouth until the palate can be repaired. A cleft lip occurs approximately once in every 1000 births and is more common in males than in females. A cleft palate occurs approximately once in every 2500 births and is more common in females than in males. A cleft lip and cleft palate may also occur in the same person.

Bones of the Skull

The **skull** is composed of 28 separate bones (see table 7.1) organized into the following groups: the auditory ossicles, the cranial vault, and the facial bones. The six **auditory ossicles,** two each of the malleus, incus, and stapes, function in hearing (see chapter 15). One set is located inside each temporal bone and cannot be observed unless the temporal bones are cut open. The remaining 22 bones of the skull, or **cranium** (krā′nē-ŭm), are roughly divided into two portions: the cranial vault and the face. The individual bones are illustrated in figure 7.13. The **cranial vault,** or **braincase,** consists of eight bones that immediately surround and protect the brain. They include the paired parietal and temporal bones and the unpaired frontal, occipital, sphenoid, and ethmoid bones.

The 14 **facial bones** form the structure of the face in the anterior skull but do not contribute to the cranial vault

(see table 7.1). They are the maxilla (two), mandible (one), zygomatic (two), palatine (two), nasal (two), lacrimal (two), vomer (one), and inferior nasal concha (two) bones. The frontal and ethmoid bones, which are part of the cranial vault, also contribute to the face. The mandible is often listed as a facial bone, even though it is not part of the intact skull.

The facial bones provide protection for the major sensory organs located in the face: the eyes, nose, and tongue. The bones of the face also provide attachment points for muscles involved in **mastication** (mas′ti-kā′shŭn, meaning chewing), facial expression, and eye movement. The jaws (mandible and maxillae) possess **alveolar** (al′vē′-o-lăr) **processes** with sockets for the attachment of the teeth. The bones of the face and their associated soft tissues, determine the unique facial features of each individual.

Hyoid

The **hyoid bone** (figure 7.14), which is unpaired, is not part of the skull (see table 7.1) and has no direct bony attachment to the skull. It is attached to the skull by muscles and ligaments and "floats" in the superior aspect of the neck just below the mandible. The hyoid bone provides an attachment for some tongue muscles, and it is also an attachment point for important neck muscles that elevate the larynx during speech or swallowing.

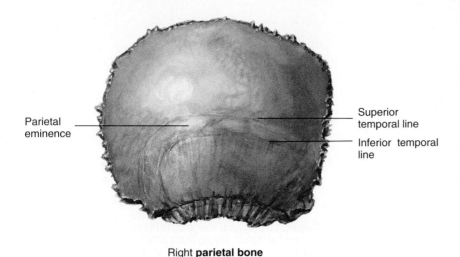

Parietal eminence

Superior temporal line

Inferior temporal line

Right **parietal bone**
(Viewed from the lateral side)

(a)

Landmarks seen on this figure:
Superior and inferior temporal lines: attachment point for temporalis muscle.
Parietal eminence: the widest part of the head is from one parietal eminence to the other.
Special feature: Forms lateral wall of skull.

Squamous portion

Zygomatic process

Mandibular fossa

External auditory meatus

Styloid process

Mastoid process

Right **temporal bone**
(Viewed from the lateral side)

(b)

Landmarks seen on this figure:
External auditory meatus: external canal of the ear; carries sound to the ear.
Mandibular fossa: articulation point between the mandible and unit skull.
Mastoid process: attachment point for muscles moving the head and for a hyoid muscle.
Squamous portion: flat, lateral portion of the temporal bone
Styloid process: attachment for muscles of the tongue, throat, and hyoid bone.
Zygomatic process: helps form the bony bridge from the cheek to just anterior to the ear; attachment for a muscle moving the mandible.
Landmarks seen in other figures:
Carotid canal: canal through which the internal carotid artery enters the cranial vault (figures 7.11 and 7.12).
Internal auditory meatus: opening through which the facial (cranial nerve VII) and vestibulocochlear (cranial nerve VIII) nerves enter the petrous portion of the temporal bone (figure 7.11).
Jugular foramen: foramen through which the internal jugular vein exits the skull (figures 7.11 and 7.12).
Middle cranial fossa: depression in the floor of the cranial vault formed by the temporal lobes of the brain (figure 7.11).
Petrous portion: thick, "rocky" portion of the temporal bone (figure 7.11).
Stylomastoid foramen: foramen through which the facial nerve (cranial nerve VII) exits the skull (figure 7.12).
Special features: Contains the middle and inner ear, and the mastoid air cells: place where the mandible articulates with the rest of the skull.

Figure 7.13 Skull Bones

(*a*) Right **parietal bone** (viewed from the lateral side). (*b*) Right **temporal bone** (viewed from the lateral side).

Frontal bone
(Viewed from in front and slightly above)

(c)

Glabella
Supraorbital foramen
Orbital plate
Supraorbital margin
Zygomatic process
Nasal spine

Landmarks seen on this figure:
Glabella: bump between the supraorbital ridges.
Nasal spine: superior part of the nasal bridge.
Orbital plate: roof of the orbit.
Supraorbital foramen: opening through which nerves and vessels exit the skull to the skin of the forehead.
Supraorbital margin: ridge forming the anterior superior border of the orbit.
Zygomatic process: connects to the zygomatic bone; helps form the lateral margin of the orbit.
Special features: Forms the forehead and roof of the orbit; contains the frontal sinus.

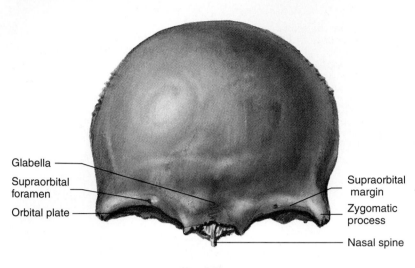

Lesser wing
Greater wing
Foramen rotundum
Foramen ovale
Foramen spinosum
Optic foramen
Superior orbital fissure
Sella turcica
Groove of carotid canal

(Superior view)

Landmarks seen on this figure:
Body: thickest part of the bone.
Foramen ovale: opening through which a branch of the trigeminal nerve (cranial nerve V) exits the cranial vault.
Foramen rotundum: opening through which a branch of the trigeminal nerve (cranial nerve V) exits the cranial vault.
Foramen spinosum: opening through which a major artery to the meninges (membranes around the brain) enters the cranial vault.
Greater wing: forms the floor of the middle cranial fossa; several foramina pass through this wing.
Lateral pterygoid plate: attachment point for muscles of mastication (chewing).
Lesser wing: superior border of the superior orbital fissure.
Medial pterygoid plate: posterolateral walls of the nasal cavity.
Optic foramen: opening through which the optic nerve (cranial nerve II) passes from the orbit to the cranial vault.
Pterygoid canal: opening through which nerves and vessels exit the cranial vault.
Pterygoid hamulus: process around which the tendon from a muscle to the soft palate passes.
Sella turcica: fossa containing the pituitary gland.
Superior orbital fissure: opening through which nerves and vessels enter the orbit from the cranial vault.
Special feature: Contains the sphenoidal sinus.

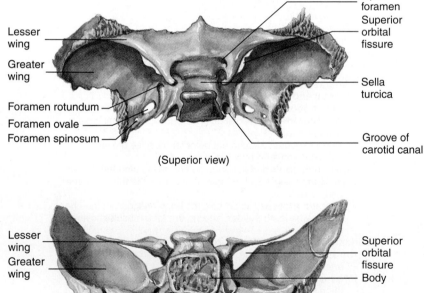
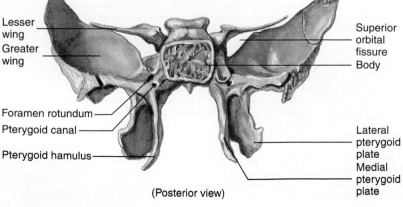

Lesser wing
Greater wing
Foramen rotundum
Pterygoid canal
Pterygoid hamulus
Superior orbital fissure
Body
Lateral pterygoid plate
Medial pterygoid plate

(Posterior view)

(d)   **Sphenoid bone**

Figure 7.13   (*continued*)

(c) **Frontal bone** (viewed from in front and slightly above). (d) **Sphenoid bone.** Superior view. **Sphenoid bone.** Posterior view.

Anterior

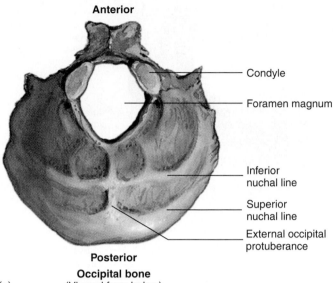

Condyle

Foramen magnum

Inferior nuchal line

Superior nuchal line

External occipital protuberance

Posterior

Occipital bone
(Viewed from below)

(e)

Landmarks seen on this figure:
    Condyle: articulation point between the skull and first vertebra.
    External occipital protuberance: attachment point for a strong ligament (nuchal ligament) in back of the neck.
    Foramen magnum: opening around the point where the brain and spinal cord connect.
    Inferior nuchal line: attachment point for neck muscles.
    Superior nuchal line: attachment point for neck muscles.
Landmarks seen in other figures:
    Hypoglossal canal: opening through which the hypoglossal nerve (cranial nerve XII) passes (figure 7.11).
    Posterior cranial fossa: depression in the posterior of the cranial vault formed by the cerebellum (figure 7.11).
Special features: Forms the base of the skull.

Anterior

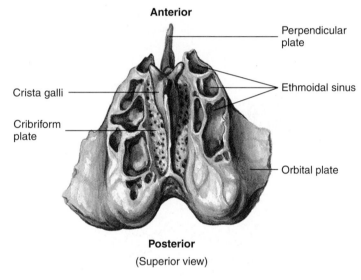

Perpendicular plate

Crista galli

Cribriform plate

Ethmoidal sinus

Orbital plate

Posterior

(Superior view)

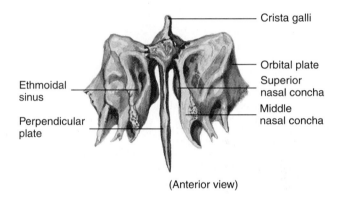

Crista galli

Orbital plate

Ethmoidal sinus

Superior nasal concha

Perpendicular plate

Middle nasal concha

(Anterior view)

Landmarks seen on this figure:
    Cribriform plate: contains numerous openings through which branches of the olfactory nerve (cranial nerve I) enter the cranial vault from the nasal cavity.
    Crista galli: attachment for meninges (membrane around brain).
    Ethmoidal sinus: spaces in the bone, help lighten the skull.
    Middle nasal concha: ridge extending into the nasal cavity, gives more surface area, helping warm and moisten air in the cavity.
    Orbital plate: forms the medial wall of the orbit.
    Perpendicular plate: forms part of the nasal septum.
    Superior nasal concha: ridge extending into the nasal cavity, gives more surface area, helping warm and moisten air in the cavity.
Landmarks seen in other figures:
    Ethmoid foramina: openings through which nerves and vessels pass from the orbit to the nasal cavity (figure 7.8).
Special features:
    Forms part of the nasal septum and part of the lateral walls and roof of the nasal cavity; contains the ethmoidal sinus, or ethmoidal air cells.

Crista galli

Ethmoidal sinus

Orbital plate

Posterior

Anterior

Middle nasal concha

Perpendicular plate

(Lateral view)

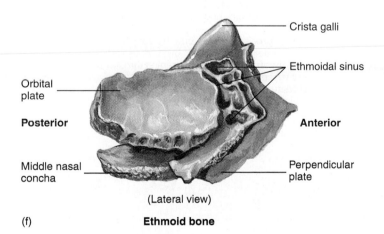

(f)   **Ethmoid bone**

Figure 7.13 *(continued)*

(e) **Occipital bone** (viewed from below) (f) **Ethmoid bone.** Superior, lateral, anterior view.

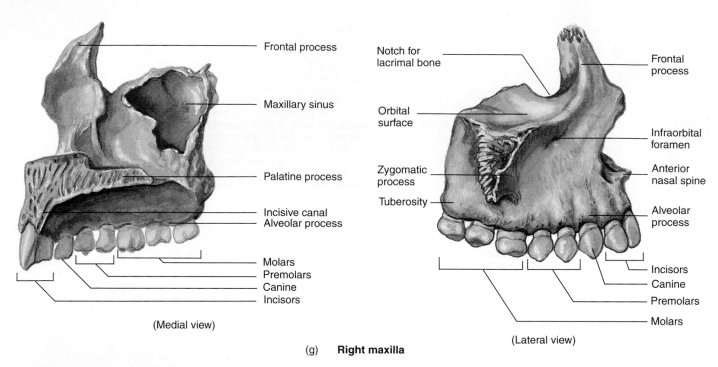

Frontal process

Maxillary sinus

Palatine process

Incisive canal
Alveolar process

Molars
Premolars
Canine
Incisors

(Medial view)

Notch for lacrimal bone

Orbital surface

Zygomatic process

Tuberosity

Frontal process

Infraorbital foramen

Anterior nasal spine

Alveolar process

Incisors
Canine
Premolars

Molars

(Lateral view)

(g)   **Right maxilla**

Landmarks seen on this figure:
  Alveolar process: ridge containing the teeth.
  Anterior nasal spine: forms part of the nasal septum.
  Frontal process: forms the sides of the nasal bridge.
  Incisive canal: opening through which a nerve exits the nasal cavity to the roof of the oral cavity.
  Infraorbital foramen: opening through which a nerve and vessels exit the skull to the face.
  Maxillary sinus: cavity in the bone, which helps lighten the skull.
  Orbital surface: forms the floor of the orbit.
  Palatine process: forms the anterior two-thirds of the hard palate.
  Tuberosity: lump posterior to the last maxillary molar tooth.
  Zygomatic process: connection to the zygomatic bone, helps form the interior margin of the orbit.
  Special feature: Contains the maxillary sinus and maxillary teeth.

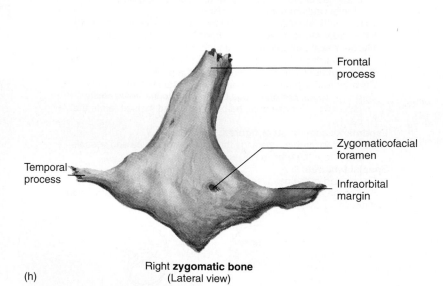

Frontal process

Zygomaticofacial foramen

Temporal process

Infraorbital margin

(h)

Right **zygomatic bone**
(Lateral view)

Landmarks seen on this figure:
  Frontal process: connection to the frontal bone, helps form the lateral margin of the orbit.
  Infraorbital margin: ridge forming the inferior border of the orbit.
  Temporal process: helps form the bony bridge from the cheek to just anterior to the ear.
  Zygomaticofacial foramen: opening through which a nerve and vessels exit the skull to the face.
  Special features: Forms the prominence of the cheek; forms the anterolateral wall of the orbit.

Figure 7.13   (*continued*)

(*g*) Right maxilla. Medial and lateral views. (*h*) Right **zygomatic bone**. Lateral view.

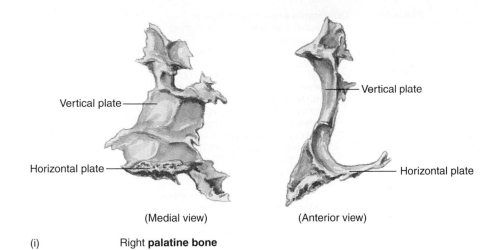

Vertical plate

Horizontal plate

(Medial view)

Vertical plate

Horizontal plate

(Anterior view)

(i)   Right **palatine bone**

Landmarks seen on this figure:
Horizontal plate: forms the posterior one-third of the hard palate.
Vertical plate: forms part of the lateral nasal wall.
Special features: Helps form the part of the hard palate and a small part of the wall of the orbit.

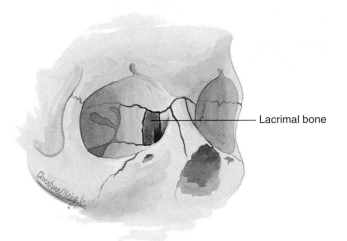

Lacrimal bone

Special feature: Forms a small portion of the orbital wall.

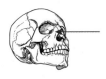

Right **lacrimal bone**
(Lateral view)

(j)

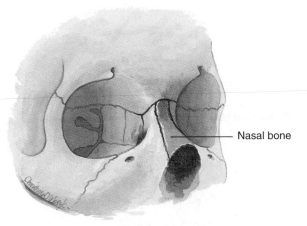

Nasal bone

Special feature: Forms the bridge of the nose.

Right **nasal bone**
(Lateral view)

(k)

Figure 7.13   (*continued*)

(*i*) Right **palatine bone.** Medial and anterior views. (*j*) Right **lacrimal bone.** Lateral view. (*k*) Right **nasal bone** (lateral view).

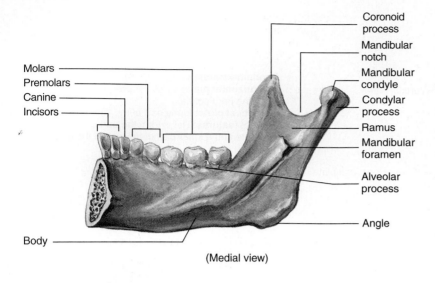

Molars
Premolars
Canine
Incisors

Coronoid process
Mandibular notch
Mandibular condyle
Condylar process
Ramus
Mandibular foramen
Alveolar process
Angle

Body

(Medial view)

Landmarks seen on this figure:

Alveolar process: ridge containing the teeth.

Angle: corner between the body and ramus.

Body: major, horizontal portion of the bone.

Condylar process: extension containing the mandibular condyle.

Coronoid process: attachment for a muscle of mastication.

Mandibular condyle: point of articulation between the mandible and unit skull.

Mandibular foramen: opening through which nerves and vessels to the mandibular teeth enter the bone.

Mandibular notch: depression between the condylar process and the coronoid process.

Mental foramen: opening through which a nerve and vessels exit the mandible to the skin of the chin.

Ramus: major, nearly vertical portion of the bone.

Special features:

The only bone in this figure that is freely movable relative to the rest of the skull bones; holds the lower teeth.

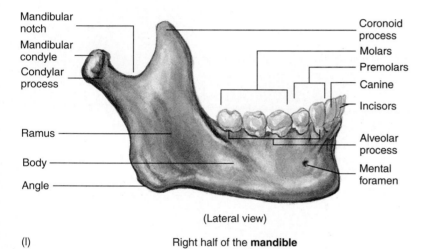

Mandibular notch
Mandibular condyle
Condylar process

Ramus

Body

Angle

Coronoid process
Molars
Premolars
Canine
Incisors
Alveolar process
Mental foramen

(Lateral view)

(l)   Right half of the **mandible**

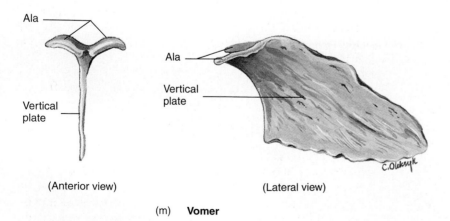

Ala

Vertical plate

(Anterior view)

Ala

Vertical plate

(Lateral view)

C.Olukryk

(m)   **Vomer**

Landmarks seen on this figure:

Ala: attachment point between the vomer and sphenoid.

Vertical plate: forms part of the nasal septum.

Special feature:

Forms most of the posterior nasal septum.

Figure 7.13 (*continued*)

(*l*) Right half of the **mandible**. Medial and lateral views. (*m*) **Vomer**. Anterior, lateral views.

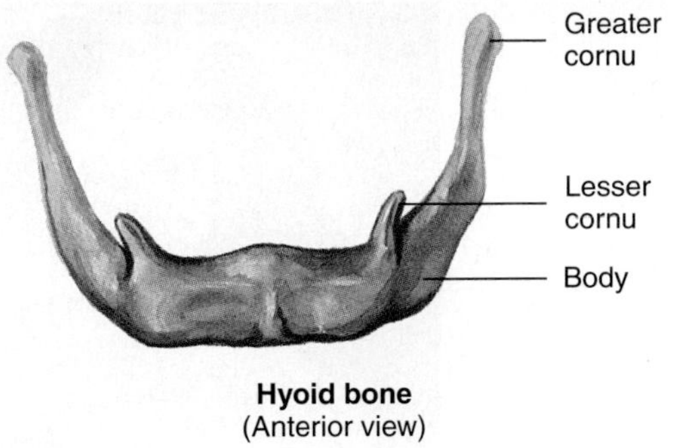

Greater
cornu

Lesser
cornu

Body

Hyoid bone
(Anterior view)

Landmarks seen on this figure:
Body: major portion of the bone.
Greater cornu: attachment point for
muscles and ligaments.
Lesser cornu: attachment point for
muscles and ligaments.
Special features:
Not actually part of the skull; one of
the few bones of the body that does
not articulate with another bone; it is
attached to the skull by muscles and
ligaments.

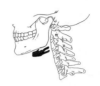

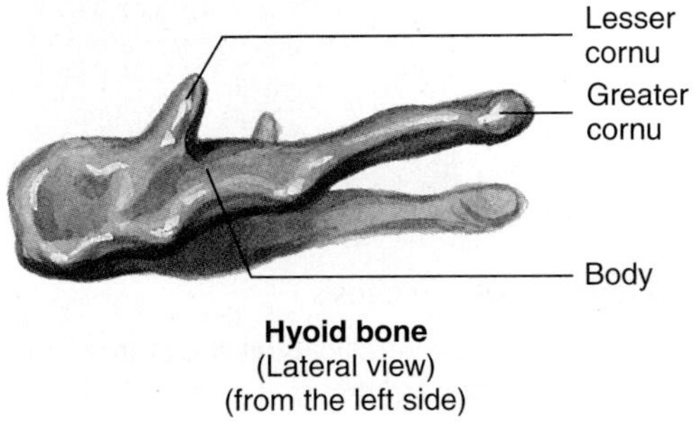

Lesser
cornu
Greater
cornu

Body

Hyoid bone
(Lateral view)
(from the left side)

Figure 7.14 Hyoid Bone
Anterior and lateral views (from the left side).

Vertebral Column

The **vertebral column** usually consists of 26 bones, which can be divided into five regions (figure 7.15). There are seven **cervical** vertebrae, 12 **thoracic** vertebrae, five **lumbar** vertebrae, one **sacral** bone, and one **coccygeal** (kok-sij′ē-ăl) bone. The developing embryo has about 34 vertebrae, but the five sacral vertebrae fuse to form one bone, and the four or five coccygeal bones usually fuse into one bone.

The adult vertebral column has four major curvatures (see figure 7.15). Two of the curves appear during embryonic development and reflect the C-shaped curve of the embryo and fetus within the uterus. When the infant raises its head in the first few months after birth, a secondary curve, which is convex anteriorly, develops in the neck. Later, when the infant learns to sit and then walk, the lumbar portion of the column also becomes convex anteriorly. Thus in the adult vertebral column the cervical region is convex anteriorly, the thoracic region is concave anteriorly, the lumbar region is convex anteriorly, and the sacral and coccygeal regions are, together, concave anteriorly.

Clinical Note

If the convex curve of the lumbar region is exaggerated, the term **lordosis** (lōr-dō′sis, meaning hollow back) is applied to the defect; and if the concave curve, especially in the thorax, is exaggerated, the term **kyphosis** (kī-fō′sis, meaning hump back) is applied. **Scoliosis** (skō′lē-ō′sis) is an abnormal bending of the spine to the side, which is often accompanied by secondary abnormal curvatures such as kyphosis (figure 7.16).

General Plan of the Vertebrae

The vertebral column performs five major functions: (1) it supports the weight of the head and trunk; (2) it protects the spinal cord; (3) it allows spinal nerves to exit the spinal cord; (4) it provides a site for muscle attachment; and (5) it permits movement of the head and trunk. The general structure of a vertebra is outlined in table 7.7. Each vertebra consists of a body, an arch, and various processes (figure 7.17). The weight-bearing portion of the vertebra is a bony disk called the **body.**

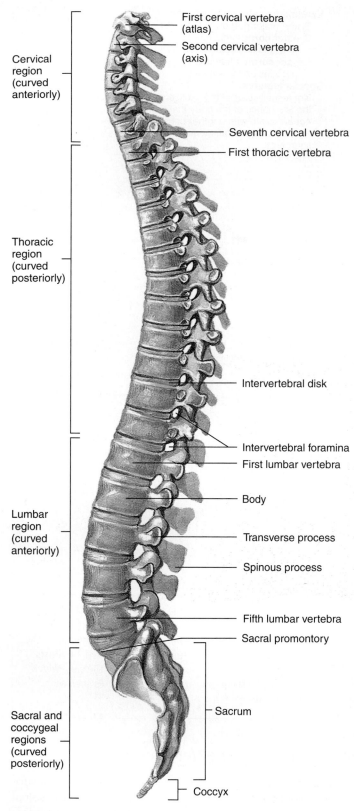

Figure 7.15   Vertebral Column

Complete column viewed from the left side.

Cervical region (curved anteriorly)

First cervical vertebra (atlas)

Second cervical vertebra (axis)

Seventh cervical vertebra

First thoracic vertebra

Thoracic region (curved posteriorly)

Intervertebral disk

Intervertebral foramina

First lumbar vertebra

Body

Lumbar region (curved anteriorly)

Transverse process

Spinous process

Fifth lumbar vertebra

Sacral promontory

Sacral and coccygeal regions (curved posteriorly)

Sacrum

Coccyx

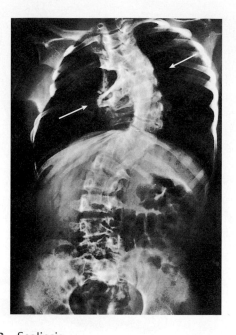

Figure 7.16   Scoliosis

Scoliosis is an abnormal lateral curvature of the spine. The abnormality is indicated by the arrows.

During life **intervertebral disks** of fibrocartilage, which are located between the bodies of adjacent vertebrae (see figures 7.15 and 7.17), provide additional support and prevent the vertebral bodies from rubbing against each other. The intervertebral disks consist of an external **anulus fibrosus** (an′ū-lŭs fī-brō′sŭs, meaning fibrous ring) and an internal gelatinous **nucleus pulposus** (pul-pō′sŭs, meaning pulp). The disk becomes more compressed with increasing age so that the distance between vertebrae and therefore the overall height of the individual decreases. The anulus fibrosus also becomes weaker with age and more susceptible to herniation.

> ### Clinical Note
>
> A **herniated,** or **ruptured, disk** results from the breakage or ballooning of the anulus fibrosus with a partial or complete release of the nucleus pulposus (figure 7.18). The herniated part of the disk may push against the spinal cord or spinal nerves, compromising their normal function and producing pain. Herniation of the inferior lumbar intervertebral disks is most common, but herniation of the inferior cervical disks is almost as common.
>
> Herniated or ruptured disks can be repaired in one of several ways. One procedure uses prolonged bed rest and is based on the tendency for the herniated part of the disk to recede and the anulus fibrosus to repair itself. In many cases, however, surgery is required, and the damaged disk is removed. To enhance the stability of the vertebral column, a piece of hipbone is sometimes inserted into the space previously occupied by the disk, and the adjacent vertebrae become fused by bone across the gap.

Protection of the spinal cord is provided by the **vertebral arch** and the dorsal portion of the body, which surround a large opening, called the **vertebral foramen.** The vertebral

Table 7.7   General Structure of a Vertebra (see figure 7.17 b and c)

Feature	Description
Body	Disk-shaped; usually largest part with flat surfaces directed superiorly and inferiorly; forms the anterior wall of the vertebral foramen; intervertebral disks are located between the bodies
Arch	Forms the lateral and posterior walls of the vertebral foramen; possesses several processes and articular surfaces
Pedicle	Foot of the arch with one on each side; forms the lateral walls of the vertebral foramen
Lamina	Posterior part of the arch; forms the posterior wall of the vertebral foramen
Transverse process	Process projecting laterally from the junction of the lamina and pedicle; a site of muscle attachment
Spinous process	Process projecting posteriorly at the point where the two laminae join; a site of muscle attachment; strengthens the vertebral column and allows for movement.
Articular processes	Superior and inferior projections containing articular facets where vertebrae articulate with each other; strengthens the vertebral column and allows for movement
Vertebral foramen	Hole in each vertebra through which the spinal cord passes; adjacent vertebral foramina form the vertebral canal
Intervertebral foramen	Opening between vertebrae through which spinal nerves exit the vertebral canal

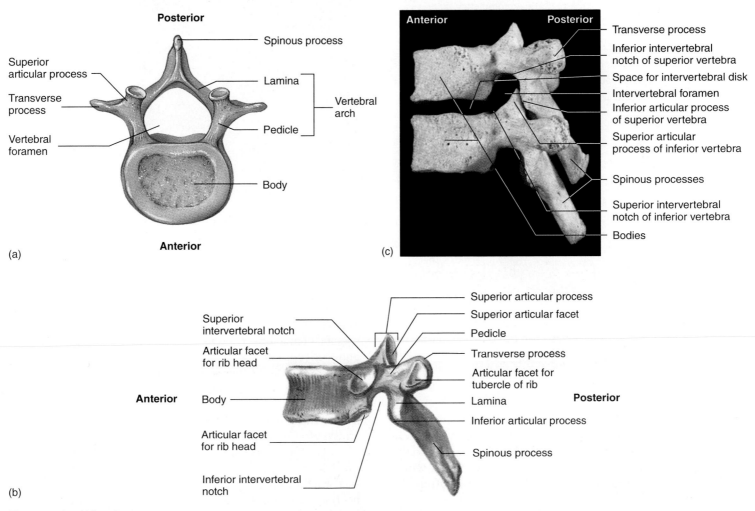

Figure 7.17   Vertebra

(*a*) Superior view. (*b*) Lateral view. (*c*) Photograph of two stacked vertebrae from a lateral view. The relationship between the inferior articular process of one vertebra and the superior articular process of the next inferior vertebra can be seen. The intervertebral foramen and the space for the intervertebral disk also can be seen.

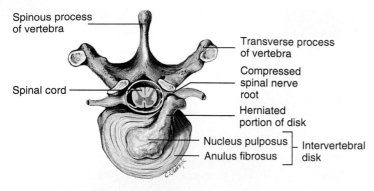

Figure 7.18   Herniated Disk

Part of the anulus fibrosus has been removed to reveal the nucleus pulposus in the center of the disk. ✗

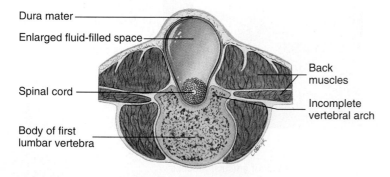

Figure 7.19   Spina Bifida

This developmental malformation occurs when two vertebral laminae fail to fuse.

foramina of adjacent vertebrae combine to form the **vertebral canal,** which contains the spinal cord. The arch can be divided into left and right halves, and each half has two parts: the **pedicle** (ped′ĭ-kl, meaning foot), which is attached to the body, and the **lamina** (lam′i-nǎ, meaning thin plate), which continues dorsally from the pedicle to join the lamina from the opposite half of the arch.

A **transverse process** extends laterally from each side of the arch between the lamina and pedicle, and a single **spinous process** is present at the point of junction between the two laminae. The spinous processes can be seen and felt as a series of lumps down the midline of the back (figure 7.20). Much vertebral movement is accomplished by the contraction of skeletal muscles that are attached to the transverse and spinous processes (see chapter 11).

Spinal nerves exit the spinal cord through the **intervertebral foramina** (see figures 7.15 and 17*c*). Each intervertebral foramen is formed by notches in the pedicles of adjacent vertebrae.

Movement and additional support of the vertebral column are made possible by the vertebral processes. Each vertebra has a **superior** and an **inferior articular process,** with the superior process of one vertebra articulating with the inferior process of the next superior vertebra. Overlap of these processes increases the rigidity of the vertebral column. The region of overlap and articulation between the superior and inferior articular processes create a smooth "little face" on each articular process called an **articular facet** (fas′et).

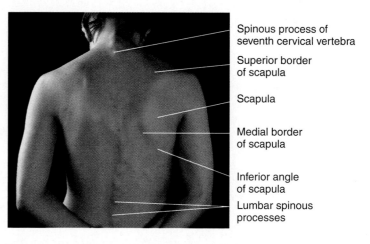

Figure 7.20   Photograph of a Person's Back Showing the Scapula and Vertebral Spinous Processes

Regional Differences in Vertebrae

The vertebrae of each region of the vertebral column have specific characteristics, which tend to blend at the boundaries between regions. The **cervical vertebrae** (see figure 7.15; figure 7.21*a–c*) have very small bodies, partly **bifid** (bī′fid, meaning split) spinous processes, and a **transverse foramen** in each transverse process through which the vertebral arteries extend toward the head. Only cervical vertebrae have transverse foramina.

The first cervical vertebra is called the **atlas** (see figure 7.21*a*) because it holds up the head, just as Atlas in classical mythology held up the world. The atlas vertebra has no body and no spinous process, but it has large superior articular facets, where it joins the occipital condyles on the base of the skull. This joint allows the head to move in a "yes" motion or to tilt from side to side. The second cervical vertebra is called the **axis** (see figure 7.21*b*) because a

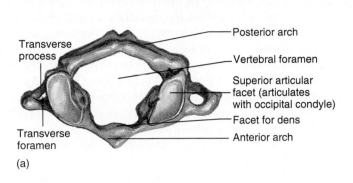

(a)

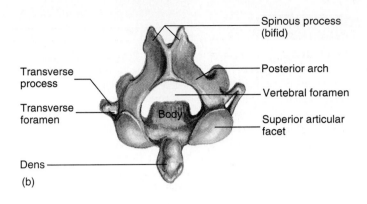

(b)

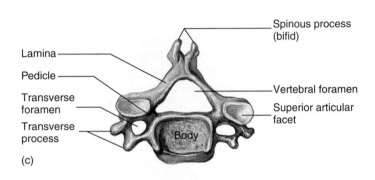

(c)

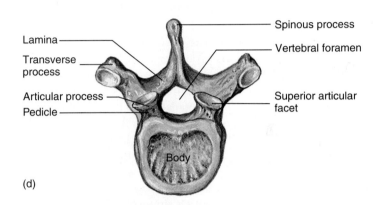

(d)

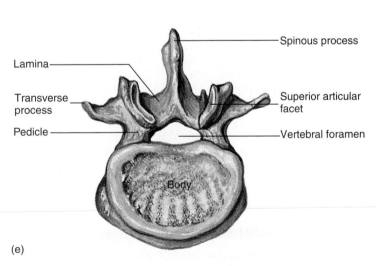

(e)

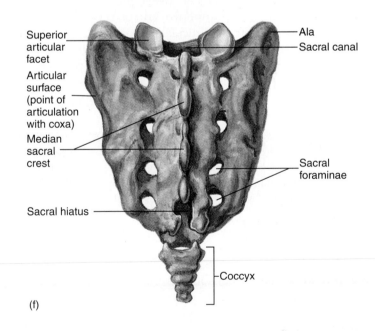

(f)

Figure 7.21   Vertebrae

(*a*) Atlas (first cervical vertebra), superior view. (*b*) Axis (second cervical vertebra), slightly posterior and superior view. (*c*) Fifth cervical vertebra, superior view. (*d*) Thoracic vertebra, superior view. (*e*) Lumbar vertebra, superior view. (*f*) Sacrum and coccyx, posterior view.

considerable amount of rotation occurs at this vertebra to produce a "no" motion of the head. The axis has a highly modified process on the superior side of its small body called the **dens,** or **odontoid process** (both dens and odontoid mean tooth-shaped). The dens fits into the en-

larged vertebral foramen of the atlas, and the latter rotates around this process. The spinous process of the seventh cervical vertebra, which is not bifid, is quite pronounced and often can be seen and felt as a lump between the shoulders (see figure 7.20).

207

Whiplash is a traumatic hyperextension of the cervical vertebrae. The head is a heavy object at the end of a flexible column, and it may become hyperextended when the head "snaps back" as a result of a sudden acceleration of the body. This commonly occurs in "rear-end" automobile accidents in which the body is quickly forced forward while the head remains stationary. Common injuries resulting from whiplash are fracture of the spinous processes of the cervical vertebrae and ruptured disks, with an anterior tear of the anulus fibrosus. These injuries can cause posterior pressure on the spinal cord or spinal nerves and strained or torn muscles, tendons, and ligaments.

The **thoracic vertebrae** (see figures 7.15 and 7.21*d*) possess long, thin spinous processes, which are directed inferiorly, and they have relatively long transverse processes. The first 10 thoracic vertebrae have articular facets on their transverse processes where they articulate with the tubercles of the ribs. Additional facets are on the superior and inferior margins of the body where the heads of the ribs articulate. The head of most ribs articulates between two vertebrae so that each vertebra has half of a facet, sometimes called a **demifacet,** at the point where the rib head articulates.

The **lumbar vertebrae** (see figures 7.15 and 7.21*e*) have large, thick bodies and heavy, rectangular transverse and spinous processes. The superior articular processes face medially, and the inferior articular processes face laterally. When the superior articular surface of one lumbar vertebra joins the inferior articulating surface of another lumbar vertebra, the resulting arrangement adds strength to the inferior portion of the vertebral column and limits rotation of the lumbar vertebrae.

3   P R E D I C T

Why is the arrangement of the superior and inferior articulating processes in the lumbar vertebrae beneficial? Why are the lumbar vertebrae more massive than the cervical vertebrae?

✔ *Answer in Appendix F*

The **sacral vertebrae** (see figure 7.15 and 7.21*f*) are highly modified in comparison with the others. These five vertebrae are fused into a single bone called the **sacrum.**

Clinical Note

The fifth lumbar vertebra or first coccygeal vertebra may become fused into the sacrum. Conversely, the first sacral vertebra may fail to fuse with the rest of the sacrum, resulting in six lumbar vertebrae.

The transverse processes of the sacral vertebrae fuse to form the **alae** (ă′lē, meaning wings) that join the sacrum to the pelvic bones. The spinous processes of the first four

sacral vertebrae are more or less separate projections on the dorsum of the bone called the **median sacral crest.** The spinous process of the fifth vertebra does not form, leaving a **sacral hiatus** at the inferior end of the sacrum, which is often the site of anesthetic injections. The intervertebral foramina are divided into dorsal and ventral foramina, the **sacral foramina,** which are lateral to the midline. The anterior edge of the body of the first sacral vertebra bulges to form the **sacral promontory** (see figure 7.15), a landmark that separates the abdominal cavity from the pelvic cavity. The sacral promontory can be felt during a vaginal examination, and it is used as a reference point during measurement of the pelvic outlet.

The **coccyx** (kok′siks, meaning shaped like a cuckoo's bill; see figures 7.15 and 7.21*f*), or tailbone, is the most inferior portion of the vertebral column and usually consists of three to five more-or-less fused vertebrae that form a triangle, with the apex directed inferiorly. The coccygeal vertebrae are greatly reduced in size relative to the other vertebrae and have neither vertebral foramina nor well-developed processes.

Clinical Note

Because the cervical vertebrae are rather delicate and have small bodies, dislocations and fractures are more common in this area than in other regions of the column. Because the lumbar vertebrae have massive bodies and carry a large amount of weight, fractures are less common, but ruptured intervertebral disks are more common in this area than in other regions of the column. The coccyx is easily broken in a fall in which a person sits down hard on a solid surface.

Thoracic Cage

The **thoracic cage,** or **rib cage,** protects the vital organs within the thorax and prevents the collapse of the lung during respiration. It consists of the thoracic vertebrae, the ribs with their associated costal (rib) cartilages, and the sternum (figure 7.22*a*).

Ribs and Costal Cartilages

The 12 pairs of **ribs** can be divided into the true and the false ribs. The superior seven pairs, called the **true,** or **vertebrosternal** (ver-tĕ′brō-ster′năl), **ribs,** articulate with the thoracic vertebrae and attach directly through their **costal cartilages** to the sternum. The inferior five pairs, or **false ribs,** articulate with the thoracic vertebrae but do not attach directly to the sternum. The false ribs consist of two groups. The eighth, ninth, and tenth ribs, the **vertebrochondral** (ver-tĕ′brō-kon′drăl) **ribs,** are joined to a common cartilage, which, in turn, is attached to the sternum. Two of the false ribs, the eleventh and twelfth ribs, are also called **floating,** or **vertebral ribs,** because they have no attachment to the ster-

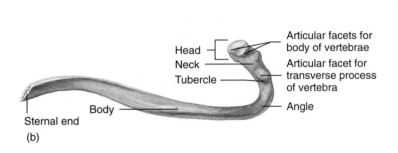

Clavicle

Seventh cervical vertebra

First thoracic vertebra

Jugular notch

1

Sternal angle

2

3

True ribs

4

Costal cartilage

5

6

Manubrium

Body — Sternum

Xiphoid process

7

8

11

T12

False ribs

9

12

L1

Floating ribs

10

(a)

Head

Neck

Tubercle

Articular facets for body of vertebrae

Articular facet for transverse process of vertebra

Body

Angle

Sternal end

(b)

Head of rib set against the inferior articular facet of the superior vertebra and the superior articular facet of the inferior vertebra

Tubercle of rib set against the articular facet on the transverse process of the inferior vertebra

Angle of rib

Body of rib

(c)

Figure 7.22 Thoracic Cage

(a) Entire thoracic cage as seen from in front. (b) Typical rib.
(c) Photograph of two thoracic vertebrae and the proximal end of
a rib as seen from the left side, showing the relationship between
the vertebra and the head and tubercle of the rib.

num. The costal cartilages are flexible and permit the thoracic
cage to expand during respiration.

Most ribs have two points of articulation with the thoracic vertebrae (figure 7.22*b* and *c*). The **head** articulates with
the bodies of two adjacent vertebrae and the intervertebral
disk between them, and the **tubercle** articulates with the
transverse process of one vertebra. The **neck** is between the
head and tubercle, and the **body,** or shaft, is the main part of
the rib. The **angle** of the rib is located just lateral to the tubercle and is the point of greatest curvature.

Clinical Note

A **separated rib** is a dislocation between a rib and its costal cartilage.
As a result of the dislocation, the rib can move, override adjacent ribs,
and cause pain. Separation of the tenth rib is the most common.

The angle is the weakest part of the rib and may be fractured in a
crushing accident, such as an automobile accident.

The transverse processes of the seventh cervical vertebra may
form separate bones called **cervical ribs.** These ribs may be just tiny
pieces of bone or may be long enough to reach the sternum. The first
lumbar vertebra may develop lumbar ribs.

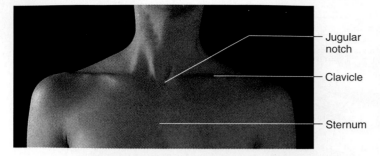

Figure 7.23   Surface Anatomy Showing Bones of the Upper Thorax

Sternum

The **sternum,** or breastbone, has been described as shaped like a sword and is divided into three parts (see figure 7.22*a*): the **manubrium** (mă-nū′brē-ŭm, meaning handle), the **body,** and the **xiphoid** (zif′oyd, meaning sword and representing the sword tip) **process.** The old term for the body was gladiolus, meaning sword, and had the same root as gladiator, a person who fights with a sword. The superior margin of the manubrium has a **jugular** (neck) **notch** in the midline, which can be easily felt at the anterior base of the neck (figure 7.23). The first rib and the clavicle articulate with the manubrium. The point at which the manubrium joins the body of the sternum can be felt as a prominence on the anterior thorax called the **sternal angle.** The cartilage of the second rib attaches to the sternum at the sternal angle, the third through seventh ribs attach to the body of the sternum, and no ribs attach to the xiphoid process.

> ### Clinical Note
>
> The sternal angle is important clinically because the second rib is found lateral to it and can be used as a starting point for counting the other ribs. Counting ribs is important because they are landmarks used to locate structures in the thorax, such as areas of the heart. The sternum often is used as a site for taking red bone marrow samples because it is readily accessible. Because the xiphoid process of the sternum is attached only at its superior end, it may be broken during cardiopulmonary resuscitation (CPR) and then may lacerate the liver.

Appendicular Skeleton

The appendicular skeleton (see figure 7.1) consists of the bones of the **upper** and **lower limbs** and the **girdles** by which they are attached to the body. The term girdle means a belt or a zone and refers to the two zones, pectoral and pelvic, where the limbs are attached to the body.

Upper Limb

The human forelimb is capable of a wide range of movements, including lifting, grasping, pulling, and touching.

Many structural characteristics of the upper limb reflect these functions. The upper limb and its girdle are attached rather loosely by muscles to the rest of the body, an arrangement that allows considerable freedom of movement of this extremity. This freedom of movement allows placement of the hand in a wide range of positions to accomplish its functions.

Pectoral Girdle

The **pectoral** (pek′tŏ-răl), or **shoulder girdle,** consists of two pairs of bones that attach the upper limb to the body: each pair is composed of a **scapula** (skap′ūlă), or shoulder blade (figure 7.24), and a **clavicle** (klav′i-kl), or collarbone (see figures 7.22 and 24*c*). The scapula is a flat, triangular bone that can easily be seen and felt in a living person (see figure 7.20). The base of the triangle, the superior border, faces superiorly; and the apex, the inferior angle, is directed inferiorly.

The large **acromion** (ă-krō′mē-on, meaning shoulder tip) **process,** which can be felt at the tip of the shoulder, has three functions: (1) to form a protective cover for the shoulder joint; (2) to form the attachment site for the clavicle; and (3) to provide attachment points for some of the shoulder muscles. The **scapular spine** extends from the acromion process across the posterior surface of the scapula and divides that surface into a small **supraspinous fossa** superior to the spine and a larger **infraspinous fossa** inferior to the spine. The deep, anterior surface of the scapula constitutes the **subscapular fossa.** The smaller **coracoid** (meaning shaped like a crow's beak) **process** provides attachments for some shoulder and arm muscles. A **glenoid** (glen′oyd) **fossa,** located in the superior lateral portion of the bone, articulates with the head of the humerus.

The clavicle (see figures 7.22 and 24*c*) is a long bone with a slight sigmoid (S-shaped) curve and is easily seen and felt in the living human (see figure 7.23). The lateral end of the clavicle articulates with the acromion process, and its medial end articulates with the manubrium of the sternum. These articulations form the only bony connections between the pectoral girdle and the axial skeleton. The clavicle holds the upper limb away from the body, facilitating the mobility of the limb.

4	P R E D I C T

A broken clavicle changes the position of the upper limb in what way?

✔ *Answer in Appendix F*

Arm

The arm, the part of the upper limb from the shoulder to the elbow, contains only one bone, the **humerus** (figure 7.25). The humeral **head** articulates with the glenoid fossa of the scapula. The **anatomical neck,** immediately distal to the

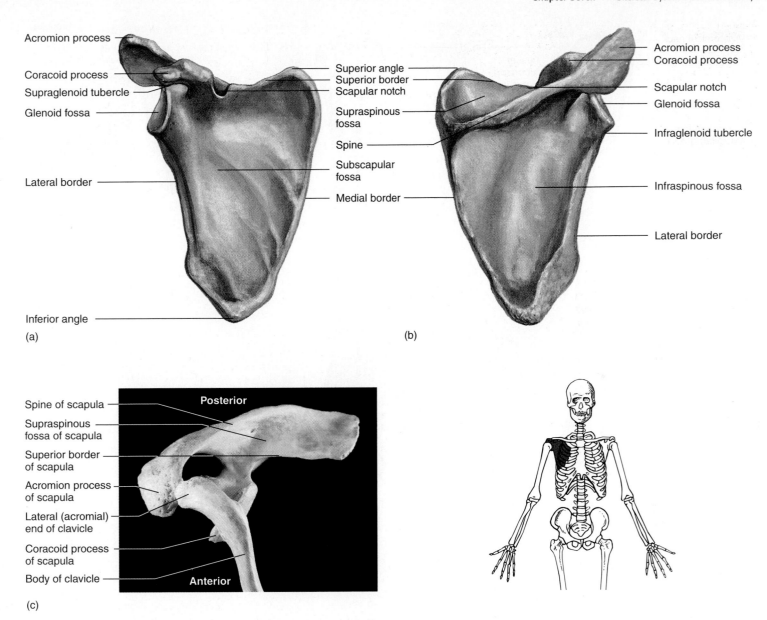

Acromion process
Coracoid process
Supraglenoid tubercle
Glenoid fossa
Lateral border
Inferior angle
(a)

Superior angle
Superior border
Scapular notch
Supraspinous fossa
Spine
Subscapular fossa
Medial border

Acromion process
Coracoid process
Scapular notch
Glenoid fossa
Infraglenoid tubercle
Infraspinous fossa
Lateral border
(b)

Spine of scapula
Supraspinous fossa of scapula
Superior border of scapula
Acromion process of scapula
Lateral (acromial) end of clavicle
Coracoid process of scapula
Body of clavicle
Posterior
Anterior
(c)

Figure 7.24 Scapula

(*a*) Right scapula, anterior view. (*b*) Right scapula, posterior view. (*c*) Photograph of the right scapula and clavicle from a superior view, showing the relationship between the distal end of the clavicle and the acromion process of the scapula.

head, is almost nonexistent; thus a surgical neck is identified. The **surgical neck** is so named because it is a common fracture site that often requires surgical repair. If it becomes necessary to remove the humeral head because of disease or injury, it is removed down to the surgical neck. The **greater** and **lesser tubercles** are located on the lateral and anterior surfaces, respectively, of the proximal end of the humerus, where they function as sites of muscle attachment. The groove between the two tubercles contains one tendon of the biceps brachii muscle and is called the **intertubercular,** or **bicipital** (bī-sip'i-tăl), **groove.** The **deltoid tuberosity** is located on the lateral surface of the humerus a little more than a third of the way along its length and is the attachment site for the deltoid muscle.

Clinical Note

Exaggerated exercise such as weight lifting increases the size of bony tubercles, including those of the humerus. Anthropologists use tubercular enlargements to estimate the occupation of the person whose bones are being examined. For example, a large deltoid tuberosity in a young skeleton suggests that the young person lifted heavier-than-normal weights. In some situations, this information may indicate that the person was a slave required to carry heavy loads.

The articular surfaces of the distal end of the humerus exhibit some unusual features where the humerus articulates with the two forearm bones. The lateral portion of the

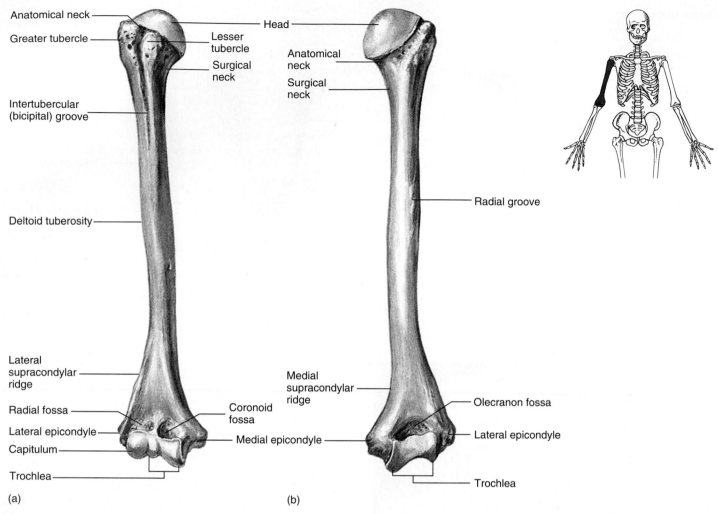

Figure 7.25   Right Humerus
(*a*) Anterior view. (*b*) Posterior view.

articular surface is very rounded, articulates with the radius, and is called the **capitulum** (kă-pit′yū-lŭm, meaning head-shaped). The medial portion somewhat resembles a spool or pulley, articulates with the ulna, and is called the **trochlea** (trok′lē-ă, meaning spool). Proximal to the capit-ulum and the trochlea are the **medial** and **lateral epi-condyles,** which function as points of muscle attachment for the muscles of the forearm.

Forearm

The forearm has two bones: the **ulna** on the medial side of the forearm, the side with the little finger, and the **radius** on the lateral, or thumb side, of the forearm (figure 7.26).

The proximal end of the ulna has a C-shaped articular surface that fits over the trochlea of the humerus. The larger, posterior process is the **olecranon** (ō-lek′ră-non, meaning the point of the elbow) **process** and can be felt as the point of the elbow (figure 7.27). Posterior arm muscles attach to the olecranon process. The smaller, anterior process is the **coro-**

noid (kōr′ŏ-noyd; also means crow's beak) **process,** and the notch between the two where the ulna articulates with the humerus is the **trochlear,** or **semilunar, notch.**

The distal end of the ulna has a small **head,** which artic-ulates with both the radius and the wrist bones (see figures 7.26 and 7.27). The head can be seen on the posterior, medial (ulnar) side of the distal forearm. The posteromedial side of the head has a small **styloid** (stī′loyd, meaning shaped like a stylus or writing instrument) **process** to which ligaments of the wrist are attached.

The proximal end of the radius is the **head.** It is concave and articulates with the capitulum of the humerus. The lateral

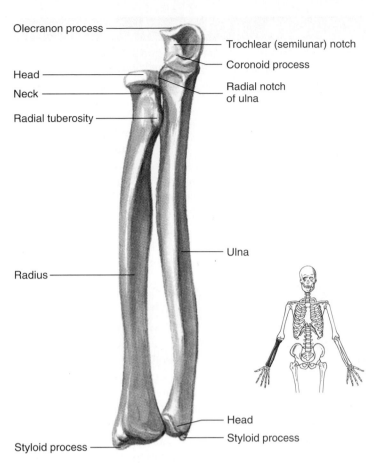

Olecranon process

Trochlear (semilunar) notch

Coronoid process

Head

Radial notch of ulna

Neck

Radial tuberosity

Ulna

Radius

Head

Styloid process

Styloid process

Figure 7.26   Ulna and Radius of the Right Forearm

surfaces of the head constitute a smooth cylinder, where the radius rotates against the **radial notch** of the ulna. As the forearm is rotated (supination and pronation; see chapter 8), the proximal end of the ulna tends to stay in place, and the radius tends to rotate. The **radial tuberosity** is the point at which a major anterior arm muscle, the biceps brachii, attaches.

The distal end of the radius, which articulates with the ulna and the carpals, is somewhat broadened, and a **styloid process** to which wrist ligaments are attached is located on the lateral side of the distal radius.

> **Clinical Note**
>
> The radius is the most commonly fractured bone in people over 50 years old. It is often fractured as the result of a fall on an outstretched hand. The most common site of fracture is 2.5 cm proximal to the wrist, and the fracture is often comminuted, or impacted.

Wrist

The wrist is a relatively short region between the forearm and hand and is composed of eight **carpal** (kar′păl) **bones,** which are arranged into two rows of four each (figure 7.28). The eight carpals, taken together, are convex posteriorly and concave anteriorly. The anterior concavity of the carpals is accentuated by the tubercle of the trapezium at the base of the thumb and the hook of the hamate at the base of the little finger. A ligament stretches across the wrist from the tubercle of the trapezium to the hook of the hamate to form a tunnel on the anterior surface of the wrist called the **carpal tunnel.** Tendons, nerves, and blood vessels pass through this tunnel to enter the hand.

Heads of metacarpals (knuckles)

Head of ulna

Lateral epicondyle

Olecranon process

Acromion process

Medial border of scapula

Olecranon process

Medial epicondyle

Figure 7.27   Surface Anatomy Showing Bones of the Pectoral Girdle and Upper Limb

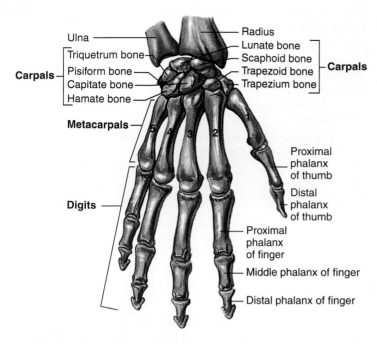

Ulna
Triquetrum bone
Pisiform bone
Capitate bone
Hamate bone

Carpals

Radius
Lunate bone
Scaphoid bone
Trapezoid bone
Trapezium bone

Carpals

Metacarpals

5 4 3 2

Digits

Proximal phalanx of thumb

Distal phalanx of thumb

Proximal phalanx of finger

Middle phalanx of finger

Distal phalanx of finger

Figure 7.28   Bones of the Right Wrist and Hand, Posterior View

Clinical Note

Because neither the bones nor the ligaments that form the walls of the carpal tunnel stretch, edema (fluid buildup) or connective tissue deposition within the carpal tunnel, caused by trauma or some other problem, may apply pressure against the nerve and vessels passing through the tunnel. This pressure causes carpal tunnel syndrome, which consists of tingling, burning, and numbness in the hand. Carpal tunnel syndrome occurs more frequently in people who use their hands a lot. The number of cases has increased in recent years and is associated with people who work long hours on computers.

Hand

Five **metacarpals** are attached to the carpal bones and constitute the bony framework of the hand (see figure 7.28). The metacarpals form a curve so that, in the resting position, the palm of the hand is concave. The distal ends of the metacarpals help form the knuckles of the hand. The spaces between the metacarpals are occupied by soft tissue.

The five **digits** of each hand include one thumb and four fingers. Each digit consists of small long bones called **phalanges** (fă-lan′jēz; the singular term phalanx refers to the Greek word, meaning a line or wedge of soldiers holding their spears, tips outward, in front of them). The thumb has two phalanges, and each finger has three. One or two **sesamoid** (ses′ă-moyd, meaning resembling a sesame seed) **bones** (not illustrated) often form near the junction between the proximal phalanx and the metacarpal of the thumb. Sesamoid bones are small bones located within tendons.

6   P R E D I C T

Explain why the dried, articulated skeleton appears to have much longer "fingers" than are seen in the hand with the soft tissue intact.

✔ *Answer in Appendix F*

Lower Limb

The general pattern of the lower limb is very similar to that of the upper limb, except that the pelvic girdle is attached much more firmly to the body than is the pectoral girdle and the bones in general are thicker, heavier, and longer than those of the upper limb. These structures reflect the function of the lower limb in support and movement of the body.

Pelvic Girdle

The **pelvis** (pel′vis, meaning basin), or **pelvic girdle,** is a ring of bones formed by the sacrum and paired bones called the **coxae** (kok′sē), or hipbones (figure 7.29). Each coxa consists of a large, concave bony plate superiorly, a slightly narrower region in the center, and an expanded bony ring inferiorly, which surrounds a large **obturator** (ob′tū-rā′tŏr, meaning to occlude or close up, indicating that the foramen is occluded by soft tissue) **foramen.** A fossa called the **acetabulum** (as-ĕ-tab′yū-lŭm, meaning a shallow vinegar cup—a common household item in ancient times) is located on the lateral surface of each coxa and is the point of articulation of the lower limb with the girdle. The articular surface of the acetabulum is crescent-shaped and occupies only the superior and lateral aspects of the fossa. The pelvic girdle is the place of attachment for the lower limbs, supports the weight of the body, and protects internal organs. In addition, the pelvic girdle in a woman protects the developing fetus and forms a passageway through which the fetus passes during delivery.

Each coxa is formed by the fusion of three bones during development: the **ilium** (il′ē-ŭm, meaning groin), the **ischium** (is′kē-ŭm, meaning hip), and the **pubis** (pyū′bis, which refers to the genital hair). All three bones join near the center of the acetabulum (figure 7.30*a*). The superior portion of the ilium is called the **iliac crest** (figure 7.30*b* and *c*). The crest ends anteriorly as the **anterior superior iliac spine** and posteriorly as the **posterior superior iliac spine.** The crest and anterior spine can be felt and even seen in thin individuals (figure 7.31). The anterior superior iliac spine is an important anatomic landmark that is used, for example, to find the correct location for giving injections in the hip muscle. A dimple overlies the posterior superior iliac spine just superior to the buttocks. The **greater ischiadic** (is-kē-ad′ik, formerly called sciatic) **notch** is on the posterior side of the ilium, just inferior to the inferior posterior iliac spine. The ischiadic nerve passes through the greater ischiadic notch. The **auricular surface** of the ilium joins the sacrum to form the **sacroiliac joint** (see figure 7.29). The medial side of the ilium consists of a large depression called the **iliac fossa.**

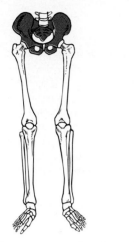

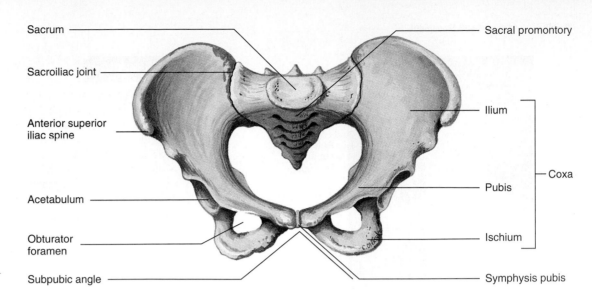

Figure 7.29 The Complete Pelvic Girdle, Anterior View

Labels for Figure 7.29: Sacrum, Sacroiliac joint, Anterior superior iliac spine, Acetabulum, Obturator foramen, Subpubic angle, Sacral promontory, Ilium, Coxa, Pubis, Ischium, Symphysis pubis

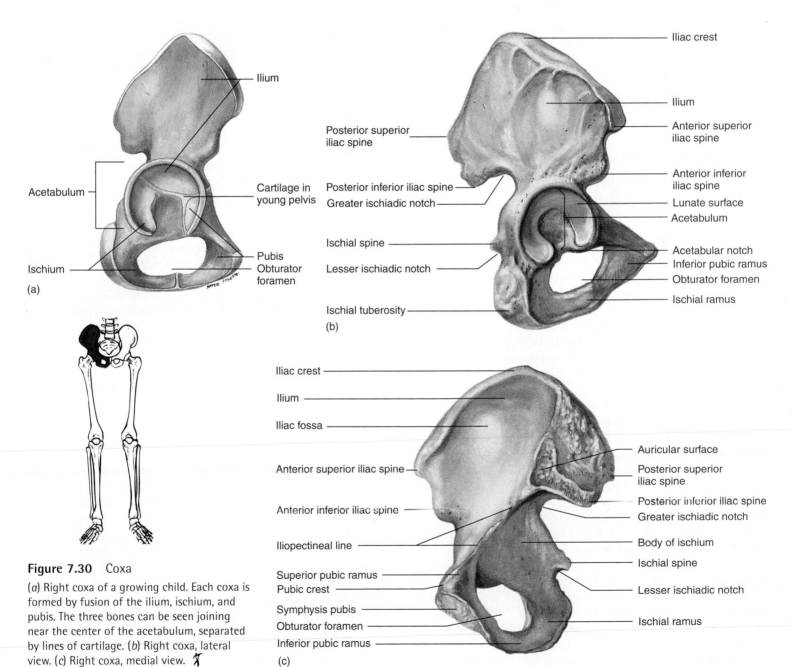

Figure 7.30 Coxa

(a) Right coxa of a growing child. Each coxa is formed by fusion of the ilium, ischium, and pubis. The three bones can be seen joining near the center of the acetabulum, separated by lines of cartilage. (b) Right coxa, lateral view. (c) Right coxa, medial view.

Labels for (a): Ilium, Acetabulum, Ischium, Cartilage in young pelvis, Pubis, Obturator foramen

Labels for (b): Iliac crest, Ilium, Anterior superior iliac spine, Anterior inferior iliac spine, Lunate surface, Acetabulum, Acetabular notch, Inferior pubic ramus, Obturator foramen, Ischial ramus, Posterior superior iliac spine, Posterior inferior iliac spine, Greater ischiadic notch, Ischial spine, Lesser ischiadic notch, Ischial tuberosity

Labels for (c): Iliac crest, Ilium, Iliac fossa, Anterior superior iliac spine, Anterior inferior iliac spine, Iliopectineal line, Superior pubic ramus, Pubic crest, Symphysis pubis, Obturator foramen, Inferior pubic ramus, Auricular surface, Posterior superior iliac spine, Posterior inferior iliac spine, Greater ischiadic notch, Body of ischium, Ischial spine, Lesser ischiadic notch, Ischial ramus

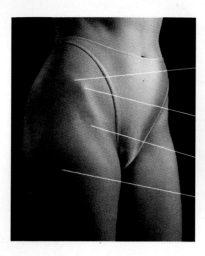

Iliac crest

Anterior
superior
iliac spine

Anterior inferior
iliac spine

Greater
trochanter

Figure 7.31   Surface Anatomy Showing an Anterior View of the Hip Bones

Clinical Note

The sacroiliac joint receives most of the weight of the upper body and is strongly supported by ligaments. Excessive strain on the joint, however, can cause slight joint movement and can stretch connective tissue and associated nerve endings in the area and cause pain. Thus is derived the expression, "My aching sacroiliac!" This problem sometimes develops in pregnant women because of the forward weight distribution of the fetus.

The ischium possesses a heavy **ischial tuberosity,** where posterior thigh muscles attach and on which a person sits (see figure 7.30b). The pubis possesses a **pubic crest,** where abdominal muscles attach (see figure 7.30c). The pubic crest can be felt anteriorly. Just inferior to the pubic crest is the point of junction, the **symphysis** (sim'fi-sis, meaning a coming together) **pubis,** or **pubic symphysis,** between the two coxae.

The pelvis can be divided into two parts by an imaginary plane passing from the sacral promontory along the **iliopectineal lines** of the ilium to the pubic crest (figure 7.32a and b). The bony boundary of this plane is the **pelvic brim.** The **false,** or **greater, pelvis** is superior to the pelvic brim and is partially surrounded by bone on the posterior and lateral sides. During life, the abdominal muscles form the anterior wall of the false pelvis. The **true pelvis** is inferior to the pelvic brim and is completely surrounded by bone. The superior opening of the true pelvis, at the level of the pelvic brim, is the **pelvic inlet.** The inferior opening of the true pelvis, bordered by the inferior margin of the pubis, the ischial spines and tuberosities, and the coccyx, is the **pelvic outlet** (see figure 7.32a).

The male pelvis usually is more massive than the female pelvis as a result of the greater weight and size of the male, and the female pelvis is broader and has a larger, more rounded pelvic inlet and outlet (see figure 7.32b and c), consistent with the need to allow the fetus to pass through these openings in the female pelvis during delivery. Table 7.8 lists additional differences between the male and female pelvis.

Clinical Note

A wide circular pelvic inlet and a pelvic outlet with widely spaced ischial spines are ideal for delivery. Variations in a woman's pelvic inlet and outlet can cause problems during delivery; thus the size of the pelvic inlet and outlet is routinely measured during prenatal pelvic examinations of pregnant women. If the pelvic outlet is too small for normal delivery, delivery can be accomplished by **cesarean section,** which is the surgical removal of the fetus through the abdominal wall.

Thigh

The thigh contains a single bone called the **femur.** The femur has a prominent rounded **head,** where it articulates with the acetabulum, and a well-defined **neck;** both are located at an oblique angle to the shaft of the femur (figure 7.33). The proximal shaft exhibits two tuberosities: a **greater trochanter** (trō-kan'ter, meaning runner) lateral to the neck and a smaller, or **lesser, trochanter** inferior and posterior to the neck. Both trochanters are attachment sites for muscles that attach the hip to the thigh. The greater trochanter and its attached muscles form a bulge that can be seen as the widest part of the hips (see figure 7.31). The distal end of the femur has **medial** and **lateral condyles,** smooth, rounded surfaces that articulate with the tibia. Located proximally to the condyles are the **medial** and **lateral epicondyles,** important sites of muscle and ligament attachment.

7	P R E D I C T

Compare the following in terms of structure and function for the upper and lower limbs: depth of sockets, size of bones, and size of tubercles and trochanters. What is the significance of these differences?

✔ *Answer in Appendix F*

The **patella** is a large sesamoid bone located within the tendon of the quadriceps femoris muscle, which is the major muscle of the anterior thigh (figure 7.34). The patella articulates with the patellar groove of the femur to create a smooth articular surface over the anterior distal end of the femur. The patella holds the tendon away from the distal end of the femur and therefore changes the angle of the tendon between the quadriceps femoris muscle and the tibia, where the tendon attaches. This change in angle increases the force that can be applied from the muscle to the tibia. As the result of this increase in applied force, less muscle contraction force is required to move the tibia.

Clinical Note

If the patella is severely fractured, the tendon from the quadriceps femoris muscle may be torn, resulting in a severe decrease in muscle function. In extreme cases, it may be necessary to remove the patella

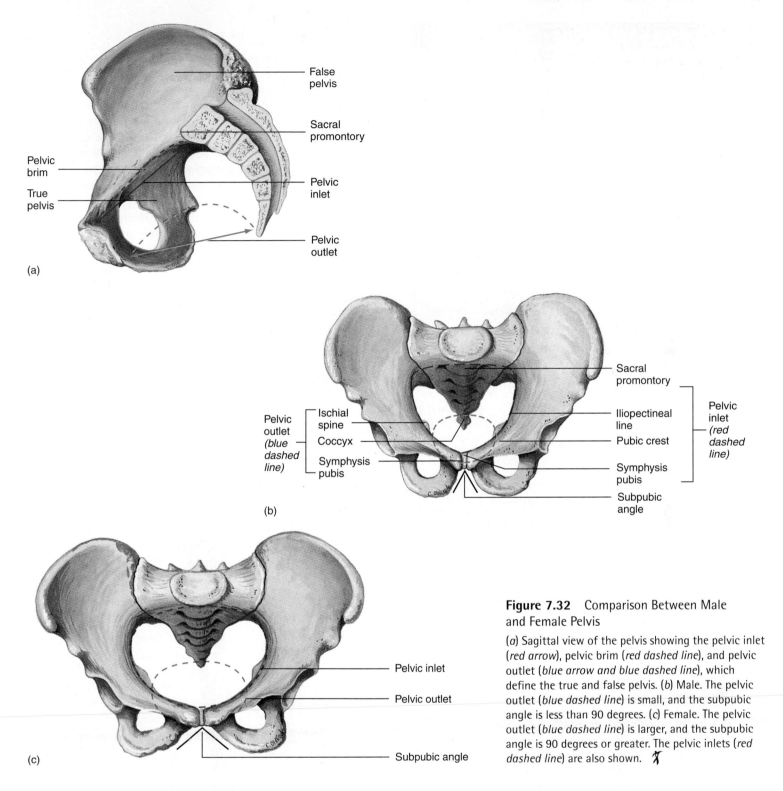

False pelvis

Sacral promontory

Pelvic brim

True pelvis

Pelvic inlet

Pelvic outlet

(a)

Pelvic outlet *(blue dashed line)*

Ischial spine

Coccyx

Symphysis pubis

Sacral promontory

Iliopectineal line

Pubic crest

Symphysis pubis

Subpubic angle

Pelvic inlet *(red dashed line)*

(b)

Pelvic inlet

Pelvic outlet

Subpubic angle

(c)

Figure 7.32   Comparison Between Male and Female Pelvis

(*a*) Sagittal view of the pelvis showing the pelvic inlet (*red arrow*), pelvic brim (*red dashed line*), and pelvic outlet (*blue arrow and blue dashed line*), which define the true and false pelvis. (*b*) Male. The pelvic outlet (*blue dashed line*) is small, and the subpubic angle is less than 90 degrees. (*c*) Female. The pelvic outlet (*blue dashed line*) is larger, and the subpubic angle is 90 degrees or greater. The pelvic inlets (*red dashed line*) are also shown.

to repair the tendon. Removal of the patella results in a decrease in the amount of power the quadriceps femoris muscle can generate at the tibia.

The patella normally tracks in the patellar groove on the anterodistal end of the femur. Abnormal tracking of the patella can become a problem in some teenage females. As the young

woman's hips widen during puberty, the angles at the joints between the hips and the tibia may change considerably. As the knee becomes located more medially relative to the hip, the patella may be forced to track more laterally than normal. This lateral tracking may result in pain in the knees of some young female athletes.

Table 7.8   Differences between the Male and Female Pelvis (see figure 7.32)

Area	Description
General	Female pelvis somewhat lighter in weight and wider laterally but shorter superiorly to inferiorly and less funnel-shaped; less obvious muscle attachment points in female than in male
Sacrum	Broader in female with the inferior part directed more posteriorly; the sacral promontory projects less far anteriorly in female
Pelvic inlet	Heart-shaped in male; oval in female
Pelvic outlet	Broader and more shallow in female
Subpubic angle	Less than 90 degrees in male; 90 degrees or more in female
Ilium	More shallow and flared laterally in female
Ischial spines	Farther apart in female
Ischial tuberosities	Turned laterally in female and medially in male

Leg

The leg, which is the part of the lower limb between the knee and the ankle, consists of two bones: the **tibia** (tib′ē-ă, or shinbone) and the **fibula** (fib′yū-lă, meaning resembling a clasp or buckle; figure 7.35). The tibia is by far the larger of the two and supports most of the weight of the leg. A **tibial tuberosity,** which is the attachment point for the quadriceps femoris muscle, can easily be seen and felt just inferior to the patella (figure 7.36). The **anterior crest** forms the shin. The proximal end of the tibia has flat **medial** and **lateral condyles,** which articulate with the condyles of the femur. Located between the condyles is the **intercondylar eminence,** which is a site of ligament attachment. The knee is unusual in that it has ligaments within the joint (see chapter 8). The distal end of the tibia is enlarged to form the **medial malleolus** (ma-lē′ō-lŭs, meaning mallet-shaped), which helps form the medial side of the ankle joint.

The fibula does not articulate with the femur but has a small proximal head where it articulates with the tibia. The distal end of the fibula is slightly enlarged as the **lateral malleolus** to create the lateral wall of the ankle joint. The lateral and medial malleoli can be felt and seen as prominent

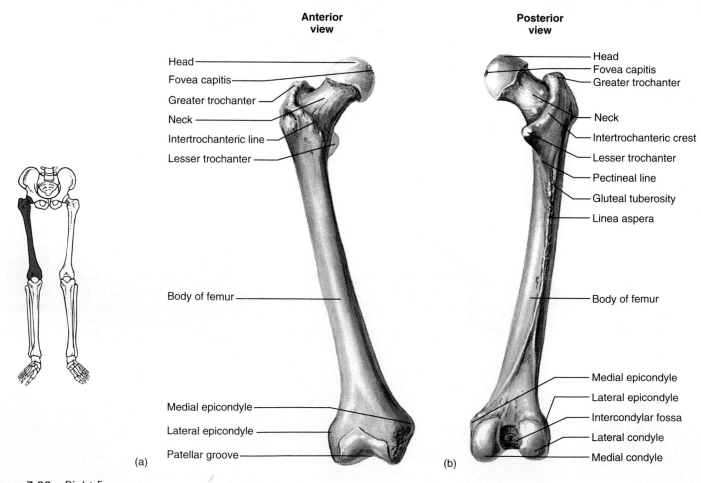

Anterior view

Posterior view

Head
Fovea capitis
Greater trochanter
Neck
Intertrochanteric line
Lesser trochanter

Body of femur

Medial epicondyle
Lateral epicondyle
Patellar groove

(a)

Head
Fovea capitis
Greater trochanter

Neck
Intertrochanteric crest
Lesser trochanter
Pectineal line
Gluteal tuberosity
Linea aspera

Body of femur

Medial epicondyle
Lateral epicondyle
Intercondylar fossa
Lateral condyle
Medial condyle

(b)

Figure 7.33   Right Femur

(a) Anterior view. (b) Posterior view.

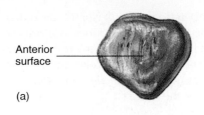

Anterior
surface

(a)

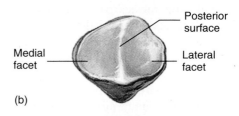

Medial
facet

Posterior
surface

Lateral
facet

(b)

Figure 7.34   Right Patella
(*a*) Anterior view. (*b*) Posterior view.

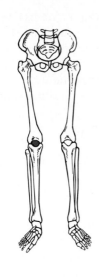

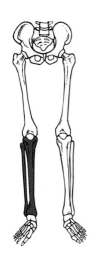

Figure 7.35   Right Tibia and
Fibula, Anterior View

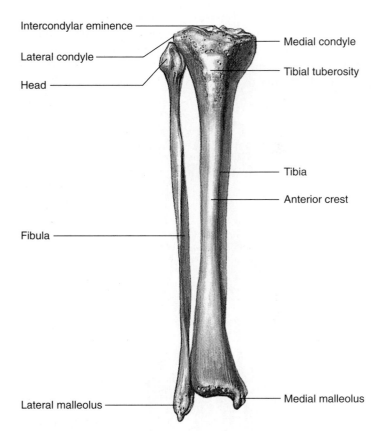

Intercondylar eminence
Lateral condyle
Head
Medial condyle
Tibial tuberosity
Tibia
Anterior crest
Fibula
Lateral malleolus
Medial malleolus

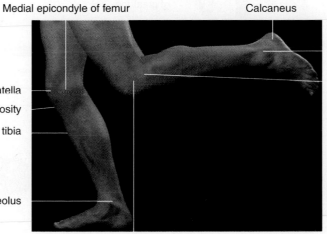

Medial epicondyle of femur
Calcaneus
Lateral malleolus
Head of fibula
Patella
Tibial tuberosity
Anterior crest of tibia
Medial malleolus
Lateral epicondyle of femur

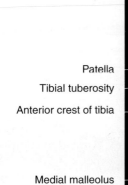

Figure 7.36   Surface Anatomy Showing
Bones of the Lower Limb

lumps on either side of the ankle (see figure 7.36). The thinnest portion of the fibula is just proximal to the lateral malleolus.

| 8 | P R E D I C T |

Explain why modern ski boots are designed with high tops that extend partway up the leg.

✔ *Answer in Appendix F*

Ankle

The ankle consists of seven **tarsal** (tar'săl, meaning the sole of the foot) **bones,** which are depicted and named in figure 7.37. The **talus** (tā'lŭs, meaning ankle bone) articulates with the tibia and the fibula to form the ankle joint. The **calcaneus**

(kal-kā'nē-us, meaning heel) is located inferior to the talus and supports that bone. The calcaneus protrudes posteriorly where the calf muscles attach to it and where it can be easily felt as the heel of the foot. The ankle is relatively much larger than the wrist and, from a functional perspective, includes the heel and the proximal third of the foot.

Foot

The **metatarsals** and **phalanges** of the foot are arranged in a manner very similar to that of the metacarpals and phalanges of the hand, with the great toe analogous to the thumb (see figure 7.37b). Small sesamoid bones often form in the tendons of muscles attached to the great toe. The ball of the foot is the junction between the metatarsals and phalanges. The foot as a unit is convex dorsally and concave ventrally to form the arches of the foot (described more fully in chapter 8).

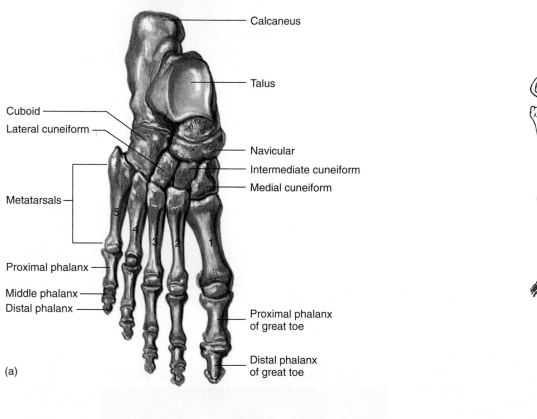

Calcaneus
Talus
Cuboid
Lateral cuneiform
Navicular
Intermediate cuneiform
Medial cuneiform
Metatarsals
Proximal phalanx
Middle phalanx
Distal phalanx
Proximal phalanx of great toe
Distal phalanx of great toe

(a)

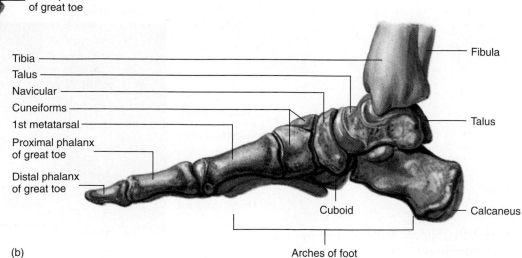

Tibia
Talus
Navicular
Cuneiforms
1st metatarsal
Proximal phalanx of great toe
Distal phalanx of great toe
Fibula
Talus
Cuboid
Calcaneus
Arches of foot

Figure 7.37   Bones of the Right Ankle and Foot

(*a*) Dorsal view. (*b*) Medial view.   ✗   (b)

Summary

The gross anatomy of the skeletal system considers the features of bone, cartilage, tendons, and ligaments that can be seen without the use of a microscope. Dried, prepared bones display the major features of bone but obscure the relationship between bone and soft tissue.

General Considerations

Bones have processes, smooth surfaces, and holes that are associated with ligaments, muscles, joints, nerves, and blood vessels.

Axial Skeleton

The axial skeleton consists of the skull, vertebral column, and thoracic cage.

Skull

The skull can be thought of as a single unit.

Superior View of the Skull

The parietal bones are joined at the midline by the sagittal suture; they are joined to the frontal bone by the coronal suture, to the occipital bone by the lambdoid suture, and to the temporal bone by the squamous suture.

Posterior View of the Skull

Nuchal lines are the points of attachment for neck muscles.

Lateral View of the Skull

1. The external auditory meatus transmits sound waves toward the eardrum.
2. Important neck muscles attach to the mastoid process.
3. The temporal lines are attachment points of the temporalis muscle.
4. The zygomatic arch, from the temporal and zygomatic bones, forms a bridge across the side of the skull.

Frontal View of the Skull

1. The orbits contain the eyes.
2. The nasal cavity is divided by the nasal septum, and the hard palate separates the nasal cavity from the oral cavity.
3. Sinuses within bone are air-filled cavities. The paranasal sinuses, which connect to the nasal cavity, are the frontal, ethmoidal, sphenoidal, and maxillary sinuses.
4. The mandible articulates with the temporal bone.

Interior of the Cranial Vault

1. The crista galli is a point of attachment for one of the meninges.
2. The olfactory nerves extend into the roof of the nasal cavity through the cribriform plate.
3. The sella turcica is occupied by the pituitary gland.
4. The spinal cord and brain are connected through the foramen magnum.

Inferior View of the Skull

1. Occipital condyles are points of articulation between the skull and the vertebral column.

2. Blood reaches the brain through the internal carotid arteries, which pass through the carotid canals, and the vertebral arteries, which pass through the foramen magnum.
3. Most blood leaves the brain through the internal jugular veins, which exit through the jugular foramina.
4. Styloid processes provide attachment points for three muscles involved in movement of the tongue, hyoid bone, and pharynx.
5. The hard palate forms the floor of the nasal cavity.

Bones of the Skull

1. The skull is composed of 28 bones.
2. The auditory ossicles, which function in hearing, are located inside the temporal bones.
3. The cranial vault protects the brain.
4. The facial bones protect the sensory organs of the head and function as muscle attachment sites (mastication, facial expression, and eye muscles).
5. The mandible and maxillae possess alveolar processes with sockets for the attachment of the teeth.

Hyoid

The hyoid bone, which "floats" in the neck, is the attachment site for throat and tongue muscles.

Vertebral Column

1. The vertebral column provides flexible support and protects the spinal cord.
2. The vertebral column has four major curvatures: cervical, thoracic, lumbar, and sacral/coccygeal. Abnormal curvatures are lordosis (lumbar), kyphosis (thoracic), and scoliosis (lateral).
3. A typical vertebra consists of a body, an arch, and various processes.
   - Adjacent bodies are separated by intervertebral disks. The disk has a fibrous outer covering (anulus fibrosus) surrounding a gelatinous interior (nucleus pulposus).
   - Vertebrae articulate with one another through the superior and inferior articular processes.
   - Part of the body and the arch (pedicle and lamina) form the vertebral foramen, which contains and protects the spinal cord.
   - Spinal nerves exit through the intervertebral foramina.
   - The transverse and spinous processes serve as points of muscle and ligament attachment.
4. Several different kinds of vertebrae can be recognized.
   - All seven cervical vertebrae have transverse foramina, and most have bifid spinous processes.
   - The 12 thoracic vertebrae are characterized by long, downward pointing spinous processes and demifacets.
   - The five lumbar vertebrae have thick, heavy bodies and processes.
   - The sacrum consists of five fused vertebrae and attaches to the pelvis.
   - The coccyx consists of four fused vertebrae attached to the sacrum.

Thoracic Cage

1. The thoracic cage (consisting of the ribs, their associated costal cartilages, and the sternum) functions to protect the thoracic

organs and to prevent the collapse of the thorax during respiration.

2. Twelve pairs of ribs attach to the thoracic vertebrae. They are divided into seven pairs of true ribs and five pairs of false ribs. Two pairs of false ribs are floating ribs.

3. The sternum is composed of the manubrium, the body, and the xiphoid process.

Appendicular Skeleton

The appendicular skeleton consists of the upper and lower limbs and the girdles that attach the limbs to the body.

Upper Limb

1. The upper limb is attached loosely and functions in grasping and manipulation.

2. The pectoral girdle consists of the scapulae and the clavicles.
   - The scapula articulates with the humerus and the clavicle. It serves as an attachment site for shoulder, back, and arm muscles.
   - The clavicle holds the shoulder away from the body, permitting free movement of the arm.

3. The arm bone is the humerus.
   - The humerus articulates with the scapula (head), the radius (capitulum), and the ulna (trochlea).
   - Sites of muscle attachment are the greater and lesser tubercles, the deltoid tuberosity, and the epicondyles.

4. The forearm consists of the ulna and the radius.
   - The ulna and the radius articulate with each other and with the humerus and the wrist bones.
   - The wrist ligaments attach to the styloid processes of the radius and the ulna.

5. There are eight carpal, or wrist, bones arranged in two rows.

6. The hand consists of five metacarpal bones.

7. The phalanges are digital bones. Each finger has three phalanges, and the thumb has two phalanges.

Lower Limb

1. The lower limb is attached solidly to the coxa and functions in support and locomotion.

2. The pelvic girdle is formed by the coxae and the sacrum. Each coxa is formed by the fusion of the ilium, the ischium, and the pubis.
   - The coxae articulate with each other (symphysis pubis) and with the sacrum (sacroiliac joint) and the femur (acetabulum).
   - Important sites of muscle attachment are the iliac crest, the iliac spines, and the ischial tuberosity.
   - The female pelvis has a larger pelvic inlet and outlet than the male pelvis.

3. The thigh bone is the femur.
   - The femur articulates with the coxa (head), the tibia (medial and lateral condyles), and the patella (patellar groove).
   - Sites of muscle attachment are the greater and lesser trochanters and the lateral and medial epicondyles.

4. The leg consists of the tibia and the fibula.
   - The tibia articulates with the femur, the fibula, and the talus. The fibula articulates with the tibia and the talus.
   - Knee ligaments attach to the intercondylar eminence, and tendons from the thigh muscles attach to the tibial tuberosity.

5. Seven tarsal bones form the ankle.

6. The foot consists of five metatarsal bones.

7. The toes have three phalanges each, except for the big toe, which has two.

Content Review

1. Define the terms axial skeleton and appendicular skeleton.
2. List the seven bones that form the orbit of the eye.
3. Name the eight bones that form the nasal cavity. Describe the bones and cartilage that form the nasal septum.
4. What is a sinus? What are the functions of the sinuses? Give the location of the paranasal sinuses.
5. Name the bones that form the hard palate. What is the function of the hard palate?
6. Through what foramen does the brainstem connect to the spinal cord? Name the foramina that contain nerves for the senses of vision (optic nerve), smell (olfactory nerves), and hearing (vestibulocochlear nerve).
7. Name the foramen through which the major blood vessels for the brain enter and exit the skull.
8. List the places where the following muscles attach to the skull: neck muscles, throat muscles, muscles of mastication, muscles of facial expression, and muscles that move the eyeballs.
9. Name the bones of the cranial vault.
10. Name the bones of the face.
11. Describe the four major curvatures of the vertebral column. Define the terms lordosis, kyphosis, and scoliosis.
12. How do the vertebrae protect the spinal cord? Where do spinal nerves exit the vertebral column?
13. Name and give the number of each type of vertebra. Describe the characteristics that distinguish the different types from one another.
14. What is the function of the thoracic cage? Distinguish between true, false, and floating ribs.
15. Describe the different parts of the sternum.
16. Name the bones that make up the pectoral girdle. What are the functions of these bones?
17. What are the functions of the acromion process and the coracoid process of the scapula?
18. Name the important sites of muscle attachment on the humerus.
19. Give the points of articulation between the scapula, humerus, ulna, radius, and wrist bones.
20. What is the function of the radial tuberosity?
21. List the eight carpal bones. What is the carpal tunnel?
22. What bones form the hand? The knuckles? How many phalanges are in each finger and in the thumb?
23. Define the term pelvis. What bones fuse to form each coxa? Where and with what bones do the coxae articulate?
24. Name the important sites of muscle attachment on the pelvis.
25. Distinguish between the true pelvis and the false pelvis.
26. Describe the difference between a male and a female pelvis.

27. What is the function of the greater trochanter and the lesser trochanter?
28. Name the bones of the leg.
29. Give the points of articulation between the pelvis, femur, leg, and ankle.
30. What is the function of the tibial tuberosity?
31. Name the seven tarsal bones. Which bones form the ankle joint? What bone forms the heel?
32. Describe the bones that constitute the foot.

Develop Your Reasoning Skills

1. A patient has an infection in the nasal cavity. Name seven adjacent structures to which the infection could spread.
2. A patient is unconscious. Radiographic films reveal that the superior articular process of the atlas has been fractured. Which of the following could have produced this condition: falling on the top of the head or being hit in the jaw with an uppercut? Explain.
3. If the vertebral column is forcefully rotated, what part of the vertebra is most likely to be damaged? In what area of the vertebral column is such damage most likely?
4. An asymmetric weakness of the back muscles can produce which of the following: scoliosis, kyphosis, or lordosis? Which could result from pregnancy? Explain.
5. What might be the consequences of a broken forearm involving both the ulna and radius when the ulna and radius fuse to each other during repair of the fracture?
6. Suppose you need to compare the length of one lower limb to the other in an individual. Using bony landmarks, suggest an easy way to accomplish the measurements.
7. A paraplegic individual develops decubitus ulcers (pressure sores) on the buttocks from sitting in a wheelchair for extended periods. Name the bony protuberance responsible.
8. Why are women knock-kneed more often than men?
9. On the basis of bone structure of the lower limb, explain why it is easier to turn the foot medially (sole of the foot facing toward the midline of the body) than laterally. Why is it easier to cock the wrist medially than laterally?
10. Justin Time leaped from his hotel room to avoid burning to death in a fire. If he landed on his heels, what bone was he likely to fracture? Unfortunately for Justin, a 240-lb fireman, Hefty Stomper, ran by and stepped heavily on the proximal part of Justin's foot (not the toes). What bones could now be broken?

Web Site Link

For a listing of the most current web sites related to this chapter, please visit the Seeley home page at:
http://www.mhhe.com/biosci/ap/seeleyap/

Chapter Eight

Articulations and Movement

Objectives

1. Define the term articulation.

2. Discuss the importance of movement to joint formation and maintenance.

3. List the general features of a fibrous joint, name the three classes of fibrous joints, list examples for each class, and describe their differences.

4. List the general features of a cartilaginous joint, name the two types of cartilaginous joints, list examples for each class, and describe the major differences between them.

5. Describe the general features of a synovial joint, including the histology of the joint capsule and the nature of synovial fluid.

6. Define the term bursa, and list three places where bursae may be located.

7. List the types of synovial joints, give an example of each type, and describe the major structural and functional features of each.

8. Describe flexion and extension, and give examples at two joints.

9. Describe abduction and adduction, and give examples at two joints.

10. Describe inversion and eversion of the foot.

11. Describe rotation, circumduction, pronation, and supination.

12. Describe the following joints, and indicate the types of movements that can occur at each: temporomandibular, shoulder, hip, knee, and ankle joints. Do the same for the foot arches.

13. List the most common types of knee injuries, and tell which part of the knee is most often damaged in each type.

14. Define the term sprain, and describe which portions of the ankle joint are most commonly damaged when it is sprained.

Part Two

We move because muscles pull on our bones, but movement required of the skeleton cannot occur without movable joints between the bones, which allow one bone to move relative to another. Without joints, movement would be severely restricted. In machines, the parts most likely to wear out are those that rub together when components of the machine are moving, and those parts require the most maintenance. The movable joints are the places in the body where the bones rub together, yet we tend to pay little attention to them, unless one is injured or not functioning properly for some other reason. Then we realize how important the movable joints are for normal function. Fortunately our joints are self-maintaining, but damage to or disease of a joint can eventually bring the whole body to a literal standstill.

An **articulation**, or **joint**, is a place where two bones come together. Joints are usually considered movable, but that is not always the case. Many joints allow only limited movement, and others seem to be immovable. The structure of a given joint is directly correlated with its degree of movement. Fibrous joints have much less movement than joints that contain fluid and have smooth articulating surfaces.

Joints develop between adjacent bones or areas of ossification, and movement is important in determining the type of joint that develops. If movement is restricted—even in a highly movable joint—at any time during an individual's life, the joint may be transformed into a nonmovable joint.

The scheme for naming joints is described first in this chapter, followed by the classification of joints, a discussion of body movements, and a more complete description of selected joints.

Naming Joints

Joints are commonly named according to the bones or portions of bones that are united at the joint, such as the temporomandibular joint between the temporal bone and the mandible.

Some joints are given the name of only one of the articulating bones, such as the humeral joint (shoulder) between the humerus and scapula. Still other joints are simply given the Greek or Latin equivalent of the common name, such as **cubital** (kyū′bi-tăl) for the elbow joint.

Classes of Joints

The three major kinds of joints are classified structurally as fibrous, cartilaginous, and synovial. In this classification scheme, joints are classified according to the major connective tissue type that binds the bones together, and whether or not there is a fluid-filled joint capsule. Joints may also be classified by function. This classification is based on the degree of motion at each joint and includes the terms synarthrosis (nonmovable joint), amphiarthrosis (slightly movable joint), and diarthrosis (freely movable joint). This functional classification is somewhat limited and is not used in this text. The structural classification scheme with its various subclasses allows for a more precise classification and is the scheme we use.

Fibrous Joints

Fibrous joints consist of two bones that are united by fibrous connective tissue, have no joint cavity, and exhibit little or no movement. Joints in this group are classified further on the basis of structure as sutures, syndesmoses, or gomphoses (table 8.1).

Sutures

Sutures (sū′chūrz) are seams between the bones of the unit skull (see figures 7.2–4 and 7.6), and some sutures may be completely immovable in adults. Sutures are seldom smooth, and the opposing bones often interdigitate (have interlocking fingerlike processes). This interdigitation adds considerable stability to sutures. The tissue between the two bones is dense, regularly arranged fibrous connective tissue, and the periosteum on the inner and outer surfaces of the adjacent bones continues over the joint. The two layers of periosteum plus the dense fibrous connective tissue in between form the **sutural ligament.**

In a newborn, wide areas between some of the sutures are called **fontanels** (fon′tă-nelz′). The fontanels make the skull flexible during the birth process and allow for growth of the head after birth (figure 8.1).

The margins of bones within sutures are sites of continuous intramembranous bone growth, and many sutures eventually become ossified. For example, ossification of the suture between the two frontal bones occurs shortly after birth so that they usually form a single frontal bone in the adult skull. In most normal adults the coronal, sagittal, and lambdoid sutures are not fused. In some very old adults, however, even these sutures may become ossified. A **synostosis** (sin′os-tō′sis) results when two bones grow together across a joint to form a single bone. This process also occurs in certain other joints such as in the epiphyseal plates.

Table 8.1   Fibrous and Cartilaginous Joints

Class and Example of Joint	Structures Joined	Movement
Fibrous Joints		
Sutures		
Coronal suture	Frontal and parietal	None
Lambdoid suture	Occipital and parietal	None
Sagittal suture	Between the two parietal bones	None
Squamous suture	Parietal and temporal	Slight
Syndesmoses		
Radioulnar syndesmosis (interosseous membrane)	Ulna and radius	Slight
Stylohyoid syndesmosis	Styloid process and hyoid bone	Slight
Stylomandibular syndesmosis	Styloid process and mandible	Slight
Tibiofibular syndesmosis (interosseous membrane)	Tibia and fibula	Slight
Gomphoses		
Dentoalveolar joint	Tooth and alveolar process	Slight
Cartilaginous Joints		
Synchrondroses		
Epiphyseal plate	Between diaphysis and epiphysis of a long bone	None
Sternocostal synchondrosis	Anterior cartilaginous part of first rib; between rib and sternum	Slight
Sphenooccipital synchondrosis	Sphenoid and occipital	None
Symphyses		
Intervertebral symphysis	Bodies of adjacent vertebrae	Slight
Manubriosternal symphysis	Manubrium and body of sternum	None
Symphysis pubis	Between the two coxae	None except during childbirth
Xiphisternal symphysis	Xiphoid process and body of sternum	None

Syndesmoses

A **syndesmosis** (sin′dez-mō′sis, meaning to fasten or bind) is a type of fibrous joint in which the bones are farther apart than in a suture and are joined by ligaments. Some movement may occur at syndesmoses because of flexibility of the ligaments, such as in the radioulnar syndesmosis, which binds the radius and ulna together (figure 8.2).

Gomphoses

Gomphoses (gom-fō′sēz) are specialized joints consisting of pegs that fit into sockets and that are held in place by fine bundles of regular collagenous connective tissue. The joints between the teeth and the sockets (alveoli) of the mandible and maxillae are gomphoses. The connective tissue bundles between the teeth and their sockets are **periodontal** (per′ē-ō-don′tăl) **ligaments** (see figure 24.6) and allow a slight amount of "give" to the teeth during mastication.

Clinical Note

The gingiva, or gums, are the soft tissues covering the alveolar ridge. Neglect of the teeth can result in **gingivitis,** an inflammation of the gingiva, often resulting from bacterial infection. Left untreated, gingivitis may spread to the tooth socket, resulting in periodontal disease, the leading cause of tooth loss in the United States. Periodontal disease involves an accumulation of plaque and bacteria, which gradually destroys the periodontal ligaments and the bone. As a result, teeth may become so loose that they come out of their sockets. Proper brushing, flossing, and professional cleaning to remove plaque can usually prevent gingivitis and periodontal disease.

Cartilaginous Joints

Cartilaginous joints unite two bones by means of either hyaline cartilage or fibrocartilage (see table 8.1). If the bones are joined by hyaline cartilage, they are called synchondroses; if they are joined by fibrocartilage, they are called symphyses.

Synchondroses

A **synchondrosis** (sin′kon-drō′sis, meaning union through cartilage) consists of two bones joined by hyaline cartilage where little or no movement occurs (figure 8.3a). The epiphyseal plates of growing bones are synchondroses (figure 8.3b). Most synchondroses are temporary and are replaced by bone to form synostoses. On the other hand, some synchondroses persist throughout life. An example is the sternocostal synchondrosis between the first rib and the sternum (figure 8.3c). The costal cartilages allow for "give," which enables the thorax to expand and contract during respiration. In fact, there is so much give between the costal

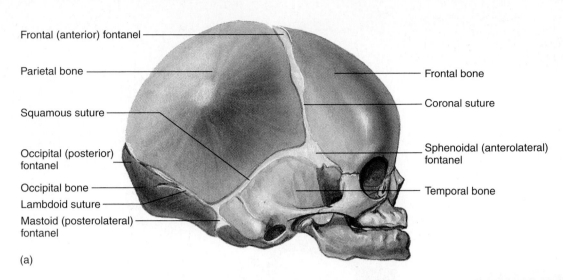

Frontal (anterior) fontanel
Parietal bone
Squamous suture
Occipital (posterior) fontanel
Occipital bone
Lambdoid suture
Mastoid (posterolateral) fontanel

Frontal bone
Coronal suture
Sphenoidal (anterolateral) fontanel
Temporal bone

(a)

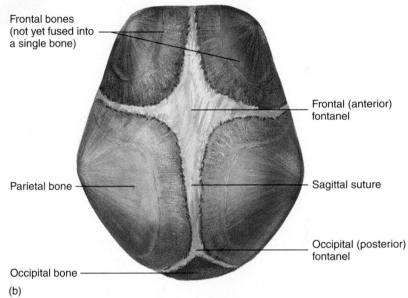

Frontal bones (not yet fused into a single bone)
Parietal bone
Occipital bone

Frontal (anterior) fontanel
Sagittal suture
Occipital (posterior) fontanel

(b)

Figure 8.1   Fetal Skull Showing Fontanels and Sutures
(a) Lateral view. (b) Superior view.

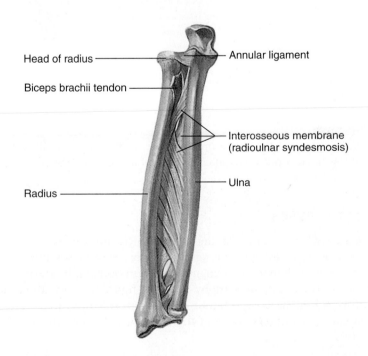

Head of radius
Biceps brachii tendon
Radius

Annular ligament
Interosseous membrane (radioulnar syndesmosis)
Ulna

Figure 8.2   Right Radioulnar Syndesmosis
(Interosseous Membrane)

Figure 8.3 Synchondroses

(*a*) Synchondroses (epiphyseal plates) between
the developing bones of the coxa. (*b*) Epiphyseal
plates. ✗

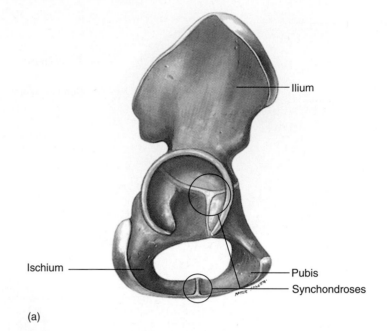

Ilium

Ischium

Pubis

Synchondroses

(a)

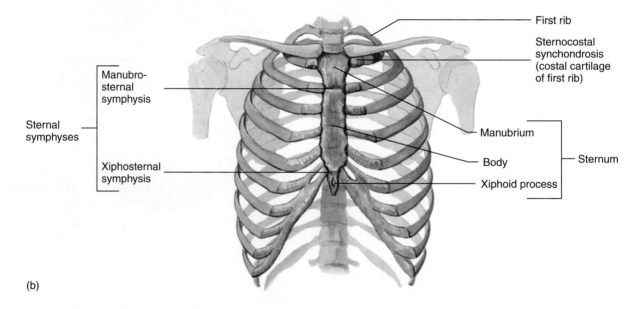

First rib

Sternocostal
synchondrosis
(costal cartilage
of first rib)

Manubro-
sternal
symphysis

Sternal
symphyses

Manubrium

Sternum

Xiphosternal
symphysis

Body

Xiphoid process

(b)

cartilages and the sternum that synovial joints usually develop between the costal cartilages of all but the first rib and
the sternum.

Symphyses

A **symphysis** (sim′fi-sis, meaning a growing together) consists of fibrocartilage uniting two bones. Symphyses include
the junction between the manubrium and body of the sternum
(see figure 8.3*b*), the symphysis pubis (figure 8.4), and the intervertebral disks (see figure 7.15). Some of these joints are
slightly movable because of the somewhat flexible nature of
fibrocartilage.

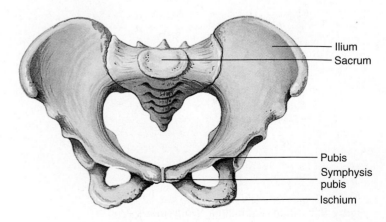

Ilium
Sacrum

Pubis
Symphysis
pubis
Ischium

Figure 8.4 Symphysis Pubis ✗

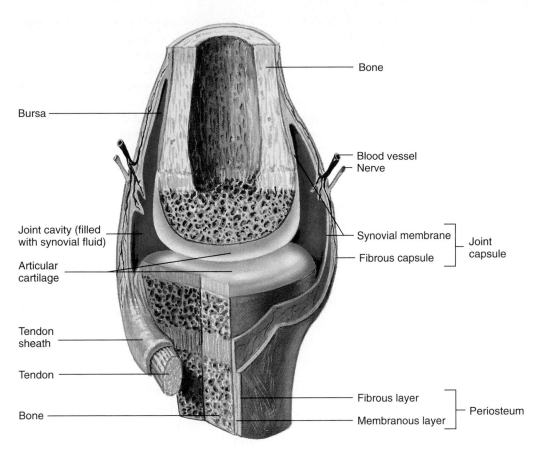

Bone

Bursa

Blood vessel
Nerve

Joint cavity (filled
with synovial fluid)

Synovial membrane
Fibrous capsule

Joint
capsule

Articular
cartilage

Tendon
sheath

Tendon

Fibrous layer
Membranous layer

Periosteum

Bone

Figure 8.5   Structure of a Synovial Joint

Synovial Joints

Synovial (si-nō′vē-ăl, meaning joint fluid; syn, coming together, ovia, resembling egg albumin) **joints,** those containing **synovial fluid,** allow considerable movement between articulating bones (figure 8.5). These joints are anatomically more complex than fibrous and cartilaginous joints. Most joints that unite the bones of the appendicular skeleton are synovial joints, reflecting the far greater mobility of the appendicular skeleton compared with that of the axial skeleton.

The articular surfaces of bones within synovial joints are covered with a thin layer of hyaline cartilage called **articular cartilage,** which provides a smooth surface where the bones meet. Additional fibrocartilage **articular disks** are associated with several synovial joints such as the knee and the temporomandibular joint. Articular disks provide extra strength and support to the joint and increase the depth of the joint cavity.

A synovial joint is enclosed by a **joint capsule,** which helps to hold the bones together while allowing for movement. The joint capsule consists of two layers: an outer **fibrous capsule** and an inner **synovial membrane** (see figure 8.5). The fibrous capsule is continuous with the fibrous layer of the periosteum that covers the bones united at the joint. Portions of the fibrous capsule may thicken to form ligaments. In addition, ligaments and tendons may be present outside the fibrous capsule, contributing to the strength and stability of the joint, while limiting movement in some directions across the joint.

The synovial membrane lines the joint cavity everywhere, except over the articular cartilage. It consists of a collection of modified connective tissue cells either intermixed with part of the fibrous capsule or separated from it by a layer of areolar tissue or adipose tissue. The membrane produces synovial fluid, which consists of a serum (blood fluid) filtrate and secretions from the synovial cells. Synovial fluid is a complex mixture of polysaccharides, proteins, fat, and cells. The major polysaccharide is hyaluronic acid, which provides much of the slippery consistency and lubricating qualities of synovial fluid. Synovial fluid forms an important thin lubricating film that covers the surfaces of a joint.

3   P R E D I C T

What would happen if a synovial membrane covered the articular cartilage?

✔ *Answer in Appendix F*

In certain synovial joints the synovial membrane may extend as a pocket, or sac, called a **bursa** (ber'să, meaning pocket) for a distance away from the rest of the joint cavity (see figure 8.5). Bursae contain synovial fluid and provide a fluid-filled cushion between structures that otherwise would rub against each other, such as tendons rubbing on bones or other tendons. Some bursae are not associated with joints, such as those located between the skin and underlying bony prominences, where friction could damage the tissues. Other bursae extend along tendons for some distance, forming **tendon sheaths. Bursitis** (ber-sī'tis) is an inflammation of a bursa and may cause considerable pain around the joint and restrict movement.

At the peripheral margin of the articular cartilage, blood vessels form a vascular circle that supplies the cartilage with nourishment, but no blood vessels penetrate the cartilage or enter the joint cavity. Additional nourishment to the articular cartilage comes from the underlying cancellous bone and from the synovial fluid covering the articular cartilage. Sensory nerves enter the fibrous capsule and, to a lesser extent, the synovial membrane. They not only supply information to the brain about pain in the joint but also furnish constant information to the brain about the position of the joint and its degree of movement (see chapters 13 and 14). Nerves do not enter the cartilage or joint cavity.

Types of Synovial Joints

Synovial joints are classified according to the shape of the adjoining articular surfaces. The six types of synovial joints are plane, saddle, hinge, pivot, ball-and-socket, and ellipsoid. Movements at synovial joints can be described as **monoaxial** (occurring around one axis), **biaxial** (occurring about two axes situated at right angles to each other), or **multiaxial** (occurring about several axes).

Plane, or **gliding, joints** consist of two opposed flat surfaces of about equal size in which a slight amount of gliding motion can occur between the bones (figure 8.6*a*). These joints are considered monoaxial because some rotation is also possible but is limited by ligaments and adjacent bone. Examples are the articular processes between vertebrae.

Saddle joints consist of two saddle-shaped articulating surfaces oriented at right angles to each other so that complementary surfaces articulate with each other (figure 8.6*b*). Saddle joints are biaxial joints. The carpometacarpal joint of the thumb is an example.

Hinge joints are monoaxial joints (figure 8.6*c*). They consist of a convex cylinder in one bone applied to a corresponding concavity in the other bone. Examples include the elbow and knee joints.

Pivot joints are monoaxial joints that restrict movement to rotation around a single axis (figure 8.6*d*). Each pivot joint consists of a relatively cylindrical bony process that rotates within a ring composed partly of bone and partly of ligament. The articulation between the dens, a process on the axis (see chapter 7), and the atlas is an example. The head of the radius articulating with the proximal end of the ulna is another example (see figure 8.2).

Ball-and-socket joints consist of a ball (head) at the end of one bone and a socket in an adjacent bone into which a portion of the ball fits (figure 8.6*e*). This type of joint is multiaxial, allowing a wide range of movement in almost any direction. Examples are the shoulder and hip joints.

Ellipsoid joints (or condyloid joints) are modified ball-and-socket joints (figure 8.6*f*). The articular surfaces are ellipsoid in shape rather than spherical as in regular ball-and-socket joints. Ellipsoid joints are biaxial, because the shape of the joint limits its range of movement almost to a hinge motion in two axes and restricts rotation. The atlantooccipital joint is an example.

Types of Movement

The movements possible at a given joint are related to its structure. Some joints are limited to only one type of movement; others can move in several directions. With few exceptions, movement is best described in relation to the anatomic position: (1) movement away from the anatomic position and (2) movement returning a structure toward the anatomic position. Most movements are accompanied by other movements in the opposite direction and therefore are listed in pairs.

Angular Movements

Angular movements are those in which one part of a linear structure, such as the body as a whole or a limb, is bent relative to another part of the structure, changing the angle between the two parts. Angular movements also involve the movement of a solid rod, such as a limb, that is attached at one end to the body, so that the angle at which it meets the body is changed. The most common angular movements are flexion and extension and abduction and adduction.

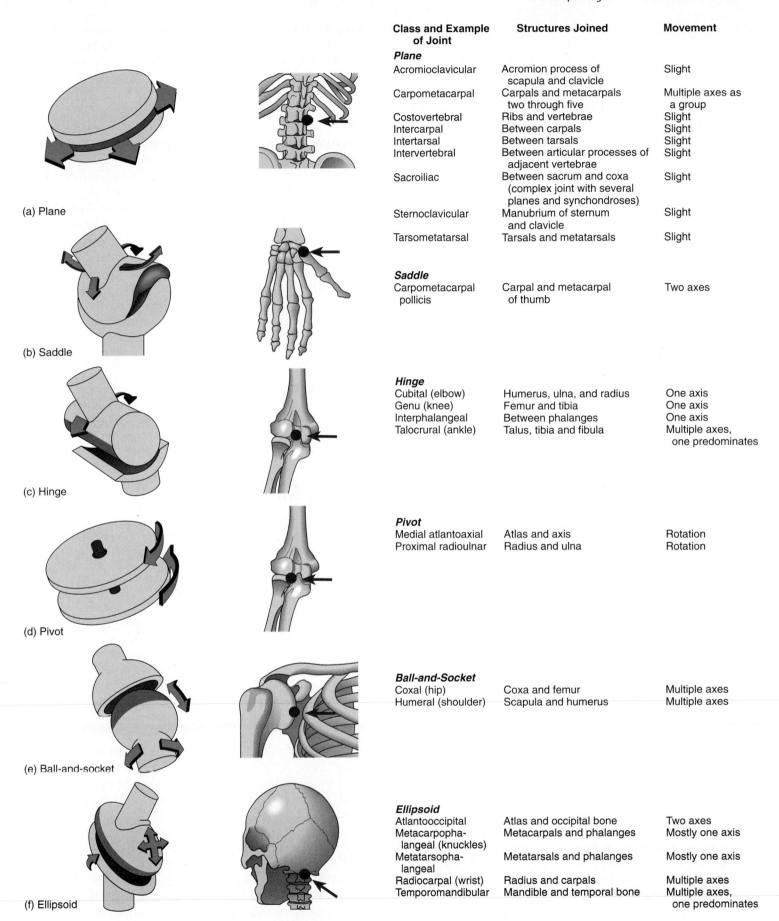

Class and Example of Joint	Structures Joined	Movement
Plane		
Acromioclavicular	Acromion process of scapula and clavicle	Slight
Carpometacarpal	Carpals and metacarpals two through five	Multiple axes as a group
Costovertebral	Ribs and vertebrae	Slight
Intercarpal	Between carpals	Slight
Intertarsal	Between tarsals	Slight
Intervertebral	Between articular processes of adjacent vertebrae	Slight
Sacroiliac	Between sacrum and coxa (complex joint with several planes and synchondroses)	Slight
Sternoclavicular	Manubrium of sternum and clavicle	Slight
Tarsometatarsal	Tarsals and metatarsals	Slight
Saddle		
Carpometacarpal pollicis	Carpal and metacarpal of thumb	Two axes
Hinge		
Cubital (elbow)	Humerus, ulna, and radius	One axis
Genu (knee)	Femur and tibia	One axis
Interphalangeal	Between phalanges	One axis
Talocrural (ankle)	Talus, tibia and fibula	Multiple axes, one predominates
Pivot		
Medial atlantoaxial	Atlas and axis	Rotation
Proximal radioulnar	Radius and ulna	Rotation
Ball-and-Socket		
Coxal (hip)	Coxa and femur	Multiple axes
Humeral (shoulder)	Scapula and humerus	Multiple axes
Ellipsoid		
Atlantooccipital	Atlas and occipital bone	Two axes
Metacarpophalangeal (knuckles)	Metacarpals and phalanges	Mostly one axis
Metatarsophalangeal	Metatarsals and phalanges	Mostly one axis
Radiocarpal (wrist)	Radius and carpals	Multiple axes
Temporomandibular	Mandible and temporal bone	Multiple axes, one predominates

(a) Plane

(b) Saddle

(c) Hinge

(d) Pivot

(e) Ball-and-socket

(f) Ellipsoid

Figure 8.6   Types of Synovial Joints and Selected Examples

Flexion and Extension

Flexion and extension can be defined in a number of ways, but in each case there are exceptions to the definition. The literal definition is to bend and straighten, respectively. We have chosen to use a definition with more utility and fewer exceptions. **Flexion** moves a part of the body in the anterior or ventral direction. **Extension** moves a part in a posterior or dorsal direction (figure 8.7*a–d*). The only exception is the knee, in which flexion moves the leg in a posterior direction and extension moves it in an anterior direction (figure 8.7*e*).

Movement of the foot toward the plantar surface, such as when standing on the toes, is commonly called **plantar flexion;** and movement of the foot toward the shin, such as when walking on the heels, is called **dorsiflexion** (figure 8.7*f*).

Clinical Note

Hyperextension is usually defined as an abnormal, forced extension of a joint beyond its normal range of motion. For example, if a person falls and attempts to break the fall by putting out her hand, the force of the fall directed into the hand and wrist may cause hyperextension of the wrist, which may result in sprained joints or broken bones. Some health professionals, however, define hyperextension as the normal movement of a structure into the space posterior to the anatomic position.

Abduction and Adduction

Abduction (meaning to take away) is movement away from the midline; **adduction** (meaning to bring together) is movement toward the midline (figure 8.7*g*). Moving the lower limbs away from the midline of the body such as in the outward half of "jumping jacks" is abduction, and bringing the lower limbs back together is adduction. Abduction of the fingers involves spreading the fingers apart, away from the midline of the hand, and adduction is bringing them back together (figure 8.7*h*). Abduction of the wrist, which is sometimes called radial deviation, is movement of the hand away from the midline of the body, and adduction of the wrist, which is sometimes called ulnar deviation, results in movement of the hand toward the midline of the body. Abduction of the head is tilting the head to one side or the other and is sometimes called lateral flexion of the neck. Bending at the waist to one side or the other is usually called lateral flexion of the vertebral column, rather than abduction.

Circular Movements

Circular movements involve the rotation of a structure around an axis or movement of the structure in an arc.

Rotation

Rotation is the turning of a structure around its long axis, such as rotation of the head, the humerus, or the entire body

(figure 8.7*i*). Medial rotation of the humerus with the forearm flexed brings the hand toward the body. Rotation of the humerus so that the hand moves away from the body is lateral rotation.

Pronation and Supination

Pronation (prō-nā′shŭn) and **supination** (sū′pi-nā′shŭn) refer to the unique rotation of the forearm (figure 8.7*j*). The word prone means lying face down; the word supine means lying face up. Pronation is rotation of the palm so that it faces posteriorly in relation to the anatomic position. The palm of the hand faces inferiorly if the elbow is flexed. Supination is rotation of the palm so that it faces anteriorly in relation to the anatomic position. The palm of the hand faces superiorly if the elbow is flexed. In pronation the radius and ulna cross; in supination they are in a parallel position. As described in chapter 7, the head of the radius rotates against the radial notch of the ulna during supination and pronation.

Circumduction

Circumduction is a combination of flexion, extension, abduction, and adduction (figure 8.7*k*). It occurs at freely movable joints such as the shoulder. In circumduction the arm moves so that it describes a cone with the shoulder joint at the apex.

Special Movements

Special movements are those movements unique to only one or two joints; they do not fit neatly into one of the other categories.

Elevation and Depression

Elevation moves a structure superiorly; **depression** moves it inferiorly (figure 8.7*l*). The mandible and scapulae are primary examples. Depression of the mandible opens the mouth, and elevation closes it. Shrugging the shoulders is an example of scapular elevation.

Protraction and Retraction

Protraction consists of moving a structure in an anterior direction (figure 8.7*m*). **Retraction** moves the structure back to the anatomic position or even more posteriorly. As with elevation and depression, the mandible and scapulae are primary examples.

Excursion

Lateral excursion refers to moving the mandible to either the right or left of the midline (figure 8.7*n*), such as in grinding the teeth or chewing. **Medial excursion** returns the mandible to the neutral position.

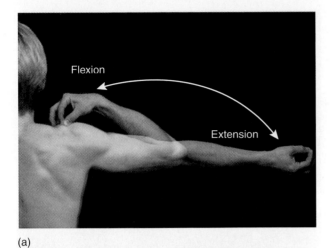

(a)

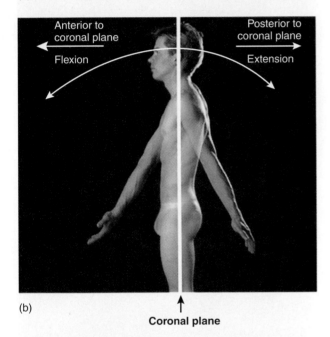

(b)

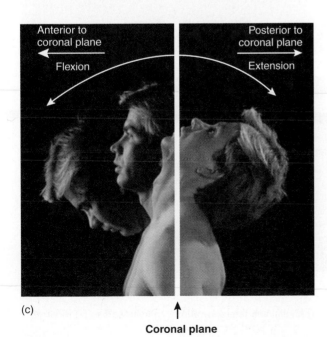

(c)

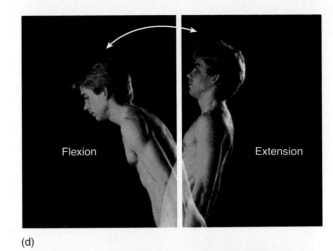

(d)

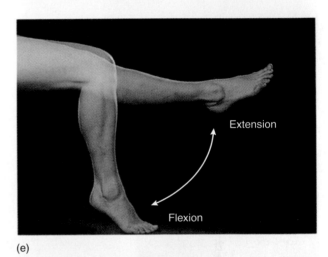

(e)

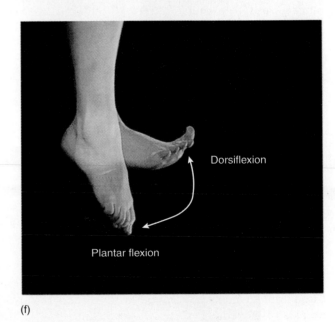

(f)

Figure 8.7   Movements

(*a*) Flexion and extension of the elbow. (*b*) Flexion and extension of the shoulder. (*c*) Flexion and extension of the neck. (*d*) Flexion and extension of the trunk. (*e*) Flexion and extension of the knee. (*f*) Plantar flexion and dorsiflexion of the foot. ✗

233

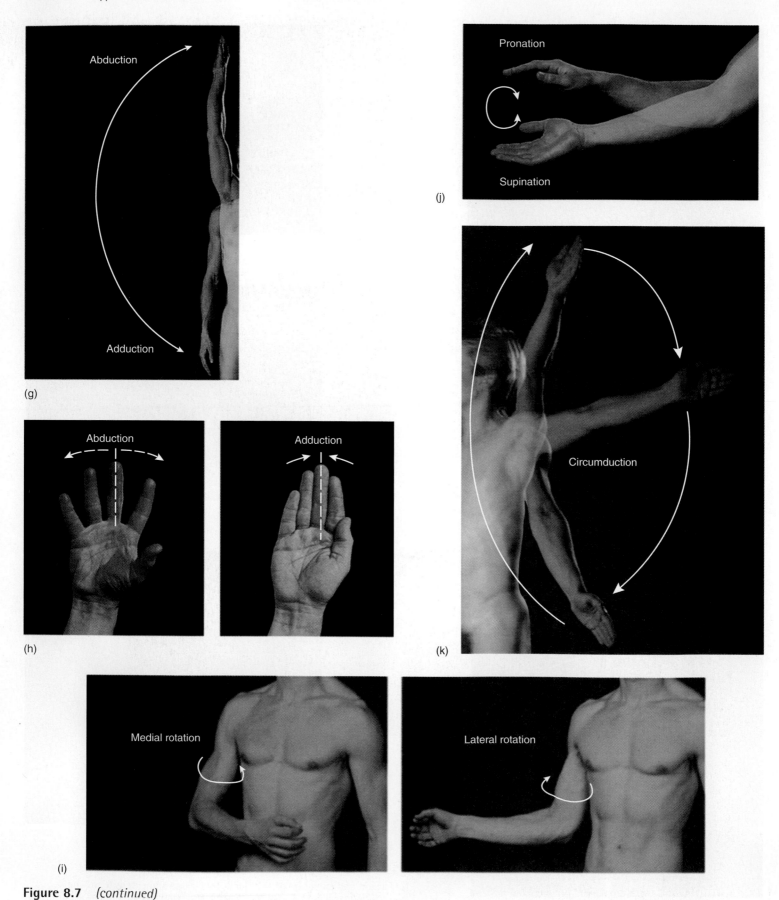

Figure 8.7   *(continued)*

(*g*) Abduction and adduction of the upper limb. (*h*) Abduction and adduction of the fingers. (*i*) Medial and lateral rotation of the humerus. (*j*) Pronation and supination. (*k*) Circumduction of the shoulder. ✗

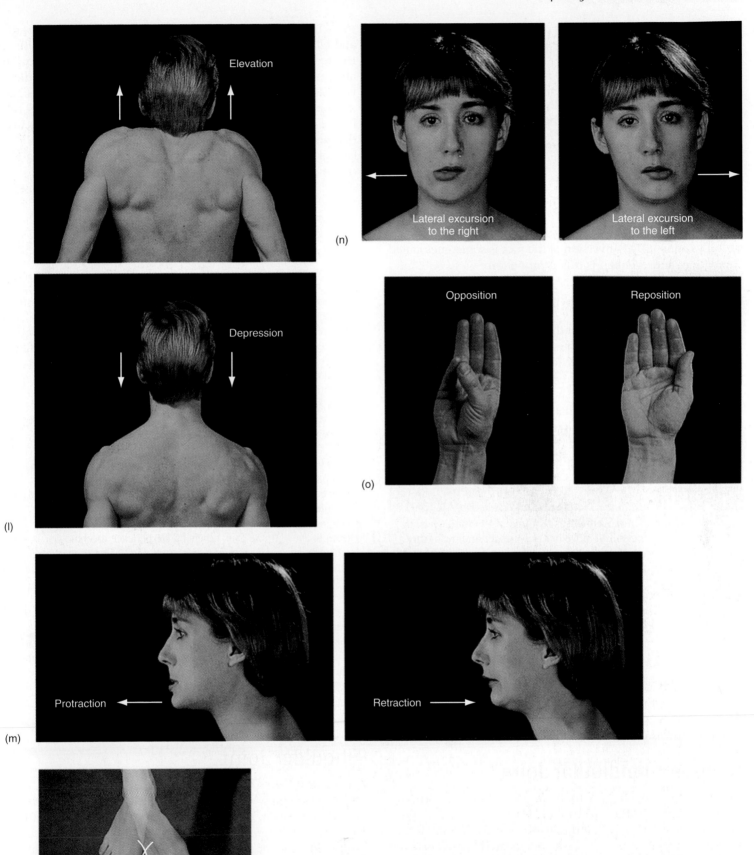

Figure 8.7 *(continued)*

(*l*) Elevation and depression of the shoulder. (*m*) Protraction and retraction of the jaw. (*n*) Lateral excursion of the jaw. (*o*) Opposition and reposition of the thumb. (*p*) Inversion and eversion of the foot.

Opposition and Reposition

Opposition is a unique movement of the thumb and little finger (figure 8.7*o*). It occurs when these two digits are brought toward each other across the palm of the hand. The thumb can also oppose the other digits. **Reposition** is the movement returning the thumb and little finger to the neutral, anatomic position.

Inversion and Eversion

Inversion consists of turning the ankle so that the plantar surface of the foot faces medially, toward the opposite foot. **Eversion** is turning the ankle so that the plantar surface faces laterally (figure 8.7*p*). Inversion of the foot is sometimes called supination, and eversion is called pronation.

Combination Movements

Most movements that occur in the course of normal activities are combinations of the movements named previously and are described by naming the individual movements involved in the combined movement. For example, if a person holds his hand straight out to his side at shoulder height and then brings the hand in front of him so that it is still at shoulder height, that movement could be considered a combination of abduction and flexion.

---

4   **P R E D I C T**

What combination of movements is required at the shoulder and elbow joints for a person to move her right upper limb from the anatomic position to touch the right side of her head with her fingertips?

✔ *Answer in Appendix F*

Description of Selected Joints

It is impossible in a limited space to describe all the joints of the body; therefore, only selected joints are described in this chapter, and they have been chosen because of their representative structure, important function, or clinical significance.

Temporomandibular Joint

The mandible articulates with the temporal bone to form the **temporomandibular joint (TMJ).** The mandibular condyle fits into the mandibular fossa of the temporal bone. A fibrocartilage articular disk is located between the mandible and the temporal bone, dividing the joint into superior and inferior cavities (figure 8.8). The joint is surrounded by a fibrous capsule to which the articular disk is attached at its margin, and it is strengthened by lateral and accessory ligaments.

The temporomandibular joint is a combination plane and ellipsoid joint, with the ellipsoid portion predominating. Depression of the mandible to open the mouth involves an anterior gliding motion of the mandibular condyle and articular disk relative to the temporal bone, which is about the same motion that occurs in protraction of the mandible; it is followed by a hinge motion that occurs between the articular disk and the mandibular head. The mandibular condyle is also capable of slight mediolateral movement, allowing excursion of the mandible.

Clinical Note

TMJ disorders are a group of conditions accounting for most chronic orofacial pain. The group includes joint noise; pain in the muscle, joint, or face; headache; and reduction in the range of joint movement. TMJ pain is often felt as referred pain in the ear. Patients may go to a physician complaining of an earache and are then referred to a dentist. As many as 65%–75% of people between ages 20 and 40 experience some of these symptoms. Symptoms appear to affect men and women about equally, but only about 10% of the symptoms are severe enough to cause people to seek medical attention. Women experience severe pain eight times more often than do men.

TMJ disorders are classified as those involving the joint, with or without pain; those involving only muscle pain; or those involving both the joint disorder and muscle pain. TMJ disorders are also classified as acute or chronic. Acute cases are usually self-limiting and have an identifiable cause. Chronic cases are not self-limiting, may be permanent, and often have no apparent cause. Chronic TMJ disorders are not easily treated, and chronic TMJ pain has much in common with other types of chronic pain. Whereas some people learn to live with the pain, others may experience psychologic problems, such as a sense of helplessness and hopelessness, high tension, and loss of sleep and appetite. Drug dependency may occur if strong drugs are used to control the pain; and relationships, lifestyle, vocation, and social interactions may be disrupted. Many of these problems may make the pain worse through positive feedback. Treatment includes teaching the patient to reduce jaw movements that aggravate the problem and to reduce stress and anxiety. Physical therapy may help to relax the muscles and restore function. Analgesic and anti-inflammatory drugs may be used, and oral splints may be helpful, especially at night.

Shoulder Joint

The **shoulder,** or **humeral, joint** is a ball-and-socket joint in which stability is reduced and mobility is increased (figure 8.9). Flexion, extension, abduction, adduction, rotation, and circumduction can all occur at the shoulder joint. The rounded head of the humerus articulates with the shallow glenoid fossa of the scapula. The rim of the glenoid fossa is built up slightly by a fibrocartilage ring, the **glenoid labrum,** to which the joint capsule is attached. A **subscapular bursa** (not shown in the figure) and a **subacromial bursa** open into the joint cavity.

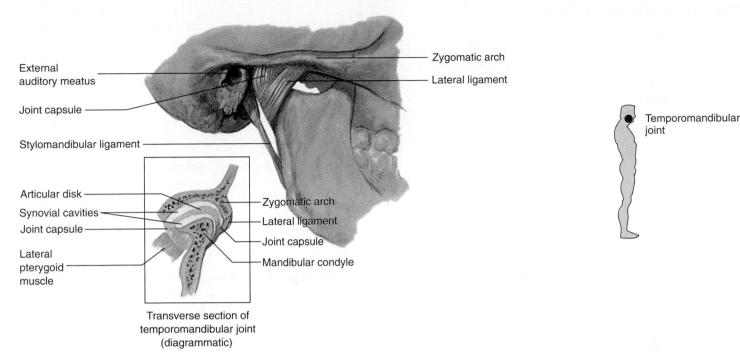

External
auditory meatus

Joint capsule

Stylomandibular ligament

Articular disk

Synovial cavities

Joint capsule

Lateral
pterygoid
muscle

Zygomatic arch

Lateral ligament

Zygomatic arch

Lateral ligament

Joint capsule

Mandibular condyle

Temporomandibular
joint

Transverse section of
temporomandibular joint
(diagrammatic)

Figure 8.8   Right Temporomandibular Joint (lateral view)

The stability of the joint is maintained primarily by three sets of ligaments and four muscles. The ligaments of the shoulder are listed in table 8.2. The four muscles, referred to collectively as the **rotator cuff,** pull the humeral head superiorly and medially toward the glenoid fossa. These muscles are discussed in more detail in chapter 11. The head of the humerus is also supported against the glenoid fossa by the tendon from the biceps brachii muscle in the anterior part of the arm. This tendon is unusual in that it passes through the articular capsule of the shoulder joint before crossing the head of the humerus and attaching to the scapula at the supraglenoid tubercle (see figure 7.24a).

Table 8.2	Ligaments of the Shoulder Joint (see figure 8.9)
Ligament	**Description**
Glenohumeral (superior, middle, and inferior)	Three slightly thickened longitudinal sets of fibers on the anterior side of the capsule; extend from the humerus to the margin of the glenoid fossa
Transverse humeral	Lateral, transverse fibrous thickening of the joint capsule; crosses between the greater and lesser tubercles and holds down the tendon from the long head of the biceps muscle
Coracohumeral	Crosses from the root of the coracoid process to the humeral neck
Coracoacromial	Crosses above the joint between the coracoid process and the acromion process; an accessory ligament

Clinical Note

The most common traumatic shoulder disorders are **dislocation** and muscle or tendon **tears.** The shoulder is the most commonly dislocated joint in the body. The major ligaments cross the superior part of the shoulder joint, and no major ligaments or muscles are associated with the inferior side. As a result, dislocation of the humerus is most likely to occur inferiorly into the axilla. Because the axilla contains some very important nerves and arteries, severe and permanent damage may result from attempts to relocate a dislocated shoulder using inappropriate techniques (see chapter 14). Chronic shoulder disorders include tendonitis, bursitis, and arthritis; they involve inflammation of tendons, bursae, or the joint, respectively. Bursitis of the subacromial bursa can become very painful when the large shoulder muscle, called the deltoid muscle, compresses the bursa during shoulder movement.

5   P R E D I C T

Separation of the shoulder consists of stretching or tearing the ligaments of the acromioclavicular joint (acromioclavicular, or AC, separation). Using figure 8.9a and your knowledge of the articulated skeleton for assistance, explain the nature of a shoulder separation, and predict the problems that may follow a separation.

✔ *Answer in Appendix F*

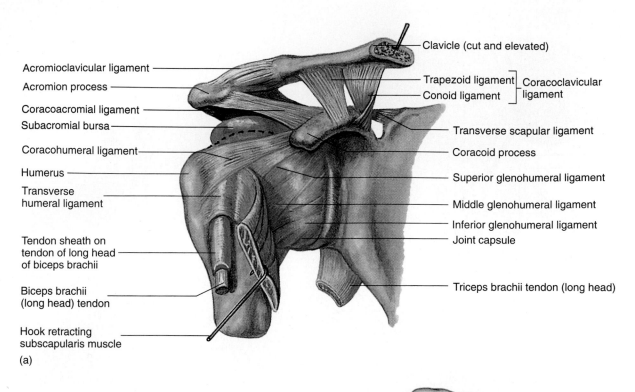

Acromioclavicular ligament

Acromion process

Coracoacromial ligament

Subacromial bursa

Coracohumeral ligament

Humerus

Transverse humeral ligament

Tendon sheath on tendon of long head of biceps brachii

Biceps brachii (long head) tendon

Hook retracting subscapularis muscle

(a)

Clavicle (cut and elevated)

Trapezoid ligament — Coracoclavicular ligament

Conoid ligament

Transverse scapular ligament

Coracoid process

Superior glenohumeral ligament

Middle glenohumeral ligament

Inferior glenohumeral ligament

Joint capsule

Triceps brachii tendon (long head)

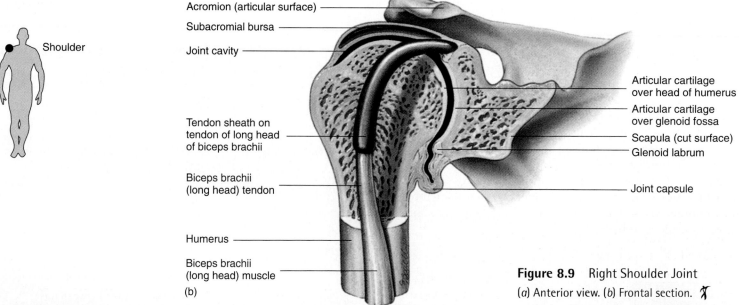

Shoulder

Acromion (articular surface)

Subacromial bursa

Joint cavity

Tendon sheath on tendon of long head of biceps brachii

Biceps brachii (long head) tendon

Humerus

Biceps brachii (long head) muscle

(b)

Articular cartilage over head of humerus

Articular cartilage over glenoid fossa

Scapula (cut surface)

Glenoid labrum

Joint capsule

Figure 8.9 Right Shoulder Joint
(*a*) Anterior view. (*b*) Frontal section.

Hip Joint

The femoral head articulates with the relatively deep, concave acetabulum of the coxa to form the **coxal,** or **hip joint** (figure 8.10). The head of the femur is more nearly a complete ball than the articulating surface of any other bone of the body. The acetabulum is deepened and strengthened by a lip of fibrocartilage called the **acetabular labrum,** which is incomplete inferiorly, and by a **transverse acetabular ligament,** which crosses the acetabular notch on the inferior edge of the acetabulum. The hip is capable of a wide range of movement, including flexion, extension, abduction, adduction, rotation, and circumduction.

An extremely strong joint capsule, reinforced by several ligaments, extends from the rim of the acetabulum to the neck of the femur (table 8.3). The **iliofemoral ligament** is especially strong. When standing, most people tend to thrust the hips anteriorly. This position is relaxing because the iliofemoral ligament supports much of the body's weight. The **ligamentum teres,** which is the ligament of the head of the femur, is located inside the hip joint between the femoral head and the acetabulum. It functions very little in strengthening the hip joint; however, it does carry a small nutrient artery to the head of the femur in about 80% of the population.

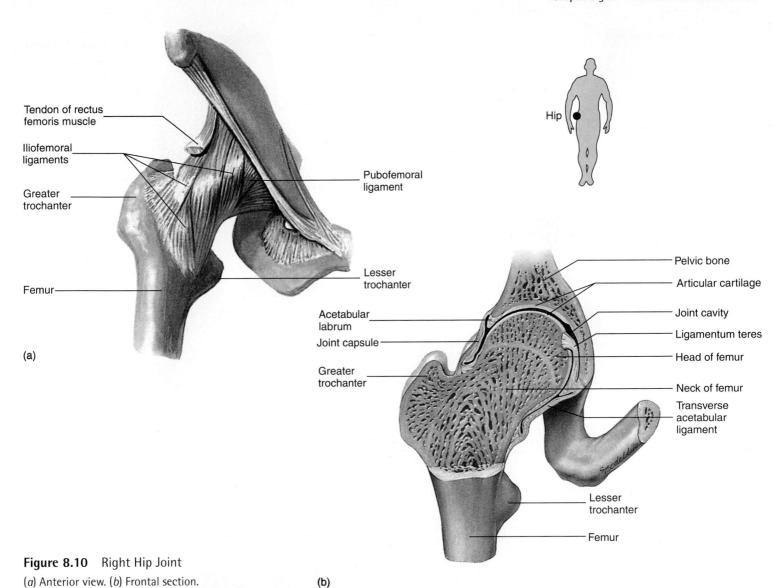

Figure 8.10   Right Hip Joint
(*a*) Anterior view. (*b*) Frontal section.

Table 8.3	Ligaments of the Hip Joint (see figure 8.10)
Ligament	**Description**
Transverse acetabular	Bridges gap in the inferior margin of the fibrocartilage acetabular labrum
Iliofemoral	Strong, thick band between the anterior inferior iliac spine and the inertrochanteric line of the femur.
Pubofemoral	Extends from the pubic portion of the acetabular rim to the inferior portion of the femoral neck
Ischiofemoral	Bridges the ischial acetabular rim and the superior portion of the femoral neck; less well defined
Ligamentum teres	Weak, flat band from the margin of the acetabular notch and the transverse ligament to a fovea in the center of the femoral head.

Clinical Note

Dislocation of the hip may occur when the hip is flexed and the femur is driven posteriorly, such as when a person sitting in an automobile is involved in an accident. The head of the femur usually dislocates posterior to the acetabulum, tearing the acetabular labrum, the fibrous capsule, and the ligaments. Fracture of the femur and the coxa often accompany hip dislocation.

Knee Joint

The **knee joint** traditionally is classified as a hinge joint located between the femur and the tibia (figure 8.11). Actually it is a complex ellipsoid joint that allows flexion, extension, and a small amount of rotation of the leg. The distal

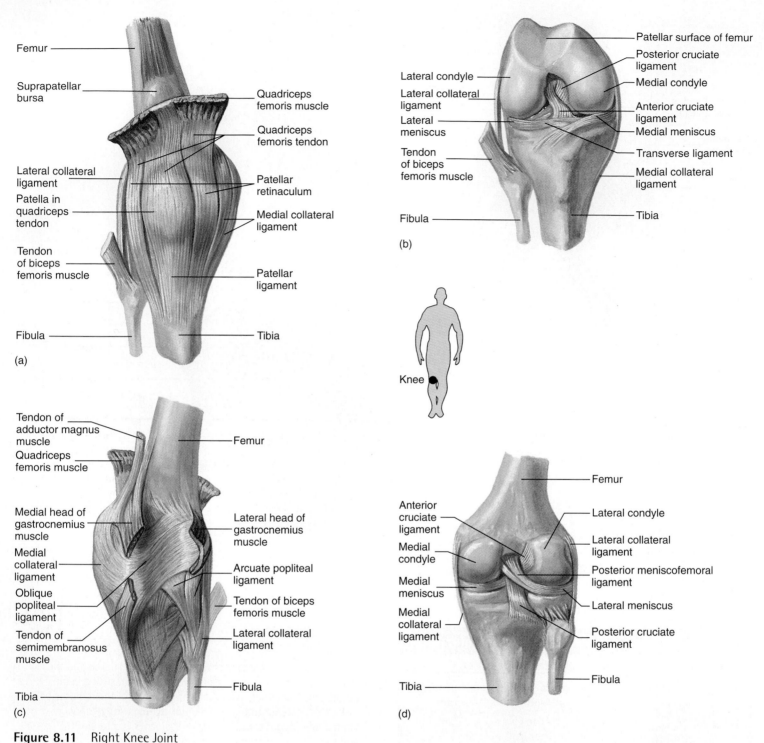

Figure 8.11   Right Knee Joint

(*a*) Anterior superficial view. (*b*) Anterior deep view (knee flexed). (*c*) Posterior superficial view. (*d*) Posterior deep view. Right knee joint.

end of the femur has two large ellipsoid surfaces and a deep fossa between them. The femur articulates with the proximal end of the tibia, which is flattened and smooth laterally, with a crest called the intercondylar eminence in the center (see figure 7.34). The margins of the tibia are built up by

thick fibrocartilage articular disks, called **menisci** (mĕ-nis'sī, meaning crescent-shaped; see figure 8.11*b* and *d*), which deepen the articular surface. The fibula does not articulate with the femur but articulates only with the lateral side of the tibia.

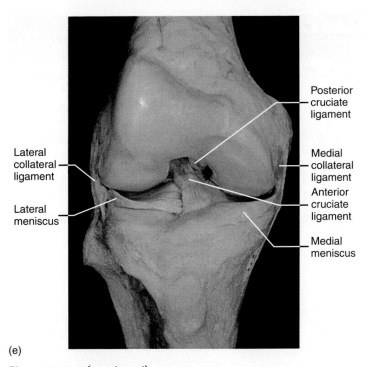

Lateral collateral ligament

Lateral meniscus

Posterior cruciate ligament

Medial collateral ligament

Anterior cruciate ligament

Medial meniscus

(e)

Quadriceps femoris tendon

Suprapatellar bursa

Subcutaneous prepatellar bursa

Patella

Fat pad

Patellar ligament

Deep infrapatellar bursa

Femur

Articular cartilage

Meniscus

Tibia

(f)

Figure 8.11   *(continued)*

(e) Photograph of anterior deep view. (*f*) Sagittal section. ✗

Clinical Note

Injuries to the medial side of the knee are much more common than injuries to the lateral side. The lateral collateral ligament strengthens the joint laterally and is stronger than the medial collateral ligament. Damage to the collateral ligaments occurs as a result of blows to the opposite side of the knee, and severe blows to the medial side of the knee, which would damage the lateral collateral ligament, are far less common than blows to the lateral side of the knee. In addition, the medial meniscus is fairly tightly attached to the medial collateral ligament and is damaged 20 times more often in a knee injury than the lateral meniscus, which is thinner and more loosely attached. A torn meniscus may result in a "clicking" sound during extension of the leg; or, if the damage is more severe, the torn piece of cartilage may move between the articulating surfaces of the tibia and femur, causing the knee to "lock" in a partially flexed position. If the knee is driven anteriorly or if the knee is hyperextended, the anterior cruciate ligament may be torn, which causes the knee joint to be very unstable. If the knee is driven posteriorly, the posterior cruciate ligament may be torn. Surgical replacement of a cruciate ligament with a transplanted or artificial ligament is now being done.

A common type of football injury results from a block or tackle to the lateral side of the knee, which can cause the knee to bend inward, opening the medial side of the joint and tearing the medial collateral ligament. Because this ligament is strongly attached to the medial meniscus, the medial meniscus often is torn as well. In severe injuries the anterior cruciate ligament, which is attached to the medial meniscus, is also damaged (figure A).

Bursitis in the subcutaneous prepatellar bursa (see figure 8.11*f*), commonly called "housemaid's knee," may result from prolonged work performed while on the hands and knees. Another bursitis, "clergyman's knee," results from excessive kneeling and affects the subcutaneous infrapatellar bursa (not illustrated). This type of bursitis is common in carpet layers and roofers.

Other common knee problems include chondromalacia, or softening of the cartilage, which results from abnormal movement of the patella within the patellar groove, and the "fat pad syndrome," which consists of an accumulation of fluid in the fat pad posterior to the patella. An acutely swollen knee appearing immediately after an injury is usually a sign of blood accumulation within the joint and is called a hemarthrosis. A slower accumulation of fluid, "water on the knee," may be caused by bursitis.

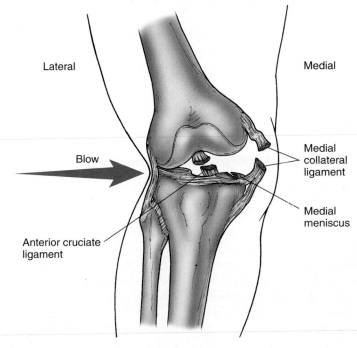

Lateral

Medial

Blow

Anterior cruciate ligament

Medial collateral ligament

Medial meniscus

Figure A   Injury to the Right Knee ✗

Two **cruciate** (krū′shē-āt, meaning crossed) **ligaments** extend between the intercondylar eminence of the tibia and the fossa of the femur (see figure 8.11*b, d,* and *e*). The anterior cruciate ligament prevents anterior displacement of the tibia relative to the femur, and the posterior cruciate ligament prevents posterior displacement of the tibia. The joint is also strengthened by **collateral** and **popliteal ligaments** and by the tendons of the thigh muscles, which extend around the knee (table 8.4).

The knee is surrounded by a number of bursae (see figure 8.11*f*). The largest is the **suprapatellar bursa,** which is a superior extension of the joint capsule and allows for movement of the anterior thigh muscles over the distal end of the femur. Other knee bursae include the subcutaneous prepatellar bursa and the deep infrapatellar bursa (see figure 8.11*f*), as well as the popliteal bursa, the gastrocnemius bursa, and the subcutaneous infrapatellar bursa (not illustrated).

Ankle Joint

The distal tibia and fibula form a highly modified hinge joint with the talus called the **ankle,** or **talocrural** (tā′lō-krū′răl), **joint** (figure 8.12). The medial and lateral malleoli of the tibia and fibula, which form the medial and lateral margins of the ankle, are rather extensive, whereas the anterior and posterior margins are almost nonexistent. As a result, a hinge joint is created from a modified ball-and-socket arrangement. A fibrous capsule surrounds the joint, with the medial and lateral parts thickened to form ligaments. Other ligaments also help stabilize the joint (table 8.5). Dorsiflexion, plantar flexion, and limited inversion and eversion can occur at this joint.

Clinical Note

The ankle is the most frequently injured major joint in the body. The most common ankle injuries result from forceful inversion of the foot. A **sprained ankle** results when the ligaments of the ankle are torn partially or completely. The calcaneofibular ligament tears most often, followed in frequency by the anterior talofibular ligament. A fibular fracture can occur with severe inversion because the talus can slide against the lateral malleolus and break it.

Arches of the Foot

The foot has three major **arches** that distribute the weight of the body between the heel and the ball of the foot during standing and walking (figure 8.13). As the foot is placed on the ground, weight is transferred from the tibia and the fibula to the talus. From there, the weight is distributed first to the heel (calcaneus) and then through the arch system along the lateral side of the foot to the ball of the foot (head of the metatarsals). This effect can be observed when a person with wet, bare feet walks across a dry surface; the print of the heel,

Table 8.4	Ligaments of the Knee Joint (see figure 8.11)
Ligament	**Description**
Patellar	Thick, heavy, fibrous band between the patella and the tibial tuberosity; actually part of the quadriceps femoris tendon
Patellar retinaculum	Thin band from the margins of the patella to the sides of the tibial condyles
Oblique popliteal	Thickening of the posterior capsule; extension of the semimembranous tendon
Arcuate popliteal	Extends from the posterior fibular head to the posterior fibrous capsule
Medial collateral	Thickening of the lateral capsule from the medial epicondyle of the femur to the medial surface of the tibia; also called the tibial collateral ligament
Lateral collateral	Round ligament extending from the lateral femoral epicondyle to the head of the fibula; also called the fibular collateral ligament
Anterior cruciate	Extends obliquely, superiorly, and posteriorly from the anterior intercondylar eminence of the tibia to the medial side of the lateral femoral condyle
Posterior cruciate	Extends superiorly and anteriorly from the posterior intercondylar eminence to the lateral side of the medial condyle
Coronary (medial and lateral)	Attaches the menisci to the tibial condyles (not illustrated)
Transverse	Connects the anterior portions of the medial and lateral menisci
Meniscofemoral (anterior and posterior)	Joins the posterior part of the lateral menisci to the medial condyle of the femur, passing anterior and posterior to the posterior cruciate ligament (not illustrated)

the lateral border of the foot, and the ball of the foot can be seen, but the middle of the plantar surface and the medial border leave no impression. The medial side leaves no mark because the arches on this side of the foot are higher than those on the lateral side. The shape of the arches is maintained by the configuration of the bones, the ligaments connecting them, and the muscles acting on the foot (see figure 8.12*a*). The ligaments of the arch serve two major functions: to hold the bones in their proper relationship as segments of the arch and to provide ties across the arch somewhat like a bowstring. As weight is transferred through the arch system, some of the ligaments are stretched, giving the foot more flexibility and allowing it to adjust to uneven surfaces. When weight is removed from the foot, the ligaments recoil and restore the arches to their unstressed shape.

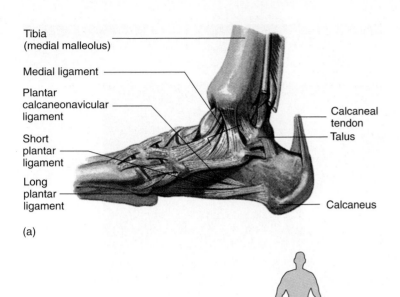

Tibia
(medial malleolus)

Medial ligament

Plantar
calcaneonavicular
ligament

Short
plantar
ligament

Long
plantar
ligament

(a)

Calcaneal
tendon

Talus

Calcaneus

Tibia

Posterior
tibiofibular
ligament

Calcaneofibular
ligament

Calcaneal
tendon

Long
plantar
ligament

Calcaneus

(b)

Fibula
(lateral malleolus)

Anterior tibiofibular ligament

Anterior talofibular ligament

Tendon of
peroneus
longus
muscle

Tendon of
peroneus
brevis
muscle

Ankle

Figure 8.12 Right Ankle Joint
(*a*) Medial view. (*b*) Lateral view.

Figure 8.13 Arches (*arrows*)
of the Right Foot

The medial longitudinal arch is
formed by the calcaneus, talus,
navicular, the cuneiforms, and the
three medial metatarsals. The
lateral longitudinal arch is formed
by the calcaneus, cuboid, and the
two lateral metatarsals. The
transverse arch is formed by the
cuboid and cuneiforms.

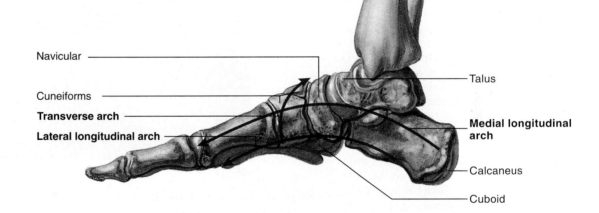

Navicular

Cuneiforms

Transverse arch

Lateral longitudinal arch

Talus

**Medial longitudinal
arch**

Calcaneus

Cuboid

Table 8.5	Ligaments of the Ankle (see figure 8.12)
Ligament	**Description**
Medial	Thickening of the medial fibrous capsule that attaches the medial malleolus to the calcaneus, navicular, and talus; also called the deltoid ligament
Calcaneofibular	Extends from the lateral malleolus to the lateral surface of the calcaneus; separate from the capsule
Anterior talofibular	Extends from the lateral malleolus to the neck of the talus; fused with the joint capsule

Clinical Note

The arches of the foot normally form early in fetal life. Failure to
form results in congenital flat feet, or fallen arches, a condition in
which the arches, primarily the medial longitudinal arch, are
depressed or collapsed. This condition is not always painful. Flat feet
may also occur when the muscles and ligaments supporting the arch
fatigue and allow the arch, usually the medial longitudinal arch, to
collapse. During prolonged standing, the plantar calcaneonavicular
ligament may stretch, flattening the medial longitudinal arch. The
transverse arch may also become flattened. The strained ligaments
can become painful.

Plantar fasciitis, inflammation of the plantar fascia, which is a
broad band of superficial connective tissue extending from the
calcaneus to the ball of the foot, can be a problem for distance
runners as a result of continuous stretching.

Clinical Focus   Joint Disorders

Arthritis

Arthritis, an inflammation of any joint, is the most common and best known of the joint disorders, affecting 10% of the world's population. There are more than 100 different types of arthritis. Classification is often based on the cause and progress of the arthritis. Causes include infectious agents, metabolic disorders, trauma, and immune disorders. Mild exercise retards joint degeneration and enhances mobility. Swimming and walking are recommended for people with arthritis; but running, tennis, and aerobics are not recommended. Therapy depends on the type of arthritis, but usually includes the use of anti-inflammatory drugs. Current research is focusing on the possible development of antibodies against the cells that initiate the inflammatory response in the joints or against cell surface markers on those cells.

Osteoarthritis (OA) is the most common type of arthritis, affecting one in ten people in the United States (85% of those over age 70). OA may begin as a molecular abnormality in articular cartilage, with heredity and normal "wear and tear" of the joint important contributing factors. Slowed metabolic rates with increased age also seem to contribute to OA. Inflammation is usually secondary in this disorder. It tends to occur in the weight-bearing joints such as the knees and is more common in overweight individuals.

Rheumatoid arthritis (RA) is the second most common type of arthritis. It affects about 3% of all women and about 1% of all men in the United States. It is a general connective tissue disorder that affects the skin, vessels, lungs, and other organs, but it is most pronounced in the joints. It is severely disabling and most commonly destroys small joints such as those in the hands and feet (figure B). The initial cause is unknown but may involve a transient infection or an autoimmune disease (an immune reaction to one's own tissues; see chapter 22) that develops against collagen. There may also be a genetic predisposition. Whatever the cause, the ultimate course appears to be immunologic. People with classic RA have a

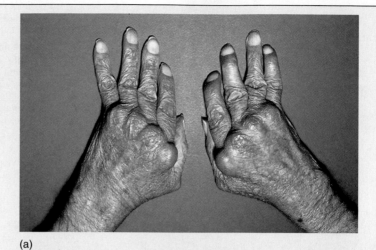

(a)

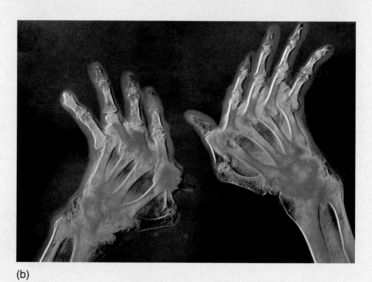

(b)

Figure B   Rheumatoid Arthritis   (*a*) Photograph of hands with rheumatoid arthritis. (*b*) Radiographs of the same hands shown in (*a*).

protein, **rheumatoid factor,** in their blood. In RA the synovial fluid and associated connective tissue cells proliferate, forming a pannus (clothlike layer), which causes the joint capsule to become thickened and which destroys the articular cartilage. In advanced stages, opposing joint surfaces can become fused. Juvenile rheumatoid arthritis is similar to the adult type in many

ways, but no rheumatoid factor is found in the serum.

Lyme Disease

Lyme disease is the result of a bacterial infection (*Borrelia burgdorferi*) transmitted to humans by a tick vector (usually *Ixodes* sp.), which affects the brain, nerves, eyes, heart, and joints. The chronic arthritis and central nervous system dysfunction that are symptoms of the disease are severely disabling but rarely fatal. The disease is named for an epidemic of childhood arthritis occurring in Lyme, Connecticut, in 1975. It has probably existed in Europe for many years and in North America before the first European colonization but was unrecognized. Humans and domestic animals are only incidental hosts to the ticks that normally infect wild mammals and birds. Deer are of particular concern. The northeastern United States was greatly deforested during the 18th and 19th centuries, and deer and other wildlife populations declined dramatically. The more recent abandonment and reforestation of farms in New England has lead to an increase in the deer and tick populations, with a resurgence of the associated joint and nervous system disease. There have been over 103,000 cases of Lyme disease reported in the United States since 1982. Although the disease is most common in the northeastern United States, cases have been reported in the northcentral states, along the West Coast, and scattered throughout the eastern and central states. Early manifestations of the disease include flulike symptoms, with localized skin rash. If untreated, the bacterium can spread to the nervous system, heart, and joints within a few weeks to months. A human vaccine against Lyme disease is currently being tested.

Suppurative (pus-forming) **arthritis** may result from a number of infectious agents. These joint infections may be transferred from some other infected site in the body or may be systemic (i.e., throughout the body). Usually only one joint, normally one of the larger joints, is affected, and the course of suppurative arthritis, if treated early, is transitory. With prolonged infection, however, the articular surfaces may degenerate. **Tuberculous arthritis** can occur as a secondary infection from pulmonary tuberculosis and is more damaging than typical suppurative arthritis. It usually affects the spine or large joints and causes ulceration of the articular cartilages and even erosion of the underlying bone. Transient arthritis of multiple joints is a common symptom of rheumatic fever, but permanent damage seldom occurs in joints with this disorder.

Hemophilic arthritis may result from bleeding into the joint cavity caused by hemophilia, a hereditary disease characterized by a deficient clotting mechanism in the blood. There is some evidence that the iron in the blood is toxic to the chondrocytes, resulting in degeneration of the articular cartilage.

Gout

Gout is a group of metabolic disorders involving joints. These disorders are largely idiopathic (of unknown cause), although some cases of gout seem to be familial (occur in families and therefore are probably genetic). Gout is more common in males than in females. The ultimate problem in gout patients is an increase in uric acid in the blood because of too much synthesis or decreased removal through the kidneys. The limited solubility of uric acid salts in the body results in precipitation of monosodium urate crystals in various tissues, including the kidneys and joint capsules.

The earliest symptom of gout is transient arthritis resulting from urate crystal accumulation and irritation in the synovial fluid. This irritation can ultimately lead to an inflammatory response in the joints, and both the crystal deposition and inflammation can become chronic. Normally only one or two joints are affected. The most commonly affected joints (85% of the cases) are the base of the great toe and other foot and leg joints to a lesser extent. Any joint may ultimately be involved, and damage to the kidneys from crystal formation occurs in almost all advanced cases. Kidney failure may occur in untreated cases. With modern medications, these complications seldom occur. Weight control and reduced alcohol consumption can help prevent gout.

Pseudogout is a disorder that causes pain and swelling similar to that seen in gout, but it is characterized by calcium hypophosphate crystal deposits in joints.

Hallux Valgus and Bunion

In people who wear pointed shoes, the great toe can be deformed and displaced laterally, a condition called **hallux valgus**. Bunions are often associated with hallux valgus. A **bunion** is a bursitis that develops over the first metatarsophalangeal joint because of pressure and rubbing by shoes.

Joint Replacement

As a result of recent advancements in biomedical technology, many joints of the body can now be replaced by artificial joints. Joint replacement, called **arthroplasty,** was first developed in the late 1950s. One of the major reasons for its use is to eliminate unbearable pain in patients near age 55 to 60 with joint disorders. Osteoarthritis is the leading disease requiring joint replacement, accounting for two-thirds of the patients. Rheumatoid arthritis accounts for more than half of the remaining cases.

The major objectives in the design of joint prostheses (artificial replacements) include the development of stable articulations, low friction, solid fixation to the bone, and normal range of motion. New synthetic replacement materials are being designed by biomedical engineers to accomplish these objectives. Prosthetic joints usually are composed of metal, such as stainless steel, titanium alloys, or cobalt–chrome alloys, in combination with modern plastics, such as high-density polyethylene, silastic, or elastomer. The bone of the articular area is removed on one side (a procedure called hemireplacement) or both sides (total replacement) of the joint, and the artificial articular areas are glued to the bone with a synthetic adhesive, such as methylmethacrylate. The smooth metal surface rubbing against the smooth plastic surface provides a low-friction contact with a range of movement that depends on the design.

The success of joint replacement depends on the joint being replaced, the age and condition of the patient, and the state of the technology. Most reports are based on examination of patients 2–10 years after joint replacement. The technology is improving constantly, so current reports do not adequately reflect the effect of the most recent improvements. Still, the current reports indicate a success rate of 80%–90% in hip replacements and 60% or more in ankle and elbow replacements. The major reason for failure of prosthetic joints is loosening of the artificial joint from the bone to which it is attached. New prostheses with porous surfaces help to overcome this problem.

Summary

An articulation, or joint, is a place where two bones come together.

Naming Joints

Joints are named according to the bones or parts of bones involved.

Classes of Joints

Joints can be classified according to function or according to the type of connective tissue that binds them together and whether there is fluid between the bones.

Fibrous Joints

1. Fibrous joints are those in which bones are connected by fibrous tissue with no joint cavity. They are capable of little or no movement.
2. Sutures involve interdigitating bones held together by dense fibrous connective tissue. They occur between most skull bones.
3. Syndesmoses are joints consisting of fibrous ligaments.
4. Gomphoses are joints in which pegs fit into sockets and are held in place by periodontal ligaments (teeth in the jaws).
5. Some sutures and other joints can become ossified (synostosis).

Cartilaginous Joints

1. Synchondroses are immovable joints in which bones are joined by hyaline cartilage. Epiphyseal plates are examples.
2. Symphyses are slightly movable joints made of fibrocartilage.

Synovial Joints

1. Synovial joints are capable of considerable movement. They consist of the following:
   - Articular cartilage on the ends of bones, which provides a smooth surface for articulation. Articular disks can provide additional support.
   - A joint capsule of fibrous connective tissue, which holds the bones together while permitting flexibility, and a synovial membrane, which produces synovial fluid that lubricates the joint.
2. Bursae are extensions of synovial joints that protect skin, tendons, or bone from structures that could rub against them.

3. Synovial joints are classified according to the shape of the adjoining articular surfaces: plane (two flat surfaces), saddle (two saddle-shaped surfaces), hinge (concave and convex surfaces), pivot (cylindrical projection inside a ring), ball-and-socket (rounded surface into a socket), and ellipsoid (ellipsoid concave and convex surfaces).

Types of Movement

1. Angular movements include flexion and extension, abduction and adduction.
2. Circular movements include rotation, pronation and supination, and circumduction.
3. Special movements include elevation and depression, protraction and retraction, excursion, opposition and reposition, and inversion and eversion.
4. Combination movements involve two or more of the above-mentioned movements.

Description of Selected Joints

1. The temporomandibular joint is a complex hinge and gliding joint between the temporal and mandibular bones. It is capable of elevation and depression, protraction and retraction, and lateral and medial excursion movements.
2. The shoulder joint is a ball-and-socket joint between the head of the humerus and the glenoid fossa of the scapula that permits a wide range of movements. It is strengthened by ligaments and the muscles of the rotator cuff. The tendon of the biceps brachii passes through the joint capsule. The shoulder joint is capable of flexion and extension, abduction and adduction, rotation, and circumduction.
3. The hip joint is a ball-and-socket joint between the head of the femur and the acetabulum of the coxa that is greatly strengthened by ligaments and that is capable of a wide range of movements.
4. The knee joint is a complex ellipsoid joint between the femur and the tibia that is supported by many ligaments. The joint allows flexion and extension and slight rotation of the leg.
5. The ankle joint is a special hinge joint of the tibia, fibula, and talus that allows dorsiflexion and plantar flexion and inversion and eversion.
6. The bony arches transfer weight from the heels to the toes and allow the foot to conform to many different positions.

Content Review

1. Define an articulation or joint.
2. On what criteria are joints named and classified? Name the three major classes of joints on the basis of their structure.
3. Define the term fibrous joints, describe the three different types, and give examples of each.
4. Define cartilaginous joints, describe two different types, and give an example of each.
5. Describe the structure of a synovial joint. How do the different parts of the joint function to permit joint movement?

6. Define the terms bursa and tendon sheath. What is their function?
7. On what basis are synovial joints classified? Describe the different types of synovial joints, and give examples of each. What movements does each type of joint allow?
8. Define the terms flexion and extension. How are they different for the upper and lower limbs? What is hyperextension?
9. Contrast abduction and adduction. Describe these movements for the head, vertebral column, upper limbs, wrist, fingers, lower limbs, and toes.

10. Distinguish among rotation, circumduction, pronation, and supination. Give an example of each.
11. Define the following jaw movements: protraction, retraction, lateral excursion, medial excursion, elevation, and depression.
12. Define the terms opposition and reposition.
13. What terms are used for flexion and extension of the foot? For turning the sole of the foot medially or laterally?
14. For each of the following joints, name the bones of the joint, the specific part of the bones that form the joint, the type of joint(s) present, and the possible movement(s) at the joint: temporomandibular, shoulder, hip, knee, and ankle joint.

Develop Your Reasoning Skills

1. What would be the result if the sternal synchondroses and the sternocostal synchondrosis of the first rib were to become synostoses?
2. Using an articulated skeleton, examine the following list of joints. Describe the type of joint and the movement(s) possible.
   a. The joint between the zygomatic bone and the maxilla
   b. The ligamentous connection between the coccyx and the sacrum
   c. The elbow joint
3. For each of the following muscles, describe the motion(s) produced when the muscle contracts. It may be helpful to use an articulated skeleton.
   a. The biceps brachii muscle attaches to the coracoid process of the scapula (one head) and the radial tuberosity of the radius. Name two movements that the muscle accomplishes in the forearm.
   b. The rectus femoris muscle attaches to the anterior superior iliac spine and the tibial tuberosity. How does contraction move the thigh? The leg?
   c. The supraspinatus muscle is located in and attached to the supraspinatus fossa of the scapula. Its tendon runs over the head of the humerus to the greater tubercle. When it contracts, what movement occurs at the humeral (shoulder) joint?
   d. The gastrocnemius muscle attaches to the medial and lateral condyles of the femur and to the calcaneus. What movement of the leg results when this muscle contracts? Of the foot?
4. Crash McBang hurt his knee in an auto accident by ramming it into the dashboard. The doctor tested the knee for ligament damage by having Crash sit on the edge of a table with his leg flexed at a 90-degree angle. The doctor attempted to pull the tibia in an anterior direction (the anterior drawer test) and then tried to push the tibia in a posterior direction (the posterior drawer test). There was no unusual movement of the tibia in the anterior drawer test, but there was during the posterior drawer test. Explain the purpose of each test, and tell Crash which ligament he has damaged.

Web Site Link

For a listing of the most current web sites related to this chapter, please visit the Seeley home page at:
http://www.mhhe.com/biosci/ap/seeleyap/

Chapter Nine

Receptor Responses and Membrane Potentials

Objectives

1. Define the term chemical signal (or ligand), and list the two main categories into which ligands are placed.

2. Describe how ligands directly alter membrane permeability.

3. Describe how ligands interact with receptors to influence G proteins, and list the ways G proteins can produce a response to a ligand.

4. Describe how ligands interact with receptors to produce intracellular mediator molecules.

5. Explain how ligands such as insulin produce a response by causing phosphorylation of intracellular proteins.

6. Explain how ligands that cross the plasma membrane can produce responses by binding to intracellular receptors.

7. Describe the concentration differences that exist between intracellular fluid and extracellular fluid.

8. Describe the factors that affect the concentration differences across the plasma membrane for proteins and for potassium (K^+), sodium (Na^+), and chloride (Cl^-) ions.

9. Define the term resting membrane potential, and explain how it is produced.

10. Predict and explain the changes that occur in the resting membrane potential as a result of changes in the K^+ ion concentration gradient across the plasma membrane, and do the same for changes in the permeability of the membrane to K^+ ions.

11. Explain how ions cross the plasma membrane.

12. List the characteristics of a local potential, and explain how a local potential gives rise to an action potential.

13. Explain the role of voltage-gated Na^+ and K^+ ion channels in the development of action potentials.

14. Describe the phases of an action potential and the events responsible for each phase.

15. Define the terms absolute and relative refractory period, and compare their effects on action potentials.

16. Describe how an action potential is propagated along a cell's membrane.

17. Define the words subthreshold, threshold, submaximal, maximal, and supramaximal stimuli.

18. Compare the effect of stimulus strength and stimulus duration on action potential frequency.

19. Define the term accommodation, and describe the effect of accommodation on action potential frequency.

Part Two

In many ways, the trillions of cells that make up the human body resemble a complex society. Extensive communication among humans is necessary to enable the thousands of activities carried on by a society. Similarly, the thousands of functions performed by the human body require intricate communication among cells. The coordinated activity of millions of cells is essential to accomplishing tasks such as observing our surroundings, moving, thinking, growing, and reproducing. Homeostasis could not be maintained without millions of cells detecting hundreds of small changes each day and producing responses to maintain a constant internal environment.

Communication among cells is accomplished by chemical signals, which are released by some cells and influence the activities of others, and by electric signals that pass from cell to cell. In addition, electric signals travel from one part of specialized cells, such as neurons, over long distances, to another part of the same cell, where the electric signals cause chemical signals to be released, which, in turn, influence the activity of other, nearby cells. Cells that produce signals are specialized to release those signals at the right time, and cells that respond to the signals are specialized to receive them and respond appropriately. The purpose of this chapter is to introduce how cells respond to chemical signals and to provide an introduction to electric signals. In later chapters specific examples of the mechanisms involving chemical and electric signals will be emphasized.

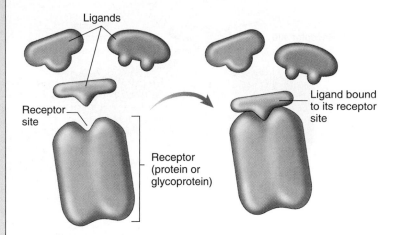

Figure 9.1   Specificity of Receptors for Ligands

The shape and chemical characteristics of receptor sites on receptors make the receptor sites very specific so that certain ligands can bind to a receptor site, but others cannot.

Ligands and Membrane-Bound Receptors

The results of ligands binding to membrane-bound receptors are to (1) directly change the permeability of the plasma membrane, (2) alter the activity of G proteins at the inner surface of the plasma membrane, (3) or alter the activity of intracellular enzymes. The changes, initiated by the combination of ligands with their receptors sites, produce specific responses in cells.

Receptors That Directly Alter Membrane Permeability

Some membrane-bound receptors are protein molecules that make up part of ion channels in the plasma membrane. When ligands bind to the receptor sites of this type of receptor, the combination alters the three-dimensional structure of the proteins of the ion channels, causing the channels either to open or close. These channels are called **ligand-gated ion channels.** The result is a change in the permeability of the plasma membrane to the specific ions passing through the ion channels (figure 9.3). For example, acetylcholine released from nerve cells is a ligand that combines with membrane-bound receptors of skeletal muscle cells. The combination of acetylcholine molecules with the receptor sites of the membrane-bound receptors for acetylcholine opens Na^+ ion channels in the plasma membrane. Consequently, Na^+ ions diffuse into the skeletal muscle cells and trigger events that cause them to contract.

Receptors and the Function of G Proteins

Many membrane-bound receptors produce responses through the action of a complex of proteins of the plasma membrane called **G proteins** (figure 9.4). G proteins consist

▊Chemical Signals

Chemical signals, commonly called **ligands** (lī′gandz), are molecules that bind to proteins or glycoproteins. The portion of each protein or glycoprotein molecule where the ligand binds is called a **binding site.** If the protein or glycoprotein molecule is a receptor, it is called a **receptor site.** The shape and chemical characteristics of each receptor site allows only a specific type of ligand to bind to it (figure 9.1). The tendency for each type of ligand to bind to a specific type of receptor site, and not to others, is called **specificity.**

Ligands can be placed into two major categories:

1. *Ligands that cannot pass through the plasma membrane.* These ligands include large molecules and water-soluble molecules. They interact with **membrane-bound receptors,** which are receptors that extend across the plasma membrane and have their receptor sites exposed to the outer surface of the plasma membrane (figure 9.2a). When a ligand binds to the receptor site on the outside of the plasma membrane, the receptor initiates a response inside the cell.

2. *Ligands that pass through the plasma membrane.* These ligands are lipid-soluble and relatively small. They bind to **intracellular receptors,** which are receptors in the cytoplasm or in the nucleus of the cell (figure 9.2b). Subsequently, the receptors, with the ligands bound to their receptor sites, interact with the DNA in the nucleus of the cell to produce a response.

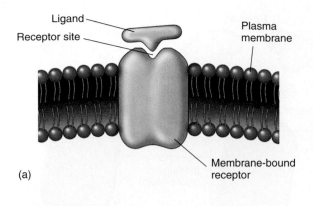

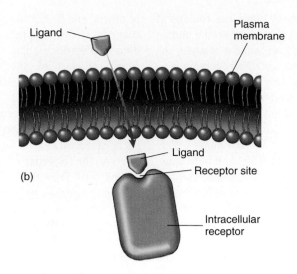

(a)

(b)

Figure 9.2   Membrane-Bound and Intracellular Receptors

(a) Membrane-Bound Receptor. A ligand combines with the receptor site of a membrane-bound receptor. The receptor site is exposed to the outside of the cell, and the receptor extends across the plasma membrane. (b) Intracellular Receptor. The small, lipid-soluble ligand diffuses through the plasma membrane and combines with the receptor site of an intracellular receptor.

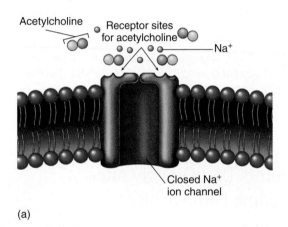

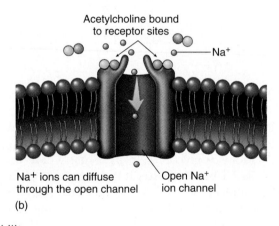

(a)

(b)

Figure 9.3   Membrane-Bound Receptors Directly Affecting Membrane Permeability

(a) The Na$^+$ ion channel has receptor sites for the ligand, acetylcholine. When the receptor sites are not occupied by acetylcholine, the Na$^+$ ion channel remains closed. (b) When two acetylcholine molecules bind to their receptor sites on the Na$^+$ ion channel, the channel opens to allow Na$^+$ ions to diffuse through the channel into the cell.

of three subunits; from the largest to smallest, they are called alpha (α), beta (β), and gamma (γ). The G proteins are so named because they bind to guanine nucleotides. In the inactive state, guanine diphosphate (GDP) molecules are bound to the α subunits of G proteins.

G proteins can bind with receptors at the inner surface of the plasma membrane. When a ligand binds to the receptor on the outside of a cell, the receptor changes shape. As a result, the receptor combines with a G protein complex on the inner surface of the cell membrane, and GDP is released from the α subunit. Guanine triphosphate (GTP), which is more abundant than GDP, binds to the α subunit, thereby activating it. The G protein complex separates from the receptor, and the activated α subunit separates from the β and γ subunits (see

figure 9.4a and b). The activated α subunit can alter the activity of molecules within the plasma membrane or inside the cell, thus producing cellular responses. After a short time, the activated α subunit is turned off because GTP is converted to GDP. The α subunit then recombines with the β and γ subunits (see figure 9.4 c and d).

The activated α subunits of G proteins can combine with ion channels, causing them to open or close (figure 9.5). For example, opening calcium ion channels in smooth muscle cells results in the movement of calcium ions into those cells. The calcium ions function as intracellular mediators. In smooth muscle cells, the calcium ions combine with **calmodulin** (kal-mod′yū-lin) molecules, and the calcium–calmodulin complexes activate enzymes that cause contraction of the

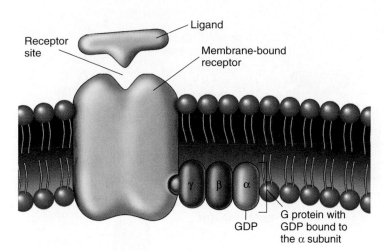

(a) The membrane-bound receptor has a receptor exposed to the outside of the cell. The portion of the receptor inside of the cell is closely associated with the G protein.

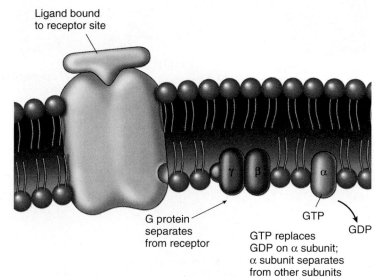

(b) The ligand binds to the receptor site of the membrane-bound receptor. The combination alters the G protein. GTP replaces GDP on the α subunit, and the α subunit separates from the γ and β subunits. The α subunit can influence ion channels in the plasma membrane or the synthesis of intracellular mediators.

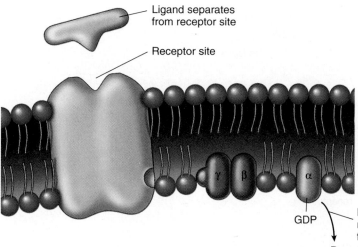

(c) When the ligand separates from the receptor site, additional G proteins are no longer activated. Inactivation of the α subunit occurs when phosphorylase removes an inorganic phosphate (P_i) from the GTP, leaving GDP bound to the α subunit.

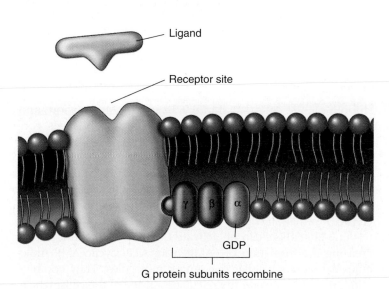

(d) The subunits of the G proteins recombine.

Figure 9.4   Membrane-Bound Receptors and G Proteins

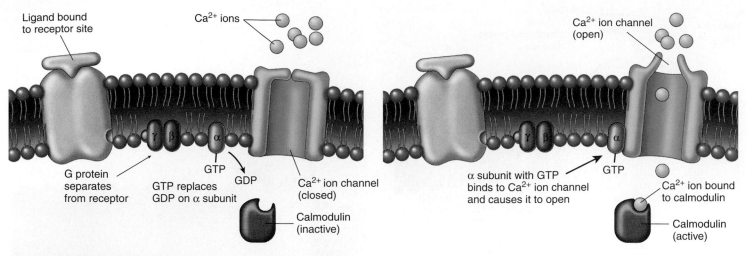

(a) A ligand binds to the receptor site of the membrane-bound receptor. The combination alters the G protein. GTP replaces GDP on the α subunit, and the α subunit separates from the γ and β subunits.

(b) The α subunit, with GTP bound to it, combines with the Ca²⁺ ion channel, and the combination causes the Ca²⁺ ion channel to open. Ca²⁺ ions diffuse into the cell and combine with calmodulin. The combination of Ca²⁺ ions with calmodulin produces the response of the cell to the ligand.

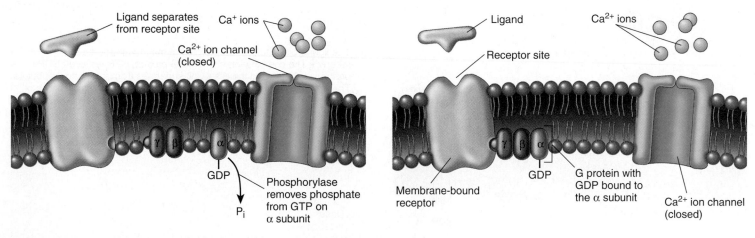

(c) Phosphorylase removes an inorganic phosphate from the GTP bound to the α subunit, leaving GDP bound to the α subunit. The α subunit can no longer stimulate a cellular response, and it separates from the Ca²⁺ ion channel and the channel closes.

(d) The α subunit recombines with γ and β subunits.

Figure 9.5   Membrane–Bound Receptors, G Proteins, and Calcium Ion Channels

smooth muscle cells (see figure 9.5*a* and *b*). After a short time, the activated α subunit is inactivated because GTP is converted to GDP. The α subunit then recombines with the β and γ subunits (see figure 9.5*c* and *d*).

Activated α subunits of G proteins can alter the activity of enzymes inside of the cell. For example, activated α subunits can influence the rate of **cyclic adenosine monophosphate (cAMP)** formation (figure 9.6). The enzyme, **adenylyl cyclase** (a-den′i-lil sī′klās), can be activated by G proteins, thereby increasing the formation of cAMP from ATP. Cyclic AMP molecules act as intracellular messenger molecules. They combine with enzymes and alter their activities inside of the cells, which, in turn, produce responses. The amount of time cAMP is pres-

ent to produce a response in a cell is limited. An enzyme in the cytoplasm, called **phosphodiesterase** (fos′fō-dī-es′ter-ās), breaks down cAMP to AMP. The response of the cell is terminated after cAMP levels are reduced below a certain level.

Cyclic AMP acts as an *intracellular mediator* in many cell types. The response in each cell type is different from responses in other cell types, however, because the enzymes activated by cAMP in each cell type are different. For example, epinephrine combines with receptors in liver cells, causing an increase in cAMP synthesis that increases the release of glucose from liver cells. In contrast, luteinizing hormone combines with receptors in cells of the ovary to increase cAMP synthesis. In ovarian cells a major response to increased cAMP is ovulation.

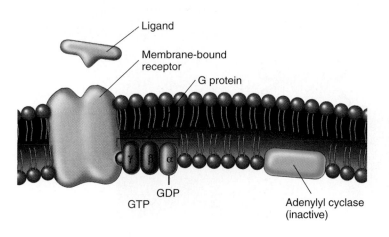

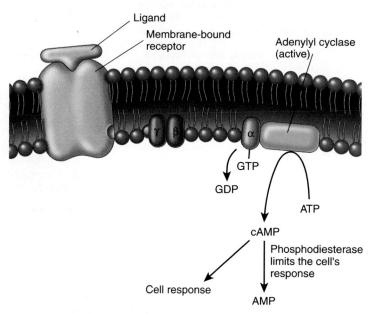

Figure 9.6   Membrane-Bound Receptors, G Proteins, and Cyclic AMP

The combination of ligands with their receptors does not always result in increased cAMP synthesis. In some cell types, the combination of certain ligands with their receptors causes the G proteins to inhibit the synthesis of cAMP. This decrease in cAMP synthesis produces a response.

G proteins can also alter the concentration of intracellular mediators other than Ca^{2+} ions or cAMP (table 9.1). For example, **diacylglycerol** (dī′as-il-glis′er-al) **(DAG)** and **inositol** (in-ō′si-tōl) **triphosphate (IP$_3$)** are intracellular mediator molecules that are influenced by G proteins (see chapter 17).

Receptors That Alter the Activity of Intracellular Enzymes

Some ligands bind to membrane-bound receptors and cause a change in the activity of intracellular enzymes. The altered enzyme activity either increases or decreases the synthesis of intracellular mediator molecules, or it results in the phosphorylation of intracellular proteins. The intracellular mediators or the phosphorylated proteins produce the responses of cells to the ligands.

Intracellular enzymes that are controlled by membrane-bound receptors can be part of the membrane-bound recep-

Table 9.1   Common Intracellular Mediators

Intracellular Mediator	Example of Cell Type	Example of Response
Cyclic guanine monophosphate (cGMP)	Kidney cells	Increases Na$^+$ ion and water excretion by the kidney
Cyclic adenosine monophosphate (cAMP)	Liver cells	Increases the breakdown of glycogen and the release of glucose into the circulatory system
Calcium ions	Smooth muscle cells	Contraction of smooth muscle cells
Inositol triphosphate (IP$_3$)	Smooth muscle cells	Contraction of certain smooth muscle cells in response to epinephrine
Diacylglycerol (DAG)	Smooth muscle cells	Contraction of certain smooth muscle cells in response to epinephrine
Nitric oxide (NO)	Smooth muscle cells	Relaxation of smooth muscle cells of blood vessels resulting in vasodilation

tor, or they may be separate molecules. The intracellular mediator molecules act as chemical signals that move from the enzymes that produce them into the cytoplasm of the cell, where they influence the activity of other molecules and produce the response of the cell.

Cyclic guanine (gwahn′ēn) **monophosphate (cGMP)** is an example of an intracellular mediator molecule that is

1. The ligand combines with the receptor site of the membrane-bound receptor.

2. The combination activates the enzyme guanylyl cyclase at the inner surface of the cell membrane. Guanylyl cyclase converts GTP to cGMP plus 2 inorganic phosphate groups.

3. Cyclic GMP is an intracellular mediator, and it functions to alter the activity of intracellular enzymes to produce a response.

4. Phosphodiesterase breaks down cGMP to GMP and limits the length of time cGMP functions in the cell.

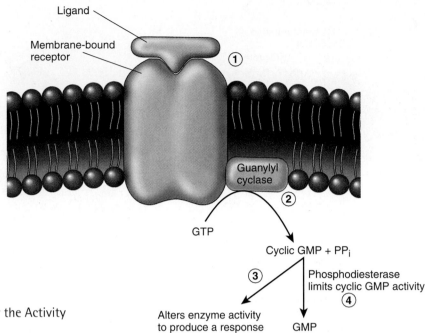

Figure 9.7 Membrane-Bound Receptors That Alter the Activity of Intracellular Enzymes; Guanylyl Cyclase

synthesized in response to a ligand binding with a membrane-bound receptor (Figure 9.7). The ligand binds to its receptor, and the combination activates an enzyme called **guanylyl cyclase** (gwahn′i-lil sī′klās) located at the inner surface of the plasma membrane. The guanylyl cyclase enzyme converts guanine triphosphate (GTP) to cGMP and two inorganic phosphate groups. The cGMP molecules then combine with specific enzymes in the cytoplasm of the cell and activate them. The activated enzymes, in turn, produce the response of the cell to the ligand.

Atrial natriuretic hormone is a ligand that combines with its receptor in the plasma membrane of kidney cells. The result is an increase in the rate of cGMP synthesis at the inner surface of the plasma membranes. Cyclic GMP influences the action of enzymes in the kidney cells, which increase the rate of sodium ion and water excretion by the kidney (see chapter 26).

The amount of time the cGMP is present to produce a response in the cell is limited. Phosphodiesterase breaks down cGMP to GMP. Consequently, the length of time a ligand increases cGMP synthesis and has an effect on a cell is brief after the ligand is no longer present.

Some ligands bind to membrane-bound receptors, and the portion of the receptor on the inner surface of the plasma membrane acts as an enzyme that adds phosphate groups, a process called **phosphorylation** (fos′fōr-i-lā-shŭn), to several specific proteins. Some of the proteins phosphorylated can be part of the membrane-bound receptor, and others can be in the cytoplasm of the cell (figure 9.8). The phosphorylated molecules influence the activity of other enzymes in the cytoplasm of the cell. For example, insulin molecules bind to their membrane-bound receptors, resulting in the phosphorylation of parts of the receptors on the inner surface of the

plasma membrane and certain other intracellular proteins. These phosphorylated proteins produce the responses of the cells to insulin.

Ligands and Intracellular Receptors

Intracellular receptors are found inside the cell, either in the cytoplasm or in the nucleus. By the process of diffusion, lipid-soluble ligands cross the plasma membrane into the cytoplasm or into the nucleus and bind to intracellular receptors (figure 9.9). After a ligand binds with an intracellular receptor, the receptor can alter the activity of enzymes in the cell, or the receptor can bind to deoxyribonucleic acid (DNA) to produce a response. Some intracellular receptors that influence the expression of DNA are located in the cytoplasm. Once a ligand binds to its receptor, the receptor diffuses into the nucleus and binds to DNA. Other intracellular receptors are located in the nucleus. A ligand diffuses into the nucleus and binds to its receptor, and the receptor then binds with DNA.

The receptor, when bound to its ligand, has specific "fingerlike" projections that interact with specific parts of a DNA molecule. The combination of the receptor and DNA increases the synthesis of specific messenger ribonucleic acid (mRNA) molecules. The mRNA molecules then move to the cytoplasm and increase the synthesis of certain proteins at the ribosomes. The newly synthesized proteins produce the cell response to the ligand. For example, testosterone from the testes and estrogen from the ovaries, stimulate the synthesis of proteins that are responsible for the secondary sex characteristics of males and females.

See table 9.2 for an overview of cell responses to hormones binding to their receptors.

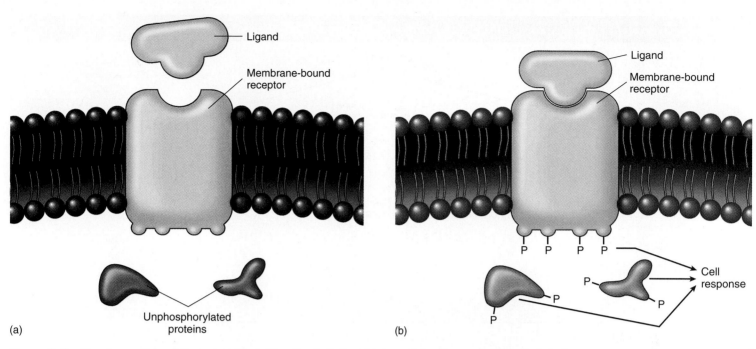

Figure 9.8   Membrane-Bound Receptors That Alter the Activity of Intracellular Enzymes; Phosphorylation

1. The ligand diffuses through the plasma membrane and enters the cytoplasm of the cell.

2. The ligand combines with the receptor in the nucleus (or in the cytoplasm).

3. The receptor with the ligand bound to it interacts with DNA and increases the synthesis of specific messenger RNA (mRNA) molecules.

4. The mRNA passes from the nucleus to the cytoplasm.

5. In the cytoplasm of the cell, mRNA combines with ribosomes. New protein molecules, which produce the response of the cell to the ligand, are synthesized.

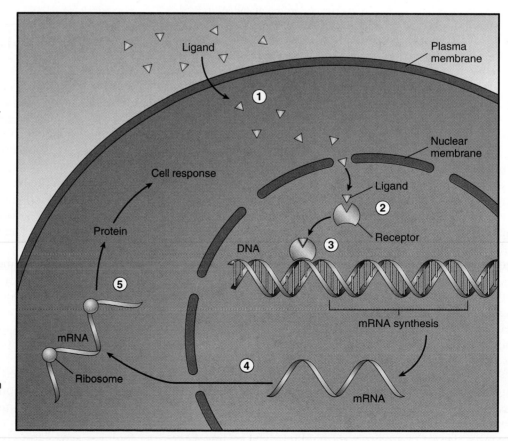

Figure 9.9   Ligands That Bind to Intracellular Receptors and Increase Protein Synthesis

Table 9.2   Overview of the Responses of Cells to Hormones Binding to Their Receptors

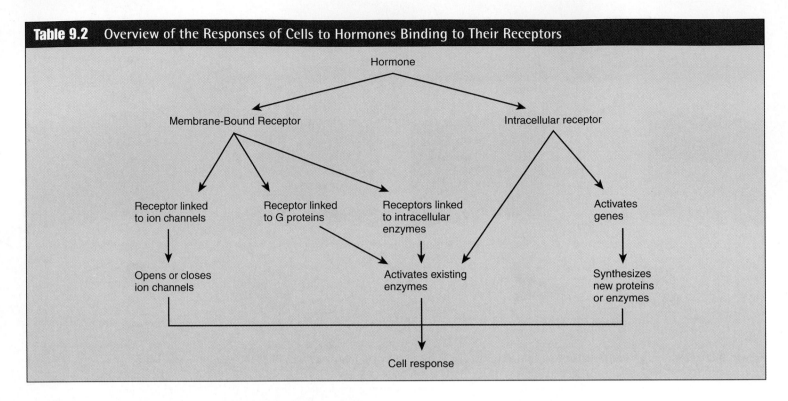

Drugs with structures similar to specific ligands may compete with those ligands for their receptor sites (see chapter 3). Depending on the exact characteristics of a drug, it may either bind to a receptor site and activate the receptor or bind to a receptor site and inhibit the action of the receptor. For example, drugs exist that compete with the ligand, epinephrine, for its receptor sites. Some of these drugs activate epinephrine receptors and others inhibit them.

Some cellular functions depend on the coordinated activity of ligands that bind to membrane-bound receptors and ligands that bind to intracellular receptors. For example, acetylcholine molecules, released from nerve cells, bind to membrane-bound receptors of endothelial cells in blood vessels, and the combination causes Ca^{2+} ion channels to open. Ca^{2+} ions then enter the endothelial cell and activate enzymes that produce nitric oxide (NO). NO is a very toxic gas that functions as a ligand, and it diffuses from the endothelial cells to smooth muscle cells. The NO binds to an intracellular receptor that is part of an enzyme called guanylyl cyclase. In re-

sponse, the guanylyl cyclase synthesizes cGMP, which causes the smooth muscle cells to relax (figure 9.10).

▌Electric Signals

There are many differences between humans and computers, but one similarity is the dependence of both on electric signals for such operations as communication and information processing. The electrical properties of many cells dramatically influence how the body functions. For example, stimuli act on specialized cells in the eye, ear, mouth, and skin to produce electric signals called **action potentials,** which are conducted from these cells to the spinal cord and brain. Within the brain, the action potentials are interpreted, causing the sensations of vision, sound, taste, and touch. Action potentials originating within the brain and spinal cord are conducted to muscles and certain glands to regulate their activities. Complex mental processes, including emotions and conscious thought, also depend on action potentials in nerve cells of the brain. Electric signals are therefore an important means by which cells transfer information from one cell to another; and interpretation of electric signals influences the ability to perceive our environment, remember, think, and act.

A basic knowledge of the electrical characteristics of cells is necessary for understanding the normal functions and many pathologies of muscle and nervous tissues. The electrical properties of cells result from the ionic concentration differences across the plasma membrane and from the perme-

1. Acetylcholine binds to the acetylcholine receptor site on the acetylcholine receptor. The combination causes a Ca²⁺ ion channel to open, allowing Ca²⁺ ions to diffuse into the endothelial cell.

2. The Ca²⁺ ion binds to nitric oxide synthase, an enzyme that acts on arginine to produce nitric oxide (NO).

3. NO diffuses out of the endothelial cell and into the smooth muscle cell.

4. NO combines with the enzyme, guanylyl cyclase, which converts GTP to cyclic GMP. Cyclic GMP causes the smooth muscle cell to relax.

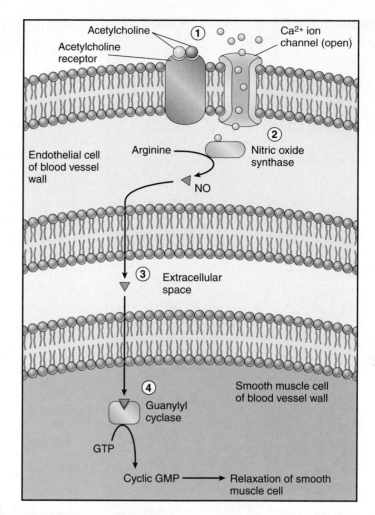

Figure 9.10 Combined Membrane-Bound and Intracellular Receptor Mechanism

The combination of a ligand with its membrane-bound receptor results in the production of nitric oxide (NO) in one cell. The NO diffuses into another cell and binds to an intracellular receptor, producing a response.

ability characteristics of the plasma membrane. This section introduces the electrical properties of cells when they are under resting conditions and in response to stimuli.

Concentration Differences Across the Plasma Membrane

Table 9.3 lists the concentration differences for positively charged ions (cations) and negatively charged ions (anions) between the intracellular and extracellular fluids. The concentration of Na⁺ and Cl⁻ ions is much greater outside of the cell than inside, and the concentration of K⁺ ions and negatively charged molecules, such as proteins and other molecules containing phosphate, are much greater inside the cell than outside. Note that there is a steep concentration gradient (see chapter 3) for Na⁺ ions from outside the cell to the inside. There is also a steep concentration gradient for K⁺ ions from the inside to the outside of the cell.

Differences in intracellular and extracellular concentrations of ions result primarily from (1) the sodium–potassium exchange pump and (2) the permeability characteristics of the plasma membrane.

The Sodium–Potassium Exchange Pump

The differences in K⁺ and Na⁺ ion concentrations across the plasma membrane are maintained primarily by the action of the **sodium–potassium exchange pump** (figure 9.11). Through active transport, the sodium–potassium exchange pump moves K⁺ and Na⁺ ions through the plasma membrane against their concentration gradients. K⁺ ions are transported into the cell, increasing the concentration of K⁺ ions inside the cell, and Na⁺ ions are transported out of the cell, increasing the concentration of Na⁺ ions outside the cell. Approximately three Na⁺ ions are transported out of the cell and two K⁺ ions are transported into the cell for each ATP molecule used.

Table 9.3	Representative Concentrations of the Principal Cations and Anions in Extracellular and Intracellular Fluids of Vertebrates	
Ions	Intracellular Fluid (mEq/L)	Extracellular Fluid (mEq/L)
Cations (Positive)		
Potassium (K^+)	148	5
Sodium (Na^+)	10	142
Calcium (Ca^{2+})	<1	5
Others	41	3
TOTAL	200	155
Anions (Negative)		
Proteins	56	16
Chloride (Cl^-)	4	103
Others	140	36
TOTAL	200	155

Permeability Characteristics of the Plasma Membrane

As noted in chapter 3, the plasma membrane is selectively permeable, allowing some, but not all, substances to pass through it. Negatively charged proteins are synthesized inside the cell, and because of their large size and their solubility characteristics, they cannot readily diffuse across the plasma membrane (figure 9.12). Negatively charged Cl^- ions are repelled by the negatively charged proteins and other negatively charged ions inside the cell. The Cl^- ions diffuse through the plasma membrane and accumulate outside it, resulting in a higher concentration of Cl^- ions outside of the cell than inside.

Ions pass through the plasma membrane through ion channels, of which there are two major types: nongated ion channels and gated ion channels.

Nongated Ion Channels

Nongated ion channels are always open and are responsible for the permeability of the plasma membrane to ions when the plasma membrane is at rest. Each ion channel is specific for one type of ion, although the specificity is not absolute. The number of each type of nongated ion channels in the plasma

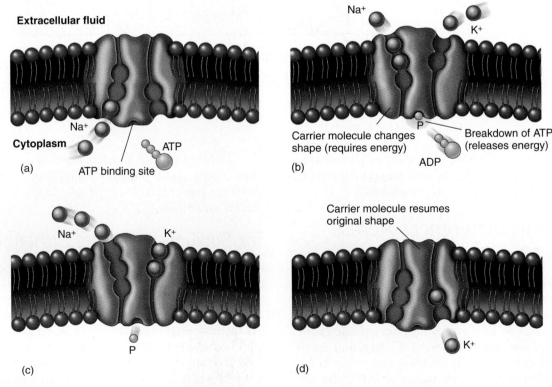

Figure 9.11   The Sodium–Potassium Exchange Pump

(a) Three Na^+ ions and an ATP bind to the carrier. (b) The ATP breaks down to ADP and energy. The carrier molecule changes shape, and the Na^+ ions are transported across the membrane. (c) The Na^+ ions diffuse away from the carrier, two K^+ ions are bound to the carrier, and the phosphate is released. (d) The carrier changes shape, transporting K^+ ions across the membrane, and the K^+ ions diffuse away from the carrier. The carrier molecule can again bind to Na^+ ions and ATP.

membrane determines the permeability characteristics of the resting plasma membrane to different types of ions. The plasma membrane is more permeable to Cl^- and K^+ ions and much less permeable to Na^+ ions because there are many more Cl^- and K^+ nongated ion channels than Na^+ nongated ion channels in the plasma membrane.

Gated Ion Channels

Gated ion channels open and close in response to stimuli. By opening and closing, the gated ion channels can change the permeability characteristics of the plasma membrane. The major types of gated ion channels are

1. *Voltage-gated ion channels.* These channels open and close in response to small voltage changes across the plasma membrane. In an unstimulated cell, the inside of the plasma membrane is negatively charged relative to the outside. This charge difference can be measured in units called **millivolts** (mV; 1 mV = 1/1000 V). When a cell is stimulated, the charge difference changes, and that causes voltage-gated ion channels to open or close. Voltage-gated channels specific for Na^+ and K^+ ions are most numerous in electrically excitable tissues, but voltage-gated Ca^{2+} channels are also important, especially in smooth muscle and cardiac muscle cells (see chapters 10 and 20).
2. *Ligand-gated ion channels.* These channels open or close in response to ligands in two ways. Ligands bind to receptor sites on ion channels, which directly results in opening or closing the ion channels, or the ligands

bind to receptors that activate G proteins or intracellular mediators, which causes ion channels to open or close. Ligand-gated ion channels are common for Na^+, K^+, Ca^{2+}, and Cl^- ions, and these ligand-gated channels are common in tissues such as nervous and muscle tissues, as well as glands.

3. *Other gated ion channels.* These ion channels are present in specialized electrically excitable tissues. Examples include touch receptors, which respond to mechanical stimulation of the skin; temperature receptors, which respond to temperature changes in the skin; and light receptors, which respond to light in the retina of the eye.

The Resting Membrane Potential

Although there are unequal concentrations of ions in the intracellular and extracellular fluids, they are nearly electrically neutral. That is, both intracellular and extracellular fluids have nearly equal numbers of positively and negatively charged ions. There is, however, an unequal distribution of charge between the immediate inside and the immediate outside of the plasma membrane. This electric charge difference across the plasma membrane, called a **potential difference,** can be measured between the inside and outside of essentially all cells. By placing the tip of one microelectrode inside a cell and another outside it, and by connecting the electrodes by wires to an appropriate measuring device such as a voltmeter or an oscilloscope, the potential difference can be measured (figure 9.13). The potential difference across the plasma membranes of skeletal muscle fibers and nerve cells is −70 to −90 mV. The potential difference is reported as a negative number, because the inside of the plasma membrane is negative compared with the outside. In an unstimulated, or resting, cell, the potential difference across the plasma membrane is called the **resting membrane potential.**

Establishment of the Resting Membrane Potential

The resting membrane potential results from the permeability characteristics of the resting plasma membrane and the difference in concentration of ions between the intracellular and the extracellular fluids. The plasma membrane is somewhat permeable to K^+ ions because of nongated K^+ ion channels. Positively charged K^+ ions can therefore diffuse down their concentration gradient from inside to just outside the cell. Negatively charged proteins and other molecules cannot diffuse through the plasma membrane with the K^+ ions. As K^+ ions diffuse out of the cell, the loss of positive charges makes the inside of the plasma membrane more negative. Because opposite charges attract, the K^+ ions are attracted back toward the cell. The K^+ ions accumulate just outside of the plasma membrane, making the outside of the plasma membrane positive relative to the inside. Thus, the tendency for K^+ ions to

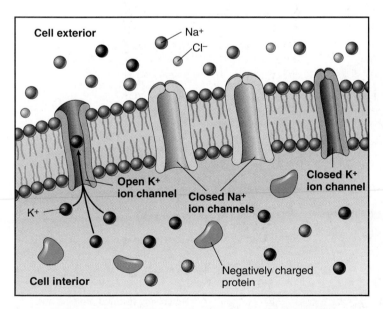

Figure 9.12   Membrane Permeability and Ion Concentrations
Concentrations of Na^+ ions, K^+ ions, Cl^- ions, and negatively charged proteins across the plasma membrane. The permeability of the membrane to K^+ ions is greater than its permeability to Na^+ ions because some K^+ channels remain open and few Na^+ channels remain open. The membrane is not permeable to the negatively charged proteins inside of the cell.

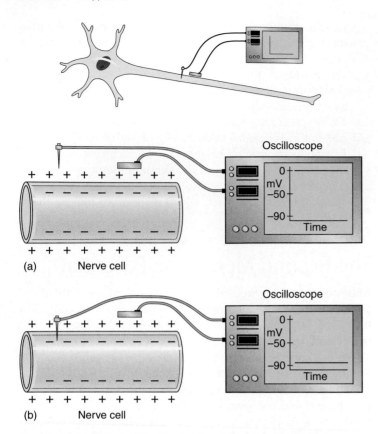

(a)   Nerve cell

(b)   Nerve cell

Oscilloscope

Oscilloscope

Figure 9.13   Measuring the Resting Membrane Potential

(*a*) Both recording (needle) and reference (block) electrodes are outside of the cell, and no potential difference (0 mV) is recorded. (*b*) The recording electrode is inside the cell, the reference electrode is outside, and a potential difference of about 85 mV is recorded, with the inside of the plasma membrane negative with respect to the outside of the plasma membrane.

diffuse from a higher concentration inside the cell to a lower concentration outside the cell is opposed by the charge difference that develops across the plasma membrane. The resting membrane potential is in equilibrium because the K^+ ion concentration gradient, which causes K^+ ions to diffuse out of the cell, is equal to the potential difference across the plasma membrane, which opposes that movement (figure 9.14).

Other ions, such as Na^+, Cl^-, and Ca^{2+} ions do have some small influence on the resting membrane potential, but the major influence on resting membrane potential is from K^+ ions. Because the resting plasma membrane is 50–100 times less permeable to Na^+ ions than to K^+ ions, very few Na^+ ions can diffuse from the outside to the inside of the resting cell. The resting plasma membrane is not very permeable to Ca^{2+} ions either. The plasma membrane is relatively permeable to Cl^- ions, but the negatively charged Cl^- ions are repelled by the negative charge inside the cell.

The resting membrane potential is proportional to the tendency for K^+ ions to diffuse out of the cell and not to the actual rate of flow for K^+ ions. At equilibrium, very few K^+ ions pass through the plasma membrane because their movement out of the cell is opposed by negative charge inside the cell. Still, some Na^+ ions and K^+ ions diffuse continuously across the plasma membrane, although at a low rate. The large concentration gradients for Na^+ and K^+ ions would eventually disappear without the continuous activity of the sodium–potassium exchange pump.

The function of the sodium–potassium exchange pump is to maintain the normal concentration gradients for Na^+ and K^+ ions across the plasma membrane; however, the sodium–potassium exchange pump is also responsible for a small portion of the resting membrane potential, usually less

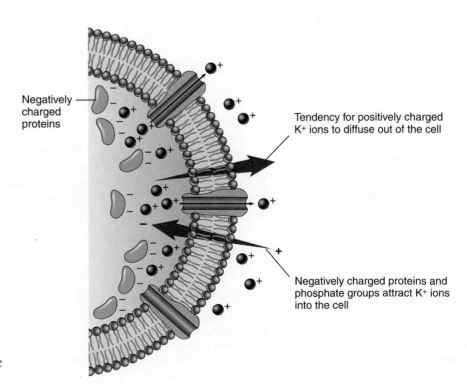

Negatively charged proteins

Tendency for positively charged K+ ions to diffuse out of the cell

Negatively charged proteins and phosphate groups attract K+ ions into the cell

Figure 9.14   Potassium Ions and the Resting Membrane Potential

At equilibrium (resting conditions) the tendency for K+ ions to diffuse out of the cell is opposed by the negative charge inside of the cell.

than 15 mV. The sodium–potassium exchange pump transports approximately three Na^+ ions out of the cell and two K^+ ions into the cell for each ATP molecule used (see figure 9.11). Because more positively charged ions are pumped out of the cell than are pumped into it, the outside is more positively charged than the inside, which contributes to the resting membrane potential.

The sodium–potassium exchange pump requires ATP. When metabolic poisons that inhibit ATP synthesis are added to electrically excitable cells, the resting membrane potential is not changed substantially as long as normal concentration gradients exist across the plasma membrane. As the normal concentration gradient for K^+ ions slowly declines, so does the resting membrane potential.

The characteristics responsible for a resting membrane potential are summarized in table 9.4.

Changing the Resting Membrane Potential

It is possible to predict how the resting membrane potential will be affected by (1) alterations in the K^+ ion concentration on either side of the plasma membrane and (2) changes in the permeability of the plasma membrane to K^+ ions. In response to each of these conditions, a new equilibrium is quickly established across the plasma membrane. For example, potassium succinate added to the extracellular fluid increases the extracellular concentration of K^+ ions and therefore decreases the normal K^+ ion concentration gradient. As a consequence, the tendency for K^+ ions to diffuse out of the cell decreases, and a smaller negative charge inside the cell is required to oppose the diffusion of K^+ ions out of the cell. At this new equilibrium, the charge difference across the plasma membrane is decreased, and the resting membrane potential is less negative (figure 9.15a). This change is called **depolarization** (dē-pō′lăr-i-zā′shŭn), or **hypopolarization** (hī′pō-pō-lăr-i-zā′shŭn) of the resting membrane potential. That is, the potential difference across the plasma membrane becomes smaller, or less polar.

A decrease in the extracellular concentration of K^+ ions, caused by reducing the potassium chloride (KCl) concentration in the extracellular fluid, increases the K^+ ion concentration gradient from the inside to the outside of the cell and increases the tendency for K^+ ions to diffuse out of the cell. As

Table 9.4	**Characteristics Responsible for the Resting Membrane Potential**

1. The number of charged molecules and ions inside and outside the cell is nearly equal.

2. The concentration of K^+ ions is higher inside than outside the cell, and the concentration of Na^+ ions is higher outside than inside the cell.

3. The plasma membrane is 50–100 times more permeable to K^+ ions than to other positively charged ions such as Na^+ ions.

4. The plasma membrane is impermeable to large intracellular negatively charged molecules such as proteins.

5. K^+ ions tend to diffuse across the plasma membrane from the inside to the outside of the cell.

6. Because negatively charged molecules cannot follow the positively charged K^+ ions, a small negative charge develops just inside the plasma membrane.

7. The negative charge inside the cell attracts positively charged K^+ ions. When the negative charge inside the cell is great enough to prevent additional K^+ ions from diffusing out of the cell through the plasma membrane, an equilibrium is established.

8. The charge difference across the plasma membrane at equilibrium is reflected as a difference in potential, which is measured in millivolts (mV).

9. The resting membrane potential is proportional to the potential for K^+ ions to diffuse out of the cell but not to the actual rate of flow for K^+ ions.

10. At equilibrium there is very little movement of K^+ ions or other ions across the plasma membrane.

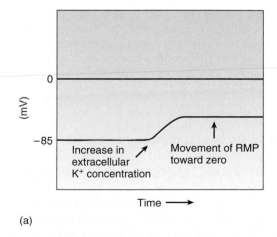

(a)

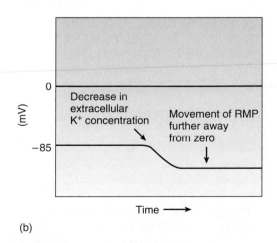

(b)

Figure 9.15   Changes in the Resting Membrane Potential Caused by Changes in Extracellular K^+ Ion Concentration

(a) Elevated extracellular K^+ ion concentration causes depolarization. (b) Decreased extracellular K^+ ion concentration causes hyperpolarization.

a result, a greater negative charge inside the cell is required to resist the diffusion of the K$^+$ ions out of the cell. Thus the resting membrane potential becomes more negative (figure 9.15*b*), a change called **hyperpolarization** (hī′per-pō′lăr-i-zā′shŭn). That is, the potential difference across the plasma membrane becomes greater, or more polar.

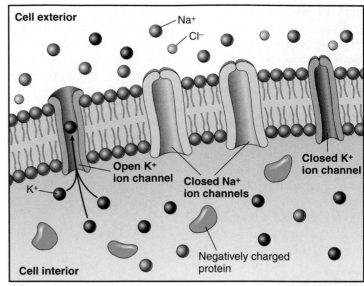

P R E D I C T
Does the resting membrane potential increase or decrease when the intracellular concentration of potassium ions is increased by the injection of a solution of potassium succinate into the cell? Explain.

✔ *Answer in Appendix F*

A change in the permeability of the membrane to K$^+$ ions also affects the resting membrane potential. The resting membrane is not freely permeable to K$^+$ ions. With an increase in the permeability of the membrane to K$^+$ ions, a new equilibrium across the plasma membrane is quickly established. The increased tendency for K$^+$ ions to diffuse out of the cell is opposed by a greater negative charge that develops inside the plasma membrane (hyperpolarization).

P R E D I C T
Explain the effect on the resting membrane potential of a reduced permeability of the plasma membrane to K$^+$ ions.

✔ *Answer in Appendix F*

Changes in the concentration of Na$^+$ ions on either side of the plasma membrane do not influence the resting membrane potential very much because the membrane is much less permeable to Na$^+$ ions than to K$^+$ ions. Large changes in the concentration gradient for sodium are required to have a significant effect on the resting membrane potential. If the permeability of the membrane to Na$^+$ ions increases, however, the resting membrane potential is affected dramatically. The concentration gradient for Na$^+$ ions is from the outside to the inside of the cell. If the permeability of the plasma membrane to Na$^+$ ions increases, Na$^+$ ions diffuse down their concentration gradient into the cell, and the inside of the plasma membrane becomes more positive, resulting in depolarization.

Electrically Excitable Cells

Electrically excitable cells, such as nerve and muscle cells, have many voltage-gated Na$^+$ and K$^+$ ion channels in their plasma membranes. Small depolarizations of the plasma membrane, regardless of the cause of the depolarization, cause voltage-gated Na$^+$ ion channels to open, increasing the plasma membranes permeability to Na$^+$ ions. The voltage-gated Na$^+$ ion channels remain open for a short time and then close (figure 9.16*a* and *b*). Depolarization of the plasma membrane also causes a greater number of voltage-gated K$^+$ ion

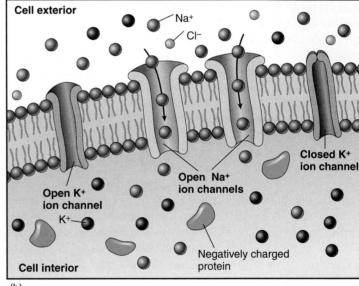

(a)

(b)

Figure 9.16   Stimuli and Membrane Permeability

(*a*) Nearly all Na$^+$ ion channels remain closed in a resting or unstimulated plasma membrane. (*b*) When a stimulus is applied to the plasma membrane that causes the plasma membrane to depolarize, the activation gates of voltage-gated Na$^+$ ion channels open. Na$^+$ ions then diffuse down their concentration gradient into the cell, causing depolarization of the plasma membrane.

channels to open, increasing the plasma membrane's permeability to K$^+$ ions. Although the voltage-gated K$^+$ ion channels begin to open at about the same time as the voltage-gated Na$^+$ ion channels, they open more slowly, remain open for a brief time, and then close.

Voltage-gated Na$^+$ ion channels are sensitive to changes in the extracellular concentration of calcium (Ca^{2+}) ions. Ca^{2+} ions in the extracellular fluid are attracted to proteins of the plasma membrane with negatively charged groups exposed to the extracellular fluid. If the extracellular concentration of

Ca^{2+} ions decreases, Ca^{2+} ions diffuse away from proteins of the plasma membrane, including the voltage-gated Na^+ ion channels, causing the voltage-gated Na^+ ion channels to open. If the extracellular concentration of Ca^{2+} ions increases, Ca^{2+} ions bind to the voltage-gated Na^+ ion channels, causing them to close. At the Ca^{2+} ion concentrations normally found in the extracellular fluid, only a small percentage of the Na^+ ion channels are open at any single moment.

5   P R E D I C T

Predict the effect on the resting membrane potential of a decrease in the extracellular concentration of calcium ions.

✔ *Answer in Appendix F*

Local Potentials

A stimulus applied at one point to the plasma membrane of a cell normally causes a change in the resting membrane potential called a **local potential,** which is confined to a small region of the plasma membrane. Local potentials can result from (1) ligands binding to their receptors, (2) mechanical stimulation, (3) temperature changes, (4) changes in the charge across the plasma membrane, or (5) spontaneous changes in membrane permeability. For example, ligands such as acetylcholine molecules bind to their receptors and cause ligand-gated Na^+ ion channels to open, resulting in a local depolarization. Local potentials are called **graded** because they vary from small to large depending on the stimulus strength or frequency. For example, weak stimuli cause smaller changes in membrane potentials, and strong stimuli cause larger changes (figure 9.17*a*). Increased frequency of stimulation can cause two or more local potentials to **summate** (sŭm-āt′) (figure 9.17*b*). That is, if a second stimulus is applied before the local potential produced by the first stimulus has returned to the resting membrane potential, a larger depolarization results than would result from a single stimulus alone. This type of summation is called **temporal summation** (summation is described in more detail in chapter 12).

Local potentials spread, or are conducted, over the plasma membrane in a decremental fashion. That is, local potentials rapidly decrease in magnitude as they spread over the surface of the plasma membrane. Normally a local potential cannot be detected more than a few micrometers (μm) from the site of stimulation. As a consequence, a local potential cannot transfer information over long distances from one part of the body to another.

Some local potentials are hyperpolarizations instead of depolarizations. An increase in the permeability of the plasma membrane to K^+ ions or Cl^- ions results in such hyperpolarizations. An increase in the permeability of the membrane to K^+ ions increases the tendency for K^+ ions to diffuse out of the cell, making the inside of the cell more negative and the outside of the cell more positive. An increase in the permeability of the plasma membrane to Cl^- ions allows Cl^- ions to diffuse into the cell, causing the inside of the cell to become

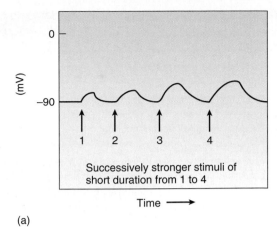

(a)

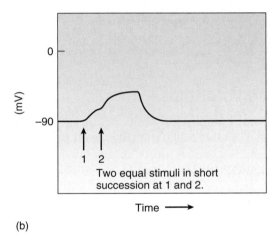

(b)

Figure 9.17   Local Potentials

(*a*) Local potentials are proportional to the stimulus strength. A weak stimulus applied briefly causes a small depolarization, which quickly returns to the resting membrane potential (1). Progressively stronger stimuli result in larger depolarizations (2 to 4). (*b*) A stimulus applied to a cell causes a small depolarization. When a second stimulus is applied before the depolarization disappears, the depolarization caused by the second stimulus is added to the depolarization caused by the first stimulus to result in a larger depolarization.

more negative. The characteristics of local potentials are summarized in table 9.5.

Local potentials occur most often in the dendrites and cell bodies of neurons and near the sites where neurons innervate muscle cells. They occur less often in the axons of neurons (see chapters 4 and 12).

6   P R E D I C T

Given two cells that are identical in all ways except that the extracellular concentration of sodium ions is greater for cell A than for cell B, how would the magnitude of the local potential in cell A differ from that in cell B if stimuli of identical strength were applied to each?

✔ *Answer in Appendix F*

Table 9.5   Characteristics of Local Potentials
1. A stimulus causes increased permeability of the membrane to Na$^+$ ions or increased permeability of the membrane to K$^+$ and Cl$^-$ ions.
2. Depolarization is a result of increased permeability of the membrane to Na$^+$ ions; hyperpolarization is a result of increased permeability of the membrane to K$^+$ or Cl$^-$ ions.
3. Local potentials are graded; that is, the size of the local potential is proportional to the strength of the stimulus. Local potentials can also summate. Thus, a local potential produced in response to several stimuli in rapid succession is larger than a local potential produced in response to a single stimulus.
4. Local potentials are conduced in a decremental fashion, meaning that their magnitude decreases as they spread over the plasma membrane. Local potentials cannot be measured a few micrometers from the point of summation.
5. A depolarizing local potential can cause an action potential.

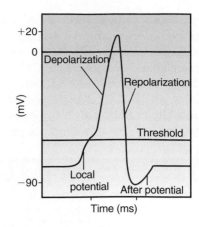

Figure 9.18   The Action Potential

The action potential consists of a depolarization phase and a repolarization phase, often followed by a short period of hyperpolarization called the afterpotential.

Action Potentials

When a local potential causes depolarization of the plasma membrane to a level called the **threshold potential,** a series of permeability changes occur that result in an **action potential** (figure 9.18). An action potential is a large change in the membrane potential that propagates, without changing its magnitude, over long distances along the plasma membrane. If the local potential causes hyperpolarization, no action potential is produced because the threshold potential is not reached. The characteristics of action potentials are summarized in table 9.6.

Action potentials occur according to the **all-or-none principle.** To understand what all-or-none implies, consider, that when a depolarization reaches threshold, all the permeability changes responsible for an action potential proceed without stopping and are constant in magnitude (the "all" part). If the local potential does not reach the threshold, few of the permeability changes occur, and the membrane potential returns to its resting level after a brief period without producing an action potential (the "none" part).

The action potential has a **depolarization phase,** in which the membrane potential moves away from the resting membrane potential and becomes more positive, and a **repolarization phase,** in which the membrane potential returns toward the resting membrane state and becomes more negative. After the repolarization phase, the plasma membrane may be slightly hyperpolarized for a short period called the **afterpotential** (see figure 9.18).

The change in charge across the plasma membrane caused by a local potential causes increasing numbers of voltage-gated Na$^+$ channels to open for a brief time. Each voltage-gated Na$^+$ ion channel has two voltage-sensitive gates, an activation gate and an inactivation gate (figure 9.19). When the plasma membrane is at rest, the activation gates of the voltage-gated Na$^+$ ion channel are closed, and the inactivation gates are open. Because the activation gates are closed,

Table 9.6   Characteristics of the Action Potential
1. Action potentials are produced when a local potential reaches threshold.
2. Action potentials are all-or-none.
3. Depolarization is a result of increased membrane permeability to Na$^+$ ions and movement of Na$^+$ ions into the cell. Activation gates of the voltage-gated Na$^+$ ion channels open.
4. Repolarization is a result of decreased membrane permeability to Na$^+$ ions and increased membrane permeability to K$^+$ ions, which stops Na$^+$ ion movement into the cell and increased K$^+$ ion movement out of the cell. The inactivation gates of the voltage-gated Na$^+$ ion channels close, and the voltage-gated K$^+$ ion channels open.
5. No action potential is produced by a stimulus, no matter how strong, during the absolute refractory period. During the relative refractory period a stronger-than-threshold stimulus can produce an action potential.
6. Action potentials are propagated, and for a given axon or muscle fiber the magnitude of the action potential is constant.
7. Stimulus strength determines the frequency of action potentials. Unless accommodation occurs, stimulus duration determines how long action potentials are produced.

Na$^+$ ions cannot pass through the channels. When the plasma membrane with voltage-gated Na$^+$ ion channels is depolarized to threshold, the change in the membrane potential causes many of the activation gates to open, and Na$^+$ ions can diffuse through the Na$^+$ ion channels into the cell. Movement of Na$^+$ ions into the cell causes the membrane to be depolarized which causes more voltage-gated Na$^+$ ion channels to open. The voltage-gated Na$^+$ ion channels continue to open until the membrane depolarizes maximally (see figure 9.19). As the membrane potential approaches its maximum depolarization, the change in the potential difference across the

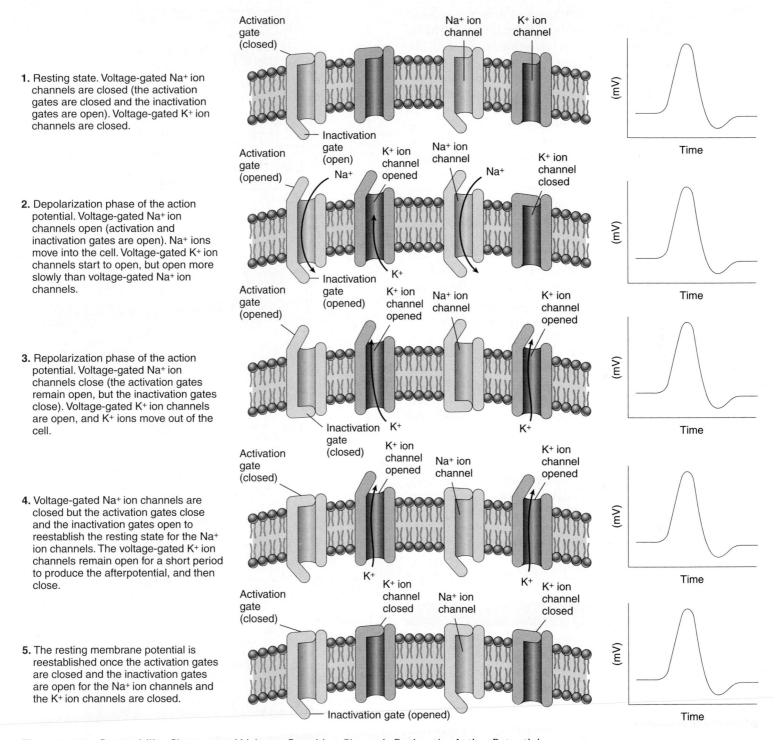

1. **Resting state.** Voltage-gated Na⁺ ion channels are closed (the activation gates are closed and the inactivation gates are open). Voltage-gated K⁺ ion channels are closed.

2. **Depolarization phase of the action potential.** Voltage-gated Na⁺ ion channels open (activation and inactivation gates are open). Na⁺ ions move into the cell. Voltage-gated K⁺ ion channels start to open, but open more slowly than voltage-gated Na⁺ ion channels.

3. **Repolarization phase of the action potential.** Voltage-gated Na⁺ ion channels close (the activation gates remain open, but the inactivation gates close). Voltage-gated K⁺ ion channels are open, and K⁺ ions move out of the cell.

4. Voltage-gated Na⁺ ion channels are closed but the activation gates close and the inactivation gates open to reestablish the resting state for the Na⁺ ion channels. The voltage-gated K⁺ ion channels remain open for a short period to produce the afterpotential, and then close.

5. The resting membrane potential is reestablished once the activation gates are closed and the inactivation gates are open for the Na⁺ ion channels and the K⁺ ion channels are closed.

Figure 9.19 Permeability Changes and Voltage-Gated Ion Channels During the Action Potential

plasma membrane causes the inactivation gates to begin closing, and the permeability of the plasma membrane to Na⁺ ions decreases. Closure of the voltage-gated Na⁺ channels is partially responsible for repolarization of the plasma membrane (see figure 9.19). It is repolarization of the plasma membrane that causes the activation gates to close and the inactivation gates to open, thus returning the voltage-gated Na⁺ channels to their resting state.

Voltage-gated K⁺ channels begin to open at the same time as the voltage-gated Na⁺ channels, but they open more slowly. The voltage-gated K⁺ channels have one gate, and once opened, these channels remain open until the resting membrane potential is reestablished. The increased movement of K⁺ ions out of the cell through the open K⁺ ion channels partially counteracts the increased movement of Na⁺ ions into the cell through the open Na⁺ channels and helps repolarize the plasma membrane during the repolarization phase of the action potential (see figure 9.19). The return of the membrane potential to its resting level causes the voltage-gated K$^+$ channels to close.

The following events are responsible for the action potential. As soon as a threshold depolarization is reached, many voltage-gated Na$^+$ ion channels begin to open. Na$^+$ ions move rapidly into the cell, and the resulting depolarization causes additional voltage-gated Na$^+$ ion channels to open. As a consequence, more Na$^+$ ions diffuse into the cell, causing a greater depolarization of the membrane, which, in turn, causes still more voltage-gated Na$^+$ ion channels to open. This is an example of a positive-feedback cycle, and it continues until most of the voltage-gated Na$^+$ ion channels in the plasma membrane are open.

Diffusion of Na$^+$ ions into the cell causes the depolarization phase of the action potential. Enough Na$^+$ ions diffuse across the plasma membrane to cause the membrane potential to become positive (i.e., the inside of the plasma membrane becomes positive relative to its outside). Although Na$^+$ ions move through the voltage-gated Na$^+$ ion channels during the depolarization phase of the action potential, the total number of Na$^+$ ions crossing the plasma membrane is small. When the positive charge reaches approximately its maximum value (between +20 and +50 mV) inside the cell, the inactivation gates of the voltage-sensitive Na$^+$ ion channels close the channels, and the influx of Na$^+$ ions slows (figure 9.20a).

During the repolarization phase of the action potential, the voltage-gated K$^+$ channels, which started to open along with the voltage-gated Na$^+$ channels, continue to open. Consequently, the permeability of the plasma membrane to Na$^+$ ions decreases, and the permeability to K$^+$ ions increases. The movement of K$^+$ ions out of the cell helps cause repolarization (figure 9.20b).

---

7   P R E D I C T

Predict the effect of a reduced extracellular concentration of Na$^+$ ions on the magnitude of the action potential in an electrically excitable cell, and explain the effect of an increased extracellular concentration of Na$^+$ ions on the magnitude of the action potential.

✔ *Answer in Appendix F*

In many cells, a period of hyperpolarization, or afterpotential, exists following each action potential. The afterpotential exists because the voltage-gated K$^+$ ion channels remain open for a short time. The increased K$^+$ ion permeability that develops during the repolarization phase of the action potential lasts slightly longer than the time required to bring the membrane potential back to its resting level. When the elevated K$^+$ ion permeability returns to normal as the voltage-gated K$^+$ ion channels close, the membrane potential resumes its resting level.

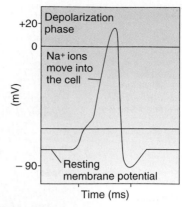

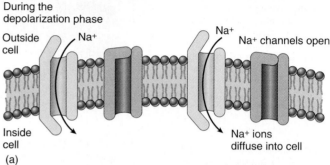

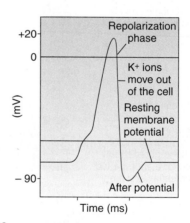

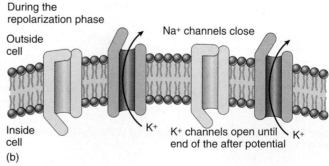

Figure 9.20   Summary of Permeability Changes During the Phases of the Action Potential

(*a*) The permeability of the membrane to Na$^+$ ions increases during depolarization, and Na$^+$ ions diffuse through the open voltage-gated Na$^+$ ion channels into the cell, causing depolarization of the plasma membrane. (*b*) The permeability of the plasma membrane to K$^+$ ions increases because the voltage-gated K$^+$ ion channels are open, and the permeability of the membrane to Na$^+$ ions decreases to its resting level as the voltage-gated Na$^+$ ion channels close during repolarization of the plasma membrane. An afterpotential occurs because of elevated K$^+$ ion permeability until the voltage-gated K$^+$ ion channels close.

Clinical Focus   Examples of Abnormal Membrane Potentials

Several important conditions provide examples of the physiology of membrane potentials and the consequence of abnormal ones. **Hypokalemia** (hī-pō-ka-lē′mē-ă) is a lower-than-normal concentration of K^+ ions in the blood or extracellular fluid. Figure 9.15b shows that reduced extracellular K^+ ion concentrations cause hyperpolarization of the resting membrane potential. Thus, a greater-than-normal stimulus is required to depolarize the membrane to its threshold level and to initiate action potentials in neurons, skeletal muscle, and cardiac muscle. Symptoms of hypokalemia include muscular weakness, an abnormal electrocardiogram, and sluggish reflexes. These symptoms are consistent with the effect of a reduced extracellular K^+ ion concentration. The symptoms result from the reduced sensitivity of the excitable tissues to stimulation. The several causes of hypokalemia include potassium depletion during starvation, alkalosis, and certain kidney diseases.

Hypocalcemia (hī-pō-kal-sē′mē-ă) is a lower-than-normal concentration of Ca^{2+} ions in blood or extracellular fluid. Symptoms of hypocalcemia include nervousness and uncontrolled contraction of skeletal muscles, called **tetany** (tet′ă-nē). The symptoms are due to an increased membrane permeability to Na^+ ions that results because low blood levels of Ca^{2+} ions cause Na^+ channels in the membrane to open. Na^+ ions diffuse into the cell, cause depolarization of the plasma membrane to threshold, and initiate action potentials. The tendency for action potentials to occur spontaneously in nervous tissue and muscles accounts for the listed symptoms. A lack of dietary calcium, a lack of vitamin D, or a reduced secretion rate of a parathyroid gland hormone are examples of conditions that cause hypocalcemia.

As long as the Na^+ and K^+ ion concentrations remain unchanged across the plasma membrane, all the action potentials produced by a cell are identical. They all take the same amount of time, and they all exhibit the same magnitude. These characteristics vary somewhat from one cell type to another, but it generally takes approximately 1–2 millisecond (ms) (1 ms = 0.001 s) for an action potential to occur.

Active transport of Na^+ and K^+ ions is not involved directly in the action potential, but it plays an important role in maintaining the concentration gradients across the plasma membrane. The sodium–potassium exchange pump functions to transport Na^+ ions from the cell and K^+ ions into the cell to maintain the normal concentrations of these ions on either side of the plasma membrane after a series of action potentials has occurred. The sodium–potassium exchange pump is too slow to have a large effect on either the depolarization or repolarization phase of individual action potentials.

Refractory Period

Once an action potential is produced at a given point on the plasma membrane, the sensitivity of that area to further stimulation decreases for a time called the **refractory** (rē-frak′tōr-ē) **period.** The first part of the refractory period, during which there is complete insensitivity to another stimulus, called the **absolute refractory period,** exists from the beginning of depolarization until shortly after repolarization is completed in some cells or during the repolarization phase in other cells (figure 9.21). The absolute refractory periods exists as long as the activation gates are open and the inactivation gates are closed in the voltage-gated Na^+ ion channels. The absolute refractory period ends when the activation gates close and the inactivation gates reopen.

The existence of the absolute refractory period guarantees that once an action potential is begun, both the depolarization and the repolarization phases will be completed, or

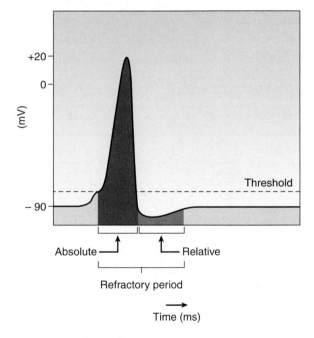

Figure 9.21   Refractory Period

The absolute and relative refractory periods of an action potential. In some cells the absolute refractory period may end during the repolarization phase of the action potential.

nearly completed, before another action potential can be started, and that a strong stimulus cannot lead to prolonged depolarization of the plasma membrane. The absolute refractory period also has important consequences for action potential propagation (see following section).

The second part of the refractory period, called the **relative refractory period,** follows the absolute refractory period. During the relative refractory period, the membrane is more permeable to K^+ ions because many voltage-gated K^+ ion channels are open. The relative refractory period ends when the voltage-gated K^+ ion channels close. A stronger-than-threshold stimulus can initiate another action potential during the relative refractory period. Thus, after the absolute refractory period, but before the relative refractory is completed, a sufficiently strong stimulus can produce another action potential.

Propagation of Action Potentials

An action potential occurs in a very small area of the plasma membrane. It does not affect the entire plasma membrane at one time. After an action potential is produced at one location, it stimulates adjacent regions of the plasma membrane

(figure 9.22). The change in the membrane potential at the point where an action potential occurs causes voltage-gated Na^+ ion channels to open in adjacent regions of the membrane. As a consequence, an action potential is produced in those adjacent regions.

Action potentials are propagated from their point of origin over the remainder of the cell. The absolute refractory period, however, makes the membrane insensitive to restimulation long enough to prevent an action potential from reinitiating another action potential at that same point and keeps an action potential from reversing its direction of propagation.

Action potentials are not propagated from one cell to an adjacent cell in the same way that they are propagated along the membrane of a single cell. Specialized structures called **synapses** (sing.; sin'aps) are the sites at which action potentials from one cell are able to produce action potentials in an adjacent cell. Most synapses are chemical, that is, an action potential causes the release of a chemical from the end of a nerve cell process (figure 9.23a). The chemical is a type of ligand, called a **neurotransmitter.** The action potential at the end of a nerve

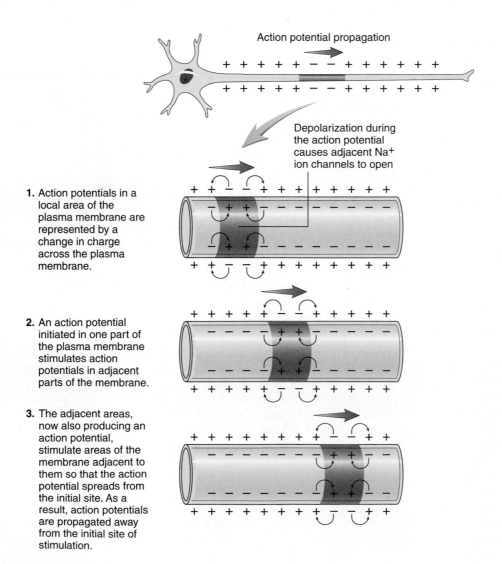

1. Action potentials in a local area of the plasma membrane are represented by a change in charge across the plasma membrane.

2. An action potential initiated in one part of the plasma membrane stimulates action potentials in adjacent parts of the membrane.

3. The adjacent areas, now also producing an action potential, stimulate areas of the membrane adjacent to them so that the action potential spreads from the initial site. As a result, action potentials are propagated away from the initial site of stimulation.

Figure 9.22   Action Potential Propagation

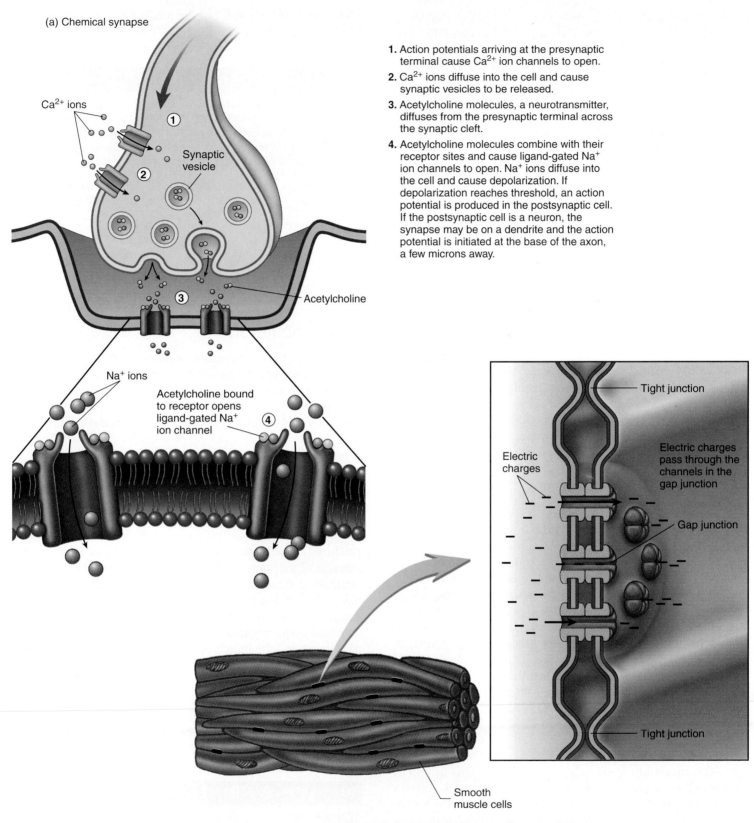

(a) Chemical synapse

1. Action potentials arriving at the presynaptic terminal cause Ca^{2+} ion channels to open.

2. Ca^{2+} ions diffuse into the cell and cause synaptic vesicles to be released.

3. Acetylcholine molecules, a neurotransmitter, diffuses from the presynaptic terminal across the synaptic cleft.

4. Acetylcholine molecules combine with their receptor sites and cause ligand-gated Na^+ ion channels to open. Na^+ ions diffuse into the cell and cause depolarization. If depolarization reaches threshold, an action potential is produced in the postsynaptic cell. If the postsynaptic cell is a neuron, the synapse may be on a dendrite and the action potential is initiated at the base of the axon, a few microns away.

Ca^{2+} ions

Synaptic vesicle

Acetylcholine

Na^+ ions

Acetylcholine bound to receptor opens ligand-gated Na^+ ion channel

Tight junction

Electric charges

Electric charges pass through the channels in the gap junction

Gap junction

Tight junction

Smooth muscle cells

Electrical charges in the form of action potentials can pass directly from one cell to another because gap junctions make an electric connection between the cells.

(b) Electrical synapse

Figure 9.23 Synapses

(a) Chemical synapse. (b) Electrical synapse.

cell process causes voltage-gated Ca^{2+} ion channels to open. Ca^{2+} ions diffuse into the cell and cause the release of the neurotransmitter molecules. The neurotransmitter molecules diffuse across the space within the synapse to bind with specific membrane receptors on the adjacent plasma membrane. The combination of the neurotransmitters with their receptors can influence ligand-gated channels in the plasma membrane. For example, some neurotransmitters bind to receptors on ligand-gated Na^+ ion channels in synapses, increasing the permeability of the plasma membrane to Na^+ ions. As a consequence, Na^+ ions diffuse into the cell and cause a local potential. Local depolarizations in synapses are sometimes called **excitatory postsynaptic potentials (EPSPs).** If the local potential exceeds threshold, an action potential results.

In some synapses, the neurotransmitter binds to receptor molecules, that open K^+ ion channels instead of Na^+ ion channels. When that occurs, K^+ ions diffuse out of the cell, and the membrane is hyperpolarized. Local hyperpolarization in synapses is sometimes called an **inhibitory postsynaptic potentials (IPSP).** Action potentials are not produced by local potentials that hyperpolarize the plasma membrane because the membrane potential is farther from threshold.

Some synapses are electrical synapses in which adjacent plasma membranes, joined at **gap junctions,** allow depolarizations in one cell to spread to the adjacent cell as though the two cells were fused (figure 9.23*b*). Gap junctions are found in cardiac muscle and in many types of smooth muscle. Synapses are discussed further in chapters 10 and 12.

Action Potential Frequency

The **action potential frequency** is the number of action potentials produced per unit of time in response to a stimulus. Action potential frequency is directly proportional to stimulus strength and to the size of the local potential. A stimulus resulting in a local potential so small that it does not reach threshold is called a **subthreshold stimulus,** and it produces no action potential (figure 9.24). A stimulus just strong enough to cause a local potential to reach threshold, a **threshold stimulus,** produces a single action potential. A stimulus strong enough to produce a maximum frequency of action potentials is a **maximal stimulus,** and a stronger stimulus is called a **supramaximal stimulus.** A **submaximal stimulus** includes stimuli between threshold and the maximal stimulus strength. For submaximal stimuli, the action potential frequency increases in proportion to the strength of the stimulus.

The maximum frequency of action potentials in an excitable cell is determined by the duration of the absolute refractory period. If the duration is 1 ms for each action potential, the maximum frequency of action potentials for that cell is approximately 1000 action potentials per second.

The length of time during which action potentials are produced in a cell in response to a given stimulus reflects how long the stimulus is applied at a strength great enough to cause a local potential to exceed threshold. Two stimuli of identical strength applied to a cell for different lengths of time normally result in identical action potential frequencies in re-

sponse to each stimulus, but the length of time the action potentials are produced is different.

9	**P R E D I C T**

Action potentials are monitored in a single pain receptor in response to three different pain stimuli. Given the following information about the action potentials produced, identify which recording was in response to the weaker stimulus, and which stimulus was applied for the shortest time.

	Total Action Potential Frequency	Number of Action Potentials
Recording 1	200/s	800
Recording 2	400/s	800
Recording 3	600/s	600

✔ *Answer in Appendix F*

Accommodation

In some cells, the local potential is maintained at a constant magnitude as long as a stimulus of a given strength is applied to the cell. In other cells, the local potential quickly returns to its resting membrane potential, even though the stimulus is applied for a long time, an adjustment called **accommodation** (ă-kom'ŏ-dā'shŭn) (figure 9.25). In cells that exhibit accommodation, a constant stimulus that is applied to the cell at first produces an action potential frequency proportional to the magnitude of the local potential. Eventually, even though the stimulus remains constant, the local potential begins to decrease in magnitude, and the action potential frequency declines.

Cells having local potentials that do not accommodate produce action potential frequencies proportional to the strength of the stimulus, and the action potential frequency is maintained as long as the stimulus is applied (see figure 9.25*a*). Nerve cells monitoring the position of the arms and legs accommodate slowly so that when a limb is moved, the action potential frequency provides information about the degree of movement and the length of time the limb is in that position.

Cells having local potentials that do accommodate rapidly are adapted for producing action potential frequencies that are proportional to the rate at which the change in stimulus strength occurs. Nerve cells monitoring the acceleration of a limb accommodate rapidly and provide action potential frequencies proportional to the rate of change in the limb movement.

10	**P R E D I C T**

Consider the following: A maximal stimulus is applied to a nerve cell and is maintained for a period of 10 s; the action potential frequency decreases from a maximum of 400/s immediately after the stimulus to a frequency of 100/s after 2 s and a frequency of 0/s after 10 s. Explain the relationship between the duration of the stimulus, the local potential, and the observed change in action potential frequency.

✔ *Answer in Appendix F*

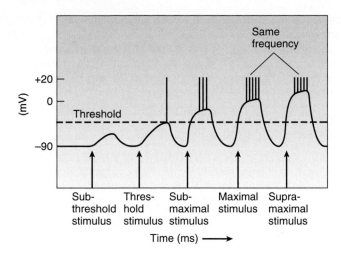

Figure 9.24   Stimuli, Local Potentials, and Action Potentials

Relationship among stimulus strength, local potential, and action potential frequency. Each stimulus in this figure is stronger than the previous one.

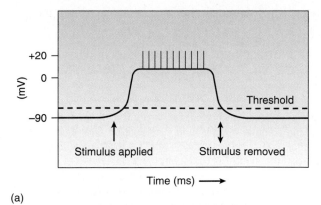

(a)

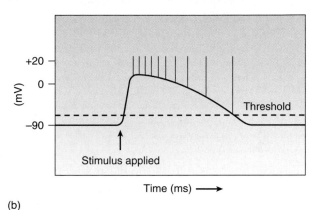

(b)

Figure 9.25   Stimulus, Action Potential Frequency, and Accommodation

(*a*) When a stimulus is applied in some neurons, the local potential remains constant as long as the stimulus is applied, so that the action potential frequency remains constant while the stimulus is applied. (*b*) The local potential is shown with action potentials superimposed. In other neurons accommodation occurs when a constant stimulus is applied. The local potential becomes smaller, and the action potential frequency decreases. Action potentials are still produced (but at a decreasing frequency) until the local potential declines below threshold, at which time the action potentials cease.

Summary

Communication between cells is accomplished by chemical and electrical signals. The purpose of this chapter is to introduce how cells respond to chemical and electric signals.

Chemical Signals

1. Chemical signals, called ligands, combine with receptor sites of receptor molecules.
2. The receptors are highly specific.
3. Drugs similar to ligands can combine with and activate or inhibit receptors.
4. There are two categories of ligands:
   - Those that cannot pass through the plasma membrane and bind to membrane-bound receptors.
   - Those that pass through the plasma membrane and bind to intracellular receptors.

Ligands and Membrane-Bound Receptors

1. Ligands can bind to receptors that are part of ion channels and alter the permeability of the ion channels.
2. Ligands can bind to receptors and activate G protein complexes inside of the plasma membrane.
   - G proteins can alter the permeability of the membrane by altering ion channels in the plasma membrane.
   - G proteins can produce intracellular mediators, such as cAMP, DAG, and IP_3.
3. Ligands can bind to receptors, and the combination alters the activity of intracellular enzymes, which synthesize intracellular mediators, or second-messenger molecules, inside of the cell such as cGMP.
   - Receptors add phosphate groups to part of the receptor inside the plasma membrane and other intracellular

271

proteins, which produces the response of the cell to the ligand.

Ligands and Intracellular Receptors

1. Intracellular receptor molecules are found in cytoplasm or in the nucleus of the cell.
2. Ligands that are small and lipid-soluble diffuse across the plasma membrane and combine with intracellular receptors.
   - The intracellular receptors in combination with ligands interact with specific portions of DNA molecules and increase the synthesis of specific proteins.
3. In some cases, membrane-bound receptors and intracellular receptors produce a response to ligands. For example, ligands can bind to receptors that cause Ca^{2+} ion channels to open, which causes nitric oxide to be synthesized in endothelial cells of blood vessels. The nitric oxide diffuses from the endothelial cells and causes smooth muscle cells to relax by increasing cGMP production in them.

Electric Signals

Electrical properties of cells result from the ionic concentration differences across the plasma membrane and from the permeability characteristics of the plasma membrane.

Concentration Differences Across the Plasma Membrane

1. The sodium–potassium exchange pump moves ions by active transport. K^+ ions are moved into the cell, and Na^+ ions are moved out of it.
2. The concentration of K^+ ions and negatively charged proteins and other molecules is higher inside, and the concentration of Na^+ and Cl^- ions is higher outside the cell.
3. Negatively charged proteins and other negatively charged ions are synthesized inside the cell and cannot diffuse out of it, and they repel negatively charged Cl^- ions.
4. Permeability of the plasma cell membrane to ions is determined by nongated and gated ion channels.
   - Nongated K^+ channels are more numerous than nongated Na^+ ion channels, thus the plasma membrane is more permeable to K^+ than to Na^+ when at rest.
   - Gated ion channels in the plasma membrane include voltage-gated ion channels, ligand-gated channels, and other gated ion channels.

The Resting Membrane Potential

1. The resting membrane potential is a charge difference across the plasma membrane when the cell is in an unstimulated condition. The inside of the cell is negatively charged to the outside of the cell.
2. The resting membrane potential is due mainly to the tendency of positively charged K^+ ions to diffuse out of the cell (carrying their positive charges with them), which is opposed by the negative charge that develops inside the plasma membrane. At equilibrium, the tendency of positive charges to diffuse out of the cell is opposed by the negative charge inside the cell, and few ions actually diffuse through the plasma membrane.
3. Depolarization is a decrease in the resting membrane potential and can result from an increase in the concentration of extracellular K^+ ions or from a decrease in membrane permeability to K^+ ions.
4. Hyperpolarization is an increase in the resting membrane potential that can result from a decrease in the concentration of extracellular K^+ ions or an increase in membrane permeability to K^+ ions.

Electrically Excitable Cells

Local Potentials

1. A local potential, either a depolarization or a hyperpolarization, is a small change in the resting membrane potential that is confined to a small area of the plasma membrane.
2. A local potential is termed graded because a stronger stimulus produces a greater potential change than a weaker stimulus.
3. A local potential decreases in magnitude as the distance from the stimulation increases.
4. A local potential can be produced in the following two ways:
   - An increase in membrane permeability to Na^+ ions can cause local depolarization.
   - An increase in membrane permeability to K^+ or Cl^- ions can result in local hyperpolarization.

Action Potentials

1. An action potential is a larger change in the resting membrane potential that spreads over the entire surface of the cell.
2. The threshold potential is the membrane potential at which a local potential depolarizes the plasma membrane sufficiently to produce an action potential.
3. Action potentials occur in an all-or-none fashion. If the action potential occurs at all, it is of the same magnitude, no matter how strong the stimulus.
4. Depolarization occurs as the inside of the membrane becomes more positive because of Na^+ ion movement into the cell through voltage-gated ion channels. Repolarization is a return of the membrane potential toward the resting membrane potential because voltage-gated Na^+ ion channels close and movement into the cell slows to resting levels and because voltage-gated K^+ ion channels continue to open and K^+ ion movement out of the cell continues for a short while before the voltage-gated K^+ ion channels close, returning the membrane permeability and the resting membrane potential to resting levels.
5. The afterpotential is a short period of hyperpolarization following repolarization to resting levels due to the voltage-gated K^+ ion channels remaining open.

Refractory Period

1. The absolute refractory period is the time during an action potential when a second stimulus, no matter how strong, cannot initiate another action potential. During the absolute refractory period, the activation gates of voltage-gated Na^+ ion channels are open.
2. The relative refractory period follows the absolute refractory period and is the time during which a stronger-than-threshold stimulus can evoke another action potential. During this time the activation gates are closing and the inactivation gates are opening in the voltage-gated Na^+ ion channels.

Propagation of Action Potentials

1. An action potential causes sodium channels in adjacent regions of the cell to open and to produce action potentials.
2. Reversal of the direction of action potential propagation is prevented by the absolute refractory period.
3. Action potentials can be transferred from cell to cell across a synapse. Transfer can result from diffusion of a chemical (neurotransmitter) or can be electrical (gap junction).

Action Potential Frequency

1. Types of stimuli
   - A subthreshold stimulus produces only a local potential.
   - A threshold stimulus causes a local potential that reaches threshold and results in a single action potential.
   - A submaximal stimulus is greater than a threshold stimulus and weaker than a maximal stimulus. The action potential frequency increases as the strength of the submaximal stimulus increases.
   - A maximal or a supramaximal stimulus produces a maximum frequency of action potentials.
2. The longer a threshold or greater stimulus is applied, the longer the action potential generated, unless accommodation occurs.

Accommodation

During accommodation, even though a long stimulus of constant strength is applied, the local potential returns to the resting membrane potential after a short time, and no action potential is produced after the local potential decreases below the threshold value.

Content Review

1. Name the two main categories into which chemical signals, or ligands, can be placed. Name two main categories of receptors.
2. Describe how a ligand can combine with a receptor and directly alter membrane permeability.
3. Explain how the combination of a ligand and its receptor can alter the G proteins at the inner surface of the plasma membrane.
4. Describe how G proteins can alter the permeability of the plasma membrane and how they can alter the synthesis of an intracellular mediator molecule such as cAMP.
5. Describe how a ligand can combine with a membrane-bound receptor and change enzyme activity inside the cell.
6. Explain how the combination of a ligand with a membrane-bound receptor alters the activity of the cell by phosphorylating proteins inside the cell.
7. Explain how a ligand that crosses the plasma membrane interacts with its receptor and how it alters the rate of protein synthesis.
8. Describe the concentration differences that exist across a plasma membrane for K^+ ions, Na^+ ions, Cl^- ions, and proteins. Explain the cause of these differences.
9. Define the term resting membrane potential. Is the outside of the plasma membrane positively or negatively charged relative to the inside?
10. Explain the role of K^+ ions in establishing the resting membrane potential.
11. Define the terms depolarization and hyperpolarization. How do changes in extracellular K^+ ion concentration or in membrane permeability to K^+ ions affect depolarization and hyperpolarization?
12. What effect does a change of extracellular sodium have on the resting membrane potential? Why is this so?
13. What effect does depolarization have on voltage-gated Na^+ ion channels?
14. Give an example of a ligand-gated Na^+ channel.
15. What effect do increases and decreases in extracellular Ca^{2+} ion concentration have on Na^+ ion channels?
16. Describe the sodium–potassium exchange pump. What effect does it have on the resting membrane potential? On Na^+ and K^+ ion concentration gradients across the plasma membrane?
17. Differentiate between a local potential and an action potential.
18. Describe two ways that a change in membrane permeability can produce a local potential.
19. What is meant by a graded potential?
20. Define the term threshold potential. What happens to a local potential that reaches threshold?
21. Discuss the all-or-none production of an action potential. Are all action potentials the same?
22. Define the depolarization and repolarization phases of an action potential. Explain how changes in membrane permeability and the movement of Na^+ and K^+ ions cause each phase.
23. Describe the afterpotential and its cause.
24. Distinguish between the absolute and relative refractory periods. Relate them to the depolarization and repolarization phases of the action potential.
25. What is the function of the sodium–potassium exchange pump after a series of action potentials?
26. What causes the propagation of the action potential? What prevents the action potential from reversing its direction of propagation?
27. Describe two ways an action potential can pass from one cell to another.
28. Define a subthreshold, threshold, submaximal, maximal, and supramaximal stimulus.
29. What determines the maximum frequency of action potential generation?
30. How does stimulus strength affect the frequency of action potential production?
31. What is accommodation? How does the length of stimulation affect the length of time that action potentials are produced?

Develop Your Reasoning Skills

1. A phosphatase enzyme functions to convert GTP to GDP when it is bound to the α subunit of the G protein complex. Predict the outcome if there was a mutation in the gene responsible for producing the phosphatase enzyme so that the enzyme was not functional.

2. Epinephrine binds to a membrane-bound receptor on liver cells, resulting in an increased rate of cAMP synthesis in liver cells. Glucagon binds to different membrane-bound receptors on the same liver cells and increases cAMP synthesis. Also, epinephrine, but not glucagon, can combine with membrane-bound receptors of smooth muscle cells in the bronchioles of the lung. Is the response of liver cells and these smooth muscle cells to epinephrine and glucagon similar or different? Explain.

3. Nitroglycerine is a drug given to certain heart patients to alleviate angina pectoris (chest pain). One of its effects is to dilate blood vessels. Apparently, nitroglycerine can increase nitric oxide synthesis, but it does not bind to membrane-bound receptors. Explain how nitroglycerine dilates the blood vessels.

4. Predict the consequence of a reduced intracellular K^+ ion concentration on the resting membrane potential.

5. Predict the effect of an elevated extracellular potassium ion concentration on nerve and muscle tissue.

6. A child eats a whole bottle of salt (NaCl) tablets. What effect would this have on action potentials?

7. Lithium ions reduce the permeability of plasma membranes to sodium ions. Predict the effect lithium ions in the extracellular fluid would have on the response of a neuron to stimuli.

8. Predict the effect of an elevated extracellular calcium concentration on nerve and muscle tissue.

9. When severe burns occur, many cells are destroyed and release their contents into the blood. Assuming that shock caused by reduced blood volume and stress is under control, explain why burn patients may suffer from ectopic contractions of heart muscle caused by action potentials originating in abnormal areas of the heart.

10. Both smooth muscle and cardiac muscle have the ability to contract spontaneously (they will contract without external stimulation). They also contract rhythmically (contractions occur at regular intervals). On the basis of what you know about membrane potentials, propose an explanation for both the spontaneous and rhythmic characteristics of these muscle tissues. Assume that an action potential in a muscle cell causes the muscle to contract.

11. Smooth muscle has some characteristics that differ from skeletal muscle or nerves. One characteristic involves the action potential. Although smooth muscle action potentials are very similar to those in skeletal muscle, the following data suggest that some differences exist:
    - A chemical compound that specifically blocks the diffusion of sodium into the cell reduces the amplitude of the action potentials but does not eliminate them in smooth muscle.
    - The amplitude of smooth muscle action potentials is reduced in a calcium-free medium.
    - Elevating the intracellular concentration of calcium reduces the amplitude of smooth muscle action potentials.
    - On the basis of the previous information, which of the following is (are) most logical?
    a. Calcium inhibits the entry of sodium into smooth muscle cells.
    b. Calcium regulates the permeability of smooth muscle plasma membranes to sodium.
    c. Calcium participates with sodium in the depolarization phase of the action potential.
    d. Calcium is responsible for the depolarization phase of the action potential.

Web Site Link

For a listing of the most current web sites related to this chapter, please visit the Seeley home page at:
http://www.mhhe.com/biosci/ap/seeleyap/

Chapter Ten

Muscular System: Histology and Physiology

Objectives

1. List the major categories of muscles, and describe their general characteristics.

2. Describe the structure of a muscle, including its connective tissue elements, blood vessels, and nerves.

3. Diagram the arrangement of myofilaments, myofibrils, sarcomeres, sarcoplasmic reticulum, and T tubules in a muscle fiber.

4. Explain the events responsible for the transmission of an action potential across the neuromuscular junction.

5. Describe the events that result in muscle contraction and relaxation in response to an action potential in a motor neuron.

6. Explain how muscle tone is maintained and how slow contraction and relaxation occur in skeletal muscle.

7. Describe how the length of a muscle influences the force of contraction.

8. Compare the mechanisms involved in psychologic fatigue, muscular fatigue, and synaptic fatigue.

9. Explain the causes of physiologic contracture and rigor mortis.

10. Describe the events that lead to an oxygen debt and recovery from it.

11. Distinguish between fast-twitch muscles and slow-twitch muscles, and explain the functions for which each type is best adapted.

12. Predict the effects of both aerobic exercise and anaerobic exercise on the structure and function of skeletal muscle.

13. Explain the events responsible for the generation of heat produced by muscle before, during, and after exercise and when shivering.

14. List the types of smooth muscle, and describe the characteristics of each.

15. Describe the relationship between the resting membrane potential, action potentials, and contraction in smooth muscle.

16. Compare the unique structural and functional characteristics of smooth muscle with those of skeletal muscle.

17. Compare the structural and functional characteristics of cardiac muscle with those of skeletal muscle.

Part Two

Muscle cells function like tiny motors to produce the forces responsible for the movement of the arms, legs, heart, and other parts of the body. The fuel for these tiny motors is extracted from nutrient molecules taken into the cells, and the nervous system coordinates their activities to produce muscle tone and coordinated movements of the body. Without muscle cells, we could not stand, walk, run, talk, or carry out most of the other tasks we do each day.

Muscle tissue is highly specialized to contract or shorten forcefully. Except for movements produced by cilia, flagella, and the effects of gravity, muscle tissue is responsible for the mechanical processes in the body. There are three types of muscle tissue: skeletal, smooth, and cardiac. Skeletal muscle moves the trunk and appendages, smooth muscle

moves food through the digestive system, and glandular secretions through ducts, and cardiac muscle moves blood through vessels. In addition, metabolism that occurs in the large mass of muscle tissue in the body produces the heat essential to maintain normal body temperature.

This chapter presents the basic structural and functional characteristics of muscle tissue. The interactions between structure and function are emphasized to illustrate the mechanical properties of muscle, its energy requirements, and the relationship of these processes to the chemical and electrical events within muscle cells. Because skeletal muscle is more abundant than other types of muscle in the body and because more is known about it, skeletal muscle is examined in the greatest detail.

General Functional Characteristics of Muscle

Muscle has four major functional characteristics: contractility, excitability, extensibility, and elasticity. **Contractility** (kon-trak-til'i-tē) refers to the capacity of muscle to contract or shorten forcefully. **Excitability** (ek-sī'tă-bil'i-tē) means that muscle responds to stimulation by nerves and hormones, making it possible for the nervous system and, in some muscle types, the endocrine system, to regulate muscle activity. **Extensibility** (eks-ten'sĭ-bil'i-tē) means that muscles can be stretched to their normal resting length and beyond to a limited degree, and **elasticity** (e-las-tis'i-tē) means that if muscles are stretched, they recoil to their original resting length.

During contraction, muscles shorten forcefully but lengthen passively. When they contract, muscles move the structures to which they are attached, but the opposite movement requires an antagonistic force such as that produced by another muscle, gravity, or the force of fluid filling a hollow organ.

The major characteristics of skeletal, smooth, and cardiac muscle are compared in table 10.1. Skeletal muscle with its associated connective tissue constitutes about 40% of the body's weight and is responsible for locomotion, facial expressions, posture, respiratory movements, and many other body movements. Its function, to a large degree, is under voluntary, or conscious, control by the nervous system. Smooth muscle is the most variable type of muscle in the body with respect to distribution and function. It is in the

Table 10.1	Comparison of Muscle Types		
Features	Skeletal Muscle	Smooth Muscle	Cardiac Muscle
Location	Attached to bones	Walls of hollow organs, blood vessels, eyes, glands, and skin	Heart
Cell shape	Very long and cylindrical (1 mm– 4 cm in length and may extend the entire length of short muscles; 10–100 μm in diameter)	Spindle-shaped (15–200 μm in length and 5–8 μm in diameter)	Cylindrical and branched (100–500 μm in length; 12–20 μm in diameter)
Nucleus	Multiple, peripherally located	Single, centrally located	Single, centrally located
Special cell-cell attachments	None	Gap junctions join some visceral smooth muscle cells together	Intercalated disks join cells to one another
Striations	Yes	No	Yes
Control	Voluntary and involuntary (reflexes)	Involuntary	Involuntary
Capable of spontaneous contraction	No	Yes (some smooth muscle)	Yes
Function	Body movement	Food movement through the digestive tract, emptying of the urinary bladder, regulation of blood vessel diameter, change in pupil size, contraction of many gland ducts, movement of hair, and many other functions	Pumps blood; contractions provide the major force for propelling blood through blood vessels

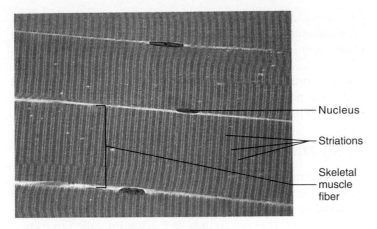

Figure 10.1   Skeletal Muscle Fibers

Skeletal muscle fibers in longitudinal section.

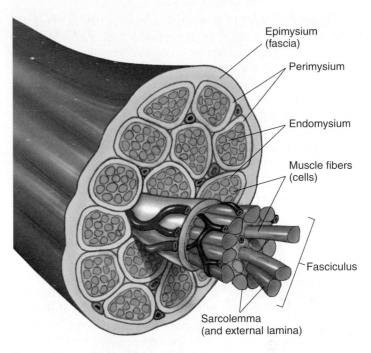

Figure 10.2   Skeletal Muscle Structure

Relationship between muscle fibers, fasciculi, and associated connective tissue layers; the epimysium, perimysium, and endomysium.

walls of hollow organs and tubes, the internal muscles of the eye, the walls of blood vessels, and other areas. Smooth muscle performs a variety of functions, including propelling urine through the urinary tract, mixing food in the stomach and intestine, dilating and constricting the pupils, and regulating the flow of blood through blood vessels. Cardiac muscle is found only in the heart, and its contractions provide the major force for moving blood through the circulatory system. Unlike skeletal muscle, cardiac muscle and many smooth muscles are autorhythmic, that is, they contract spontaneously at somewhat regular intervals, and nervous or hormonal stimulation is not always required for them to contract. Furthermore, smooth muscle and cardiac muscle are not under direct conscious control but instead are innervated and, in part, regulated unconsciously, or involuntarily, by the autonomic nervous system and the endocrine system (see chapters 16 and 18).

Skeletal Muscle: Structure

Skeletal muscles are composed of skeletal **muscle fibers** associated with smaller amounts of connective tissue, blood vessels, and nerves. Skeletal muscle fibers are skeletal muscle cells. Each skeletal muscle fiber is a single cylindrical cell containing several nuclei located around the periphery of the fiber near the plasma membrane (figure 10.1). The muscle fibers develop from less mature multinucleated cells called **myoblasts** (mī′ō-blasts). Their multiple nuclei result from the fusion of myoblast precursor cells and not from the division of nuclei within myoblasts. Myoblasts are converted to muscle fibers as contractile proteins accumulate within their cytoplasm. Shortly after the myoblasts form, nerves grow into the area and innervate the developing muscle fibers.

The number of skeletal muscle fibers remains relatively constant after birth. Enlargement, or hypertrophy, of muscles after birth therefore results not from a significant increase in the number of muscle fibers but from an increase in their size. Similarly, hypertrophy of muscles in response to exercise is due mainly to an increase in muscle fiber size rather than number.

As seen in longitudinal section, alternating light and dark bands give the muscle fiber a **striated** (strī′āt-ĕd, meaning banded), or striped, appearance (see figure 10.1). A single fiber can extend from one end of a small muscle to the other, but several muscle fibers arranged end to end are required to extend the full length of most longer muscles. Muscle fibers range from approximately 1 mm to about 4 cm in length and from 10–100 μm in diameter. Large muscles contain large-diameter fibers, whereas small, delicate muscles have small-diameter fibers. All the muscle fibers in a given muscle have similar dimensions.

Connective Tissue

Surrounding each muscle fiber is a delicate **external lamina** (lam′i-nă) composed primarily of reticular fibers. This external lamina is produced by the muscle fiber and, when observed through the light microscope, cannot be distinguished from the cell membrane of the muscle fiber, the **sarcolemma** (sar′kō-lem′ă; figure 10.2). The prefix *sarco-* refers to muscle and is used to rename some structures found in muscle cells. The **endomysium** (en′dō-miz′ē-ŭm; en′dō-mis′ē-ŭm), a delicate network of loose connective tissue with numerous reticular fibers, surrounds each muscle fiber outside the external lamina. A bundle of muscle fibers with their endomysium is surrounded by another, heavier connective tissue layer called the **perimysium** (per′ĭ-mis′ē-ŭm; per′ĭ-miz′ē-ŭm). Each bundle ensheathed by perimysium is a muscle **fasciculus** (fă-sik′yū-lus). A muscle consists of many fasciculi grouped together and surrounded by a third and heavier layer, the **epimysium** (ep-ĭ-mis′ē-ŭm), which is composed of dense, collagenous connective tissue and covers the entire surface of the muscle.

Fascia (fash′ē-ă) is connective tissue that covers the body by forming a sheet of tissue under the skin; it also surrounds individual muscles or groups of muscles. The fascia around an individual muscle is also called epimysium. The connective tissue components of muscles are continuous with one another. At the end of muscles, the connective tissue components of muscle are continuous with the connective tissue of tendons or the periosteum of bone (see chapter 6). The connective tissue of muscle holds the muscle cells together and attaches muscles to tendons or bones.

Muscle Fibers

The many nuclei of each muscle fiber lie just inside the sarcolemma, whereas most of the interior of the fiber is filled with myofibrils. Other organelles, such as the numerous mitochondria and glycogen granules, are packed between the myofibrils. The cytoplasm without the myofibrils is called **sarcoplasm** (sar′kō-plazm). Each **myofibril** (mĭ′-ō-fī′bril) is a threadlike structure approximately 1–3 μm in diameter that extends from one end of the muscle fiber to the other (figure 10.3). Myofibrils are composed of two kinds of protein filaments called **myofilaments** (mī-ō-fil′ă-mentz). **Actin** (ak′-tin) **myofilaments,** or thin myofilaments, are approximately 8 nanometers (nm) in diameter and 1000 nm in length, whereas **myosin** (mī′ō-sin) **myofilaments,** or thick myofilaments, are approximately 12 nm in diameter and 1800 nm in length.

Sarcomeres

The actin and myosin myofilaments are organized in highly ordered units called **sarcomeres** (sar′kō-mērz), which are joined end to end to form the myofibrils (see figure 10.3; figure 10.4).

Each sarcomere extends from one Z disk to an adjacent Z disk. A **Z disk** is a filamentous network of protein forming a disklike structure for the attachment of actin myofilaments (figure 10.5). The arrangement of the actin myofilaments and myosin myofilaments gives the myofibril a banded, or striated, appearance when viewed longitudinally. Each **isotropic** (ī-sō-trop′ik) (light band), or **I, band** includes a Z disk and extends from either side of the Z disk to the ends of the myosin myofilaments. When seen in longitudinal and cross sections, the I band on either side of the Z disk consists only of actin myofilaments. Each **anisotropic** (an-ī-sō-trop′ik) (dark band), or **A, band** extends the length of the myosin myofilaments within a sarcomere. The actin and myosin myofilaments overlap for some distance at both ends of the A band. In a cross section of the A band in the area where actin and myosin myofilaments overlap, each myosin myofilament is visibly surrounded by six actin myofilaments. In the center of each A band is a smaller band called the **H zone,** where the actin and myosin myofilaments do not overlap and only myosin myofilaments are present. A dark band called the **M line** is in the middle of the H zone and consists of delicate filaments that attach to the center of the myosin myofilaments. The M line helps to hold the myosin myofilaments in place similar to the way the Z disk holds actin myofilaments in place. Several less visible structural

protein molecules also function to hold the actin and myosin myofilaments in place and are responsible for a large part of the elastic properties of skeletal muscle fibers. One of these proteins extends from the Z disk to the M line. The numerous myofibrils are oriented within each muscle fiber so that A bands and I bands of parallel myofibrils are aligned and thus produce the striated pattern seen through the microscope.

Actin and Myosin Myofilaments

Each actin myofilament is composed of two strands of **fibrous actin (F actin),** a series of **tropomyosin** (trō-pō-mī′ō-sin) **molecules,** and a series of **troponin molecules** (trō′pō-nin) (figure 10.6a). The two strands of F actin are coiled to form a double helix that extends the length of the actin myofilament. Each F actin strand is a polymer of approximately 200 small globular units called **globular actin (G actin)** monomers. Each G actin monomer has an active site to which myosin molecules can bind during muscle contraction. Tropomyosin is an elongated protein that winds along the groove of the F actin double helix. Each tropomyosin molecule is sufficiently long to cover seven G actin active sites. Troponin is composed of three subunits: one that binds to actin; the second, which binds to tropomyosin; and the third, which binds to calcium ions. The troponin molecules are spaced between the ends of the tropomyosin molecules in the groove between the F actin strands. The complex of tropomyosin and troponin regulates the interaction between active sites on G actin and myosin.

Myosin myofilaments are composed of many elongated **myosin molecules,** which are shaped like golf clubs. Each myosin molecule consists of two **heavy myosin molecules,** which are wound together to form a rodlike portion lying parallel to the myosin myofilament, and two heads which extend laterally (figure 10.6b). Four molecules of light myosin are attached to the heads of each myosin molecule (see figure 10.6b). Each myosin myofilament consists of about 300 myosin molecules arranged so that about 150 myosin molecules have their heads projecting toward each end. The centers of the myosin myofilaments consist of only the rodlike portions of the myosin molecules and cannot form cross-bridges (see figures 10.5 and 10.6c).

The heads of each myosin molecule are attached to the rodlike portion of the myosin molecule by a hingelike area that can bend and straighten during contraction. The heads of myosin molecules have ATPase activity, the enzymatic activity that breaks down adenosine triphosphate (ATP), releasing energy. The heads of myosin molecules also have proteins that bind the heads of the myosin molecules to active sites on the actin molecules. The combination of myosin heads with the active sites of actin molecules is called a **cross-bridge.**

T Tubules and Sarcoplasmic Reticulum

The sarcolemma has along its surface many tubelike invaginations called **transverse,** or **T, tubules,** which are regularly arranged and project into the muscle fiber and wrap around sarcomeres in the region where actin myofilaments and

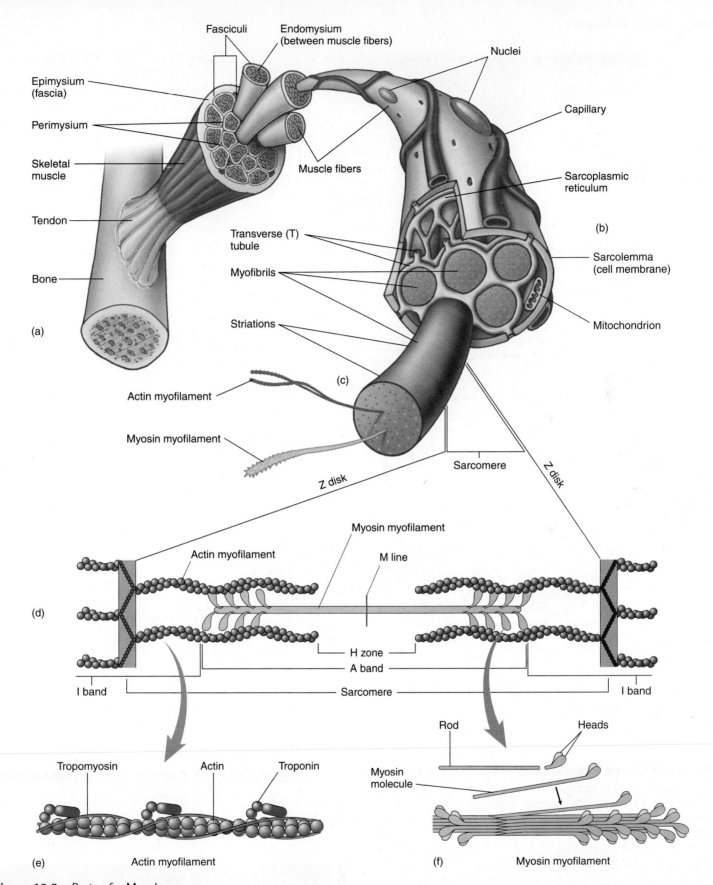

Figure 10.3 Parts of a Muscle

(a) Part of a muscle attached by a tendon to a bone. A muscle is composed of muscle fasciculi, each surrounded by perimysium. The fasciculi are composed of bundles of individual muscle fibers (muscle cells), each surrounded by endomysium. (b) Enlargement of one muscle fiber. The muscle fiber contains several myofibrils. (c) A myofibril extended out the end of the muscle fiber. The banding patterns of the sarcomeres are shown in the myofibril. (d) A single sarcomere forms a myofibril, which is composed of actin myofilaments and myosin myofilaments. The Z disk anchors the actin myofilaments, and the M line anchors the myosin myofilaments. The A band is the length of the myosin myofilaments, the H zone is the area where only myosin myofilaments occur, with no overlapping actin myofilaments, and the I band is the area where only actin myofilaments occur with no overlapping myosin myofilaments. (e) Actin myofilaments are made up of actin, troponin, and tropomyosin molecules. (f) Myosin myofilaments are made up of myosin molecules.

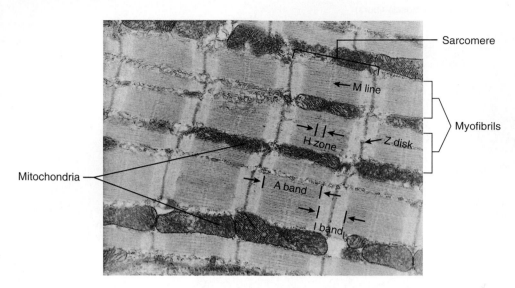

Figure 10.4   Electron Micrograph of a Skeletal Muscle Fiber

Electron micrograph of longitudinal section of a skeletal muscle fiber showing several sarcomeres, with A bands, I bands, Z disks, H zones, and M lines.

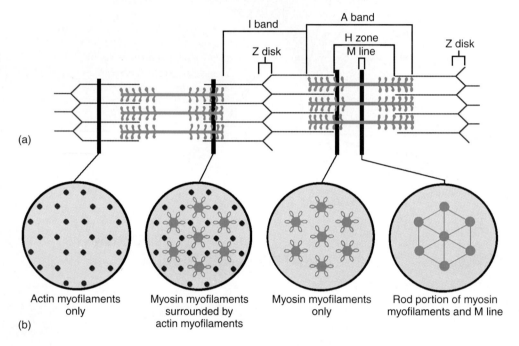

Figure 10.5   Components of a Sarcomere

(*a*) The arrangement of I and A bands, H zones, Z disks, and M lines in sarcomeres. (*b*) Cross sections through regions of the sarcomeres are indicated by black bars.

myosin myofilaments overlap (figure 10.7). The lumen of each T tubule is filled with extracellular fluid and is continuous with the exterior of the muscle fiber. Suspended in the sarcoplasm between the T tubules is a highly specialized, smooth endoplasmic reticulum called the **sarcoplasmic reticulum** (sar′kō-plaz′mik re-tik′yū-lŭm). Near the T tubules, the sarcoplasmic reticulum is enlarged to form **terminal cisternae** (sis-ter′nē). A T tubule and the two adjacent terminal cisternae, together are called a **triad** (trī′ad). The sarcoplasmic reticulum membrane actively transports calcium ions from the sarcoplasm into its lumen, where calcium ions are stored in very high concentrations compared with the sarcoplasm.

▣ Sliding Filament Model

The **sliding filament model** of muscle contraction includes all of the events that result in actin myofilaments sliding over

myosin myofilaments to shorten the sarcomeres of muscle fibers. Actin and myosin myofilaments do not change length during contraction of skeletal muscle. Instead, the actin and myosin myofilaments slide past one another in a way that causes the sarcomeres to shorten (figure 10.8).

Muscle relaxation is a passive process. During relaxation, the cross-bridges are released, and the sarcomeres can lengthen. For this to happen, some force must be applied to a muscle to cause the sarcomeres to lengthen, such as forces produced by other muscles or by gravity.

1   P R E D I C T

Explain the events that influence the width of each band of a sarcomere when a muscle goes through the sequence of being stretched, contracting, and relaxing.

✔ *Answer in Appendix F*

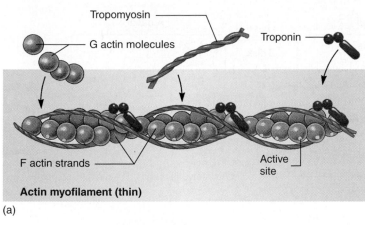

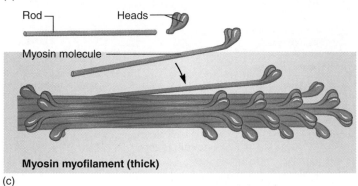

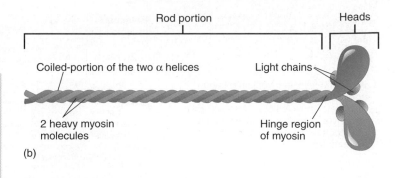

Figure 10.6   Structure of Actin and Myosin

(*a*) Actin myofilaments are composed of individual globular actin (G actin) molecules (*purple spheres*), filamentous tropomyosin molecules (*blue strands*), and globular troponin molecules (*red spheres*) assembled into a single filament. (*b*) A myosin molecule (*green*) is a golf-club-shaped structure composed of two molecules of heavy myosin wound together to form the rod portion and a double globular head. Four smaller light myosin molecules are located on the heads of the myosin molecule. (*c*) Myosin myofilaments are composed of many individual golf-club-shaped myosin molecules. The rod portions are in a parallel arrangement, with all the heads pointing in the same direction at one end and in the opposite direction at the other end of the myosin myofilament.

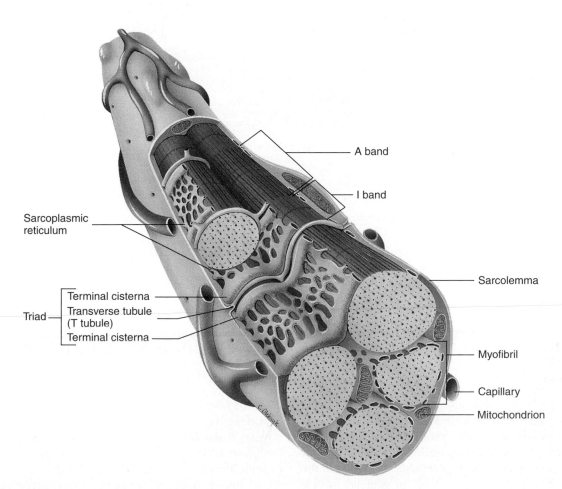

Figure 10.7   T Tubules and Sarcoplasmic Reticulum

A T tubule and the sarcoplasmic reticulum on either side of the T tubule (triad).

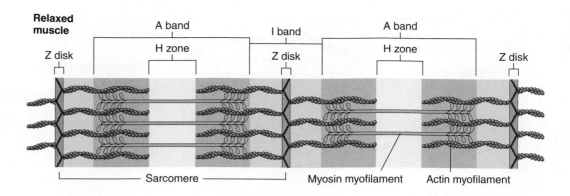

Relaxed muscle

1. Actin and myosin myofilaments do not change length during contraction of skeletal muscle fibers. Instead, actin and myosin myofilaments slide past one another in a way that causes the sarcomeres to shorten.

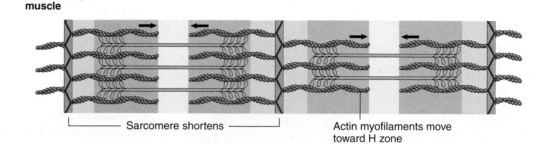

Contracting muscle

2. During contraction, actin myofilaments at each end of the sarcomere slide past the myosin myofilaments toward the H zone, and the I bands (blue) shorten, but the A bands (pink) do not. The H zone (yellow) narrows or even disappears as the actin myofilaments meet at the center of the sarcomere.

Fully contracted muscle

3. As the actin myofilaments slide over the myosin myofilaments, the Z disks are brought closer together, and the sarcomere is shortened.

A band: pink
I band: blue
H zone: yellow

Figure 10.8   Sarcomere Shortening

Note that the I bands (*blue*) shorten, but the A bands (*pink*) do not. The H zone (*yellow*) narrows or even disappears as the actin myofilaments meet at the center of the sarcomere.

Physiology of Skeletal Muscle Fibers

Skeletal muscle contracts in response to electric signals, called action potentials, that are conducted along axons of nerve cells to the synapses between the axons and muscle fibers. Nerve cells regulate the function of skeletal muscle fibers by controlling the frequency of action potentials produced in the muscle cell membrane. Action potentials in skeletal muscle fibers trigger a series of chemical events that result in muscle contraction.

Neuromuscular Junction

Motor neurons are specialized nerve cells with axons that propagate action potentials from the brainstem and spinal cord to skeletal muscle fibers at a relatively high velocity. The connective tissue of muscle provides a passageway for blood vessels and motor neurons to reach the individual muscle cells (figure 10.9). When the axons reach the level of the perimysium, they branch repeatedly, each branch projecting toward one muscle fiber and forming a **neuromuscular junction,** or **synapse** (sin′aps), near the center of the muscle fiber (see figure 10.9; figure 10.10). Thus, each muscle fiber receives a

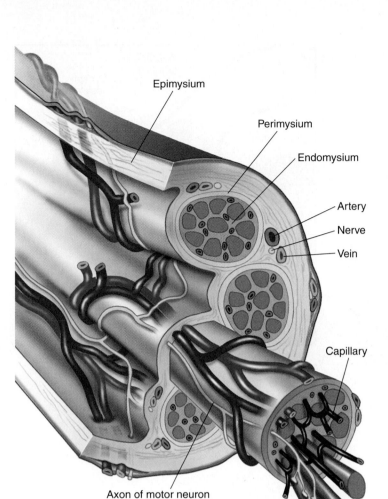

Figure 10.9   Innervation and Blood Supply of a Muscle

Arteries, veins, and nerves course together through the connective tissue of muscles. They branch frequently as they approach individual muscle fibers. 🏃

branch of an axon, and each axon innervates more than one muscle fiber.

Near the muscle fiber it innervates, each axon branch forms a cluster of enlarged axon terminals that rest in an invagination of the sarcolemma. The neuromuscular junction consists of the axon terminals and the area of the muscle fiber they innervate. Each axon terminal is the **presynaptic** (prē′si-nap′tik) **terminal,** the space between an axon terminal and the muscle fiber is the **synaptic** (si-nap′tik) **cleft,** and the muscle cell membrane in the area of the junction is the **postsynaptic** (pōst-si-nap′tik) **membrane,** or **motor end-plate** (see figure 10.10).

Each presynaptic terminal contains many small, spherical sacs approximately 45 μm in diameter, called **synaptic vesicles,** and numerous mitochondria. The vesicles contain **acetylcholine** (as′e-til-kō′lēn)—an organic molecule composed of acetic acid and choline, which functions as a neuro-

transmitter. A **neurotransmitter** (nūr′ō-trans-mit′er) is a substance released from a presynaptic membrane that diffuses across the synaptic cleft and stimulates (or inhibits) the production of an action potential in the postsynaptic membrane.

When an action potential reaches the presynaptic terminal, it causes voltage-gated calcium (Ca^{2+}) channels in the cell membrane of the axon to open, and as a result Ca^{2+} ions diffuse into the cell (figure 10.11a). Once inside the cell, the Ca^{2+} ions cause the contents of a few synaptic vesicles to be secreted by exocytosis from the presynaptic terminal into the synaptic cleft. The acetylcholine molecules released from the synaptic vesicles then diffuse across the cleft and bind to receptor molecules located within the postsynaptic membrane. This causes ligand-gated sodium channels to open, increasing the permeability of the membrane to sodium (Na^+) ions. Na^+ ions then diffuse into the cell, producing a local potential. In skeletal muscle, the local potential caused by each action potential in the motor neuron exceeds threshold, and an action potential in the muscle fiber is produced.

---

2   P R E D I C T

Predict the consequence if presynaptic action potentials in an axon could not release sufficient acetylcholine to cause depolarization to threshold in a skeletal muscle fiber.

✔ *Answer in Appendix F*

Acetylcholine released into the synaptic cleft is rapidly broken down to acetic acid and choline by the enzyme **acetylcholinesterase** (as′e-til-kō-lin-es′ter-ās; figure 10.11b). Acetylcholinesterase keeps acetylcholine from accumulating within the synaptic cleft, where it would act as a constant stimulus at the postsynaptic terminal. The release of acetylcholine and its rapid degradation in the synaptic cleft ensures that one presynaptic action potential yields only one postsynaptic action potential. Choline molecules are actively reabsorbed by the presynaptic terminal and then combined with the acetic acid produced within the cell to form acetylcholine. Recycling choline molecules requires less energy and is more rapid than completely synthesizing new acetylcholine molecules each time they are released from the presynaptic terminal. Acetic acid is an intermediate in the process of glucose metabolism (see chapter 25) and can be taken up and used by a variety of cells after it diffuses away from the area of the neuromuscular junction.

Excitation–Contraction Coupling

When an action potential is produced in a skeletal muscle fiber, it leads to contraction of the muscle fiber. The mechanism by which an action potential causes contraction of a muscle fiber is called **excitation–contraction coupling.** The action potential is propagated along the sarcolemma of the muscle fiber. When the action potential reaches the T tubules, the membranes of the T tubules undergo depolarization, because the T tubules are invaginations of the sarcolemma. The

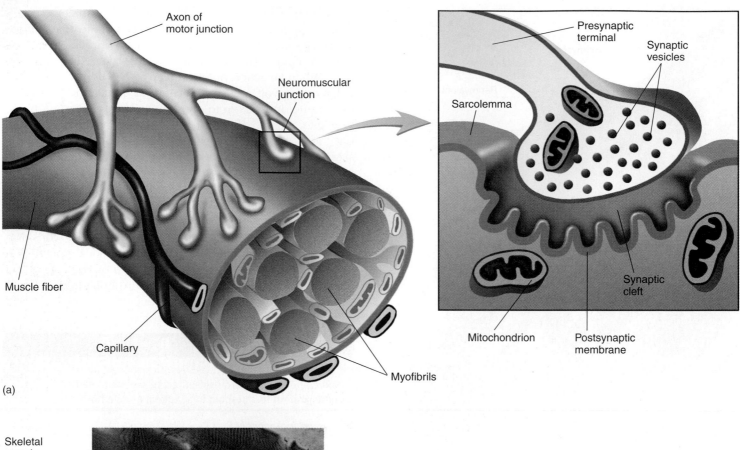

Axon of motor junction

Neuromuscular junction

Muscle fiber

Capillary

Myofibrils

(a)

Presynaptic terminal

Synaptic vesicles

Sarcolemma

Synaptic cleft

Mitochondrion

Postsynaptic membrane

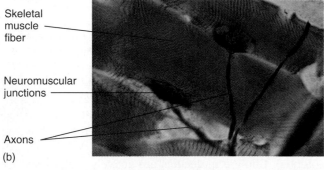

Skeletal muscle fiber

Neuromuscular junctions

Axons

(b)

Figure 10.10   Neuromuscular Junction

(*a*) Diagram showing the neuromuscular junction. Several branches of an axon form the neuromuscular junction with a single muscle fiber. (*b*) Photomicrograph of a neuromuscular junction.

T tubules carry the depolarizations into the interior of the muscle fiber, and the depolarizations of the T tubules of the triads cause voltage-gated Ca^{2+} ion channels in the sarcoplasmic reticulum to open. The sarcoplasmic reticulum actively transports Ca^{2+} ions into its lumen; thus the concentration of Ca^{2+} ions is approximately 2000 times higher within the sarcoplasmic reticulum than in the sarcoplasm of a resting muscle. When the voltage-gated Ca^{2+} channels of the sarcoplasmic reticulum open, Ca^{2+} ions rapidly diffuse the short distance from the sarcoplasmic reticulum into the sarcoplasm surrounding the myofibrils (figure 10.12).

Clinical Note

Anything that affects the production, release, and degradation of acetylcholine or its ability to bind to its receptor molecule also affects the transmission of action potentials across the neuromuscular junction. For example, some insecticides contain

organophosphates that bind to and inhibit the function of acetylcholinesterase. As a result, acetylcholine is not degraded and accumulates in the synaptic cleft, where it acts as a constant stimulus to the muscle fiber. Insects exposed to the insecticide die, partly because their muscles contract and cannot relax—a condition called **spastic paralysis** (spas′tik pă-ral′i-sis), which is followed by fatigue of the muscles. In humans a similar response is seen. The skeletal muscles responsible for respiration cannot undergo their normal cycle of contraction and relaxation. Instead they remain in a state of spastic paralysis until they become fatigued. Victims die of respiratory failure. Other organic poisons such as curare bind to the acetylcholine receptors, preventing acetylcholine from binding to them. Curare does not allow activation of the receptors, and therefore the muscle is incapable of contracting in response to nervous stimulation—a condition called **flaccid** (flas′id) **paralysis.** Curare is not a poison to which people are commonly exposed, but it has been used to investigate the role of acetylcholine in the neuromuscular synapse and is sometimes used in small doses to relax muscles during certain kinds of surgery. **Myasthenia gravis**

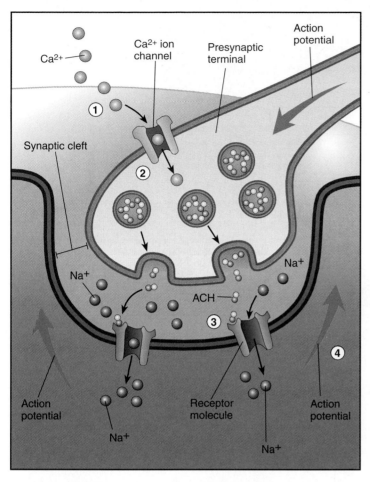

1. An action potential arrives at the presynaptic terminal causing voltage-gated Ca²⁺ ion channels to open, increasing the Ca²⁺ ion permeability of the presynaptic terminal.

2. Ca²⁺ ions enter the presynaptic terminal and initiate the release of a neurotransmitter, acetylcholine (ACH), from synaptic vesicles in the presynaptic terminal.

3. Diffusion of ACH across the synaptic cleft and binding of ACH to ACH receptors on the postsynaptic muscle fiber membrane causes an increase in the permeability of ligand-gated Na⁺ ion channels.

4. The increase in Na⁺ ion permeability results in depolarization of the postsynaptic membrane; once threshold has been reached a postsynaptic action potential results.

(a)

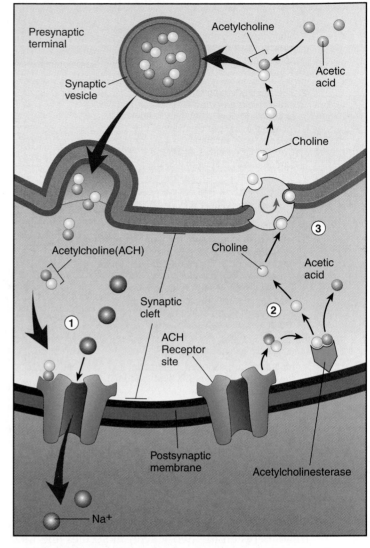

1. Once acetylcholine is released into the synaptic cleft it binds to the receptors for acetylcholine on the postsynaptic membrane and causes Na⁺ ion channels to open.

2. Acetylcholine is rapidly broken down in the synaptic cleft by acetylcholinesterase to acetic acid and choline.

3. The choline is reabsorbed by the presynaptic terminal and combined with acetic acid to form more acetylcholine, which enters synaptic vesicles. Acetic acid is taken up by many cell types.

(b)

Figure 10.11   Function of the Neuromuscular Junction

(a) Release of acetylcholine in response to an action potential at the neuromuscular junction. (b) Breakdown of acetylcholine in the neuromuscular junction.

(mī-as-thē′nē-a grăv′is) results from the production of antibodies that bind to acetylcholine receptors, eventually causing the destruction of the receptor and thus reducing the number of receptors. As a consequence, muscles exhibit a degree of flaccid paralysis or are extremely weak. A class of drugs that includes neostigmine partially blocks the action of acetylcholinesterase and sometimes is used to treat myasthenia gravis. The drugs cause acetylcholine levels to increase in the synaptic cleft and combine more effectively with the remaining acetylcholine receptor sites.

3   P R E D I C T

Predict the specific cause of death resulting from a lethal dose of (a) organophosphate poison or (b) curare.

✔ *Answer in Appendix F*

The Ca²⁺ ions bind to troponin of the actin myofilaments (see figure 10.12). The combination of Ca²⁺ ions with troponin causes the troponin–tropomyosin complex to move deeper

1. An action potential is propagated along the sarcolemma of the skeletal muscle, causing a depolarization to spread along the membrane of the T tubules.

2. The depolarization of the T tubule causes voltage-gated Ca²⁺ ion channels to open, resulting in an increase in the permeability of the sarcoplasmic reticulum to Ca²⁺ ions. Ca²⁺ ions then diffuse from the sarcoplasmic reticulum into the sarcoplasm.

3. Ca²⁺ ions released from the sarcoplasmic reticulum bind to troponin molecules in the actin myofilament. Consequently, the troponin molecules bound to G actin molecules are released. This causes tropomyosin molecules to move exposing active sites on the G actin molecules.

4. Once active sites on G actin molecules are exposed, the heads of the myosin myofilaments bind to them to form cross bridges.

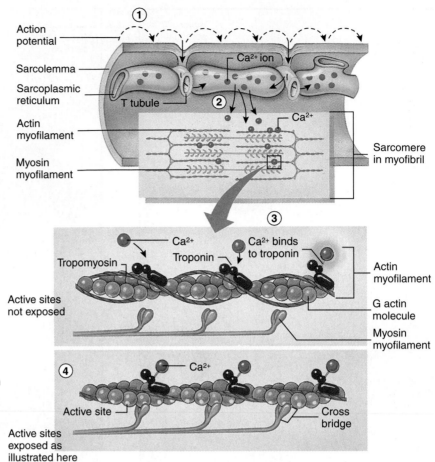

Figure 10.12   Action Potentials and Muscle Contraction

into the groove between the two F actin molecules and thus expose active sites on the actin myofilaments. These exposed active sites bind to the heads of the myosin molecules to form cross-bridges (see figure 10.12). When the heads of the myosin molecules bind to actin, a cycle of events resulting in contraction proceeds very rapidly. The heads of myosin molecules move at their hinged area, forcing the actin myofilament, to which the heads of the myosin molecules are attached, to slide over the surface of the myosin myofilament. After movement, each myosin head releases from the actin and returns to its original position. It can then form another cross-bridge at a different site on the actin myofilament, followed by movement, release of the cross-bridge, and return to its original position. During a single contraction, each myosin molecule undergoes the cycle of cross-bridge formation, movement, release, and return to its original position many times.

Energy Requirements for Muscle Contraction

The energy from one ATP molecule is required for each cycle of cross-bridge formation, movement, and release. After a cross-bridge has formed and movement has occurred, release of the myosin head from actin requires ATP to bind to the head of the myosin molecule. The ATP is broken down to adenosine diphosphate (ADP) and a phosphate molecule by ATPase in the head of the myosin myofilament, and energy is stored in the head of the myosin molecule. Both ATP and phosphate remain bound to the myosin head. As a result of ATP being broken down, the cross-bridge is released, and the myosin head is restored to its original position (figure 10.13). Then the myosin molecule binds to another actin active site to form another cross-bridge, and the phosphate is released from the myosin head. Much of the stored energy is used for cross-bridge formation and movement, and the ADP molecule is then released from the myosin head (see figure 10.13). Before the cross-bridge can be released for another cycle, an ATP molecule must once again bind to the head of the myosin molecule (see figure 10.13).

Movement of the myosin molecule while the cross-bridge is attached is a **power stroke,** whereas return of the myosin head to its original position after cross-bridge release is a **recovery stroke.** Many cycles of power and recovery strokes occur during each muscle contraction. While muscle is relaxed, energy stored in the heads of the myosin molecules is held in reserve until the next contraction. When calcium is released from the sarcoplasmic reticulum in response

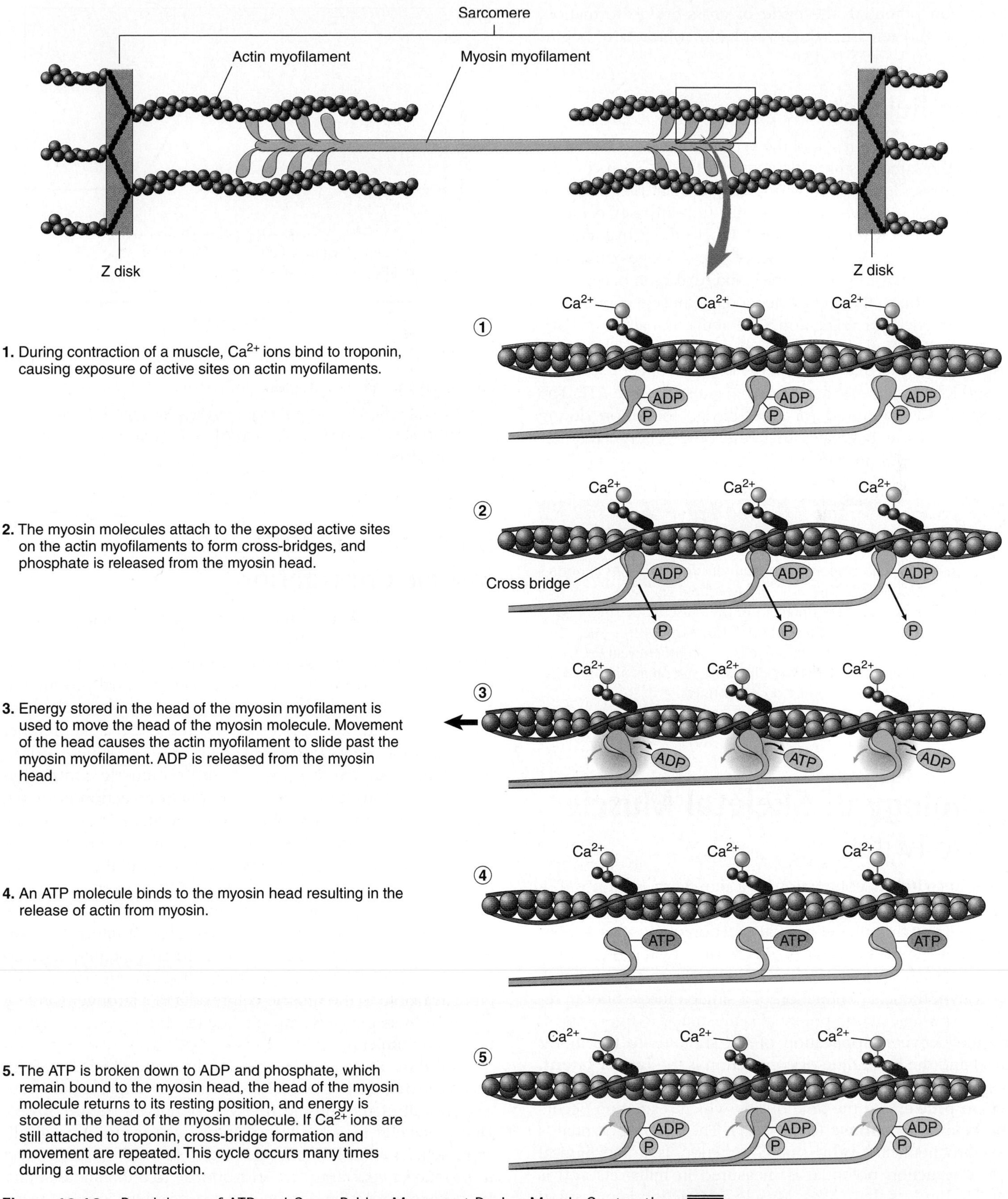
Sarcomere
Actin myofilament
Myosin myofilament
Z disk
Z disk
1. During contraction of a muscle, Ca^{2+} ions bind to troponin, causing exposure of active sites on actin myofilaments.
2. The myosin molecules attach to the exposed active sites on the actin myofilaments to form cross-bridges, and phosphate is released from the myosin head.
3. Energy stored in the head of the myosin myofilament is used to move the head of the myosin molecule. Movement of the head causes the actin myofilament to slide past the myosin myofilament. ADP is released from the myosin head.
4. An ATP molecule binds to the myosin head resulting in the release of actin from myosin.
5. The ATP is broken down to ADP and phosphate, which remain bound to the myosin head, the head of the myosin molecule returns to its resting position, and energy is stored in the head of the myosin molecule. If Ca^{2+} ions are still attached to troponin, cross-bridge formation and movement are repeated. This cycle occurs many times during a muscle contraction.
Ca^{2+}
ADP
P
Cross bridge
ATP

Figure 10.13 Breakdown of ATP and Cross-Bridge Movement During Muscle Contraction

to an action potential, the cycle of cross-bridge formation, movement, and release, which results in contraction, begins (see figures 10.12 and 10.13).

Muscle Relaxation

Relaxation occurs as a result of the active transport of Ca^{2+} ions back into the sarcoplasmic reticulum. As the Ca^{2+} ion concentration decreases in the sarcoplasm, Ca^{2+} ions diffuse away from the troponin molecules. The troponin–tropomyosin complex then reestablishes its position, which blocks the active sites on the actin molecules. As a consequence, cross-bridges cannot reform once they have been released, and relaxation occurs.

In addition to the energy needed for muscle contraction, energy is needed for relaxation. The active transport of Ca^{2+} ions into the sarcoplasmic reticulum requires ATP. The active transport processes that maintain the normal concentrations of Na^+ and K^+ ions across the sarcolemma also require ATP. The amount of ATP required for cross-bridge formation during contraction is much greater than the other energy requirements in a skeletal muscle.

4	P R E D I C T

Predict the consequences of having the following conditions develop in a muscle in response to a stimulus: (a) Na^+ ions cannot enter the skeletal muscle through voltage-gated Na^+ ion channels; (b) very little ATP is present in the muscle fiber before a stimulus is applied; and (c) adequate ATP is present within the muscle fiber, but action potentials occur at a frequency so great that calcium is not transported back into the sarcoplasmic reticulum between individual action potentials.

✔ *Answer in Appendix F*

Physiology of Skeletal Muscle
Muscle Twitch

A **muscle twitch** is contraction of a muscle in response to a stimulus that causes an action potential in one or more muscle fibers. Even though the normal function of muscles is more complex, an understanding of the muscle twitch makes the function of muscles in living organisms easier to comprehend.

A hypothetical contraction of a single muscle fiber in response to a single action potential is illustrated in figure 10.14. The time between application of the stimulus to the motor neuron and the beginning of contraction is the **lag,** or **latent, phase;** the time during which contraction occurs is the **contraction phase;** and the time during which relaxation occurs is the **relaxation phase** (table 10.2). The action potential is an electrochemical event, but contraction is a mechanical event. The action potential is measured in millivolts and is completed in less than 2 ms. Muscle contraction is measured as a force, also called tension, and is reported as the number of grams lifted, or the distance the muscle shortens, and requires up to 1 s to occur.

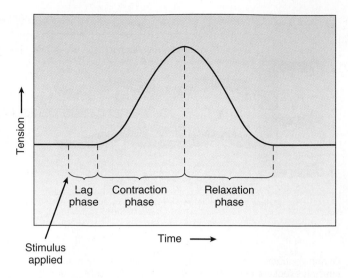

Figure 10.14   Phases of a Muscle Twitch

Hypothetical muscle twitch in a single muscle fiber. There is a short lag phase after stimulus application, followed by a contraction phase and a relaxation phase.

Stimulus Strength and Muscle Contraction

An isolated skeletal muscle fiber produces contractions of equal force in response to each action potential. This is called the **all-or-none law of skeletal muscle contraction** and can be explained on the basis of action potential production in the skeletal muscle fiber. When brief electric stimuli of increasing strength are applied to the muscle cell membrane, the following events occur: (1) a subthreshold stimulus does not produce an action potential, and no muscle contraction occurs; (2) a threshold stimulus produces an action potential and results in contraction of the muscle cell; or (3) a stronger-than-threshold stimulus produces an action potential of the same magnitude as the threshold stimulus and therefore produces an identical contraction. Thus, for a given condition, once an action potential is generated, the skeletal muscle fiber contracts to produce a constant force. If internal conditions change, it is possible for the force of contraction to change as well. For example, increasing the amount of calcium available to the muscle cell results in a stronger force of contraction; conversely, muscle fatigue can result in a weaker force of contraction.

Within a skeletal muscle, skeletal muscle fibers form **motor units,** each of which consists of a single motor neuron and all of the muscle fibers it innervates (figure 10.15). Like individual muscle fibers, motor units respond in an all-or-none fashion. All the muscle fibers of a motor unit contract to produce a constant force in response to a threshold stimulus because an action potential in a motor neuron initiates action potentials in all of the muscle fibers it innervates.

Whole muscles exhibit characteristics that are more complex than those of individual muscle fibers or motor units.

Table 10.2 Events That Occur During Each Phase of a Muscle Twitch*

Lag Phase

An action potential is propagated to the presynaptic terminal of the motor neuron.

The action potential causes the permeability of the presynaptic terminal to increase.

Calcium ions diffuse into the presynaptic terminal, causing acetylcholine contained within several synaptic vesicles to be released by exocytosis into the synaptic cleft.

Acetylcholine released from the presynaptic terminal diffuses across the synaptic cleft and binds to acetylcholine receptor molecules in the postsynaptic membrane of the sarcolemma.

The binding of acetylcholine to its receptor site causes ligand-gated Na^+ ion channels to open, and the postsynaptic membrane becomes more permeable to Na^+ ions.

Na^+ ions diffuse into the muscle fiber, causing a local depolarization that exceeds threshold and produces an action potential.

Acetycholine is rapidly degraded in the synaptic cleft to acetic acid and choline by acetylcholinesterase, thus limiting the length of time acetylcholine is bound to its receptor site. The result is that one presynaptic action potential produces one postsynaptic action potential in each muscle fiber.

The action potential produced in a muscle fiber is propagated from the postsynaptic membrane near the middle of the fiber toward both ends and into the T tubules.

The depolarization that occurs in the T tubule in response to the action potential causes voltage-gated Ca^{2+} channels of the membrane of the sarcoplasmic reticulum to open, and the membrane of the sarcoplasmic reticulum becomes very permeable to Ca^{2+} ions.

Calcium ions diffuse from the sarcoplasmic reticulum into the sarcoplasm.

Calcium ions bind to troponin; the troponin–tropomyosin complex changes its position and exposes the active site on the actin myofilaments.

Contraction Phase

Cross-bridges between actin molecules and myosin molecules form, move, release, and re-form many times, causing the sarcomeres to shorten. Energy stored in the head of the myosin molecule allows cross-bridge formation and movement. After cross-bridge movement has occurred, ATP must bind to the myosin head. The ATP is broken down to ADP, and some of the energy is used to release the cross-bridge and cause the head of the myosin molecule to move back to its resting position, where it is ready to form another cross-bridge. Some of the energy from the ATP is stored in the myosin head and is used for the next cross-bridge formation and movement (see figure 10.13). Energy is also released as heat.

Relaxation Phase

Calcium ions are actively transported into the sarcoplasmic reticulum.

The troponin–tropomyosin complexes inhibit cross-bridge formation.

The muscle fibers lengthen passively.

*Assuming that the process begins with a single action potential in the motor neuron.

A muscle is composed of many motor units, and the axons of the motor units combine to form a nerve. If brief electric stimuli of increasing strength are applied to the nerve, the muscle responds in a graded rather than an all-or-none fashion (figure 10.16). A **subthreshold stimulus** is not strong enough to cause an action potential in any of the motor neuron axons and causes no contraction. As the stimulus increases in strength, however, it eventually becomes a **threshold stimulus.** At threshold, the stimulus is strong enough to produce an action potential in the axon of a single motor neuron, and all of the muscle fibers of that motor unit contract. Progressively stronger stimuli called **submaximal stimuli** activate additional motor units. All of the motor units are activated by a **maximal stimulus,** at which point a greater stimulus strength, a **supramaximal stimulus,** has no additional effect. As the stimulus strength increases between threshold and maximum values, motor units are **recruited,** which means

that the number of motor units responding to the stimulus increases and the force of contraction produced by the muscle increases in a graded fashion. This relationship is called **multiple motor unit summation.** A whole muscle contracts with either a small force or a large force, depending on the number of motor units recruited. Each motor unit, however, responds to every action potential by producing contractions of equal magnitude.

Motor units in different muscles do not always contain the same number of muscle fibers. Muscles performing delicate and precise movements have motor units with a small number of muscle fibers, whereas muscles performing more powerful but less precise contractions have motor units with many muscle fibers. For example, in very delicate muscles, such as those that move the eye, the number of muscle fibers per motor unit can be fewer than 10, whereas in the heavy muscles of the leg the number can be several hundred.

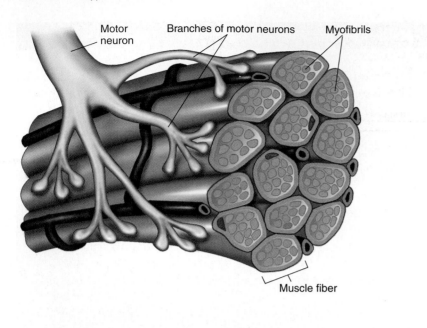

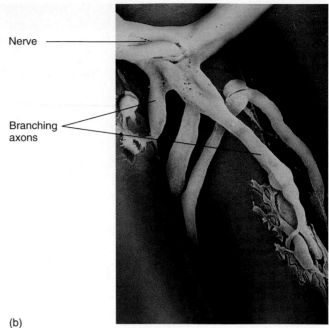

(a)

(b)

Figure 10.15 The Motor Unit

(*a*) A motor unit consists of a single neuron and all the muscle fibers its branches innervate. (*b*) Photomicrograph of a motor unit.

5   **P R E D I C T**

In patients with poliomyelitis (pō'lē-ō-mī'ĕ-lī'tis), motor neurons are destroyed, causing loss of muscle function and even flaccid paralysis. Sometimes recovery occurs because of the formation of axon branches from the remaining motor neurons. These branches innervate the paralyzed muscle fibers to produce motor units with many more muscle fibers than usual, resulting in recovery of muscle function. What effect would this reinnervation of muscle fibers have on the degree of muscle control in a person who has recovered from poliomyelitis?

✔ *Answer in Appendix F*

Stimulus Frequency and Muscle Contraction

A single muscle fiber contracts in response to an action potential. Although an action potential triggers contraction of a muscle fiber, the action potential and its refractory period (see chapter 9) are complete long before the contraction and relaxation phases are complete. In addition, the contractile mechanism in a muscle fiber exhibits no refractory period. Relaxation of a muscle fiber is therefore not required before a second action potential can stimulate a second contraction. As the frequency of action potentials in a skeletal muscle fiber increases, the frequency of contraction also increases (figure 10.17). In **incomplete tetanus** (tet'ă-nŭs) muscle fibers partially relax between contractions, but in **complete tetanus** action potentials occur so rapidly that no muscle relaxation can happen between the action potentials.

Tetanus of a muscle caused by stimuli of increasing frequency can be explained by the effect of the action potentials on Ca^{2+} ion release from the sarcoplasmic reticulum. The first action potential causes Ca^{2+} ion release from the sarcoplasmic reticulum, the Ca^{2+} ions diffuse to the myofibrils, and contraction occurs. Relaxation begins as the Ca^{2+} ions are pumped back into the sarcoplasmic reticulum. If the next action potential occurs before relaxation is complete, however, two things happen. First, because there has not been enough time for all the Ca^{2+} ions to reenter the sarcoplasmic reticulum, Ca^{2+} ion levels around the myofibrils remain elevated. Second, the next action potential causes the release of additional Ca^{2+} ions from the sarcoplasmic reticulum. Thus, the elevated Ca^{2+} ion levels in the sarcoplasm produce continued contraction of the muscle fiber. Action potentials at a high frequency can increase Ca^{2+} ion concentrations in the sarcoplasm to an extent that the muscle fiber is contracted completely and does not relax at all.

The tension produced by a muscle increases as the stimulus frequency increases. The increased tension is called **multiple-wave summation,** which is apparent when a muscle is exhibiting incomplete or complete tetanus.

At least two factors play a role in the increased tension observed during multiple-wave summation. First, as the action potential frequency increases, the concentration of Ca^{2+} ions around the myofibrils becomes greater than during a single muscle twitch, causing a greater degree of contraction. The additional Ca^{2+} ions cause the exposure of additional active sites on the actin myofilaments. Second, the sarcoplasm and the connective tissue components of muscle have some elasticity. During each separate muscle twitch, some tension pro-

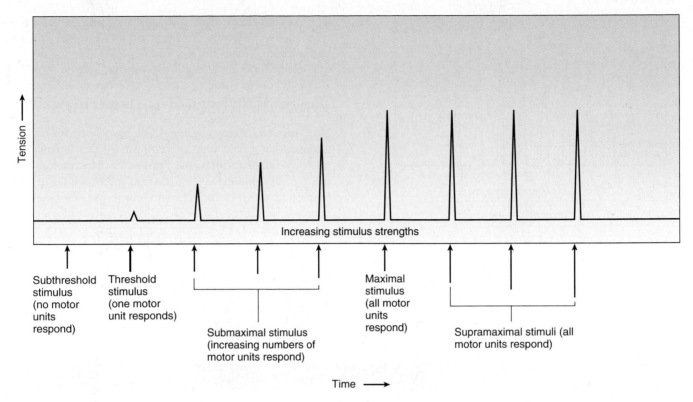

Figure 10.16   Multiple Motor Unit Summation

Multiple motor unit summation occurs as stimuli of increasing strength are applied to a nerve that innervates a muscle. The amount of tension (height of peaks) is influenced by the number of motor units responding.

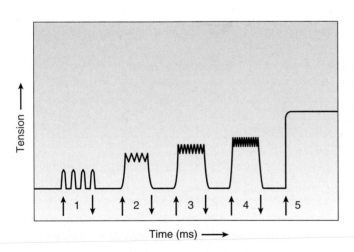

Figure 10.17   Multiple-Wave Summation

Multiple-wave summation caused by stimuli of increased frequency (1–5): complete relaxation between stimuli (1), incomplete tetanus—partial relaxation between stimuli (2–4), and complete tetanus—no relaxation between stimuli (5).

duced by the contracting muscle fibers is used to stretch those elastic elements, and the remaining tension is applied to the load to be lifted. In a single muscle twitch, relaxation begins before the elastic components are totally stretched. The maximum tension produced during a single muscle twitch is therefore not applied to the load to be lifted. In a muscle stimulated

at a high frequency, the elastic elements are stretched during the early part of the prolonged contraction. The stretching allows all of the tension produced by the muscle to be applied to the load to be lifted, and the observed tension produced by the muscle increases.

Another example of a graded response is **treppe** (trep'eh, meaning staircase), which occurs in muscle that has rested for a prolonged period (figure 10.18). If the muscle is stimulated with a maximal stimulus at a low frequency, which allows complete relaxation between the stimuli, the contraction triggered by the second stimulus produces a slightly greater tension than the first. The contraction triggered by the third stimulus produces a contraction with a greater tension than the second. After only a few stimuli, the tension produced by all the contractions is equal.

A possible explanation for treppe is an increase in Ca^{2+} ion levels around the myofibrils. The Ca^{2+} ions released in response to the first stimulus are not taken up completely by the sarcoplasmic reticulum before the second stimulus causes the release of additional Ca^{2+} ions, even though the muscle completely relaxes between the muscle twitches. As a consequence, during the first few contractions of the muscle, the Ca^{2+} ion concentration in the sarcoplasm increases slightly, making contraction more efficient because of the increased number of Ca^{2+} ions available to bind to troponin. Treppe achieved during warm-up exercises can contribute to improved muscle efficiency during athletic events. Factors such

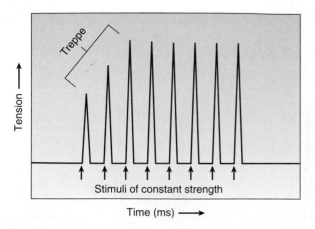

Figure 10.18 Treppe

When a rested muscle is stimulated with maximal stimuli at a frequency that allows complete relaxation between stimuli, the second contraction produces a slightly greater tension than the first, and the third contraction produces greater tension than the second. After a few contractions, the tension produced by all contractions is equal.

as increased blood flow to the muscle and increased muscle temperature probably are involved as well. Increased muscle temperature causes the enzymes responsible for muscle contraction to function at a more rapid rate.

Types of Muscle Contractions

Muscle contractions are classified based on the type of contraction that predominates (table 10.3). In **isometric** (ī-sō-met′rik) **contractions** the length of the muscle does not change, but the amount of tension increases during the contraction process. Isometric contractions are responsible for the constant length of the postural muscles of the body such as muscles that hold the spine erect while a person is sitting or standing. In **isotonic** (ī-sō-ton′ik) **contractions** the amount of tension produced by the muscle is constant during contraction, but the length of the muscle changes. Movements of the arms or fingers are predominantly isotonic contractions. Examples include waving or using a computer keyboard. Most muscle contractions are not strictly isometric or isotonic con-

Table 10.3 Types of Muscle Contractions	
Contraction Types	**Characteristics**
Multiple motor unit summation	Each motor unit responds in an all-or-none fashion.
	A whole muscle is capable of producing an increasing amount of tension as the number of motor units stimulated increases.
Multiple-wave summation	Summation results when many action potentials are produced in a muscle fiber.
	Contraction occurs in response to the first action potential, but there is not enough time for relaxation to occur between action potentials.
	Because each action potential causes the release of calcium ions from the sarcoplasmic reticulum, calcium ions remain elevated in the sarcoplasm to produce a tetanic contraction.
	The tension produced as a result of multiple-wave summation is greater than the tension produced by a single muscle twitch. The increased tension results from the greater concentration of calcium in the sarcoplasm and the stretch of the elastic components of the muscle early in contraction.
Tetanus of muscles	Tetanus of muscles results from multiple-wave summation.
	Incomplete tetanus occurs when the action potential frequency is low enough to allow partial relaxation of the muscle fibers.
	Complete tetanus occurs when the action potential frequency is high enough that no relaxation of the muscle fibers occurs.
Treppe	The tension produced increases for the first few contractions in response to a maximal stimulus at a low frequency in a muscle that has been at rest for some time.
	The increased tension may result from the accumulation of small amounts of calcium in the sarcoplasm for the first few contractions or from an increasing rate of enzyme activity.
Isotonic contraction	The muscle produces a constant tension during contraction.
	The muscle shortens during contraction.
	This type is characteristic of finger and hand movements.
Isometric contraction	The muscle produces an increasing tension during contraction.
	The length of the muscle remains constant during contraction.
	This type is characteristic of postural muscles that maintain a constant tension without changing their length.
Concentric contractions	The muscle produces increasing tension as the muscle shortens.
Eccentric contractions	The muscle produces tension, but the length of the muscle is increasing.

tractions. For example, both the length and tension of muscles change when a person walks or opens a heavy door. Although there are some mechanical differences, both types of contractions result from the same contractile process within muscle cells.

Concentric contractions (kon-sen′trik) are isotonic contractions in which tension in the muscle is great enough to overcome the opposing resistance, and the muscle shortens. Concentric contractions include contractions that result in an increasing tension as the muscle shortens. A large percentage of the movements performed by muscle contractions are concentric contractions. **Eccentric contractions** (ek-sen′trik) are isotonic contractions in which tension is maintained in a muscle, but the opposing resistance is great enough to cause the muscle to increase in length (see table 10.3). Eccentric contractions are performed when a person lets a heavy weight down slowly. During eccentric contractions substantial force is produced by muscles, and eccentric contractions are of clinical interest because repetitive eccentric contractions, such as seen in the legs of people who run downhill for long distances, tend to injure muscle fibers and the connective tissue of muscles.

Muscle tone refers to the constant tension produced by muscles of the body for long periods of time. Muscle tone is responsible for keeping the back and legs straight, the head upright, and the abdomen flat. Muscle tone depends on a small percentage of all the motor units contracting out of phase with one another at any point in time. The same motor units are not contracting all the time, however. A small percentage of all motor units are stimulated with a frequency of nerve impulses that causes incomplete tetanus for short periods. The motor units that are contracting are stimulated in such a way that the tension produced by the whole muscle remains constant.

---

6   P R E D I C T

Marty Myosin overheard an argument between two students who could not decide if a weight lifter who lifts a weight above his head and then holds it there before lowering it is using isometric, isotonic, concentric, or eccentric muscle contractions. Marty was an expert on muscle contractions, so he settled the debate. What was his explanation?

✔ *Answer in Appendix F*

Movements of the body are usually smooth and occur at widely differing rates—some very slow and others quite rapid. All movements are produced by muscle contractions, but very few of the movements resemble the rapid contractions of individual muscle twitches. Smooth, slow contractions result from an increasing number of motor units contracting out of phase as the muscles shorten, and from a decreasing number of motor units contracting out of phase as muscles lengthen. Each individual motor unit exhibits either incomplete or complete tetanus, but because the contractions are out of phase and because the number of motor units activated varies at

each point in time, a smooth contraction results. Consequently, muscles are capable of contracting either slowly or rapidly, depending on the number of motor units stimulated and the rate at which that number increases or decreases.

Length Versus Tension

Active tension is the force applied to an object to be lifted when a muscle contracts. The initial length of a muscle has a strong influence on the amount of active tension it produces. As the length of a muscle increases, its active tension also increases, to a point. If the muscle is stretched farther than that optimum length, the active tension it produces begins to decline. The muscle length plotted against the tension produced by the muscle in response to maximal stimuli is the **active tension curve** (figure 10.19).

If a muscle is stretched so that the actin and myosin myofilaments within the sarcomeres do not overlap or overlap to a very small extent, the muscle produces very little active tension when it is stimulated. Also, if the muscle is not stretched at all, the myosin myofilaments touch each of the Z disks in each sarcomere, and very little contraction of the sarcomeres can occur. If the muscle is stretched to its optimum length, there is optimal overlap of the actin and myosin myofilaments. When the muscle is stimulated, cross-bridge formation results in maximal contraction.

Passive tension is the tension applied to the load when a muscle is stretched, but not stimulated. It is similar to the tension produced if the muscle is replaced with an elastic band. Passive tension exists because the muscle and its connective tissue have some elasticity. The sum of active and passive tension is **total tension.**

Weight lifters and others who lift heavy objects usually assume positions so that their muscles are stretched close to their optimum length before lifting. For example, the position a weight lifter assumes before power lifting stretches the arm and leg muscles to a near-optimum length for muscle contraction, and the stance a lineman assumes in a football game stretches most muscle groups in the legs so they are near their optimum length for suddenly moving the body forward.

Fatigue

Fatigue (fă-tēg′) is the decreased capacity to do work and the reduced efficiency of performance that normally follows a period of activity. The rate at which individuals develop fatigue is highly variable, but it is a phenomenon that everyone has experienced. Fatigue can develop at three possible sites: the nervous system, the muscles, and the neuromuscular junction.

Psychologic fatigue, the most common type of fatigue, involves the central nervous system. The muscles are capable of functioning, but the individual "perceives" that additional muscular work is not possible. A burst of activity in a tired athlete in response to encouragement from spectators is an illustration of how psychologic fatigue can be overcome. The onset and duration of psychologic fatigue vary greatly and depend on the emotional state of the individual.

There is an optimal muscle length at which the muscle produces a maximal tension in response to a maximal stimulus.

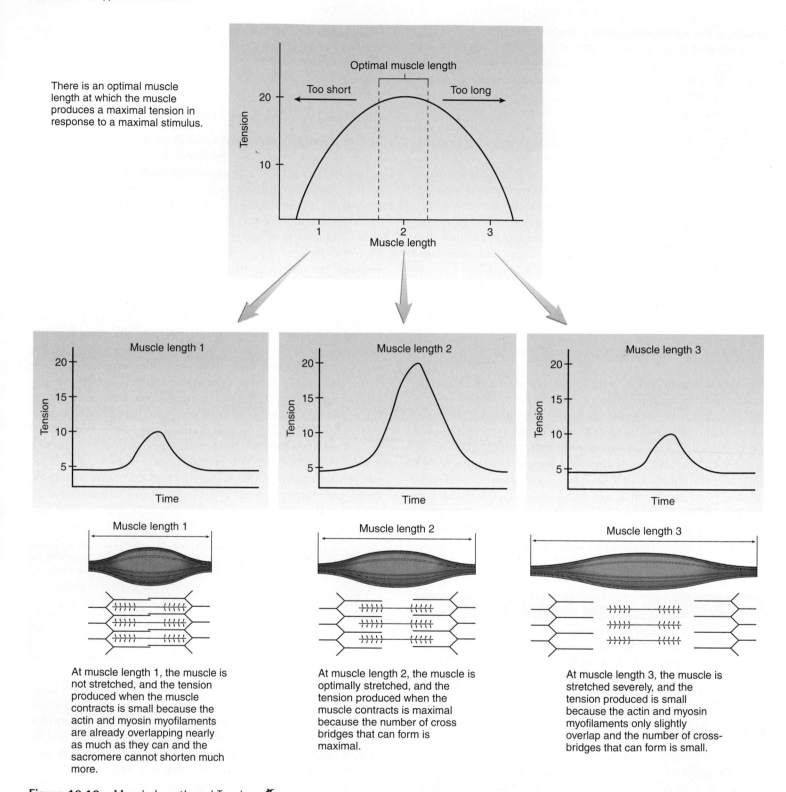

At muscle length 1, the muscle is not stretched, and the tension produced when the muscle contracts is small because the actin and myosin myofilaments are already overlapping nearly as much as they can and the sacromere cannot shorten much more.

At muscle length 2, the muscle is optimally stretched, and the tension produced when the muscle contracts is maximal because the number of cross bridges that can form is maximal.

At muscle length 3, the muscle is stretched severely, and the tension produced is small because the actin and myosin myofilaments only slightly overlap and the number of cross-bridges that can form is small.

Figure 10.19   Muscle Length and Tension

The second most common type of fatigue occurs in the muscle fiber. **Muscular fatigue** results from ATP depletion. Without adequate ATP levels in muscle fibers, cross-bridges cannot function normally. As a consequence, the tension that a muscle is capable of producing declines. Fatigue in the legs of marathon runners or in the arms and legs of swimmers are examples.

The least common type of fatigue, called **synaptic fatigue,** occurs in the neuromuscular junction. If the action potential frequency in motor neurons is great enough, the release of acetylcholine from the presynaptic terminals is greater than the rate of acetylcholine synthesis. As a result, the synaptic vesicles become depleted, and insufficient acetylcholine is released to stimulate the muscle fibers. Under normal physiologic conditions, fatigue of neuromuscular junctions is rare; however, it may occur under conditions of extreme exertion.

Pain frequently develops after 1 or 2 days in muscles that are vigorously exercised, and the pain can last for several days. The pain is more common in untrained people who exercise vigorously. In addition, highly repetitive eccentric contractions of muscles produce pain more readily than concentric contractions. The pain is associated with damage to skeletal muscle fibers and with connective tissue surrounding the skeletal muscle fibers. In people with muscle soreness induced by exercise, enzymes that are normally found inside muscle fibers can be detected in the extracellular fluid. In addition, fragments of collagen molecules can be found in the extracellular fluid of muscles. These observations indicate that injury occurs to both muscle fibers and the connective tissue of muscles. The pain produced appears to be the result of inflammation resulting from damage to the muscle fibers and connective tissue of muscles.

Physiologic Contracture and Rigor Mortis

As a result of extreme muscular fatigue, muscles occasionally become incapable of either contracting or relaxing—a condition called **physiologic contracture** (kon-trak′chūr), which is caused by a lack of ATP within the muscle fibers. ATP can decline to very low levels when a muscle is stimulated strongly, such as under conditions of extreme exercise. When ATP levels are very low, active transport of Ca^{2+} ions into the sarcoplasmic reticulum slows, Ca^{2+} ions accumulate within the sarcoplasm, and ATP is unavailable to bind to the myosin molecules that have formed cross-bridges with the actin myofilaments. As a consequence, the previously formed cross-bridges cannot release, resulting in physiologic contracture.

Rigor mortis (rig′er mōr′tĭs) is the development of rigid muscles several hours after death and is similar to physiologic contracture. ATP production stops shortly after death, and ATP levels within muscle fibers decline. Because of low ATP levels, active transport of Ca^{2+} ions into the sarcoplasmic reticulum stops, and Ca^{2+} ions leak from the sarcoplasmic reticulum into the sarcoplasm. Ca^{2+} can also leak from the sarcoplasmic reticulum as a result of the breakdown of the sarcoplasmic reticulum membrane after cell death. As calcium levels increase in the sarcoplasm, cross-bridges form. Too little ATP is available to bind to the myosin molecules, however, so the cross-bridges are unable to release and re-form in a cyclic fashion to produce contractions. As a consequence, the muscles remain stiff until tissue degeneration occurs.

Energy Sources

ATP provides the immediate source of energy for muscle contractions. As long as adequate amounts of ATP are present, muscles can contract repeatedly for a long time. ATP must be synthesized continuously to sustain muscle contractions, and ATP synthesis must be equal to ATP breakdown because only small amounts of ATP are stored in the muscle fibers. The energy required to produce ATP comes from three sources: (1) creatine phosphate, (2) anaerobic respiration, and (3) aerobic respiration. Only the main points of anaerobic respiration and aerobic respiration are considered here (a more detailed discussion of anaerobic and aerobic respiration can be found in chapter 25).

Creatine Phosphate

During resting conditions, energy from aerobic respiration (see later section) is used to synthesize **creatine** (krē′ă-tēn) **phosphate.** Creatine phosphate accumulates in muscle cells and functions to store energy, which can be used to synthesize ATP. As ATP levels begin to fall, ADP reacts with creatine phosphate to produce ATP and creatine.

$$ADP + creatine\ phosphate \rightarrow creatine + ATP$$

The reaction occurs very rapidly and is able to maintain ATP levels as long as creatine phosphate is available in the cell. During intense muscular contraction, however, creatine phosphate levels are quickly exhausted. ATP and creatine phosphate present in the cell provide enough energy to sustain vigorous contractions for about 10–15 s.

Anaerobic Respiration

Anaerobic (an-ār-ō′bik) **respiration** occurs in the absence of oxygen and results in the breakdown of glucose to yield ATP and lactic acid. For each molecule of glucose metabolized, there is a net production of two ATP molecules and two molecules of lactic acid. The first part of anaerobic metabolism and aerobic metabolism are common to each other. In both cases, each glucose molecule is broken down to two molecules of pyruvic acid. Two molecules of ATP are used in this process, but four molecules of ATP are produced, resulting in a net gain of two ATP molecules for each glucose molecule metabolized. In anaerobic metabolism, the pyruvic acid is then converted to lactic acid. Unlike pyruvic acid, much of the lactic acid diffuses out of the muscle fibers into the bloodstream.

Anaerobic respiration is less efficient than aerobic respiration, but it is faster, especially when oxygen availability limits aerobic respiration. By using many glucose molecules, anaerobic respiration can rapidly produce ATP for a short time. During short periods of intense exercise, such as sprinting, anaerobic respiration combined with the breakdown of creatine phosphate provides enough ATP to support intense muscle contraction for up to 3 min. ATP formation from creatine phosphate and anaerobic metabolism are limited by depletion of creatine phosphate and glucose and the buildup of lactic acid within muscle fibers.

Aerobic Respiration

Aerobic (ār-ō′bik) **respiration** requires oxygen and breaks down glucose to produce ATP, carbon dioxide, and water.

Compared with anaerobic respiration, aerobic respiration is much more efficient. The metabolism of a glucose molecule by anaerobic respiration produces a net gain of two ATP molecules for each glucose molecule. In contrast, aerobic respiration can produce about 38 ATP molecules for each glucose molecule. In addition, aerobic respiration uses a greater variety of molecules as energy sources, such as fatty acids and amino acids. Some glucose is used as an energy source in skeletal muscles, but fatty acids provide a more important source of energy during sustained exercise and during resting conditions.

In aerobic respiration, pyruvic acid is metabolized by chemical reactions within mitochondria. Two closely coupled sequences of reactions in mitochondria, called the citric acid cycle and the electron transport chain, produce many ATP molecules. Carbon dioxide molecules are produced, and in the last step, oxygen atoms are combined with hydrogen atoms to form water. Thus carbon dioxide, water, and ATP are major end products of aerobic metabolism. The following equation represents aerobic respiration of one molecule of glucose:

$$\text{Glucose} + 6\,O_2 + 38\,\text{ADP} + 38P \rightarrow$$
$$6\,CO_2 + 6\,H_2O + \text{about 38 ATP}$$

Although aerobic metabolism produces many more ATP molecules for each glucose molecule metabolized than does anaerobic metabolism, the rate at which the ATP molecules are produced is slower. Resting muscles or muscles undergoing long-term exercise, such as long-distance running or other endurance exercises, depend primarily on aerobic respiration for ATP synthesis.

Oxygen Debt

After intense exercise, the rate of aerobic metabolism remains elevated for a time. The oxygen taken in by the body above that required for resting metabolism after exercise is called the **oxygen debt.** It represents the difference between the amount of oxygen needed for aerobic respiration during muscle activity and the amount that actually was used. ATP produced by anaerobic sources and used during muscle activity contributes to the oxygen debt. The increased aerobic metabolism after exercise reestablishes normal ATP and creatine phosphate levels in muscle fibers. It also converts excess lactic acid to pyruvic acid and then to glucose, primarily in the liver. The glucose is used to help restore glycogen levels in muscle fibers and in liver cells.

Clinical Note

During brief, but intense exercise such as during a sprint, much of the ATP used by exercising muscles comes from the conversion of creatine phosphate to creatine and from anaerobic respiration. Glycogen is broken down to glucose in the skeletal muscle fibers and in the liver. Glycogen is released from the liver into the circulatory system and can be taken up by skeletal muscle fibers. Anaerobic respiration converts the glucose molecules to ATP and lactic acid. Heavy breathing and elevated aerobic respiration after the race results from the oxygen debt. The increased aerobic respiration pays back the oxygen debt by converting creatine to creatine phosphate and converting the excess lactic acid to glucose, which is then stored as glycogen in muscles and in the liver once again.

The magnitude of the oxygen debt depends on the intensity of the exercise, the length of time it was sustained, and the physical condition of the individual. Those who are in poor physical condition do not have as great a capacity to perform aerobic metabolism as well-trained athletes.

7   P R E D I C T

After a 10-km run with a sprint at the end, a runner continues to breathe heavily for a time. Compare the function of the elevated metabolic processes during the run, near the end, and shortly after the run.

✔ *Answer in Appendix F*

Slow and Fast Fibers

Not all skeletal muscles have identical functional capabilities. Slow-twitch muscle fibers contract more slowly and are more resistant to fatigue, whereas fast-twitch muscle fibers contract quickly and fatigue quickly. The proportion of muscle fiber types differs within individual muscles.

Slow-Twitch, or High-Oxidative, Muscle Fibers

Slow-twitch, or **high-oxidative, muscle fibers** contract more slowly, are smaller in diameter, have a better developed blood supply, have more mitochondria, and are more fatigue-resistant than fast-twitch muscle fibers. Slow-twitch muscle fibers respond relatively slowly to nervous stimulation and break down ATP at a limited rate within the heads of their myosin molecules. Aerobic respiration is the primary source for ATP synthesis in slow-twitch muscles, and their capacity to perform aerobic respiration is enhanced by a plentiful blood supply and the presence of numerous mitochondria. They are sometimes called high-oxidative muscle fibers because of their enhanced capacity to carry out aerobic respiration. Slow-twitch fibers also contain large amounts of **myoglobin** (mī-ō-glō′bin), a dark pigment similar to hemoglobin, that binds oxygen and acts as a reservoir for it when the blood does not supply an adequate amount. Myoglobin thus enhances the capacity of the cell to perform aerobic respiration.

Fast-Twitch, or Low-Oxidative, Muscle Fibers

Fast-twitch, or **low-oxidative, muscle fibers** respond to nervous stimulation and contain myosin molecules that break down ATP more rapidly than do slow-twitch muscle fibers. This allows their cross-bridges to form, release, and re-form more rapidly than those in slow-twitch muscles. Muscles containing these fibers have a less well-developed blood supply than that of slow-twitch muscles. In addition, fast-twitch muscles have very little myoglobin and fewer and smaller mitochondria. Fast-twitch muscles have large deposits of glycogen and are well adapted to perform anaerobic respiration. The anaerobic respiration of fast-twitch muscles, however, is not adapted for supplying a large amount of ATP for a prolonged period. The muscles tend to contract rapidly for a shorter time and fatigue relatively quickly.

Distribution of Fast-Twitch and Slow-Twitch Muscle Fibers

The muscles of many animals are composed primarily of either fast-twitch or slow-twitch muscle fibers. The white meat of a chicken or pheasant breast, which is composed mainly of fast-twitch fibers, appears whitish because of its relatively poor blood supply and lack of myoglobin. The muscles are adapted to contract rapidly for a short time but fatigue quickly. The red, or dark, meat of a chicken leg or of a duck breast is composed of slow-twitch fibers and appears darker because of the relatively well-developed blood supply and a large amount of myoglobin. These muscles are adapted to contract slowly for a longer time and to fatigue slowly. The distribution of slow-twitch and fast-twitch muscle fibers is consistent with the behavior of these animals. For example, pheasants can fly relatively fast for short distances, and ducks fly more slowly for long distances.

Humans exhibit no clear separation of slow-twitch and fast-twitch muscle fibers in individual muscles. Most muscles have both types of fibers, although the number of each varies in a given muscle. The large postural muscles contain more slow-twitch fibers, whereas muscles of the upper limbs contain more fast-twitch fibers.

The distribution of slow- and fast-twitch muscle fibers in a given muscle is constant for each individual and apparently is established developmentally. People who are good sprinters have a greater percentage of fast-twitch muscle fibers, whereas good long-distance runners have a higher percentage of slow-twitch fibers in their leg muscles. Athletes who are able to perform a variety of anaerobic and aerobic exercises tend to have a more balanced mixture of fast-twitch and slow-twitch muscle fibers.

Effects of Exercise

Neither fast-twitch nor slow-twitch muscle fibers can be converted to muscle fibers of the other type. Nevertheless, training can increase the capacity of both types of muscle fibers to perform more efficiently. Intense exercise resulting in anaerobic metabolism increases muscular strength and mass and has a greater effect on fast-twitch than on slow-twitch muscle fibers. On the other hand, aerobic exercise increases the vascularity of

muscle and causes enlargement of slow-twitch muscle fibers. Aerobic metabolism also can convert fast-twitch muscle fibers that fatigue readily to fast-twitch muscle fibers that resist fatigue, by increasing the number of mitochondria in the muscle cells and increasing their blood supply. Trained fast-twitch muscles are called **fatigue-resistant fast-twitch muscles.** Through training, a person with more fast-twitch muscle fibers can run long distances, and a person with more slow-twitch muscle fibers can increase the speed at which he or she runs.

8	P R E D I C T

What kind of exercise regimen is appropriate for people who are training to be endurance runners? What effect will the composition of their muscles, in terms of muscle fiber type, have on their ability to perform in an endurance race?

✔ *Answer in Appendix F*

A muscle increases in size, or **hypertrophies** (hī-per′trō-fēz), and increases in strength and endurance in response to exercise. Conversely, a muscle that is not used decreases in size, or **atrophies** (at′rō-fēz). The muscular atrophy that occurs in limbs placed in casts for several weeks is an example. Because muscle cell numbers do not change appreciably during a person's life, atrophy and hypertrophy of muscles result from changes in the size of individual muscle fibers. As fibers increase in size, the number of myofibrils and sarcomeres increases within each muscle fiber. Other elements such as blood vessels, connective tissue, and mitochondria also increase. Atrophy involves a decrease in these elements without a decrease in muscle fiber number. Severe atrophy such as occurs in elderly people who cannot readily move their limbs, however, does involve an irreversible decrease in the number of muscle fibers and can lead to paralysis.

The increased strength of trained muscle is greater than would be expected if it were based only on the change in muscle size. Part of the increase in strength results from the ability of the nervous system to recruit a large number of the motor units simultaneously in a trained person to perform movements with better neuromuscular coordination. In addition, trained muscles usually are restricted less by excess fat. Metabolic enzymes increase in hypertrophied muscle fibers, resulting in a greater capacity for nutrient uptake and ATP production. Improved endurance in trained muscles is in part a result of improved metabolism, increased circulation to the exercising muscles, increased numbers of capillaries, more efficient respiration, and a greater capacity for the heart to pump blood.

Clinical Note

Some people take synthetic hormones called **anabolic steroids** (an-ă-bol′ik ster′oydz) to increase the size and strength of their muscles. The anabolic steroids are related to testosterone, a reproductive hormone secreted by the testes, except that they have been altered so that the reproductive effects of these compounds are minimized, but their effect on skeletal muscles is maintained. Testosterone and anabolic steroids cause skeletal muscle tissue to hypertrophy. People who take large doses of an anabolic steroid exhibit an increase in body weight and total skeletal muscle mass, and many athletes believe that anabolic steroids improve performance that depends on strength. Unfortunately, evidence indicates that harmful side effects are associated with taking anabolic steroids, including periods of irritability, testicular atrophy and sterility, cardiovascular diseases such as heart attack or stroke, and abnormal liver function. Most athletic organizations prohibit the use of anabolic steroids, and some even analyze urine samples either randomly or periodically for evidence of their use. Penalties exist for athletes who have evidence of anabolic steroid metabolites in their urine.

Growth hormone is also used inappropriately to increase muscle size by some individuals. Growth hormone increases protein synthesis in muscle tissue although it does not produce the same kinds of side effects as those produced by anabolic steroids. The large doses of growth hormone used by athletes, however, can cause harmful side effects if taken over a long period (see chapter 18).

✔ *Answer in Appendix F*

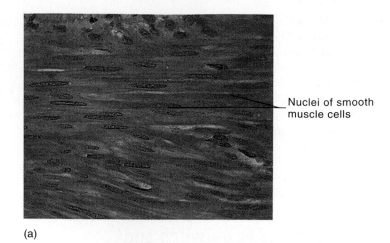

Nuclei of smooth muscle cells

(a)

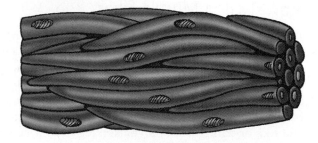

(b)

Figure 10.20   Smooth Muscle Histology

Heat Production

The rate of metabolism in skeletal muscle differs before, during, and after exercise. As chemical reactions occur within cells, some energy is released in the form of heat. Normal body temperature results in large part from this heat. Because the rate of chemical reactions increases in muscle fibers during contraction, the rate of heat production also increases, causing an increase in body temperature. After exercise, elevated metabolism resulting from the oxygen debt helps keep the body temperature elevated. If the body temperature increases as a result of increased contraction of skeletal muscle, vasodilation of blood vessels in the skin and sweating function to speed heat loss and keep the body temperature within its normal range (see chapter 25).

When the body temperature declines below a certain level, the nervous system responds by inducing **shivering,** which involves rapid skeletal muscle contractions that produce shaking rather than coordinated movements. The muscle movement increases heat production up to 18 times that of resting levels, and the heat produced during shivering can exceed that produced during moderate exercise. The elevated heat production during shivering helps raise the body temperature to its normal range.

Smooth Muscle

Smooth muscle is distributed widely throughout the body and is more variable in function than other muscle types. Smooth muscle cells (figure 10.20) are smaller than skeletal muscle cells, ranging from 15 to 200 μm in length and from 5 to 10 μm in diameter. They are spindle-shaped, with a sin-

gle nucleus located in the middle of the cell. Compared with skeletal muscle, fewer actin and myosin myofilaments are present. Although the myofilaments approximate a longitudinal, or spiral, orientation within the smooth muscle cell, they are not organized into sarcomeres. Consequently, smooth muscle does not have a striated appearance. Smooth muscle cells contain noncontractile **intermediate filaments.** These attach to **dense bodies** that are scattered through the cell and are occasionally attached to the plasma membrane. Myofilaments containing actin and myosin are attached by dense bodies to the intermediate filaments, which are considered to be equivalent to the Z disks in skeletal muscle and to dense bodies of the plasma membrane. The myofilaments, intermediate filaments, and dense bodies form an intracellular cytoskeleton. The myofilaments are obliquely arranged so the cell shortens when the actin and myosin slide over one another, causing the myofilaments to shorten during contraction (figure 10.21).

Sarcoplasmic reticulum is not as well developed in smooth muscle cells as it is in skeletal muscle fibers, and there is no T tubule system in smooth muscle. Some shallow invaginated areas called **caveolae** (kav′ē-ō-lē) are along the surface of the cell membrane. The function of caveolae is not well known, but it may be similar to that of both the T tubules and the sarcoplasmic reticulum of skeletal muscle.

The Ca^{2+} ions required to initiate contractions in smooth muscle enter the cell from the extracellular fluid, and Ca^{2+} ions enter the sarcoplasm from the smooth endoplasmic retic-

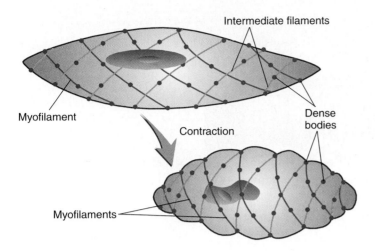

Figure 10.21   Contractile Proteins in a Smooth Muscle Cell

Bundles of contractile myofilaments containing actin and myosin are anchored at one end to dense bodies in the plasma membrane and at the other end, through dense bodies, to intermediate filament. The contractile myofilaments are oriented obliquely to the long axis of the cell, and when actin and myosin slide over one another during contraction, the myofilaments shorten and the cell shortens.

ulum. The distance that Ca^{2+} ions must diffuse, the rate at which action potentials are propagated between smooth muscle cells, and the smaller number of actin and myosin filaments are partially responsible for the slower contraction of smooth muscle compared with skeletal muscle.

Ca^{2+} ions bind to a protein called **calmodulin** (kal-mod'yū-lin) in smooth muscle cells. Calmodulin molecules with calcium ions bound to them activate an enzyme called **myosin kinase** (kī'nās), which transfers a phosphate group from ATP to light myosin molecules on the heads of myosin molecules. Cross-bridge formation occurs when myosin filaments have phosphate groups bound to them. The enzymes responsible for cross-bridge cycling function more slowly in smooth muscle cells than in skeletal muscle, which, in part, accounts for the slow rate of contraction in smooth muscle cells. Another enzyme called **myosin phosphatase** (fos'fă-tās) removes the phosphate group from the myosin molecules (figure 10.22). If the phosphate is removed from myosin while the cross-bridges are attached, the cross-bridges release very slowly. This explains how smooth muscle is able to sustain tension for long periods without rapid cross-bridge cycling. A sustained concentration of Ca^{2+} ions in the sarcoplasm of smooth muscle cells results in the activation of myosin molecules and cross-bridge formation. The action of myosin phosphatase results in a high percentage of myosin molecules having their phosphates removed while bound to actin. This process favors sustained contractions and a low rate of energy consumption because of the slow release of cross-bridges. As long as Ca^{2+} ions are present, cross-bridges re-form quickly after they are released. Consequently, many cross-bridges are intact at a give time in contracted smooth muscle.

Types of Smooth Muscle

Smooth muscle can be either visceral or multiunit. **Visceral** (vis'er-ăl), or **unitary, smooth muscle** is more common than multiunit smooth muscle. It normally occurs in sheets and includes smooth muscle of the digestive, reproductive, and urinary tracts. Visceral smooth muscle exhibits numerous gap junctions (see chapter 4), which allow action potentials to pass directly from one cell to another. As a consequence, sheets of smooth muscle cells function as a single unit, and a wave of contraction traverses the entire smooth muscle sheet. Visceral smooth muscle is often autorhythmic, but some contracts only when stimulated. For example, visceral smooth muscles of the digestive tract contract spontaneously and at relatively regular intervals, whereas the visceral smooth muscle of the urinary bladder contracts when stimulated by the nervous system.

Multiunit smooth muscle occurs as sheets such as in the walls of blood vessels, in small bundles such as in the arrector pili muscles and the iris of the eye, or as single cells such as in the capsule of the spleen. Multiunit smooth muscle has fewer gap junctions than multiunit smooth muscle cells, and cells or groups of cells act as independent units. It normally contracts only when stimulated by nerves or hormones.

Electrical Properties of Smooth Muscle

The resting membrane potential (RMP) (see chapter 9) of smooth muscle cells ranges from 55 to 60 mV, in contrast to the RMP of 85 mV in skeletal muscle. Furthermore, the RMP fluctuates, with slow depolarization and repolarization phases occurring in many visceral smooth muscle cells. These slow waves of depolarization and repolarization are propagated from cell to cell for short distances and cause contractions (figure 10.23a). More "classic" action potentials can be triggered by the slow waves of depolarization and usually are propagated for longer distances (figure 10.23b). The slow waves in the RMP result from a spontaneous and progressive increase in the permeability of the cell membrane to Na^+ and Ca^{2+} ions. Both types of ions diffuse into the cell through their respective channels and produce the depolarization.

Smooth muscle does not respond in an all-or-none fashion to action potentials. A series of action potentials in smooth muscle can result in a single, slow contraction followed by slow relaxation, instead of individual contractions in response to each action potential as occurs in skeletal muscle. A slow wave of depolarization that has one to several more classic-appearing action potentials superimposed on it is common in many types of smooth muscle. After the wave of depolarization, the smooth muscle undergoes a wave of contraction.

Spontaneously generated action potentials that lead to contractions are characteristic of visceral smooth muscle in the uterus, the ureter, and the digestive tract. Certain smooth muscle cells in these organs function as **pacemaker cells,** which

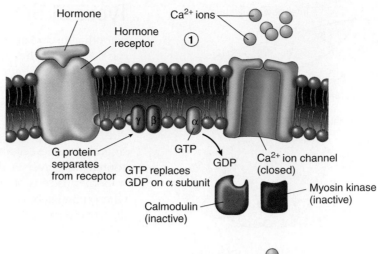

1. Either a hormone combines with a hormone receptor, or depolarization of the cell membrane activates the G protein.

2. An α-subunit opens the Ca²⁺ ion channel in the plasma membrane and Ca²⁺ ions diffuse through Ca²⁺ ion channel and combine with calmodulin.

3. Calmodulin with a Ca²⁺ ion bound to it binds with myosin kinase and activates it.

4. Activated myosin kinase attaches phosphate from ATP to myosin heads to activate the contractile process.

5. A cycle of cross-bridge formation, movement, detachment, and cross-bridge formation occurs.

6. Relaxation occurs when myosin phosphatase removes phosphate from myosin.

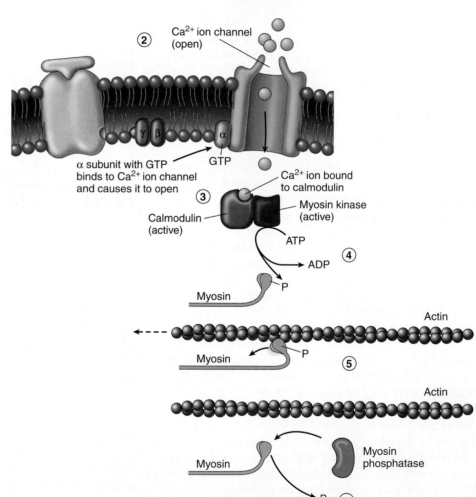

Figure 10.22 Ca²⁺ Ions in Smooth Muscle

The role of Ca²⁺ ions in smooth muscle contractions.

tend to develop action potentials more rapidly than other cells. Also, hormones can bind to hormone receptors on some smooth muscle cell membranes. The combination of the hormone with the receptor causes ligand-gated Ca²⁺ ion channels in the cell membrane to open. Ca²⁺ ions then enter the cell and cause smooth muscle contractions to occur without a major change in the membrane potential. For example, some smooth muscles contract when exposed to the hormone epinephrine because epinephrine combines with epinephrine receptors. Epinephrine combined with its receptors activates G proteins in the plasma membrane (see chapter 9). The G protein molecules can produce intracellular mediator molecules, which open the ligand-gated Ca²⁺ ion channels in the plasma membrane or sarcoplasmic reticulum.

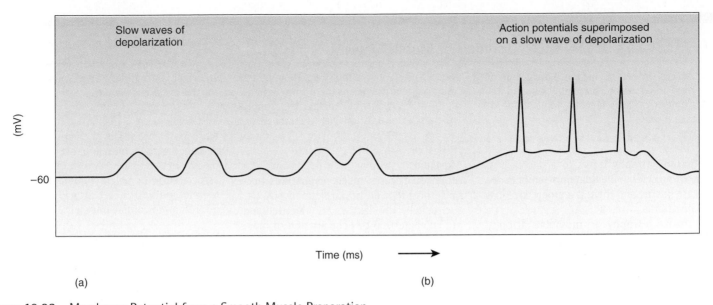

Figure 10.23   Membrane Potential from a Smooth Muscle Preparation

(*a*) Slow waves of depolarization. (*b*) Action potentials in smooth muscle superimposed on a slow wave of depolarization.

9   P R E D I C T

Explain how a ligand could bind to a membrane-bound receptor in a smooth muscle cell and cause a sustained contraction of the smooth muscle cell for a prolonged period without a large increase in ATP breakdown.

✔ *Answer in Appendix F*

Functional Properties of Smooth Muscle

Smooth muscle has four functional properties not seen in skeletal muscle: (1) some visceral smooth muscle has autorhythmic contractions; (2) smooth muscle tends to contract in response to a sudden stretch but not to a slow increase in length; (3) smooth muscle exhibits a relatively constant tension, called **smooth muscle tone,** over a long period and maintains that same tension in response to a gradual increase in the smooth muscle length; (4) the amplitude of contraction produced by smooth muscle also remains constant, although the muscle length varies. Smooth muscle is therefore well adapted for lining the walls of hollow organs such as the stomach and the urinary bladder. As the volume of the stomach or urinary bladder increases, only a small increase develops in the tension applied to their contents. Also, as the volume of the large and small intestines increases, the contractions that move food through them do not change dramatically in amplitude.

The metabolism of smooth muscle cells is similar to that of skeletal muscle fibers. They are poorly adapted to perform anaerobic metabolism, however. An oxygen debt does not develop in smooth muscle, and fatigue occurs quickly in the absence of an adequate oxygen supply.

Regulation of Smooth Muscle

Smooth muscle is innervated by nerves of the autonomic nervous system (see chapter 16), whereas skeletal muscle is innervated by the somatic motor nervous system (see chapter 14). The regulation of smooth muscle is therefore involuntary, and the regulation of skeletal muscle is voluntary.

Hormones are also important in regulating smooth muscle. Epinephrine, a hormone from the adrenal medulla, stimulates some smooth muscles, such as those in the blood vessels of the intestine, and inhibits other smooth muscles, such as those in the intestine itself. Oxytocin stimulates contractions of uterine smooth muscle, especially during delivery of a baby. These and other hormones are discussed more thoroughly in chapters 17 and 18. Other chemical substances, such as histamine and prostaglandins, also influence smooth muscle function.

Cardiac Muscle

Cardiac muscle is found only in the heart and is discussed in detail in chapter 20. Cardiac muscle tissue is striated like skeletal muscle, but each cell usually contains one nucleus located near the center. Adjacent cells join together to form branching fibers by specialized cell-to-cell attachments called **intercalated** (in-ter′kă-lā-ted) **disks,** which have gap junctions that allow action potentials to pass from cell to cell. Cardiac muscle cells are autorhythmic, and one part of the heart normally acts as the pacemaker. The action potentials of cardiac muscle are similar to those in nerve and skeletal muscle but have a much longer duration and refractory period. The depolarization of cardiac muscle results from the influx of both Na^+ and K^+ ions across the cell membrane. Regulation of contraction in cardiac muscle by Ca^{2+} ions is similar to that of skeletal muscle.

Clinical Focus   Disorders of Muscle Tissue

Muscle disorders are caused by disruption of normal innervation, degeneration and replacement of muscle cells, injury, lack of use, or disease.

Atrophy

Muscular atrophy refers to a decrease in size of muscles. Individual muscle fibers decrease in size, and there is a progressive loss of myofibrils.

Disuse atrophy is muscular atrophy that results from a lack of muscle use. Bedridden people, people with limbs in casts, or those who are inactive for other reasons experience disuse atrophy in the muscles that are not used. Disuse atrophy is temporary if a muscle is exercised after it is taken out of a cast. Extreme disuse of a muscle, however, results in muscular atrophy in which skeletal muscle fibers are permanently lost and replaced by connective tissue. Immobility that occurs in bedridden elderly people can lead to permanent and severe muscular atrophy.

Denervation atrophy (dē-ner-vā'shŭn) results when nerves that supply skeletal muscles are severed. When motor neurons innervating skeletal muscle fibers are severed, the result is flaccid paralysis. If the muscle is reinnervated, muscle function is restored, and atrophy is stopped. If skeletal muscle is permanently denervated, however, it atrophies and exhibits permanent flaccid paralysis. Eventually muscle fibers are replaced by connective tissue, and the condition cannot be reversed.

Transcutaneous stimulators are used to supply electric stimuli to muscles that have had their nerves temporarily damaged or to muscles that are put in casts for a prolonged period. The electric stimuli keep the muscles functioning and prevent them from permanently atrophying while the nerves resupply the muscles or until the cast is removed.

Muscular Dystrophy

Muscular dystrophy (dis'trō-fē) is one of a group of diseases called **myopathies** (mī-op'ă-thēz) that destroy skeletal muscle tissue. Usually the diseases are inherited and are characterized by degeneration of muscle cells, leading to atrophy and eventual replacement by connective tissue.

Duchenne's muscular dystrophy is an inherited sex-linked (X-linked) recessive disorder that almost exclusively affects males. As muscles atrophy and are replaced by connective tissue, they shorten, causing immobility of joints and postural abnormalities such as scoliosis. By early adolescence affected individuals are usually confined to wheelchairs (see Systems Pathology).

Facioscapulohumoral (fa'sĭ-o-skap-u-lo-hu'mor-al) muscular dystrophy is generally less severe, and it affects both sexes later in life. The muscles of the face and shoulder girdle are primarily involved. Facioscapulohumoral muscular dystrophy appears to be inherited as an autosomal-dominant condition. Both types of muscular dystrophy are inherited and progressive, and no drugs can prevent the progression of the disease. Therapy primarily involves exercises. Braces and corrective surgery sometimes help correct abnormal posture caused by the advanced disease.

Research is directed at identifying the genes responsible for all types of muscular dystrophy, exploring the mechanism that leads to the disease condition, and finding an effective treatment once the mechanism for the disease is known.

Fibrosis

Fibrosis (fī-brō'sis) is the replacement of damaged cardiac muscle or skeletal muscle by connective tissue. Fibrosis, or scarring, is associated with severe trauma to skeletal muscle and with heart attack (myocardial infarction) in cardiac muscle.

Fibrositis

Fibrositis (fī-brō-sī'tis) is an inflammation of fibrous connective tissue, resulting in stiffness, pain, or soreness. It is not progressive, nor does it lead to tissue destruction. Fibrositis can be caused by repeated muscular strain or prolonged muscular tension.

Cramps

Cramps are painful, spastic contractions of muscles that usually result from an irritation within a muscle that causes a reflex contraction (see chapter 13). Local inflammation resulting from a buildup of lactic acid and fibrositis causes reflex contraction of muscle fibers surrounding the irritated region.

Fibromyalgia (fī-brō-mī-al'jē-ă), or chronic muscle pain syndrome, has muscle pain as its main symptom. Fibromyalgia has no known cure, but it is not progressive, crippling, or life-threatening. The pain occurs in muscles or where muscles join their tendons, but not in joints. The pain is chronic and widespread, and it is distinguished from other causes of chronic pain by the identification of tender points in muscles, by the length of time the pain persists, and by failure to identify any other cause of the condition.

Systems Pathology
duchenne's muscular dystrophy

DUCHENNE'S MUSCULAR DYSTROPHY

A couple became concerned about their 3-year-old boy when they noticed that he was much weaker than other boys his age, and the differences appeared to become more obvious as time passed. He had difficulty sitting, standing, and walking. He seemed clumsy and he fell often. He had difficulty climbing stairs, and he often got from a sitting position on the floor to a standing position by using his hands and arms to climb up his legs. His muscles appeared to be poorly developed. The couple took their son to a physician to have him exam-ined. After several kinds of tests, they were informed that their son had Duchenne's muscular dystrophy.

BACKGROUND INFORMATION

Duchenne's muscular dystrophy (DMD) is usually identi-fied in children at around 3 years of age when the parents no-tice slow motor development with progressive weakness and muscle wasting (figure A). Typically, muscular weakness be-gins in the pelvic girdle, causing a waddling gait. Temporary enlargement of the calf muscles is apparent in 80% of cases. Rising from the floor by "climbing up the legs" is characteris-tic and is caused by weakness of the lumbar and gluteal mus-cles. Within 3–5 years, muscles of the shoulder girdle become involved. Wasting of the muscles contributes to muscular at-rophy and deformity of the skeleton. People with DMD are usually unable to walk by 10–12 years of age, and few live be-yond age 20. There is no effective treatment to prevent the progressive deterioration of muscles in DMD.

Duchenne's muscular dystrophy results from an abnor-mal gene located on the X chromosome, at a position called Xp21, and is therefore a sex-linked (X-linked) condition. Al-though the gene is carried by females, DMD affects males al-most exclusively. This position, or gene locus, is responsible for producing a protein called dystrophin, which plays a role in attaching myofibrils to and regulating the activity of other proteins in the plasma membrane. Dystrophin is thought to protect muscles' cells against mechanical stress in the normal individual. In DMD, part of the gene at Xp21 is missing, and the protein it produces malfunctions, resulting in abnormal contractions and progressive muscular weakness.

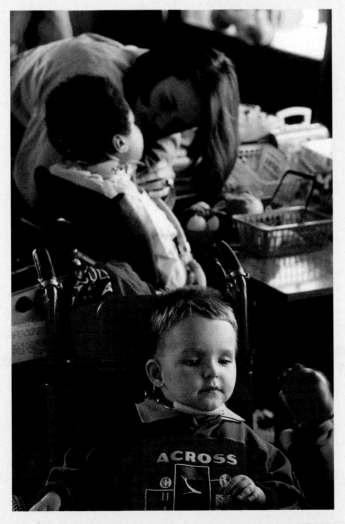

Figure A Young children with Duchenne's muscular dystrophy.

---

10   P R E D I C T

A boy with Duchenne's muscular dystrophy developed pulmonary edema and then pneumonia. His physician diagnosed the condition in the following way: the pulmonary edema was the result of heart failure and the increased fluid in the lungs acted as a site where bacteria invaded and grew. The fact that the boy could not breath deeply or cough effectively made the condition worse. Explain how a boy with DMD might develop heart failure and ineffective respiratory movements.

✔ *Answer in Appendix F*

Systemic Interactions

System	Interaction
Skeletal	Replacement of muscles by connective tissue results in shortened inflexible muscles, causing severe deformities of the skeletal system. The shortened muscles are referred to as contractures. Kyphoscoliosis, severe curvature of the spinal column laterally and anteriorly, can be so severe that normal respiratory movements are impaired. Deformities of the limbs result from the contractures. Surgery is sometimes required to prevent contractures from making it impossible for the individual to sit in a wheelchair.
Nervous	Some degree of mental retardation occurs in a large percentage of people with DMD.
Cardiovascular	Cardiac muscle is affected by DMD. Consequently, heart failure occurs in a large number of people with advanced DMD. Heart and respiratory muscles are affected, and death caused by respiratory or cardiac failure usually occurs before age 20. Cardiac involvement becomes serious in as much as 95% of cases.
Lymphatic and Immune	There are no obvious direct effects on the lymphatic system, but phagocytosis of muscle fibers is accomplished mainly by macrophages.
Respiratory	Deformity of the thorax and increasing weakness of the respiratory muscles results in inadequate respiratory movements and an increase in respiratory infections such as pneumonia. Inadequate respiratory movements due to weak respiratory muscles is a major factor in many deaths.
Digestive	Smooth muscle tissue is influenced by muscular dystrophy. The reduced ability of smooth muscle to contract can result in abnormalities of the digestive system such as an enlarged colon diameter, a twisting of the intestine resulting in increased intestinal obstruction, cramping, and reduced absorption of nutrients.
Urinary	Reduced smooth muscle function and being wheelchair-dependent increase the frequency of urinary tract infections.

Summary

General Functional Characteristics of Muscle

1. Muscle exhibits contractility (shortens forcefully), excitability (responds to stimuli), extensibility (can be stretched), and elasticity (recoils to resting length).
2. The three types of muscle are skeletal, smooth, and cardiac.

Skeletal Muscle: Structure

Muscle fibers are multinucleated and appear striated.

Connective Tissue

1. Endomysium surrounds each muscle fiber.
2. Muscle fibers are covered by the external lamina and the endomysium.
3. Muscle fasciculi, bundles of muscle fibers, are covered by the perimysium.
4. Muscle consisting of fasciculi is covered by the epimysium, or fascia.
5. The connective tissue of muscle is bound firmly to the connective tissue of tendons and bone.

Muscle Fibers

1. A muscle fiber is a single cell consisting of a cell membrane (sarcolemma), cytoplasm (sarcoplasm), several nuclei, and myofibrils.
2. Myofibrils are composed of many adjoining sarcomeres.
   - Sarcomeres are bound by Z disks that hold actin myofilaments.
   - Six actin myofilaments (thin filaments) surround a myosin myofilament (thick filament).
   - Myofibrils appear striated because of A bands and I bands.

3. Actin myofilaments consist of a double helix of F actin (composed of G actin monomers), tropomyosin, and troponin.
4. Myosin molecules, consisting of two globular heads and a rodlike portion, constitute myosin myofilaments.
5. A cross-bridge is formed when the myosin binds to the actin.
6. Invaginations of the sarcolemma form T tubules that wrap around the sarcomeres.
7. A triad is a T tubule and two terminal cisternae (an enlarged area of sarcoplasmic reticulum).

Sliding Filament Model

1. Actin and myosin myofilaments do not change in length during contraction.
2. Actin and myosin myofilaments slide past one another in a way that causes sarcomeres to shorten.
3. The I band and H zones become narrower during contraction, and the A band remains constant in length.

Physiology of Skeletal Muscle Fibers
Neuromuscular Junction

1. The presynaptic terminal of the axon is separated from the postsynaptic membrane of the muscle fiber by the synaptic cleft.
2. Acetylcholine released from the presynaptic terminal binds to receptors of the postsynaptic membrane, thereby changing membrane permeability and producing an action potential.
3. After an action potential occurs, acetylcholinesterase splits acetylcholine into acetic acid and choline. Choline is reabsorbed into the presynaptic terminal to re-form acetylcholine.

Excitation–Contraction Coupling

1. Action potentials move into the T tubule system, causing voltage-gated Ca^{2+} ion channels to open to release Ca^{2+} ions from the sarcoplasmic reticulum.
2. Ca^{2+} ions diffuse to the myofilaments and bind to troponin, causing tropomyosin to move and expose actin to myosin.
3. Contraction occurs when actin and myosin bind, myosin changes shape, and actin is pulled past the myosin.
4. Relaxation occurs when calcium is taken up by the sarcoplasmic reticulum, ATP binds to myosin, and tropomyosin moves back so actin is no longer exposed to myosin.

Energy Requirements for Muscle Contraction

1. One ATP molecule is required for each cycle of cross-bridge formation, movement, and release.
2. ATP is also required to transport Ca^{2+} ions into the sarcoplasmic reticulum and to maintain normal concentration gradients across the cell membrane.

Muscle Relaxation

1. Ca^{2+} ions are transported into the sarcoplasmic reticulum.
2. Ca^{2+} ions diffuse away from troponin, preventing further cross-bridge formation.

Physiology of Skeletal Muscle
Muscle Twitch

1. A muscle twitch is the contraction of a single muscle fiber or a whole muscle in response to a stimulus.
2. A muscle twitch has a lag, contraction, and relaxation phase.

Stimulus Strength and Muscle Contraction

1. For a given condition, a muscle fiber or motor unit contracts with a consistent force in response to each action potential, which is called the all-or-none law of muscle contraction.
2. For a whole muscle, a stimulus of increasing magnitude results in graded contractions of increased force as more motor units are recruited (multiple motor unit summation).

Stimulus Frequency and Muscle Contraction

1. A stimulus of increasing frequency increases the force of contraction (multiple-wave summation).
2. Incomplete tetanus is partial relaxation between contractions, and complete tetanus is no relaxation between contractions.
3. The force of contraction of a whole muscle increases with increased frequency of stimulation because of an increasing concentration of Ca^{2+} ions around the myofibrils and because of complete stretching of muscle elastic elements.
4. Treppe is an increase in the force of contraction during the first few contractions of a rested muscle.

Types of Muscle Contractions

1. Isometric contractions cause a change in muscle tension but no change in muscle length.
2. Isotonic contractions cause a change in muscle length but no change in muscle tension.
3. Asynchronous contractions of motor units produce smooth, steady muscle contractions.
4. Muscle tone is maintenance of a steady tension for long periods.
5. Concentric contractions cause muscles to shorten and tension to increase.
6. Eccentric contractions cause muscles to increase in length and the tension to gradually decrease.

Length Versus Tension

Muscle contracts with less-than-maximum force if its initial length is shorter or longer than optimum.

Fatigue

Fatigue is the decreased ability to do work and can be caused by the central nervous system, depletion of ATP in muscles, or depletion of acetylcholine in the neuromuscular synapse.

Physiologic Contracture and Rigor Mortis

Physiologic contracture (inability of muscles to contract or relax) and rigor mortis (stiff muscles after death) result from inadequate amounts of ATP.

Energy Sources

1. Energy for muscle contraction comes from ATP.
2. ATP can be synthesized when ADP reacts with creatine phosphate to form creatine and ATP. ATP from this source provides energy for a short time during intense exercise.
3. ATP is synthesized by anaerobic respiration and is used to provide energy for a short time during intense exercise. Anaerobic respiration produces ATP less efficiently but more rapidly than aerobic respiration. Lactic acid levels increase because of anaerobic respiration.
4. ATP is synthesized by aerobic respiration. Although ATP is produced more efficiently, it is produced more slowly. Aerobic respiration produces energy for muscle contractions under resting conditions or during exercises such as long-distance running.

Oxygen Debt

After anaerobic respiration, aerobic respiration is higher than normal, restoring creatine phosphate levels and converting lactic acid to glucose.

Slow and Fast Fibers

1. Slow-twitch fibers split ATP slowly and have a well-developed blood supply, many mitochondria, and myoglobin.
2. Fast-twitch fibers split ATP rapidly.
   - Fast-twitch, fatigable fibers have large amounts of glycogen, a poor blood supply, fewer mitochondria, and little myoglobin.
   - Fast-twitch, fatigue-resistant fibers have a well-developed blood supply, more mitochondria, and more myoglobin.
3. People who are good sprinters have a greater percentage of fast-twitch muscle fibers, and people who are good long-distance runners have a higher percentage of slow-twitch muscle fibers in their leg muscles.

Effects of Exercise

1. Muscles increase (hypertrophy) or decrease (atrophy) in size because of a change in the size of muscle fibers.
2. Anaerobic exercise develops fast-twitch, fatigable fibers. Aerobic exercise develops slow-twitch fibers and changes fast-twitch, fatigable fibers into fast-twitch, fatigue-resistant fibers.

Heat Production

1. Heat is produced as a by-product of chemical reactions in muscles.
2. Shivering produces heat to maintain body temperature.

Smooth Muscle

1. Smooth muscle cells are spindle-shaped with a single nucleus. They have actin myofilaments and myosin myofilaments but are not striated.
2. The sarcoplasmic reticulum is poorly developed, and caveolae may function as a T tubule system.
3. Ca^{2+} ions enter the cell to initiate contraction; calmodulin binds to Ca^{2+} ions and activates an enzyme that transfers a phosphate group from ATP to myosin. When phosphate groups are attached to myosin, cross-bridges form.

Types of Smooth Muscle

1. Visceral smooth muscle fibers contract slowly, have gap junctions (and thus function as a single unit), and can be autorhythmic.
2. Multiunit smooth muscle fibers contract rapidly in response to stimulation by neurons and function independently.

Electrical Properties of Smooth Muscle

1. Spontaneous contractions result from Na^+ and Ca^{2+} ion leakage into cells. Na^+ ion and Ca^{2+} ion movement into the cell is involved in depolarization.

2. The autonomic nervous system and hormones can inhibit or stimulate action potentials (and thus contractions). Hormones can also stimulate or inhibit contractions without affecting membrane potentials.

Functional Properties of Smooth Muscle

1. Smooth muscle can contract autorhythmically in response to stretch or when stimulated by the autonomic nervous system or hormones.
2. Smooth muscle maintains a steady tension for long periods.
3. The force of smooth muscle contraction remains nearly constant, despite changes in muscle length.
4. Smooth muscle does not develop an oxygen debt.

Regulation of Smooth Muscle

1. Smooth muscle is innervated by the autonomic nervous system and is involuntary.
2. Hormones are important in regulatory smooth muscle.

Cardiac Muscle

Cardiac muscle fibers are striated, have a single nucleus, are connected by intercalated disks (thus function as a single unit), and are capable of autorhythmicity.

Content Review

1. Compare the structure, function, location, and control of the three major muscle types.
2. Name the connective tissue structures that surround muscle fibers, muscle fasciculi, and whole muscles.
3. Define the terms sarcolemma, sarcoplasm, myofibril, and sarcomere.
4. What are Z disks and M lines, and what are their functions?
5. Explain how the arrangement of actin myofilaments and myosin myofilaments produce I bands, A bands, and H zones.
6. How do G actin, tropomyosin, and troponin combine to form an actin myofilament?
7. What is the T tubule system? What is a triad?
8. Describe the blood and nerve supply of a muscle fiber.
9. Describe the neuromuscular junction. How does an action potential in the neuron produce an action potential in the muscle cell?
10. How does an action potential produced in the postsynaptic terminal of the neuromuscular junction eventually result in contraction of the muscle fiber?
11. Where in the contraction and relaxation processes is ATP required?
12. Describe the phases of a muscle twitch and the events that occur in each phase.
13. Why does a single muscle fiber either not contract or contract with the same force in response to stimuli of different magnitudes?
14. How does increasing the magnitude of a stimulus cause a whole muscle to respond in a graded fashion?
15. Explain why increasing the frequency of stimulation increases the force of contraction of a single muscle fiber.

16. Define isometric, concentric, and eccentric contractions. What is muscle tone, and how is it maintained?
17. How are smooth contractions produced in muscles?
18. Draw an active tension curve. How does the overlap of actin and myosin explain the shape of the curve?
19. Define the term fatigue, and list three locations in which fatigue can develop.
20. Define and explain the cause of physiologic contracture and rigor mortis.
21. Contrast the efficiency of aerobic and anaerobic respiration. When is each type used by cells?
22. What is the function of creatine phosphate? When does lactic acid production increase in a muscle cell?
23. Contrast the structural and functional differences between slow and fast fibers.
24. What factors contribute to an increase in muscle strength and endurance? How does anaerobic versus aerobic exercise affect muscles?
25. How do muscles contribute to the heat responsible for body temperature before, during, and after exercise? What is accomplished by shivering?
26. Describe a typical smooth muscle cell. How does it differ from a skeletal muscle cell or a cardiac muscle cell?
27. Compare visceral smooth muscle to multiunit smooth muscle. Explain why visceral smooth muscle contracts as a single unit.
28. How are spontaneous contractions produced in smooth muscle?
29. How do the nervous system and hormones regulate smooth muscle activity?

Develop Your Reasoning Skills

1. Bob Canner improperly canned some homegrown vegetables. As a result, he contracted botulism poisoning after eating the vegetables. Symptoms included difficulty in swallowing and breathing. Eventually he died of respiratory failure (his respiratory muscles relaxed and would not contract). Assuming that botulism toxin affects the neuromuscular synapse, propose the ways that botulism toxin could produce the observed symptoms.

2. A patient is thought to be suffering from either muscular dystrophy or myasthenia gravis. How would you distinguish between the two conditions?

3. Under certain circumstances, the actin and myosin myofilaments can be extracted from muscle cells and placed in a beaker. They subsequently bind together to form long filaments of actin and myosin. Addition of what cell organelle or molecule to the beaker would make the actin and myosin myofilaments unbind?

4. Explain the effect of a lower-than-normal temperature on each of the processes that occur in the lag (latent) phase of muscle contraction.

5. Design an experiment to test the following hypothesis: muscle A has the same number of motor units as muscle B. Assume you could stimulate the nerves that innervate skeletal muscles with an electronic stimulator and monitor the tension produced by the muscles.

6. Compare the differences that occur when a muscle such as the biceps slowly lifts and lowers a weight and when a muscle twitches.

7. Predict the shape of an active tension curve for visceral smooth muscle. How does it differ from the active tension curve for skeletal muscle?

8. A researcher is investigating the composition of muscle tissue in the gastrocnemius muscles (in the calf of the leg) of athletes. A needle biopsy is taken from the muscle, and the concentration (or enzyme activity) of several substances is determined. Describe the major differences this researcher sees when comparing the muscles from athletes who perform in the following events: 100-m dash, weight lifting, and 10,000-m run.

9. Harvey Leche milked cows by hand each morning before school. One morning he slept later than usual and had to hurry to get to school on time. As he was milking the cows as fast as he could, his hands became very tired, and for a short time he could neither release his grip nor squeeze harder. Explain what happened.

10. Blood vessels that supply oxygen to smooth muscle undergo constriction. Explain how this phenomenon affects the ability of smooth muscle to contract.

11. Shorty McFleet noticed that his rate of respiration was elevated after running a 100 m race but was not as elevated after running slowly for a much longer distance. Because you studied muscle physiology, he asked you for an explanation. What would you say?

12. It is known that high blood K^+ ion concentrations cause depolarization of the resting membrane potential. Predict the effect of high blood K^+ ion levels on smooth muscle function. Explain.

13. Predict and explain the response if the ATP concentration in a muscle that was exhibiting rigor mortis could be instantly increased.

14. A hormone stimulates smooth muscle from a blood vessel to contract. The hormone only causes a small change in the membrane potential, however, even though the smooth muscle tissue contracts substantially. Explain.

Web Site Link

For a list of the most current web sites related to this chapter, please visit the Seeley home page at:
http://www.mhhe.com/biosci/ap/seeleyap/

Chapter Eleven

Muscular System: Gross Anatomy

Objectives

1. Discuss what is meant by the terms origin and insertion of a muscle.

2. Define the following terms, and give an example of each: synergist, antagonist, prime mover, and fixator.

3. List the major muscle shapes, and indicate how each relates to function.

4. List and describe the three lever classes, and give an example of each. Which lever class is most common in the body?

5. Describe the major movements of the head, and list the muscles involved in each movement.

6. Describe various facial expressions, and list the major muscles causing them.

7. List the muscles of mastication, and indicate the effect of each muscle on mandibular movement.

8. Explain the location of and functional differences between extrinsic and intrinsic tongue muscles.

9. Describe the process of swallowing, and explain the action of each muscle in this process.

10. Describe the muscles of the eye and how each affects eye movement.

11. Describe movements of the vertebral column, and list the muscles involved.

12. Describe the placement, fascicular orientation, and function of the muscles of the thorax and abdominal wall.

13. Describe the pelvic floor and perineum, and list the muscles forming them.

14. List the muscles forming the rotator cuff, and describe its function.

15. Describe the movements of the arm and the muscles involved.

16. Describe the forearm muscles in terms of their functional groupings and the movements they produce.

17. Explain the difference between extrinsic and intrinsic hand muscles.

18. Describe the movements of the thigh, and list the muscles involved in each movement.

19. Describe the leg in terms of compartments, list the muscles contained in each compartment, and indicate the function of each muscle.

f it were not for our skeletal muscles, we would not be able to hold this book; we wouldn't even be able to sit or stand. Without our muscular system, we would be restricted to one position in one place. We wouldn't be able to blink so our eyes would dry out. None of these inconveniences would bother us for long because we wouldn't be able to breathe either.

We use our skeletal muscles all the time. Even when we aren't "moving," the postural muscles are constantly contracting to keep us sitting or standing upright. Respiratory muscles are constantly functioning to keep us breathing, and muscles in our eyelids allow us to blink. While we sleep, we continue to breathe, and most people change positions 40–70 times in a single night of sleep. Communication of any kind requires skeletal muscles; whether we are writing, typing, or speaking. Silent communication by hand signals or facial expression also requires skeletal muscle function.

General Principles

This chapter is devoted entirely to the description of the major named skeletal muscles. The structure and function of cardiac and smooth muscle are considered in other chapters. Most skeletal muscles extend from one bone to another and cross at least one joint. Muscle contractions usually cause movement by pulling one bone toward another across a movable joint. Some muscles of the face, however, are not attached to bone at both ends but attach to the connective tissue of skin and move the skin when they contract.

Muscles are attached to bones and other connective tissue by **tendons.** A very broad tendon is called an **aponeurosis** (ap′ō-nū-rō′sis). The points of attachment for each muscle are the origin and insertion. The **origin,** also called the **head,** is normally that end of the muscle attached to the more stationary of the two bones, and the **insertion** is the end of the muscle attached to the bone undergoing the greatest movement. The largest portion of the muscle, between the origin and the insertion, is the **belly.** Some muscles have multiple origins and a common insertion and are said to have multiple heads (such as the biceps with two heads).

Most muscles function as members of a group to accomplish specific movements. Furthermore, many muscles are members of more than one group, depending on the type of movement being considered. For example, the anterior part of the deltoid muscle functions with the flexors of the arm, whereas the posterior part functions with the extensors of the arm. Muscles that work together to cause a movement are **synergists** (sin′er-jists), and a muscle working in opposition to another muscle, moving a structure in the opposite direction, is an **antagonist.** The brachialis and biceps brachii are synergists in flexing the forearm; the triceps brachii is the antagonist and extends the forearm. Among a group of synergists, if one muscle plays the major role in accomplishing the desired movement, it is called the **prime mover.**

Other muscles, called **fixators** (fik′sā-ters), may stabilize one or more joints crossed by the prime mover. The extensor digitorum is the prime mover in finger extension. The flexor carpi radialis and flexor carpi ulnaris are the fixators that keep the wrist from extending as the fingers are extended.

Muscle Shapes

Muscles come in a wide variety of shapes, the shape of a given muscle determining the degree to which it can contract and the amount of force it can generate. The large number of muscular shapes can be grouped into four classes according to the orientation of the muscle fasciculi: pennate, parallel, convergent, and circular. Some muscles have their fasciculi arranged like the barbs of a feather along a common tendon and are therefore called **pennate** (pen′āt; *pennatus* is Latin meaning feather) muscles. A muscle with fasciculi on one side of the tendon only is **unipennate,** one with fasciculi on both sides is **bipennate,** and a muscle with fasciculi arranged at many places around the central tendon is **multipennate** (figure 11.1*a*). The pennate arrangement allows a large number of fasciculi to attach to a single tendon with the force of contraction concentrated at the tendon. The muscles that extend the leg are examples of multipennate muscles (see table 11.20). In other muscles, called **parallel** muscles, fasciculi are organized parallel to the long axis of the muscle (figure 11.1*b*). As a consequence, the muscles shorten to a greater degree than do pennate muscles because the fasciculi are in a direct line with the tendon; however, they contract with less force because fewer total fascicles are attached to the tendon. The infrahyoid muscles are an example of parallel muscles (see table 11.4). In **convergent** muscles, such as the deltoid muscle, the base is much wider than the insertion, giving the muscle a triangular shape and allowing it to contract with more force than could occur in a parallel muscle. **Circular** muscles, such as the orbicularis oris and orbicularis oculi (see table 11.2) have their fasciculi arranged in a circle around an opening and act as sphincters to close the opening.

Muscles may have specific shapes, such as quadrangular, triangular, rhomboidal, or fusiform (figure 11.1*c*). Muscles also may have multiple components, such as two bellies or two heads. A digastric muscle has two bellies separated by a tendon, whereas a bicipital muscle has two origins (heads) and a single insertion (figure 11.1*d*).

Nomenclature

Muscles are named according to their location, size, shape, orientation of fasciculi, origin and insertion, number of heads, or function. Recognizing the descriptive nature of muscle names makes learning those names much easier.

1. *Location.* Some muscles are named according to their location. For example, a pectoralis (chest) muscle is located in the chest, a gluteus (buttock) muscle is located in the buttock, and a brachial (arm) muscle is located in the arm.

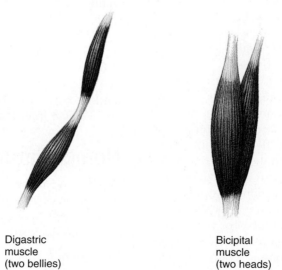

Unipennate
muscle

(a)

Bipennate
muscle

Multipennate
muscle

Parallel
muscle

Circular
muscle

Convergent
muscle

(b)

Quadrangular
muscle

Trapezoidal
muscle

Triangular
muscle

Rhomboidal
muscle

Fusiform muscle

(c)

Digastric
muscle
(two bellies)

Bicipital
muscle
(two heads)

(d)

Figure 11.1   Examples of Muscle Types

(*a*) Muscles with various pennate arrangements. (*b*) Muscles with various fascicular orientations. (*c*) Muscles with various shapes. (*d*) Muscles with various components.

2. *Size.* Muscle names may also refer to the relative size of the muscle. For example, the gluteus maximus (large) is the largest muscle of the buttock, and the gluteus minimus (small) is the smallest. A longus (long) muscle is longer than a brevis (short) muscle.

3. *Shape.* Some muscles are named according to their shape. The deltoid (triangular) muscle is triangular, a quadratus (quadrangular) muscle is rectangular, and a teres (round) muscle is round.

4. *Orientation.* Muscles are also named according to their fascicular orientation. A rectus (straight) muscle has muscle fasciculi running straight down the body, whereas the fasciculi of an oblique muscle lie oblique to the longitudinal axis of the body.

5. *Origin and insertion.* Muscles may be named according to the origin and insertion of the muscle. The sternocleidomastoid originates on the sternum and clavicle and inserts onto the mastoid process of the temporal bone. The brachioradialis originates in the arm (brachium) and inserts onto the radius.

6. *Number of heads.* The number of heads (origins) a muscle has may also be used in naming it. A biceps muscle has two heads, and a triceps muscle has three heads.

7. *Function.* Muscles are also named according to their function. An abductor moves a structure away from the midline, and an adductor moves a structure toward the midline. The masseter (a chewer) is a chewing muscle.

Movements Accomplished by Muscles

When muscles contract, the **pull** (P), or force, of muscle contraction is applied to levers, such as bones, resulting in movement of the levers (figure 11.2). A **lever** is a rigid shaft capable of turning about a pivot point called a **fulcrum** (F) and transferring a force applied at one point along the lever to a **weight** (W), or resistance, placed at some other point along the lever. The joints function as fulcrums, the bones function as levers, and the muscles provide the pull to move the levers. The relative positions of levers, weights, fulcrums, and forces make up three classes of levers.

Class I Lever

In a **class I lever system** the fulcrum is located between the force and the weight (figure 11.2*a*). An example of this type of lever is a child's seesaw. The children on the seesaw alternate between being the weight and the pull across a fulcrum in the center of the board. An example in the body is the head. The atlantooccipital joint is the fulcrum, the posterior neck muscles provide the pull depressing the back of the head, and the face, which is elevated, is the weight. With the weight balanced over the fulcrum, only a small amount of pull is required to lift a weight. For example, only a very small shift in

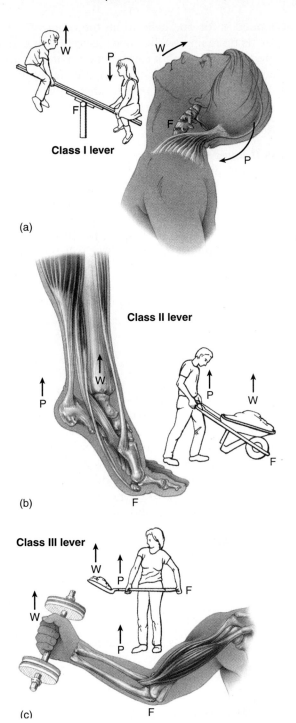

Figure 11.2   Lever Classes

(*a*) Class I: The fulcrum (F) is located between the weight (W) and the force or pull (P). The pull is directed downward, and the weight, on the opposite side of the fulcrum, is lifted. (*b*) Class II: The weight (W) is located between the fulcrum (F) and the force or pull (P). The upward pull lifts the weight. (*c*) Class III: The force or pull (P) is located between the fulcrum (F) and the weight (W). The upward pull lifts the weight. 🏃

weight is needed for one child to lift the other on a seesaw. This system is quite limited, however, as to how much weight can be lifted and how high it can be lifted. For example, consider what happens when the child on one end of the seesaw is much larger than the child on the other end.

Class II Lever

In a **class II lever system** the weight is located between the fulcrum and the pull (figure 11.2*b*). An example is a wheelbarrow, where the wheel is the fulcrum and the person lifting on the handles provides the pull. The weight, or load, carried in the wheelbarrow is placed between the wheel and the operator. In the body, an example of a class II lever is the foot of a person standing on her toes. The calf muscles pulling (force) on the calcaneus (end of the lever) elevate the foot and the weight of the entire body, with the ball of the foot acting as the fulcrum. A considerable amount of weight can be lifted by using this type of lever system, but the weight usually isn't lifted very high.

Class III Lever

In a **class III lever system,** the most common type in the body, the pull is located between the fulcrum and the weight (figure 11.2*c*). An example is a person using a shovel. The hand placed on the part of the handle closest to the blade provides the pull to lift the weight, such as a shovel full of dirt, and the hand placed near the end of the handle acts as the fulcrum. In the body, the action of the biceps brachii muscle (force) pulling on the radius (lever) to flex the elbow (fulcrum) and elevate the hand (weight) is an example of a class III lever. This type of lever system does not allow as great a weight to be lifted, but the weight can be lifted a greater distance.

Muscle Anatomy

An overview of the superficial skeletal muscles is presented in figure 11.3.

Head Muscles
Head Movement

Most of the flexors of the head and neck (table 11.1 and figure 11.4*a*) lie deep within the neck along the anterior margins of the vertebral bodies. Extension of the head is accomplished by posterior neck muscles that attach to the occipital bone (figure 11.4*b* and *c*) and function as the force of a class I lever system.

The muscular ridge seen superficially in the posterior part of the neck and lateral to the midline is composed of the trapezius muscle overlying the splenius capitis (figure 11.5*a*). The fasciculi of the trapezius muscles are shorter at the base

of the neck, leaving a diamond-shaped area over the inferior cervical and superior thoracic vertebral spines.

Rotation and abduction of the head are accomplished by muscles of both the lateral and posterior groups (see table 11.1). The **sternocleidomastoid** (ster′nō-klī′dō-mas′toyd) muscle is the prime mover of the lateral group. It is easily seen on the anterior and lateral sides of the neck, especially if the head is extended slightly and rotated to one side (see figure 11.5*b*). If the sternocleidomastoid muscle on only one side of the neck contracts, the head is rotated toward the opposite side. If both contract together, they flex the neck. Adduction of the head (moving the head back to the midline after it has been tilted to one side or the other) is accomplished by the abductors of the opposite side.

1   P R E D I C T

Shortening of the right sternocleidomastoid muscle rotates the head in which direction?

✔ *Answer in Appendix F*

Clinical Note

Torticollis (tōr′ti-kol′is, meaning a twisted neck), or wry neck, may result from injury to one of the sternocleidomastoid muscles. It sometimes is caused by damage to an infant's neck muscles during a difficult birth and usually can be corrected by exercising the muscle.

Facial Expression

The skeletal muscles of the face (table 11.2 and figure 11.6) are cutaneous muscles attached to the skin. Many animals have cutaneous muscles over the trunk that allow the skin to twitch to remove irritants such as insects. In humans, facial expressions are important components of nonverbal communication, and the cutaneous muscles are confined primarily to the face and neck.

Several muscles act on the skin around the eyes and eyebrows (figure 11.7). The **occipitofrontalis** (ok-sip′i-tō-frŭn-tā′lis) raises the eyebrows and furrows the skin of the forehead. The **orbicularis oculi** (ōr-bik′yū-lā′ris ok′yū-lī) closes the eyelids and causes "crow's-feet" wrinkles in the skin at the lateral corners of the eyes. The **levator palpebrae** (lē′vā-tor pal-pē′brē; the palpebral fissure is the opening between the eyelids) **superioris** raises the upper lids (see figure 11.7*a*). A droopy eyelid on one side, called **ptosis** (tō′sis), usually indicates that the nerve to the levator palpebrae superioris has been damaged. The **corrugator supercilii** (kōr′ŭ-gā′tor sū′per-sil′ē-ī) draws the eyebrows inferiorly and medially, producing vertical corrugations (furrows) in the skin between the eyes (see figures 11.6 and 11.7*c*).

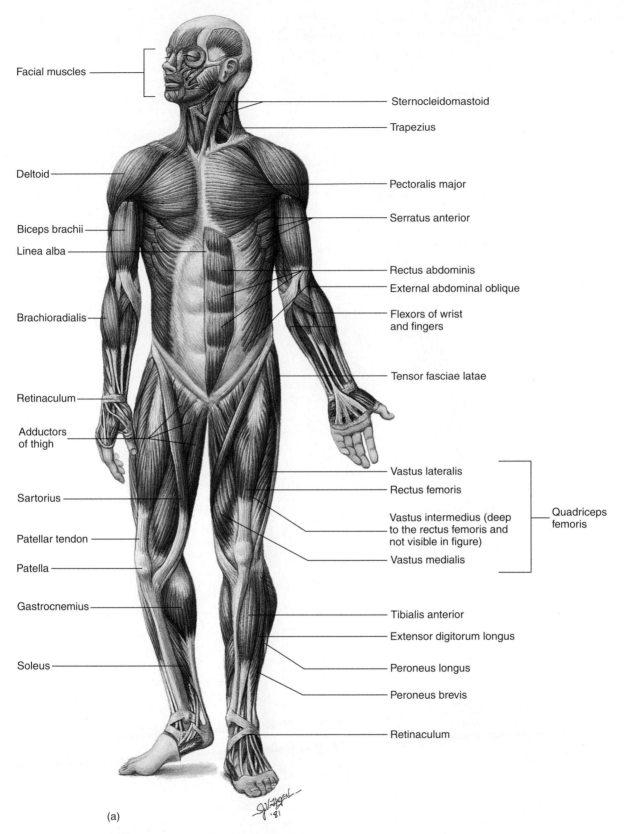

Facial muscles

Sternocleidomastoid

Trapezius

Deltoid

Pectoralis major

Serratus anterior

Biceps brachii

Linea alba

Rectus abdominis

External abdominal oblique

Brachioradialis

Flexors of wrist
and fingers

Tensor fasciae latae

Retinaculum

Adductors
of thigh

Vastus lateralis

Rectus femoris

Sartorius

Vastus intermedius (deep
to the rectus femoris and
not visible in figure)

Quadriceps
femoris

Patellar tendon

Vastus medialis

Patella

Gastrocnemius

Tibialis anterior

Extensor digitorum longus

Soleus

Peroneus longus

Peroneus brevis

Retinaculum

(a)

Figure 11.3   General Overview of the Superficial Body Musculature
(a) Anterior view. ✗

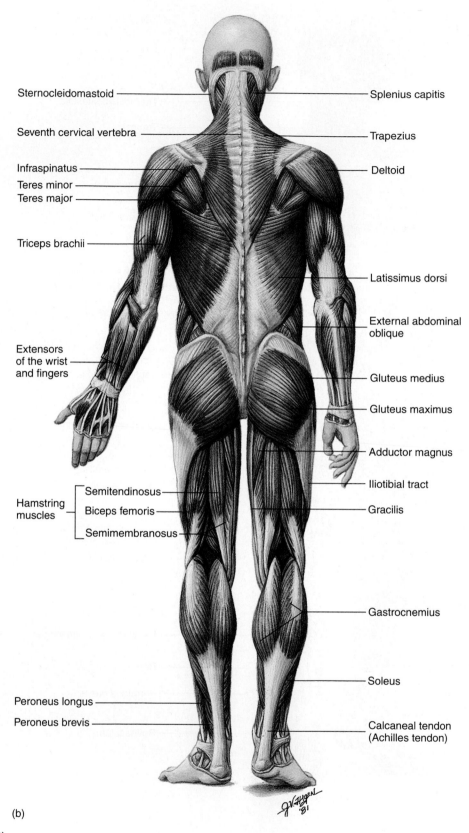

Sternocleidomastoid

Seventh cervical vertebra

Infraspinatus

Teres minor

Teres major

Triceps brachii

Extensors
of the wrist
and fingers

Hamstring
muscles

Semitendinosus

Biceps femoris

Semimembranosus

Peroneus longus

Peroneus brevis

Splenius capitis

Trapezius

Deltoid

Latissimus dorsi

External abdominal
oblique

Gluteus medius

Gluteus maximus

Adductor magnus

Iliotibial tract

Gracilis

Gastrocnemius

Soleus

Calcaneal tendon
(Achilles tendon)

(b)

Figure 11.3   *(continued)*
(*b*) Posterior view.

Table 11.1* Muscles Moving the Head (see figure 11.4)

Muscle	Origin	Insertion	Nerve	Function
Anterior				
Longus capitis (lon'gŭs ka'pi-tis) (not illustrated)	C3–C6	Occipital bone	C1–C3	Flexes head
Rectus capitis anterior (rek'tŭs ka'pi-tis) (not illustrated)	Atlas	Occipital bone	C1–C2	Flexes head
Posterior				
Longissimus capitis (lon-gis'ĭ-mŭs kă'pĭ-tis) (see figure 11.13)	Upper thoracic and lower cervical vertebrae	Mastoid process	Dorsal rami of cervical nerves	Extends, rotates, and laterally flexes head
Oblique capitis superior (ka'pi-tis) (figure 11.4c)	Atlas	Occipital bone (inferior nuchal line)	Dorsal ramus of C1	Extends and laterally flexes head
Rectus capitis posterior (rek'tŭs ka'pi-tis) (figure 11.4c)	Axis, atlas	Occipital bone	Dorsal ramus of C1	Extends and rotates head
Semispinalis capitis (figures 11.4b, c; 11.13)	C4–T6	Occipital bone	Dorsal rami of cervical nerves	Extends and rotates head
Splenius capitis	C4–T6	Superior nuchal line and mastoid process	Dorsal rami of cervical nerves	Extends, rotates, and laterally flexes head
Trapezius	Occipital protuberance, nuchal ligament, spinous processes of C7–T12	Clavicle, acromion process, and scapular spine	Accessory	Extends and laterally flexes head
(figures 11.3a, b; 11.4a, b; 11.5; 11.6a; 11.9; 11.19a; 11.21a–d)				
Lateral				
Rectus capitis lateralis (not illustrated)	Atlas	Occipital bone	C1	Laterally flexes head
Sternocleidomastoid	Manubrium and medial clavicle	Mastoid process and superior nuchal line	Accessory	One contracting alone: rotates and extends head; Both contracting together: flex head
(figures 11.3a, b; 11.4a, b; 11.5a, b; 11.6a; 11.9; 11.21a, b)				

*The tables in this chapter are to be used as references. As you study the muscular system, first locate the muscle on the figure, and then find its description in the corresponding table.

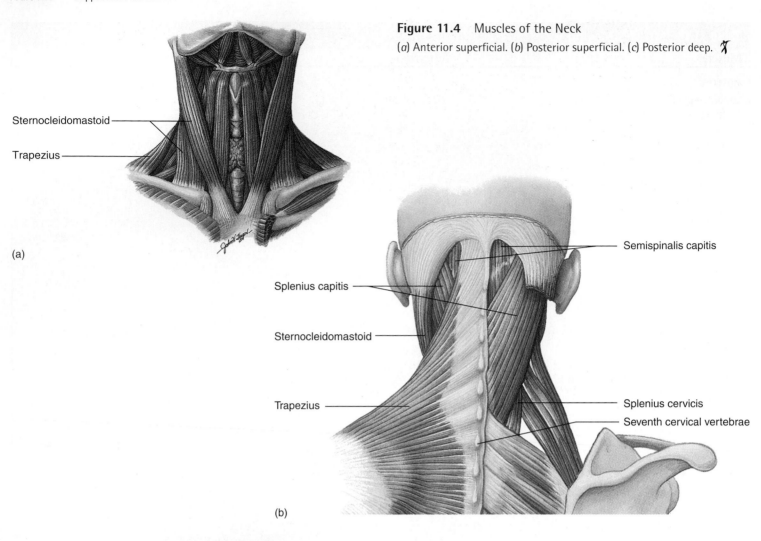

Figure 11.4   Muscles of the Neck

(*a*) Anterior superficial. (*b*) Posterior superficial. (*c*) Posterior deep.

Sternocleidomastoid

Trapezius

(a)

Semispinalis capitis

Splenius capitis

Sternocleidomastoid

Trapezius

Splenius cervicis

Seventh cervical vertebrae

(b)

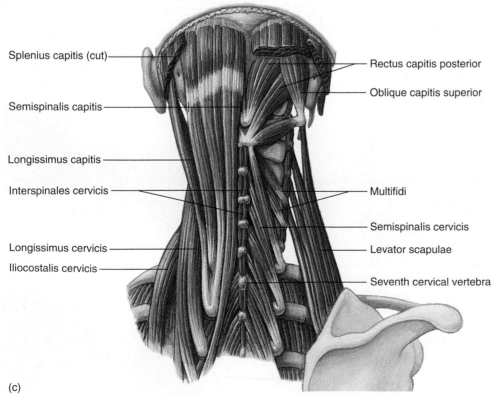

Splenius capitis (cut)

Semispinalis capitis

Longissimus capitis

Interspinales cervicis

Longissimus cervicis

Iliocostalis cervicis

Rectus capitis posterior

Oblique capitis superior

Multifidi

Semispinalis cervicis

Levator scapulae

Seventh cervical vertebra

(c)

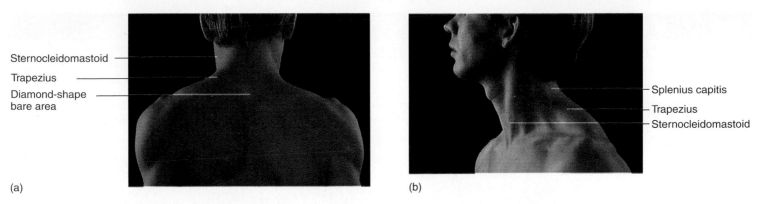

Sternocleidomastoid

Trapezius

Diamond-shape
bare area

Splenius capitis

Trapezius

Sternocleidomastoid

(a)

(b)

Figure 11.5   Surface Anatomy, Muscles of the Neck

(*a*) Posterior view. (*b*) Lateral view.

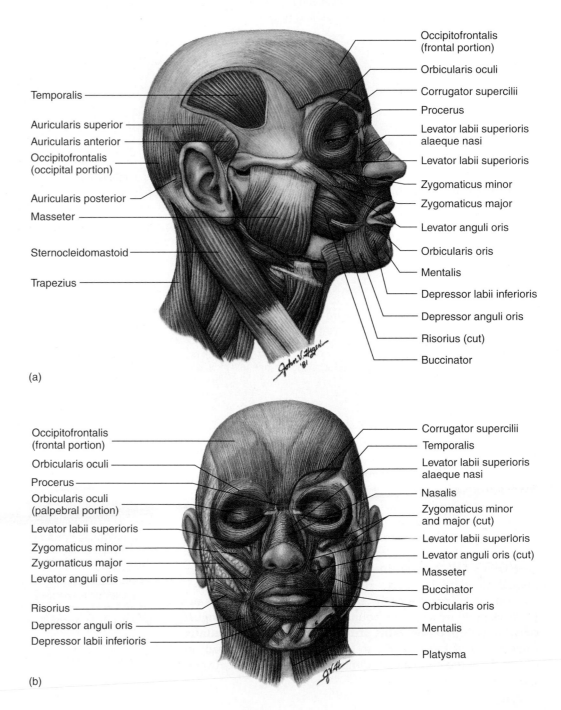

Occipitofrontalis
(frontal portion)

Orbicularis oculi

Corrugator supercilii

Procerus

Temporalis

Levator labii superioris
alaeque nasi

Auricularis superior

Auricularis anterior

Levator labii superioris

Occipitofrontalis
(occipital portion)

Zygomaticus minor

Zygomaticus major

Auricularis posterior

Levator anguli oris

Masseter

Orbicularis oris

Sternocleidomastoid

Mentalis

Depressor labii inferioris

Trapezius

Depressor anguli oris

Risorius (cut)

Buccinator

(a)

Occipitofrontalis
(frontal portion)

Corrugator supercilii

Temporalis

Orbicularis oculi

Levator labii superioris
alaeque nasi

Procerus

Nasalis

Orbicularis oculi
(palpebral portion)

Zygomaticus minor
and major (cut)

Levator labii superioris

Levator labii superloris

Zygomaticus minor

Levator anguli oris (cut)

Zygomaticus major

Masseter

Levator anguli oris

Buccinator

Risorius

Orbicularis oris

Depressor anguli oris

Mentalis

Depressor labii inferioris

Platysma

Figure 11.6   Muscles of
Facial Expression

(*a*) Lateral view. (*b*) Anterior view.

(b)

317

Table 11.2 Muscles of Facial Expression (see figure 11.6)

Muscle	Origin	Insertion	Nerve	Function
Auricularis (aw-rik′ū-lăr′is)				
Anterior (figure 11.6a)	Aponeurosis over head	Cartilage of auricle	Facial	Draws auricle superiorly and anteriorly
Posterior (figure 11.6a)	Mastoid process	Posterior root of auricle	Facial	Draws auricle posteriorly
Superior (figure 11.6a)	Aponeurosis over head	Cartilage of auricle	Facial	Draws auricle superiorly and posteriorly
Buccinator (buk′sĭ-nā′tōr) (figures 11.6; 11.7d; 11.8a; 11.11b)	Mandible and maxilla	Orbicularis oris at angle of mouth	Facial	Retracts angle of mouth; flattens cheek
Corrugator supercilii (kōr′ŭ-gā′tōr sū′per-sil′ē-ī) (figures 11.6; 11.7c)	Nasal bridge and orbicularis oculi	Skin of eyebrow	Facial	Depresses medial portion of eyebrow and draws eyebrows together as in frowning
Depressor anguli oris (dē-pres′ōr an′gū-lī ōr′ŭs) (figures 11.6; 11.7c)	Lower border of mandible	Lip near angle of mouth	Facial	Depresses angle of mouth
Depressor labii inferioris (dē-pres′ōr lā′bē-ī in-fēr′ē-ōr-is) (figures 11.6; 11.7f)	Lower border of mandible	Skin of lower lip and orbicularis oris	Facial	Depresses lower lip
Levator anguli oris (le-vā′ter an′gū-lī ōr′-ŭs) (figures 11.6; 11.7e)	Maxilla	Skin at angle of mouth and orbicularis oris	Facial	Elevates angle of mouth
Levator labii superioris (le-vā′ter lā′bē-ī sū-pēr′ē-ōr-is) (figures 11.6; 11.7b)	Maxilla	Skin and orbicularis oris of upper lip	Facial	Elevates upper lip
Levator labii superioris alaeque nasi (le-vā′ter lā′bē-ī sū-pēr′ē-ōr-ĭs ă-lak′ă nā′zī) (figures 11.6; 11.7b, f)	Maxilla	Ala at nose and upper lip	Facial	Elevates ala of nose and upper lip
Levator palpebrae superioris (le-vā′ter pal-pē′brē sū-pēr′ē-ōr-is) (figures 11.7a; 11.12a, b)	Lesser wing of sphenoid	Skin of eyelid	Oculomotor	Elevates upper eyelid

Several muscles function in moving the lips and the skin surrounding the mouth. The **orbicularis oris** (ōr-bik′yū-lā′ris ōr′is) and **buccinator** (buk′si-nā-tōr), the kissing muscles, pucker the mouth. Smiling is accomplished by the **zygomaticus** (zī′gō-mat′i-kŭs) **major** and **minor**, the **levator anguli** (ang′gyū-lī) **oris**, and the **risorius** (rī-sōr′ē-ŭs). Sneering is accomplished by the **levator labii** (lā′bē-ī) **superioris**; and frowning or pouting by the **depressor anguli oris**, the **depressor labii inferioris**, and the **mentalis** (men-tā′lis). If the mentalis muscles are well developed on each side of the chin, a chin dimple may be located between the two muscles.

2   **P R E D I C T**

Harry Wolf, a notorious flirt, on seeing Sally Gorgeous raises his eyebrows, winks, whistles, and smiles. Name the facial muscles he uses to carry out this communication. Sally, thoroughly displeased with this exhibition, frowns and flares her nostrils in disgust. What muscles does she use?

✔ *Answer in Appendix F*

Table 11.2   Muscles of Facial Expression (see figure 11.6)—cont'd

Muscle	Origin	Insertion	Nerve	Function
Mentalis (men-tā′lis) (figures 11.6; 11.7c)	Mandible	Skin of chin	Facial	Elevates and wrinkles skin over chin; elevates lower lip
Nasalis (nā′ză-lis) (figures 11.6b; 11.7d)	Maxilla	Bridge and ala of nose	Facial	Dilates nostril
Occipitofrontalis (ok-sip′i-tō-frŏn′tā′lis) (figures 11.6; 11.7a)	Occipital bone	Skin of eyebrow and nose	Facial	Moves scalp; elevates eyebrows
Orbicularis oculi (ōr-bik′yū-lā′ris ok′yū-lī) (figures 11.6; 11.7b)	Maxilla and frontal bones	Circles orbit and inserts near origin	Facial	Closes eye
Orbicularis oris (ōr-bik′yū-lā′ris ōr′is) (figures 11.6; 11.7d)	Nasal septum, maxilla, and mandible	Fascia and other muscles of lips	Facial	Closes lip
Platysma (plă-tiz′mă) (figures 11.6b; 11.7d)	Fascia of deltoid and pectoralis major	Skin over inferior border of mandible	Facial	Depresses lower lip; wrinkles skin of neck and upper chest
Procerus (prō-se′rŭs) (figures 11.6; 11.7b, c)	Bridge of nose	Frontalis	Facial	Creates horizontal wrinkle between eyes, as in frowning
Risorius (rī-sō′rē-ŭs) (figures 11.6; 11.7e, f)	Platysma and masseter fascia	Orbicularis oris and skin at corner of mouth	Facial	Abducts angle of mouth
Zygomaticus major (zī′gō-mat′i-kŭs) (figures 11.6; 11.7e, f)	Zygomatic bone	Angle of mouth	Facial	Elevates and abducts upper lip
Zygomaticus minor (zī′gō-mat′i-kŭs) (figures 11.6; 11.7e, f)	Zygomatic bone	Orbicularis oris of upper lip	Facial	Elevates and abducts upper lip

Mastication

Chewing, or **mastication** (mas′ti-kā′shŭn), involves force-fully closing the mouth (elevating the mandible) and grinding the food between the teeth (medial and lateral excursion of the mandible). The **muscles of mastication** and the **hyoid muscles** move the mandible (tables 11.3 and 11.4; figures 11.8 and 11.9). The elevators of the mandible are some of the strongest muscles of the body, bringing the mandibular teeth forcefully against the maxillary teeth to crush food. Slight mandibular depression involves relaxation of the mandibular elevators and the pull of gravity. Opening the mouth wide requires the action of the depressors of the mandible; and even though the muscles of the tongue and the buccinator (see table 11.2; table 11.5) are not involved in the actual process of chewing, they help move the food in the mouth and hold it in place between the teeth.

Tongue Movements

The tongue is very important in mastication and speech: (1) it moves the food around in the mouth; (2) with the buccinator it holds the food in place while the teeth grind it; and (3) it pushes the food up to the palate and back toward the pharynx to initiate swallowing. The tongue consists of a mass of **intrinsic muscles** (entirely within the tongue), which are involved in changing the shape of the tongue, and **extrinsic muscles** (outside of the tongue but attached to it), which help change the shape and move the tongue (see table 11.5; figure 11.10).

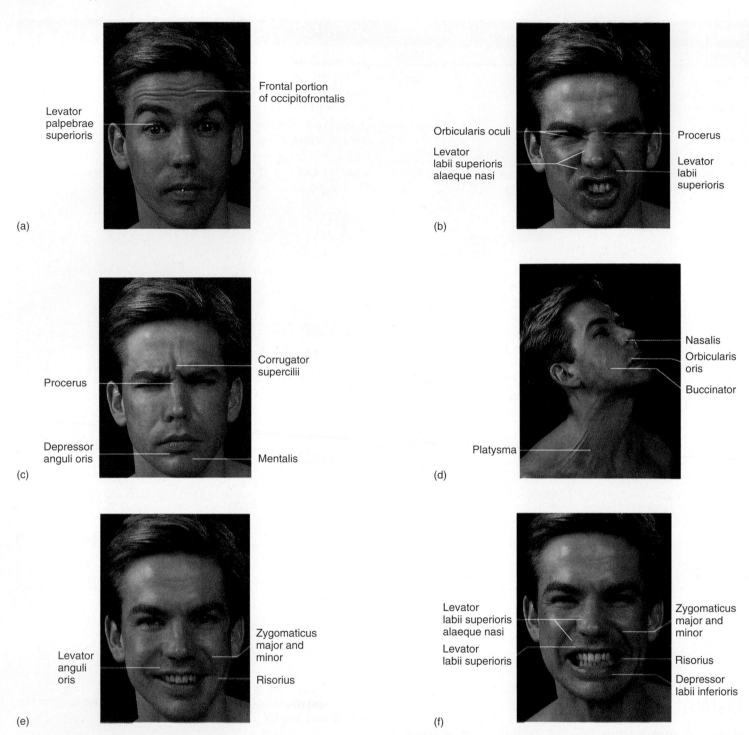

(a)

Levator
palpebrae
superioris

Frontal portion
of occipitofrontalis

(b)

Orbicularis oculi

Levator
labii superioris
alaeque nasi

Procerus

Levator
labii
superioris

(c)

Procerus

Depressor
anguli oris

Corrugator
supercilii

Mentalis

(d)

Platysma

Nasalis

Orbicularis
oris

Buccinator

(e)

Levator
anguli
oris

Zygomaticus
major and
minor

Risorius

(f)

Levator
labii superioris
alaeque nasi

Levator
labii superioris

Zygomaticus
major and
minor

Risorius

Depressor
labii inferioris

Figure 11.7 Surface Anatomy, Muscles of Facial Expression

Everyone can change the shape of her tongue, but not everyone can roll her tongue into the shape of a tube. The ability to accomplish such movements apparently is partially controlled genetically, but apparently other factors are involved. In some cases one of a pair of identical twins can roll the tongue but the other twin cannot. It is not known exactly what tongue muscles are involved in tongue rolling, and no anatomic differences are reported to exist between tongue rollers and nonrollers.

Swallowing and the Larynx

The hyoid muscles (see table 11.4 and figure 11.9) are divided into a **suprahyoid group** superior to the hyoid bone and an **infrahyoid group** inferior to the hyoid. When the hyoid bone is fixed by the infrahyoid muscles so that the bone is stabilized from below, the suprahyoid muscles can help depress the mandible. If the suprahyoid muscles fix the hyoid and thus stabilize it from above, the thyrohyoid muscle (an infrahyoid muscle) can elevate the larynx. To observe this effect, place your hand on your larynx (Adam's apple) and swallow.

Table 11.3   Muscles of Mastication (see figures 11.6 and 11.8)

Muscle	Origin	Insertion	Nerve	Function
Temporalis (tem′pŏ-rā′lis) (figures 11.6; 11.8a)	Temporal fossa	Anterior portion of mandibular ramus and coronoid process	Mandibular division of trigeminal	Elevates and retracts mandible; involved in excursion
Masseter (ma′se-ter) (figures 11.6; 11.8a)	Zygomatic arch	Lateral side of mandibular ramus	Mandibular division of trigeminal	Elevates and protracts mandible; involved in excursion
Pterygoids (ter′i-goydz)				
Lateral (figure 11.8b)	Pterygoid process and greater wing of sphenoid	Condylar process of mandible and articular disk	Mandibular division of trigeminal	Protracts and depresses mandible; involved in excursion
Medial (figure 11.8b)	Pterygoid process of sphenoid and tuberosity of maxilla	Medial surface of mandible	Mandibular division of trigeminal	Protracts and elevates mandible; involved in excursion

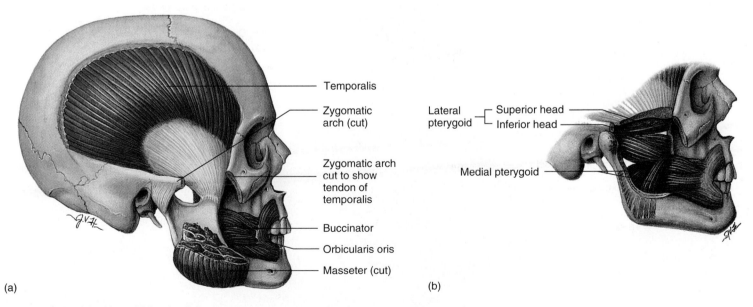

Temporalis

Zygomatic arch (cut)

Zygomatic arch cut to show tendon of temporalis

Buccinator

Orbicularis oris

Masseter (cut)

(a)

Lateral pterygoid — Superior head — Inferior head —

Medial pterygoid —

(b)

Figure 11.8   Muscles of Mastication

(a) Lateral (superficial) view. Masseter and zygomatic arch have been cut away to expose the temporalis. (b) Lateral (deep) view. Masseter and temporalis muscles have been removed, and the zygomatic arch and part of the mandible have been cut away to reveal the deeper muscles.

Table 11.4 Hyoid Muscles (see figures 11.9 and 11.10)

Muscle	Origin	Insertion	Nerve	Function
Suprahyoid Muscles				
Digastric (dī-gas′trik) (figure 11.9)	Mastoid process (posterior belly)	Mandible near midline (anterior belly)	Posterior belly—facial; anterior belly—mandibular division of trigeminal	Depresses and retracts mandible; elevates hyoid
Geniohyoid (je′nē-ō-hī′oyd) (figure 11.10)	Genu of mandible	Body of hyoid	Fibers of C1 and C2 with hypoglossal	Protracts hyoid; depresses mandible
Mylohyoid (mī′lō-hī′oyd) (figures 11.9; 11.11b)	Body of mandible	Hyoid	Mandibular division of trigeminal	Elevates floor of mouth and tongue; depresses mandible when hyoid is fixed
Stylohyoid (stī′lō-hī′oyd) (figures 11.9; 11.10)	Styloid process	Hyoid	Facial	Elevates hyoid
Infrahyoid Muscles				
Omohyoid (ō′mō-hī′oyd) (figure 11.9)	Superior border of scapula	Hyoid	Upper cervical through ansa cervicalis	Depresses hyoid; fixes hyoid in mandibular depression
Sternohyoid (ster′nō-hī′oyd) (figure 11.9)	Manubrium and first costal cartilage	Hyoid	Upper cervical through ansa cervicalis	Depresses hyoid; fixes hyoid in mandibular depression
Sternothyroid (ster′nō-thī′royd) (figure 11.9)	Manubrium and first or second costal cartilage	Thyroid cartilage	Upper cervical through ansa cervicalis	Depresses larynx; fixes hyoid in mandibular depression
Thyrohyoid (thī′rō-hī′oyd) (figure 11.9)	Thyroid cartilage	Hyoid	Upper cervical, passing with hypoglossal	Depresses hyoid and elevates thyroid cartilage of larynx; fixes hyoid in mandibular depression

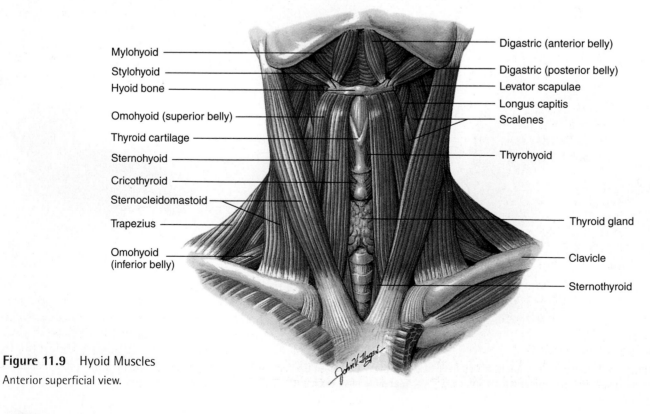

Figure 11.9 Hyoid Muscles

Anterior superficial view.

Table 11.5   Tongue Muscles (see figure 11.10)

Muscle	Origin	Insertion	Nerve	Function
Intrinsic Muscles				
Longitudinal, transverse, and vertical (not illustrated)	Within tongue	Within tongue	Hypoglossal	Change tongue shape
Extrinsic Muscles				
Genioglossus (jē′nē-ō-glos′ŭs) (figure 11.10)	Genu of mandible	Tongue	Hypoglossal	Depresses and protrudes tongue
Hyoglossus (hī′ō-glos′ŭs) (figures 11.10; 11.11b)	Hyoid	Side of tongue	Hypoglossal	Retracts and depresses side of tongue
Styloglossus (stī′lō-glos′ŭs) (figures 11.10; 11.11b)	Styloid process of temporal bone	Tongue (lateral and inferior)	Hypoglossal	Retracts tongue
Palatoglossus (pal′ă-tō-glos′ŭs) (figures 11.10; 11.11a)	Soft palate	Tongue	Pharyngeal plexus	Elevates posterior tongue

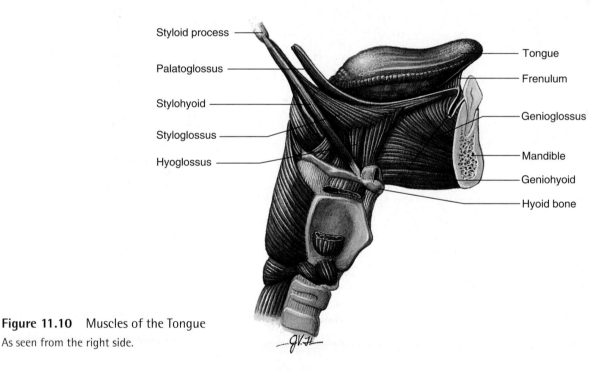

Figure 11.10   Muscles of the Tongue
As seen from the right side.

The soft palate, pharynx, and larynx contain several muscles involved in swallowing and speech (table 11.6 and figure 11.11). The muscles of the soft palate close the posterior opening to the nasal cavity during swallowing.

Swallowing (see chapter 24) is accomplished by elevation of the pharynx, which in turn is accomplished by elevation of the larynx, to which the pharynx is attached, and constriction of the **palatopharyngeus** (pal′ă-tō-far-in-jē′ŭs) and **salpingopharyngeus** (sal-pin′gō-far-in-jē′ŭs; *salpingo* means trumpet and refers to the trumpet-shaped opening of the auditory, or eustachian, tube). The pharyngeal constrictor muscles then constrict from superior to inferior, forcing the food into the esophagus.

The salpingopharyngeus also opens the auditory tube, which connects the middle ear with the pharynx. Opening the auditory tube equalizes the pressure between the middle ear and the atmosphere; this is why it is sometimes helpful to chew gum or swallow when ascending or descending a mountain in a car or when changing altitudes in an airplane.

Table 11.6   Muscles of Swallowing and the Larynx (see figure 11.11)

Muscle	Origin	Insertion	Nerve	Function
Larynx				
Arytenoids (ăr-i-tĕ′noydz)				
Oblique (not illustrated)	Arytenoid cartilage	Opposite arytenoid cartilage	Recurrent laryngeal	Narrows opening to larynx
Transverse (not illustrated)	Arytenoid cartilage	Opposite arytenoid cartilage	Recurrent laryngeal	Narrows opening to larynx
Cricoarytenoids (kri′kō-ăr-i-tĕ′noydz)				
Lateral (not illustrated)	Lateral side of cricoid cartilage	Arytenoid cartilage	Recurrent laryngeal	Narrows opening to larynx
Posterior (not illustrated)	Posterior side of cricoid cartilage	Arytenoid cartilage	Recurrent laryngeal	Widens opening of larynx
Cricothyroid (kri′kō-thī′-royd) (figures 11.9a; 11.11b)	Anterior cricoid cartilage	Thyroid cartilage	Superior laryngeal	Tenses vocal cords
Thyroarytenoid (thī′rō-ăr-i-tĕ′noyd) (not illustrated)	Thyroid cartilage	Arytenoid cartilage	Recurrent laryngeal	Shortens vocal cords
Vocalis (vō-kal′ĭs) (not illustrated)	Thyroid cartilage	Arytenoid cartilage	Recurrent laryngeal	Shortens vocal cords
Soft Palate				
Levator veli palatini (le-vā′ter vel′ī pal′ă-tē′nī) (figure 11.11a, b)	Temporal bone and auditory tube	Soft palate	Pharyngeal plexus	Elevates soft palate
Palatoglossus (pal′ă-tō-glos′ŭs) (figures 11.10; 11.11a)	Soft palate	Tongue	Pharyngeal plexus	Narrows fauces; elevates posterior tongue

The muscles of the larynx are listed in table 11.6 and are illustrated in figure 11.11b. Most of the laryngeal muscles help to narrow or close the laryngeal opening so food does not enter the larynx when a person swallows. The remaining muscles shorten the vocal cords to raise the pitch of the voice.

Movements of the Eyeball

The eyeball rotates within the orbital fossa, allowing vision in a wide range of directions. The movements of each eye are accomplished by six muscles named for the orientation of their fasciculi relative to the spherical eye (table 11.7; figure 11.12).

Each rectus muscle (so named because the fibers are nearly straight with the axis of the eye) attaches to the globe of the eye anterior to the center of the sphere. The superior rectus rotates the anterior portion of the globe superiorly so that the pupil, and thus the gaze, are directed superiorly (looking up). The inferior rectus depresses the gaze, the lateral rectus laterally deviates the gaze (looking to the side), and the medial rectus medially deviates the gaze (looking toward the nose). The superior rectus and inferior rectus are not completely straight in their orientation to the eye; thus they also medially deviate the gaze as they contract.

The oblique muscles (so named because their fibers are oriented obliquely to the axis of the eye) insert onto the posterolateral margin of the globe so that both muscles laterally de-

Table 11.6   Muscles of Swallowing and the Larynx (see figure 11.11)—cont'd

Muscle	Origin	Insertion	Nerve	Function
Soft Palate—cont'd				
Palatopharyngeus (pal'ă-tō-far-in-jē'ŭs) (figure 11.11a)	Soft palate	Pharynx	Pharyngeal plexus	Narrows fauces; depresses palate; elevates pharynx
Tensor veli palatini (ten'sor vel'ī pal'ă-tē'nī) (figure 11.11)	Sphenoid and auditory tube	Soft palate division of auditory tube	Mandibular, division of trigeminal	Tenses soft palate; opens auditory tube
Uvulae (ū'vū-lē) (figure 11.11a)	Posterior nasal spine	Uvula	Pharyngeal plexus	Elevates uvula
Pharynx				
Pharyngeal constrictors (far'in-jē'ăl)				
Inferior (figure 11.11b)	Thyroid and cricoid cartilages	Pharyngeal raphe	Pharyngeal plexus and external laryngeal nerve	Narrows lower pharynx in swallowing
Middle (figure 11.11b)	Stylohyoid ligament and hyoid	Pharyngeal raphe	Pharyngeal plexus	Narrows pharynx in swallowing
Superior (figure 11.11b)	Medial pterygoid plate, mandible, floor of mouth, and side of tongue	Pharyngeal raphe	Pharyngeal plexus	Narrows pharynx in swallowing
Salpingopharyngeus (sal-pin'gō-far'in-jē'ŭs) (figure 11.11a)	Auditory tube	Pharynx	Pharyngeal plexus	Elevates pharynx; opens auditory tube in swallowing
Stylopharyngeus (stī'lō-far'in-jē'ŭs) (figure 11.11b)	Styloid process	Pharynx	Glossopharyngeus	Elevates pharynx

viate the gaze as they contract. The superior oblique elevates the posterior part of the eye, thus directing the pupil inferiorly and depressing the gaze. The inferior oblique elevates the gaze.

3   P R E D I C T

Strabismus (stră-biz'mus) is a condition in which one or both eyes deviate in a medial or lateral direction. In some cases the condition may be caused by a weakness in either the medial or lateral rectus muscle. If the lateral rectus of the right eye is weak, in which direction would the eye deviate?

✔ *Answer in Appendix F*

Trunk Muscles
Muscles Moving the Vertebral Column

The muscles that extend the spine and abduct and rotate the vertebral column can be divided into deep and superficial groups (table 11.8). In general, the muscles of the deep group extend from vertebra to vertebra, whereas the muscles of the superficial group extend from the vertebrae to the ribs. In humans, these back muscles are very strong to maintain erect posture, but comparable muscles in cattle are relatively delicate, constituting the area from which New York (top loin) steaks are cut. The **erector** (ĕ-rek'tōr) **spinae** (spī'nē) group of muscles on each side of the back consists of three subgroups: the **iliocostalis** (il'ē-ō-kos-tā'lĭs), the **longissimus** (lon-gis'i-mŭs), and the **spinalis** (spī-nā'lĭs). The longissimus group accounts for most of the muscle mass in the lower back (figure 11.13).

Clinical Note

Low back pain can result from poor posture, being overweight, or from having a poor fitness level. A few changes may help: sitting and standing up straight, using a low back support when sitting, losing weight, exercising, especially the back and abdominal muscles, and sleeping on your side on a firm mattress. Sleeping on your side all night, however, may be difficult because most people change position over 40 times during the night.

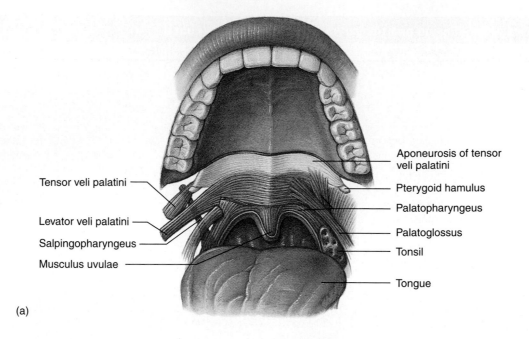

(a)

Tensor veli palatini

Levator veli palatini

Salpingopharyngeus

Musculus uvulae

Aponeurosis of tensor veli palatini

Pterygoid hamulus

Palatopharyngeus

Palatoglossus

Tonsil

Tongue

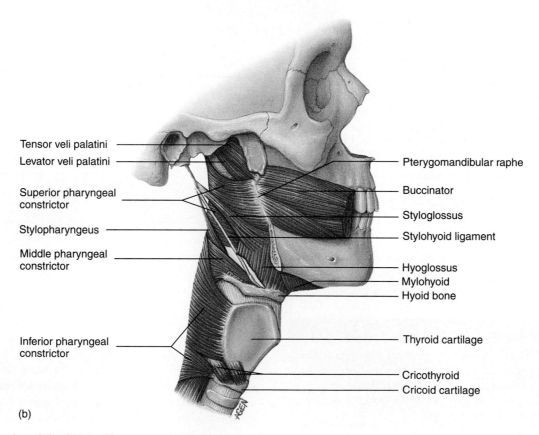

(b)

Tensor veli palatini

Levator veli palatini

Superior pharyngeal constrictor

Stylopharyngeus

Middle pharyngeal constrictor

Inferior pharyngeal constrictor

Pterygomandibular raphe

Buccinator

Styloglossus

Stylohyoid ligament

Hyoglossus

Mylohyoid

Hyoid bone

Thyroid cartilage

Cricothyroid

Cricoid cartilage

Figure 11.11   Muscles of the Palate, Pharynx, and Larynx

(*a*) Inferior view of the palate. Palatoglossus and part of the palatopharyngeus muscles have been cut on one side to reveal the deeper muscles.
(*b*) Lateral view of the palate, pharynx, and larynx. Part of the mandible has been removed to reveal the deeper structures.

Table 11.7   Muscles Moving the Eye (see figure 11.12)

Muscle	Origin	Insertion	Nerve	Function
Oblique				
Inferior (figure 11.7)	Orbital plate of maxilla	Sclera of eye	Oculomotor	Elevates and laterally deviates gaze
Superior (figure 11.7)	Fibrous ring	Sclera of eye	Trochlear	Depresses and laterally deviates gaze
Rectus				
Inferior (figure 11.7)	Fibrous ring	Sclera of eye	Oculomotor	Depresses and medially deviates gaze
Lateral (figure 11.7)	Fibrous ring	Sclera of eye	Abducens	Laterally deviates gaze
Medial (figure 11.7)	Fibrous ring	Sclera of eye	Oculomotor	Medially deviates gaze
Superior (figure 11.7)	Fibrous ring	Sclera of eye	Oculomotor	Elevates and medially deviates gaze

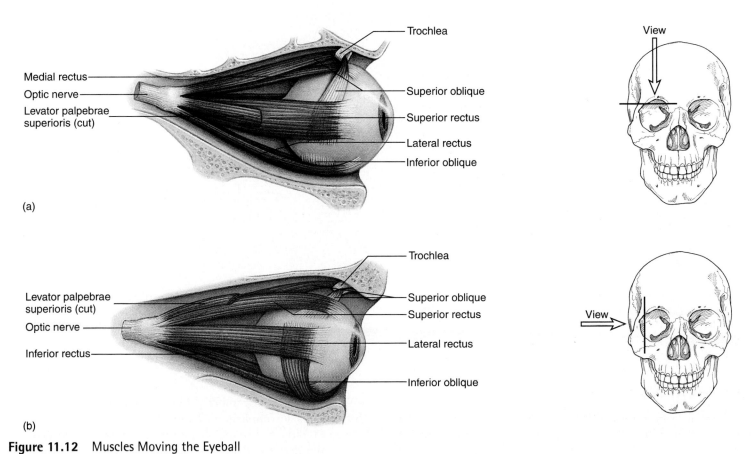

Figure 11.12   Muscles Moving the Eyeball

(*a*) Superior view of the right eyeball. (*b*) Lateral view of the right eyeball.

Table 11.8   Muscles Acting on the Vertebral Column (see figures 11.4 and 11.13)

Muscle	Origin	Insertion	Nerve	Function
Superficial				
Erector spinae (ĕ-rek′tōr spī′nē) (divides into three columns)				
Iliocostalis (il′ē-ō-kos-tā′lis)	Sacrum, ilium, and lumbar spines	Ribs and vertebrae	Dorsal rami of spinal nerves	Extends vertebral column
Cervicis (ser′vi-sis) (figures 11.4c; 11.13)	Superior six ribs	Middle cervical vertebrae	Dorsal rami of thoracic nerves	Extends, laterally flexes, and rotates vertebral column
Thoracis (thō-ra′sis) (figure 11.13)	Inferior six ribs	Superior six ribs	Dorsal rami of thoracic nerves	Extends, laterally flexes, and rotates vertebral column
Lumborum, (lum-bōr′ŭm) (figure 11.13)	Sacrum, ilium, and lumbar vertebrae	Inferior six ribs	Dorsal rami of thoracic and lumbar nerves	Extends, laterally flexes, and rotates vertebral column
Longissimus (lon-gis′i-mŭs)				
Capitis, (ka′pĭ-tis) (figures 11.4c; 11.13)	Upper thoracic and lower cervical vertebrae	Mastoid process	Dorsal rami of cervical nerves	Extends head
Cervicis (ser′vĭ-sis) (figures 11.4c; 11.13)	Upper thoracic vertebrae	Upper cervical vertebrae	Dorsal rami of cervical nerves	Extends neck
Thoracis (thō-ra′sis) (figure 11.13)	Ribs and lower thoracic vertebrae	Upper lumbar vertebrae and ribs	Dorsal rami of thoracic and lumbar nerves	Extends vertebral column
Spinalis (spī-nā′lis)				
Cervicis (ser′vĭ-sis) (not illustrated)	C6–C7	C2–C3	Dorsal rami of cervical nerves	Extends neck
Thoracis (thō-ra′sis) (figure 11.13)	T11–L2	Middle and upper thoracic vertebrae	Dorsal rami of thoracic nerves	Extends vertebral column

Thoracic Muscles

The muscles of the thorax are involved almost entirely in the process of breathing (see chapter 23). Four major groups of muscles are associated with the rib cage (table 11.9 and figure 11.14). The **scalene** (skā′lēn) muscles elevate the first two ribs during inspiration. The **external intercostals** (in′ter-kos′tulz) also elevate the ribs during inspiration. The **internal intercostals** and **transversus thoracis** (thō-ra′sis) muscles contract during forced expiration.

The major movement produced during quiet breathing, however, is accomplished by the **diaphragm** (dī′ă-fram; see figure 11.14a). It is dome-shaped when relaxed; when it contracts, the dome is flattened, causing the volume of the thoracic cavity to increase, resulting in inspiration. If this dome of skeletal muscle or the phrenic nerve supplying it is severely damaged, the amount of air exchanged in the lungs may be so small that the individual is likely to die unless connected to an artificial respirator.

Table 11.8   Muscles Acting on the Vertebral Column (see figures 11.4 and 11.13)—cont'd

Muscle	Origin	Insertion	Nerve	Function
Superficial—cont'd				
Erector spinae—cont'd (ĕ-rek'tōr spī'nē) (divides into three columns)				
Longus colli (lon'gŭs kō'lī) (not illustrated)	C3–T3	C1–C6	Ventral rami of cervical nerves	Rotates and flexes neck
Splenius cervicis (sple'nē-ŭs ser'vĭ-sis) (figure 11.4b)	C3–C5	C1–C3	Dorsal rami of cervical nerves	Rotates and extends neck
Deep				
Interspinales (in'ter-spī-nā'lēz) (figures 11.4c; 11.13)	Spinous processes of all vertebrae	Next superior spinous process	Dorsal rami of spinal nerves	Extends back and neck
Intertransversarii (in'ter-trans'ver-săr'ē-ī) (figure 11.13)	Transverse processes of all vertebrae	Next superior transverse process	Dorsal rami of spinal nerves	Laterally flexes vertebral column
Multifidus (mul-tif'ĭ-dŭs) (figures 11.4c; 11.13)	Transverse processes of vertebrae, posterior surface of sacrum and ilium	Spinous processes of next superior vertebrae	Dorsal rami of spinal nerves	Extends and rotates vertebral column
Psoas minor (sō'ŭs mī'nor) (figure 11.27a)	T12–L1	Near pubic crest	L1	Flexes vertebral column
Rotatores (rō-tā'tōrz) (not illustrated)	Transverse processes of all vertebrae	Base of spinous process of superior vertebrae	Dorsal rami of spinal nerves	Extends and rotates vertebral column
Semispinalis (sem'ē-spī-nā'lis)				
Cervicis (ser'vĭ-sis) (figures 11.4c; 11.13)	Transverse processes of T2–T5	Spinous processes of C2–C5	Dorsal rami of cervical nerves	Extends neck
Thoracis (thō-ra'sis) (figure 11.13)	Transverse processes of T5–T11	Spinous processes of C5–T4	Dorsal rami of thoracic nerves	Extends vertebral column

Abdominal Wall

The muscles of the anterior abdominal wall (table 11.10) flex and rotate the vertebral column. Contraction of the abdominal muscles when the vertebral column is fixed decreases the volume of the abdominal cavity and the thoracic cavity and can aid in such functions as forced expiration, vomiting, defecation, urination, and childbirth. The crossing pattern of the abdominal muscles creates a strong anterior wall that holds in and protects the abdominal viscera.

In a relatively muscular person with little fat, a vertical line is visible, extending from the area of the xiphoid process of the sternum through the navel to the pubis. This tendinous area of the abdominal wall is devoid of muscle; the **linea alba** (lin'ē-ă al'bă), or white line, is so named because it consists of white connective tissue rather than muscle (figure 11.15). On each side of the linea alba is the **rectus abdominis** (figures 11.16 and 11.17). **Tendinous intersections** (tendinous inscriptions) transect the rectus abdominis at three, or sometimes more, locations, causing the abdominal wall of a well-muscled person to appear segmented. Lateral to the rectus abdominis is the **linea semilunaris** (sem'ē-lū-nar'is, meaning a crescent- or half-moon-shaped line); lateral to it are three layers of muscle (see figures 11.15 through 11.17). From superficial to deep, these muscles are the **external abdominal oblique, internal abdominal oblique,** and **transversus abdominis.**

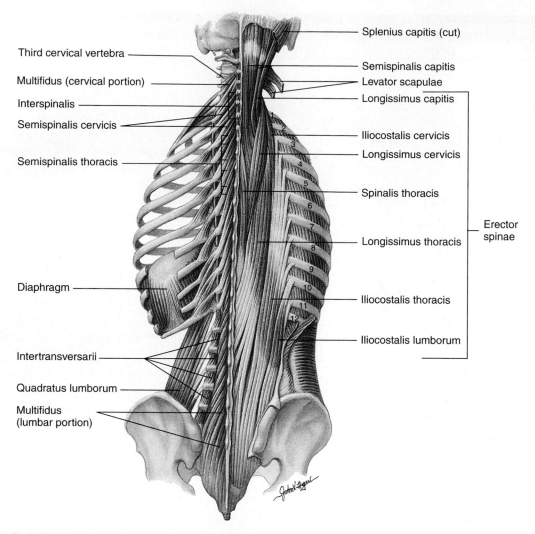

Third cervical vertebra
Multifidus (cervical portion)
Interspinalis
Semispinalis cervicis
Semispinalis thoracis
Diaphragm
Intertransversarii
Quadratus lumborum
Multifidus (lumbar portion)

Splenius capitis (cut)
Semispinalis capitis
Levator scapulae
Longissimus capitis
Iliocostalis cervicis
Longissimus cervicis
Spinalis thoracis
Longissimus thoracis
Iliocostalis thoracis
Iliocostalis lumborum

Erector spinae

Figure 11.13   Deep Back Muscles

On the right, the erector spinae group of muscles is demonstrated. On the left, these muscles have been removed to reveal the deeper back muscles.

Pelvic Floor and Perineum

The pelvis is a ring of bone (see chapter 7) with an inferior opening that is closed by a muscular wall through which the anus and the urogenital openings penetrate (table 11.11). Most of the pelvic floor is formed by the **coccygeus** (kok-si'jē-ŭs) muscle and the **levator ani** (a'nī) muscle, referred to jointly as the **pelvic diaphragm.** The area inferior to the pelvic floor is the **perineum** (per'i-nē'ŭm), which is somewhat diamond-shaped (figure 11.18). The anterior half of the diamond is the urogenital triangle, and the posterior half is the anal triangle (see chapter 28). The urogenital triangle contains the **urogenital diaphragm,** which forms a "subfloor" to the pelvis in that area and consists of the **deep transverse peroneus** (pĕr'ĭ-nē'us) muscle and the **sphincter urethrae** (ū-rē'thrē) muscle. During pregnancy the muscles of the pelvic diaphragm and urogenital diaphragm may be stretched by the extra weight of the fetus, and specific exercises are designed to strengthen them.

Upper Limb Muscles

The muscles of the upper limb include the ones that attach the limb and girdle to the body, and those that are in the arm, forearm, and hand.

Scapular Movements

The major connection of the upper limb to the body is accomplished by muscles (table 11.12 and figure 11.19). The muscles attaching the scapula to the thorax include the **trapezius, levator scapulae** (skap'yū-lē), **rhomboideus** (rom-bō-id'ē-ŭs) **major** and **minor, serratus** (sĕr-ā'tŭs) **anterior,** and **pectoralis** (pek'tō-ra'lis) **minor.** These muscles move the scapula, permitting a wide range of movements of the upper limb, or act as fixators to hold the scapula firmly in position when the muscles of the arm contract. The superficial muscles that act on the scapula can be easily seen on a

Table 11.9 Muscles of the Thorax (see figure 11.14)

Muscle	Origin	Insertion	Nerve	Function
Diaphragm (figure 11.14a)	Interior of ribs, sternum, and lumbar vertebrae	Central tendon of diaphragm	Phrenic	Inspiration; depresses floor of thorax
Intercostalis (in'ter-kos-ta'lis)				
External (figure 11.14a, b)	Inferior margin of each rib	Superior border of next rib below	Intercostal	Inspiration; elevates ribs
Internal (figure 11.14a, b)	Superior margin of each rib	Inferior border of next rib above	Intercostal	Expiration; depresses ribs
Scalenus (skā-lē'nŭs) (figure 11.9)				
Anterior	C3–C6	First rib	Cervical plexus	Elevates first rib
Medial	C2–C6	First rib	Cervical plexus	Elevates first rib
Posterior	C4–C6	Second rib	Cervical and brachial plexuses	Elevates second rib
Serratus posterior (sĕr-ā'tŭs)				
Inferior (not illustrated)	T11–L2	Inferior four ribs	Ninth to twelfth intercostals	Depresses inferior ribs and extends back
Superior (not illustrated)	C6–T2	Second to fifth ribs	First to fourth intercostals	Elevates superior ribs
Transversus thoracis (trans-ver'sus thō-ra'sis) (not illustrated)	Sternum and xiphoid process	Second to sixth costal cartilages	Intercostal	Decreases diameter of thorax

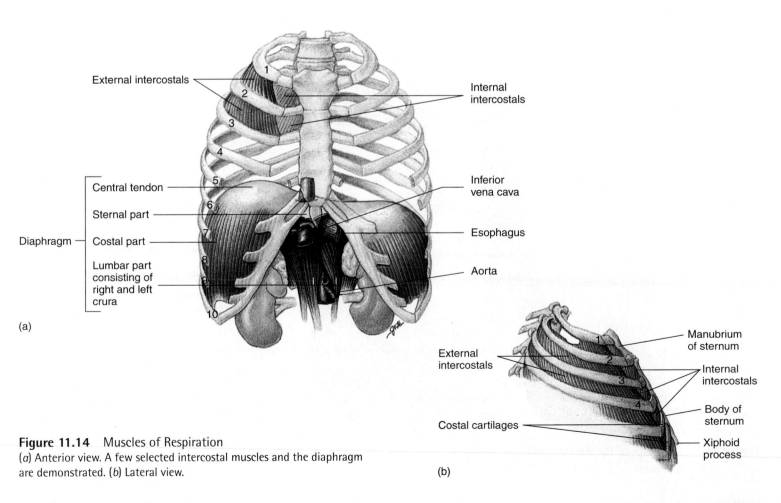

Figure 11.14 Muscles of Respiration
(a) Anterior view. A few selected intercostal muscles and the diaphragm are demonstrated. (b) Lateral view.

Table 11.10   Muscles of the Abdominal Wall (see figures 11.3, 11.16, and 11.17)

Muscle	Origin	Insertion	Nerve	Function
Anterior				
Rectus abdominis (rek'tŭs ab-dom'i-nis) (figures 11.3a; 11.15; 11.16; 11.17)	Pubic crest and symphysis pubis	Xiphoid process and inferior ribs	Branches of lower thoracic	Flexes vertebral column; compresses abdomen
External abdominal oblique (figures 11.3a, b; 11.15; 11.16; 11.19b)	Fifth to twelfth ribs	Iliac crest, inguinal ligament, and rectus sheath	Branches of lower thoracic	Flexes and rotates vertebral column; compresses abdomen; depresses thorax
Internal abdominal oblique (figures 11.15; 11.16)	Iliac crest, inguinal ligament, and lumbar fascia	Tenth to twelfth ribs and rectus sheath	Lower thoracic	Flexes and rotates vertebral column; compresses abdomen; depresses thorax
Transversus abdominis (trans-ver'sŭs ab-dom'i-nis) (figures 11.15; 11.16)	Seventh to twelfth costal cartilages, lumbar fascia, iliac crest, and inguinal ligament	Xiphoid process, linea alba, and pubic tubercle	Lower thoracic	Compresses abdomen
Posterior				
Quadratus lumborum (kwah-drā'tŭs lŭm-bōr'ŭm) (figure 11.13)	Iliac crest and lower lumbar vertebrae	Twelfth rib and upper lumbar vertebrae	Upper lumbar	Laterally flexes vertebral column and depresses twelfth rib

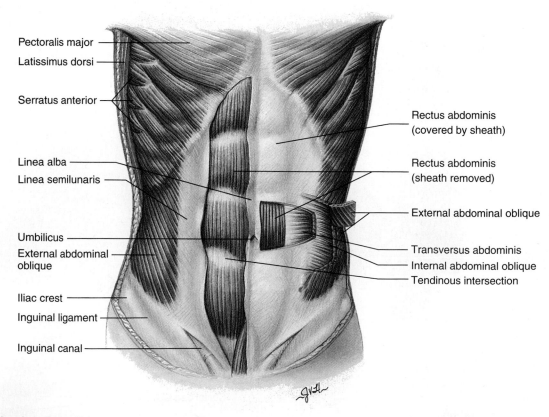

Figure 11.15   Muscles of the Anterior Abdominal Wall

Windows have been made in the side to reveal the various muscle layers.

Figure 11.16
Muscles of the Anterior Abdominal Wall

Cross section superior to the umbilicus.

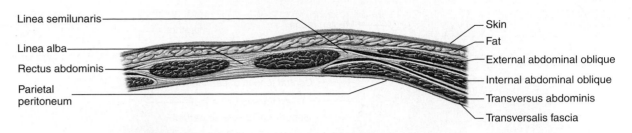

Linea semilunaris
Linea alba
Rectus abdominis
Parietal peritoneum

Skin
Fat
External abdominal oblique
Internal abdominal oblique
Transversus abdominis
Transversalis fascia

Figure 11.17
Surface Anatomy, Muscles of the Anterior Abdominal Wall

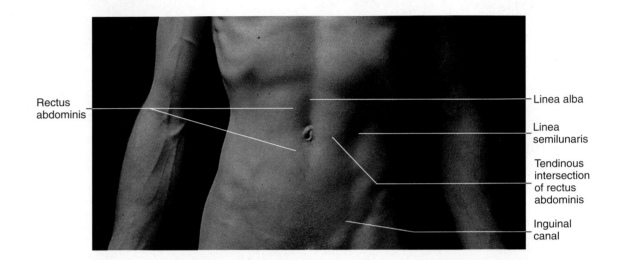

Rectus abdominis

Linea alba
Linea semilunaris
Tendinous intersection of rectus abdominis
Inguinal canal

Table 11.11	**Muscles of the Pelvic Floor and Perineum (see figure 11.18)**			
Muscle	**Origin**	**Insertion**	**Nerve**	**Function**
Bulbospongiosus (bul'bō-spŭn'jē-ō'sŭs) (figure 11.18)	Male—central tendon of perineum and median raphe of penis	Dorsal surface of penis and bulb of penis	Pudendal	Constricts urethra; erects penis
	Female—central tendon of perineum	Base of clitoris	Pudendal	Erects clitoris
Coccygeus (kok-si'jē-ŭs) (not illustrated)	Ischial spine	Coccyx	S3 and S4	Elevates and supports pelvic floor
Ischiocavernosus (ish'ē-ō-kav'er-nō'sŭs) (figure 11.18)	Ischial ramus	Corpus cavernosum	Perineal	Compresses base of penis or clitoris
Levator ani (le-vā'ter ā'nī) (figure 11.18)	Posterior pubis and ischial spine	Sacrum and coccyx	Fourth sacral	Elevates anus; supports pelvic viscera
Sphincter ani externus (sfing'ter ā'nī ex-ter'nŭs) (figure 11.18)	Coccyx	Central tendon of perineum	Fourth sacral and pudenda	Keeps orifice of anal canal closed
Sphincter urethrae (sfingk'ter ū-rē'thrē) (not illustrated)	Pubic ramus	Median raphe	Pudendal	Constricts urethra
Transverse perinei (pĕr'i-nē'ī)				
Deep (figure 11.18)	Ischial ramus	Median raphe	Pudendal	Supports pelvic floor
Superficial (figure 11.18a)	Ischial ramus	Central perineal tendon	Pudendal	Fixes central tendon

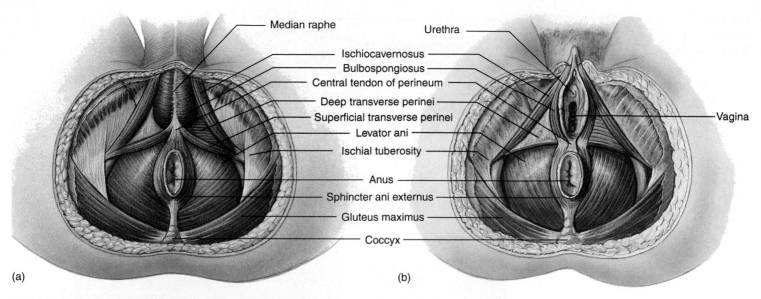

Figure 11.18 Muscles of the Pelvic Floor and Perineum. Inferior view.
(a) Male. (b) Female.

Table 11.12	Muscles Acting on the Scapula (see figure 11.19)			
Muscle	**Origin**	**Insertion**	**Nerve**	**Function**
Levator scapulae (le-vā′ter skap′ū-lē) (figures 11.4c; 11.19a; 11.20b)	C1–C4	Superior angle of scapula	Dorsal scapular	Elevates, retracts, and rotates scapula; laterally flexes neck
Pectoralis minor (pek′tō-ra′lis) (figure 11.19b)	Third to fifth ribs	Coracoid process of scapula	Anterior thoracic	Depresses scapula or elevates ribs
Rhomboideus (rom-bō-id′ē-ŭs)				
Major (figures 11.19a; 11.20b)	T1–T4	Medial border of scapula	Dorsal scapular	Retracts, rotates, and fixes scapula
Minor (figures 11.19a; 11.20b)	C6–C7	Medial border of scapula	Dorsal scapular	Retracts, slightly elevates, rotates, and fixes scapula
Serratus anterior (ser-ā′tŭs) (figures 11.3a; 11.19b; 11.20a; 11.21a; 11.22a)	First to ninth ribs	Medial border of scapula	Long thoracic	Rotates and protracts scapula; elevates ribs
Subclavius (sŭb-klā′vē-ŭs) (figure 11.19b)	First rib	Clavicle	Subclavian	Fixes clavicle or elevates first rib
Trapezius (tra-pē′zē-ŭs) (figures 11.3a, b; 11.19a; 11.21a–d)	External occipital protuberance, ligamentum nuchae, and C7–T12	Clavicle, acromion process, and scapular spine	Accessory and cervical plexus	Elevates, depresses, retracts, rotates, and fixes scapula; extends neck

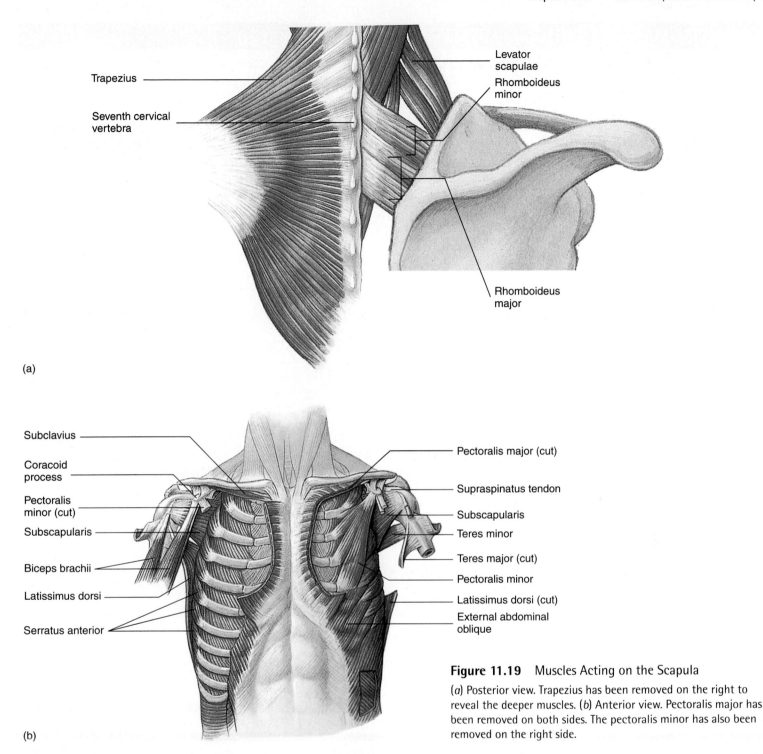

(a)

(b)

Figure 11.19   Muscles Acting on the Scapula

(*a*) Posterior view. Trapezius has been removed on the right to reveal the deeper muscles. (*b*) Anterior view. Pectoralis major has been removed on both sides. The pectoralis minor has also been removed on the right side.

living person (see figure 11.21): the trapezius forms the upper line from each shoulder to the neck, and the origin of the serratus anterior from the first eight or nine ribs can be seen along the lateral thorax.

Arm Movements

The arm is attached to the thorax by the **pectoralis major** and the **latissimus dorsi** (lă-tis′i-mŭs dōr′sī) muscles (table

11.13 and figure 11.20; see figure 11.19*b*). Notice that the pectoralis major muscle is listed in table 11.13 as both a flexor and extensor. The muscle flexes the extended arm and extends the flexed arm. Try these movements yourself and notice the position and action of the muscle. The **deltoid** (deltoideus) muscle also is listed in table 11.13 as a flexor and extensor. The deltoid muscle is like three muscles in one: the anterior fibers flex the arm; the lateral fibers abduct the arm; and the posterior fibers extend the arm. The deltoid muscle is part of the

Table 11.13   Muscles Acting on the Arm (see figures 11.19, 11.20, 11.21, and 11.22)

Muscle	Origin	Insertion	Nerve	Function
Coracobrachialis (kōr′ă-kō-brā-kē-a′lis) (figures 11.20a; 11.22c)	Coracoid process of scapula	Midshaft of humerus	Musculocutaneous	Adducts and flexes arm
Deltoid (del′toyd) (figures 11.3a, b; 11.20a; 11.21; 11.22a, b; 11.25)	Clavicle, acromion process, and scapular spine	Deltoid tuberosity	Axillary	Abducts, flexes, extends, and medially and laterally rotates arm
Latissimus dorsi (lă-tis′i-mŭs dōr′sī) (figures 11.3b; 11.19b; 11.20b; 11.21b)	T7–L5, sacrum and iliac crest	Medial crest of intertubercular groove	Thoracodorsal	Adducts, medially rotates, and extends arms
Pectoralis major (pek′tō-rā′lis) (figures 11.3a; 11.20a; 11.21a, b; 11.22a)	Clavicle, sternum, and abdominal aponeurosis	Lateral crest of intertubercular groove	Anterior thoracic	Adducts, flexes, and medially rotates arm; extends arm from flexed position
Teres major (te′rēz) (figures 11.3b; 11.19b; 11.20b; 11.21c, d; 11.22c)	Lateral border of scapula	Medial crest of intertubercular groove	Subscapular C5 and C6	Adducts, extends, and medially rotates arm
Rotator Cuff				
Infraspinatus (in′fră-spī-nā′tŭs) (figures 11.3b; 11.20b; 11.21c, d)	Infraspinous fossa of scapula	Greater tubercle of humerus	Suprascapular C5 and C6	Extends and laterally rotates arm
Subscapularis (sŭb′skap-yū-lā′ris) (figures 11.19b; 11.20c)	Subscapular fossa	Lesser tubercle of humerus	Subscapular C5 and C6	Extends and medially rotates arm
Supraspinatus (sū′pră-spī-nā′tŭs) (figures 11.19b; 11.20b, c)	Supraspinous fossa	Greater tubercle of humerus	Suprascapular C5 and C6	Abducts arm
Teres minor (te′rēz) (figures 11.3b; 11.19b; 11.20b, c)	Lateral border of scapula	Greater tubercle of humerus	Axillary C5 and C6	Adducts, extends, and laterally rotates arm

group of muscles that binds the humerus to the scapula. The primary muscles holding the head of the humerus in the glenoid fossa, however, are called the **rotator cuff muscles** (listed separately in table 11.13) because they form a cuff or cap over the proximal humerus (figure 11.20c). A rotator cuff injury involves damage to one or more of these muscles or their tendons, usually the supraspinatus muscle. The muscles moving the arm are involved in flexion, extension, abduction, adduction, rotation, and circumduction (table 11.14).

Abduction of the arm involves the deltoid, rotator cuff muscles, and the trapezius. Abduction from the anatomic position through the first 90 degrees (to the point at which the hand is level to the shoulder) is accomplished almost entirely by the deltoid muscle. Place your hand on your deltoid and feel it contract as you abduct 90 degrees. Abduction from 90 degrees to 180 degrees, so that the hand is held high above the head, primarily involves rotation of the scapula, which is accomplished by the trapezius. Feel the inferior angle of your scapula as you abduct to 90 degrees and then to 180 degrees. There's a big difference. Abduction from 90 degrees to 180 degrees, however, cannot occur unless the head of the humerus is held tightly in the glenoid fossa by the rotator cuff muscles. Damage to the supraspinatus muscle can prevent abduction past 90 degrees.

4   P R E D I C T

A tennis player complains of pain in the shoulder when attempting to serve or when attempting an overhead volley (extreme abduction). What rotator cuff muscle is probably damaged?

✔ *Answer in Appendix F*

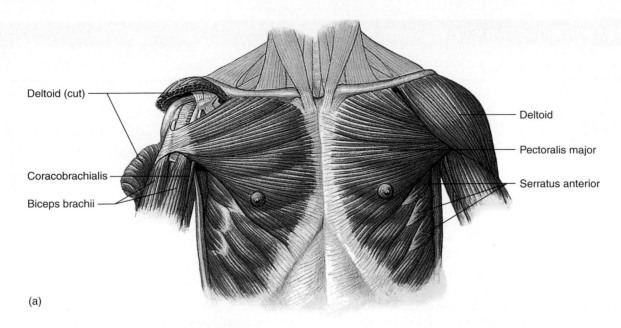

Deltoid (cut)

Coracobrachialis

Biceps brachii

Deltoid

Pectoralis major

Serratus anterior

(a)

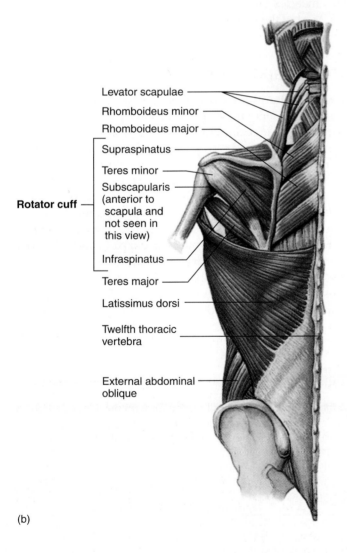

Levator scapulae

Rhomboideus minor

Rhomboideus major

Supraspinatus

Teres minor

Subscapularis (anterior to scapula and not seen in this view)

Infraspinatus

Teres major

Latissimus dorsi

Twelfth thoracic vertebra

External abdominal oblique

Rotator cuff

(b)

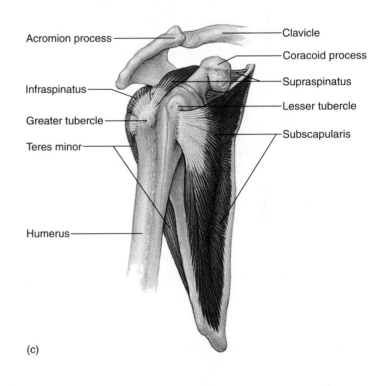

Acromion process

Infraspinatus

Greater tubercle

Teres minor

Humerus

Clavicle

Coracoid process

Supraspinatus

Lesser tubercle

Subscapularis

(c)

Figure 11.20   Muscles Attaching the Upper Limb to the Body

(a) Anterior view. (b) Posterior view. (c) Anterior view of the rotator cuff, showing the teres minor, infraspinatus, supraspinatus, and subscapularis muscles.

Table 11.14   Summary of Muscle Actions on the Arm

Flexion	Extension	Abduction	Adduction	Medial Rotation	Lateral Rotation
Deltoid	Deltoid	Deltoid	Pectoralis major	Pectoralis major	Deltoid
Pectoralis major	Teres major	Supraspinatus	Latissimus dorsi	Teres major	Infraspinatus
Coracobrachialis	Lattissimus dorsi		Teres major	Lattissimus dorsi	Teres minor
Biceps brachii	Pectoralis major		Teres minor	Deltoid	
	Triceps brachii		Triceps brachii	Subscapularis	
			Coracobrachialis		

Several muscles acting on the arm can be seen very clearly in the living individual (figure 11.21). The pectoralis major forms the upper chest, and the deltoids are prominent over the shoulders. The deltoid is a common site for administering injections.

Forearm Movements

The surface anatomy of the arm muscles is illustrated in figure 11.21. The triceps constitute the main mass visible on the posterior aspect of the arm (see figure 11.25). The biceps brachii is readily visible on the anterior aspect of the arm. The brachialis lies deep to the biceps and can be seen only as a small mass on the medial and lateral sides of the arm. The brachioradialis forms a bulge on the anterolateral side of the forearm just distal to the elbow. If the elbow is forcefully flexed in the midprone position (midway between pronation and supination), the brachioradialis stands out clearly on the forearm.

Flexion and Extension of the Forearm

Extension of the forearm is accomplished by the **triceps brachii** (brā′kē-ī) and **anconeus** (ang-kō′nē-ŭs); flexion of the forearm is accomplished by the **brachialis** (brā′kē-al′is), **biceps brachii,** and **brachioradialis** (brā′kē-ō-rā′dē-al′is; table 11.15; figure 11.22).

Supination and Pronation

Supination of the forearm is accomplished by the **supinator** and the **biceps brachii** (see figure 11.22b; figure 11.23c and d). Pronation is a function of the **pronator quadratus** (kwah-drā′tŭs) and the **pronator teres** (te′rēz) (figure 11.23a and c).

5   P R E D I C T

Explain the difference between doing chin-ups with the forearm supinated versus pronated. Which muscle or muscles are used in each type of chin-up? Which type is easier? Why?

✔ *Answer in Appendix F*

Wrist, Hand, and Finger Movements

The forearm muscles can be divided into anterior and posterior groups (table 11.16; see figure 11.23). Most of the anterior forearm muscles are responsible for flexion of the wrist and fingers. Most of the posterior forearm muscles cause extension of the wrist and fingers.

Extrinsic Hand Muscles

The **extrinsic hand muscles** are in the forearm but have tendons that extend into the hand. A strong band of fibrous connective tissue, the **retinaculum** (ret-i-nak′y ū-lŭm, meaning bracelet), covers the flexor and extensor tendons and holds them in place around the wrist so that they do not "bowstring" during muscle contraction (see figures 11.23e).

Two major anterior muscles, the **flexor carpi** (kar′pī) **radialis** (rā-dī-ă′lis) and the **flexor carpi ulnaris** (ŭl-nā′ris), flex the wrist; and three posterior muscles, the **extensor carpi radialis longus,** the **extensor carpi radialis brevis,** and the **extensor carpi ulnaris,** extend the wrist. The wrist flexors and extensors are visible on the anterior and posterior surfaces of the forearm. The tendon of the flexor carpi radialis is an important landmark because the radial pulse can be felt just lateral to the tendon (see figure 11.23a).

Clinical Note

Forceful repeated pronation and supination, with the wrist extended and the fingers flexed, as may occur when grasping a tennis racket and participating in a tennis match, can result in inflammation and pain where the extensor muscles originate on the lateral humeral epicondyle. This condition, referred to as **"tennis elbow,"** may occur as a result of any activity involving a similar motion, such as shoveling snow.

Flexion of the four medial digits is a function of the **flexor digitorum** (dij′i-tor′ŭm) **superficialis** and **flexor digitorum profundus** (prō-fŭn′dŭs meaning deep). Extension is

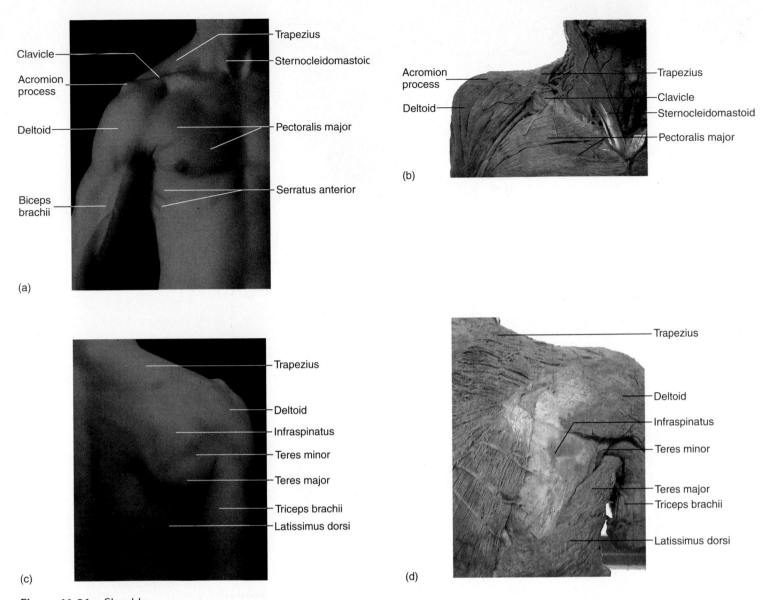

Figure 11.21   Shoulder

(*a*) Surface anatomy of the anterior shoulder. (*b*) Photograph showing a dissection of the anterior shoulder. (*c*) Surface anatomy of the posterior shoulder. (*d*) Photograph showing a dissection of the posterior shoulder.

accomplished by the **extensor digitorum.** The tendons of this muscle are very visible on the dorsum of the hand (see figure 11.25*b*). The little finger has an additional extensor, the **extensor digiti minimi** (di′ji-tī min′i-mī). The index finger also has an additional extensor, the **extensor indicis** (in′di-sis).

Movement of the thumb is caused in part by the **abductor pollicis** (pol′i-sis) **longus,** the **extensor pollicis longus,** and the **extensor pollicis brevis.** These tendons form the sides of a depression on the posterolateral side of the wrist called the "anatomical snuffbox" (see figure 11.25*b*). When snuff was in use, a small pinch could be placed into the anatomical snuffbox and inhaled through the nose.

Intrinsic Hand Muscles

The **intrinsic hand muscles** are entirely within the hand (table 11.17 and figure 11.24). Abduction of the fingers is accomplished by the **interossei** (in-ter-os′ē-ī) **dorsales** (dōr-sa′lēz) and the **abductor digiti minimi,** whereas adduction is a function of the **interossei palmares** (pal-măr′ēz).

The **flexor pollicis brevis,** the **abductor pollicis brevis,** and the **opponens pollicis** form a fleshy prominence at the base of the thumb called the **thenar** (thē′nar) **eminence** (see figures 11.24 and 11.25*a*). The **abductor digiti minimi, flexor digiti minimi brevis,** and **opponens digiti minimi**

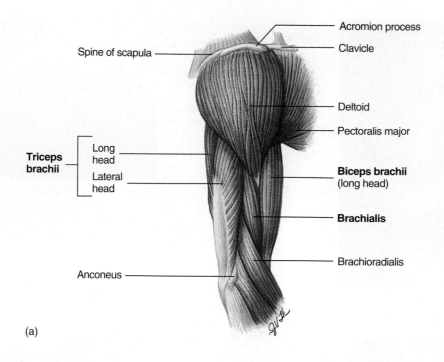

Acromion process
Clavicle
Spine of scapula
Deltoid
Pectoralis major
Triceps brachii — Long head
Lateral head
Biceps brachii (long head)
Brachialis
Brachioradialis
Anconeus

(a)

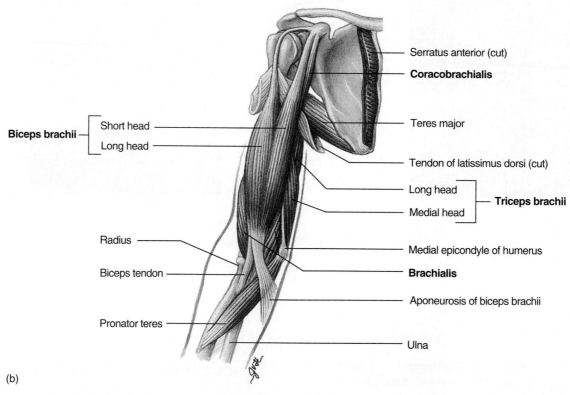

Serratus anterior (cut)
Coracobrachialis
Biceps brachii — Short head
Long head
Teres major
Tendon of latissimus dorsi (cut)
Long head
Medial head — **Triceps brachii**
Radius
Biceps tendon
Medial epicondyle of humerus
Brachialis
Aponeurosis of biceps brachii
Pronator teres
Ulna

(b)

Figure 11.22   Muscles of the Arm

(*a*) Lateral view of the right shoulder and arm. (*b*) Anterior view of the right shoulder and arm (deep). Deltoid, pectoralis major, and pectoralis minor muscles have been removed to reveal deeper structures.

Table 11.15   Muscles Acting on the Forearm (see figures 11.22 and 11.23)

Muscle	Origin	Insertion	Nerve	Function
Arm				
Biceps brachii (bī′seps brā′kē-ī) (figures 11.3a; 11.19b; 11.20a; 11.21a; 11.22; 11.25)	Long head—supraglenoid tubercle; Short head— coracoid process	Radial tuberosity	Musculocutaneous	Flexes and supinates forearm; flexes arm
Brachialis (brā′kē-al′is) (figures 11.22; 11.25)	Humerus	Coronoid process of ulna	Musculocutaneous and radial	Flexes forearm
Triceps brachii (trī′seps brā′kē-ī) (figures 11.3b; 11.21c, d; 11.22; 11.25)	Long head—lateral border of scapula; Lateral head—lateral and posterior surface of humerus; Medial head— posterior humerus	Olecranon process of ulna	Radial	Extends forearm; extends and adducts arm
Forearm				
Anconeus (ang-kō′nē-ŭs) (figure 11.23d)	Lateral epicondyle of humerus	Olecranon process and posterior ulna	Radial	Extends forearm
Brachioradialis (brā′kē-ō-rā′dē-al′is) (figures 11.3a; 11.22a; 11.23b, e; 11.25)	Lateral supracondylar ridge of humerus	Styloid process of radius	Radial	Flexes forearm
Pronator quadratus (prō′nā-tōr kwah-drā′tŭs) (figure 11.23c)	Distal ulna	Distal radius	Anterior interosseous	Pronates forearm
Pronator teres (prō′nā-tōr te′rēz) (figures 11.22c; 11.23a)	Medial epicondyle of humerus and coronoid process of ulna	Radius	Median	Pronates forearm
Supinator (sū′pi-nā′tōr) (figure 11.23c, d)	Lateral epicondyle of humerus and ulna	Radius	Radial	Supinates forearm

constitute the **hypothenar eminence** on the ulnar side of the hand. The thenar and hypothenar muscles are involved in the control of the thumb and little finger.

Lower Limb Muscles
Thigh Movements

Several hip muscles originate on the coxa and insert onto the femur (table 11.18 and figures 11.26 through 11.28). These muscles can be divided into three groups: anterior, posterolateral, and deep.

The anterior muscles, the **iliacus** (il-ē′ā-kŭs) and the **psoas** (sō′as) **major,** flex the thigh. Because these muscles share a common insertion and produce the same movement,

they often are referred to as the **iliopsoas** (il-ē-ō-sō′as). When the thigh is fixed, the iliopsoas flexes the trunk on the thigh. For example, the iliopsoas actually does most of the work when a person does sit-ups.

The posterolateral hip muscles consist of the gluteal muscles and the **tensor fasciae** (fash′ē-ē) **latae** (lā′tē). The **gluteus maximus** (glū′tē-ŭs) contributes most of the mass that can be seen as the buttocks, and the **gluteus medius,** a common site for injections, creates a smaller mass just superior and lateral to the maximus. The gluteus maximus functions at its maximum force in extension of the thigh when the hip is flexed at a 45-degree angle so that the muscle is optimally stretched, which accounts for both the sprinter's stance and the bicycle racing posture.

The deep hip muscles function as lateral thigh rotators (see table 11.18).

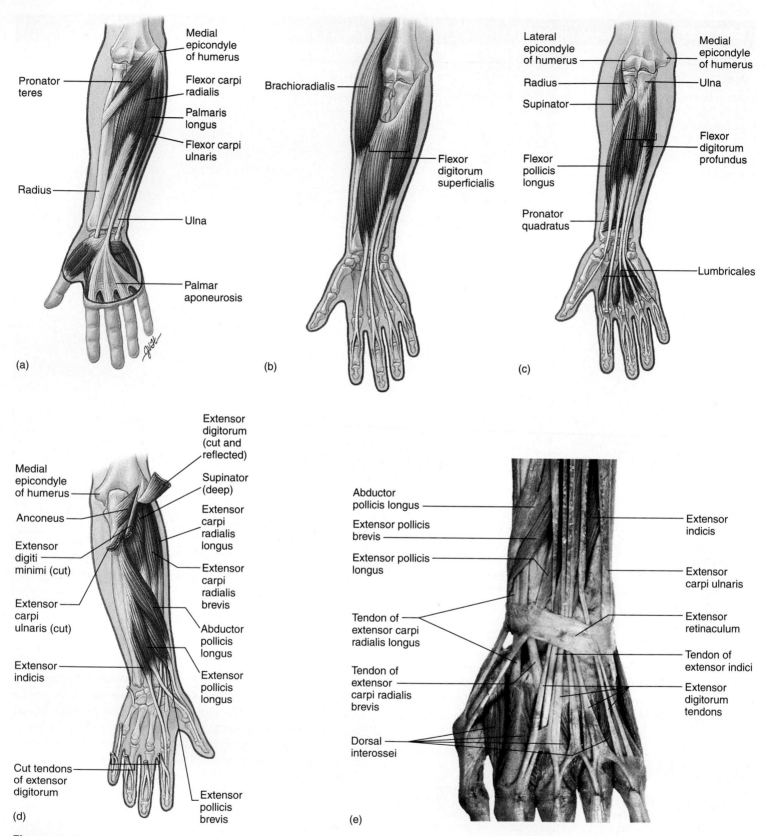

Figure 11.23   Muscles of the Forearm

(*a*) Anterior view of the right forearm (superficial). Brachioradialis muscle has been removed. (*b*) Anterior view of the right forearm (deeper than *a*). Pronator teres, flexor carpi radialis and ulnaris, and palmaris longus muscles have been removed. (*c*) Anterior view of the right forearm (deeper than *a* or *b*). Brachioradialis, pronator teres, flexor carpi radialis and ulnaris, palmaris longus, and flexor digitorum superficialis muscles have been removed. (*d*) Deep muscles of the right posterior forearm. Extensor digitorum, extensor digiti minimi, and extensor carpi ulnaris muscles have been cut to reveal deeper muscles. (*e*) Photograph showing dissection of the posterior right forearm and hand.

Table 11.16 Muscles of the Forearm Acting on the Wrist, Hand, and Fingers (see figure 11.23)

Muscle	Origin	Insertion	Nerve	Function
Anterior Forearm				
Flexor carpi radialis (kar'pī rā'dē-a-lĭs) (figures 11.23a; 11.25a)	Medial epicondyle of humerus	Second and third metacarpals	Median	Flexes and abducts wrist
Flexor carpi ulnaris (kar'pī ŭl-nā'ris) (figure 11.23a)	Medial epicondyle of humerus, and ulna	Pisiform	Ulnar	Flexes and adducts wrist
Flexor digitorum profundus (dij'i-tōr'ŭm prō-fŭn'dŭs) (figure 11.23c)	Ulna	Distal phalanges of digits two through five	Ulnar and median	Flexes fingers and wrist
Flexor digitorum superficialis (dij'i-tōr'ŭm sū'per-fish'ē-a'lis) (figure 11.23b)	Medial epicondyle of humerus, coronoid process, and radius	Middle phalanges of digits two through five	Median	Flexes fingers and wrist
Flexor pollicis longus (pol'i-sis lon'gŭs) (figure 11.23c)	Radius	Distal phalanx of thumb	Median	Flexes thumb and wrist
Palmaris longus (pal-măr'is lon'gŭs) (figures 11.23a; 11.25a)	Medial epicondyle of humerus	Palmar fascia	Median	Tenses palmar fascia; flexes wrist
Posterior Forearm				
Abductor pollicis longus (pol'i-sis lon'gŭs) (figure 11.23d, e)	Posterior ulna and radius and interosseous membrane	Base of first metacarpal	Radial	Abducts and extends thumb; abducts wrist
Extensor carpi radialis brevis (kar'pī rā'dē-a'lis brev'is) (figure 11.23d, e)	Lateral epicondyle of humerus	Base of third metacarpal	Radial	Extends and abducts wrist

continued next page

In addition to the hip muscles, some of the muscles located in the thigh originate on the coxa and can cause movement of the thigh (table 11.19; table 11.20). There are three groups of thigh muscles, based on their location in the thigh: the anterior, which flex; the posterior, which extend; and the medial, which adduct the thigh.

Leg Movements

The anterior thigh muscles are the **quadriceps femoris** (fem'ŏ-ris) and the **sartorius** (sar-tōr'ē-ŭs) (see table 11.20 and figure 11.27a). The quadriceps femoris is actually four muscles: the **rectus femoris,** the **vastus lateralis,** the **vastus medialis,** and the **vastus intermedius.** The vastus lateralis sometimes is used as an injection site, especially in infants who may not have well-developed deltoid or gluteal muscles. The muscles of the quadriceps femoris have a common insertion, the patellar tendon, on and around the patella. The patel-

lar ligament is an extension of the patellar tendon onto the tibial tuberosity. The patellar ligament is the point that is tapped with a rubber hammer when testing the knee-jerk reflex in a physical examination.

The sartorius is the longest muscle of the body, crossing from the lateral side of the hip to the medial side of the knee. As the muscle contracts, it flexes the thigh and leg and laterally rotates the thigh. This movement is the action required for crossing the legs.

Clinical Note

The term sartorius means tailor. The sartorius muscle is so named because its action is to cross the legs, a common position traditionally preferred by tailors because they can hold their sewing in their lap as they sit and sew by hand.

Table 11.16 Muscles of the Forearm Acting on the Wrist, Hand, and Fingers (see figure 11.23)—cont'd

Muscle	Origin	Insertion	Nerve	Function
Posterior Forearm—cont'd				
Extensor carpi radialis longus (kar′pī rā′dē-a′lis lon′gus) (figures 11.23*d, e*)	Lateral supracondylar ridge of humerus	Base of second metacarpal	Radial	Extends and abducts wrist
Extensor carpi ulnaris (kar′pī ŭl-nā′ris) (figures 11.23*d, e;* 11.25*b*)	Lateral epicondyle of humerus; ulna	Base of fifth metacarpal	Radial	Extends and adducts wrist
Extensor digiti minimi (dij′i-tī mi′nĭ-mī) (figure 11.23*d, e*)	Lateral epicondyle of humerus	Phalanges of fifth digit	Radial	Extends little finger and wrist
Extensor digitorum (dij′i-tōr′ŭm) (figures 11.23*d, e;* 11.25*b*)	Lateral epicondyle of humerus	Bases of phalanges of digits two through five	Radial	Extends fingers and wrist
Extensor indicis (in′di-sis) (figure 11.23*d, e*)	Ulna	Second digit	Radial	Extends forefinger and wrist
Extensor pollicis brevis (pol′i-sis brev′is) (figure 11.23*d, e*)	Radius	Proximal phalanx of thumb	Radial	Extends and abducts thumb; abducts wrist
Extensor pollicis longus (pol′i-sis lon′gŭs) (figure 11.23*d, e*)	Ulna	Distal phalanx of thumb	Radial	Extends thumb

The medial group of muscles is involved primarily in adduction of the thigh (figure 11.27*b* and *c*).

The posterior thigh muscles are collectively called the hamstring muscles and consist of the **biceps femoris, semimembranosus** (se′mē-mem′bră-nō′sŭs), and **semitendinosus** (se′mē-ten′di-nō′sŭs) (see table 11.20; figure 11.28). Their tendons are easily felt and seen on the medial and lateral posterior aspect of a slightly bent knee (see figure 11.30).

> **Clinical Note**
>
> The hamstrings are so named because in pigs these tendons can be used to suspend hams during curing. Some animals such as wolves often bring down their prey by biting through the hamstrings; therefore, "to hamstring" someone is to render him helpless. A "pulled hamstring" results from tearing one or more of these muscles or their tendons, usually near the origin of the muscle.

Ankle, Foot, and Toe Movements

Muscles of the leg that move the ankle and the foot are listed in table 11.21 and are illustrated in figures 11.29 and 11.30. These **extrinsic foot muscles** can be divided into three groups, each located within a separate compartment of the leg (figure 11.31): anterior, posterior, and lateral. The anterior muscles are extensor muscles involved in dorsiflexion of the foot and extension of the toes.

> **Clinical Note**
>
> The term **shin-splints** is a catchall term involving any one of the following four conditions associated with pain in the anterior portion of the leg.
>
> - Excessive stress on the tibialis posterior, resulting in pain along the origin of the muscle.
> - Tibial periostitis, an inflammation of the tibial periosteum.
> - Anterior compartment syndrome. During hard exercise the anterior compartment muscles may swell with blood. The overlying fascia is very tough and does not expand; thus the nerves and vessels are compressed, causing pain.
> - Stress fracture of the tibia 2–5 cm distal to the knee.
>
> The best treatment for any of these types of shin-splints is to rest the leg for 1–4 weeks, depending on the type of shin-splint.

The superficial muscles of the posterior compartment, the **gastrocnemius** (gas′trok-nē′mē-ŭs) and **soleus,** form the bulge of the calf (posterior leg) (see figures 11.29 and 11.30). They join with the small **plantaris** muscle to form the common **calcaneal** (kal-kā′nē-al), or **Achilles, tendon** (see figure 11.29*c*). These muscles are involved in plantar flexion of

Table 11.17   Intrinsic Hand Muscles (see figure 11.24)

Muscle	Origin	Insertion	Nerve	Function
Midpalmar Muscles				
Interossei (in′ter-os′ē-ī)				
Dorsales (dōr-sā′lēz) (figure 11.24)	Sides of metacarpal bones	Proximal phalanges of second, third, and fourth digits	Ulnar	Abducts second, third, and fourth digits
Palmares (pal-măr′ēz) (figure 11.24)	Second, fourth, and fifth metacarpals	Second, fourth, and fifth digits	Ulnar	Adducts second, fourth, and fifth digits
Lumbricales (lum′bră-ka′lēz) (figures 11.23c; 11.24)	Tendons of flexor digitorum profundis	Second through fifth digits	Two on radial side—median; two on ulnar side—ulnar	Flexes proximal and extends middle and distal phalanges
Thenar Muscles				
Abductor pollicis brevis (ab′dŭk-ter pol′i-sis brev′is) (figure 11.24)	Trapezium	Proximal phalanx of thumb	Median	Abducts thumb
Adductor pollicis (a′dŭk-ter pol′i-sis) (figure 11.24)	Third metacarpal, second metacarpal, trapezoid, and capitate	Proximal phalanx of thumb	Ulnar	Adducts thumb
Flexor pollicis brevis (pol′i-sis brev′is) (figure 11.24)	Flexor retinaculum and first metacarpal	Proximal phalanx of thumb	Median and ulnar	Flexes thumb
Opponens pollicis (ō-pō′nenz pol′i-sis) (figure 11.24)	Trapezium and flexor retinaculum	First metacarpal	Median	Opposes thumb
Hypothenar Muscles				
Abductor digiti minimi (ab′duk-ter dij′i-tī min′imī) (figure 11.24)	Pisiform	Base of fifth digit	Ulnar	Abducts and flexes little finger
Flexor digiti minimi brevis (dij′i-tī min′ĭ-mī brev′is) (figure 11.24)	Hamate	Middle and proximal phalanx of fifth digit	Ulnar	Flexes little finger
Opponens digiti minimi (ō-pō′nenz dij′i-tī min′i-mī) (figure 11.24)	Hamate and flexor retinaculum	Fifth metacarpal	Ulnar	Opposes little finger

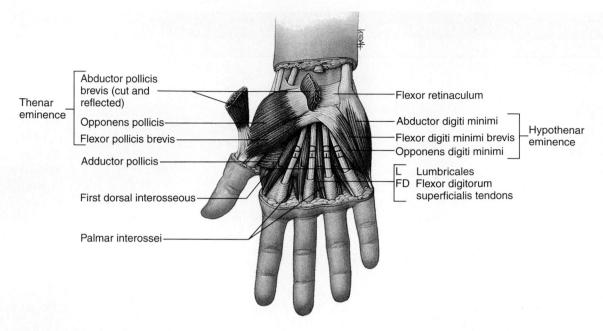

Thenar eminence

Abductor pollicis brevis (cut and reflected)

Opponens pollicis

Flexor pollicis brevis

Adductor pollicis

First dorsal interosseous

Palmar interossei

Flexor retinaculum

Abductor digiti minimi

Flexor digiti minimi brevis

Opponens digiti minimi

Hypothenar eminence

L   Lumbricales
FD Flexor digitorum superficialis tendons

Figure 11.24 Hand

Palmar surface of the right hand. Abductor pollicis brevis has been cut.

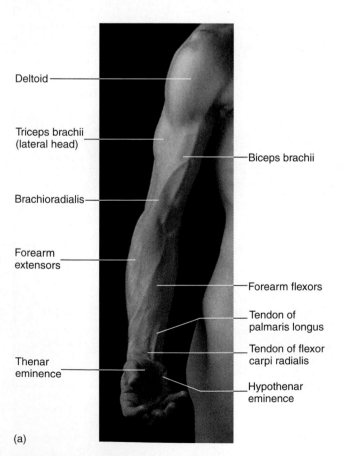

Deltoid

Triceps brachii (lateral head)

Brachioradialis

Forearm extensors

Thenar eminence

Biceps brachii

Forearm flexors

Tendon of palmaris longus

Tendon of flexor carpi radialis

Hypothenar eminence

(a)

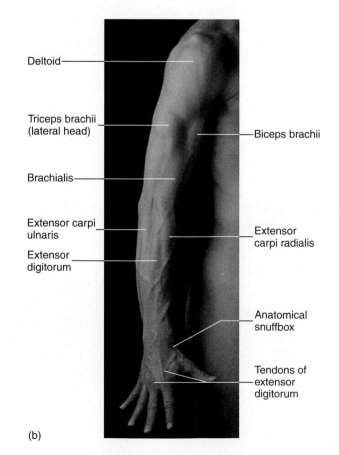

Deltoid

Triceps brachii (lateral head)

Brachialis

Extensor carpi ulnaris

Extensor digitorum

Biceps brachii

Extensor carpi radialis

Anatomical snuffbox

Tendons of extensor digitorum

(b)

Figure 11.25 Surface Anatomy, Muscles of the Upper Limb

(*a*) Anterior view. (*b*) Lateral and posterior view.

Table 11.18   Muscles Acting on the Thigh (see figure 11.26)

Muscle	Origin	Insertion	Nerve	Function
Anterior				
Iliopsoas (il'ē-ō-sō'as)				
Iliacus (il'ē-ā'kŭs) (figure 11.27a)	Iliac fossa	Lesser trochanter of femur and capsule of hip joint	Lumbar plexus	Flexes and medially rotates thigh
Psoas major (sō'ŭs) (figure 11.27a)	T12–L5	Lesser trochanter of femur	Lumbar plexus	Flexes thigh
Posterior and Lateral				
Gluteus maximus (glū'tē-ŭs mak'si-mŭs) (figures 11.3b; 11.26; 11.29b)	Ilium, sacrum, and coccyx	Gluteal tuberosity of femur and the fascia lata	Inferior gluteal	Extends, abducts, and laterally rotates thigh
Gluteus medius (glū'tē-ŭs mē'dē-ŭs) (figures 11.3b; 11.26; 11.29b)	Ilium	Greater trochanter of femur	Superior gluteal	Abducts and medially rotates thigh
Gluteus minimus, (glū'tē-ŭs min'i-mŭs) (figure 11.26b)	Ilium	Greater trochanter of femur	Superior gluteal	Abducts and medially rotates thigh
Tensor fasciae latae (ten'sōr fa'shē-ē lā'tē) (figures 11.3a; 11.27a; 11.30a)	Anterior superior iliac spine	Through iliotibial tract to lateral condyle of tibia	Superior gluteal	Tenses lateral fascia; flexes, abducts, and medially rotates thigh
Deep Thigh Rotators				
Gemellus (jĕ-mēl'ŭs)				
Inferior (figure 11.26b)	Ischial tuberosity	Obturator internus tendon	L5 and S1	Laterally rotates and abducts thigh
Superior (figure 11.26b)	Ischial spine	Obturator internus tendon	L5 and S1	Laterally rotates and abducts thigh
Obturator (ob'tūr-ā'tōr)				
Externus (ex-ter'nŭs) (figure 11.26b)	Inferior margin of obturator foramen	Greater trochanter of femur	Obturator	Laterally rotates thigh
Internus (in-ter'nŭs) (figure 11.26b)	Margin of obturator foramen	Greater trochanter of femur	Ischiadic plexus*	Laterally rotates and abducts thigh
Piriformis (pir'i-fōr'mis) (figure 11.26b)	Sacrum and ilium	Greater trochanter of femur	Ischiadic plexus*	Laterally rotates and abducts thigh
Quadratus femoris (kwah'-drā'tŭs fem'ō-ris) (figure 11.26b)	Ischial tuberosity	Intertrochanteric ridge of femur	Ischiadic plexus*	Laterally rotates thigh

*Formerly referred to as the sciatic nerve.

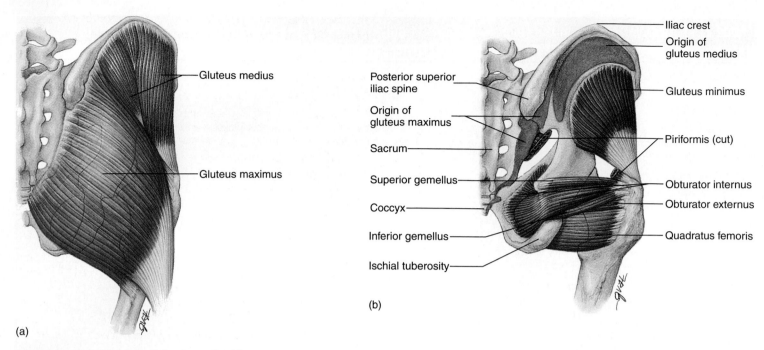

Figure 11.26   Muscles of the Posterior Hip

(*a*) Posterior view of the right hip, superficial. (*b*) Posterior view of the right hip, deep. Gluteus maximus and medius have been removed to reveal deeper muscles. The piriformis has been cut.

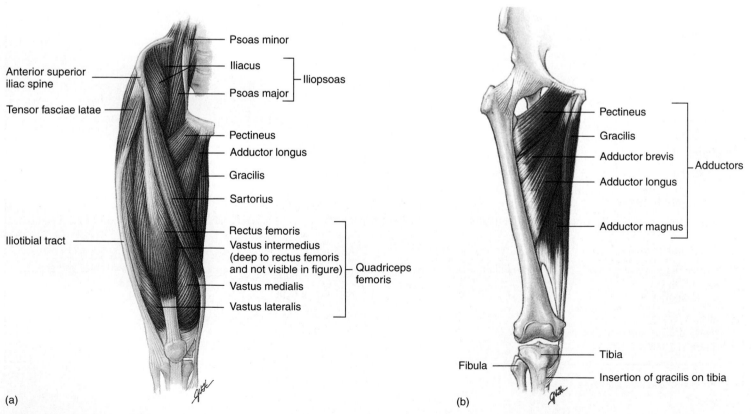

Figure 11.27   Muscles of the Anterior Thigh

(*a*) Anterior view of the right thigh. (*b*) Adductor region of the right thigh. Tensor fasciae latae, sartorius, and quadriceps femoris muscles have been removed.

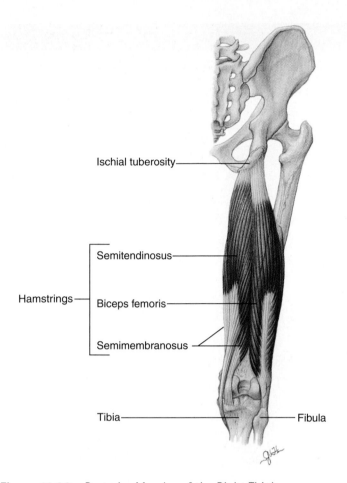

Figure 11.28   Posterior Muscles of the Right Thigh
Hip muscles have been removed.

the foot. The deep muscles of the posterior compartment plantar flex and invert the foot and flex the toes.

> ### Clinical Note
>
> The Achilles tendon derives its name from a hero of Greek mythology. As a baby, Achilles was dipped into magic water, which made him invulnerable to harm everywhere the water touched his skin. But his mother held him by the heel and failed to submerge this part of his body under the water. Consequently, his heel was vulnerable and proved to be his undoing; he was shot in the heel with an arrow at the battle of Troy and died. Thus, saying that someone has an "Achilles' heel" means that he has a weak spot that can be attacked.

The lateral muscles are primarily everters of the foot, but they also aid plantar flexion.

Intrinsic foot muscles, located within the foot itself (table 11.22; figure 11.32), flex, extend, abduct, and adduct the toes. They are arranged in a manner similar to that of the intrinsic muscles of the hand.

Table 11.19	Summary of Muscle Actions on the Thigh				
Flexion	**Extension**	**Abduction**	**Adduction**	**Medial Rotation**	**Lateral Rotation**
Iliopsoas	Gluteus maximus	Gluteus maximus	Adductor magnus	Iliopsoas	Gluteus maximus
Tensor fasciae latae	Semitendinosus	Gluteus medius	Adductor longus	Tensor fasciae latae	Obturator internus
Rectus femoris	Semimembranosus	Gluteus minimus	Adductor brevis	Gluteus medius	Obturator externus
Sartorius	Biceps femoris	Tensor fasciae latae	Pectineus	Gluteus minimus	Superior gemellus
Adductor longus	Adductor magnus	Obturator internus	Gracilis		Inferior gemellus
Adductor brevis		Gemellus superior and inferior			Quadratus femoris
Pectineus		Piriformis			Piriformis
					Adductor magnus
					Adductor longus
					Adductor brevis

Table 11.20 Muscles of the Thigh (see figures 11.27 and 11.28)

Muscle	Origin	Insertion	Nerve	Function
Anterior Compartment				
Quadriceps femoris (kwah'-dri-seps fem'ō-ris) (figures 11.3a; 11.27a, b; 11.29)	Rectus femoris— anterior inferior iliac spine; Vastus lateralis—femur; Vastus intermedius— femur; Vastus medialis—linea aspera	Patella and onto tibial tuberosity through patellar ligament	Femoral	Extends leg: rectus femoris also flexes thigh
Sartorius (sar-tōr'ē-ŭs) (figures 11.3a; 11.27a; 11.29a)	Anterior superior iliac spine	Medial side of tibial tuberosity	Femoral	Flexes thigh and leg: rotates leg medially and thigh laterally
Medial Compartment				
Adductor brevis (a'dŭk-ter brev'is) (figure 11.27c)	Pubis	Femur	Obturator	Adducts, flexes, and laterally rotates thigh
Adductor longus (a'dŭk-ter lon'gŭs) (figure 11.27)	Pubis	Femur	Obturator	Adducts, flexes, and laterally rotates thigh
Adductor magnus (a'dŭk-ter mag'nŭs) (figures 11.3b; 11.27b, c)	Pubis and ischium	Femur	Obturator and tibial	Adducts, extends, and laterally rotates thigh
Gracilis (gras'i-lis) (figures 11.3b; 11.27)	Pubis near symphysis	Tibia	Obturator	Adducts thigh; flexes leg
Pectineus (pek'ti-nē'ŭs) (figure 11.27a, c)	Pubic crest	Pectineal line of femur	Femoral and obturator	Adducts and flexes thigh
Posterior Compartment				
Biceps femoris (bī'seps fem'ō-ris) (figures 11.3b; 11.28; 11.29b)	Long head—ischial tuberosity; Short head—femur	Head of fibula	Long head—tibial; Short head—common fibular	Flexes and laterally rotates leg; extends thigh
Semimembranosus (se'mē-mem'bră-nō'sŭs) (figures 11.3b; 11.28; 11.29b)	Ischial tuberosity	Medial condyle of tibia and collateral ligament	Tibial	Flexes and medially rotates leg; tenses capsule of knee joint; extends thigh
Semitendinosus (se'mē-ten'di-nō'sŭs) (figures 11.3b; 11.28; 11.29b)	Ischial tuberosity	Tibia	Tibial	Flexes and medially rotates leg; extends thigh

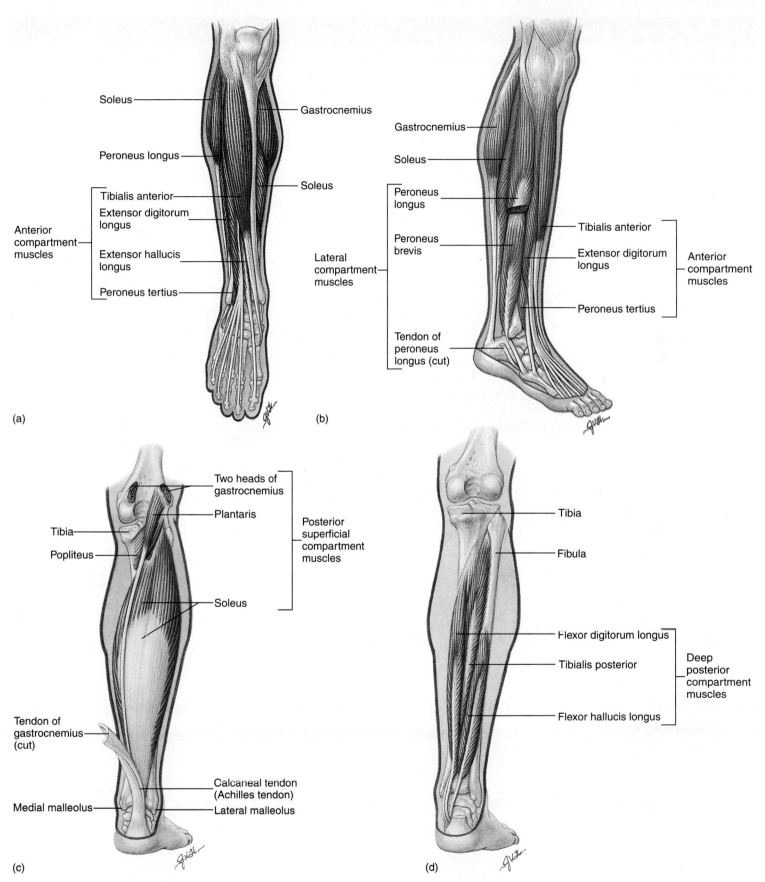

Soleus

Gastrocnemius

Peroneus longus

Soleus

Tibialis anterior

Extensor digitorum longus

Anterior compartment muscles

Extensor hallucis longus

Peroneus tertius

(a)

Gastrocnemius

Soleus

Peroneus longus

Peroneus brevis

Lateral compartment muscles

Tendon of peroneus longus (cut)

Tibialis anterior

Extensor digitorum longus

Anterior compartment muscles

Peroneus tertius

(b)

Two heads of gastrocnemius

Plantaris

Posterior superficial compartment muscles

Tibia

Popliteus

Soleus

Tendon of gastrocnemius (cut)

Medial malleolus

Calcaneal tendon (Achilles tendon)

Lateral malleolus

(c)

Tibia

Fibula

Flexor digitorum longus

Tibialis posterior

Deep posterior compartment muscles

Flexor hallucis longus

(d)

Figure 11.29   Muscles of the Leg

(a) Anterior view of the right leg. (b) Lateral view of the right leg. (c) Posterior view of the right calf, superficial. Gastrocnemius has been removed.
(d) Posterior view of the right calf, deep. Gastrocnemius, plantaris, and soleus muscles have been removed.

Table 11.21   Muscles of the Leg Acting on the Leg, Ankle, and Foot (see figures 11.29 and 11.31)

Muscle	Origin	Insertion	Nerve	Function
Anterior Compartment				
Extensor digitorum longus (dij'i-tōr-ŭm lon'gŭs) (figures 11.3a; 11.29a, b)	Lateral condyle of tibia and fibula	Four tendons to phalanges of four lateral toes	Deep fibular*	Extends four lateral toes; dorsiflexes and everts foot
Extensor hallucis longus (hal'i-sis lon'gŭs) (figure 11.29a)	Middle fibula and interosseous membrane	Distal phalanx of great toe	Deep fibular*	Extends great toe; dorsiflexes and inverts foot
Tibialis anterior (tib'ē-a'lis) (figures 11.3a; 11.29a, b)	Tibia and interosseous membrane	Medial cuneiform and first metatarsal	Deep fibular*	Dorsiflexes and inverts foot
Fibularis tertius (peroneus tertius) (per'ō-nē'ŭs ter'shē-ŭs) (figure 11.29b)	Fibula and interosseous membrane	Fifth metatarsal	Deep fibular*	Dorsiflexes and everts foot
Posterior Compartment				
Superficial				
Gastrocnemius (gas'trok-nē'mē-ŭs) (figures 11.3a, b; 11.29 a, b; 11.30b)	Medial and lateral epicondyles of femur	Through calcaneal (Achilles) tendon to calcaneus	Tibial	Plantar flexes foot; flexes leg
Plantaris (plan-tār'is) (figure 11.29c)	Femur	Through calcaneal tendon to calcaneus	Tibial	Plantar flexes foot; flexes leg
Soleus (sō'lē-ŭs) (figures 11.3a, b; 11.29 a, b, c; 11.30b)	Fibula and tibia	Through calcaneal tendon to calcaneus	Tibial	Plantar flexes foot
Deep				
Flexor digitorum longus (dij'i-tōr'ŭm lon'gŭs) (figure 11.30 b, d)	Tibia	Four tendons to distal phalanges of four lateral toes	Tibial	Flexes four lateral toes; plantar flexes and inverts foot
Flexor hallucis longus (hal'i-sis lon'gŭs) (figure 11.29d)	Fibula	Distal phalanx of great toe	Tibial	Flexes great toe; plantar flexes and inverts foot
Popliteus (pop'li'-tē'ŭs) (figure 11.29c)	Lateral femoral condyle	Posterior tibia	Tibial	Flexes and medially rotates leg
Tibialis posterior (tib'ē-a'lis) (figure 11.29d)	Tibia, interosseous membrane, and fibula	Navicular, cuneiforms, cuboid, and second through fourth metatarsals	Tibial	Plantar flexes and inverts foot

Table 11.21 Muscles of the Leg Acting on the Leg, Ankle, and Foot (see figures 11.29 and 11.31)—cont'd

Muscle	Origin	Insertion	Nerve	Function
Lateral Compartment				
Fibularis brevis (peroneus brevis) (fib-yū-lā'ris brev'is) (figures 11.3a, b; 11.29b)	Fibula	Fifth metatarsal	Superficial fibular*	Everts and plantar flexes foot
Fibularis longus (peroneus longus) (fib-yū-lā'ris lon'gŭs) (figures 11.3a, b; 11.29a, b)	Fibula	Medial cuneiform and first metatarsal	Superficial fibular*	Everts and plantar flexes foot

*Formerly referred to as the peroneal nerve.

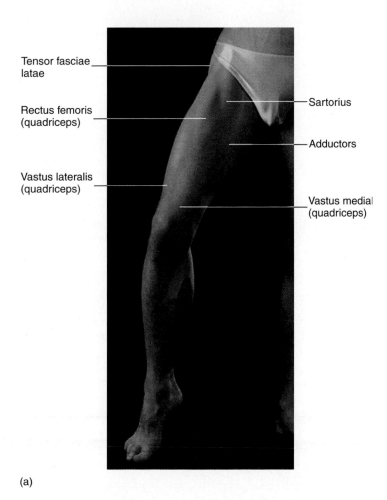

Tensor fasciae latae

Rectus femoris (quadriceps)

Vastus lateralis (quadriceps)

Sartorius

Adductors

Vastus medial (quadriceps)

(a)

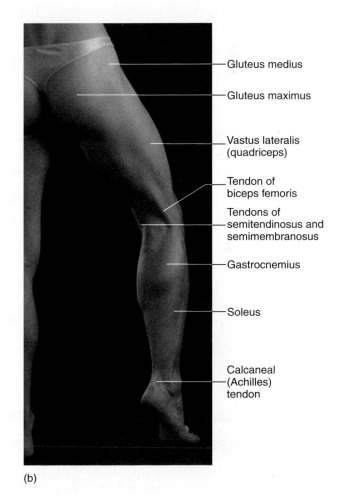

Gluteus medius

Gluteus maximus

Vastus lateralis (quadriceps)

Tendon of biceps femoris

Tendons of semitendinosus and semimembranosus

Gastrocnemius

Soleus

Calcaneal (Achilles) tendon

(b)

Figure 11.30 Surface Anatomy, Muscles of the Lower Limb
(a) Anterior view. (b) Posterior view.

Table 11.22 Intrinsic Muscles of the Foot (see figure 11.32)

Muscle	Origin	Insertion	Nerve	Function
Abductor digiti minimi (ab′dŭk-ter dij′i-tī min′ĭ-mī) (figure 11.32)	Calcaneus	Proximal phalanx of fifth toe	Lateral plantar	Abducts and flexes little toe
Abductor hallucis (ab′dŭk-ter hal′i-sis) (figure 11.32)	Calcaneus	Great toe	Medial plantar	Abducts great toe
Adductor hallucis (a′dŭk-ter hal′i-sis) (not illustrated)	Lateral four metatarsals	Proximal phalanx of great toe	Lateral plantar	Adducts great toe
Extensor digitorum brevis (dij′i-tōr′ŭm brev′is) (not illustrated)	Calcaneus	Four tendons fused with tendons of extensor digitorum longus	Deep fibular*	Extends toes
Flexor digiti minimi brevis (dij′i-tī min′ĭ-mī brev′is) (figure 11.32)	Fifth metatarsal	Proximal phalanx of fifth digit	Lateral plantar	Flexes little toe (proximal phalanx)
Flexor digitorum brevis (dij′i-tōr′ŭm brev′is) (figure 11.32)	Calcaneus and plantar fascia	Four tendons to middle phalanges of four lateral toes	Medial plantar	Flexes lateral four toes
Flexor hallucis brevis (hal′i-sis brev′is) (figure 11.32)	Cuboid; medial and lateral cuneiforms	Two tendons to proximal phalanx of great toe	Medial and lateral plantar	Flexes great toe
Dorsal interossei (in′ter-os′ē-ī) (not illustrated)	Metatarsal bones	Proximal phalanges of second, third, and fourth digits	Lateral plantar	Abduct second, third, and fourth toes; adduct second toe
Plantar interossei (plan′tăr in′ter-os′ē-ī) (not illustrated)	Third, fourth, and fifth metatarsals	Proximal phalanges of third, fourth, and fifth digits	Lateral plantar	Adduct third, fourth, and fifth toes
Lumbricales (lum′bri-kā-lēz) (figure 11.32)	Tendons of flexor digitorum longus	Second through fifth digits	Lateral and medial plantar	Flex proximal and extend middle and distal phalanges
Quadratus plantae (kwah′drā′tŭs plan′tē) (not illustrated)	Calcaneus	Tendons of flexor digitorum longus	Lateral plantar	Flexes toes

*Formerly referred to as the peroneal nerve.

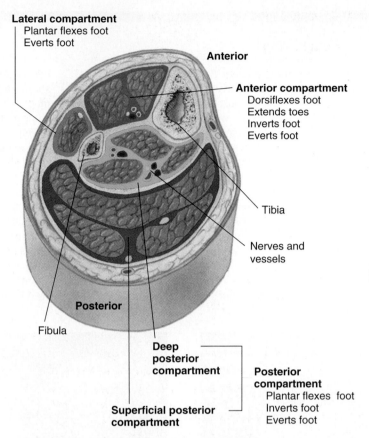

Lateral compartment
  Plantar flexes foot
  Everts foot

Anterior

Anterior compartment
  Dorsiflexes foot
  Extends toes
  Inverts foot
  Everts foot

Tibia

Nerves and
vessels

Posterior

Fibula

**Deep
posterior
compartment**

**Posterior
compartment**
  Plantar flexes foot
  Inverts foot
  Everts foot

**Superficial posterior
compartment**

Figure 11.31   Cross Section Through the Right Leg

Drawing of the muscular compartments. ✗

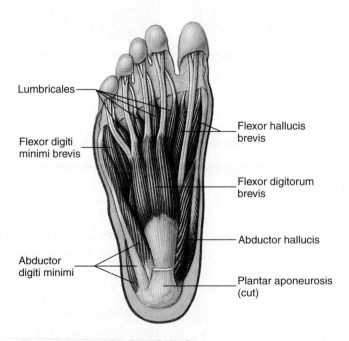

Lumbricales

Flexor hallucis
brevis

Flexor digiti
minimi brevis

Flexor digitorum
brevis

Abductor hallucis

Abductor
digiti minimi

Plantar aponeurosis
(cut)

Figure 11.32   Muscles of the Foot

Plantar view of the right foot.

Clinical Focus Bodybuilding

Bodybuilding has become a popular sport worldwide. Once considered only for men, it is now enjoyed by thousands of women as well. Participants in this sport combine diet and specific weight training to develop maximum muscle mass and minimum body fat, with their major goal being a well-balanced, complete physique. An uninformed, untrained muscle builder can build some muscles and ignore others; the result is a disproportioned body. Skill, training, and concentration are required to build a well-proportioned, muscular body and to know which exercises build a large number of muscles and which are specialized to build certain parts of the body. Is

the old adage "no pain, no gain" correct? Not really. Overexercising can cause small tears in muscles, causing soreness. Torn muscles are weaker, and it may take up to 3 weeks to repair the damage, even though the soreness may only last 5–10 days.

Bodybuilders concentrate on increasing skeletal muscle mass. Endurance tests conducted several years ago demonstrated that the cardiovascular and respiratory abilities of bodybuilders were similar to those abilities in normal, healthy persons untrained in a sport. More recent studies, however, indicate that the cardiorespiratory fitness of bodybuilders is similar to that of other well-trained athletes. The

difference between the results of the new studies and the older ones is attributed to modern bodybuilding techniques that include aerobic exercise and running, as well as "pumping iron."

Bodybuilding has its own language. Bodybuilders refer to the "lats," "traps," and "delts" rather than the latissimus dorsi, trapezius, and deltoids. The exercises also have special names such as "lat pulldowns," "preacher curls," and "triceps extensions."

Photographs of bodybuilders are very useful in the study of anatomy because they enable easy identification of the surface anatomy of muscles that cannot usually be seen in untrained people (figure A).

 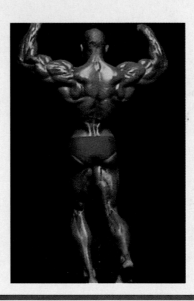

Figure A   Bodybuilders

Summary

Body movements result from the contraction of skeletal muscles.

General Principles

1. The less movable end of a muscle attachment is the origin; the more movable end is the insertion.
2. Synergists are muscles that function together to produce movement. Antagonists oppose or reverse the movement of another muscle.
3. Prime movers are mainly responsible for a movement. Fixators stabilize the action of prime movers.

Muscle Shapes

Muscle shape is determined primarily by the arrangement of muscle fasciculi.

Nomenclature

Muscles are named according to their shape, origin and insertion, location, size, orientation of fasciculi, or function.

Movements Accomplished by Muscles

Contracting muscles generate a force that acts on bones (levers) across joints (fulcrums) to create movement. There are three classes of levers.

Head Muscles
Head Movement

Origins of these muscles are mainly on the cervical vertebrae (except for the sternocleidomastoid); insertions are on the occipital bone or mastoid process. They cause flexion, extension, rotation, abduction, and adduction of the head.

Facial Expression

Origins of facial muscles are on skull bones or fascia; insertions are into the skin, causing movement of the facial skin, lips, and eyelids.

Mastication

Three pairs of muscles close the jaw; gravity opens the jaw. Forced opening is caused by the lateral pterygoids and the hyoid muscles.

Tongue Movement

Intrinsic tongue muscles change the shape of the tongue; extrinsic tongue muscles move the tongue.

Swallowing and the Larynx

1. Hyoid muscles can depress the jaw and assist in swallowing.
2. Muscles open and close the openings to the nasal cavity, auditory tubes, and larynx.

Movements of the Eyeball

Six muscles with their origins on the orbital bones insert on the eyeball and cause it to move within the orbit.

Trunk Muscles
Muscles Moving the Vertebral Column

1. These muscles extend, abduct, rotate, or flex the vertebral column.
2. A deep group of muscles connects adjacent vertebrae.
3. A more superficial group of muscles runs from the pelvis to the skull, extending from the vertebrae to the ribs.

Thoracic Muscles

1. Most respiratory movement is caused by the diaphragm.
2. Muscles attached to the ribs aid in respiration.

Abdominal Wall

Abdominal wall muscles hold and protect abdominal organs and cause flexion, rotation, and lateral flexion of the vertebral column.

Pelvic Floor and Perineum

These muscles support the abdominal organs inferiorly.

Upper Limb Muscles
Scapular Movements

Six muscles attach the scapula to the trunk, enabling the scapula to function as an anchor point for the muscles and bones of the arm.

Arm Movements

Seven muscles attach the humerus to the scapula. Two additional muscles attach the humerus to the trunk. These muscles cause flexion, extension, abduction, adduction, rotation, and circumduction of the arm.

Forearm Movements

1. Flexion and extension of the forearm are accomplished by three muscles located in the arm and two in the forearm.
2. Supination and pronation are accomplished primarily by forearm muscles.

Wrist, Hand, and Finger Movements

1. Forearm muscles that originate on the medial epicondyle are responsible for flexion of the wrist and fingers. Muscles extending the wrist and fingers originate on the lateral epicondyle.
2. Extrinsic hand muscles are in the forearm. Intrinsic hand muscles are in the hand.

Lower Limb Muscles
Thigh Movements

1. Anterior pelvic muscles cause flexion of the thigh.
2. Muscles of the buttocks are responsible for extension, abduction, and rotation of the thigh.
3. The thigh can be divided into three compartments.
   - The medial compartment muscles adduct the thigh.
   - The anterior compartment muscles flex the thigh.
   - The posterior compartment muscles extend the thigh.

Leg Movements

Some muscles of the thigh also act on the leg. The anterior thigh muscles extend the leg, and the posterior thigh muscles flex the leg.

Ankle, Foot, and Toe Movements

1. The leg can be divided into three compartments.
   - Muscles in the anterior compartment cause dorsiflexion, inversion, or eversion of the foot and extension of the toes.
   - Muscles of the lateral compartment plantar flex and evert the foot.
   - Muscles of the posterior compartment flex the leg, plantar flex and invert the foot, and flex the toes.
2. Intrinsic foot muscles flex or extend and abduct or adduct the toes.

Content Review

1. Define the terms origin and insertion, synergist and antagonist, prime mover and fixator.
2. Describe the different shapes of muscles. How are the shapes related to the force of contraction of the muscle and the range of movement the contraction produces?
3. List the different criteria used to name muscles, and give an example of each.
4. Using the terms fulcrum, lever, and force, explain how contraction of a muscle results in movement. Define the three classes of levers, and give an example of each in the body.

5. Name the movements of the head and neck caused by contraction of the sternocleidomastoid muscle. What is wry neck?
6. What is unusual about the insertion (and sometimes origin) of facial muscles?
7. Which muscles are responsible for moving the ears, the eyebrows, the eyelids, and the nose? For puckering the lips, smiling, sneering, and frowning?
8. What causes a dimple on the chin?
9. Name the muscles responsible for closing and opening the jaw and for lateral and medial excursion of the jaw.
10. Contrast the movements produced by the extrinsic and intrinsic tongue muscles.
11. Which muscles open and close the openings to the auditory tube and larynx?
12. Describe the muscles of the eye and the movements that they cause.
13. Describe the group of muscles that attaches to the vertebrae or ribs (or both) and causes movement of the spine.
14. Name the muscle that is mainly responsible for respiratory movements. How do other muscles aid this movement?
15. Explain the anatomic basis for the lines seen on a well-muscled individual's abdomen. What are the functions of the abdominal muscles?
16. What openings penetrate the pelvic floor muscles?
17. What two muscles attach the humerus directly to the trunk? Name seven muscles that attach the humerus to the scapula.
18. What muscles cause extension and flexion of the arm? Abduction and adduction? Rotation?
19. List the muscles that cause flexion and extension of the forearm. Where are these muscles located?
20. Supination and pronation of the forearm are produced by what muscles? Where are these muscles located?
21. Describe the muscles that cause flexion and extension of the wrist.
22. Contrast the location and actions of the extrinsic and intrinsic hand muscles. What are the thenar and hypothenar eminence?
23. Describe the muscles that move the thumb. The tendons of what muscles form the anatomical snuffbox?
24. What muscle is the prime mover for flexion of the thigh? What muscles act as synergists to this muscle?
25. Describe the movements produced by the buttock muscles.
26. Name the muscle compartments of the thigh and the thigh movements produced by the muscles of each compartment. List the muscles of each compartment and the individual function of each muscle.
27. How is it possible for thigh muscles to move both the thigh and the leg? Name at least four muscles that can do this.
28. What movements are produced by the three muscle compartments of the leg? Name the muscles of each compartment, and describe the movements for which each muscle is responsible.
29. What movement do the peroneus muscles have in common? The tibialis muscles?
30. Name the leg muscles that flex the leg. Which of them can also flex the foot?
31. Describe the intrinsic foot muscles and their functions.

Develop Your Reasoning Skills

1. For each of the following muscles, (1) describe the movement that the muscle produces, and (2) name the muscles that act as synergists and antagonists for them: longus capitis, erector spinae, coracobrachialis.
2. Propose an exercise that would benefit each of the following muscles specifically: biceps brachii, triceps brachii, deltoid, rectus abdominis, quadriceps femoris, and gastrocnemius.
3. Consider only the effect of the brachioradialis muscle for this question. If a weight is held in the hand and the forearm is flexed, what type of lever system is in action? If the weight is placed on the forearm? Which system can lift more weight, and how far?
4. A patient was involved in an automobile accident in which the car was "rear-ended," resulting in whiplash injury of the head (hyperextension). What neck muscles might be injured in this type of accident? What is the easiest way to prevent such injury in an automobile accident?
5. During surgery, a branch of the patient's facial nerve was accidentally cut on one side of the face. As a result, after the operation, the lower eyelid and the corner of the patient's mouth drooped on that side of the face. What muscles were apparently affected?
6. When a person becomes unconscious, the tongue muscles relax, and there is a tendency for the tongue to retract or fall back and obstruct the airway. Which tongue muscle is responsible? How can this be prevented or reversed?
7. The mechanical support of the head of the humerus in the glenoid fossa is weakest in the inferior direction. What muscles help prevent dislocation of the shoulder when a heavy weight such as a suitcase is carried?
8. How would paralysis of the quadriceps femoris of the left leg affect a person's ability to walk?
9. Speedy Sprinter started a 200 m dash and fell to the ground in pain. Examination of her right leg revealed the following symptoms: inability to plantar flex the foot against resistance, normal ability to evert the foot, dorsiflexion of the foot more than normal, and abnormal bulging of the calf muscles. Explain the nature of her injury.
10. What muscles are required to turn this page?

Web Site Link

For a listing of the most current web sites related to this chapter, please visit the Seeley home page at:
http://www.mhhe.com/biosci/ap/seeleyap/

Chapter Twelve

Functional Organization of Nervous Tissue

Objectives

1. List the divisions of the nervous system, and describe the characteristics of each.

2. Describe the structure of neurons and the function of their components.

3. Describe the three basic neuron shapes, and give an example of where each one is found.

4. Describe the location, relative number, structure, and general function of neuroglial cells.

5. Explain the role of the myelin sheath in saltatory conduction of action potentials.

6. Define the terms nerve, nerve tract, nucleus, and ganglion.

7. Describe the structure of a nerve, including its connective tissue components.

8. Describe the structure and function of the synapse.

9. Describe the release and metabolism of acetylcholine and norepinephrine in the synaptic cleft.

10. Explain how changes in the permeability of the neuron membrane result in the development of IPSPs and EPSPs.

11. Describe presynaptic inhibition and facilitation.

12. Define the terms spatial summation and temporal summation.

13. Describe the role of the synapse in eliminating or enhancing information.

14. List the components and characteristics of a reflex.

15. Diagram a convergent pathway, a divergent pathway, and an oscillating circuit, and describe what is accomplished in each.

Part Three

As a hungry person prepares to drink a cup of hot soup, he smells the aroma and anticipates the taste of the soup. Feeling the warmth of the cup in his hand, he carefully raises the cup to his lips and takes a sip. The soup is so hot that he "burns" his tongue, quickly jerks the cup away from his lips, and gasps with pain. All of these sensations and activities are monitored and controlled by the nervous system, which consists of the brain, spinal cord, nerves, and sensory receptors.

Homeostasis is maintained to a large degree by the regulatory and coordinating activities of the nervous system and depend on its ability to detect, interpret, and respond to changes. Sensory receptors monitor temperature, touch, smell, pain, sound, light, blood pressure, pH of the body fluids, the relative position of body parts, and other conditions. This information is transmitted from the sensory receptors by nerves to the brain and the spinal cord. On the basis of this information, the brain and the spinal cord assess conditions outside and inside the body. The information may produce a response, be stored for later use, or be ignored.

The nervous system is also responsible for mental activity, including consciousness, memory, and thinking. The abilities to solve problems, communicate using symbols, develop values, and exhibit a wide variety of emotions cannot occur without the nervous system.

Divisions of the Nervous System

There is only one nervous system, even though some of its subdivisions are referred to as separate systems. Thus the central nervous system and the peripheral nervous system are subdivisions of the nervous system, instead of separate organ systems as their names suggest (figure 12.1). Each subdivision has structural and functional features that separate it from the other subdivisions.

The **central nervous system (CNS)** consists of the brain and the spinal cord, which are protected by surrounding bone. The brain is located within the cranial vault of the skull, and the spinal cord is located within the vertebral canal, formed by the vertebrae (see chapter 7). The brain and spinal cord are continuous with each other at the foramen magnum.

The **peripheral nervous system (PNS)** consists of nerves and ganglia. **Nerves** are bundles of axons and their sheaths that extend from the CNS to peripheral structures, such as muscles and glands, and from sensory organs to the CNS. Forty-three pairs of nerves originate from the CNS to form the PNS. Twelve pairs of **cranial nerves** originate from the brain, and 31 pairs of **spinal nerves** originate from the spinal cord. **Ganglia** (gang′glē-ă; sing. ganglion, gang′glē-on, meaning knots) are collections of neuron cell bodies located outside the CNS.

The PNS has two subcategories: the afferent and the efferent divisions (figure 12.2). The **afferent, or sensory division,** transmits action potentials from the sensory organs to the CNS. The cell bodies of these neurons are located in ganglia near the spinal cord or near the origin of certain cranial nerves (figure 12.3a). The **efferent, or motor division,** transmits action potentials from the CNS to effector organs, such as muscles and glands.

The efferent division of the nervous system is divided into two subdivisions: the **somatic motor nervous system** and the **autonomic nervous system (ANS).** The somatic motor nervous system transmits action potentials from the CNS to skeletal muscle. Its neuron cell bodies are located within the CNS, and their axons extend through nerves to neuromuscular junctions (figure 12.3b), which are the only somatic motor nervous system synapses outside of the CNS.

The ANS transmits action potentials from the CNS to smooth muscle, cardiac muscle, and certain glands. It sometimes is called the involuntary nervous system because control of its target tissues occurs subconsciously. The ANS has two sets of neurons that exist in a series between the CNS and the effector organs. Cell bodies of the first neurons are within the CNS and send their axons to autonomic ganglia, where neuron cell bodies of the second neurons are located. Synapses exist between the first and second neurons within the autonomic ganglia, and the axons of the second neurons extend from the autonomic ganglia to the effector organs (figure 12.3c).

The ANS is subdivided into the **sympathetic** and the **parasympathetic divisions.** In general the sympathetic division prepares the body for physical activity when activated, whereas the parasympathetic division regulates resting or vegetative functions, such as digesting food or emptying the urinary bladder.

The CNS is the major site for processing information, initiating responses, and integrating mental processes. It is analogous to a highly sophisticated computer with the ability to receive input, process and store information, and generate responses. It also can produce ideas, emotions, and other mental processes that are not the automatic consequences of information input. The PNS functions primarily to detect stimuli and transmit information in the form of action potentials to and from the CNS. The PNS, however, does perform some limited integration at the sensory receptors and in some ganglia.

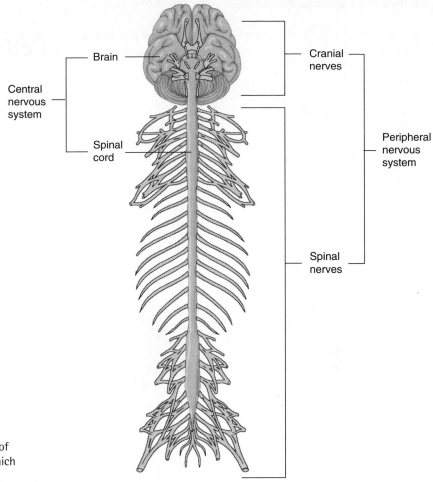

Figure 12.1 The Nervous System

The CNS consists of the brain and spinal cord. The PNS consists of cranial nerves, which arise from the brain, and spinal nerves, which arise from the spinal cord.

Figure 12.2
Organization of the Nervous System

The afferent division of the peripheral nervous system (PNS) detects stimuli and conducts action potentials to the central nervous system (CNS). The CNS interprets incoming action potentials and initiates action potentials that are conducted through the efferent division to produce a response. The efferent division is divided into the somatic motor nervous system and autonomic nervous system.

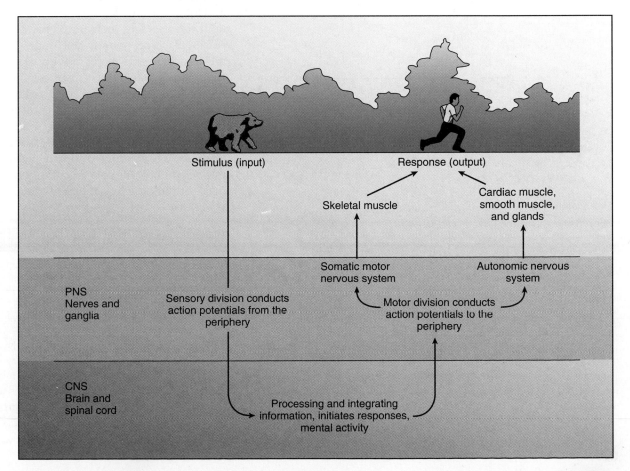

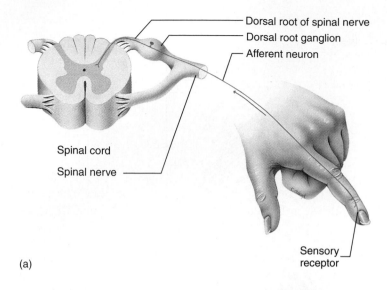

Dorsal root of spinal nerve
Dorsal root ganglion
Afferent neuron

Spinal cord

Spinal nerve

Sensory receptor

(a)

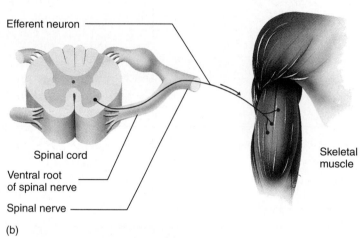

Efferent neuron

Spinal cord

Ventral root of spinal nerve

Spinal nerve

Skeletal muscle

(b)

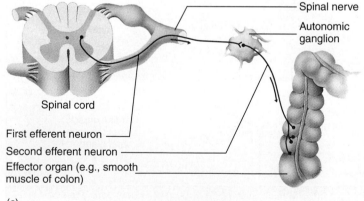

Spinal nerve

Autonomic ganglion

Spinal cord

First efferent neuron

Second efferent neuron

Effector organ (e.g., smooth muscle of colon)

(c)

Figure 12.3   Divisions of the Peripheral Nervous System

(*a*) **Afferent division.** A neuron with its cell body in a dorsal root ganglion. (*b*) **Efferent division** (somatic motor nervous system). The neuron extends from the CNS to effector cells (skeletal muscle). (*c*) **Efferent division** (ANS). Two neurons are in series between the CNS and the effector cells (smooth muscle). The first neuron has its cell body in the CNS, and the second neuron has its cell body in an autonomic ganglion.

Cells of the Nervous System

Cells of the nervous system include nonneural cells and neurons. Nonneural cells are **neuroglia** (nū-rog′lē-ă, meaning nerve glue), or **glial** (glī-ăl) **cells,** which support and protect neurons and perform other functions, whereas neurons receive stimuli and conduct action potentials.

Neurons

Neurons, or **nerve cells,** receive stimuli and transmit action potentials to other neurons or to effector organs. They are organized to form complex networks that perform the functions of the nervous system. Each neuron consists of a cell body and two types of processes. The cell body is called the **neuron cell body,** or **soma** (sō′mă, meaning body), and the processes are called **dendrites** (den′drītz, meaning tree), which refers to the branching organization of dendrites, and **axons** (ak′sonz, meaning axle), which refers to the straight alignment and uniform diameter of most axons (figure 12.4). Axons are also referred to as **nerve fibers.**

Neuron Cell Body

Each neuron cell body contains a single relatively large and centrally located nucleus with a prominent nucleolus. Extensive rough endoplasmic reticulum and Golgi apparatuses surround the nucleus, and a moderate number of mitochondria and other organelles are also present. Randomly arranged lipid droplets and melanin pigments accumulate in the cytoplasm of some neuron cell bodies. The lipid droplets and melanin pigments increase with age, but their functional significance is unknown. Large numbers of intermediate filaments (neurofilaments) and microtubules form bundles that course through the cytoplasm in all directions. The neurofilaments separate areas of rough endoplasmic reticulum called **chromatophilic** (krō-mă-tō-fil′ik) **substance,** or **Nissl** (nis′l) **bodies.** The presence of organelles such as rough endoplasmic reticulum indicates that the neuron cell body is the primary site of protein synthesis within neurons.

1	P R E D I C T

Predict the effect on the part of a severed axon that is no longer connected to its neuron cell body. Explain your prediction.

✔ *Answer in Appendix F*

Dendrites

Dendrites are short, often highly branched cytoplasmic extensions that are tapered from their bases at the neuron cell body to their tips (see figure 12.4). The surface of many dendrites has small extensions called **dendritic spines,** with which axons of other neurons form synapses with the dendrites. Den-

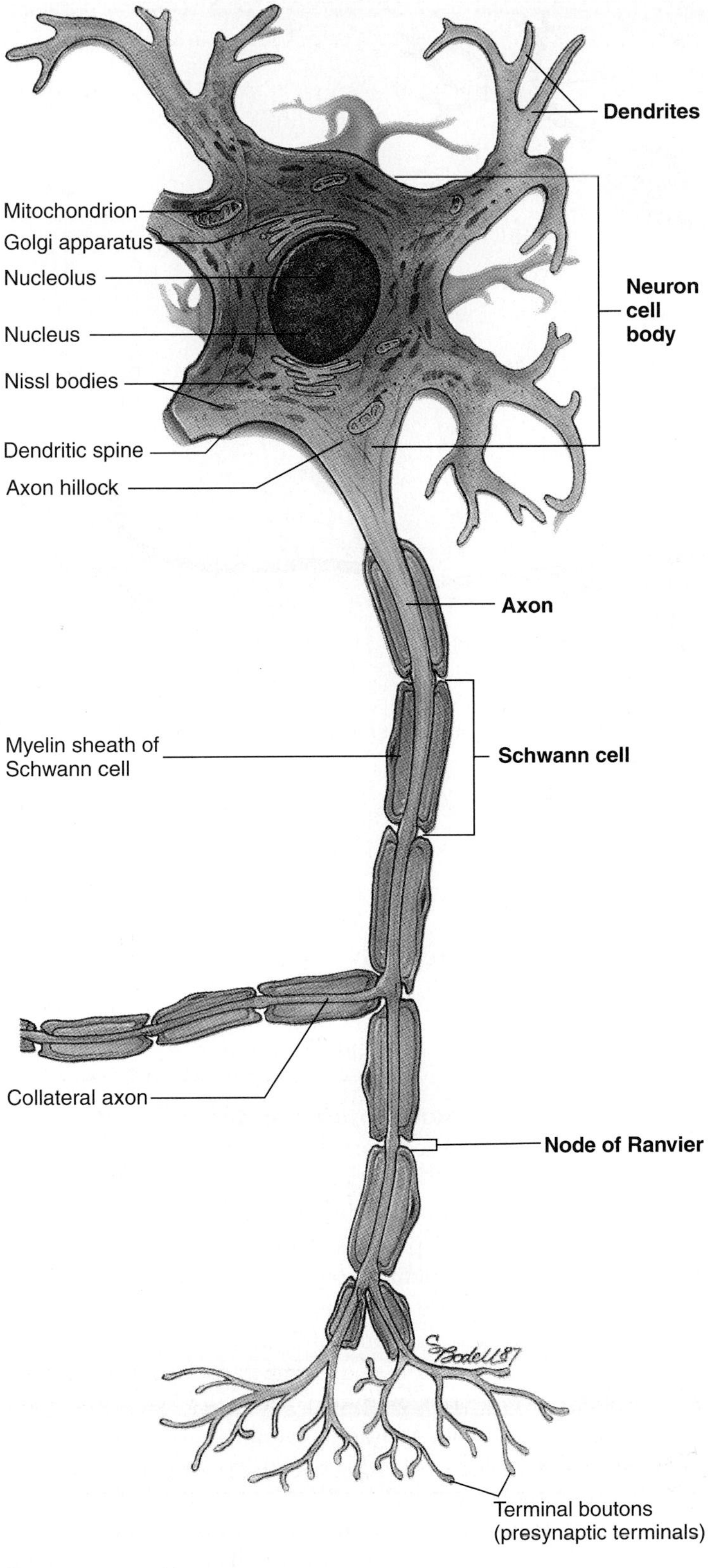

Figure 12.4   Neuron

Structural features of a neuron include a cell body and two types of cell processes: dendrites and an axon. ✗

Labels on figure:
Dendrites
Mitochondrion
Golgi apparatus
Nucleolus
Nucleus
Nissl bodies
Dendritic spine
Axon hillock
Neuron cell body
Myelin sheath of Schwann cell
Axon
Schwann cell
Collateral axon
Node of Ranvier
Terminal boutons (presynaptic terminals)

drites respond to **neurotransmitter** substances released from the axons of other neurons by producing local potentials. Most dendrites do not conduct action potentials. Functionally, dendrites have traditionally been classified as processes that conduct electric signals toward the cell body.

Axons

In most neurons a single axon arises from a cone-shaped area of the neuron cell body called the **axon hillock.** An axon can remain as a single structure or can branch to form collateral axons or side branches (see figure 12.4). Each axon has a constant diameter and can vary in size from a few millimeters to more than 1 m in length. The cytoplasm of the axon is called **axoplasm,** and its cell membrane is called the **axolemma** (*lemma* is Greek, meaning husk or sheath). Axons terminate by branching to form small extensions with enlarged ends called **presynaptic terminals,** or **terminal boutons** (bū-tonz', meaning buttons). Numerous small vesicles that contain neurotransmitters are present in the presynaptic terminals. Functionally, axons conduct action potentials from the neuron cell body to the presynaptic terminals.

Axon transport mechanisms can move cytoskeletal proteins (see chapter 3), organelles such as mitochondria, and vesicles containing neurohormones to be secreted (see chapter 17) down the axon to the presynaptic terminals. In addition, damaged organelles, recycled plasma membrane, and substances taken in by endocytosis can be transported up the axon to the neuron cell body. Although the movement of materials within the axon is necessary for normal function, it also provides a way for infectious agents and harmful substances to enter neurons. For example, rabies and herpes viruses enter the axon endings of damaged skin and are transported to the CNS.

Types of Neurons

Neurons can be classified by function or by structure. The functional classification considers the direction in which action potentials are conducted. **Afferent,** or **sensory, neurons** conduct action potentials toward the CNS, and **efferent,** or **motor, neurons** conduct action potentials from the CNS to muscles or glands. **Association neurons,** or **interneurons,** conduct action potentials from one neuron to another.

The structural classification scheme considers the number of processes that extend from the neuron cell body. The three major categories of neurons are multipolar, bipolar, and unipolar. **Multipolar neurons** have many processes, which consist of many dendrites and a single axon. The dendrites vary in number and in their degree of branching (figure 12.5*a*). Association and motor neurons are multipolar.

Bipolar neurons have two processes: a dendrite and an axon (figure 12.5*b*). The dendrite often is specialized to receive the stimulus, and the axon conducts action potentials to the CNS. Bipolar neurons are sensory neurons that are components of certain specialized sensory organs such as the rods and cones of the eye.

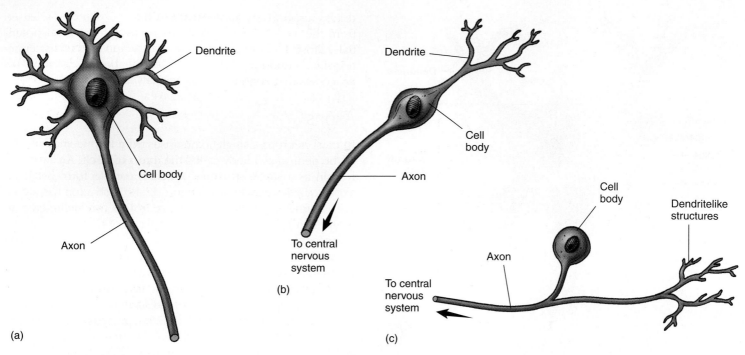

Figure 12.5 Types of Neurons

(*a*) A multipolar neuron has many dendrites and one axon. (*b*) A bipolar neuron has a dendrite and an axon. (*c*) A unipolar neuron has an axon and no dendrites.

Unipolar neurons have one process, an axon (figure 12.5*c*). Most sensory neurons are unipolar. The peripheral ends of their axons have dendritelike processes that respond to stimuli, producing action potentials that are conducted by the axon to the CNS. The branch of a unipolar neuron that extends from the periphery to the neuron cell body conducts action potentials toward the neuron cell body, and, according to a functional definition of a dendrite, it could be classified as a dendrite. It often is referred to as an axon, however, for two reasons: it cannot be distinguished from an axon on the basis of its structure, and it conducts action potentials in the same fashion as an axon (see the section Axon Sheaths later in the chapter).

Neuroglia

Neuroglia are far more numerous than neurons and account for more than half of the brain's weight. They are the major supporting cells in the CNS, participate in the formation of a permeability barrier between the blood and the neurons, phagocytize foreign substances, produce cerebrospinal fluid, and form myelin sheaths around axons. Each of the five types of neuroglial cells has unique structural and functional characteristics.

Astrocytes

Astrocytes (as′trō-sītz, the word *aster* is Greek, meaning star) are neuroglia that are star-shaped because of cytoplas-

mic processes that extend from the cell body. The processes of the astrocytes extend to and cover the surfaces of blood vessels, neurons (figure 12.6), and the pia mater. (The pia mater is a membrane covering the outside of the brain and spinal cord.) Astrocytes are a nonrigid supporting matrix and help to regulate the composition of the extracellular fluid around neurons.

Because of the **blood–brain barrier,** only certain substances can pass from the blood into the nervous tissue of the brain and spinal cord. The blood–brain barrier protects neurons from toxic substances in the blood, allows the exchange of nutrients and waste products between neurons and the blood, and prevents fluctuations in the composition of the blood from affecting the functions of the brain. Endothelial cells of the blood vessels, which are joined by tight junctions (see chapter 4), form the blood–brain barrier. Consequently, substances do not pass between the cells but must pass through the cells. Lipid-soluble substances, such as nicotine, ethanol, and heroin, can diffuse through the phospholipid membrane of the endothelial cells and enter the brain. Water-soluble molecules such as amino acids and glucose move across the blood–brain barrier by mediated transport (see chapter 3).

Astrocytes play a role in regulating the extracellular composition of brain fluid. They influence the formation of tight junctions between endothelial cells and the types of molecules transported by the endothelial cells. In addition, after substances pass through the blood–brain barrier, astrocytes function to remove or process molecules and ions.

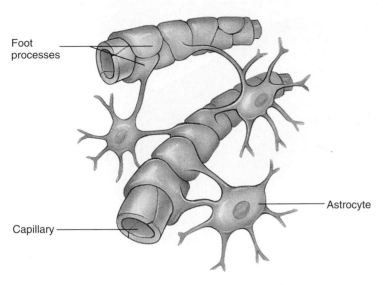

Foot processes

Capillary

Astrocyte

Figure 12.6   Astrocytes

Astrocyte processes form feet that cover the surfaces of neurons and blood vessels. The astrocytes provide structural support and play a role in regulating what substances from the blood reach the neurons.

Clinical Note

The permeability characteristics of the blood–brain barrier must be considered when developing drugs designed to affect the CNS. For example, Parkinson's disease is caused by a lack of the neurotransmitter dopamine, which normally is produced by certain neurons of the brain. This lack results in decreased muscle control and shaking movements. Administering dopamine is not helpful because dopamine cannot cross the blood–brain barrier. Levodopa (L-dopa), a precursor to dopamine, is administered instead because it can cross the blood–brain barrier. CNS neurons then convert levodopa to dopamine, which reduces the symptoms of Parkinson's disease.

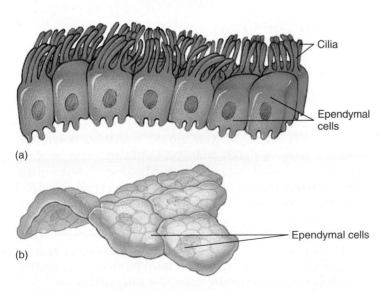

Cilia

Ependymal cells

(a)

Ependymal cells

(b)

Figure 12.7   Ependymal Cells

(a) Ciliated ependymal cells lining the ventricles of the brain help to move cerebrospinal fluid. (b) Ependymal cells on the surface of the choroid plexus secrete cerebrospinal fluid. 🏃 ▭

Ependymal Cells

Ependymal (ep-en'di-măl) cells line the ventricles (cavities) of the brain and the central canal of the spinal cord (figure 12.7a). Specialized ependymal cells within certain regions of the ventricles, the **choroid plexuses** (ko'royd plek'sŭs-ez) (figure 12.7b), secrete the cerebrospinal fluid that circulates through the ventricles of the brain (see chapter 13). The free surface of the ependymal cells frequently has patches of cilia that assist in moving cerebrospinal fluid through the cavities of the brain. Ependymal cells also have long processes at their basal surfaces that extend deep into the brain and the spinal cord and seem, in some cases, to have astrocytelike functions.

Microglia

Microglia (mī-krog'lē-ă) are specialized macrophages in the CNS that become mobile and phagocytic in response to inflammation, phagocytizing necrotic tissue, microorganisms, and foreign substances that invade the CNS (figure 12.8).

Clinical Note

Numerous microglia migrate to areas damaged by infection, trauma, or stroke and perform phagocytosis. A pathologist can identify these damaged areas in the CNS during an autopsy because large numbers of microglia are found in them.

Oligodendrocytes

Oligodendrocytes (ol'i-gō-den'drō-sītz) have cytoplasmic extensions that form sheaths around axons in the CNS (figure 12.9). A single oligodendrocyte can surround portions of

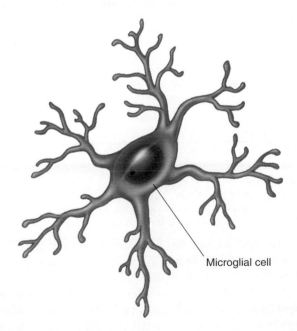

Microglial cell

Figure 12.8   Microglia

Microglia within the central nervous system are similar to macrophages.

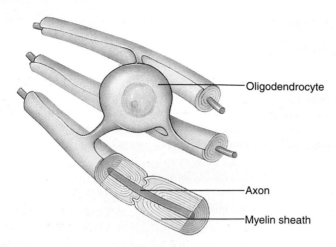

Figure 12.9   Oligodendrocyte

Extensions from the oligodendrocyte form the myelin sheaths of axons within the central nervous system.

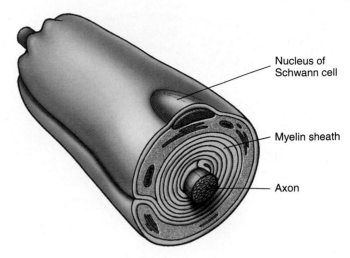

Figure 12.10   Schwann Cell

Extension from the Schwann cell forms the myelin sheath of an axon within the peripheral nervous system. 🏃 📼

several axons. Oligodendrocyte processes are modified to form coverings called **myelin** (mī′ĕ-lin) **sheaths** around parts of more than one axon in the CNS. The normal rate of action potential propagation along axons depends on myelin sheaths.

Schwann Cells

Schwann cells, or **neurolemmocytes** (nūr-ō-lem′mō-sītz), are neuroglial cells in the PNS that form myelin sheaths around axons. Unlike oligodendrocytes, however, each Schwann cell forms a myelin sheath around a portion of only one axon (figure 12.10).

Satellite cells, which are specialized neurolemmocytes, surround neuron cell bodies in ganglia, provide support, and can provide nutrients to the neuron cell bodies (figure 12.11).

Axon Sheaths

Cytoplasmic extensions of the oligodendrocytes in the CNS and of the Schwann cells in the PNS surround axons to form either unmyelinated or myelinated axons. The cytoplasmic extensions protect and electrically insulate the axons from one another. In addition, action potentials are propagated along unmyelinated and myelinated axons in different ways and at different rates.

Unmyelinated Axons

Unmyelinated axons rest in an invagination of the oligodendrocytes or the Schwann cells (figure 12.12a). The axons are surrounded by an extension of the cell's cytoplasm, but they are not actually inside the cell. Thus each axon is surrounded by a series of cells, and each cell can simultaneously surround more than one unmyelinated axon. In unmyelinated axons, action potentials are propagated along the entire axon membrane (see figure 9.22).

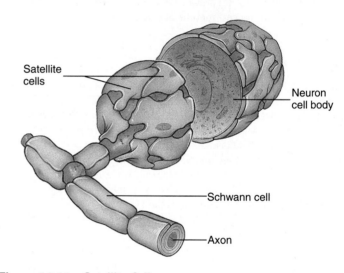

Figure 12.11   Satellite Cells

Neuron cell bodies within ganglia are surrounded by satellite cells.

Myelinated Axons

In **myelinated axons,** extensions from oligodendrocytes or Schwann cells repeatedly wrap around a segment of an axon to form a series of tightly wrapped membranes rich in phospholipids with little cytoplasm sandwiched between the membrane layer (see figure 12.9; figure 12.12b). The tightly wrapped membranes constitute the **myelin sheath,** giving myelinated axons a white appearance because of the high lipid concentration. When viewed longitudinally, gaps can be seen every 0.1–1.5 mm between the individual cells. These interruptions in the myelin sheath are the **nodes of Ranvier** (ron′vē-ā), and the areas between the nodes are called the **internodes.**

Conduction of action potentials from one node of Ranvier to another in myelinated neurons is called **saltatory conduction** (L. *saltare,* meaning to leap). The reversal of the electric charge across the cell membrane, which occurs during an

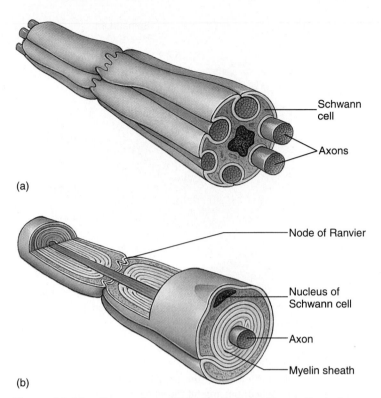

(a)

(b)

Schwann cell

Axons

Node of Ranvier

Nucleus of Schwann cell

Axon

Myelin sheath

Figure 12.12   Comparison of Myelinated and Unmyelinated Axons

(*a*) Unmyelinated axons with two Schwann cells surrounding several axons in parallel formation. Each Schwann cell surrounds part of several axons. (*b*) Myelinated axon with two Schwann cells forming the myelin sheath around a single axon. Each Schwann cell surrounds part of one axon.

ducted from node to node without being propagated along the entire length of the axon.

The speed of action potential conduction along an axon depends on the myelination of the axon. Action potentials are conducted more rapidly in myelinated than unmyelinated axons because action potentials are formed quickly at each successive node of Ranvier instead of being propagated more slowly through every part of the axon's membrane. Action potential conduction in a myelinated fiber is like a grasshopper jumping, whereas action potential conduction in an unmyelinated axon is like a grasshopper walking. The grasshopper (action potential) moves more rapidly by jumping.

In addition to myelination, the diameter of axons affects the speed of action potential conduction. Large-diameter axons conduct action potentials more rapidly than small-diameter axons because large-diameter axons provide less resistance to action potential propagation.

Nerve fibers (axons) can be classified on the basis of their size and myelination. Not surprisingly, the structure of nerve fibers reflects their functions. Type A fibers are large-diameter, myelinated axons that conduct action potentials at 15–120 m/s. Motor neurons supplying skeletal muscles and most sensory neurons have type A fibers. Rapid response to the external environment is possible because of the rapid input of sensory information to the CNS and rapid output of action potentials to skeletal muscles.

Type B fibers are medium-diameter, lightly myelinated axons that conduct action potentials at 3–15 m/s, and type C fibers are small-diameter, unmyelinated axons that conduct action potentials at 2 m/s or less. Type B and C fibers are primarily part of the ANS, which supplies internal organs such as the stomach, intestines, and heart. The responses necessary to maintain internal homeostasis such as digestion need not be as rapid as responses to the external environment.

2	P R E D I C T

What is the advantage of having small-diameter axons that conduct action potentials rapidly? (*Hint:* Consider what an animal with only unmyelinated axons would be like.)

✔ *Answer in Appendix F*

action potential at one node of Ranvier, causes electric current to flow across the membrane at the adjacent node almost instantaneously (figure 12.13). The lipid within the membranes of the myelin sheath acts as a layer of insulation, forcing the electric current to flow to the adjacent node and preventing flow across the membrane within the internode. The flow of the electric current acts as a stimulus and initiates an action potential at the adjacent node. Action potentials are thus con-

Figure 12.13   Saltatory Conduction

(*a*) to (*c*) shows the flow of electric charge from node to node during saltatory conduction, producing action potentials at the nodes of Ranvier in myelinated neurons.

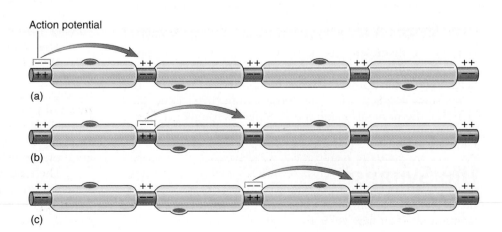

Action potential

(a)

(b)

(c)

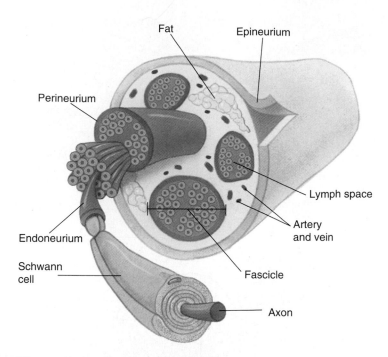

Figure 12.14   Nerve

Nerve structure illustrating axons surrounded by various layers of connective tissue: epineurium around the whole nerve, perineurium around nerve fascicles, and endoneurium around Schwann cells and axons.

Organization of Nervous Tissue

Nervous tissue is organized so that axons form bundles, and neuron cell bodies and their relatively short dendrites are organized into groups. Bundles of parallel axons with their associated myelin sheaths are whitish in color, which accounts for their name, **white matter.** Collections of neuron cell bodies and unmyelinated axons are more gray in color and are called **gray matter.**

The axons that make up the white matter of the CNS form **nerve tracts,** which propagate action potentials from one area in the CNS to another. The gray matter of the CNS performs integrative functions or acts as relay areas in which axons synapse with the cell bodies of neurons. The central area of the spinal cord is gray matter, and the outer surface of much of the brain consists of gray matter called **cortex.** Within the brain are other collections of gray matter called **nuclei.**

In the PNS, bundles of axons and their sheaths form **nerves,** which conduct action potentials to and from the CNS. Most nerves contain myelinated axons, but some nerves consist of unmyelinated axons. Collections of neuron cell bodies in the PNS are called **ganglia.**

Peripheral nerves consist of axon bundles (figure 12.14). Each axon and its Schwann cell sheath are surrounded by a delicate connective tissue layer, the **endoneurium** (en-dō-nū′rē-ŭm). Groups of axons are surrounded by a heavier connective tissue layer, the **perineurium** (per-i-nū′rē-ŭm), to form **nerve fascicles** (fas′i-klz). A third layer of connective tissue, the **epineurium** (ep-i-nū′rē-ŭm), binds the nerve fascicles together to form a nerve. The connective tissue of the epineurium merges with the loose connective tissue surrounding the nerve to make the PNS nerves tougher than the nerve tracts of the CNS.

The Synapse

Just as the fire from one lit torch can light another torch, action potentials in one cell can result in action potentials being produced in another cell, allowing communication between them. For example, if you touch a hot pan, action potentials produced in temperature and pain sensory neurons are sent to the CNS. Thus information in the form of action potentials is carried by nerve fibers from the finger toward the CNS. For the CNS to get this information, the action potentials of the sensory neurons must cause the production of action potentials in the CNS neurons. After the CNS has received the information, it can produce a response. One response is to remove the finger from the hot object by causing the appropriate skeletal muscles to contract. CNS action potentials cause the production of action potentials in motor neurons, which are transmitted by the motor neurons toward skeletal muscles. The action potentials of the motor neuron result in the production of skeletal muscle action potentials, which are stimuli that cause the muscle fibers to contract (see chapter 10).

The action potentials in one cell can cause the production of action potentials in another cell at the **synapse** (sin′aps), which is a junction between the two cells. In this sequence, the first cell to produce actions potentials is called the **presynaptic cell,** and the second cell, which responds to the first cell across the synapse, is called the **postsynaptic cell.** Presynaptic cells are typically neurons, and postsynaptic cells are typically other neurons, muscle cells, or gland cells.

The essential components of a synapse are the presynaptic terminal, the synaptic cleft, and the postsynaptic membrane (figure 12.15a). The **presynaptic terminal** is formed from the end of an axon, and the space separating the axon

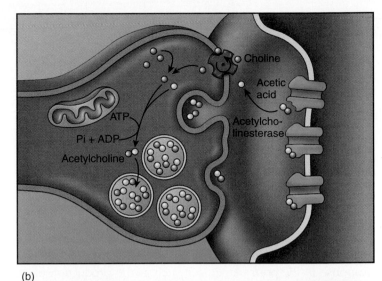

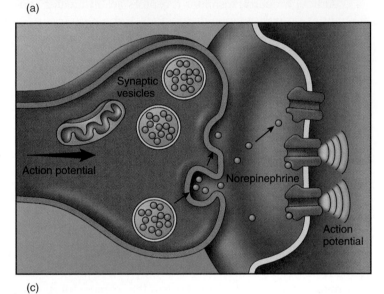

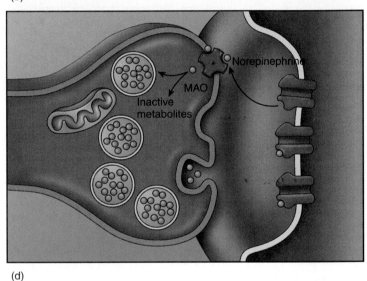

(a)   (b)   (c)   (d)

Figure 12.15   Synaptic Transmission

The end of the axon, the presynaptic terminal, is separated by a space, the synaptic cleft, from the postsynaptic membrane. (a) The neurotransmitter acetylcholine diffuses from the presynaptic terminal across the synaptic cleft to the receptors in the postsynaptic membrane. (b) Acetylcholinesterase acts on the acetylcholine to break it down into acetic acid and choline. Choline is transported back to the presynaptic terminal, where it is used to resynthesize acetylcholine. (c) The neurotransmitter norepinephrine diffuses from the presynaptic terminal across the synaptic cleft to the receptors in the postsynaptic membrane. (d) Norepinephrine is transported back into the presynaptic terminal and packaged into synaptic vesicles for reuse, or the enzyme monoamine oxidase (MAO) alters its structure and inactivates it.

ending and the cell with which it synapses is the **synaptic cleft.** The membrane of the postsynaptic cell opposed to the presynaptic terminal is the **postsynaptic membrane.**

Action potentials at the presynaptic terminal can cause the production of action potentials in the postsynaptic cell. Action potentials do not directly pass from the presynaptic terminal to the postsynaptic membrane, however. Instead, the action potentials in the presynaptic terminal cause the release of a chemical called a **neurotransmitter** from the presynaptic terminal. The neurotransmitter is a ligand (see chapter 9) that can diffuse across the synaptic cleft and bind to a receptor in the postsynaptic membrane. As a result, local potential changes are produced in the postsynaptic membrane. The lo-

cal potential changes can stimulate or inhibit action potential production in the postsynaptic cell.

Presynaptic terminals are specialized to produce and release neurotransmitters. The major cytoplasmic organelles within presynaptic terminals are mitochondria and numerous membrane-bound **synaptic vesicles,** which contain neurotransmitter. Each action potential arriving at the presynaptic terminal initiates a series of specific events that result in the release of neurotransmitter. In response to an action potential, voltage-gated calcium (Ca^{2+}) ion channels open, and Ca^{2+} ions diffuse into the presynaptic terminal. Ca^{2+} ions cause synaptic vesicles to fuse with the presynaptic membrane and release their neurotransmitter by exocytosis into the synaptic cleft.

Once the neurotransmitter is released from the presynaptic terminal, it diffuses rapidly across the synaptic cleft, which is about 20 nm wide, and binds in a reversible fashion to specific receptors in the postsynaptic membrane. This binding causes a response in the postsynaptic cell such as the production of an action potential. The interaction between the neurotransmitter substance and the receptor represents an equilibrium.

Neurotransmitter + Receptor ⇄ Neurotransmitter–receptor complex

When the neurotransmitter concentration in the synaptic cleft is high, many of the receptor molecules have neurotransmitter molecules bound to them, and when the neurotransmitter concentration declines, the neurotransmitter molecules diffuse away from the receptor molecules.

Neurotransmitters have short-term effects on postsynaptic membranes because the neurotransmitter is rapidly destroyed or removed from the synaptic cleft. For example, the neurotransmitter acetylcholine is broken down by the enzyme **acetylcholinesterase** (as'e-til-kō-lin-es'ter-ās) to acetic acid and choline. Choline is then transported back into the presynaptic terminal and is used to resynthesize acetylcholine (figure 12.15*b*). On the other hand, when the neurotransmitter norepinephrine is released into the synaptic cleft, most of it is actively transported back into the presynaptic terminal, where the norepinephrine is repackaged into synaptic vesicles for reuse. Within the presynaptic terminal, some of the norepinephrine is inactivated by the enzyme **monoamine oxidase** (mon'ō-am'īn ok'si-dās) (figure 12.15*c* and *d*).

Diffusion of neurotransmitter molecules away from the synapse and into the extracellular fluid also limits the length of time the neurotransmitter molecules remain bound to their receptors. Norepinephrine in the circulation is taken up primarily by liver and kidney cells, where the enzymes monoamine oxidase and **catechol-*O*-methyltransferase** (kat'ĕ-kol-ō-meth'il-trans'fer-ās) convert it into inactive metabolites.

Receptor Molecules in Synapses

Receptor molecules in synapses are membrane-bound, ligand-activated receptors with highly specific receptor sites (see chapter 9). Consequently, only neurotransmitter molecules or very closely related substances normally bind to their receptors. For example, acetylcholine receptors bind to acetylcholine but not to other neurotransmitters. More than one type of receptor molecule exists for some neurotransmitters. A single neurotransmitter can therefore bind to one type of specific receptor to cause depolarization in one synapse and to another type of specific receptor to cause hyperpolarization in another synapse. For example, norepinephrine is either inhibitory or stimulatory, depending on the type of norepinephrine receptor to which it binds and on the effect that receptor has on the permeability of the postsynaptic membrane.

Although neurotransmitter receptors are in greater concentrations on postsynaptic membranes, some receptors exist on presynaptic membranes. For example, norepinephrine released from the presynaptic membrane binds to receptors on both the presynaptic and postsynaptic membranes. Its binding to the presynaptic membrane decreases the release of additional synaptic vesicles. Norepinephrine can therefore modify its own release by binding to presynaptic receptors. A high frequency of presynaptic action potentials results in the release of fewer synaptic vesicles in response to later action potentials.

Neurotransmitters and Neuromodulators

Several substances have been identified as neurotransmitters, and others are suspected neurotransmitters. It was once thought that each neuron contains only one type of neurotransmitter; however, it is now known that some neurons can secrete more than one type. If a neuron does produce more than one neurotransmitter, it secretes all of them from each of its presynaptic terminals. The physiologic significance of presynaptic terminals that secrete more than one type of neurotransmitter has not been clearly established.

Neuromodulators are substances released from neurons that can presynaptically or postsynaptically influence the likelihood that an action potential in the presynaptic terminal will result in the production of an action potential in the postsynaptic cell. For example, a neuromodulator that decreases the release of an excitatory neurotransmitter from a presynaptic terminal decreases the likelihood of action potential production in the postsynaptic cell. A list of neurotransmitters and neuromodulators is presented in table 12.1.

Table 12.1	Substances That Are Neurotransmitters or Neuromodulators (or both)		
Substance	**Location**	**Effect**	**Clinical Example**
Acetylcholine	Many nuclei scattered throughout the brain and spinal cord. Nerve tracts from the nuclei extend to many areas of the brain and spinal cord. Also found in the neuromuscular junction of skeletal muscle and many ANS synapses	Excitatory or inhibitory	Alzheimer's disease (a type of senile dementia) is associated with a decrease in acetylcholine-secreting neurons. Myasthenia gravis (weakness of skeletal muscles) results form a reduction in acetylcholine receptors.
Monoamines			
Norepinephrine	A small number of small-sized nuclei in the brainstem. Nerve tracts extend from the nuclei to many areas of the brain and spinal cord. Also in some ANS synapses.	Excitatory or inhibitory	Cocaine and amphetamines increase the release and block the reuptake of norepinephrine, resulting in overstimulation of postsynaptic neurons.
Serotonin	A small number of small-sized nuclei in the brainstem. Nerve tracts extend from the nuclei to many areas of the brain and spinal cord.	Generally inhibitory	Involved with mood, anxiety, and sleep induction. Levels of serotonin are elevated in schizophrenia (delusions, hallucinations, and withdrawal).
Dopamine	Confined to a small number of nuclei and nerve tracts. Distribution is more restricted than that of norepinephrine or serotonin. Also found in some ANS synapses.	Generally excitatory	Parkinson's disease (depression of voluntary motor control) results from destruction of dopamine-secreting neurons. Drugs used to increase dopamine production induce vomiting and schizophrenia.
Histamine	Hypothalamus, with nerve tracts to many parts of the brain and spinal cord.	Generally inhibitory	No clear indication of histamine-associated pathologies. Histamine apparently is involved with arousal from sleep, pituitary hormone secretion, control of cerebral circulation, and thermoregulation.
Amino Acids			
Gamma-aminobutyric acid (GABA)	GABA-secreting neurons mostly control activities in their own area and are not usually involved with transmission from one part of the CNS to another. Most neurons of the CNS have GABA receptors.	Majority of postsynaptic inhibition in the brain; some presynaptic inhibition in the spinal cord	Drugs that increase GABA function have been used to treat epilepsy (excessive discharge of neurons).
Glycine	Spinal cord and brain. Like GABA, glycine predominantly produces local effects.	Most postsynaptic inhibition in the spinal cord	Glycine receptors are inhibited by the poison strychnine.
Glutamate and aspartate	Widespread in the brain and spinal cord, especially in nerve tracts that ascend or descend the spinal cord or in tracts that project from one part of the brain to another.	Excitatory	Drugs that block glutamate or aspartate are under development. These drugs might prevent seizures and neural degeneration from overexcitation.

continued next page

Table 12.1   Substances That Are Neurotransmitters or Neuromodulators (or both)—cont'd

Substance	Location	Effect	Clinical Example
Nitric Oxide	Brain, spinal cord, adrenal gland, intramural plexus, nerves to penis.	Excitatory	Blocking nitric oxide production may prevent stroke damage. Stimulating nitric oxide release is used to treat impotence.
Neuropeptides			
Endorphins and enkephalins	Widely distributed in the CNS and PNS.	Generally inhibitory	The opiates morphine and heroin bind to endorphin and enkephalin receptors on presynaptic neurons and reduce pain by blocking the release of neurotransmitter.
Substance P	Spinal cord, brain, and sensory neurons associated with pain.	Generally excitatory	Substance P is a neurotransmitter in pain transmission pathways. Blocking the release of substance P by morphine reduces pain.

glutamate binds to postsynaptic neurons and stimulates them to release nitric oxide (NO), which in high concentrations can be toxic to cells. The nitric oxide diffuses from the postsynaptic neurons and causes damage to surrounding cells. It is possible that stroke damage may be reduced by developing drugs that block glutamate receptors or inhibit the production of NO.

Excitatory and Inhibitory Postsynaptic Potentials

The combination of neurotransmitters with their specific receptors causes either depolarization or hyperpolarization of the postsynaptic membrane. When depolarization occurs, the response is stimulatory, and the local depolarization is an **excitatory postsynaptic potential (EPSP)** (figure 12.16*a*). Neurons releasing neurotransmitter substances that cause EPSPs are **excitatory neurons.** In general, an EPSP occurs because of an increase in the permeability of the membrane to sodium (Na^+) ions. For example, glutamate in the brain and acetylcholine in skeletal muscle can bind to their receptors, causing Na^+ ion channels to open (see chapter 9). Because the concentration gradient is large for Na^+ ions and because the negative charge inside the cell attracts the positively charged Na^+ ions, they diffuse into the cell and cause depolarization. If depolarization reaches threshold, an action potential is produced.

When the combination of a neurotransmitter with its receptor results in hyperpolarization of the postsynaptic membrane, the response is inhibitory, and the local hyperpolarization is an **inhibitory postsynaptic potential (IPSP)** (figure 12.16*b*). Neurons releasing neurotransmitter substances that cause IPSPs are **inhibitory neurons.** The IPSP is the result of an increase in the permeability of the cell membrane to chloride (Cl^-) or potassium (K^+) ions. For example, in the spinal cord, glycine binds to its receptors, directly causing Cl^- ion channels to open. Because Cl^- ions are more concentrated outside the cell than inside, when the permeability of the membrane to Cl^- ions increases, they diffuse into the cell, causing the inside of the cell to become more negative and resulting in hyperpolarization. Acetylcholine can bind to its receptors in the heart, causing G protein-mediated opening of K^+ ion channels. The concentration of K^+ ions is greater inside the cell than outside, and increased permeability of the membrane to K^+ results in diffusion of K^+ ions out of the cell. Consequently, the outside of the cell becomes more positive than the inside, resulting in hyperpolarization.

Presynaptic Inhibition and Facilitation

Many of the synapses of the CNS are **axoaxonic synapses,** meaning that the axon of one neuron synapses with the presynaptic terminal (axon) of another (figure 12.17). The axoaxonic synapse does not initiate an action potential in the presynaptic terminal. When an action potential reaches the presynaptic terminal, however, neuromodulators released in the axoaxonic synapse can alter the amount of neurotransmitter released from the presynaptic terminal.

In **presynaptic inhibition** the amount of neurotransmitter released from the presynaptic terminal decreases, and in **presynaptic facilitation** the amount released from the presynaptic terminal increases (see figure 12.17). The amount

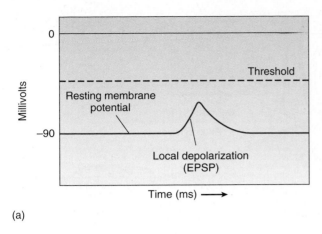

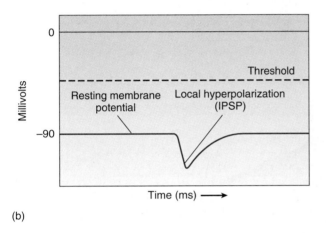

Figure 12.16   Postsynaptic Potentials

(*a*) Excitatory postsynaptic potential (EPSP) is closer to threshold.
(*b*) Inhibitory postsynaptic potential (IPSP) is further from threshold.

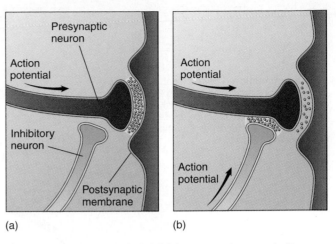

Figure 12.17   Presynaptic Inhibition at an Axoaxonic Synapse

(*a*) The inhibitory neuron of the axoaxonic synapse is inactive and has no effect on the release of neurotransmitter from the presynaptic terminal. (*b*) Release of a neuromodulator from the inhibitory neuron of the axoaxonic synapse reduces the amount of neurotransmitter released from the presynaptic terminal.

Spatial and Temporal Summation

Depolarizations produced in postsynaptic membranes are local depolarizations. Within the CNS and in many PNS synapses a single presynaptic action potential does not cause a local depolarization in the postsynaptic membrane sufficient to reach threshold and produce an action potential (see chapter 9). Instead, a series of presynaptic action potentials causes a series of local potentials in the postsynaptic neuron. The local potentials combine in a process called **summation** at the axon hillock of the postsynaptic neuron, which is the normal site of action potential generation for most neurons. If summation results in a local potential that exceeds threshold at the axon hillock, an action potential is produced.

Two types of summation, called spatial summation and temporal summation, are possible. The simplest type of **spatial summation** occurs when two action potentials arrive simultaneously at two different presynaptic terminals that synapse with the same postsynaptic neuron. In the postsynaptic neuron each action potential causes a local depolarization that undergoes summation at the axon hillock. If the summated depolarization reaches threshold, an action potential is produced (figure 12.18*a*).

Temporal summation results when two action potentials arrive in very close succession at a single presynaptic terminal. The first action potential causes a local depolarization in the postsynaptic membrane that remains for a few milliseconds before it disappears, although its magnitude decreases through time. Before the local depolarization caused by the first action potential repolarizes to its resting value, a second action potential initiates a second local depolarization. Temporal summation results when the second local depolarization summates with the remainder of the first local depolarization. If the summated local depolarization reaches

of neurotransmitter released by the presynaptic terminal affects the response produced in the postsynaptic membrane. The greater the amount of neurotransmitter, the larger the IPSP or EPSP produced. Consequently, presynaptic inhibition or facilitation can decrease or increase the likelihood of producing action potentials in the postsynaptic cell. For example, enkephalins and endorphins in the brain and spinal cord produce presynaptic inhibition of neurons transmitting pain sensations. A reduction in neurotransmitter release can reduce or prevent the production of action potentials by postsynaptic neurons. This interference with the transmission of the pain signal results in reduction or elimination of the awareness of pain.

An example of presynaptic facilitation involves the neurotransmitters **glutamate** and **nitric oxide.** A presynaptic neuron releases glutamate, which binds to glutamate receptors on the postsynaptic membrane and stimulates the postsynaptic neuron to produce nitric oxide. The nitric oxide diffuses out of the postsynaptic neuron, crosses the synaptic cleft, diffuses into the presynaptic neuron, and stimulates the release of additional glutamate from the presynaptic neuron.

(a) **Spatial summation.** The two local depolarizations produced at 1 and 2 by action potentials that arrive simultaneously summate at the axon hillock to produce a local depolarization that exceeds threshold, resulting in an action potential.

(b) **Temporal summation.** Two action potentials arrive in close succession at the presynaptic membrane. Before the first local depolarization returns to threshold, the second is produced. They summate to exceed threshold and produce an action potential.

(c) Combined spatial and temporal summation with both excitatory postsynaptic potentials and inhibitory postsynaptic potentials. The outcome, which is the product of summation, is determined by which influence is greater.

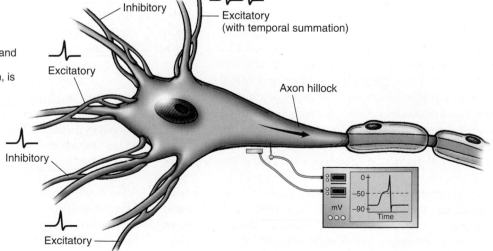

Figure 12.18   Summation

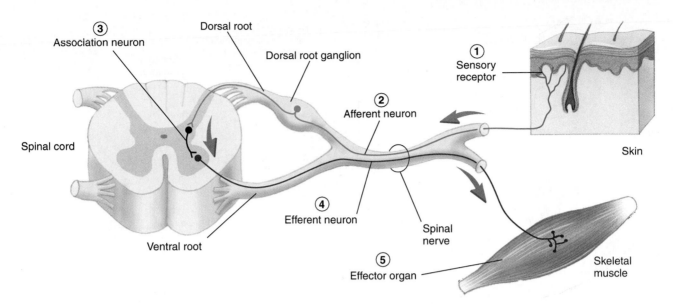

Figure 12.19   Reflex Arc

The parts of a reflex arc are labeled in the order in which action potentials pass through them. The five components are the sensory receptor, afferent neuron, association neuron, efferent neuron, and effector organ. ✗ ▭

threshold at the axon hillock, an action potential is produced in the postsynaptic neuron (figure 12.18*b*).

Excitatory and inhibitory neurons can synapse with a single postsynaptic neuron. Summation of EPSPs and IPSPs occurs in the postsynaptic neuron, and the IPSPs tend to cancel the EPSPs. Whether a postsynaptic action potential is initiated or not depends on which type of local potential has the greatest influence on the postsynaptic membrane potential (figure 12.18*c*).

The synapse is an essential structure for the process of integration carried out by the CNS. For example, action potentials propagated along axons from sensory organs to the CNS can produce a sensation, or they can be ignored. To produce a sensation, action potentials must be transmitted across synapses as they travel through the CNS to the cerebral cortex where information is interpreted. Stimuli that do not result in action potential transmission across synapses are ignored because information never reaches the cerebral cortex. The brain can ignore large amounts of sensory information as a result of complex integration.

▐ Reflexes

The basic structural unit of the nervous system is the neuron. The **reflex arc** is the basic functional unit of the nervous system and is the smallest, simplest portion capable of receiving a stimulus and yielding a response. It has five basic components: (1) a sensory receptor, (2) an afferent or sensory neuron, (3) association neurons, (4) an efferent or motor neuron, and (5) an effector organ (figure 12.19).

Action potentials initiated in sensory receptors are propagated along afferent axons within the PNS to the CNS, where they usually synapse with association neurons. Association neurons synapse with efferent (motor) neurons, which send axons out of the spinal cord and through the PNS to muscles or glands, where the action potentials of the efferent neurons cause effector organs to respond. The response produced by the reflex arc is called a **reflex.** It is an automatic response to a stimulus that occurs without conscious thought.

Reflexes are homeostatic. Some function to remove the body from painful stimuli that would cause tissue damage, and others function to keep the body from suddenly falling or moving because of external forces. A number of reflexes are responsible for maintaining relatively constant blood pressure, body fluid pH, blood carbon dioxide levels, and water intake. Specific reflexes are described in chapter 13.

Individual reflexes vary in their complexity. Some involve simple neuronal pathways and few association neurons, whereas others involve complex pathways and integrative centers. Many are integrated within the spinal cord, and others are integrated within the brain. In addition, higher brain centers influence reflexes by either suppressing or exaggerating them.

Reflexes do not operate as isolated entities within the nervous system because branches of the afferent neurons or

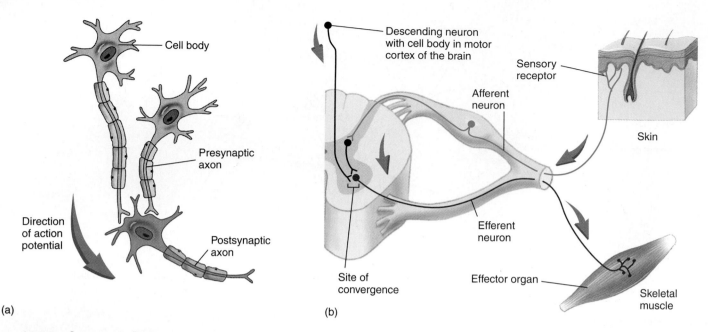

Figure 12.20   Convergent Pathways

(*a*) General model of a convergent pathway, showing two neurons converging on one neuron. (*b*) Example of a convergent pathway in the spinal cord. An association neuron that is part of a reflex arc and a descending neuron from the brain converge on a single motor neuron. 🏃

association neurons send information along nerve tracts to the brain. A pain stimulus, for example, not only initiates a response that removes the affected part of the body from the painful stimulus but also causes perception of the pain sensation as a result of action potentials sent to the brain.

Neuronal Pathways and Circuits

The organization of neurons within the CNS varies from relatively simple to extremely complex patterns. The axon of a neuron can branch repeatedly to form synapses with many other neurons, and hundreds or even thousands of axons can synapse with the cell body and dendrites of a single neuron. Although their complexity varies, three basic patterns can be recognized: convergent pathways, divergent pathways, and oscillating circuits.

Convergent Pathways

Convergent pathways have many neurons that converge and synapse with a smaller number of neurons (figure 12.20*a*). The simplest convergent pathway occurs when two presynaptic neurons synapse with a single postsynaptic neuron, the activity of which is influenced by spatial summation. If action potentials in one presynaptic neuron cause a subthreshold depolarization in the postsynaptic neuron, no postsynaptic action potential occurs. That subthreshold depolarization, however, facilitates the response to action potentials from other presynaptic neurons. Also, if some presynaptic neurons are inhibitory and others are excitatory, the response of the postsy-

naptic neuron depends on the summation of both the EPSPs and the IPSPs.

An example of a convergent pathway is the motor neurons of the spinal cord that control muscle movements (figure 12.20*b*). Afferent neurons from pain receptors carry action potentials to the spinal cord and synapse with association neurons, which, in turn, synapse with motor neurons. Stimulation of the pain receptors causes a reflex response that results in stimulation of the motor neurons. Neurons with their cell bodies located within the cerebrum also synapse with the motor neurons, however. Conscious movements are controlled by the cerebrum by sending action potentials through nerve tracts that synapse with motor neurons in the spinal cord. Inhibitory axons also descend within the spinal cord and synapse either directly or through association neurons on the motor neurons. The activity of motor neurons in the spinal cord thus depends on the activity in at least these three different types of presynaptic neurons.

Divergent Pathways

In **divergent pathways** a smaller number of presynaptic neurons synapse with a larger number of postsynaptic neurons to allow information transmitted in one neuronal pathway to diverge into two or more pathways (figure 12.21*a*). The simplest divergent pathway occurs when a single presynaptic neuron branches to synapse with two postsynaptic neurons. An example of a divergent pathway is found within the spinal cord (figure 12.21*b*). Afferent neurons carrying action potentials from pain receptors synapse within the spinal cord with association neurons that, in turn, induce a reflex response. In addition to synapsing with association neurons, collateral axons

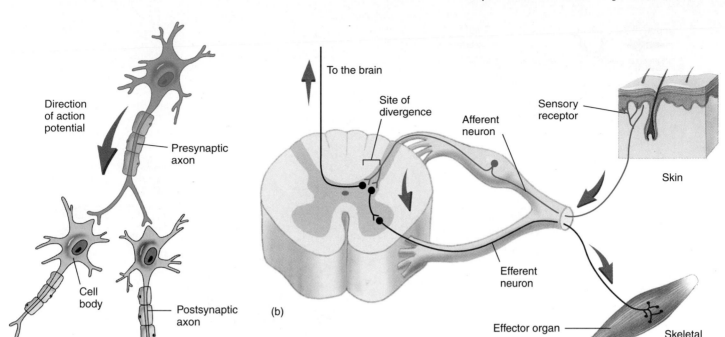

Figure 12.21   Divergent Pathways

(*a*) General model of a divergent pathway, showing one neuron diverging onto two neurons. (*b*) Divergent pathway in the spinal cord. An association neuron that is part of a reflex arc sends information to a motor neuron and to an ascending neuron to the brain. ✗

synapse with ascending neurons that carry action potentials toward the brain. The reflex response and the conscious sensation of pain are possible because of divergent pathways.

4   P R E D I C T

Ima Player taps her foot to the sound of the music while playing the trumpet. To accomplish these activities, she must read and understand the music sheets, produce the muscle movements necessary to move her foot and play the trumpet, and listen to and analyze the sounds produced. The ability of the nervous system to perform many activities at the same time is called parallel processing. In general terms, explain how parallel processing is possible.

✔ *Answer in Appendix F*

Oscillating Circuits

Oscillating circuits have neurons arranged in a circular fashion, which allows action potentials entering the circuit to cause a neuron farther along in the circuit to produce an action potential more than once (figure 12.22). This response is called **after-discharge,** and its effect is to prolong the response to a stimulus. Oscillating circuits are similar to positive-feedback systems. Once an oscillating circuit is stimulated, it continues to discharge until the synapses involved become fatigued or until they are inhibited by other neurons. Figure 12.22*a* illustrates a simple circuit in which a collateral axon stimulates its own cell body; figure 12.22*b* shows a more complex circuit. Oscillating circuits play a role in neuronal circuits that are periodically active. Respiration may be controlled by an oscillating circuit that controls inspiration and another that controls expiration.

Neurons that spontaneously produce action potentials are common in the CNS and may activate oscillating circuits, which remain active a while. The cycle of wakefulness and sleep may involve circuits of this type. Spontaneously active neurons are also capable of influencing the activity of other circuit types. The complex functions carried out by the CNS are affected by the numerous circuits operating together and influencing the activity of one another.

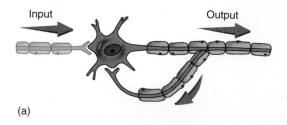

(a)

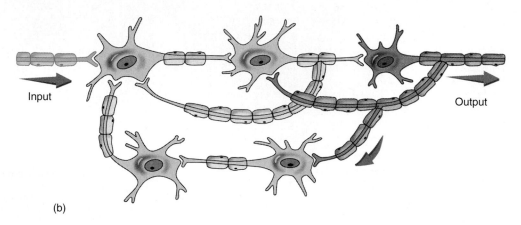

(b)

Figure 12.22   Oscillating Circuits

(*a*) A single neuron stimulates itself. (*b*) A more complex oscillating circuit in which the input neuron is stimulated by two other neurons.

Summary

Divisions of the Nervous System

1. The nervous system has two anatomic divisions.
    *   The central nervous system (CNS) consists of the brain and spinal cord and is encased in bone.
    *   The peripheral nervous system (PNS), the nervous tissue outside of the CNS, consists of nerves and ganglia.
2. The anatomic divisions perform different functions.
    *   The CNS processes, integrates, stores, and responds to information from the PNS.
    *   The PNS detects stimuli and transmits information to and receives information from the CNS.
3. The PNS has two divisions.
    *   The afferent division transmits action potentials to the CNS and usually consists of single neurons that have their cell bodies in ganglia.
    *   The efferent division carries action potentials away from the CNS in cranial or spinal nerves.
4. The efferent division has two subdivisions.
    *   The somatic motor nervous system innervates skeletal muscle and is mostly under voluntary control. It consists of single neurons that have their cell bodies located within the CNS.
    *   The autonomic nervous system (ANS) innervates cardiac muscle, smooth muscle, and glands. It has two sets of neurons between the CNS and effector organs. The first set has its cell bodies within the CNS, and the second set has its cell bodies within autonomic ganglia.

Cells of the Nervous System
Neurons

1. Neurons receive stimuli and transmit action potentials.
2. Neurons have three components.
    *   The cell body is the primary site of protein synthesis.
    *   Dendrites are short, branched cytoplasmic extensions of the cell body that usually conduct electric signals toward the cell body.
    *   Axons are cytoplasmic extensions of the cell body that transmit action potentials to other cells.

Types of Neurons

1. Multipolar neurons have several dendrites and a single axon. Association and motor neurons are multipolar.
2. Bipolar neurons have a single axon and dendrite and are found as components of sensory organs.
3. Unipolar neurons have a single axon. Most sensory neurons are unipolar.

Neuroglia

1. Neuroglia are nonneural cells that support and aid the neurons of the CNS and PNS.
2. CNS neuroglia
    *   Astrocytes provide structural support for neurons and blood vessels. The endothelium of blood vessels forms the blood–brain barrier, which regulates the movement of

Clinical Focus   Nervous Tissue Response to Injury

When a nerve is cut, either healing or permanent interruption of the neural pathways occurs. The final outcome depends on the severity of the injury and on its treatment.

Several degenerative changes result when a nerve is cut (figure A). Within about 3–5 days, the axons in the part of the nerve distal to the cut break into irregular segments and degenerate. This occurs because the neuron cell body produces the substances essential to maintain the axon and these substances have no way of reaching parts of the axon distal to the point of damage. Eventually the distal part of the axon completely degenerates. At the same time the axons are degenerating, the myelin part of the Schwann cells around them also degenerates, and macrophages invade the area to phagocytize the myelin. The Schwann cells then enlarge, undergo mitosis, and finally form a column of cells along the regions once occupied by the axons. The columns of Schwann cells are essential for the growth of new axons. If the ends of the regenerating axons encounter a Schwann cell column, their rate of growth increases, and reinnervation of peripheral structures is likely. If the ends of the axons do not encounter the columns, they fail to reinnervate the peripheral structures.

The part of the axon proximal to the cut degenerates for a distance up to several Schwann cells in length and then begins regenerative processes that lead to growth from the end of the severed axon. The end of each regenerating axon forms bulbous enlargements and several axonal sprouts. It normally takes about 2 weeks for the axonal sprouts to grow across the scar that develops in the area in which the nerve was cut and to enter the Schwann cell columns. Only one of the sprouts from each severed neuron forms an axon, however. The other branches degenerate. After the axons grow through the Schwann cell columns, new myelin sheaths are formed, and the neurons reinnervate the structures they previously supplied.

Treatment strategies that increase the probability of reinnervation include bringing the ends of the severed nerve close together surgically. In some cases in which sections of nerves are destroyed as a result of trauma, nerve transplants are performed to replace damaged segments. The transplanted nerve eventually degenerates, but it does provide Schwann cell columns through which axons can grow.

Regeneration of damaged nerve tracts within the CNS is very limited and is poor in comparison with regeneration of nerves in the PNS. In part, the difference may result from the oligodendrocytes, which exist only in the CNS. Each oligodendrocyte has several processes, each of which forms part of a myelin sheath. The cell bodies of the oligodendrocytes are a short distance from the axons they ensheathe, and there are fewer oligodendrocytes than Schwann cells. Consequently, when the myelin degenerates following damage, no column of cells remains in the CNS to act as a guide for the growing axons.

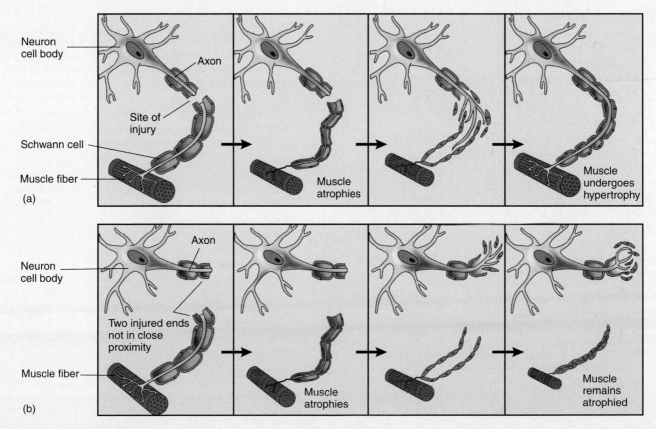

Figure A   Changes That Occur in an Injured Nerve Fiber   (*a*) When the two ends of the injured nerve fiber are aligned in close proximity, healing and regeneration of the axon are likely to occur. Without stimulation from the nerve, the muscle is paralyzed and atrophies (shrinks in size). After reinnervation the muscle can become functional and hypertrophy (increase in size). (*b*) When the two ends of the injured nerve fiber are not aligned in close proximity, regeneration is unlikely to occur. Without innervation from the nerve, muscle function is completely lost, and the muscle remains atrophied.

substances between the blood and the CNS. Astrocytes influence the functioning of the blood–brain barrier and process substances that pass through it.
- Microglia are macrophages that phagocytize microorganisms, foreign substances, or necrotic tissue.
- Ependymal cells line the ventricles and the central canal of the spinal cord. Some are specialized to produce cerebrospinal fluid.
- Oligodendrocytes form myelin sheaths around the axons of neurons of the CNS.
3. Schwann cells (PNS neuroglia)
- Schwann cells, or neurolemmocytes, form myelin sheaths around the axons of neurons of the PNS.
- Satellite cells support and nourish neuron cell bodies within ganglia.

Axon Sheaths

1. Unmyelinated axons rest in invaginations of oligodendrocytes (CNS) or Schwann cells (PNS). They conduct action potentials slowly.
2. Myelinated axons are wrapped by several layers of cell membrane from oligodendrocytes (CNS) or Schwann cells (PNS). Spaces between the wrappings are the nodes of Ranvier, and action potentials are conducted rapidly by saltatory conduction from one node of Ranvier to the next.

Organization of Nervous Tissue

1. Nervous tissue can be grouped into white and gray matter.
- White matter is myelinated axons and functions to propagate action potentials.
- Gray matter is collections of neuron cell bodies or unmyelinated axons. Axons synapse with neuron cell bodies, which is functionally the site of integration in the nervous system.
- White matter forms nerve tracts in the CNS and nerves in the PNS. Gray matter forms cortex and nuclei in the CNS and ganglia in the PNS.
2. In the PNS, individual axons are surrounded by the endoneurium. Groups of axons, fascicles, are bound together by the perineurium. The fascicles form the nerve and are held together by the epineurium.

The Synapse

1. Anatomically the synapse has three components.
- The enlarged ends of the axon are the presynaptic terminals containing synaptic vesicles.
- The postsynaptic membranes contain receptors for the neurotransmitter.
- The synaptic cleft, a space, separates the presynaptic and postsynaptic membranes.
2. An action potential arriving at the presynaptic terminal causes the release of a neurotransmitter, which diffuses across the synaptic cleft and binds to the receptors of the postsynaptic membrane.
3. The effect of the neurotransmitter on the postsynaptic membrane can be stopped in several ways.
- The neurotransmitter is broken down by an enzyme.
- The neurotransmitter is taken up by the presynaptic terminal.
- The neurotransmitter diffuses out of the synaptic cleft.

Receptor Molecules in Synapses

1. Neurotransmitters are specific for their receptors.
2. A neurotransmitter can be stimulatory in one synapse and inhibitory in another, depending on the type of receptor present.
3. Some presynaptic terminals have receptors.

Neurotransmitters and Neuromodulators

Neuromodulators influence the likelihood that an action potential in a presynaptic terminal will result in an action potential in a postsynaptic cell.

Excitatory and Inhibitory Postsynaptic Potentials

1. Depolarization of the postsynaptic membrane caused by an increase in membrane permeability to sodium ions is an excitatory postsynaptic potential (EPSP).
2. Hyperpolarization of the postsynaptic membrane caused by an increase in membrane permeability to chloride ions or potassium ions is an inhibitory postsynaptic potential (IPSP).

Presynaptic Inhibition and Facilitation

1. Presynaptic inhibition decreases neurotransmitter release.
2. Presynaptic facilitation increases neurotransmitter release.

Spatial and Temporal Summation

1. Presynaptic action potentials through neurotransmitters produce local potentials in postsynaptic neurons. The local potential can summate to produce an action potential at the axon hillock.
2. Spatial summation occurs when two or more presynaptic terminals simultaneously stimulate a postsynaptic neuron.
3. Temporal summation occurs when two or more action potentials arrive in succession at a single presynaptic terminal.
4. Inhibitory and excitatory presynaptic neurons can converge on a postsynaptic neuron. The activity of the postsynaptic neuron is determined by the integration of the EPSPs and IPSPs produced in the postsynaptic neuron.

Reflexes

1. A reflex arc is the functional unit of the nervous system.
- Sensory receptors respond to stimuli and produce action potentials in afferent neurons.
- Afferent neurons propagate action potentials to the CNS.
- Association neurons in the CNS synapse with afferent neurons and with efferent neurons.
- Efferent neurons carry action potentials from the CNS to effector organs.
- Effector organs such as muscles or glands respond to the action potentials.
2. Reflexes do not require conscious thought, and they produce a consistent and predictable result.
3. Reflexes are homeostatic.
4. Reflexes are integrated within the brain and spinal cord. Higher brain centers can suppress or exaggerate reflexes.

Neuronal Pathways and Circuits

1. Convergent pathways have many neurons synapsing with a few neurons.
2. Divergent pathways have a few neurons synapsing with many neurons.
3. Oscillating circuits have collateral branches of postsynaptic neurons synapsing with presynaptic neurons.

Content Review

1. Describe the CNS and PNS anatomically and functionally.
2. Define the afferent and efferent divisions of the PNS.
3. Contrast the afferent, somatic motor system, and autonomic nervous systems in terms of the number of neurons, the location of neuron cell bodies, and the structures innervated.
4. What are the functions of neurons? Name the three parts of a neuron, and describe their functions.
5. Describe the three types of neurons on the basis of their structure, and give an example of where each type is found.
6. Define the term neuroglia. Name and describe the functions of the different kinds of neuroglia.
7. What are the differences between unmyelinated and myelinated axons with regard to the arrangement of the cells that cover their axons? What are the nodes of Ranvier?
8. Do unmyelinated or myelinated neurons propagate action potentials more rapidly? Describe what occurs during saltatory conduction.
9. For nerve tract, nerve, nucleus, and ganglion, name the cells or parts of cells found in each, state if they are white or gray matter, and name the part (CNS or PNS) of the nervous system in which they are found.
10. Describe the layers of connective tissue found in nerves.
11. Describe the operation of the synapse, starting with an action potential in the presynaptic neuron and ending with the generation of an action potential in the postsynaptic neuron.
12. Describe the specific, reversible reaction that occurs between a neurotransmitter and its receptor.
13. Name three ways to stop the effect of a neurotransmitter on the postsynaptic membrane. Give an example of each way.
14. How can a neurotransmitter cause depolarization in one synapse but hyperpolarization in another?
15. What is a neuromodulator?
16. Define and explain the production of EPSPs and IPSPs. Why are they important?
17. What are presynaptic inhibition and facilitation?
18. In what part of the postsynaptic neuron are local potentials produced? Where do they summate? What happens when they summate?
19. Contrast spatial and temporal summation. Give an example of each.
20. Explain how inhibitory and excitatory presynaptic neurons can influence the activity of a postsynaptic neuron.
21. At the cell level, where does integration take place in the CNS?
22. Name the five components of a reflex arc. Describe the operation of a reflex arc, starting with a stimulus and ending with the reflex response.
23. Define a reflex. What is the relationship between a reflex response and awareness of the stimuli that caused the reflex? What effects can higher brain centers have on reflexes?
24. Define the terms convergent pathway, divergent pathway, and oscillating circuit, and give an example of each.

Develop Your Reasoning Skills

1. Assume that you have two nerve fibers of the same diameter, but one nerve fiber is myelinated and the other is unmyelinated. Along which type of fiber is the conduction of an action potential most energy-efficient (*Hint:* ATP).
2. Explain the consequences when an inhibitory neuromodulator is released from a presynaptic terminal and a stimulatory neurotransmitter is released from another presynaptic terminal, both of which synapse with the same neuron.
3. With aging, the speed of reflex responses slows down. The decreased rate of response is believed to result from many age-related changes in the nervous system. List possible explanations for slower reflexes in the elderly.
4. Students in a veterinary school were given the following hypothetical problem. A dog ingests organophosphate poison, and the students are responsible for saving the animal's life. Organophosphate poisons bind to and inhibit acetylcholinesterase. Several substances they could inject include the following: acetylcholine, curare (which blocks acetylcholine receptors), and potassium chloride. If you were a student in the class, what would you do to save the animal?
5. Strychnine blocks receptor sites for inhibitory neurotransmitter substances in the central nervous system. Explain how strychnine could produce tetany in skeletal muscles.
6. Describe how stimulation of a neuron that has its cell body in the cerebrum could inhibit a reflex that is integrated within the spinal cord.

Web Site Link

For a listing of the most current web sites related to this chapter, please visit the Seeley home page at:
http://www.mhhe.com/biosci/ap/seeleyap/

Chapter Thirteen

Central Nervous System: Brain and Spinal Cord

Physiology

Human Anatomy

Objectives

1. Describe the formation of the neural tube, and list the structures that develop from its various parts.

2. List the parts of the brain.

3. Define what is included in the brainstem, and describe its major features.

4. Describe the major function of the reticular formation.

5. List the regions of the diencephalon, and indicate their major functions.

6. Describe the major functional areas of the cerebral cortex, and explain their interactions.

7. Describe and explain the pathway for speech.

8. Describe the basic brain waves, and correlate them with brain function.

9. Explain how sensory, short-term, and long-term memory work.

10. Describe the major functions of the basal nuclei.

11. Describe the components and functions of the limbic system.

12. Describe the major functions of the cerebellum, and explain its comparator function.

13. Describe the spinal cord in cross section, explaining the functions of each area.

14. Diagram the stretch reflex, Golgi tendon reflex, withdrawal reflex, reciprocal innervation, and the crossed extensor reflex. Explain the function of each of these reflexes.

15. Describe the course of the fibers associated with the spinothalamic and dorsal-column/medial-lemniscal systems.

16. Outline the course and describe the function of the corticospinal, corticobulbar, and indirect tracts.

17. Describe the three meningeal layers surrounding the central nervous system.

18. Name the four ventricles of the brain, and describe their locations and the connections between them.

19. Describe the origin, composition, and circulation of the cerebrospinal fluid.

The brain is involved, in some way, in almost all bodily functions. Although humans have larger, more complex brains than other animals, many human brain functions are similar to those of other animals. The sensory input we receive and most of the ways we respond to that input do not require uniquely human brain functions. Yet, the human brain is also capable of complex functions, such as recording history, reasoning, and planning, which are unparalleled in the animal kingdom. Many of these functions can only be studied in humans. That is why much of human brain function remains elusive and why an understanding of the human brain remains one of the most challenging frontiers of anatomy and physiology.

The **central nervous system (CNS)** consists of the brain and the spinal cord (figure 13.1), with the division between these two parts of the CNS placed somewhat arbitrarily at the level of the foramen magnum. The **brain** is that part of the CNS housed within the cranial vault. The **spinal cord** is contained within the vertebral column. The anatomic features and some basic functional features of the brain and spinal cord are presented in this chapter, followed by a description of the ascending and descending pathways. The pathways are described last because understanding the basic anatomy and physiology of both the brain and spinal cord makes the pathways easier to comprehend.

13.2a). The lateral sides of the neural plate become elevated as waves, called **neural folds.** The crest of each fold is called a **neural crest,** and the center of the neural plate becomes the **neural groove.** The neural folds move toward each other in the midline, and the crests fuse to create a **neural tube** (figure 13.2b). The cephalic portion of the neural tube becomes the brain, and the caudal portion becomes the spinal cord. **Neural crest cells** separate from the neural crests and give rise to part of the peripheral nervous system (see chapter 14).

Figure 13.1   Brain and Spinal Cord

Computer-generated 3-D image of the brain and spinal cord (*yellow*) from a computed tomographic (CT) scan.

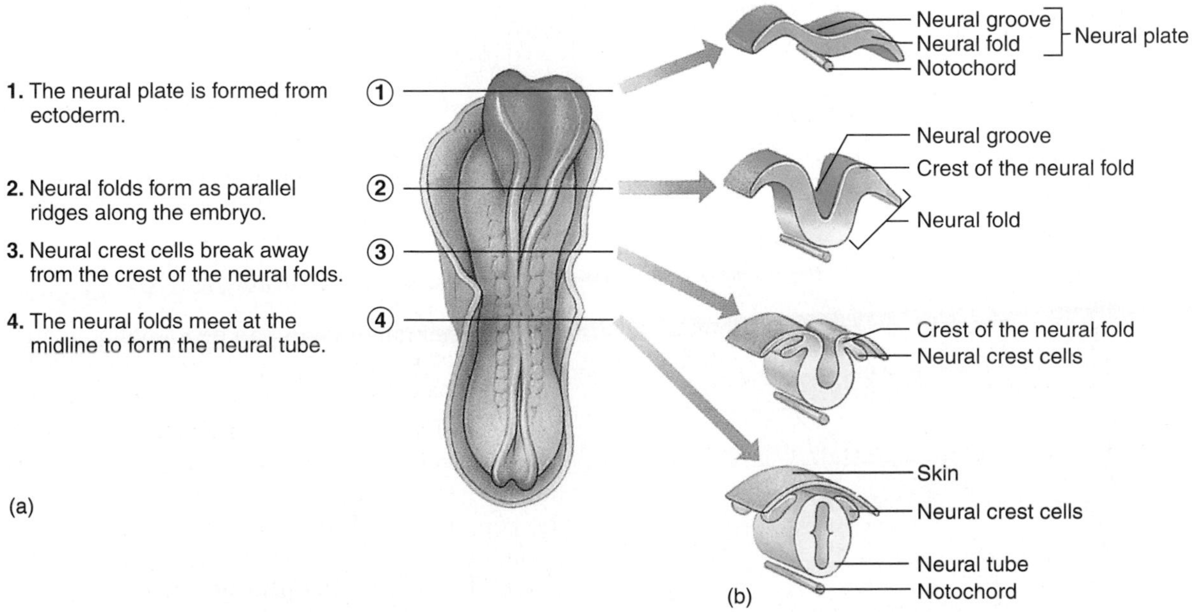

Development

The CNS develops from a flat plate of tissue, the **neural plate,** on the upper surface of the embryo, as a result of the influence of the underlying rod-shaped **notochord** (figure

1. The neural plate is formed from ectoderm.

2. Neural folds form as parallel ridges along the embryo.

3. Neural crest cells break away from the crest of the neural folds.

4. The neural folds meet at the midline to form the neural tube.

Neural groove
Neural fold — Neural plate
Notochord

Neural groove
Crest of the neural fold
Neural fold

Crest of the neural fold
Neural crest cells

Skin
Neural crest cells

Neural tube
Notochord

(a)

(b)

Figure 13.2   Formation of the Neural Tube

(*a*) A 21-day-old human embryo. (*b*) Cross sections through the embryo. The level of each section is indicated by a line in part (*a*).

Table 13.1   Development of the Central Nervous System (see figure 13.3)

Early Embryo	Late Embryo	Adult	Cavity	Function
Prosencephalon (forebrain)	Telencephalon	Cerebrum	Lateral ventricles	Higher brain functions
	Diencephalon	Diencephalon (thalamus, subthalamus, epithalamus, hypothalamus)	Third ventricle	Relay center, autonomic nerve control, endocrine control
Mesencephalon (midbrain)	Mesencephalon	Mesencephalon (midbrain)	Cerebral aqueduct	Nerve pathways, reflex centers
Rhombencephalon (hindbrain)	Metencephalon	Pons and cerebellum	Fourth ventricle	Nerve pathways, reflex centers, muscle coordination, and balance
	Myelencephalon	Medulla oblongata	Central canal	Nerve pathways, reflex centers

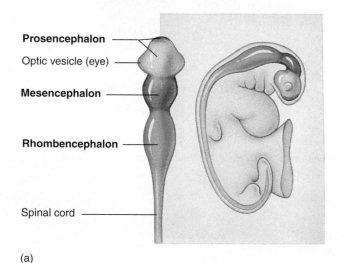

(a)

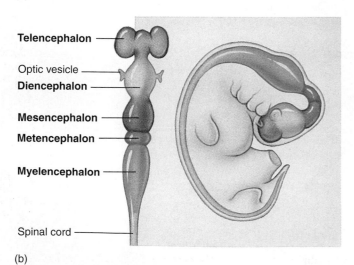

(b)

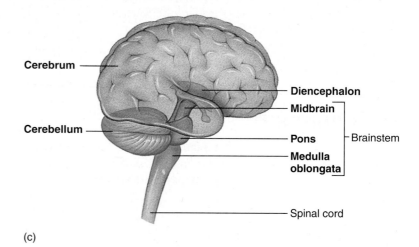

(c)

Figure 13.3   Development of the Brain Segments and Ventricles (a) Young embryo. (b) Older embryo. (c) Adult.

The part of the neural tube that becomes the brain forms a series of pouches (table 13.1 and figure 13.3). The pouch walls become the various portions of the adult brain, and the cavities become fluid-filled **ventricles** (ven'tri-klz). The ventricles are continuous with the **central canal** of the spinal cord, which is also formed from the hollow center of the neural tube. The neural tube develops flexures that cause the brain to be oriented almost 90 degrees to the spinal cord.

Three brain regions can be identified in the early embryo (see table 13.1 and figure 13.3a): a forebrain, or **prosencephalon** (pros-en-sef'ă-lon); a midbrain, or **mesencephalon** (mes-en-sef'ă-lon); and a hindbrain, or **rhombencephalon** (romb-en-sef'ă-lon). During development, the forebrain divides into the **telencephalon** (tel-en-sef'ă-lon), which becomes the cerebrum, and the **diencephalon** (dī-en-sef'ă-lon). The midbrain remains as a single structure, but the hindbrain divides into the **metencephalon** (met-en-sef'ă-lon), which becomes the pons and cerebellum, and the **myelencephalon** (mī'el-en-sef'ă-lon), which becomes the medulla oblongata (see figure 13.3b and c).

Table 13.2   Divisions and Functions of the Central Nervous System

Brainstem 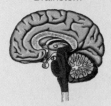	Connects the spinal cord to the cerebrum; several important functions (see below); location of cranial nerve nuclei	**Diencephalon** 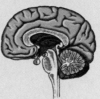	
Medulla oblongata 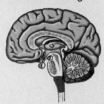	Pathway for ascending and descending nerve tracts; center for several important reflexes (e.g., heart rate, breathing, swallowing, vomiting)	**Thalamus**	Major sensory relay center; influences mood and movement
		Subthalamus	Contains nerve tracts and nuclei
		Epithalamus	Contains nuclei responding to olfactory stimulation and contains pineal body
Pons 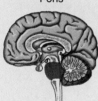	Contains ascending and descending nerve tracts; relay between cerebrum and cerebellum; reflex center	**Hypothalamus**	Major control center for maintaining homeostasis and regulating endocrine function
		Cerebrum	Conscious perception, thought, and conscious motor activity; can override most other systems
Midbrain 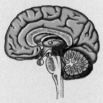	Contains ascending and descending nerve tracts; visual reflex center; part of auditory pathway	**Basal nuclei**	Control of muscle activity and posture; largely inhibit unintentional movement
		Limbic system	Autonomic response to smell, emotion, mood, and other such functions
Reticular formation 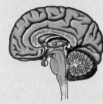	Scattered throughout brainstem; controls cyclic activities such as the sleep–wake cycle	**Cerebellum**	Control of muscle movement and tone; regulates extent of intentional movement

Brainstem

The major regions of the adult brain are the cerebrum, diencephalon (thalamus and hypothalamus), midbrain (or mesencephalon), pons, cerebellum, and medulla oblongata (table 13.2 and figure 13.4; see figure 13.3c).

The medulla oblongata, pons, and midbrain constitute the **brainstem** (figure 13.5). The brainstem connects the spinal cord to the remainder of the brain and is responsible for many essential functions. Damage to small brainstem areas often causes death because reflexes essential for survival are integrated in the brainstem, whereas relatively large areas of the cerebrum or cerebellum may be damaged without being life-threatening. All but 2 of the 12 cranial nerves enter or exit the brain through the brainstem (see chapter 14).

Medulla Oblongata

The **medulla oblongata** (ob′long-gah′tă), often called the medulla, is about 3 cm long, is the most inferior part of the brainstem, and is continuous inferiorly with the spinal cord. Superficially, the spinal cord blends into the medulla, but

Figure 13.4   Regions of the
Right Half of the Brain

(as seen in a midsagittal section) ✗

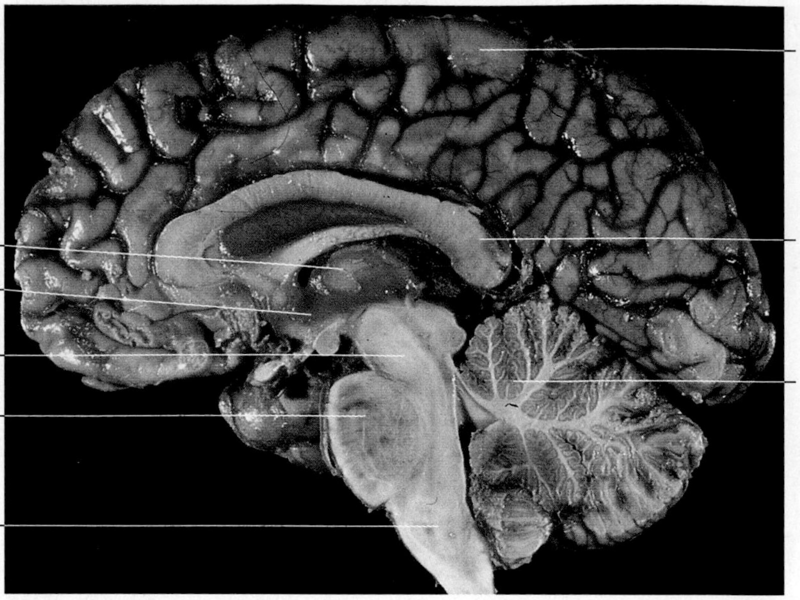

internally there are several differences. Discrete **nuclei,** clusters of gray matter composed mostly of cell bodies, with specific functions, are found in the medulla oblongata, whereas the gray matter of the spinal cord extends as a continuous mass in the center of the cord. In addition, the nerve tracts that pass through the medulla do not have the same organization as those of the spinal cord.

On the anterior surface of the medulla are two prominent enlargements, called **pyramids** because they are broader near the pons and taper toward the spinal cord (see figure 13.5a). The pyramids are descending nerve tracts involved in the conscious control of skeletal muscles. Near their inferior ends, most of the fibers of the descending nerve tracts cross to the opposite side, or **decussate** (dē′kŭ-sāt; the Latin word *decussatus* means to form an X, as in the Roman numeral X). This decussation accounts, in part, for the fact that each half of the brain controls the opposite half of the body.

Two rounded, oval structures, called **olives,** protrude from the anterior surface of the medulla oblongata just lateral to the superior margins of the pyramids (see figure 13.5a and b). The olives are nuclei involved in functions such as balance, coordination, and modulation of sound impulses from the inner ear (see chapter 15). The nuclei of cranial nerves V (trigeminal), IX (glossopharyngeal), X (vagus), XI (accessory), and XII (hypoglossal) also are located within the medulla (figure 13.5c).

Functionally, the medulla oblongata acts as a conduction pathway for both ascending and descending nerve tracts. Its role as a conduction pathway is discussed in the description of ascending and descending nerve tracts. Various medullary nuclei also function as centers for several reflexes, such as those involved in the regulation of heart rate, blood vessel diameter, breathing, swallowing, vomiting, coughing, and sneezing.

Pons

The part of the brainstem just superior to the medulla oblongata is the **pons** (see figure 13.5a), which contains ascending

and descending nerve tracts and several nuclei. The pontine nuclei, located in the anterior portion of the pons, relay information from the cerebrum to the cerebellum.

The nuclei for cranial nerves V (trigeminal), VI (abducens), VII (facial), VIII (vestibulocochlear), and IX (glossopharyngeal) are contained within the posterior pons. Other important pontine areas include the pontine sleep center and the respiratory centers. These centers function with the respiratory centers in the medulla to help control respiratory movements (see chapter 23).

Midbrain

The **midbrain,** or **mesencephalon,** is the smallest region of the brainstem (see figure 13.5b). It is just superior to the pons and contains the nuclei of cranial nerves III (oculomotor), IV (trochlear), and V (trigeminal).

The **tectum** (tek′tŭm, meaning roof) of the midbrain consists of four nuclei that form mounds on the dorsal surface, collectively called **corpora** (kōr′pōr-ă, meaning bodies) **quadrigemina** (kwah′dri-jem′i-nă, meaning four twins). Each mound is called a **colliculus** (ko-lik′yū-lŭs, meaning hill); the two superior mounds are called **superior colliculi,** and the two inferior mounds are called **inferior colliculi.** The inferior colliculi are involved in hearing and are an integral part of the auditory pathways in the CNS. Neurons conducting impulses from the structures of the inner ear (see chapter 15) to the brain synapse in the inferior colliculi. The superior colliculi are involved in visual reflexes, and they receive input from the eyes, the inferior colliculi, the skin, and the cerebrum. Nerve fibers from the superior colliculi project to the oculomotor, trochlear, and abducens cranial nerve nuclei and to the superior cervical part of the spinal cord, where they stimulate motor neurons involved in turning the eyes and the head. Impulses reaching the superior colliculi from the cerebrum are involved in the visual tracking of moving objects (see chapter 15).

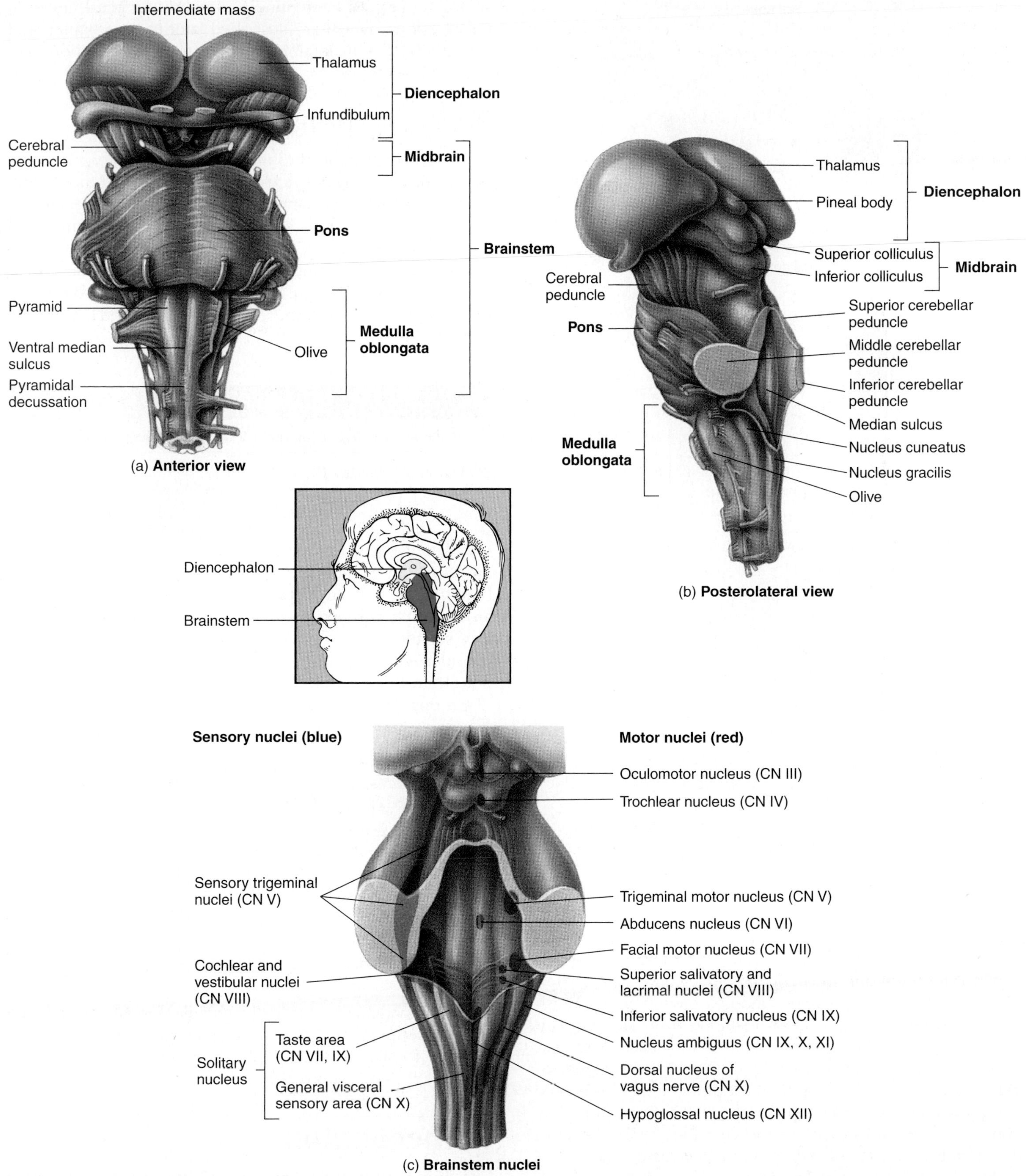

Intermediate mass

Thalamus

Diencephalon

Infundibulum

Cerebral peduncle

Midbrain

Pons

Brainstem

Pyramid

Medulla oblongata

Ventral median sulcus

Olive

Pyramidal decussation

(a) **Anterior view**

Thalamus

Pineal body

Diencephalon

Superior colliculus

Cerebral peduncle

Inferior colliculus

Midbrain

Superior cerebellar peduncle

Pons

Middle cerebellar peduncle

Inferior cerebellar peduncle

Median sulcus

Medulla oblongata

Nucleus cuneatus

Nucleus gracilis

Olive

(b) **Posterolateral view**

Diencephalon

Brainstem

Sensory nuclei (blue)

Motor nuclei (red)

Oculomotor nucleus (CN III)

Trochlear nucleus (CN IV)

Sensory trigeminal nuclei (CN V)

Trigeminal motor nucleus (CN V)

Abducens nucleus (CN VI)

Facial motor nucleus (CN VII)

Cochlear and vestibular nuclei (CN VIII)

Superior salivatory and lacrimal nuclei (CN VIII)

Inferior salivatory nucleus (CN IX)

Nucleus ambiguus (CN IX, X, XI)

Taste area (CN VII, IX)

Dorsal nucleus of vagus nerve (CN X)

Solitary nucleus

General visceral sensory area (CN X)

Hypoglossal nucleus (CN XII)

(c) **Brainstem nuclei**

Figure 13.5   Brainstem and Diencephalon

(*a*) Anterior view. (*b*) Posterolateral view. (*c*) Brainstem nuclei. The sensory nuclei are shown on the left (*blue*). The motor nuclei are shown on the right (*red*). Even though the nuclei are shown on only one side, each half of the brainstem has both sensory and motor nuclei. The inset shows the location of the diencephalon (*yellow*) and brainstem (*green*).

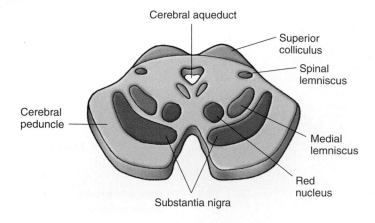

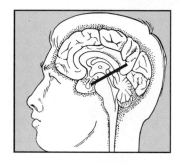

Figure 13.6   Cross Section Through the Midbrain
Inset shows the level of section.

The **tegmentum** (teg-men′tŭm, meaning floor) of the midbrain largely consists of ascending tracts from the spinal cord to the brain and also contains the paired **red nuclei.** The red nuclei are so named because they have a pinkish color in fresh brain specimens, resulting from an abundant blood supply. The red nuclei aid in the unconscious regulation and coordination of motor activities. **Cerebral peduncles** (pe-dŭng′klz, meaning the foot of a column) constitute that portion of the midbrain inferior to the tegmentum. They consist primarily of descending tracts from the cerebrum to the spinal cord and constitute one of the major CNS motor pathways. The **substantia nigra** (nī′gră, meaning black substance) is a nuclear mass between the tegmentum and cerebral peduncles, containing cytoplasmic melanin granules that give it a dark gray or black color (figure 13.6). The

substantia nigra is interconnected with other basal nuclei of the cerebrum, which are described later in this chapter, and it is involved in maintaining muscle tone and in coordinating movements.

Reticular Formation

Scattered like a cloud throughout most of the length of the brainstem is a group of nuclei collectively called the **reticular formation** (not illustrated), which receives afferent axons from a large number of sources and especially from nerves that innervate the face. These axons play an important role in arousing and maintaining consciousness. The reticular formation and its connections constitute a system, called the **reticular activating system,** which is involved with the sleep–wake cycle.

Visual and acoustical stimuli and mental activities can stimulate the reticular activating system to maintain alertness and attention. Stimuli such as a ringing alarm clock, sudden bright lights, or cold water being splashed on the face can arouse consciousness. Conversely, removal of visual or auditory stimuli may lead to drowsiness or sleep. For example, consider what happens to many students during a monotonous lecture in a dark lecture hall. Damage to cells of the reticular formation can result in coma.

Descending fibers from the reticular formation constitute one of the most important motor pathways. Fibers from the reticular formation are critical in controlling vital functions such as respiratory movements and cardiac rhythms.

Diencephalon

The **diencephalon** (dī-en-sef′ă-lon) is the part of the brain between the brainstem and the cerebrum (see figure 13.4; figure 13.7*a*). Its main components are the thalamus, subthalamus, hypothalamus, and epithalamus.

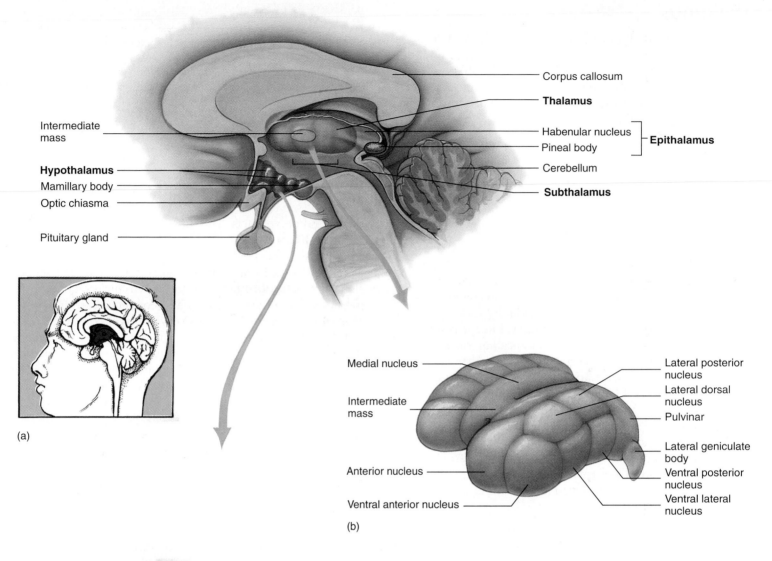

Corpus callosum

Thalamus

Intermediate mass

Habenular nucleus

Pineal body

Epithalamus

Hypothalamus

Cerebellum

Mamillary body

Subthalamus

Optic chiasma

Pituitary gland

(a)

Medial nucleus

Lateral posterior nucleus

Lateral dorsal nucleus

Intermediate mass

Pulvinar

Lateral geniculate body

Anterior nucleus

Ventral posterior nucleus

Ventral anterior nucleus

Ventral lateral nucleus

(b)

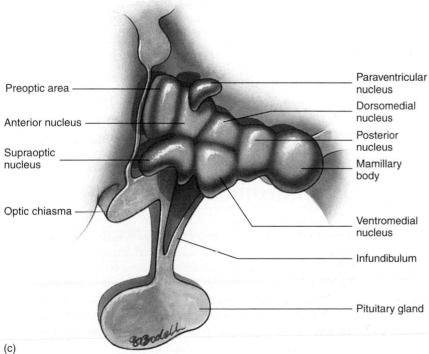

Preoptic area

Paraventricular nucleus

Dorsomedial nucleus

Anterior nucleus

Posterior nucleus

Supraoptic nucleus

Mamillary body

Optic chiasma

Ventromedial nucleus

Infundibulum

Pituitary gland

(c)

Figure 13.7   Diencephalon

(*a*) General overview of the right half of the diencephalon as seen in a midsagittal section. (*b*) Thalamus showing the nuclei. (*c*) Hypothalamus showing the nuclei and right half of the pituitary.

Thalamus

The **thalamus** (thal'ă-mŭs) (see figure 13.7*a* and *b*) is by far the largest part of the diencephalon, constituting about four-fifths of its weight. It is a cluster of nuclei shaped somewhat like a yo-yo, with two large, lateral portions connected in the center by a small stalk called the **intermediate mass.** The space surrounding the intermediate mass and separating the two large portions of the thalamus is the third ventricle of the brain.

Most sensory input projects to the thalamus, where afferent neurons synapse with thalamic neurons, which send projections from the thalamus to the cerebral cortex. Axons carrying auditory information synapse in the **medial geniculate** (je-nik'yū-lāt; L. *genu,* meaning bent like a knee) **nucleus** of the thalamus, axons carrying visual information synapse in the **lateral geniculate nucleus,** and most other sensory impulses synapse in the **ventral posterior nucleus.**

The thalamus also influences mood and general body movements associated with strong emotions such as fear or rage. The **ventral anterior** and **ventral lateral nuclei** are involved in motor functions, communicating between the basal nuclei and the motor cortex (these areas are described later in this chapter). The **anterior** and **medial nuclei** are connected to the limbic system and to the prefrontal cortex (described later in this chapter) and are involved in mood modification. The **lateral dorsal nucleus** is connected to other thalamic nuclei and to the cerebral cortex and is involved in regulating emotions. The **lateral posterior nucleus** and the **pulvinar** (pŭl-vī'năr, meaning pillow) also have connections to other thalamic nuclei and are involved in sensory integration.

Subthalamus

The **subthalamus** is a small area immediately inferior to the thalamus (see figure 13.7*a*) that contains several nerve tracts and the **subthalamic nuclei.** A small portion of the red nucleus and substantia nigra of the midbrain extend into this area.

The subthalamic nuclei are associated with the basal nuclei and are involved in controlling motor functions.

Epithalamus

The **epithalamus** is a small area superior and posterior to the thalamus (see figure 13.7*a*). It consists of habenular nuclei and the pineal body. The **habenular** (hă-ben'yū-lăr) **nuclei** are influenced by the sense of smell and are involved in emotional and visceral responses to odors. The **pineal** (pin'ē-ăl) **body** is shaped somewhat like a pinecone, from which the name pineal is derived. It appears to play a role in controlling the onset of puberty, but data are inconclusive, so active research continues in this field. The pineal body also may be involved in the sleep–wake cycle.

Clinical Note

In about 75% of adults, the pineal body contains granules of calcium and magnesium salts called "brain sand." These granules can be seen on radiographs and are useful as a landmark in determining whether or not the pineal body has been displaced by a pathologic enlargement of a part of the brain, such as a tumor or a hematoma.

Hypothalamus

The **hypothalamus** is the most inferior portion of the diencephalon (see figure 13.7*a* and *c*) and contains several small nuclei and nerve tracts. The most conspicuous nuclei, called the **mamillary bodies,** appear as bulges on the ventral surface of the diencephalon. They are involved in olfactory reflexes and emotional responses to odors. A funnel-shaped stalk, the **infundibulum** (in-fŭn-dib'yū-lūm), extends from the floor of the hypothalamus and connects it to the **posterior pituitary gland,** or **neurohypophysis** (nūr'ō-hī-pof'i-sis). The hypothalamus plays an important role in controlling the endocrine system because it regulates the pituitary gland's secretion of hormones, which influence functions as diverse as metabolism, reproduction, responses to stressful stimuli, and urine production (see chapter 18).

Afferent fibers that terminate in the hypothalamus provide input from: (1) visceral organs; (2) taste receptors of the tongue; (3) the limbic system, which is involved in responses to smell; (4) specific cutaneous areas such as the nipples and external genitalia; and (5) the prefrontal cortex of the cerebrum carrying information relative to "mood" through the thalamus. Efferent fibers from the hypothalamus extend into the brainstem and the spinal cord, where they synapse with neurons of the autonomic nervous system (see chapter 16). Other fibers extend through the infundibulum to the posterior portion of the pituitary gland (see chapter 18); some extend to trigeminal and facial nerve nuclei (see chapter 14) to help control the head muscles involved in swallowing; and some extend to motor neurons of the spinal cord to stimulate shivering.

The hypothalamus is very important in a number of functions, all of which have emotional and mood relationships (table 13.3). Sensations such as sexual pleasure, feeling relaxed and "good" after a meal, rage, and fear are related to hypothalamic functions.

Cerebrum

The cerebrum is the part of the brain that most people think of when the term brain is mentioned. It is the largest portion of the brain, weighing about 1200 g in females and 1400 g in males. Brain size is related to body size; larger brains are associated with larger bodies, not with greater intelligence.

The cerebrum is divided into left and right hemispheres by a **longitudinal fissure** (figure 13.8*a*). The most conspicuous features on the surface of each hemisphere are numerous

Table 13.3	Hypothalamic Functions
Function	**Description**
Autonomic	Helps control heart rate, urine release from the bladder, movement of food through the digestive tract, and blood vessel diameter
Endocrine	Helps regulate pituitary gland secretions and influences metabolism, ion balance, sexual development, and sexual functions
Muscle control	Controls muscles involved in swallowing and stimulates shivering in several muscles
Temperature regulation	Promotes heart loss when the hypothalamic temperature increases by increasing sweat production (anterior hypothalamus) and promotes heat production when the hypothalamic temperature decreases by promoting shivering (posterior hypothalamus)
Regulation of food and water intake	Hunger center promotes eating and satiety center inhibits eating; thirst center promotes water intake
Emotions	Large range of emotional influences over body functions; directly involved in stress-related and psychosomatic illnesses and with feelings of fear and rage
Regulation of the sleep–wake cycle	Coordinates responses to the sleep–wake cycle with the other areas of the brain (e.g., the reticular activating system)

folds called **gyri** (jī′rī; sing., gyrus), which greatly increase the surface area of the cortex. The intervening grooves between the gyri are called **sulci** (sŭl′sī; sing., sulcus; see figure 13.8). A **central sulcus,** which runs in the lateral surface of the cerebrum from superior to inferior, is located about midway along the length of the brain. The central sulcus is located between the **precentral gyrus** anteriorly and a **postcentral gyrus** posteriorly. The general pattern of the gyri is similar in all normal human brains, but some variation exists between individuals and even between the two hemispheres of the same cerebrum.

Each cerebral hemisphere is divided into lobes, which are named for the skull bones overlying each one (figure 13.8*b*). The **frontal lobe** is important in voluntary motor function, motivation, aggression, the sense of smell, and mood. The **parietal lobe** is the major center for the reception and evaluation of sensory information, except for smell, hearing, and vision. The frontal and parietal lobes are separated by the central sulcus. The **occipital lobe** functions in the reception and integration of visual input and is not distinctly separate from the other lobes. The **temporal lobe** receives and evaluates input for smell and hearing and plays an important role in memory. Its anterior and inferior portions are referred to as the "psychic cortex," and they are associated with such brain functions as abstract thought and judgment. The temporal lobe is separated from the rest of the cerebrum by a **lateral fissure,** and deep

within the fissure is the **insula** (in′sū-lă, meaning island), often referred to as a fifth lobe.

The gray matter on the outer surface of the cerebrum is the **cortex,** and clusters of gray matter deep inside the brain are **nuclei** (figure 13.9). The white matter of the brain between the cortex and nuclei is the **cerebral medulla.** This term should not be confused with the medulla oblongata; medulla is a general term meaning the center of a structure, or marrow. The cerebral medulla consists of nerve tracts that connect the cerebral cortex to other areas of cortex or other parts of the CNS. These tracts fall into three main categories: (1) **association fibers,** which connect areas of the cerebral cortex within the same hemisphere; (2) **commissural fibers,** which connect one cerebral hemisphere to the other; and (3) **projection fibers,** which are between the cerebrum and other parts of the brain and spinal cord (see figure 13.9).

Cerebral Cortex

Figure 13.10 depicts a lateral view of the left cerebral cortex with some of the functional areas labeled. Sensory pathways project to specific regions of the cerebral cortex, called **primary sensory areas,** where these sensations are perceived.

Most of the postcentral gyrus is called the **primary somatic sensory cortex,** or **general sensory area.** The terms area and cortex are often used interchangeably for the same functional region of the cerebral cortex. Afferent fibers carrying general sensory input such as pain, pressure, and temperature synapse in the thalamus, and thalamic neurons relay the information to the primary somatic sensory cortex.

The somatic sensory cortex is organized topographically relative to the general plan of the body (figure 13.11*a*). Sensory impulses conducting input from the feet project to the most superior portion of the somatic sensory cortex, and sensory impulses from the face project to the most inferior portion of the somatic sensory cortex. The pattern of the somatic sensory cortex in each hemisphere is arranged in the form of an upside down half homunculus (hō-mungk′yū-lŭs, meaning a little human) representing the opposite side of the body, with the feet located superiorly and the head located inferiorly. The size of various regions of the somatic sensory cortex is relative to the number of sensory receptors in the associated regions of the body. The density of sensory receptors is much greater in the face than in the legs; therefore, a greater area of the somatic sensory cortex contains sensory neurons associated with the face, and the homunculus has a disproportionately large face.

Other primary sensory areas are the **taste area,** where taste sensations are consciously perceived in the cortex; the **olfactory cortex** (not shown in figure 13.10), which is on the inferior surface of the frontal lobe and is the area in which both conscious and unconscious responses to odor are initiated (see chapter 15); the **primary auditory cortex,** where auditory stimuli are processed by the brain; and the **visual cortex,** where portions of visual images are processed. In the visual cortex, color, shape, and movement are processed separately rather than as a complete "color motion picture."

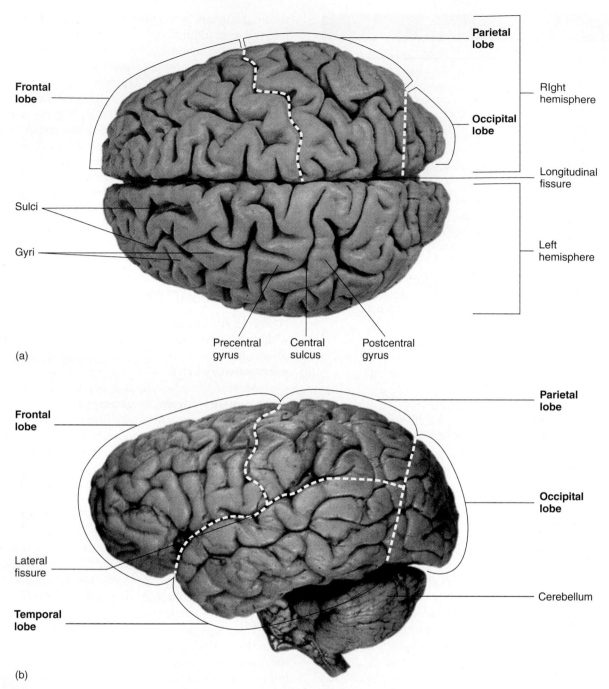

Figure 13.8   The Brain

(a) Superior view. (b) Lateral view of the left cerebral hemisphere. ✗

The primary sensory areas of the cerebral cortex must be intact for conscious perception, localization, and identification of a stimulus. Cutaneous sensations, although integrated within the cerebrum, are perceived as though they were on the surface of the body. This is called **projection** and indicates that the brain refers a cutaneous sensation to the superficial site at which the stimulus interacts with the sensory receptors.

Cortical areas immediately adjacent to the primary sensory centers, called **association areas,** are involved in the process of recognition. The **somatic sensory association area** is posterior to the primary somatic sensory cortex, and the

visual association area is anterior to the visual cortex (see figure 13.10). Afferent action potentials originating in the retina of the eye reach the visual cortex, where the image is "perceived." Action potentials then pass from the visual cortex to the visual association area, where the present visual information is compared with past visual experience ("Have I seen this before?"). On the basis of this comparison, the visual association area "decides" whether or not the visual input is recognized and passes judgment concerning the significance of the input. For example, we pay less attention to a person in a crowd that we have never seen before than to someone we know.

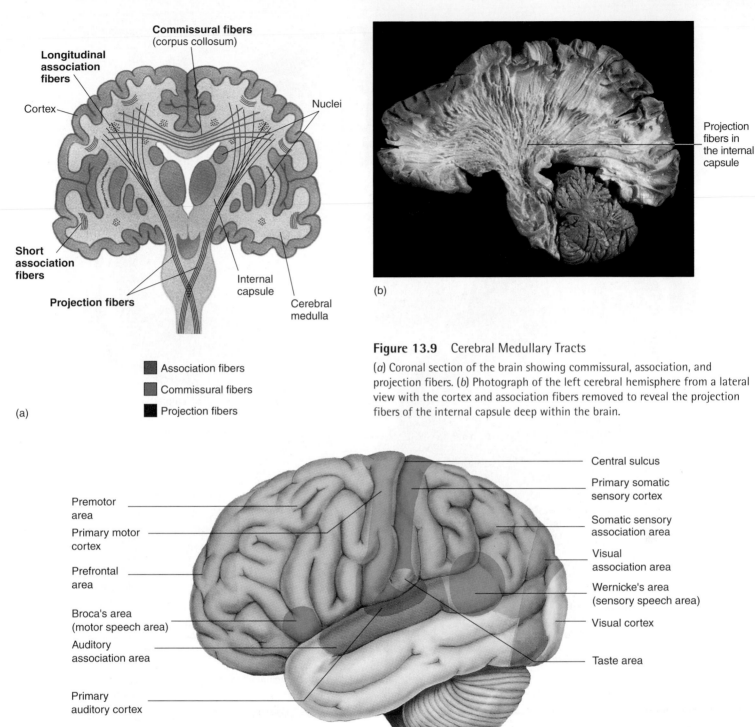

Longitudinal
association
fibers

Commissural fibers
(corpus collosum)

Cortex

Nuclei

**Short
association
fibers**

Projection fibers

Internal
capsule

Cerebral
medulla

Projection
fibers in
the internal
capsule

(b)

■ Association fibers

■ Commissural fibers

■ Projection fibers

(a)

Figure 13.9   Cerebral Medullary Tracts

(*a*) Coronal section of the brain showing commissural, association, and projection fibers. (*b*) Photograph of the left cerebral hemisphere from a lateral view with the cortex and association fibers removed to reveal the projection fibers of the internal capsule deep within the brain.

Premotor
area

Primary motor
cortex

Prefrontal
area

Broca's area
(motor speech area)

Auditory
association area

Primary
auditory cortex

Central sulcus

Primary somatic
sensory cortex

Somatic sensory
association area

Visual
association area

Wernicke's area
(sensory speech area)

Visual cortex

Taste area

Figure 13.10   Some Functional Areas of the Lateral Side of the Left Cerebral Cortex

The visual association area, like other association areas of the cortex, has reciprocal connections with other parts of the cortex, which influence decisions. For example, the visual association area has input from the frontal lobe, where emotional value is placed on the visual input. Because of these numerous connections, visual information is judged several times as it passes beyond the visual association area. This may be one of the reasons why two people who witness the same event can present somewhat different versions of what happened.

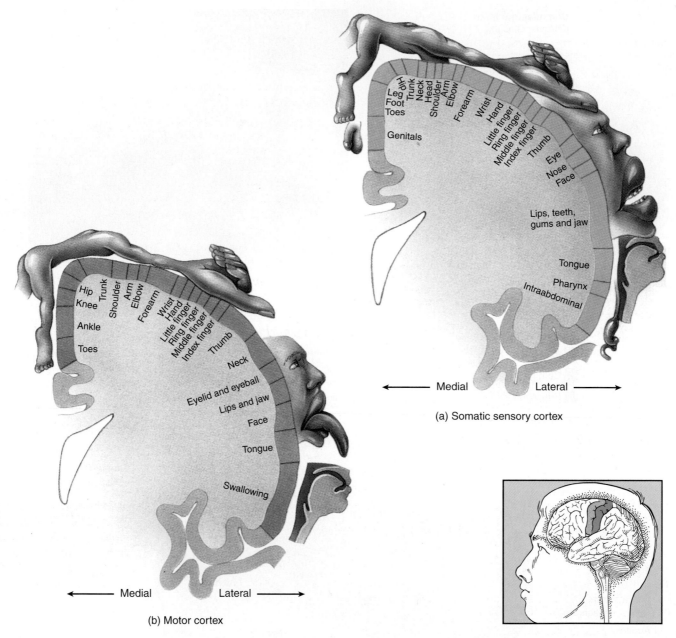

Figure 13.11   Topography of the Somatic Sensory and Motor Cortex

Cerebral cortex seen in coronal section on the left side of the brain. The figure of the body (homunculus) depicts the nerve distributions; the size of each body region shown indicates relative innervation. Each cortex occurs on both sides of the brain but appears on only one side in this illustration. The inset shows the motor and somatic sensory regions of the left hemisphere. (a) Somatic sensory cortex. (b) Motor cortex.

---

2   PREDICT

Using the visual association area as an example, explain the general functions of the association areas around the other primary cortical areas (see figure 13.10).

✔ *Answer in Appendix F*

The **precentral gyrus** is also called the **primary motor cortex,** or **primary motor area.** Efferent action poten-

tials initiated in this region control many voluntary movements, especially the fine motor movements of the hands. Cortical neurons that control skeletal muscles are called **upper motor neurons.** Upper motor neurons are not confined to only the precentral gyrus. Only about 30% of them are located in the precentral gyrus. Another 30% are in the premotor area, and the rest are in the somatic sensory cortex.

The cortical functions of the precentral gyrus are arranged topographically according to the general plan of the body—similar to that of the postcentral gyrus (figure 13.11b). The neuron cell bodies providing motor function to the feet are in the most superior and medial portions, whereas those for the face are in the inferior region. Muscle groups that have many motor units and therefore greater innervation are represented by a relatively larger area of the motor cortex. For example, muscles of the hands and mouth are represented by a larger area in the motor cortex than the muscles of the thighs and legs because, even though the muscles in the hands and mouth are small, they contain far more motor units than do the larger muscles of the thighs and legs.

The **premotor area,** located anterior to the primary motor cortex (see figure 13.10), is the staging area in which motor functions are organized before they are initiated in the motor cortex. For example, if a person decides to take a step, the neurons of the premotor area are stimulated first. The determination is made in the premotor area as to which muscles must contract, in what order, and to what degree. Impulses are then passed to the upper motor neurons in the motor cortex, which actually initiate the planned movements.

Clinical Note

The premotor area must be intact for a person to carry out complex, skilled, or learned movements, especially ones related to manual dexterity. Impairment in the performance of learned movements, called **apraxia** (ă-prak′sē-ă), can result from a lesion in the premotor area. Apraxia is characterized by hesitancy in performing these movements.

The motivation and foresight to plan and initiate movements occur in the next most anterior portion of the brain, the **prefrontal area,** an association area that is well developed only in primates and especially in humans. It is involved in motivation and regulation of emotional behavior and mood. The large size of this area in humans may account for our relatively well-developed forethought and motivation and for our emotional complexity.

Clinical Note

In relation to its involvement in motivation, the prefrontal area is also thought to be the functional center for aggression. Beginning in 1935, one method used to eliminate uncontrollable aggression or anxiety in psychiatric hospital patients was to surgically remove or destroy the prefrontal regions of the brain, a procedure called a **prefrontal,** or **frontal, lobotomy.** This operation was sometimes successful in eliminating aggression, but its effect was often only temporary, and some patients developed epilepsy or abnormal personality changes, such as lack of inhibition or a lack of initiative and drive. Later studies failed to confirm the usefulness of lobotomies, and the practice was largely discontinued in the late 1950s.

Speech

In most people, the speech area is in the left cortex. Two major cortical areas are involved in speech: **Wernicke's area** (sensory speech area), a portion of the parietal lobe, and **Broca's area** (motor speech area) in the inferior part of the frontal lobe (see figure 13.10). Wernicke's area is necessary for understanding and formulating coherent speech. Broca's area initiates the complex series of movements necessary for speech.

The following pathway must function for someone to repeat a word that he hears. Action potentials from the ear reach the primary auditory cortex, where the word is heard. The word is then recognized in the auditory association area and comprehended in parts of Wernicke's area. Then action potentials representing the word are conducted through association fibers that connect Wernicke's and Broca's areas. In Broca's area, the word is formulated as it will be repeated. Impulses then go to the premotor area, where the movements are programmed, and finally to the primary motor cortex, where the proper movements are triggered (figure 13.12).

Speaking a written word is somewhat similar (see figure 13.12). The information passes from the eyes to the visual cortex, then passes to the visual association area where the word is recognized, and continues to Wernicke's area where the word is understood and formulated as it will be spoken. From Wernicke's area it follows the same route as followed for repeating audibly received words.

3 PREDICT

Propose the pathway needed for a blindfolded person to name an object placed in her right hand.

✔ *Answer in Appendix F*

Clinical Note

Aphasia (ă-fā′zē-ă), absent or defective speech or language comprehension, results from a lesion in the language areas of the cortex. The several types of aphasia depend on the site of the lesion. **Receptive aphasia** (Wernicke's aphasia), which includes defective auditory and visual comprehension of language, defective naming of objects, and repetition of spoken sentences, is caused by a lesion in Wernicke's area. Both **jargon aphasia,** in which a person may speak fluently but unintelligibly, and **conduction aphasia,** in which a person has poor repetition but relatively good comprehension, can result from a lesion in the tracts between Wernicke's and Broca's areas. **Anomic** (ă-nō′mik) **aphasia,** caused by the isolation of Wernicke's area from the parietal or temporal association areas, is characterized by fluent but circular speech resulting from poor word-finding ability. **Expressive aphasia** (Broca's aphasia), caused by a lesion in Broca's area, is characterized by hesitant and distorted speech.

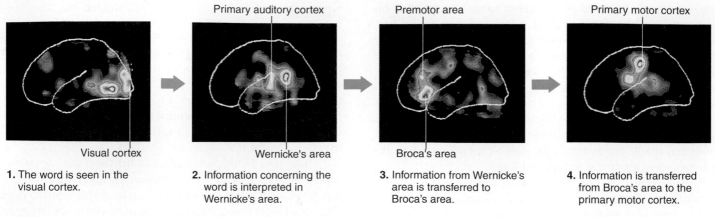

1. The word is seen in the visual cortex.

2. Information concerning the word is interpreted in Wernicke's area.

3. Information from Wernicke's area is transferred to Broca's area.

4. Information is transferred from Broca's area to the primary motor cortex.

Figure 13.12   Demonstration of Cortical Activities During Speech

The figures, beginning on the left and following the arrows, show the pathway for reading and naming something that is seen, such as reading aloud. PET scans show the areas of the brain that are most active during various phases of speech. Red indicates the most active areas; blue indicates the least active areas.

(a)

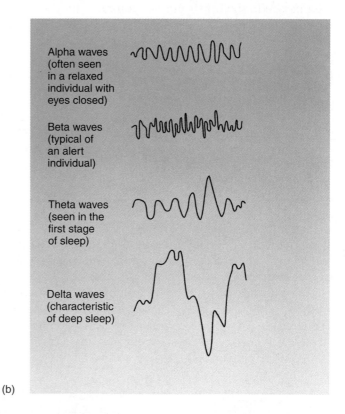

(b)

Figure 13.13   Electroencephalograms (EEGs) Showing Brain Waves

(*a*) Photo shows a person with EEG leads attached to the head.
(*b*) Tracings from EEGs.

Brain Waves

Electrodes placed on a person's scalp and attached to a recording device can record the electrical activity of the brain, producing an **electroencephalogram** (ē-lek′trō-en-sef′ă-lō-gram) (**EEG**) (figure 13.13). These electrodes are not sensitive enough to detect individual action potentials, but they can detect the simultaneous action potentials in large numbers of neurons. As a result, the EEG displays wavelike patterns known as **brain waves.** Brain waves are produced continuously, but their intensity and frequency differ from time to time based on the state of brain activity.

Most of the time EEG patterns from a given individual are irregular with no particular pattern because, although the normal brain is active, most of its electrical activity is not synchronous. At other times, however, specific patterns can be detected. These regular patterns are classified as alpha, beta, theta, or delta waves (see figure 13.13). **Alpha waves** are observed in a normal person who is awake but in a quiet, resting state with the eyes closed. **Beta waves** have a higher frequency than alpha waves and occur during intense mental activity. **Theta waves** usually occur in children, but they can also occur in adults who are experiencing frustration or who have certain brain disorders. **Delta**

waves occur in deep sleep, in infancy, and in patients with severe brain disorders.

Distinct types of EEG patterns can be detected in patients with specific brain disorders such as epileptic seizures. Neurologists use these patterns to diagnose the disorders and determine the treatment for them.

Memory

Memory can be divided into at least three types: sensory, short-term (or primary), and long-term. **Sensory memory** is the very short-term retention of sensory input received by the brain while something is scanned, evaluated, and acted on. This type of memory lasts less than a second and apparently involves transient changes in membrane potentials.

If a given piece of data held in sensory memory is considered valuable enough, it is moved into **short-term memory,** where information is retained for a few seconds to a few minutes. This memory is limited primarily by the number of bits of information (usually about seven) that can be stored at any one time, although the amount varies from person to person. Have you ever wondered why telephone numbers are seven digits long? More bits can be stored when the numbers are grouped into specific segments separated by spaces such as when adding an area code. When new information is presented, old information previously stored in short-term memory is eliminated; therefore, if a person is given a second telephone number or if the person's attention is drawn to something else, the first number usually is forgotten.

Several physiologic explanations have been proposed for short-term memory, most of which involve short-term changes in membrane potentials. The changes in membrane potentials are transitory but are longer than those involved in sensory memory, and they can be eliminated by new signals reaching the cells.

Certain pieces of information are transferred from short-term to **long-term memory.** Long-term memory may involve a physical change in neuron shape, called **long-term potentiation,** which facilitates future transmission of impulses. Long-term memory storage in a single neuron involves a calcium influx into the cell that activates an enzyme called **calpain** (kal'pān). Calpain, in turn, partially degrades the dendritic cytoskeleton of the neuron, changing the shape of the dendrite. The change in shape is stabilized by the creation of a new cytoskeleton, and the memory becomes more-or-less permanent.

A whole series of neurons and their pattern of activity, called a **memory engram,** or memory trace, probably are involved in the long-term retention of a given piece of information, a thought, or an idea. Repetition of the information and association of the new information with existing memories assist in the transfer of information from short-term to long-term memory.

There are two types of long-term memory: declarative and procedural. **Declarative memory** involves the retention of facts, such as names, dates, and places. Declarative memory is localized in a part of the temporal lobe called the **hip-pocampus** (hip-ō-kam'pŭs, meaning shaped like a seahorse) (see figure 13.15) and the **amygdaloid** (ă-mig'dă-loyd, meaning almond-shaped) **nucleus.** The hippocampus is involved in the actual memory, such as recalling a person's name; and the amygdala is involved in the emotional overtones of that memory, such as feelings of like or dislike, and the recollection of good or bad memories associated with that person. A lesion in the temporal lobe affecting the hippocampus can prevent the brain from moving information from short-term to long-term memory. Emotion and mood apparently serve as gates in the brain, determining what is or is not stored in long-term declarative memory. The amygdaloid nucleus is also a key to the development of fear, which also involves the prefrontal cortex and the hypothalamus.

Clinical Note

Some aspects of fearful responses appear to be "hard-wired" in the brain and do not require learning. For example, infant rodents are terrified when exposed to a cat, even though they have never seen a cat. Loud sounds seem to be particularly effective in eliciting fear responses. There is a direct collateral branch from the auditory pathway to the amygdala, which does not involve the cerebral cortex. Fear can be evoked by a loud sound acting directly on the amygdala. Overcoming fear, however, requires the involvement of the cerebral cortex; therefore, the stimulation of fear appears to involve one process, and its suppression another. Flaws in either process could result in fear-related disorders, such as anxiety, depression, panic, phobias, and posttraumatic stress disorder.

Procedural memory, also called **reflexive memory,** involves the development of skills such as riding a bicycle or playing a piano. Procedural memory is stored primarily in the cerebellum and the premotor area of the cerebrum. Conditioned, or pavlovian, reflexes are also procedural and can be eliminated in experimental animals by producing cerebellar lesions in the animals. The most famous example of a conditioned reflex is that of Ivan Pavlov's experiments with dogs. Each time he fed the dogs, a bell was rung; soon the dogs would salivate when the bell rang, even if no food was presented.

Right and Left Cortex

The cortex of the right cerebral hemisphere controls muscular activity in and receives sensory input from the left half of the body. The left cerebral hemisphere controls muscles in and receives sensory input from the right half of the body. Sensory information received by the cortex of one hemisphere is shared with the other through connections between the two hemispheres called **commissures** (kom'i-shyūrz, meaning a joining together). The largest of these commissures is the **corpus callosum** (kōr'pŭs kă-lō'sŭm, meaning callous body), which is a broad band of nerve tracts at the base of the longitudinal fissure (see figures 13.4 and 13.9).

Language and perhaps other functions such as artistic activities are not shared equally between the left and right

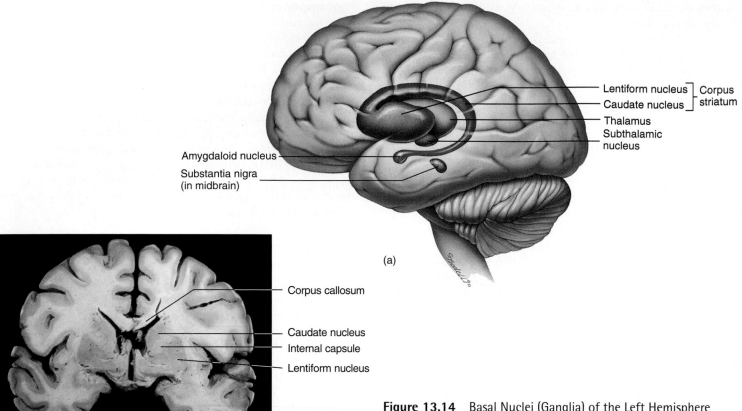

Figure 13.14 Basal Nuclei (Ganglia) of the Left Hemisphere

(*a*) A "transparent 3-D" drawing of the basal nuclei inside the left hemisphere. (*b*) Photograph of a frontal section of the brain showing the basal nuclei and other structures.

cerebral hemispheres. The left hemisphere is more involved in such skills as mathematics and speech. The right hemisphere is involved in functions such as three-dimensional or spatial perception, recognition of faces, and musical ability.

Clinical Note

Dominance for most functions is probably not very important in most people because the two hemispheres are in constant communication through the corpus callosum, literally allowing the right hand to know what the left hand is doing. Surgical cutting of the corpus callosum has been successful in treating a limited number of epilepsy cases. Under certain conditions, however, interesting functional defects can be seen in people whose corpus callosum has been severed. For example, if a patient with a severed corpus callosum is asked to reach behind a screen to touch one of several items with one hand without being able to see it and then is asked to point out the same object with the other hand, he cannot do it. Tactile information from the left hand enters the right somatic sensory cortex but is not transferred to the left hemisphere, which controls the right hand. As a result, the left hemisphere cannot direct the right hand to the correct object.

A person suffering a stroke in the right parietal lobe may lose the ability to recognize faces while retaining essentially all other brain functions. In more severe lesion cases, a person may lose the ability to identify simple objects. This defect is called **amorphosynthesis**

(ă'mōr'fō-sin'thĕ-sis). Other people with a similar lesion may tend to ignore the left half of the world, including the left half of their own bodies. Such people may completely ignore a person who is to their left but react normally when the person moves to their right. They may also fail to dress the left half of the body or eat the food on the left half of the plate.

Basal Nuclei

The **basal nuclei,** or basal ganglia, are a group of functionally related nuclei located bilaterally in the inferior cerebrum, diencephalon, and midbrain (figure 13.14). The **subthalamic nucleus** is located in the diencephalon, and the **substantia nigra** is located in the midbrain. The nuclei in the cerebrum are collectively called the **corpus striatum** (kōr'pŭs strī-ā'tŭm, meaning striped body) and include the **caudate** (kaw'-dāt, meaning having a tail) **nucleus** and **lentiform** (len'ti-fōrm, meaning lens-shaped) **nucleus.** They are the largest nuclei of the brain and occupy a large part of the cerebrum.

The basal nuclei are important in organizing and coordinating motor movements and posture. Complex neural connections link the basal nuclei with the cerebral cortex. The major effect of the basal nuclei is to decrease muscle tone and inhibit unwanted muscular activity. Disorders of the basal nuclei result in increased muscle tone and exaggerated, uncon-

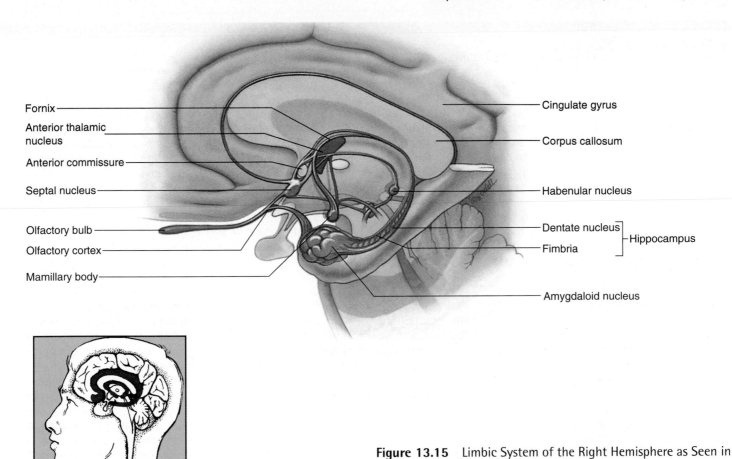

Figure 13.15   Limbic System of the Right Hemisphere as Seen in a Midsagittal Section

Limbic System

Parts of the cerebrum and diencephalon are grouped together under the title **limbic** (lim′bik) **system** (figure 13.15). *Limbus* means border, and the term limbic refers to deep portions of the cerebrum that form a ring around the brainstem. The limbic lobe is thought to be the oldest and most primitive part of the brain. Structurally the limbic system consists of (1) certain cerebral cortical areas, including the **cingulate** (sin′gyū-lāt, meaning to surround) **gyrus,** located along the inner surface of the longitudinal fissure just above the corpus callosum, and the **hippocampus;** (2) various nuclei such as anterior nuclei of the thalamus and the **habenular** (ha-ben′yū-lăr) **nuclei** in the epithalamus; (3) parts of the **basal nuclei;** (4) the hypothalamus, especially the **mamillary bodies;** (5) the **olfactory cortex;** and (6) tracts connecting the various cortical areas and ganglia, such as the **fornix,** which connects the hippocampus to the thalamus and mamillary bodies.

The limbic system influences emotions, the visceral responses to emotions, motivation, mood, and sensations of pain and pleasure. This system is associated with basic sur-

vival instincts: the acquisition of food and water, as well as reproduction. One of the major sources of sensory input into the limbic system is the olfactory nerves. The smell or thought of food stimulates the sense of hunger in the hypothalamus, which motivates us to seek food. Many animals can also smell water, even over great distances. In animals such as dogs and cats, olfactory detection of **pheromones** (fer′ō-mōnz) is important in reproduction. Pheromones are molecules released into the air by one animal that attract another animal of the same species, usually of the opposite sex.

Apparently the cingulate gyrus is a "satisfaction center" for the brain and associated with the feeling of satisfaction after a meal or after sexual intercourse. The relationship of the hippocampus with the limbic system and with memory is probably very important to survival. For example, it is very important for an animal to remember where to obtain food. Once a person has eaten, the satiety center in the hypothalamus is stimulated, the hunger center is inhibited, and the person feels satiated. The hypothalamus interacts with the cingulate gyrus and other parts of the limbic system, causing a sense of satisfaction associated with the satiation.

Lesions in the limbic system can result in a voracious appetite, increased sexual activity, which is often inappropriate, and docility, including the loss of normal fear and anger responses. Because the hippocampus is part of the temporal lobe, damage to that portion can also result in a loss of memory.

trolled movements, especially when a person is at rest. A specific feature of basal nuclei disorders is a "resting tremor," a slight shaking of the hands when a person is not performing a task.

Clinical Focus   Dyskinesias

Dyskinesias (dis-ki-nē'sē-ăs) are a group of disorders often involving the basal nuclei, in which unwanted, superfluous movements occur. Defects in the basal nuclei may result in brisk, jerky, purposeless movements that resemble fragments of voluntary movements. **Sydenham's chorea** (kōr-ē'ă; also called St. Vitus' dance) is a disease usually associated with a toxic or infectious disorder that apparently causes temporary dysfunction of the corpus striatum and usually affects children. **Huntington's chorea** is a dominant hereditary disorder that begins in middle life and causes mental deterioration and progressive degeneration of the corpus striatum in affected individuals.

Cerebral palsy (pawl'zē) is a general term referring to defects in motor functions or coordination resulting from several types of brain damage, which may be caused by abnormal brain development or birth-related injury. Some symptoms of cerebral palsy are related to basal nuclei dysfunction. **Athetosis** (ath-ĕ-tō'sis), often one of the features of cerebral palsy, is characterized by slow, sinuous, aimless movements. When the face, neck, and tongue muscles are involved, grimacing, protrusion, and writhing of the tongue, and difficulty in speaking and swallowing are characteristics.

Damage to the subthalamic nucleus can result in **hemiballismus** (hem'ē-bă-liz'mŭs), an uncontrolled, purposeless, and forceful throwing or flailing of the arm. Forceful twitching of the face and neck may also result from subthalamic nuclear damage.

Parkinson's disease, characterized by muscular rigidity, loss of facial expression, tremor, a slow, shuffling gait, and general lack of movement, is caused by a dysfunction in the substantia nigra. The disease usually occurs after age 55 and is not contagious or inherited. A resting tremor called "pill-rolling" is characteristic of Parkinson's disease and consists of circular movement of the opposed thumb and index fingertips. The increased muscular rigidity in Parkinson's disease results from defective inhibition of some of the basal nuclei by the substantia nigra. In this disease, dopamine, an inhibitory neurotransmitter substance, is deficient. The melanin-containing cells of the substantia nigra degenerate, resulting in a loss of pigment.

Parkinson's disease can be treated with levodopa (L-dopa), a precursor to dopamine, or, more effectively, with Sinemet, a combination of L-dopa and carbidopa. Carbidopa prevents L-dopa from being absorbed by tissues other than the brain. A protein called

glial cell line-derived neurotrophic factor (GDNF) has been discovered that selectively promotes the survival of dopamine-secreting neurons. Experimental treatment of the disorder by transplanting fetal tissues capable of producing dopamine is also under investigation.

Cerebellar lesions result in a spectrum of characteristic functional disorders. Movements tend to be **ataxic** (jerky) and **dysmetric** (overshooting—for example, pointing past or deviating from a mark that one tries to touch with the finger). Alternating movements such as supination and pronation are performed in a clumsy manner. **Nystagmus,** (nis-tag'mŭs), which is a constant motion of the eyes, may also occur. A cerebellar tremor is an intention tremor (i.e., the more carefully one tries to control a given movement, the greater the tremor becomes). For example, when a person with a cerebellar tremor attempts to drink a glass of water, the closer the glass comes to the mouth, the shakier the movement becomes. This type of tremor is in direct contrast to basal nuclei tremors described previously in which the resting tremor largely or completely disappears during purposeful movement.

Cerebellum

The term **cerebellum** (ser-e-bel'ŭm; figure 13.16) means little brain. It communicates with other regions of the CNS through three large nerve tracts: the superior, middle, and inferior **cerebellar peduncles** (pe-dŭng'klz). The cerebellum has a gray cortex and nuclei, with white medulla in between. The cerebellar cortex has ridges called **folia.**

The cerebellum consists of three parts: a small anterior part, the **flocculonodular** (flok'yū-lō-nod'yū-lăr; floccular, meaning a tuft of wool) **lobe;** a narrow central **vermis** (meaning worm-shaped); and two large **lateral hemispheres** (see figure 13.16). The flocculonodular lobe is the simplest part of the cerebellum and is involved in balance and maintaining muscle tone. The anterior part of the vermis is involved in gross motor coordination and muscle tone, and the posterior vermis and lateral hemispheres are involved in fine motor coordination, producing smooth, flowing movements.

The cerebellum accomplishes fine motor coordination by means of its **comparator** function. Action potentials from the motor cortex descend into the spinal cord to initiate voluntary movements, and at the same time action potentials

from the motor cortex to the cerebellum give the cerebellar neurons information representing the intended movement (figure 13.17). Simultaneously, action potentials from proprioceptive neurons ascend to the cerebellum. Proprioceptive neurons innervate the joints and tendons of the structure being moved, such as the elbow or knee, and provide information about the position of the body or body parts. These action potentials give the cerebellar neurons information from the periphery about the actual movements. The cerebellum compares the action potentials from the motor cortex with those from the moving structures. That is, it compares the intended movement with the actual movement, and if a difference is detected, the cerebellum sends action potentials to the motor cortex and the spinal cord to correct the discrepancy. The result is smooth and coordinated movements.

With training, a person can develop highly skilled and rapid movements that are accomplished more rapidly than can be accounted for by the comparator function of the cerebellum. In these cases, the cerebellum can "learn" highly specialized motor functions through specific, repeated comparator activities. There is some evidence that the cerebellum may also function in some sensory perception.

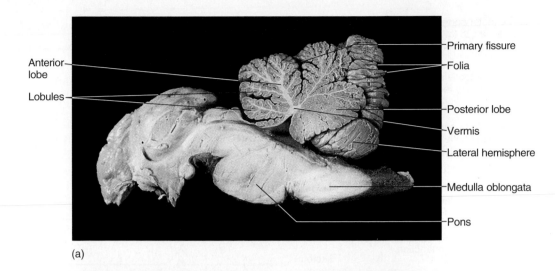

Anterior lobe — Primary fissure

Lobules — Folia

Posterior lobe

Vermis

Lateral hemisphere

Medulla oblongata

Pons

(a)

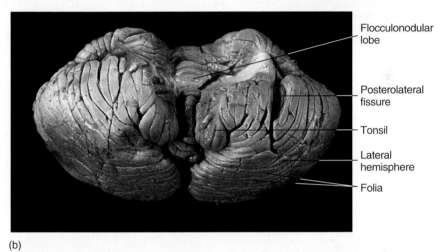

Flocculonodular lobe

Posterolateral fissure

Tonsil

Lateral hemisphere

Folia

(b)

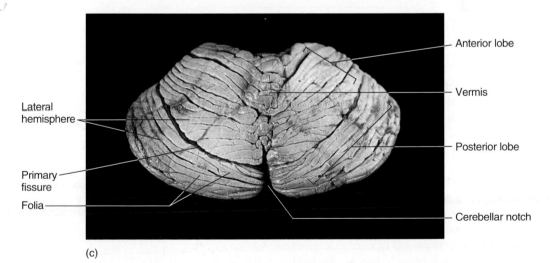

Anterior lobe

Vermis

Lateral hemisphere

Posterior lobe

Primary fissure

Folia

Cerebellar notch

(c)

Figure 13.16 Cerebellum

(a) Right half of the cerebellum as seen in a midsagittal section.
(b) Inferior view of the cerebellum.
(c) Superior view of the cerebellum. ✗

Clinical Note

Cerebellar dysfunction results in (1) decreased muscle tone, (2) balance impairment, (3) a tendency to overshoot when reaching for or touching an object, and (4) an intention tremor, which is a shaking in the hands that occurs only while attempting to perform a task. Notice that although the cerebellum and basal nuclei both control motor functions, they have opposite effects, and many symptoms associated with dysfunction are also opposite. For example, cerebellar dysfunction results in decreased motor tone and an intention tremor, whereas basal nuclear dysfunction results in increased motor tone and a resting tremor.

1. The motor cortex sends action potentials to lower motor neurons in the spinal cord.

2. Action potentials from the motor cortex inform the cerebellum of the intended movement.

3. Lower motor neurons in the spinal cord send action potentials to skeletal muscles, causing them to contract.

4. Proprioceptive signals from the skeletal muscles and joints to the cerebellum convey information concerning the status of the muscles and the structure being moved during contraction.

5. The cerebellum compares the information from the motor cortex to the proprioceptive information from the skeletal muscles and joints.

6. Action potentials from the cerebellum to the spinal cord modify the stimulation from the motor cortex to the lower motor neurons.

7. Action potentials from the cerebellum are sent to the motor cortex, which modify its motor activity.

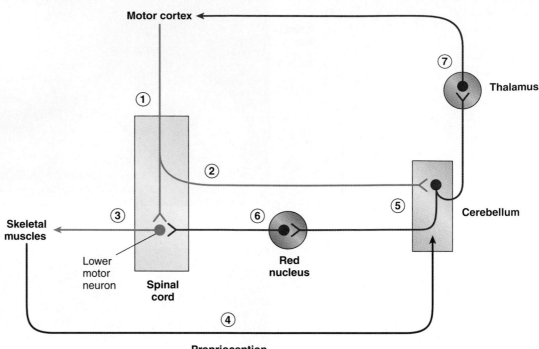

Figure 13.17   Cerebellar Comparator Function

Spinal Cord

The spinal cord is extremely important to the overall function of the nervous system. It is the communication link between the brain and the **peripheral nervous system (PNS)** inferior to the head, integrating incoming information and producing responses through reflex mechanisms.

General Structure

The **spinal cord** (figure 13.18) extends from the foramen magnum to the level of the second lumbar vertebra. It is considerably shorter than the vertebral column because it does not grow as rapidly as the vertebral column during development. It is composed of cervical, thoracic, lumbar, and sacral segments, which are named according to the area of the vertebral column from which their nerves enter and exit. Thirty-one pairs of spinal nerves exit the spinal cord and pass out of the vertebral column through the intervertebral foramina (see chapter 14). Because the spinal cord is shorter than the vertebral column, the nerves do not always exit the vertebral column at the same level that they exit the spinal cord.

The spinal cord is not uniform in diameter throughout its length. There is a general decrease in diameter superiorly to inferiorly, and there are two enlargements where nerves supplying the extremities enter and leave the cord (see figure 13.18). The **cervical enlargement** in the inferior cervical region corresponds to the location where nerves that supply the upper limbs enter or exit the cord, and the **lumbar enlargement** in the inferior thoracic and superior lumbar regions is the site where the nerves supplying the lower limbs enter or exit.

4   **P R E D I C T**

Why is the cord enlarged in the cervical and lumbar areas?

✔ *Answer in Appendix F*

Immediately inferior to the lumbar enlargement, the spinal cord tapers to form a conelike region called the **conus medullaris.** Its tip is at the level of the second lumbar vertebra and is the inferior end of the spinal cord. A connective tissue filament, the **filum terminale** (fī′lŭm ter′mi-nal′ē), extends inferiorly from the apex of the conus medullaris to the coccyx and anchors the cord to the coccyx. The nerves sup-

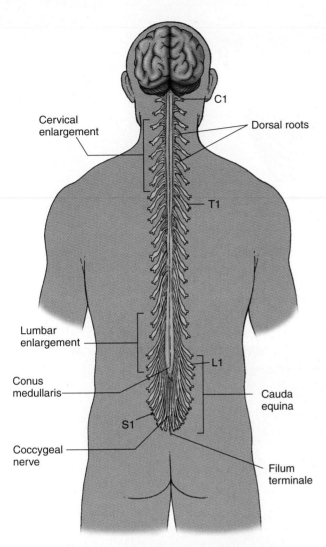

Figure 13.18 Spinal Cord and Spinal Nerve Roots

plying the legs and other inferior structures of the body, the second lumbar (L2) to the fifth sacral nerves (S5), exit the lumbar enlargement and conus medullaris, course inferiorly through the vertebral canal, and exit through the intervertebral foramina from L2 to S5. The conus medullaris and the numerous nerves extending inferiorly from it resemble a horse's tail and are therefore called the **cauda equina** (kaw'dă ē-kwī'nă; see figure 13.18).

Cross Section

A cross section of the spinal cord reveals that the cord consists of a central gray portion and a peripheral white portion (figure 13.19). The white matter consists of nerve tracts, and the gray matter consists of neuron cell bodies and dendrites. An **anterior median fissure** and a **posterior median sulcus** are deep clefts partially separating the two halves of the cord. The white matter in each half of the spinal cord is organized into three **columns,** or **funiculi**

(fyū-nik'yū-lī), called the **ventral** (anterior), **dorsal** (posterior), and **lateral columns.** Each funiculus is subdivided into **fasciculi** (fă-sik'yū-lī), or **nerve tracts.** Individual axons carrying action potentials to (ascending) or from (descending) the brain are usually grouped together within the fasciculi, and axons within a given fasciculus carry basically the same type of information, although fasciculi may overlap to some extent.

The central gray matter is organized into horns. Each half of the central gray matter of the spinal cord consists of a relatively thin **posterior** (dorsal) **horn** and a larger **anterior** (ventral) **horn.** Small **lateral horns** also exist in levels of the cord associated with the autonomic nervous system. The axons of sensory neurons synapse with cell bodies of association neurons in the posterior horn; the cell bodies of somatic motor neurons are in the anterior horn, or motor horn; and the cell bodies of autonomic neurons are in the lateral horns. The two halves of the spinal cord are connected by **gray** and **white commissures** (figure 13.19a). The central canal is in the center of the gray commissure.

Dorsal (posterior) and ventral (anterior) roots exit the spinal cord near the dorsal and ventral horns (figure 13.19b). The **dorsal root** carries afferent action potentials to the cord, and the **ventral root** carries efferent action potentials away from the cord. The dorsal roots possess **dorsal root ganglia** (gang'glē-ă, meaning a swelling or knot; also called spinal ganglia), which contain the cell bodies of sensory neurons. The axons of these neurons form the dorsal root and project into the posterior horn, where they synapse with other neurons or ascend or descend in the spinal cord. The ventral root is formed by the axons of neurons in the anterior and lateral horns. The dorsal and ventral roots unite to form the spinal nerves.

5	P R E D I C T

Explain why the dorsal root ganglia are larger in diameter than the dorsal roots.

✔ *Answer in Appendix F*

Spinal Reflexes

Automatic reactions to stimuli that occur without conscious thought are called **reflexes** (see chapter 12). Because they occur without conscious thought, reflexes are considered involuntary, even though they often involve skeletal muscles.

Reflexes are integrated both in the brainstem and in the spinal cord. Many of the reflexes occurring in the brainstem are autonomic, or visceral, reflexes (see chapter 16) and include such reactions as constriction of the pupil in response to increased light or an increased heart rate in response to reduced blood pressure. Major spinal cord reflexes include the stretch reflex, the Golgi tendon reflex, and the withdrawal reflex.

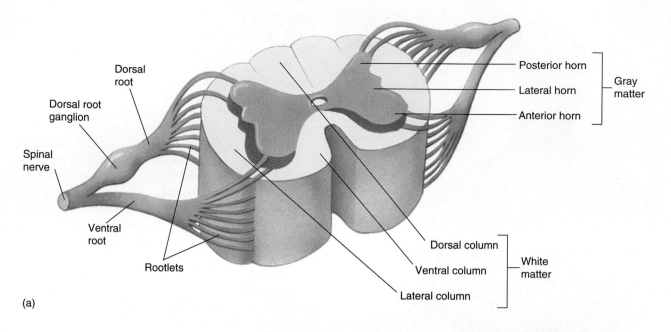

(a)

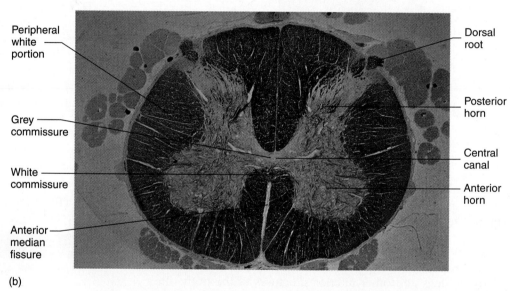

(b)

Figure 13.19   Cross Section of the Spinal Cord

(*a*) A 3-D drawing of a segment of the spinal cord showing one dorsal and one ventral root on each side and the rootlets that form them. (*b*) Photograph of a cross section through the midlumbar region. The darker-colored areas are white matter, where tracts are located. The lighter area is gray matter, where neuron cell bodies are located. ↗

Stretch Reflex

The simplest reflex is the **stretch reflex** (figure 13.20*a*), a reflex in which muscles contract in response to a stretching force applied to them. The sensory receptor of the reflex is the **muscle spindle,** which consists of 3–10 small specialized skeletal muscle cells. The cells are contractile only at their ends and are innervated by specific efferent fibers called **gamma motor neurons,** originating from the spinal cord and controlling contraction of the ends of the muscle spindle cells. The noncontractile centers of the muscle spindle cells are innervated by afferent neurons that carry impulses into the spinal cord, where they synapse directly with motor neurons called **alpha motor neurons,** which in turn innervate the muscle in which the muscle spindle is embedded. The stretch reflex is unique in that it does not require an association neuron between the afferent and efferent neurons.

Stretching a muscle also stretches the muscle spindle located among the muscle fibers. The stretch stimulates the afferent neurons that innervate the center of the muscle spindle. The increased frequency of action potentials in the afferent neurons stimulates the alpha motor neurons in the spinal cord. The alpha motor neurons transmit action potentials to skeletal muscle, causing a rapid contraction of the stretched muscle, which opposes the stretch of the muscle. The postural muscles demonstrate the adaptive nature of this reflex. If a person is standing upright and then bends slightly to one side, the postural muscles associated with the vertebral column on the other side are stretched. As a result, stretch reflexes are initiated in those muscles, which cause them to contract and reestablish normal posture.

Collateral axons from the afferent neurons of the muscle spindles also synapse with ascending nerve tracts, which enable the brain to perceive that a muscle has been stretched.

1. Muscle spindles detect stretch of the muscle.

2. Afferent neurons conduct action potentials to the spinal cord.

3. Afferent neurons synapse with alpha motor neurons.

4. Stimulation of the alpha motor neurons causes the muscle to contract and resist being stretched.

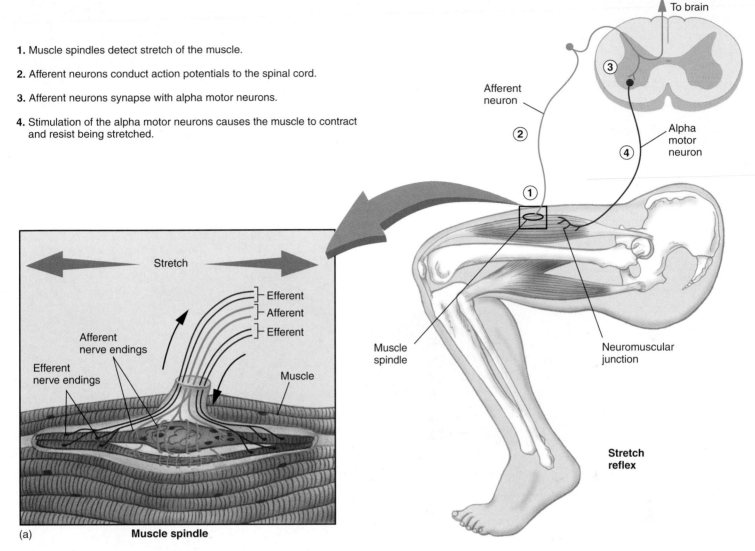

Figure 13.20   Spinal Cord Reflexes

(*a*) Stretch reflex.

Descending neurons within the spinal cord synapse with the neurons of the stretch reflex and modulate their activity. This activity is important in maintaining posture and in coordinating muscular activity.

Gamma motor neurons are responsible for regulating the sensitivity of the muscle spindle. As a skeletal muscle contracts, the tension on the muscle spindles within the muscle is relaxed, causing the muscle spindles to be less sensitive. Action potentials carried by gamma motor neurons stimulate the ends of the muscle spindle cells and cause them to shorten, along with the skeletal muscle. Contraction of the muscle spindles stimulates the afferent neurons from the center of the muscle spindle cells and maintains their sensitivity to stretching. Action potentials from the brain that stimulate alpha motor neurons to the skeletal muscle and result in contraction of the muscle also stimulate gamma motor neurons to the muscle spindle, enhancing the activity of the muscle spindle, which in turn helps

control and coordinate muscular activity, such as posture, muscle tension, and muscle length.

Clinical Note

The **knee-jerk reflex,** or **patellar reflex,** is a classic example of the stretch reflex and is used by clinicians to determine if the higher CNS centers that normally influence this reflex are functional. When the patellar ligament is tapped, the quadriceps femoris muscle tendon and the muscles themselves are stretched. The muscle spindle fibers within these muscles are also stretched, and the stretch reflex is activated. Consequently, contraction of the muscles extends the leg, producing the characteristic knee-jerk response. When the stretch reflex is greatly exaggerated, it indicates that the neurons within the brain that normally innervate the gamma motor neurons and enhance the stretch reflex are overly active. On the other hand, if the facilitatory action potentials to the gamma motor neurons are depressed, the stretch reflex is greatly suppressed or absent.

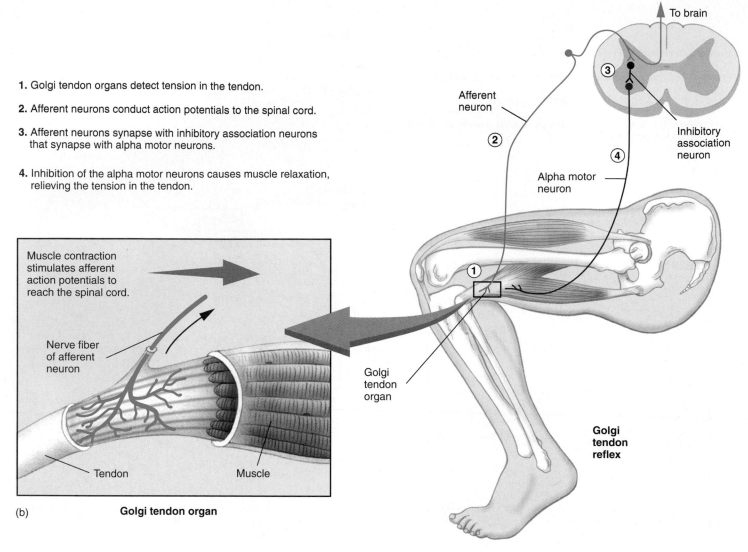

1. Golgi tendon organs detect tension in the tendon.

2. Afferent neurons conduct action potentials to the spinal cord.

3. Afferent neurons synapse with inhibitory association neurons that synapse with alpha motor neurons.

4. Inhibition of the alpha motor neurons causes muscle relaxation, relieving the tension in the tendon.

Muscle contraction stimulates afferent action potentials to reach the spinal cord.

Nerve fiber of afferent neuron

Tendon

Muscle

(b)   **Golgi tendon organ**

To brain

Afferent neuron

② ④

③

Inhibitory association neuron

Alpha motor neuron

Golgi tendon organ

①

Golgi tendon reflex

Figure 13.20   (*continued*)

(*b*) Golgi tendon reflex.

Golgi Tendon Reflex

The **Golgi tendon reflex** prevents the production of excessive tension in a muscle. **Golgi tendon organs** are encapsulated nerve endings that have at their ends numerous terminal branches with small swellings that are associated with bundles of collagen fibers in tendons. The Golgi tendon organs are located within tendons near the muscle–tendon junction and are stimulated as the tendon is stretched during muscle contraction (figure 13.20*b*). As a muscle contracts, the attached tendons are stretched, resulting in increased tension in the tendon. The increased tension stimulates action potentials in the afferent fibers from the Golgi tendon organs.

The afferent neurons of the Golgi tendon organs pass through the dorsal root to the spinal cord and enter the posterior gray matter, where they branch and synapse with inhibitory association neurons (see chapter 12). The association neurons synapse with alpha motor neurons that innervate the muscle to which the Golgi tendon organ is attached. When a great amount of tension is applied to the tendon, this reflex inhibits the motor neurons of the associated muscle and causes it to relax, thus protecting muscles and tendons from damage caused by excessive tension. The sudden relaxation of the muscle reduces the tension applied to the muscle and tendons. A weight lifter who suddenly drops a heavy weight after straining to lift it does so, in part, because of the effect of the Golgi tendon reflex.

Tremendous amounts of tension can be applied to muscles and tendons in the legs. Frequently an athlete's Golgi tendon reflex is inadequate to protect muscles and tendons from excessive tension. The large muscles and sudden movements of football players and sprinters can make them vulnerable to relatively frequent hamstring pulls and Achilles (calcaneal) tendon injuries.

1. Pain receptors detect a painful stimulus.

2. Afferent neurons conduct action potentials to the spinal cord.

3. Afferent neurons synapse with excitatory association neurons that synapse with alpha motor neurons.

4. Excitation of the alpha motor neurons results in contraction of the flexor muscles and withdrawal of the limb from the painful stimulus.

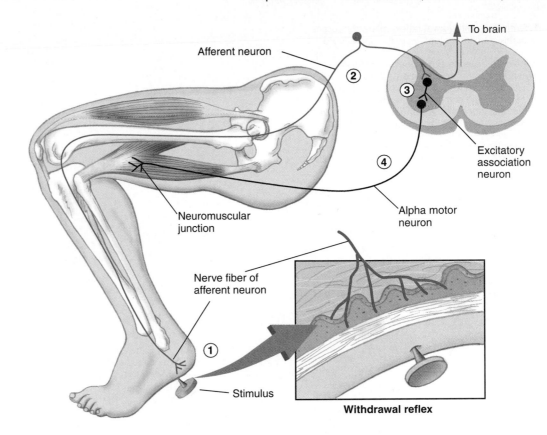

(c)

1. During the withdrawal reflex, afferent neurons conduct action potentials to the spinal cord.

2. Afferent neurons synapse with excitatory association neurons that are part of the withdrawal reflex.

3. Collateral branches also synapse with inhibitory association neurons that are part of reciprocal innervation.

4. Inhibition of the alpha motor neurons supplying the extensor muscles causes them to relax and not oppose the flexor muscles of the withdrawal reflex.

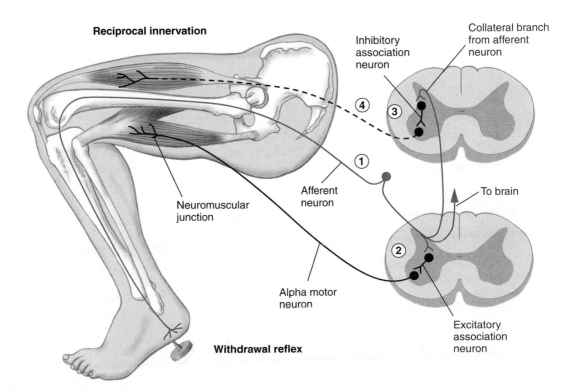

(d)

Figure 13.20   (*continued*)

(*c*) Withdrawal reflex. (*d*) Reciprocal innervation.

1. During the withdrawal reflex, afferent neurons conduct action potentials to the spinal cord.

2. Afferent neurons synapse with excitatory association neurons that are part of the withdrawal reflex. Collateral branches also synapse with excitatory association neurons that cross over to the opposite side of the spinal cord as part of the crossed extensor reflex.

3. Stimulation of the alpha motor neurons supplying extensor muscles on the opposite limb causes them to contract and support body weight during the withdrawal reflex.

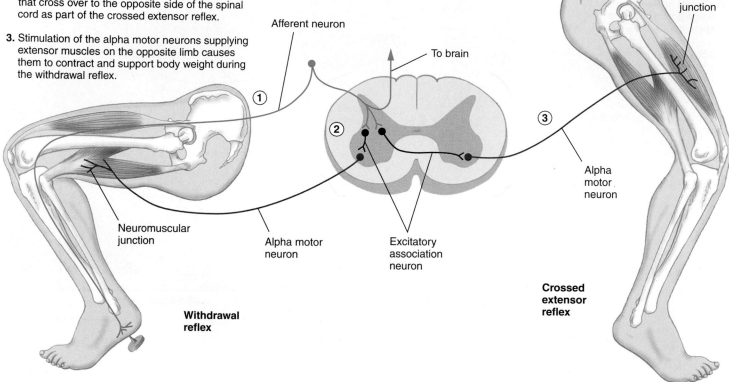

Figure 13.20 (*continued*)
(*e*) Cross extensor reflex.

Withdrawal Reflex

The function of the **withdrawal,** or **flexor reflex,** is to remove a limb or other body part from a painful stimulus. The sensory receptors are pain receptors (see chapter 15). Action potentials from painful stimuli are conducted by afferent neurons through the dorsal root to the spinal cord, where they synapse with excitatory association neurons, which in turn synapse with alpha motor neurons (figure 13.20*c*). The alpha motor neurons stimulate muscles, usually flexor muscles, that remove the limb from the source of the painful stimulus. Collateral branches of the afferent neurons synapse with ascending fibers to the brain, providing conscious awareness of the painful stimuli.

Reciprocal Innervation

Reciprocal innervation is associated with the withdrawal reflex and reinforces its efficiency (figure 13.20*d*). Collateral axons of afferent neurons that carry action potentials from pain receptors innervate inhibitory association neurons, which inhibit alpha motor neurons of the extensor (antago-

nist) muscles. When the withdrawal reflex is initiated, the flexor muscles usually contract, and reciprocal innervation causes relaxation of the extensor muscles, which reduces the resistance to movement that the extensor muscles would otherwise generate.

Reciprocal innervation is also involved in the stretch reflex. When the stretch reflex causes a muscle to contract, reciprocal innervation causes opposing muscles to be inhibited. In the patellar reflex, for example, the leg flexors are inhibited when the leg extensors are stimulated.

Crossed Extensor Reflex

The **crossed extensor reflex** is another reflex associated with the withdrawal reflex (figure 13.20*e*). Association neurons that stimulate alpha motor neurons, resulting in withdrawal of a limb, send collateral axons through the white commissure to the opposite side of the spinal cord to stimulate alpha motor neurons that innervate extensor muscles in the opposite side of the body. If a withdrawal reflex is initiated in one leg, the crossed extensor reflex causes extension of the opposite leg.

The crossed extensor reflex is adaptive in that it prevents falls by shifting the weight of the body from the affected to the unaffected leg. It is exemplified by a person's reaction to stepping on a sharp object. The withdrawal reflex occurs in response to the painful stimulus in the affected leg, and the crossed extensor reflex occurs in the opposite leg. Initiating a withdrawal reflex in both legs at the same time would cause one to fall.

Spinal Pathways

The names of most ascending and descending pathways, or tracts, in the CNS indicate their origin and termination (tables 13.4 and 13.5 and figure 13.21). Each pathway usually is given a composite name in which the first half of the word indicates its origin and the second half indicates its termination. Ascending pathways therefore usually begin with the prefix spino-, indicating that they originate in the spinal cord. For example, a spinothalamic tract is one that originates in the spinal cord and terminates in the thalamus. An exception to this rule of nomenclature is the dorsal-column/medial-lemniscal system, whose name is a combination of the pathway names in the spinal cord and brainstem. Descending pathways usually begin with the prefix cortico-, indicating that they begin in the cerebral cortex. The corticospinal tract is a descending tract that originates in the cerebral cortex and terminates in the spinal cord. The specific function of each ascending or descending tract, however, is not suggested by its name.

Ascending Pathways

The major ascending pathways or tracts involved in the conscious perception of external stimuli are the spinothalamic pathways and the dorsal-column/medial-lemniscal system (see table 13.4). Those carrying sensations about which we are not consciously aware are the spinocerebellar, spinoolivary, spinotectal, and spinoreticular tracts.

Spinothalamic System

The spinothalamic system is the least discriminative of the two systems that convey cutaneous sensory information to the brain. Pain and temperature information are carried primarily by the **lateral spinothalamic** (spī′nō-tha-lam′ik) **tracts** (figure 13.22*a*). Light touch, pressure, tickle, and itch sensations are carried by the **anterior spinothalamic tracts** (figure 13.22*b*).

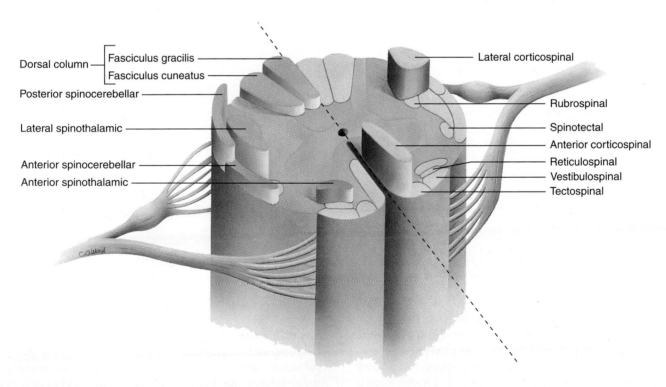

Figure 13.21   Cross Section of the Spinal Cord Depicting the Pathways

Ascending pathways are labeled on the left side of the figure (*blue*); descending pathways are labeled on the right side (*pink*). ✗

Table 13.4   Ascending Spinal Pathways

Pathway	Modality (information transmitted)	Origin	Termination
Spinothalamic		Cutaneous receptors	Cerebral cortex
Lateral	Pain and temperature		
Anterior	Light touch, pressure, tickle, and itch sensation		
Dorsal-column/ medial-lemniscal system	Proprioception, two-point discrimination, pressure, and vibration	Cutaneous receptors, joints	Cerebral cortex and cerebellum
Spinocerebellar	Proprioception to cerebellum	Joints, tendons	Cerebellum
Posterior			
Anterior			
Spinoolivary	Proprioception relating to balance	Joints, tendons	Accessory olivary nucleus, then to cerebellum
Spinotectal	Tactile stimulation causing visual reflexes	Cutaneous receptors	Superior colliculus
Spinoreticular	Tactile stimulation arousing consciousness	Cutaneous receptors	Reticular formation

Table 13.5   Descending Spinal Pathways

Pathway	Functions Controlled	Origin	Termination	Crossover
Pyramidal	Muscle tone and conscious skilled movements, especially of the hands			
Croticospinal	Movements, especially of the hands	Cerebral cortex (upper motor neuron)	Anterior horn of spinal cord (lower motor neuron)	
Lateral				Inferior end of medulla oblongata
Anterior				At level of lower motor neuron
Corticobulbar	Facial and head movements	Cerebral cortex (upper motor neuron)	Cranial nerve nuclei in brainstem (lower motor neuron)	Varies for the various cranial nerves

Light touch is also called crude touch (poorly localized); although the receptors of these nerves respond to very light touch, the stimulus is not well localized.

Three neurons in sequence—the primary, secondary, and tertiary—are involved in the pathway from the peripheral receptor to the cerebral cortex. The **primary neuron** cell bodies of the spinothalamic system are in the dorsal root ganglia. The primary neurons pick up sensory input from the periphery and relay it to the posterior horn of the spinal cord, where they synapse with association neurons. The association neurons, which are not specifically named in the three-neuron sequence, synapse with secondary neurons. Axons from the **secondary neurons** cross to the opposite side of the spinal cord through the anterior portion of the

Table 13.4   Ascending Spinal Pathways—cont'd

Primary Cell Body	Secondary Cell Body	Tertiary Cell Body	Crossover
Dorsal root ganglion	Posterior horn of spinal cord	Thalamus	
			Level at which primary neuron enters cord
			Eight to 10 segments from where primary neuron entered cord; many collaterals
Dorsal root ganglion	Medulla oblongata	Thalamus	Medulla oblongata
Dorsal root ganglion	Posterior horn of spinal cord	Cerebellum	
			Uncrossed
			Some uncrossed; some cross at point of origin and recross in cerebellum
Dorsal root ganglion	Posterior horn of spinal cord	Accessory olivary nucleus	At point of origin; recross to reach cerebellum
Dorsal root ganglion	Posterior horn of spinal cord	Superior colliculus	At point of origin
Dorsal root ganglion	Posterior horn of spinal cord	Reticular formation	Some uncrossed; some cross spinal cord at point of entry

Table 13.5   Descending Spinal Pathways—cont'd

Pathway	Functions Controlled	Origin	Termination	Crossover
Extrapyramidal	Unconscious movements			
Rubrospinal	Movement coordination	Red nucleus	Anterior horn of spinal cord	Midbrain
Vestibulospinal	Posture, balance	Vestibular nucleus	Anterior horn of spinal cord	Uncrossed
Reticulospinal	Posture adjustment, especially during movement	Reticular formation	Anterior horn of spinal cord	Some uncrossed; some cross at level of termination
Tectospinal	Movement of head and neck in response to visual reflexes	Superior colliculus	Cranial nerve nucleus in medulla oblongata and anterior horn of upper levels of spinal cord (lower motor neurons that turn head and neck)	Midbrain

gray and white commissures and enter the spinothalamic tract, where they ascend to the thalamus. As these fibers pass through the brainstem, they are joined by fibers of the **trigeminothalamic tract** (trigeminal nerve, or cranial nerve V; see chapter 14), which carries pain and temperature impulses from the face and teeth. This input is much like that from the spinal nerves in that primary neurons from one side

of the face synapse with secondary neurons, which cross to the opposite side of the brainstem. Collateral branches, especially from this tract, project to the reticular formation, where they stimulate wakefulness and consciousness. The secondary neurons synapse with cell bodies of tertiary neurons in the thalamus. **Tertiary neurons** from the thalamus project to the somatic sensory cortex.

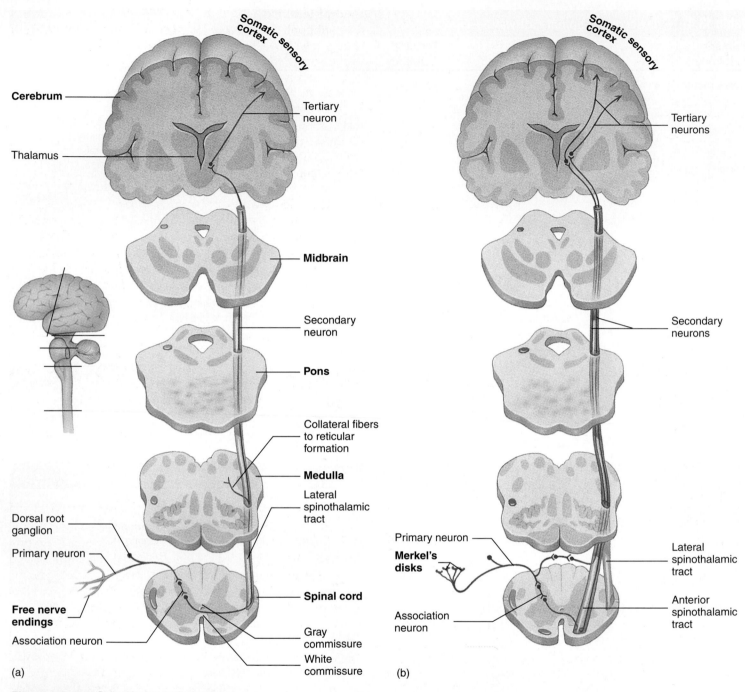

Figure 13.22 Spinothalamic System

(*a*) The lateral spinothalamic tract, which transmits action potentials for pain and temperature. Lines on the inset indicate levels of section. (*b*) The anterior spinothalamic tract, which transmits action potentials for light touch.

Primary neurons contributing to the lateral spinothalamic tract (pain and temperature) ascend or descend only one or two segments before synapsing with secondary neurons, whereas those entering the anterior spinothalamic tract (light touch and pressure) may ascend or descend for 8–10 segments before synapsing. Throughout this distance the primary neurons of the anterior spinothalamic system send out collateral branches that synapse with secondary neurons at several intermediate levels. Thus collateral branches from a number of sensory neurons,

each conducting information from a different patch of skin, may converge on a single secondary neuron in the spinal cord.

7   P R E D I C T

Explain why light touch is very sensitive but not highly discriminative in relation to the exact point of stimulation.

✔ *Answer in Appendix F*

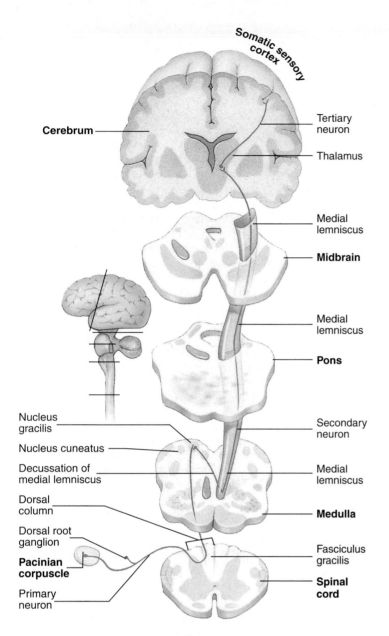

Figure 13.23 Dorsal-Column/Medial-Lemniscal System

The fasciculus gracilis and fasciculus cuneatus convey proprioception and two-point discrimination. Only the fasciculus gracilis pathway is shown. Lines on the inset indicate levels of section.

Lesions on one side of the spinal cord that sever the lateral spinothalamic tract eliminate pain and temperature sensation below that level on the opposite side of the body. Lesions on one side of the spinal cord that sever the anterior spinothalamic tract, however, do not eliminate all of the light touch and pressure sensations below the level of the lesion because of the large number of collateral branches crossing the cord at various levels.

Dorsal-Column/Medial-Lemniscal System

The sensations of two-point discrimination, proprioception, pressure, and vibration are carried by the **dorsal-column/**

medial-lemniscal (lem-nis′kăl; ribbon; the fibers form a thin, ribbonlike pathway through the brainstem) **system** (figure 13.23). This system is named for the dorsal column of the spinal cord (see figure 13.21) and the medial lemniscus, which is the continuation of the dorsal column in the brainstem.

Two-point discrimination (fine touch) is the ability to detect simultaneous stimulation at two points on the skin. The distance between two points that a person can detect as separate points of stimulation differs for various regions of the body (figure 13.24). This sensation is important in evaluating the texture of objects.

Proprioception (prō′prē-ō-sep′shŭn, meaning perception of position) provides information about the precise position and the rate of movement of various body parts, the weight of an object being held in the hand, and the range of movement of a joint. This information is involved in activities such as walking, climbing stairs, shooting a basketball, driving a car, eating, or writing. Receptors for this system are located around joints and in muscles.

---

8   P R E D I C T

A man has constipation, which causes distention and painful cramping in his colon. What kind of pain would he experience (local or diffuse) and where would it be perceived? Explain.

✔ *Answer in Appendix F*

The cell bodies of the afferent spinal neurons of the dorsal-column/medial-lemniscal system are the largest in the dorsal root ganglia, especially ones for discriminative touch. The primary neurons of the dorsal-column/medial-lemniscal system ascend the entire length of the spinal cord without crossing to its opposite side and synapse with secondary neurons located in the medulla oblongata.

In the spinal cord the dorsal-column/medial-lemniscal system can be divided into two separate tracts (see figure 13.21) based on the source of the stimulus. The **fasciculus gracilis** (gras′i-lis, meaning thin) conveys sensations from nerve endings below the midthoracic level, and the **fasciculus cuneatus** (kyū′nē-ā′tŭs, meaning wedge-shaped) conveys impulses from nerve endings above the midthorax. The fasciculus gracilis terminates by synapsing with secondary neurons in the **nucleus gracilis** and with fibers of the posterior spinocerebellar tracts. The fasciculus cuneatus terminates by synapsing with secondary neurons in the **nucleus cuneatus** (see figure 13.5). Both the nucleus gracilis and the nucleus cuneatus are in the medulla oblongata. The secondary neurons then exit the nucleus gracilis and the nucleus cuneatus, cross to the opposite side of the medulla through the decussations of the medial lemniscus, and ascend through the medial lemniscus to terminate in the thalamus. Tertiary neurons from the thalamus project to the somatic sensory cortex.

Clinical Focus Pain

Pain is a sensation characterized by a group of unpleasant perceptual and emotional experiences that trigger autonomic, psychologic, and somatomotor responses. Pain sensation consists of two portions: (1) rapidly conducted action potentials carried by large-diameter, myelinated axons, resulting in sharp, well-localized, pricking, or cutting pain, followed by (2) more slowly propagated action potentials, carried by smaller, less heavily myelinated axons, resulting in diffuse burning or aching pain. Research indicates that pain receptors have very uniform sensitivity that does not change dramatically from one instant to another. Variations in pain sensation result from the differences in integration of action potentials from the pain receptors and the mechanisms by which pain receptors are stimulated.

Although the dorsal-column/medial-lemniscal system contains no pain fibers, tactile and mechanoreceptors are often activated by the same stimuli that affect pain receptors. Action potentials from the tactile receptors help localize the source of pain and monitor changes in the stimuli. Superficial pain is highly localized because of the simultaneous stimulation of pain receptors and mechanoreceptors in the skin. Deep or visceral pain is not highly localized because of the absence of numerous mechanoreceptors in the deeper structures, and it is normally perceived as a diffuse pain.

Dorsal-column/medial-lemniscal system neurons are involved in what is called the gate-control theory of pain control. Primary neurons of the dorsal-column/medial-lemniscal system send out collateral branches that synapse with association neurons in the posterior horn of the spinal cord. The association neurons have an inhibitory effect on the secondary neurons of the lateral spinothalamic tract. Thus pain action potentials traveling through the lateral spinothalamic tract can be suppressed by action potentials that originate in neurons of the dorsal-column/medial-lemniscal system. These neurons may act as a "gate" for pain action potentials transmitted in the lateral spinothalamic tract. Increased activity in the dorsal-column/medial-lemniscal system tends to close the gate, reducing pain action potentials transmitted in the lateral spinothalamic tract.

The gate-control theory may explain the physiologic basis for the following methods that have been used to reduce the intensity of chronic pain: electric stimulation of the dorsal-column/medial-lemniscal neurons, transcutaneous electric stimulation (applying a weak electric stimulus to the skin), acupuncture, massage, and exercise. The frequency of action potentials that are transmitted in the dorsal-column/medial-lemniscal system is increased when the skin is rubbed vigorously and when the limbs are moved and may explain why vigorously rubbing a large area around a source of pricking pain tends to reduce the intensity of the painful sensation. Exercise normally decreases the sensation of pain, and exercise programs are important components in the clinical management of chronic pain not associated with illness. Action potentials initiated by acupuncture procedures may also inhibit the action potentials in neurons that transmit pain action potentials upward in the spinal cord by influencing afferent cells of the posterior horn.

Referred Pain

Referred pain is a painful sensation in a region of the body that is not the source of the pain stimulus. Most commonly, referred pain is sensed in the skin or other superficial structures when internal organs are damaged or inflamed. This sensation usually occurs because both the area to which the pain is referred and the area where the actual damage occurs are innervated by neurons from the same spinal segment.

Many cutaneous afferent neurons and visceral afferent neurons that transmit pain action potentials converge on the same ascending neurons; however, the brain cannot distinguish between the two sources of painful stimuli, and the painful sensation is referred to the most superficial structures innervated by the converging neurons. This referral may occur because the number of receptors is much greater in superficial structures than in deep structures and the brain is more "accustomed" to dealing with superficial stimuli.

Referred pain is clinically useful in diagnosing the actual cause of the painful stimulus. Heart attack victims often feel cutaneous pain radiating from the left shoulder down the arm. Other examples of referred pain are shown in figure A.

Phantom Pain

Phantom pain occurs in people who have had appendages amputated or a structure such as a tooth removed. Frequently these people perceive pain, which can be intense, in the amputated structure as if it were still in place. If a neuron pathway that transmits action potentials is stimulated at any point along that pathway, action potentials are initiated and propagated toward the CNS. Integration results in the perception of pain that is projected to the site of the sensory receptors, even if those sensory receptors are no longer present. A similar phenomenon can be easily demonstrated by bumping the ulnar nerve as it crosses the elbow (the funny bone). A sensation of pain is often felt in the fourth and fifth digits, even though the neurons were stimulated at the elbow.

A factor that may be very important in phantom pain results from the lack of touch, pressure, and proprioceptive impulses from the amputated limb. Those action potentials suppress the transmission of pain action potentials in the pain pathways. When a limb is amputated, the inhibitory effect of sensory information, which is normally transmitted through the dorsal-column/medial-lemniscal system, on the ascending pain pathways is removed. As a consequence, the intensity of phantom pain may be increased. Another factor in phantom pain may be that the brain retains an image of the amputated body part and creates an impression that the part is still there.

Chronic Pain

Although pain is important in warning us of potentially injurious conditions, the pain itself can become a problem. Chronic pain, such as migraine headaches, localized facial pain, or back pain, can be very debilitating and loses its value of providing information about the condition of the body. People suffering from chronic pain often feel helpless and hopeless, and they may become dependent on drugs. The pain can interfere with vocational pursuits, and the victims are often unemployed or even housebound and socially isolated. They are easily frustrated or angered, and they suffer symptoms of major depression. These qualities are associated with what is called chronic pain syndrome. Over 2 million people in the United States at any given time suffer chronic pain sufficient to impair activity. Chronic pain may originate with acute pain associated with an injury or may develop for no apparent reason. How afferent signals are

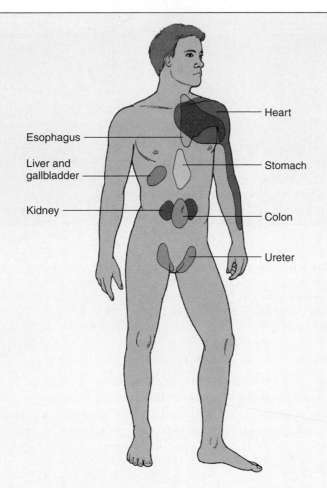

Figure A   Regions of Referred Pain

Esophagus

Liver and gallbladder

Kidney

Heart

Stomach

Colon

Ureter

Sensitization in Chronic Pain

Tissue damage within an area of injury, such as the skin, can cause an increase in the sensitivity of nerve endings in the area of damage, a condition called **peripheral sensitization.** Research has also revealed a novel class of pain receptors which are not activated by traditional noxious stimuli but are recruited only when tissues become inflamed. These receptors, once activated, add to the total barrage of afferent signals to the brain and intensify the sensation of pain. The CNS may also respond to tissue damage by decreasing its threshold and increasing its sensitivity to pain. This condition is called **central sensitization.** Under this condition, neurons in the CNS release the excitatory amino acids, glutamate and aspartate. Central sensitization apparently results from a specific subset of aspartate receptors, which have little function in normal sensation. These receptors are only recruited during repetitive neuron firing, such as when intense pain sensations are experienced. These receptors open Ca^{2+} ion channels, which results in the production of nitric oxide and the maintenance of a hyperexcitable state in the CNS cells. This chronic hyperexcitable state results in persistent, chronic pain states. This information concerning peripheral and central sensitization, and the knowledge that sensitization involves neuronal and chemical receptors not normally involved in sensation, may lead to the discovery of new drugs for treating chronic pain. Rather than searching for new analgesics, which may decrease a broad range of sensations, there is now an opportunity to develop a new class of drugs, the "antihyperalgesics," which may block sensitization without diminishing other sensations, including that to normal pain.

processed in the thalamus and cerebrum may determine if the input is evaluated as only a discomfort, a minor pain, or a severe pain and how much distress is associated with the sensation. The brain actively regulates the amount of pain information that gets through to the level of perception, suppressing much of the input. If this dampening system becomes less functional, pain perception may increase. Other nervous system factors, such as a loss of some sensory modalities from an area, or habituation of pain transmission, which may remain even after the stimulus is removed, may actually intensify otherwise normal pain sensations. The depression, anxiety, and stress associated with chronic pain syndrome can also perpetuate the pain sensations. Treatment often requires a multidisciplinary approach, including such interventions as surgery or psychotherapy. Some sufferers respond well to drug therapy, but some drugs, such as opiates, have a diminishing effect and may become addictive.

Two people, Bill and Mary, were each involved in an accident and each experienced a loss of proprioception, fine touch, and vibration on the left side of the body below the waist. It was determined that Bill had damage to his spinal cord as a result of the accident and that Mary had damage to her brainstem. Explain which side of the spinal cord was damaged in Bill and which side was damaged in Mary.

✔ *Answer in Appendix F*

Spinocerebellar System and Other Tracts

The spinocerebellar tracts (see figure 13.21) carry proprioceptive information to the cerebellum so that information concerning actual movements can be monitored and compared with cerebral information representing intended movements.

Two spinocerebellar tracts extend through the spinal cord: (1) the **posterior spinocerebellar tract** (figure 13.25), which originates in the thoracic and upper lumbar regions and contains uncrossed nerve fibers that enter the cerebellum

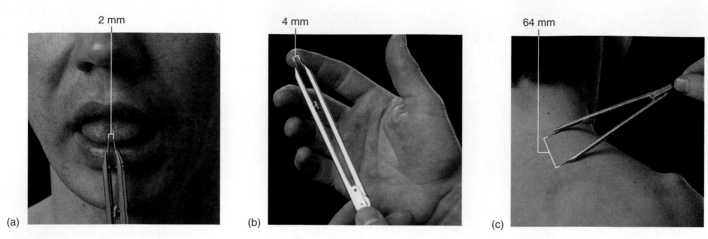

(a)   (b)   (c)

Figure 13.24   Two-Point Discrimination

Two-point discrimination can be demonstrated by touching a person's skin with the two points of a compass. When the two points are close together, the individual perceives only one point. When the points of the compass are opened wider, the person becomes aware of two points. (*a*) Awareness of two-point discrimination occurs when there is at least 2 mm between the points on the tip of the tongue, the most sensitive area of the body. (*b*) Awareness of two-point discrimination occurs when there is at least 4 mm between the points for the fingertips. (*c*) Awareness of two-point discrimination occurs when there is at least 64 mm distance between the points for the back.

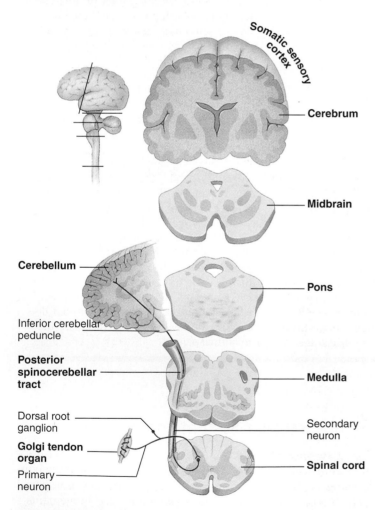

Figure 13.25   Posterior Spinocerebellar Tract

This tract transmits proprioceptive information from the thorax, upper limbs, and upper lumbar region to the cerebellum. Lines on the inset indicate levels of section.

through the inferior cerebellar peduncles; and (2) the **anterior spinocerebellar tract,** which carries information from the lower trunk and lower limbs and contains both crossed and uncrossed nerve fibers that enter the cerebellum through the superior cerebellar peduncle. The crossed fibers recross in the cerebellum. Both spinocerebellar tracts transmit proprioceptive information to the cerebellum from the same side of the body as the cerebellar hemisphere to which they project. Why the anterior spinocerebellar tract crosses twice to accomplish this feat is unknown. Much of the proprioceptive information carried from the legs by the fasciculus gracilis of the dorsal-column/medial-lemniscal system is transferred by synapses in the inferior thorax to the spinocerebellar system and enters the cerebellum as unconscious proprioceptive information. In addition, the spinocerebellar tracts convey no information from the arms to the cerebellum. This input enters the cerebellum through the inferior peduncle from the cuneate nucleus of the dorsal-column/medial-lemniscal system. The dorsal-column/medial-lemniscal system, therefore, is involved not only in conscious awareness of proprioception but also unconscious neuromuscular functions.

10   **P R E D I C T**

Most of the neurons from the fasciculus gracilis synapse in the inferior thorax and enter the spinocerebellar system, whereas most of the neurons from the fasciculus cuneatus synapse in the nucleus cuneatus and then continue to the thalamus and cerebrum. It can therefore be deduced that most of the proprioception from the lower limbs is unconscious and most of the proprioception from the upper limbs is conscious. Explain why this difference in the two sets of limbs is of value.

✔ *Answer in Appendix F*

The **spinoolivary tracts** project to the accessory olivary nucleus and to the cerebellum, where action potentials carried by these tracts contribute to coordination of movement associated primarily with balance. The **spinotectal** (spī′nō-tek′tăl) **tracts** end in the superior colliculi of the midbrain and transmit action potentials involved in reflexively turning the head and eyes toward a point of cutaneous stimulation. The **spinoreticular tracts** transmit action potentials involved in arousing consciousness in the reticular activating system through cutaneous stimulation.

Descending Pathways

Most of the descending pathways are involved in the control of motor functions (see table 13.5). Some descending fibers, however, are part of the sensory system and modulate the transmission of sensory information from the spinal cord to the somatic sensory cortex (see discussion at the end of this section).

The voluntary motor system consists of two primary groups of neurons: upper motor neurons and lower motor neurons. **Upper motor neurons** originate in the cerebral cortex, cerebellum, and brainstem and modulate the activity of the lower motor neurons. Upper motor neuron fibers constitute the descending motor pathways. **Lower motor neurons** are either in the anterior horn of the spinal cord central gray matter or in the cranial nerve nuclei of the brainstem, and the axons of both groups extend through peripheral nerves to skeletal muscles.

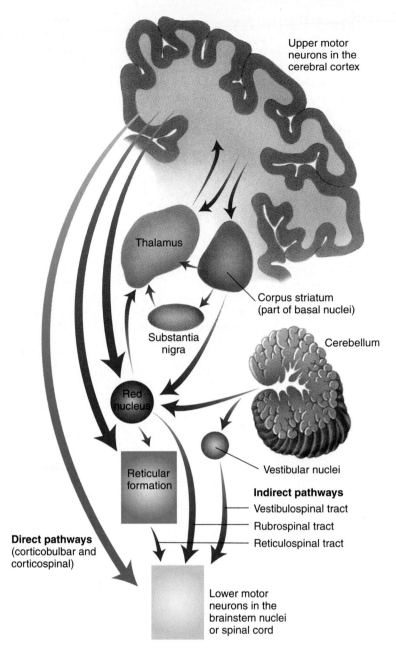

Figure 13.26 Descending Pathways

The direct pathways (corticobulbar and corticospinal) are indicated by the blue arrow. The indirect pathways and their interconnections are indicated by the red arrows.

Clinical Note

Amyotrophic (ă-mī-ō-trō′fik) **lateral sclerosis** (**ALS**), also called Lou Gehrig's disease, usually affects people between the ages of 40 and 70. It begins with weakness and clumsiness and progresses within 2–5 years to loss of muscle control. The disease selectively destroys both upper and lower motor neurons. About 10% of the cases of ALS are inherited. The inherited form of ALS apparently results from a mutation in the DNA coding for the enzyme **superoxide dismutase** (**SOD**) and is located on chromosome 21. SOD is involved in eliminating free radicals from the body. Free radicals are molecules with an odd number of electrons in their outer shells, which makes them highly reactive. They can strip electrons from proteins, lipids, or nucleic acids, destroying their functions and resulting in cell dysfunction or death. Free radical damage has been implicated in ALS, arteriosclerosis, arthritis, cancer, and aging. Superoxide is one of the most important free radicals, and it forms as the result of oxygen reacting with other free radicals. Although oxygen is critical for aerobic metabolism, it is also dangerous to tissues. Superoxide is very toxic. SOD catalyzes the conversion of superoxide to hydrogen peroxide, which is then converted by catalase to oxygen and water. Apparently, if SOD is defective, superoxide is not degraded and can destroy cells. Motor neurons appear to be particularly sensitive to superoxide attack.

The descending motor fibers are divided into two groups: direct pathways and indirect pathways (figure 13.26). The **direct pathways,** also called the **pyramidal** (pi-ram′i-dal) **system,** are involved in the maintenance of muscle tone and in controlling the speed and precision of skilled movements, primarily fine movements involved in functions such as dexterity. The **indirect pathways,** sometimes called the **extrapyramidal system,** are involved in less precise control of motor functions, especially ones associated with overall body coordination and cerebellar function. The indirect pathways are phylogenetically older and control more "primitive" trunk and proximal limb movements. The direct pathways, which exist only in mammals, may be thought of as overlying the indirect pathways and are more involved in finely controlled movements of the face and distal limbs.

Direct Pathways

Direct pathways are so named because upper motor neurons in the cerebral cortex, whose axons form these pathways, synapse directly with lower motor neurons in the brainstem or spinal cord. They are also called the pyramidal system because the fibers of these pathways primarily pass through the medullary **pyramids** (see figure 13.5*a*). They include groups of nerve fibers arrayed into two tracts: the **corticospinal tract,** which is involved in direct cortical control of movements below the head, and the **corticobulbar tract,** which is involved in direct cortical control of movements in the head and neck.

Axons constituting the corticospinal tracts originate from neuron cell bodies in the primary motor and premotor areas of the frontal lobes and from the somatic sensory part of the parietal lobe. They descend through the **internal capsule,** which is the major entrance and exit pathway of the cerebrum; the cerebral peduncles of the midbrain; and the pyramids of the medulla oblongata. At the inferior end of the medulla 75%–85% of the corticospinal fibers cross to the opposite side of the CNS through the **pyramidal decussation,** which is visible on the anterior surface of the inferior medulla (see figure 13.5*a*). The crossed fibers descend in the **lateral corticospinal tract** of the spinal cord (figure 13.27). The re-

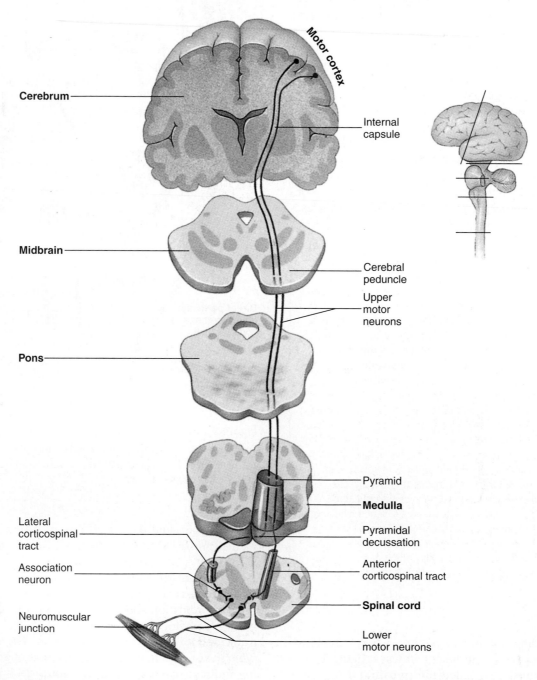

Figure 13.27 Direct Pathways

Lateral corticospinal tract, which is responsible for movement below the head. Lines on the inset indicate levels of section.

maining 15%–25% descend uncrossed in the **anterior corti-cospinal tract** and decussate near the level where they synapse with lower motor neurons. The anterior corticospinal tracts supply the neck and upper limbs, and the lateral corticospinal tracts supply all levels of the body.

Most of the corticospinal fibers synapse with association neurons in the lateral portions of the spinal cord central gray matter. The association neurons, in turn, synapse with the lower motor neurons of the anterior horn that innervate primarily distal limb muscles.

Damage to the corticospinal tracts results in reduced muscle tone, clumsiness, and weakness but not in complete paralysis, even if the damage is bilateral. Experiments with monkeys have demonstrated that bilateral sectioning of the medullary pyramids results in (1) loss of contact-related activities such as tactile placing of the foot and grasping; (2) defective fine movements; and (3) hypotonia (reduced tone). On the basis of these and other experimental data, it has been concluded that the corticospinal system is superimposed over the older indirect pathways and has many parallel functions. It is proposed that the main function of the direct pathways is to add speed and agility to conscious movements, especially of the hands, and to provide a high degree of fine motor control such as in movements of individual fingers. Most spinal cord lesions, however, affect both the direct and indirect pathways and result in complete paralysis.

The **corticobulbar tracts** are analogous to the corticospinal tracts. The former innervate the head, and the latter innervate the rest of the body. Cells that contribute to the corticobulbar tracts are in regions of the cortex similar to those of the corticospinal tracts, except that they are more laterally and inferiorly located on the cortex. Corticobulbar tracts follow the same basic route as the corticospinal system down to the level of the brainstem. At that point most corticobulbar fibers terminate in the reticular formation near the cranial nerve nuclei. Association neurons from the reticular formation then enter the **cranial nerve nuclei,** where they synapse with lower motor neurons. These nuclei give rise to nerves that control eye and tongue movements; mastication; facial expression; and palatine, pharyngeal, and laryngeal movements.

Indirect Pathways

The indirect pathways (figure 13.28) include all the upper motor neurons whose axons synapse in some intermediate nucleus rather than directly with lower motor neurons. Axons from these nuclei do not pass through the pyramids or through the corticobulbar tracts and, therefore, are sometimes called extrapyramidal. The major tracts are the rubrospinal, vestibulospinal, and reticulospinal tracts. There are many interconnections and feedback loops in this system.

The **rubrospinal tract** begins in the red nucleus, decussates in the midbrain, and descends in the lateral funiculus. The red nucleus receives input from both the motor cortex and the cerebellum. Lesions in the red nucleus result in intention, or action, tremors similar to those seen in cerebellar lesions (see the Clinical Focus on Dyskinesias). Its function therefore is related closely to cerebellar function. Damage to

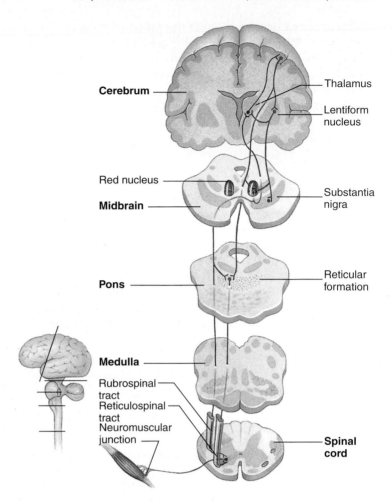

Figure 13.28 Indirect Pathways

Examples of indirect pathways: rubrospinal and reticulospinal tracts. Lines on the inset indicate levels of section.

the rubrospinal tract impairs distal arm and hand movements but does not greatly affect general body movements. The rubrospinal tract is the one indirect tract that is very closely related to the direct, corticospinal tract. It is present only in mammals, it terminates in the lateral portion of the spinal cord central gray matter with the corticospinal tract, and it transmits action potentials involved in the cerebellar comparator function and, as a result, fine motor control of distal limb muscles.

The **vestibulospinal tracts** (see figure 13.21) originate in the vestibular nuclei, descend in the anterior funiculus, and synapse with lower motor neurons in the ventromedial portion of the spinal cord central gray matter. Their fibers preferentially influence neurons innervating extensor muscles in the trunk and proximal limbs and are involved primarily in the maintenance of upright posture. The vestibular nuclei receive major input from the vestibular nerve (see chapter 15) and the cerebellum.

The **reticulospinal tract** originates in the reticular formation of the pons and medulla oblongata, descends in the anterior portion of the lateral funiculus, and synapses with lower motor neurons in the ventromedial portion of the spinal cord central gray matter. The function of this tract involves the maintenance of posture through the action of trunk and proximal limb muscles during certain movements. For example, when a person who is standing lifts one foot off the ground, the weight of the body is shifted over to the other limb. The reticulospinal tract apparently enhances the functions of the crossed extensor reflex during this type of movement so that balance is maintained.

Another major portion of the indirect pathways involves the basal nuclei (see figure 13.26). The basal nuclei also have a number of connections with other indirect pathways by which they modulate motor functions.

Descending Pathways Modulating Sensation

The corticospinal and other descending pathways send collateral axons to the thalamus, reticular formation, trigeminal nuclei, and spinal cord; and the axons function to modulate pain impulses in ascending tracts. Through this route the cerebral cortex or other brain regions may reduce the conscious perception of sensation. The descending pathways reduce the transmission of pain impulses by secreting **endorphins** (natural analgesics).

Clinical Note

About 10,000 new cases of **spinal cord injury** occur each year in the United States. Automobile and motorcycle accidents are the leading cause, followed by gunshot wounds, falls, and swimming accidents. Spinal cord injury is classified according to the vertebral level at which the injury occurred, whether the entire cord is damaged at that level or only a portion of the cord, and the mechanism of injury. Most spinal cord injuries occur in the cervical region or at the thoracolumbar junction and are incomplete. The mechanisms include concussion, contusion, or laceration and involve excessive flexion, extension, rotation, or compression of the vertebral column. The majority of spinal cord injuries are acute contusions of the cord due to bone or disk displacement into the cord and involve a combination of excessive directional movements, such as simultaneous flexion and compression.

At the time of spinal cord injury, two types of tissue damage occur: (1) primary mechanical damage, and (2) secondary tissue damage extending into a much larger region of the cord than the primary damage. This region of secondary tissue damage extends the domain of the injury and is the focal point of current research in spinal cord injury. The only treatment for primary damage is prevention, such as wearing seat belts when riding in automobiles and not diving in shallow water. Once an accident occurs, however, little can be done at present about the primary damage. On the other hand, it is now known that much of the secondary damage can be prevented or reversed. Secondary spinal cord damage, which begins within minutes of the primary damage, is caused by ischemia, edema, ion imbalances, the release of "excitotoxins" such as glutamate, and inflammatory cell invasion.

Until the 1950s, spinal cord injuries were often ultimately fatal. Now, with quick treatment, directed at the mechanisms of secondary tissue damage, much of the total damage to the spinal cord can be prevented. Treatment of the damaged spinal cord with large doses of methylprednisolone, a synthetic steroid, within 8 h of the injury, can dramatically reduce the secondary damage to the cord. Current treatment includes anatomic realignment and stabilization of the vertebral column, decompression of the spinal cord, and administration of methylprednisolone. Rehabilitation is based on retraining the patient to use whatever residual connections exist across the site of damage.

It had long been thought that the spinal cord is incapable of regeneration following severe damage. It is now known that following injury, most neurons of the adult spinal cord survive and begin to regenerate, growing about 1 mm into the site of damage, but then they regress to an inactive, atrophic state. In addition, fetuses and newborns exhibit considerable regenerative ability and functional improvement. The major block to adult spinal cord regeneration is the formation of a scar, consisting mainly of astrocytes, at the site of injury. Myelin in the scar is apparently the primary inhibitor of regeneration. Implantation of peripheral nerves, Schwann cells, or fetal CNS tissue can bridge the scar and stimulate some regeneration. Certain growth factors can also stimulate some regeneration. Current research continues to look for the right combination of chemicals and other factors to stimulate regeneration of the spinal cord following injury.

Meninges and Cerebrospinal Fluid

Meninges

Three connective tissue layers, the **meninges** (mĕ-nin′jēz), surround and protect the brain and spinal cord (figure 13.29). The most superficial and thickest layer is the **dura mater** (dū′ră mā′ter, meaning tough mother). Three dural folds, the **falx cerebri** (falks se-rē′brī, meaning sickle-shaped), the **tentorium cerebelli** (ten-tō′rē-ŭm ser′ĕ-bel′ī, meaning tent), and the **falx cerebelli** extend into the major brain fissures. The falx cerebri is located between the two central hemispheres in the longitudinal fissure, the tentorium cerebelli is between the cerebrum and cerebellum, and the falx cerebelli lies between the two cerebellar hemispheres.

The dura mater surrounding the brain is tightly attached to and continuous with the periosteum of the cranial vault, forming a single functional layer, whereas the dura mater of the spinal cord is separated from the periosteum of the vertebral canal by the **epidural space.** This is a true space around the spinal cord that contains spinal nerves, blood vessels, areolar connective tissue, and fat. **Epidural anesthesia** of the spinal nerves is induced by injecting anesthetics into this space.

The two layers of the dura mater around the brain are fused but separate in several places, primarily at the bases of the three dural folds, to form venous **dural sinuses.** The dural sinuses collect most of the blood that returns from the brain, as well as cerebrospinal fluid (CSF) from around the brain (see next section). The sinuses then empty into the veins that exit the skull (see chapter 21).

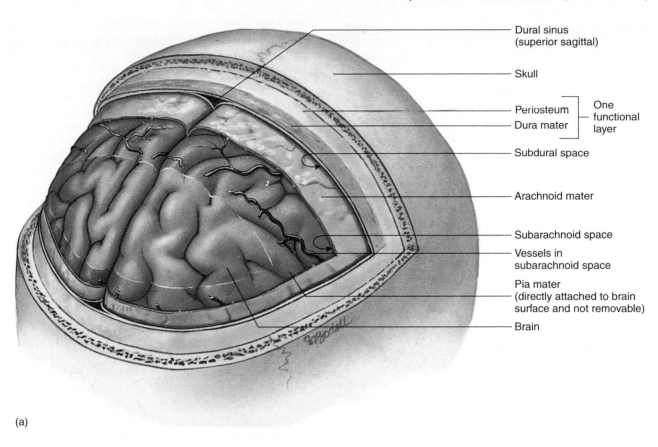

(a)

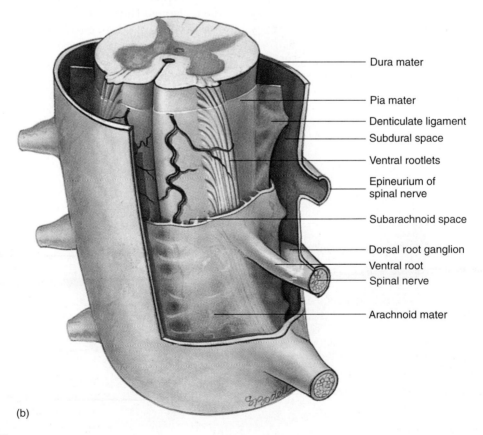

(b)

Figure 13.29   Meninges

(*a*) Meningeal coverings of the brain. (*b*) Meningeal coverings of the spinal cord.

The next meningeal layer is a very thin, wispy **arachnoid** (ă-rak'noyd, meaning spiderlike; that is, cobwebs) **mater,** or arachnoid layer. The space between this layer and the dura mater is the **subdural space** and contains only a very small amount of serous fluid. The third meningeal layer, the **pia** (pē'ă, meaning affectionate) **mater** is bound very tightly to the surface of the brain and spinal cord. Between the arachnoid mater and the pia mater is the **subarachnoid space,** which contains weblike strands of the arachnoid mater and blood vessels and is filled with CSF.

Clinical Note

Damage to the venous dural sinuses can cause bleeding into the subdural space, resulting in a **subdural hematoma,** which can cause pressure on the brain.

The dura mater and dural folds help hold the brain in place within the skull and keep it from moving around too freely. The spinal cord is held in place within the vertebral canal by a series of connective tissue strands connecting the pia mater to the dura mater, which cause the arachnoid mater to form points between each of the nerves. Because the points create a "toothed" appearance, these attachments are called **denticulate** (den-tik'yū-lāt) **ligaments** (figure 13.29b).

Clinical Note

Several clinical procedures involve the insertion of a needle into the subarachnoid space inferior to the level of L2. Because the spinal cord extends only to approximately level L2 of the vertebral column and the meninges extend to the end of the vertebral column, the needle does not damage the spinal cord. The nerves of the cauda equina are pushed aside quite easily by the needle during these procedures and normally are not damaged. In **spinal anesthesia,** or spinal block, drugs that block action potential transmission are introduced into the subarachnoid space, preventing pain sensations from the lower half of the body. Women are often given epidural anesthesia during childbirth. A **spinal tap** is the removal of CSF from the subarachnoid space. A spinal tap may be performed to examine the CSF for infectious agents (meningitis), for the presence of blood (hemorrhage), or for CSF pressure. A radiopaque substance may also be injected into this area, and a **myelogram** (radiograph of the spinal cord) may be taken to visualize spinal cord defects or damage.

Ventricles

As already stated, the CNS is formed as a hollow tube that may be quite reduced in some areas of the adult CNS and expanded in other areas (see figure 13.3c). It is lined with a single layer of epithelial cells called **ependymal cells** (ep-en'di-măl; see chapter 12). Each cerebral hemisphere contains a relatively large cavity, the **lateral ventricle** (figure 13.30). The lateral ventricles are separated from each other by thin **septa pellucida** (sep'tă pe-lū'sid-ă,; sing. septum pellucidum, meaning translucent walls), which lie in the midline just inferior to the corpus callosum and usually are fused with each other. A smaller midline cavity, the **third ventricle,** is located in the center of the diencephalon between the two halves of the thalamus. The two lateral ventricles communicate with the third ventricle through two **interventricular foramina** (foramina of Monro). The lateral ventricles can be thought of as the first and second ventricles in the numbering scheme, but they are not designated as such. The **fourth ventricle** is

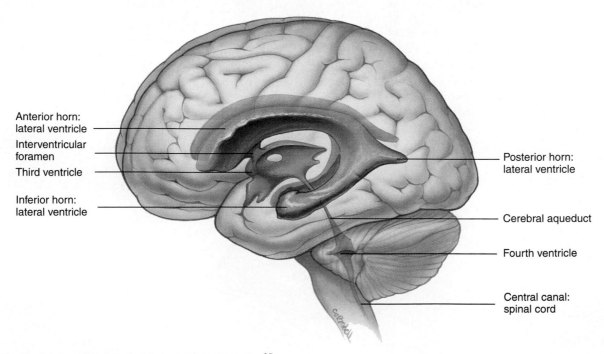

Anterior horn:
lateral ventricle

Interventricular foramen

Third ventricle

Inferior horn:
lateral ventricle

Posterior horn:
lateral ventricle

Cerebral aqueduct

Fourth ventricle

Central canal:
spinal cord

Figure 13.30 Ventricles of the Brain Viewed from the Left

in the inferior part of the pontine region and the superior region of the medulla oblongata at the base of the cerebellum. The third ventricle communicates with the fourth ventricle through a narrow canal, the **cerebral aqueduct** (aqueduct of Sylvius), which passes through the midbrain. The fourth ventricle is continuous with the central canal of the spinal cord, which extends nearly the full length of the cord. The fourth ventricle is also continuous with the subarachnoid space through foramina in its walls and roof.

Cerebrospinal Fluid

Cerebrospinal (ser-ĕ′brō-spī′năl; sĕ-rē′brō-spī′năl) **fluid** is a fluid similar to plasma and interstitial fluid that bathes the brain and the spinal cord and provides a protective cushion around the CNS. It also provides some nutrients to CNS tissues. About 80%–90% of the CSF is produced by specialized ependymal cells within the lateral ventricles, with the remainder produced by similar cells in the third and fourth ventricles. These specialized ependymal cells, their support tissue, and the associated blood vessels together are called **choroid** (kō′royd, meaning lacy) **plexuses** (plek′sŭs-ēz) (figure 13.31). The choroid plexuses are formed by invaginations of the vascular pia mater into the ventricles, producing a vascular connective tissue core covered by ependymal cells.

Clinical Note

In skull fractures in which the meninges are torn, CSF may leak from the nose if the fracture is in the frontal area, or from the ear if the fracture is in the temporal area. Leakage of CSF indicates serious damage to the head and presents a risk of meningitis, because bacteria may pass from the nose or ear through the tear and into the meninges.

Production of the CSF by the choroid plexuses is not fully understood. Some portions of the blood plasma cross the membranes of the plexus by both simple and facilitated diffusion, whereas other portions require active transport.

CSF fills the ventricles, the subarachnoid space of the brain and spinal cord, and the central canal of the spinal cord. Approximately 23 mL of fluid fills the ventricles, and 117 mL fills the subarachnoid space. The route taken by the CSF from its origin in the choroid plexuses to its return to the circulation is depicted in figure 13.31. The flow rate of CSF from its origin to the point at which it enters the bloodstream is about 0.4 mL/min. CSF passes from the lateral ventricles through the interventricular foramina into the third ventricle and then through the cerebral aqueduct into the fourth ventricle. It can exit the interior of the brain only through the wall of the fourth ventricle. One **median foramen** (foramen of Magendie), which opens through the roof of the fourth ventricle, and two **lateral foramina** (foramina of Luschka) allow the CSF to pass from the fourth ventricle to the subarachnoid space. Masses of arachnoid tissue, **arachnoid granulations,** penetrate into the superior sagittal sinus, and CSF passes into the dural sinuses through these granulations. The sinuses are blood-filled; thus it is within these dural sinuses that the CSF reenters the bloodstream. From the venous dural sinuses the blood flows to veins of the general circulation.

Clinical Note

If the foramina of the fourth ventricle or the cerebral aqueduct are blocked, CSF can accumulate within the ventricles. This condition is called internal hydrocephalus and is the result of increased CSF pressure. The production of CSF continues, even when the passages that normally allow it to exit the brain are blocked. Consequently, fluid builds inside the brain, causing pressure that compresses the nervous tissue and dilates the ventricles. Compression of the nervous tissue usually results in irreversible brain damage. If the skull bones are not completely ossified when the hydrocephalus occurs, the pressure may also severely enlarge the head. The cerebral aqueduct may be blocked at the time of birth or may become blocked later in life because of a tumor growing in the brainstem.

A subarachnoid hemorrhage may block the return of CSF to the circulation. Hydrocephalus can be successfully treated by placing a drainage tube (shunt) between the brain ventricles and one of the body cavities to eliminate the high internal pressures. Infection may be introduced into the brain through these shunts, however, and the shunts must be replaced as the person grows. If CSF accumulates in the subarachnoid space, the condition is called external hydrocephalus. In this condition, pressure is applied to the brain externally, compressing neural tissues and causing brain damage.

Clinical Focus   General CNS Disorders

Infections

Encephalitis (en-sef-ă-lī'tis) is an inflammation of the brain most often caused by a virus and less often by bacteria or other agents. A large variety of symptoms may result, including fever, paralysis, coma, or even death.

Myelitis (mī-e-lī'tis) is an inflammation of the spinal cord with causes and symptoms similar to encephalitis.

Meningitis (men-in-jī'tis) is an inflammation of the meninges. It may be virally induced but is more often bacterial. Symptoms usually include stiffness in the neck, headache, and fever. Pus may accumulate in the subarachnoid space, block CSF flow, and result in hydrocephalus. In severe cases meningitis may also cause paralysis, coma, or death.

Reye's syndrome may develop in children following a viral infection, especially influenza or chicken pox. The use of aspirin in cases of viral infection has been linked to development of the syndrome in the United States. There may also be a predisposing disorder in fat metabolism in some cases. In children affected by the syndrome, the brain cells swell, and the liver and kidneys accumulate fat. Symptoms include vomiting, lethargy, and loss of consciousness and may progress to coma and death or to permanent brain damage.

Rabies is a viral disease transmitted by the bite of an infected mammal. The rabies virus infects the brain, salivary glands (through which it is transmitted), muscles, and connective tissue. When the patient attempts to swallow, the effort can produce pharyngeal muscle spasms; sometimes even the thought of swallowing water or the sight of water can induce pharyngeal spasms. Thus the term hydrophobia, fear of water, is applied to the disease. The virus also infects the brain and results in abnormal excitability, aggression, and, in later stages, paralysis and death.

Tabes dorsalis (tā'bēs dōr-sā'lis) is a progressive disorder occurring as a result of untreated syphilis. Tabes means a wasting away, and dorsalis refers to a degeneration of the dorsal roots and dorsal columns of the spinal cord. The symptoms include ataxia, resulting from lack of proprioceptive input; anesthesia, resulting from dorsal root damage; and eventually paralysis as the infection spreads.

Multiple sclerosis (MS), although of unknown cause, possibly involves an autoimmune response to a viral infection. It results in localized brain lesions and demyelination of neurons in the brain and spinal cord, in which the myelin sheaths become sclerotic, or hard—thus the name, resulting in poor conduction of action potentials. Its symptomatic periods are separated by periods of apparent remission. With each recurrence, however, many neurons are permanently damaged so that the progressive symptoms of the disease include exaggerated reflexes, tremor, nystagmus, and speech defects.

Other Disorders

Tumors of the brain develop from neuroglial cells. Symptoms vary widely, depending on the location of the tumor but may include headaches, neuralgia (pain along the distribution of a peripheral nerve), paralysis, seizures, coma, and death. Meningiomas (mě-nin'jē-ō'măz), tumors of the meninges, account for 25% of all primary intracranial tumors.

Stroke is a term meaning a blow or sudden attack, suggesting the speed with which this type of defect can occur. It is also referred to clinically as a cerebrovascular accident (CVA) and is caused by hemorrhage, thrombosis, embolism, or vasospasm of the cerebral blood vessels, which result in an infarct, a local area of neuronal cell death caused by a lack of blood supply. Symptoms depend on the location but include anesthesia or paralysis on the side of the body opposite the cerebral infarct. Each year 75,000 Americans suffer strokes. Cigarette smokers are 2.5 times more likely to suffer strokes than are nonsmokers. A daily dose of aspirin may reduce a person's risk of stroke by 50%–80% through its ability to interfere with blood clotting.

An aneurysm (an'yū-rizm) is a dilation, or ballooning, of an artery. The arteries around the brain are common sites for aneurysms, and hypertension can cause one of these "balloons" to burst or leak, causing a hemorrhage around the brain. With hemorrhaging, blood may enter the epidural space (epidural hematoma), subdural space (subdural hematoma), subarachnoid space, or the brain tissue. Blood in the subdural or subarachnoid space can apply pressure to the brain, causing damage to brain tissue. Blood is toxic to brain tissue, so that blood entering the brain can directly damage brain tissue.

Cerebral compression may occur as the result of hematomas, hydrocephalus, tumors, or edema of the brain, which can occur as the result of a severe blow to the head. The intracranial pressure increases, which may directly damage brain tissue. The cerebellum may compress the fourth ventricle, blocking the foramina, and causing internal hydrocephalus, which further increases intracranial pressure. The greatest problem comes from compression of the brainstem. Compression of the midbrain can kink the oculomotor nerves, resulting in dilation of the pupils with no light response. Compression of the medulla oblongata may disrupt cardiovascular and respiratory centers, which can cause death. Compression of any part of the CNS that results in ischemia for as little as 3–5 min can result in local neuronal cell death. This is a major problem in spinal cord injuries.

Syringomyelia (sĭ-ring'gō-mī-ē'lē-ă) is a degenerative cavitation of the central canal of the spinal cord, often caused by a cord tumor. Symptoms include neuralgia, paresthesia (increased sensitivity to pain), specific loss of pain and temperature sensation, and paresis. This defect is unusual in that it occurs in a distinct band that includes both sides of the body because commissural tracts are destroyed.

Alzheimer's disease is a severe type of mental deterioration, or dementia, usually affecting older people, but occasionally affecting people younger than 60. It accounts for half of all dementias; the other half result from drug and alcohol abuse, infections, or CVAs. Alzheimer's disease is estimated to affect 10% of all people older than 65 and nearly half of those older than 85.

Alzheimer's disease involves a general decrease in brain size resulting from loss of neurons in the cerebral cortex. The gyri become narrower, and the sulci widen. The frontal lobes and specific regions of the temporal lobes are affected most severely. Symptoms include general intellectual deficiency, memory loss, short attention span, moodiness, disorientation, and irritability.

Amyloid plaques and neurofibrillary tangles, both containing aluminum accumulations, form in the cortex of patients with Alzheimer's disease. Amyloid (am'i-loyd) plaques are localized axonal enlargements of degenerating nerve fibers, containing large amounts of β-amyloid protein, and neurofibrillary tangles, which are filaments inside the cell bodies of the dead or dying neurons.

There is some evidence that Alzheimer's disease may have some characteristics of a chronic inflammatory disease, similar to arthritis, and anti-inflammatory drug therapy has had some affect in slowing its progress. Estrogen treatment may decrease or postpone symptoms in women.

The gene for β-amyloid protein has been mapped to chromosome 21; however, it is thought that only the rare, inherited, early-onset (beginning before age 60) form of Alzheimer's maps to chromosome 21. The more common, late-onset form (beginning after age 65), which makes up more than three-fourths of all cases, maps to chromosome 19. It is noteworthy that people with Down's syndrome, or trisomy 21, which means that a person has three copies of chromosome 21, exhibit the cortical and other changes associated with Alzheimer's disease.

Another protein, **apolipoprotein E** (ap′ō-lip-ō-prō′tēn; **apo E**), which binds to β-amyloid protein and is known to transport cholesterol in the blood, has also been associated with Alzheimer's disease. The protein has been found in the plaques and tangles and has been mapped to the same region of chromosome 19 as the late-onset form of Alzheimer's. People with two copies of the apo E-IV gene are eight times more likely to develop the disease than people with no copies of the defective gene. Apo E-IV apparently binds to β-amyloid more rapidly and more tightly than does apo E-III, which is the normal form of the protein.

Apo E may also be involved with regulating phosphorylation of another protein, called τ (tau), which, in turn, is involved in microtubule formation inside neurons. If τ is overphosphorylated, microtubules are not properly constructed, and the τ proteins intertwine to form neurofibrillary tangles. It has been demonstrated that apo E-III interacts with τ but that apo E-IV does not. It may be that the less stable microtubules, formed with a decreased τ involvement, begin to eventually break down, resulting in neuronal dysfunction. The neurofibrillary tangles of τ proteins may also clog up the cell, further decreasing cell function.

Tay-Sachs disease is a hereditary disorder of infants involving abnormal sphingolipid (lipids with long base chains) metabolism that results in severe brain dysfunction. Symptoms include paralysis, blindness, and death, usually before age 5.

Chronic mercury poisoning can cause brain disorders such as intention tremor, exaggerated reflexes, and emotional instability.

Lead poisoning is a serious problem, particularly among urban children. Lead is taken into the body from contaminated air, food, and water. Flaking lead paint in older houses and soil contamination can be major sources of lead poisoning in children. Lead usually accumulates slowly in the body until toxic levels are reached.

Brain damage caused by lead poisoning includes edema, demyelination, and cortical neuron necrosis with astrocyte proliferation. This damage appears to be permanent and can result in reduced intelligence, learning disabilities, poor psychomotor development, and blindness. In severe cases, psychoses, seizures, coma, or death may occur. Adults exhibit more mild PNS symptoms, including demyelination with decreased neuromuscular function.

Epilepsy is a group of brain disorders that have seizure episodes in common. The seizure, a sudden massive neuronal discharge, can be either partial or complete, depending on the amount of brain involved and whether or not consciousness is impaired. Normally there is a balance between excitation and inhibition in the brain. When this balance is disrupted by increased excitation or decreased inhibition, a seizure may result. The neuronal discharges may stimulate muscles innervated by the neurons involved, resulting in involuntary muscle contractions, or convulsions.

Headaches have a variety of causes that can be grouped into two basic classes: extracranial and intracranial. Extracranial headaches can be caused by inflammation of the sinuses, dental irritations, temperomandibular joint disorders, ophthalmologic disorders, or tension in the muscles moving the head and neck. Intracranial headaches may result from inflammation of the brain or meninges, vascular problems, mechanical damage, or tumors.

Tension headaches are extracranial muscle tension, stress headaches, consisting of a dull, steady pain in the forehead, temples, neck, or throughout the head. Tension headaches are associated with stress, fatigue, and posture.

Migraine headaches (migraine means half a skull) occur in only one side of the head and appear to involve the abnormal dilation and constriction of blood vessels. They often start with distorted vision, shooting spots, and blind spots. Migraines consist of severe throbbing, pulsating pain. About 80% of migraine sufferers have a family history of the disorder, and women are affected four times more often than men. Those suffering migraines are usually women younger than 35. The severity and frequency usually decrease with age.

A **concussion** is a blow to the head producing momentary loss of consciousness without immediate detectable damage to the brain. Often there are no more problems after the person regains consciousness; however, in some cases, **postconcussion syndrome** may occur a short time after the injury. The syndrome includes increased muscle tension or migraine headaches; reduced alcohol tolerance; difficulty in learning new things; reduction in creativity; and motivation, fatigue, and personality changes. The symptoms may be gone in a month or may persist for as much as a year. In some cases, postconcussion syndrome may be the result of a slowly occurring subdural hematoma that may be missed by an early examination. The blood may accumulate from small leaks in the dural sinuses.

Alexia (ă-lek′sē-ă), loss of the ability to read, may result from a lesion in the visual association cortex. **Dyslexia** (dĭs-lek′sē-ă) is a defect in which an individual's reading level is below that expected on the basis of his overall intelligence. Most people with dyslexia have normal or above-normal intelligence quotients. The term means reading deficiency and is also called partial alexia. It is three times more common in males than females. As many as 10% of American males suffer from the disorder. The symptoms vary considerably from person to person and include transposition of letters in a word, confusion between the letters "b" and "d," and lack of orientation in three-dimensional space. The brains of dyslexics are physically different from other brains, having abnormal cellular arrangements, including cortical disorganization and the appearance of bits of gray matter in medullary areas. Dyslexia apparently results from abnormal brain development.

Children with **attention deficit disorder (ADD)** are easily distractible, have short attention spans, and may shift from one uncompleted task to another. Children with **attention deficit/hyperactivity disorder (ADHD)** exhibit the characteristics of ADD, but they are also fidgety, have difficulty remaining seated and waiting their turn, engage in excessive talking, and commonly interrupt others. About 3% of all children exhibit ADHD, more boys than girls. Symptoms usually occur before age 7. The neurologic basis of both ADD and ADHD is as yet unknown.

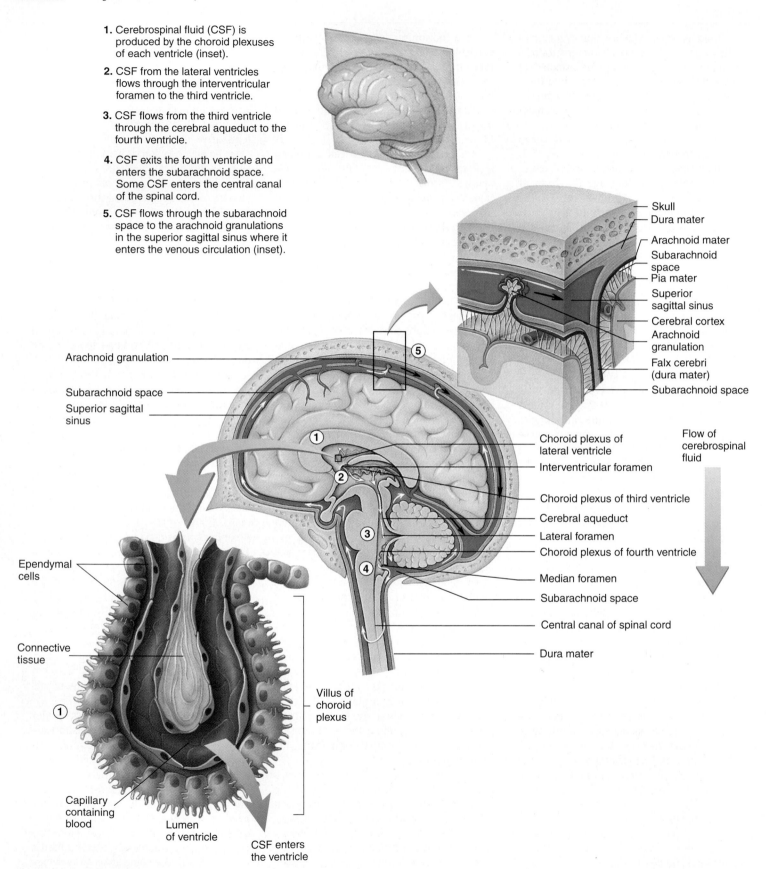

1. Cerebrospinal fluid (CSF) is produced by the choroid plexuses of each ventricle (inset).

2. CSF from the lateral ventricles flows through the interventricular foramen to the third ventricle.

3. CSF flows from the third ventricle through the cerebral aqueduct to the fourth ventricle.

4. CSF exits the fourth ventricle and enters the subarachnoid space. Some CSF enters the central canal of the spinal cord.

5. CSF flows through the subarachnoid space to the arachnoid granulations in the superior sagittal sinus where it enters the venous circulation (inset).

Skull
Dura mater
Arachnoid mater
Subarachnoid space
Pia mater
Superior sagittal sinus
Cerebral cortex
Arachnoid granulation
Falx cerebri (dura mater)
Subarachnoid space

Arachnoid granulation
Subarachnoid space
Superior sagittal sinus

Choroid plexus of lateral ventricle
Interventricular foramen
Choroid plexus of third ventricle
Cerebral aqueduct
Lateral foramen
Choroid plexus of fourth ventricle
Median foramen
Subarachnoid space
Central canal of spinal cord
Dura mater

Flow of cerebrospinal fluid

Ependymal cells
Connective tissue
Capillary containing blood
Lumen of ventricle
Villus of choroid plexus
CSF enters the ventricle

Figure 13.31 Cerebrospinal Fluid (CSF) Circulation

The white arrows represent the route of the CSF. The black arrows represent the route of blood flow.

Systems Pathology

stroke

▌STROKE

Mr. S, who is approaching middle age, is somewhat over-weight, and has high blood pressure, was seated on the edge of his couch, at least most of the time, when he was not jump-ing to his feet and shouting at the referees for an obviously bad call. He was surrounded by empty pizza boxes, bowls of chips and salsa, empty beer cans, and full ashtrays (figure B). As he cheered on his favorite team in a hotly contested big game, which they would be winning easily if it weren't for the lousy officiating, he noticed that he felt drowsy and that the television screen seemed blurry. He began to feel dizzy. As he tried to stand up, he suddenly vomited and collapsed to the floor, unconscious.

Mr. S was rushed to the local hospital where the follow-ing signs and symptoms were observed. He exhibited weak-ness in his limbs, especially on the right, and ataxia (inability to walk). He had loss of pain and temperature sensation in his right lower limb and the left side of his face. The dizziness persisted and he appeared disoriented and lacked attentive-ness. He also exhibited dysphagia (the inability to swallow) and hoarseness. He had nystagmus (rhythmic oscillation of the eyes). His pupils were slightly dilated, his respiration was short and shallow, and his pulse rate and blood pressure were elevated.

BACKGROUND INFORMATION:

Mr. S suffered a "stroke," also referred to as a cerebrovascular accident (CVA). The term **stroke** describes a heterogeneous group of conditions involving death of brain tissue resulting from disruption of its vascular supply. There are two types of stroke: **hemorrhagic stroke,** which results from bleeding of arteries supplying brain tissue, and **ischemic stroke,** which results from blockage of arteries supplying brain tissue. The blockage in ischemic stroke can result from a thrombus (a clot that develops in place within an artery) or an embolism (a plug, composed of a detached thrombus or other foreign body, such as a fat globule or gas bubble, which becomes lodged in an artery, blocking it). Mr. S was at high risk for de-veloping a stroke. He was approaching middle age, was over-weight, did not exercise enough, smoked, was under stress, and had a poor diet.

The combination of motor loss, which was seen as weakness in his limbs, and sensory loss, seen as loss of pain and temperature sensation in his right lower limb and loss of all sensation in the left side of his face; along with the ataxia, dizziness, nystagmus, and hoarseness, suggest that the stroke affected the brainstem and cerebellum. Blockage of the verte-bral artery, a major artery supplying the brain, or its branches can result in what is called a lateral medullary infarction (an area of dead tissue resulting from a loss of blood supply to an area). Damage to the descending motor pathways in that area, above the medullary decussation, results in muscle weakness and damage to ascending pathways can result in loss of pain and temperature sensation (or other sensory modalities de-pending on the affected tract). Damage to cranial nerve nuclei results in the loss of pain and temperature sensation in the face, dizziness, blurred vision, nystagmus, vomiting, and hoarseness. These signs and symptoms are not observed un-less the lesion is in the brainstem, where these nuclei are lo-cated. Some damage to the cerebellum, also supplied by branches of the vertebral artery can account for the ataxia.

Drowsiness, disorientation, inattentiveness, and loss of consciousness are examples of generalized neurologic re-sponse to damage. Seizures may also result from severe neu-rologic damage. Depression from neurologic damage or from discouragement is also common. Slight dilation of the pupils; short, shallow respiration, and increased pulse rate and blood pressure are all signs of Mr. S's anxiety, not about the outcome of the game, but about his current condition and his immediate future. With a loss of consciousness, Mr. S would not remember the last few minutes of what he saw in the game he was watching. People in these circumstances are often worried about how they are going to deal with work tomorrow. They often have no idea that they may be permanently debilitated in that the motor and sensory losses may be permanent, or that they will have a long term of ther-apy ahead.

---

11 **P R E D I C T**

Given that Mr. S exhibited weakness in his right limbs, loss of pain and temperature sensation in his right lower limb and the left side of his face, state which side of the brainstem was most severely affected by the stroke. Explain your answer.

✔ *Answer in Appendix F*

Figure B   Sitting for a Stroke 🏃

Systemic Interactions	
System	**Interactions**
Integumentary	Decubitus ulcers (bedsores) from immobility; loss of motor function following a stroke leads to immobility.
Skeletal	Loss of bone mass, if muscles are dysfunctional for a prolonged time; in the absence of muscular activity, the bones to which those muscle are attached begin to be resorbed by osteoclasts.
Muscular	Major area of effect; absence of stimulation due to damaged pathways or neurons leads to decreased motor function and may result in muscle atrophy.
Endocrine	Strokes in other parts of the brain could involve the hypothalamus, pineal body, or pituitary gland functions.
Cardiovascular	Risks: Phlebothrombosis (blood clot in a vein) can occur from inactivity. Edema around the brain could apply pressure to the cardioregulatory and vasomotor centers of the brain. This pressure could stimulate these centers, which would result in elevated blood pressure, and congestive heart failure could result. If the cardioregulatory center in the brain is damaged, death may occur rapidly. Bleeding is due to the use of anticoagulants. Hypotension results from use of antihypertensives.
Respiratory	Pneumonia from aspiration of the vomitus or hypoventilation results from decreased function in the respiratory center. If the respiratory center is severely damaged, death may occur rapidly.
Digestive	Vomiting, dysphagia (difficulty swallowing); hypovolemia (decreased blood volume) result from decreased fluid intake; occurs because of dysphagia; may be a loss of bowel control.
Urinary	Control of the micturition reflex may be inhibited. Urinary tract infection results from catheter implantation or from urinary bladder distension.
Reproductive	Loss of libido; innervation of the reproductive organs is often affected.

Summary

Development

The brain and spinal cord develop from the neural tube. The ventricles and central canal develop from the lumen of the neural tube.

Brainstem

1. The medulla oblongata is continuous with the spinal cord and contains ascending and descending nerve tracts.
   - The pyramids are nerve tracts controlling voluntary muscle movement.
   - The olives are nuclei that function in equilibrium, coordination, and modulation of sound from the inner ear.
   - Medullary nuclei regulate the heart, blood vessels, respiration, swallowing, vomiting, coughing, sneezing, and hiccuping. The nuclei of cranial nerves V and IX–XII are in the medulla.
2. The pons is superior to the medulla.
   - Ascending and descending nerve tracts pass through the pons.
   - Pontine nuclei regulate sleep and respiration. The nuclei of cranial nerves V–IX are in the pons.
3. The midbrain is superior to the pons.
   - The midbrain contains the nuclei for cranial nerves III, IV, and V.
   - The tectum consists of four colliculi. The two inferior colliculi are involved in hearing, and the two superior colliculi in visual reflexes.
   - The tegmentum contains ascending tracts and the red nuclei, which are involved in motor activity.
   - The cerebral peduncles are the major descending motor pathway.
   - The substantia nigra connects to the basal nuclei and is involved with muscle tone and movement.
4. The reticular formation consists of nuclei scattered throughout the brainstem. The reticular activating system extends to the thalamus and cerebrum and maintains consciousness.

Diencephalon

1. The diencephalon is located between the brainstem and the cerebrum.
2. The thalamus consists of two lobes connected by the intermediate mass. The thalamus functions as an integration center.
   - Most sensory input synapses in the thalamus.
   - The thalamus also has some motor functions.
3. The subthalamus is inferior to the thalamus and is involved in motor function.
4. The epithalamus is superior and posterior to the thalamus and contains the habenular nuclei, which influence emotions through the sense of smell. The pineal body may play a role in the onset of puberty.
5. The hypothalamus, the most inferior portion of the diencephalon, contains several nuclei and tracts.
   - The mamillary bodies are reflex centers for olfaction.
   - The hypothalamus regulates many endocrine functions (e.g., metabolism, reproduction, response to stress, and urine production). The pituitary gland attaches to the hypothalamus.
   - The hypothalamus regulates body temperature, hunger, thirst, satiety, swallowing, and emotions.

Cerebrum

1. The cortex of the cerebrum is folded into ridges called gyri and grooves called sulci, or fissures. Nerve tracts connect areas of the cortex within the same hemisphere (association fibers), between different hemispheres (commissural fibers), and with other parts of the brain and the spinal cord (projection fibers).
2. The longitudinal fissure divides the cerebrum into left and right hemispheres. Each hemisphere has five lobes.
   - The frontal lobes are involved in smell, voluntary motor function, motivation, aggression, and mood.
   - The parietal lobes contain the major sensory areas receiving general sensory input, taste, and balance.
   - The occipital lobes contain the visual centers.
   - The temporal lobes receive olfactory and auditory input, and are involved in memory, abstract thought, and judgment.

Cerebral Cortex

1. Sensory pathways project to primary sensory areas in the cerebral cortex. Association areas interpret input from the primary sensory areas.
2. The frontal lobe contains areas that deal with voluntary muscle movement.
   - The primary motor area controls many muscle movements.
   - The premotor area is necessary for complex, skilled, and learned movements.
   - The prefrontal area is involved with the motivation and foresight associated with movement.
3. Cortical functions are arranged topographically on the cortex.
4. Speech is located only in the left cortex in most people.
   - Wernicke's area comprehends and formulates speech.
   - Broca's area receives input from Wernicke's area and sends impulses to the premotor and motor areas, which cause the muscle movements required for speech.
5. Electroencephalograms (EEGs) record the electrical activity of the brain as alpha, beta, theta, and delta waves. Some brain disorders can be detected with EEGs.
6. There are at least three kinds of memory: sensory, short-term, and long-term.
7. Each cerebral hemisphere controls and receives input from the opposite side of the body.
   - The right and left hemispheres are connected by commissures. The largest commissure is the corpus callosum, which allows sharing of information between hemispheres.
   - In most people the left hemisphere is dominant, controlling speech and analytic skills. The right hemisphere controls spatial and musical abilities.

Basal Nuclei

1. Basal nuclei include the subthalamic nuclei, substantia nigra, and corpus striatum.

2. The basal nuclei are important in coordinating motor movements and posture. They mainly have an inhibitory effect.

Limbic System

1. The limbic system includes parts of the cerebral cortex, basal nuclei, thalamus, hypothalamus, and the olfactory cortex.
2. The limbic system controls visceral functions through the autonomic nervous system and the endocrine system and is also involved in emotions and memory.

Cerebellum

1. The cerebellum has three parts that control balance, gross motor coordination, and fine motor coordination.
2. The cerebellum functions to correct discrepancies between intended movements and actual movements.
3. The cerebellum can "learn" highly specific complex motor activities.

Spinal Cord
General Structure

1. Thirty-one pairs of spinal nerves exit the spinal cord. The spinal cord has cervical and lumbar enlargements where nerves of the limbs enter and exit.
2. The spinal cord is shorter than the vertebral column. Nerves from the end of the spinal cord form the cauda equina.

Cross Section

1. The cord consists of peripheral white matter and central gray matter.
2. White matter is organized into funiculi, which are subdivided into fasciculi, or nerve tracts, which carry action potentials to and from the brain.
3. Gray matter is divided into horns.
   - The dorsal horns contain sensory axons that synapse with association neurons. The ventral horns contain the neuron cell bodies of somatic motor neurons, and the lateral horns contain the neuron cell bodies of autonomic neurons.
   - The gray and white commissures connect each half of the spinal cord.
4. The dorsal root conveys sensory input into the spinal cord, and the ventral root conveys motor output away from the spinal cord.

Spinal Reflexes
Stretch Reflex

Muscle spindles detect stretch of skeletal muscles and cause the muscle to shorten reflexively.

Golgi Tendon Reflex

Golgi tendon organs respond to increased tension within tendons and cause skeletal muscles to relax.

Withdrawal Reflex

1. Activation of pain receptors causes contraction of muscles and the removal of some part of the body from a painful stimulus.
2. Reciprocal innervation causes relaxation of muscles that would oppose the withdrawal movement.

3. In the crossed extensor reflex, during flexion of one limb caused by the withdrawal reflex, the opposite limb is stimulated to extend.

Spinal Pathways
Ascending Pathways

1. Ascending pathways carry conscious and unconscious sensations.
2. Spinothalamic system
   - The lateral spinothalamic tracts carry pain and temperature sensations. The anterior spinothalamic tracts carry light touch, pressure, tickle, and itch sensations.
   - Both tracts are formed by primary neurons that enter the spinal cord and synapse with secondary neurons. The secondary neurons cross the spinal cord and ascend to the thalamus, where they synapse with tertiary neurons that project to the somatic sensory cortex.
3. Dorsal-column/medial-lemniscal system
   - The dorsal-column/medial-lemniscal system carries the sensations of two-point discrimination, proprioception, pressure, and vibration.
   - Primary neurons enter the spinal cord and ascend to the medulla, where they synapse with secondary neurons. The secondary neurons cross over and project to the thalamus, where they synapse with tertiary neurons that extend to the somatic sensory cortex.
4. Spinocerebellar system and other tracts
   - The spinocerebellar tracts carry unconscious proprioception to the cerebellum from the same side of the body.
   - Neurons of the dorsal-column/medial-lemniscal system synapse with the neurons that carry proprioception information to the cerebellum.
   - The spinoolivary tract contributes to coordination of movement, the spinotectal tract to eye reflexes, and the spinoreticular tract to arousing consciousness.

Descending Pathways

1. Upper motor neurons are located in the cerebral cortex, cerebellum, and brainstem. Lower motor neurons are found in the cranial nuclei or the ventral horn of the spinal cord gray matter.
2. The direct pathways maintain muscle tone and controls fine, skilled movements in the face and distal limbs. The indirect pathways control conscious and unconscious muscle movements in the trunk and proximal limbs.
3. The corticospinal tracts control muscle movements below the head.
   - About 75%–85% of the upper motor neurons of the corticospinal tracts cross over in the medulla to form the lateral corticospinal tracts in the spinal cord.
   - The remaining upper motor neurons pass through the medulla to form the anterior corticospinal tracts, which cross over in the spinal cord.
   - The upper motor neurons of both tracts synapse with association neurons that then synapse with lower motor neurons in the spinal cord.
4. The corticobulbar tracts innervate the head muscles. Upper motor neurons synapse with association neurons in the

reticular formation that, in turn, synapse with lower motor neurons in the cranial nerve nuclei.

5. The indirect pathways include the rubrospinal, vestibulospinal, and reticulospinal tracts and fibers from the basal nuclei.

6. The indirect pathways are involved in conscious and unconscious trunk and proximal limb muscle movements, posture, and balance.

7. Some axons from the somatic sensory cortex synapse with secondary and tertiary neurons of the ascending sensory system and modify their activity.

Meninges and Cerebrospinal Fluid
Meninges

1. The brain and spinal cord are covered by the dura, arachnoid, and pia mater.

2. The dura mater attaches to the skull and has two layers that can separate to form dural sinuses.

3. Beneath the arachnoid mater the subarachnoid space contains CSF that helps cushion the brain.

4. The pia mater attaches directly to the brain.

Ventricles

1. The lateral ventricles in the cerebrum are connected to the third ventricle in the diencephalon by the interventricular foramen.

2. The third ventricle is connected to the fourth ventricle in the pons by the cerebral aqueduct. The central canal of the spinal cord is connected to the fourth ventricle.

Cerebrospinal Fluid

1. CSF is produced from the blood in the choroid plexus of each ventricle. CSF moves from the lateral to the third and then to the fourth ventricle.

2. From the fourth ventricle CSF enters the subarachnoid space through three foramina.

3. CSF leaves the subarachnoid space through arachnoid granulations and returns to the blood in the dural sinuses.

Content Review

1. Describe the formation of the neural tube. Name the five divisions of the neural tube and the parts of the brain that each division becomes.
2. Name the parts of the brainstem, and describe their functions.
3. What are the reticular formation and the reticular activating systems?
4. Name the four main components of the diencephalon.
5. Describe the functions of the thalamus and the hypothalamus.
6. Name the five lobes of the cerebrum, and describe their location and functions.
7. Describe in the cerebral cortex, the locations of the special and general senses and their association areas.
8. How do the primary and association areas interact to perceive a sensation?
9. Describe the topographical arrangement of the sensory and motor areas of the cerebral cortex.
10. Starting with hearing or reading a word, name the areas of the brain that are involved with receiving the word stimulus and the areas that eventually cause the word to be spoken aloud.
11. What is an EEG? What four conditions produce alpha, beta, theta, and delta waves, respectively?
12. Name the three different types of memory, and describe the processes that eventually result in the ability to remember something for a long time.
13. Name the pathways that connect the right and left cerebral hemispheres.
14. Name the basal nuclei, and state where they are located. What are their functions?
15. Name the parts of the limbic system. What does the limbic system do?
16. Describe the comparator activities of the cerebellum. Describe the role of the cerebellum in rapid and skilled motor movements such as playing the piano.
17. Define the cervical and lumbar enlargements of the spinal cord, the conus medullaris, and the cauda equina.
18. Describe the spinal cord gray matter. Where are sensory, autonomic, and somatic motor neurons located in the gray matter?
19. Contrast and describe a stretch reflex and a Golgi tendon reflex.
20. What is a withdrawal reflex? How do reciprocal innervation and the crossed extensor reflex assist the withdrawal reflex?
21. Describe the operation of the gamma motor system. What does it accomplish?
22. What are the functions of the lateral and anterior spinothalamic tracts and the dorsal-column/medial-lemniscal system? Describe where the neurons of these tracts cross over and synapse.
23. What kind of information is carried in the spinocerebellar tracts?
24. What are the functions of the spinoolivary, spinotectal, and spinoreticular tracts?
25. Distinguish between upper and lower motor neurons.
26. What two tracts form the direct pathways? What area of the body is supplied by each tract? Describe the location of the neurons in each tract, where they cross over, and where they synapse.
27. Name the tracts and structures that form the indirect pathways. What functions do they control? Contrast them with the functions of the direct pathways.
28. Describe the three meninges that surround the CNS. What are the falx cerebri, tentorium cerebelli, and falx cerebelli?
29. What space between what dural layers contains the CSF?
30. Describe the production and circulation of the CSF. Where does the CSF return to the blood?

Develop Your Reasoning Skills

1. Woody Knothead was accidentally struck in the head with a baseball bat. He fell to the ground unconscious. Later, when he regained consciousness, he was not able to remember any of the events that happened 10 min before the accident. Explain. What complications might be looked for at a later time?

2. A patient suffered brain damage in an automobile accident. It was suspected that the cerebellum was the part of the brain that was affected. On the basis of what you know about cerebellar function, how could you determine that the cerebellum was involved?

3. If there were a decrease in the frequency of action potentials in gamma motor fibers to a muscle, would the muscle tend to relax or contract?

4. A patient is suffering from the loss of two-point discrimination and proprioceptive sensations on the right side of the body resulting from a lesion in the pons. What tract is affected, and which side of the pons is involved?

5. A patient suffered a lesion in the central core of the spinal cord. It was suspected that the fibers that decussate and that are associated with the lateral spinothalamic tracts were affected in the area of the lesion. What observations would be consistent with that diagnosis?

6. A person in a car accident exhibited the following symptoms: extreme paresis on the right side, including the arm and leg, reduction of pain sensation on the left side, and normal tactile sensation on both sides. Which nerve tracts were damaged? Where did the patient suffer nerve tract damage?

7. If the right side of the spinal cord is completely transected, what symptoms do you expect to observe with regard to motor function, two-point discrimination, light touch, and pain perception?

8. A patient with a cerebral lesion exhibited a loss of fine motor control of the left hand, arm, forearm, and shoulder. All other motor and sensory functions appeared to be intact. Describe the location of the lesion as precisely as possible.

9. A baby is born with enlarged lateral and third ventricles but a normal fourth ventricle. Describe the defect and its location.

10. Near the end of the fall semester, a female nursing student developed a throbbing headache, which lasted about 30 minutes and then faded away. Similar headaches recurred several times over the next couple of weeks and during final examinations. As she was loading her car to go home for the semester break, she suddenly experienced a sudden, severe headache, which was accompanied by nausea, vomiting, and a general feeling of weakness. Some of her fellow students took her to the hospital emergency room, where she was examined. The physical examination revealed that the student had a stiff neck and an elevated blood pressure. An examination of the optic fundus (back of the eye) through an ophthalmoscope revealed subhyaloid hemorrhages (bleeding between the retina and the vitreous body of the eye). Her reflexes and sensory modalities appeared normal. She had no history of headache other than the recent episodes. A spinal tap was ordered, which revealed blood in the CSF. Present a possible tentative diagnosis.

Web Site Link

For a listing of the most current web sites related to this chapter, please visit the Seeley home page at:
http://www.mhhe.com/biosci/ap/seeleyap/

Chapter Fourteen

Peripheral Nervous System:
Cranial Nerves and Spinal Nerves

Objectives

1. Describe the distribution and function of each cranial nerve.

2. List the sensory cranial nerves and their functions.

3. List the somatic motor cranial nerves and their functions.

4. List the combination somatic motor and sensory cranial nerves and their functions.

5. List the combination somatic motor and parasympathetic cranial nerves and their functions.

6. List the combination somatic motor, sensory, and parasympathetic cranial nerves and their functions.

7. List, by letter and number, the spinal nerves existing at each vertebral level.

8. Explain what is meant by a dermatomal map.

9. Outline the pattern and distribution of the simplest spinal nerves, which do not participate in the formation of plexuses.

10. Explain the difference between ventral and dorsal rami.

11. Define the term plexus.

12. Describe the structure, distribution, and function of the cervical plexus.

13. Describe the general structure, distribution, and function of the nerves derived from the brachial plexus.

14. Explain the structural and functional basis of deficits resulting from damage to each upper limb nerve.

15. Describe the structure, distribution, and function of the obturator, femoral, tibial, and common fibular nerves.

16. Discuss the parts of the ischiadic (sciatic) nerve.

17. List the structures innervated by the coccygeal plexus.

Part Three

ven though the central nervous system (CNS) receives sensory information, evaluates that information, and initiates actions, without the peripheral nervous system (PNS), the CNS would remain isolated from both the body and the rest of the world. The PNS collects information from numerous sources both inside and outside the body and relays it by way of afferent fibers to the CNS. Efferent fibers in the PNS relay information from the CNS to various parts of the body, primarily to muscles and glands, regulating activity in those structures. Without the PNS, the CNS would receive no sensory information and could produce no observable responses. Even thoughts and emotions could not be expressed because of the CNS's isolation.

The PNS can be divided into two parts: a cranial part, consisting of 12 pairs of nerves, and a spinal part, consisting of 31 pairs of nerves. The cranial part of the PNS is discussed first in this chapter; discussion of the spinal nerves follows.

tal muscles also contain proprioceptive afferent fibers, which convey impulses to the CNS from those muscles. Because proprioception is the only sensory function of several otherwise somatic motor cranial nerves, however, that function is usually ignored, and the nerves are designated by convention as motor only. **Parasympathetic** function involves the regulation of glands, smooth muscles, and cardiac muscle.

These functions are part of the autonomic nervous system and are discussed in chapter 16. Table 14.2 lists the general organization of the cranial nerves by function. A given cranial nerve may have one or more of the three functions. Several of the cranial nerves have ganglia associated, and these ganglia are of two types: parasympathetic and sensory.

The **olfactory** (I) and **optic** (II) **nerves** are exclusively sensory and are involved in the special senses of smell and vision, respectively. These nerves are discussed in chapter 15.

The **oculomotor nerve** (III) innervates four of the six muscles that move the eyeball and the levator palpebrae superioris muscle, which raises the superior eyelid. In addition, parasympathetic nerve fibers in the oculomotor nerve innervate smooth muscles in the eye and regulate the size of the pupil and the shape of the lens of the eye.

Cranial Nerves

The 12 **cranial nerves** are listed and illustrated in table 14.1 and are illustrated in figure 14.1. By convention, the cranial nerves are indicated by Roman numerals (I–XII) from anterior to posterior. The three general categories of cranial nerve function are (1) sensory, (2) somatic motor, and (3) parasympathetic. **Sensory** functions include the special senses such as vision and the more general senses such as touch and pain. **Somatic** (sō-mat'ik) **motor** functions refer to the control of skeletal muscles through motor neurons. **Proprioception** (prō-prē-ō-sep'shun) informs the brain about the position of various body parts, including joints and muscles. The cranial nerves innervating skele-

1 P R E D I C T

A drooping upper eyelid on one side of the face is a sign of possible oculomotor nerve damage. Describe how this could possibly be evaluated by examining other oculomotor nerve functions. Describe the movements of the eye that would distinguish among oculomotor, trochlear, and abducens nerve damage.

✔ *Answer in Appendix F*

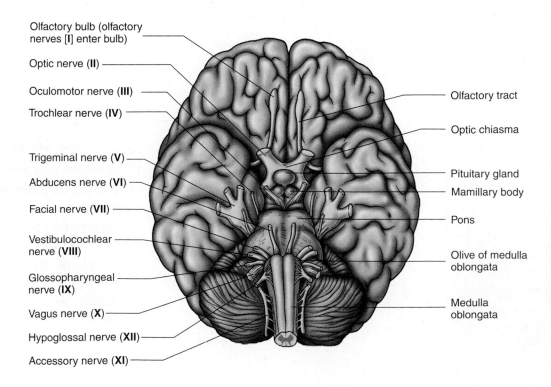

Olfactory bulb (olfactory nerves [I] enter bulb)

Optic nerve (II)

Oculomotor nerve (III)

Trochlear nerve (IV)

Trigeminal nerve (V)

Abducens nerve (VI)

Facial nerve (VII)

Vestibulocochlear nerve (VIII)

Glossopharyngeal nerve (IX)

Vagus nerve (X)

Hypoglossal nerve (XII)

Accessory nerve (XI)

Olfactory tract

Optic chiasma

Pituitary gland

Mamillary body

Pons

Olive of medulla oblongata

Medulla oblongata

Figure 14.1 Inferior Surface of the Brain Showing the Origin of the Cranial Nerves

Table 14.1   Cranial Nerves and Their Functions

Cranial Nerve	Foramen or Fissure*	Function

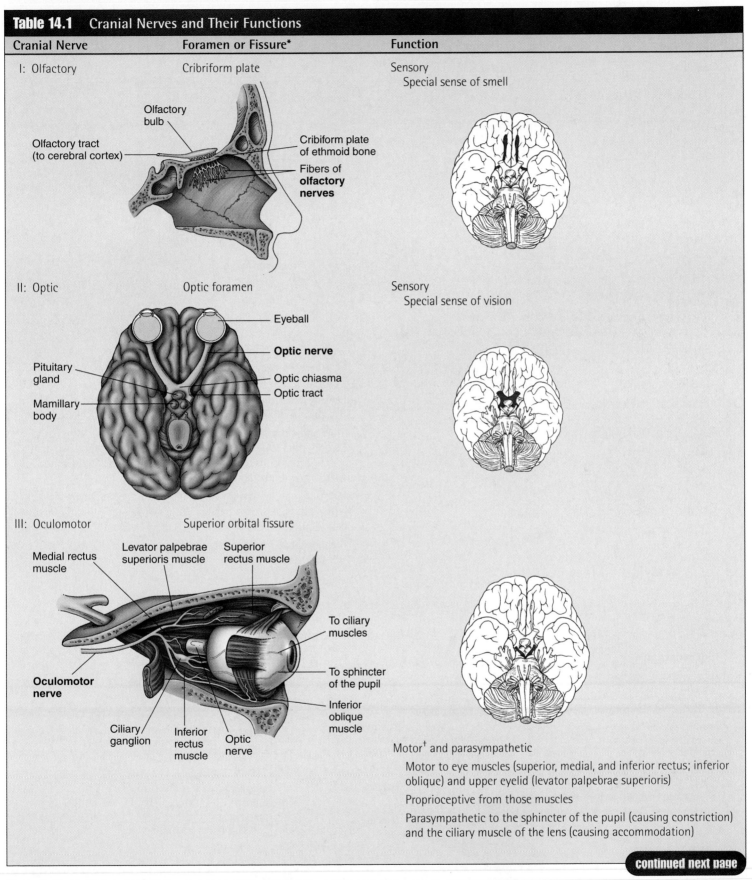

I: Olfactory — Cribriform plate

Sensory
  Special sense of smell

Olfactory bulb

Olfactory tract (to cerebral cortex)

Cribiform plate of ethmoid bone

Fibers of **olfactory nerves**

II: Optic — Optic foramen

Sensory
  Special sense of vision

Eyeball

Optic nerve

Pituitary gland

Optic chiasma

Optic tract

Mamillary body

III: Oculomotor — Superior orbital fissure

Medial rectus muscle

Levator palpebrae superioris muscle

Superior rectus muscle

To ciliary muscles

To sphincter of the pupil

Oculomotor nerve

Inferior oblique muscle

Ciliary ganglion

Inferior rectus muscle

Optic nerve

Motor† and parasympathetic
  Motor to eye muscles (superior, medial, and inferior rectus; inferior oblique) and upper eyelid (levator palpebrae superioris)
  Proprioceptive from those muscles
  Parasympathetic to the sphincter of the pupil (causing constriction) and the ciliary muscle of the lens (causing accommodation)

continued next page

*Route of entry or exit from the skull.

†Proprioception is a sensory function, not a motor function; however, motor nerves to muscles also contain some proprioceptive afferent fibers from those muscles. Because proprioception is the only sensory information carried by some cranial nerves, these nerves still are considered "motor."

Table 14.1   Cranial Nerves and Their Functions—cont'd

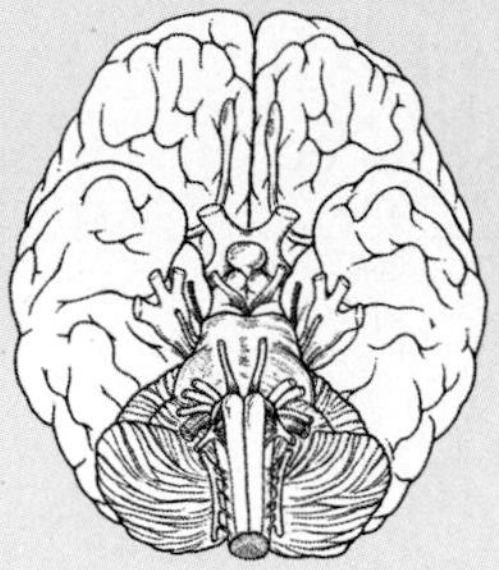

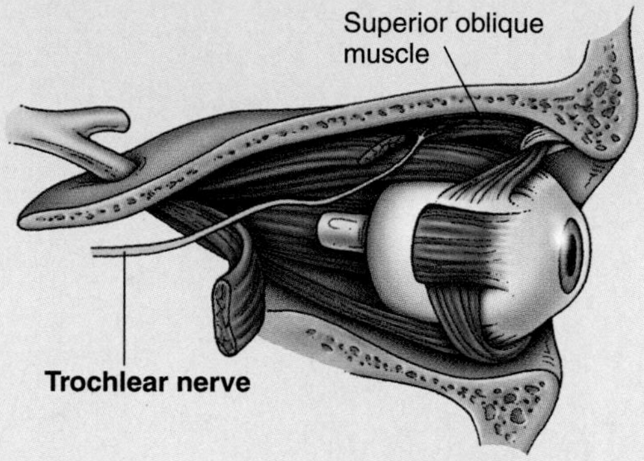

Cranial Nerve	Foramen or Fissure*	Function
IV: Trochlear	Superior orbital fissure	Motor[†] Motor to one eye muscle (superior oblique) Proprioceptive from that muscle

Trochlear nerve

Superior oblique muscle

V: Trigeminal
The trigeminal nerve is divided into three branches: the ophthalmic (V_1), the maxillary (V_2), and the mandibular (V_3)

Ophthalmic branch (V_1)	Superior orbital fissure	Sensory Sensory from scalp, forehead, nose, upper eyelid, and cornea
Maxillary branch (V_2)	Foramen rotundum	Sensory Sensory from palate, upper jaw, upper teeth and gums, nasopharynx, nasal cavity, skin of cheek, lower eyelid, and upper lip
Mandibular branch (V_3)	Foramen ovale	Sensory and motor[†] Sensory from lower jaw, lower teeth and gums, anterior two-thirds of tongue, mucous membrane of cheek, lower lip, skin of chin, auricle, and temporal region Motor to muscles of mastication (masseter, temporalis, medial and lateral pterygoids), soft palate (tensor veli palatini), throat (anterior belly of digastric, mylohyoid), and middle ear (tensor tympani) Proprioceptive from those muscles

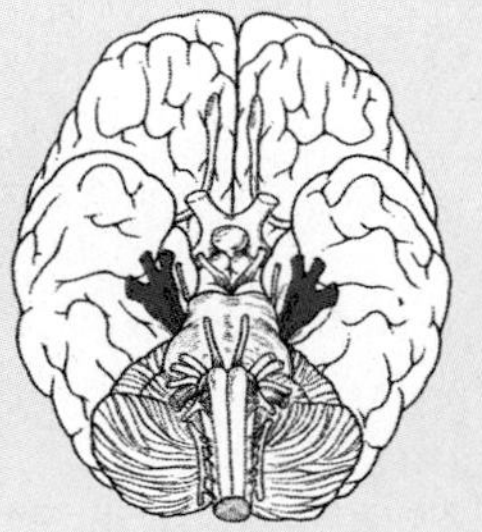

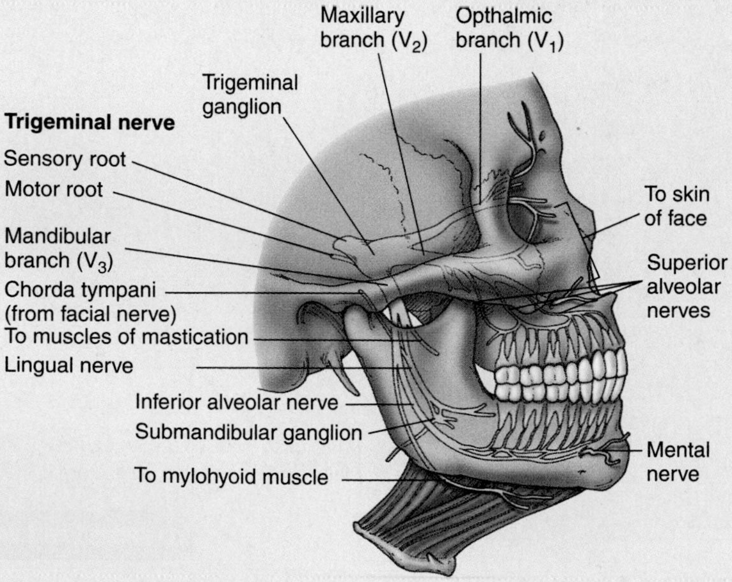

Maxillary branch (V_2)
Opthalmic branch (V_1)
Trigeminal ganglion
Trigeminal nerve
Sensory root
Motor root
Mandibular branch (V_3)
Chorda tympani (from facial nerve)
To muscles of mastication
Lingual nerve
Inferior alveolar nerve
Submandibular ganglion
To mylohyoid muscle
To skin of face
Superior alveolar nerves
Mental nerve

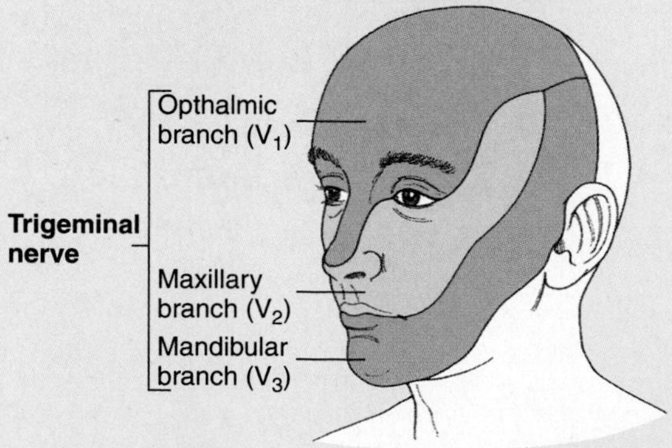

Trigeminal nerve
Opthalmic branch (V_1)
Maxillary branch (V_2)
Mandibular branch (V_3)

continued next page

Table 14.1 Cranial Nerves and Their Functions—cont'd

Cranial Nerve	Foramen or Fissure*	Function
VI: Abducens	Superior orbital fissure	Motor[†] Motor to one eye muscle (lateral rectus) Proprioceptive from that muscle
VII: Facial	Internal auditory meatus Stylomastoid foramen	Sensory, motor,[†] and parasympathetic Sense of taste from anterior two-thirds of tongue, sensory from some of external ear and palate Motor to muscles of facial expression, throat (posterior belly of digastric, stylohyoid), and middle ear (stapedius) Proprioceptive from those muscles Parasympathetic from submandibular and sublingual salivary glands, lacrimal gland, glands of the nasal cavity and palate

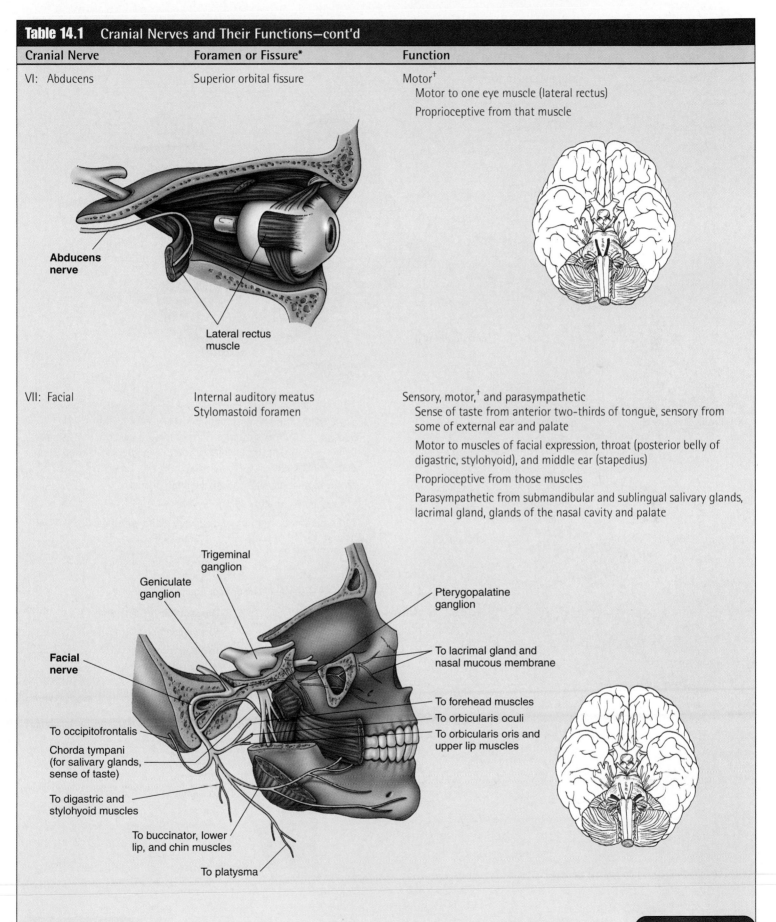

Abducens nerve

Lateral rectus muscle

Geniculate ganglion

Trigeminal ganglion

Facial nerve

Pterygopalatine ganglion

To lacrimal gland and nasal mucous membrane

To occipitofrontalis

Chorda tympani (for salivary glands, sense of taste)

To digastric and stylohyoid muscles

To buccinator, lower lip, and chin muscles

To platysma

To forehead muscles

To orbicularis oculi

To orbicularis oris and upper lip muscles

continued next page

Table 14.1   Cranial Nerves and Their Functions—cont'd

Cranial Nerve	Foramen or Fissure*	Function
VIII: Vestibulocochlear	Internal auditory meatus	Sensory Special senses of hearing and balance

Vestibular ganglion

Vestibular nerve

Vestibulocochlear nerve

Cochlear nerve

Spiral ganglion of cochlea

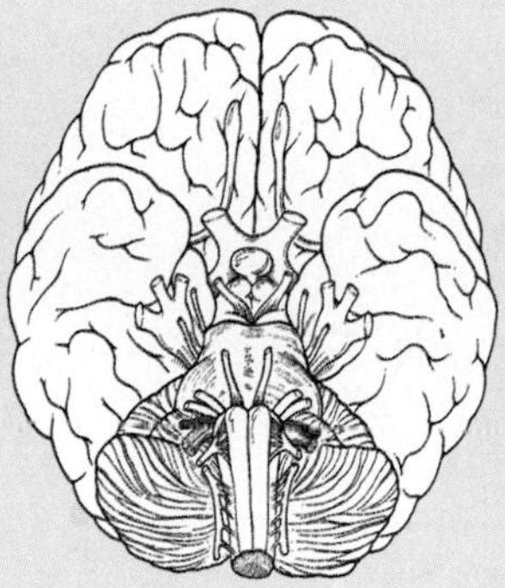

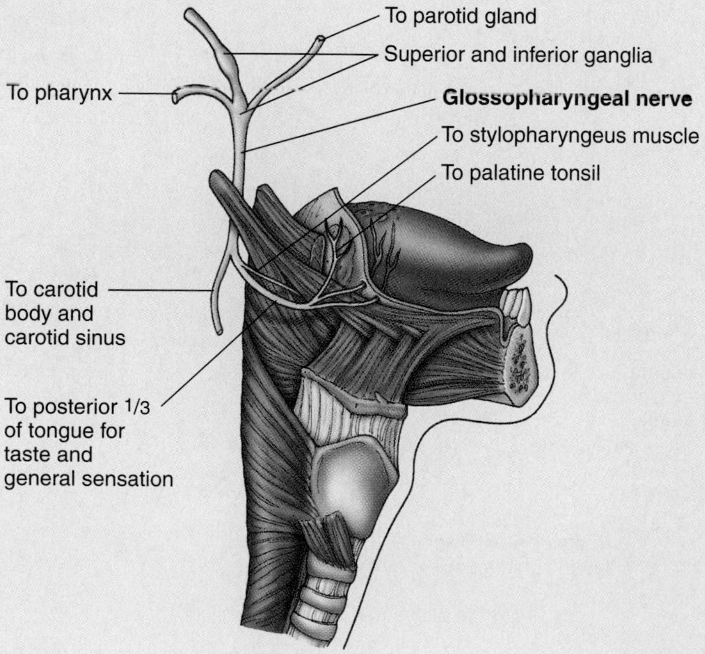

Cranial Nerve	Foramen or Fissure*	Function
IX: Glossopharyngeal	Jugular foramen	Sensory, motor,[†] and parasympathetic Sense of taste from posterior third of tongue, sensory from pharynx, palatine tonsils, posterior third of tongue, middle ear, carotid sinus and carotid body Motor to pharyngeal muscle (stylopharyngeus) Proprioceptive from that muscle Parasympathetic from parotid salivary gland and the glands of the posterior third of tongue

To parotid gland

Superior and inferior ganglia

To pharynx

Glossopharyngeal nerve

To stylopharyngeus muscle

To palatine tonsil

To carotid body and carotid sinus

To posterior 1/3 of tongue for taste and general sensation

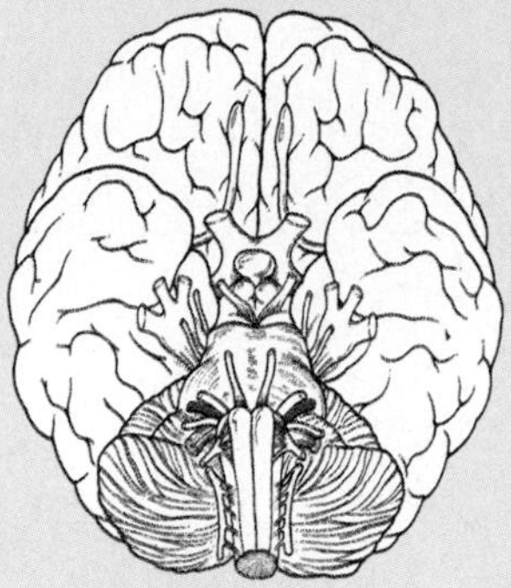

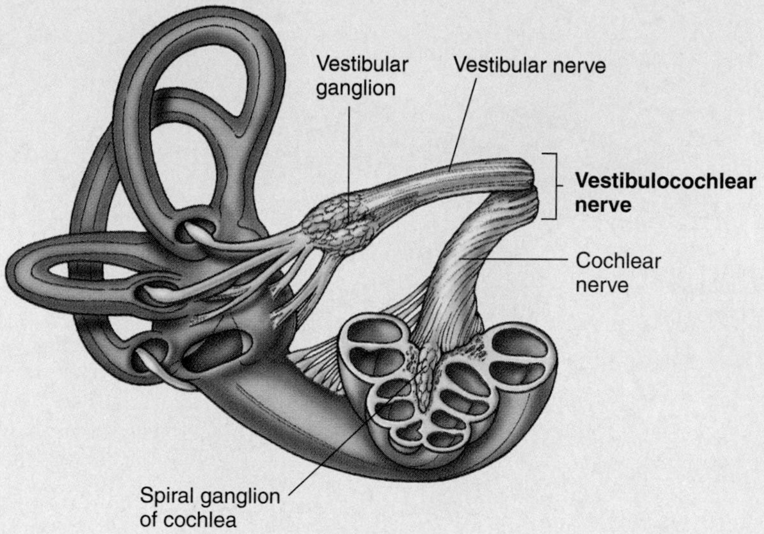

continued next page

Table 14.1   Cranial Nerves and Their Functions—cont'd

Cranial Nerve	Foramen or Fissure*	Function
X: Vagus	Jugular foramen	Sensory, motor,[†] and parasympathetic Sensory from inferior pharynx, larynx, thoracic and abdominal organs, sense of taste from posterior tongue Motor to soft palate, pharynx, intrinsic laryngeal muscles (voice production), and an extrinsic tongue muscle (palatoglossus) Proprioceptive from those muscles Parasympathetic to thoracic and abdominal viscera
XI: Accessory	Foramen magnum Jugular foramen	Motor[†] Motor to soft palate, pharynx, sternocleidomastoid, and trapezius Proprioceptive from those muscles

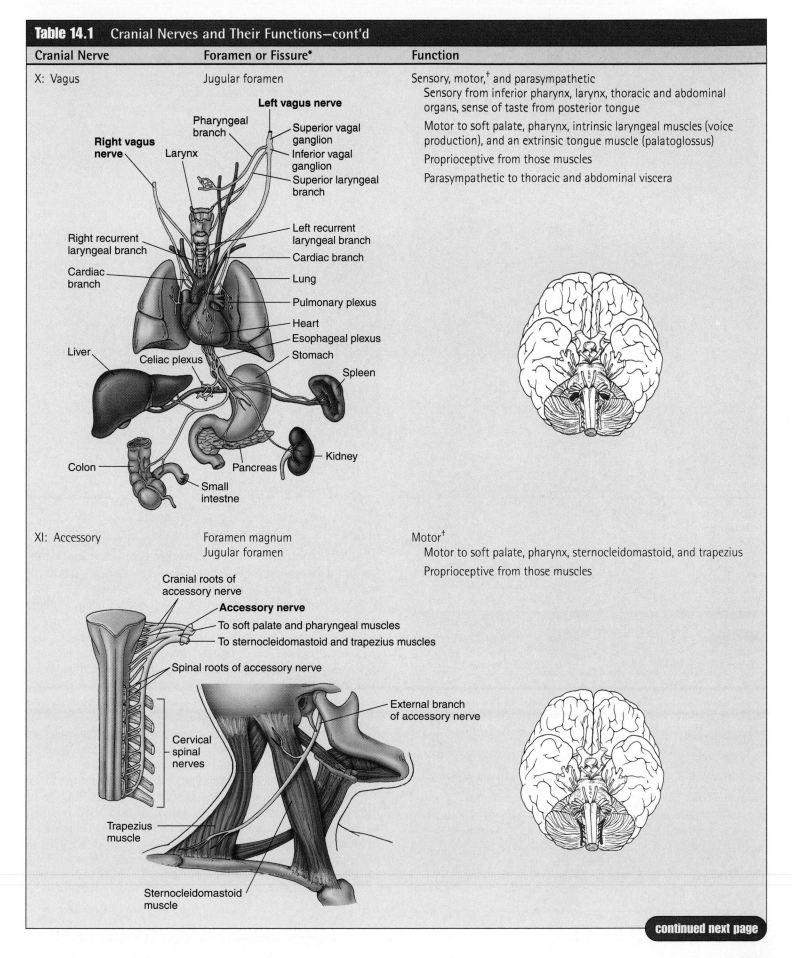

continued next page

Table 14.1   Cranial Nerves and Their Functions—cont'd

Cranial Nerve	Foramen or Fissure*	Function
XII: Hypoglossal	Hypoglossal canal	Motor[†] Motor to intrinsic and extrinsic tongue muscles (styloglossus, hypoglossus, genioglossus), and throat muscles (thyrohyoid and geniohyoid) Proprioceptive from those muscles

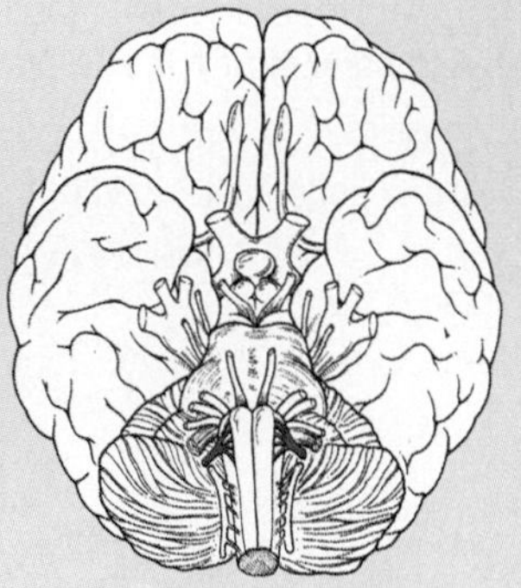

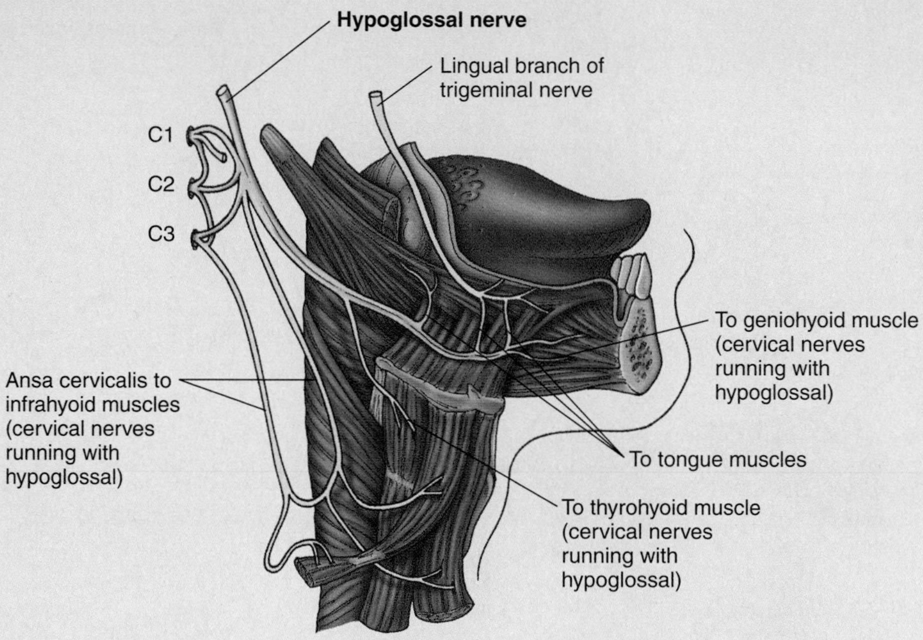

Hypoglossal nerve

Lingual branch of trigeminal nerve

C1
C2
C3

Ansa cervicalis to infrahyoid muscles (cervical nerves running with hypoglossal)

To geniohyoid muscle (cervical nerves running with hypoglossal)

To tongue muscles

To thyrohyoid muscle (cervical nerves running with hypoglossal)

Table 14.2   Functional Organization of the Cranial Nerves

Nerve Function	Cranial Nerve	
Sensory	I	Olfactory
	II	Optic
	VIII	Vestibulocochlear
Somatomotor/proprioception	IV	Trochlear
	VI	Abducens
	XI	Accessory
	XII	Hypoglossal
Somatomotor/proprioception and sensory	V	Trigeminal
Somatomotor/proprioception and parasympathetic	III	Oculomotor
Somatomotor/proprioception, sensory, and parasympathetic	VII	Facial
	IX	Glossopharyngeal
	X	Vagus

The **trochlear** (trōk′lē-ar) **nerve** (IV) is a somatic motor nerve that innervates one of the six eye muscles responsible for moving the eyeball.

The **trigeminal** (trī-jem′i-năl) **nerve** (V) has somatic motor, proprioceptive, and cutaneous sensory functions. It supplies motor innervation to the muscles of mastication, one middle ear muscle, one palatine muscle, and two throat muscles. In addition to proprioception associated with its somatic motor functions, the trigeminal nerve also supplies proprioception to the temporomandibular joint. Damage to the trigeminal nerve may impede chewing.

The trigeminal nerve has the greatest general sensory function of all the cranial nerves and is the only cranial nerve involved in **sensory cutaneous innervation.** All other cutaneous innervation comes from spinal nerves (see figure 14.4). Trigeminal means three twins, and the sensory distribution of the trigeminal nerve in the face is divided into three regions, each supplied by a branch of the nerve. The three branches—ophthalmic, maxillary, and mandibular—arise directly from the trigeminal ganglion, which serves the same function as the dorsal root ganglia of the spinal nerves.

In addition to these cutaneous functions, the maxillary and mandibular branches are important in dentistry. The max-

Clinical Focus   Peripheral Nervous System Disorders

General Types of PNS Disorders

Anesthesia is the loss of sensation (the Greek word *esthesis* means sensation). It may be a pathologic condition if it happens spontaneously, or it may be induced to facilitate surgery or some other medical treatment.

Hyperesthesia is an abnormal acuteness to sensation, especially an increased sensitivity to pain, pressure, or light.

Paresthesia is an abnormal spontaneous sensation, such as tingling, prickling, or burning.

Neuralgia (nū-ral′jē-ă) consists of severe spasms of throbbing or stabbing pain along the pathway of a nerve. Two types of neuralgia are described here.

Trigeminal neuralgia, also called tic douloureux, involves one or more of the trigeminal nerve branches and consists of sharp bursts of pain in the face. This disorder often has a trigger point in or around the mouth, which, when touched, elicits the pain response in some other part of the face. The cause of trigeminal neuralgia is unknown.

Ischiadica (is′kē-ad′i-kă), or **sciatica,** is a neuralgia of the ischiadic nerve, with pain radiating down the back of the thigh and leg. The most common cause is a herniated lumbar disk, resulting in pressure on the spinal nerves contributing to the lumbar plexus. Ischiadica may also be produced by ischiadic neuritis arising from a number of causes, including mechanical stretching during exertion, vitamin deficiency, or metabolic disorders (such as gout or diabetes).

Neuritis (nū-rī′tis) is a general term referring to inflammation of a nerve that has a wide variety of causes, including mechanical injury or pressure, viral or bacterial infection, poisoning by drugs or other chemicals, and vitamin deficiencies. Neuritis in sensory nerves is characterized by neuralgia or may result in anesthesia and loss of reflexes in the affected area. Neuritis in motor nerves results in loss of motor function.

Facial palsy (called Bell's palsy) is a unilateral paralysis of the facial muscles. The affected side of the face droops because of the absence of muscle tone. Facial palsy involves the facial nerve and may result from facial nerve neuritis.

Infections

Herpes is a family of diseases characterized by skin lesions, which are caused by a group of closely related viruses (the herpesviruses). The term is derived from the Greek word *herpo* meaning to creep and indicates a spreading skin eruption. The viruses apparently reside in the ganglia of sensory nerves and cause lesions along the course of the nerve.

Herpes simplex I is usually characterized by one or more lesions on the lips or nose. The virus apparently resides in the trigeminal ganglion. Eruptions are usually recurrent and often occur in times of reduced resistance such as during a case of the common cold. For this reason they are called cold sores or fever blisters. A different herpesvirus, **herpes simplex II,** or genital herpes, is usually responsible for a sexually transmitted disease causing lesions on the external genitalia.

The varicella-zoster virus causes the diseases chicken pox in children and shingles in older adults, a disease also called **herpes zoster.** Normally, this virus first enters the body in childhood to cause chicken pox. The virus then lies dormant in the spinal ganglia for many years and can become active during a time of reduced resistance to cause shingles, a unilateral patch of skin blisters and discoloration along the path of one or more spinal nerves, most commonly around the waist. The symptoms can persist for 3–6 months.

Poliomyelitis (pō′lē-ō-mī′ĕ-lī′tis) ("polio" or infantile paralysis; the Greek word *polio* means gray matter) is a disease caused by an *Enterovirus.* It is actually a CNS infection, but its major effect is on the peripheral nerves and the muscles they supply. The virus infects the motor neurons in the anterior horn of the central gray matter of the spinal cord. The infection causes degeneration of the motor neurons, which results in paralysis and atrophy of the muscles innervated by those nerves.

Anesthetic leprosy is a bacterial infection of the peripheral nerves caused by *Mycobacterium leprae.* The infection results in anesthesia, paralysis, ulceration, and gangrene.

Genetic and Autoimmune Disorders

Myotonic dystrophy is an autosomal-dominant hereditary disease characterized by muscle weakness, dysfunction, and atrophy and by visual impairment as a result of nerve degeneration.

Myasthenia (mī-as-thē′nē-ă) **gravis** is an autoimmune disorder resulting in a reduction in the number of functional acetylcholine receptors in neuromuscular junctions. T cells of the immune system break down acetylcholine receptor proteins into two fragments, which trigger antibody production by the immune system. Myasthenia gravis results in fatigue and progressive muscular weakness because of the neuromuscular dysfunction.

Neurofibromatosis (nūr′ō-fī-brō-mă-tō′sis) is a genetic disorder in which small skin lesions appear in early childhood followed by the development of multiple subcutaneous neurofibromas, which are benign tumors resulting from Schwann cell (neurolemmocyte) proliferation. The neurofibromas may slowly increase in size and number over several years and cause extreme disfiguration.

---

illary nerve supplies sensory innervation to the maxillary teeth, palate, and gingiva (jin′jĭ-vah, meaning gum). The mandibular branch supplies sensory innervation to the mandibular teeth, tongue, and gingiva. The various nerves innervating the teeth are referred to as **alveolar** (al′vē-ō′lăr, refers to the sockets in which the teeth are located). The **superior alveolar nerves** to the maxillary teeth are derived from the maxillary branch of the trigeminal nerve, and the **inferior alveolar nerves** to the mandibular teeth are derived from the mandibular branch of the trigeminal nerve.

Dentists inject anesthetic to block sensory transmission by the alveolar nerves. The superior alveolar nerves are not usually anesthetized directly because they are difficult to approach with a needle. For this reason the maxillary teeth usually are anesthetized locally by inserting the needle beneath the oral mucosa surrounding the teeth. The inferior alveolar nerve probably is anesthetized more often than any other nerve in the body. To anesthetize this nerve, the dentist inserts the needle somewhat posterior to the patient's last molar.

Several nondental nerves are usually anesthetized during an inferior alveolar block. The mental nerve, which is cutaneous to the anterior lip and chin, is a distal branch of the inferior alveolar nerve; therefore, when the inferior alveolar nerve is blocked, the mental nerve is blocked also, resulting in a numb lip and chin. Nerves lying near the point where the inferior alveolar nerve enters the mandible often are also anesthetized during inferior alveolar anesthesia. For example, the lingual nerve can be anesthetized to produce a numb tongue. The facial nerve lies some distance from the inferior alveolar nerve, but in rare cases anesthetic can diffuse far enough posteriorly to anesthetize that nerve. The result is a temporary facial palsy, with the injected side of the face drooping because of flaccid muscles, which disappears when the anesthesia wears off. If the facial nerve is cut by an improperly inserted needle, permanent facial palsy may occur.

The **abducens** (ab-dū′senz) **nerve** (VI), like the trochlear nerve, is a somatic motor nerve that innervates one of the six eye muscles responsible for moving the eyeball.

The **facial nerve** (VII) is somatic motor, sensory, and parasympathetic. It controls all the muscles of facial expression, a small muscle in the middle ear, and two throat muscles. It is sensory for the sense of taste in the anterior two-thirds of the tongue (see chapter 15). The facial nerve supplies parasympathetic innervation to the submandibular and sublingual salivary glands and to the lacrimal glands.

The **vestibulocochlear** (ves-tib′yū-lō-kok′lē-ăr) **nerve** (VIII), like the olfactory and optic nerves, is exclusively sensory and transmits action potentials from the inner ear responsible for the special senses of hearing and balance (see chapter 15).

The **glossopharyngeal** (glos′ō-fă-rin′jē-ăl) **nerve** (IX), like the facial nerve, is somatic motor, sensory, and parasympathetic and has both sensory and parasympathetic ganglia. The glossopharyngeal nerve is somatic motor to one muscle of the pharynx and supplies parasympathetic innervation to the parotid salivary glands. The glossopharyngeal nerve is sensory for the sense of taste in the posterior third of the tongue. It also supplies tactile sensory innervation from the posterior tongue, middle ear, and pharynx and transmits sensory stimulation from receptors in the carotid arteries and the aortic arch, which monitor blood pressure and blood carbon dioxide, blood oxygen, and blood pH levels (see chapter 21).

The **vagus** (vā′gŭs) **nerve** (X), like the facial and glossopharyngeal nerves, is somatic motor, sensory, and parasympathetic and has both sensory and parasympathetic ganglia. Most muscles of the soft palate, pharynx, and larynx are innervated by the vagus nerve. Damage to the laryngeal branches of the vagus nerve can interfere with normal speech. The vagus nerve is sensory for taste from the root of the tongue (see chapter 15). It is sensory for the inferior pharynx and the larynx and assists the glossopharyngeal nerve in transmitting sensory stimulation from receptors in the carotid arteries and the aortic arch, which monitor blood pressure and carbon dioxide, oxygen, and pH levels in the blood (see chapter 21). In addition, the vagus nerve conveys sensory information from the thoracic and abdominal organs. The parasympathetic part of the vagus nerve is very important in regulating the functions of the thoracic and abdominal organs. It carries parasympathetic fibers to the heart and lungs in the thorax and to the digestive organs, spleen, and kidneys in the abdomen.

The **accessory** (XI), and **hypoglossal** (XII) **nerves** are somatic motor nerves. The accessory nerve has both a cranial and a spinal component. The cranial component joins the vagus nerve (hence the name accessory) and participates in its function. The spinal component of the accessory nerve provides the major innervation to the sternocleidomastoid and trapezius muscles of the neck and shoulder. The hypoglossal nerve supplies the intrinsic tongue muscles, three of the four extrinsic tongue muscles, and the thyrohyoid and the geniohyoid muscles.

2   P R E D I C T

Injury to the spinal portion of the accessory nerve may result in sternocleidomastoid muscle dysfunction, a condition called "wry neck." If the head of a person with wry neck is turned to the left, would this position indicate injury to the left or right spinal component of the accessory nerve?

✔ *Answer in Appendix F*

3   P R E D I C T

Unilateral damage to the hypoglossal nerve results in loss of tongue movement on one side, which is most obvious when the tongue is protruded. If the tongue is deviated to the right, is the left or right hypoglossal nerve damaged?

✔ *Answer in Appendix F*

Spinal Nerves

The **spinal nerves** arise through numerous rootlets along the dorsal and ventral surfaces of the spinal cord (figure 14.2). About six to eight of these rootlets combine to form a **ventral root** on the ventral (anterior) side of the spinal cord, and

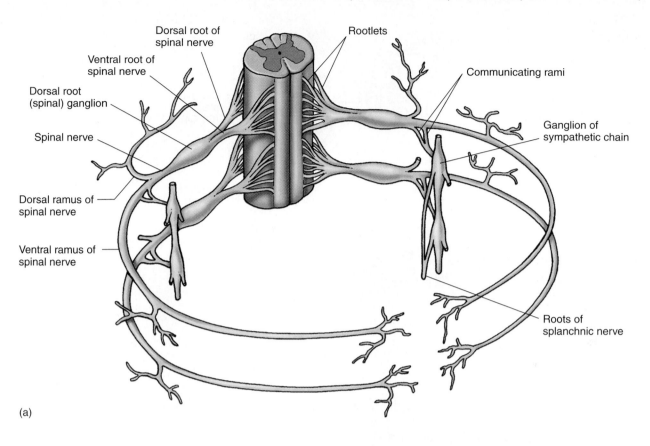

(a)

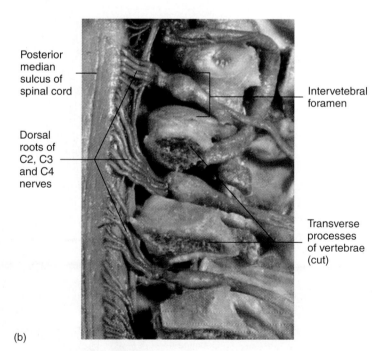

(b)

Figure 14.2   Spinal Nerves

(a) Typical thoracic spinal nerves. *(b)* Photograph of four dorsal roots in place along the vertebral column.

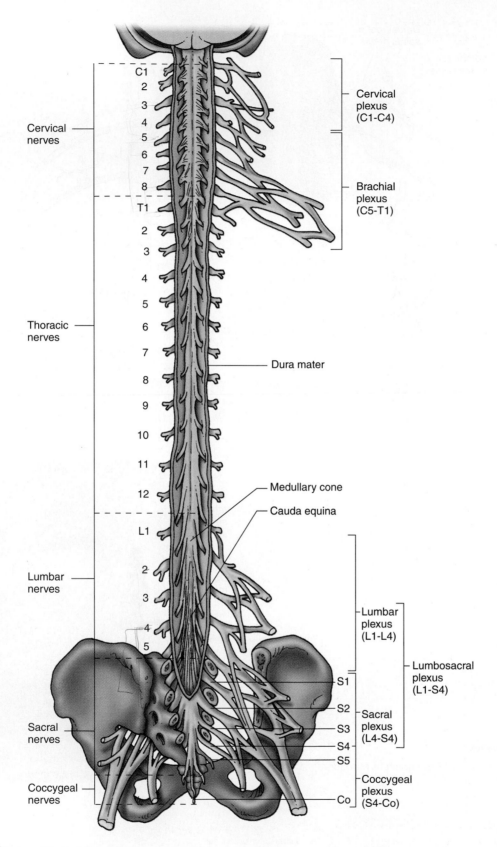

Figure 14.3   Spinal Cord and Spinal Nerves

another six to eight form a **dorsal root** on the dorsal (posterior) side of the cord at each segment. The ventral root contains efferent (motor) fibers, and the dorsal root contains afferent (sensory) fibers. The dorsal and ventral roots join one another just lateral to the spinal cord to form the spinal nerve. The dorsal root contains a ganglion, called the **dorsal root, or spinal, ganglion,** near where it joins the ventral root.

All of the 31 pairs of spinal nerves, except the first pair and those in the sacrum, exit the vertebral column through an intervertebral foramen located between adjacent vertebrae. The first pair of spinal nerves exits between the skull and the first cervical vertebra. The nerves of the sacrum exit from the single bone of the sacrum through the sacral foramina (see chapter 7). Eight spinal nerve pairs exit the vertebral column in the cervical region, 12 in the thoracic region, 5 in the lumbar region, 5 in the sacral region, and 1 in the coccygeal region (figure 14.3). For convenience, each of the spinal nerves is designated by a letter and number. The letter indicates the region of the vertebral column from which the nerve emerges: C, cervical; T, thoracic; L, lumbar; and S, sacral. The single coccygeal nerve is often not designated, but when it is, the symbol often used is Co. The number indicates the location in each region where the nerve emerges from the vertebral column, with the smallest number always representing the most superior origin. For example, the most superior nerve exiting from the thoracic region of the vertebral column is designated T1. The cervical nerves are designated C1–C8, the thoracic nerves T1–T12, the lumbar nerves L1–L5, and the sacral nerves S1–S5.

Each of the spinal nerves except C1 has a specific cutaneous sensory distribution. Figure 14.4 depicts the **dermatomal** (der-mă-tō′măl) **map** for the sensory cutaneous distribution of the spinal nerves. A **dermatome** is the area of skin supplied with sensory innervation by a pair of spinal nerves.

---

4 **P R E D I C T**

The dermatomal map is important in clinical considerations of nerve damage. Loss of sensation in a dermatomal pattern can provide valuable information about the location of nerve damage. Predict the possible site of nerve damage for a patient who suffered whiplash in an automobile accident and subsequently developed anesthesia (no sensations) in the left arm, forearm, and hand (see figure 14.4 for help).

✔ *Answer in Appendix F*

Figure 14.2*a* depicts an idealized section through the trunk. Each spinal nerve has a dorsal and a ventral **ramus** (rā′mŭs, meaning branch). Additional rami (rā′mī), called communicating rami, from the thoracic and upper lumbar spinal cord regions carry axons associated with the sympathetic nervous system (see chapter 16). The **dorsal rami** (rā′mī) innervate most of the deep muscles of the dorsal trunk responsible for movement of the vertebral column. They also innervate the connective tissue and skin near the midline of the back.

The **ventral rami** are distributed in two ways. In the thoracic region the ventral rami form **intercostal** (meaning between ribs) **nerves,** which extend along the inferior margin of each rib and innervate the intercostal muscles and the skin over the thorax. The ventral rami of the remaining spinal nerves form five **plexuses** (plek′sŭs-ēz). The term plexus means braid and describes the organization produced by the intermingling of the nerves. The ventral rami of different spinal nerves, called the **roots** of the plexus, join with each other to form a plexus. These roots should not be confused with the dorsal and ventral roots from the spinal cord, which are more medial. The axons from the spinal nerves mix; thus the nerves that arise from plexuses usually have axons from more than one level of the spinal cord. The ventral rami of spinal nerves C1–C4 form the cervical plexus; C5–T1 form the brachial plexus; L1–L4 form the lumbar plexus; L4–S4 form the sacral plexus; and S4, S5, and the coccygeal nerve (Co) form the coccygeal plexus.

Several smaller somatic plexuses, such as the pudendal plexus in the pelvis, are derived from more distal branches of the spinal nerves. Some of the somatic plexuses are mentioned where appropriate in this chapter.

Cervical Plexus

The **cervical plexus** is a relatively small plexus originating from spinal nerves C1–C4 (figure 14.5). Branches derived from this plexus innervate superficial neck structures, including several of the muscles attached to the hyoid bone. The cervical plexus innervates the skin of the neck and posterior portion of the head (see figure 14.4).

One of the most important derivatives of the cervical plexus is the **phrenic** (fren′ik) **nerve,** which originates from spinal nerves C3–C5, derived from both the cervical and brachial plexus. The phrenic nerves descend along each side of the neck to enter the thorax. They descend along the side of the mediastinum to reach the diaphragm, which they innervate. Contraction of the diaphragm is largely responsible for the ability to breathe.

---

5 **P R E D I C T**

The phrenic nerve may be damaged where it descends along the neck. Because it descends along the mediastinum, care must be taken not to damage the nerve during open thoracic surgery or open heart surgery. Cancer of the bronchus is the most common type of cancer in men, accounting for about 30% of all male cancers and most often occurs in men who smoke cigarettes. Tumors at the base of the lung can compress the phrenic nerve. Explain how damage to or compression of the right phrenic nerve would affect the diaphragm. Describe the effect on breathing of a completely severed spinal cord at the C2 level versus at the C6 level.

✔ *Answer in Appendix F*

Brachial Plexus

The **brachial plexus** originates from spinal nerves C5–T1 (figure 14.6). There is also a connection from the brachial

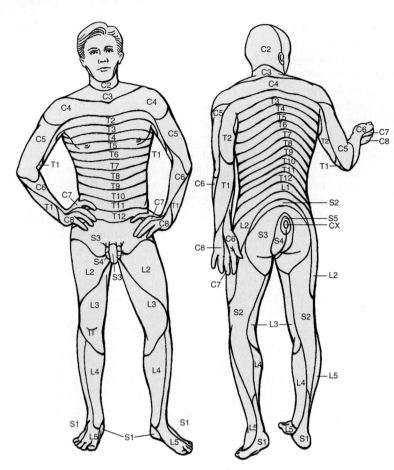

Figure 14.4   Dermatomal Map
Letters and numbers indicate the spinal nerves innervating a given region of skin.

plexus to C4 of the cervical plexus. The five ventral rami that constitute the brachial plexus join to form three **trunks,** which separate into six **divisions,** and then join again to create three **cords** (posterior, lateral, and medial) from which five **branches,** or nerves of the upper limb, emerge.

The five major nerves emerging from the brachial plexus to supply the upper limb are the axillary, radial, musculocutaneous, ulnar, and median nerves. The axillary nerve innervates part of the shoulder; the radial nerve innervates the posterior arm, forearm, and hand; the musculocutaneous nerve innervates the anterior arm; and the ulnar and median nerves innervate the anterior forearm and hand. Additional brachial plexus nerves innervate the shoulder and pectoral muscles.

> **Clinical Note**
>
> Sometimes it is necessary to anesthetize the entire upper limb; in this case, an anesthetic can be injected near the brachial plexus. The injection, called **brachial anesthesia,** is made between the neck and the shoulder posterior to the clavicle.

Axillary Nerve

The **axillary** (ak′sil-ār′ē) **nerve** innervates the deltoid and teres minor muscles (figure 14.7). It also provides sensory innervation to the shoulder joint and to the skin over part of the shoulder.

Radial Nerve

The **radial nerve** emerges from the posterior cord of the brachial plexus and descends within the deep aspect of the posterior arm (figure 14.8). About midway down the shaft of the humerus it lies against the bone in the radial groove. The radial nerve innervates all of the extensor muscles of the upper limb, the supinator muscle, and two muscles that flex the forearm. Its cutaneous sensory distribution is to the posterior portion of the upper limb, including the posterior surface of the hand.

> **Clinical Note**
>
> Because the radial nerve lies near the humerus in the axilla, it can be damaged if it is compressed against the humerus. Improper use of crutches (i.e., when the crutch is pushed tightly into the axilla) can result in **"crutch paralysis."** In this disorder, the radial nerve is compressed between the top of the crutch and the humerus. As a result, the radial nerve is damaged, and the muscles it innervates lose their function. The major symptom of crutch paralysis is **"wrist drop"** in which the extensor muscles of the wrist and fingers, which are innervated by the radial nerve, fail to function; as a result, the elbow, wrist, and fingers are constantly flexed.

---

6   P R E D I C T

Wrist drop can also result from a compound fracture of the humerus. Explain how and where damage to the nerve may occur.

✔ *Answer in Appendix F*

Musculocutaneous Nerve

The **musculocutaneous** (mŭs′kyū-lō-kyū-tā′nē-ŭs) **nerve** provides motor innervation to the anterior muscles of the arm as well as cutaneous sensory innervation for part of the forearm (figure 14.9).

Ulnar Nerve

The **ulnar nerve** innervates two forearm muscles plus most of the intrinsic hand muscles, except some associated with the thumb. Its sensory distribution is to the ulnar side of the hand (figure 14.10).

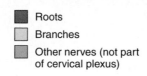

■ Roots
□ Branches
■ Other nerves (not part
   of cervical plexus)

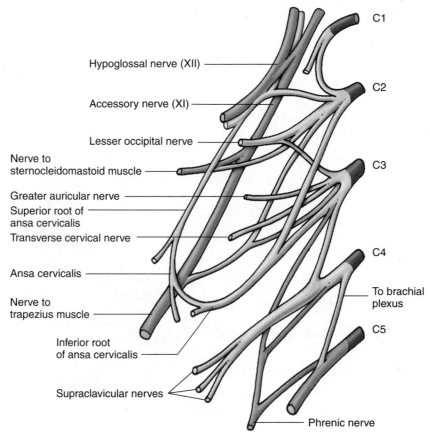

C1

Hypoglossal nerve (XII)

C2

Accessory nerve (XI)

Lesser occipital nerve

C3

Nerve to
sternocleidomastoid muscle

Greater auricular nerve

Superior root of
ansa cervicalis

Transverse cervical nerve

C4

Ansa cervicalis

To brachial
plexus

Nerve to
trapezius muscle

C5

Inferior root
of ansa cervicalis

Supraclavicular nerves

Phrenic nerve

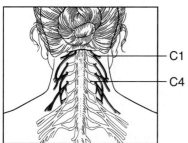

C1

C4

Figure 14.5   Cervical Plexus
The roots of the plexus are formed by the ventral rami of the spinal nerves C1–C4.

Median Nerve

The **median nerve** innervates all but one of the flexor muscles of the forearm and most of the hand muscles near the thumb, called the thenar area of the hand. Its cutaneous sensory distribution is to the radial portion of the palm of the hand (figure 14.11).

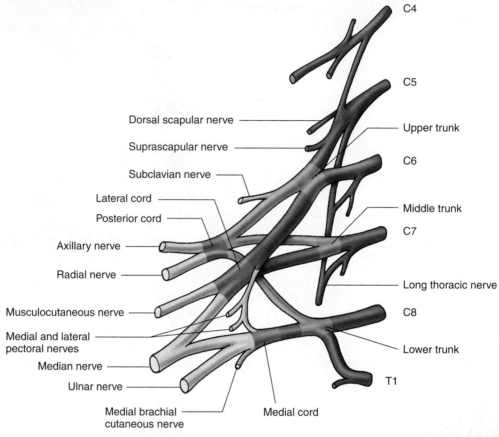

Roots: C5, C6, C7, C8, T1
Trunks: upper, middle, lower
Anterior divisions
Posterior divisions
Cords: posterior, lateral, medial
Branches: Axillary nerve
              Radial nerve
              Musculocutaneous nerve
              Median nerve
              Ulnar nerve

C4
C5
Dorsal scapular nerve
Suprascapular nerve — Upper trunk
C6
Subclavian nerve
Lateral cord
Posterior cord — Middle trunk
C7
Axillary nerve
Radial nerve
Long thoracic nerve
Musculocutaneous nerve
C8
Medial and lateral
pectoral nerves — Lower trunk
Median nerve
Ulnar nerve
T1
Medial brachial
cutaneous nerve — Medial cord

Figure 14.6   Brachial Plexus

The roots of the plexus are formed by the ventral rami of the spinal nerves C5–T1 and join to form an upper, middle, and lower trunk. Each trunk divides into anterior and posterior divisions. The divisions join together to form the posterior, lateral, and medial cords from which the major brachial plexus nerves arise.

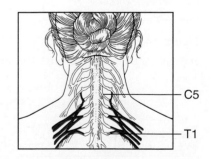

C5

T1

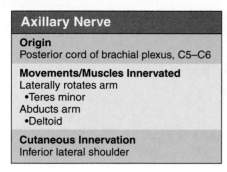

Axillary Nerve

Origin
Posterior cord of brachial plexus, C5–C6

Movements/Muscles Innervated
Laterally rotates arm
•Teres minor
Abducts arm
•Deltoid

Cutaneous Innervation
Inferior lateral shoulder

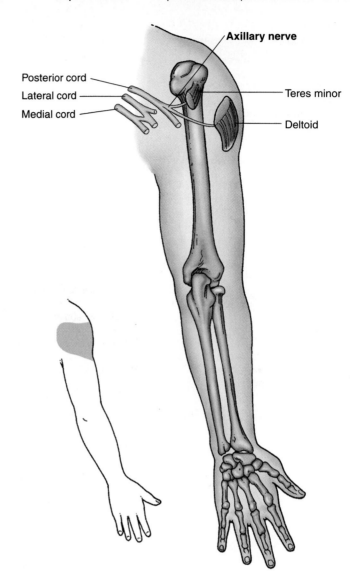

Figure 14.7 Axillary Nerve
Route of the axillary nerve and the muscles it innervates. The inset depicts the cutaneous distribution of the nerve (*shaded area*).

Clinical Focus   Nerve Replacement

Patients paralyzed by strokes or spinal cord lesions are now able to regain certain functions. Microcomputers are being perfected that stimulate certain programmed activities such as grasping and walking. Fine wire leads convey an electric impulse initiated by the microcomputer to either peripheral nerves or directly to the muscles responsible for the desired movement. The program is initiated by the subtle movement of muscles not affected by the paralysis. Sensors connected to the microcomputer are attached to the skin overlying functional muscles and are able to detect electrical activity associated with movement of the underlying muscles. For example, a person with both legs paralyzed may have such a sensor attached to the abdomen. The abdominal muscles, normally involved in walking, are stimulated by descending tracts when walking is initiated by CNS centers. The resultant abdominal muscle activity is detected by the sensor, which activates the program that stimulates the appropriate sequence of muscles, and the paralyzed person walks. Similarly, a quadriplegic can initiate certain grasping actions by subtle movements of the shoulder, neck, or face, where specific sensors can be placed.

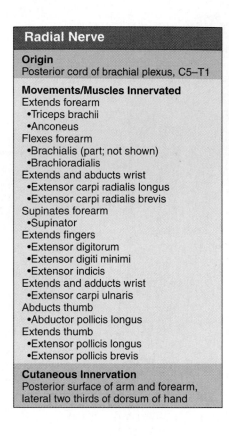

Radial Nerve

Origin
Posterior cord of brachial plexus, C5–T1

Movements/Muscles Innervated
Extends forearm
 •Triceps brachii
 •Anconeus
Flexes forearm
 •Brachialis (part; not shown)
 •Brachioradialis
Extends and abducts wrist
 •Extensor carpi radialis longus
 •Extensor carpi radialis brevis
Supinates forearm
 •Supinator
Extends fingers
 •Extensor digitorum
 •Extensor digiti minimi
 •Extensor indicis
Extends and adducts wrist
 •Extensor carpi ulnaris
Abducts thumb
 •Abductor pollicis longus
Extends thumb
 •Extensor pollicis longus
 •Extensor pollicis brevis

Cutaneous Innervation
Posterior surface of arm and forearm, lateral two thirds of dorsum of hand

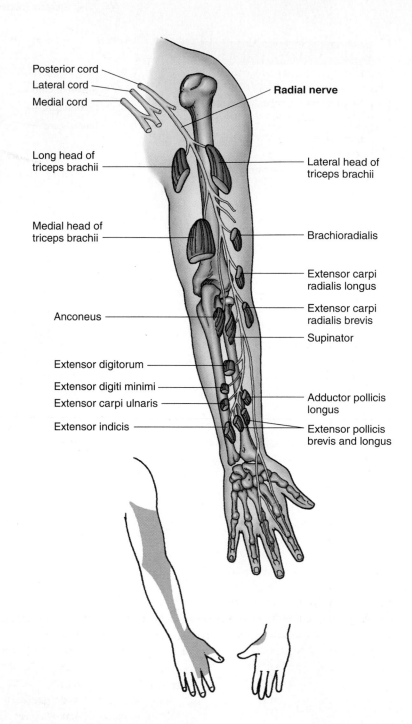

Figure 14.8 Radial Nerve
Route of the radial nerve and the muscles it innervates. The inset depicts the cutaneous distribution of the nerve (*shaded area*).

Other Nerves of the Brachial Plexus

Several nerves, other than the five just described, arise from the brachial plexus (see figure 14.6). They supply most of the muscles acting on the scapula and arm and include the pectoral, long thoracic, thoracodorsal, subscapular, and supra-scapular nerves. In addition, brachial plexus nerves supply the cutaneous innervation of the medial arm and forearm.

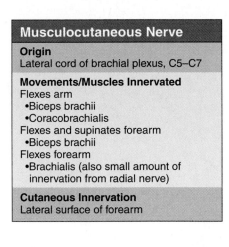

Musculocutaneous Nerve

Origin
Lateral cord of brachial plexus, C5–C7

Movements/Muscles Innervated
Flexes arm
 •Biceps brachii
 •Coracobrachialis
Flexes and supinates forearm
 •Biceps brachii
Flexes forearm
 •Brachialis (also small amount of
  innervation from radial nerve)

Cutaneous Innervation
Lateral surface of forearm

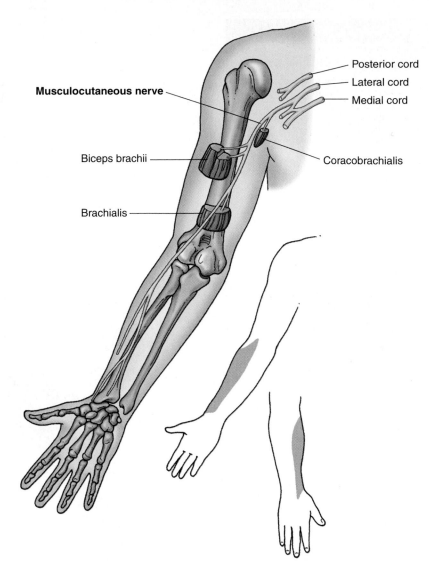

Figure 14.9   Musculocutaneous Nerve

Route of the musculocutaneous nerve and the muscles it innervates. The inset depicts the cutaneous distribution of the nerve (*shaded area*).

Clinical Note

Damage to the median nerve occurs most commonly where it enters the wrist through the **carpal tunnel.** This tunnel is created by the concave organization of the carpal bones and the flexor retinaculum on the anterior surface of the wrist. None of the connective tissue components expands readily. Inflammation in the wrist or an increase in the size of the tendons in the carpal tunnel can produce pressure within the tunnel, compressing the median nerve and resulting in numbness, tingling, and pain in the fingers. This condition is referred to as **carpal tunnel syndrome.** Surgery is often required to relieve the pressure.

Persons attempting suicide by cutting the wrists commonly cut the median nerve proximal to the carpal tunnel.

Lumbar and Sacral Plexuses

The **lumbar plexus** originates from the ventral rami of spinal nerves L1–L4, and the **sacral plexus** from L4–S4. Because of their close, overlapping relationship and their similar distribution, however, the two plexuses often are considered together as a single **lumbosacral plexus** (L1–S4; figure 14.12). Four major nerves exit the lumbosacral plexus and enter the lower limb: the obturator, femoral, tibial, and common fibular (peroneal). The obturator nerve innervates the medial thigh; the femoral nerve innervates the anterior thigh; the tibial nerve innervates the posterior thigh and leg; and the common fibular nerve innervates the posterior thigh, the anterior and lateral leg, and the foot. Other lumbosacral nerves supply the lower back, the hip, and the lower abdomen.

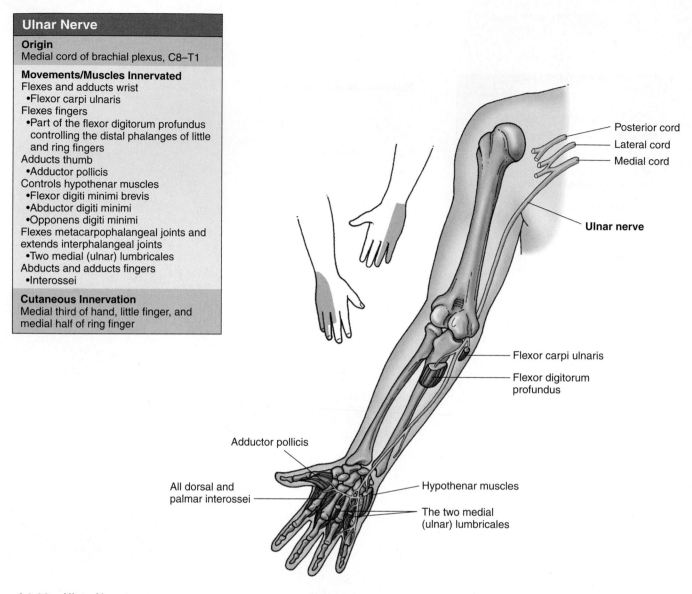

Ulnar Nerve

Origin
Medial cord of brachial plexus, C8–T1

Movements/Muscles Innervated
Flexes and adducts wrist
 •Flexor carpi ulnaris
Flexes fingers
 •Part of the flexor digitorum profundus
  controlling the distal phalanges of little
  and ring fingers
Adducts thumb
 •Adductor pollicis
Controls hypothenar muscles
 •Flexor digiti minimi brevis
 •Abductor digiti minimi
 •Opponens digiti minimi
Flexes metacarpophalangeal joints and
extends interphalangeal joints
 •Two medial (ulnar) lumbricales
Abducts and adducts fingers
 •Interossei

Cutaneous Innervation
Medial third of hand, little finger, and
medial half of ring finger

Posterior cord
Lateral cord
Medial cord

Ulnar nerve

Flexor carpi ulnaris

Flexor digitorum
profundus

Adductor pollicis

Hypothenar muscles

All dorsal and
palmar interossei

The two medial
(ulnar) lumbricales

Figure 14.10   Ulnar Nerve
Route of the ulnar nerve and the muscles it innervates. The inset depicts the cutaneous distribution of the nerve (*shaded area*).

Obturator Nerve

The **obturator** (ob′tū-rā′tŏr) **nerve** supplies the muscles that adduct the thigh. Its cutaneous sensory distribution is to the medial side of the thigh (figure 14.13).

Femoral Nerve

The **femoral nerve** innervates the iliopsoas and sartorius muscles and the quadriceps femoris group. Its cutaneous sensory distribution is the anterior and lateral thigh and the medial leg and foot (figure 14.14).

Tibial and Common Fibular Nerves

The **tibial** and **common fibular** (**peroneal**) (per-o-nē′ăl) **nerves** originate from spinal segments L4–S3 and are bound together within a connective tissue sheath for the length of the thigh (figure 14.15). These two nerves, combined within the same sheath, are referred to jointly as the **ischiadic** (is-kē-ad′ik) **nerve** (see figure 14.12); formerly called the **sciatic** (sī-at′ik) **nerve.** The term sciatic originated as a degenerate form of ischiadic, and the International Conference of Anatomists has recently decided to begin using the correct term. The ischiadic nerve, by far the largest peripheral

Median Nerve

Origin
Medial and lateral cords of brachial plexus, C5–T1

Movements/Muscles Innervated
Pronates forearm
•Pronator teres
•Pronator quadratus
Flexes and abducts wrist
•Flexor carpi radialis
Flexes wrist
•Palmaris longus
Flexes fingers
•Part of flexor digitorum profundus controlling the distal phalanx of the middle and index fingers
•Flexor digitorum superficialis
Controls thumb muscle
•Flexor pollicis longus
Controls thenar muscles
•Abductor pollicis brevis
•Opponens pollicis
•Flexor pollicis brevis
Flexes metacarpophalangeal joints and extends interphalangeal joints
•Two lateral (radial) lumbricales

Cutaneous Innervation
Lateral two thirds of palm of hand, thumb, index and middle fingers, and the lateral half of ring finger and dorsal tips of the same fingers

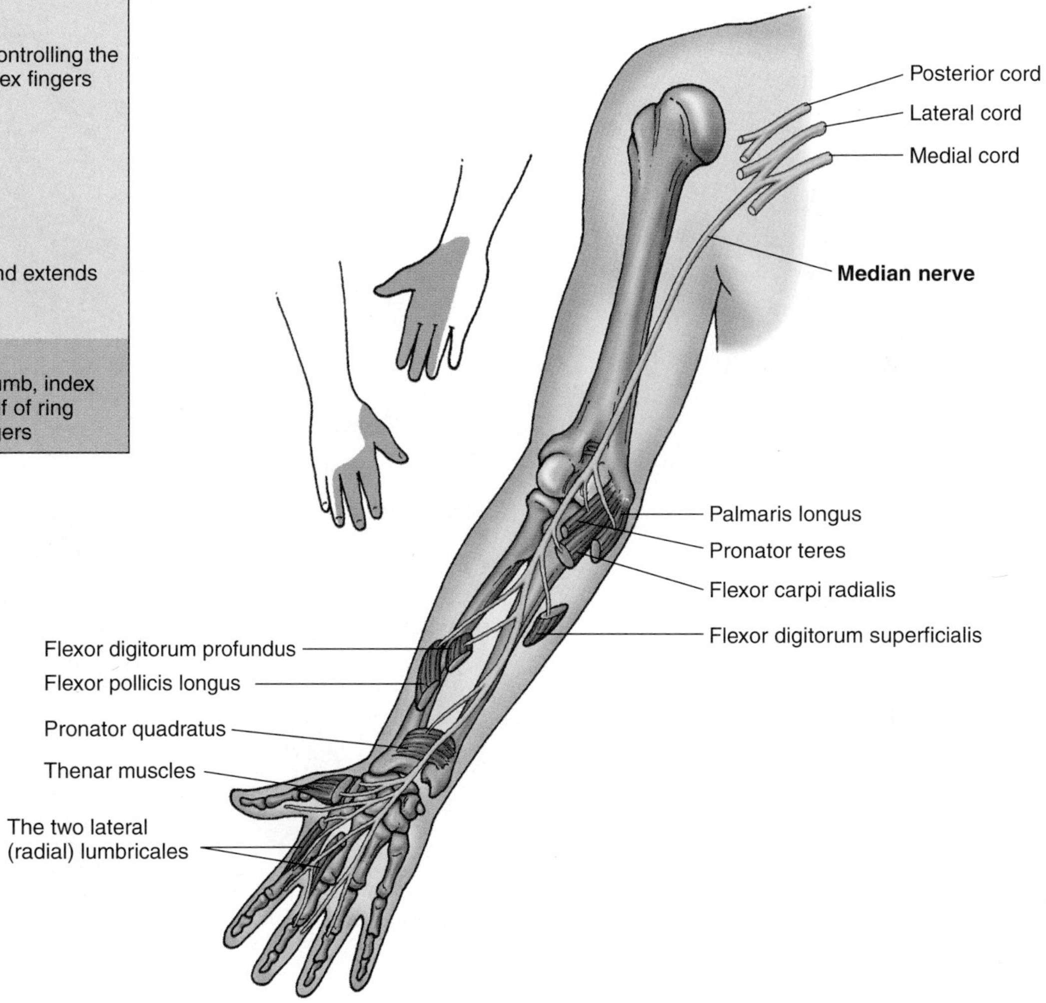

Figure 14.11   Median Nerve
Route of the median nerve and the muscles it innervates. The inset depicts the cutaneous distribution of the nerve (*shaded area*).

nerve in the body, passes through the greater ischiadic notch in the pelvis and descends in the posterior thigh to the popliteal fossa, where the two portions of the ischiadic nerve separate.

The tibial nerve innervates most of the posterior thigh and leg muscles. It branches in the foot to form the **medial** and **lateral plantar** (plan′tăr) **nerves** that innervate the plantar muscles of the foot and the skin over the sole of the foot. Another branch, the **sural** (sū′răl) **nerve,** supplies part of the cutaneous innervation over the calf of the leg and the plantar surface of the foot (see figure 14.15).

Clinical Note

If a person sits on a hard surface for a considerable time, the ischiadic (sciatic) nerve may be compressed against the ischial portion of the coxa. When the person stands up, a tingling sensation described as "pins and needles" can be felt throughout the lower limb, and the limb is said to have "gone to sleep."

The ischiadic nerve may be seriously injured in a number of ways. A ruptured disk or pressure from the uterus during pregnancy may compress the roots of the ischiadic nerve. Other possibilities for causing ischiadic nerve damage include hip injury or an improperly administered injection in the hip region.

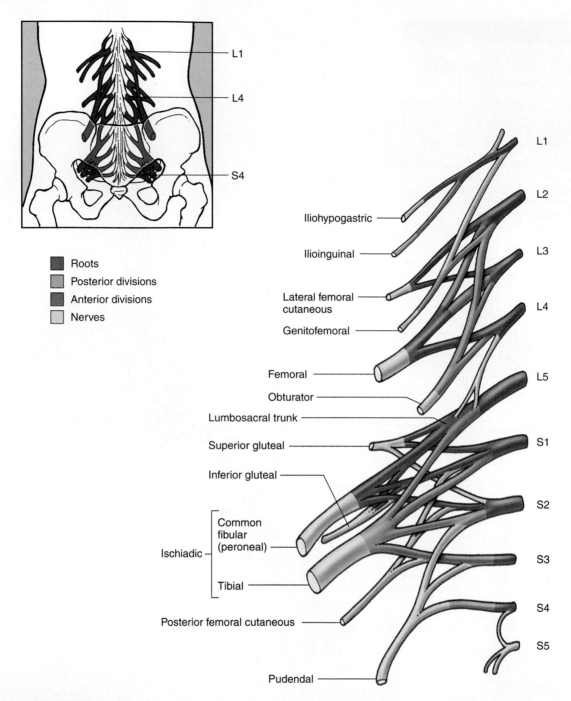

L1

L4

S4

Roots
Posterior divisions
Anterior divisions
Nerves

L1

L2

Iliohypogastric

Ilioinguinal

L3

Lateral femoral
cutaneous

L4

Genitofemoral

Femoral

L5

Obturator

Lumbosacral trunk

Superior gluteal

S1

Inferior gluteal

S2

Common
fibular
(peroneal)

Ischiadic

S3

Tibial

Posterior femoral cutaneous

S4

S5

Pudendal

Figure 14.12   Lumbosacral Plexus

The roots of the plexus are formed by the ventral rami of the spinal nerves L1–S4 and form anterior and posterior divisions, which give rise to the lumbrosacral nerves. The lumbo sacral trunk joins the lumbar and sacral plexuses.

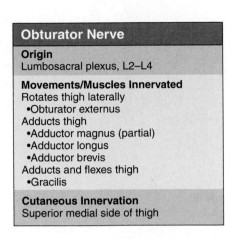

Obturator Nerve

Origin
Lumbosacral plexus, L2–L4

Movements/Muscles Innervated
Rotates thigh laterally
•Obturator externus
Adducts thigh
•Adductor magnus (partial)
•Adductor longus
•Adductor brevis
Adducts and flexes thigh
•Gracilis

Cutaneous Innervation
Superior medial side of thigh

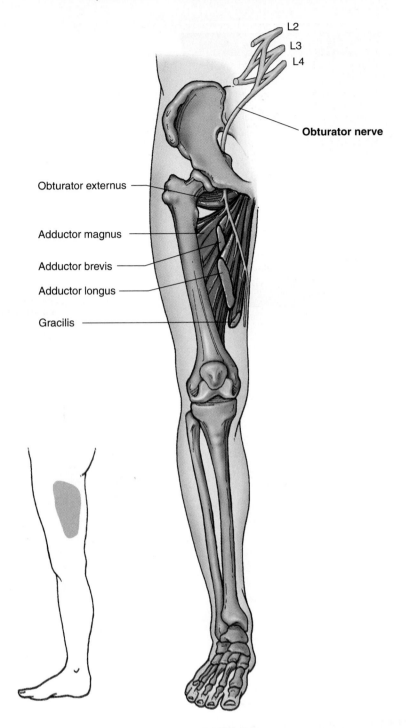

Figure 14.13   Obturator Nerve
Route of the obturator nerve and the muscles it innervates. The inset depicts the cutaneous distribution of the nerve (*shaded area*).

The common fibular nerve divides into the **deep** and **superficial fibular (peroneal) nerves.** These branches innervate the anterior and lateral muscles of the leg and foot. The cutaneous distribution of the common fibular nerve and its branches is the lateral and anterior leg and the dorsum of the foot (figure 14.16).

Other Lumbosacral Plexus Nerves

In addition to the nerves just described, the lumbosacral plexus gives rise to nerves that supply the lower abdominal muscles (iliohypogastric nerve), the hip muscles that act on the femur (gluteal nerves), and the muscles of the abdominal

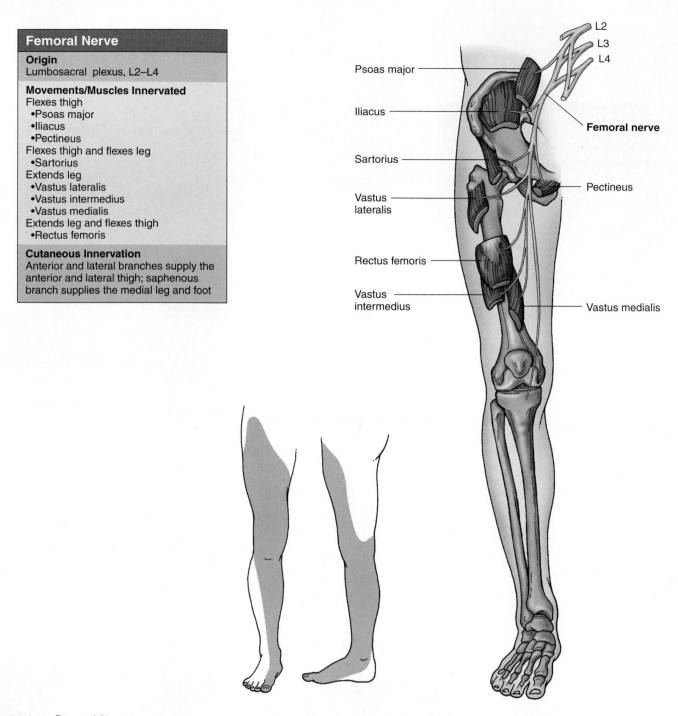

Femoral Nerve

Origin
Lumbosacral plexus, L2–L4

Movements/Muscles Innervated
Flexes thigh
  • Psoas major
  • Iliacus
  • Pectineus
Flexes thigh and flexes leg
  • Sartorius
Extends leg
  • Vastus lateralis
  • Vastus intermedius
  • Vastus medialis
Extends leg and flexes thigh
  • Rectus femoris

Cutaneous Innervation
Anterior and lateral branches supply the anterior and lateral thigh; saphenous branch supplies the medial leg and foot

Figure 14.14   Femoral Nerve

Route of the femoral nerve and the muscles it innervates. The inset depicts the cutaneous distribution of the nerve (*shaded area*).

floor (pudendal nerve; see figure 14.12). The **iliohypogastric** (il′ē-ō-hī-pō-gas′trik), **ilioinguinal** (il′ē-ō-ing′gwi-năl), **genitofemoral** (jen′i-tō-fem′ŏ-răl), **cutaneous femoral,** and **pudendal** (pyū-den′dăl) nerves innervate the skin of the suprapubic area, the external genitalia, the superior medial thigh, and the posterior thigh. The pudendal nerve plays a vital role in sexual stimulation and response.

Clinical Note

Branches of the pudendal nerve are anesthetized before a doctor performs an episiotomy for childbirth. An **episiotomy** (e-piz-ē-ot′ō-mē) is a cut in the perineum that makes the opening of the birth canal larger.

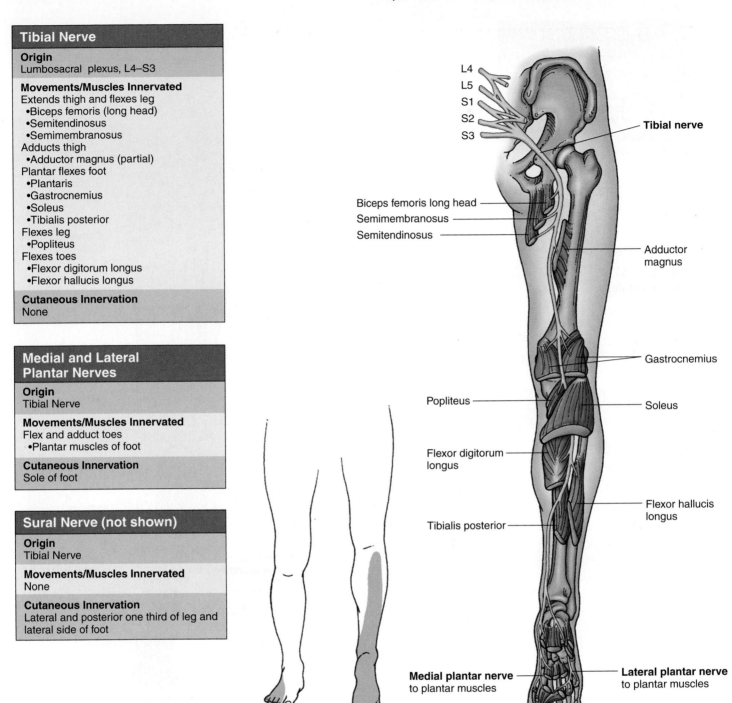

Tibial Nerve

Origin
Lumbosacral plexus, L4–S3

Movements/Muscles Innervated
Extends thigh and flexes leg
•Biceps femoris (long head)
•Semitendinosus
•Semimembranosus
Adducts thigh
•Adductor magnus (partial)
Plantar flexes foot
•Plantaris
•Gastrocnemius
•Soleus
•Tibialis posterior
Flexes leg
•Popliteus
Flexes toes
•Flexor digitorum longus
•Flexor hallucis longus

Cutaneous Innervation
None

Medial and Lateral Plantar Nerves

Origin
Tibial Nerve

Movements/Muscles Innervated
Flex and adduct toes
•Plantar muscles of foot

Cutaneous Innervation
Sole of foot

Sural Nerve (not shown)

Origin
Tibial Nerve

Movements/Muscles Innervated
None

Cutaneous Innervation
Lateral and posterior one third of leg and lateral side of foot

Figure 14.15   Tibial Nerve
Route of the tibial nerve and the muscles it innervates. The inset depicts the cutaneous distribution of the nerve (*shaded area*).

Coccygeal Plexus

The **coccygeal** (kok-sij′ē-ăl) **plexus** is a very small plexus formed from the ventral rami of spinal nerves S4, S5, and the coccygeal nerve. This small plexus supplies motor innervation to muscles of the pelvic floor and sensory cutaneous innervation to the skin over the coccyx. Some skin over the coccyx is innervated by the dorsal rami of the coccygeal nerves.

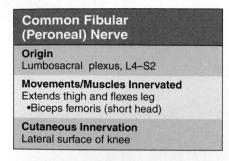

Common Fibular (Peroneal) Nerve

Origin
Lumbosacral plexus, L4–S2

Movements/Muscles Innervated
Extends thigh and flexes leg
 •Biceps femoris (short head)

Cutaneous Innervation
Lateral surface of knee

Deep Fibular (Peroneal) Nerve

Origin
Common fibular (peroneal) nerve

Movements/Muscles Innervated
Dorsiflexes foot
 •Tibialis anterior
 •Peroneus tertius
Extends toes
 •Extensor digitorum longus
 •Extensor hallucis longus

Cutaneous Innervation
Great and second toe

Superficial Fibular (Peroneal) Nerve

Origin
Common fibular (peroneal) nerve

Movements/Muscles Innervated
Plantar flexes and everts foot
 •Peroneus longus
 •Peroneus brevis
Extends toes
 •Extensor digitorum brevis

Cutaneous Innervation
Dorsal anterior third of leg and dorsum of foot

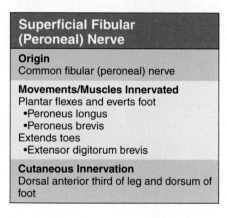

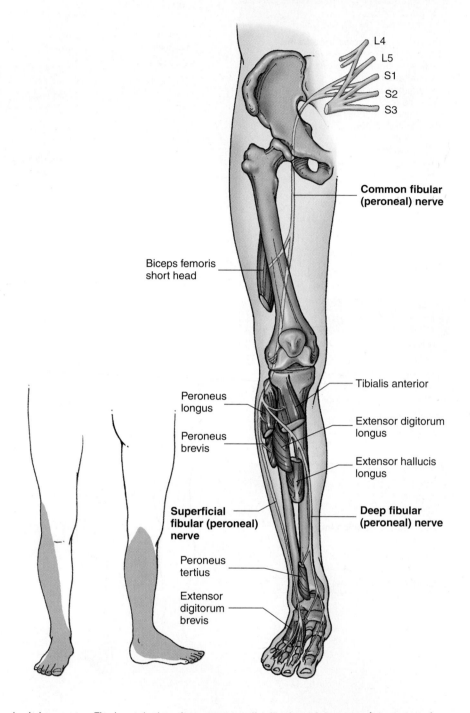

Figure 14.16   Fibular Nerve

Route of the common fibular (peroneal) nerve and the muscles it innervates. The inset depicts the cutaneous distribution of the nerve (*shaded area*).

Summary

Cranial Nerves

1. Cranial nerves perform sensory, somatic motor, proprioceptive, and parasympathetic functions.

2. The olfactory (I) and optic (II) nerves are involved in the sense of smell and vision.

3. The oculomotor nerve (III) innervates four of six extrinsic eye muscles and the upper eyelid. The oculomotor nerve also provides parasympathetic supply to the iris and lens of the eye.
4. The trochlear nerve (IV) controls an extrinsic eye muscle.
5. The trigeminal nerve (V) supplies the muscles of mastication, as well as a middle ear muscle, a palatine muscle, and two throat muscles. The trigeminal nerve has the greatest cutaneous sensory distribution of any cranial nerve. There are three branches of the trigeminal nerve. Two of the three trigeminal nerve branches innervate the teeth.
6. The abducens nerve (VI) controls an extrinsic eye muscle.
7. The facial nerve (VII) supplies the muscles of facial expression, an inner ear muscle, and two throat muscles. It is involved in the sense of taste. It is parasympathetic to two sets of salivary glands and to the lacrimal glands.
8. The vestibulocochlear nerve (VIII) is involved in the sense of hearing and balance.
9. The glossopharyngeal nerve (IX) is involved in taste and supplies tactile sensory innervation from the posterior tongue, middle ear, and pharynx. It is also sensory for receptors that monitor blood pressure and gas levels in the blood. The glossopharyngeal nerve is parasympathetic to the parotid salivary glands.
10. The vagus nerve (X) innervates the muscles of the pharynx, palate, and larynx. It is also involved in the sense of taste. The vagus nerve is sensory for the pharynx and larynx and for receptors that monitor blood pressure and gas levels in the blood. The vagus nerve is sensory for thoracic and abdominal organs. The vagus nerve provides parasympathetic innervation to the thoracic and abdominal organs.
11. The accessory nerve (XI) has a cranial and a spinal component. The cranial component joins the vagus nerve. The spinal component supplies the sternocleidomastoid and trapezius muscles.
12. The hypoglossal nerve (XII) supplies the intrinsic tongue muscles, three of four extrinsic tongue muscles, and two throat muscles.

Spinal Nerves

1. Nerve rootlets from the spinal cord combine to form the ventral (efferent) and dorsal (afferent) roots, which join to form spinal nerves.
2. There are 8 cervical, 12 thoracic, 5 lumbar, 5 sacral pairs, and 1 coccygeal pair of spinal nerves.
3. Spinal nerves have specific cutaneous distributions called dermatomes.
4. Spinal nerves branch to form rami.
   - The dorsal rami supply the muscles and skin near the midline of the back.
   - The ventral rami in the thoracic region form intercostal nerves that supply the thorax and upper abdomen. The remaining ventral rami join to form plexuses (see following

summary sections). Communicating rami supply sympathetic nerves (see chapter 16).

Cervical Plexus

Spinal nerves C1–C4 form the cervical plexus that supplies some muscles and the skin of the neck and shoulder. The phrenic nerves innervate the diaphragm.

Brachial Plexus

1. Spinal nerves C5–T1 form the brachial plexus, which supplies the upper limb.
2. The axillary nerve innervates the deltoid and teres minor muscles and the skin of the shoulder.
3. The radial nerve supplies the extensor muscles of the arm and forearm and the skin of the posterior surface of the arm, forearm, and hand.
4. The musculocutaneous nerve supplies the anterior arm muscles and the skin of the lateral surface of the forearm.
5. The ulnar nerve innervates most of the intrinsic hand muscles and the skin on the ulnar side of the hand.
6. The median nerve innervates the pronator and most of the flexor muscles of the forearm, most of the thenar muscles, and the skin of the radial side of the palm of the hand.
7. Other nerves supply most of the muscles that act on the arm, the scapula, and the skin of the medial arm and forearm.

Lumbar and Sacral Plexuses

1. Spinal nerves L1–S4 form the lumbosacral plexus.
2. The obturator nerve supplies the muscles that adduct the thigh and the skin of the medial thigh.
3. The femoral nerve supplies the muscles that flex the thigh and extend the leg and the skin of the anterior and lateral thigh and the medial leg and foot.
4. The tibial nerve innervates the muscles that extend the thigh and flex the leg and the foot. It also supplies the plantar muscles and the skin of the posterior leg and the sole of the foot.
5. The common fibular nerve supplies the short head of the biceps femoris, the muscles that dorsiflex and plantar flex the foot, and the skin of the lateral and anterior leg and the dorsum of the foot.
6. In the thigh the tibial nerve and the common fibular nerve are combined as the ischiadic (sciatic) nerve.
7. Other lumbosacral nerves supply the lower abdominal muscles, the hip muscles, and the skin of the suprapubic area, external genitalia, and upper medial thigh.

Coccygeal Plexus

Spinal nerves S4, S5, and Co form the coccygeal plexus, which supplies the muscles of the pelvic floor and the skin over the coccyx.

Content Review

1. What are the three major functions of the cranial nerves?
2. Which cranial nerves are sensory only? With what sense is each of these nerves associated?

3. Name the cranial nerves that are somatic motor and proprioceptive only. What muscles or muscle groups does each nerve supply?

4. The sensory cutaneous innervation of the face is provided by what cranial nerve? How is this nerve important in dentistry? Name the muscles that would no longer function if this nerve were damaged.

5. Which four cranial nerves have a parasympathetic function? Describe the functions of each of these nerves.

6. Name the cranial nerves that control movement of the eyeball.

7. Which cranial nerves are involved in the sense of taste? What part of the tongue does each nerve supply?

8. Speech production involves what cranial nerves? Describe the branches of these nerves.

9. Differentiate between rootlet, dorsal root, ventral root, and spinal nerve. Which of them contain sensory fibers or motor fibers?

10. Describe all the spinal nerves by name and number. Where do they exit the vertebral column?

11. What is a dermatome? Why are dermatomes clinically important?

12. Contrast dorsal, ventral, and communicating rami of spinal nerves. What muscles do the dorsal rami innervate?

13. Describe the distribution of the ventral rami of the thoracic region.

14. What is a plexus? What happens to the axons of spinal nerves as they pass through a plexus?

15. Name the main spinal plexuses and the spinal nerves associated with each one.

16. Name the structures innervated by the cervical plexus. Describe the innervation of the phrenic nerve.

17. Name the five major nerves that emerge from the brachial plexus. List the muscles they innervate and the areas of the skin they supply. In addition to these five nerves, name the muscles and skin areas supplied by the remaining brachial plexus nerves.

18. Name the two principle nerves that arise from the lumbosacral plexus, and describe the muscles and skin areas they supply. Describe the structures innervated by the remaining lumbosacral nerves.

19. What structures are innervated by the coccygeal plexus?

Develop Your Reasoning Skills

1. Injury to which cranial nerve produces the following symptoms?
   a. A patient has strabismus, the left eye is turned inferiorly and laterally.
   b. A patient is unable to move the eyeball medially.
   c. The upper left eyelid is drooping (ptosis), and the left pupil is dilated.

2. Damage to which cranial nerve produces each of the following symptoms?
   a. Vertigo (a balance disorder in which the patient feels as if she is spinning)
   b. Tinnitus (a ringing sound in the ear)
   c. Anosmia (loss of the sense of smell)
   d. Blindness

3. Wendy Frost went cross-country skiing on a very cold day. Afterward she was unable to close her right eye or raise her right eyebrow. Although she could move her jaw, she had difficulty chewing because food would drool out of her mouth. What nerve was affected? Explain the observed symptoms.

4. Red Blister has herpes zoster, a viral infection. The virus lies dormant in nervous tissue and sporadically becomes active, causing lesions in the skin supplied by the nerves that it infects. Red exhibited lesions on the scalp, the forehead, and the cornea of the eye. Name the nerve that was infected by the herpesvirus. Be as specific as you can.

5. The act of swallowing involves two components. The voluntary portion involves the movement of food to the superior part of the pharynx. There the food stimulates tactile receptors that initiate the second component, an involuntary swallowing reflex. Sensory impulses from the tactile receptors are transmitted to the medulla oblongata. From the medulla, motor impulses are transmitted back to the muscles of the soft palate, pharynx, larynx, and throat, and the food is swallowed. Name the two cranial nerves that convey the sensory impulses and the five cranial nerves that carry the motor impulses.

6. A cancer patient has his left lung removed. To reduce the space remaining where the lung was removed, the diaphragm on the left side was paralyzed, allowing the abdominal viscera to push the diaphragm upward. What nerve would be cut? Where would be a good place to cut it, and when would the surgery be done?

7. Based on sensory response to pain in the skin of the hand, how could you distinguish between damage to the ulnar, median, and radial nerves?

8. During a difficult delivery the baby's arm delivered first. The attending physician grasped the arm and forcefully pulled it. Later a nurse observed that the baby could not abduct or adduct the medial four fingers and flexion of the wrist was impaired. What nerve was damaged?

9. Two patients were admitted to the hospital. According to their charts, both had herniated disks that were placing pressure on the roots of the ischiadic nerve. One patient had pain in the buttocks and the posterior aspect of the thigh. The other patient experienced pain in the posterior and lateral aspects of the leg and the lateral part of the ankle and foot. Explain how the same condition, a herniated disk, could produce such different symptoms.

10. In an automobile accident a woman suffered a crushing hip injury. For each of the conditions given here, state what nerve was damaged.
    a. Unable to adduct the thigh
    b. Unable to extend the leg
    c. Unable to flex the leg
    d. Loss of sensation from the skin of the anterior thigh
    e. Loss of sensation from the skin of the medial thigh

Web Site Link

For a listing of the most current web sites related to this chapter, please visit the Seeley home page at:
http://www.mhhe.com/biosci/ap/seeleyap/

Chapter Fifteen

The Senses

Objectives

1. Define the term sensation. Explain the differences between somatic, visceral, and special senses, and give examples of each.

2. Describe the major sensory nerve endings, their locations, and their functions.

3. Describe the histologic structure and function of the olfactory epithelium and the olfactory bulb.

4. Describe the central nervous system connections for smell, and explain how these connections elicit various visceral and conscious responses to smell.

5. Explain adaptation to odor, and describe various levels at which it can occur.

6. Describe the histology and function of a typical taste bud.

7. Describe the central nervous system pathways and cortical locations for taste.

8. List the accessory structures of the eye, and explain their functions.

9. Describe the tunics of the eye, and give the function of each.

10. Describe the internal structures of the eye, including the lens, ciliary body, iris, macula lutea, and optic disc, and explain the function of each.

11. Describe the compartments and chambers of the eye, and explain the function of the canal of Schlemm.

12. Explain light refraction and reflection and how they relate to eye function.

13. Name and describe the structure and function of the layers of the retina.

14. Describe the chemical reaction in rhodopsin as a result of light stimulation.

15. Outline the central nervous system pathway for visual input, and describe what happens to images from each half of the visual fields.

16. Describe the structures of the outer and middle ears, and state their functions.

17. Describe the microanatomy of the cochlea, and explain how sounds are detected.

18. Describe the central nervous system pathway for the appreciation of hearing, the pathway for determining pitch, and the reflex pathway for dampening sound.

19. Explain how the static and kinetic labyrinths function in balance.

20. Describe the central nervous system pathways for balance.

Most people are familiar with the story of Helen Keller, who, because she was both deaf and blind, found it very difficult to learn to interact with other people. Yet she could be reached, and taught, by means of her other senses, such as the sense of touch. Imagine how difficult interactions would be if none of the senses were functioning. Through our senses, our brains establish and maintain contact with the world around us.

Historically, five senses were recognized: smell, taste, sight, hearing, and touch. Today we recognize many more. Some specialists propose that there are at least 20, perhaps as many as 40, different senses. The senses may be classified into two major groups: general and special. Touch, pressure, pain, temperature, vibration, and proprioception (knowing where the body is located in space) are referred to as general senses. Smell, taste, sight, hearing, and balance are referred to as special senses.

Usually receptors are quite specific and are most sensitive to only one type of stimulus. **Mechanoreceptors** respond to mechanical stimuli, such as compression, bending, or stretching of cells. **Chemoreceptors** respond to chemicals that become attached to receptors on their membranes. **Photoreceptors** respond to light striking the receptor cells. **Thermoreceptors** respond to changes in temperature at the site of the receptor. **Nociceptors** (nō-si-sep′ters; L. *noceo* means hurt) respond to painful mechanical, chemical, or thermal stimuli.

The general senses of touch, pressure, and proprioception all depend on a variety of mechanoreceptors, as do the special senses of sound and balance. The general senses of pain and temperature depend on nociceptors and thermoreceptors, respectively. The special senses of smell and taste depend on chemoreceptors, and the special sense of vision depends on photoreceptors.

Classification of the Senses

The general senses can be divided into two groups: the somatic senses, which provide general sensory information about the body and the environment, and the visceral senses, which provide information about various internal organs. The special senses are those with highly localized receptors that provide specific information about the environment (table 15.1). Modalities of sensation refer to the form of the sensation. For example, somatic modalities include touch, pressure, temperature, proprioception, and pain. Visceral modalities consist primarily of pain and pressure. Modalities of the special senses are smell, taste, sight, sound, and balance.

Sensation

Sensation, or **perception,** is the conscious awareness of stimuli received by sensory receptors. The brain constantly receives a wide variety of stimuli from both inside and outside the body. To be perceived as a conscious sensation, action potentials generated by receptors must reach the cerebral cortex. Some action potentials reach other areas of the brain such as the cerebellum, where they are not consciously perceived. In addition, the cerebral cortex screens much of what it receives, ignoring a large part of the action potentials that reach it. Because we ignore much of the information carried by these action potentials, they remain subconscious.

Table 15.1	Classification of the Senses	
Type of Sense	**Receptor Type**	**Initiation of Response**
Somatic		
Touch	Mechanoreceptors	Compression of receptors
Pressure	Mechanoreceptors	Compression of receptors
Temperature	Thermoreceptors	Temperature around nerve endings
Proprioception	Mechanoreceptors	Compression of receptors
Pain	Nociceptors	Irritation of nerve endings (e.g., mechanical, chemical, or thermal)
Visceral		
Pain	Nociceptors	Irritation of nerve endings
Pressure	Mechanoreceptors	Compression of receptors
Special		
Smell	Chemoreceptors	Binding of molecules to membrane receptors
Taste	Chemoreceptors	Binding of molecules to membrane receptors
Sight	Photoreceptors	Chemical change in receptors initiated by sight
Sound	Mechanoreceptors	Bending of microvilli on receptor cells
Balance	Mechanoreceptors	Bending of microvilli on receptor cells

Sensation requires the following steps:

1. There must be a **stimulus** originating either inside or outside of the body.
2. There must be a **receptor** capable of detecting the stimulus and of converting the stimulus into action potentials.
3. Action potentials generated by the receptor must be **conducted** to the central nervous system (CNS) through afferent nerves.
4. Action potentials reaching the CNS must be **translated** within the brain before the person is **aware** of the stimulus.

Because sensations are an awareness of a stimulus, they can occur only in the cerebral cortex, where action potentials are translated. Receptors can only generate local potentials or action potentials in response to a stimulus. Action potentials from a given receptor are carried to a specific part of the cerebral cortex, where they are interpreted. Sensation results from the interpretation of these action potentials. In the case of touch, the sensation is **projected** to the site of origin of the stimulus, that is, the sensation of touch is perceived to be at the site at which the sensory receptor was stimulated, such as at the tip of the finger, rather than in the cerebral cortex, where the sensation actually occurs.

Clinical Note

The CNS cannot be consciously aware of all stimuli. Much information is processed at an unconscious level. For example, receptors that detect changes in blood pH or blood pressure do not produce a conscious sensation.

In addition, humans exhibit selective awareness. That is, we are more aware of sensations on which we have our attention focused than on other sensations. If we were simultaneously aware of all the stimuli with which the brain is constantly bombarded, it is unlikely that we would be able to function. Being aware of so many stimuli would require us to constantly make conscious decisions about the stimuli to which we should respond. We might not be able to focus our attention on a single task and might be incapable of performing simple functions.

For example, as you read this paragraph, are you aware of the weight of the book in your hands if you are holding it, or the weight of your arms on the desk or on your lap if you are reading at a desk? Are you aware of the small noises around you? You are probably not aware of your clothes touching your body until your attention is drawn to them.

An example of projection is a person's finger touching an object. Touch receptors are stimulated, and action potentials are conducted to the primary somatic sensory area of the cerebral cortex, where the action potentials are interpreted as touch in the finger. Remember that the finger maps to a specific part of the primary somatic sensory cortex (see chapter 13). If an afferent neuron from a touch receptor is electrically stimulated anywhere along its length and action potentials are produced in that afferent neuron, the action potentials are conducted to the finger region of the primary somatic sensory cortex and are interpreted as touch in the finger. Phantom pain (see chapter 13) is another example of projection.

1   P R E D I C T

What would be the result of directly stimulating a neuron in the "finger region" of the somatic sensory cortex?

✔ *Answer in Appendix F*

Some sensations have the quality of **adaptation,** or **accommodation,** a decreased sensitivity to a continued stimulus. After exposure to a stimulus for a time, the response of the receptors or the afferent pathways to a certain stimulus strength lessens from that which occurs when the stimulus was first applied. For example, when a person first gets dressed, tactile receptors and pathways relay information to the brain that create an awareness that the clothes are touching the skin. After a time, the action potentials decrease, and the clothes are ignored.

Another way that sensations change through time occurs in proprioception. Two types of proprioceptors are involved in providing positional information: tonic receptors and phasic receptors. **Tonic** receptors generate action potentials as long as a stimulus is applied and accommodate very slowly. Information from tonic proprioceptors allows a person to know, for example, where the little finger is at all times without having to look for it. **Phasic** receptors, by contrast, accommodate rapidly and are most sensitive to changes in stimuli. For example, information from phasic proprioceptors allows us to know where our hand is as it moves, allowing us to control its movement through space and to predict where it will be in the next moment.

We are usually not conscious of tonic or phasic input, but through selective awareness we can call up the information when we wish. For example, where is the thumb of your right hand at this moment? Were you aware of its position a few seconds ago?

Types of Afferent Nerve Endings

At least eight major types of sensory nerve endings are involved in general sensation: free nerve endings, Merkel's disks, hair follicle receptors, Pacinian corpuscles, Meissner's corpuscles, Ruffini's end organs, Golgi tendon apparatuses, and muscle spindles (table 15.2 and figure 15.1). Many of these nerve endings are associated with the skin; others are associated with deeper structures, such as tendons, ligaments, and muscles; and some can be found in both the skin and deeper structures. In general, sensory nerve endings are classified into three groups: **exteroreceptors, visceroreceptors,** or **proprioceptors.** Exteroreceptors (cutaneous receptors) are associated with the skin, visceroreceptors are associated with the viscera or organs, and proprioceptors are associated with joints, tendons, and other connective tissue. Exteroreceptors provide information about

Table 15.2   Afferent Nerve Endings

Type of Nerve Ending	Structure	Function
Free nerve endings	Branching, no capsule	Pain, itch, tickle, temperature, joint movement, and proprioception
Merkel's disks	Flattened expansions at the end of axons; each expansion associated with a Merkel's cell	Light touch and superficial pressure
Hair follicle	Wrapped around hair follicles or extending along the hair axis, each axon supplies several hairs, and each hair receives branches from several neurons, resulting in considerable overlap	Light touch; responds to very slight bending of the hair
Pacinian corpuscle	Onion-shaped capsule of several cell layers with a single central nerve process	Deep cutaneous pressure, vibration, and proprioception
Meissner's corpuscles	Several branches of a single axon associated with wedge-shaped epitheloid cells and surrounded by a connective tissue capsule	Two-point discrimination
Ruffini's end organs	Branching axon with numerous small, terminal knobs surrounded by a connective tissue capsule	Continuous touch or pressure; respond to depression or stretch of the skin
Golgi tendon organs	Surrounds a bundle of tendon fascicles and is enclosed by a delicate connective tissue capsule; nerve terminations are branched with small swellings applied to individual tendon fascicles	Proprioception associated with the stretch of a tendon; important in the control of muscle contraction
Muscle spindle	Three to 10 striated muscle fibers enclosed by a loose connective tissue capsule, striated only at the ends, with sensory nerve endings in the center	Proprioception associated with detection of muscle stretch; important for control of muscle tone

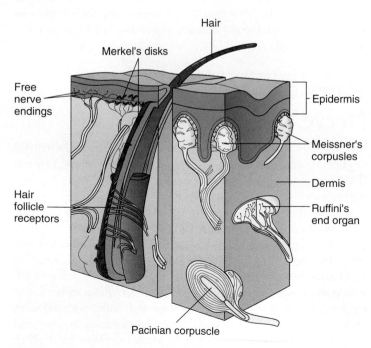

Figure 15.1   Sensory Cutaneous Nerve Endings

the external environment, visceroreceptors provide information about the internal environment, and proprioceptors provide information about body position, movement, and the extent of stretch or the force of muscular contractions.

The simplest and most common sensory nerve endings are the **free nerve endings** (see figure 15.1), which are distributed throughout almost all parts of the body. Most visceroreceptors consist of free nerve endings, which are responsible for a number of sensations, including pain, temperature, itch, and movement. The free nerve endings responsible for temperature detection are of three types. One type, the **cold receptors,** increases its rate of action potential firing as the skin is cooled. The second type, **warm receptors,** increases its rate of action potential firing as skin temperature increases. Both cold and warm receptors are phasic receptors and therefore respond most strongly to changes in temperature. Cold receptors are 10–15 times more numerous in any given area of skin than warm receptors. The third type is a **pain receptor,** which is stimulated in extreme cold or heat. At very cold temperatures (0°–12°C), only pain fibers are stimulated. The pain sensation ends as the temperature exceeds 15°C. Between 12° and 35°C, cold fibers are stimulated. Nerve fibers from warm receptors are stimulated between 25° and 47°C. "Comfortable" temperatures, between 25° and 35°C,

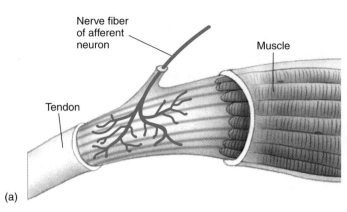

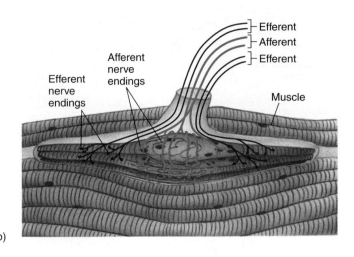

Figure 15.2   Nerve Endings Associated with Muscles

(*a*) Golgi tendon organ. (*b*) Muscle spindle.

therefore stimulate both warm and cold receptors. Temperatures above 47°C not only no longer stimulate warm receptors but actually stimulate cold receptors and pain receptors.

2   **P R E D I C T**

How might a very cold object placed in the hand be misperceived as being hot?

✔ *Answer in Appendix F*

Merkel's (mer'kĕlz), or **tactile, disks,** are more complex than free nerve endings (see figure 15.1) and consist of axonal branches that end as flattened expansions, each associated with a specialized epithelial cell. They are distributed throughout the basal layers of the epidermis just superficial to the basement membrane and are associated with dome-shaped mounds of thickened epidermis in hairy skin. Merkel's disks are involved with the sensations of light touch and superficial pressure. These receptors can detect a skin displacement of 1/25,000th of an inch.

Hair follicle receptors, or **hair end organs,** respond to very slight bending of the hair and are involved in light touch (see figure 15.1). Even though these nerve endings are extremely sensitive, requiring very little stimulation to elicit a response, they are not very discriminative (the sensation is not very well localized). The dendritic tree at the distal end of a sensory axon has several hair follicle receptors. The field of hairs innervated by these receptors overlaps with the fields of hair follicle receptors of adjacent axons. The considerable overlap that exists in the sensory endings of afferent neurons helps explain why light touch is not very discriminative, yet because of converging signals within the CNS, it is very sensitive (see chapter 13).

Pacinian (pa-sin'ē-an, pa-chin'-ē-an), or **lamellated, corpuscles** are very complex nerve endings resembling an onion (see figure 15.1). A single dendrite extends to the center of each lamellated corpuscle. The corpuscles are located within the deep dermis or hypodermis, where they are re-

sponsible for deep cutaneous pressure and vibration. Pacinian corpuscles associated with the joints help relay proprioceptive information about joint positions.

Meissner's (mīs'nerz), or **tactile, corpuscles** are distributed throughout the dermal papillae (see figure 15.1 and chapter 5) and are involved in two-point discrimination touch. Meissner's corpuscles are numerous and close together in the tongue (about 2 mm apart) and fingertips (about 4 mm apart) but are less numerous and more widely separated in other areas such as the back (about 64 mm apart).

Ruffini's (rū-fē'nēz) **end organs** are located in the dermis of the skin (see figure 15.1), primarily in the fingers. They respond to pressure on the skin directly superficial to the receptor and to stretch of adjacent skin. These nerve endings are important in responding to continuous touch or pressure.

Golgi tendon organs are proprioceptive nerve endings associated with the fibers of a tendon at the muscle–tendon junction (figure 15.2*a*). They are activated by an increase in tendon tension, whether it is caused by contraction of the muscle or by passive stretch of the tendon. When a great amount of tension is applied to the tendon, afferent fibers from the Golgi tendon organs inhibit the motor neuron of the associated muscle and cause it to relax (Golgi tendon reflex), preventing muscle and tendon damage caused by excessive tension (see chapter 13).

Muscle spindles consist of 3–10 muscle fibers that are striated only on the ends so that only the ends of each cell can contract. Sensory nerve endings are wrapped around the center of the muscle fibers, and gamma motor neurons supply the striated ends (figure 15.2*b*). When the skeletal muscle is stretched or when the ends of the muscle spindle fibers contract, the center of the muscle spindle is stretched, stimulating the sensory neurons of the muscle spindle. The sensory neurons synapse with alpha motor neurons in the spinal cord. When alpha motor neurons are stimulated, they cause a rapid contraction of the stretched muscle (the stretch reflex; see chapter 13).

Muscle spindles are important to the control and tone of postural muscles. Brain centers act through descending tracts

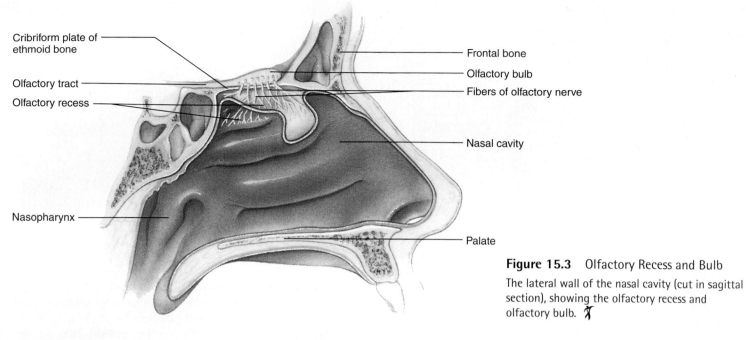

Cribriform plate of ethmoid bone

Olfactory tract

Olfactory recess

Nasopharynx

Frontal bone

Olfactory bulb

Fibers of olfactory nerve

Nasal cavity

Palate

Figure 15.3   Olfactory Recess and Bulb

The lateral wall of the nasal cavity (cut in sagittal section), showing the olfactory recess and olfactory bulb.

to either increase or decrease action potentials in gamma motor fibers. Stimulation of the gamma motor system activates the stretch reflex, which in turn increases the tone of the muscles involved.

Olfaction

Olfaction (ol-fak'shŭn), the sense of smell, occurs in response to odors that stimulate sensory receptors located in the extreme superior region of the nasal cavity, called the **olfactory recess** (figure 15.3). Most of the nasal cavity is involved in respiration, with only a small superior part devoted to olfaction. During normal respiration, air passes through the nasal cavity without much of it entering the olfactory recess. The major anatomic features of the nasal cavity are described in chapter 23 in relation to respiration. The specialized nasal epithelium of the olfactory recess is called the **olfactory epithelium.**

---

3   **P R E D I C T**

Explain why it sometimes helps to inhale slowly and deeply through the nose when trying to identify an odor.

✔ *Answer in Appendix F*

Olfactory Epithelium and Bulb

There are 10 million **olfactory neurons** within the olfactory epithelium (figure 15.4). The axons of these bipolar neurons project through numerous small foramina of the bony cribriform plate (see chapter 7) to the **olfactory bulbs** (see figures 15.3 and 15.4). **Olfactory tracts** project from the bulbs to cerebral cortex.

The dendrites of olfactory neurons extend to the epithelial surface of the nasal cavity, and their ends are modified into bulbous enlargements called **olfactory vesicles** (see figure 15.4). These vesicles possess cilia called **olfactory hairs,** which lie in a thin mucous film on the epithelial surface.

Airborne molecules enter the nasal cavity and are dissolved in the fluid covering the olfactory epithelium. They interact with chemoreceptor molecules of the olfactory hair membranes. Although the exact nature of this interaction is not yet fully understood, it appears that chemoreceptors are membrane receptor molecules that bind to certain molecules. Once an odor-producing molecule has become bound to a receptor, the cilia of the olfactory neurons react by depolarizing and initiating action potentials in the olfactory neurons.

The mechanism of olfactory discrimination is not completely known. Most physiologists believe that the wide variety of detectable smells, which is about 4000 for the average person, are actually combinations of a smaller number of primary odors. Seven primary classes of odors have been proposed: (1) camphoraceous, (2) musky, (3) floral, (4) pepperminty, (5) ethereal, (6) pungent, and (7) putrid. It is very unlikely, however, that this list is an accurate representation of all primary odors, and some studies point to the possibility of as many as 50 primary odors.

The threshold for the detection of odors is very low, so very few molecules are required to trigger the response. Apparently there is rather low specificity in the olfactory epithelium in that a given receptor may react to more than one type of airborne molecule.

Clinical Note

Methylmercaptan, which has a nauseating odor similar to that of rotten cabbage, is added at a concentration of about 1 part per million to natural gas. A person can detect the odor of about 1/25 billionth of a milligram of the substance and therefore is aware of the presence of the more dangerous but odorless natural gas.

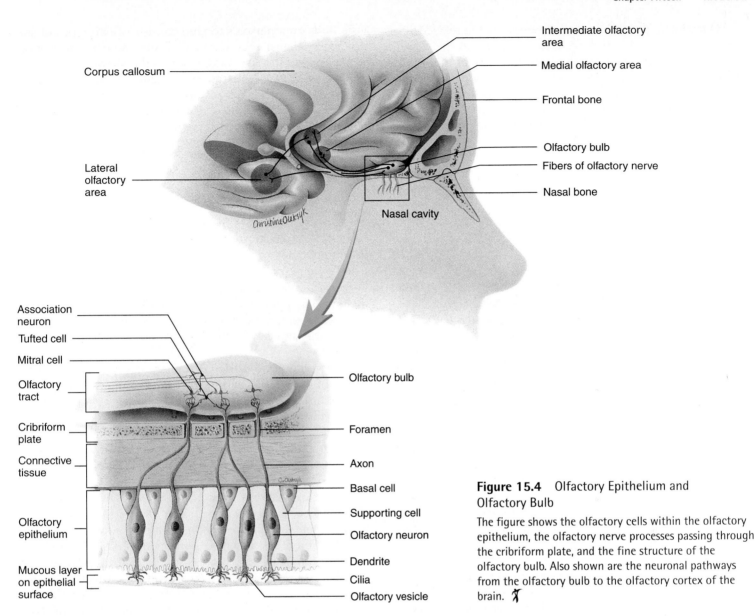

Corpus callosum

Intermediate olfactory area

Medial olfactory area

Frontal bone

Olfactory bulb

Fibers of olfactory nerve

Nasal bone

Lateral olfactory area

Nasal cavity

Association neuron

Tufted cell

Mitral cell

Olfactory tract

Cribriform plate

Connective tissue

Olfactory epithelium

Mucous layer on epithelial surface

Olfactory bulb

Foramen

Axon

Basal cell

Supporting cell

Olfactory neuron

Dendrite

Cilia

Olfactory vesicle

Figure 15.4 Olfactory Epithelium and Olfactory Bulb

The figure shows the olfactory cells within the olfactory epithelium, the olfactory nerve processes passing through the cribriform plate, and the fine structure of the olfactory bulb. Also shown are the neuronal pathways from the olfactory bulb to the olfactory cortex of the brain. 𝑋

The primary olfactory neurons have the most exposed nerve endings of any neurons, and they are constantly being replaced. The entire olfactory epithelium, including the neurosensory cells, is lost about every 2 months as the olfactory epithelium degenerates and is lost from the surface. Lost ol-factory cells are replaced by proliferation of **basal cells** in the olfactory epithelium. This replacement of olfactory neurons is unique among neurons, most of which are permanent cells that do not replicate and are not replaced by other neurons (see chapter 4).

Neuronal Pathways for Olfaction

Axons from the olfactory neurons (cranial nerve I) enter the olfactory bulb (see figure 15.4), where they synapse with **mitral** (mi′trăl; triangular cells; shaped like a bishop's miter or hat) **cells** or **tufted cells.** The mitral and tufted cells relay olfactory information to the brain through the olfactory tracts and synapse with **association neurons** in the olfactory bulb. Association neurons also receive input from nerve cell processes entering the olfactory bulb from the brain. As a result of input from both mitral cells and the brain, association neurons can modify olfactory information before it leaves the olfactory bulb.

Olfaction is the only major sensation that is relayed directly to the cerebral cortex without first going to the thalamus. Each olfactory tract terminates in an area of the brain called the **olfactory cortex.** The olfactory cortex is in the frontal lobe, within the lateral fissure of the cerebrum and can be divided structurally and functionally into three areas: lateral, intermediate, and medial. The **lateral olfactory area** is involved in the conscious perception of smell. The **medial olfactory area** is responsible for visceral and emotional reactions to odors and has connections to the limbic system, through which it connects to the hypothalamus. Axons extend from the **intermediate olfactory area** along the olfactory tract to the bulb, synapse with the association neurons, and thus constitute a major mechanism by which sensory information is modulated within the olfactory bulb.

---

4   P R E D I C T

The olfactory system quickly adapts to continued stimulation, and a particular odor becomes unnoticed before very long, even though the odor molecules are still present in the air. Describe as many sites as you can in the olfactory pathways where such adaptation can occur.

✔ *Answer in Appendix F*

▊Taste

The sensory structures that detect **gustatory, or taste,** stimuli are the **taste buds.** Most taste buds are associated with specialized portions of the tongue called **papillae** (pă-pil'ē). Taste buds, however, are also located on other areas of the tongue, the palate, and even the lips and throat, especially in children. There are four major types of papillae, named according to their shape (figure 15.5): **circumvallate** (ser'kŭm-val'āt, meaning surrounded by a groove or valley), **fungiform** (fŭn'ji-fōrm, meaning mushroom-shaped), **foliate** (fō'lē-āt, meaning leaf-shaped), and **filiform** (fil'i-fōrm, meaning filament-shaped). Taste buds (figure 15.5b–d) are associated with circumvallate, fungiform, and foliate papillae. Filiform papillae are the most numerous papillae on the surface of the tongue but have no taste buds.

Circumvallate papillae are the largest but least numerous of the papillae. Eight to 12 of these papillae form a V-shaped row along the border between the anterior and posterior parts of the tongue (see figure 15.5a). Fungiform papillae are scattered irregularly over the entire dorsal surface of the tongue and appear as small red dots interspersed among the far more numerous filiform papillae. Foliate papillae are distributed over the sides of the tongue and contain the most sensitive of the taste buds. They are most numerous in young children and decrease with age, becoming rare in adults.

Histology of Taste Buds

Taste buds are oval structures embedded in the epithelium of the tongue and mouth (see figure 15.5f). Each of the 10,000 taste buds on a person's tongue consists of two types of specialized epithelial cells. One type forms the exterior supporting capsule of the taste bud, whereas the interior of each bud consists of about 50 **gustatory, or taste, cells.** Like olfactory cells, cells of the taste buds are replaced continuously, each having a normal life span of about 10 days. Each gustatory cell has several microvilli, called **gustatory hairs,** extending from its apex into a tiny opening in the epithelium called the **gustatory, or taste, pore.**

Function of Taste

Substances dissolved in the saliva enter the taste pore and apparently become attached to chemoreceptor molecules on the cell membranes of the gustatory hairs, causing a change in membrane permeability and subsequent depolarization of the taste cells. These cells have no axons and do not generate their own action potentials. Neurotransmitters apparently are released from the gustatory cells and stimulate action potentials in the axons of gustatory neurons associated with the taste cells.

Hot or cold food temperatures may interfere with the ability of the taste buds to function in tasting food. If a cold fluid is held in the mouth, the fluid becomes warmed by the body, and the taste becomes enhanced. On the other hand, adaptation is very rapid for taste. This adaptation apparently occurs both at the level of the taste bud and within the CNS. Adaptation may begin within 1 or 2 s after a taste sensation is perceived, and complete adaptation may occur within 5 min.

The basic tastes detected by the taste buds can be divided into four types: sour, salty, bitter, and sweet. Even though there are only four primary tastes, a fairly large number of different tastes can be perceived, presumably by combining the four basic taste sensations. As with olfaction, the specificity of the receptor molecules is not perfect. For example, artificial sweeteners have different chemical structures than the sugars they are designed to replace and are often many times more powerful than natural sugars in stimulating taste sensations.

Many of the sensations thought of as being taste are strongly influenced by olfactory sensations. This phenomenon can be demonstrated by pinching one's nose to close the nasal passages, while trying to taste something. With olfaction blocked, it is difficult to distinguish between the taste of a piece of apple and a piece of potato. Much of the "taste" is lost by this action. Although all taste buds are able to detect all four of the basic tastes, each taste bud is usually most sensitive to one. The stimulus type to which each taste bud responds most strongly is related more to its position on the tongue than to the type of papilla with which it is associated (figure 15.6). The tip of the tongue commonly reacts more strongly to sweet and salty tastes, the back of the tongue to bitter taste, and the sides of the tongue to sour taste.

Thresholds vary for the four primary tastes. Sensitivity for bitter substances is the highest; sensitivities for sweet and salty tastes are the lowest. Sugars, some other carbohydrates, and some proteins produce sweet tastes; acids produce sour

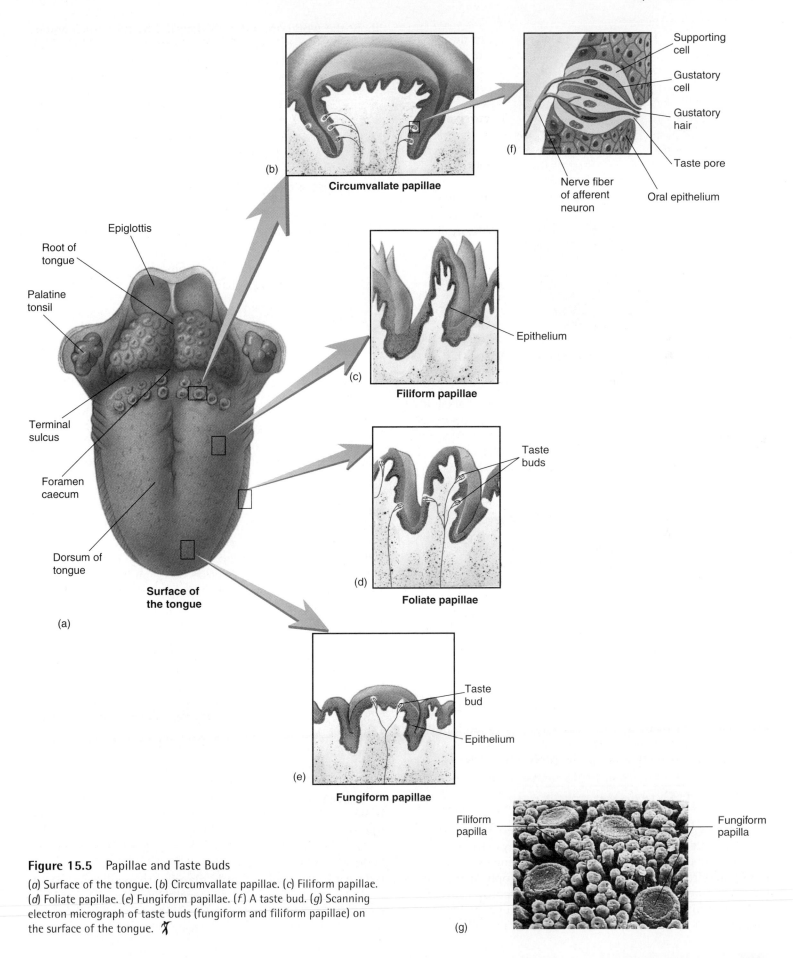

Figure 15.5   Papillae and Taste Buds

(*a*) Surface of the tongue. (*b*) Circumvallate papillae. (*c*) Filiform papillae. (*d*) Foliate papillae. (*e*) Fungiform papillae. (*f*) A taste bud. (*g*) Scanning electron micrograph of taste buds (fungiform and filiform papillae) on the surface of the tongue.

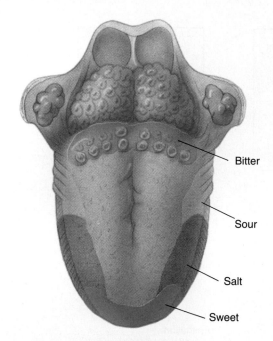

Figure 15.6   Regions of the Tongue Sensitive to Various Tastes

tastes; metal ions tend to produce salty tastes; and alkaloids (bases) produce bitter tastes. Many alkaloids are poisonous; thus the high sensitivity for bitter tastes may be protective. On the other hand, humans tend to crave sweet and salty tastes, perhaps in response to the body's need for sugars, carbohydrates, proteins, and minerals.

Neuronal Pathways for Taste

Taste from the anterior two-thirds of the tongue, except from the circumvallate papillae, is carried by means of a branch of the facial nerve (VII) called the **chorda tympani** (kōr'dă tim'pă-nē; so named because it crosses over the surface of the tympanic membrane of the middle ear). Taste from the posterior third of the tongue, the circumvallate papillae, and the superior pharynx is carried by means of the glossopharyngeal nerve (IX). In addition to these two major nerves, the vagus nerve (X) carries a few fibers for taste sensation from the epiglottis.

These nerves extend from the taste buds to the tractus solitarius of the medulla oblongata (figure 15.7). Fibers from this nucleus decussate and extend to the thalamus. Neurons from the thalamus project to the taste area of the cortex, which is at the extreme inferior end of the postcentral gyrus.

▌Visual System

The visual system includes the eyes, the accessory structures, and the optic nerves (II), tracts, and pathways. The eyes respond to light and initiate afferent action potentials, which are transmitted from the eyes to the brain by the optic nerves and tracts. The accessory structures, such as eyebrows, eyelids, eyelashes, and tear glands, help protect the eyes from direct sunlight and damaging particles. Much of the information about the world around us is detected by the visual system. Our education is largely based on visual input and depends on our ability to read words and numbers. Visual input includes information about light and dark, color and hue.

Accessory Structures

Accessory structures protect, lubricate, move, and in other ways aid in the function of the eye (figures 15.8 through 15.12). They include the eyebrows, eyelids, conjunctiva, lacrimal apparatus, and extrinsic eye muscles.

Eyebrows

The **eyebrows** protect the eyes by preventing perspiration, which can irritate the eyes, from running down the forehead and into them, and they help shade the eyes from direct sunlight.

Eyelids

The **eyelids,** also called **palpebrae** (pal-pē'brē), with their associated lashes, protect the eyes from foreign objects. The space between the two eyelids is called the **palpebral fissure,** and the angles where the eyelids join at the medial and lateral margins of the eye are called **canthi** (kan'thī, meaning corners of the eye). The medial canthus contains a small reddish-pink mound called the **caruncle** (kar'ŭng-kl, meaning a mound of tissue). The caruncle contains some modified sebaceous and sweat glands.

The eyelids consist of five layers of tissue. From the outer to the inner surface, they are (1) a thin layer of integument on the external surface; (2) a thin layer of areolar connective tissue; (3) a layer of skeletal muscle consisting of the orbicularis oculi and levator palpebrae superioris muscles; (4) a crescent-shaped layer of dense connective tissue called the **tarsal** (tar'săl) **plate,** which helps maintain the shape of the eyelid; and (5) the palpebral conjunctiva (described in the next section), which lines the inner surface of the eyelid and lies against the surface of the eyeball.

If an object suddenly approaches the eye, the eyelids protect the eye by rapidly closing and then opening (blink reflex). Blinking, which normally occurs about 25 times per minute, also helps keep the eye lubricated by spreading tears over the surface of the eye. Movements of the eyelids are a function of skeletal muscles. The orbicularis oculi muscle closes the lids, and the levator palpebrae superioris elevates the upper lid (see chapter 11). The eyelids also help regulate the amount of light entering the eye.

Eyelashes (figure 15.9) are attached as a double or triple row of hairs to the free edges of the eyelids. **Ciliary glands** are modified sweat glands that open into the follicles of the eyelashes, keeping them lubricated. When one of these glands becomes inflamed, it is called a **sty. Meibomian** (mī-bō'mē-an; also called tarsal) **glands** are sebaceous glands near the inner margins of the eyelids and produce **sebum** (sē'bŭm; an oily semifluid substance), which lubricates the lids and restrains tears from flowing over the margin of the

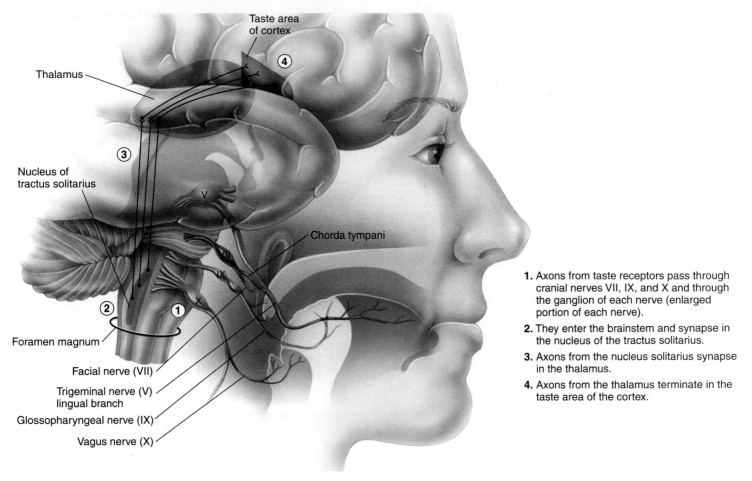

1. Axons from taste receptors pass through cranial nerves VII, IX, and X and through the ganglion of each nerve (enlarged portion of each nerve).
2. They enter the brainstem and synapse in the nucleus of the tractus solitarius.
3. Axons from the nucleus solitarius synapse in the thalamus.
4. Axons from the thalamus terminate in the taste area of the cortex.

Figure 15.7   Pathways for the Sense of Taste

The facial nerve (anterior two-thirds of the tongue), glossopharyngeal nerve (posterior third of the tongue), and vagus nerve (root of the tongue) all carry taste sensations. The trigeminal nerve is also shown. It carries tactile sensations from the anterior two-thirds of the tongue. The chorda tympani from the facial nerve (carrying taste input) joins the trigeminal nerve. ✗

eyelids. An infection or blockage of a meibomian gland is called a **chalazion** (ka-lā′zē-on), or **meibomian cyst.**

Conjunctiva

The **conjunctiva** (kon-jŭnk-tī′vă) (see figure 15.9) is a thin, transparent mucous membrane. The **palpebral conjunctiva** covers the inner surface of the eyelids, and the **bulbar conjunctiva** covers the anterior surface of the eye. The points at which the palpebral and bulbar conjunctivae meet are the superior and inferior **conjunctival fornices.**

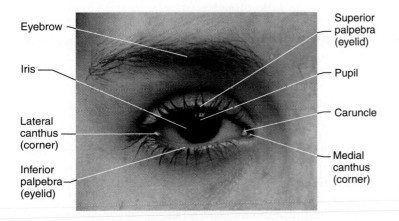

Figure 15.8   The Right Eye and Its Accessory Structures

Clinical Note

Conjunctivitis is an inflammation of the conjunctiva caused by infection or some other irritation. An example of conjunctivitis caused by a bacterium is **acute contagious conjunctivitis,** also called **pinkeye.**

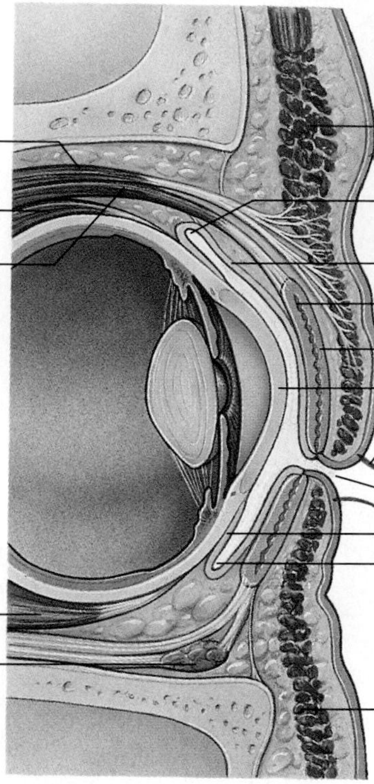

Figure 15.9   Sagittal Section Through the Eye Showing Its Accessory Structures 𝒳

Eyebrow
Orbicularis oculi muscle
Levator palpebrae superioris muscle
Superior conjunctival fornix
Superior rectus muscle
Palpebral conjunctiva
Smooth muscle to tarsal plate
Tarsal (meibomian) gland
Tarsal plate
Cornea
Eyelash
Palpebral fissure
Bulbar conjunctiva
Inferior conjunctival fornix
Inferior rectus muscle
Inferior oblique muscle
Orbicularis oculi muscle

Lacrimal Apparatus

The **lacrimal** (lak′ri-măl) **apparatus** (figure 15.10) consists of a lacrimal gland situated in the superolateral corner of the orbit and a nasolacrimal duct beginning in the inferomedial corner of the orbit. The **lacrimal gland** is innervated by parasympathetic fibers from the facial nerve (VII). The gland produces tears, which leave the gland through several ducts and pass over the anterior surface of the eyeball. Tears are produced constantly by the gland at the rate of about 1 mL/day to moisten the surface of the eye, lubricate the eyelids, and wash away foreign objects. Tears are mostly water, with some salts, mucus, and lysozyme, an enzyme that kills certain bacteria. Most of the fluid produced by the lacrimal glands evaporates from the surface of the eye, but excess tears are collected in the medial corner of the eye by the **lacrimal canaliculi.** The opening of each lacrimal canaliculus is called a **punctum** (pŭngk′tŭm). The upper and lower eyelids each have a punctum near the medial canthus. Each punctum is located on a small lump called the **lacrimal papilla.** The lacrimal canaliculi open into a **lacrimal sac,** which in turn continues into the **nasolacrimal duct** (see figure 15.10). The nasolacrimal duct opens into the inferior meatus of the nasal cavity beneath the inferior nasal concha (see chapter 23).

Clinical Note

Facial nerve damage results in the inability to close the eyelid on the affected side. With the ability to blink being lost, tears cannot be washed across the eye, and the cornea becomes dry. A dry cornea may become ulcerated, and, if not treated, eyesight may be lost.

5   P R E D I C T

Explain why it is often possible to "taste" medications, such as eyedrops, that have been placed into the eyes. Why does a person's nose "run" when he or she cries?

✔ *Answer in Appendix F*

Extrinsic Eye Muscles

Movement of each eyeball is accomplished by six muscles, the **extrinsic muscles** of the eye (see figures 15.11 and 15.12; also see chapter 11). Four of these muscles run more or less straight anteroposteriorly. They are the superior, inferior, medial, and lateral **rectus muscles.** Two muscles, the superior and inferior **oblique muscles,** are placed at an angle to the globe of the eye.

The movements of the eye can be described by a figure resembling the letter "H." The clinical test for normal eye movement is therefore called the **H test.** A person's inability to move his eye toward one part of the "H" may indicate dysfunction of an extrinsic eye muscle.

The superior oblique muscle is innervated by the trochlear nerve (IV). The nerve is so named because the superior oblique muscle goes around a little pulley, or trochlea, in the superomedial corner of the orbit. The lateral rectus muscle is innervated by the abducens nerve (VI), so named because the lateral rectus muscle abducts the eye. The other four extrinsic eye muscles are innervated by the oculomotor nerve (III).

Figure 15.10 Lacrimal Structures

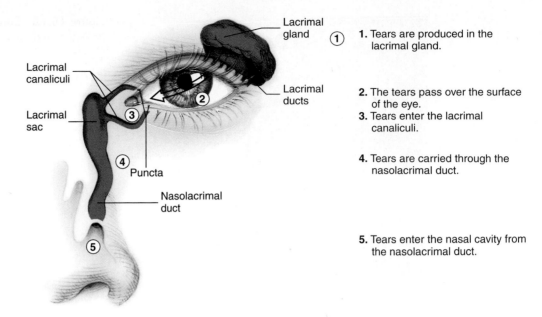

1. Tears are produced in the lacrimal gland.

2. The tears pass over the surface of the eye.
3. Tears enter the lacrimal canaliculi.

4. Tears are carried through the nasolacrimal duct.

5. Tears enter the nasal cavity from the nasolacrimal duct.

Figure 15.11
Extrinsic Muscles of the Eye

(*a*) Superior view.
(*b*) Lateral view.

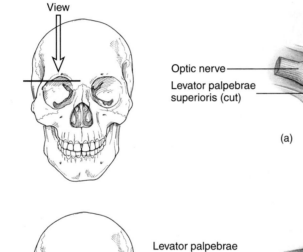
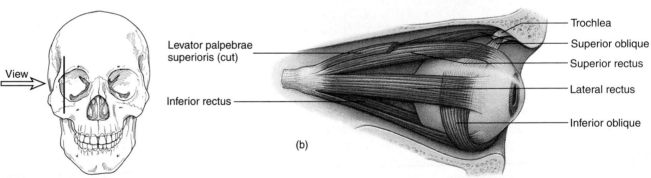

Anatomy of the Eye

The eye is composed of three coats, or tunics (figure 15.13). The outer, or **fibrous, tunic** consists of the sclera and cornea; the middle, or **vascular, tunic** consists of the choroid, ciliary body, and iris; and the inner, or **nervous, tunic** consists of the retina.

Fibrous Tunic

The **sclera** (sklĕr′ă) is the firm, opaque, white, outer layer of the posterior five-sixths of the eye. It consists of dense col-

lagenous connective tissue with elastic fibers. The sclera helps maintain the shape of the eye, protects its internal structures, and provides an attachment point for the muscles that move it. Usually, a small portion of the sclera can be seen as the "white of the eye" when the eye and its surrounding structures are intact (see figure 15.8).

The sclera is continuous anteriorly with the cornea. The **cornea** (kōr′nē-ă) is an avascular, transparent structure that permits light to enter the eye and bends, or refracts, that light as part of the focusing system of the eye. The cornea consists of a connective tissue matrix containing collagen, elastic fibers, and proteoglycans, with a layer of stratified squamous

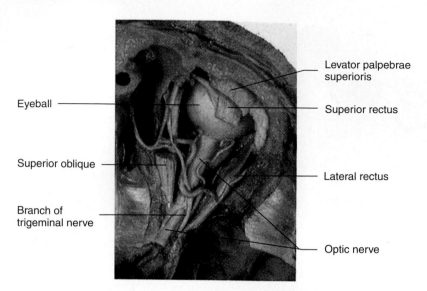

Figure 15.12   Superior Photographic View of the Eye and Its Associated Structures

Levator palpebrae superioris

Eyeball

Superior rectus

Superior oblique

Lateral rectus

Branch of trigeminal nerve

Optic nerve

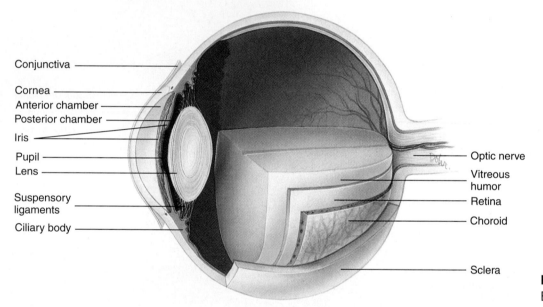

Conjunctiva

Cornea
Anterior chamber
Posterior chamber
Iris
Pupil
Lens

Suspensory ligaments
Ciliary body

Optic nerve
Vitreous humor
Retina
Choroid
Sclera

Figure 15.13   Sagittal Section of the Eye Demonstrating Its Layers

epithelium covering the outer surface and a layer of simple squamous epithelium on the inner surface. Large collagen fibers are white, whereas smaller collagen fibers and proteo glycans are transparent. The cornea is transparent, rather than white like the sclera, in part, because there are fewer large collagen fibers and more proteoglycans in the cornea than in the sclera. The transparency of the cornea also results from its low water content. In the presence of water, proteoglycans trap water and expand, which scatters light. In the absence of water, the proteoglycans decrease in size and do not interfere with the passage of light through the matrix.

Clinical Note

The central part of the cornea receives oxygen from the outside air. Soft plastic contact lenses worn for long periods must therefore be permeable to air so that air can reach the cornea.

The most common eye injuries are cuts or tears of the cornea caused by foreign objects such as stones or sticks hitting the cornea.

The cornea was one of the first organs transplanted. Several characteristics make it relatively easy to transplant: it is easily accessible and relatively easily removed; it is avascular and therefore does not require as extensive circulation as do other tissues; and it is less immunologically active and therefore less likely to be rejected than other tissues.

6   P R E D I C T

Predict the effect of inflammation of the cornea on vision.

✔ *Answer in Appendix F*

Vascular Tunic

The middle tunic of the eyeball is called the vascular tunic because it contains most of the blood vessels of the eyeball (see

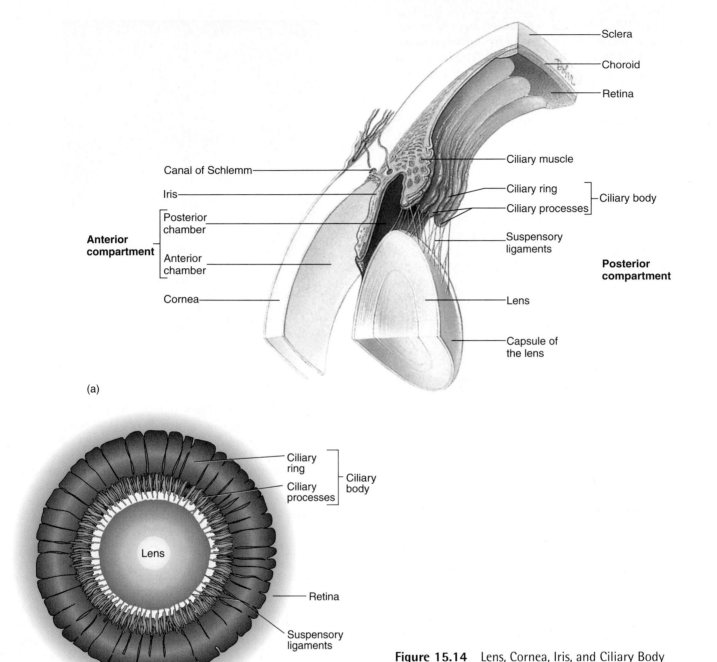

Figure 15.14   Lens, Cornea, Iris, and Ciliary Body

(*a*) The orientation is the same as in figure 15.13. (*b*) The lens and ciliary body seen from an anterior view.

figure 15.13). The arteries of the vascular tunic are derived from a number of arteries called **short ciliary arteries,** which pierce the sclera in a circle around the optic nerve. These arteries are branches of the **ophthalmic** (of-thal′mik) **artery,** which is a branch of the internal carotid artery. The vascular tunic contains a large number of melanin-containing pigment cells and appears black in color. The portion of the vascular tunic associated with the sclera of the eye is the **choroid** (ko′royd). The term choroid means membrane and suggests that this layer is relatively thin (0.1–0.2 mm thick). Anteriorly the vascular tunic consists of the ciliary body and iris.

The **ciliary** (sil′ē-ar-ē) **body** is continuous with the choroid, and the **iris** is attached at its lateral margins to the cil-

iary body (figure 15.14). The ciliary body consists of an outer **ciliary ring** and an inner group of **ciliary processes,** which are attached to the lens by **suspensory ligaments.** The ciliary body contains smooth muscles called the **ciliary muscles,** which are arranged somewhat like a pinwheel, with the outer muscle fibers oriented radially and the central fibers oriented circularly. The ciliary muscles function as a sphincter, and contraction of the ciliary muscles can change the shape of the lens. (This function is described in more detail later in this chapter.) The ciliary processes are a complex of capillaries and cuboidal epithelium that produce aqueous humor.

The **iris** is the "colored part" of the eye, and its color differs from person to person. Brown eyes have brown melanin pigment in the iris. Blue eyes are not caused by a blue pigment but result from the scattering of light by the tissue of the iris, overlying a deeper layer of black pigment. The blue color is produced in a fashion similar to the scattering of light as it passes through the atmosphere to form the blue skies from the black background of space.

The iris is a contractile structure consisting mainly of smooth muscle and surrounding an opening called the **pupil.** Light enters the eye through the pupil, and the iris regulates the amount of light by controlling the size of the pupil. The iris contains two groups of smooth muscles: a circular group called the **sphincter pupillae** (pyū-pil'ē), and a radial group called the **dilator pupillae.** The sphincter pupillae are innervated by parasympathetic fibers from the oculomotor nerve (III) and contract the iris, decreasing or constricting the size of the pupil. The dilator pupillae are innervated by sympathetic fibers and dilate the pupil. The ciliary muscles, sphincter pupillae, and dilator pupillae are sometimes referred to as the intrinsic eye muscles.

Retina

The **retina** is the innermost tunic of the eye (see figure 15.13). It consists of the outer **pigmented retina,** which is pigmented simple cuboidal epithelium, and the inner **sensory retina,** which responds to light. The sensory retina contains 120 million photoreceptor cells called **rods** and another 6 or 7 million **cones,** as well as numerous relay neurons. The visual portion of the retina covers the inner surface of the eye posterior to the ciliary body. A more detailed description of the histology and function of the retina is presented later in this chapter.

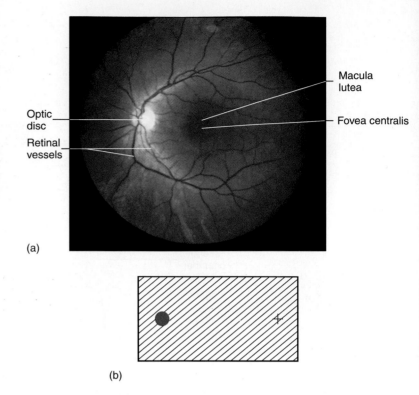

(a)

(b)

Figure 15.15 Ophthalmoscopic View of the Left Retina

(*a*) The posterior wall of the retina as seen when looking through the pupil. Notice the vessels entering the eye through the optic disc (the optic nerve), and notice the macula lutea with the fovea (the part of the retina with the greatest visual acuity). (*b*) Demonstration of the blind spot. Close your right eye. Hold the figure in front of your left eye and stare at the +. Move the figure toward your eye. At a certain point, when the image of the spot is over the optic disc, the red spot seems to disappear.

tralis, which is normally the point where light is focused. The fovea is the portion of the retina with the greatest visual acuity, the ability to see fine images, because the photoreceptor cells are more tightly packed in that portion of the retina than anywhere else. Just medial to the macula lutea is a white spot, the **optic disc,** through which blood vessels enter the eye and spread over the surface of the retina. This is also the spot where nerve processes from the sensory retina meet, pass through the outer two tunics, and exit the eye as the optic nerve. The optic disc contains no photoreceptor cells and does not respond to light; therefore it is called the **blind spot** of the eye.

Clinical Note

The pupil appears black when you look into a person's eye because of the pigment in the choroid and the pigmented portion of the retina. The eye is a closed chamber, which allows light to enter only through the pupil. Light is absorbed by the pigmented inner lining of the eye; thus looking into it is like looking into a dark room. If a bright light is directed into the pupil, however, the reflected light is red because of the blood vessels on the surface of the retina, which is why the pupils in the eyes of a person looking directly at a flash camera are red in a photograph. People with albinism lack the pigment melanin, and the pupil always appears red because there is no pigment to absorb light and prevent it from being reflected from the back of the eye. The diffusely lighted blood vessels in the interior of the eye contribute to the red color of the pupil.

Clinical Note

Ophthalmoscopic examination of the posterior retina can reveal some general disorders of the body. **Hypertension,** or high blood pressure, results in "nicking" (compression) of the retinal veins where the abnormally pressurized arteries cross them. **Increased cerebrospinal fluid (CSF) pressure** associated with hydrocephalus may cause swelling of the optic disc. This swelling is referred to as **papilledema** (pă-pil-e-dē'mă).

When the posterior region of the retina is examined with an **ophthalmoscope** (of-thal'mō-skōp) (figure 15.15), several important features can be observed. Near the center of the posterior retina is a small yellow spot approximately 4 mm in diameter, the **macula lutea** (mak'yū-lă lū'tē-ă). In the center of the macula lutea is a small pit, the **fovea** (fō'vē-ă) **cen-**

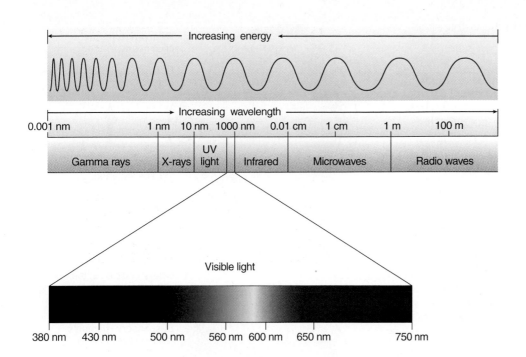

Figure 15.16   The Electromagnetic Spectrum

The spectrum of visible light is pulled out and expanded. The wavelengths of the various colors are also depicted.

Compartments of the Eye

Two major compartments exist within the eye, a larger compartment posterior to the lens and a much smaller compartment anterior to the lens (see figure 15.13). The **anterior compartment** is divided into two chambers: the **anterior chamber** lies between the cornea and iris, and a smaller **posterior chamber** lies between the iris and lens (see figure 15.14). These two chambers are filled with **aqueous humor,** which helps maintain intraocular pressure. The pressure within the eye keeps the eye inflated and is largely responsible for maintaining the shape of the eye. The aqueous humor also refracts light and provides nutrition for the structures of the anterior chamber, such as the cornea, which has no blood vessels. Aqueous humor is produced by the ciliary processes as a blood filtrate and is returned to the circulation through a venous ring at the base of the cornea called the **canal of Schlemm** (shlem), or the **scleral venous sinus** (see figure 15.14). The production and removal of aqueous humor results in "circulation" of aqueous humor and maintenance of a constant intraocular pressure. If circulation of the aqueous humor is inhibited, a defect called **glaucoma** (glaw-kō′mă), which is an abnormal increase in intraocular pressure, can result (see the following Clinical Focus).

The posterior compartment of the eye is much larger than the anterior compartment. It is surrounded almost completely by the retina and is filled with a transparent jellylike substance, the **vitreous** (vit′rē-ŭs) **humor.** The vitreous humor is not produced on a regular basis as is the aqueous humor, and its turnover is extremely slow. The vitreous humor helps maintain intraocular pressure and therefore the shape of the eyeball, and it holds the lens and the retina in place. It also functions in the refraction of light in the eye.

Lens

The **lens** is an unusual biologic structure. Transparent and biconvex, with the greatest convexity on its posterior side, the lens consists of a layer of cuboidal epithelial cells on its anterior surface and a posterior region of very long columnar epithelial cells called **lens fibers.** Cells from the anterior epithelium proliferate and give rise to the lens fibers at the equator of the lens. The lens fibers lose their nuclei and other cellular organelles and accumulate a special set of proteins called **crystallines.** This crystalline lens is covered by a highly elastic transparent **capsule.**

The lens is suspended between the two eye compartments by the suspensory ligaments of the lens, which are connected from the ciliary body to the lens capsule.

Functions of the Complete Eye

The eye functions much like a camera. The iris allows light into the eye, and the light is focused by the lens, cornea, and humors onto the retina. The light striking the retina is converted into action potentials that are relayed to the brain.

Light

The electromagnetic spectrum is the entire range of wavelengths or frequencies of electromagnetic radiation from very short gamma waves at one end of the spectrum to the longest radio waves at the other end (figure 15.16). **Visible light** is the portion of the electromagnetic spectrum that can be detected by the human eye. Light has characteristics of both particles (photons) and waves, with a wavelength between 400 and 700 nm. This range sometimes is called the range of visible light or, more correctly, the **visible spectrum.** Within the visible spectrum each color has a different wavelength.

Light Refraction and Reflection

An important characteristic of light is that it can be refracted (bent). As light passes from air to a denser substance such as glass or water, its speed is reduced. If the surface of that substance is at an angle other than 90 degrees to the direction the light rays are traveling, the rays are bent as a result of variation in the speed of light as it encounters the new medium. This bending of light is called **refraction.**

If the surface of a lens is concave, with the lens thinnest in the center, the light rays diverge as a result of refraction. If the surface is convex, with the lens thickest in the center, the light rays tend to converge. As light rays converge, they finally reach a point at which they cross. This point is called the **focal point,** and causing light to converge is called **focusing.** No image is formed exactly at the focal point, but an inverted, focused image can form on a surface located some distance past the focal point. How far past the focal point the focused image forms depends on a number of factors. A biconvex lens causes light to focus closer to the lens than does a lens with a single convex surface. Furthermore, the more nearly spherical the lens, the closer to the lens the light is focused; the more flattened the biconcave lens, the more distant is the point where the light is focused.

If light rays strike an object that is not transparent, they bounce off the surface. This phenomenon is called **reflection.** If the surface is very smooth, such as the surface of a mirror, the light rays bounce off in a specific direction. If the surface is rough, the light rays are reflected in several directions, producing a more diffuse reflection. We can see most solid objects because of the light reflected from their surfaces.

Focusing of Images on the Retina

The focusing system of the eye projects a clear image on the retina. Light rays converge as they pass from the air through the convex cornea. Additional convergence occurs as light encounters the aqueous humor, lens, and vitreous humor. The greatest contrast in media density is between the air and the cornea; therefore, the greatest amount of convergence occurs at that point. The shape of the cornea and its distance from the retina are fixed, however, so that no adjustment in the location of the focal point can be made by the cornea. Fine adjustment in focal point location is accomplished by changing the shape of the lens. In general, focusing can be accomplished in two ways. One is to keep the shape of the lens constant and move it nearer or farther from the point at which the image will be focused, such as occurs in a camera, microscope, or telescope. The second way is to keep the distance constant and to change the shape of the lens, which is the technique used in the eye.

As light rays enter the eye and are focused, the image formed just past the focal point is inverted (figure 15.17). Action potentials that represent the inverted image are passed to the visual cortex of the cerebrum, where they are interpreted by the brain as being right side up.

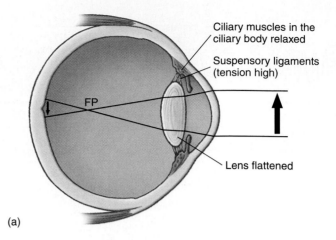

(a)

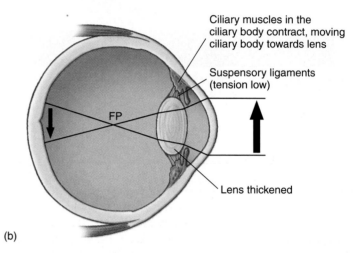

(b)

Figure 15.17   Ability of the Lens to Focus Images on the Retina

The focal point (FP) is where light rays cross. (*a*) Distant object: the lens is flattened, and the image is focused on the retina. (*b*) Close object: the lens is more rounded, and the image is focused on the retina. 𝕏

> ## Clinical Note
>
> Because the visual image is inverted when it reaches the retina, the image of the world focused on the retina is upside down. The brain processes information from the retina so that the world is perceived the way "it really is." If a person wears glasses that invert the image entering the eye, he will see the world upside down for a few days, after which time the brain adjusts to the new input to set the world right side up again. If the glasses are then removed, another adjustment period is required before the world is made right by the brain.

When the ciliary muscles are relaxed, the suspensory ligaments of the ciliary body maintain elastic pressure on the lens, keeping it relatively flat and allowing for distant vision (figure 15.17*a*). The condition in which the lens is flattened so that nearly parallel rays from a distant object are focused on the retina is referred to as **emmetropia** (em-ĕ-trō′pē-ă, meaning measure) and is the normal resting condition of the lens. The point at which the lens does not have to thicken for focusing to occur is called the **far point of vision** and normally is 20 feet or more from the eye.

When an object is brought closer than 20 feet to the eye, three events occur to bring the image into focus on the retina: accommodation by the lens, constriction of the pupil, and convergence of the eyes.

---

7   P R E D I C T

Explain how several hours of reading can cause eyestrain, or eye fatigue. Describe what structures are involved.

✔ *Answer in Appendix F*

---

1. *Accommodation.* When focusing on a nearby object, the ciliary muscles contract as a result of parasympathetic stimulation from the oculomotor nerve (III). This sphincterlike contraction pulls the choroid toward the lens, reducing the tension on the suspensory ligaments. This allows the lens to assume a more spherical form because of its own elastic nature (figure 15.17*b*). The spherical lens then has a more convex surface, causing greater refraction of light. This process is called accommodation.

   As light strikes a solid object, the rays are reflected in every direction from the surface of the object. Only a small portion of the light rays reflected from a solid object, however, pass through the pupil and enter the eye of any given person (see figure 15.17). An object far away from the eye appears small compared with a nearby object because only nearly parallel light rays enter the eye from a distant object (see figure 15.17*a*). Converging rays leaving an object closer to the eye can also enter the eye (see figure 15.17*b*), and the object appears larger.

   When rays from a distant object reach the lens, they do not have to be refracted to any great extent to be focused on the retina, and the lens can remain fairly flat. When an object is closer to the eye, the more obliquely directed rays must be refracted to a greater extent to be focused on the retina.

   As an object is brought closer and closer to the eye, accommodation becomes more and more difficult because the lens cannot become any more convex. At some point, the eye no longer can focus the object, and it is seen as a blur. The point at which this blurring occurs is called the **near point of vision,** which is usually about 2–3 inches from the eye for children, 4–6 inches for a young adult, 20 inches for a 45-year-old adult, and 60 inches for an 80-year-old adult. This increase in the near point of vision occurs because the lens becomes more rigid with increasing age, which is primarily why some older people say they could read with no problem if they only had longer arms.

Clinical Note

When a person's vision is tested, a chart is placed 20 feet from the eye, and the person is asked to read a line of letters that has been standardized for normal vision. If the person can read the line, the vision is considered to be 20/20, which means that the person can see at 20 feet what people with normal vision can see at 20 feet. If, on the other hand, the person can see words only at 20 feet that people with normal vision can see at 40 feet, the vision is considered 20/40.

---

2. *Pupil constriction.* Another factor involved in focusing is the **depth of focus,** which is the greatest distance through which an object can be moved and still remain in focus on the retina. The main factor affecting the depth of focus is the size of the pupil. If the pupillary diameter is small, the depth of focus is greater than if the pupillary diameter is large. With a smaller pupillary opening an object may therefore be moved slightly nearer or farther from the eye without disturbing its focus. This is particularly important when viewing an object at close range because the interest in detail is much greater, and therefore the acceptable margin for error is smaller. When the pupil is constricted, the light entering the eye tends to pass more nearly through the center of the lens and is more accurately focused than light passing through the edges of the lens. Pupillary diameter also regulates the amount of light entering the eye. The dimmer the light, the greater the pupil diameter must be. As the pupil constricts during close vision, therefore, more light is required on the object being observed.

3. *Convergence.* Because the light rays entering the eyes from a distant object are nearly parallel, both pupils can pick up the light rays when the eyes are directed more or less straight ahead. As an object moves closer, however, the eyes must be rotated medially so that the object is kept focused on corresponding areas of each retina. Otherwise the object will appear blurry. This medial rotation of the eyes is accomplished by reflexive stimulation of the medial rectus muscle of each eye and is called convergence. Convergence can easily be observed. Have someone stand facing you. Have him reach out one hand and extend his index finger as far in front of his face as he can. While he keeps his gaze fixed on his finger, have him slowly bring his finger in toward his nose until he finally touches it. Notice the movement of his pupils as he does this. What happens?

Structure and Function of the Retina

The retina of each eye is about the size and thickness of a postage stamp, yet with only two tiny, postage stamp-sized

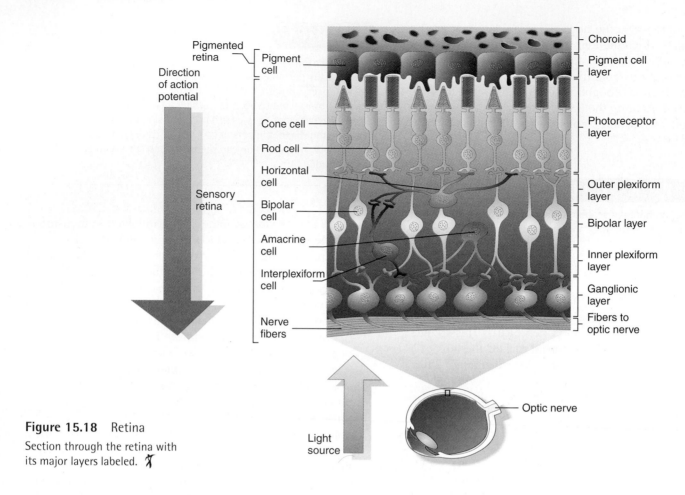

Figure 15.18   Retina

Section through the retina with its major layers labeled. ⚉

windows we have the potential to see the whole world. In speaking of the eye, Leonardo da Vinci said, "Who would believe that so small a space could contain the images of all the universe?"

The retina consists of a pigmented retina and a sensory retina. The **sensory retina** contains three layers of neurons: photoreceptor, bipolar, and ganglionic. The cell bodies of these neurons form nuclear layers separated by plexiform layers, where the neurons of adjacent layers synapse with each other (figure 15.18). The outer plexiform (plexuslike) layer is between the photoreceptor and bipolar cell layers. The inner plexiform layer is between the bipolar and ganglionic cell layers.

The **pigmented retina,** or pigmented epithelium, consists of a single layer of cells. This layer of cells is filled with melanin pigment and, together with the pigment in the choroid, provides a black matrix, which enhances visual acuity by isolating individual photoreceptors and reducing light scattering. Pigmentation is not strictly necessary for vision, however. People with albinism (lack of pigment) can see, although their visual acuity is reduced because of some light scattering.

The layer of the sensory retina nearest the pigmented retina is the layer of rods and cones. The rods and cones are photoreceptor cells, which are sensitive to stimulation from

"visible" light. The light-sensitive portion of each photoreceptor cell is adjacent to the pigmented layer.

Rods

Rods are bipolar photoreceptor cells involved in noncolor vision and are responsible for vision under conditions of reduced light (table 15.3). The modified, dendritic, light-sensitive part of rod cells is cylindrical, with no taper from base to apex (figure 15.19a). This rod-shaped photoreceptive part of the rod cell contains about 700 double-layered membranous discs. The discs contain **rhodopsin** (rō-dop'sin), which consists of the protein **opsin** in loose chemical combination with a pigment called **retinal** (derived from vitamin A).

Function of Rhodopsin

Figure 15.20 depicts the changes that rhodopsin undergoes in response to light. In the resting state the shape of opsin and retinal keeps the retinal tightly bound to the opsin surface. In a process called **bleaching,** retinal separates from opsin when rhodopsin is exposed to light. As light is absorbed, retinal changes shape and begins to detach from the opsin molecule. Because of the retinal separation, opsin "opens up" with a release of energy. This reaction is somewhat like a

Table 15.3   Rods and Cones

Photoreceptive End	Photoreceptive Molecule	Function	Location
Rod			
Cylindrical	Rhodospin	Noncolor vision; vision under conditions of low light	Over most of retina; none in fovea
Cone			
Conical	Iodopsin	Color vision; visual acuity	Numerous in fovea and macula lutea; sparse over rest of retina

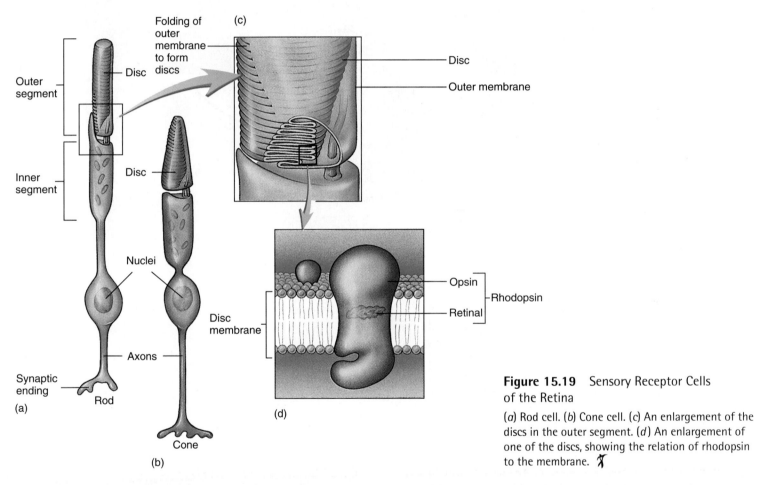

Figure 15.19   Sensory Receptor Cells of the Retina

(*a*) Rod cell. (*b*) Cone cell. (*c*) An enlargement of the discs in the outer segment. (*d*) An enlargement of one of the discs, showing the relation of rhodopsin to the membrane.

spring (opsin) that is held by a trigger (retinal). Light simply activates the trigger, which, when released, allows the spring to uncoil forcefully. It is thought that the separation of opsin and retinal exposes some active sites that change the membrane potential of the rod cell, which results in hyperpolarization of the cell (figure 15.20*b*).

This hyperpolarization in the photoreceptor cells is somewhat remarkable, because most neurons respond to stimuli by depolarizing. When photoreceptor cells are not exposed to light and are in a resting, nonactivated state, some of the Na^+ ion channels in their membranes are open, and Na^+ ions flow into the cell. This influx of Na^+ ions causes the photoreceptor cells to release the neurotransmitter glutamate from their presynaptic terminals. Glutamate binds to receptors on the postsynaptic membranes of the bipolar cells of the retina, causing them to hyperpolarize. Thus, glutamate causes an inhibitory postsynaptic potential (IPSP) in the bipolar cells.

When photoreceptor cells are exposed to light, the Na^+ ion channels close, fewer Na^+ ions enter the cell, and the amount of glutamate released from the postsynaptic terminals decreases. As a result, the hyperpolarization, or IPSP in the postsynaptic, bipolar cells, decreases, the cells depolarize, and an action potential is generated in the bipolar cells. The number of Na^+ ion channels that close and the degree to which they close is proportional to the amount of light exposure.

At the final stage of this light-initiated reaction, retinal is completely released from the opsin. This free retinal may

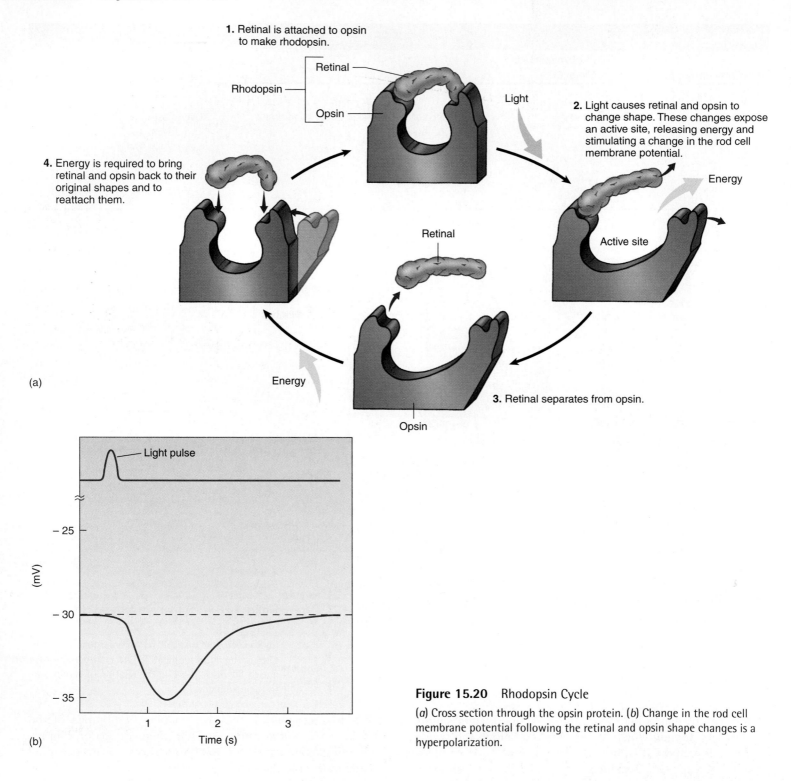

1. Retinal is attached to opsin to make rhodopsin.

Rhodopsin
Retinal
Opsin

Light

2. Light causes retinal and opsin to change shape. These changes expose an active site, releasing energy and stimulating a change in the rod cell membrane potential.

Energy

Active site

Retinal

4. Energy is required to bring retinal and opsin back to their original shapes and to reattach them.

(a)

Energy

Opsin

3. Retinal separates from opsin.

Light pulse

(mV)

− 25

− 30

− 35

1   2   3

Time (s)

(b)

Figure 15.20   Rhodopsin Cycle

(*a*) Cross section through the opsin protein. (*b*) Change in the rod cell membrane potential following the retinal and opsin shape changes is a hyperpolarization.

then be converted back to vitamin A, from which it was originally derived. The total vitamin A/retinal pool is in equilibrium so that under normal conditions the amount of free retinal is relatively constant. To create more rhodopsin, the altered retinal must be converted back to its original shape, a reaction that requires energy. Once the retinal resumes its original shape, its recombination with opsin is spontaneous, and the newly formed rhodopsin can again respond to light.

Light and **dark adaptation** is the adjustment of the eyes to changes in light. Adaptation to light or dark conditions, which occurs when a person comes out of a darkened building into the sunlight or vice versa, is accomplished by changes in the amount of available rhodopsin. In bright light excess rhodopsin is broken down so that not as much is available to initiate action potentials, and the eyes become "adapted" to bright light. Conversely, in a dark room more rhodopsin is produced, making the retina more light-sensitive.

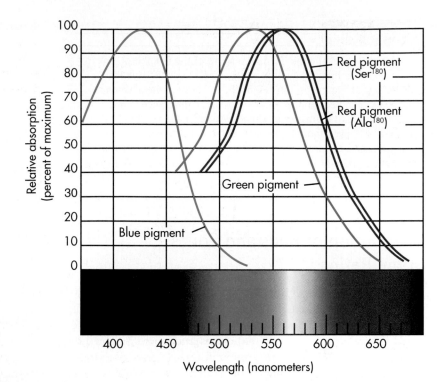

Figure 15.21   Wavelengths to Which Each of the Three Visual Pigments are Sensitive: Blue, Green, Red

There are actually two forms of the red pigment. One, found in 60% of the population, has a serine at position 180; and the other, found in 40% of the population, has an alanine at position 180. Each red pigment has a slightly different wavelength sensitivity.

---

8   P R E D I C T

If breakdown of rhodopsin occurs rapidly and production is slow, do eyes adapt more rapidly to light or dark conditions?

✔ *Answer in Appendix F*

Light and dark adaptation also involves pupil reflexes. The pupil enlarges in dim light and contracts in bright light to allow more or less light into the eye, respectively. In addition, rod function decreases and cone function increases in light conditions, and vice versa during dark conditions. This occurs because rod cells are more sensitive to light than cone cells and because rhodopsin is depleted more rapidly in rods than in cones.

Cones

Color vision and visual acuity are functions of cone cells. Color is a function of the wavelength of light, and each color can be assigned a certain wavelength within the visible spectrum. Even though rods are very sensitive to light, they cannot detect color, and afferent signals that ultimately reach the brain from these cells are interpreted by the brain as shades of gray. Cones require relatively bright light to function. As a result, as the light decreases, so does the color of objects that can be seen until, under conditions of very low illumination, the object appears gray. This occurs because as the light decreases, fewer cone cells respond to the dim light.

Cones are bipolar photoreceptor cells with a conical light-sensitive part that tapers slightly from base to apex (figure 15.19b). The outer segments of the cone cells, like those of the rods, consist of double-layered discs. The discs are slightly more numerous and more closely stacked in the cones than in the rods. Cone cells contain a visual pigment, **iodopsin** (ī-ō-dop'sin), which consists of retinal combined with a photopigment opsin protein. There are three major types of color-sensitive opsin: blue, red, and green; each closely resembles the opsin proteins of rod cells, but with somewhat different amino acid sequences. These color photopigments function in much the same manner as rhodopsin, but whereas rhodopsin responds to the entire spectrum of visible light, each iodopsin is sensitive to a much narrower spectrum.

Most people have one red pigment gene and one or more green pigment genes located in a tandem array on each X chromosome. An enhancer gene on the X chromosome apparently determines that only one color opsin gene is expressed in each cone cell. Only the first or second gene in the tandem array is expressed in each cone cell, so that some cone cells express only the red pigment gene and others express only one of the green pigment genes.

As can be seen in figure 15.21, although there is considerable overlap in the wavelength of light to which these pigments are sensitive, each pigment absorbs light of a certain range of wavelengths. As light of a given wavelength, representing a certain color, strikes the retina, all cone cells containing photopigments capable of responding to that wavelength generate action potentials. Because of the overlap among the three types of cones, especially between the green and red pigments, different proportions of cone cells respond to each wavelength, thus allowing color perception over a wide range. Color is interpreted in the visual cortex as combinations of afferent signals originating from cone cells. For example, when orange light strikes the retina, 99% of the red-sensitive cones respond, 42% of the green-sensitive cones respond, and no blue cones respond. When yellow light strikes the retina, the response is shifted so that a greater number of green-sensitive cones respond. The variety of combinations created allows humans to distinguish several million gradations of light and shades of color.

Distribution of Rods and Cones in the Retina

Cones are involved in visual acuity, in addition to their role in color vision. The fovea centralis is used when visual acuity is required, such as for focusing on the words of this page. The fovea centralis has about 35,000 cones and no rods. The 120 million rods are 20 times more plentiful than cones over most of the remaining retina, however. They are more highly concentrated away from the fovea and are more important in low-light conditions.

9 P R E D I C T

Explain why at night a person may notice a movement "out of the corner of her eye," but, when she tries to focus on the area where she noticed the movement, it appears as though nothing is there.

✔ *Answer in Appendix F*

Inner Layers of the Retina

The middle and inner nuclear layers of the retina consist of two major types of neurons: bipolar and ganglion cells. The rod and cone photoreceptor cells synapse with **bipolar cells,** which in turn synapse with **ganglion cells.** Axons from the ganglion cells pass over the inner surface of the retina (see figure 15.18), except in the area of the fovea centralis, converge at the **optic disc,** and exit the eye as the **optic nerve** (II). The fovea centralis is devoid of ganglion cell processes, resulting in a small depression in this area; thus the name fovea, meaning small pit. As a result of the absence of ganglion cell processes in addition to the concentration of cone cells mentioned previously, visual acuity is further enhanced in the fovea centralis because light rays do not have to pass through as many tissue layers before reaching the photoreceptor cells.

Rod and cone cells differ in the way they interact with bipolar and ganglion cells. One bipolar cell receives input from numerous rods, and one ganglion cell receives input from several bipolar cells so that spatial summation of the signal occurs and the signal is enhanced, allowing awareness of stimulus from very dim light sources but decreasing visual acuity in these cells. Cones, on the other hand, exhibit little or no convergence on bipolar cells so that one cone cell may synapse with only one bipolar cell. This system reduces light sensitivity but enhances visual acuity.

Within the inner layers of the retina, there are also **association neurons,** which modify the signals from the photoreceptor cells before the signal ever leaves the retina (see figure 15.18). **Horizontal cells** form the outer plexiform layer and synapse with photoreceptor cells and bipolar cells. **Amacrine** (am′ă-krin) **cells** form the inner plexiform layer and synapse with bipolar and ganglion cells. **Interplexiform cells** form the bipolar layer and synapse with amacrine, bipolar, and horizontal cells, forming a feedback loop. Association neurons are either excitatory or inhibitory on the cells with which they synapse. These association cells enhance borders and contours, increasing the intensity at boundaries, such as the edge of a dark object against a light background.

Neuronal Pathways for Vision

The optic nerve (II) (figure 15.22) leaves the eye and exits the orbit through the optic foramen to enter the cranial vault. Just inside the vault and just anterior to the pituitary, the optic nerves are connected to each other at the **optic chiasma** (kī′az-mă). Ganglion cell axons from the nasal retina (the medial portion of the retina) cross through the optic chiasma and project to the opposite side of the brain. Ganglion cell axons from the temporal retina (the lateral portion of the retina) pass through the optic nerves and project to the brain on the same side of the body without crossing.

Beyond the optic chiasma, the route of the ganglionic axons is called the **optic tract** (see figure 15.22). Most of the optic tract axons terminate in the **lateral geniculate nucleus** of the thalamus. Some axons do not terminate in the thalamus but separate from the optic tract to terminate in the **superior colliculi,** the center for visual reflexes (see chapter 13). Neurons of the lateral geniculate ganglion form the fibers of the **optic radiations,** which project to the **visual cortex** in the **occipital lobe** (see figure 15.22). Neurons of the visual cortex integrate the messages coming from the retina into a single message, translate that message into a mental image, and then transfer the image to other parts of the brain, where it is evaluated and either ignored or acted on.

The projections of ganglion cells from the retina can be related to the **visual fields** (see figure 15.22). The visual field of one eye can be evaluated by closing the other eye. Everything that can be seen with the one open eye is the visual field of that eye. The visual field of each eye can be divided into a temporal half (lateral) and a nasal half (medial). In each eye the temporal

1. Each visual field is divided into a temporal and nasal half.

2. After passing through the lens, each half of a visual field projects to the opposite side of the retina.

3. An optic tract consists of axons extending from the retina to the brain.

4. In the optic chiasma, axons from the nasal half of each retina cross to the opposite side of the brain. Axons from the temporal half of each retina do not cross.

5. An optic tract consists of axons that have passed through the optic chiasma (with or without crossing).

6. The axons synapse in the lateral geniculate nucleus of the thalamus. Collateral branches of the lateral geniculate nucleus of the axons synapse in the superior colliculi.

7. An optic radiation consists of thalamic neurons that project to the visual cortex.

8. The right half of each visual field projects to the left side of the brain, and the left half of each visual field projects to the right side of the brain (see part b).

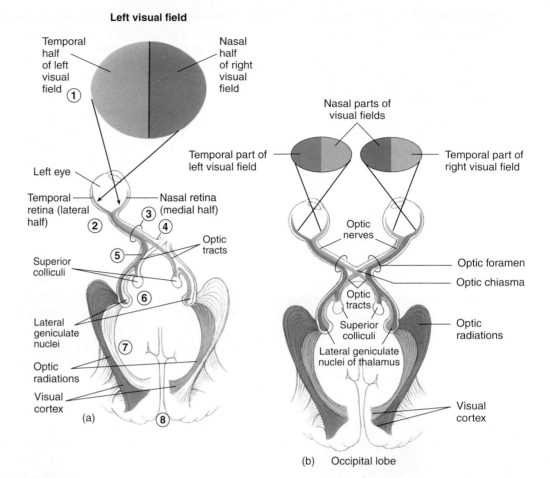

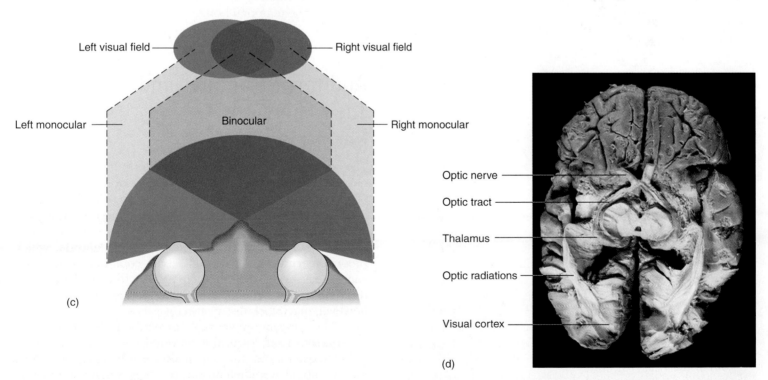

Figure 15.22 Visual Pathways

(*a*) Left visual field and the left visual CNS pathways (superior view). (*b*) Right and left visual fields and visual CNS pathways (superior view). The visual fields are shown separately to help in understanding each field. (*c*) The right and left visual fields are superimposed to show the area of binocular vision (*blue-green area*). (*d*) Photo of the optic radiations.

half of the visual field projects onto the nasal retina, whereas the nasal half of the visual field projects to the temporal retina. The projections and nerve pathways are arranged in such a way that images entering the eye from the right half of each visual field project to the left half of the brain. Conversely, the left half of each visual field projects to the right side of the brain.

Clinical Note

Because the optic chiasma lies just anterior to the pituitary, a pituitary tumor can put pressure on the optic chiasma and may result in visual defects. Because the nerve fibers crossing in the optic chiasma are carrying information from the temporal halves of the visual fields, a person with optic chiasma damage cannot see objects in the temporal halves of the visual fields, a condition called **tunnel vision.** Tunnel vision is often an early sign of a pituitary tumor.

10 P R E D I C T

The figure depicts examples of two lesions in the visual pathways. In the first example, the effect of a lesion at A in the optic radiations on the visual fields is depicted (with the right and left fields separated). The darkened areas indicate what parts of the visual fields are defective. In the second example, a lesion at B in the right optic nerve is depicted. Describe the effect that lesion B would have on the visual fields.

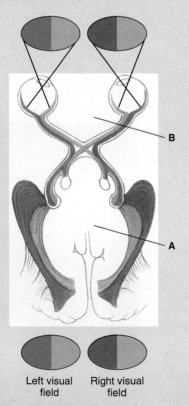

B

A

Left visual   Right visual
field   field

Lesions of the Visual Pathways
The lines at A and B represent lesions. The oval insets depict the effects of the lesion at A, in the optic radiations, on the visual fields. The darkened areas indicate lack of vision in the area.

✔ *Answer in Appendix F*

The visual fields of the eyes partially overlap (see figure 15.22). The region of overlap is the area of **binocular vision,** seen with two eyes at the same time, and it is responsible for **depth perception,** the ability to distinguish between near and far objects and to judge their distance. Because humans see the same object with both eyes, the image of the object reaches the retina of one eye at a slightly different angle from that of the other. With experience, the brain can interpret these differences in angle so that distance can be judged quite accurately.

Hearing and Balance

The organs of hearing and balance can be divided into three parts: external, middle, and inner ears (figure 15.23). The external and middle ears are involved in hearing only, whereas the inner ear functions in both hearing and balance.

The **external ear** includes the **auricle** (aw'ri-kl, meaning ear) and the **external auditory meatus** (mē-ā'tŭs, the passageway from the outside to the eardrum). The external ear terminates medially at the **eardrum,** or **tympanic** (tim-pan'ik) **membrane.** The **middle ear** is an air-filled space within the petrous portion of the temporal bone, which contains the **auditory ossicles.** The **inner ear** contains the sensory organs for hearing and balance. It consists of interconnecting fluid-filled tunnels and chambers within the petrous portion of the temporal bone.

Auditory Structures and Their Functions
External Ear

The auricle, or **pinna** (pin'ă), is the fleshy part of the external ear on the outside of the head and consists primarily of elastic cartilage covered with skin (figure 15.24). Its shape helps to collect sound waves and direct them toward the external auditory meatus. The external auditory meatus is lined with **hairs** and **ceruminous** (sĕ-rū'mi-nŭs) **glands,** which produce **cerumen,** a modified sebum commonly called earwax. The hairs and cerumen help prevent foreign objects from reaching the delicate eardrum. Overproduction of cerumen, however, may block the meatus.

The tympanic membrane, or eardrum, is a thin, semitransparent, nearly oval, three-layered membrane that separates the external ear from the middle ear. It consists of a low, simple cuboidal epithelium on the inner surface and a thin stratified squamous epithelium on the outer surface, with a layer of connective tissue between. Sound waves reaching the tympanic membrane through the external auditory meatus cause it to vibrate.

Clinical Focus   Eye Disorders

Myopia

Myopia (mī-ō′pē-ă), or nearsightedness, is the ability to see close objects clearly, but distant objects appear blurry. Myopia is a defect of the eye in which the focusing system, the cornea and lens, is optically too powerful, or the eyeball is too long (axial myopia). As a result, the focal point is too near the lens, and the image is focused in front of the retina (figure A*a*).

Myopia is corrected by a concave lens that counters the refractive power of the eye. Concave lenses cause the light rays coming to the eye to diverge and are therefore called "minus" lenses (figure A*b*).

Another technique for correcting myopia is **radial keratotomy** (ker′ă-tot′ō-mē), which consists of making a series of four to eight radiating cuts in the cornea. The cuts are intended to slightly weaken the dome of the cornea so that it becomes more flattened and eliminates the myopia. One problem with the technique is that it is difficult to predict exactly how much flattening will occur. In one study of 400 patients 5 years after the surgery, 55% had normal vision, 28% were still somewhat myopic, and 17% had become hyperopic. Another problem is that some patients are bothered by glare following radial keratotomy because the slits apparently don't heal evenly.

An alternative procedure that is currently being investigated is **laser corneal sculpturing,** in which a thin portion of the cornea is etched away to make the cornea less convex. The advantage of this procedure is that the results can be more accurately predicted than those from radial keratotomy.

Hyperopia

Hyperopia (hī-per-ō′pē-ă), or farsightedness, is the ability to see distant objects clearly, but close objects appear blurry. Hyperopia is a disorder in which the cornea and lens system is optically too weak or the eyeball is too short. The image is focused behind the retina (figure A*c*).

Hyperopia can be corrected by convex lenses that cause light rays to converge as they approach the eye (figure A*d*). Such lenses are called "plus" lenses.

Presbyopia

Presbyopia (prez-bē-ō′pē-ă) is the normal, presently unavoidable, degeneration of the accommodation power of the eye that occurs as a consequence of aging. It

occurs because the lens becomes sclerotic and less flexible. The eye is presbyopic when the near point of vision has increased beyond 9 inches. The average age

for onset of presbyopia is the midforties. Avid readers or people engaged in fine, close work may develop the symptoms earlier.

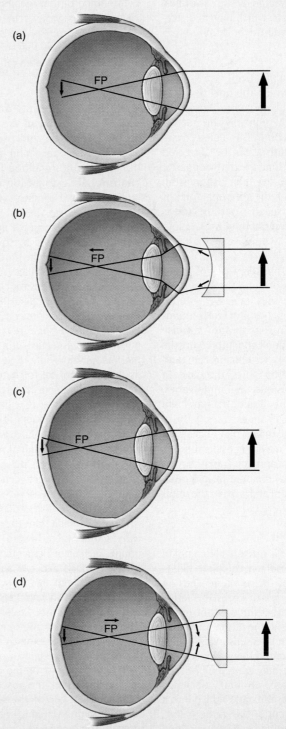

Figure A   Visual Disorders and Their Correction by Various Lenses. FP is the focal point.   (*a*) Myopia (nearsightedness). (*b*) Correction of myopia with a concave lens. (*c*) Hyperopia (farsightedness). (*d*) Correction of hyperopia with a convex lens.

Presbyopia can be corrected by the use of "reading glasses" that are worn only for close work and are removed when the person wants to see at a distance. It is sometimes annoying to keep removing and replacing glasses because reading glasses hamper vision of only a few feet away. This problem may be corrected by the use of half glasses, or by **bifocals,** which have a different lens in the top and the bottom.

Astigmatism

Astigmatism (ă-stig′mă-tizm) is a type of refractive error in which the quality of focus is affected. If the cornea or lens is not uniformly curved, the light rays do not focus at a single point but fall as a blurred circle. Regular astigmatism can be corrected by glasses that are formed with the opposite curvature gradation. Irregular astigmatism is a situation in which the abnormal form of the cornea fits no specific pattern and is very difficult to correct with glasses.

Strabismus

Strabismus (stra-biz′mŭs) is a lack of parallelism of light paths through the eyes. Strabismus can involve only one eye or both eyes, and the eyes may turn in (convergent) or out (divergent). In **concomitant strabismus,** the most common congenital type, the angle between visual axes remains constant, regardless of the direction of the gaze. In **noncomitant strabismus,** the angle varies, depending on the direction of the gaze, and deviates as the gaze changes.

In some cases, the image that appears on the retina of one eye may be considerably different from that appearing on the other eye. This problem is called **diplopia** (di-plō′pē-ă, meaning double vision) and is often the result of weak or abnormal eye muscles.

Retinal Detachment

Retinal detachment is a relatively common problem that can result in complete blindness. The integrity of the retina depends on the vitreous humor, which keeps the retina pushed against the other tunics of the eye. If a hole or tear occurs in the retina, fluid may accumulate between the sensory and pigmented retina, separating them. This separation, or detachment, may continue until the sensory retina has become totally detached from the pigmented retina and folded into a funnellike form around the optic nerve. When the sensory retina becomes separated from its nutrient supply in the choroid, it degenerates, and blindness follows. Causes of retinal detachment include a severe blow to the eye or head; a shrinking of the vitreous humor, which may occur with aging; or diabetes. The space between the sensory and pigmented retina, called the subretinal space, is also important in keeping the retina from detaching, as well as in maintaining the health of the retina. The space contains a gummy substance that glues the sensory retina to the pigmented retina.

Color Blindness

Color blindness results from the disfunction of one or more of the three photopigments involved in color vision. If one pigment is disfunctional and the other two are functional, the condition is called **dichromatism.** An example of dichromatism is red-green color blindness (figure B).

The genes for the red and green photopigments are arranged in tandem on the X chromosome, which explains why color blindness is over eight times more common in males than in females (see chapter 29).

There are six exons for each gene. The red and green genes are 96%–98% identical and, as a result, the exons may be shuffled to form hybrid genes in some people. Some of the hybrid genes produce proteins with nearly normal function, but others do not. Exon five is the most critical for determining normal red-green function. If a red pigment gene with the fifth exon is replaced by the fifth exon from a green gene, the protein made from the gene responds to wavelengths more toward the green pigment range. The person has a red perception deficiency and is not able to distinguish between red and green. If a green pigment gene with the fifth exon is replaced by the fifth exon from a red gene, the protein made from the gene responds to wavelengths more toward the red pigment range. The person has a green perception deficiency and is also not able to distinguish between red and green.

Apparently only about 3 of the over 360 amino acids in the color opsin proteins (those at positions 180 in exon 3 and those at 277 and 285 in exon 5) are key to determining their wavelength absorption characteristics. If those amino acids are altered by hydroxylation, the absorption shifts toward the red end of the spectrum. If they are not hydroxylated, the absorption shifts toward the green end.

Night Blindness

Everyone sees less clearly in the dark than in the light. A person with **night blindness,** however, may not see well enough in a dimly lit environment to function adequately. **Progressive** night blindness results from general retinal degeneration. **Stationary** night blindness results from nonprogressive abnormal rod function. Temporary night blindness can result from a vitamin A deficiency.

Patients with night blindness can now be helped with special electronic optical devices. These include monocular pocket scopes and binocular goggles that electronically amplify light.

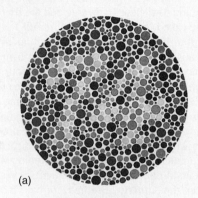

(a)

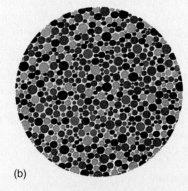

(b)

Figure B Color Blindness Charts

(*a*) A person with normal color vision can see the number 74, whereas a person with red-green color blindness sees the number 21. (*b*) A person with normal color vision can see the number 42. A person with red color blindness sees the number 2, and a person with green color blindness sees the number 4.

The above has been reproduced from Ishihara's *Tests for Colour Blindness* published by Kanehara & Co., Ltd., Tokyo, Japan, but tests for color blindness cannot be conducted with this material. For accurate testing, the original plates should be used.

Glaucoma

Glaucoma (glaw-kō′mă) (figure C*a*) is a disease of the eye involving increased intraocular pressure caused by a buildup of aqueous humor. It usually results from blockage of the aqueous veins or the canal of Schlemm, restricting drainage of the aqueous humor, or from overproduction of aqueous humor. If untreated, glaucoma can lead to retinal, optic disc, and optic nerve damage. The damage results from the increased intraocular pressure, which is sufficient to close off the blood vessels, causing starvation and death of the retinal cells.

Glaucoma is one of the leading causes of blindness in the United States, affecting 2% of people over age 35, and accounting for 15% of all blindness. Fifty thousand people in the United States are blind as the result of glaucoma, and it occurs three times more often in black people than in white people. The symptoms include a slow closing in of the field of vision. There is no pain or redness, and there are no light flashes.

Glaucoma has a strong hereditary tendency but may develop after surgery or with the use of certain eyedrops containing cortisone. Everyone older than 40 should be checked every 2–3 years for glaucoma; those older than 40 who have relatives with glaucoma should have an annual checkup. During a checkup, the field of vision is checked, and the optic nerve is examined. Ocular pressures can also be measured. Glaucoma is usually treated with eyedrops, which do not cure the problem but keep it from advancing. In some cases, laser or conventional surgery may be used.

Cataract

Cataract (figure C*b*) is a clouding of the lens resulting from a buildup of proteins. The lens relies on the aqueous humor for its nutrition. Any loss of this nutrient source leads to degeneration of the lens and, ultimately, opacity of the lens (i.e., a cataract). A cataract may occur with advancing age, infection, or trauma.

A certain amount of lens clouding occurs in 65% of patients older than 50 and 95% of patients older than 65. The decision of whether to remove the cataract depends on the extent to which light passage is blocked. Over 400,000 cataracts are removed in the United States each year. Surgery to remove a cataract is actually the removal of the lens. The posterior portion of the lens capsule is left intact. Although light convergence is still accomplished by the cornea, with the lens gone, the rays cannot be focused as well, and an artificial lens must be supplied to help accomplish focusing. In most cases, an artificial lens is implanted into the remaining portion of the lens capsule at the time that the natural lens is removed. The implanted lens helps to restore normal vision, but glasses may be required for near vision.

Macular Degeneration

Macular degeneration (figure C*c*) is very common in older people. It does not cause total blindness but results in the loss of acute vision. This degeneration has a variety of causes, including hereditary disorders, infections, trauma, tumor, or most often, poorly understood degeneration associated with aging. Because no satisfactory medical treatment has been developed, optical aids, such as magnifying glasses, are used to improve visual function.

Diabetes

Loss of visual function is one of the most common consequences of diabetes because a major complication of the disease is disfunction of the peripheral circulation. Defective circulation to the eye may result in retinal degeneration or detachment. Diabetic retinal degeneration (figure C*d*) is one of the leading causes of blindness in the United States.

Infections

Trachoma (tră-kō′mă) is the leading cause of blindness worldwide. It is caused by an intracellular microbial infection (*Chlamydia trachomatis*) of the corneal epithelial cells, resulting in scar tissue formation in the cornea. The bacteria are spread from one eye to another eye by towels, fingers, and other objects. There are 500 million cases of trachoma in the world, and 7 million people are blind or visually impaired as a result of it.

Neonatal gonorrheal ophthalmia (of-thal′mē-ă) is a bacterial infection (*Neisseria gonorrhoeae*) of the eye that causes blindness. If the mother has gonorrhea, which is a sexually transmitted disease of the reproductive tract, the bacteria can infect the newborn during delivery. The disease can be prevented by treating the infant's eyes with silver nitrate, tetracycline, or erythromycin drops.

(a)

(b)

(c)

(d)

Figure C   Defects in Vision
Visual images as seen with various defects in vision. (*a*) Glaucoma. (*b*) Cataract. (*c*) Macular degeneration. (*d*) Diabetic retinopathy.

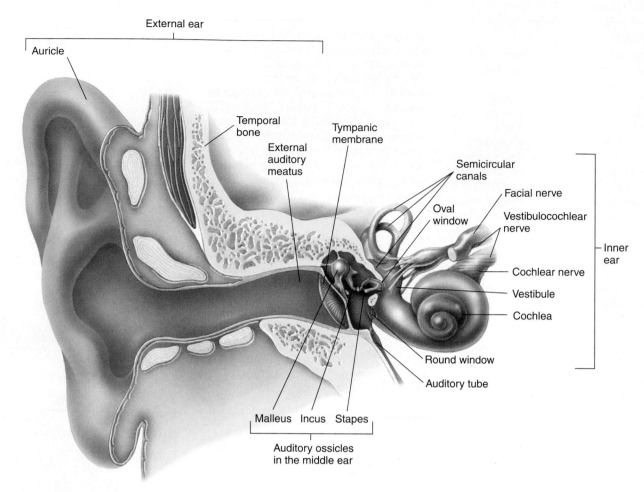

Figure 15.23   External, Middle, and Inner Ear ✗

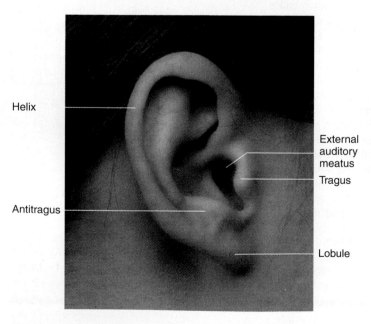

Figure 15.24   Structures of the Auricle (the Right Ear)

Middle Ear

Medial to the tympanic membrane is the air-filled cavity of the middle ear (see figure 15.23). Two covered openings, the round and oval windows, on the medial side of the middle ear separate it from the inner ear. Two openings provide air passages from the middle ear. One passage opens into the **mastoid air cells** in the mastoid process of the temporal bone. The other passageway, the **auditory,** or **eustachian** (yū-stā′shŭn) **tube,** opens into the pharynx and equalizes air pressure between the outside air and the middle ear cavity.

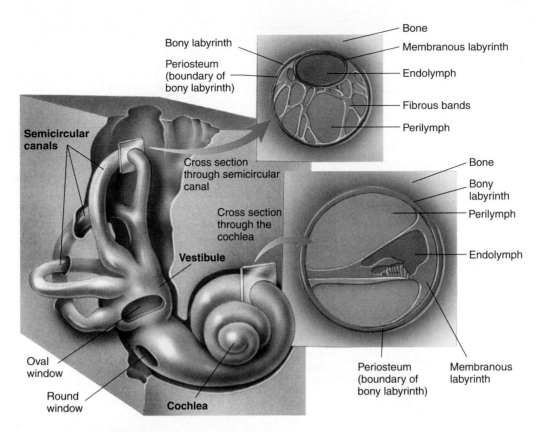

Figure 15.25   The Inner Ear: Bony and Membranous Labyrinth
The cross sections are taken through a semicircular canal and the cochlea to show the relationship between the bony and membranous labyrinth.

Unequal pressure between the middle ear and the outside environment can distort the eardrum, dampen its vibrations, and make hearing difficult. Distortion of the eardrum, which occurs under these conditions, also stimulates pain fibers associated with it. Because of this distortion, when a person changes altitude, sounds seem muffled, and the eardrum may become painful. These symptoms can be relieved by opening the auditory tube, allowing air to pass through the auditory tube to equalize air pressure. Swallowing, yawning, chewing, and holding the nose and mouth shut while gently trying to force air out of the lungs are methods used to open the auditory tube.

The middle ear contains three auditory ossicles: the **malleus** (mal′ē-ŭs, meaning hammer), **incus** (ing′kŭs, meaning anvil), and **stapes** (stā′pēz, meaning stirrup), which transmit vibrations from the tympanic membrane to the **oval window.** The handle of the malleus is attached to the inner surface of the tympanic membrane, and vibration of the membrane causes the malleus to vibrate as well. The head of the malleus is attached by a very small synovial joint to the incus, which in turn is attached by a small synovial joint to the stapes. The foot plate of the stapes fits into the oval window and is held in place by a flexible **annular ligament.**

Clinical Note

A structure that students might be somewhat surprised to find in the middle ear is the **chorda tympani.** It is a branch of the facial nerve carrying taste impulses from the anterior two-thirds of the tongue. It crosses over the inner surface of the tympanic membrane (see figures 15.23 and 15.30). The chorda tympani has nothing to do with hearing but is just passing through. This nerve can be damaged, however, during ear surgery or by a middle ear infection, resulting in loss of taste sensation carried by that nerve.

Inner Ear

The tunnels and chambers inside the temporal bone are called the **bony labyrinth** (lab′i-rinth, meaning a maze) (figure 15.25). Because the bony labyrinth consists of tunnels within the bone, it cannot easily be removed and examined separately. The bony labyrinth is lined with periosteum, and when the inner ear is shown separately (figure 15.26a), the periosteum is what is depicted. Inside the bony labyrinth is a similarly shaped but

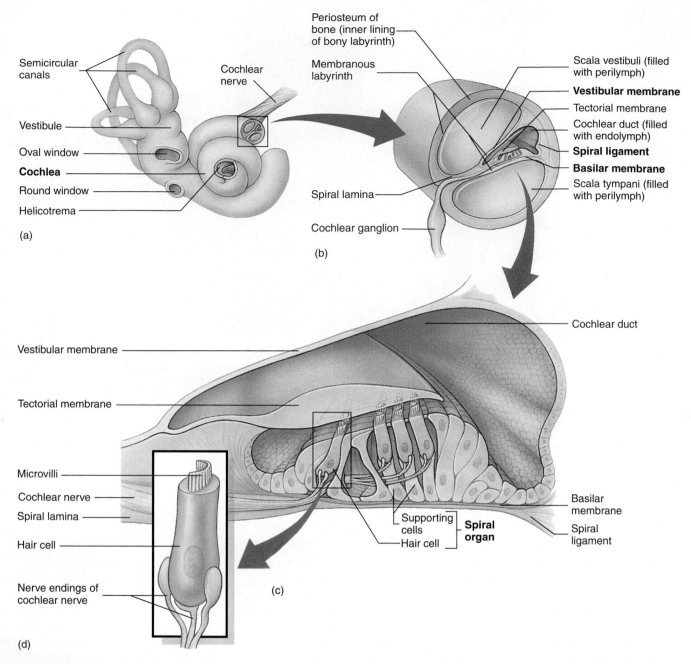

Figure 15.26   Structure of the Cochlea

(*a*) **The inner ear.** The outer surface (*gray*) is the periosteum lining the inner surface of the bony labyrinth. (*b*) **A cross section of the cochlea.** The outer layer is the periosteum lining the inner surface of the bony labyrinth. The membranous labyrinth is very small in the cochlea and consists of the vestibular and basilar membranes. The space between the membranous and bony labyrinth consists of two parallel tunnels: the scala vestibuli and scala tympani. (*c*) An enlarged section of the cochlear duct (membranous labyrinth). (*d*) A greatly enlarged individual sensory hair cell.

smaller set of membranous tunnels and chambers called the **membranous labyrinth.** The membranous labyrinth is filled with a clear fluid called **endolymph,** and the space between the membranous and bony labyrinth is filled with a fluid called **perilymph.** Perilymph is very similar to CSF, but endolymph has a high concentration of potassium and a low

concentration of sodium, which is opposite from perilymph and CSF.

The bony labyrinth can be divided into three regions: cochlea, vestibule, and semicircular canals. The **vestibule** (ves′ti-būl) and **semicircular canals** are involved primarily in balance, and the **cochlea** (kok′lē-ă) is involved in hearing.

The cochlea is divided into three parts: the scala vestibuli, the scala tympani, and the cochlear duct.

The oval window communicates with the vestibule of the inner ear, which in turn communicates with a cochlear chamber, the **scala** (skā′lă) **vestibuli** (ves-tib′yū-lē) (figure 15.26*a*). The scala vestibuli extends from the oval window to the **helicotrema** (hel′i-kō-trē′mă, meaning a hole at the end of a helix or spiral) at the apex of the cochlea; a second cochlear chamber, the **scala tympani** (tim′pă-nē), extends from the helicotrema, back from the apex, parallel to the scala vestibuli, to the membrane of the **round window.**

The scala vestibuli and the scala tympani are the perilymph-filled spaces between the walls of the bony and membranous labyrinths. The bony walls of each of these chambers are covered by a layer of simple squamous epithelium that is attached to the periosteum of the bone. The wall of the membranous labyrinth that bounds the scala vestibuli is called the **vestibular membrane** (Reissner's membrane); the wall of the membranous labyrinth bordering the scala tympani is the **basilar membrane** (figure 15.26*b* and *c*). The space between the vestibular membrane and the basilar membrane is the interior of the membranous labyrinth and is called the **cochlear duct** or **scala media,** which is filled with endolymph.

The vestibular membrane consists of a double layer of squamous epithelium and is the simplest region of the membranous labyrinth. The vestibular membrane is so thin that it has little or no mechanical effect on the transmission of sound waves through the inner ear; therefore the perilymph and endolymph on the two sides of the vestibular membrane can be thought of mechanically as one fluid. The role of the vestibular membrane is to separate the two chemically different fluids. The basilar membrane is somewhat more complex and is of much greater physiologic interest in relation to the mechanics of hearing. It consists of an acellular portion with collagen fibers, ground substance, and sparsely dispersed elastic fibers and a cellular part with a thin layer of vascular connective tissue that is overlaid with simple squamous epithelium.

The basilar membrane is attached at one side to the bony **spiral lamina,** which projects from the sides of the **modiolus** (mō′dī′ō-lus), the bony core of the cochlea, like the threads of a screw, and at the other side to the lateral wall of the bony labyrinth by the **spiral ligament,** a local thickening of the periosteum. The distance between the spiral lamina and the spiral ligament (i.e., the width of the basilar membrane) increases from 0.04 mm near the oval window to 0.5 mm near the helicotrema. The collagen fibers of the basilar membrane are oriented across the membrane between the spiral lamina and the spiral ligament, somewhat like the strings of a piano. The collagen fibers near the oval window are both shorter and thicker than those near the helicotrema. The diameter of the collagen fibers in the membrane decreases as the basilar membrane widens. As a result, the basilar membrane near the oval window is short and stiff, and responds to high-frequency vibrations, whereas that part near

Figure 15.27   Scanning Electron Micrograph of Cochlear Hair Cell Microvilli

the helicotrema is wide and limber and responds to low-frequency vibrations.

The cells inside the cochlear duct are highly modified to form a structure called the **spiral organ,** or the **organ of Corti** (see figure 15.26*b* and *c*). The spiral organ contains supporting epithelial cells and specialized sensory cells called **hair cells,** which have specialized hairlike projections at their apical ends. In children, these projections consist of one cilium (kinocilium) and about 80 very long microvilli, often referred to as **stereocilia;** but in adults the cilium is absent from most hair cells (see figures 15.26*d* and 15.27). The hair cells are arranged in four long rows extending the length of the cochlear duct. The tips of the hairs are embedded within an acellular gelatinous shelf called the **tectorial** (tek-tōr′ē-ăl) **membrane,** which is attached to the spiral lamina.

Hair cells have no axons, but the basilar regions of each hair cell are covered by synaptic terminals of sensory neurons, the cell bodies of which are located within the cochlear modiolus and are grouped into a **cochlear,** or **spiral ganglion** (see figures 15.26*b* and 15.32). Afferent fibers of these neurons join to form the **cochlear nerve.** This nerve then joins the vestibular nerve to become the **vestibulocochlear nerve** (VIII), which traverses the internal auditory meatus and enters the cranial vault.

Auditory Function

Sound is created by the vibration of matter such as air, water, or a solid material. There is no sound in a vacuum. When a person speaks, the vocal cords vibrate, causing the air passing out of the lungs to vibrate. The vibrations consists of bands of compressed air followed by bands of less compressed air (figure 15.28*a*). These vibrations are propagated through the air as sound waves, somewhat like ripples are propagated over the surface of water. **Volume,** or loudness, is a function of wave

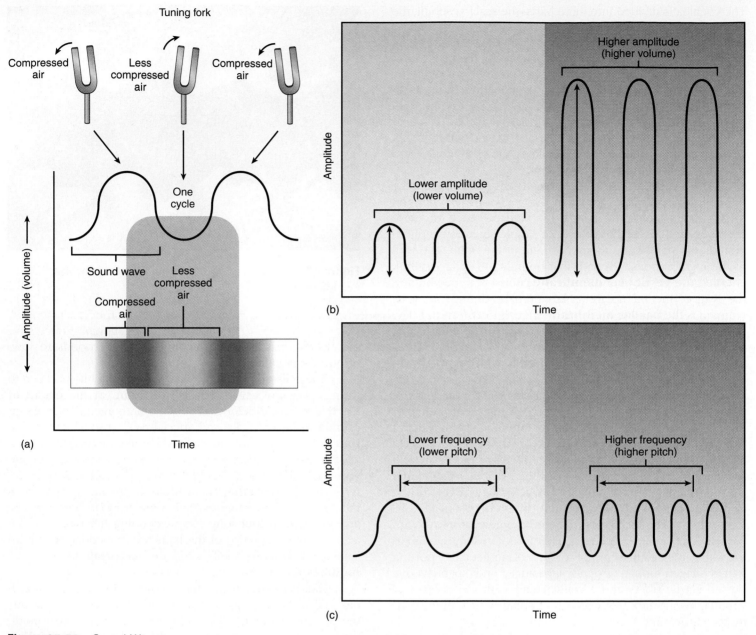

Figure 15.28   Sound Waves

(*a*) Each sound wave consists of a region of compressed air between two regions of less compressed air (*blue bars*). The sigmoid waves depicted above correspond to the regions of more compressed air (peaks) and less compressed air (troughs). The green shadowed area represents the width of one cycle (distance between peaks). When something like a tuning fork (shown in the figure) or vocal cords vibrate, the movements of the object alternate between compressing the air and decompressing the air, or making the air less compressed, thus producing sound. (*b*) Depicts low- and high-volume sound waves. Compare the relative lengths of the arrows indicating the wave height (amplitude). (*c*) Depicts lower and higher pitch sound. Compare the relative number of peaks (frequency) within a given time interval (between arrows).

amplitude, or height, measured in decibels (figure 15.28*b*). The greater the amplitude, the louder is the sound. **Pitch** is a function of the wave frequency (i.e., the number of waves or cycles per second) measured in hertz (Hz) (figure 15.28*c*). The higher the frequency, the higher the pitch. The normal range of human hearing is 20–20,000 Hz and 0 or more decibels (db). Sounds louder than 125 db are painful to the ear.

Clinical Note

The range of normal human speech is 250–8000 Hz. This is the range that is tested for the possibility of hearing impairment because it is the most important for communication.

Table 15.4   Steps Involved in Hearing

1. Sound waves are collected by the auricle and are conducted through the external auditory meatus to the tympanic membrane, causing it to vibrate.
2. The vibrating tympanic membrane causes the malleus, incus, and stapes to vibrate.
3. Vibration of the stapes produces vibration in the perilymph of the scala vestibuli.
4. The vibration of the perilymph produces simultaneous vibration of the vestibular membrane and the endolymph in the cochlear duct.
5. Vibration of the endolymph causes the basilar membrane to vibrate.
6. As the basilar membrane vibrates, the hair cells attached to the membrane move relative to the tectorial membrane, which remains stationary.
7. The hair cell microvilli, embedded in the tectorial membrane, become bent.
8. Bending of the microvilli causes depolarization of the hair cells.
9. The hair cells induce action potentials in the cochlear neurons.
10. The action potentials generated in the cochlear neurons are conducted to the CNS.
11. The action potentials are translated in the cerebral cortex and are perceived as sound.

Timbre (tam′br or tim′br) is the resonance quality or overtones of a sound. A smooth sigmoid curve is the image of a "pure" sound wave, but such a wave almost never exists in nature. The sounds made by musical instruments or the human voice are not smooth sigmoid curves but rather are rough, jagged curves formed by numerous, superimposed curves of various amplitudes and frequencies. The roughness of the curve accounts for the timbre. Timbre allows one to distinguish between, for example, an oboe and a French horn playing a note at the same pitch and volume. The steps involved in hearing are listed in table 15.4 and are illustrated in figure 15.29.

External Ear

Sound waves are collected by the auricle and are conducted through the external auditory meatus toward the tympanic membrane. Sound waves travel relatively slowly in air, 332 m/s, and a significant time interval may elapse between the time a sound wave reaches one ear and the time that it reaches the other. The brain can interpret this interval to determine the direction from which a sound is coming.

Middle Ear

Sound waves strike the tympanic membrane and cause it to vibrate. This vibration causes vibration of the three ossicles of the middle ear, and by this mechanical linkage vibration is transferred to the oval window. More force is required to cause vibration in a liquid such as the perilymph of the inner ear than is required in air; thus the vibrations reaching the perilymph must be amplified as they cross the middle ear. The footplate of the stapes and its annular ligament, which occupy the oval window, are much smaller than the tympanic membrane. Because of this size difference, the mechanical force of vibration is amplified about 20-fold as it passes from the tympanic membrane, through the ossicles, and to the oval window.

Two small skeletal muscles are attached to the ear ossicles and reflexively dampen excessively loud sounds (figure 15.30). This **sound attenuation reflex** protects the delicate ear structures from damage by loud noises. The **tensor tympani** (ten′sŏr tim′pănē) muscle is attached to the malleus and is innervated by the trigeminal nerve (V). The **stapedius** (stā-pē′dē-ŭs) muscle is attached to the stapes and is supplied by the facial nerve (VII). The sound attenuation reflex responds most effectively to low-frequency sounds and can reduce by a factor of 100 the energy reaching the eardrum. The reflex is too slow to prevent damage from a sudden noise, such as a gunshot, and it cannot function effectively for longer than about 10 min, in response to prolonged noise.

11   P R E D I C T

What effect does facial nerve damage have on hearing?

✔ *Answer in Appendix F*

Inner Ear

As the stapes vibrates, it produces waves in the perilymph of the scala vestibuli (see figure 15.29). Vibrations of the perilymph are transmitted through the thin vestibular membrane and cause simultaneous vibrations of the endolymph. The mechanical effect is as though the perilymph and endolymph were a single fluid. Vibration of the endolymph causes distortion of the basilar membrane. Waves in the perilymph of the scala vestibuli are transmitted also through the helicotrema and into the scala tympani. Because the helicotrema is very small, however, this transmitted vibration is probably of little consequence. Distortions of the basilar membrane, together with weaker waves coming through the helicotrema, cause waves in the scala tympani perilymph and ultimately result in vibration of the membrane of the round window. Vibration of the round window membrane is important to hearing because it acts as a mechanical release for waves from within the cochlea. If this window were solid, it would reflect the waves, which would interfere with and dampen later waves. The round window also allows relief of pressure in the perilymph because fluid is not compressible, preventing compression damage to the spiral organ.

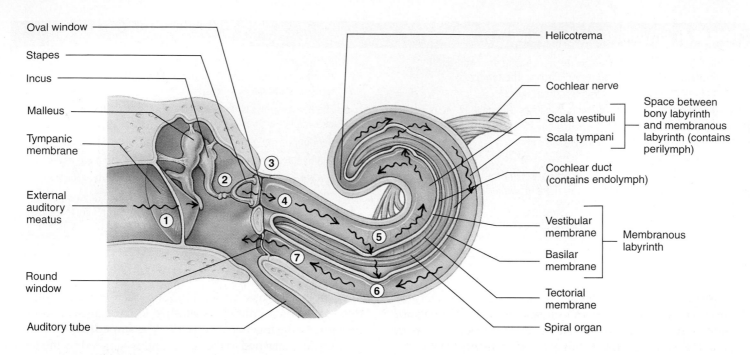

Oval window

Stapes

Incus

Malleus

Tympanic membrane

External auditory meatus

Round window

Auditory tube

Helicotrema

Cochlear nerve

Scala vestibuli

Scala tympani

Space between bony labyrinth and membranous labyrinth (contains perilymph)

Cochlear duct (contains endolymph)

Vestibular membrane

Basilar membrane

Membranous labyrinth

Tectorial membrane

Spiral organ

1. Sound waves strike the tympanic membrane and cause it to vibrate.
2. Vibration of the tympanic membrane causes the three bones of the middle ear to vibrate.
3. The foot plate of the stapes vibrates in the oval window.
4. Vibration of the foot plate causes the perilymph in the scala vestibuli to vibrate.
5. Vibration of the perilymph causes displacement of the basilar membrane. Short waves (high pitch) cause displacement of the basilar

membrane near the oval window, and longer waves (low pitch) cause displacement of the basilar membrane some distance from the oval window. Movement of the basilar membrane is detected in the hair cells of the spiral organ, which are attached to the basilar membrane.
6. Vibrations of the perilymph in the scala vestibuli and of the endolymph in the cochlear duct are transferred to the perilymph of the scala tympani.
7. Vibrations in the perilymph of the scala tympani are transferred to the round window, where they are dampened.

Figure 15.29   Effect of Sound Waves on Cochlear Structures 🏃

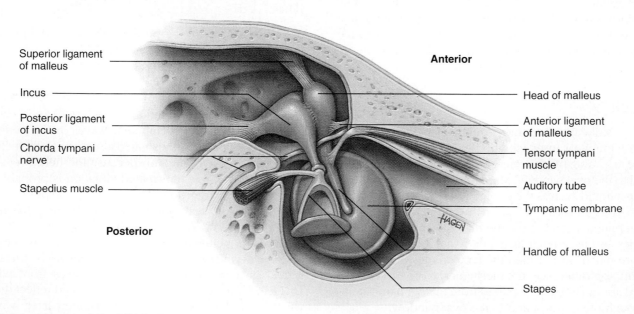

Superior ligament of malleus

Incus

Posterior ligament of incus

Chorda tympani nerve

Stapedius muscle

Posterior

Anterior

Head of malleus

Anterior ligament of malleus

Tensor tympani muscle

Auditory tube

Tympanic membrane

Handle of malleus

Stapes

HAGEN

Figure 15.30   Muscles of the Middle Ear

Medial view of the middle ear (as though viewed from the inner ear), showing the three ear ossicles with their ligaments and the two muscles of the middle ear: the tensor tympani and the stapedius. 🏃

The distortion of the basilar membrane is most important to hearing. As this membrane distorts, the hair cells resting on the basilar membrane move relative to the tectorial membrane, which remains stationary. The hair cell microvilli, which are embedded in the tectorial membrane, become bent, causing depolarization of the hair cells. The hair cells then induce action potentials in the cochlear neurons that synapse on the hair cells, apparently by direct electrical excitation through electrical synapses rather than by neurotransmitters.

The hairs of the hair cells are bathed in endolymph. Because of the difference in the potassium and sodium ion concentrations between the perilymph and endolymph, there is approximately an 80-mV potential across the vestibular membrane between the two fluids. This is called the **endocochlear potential.** Because the hair cell hairs are surrounded by endolymph, the hairs have a greater electric potential than if they were surrounded by perilymph. It is believed that this potential difference makes the hair cells much more sensitive to slight movement than they would be if surrounded by perilymph.

The part of the basilar membrane that distorts as a result of endolymph vibration depends on the pitch of the sound that created the vibration and, as a result, on the vibration frequency within the endolymph. The width of the basilar membrane and the length and diameter of the collagen fibers stretching across the membrane at each level along the cochlear duct determine the location of the optimum amount of basilar membrane vibration produced by a given pitch (figure 15.31). Higher pitched tones cause optimal vibration near the base, and lower pitched tones cause optimal vibration near the apex of the basilar membrane. As the basilar membrane vibrates, hair cells along a large part of the basilar membrane are stimulated. In areas of minimum vibration, the amount of stimulation may not reach threshold. In other areas, a low frequency of afferent action potentials may be transmitted, whereas in the optimally vibrating regions of the basilar membrane, a high frequency of action potentials is initiated.

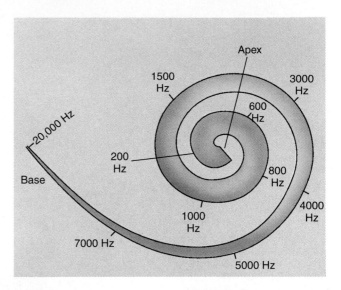

Figure 15.31   Effect of Sound Waves on Points Along the Basilar Membrane

Points of maximum vibration along the basilar membrane resulting from stimulation by sounds of various frequencies (in hertz).

Afferent action potentials conducted by cochlear nerve fibers from all along the spiral organ terminate in the **superior olivary nucleus** in the medulla oblongata (figure 15.32 and chapter 13). These action potentials are compared with one another, and the strongest action potential, corresponding to the area of maximum basilar membrane vibration, is taken as standard. Efferent action potentials then are sent from the superior olivary nucleus back to the spiral organ to all regions where the maximum vibration did not occur. These action potentials inhibit the hair cells from initiating additional action potentials in the afferent neurons. Thus, only action potentials from regions of maximum vibration are received by the cortex, where they become consciously perceived.

By this process tones are localized along the cochlea. As a result of this localization, neurons along a given portion of the cochlea send action potentials only to the cerebral cortex in response to specific pitches. Action potentials near the base of the basilar membrane stimulate neurons in a certain part of the auditory cortex, which interpret the stimulus as a high-pitched sound, whereas action potentials from the apex stimulate a different part of the cortex, which interprets the stimulus as a low-pitched sound.

12   **P R E D I C T**

Suggest some possible sites and mechanisms to explain why certain people have "perfect pitch" and other people are "tone deaf."

✔ *Answer in Appendix F*

Sound volume, or loudness, is a function of sound wave amplitude. As high-amplitude sound waves reach the ear, the

1. Afferent axons from the cochlear ganglion terminate in the cochlear nucleus in the brainstem.

2. Axons from the neurons in the cochlear nucleus project to the superior olivary nucleus or to the inferior colliculus.

3. Axons from the inferior colliculus project to the medial geniculate nucleus of the thalamus.

4. Thalamic neurons project to the auditory cortex.

5. Neurons in the superior olivary nucleus send axons to the inferior colliculus, back to the inner ear, or to motor nuclei in the brainstem that send efferent fibers to the middle ear muscles.

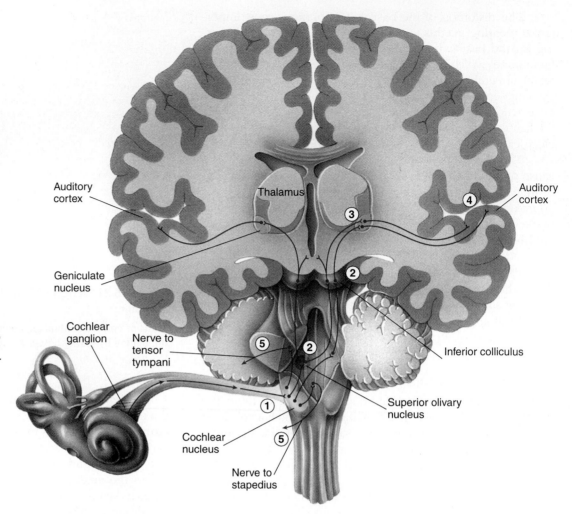

Figure 15.32   Central Nervous System Pathways for Hearing

perilymph, endolymph, and basilar membrane vibrate more intensely, and the hair cells are stimulated more intensely. As a result of the increased stimulation, more hair cells send action potentials at a higher frequency to the cerebral cortex, where this information is perceived as a greater sound volume.

---

13   P R E D I C T

Explain why it is much easier to perceive subtle musical tones when music is played somewhat softly as opposed to very loudly.

✔ *Answer in Appendix F*

Neuronal Pathways for Hearing

The special senses of hearing and balance are both transmitted by the vestibulocochlear (VIII) nerve. The term vestibular refers to the vestibule of the inner ear, which is involved in balance. The term cochlear refers to the cochlea and is that portion of the inner ear involved in hearing. The vestibulocochlear nerve functions as two separate nerves carrying information from two separate but closely related structures.

The auditory pathways within the CNS are very complex, with both crossed and uncrossed tracts (see figure 15.32). Unilateral CNS damage therefore usually has little effect on hearing. The neurons from the cochlear ganglion synapse with CNS neurons in the dorsal or ventral **cochlear nucleus** in the superior medulla near the inferior cerebellar peduncle. These neurons in turn either synapse in or pass through the superior olivary nucleus. Neurons terminating in this nucleus may synapse with efferent neurons returning to the cochlea to modulate pitch perception. Nerve fibers from the superior olivary nucleus also project to the trigeminal (V) and facial (VII) nuclei, controlling the tensor tympani and stapedius muscles, respectively. This reflex pathway dampens loud sounds by initiating contractions of these muscles. This is the sound attenuation reflex described previously. Neurons synapsing in the superior olivary nucleus may also join other ascending neurons to the cerebral cortex.

Clinical Focus   Deafness and Functional Replacement of the Ear

Deafness can have many causes. In general, there are two categories of deafness: conduction and sensorineural (or nerve) deafness. **Conduction deafness** involves a mechanical deficiency in transmission of sound waves from the external ear to the spiral organ and may often be corrected surgically. Hearing aids help people with such hearing deficiencies by boosting the sound volume reaching the ear. **Sensorineural deafness** involves the spiral organ or nerve pathways and is more difficult to correct.

Research is currently being conducted on ways to replace the hearing pathways with electric circuits. One approach involves the direct stimulation of nerves by electric impulses. There has been considerable success in the area of cochlear nerve stimulation. Certain types of sensorineural deafness in which the hair cells of the spiral organ are impaired can now be partially corrected. Prostheses are available that consist of a microphone for picking up the initial sound waves; a microelectronic processor for con-

verting the sound into electric signals; a transmission system for relaying the signals to the inner ear; and a long, slender electrode that is threaded into the cochlea. This electrode delivers electric signals directly to the endings of the cochlear nerve (figure D). High-frequency sounds are picked up by the microphone and transmitted through specific circuits to terminate near the oval window, whereas low-frequency sounds are transmitted farther up the cochlea to cochlear nerve endings near the helicotrema.

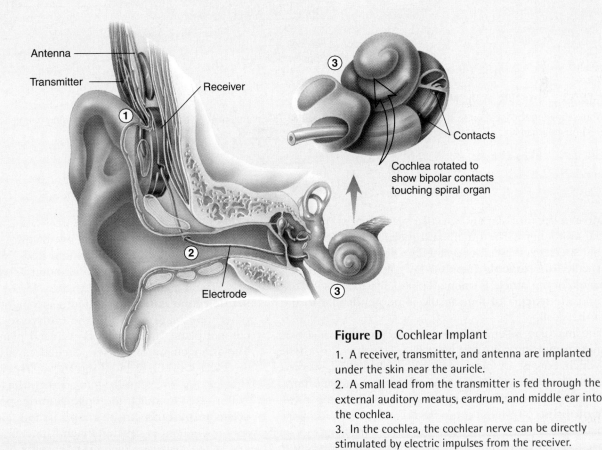

Figure D   Cochlear Implant

1. A receiver, transmitter, and antenna are implanted under the skin near the auricle.
2. A small lead from the transmitter is fed through the external auditory meatus, eardrum, and middle ear into the cochlea.
3. In the cochlea, the cochlear nerve can be directly stimulated by electric impulses from the receiver.

Ascending neurons from the superior olivary nucleus travel in the **lateral lemniscus.** All ascending fibers synapse in the **inferior colliculi,** and neurons from there project to the **medial geniculate nucleus** of the **thalamus,** where they synapse with neurons that project to the cortex. These neurons terminate in the **auditory cortex** in the dorsal portion of the temporal lobe within the lateral fissure and, to a lesser extent, on the superolateral surface of the temporal lobe (see chapter 13). Neurons from the inferior colliculus also project to the **superior colliculus,** where reflexes that turn the head and eyes in response to loud sounds are initiated.

Balance

The organs of balance can be divided structurally and functionally into two parts. The first, the **static labyrinth,** consists of the **utricle** (yū′tri-kl) and **saccule** (sak′yūl) of the vestibule and is primarily involved in evaluating the position of the head relative to gravity, although the system also responds to linear acceleration or deceleration, such as when a person is in a car that is increasing or decreasing speed. The second, the **kinetic labyrinth,** is associated with the semicircular canals and is involved in evaluating movements of the head.

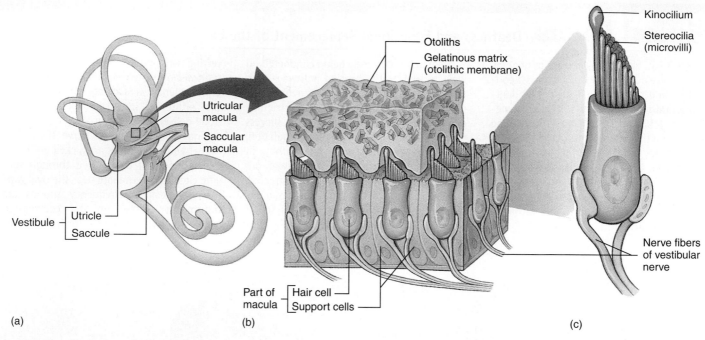

Figure 15.33   Structure of the Macula

(*a*) Vestibule showing the location of the utricular and saccular maculae. (*b*) Enlargement of the utricular macula, showing hair cells and otoliths in the macula. (*c*) An enlarged hair cell, showing the kinocilium and stereocilia.

Most of the utricular and saccular walls consist of simple cuboidal epithelium. The utricle and saccule, however, each contain a specialized patch of epithelium about 2–3 mm in diameter called the **macula** (mak′yū-lă; figure 15.33*a* and *b*). The macula of the utricle is oriented parallel to the base of the skull, and the macula of the saccule is perpendicular to the base of the skull.

The maculae resemble the spiral organ and consist of columnar supporting cells and hair cells. The "hairs" of these cells, which consist of numerous microvilli, called **stereocilia,** and one cilium, called a **kinocilium** (kī-nō-sil′ē-ŭm), are embedded in a **gelatinous mass** weighted by the presence of **otoliths** (ō′tō-liths) composed of protein and calcium carbonate (see figure 15.33*b*). The gelatinous mass moves in response to gravity, bending the hair cells and initiating action potentials in the associated neurons. Deflection of the hairs toward the kinocilium results in depolarization of the hair cell, whereas deflection of the hairs away from the kinocilium results in hyperpolarization of the hair cell. If the head is tipped, otoliths move in response to gravity and stimulate certain hair cells (figure 15.34). The hair cells are constantly being stimulated at a low level by the presence of the otolith-weighted covering of the macula; but as this covering moves in response to gravity, the pattern of intensity of hair cell stimulation changes. This pattern of stimulation and the subsequent pattern of action potentials from the numerous hair cells of the maculae can be translated by the brain into specific information about head position or acceleration. Much of this infor-

mation is not perceived consciously but is dealt with subconsciously. The body responds by making subtle tone adjustments in muscles of the back and neck, which are intended to restore the head to its proper neutral, balanced position.

The kinetic labyrinth (figure 15.35) consists of three **semicircular canals** placed at nearly right angles to one another, one lying nearly in the transverse plane, one in the coronal plane, and one in the sagittal plane (see chapter 1). The arrangement of the semicircular canals enables a person to detect movement in all directions. The base of each semicircular canal is expanded into an **ampulla** (see figure 15.35*a*). Within each ampulla, the epithelium is specialized to form a **crista ampullaris** (kris′tă am-pū-lar′ŭs). This specialized sensory epithelium is structurally and functionally very similar to that of the maculae. Each crista consists of a ridge or crest of epithelium with a curved gelatinous mass, the **cupula** (kū′pū-lă), suspended over the crest. The hairlike processes of the crista hair cells, similar to those in the maculae, are embedded in the cupula (see figure 15.35*b*). The cupula contains no otoliths and therefore does not respond to gravitational pull. Instead, the cupula is a float that is displaced by fluid movements within the semicircular canals. Endolymph movement within each semicircular canal moves the cupula, bends the hairs, and initiates action potentials (figure 15.36).

As the head begins to move in a given direction, the endolymph does not move at the same rate as the semicircular canals (see figure 15.36). This difference causes displacement of the cupula in a direction opposite to that of the movement

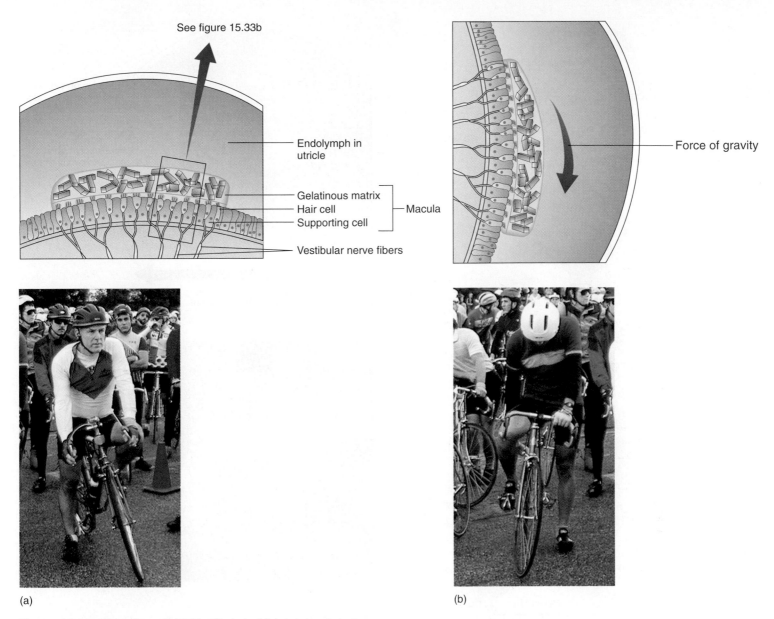

See figure 15.33b

Endolymph in utricle

Gelatinous matrix
Hair cell — Macula
Supporting cell

Vestibular nerve fibers

Force of gravity

(a)   (b)

Figure 15.34   Function of the Vestibule in Maintaining Balance

(*a*) In an upright position, the maculae do not move. (*b*) As the position of the head changes, such as when a person bends over, the maculae respond to changes in position of the head relative to gravity by moving in the direction of gravity.

of the head, resulting in relative movement between the cupula and the endolymph. As movement continues, the fluid of the semicircular canals begins to move and "catches up" with the cupula, and stimulation is stopped. As movement of the head ceases, the endolymph continues to move because of its momentum, causing displacement of the cupula in the same direction as the head had been moving. Because displacement of the cupula is most intense when the rate of head movement changes, this system detects changes in the rate of movement rather than movement alone. As with the static labyrinth, the information obtained by the brain from the kinetic labyrinth is largely subconscious.

Clinical Note

Space sickness is a balance disorder occurring in zero gravity and resulting from unfamiliar sensory input to the brain. The brain must adjust to these unusual signals, or severe symptoms may result such as headaches and dizziness. Space sickness is unlike motion sickness in that motion sickness results from an excessive stimulation of the brain, whereas space sickness results from too little stimulation as a result of weightlessness.

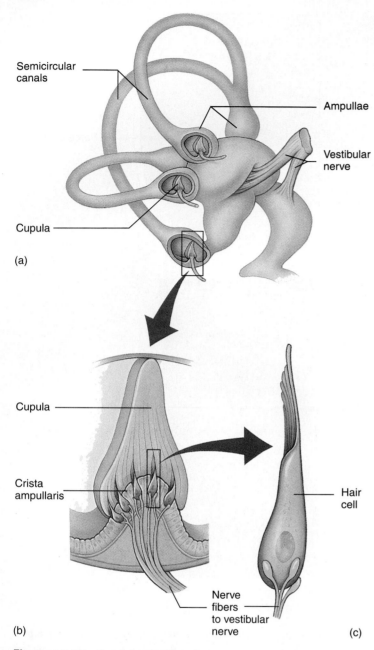

Semicircular canals

Ampullae

Vestibular nerve

Cupula

(a)

Cupula

Crista ampullaris

Hair cell

Nerve fibers to vestibular nerve

(b)

(c)

Figure 15.35 Semicircular Canals

(a) Semicircular canals showing location of the crista ampullaris in the ampullae of the semicircular canals. (b) Enlargement of the crista ampullaris, showing the cupula and hair cells. (c) Enlargement of a hair cell.

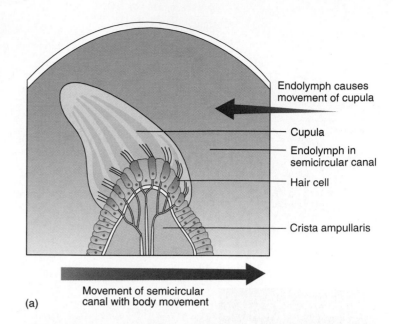

Endolymph causes movement of cupula

Cupula

Endolymph in semicircular canal

Hair cell

Crista ampullaris

Movement of semicircular canal with body movement

(a)

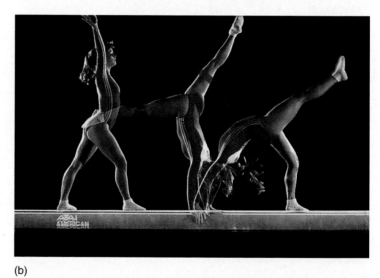

(b)

Figure 15.36 Function of the Semicircular Canals

The crista ampullaris responds to fluid movements within the semicircular canals. (a) When a person is at rest, the crista ampullaris does not move. (b) As a person begins to move in a given direction, the semicircular canals begin to move with the body (*blue arrow*), but the endolymph tends to remain stationary relative to the movement (momentum force; *red arrow* pointing in the opposite direction of body and semicircular canal movement), and the crista ampullaris is displaced by the endolymph in a direction opposite to the direction of movement.

Neuronal Pathways for Balance

Neurons synapsing on the hair cells of the maculae and cristae ampullares converge into the **vestibular ganglion,** where their cell bodies are located (figure 15.37). Afferent fibers from these neurons join afferent fibers from the cochlear ganglion to form the vestibulocochlear nerve (VIII) and terminate in the **vestibular nucleus** within the medulla oblongata. Axons run from this nucleus to numerous areas of the CNS, such as the spinal cord, cerebellum, cerebral cortex, and the nuclei controlling extrinsic eye muscles.

Balance is a complex process not simply confined to one type of input. In addition to vestibular sensory input, the vestibular nucleus receives input from proprioceptive neurons throughout the body, and from the visual system. People are asked to close their eyes while balance is evaluated in a sobriety test because alcohol affects the proprioceptive and

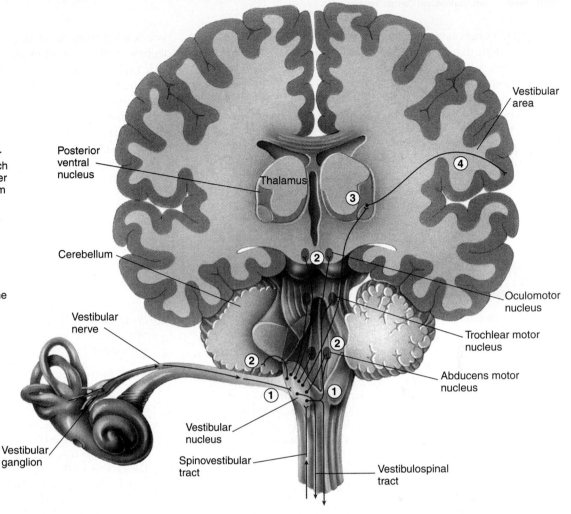

1. Afferent axons from the vestibular ganglion pass through the vestibular nerve to the vestibular nucleus, which also receives input from several other sources, such as proprioception from the legs.

2. Vestibular neurons send projections to the cerebellum, which controls postural muscles, and to the motor nuclei (oculomotor, trochlear, and abducens), which control extrinsic eye muscles.

3. Vestibular neurons also project to the posterior ventral nucleus of the thalamus.

4. Thalamic neurons project to the vestibular area of the cortex.

Figure 15.37 Central Nervous System Pathways for Balance

Clinical Focus — Ear Disorders

Otosclerosis

Otosclerosis (ō′tō-sklē-rō′sis) is an ear disorder in which spongy bone grows over the oval window and immobilizes the stapes, leading to progressive loss of hearing. This disorder can be surgically corrected by breaking away the bony growth and the immobilized stapes. During surgery, the stapes is replaced by a small rod connected by a fat pad or a synthetic membrane, to the oval window at one end and to the incus at the other end.

Tinnitus

Tinnitus (ti-nī′tŭs) consists of noises such as ringing, clicking, whistling, or booming in the ears. These noises may occur as a result of disorders in the middle or inner ear, or along the central neuronal pathways.

Motion Sickness

Motion sickness consists of nausea, weakness, and other dysfunctions caused by stimulation of the semicircular canals during motion, such as in a boat, automobile, airplane, swing, or amusement park ride. It may progress to vomiting and incapacitation. Antiemetics such as anticholinergic or antihistamine medications can be taken to counter the nausea and vomiting associated with motion sickness. Scopolamine is an anticholinergic drug that reduces the excitability of vestibular receptors. Cyclizine (Marezine), dimenhydrinate (Dramamine), and diphenhydramine (Benadryl) are antihistamines that affect the neural pathways from the vestibule. Scopolamine can be administered transdermally in the form of a patch placed on the skin behind the ear (Transdermal-Scop), which lasts about 3 days.

Otitis Media

Infections of the middle ear, called **otitis media**, are quite common in young children. These infections usually result from the spread of infection from the mucous membrane of the pharynx through the auditory tube to the mucous lining of the middle ear. The symptoms of otitis media, consisting of low-grade fever, lethargy, and irritability, are often not easily recognized by the parent as signs of middle ear infection. The infection can also cause a temporary decrease or loss of hearing because fluid buildup has dampened the tympanic membrane or ossicles.

Earache

Earache can result from otitis media, otitis externa (inflammation of the external auditory meatus), dental abscesses, or temporomandibular joint pain.

vestibular components of balance (cerebellar function) to a greater extent than it does the visual portion.

Reflex pathways exist between the kinetic part of the vestibular system and the nuclei controlling the extrinsic eye muscles (oculomotor, trochlear, and abducens). A reflex pathway allows maintenance of visual fixation on an object while the head is in motion. This function can be demonstrated by spinning a person around about 10 times in 20 seconds, stopping him, and observing his eye movements. The reaction is most pronounced if the individual's head is tilted forward about 30 degrees while he is spinning, thus bringing the lateral semicircular canals into the horizontal plane. There is slight oscillatory movement of the eyes. The eyes track in the direction of motion and return with a rapid recovery movement before repeating the tracking motion. This oscillation of the eyes is called **nystagmus** (nis-tag′mŭs). If asked to walk in a straight line, the individual deviates in the direction of rotation, and if asked to point to an object, his finger deviates in the direction of rotation.

Summary

The senses include general senses and special senses.

Classification of the Senses

1. Somatic modalities include touch, pressure, temperature, proprioception, and pain.
2. Visceral modalities are primarily pain and pressure.
3. Special modalities are smell, taste, sight, sound, and balance.
4. Receptors include mechanoreceptors, chemoreceptors, photoreceptors, thermoreceptors, and nociceptors.

Sensation

1. Sensation or perception is the conscious awareness of stimuli received by sensory receptors.
2. Sensation requires a stimulus, a receptor, conduction of an action potential to the CNS, translation of the action potential, and processing of the action potential in the CNS so that the person is aware of the sensation.

Types of Afferent Nerve Endings

1. Free nerve endings detect light touch, pain, itch, tickle, and temperature.
2. Merkel's disks respond to light touch and superficial pressure.
3. Hair follicle receptors wrap around the hair follicle and are involved in the sensation of light touch when the hair is bent.
4. Pacinian corpuscles, located in the dermis and hypodermis, detect pressure. In joints, they serve a proprioceptive function.
5. Meissner's corpuscles, located in the dermis, are responsible for two-point discriminative touch.
6. Ruffini's end organs are involved in continuous touch or pressure.
7. Golgi tendon organs, embedded in tendons, respond to changes in tension.
8. Muscle spindles, located in skeletal muscle, are proprioceptors.

Olfaction

Olfaction is the sense of smell.

Olfactory Epithelium and Bulb

1. Olfactory neurons in the olfactory epithelium are bipolar neurons. Their distal ends are enlarged as olfactory vesicles, which have long cilia. The cilia have receptors that respond to dissolved substances.

2. There are at least seven (perhaps 50) primary odors. The olfactory neurons have a very low threshold and accommodate rapidly.

Neuronal Pathways of Olfaction

1. Axons from the olfactory neurons extend as olfactory nerves to the olfactory bulb, where they synapse with mitral and tufted cells. Axons from these cells form the olfactory tracts. Association neurons in the olfactory bulbs can modulate output to the olfactory tracts.
2. The olfactory tracts terminate in the olfactory cortex. The lateral olfactory area is involved in the conscious perception of smell, the intermediate area with modulating smell, and the medial area with visceral and emotional responses to smell.

Taste

Taste buds usually are associated with papillae.

Histology of Taste Buds

1. The papillae are the circumvallate, fungiform, foliate, and filiform.
2. Taste buds consist of support and gustatory cells.
3. The gustatory cells have gustatory hairs that extend into taste pores.

Function of Taste

1. Receptors on the hairs detect dissolved substances.
2. There are four basic types of taste: sour, salty, bitter, and sweet.

Neuronal Pathways for Taste

1. The facial nerve carries taste sensations from the anterior two-thirds of the tongue, the glossopharyngeal nerve from the posterior third of the tongue, and the vagus nerve from the epiglottis.
2. The neural pathways for taste extend from the medulla oblongata to the thalamus and to the cerebral cortex.

Visual System
Accessory Structures

1. The eyebrows prevent perspiration from entering the eyes and help shade the eyes.

2. The eyelids consist of five tissue layers. They protect the eyes from foreign objects and help lubricate the eyes by spreading tears over their surface.
3. The conjunctiva covers the inner eyelid and the anterior part of the eye.
4. Lacrimal glands produce tears that flow across the surface of the eye. Excess tears enter the lacrimal canaliculi and reach the nasal cavity through the nasolacrimal canal. Tears lubricate and protect the eye.
5. The extrinsic eye muscles move the eyeball.

Anatomy of the Eye

1. The fibrous tunic is the outer layer of the eye. It consists of the sclera and cornea.
   - The sclera is the posterior four-fifths of the eye. It is white connective tissue that maintains the shape of the eye and provides a site for muscle attachment.
   - The cornea is the anterior one-fifth of the eye. It is transparent and refracts light that enters the eye.
2. The vascular tunic is the middle layer of the eye.
   - The iris is smooth muscle regulated by the autonomic nervous system. It controls the amount of light entering the pupil.
   - The ciliary muscles control the shape of the lens. They are smooth muscles regulated by the autonomic nervous system. The ciliary process produces aqueous humor.
3. The retina is the inner layer of the eye and contains neurons sensitive to light.
4. The eye has two compartments.
   - The anterior compartment is filled with aqueous humor, which circulates and leaves by way of the canal of Schlemm.
   - The posterior compartment is filled with vitreous humor.
5. The lens is held in place by the suspensory ligaments, which are attached to the ciliary muscles.
6. The macula lutea (fovea centralis) is the area of greatest visual acuity.
7. The optic disc is the location through which nerves exit and blood vessels enter the eye. It has no photosensory cells and is therefore the blind spot of the eye.

Functions of the Complete Eye

1. Light is that portion of the electromagnetic spectrum that humans can see.
2. When light travels from one medium to another, it can bend or refract. Light striking a concave surface refracts outward (divergence). Light striking a convex surface refracts inward (convergence).
3. Converging light rays meet at the focal point and are said to be focused.
4. The cornea, aqueous humor, lens, and vitreous humor all refract light. The cornea is responsible for most of the convergence, whereas the lens can adjust the focal point by changing shape.
   - Relaxation of the ciliary muscles causes the lens to flatten, producing the emmetropic eye.
   - Contraction of the ciliary muscles causes the lens to become more spherical. This change in lens shape enables the eye to focus on objects that are less than 20 feet away, a process called accommodation.
5. The far point of vision is the distance at which the eye no longer has to change shape to focus on an object. The near point of vision is the closest an object can come to the eye and still be focused.

6. The pupil becomes smaller during accommodation, increasing the depth of focus.

Structure and Function of the Retina

1. The pigmented retina provides a black backdrop for increasing visual acuity.
2. Rods are responsible for vision in low illumination (night vision).
   - A pigment, rhodopsin, is split by light into retinal and opsin, producing an action potential in the rod.
   - Light adaptation is caused by a reduction of rhodopsin; dark adaptation is caused by rhodopsin production.
3. Cones are responsible for color vision and visual acuity.
   - There are three types of cones, each with a different photopigment. The pigments are most sensitive to blue, red, and green lights.
   - Perception of many colors results from mixing the ratio of the different types of cones that are active at a given moment.
4. Most visual images are focused on the fovea centralis, which has a very high concentration of cones. Moving away from the fovea, there are fewer cones (the macula lutea); mostly rods are in the periphery of the retina.
5. The rods and the cones synapse with bipolar cells that in turn synapse with ganglion cells, which form the optic nerves.
6. Association neurons in the retina can modify information sent to the brain.

Neuronal Pathways for Vision

1. Ganglia cell axons extend to the lateral geniculate ganglion of the thalamus, where they synapse. From there neurons form the optic radiations that project to the visual cortex.
2. Neurons from the nasal visual field (temporal retina) of one eye and the temporal visual field (nasal retina) of the opposite eye project to the same cerebral hemisphere. Axons from the nasal retina cross in the optic chiasma, and axons from the temporal retina remain uncrossed.
3. Depth perception is the ability to judge relative distances of an object from the eyes and is a property of binocular vision. Binocular vision results because a slightly different image is seen by each eye.

Hearing and Balance

The osseous labyrinth is a canal system within the temporal bone that contains perilymph and the membranous labyrinth. Endolymph is inside the membranous labyrinth.

Auditory Structures and Their Functions

1. The external ear consists of the auricle and external auditory meatus.
2. The middle ear connects the external and inner ears.
   - The tympanic membrane is stretched across the external auditory meatus.
   - The malleus, incus, and stapes connect the tympanic membrane to the oval window of the inner ear.
   - The auditory tube connects the middle ear to the pharynx and functions to equalize pressure.
   - The middle ear is connected to the mastoid air cells.
3. The inner ear has three parts: the semicircular canals; the vestibule, which contains the utricle and the saccule; and the cochlea.

4. The cochlea is a spiral-shaped canal within the temporal bone.
   • The cochlea is divided into three compartments by the vestibular and basilar membranes. The scala vestibuli and scala tympani contain perilymph. The cochlear duct contains endolymph and the spiral organ (organ of Corti).
   • The spiral organ consists of hair cells that attach to the tectorial membrane.

Auditory Function

1. Sound waves are funneled by the auricle down the external auditory meatus, causing the tympanic membrane to vibrate.
2. The tympanic membrane vibrations are passed along the ossicles to the oval window of the inner ear.
3. Movement of the stapes in the oval window causes the perilymph, vestibular membrane, and endolymph to vibrate, producing movement of the basilar membrane. Movement of the basilar membrane causes displacement of the hair cells in the spiral organ and the generation of action potentials, which travel along the vestibulocochlear nerve.
4. Some vestibulocochlear nerve axons synapse in the superior olivary nucleus. Efferent neurons from this nucleus project back to the cochlea, where they regulate the perception of pitch.
5. The round window protects the inner ear from pressure buildup and dissipates waves.

Neuronal Pathways for Hearing

1. Axons from the vestibulocochlear nerve synapse in the medulla. Neurons from the medulla project axons to the inferior colliculi, where they synapse. Neurons from this point project to the thalamus and synapse. Thalamic neurons extend to the auditory cortex.
2. Efferent neurons project to cranial nerve nuclei responsible for controlling muscles that dampen sound in the middle ear.

Balance

1. Static balance evaluates the position of the head relative to gravity and detects linear acceleration and deceleration.
   • The utricle and saccule in the inner ear contain maculae. The maculae consist of hair cells with the hairs embedded in a gelatinous mass that contains otoliths.
   • The gelatinous mass moves in response to gravity.
2. Kinetic balance evaluates movements of the head.
   • There are three semicircular canals at right angles to one another in the inner ear. The ampulla of each semicircular canal contains the crista ampullaris, which has hair cells with hairs embedded in a gelatinous mass, the cupula.
   • The cupula is moved by endolymph within the semicircular canal when the head moves.

Neuronal Pathways for Balance

1. Axons from the maculae and the cristae ampullares extend to the vestibular nucleus of the medulla. Fibers from the medulla run to the spinal cord, cerebellum, cortex, and nuclei that control the extrinsic eye muscles.
2. Balance also depends on proprioception and visual input.

Content Review

1. Define the terms somatic, visceral, and special sense.
2. List the types of receptors.
3. How is a stimulus perceived as a sensation?
4. Define adaptation.
5. List the eight major types of afferent nerve endings, indicate where they are located, and state the functions they perform.
6. Describe the initiation of an action potential in an olfactory neuron. Name all the structures and cells that the action potential would encounter on the way to the olfactory cortex.
7. Name the three areas of the olfactory cortex, and give their functions.
8. How is the sense of smell modified in the olfactory bulb?
9. What is a primary odor? Name seven possible examples. How do the primary odors relate to our ability to smell many different odors?
10. How is the sense of taste related to the sense of smell?
11. Name and describe the four kinds of papillae found on the tongue. Which ones have taste buds associated with them?
12. Starting with a gustatory hair, name the structures and cells that an action potential would encounter on the way to the taste area of the cerebral cortex.
13. What are the four primary tastes? Where are they concentrated on the tongue? How do they produce many different kinds of taste sensations?
14. Describe the following structures, and state their functions: eyebrows, eyelids, conjunctiva, lacrimal apparatus, and extrinsic eye muscles.
15. Name the three layers (tunics) of the eye. For each layer describe the parts or structures it forms, and explain their functions.
16. What is the blind spot?
17. Name the two compartments of the eye and the substances that fill each compartment.
18. What is the function of the canal of Schlemm and the ciliary processes?
19. Describe the lens of the eye, and explain how the lens is held in place.
20. How does the pupil constrict? How does it dilate?
21. What causes light to refract? What is a focal point?
22. Describe the changes that occur in the lens, pupil, and extrinsic eye muscles as an object moves from 25 feet away to 6 inches away. What is meant by the terms near point and far point of vision?
23. Starting with a rod or a cone, name the cells or structures that an action potential would encounter while traveling to the visual cortex.
24. What is the function of the pigmented retina and of the choroid?
25. Describe the breakdown of rhodopsin by light. How does it re-form?
26. Describe the arrangement of cones and rods in the fovea, the macula lutea, and the periphery of the eye.
27. What is a visual field? How do the visual fields project to the brain?

28. What is depth perception? How does it occur?
29. Name the three regions of the ear, and list each region's parts.
30. Describe the relationship between the tympanic membrane, the ear ossicles, and the oval window of the inner ear.
31. What is the function of the external auditory meatus and of the auditory tube?
32. Explain how the cochlear duct is divided into three compartments. What is found in each compartment?
33. Starting with the auricle, trace a sound wave into the inner ear to the point at which action potentials are generated in the vestibulocochlear nerve.

34. Describe the neural pathways for hearing from the vestibulocochlear nerve to the cerebral cortex.
35. What are the functions of the saccule and the utricle? Describe the macula and its function.
36. What is the function of the semicircular canals? Describe the crista ampullaris and its mode of operation.
37. Describe the neural pathways for balance.

Develop Your Reasoning Skills

1. Describe all the sensations involved when a woman picks up an apple and bites into it. Explain which of those sensations are special and which are general. What types of receptors are involved? Which aspects of the taste of the apple are actually taste and which are olfaction?
2. An elderly man with normal vision developed cataracts. He was surgically treated by removing the lenses of his eyes. What kind of glasses would you recommend he wear to compensate for the removal of his lenses?
3. Some animals have a reflective area in the choroid called the tapetum lucidum. Light entering the eye is reflected back instead of being absorbed by the choroid. What would be the advantage of this arrangement? The disadvantage?
4. Perhaps you have heard someone say that eating carrots is good for the eyes. What is the basis for this claim?
5. On a camping trip Jean Tights ripped her pants. That evening she was going to repair the rip. As the sun went down, there was less and less light. When she tried to thread the needle, it was obvious that she was not looking directly at the needle but was looking a few inches to the side. Why did she do this?
6. A man stared at a black clock on a white wall for several minutes. Then he shifted his view and looked at only the blank white wall. Although he was no longer looking at the clock, he saw a light clock against a dark background. Explain what happened.
7. Describe the results of a lesion of the optic chiasma.
8. Persistent exposure to loud noise can cause loss of hearing, especially for high-frequency sounds. What part of the ear is probably damaged? Be as specific as possible.

9. Professional divers are subject to increased pressure as they descend to the bottom of the ocean. Sometimes this pressure can lead to damage to the ear and loss of hearing. Describe the normal mechanisms that adjust for changes in pressure, suggest some conditions that might interfere with pressure adjustment, and explain how the increased pressure might cause loss of hearing.
10. If a vibrating tuning fork is placed against the mastoid process of the temporal bone, the vibrations will be perceived as sound, even if the external auditory meatus is plugged. Explain how this could happen.
11. Some student nurses are at a party. Because they love anatomy and physiology so much, they are discussing adaptation of the special senses. They make the following observations:
    a. When entering a room, an odor such as brewing coffee is easily noticed. A few minutes later the odor might be barely, if at all, detectable, no matter how hard one tries to smell it.
    b. When entering a room, the sound of a ticking clock can be detected. Later the sound is not noticed until a conscious effort is made to hear it. Then it is easily heard. Explain the basis for each of the above observations.

Web Site Link

For a listing of the most current web sites related to this chapter, please visit the Seeley home page at:
http://www.mhhe.com/biosci/ap/seeleyap/

Chapter Sixteen

Autonomic Nervous System

Objectives

1. Compare the structural differences between the autonomic nervous system and the somatic motor nervous system.

2. Define the terms preganglionic neuron, postganglionic neuron, afferent neuron, somatic motor neuron, autonomic ganglion, and effector.

3. For both divisions of the autonomic nervous system, describe the location of the preganglionic and postganglionic neurons, the location of ganglia, the relative length of preganglionic and postganglionic axons, and the ratio of preganglionic to postganglionic neurons.

4. Describe the four pathways by which the sympathetic neurons extend from the sympathetic chain ganglia to target organs.

5. List the neurotransmitter substances for the preganglionic and postganglionic neurons for both the parasympathetic and sympathetic divisions.

6. Describe receptor types within autonomic synapses that respond to acetylcholine and norepinephrine, and describe their location.

7. Compare the autonomic nervous system's response to nicotine and muscarine.

8. Using examples, describe how autonomic reflexes help maintain homeostasis.

9. List the generalizations that can be made about the autonomic nervous system, and describe the limitations of each generalization.

10. Give an example for each category of drugs that affect the autonomic nervous system, and explain the general influence of the drug on the autonomic nervous system.

It is a sunny spring day and you are on a picnic. As you concentrate on the pleasant surroundings and the delicious food, you feel that everything is in balance and harmony. In physiologic terms, balance results from the maintenance of homeostasis. Your autonomic nervous system (ANS) keeps your body temperature at a constant level by controlling the activity of your sweat glands and the amount of blood flowing through your skin. After lunch, you are unaware of all the activities that are controlled by the ANS as your meal is digested. Furthermore, the movement of digested nutrients to tissues is possible because the ANS controls heart rate, which helps to maintain the blood pressure necessary to deliver blood to tissues. You would probably be overwhelmed by all the activities necessary to maintain homeostasis if you had to consciously control them. Fortunately, these activities are controlled on an unconscious level by the ANS.

The major anatomic and physiologic characteristics of the ANS are described in this chapter. A functional knowledge of the ANS enables you to predict general responses to a variety of stimuli, explain responses to changes in environmental conditions, comprehend symptoms that result from abnormal autonomic functions, and understand how drugs affect the ANS.

Contrasting the Somatic Motor and Autonomic Nervous Systems

The peripheral nervous system is composed of afferent and efferent neurons, with axons that course through the same nerves. Afferent neurons carry action potentials from the periphery to the central nervous system (CNS), and efferent neurons carry action potentials from the CNS to the periphery. The efferent neurons belong to either the somatic motor nervous system, which innervates skeletal muscle, or the ANS, which innervates smooth muscle, cardiac muscle, and glands.

Although axons of autonomic, somatic motor, and afferent neurons are found within the same nerves, the proportion varies from nerve to nerve. For example, nerves innervating smooth muscle, cardiac muscle, and glands consist primarily of autonomic neurons; and nerves innervating skeletal muscles consist primarily of somatic motor neurons. Cranial nerves such as the optic, vestibulocochlear, and trigeminal nerves are composed entirely or mainly of afferent neurons.

Unlike efferent neurons, afferent neurons are not divided into functional groups. Afferent neurons propagate action potentials from sensory receptors to the CNS and provide information for reflexes mediated through the somatic motor system or the ANS. For example, stimulation of pain receptors can initiate both somatic motor and autonomic reflexes such as the withdrawal reflex and an increase in heart rate, respectively. Although some afferent neurons primarily affect autonomic functions and others primarily influence somatic motor functions, functional overlap makes the classification of afferent neurons as autonomic or somatic misleading.

In contrast to afferent neurons, the efferent neurons are separated into the somatic motor system and the ANS, which differ structurally and functionally. Axons of somatic motor neurons extend from the CNS to skeletal muscle. The ANS, on the other hand, has two neurons in a series extending between the CNS and the organs innervated (figure 16.1). The first neurons of the series are called **preganglionic neurons.** Their cell bodies are located within either the brainstem or the spinal cord, and their axons extend through nerves to autonomic ganglia located outside the CNS. The **autonomic ganglia** contain the cell bodies of the second neurons of the series, which are called **postganglionic neurons.** The preganglionic neurons synapse with the postganglionic neurons in the autonomic ganglia. The axons of the postganglionic neurons extend to effector organs, where they synapse with their target tissues.

Many movements controlled by the somatic motor system are conscious, whereas ANS functions are unconsciously controlled. The effect of somatic motor neurons on skeletal muscle is always excitatory, but the effect of the ANS on target tissues can be excitatory or inhibitory. For example, after a meal the ANS can stimulate stomach activities, but during exercise, the ANS can inhibit those activities. A comparison of the somatic motor nervous system and the ANS is summarized in table 16.1.

Divisions of the Autonomic Nervous System: Structural Features

The ANS is composed of **sympathetic** and **parasympathetic divisions,** each with unique structural and functional features. Structurally these divisions differ in (1) the location of their preganglionic neuron cell bodies within the CNS, (2) the location of their autonomic ganglia, (3) the relative lengths of their preganglionic and postganglionic axons, and (4) the ratio of preganglionic and postganglionic neurons. Table 16.2 summarizes the structural differences between the sympathetic and parasympathetic divisions.

Sympathetic Division

Cell bodies of sympathetic preganglionic neurons are in the lateral horns of the spinal cord gray matter between the first thoracic (T1) and the second lumbar (L2) segments (figure 16.2). Because of the location of the preganglionic cell bodies, the sympathetic division is sometimes called the **thoracolumbar division.** The axons of the preganglionic neurons pass through the ventral roots of spinal nerves T1–L2, course through the spinal nerves for a short distance, leave the spinal nerves, and project to autonomic ganglia. These ganglia, called **sympathetic chain ganglia,** are on either side of the vertebral column behind the epithelial linings of the

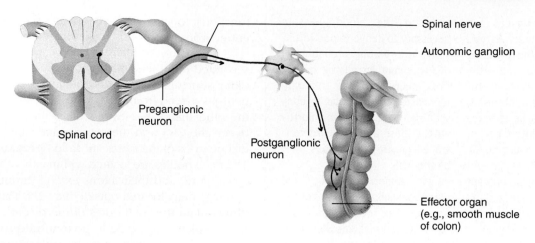

Figure 16.1   Organization of Autonomic Nervous System Neurons

The cell body of the preganglionic neuron is in the CNS, and its axon extends to the autonomic ganglion and synapses with the postganglionic neuron. The postganglionic neuron extends to and synapses with its effector organ.

Table 16.1	Comparison of the Autonomic and Somatic Motor	
Features	**Somatic Motor Nervous System**	**Autonomic Nervous System**
Target tissues	Skeletal muscle	Smooth muscle, cardiac muscle, and glands
Regulation	Controls all conscious and unconscious movements of skeletal muscle	Unconscious regulation, although influenced by conscious mental functions
Response to stimulation	Skeletal muscle contracts	Target tissues are stimulated or inhibited
Neuron arrangement	One neuron extends from the central nervous system (CNS) to skeletal muscle	Two neurons in series; the preganglionic neuron extends from the CNS to an autonomic ganglion, and the postganglionic neuron extends from the autonomic ganglion to the target tissue
Neuron cell body location	Neuron cell bodies are in motor nuclei of the cranial nerves and in the ventral horn of the spinal cord	Preganglionic neuron cell bodies are in autonomic nuclei of the cranial nerves and in the lateral part of the spinal cord; postganglionic neuron cell bodies are in autonomic ganglia
Number of synapses	One synapse between the somatic motor neuron and the skeletal muscle	Two synapses; first is in the autonomic ganglia; second is at the target tissue
Axon sheaths	Myelinated	Preganglionic axons are myelinated; postganglionic axons are unmyelinated
Neurotransmitter substance	Acetylcholine	Acetylcholine is released by preganglionic neurons; either acetylcholine or norepinephrine is released by postganglionic neurons
Receptor molecules	Receptor molecules for acetylcholine are nicotinic	In autonomic ganglia, receptor molecules for acetylcholine are nicotinic; in target tissues, receptor molecules for acetylcholine are muscarinic, whereas receptor molecules for norepinephrine are either alpha- or beta-adrenergic

pleural and peritoneal cavities. The ganglia are connected to one another and form a chain along both sides of the spinal cord. Although only ganglia from T1 to L2 receive preganglionic axons from the spinal cord, the sympathetic chain extends into the cervical and sacral regions so that one pair of ganglia is associated with nearly every pair of spinal nerves. In the cervical region the ganglia usually fuse during fetal development so only two or three pairs occur in the adult.

The axons of the preganglionic neurons are small in diameter and myelinated. The short connection between a spinal nerve and a sympathetic chain ganglion through which the preganglionic axons pass is called the **white ramus communicans** (rā′mŭs kŏ-myū′ni-kans; pl., rami communicantes, rā′mī kŏ-myū-ni-kan′tēz) because of the whitish color of the myelinated axons (figure 16.3).

Sympathetic axons exit the sympathetic chain ganglia by four different routes.

Table 16.2   Comparison of the Sympathetic and Parasympathetic Divisions

Feature	Sympathetic Division	Parasympathetic Division
Location of preganglionic cell body	Lateral horns of spinal cord gray matter (T1–L2)	Brainstem and lateral parts of spinal cord gray matter (S2–S4)
Outflow from central nervous system	Spinal nerves Sympathetic nerves Splanchnic nerves	Cranial nerves Pelvic nerves
Ganglia	Sympathetic chain ganglia along spinal cord for spinal and sympathetic nerves; collateral ganglia for splanchnic nerves	Terminal ganglia near or on effector organ
Number of postganglionic neurons for each preganglionic neuron	Many	Few
Relative length of neurons	Short preganglionic Long postganglionic	Long preganglionic Short postganglionic

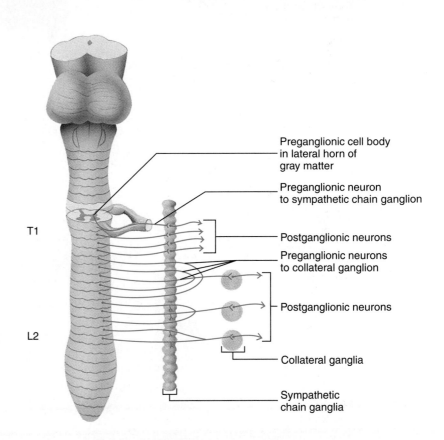

T1

L2

Preganglionic cell body in lateral horn of gray matter

Preganglionic neuron to sympathetic chain ganglion

Postganglionic neurons

Preganglionic neurons to collateral ganglion

Postganglionic neurons

Collateral ganglia

Sympathetic chain ganglia

Figure 16.2   Sympathetic Division

Location of the preganglionic and postganglionic cell bodies of the sympathetic neurons. The preganglionic cell bodies are in the lateral gray matter of the thoracic and lumbar parts of the spinal cord. The cell bodies of the postganglionic neurons are primarily within the sympathetic chain ganglia. Some postganglionic cell bodies are within collateral ganglia that lie outside the sympathetic chain ganglia.

1. *Spinal nerves* (see figure 16.3*a*). The preganglionic axons synapse with postganglionic neurons in a sympathetic chain ganglion at the same level that the preganglionic axons enter the sympathetic chain. Alternatively, the preganglionic axons pass either superiorly or inferiorly through one or more ganglia and synapse with postganglionic neurons in a sympathetic chain ganglion at a different level. The axons of the postganglionic neurons pass through a **gray ramus communicans** and reenter a spinal nerve. The postganglionic axons are not myelinated, giving the gray

ramus communicans its grayish color. The postganglionic axons then project through the spinal nerve to the organs they innervate. These structures include sweat glands in the skin, smooth muscle in skeletal and skin blood vessels, and the smooth muscle of the arrector pili.

2. *Sympathetic nerves* (see figure 16.3*b*). The preganglionic axons enter the sympathetic chain and synapse in a sympathetic chain ganglion at the same or a different level with the postganglionic neuron. The postganglionic axons leave the sympathetic chain

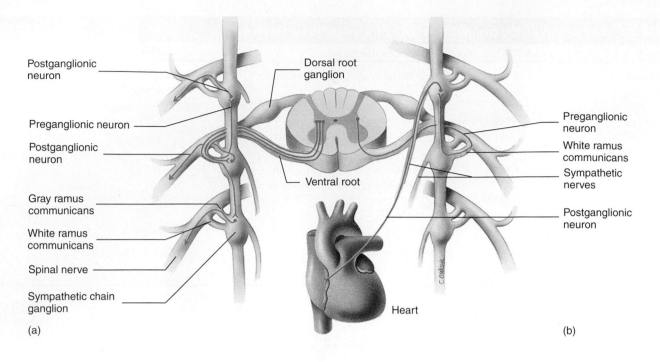

Postganglionic neuron

Preganglionic neuron

Postganglionic neuron

Gray ramus communicans

White ramus communicans

Spinal nerve

Sympathetic chain ganglion

(a)

Dorsal root ganglion

Ventral root

Preganglionic neuron

White ramus communicans

Sympathetic nerves

Postganglionic neuron

Heart

(b)

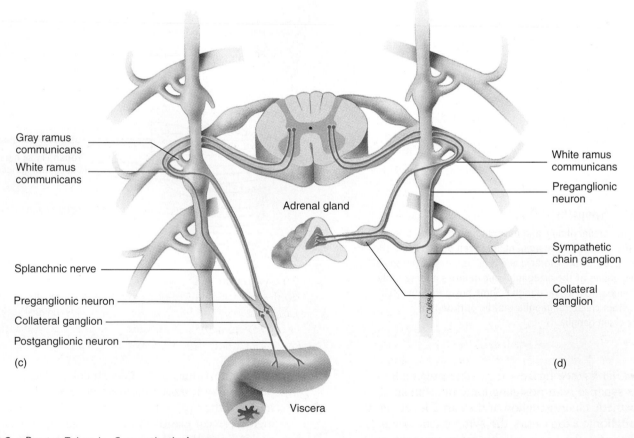

Gray ramus communicans

White ramus communicans

Splanchnic nerve

Preganglionic neuron

Collateral ganglion

Postganglionic neuron

(c)

Adrenal gland

Viscera

White ramus communicans

Preganglionic neuron

Sympathetic chain ganglion

Collateral ganglion

(d)

Figure 16.3   Routes Taken by Sympathetic Axons

(a) Preganglionic axons enter a sympathetic chain ganglion through a white ramus communicans. Some axons synapse with a postganglionic neuron at the level of entry; other axons ascend or descend to other levels before synapsing. Postganglionic axons exit the sympathetic chain ganglia through gray rami communicantes and enter spinal nerves. (b) Like part (a), except that postganglionic axons exit through a sympathetic nerve (only an ascending axon is illustrated). (c) Preganglionic neurons do not synapse in the sympathetic chain ganglia but exit in splanchnic nerves and extend to collateral ganglia where they synapse with postganglionic neurons. (d) Like part (c), except that preganglionic axons extend to the adrenal medulla, where they synapse. There are no postganglionic neurons.

ganglion in a sympathetic nerve. Many of the sympathetic nerves innervate the thoracic organs, including cardiac muscle, smooth muscle in thoracic blood vessels, and smooth muscle in the esophagus and lungs. In the cervical region, sympathetic nerves form plexuses around the carotid arteries and project to the organs of the head. These sympathetic axons innervate areas of the head and neck not innervated through the spinal nerves. These structures include sweat glands in the skin; salivary glands in the mouth; and smooth muscle in blood vessels, the eye, and the arrector pili.

3. *Splanchnic* (splangk'nik) *nerves* (see figure 16.3*c*). Some preganglionic axons that originate between T5 and T12 of the spinal cord enter the sympathetic chain ganglia and, without synapsing, exit at the same or a different level. The preganglionic axons then pass through splanchnic nerves to **collateral,** or **prevertebral, ganglia,** where they synapse with postganglionic neurons. Axons of the postganglionic neurons leave from the collateral ganglia through small nerves that extend to target organs. The collateral ganglia are in the abdomen close to sites where major arteries branch from the abdominal aorta and are named after the blood vessels near which they are located. The three major collateral ganglia are the celiac (sē'lē-ak), superior mesenteric (mez-en-ter'ik), and inferior mesenteric ganglia. The splanchnic nerves innervate the abdominopelvic organs: smooth muscle in blood vessels or the walls of organs and glands such as the pancreas, liver, and prostate gland.

4. *Innervation to the adrenal gland* (see figure 16.3*d*). The splanchnic nerve innervation to the adrenal glands is different from other ANS nerves because it consists of only preganglionic neurons. The axons of the preganglionic neurons do not synapse in the sympathetic chain ganglia or in collateral ganglia. Instead the preganglionic axons pass through the sympathetic chain ganglia and collateral ganglia and synapse with cells in the adrenal medulla. The **adrenal medulla** (me-dūl'ă) is the inner portion of the adrenal gland and consists of specialized cells derived during development from the same cells that give rise to the postganglionic cells of the ANS. These specialized cells are round in shape, have no axons or dendrites, and are divided into two populations of cells. About 80% of the cells secrete **epinephrine** (ep'i-nef'rin), also called **adrenaline** (ă-dren'ă-lin), and about 20% secrete **norepinephrine** (nōr'ep-i-nef'rin), also called **noradrenaline** (nōr-ă-dren'ă-lin). Stimulation of these cells by the preganglionic axons causes the release of epinephrine and norepinephrine. These substances circulate in the blood and affect all tissues having receptors to which they can bind. The general response to epinephrine and norepinephrine released from the adrenal medulla is to prepare the individual for physical activity. Secretions of the adrenal medulla are considered hormones because they are released into the general circulation and travel some distance to the tissues in which they have their effect (see chapters 17 and 18).

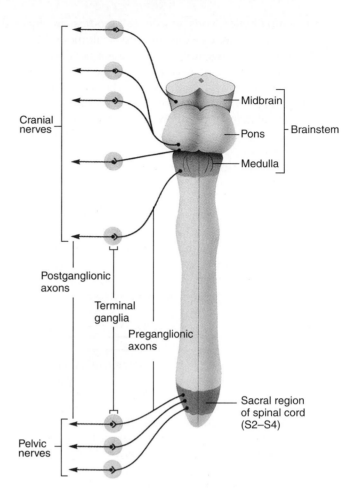

Figure 16.4   Parasympathetic Division

Location of the preganglionic and postganglionic neuron cell bodies of the parasympathetic division. The preganglionic neuron cell bodies are in the brainstem and the lateral gray matter of the sacral part of the spinal cord, and the postganglionic neuron cell bodies are within terminal ganglia.

Parasympathetic Division

Parasympathetic preganglionic neuron cell bodies are found both superior and inferior to the area of the CNS where preganglionic neuron cell bodies of the sympathetic division are found. Preganglionic cell bodies of the parasympathetic division are either within brainstem nuclei or within the lateral parts of the gray matter in the sacral region of the spinal cord from S2 to S4 (figure 16.4). For that reason the parasympathetic division is sometimes called the **craniosacral** (krā'nē-ō-sā'krăl) **division.**

Axons of the preganglionic neurons course through cranial or pelvic nerves to **terminal ganglia** either near or embedded within the walls of the organs innervated by the parasympathetic neurons. Many of the parasympathetic ganglia are small in size, but some, such as those in the wall of the digestive tract, are extensive. The axons of the postganglionic neurons extend relatively short distances from the terminal ganglia to the target organs.

Parasympathetic axons whose cell bodies are located within brainstem nuclei exit through the oculomotor (III), facial (VII), glossopharyngeal (IX), and vagus nerves (X). The oculomotor nerves carry parasympathetic fibers that innervate smooth muscle cells within the eyes; the facial and glossopharyngeal nerves carry parasympathetic fibers to the salivary and lacrimal glands; and the vagus nerves carry parasympathetic fibers to most thoracic and abdominal viscera, including the heart, lungs, esophagus, stomach, pancreas, liver, intestine, and upper colon. Approximately 75% of all parasympathetic axons course through the vagus nerves. Parasympathetic axons in the head leave the oculomotor, facial, and glossopharyngeal nerves and travel with branches of the trigeminal nerve to their target organs.

Parasympathetic preganglionic axons whose cell bodies are in the sacral region of the spinal cord course through pelvic nerves that innervate the urinary bladder, lower colon, rectum, and organs of the reproductive system.

Clinical Note

Spinal cord injury can damage nerve tracts, interrupting control of autonomic preganglionic neurons by ANS centers in the brain. For the parasympathetic division, effector organs innervated through the sacral region of the spinal cord are affected, but most effector organs still have normal parasympathetic function because they are innervated by the vagus nerve. For the sympathetic division, brain control of sympathetic preganglionic neurons is lost below the site of the injury. The higher the level of injury, the greater the number of body parts affected.

Physiology of the Autonomic Nervous System
Neurotransmitters

The sympathetic and parasympathetic nerve endings secrete one of two neurotransmitters. If the neuron secretes acetylcholine, it is a **cholinergic** (kol-in-er′jik) **neuron,** and if it secretes norepinephrine, it is an **adrenergic** (ad-rĕ-ner′jik) **neuron.** All preganglionic neurons of the sympathetic and parasympathetic divisions and all postganglionic neurons of the parasympathetic division are cholinergic. Almost all postganglionic neurons of the sympathetic division are adrenergic, but a few postganglionic neurons that innervate thermoregulatory sweat glands are cholinergic (figure 16.5).

In recent years, substances in addition to the regular neurotransmitters have been extracted from ANS neurons. These substances include fatty acids, such as prostaglandins, and peptides, such as gastrin, somatostatin, cholecystokinin, vasoactive intestinal peptide, enkephalins, substance P, and dopamine. The specific role that each of these compounds plays in the regulation of the ANS is unclear, but they can function as either neurotransmitters or neuromodulator substances (see chapter 12).

Receptors

Receptors in the cell membrane of certain cells can combine with either acetylcholine or norepinephrine. The combination of neurotransmitter and receptor functions as a signal to cells, causing them to respond. Depending on the type of cell, the response can be excitatory or inhibitory.

Cholinergic Receptors

In a cholinergic synapse, acetylcholine molecules released from the presynaptic terminal diffuse across the synaptic cleft and combine with receptor molecules within the postsynaptic membrane. These receptors are called **cholinergic receptors,** and they have two major structurally different forms. **Nicotinic** (nik-ō-tin′ik) **receptors** bind to nicotine, an alkaloid substance found in tobacco; and **muscarinic** (mŭs-kă-rin′ik) **receptors** bind to muscarine, an alkaloid extracted from some poisonous mushrooms. Although nicotine and muscarine are not naturally in the human body, they demonstrate differences in the two classes of cholinergic receptors. Nicotine binds to nicotinic receptors but not to muscarinic receptors, whereas muscarine binds to muscarinic receptors but not to nicotinic receptors. On the other hand, nicotinic and muscarinic receptors are very similar because acetylcholine binds to and activates both types of receptors.

The membranes of all postganglionic neurons in autonomic ganglia and the membranes of skeletal muscle cells have nicotinic receptors. The membranes of effector cells that respond to acetylcholine released from postganglionic neurons have muscarinic receptors.

1   PREDICT
Would structures innervated by the sympathetic division or the parasympathetic division be stimulated after the consumption of nicotine? After the consumption of muscarine? Explain.

✔ *Answer in Appendix F*

Acetylcholine binding to nicotinic receptors results in the direct opening of Na$^+$ ion channels and the production of action potentials. When acetylcholine binds to muscarinic receptors, the cell's response is mediated through G proteins (see chapter 9). The response can be either excitatory or inhibitory, depending on the target tissue in which the receptors are found. For example, acetylcholine binds to muscarinic receptors in cardiac muscle, reducing the heart rate; and acetylcholine binds to muscarinic receptors in smooth muscle cells of the stomach, increasing the rate of contraction.

Adrenergic Receptors

Norepinephrine is released from adrenergic postganglionic neurons of the sympathetic division (see figure 16.5), diffuses across the synapse, and binds to receptor molecules within the cell membranes of the effector organ. These receptors are called **adrenergic receptors,** and they are subdivided into

Sympathetic division

Most target tissues innervated by the sympathetic division have adrenergic receptors. When norepinephrine binds to adrenergic receptors, some target tissues are stimulated, and others are inhibited. For example, blood vessels are stimulated to constrict, and stomach glands are inhibited.

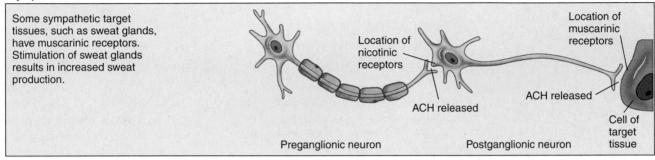

Sympathetic division

Some sympathetic target tissues, such as sweat glands, have muscarinic receptors. Stimulation of sweat glands results in increased sweat production.

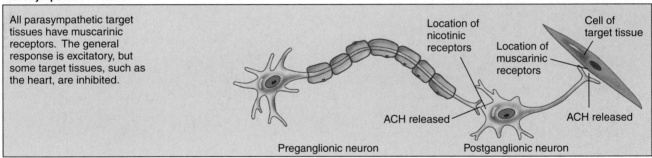

Parasympathetic division

All parasympathetic target tissues have muscarinic receptors. The general response is excitatory, but some target tissues, such as the heart, are inhibited.

Figure 16.5   Location of ANS Receptors

Nicotinic receptors are on the cell bodies of both sympathetic and parasympathetic postganglionic cells in the autonomic ganglia. *Abbreviations: NE,* norepinephrine; *ACH,* acetylcholine.

two major structural categories: **alpha (α)-receptors** and **beta (β)-receptors.** Effector cells can have alpha receptors, beta receptors, or both alpha and beta receptors (table 16.3).

When norepinephrine binds to adrenergic receptors, the cell's response is mediated through G proteins. Norepinephrine binding to alpha or beta receptors in some tissues produces an excitatory response, whereas in other tissues it produces an inhibitory response. For example, norepinephrine binding to the beta receptors in cardiac muscle produces an excitatory response that increases the force of contraction of cardiac muscle cells. Norepinephrine binding to the beta receptors in stomach smooth muscle, however, produces an inhibitory response that relaxes the smooth muscle cells.

Epinephrine and norepinephrine released from the adrenal medulla can also bind to alpha and beta receptors. Epinephrine stimulates both types of receptors nearly equally,

but norepinephrine stimulates alpha receptors more than beta receptors.

Clinical Note

Norepinephrine is produced from a precursor molecule called dopamine. Certain preganglionic and postganglionic neurons can release dopamine, which binds to specific dopamine receptors. Because dopamine is structurally similar to norepinephrine, it also can bind to beta receptors. Dopamine hydrochloride has been used successfully to treat circulatory shock because it can bind to dopamine receptors in kidney blood vessels. The resulting vasodilation increases blood flow to the kidneys and prevents damage to the kidneys. At the same time, dopamine can bind to beta receptors in the heart, causing increased force of contraction.

Table 16.3   Autonomic Innervation of Target Tissues

Organ	Effect of Sympathetic Stimulation	Effect of Parasympathetic Stimulation
Heart		
Muscle	Increased rate and force (b)	Slowed rate (c)
Coronary arteries	Dilated (b), constricted (a)*	Dilated (c)
Systemic blood vessels		
Abdominal	Constricted (a)	None
Skin	Constricted (a)	None
Muscle	Dilated (b), constricted (a)	None
Lungs		
Bronchi	Dilated (b)	Constricted (c)
Liver	Glucose released into blood (b)	None
Skeletal muscles	Breakdown of glycogen to glucose (b)	None
Metabolism	Increased up to 100% (a, b)	None
Glands		
Adrenal	Release of epinephrine and norepinephrine (c)	None
Salivary	Constriction of blood vessels and slight production of a thick, viscous secretion (a)	Dilation of blood vessels and thin, copious secretion (c)
Gastric	Inhibition (a)	Stimulation (c)
Pancreas	Decreased insulin secretion (a)	Increased insulin secretion (c)
Lacrimal	None	Secretion (c)
Sweat		
Merocrine	Copious, watery secretion (c)	None
Apocrine	Thick, organic secretion (c)	None
Gut		
Wall	Decreased tone (b)	Increased motility (c)
Sphincter	Increased tone (a)	Decreased tone (c)
Gallbladder and bile ducts	Relaxed (b)	Contracted (c)
Urinary bladder		
Wall	Relaxed (b)	Contracted (c)
Sphincter	Contracted (a)	Relaxed (c)
Eye		
Ciliary muscle	Relaxed for far vision (b)	Contracted for near vision (c)
Pupil	Dilated (a)	Constricted (c)
Arrector pili muscles	Contraction (a)	None
Blood	Increased coagulation (a)	None
Sex organs	Ejaculation (a)	Erection (c)

(a) Mediated by alpha-adrenergic receptors; (b) mediated by beta-adrenergic receptors; (c) mediated by cholinergic receptors.

*Normally blood flow increases through coronary arteries as a result of sympathetic stimulation of the heart because of increased demand by cardiac tissue for oxygen (local control of blood flow is discussed in chapter 21). In experiments that isolate the coronary arteries, however, sympathetic nerve stimulation, acting through alpha-adrenergic receptors, causes vasoconstriction. The beta-adrenergic receptors are relatively insensitive to sympathetic nerve stimulation but can be activated by epinephrine released from the adrenal gland and by drugs.

Clinical Focus   The Influence of Drugs on the Autonomic Nervous System

Drugs that affect the ANS can have important therapeutic value, and they can be used to treat certain diseases because they can increase or decrease activities normally controlled by the ANS. Drugs that affect the ANS can also be found in medically hazardous substances such as tobacco and insecticides.

Direct-acting drugs bind to ANS receptors to produce their effects. For example, **stimulating agents** bind to specific receptors and activate them, and **blocking agents** bind to specific receptors and prevent them from being activated. The main topic of this essay is direct-acting drugs. It should be noted, however, that indirect-acting drugs can also influence the ANS. For example, some drugs indirectly produce a stimulatory effect by causing the release of neurotransmitters or by preventing the metabolic breakdown of neurotransmitters. Other drugs indirectly produce an inhibitory effect by preventing the biosynthesis or release of neurotransmitters.

Drugs That Bind to Nicotinic Receptors

Drugs that bind to nicotinic receptors and activate them are **nicotinic agents**. Although these agents have little therapeutic value and are mainly of interest to researchers, nicotine is medically important because of its presence in tobacco. Nicotinic agents bind to the nicotinic receptors on all postganglionic neurons within the autonomic ganglia and produce stimulation. Responses to nicotine are variable and depend on the amount taken into the body. Because nicotine stimulates the postganglionic neurons of both the sympathetic and parasympathetic divisions, much of the variability of its effects results from the opposing actions of these divisions. For example, in response to the nicotine contained in a cigarette, the heart rate may either increase or decrease; and its rhythm tends to become less regular as a result of the simultaneous actions on the sympathetic division, which increases the heart rate, and the parasympathetic division, which decreases the heart rate. Blood pressure tends to increase because of the constriction of blood vessels, which are almost exclusively innervated by sympathetic neurons. In addition to its influence on the ANS, nicotine also affects the central nervous system; therefore, not all of its effects can be explained on the basis of action on the ANS. Nicotine is extremely toxic, and small amounts can be lethal.

Drugs that bind to and block nicotinic receptors are called **ganglionic blocking agents** because they block the effect of acetylcholine on both parasympathetic and sympathetic postganglionic neurons. The effect of these substances on the sympathetic division, however, overshadows the effect on the parasympathetic division. For example, trimethaphan camsylate (trī-meth′ă-fan kam′sil-āt) can be used to treat high blood pressure. It blocks sympathetic stimulation of blood vessels, causing the blood vessels to dilate, which decreases blood pressure. Ganglionic blocking agents have limited uses because they affect both sympathetic and parasympathetic ganglia. Whenever possible, more selective drugs are now used.

Drugs That Bind to Muscarinic Receptors

Drugs that bind to and activate muscarinic receptors are **muscarinic, or parasympathomimetic** (par-ă-sim′pă-thō-mi-met′ik), **agents.** These drugs activate the muscarinic receptors of target tissues of the parasympathetic division and the muscarinic receptors of sweat glands, which are innervated by the sympathetic division. Muscarine causes increased sweating; increased secretion of glands in the digestive system; decreased heart rate; constriction of the pupils; and contraction of respiratory, digestive, and urinary system smooth muscles. Bethanechol (be-than′ĕ-kol) chloride is a parasympathomimetic agent used to stimulate the urinary bladder following surgery, because the general anesthetics used for surgery can temporarily inhibit a person's ability to urinate.

Drugs such as atropine that bind to and block the action of muscarinic receptors are **muscarinic, or parasympathetic, blocking agents.** These drugs dilate the pupil of the eye and are used during eye examinations to help the examiner to see the retina through the pupil. They also decrease salivary secretion and are used during surgery to prevent patients from choking on excess saliva while they are anesthetized.

Drugs That Bind to Alpha and Beta Receptors

Drugs that activate adrenergic receptors are **adrenergic, or sympathomimetic** (sim′pă-thō-mi-met′ik) **agents.** Drugs such as phenylephrine (fen-il-ef′rin) stimulate alpha receptors, which are numerous in the smooth muscle cells of certain blood vessels, especially in the digestive tract and the skin. These drugs increase blood pressure by causing vasoconstriction. On the other hand, albuterol (al-byū′ter-ol) is a drug that selectively activates beta receptors, that are found in cardiac muscle and bronchiolar smooth muscle. Beta-adrenergic-stimulating agents are sometimes used to dilate bronchioles in respiratory disorders such as asthma and are occasionally used as cardiac stimulants.

Drugs that bind to and block the action of alpha receptors are **alpha-adrenergic-blocking agents.** For example, prazosin (pra′zō-sin) hydrochloride is used to treat hypertension. By binding to alpha receptors in the smooth muscle of blood vessel walls, prazosin hydrochloride blocks the normal effects of norepinephrine released from sympathetic postganglionic neurons. Thus, the blood vessels relax, and blood pressure decreases.

Propranolol (prō-pran′ō-lōl) is an example of a **beta-adrenergic-blocking agent.** These drugs are sometimes used to treat high blood pressure, some types of cardiac arrhythmias, and patients recovering from heart attacks. Blockage of the beta receptors within the heart prevents sudden increases in the heart rate and thus decreases the probability of arrhythmic contractions.

Future Research

Our present knowledge of the ANS is more complicated than the broad outline presented here. In fact, there are subtype receptors for each of the major receptor types. For example, alpha receptors are subdivided into the following subgroups: α_{1A}-, α_{1B}-, α_{2A}-, and α_{2B}-receptors. The exact number of subtypes in humans is not yet known; however, their existence suggests the possibility of designing drugs that affect only one subtype. For example, a drug that affects the blood vessels of the heart but not other blood vessels might be developed. Such drugs could produce specific effects yet would not produce undesirable side effects because they would act only on specific target tissues.

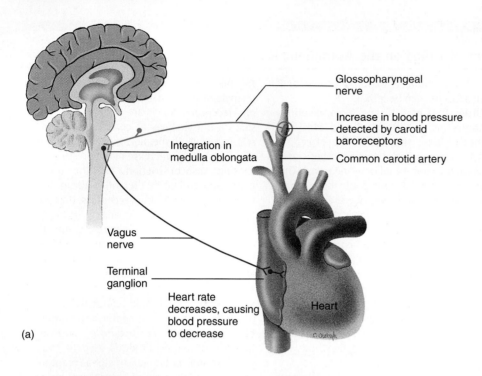

(a)

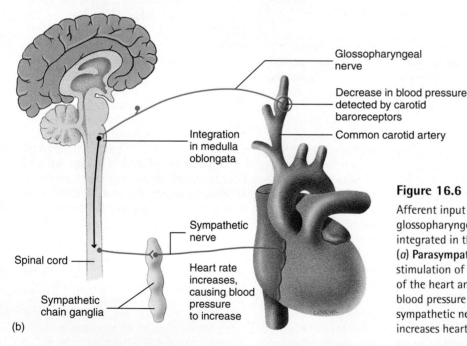

(b)

Figure 16.6   Autonomic Reflexes

Afferent input from the carotid baroreceptors are sent along the glossopharyngeal nerves to the medulla oblongata. The input is integrated in the medulla, and efferent impulses are sent to the heart. (*a*) **Parasympathetic reflex.** Increased blood pressure results in increased stimulation of the heart by the vagus nerves, which increases inhibition of the heart and lowers heart rate. (*b*) **Sympathetic reflex.** Decreased blood pressure results in increased stimulation of the heart by sympathetic nerves, which, in turn, increases stimulation of the heart and increases heart rate and the force of contraction.

Regulation of the Autonomic Nervous System

Much of the regulation of structures by the ANS occurs through autonomic reflexes, but input from the cerebrum, hypothalamus, and other areas of the brain allows conscious thoughts and actions, emotions, and other CNS activities to influence autonomic functions. Without the regulatory activity of the ANS, an individual has limited ability to maintain homeostasis.

Autonomic reflexes, like other reflexes, involve sensory receptors; afferent, association, and efferent neurons; and effector cells (figure 16.6; see chapter 12). For example, **baroreceptors** (stretch receptors) in the walls of large arteries near the heart detect changes in blood pressure, and afferent neurons transmit information from the baroreceptors through the glossopharyngeal and vagus nerves to the medulla oblongata. In the medulla oblongata, association neurons integrate the information, and action potentials are produced in autonomic neurons that extend to the heart. If baroreceptors detect a change in blood pressure, autonomic reflexes change heart rate, which returns blood pressure to

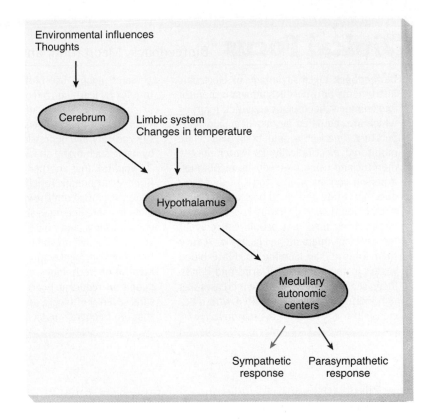

Environmental influences
Thoughts

Cerebrum

Limbic system
Changes in temperature

Hypothalamus

Medullary
autonomic
centers

Sympathetic   Parasympathetic
response   response

Figure 16.7 Influence of Higher Parts of the Brain on Autonomic Functions

The influence of the hypothalamus and the cerebrum on the ANS. Neural pathways extend from the cerebrum to the hypothalamus and from the hypothalamus to neurons of the ANS.

normal. A sudden increase in blood pressure initiates a parasympathetic reflex that inhibits cardiac muscle cells and reduces the heart rate, thus bringing blood pressure down toward its normal value. Conversely, a sudden decrease in blood pressure initiates a sympathetic reflex, which stimulates the heart to increase its rate and force of contraction, thus increasing blood pressure.

PREDICT

Sympathetic neurons stimulate sweat glands in the skin. Predict how they function to control body temperature during exercise and during exposure to cold temperatures.

✔ *Answer in Appendix F*

Other autonomic reflexes also participate in the regulation of blood pressure. For example, numerous sympathetic neurons transmit a low but relatively constant frequency of action potentials that stimulate blood vessels throughout the body, keeping them partially constricted. If the vessels constrict further, blood pressure increases; and if they dilate, blood pressure decreases. Thus altering the frequency of action potentials delivered to blood vessels along sympathetic neurons can either raise or lower blood pressure.

PREDICT

How do sympathetic reflexes that control blood vessels respond to a sudden decrease and a sudden increase in blood pressure?

✔ *Answer in Appendix F*

Specific areas of the CNS, including the spinal cord, medulla oblongata, and hypothalamus, integrate a large number of autonomic reflexes (figure 16.7). Areas in which autonomic reflexes are integrated, however, are influenced by other areas of the CNS. For example, the hypothalamus integrates responses to temperature changes and coordinates responses to stress, rage, and other emotions. The limbic system, which plays an important role in emotions and their expression, and the cerebral cortex affect autonomic functions by influencing the hypothalamus.

Higher centers of the brain also affect autonomic functions. Emotions such as anger increase blood pressure by increasing heart rate and constricting blood vessels through sympathetic stimulation. Pleasant thoughts of a delicious banquet initiate increased secretion by salivary glands and by glands within the stomach and increased smooth muscle contractions within the digestive system, all of which are controlled by parasympathetic neurons.

Functional Generalizations About the Autonomic Nervous System

Generalizations can be made about the function of the ANS on effector organs, but most have exceptions.

Stimulatory Versus Inhibitory Effects

Both divisions of the ANS produce stimulatory and inhibitory effects. For example, the parasympathetic division stimulates

Clinical Focus   Biofeedback, Meditation, and the Fight-or-Flight Response

Biofeedback takes advantage of electronic instruments or other techniques to monitor and change subconscious activities, many of which are regulated by the ANS. Skin temperature, heart rate, and brain waves are monitored electronically. By watching the monitor and using biofeedback techniques, a person can learn how consciously to reduce heart rate and blood pressure and regulate blood flow in the limbs. For example, it has been claimed that people can prevent the onset of migraine headaches or reduce their intensity by learning to dilate blood vessels in the skin of their arms and hands. Increased blood vessel dilation increases skin temperature, which is correlated with a decrease in the severity of the migraine.

Some people use biofeedback methods to relax by learning to reduce the heart rate or change the pattern of brain waves. The severity of stomach ulcers, high blood pressure, anxiety, and depression may be reduced by using such biofeedback techniques.

Meditation is another technique that influences autonomic functions. Although numerous claims about the value of meditation include improving one's spiritual well-being, consciousness, and holistic view of the universe, it has been established that meditation does influence autonomic functions. Meditation techniques are useful in some people in reducing heart rate, blood pressure, severity of ulcers, and other symptoms that are frequently associated with stress.

The **fight-or-flight response** can occur when an individual is subjected to severe stress such as a threatening situation or a repugnant event. The response may be confrontation or avoidance. The response involves all parts of the nervous system, as well as the endocrine system, and can be consciously or unconsciously mediated. The autonomic part of the fight-or-flight response results in a general increase in sympathetic activity, including heart rate, blood pressure, sweating, and other responses, that prepare the individual for physical activity. The fight-or-flight response is adaptive because it enables the individual to resist or move away from a threatening situation.

---

contraction of the urinary bladder and inhibits the heart, causing a decrease in heart rate. The sympathetic division causes vasoconstriction by stimulating smooth muscle contraction in blood vessel walls and produces dilation of lung air passageways by inhibiting smooth muscle contraction in the walls of the passageways. Thus, it is *not* true that one division of the ANS is always stimulatory and the other is always inhibitory.

Dual Innervation

Most organs that receive autonomic neurons are innervated by both the parasympathetic and the sympathetic divisions (figure 16.8). The gastrointestinal tract, heart, urinary bladder, and reproductive tract are examples (see table 16.3). Dual innervation of organs by both divisions of the ANS is not universal, however. For example, sweat glands and blood vessels are innervated by sympathetic neurons almost exclusively. In addition, most structures receiving dual innervation are not regulated equally by both divisions. For example, parasympathetic innervation of the gastrointestinal tract is more extensive and exhibits a greater influence than does sympathetic innervation.

Opposite Effects

When a *single* structure is innervated by both autonomic divisions, the two divisions usually produce opposite effects on the structure. As a consequence, the ANS is capable of both increasing and decreasing the activity of the structure, resulting in an efficient control system. For example, in the gastrointestinal tract, parasympathetic stimulation increases secretion from glands, whereas sympathetic stimulation decreases secretion. In a few instances, however, the effect of the two divisions is not clearly opposite. For example, both

divisions of the ANS increase salivary secretion: the parasympathetic division initiates the production of a large volume of thin, watery saliva, and the sympathetic division causes the secretion of a small volume of viscous saliva.

Cooperative Effects

One autonomic division alone or both divisions acting together can coordinate the activities of *different* structures. For example, the parasympathetic division stimulates the pancreas to release digestive enzymes into the small intestine. At the same time the parasympathetic division stimulates contractions of the small intestine to mix the digestive enzymes with food within the small intestine, resulting in increased digestion and absorption of the food.

Both divisions cooperate to achieve normal reproductive function. The parasympathetic division initiates erection of the penis, and the sympathetic division stimulates the release of secretions from male reproductive glands and helps initiate ejaculation in the male reproductive tract.

General Versus Localized Effects

The sympathetic division has a more general effect than the parasympathetic division because activation of the sympathetic division often causes secretion of both epinephrine and norepinephrine from the adrenal medulla. These hormones circulate in the blood and stimulate effector organs throughout the body. Because circulating epinephrine and norepinephrine can persist for a few minutes before being broken down, they can also produce an effect for a longer time than direct stimulation of effector organs by postganglionic sympathetic axons.

The sympathetic division diverges more than the parasympathetic division. Each sympathetic preganglionic

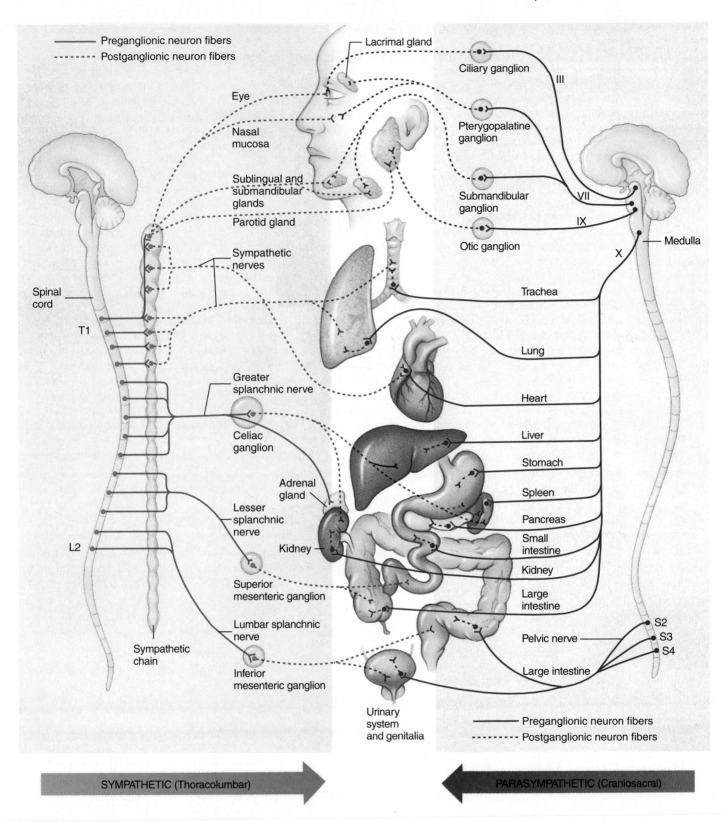

Figure 16.8   Innervation of Organs by the ANS

Preganglionic fibers are indicated by solid lines, and postganglionic fibers are indicated by dashed lines.

Clinical Focus   Disorders of the Autonomic Nervous System

Normal function of all components of the ANS is not required to maintain life, as long as environmental conditions are constant and optimum. Abnormal autonomic functions, however, markedly affect the individual's ability to respond to changing conditions. This can be demonstrated by **sympathectomy,** the removal of sympathetic ganglia. The normal regulation of body temperature is lost following sympathectomy. In a hot environment, the ability to lose heat by increasing blood flow to the skin and by sweating is decreased. When exposed to the cold, the ability to reduce blood flow to the skin and conserve heat is decreased. Sympathectomy also results in low blood pressure caused by dilation of peripheral blood vessels and results in the inability to increase blood pressure during periods of physical activity.

Orthostatic hypotension is a drop in blood pressure that occurs when a person who was sitting or lying down suddenly stands up. It is sometimes caused by disorders, such as diabetes mellitus, that decreases the frequency of action potentials in sympathetic nerves innervating blood vessels. Consequently, on standing, blood pools in dilated blood vessels in the lower extremities, less blood returns to the heart, and the amount of blood the heart pumps decreases. Blood pressure decreases, resulting in reduced blood flow to the brain, which causes fainting because of a lack of oxygen.

Raynaud's disease involves the spasmodic contraction of blood vessels in the periphery of the body, especially in the digits, and results in pale, cold hands that are prone to ulcerations and gangrene because of poor circulation. This condition can be caused by exaggerated sensitivity of blood vessels to sympathetic innervation. Preganglionic denervation (cutting the preganglionic neurons) is occasionally performed to alleviate the condition.

Hyperhidrosis (hī′per-hī-drō′sis), or excessive sweating, is caused by exaggerated sympathetic innervation of the sweat glands.

Achalasia (ak′ă-lā′zē-ă) is characterized by difficulty in swallowing and in controlling contraction of the esophagus where it enters the stomach, therefore interrupting normal peristaltic contractions of the esophagus. The swallowing reflex is controlled partly by somatic motor reflexes and partly by parasympathetic reflexes. The cause of achalasia can be abnormal parasympathetic regulation of the swallowing reflex. The condition is aggravated by emotions.

Dysautonomia (dis′aw-tō-nō′mē-ă), an inherited condition involving an autosomal-recessive gene, causes reduced tear gland secretion, poor vasomotor control, trouble in swallowing, and other symptoms. It is the result of poorly controlled autonomic reflexes.

Hirschsprung's disease, or **megacolon,** is caused by a functional obstruction in the lower colon and rectum. Ineffective parasympathetic innervation and a predominance of sympathetic innervation of the colon inhibit peristaltic contractions, causing feces to accumulate above the inhibited area. The resulting dilation of the colon can be so great that surgery is required to alleviate the condition.

---

neuron synapses with many postganglionic neurons, whereas parasympathetic preganglionic neurons synapse with about two postganglionic neurons. Consequently, stimulation of sympathetic preganglionic neurons can result in greater stimulation of an effector organ.

Sympathetic stimulation often activates many different kinds of effector organs at the same time as a result of CNS stimulation or epinephrine and norepinephrine release from the adrenal medulla. It is possible, however, for the CNS to selectively activate effector organs. For example, vasoconstriction of cutaneous blood vessels in a cold hand is not always associated with an increased heart rate or other responses controlled by the sympathetic division.

Functions at Rest Versus Activity

In cases in which both parasympathetic and sympathetic neurons innervate a single organ, the parasympathetic division tends to have a greater influence under resting conditions, whereas the sympathetic division has a major influence under conditions of physical activity or stress. Increased sympathetic activity results in increased nervous stimulation of effector organs and increased epinephrine and norepinephrine release from the adrenal medulla. Consequently, the pumping effectiveness of the heart increases, blood vessels in skeletal muscle dilate, and blood vessels in visceral structures and the skin constrict. There is also decreased activity of the gastrointestinal tract, increased glucose release from the liver, and a large increase in metabolism, especially in skeletal muscle. In general, the sympathetic division decreases the activity of organs not essential for the maintenance of physical activity and shunts blood and nutrients to structures that are active during physical exercise. This is sometimes referred to as the fight-or-flight response (see Clinical Focus "Biofeedback Meditation, and the Fight-or-Flight Response"). The sympathetic division, however, also plays a major role during resting conditions by maintaining blood pressure and body temperature.

Increased activity of the parasympathetic division is generally consistent with resting conditions during which eating, digestion, urination, defecation, and other vegetative functions are emphasized. Many of the reflexes that regulate the digestive, urinary, and reproductive systems are mediated by the parasympathetic division (see table 16.2).

4	P R E D I C T

Make a list of the responses controlled by the ANS in (a) a person who is extremely angry and (b) a person who has just finished eating and is relaxing.

✔ *Answer in Appendix F*

Summary

Contrasting the Somatic Motor and Autonomic Nervous Systems

1. The cell bodies of somatic motor neurons are located in the CNS, and their axons extend to skeletal muscles, where they have an excitatory effect that usually is controlled consciously.
2. The cell bodies of the preganglionic neurons of the ANS are located in the CNS and extend to ganglia, where they synapse with postganglionic neurons. The postganglionic axons extend to smooth muscle, cardiac muscle, or glands and have an excitatory or inhibitory effect that usually is controlled unconsciously.

Divisions of the Autonomic Nervous System: Structural Features

Sympathetic Division

1. Preganglionic cell bodies are in the lateral horns of the spinal cord gray matter from T1 to L2.
2. Preganglionic axons pass through the ventral roots to the white rami communicantes to the sympathetic chain ganglia. From there four courses are possible:
   - Preganglionic axons synapse (at the same or a different level) with postganglionic neurons, which exit the ganglia through the gray rami communicantes and enter spinal nerves.
   - Preganglionic axons synapse (at the same or a different level) with postganglionic neurons, which exit the ganglia through sympathetic nerves.
   - Preganglionic axons pass through the chain ganglia without synapsing to form splanchnic nerves. Preganglionic axons then synapse with postganglionic neurons in collateral ganglia.
   - In the case of the adrenal gland, the preganglionic axons synapse with the cells of the adrenal medulla.

Parasympathetic Division

1. Preganglionic cell bodies are in nuclei in the brainstem or the lateral parts of the spinal cord gray matter from S2 to S4.
   - Preganglionic axons from the brain pass to ganglia through cranial nerves III, VII, IX, and X.
   - Preganglionic axons from the sacral region pass through ventral roots of the pelvic nerves to the ganglia.

2. Preganglionic axons pass to terminal ganglia within the wall of or near the organ that is innervated.

Physiology of the Autonomic Nervous System

Neurotransmitters

1. Acetylcholine is released by cholinergic neurons (all preganglionic neurons, all parasympathetic postganglionic neurons, and some sympathetic postganglionic neurons).
2. Norepinephrine is released by adrenergic neurons (most sympathetic postganglionic neurons).

Receptors

1. Acetylcholine binds to nicotinic receptors (found in all postganglionic neurons) and muscarinic receptors (found in all parasympathetic and some sympathetic effector organs).
2. Norepinephrine binds to alpha and beta receptors (found in most sympathetic effector organs).
3. Activation of nicotinic receptors is excitatory, whereas activation of the other receptors can be excitatory or inhibitory.

Regulation of the Autonomic Nervous System

1. Autonomic reflexes control most of the activity of visceral organs, glands, and blood vessels.
2. Autonomic reflex activity can be influenced by the hypothalamus and higher brain centers.

Functional Generalizations About the Autonomic Nervous System

1. Both divisions of the ANS produce stimulatory and inhibitory effects.
2. Most organs are innervated by both divisions. Usually each division produces an opposite effect on a given organ.
3. Either division alone or both working together can coordinate the activities of different structures.
4. The sympathetic division produces more generalized effects than the parasympathetic division.
5. Sympathetic activity generally prepares the body for physical activity, whereas parasympathetic activity is more important for vegetative functions.

Content Review

1. Define the terms preganglionic neuron, postganglionic neuron, autonomic ganglia, and effector organ.
2. Contrast the somatic motor nervous system and the ANS for each of the following:
   a. The number of neurons between the CNS and the effector organ
   b. The location of neuron cell bodies
   c. The structures each innervates
   d. Inhibitory or excitatory effects
   e. Conscious or unconscious control
3. Contrast the sympathetic and parasympathetic divisions with regard to the following:
   a. Location of preganglionic cell bodies
   b. Location of ganglia
   c. Length of preganglionic and postganglionic axons
   d. Number of preganglionic and postganglionic neurons
4. Describe four ways in which efferent neurons of the sympathetic division exit the CNS and extend to effector organs. Describe two ways for the parasympathetic division.
5. Why is the adrenal medulla considered a part of the sympathetic division? What substances does it release, and what effects do these substances have? Are these substances neurotransmitters or hormones?
6. What kinds of neurons (preganglionic or postganglionic, myelinated or unmyelinated) are found in the following:
   a. Cranial nerves
   b. Spinal nerves
   c. Sympathetic nerves
   d. Splanchnic nerves
   e. Pelvic nerves
7. Describe three types (locations) of autonomic ganglia.
8. Generally speaking, what structures are innervated by autonomic fibers in cranial nerves, spinal nerves, sympathetic nerves, splanchnic nerves, and pelvic nerves?
9. What neurotransmitter is released by cholinergic neurons? Which neurons of the ANS are cholinergic?
10. What neurotransmitter is released by adrenergic neurons? Which neurons of the ANS are adrenergic?
11. With what type of receptors does acetylcholine bind? Where are these receptors found? Is the effect excitatory or inhibitory?
12. With what type of receptors does norepinephrine bind? Where are these receptors found? Is the effect excitatory or inhibitory?
13. Name the components of an autonomic reflex. Describe the autonomic reflex that maintains blood pressure by altering heart rate or the diameter of blood vessels.
14. In what area of the CNS are autonomic reflexes integrated? What role do other areas of the CNS have in modifying autonomic reflexes resulting from emotions or stress?
15. Most organs of the body are innervated by both divisions of the ANS. Give an exception.
16. For organs innervated by both the parasympathetic and sympathetic divisions, list four organs for which the parasympathetic division has an excitatory effect and four organs for which it has an inhibitory effect. For each set of organs listed, describe the effect of the sympathetic division on that set of organs. What conclusions can be made about the following?
    a. The ability of the parasympathetic or sympathetic divisions to have an excitatory or inhibitory effect
    b. The effect of the parasympathetic and sympathetic divisions on an organ they both innervate
17. To help you remember whether or not the sympathetic or parasympathetic division has an excitatory or inhibitory effect on a particular organ, what generalization is useful?

Develop Your Reasoning Skills

1. When a person is startled or when she sees a "pleasurable" object, the pupils of the eyes can dilate. What division of the ANS is involved in this reaction? Describe the nerve pathway involved.

2. Reduced secretion from salivary and lacrimal glands could indicate damage to what nerve?

3. In a patient with Raynaud's disease, blood vessels in the skin of the hand can become chronically constricted, reducing blood flow and producing gangrene. These vessels are supplied by nerves that originate at levels T2 and T3 of the spinal cord and eventually exit through the first thoracic and inferior cervical sympathetic ganglia. Surgical treatment for Raynaud's disease severs this nerve supply. At which of the following locations would you recommend that the cut be made: white rami of T2–T3, gray rami of T2–T3, spinal nerves T2–T3, or spinal nerves C1–T1? Explain.

4. Patients with diabetes mellitus can develop autonomic neuropathy, which is damage to parts of the autonomic nerves. Given the following parts of the ANS—vagus nerve, splanchnic nerve, pelvic nerve, cranial nerve, outflow of gray ramus—match the part with the symptom it would produce if the part were damaged:
   a. Impotence
   b. Subnormal sweat production
   c. Gastric atony and delayed emptying of the stomach
   d. Diminished pupil reaction (constriction) to light
   e. Bladder paralysis with urinary retention

5. Explain why methacholine, a drug that acts like acetylcholine, is effective for treating tachycardia (heart rate faster than normal). Which of the following side effects would you predict: increased salivation, dilation of the pupils, sweating, and difficulty in breathing?

6. A patient has been exposed to the organophosphate pesticide malathion, which inactivates acetylcholinesterase. Which of the following symptoms would you predict: blurring of vision, excess tear formation, frequent or involuntary urination, pallor (pale skin), muscle twitching, or cramps? Would atropine be an effective drug to treat the symptoms? Explain.

7. Epinephrine is routinely mixed with local anesthetic solutions. Why?

8. A drug blocks the effect of the sympathetic division on the heart. Careful investigation reveals that, after administration of the drug, normal action potentials are produced in the sympathetic preganglionic and postganglionic neurons. Also, injection of norepinephrine produces a normal response in the heart. Explain, in as many ways as you can, the mode of action of the unknown drug.

9. A drug is known to decrease heart rate. After cutting the white rami of T1–T4, the drug still causes heart rate to decline. After cutting the vagus nerves, the drug no longer affects heart rate. Which division of the ANS does the drug affect? Does the drug have its effect at the synapse between preganglionic and postganglionic neurons, at the synapse between postganglionic neurons and effector organs, or in the CNS? Is the effect of the drug excitatory or inhibitory?

Web Site Link

For a listing of the most current web sites related to this chapter, please visit the Seeley home page at:
http://www.mhhe.com/biosci/ap/seeleyap/

Chapter Seventeen

Functional Organization of the Endocrine System

Objectives

1. Define the terms endocrine gland, hormone, and endocrine system.

2. Explain why a simple definition for hormone is difficult to create.

3. Explain why the endocrine system is primarily an amplitude-modulated system.

4. Describe the functional relationship between the nervous system and the endocrine system.

5. Explain how the regulation of hormone secretion is achieved.

6. Define the term half-life, and explain how the combination of hormones with plasma proteins affects their half-life.

7. Describe the means by which hormones are metabolized and excreted.

8. Explain how the sensitivity of target tissues to hormones can change.

9. Compare the relationship between hormones and their receptor molecules for membrane-bound and intracellular receptors.

10. List the categories of responses that can occur following the combination of hormones with membrane-bound receptors.

11. List some of the major hormones that bind to membrane-bound receptors and alter the permeability of the plasma membrane, alter G protein function, or activate intracellular enzymes.

12. Explain how the combination of a hormone with its membrane-bound receptor rapidly activates many intracellular enzymes to produce a cellular response.

13. Using diagrams, explain the intracellular mediator model of hormone action, and describe the characteristics of the responses that are produced.

14. List some of the intracellular mediator molecules produced in response to hormones binding to membrane-bound receptors.

Part Three

The nervous and endocrine systems are the two major regulatory systems of the body, and together they regulate and coordinate the activity of essentially all other body structures. The nervous system transmits information in the form of action potentials along the axons of nerve cells. Chemical signals in the form of neurotransmitters are released at synapses between neurons and the tissues they control. The endocrine system sends information to the tissues it controls in the form of chemical signals released from endocrine glands. The chemical signals are released into the circulatory system and carried to all parts of the body. Body structures that are able to recognize the chemical signals respond to them. Thus the nervous system functions some-thing like telephone messages sent along telephone wires to their destination, whereas the endocrine system is more like radio signals broadcast widely that everyone with radios tuned to the proper channel can receive.

The basic features of the nervous system are presented in chapters 12 through 16. This chapter introduces the general characteristics of the endocrine system, compares some of the functions of the nervous and endocrine systems, emphasizes the role of the endocrine system in the maintenance of homeostasis, and illustrates the means by which the endocrine system regulates the functions of cells. The structure and function of each endocrine gland, its secretory products, and the means by which its activity is regulated are described in chapter 18.

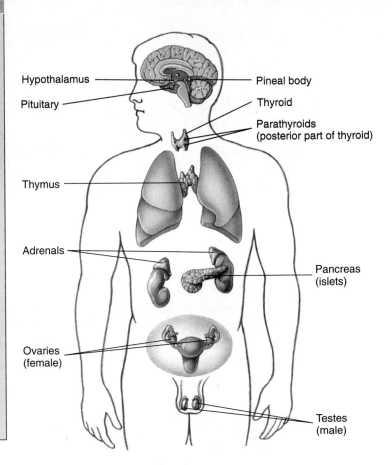

Figure 17.1 Endocrine Glands

The location of major endocrine glands (*yellow*) in the human body. ✗

▋General Characteristics of the Endocrine System

The term **endocrine** (en'dō-krin) is derived from the Greek words *endo,* meaning within, and *crino,* to separate. The term implies that cells of endocrine glands secrete chemical signals internally and influence tissues that are separated by some distance from them. The endocrine system is composed of glands that secrete chemical signals into the circulatory system (figure 17.1). In contrast, exocrine glands have ducts that carry their secretions to surfaces (see chapter 4). The secretory products of endocrine glands are **hormones** (hōr'mōnz), a term derived from the Greek word *hormon,* meaning to set into motion. Traditionally a hormone is defined as a chemical signal that (1) is produced in minute amounts by a collection of cells, (2) is secreted into the interstitial spaces, (3) enters the circulatory system, where it is transported some distance, and (4) acts on specific tissues called **target tissues** at another site in the body to influence the activity of those tissues in a specific fashion. All hormones exhibit most components of this definition, but some components do not apply to every hormone.

Both the endocrine system and the nervous system regulate the activities of structures in the body, but they do so in different ways. For example, hormones secreted by most endocrine glands can be described as **amplitude-modulated signals** (am'pli-tūd mod-yū-lāt'ed), which consist mainly of increases or decreases in the concentration of hormones in the body fluids (figure 17.2*a*). The effects produced by the hormones either increase or decrease responses as a function of the hormone concentration. On the other hand, the all-or-none action potentials carried along axons can be described as **frequency-modulated signals** (figure 17.2*b*), which vary in frequency but not in amplitude. Weak signals are represented by a lower frequency of action potentials, whereas strong signals are represented by a higher frequency of action potentials (see chapter 9). The responses of the endocrine system are usually slower and of longer duration and its effects are usually more generally distributed than those of the nervous system.

Although the stated differences between the endocrine and nervous systems are generally true, exceptions exist. For example, some endocrine responses are more rapid than some neural responses, and some endocrine responses have a shorter duration than some neural responses. In addition, some hormones act as both amplitude- and frequency-modulated signals, in which the concentrations of the hormones and the frequencies at which the increases in hormone concentrations occur are important.

At one time, the endocrine system was believed to be relatively independent and different from the nervous system. An intimate relationship between these systems is now recognized,

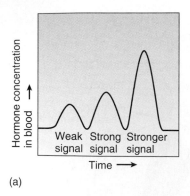

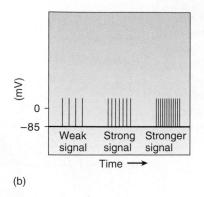

Figure 17.2   Regulatory Systems

(*a*) Amplitude-modulated system. The concentration of the hormone determines the strength of the signal and the magnitude of the response. For most hormones, a small concentration of a hormone is a weak signal and produces a small response, whereas a larger concentration is a stronger signal and results in a greater response. (*b*) Frequency-modulated system. The strength of the signal depends on the frequency, not the size, of the action potentials. A low frequency of action potentials is a weak stimulus, and a higher frequency is a stronger stimulus.

and the two systems cannot be separated completely either anatomically or functionally. Some neurons secrete into the circulatory system chemical signals called **neurohormones** (nūr-ō-hōr′mōnz), which function like hormones. Also, some neurons directly innervate endocrine glands and influence their secretory activity. Neurons release chemical signals at synapses in the form of neurotransmitters and neuromodulators, and the membrane potentials of some endocrine glands undergo depolarization or hyperpolarization, which results in either an increase or a decrease in the rate of hormone secretion. Conversely, some hormones secreted by endocrine glands affect the nervous system and markedly influence its activity.

Both the nervous and endocrine systems rely on chemical signals. **Intercellular chemical signals** allow one cell to communicate with another. These signals coordinate and regulate the activities of most cells. Neurotransmitters and neuromodulators are intercellular chemical signals that play important roles in the function of the nervous system (see chapter 12). Hormones are intercellular chemical signals secreted by endocrine glands.

Autocrine (aw′tō-krin) **chemical signals** are released by cells and have a local effect on the same cell type from which the chemical signals are released. Examples include prostaglandinlike chemicals released from smooth muscle cells and platelets in response to inflammation. These chemicals cause the relaxation of blood vessel smooth muscle cells and the aggregation of platelets. As a result the blood vessels dilate and blood clots.

Paracrine (par′ă-krin) **chemical signals** are released by cells and affect other cell types locally without being transported in blood. For example, a peptide called somatostatin is released by cells in the pancreas and functions locally to inhibit the secretion of insulin from other cells of the pancreas (see chapter 18).

Pheromones (fer′ō-mōnz) are chemical signals secreted into the environment that modify the behavior and the physiology of other individuals. For example, pheromones released in the urine of cats and dogs at certain times are olfactory signals that indicate fertility, and evidence supports the existence of pheromones produced by women that influence the length of menstrual cycles of other women (table 17.1).

Many intercellular chemical signals consistently fit one specific definition, but others do not. For example, norepinephrine functions both as a neurotransmitter and as a neurohormone; and prostaglandins function as neurotransmitters, neuromodulators, parahormones, and autocrine chemical signals. The schemes used to classify chemicals on the basis of their functions are useful, but they do not indicate that a specific molecule always performs as the same type of chemical signal. For that reason, the study of endocrinology often includes the study of autocrine and paracrine chemical signals in addition to hormones.

Chemical Structure of Hormones

Hormones, including neurohormones, are proteins, short sequences of amino acids called polypeptides, derivatives of amino acids, or lipids. Some protein hormones, called glycoprotein hormones, are composed of one or more polypeptide chains and carbohydrate molecules. The lipid hormones are either steroids or derivatives of fatty acids. Table 17.2 and figure 17.3 provide information concerning the chemical structure of the major hormones.

Hormones that are soluble in lipids, such as steroids, can diffuse through the plasma membrane of cells and bind to intracellular receptors in the cytoplasm or in the nucleus. Large, water-soluble hormones, such as the protein hormones, cannot diffuse through the plasma membrane. Protein hormones, therefore, bind to membrane-bound receptors on the cell surface (see chapter 9).

Control of Secretion Rate

Most hormones are not secreted at a constant rate. Instead, most endocrine glands increase and decrease their secretory activity dramatically over time. The specific mechanisms that regulate the secretion rates for each hormone are presented in chapter 18, but the general patterns of regulation are introduced in this chapter. Hormones function to regulate the rates of many activities in the body. The secretion rate of each hormone is controlled by negative feedback mechanisms (see chapter 1), so that the function it regulates is maintained within a normal range and homeostasis is maintained.

Hormones have three major patterns of regulation. One method involves the action of a substance other than a hormone on the endocrine gland. Figure 17.4 describes the influence of blood glucose on insulin secretion from the pan-

Table 17.1   Functional Classification of Intercellular Chemical Signals

Intercellular Chemical Signal	Description	Example	
Autocrine	Secreted by cells in a local area and influences the activity of the same cell type from which it was secreted	Prostaglandins	
Paracrine	Produced by a wide variety of tissues and secreted into tissue spaces; usually has a localized effect on other tissues	Histamine prostaglandins	
Hormone	Secreted into the blood by specialized cells; travels some distance to target tissues; influences specific activities	Thyroxine, insulin	
Neurohormone	Produced by neurons and functions like hormones	Oxytocin, antidiuretic hormone	
Neurotransmitter or neurohumor	Produced by neurons and secreted into extracellular spaces by presynaptic nerve terminals; travels short distances; influences postsynaptic cells	Acetylcholine, epinephrine	
Pheromone	Secreted into the environment; modifies physiology and behavior of other individuals	Sex pheromones are released by humans and many other animals. They are released in the urine of animals, such as dogs and cats. Pheromones produced by women influence the length of the menstrual cycle of other women.	

Table 17.2   Structural Categories of Hormones

Structural Category	Examples	Structural Category	Examples
Proteins	Growth hormone Prolactin Insulin	Amino acid derivatives	Epinephrine Norepinephrine Thyroid hormones (both T_4 and T_3) Melatonin
Glycoproteins (protein and carbohydrate)	Follicle-stimulating hormone Luteinizing hormone Thyroid-stimulating hormone Parathyroid hormone	Lipids Steroids (Cholesterol is a precursor for all steroids)	Estrogens Progestins (progesterone) Testosterone Mineralocorticoids (aldosterone) Glucocorticoids (cortisol)
Polypeptides	Thyrotropin-releasing hormone Oxytocin Antidiuretic hormone Calcitonin Glucagon Adrenocorticotropic hormone Endorphins Thymosin Melanocyte-stimulating hormones Hypothalamic hormones Lipotropins Somatostatin	Fatty acids	Prostaglandins Thromboxanes Prostacyclins Leukotrienes

Abbreviations: T_4 = tetraiodothyronine or thyroxine; T_3 = triiodothyronine.

creas. An increase in blood sugar levels causes increased insulin secretion from the pancreas. Insulin increases glucose movement into tissues, resulting in a decrease in blood glucose. As blood glucose levels decrease, insulin levels decrease, thus insulin levels increase and decrease in response to blood glucose.

A second pattern of hormone regulation involves neural control of the endocrine gland. Neurons synapse with the cells that produce the hormone; and, when action potentials result, the neurons release a neurotransmitter. In some cases, the neurotransmitter is stimulatory and causes the cells to increase hormone secretion. In other cases the neurotransmitter is inhibitory and decreases hormone secretion. Thus sensory input and emotions acting through the nervous system can influence hormone secretion. Figure 17.5 illustrates the neural control of epinephrine and norepinephrine secretion from the adrenal gland. In response to stimuli such as stress or exercise, the nervous system stimulates the adrenal gland to secrete epinephrine and norepinephrine, which helps the body respond to the stimuli. When the stimuli are no longer present, secretion of epinephrine and norepinephrine decreases.

A third pattern of hormone regulation involves the control of the secretory activity of one endocrine gland by a hormone or a neurohormone secreted by another endocrine gland. Figure 17.6 illustrates how thyroid-releasing hormone (TRH) from the hypothalamus of the brain stimulates the secretion of thyroid-stimulating hormone (TSH) from the anterior pituitary gland, which in turn, stimulates the secretion of thyroid hormones from the thyroid gland. Because thyroid hormones can inhibit the secretion of TRH and TSH, a negative feedback mechanism for regulating hormone levels is established. As TRH levels increase, TSH levels and then thyroid hormone levels increase. Through negative feedback, the thyroid hormones cause a decrease in TRH and TSH, which results in a decrease in thyroid hormone levels. Thus, the concentrations of TRH, TSH, and thyroid hormone increase and decrease within a normal range.

When action potentials in parasympathetic neurons that innervate the pancreas increase, the neurotransmitter acetylcholine is released. Acetylcholine causes depolarization of pancreatic cells, and insulin is secreted. When action potentials in sympathetic neurons that innervate the pancreas increase, the neurotransmitter norepinephrine is released. Norepinephrine causes hyperpolarization of pancreatic cells, and insulin secretion decreases. Thus, nervous stimulation of the pancreas can either increase or decrease insulin secretion.

---

1   P R E D I C T

Thyroid-stimulating hormone (TSH) is secreted from the anterior pituitary gland and acts on the thyroid gland to increase the secretion of thyroid hormones. Thyroid hormones, on the other hand, have a negative-feedback effect on TSH secretion (see figure 17.6). For a person having normal thyroid function, the rate at which TSH and thyroid hormones are secreted remains within a normal range of concentrations. In some people, however, the immune system begins to produce an abnormal substance that functions like TSH. Predict what that substance will do to the rate of thyroid-stimulating hormone secretion and the rate of thyroid hormone secretion.

✔ *Answer in Appendix F*

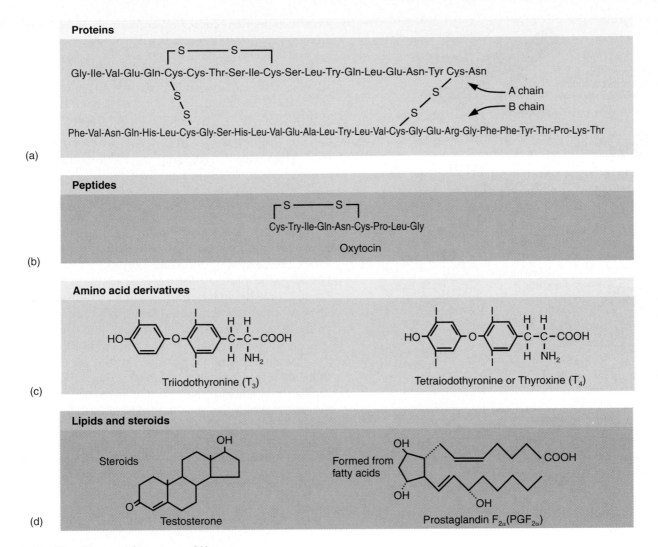

Figure 17.3   The Chemical Structure of Hormones

(*a*) Insulin is an example of a protein hormone. (*b*) Oxytocin is an example of a peptide hormone. (*c*) The thyroid hormones, triiodothyronine (T_3) and tetraiodothyronine (T_4), are examples of modified amino acid hormones. (*d*) Testosterone, a steroid, and prostaglandin $F_{2\alpha}$ are examples of lipid hormones.

One of the three major patterns by which hormone secretion is regulated applies to each hormone, but the complete picture is not quite so simple. The regulation of hormone secretion often involves more than one mechanism. In some cases the nervous system and hormones from other endocrine glands regulate the rate at which a hormone is secreted. For example, both the concentration of blood glucose and the autonomic nervous system influence insulin secretion from the pancreatic islets.

There are a few examples of positive-feedback regulation in the endocrine system. In each instance, however, negative-feedback mechanisms limit the positive-feedback process. The secretion of luteinizing hormone (LH) before ovulation (figure 17.7) and the role of oxytocin in delivery of an infant are often cited as examples (see chapters 28 and 29).

Some hormones are in the circulatory system at relatively constant levels, some change suddenly in response to certain stimuli, and others change in relatively constant cycles (figure 17.8). For example, thyroid hormones in the blood

vary within a small range of concentrations, which remain relatively constant. Epinephrine is released in large amounts in response to stress or physical exercise; thus its concentration can change suddenly. Reproductive hormones increase and decrease in a cyclic fashion in women during their reproductive years.

Transport and Distribution in the Body

Hormones are dissolved in blood plasma and transported either in a free form or bound to plasma proteins. Hormones that are free in the plasma can diffuse from capillaries into interstitial spaces. As the concentration of free hormone molecules increases in the blood, more hormone molecules diffuse from the capillaries into the interstitial spaces and bind to target cells. As the concentration of the free hormone molecules

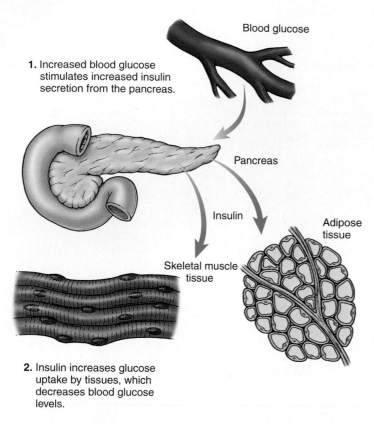

1. Increased blood glucose stimulates increased insulin secretion from the pancreas.

Blood glucose

Pancreas

Insulin

Adipose tissue

Skeletal muscle tissue

2. Insulin increases glucose uptake by tissues, which decreases blood glucose levels.

Figure 17.4   Nonhormonal Regulation of Hormone Secretion

Glucose, which is not a hormone, regulates the secretion of insulin from the pancreas.

decreases in the blood, fewer hormone molecules diffuse from the capillaries into the interstitial spaces and bind to target cells (figure 17.9).

Hormones that bind to plasma proteins do so in a reversible fashion. An equilibrium is established between the free plasma hormones and hormones bound to the plasma proteins.

$$
\begin{array}{ccccc}
\text{H} & + & \text{BP} & \leftrightarrow & \text{HBP} \\
\text{Hormone} & & \text{Binding} & & \text{Hormone bound} \\
& & \text{protein} & & \text{to binding protein}
\end{array}
$$

Many hormones bind only to certain types of plasma proteins. For example, a specific type of plasma protein binds to thyroid hormones, and a different type of plasma protein binds to sex hormones such as testosterone. An equilibrium exists between the unbound hormone and the hormone bound to the plasma proteins. The equilibrium is important because only the free hormone is able to diffuse through capillary walls and bind to target tissues. A large increase or decrease in the plasma protein concentration can influence the concentration of free hormone in the blood (figure 17.10).

Because hormones circulate in the blood, they are distributed quickly throughout the body. They diffuse through the capillary endothelium and enter the interstitial spaces, although the rate at which this movement occurs varies from one hormone to the next. Lipid-soluble hormones readily diffuse through the walls of all capillaries. In contrast, water-soluble hormones such as proteins must pass through pores in the capillary endothelium. The capillary endothelia of organs that are regulated by protein hormones have large pores.

1. Stimuli such as stress or exercise activate the sympathetic division of the autonomic nervous system.

2. Sympathetic neurons stimulate the release of epinephrine and smaller amounts of norepinephrine from the adrenal medulla. Epinephrine and norepinephrine prepare the body to respond to stressful conditions.

Once the stressful stimuli are removed, less epinephrine is released as a result of decreased stimulation from the autonomic nervous system.

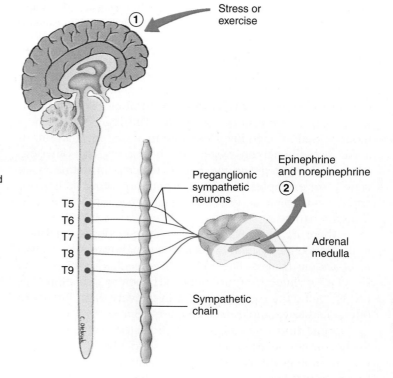

Stress or exercise

Preganglionic sympathetic neurons

Epinephrine and norepinephrine

Adrenal medulla

T5
T6
T7
T8
T9

Sympathetic chain

Figure 17.5   Nervous System Regulation of Hormone Secretion

The sympathetic division of the autonomic nervous system stimulates the adrenal gland to secrete epinephrine and norepinephrine.

1. Thyroid-releasing hormone (TRH) is released from neurons in the hypothalamus and travels to the anterior pituitary gland.

2. TRH stimulates the release of thyroid stimulating hormone (TSH) from the anterior pituitary gland. TSH travels to the thyroid gland.

3. TSH stimulates the secretion of thyroid hormones from the thyroid gland.

4. Thyroid hormones act on tissues to produce the usual response to thyroid hormones.

5. Thyroid hormones also act on the hypothalamus and the anterior pituitary to inhibit both TRH secretion and TSH secretion.

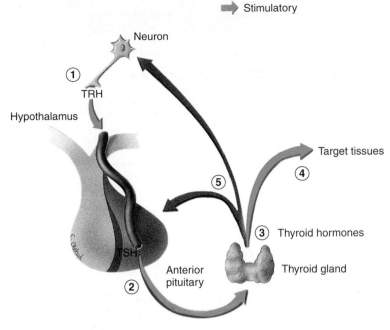

Figure 17.6 Hormonal Regulation of Hormone Secretion

Hormones can stimulate or inhibit the secretion of other hormones.

1. During the menstrual cycle, before ovulation, small amounts of estrogen are secreted from the ovary.

2. Estrogen stimulates the release of gonadotropin-releasing hormone (GnRH) from the hypothalamus and luteinizing hormone (LH) from the anterior pituitary.

3. GnRH also stimulates the release of LH from the anterior pituitary.

4. LH causes the release of additional estrogen from the ovary. Consequently, the blood levels of GnRH and LH increase because of this positive-feedback effect.

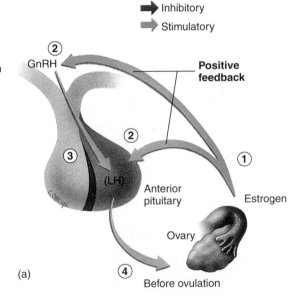

(a)

Figure 17.7 Positive and Negative Feedback

(a) The menstrual cycle, before ovulation, is often cited as an example of positive-feedback regulation of hormone secretion. (b) The menstrual cycle, after ovulation, is an example of negative-feedback regulation of hormone secretion.

1. During the menstrual cycle, after ovulation, the ovary begins to secrete progesterone in response to LH.

2. Progesterone inhibits the release of GnRH from the hypothalamus and LH from the anterior pituitary.

3. Decreased GnRH release from the hypothalamus reduces LH secretion from the anterior pituitary. Consequently, GnRH and LH levels in the blood decrease because of this negative-feedback effect.

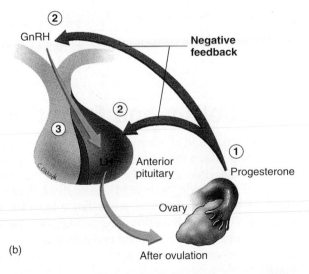

(b)

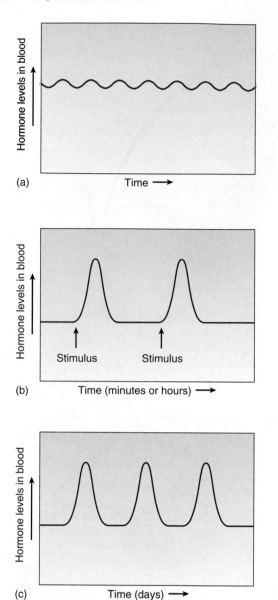

Figure 17.8   Changes in Hormone Secretion Through Time

At least three basic patterns of hormone secretion exist. (*a*) Chronic hormone regulation—the maintenance of a relatively constant concentration of hormone in the circulating blood over a relatively long period. (*b*) Acute hormone regulation—a hormone rapidly increases in the blood for a short time in response to a stimulus. (*c*) Cyclic hormone regulation—a hormone is regulated so that it increases and decreases in the blood at a relatively constant time and to roughly the same amount.

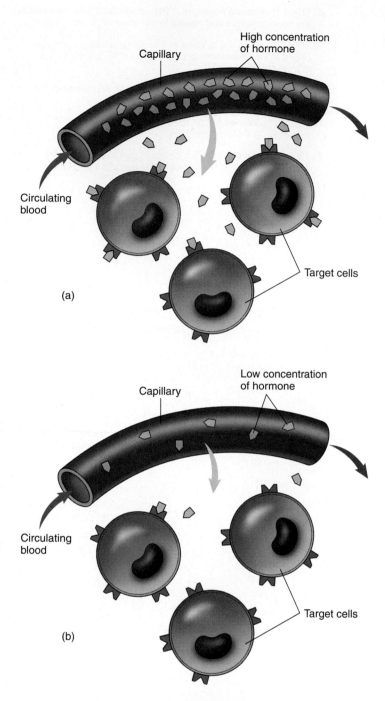

Figure 17.9   Hormone Concentrations at the Target Cell

Hormone diffuses from the blood through the walls of the capillaries into the interstitial spaces. Once within the interstitial spaces, hormone diffuses to the target cells. (*a*) As the concentration of free hormone increases in the blood, more hormone diffuses from the capillary to the target cells. (*b*) As the concentration of free hormone decreases in the blood, less hormone diffuses from the capillary to the target cells.

Metabolism and Excretion

The destruction and elimination of hormones limit the length of time during which they are active, and regulation of body activities is more precise when hormones are secreted and remain active for only short periods. The length of time it takes for elimination of half a dose of a substance from the circulatory system is called its **half-life.** The half-life of a hormone is a standard measurement used by endocrinologists because it allows them to predict the rate at which hormones are eliminated

from the body. The length of time required for total removal of a hormone from the body is not as useful because that measurement is influenced dramatically by the starting concentration. Water-soluble hormones, such as proteins, glycoproteins, epinephrine, and norepinephrine, have relatively short half-lives because they are degraded rapidly by enzymes within the circulatory system or organs, such as the kidneys, liver, or lungs.

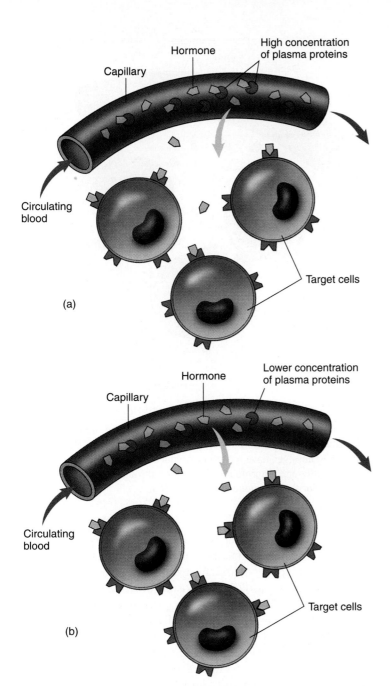

Figure 17.10 Effect of Changes in Plasma Protein Concentration on the Concentration of Free Hormone

(a) An equilibrium exists between free hormone molecules and hormone molecules bound to plasma proteins. The free hormone molecules can diffuse from the capillaries to the interstitial spaces. (b) A decrease in plasma protein concentration reduces the number of hormone molecules bound to plasma proteins. This increases the rate at which free hormone molecules diffuse from the capillaries and combine with target cells. In addition, hormones that diffuse from capillaries are eliminated from the blood by the kidney and liver. The result is that the hormone concentration in the body is reduced and fewer hormone molecules are available to bind to receptors.

Hormones with short half-lives normally have concentrations that increase and decrease rapidly within the blood. They generally regulate activities that have a rapid onset and a short duration.

Table 17.3   Factors That Influence the Half-Life of Hormones

A. Means by which hormones are eliminated from the circulatory system

1. **Excretion**
   Hormones are excreted by the kidney into the urine or excreted by the liver into the bile.

2. **Metabolism**
   Hormones are enzymatically degraded in the blood, liver, kidney, lungs, or target tissues. End products of metabolism are either excreted in urine or bile or used in other metabolic processes by cells in the body.

3. **Active Transport**
   Some hormones are actively transported into cells and are used again as either hormones or neurotransmitter substances.

4. **Conjugation**
   Substances such as sulfate or glucuronic acid groups are attached to hormones primarily in the liver, normally making them less active as hormones and increasing the rate at which they are excreted in the urine or bile.

B. Means by which the half-life of hormones is prolonged

1. Some hormones are protected from rapid excretion or metabolism by binding reversibly with plasma proteins.

2. Some hormones are protected by their structure. The carbohydrate components of the glycoprotein hormones protect them from proteolytic enzymes in the circulatory system.

Hormones that are lipid-soluble, such as steroids and thyroid hormones, circulate in the blood in combination with the plasma proteins. The combination reduces the rate at which they diffuse through the wall of the blood vessels and increases their half-life. Hormones with a long half-life have blood levels that are maintained at a relatively constant level through time. Table 17.3 outlines the ways hormone half-life is shortened or lengthened.

Hormones are removed from the blood in four major ways: excretion, metabolism, active transport, and conjugation. Hormones are excreted by the kidney into the urine or by the liver into the bile. Some hormones are metabolized or are chemically modified by enzymes in the blood or in tissues such as the liver, kidney, lungs, or their target cells. The end products can be excreted in the urine or bile, or they can be taken up by cells and used in metabolic processes. For example, epinephrine is modified enzymatically and then excreted by the kidney. Protein hormones are broken down to their amino acid building blocks. The amino acids can then be taken up by cells and used to synthesize new proteins. Some hormones can be actively transported into cells and recycled. For example, both epinephrine and norepinephrine can be actively transported into cells and secreted again.

Some hormones are conjugated by the liver. **Conjugation** (kon-jū-gā′shŭn) is accomplished when the liver attaches water-soluble molecules to the hormone. These substances are usually sulfate or glucuronic acid. Once they are conjugated, hormones are excreted by the kidney and liver at a greater rate.

Interaction of Hormones with Their Target Tissues

Hormones bind to receptors in their target tissues and alter the rate at which specific activities occur. They do not cause cells to do new things, but they affect the rate at which target cells perform processes they can already do. Hormones can activate or inactivate enzymes that already exist in the cytoplasm of target cells, alter the rate at which specific molecules are synthesized within cells, or alter membrane permeability.

Hormone receptors are either protein or glycoprotein molecules that exist in specific three-dimensional shapes. The unique shape and chemical composition allow the receptor sites of hormone receptors to be highly specific. That is, each receptor site binds only to a single type of hormone or closely related substance. Insulin therefore binds to insulin receptors but not to receptors for growth hormone. A hormone can bind, however, to a number of different receptors that are closely related. For example, epinephrine can bind to more than one type of epinephrine receptor.

Hormones are secreted and distributed throughout the body by the circulatory system, but the presence or absence of specific receptor molecules in cells determines which cells will or will not respond to each hormone (figure 17.11). For example, there are receptors for TSH in cells of the thyroid gland, but there are no such receptors in most other cells of the body.

The response to a given concentration of a hormone is constant in some cases but variable in others. In some tissues the response rapidly decreases through time. Fatigue of the target tissues after prolonged stimulation explains some decreases in responsiveness. In many tissues the number of hormone receptors rapidly decreases after exposure to certain hormones—a phenomenon called **down-regulation** (figure 17.12*a*). Two known mechanisms are responsible for the decrease in the number of receptors. First, the rate at which receptors are synthesized decreases in some tissues after the cells are exposed to a hormone. Because most receptor molecules are degraded after a time, a decrease in synthesis rate reduces the total number of receptor molecules in a cell. Second, the combination of hormones and receptors can increase the rate at which receptor molecules are degraded. In some cases, when a hormone binds to a receptor, both the hormone and the receptor are taken into the cell by phagocytosis. Once inside the cell, the hormone and receptor can be broken down by the cell.

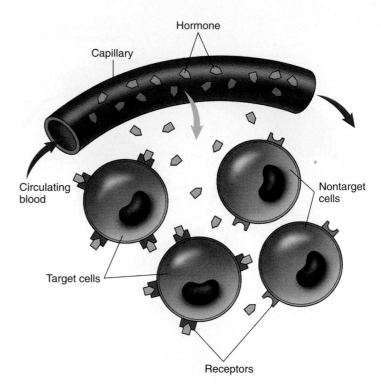

Figure 17.11   Response of Target Cells to Hormones

Hormones are secreted into the blood and distributed throughout the body, where they diffuse from the blood into the interstitial fluid. Only target cells, however, have receptors to which hormones can bind; therefore, although a hormone is distributed throughout the body, only target cells for that hormone can respond to it.

Gonadotropin-releasing hormone (GnRH), which is released from the hypothalamus of the brain, causes the secretion of LH and follicle-stimulating hormone (FSH) from the anterior pituitary. In addition, exposure of the anterior pituitary to GnRH causes the number of receptor molecules for GnRH in the pituitary gland to dramatically decrease several hours after exposure to GnRH. The down-regulation of GnRH receptors causes the pituitary gland to become less sensitive to additional GnRH. The normal response of the pituitary gland to GnRH therefore depends on periodic rather than constant exposure of the gland to the hormone.

In general, tissues that exhibit down-regulation of receptor molecules are adapted to respond to short-term increases in hormone concentrations, and tissues that respond to hormones maintained at constant levels normally do not exhibit down-regulation.

In addition to down-regulation, periodic increases in the sensitivity of some tissues to certain hormones also occur. This is called **up-regulation,** and it results from an increase in the rate of receptor molecule synthesis (figure 17.12*b*). An example of up-regulation is the increased number of receptor molecules for LH in ovarian tissues during each menstrual cycle. FSH secreted by the pituitary gland increases the rate of LH receptor molecule synthesis. Thus exposure of a tissue to one hormone can increase its sensitivity to a second by causing up-regulation in the number of hormone receptors.

Down regulation

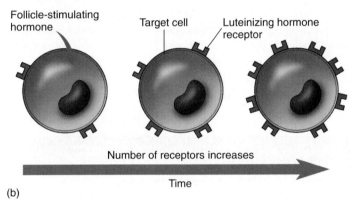

GnRH

Target cell   GnRH receptor

Number of receptors decreases

Time

(a)

Up regulation

Follicle-stimulating
hormone

Target cell   Luteinizing hormone
receptor

Number of receptors increases

Time

(b)

Figure 17.12   Down-Regulation and Up-Regulation

(a) Down-regulation occurs when the number of receptors for a hormone decreases within target cells. For example, gonadotropin-releasing hormone (GnRH) released from the hypothalamus binds to GnRH receptors in the anterior pituitary. The combination of the GnRH bound to its receptors causes the target cell to secrete luteinizing hormone (LH) and follicle-stimulating hormone (FSH); it also causes down-regulation of the GnRH receptors so that eventually the target cells become less sensitive to the GnRH. (b) Up-regulation occurs when some stimulus causes the number of receptors for a hormone to increase within a target cell. For example, FSH acts on cells of the ovary to up-regulate the number of receptors for LH. Thus the ovary becomes more sensitive to the effect of LH.

---

3   P R E D I C T

Estrogen is a hormone secreted by the ovary. It is secreted in greater amounts after menstruation and a few days before ovulation. Among its many effects is causing up-regulation of receptors in the uterus for another hormone secreted by the ovary called progesterone. Progesterone is secreted after ovulation. A major effect of progesterone is to cause the uterus to become ready for the embryo to attach to its wall following ovulation. Pregnancy cannot occur unless the embryo attaches to the wall of the uterus. Predict the consequence if the ovary secretes too little estrogen.

✔ *Answer in Appendix F*

Classes of Hormone Receptors

The two major classes of hormone receptors are **membrane-bound receptors** and **intracellular receptors.** Membrane-bound receptors extend across the plasma membranes of cells. The part of the receptor exposed to the outside of the cell binds to a specific hormone, and the part of the receptor that extends to the interior of the cell produces a response when a hormone is bound to the receptor (see chapter 9). Hormones that are water-soluble or that have a large molecular weight and therefore cannot pass through the plasma membrane bind to membrane-bound receptor molecules. These hormones include proteins, glycoproteins, polypeptides, epinephrine, and norepinephrine. Intracellular receptors are found in the cytoplasm or in the nucleus of cells. Lipid-soluble hormones, including steroids and thyroid hormones, readily diffuse through plasma membranes, enter the cytoplasm, and bind to intracellular receptor molecules. When bound to a hormone, intracellular receptor molecules produce a response (see chapter 9).

Membrane-Bound Hormone Receptors

Hormones bind in a reversible fashion to the receptor sites of membrane-bound receptor molecules that are exposed to the extracellular fluid. Hormone receptor molecules have peptide chains that cross the membrane once, in the case of some receptors, and several times for other receptors. After a hormone binds to its receptor site, the intracellular part of the receptor initiates events that lead to a response. The mechanisms by which all membrane-bound receptors produce an intracellular response is not known, but evidence exists for at least three major mechanisms. The intracellular portion of membrane-bound receptors can (1) directly control membrane channels, (2) activate G proteins on the inner surface of the cell membrane, which can result in alterations of membrane channels or activate intracellular mediator molecules, or (3) alter the activity of intracellular enzymes, which, in turn, catalyze the synthesis of intracellular mediator molecules or attach phosphate groups to specific proteins in the membrane and cytoplasm of the cell (see chapter 9).

Receptors That Directly Control Membrane Channels

The combination of some hormones with their membrane-bound receptors directly changes the structure of membrane channels (see chapter 9). For example, serotonin binds to serotonin receptor sites that are part of the ligand-gated Na^+ ion channels and causes them to open (figure 17.13). As a result, Na^+ ions and other small cations diffuse into the cell and cause depolarization of the cell membrane. Depolarization of the target cell leads to the response of the target cell to the

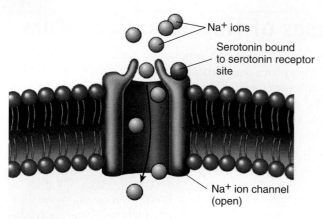

Figure 17.13   Receptors That Directly Control Membrane Channels

Membrane-bound receptors for serotonin are part of the Na$^+$ ion channel. When serotonin binds to its receptor site on the serotonin receptor, the Na$^+$ ion channel opens and the permeability of the membrane to Na$^+$ ions increases. Na$^+$ ions diffuse through the channels into the cell.

chemical signal. Table 17.4 lists some examples of ligand-gated ion channels. Many of these channels respond to neurotransmitters and not hormones, but some play important roles in regulating hormone secretion or mediating responses to paracrine chemical signals.

Hormone Receptors That Activate G Proteins

Several hormones combine with their membrane-bound receptors and alter the activity of G proteins at the inner surface of the plasma membrane (Table 17.5 and see chapter 9). When a hormone is bound to its hormone receptor, the intracellular portion of the hormone receptor causes the G protein subunits to separate from the membrane-bound receptor, and the alpha subunit of the G protein separates from the beta and gamma subunits. Guanosine triphosphate (GTP) then binds to the alpha subunit and replaces guanosine diphosphate (GDP), which was previously bound to the alpha subunit. The alpha subunit, bound to GTP, typically alters the activity of the cell by (1) changing membrane permeability, or (2) increasing or decreasing the concentration of intracellular mediator molecules inside the cell. The ac-

Table 17.4   Chemical Signals, Including Paracrine, That Bind to Receptors and Directly Control Ion Channels

Ligand	Channel Type	Response
Acetylcholine	Cation channel (primarily Na$^+$ ion channels)	Excitatory
Serotonin	Cation channel (primarily Na$^+$ ion channels)	Excitatory
Glutamate	Cation channel (primarily Na$^+$ ion channels)	Excitatory
Glycine	Cl$^-$ ion channels	Inhibitory
GABA	Cl$^-$ ion channels	Inhibitory

Abbreviation: GABA = gamma-aminobutyric acid.

Table 17.5   Examples of Hormones That Bind to Membrane-Bound Receptors and Activate G Proteins

Hormone	Source	Target Tissue
Luteinizing hormone	Anterior pituitary	Ovary or testis
Follicle-stimulating hormone	Anterior pituitary	Ovary or testis
Prolactin	Anterior pituitary	Ovary or testis
Thyroid-stimulating hormone	Anterior pituitary	Thyroid gland
Adrenocorticotropic hormone	Anterior pituitary	Adrenal cortex
Oxytocin	Posterior pituitary	Uterus
Vasopressin	Posterior pituitary	Kidney
Calcitonin	Thyroid gland (parafollicular cells)	Osteoclasts and osteocytes
Parathyroid hormone	Parathyroid gland	Osteoclasts
Glucagon	Pancreas	Liver
Epinephrine	Medulla of adrenal gland	Cardiac muscle
Atrial natriuretic hormone	Atrium of the heart	Kidney cells

tivity of the alpha subunit is terminated when the GTP is broken down to GDP by an enzyme called **phosphodiesterase** (fos′fō-dī-es′ter-ās). The inactive alpha subunit then recombines with the beta and gamma subunits.

1. *G proteins regulate membrane channels.* The alpha subunits of G proteins, bound to GTP, can combine with membrane channels and alter their permeability. The G proteins of some cells can open Ca^{2+} ion channels. Ca^{2+} ions then diffuse through the opened ion

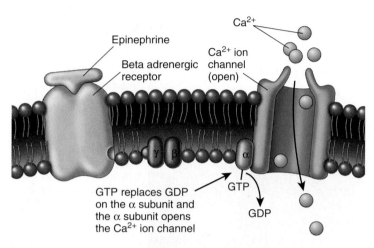

Figure 17.14 Hormone Receptors That Activate G Proteins and Open Ca^{2+} Channels

Beta-adrenergic receptors in heart muscle are associated with G proteins, that regulate Ca^{2+} ion channels. Epinephrine binds to the beta-adrenergic receptor and causes the G proteins to dissociate. GDP on the alpha subunit is replaced by GTP, and the alpha subunit then combines with the Ca^{2+} ion channel, causing it to open. Ca^{2+} ions then diffuse into the cell and increase the force of contraction of cardiac muscle.

channels, increasing the intracellular concentration of Ca^{2+} ions. The Ca^{2+} ions, acting as intracellular mediators, alter the activity of cells. For example, epinephrine binds to certain beta-adrenergic receptors in the heart, and the alpha subunit of a G protein then interacts with the Ca^{2+} ion channels to increase the permeability of the membrane to Ca^{2+} ions. The increased intracellular concentration of Ca^{2+} ions in heart muscle increases the rate and force of contraction (figure 17.14).

2. *G proteins regulate the synthesis of intracellular mediators.* The alpha subunits of G proteins can alter the activity of enzymes at the inner surface of the plasma membrane causing them to either increase or decrease the synthesis of intracellular mediator molecules (see chapter 9). For example, G proteins can alter the activity of an enzyme called **adenylyl** (a-den′i-lil) **cyclase,** which produces **cyclic adenosine** (ă-den′ō-sēn) **monophosphate (cAMP),** or an enzyme called guanylyl (gwahn′i-lil) cyclase that produces **cyclic guanosine** (gwahn′ō-sēn) **monophosphate (cGMP).** The combination of a hormone with its receptor causes the alpha subunit of the G proteins to be released from its beta and gamma subunits. GTP binds to the alpha subunit, and then the alpha subunit of the G protein binds to adenylyl cyclase or guanylyl cyclase. Depending on the type of G protein, an increase in cAMP or cGMP synthesis or a reduction in cAMP or cGMP can result (figure 17.15).

Cyclic AMP and cGMP bind to specific enzymes called **protein kinases** (kī′nās-sez), or **phosphokinases** (fos-fō-kī′nās-sez). Protein kinases attach phosphate groups to enzymes within the cytoplasm of cells, which alters their activity.

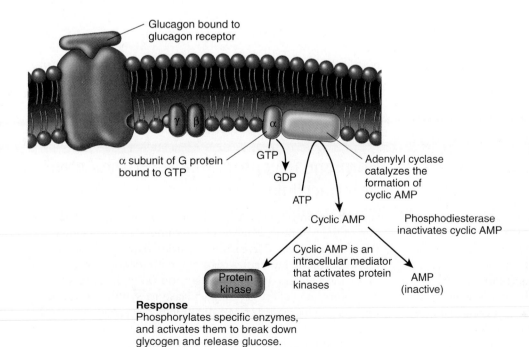

Figure 17.15 Hormone Receptors That Activate G Proteins and Increase Second Messenger Synthesis

Membrane-bound receptors for glucagon are associated with G proteins in liver cells. When glucagon binds to glucagon receptors, the alpha subunit of the G proteins dissociates from the other subunits and GTP binds to it. The alpha subunit then binds to adenylyl cyclase and activates it. The increase in cyclic AMP that results activates protein kinase enzymes, which phosphorylate other specific enzymes that break down glycogen and release glucose from the liver cells.

539

For example, glucagon binds to glucagon receptors within the plasma membrane of liver cells. The occupied receptors, through G proteins, activate adenylyl cyclase enzymes on the inner surface of the plasma membrane to increase cAMP synthesis. From the plasma membrane, cAMP molecules diffuse throughout the cytoplasm of the cell and bind to and activate specific protein kinase enzymes, which add phosphate groups to other enzymes and alter their activity. The final result is the breakdown of glycogen to individual glucose molecules by liver enzymes. The glucose is released into the circulatory system. Finally, phosphodiesterase molecules within the cell break down the cAMP molecules; and as cAMP levels decrease, the activity of enzymes return to their previous rates.

4	P R E D I C T

When smooth muscle cells in the airways of the lungs contract, breathing becomes very difficult, whereas breathing is easy if the smooth muscle cells are relaxed. During asthma attacks, the smooth muscle cells in the airways of the lungs contract. Cyclic AMP causes the smooth muscle cells of airways to relax. Some of the drugs used to treat asthma increase cAMP in smooth muscle cells. Explain as many ways as possible how these drugs might work.

✔ *Answer in Appendix F*

Intracellular mediator molecules mediate the effect of numerous hormones on target tissues. Examples of hormones for which cAMP functions as an intracellular mediator molecule include the gonadotropins, which control events in the ovary and testes; adrenocorticotropic hormone, which regulates secretions from the adrenal cortex; and thyroid-stimulating hormone (TSH), which controls the rate of secretion from the thyroid glands.

Cyclic AMP functions as an intracellular mediator molecule in many cells, but for each cell type, it stimulates a different set of enzymes or other processes and produces a different type of response. For example, cAMP stimulates one type of response in liver cells but another type of response in the ovary.

The alpha subunit of the G proteins can increase the activity of an enzyme called phospholipase C, which acts on a membrane lipid called **phosphoinositol** (fos′fō-in-ō′sĭ-tōl) **(PIP$_2$),** which then leads to the formation of two intracellular mediators from PIP$_2$ called **diacylglycerol** (dī-as′ĭl-glis′er-ol) **(DAG)** and **inositol** (in-ō′sĭ-tōl) **triphosphate (IP$_3$).** IP$_3$ causes calcium to be released from the endoplasmic reticulum of the cell, or it may cause calcium ion channels in the cell membrane to open. DAG and Ca^{2+} ions can alter the activity of specific protein kinase enzymes, which in turn, can alter the activity of the cell by increasing calcium permeability of the plasma membrane or by altering the synthesis of other chemicals within the cell such as prostaglandins (figure 17.16). For example, IP$_3$ is produced in some smooth muscle cells in response to the combination of epinephrine with certain receptors for epinephrine. The IP$_3$ then stimulates calcium release from the endoplasmic reticulum or opens calcium ion chan-

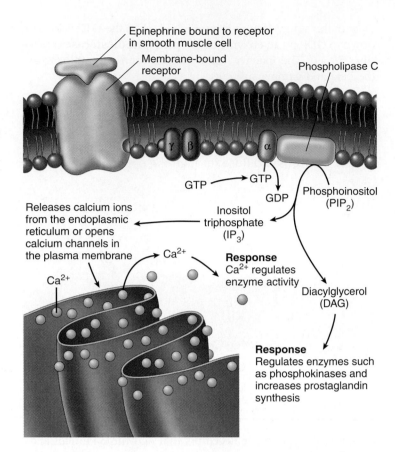

Figure 17.16   Membrane-Bound Receptors That Activate G Proteins and Increase the Synthesis of Intracellular Mediators

Epinephrine receptors in some smooth muscle cells are associated with G proteins. When epinephrine binds to the receptor, the G proteins dissociate and the alpha subunit binds to GTP. The alpha subunit then binds with phospholipase C, which acts on phosphoinositol (PIP$_2$) and produces inositol triphosphate (IP$_3$) and diacylglycerol (DAG). IP$_3$ releases Ca^{2+} ions from the endoplasmic reticulum, and DAG regulates enzymes such as those that synthesize prostaglandin synthesis. These responses increase smooth muscle contraction.

nels in the cell membrane. The calcium ions bind with a protein called **calmodulin** (kal-mod′yū-lin). Calmodulin then binds to an enzyme that phosphorylates myosin molecules in the smooth muscle cells, which stimulates cross-bridge formation and contractions (see chapter 10).

Hormone Receptors That Alter the Activity of Intracellular Enzymes

Some hormones bind with receptors, and the intracellular portion of the receptors catalyzes the synthesis of intracellular mediator molecules. For example, atrial natriuretic (nā′trē-yū-ret′ik) hormone molecules combine with receptor sites on cells of the kidney, and the intracellular portion of the receptors acts as an enzyme called guanylyl cyclase to synthesize cGMP. In response to increased cGMP, the kidney cells cause an increase in the number of Na$^+$ ions excreted and an increase in the volume of water eliminated as urine (figure 17.17).

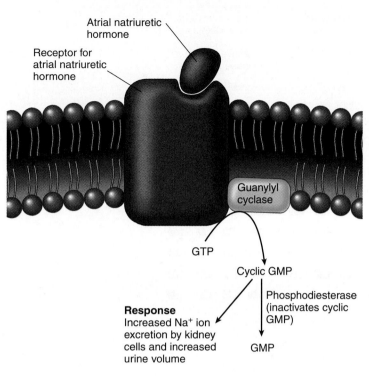

Figure 17.17   Hormone Receptor That Directly Synthesizes an Intracellular Mediator

Atrial natriuretic hormone binds with its receptor site. At the inner surface of the plasma membrane, guanylyl cyclase is activated to produce cyclic GMP from GTP. Cyclic GMP is an intracellular mediator that mediates the response of the cell. Phosphodiesterase is an enzyme that breaks down cyclic GMP to inactive GMP.

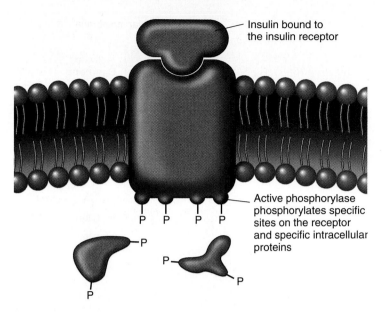

Figure 17.18   Hormone Receptors That Phosphorylate Intracellular Proteins

Insulin receptors are membrane-bound receptors. When insulin binds to the insulin receptor, the receptor acts as a phosphorylase enzyme and attaches phosphate groups from ATP to specific sites on the receptor and on intracellular proteins. The phosphorylated proteins produce the normal response to insulin.

Table 17.6 Hormones That Bind to Receptors That Phosphorylate Intracellular Proteins		
Hormone	**Source**	**Target Tissue and Effect**
Insulin	Pancreatic islets	Most cells, increases glucose and amino acid uptake
Growth hormone	Anterior pituitary gland	Most cells; increases protein synthesis and resists protein breakdown
Prolactin	Anterior pituitary gland	Mammary glands and ovary; initiates milk production following pregnancy and helps maintain the corpus luteum
Growth factors	Various tissues	Stimulate growth in certain cell types
Some intercellular immune signal molecules	Cells of the immune system	Immune-competent cells; help mediate responses of the immune system

Some membrane-bound receptors have intracellular enzymes as part of the hormone receptor or are closely associated with enzymes that phosphorylate proteins. When a hormone is bound to the receptor site, the intracellular enzyme can phosphorylate specific proteins of the receptor and specific intracellular proteins. For example, when insulin binds to insulin receptors, the enzyme activity of the insulin receptors results in phosphate groups being attached to part of the receptor and to other specific intracellular proteins. Major hormones and intercellular messenger molecules that alter enzyme activity by attaching phosphate groups to proteins inside of the cell are listed in table 17.6 (figure 17.18).

Hormones that stimulate the synthesis of an intracellular mediator molecule often produce rapid responses because the mediator influences already existing enzymes and causes a **cascade effect,** which results when a few mediator molecules activate several enzymes and each of the activated enzymes in turn activates several other enzymes that produce the final response. Thus an amplification system exists in which a few molecules, such as cAMP, can affect the activity of many enzymes within a cell (figure 17.19).

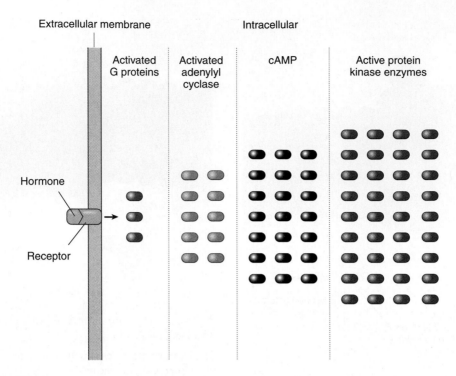

Figure 17.19 The Cascade Effect

The combination of a hormone with a membrane-bound receptor activates several G proteins. The G proteins, in turn, activate adenylyl cyclase enzymes, which cause the synthesis of a large number of cAMP molecules. The cAMP molecules, in turn, activate many protein kinase enzymes, which produce a rapid and amplified response.

Intracellular Hormone Receptors

Intracellular receptors consist of protein molecules inside of cells. Some of the receptors float freely in the cytoplasm of the target cells, whereas others are located in the nucleus (see chapter 9). By the process of diffusion, lipid-soluble hormones cross the plasma membrane into the cytoplasm. Some hormones bind to their receptors in the cytoplasm and activate enzymes (see figure 9.10) or the receptor–hormone complex moves into the nucleus. Hormones for which there are no cytoplasmic receptors can diffuse directly into the nucleus and form receptor–hormone complexes in the nucleus (see figure 9.9).

The nuclear receptors have specific "fingerlike" projections that can interact with specific parts of the DNA molecule when bound to the hormone. The receptors then function to increase the synthesis of specific **messenger ribonucleic acid (mRNA)** molecules. The mRNA moves to the cytoplasm and initiates the synthesis of new proteins at the ribosomes. The newly synthesized proteins produce the cell response to the hormone. For example, the effect of the steroid aldosterone on its target cells within the kidney is to stimulate the synthesis of proteins that increase the rate of sodium chloride transport (figure 17.20). The result is an increase in the reabsorption of sodium chloride from filtrate in the kidney and a reduction in the amount of sodium chloride lost in the urine.

Cells that synthesize new protein molecules in response to a hormonal stimulus normally have a latent period of several hours between the time the hormone binds to its receptor and the time a response is observed. During this latent period, mRNA and new proteins are synthesized. Receptor–hormone complexes normally are degraded within the cell, limiting the length of time the hormone influences the activity of the cell, and the cell slowly returns to its previous functional state.

---

5 **P R E D I C T**

Of membrane-bound receptors and intracellular receptors, which is better adapted for mediating a response that lasts a considerable length of time and which is better for mediating a response with a rapid onset and a short duration? Explain why.

✔ *Answer in Appendix F*

Hormones that produce responses through intracellular receptor mechanisms include the steroid hormones, thyroid hormones, and vitamin D (see table 17.7).

1. Aldosterone is a lipid-soluble hormone and can easily diffuse through the plasma membrane.

2. Aldosterone, once inside of the cell, binds with an aldosterone receptor molecule in the cytoplasm.

3. The aldosterone–receptor complex moves into the nucleus and binds to DNA.

4. The synthesis of messenger RNA (mRNA) that codes for specific proteins.

5. The mRNA leaves the nucleus, passes into the cytoplasm of the cell, and binds to ribosomes, where it directs the synthesis of the specific proteins.

6. The proteins synthesized on the ribosomes produce the response of the cell to aldosterone.

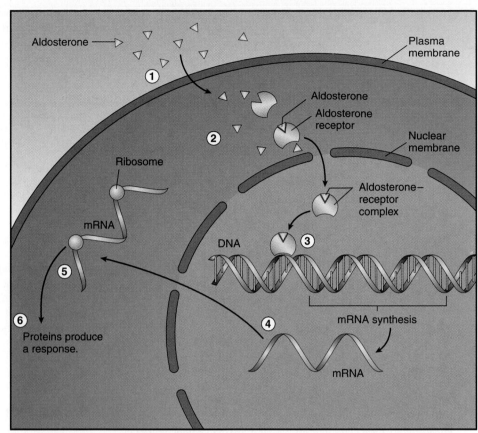

Figure 17.20   Intracellular Receptor Model

Table 17.7	**Major Hormones That Combine with Intracellular Receptors**		
Category of Hormone	**Hormone**	**Source**	**Target Tissue and Effect**
Sex steroids	Testosterone	Testis	Responsible for development of the reproductive structures and development of male secondary sex characteristics
	Progesterone	Ovary	Causes increased size of cells lining the uterus
	Estrogen	Ovary	Causes increased cell division in the lining of the uterus
Mineralocorticoid steroids	Aldosterone	Adrenal cortex	Increased reabsorption of Na^+ ions and increased secretion of K^+ ions in the kidney
Glucocorticoid steroid hormones	Cortisol	Adrenal cortex	Increased breakdown of proteins and fats and increased blood levels of glucose
Thyroid hormones	Triiodothyronine (T_3)	Thyroid gland	Regulate development and metabolism
Vitamin D	1,25-Dihydroxycholecalciferol	Combination of the skin, liver, and kidney	Increased reabsorption of Ca^{2+} ions in the kidney and absorption of Ca^{2+} ions in the gastrointestinal tract

Summary

General Characteristics of the Endocrine System

1. Endocrine glands produce hormones that are released into the interstitial fluid, diffuse into the blood, and travel to target tissues, where they cause a specific response.
2. Other chemical messengers produced by endocrine glands include neurohormones, neurotransmitters, neuromodulators, parahormones, and pheromones.

3. Generalizations about the differences between the endocrine and nervous system include the following: (a) the endocrine system is amplitude-modulated, whereas the nervous system is frequency-modulated; and (b) the response of target tissues to hormones is usually slower and of longer duration than that to neurons.

Chemical Structure of Hormones

Hormones are proteins, glycoproteins, polypeptides, derivatives of amino acids, or lipids (steroids or derivatives of fatty acids).

Control of Secretion Rate

1. Most hormones are not secreted at a constant rate.
2. Most hormone secretion is controlled by negative-feedback mechanisms that function to maintain homeostasis.
3. Hormone secretion from an endocrine tissue is regulated by one or more of three mechanisms: (a) a nonhormone substance; (b) stimulation by the nervous system; or (c) a hormone from another endocrine tissue.

Transport and Distribution in the Body

Hormones are dissolved in plasma or bind to plasma proteins. The blood quickly distributes hormones throughout the body.

Metabolism and Excretion

1. Nonpolar, readily diffusible hormones bind to plasma proteins and have an increased half-life.
2. Water-soluble hormones, such as proteins, epinephrine, and norepinephrine, do not bind to plasma proteins or readily diffuse out of the blood. Instead, they are broken down by enzymes or are taken up by tissues. They have a short half-life.
3. Hormones with a short half-life regulate activities that have a rapid onset and a short duration.
4. Hormones with a long half-life regulate activities that remain at a constant rate through time.
5. Hormones are eliminated from the blood by excretion from the kidneys and liver, enzymatic degradation, conjugation, or active transport.

Interaction of Hormones with Their Target Tissues

1. Target tissues have receptor molecules that are specific for a particular hormone.
2. Hormones bound with receptors affect the rate at which already existing processes occur.

3. Down-regulation is a decrease in the number of receptor molecules in a target tissue, and up-regulation is an increase in the number of receptor molecules.

Classes of Hormone Receptors

1. Membrane-bound receptors bind to water-soluble or large-molecular-weight hormones.
2. Intracellular receptors bind to lipid-soluble hormones.

Membrane-Bound Hormone Receptors

1. Membrane-bound receptors are proteins or glycoproteins that have polypeptide chains that are folded to cross the cell several times.
2. When a hormone binds to a membrane-bound receptor:
   - A change in the structure of membrane channels can result in a change in permeability of the plasma membrane to ions.
   - G proteins are activated. The alpha subunit of the G protein can bind to ion channels and cause them to open or change the rate of synthesis of intracellular mediator molecules, such as cAMP, cGMP, IP_3, and DAG.
   - Intracellular enzymes can be directly activated, which in turn synthesizes intracellular mediators, such as cGMP, or adds a phosphate group to intracellular enzymes, which alters their activity.
3. Intracellular mediator mechanisms are rapid-acting because they act on already existing enzymes and produce a cascade effect.

Intracellular Hormone Receptors

1. Intracellular receptors are proteins in the cytoplasm or nucleus.
2. Hormones bind with the intracellular receptor, and the receptor–hormone complex activates genes. Consequently, DNA is activated to produce mRNA. The mRNA initiates the production of certain proteins (enzymes) that produce the response of the target cell to the hormone.
3. Intracellular receptor mechanisms are slow-acting because time is required to produce the mRNA and the protein.
4. Intracellular receptor-activated processes are limited by the breakdown of the receptor–hormone complex.

Content Review

1. Define the terms endocrine gland, endocrine system, and hormone.
2. Name and describe five chemical messengers, other than hormones, produced by endocrine glands.
3. Contrast the endocrine system and the nervous system for the following:
   - Amplitude versus frequency modulation
   - Speed and duration of target cell response
4. List the categories of hormones on the basis of chemical structure, and give an example of each.
5. Describe the ways hormone secretion is regulated. Give examples of three patterns of hormone secretion.
6. Define the half-life of a hormone. What happens to this half-life when a hormone binds to a plasma protein? What kinds of hormones bind to plasma proteins?
7. What kinds of activities do hormones with a short half-life regulate? With a long half-life?

8. List and describe the ways that hormones are eliminated from the blood.
9. Many different hormones circulate in the blood. What determines to which hormone a tissue will respond?
10. When a hormone combines with a receptor, does it cause the cell to do new things, alter already existing processes, or both?
11. Contrast membrane-bound receptors and intracellular receptors for the following:
    - Their location in the cell
    - The characteristics of hormone to which they bind
12. Describe how membrane permeability can be changed when a hormone binds to a membrane-bound receptor. Give an example.
13. Describe how G proteins can alter membrane permeability and how intracellular mediator molecules can be produced within cells in response to the binding of hormones to their membrane-bound receptors. Give examples.

14. Describe how hormones can bind to membrane-bound receptors, directly increase intracellular enzyme activity, and increase phosphorylation of intracellular molecules. Give examples.
15. What limits the activity of intracellular mediator molecules such as cAMP and phosphorylated proteins.
16. What finally limits the processes activated by cAMP?

17. Explain what is meant by a cascade effect for the intracellular mediator model of hormone action. Does the intracellular mediator mechanism produce a slow or a rapid response?
18. Describe the intracellular model for hormone action. Compared with the intracellular mediator mechanism, is it fast- or slow-acting? Explain.
19. What finally limits the processes activated by the intracellular receptor mechanism?

Develop Your Reasoning Skills

1. Consider a hormone that is secreted in large amounts at a given interval, modified chemically by the liver, and excreted by the kidney at a rapid rate, making the half-life of the hormone in the circulatory system very short. The hormone therefore rapidly increases in the blood and then decreases rapidly. Predict the consequences of liver and kidney disease on the blood levels of that hormone.
2. Consider a hormone that controls the concentration of some substance in the circulatory system. If a tumor begins to produce that substance in large amounts in an uncontrolled fashion, predict the effect on the secretion rate for the hormone.
3. How could you determine whether or not a hormone-mediated response resulted from the intracellular mediator mechanism or the intracellular receptor mechanism?
4. If the effect of a hormone on a target tissue is through a membrane-bound receptor that has a G protein associated with it, predict the consequences if a genetic disease causes the alpha unit of the G protein to have a structure that prevents it from binding to GTP.
5. Prostaglandins are a group of hormones produced by many cells of the body. Unlike other hormones, prostaglandins do not circulate but usually have their effect at or very near their site of production. Prostaglandins apparently affect many body functions, including blood pressure, inflammation, induction of labor, vomiting, fever, and inhibition of the clotting process. Prostaglandins also influence the formation of cAMP. Explain how an inhibitor of prostaglandin synthesis could be used as a therapeutic agent. Inhibitors of prostaglandin synthesis can produce side effects. Why?

6. For a hormone that binds to a membrane-bound receptor and has cAMP as the intracellular mediator, predict and explain the consequences if a drug is taken that strongly inhibits phosphodiesterase.
7. When an individual is confronted with a potentially harmful or dangerous situation, epinephrine (adrenaline) is released from the adrenal gland. Epinephrine prepares the body for action by increasing the heart rate and blood sugar levels. Explain the advantages or disadvantages associated with a short half-life for epinephrine and those associated with a long half-life.
8. Thyroid hormones are important in regulating the basal metabolic rate of the body. What are the advantages or disadvantages of
   (a) A long half-life for thyroid hormones
   (b) A short half-life
9. An increase in thyroid hormones causes an increase in metabolic rate. If liver disease results in reduced production of the plasma proteins to which thyroid hormones normally bind, what is the effect on metabolic rate? Explain.
10. Predict the effect on LH and FSH secretion if a small tumor in the hypothalamus of the brain secretes large concentrations of GnRH continuously. Given that LH and FSH regulate the function of the male and female reproductive systems, predict whether the condition increases or decreases the activity of the reproductive system.

Web Site Link

For a listing of the most current web sites related to this chapter, please visit the Seeley home page at:
http://www.mhhe.com/biosci/ap/seeleyap/

Chapter Eighteen

Endocrine Glands

Objectives

1. List the most important information relating to endocrine glands and their secretions.

2. Describe the embryonic development, anatomy, and location of the pituitary gland, and describe the functional and structural relationships between the hypothalamus and the pituitary gland.

3. Describe the secretory cells of the posterior pituitary, including the location of their cell bodies, the site of hormone synthesis, the transport of hormones to the posterior pituitary, and hormone secretion.

4. Outline the means by which anterior pituitary hormone secretion is regulated.

5. Describe the target tissues, regulation, and responses to each of the posterior and anterior pituitary hormones.

6. Describe the structure and location of the thyroid gland.

7. Describe the response of target tissues to thyroid hormones, and outline the regulation of thyroid hormone secretion.

8. Describe the regulation of calcitonin secretion, and describe its function.

9. Explain the function of parathyroid hormone, and describe the means by which its secretion is regulated.

10. Explain the relationship between parathyroid hormone and vitamin D.

11. Describe the structure and embryologic development of the adrenal glands, and describe the response of the target tissues to each of the adrenal hormones.

12. Describe the means by which the adrenal hormones are regulated.

13. Describe the position and structure of the pancreas, and list the substances secreted by the pancreas and their functions.

14. Explain the regulation of insulin and glucagon secretion, and describe how blood-nutrient levels are regulated by hormones after a meal and during exercise.

15. Describe the structure and location of the pineal body, the products it secretes, and the functions of these products.

Part Three

Early in the 1900s, people who developed insulin-dependent diabetes or Addison's disease died, almost without exception. There were few effective treatments for these and other diseases of the endocrine system, such as diabetes insipidis, Cushing's syndrome, and many reproductive abnormalities. Although these conditions are still serious and not all can be completely reversed, advances made in understanding the endocrine system have improved the outlook for people with these and other endocrine diseases.

Understanding how body functions are controlled requires some knowledge of the endocrine system. This system is one of the two major control systems of the body, and it is involved in so many functions that homeostasis cannot be maintained without its normal function. Diseases caused by disorders of the endocrine system emphasize its importance. Yet the endocrine system consists of several small glands distributed throughout the body, which could escape notice, if it were not for the importance of the small amounts of hormones they secrete.

Several pieces of information are needed for a thorough understanding of the role endocrine glands and their secretions play in the body. They are

1. The anatomy of each gland and its location
2. The hormone secreted by each gland
3. The target tissues and the response of target tissues to each hormone
4. The means by which the secretion of each hormone is regulated
5. The consequences and causes, if known, of hypersecretion and hyposecretion of the hormone

This information is provided for each of the endocrine glands discussed in this chapter. Certain hormones, such as those that regulate digestion and reproduction, are mentioned only briefly because they are explained more fully in later chapters.

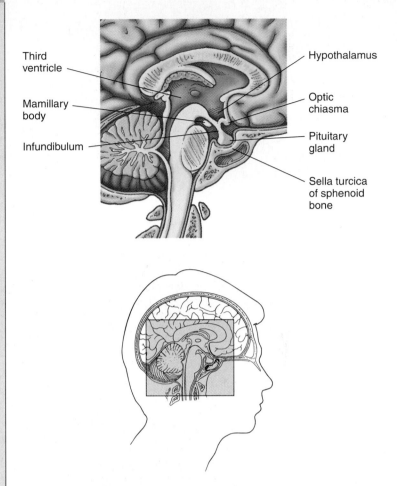

Figure 18.1   The Hypothalamus and Pituitary Gland

A midsagittal section of the head through the pituitary gland showing the location of the hypothalamus and the pituitary. The pituitary gland is in a depression called the sella turcica in the floor of the skull. It is connected to the hypothalamus of the brain by the infundibulum. ✗

Pituitary Gland and Hypothalamus

The **pituitary** (pi-tū′i-tăr-rē) **gland**, or **hypophysis** (hī-pof′ĭ-sis, meaning an undergrowth), secretes nine major hormones that regulate numerous body functions and the secretory activity of several other endocrine glands.

The **hypothalamus** (hī′pō-thal′ă-mŭs) of the brain and pituitary gland are major sites where the nervous and endocrine systems interact (figure 18.1). The hypothalamus regulates the secretory activity of the pituitary gland. Indeed, the posterior pituitary is an extension of the hypothalamus. The activity of the hypothalamus, in turn, is influenced by hormones, by sensory information that enters the central nervous system, and by emotions.

Structure of the Pituitary Gland

The pituitary gland is roughly 1 cm in diameter, weighs 0.5–1 g, and rests in the sella turcica of the sphenoid bone (see figure 18.1). It is located inferior to the hypothalamus and is connected to it by a stalk of tissue called the **infundibulum** (in-fŭn-dib′yū-lŭm).

The pituitary gland is divided functionally into two parts: the **posterior pituitary,** or **neurohypophysis** (nū′rō-hī-pof′i-sis), and the **anterior pituitary,** or **adenohypophysis** (ad′ĕ-nō-hī-pof′i-sis).

Posterior Pituitary, or Neurohypophysis

The posterior pituitary is called the neurohypophysis because it is continuous with the brain (*neuro-* refers to the nervous system). It is formed during embryonic development from an outgrowth of the inferior part of the brain in the area of the hypothalamus (see chapter 29). The outgrowth of the brain forms the infundibulum, and the distal end of the infundibulum enlarges to form the posterior pituitary (figure 18.2). Secretions of the posterior pituitary are **neurohormones** (nūr-ō-hōr′-mōnz) because the posterior pituitary is an extension of the nervous system.

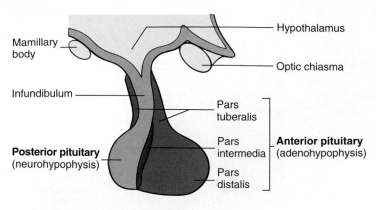

Figure 18.2   Subdivisions of the Pituitary Gland

The pituitary gland is divided into the anterior pituitary, or adenohypophysis, and the posterior pituitary, or neurohypophysis. The anterior pituitary is subdivided further into the pars distalis, pars intermedia, and pars tuberalis. The posterior pituitary consists of the enlarged distal end of the infundibulum, which connects the posterior pituitary to the hypothalamus.

Anterior Pituitary, or Adenohypophysis

The anterior pituitary, or adenohypophysis (*adeno*- means gland), arises as an outpocketing of the roof of the embryonic oral cavity called Rathke's pouch, which grows toward the posterior pituitary. As it nears the posterior pituitary, Rathke's pouch loses its connection with the oral cavity and becomes the anterior pituitary. The anterior pituitary is subdivided into three areas with indistinct boundaries: the pars tuberalis, the pars distalis, and the pars intermedia (see figure 18.2). The hormones secreted from the anterior pituitary, in contrast to those from the posterior pituitary, are not neurohormones because the anterior pituitary is derived from epithelial tissue of the embryonic oral cavity and not from neural tissue.

Relationship of the Pituitary to the Brain

Portal vessels are blood vessels that begin and end in a capillary network. The **hypothalamohypophyseal** (hī′pō-thal′ă-mō-hī′pō-fiz′ē-ăl) **portal system** extends from a part of the hypothalamus to the anterior pituitary (figure 18.3*a*). The primary capillary network in the hypothalamus is supplied with blood from arteries that deliver blood to the hypothalamus. From the primary capillary network, the hypothalamohypophyseal portal vessels carry blood to a secondary capillary network in the anterior pituitary. Veins from the secondary capillary network eventually merge with the general circulation.

Neurohormones, produced and secreted by the hypothalamus, enter the primary capillary network and are carried to the secondary capillary network. There the neurohormones leave the blood and act on the cells of the anterior pituitary. They act either as **releasing hormones,** increasing the secretion of anterior pituitary hormones, or as **inhibiting hormones,** decreasing the secretion of anterior pituitary hormones. Each releasing hormone stimulates and each inhibiting hormone inhibits the production and secretion of a specific hormone by the anterior pituitary. In response to the releasing hormones, anterior pituitary cells secrete hormones that enter the secondary capillary network and are carried by the general circulation to their target tissues. Thus the hypothalamohypophyseal portal system provides a means by which the hypothalamus, using neurohormones as chemical signals, regulates the secretory activity of the anterior pituitary (figure 18.4 and table 18.1).

In contrast, there is no portal system that carries hypothalamic neurohormones to the posterior pituitary. Neurohormones released from the posterior pituitary are produced by neurosecretory cells with their cell bodies in the hypothalamus. The axons of these cells extend from the hypothalamus through the infundibulum into the posterior pituitary and constitute a nerve tract called the **hypothalamohypophyseal tract** (figure 18.3*b*). Neurohormones produced in the hypothalamus pass down these axons in tiny vesicles and are stored in secretory granules in the enlarged ends of the axons. Action potentials originating in the neuron cell bodies in the hypothalamus are propagated along the axons to the axon terminals in the posterior pituitary. The action potentials cause the release of neurohormones from the axon terminals, and they enter the circulatory system (see figure 18.4).

1   P R E D I C T

Surgical removal of the posterior pituitary in experimental animals results in marked symptoms, but these symptoms associated with hormone shortage are temporary. Explain these results.

✔ *Answer in Appendix F*

Hormones of the Pituitary Gland

This section describes the hormones secreted from the pituitary gland (table 18.2), their effects on the body, and the mechanisms that regulate their secretion rate. In addition, some major consequences of abnormal hormone secretion are stressed.

Posterior Pituitary Hormones

The posterior pituitary stores and secretes two polypeptide neurohormones called antidiuretic hormone and oxytocin. Each hormone is secreted by a separate population of cells.

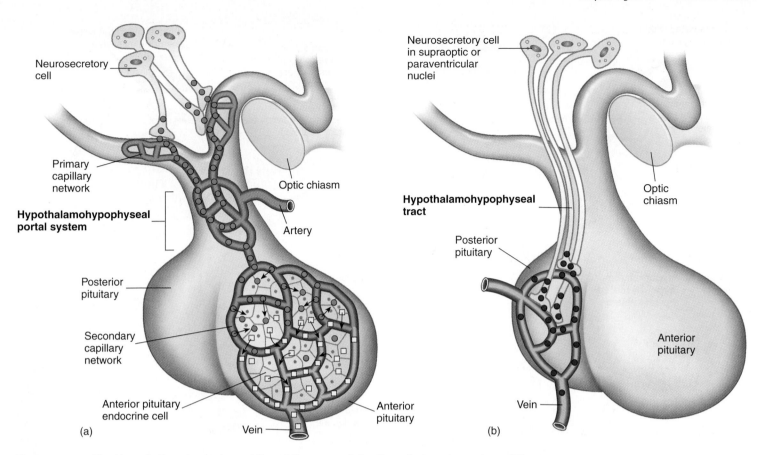

Figure 18.3   The Hypothalamohypophyseal Portal System and the Hypothalamohypophyseal Tract

(*a*) The hypothalamohypophyseal portal system originates from a primary capillary network in the hypothalamus and extends to a secondary capillary network in the anterior pituitary. (*b*) The hypothalamohypophyseal tract originates in the hypothalamus with the cell bodies of neurosecretory cells whose axons extend to the posterior pituitary. 🏃

Antidiuretic Hormone

Antidiuretic (an′tē-dī-yū-ret′ik) **hormone (ADH)** is so named because it prevents (anti-) the output of large amounts of urine (diuresis). ADH is also called **vasopressin** (vā-sō-pres′in) because it constricts blood vessels and raises blood pressure when present in high concentrations. ADH is synthesized by neuron cell bodies in the supraoptic nuclei of the hypothalamus and transported within the axons of the hypothalamohypophyseal tract to the posterior pituitary, where it is stored in axon terminals. ADH is released from these endings into the blood and carried to its primary target tissue, the kidneys, where it promotes the retention of water and reduces urine volume (see chapter 26).

The secretion rate for ADH changes in response to alterations in blood osmolality and blood volume. The osmolality of a solution increases as the concentration of solutes in the solution increases. Specialized neurons, called **osmoreceptors** (os′mō-rē-sep′terz), synapse with the ADH neurosecretory cells in the hypothalamus. When blood osmolality increases, the frequency of action potentials in the osmoreceptors increases, resulting in a greater frequency of action potentials in the neurosecretory cells. As a consequence, ADH secretion increases. Alternatively, the ADH neurosecretory cells can be stimulated directly by an increase in blood osmolality. Because ADH promotes water retention by the kidneys, it functions to reduce blood osmolality and resists any further increase in the osmolality of body fluids.

As the osmolality of the blood decreases, the action potential frequency in the osmoreceptors and the neurosecretory cells decreases. Thus less ADH is secreted from the posterior pituitary gland, and the volume of water eliminated in the form of urine increases.

Sensory receptors that detect changes in blood pressure send action potentials through afferent vagal nerve fibers that eventually synapse with the ADH neurosecretory cells. A drop in blood pressure, which normally accompanies a drop in blood volume, causes an increased action potential frequency in the neurosecretory cells and increased ADH secretion, which promotes water retention by the kidneys. Because the water in urine is derived from blood as it passes through the kidney, ADH slows any further reduction in blood volume.

An increase in blood pressure decreases the action potential frequency in neurosecretory cells. This leads to the

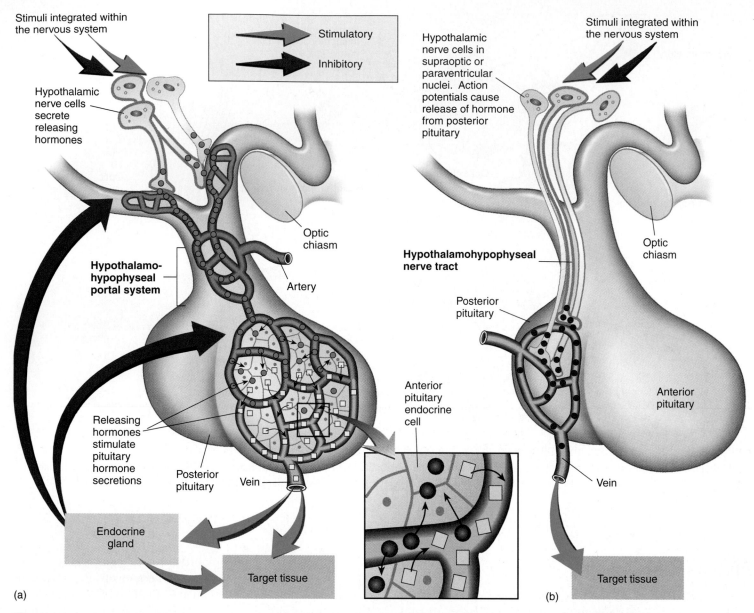

Figure 18.4 General Relationship Between the Hypothalamus, the Pituitary, and Target Tissues

(*a*) Substances called releasing hormones are secreted from the hypothalamic neurons as a result of certain stimuli. They pass through the hypothalamohypophyseal portal system to the anterior pituitary. The releasing hormones can either stimulate or inhibit the secretion of anterior pituitary hormones. Hormones secreted from cells within the anterior pituitary pass through the blood and influence the activity of their target tissues. (*b*) In response to stimulation of hypothalamic neurosecretory cells, action potentials pass along the axons of the neurosecretory cells to the posterior pituitary. The action potentials cause the release from the posterior pituitary of neurohormones that pass through the blood to target tissues. ✗

secretion of less ADH from the posterior pituitary. As a result, the volume of urine produced by the kidneys increases (figure 18.5).

Clinical Note

A lack of ADH secretion is one cause of diabetes insipidus and leads to the production of a large amount of dilute urine, which can approach 20 L/day. The loss of many liters of water in the form of urine causes an increase in the osmolality of the body fluids, but negative-feedback mechanisms fail to stimulate ADH release. The volume of urine produced each day increases rapidly as the rate of ADH secretion becomes less than 50% of normal. Diabetes insipidus also results from either damage to the kidney or a genetic disorder that makes the kidney incapable of responding to ADH. The consequences of diabetes insipidus are not obvious until the condition becomes severe. When the condition is severe, dehydration and death can result unless the intake of water is adequate to accommodate its loss.

Table 18.1   Hormones of the Hypothalamus

Hormones	Structure	Target Tissue	Response
Growth hormone-releasing hormone (GHRH)	Small peptide	Anterior pituitary cells that secrete growth hormone	Increased growth hormone secretion
Growth hormone-inhibiting hormone (GHIH), or somatostatin	Small peptide	Anterior pituitary cells that secrete growth hormone	Decreased growth hormone secretion
Corticotropin-releasing hormone (CRH)	Peptide	Anterior pituitary cells that secrete adrenocorticotropic hormone	Increased adrenocorticotropic hormone secretion
Gonadotropin-releasing hormone (GnRH)	Small peptide	Anterior pituitary cells that secrete luteinizing hormone and follicle-stimulating hormone	Increased secretion of luteinizing hormone and follicle-stimulating hormone
Prolactin-inhibiting hormone (PIH)	Unknown (possibly dopamine)	Anterior pituitary cells that secrete prolactin	Decreased prolactin secretion
Prolactin-releasing hormone (PRH)	Unknown	Anterior pituitary cells that secrete prolactin	Increased prolactin secretion

Table 18.2   Hormones of the Pituitary Gland

Hormones	Structure	Target Tissue	Response
Posterior Pituitary (Neurohypophysis)			
Antidiuretic hormone (ADH)	Small peptide	Kidney	Increased water reabsorption (less water is lost in the form of urine)
Oxytocin	Small peptide	Uterus; mammary glands	Increased uterine contractions; increased milk expulsion from mammary glands
Anterior Pituitary (Adenohypophysis)			
Growth hormone (GH), or somatotropin	Protein	Most tissues	Increased growth in tissues; increased amino acid uptake and protein synthesis; increased breakdown of lipids and release of fatty acids from cells; increased glycogen synthesis and increased blood glucose levels; increased somatomedin production
Thyroid-stimulating hormone (TSH)	Glycoprotein	Thyroid gland	Increased thyroid hormone secretion
Adrenocorticotropic hormone (ACTH)	Peptide	Adrenal cortex	Increased glucocorticoid hormone secretion
Lipotropins	Peptides	Fat tissues	Increased fat breakdown
β Endorphins	Peptides	Brain, but not all target tissues are known	Analgesia in the brain; inhibition of gonadotropin-releasing hormone secretion
Melanocyte-stimulating hormone (MSH)	Peptide	Melanocytes in the skin	Increased melanin production in melanocytes to make the skin darker in color
Luteinizing hormone (LH)	Glycoprotein	Ovaries in females; testes in males	Ovulation and progesterone production in ovaries, testosterone synthesis and support for sperm cell production in testes
Follicle-stimulating hormone (FSH)	Glycoprotein	Follicles in ovaries in females; seminiferous tubes in males	Follicle maturation and estrogen secretion in ovaries, sperm cell production in testes
Prolactin	Protein	Ovaries and mammary glands in females	Milk production in lactating women; increased response of follicle to LH and FSH; unclear function in males

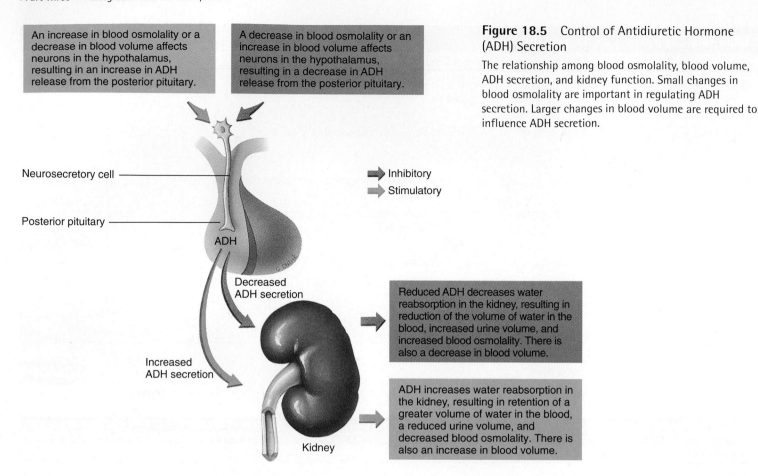

An increase in blood osmolality or a decrease in blood volume affects neurons in the hypothalamus, resulting in an increase in ADH release from the posterior pituitary.

A decrease in blood osmolality or an increase in blood volume affects neurons in the hypothalamus, resulting in a decrease in ADH release from the posterior pituitary.

Neurosecretory cell

Posterior pituitary

ADH

➡ Inhibitory
➡ Stimulatory

Decreased ADH secretion

Increased ADH secretion

Reduced ADH decreases water reabsorption in the kidney, resulting in reduction of the volume of water in the blood, increased urine volume, and increased blood osmolality. There is also a decrease in blood volume.

ADH increases water reabsorption in the kidney, resulting in retention of a greater volume of water in the blood, a reduced urine volume, and decreased blood osmolality. There is also an increase in blood volume.

Kidney

Figure 18.5   Control of Antidiuretic Hormone (ADH) Secretion

The relationship among blood osmolality, blood volume, ADH secretion, and kidney function. Small changes in blood osmolality are important in regulating ADH secretion. Larger changes in blood volume are required to influence ADH secretion.

Oxytocin

Oxytocin (ok-sē-tō′sin) is synthesized by neuron cell bodies in the paraventricular nuclei of the hypothalamus and then is transported through axons to the posterior pituitary, where it is stored in the axon terminals.

Oxytocin stimulates the smooth muscle cells of the uterus. Oxytocin plays an important role in the expulsion of the fetus from the uterus during delivery by stimulating uterine smooth muscle contraction. It also causes contraction of the uterine smooth muscle cells in nonpregnant women, primarily during menses and sexual intercourse. The uterine contractions play a role in the expulsion of the uterine epithelium and small amounts of blood during menses and can participate in the movement of sperm cells through the uterus after sexual intercourse. Oxytocin is also responsible for milk ejection in lactating females by promoting contraction of smooth musclelike cells surrounding the alveoli of the mammary glands. Little is known about the effect of oxytocin in males.

Stretch of the uterus, or mechanical stimulation of the cervix, or stimulation of the nipples of the breast when a baby nurses activate nervous reflexes that stimulate oxytocin release. Action potentials are carried from the uterus and from the nip-

ples to the spinal cord and then up the spinal cord to the hypothalamus of the brain. Action potentials in the oxytocin-secreting neurons pass along the axons to the posterior pituitary, where they cause axon terminals to release oxytocin. The role of oxytocin in the reproductive system is described in greater detail in chapter 29.

Anterior Pituitary Hormones

The anterior pituitary secretions are influenced by releasing and inhibiting hormones that pass from the hypothalamus through the hypothalamohypophyseal portal system to the anterior pituitary. For some anterior pituitary hormones, the hypothalamus produces both releasing hormones and inhibiting hormones (see table 18.1).

All of the hormones released from the anterior pituitary are proteins, glycoproteins, or polypeptides. They are transported in the circulatory system, have a half-life measured in terms of minutes, and bind to membrane-bound receptor molecules on their target cells. For the most part, each hormone is secreted by a separate cell type. Adrenocorticotropic hormone and lipotropin are exceptions because both of these hormones are derived from the same precursor protein.

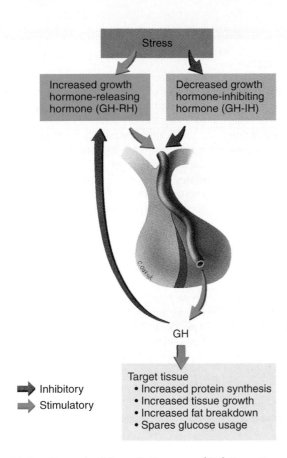

Figure 18.6 Control of Growth Hormone (GH) Secretion

Secretion of GH is controlled by two neurohormones released from the hypothalamus: growth hormone-releasing hormone (GH-RH), which stimulates GH secretion, and growth hormone-inhibiting hormone (GH-IH), which inhibits GH secretion. Stress increases GH-RH secretion and inhibits GH-IH secretion. High levels of GH have a negative-feedback effect on the production of GH-RH by the hypothalamus.

Some anterior pituitary gland hormones are called **tropic** (trō'pik) **hormones** because as they are released from the anterior pituitary gland they regulate the secretion of hormones from other endocrine glands. Tropic hormones include thyroid-stimulating hormone, luteinizing hormone, follicle-stimulating hormone, adrenocorticotropic hormone, and prolactin.

Growth Hormone

Growth hormone (GH), sometimes called **somatotropin** (sō'mă-to-trō'pin), stimulates growth in most tissues and is one of the major regulators of metabolism. It increases the number of amino acids entering cells and favors their incorporation into proteins. GH increases lipolysis, or the breakdown of lipids, and the release of fatty acids from fat cells. The fatty acids then can be used as energy sources to drive chemical reactions, including anabolic reactions, by other cells. GH increases glycogen synthesis and storage in tissues, and the increased use of fats as an energy source spares glucose. GH

plays an important role in regulating blood nutrient levels after a meal and during periods of fasting.

GH binds directly to membrane-bound receptors on target cells (see chapter 17), such as fat cells, to produce a response. These responses are called the direct effects of GH and include the increased breakdown of lipids and decreased use of glucose as an energy source.

GH also has indirect effects on some tissues, because GH increases the production of a number of polypeptides, primarily by the liver, but also by skeletal muscle and other tissues. These polypeptides, called **somatomedins** (sō-mă'tō-mē'dinz), circulate in the blood and bind to receptors on target tissues. The best understood effects of the somatomedins is the stimulation of growth in cartilage and bone and increased synthesis of protein in skeletal muscles. The best known somatomedins are two polypeptide hormones produced by the liver called **insulinlike growth factor I** and **II** because of the similarity of their structure to insulin. Growth hormone, growth factors such as somatomedins, and prolactin bind to membrane-bound receptors that phosphorylate intracellular proteins (see chapter 17).

Two neurohormones released from the hypothalamus regulate the secretion of GH (figure 18.6). One factor, **growth hormone-releasing hormone (GHRH),** stimulates the secretion of GH, and the other, **growth hormone-inhibiting hormone (GHIH),** or **somatostatin** (sō'mă-tō-stat'in), inhibits the secretion of GH. Note in figure 18.6 that the stimuli that influence GH secretion act on the hypothalamus to increase or decrease the secretion of the releasing and inhibiting hormones. Secretion of GH is stimulated by low blood glucose levels and stress and is inhibited by high blood glucose levels. Rising blood levels of certain amino acids also increase GH secretion.

In most people a rhythm of GH secretion occurs with daily peak levels correlated with deep sleep. There is not a chronically elevated blood GH level during periods of rapid growth, although children tend to have somewhat higher blood levels of GH than adults. In addition to GH, factors such as genetics, nutrition, and sex hormones influence growth.

2   P R E D I C T

Mr. Hoops has a son who wants to be a basketball player almost as much as Mr. Hoops wants him to be one. Mr. Hoops knows a little bit about growth hormone and asked his son's doctor if he would prescribe some for his son, so he can grow tall. What do you think the doctor tells Mr. Hoops?

✔ *Answer in Appendix F*

Thyroid-Stimulating Hormone

Thyroid-stimulating hormone   (TSH), also called **thyrotropin** (thī-rō-trō'-pin), stimulates the synthesis and secretion of thyroid hormones from the thyroid gland.

Clinical Focus   Growth Hormone and Growth Disorders

Several pathologic conditions are associated with abnormal GH secretion. In general, the causes for hypersecretion or hyposecretion of GH involve tumors in the hypothalamus or the pituitary, the synthesis of structurally abnormal GH, the inability of the liver to produce somatomedins, or the lack of receptor molecules in the target cells.

Chronic hyposecretion of GH in infants and children leads to **dwarfism** (dwōrf′izm), or short stature due to delayed bone growth. The bones usually have a normal shape, however. In contrast to the dwarfism caused by hyposecretion of thyroid hormones, these dwarfs exhibit normal intelligence. Other symptoms resulting from the lack of GH include mild obesity and retarded development of adult reproductive functions. Two types of dwarfism result from a lack of GH secretion: (1) In approximately two-thirds of the cases, GH and other anterior pituitary hormones are secreted in reduced amounts; (2) in the

remaining approximately one-third of cases, a reduced amount of GH is observed, and normal reproduction can occur. No obvious pathology is associated with hyposecretion of GH in adults, although some evidence indicates that lack of GH can lead to reduced bone mineral content in adults.

The gene responsible for determining the structure of GH has been transferred successfully from human cells to bacterial cells, which produce GH that is identical to human GH. The GH produced in this fashion is available to treat patients who suffer from a lack of GH secretion.

Chronic hypersecretion of GH leads to **giantism** (jī′an-tizm), or **acromegaly** (ak-rō-meg′ă-lē), depending on whether the hypersecretion occurs before or after complete ossification of the epiphyseal plates in the skeletal system. Chronic hypersecretion of GH before the epiphyseal plates have ossified causes exaggerated and prolonged

growth in long bones, resulting in giantism. Some individuals thus affected have grown to 8 feet tall or more.

In adults, chronically elevated GH levels result in acromegaly. No increase in height occurs because of the ossified epiphyseal plates. The condition does result in an increased diameter of fingers, toes, hands, and feet; the deposition of heavy bony ridges above the eyes; and a prominent jaw. The influence of GH on soft tissues results in a bulbous or broad nose, an enlarged tongue, thickened skin, and sparse subcutaneous adipose tissue. Nerves frequently are compressed as a result of the proliferation of connective tissue. Because GH spares glucose usage, chronic hyperglycemia results, frequently leading to diabetes mellitus and the development of severe atherosclerosis. Treatment for chronic hypersecretion of GH often involves surgical removal or irradiation of a GH-producing tumor.

Adrenocorticotropic Hormone and Related Substances

Adrenocorticotropic (ă-drē′nō-kōr′ti-kō-trō′pik) **hormone (ACTH)** is one of several anterior pituitary hormones derived from a precursor molecule called **proopiomelanocortin** (prō-ō′-pē-ō-mel′ă-nō-kōr′tin). The major products of this large molecule are ACTH, lipotropins, β endorphin, and melanocyte-stimulating hormone.

ACTH functions to increase the secretion of hormones, primarily cortisol, from the adrenal cortex. ACTH and melanocyte-stimulating hormone also bind to melanocytes in the skin and increase skin pigmentation (see chapter 5). In pathologic conditions such as Addison's disease, blood levels of ACTH and related hormones are chronically elevated, and the skin becomes markedly darker.

The **lipotropins** (li-pō-trō′pinz) secreted from the anterior pituitary bind to membrane-bound receptor molecules on adipose tissue cells. They cause fat breakdown and the release of fatty acids into the circulatory system.

The **beta endorphins** (en′-dōr-phinz) have the same effects as opiate drugs such as morphine, and they can play a role in analgesia in response to stress and exercise. Other functions have been proposed for the β endorphins, including regulation of body temperature, food intake, and water balance. Both ACTH and β-endorphin secretions increase in response to stress and exercise.

Melanocyte-stimulating hormone (MSH) binds to membrane-bound receptors on skin melanocytes and stimu-

lates increased melanin deposition in the skin. The regulation of MSH secretion and its function in humans is not well understood, although it is an important regulator of skin pigmentation in some other vertebrates.

Gonadotropins and Prolactin

Gonadotropins (gō′năd-ō-trō′pinz) are hormones capable of promoting growth and function of the **gonads,** which include the ovaries and testes. The two major gonadotropins secreted from the anterior pituitary are **luteinizing** (lū′tē-ĭ-nīz-ing) **hormone (LH),** and **follicle-stimulating hormone (FSH).** LH, FSH, and another anterior pituitary hormone called **prolactin** (prō-lak′tin) play important roles in regulating reproduction.

LH and FSH secreted into the blood bind to membrane-bound receptors, increase the intracellular synthesis of cyclic adenosine monophosphate (cAMP) through G protein mechanisms, and stimulate the production of **gametes** (gam-ēts—sperm cells in the testes and oocytes in ovaries. LH and FSH also control the production of reproductive hormones, called estrogens and progesterone in the ovaries and testosterone in the testes.

LH and FSH are released from the anterior pituitary cells under the influence of a single hypothalamic-releasing hormone called **gonadotropin-releasing hormone (GnRH).** GnRH is also called **luteinizing hormone-releasing hormone (LHRH).**

Prolactin plays an important role in milk production in the mammary gland of lactating females. It can increase the

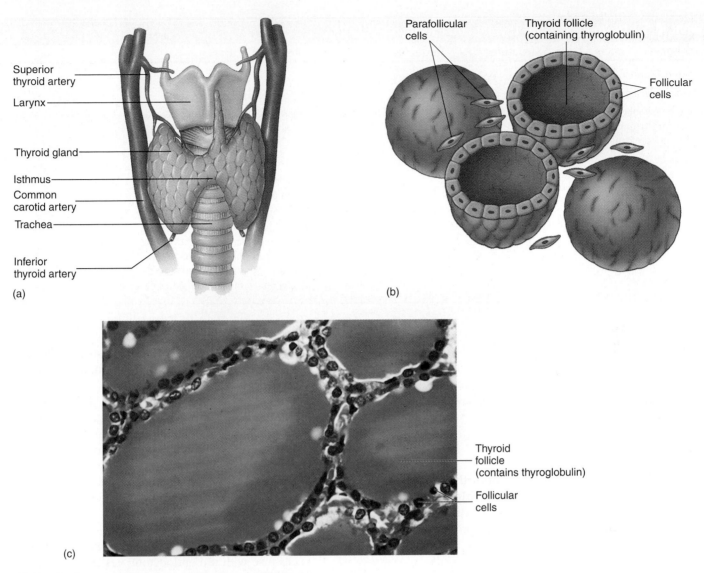

Figure 18.7   Anatomy and Histology of the Thyroid Gland

(*a*) Frontal view of the thyroid gland. (*b*) Histology of the thyroid gland. The thyroid gland is made up of many spheric thyroid follicles containing thyroglobulin. Parafollicular cells are in the tissue between the thyroid follicles. (*c*) Low-power photomicrograph of thyroid follicles.

number of receptor molecules for FSH and LH in the ovaries, and prolactin therefore has a permissive effect for FSH and LH on the ovary. It can also enhance progesterone secretion of the ovary after ovulation. No role for prolactin has been clearly established in males. Several hypothalamic neurohormones can be involved in the complex regulation of prolactin secretion. One neurohormone is **prolactin-releasing hormone (PRH),** and another is **prolactin-inhibiting hormone (PIH).** The regulation of gonadotropin and prolactin secretion and their specific effects are explained more fully in chapter 28.

Thyroid Gland

The **thyroid gland** is composed of two lobes connected by a narrow band of thyroid tissue called the **isthmus.** The lobes are lateral to the upper portion of the trachea just inferior to

the larynx, and the isthmus extends across the anterior aspect of the trachea (figure 18.7*a*). The thyroid gland is one of the largest endocrine glands, with a weight of approximately 20 g. It is highly vascular and appears more red than its surrounding tissues.

Histology

The thyroid gland contains numerous **follicles,** which are small spheres whose walls are composed of a single layer of cuboidal epithelial cells (figure 18.7*b*). The center, or lumen, of each thyroid follicle is filled with a protein called **thyroglobulin** (thī-rō-glob′yū-lin) to which thyroid hormones are bound. The thyroglobulin stores large amounts of thyroid hormone.

Between the follicles a delicate network of loose connective tissue contains numerous capillaries. Scattered

Table 18.3   Hormones of the Thyroid and Parathyroid Glands

Hormones	Structure	Target Tissue	Response
Thyroid Gland			
Thyroid Follicles			
Thyroid hormones (triiodothyronine and tetraiodothyronine)	Amino acid derivative	Most cells of the body	Increased metabolic rate; essential for normal process of growth and maturation
Parafollicular Cells			
Calcitonin	Polypeptide	Bone	Decreased rate of breakdown of bone by osteoclasts; prevention of a large increase in blood calcium levels
Parathyroid			
Parathyroid hormone	Peptide	Bone; kidney; small intestine	Increased rate of breakdown of bone by osteoclasts; increased reabsorption of calcium in kidneys, increased absorption of calcium from the small intestine, increased vitamin D synthesis, increased blood calcium levels

parafollicular (par'ă-fo-lik'yū-lăr) **cells** are found between the follicles and among the cells that make up the walls of the follicle. **Calcitonin** (kal-si-tō'nin) is secreted from the parafollicular cells and plays a role in reducing the concentration of calcium in the body fluids when calcium levels become elevated.

Thyroid Hormones

The thyroid hormones include both **triiodothyronine** (trī-ī'ō-dō-thī'rō-nēn; T_3) and **tetraiodothyronine** (tet'ră-ī'ō-dō-thī'rō-nēn; T_4). T_4 is also called **thyroxine** (thī-rok'sin). These substances constitute the major secretory products of the thyroid gland, consisting of 10% T_3 and 90% T_4 (table 18.3).

Thyroid Hormone Synthesis

TSH from the anterior pituitary must be present to maintain thyroid hormone synthesis and secretion. TSH binds to membrane-bound receptors on cells of the thyroid follicles, and increases cAMP synthesis through a G protein mechanism. The cAMP mediates the response of thyroid cells to TSH (see chapter 17). An adequate amount of iodine in the diet also is required. The following events in the thyroid follicles result in thyroid hormone synthesis and secretion:

1. Iodide (I^-) ions are actively absorbed by the cells of the thyroid follicle. The transport of the I^- ions is against a concentration gradient of approximately 30-fold in healthy individuals.
2. Large proteins called **thyroglobulins** (thī-rō-glob'yū-linz), which contain numerous tyrosine amino acid molecules, are synthesized within the cells of the follicle.
3. Nearly simultaneously, the I^- ions are oxidized to form iodine (I) and one or two iodine atoms are bound chemically to tyrosine molecules of thyroglobulin. This occurs close to the time the thyroglobulin molecules are secreted by the process of exocytosis into the lumen of the follicle. The secreted thyroglobulin contains the iodinated tyrosines.
4. In the lumen of the follicle two diiodotyrosine molecules are combined to form tetraiodothyronine (T_4), and one monoiodotyrosine and one diiodotyrosine molecule are combined to form triiodothyronine (T_3). Large amounts of T_3 and T_4 are stored within the thyroid follicles as components of thyroglobulin. A reserve sufficient to supply thyroid hormones for approximately 2 weeks is stored in this form.
5. Thyroglobulin is taken into the thyroid cells by endocytosis, in which lysosomes within the follicular cells fuse with the endocytotic vesicles.
6. Proteolytic enzymes break down thyroglobulin to release T_3 and T_4, which then diffuse from the follicular cells into the interstitial spaces and finally into the capillaries of the thyroid gland. The remaining amino acids of thyroglobulin are used to synthesize more thyroglobulin.

The biosynthesis of the thyroid hormones is outlined in figure 18.8.

Transport in the Blood

Thyroid hormones are transported in combination with plasma proteins in the circulatory system. Approximately 70%–75% of the circulating T_3 and T_4 are bound to **thyroxine-binding globulin (TBG),** which is synthesized by the liver. T_3 and T_4, bound to plasma protein, form a large reservoir of circulating thyroid hormones, and the half-life of these hormones is increased greatly because of this binding. After thyroid gland removal in experimental animals, it takes approximately 1 week for the T_3 and T_4 levels in the blood to decrease by 50%. As free T_3 and T_4 levels decrease in the interstitial spaces, additional T_3 and T_4 dissociate from

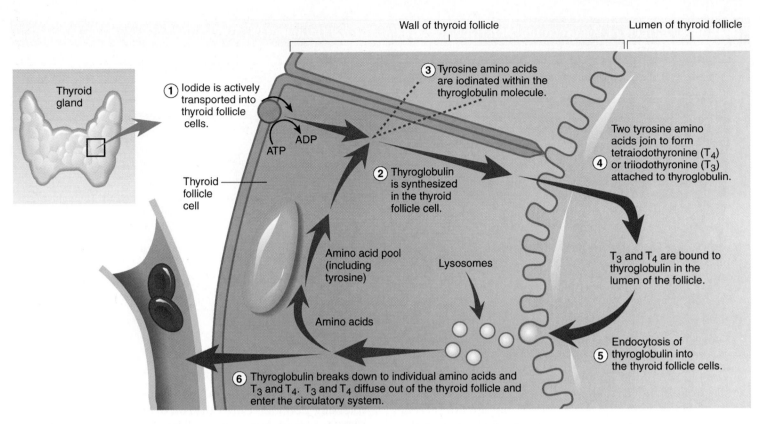

Figure 18.8 Biosynthesis of Thyroid Hormones

The numbered steps describe the synthesis and the secretion of thyroid hormones from the thyroid gland. See text for details of each numbered step.

plasma proteins to maintain the levels in the tissue spaces. When sudden secretion of T_3 and T_4 occurs, the excess is taken up by the plasma proteins. As a consequence, the concentration of thyroid hormones in the tissue spaces fluctuates very little.

Approximately 33%–40% of the T_4 is converted to T_3 in the body tissues. This conversion can be important in the action of thyroid hormones on their target tissues because T_3 is the major hormone that interacts with the target cells. In addition, T_3 is several times more potent than T_4.

Much of the circulating T_4 is eliminated from the body by being converted to tetraiodothyroacetic acid and then excreted in the urine or bile. In addition, a large amount is converted to an inactive form of T_3 and rapidly metabolized and excreted.

Mechanism of Action of Thyroid Hormones

Thyroid hormones interact with their target tissues in a fashion similar to that of the steroid hormones. They readily diffuse through cell membranes into the cytoplasm of cells. Within cells they bind to receptor molecules in the nuclei. Thyroid hormones combined with their receptor molecules interact with deoxyribonucleic acid in the nucleus to influence regulatory genes and initiate new protein synthesis. The newly synthesized proteins within the target cells mediate the response of the cells to thyroid hormones. It takes up to a week after the administration of thyroid hormones for a maximal response to develop, and new protein synthesis occupies much of that time.

Effects of Thyroid Hormones

Thyroid hormones affect nearly every tissue in the body, but not all tissues respond identically. Metabolism is primarily affected in some tissues, and growth and maturation are influenced in others.

The normal rate of metabolism for an individual depends on an adequate supply of thyroid hormone, which increases the rate at which glucose, fat, and protein are metabolized, which in turn increases body temperature. Thyroid hormones can alter the function of mitochondria, resulting in greater adenosine triphosphate production and a greater rate of heat production. Low levels of thyroid hormones lead to the opposite effect. Normal body temperature depends on an adequate amount of thyroid hormone.

Normal growth and maturation of organs also depend on thyroid hormones. For example, bone, hair, teeth, connective tissue, and nervous tissue require thyroid hormone for normal growth and development. Also, T_3 and T_4 play a

Table 18.4   Effects of Hyposecretion and Hypersecretion of Thyroid Hormones

Hypothyroidism	Hyperthyroidism
Decreased metabolic rate, low body temperature, cold intolerance	Increased metabolic rate, high body temperature, heat intolerance
Weight gain, reduced appetite	Weight loss, increased appetite
Reduced activity of sweat and sebaceous glands, dry and cold skin	Copious sweating, warm and flushed skin
Reduced heart rate, reduced blood pressure, dilated and enlarged heart	Rapid heart rate, elevated blood pressure, abnormal electrocardiogram
Weak, flabby skeletal muscles, sluggish movements	Weak skeletal muscles that exhibit tremors, quick movements with exaggerated reflexes
Constipation	Bouts of diarrhea
Myxedema (swelling of the face and body) as a result of mucoprotein deposits	Exophthalmos (protruding of the eyes) as a result of mucoprotein and other deposits behind the eye
Apathetic, somnolent	Hyperactivity, insomnia, restlessness, irritability, short attention span
Coarse hair, rough and dry skin	Soft, smooth hair and skin
Decreased iodide uptake	Increased iodide uptake
Possible goiter (enlargement of the thyroid gland)	Almost always develops goiter

permissive role for GH, and GH does not have its normal effect on target tissues if the thyroid hormones are not present.

The specific effects of hyposecretion and hypersecretion of thyroid hormones are outlined in table 18.4. Hypersecretion of thyroid hormone increases the rate of metabolism. High body temperature, weight loss, increased appetite, rapid heart rate, elevated blood pressure, and an enlarged thyroid gland are major symptoms.

Hyposecretion of thyroid hormone decreases the rate of metabolism. Low body temperature, weight gain, reduced appetite, reduced heart rate, reduced blood pressure, weak skeletal muscles, and apathy are major symptoms of hyposecretion of thyroid hormones. If hyposecretion of thyroid hormones occurs during development in addition to a decreased rate of metabolism, abnormal nervous system development, abnormal growth, and abnormal maturation of tissues occur. The consequence is a mentally retarded person of short stature and distinctive form called a **cretin** (krē'tin).

Regulation of Thyroid Hormone Secretion

The most important regulator of thyroid hormone secretion is TSH. Small fluctuations occur in blood TSH levels on a daily basis, with a small nocturnal increase. The effects of increased TSH levels are increased synthesis and secretion of T_3 and T_4 and hypertrophy (increased cell size) and hyperplasia (increased cell number) of the thyroid gland. Decreased blood levels of TSH lead to decreased T_3 and T_4 secretion and thyroid gland atrophy. Figure 18.9 illustrates the regulation of thyroid hormone secretion by TSH and the negative-feedback mechanisms that influence TSH secretion. Note that T_3 and T_4 inhibit TSH secretion by acting on the anterior pituitary and hypothalamus. If the thyroid gland is removed or if T_3 and T_4 secretion declines, TSH levels increase dramatically.

The secretion of TSH is initiated by a hypothalamic-releasing hormone called **thyrotropin-releasing hormone (TRH).** Several phenomena affect TRH release from the hypothalamus. Exposure to cold and stress cause increased TRH secretion. As a consequence, TSH and thyroid hormone levels increase. In contrast, prolonged fasting decreases TRH secretion, which leads to a decrease in thyroid hormone levels in the circulation. The thyroid hormones play a role in acclimation to chronic exposure to cold, stress, and food deprivation. In all of these conditions the response is most important if the condition is relatively chronic because approximately 1 week is required before a maximum change in metabolism occurs in response to the thyroid hormones.

Abnormal thyroid conditions are outlined in table 18.5.

3   P R E D I C T

An enlargement of the thyroid gland, called a goiter, develops when there is too little iodine in the diet. Explain why the thyroid gland enlarges in response to iodine deficiency in the diet.

✔ *Answer in Appendix F*

Calcitonin

The parafollicular cells of the thyroid gland, which secrete calcitonin, are dispersed between the thyroid follicles throughout the thyroid gland. The major stimulus for increased calcitonin secretion is an increase in calcium levels in the body fluids.

The primary target tissue for calcitonin is bone (see chapter 6). Calcitonin binds to membrane-bound receptors, decreases osteoclast activity, and lengthens the life span of osteoblasts. The result is a decrease in blood calcium and phosphate levels caused by increased bone deposition.

1. Thyroid-releasing hormone (TRH) is released from neurons within the hypothalamus and passes through the hypothalamohypophyseal portal blood vessels to the anterior pituitary.

2. TRH causes cells of the anterior pituitary to secrete thyroid-stimulating hormone (TSH).

3. TSH passes through the general circulation to the thyroid gland, where it causes both increased synthesis and secretion of thyroid hormones (T_3 and T_4).

4. T_3 and T_4 have an inhibitory effect on the secretion of TRH from the hypothalamus and TSH from the anterior pituitary.

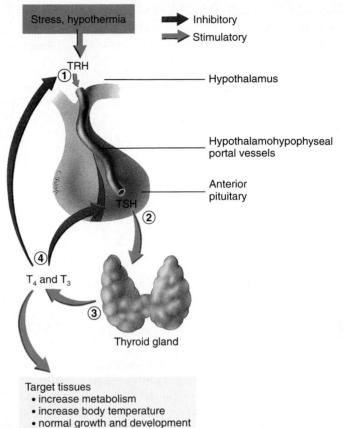

Stress, hypothermia

➡ Inhibitory
➡ Stimulatory

TRH
①
Hypothalamus

Hypothalamohypophyseal portal vessels

Anterior pituitary

TSH
②

T_4 and T_3
④

③
Thyroid gland

Target tissues
• increase metabolism
• increase body temperature
• normal growth and development

Figure 18.9   Regulation of Thyroid Hormone (T_3 and T_4) Secretion

Table 18.5	Abnormal Thyroid Conditions
Cause	**Description**
Hypothyroidism	
Iodine deficiency	Causes inadequate thyroid hormone synthesis, which results in elevated thyroid-stimulating hormone (TSH) secretion; thyroid gland enlarges (goiter) as a result of TSH stimulation; thyroid hormones frequently remain in the low-to-normal range
Goiterogenic substances	Found in certain drugs and in small amounts in certain plants such as cabbage; inhibit thyroid hormone synthesis
Cretinism	Caused by maternal iodine deficiency or congenital errors in thyroid hormone synthesis; results in mental retardation and a short, grotesque appearance
Lack of thyroid gland	Removed surgically or destroyed as a treatment for Graves' disease (hyperthyroidism)
Pituitary insufficiency	Results from lack of TSH secretion; often associated with inadequate secretion of other adenohypophyseal hormones
Hashimoto's disease	Autoimmune disease in which thyroid function is normal or depressed
Hyperthyroidism	
Graves' disease	Characterized by goiter and exophthalmos; apparently an autoimmune disease; most patients have long-acting thyroid stimulator, a TSH-like immune globulin, in their plasma
Tumors—benign adenoma or cancer	Result in either normal secretion or hypersecretion of thyroid hormones (rarely hyposecretion)
Thyroiditis—a viral infection	Produces painful swelling of the thyroid gland with normal or slightly increased thyroid hormone production
Elevated TSH levels	Result from a pituitary tumor
Thyroid storm	Sudden release of large amounts of thyroid hormones; caused by surgery, stress, infections, and unknown reasons

The importance of calcitonin in the regulation of blood calcium levels is unclear. Its rate of secretion increases in response to elevated blood calcium levels, and it may function to prevent large increases in blood calcium levels following a meal. Blood levels of calcitonin decrease with age to a greater extent in females than males. Osteoporosis increases with age and occurs to a greater degree in females than males. Complete thyroidectomy does not result in high blood calcium levels, however. It is possible that the regulation of blood calcium levels by other hormones, such as parathyroid hormone and vitamin D, compensates for the loss of calcitonin in individuals who have undergone a thyroidectomy. No pathologic condition is associated directly with a lack of calcitonin secretion.

Parathyroid Glands

The **parathyroid** (par-ă-thī′royd) **glands** are usually embedded in the posterior part of each lobe of the thyroid gland. There are usually four parathyroid glands, with their cells organized in densely packed masses or cords rather than in follicles (figure 18.10).

The parathyroid glands secrete **parathyroid hormone (PTH),** a polypeptide hormone that is important in the regulation of calcium levels in body fluids (see table 18.3). Bone, the kidneys, and the intestine are its major target tissues. Parathyroid hormone binds to membrane-bound receptors, activating a G protein mechanism that increases intracellular cAMP levels in target tissues. Without functional parathyroid glands, the ability to adequately regulate blood calcium levels is lost.

PTH stimulates osteoclast activity in bone and can cause the number of osteoclasts to increase. The increased osteoclast activity results in bone resorption and the release of calcium and phosphate, causing an increase in blood calcium levels. There are no PTH receptors on osteoclasts, but PTH receptors are present on osteoblasts and on cells giving rise to osteoblasts. Calcium resorption from bone may depend on the response of osteoblasts to PTH, because PTH does not cause the resorption of bone unless osteoblasts are present. The increase in osteoclast activity may be an indirect effect, possibly resulting from the release of substances from osteoblasts in response to exposure to PTH.

PTH induces calcium reabsorption within the kidneys so that less calcium leaves the body in urine. It also increases the enzymatic formation of active vitamin D in the kidneys. Calcium is actively absorbed by the epithelial cells of the small intestine, and the synthesis of transport proteins in the intestinal cells requires active vitamin D. PTH increases the rate of active vitamin D synthesis, which in turn increases the rate of calcium and phosphate absorption in the intestine, elevating blood levels of calcium.

Although PTH increases the release of phosphate ions from bone and increases phosphate ion absorption in the gut, it increases phosphate ion excretion in the kidney. The overall effect of PTH is to decrease blood phosphate levels.

The regulation of PTH secretion is outlined in figure 18.11. The primary stimulus for the secretion of PTH is a decrease in

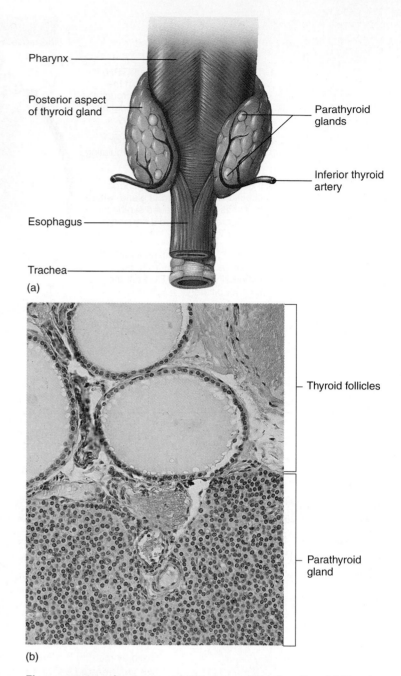

(a)

(b)

Figure 18.10   Anatomy and Histology of the Parathyroid Glands

(*a*) The parathyroid glands are embedded in the posterior part of the thyroid gland. (*b*) The parathyroid glands are composed of densely packed cords of cells.

plasma calcium levels, whereas elevated plasma calcium levels inhibit PTH secretion. This regulation keeps calcium blood levels fluctuating within a normal range of values. Both hypersecretion and hyposecretion of PTH cause serious symptoms (table 18.6).

4   P R E D I C T

Predict the effect of an inadequate dietary intake of calcium on PTH secretion and on target tissues for PTH.

✔ *Answer in Appendix F*

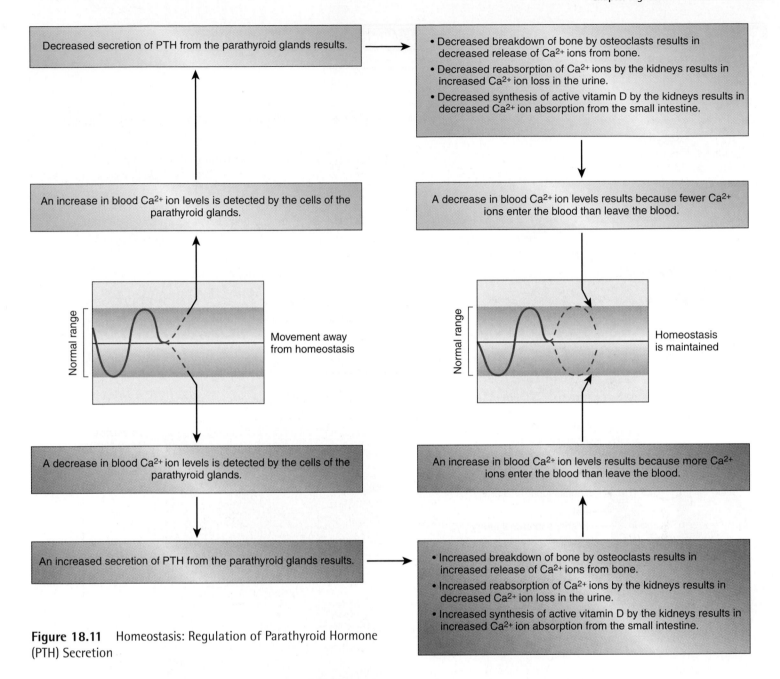

Decreased secretion of PTH from the parathyroid glands results.

- Decreased breakdown of bone by osteoclasts results in decreased release of Ca^{2+} ions from bone.
- Decreased reabsorption of Ca^{2+} ions by the kidneys results in increased Ca^{2+} ion loss in the urine.
- Decreased synthesis of active vitamin D by the kidneys results in decreased Ca^{2+} ion absorption from the small intestine.

An increase in blood Ca^{2+} ion levels is detected by the cells of the parathyroid glands.

A decrease in blood Ca^{2+} ion levels results because fewer Ca^{2+} ions enter the blood than leave the blood.

Normal range

Movement away from homeostasis

Normal range

Homeostasis is maintained

A decrease in blood Ca^{2+} ion levels is detected by the cells of the parathyroid glands.

An increase in blood Ca^{2+} ion levels results because more Ca^{2+} ions enter the blood than leave the blood.

An increased secretion of PTH from the parathyroid glands results.

- Increased breakdown of bone by osteoclasts results in increased release of Ca^{2+} ions from bone.
- Increased reabsorption of Ca^{2+} ions by the kidneys results in decreased Ca^{2+} ion loss in the urine.
- Increased synthesis of active vitamin D by the kidneys results in increased Ca^{2+} ion absorption from the small intestine.

Figure 18.11   Homeostasis: Regulation of Parathyroid Hormone (PTH) Secretion

Inactive parathyroid glands result in hypocalcemia. Reduced extracellular calcium levels cause voltage-gated Na^+ ion channels in cell membranes to open, which increases the permeability of cell membranes to Na^+ ions. As a consequence, Na^+ ions diffuse into cells and cause depolarization (see chapter 9). Symptoms of hypocalcemia are nervousness, muscle spasms, cardiac arrhythmias, and convulsions. In extreme cases tetany of the respiratory muscles can cause death.

5   **P R E D I C T**

A patient with a malignant tumor had his thyroid gland removed. What effect would this removal have on blood levels of thyroid hormone, TRH, TSH, and calcitonin? What would result if the parathyroid glands were inadvertently removed during surgery?

✔ *Answer in Appendix F*

Adrenal Glands

The **adrenal** (ă-drē′năl) **glands,** also called the **suprarenal** (sū′pră-rē′năl) **glands,** are near the superior pole of each kidney. Like the kidneys, they are retroperitoneal, and they are surrounded by abundant adipose tissue. The adrenal glands are enclosed by a connective tissue capsule and have a well-developed blood supply (figure 18.12*a*).

The adrenal glands are composed of an inner **medulla** and an outer **cortex,** which are derived from two separate embryonic tissues. The adrenal medulla arises from neural crest cells, which also give rise to postganglionic neurons of the sympathetic division of the autonomic nervous system (see chapter 16). Unlike most glands of the body, which develop from invaginations of epithelial tissue, the adrenal cortex is derived from mesoderm.

Table 18.6 Causes and Symptoms of Hypersecretion and Hyposecretion of Parathyroid Hormone

Hypoparathyroidism	Hyperparathyroidism
Causes	
Accidental removal during thyroidectomy	Primary hyperparathyroidism: a result of abnormal parathyroid function —adenomas of the parathyroid gland (90%), hyperplasia of parathyroid Idiopathic (unknown cause) cells (9%), and carcinomas (1%) Secondary hyperparathyroidism: caused by conditions that reduce blood calcium levels, such as inadequate calcium in the diet, inadequate levels of vitamin D, pregnancy, or lactation
Symptoms	
Hypocalcemia	Hypercalcemia or normal blood calcium levels; calcium carbonate salts may be deposited throughout the body, especially in the renal tubules (kidney stones), lungs, blood vessels, and gastric mucosa
Normal bone structure	Bones weak and eaten away as a result of resorption; some cases are first diagnosed when a radiograph is taken of a broken bone
Increased neuromuscular excitability; tetany, laryngospasm, and death from asphyxiation can result	Neuromuscular system less excitable; muscular weakness may be present
Flaccid heart muscle; cardiac arrhythmia may develop	Increased force of contraction of cardiac muscle; at very high levels of calcium, cardiac arrest during contraction is possible
Diarrhea	Constipation

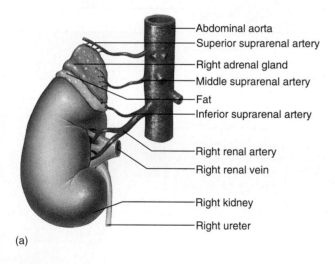

(a)

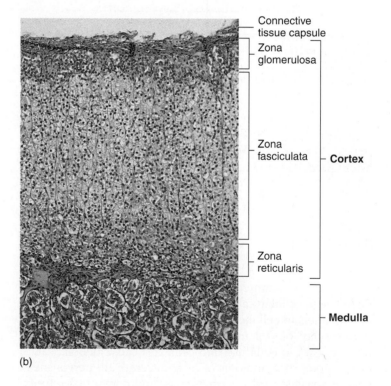

(b)

Figure 18.12 Anatomy and Histology of the Adrenal Gland

(*a*) An adrenal gland is at the superior pole of each kidney. (*b*) The adrenal glands have an outer cortex and an inner medulla. The cortex is surrounded by a connective tissue capsule and consists of three layers: the zona glomerulosa, the zona fasciculata, and the zona reticularis. 𝕏

Histology

Trabeculae of the connective tissue capsule penetrate into the adrenal gland in several locations, and numerous small blood vessels course with them to supply the gland. The medulla consists of closely packed polyhedral cells centrally located in the gland (figure 18.12*b*). The cortex is composed of smaller cells and forms three indistinct layers: the **zona glomerulosa** (glō-mār′yū-lōs-ă), the **zona fasciculata** (fă-sik′yū-lă-tă), and the **zona reticularis** (re-tik′yū-lăr′is). These three layers are functionally and structurally specialized. The zona glomerulosa is immediately beneath the capsule and is composed of small clusters of cells. Beneath the zona glomerulosa is the thickest part of the adrenal cortex,

Table 18.7   Hormones of the Adrenal Gland

Hormones	Structure	Target Tissue	Response
Adrenal Medulla			
Epinephrine primarily; norepinephrine	Amino acid derivatives	Heart, blood vessels, liver, fat cells	Increased cardiac output; increased blood flow to skeletal muscles and heart; increased release of glucose and fatty acids into blood; in general, preparation for physical activity
Adrenal Cortex			
Cortisol	Steroid	Most tissues	Increased protein and fat breakdown; increased glucose production; inhibition of immune response
Aldosterone	Steroid	Kidney	Increased sodium ion reabsorption, and potassium and hydrogen ion excretion
Sex steroids (primarily androgens)	Steroids	Many tissues	Minor importance in males; in females, development of some secondary sexual characteristics such as axillary and pubic hair

the zona fasciculata. In this layer the cells form long columns, or fascicles, of cells that extend from the surface toward the medulla of the gland. The deepest layer of the adrenal cortex is the zona reticularis, which is a thin layer of irregularly arranged cords of cells.

Hormones of the Adrenal Medulla

The adrenal medulla secretes two major hormones: **epinephrine** (**adrenaline**; ă-dren′ă-lin), 80%, and **norepinephrine** (**noradrenaline;** nor-ă-dren′ă-lin), 20% (table 18.7). Epinephrine and norepinephrine are closely related to each other. In fact, norepinephrine is a precursor to the formation of epinephrine. Because the adrenal medulla consists of cells derived from the same cells that give rise to postganglionic sympathetic neurons, its secretory products are neurohormones.

Epinephrine increases blood levels of glucose. It combines with membrane-bound receptors in the liver cells and activates cAMP synthesis within the cells. Cyclic AMP, in turn, activates enzymes that catalyze the breakdown of glycogen to glucose, causing its release into the blood. Epinephrine also increases glycogen breakdown, the intracellular metabolism of glucose in skeletal muscle cells, and the breakdown of fats in adipose tissue. Epinephrine and norepinephrine increase the heart's rate and force of contraction and cause blood vessels to constrict in the skin, kidneys, gastrointestinal tract, and other viscera. Also, epinephrine causes dilation of blood vessels in skeletal muscles and cardiac muscle.

Secretion of adrenal medullary hormones prepares the individual for physical activity and is a major component of the fight-or-flight response (see chapter 16). The response results in reduced activity in organs not essential for physical activity and in increased blood flow and metabolic activity in organs that participate in physical activity. In addition, it mobilizes nutrients that can be used to sustain physical exercise.

The effects of epinephrine and norepinephrine are short-lived because they are rapidly metabolized, excreted, or taken up by tissues. Their half-life in the circulatory system is measured in minutes.

Regulation

The release of adrenal medullary hormones primarily occurs in response to stimulation by sympathetic neurons because the adrenal medulla is a specialized part of the autonomic nervous system. Several conditions, including emotional excitement, injury, stress, exercise, and low blood glucose levels, lead to the release of adrenal medullary neurohormones (figure 18.13).

Clinical Note

The two major disorders of the adrenal medulla are tumors: pheochromocytoma (fē′ō-krō′mō-sī-tō′mă), a benign tumor, and neuroblastoma (nūr′ō-blas-tō′mă), a malignant tumor. Symptoms result from the release of large amounts of epinephrine and norepinephrine and include hypertension (high blood pressure), sweating, nervousness, pallor, and tachycardia (rapid heart rate). The high blood pressure results from the effect of these hormones on the heart and blood vessels and is correlated with an increased chance of heart disease and stroke.

Hormones of the Adrenal Cortex

The adrenal cortex secretes three hormone types: **mineralocorticoids** (min′er-al-ō-kōr′ti-koydz), **glucocorticoids** (glū′kō-kōr′ti-koydz), and **androgens** (an′drō-jenz) (see table 18.7). All are similar in structure in that they are steroids, highly specialized lipids that are derived from cholesterol. Because they are lipid-soluble, they are not stored in the adrenal gland cells but diffuse from the cells as they are synthesized. Adrenal cortical hormones are transported in the blood in combination with specific plasma proteins; they are metabolized in the liver and excreted in the bile and urine. The hormones of the adrenal cortex bind to intracellular receptors, stimulating the synthesis of specific proteins that are responsible for producing the cell's responses.

Mineralocorticoids

The major secretory products of the zona glomerulosa are the mineralocorticoids. **Aldosterone** (al-dos′ter-ōn) is produced

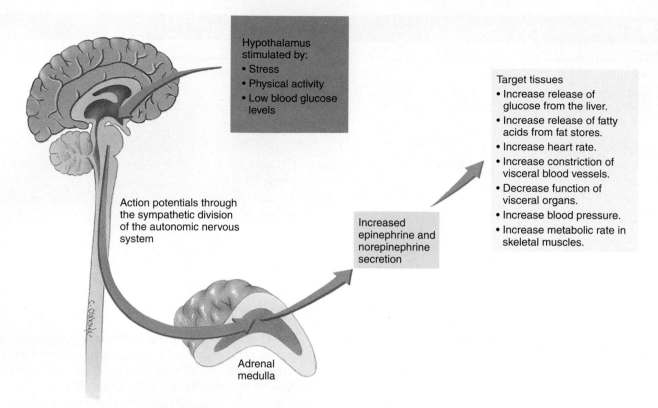

Figure 18.13   Regulation of Adrenal Medullary Secretions

Stress, physical exercise, and low blood glucose levels cause increased activity of the sympathetic nervous system, which increases epinephrine and norepinephrine secretion from the adrenal medulla.

in the greatest amounts, although other closely related mineralocorticoids are also secreted. Aldosterone increases the rate of sodium reabsorption by the kidneys, thereby increasing blood levels of sodium. Sodium reabsorption results in increased water reabsorption by the kidneys and an increase in blood volume. Aldosterone increases potassium excretion into the urine by the kidneys, thereby decreasing blood levels of potassium. It also increases the rate of hydrogen ion excretion into the urine, and, when present in high concentrations, it can result in alkalosis (elevated pH of body fluids). The details of the effects of aldosterone and the mechanisms controlling aldosterone secretion are discussed along with kidney functions in chapters 26 and 27 and with the cardiovascular system in chapter 21.

Glucocorticoids

The zona fasciculata of the adrenal cortex primarily secretes glucocorticoid hormones, and the major one is **cortisol** (kōr′ti-sol). The target tissues and responses to the glucocorticoids are numerous (table 18.8). The responses are classified as metabolic, developmental, or anti-inflammatory. Glucocorticoids increase fat catabolism, decrease glucose and amino acid uptake in skeletal muscle, increase **gluconeogenesis** (glū′kō-nē-ō-jen′ě-sis, which is the synthesis of glucose from precursor molecules such as amino acids in the liver), and increase protein degradation. Thus some major effects of glucocorticoids are an increase in fat and protein metabolism and an increase in blood glucose levels and glycogen deposits in cells. As a result, a reservoir of molecules that can be metabolized rapidly is available to cells. Glucocorticoids are also required for the maturation of tissues such as fetal lungs and for the development of receptor molecules in target tissues for epinephrine and norepinephrine. Glucocorticoids decrease the intensity of the inflammatory response by decreasing both the number of white blood cells and the secretion of inflammatory chemicals from tissues. This anti-inflammatory effect is most important under conditions of stress when the rate of glucocorticoid secretion is relatively high.

ACTH is required to maintain the secretory activity of the adrenal cortex, which rapidly atrophies without this hormone.

Table 18.8 Target Tissues and Their Responses to Glucocorticoid Hormones

Target Tissues	Responses
Peripheral tissues such as skeletal muscle, liver, and adipose tissue	Inhibits glucose use; stimulates formation of glucose from amino acids and, to some degree, from fats (gluconeogenesis) in the liver, which results in elevated blood glucose levels; stimulates glycogen synthesis in cells; mobilizes fats by increasing lipolysis, which results in the release of fatty acids into the blood and an increased rate of fatty acid metabolism; increases protein breakdown and decreases protein synthesis
Immune tissues	Anti-inflammatory—depresses antibody production, white blood cell production, and the release of inflammatory components in response to injury
Target cells for epinephrine	Receptor molecules for epinephrine and norepinephrine decrease without adequate amounts of glucocorticoid hormone

Clinical Focus Hormone Pathologies of the Adrenal Cortex

Several pathologies are associated with abnormal secretion of adrenal cortex hormones.

Addison's disease results from abnormally low levels of aldosterone and cortisol. The cause of many cases of Addison's disease is unknown, but it is a suspected autoimmune disease in which the body's defense mechanisms inappropriately destroy the adrenal cortex. Some cases of Addison's disease are caused by the destruction of the adrenal cortex by bacteria, such as tuberculosis bacteria, acquired immune deficiency syndrome (AIDS), fungal infections, adrenal hemorrhage, or cancer. It can also be caused by suppression of pituitary gland function by prolonged treatment with glucocorticoids or can result from neoplasms that damage the hypothalamus. Symptoms of Addison's disease include weakness, fatigue, weight loss, anorexia, and in many cases increased pigmentation of the skin. Reduced blood pressure results from the loss of sodium ions and water through the kidney. Reduced blood pressure is the most critical manifestation and requires immediate treatment. Low blood levels of sodium ions, high blood levels of potassium ions, and reduced blood pH are consistent with the condition.

Aldosteronism (al-dos′ter-on-izm) is caused by excess production of aldosterone. Primary aldosteronism results from an adrenal cortex tumor, and secondary aldosteronism occurs when some extraneous factor such as overproduction of renin, a sub-stance produced by the kidney, increases aldosterone secretion. Major symptoms of aldosteronism include reduced blood levels of potassium ions, increased blood pH, and elevated blood pressure. Elevated blood pressure is a result of the retention of water and sodium ions by the kidneys.

Cushing's syndrome (figure A) is a disorder characterized by hypersecretion of cor-

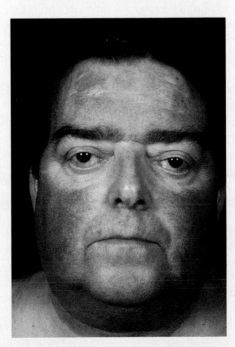

Figure A Male Patient with Cushing's Syndrome

tisol and androgens and possibly by excess aldosterone production. The majority of cases are caused by excess ACTH production by nonpituitary tumors, which usually result from a type of lung cancer, or by pituitary tumors. Sometimes adrenal tumors or unidentified causes can be responsible for hypersecretion of the adrenal cortex without increases in ACTH secretion. Elevated secretion of glucocorticoids results in muscle wasting, the accumulation of adipose tissue in the face and trunk of the body, and increased blood glucose levels.

Hypersecretion of androgens from the adrenal cortex causes a condition called **adrenogenital** (ă-drē′nō-jen′i-tăl) **syndrome,** in which secondary sexual characteristics develop early in male children, and female children are masculinized. If the condition develops before birth in females, the external genitalia can be masculinized to the extent that the infant's reproductive structures can be neither clearly female nor male. Hypersecretion of adrenal androgens in male children before puberty results in rapid and early development of the reproductive system. If not treated, early sexual development and a short stature result. The short stature results from the effect of androgens on skeletal growth. In adult females partial development of male secondary sexual characteristics such as facial hair and a masculine voice occurs.

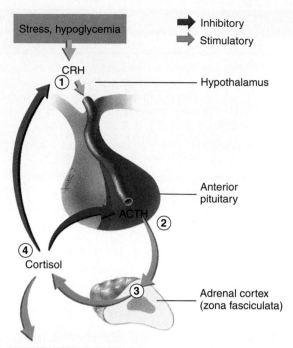

Figure 18.14   Regulation of Cortisol Secretion

→ Inhibitory
→ Stimulatory

Stress, hypoglycemia

CRH
① ——— Hypothalamus

Anterior pituitary

ACTH ②

③ Adrenal cortex (zona fasciculata)

④ Cortisol

Target tissue
• Increase fat and protein breakdown.
• Increase blood glucose levels.
• Have antiinflammatory effects.

1. Cortiocotropin-releasing hormone (CRH) is released from hypothalamic neurons in response to stress or hypoglycemia and passes, by way of the hypothalamohypophyseal portal blood vessels, to the anterior pituitary.

2. In the anterior pituitary CRH binds to and stimulates cells that secrete adrenocorticotropic hormone (ACTH).

3. ACTH binds to membrane-bound receptors on cells of the adrenal cortex and stimulates the secretion of glucocorticoids, primarily cortisol.

4. Cortisol inhibits CRH and ACTH secretion.

ACTH acts on the zona fasiculata to increase aldosterone secretion. The regulation of ACTH and cortisol secretion is outlined in figure 18.14. **Corticotropin-releasing hormone (CRH)** is released from the hypothalamus and stimulates the anterior pituitary to secrete ACTH. ACTH and cortisol inhibit CRH secretion from the hypothalamus and thus constitute a negative-feedback influence on CRH secretion. In addition, high concentrations of cortisol in the blood inhibit ACTH secretion from the anterior pituitary, and low concentrations stimulate it. This negative-feedback loop is important in maintaining blood cortisol levels within a narrow range of concentrations. In response to stress or hypoglycemia, blood levels of cortisol increase rapidly because these stimuli trigger a large increase in CRH release from the hypothalamus. Table 18.9 outlines several abnormalities associated with hypersecretion and hyposecretion of adrenal hormones.

7	P R E D I C T

Cortisone, a drug similar to cortisol, is sometimes given to people who have severe allergies. Taking this substance chronically can damage the adrenal cortex. Explain how this damage can occur.

✔ *Answer in Appendix F*

Adrenal Androgens

Some adrenal steroids, including **androstenedione** (an-drō-stēn′dī-ōn), are weak **androgens.** They are secreted by the zona reticularis and converted by peripheral tissues to the more potent androgen, testosterone. Adrenal androgens stimulate pubic and axillary hair growth and sexual drive in females. Their effects in males are negligible in comparison with testosterone secreted by the testes. Chapter 28 presents additional information about androgens.

█ Pancreas

The **pancreas** (pan′krē-us) lies behind the peritoneum between the greater curvature of the stomach and the duodenum. It is an elongated structure approximately 15 cm long, weighing approximately 85–100 g. The head of the pancreas lies near the duodenum, and its body and tail extend toward the spleen.

Histology

The pancreas is both an exocrine gland and an endocrine gland. The exocrine portion consists of **acini** (as′ĭ-nī), which produce pancreatic juice, and a duct system, which carries pancreatic juice to the small intestine (see chapter 24). The endocrine part, consisting of **pancreatic islets** (islets of Langerhans), (figure 18.15) produces hormones that enter the circulatory system.

Between 500,000 and 1,000,000 pancreatic islets are dispersed among the ducts and acini of the pancreas. Each islet

Table 18.9 Symptoms of Hyposecretion and Hypersecretion of Adrenal Cortex Hormones

Hyposecretion	Hypersecretion
Aldosterone	
Hyponatremia (low blood levels of sodium)	Slight hypernatremia (high blood levels of sodium)
Hyperkalemia (high blood levels of potassium)	Hypokalemia (low blood levels of potassium)
Acidosis	Alkalosis
Low blood pressure	High blood pressure
Tremors and tetany of skeletal muscles	Weakness of skeletal muscles
Polyuria	Acidic urine
Cortisol	
Hypoglycemia (low blood glucose levels)	Hyperglycemia (high blood glucose levels; adrenal diabetes)—leads to diabetes mellitus
Depressed immune system	Depressed immune system
Protein and fats from diet are unused, resulting in weight loss	Destruction of tissue proteins, causing muscle atrophy and weakness, osteoporosis, weak capillaries (easy bruising), thin skin, and impaired wound healing; mobilization and redistribution of fats, causing depletion of fat from limbs and deposition in face (moon face), neck (buffalo hump), and abdomen
Loss of appetite, nausea, and vomiting	Emotional effects, including euphoria and depression
Increased skin pigmentation (caused by elevated ACTH)	
Androgens	
In women reduction of pubic and axillary hair	In women hirsuitism (excessive facial and body hair), acne, increased sex drive, regression of breast tissue, and loss of regular menses

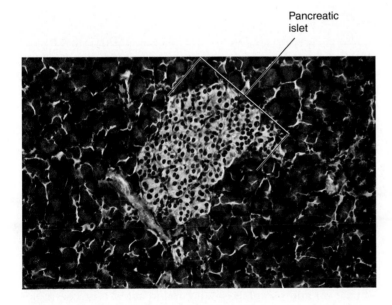

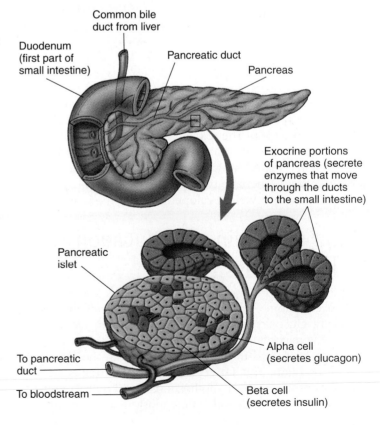

Figure 18.15 Histology of the Pancreatic Islets

A pancreatic islet consists of clusters of specialized cells among the acini of the exocrine portion of the pancreas. The stain used for this slide does not distinguish between alpha and beta cells. 𝒳

Table 18.10 Pancreatic Hormones

Cells In Islets	Hormone	Structure	Target Tissue	Response
Beta (β)	Insulin	Protein	Especially liver, skeletal muscle, fat tissue	Increased uptake and use of glucose and amino acids
Alpha (α)	Glucagon	Polypeptide	Liver primarily	Increased breakdown of glycogen; release of glucose into the circulatory system
Delta (δ)	Somatostatin	Peptide	Alpha and beta cells (some somatostatin is produced in the hypothalamus)	Inhibition of insulin and glucagon secretion

Table 18.11 Effect of Insulin and Glucagon on Target Tissues

Target Tissue	Response to Insulin	Response to Glucagon
Skeletal muscle, cardiac muscle, cartilage, bone, fibroblasts, leukocytes, and mammary glands	Increased glucose uptake and glycogen synthesis; increased uptake of certain amino acids	Little effect
Liver	Increased glycogen synthesis; increased use of glucose for energy (glycolysis)	Causes rapid increase in the breakdown of glycogen to glucose (glycogenolysis) and release of glucose into the blood Increased formation of glucose (gluconeogenesis) from amino acids and, to some degree, from fats Increased metabolism of fatty acids, resulting in increased ketones in the blood
Adipose cells	Increased glucose uptake, glycogen synthesis, fat synthesis, and fatty acid uptake; increased glycolysis	High concentrations cause breakdown of fats (lipolysis); probably unimportant under most conditions
Nervous system	Little effect except to increase glucose uptake in the satiety center	No effect

is composed of **alpha (α) cells** (20%), which secrete glucagon, **beta (β) cells** (75%), which secrete insulin, and other cell types (5%). The remaining cells are either immature cells of questionable function or **delta (δ) cells,** which secrete somatostatin. Nerves from both divisions of the autonomic nervous system innervate the pancreatic islets, and each islet is surrounded by a well-developed capillary network.

Effect of Insulin and Glucagon on Their Target Tissues

The pancreatic hormones play an important role in regulating the concentration of critical nutrients in the circulatory system, especially glucose, or blood sugar, and amino acids (table 18.10). The major target tissues of insulin are the liver, adipose tissue, muscles, and the satiety center within the hypothalamus of the brain. The **satiety** (sa′-tī-ĕ-tē) **center** is a collection of neurons in the hypothalamus that controls appetite, but insulin does not directly affect most areas of the nervous

system. The specific effects of insulin on these target tissues are listed in table 18.11.

Insulin molecules bind to a membrane-bound receptor on its target cells. Before the cells exhibit a response to insulin, specific proteins in the membrane become phosphorylated. Part of the cell's response to glucose is to increase the number of active transport proteins in the membrane of cells for glucose and amino acids. Subsequently, the insulin and receptor molecules are taken through endocytosis into the cell, and the insulin is released from the insulin receptor and broken down.

In general, insulin increases the ability of its target tissue to take up and use glucose and amino acids. Glucose molecules that are not needed immediately as an energy source to maintain cell metabolism are stored as glycogen in skeletal muscle, the liver, and other tissues and are converted to fat in adipose tissue. Amino acids can be broken down and used as an energy source or to synthesize glucose, or they can be converted to protein. Without insulin, the ability of these tissues to accept glucose and amino acids and use them is minimal.

When too much insulin is present, target tissues rapidly take up glucose from the blood, causing blood levels of glucose to decline to very low levels. Although the nervous system, except for cells of the satiety center, is not a target tissue for insulin, the nervous system depends primarily on blood glucose for a nutrient source. Consequently, low blood glucose levels cause the central nervous system to malfunction.

In the absence of insulin, the movement of glucose and amino acids into cells declines dramatically, even though blood levels of these substances can be very high. The satiety center requires insulin to take up glucose. In the absence of insulin, the satiety center cannot detect the presence of glucose in the extracellular fluid even when high levels are present. The result is an intense sensation of hunger in spite of high blood glucose levels.

Glucagon primarily influences the liver, although it has some effect on skeletal muscle and adipose tissue (see table 18.11). In general, glucagon causes the breakdown of glycogen and increased glucose synthesis in the liver. It also increases the breakdown of fats. The amount of glucose released from the liver into the blood increases dramatically after glucagon secretion increases. Because glucagon is secreted into the hepatic portal vein, which carries blood from the intestine and pancreas to the liver, it is delivered in a relatively high concentration to the liver, where it is metabolized rapidly. Thus it has less of an effect on skeletal muscles and adipose tissue.

Regulation of Pancreatic Hormone Secretion

The secretion of insulin is controlled by blood levels of nutrients, neural stimulation, and hormones. Hyperglycemia, or elevated blood levels of glucose, directly affects the beta cells and stimulates insulin secretion. Hypoglycemia, or low blood levels of glucose, directly inhibits insulin secretion. Thus blood glucose levels play a major role in the regulation of insulin secretion. Certain amino acids also stimulate insulin secretion by acting directly on the beta cells. After a meal when glucose and amino acid levels increase in the circulatory system, insulin secretion increases. During periods of fasting when blood glucose levels are low, the rate of insulin secretion declines (figure 18.16).

The autonomic nervous system also controls insulin secretion. Parasympathetic stimulation is associated with food intake, and its stimulation acts with the elevated blood glucose levels to increase insulin secretion. Sympathetic innervation inhibits insulin secretion and helps prevent a rapid fall in blood glucose levels. Because most tissues, except nervous tissue, require insulin to take up glucose, sympathetic stimulation maintains blood glucose levels in a normal range during periods of physical activity or excitement. This response is important for maintaining normal functioning of the nervous system.

Gastrointestinal hormones involved with the regulation of digestion, such as gastrin, secretin, and cholecystokinin (see chapter 24), increase insulin secretion. Somatostatin inhibits insulin and glucagon secretion, but the factors that regulate somatostatin secretion are not clear. It can be released in response to food intake, in which case somatostatin may prevent oversecretion of insulin.

8	P R E D I C T

Explain why the increase in insulin secretion in response to parasympathetic stimulation and gastrointestinal hormones is consistent with the maintenance of blood glucose levels in the circulatory system.

✔ *Answer in Appendix F*

Low blood glucose levels stimulate glucagon secretion, and high blood glucose levels inhibit it. Certain amino acids and sympathetic stimulation also increase glucagon secretion. After a high-protein meal, amino acids increase both insulin and glucagon secretion. Insulin causes target tissues to accept the amino acids for protein synthesis, and glucagon increases the process of glucose synthesis from amino acids in the liver (gluconeogenesis). Both protein synthesis and the use of amino acids to maintain blood glucose levels result from the low, but simultaneous, secretion of insulin and glucagon induced by a high-protein intake.

9	P R E D I C T

Compare the regulation of glucagon and insulin secretion after a meal high in carbohydrates, after a meal low in carbohydrates but high in proteins, and during physical exercise.

✔ *Answer in Appendix F*

Hormonal Regulation of Nutrients

Two different situations—after a meal and during exercise—can illustrate how several hormones function together to regulate blood nutrient levels.

After a meal and under resting conditions, secretion of glucagon, cortisol, GH, and epinephrine is reduced (figure 18.17*a*). The high blood glucose levels and parasympathetic stimulation elevate insulin secretion, increasing the uptake of glucose, amino acids, and fats by target tissues. Substances not immediately used for cell metabolism are stored. Glucose is converted to glycogen in skeletal muscle and the liver and is used for fat synthesis in adipose tissue and the liver. The rapid uptake and storage of glucose prevent too large an increase in blood glucose levels. Amino acids are incorporated into protein, and fats that were ingested as part of the meal are stored in adipose tissue and the liver. If the meal is high in protein, a small amount of glucagon is secreted, increasing the rate at which the liver uses amino acids to form glucose.

Within 1–2 h after the meal, absorption of digested materials from the gastrointestinal tract declines, and blood sugar levels decline (figure 18.17*b*). As a result, secretion of

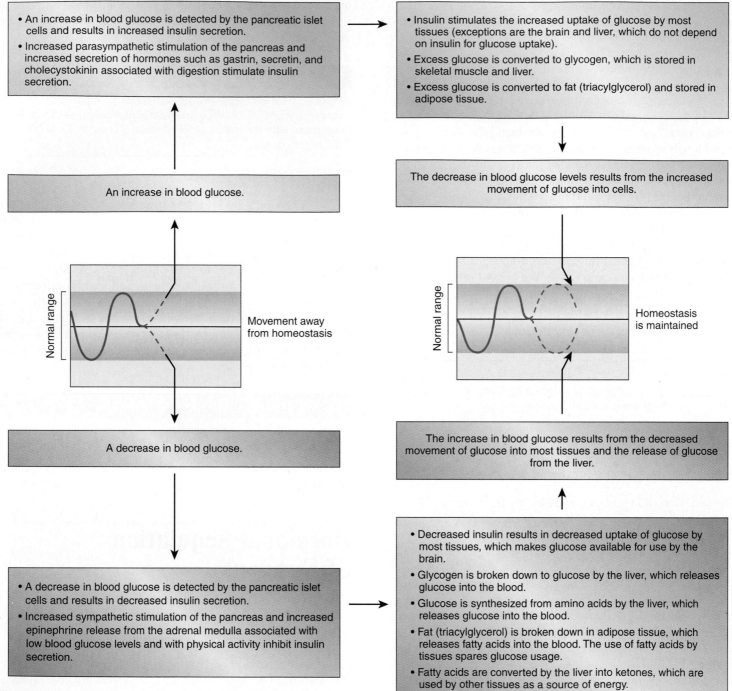

- An increase in blood glucose is detected by the pancreatic islet cells and results in increased insulin secretion.
- Increased parasympathetic stimulation of the pancreas and increased secretion of hormones such as gastrin, secretin, and cholecystokinin associated with digestion stimulate insulin secretion.

- Insulin stimulates the increased uptake of glucose by most tissues (exceptions are the brain and liver, which do not depend on insulin for glucose uptake).
- Excess glucose is converted to glycogen, which is stored in skeletal muscle and liver.
- Excess glucose is converted to fat (triacylglycerol) and stored in adipose tissue.

An increase in blood glucose.

The decrease in blood glucose levels results from the increased movement of glucose into cells.

Normal range

Movement away from homeostasis

Normal range

Homeostasis is maintained

A decrease in blood glucose.

The increase in blood glucose results from the decreased movement of glucose into most tissues and the release of glucose from the liver.

- A decrease in blood glucose is detected by the pancreatic islet cells and results in decreased insulin secretion.
- Increased sympathetic stimulation of the pancreas and increased epinephrine release from the adrenal medulla associated with low blood glucose levels and with physical activity inhibit insulin secretion.

- Decreased insulin results in decreased uptake of glucose by most tissues, which makes glucose available for use by the brain.
- Glycogen is broken down to glucose by the liver, which releases glucose into the blood.
- Glucose is synthesized from amino acids by the liver, which releases glucose into the blood.
- Fat (triacylglycerol) is broken down in adipose tissue, which releases fatty acids into the blood. The use of fatty acids by tissues spares glucose usage.
- Fatty acids are converted by the liver into ketones, which are used by other tissues as a source of energy.

Figure 18.16   Homeostasis: Regulation of Insulin Secretion

glucagon, cortisol, GH, and epinephrine increases, stimulating the release of glucose from tissues. Insulin levels decrease, the rate of glucose entry into the target tissues for insulin decreases, and glycogen is converted back to glucose and released into the blood. The decreased uptake of glucose by most tissues, combined with its release from the liver, helps maintain blood glucose at levels necessary for normal brain function. Cells that use less glucose start using more fats and proteins. Adipose tissue releases fatty acids, and the liver releases triacylglycerol (in lipoproteins) and ketones into the blood. Tissues take up these substances from the blood and

use them for energy. Fat molecules are a major source of energy for most tissues when blood glucose levels are low.

The interactions of insulin, GH, glucagon, epinephrine, and cortisol are excellent examples of negative-feedback mechanisms. When blood sugar levels are high, these hormones cause rapid uptake and storage of glucose, amino acids, and fats. When blood sugar levels are low, they cause release of glucose and a switch to fat and protein metabolism as a source of energy for most tissues.

During exercise, skeletal muscles require energy to support the contraction process (see chapter 10). Although me-

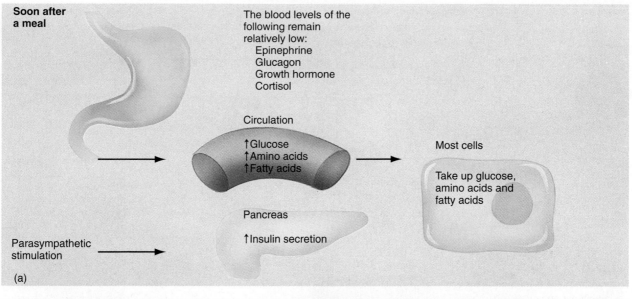

Soon after a meal

The blood levels of the following remain relatively low:
    Epinephrine
    Glucagon
    Growth hormone
    Cortisol

Circulation

↑Glucose
↑Amino acids
↑Fatty acids

Most cells

Take up glucose, amino acids and fatty acids

Pancreas

↑Insulin secretion

Parasympathetic stimulation

(a)

Several hours after a meal

Epinephrine, growth hormone, and cortisol secretion increase

Most cells

Glucose uptake decreases and switch to fat and protein metabolism

Circulation

↓Glucose
↓Amino acids
↓Fatty acids

Liver

Releases glucose, ketones, and triacylglycerols into circulation

Pancreas

↓Insulin secretion
↑Glucagon secretion

Adipose tissue

Releases fatty acids into circulation

Sympathetic stimulation

(b)

Figure 18.17 Regulation of Blood Nutrient Levels After a Meal

(*a*) Soon after a meal, glucose, amino acids, and fatty acids enter the bloodstream from the intestinal tract. Glucose and amino acids stimulate insulin secretion. In addition, parasympathetic stimulation increases insulin secretion. Cells take up the glucose and amino acids and use them in their metabolism. (*b*) Several hours after a meal, absorption from the intestinal tract decreases, and blood levels of glucose, amino acids, and fatty acids decrease. As a result, insulin secretion decreases, and glucagon, epinephrine, and GH secretion increase. Cell uptake of glucose decreases, and usage of fats and proteins increases.

tabolism of intracellular nutrients can sustain muscle contraction for a short time, additional energy sources are required during prolonged activity. Sympathetic nervous system activity, which increases during exercise, stimulates the release of epinephrine from the adrenal medulla and of glucagon from the pancreas (figure 18.18). These hormones induce the conversion of glycogen to glucose in the liver and the release of glucose into the blood, thus providing skeletal muscles with a source of energy. Because epinephrine and glucagon have short half-lives, they can rapidly adjust blood sugar levels for varying conditions of activity.

During sustained activity, glucose release from the liver and other tissues is not adequate to support muscle activity, and there is a danger that blood glucose levels will become too low to support brain function. A decrease in insulin prevents uptake of glucose by most tissues, thus conserving glucose for the brain. Epinephrine, glucagon, cortisol, and GH cause an increase of fatty acids, triacylglycerols, and ketones in the blood. GH also inhibits the breakdown of proteins, preventing muscles from using themselves as an energy source. Consequently, glucose metabolism decreases, and fat metabolism in skeletal muscles increases. At the end of a long race,

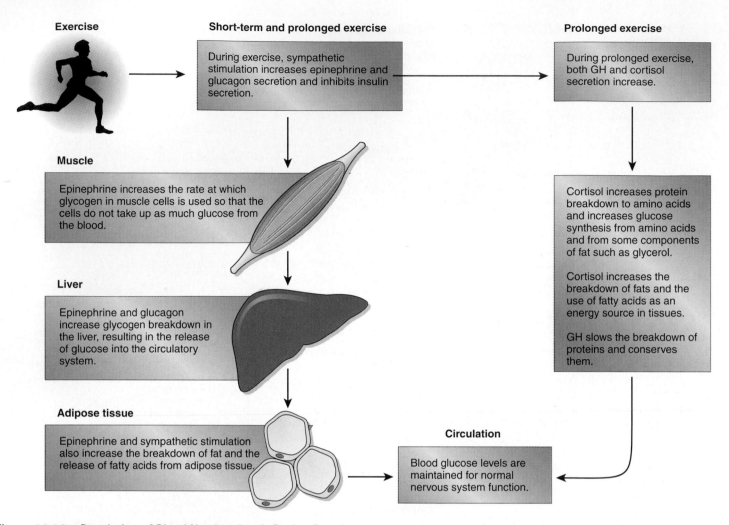

Exercise

Short-term and prolonged exercise

During exercise, sympathetic stimulation increases epinephrine and glucagon secretion and inhibits insulin secretion.

Prolonged exercise

During prolonged exercise, both GH and cortisol secretion increase.

Muscle

Epinephrine increases the rate at which glycogen in muscle cells is used so that the cells do not take up as much glucose from the blood.

Liver

Epinephrine and glucagon increase glycogen breakdown in the liver, resulting in the release of glucose into the circulatory system.

Cortisol increases protein breakdown to amino acids and increases glucose synthesis from amino acids and from some components of fat such as glycerol.

Cortisol increases the breakdown of fats and the use of fatty acids as an energy source in tissues.

GH slows the breakdown of proteins and conserves them.

Adipose tissue

Epinephrine and sympathetic stimulation also increase the breakdown of fat and the release of fatty acids from adipose tissue.

Circulation

Blood glucose levels are maintained for normal nervous system function.

Figure 18.18   Regulation of Blood Nutrient Levels During Exercise

for example, muscles rely to a large extent on fat metabolism for energy.

Reproductive Hormones

Reproductive hormones are secreted primarily from the ovaries, testes, placenta, and pituitary gland (table 18.12). These hormones are discussed in chapter 28.

Hormones of the Pineal Body, Thymus Gland, and Others

The **pineal** (pin′ē-ăl) **body** in the epithalamus of the brain secretes hormones that act on the hypothalamus or the gonads

to inhibit reproductive functions. Two substances have been proposed as secretory products: **melatonin** (mel′ă-tōn′in) and **arginine vasotocin** (ar′ji-nēn vā-sō-tō′sin) (table 18.13). Melatonin can decrease GnRH secretion from the hypothalamus and may inhibit reproductive functions through this mechanism. It may also help regulate sleep cycles by increasing the tendency to sleep.

The **photoperiod,** which is the daily amount of daylight and dark that occurs each day, and changes with the seasons of the year. In some animals, the photoperiod regulates pineal secretions (figure 18.19). For example, increased daylight initiates action potentials in the retina of the eye that are propagated to the brain and cause a decrease in the action potentials sent first to the spinal cord and then through sympathetic neurons to the pineal body. Decreased pineal secretion results. In the dark, action potentials delivered by sympathetic neurons to the pineal body increase, stimulating the secretion of pineal hormones. Humans secrete larger amounts of melatonin at night than in the daylight. In animals that breed in the spring, the increased day length decreases pineal secretions. Because pineal secretions inhibit reproductive functions in these species, the increased day length results in hypertrophy of the reproductive structures.

Clinical Focus   Diabetes Mellitus

Diabetes mellitus results primarily from inadequate secretion of insulin or the inability of tissues to respond to insulin. **Insulin-dependent diabetes mellitus (IDDM)**, also called **type I diabetes mellitus,** affects approximately 3% of people with diabetes mellitus, and results from diminished insulin secretion. It develops as a result of autoimmune destruction of the pancreatic islets, and symptoms appear after approximately 90% of the islets are destroyed. IDDM most commonly develops in young people. Heredity may play some role in the condition, although initiation of pancreatic islet destruction may involve a viral infection of the pancreas (see the Systems Pathology essay at the end of this chapter).

Noninsulin-dependent diabetes mellitus (NIDDM), also called **type II diabetes mellitus,** results from the inability of the tissues to respond to insulin. NIDDM usually develops in people older than 40–45 years of age, although the age of onset varies considerably. There is a strong genetic component to the disease, but its actual cause is unknown. In some cases, abnormal receptors for insulin or antibodies appear to bind to and damage insulin receptors, but, in other cases, abnormalities may occur in the mechanisms activated by the insulin receptors.

NIDDM is more common than IDDM. Approximately 97% of people who have diabetes mellitus have NIDDM. The reduced number of functional receptors for insulin make the uptake of glucose by cells very slow, which results in elevated blood glucose levels after a meal. Obesity is common, although not universal, in patients with NIDDM. Elevated blood glucose levels cause fat cells to convert glucose to fat, even though the rate at which adipose cells take up glucose is impaired. Increased blood glucose and increased urine production lead to hyperosmolality of blood and dehydration of cells. The poor use of nutrients and dehydration of cells leads to lethargy, fatigue, and periods of irritability. The elevated blood glucose levels lead to recurrent infections and prolonged wound healing.

Patients with NIDDM do not suffer sudden large increases of blood glucose and severe tissue wasting because a slow rate of glucose uptake does occur, even though the insulin receptors are defective. In some people with NIDDM, insulin production eventually decreases because pancreatic islet cells atrophy and IDDM develops. Approximately 25%–30% of patients with NIDDM take insulin, 50% take oral medication to increase insulin secretion and increase the efficiency of glucose utilization, and the remainder control blood glucose levels with exercise and diet.

Glucose tolerance tests are used to diagnose diabetes mellitus. In general, the test involves feeding the patient a large amount of glucose after a period of fasting. Blood samples are collected for a few hours, and a sustained increase in blood glucose levels strongly indicates that the person is suffering from diabetes mellitus.

Too much insulin relative to the amount of glucose ingested leads to **insulin shock.** The high levels of insulin cause target tissues to take up glucose at a very high rate. As a result, blood glucose levels rapidly fall to a low level. Because the nervous system depends on glucose as its major source of energy, neurons malfunction because of a lack of metabolic energy. The result is a series of nervous system responses that include disorientation, confusion, and convulsions. Taking too much insulin, too little food intake after an injection of insulin, or increased metabolism of glucose due to excess exercise by a diabetic patient can cause insulin shock.

It appears that damage to blood vessels and reduced nerve function can be reduced in diabetic patients suffering from either IDDM or NIDDM by keeping the blood glucose well within normal levels at all times. Doing so, however, requires increased attention to diet, frequent blood glucose testing, and increased chance of suffering from low blood glucose levels, which leads to symptoms of insulin shock. A strict diet and routine exercise are often effective components of a treatment strategy for diabetes mellitus, and in many cases diet and exercise are adequate to control NIDDM.

Clinical Focus   Stress

The adrenal cortex and the adrenal medulla play major roles in response to stress.

In general, stress activates nervous and endocrine responses that prepare the body for physical activity, even when physical activity is not the most appropriate response to the stressful conditions, such as during an examination or other mentally stressful situations. The endocrine response to stress involves increased CRH release from the hypothalamus and increased sympathetic stimulation of the adrenal medulla. CRH stimulates ACTH secretion from the anterior pituitary, which in turn stimulates cortisol from the adrenal cortex. Increased sympathetic stimulation of the adrenal medulla increases epinephrine and norepinephrine secretion.

Together epinephrine and cortisol increase blood glucose levels and the release of fatty acids from adipose tissue and the liver. Sympathetic innervation of the pancreas decreases insulin secretion. Consequently, most tissues do not readily take up and use glucose. Thus glucose is available primarily to the nervous system; and fatty acids are used by skeletal muscle, cardiac muscle, and other tissues.

Epinephrine and sympathetic stimulation also increase cardiac output, increase blood pressure, and act on the central nervous system to increase alertness and aggressiveness. Cortisol also decreases the initial inflammatory response.

Responses to stress illustrate the close relationship between the nervous and endocrine systems and provide an example of their integrated functions. Our ability to respond to stressful conditions depends on the nervous and endocrine responses to stress.

Although responses to stress are adaptive under many circumstances, they can become harmful. For example, if stress is chronic, the elevated secretion of cortisol and epinephrine produces harmful effects.

Table 18.12 Hormones of the Reproductive Organs

Hormones	Structure	Target Tissue	Response
Testis			
Testosterone	Steroid	Most cells	Aids in spermatogenesis; maintenance of functional reproductive organs; secondary sexual characteristics; sexual behavior
Ovary			
Estrogens	Steroids	Most cells	Uterine and mammary gland development and function; external genitalia structure; secondary sexual characteristics; sexual behavior and menstrual cycle
Progesterone	Steroid	Most cells	Uterine and mammary gland development and function; external genitalia structure; secondary sexual characteristics; menstrual cycle

Table 18.13 Other Hormones and Hormonelike Substances

Chemical Signal	Structure	Target tissue	Response
Pineal Body			
Melatonin	Amino acid derivative	At least the hypothalamus	Inhibition of gonadotropin-releasing hormone secretion, thereby inhibiting reproduction; significance is not clear in humans
Arginine vasotocin	Amino acid derivative	Possibly the hypothalamus	Possible inhibition of gonadotropin-releasing hormone secretion
Thymus Gland			
Thymosin	Peptide	Immune tissues	Development and function of the immune system
Several Tissues (autocrine and paracrine regulatory substances)			
Prostaglandins	Modified fatty acid	Most tissues	Mediation of the inflammatory response, increased uterine contractions; ovulation, possible inhibition of progesterone synthesis; blood coagulation; and other functions
Prostacyclins	Modified fatty acid	Most tissues	Mediation of the inflammatory response, and other functions
Thromboxanes	Modified fatty acid	Most tissues	Mediation of the inflammatory response, and other functions
Leukotrienes	Modified fatty acid	Most tissues	Mediation of the inflammatory response, and other functions
Enkephalins and endorphins	Peptides	Nervous system	Reduction of pain sensation, and other functions
Epidermal growth factor	Protein	Many tissues	Stimulates division in many cell types and plays a role in embryonic development
Fibroblast growth factor	Protein	Many tissues	Stimulates cell division in many cell types and plays a role in embryonic development
Interleukin-2	Protein	Certain immune competent cells	Stimulates cell division of T lymphocytes

The function of the pineal body in humans is not clear, but tumors that destroy the pineal body correlate with early sexual development, and tumors that result in pineal hormone secretion correlate with retarded development of the reproductive system. It is not clear, however, if the pineal body controls the onset of puberty.

Arginine vasotocin works with melatonin to regulate the function of the reproductive system in some animals. Evidence for the role of melatonin is more extensive, however.

The **thymus gland** (thī′mŭs) is in the neck and superior to the heart in the thorax, and secretes a hormone called **thymosin** (thī′mō-sin) (see table 18.13). Both the thymus

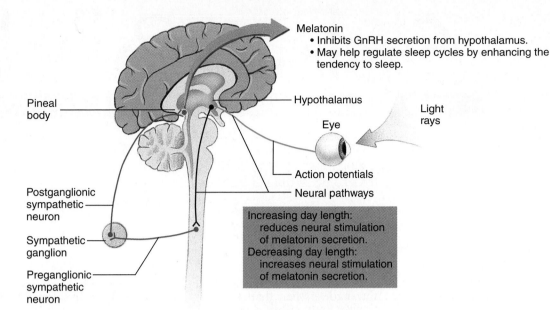

Figure 18.19   Regulation of Melatonin Secretion from the Pineal Body

Light entering the eye inhibits and dark stimulates neural stimulation of melatonin secretion from the pineal body.

Labels within figure:

Pineal body

Postganglionic sympathetic neuron

Sympathetic ganglion

Preganglionic sympathetic neuron

Melatonin
• Inhibits GnRH secretion from hypothalamus.
• May help regulate sleep cycles by enhancing the tendency to sleep.

Hypothalamus

Eye

Light rays

Action potentials

Neural pathways

Increasing day length: reduces neural stimulation of melatonin secretion.
Decreasing day length: increases neural stimulation of melatonin secretion.

---

gland and thymosin play an important role in the development of the immune system and are discussed in chapter 22.

Several hormones are released from the gastrointestinal tract. They regulate digestive functions by influencing the activity of the stomach, intestines, liver, and pancreas. They are discussed in chapter 24.

Hormonelike Substances

Autocrine chemical signals are chemicals released from cells that influence the cell type from which they are released. **Paracrine chemical signals** are chemicals released from cells near their target cells that reach their targets by diffusion. Autocrine and paracrine chemical signals differ from hormones in that they are not secreted from discrete endocrine glands, they have local effects rather than systemic effects, or they have functions that are not understood adequately to explain their role in the body. Examples of autocrine chemical signals include chemical mediators of inflammation derived from the fatty acid **arachidonic** (ă-rak-i-don′ik) **acid.** These include **prostaglandins,** (pros′stă-glan′dinz), **thromboxanes** (throm′box-zānz), **prostacyclins** (pros-tă-sī′klinz), and **leukotrienes** (lū′kō-trī-ēnz). Paracrine chemical signals include substances that play a role in modulating the sensation of pain, such as **endorphins** (en-dōr′phinz) and **enkephalins** (en-kef′ă-linz); and several peptide growth factors, such as **epidermal growth factor, fibroblast growth factor,** and **interleukin-2** (in-ter-lū′kin) (see table 18.13).

Prostaglandins, thromboxanes, prostacyclins, and leukotrienes are released from injured cells and are responsible for initiating some of the symptoms of inflammation (see chapter 22), in addition to being released from certain healthy cells. For example, prostaglandins are involved in the regulation of uterine contractions during menstruation and childbirth, the process of ovulation, the inhibition of progesterone synthesis by the corpus luteum, the regulation of coagulation,

kidney function, and modification of the effect of other hormones on their target tissues. They are paracrine regulatory substances in these examples because they are synthesized and secreted by the cells near their target cells. Once prostaglandins enter the circulatory system, they are metabolized rapidly.

Pain receptors are stimulated directly by prostaglandins and other inflammatory compounds, or prostaglandins cause vasodilation of blood vessels, which is associated with headaches. Anti-inflammatory drugs such as aspirin inhibit prostaglandin synthesis and, as a result, reduce inflammation and pain.

Three classes of peptide molecules bind to the same receptor molecules as morphine. They include enkephalins, endorphins, and **dynorphins** (dī′nōr-fin); and they are produced in several sites in the body, such as parts of the brain, pituitary, spinal cord, and gut. These substances are endogenously produced analgesics. They act as neurotransmitters in some neurons of both the central and peripheral nervous systems and as hormones or paracrine regulatory substances. In general they moderate the sensation of pain (see chapter 13). Decreased sensitivity to painful stimuli during exercise and stress may result from the increased secretion of these substances.

Several proteins can be classified as growth factors. They generally function as paracrine chemical signals because they are secreted near their target tissues. Epidermal growth factor stimulates cell divisions in a number of tissues and plays an important role in embryonic development. Interleukin-2 stimulates the proliferation of T lymphocytes and plays a very important role in immune responses (see chapter 22). The number of hormonelike substances in the body is large, and only a few of them have been mentioned here. Chemical communication among cells in the body is complex, well developed, and necessary for maintenance of homeostasis. Investigations of chemical regulation increase our knowledge of body functions—knowledge that can be used in the development of techniques for the treatment of pathologic conditions.

Systems Pathology

INSULIN-DEPENDENT DIABETES MELLITUS

Billy, a 10-year-old boy, was diagnosed as having insulin-dependent diabetes mellitus (IDDM). Billy's mother took him to a physician after noticing that he was constantly hungry and was losing weight rapidly in spite of his unusually large food intake. More careful observation made it clear that Billy was constantly thirsty and that he urinated frequently. In addition, he felt weak and lethargic, and his breath occasionally had a distinctive sweet, or acetone, odor. Diagnostic tests confirmed that he had IDDM.

BACKGROUND INFORMATION

IDDM is caused by diminished insulin secretion. In patients with IDDM, nutrients are absorbed from the intestine after a meal, but skeletal muscle, adipose tissue, and other target tissues do not readily take glucose into their cells, and liver cells cannot convert glucose to glycogen. Consequently, blood levels of glucose increase dramatically. Glucagon and glucocorticoid secretion increase because the glucose in the blood cannot enter the cells that produce these hormones, so their rate of secretion is similar to when blood glucose levels are low. Epinephrine secretion also increases. In response to these hormones, glycogen, fats, and proteins are broken down and metabolized to produce the adenosine triphosphate (ATP) required by cells.

When blood glucose levels are very high, glucose is excreted in the urine, which results in an increase in urine volume. The rapid loss of water in the urine increases the osmotic concentration of blood, which increases the sensation of thirst. The increased osmolality of blood and the ionic imbalances caused by the loss of Ca^{2+} and K^+ ions in the large amount of urine produced, cause neurons to malfunction and result in diabetic coma in severe cases. When insulin levels in the blood are low and cells of the nervous system that control appetite appear to be unable to take up glucose even when blood glucose levels are high, the result is an increased appetite. **Polyurea** (pol-ē-yū′rē-ă) (increased urine volume), **polydipsia** (pol-ē-dip′sē-ă) (increased thirst), and **polyphagia** (pol-ē-fā′jē-ă) (increased appetite) are major symptoms of IDDM. Acidosis is caused by rapid fat catabolism that results in increased levels of **acetoacetic** (as′e-tō-a-sē′tik) **acid,** which is converted to **acetone** (as′e-tōn) and **β-hydroxybutyric** (bā′tă hī-drōk′sē-byū-tir′ik) **acid.** These

three substances collectively are referred to as **ketone** (kē′tōn) bodies. The presence of excreted ketone bodies in urine and in expired air ("acetone breath") suggest that the person has diabetes mellitus.

Billy's physician explained that prior to the late 1920s people with his condition always died in a relatively short time. They suffered from massive weight loss and appeared to starve to death in spite of eating a large amount of food. His physician explained that because of the discovery of insulin,

Figure B A 10-year-old boy giving himself an insulin injection. 🏃

System Interactions

System	Interactions
Muscular	Untreated diabetes mellitus, especially IDDM, results in severe muscle atrophy because glycogen, stored fat, and proteins of muscles are broken down and used as energy sources. Ionic imbalances can also lead to muscular weakness.
Nervous	Untreated IDDM can have dramatic effects on the nervous system. When the blood glucose reaches very high levels, the osmolality of the extracellular fluid is increased. Thus, water diffuses from the neurons of the brain. In addition, acidosis develops because of the rapid metabolism of fats. As a result, the nervous system cannot function normally, and diabetic coma can result. A long-term effect is the degeneration of the myelin sheaths of neurons, resulting in abnormal nerve functions.
Cardiovascular	Atherosclerosis develops more rapidly in diabetics than in the normal population. Changes in the capillary structure and high blood glucose levels increase the probability of reduced circulation and gangrene.
Lymphatic and immune	The tendency to develop infections increases, and the rate of healing is slower. In some cases, there is an allergic reaction to the injected insulin.
Respiratory	Acidosis causes hyperventilation, which increases blood pH back toward normal levels by decreasing blood CO_2 levels.
Urinary	High blood glucose levels cause polyuria, the urine contains glucose and urine has a high osmolality, and people with diabetes are more likely to develop urinary tract infections.
Reproductive	Pregnant women with diabetes mellitus may have babies with a larger than normal birth weight because the blood glucose levels may be high in the mother and fetus, and the fetus's pancreas produces insulin. Glucose is therefore taken up by cells of the fetus where it is converted to fat.

many people with his type of diabetes mellitus are able to live nearly normal lives. Taking insulin injections, monitoring blood glucose levels, and following a strict diet to keep blood glucose levels within a normal range of values are the major treatments for IDDM.

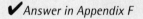

11   **P R E D I C T**

After Billy was diagnosed with diabetes mellitus, he followed a strict diet and took insulin for a few months. He began to feel much better than before. In fact, he felt so well that he began to sneak candy and soft drinks when his parents were not around. Predict the consequences of his actions on his health.

✔ *Answer in Appendix F*

Summary

Pituitary Gland and Hypothalamus

1. The pituitary gland secretes at least nine hormones that regulate numerous body functions and other endocrine glands.
2. The hypothalamus regulates pituitary gland activity through neurohormones and action potentials.

Structure of the Pituitary Gland

1. The posterior pituitary develops from the floor of the brain and consists of the infundibulum and pars nervosa.
2. The anterior pituitary develops from the roof of the mouth and consists of the pars distalis, pars intermedia, and pars tuberalis.

Relationship of the Pituitary to the Brain

1. The hypothalamohypophyseal portal system connects the hypothalamus and the anterior pituitary.
   - Neurohormones are produced in hypothalamic neurons.
   - Through the portal system the neurohormones inhibit or stimulate hormone production in the anterior pituitary.

2. The hypothalamohypophyseal nerve tract connects the hypothalamus and the posterior pituitary.
   - Neurohormones are produced in hypothalamic neurons.
   - The neurohormones move down the axons of the nerve tract and are secreted from the posterior pituitary.

Hormones of the Pituitary Gland
Posterior Pituitary Hormones

1. ADH promotes water retention by the kidneys.
2. Oxytocin promotes uterine contractions during delivery and causes milk ejection in lactating women.

Anterior Pituitary Hormones

1. GH, or somatotropin
   - GH stimulates the uptake of amino acids and their conversion into proteins and stimulates the breakdown of fats and glycogen.

- GH stimulates the production of somatomedins; together they promote bone and cartilage growth.
- GH secretion increases in response to an increase in blood amino acids, low blood glucose, or stress.
- GH is regulated by GHRH and GHIH, or somatostatin.

2. TSH, or thyrotropin, causes the release of thyroid hormones.
3. ACTH
   - ACTH is derived from proopiomelanocortin.
   - ACTH stimulates cortisol secretion from the adrenal cortex and increases skin pigmentation.
4. Several hormones in addition to ACTH are derived from proopiomelanocortin.
   - Lipotropins cause fat breakdown.
   - Beta-endorphins play a role in analgesia.
   - MSH increases skin pigmentation.
5. LH and FSH
   - Both hormones regulate the production of gametes and reproductive hormones (testosterone in males; estrogen and progesterone in females).
   - GnRH from the hypothalamus stimulates LH and FSH secretion.
6. Prolactin
   - Prolactin stimulates milk production in lactating females.
   - Prolactin-releasing hormone and prolactin-inhibiting hormone from the hypothalamus affect prolactin secretion.

Thyroid Gland

The thyroid gland is just inferior to the larynx.

Histology

1. The thyroid gland is composed of small, hollow balls of cells called follicles, which contain thyroglobulin.
2. Parafollicular cells are scattered throughout the thyroid gland.

Thyroid Hormones

1. Thyroid hormone synthesis
   - Iodide ions are taken into the follicles by active transport, are oxidized, and are bound to tyrosine molecules in thyroglobulin.
   - Thyroglobulin is secreted into the follicle lumen. Tyrosine molecules with iodine combine to form T_3 and T_4, thyroid hormones.
   - Thyroglobulin is taken into the follicular cells and is broken down; T_3 and T_4 diffuse from the follicles to the blood.
2. Transport in the blood
   - T_3 and T_4 bind to thyroxine-binding globulin and other plasma proteins.
   - The plasma proteins prolong the half-life of T_3 and T_4 and regulate the levels of T_3 and T_4 in the blood.
   - Approximately one-third of the T_4 is converted into functional T_3.
3. Mechanism of action of thyroid hormones
   - Thyroid hormones bind with intracellular receptor molecules and initiate new protein synthesis.
4. Effects of thyroid hormones
   - Thyroid hormones increase the rate of glucose, fat, and protein metabolism in many tissues, thus increasing body temperature.
   - Normal growth of many tissues is dependent on thyroid hormones.

5. Regulation of thyroid hormone secretion
   - Increased TSH from the anterior pituitary increases thyroid hormone secretion.
   - TRH from the hypothalamus increases TSH secretion. TRH increases as a result of chronic exposure to cold, food deprivation, and stress.
   - T_3 and T_4 inhibit TSH and TRH secretion.

Calcitonin

1. The parafollicular cells secrete calcitonin.
2. An increase in blood calcium levels stimulates calcitonin secretion.
3. Calcitonin decreases blood calcium and phosphate levels by inhibiting osteoclasts.

Parathyroid Glands

1. The parathyroid glands are embedded in the thyroid glands.
2. PTH increases blood calcium levels.
   - PTH stimulates osteoclasts.
   - PTH promotes calcium reabsorption by the kidneys and the formation of active vitamin D by the kidneys.
   - Active vitamin D increases calcium absorption by the intestine.
3. A decrease in blood calcium levels stimulates PTH secretion.

Adrenal Glands

1. The adrenal glands are near the superior pole of each kidney.
2. The adrenal medulla arises from neural crest cells and functions as part of the sympathetic nervous system. The adrenal cortex is derived from mesoderm.

Histology

1. The medulla is composed of closely packed cells.
2. The cortex is divided into three layers: the zona glomerulosa, the zona fasciculata, and the zona reticularis.

Hormones of the Adrenal Medulla

1. Epinephrine accounts for 80% and norepinephrine for 20% of the adrenal medulla hormones.
   - Epinephrine increases blood glucose levels, use of glycogen and glucose by skeletal muscle, and heart rate and force of contraction and causes vasoconstriction in the skin and viscera and vasodilation in skeletal and cardiac muscle.
   - Norepinephrine stimulates cardiac muscle and causes constriction of most peripheral blood vessels.
2. The adrenal medulla hormones prepare the body for physical activity.
3. Release of adrenal medulla hormones is mediated by the sympathetic nervous system in response to emotions, injury, stress, exercise, and low blood glucose levels.

Hormones of the Adrenal Cortex

1. The zona glomerulosa secretes the mineralocorticoids, especially aldosterone. Aldosterone acts on the kidneys to increase sodium and to decrease potassium and hydrogen levels in the blood.
2. The zona fasciculata secretes glucocorticoids, especially cortisol.
   - Cortisol increases fat and protein breakdown, increases glucose synthesis from amino acids, decreases the

inflammatory response, and is necessary for the development of some tissues.
  • ACTH from the anterior pituitary stimulates cortisol secretion. CRH from the hypothalamus stimulates ACTH release. Low blood sugar levels or stress stimulate CRH secretion.
3. The zona reticularis secretes androgens. In females androgens stimulate axillary and pubic hair growth and sexual drive.

Pancreas

The pancreas is located along the small intestine and the stomach. It is both an exocrine and an endocrine gland.

Histology

1. The exocrine portion of the pancreas consists of a complex duct system that ends in small sacs called acini that produce pancreatic digestive juices.
2. The endocrine portion consists of the pancreatic islets. Each islet is composed of alpha cells that secrete glucagon, beta cells that secrete insulin, and delta cells that secrete somatostatin.

Effect of Insulin and Glucagon on Their Target Tissues

1. Insulin
  • Insulin's target tissues are the liver, adipose tissue, muscle, and the satiety center in the hypothalamus. The nervous system is not a target tissue, but it does rely on blood glucose levels maintained by insulin.
  • Insulin increases the uptake of glucose and amino acids by cells. Glucose is used for energy or is stored as glycogen. Amino acids are used for energy or are converted to glucose or proteins.
2. Glucagon
  • Glucagon's target tissue is mainly the liver.
  • Glucagon causes the breakdown of glycogen and fats for use as an energy source.

Regulation of Pancreatic Hormone Secretion

1. Insulin secretion increases because of elevated blood glucose levels, an increase in some amino acids, parasympathetic stimulation, and gastrointestinal hormones. Sympathetic stimulation decreases insulin secretion.
2. Glucagon secretion is stimulated by low blood glucose levels, certain amino acids, and sympathetic stimulation.
3. Somatostatin inhibits insulin and glucagon secretion.

Hormonal Regulation of Nutrients

1. After a meal the following events take place:
  • Glucagon, cortisol, GH, and epinephrine are inhibited by high blood glucose levels, reducing the release of glucose from tissues.
  • Insulin secretion increases as a result of the high blood glucose levels, increasing the uptake of glucose, amino acids, and fats, which are used for energy or are stored.
  • Sometime after the meal, blood glucose levels drop. Glucagon, cortisol, GH, and epinephrine levels increase, insulin levels decrease, and glucose is released from tissues.
  • Adipose tissue releases fatty acids, triacylglycerols, and ketones, which are used for energy by most tissues.
2. During exercise the following events occur:
  • Sympathetic activity increases epinephrine and glucagon secretion, causing a release of glucose into the blood.
  • Low blood sugar levels, caused by uptake of glucose by skeletal muscles, stimulates epinephrine, glucagon, GH, and cortisol secretion, causing an increase in fatty acids, triacylglycerols, and ketones in the blood, all of which are used for energy.

Reproductive Hormones

Reproductive hormones are secreted by the ovaries, testes, placenta, and pituitary gland.

Hormones of the Pineal Body, Thymus Gland, and Others

1. The pineal body produces melatonin and arginine vasotocin, which can inhibit reproductive maturation.
2. The thymus gland produces thymosin, which is involved in the development of the immune system.
3. Several hormones produced by the gastrointestinal tract regulate digestive functions.

Hormonelike Substances

1. Autocrine and paracrine chemical signals are produced by many cells of the body and usually have a local effect. They affect many body functions.
2. Prostaglandins, prostacyclins, thromboxanes, and leukotrienes are derived from fatty acids and mediate inflammation and other functions. Endorphins, enkephalins, and dynorphins are analgesic substances. Growth factors influence cell division and growth in many tissues, and interleukin-2 influences cell division in T cells of the immune system.

Content Review

1. To understand the role of an endocrine gland and its secretions in the body, what five things should be kept in mind?
2. Where is the pituitary gland located? Contrast the embryonic origin of the posterior pituitary and the anterior pituitary.
3. Name the parts of the pituitary gland and the function of each part.
4. Describe the hypothalamohypophyseal portal system. How does the hypothalamus regulate the hormone secretion of the anterior pituitary?
5. Describe the production of a neurohormone in the hypothalamus and its secretion in the posterior pituitary.
6. Where is ADH produced, where is it secreted, and what is its target tissue? What happens when ADH levels increase?

7. Where is oxytocin produced and secreted, and what effects does it have on its target tissues?

8. Structurally, what kinds of hormones are released from the posterior pituitary and the anterior pituitary? Do these hormones bind to plasma proteins, how long is their half-life, and how do they activate their target tissues?

9. For each of the following hormones secreted by the anterior pituitary—GH, TSH, ACTH, LH, FSH, and prolactin—name their target tissues and the effect of the hormone on its target tissue.

10. What effect do amino acids, glucose, and stress have on GH secretion?

11. What stimulates somatomedin production, where is it produced, and what are its effects?

12. How are ACTH, MSH, lipotropins, and β endorphins related? What are the functions of these hormones?

13. Where is the thyroid gland located? Describe the follicles and the parafollicular cells within the thyroid. What hormones do they produce?

14. Starting with the uptake of iodide by the follicles, describe the production and secretion of thyroid hormones.

15. How are the thyroid hormones transported in the blood? What effect does this transportation have on their half-life?

16. What are the target tissues of thyroid hormone? By what mechanism do thyroid hormones alter the activities of their target tissues? What effects are produced?

17. Starting in the hypothalamus, explain how chronic exposure to the cold, food deprivation, or stress can affect thyroid hormone production.

18. Diagram two negative-feedback mechanisms involving hormones that function to regulate the production of thyroid hormones.

19. What effect does calcitonin have on osteoclasts, osteoblasts, and blood calcium levels? What stimulus can cause an increase in calcitonin secretion?

20. Where are the parathyroid glands located, and what hormone do they produce?

21. What effect does PTH have on osteoclasts, osteoblasts, the kidneys, the small intestine, and blood calcium and phosphate levels? What stimulus can cause an increase in PTH secretion?

22. Where are the adrenal glands located? Describe the embryonic origin of the adrenal medulla and the adrenal cortex.

23. Name two hormones secreted by the adrenal medulla, and list the effects of these hormones.

24. List several conditions that can stimulate the production of adrenal medullary hormones. What role does the nervous system play in the release of adrenal medullary hormones? How does this role relate to the embryonic origin of the adrenal medulla?

25. Describe the three layers of the adrenal cortex, and name the hormones produced by each layer.

26. Name the target tissue of aldosterone, and list the effects of an increase in aldosterone secretion on the concentration of ions in the blood.

27. Describe the effects produced by an increase in cortisol secretion. Starting in the hypothalamus, describe how stress or low blood sugar levels can stimulate cortisol release.

28. What effects do adrenal androgens have on males and females?

29. Where is the pancreas located? Describe the exocrine and endocrine parts of this gland and the secretions produced by each portion.

30. Name the target tissues for insulin and glucagon, and list the effects they have on their target tissues.

31. How does insulin affect the nervous system in general and the satiety center in the hypothalamus in particular?

32. What effect do blood glucose levels, blood amino acid levels, the autonomic nervous system, and somatostatin have on insulin and glucagon secretion?

33. Describe the hormonal effects after a meal that result in the movement of nutrients into cells and their storage. Describe the hormonal effects that later cause the release of stored materials for use as energy.

34. During exercise, how does sympathetic nervous system activity regulate blood sugar levels? Name five hormones that interact to ensure that both the brain and muscles have adequate energy sources.

35. Where is the pineal body located? Name the hormones it produces and their possible effects.

36. Where is the thymus gland located, what hormone does it produce, and what effects does the hormone have?

37. Describe the site of production and actions of prostaglandins, endorphins, enkephalins, and dynorphins.

Develop Your Reasoning Skills

1. The hypothalamohypophyseal portal system connects the hypothalamus with the anterior pituitary. Why is such a special circulatory system advantageous?

2. The secretion of ADH can be affected by exposure to hot or cold environmental temperatures. Predict the effect of a hot environment on ADH secretion, and explain why it is advantageous. Propose a mechanism by which temperature produces a change in ADH secretion.

3. A patient exhibited polydipsia (thirst), polyuria (excess urine production), and urine with a low specific gravity (contains few ions and no glucose). If you wanted to reverse the symptoms, would you administer insulin, glucagon, ADH, or aldosterone? Explain.

4. A patient complains of headaches and visual disturbances. A casual glance reveals that the patient's finger bones are enlarged in diameter, there is a heavy deposition of bone over the eyes, and the patient has a prominent jaw. The doctor tells you that the headaches and visual disturbances result from increased pressure within the skull and that the patient is suffering from a pituitary tumor that is affecting hormone secretion. Name the hormone that is causing the problem, and explain why there is an increase in pressure within the skull.

5. Most laboratories have the ability to determine blood levels of TSH, T_3, and T_4. Given that ability, design a method of determining whether hyperthyroidism in a patient results from

a pituitary abnormality or from the production of a nonpituitary thyroid stimulatory substance.

6. An anatomy and physiology instructor asked two students to predict a patient's response to chronic vitamin D deficiency. One student claimed that the person would suffer from hypocalcemia and the symptoms associated with that condition. The other student claimed that calcium levels would remain within their normal range, although at the low end of the range, and bone resorption would occur to the point that advanced osteomalacia might be seen. With whom do you agree, and why?

7. Given the ability to measure blood glucose levels, design an experiment that distinguishes between a person with diabetes, a healthy person, and a person who has a pancreatic tumor that secretes large amounts of insulin.

8. A patient arrives in an unconscious condition. A medical emergency bracelet reveals that he has diabetes. The patient can be in diabetic coma or insulin shock. How could you tell which, and what treatment would you recommend for each condition?

9. Diabetes mellitus can result from a lack of insulin that results in hyperglycemia. Adrenal diabetes and pituitary diabetes also produce hyperglycemia. What hormones produce the last two conditions?

10. Predict some of the consequences of exposure to intense and prolonged stress.

Web Site Link

For a listing of the most current web sites related to this chapter, please visit the Seeley home page at:
http://www.mhhe.com/biosci/ap/seeleyap/

Chapter Nineteen

Cardiovascular System: Blood

Physiology

Human Anatomy

Objectives

1. List the functions of blood.

2. List the components of blood plasma, and explain their functions.

3. Define the term formed element, and list the three types of formed elements.

4. Describe the origin and production of the formed elements.

5. Describe the structure and function of erythrocytes.

6. Explain the role of iron in blood.

7. Define erythropoiesis, and name the cells produced during erythropoiesis.

8. Describe the removal of damaged or "worn-out" erythrocytes from the circulation, and describe the production and fate of bilirubin.

9. Describe the structures and functions of the five types of leukocytes.

10. Describe the structure, origin, and function of platelets.

11. Name and describe the three stages of hemostasis.

12. Compare the extrinsic and intrinsic pathways of coagulation.

13. Explain the importance of the balance between coagulation factors and anticoagulants.

14. Describe how a clot functions in wound healing and how the clot is removed.

15. Explain the basis of ABO and Rh incompatibilities.

16. Describe diagnostic blood tests and the normal values for the tests, and give examples of disorders that produce abnormal test values.

Blood has fascinated humans for thousands of years. Blood was considered to be the "essence of life" because the uncontrolled loss of it can result in death. Blood was also thought to define our character and emotions. People of a noble bloodline were described as "blue bloods," whereas criminals were considered to have "bad" blood. It was said that anger caused the blood to "boil," and fear resulted in blood "curdling". The scientific study of blood reveals a story as fascinating as any of these speculations. Blood performs many functions essential to life and often can reveal much about our health.

Blood is a type of connective tissue, consisting of cells and cell fragments surrounded by a liquid matrix, which circulates through the heart and blood vessels. The cells and cell fragments are the formed elements, and the liquid is the plasma. The formed elements make up about 45%, and plasma makes up about 55% of the total blood volume (figure 19.1). The total blood volume in the average adult is about 4–5 L in females and 5–6 L in males. Blood makes up about 8% of the total weight of the body.

Cells require constant nutrition and waste removal because they are metabolically active. Most cells are located some distance from nutrient sources such as the digestive tract and sites of waste disposal such as the kidneys. The cardiovascular system, which consists of the heart, blood vessels, and blood, connects the various tissues of the body. The heart pumps blood through blood vessels, which extend throughout the body, and the blood delivers nutrients and picks up waste products.

Functions

The functions of blood can be categorized as transportation, maintenance, and protection. Many of these functions, however, can be placed in more than one category. For example, for blood cells to protect against microorganisms, they must be transported to sites of infection.

Transportation

Blood is the primary transport medium of the body. Oxygen enters blood in the lungs and is carried to cells; and carbon dioxide, produced by cells, is carried in blood to the lungs from which it is expelled. Ingested nutrients, electrolytes, and water are transported by the blood from the digestive tract to cells, and waste products are transported from cells to the kidneys for elimination in urine. In addition to transporting gases, nutrients, and waste products, blood transports other substances. For example, the precursor to vitamin D is produced in the skin (see chapter 5) and transported by the blood to the liver and then to the kidneys for processing into active vitamin D. The active vitamin D is transported in blood to the small intestines, where it promotes the uptake of calcium. Another example is lactic acid produced by skeletal muscles during anaerobic respiration (see chapter 10). The lactic acid is carried in blood to the liver and converted into glucose. Finally, many of the substances necessary for maintenance and protection must be transported throughout the body.

Maintenance

Blood plays a crucial role in maintaining homeostasis. Many of the hormones and enzymes that regulate body processes are found in blood, as are buffers (see chapter 2), which help keep the blood's pH within its normal limits of 7.35–7.45. The osmotic composition of blood is also critical for maintaining the normal fluid and electrolyte balance. Because blood can hold heat, it is involved with temperature regulation, transporting heat from the interior to the surface of the body, where the heat is released. When blood vessels are damaged, the blood clots that form are the first step in tissue repair and the restoration of normal function.

Protection

Cells and chemicals of the blood constitute an important part of the immune system, protecting against foreign substances, such as microorganisms and toxins. Blood clotting also provides protection against excessive fluid and cell loss when blood vessels are damaged.

Plasma

Plasma (plaz'mă) is a pale yellow fluid that consists of about 91% water and 9% other substances, such as proteins, ions, nutrients, gases, and waste products (table 19.1). Plasma is a **colloidal** (ko-loyd'ăl) **solution,** which is a liquid containing suspended substances that do not settle out of solution. Most of the suspended substances are plasma proteins, which include albumin, globulins, and fibrinogen. **Albumin** (al-byū'min) makes up 58% of the plasma proteins and is important in the regulation of water movement between tissues and blood. Because albumin does not easily pass from the blood into tissues, it plays an important role in maintaining the osmotic concentration of blood (see chapters 3 and 26). **Globulins** (glob'yū-linz) account for 38% of the plasma proteins. Some globulins, such as antibodies and complement, are part of the immune system (see chapter 22), whereas others function as transport molecules (see chapter 17). **Fibrinogen** (fī-brin'ō-jen) constitutes 4% of the plasma proteins and is responsible for the formation of blood clots (see discussion later in the chapter).

In addition to the suspended molecules, plasma contains a number of dissolved components, such as ions, nutrients, waste products, gases, and regulatory substances.

Plasma volume remains relatively constant. Normally, water intake through the digestive tract closely matches water loss through the kidneys, lungs, digestive tract, and skin. Oxygen enters blood in the lungs and carbon dioxide enters blood from the tissues. Other suspended or dissolved substances in the blood come from the liver, kidneys, intestines, endocrine

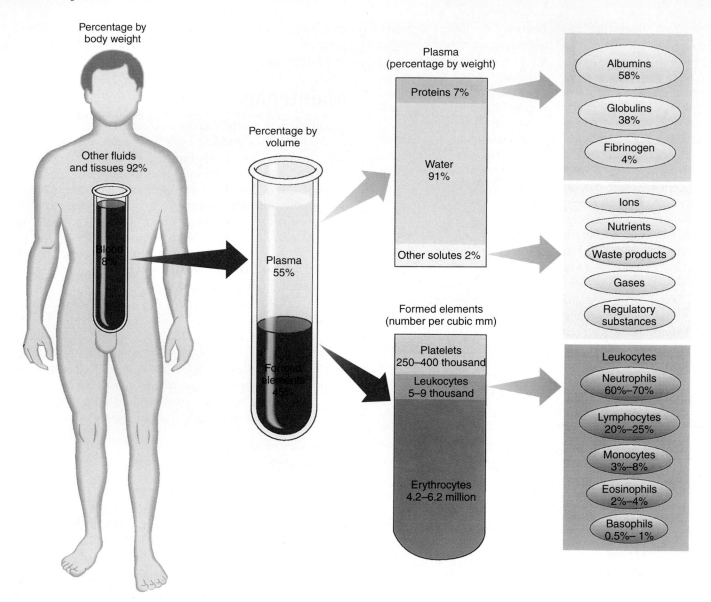

Figure 19.1   Composition of Blood

Approximate values for the components of blood in a normal adult.

glands, and immune tissues such as the spleen. These other substances are also regulated and maintained within narrow limits.

Formed Elements

About 95% of the volume of the **formed elements** consists of erythrocytes (ĕ-rith′rō-sītz), or red blood cells. The remaining 5% consists of leukocytes (lū′kō-sītz), or white blood cells, and cell fragments called platelets, or thrombocytes (throm′bō-sītz). The formed elements of the blood are outlined and illustrated in table 19.2. In healthy adults, leukocytes are the only formed elements possessing nuclei, whereas erythrocytes and platelets have few organelles and lack nuclei.

Leukocytes are named according to their appearance in stained preparations. Leukocytes containing large cytoplasmic granules are **granulocytes** (gran′yū-lō-sītz), and those with very small granules that cannot be seen easily with the light microscope are **agranulocytes** (see table 19.2). The three types of granulocytes are named according to the staining characteristics of their cytoplasm: neutrophils (nū′trō-filz), eosinophils (ē-ō-sin′ō-filz), and basophils (bā′sō-filz). There are two types of agranulocytes: monocytes (mon′ō-sītz), named according to nuclear morphology, and lymphocytes (lim′fō-sītz), named according to a major site of proliferation.

Production of Formed Elements

The process of blood cell production, called **hematopoiesis** (hē′mă-tō-poy-ē′sis) or **hemopoiesis** (hē′mō-poy-ē′sis), occurs

Table 19.1   Composition of Plasma

Plasma Components	Function
Water	Acts as a solvent and suspending medium for blood components
Plasma proteins	
Albumin	Partly responsible for blood viscosity and osmotic pressure; acts as a buffer; transport of fatty acids, free bilirubin, and thyroid hormones
Globulins	Transport of lipids, carbohydrates, hormones, and ions such as iron and copper; antibodies and complement are involved in immunity
Fibrinogen	Functions in blood clotting
Ions	
Sodium, potassium, calcium, magnesium, chloride, iron, phosphate, hydrogen, hydroxide, bicarbonate	Involved in osmosis, membrane potentials, and acid–base balance
Nutrients	
Glucose, amino acids, triacylglycerol, cholesterol	Source of energy and basic "building blocks" of more complex molecules
Vitamins	Promote enzyme activity
Waste products	
Urea, uric acid, creatinine, ammonia salts	Breakdown products of protein metabolism; excreted by the kidneys
Bilirubin	Breakdown product of erythrocytes; excreted as part of the bile from the liver into the intestine
Lactic acid	End product of anaerobic respiration; converted to glucose by the liver
Gases	
Oxygen	Necessary for aerobic respiration; terminal electron acceptor in electron transport chain
Carbon dioxide	Waste product of aerobic respiration; as bicarbonate, helps buffer blood
Nitrogen	Inert
Regulatory substances	Enzymes catalyze chemical reactions; hormones stimulate or inhibit many body functions

in the embryo and fetus in tissues, such as the yolk sac, liver, thymus, spleen, lymph nodes, and red bone marrow. After birth, hematopoiesis is confined primarily to red bone marrow, with some lymphoid tissue helping in the production of lymphocytes (see chapter 22). In young children, nearly all the marrow is red bone marrow. In adults, however, red marrow is confined to the skull, ribs, sternum, vertebrae, pelvis, proximal femur, and proximal humerus. The red marrow in other locations is replaced by yellow marrow (see chapter 6).

All the formed elements of the blood are derived from a single population of stem cells located in the red bone marrow. **Stem cells** are precursor cells capable of dividing to produce daughter cells that can differentiate into other cell types. The hemopoietic stem cells produce the cells that give rise to the various types of blood cells (figure 19.2): **proerythroblasts** (prō-ĕ-rith′rō-blastz), from which erythrocytes develop; **myeloblasts** (mī′ĕ-lō-blastz), from which granulocytes develop; **lymphoblasts** (lim′fō-blastz), from which lymphocytes develop; **monoblasts** (mon′ō-blastz), from which monocytes develop; and **megakaryoblasts** (meg-ă-kar′ē-ō-blastz), from which platelets develop. The develop-

ment of each cell line is regulated by a specific growth factor. That is, the type of formed element derived from the stem cells and how many formed elements are produced are determined by the growth factor.

Clinical Note

Many cancer therapies affect dividing cells such as those found in tumors. An undesirable side effect of such therapies, however, can be the destruction of nontumor cells that are dividing, such as the cells in red bone marrow. After treatment for cancer, growth factors are used to stimulate the rapid regeneration of the red bone marrow. Although not a cure for cancer, the use of growth factors can speed recovery from the cancer therapy.

Some types of leukemia and genetic immune deficiency diseases can be treated with a bone marrow transplant containing blood stem cells. To avoid problems of tissue rejection, families with a history of these disorders can freeze the umbilical cord blood of their newborn children. The cord blood contains many stem cells and can be used instead of a bone marrow transplant.

Table 19.2   Formed Elements of the Blood

Cell Type	Illustration	Description	Function
Erythrocyte		Biconcave disk; no nucleus; contains hemoglobin, which colors the cell red; 7.5 μm in diameter	Transports oxygen and carbon dioxide
Leukocyte *Granulocytes*		Spherical cell with a nucleus; white in color because it lacks hemoglobin	Five types of leukocytes, each with specific functions
Neutrophil		Nucleus with two to four lobes connected by thin filaments; cytoplasmic granules stain a light pink or reddish purple; 10–12 μm in diameter	Phagocytizes microorganisms and other substances
Basophil		Nucleus with two indistinct lobes; cytoplasmic granules stain blue-purple; 10–12 μm in diameter	Releases histamine, which promotes inflammation, and heparin, which prevents clot formation
Eosinophil		Nucleus often bilobed; cytoplasmic granules stain orange-red or bright red; 11–14 μm in diameter	Releases chemicals that reduce inflammation; attacks certain worm parasites
Agranulocytes Lymphocyte		Round nucleus; cytoplasm forms a thin ring around the nucleus; 6–14 μm in diameter	Produces antibodies and other chemicals responsible for destroying microorganisms; contributes to allergic reactions, graft rejection, tumor control, and regulation of the immune system
Monocyte		Nucleus round, kidney, or horseshoe-shaped; contains more cytoplasm than does lymphocyte; 12–20 μm in diameter	Phagocytic cell in the blood; leaves the blood and becomes a macrophage, which phagocytizes bacteria, dead cells, cell fragments, and other debris within tissues
Platelet		Cell fragment surrounded by a plasma membrane and containing granules; 2–4 μm in diameter	Forms platelet plugs; releases chemicals necessary for blood clotting

Erythrocytes

Erythrocytes, or **red blood cells** (figure 19.3) are about 700 times more numerous than leukocytes and 17 times more numerous than platelets. Males have about 5.2 million erythrocytes per cubic millimeter of blood (range: 4.2–5.8 million), whereas females have about 4.5 million/mm³ (range: 3.6–5.2 million). Erythrocytes cannot move of their own accord and are passively moved by forces that cause the blood to circulate.

Structure

Normal erythrocytes are biconcave disks about 7.5 μm in diameter with edges that are thicker than the center of the cell. Compared with a flat disk of the same size, the biconcave shape increases the surface area of the erythrocyte. The greater surface area makes the movement of gases into and out of the erythrocyte more rapid. In addition, the erythrocyte can bend or fold around its thin center, decreasing its size and enabling it to pass more easily through small blood vessels.

Erythrocytes are highly specialized cells that lose their nuclei and nearly all their cellular organelles during maturation. The main component of the erythrocyte is the pigmented protein **hemoglobin** (hē-mō-glō′bin), which occupies about one-third of the total cell volume and accounts for its red color. Other erythrocyte contents include lipids, adenosine triphosphate (ATP), and the enzyme carbonic anhydrase.

Function

The primary functions of erythrocytes are to transport oxygen from the lungs to the various tissues of the body and to transport carbon dioxide from the tissues to the lungs. Approximately 98.5% of the oxygen transported in the blood from the lungs to the tissues is transported in combination with the hemoglobin in the erythrocytes, and the remaining 1.5% is dissolved in the water part of the plasma. If erythrocytes rupture, the hemoglobin leaks out into the plasma and becomes nonfunctional because the shape of the molecule changes as a result of denaturation (see chapter 2). Erythrocyte rupture followed by hemoglobin release is called **hemolysis** (hē-mol′-i-sis).

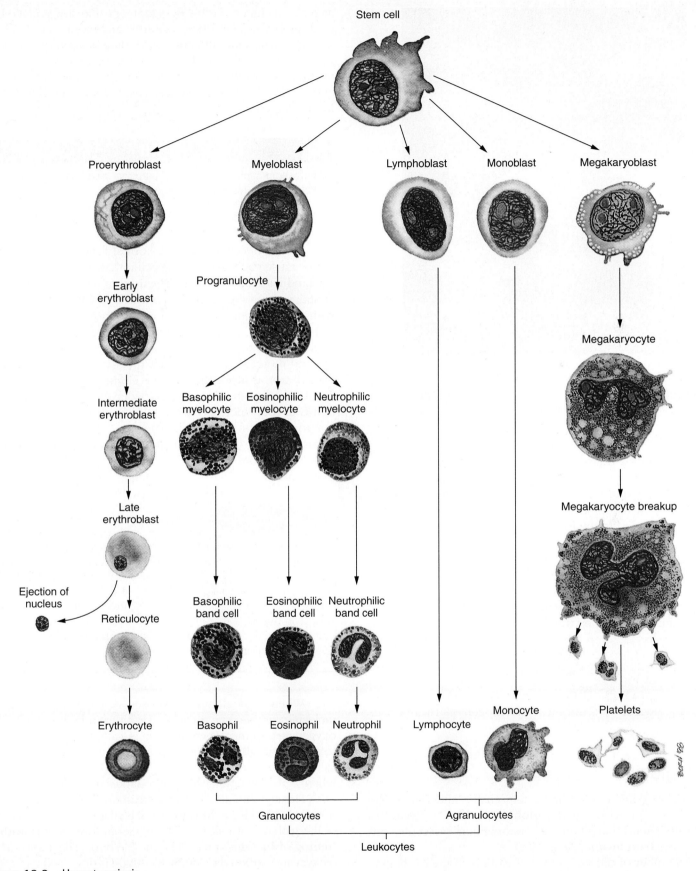

Stem cell

Proerythroblast   Myeloblast   Lymphoblast   Monoblast   Megakaryoblast

Early
erythroblast

Progranulocyte

Intermediate
erythroblast

Basophilic
myelocyte

Eosinophilic
myelocyte

Neutrophilic
myelocyte

Megakaryocyte

Late
erythroblast

Megakaryocyte breakup

Ejection of
nucleus

Basophilic
band cell

Eosinophilic
band cell

Neutrophilic
band cell

Reticulocyte

Erythrocyte   Basophil   Eosinophil   Neutrophil   Lymphocyte   Monocyte   Platelets

Granulocytes   Agranulocytes

Leukocytes

Figure 19.2   Hematopoiesis

All blood cells are derived from stem cells, which give rise to the progenitors of each cell type.

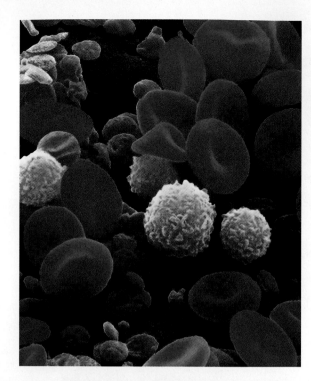

Figure 19.3   Erythrocytes and Leukocytes

Scanning electron micrograph of formed elements: erythrocytes (*red doughnut-shapes*) and leukocytes (yellow).

Carbon dioxide is transported in the blood in three major ways: approximately 7% is transported as carbon dioxide dissolved in the plasma, approximately 23% is transported in combination with blood proteins (mostly hemoglobin), and 70% is transported in the form of bicarbonate ions. The bicarbonate ions (HCO_3^-) are produced when carbon dioxide (CO_2) and water (H_2O) combine to form carbonic acid (H_2CO_3), which dissociates to form hydrogen (H^+) and bicarbonate ions. The combination of carbon dioxide and water is catalyzed by an enzyme, **carbonic anhydrase,** which is located primarily within erythrocytes.

$$\underset{\substack{\text{Carbon}\\\text{dioxide}}}{CO_2} + \underset{\text{Water}}{H_2O} \rightleftarrows \overset{\substack{\text{Carbonic}\\\text{anhydrase}}}{\underset{\substack{\text{Carbonic}\\\text{acid}}}{H_2CO_3}} \rightleftarrows \underset{\substack{\text{Hydrogen}\\\text{ion}}}{H^+} + \underset{\substack{\text{Bicarbonate}\\\text{ion}}}{HCO_3^-}$$

Hemoglobin

Hemoglobin consists of four protein chains and four heme groups. Each protein, called a **globin** (glō'bin), is bound to one **heme** (hēm). Each heme is a red-pigment molecule containing one **iron atom** (figure 19.4).

A number of different types of globin exist, each having a slightly different amino acid composition. The four globins in normal adult hemoglobin consist of two alpha (α) chains and two beta (β) chains. Embryonic and fetal globins appear at different times during development and are replaced by adult globin near the time of birth. Embryonic and fetal hemoglobins are more effective at binding oxygen than is adult hemoglobin. Abnormal hemoglobins are less effective at attracting oxygen than is normal hemoglobin and can result in anemia (see the Clinical Focus "Disorders of the Blood" later in the chapter).

1   P R E D I C T

What would happen to a fetus if maternal blood had an equal or greater affinity for oxygen than fetal blood?

✔ *Answer in Appendix F*

Iron is necessary for the normal function of hemoglobin because each oxygen molecule that is transported is associated with an iron atom. The adult human body normally contains about 4 g of iron, two-thirds of which is associated with hemoglobin. Small amounts of iron are regularly lost from the body in waste products such as urine and feces. Females lose additional iron as a result of menstrual bleeding and therefore require more dietary iron than do males. Dietary iron is absorbed into the circulation from the upper part of the intestinal tract. Acid from the stomach and vitamin C in food increase the solubility of iron in the alkaline environment of the small intestine, thus facilitating the absorption of iron in the small intestine. Iron absorption is regulated according to need, and iron deficiency can result in anemia.

Clinical Note

Various types of poisons affect the hemoglobin molecule. Carbon monoxide (CO), such as occurs in incomplete combustion of gasoline, binds to the iron of hemoglobin, forming the relatively stable compound carboxyhemoglobin (kar-bok'sē-hē-mō-glō'bin). As a result of the stable binding of carbon monoxide, hemoglobin cannot transport oxygen, and death may occur. Cigarette smoke also produces carbon monoxide, and the blood of smokers can contain 5%–15% carboxyhemoglobin.

When hemoglobin is exposed to oxygen, one oxygen molecule can become associated with each heme group. This oxygenated form of hemoglobin is called **oxyhemoglobin** (ok'sē-hē-mō-glō'bin). Hemoglobin containing no oxygen is called **deoxyhemoglobin.** Oxyhemoglobin is bright red, whereas deoxyhemoglobin has a darker red color.

Hemoglobin also transports carbon dioxide, which does not combine with the iron atoms but is attached to amino groups of the globin molecule. This hemoglobin form is **carbamino-hemoglobin** (kar-bam'i-nō-hē-mō-glō'bin). The transport of oxygen and carbon dioxide by the blood is discussed more fully in chapter 23.

A recently discovered function of hemoglobin is the transport of nitric oxide, which is produced by the endothelial

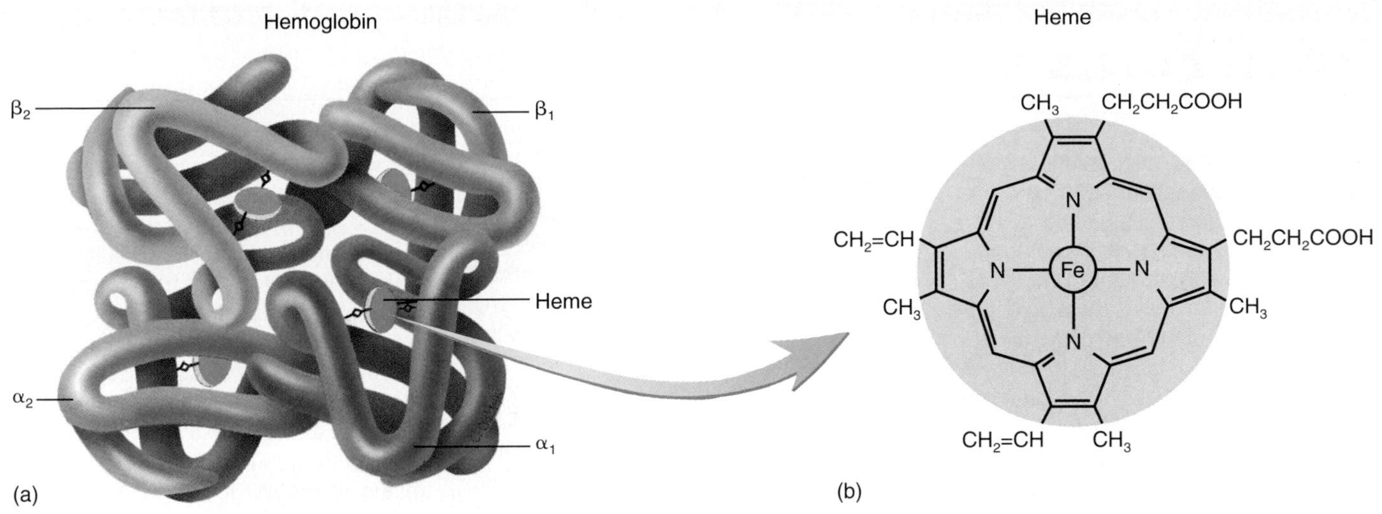

Hemoglobin

Heme

(a)

(b)

Figure 19.4   Hemoglobin

(*a*) Four protein chains, each with a heme, form a hemoglobin molecule. (*b*) Each heme contains one iron atom.

cells lining blood vessels. In the lungs, at the same time that heme picks up oxygen, in each β-globin a sulfur-containing amino acid, cysteine, picks up a nitric oxide molecule to form *S*-nitrosothiol (nī-trōs'ō-thī-ol; SNO). When oxygen is released in tissues so is the nitric oxide, which functions as a chemical signal that induces the smooth muscle of blood vessels to relax. By affecting the amount of nitric oxide in tissues, hemoglobin may play a role in regulating blood pressure, because relaxation of blood vessels results in a decrease in blood pressure (see chapter 21).

Clinical Note

Current research is being conducted in an attempt to develop artificial hemoglobin. One chemical that has been used in clinical trials is fluosol-DA, a white liquid with a high oxygen affinity. Although the usefulness of hemoglobin substitutes is currently limited because artificial hemoglobin is destroyed fairly quickly in the body, future work may uncover more successful substitutes that can provide long-term relief for patients with blood disorders.

The use of artificial hemoglobin could eliminate some of the disadvantages of using blood for blood transfusions. The use of artificial hemoglobin would prevent the possibility of transferring diseases such as hepatitis or AIDS or of causing transfusion reactions because of mismatched blood. In addition, artificial hemoglobin could be used when blood is not available.

Life History of Erythrocytes

Under normal conditions about 2.5 million erythrocytes are destroyed every second. This loss seems staggering until you realize that it represents only 0.00001% of the total 25 trillion erythrocytes contained in the normal adult circulation. Furthermore, these 2.5 million erythrocytes are being replaced by the production of an equal number of erythrocytes every second, thus maintaining homeostasis.

The process by which new erythrocytes is produced is called **erythropoiesis** (ĕ-rith'rō-poy-ē'sis; see figure 19.2), and the time required for the production of a single erythrocyte is about 4 days. Stem cells, from which all blood cells originate, give rise to proerythroblasts. After several mitotic divisions, proerythroblasts become **early (basophilic) erythroblasts** (ĕ-rith'rō-blastz), which stain with a basic dye. The dye stains the cytoplasm a purplish color because it binds to the large numbers of ribosomes, which are sites of synthesis for the protein hemoglobin. Early erythroblasts give rise to **intermediate (polychromatic) erythroblasts,** which stain different colors with basic and acidic dyes. As hemoglobin is synthesized and accumulates in the cytoplasm, it is stained a reddish color by an acidic dye. Intermediate erythroblasts continue to produce hemoglobin, and then most of their ribosomes and other organelles degenerate. The resulting **late erythroblasts** have a reddish color because about one-third of the cytoplasm is now hemoglobin.

The late erythroblasts lose their nuclei by a process of extrusion to become immature erythrocytes, which are called **reticulocytes** (re-tik'yū-lō-sītz), because a reticulum, or network, can be observed in the cytoplasm using a special staining technique. It is now known that the reticulum is artificially produced by the reaction of the dye with the few remaining ribosomes in the reticulocyte. The reticulocytes become mature erythrocytes when these remaining ribosomes degenerate. Mature erythrocytes and reticulocytes are released from the bone marrow into the circulating blood, which normally consists of mature erythrocytes and 1%–3% reticulocytes.

2   P R E D I C T

What does an elevated reticulocyte count indicate? Would a person's reticulocyte count change during the week after he had donated a unit (about 500 mL) of blood?

✔ *Answer in Appendix F*

Clinical Focus   Disorders of the Blood

Polycythemia

Polycythemia (pol'ē-sī-thē'mē-ǎ) is characterized by an overabundance of erythrocytes, resulting in increased blood viscosity, reduced flow rates, and, if severe, plugging of the capillaries. **Polycythemia vera** (ve'rǎ) is a chronic form of polycythemia of unknown cause. **Secondary polycythemia** results from a decreased oxygen supply, such as occurs at high altitudes, in chronic obstructive pulmonary disease, or congestive heart failure. The resulting decrease in oxygen delivery to the kidneys stimulates erythropoietin secretion, causing an increase in erythrocyte production.

Anemia (ǎ-nē'mē-ǎ) is a deficiency of hemoglobin in the blood. It can result from a decrease in the number of erythrocytes, a decrease in the amount of hemoglobin in each erythrocyte, or both. The decreased hemoglobin reduces the ability of the blood to transport oxygen. Anemic patients suffer from a lack of energy and feel excessively tired and listless. They can appear pale and quickly become short of breath with only slight exertion.

One general cause of anemia is insufficient production of erythrocytes. **Aplastic anemia** is caused by an inability of the red bone marrow to produce erythrocytes. It is usually acquired as a result of damage to the red marrow by chemicals (e.g., benzene), drugs (e.g., certain antibiotics and sedatives), or radiation.

Erythrocyte production can also be reduced because of nutritional deficiencies. **Iron-deficiency anemia** results from a deficient intake or absorption of iron or from excessive iron loss. Consequently, not enough hemoglobin is produced, and the erythrocytes are smaller than normal. **Folate deficiency** can also cause anemia. Inadequate amounts of folate in the diet is the usual cause of folate deficiency, with the disorder developing most often in the poor, in pregnant women, and in chronic alcoholics. Because folate helps in the synthesis of DNA, a folate deficiency results in fewer cell divisions and, therefore, decreased erythrocyte production. Another type of nutritional anemia is **pernicious** (per-nish'ŭs) **anemia**, which is caused by inadequate amounts of vitamin B_{12}. Because vitamin B_{12} is important for folate synthesis, inadequate amounts of vitamin B_{12} can also result in decreased erythrocyte production. Although inadequate levels of vitamin

B_{12} in the diet can cause pernicious anemia, the usual cause is insufficient absorption of the vitamin. Normally the stomach produces intrinsic factor, a protein that binds to vitamin B_{12}. The combined molecules pass into the lower intestine, where intrinsic factor facilitates the absorption of the vitamin. Without adequate levels of intrinsic factor, insufficient vitamin B_{12} is absorbed, and pernicious anemia develops. Present evidence suggests that the inability to produce intrinsic factor is an autoimmune disease in which the body's immune system damages the cells in the stomach that produce intrinsic factor.

Another general cause of anemia is loss or destruction of erythrocytes. **Hemorrhagic** (hem-ŏ-raj'ik) **anemia** results from a loss of blood, such as can result from trauma, ulcers, or excessive menstrual bleeding. Chronic blood loss, in which small amounts of blood are lost over time, can result in iron-deficiency anemia. **Hemolytic** (hē-mō-lit'ik) **anemia** is a disorder in which erythrocytes rupture or are destroyed at an excessive rate. It can be caused by inherited defects within the erythrocytes. For example, one kind of inherited hemolytic anemia results from a defect in the cell membrane that causes erythrocytes to rupture easily. Many kinds of hemolytic anemia result from unusual damage to the erythrocytes by drugs, snake venom, artificial heart valves, autoimmune disease, or hemolytic disease of the newborn.

Some anemias result from inadequate or defective hemoglobin production. **Thalassemia** (thal-ǎ-sē'mē-ǎ) is a hereditary disease found predominately in people of Mediterranean, Asian, and African ancestry. It is caused by insufficient production of the globin part of the hemoglobin molecule. The major form of the disease results in death by age 20, and the minor form in a mild anemia. **Sickle cell anemia** is a hereditary disease found mostly in people of African ancestry but also occasionally among people of Mediterranean heritage. It results in the formation of an abnormal hemoglobin, in which the erythrocytes assume a rigid sickle shape and plug up small blood vessels. They are also more fragile than normal erythrocytes. In its severe form, sickle cell anemia is usually fatal before the person is 30 years of age, whereas in its minor form, sickle cell trait, there are usually no symptoms.

Von Willibrand's Disease

Von Willibrand's disease is the most common inherited bleeding disorder, occurring as frequently as 1 in 1000 individuals. Von Willibrand factor (vWF) facilitates platelet adhesion and is the plasma carrier for factor VIII (see discussion on Coagulation and table 19.3 later in the chapter for more on factors). One treatment for von Willebrand's disease involves injections of vWF or concentrates of factor VIII to which vWF is attached. Another therapeutic approach is to administer a drug that increases vWF levels in the blood.

Hemophilia

Hemophilia (hē-mō-fil'ē-ǎ) is a genetic disorder in which clotting is abnormal or absent. It is most often found in people from northern Europe and their descendants. Because hemophilia is an X-linked trait (see chapter 29), it occurs almost exclusively in males. **Hemophilia A** (classic hemophilia) results from a deficiency of plasma coagulation factor VIII, and **hemophilia B** is caused by a deficiency in plasma factor IX. Hemophilia A occurs in approximately 1 in 10,000 male births, and hemophilia B occurs in approximately 1 in 100,000 male births. Treatment of hemophilia involves injection of the missing clotting factor taken from donated blood.

Thrombocytopenia

Thrombocytopenia (throm'bō-sī-tō-pē'nē-ǎ) is a condition in which the platelet number is greatly reduced, resulting in chronic bleeding through small vessels and capillaries. There are several causes of thrombocytopenia, including increased platelet destruction, caused by autoimmune disease (see chapter 22) or infections, or decreased platelet production, resulting from hereditary disorders, pernicious anemia, drug therapy, or radiation therapy.

Leukemia

The **leukemias** (lū-kē-mē'ǎs) are cancers of the red bone marrow in which abnormal production of one or more of the leukocyte types occurs. Because these cells are usually immature or abnormal and lack their normal immunologic functions, patients are very susceptible to infections. The excess production of leukocytes in the red marrow can also interfere with erythrocyte and platelet formation and thus lead to anemia and bleeding.

Infectious Diseases of the Blood

Microorganisms do not normally survive in the blood. Blood can transport microorganisms, however, and they can multiply in the blood. Microorganisms can enter the body and be transported by the blood to the tissues they infect. For example, the poliomyelitis virus enters through the gastrointestinal tract and is carried to nervous tissue. After microorganisms are established at a site of infection, some can enter the blood. They can then be transported to other locations in the body, multiply within the blood, or be eliminated by the body's immune system.

Septicemia (sep-ti-sē'mē-ă), or blood poisoning, is the spread of microorganisms and their toxins by the blood. Often septicemia results from the introduction of microorganisms by a medical procedure such as the insertion of an intravenous tube into a blood vessel. The release of toxins by microorganisms can cause septic shock, which is a decrease in blood pressure that can result in death.

In a few diseases, microorganisms actually multiply within blood cells. **Malaria** (mă-lār'ē-ă) is caused by a protozoan (*Plasmodium*) that is introduced into the blood by the bite of the *Anopheles* mosquito. Part of the development of the protozoan occurs inside erythrocytes. The symptoms of chills and fever in malaria are produced by toxins released when the protozoan causes the erythrocytes to rupture. **Infectious mononucleosis** (mon'ō-nū-klē-ō'sis) is caused by a virus (Epstein-Barr virus) that infects the salivary glands and lymphocytes. The lymphocytes are altered by the virus, and the immune system attacks and destroys the lymphocytes. The immune system response is believed to produce the symptoms of fever, sore throat, and swollen lymph nodes. The **acquired immune deficiency syndrome (AIDS)** is caused by the human immunodeficiency virus (HIV), which infects lymphocytes and suppresses the immune system (see chapter 22).

The presence of microorganisms in the blood is a concern when transfusions are made, because it is possible to infect the blood recipient. Blood is routinely tested in an effort to eliminate this risk, especially for AIDS and hepatitis. **Hepatitis** (hep-ă-tī'tis) is an infection of the liver caused by several different kinds of viruses. After recovering, hepatitis victims can become carriers. Although they show no signs of the disease, they release the virus into their blood or bile. To prevent infection of others, anyone who has had hepatitis is asked not to donate blood products.

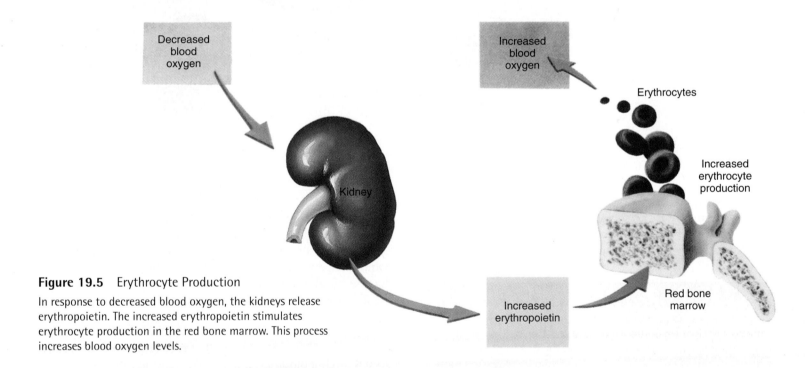

Figure 19.5   Erythrocyte Production

In response to decreased blood oxygen, the kidneys release erythropoietin. The increased erythropoietin stimulates erythrocyte production in the red bone marrow. This process increases blood oxygen levels.

Cell division requires the vitamins folate and B_{12}, which are necessary for the synthesis of deoxyribonucleic acid (see chapter 3). Hemoglobin production requires iron. Consequently, lack of folate, vitamin B_{12}, or iron can interfere with normal erythrocyte production.

Erythrocyte production is stimulated by low blood oxygen levels, typical causes of which are decreased numbers of erythrocytes, decreased or defective hemoglobin, diseases of the lungs, high altitude, inability of the cardiovascular system to deliver blood to tissues, and increased tissue demands for oxygen such as during endurance exercises.

Low blood oxygen levels stimulate erythrocyte production by increasing the formation of the glycoprotein **erythropoietin** (ě-rith-rō-poy'ě-tin) by the kidneys (figure 19.5). Erythropoietin stimulates red bone marrow to produce more erythrocytes by increasing the number of proerythroblasts formed and by decreasing the time required for erythrocytes to mature. Thus, when oxygen levels in the blood decrease, erythropoietin production increases, which increases erythrocyte production. The increased number of erythrocytes increases the ability of the blood to transport oxygen. This mechanism returns blood oxygen levels to normal and

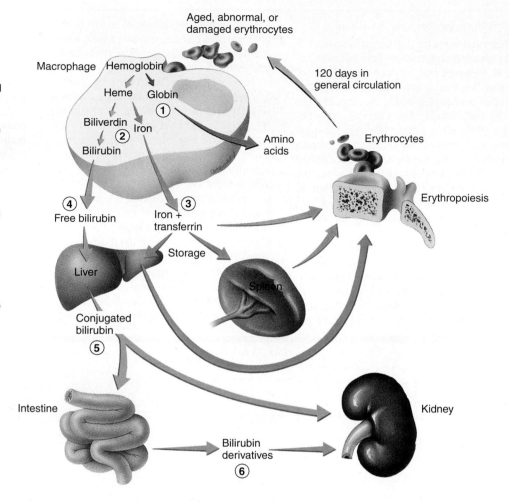

1. The globin chains of hemoglobin are broken down to individual amino acids (*lavender arrow*) and are metabolized or used to build new proteins.

2. Iron is released from the heme of hemoglobin. The heme is converted into biliverdin, which is converted into bilirubin.

3. Iron is distributed to various tissues for storage or transported to the red bone marrow and used in the production of new hemoglobin (*green arrows*).

4. Free bilirubin (*blue arrow*) is transported to the liver.

5. Most conjugated bilirubin is excreted as part of the bile; some is excreted in the urine.

6. Bilirubin derivatives contribute to the color of feces or are reabsorbed and excreted in the urine.

Figure 19.6   Hemoglobin Breakdown

Hemoglobin is broken down in macrophages, and the breakdown products are used or excreted.

---

maintains homeostasis by increasing the delivery of oxygen to tissues. Conversely, if blood oxygen levels increase, less erythropoietin is released, and erythrocyte production decreases.

3   P R E D I C T

Cigarette smoke produces carbon monoxide. If a nonsmoker smoked a pack of cigarettes a day for a few weeks, what would happen to the number of erythrocytes in the person's blood? Explain.

✔ *Answer in Appendix F*

Erythrocytes normally stay in the circulation for about 120 days in males and 110 days in females. These cells have no nuclei and therefore cannot produce new proteins. As their existing proteins, enzymes, cell membrane components, and other structures degenerate, the erythrocytes become old and abnormal in form and function. Erythrocytes can also be damaged in various ways while passing through the circulation.

Old, damaged, or defective erythrocytes are removed from the blood by **macrophages** located in the spleen, liver,

and other lymphatic tissues (figure 19.6). Within the macrophage, lysosomal enzymes break open erythrocytes and begin to digest hemoglobin. Globin is broken down into its component amino acids, most of which are reused in the production of other proteins. Iron atoms also are released for recycling. The heme groups are converted to **biliverdin** (bil-i-ver′din) and then to **bilirubin** (bil-i-rū′bin), which is released into the plasma. Bilirubin binds to albumin and is transported to liver cells. This bilirubin is called **free (indirect) bilirubin** even though it binds to albumin. The free bilirubin is taken up by the liver cells and is conjugated, or joined, to glucuronic acid to form **conjugated (direct) bilirubin,** which is more water-soluble than free bilirubin. Some of the conjugated bilirubin escapes back into the blood and is excreted by the kidneys. Most of the conjugated bilirubin becomes part of the **bile,** which is the fluid secreted from the liver into the small intestine. In the intestines, bacteria convert bilirubin into the pigments that give the feces its characteristic brownish color. Some of these pigments are absorbed from the intestine, modified in the kidneys, and excreted in the urine, contributing to the characteristic yellowish color of urine. A yellowish staining of the skin and sclerae by bile pigments, associated with

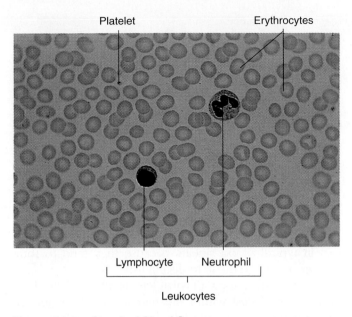

Figure 19.7   Standard Blood Smear

The erythrocytes are pink with whitish centers. The centers appear whitish because light more readily shines through the thin center of the disk than through the thicker edges. The leukocytes have been stained and have pink-colored cytoplasm and purple-colored nuclei.

a buildup of bilirubin in the circulation and interstitial spaces, is known as **jaundice** (jawn'dis).

Leukocytes

Leukocytes, or **white blood cells,** are clear or whitish colored cells that lack hemoglobin but have a nucleus. In stained preparations, leukocytes attract stain, whereas erythrocytes remain relatively unstained (figure 19.7; see table 19.2).

Leukocytes protect the body against invading microorganisms and remove dead cells and debris from the body. Most leukocytes are motile, exhibiting **ameboid movement,** which is the ability to move like an ameba by putting out irregular cytoplasmic projections. Leukocytes leave the circulation and enter tissues by **diapedesis** (dī'ă-pĕ-dē'sis), a process in which they become thin and elongated and slip between or in some cases through the cells of blood vessel walls. The leukocytes can then be attracted to foreign materials or dead cells within the tissue by **chemotaxis** (kem-ō-tak'sis). At the site of an infection, leukocytes accumulate and phagocytize bacteria, dirt, and dead cells; then they die. The accumulation of dead leukocytes and bacteria, along with fluid and cell debris, is called **pus.**

The five types of leukocytes are neutrophils, eosinophils, basophils, lymphocytes, and monocytes.

Neutrophils

Neutrophils (see table 19.2), the most common type of leukocytes in the blood, have small cytoplasmic granules that stain with both acidic and basic dyes. Their nuclei are commonly trilobed, but the number of lobes varies from two to five. Neutrophils often are called **polymorphonuclear** (pol'ē-mōr-fō-nū'klē-ăr) **neutrophils,** or PMNs, to indicate that their nuclei can occur in more than one (poly) form (morph). Neutrophils usually remain in the circulation for about 10–12 h, and then move into other tissues, where they become motile and seek out and phagocytize bacteria, antigen–antibody complexes (antigens and antibodies bound together), and other foreign matter. Neutrophils also secrete a class of enzymes called **lysozymes** (lī'sō-zīmz), which are capable of destroying certain bacteria. Neutrophils usually survive for 1–2 days after leaving the circulation.

Eosinophils

Eosinophils (see table 19.2) contain cytoplasmic granules that stain bright red with eosin, an acidic stain. They are motile cells that leave the circulation to enter the tissues during an inflammatory reaction. They are most common in tissues undergoing an allergic response, and their numbers are elevated in the blood of people with allergies or certain parasitic infections. Eosinophils apparently reduce the inflammatory response by producing enzymes that destroy inflammatory chemicals such as histamine. Eosinophils also phagocytize antigen–antibody complexes formed during the allergic response, although they are not as important in this function as neutrophils.

Basophils

Basophils (see table 19.2), the least common of all leukocytes, contain large cytoplasmic granules that stain blue or purple with basic dyes. Basophils, like other granulocytes, leave the circulation and migrate through the tissues, where they play a role in both allergic and inflammatory reactions. Basophils contain large amounts of **histamine,** which they release within tissues to increase inflammation. They also release **heparin,** which inhibits blood clotting.

Lymphocytes

The smallest leukocytes are **lymphocytes,** most of which are slightly larger in diameter than erythrocytes (see table 19.2). The lymphocytic cytoplasm consists of only a thin, sometimes imperceptible, ring around the nucleus. Although lymphocytes originate in bone marrow, they migrate through the blood to lymphatic tissues where they can proliferate and produce more lymphocytes. The majority of the body's total lymphocyte population is in the lymphatic tissues: the lymph nodes, spleen, tonsils, lymph nodules, and thymus.

Although they cannot be identified using standard microscopic examination, a number of different kinds of lymphocytes play important roles in immunity (see chapter 22 for details). For example, **B cells** can be stimulated by bacteria or toxins to divide, forming cells that produce proteins called **antibodies.** The antibodies can attach to the bacteria

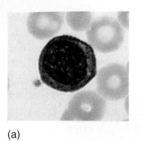
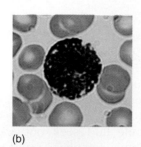
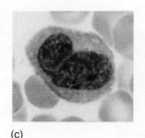
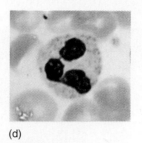
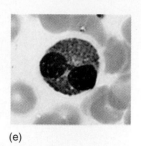

(a)   (b)   (c)   (d)   (e)

Figure 19.8   Identification of Leukocytes

See Predict question 4.

and activate mechanisms that result in destruction of the bacteria. **T cells** protect against viruses by attacking and destroying cells in which viruses are reproducing. In addition, T cells are involved with the destruction of tumor cells and tissue graft rejections.

Monocytes

Monocytes are typically the largest of the leukocytes (see table 19.2). Monocytes normally remain in the circulation for about 3 days, leave the circulation, become transformed into macrophages, and migrate through the various tissues. They phagocytize bacteria, dead cells, cell fragments, and other debris within the tissues. An increase in the number of monocytes is often associated with chronic infections. In addition, macrophages can break down phagocytized foreign substances and present the processed substances to lymphocytes, which results in activation of the lymphocytes (see chapter 22).

4	P R E D I C T

Based on their morphology, identify each of the leukocytes shown in figure 19.8.

✔ *Answer in Appendix F*

Platelets

Platelets, or **thrombocytes** (see figure 19.7 and table 19.2), are minute fragments of cells consisting of a small amount of cytoplasm surrounded by a plasma membrane. The surface of platelets has glycoproteins and proteins that allow platelets to attach to other molecules such as collagen in connective tissue. Some of these surface molecules, as well as molecules released from granules in the platelet cytoplasm, play important roles in controlling blood loss. The platelet cytoplasm also contains actin and myosin that can cause contraction of the platelet (see section on Clot Retraction and Dissolution later in this chapter).

Platelets are roughly disk-shaped and average about 3 μm in diameter. Even though they are about 40 times more common in the blood than leukocytes, platelets often are not

counted in typical blood smears because they tend to form clumps and become difficult to distinguish.

The life expectancy of platelets is about 5–9 days. Platelets are produced within the marrow and are derived from **megakaryocytes** (meg-ă-kar′ē-ō-sītz), which are extremely large cells with diameters up to 100 μm. Small fragments of these cells break off and enter the circulation as platelets.

Platelets play an important role in preventing blood loss by (1) forming platelet plugs, which seal holes in small vessels and (2) by forming clots, which help seal off larger wounds in the vessels.

Hemostasis

Hemostasis (hē′mō-stā-sis), the arrest of bleeding, is very important to the maintenance of homeostasis. If not stopped, excessive bleeding from a cut or torn blood vessel can result in a positive-feedback pathway, consisting of ever-decreasing blood volume and blood pressure, leading away from homeostasis, and resulting in death. Fortunately, when a blood vessel is damaged, a number of events occur that help prevent excessive blood loss. Hemostasis can be divided into three stages: vascular spasm, platelet plug formation, and coagulation.

Vascular Spasm

Vascular spasm is an immediate but temporary closure of a blood vessel resulting from contraction of smooth muscle within the wall of the blood vessel. In small vessels, this constriction can close the vessels completely and stop the flow of blood through the vessels. Vascular spasm is produced by nervous system reflexes and by chemicals. For example, during the formation of a platelet plug, platelets release **thromboxanes** (throm′bok-zānz), which are derived from prostaglandins, and endothelial cells release the peptide **endothelin** (en-dō′thē-lin).

Platelet Plug Formation

A **platelet plug** is an accumulation of platelets that can seal up small breaks in blood vessels. Platelet plug formation is

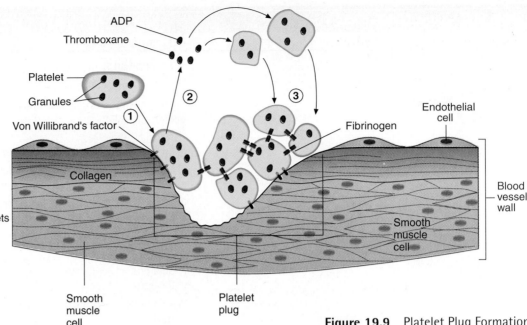

1. Platelet adhesion to collagen is mediated by Von Willibrand's factor and platelet surface receptors.

2. The platelet release reaction is the release of ADP, thromboxanes, and other chemicals that activate other platelets.

3. Platelet aggregation occurs as activated platelets produce surface receptors that bind to fibrinogen, connecting the platelets to one another. A platelet plug is formed by the accumulating mass of platelets.

Figure 19.9   Platelet Plug Formation

very important in maintaining the integrity of the circulatory system because small tears occur in the smaller vessels and capillaries many times each day, and platelet plug formation quickly closes them. People who lack the normal number of platelets tend to develop numerous small hemorrhages in their skin and internal organs.

The formation of a platelet plug can be described as a series of steps, but in actuality many of the events occur simultaneously (figure 19.9).

1. **Platelet adhesion** occurs when platelets bind to collagen exposed by blood vessel damage. Von Willibrand's factor, produced and secreted by blood vessel endothelial cells, binds to platelet surface receptors and collagen, causing the platelets to adhere to the collagen. In addition, other platelet surface receptors can bind directly to collagen.

2. After platelets adhere to collagen, they become activated, and in the **platelet release reaction,** adenosine diphosphate (ADP), thromboxanes, and other chemicals are extruded from the platelets by exocytosis. The ADP and thromboxanes stimulate other platelets to become activated and release additional chemicals, producing a cascade of chemical release by the platelets. Thus more and more platelets become activated.

3. As platelets become activated they also express surface receptors that can bind to fibrinogen, a plasma protein. In **platelet aggregation,** the fibrinogen forms a bridge between the surface receptors of different platelets, resulting in the formation of a platelet plug.

4. Activated platelets express phospholipids (platelet factor III) and coagulation factor V, which are an important part of clot formation (see following section on Coagulation).

Clinical Note

The production of prostaglandins is very important to platelet plug formation, as is demonstrated by the effect of aspirin. Because aspirin inhibits prostaglandin synthesis, the production of thromboxanes, which are derived from prostaglandins, is also inhibited, resulting in reduced platelet activation. If an expectant mother ingests aspirin near the end of pregnancy, prostaglandin synthesis is inhibited and several effects are possible. Two of these effects are (1) the mother can experience excessive postpartum hemorrhage because of decreased platelet function, and (2) the baby can exhibit numerous localized hemorrhages called **petechiae** over the surface of its body as a result of decreased platelet function. If the quantity of ingested aspirin is large, the infant, mother, or both may die as a result of hemorrhage. On the other hand, in a stroke or heart attack, platelet plugs and clots can form in vessels and threaten the life of the individual. Studies of individuals who are at risk because of the development of clots, such as people who have had a previous heart attack, indicate that taking small amounts of aspirin daily can reduce the likelihood of clot formation and another heart attack. That everyone should take aspirin daily, however, is not currently recommended.

Coagulation

Vascular spasms and platelet plugs alone are not sufficient to close large tears or cuts. When a blood vessel is severely damaged, **coagulation** (kō-ag-yū-lā′shŭn), or **blood clotting,** results in the formation of a clot. A **blood clot** is a network of threadlike protein fibers, called **fibrin,** that traps blood cells, platelets, and fluid.

The formation of a blood clot depends on a number of proteins found within plasma called **coagulation factors** (table 19.3). Normally the coagulation factors are in an inactive state and do not cause clotting. After injury, the clotting factors

Table 19.3   Coagulation Factors

Factor Number	Name (synonym)	Description and Function
I	Fibrinogen	Plasma protein synthesized in liver; converted to fibrin in stage 3
II	Prothrombin	Plasma protein synthesized in liver (requires vitamin K); converted to thrombin in stage 2
III	Tissue factor (thromboplastin)	Mixture of lipoproteins released from damaged tissue; required in extrinsic stage 1
IV	Calcium ion	Required throughout entire clotting sequence
V	Proaccelerin (labile factor)	Plasma protein synthesized in liver; activated form functions in stages 1 and 2 of both intrinsic and extrinsic clotting pathways
VI		Once thought to be involved but no longer accepted as playing a role in coagulation; apparently the same as activated factor V
VII	Serum prothrombin conversion accelerator (stable factor, proconvertin)	Plasma protein synthesized in liver (requires vitamin K); functions in extrinsic stage 1
VIII	Antihemophilic factor (antihemophilic globulin)	Plasma protein synthesized in megakaryocytes and endothelial cells; required for intrinsic stage 1
IX	Plasma thromboplastin component (Christmas factor)	Plasma protein synthesized in liver (requires vitamin K); required for intrinsic stage 1
X	Stuart factor (Stuart-Prower factor)	Plasma protein synthesized in liver (requires vitamin K); required in stages 1 and 2 of both intrinsic and extrinsic clotting pathways
XI	Plasma thromboplastin antecedent	Plasma protein synthesized in liver; required for intrinsic stage 1
XII	Hageman factor	Plasma protein required for intrinsic stage 1
XIII	Fibrin-stabilizing factor	Protein found in plasma and platelets; required for stage 3
Platelet Factors		
I	Platelet accelerator	Same as plasma factor V
II	Thrombin accelerator	Accelerates thrombin (intrinsic clotting pathway) and fibrin production
III		Phospholipids necessary for the intrinsic and extrinsic clotting pathways
IV		Binds heparin, which prevents clot formation

are activated to produce a clot. This activation is a complex process involving many chemical reactions, some of which require calcium (Ca^{2+}) ions and molecules on the surface of activated platelets, such as phospholipids and coagulation factor V.

5   P R E D I C T

Why is it advantageous for clot formation to involve molecules on the surface of activated platelets.

✔ *Answer in Appendix F*

The activation of clotting proteins can be summarized in three main stages (figure 19.10). **Stage 1** consists of the formation of **prothrombinase, stage 2** is the conversion of **prothrombin** to **thrombin** by prothrombinase, and **stage 3** consists of the conversion of soluble **fibrinogen** to insoluble **fibrin** by thrombin.

Depending on how prothrombinase is formed in stage 1, two separate pathways for coagulation have been described: the **extrinsic clotting pathway** and the **intrinsic clotting pathway** (see figure 19.10).

Extrinsic Clotting Pathway

The extrinsic clotting pathway is so named because it begins with chemicals that are outside of, or extrinsic to, the blood (see figure 19.10). In stage 1, damaged tissues release a mixture of lipoproteins and phospholipids called **tissue factor (TF),** also known as thromboplastin (throm-bō-plas′tin), or factor III. Tissue factor, in the presence of calcium (Ca^{2+}) ions, forms a complex with factor VII, which activates factor X. On the surface of platelets, activated factor X, factor V, platelet phospholipids, and Ca^{2+} ions complex to form prothrombinase. In stage 2, the soluble plasma protein prothrombin is converted to the enzyme thrombin by prothrombinase. During stage 3, the soluble plasma protein fibrinogen is converted to the insoluble protein fibrin by thrombin. The fibrin forms the fibrous network of the clot. Thrombin also stimulates factor XIII activation, which is necessary to stabilize the clot.

Intrinsic Clotting Pathway

The intrinsic clotting pathway is so named because it begins with chemicals that are inside, or intrinsic to, the blood (see

Stage 1 can be activated in two ways:

Extrinsic pathway starts with tissue factor, which is released outside of the plasma in damaged tissue.

Intrinsic pathway starts when inactive factor XII, which is in the plasma, is activated by coming into contact with a damaged blood vessel.

Stage 1: Damage to tissue or blood vessels activates clotting factors that activate other clotting factors, which leads to the production of prothrombinase. The activated factors are within white ovals, whereas the inactive precursors are shown as yellow ovals.

Stage 2: Prothrombin is activated by prothrombinase to form thrombin.

Stage 3: Fibrinogen is activated by thrombin to form fibrin, which forms the clot.

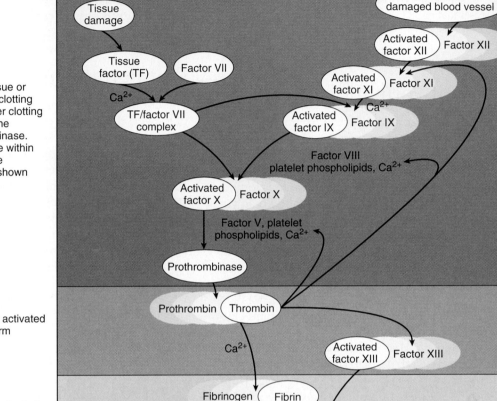

Figure 19.10   Clot Formation

figure 19.10). In stage 1, damage to blood vessels can expose collagen in the connective tissue beneath the epithelium lining the blood vessel. When plasma factor XII comes into contact with collagen, it is activated and stimulates factor XI, which in turn activates factor IX. Activated factor IX joins with factor VIII, platelet phospholipids, and Ca^{2+} ions to activate factor X. On the surface of platelets, activated factor X, factor V, platelet phospholipids, and Ca^{2+} ions complex to form prothrombinase. Stages 2 and 3 then are activated, and a clot results.

Although once considered distinct pathways, it is now known that the extrinsic pathway can activate the clotting proteins in the intrinsic pathway. The TF–VII complex from the extrinsic pathway can stimulate the formation of activated factors IX and X in the intrinsic pathway. When tissues are damaged, tissue factor also rapidly leads to the production of thrombin, which can activate many of the clotting proteins in the intrinsic pathway, such as factor XI and prothrombinase.

Thus thrombin is part of a positive-feedback system in which thrombin production stimulates the production of additional thrombin. Thrombin also has a positive-feedback effect on coagulation by stimulating platelet activation.

Clinical Note

Many of the factors involved in clot formation require vitamin K for their production (see table 19.3). Humans rely on two sources of vitamin K. About half comes from the diet, and half from bacteria within the large intestine. Antibiotics taken to fight bacterial infections sometimes kill these intestinal bacteria, reducing vitamin K levels and resulting in bleeding problems. Vitamin K supplements may be necessary for patients on prolonged antibiotic therapy. Newborns lack these intestinal bacteria, and a vitamin K injection is routinely given to infants at birth. Infants can also obtain vitamin K from food such as milk. Because cow's milk contains more vitamin K than does human milk, breast-fed infants are more susceptible to hemorrhage than bottle-fed infants.

Vitamin K is absorbed from the small intestine into the circulation and is necessary for the synthesis of several clotting factors by the liver. The absorption of vitamin K from the intestine requires the presence of bile. Certain disorders such as obstruction of bile flow to the intestine can interfere with vitamin K absorption and lead to insufficient clotting. Liver diseases that result in the decreased synthesis of clotting factors also can lead to insufficient clot formation.

of anticoagulants such as heparin, which acts rapidly, or warfarin, which acts more slowly than heparin. Warfarin (Coumadin) prevents clot formation by suppressing the production of vitamin K-dependent coagulation factors (II, VII, IX, and X) by the liver. Interestingly, warfarin was first used as a rat poison, causing rats to bleed to death. In small doses warfarin is a proven, effective anticoagulant in humans. Caution is necessary with anticoagulant treatment, however, because the patient can hemorrhage internally or bleed excessively when cut.

Control of Clot Formation

Without control, coagulation would spread from the point of initiation to the entire circulatory system. Furthermore, vessels in a normal person contain rough areas that can stimulate clot formation, and small amounts of prothrombin are constantly being converted into thrombin. To prevent unwanted clotting, the blood contains several **anticoagulants** (an'tē-kō-ag'yū-lantz), which prevent coagulation factors from initiating clot formation. Only when coagulation factor concentrations exceed a given threshold does coagulation occur. At the site of injury so many coagulation factors are activated that the anticoagulants are unable to prevent clot formation. Away from the injury site, however, the activated coagulation factors are diluted in the blood, anticoagulants neutralize them, and clotting is prevented.

Examples of anticoagulants in the blood are antithrombin, heparin, and prostacyclin. **Antithrombin,** a plasma protein produced by the liver, slowly inactivates thrombin. **Heparin,** produced by basophils and endothelial cells, increases the effectiveness of antithrombin because heparin and antithrombin together rapidly inactivate thrombin. **Prostacyclin** (pros-tă-sī′klin) is a prostaglandin derivative produced by endothelial cells. It counteracts the effects of thrombin by causing vasodilation and by inhibiting the release of coagulation factors from platelets.

Anticoagulants are also important outside the body, preventing the clotting of blood used in transfusions and laboratory blood tests. Examples include heparin, **ethylenediaminetetraacetic** (eth'il-ēn-dī'ă-mēn-tet-ră-ă-sē'tik) **acid (EDTA),** and sodium citrate. EDTA and sodium citrate prevent clot formation by binding to calcium ions, making them inaccessible for clotting reactions.

> ### Clinical Note
>
> When platelets encounter damaged or diseased areas on the walls of blood vessels or the heart, an attached clot called a **thrombus** (throm'bŭs) can form. A thrombus that breaks loose and begins to float through the circulation is called an **embolus** (em'bō-lŭs). Both thrombi and emboli can result in death if they block vessels that supply blood to essential organs such as the heart, brain, or lungs. Abnormal coagulation can be prevented or hindered by the injection

Clot Retraction and Dissolution

The fibrin meshwork constituting the clot adheres to the walls of the vessel. Once the clot has formed, it begins to condense into a denser, compact structure through a process known as **clot retraction.** Platelets contain contractile proteins, actin and myosin, which operate in a similar fashion to the actin and myosin in smooth muscle (see chapter 10). Platelets form small extensions that attach to fibrin. Contraction of the extensions pulls on the fibrin and is responsible for clot retraction. As the clot condenses, a fluid called **serum** (sēr'ŭm) is squeezed out of it. Serum is plasma from which fibrinogen and some of the clotting factors have been removed.

Consolidation of the clot pulls the edges of the damaged vessel together, which can help to stop the flow of blood, reduce infection, and enhance healing. The damaged vessel is repaired by the movement of fibroblasts into the damaged area and the formation of new connective tissue. In addition, epithelial cells around the wound proliferate and fill in the torn area.

The clot usually is dissolved within a few days after clot formation by a process called **fibrinolysis** (fī-bri-nol'-i-sis), which involves the activity of **plasmin** (plaz'min), an enzyme that hydrolyzes fibrin. Plasmin is formed from inactive plasminogen, which is a normal blood protein. It is activated by thrombin, factor XII, tissue plasminogen activator (t-PA), urokinase, and lysosomal enzymes released from damaged tissues (figure 19.11). In disorders that are caused by blockage of a vessel by a clot, such as a heart attack, dissolving the clot can restore blood flow and reduce damage to tissues. For example, streptokinase (a bacterial enzyme), tissue plasminogen activator, or urokinase can be injected into the blood or introduced at the clot site by means of a catheter. These substances convert plasminogen to plasmin, which breaks down the clot.

Blood Grouping

If large quantities of blood are lost during surgery or in an accident, the blood volume must be increased, or the patient can go into shock and die. A **transfusion** is the transfer of blood or other solutions into the blood of the patient. In many cases the return of blood volume to normal levels is all that is necessary. This can be accomplished by the transfusion of plasma or plasma expanders, which are prepared solutions having the

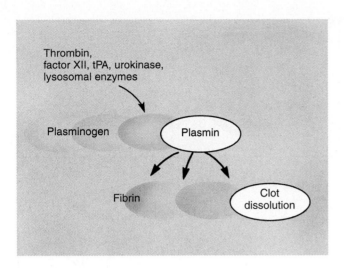

Figure 19.11   Fibrinolysis

Plasminogen is converted by thrombin, factor XII, tissue plasminogen activator (t-PA), urokinase, or lysosomal enzymes to the active enzyme plasmin. Plasmin breaks the fibrin molecules and therefore the clot into smaller pieces, which are washed away in the blood or are phagocytized.

proper amounts of solutes. When large quantities of blood are lost, however, erythrocytes must also be replaced so that the oxygen-carrying capacity of the blood is restored.

Early attempts to transfuse blood from one person to another were often unsuccessful because they resulted in transfusion reactions, which included clotting within blood vessels, kidney damage, and death. It is now known that transfusion reactions are caused by interactions between antigens and antibodies (see chapter 22). In brief, the surfaces of erythrocytes have molecules called **antigens** (an'ti-jenz), and in the plasma there are molecules called **antibodies.** Antibodies are very specific, meaning that each antibody can combine only with a certain antigen. When the antibodies in the plasma bind to the antigens on the surfaces of the erythrocytes, they form molecular bridges that connect the erythrocytes. As a result, **agglutination** (ă-glū-ti-nā'shŭn), or clumping, of the cells occurs. The combination of the antibodies with the antigens can also initiate reactions that cause **hemolysis,** or rupture of the erythrocytes. Because the antigen–antibody combinations can cause agglutination, the antigens are often called **agglutinogens** (ă-glū-tin'ō-jenz), and the antibodies are called **agglutinins** (ă-glū'ti-ninz).

The antigens on the surface of erythrocytes have been categorized into **blood groups,** and more than 35 blood groups, most of which are rare, have been identified. For transfusions, the ABO and Rh blood groups are among the most important. Other well-known groups include the Lewis, Duffy, MNSs, Kidd, Kell, and Lutheran groups.

ABO Blood Group

In the **ABO blood group,** type A blood has type A antigens, type B blood has type B antigens, type AB blood has both

types of antigens, and type O blood has neither A nor B antigens (figure 19.12). In addition, plasma from type A blood contains type B antibodies, which act against type B antigens, whereas plasma from type B blood contains type A antibodies, which act against type A antigens. Type AB blood has neither type of antibody, and type O blood has both A and B antibodies.

The ABO blood types are not found in equal numbers. In white people in the United States the distribution is type O, 47%; type A, 41%; type B, 9%; and type AB, 3%. Among black people in the United States the distribution is type O, 46%; type A, 27%; type B, 20%; and type AB, 7%.

The presence of A and B antibodies in blood is not clearly understood. Antibodies normally do not develop against an antigen unless the body is exposed to that antigen. This means, for example, that a person with type A blood should not have type B antibodies unless he or she has received a transfusion of type B blood, which contains type B antigens. People with type A blood do have type B antibodies, however, even though they have never received a transfusion of type B blood. One possible explanation is that type A or B antigens on bacteria or food in the digestive tract stimulate the formation of antibodies against antigens that are different from one's own antigens. Thus a person with type A blood would produce type B antibodies against the B antigens on the bacteria or food. In support of this hypothesis is the observation that A and B antibodies are not found in the blood until about 2 months after birth.

A blood **donor** gives blood, and a **recipient** receives blood. Usually a donor can give blood to a recipient if they both have the same blood type. For example, a person with type A blood could donate to another person with type A blood. There would be no ABO transfusion reaction because the recipient has no antibodies against the type A antigen. On the other hand, if type A blood were donated to a person with type B blood, there would be a transfusion reaction because the person with type B blood has antibodies against the type A antigen, and agglutination would result (figure 19.13).

Historically, people with type O blood have been called universal donors because they usually can give blood to the other ABO blood types without causing an ABO transfusion reaction. Their erythrocytes have no ABO surface antigens and therefore do not react with the recipient's A or B antibodies. For example, if type O blood is given to a person with type A blood, the type O erythrocytes do not react with the type B antibodies in the recipient's blood. In a similar fashion, if type O blood is given to a person with type B blood, there would be no reaction with the recipient's type A antibodies.

The term universal donor is misleading, however. Transfusion of type O blood can produce a transfusion reaction for two reasons. First, other blood groups can cause a transfusion reaction. Second, antibodies in the blood of the donor can react with antigens in the blood of the recipient. For example, type O blood has type A and B antibodies. If type O blood is transfused into a person with type A blood, the A antibodies (in the type O blood) react against the A antigens (in the type A blood). Usually such reactions are not

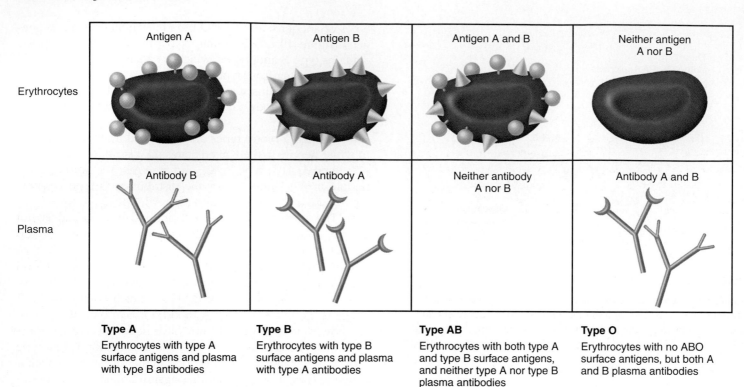

Type A
Erythrocytes with type A surface antigens and plasma with type B antibodies

Type B
Erythrocytes with type B surface antigens and plasma with type A antibodies

Type AB
Erythrocytes with both type A and type B surface antigens, and neither type A nor type B plasma antibodies

Type O
Erythrocytes with no ABO surface antigens, but both A and B plasma antibodies

Figure 19.12   ABO Blood Groups

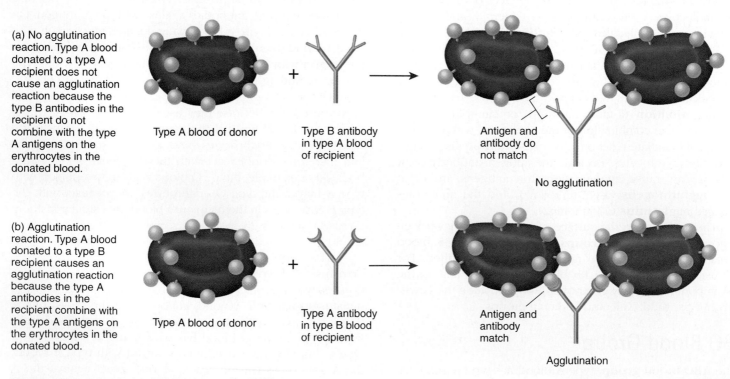

(a) No agglutination reaction. Type A blood donated to a type A recipient does not cause an agglutination reaction because the type B antibodies in the recipient do not combine with the type A antigens on the erythrocytes in the donated blood.

Type A blood of donor

Type B antibody in type A blood of recipient

Antigen and antibody do not match

No agglutination

(b) Agglutination reaction. Type A blood donated to a type B recipient causes an agglutination reaction because the type A antibodies in the recipient combine with the type A antigens on the erythrocytes in the donated blood.

Type A blood of donor

Type A antibody in type B blood of recipient

Antigen and antibody match

Agglutination

Figure 19.13   Agglutination Reaction

serious because the antibodies in the donor's blood are diluted in the blood of the recipient, and few reactions take place. Because type O blood sometimes causes transfusion reactions, it is given to a person with another blood type only in life-or-death emergency situations.

6 P R E D I C T

Historically, people with type AB blood were called universal recipients. What is the rationale for this term? Explain why the term is misleading.

✔ *Answer in Appendix F*

Rh Blood Group

Another important blood group is the **Rh blood group,** so named because it was first studied in the rhesus monkey. People are Rh-positive if they have a certain Rh antigen (the D antigen) on the surface of their erythrocytes, and people are Rh-negative if they do not have this Rh antigen. About 85% of white people and 88% of black people in the United States are Rh-positive. The ABO blood type and the Rh blood type usually are designated together. For example, a person designated as A positive is type A in the ABO blood group and Rh-positive. The rarest combination in the United States is AB negative, which occurs in less than 1% of all Americans.

Antibodies against the Rh antigen do not develop unless an Rh-negative person is exposed to Rh-positive blood. This can occur through a transfusion or by transfer of blood between a mother and her fetus across the placenta. When an Rh-negative person receives a transfusion of Rh-positive blood, the recipient becomes sensitized to the Rh antigen and produces Rh antibodies. If the Rh-negative person is unfortunate enough to receive a second transfusion of Rh-positive blood after becoming sensitized, a transfusion reaction results.

Rh incompatibility can pose a major problem in some pregnancies when the mother is Rh-negative and the fetus is Rh-positive (figure 19.14). If fetal blood leaks through the placenta and mixes with the mother's blood, the mother becomes sensitized to the Rh antigen. The mother produces Rh antibodies that cross the placenta and cause agglutination and hemolysis of fetal erythrocytes. This disorder is called **hemolytic disease of the newborn (HDN),** or **erythroblastosis fetalis** (ĕ-rith′rō-blas-tō′sis fē-ta′lis), and it may be fatal to the fetus. In the woman's first pregnancy, however, there is usually no problem. The leakage of fetal blood is usually the result of a tear in the placenta that takes place either late in the pregnancy or during delivery. Thus there is not enough time for the mother to produce enough Rh antibodies to harm the fetus. In later pregnancies, however, a problem can arise because the mother has already been sensitized to the Rh antigen. Consequently, if the fetus is Rh-positive and if there is any leakage of fetal blood into the mother's blood, she rapidly produces large amounts of Rh antibodies, and HDN develops.

HDN can be prevented if the Rh-negative woman is given an injection of a specific type of antibody preparation, called Rh$_0$(D) immune globulin (RhoGAM). The injection can be administered during the pregnancy or before or immediately after each delivery or abortion. The injection contains antibodies against Rh antigens. The injected antibodies bind to the Rh antigens of any fetal erythrocytes that may have entered the mother's blood. This treatment inactivates the fetal Rh antigens and prevents sensitization of the mother.

If HDN develops, treatment consists of slowly removing the newborn's blood and replacing it with Rh-negative blood. The newborn can also be exposed to fluorescent light, because the light helps to break down the large amounts of bilirubin formed as a result of erythrocyte destruction. High levels of bilirubin are toxic to the nervous system and can damage brain tissue.

Diagnostic Blood Tests
Type and Crossmatch

To prevent transfusion reactions the blood is typed, and a **crossmatch** is made. **Blood typing** determines the ABO and Rh blood groups of the blood sample. Typically, the cells are separated from the serum. The cells are tested with known antibodies to determine the type of antigen on the cell surface. For example, if a patient's blood cells agglutinate when mixed with type A antibodies but do not agglutinate when mixed with type B antibodies, it is concluded that the cells have type A antigen. In a similar fashion the serum is mixed with known cell types (antigens) to determine the type of antibodies in the serum.

Normally donor blood must match the ABO and Rh type of the recipient. Because other blood groups can also cause a transfusion reaction, however, a crossmatch is performed. In a crossmatch the donor's blood cells are mixed with the recipient's serum, and the donor's serum is mixed with the recipient's cells. The donor's blood is considered safe for transfusion only if there is no agglutination in either match.

Complete Blood Count

The **complete blood count (CBC)** is an analysis of the blood that provides much information. It consists of a red blood cell count, hemoglobin and hematocrit measurements, and a white blood cell count.

Red Blood Cell Count

Blood cell counts usually are done electronically with a machine, but they can also be done manually with a microscope. The normal range for a **red blood cell count (RBC)** (expressed in millions of erythrocytes per cubic millimeter of blood) is 4.2–5.8 million/mm^3 for a male, and 3.6–5.2 million/mm^3 of blood for a female. **Polycythemia** (pol′ē-sī-thē′mē-ă) is an overabundance of erythrocytes. It can result

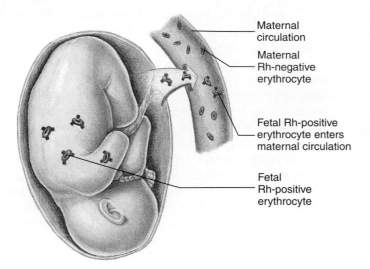

Maternal circulation

Maternal Rh-negative erythrocyte

Fetal Rh-positive erythrocyte enters maternal circulation

Fetal Rh-positive erythrocyte

1. Before or during delivery, Rh-positive erythrocytes from the fetus enter the blood of an Rh-negative woman through a tear in the placenta.

Maternal circulation

Maternal Rh-negative erythrocyte

Rh antibodies

2. The mother is sensitized to the Rh antigen and produces Rh antibodies. Because this usually happens after delivery, there is no effect on the fetus in the first pregnancy.

3. During a subsequent pregnancy with an Rh-positive fetus, Rh-positive erythrocytes cross the placenta, enter the maternal circulation, and stimulate the mother to produce antibodies against the Rh antigen. Antibody production is rapid because the mother has been sensitized to the Rh antigen. The Rh antibodies from the mother cross the placenta, causing agglutination and hemolysis of fetal erythrocytes, and hemolytic disease of the newborn develops.

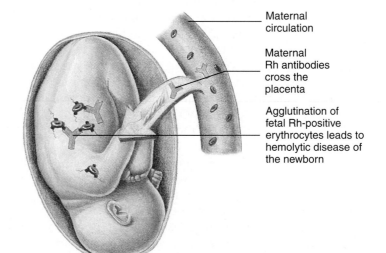

Maternal circulation

Maternal Rh antibodies cross the placenta

Agglutination of fetal Rh-positive erythrocytes leads to hemolytic disease of the newborn

Figure 19.14   Hemolytic Disease of the Newborn (HDN)

from a decreased oxygen supply, which stimulates erythropoietin production, or from red bone marrow tumors. Because erythrocytes tend to stick to one another, increasing the number of erythrocytes makes it more difficult for blood to flow. Consequently, polycythemia increases the workload of the heart. It also can reduce blood flow through tissues and, if severe, can result in plugging of small blood vessels (capillaries).

Hemoglobin Measurement

The **hemoglobin measurement** determines the amount of hemoglobin in a given volume of blood, usually expressed as grams of hemoglobin per 100 mL of blood. The normal hemoglobin count for a male is 14–18 g/100 mL of blood, and for a female it is 12–16 g/100 mL of blood. Abnormally low hemoglobin is an indication of anemia (ă-nē′mē-ă), which is a reduced number of erythrocytes per 100 mL of blood or a reduced amount of hemoglobin in each erythrocyte.

Hematocrit Measurement

The percentage of total blood volume composed of erythrocytes is the **hematocrit** (hē′mă-tō-krit). One way to determine hematocrit is to place blood in a tube and spin the tube in a centrifuge. The formed elements are heavier than the plasma and are forced to one end of the tube (figure 19.15). The erythrocytes account for 44%–54% of the total blood volume in males and 38%–48% in females. Because the hematocrit measurement is based on volume, it is affected by the number and size of erythrocytes. For example, a decreased hematocrit can result from a decreased number of normal-sized erythrocytes or a normal number of small-sized erythrocytes. The average volume of an erythrocyte is calculated by dividing the hematocrit by the red blood cell count. A number of disorders cause erythrocytes to be smaller or larger than normal. For example, inadequate iron in the diet can impair hemoglobin production. Consequently, during their formation erythrocytes do not fill with hemoglobin, and they remain smaller than normal.

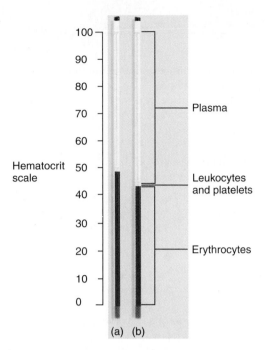

Figure 19.15 Hematocrit

Normal hematocrits of (*a*) male and (*b*) female. Blood is separated into plasma, erythrocytes, and a small amount of leukocytes and platelets, which rest on the erythrocytes. The hematocrit measurement includes only the erythrocytes and does not measure the leukocytes and platelets.

White Blood Cell Count

A **white blood cell count (WBC)** measures the total number of leukocytes in the blood. There are normally 5000–10,000 leukocytes per cubic millimeter of blood. **Leukopenia** (lū-kō-pē′nē-ă) is a lower-than-normal WBC count and can indicate depression or destruction of the red marrow by radiation, drugs, tumor, or a deficiency of vitamin B_{12} or folate. **Leukocytosis** (lū′kō-sī-tō′sis) is an abnormally high white blood cell count. **Leukemia** (lū-kē′mē-ă) (a cancer of the red marrow) and bacterial infections often cause leukocytosis.

White Blood Cell Differential Count

A **white blood cell differential count** determines the percentage of each of the five kinds of leukocytes in the white blood cell count. Normally neutrophils account for 60%–70%; lymphocytes, 20%–30%; monocytes, 2%–8%; eosinophils, 1%–4%; and basophils, 0.5%–1%. A white blood cell differential count can provide much insight about a patient's condition. For example, in patients with bacterial infections the neutrophil count is often greatly increased, whereas in patients with allergic reactions the eosinophil and basophil counts are elevated.

Clotting

Two measurements that test the ability of the blood to clot are the platelet count and the prothrombin time.

Platelet Count

A normal **platelet count** is 250,000–400,000 platelets per cubic millimeter of blood. **Thrombocytopenia** (throm′bō-sī-tō-pē′nē-ă) is a condition in which the platelet count is greatly reduced, resulting in chronic bleeding through small vessels and capillaries. It can be caused by decreased platelet production as a result of hereditary disorders, lack of vitamin B_{12} (pernicious anemia), drug therapy, or radiation therapy.

Prothrombin Time Measurement

Prothrombin time measurement is a measure of how long it takes for the blood to start clotting, which normally is 9–12 s. Because many clotting factors must be activated to form prothrombin, a deficiency of any one of them can cause an abnormal prothrombin time. Vitamin K deficiency, certain liver diseases, and drug therapy can cause an increased prothrombin time.

Blood Chemistry

The composition of materials dissolved or suspended in the plasma can be used to assess the functioning of many of the body's systems. For example, high blood glucose levels can indicate that the pancreas is not producing enough insulin; high blood urea nitrogen is a sign of reduced kidney function; increased bilirubin can indicate liver dysfunction; and high cholesterol levels can indicate an increased risk of developing cardiovascular disease. A number of blood chemistry tests are routinely done when a blood sample is taken, and additional tests are available.

7   P R E D I C T

When a patient complains of acute pain in the abdomen, the physician suspects appendicitis, which is a bacterial infection of the appendix. What blood test should be done to support the diagnosis?

✔*Answer in Appendix F*

Summary

Functions

1. Blood transports gases, nutrients, waste products, and hormones.
2. Blood is involved in the regulation of homeostasis and the maintenance of pH, body temperature, fluid balance, and electrolyte levels.
3. Blood protects against disease and blood loss.

Plasma

1. Plasma is mostly water (91%) and contains proteins such as albumin (maintains osmotic pressure), globulins (function in transport and immunity), fibrinogen (involved in clot formation), and hormones and enzymes (involved in regulation).
2. Plasma also contains ions, nutrients, waste products, and gases.

Formed Elements

The formed elements include erythrocytes (red blood cells), leukocytes (white blood cells), and platelets (cell fragments).

Production of Formed Elements

1. In the embryo and fetus the formed elements are produced in a number of locations.
2. After birth, red bone marrow becomes the source of the formed elements.
3. All formed elements are derived from stem cells.

Erythrocytes

1. Erythrocytes are biconcave disks containing hemoglobin and carbonic anhydrase.
   - A hemoglobin molecule consists of four heme and four globin molecules. The heme molecules transport oxygen, and the globin molecules transport carbon dioxide and nitric oxide. Iron is required for oxygen transport.
   - Carbonic anhydrase is involved with the transport of carbon dioxide.
2. Erythropoiesis is the production of erythrocytes.
   - Stem cells in red bone marrow eventually give rise to late erythroblasts, which lose their nuclei and are released into the blood as reticulocytes. Loss of the endoplasmic reticulum by a reticulocyte produces an erythrocyte.
   - In response to low blood oxygen the kidneys produce erythropoietin, which stimulates erythropoiesis.
3. Worn out erythrocytes are phagocytized by macrophages. Hemoglobin is broken down, and heme becomes bilirubin, which is secreted in bile.

Leukocytes

1. Leukocytes protect the body against microorganisms and remove dead cells and debris.
2. There are five types of leukocytes:
   - Neutrophils are small phagocytic cells.
   - Eosinophils function to reduce inflammation.
   - Basophils release histamine and are involved with increasing the inflammatory response.

- Lymphocytes are important in immunity, including the production of antibodies.
- Monocytes leave the blood, enter tissues, and become large phagocytic cells called macrophages.

Platelets

Platelets, or thrombocytes, are cell fragments pinched off from megakaryocytes in the red bone marrow.

Hemostasis
Vascular Spasm

Vasoconstriction of damaged blood vessels reduces blood loss.

Platelet Plug Formation

1. Platelets repair minor damage to blood vessels by forming platelet plugs.
   - In platelet adhesion, platelets bind to collagen in damaged tissues.
   - In the platelet release reaction, platelets release chemicals that activate additional platelets.
   - In platelet aggregation, platelets bind to one another to form a platelet plug.
2. Platelets also release chemicals involved with coagulation.

Coagulation

1. Coagulation is the formation of a blood clot.
2. Coagulation consists of three stages.
   - Activation of prothrombinase
   - Conversion of prothrombin to thrombin by prothrombinase
   - Conversion of fibrinogen to fibrin by thrombin. The insoluble fibrin forms the clot
3. The first stage of coagulation occurs through the extrinsic or intrinsic clotting pathway. Both pathways end with the production of prothrombinase.
   - The extrinsic clotting pathway begins with the release of tissue factor from damaged tissues.
   - The intrinsic clotting pathway begins with the activation of factor XII.

Control of Clot Formation

1. Heparin and antithrombin inhibit thrombin activity. Fibrinogen is therefore not converted to fibrin, and clot formation is inhibited.
2. Prostacyclin counteracts the effects of thrombin.

Clot Retraction and Dissolution

1. Clot retraction results from the contraction of platelets, which pull the edges of damaged tissue closer together.
2. Serum, plasma minus fibrinogen and some clotting factors, is squeezed out of the clot.
3. Factor XII, thrombin, tissue plasminogen activator, and urokinase activate plasmin, which dissolves fibrin (the clot).

Blood Grouping

1. Blood groups are determined by antigens on the surface of erythrocytes.

2. Antibodies can bind to erythrocyte antigens, resulting in agglutination or hemolysis of erythrocytes.

ABO Blood Group

1. Type A blood has A antigens, type B blood has B antigens, type AB blood has A and B antigens, and type O blood has neither A or B antigens.
2. Type A blood has B antibodies, type B blood has A antibodies, type AB blood has neither A or B antibodies, and type O blood has both A and B antibodies.
3. Mismatching the ABO blood group is responsible for transfusion reactions.

Rh Blood Group

1. Rh-positive blood has a certain Rh antigen (the D antigen), whereas Rh-negative blood does not.
2. Antibodies against the Rh antigen are produced by an Rh-negative person when the person is exposed to Rh-positive blood.
3. The Rh blood group is responsible for HDN.

Diagnostic Blood Tests

Type and Crossmatch

Blood typing determines the ABO and Rh blood groups of a blood sample. A crossmatch tests for agglutination reactions between donor and recipient blood.

Complete Blood Count

The complete blood count consists of the following: red blood cell count, hemoglobin measurement (grams of hemoglobin per 100 mL of blood), hematocrit measurement (percent volume of erythrocytes), and white blood cell count.

White Blood Cell Differential Count

The white blood cell differential count determines the percentage of each type of leukocyte.

Clotting

Platelet count and prothrombin time measure the ability of the blood to clot.

Blood Chemistry

The composition of materials dissolved or suspended in plasma (e.g., glucose, urea nitrogen, bilirubin, and cholesterol) can be used to assess the functioning and status of the body's systems.

Content Review

1. Describe the three major functions performed by blood, and give examples for each function.
2. Define the term plasma. What is the function of albumin, globulins, and fibrinogen in plasma? What other substances are found in plasma?
3. Name the three general types of formed elements in blood.
4. Define the word hematopoiesis. What is a stem cell?
5. Describe the two basic parts of a hemoglobin molecule. Which part is associated with iron? What gases are transported by each part?
6. What is erythropoiesis? Where does it occur?
7. Describe the formation of erythrocytes, starting with the stem cell in red bone marrow.
8. What is erythropoietin, where is it produced, what causes it to be produced, and what effect does it have on erythrocyte production?
9. Where are erythrocytes mainly broken down? List the three breakdown products of hemoglobin, and explain what happens to them.
10. What are the two major functions of leukocytes?
11. Name the three types of granulocytes and the two types of agranulocytes. Describe the morphology and function of each type.
12. Name the two leukocytes that function primarily as phagocytic cells.
13. Which leukocyte reduces the inflammatory response? Which leukocyte releases histamine and promotes inflammation?
14. What is the function of a platelet plug? Describe the process of platelet plug formation.

15. What is a clot? What is the function of a clot?
16. Clotting is divided into three stages. Describe the final event that occurs in each stage.
17. What is the difference between extrinsic and intrinsic activation of clotting?
18. What is the function of anticoagulants in blood? Name three anticoagulants in blood, and explain how they prevent clot formation.
19. Define the terms thrombus and embolus, and explain why they are dangerous.
20. Describe clot retraction and clot dissolution. What do they accomplish? What is serum?
21. What are blood groups, and how do they cause transfusion reactions? List the four ABO blood types. Why is a person with type O blood considered a universal donor?
22. What is meant by the term Rh-positive? How can Rh incompatibility affect a pregnancy?
23. For each of the following tests, define the test and give an example of a disorder that would cause an abnormal test result.
    - Type and crossmatch
    - Red blood cell count
    - Hemoglobin measurement
    - Hematocrit measurement
    - White blood cell count
    - White blood cell differential count
    - Platelet count
    - Prothrombin time measurement
    - Blood chemistry tests

Develop Your Reasoning Skills

1. In hereditary hemolytic anemia massive destruction of erythrocytes occurs. Would you expect the reticulocyte count to be above or below normal? Explain why one of the symptoms of the disease is jaundice. In 1910 it was discovered that hereditary hemolytic anemia could be successfully treated by removing the spleen. Explain why this treatment is effective.

2. Red Packer, a physical education major, wanted to improve his performance in an upcoming marathon race. About 6 weeks before the race, 500 mL of blood was removed from his body, and the formed elements were separated from the plasma. The formed elements were frozen, and the plasma was reinfused into his body. Just before the competition, the formed elements were thawed and injected into his body. Explain why this procedure, called blood doping or blood boosting, would help Red's performance. Can you suggest any possible bad effects?

3. Chemicals such as benzene and chloramphenicol can destroy red bone marrow, causing aplastic anemia. What symptoms develop as a result of the lack of (1) erythrocytes, (2) platelets, and (3) leukocytes?

4. Some people habitually use barbiturates to depress feelings of anxiety. Barbiturates cause hypoventilation, which is a slower than normal rate of breathing because they suppress the respiratory centers in the brain. What happens to the red blood cell count of a habitual user of barbiturates? Explain.

5. What blood problems would you expect to observe in a patient after total gastrectomy (removal of the stomach)? Explain.

6. According to the old saying, "Good food makes good blood." Name three substances in the diet that are essential for "good blood." What blood disorders develop if these substances are absent from the diet?

Web Site Link

For a listing of the most current web sites related to this chapter, please visit the Seeley home page at:
http://www.mhhe.com/biosci/ap/seeleyap/

Cardiovascular System: The Heart

Objectives

1. Describe the size, shape, and location of the heart.
2. Describe the structure and function of the pericardium.
3. Describe the histology of the three major layers of the heart.
4. Describe the external and internal anatomy of the heart.
5. Discuss the similarities and differences between cardiac muscle and skeletal muscle.
6. Describe the conducting system of the heart.
7. Explain why Purkinje fibers conduct action potentials more rapidly than other cardiac muscle cells.
8. State the differences between slow and fast channels in cardiac muscle, and describe action potentials in cardiac muscle.
9. Define the term autorhythmicity, and explain how the sinoatrial node functions as the pacemaker.
10. Explain the importance of a refractory period in cardiac muscle.
11. Explain the various features of an electrocardiogram and the events that those features reflect.
12. Explain the various components of the cardiac cycle, including systole, diastole, and the heart sounds.
13. Explain the bases of the major heart sounds.
14. Describe the aortic pressure curve.
15. List the major factors involved in the intrinsic regulation of the heart, and describe the major functions of each factor.
16. List the major factors involved in the extrinsic regulation of the heart, and describe the major functions of each factor.
17. Describe the differences between sympathetic and parasympathetic stimulation on heart function.
18. Discuss the role of the heart in homeostasis and how changes in blood pressure, pH, carbon dioxide, oxygen, ions, and temperature affect the heart.

The heart functions as a pump and is responsible for the circulation of the blood through the blood vessels. The heart produces the pressure responsible for making blood flow through the blood vessels by contracting forcefully. The heart is actually two pumps in one. The right side of the heart receives blood from the body and pumps blood through the pulmonary (pŭl′mō-nār-rē) circulation, which carries blood to the lungs and returns it to the left side of the heart. In the lungs carbon dioxide is exchanged for oxygen. Carbon dioxide diffuses from the blood into the lungs, and oxygen diffuses from the lungs into the blood. The left side of the heart pumps blood through the systemic circulation, which carries blood to all remaining tissues of the body and returns it to the right side of the heart (figure 20.1).

The heart of a healthy 70-kg person pumps approximately 7200 L of blood each day at a rate of 5 L/min. For most people, the heart continues to pump near that rate for more than 75 years. During short periods of vigorous exercise, the amount of blood pumped per minute increases severalfold. If the heart loses its ability to pump blood for even a few minutes, however, the life of the individual is in danger.

In this chapter, the location and the anatomy of the heart are described first, followed by a discussion of its function and regulation. The structure and functional characteristics of the blood vessels and the integrated function of the heart and blood vessels are presented in chapter 21.

Size, Form, and Location of the Heart

The adult heart has the shape of a blunt cone and is approximately the size of a closed fist. The blunt, rounded point of the cone is the **apex**; and the larger, flat part at the opposite end of the cone is the **base.**

The heart is in the thoracic cavity between the lungs. The heart, trachea, esophagus, and associated structures form a midline partition, called the **mediastinum** (mē′dē-as-tī′nŭm) (see figure 1.14).

The heart lies obliquely in the mediastinum, with its base directed posteriorly and slightly superiorly and the apex directed anteriorly and slightly inferiorly. The apex is also directed to the left so that approximately two-thirds of the heart's mass lies to the left of the midline of the sternum (figure 20.2). It is important for clinical reasons to know the location of the heart in the thoracic cavity. Positioning a stethoscope to hear the heart sounds and positioning electrodes to record an **electrocardiogram** (ē-lek-trō-kar′dē-ō-gram; **ECG** or **EKG**) from chest leads depend on this knowledge. Effective cardiopulmonary resuscitation (kar′dē-ō-pŭl′mo-nār-ē rē-sŭs′i-tā-shŭn; CPR) also depends on a reasonable knowledge of the position and form of the heart.

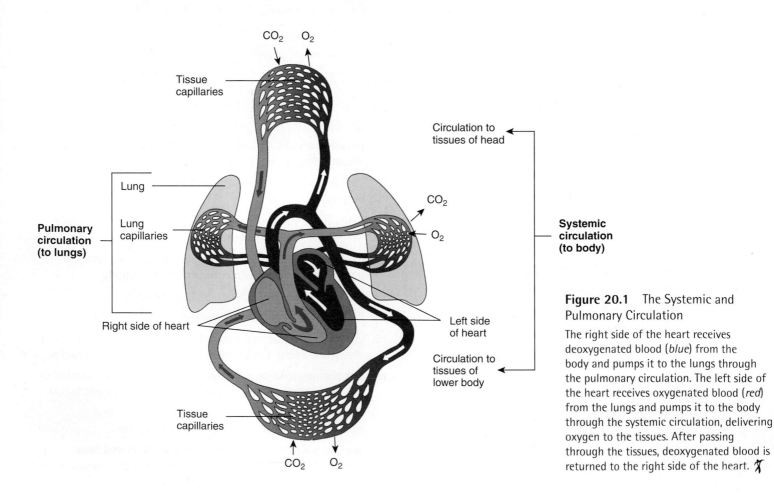

Figure 20.1 The Systemic and Pulmonary Circulation

The right side of the heart receives deoxygenated blood (*blue*) from the body and pumps it to the lungs through the pulmonary circulation. The left side of the heart receives oxygenated blood (*red*) from the lungs and pumps it to the body through the systemic circulation, delivering oxygen to the tissues. After passing through the tissues, deoxygenated blood is returned to the right side of the heart. 🏃

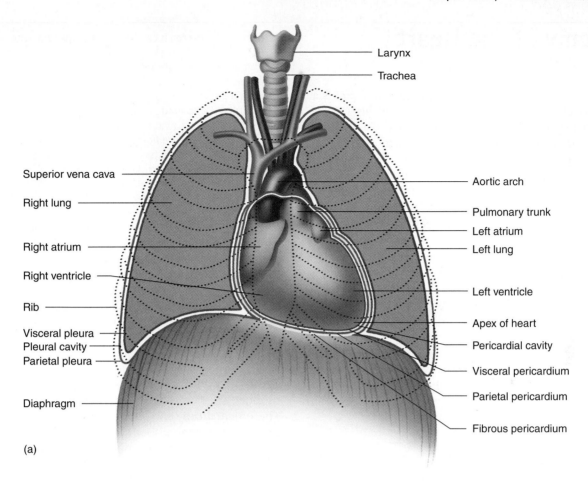

Larynx

Trachea

Superior vena cava

Right lung

Right atrium

Right ventricle

Rib

Visceral pleura
Pleural cavity
Parietal pleura

Diaphragm

Aortic arch

Pulmonary trunk
Left atrium
Left lung

Left ventricle

Apex of heart

Pericardial cavity

Visceral pericardium

Parietal pericardium

Fibrous pericardium

(a)

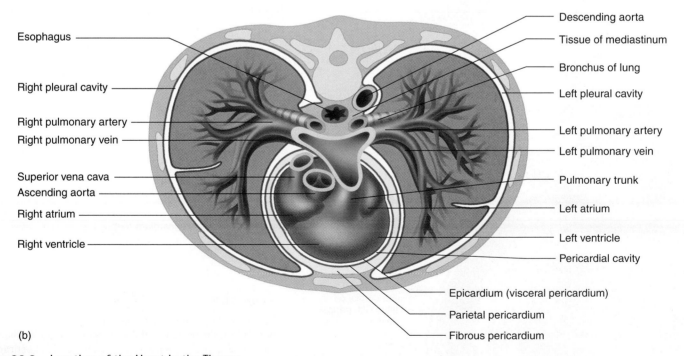

Esophagus

Right pleural cavity

Right pulmonary artery
Right pulmonary vein

Superior vena cava
Ascending aorta

Right atrium

Right ventricle

Descending aorta

Tissue of mediastinum

Bronchus of lung

Left pleural cavity

Left pulmonary artery
Left pulmonary vein

Pulmonary trunk

Left atrium

Left ventricle

Pericardial cavity

Epicardium (visceral pericardium)

Parietal pericardium

Fibrous pericardium

(b)

Figure 20.2 Location of the Heart in the Thorax

(a) The heart lies deep and slightly to the left of the sternum. The base of the heart, located deep to the sternum, extends to the second intercostal space, and the apex of the heart is in the fifth intercostal space, approximately 9 cm to the left of the midline. (b) Cross section of the thorax showing the position of the heart in the mediastinum and its relationship to other structures.

Anatomy of the Heart
Pericardium

The **pericardium** (per-i-kar′dē-ŭm), or **pericardial sac,** is a double-layered closed sac that surrounds the heart (figure 20.3). It consists of a tough, fibrous connective tissue outer layer called the **fibrous pericardium** and a thin, transparent inner layer of simple squamous epithelium called the **serous pericardium.** The fibrous pericardium prevents overdistention of the heart and anchors it within the mediastinum. Superiorly, the fibrous pericardium is continuous with the connective tissue coverings of the great vessels, and inferiorly it is attached to the surface of the diaphragm.

The part of the serous pericardium lining the fibrous pericardium is the **parietal pericardium,** and that part covering the heart surface is the **visceral pericardium,** or **epicardium** (ep-i-kar′dē-ŭm) (see figure 20.3). The parietal and visceral portions of the serous pericardium are continuous with each other where the great vessels enter or leave the heart. The **pericardial cavity,** between the visceral and parietal pericardia, is filled with a thin layer of serous **pericardial fluid** that helps reduce friction as the heart moves within the pericardial sac.

Clinical Note

Pericarditis (per′i-kar-dī′tis) is an inflammation of the serous pericardium. The cause is frequently unknown, but it can result from infection, diseases of connective tissue, or damage due to radiation treatment for cancer. It can be extremely painful, with sensations of pain referred to the back and to the chest, which can be confused with the pain of a myocardial infarction (heart attack). Pericarditis can result in a small amount of fluid accumulation within the pericardial sac.

Cardiac tamponade (tam-pŏ-nād′) is a potentially fatal condition in which a large volume of fluid or blood accumulates in the pericardial sac. The fluid compresses the heart from the outside. Although the heart is a powerful muscle, it relaxes passively. When it is compressed by fluid within the pericardial sac, it cannot dilate when the cardiac muscle relaxes. Consequently, it cannot fill with blood during relaxation, which makes it impossible for it to pump blood. Cardiac tamponade can cause a person to die quickly unless the fluid is removed. Causes of cardiac tamponade include rupture of the heart wall following a myocardial infarction, rupture of blood vessels in the pericardium after a malignant tumor invades the area, damage to the pericardium resulting from radiation therapy, and trauma (e.g., traffic accident).

Heart Wall

The heart wall is composed of three layers of tissue: the epicardium, the myocardium, and the endocardium (figure 20.4). The **epicardium** (ep-i-kar′dē-ŭm) is a thin serous membrane that constitutes the smooth outer surface of the heart. The epicardium and the visceral pericardium are two names for the same structure. The serous pericardium is called the epicardium when considered a part of the heart and the visceral pericardium when considered a part of the pericardium. The thick middle layer of the heart, the **myocardium** (mī-ō-kar′dē-ŭm), is

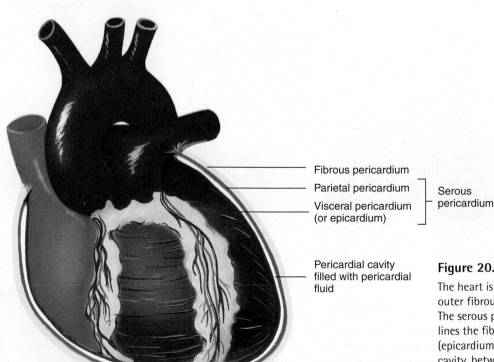

Fibrous pericardium
Parietal pericardium
Visceral pericardium (or epicardium)
Serous pericardium
Pericardium

Pericardial cavity filled with pericardial fluid

Figure 20.3   Heart in the Pericardium

The heart is located in the pericardium, which consists of an outer fibrous pericardium and an inner serous pericardium. The serous pericardium has two parts: the parietal pericardium lines the fibrous pericardium, and the visceral pericardium (epicardium) covers the surface of the heart. The pericardial cavity, between the parietal and visceral pericardium, is filled with a small amount of pericardial fluid. 🐍

composed of cardiac muscle cells and is responsible for the ability of the heart to contract. The smooth inner surface of the heart chambers is the **endocardium** (en-dō-kar′de-ŭm), which consists of simple squamous epithelium over a layer of connective tissue. The smooth inner surface allows blood to move easily through the heart. The heart valves are formed by a fold of the endocardium, making a double layer of endocardium with connective tissue in between.

The interior surfaces of the atria are mainly flat, but the interior of both auricles and a part of the right atrial wall are modified by muscular ridges called **musculi pectinati** (pek′ti-nah′tĕ; meaning, hair comb). The musculi pectinati of the right atrium are separated from the larger, smooth portions of the atrial wall by a ridge called the **crista terminalis** (kris′tă ter′mi-nal′is; meaning, terminal crest). The interior walls of the ventricles are modified by ridges and columns called **trabeculae** (tră-bek′yū-lē; meaning, beams) **carneae** (kar′nē ē; meaning, flesh).

External Anatomy

The heart consists of four chambers: two **atria** (ā′trē-ă; meaning, entrance chamber) and two **ventricles** (ven′tri-klz; meaning, belly). The thin-walled atria form the superior and posterior parts of the heart, and the thick-walled ventricles form the anterior and inferior portions (figure 20.5). Flaplike **auricles** (aw′ri-klz; meaning, ears) are extensions of the atria that can be seen anteriorly between each atrium and ventricle. The entire atrium used to be called the auricle, and some medical personnel still refer to it as such.

Several large veins carry blood to the heart. The **superior vena cava** (vēnă kā′vă) and the **inferior vena cava** carry blood from the body to the right atrium, and four **pulmonary veins** carry blood from the lungs to the left atrium. In addition, the smaller coronary sinus carries blood from the walls of the heart to the right atrium.

Two arteries, the **aorta** and the **pulmonary trunk,** exit the heart. The aorta carries blood from the left ventricle to the body, and the pulmonary trunk carries blood from the right ventricle to the lungs.

A large **coronary** (kōr′o-nār-ē; meaning, circling like a crown) **sulcus** (sūl′kŭs; meaning, ditch) runs obliquely around the heart, separating the atria from the ventricles. Two more sulci extend inferiorly from the coronary sulcus, indicating the division between the right and left ventricles. The **anterior interventricular sulcus,** or **groove,** is on the anterior surface of the heart, and the **posterior interventricular sulcus,** or **groove,** is on the posterior surface of the heart. In the normal, intact heart the sulci are covered by fat, and only after this fat is removed can the actual sulci be seen.

The major arteries supplying blood to the tissue of the heart lie within the coronary sulcus and interventricular sulci on the surface of the heart. The **right** and **left coronary arteries** exit the aorta just above the point where the aorta leaves the heart and lie within the coronary sulcus (figure 20.6a). The right coronary artery is usually smaller than the left one, and it does not supply as much of the heart with blood.

A major branch of the left coronary artery, called the **anterior interventricular artery,** or the **anterior descending artery,** extends inferiorly in the anterior interventricular sulcus

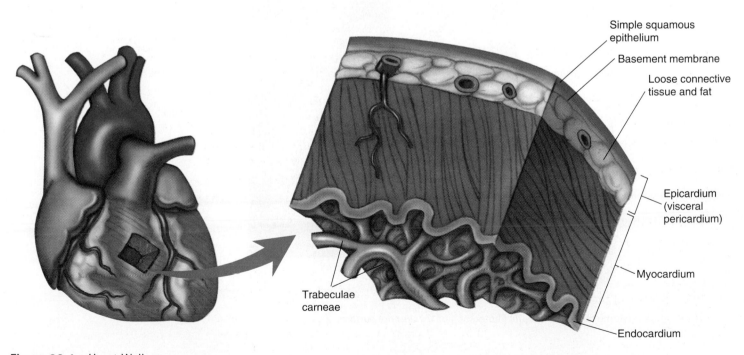

Figure 20.4   Heart Wall

Part of the wall of the heart has been removed to show its structure. The enlarged section illustrates the epicardium, the myocardium, and the endocardium. 🎋

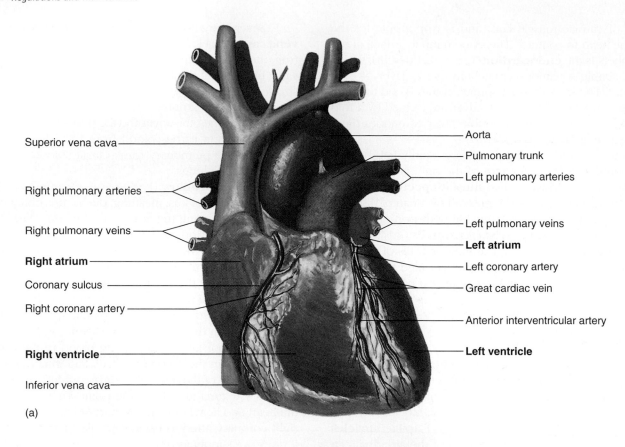

Superior vena cava

Right pulmonary arteries

Right pulmonary veins

Right atrium

Coronary sulcus

Right coronary artery

Right ventricle

Inferior vena cava

Aorta

Pulmonary trunk

Left pulmonary arteries

Left pulmonary veins

Left atrium

Left coronary artery

Great cardiac vein

Anterior interventricular artery

Left ventricle

(a)

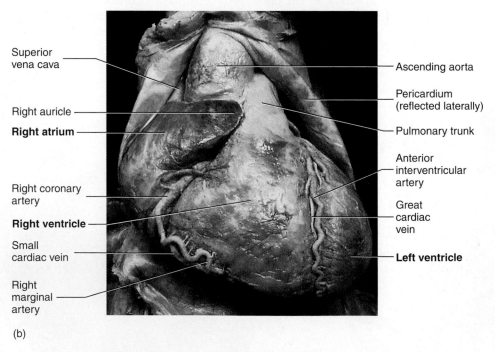

Superior vena cava

Right auricle

Right atrium

Right coronary artery

Right ventricle

Small cardiac vein

Right marginal artery

Ascending aorta

Pericardium (reflected laterally)

Pulmonary trunk

Anterior interventricular artery

Great cardiac vein

Left ventricle

(b)

Figure 20.5   Surface of the Heart

(a) View of the anterior (sternocostal) surface. (b) Photograph of the anterior surface.

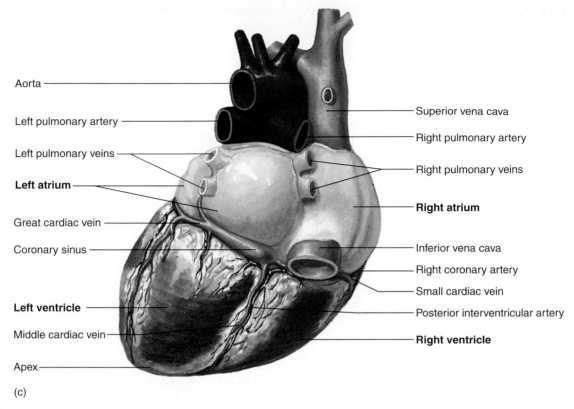

Aorta

Left pulmonary artery

Left pulmonary veins

Left atrium

Great cardiac vein

Coronary sinus

Left ventricle

Middle cardiac vein

Apex

(c)

Superior vena cava

Right pulmonary artery

Right pulmonary veins

Right atrium

Inferior vena cava

Right coronary artery

Small cardiac vein

Posterior interventricular artery

Right ventricle

Figure 20.5   (continued)

(c) View of the posterior (base) and inferior (diaphragmatic) surfaces of the heart.

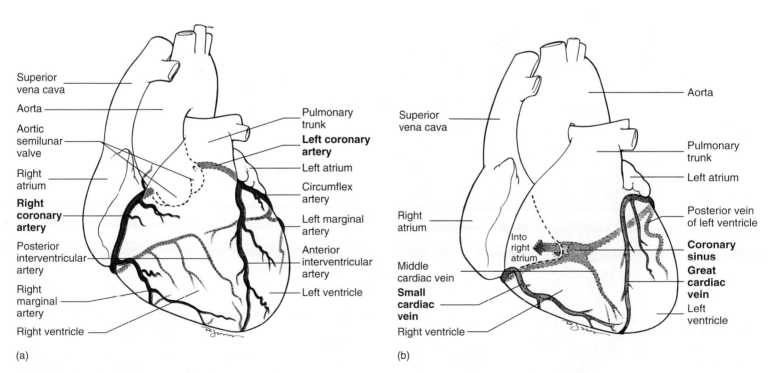

Superior vena cava

Aorta

Aortic semilunar valve

Right atrium

Right coronary artery

Posterior interventricular artery

Right marginal artery

Right ventricle

(a)

Pulmonary trunk

Left coronary artery

Left atrium

Circumflex artery

Left marginal artery

Anterior interventricular artery

Left ventricle

Superior vena cava

Right atrium

Middle cardiac vein

Small cardiac vein

Right ventricle

(b)

Aorta

Pulmonary trunk

Left atrium

Posterior vein of left ventricle

Coronary sinus

Great cardiac vein

Left ventricle

Into right atrium

Figure 20.6   Circulation to the Heart

(a) Arteries supplying blood to the heart. The arteries of the anterior surface are seen directly and are darker in color; the arteries of the posterior surface are seen through the heart and are outlined with dashed lines. (b) Veins draining blood from the heart. The veins of the anterior surface are seen directly and are darker in color; the veins of the posterior surface are seen through the heart and are outlined with dashed lines.

and supplies blood to most of the anterior part of the heart. A **marginal branch** of the left coronary artery supplies blood to the lateral wall of the left ventricle. The **circumflex** (ser'kŭm-fleks) **branch** of the left coronary artery extends around to the posterior side of the heart in the coronary sulcus. Its branches supply blood to much of the posterior wall of the heart.

The right coronary artery lies within the coronary sulcus and extends from the aorta around to the posterior part of the heart. Branches of the right coronary artery supply blood to the lateral wall of the right ventricle, and a branch called the **posterior interventricular artery** lies in the posterior interventricular sulcus and supplies blood to the posterior and inferior part of the heart.

1	**P R E D I C T**

Predict the effect on the heart if blood flow through a coronary artery, such as the anterior interventricular artery, is restricted or completely blocked.

✔ *Answer in Appendix F*

The major vein draining the tissue on the left side of the heart is the **great cardiac vein,** and a **small cardiac vein** drains the right margin of the heart (figure 20.6*b*). These veins converge toward the posterior part of the coronary sulcus and empty into a large venous cavity called the **coronary sinus,** which in turn empties into the right atrium. A number of smaller veins empty into the cardiac veins, the coronary sinus, or directly into the right atrium.

Heart Chambers and Valves

Right and Left Atria

The **right atrium** has three major openings that receive blood vessels from the body: the superior vena cava, the inferior vena cava, and the coronary sinus. The **left atrium** has four relatively uniform openings that receive the four pulmonary veins. The two atria are separated from each other by the **interatrial septum.** A slight oval depression, the **fossa ovalis** (fos'ă ō-va'lis), on the right side of the septum marks the former location of the **foramen ovale** (ō-va'lē), an opening between the right and left atria in the embryo and the fetus (see chapter 29).

Right and Left Ventricles

The atria open into the ventricles through **atrioventricular canals** (figure 20.7). Each ventricle has one large, superiorly placed outflow route near the midline of the heart. The **right ventricle** opens into the pulmonary trunk, and the **left ventricle** opens into the aorta. The two ventricles are separated from each other by the **interventricular septum,** which has a thick muscular part toward the apex and a thin membranous part toward the atria.

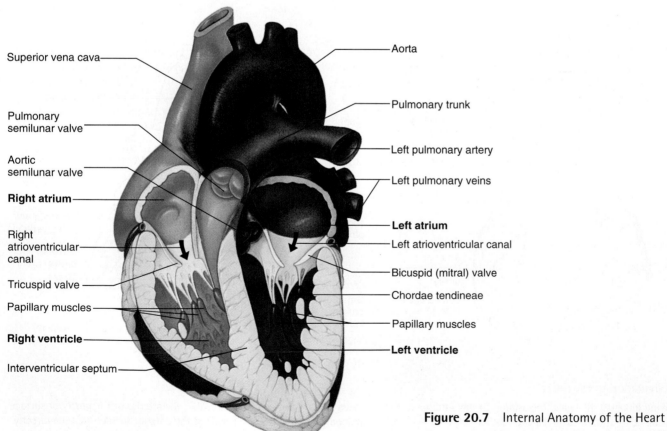

Figure 20.7   Internal Anatomy of the Heart

The heart is cut in a frontal plane to show the internal anatomy.

Clinical Focus   Angina, Infarctions, and Treatment of Blocked Coronary Arteries

Angina (an'ji-nă, an'jī-nă) **pectoris** (pek'tō-ris) is pain that results from a reduction in blood supply to cardiac muscle.

The pain is temporary, and, if blood flow is restored, little permanent change or damage results. Angina pectoris is characterized by chest discomfort deep to the sternum, often described as heaviness, pressure, or moderately severe pain; and it is often mistaken for indigestion. The pain can also be referred to the neck, lower jaw, left arm, and left shoulder. Most often, angina pectoris results from narrowed and hardened coronary arterial walls. The reduced blood flow results in a reduced supply of oxygen to cardiac muscle cells. As a consequence, the limited anaerobic metabolism of cardiac muscle results in a buildup of lactic acid and reduced pH in affected areas of the heart. Pain receptors are stimulated by the lactic acid. The pain is predictably associated with exercise because the increased pumping activity of the heart requires more oxygen and the narrowed blood vessels cannot supply it. Angina pectoris is frequently relieved by rest and by drugs such as nitroglycerin. Nitroglycerin dilates the blood vessels, including the dilation of coronary arteries. Consequently, it increases the oxygen supply to cardiac muscle and reduces the workload of the heart and the need for oxygen because the heart has to pump blood against a smaller pressure. The heart pumps less blood because the blood tends to remain in the dilated blood vessels and less blood is returned to the heart.

Myocardial infarction (mī-ō-kar'dē-ăl in-fark'shŭn) results from a prolonged lack of blood flow to a part of the cardiac muscle, resulting in a lack of oxygen and cellular death. Myocardial infarctions vary with the amount of cardiac muscle and the part of the heart that is affected. If blood supply to cardiac muscle is reestablished within 20 minutes, no permanent damage occurs. If the lack of oxygen lasts longer, cell death results. Within 30–60 s after blockage of a coronary blood vessel, however, functional changes are obvious. The electrical properties of the cardiac muscle are altered, and the ability of the heart to function properly is lost. The most common cause of myocardial infarction is thrombus formation that blocks a coronary artery. Coronary arteries narrowed by **atherosclerotic** (ath'er-ō-skler-ot'ik) lesions provide one of the conditions that increase the chances of myocardial infarction. Atherosclerotic lesions partially block blood vessels, resulting in turbulent blood flow, and the surfaces of the lesions are rough. These changes increase the probability of thrombus formation.

Angioplasty (an'jē-ō-plas-tē) is a process whereby a small balloon is threaded through the aorta and into a coronary artery. After the balloon has entered the partially occluded coronary artery, it is inflated, flattening the atherosclerotic deposits against the vessel walls and opening the occluded blood vessel. This technique improves the function of cardiac muscle in patients suffering from an inadequate blood flow to the cardiac muscle through the coronary arteries. Some controversy exists about its effectiveness, at least in some patients, because dilation of the coronary arteries can be reversed within a few weeks or months and because blood clots can form in coronary arteries following angioplasty. To help prevent future blockage, a metal-mesh tube called a **stent** can be inserted into the vessel. Although the stent is better able to hold the vessel open, it too can eventually become blocked. Small rotating blades and lasers are also used to remove lesions from coronary vessels.

A **coronary bypass** is a surgical procedure that relieves the effects of obstructions in the coronary arteries. The technique involves taking healthy segments of blood vessels from other parts of the patient's body and using them to bypass obstructions in the coronary arteries. The technique is common for those who suffer from severe occlusion of parts of the coronary arteries.

Special enzymes are used to break down blood clots that form in the coronary arteries and cause heart attacks. The major enzymes used are **streptokinase** (strep-tō-kī'nās), **tissue plasminogen** (plaz-min'ō-jen) **activator** (tPA), or sometimes **urokinase** (yūr-ō-kī'nās). These enzymes function to activate plasminogen, which is an inactive form of an enzyme in the body that breaks down the fibrin of clots. The strategy is to administer these drugs to people suffering from myocardial infarctions as soon as possible following the onset of symptoms. Removal of the occlusions produced by clots reestablishes blood flow to the cardiac muscle and reduces the amount of cardiac muscle permanently damaged by the occlusion.

Atrioventricular Valves

An **atrioventricular valve** is in each atrioventricular canal and is composed of cusps, or flaps. These valves allow blood to flow from the atria into the ventricles but prevent blood from flowing back into the atria. The atrioventricular valve between the right atrium and the right ventricle has three cusps and is therefore called the **tricuspid valve** (trī-kŭs'pid). The atrioventricular valve between the left atrium and left ventricle has two cusps and is therefore called the **bicuspid** (bī-kŭs'pid), or **mitral** (mī'trăl; meaning, resembling a bishop's miter, a two-pointed hat), **valve.**

Each ventricle contains cone-shaped muscular pillars called **papillary** (pap'i-lār'ē; meaning, nipple or pimple-shaped) **muscles.** These muscles are attached by thin, strong connective tissue strings called **chordae tendineae** (kōr'dē ten'di-nē-ē; meaning, heart strings) to the cusps of the atrioventricular valves (see figure 20.7; figure 20.8a). The papillary muscles contract when the ventricles contract and prevent the valves from opening into the atria by pulling on the chordae tendineae attached to the valve cusps. Blood flowing from the atrium into the ventricle pushes the valve open into the ventricle, but, when the ventricle contracts, blood pushes the valve back toward the atrium. The atrioventricular canal is closed as the valve cusps meet (figure 20.9).

Semilunar Valves

Within the aorta and pulmonary trunk are **aortic** and **pulmonary semilunar** (sem-ē-lū'năr; meaning, halfmoon-shaped)

valves. Each valve consists of three pocketlike semilunar cusps, the free inner borders of which meet in the center of the artery to block blood flow (see figures 20.7 and 20.8b). Blood flowing out of the ventricles pushes against each valve, forcing it open, but when blood flows back from the aorta or pulmonary trunk toward the ventricles, it enters the pockets of the cusps, causing them to meet in the center of the aorta or pulmonary trunk, thus closing them and keeping blood from flowing back into the ventricles (see figure 20.9).

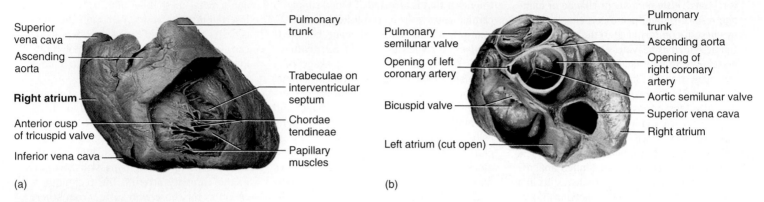

Figure 20.8   Heart Valves

(a) View of the tricuspid valve, the chordae tendineae, and the papillary muscles. (b) A superior view of the heart valves. Note the three cusps of each semilunar valve meeting to prevent the backflow of blood.

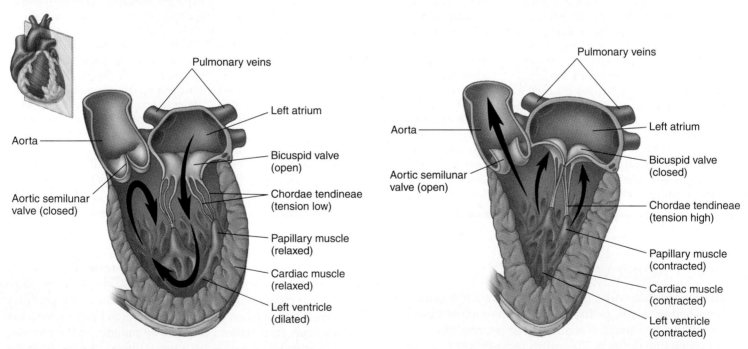

(a) When the bicuspid valve is open, the cusps of the valve are pushed by blood into the ventricle. The tension on the chordae tendineae is low, and the papillary muscles are relaxed. Blood flows from the left atrium into the left ventricle. When the aortic semilunar valve is closed, the cusps of the valve overlap as they are pushed by the blood in the aorta toward the ventricle. There is no blood flow from the aorta into the ventricle.

(b) When the bicuspid valve is closed, the cusps of the valves overlap as they are pushed by the blood toward the left atrium. There is no blood flow from the ventricle into the atrium. The tension on the chordae tendineae is increased, and the papillary muscles are contracted. When the aortic semilunar valve is open, the cusps of the valve are pushed by the blood toward the aorta. Blood then flows from the left ventricle into the aorta.

Figure 20.9   Function of the Heart Valves

(a) Valve positions when blood is flowing into the left ventricle. (b) Valve positions when blood is flowing out of the left ventricle.

█Route of Blood Flow Through the Heart

Blood flow through the heart is depicted in figure 20.10. Even though it is more convenient to discuss blood flow through the heart one side at a time, it is important to understand that both atria contract at about the same time and both ventricles contract at about the same time. This concept is particularly important when considering electrical activity, pressure changes, and heart sounds.

Blood enters the right atrium from the systemic circulation, which returns blood from all the tissues of the body. Blood flows from an area of higher pressure in the systemic circulation to the right atrium, which has a lower pressure. Most of the blood in the right atrium then passes into the right ventricle as the ventricle relaxes following the previous contraction. The right atrium then contracts, and most of the blood remaining in the atrium is pushed into the ventricle to complete right ventricular filling.

Contraction of the right ventricle pushes blood against the tricuspid valve, forcing it closed, and against the pulmonary

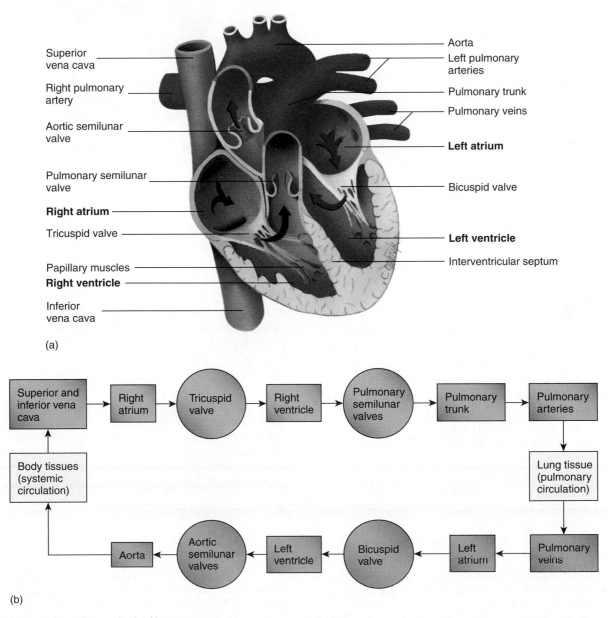

Figure 20.10   Blood Flow Through the Heart

(a) Frontal section of the heart revealing the four chambers and the direction of blood flow through the heart. (b) Diagram listing in order the structures through which blood flows in the systemic and pulmonary circulations. The heart valves are indicated by circles, deoxygenated blood (*blue*); oxygenated blood (*red*).

semilunar valve, forcing it open, thus allowing blood to enter the pulmonary trunk.

The pulmonary trunk branches to form the **pulmonary arteries** (see figure 20.5), which carry blood to the lungs, where carbon dioxide is released and oxygen is picked up (see chapters 21 and 23). Blood returning from the lungs enters the left atrium through the four pulmonary veins. The blood passing from the left atrium to the left ventricle opens the bicuspid valve, and contraction of the left atrium completes left ventricular filling.

Contraction of the left ventricle pushes blood against the bicuspid valve, closing it, and against the aortic semilunar valve, opening it and allowing blood to enter the aorta. Blood flowing through the aorta is distributed to all parts of the body except to the parts of the lungs supplied by the pulmonary blood vessels (see chapter 23).

▮Histology
Heart Skeleton

The **skeleton of the heart** consists of a plate of fibrous connective tissue between the atria and ventricles. This connective tissue plate forms **fibrous rings** around the atrioventricular and semilunar valves and provides a solid support for them (figure 20.11). The fibrous connective tissue plate also serves as electrical insulation between the atria and the ventricles and provides a rigid site for attachment for the cardiac muscles.

Cardiac Muscle

Cardiac muscle cells are elongated, branching cells that contain one or occasionally two centrally located nuclei. Cardiac muscle cells contain actin and myosin myofilaments organized to form sarcomeres, which join end to end to form myofibrils (see chapter 4). The actin and myosin myofilaments are responsible for muscle contraction, and their organization gives cardiac muscle a striated (banded) appearance. The striations are less regularly arranged and less numerous than in skeletal muscle (figure 20.12a and b).

Cardiac muscle has a **smooth sarcoplasmic reticulum,** but it is neither as regularly arranged nor as abundant as in skeletal muscle fibers, and no dilated cisternae are present, as occurs in skeletal muscle. The sarcoplasmic reticulum comes into close association at various points with membranes of **transverse tubules (T tubules).** This loose association between the sarcoplasmic reticulum and the T tubules is partly responsible for the slow onset of contraction and the prolonged contraction phase in cardiac muscle. Depolarizations of the cardiac muscle cell membrane are not carried from the surface of the cell to the sarcoplasmic reticulum as efficiently as they are in skeletal muscles, and calcium must diffuse a greater distance from the sarcoplasmic reticulum to the actin myofilaments.

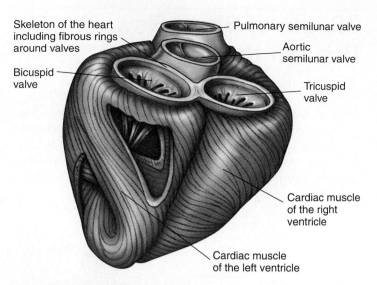

Figure 20.11   Skeleton of the Heart

The skeleton of the heart consists of fibrous connective tissue rings that surround the heart valves and separates the atria from the ventricles. Cardiac muscle attaches to the fibrous connective tissue. The muscle fibers are arranged so that when the ventricles contract a wringing motion is produced and the distance between the apex and base of the heart shortens.

Adenosine triphosphate (ATP) provides the energy for cardiac muscle contraction, and, as in other tissues, ATP production depends on oxygen availability. Cardiac muscle, however, cannot develop a large oxygen debt, a characteristic that is consistent with the function of the heart. Development of a large oxygen debt would result in muscular fatigue and cessation of cardiac muscle contraction. Cardiac muscle cells are rich in mitochondria, which perform oxidative metabolism at a rate rapid enough to sustain normal myocardial energy requirements. An extensive capillary network provides an adequate oxygen supply to the cardiac muscle cells.

2   P R E D I C T

Under resting conditions most ATP produced in cardiac muscle is derived from the metabolism of fatty acids. During periods of heavy exercise, however, cardiac muscle cells use lactic acid as an energy source. Explain why this arrangement is an advantage.

✔ *Answer in Appendix F*

Cardiac muscle cells are organized in spiral bundles or sheets. The cells are bound end to end and laterally to adjacent cells by specialized cell-to-cell contacts called **intercalated** (in-ter′kă-lā-ted) **disks** (see figure 20.12). The membranes of the intercalated disks have folds, and the adjacent cells fit together, thus greatly increasing contact between them. Specialized cell membrane structures called **desmosomes** (dez′mō-sōmz) hold the cells together, and **gap junctions** function as areas of low electric resistance between the cells, allowing action potentials to pass from one cell to adja-

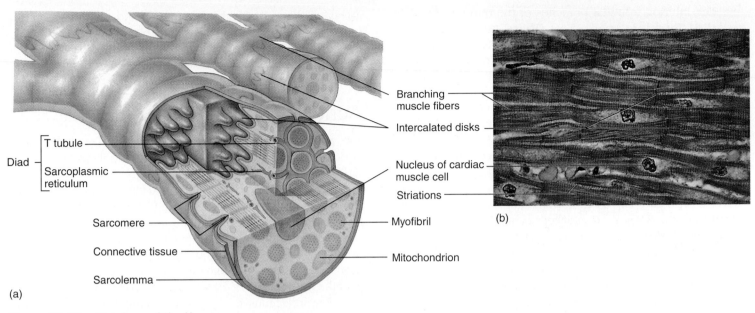

Figure 20.12   Histology of the Heart

(*a*) Heart muscle demonstrating the structure and arrangement of the individual muscle fibers. (*b*) Photomicrograph of heart muscle.

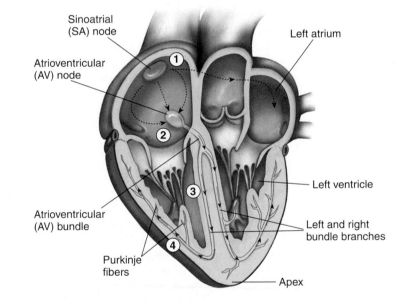

1. Action potentials originate in the sinoatrial (SA) node and travel across the wall of the atrium (*arrows*) from the SA node to the atrioventricular (AV) node.

2. Action potentials pass through the AV node and along the atrioventricular (AV) bundle, which extends from the AV node, through the fibrous skeleton, into the interventricular septum.

3. The AV bundle divides into right and left bundle branches, and action potentials descend to the apex of each ventricle along the bundle branches.

4. Action potentials are carried by the Purkinje fibers from the bundle branches to the ventricular walls.

Figure 20.13   Conducting System of the Heart

cent cells (see figure 4.3). Electrically the cardiac muscle cells behave as a single unit, and the highly coordinated contractions of the heart depend on this functional characteristic.

Conducting System

The conducting system of the heart, which relays electric action potentials through the heart, consists of modified cardiac muscle cells that form two nodes (meaning, a knot or lump) and a conducting bundle (figure 20.13). The two nodes are contained within the walls of the right atrium and are named according to their position in the atrium. The **sinoatrial (SA) node** is medial to the opening of the superior vena cava, and the **atrioventricular (AV) node** is medial to the right atrioventricular valve. The AV node gives rise to a conducting bundle of the heart, the **atrioventricular bundle.** This bundle passes through a small opening in the fibrous skeleton to reach the interventricular septum, where it divides to form the **right** and **left bundle branches,** which extend beneath the endocardium on either side of the interventricular septum to the apices of the right and left ventricles, respectively.

The inferior, terminal branches of the bundle branches are called **Purkinje** (per-kin′jē) **fibers,** which are large-diameter cardiac muscle fibers. They have fewer myofibrils than most cardiac muscle cells and do not contract as forcefully. Intercalated disks are well developed between the Purkinje fibers and contain numerous gap junctions. As a result of these structural modifications, action potentials travel along the Purkinje fibers much more rapidly than through other cardiac muscle tissue.

Cardiac muscle cells have the capacity to generate spontaneous action potentials, but cells of the SA node do so at a greater frequency. As a result, the SA node is called the **pacemaker** of the heart. Once action potentials are produced, they spread from the SA node to adjacent cardiac muscle fibers of the atrium. Preferential pathways conduct action potentials from the SA node to the AV node at a greater velocity than they are transmitted in the remainder of the atrial muscle fibers, although such pathways cannot be distinguished structurally from the remainder of the atrium.

When the heart beats under resting conditions, approximately 0.04 s is required for action potentials to travel from the SA node to the AV node. Within the AV node action potentials are propagated slowly compared with the remainder of the conducting system. As a consequence, there is a delay of 0.11 s from the time action potentials reach the AV node until they pass to the AV bundle. The total delay of 0.15 s allows completion of the atrial contraction before ventricular contraction begins.

After action potentials pass from the AV node to the highly specialized conducting bundles, the velocity of conduction increases dramatically. The action potentials pass through the left and right bundle branches and through the individual Purkinje fibers that penetrate into the myocardium of the ventricles (see figure 20.13).

Because of the arrangement of the conducting system, the first part of the myocardium that is stimulated is the inner wall of the ventricles near the apex. Thus ventricular contraction begins at the apex and progresses throughout the ventricles. Once stimulated, the spiral arrangement of muscle layers in the wall of the heart results in a wringing action that proceeds from the apex toward the base of the heart. During the process, the distance between the apex and the base of the heart decreases.

3   P R E D I C T

Explain why it is more efficient for contraction of the ventricles to begin at the apex of the heart than at the base.

✔ *Answer in Appendix F*

Electrical Properties

Cardiac muscle cells, like other electrically excitable cells such as neurons and skeletal muscle fibers, have a **resting membrane potential (RMP).** The RMP depends on a low perme-

ability of the cell membrane to Na^+ and Ca^{2+} ions, and a higher permeability to K^+ ions. When neurons, skeletal muscle cells, and cardiac muscle cells are depolarized to their threshold level, action potentials result (see chapter 9).

Action Potentials

Like action potentials in skeletal muscle, those in cardiac muscle exhibit depolarization followed by repolarization of the RMP. Alterations in membrane channels are responsible for the changes in the permeability of the cell membrane that produce the action potentials. Action potentials in cardiac muscle last longer than those in skeletal muscle, and the membrane channels differ from those in skeletal muscle. In contrast to action potentials in skeletal muscle, which take less than 2 milliseconds (ms) to complete, action potentials in cardiac muscle take approximately 200–500 ms to complete.

In cardiac muscle, the action potential consists of a rapid **depolarization phase,** followed by rapid, but partial, **early repolarization.** Then a prolonged period of slow repolarization occurs, called the **plateau phase.** At the end of the plateau, a more rapid **final repolarization phase** takes place, during which the membrane potential returns to its resting level (figure 20.14).

The depolarization phase of the action potential is brought about by the opening of membrane channels, called **voltage-gated Na^+ ion channels,** or **sodium fast channels** (or **fast channels**). As the voltage-gated Na^+ ion channels open, Na^+ ions diffuse into the cell, causing rapid depolarization until the cell is depolarized to approximately +20 millivolts (mV).

The voltage change occurring during depolarization affects other ion channels in the cell membrane. There are several different types of **voltage-gated K^+ ion channels,** each of which open and close at different membrane potentials, causing changes in membrane permeability to K^+ ions. For example, at rest, the movement of K^+ ions through open voltage-gated K^+ ion channels is primarily responsible for establishing the resting membrane potential in cardiac muscle cells. Depolarization causes these voltage-gated K^+ ion channels to close, decreasing membrane permeability to K^+ ions. Depolarization also causes membrane channels called **voltage-gated Ca^{2+} ion,** or **calcium slow channels** (or **slow channels**) to begin to open. Compared with sodium fast channels, the calcium slow channels open and close slowly.

Repolarization is the result of changes in membrane permeability to Na^+, K^+, and Ca^{2+} ions. Early repolarization occurs when the voltage-gated Na^+ ion channels close and a small number of voltage-gated K^+ ion channels open. Na^+ ion movement into the cell stops, and K^+ ions move out of the cell. The plateau phase occurs as voltage-gated Ca^{2+} ion channels continue to open, and the movement of Ca^{2+} ions into the cell counteracts the potential change produced by the movement of K^+ ions out of the cell. The plateau phase ends and final repolarization begins as the voltage-gated Ca^{2+} ion

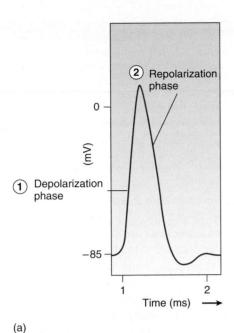

(a)

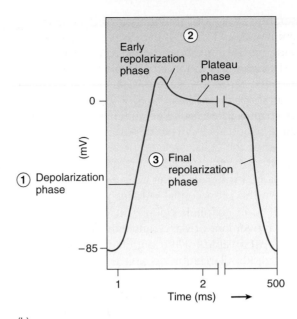

(b)

Permeability changes in skeletal muscle:
1. **Depolarization phase**
   - Voltage-gated Na$^+$ ion channels open.
   - Voltage-gated K$^+$ ion channels begin to open.

2. **Repolarization phase**
   - Voltage-gated Na$^+$ ion channels close.
   - Voltage-gated K$^+$ ion channels continue to open.
   - Voltage-gated K$^+$ ion channels close at the end of repolarization and return the membrane potential to its resting value.

Permeability changes in cardiac muscle:
1. **Depolarization phase**
   - Voltage-gated Na$^+$ ion channels open.
   - Voltage-gated K$^+$ ion channels close.
   - Voltage-gated Ca^{2+} ion channels begin to open.

2. **Early repolarization and plateau phases**
   - Voltage-gated Na$^+$ ion channels close.
   - Some voltage-gated K$^+$ ion channels open, causing early repolarization.
   - Voltage-gated Ca^{2+} ion channels are open, producing the plateau by slowing further repolarization.

3. **Final repolarization phase**
   - Voltage-gated Ca^{2+} ion channels close.
   - Many voltage-gated K$^+$ ion channels open.

Figure 20.14   Comparison of Action Potentials in Skeletal and Cardiac Muscle

(*a*) An action potential in skeletal muscle (*red line*) consists of depolarization and repolarization phases. (*b*) An action potential in cardiac muscle (*blue line*) consists of depolarization, early repolarization, plateau, and final repolarization phases. Cardiac muscle does not repolarize as rapidly as skeletal muscle (indicated by the break in the curve) because of the plateau phase.

channels close and many voltage-gated K$^+$ ion channels open. Thus Ca^{2+} ions stop diffusing into the cell, and the tendency for K$^+$ ions to diffuse out of the cell increases. These permeability changes cause the membrane potential to return to its resting level.

Action potentials in cardiac muscle are conducted from cell to cell, whereas action potentials in skeletal muscle fibers are conducted along the length of a single muscle fiber, but not from fiber to fiber. Also, the rate of action potential propagation is slower in cardiac muscle than in skeletal muscle because cardiac muscle cells are smaller in diameter and much shorter than skeletal muscle fibers. Although the gap junctions of intercalated disks allow transfer of action potentials between cardiac muscle cells, they do slow the rate of action potential conduction between the cardiac muscle cells.

Autorhythmicity of Cardiac Muscle

The heart is said to be **autorhythmic** (aw′tō-rith′mik) because it stimulates itself (auto) to contract at regular intervals (rhythmic). If the heart is removed from the body and maintained under physiologic conditions with the proper nutrients and temperature, it will continue to beat autorhythmically for a long time.

In the SA node, specialized cardiac muscle cells, called **pacemaker cells,** generate action potentials spontaneously and at regular intervals. These action potentials spread through the conducting system of the heart to other cardiac muscle cells, causing voltage-gated Na$^+$ ion channels to open. As a result, action potentials are produced and the cardiac muscle cells contract.

The generation of action potentials in the SA node results when a spontaneously developing local potential, called the **prepotential,** reaches threshold (figure 20.15). The prepotential is caused by changes in ion movement into and out of the pacemaker cells. Na^+ ions cause depolarization by moving into the cells through specialized Na^+ ion channels different from the voltage-gated Na^+ ion channels. A decreasing permeability to K^+ ions also causes depolarization as fewer K^+ ions move out of the cells. As a result of the depolarization, voltage-gated Ca^{2+} ion channels open, and the movement of Ca^{2+} ions into the pacemaker cells causes further depolarization. When the prepotential reaches threshold, many voltage-gated Ca^{2+} ion channels open. Unlike other cardiac muscle cells, the movement of Ca^{2+} ions into the pacemaker cells is primarily responsible for the depolarization phase of the action potential. Repolarization occurs, as in other cardiac muscle cells, when the voltage-gated Ca^{2+} ion channels close and the voltage-gated K^+ ion channels open. After the RMP is reestablished, production of another prepotential starts the generation of the next action potential.

Clinical Note

Various chemical agents such as manganese ions (Mn^{2+}) and verapamil (ver-ap′ă-mil) block the voltage-gated Ca^{2+} ion channels. Voltage-gated Ca^{2+} ion channel-blocking agents prevent the movement of Ca^{2+} ions through voltage-gated Ca^{2+} ion channels into the cell and, for that reason, are called **calcium channel blockers.** Some calcium channel blockers are widely used clinically in the treatment of various cardiac disorders, including tachycardia and certain arrhythmias. Calcium channel blockers slow the development of the prepotential and thus reduce the heart rate. If action potentials arise prematurely within the SA node or other areas of the heart, calcium channel blockers reduce that tendency. Calcium channel blockers also reduce the amount of work performed by the heart because less calcium enters cardiac muscle cells to activate the contractile mechanism. On the other hand, epinephrine and norepinephrine increase the heart rate and its force of contraction by opening the voltage-gated Ca^{2+} ion channels.

Although most cardiac muscle cells respond to action potentials produced by the SA node, some cardiac muscle cells in the conducting system can generate spontaneous action potentials. Normally, the SA node controls the rhythm of the heart because its pacemaker cells generate action potentials at a faster rate than other potential pacemaker cells, producing a heart rate of 70–80 beats per minute (bpm). An **ectopic** (ek-top′ik) **focus** (fō′kŭs; pl. foci, fō′sī) is any part of the heart other than the SA node that generates a heartbeat. For example, if the SA node does not function properly, the part of the heart to produce action potentials at the next highest frequency is the AV node, which produces a heart rate of 40–60 bpm. Another cause of an ectopic focus is blockage of the conducting pathways between the SA node and other parts of the heart. For example, if action potentials do not pass through the AV node, an ectopic focus can develop in an AV bundle, resulting in a heart rate of 30 bpm.

Ectopic foci can also appear when the rate of action potential generation in the ectopic focus becomes enhanced. For example, when cells are injured their plasma membranes become more permeable, resulting in depolarization. These injured cells can be the source of ectopic action potentials.

Permeability changes in pacemaker cells:
1. **Prepotential**
   - A small number of Na^+ ion channels are open.
   - Voltage-gated K^+ ion channels that opened in the repolarization phase of the previous action potential are closing.
   - Voltage-gated Ca^{2+} ion channels begin to open.

2. **Depolarization phase**
   - Voltage-gated Ca^{2+} ion channels are open.
   - Voltage-gated K^+ ion channels are closed.

3. **Repolarization phase**
   - Voltage-gated Ca^{2+} ion channels close.
   - Voltage-gated K^+ ion channels open.

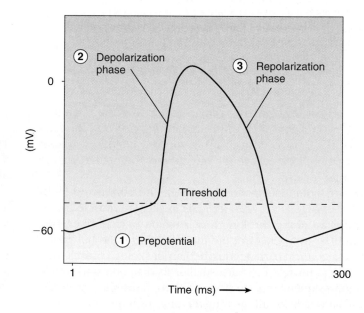

Figure 20.15   Sinoatrial Node Action Potential

The production of action potentials by the sinoatrial (SA) node is responsible for the autorhythmicity of the heart.

P R E D I C T

Predict the consequences for the pumping effectiveness of the heart if numerous ectopic foci in the ventricles produce action potentials at the same time.

✔ *Answer in Appendix F*

Refractory Period of Cardiac Muscle

Cardiac muscle, like skeletal muscle, has a **refractory** (rē-frak′tōr-ē) **period** associated with the action potential. During the **absolute refractory period** the cardiac muscle cell is completely insensitive to further stimulation, and during the **relative refractory period** the cell exhibits reduced sensitivity to additional stimulation. Because the plateau phase of the action potential in cardiac muscle delays repolarization to the RMP, the refractory period is prolonged. The long refractory period ensures that, after contraction, relaxation is nearly complete before another action potential can be initiated, thus preventing tetanic contractions in cardiac muscle.

5 **P R E D I C T**

Predict the consequences if cardiac muscle could undergo tetanic contraction.

✔ *Answer in Appendix F*

▌ Electrocardiogram

The conduction of action potentials through the myocardium during the cardiac cycle produces electric currents that can be measured at the surface of the body. Electrodes placed on the surface of the body and attached to an appropriate recording device can detect small voltage changes resulting from action potentials of the cardiac muscle. The electrodes detect a summation of all the action potentials that are transmitted through the heart at a given time. Electrodes do not detect individual action potentials. The summated record of the cardiac action potentials is an **electrocardiogram (ECG** or **EKG).**

The ECG is not a direct measurement of mechanical events in the heart, and neither the force of contraction nor blood pressure can be determined from it. Each deflection in the ECG record, however, indicates an electrical event within the heart and correlates with a subsequent mechanical event. Consequently, it is an extremely valuable diagnostic tool in identifying a number of cardiac abnormalities (table 20.1), particularly because it is painless, easy to record, and noninvasive (meaning that it does not require surgical procedures). Abnormal heart rates or rhythms, abnormal conduction pathways, hypertrophy or atrophy of portions of the heart, and the approximate location of damaged cardiac muscle can be determined from analysis of an ECG.

The normal ECG consists of a P wave, a QRS complex, and a T wave (figure 20.16). The **P wave,** which is the result of action potentials that cause depolarization of the atrial myocardium, signals the onset of atrial contraction. The **QRS com-** plex is composed of three individual waves: the Q, R, and S waves. The QRS complex results from ventricular depolarization and signals the onset of ventricular contraction. The **T wave** represents repolarization of the ventricles and precedes ventricular relaxation. A wave representing repolarization of the atria cannot be seen because it occurs during the QRS complex.

The time between the beginning of the P wave and the beginning of the QRS complex is the **PQ interval,** commonly called the **PR interval** because the Q wave is often very small. During the PR interval, which lasts approximately 0.16 second, the atria contract and begin to relax. The ventricles begin to depolarize at the end of the PR interval. The **QT interval** extends from the beginning of the QRS complex to the end of the T wave, lasts approximately 0.36 s, and represents the approximate length of time required for the ventricles to contract and begin to relax.

Clinical Note

Elongation of the PR interval can result from (1) a delay of action potential conduction through the atrial muscle because of damage such as that caused by **ischemia** (is-kē′mē-ă), which is the obstruction of the blood supply to the walls of the heart, (2) a delay of action potential conduction through atrial muscle because of a dilated atrium, or (3) a delay of action potential conduction through the AV node and bundle because of ischemia, compression, or necrosis of the AV node or bundle. These conditions result in slow conduction of action potentials through the bundle branches. An unusually long QT interval reflects the abnormal conduction of action potentials through the ventricles, which can result from myocardial infarctions or from an abnormally enlarged left or right ventricle.

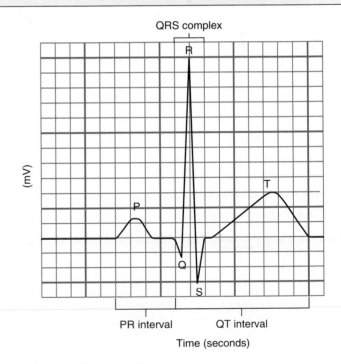

Figure 20.16   Electrocardiogram

The major waves and intervals of an electrocardiogram are labeled. Each thin horizontal line on the ECG recording represents 1 mV, and each thin vertical line represents 0.04 s. ✗ ▭

Table 20.1   Major Cardiac Arrhythmias

Conditions	Symptoms	Possible Causes
Abnormal Heart Rhythms		
Tachycardia	Heart rate in excess of 100 bpm	Elevated body temperature; excessive sympathetic stimulation; toxic conditions
Paroxysmal atrial tachycardia	Sudden increase in heart rate to 95–150 bpm for a few seconds or even for several hours; P wave precedes every QRS complex; P wave inverted and superimposed on T wave	Excessive sympathetic stimulation; abnormally elevated permeability of slow channels
Ventricular tachycardia	Frequently causes fibrillation	Often associated with damage to AV node or ventricular muscle
Abnormal Rhythms Resulting from Ectopic Action Potentials		
Atrial flutter	300 P waves/min; 125 QRS complexes/min resulting in two or three P waves (atrial contraction) for every QRS complex (ventricular contraction)	Ectopic action potentials in the atria
Atrial fibrillation	No P waves; normal QRS complexes; irregular timing; ventricles constantly stimulated by atria; reduced pumping effectiveness and filling time	Ectopic action potentials in the atria
Ventricular fibrillation	No QRS complexes; no rhythmic contraction of the myocardium; many patches of asynchronously contracting ventricular muscle	Ectopic action potentials in the ventricles
Bradycardia	Heart rate less than 60 bpm	Elevated stroke volume in athletes; excessive vagal stimulation; carotid sinus syndrome
Sinus Arrhythmia	Heart rate varies 5% during respiratory cycle and up to 30% during deep respiration	Cause not always known; occasionally caused by ischemia or inflammation or associated with cardiac failure
SA Node Block	Cessation of P wave; new low heart rate due to AV node acting as pacemaker; normal QRS complex and T wave	Ischemia; tissue damage due to infarction; causes unknown
AV Node Block		
First-degree	PR interval greater than 0.2 s	Inflammation of AV bundle
Second-degree	PR interval 0.25–0.45 s; some P waves trigger QRS complexes and others do not; 2:1, 3:1, and 3:2 P wave/QRS complex ratios may occur	Excessive vagal stimulation
Complete heart block	P wave dissociated from QRS complex; atrial rhythm approximately 100 bpm; ventricular rhythm less than 40 bpm	Ischemia of AV nodal fibers or compression of AV bundle
Premature Atrial Contractions	Occasional shortened intervals between one contraction and the succeeding contraction; frequently occurs in healthy people	Excessive smoking; lack of sleep; too much coffee; alcoholism
	P wave superimposed on QRS complex	
Premature Ventricular Contractions (PVCs)	Prolonged QRS complex; exaggerated voltage because only one ventricle may depolarize; inverted T wave; increased probability of fibrillation	Ectopic foci in ventricles; lack of sleep; too much coffee; irritability; occasionally occurs with coronary thrombosis

Abbreviations: SA = sinoatrial; AV = atrioventricular.

Cardiac Cycle

The heart can be viewed as two separate pumps represented by its right and left halves. Each pump consists of a primer pump—the atrium—and a power pump—the ventricle. Both atrial primer pumps complete the filling of the ventricles with blood, and both ventricular power pumps produce the major force that causes blood to flow through the pulmonary and systemic arteries. The term **cardiac cycle** refers to the repetitive pumping process that begins with the onset of cardiac muscle contraction and ends with the beginning of the next contraction (figures 20.17 and 20.18). Pressure changes produced within the heart chambers as a result of cardiac muscle contraction are responsible for blood movement because blood moves from areas of higher pressure to areas of lower pressure.

The duration of the cardiac cycle varies considerably among humans and also during an individual's lifetime. It can be as short as 0.25–0.3 s in a newborn infant or as long as 1 or more seconds in a well-trained athlete. The normal cardiac cycle of 0.7–0.8 s depends on the capability of cardiac muscle to contract and on the functional integrity of the conducting system.

The term **systole** (sis′tō-lē) means to contract, and **diastole** (dī-as′tō-lē) means to dilate. **Atrial systole** is contraction of the atrial myocardium, and **atrial diastole** is relaxation of the atrial myocardium. Similarly, **ventricular systole** is contraction of the ventricular myocardium, and **ventricular diastole** is relaxation of the ventricular myocardium. When the terms systole and diastole are used without reference to specific chambers, however, they mean ventricular systole or diastole.

Before considering the details of the cardiac cycle, an overview of the main events is helpful. Just before systole begins, the atria and ventricles are relaxed, the ventricles are filled with blood, the semilunar valves are closed, and the AV valves are open. As systole begins, contraction of the ventricles increases ventricular pressures, causing blood to flow toward the atria and close the AV valves. As contraction proceeds, ventricular pressures continue to rise, but no blood flows from the ventricles because all the valves are closed. This brief interval is called the **period of isovolumic** (ī-sō-vol-yū′mik) **contraction** because the volume of blood in the ventricles does not change, even though the ventricles are contracting (see figure 20.17a). As the ventricles continue to contract, ventricular pressures become greater than the pressures in the pulmonary trunk and aorta. As a result, during the **period of ejection,** the semilunar valves are pushed open and blood flows from the ventricles into those arteries (see figure 20.17b).

As diastole begins, the ventricles relax and ventricular pressures decrease below the pressures in the pulmonary trunk and aorta. Consequently, blood begins to flow back toward the ventricles, causing the semilunar valves to close (see figure 20.17c). With closure of the semilunar valves, all the heart valves are closed and no blood flows into the relaxing ventricles during the **period of isovolumic relaxation.**

Throughout ventricular systole and the period of isovolumic relaxation, the atria relax and blood flows into them from the veins. As the ventricles continue to relax, ventricular pressures become lower than atrial pressures, the AV valves open, and blood flows from the atria into the relaxed ventricles (see figure 20.17d). At rest, most ventricular filling is a passive process resulting from the greater pressure of blood in the veins and atria than in the completely relaxed ventricles. Completion of ventricular filling is an active process resulting from increased atrial pressure produced by contraction of the atria (see figure 20.17e). During exercise, atrial contraction is more important for ventricular filling because, as heart rate increases, there is less time for passive ventricular filling.

Events Occurring During Ventricular Systole

Figure 20.18 displays the main events of the cardiac cycle in graphic form and should be examined from top to bottom for each period of the cardiac cycle. The ECG indicates the electrical events that cause contraction and relaxation of the atria and ventricles. The pressure graph shows the pressure changes within the left atrium, left ventricle, and aorta resulting from atrial and ventricular contraction and relaxation. Although pressure changes in the right side of the heart are not shown, they are similar to those in the left side, only lower. The volume graph presents the changes in left ventricular volume as blood flows into and out of the left ventricle as a result of the pressure changes. The sound graph records the closing of valves caused by blood flow. See also figure 20.17 for illustrations of the valves and blood flow and table 20.2 for a summary of the events occurring during each period.

Period of Isovolumic Contraction

Completion of the QRS complex initiates contraction of the ventricles. Ventricular pressure rapidly increases, resulting in closure of the AV valves. During the previous ventricular diastole, the ventricles were filled with 120–130 mL of blood, which is called the **end-diastolic volume.** Ventricular volume does not change during the period of isovolumic contraction because all the heart valves are closed at this time.

6	P R E D I C T

Is the cardiac muscle contracting isotonically or isometrically during the period of isovolumic contraction?

✔ *Answer in Appendix F*

Period of Ejection

As soon as ventricular pressures exceed the pressures in the aorta and pulmonary trunk, the semilunar valves open. The aortic semilunar valve opens at approximately 80 mm Hg ventricular pressure, whereas the pulmonary semilunar valve

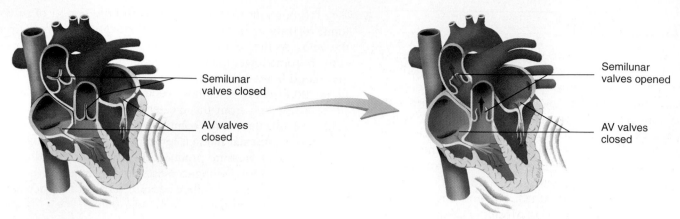

(a) Systole: Period of Isovolumic Contraction. Ventricular contraction causes the AV valves to close, which is the beginning of ventricular systole. The semilunar valves were closed in the previous diastole.

Semilunar valves closed

AV valves closed

(b) Systole: Period of Ejection. Continued ventricular contraction pushes blood out of the ventricles, causing the semilunar valves to open.

Semilunar valves opened

AV valves closed

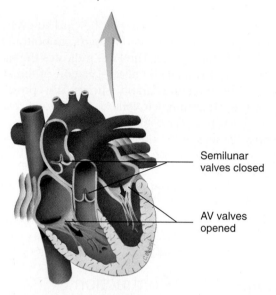

Semilunar valves closed

AV valves opened

(e) Diastole: Active Ventricular Filling. The atria contract and complete ventricular filling.

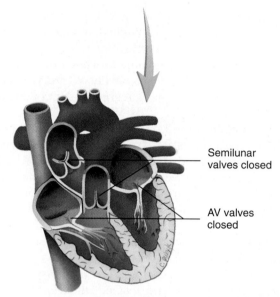

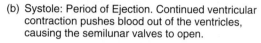

Semilunar valves closed

AV valves closed

(c) Diastole: Period of Isovolumic Relaxation. Blood flowing back toward the relaxed ventricles causes the semilunar valves to close, which is the beginning of ventricular diastole.

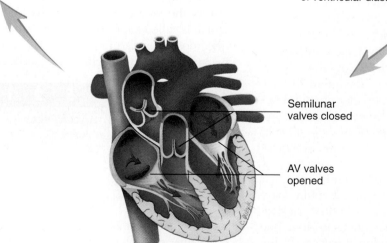

Semilunar valves closed

AV valves opened

(d) Diastole: Passive Ventricular Filling. The AV valves open and blood flows into the relaxed ventricles, accounting for most of the ventricular filling.

Figure 20.17   The Cardiac Cycle

The cardiac cycle is a repeating series of contraction and relaxation that moves blood through the heart. See figure 20.18 and table 20.2 for additional details and explanations. (AV = atrioventricular)

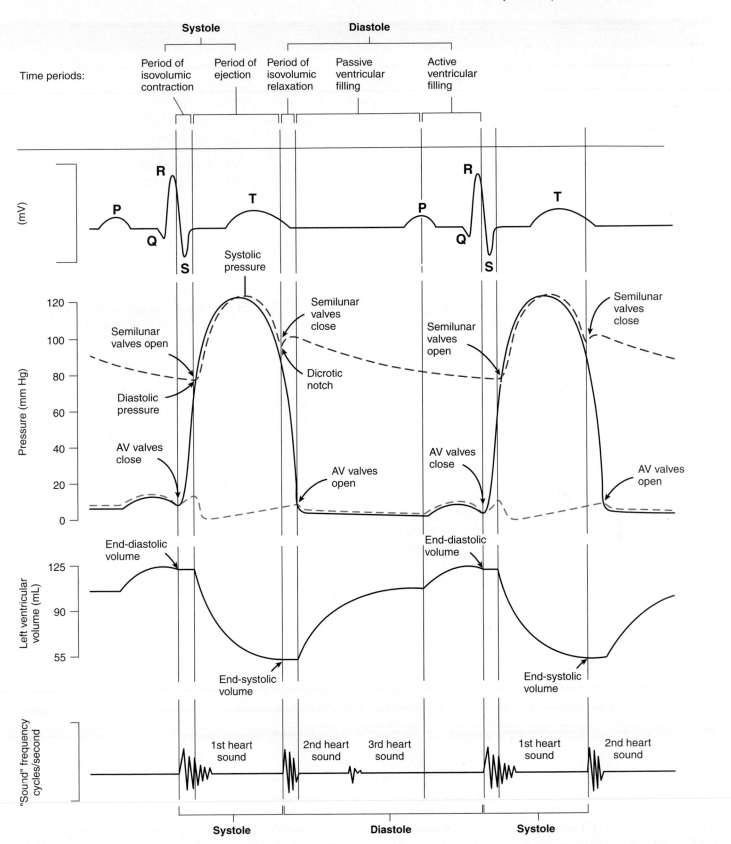

Figure 20.18   Events Occurring During the Cardiac Cycle

The cardiac cycle is divided into five periods (see top of figure). Within these periods, four graphs are presented. From top to bottom, the electrocardiograph; pressure changes for the left atrium (*blue line*), left ventricle (*black line*), and aorta (*red line*); left ventricular volume curve; and heart sounds are illustrated. See table 20.2 for explanations of events during each period and figure 20.17 for illustrations of the valves and blood flow movement.

Table 20.2   Summary of the Events of the Cardiac Cycle

	Ventricular Systole	
	Period of Isovolumic Contraction	**Period of Ejection**
Time period	The ventricles begin to contract, but ventricular volume does not change.	The ventricles continue to contract and blood is pumped out of the ventricles.
Condition of valves	Semilunar valves closed. AV valves closed. (see figure 20.17a)	Semilunar valves opened. AV valves closed. (see figure 20.17b)
ECG	The QRS complex is completed and the ventricles are depolarized. As a result, the ventricles begin to contract. Atrial repolarization is masked by the QRS complex. The atria are relaxed (atrial diastole).	The **T wave** results from ventricular repolarization.
Atrial pressure graph	Atrial pressure decreases in the relaxed atria. When atrial pressure is less than venous pressure, blood flows into the atria. Atrial pressure increases briefly as the contracting ventricles push blood back toward the atria.	Atrial pressure increases gradually as blood flows from the veins into the relaxed atria.
Ventricular pressure graph	Ventricular contraction causes an increase in ventricular pressure, which causes blood to flow toward the atria, closing the AV valves. Ventricular pressure increases rapidly.	Ventricular pressure becomes greater than pressure in the aorta as the ventricles continue to contract. The semilunar valves are pushed open as blood flows out of the ventricles. Ventricular pressure peaks as the ventricles contract maximally, then pressure decreases as blood flows out of the ventricles decreases.
Aortic pressure graph	Just before the semilunar valves open, pressure in the aorta decreases to its lowest value, called the **diastolic pressure** (approximately 80 mm Hg).	As blood is pushed into the aorta by ventricular contraction, pressure in the aorta increases to its highest value, called the **systolic pressure** (approximately 120 mm Hg).
Volume graph	During the **period of isovolumic contraction** ventricular volume does not change because the semilunar and AV valves are closed.	After the semilunar valves open, blood volume decreases as blood flows out of the ventricles during the **period of ejection.** The amount of blood left in a ventricle at the end of the period of ejection is called the **end-systolic volume.**
Sound graph	Blood flowing from the ventricles toward the atria closes the AV valves. Vibrations of the valves and the turbulent flow of blood produce the **first heart sound,** which marks the beginning of ventricular systole.	

opens at approximately 8 mm Hg. Although the pressures are different, both valves open at nearly the same time.

As blood flows from the ventricles during the period of ejection, the left ventricular pressure continues to climb to approximately 120 mm Hg, and the right ventricular pressure increases to approximately 25 mm Hg. The larger left ventricular pressure causes blood to flow throughout the body (systemic circulation), whereas the lower right ventricle pressure causes blood to flow through the lungs (pulmonary circuit). Even though the pressure generated by the left ventricle is much higher than that of the right ventricle, the amount of blood pumped by each is almost the same.

Ventricular Diastole		
Period of Isovolumic Relaxation	**Passive Ventricular Filling**	**Active Ventricular Filling**
The ventricles relax, but ventricular volume does not change.	Blood flows into the ventricles because blood pressure is higher in the veins and atria than in the relaxed ventricles.	Contraction of the atria pumps blood into the ventricles.
Semilunar valves closed. AV valves closed. (see figure 20.17c)	Semilunar valves closed. AV valves opened. (see figure 20.17d)	Semilunar valves closed. AV valves opened. (see figure 20.17e)
The T wave is completed and the ventricles are repolarized. The ventricles relax.	The **P wave** is produced when the SA node generates action potentials and a wave of depolarization begins to propagate across the atria.	The P wave is completed and the atria are stimulated to contract. Action potentials are delayed in the AV node for 0.11 sec allowing time for the atria to contract.

The **QRS complex** begins as action potentials are propagated from the AV node to the ventricles. |
| Atrial pressure continues to increase gradually as blood flows from the veins into the relaxed atria. | After the AV valves open, atrial pressure decreases as blood flows out of the atria into the relaxed ventricles. | Atrial contraction (systole) causes an increase in atrial pressure and blood is forced to flow from the atria into the ventricles. |
| Elastic recoil of the aorta pushes blood back toward the heart, causing the semilunar valves to close.

After closure of the semilunar valves, the pressure in the relaxing ventricles rapidly decreases. | No significant change occurs in ventricular pressure during this time period. | Atrial contraction (systole) and the movement of blood into the ventricles cause a slight increase in ventricular pressure. |
| After the semilunar valves close, elastic recoil of the aorta causes a slight increase in aortic pressure, producing the **dicrotic notch,** or **incisura.** | Aortic pressure gradually decreases as blood runs out of the aorta into other systemic blood vessels. | Aortic pressure gradually decreases as blood runs out of the aorta into other systemic blood vessels. |
| During the **period of isovolumic relaxation** ventricular volume does not change because the semilunar and AV valves are closed. | After the AV valves open, blood flows from the atria and veins into the ventricles because of pressure differences. Most ventricular filling occurs during the first third of diastole.

Little ventricular filling occurs during the middle third of diastole. | Atrial contraction (systole) completes ventricular filling during the last third of diastole. |
| Blood flowing from the ventricles toward the aorta and pulmonary trunk closes the semilunar valves. Vibrations of the valves and the turbulent flow of blood produce the **second heart sound,** which marks the beginning of ventricular diastole. | Sometimes a **third heart sound** is produced by the turbulent flow of blood into the ventricles. | The amount of blood in a ventricle at the end of ventricular diastole is called the **end–diastolic volume.** |

Abbreviation: AV = atrioventricular.

---

7   P R E D I C T

Which ventricle has the thickest wall? Why is it important for each ventricle to pump approximately the same volume of blood?

✔ *Answer in Appendix F*

During the first part of ejection, blood flows rapidly out of the ventricles. Toward the end of ejection, there is very little blood flow, which causes the ventricular pressure to decrease despite continued ventricular contraction. At the end of ejection, the volume has decreased to 50–60 mL, which is called the **end-systolic volume.**

Clinical Focus   Abnormal Heart Sounds

To clinicians, heart sounds provide important information about the normal function of the heart and assist in diagnosing cardiac abnormalities. Abnormal heart sounds are called **murmurs** (mer'merz), and certain murmurs are important indicators of specific cardiac abnormalities. For example, an **incompetent valve** leaks significantly. After closure of an incompetent valve, blood flows through it in a reverse direction. The movement of blood in a direction opposite to normal results in turbulence, which causes a gurgling or swishing sound immediately after closure of the valve. An incompetent tricuspid valve or bicuspid valve exhibits a swish sound immediately after the first heart sound, and the first heart sound can be muffled. An incompetent aortic or pulmonary semilunar valve results in a swish sound immediately after the second heart sound.

Stenosed (sten'ōzd) valves have an abnormally narrow opening and also produce abnormal heart sounds. Blood flows through stenosed valves in a very turbulent fashion, producing a rushing sound before the valve closes. A stenosed atrioventricular valve therefore results in a rushing sound immediately before the first heart sound, and a stenosed semilunar valve results in a rushing sound immediately before the second heart sound.

Inflammation of the heart valves, resulting from conditions such as rheumatic fever, can cause valves to become either incompetent or stenosed. In addition, myocardial infarctions that cause papillary muscles to become nonfunctional can cause bicuspid or tricuspid valves to be incompetent.

Events Occurring During Ventricular Diastole
Period of Isovolumic Relaxation

Completion of the T wave results in ventricular repolarization and relaxation. The already decreasing ventricular pressure falls very rapidly as the ventricles suddenly relax. When the ventricular pressures fall below the pressures in the aorta and the pulmonary trunk, the recoil of the elastic arterial walls, which were stretched during the period of ejection, forces the blood to flow back toward the ventricles, closing the semilunar valves. Ventricular volume does not change during the period of isovolumic relaxation because all the heart valves are closed at this time.

Passive Ventricular Filling

During ventricular systole and the period of isovolumic relaxation, the relaxed atria filled with blood. As the ventricular pressure drops below the atrial pressure, the atrioventricular valves open, allowing blood to flow from the atria into the ventricles. Blood flows from the area of higher pressure in the veins and atria toward the area of lower pressure in the relaxed ventricles. Most ventricular filling occurs during the first third of ventricular diastole. At the end of passive ventricular filling the ventricles are approximately 70% filled.

---

8   P R E D I C T

Fibrillation is abnormal, rapid contractions of different parts of the heart that prevent the heart muscle from contracting as a single unit. Explain why atrial fibrillation does not immediately cause death, but ventricular fibrillation does.

✔ *Answer in Appendix F*

Active Ventricular Filling

Depolarization of the SA node generates action potentials that spread over the atria, producing the P wave and stimulating both atria to contract (atrial systole). The atria contract during the last third of ventricular diastole and complete ventricular filling.

Under most conditions the atria function primarily as reservoirs, and the ventricles can pump sufficient blood to maintain homeostasis even if the atria do not contract at all. During exercise, however, the heart pumps 300–400% more blood than during resting conditions. It is under these conditions that the pumping action of the atria becomes important in maintaining the pumping efficiency of the heart.

⬛ Heart Sounds

When a stethoscope is used to listen to the heart sounds, distinct sounds normally are heard (figure 20.19 and see figure 20.18). The **first heart sound** is a low-pitched sound, often described as a "lubb" sound. It is caused by vibration of the atrioventricular valves and surrounding fluid as the atrioventricular valves close at the beginning of ventricular systole. The **second heart sound** is a higher pitched sound often described as a "dupp" sound. It results from closure of the aortic and pulmonary semilunar valves, at the beginning of ventricular diastole. Systole is therefore approximately the time between the first and second heart sounds. Diastole, which lasts somewhat longer, is approximately the time between the second heart sound and the next first heart sound.

Occasionally a **third heart sound,** caused by blood flowing in a turbulent fashion into the ventricles, can be detected near the end of the first third of diastole. The third heart sound is normal, although faint, and is detected most easily in thin, young people.

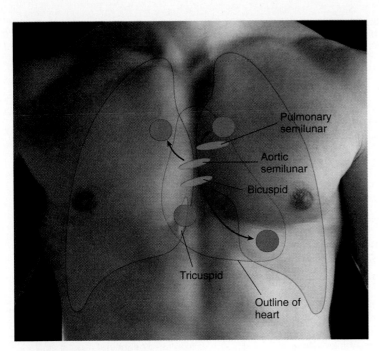

Figure 20.19   Location of the Heart Valves in the Thorax

Surface markings of the heart in the male. The positions of the four heart valves are indicated by blue ellipses, and the sites where the sounds of the valves are best heard with the stethoscope are indicated by pink circles. 𝄐

Aortic Pressure Curve

The elastic walls of the aorta are stretched as blood is ejected into the aorta from the left ventricle. The aortic pressure remains slightly below the ventricular pressure during this period of ejection. As the ventricular pressure drops below that in the aorta, blood flows back toward the ventricle because of the elastic recoil of the aorta. Consequently, the aortic semilunar valve closes, and pressure within the aorta increases slightly producing a **dicrotic** (dī-krot′ik) **notch** in the aortic pressure curve (see figure 20.18). The term dicrotic means double-beating; when increased pressure caused by recoil is large, a double pulse can be felt. The dicrotic notch is also called the **incisura** (in′sī-sū′ră; a cutting into). The aortic pressure then gradually falls throughout the rest of ventricular diastole as blood flows through the peripheral vessels. By the time the aortic pressure has fallen to approximately 80 mm Hg, the ventricles again contract, forcing blood once more into the aorta.

Blood pressure measurements performed for clinical purposes reflect the pressure changes that occur in the aorta rather than in the left ventricle (see chapter 21). The blood pressure in the aorta fluctuates between the systolic pressure, which is about 120 mm Hg, and the diastolic pressure, which is about 80 mm Hg for the average young adult at rest.

Mean Arterial Blood Pressure

Blood pressure is responsible for blood movement and therefore is critical to the maintenance of homeostasis in the body.

Blood flows from areas of higher to areas of lower pressure. For example, during one cardiac cycle, blood flows from the higher pressure in the aorta toward the lower pressure in the relaxed left ventricle.

Mean arterial pressure (MAP) is the average blood pressure between systolic and diastolic pressure in the aorta. It is proportional to the **cardiac output (CO)** times the **peripheral resistance (PR).** Cardiac output, or **minute volume,** is the amount of blood pumped by the heart per minute, and the peripheral resistance is the total resistance against which blood must be pumped.

$$MAP = CO \times PR$$

Mean arterial pressure can be altered by changes in cardiac output and peripheral resistance (figure 20.20). Cardiac output is discussed in this chapter, and peripheral resistance is considered in chapter 21.

Cardiac output is equal to heart rate times stroke volume. **Heart rate (HR)** is the number of times the heart beats (contracts) per minute. **Stroke volume (SV),** which is the volume of blood pumped during each heartbeat (cardiac cycle), is equal to the end-diastolic volume minus the end-systolic volume. During diastole, blood flows from the atria into the ventricles, and end-diastolic volume normally increases to approximately 125 mL. After the ventricles partially empty during systole, the end-systolic volume decreases to approximately 55 mL. The **stroke volume** is therefore equal to 70 mL (125 − 55).

To better understand stroke volume, imagine that you are rinsing out a sponge under a running water faucet. As you relax your hand, the sponge fills with water; as your fingers contract, water is squeezed out of the sponge; and, after you have squeezed it, some water is left within the sponge. In this analogy, the amount of water you squeeze out of the sponge (stroke volume) is the difference between the amount of water in the sponge when your hand is relaxed (end-diastolic volume) and the amount that is left in the sponge after you squeeze it (end-systolic volume).

Stroke volume can be increased by increasing end-diastolic volume or by decreasing end-systolic volume (see figure 20.20). During exercise, end-diastolic volume increases because of an increase in **venous return,** which is the amount of blood returning to the heart. End-systolic volume decreases because the heart contracts more forcefully. For example, stroke volume could increase from a resting value of 70 mL to an exercising value of 115 mL by increasing end-diastolic volume to 145 mL and decreasing end-systolic volume to 30 mL.

Under resting conditions the heart rate is approximately 72 bpm, and the stroke volume is approximately 70 mL/beat, although these values can vary considerably from person to person. The cardiac output is therefore:

$$CO = HR \times SV$$
$$= 72 \text{ bpm} \times 70 \text{ mL/beat}$$
$$= 5040 \text{ mL/min (approximately 5 L/min)}$$

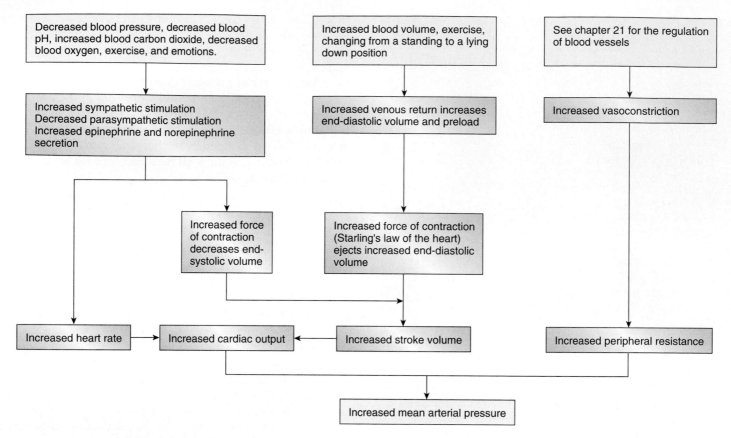

Figure 20.20   Factors Affecting Mean Arterial Pressure
Mean arterial pressure is regulated by controlling cardiac output and peripheral resistance.

During exercise, the heart rate can increase to 190 bpm, and the stroke volume can increase to 115 mL. Consequently, cardiac output is:

$$CO = 190 \text{ bpm} \times 115 \text{ mL/beat}$$
$$= 21{,}850 \text{ mL/min (approximately 22 L/min)}$$

The difference between the cardiac output when a person is at rest and the maximum cardiac output is called the **cardiac reserve.** The greater a person's cardiac reserve, the greater his or her capacity for doing exercise. Lack of exercise and cardiovascular diseases can reduce cardiac reserve and affect a person's quality of life. Exercise training can greatly increase cardiac reserve by increasing cardiac output. In well-trained athletes, stroke volume during exercise can increase to over 200 mL/beat, resulting in cardiac outputs of 40 L/min or more.

Regulation of the Heart

To maintain homeostasis, the amount of blood pumped by the heart must vary dramatically. For example, during exercise the cardiac output can increase several times over resting values. The cardiac output is controlled by regulatory mechanisms that can be classified as either intrinsic or extrinsic. **Intrinsic**

regulation results from the normal functional characteristics of the heart and does not depend on either neural or hormonal regulation. It functions when the heart is in place in the body or is removed and maintained outside the body under proper conditions. On the other hand, **extrinsic regulation** involves neural and hormonal control. Neural regulation of the heart results from sympathetic and parasympathetic reflexes, and the major hormonal regulation comes from epinephrine and norepinephrine secreted from the adrenal medulla.

Intrinsic Regulation

The amount of blood that flows into the right atrium from the veins during diastole is called the venous return. As venous return increases, the end-diastolic volume of the heart increases (see figure 20.20). The greater the end-diastolic volume, the greater the stretch of the ventricular walls. The extent to which the ventricular walls are stretched is sometimes called the **preload.** An increased preload causes an increase in cardiac output, and a decreased preload causes a decrease in cardiac output.

Cardiac muscle exhibits a length-versus-tension relationship similar to that of skeletal muscle. Skeletal muscle, however, is stretched to nearly its optimal length before contraction, whereas cardiac muscle fibers are not stretched to the

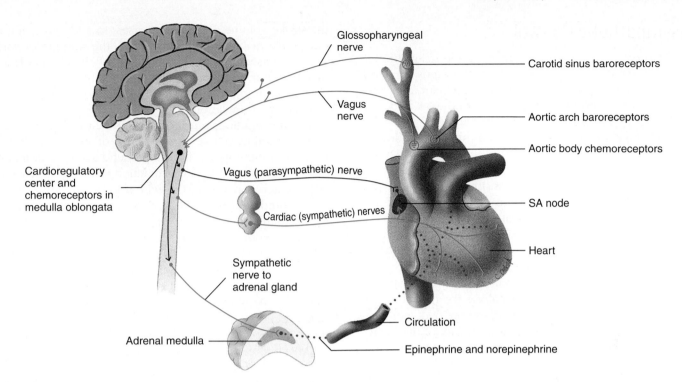

Figure 20.21   Baroreceptor and Chemoreceptor Reflexes

Afferent (*green*) nerves carry action potentials from sensory receptors to the medulla oblongata. Sympathetic (*blue*) and parasympathetic (*red*) nerves exit the spinal cord or medulla and extend to the heart to regulate its function. Hormonal influences such as epinephrine and norepinephrine from the adrenal gland also help regulate the heart's action. (SA = sinoatrial)

point at which they contract with a maximal force (see chapter 10). An increased preload therefore causes the cardiac muscle fibers to contract with a greater force and produce a greater stroke volume. This relationship between preload and stroke volume is commonly referred to as **Starling's law of the heart,** which describes the relationship between changes in the pumping effectiveness of the heart and changes in preload (see figure 20.20). Venous return can decrease to a value as low as 2 L/min or increase to as much as 24 L/min, which has a major effect on the preload.

Afterload is the pressure the contracting ventricles must produce to overcome the pressure in the aorta and move blood into the aorta. Although the pumping effectiveness of the heart is greatly influenced by relatively small changes in the preload, it is very insensitive to large changes in afterload. Aortic blood pressure must increase to more than 170 mm Hg before it hampers the ability of the ventricles to pump blood.

During physical exercise, blood vessels in exercising skeletal muscles dilate, allowing increased flow of blood through the blood vessels. The increased blood flow increases oxygen and nutrient delivery to skeletal muscles. In addition, muscle contractions repeatedly compress veins and cause an increased rate of blood flow toward the heart. As the blood rapidly flows through skeletal muscles and back to the heart, venous return to the heart increases, resulting in an increased preload. The increased preload causes an increased force of contraction, which increases stroke volume. The increase in stroke volume results in an increased cardiac output, and the volume of blood flowing to the exercising muscles increases. When at rest, venous return to the heart decreases because blood vessels in the skeletal muscles constrict and because muscular contractions no longer repeatedly compress the veins. As a result, the preload decreases, and the cardiac output declines, resulting in reduced blood flow to the resting muscles.

Extrinsic Regulation

The heart is innervated by both **parasympathetic** and **sympathetic** nerve fibers (figure 20.21). They influence the pumping action of the heart by affecting both the heart rate and stroke volume. The influence of parasympathetic stimulation on the heart is much less than that of sympathetic stimulation. Sympathetic stimulation can increase cardiac output by 50–100% over resting values, whereas parasympathetic stimulation can cause only a 10–20% decrease.

Extrinsic regulation of the heart functions to keep blood pressure, blood oxygen levels, blood carbon dioxide levels, and blood pH within their normal ranges of values. For example, if the blood pressure suddenly decreases, extrinsic mechanisms detect the decrease and initiate responses that increase the cardiac output to bring blood pressure back to its normal range.

Parasympathetic Control

Parasympathetic nerve fibers are carried to the heart through the **vagus nerves.** Preganglionic fibers of the vagus nerve extend to terminal ganglia within the wall of the heart, and postganglionic fibers extend from the ganglia to the SA node, AV node, coronary vessels, and atrial myocardium.

Parasympathetic stimulation has an inhibitory influence on the heart, primarily by decreasing the heart rate. During resting conditions continuous parasympathetic stimulation inhibits the heart to a small degree. An increase in heart rate during exercise results, in part, from decreased parasympathetic stimulation. Strong parasympathetic stimulation can decrease the heart rate 20–30 bpm. Parasympathetic stimulation has little effect on the stroke volume. In fact, if the venous return remains constant while the heart is inhibited by parasympathetic stimulation, the stroke volume actually can increase. The longer time between heartbeats allows the heart to fill to a greater capacity, resulting in an increased preload, which increases stroke volume because of Starling's law of the heart.

Acetylcholine, the neurotransmitter produced by the postganglionic parasympathetic neurons, binds to ligand-gated channels that cause the cardiac cell membranes to become more permeable to K^+ ions. As a consequence, the RMP hyperpolarizes. Heart rate decreases because it takes longer for the depolarization to cause an action potential.

Sympathetic Control

Sympathetic nerve fibers originate in the thoracic region of the spinal cord as preganglionic neurons. These neurons synapse with postganglionic neurons of the **cervical** and **upper thoracic sympathetic chain ganglia,** which project to the heart as **cardiac nerves** (see figure 20.21 and chapter 16). The postganglionic sympathetic nerve fibers innervate the SA and AV nodes, the coronary vessels, and the atrial and ventricular myocardium.

Sympathetic stimulation increases both the heart rate and the force of muscular contraction. In response to strong sympathetic stimulation, the heart rate can increase to 250 or occasionally 300 bpm. Stronger contractions also can increase stroke volume. The increased force of contraction resulting from sympathetic stimulation causes a lower end-systolic volume in the heart; therefore, the heart empties to a greater extent (see figure 20.20).

9	P R E D I C T

What effect does sympathetic stimulation have on stroke volume if the venous return remains constant? Sympathetic stimulation of the heart also results in dilation of the coronary blood vessels. Explain the functional advantage of that effect.

✔ *Answer in Appendix F*

There are limitations, however, to the relationship between increased heart rate and cardiac output. If the heart rate becomes too fast, diastole is not long enough to allow complete ventricular filling, end-diastolic volume decreases, and the stroke volume actually decreases. In addition, if the heart rate increases beyond a critical level, the strength of contraction decreases, probably as a result of the accumulation of metabolites in the cardiac muscle cells. The limit of the heart's ability to pump blood is 170–250 bpm in response to intense sympathetic stimulation.

Sympathetic stimulation of the ventricular myocardium plays a significant role in regulation of its contraction force during resting conditions. Sympathetic stimulation maintains the strength of ventricular contraction at a level approximately 20% greater than it would be with no sympathetic stimulation.

Norepinephrine, the postganglionic sympathetic neurotransmitter, increases the rate and degree of cardiac muscle depolarization so that both the frequency and the amplitude of the action potentials are increased. The effect of norepinephrine on the heart involves the association between norepinephrine and cell surface β-adrenergic receptors. The combination of norepinephrine molecules with beta-adrenergic receptors causes a G protein-mediated synthesis and accumulation of cyclic adenosine monophosphate (cAMP) in the cytoplasm of cardiac muscle cells. Cyclic-AMP increases the permeability of the cell membrane to Ca^{2+} ions, primarily by opening calcium slow channels in the cell membrane.

Hormonal Control

Epinephrine and norepinephrine released from the adrenal medulla can markedly influence the pumping effectiveness of the heart. Epinephrine has essentially the same effect on cardiac muscle as norepinephrine and therefore increases the rate and force of heart contractions (see figure 20.20). The secretion of epinephrine and norepinephrine from the adrenal medulla is controlled by sympathetic stimulation and occurs in response to increased physical activity, emotional excitement, or stressful conditions. Many stimuli that increase sympathetic stimulation of the heart also increase release of epinephrine and norepinephrine from the adrenal gland (see chapter 18). Epinephrine and norepinephrine are transported in the blood through the vessels of the heart to the cardiac muscle cells, where they bind to beta-adrenergic receptors and stimulate cAMP synthesis. Epinephrine takes a longer time to act on the heart than sympathetic stimulation does, but the effect lasts longer.

Heart and Homeostasis

The pumping efficiency of the heart plays an important role in the maintenance of homeostasis. Blood pressure in the systemic vessels must be maintained at a level at which blood flow is sufficient to achieve nutrient and waste product exchange across the walls of the capillaries at a rate that meets metabolic demands. The activity of the heart must be regulated because the metabolic activities of the tissues change under such conditions as exercise and rest.

Effect of Blood Pressure

The **baroreceptor** (bar′ō-rē-sep′tor) **reflexes** detect changes in blood pressure and result in changes in heart rate and in the force of contraction of the heart. The sensory receptors of the baroreceptor reflexes are stretch receptors. They are in the walls of certain large arteries such as the internal carotid arteries and the aorta, and they function to measure blood pressure (see figure 20.21). The anatomy of these sensory structures and their afferent pathways are described in chapter 21.

Afferent neurons project primarily through the glossopharyngeal (cranial nerve IX) and vagus (cranial nerve X) nerves from the baroreceptors to an area in the medulla oblongata called the **cardioregulatory center,** where sensory action potentials are integrated (see figure 20.21). The part of the cardioregulatory center that functions to increase heart rate is called the **cardioacceleratory center,** and the part that functions to decrease heart rate is called the **cardioinhibitory center.** Efferent action potentials then are sent from the cardioregulatory center to the heart through both the sympathetic and parasympathetic divisions of the autonomic nervous system.

Increased blood pressure within the internal carotid arteries and aorta causes their walls to stretch, stimulating an increase in action potential frequency in the baroreceptors (figure 20.22). At normal blood pressures (80–120 mm Hg), afferent action potentials are sent from the baroreceptors to the medulla oblongata at a relatively constant frequency. When blood pressure increases, the arterial walls are stretched further, and the afferent action potential frequency increases. When blood pressure decreases, the arterial walls are stretched to a lesser extent, and the afferent

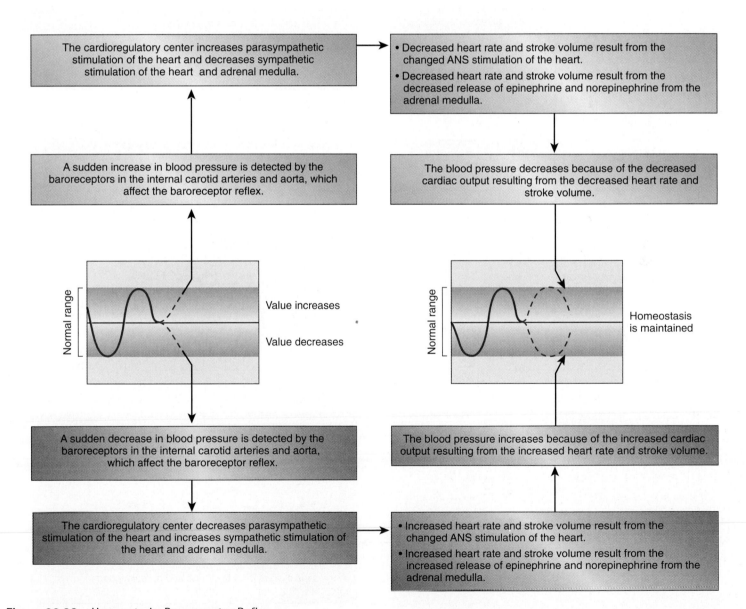

Figure 20.22   Homeostasis: Baroreceptor Reflex.

The baroreceptor reflex maintains homeostasis in response to changes in blood pressure. (ANS = autonomic nervous system)

action potential frequency decreases. In response to increased blood pressure, the baroreceptor reflexes decrease sympathetic stimulation and increase parasympathetic stimulation of the heart, causing the heart rate to decrease. Decreased blood pressure causes decreased parasympathetic and increased sympathetic stimulation of the heart, resulting in an increased heart rate and force of contraction (see figure 20.22). Withdrawal of parasympathetic stimulation is primarily responsible for increases in heart rate up to approximately 100 bpm. Larger increases in heart rate, especially during exercise, result from sympathetic stimulation. The baroreceptor reflexes are homeostatic because they keep the blood pressure within a narrow range of values, which is adequate to maintain blood flow to the tissues.

Effect of pH, Carbon Dioxide, and Oxygen

Chemoreceptor (kem′ō-rē-sep′tŏr) **reflexes** help regulate the activity of the heart. Chemoreceptors sensitive to changes in pH and carbon dioxide levels exist within the medulla oblongata. A decrease in pH and an increase in carbon dioxide decrease parasympathetic and increase sympathetic stimulation of the heart, resulting in increased heart rate and force of contraction (figure 20.23).

The increased cardiac output causes greater blood flow through the lungs, where carbon dioxide is eliminated from the body. The increased blood flow through the lungs helps

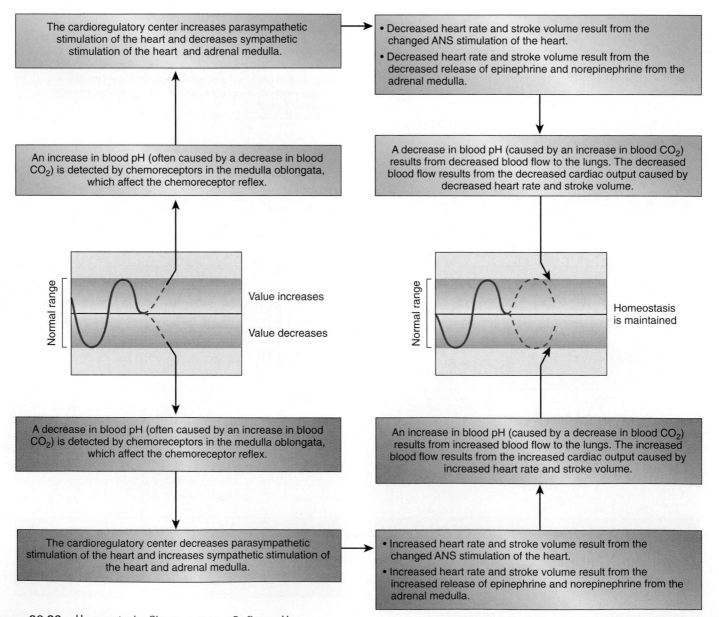

Figure 20.23 Homeostasis: Chemoreceptor Reflex—pH.

The chemoreceptor reflex maintains homeostasis in response to changes in blood concentrations of CO_2 and H^+ ions. (ANS = autonomic nervous system)

bring the blood carbon dioxide level down to its normal range of values and helps to increase the blood pH.

Chemoreceptors primarily sensitive to blood oxygen levels are found in the carotid and aortic bodies. These small structures are located by large arteries near the heart and monitor blood flowing to the brain and to the rest of the body. The carotid and aortic body chemoreceptor reflexes are activated when blood oxygen levels decrease dramatically such as during asphyxiation. In carefully controlled experiments, it is possible to isolate the effects of the carotid and aortic body chemoreceptor reflexes from other reflexes such as the medullary chemoreceptor reflexes. These experiments indicate that a decrease in blood oxygen results in a decrease in heart rate and an increase in vasoconstriction. The increased vasoconstriction causes blood pressure to rise, which promotes blood delivery despite the decrease in heart rate. The carotid and aortic body chemoreceptor reflexes may protect the heart for a short time by slowing the heart and thereby reducing its need for oxygen. The carotid and aortic body chemoreceptor reflexes normally do not function independently of other regulatory mechanisms. When all regulatory mechanisms function together, the effect of large, prolonged decreases in blood oxygen levels is to increase the heart rate. Low blood oxygen levels result in increased stimulation of respiratory movements (see chapter 23). Increased inflation of the lungs stimulates stretch receptors in the lung. Afferent action potentials from these stretch receptors influence the cardioregulatory center, which causes an increase in the heart rate. The reduced oxygen levels at high altitudes can cause an increase in the heart rate even when blood carbon dioxide levels remain low. The carotid and aortic body chemoreceptor reflexes are more important in the regulation of respiration (see chapter 23) and blood vessel constriction (see chapter 21) than in the regulation of heart rate.

Effect of Extracellular Ion Concentration

Ions that affect cardiac muscle function are the same ions (potassium, calcium, and sodium) that influence membrane potentials in other electrically excitable tissues. Some differences do exist, however, between the response of cardiac muscle and that of nerve or muscle tissue to these ions. For example, the extracellular levels of Na^+ ions rarely deviate enough from the normal value to affect the function of cardiac muscle significantly.

Excess K^+ ions in cardiac tissue cause the heart rate and stroke volume to decrease. A twofold increase in extra-cellular K^+ ions results in **heart block,** which is loss of the functional conduction of action potentials through the conducting system of the heart. The excess potassium in the extracellular fluid causes partial depolarization of the RMP, resulting in decreased amplitude of action potentials and decreased rate at which action potentials are conducted along muscle fibers. As the conduction rates decrease, ectopic action potentials can occur. The reduced action potential amplitude also results in less calcium entering the sarcoplasm of the cell; thus the strength of cardiac muscle contraction decreases.

Although the extracellular concentration of K^+ ions normally is small, a decrease in extracellular K^+ ions results in a decrease in the heart rate because the RMP is hyperpolarized; as a consequence, it takes longer for the membrane to depolarize to threshold. The force of contraction is not affected, however.

An increase in the extracellular concentration of Ca^{2+} ions produces an increase in the force of cardiac contraction because of a greater influx of Ca^{2+} ions into the sarcoplasm during action potential generation. Elevated plasma Ca^{2+} ions levels have an indirect effect on heart rate because they reduce the frequency of action potentials in nerve fibers, thus reducing sympathetic and parasympathetic stimulation of the heart (see chapter 9). Generally, elevated blood calcium levels reduce the heart rate.

A low blood calcium level increases the heart rate, although the effect is imperceptible until blood calcium levels are reduced to approximately one-tenth their normal value. The reduced extracellular Ca^{2+} ion levels cause Na^+ ion channels to open, which allows Na^+ ions to diffuse more readily into the cell, resulting in depolarization and action potential generation. Reduced Ca^{2+} ion levels, however, usually cause death as a result of tetany of skeletal muscles before they decrease enough to markedly influence the heart's function.

Effect of Body Temperature

Under resting conditions the temperature of cardiac muscle normally does not change dramatically in humans, although alterations in temperature influence the heart rate. Small increases in cardiac muscle temperature cause the heart rate to increase, and decreases in temperature cause the heart rate to decrease. For example, during exercise or fever, increased heart rate and force of contraction accompany temperature increases, but the heart rate decreases under conditions of hypothermia. During heart surgery the body temperature sometimes is reduced dramatically to slow the heart rate and other metabolic functions in the body.

Inflammation of Heart Tissues

Endocarditis (en'dō-kar-dī'tis) is inflammation of the endocardium. It affects the valves more severely than other areas of the heart and can lead to deposition of scar tissue, causing valves to become stenosed or incompetent.

Myocarditis (mī'ō-kar-dī'tis) is inflammation of the myocardium and can lead to heart failure.

Pericarditis (per'i-kar-dī'tis) is inflammation of the pericardium. Pericarditis can result from bacterial or viral infections and can be extremely painful.

Rheumatic (rū-mat'ik) heart disease can result from a streptococcal infection in young people. Toxin produced by the bacteria can cause an immune reaction called rheumatic fever about 2–4 weeks after the infection. The immune reaction can cause inflammation of the endocardium, called rheumatic endocarditis. The inflamed valves, especially the bicuspid valve, can become stenosed or incompetent. The effective treatment of streptococcal infections with antibiotics has reduced the frequency of rheumatic heart disease.

Reduced Blood Flow to Cardiac Muscle

Coronary heart disease reduces the amount of blood that the coronary arteries are able to deliver to the myocardium. The reduction in blood flow damages the myocardium. The degree of damage depends on the size of the arteries involved, whether occlusion (blockage) is partial or complete, and whether occlusion is gradual or sudden. As the walls of the arteries thicken and harden with age, the volume of blood they can supply to the heart muscle declines, and the ability of the heart to pump blood decreases. Inadequate blood flow to the heart muscle can result in angina (an'ji-nă, an-jī'nă) pectoris (pek'tō-ris), which is a poorly localized sensation of pain in the region of the chest, left arm, and left shoulder.

Degenerative changes in the artery wall can cause the inside surface of the artery to become roughened. The chance of platelet aggregation increases at the rough surface, which increases the chance of coronary thrombosis (throm-bō'sis; meaning, formation of a blood clot in a coronary vessel). Inadequate blood flow can cause an infarct (in'farkt), an area of damaged cardiac tissue. A heart attack is often referred to as a coronary thrombosis or a myocardial infarct. The outcome of coronary thrombosis depends on the extent of the damage to heart muscle caused by inadequate blood flow and whether other blood vessels can supply enough blood to maintain the heart's function. Death can occur swiftly if the infarct is large; if the infarct is small, the heart can continue to function. In most cases, scar tissue replaces damaged cardiac muscle in the area of the infarct.

People who survive infarctions often lead fairly normal lives if they take precautions. Most cases call for moderate exercise, adequate rest, a disciplined diet, and reduced stress.

Congenital Conditions Affecting the Heart

Congenital heart disease is the result of abnormal development of the heart. The following conditions are common congenital defects:

Septal defect is a hole in a septum between the left and right sides of the heart. The hole may be in the interatrial or interventricular septum. These defects allow blood to flow from one side of the heart to the other and as a consequence, greatly reduce the pumping effectiveness of the heart.

Patent ductus arteriosus (dŭk'tŭs ar-tēr'ē-ō-sŭs) results when a blood vessel called the ductus arteriosus, which is present in the fetus, fails to close after birth. The ductus arteriosus extends between the pulmonary trunk and the aorta. It allows blood to pass from the pulmonary trunk to the aorta, thus bypassing the lungs. This is normal before birth because the lungs are not functioning. If the ductus arteriosus fails to close after birth, blood flows in the opposite direction, from the aorta to the pulmonary trunk. As a consequence, blood flows through the lungs under a higher pressure and damages them. In addition, the amount of work required of the left ventricle to maintain an adequate systemic blood pressure increases.

Stenosis (ste-nō'sis) of a heart valve is a narrowed opening through one of the heart valves. In aortic or pulmonary valve stenosis, the workload of the heart is increased because the ventricles must contract with a much greater force to pump blood from the ventricles. Stenosis of the bicuspid valve prevents the flow of blood into the left ventricle, causing blood to back up in the left atrium and in the lungs, resulting in congestion of the lungs. Stenosis of the tricuspid valve causes blood to back up in the right atrium and systemic veins, causing swelling in the periphery.

An incompetent heart valve is one that leaks. Blood therefore flows through the valve when it is closed. The workload of the heart is increased because incompetent valves reduce the pumping efficiency of the heart. For example, an incompetent aortic semilunar valve allows blood to flow from the aorta into the left ventricle during diastole. Thus, the left ventricle fills with blood to a greater degree than normal. The increased filling of the left ventricle results in a greater stroke volume because of Starling's law of the heart. The pressure produced by the contracting ventricle and the pressure in the aorta is greater than normal during ventricular systole. The pressure in the aorta, however, decreases very rapidly as blood leaks into the left ventricle during diastole.

An incompetent bicuspid valve allows blood to flow back into the left atrium from the left ventricle during ventricular systole. This increases the pressure in the left atrium and pulmonary veins, which results in pulmonary edema. Also, the stroke volume of the left ventricle is reduced, which causes a decrease in systemic blood pressure. Similarly, an incompetent tricuspid valve allows blood to flow back into the right atrium and systemic veins causing edema in the periphery.

Cyanosis (sī-ă-nō'sis) is a symptom of inadequate heart function in babies suffering from congenital heart disease. The term "blue baby" is sometimes used to refer to infants with cyanosis. The blueness of the skin is caused by low oxygen levels in the blood in peripheral blood vessels.

Conditions Associated with Aging

Several heart diseases develop as people age and gradually become more severe as they grow older. The following conditions are common in elderly people.

Heart failure is the result of progressive weakening of the heart muscle and the failure of the heart to pump blood effectively. Hypertension (high blood pressure) increases the afterload on the heart, can produce significant enlargement of the heart, and can finally result in heart failure. Advanced age, malnutrition, chronic infections, toxins, severe anemias, or hyperthyroidism can cause degeneration of the heart muscle, resulting in heart failure. Heredity factors can also be responsible for increased susceptibility to heart failure.

Heart Medications

Digitalis (dij-i-tal′is): Slows and strengthens contractions of the heart muscle. This drug is frequently given to people who suffer from heart failure, although it also can be used to treat atrial tachycardia.

Nitroglycerin (nī-trō-glis′er-in): Causes dilation of all of the veins and arteries, including coronary arteries, without an increase in heart rate or stroke volume. When all blood vessels dilate, a greater volume of blood pools in the dilated blood vessels, causing a decrease in the venous return to the heart. The flow of blood through coronary arteries also increases. The reduced preload causes cardiac output to decrease, resulting in a decreased amount of work performed by the heart. Nitroglycerin is frequently given to people who suffer from coronary artery disease, which restricts coronary blood flow. The decreased work performed by the heart reduces the amount of oxygen required by the cardiac muscle. Consequently, the heart does not suffer from a lack of oxygen, and angina pectoris does not develop.

Beta-adrenergic-blocking agents: Reduce the rate and strength of cardiac muscle contractions, thus reducing the heart's demand for oxygen. Beta-adrenergic-blocking agents bind to receptors for norepinephrine and epinephrine and prevent these substances from having their normal effects. These blocking agents are often used to treat people who suffer from rapid heart rates, certain types of arrhythmias, and hypertension.

Calcium channel blockers: Reduce the rate at which Ca^{2+} ions diffuse into cardiac muscle cells and smooth muscle cells. Because the action potentials that produce cardiac muscle contractions depend in part on the flow of Ca^{2+} ions into the cardiac muscle cells, the calcium channel blockers can be used to control the force of heart contractions and reduce arrhythmia, tachycardia, and hypertension. Because entry of calcium into smooth muscle cells causes contraction, calcium channel blockers cause dilation of coronary blood vessels and can be used to treat angina pectoris.

Antihypertensive (an′tē-hī-per-ten′siv) **agents:** Several drugs are used specifically to treat hypertension. These drugs reduce blood pressure and therefore reduce the work required by the heart to pump blood. In addition, the reduction of blood pressure reduces the risk of heart attacks and strokes. Drugs used to treat hypertension include drugs that reduce the activity of the sympathetic nervous system, drugs that dilate arteries and veins, drugs that increase urine production (diuretics), and drugs that block the conversion of angiotensinogen to angiotensin I.

Anticoagulants (an′tē-kō-ag′yū-lantz): Prevent clot formation in persons with damage to heart valves or blood vessels or in persons who have had a myocardial infarction. Aspirin functions as a weak anticoagulant.

Instruments and Selected Procedures

Artificial pacemaker: An instrument placed beneath the skin, equipped with an electrode that extends to the heart. An artificial pacemaker provides an electric stimulus to the heart at a set frequency. Artificial pacemakers are used in patients in whom the natural pacemaker of the heart does not produce a heart rate high enough to sustain normal physical activity. Modern electronics has made it possible to design artificial pacemakers that can increase the heart rate as increases in physical activity occur. Pacemakers can also detect cardiac arrest, extreme arrythmias, or fibrillation. In response, strong stimulation of the heart by the pacemaker may restore heart function.

Heart lung machine: A machine that serves as a temporary substitute for the patient's heart and lungs. It pumps blood throughout the body and oxygenates and removes carbon dioxide from the blood. It has made possible many surgeries on the heart and lungs.

Heart valve replacement or repair: A surgical procedure performed on those who have diseased valves that are so deformed and scarred from conditions such as endocarditis that the valves are severely incompetent or stenosed. Substitute valves made of synthetic materials such as plastic or Dacron are effective; valves transplanted from pigs are also used.

Heart transplants: A surgical procedure made possible when the immune characteristics of a donor and the recipient are closely matched. The heart of a recently deceased donor is transplanted to the recipient, and the diseased heart of the recipient is removed. People who have received heart transplants must continue to take drugs that suppress their immune responses for the rest of their lives. Unless they do so, their immune system rejects the transplanted heart.

Artificial heart: Replacement of the heart with an artificial heart. This mechanical pump is still experimental and cannot be viewed as a permanent substitute for the heart. It has been used to keep a patient alive until a donor heart can be found.

Cardiac assistance: A temporarily implanted mechanical device that assists the heart in pumping blood. In some cases, the decreased workload on the heart provided by the mechanical device appears to promote recovery of failing hearts, and the mechanical device has been successfully removed. In **cardiomyoplasty** a piece of a back muscle (latissimus dorsi) is wrapped around the heart and stimulated to contract in synchrony with the heart.

Prevention of Heart Disease

Proper nutrition is important in reducing the risk of heart disease. A recommended diet is low in fats, especially saturated fats and cholesterol, and low in refined sugar. Diets should be high in fiber, whole grains, fruits, and vegetables. Total food intake should be limited to avoid obesity, and sodium chloride intake should be reduced.

Tobacco and excessive use of alcohol should be avoided. Smoking increases the risk of heart disease by at least 10-fold, and excessive use of alcohol also substantially increases the risk of heart disease.

Chronic stress, frequent emotional upsets, and a lack of physical exercise can increase the risk of cardiovascular disease. Remedies include relaxation techniques and aerobic exercise programs involving gradual increases in duration and difficulty in activities such as swimming, jogging, or aerobic dancing.

Hypertension (hī′per-ten′shŭn) is abnormally high systemic blood pressure. Hypertension affects about one-fifth of the U.S. population. Regular blood pressure measurements are important because hypertension does not produce obvious symptoms. If hypertension cannot be controlled by diet and exercise, it is important to treat the condition with prescribed drugs. The cause of hypertension in the majority of cases is unknown.

Some data suggest that taking an aspirin daily reduces the chance of a heart attack. Aspirin inhibits the synthesis of prostaglandins in platelets, thereby helping to prevent clot formation.

Aging

Although the age at which the heart becomes less efficient varies considerably and depends on many factors, by the age of 70 cardiac output often decreases by about one-third. Because of the decrease in reserve strength of the heart, many elderly people are often limited in their ability to respond to emergencies, infections, blood loss, or stress.

Systems Pathology
myocardial infarction

MYOCARDIAL INFARCTION

Mr. P. was an overweight, out-of-shape executive who regularly smoked and consumed food with a high fat content. He viewed his job as frustrating because he was frequently confronted with stressful deadlines. He had not had a physical examination for several years so he was not aware that his blood pressure was high. One evening Mr. P was walking to his car after work when he began to feel chest pain that radiated down his left arm. Shortly after the onset of pain he felt out of breath, developed marked pallor, became dizzy, and had to lie down on the sidewalk. The pain in his chest and arm was poorly localized, but intense and he became anxious and then disoriented. Mr. P lost consciousness, although he did not stop breathing. After a short delay, one of his coworkers noticed him and called for help. When paramedics arrived they determined that Mr. P's blood pressure was low and he exhibited arrhythmia and tachycardia. The paramedics transmitted the electrocardiogram they took to a physician by way of their electronic communications system, and they discussed Mr. P's symptoms with the physician who was at the hospital. The paramedics were directed to administer oxygen and medication to control arrhythmias and transport him to the hospital. At the hospital, tissue plasminogen activator (tPA) was administered, which improved blood flow to the damaged area of the heart by activating plasminogen, which dissolves blood clots. Blood levels of enzymes such as creatine phosphokinase increased in Mr. P's blood over the next few days, which confirmed that damage to cardiac muscle resulted from an infarction.

In the hospital, Mr. P began to experience shortness of breath because of pulmonary edema, and after a few days in the hospital he developed pneumonia. He was treated for pneumonia and gradually improved over the next few weeks. An angiogram performed several days after Mr. P's infarction indicated that he had suffered damage to a significant part of the lateral wall of his left ventricle and that neither angioplasty nor bypass surgery were necessary, although Mr. P has some serious restrictions to blood flow in his coronary arteries.

BACKGROUND INFORMATION

Mr. P experienced a myocardial infarction. A thrombosis in one of the branches of the left coronary artery reduces blood supply to the lateral wall of the left ventricle, resulting in ischemia of the left ventricle wall. That tPA is effective in treating a heart attack is consistent with the conclusion that the in-

farction was caused by a thrombosis. An ischemic area of the heart wall is not able to contract normally and, therefore, the pumping effectiveness of the heart is dramatically reduced. The reduced pumping capacity of the heart is responsible for the low blood pressure, which causes the blood flow to the brain to decrease resulting in confusion, disorientation, and unconsciousness.

Low blood pressure, increasing blood carbon dioxide levels, pain, and anxiousness increase sympathetic stimulation of the heart and adrenal glands. Increased sympathetic stimulation of the adrenal medulla results in release of epinephrine from the adrenal medulla. Increased parasympathetic stimulation of the heart results from pain sensations. In such cases, the heart is periodically arrhythmic due to the combined effects of parasympathetic stimulation, epinephrine and norepinephrine from the adrenal gland, and sympathetic stimulation. In addition, ectopic beats are produced by the ischemic areas of the left ventricle.

Pulmonary edema results from the increased pressure in the pulmonary veins because of the inability of the left ventricle to pump blood. The edema allows bacteria to infect the lungs and cause pneumonia.

The heart begins to beat rhythmically in response to medication because the infarction did not damage the conducting system of the heart which is an indication that there are no permanent arrhythmias. Permanent arrhythmias are indications of damage done to cardiac muscle specialized to conduct action potentials in the heart.

Analysis of the electrocardiogram, blood pressure measurements, and the angiogram (figure A) indicate that the infarction, in this case, is located on the left side of Mr. P's heart. Mr. P exhibited several characteristics that are correlated with an increased probability of myocardial infarction: lack of physical exercise, being overweight, smoking, and stress.

Mr. P's physician made it very clear to him that he was lucky to have survived a myocardial infarction, and the physician recommended a weight-loss program, a low-sodium and low-fat diet, and that Mr. P should stop smoking. He explained that Mr. P would have to take medication for high blood pressure if his blood pressure did not decrease in response to the recommended changes. After a period of recovery the physician recommended an aerobic exercise program, and he recommended that Mr. P seek ways to reduce the stress associated

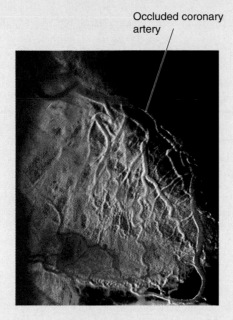

Occluded coronary
artery

Figure A Angiogram

An angiogram (an'jē-ō-gram) is a picture of a blood vessel. It is usually obtained by placing a catheter into a blood vessel and injecting a dye that can be detected with x-rays. Note the occluded (blocked) coronary blood vessel in this angiogram, which has been computer-enhanced to show colors.

with his job. His physician also recommended that Mr. P regularly take a small amount of aspirin. The aspirin is prescribed to reduce the probability of thrombosis. Because aspirin inhibits prostaglandin synthesis it reduces the tendency for blood to clot. Mr. P followed the doctor's recommendations, and after several months he began to feel better than he had in years and his blood pressure was normal.

10   P R E D I C T

Severe ischemia in the wall of a ventricle can result in the death of cardiac muscle cells. Inflammation around the necrotic tissue results, and macrophages invade the necrotic tissue and phagocytize dead cells. At the same time, blood vessels and connective tissue grow into the necrotic area and begin to deposit connective tissue to replace the necrotic tissue. Assume that Mr. P had a myocardial infarction and was recovering. After about a week, however, his blood pressure suddenly decreased to very low levels and he died within a very short time. At autopsy, a large amount of blood was found in the pericardial sac, and the wall of the left ventricle was ruptured. Explain.

✔ *Answer in Appendix F*

System Interactions

System	Interaction
Integumentary	Pallor of the skin resulted from intense constriction of peripheral blood vessels, including those in the skin.
Muscular	Reduced skeletal muscle activity required for activities such as walking results from lack of blood flow to the brain and because blood is shunted from blood vessels that supply skeletal muscles to those that supply the heart and brain.
Nervous	Decreased blood flow to the brain, decreased blood pressure, and pain due to ischemia of heart muscle result in increased sympathetic and decreased parasympathetic stimulation of the heart. Loss of consciousness occurs when the blood flow to the brain decreases enough to result in too little oxygen to maintain normal brain function, especially in the reticular activating system.
Endocrine	When blood pressure decreases to low values, antidiuretic hormone (ADH) is released from the posterior pituitary gland and renin, released from the kidney, activates the renin-angiotensinogen-aldosterone mechanism. ADH, secreted in large amounts, and angiotensin II cause vasoconstriction of peripheral blood vessels. ADH and aldosterone act on the kidneys to retain water and electrolytes. An increased blood volume increases venous return, which results in an increased stroke volume of the heart and an increase in blood pressure unless damage to the heart is very severe.
Lymphatic or Immune	White blood cells, including macrophages, move to the area of cardiac muscle damaged and phagocytize any dead cardiac muscle cells.
Respiratory	Decreased blood pressure results in a decreased blood flow to the lungs. The decrease in gas exchange results in increased blood CO_2 levels, acidosis, and decreased blood O_2 levels. Initially, respiration becomes deep and labored because of the elevated CO_2 levels, decreased blood pH, and depressed O_2 levels. If the blood O_2 levels decrease too much, the person loses consciousness. Pulmonary edema can result when the pumping effectiveness of the left ventricle is substantially reduced.
Digestive	Decreased blood flow to the digestive system to very low levels often results in increased nausea and vomiting.
Urinary	Blood flow to the kidney decreases dramatically in response to sympathetic stimulation. If the kidney becomes ischemic, damage to the kidney tubules can occur, resulting in acute renal failure. Acute renal failure reduces urine production. Increased blood urea nitrogen, increased blood levels of K^+ ions, and edema are indications that the kidneys cannot eliminate waste products and excess water. If damage is not too great, the period of reduced urine production may last up to 3 weeks and then the rate of urine production slowly returns to normal as the kidney tubules heal.

Summary

The heart produces the force that causes blood circulation.

Size, Form, and Location of the Heart

The heart is approximately the size of a closed fist and is shaped like a blunt cone. It is in the mediastinum.

Anatomy of the Heart

The heart consists of two atria and two ventricles.

Pericardium

1. The pericardium is a sac that surrounds the heart and consists of the fibrous pericardium and the serous pericardium.
2. The fibrous pericardium helps hold the heart in place.
3. The serous pericardium reduces friction as the heart beats. It consists of the following parts:
   - The parietal pericardium lines the fibrous pericardium.
   - The visceral pericardium lines the exterior surface of the heart.
   - The pericardial cavity lies between the parietal and visceral pericardium and is filled with pericardial fluid.

Heart Wall

1. The heart wall has three layers:
   - The outer epicardium (visceral pericardium) provides protection against the friction of rubbing organs.
   - The middle myocardium is responsible for contraction.
   - The inner endocardium reduces the friction resulting from blood's passing through the heart.
2. The inner surfaces of the atria are mainly smooth. The auricles have raised areas called musculi pectinati.
3. The ventricles have ridges called trabeculae carneae.

External Anatomy

1. Each atrium has a flap called the auricle.
2. The atria are separated from the ventricles by the coronary sulcus. The right and left ventricles are separated by the interventricular grooves.
3. The inferior and superior venae cavae and the coronary sinus enter the right atrium. The four pulmonary veins enter the left atrium.
4. The pulmonary trunk exits the right ventricle, and the aorta exits the left ventricle.
5. Coronary arteries branch off the aorta to supply the heart. Blood returns from the heart tissues to the right atrium through the coronary sinus and cardiac veins.

Heart Chambers and Valves

1. The atria are separated from each other by the interatrial septum, and the ventricles are separated by the interventricular septum.
2. The right atrium and ventricle are separated by the tricuspid valve. The left atrium and ventricle are separated by the bicuspid valve. The papillary muscles attach by the chordae tendineae to the atrioventricular valves.
3. The aorta and pulmonary trunk are separated from the ventricles by the semilunar valves.

Route of Blood Flow Through the Heart

1. Blood from the body flows through the right atrium into the right ventricle and then to the lungs.
2. Blood returns from the lungs to the left atrium, enters the left ventricle, and is pumped back to the body.

Histology
Heart Skeleton

The fibrous heart skeleton supports the openings of the heart, electrically insulates the atria from the ventricles, and provides a point of attachment for heart muscle.

Cardiac Muscle

1. Cardiac muscle cells are branched and have a centrally located nucleus. Actin and myosin are organized to form sarcomeres. The sarcoplasmic reticulum and T tubules are not as organized as in skeletal muscle.
2. Cardiac muscle cells are joined by intercalated disks, which allow action potentials to move from one cell to the next. Thus cardiac muscle cells function as a unit.
3. Cardiac muscle cells have a slow onset of contraction and a prolonged contraction time caused by the length of time required for calcium to move to and from the myofibrils.
4. Cardiac muscle is well supplied with blood vessels that support aerobic respiration.
5. Cardiac muscle aerobically uses glucose, fatty acids, and lactic acid to produce ATP for energy. Cardiac muscle does not develop a significant oxygen debt.

Conducting System

1. The SA node and the AV node are in the right atrium.
2. The AV node is connected to the bundle branches in the interventricular septum by the AV bundle.
3. The bundle branches give rise to Purkinje fibers, which supply the ventricles.
4. The SA node initiates action potentials, which spread across the atria and cause them to contract.
5. Action potentials are slowed in the AV node, allowing the atria to contract and blood to move into the ventricles. Then the action potentials travel through the AV bundles and bundle branches to the Purkinje fibers, causing the ventricles to contract, starting at the apex.

Electrical Properties
Action Potentials

1. After depolarization and partial repolarization there is a plateau, during which the membrane potential only slowly repolarizes.
2. Depolarization is caused by the movement of sodium ions through the voltage-gated Na^+ ion channels.
3. During depolarization voltage-gated K^+ ion channels close and voltage-gated Ca^{2+} ion channels begin to open.
4. Early repolarization results from closure of the voltage-gated Na^+ ion channels and the opening of some voltage-gated K^+ ion channels.

5. The plateau exists because voltage-gated Ca^{2+} ion channels remain open.
6. The rapid phase of repolarization results from closure of the voltage-gated Ca^{2+} ion channels and the opening of many voltage-gated K^+ ion channels.

Autorhythmicity of Cardiac Muscle

1. Cardiac pacemaker muscle cells are autorhythmic, because of the spontaneous development of a prepotential.
2. The prepotential results from the movement of Na^+ and Ca^{2+} ions into the pacemaker cells.
3. Ectopic foci are areas of the heart that regulate heart rate under abnormal conditions.

Refractory Period of Cardiac Muscle

Cardiac muscle has a prolonged depolarization and thus a prolonged refractory period, which allows time for the cardiac muscle to relax before the next action potential causes a contraction.

Electrocardiogram

1. The ECG records only the electrical activities of the heart.
   - Depolarization of the atria produces the P wave.
   - Depolarization of the ventricles produces the QRS complex. Repolarization of the atria occurs during the QRS complex.
   - Repolarization of the ventricles produces the T wave.
2. Based on the magnitude of the ECG waves and the time between waves, ECGs can be used to diagnose heart abnormalities.

Cardiac Cycle

1. The cardiac cycle is repetitive contraction and relaxation of the heart chambers.
2. Blood moves through the circulatory system from areas of higher pressure to areas of lower pressure. Contraction of the heart produces the pressure.
3. The cardiac cycle can be divided into five periods.
   - Although the heart is contracting, during the period of isovolumic contraction ventricular volume does not change because all the heart valves are closed.
   - During the period of ejection, the semilunar valves open and blood is ejected from the heart.
   - Although the heart is relaxing, during the period of isovolumic relaxation ventricular volume does not change because all the heart valves are closed.
   - Passive ventricular filling results when blood flows from the higher pressure in the veins and atria to the lower pressure in the relaxed ventricles.
   - Active ventricular filling results when the atria contract and pump blood into the ventricles.

Events Occurring During Ventricular Systole

1. Contraction of the ventricles closes the AV valves, opens the semilunar valves, and ejects blood from the heart.
2. The volume of blood in a ventricle just before it contracts is the end-diastolic volume. The volume of blood after contraction is the end-systolic volume.

Events Occurring During Ventricular Diastole

1. Relaxation of the ventricles results in closing of the semilunar valves, opening of the AV valves, and the movement of blood into the ventricles.
2. Most ventricular filling occurs when blood flows from the higher pressure in the veins and atria to the lower pressure in the relaxed ventricles.
3. Contraction of the atria completes ventricular filling.

Heart Sounds

1. The first heart sound is produced by closure of the atrioventricular valves.
2. The second heart sound is produced by closure of the semilunar valves.

Aortic Pressure Curve

1. Contraction of the ventricles forces blood into the aorta, thus producing the peak systolic pressure.
2. Blood pressure in the aorta falls to the diastolic level as blood flows out of the aorta.
3. Elastic recoil of the aorta maintains pressure in the aorta and produces the dicrotic notch.

Mean Arterial Blood Pressure

1. Mean arterial pressure is the average blood pressure in the aorta. Adequate blood pressure is necessary to ensure delivery of blood to the tissues.
2. Mean arterial pressure is proportional to cardiac output (amount of blood pumped by the heart per minute) times peripheral resistance (total resistance to blood flow through blood vessels).
3. Cardiac output is equal to stroke volume times heart rate.
4. Stroke volume, the amount of blood pumped by the heart per beat, is equal to end-diastolic volume minus end-systolic volume.
   - Venous return is the amount of blood returning to the heart. Increased venous return increases stroke volume by increasing end-diastolic volume.
   - Increased force of contraction increases stroke volume by decreasing end-systolic volume.
5. Cardiac reserve is the difference between resting and exercising cardiac output.

Regulation of the Heart
Intrinsic Regulation

1. Venous return is the amount of blood that returns to the heart during each cardiac cycle.
2. Starling's law of the heart describes the relationship between preload and the stroke volume of the heart. An increased preload causes the cardiac muscle fibers to contract with a greater force and produce a greater stroke volume.

Extrinsic Regulation

1. The cardioregulatory center in the medulla oblongata regulates the parasympathetic and sympathetic nervous control of the heart.
2. Parasympathetic control.
   - Parasympathetic stimulation is supplied by the vagus nerve.
   - Parasympathetic stimulation decreases heart rate.
   - Postganglionic neurons secrete acetylcholine, which increases membrane permeability to K^+ ions, producing hyperpolarization of the membrane.

3. Sympathetic control
   - Sympathetic stimulation is supplied by the cardiac nerves.
   - Sympathetic stimulation increases heart rate and the force of contraction (stroke volume).
   - Postganglionic neurons secrete norepinephrine, which increases membrane permeability to Na^+ and Ca^{2+} ions, producing depolarization of the membrane.
4. Epinephrine and norepinephrine are released into the blood from the adrenal medulla as a result of sympathetic stimulation.
   - The effects of epinephrine and norepinephrine on the heart are long lasting compared with those of neural stimulation.
   - Epinephrine and norepinephrine increase the rate and force of heart contraction.

Heart and Homeostasis
Effect of Blood Pressure

1. Baroreceptors monitor blood pressure.
2. In response to a decrease in blood pressure, the baroreceptor reflexes increase sympathetic stimulation and decrease parasympathetic stimulation of the heart, resulting in an increase in heart rate and force of contraction.

Effect of pH, Carbon Dioxide, and Oxygen

1. Chemoreceptors monitor blood carbon dioxide, pH, and oxygen levels.

2. In response to increased carbon dioxide and decreased pH, medullary chemoreceptor reflexes increase sympathetic stimulation and decrease parasympathetic stimulation of the heart.
3. Carotid body chemoreceptor receptors stimulated by low oxygen levels result in a decreased heart rate and vasoconstriction.
4. All regulatory mechanisms functioning together in response to low blood pH, high blood carbon dioxide, and low blood oxygen levels usually produce an increase in heart rate and vasoconstriction. Decreased oxygen levels stimulate an increase in heart rate indirectly by stimulating respiration, and the stretch of the lungs activates a reflex that increases sympathetic stimulation of the heart.

Effect of Extracellular Ion Concentration

1. An increase or decrease in extracellular K^+ ions decreases heart rate.
2. Increased extracellular Ca^{2+} ions increase the force of contraction of the heart and decrease the heart rate. Decreased Ca^{2+} ion levels produce the opposite effect.

Effect of Body Temperature

Heart rate increases when body temperature increases, and it decreases when body temperature decreases.

Content Review

1. Give the approximate size and shape of the heart. Where is it located?
2. What is the pericardium? Name its parts and their functions.
3. Describe the three layers of the heart, and state their functions.
4. Name the major blood vessels that enter and leave the heart. Which chambers of the heart do they enter or exit?
5. What structure separates the atria from each other? What structure separates the ventricles from each other?
6. Name the valves that separate the right atrium from the right ventricle and the left atrium from the left ventricle.
7. What are the functions of the papillary muscles and the chordae tendineae?
8. Describe the flow of blood through the heart.
9. What is the skeleton of the heart? Give three of its functions.
10. How does the structure of cardiac muscle differ from skeletal muscle?
11. Why does cardiac muscle have a slow onset of contraction and a prolonged contraction?
12. What anatomic features are responsible for the ability of cardiac muscle cells to contract as a unit?
13. What substances are used by cardiac muscle as an energy source? Do cardiac muscle cells develop an oxygen debt?
14. List the parts of the conducting system of the heart. Explain how the conducting system coordinates contraction of the atria and ventricles.
15. Describe ion movement during depolarization in cardiac muscle. What is the plateau phase?
16. Why is cardiac muscle referred to as autorhythmic? What are ectopic foci?

17. Why does cardiac muscle have a prolonged refractory period? What is the advantage of having a prolonged refractory period?
18. What does an ECG measure? Name the waves produced by an ECG, and state what events occur during each wave.
19. Define systole and diastole.
20. State whether or not the AV and semilunar valves are open or closed during each of the five periods of the cardiac cycle.
21. Define end-diastolic volume and end-systolic volume.
22. What produces the first heart sound, the second heart sound, and the third heart sound?
23. Explain the production in the aorta of systolic pressure, diastolic pressure, and the dicrotic notch or incisura.
24. Define mean arterial pressure, cardiac output, and peripheral resistance. Explain the role of mean arterial pressure in causing blood flow.
25. Define stroke volume, and state two ways to increase stroke volume.
26. Define the term venous return, and explain how it affects preload. How does preload affect cardiac output? State Starling's law of the heart.
27. Define the term afterload, and describe its effect on heart function.
28. What part of the brain regulates the heart? Describe the autonomic nerve supply to the heart.
29. What effect do parasympathetic stimulation and sympathetic stimulation have on heart rate, force of contraction, and stroke volume?
30. What neurotransmitters are released by the parasympathetic and sympathetic postganglionic neurons of the heart? What

effects do they have on membrane permeability and excitability?

31. Name the two main hormones that affect the heart. Where are they produced, what causes their release, and what effects do they have on the heart?

32. How does the nervous system detect and respond to the following: (1) a decrease in blood pressure, (2) an increase in blood carbon dioxide levels, (3) a decrease in blood pH, and (4) a decrease in blood oxygen levels?

33. Describe the baroreceptor reflex and the response of the heart to an increase in venous return.

34. What effect does an increase or decrease in extracellular potassium, calcium, and sodium ions have on heart rate and the force of contraction of the heart?

35. What effect does temperature have on heart rate?

Develop Your Reasoning Skills

1. Explain why the walls of the ventricles are thicker than the walls of the atria.

2. In most tissues, peak blood flow occurs during systole and decreases during diastole. In heart tissue, however, the opposite is true, and peak blood flow occurs during diastole. Explain why this difference occurs.

3. A patient has tachycardia. Would you recommend a drug that prolongs or shortens the plateau of cardiac muscle cell action potentials?

4. Endurance-trained athletes often have a decreased heart rate compared with that of a nonathlete when both are resting. Explain why an endurance-trained athlete's heart rate decreases rather than increases.

5. A doctor lets you listen to a patient's heart with a stethoscope at the same time that you feel the patient's pulse. Once in a while you hear two heartbeats very close together, but you feel only one pulse beat. Later the doctor tells you that the patient has an ectopic focus in the right atrium. Explain why you hear two heartbeats very close together. The doctor also tells you that the patient exhibits a pulse deficit (i.e., the number of pulse beats felt is fewer than the number of heartbeats heard). Explain why a pulse deficit occurs.

6. Heart rate and cardiac output were measured in a group of nonathletic students. After 2 months of aerobic exercise training their measurements were repeated. It was found that heart rate had decreased, but cardiac output remained the same for many activities. Explain these findings.

7. Explain why it is sufficient to replace the ventricles, but not the atria, in artificial heart transplantation.

8. During an experiment in a physiology laboratory a student named Cee Saw was placed on a table that could be tilted. The instructor asked the students to predict what would happen to Cee Saw's heart rate if the table were tilted so that her head was lower than her feet. Some students predicted an increase in heart rate, and others claimed it would decrease. Can you explain why both predictions might be true?

9. After Cee Saw is tilted so that her head is lower than her feet for a few minutes, the table is tilted so that her head is higher than her feet. Predict the effect this change would have on Cee Saw's heart rate.

Web Site Link

For a listing of the most current web sites related to this chapter, please visit the Seeley home page at:
http://www.mhhe.com/biosci/ap/seeleyap/

Chapter Twenty-One

Cardiovascular System: Peripheral Circulation and Regulation

Objectives

1. Describe the structure and function of capillaries, arteries, and veins.

2. Describe the structural and functional changes that occur in arteries as they age.

3. List the blood vessels of the pulmonary circulation, and describe their function.

4. List the major arteries that supply each of the major body areas, and describe their functions.

5. List the major veins that carry blood from each of the major body areas, and describe their functions.

6. Explain how lymph is formed and transported.

7. List the major lymph vessels, and describe their function.

8. Describe the significance of each of the following for the circulation of blood: viscosity, laminar and turbulent flow, blood pressure, rate of blood flow, Poiseuille's law, critical closing pressure, LaPlace's law, and vascular compliance.

9. Explain how blood pressure can be measured.

10. Explain how the total cross-sectional area of blood vessels, blood pressure, and resistance to flow change as blood flows through the aorta, small arteries, arterioles, capillaries, venules, small veins, and venae cavae.

11. Describe how the exchange of materials across the capillary occurs, and describe how edema can result from decreases in plasma protein and increases in the permeability of the capillary.

12. Describe the functional characteristics of veins.

13. Describe the mechanisms responsible for the local control of blood flow through tissues, and explain under what conditions nervous control of blood flow through tissues is important.

14. Describe the short-term and long-term mechanisms that regulate the mean arterial pressure.

15. Define hypertension, and explain its effect on the circulatory system.

16. Describe how the circulatory system responds to exercise and shock.

Part Four

The heart provides the major force that causes blood to circulate, but the blood vessels carry blood to all tissues of the body and back to the heart. In addition, the blood vessels participate in the regulation of blood pressure and help to direct blood flow to tissues that are most active. The intricacy and coordinated function of blood vessels make the design of complex urban water systems seem rather simple in comparison.

The peripheral circulatory system can be divided into two sets of blood vessels. The **systemic vessels** transport blood through essentially all parts of the body from the left ventricle and back to the right atrium. The **pulmonary vessels** transport blood from the right ventricle through the lungs and back to the left atrium (see figure 20.1). Both the blood vessels and the heart are regulated to ensure that the blood pressure is high enough to cause blood flow in sufficient quantities to meet the metabolic needs of the tissues. The cardiovascular system ensures the survival of each tissue type in the body by supplying nutrients to and removing waste products from tissues.

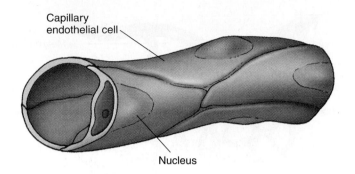

Figure 21.1 Capillary

Section of a capillary showing that it is composed of flattened endothelial cells. ⚡

▌General Features of Blood Vessel Structure

Blood is pumped from the ventricles of the heart into large elastic arteries that branch repeatedly to form many progressively smaller arteries. As they become smaller, the arteries undergo a gradual transition from having walls that contain a large amount of elastic tissue and a smaller amount of smooth muscle to having walls with a smaller amount of elastic tissue and a relatively large amount of smooth muscle. Although the arteries form a continuum from the largest to the smallest branches, they normally are classified as (1) elastic arteries, (2) muscular arteries, or (3) arterioles.

Blood flows from the arterioles into the capillaries. Most of the exchange that occurs between the interstitial spaces and the blood occurs across the walls of the capillaries. Their walls are the thinnest of all the blood vessels, blood flows through them slowly, and there is a greater number of them than of any other blood vessel type.

From the capillaries blood flows into the venous system. When compared with arteries, the walls of the veins are thinner and contain less elastic tissue and fewer smooth muscle cells. The veins increase in diameter and decrease in number, and their walls increase in thickness as they project toward the heart. They are classified as (1) venules, (2) small veins, or (3) medium or large veins.

Capillaries

All blood vessels have an internal lining of simple squamous epithelial cells called the **endothelium** (en-dō-thē′lē-ŭm), which is continuous with the endocardium of the heart.

The capillary wall consists primarily of endothelial cells (figure 21.1), which rest on a basement membrane. Outside the basement membrane is a delicate layer of loose connective tissue that merges with the connective tissue surrounding the capillary.

Along the length of the capillary are some scattered cells that are closely associated with the endothelial cells. These scattered cells lie between the basement membrane and the endothelial cells and are called **pericapillary cells.** They are apparently fibroblasts, macrophages, or undifferentiated smooth muscle cells.

Most capillaries range from 7 to 9 μm in diameter, and they branch without a change in their diameter. Capillaries are variable in length, but in general, they are approximately 1 mm long. Red blood cells flow through most capillaries in a single file and frequently are folded as they pass through the smaller-diameter capillaries.

Types of Capillaries

Capillaries can be classified as continuous, fenestrated, or sinusoidal, depending on their diameter and permeability characteristics.

Continuous capillaries are approximately 7–9 μm in diameter, and their walls exhibit no gaps between the endothelial cells. Continuous capillaries are less permeable to large molecules than are other capillary types and occur in muscle, nervous tissue, and many other locations.

In **fenestrated** (fen′es-trāt′ed) capillaries endothelial cells have numerous fenestrae. The **fenestrae** (fe-nes′trē, meaning windows) are areas approximately 70–100 μm in diameter in which the cytoplasm is absent and the cell membrane consists of a porous diaphragm that is thinner than the normal cell membrane. Fenestrated capillaries are in tissues where capillaries are highly permeable, such as in the intestinal villi, ciliary process of the eye, choroid plexuses of the central nervous system, and glomeruli of the kidney.

Sinusoidal (sī-nŭ-soy′dăl) **capillaries** are larger in diameter than either continuous or fenestrated capillaries, and their basement membrane is less prominent. Their fenestrae are larger than those in fenestrated capillaries. The sinusoidal

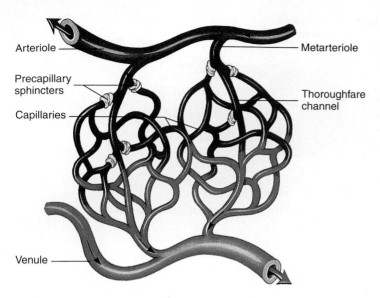

Figure 21.2   Capillary Network

The metarteriole giving rise to the network feeds directly from an arteriole into the thoroughfare channel, which feeds into the venule. The network forms numerous branches that transport blood from the thoroughfare channel and can return to the thoroughfare channel. Smooth muscle cells, called precapillary sphincters, regulate blood flow through the capillaries. Blood flow decreases when the precapillary sphincters constrict and increases when they dilate. ✗

capillaries occur in such places as endocrine glands, where large molecules cross their walls.

Sinusoids are large-diameter sinusoidal capillaries. Their basement membrane is sparse and often missing, and their structure suggests that large molecules and sometimes cells can move readily across their walls between the endothelial cells. Sinusoids are common in the liver and the bone marrow. Macrophages are closely associated with the endothelial cells of the liver sinusoids. **Venous sinuses** are similar in structure to the sinusoidal capillaries but are even larger in diameter. They occur primarily in the spleen, and they have large gaps between the endothelial cells that make up their walls.

Substances cross capillary walls by diffusing through the endothelial cells, through fenestrae, and between the endothelial cells. Lipid-soluble substances, such as oxygen and carbon dioxide, and small water-soluble molecules readily diffuse through the cell membrane. Larger water-soluble substances must pass through the fenestrae or gaps between the endothelial cells. In addition, transport by pinocytosis occurs, but little is known about its role in the capillaries. Because red blood cells and large water-soluble molecules, such as proteins, cannot readily pass through the walls of capillaries, they are effective permeability barriers.

Capillary Network

Arterioles supply blood to each capillary network (figure 21.2). Blood then flows through the capillary network and into the venules. The ends of capillaries closest to the arterioles are **ar-terial capillaries,** and the ends closest to venules are **venous capillaries.**

Blood flows from arterioles through **metarterioles** (met′ar-tēr′ē-ōlz), which have isolated smooth muscle cells along their walls. From a metarteriole blood flows into a **thoroughfare channel** that extends in a relatively direct fashion from a metarteriole to a venule. Blood flow through thoroughfare channels is relatively continuous. Several capillaries branch from the thoroughfare channels, and in these branches blood flow is intermittent. Flow in these capillaries is regulated by smooth muscle cells called **precapillary sphincters,** which are located at the origin of the branches (see figure 21.2).

Capillary networks are more numerous and more extensive in highly metabolic tissues, such as the lung, liver, kidney, skeletal muscle, and cardiac muscle. Capillary networks in the skin have many more thoroughfare channels than capillary networks in cardiac or skeletal muscle. Capillaries in the skin function in thermoregulation, and heat loss results from the flow of a large volume of blood through them. In muscle, however, nutrient and waste product exchange is the major function of the capillaries.

Structure of Arteries and Veins
General Features

Except for the capillaries and the venules, the blood vessel walls consist of three relatively distinct layers, which are most apparent in the muscular arteries and least apparent in the veins. From the lumen to the outer wall of the blood vessels the layers, or **tunics** (tū′niks), are (1) the tunica intima; (2) the tunica media; (3) and the tunica adventitia, or tunica externa (figure 21.3).

The **tunica intima** consists of endothelium, a delicate connective tissue basement membrane, a thin layer of connective tissue called the lamina propria, and a fenestrated layer of elastic fibers called the **internal elastic membrane.** The internal elastic membrane separates the tunica intima from the next layer, the tunica media.

The **tunica media,** or middle layer, consists of smooth muscle cells arranged circularly around the blood vessel. The amount of blood flowing through a blood vessel can be regulated by contraction or relaxation of the smooth muscle in the tunica media. A decrease in blood flow results from **vasoconstriction** (vā′sō-kon-strik′shŭn), a decrease in blood vessel diameter caused by smooth muscle contraction, whereas an increase in blood flow is produced by **vasodilation** (vā′sō-dī-lā′shŭn), an increase in blood vessel diameter because of smooth muscle relaxation. The tunica media also contains variable amounts of elastic and collagen fibers, depending on the size of the vessel. At the outer border of the tunica media in some arteries, an external elastic membrane, which separates the tunica media from the tunica adventitia, can be identified. A few longitudinally oriented smooth muscle cells occur in some arteries near the tunica intima.

The **tunica adventitia** (tū′ni-kă ad-ven-tish′ă) is composed of connective tissue, which varies from dense connective tissue near the tunica media to loose connective tissue

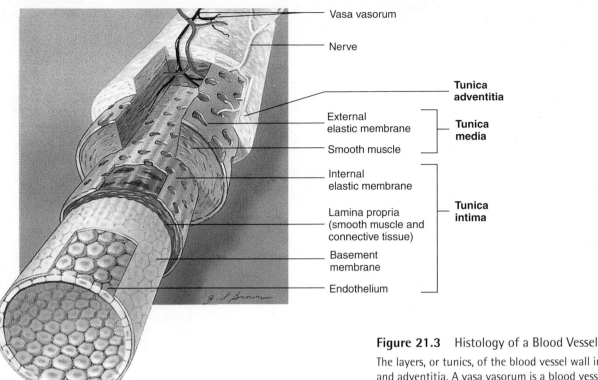

Vasa vasorum

Nerve

Tunica adventitia

External elastic membrane

Tunica media

Smooth muscle

Internal elastic membrane

Tunica intima

Lamina propria (smooth muscle and connective tissue)

Basement membrane

Endothelium

Figure 21.3   Histology of a Blood Vessel

The layers, or tunics, of the blood vessel wall include the intima, media, and adventitia. A vasa vasorum is a blood vessel that supplies blood to the wall of the blood vessel.

that merges with the connective tissue surrounding the blood vessels.

The relative thickness and composition of each layer varies with the diameter of the blood vessel and its type. The transition from one artery type or from one vein type to another is gradual, as are the structural changes.

Large Elastic Arteries

Elastic arteries have the largest diameters (figure 21.4a) and often are called **conducting arteries.** The pressure is relatively high in these vessels, and it fluctuates between systolic and diastolic values. A greater amount of elastic tissue and a smaller amount of smooth muscle occur in their walls compared with the elastic tissue and smooth muscle of other arteries. The elastic fibers are responsible for the elastic characteristics of the blood vessel wall, but collagenous connective tissue determines the degree to which the arterial wall can be stretched.

The tunica intima is relatively thick. The elastic fibers of the internal and external elastic membranes merge and are not recognizable as distinct layers. The tunica media consists of a meshwork of elastic fibers with interspersed circular smooth muscle cells and some collagen fibers. The tunica adventitia is relatively thin.

Muscular Arteries

The larger muscular arteries, often called medium arteries, can be observed in a gross dissection. They include most of the smaller arteries that are not elastic arteries with names. Their walls are relatively thick compared with their diameter, mainly

because the tunica media contains 25–40 layers of smooth muscle (figure 21.4b). The tunica intima of the medium arteries has a well-developed internal elastic membrane. The tunica adventitia is composed of a relatively thick layer of collagenous connective tissue that blends with the surrounding connective tissue. Medium arteries frequently are called **distributing arteries** because the smooth muscle cells allow these vessels to partially regulate blood supply to different regions of the body by either constricting or dilating.

Smaller muscular arteries range from 40 to 300 μm in diameter, and those that are 40 μm in diameter have approximately three or four layers of smooth muscle in their tunica media, whereas arteries that are 300 μm across have essentially the same structure as the larger muscular arteries. The small muscular arteries are adapted for vasodilation and vasoconstriction.

Arterioles

The **arterioles** (ar-tēr′ē-ōl) transport blood from small arteries to capillaries and are the smallest arteries in which the three tunics can be identified. The tunica intima has no observable internal elastic membrane, and the tunica media consists of one or two layers of circular smooth muscle cells. The arterioles, like the small arteries, are capable of vasodilation and vasoconstriction.

Venules and Small Veins

Venules (ven′yūlz), with a diameter of 40–50 μm, are tubes composed of endothelium resting on a delicate basement membrane. Their structure, except for their diameter, is very

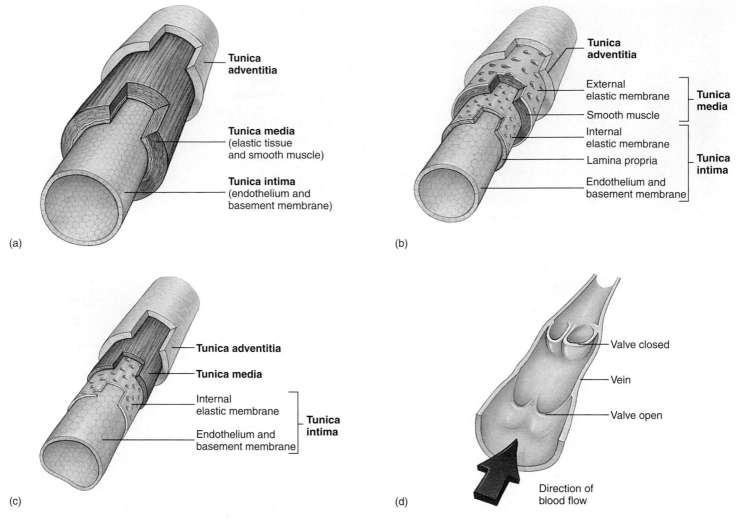

Figure 21.4   Structural Comparison of Blood Vessel Types

(*a*) An elastic artery. Large-diameter arteries with thick walls that contain a large amount of elastic connective tissue in the tunica media. (*b*) A muscular artery. Muscular arteries have a distinctive layer of smooth muscle cells in the tunica media, and they are capable of constriction and dilation. (*c*) A medium vein. Veins have thinner walls. The tunica media is thinner than the tunica media in arteries and contains fewer smooth muscle cells. The dominant layer in the veins is the tunica adventitia. (*d*) Valves in veins. The valves in veins are folds in the endothelium that allow blood to flow toward the heart but not in the opposite direction.

similar to that of capillaries. A few isolated smooth muscle cells exist outside the endothelial cells, especially in the larger venules. As the vessels increase to 0.2–0.3 mm in diameter, the smooth muscle cells form a continuous layer; the vessels then are called **small veins.** The small veins also have a tunica adventitia composed of collagenous connective tissue.

The venules collect blood from the capillaries and transport it to the small veins, which in turn transport it to the medium-sized veins. Nutrient exchange occurs across the walls of the venules, but, as the walls of the small veins increase in thickness, the degree of nutrient exchange decreases.

Medium and Large Veins

Most of the veins observed in gross anatomic dissections, except for the large veins, are **medium veins.** They collect blood from small veins and deliver it to large veins. The **large veins** transport blood from the medium veins to the heart. Their tunica intima is thin and consists of endothelial cells, a relatively thin layer of collagenous connective tissue, and a few scattered elastic fibers. The tunica media is also thin and is composed of a thin layer of circularly arranged smooth muscle cells, collagen fibers, and a few sparsely distributed elastic fibers. The tunica adventitia, which is composed of collagenous connective tissue, is the predominant layer (figure 21.4*c*).

Valves

Veins having diameters greater than 2 mm contain **valves** that allow blood to flow toward the heart but not in the opposite direction (figure 21.4*d*). The valves consist of folds in the tunica intima that form two flaps that are shaped and function like the semilunar valves of the heart. The two folds overlap in the middle of the vein so that, when blood attempts to flow in a reverse direction, the valves occlude the vessel. There are many valves in the medium veins, and the number is greater in veins of the lower extremities than in veins of the upper extremities.

Vasa Vasorum

For arteries and veins greater than 1 mm in diameter, nutrients cannot diffuse from the lumen of the vessel to all of the layers of the wall. Nutrients are therefore supplied to their walls by way of small blood vessels called **vasa vasorum** (vā′să vā′sor-ŭm), which penetrate from the exterior of the vessel to form a capillary network in the tunica adventitia and the tunica media (see figure 21.3).

Arteriovenous Anastomoses

Arteriovenous anastomoses (ă-nas′tō-mō′sez) allow blood to flow from arteries to veins without passing through capillaries. The arterioles directly enter the small veins without an intermediate capillary. A **glomus** (glō′mŭs) is an arteriovenous anastomosis that consists of arterioles arranged in a convoluted fashion surrounded by collagenous connective tissue.

Naturally occurring arteriovenous anastomoses are present in large numbers in the sole of the foot, the palm of the hand, the terminal phalanges, and the nail beds. They function in temperature regulation. **Pathologic arteriovenous anastomoses** can result from injury or tumors. They cause the direct flow of blood from arteries to veins and can, if they are sufficiently severe, lead to heart failure because of the tremendous increase in venous return to the heart.

Nerves

The walls of most blood vessels are richly supplied by unmyelinated sympathetic nerve fibers (see figure 21.3). Some blood vessels, such as those in the penis or clitoris, are innervated by parasympathetic fibers. The nerve fibers project among the smooth muscle cells of the tunica media and form synapses consisting of enlargements of the nerve fibers. The small arteries and arterioles are innervated to a greater extent than other blood vessel types. The response of the blood vessels to sympathetic stimulation is vasoconstriction. Parasympathetic stimulation of blood vessels in the penis or clitoris results in vasodilation.

The smooth muscle cells of blood vessels act to some extent as a functional unit. Frequent gap junctions occur between adjacent smooth muscle cells, and as a consequence, stimulation of a few smooth muscle cells in the vessel wall results in constriction of a relatively large segment of the blood vessel.

A few myelinated sensory neurons also innervate some blood vessels and function as baroreceptors. They monitor stretch in the blood vessel wall and detect changes in blood pressure.

Aging of the Arteries

The walls of all arteries undergo changes as they age, although some arteries change more rapidly than others and some individuals are more susceptible to change than others. The most significant changes occur in the large elastic arteries such as the aorta, large arteries that carry blood to the brain, and the coronary arteries. The age-related changes described here refer to these blood vessel types. Changes in muscular arteries do occur, but they are less dramatic and often do not result in disruption of normal blood vessel function.

Degenerative changes in arteries that make them less elastic are referred to collectively as **arteriosclerosis** (ar-tēr′ē-ō-skler-ō′sis, meaning hardening of the arteries). These changes occur in many individuals and become more severe with advancing age. A related term, **atherosclerosis** (ath′er-ō-skler-ō′sis), refers to the deposition of material in the walls of arteries to form plaques. The material is a fatlike substance containing cholesterol (figure 21.5). The fatty material can be replaced later with dense connective tissue and calcium deposits. The initial signs of arteriosclerosis have been identified in the arteries of people in their teens, and it develops earlier and progresses more rapidly in some individuals than in others.

Arteriosclerosis is characterized by a thickening of the tunica intima and a chemical change in the elastic fibers of the tunica media, making the tunica media less elastic. Fat gradually accumulates between the elastic and collagen fibers to produce a lesion that protrudes into the lumen of the vessel, which can eventually hamper normal blood flow. In advanced forms of arteriosclerosis, calcium deposits, primarily in the form of calcium carbonate, accumulate in the walls of the blood vessels.

Arteriosclerosis greatly increases resistance to blood flow. Advanced arteriosclerosis, as a consequence, adversely affects the normal circulation of blood and greatly increases the work performed by the heart.

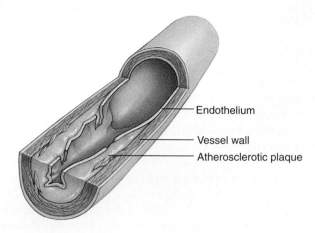

Figure 21.5   Atherosclerotic Plaque in an Artery
Atherosclerotic plaques develop within the tissue of the artery wall. ✗

Some investigators think that arteriosclerosis may not be a pathologic process. Instead, they think it may be simply an aging or wearing out process. Evidence also suggests that arteriosclerosis may result from inflammation which, in some cases may the result of an autoimmune disease. In either case several factors increase the rate at which it develops. Obesity, high dietary cholesterol and other fat consumption, and smoking are some of the factors correlated with the premature development of arteriosclerosis.

Pulmonary Circulation

Blood from the right ventricle is pumped into the **pulmonary** (pŭl′mō-năr-ē, meaning relating to the lungs) trunk. This short vessel, 5 cm long, branches into the right and left **pulmonary arteries,** one transporting blood to each lung (figure 21.6). Within the lungs, gas exchange occurs between air in the lungs and the blood. Two **pulmonary veins** exit each lung and enter the left atrium (see figure 20.6).

▉Systemic Circulation: Arteries

Oxygenated blood entering the heart from the pulmonary veins passes through the left atrium into the left ventricle and from the left ventricle into the aorta. Blood is distributed from the aorta to all parts of the body (see figure 21.6).

Aorta

All **arteries** of the systemic circulation are derived either directly or indirectly from the **aorta** (ā-ōr′t˘a), which usually is divided into three general parts: the ascending aorta, the aortic arch, and the descending aorta. The descending aorta is divided further into a thoracic aorta and an abdominal aorta (see figure 21.12).

At its origin from the left ventricle, the aorta is approximately 2.8 cm in diameter. Because it passes superiorly from

the heart, this part is called the **ascending aorta.** It is approximately 5 cm long and has only two arteries branching from it: the right and left **coronary arteries,** which supply blood to the cardiac muscle.

The aorta then arches posteriorly and to the left as the **aortic arch.** Three major branches, which carry blood to the head and upper limbs, originate from the aortic arch: the brachiocephalic artery, the left common carotid artery, and the left subclavian artery.

> ### Clinical Note
>
> Trauma that ruptures the aorta is almost immediately fatal. Trauma can also lead to an **aneurysm** (an′yū-rizm), however, a bulge caused by a weakened spot in the aortic wall. If the weakened aortic wall leaks blood slowly into the thorax, the aneurysm must be corrected surgically. The majority of traumatic aortic arch ruptures occur in automobile accidents and result from the great force with which the body is thrown into the steering wheel, dashboard, or other objects. Waist-type safety belts alone do not prevent this type of injury as effectively as shoulder-type safety belts.

The next part of the aorta is the **descending aorta.** It is the longest part of the aorta and extends through the thorax in the left side of the mediastinum and through the abdomen to the superior margin of the pelvis. The **thoracic aorta** is that portion of the descending aorta located in the thorax. It has several branches that supply various structures between the aortic arch and the diaphragm. The **abdominal aorta** is that part of the descending aorta between the diaphragm and the point at which the aorta ends by dividing into the two **common iliac** (il′ē-ak, meaning relating to the flank area) **arteries.** The abdominal aorta has several branches that supply the abdominal wall and organs. Its terminal branches, the common iliac arteries, supply blood to the pelvis and the lower limbs.

Coronary Arteries

The **coronary** (kōr′o-năr-ē, meaning encircling the heart like a crown) **arteries,** which are the only branches of the ascending aorta, are described in chapter 20.

Arteries to the Head and the Neck

The first vessel to branch from the aortic arch is the **brachiocephalic** (brā′kē-ō-se-fal′ik, meaning vessel to the arm and head) **artery.** It is a very short artery and branches at the level of the clavicle to form the **right common carotid** (ka-rot′id) **artery,** which transports blood to the right side of the head and the neck, and the **right subclavian** (sŭb-klā′vē-an, meaning below the clavicle) **artery,** which transports blood to the right upper limb (figures 21.6, 21.7, 21.9, and 21.10).

The second and third branches of the aortic arch are the **left common carotid artery,** which transports blood to the

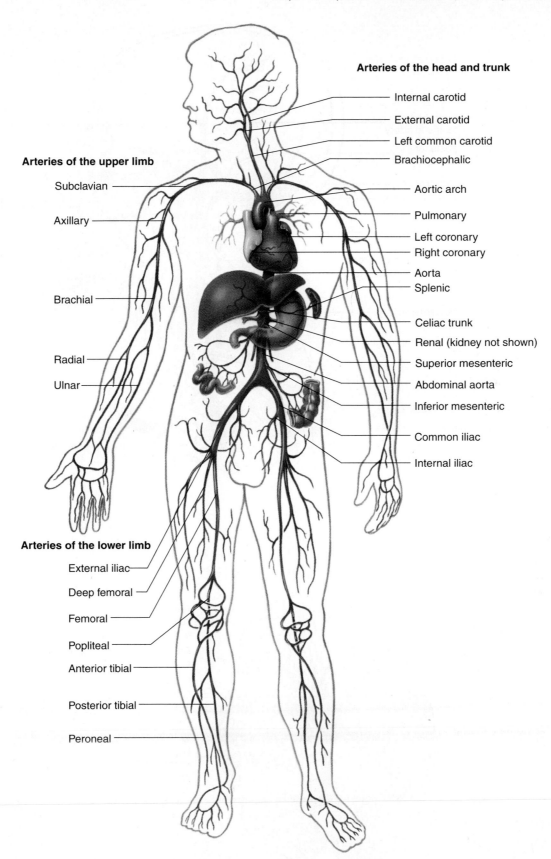

Arteries of the head and trunk

Internal carotid
External carotid
Left common carotid
Brachiocephalic
Aortic arch
Pulmonary
Left coronary
Right coronary
Aorta
Splenic
Celiac trunk
Renal (kidney not shown)
Superior mesenteric
Abdominal aorta
Inferior mesenteric
Common iliac
Internal iliac

Arteries of the upper limb

Subclavian
Axillary
Brachial
Radial
Ulnar

Arteries of the lower limb

External iliac
Deep femoral
Femoral
Popliteal
Anterior tibial
Posterior tibial
Peroneal

Figure 21.6   The Major Arteries
The arteries carry blood from the heart to the tissues of the body.

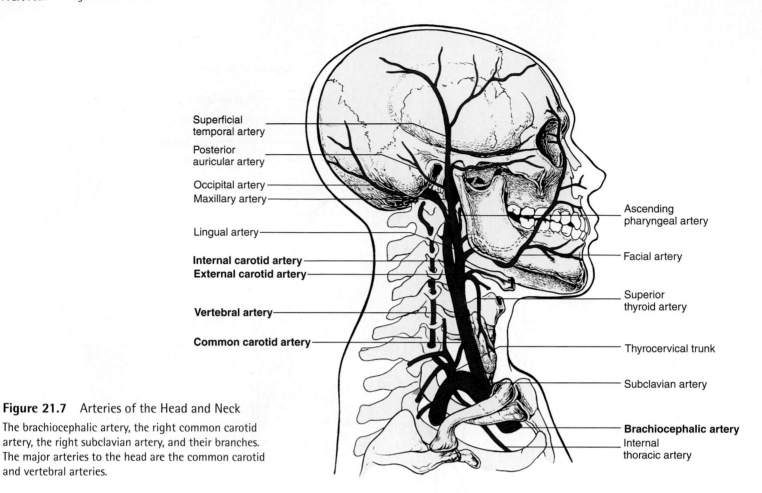

Figure 21.7 Arteries of the Head and Neck

The brachiocephalic artery, the right common carotid artery, the right subclavian artery, and their branches. The major arteries to the head are the common carotid and vertebral arteries.

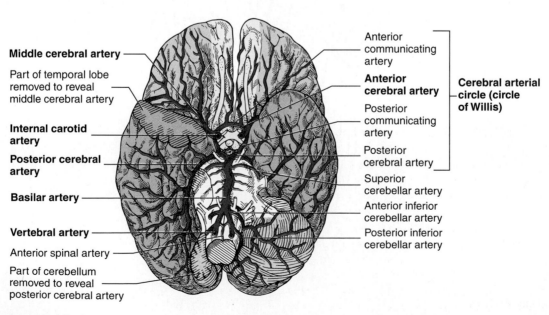

Figure 21.8 Arteries of the Brain

Inferior view of the brain showing the vertebral, basilar, and internal carotid arteries and their branches. (Colors indicate brain regions supplied by various arteries: *yellow*, anterior cerebral; *pink*, middle cerebral; *purple*, posterior cerebral; *blue*, cerebellar arteries; *white*, arteries to brainstem.)

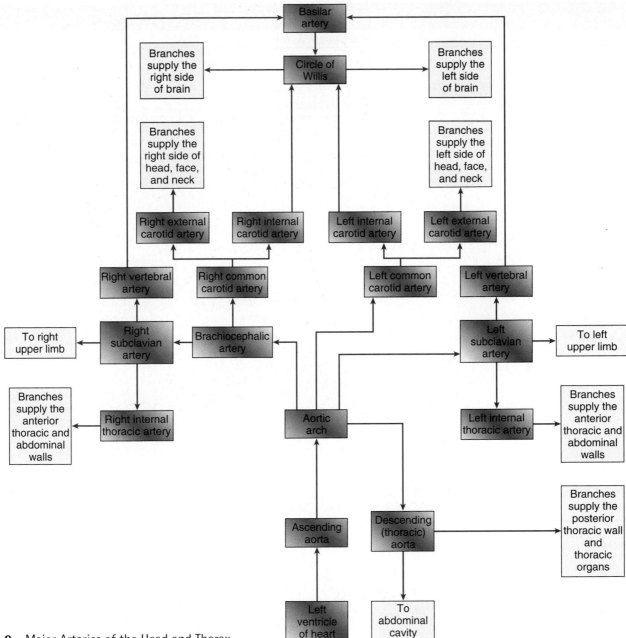

Figure 21.9   Major Arteries of the Head and Thorax

left side of the head and the neck, and the **left subclavian artery,** which transports blood to the left upper limb.

The common carotid arteries extend superiorly, without branching, along either side of the neck from their base to the inferior angle of the mandible, where each common carotid artery branches into **internal** and **external carotid arteries** (see figures 21.7 and 21.9). At the point of bifurcation on each side of the neck, the common carotid artery and the base of the internal carotid artery are dilated slightly to form the **carotid sinus,** which is important in monitoring blood pressure (baroreceptor reflex). The external carotid arteries have several branches that supply the structures of the neck and face (table 21.1; see figures 21.7 and 21.9). The internal carotid arteries, together with the vertebral arteries, which are branches of the subclavian arteries, supply the brain (see table 21.1; and see figures 21.7, 21.8 and 21.9).

Branches of the subclavian arteries, the **left** and **right vertebral arteries,** enter the cranial vault through the foramen magnum, give off arteries to the cerebellum, and then

Table 21.1   Arteries of the Head and Neck (see figures 21.7, 21.8 and 21.9)

Arteries	Tissues Supplied
Common Carotid Arteries	Head and neck by branches listed below
External Carotid	
Superior thyroid	Neck, larynx, and thyroid gland
Lingual	Tongue, mouth, and submandibular and sublingual glands
Facial	Mouth, pharynx, and face
Occipital	Posterior head and neck and meninges around posterior brain
Posterior auricular	Middle and inner ear, head, and neck
Ascending pharyngeal	Deep neck muscles, middle ear, pharynx, soft palate, and meninges around posterior brain
Superficial temporal	Temple, face, and anterior ear
Maxillary	Middle and inner ears, meninges, lower jaw and teeth, upper jaw and teeth, temple, external eye structures, face, palate, and nose
Internal Carotid	
Posterior communicating	Joins the posterior cerebral artery
Anterior cerebral	Anterior portions of the cerebrum and forms the anterior communicating arteries
Middle cerebral	Most of the lateral surface of the cerebrum
Vertebral Arteries	
(branches of the subclavian arteries)	
Anterior spinal	Anterior spinal cord
Posterior inferior cerebellar	Cerebellum and fourth ventricle
Basilar Artery	
(formed by junction of vertebral arteries)	
Anterior inferior cerebellar	Cerebellum
Superior cerebellar	Cerebellum and midbrain
Posterior cerebral	Posterior portions of the cerebrum

unite to form a single, midline **basilar** (bas′i-lăr) **artery** (see figures 21.8, figure 21.9 and table 21.1). The basilar artery gives off branches to the pons and the cerebellum and then branches to form the **posterior cerebral arteries,** which supply the posterior part of the cerebrum (see figure 21.8).

The internal carotid arteries enter the cranial vault through the carotid canals and terminate by forming the **middle cerebral arteries,** which supply large parts of the lateral cerebral cortex (see figure 21.8). Posterior branches of these arteries, the **posterior communicating arteries,** unite with the posterior cerebral arteries; and anterior branches, the **anterior cerebral arteries,** supply blood to the frontal lobes of the brain. The anterior cerebral arteries are in turn connected by an **anterior communicating artery,** which completes a circle around the pituitary gland and the base of the brain called the **cerebral arterial circle** (circle of Willis) (see figures 21.8 and 21.9).

Clinical Note

A **stroke** is a sudden neurologic disorder often caused by a decreased blood supply to a part of the brain. It can occur as a result of a **thrombosis** (throm-bō′sis; a stationary clot), an **embolism** (em′bō-lizm; a floating clot that becomes lodged in smaller vessels), or a **hemorrhage** (hem′ŏ-rij; rupture or leaking of blood from vessels). Any one of these conditions can result in a loss of blood supply or in trauma to a part of the brain. As a result, the tissue normally supplied by the arteries becomes **necrotic** (nĕ-krot′ik, meaning dead). The affected area is called an **infarct** (in′farkt, meaning to stuff into, an area of cell death). The neurologic results of a stroke are described in chapter 13.

Arteries of the Upper Limb

The three major arteries of the upper limb, called the **subclavian, axillary,** and the **brachial arteries,** are a continuum rather than a branching system. The axillary artery is the continuation of the subclavian artery, and the brachial artery is the continuation of the axillary artery. The subclavian artery is located deep to the clavicle, the axillary artery is within the axilla, and the brachial artery lies within the arm itself (table 21.2; figures 21.10 and 21.11).

Table 21.2 Arteries of the Upper Limbs (see figure 21.10)

Arteries	Tissues Supplied
Subclavian Arteries	
(right subclavian originates from the brachiocephalic artery, and left subclavian originates directly from the aorta)	
Vertebral	Spinal cord and cerebellum (see table 21.1) forms basilar artery
Internal thoracic	Diaphragm, mediastinum, pericardium, anterior thoracic wall, and anterior abdominal wall
Thyrocervical trunk	Inferior neck and shoulder
Axillary Arteries	
(continuation of subclavian)	
Thoracoacromial	Pectoral region and shoulder
Lateral thoracic	Pectoral muscles, mammary gland, and axilla
Subscapular	Scapular muscles
Brachial Arteries	
(continuation of axillary arteries)	
Deep brachial	Arm and humerus
Radial	Forearm
Deep palmar arch	Hand and fingers
Digital arteries	Fingers
Ulnar	Forearm
Superficial palmar arch	Hand and fingers
Digital arteries	Fingers

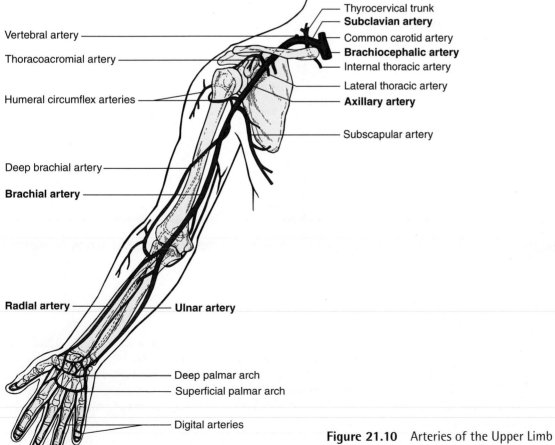

Figure 21.10 Arteries of the Upper Limb

The right brachiocephalic, subclavian, axillary, and brachial arteries and their branches.

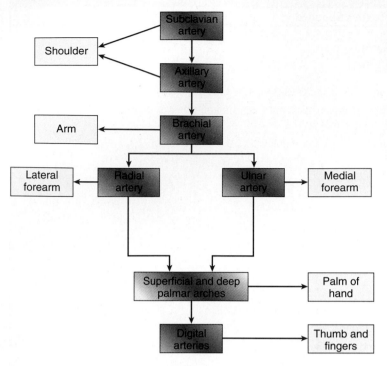

Figure 21.11   Major Arteries of the Shoulder and Upper Limb

Table 21.3	Thoracic and Abdominal Aorta (see figures 21.12 and 21.13)
Arteries	**Tissues Supplied**
Thoracic Aorta	
Visceral Branches	
Bronchial	Lung tissue
Esophageal	Esophagus
Parietal Branches	
Intercostal	Thoracic wall
Superior phrenic	Superior surface of diaphragm
Abdominal Aorta	
Visceral Branches	
Unpaired	
Celiac trunk	
Left gastric	Stomach and esophagus
Common hepatic	
Gastroduodenal	Stomach, duodenum
Right gastric	Stomach
Hepatic	Liver
Splenic	Spleen and pancreas
Left gastroepiploic	Stomach
Superior mesenteric	Pancreas, small intestine, and colon
Inferior mesenteric	Descending colon and rectum
Paired	
Suprarenal	Adrenal gland
Renal	Kidney
Gonadal	
Testicular (male)	Testis and ureter
Ovarian (female)	Ovary, ureter, and uterine tube
Parietal Branches	
Inferior phrenic	Adrenal gland and inferior surface of diaphragm
Lumbar	Lumbar vertebrae and back muscles
Median sacral	Inferior vertebrae
Common iliac	
External iliac	Lower Limb (see table 21.5)
Internal iliac	Lower back, hip, pelvis, urinary bladder, vagina, uterus, rectum, and external genitalia (see table 21.4)

The brachial artery divides into **ulnar** and **radial arteries,** which form two arches within the palm of the hand, referred to as the superficial and deep palmar arches. The **superficial palmar arch** is formed by the ulnar artery and is completed by anastomosing with the radial artery. The **deep palmar arch** is formed by the radial artery and is completed by anastomosing with the ulnar artery. This arch is not only deep to the superficial arch but is proximal as well.

Digital (dij′i-tăl, meaning relating to the digits—the fingers and the thumb) **arteries** branch from each of the two palmar arches and unite to form single arteries on the medial and lateral sides of each digit.

Thoracic Aorta and Its Branches

The branches of the thoracic aorta can be divided into two groups: the **visceral branches** supplying the thoracic organs, and the **parietal branches** supplying the thoracic wall (table 21.3 and figure 21.12*a*). The visceral branches supply the lungs, esophagus, and pericardium. Even though a large quantity of blood flows through the lungs, the lung tissue requires a separate oxygenated blood supply from the left ventricle through small bronchial branches from the thoracic aorta.

The walls of the thorax are supplied with blood by the **intercostal** (in-ter-kos′tăl, meaning between the ribs) **arteries,** which consist of two sets: the anterior intercostals and the posterior intercostals. The **anterior intercostals** are derived from the **internal thoracic arteries,** which are branches of the subclavian arteries and lie on the inner surface of the an-

terior thoracic wall (see table 21.3 and figure 21.12*a*). The **posterior intercostals** are derived as bilateral branches directly from the descending aorta. The anterior and posterior intercostal arteries lie along the inferior margin of each rib and anastomose with each other approximately midway between the ends of the ribs. **Superior phrenic** (fren′ik, meaning to the diaphragm) **arteries** supply blood to the diaphragm.

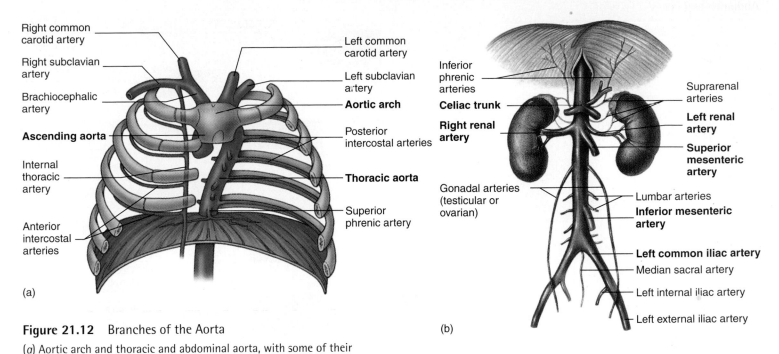

Figure 21.12   Branches of the Aorta

(*a*) Aortic arch and thoracic and abdominal aorta, with some of their branches. (*b*) Abdominal aorta and its deep branches.

Abdominal Aorta and Its Branches

The branches of the abdominal aorta, like those of the thoracic aorta, can be divided into visceral and parietal parts (see table 21.3 and figures 21.12*b* and 21.13). The visceral arteries can in turn be divided into paired and unpaired branches. There are three major unpaired branches: the **celiac** (sē′lē-ak, meaning belly) **trunk,** the **superior mesenteric** (mes′en-ter′ik, meaning relating to the mesenteries), and **inferior mesenteric artery** (see figure 21.12). Each has several major branches supplying the abdominal organs.

The paired visceral branches of the abdominal aorta supply the kidneys, adrenal glands, and gonads (testes or ovaries). The parietal arteries of the abdominal aorta supply the diaphragm and abdominal wall. The arteries of the abdomen and the areas they supply are shown schematically in figure 21.13.

Arteries of the Pelvis

The abdominal aorta divides at the level of the fifth lumbar vertebra into two **common iliac arteries.** They divide to form the **external iliac arteries,** which enter the lower limbs, and the **internal iliac arteries,** which supply the pelvic area. Visceral branches supply the pelvic organs, such as the urinary bladder, rectum, uterus, and vagina; and parietal branches supply blood to the walls and floor of the pelvis; the lumbar, gluteal, and proximal thigh muscles; and the external genitalia (table 21.4 and figures 21.13 and 21.14).

Table 21.4	Arteries of the Pelvis (see figures 21.13 and 21.14)
Arteries	**Tissues Supplied**
Internal Iliac	Pelvis through the branches listed below
Visceral Branches	
Middle rectal	Rectum
Vaginal	Vagina and uterus
Uterine	Uterus, vagina, uterine tube, and ovary
Parietal Branches	
Lateral sacral	Sacrum
Superior gluteal	Muscles of the gluteal region
Obturator	Pubic region, deep groin muscles, and hip joint
Internal pudendal	Rectum, external genitalia, and floor of pelvis
Inferior gluteal	Inferior gluteal region, coccyx, and proximal thigh

Figure 21.13 Major Arteries of the Abdomen and Pelvis

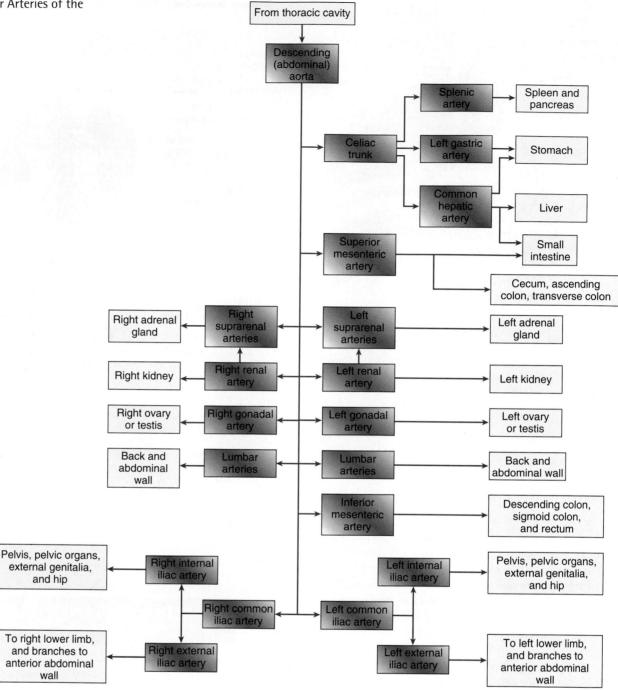

Arteries of the Lower Limb

The arteries of the lower limb form a continuum similar to that of the arteries of the upper limb. The **external iliac artery** becomes the **femoral** (fem′ŏ-răl, meaning relating to the thigh) **artery** in the thigh, which becomes the **popliteal** (pop-lit′ē-ăl, meaning ham, the hamstring area posterior to the knee) **artery** in the popliteal space. The popliteal artery gives off the **anterior tibial artery** just inferior to the knee and then continues as the **posterior tibial artery.** The anterior tibial artery becomes the **dorsalis pedis artery** at the foot. The posterior tibial artery gives off the **fibular,** or **peroneal, artery** and then gives rise to **medial** and **lateral plantar** (plan′tar, meaning the sole of the foot) **arteries,** which in turn give off **digital branches** to the toes. The arteries of the lower limb are listed in table 21.5 and illustrated in figures 21.14 and 21.15.

▌Systemic Circulation: Veins

Three major veins return blood from the body to the right atrium: the **coronary sinus,** returning blood from the walls of the heart (see figures 20.6c and 20.7b); the **superior vena cava** (vē′nă kă′vă, meaning venous cave), returning blood from the head, neck, thorax, and upper limbs; and the **inferior vena cava,** returning blood from the abdomen, pelvis, and lower limbs (figure 21.16).

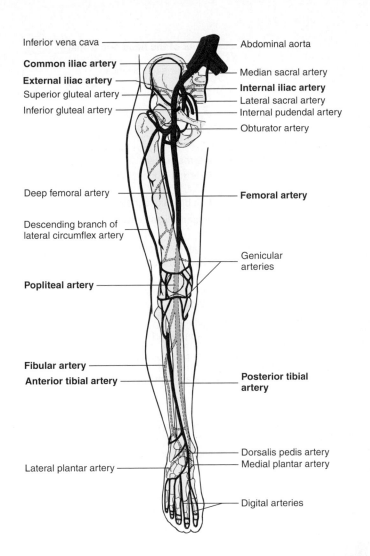

Inferior vena cava

Common iliac artery

External iliac artery

Superior gluteal artery

Inferior gluteal artery

Deep femoral artery

Descending branch of lateral circumflex artery

Popliteal artery

Fibular artery

Anterior tibial artery

Lateral plantar artery

Abdominal aorta

Median sacral artery

Internal iliac artery

Lateral sacral artery

Internal pudendal artery

Obturator artery

Femoral artery

Genicular arteries

Posterior tibial artery

Dorsalis pedis artery

Medial plantar artery

Digital arteries

Table 21.5	Arteries of the Lower Limb (see figures 21.14 and 21.15)
Arteries	**Tissues Supplied**
Femoral	Thigh, external genitalia, anterior abdominal wall
Deep femoral	Thigh, knee, and femur
Popliteal (continuation of the femoral artery)	
Posterior tibial	Knee and leg
Fibular (peroneal)	Calf and peroneal muscles and ankle
Medial plantar	Plantar region of foot
Digital arteries	Digits of foot
Lateral plantar	Plantar region of foot
Digital arteries	Digits of foot
Anterior tibial	Knee and leg
Dorsalis pedis	Dorsum of foot
Digital arteries	Digits of foot

Figure 21.14   Arteries of the Pelvis and Lower Limb

The internal and external iliac arteries and their branches are shown. The internal iliac artery supplies the pelvis and hip, and the external iliac artery supplies the lower limb through the femoral artery.

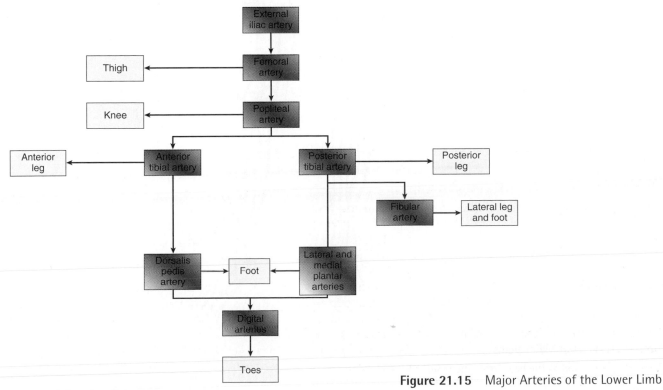

Figure 21.15   Major Arteries of the Lower Limb

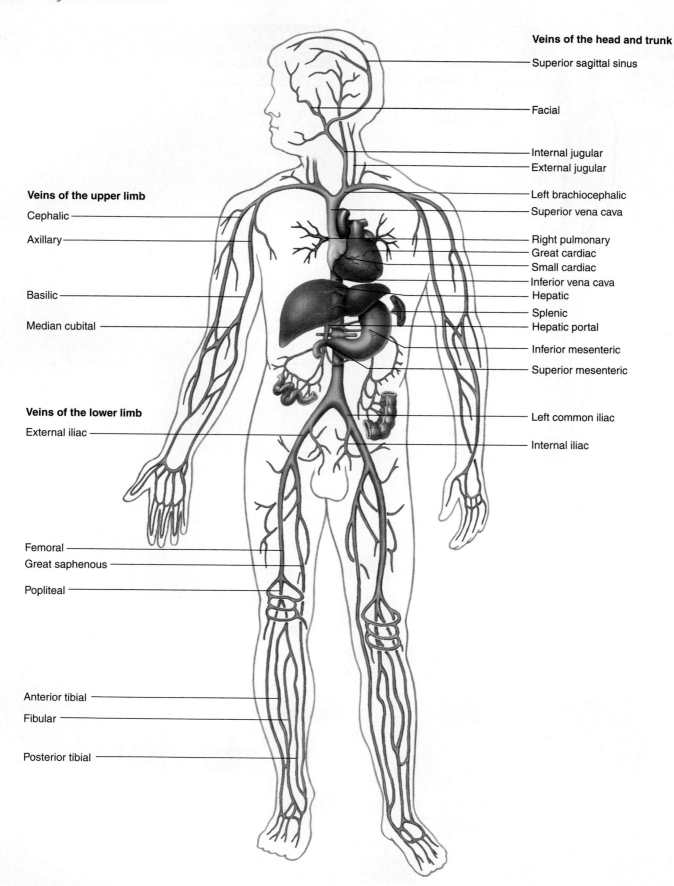

Veins of the head and trunk

Superior sagittal sinus

Facial

Internal jugular
External jugular

Left brachiocephalic
Superior vena cava

Right pulmonary
Great cardiac
Small cardiac
Inferior vena cava
Hepatic
Splenic
Hepatic portal
Inferior mesenteric
Superior mesenteric

Left common iliac

Internal iliac

Veins of the upper limb

Cephalic

Axillary

Basilic

Median cubital

Veins of the lower limb

External iliac

Femoral
Great saphenous
Popliteal

Anterior tibial
Fibular
Posterior tibial

Figure 21.16   The Major Veins

The veins carry blood to the heart from the tissues of the body.

In a very general way, the smaller veins follow the same course as the arteries and often are given the same names. The veins, however, are more numerous and more variable. The larger veins often follow a very different course and have names different from the arteries.

There are three major types of veins: superficial veins, deep veins, and sinuses. The superficial veins of the limbs are, in general, larger than the deep veins, whereas in the head and trunk the opposite is the case. Venous sinuses occur primarily in the cranial vault and the heart.

Veins Draining the Heart

The **cardiac veins,** which transport blood from the walls of the heart and return it through the coronary sinus to the right atrium, are described in chapter 20.

Veins of the Head and Neck

The two pairs of major veins that drain blood from the head and neck are the **external** and **internal jugular** (jŭg′yū-lar, meaning neck) **veins.** The external jugular veins are the more superficial of the two sets, and they drain blood primarily from the posterior head and neck. The external jugular vein usually drains into the subclavian vein. The internal jugular veins are much larger and deeper than the external jugular veins. They drain blood from the cranial vault and the anterior head, face, and neck.

The internal jugular vein is formed primarily as the continuation of the **venous sinuses** of the cranial vault. The venous sinuses are actually spaces within the dura mater surrounding the brain (chapter 13). They are depicted in figure 21.17 and are listed in table 21.6.

Clinical Note

Because venous communication exists between the facial veins and venous sinuses through the ophthalmic veins, infections can potentially be introduced into the cranial vault through this route. A superficial infection of the face on either side of the nose can enter the facial vein. The infection can then pass through the ophthalmic veins to the venous sinuses and result in meningitis. For this reason people are warned not to aggravate pimples or boils on the face on either side of the nose.

Table 21.6   Venous Sinuses of the Cranial Vault (see figure 21.17)

Veins	Tissues Drained
Internal Jugular Vein	
Sigmoid sinus	
Superior and inferior petrosal sinuses	Anterior portion of cranial vault
Cavernous sinus	
Ophthalmic veins	Orbit
Transverse sinus	
Occipital sinus	Central floor of posterior fossa of skull
Superior sagittal sinus	Superior portion of cranial vault and brain
Straight sinus	
Inferior sagittal sinus	Deep portion of longitudinal fissure

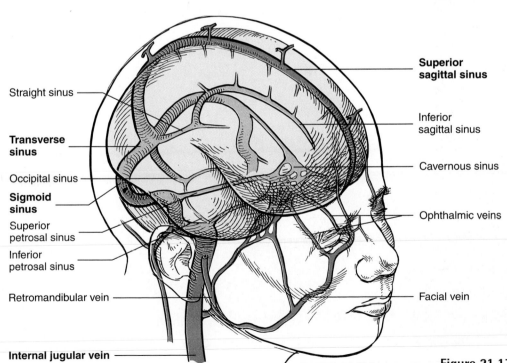

Straight sinus

Transverse sinus

Occipital sinus

Sigmoid sinus

Superior petrosal sinus

Inferior petrosal sinus

Retromandibular vein

Internal jugular vein

Superior sagittal sinus

Inferior sagittal sinus

Cavernous sinus

Ophthalmic veins

Facial vein

Figure 21.17   Venous Sinuses Associated with the Brain

Once the internal jugular veins exit the cranial vault, they receive several venous tributaries that drain the external head and face (table 21.7 and figures 21.18 and 21.19). The **internal jugular veins** join the **subclavian veins** on each side of the body to form the **brachiocephalic veins.**

Table 21.7	Veins Draining the Head and Neck (see figures 21.18 and 21.19)
Veins	**Tissues Drained**
Brachiocephalic	
Internal jugular	Brain
Lingual	Tongue and mouth
Superior thyroid	Thyroid and deep posterior facial structures (also empties into external jugular)
Facial	Superficial and anterior facial structures
External jugular	Superficial surface of posterior head and neck

Veins of the Upper Limb

The **cephalic** (se-fal′ik, meaning toward the head), **basilic** (ba-sil′ik), and **brachial veins** are responsible for draining most of the blood from the upper limbs (table 21.8 and figures 21.20 and 21.21). Many of the tributaries of the cephalic and basilic veins in the forearm and hand can be seen through the skin. Because of the considerable variation in the tributary veins of the forearm and hand, they often are left unnamed. The basilic vein of the arm becomes the **axillary vein** as it courses through the axillary region. The axillary vein then becomes the **subclavian vein** at the margin of the first rib. The cephalic vein enters the axillary vein.

The **median cubital** (kyū′bi-tăl, meaning pertaining to the elbow) **vein** is a variable vein that usually connects the cephalic vein or its tributaries with the basilic vein. In many people this vein is quite prominent on the anterior surface of the upper limb at the level of the elbow (cubital fossa) and is therefore often used as a site for drawing blood from a patient.

Figure 21.18   Veins of the Head and Neck

The right brachiocephalic vein and its tributaries. The major veins draining the head and neck are the internal and external jugular veins.

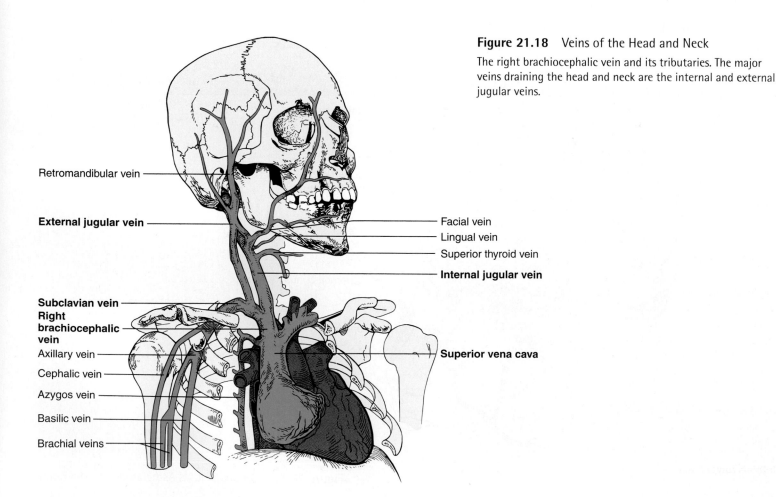

Retromandibular vein

External jugular vein

Subclavian vein
Right brachiocephalic vein

Axillary vein

Cephalic vein

Azygos vein

Basilic vein

Brachial veins

Facial vein
Lingual vein
Superior thyroid vein
Internal jugular vein

Superior vena cava

Figure 21.19 Major Veins of the Head and Thorax

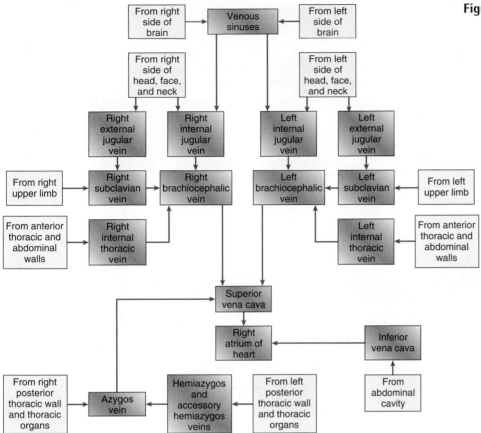

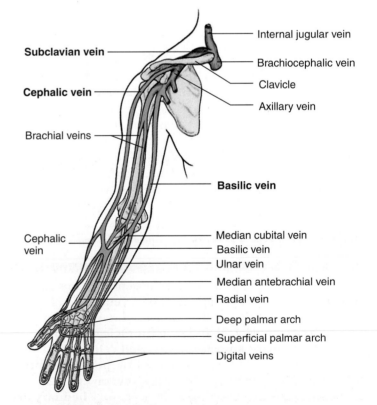

Figure 21.20 Veins of the Upper Limb

The subclavian vein and its tributaries. The major veins draining the superficial structures of the limb are the cephalic and basilic veins. The brachial veins drain the deep structures.

Table 21.8	Veins of the Upper Limb (see figures 21.20 and 21.21)
Veins	**Tissues Drained**
Subclavian (continuation of the axillary vein)	
Axillary (continuation of the basilic vein)	
Cephalic	Lateral arm, forearm, and hand (superficial veins of the forearm and hand are variable)
Brachial (paired, deep veins)	Deep structures of the arm
Radial vein	Deep forearm
Ulnar vein	Deep forearm
Basilic	Medial arm, forearm, and hand (superficial veins of the forearm and hand are variable)
Median cubital	Connects basilic and cephalic veins
Deep and superficial palmar venous arches	Drain into superficial and deep veins of the forearm
Digital veins	Fingers

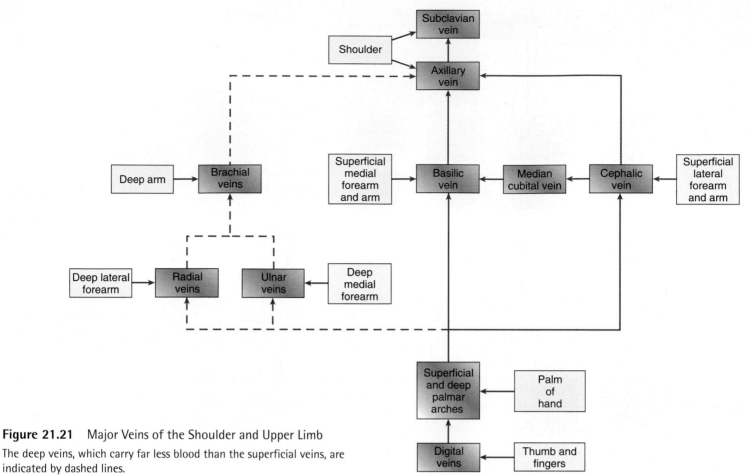

Figure 21.21   Major Veins of the Shoulder and Upper Limb

The deep veins, which carry far less blood than the superficial veins, are indicated by dashed lines.

The deep veins draining the upper limb follow the same course as the arteries. The **radial** and **ulnar veins** therefore are named for the arteries they attend. They usually are paired, with one small vein lying on each side of the artery, and they have numerous connections with one another and with the superficial veins. The radial and ulnar veins empty into the **brachial veins,** which accompany the brachial artery and empty into the axillary vein (see figures 21.20 and 21.21).

Veins of the Thorax

Three major veins return blood from the thorax to the superior vena cava: the right and left brachiocephalic veins and the **azygos** (az'ī-gos, meaning unpaired) **vein.** The thoracic drainage to the brachiocephalic veins is through the anterior thoracic wall by way of the **internal thoracic veins.** They receive blood from the **anterior intercostal veins.** Blood from the posterior thoracic wall is collected by **posterior intercostal veins** that drain into the azygos vein on the right and the **hemiazygos** (hem'ē-az'ī-gos) or **accessory hemiazygos vein** on the left. The hemiazygos and accessory hemiazygos veins empty into the azygos vein, which drains into the superior vena cava. The thoracic veins are listed in table 21.9 and illustrated in figure 21.22 (see also figure 21.19).

Table 21.9 Veins of the Thorax (see figure 21.22)	
Veins	**Tissues Drained**
Superior Vena Cava	
Brachiocephalic	
Azygos vein	Right side, posterior thoracic wall and posterior abdominal wall; esophagus, bronchi, pericardium, and mediastinum
Hemiazygos	Left side, inferior posterior thoracic wall and posterior abdominal wall; esophagus and mediastinum
Accessory hemiazygos	Left side, superior posterior thoracic wall

Veins of the Abdomen and Pelvis

Blood from the posterior abdominal wall drains into the **ascending lumbar veins.** These veins are continuous superiorly with the hemiazygos on the left and the azygos on the right. Blood from the rest of the abdomen, pelvis, and lower limbs returns to the heart through the inferior vena cava. The gonads (testes or ovaries), kidneys, and adrenal glands are the

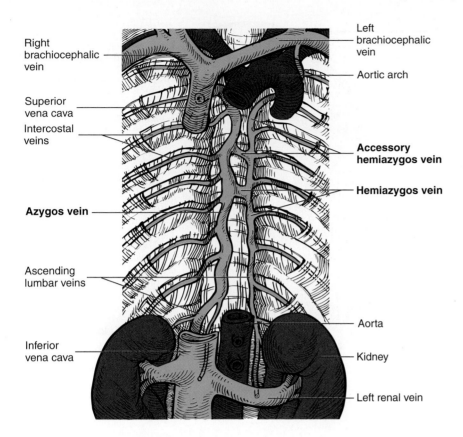

Right
brachiocephalic
vein

Superior
vena cava

Intercostal
veins

Azygos vein

Ascending
lumbar veins

Inferior
vena cava

Left
brachiocephalic
vein

Aortic arch

**Accessory
hemiazygos vein**

Hemiazygos vein

Aorta

Kidney

Left renal vein

Figure 21.22   Veins of the Thorax
The azygos and hemiazygos veins and their tributaries.

only abdominal organs outside the pelvis that drain directly into the inferior vena cava. The **internal iliac veins** drain the pelvis and join the **external iliac veins** from the lower limbs to form the **common iliac veins,** which unite to form the inferior vena cava. The major abdominal and pelvic veins are listed in table 21.10 and illustrated in figures 21.23 and 21.25.

Hepatic Portal System

Blood from the capillaries within most of the abdominal viscera, such as the stomach, intestines, and spleen, drains through a specialized system of blood vessels to the liver. Within the liver the blood flows through a series of dilated capillaries called **si-nusoids.** A **portal** (pŏr′tăl, meaning door) system is a vascular system that begins and ends with capillary beds and has no pumping mechanism such as the heart between the capillary beds. The portal system that begins with capillaries in the viscera and ends with the sinusoidal capillaries in the liver is the **hepatic** (he-pat′ik, meaning relating to the liver) **portal system** (table 21.11 and figures 21.24 and 21.25). The **hepatic portal vein,** the largest vein of the system, is formed by the

Table 21.10	Veins Draining the Abdomen and Pelvis (see figures 21.23 and 21.25)
Veins	**Tissues Drained**
Inferior Vena Cava	
Hepatic veins	Liver (see hepatic portal system)
Common iliac	
External iliac	Lower limb (see table 21.12)
Internal iliac	Pelvis and its viscera
Ascending lumbar	Posterior abdominal wall (empties into common iliac, azygos, and hemiazygos veins)
Renal	Kidney
Suprarenal	Adrenal gland
Gonadal	
Testicular (male)	Testis
Ovarian (female)	Ovary
Phrenic	Diaphragm

Table 21.11	Hepatic Portal System (see figures 21.24 and 21.25)
Veins	**Tissues Drained**
Hepatic Portal	
Superior mesenteric	Small intestine and most of the colon
Splenic	Spleen
Inferior mesenteric	Descending colon and rectum
Pancreatic	Pancreas
Left gastroepiploic	Stomach
Gastric	Stomach
Cystic	Gallbladder

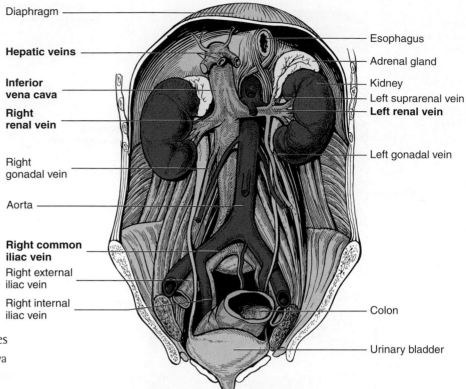

Diaphragm

Esophagus
Adrenal gland
Kidney
Left suprarenal vein
Left renal vein

Hepatic veins

**Inferior
vena cava**

**Right
renal vein**

Left gonadal vein

Right
gonadal vein

Aorta

**Right common
iliac vein**

Right external
iliac vein

Right internal
iliac vein

Colon

Urinary bladder

Figure 21.23   Inferior Vena Cava and Its Tributaries

The hepatic veins transport blood to the inferior vena cava
from the hepatic portal system, which ends as a series of
blood sinusoids in the liver (see figure 21.24).

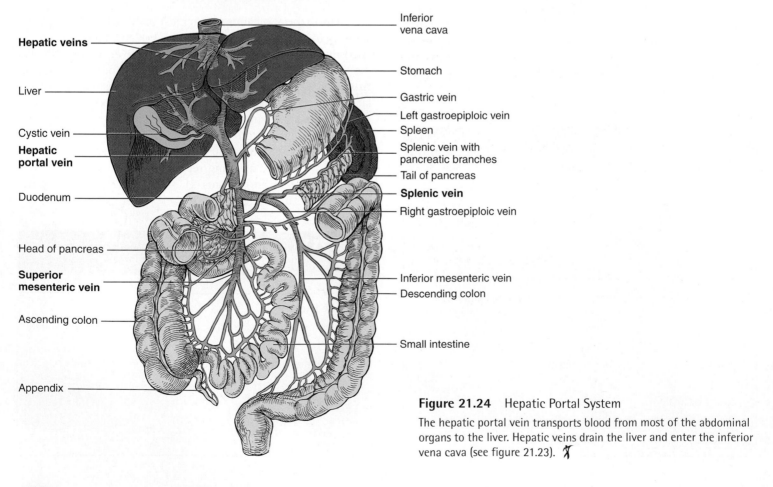

Inferior
vena cava

Hepatic veins

Stomach

Liver

Gastric vein
Left gastroepiploic vein
Spleen
Splenic vein with
pancreatic branches
Tail of pancreas

Cystic vein

**Hepatic
portal vein**

Splenic vein

Duodenum

Right gastroepiploic vein

Head of pancreas

Inferior mesenteric vein
Descending colon

**Superior
mesenteric vein**

Ascending colon

Small intestine

Appendix

Figure 21.24   Hepatic Portal System

The hepatic portal vein transports blood from most of the abdominal
organs to the liver. Hepatic veins drain the liver and enter the inferior
vena cava (see figure 21.23). ✗

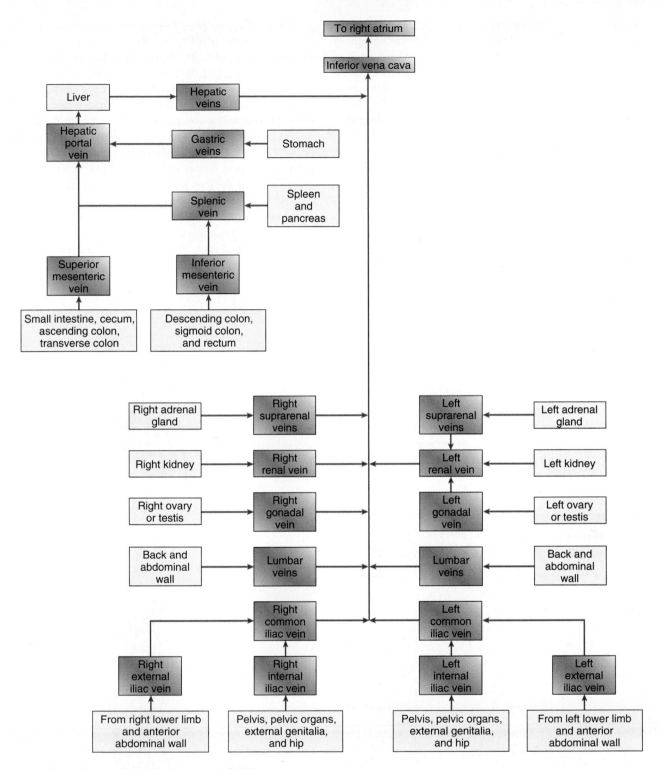

Figure 21.25   Major Veins of the Abdomen and Pelvis

union of the **superior mesenteric vein,** which drains the small intestine, and the **splenic vein,** which drains the spleen. The splenic vein receives the **inferior mesenteric** and **pancreatic veins,** which drain the large intestine and pancreas, respectively. The hepatic portal vein also receives gastric veins before entering the liver.

Blood from the liver sinusoids is collected into **central veins,** which empty into **hepatic veins.** Blood from the cystic veins also enters the hepatic veins. The hepatic veins join the inferior vena cava. The blood entering the liver through the hepatic portal vein is rich with nutrients collected from the intestines, but it also can contain a number of toxic substances harmful to the tissues of the body. Within the liver the nutrients are either taken up and stored or are modified chemically and used by other cells of the body (see chapter 24). The cells of the liver also help remove toxic substances by altering their structure or making them water-soluble, a process called **biotransformation.** The water-soluble substances can then be transported in the blood to the kidneys, from which they are excreted in the urine (see chapter 26).

Veins of the Lower Limb

The veins of the lower limb, like those of the upper limb, consist of superficial and deep groups. The distal deep veins are paired and follow the same path as the arteries, whereas the proximal deep veins are unpaired. The **anterior** and **posterior tibial veins** are paired and accompany the anterior and posterior tibial arteries. They unite just inferior to the knee to form the single **popliteal vein,** which ascends through the thigh and becomes the **femoral vein.** The femoral vein becomes the external iliac vein. **Fibular,** or **peroneal** (per-ō-nē′ăl) **veins,** also are paired in each leg and accompany the fibular arteries. They empty into the posterior tibial veins just before those veins contribute to the popliteal vein.

The superficial veins consist of the great and small saphenous veins. The **great saphenous** (să-fē′nŭs, meaning visible) **vein,** the longest vein of the body, originates over the dorsal and medial side of the foot and ascends along the medial side of the leg and thigh to empty into the femoral vein. The **small saphenous vein** begins over the lateral side of the foot and ascends along the posterior leg to the popliteal space, where it empties into the popliteal vein. The veins of the lower limb are illustrated in figures 21.26 and 21.27 and are listed in table 21.12.

Table 21.12	Veins of the Lower Limb (see figures 21.26 and 21.27)
Veins	**Tissues Drained**
External Iliac Vein	
(continuation of the femoral vein)	
Femoral	Thigh
(continuation of the popliteal vein)	
Popliteal	
Anterior tibial	Deep anterior leg
Dorsal vein of foot	Dorsum of foot
Posterior tibial	Deep posterior leg
Plantar veins	Plantar region of foot
Fibular (peroneal)	Deep lateral leg and foot
Small saphenous	Superficial posterior leg and lateral side of foot
Great saphenous	Superficial anterior and medial leg, thigh, and dorsum of foot
Dorsal vein of foot	Dorsum of foot
Dorsal venous arch	Foot
Digital veins	Toes

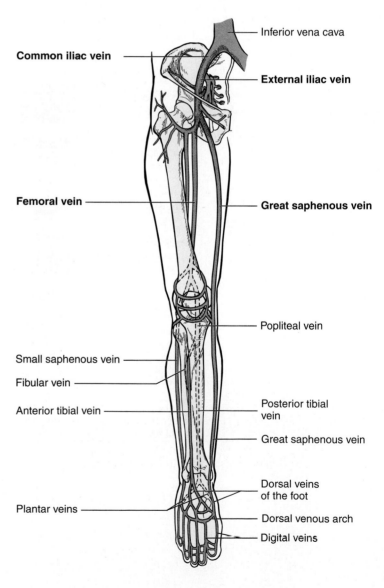

Figure 21.26   Veins of the Pelvis and Lower Limb

The right common iliac vein and its tributaries.

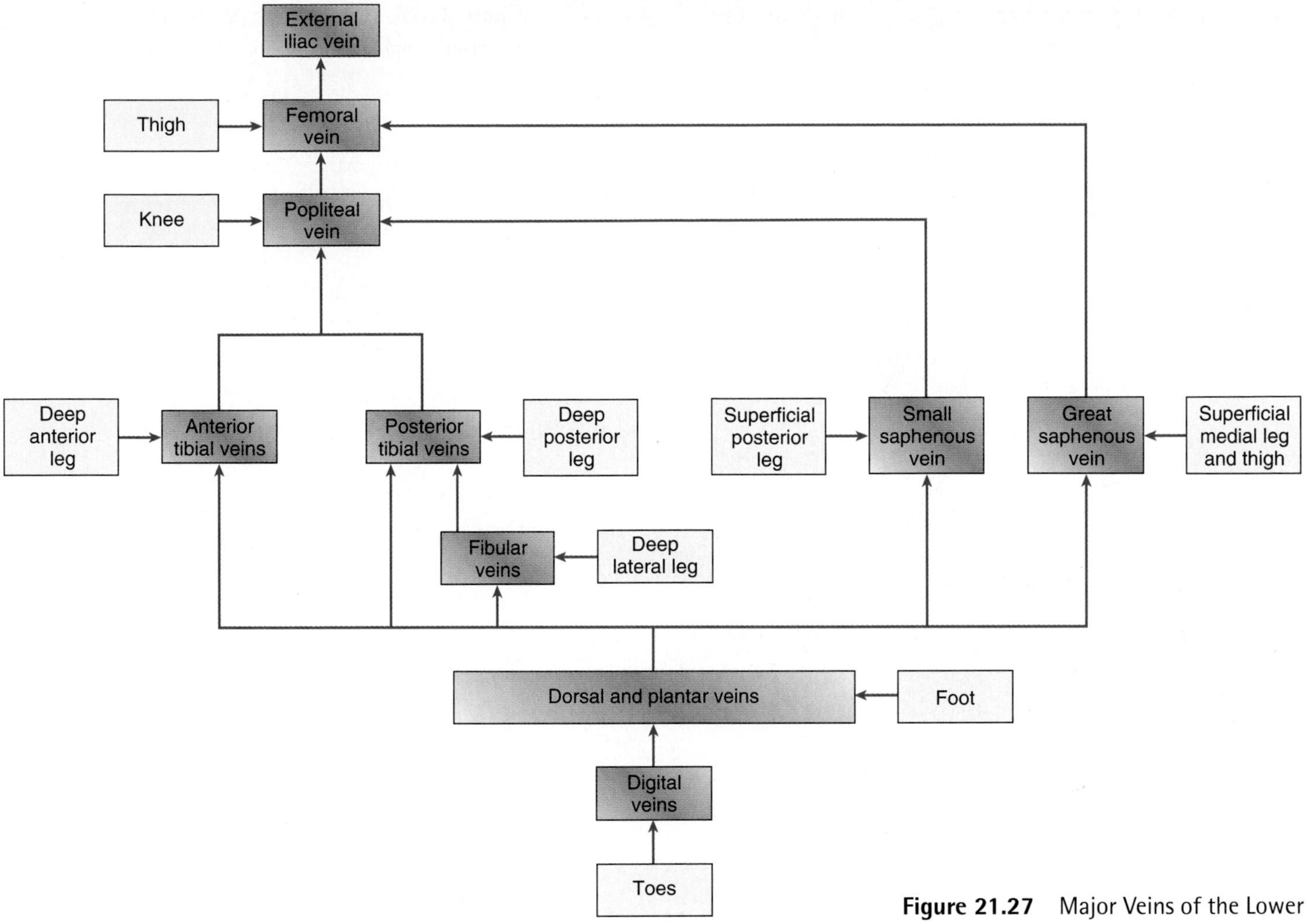

Figure 21.27   Major Veins of the Lower Limb

Lymphatic Vessels

The lymphatic system (figure 21.28), unlike the circulatory system, only carries fluid away from the tissues. The lymphatic system begins in the tissues as **lymph capillaries,** which differ from blood capillaries in that they lack a basement membrane and the cells of the simple squamous epithelium slightly overlap and are attached loosely to one another (figure 21.29). Two things occur as a result of this structure. First, the lymph capillaries are far more permeable than blood capillaries, and nothing in the interstitial fluid is excluded from the lymph capillaries. Second, the lymph capillary epithelium functions as a series of one-way valves that allow fluid to enter the capillary but prevent it from passing back into the interstitial spaces.

Lymph capillaries are in almost all tissues of the body, with the exception of the central nervous system; the bone marrow; and the tissues without blood vessels, such as cartilage, epidermis, and the cornea. A superficial group of lymph capillaries is in the dermis of the skin and the hypodermis. A deep group of lymph capillaries drains the muscles, joints, viscera, and other deep structures. Fluids tend to move out of

blood capillaries into tissue spaces and then out of the tissue spaces into lymph capillaries (figure 21.30). The fluid moving into the lymph capillaries is called **lymph.**

The lymph capillaries join to form larger **lymph vessels** that resemble small veins. The inner layer of the lymph vessel consists of endothelium surrounded by an elastic membrane, the middle layer consists of smooth muscle cells and elastic fibers, and the outer layer is a thin layer of fibrous connective tissue.

Small lymphatic vessels have a beaded appearance because of the presence of one-way valves along their lengths that are similar to the valves of veins (see figure 21.29). When a lymphatic vessel is squeezed shut, backward movement of lymph is prevented by the valves; as a consequence, the lymph moves forward through the lymphatic vessel. Three factors are believed responsible for the compression of lymphatic vessels: (1) contraction of surrounding skeletal muscles during activity, (2) contraction of the smooth muscles in the lymphatic vessel wall, and (3) pressure changes in the thorax during respiration.

Lymph nodes are round, oval, or bean-shaped bodies distributed along the various lymphatic vessels. The lymph nodes function to filter lymph, and most lymph passes

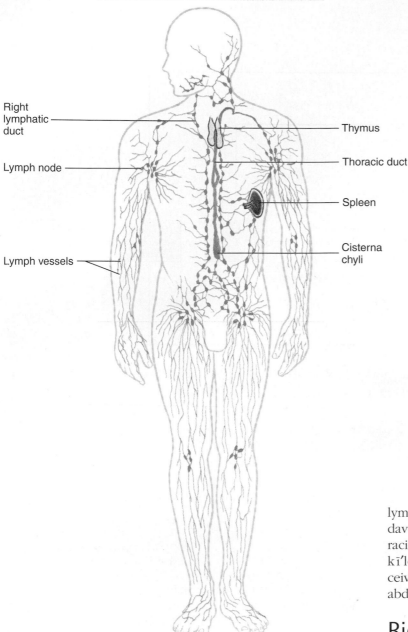

Figure 21.28   Lymphatic System
The major lymphatic organs and vessels.

Right lymphatic duct

Lymph node

Lymph vessels

Thymus

Thoracic duct

Spleen

Cisterna chyli

through at least one lymph node before entering the blood. See chapter 22 for more information on lymph nodes.

After passing through superficial or deep lymph nodes, the lymphatic vessels converge toward either the right or the left subclavian vein. Vessels from the upper right limb and the right side of the head and the neck enter the right lymphatic duct. Lymphatic vessels from the rest of the body enter the larger thoracic duct (see figures 21.28 and 21.31).

Thoracic Duct

The **thoracic duct** drains the lower limbs, abdomen, the left thorax, the left upper extremity, and the left side of the head and neck (see figure 21.31). The duct ends by entering the left subclavian vein. Although the thoracic duct is the largest

lymph vessel, it is still so small that it is difficult to see in cadavers. At the level of the superior abdominal cavity the thoracic duct is expanded to form the **cisterna chyli** (sis-ter′nă kī′lē, meaning a cistern or tank that contains juice), which receives several lymph vessels from the lower limbs and from the abdomen, especially from the digestive tract (see figure 21.28).

Right Lymphatic Duct

The **right lymphatic duct** is much shorter and smaller in diameter than the thoracic duct. It drains the right thorax, right upper limb, and right side of the head and neck and opens into the right subclavian vein. The right lymphatic duct can consist of a single duct, but more commonly two or three separate right lymphatic ducts open into the right subclavian vein, the right internal jugular vein, and the right brachiocephalic vein.

2   P R E D I C T

During radical cancer surgery malignant lymph nodes are often removed, and the lymph vessels to them are tied off to prevent metastasis, or spread, of the cancer. Predict the consequences of tying off the lymph vessels.

✔ *Answer in Appendix F*

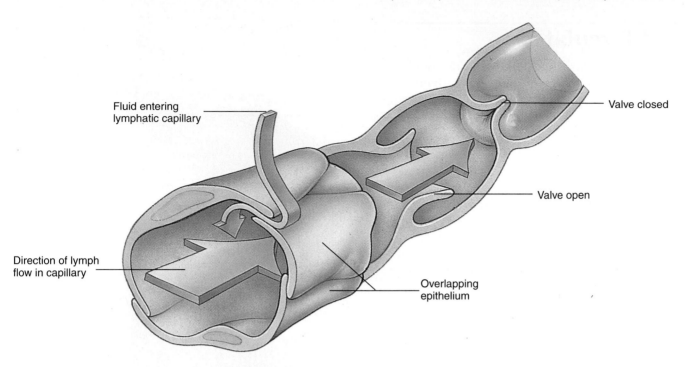

Figure 21.29 One-Way Flow of Lymph in a Lymph Capillary

The overlap of the epithelial cells of the lymph capillary allows easy entry of interstitial fluid but prevents movement back into the tissue. Valves located along the lymph capillary also ensure one-way flow of lymph. 🏃

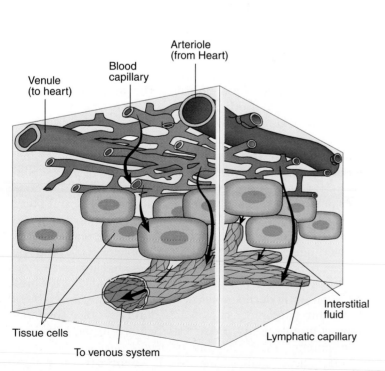

Figure 21.30 Movement of Fluid Between Blood and Lymphatic Capillaries 🏃

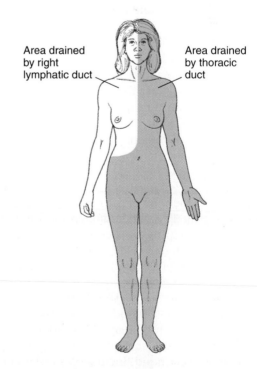

Figure 21.31 Overall Lymphatic Drainage

Lymph from the colored area drains through the thoracic duct. Lymph from the white area enters the right lymphatic duct.

Physics of Circulation

The basic physical characteristics of blood and the principles affecting the flow of liquids through vessels dramatically influence the circulation of blood. The interrelationships between pressure, flow, resistance, and the control mechanisms that regulate blood pressure and blood flow through vessels play a critical role in the function of the circulatory system.

Laminar and Turbulent Flow in Vessels

Fluid, including blood, tends to flow through long, smooth-walled tubes in a streamlined fashion called **laminar flow** (figure 21.32*a*). Fluid behaves as if it is composed of a large number of concentric layers. The layer nearest the wall of the tube experiences the greatest resistance to flow because it moves against the stationary wall. The innermost layers slip over the surface of the outermost layers and experience less resistance to movement. Thus flow in a vessel consists of movement of concentric layers, with the outermost layer moving slowest and the layer at the center moving fastest.

Laminar flow is interrupted and becomes **turbulent flow** when the rate of flow exceeds a critical velocity or when the fluid passes a constriction, a sharp turn, or a rough surface (figure 21.32*b*). Vibrations of the liquid and of the blood vessel walls during turbulent flow cause the sounds produced when blood pressure is measured using a blood pressure cuff. Turbulent flow is also common as blood flows past the valves in the heart and is partially responsible for the heart sounds.

Turbulent flow of blood through vessels occurs primarily in the heart and to a lesser extent where arteries branch. Sounds caused by turbulent blood flow in arteries are not normal and usually indicate that the blood vessel is constricted abnormally. In addition, turbulent flow in abnormally constricted arteries increases the probability that thromboses will develop in the area of turbulent flow.

Blood Pressure

Blood pressure is a measure of the force blood exerts against the blood vessel walls. The standard reference for blood pressure is the **mercury** (Hg) **manometer,** which measures pressure in millimeters of mercury (mm Hg). If the blood pressure is 100 mm Hg, the pressure is great enough to lift a column of mercury 100 mm.

Blood pressure can be measured directly by inserting a **cannula** (or tube) into a blood vessel and connecting a manometer or an electronic pressure transducer to it. Electronic transducers are very sensitive to changes in pressure and can precisely detect rapid fluctuation in pressure.

Placing catheters in blood vessels or in chambers of the heart to monitor pressure changes is possible, but these procedures are not appropriate for routine clinical determina-

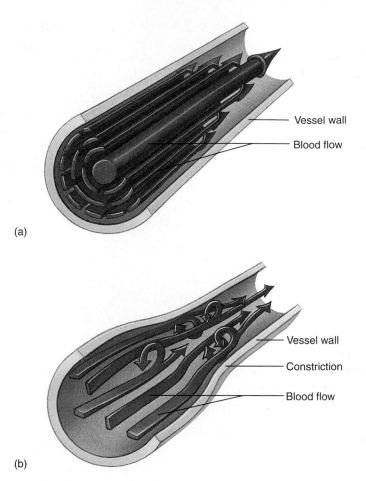

(a)

(b)

Figure 21.32   Laminar and Turbulent Flow

(*a*) Laminar flow. Fluid flows in long smooth-walled tubes as if it is composed of a large number of concentric layers. (*b*) Turbulent flow. Turbulent flow is caused by numerous small currents flowing crosswise or obliquely to the long axis of the vessel, resulting in flowing whorls and eddy currents.

tions of systemic blood pressure. The **auscultatory** (aws-kŭl′tăh-tō′rē) **method** can be used to measure blood pressure without surgical procedures or causing discomfort, so it is used under most clinical conditions. A blood pressure cuff connected to a **sphygmomanometer** (sfig′mō-mă-nom′ĕ-ter) is placed around the patient's upper arm, and a stethoscope is placed over the brachial artery (figure 21.33). Some sphygmomanometers have mercury manometers, and others have electronic manometers, but they all measure pressure in terms of millimeters of mercury. The blood pressure cuff is inflated until the brachial artery is completely collapsed. Because no blood flows through the constricted area, no sounds can be heard. The pressure in the cuff is gradually lowered. As soon as it declines below the systolic pressure, blood flows through the constricted area during systole. The blood flow is turbulent and produces vibrations in the blood and surrounding tissues that can be heard through the stethoscope. These sounds are called **Korotkoff** (kō-rot′kof) **sounds,** and the pressure at which a Korotkoff sound is first heard represents the **systolic pressure.**

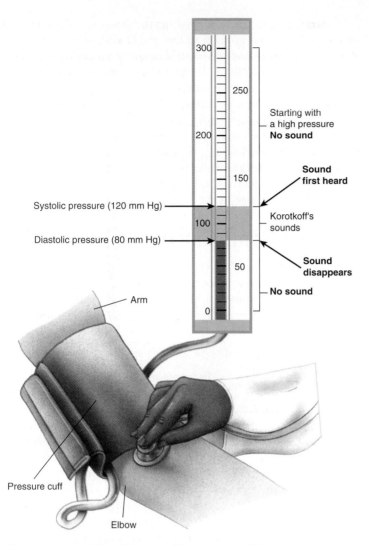

Figure 21.33   Blood Pressure Measurement Using a Sphygmomanometer

The blood pressure cuff is inflated to a high pressure until the brachial artery is collapsed and there is no blood flow, and then the pressure is decreased slowly. The systolic pressure is the pressure measured by the sphygmomanometer at which blood is forced through the collapsed blood vessel and a Korotkoff sound can first be heard. The Korotkoff sound can be heard because the blood flow is turbulent. The pressure at which sounds disappear, and blood flow becomes laminar, is the diastolic blood pressure.

As the pressure in the blood pressure cuff is lowered still more, the Korotkoff sounds change tone and loudness. When the pressure has dropped until the sound disappears completely, continuous laminar blood flow is reestablished. The pressure at which continuous laminar flow is reestablished is the **diastolic pressure.** This method for determining systolic and diastolic pressures is not entirely accurate, but its results are within 10% of methods that are more direct.

Rate of Blood Flow

The **rate** at which blood or any other liquid flows through a tube is expressed as the volume that passes a specific point per unit of time. Blood flow usually is reported in either mil-

liters (ml) per minute or liters (L) per minute. For example, when a person is resting, the **cardiac output** of the heart is approximately 5 L/min; thus blood flow through the aorta is approximately 5 L/min.

Blood flow in a vessel is proportional to the pressure difference in that vessel. For example, if the pressures at point 1 (P_1) and point 2 (P_2) in a vessel are the same, no flow occurs. If, however, the pressure at P_1 is greater than that at P_2, flow proceeds from P_1 toward P_2, and the greater the pressure difference, the greater is the rate of flow. If P_2 is greater than P_1, flow proceeds from P_2 toward P_1. Flow always occurs from a higher to a lower pressure.

The flow of blood resulting from a pressure difference in a vessel is opposed by a **resistance** (R) to blood flow. As the resistance increases, blood flow decreases, and as the resistance decreases, blood flow increases.

The effect of pressure differences and resistance to blood flow can be expressed mathematically:

$$\text{Flow} = \frac{P_1 - P_2}{R}$$

Poiseuille's Law

Several factors affect resistance to blood flow and are expressed individually in **Poiseuille's** (pwah-zuh'yez) **law.**

Poiseuilles's law is expressed by the following formula:

$$\text{Flow} = \frac{(P_1 - P_2)}{8vl/r^4} \quad \text{or} \quad \text{Flow} = \frac{(P_1 - P_2)r^4}{8vl}$$

where v equals viscosity of blood, l equals length of the vessel, P equals pressure, and r equals blood vessel radius.

According to Poiseuille's law, flow decreases when resistance increases. Resistance to flow dramatically decreases when the blood vessel diameter increases because flow is proportional to the fourth power of the blood vessel's radius. On the other hand, an increase in resistance caused by a small decrease in the blood vessel's radius results in a dramatic decrease in flow. In addition, either an increase in blood viscosity (see following section on Viscosity) or an increase in blood vessel length reduces flow.

During exercise the heart contracts with a greater force, and the blood pressure increases in the aorta. In addition, blood vessels in skeletal muscles dilate, making their radii larger and the resistance to blood flow smaller. As a consequence, the rate of flow can increase from 5 L/min in the aorta to several times that value.

Viscosity

Viscosity (vis-kos'i-tē) is a measure of the resistance of a liquid to flow. As the viscosity of a liquid increases, the pressure required to force it to flow increases. A common means for reporting the viscosity of liquids is to consider the viscosity of distilled water as 1 and to compare the viscosity of other liquids with it. Using this procedure, whole blood has a viscosity of 3–4.5, which means that about three times as much pressure is required to force whole blood to flow through a given tube at the same rate as water.

The viscosity of blood is influenced largely by **hematocrit** (hē′mă-tō-krit), which is the percent of the total blood volume composed of erythrocytes (see chapter 19). As the hematocrit increases, the viscosity of blood increases logarithmically. Blood with a hematocrit of 45% has a viscosity about three times that of water, whereas blood with a very high hematocrit of 65% has a viscosity about seven to eight times that of water. The plasma proteins have only a minor effect on the viscosity of blood. Dehydration or uncontrolled production of erythrocytes can increase hematocrit and the viscosity of blood substantially. Viscosity above its normal range of values increases the workload on the heart, and, if this workload is great enough, heart failure can result.

3	**P R E D I C T**

Predict the effect of each of the following conditions on blood flow: (a) vasoconstriction of blood vessels in the skin in response to cold exposure; (b) vasodilation of the blood vessels in the skin in response to an elevated body temperature; (c) polycythemia vera, which results in a greatly increased hematocrit.

✔ *Answer in Appendix F*

Critical Closing Pressure and Laplace's Law

Each blood vessel exhibits a **critical closing pressure,** the pressure below which the vessel collapses and blood flow through the vessel stops. Under conditions of shock, blood pressure can decrease below the critical closing pressure in vessels (see the Clinical Focus "Shock" later in the chapter). As a consequence, the blood vessels collapse, and flow ceases. Tissues supplied by these vessels can become necrotic because of the lack of blood supply.

Laplace's (la-plas′ez) **law** states that the force that stretches the vascular wall is proportional to the diameter of the vessel times the blood pressure. Laplace's law helps explain the critical closing pressure. As the pressure in a vessel decreases, the force that stretches the vessel wall also decreases. Some minimum force is required to keep the vessel open. If the pressure decreases so that the force is below that minimum requirement, the vessel will close. As the pressure in a vessel increases, the force that stretches the vessel wall also increases.

Laplace's law is expressed by the following formula:

$$F = D \times P$$

where F is force, D is vessel diameter, and P is pressure.

According to Laplace's law, as the diameter of the vessel increases, the force applied to the vessel wall increases, even if the pressure remains constant. If a part of an arterial wall becomes weakened so that a bulge forms in it, the force applied to the weakened part is greater than at other points along the blood vessel because its diameter is greater. The greater force causes the weakened vessel wall to bulge even more, further increasing the force applied to it. This series of events can proceed until the vessel finally ruptures. As the bulges in weakened blood vessel walls, called aneurysms, enlarge, the danger of their rupturing increases. Ruptured aneurysms in the blood vessels of the brain or in the aorta often result in death.

Vascular Compliance

Compliance (kom-plī′ans) is the tendency for blood vessel volume to increase as the blood pressure increases. The more easily the vessel wall stretches, the greater is its compliance. The less easily the vessel wall stretches, the smaller is its compliance.

Compliance is expressed by the following formula:

$$\text{Compliance} = \frac{\text{Increase in volume (mL)}}{\text{Increase in pressure (mm Hg)}}$$

Vessels with a large compliance exhibit a large increase in volume when the pressure increases a small amount. Vessels with a small compliance do not show a large increase in volume when the pressure increases.

Venous compliance is approximately 24 times greater than the compliance of arteries. As venous pressure increases, the volume of the veins increases greatly. Consequently, veins act as storage areas for blood because their large compliance allows them to hold much more blood than other areas of the vascular system (table 21.13).

Physiology of Systemic Circulation

The anatomy of the circulatory system, the physics of blood flow, and the regulatory mechanisms that control the heart and blood vessels determine the physiologic characteristics of the circulatory system. The entire circulatory system functions to maintain adequate blood flow to all tissues.

Approximately 84% of the total blood volume is contained in the systemic circulatory system. Most of the blood volume is in the veins, which are the vessels with the greatest compliance. Smaller volumes of blood are in the arteries and capillaries (see table 21.13).

Cross-Sectional Area of Blood Vessels

If the cross-sectional area of each blood vessel type is determined and multiplied by the number of each type of blood

Table 21.13	Distribution of Blood Volume in Blood Vessels	
Vessels		**Total Blood Volume (%)**
Systemic		
Veins		64
Large veins	(39%)	
Small veins	(25%)	
Arteries		15
Large arteries	(8%)	
Small arteries	(5%)	
Arterioles	(2%)	
Capillaries		5
	TOTAL IN SYSTEMIC VESSELS	84
Pulmonary vessels		9
Heart		7
	TOTAL BLOOD VOLUME	100

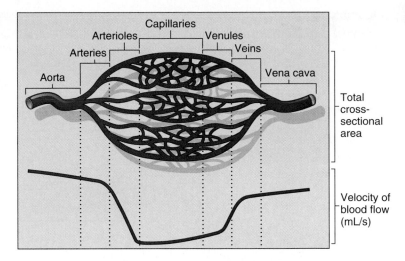

Figure 21.34   Blood Vessel Types and Velocity of Blood Flow

Total cross-sectional area for each of the major blood vessel types is illustrated. The cross-sectional area of each blood vessel is the space through which blood flows, measured in square centimeters. The cross-sectional area of the aorta is about 5 cm^2. The cross-sectional area of each capillary is much smaller, but there are so many that the total cross-sectional area of all capillaries is much greater (2500 cm^2) than the cross-sectional area of the aorta. The line at the bottom of the graph shows that blood velocity drops dramatically in arterioles, capillaries, and venules. As the total cross-sectional area increases the velocity of blood flow decreases.

vessel, the result is the total cross-sectional area for each blood vessel type. For example, there is only one aorta, and it has a cross-sectional area of 5 square centimeters (cm^2). On the other hand, there are millions of capillaries, and each has a very small cross-sectional area. The total cross-sectional area of all capillaries, however, is 2500 cm^2, which is much greater than the cross-sectional area of the aorta (figure 21.34).

The velocity of blood flow is greatest in the aorta, but the total cross-sectional area is small. In contrast, the total cross-sectional area for the capillaries is large, but the velocity of blood flow is low. As the veins become larger in diameter, their total cross-sectional area decreases, and the velocity of blood flow increases. The relationship between blood vessel diameter and velocity of blood flow is much like a stream that flows rapidly through a narrow gorge, but flows slowly through a broad plane.

Pressure and Resistance

The left ventricle of the heart forcefully ejects blood from the heart into the aorta. Because the pumping action of the heart is pulsatile, the aortic pressure fluctuates between a systolic pressure of 120 mm Hg and a diastolic pressure of 80 mm Hg (table 21.14 and figure 21.35). As blood flows from arteries through the capillaries and the veins, the pressure falls progressively to approximately 0 mm Hg or even slightly lower by the time it returns to the right atrium.

The decrease in arterial pressure in each part of the systemic circulation is directly proportional to the resistance to blood flow. There is little resistance in the aorta, so that the average pressure at the end of the aorta is nearly 100 mm Hg. The resistance in medium arteries, which are as small as 3 mm in diameter, is also small, so that their average pressure is still near 95 mm Hg. In the smaller arteries, however, the resistance to blood flow is greater; by the time blood reaches the arteri-

oles, the mean pressure is approximately 85 mm Hg. Within the arterioles the resistance to flow is higher than in any other part of the systemic circulation, and at their ends the mean pressure is only approximately 30 mm Hg. The resistance is also fairly high in the capillaries. The blood pressure at the arterial end of the capillaries is approximately 30 mm Hg, and it decreases to approximately 10 mm Hg at the venous end. Resistance to blood flow in the veins is low because of their relatively large diameter; by the time the blood reaches the right atrium in the venous system, the mean pressure has decreased from 10 mm Hg to approximately 0 mm Hg.

The muscular arteries and arterioles are capable of constricting or dilating in response to autonomic and hormonal stimulation. If constriction occurs, the resistance to blood flow increases, less blood flows through the constricted blood vessels, and blood is shunted to other, nonconstricted areas of the body. Muscular arteries help control the amount of blood flowing to each region of the body, and arterioles regulate blood flow through specific tissues. Constriction of an arteriole decreases blood flow through the local area it supplies, and vasodilation increases the blood flow.

Pulse Pressure

The difference between the systolic and diastolic pressures is called the **pulse pressure** (see figure 21.35). In a healthy young adult at rest the systolic pressure is approximately 120 mm Hg, and the diastolic pressure is approximately 80 mm Hg; thus the pulse pressure is approximately 40 mm Hg. Two major factors influence the pulse pressure: stroke volume of the heart and vascular compliance. When the stroke volume decreases, the pulse

Table 21.14   Blood Pressure Classifications

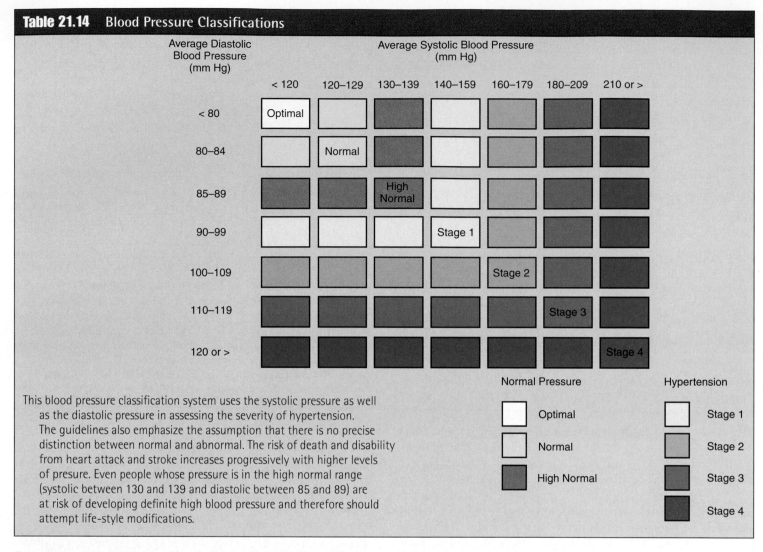

Average Diastolic Blood Pressure (mm Hg)	Average Systolic Blood Pressure (mm Hg)						
	< 120	120–129	130–139	140–159	160–179	180–209	210 or >
< 80	Optimal						
80–84		Normal					
85–89			High Normal				
90–99				Stage 1			
100–109					Stage 2		
110–119						Stage 3	
120 or >							Stage 4

Normal Pressure: Optimal, Normal, High Normal
Hypertension: Stage 1, Stage 2, Stage 3, Stage 4

This blood pressure classification system uses the systolic pressure as well as the diastolic pressure in assessing the severity of hypertension. The guidelines also emphasize the assumption that there is no precise distinction between normal and abnormal. The risk of death and disability from heart attack and stroke increases progressively with higher levels of presure. Even people whose pressure is in the high normal range (systolic between 130 and 139 and diastolic between 85 and 89) are at risk of developing definite high blood pressure and therefore should attempt life-style modifications.

Source: National High Blood Pressure Education Program, National Institutes of Health, Bethesda, MD.

pressure also decreases; and when the stroke volume increases, the pulse pressure increases. The compliance of blood vessels decreases as arteries age. Arteries in older people become less elastic, or arteriosclerotic, and the resulting decrease in compliance causes the pressure in the aorta to rise more rapidly and to a greater degree during systole and to fall more rapidly to its diastolic value. Thus, for a given stroke volume, the systolic pressure and the pulse pressure are higher as vascular compliance decreases.

4   P R E D I C T

Explain the consequences of arteriosclerosis, which is getting progressively more severe, on a large aortic aneurysm.

✔ *Answer in Appendix F*

The pulse pressure caused by the ejection of blood from the left ventricle into the aorta produces a pressure wave, or pulse, that travels rapidly along the arteries. Its rate of transmission is approximately 15 times greater in the aorta (7–10 m/s) and 100 times greater (15–35 m/s) in the distal arteries than the velocity of blood flow. The pulse is monitored frequently, especially in the radial artery, where it is called the **radial pulse,** to determine heart rate and rhythm. Also, weak pulses usually indicate a decreased stroke volume or increased constriction of the arteries as a result of intense sympathetic stimulation of the arteries. As the pulse passes through the smallest arteries and arterioles, it is gradually damped so that it is almost absent in the capillaries (see figure 21.35).

5   P R E D I C T

Explain each of the following: weak pulses in response to ectopic and premature beats of the heart, strong bounding pulses in a person who had received too much saline solution intravenously, weak pulses in a person who is suffering from hemorrhagic shock.

✔ *Answer in Appendix F*

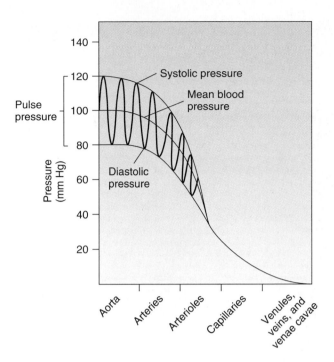

Figure 21.35 Blood Pressure in Major Blood Vessel Types

Blood pressure fluctuations between systole and diastole are damped in small arteries and arterioles. There are no large fluctuations in blood pressure in capillaries and veins. ✗

Capillary Exchange

There are approximately 10 billion capillaries in the body. The major means by which nutrients and waste products are exchanged is by the process of diffusion. Nutrients diffuse from the capillary into the interstitial spaces, and waste products diffuse in the opposite direction. In addition, a small amount of fluid moves out of the capillary at the arteriolar end, and most, but not all, of that fluid reenters the capillary at its venous end.

At the arterial end of the capillary, blood pressure and a small negative pressure in the interstitial spaces move fluid from the capillary. The small negative pressure exists in the interstitial spaces because fluid moves into lymph capillaries when tissues are compressed during movement. When the tissues are no longer compressed, the volume of the tissue increases slightly, and the interstitial pressure decreases. One-way valves in the lymphatic vessels prevent the fluid from passing back from the lymph capillaries into the interstitial spaces.

Fluid is attracted into the capillary by osmosis (figure 21.36). The concentration of small solute molecules is approximately the same in blood and in the interstitial fluid. The concentration of proteins in the interstitial fluid, however, is much lower than in the plasma, and the proteins in the plasma are too large to pass through the wall of the capillary. The osmotic pressure caused by the plasma proteins is called the **blood colloid osmotic pressure,** and it is much higher than the osmotic pressure of the interstitial fluid.

Thus, at the arteriolar end of the capillary, the forces moving fluid out of the capillary are the blood pressure and the negative pressure in the interstitial spaces, whereas the force moving fluid into the capillary results from osmosis. There is a net movement of fluid from the capillaries into the interstitial spaces because the forces moving fluid out of the capillary are greater than the forces moving fluid into it.

Between the arterial end of the capillary and its venous end, the blood pressure decreases from about 30 to 10 mm Hg, which reduces the force moving fluid out of the capillary. The concentration of proteins within the capillary increases slightly at the venous end of the capillary because of the movement of fluid out of its arteriolar end. As a consequence, the plasma protein concentration and the blood colloid osmotic pressure are greater, which slightly increases the osmotic force moving fluid into the capillary. Because of these changes, the forces moving fluid out of the capillary at its venous end are now smaller than the forces moving fluid into it. As a result, about nine-tenths of the fluid that leaves the capillary at its arterial end reenters the capillary at its venous end. The remaining one-tenth enters the lymphatic capillaries and eventually is returned to the general circulation.

Exchange of fluid across the capillary wall also can result from the cyclic dilation and constriction of the precapillary sphincter. When the precapillary sphincter dilates, the pressure rises in the capillary, forcing fluid to move into the interstitial spaces. When the precapillary sphincter constricts, the pressure in the capillary drops, and fluid moves into the capillary. This may be the primary means by which fluids are exchanged across the walls of capillaries in tissues not compressed by movements.

✔ *Answer in Appendix F*

Functional Characteristics of Veins

Cardiac output depends on the preload, which is determined by the volume of blood that enters the heart from the veins (see chapter 20). The factors that affect flow in the veins are therefore of great importance to the overall function of the cardiovascular system. If the volume of blood is increased because of a rapid transfusion, the amount of blood flow to the heart through the veins increases. This increases the preload, which causes the cardiac output to increase because of Starling's law of the heart. On the other hand, a rapid loss of a

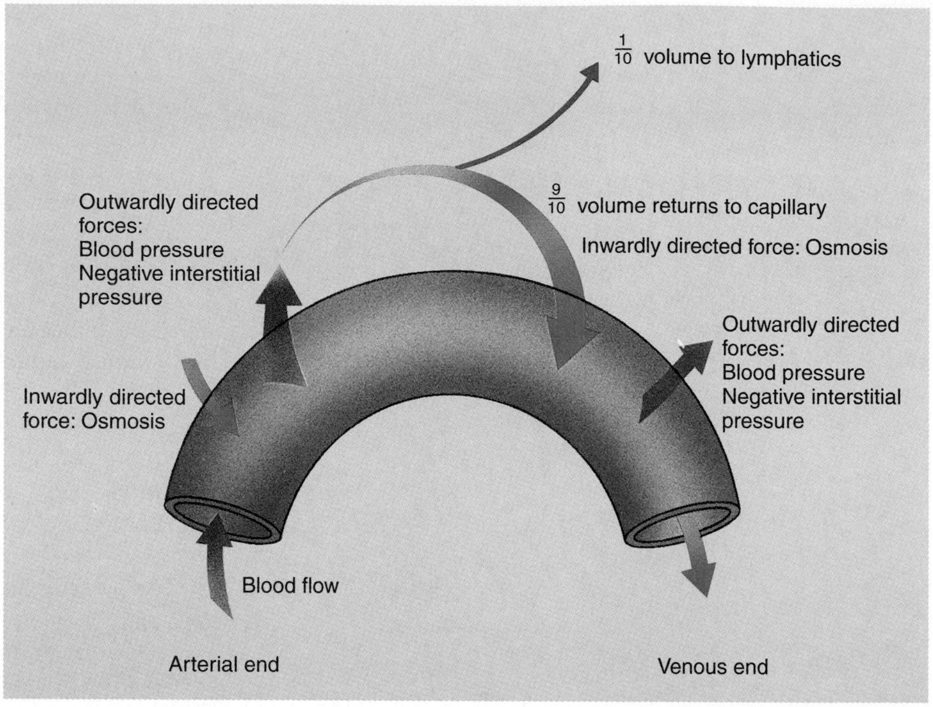

Figure 21.36 Fluid Exchange Across the Walls of Capillaries

The total pressure differences between the inside and the outside of the capillary at its arterial and venous ends are illustrated. At the arterial end, the sum of the forces causes fluid to move from the capillaries into the tissue. At the venous end, the sum of the forces causes fluid to move into the capillary. About nine-tenths of the fluid that leaves the capillary at its arterial end reenters the capillary at its venous end. About one-tenth of the fluid passes into the lymph capillaries.

large blood volume decreases venous return to the heart, which decreases the preload and cardiac output.

Venous tone is a continual state of partial contraction of the veins as a result of sympathetic stimulation. Increased sympathetic stimulation increases venous tone by causing constriction of the veins, which forces the large venous volume to flow toward the heart. Consequently, venous return and preload increase, causing an increase in cardiac output. Conversely, decreased sympathetic stimulation decreases venous tone, allowing veins to relax and dilate. As the veins fill with blood, venous return to the heart, preload, and cardiac output decrease.

The periodic muscular compression of veins forces blood to flow through them toward the heart more rapidly. The valves in the veins prevent flow away from the heart so that when veins are compressed, blood is forced to flow toward the heart. The combination of arterial dilation and compression of the veins by muscular movements during exercise causes blood to return to the heart more rapidly than under conditions of rest.

Blood Pressure and the Effect of Gravity

Blood pressure is approximately 0 mm Hg in the right atrium, and it averages approximately 100 mm Hg in the aorta. The pressure in vessels above and below the heart, however, is affected by gravity. While a person is standing, the venous pressure in the feet is influenced greatly by the force of gravity. Instead of its usual 10 mm Hg pressure at the venules, the pressure can be as much as 90 mm Hg. The arterial pressure is influenced by gravity to the same degree; thus the arteriolar ends of the capillaries can have a pressure of 110 mm Hg rather than 30 mm Hg. The normal pressure difference between the

arterial and the venous ends of capillaries still remains the same, so that flow continues through the capillaries. The major effect of the high pressure in the feet and legs when a person stands for a prolonged time without moving is edema. Without muscular movement the pressure at the venous end of the capillaries increases. Up to 15%–20% of the total blood volume can pass through the walls of the capillaries into the interstitial spaces of the legs during 15 min of standing still.

7   P R E D I C T

Explain why people who are suffering from edema in the legs are told to keep them elevated.

✔ *Answer in Appendix F*

Control of Blood Flow in Tissues

Blood flow provided to the tissues by the cardiovascular system is highly controlled and matched closely to the metabolic needs of tissues. Mechanisms that control blood flow through tissues are classified as (1) local control and (2) nervous and hormonal control.

Local Control of Blood Flow by the Tissues

Blood flow is much greater in some organs than in others. For example, blood flow through the brain, kidneys, and

Clinical Focus   Pulse

A pulse can be felt at locations where large arteries are close to the surface of the body. It is helpful to know where the major pulses can be detected because monitoring the pulse is important clinically. The heart rate, rhythmicity, and other characteristics can be determined by feeling the pulse.

A pulse can be felt at three major locations on each side of the head and neck. One site is the common carotid artery at the point where it divides into internal and external carotid arteries. A second is the superficial temporal artery immediately anterior to the ear. A third is in the facial artery at the point where it crosses the inferior border of the mandible approximately midway between the angle and the genu (figure A).

A pulse can be felt at three major points in the upper limb: in the axilla, in the brachial artery on the medial side of the arm slightly proximal to the elbow, and in the radial artery on the lateral side of the anterior forearm just proximal to the wrist. The radial artery is by tradition the most common site for detecting the pulse of a patient because it is the most easily accessible pulse in the body.

A pulse may be felt at the femoral artery in the groin, the popliteal artery just proximal to the knee, and the dorsalis pedis artery and the posterior tibial artery at the ankle (see figure A).

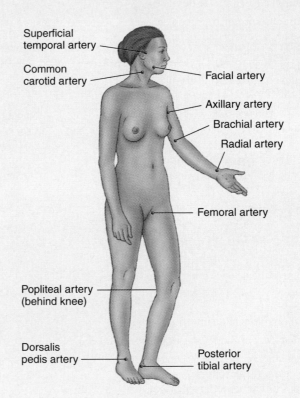

Figure A   Location of Major Points at Which the Pulse Can Be Monitored
Each pulse point is named after the artery on which it occurs.

---

liver is relatively high. The muscle mass of the body is large so that flow through resting skeletal muscles, although not high, is greater than that through other tissue types because skeletal muscle constitutes 35%–40% of the total body mass. Flow through exercising skeletal muscles can increase up to 20-fold, however, and the blood flow through the viscera, including the kidneys and liver, either remains the same or decreases.

In most tissues blood flow is proportional to the metabolic needs of the tissue; therefore, as the activity of skeletal muscle increases, blood flow increases to supply the increased need for oxygen and other nutrients. Blood flow also increases in response to a buildup of metabolic end products.

In some tissues, however, blood flow serves purposes other than the delivery of nutrients and the removal of waste products. In the skin, blood flow also dissipates heat from the body. In the kidneys it eliminates metabolic waste products, regulates water balance, and controls the pH of body fluids. Among other functions, blood flow through the liver delivers nutrients that have entered the blood from the small intestine en route to the liver for processing.

Functional Characteristics of the Capillary Bed

The innervation of the metarterioles and the precapillary sphincters in capillary beds is sparse (see figure 21.2; table 21.15). These structures are regulated primarily by local factors. As the rate of metabolism increases in a tissue, blood flow through its capillaries increases. The precapillary sphincters relax, allowing blood to flow into the local capillary bed. Blood flow can increase sevenfold to eightfold as a result of vasodilation of the metarterioles and the precapillary sphincters in response to an increased rate of metabolism.

Vasodilator substances are produced as the rate of metabolism increases. The vasodilator substances then diffuse from the tissues supplied by the capillary to the area of the precapillary sphincter, the metarterioles, and the arterioles to cause vasodilation (figure 21.37*a*). Several chemicals, including carbon dioxide, lactic acid, adenosine, adenosine monophosphate, adenosine diphosphate, endothelium-derived relaxation factor (EDRF), potassium ions, and hydrogen ions, cause vasodilation, and they increase in concentration in the extracellular fluid as the rate of metabolism in tissues increases.

Table 21.15   Homeostasis: Local Control of Blood Flow

Stimulus	Response
Regulation by Metabolic Need of Tissues	
Increased vasodilator substances (e.g., carbon dioxide, lactic acid, adenosine, adenosine monophosphate, adenosine diphosphate, endothelium-derived relaxation factor, K^+ ions, decreased pH) or decreased nutrients (e.g., oxygen, glucose, amino acids, fatty acids, and other nutrients) as a result of increased metabolism	Relaxation of precapillary sphincters and subsequent increase in blood flow through capillaries
Decreased vasodilator substances and a reduced need for O_2 and other nutrients	Contraction of precapillary sphincters and subsequent decrease in blood flow through capillaries
Regulation by Nervous Mechanisms	
Increased physical activity or increased sympathetic activity	Constriction of blood vessels in skin and viscera
Increased body temperature detected by neurons of the hypothalamus	Dilation of blood vessels in skin (see chapter 5)
Decreased body temperature detected by neurons of the hypothalamus	Constriction of blood vessels in skin (see chapter 5)
Decrease in skin temperature below a critical value	Dilation of blood vessels in skin (protects skin from extreme cold)
Anger or embarrassment	Dilation of blood vessels in skin of face and upper thorax
Regulation by Hormonal Mechanisms	
(reinforces increased activity of the sympathetic nervous system)	
Increased physical activity and increased sympathetic activity causing release of epinephrine and small amounts of norepinephrine from the adrenal medulla	Constriction of blood vessels in skin and viscera; dilation of blood vessels in skeletal and cardiac muscle
Autoregulation	
Increased blood pressure	Contraction of precapillary sphincters to maintain constant capillary blood flow
Decreased blood pressure	Relaxation of precapillary sphinters to maintain constant capillary blood flow
Long-Term Local Blood Flow	
Increased metabolic activity of tissues over a long period	Increased diameter and number of capillaries
Decreased metabolic activity of tissues over a long period	Decreased diameter and number of capillaries

Lack of nutrients can also be important in regulating local blood flow. For example, oxygen and other nutrients are required to maintain vascular smooth muscle contraction. An increased rate of metabolism decreases the amount of oxygen and other nutrients in the tissues. Smooth muscle cells of the precapillary sphincter relax in response to a lack of oxygen and other nutrients, resulting in vasodilation (see figure 21.37a).

Blood flow through capillaries is not continuous but cyclic. The cyclic fluctuation is the result of periodic contraction and relaxation of the precapillary sphincters called **vasomotion** (vā-sō-mō′shŭn). Blood flows through the capillaries until the by-products of metabolism are reduced in concentration and until nutrient supplies to precapillary smooth muscles are replenished. Then the precapillary sphincters constrict and remain constricted until the by-products of metabolism increase and nutrients decrease (figure 21.37b).

Autoregulation of Blood Flow

Arterial pressure can change over a wide range, whereas blood flow through tissues remains relatively constant. The maintenance of blood flow by tissues is called **autoregulation** (aw′tō-reg′yū-lā′shŭn). Between arterial pressures of approximately 75 mm Hg and 175 mm Hg, blood flow through tissues remains within 10%–15% of its normal value. The mechanisms responsible for autoregulation are the same as those for vasomotion. The need for nutrients and the buildup of metabolic by-products cause precapillary sphincters to di-

Clinical Focus   Hypertension

Hypertension, or high blood pressure, affects approximately 20% of the human population at sometime in their lives. Generally a person is considered hypertensive if the systolic blood pressure is greater then 140mm Hg and the diastolic pressure is greater than 90 mm Hg. Current methods of evaluation, however, take into consideration combinations of diastolic and systolic blood pressure in determining if a person is suffering from hypertension (see table 21.14). In addition, normal blood pressure is age-dependent, so classification of an individual as hypertensive depends on the person's age.

Chronic hypertension has an adverse effect on the function of both the heart and the blood vessels. Hypertension requires the heart to work harder than normal. This extra work leads to hypertrophy of the cardiac muscle, especially in the left ventricle,. and can lead to heart failure. Hypertension also increases the rate at which arteriosclerosis develops. Arteriosclerosis, in turn, increases

the probability that blood clots, or thromboemboli (throm′bō-em′bō-lī), may form and that blood vessels will rupture. Common medical problems associated with hypertension are cerebral hemorrhage, coronary infarction, hemorrhage of renal blood vessels, and poor vision caused by burst blood vessels in the retina.

Some conditions leading to hypertension include a decrease in functional kidney mass, excess aldosterone or angiotensin production, and increased resistance to blood flow in the renal arteries. All of these conditions cause an increase in the total blood volume, which causes the cardiac output to increase. The increased cardiac output forces blood to flow through tissue capillaries, causing the precapillary sphincters to constrict. Thus increased blood volume increases the cardiac output and the peripheral resistance, both of which result in a greater blood pressure.

Although these conditions result in hypertension, roughly 90% of the diagnosed

cases of hypertension are called **idiopathic,** or **essential, hypertension,** which means the cause of the condition is unknown. Drugs that dilate blood vessels (called vasodilators), drugs that increase the rate of urine production (called diuretics), or drugs that decrease cardiac output normally are used to treat essential hypertension. The vasodilator drugs increase the rate of blood flow through the kidneys and thus increase urine production, and the diuretics also increase urine production. Increased urine production reduces the blood volume, which reduces the blood pressure. Substances that decrease cardiac output, such as beta-adrenergic-blocking agents, decrease the heart rate and the force of contraction. In addition to these treatments, low-sale diets normally are recommended to reduce the amount of sodium chloride and water absorbed from the intestine into the bloodstream.

late, and blood flow through tissues increases if a minimum blood pressure exists. On the other hand, once the supply of nutrients and oxygen to tissues is adequate, the precapillary sphincters constrict, and blood flow through the tissues decreases, even if blood pressure is very high.

The availability of oxygen to a tissue can be a major factor in determining the adjustment of the vascularity of a tissue to its long-term metabolic needs. If there is a lack of oxygen, capillaries increase in diameter and in number, and if the oxygen levels remain elevated in a tissue, the vascularity decreases.

8 PREDICT

When blood flow to a tissue has been blocked for a short time, the blood flow through that tissue increases to as much as five times its normal value after the removal of the blockade. The response is called reactive hyperemia. Create a reasonable explanation for this phenomenon on the basis of what you know about the local control of blood flow.

✔ *Answer in Appendix F*

Long-Term Local Blood Flow

The long-term regulation of blood flow through tissues is matched closely to the metabolic requirements of the tissue. If the metabolic activity of a tissue increases and remains elevated for an extended period, the diameter and the number of capillaries in the tissue increase, and local blood flow increases. The increased density of capillaries in the well-trained skeletal muscles of athletes compared with that in poorly trained skeletal muscles is an example.

Clinical Note

Blockage, or occlusion, of a blood vessel leads to an increase in the diameter of smaller blood vessels that bypass the occluded vessel. In many cases the development of these collateral vessels is marked. For example, if a vessel such as the femoral artery becomes occluded, the small vessels that bypass the occluded vessel become greatly enlarged. An adequate blood supply to the leg is often reestablished over a period of weeks. If the occlusion is sudden and so complete that tissues supplied by a blood vessel suffer from ischemia (lack of blood flow), cell death can occur. In this instance, collateral circulation does not have a chance to develop before necrosis occurs.

Nervous and Hormonal Regulation of Local Circulation

Nervous control of arterial blood pressure is important in minute-to-minute regulation while at rest, during exercise, or

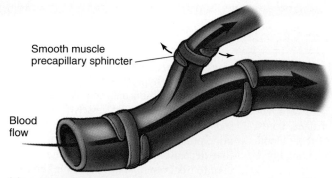

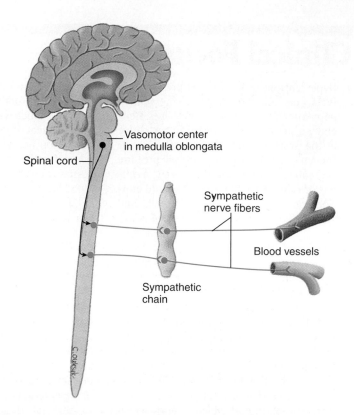

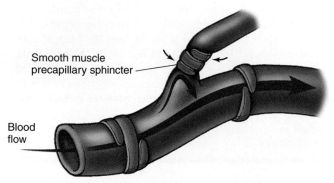

(a)   Precapillary sphincters relax due to an increase in vasodilator substances such as CO_2, lactic acid, adenosine, adenosine monophosphate, adenosine diphosphate, endothelium-derived relaxation factor, K^+ ions, and H^+ ions.

The need for O_2, glucose, amino acids, fatty acids, and other nutrients cause precapillary sphincters to relax.

(b)   Removal of vasodilator substances and a reduced need for O_2 and other nutrients cause precapillary sphincters to contact.

Figure 21.37   Control of Local Blood Flow Through Capillary Beds
(*a*) Dilation of precapillary sphincters. (*b*) Constriction of precapillary sphincters.

Figure 21.38   Nervous Regulation of Blood Vessels
Most blood vessels are innervated by sympathetic nerve fibers. The vasomotor center within the medulla oblongata plays a major role in regulating the frequency of action potentials in nerve fibers that innervate blood vessels.

in response to circulatory shock, during which blood pressure decreases to a very low value. For example, during exercise increased arterial blood pressure is needed to cause blood to flow through the capillaries of skeletal muscles at a rate great enough to supply their oxygen need.

Nervous regulation also provides a means by which blood can be shunted from one large area of the peripheral circulatory system to another. For example, in response to blood loss, blood flow to the viscera and the skin is reduced dramatically. This helps maintain the arterial blood pressure within a range sufficient to allow adequate blood flow through the capillaries of the brain and cardiac muscle.

Nervous regulation, by the autonomic nervous system, can function rapidly (within 1–30 s). The most important part of the autonomic nervous system for this regulation is the sympathetic division. Sympathetic vasomotor fibers innervate all blood vessels of the body except the capillaries, precapillary sphincters, and most metarterioles (figure 21.38). The innervation of the small arteries and arterioles allows the sympathetic nervous system to increase or decrease resistance to blood flow.

9   PREDICT

A strong athlete just finished a 1-mile run and sat down to have a drink with her friends. Her blood pressure was not dramatically elevated during the run, but her cardiac output was greatly increased. After the run, her cardiac output decreased dramatically, but her blood pressure only decreased to its resting level. Predict how sympathetic stimulation of her large veins, arteries in her digestive system, and arteries in her skeletal muscles changed while she was relaxing. Explain why this is consistent with the decrease in her cardiac output.

✔ *Answer in Appendix F*

Sympathetic vasoconstrictor fibers extend to most parts of the circulatory system, but they are less prominent in skeletal muscle, cardiac muscle, and the brain and more prominent in the kidneys, gut, spleen, and skin.

An area of the lower pons and upper medulla oblongata, called the **vasomotor** (vā-sō-mō′ter) **center** (see figure 21.38), is tonically active. A low frequency of action potentials is transmitted continually in the sympathetic vasoconstrictor fibers. As a consequence, the peripheral blood vessels are partially constricted, a condition called **vasomotor tone.**

Part of the vasomotor center inhibits vasomotor tone. Thus the vasomotor center consists of an excitatory part,

Clinical Focus   Blood Flow Through Tissues During Exercise

Blood flow through tissues is matched with the metabolic needs of the tissues. During exercise, blood flow through tissues is changed dramatically. Its rate of flow through exercising skeletal muscles can be 15–20 times greater than through resting muscles. The increased blood flow is the product of local, nervous, and hormonal regulatory mechanisms. When skeletal muscle is resting, only 20%–25% of the capillaries are open, whereas during exercise 100% of the capillaries are open.

Low oxygen tensions resulting from greatly increased muscular activity or the release of vasodilator substances, such as lactic acid, carbon dioxide, and potassium ions, causes dilation of precapillary sphincters. Increased sympathetic stimulation and epinephrine released from the adrenal medulla cause some vasoconstriction in the blood vessels of the skin and viscera and some vasodilation of blood vessels in skeletal muscles. Consequently, resistance to blood flow in skeletal muscle decreases, and resistance to blood flow in the skin and viscera increases somewhat. Blood is therefore shunted from the viscera and the skin through the vessels in skeletal muscles.

The movement of skeletal muscles that compresses veins in a cyclic fashion and the constriction of veins greatly increase the venous return to the heart. The resulting increase in the preload and increased sympathetic stimulation of the heart result in elevated heart rate and stroke volume, which increases the cardiac output. As a consequence, the blood pressure usually increases by 20–60 mm Hg, which helps sustain the increased blood flow through skeletal muscle blood vessels.

In response to sympathetic stimulation, some decrease in the blood flow through the skin can occur at the beginning of exercise. As the body temperature increases in response to the increased muscular activity, however, temperature receptors in the hypothalamus are stimulated. As a result, action potentials in sympathetic nerve fibers causing vasoconstriction decrease, resulting in vasodilation of blood vessels in the skin. As a consequence, the skin turns a red or pinkish color, and a great deal of excess heat is lost as blood flows through the dilated blood vessels.

The overall effect of exercise on circulation is to greatly increase the blood flow through exercising muscles and to keep blood flow through other organs at a value just adequate to supply their metabolic needs.

---

which is tonically active, and an inhibitory part, which can induce vasodilation. Vasoconstriction results from an increase and vasodilation from a decrease in vasomotor tone.

Areas throughout the pons, midbrain, and diencephalon can either stimulate or inhibit the vasomotor center. For example, the hypothalamus can exert either strong excitatory or inhibitory effects on the vasomotor center. Increased body temperature detected by temperature receptors in the hypothalamus causes vasodilation of blood vessels in the skin (see chapter 5). The cerebral cortex also can either excite or inhibit the vasomotor center. For example, action potentials that originate in the cerebral cortex during periods of emotional excitement activate hypothalamic centers, which in turn increase vasomotor tone (see table 21.15).

The neurotransmitter for the vasoconstrictor fibers is norepinephrine, which binds to alpha-adrenergic receptors on vascular smooth muscle cells to cause vasoconstriction. Sympathetic action potentials also cause the release of epinephrine and norepinephrine into the blood from the adrenal medulla. These hormones are transported in the blood to all parts of the body. In most vessels they cause vasoconstriction, but in some vessels, especially those in skeletal muscle, epinephrine binds to beta-adrenergic receptors, which are present in large numbers, and causes the skeletal muscle blood vessels to dilate.

Regulation of Mean Arterial Pressure

The **mean arterial pressure** is slightly less than the average of the systolic and diastolic pressures because diastole lasts longer than systole. The mean arterial pressure is approximately 70 mm Hg at birth, is approximately 100 mm Hg from adolescence to middle age, and reaches 110 mm Hg in the healthy older person, but it can be as high as 130 mm Hg. The range of normal systolic and diastolic blood pressures for adults is presented in table 21.14.

Blood flow through the entire circulatory system is determined by the cardiac output (CO), which is equal to the heart rate (HR) times the stroke volume (SV) and **peripheral resistance (PR),** which is the resistance to blood flow in all of the blood vessels. The **mean arterial pressure (MAP)** in the body is proportional to the cardiac output times the peripheral resistance:

$$MAP = CO \times PR \qquad \text{or} \qquad MAP = HR \times SV \times PR$$

This equation expresses the effect of heart rate, stroke volume, and peripheral resistance on blood pressure. An increase in any one of them results in an increase in blood pressure. Conversely, a decrease in any one of them produces a decrease in blood pressure. The mechanisms that control blood pressure do so by changing peripheral resistance, heart rate, or stroke volume. Because stroke volume depends on the amount of blood entering the heart, regulatory mechanisms that control blood volume also affect blood pressure. For example, an increase in blood volume increases venous return, which increases the preload, and the increased preload increases the stroke volume.

When blood pressure suddenly drops because of hemorrhage or some other cause, the control systems respond by increasing the blood pressure to a value consistent with life and by increasing the blood volume to its normal value. Two major types of control systems operate to achieve these responses: (1) those that respond in the short-term and (2) those that respond in the long-term.

1. Baroreceptors in the carotid sinus and aortic arch monitor blood pressure.

2. Action potentials are conducted to the cardioregulatory and vasomotor centers in the medulla oblongata.

3. Increased parasympathetic stimulation of the heart decreases the heart rate.

4. Increased sympathetic stimulation of the heart increases the heart rate and stroke volume.

5. Increased sympathetic stimulation of blood vessels increases vasoconstriction.

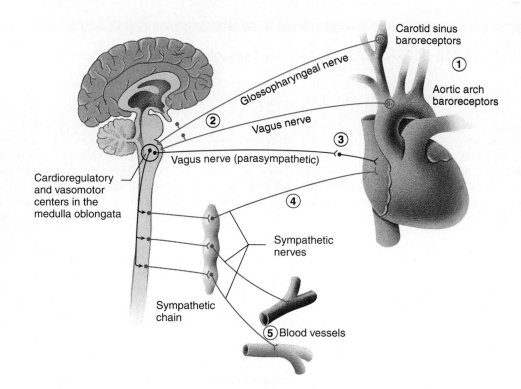

Figure 21.39   Baroreceptor Reflex Control of Blood Pressure

An increase in blood pressure increases parasympathetic stimulation of the heart and decreases sympathetic stimulation of the heart and blood vessels, resulting in a decrease in blood pressure. A decrease in blood pressure decreases parasympathetic stimulation of the heart and increases sympathetic stimulation of the heart and blood vessels, resulting in an increase in blood pressure.

The regulatory mechanisms that control pressure on a short-term basis begin to lose their capacity to regulate blood pressure a few hours to a few days after blood pressure is maintained at higher or lower values. This occurs because sensory receptors adapt to the altered pressures. Long-term regulation of blood pressure is controlled primarily by mechanisms that influence kidney function, and those mechanisms do not adapt rapidly to altered blood pressures.

Short-Term Regulation of Blood Pressure

The short-term, rapidly acting mechanisms controlling blood pressure include the baroreceptor reflexes, chemoreceptor reflexes, the central nervous system ischemic response, and the adrenal medullary mechanism.

Baroreceptor Reflexes

Baroreceptors, or **pressoreceptors,** are sensory receptors sensitive to stretch. They are scattered along the walls of most of the large arteries of the neck and the thorax and are most numerous in the area of the carotid sinus at the base of the internal carotid artery and in the walls of the aortic arch (figure 21.39). Action potentials are transmitted from the carotid sinus baroreceptors through the glossopharyngeal nerves to the cardioregulatory and vasomotor centers in the

medulla oblongata and from the aortic arch through the vagus nerves to the medulla oblongata. Stimulation of baroreceptors in the carotid sinus activates the **carotid sinus reflex,** and stimulation of baroreceptors in the aortic arch activates the **aortic arch reflex.** These reflexes function to control blood within a narrow range of values.

In the carotid sinus and the aortic arch, normal blood pressure partially stretches the arterial wall so that a constant, but low, frequency of action potentials is produced by the baroreceptors. Increased pressure in the blood vessels stretches the vessel walls and causes the baroreceptors to increase the frequency of action potentials. Conversely, a decrease in blood pressure reduces the stretch of the arterial wall and causes the baroreceptors to decrease the frequency of action potentials.

The increased frequency of action potentials produced in the baroreceptors by an increase in blood pressure stimulates the cardioregulatory and vasomotor centers of the medulla oblongata. The vasomotor center responds by causing vasodilation of blood vessels, and the cardioregulatory center responds by increasing parasympathetic stimulation of the heart. As a result, increased systemic blood pressure causes both dilation of peripheral blood vessels and a decreased heart rate, resulting in decreased blood pressure.

Sudden decreases in arterial pressure result in a decreased frequency of action potentials produced by the baroreceptors. As a consequence, the vasomotor center responds by increasing peripheral vasoconstriction. In addition, an increase in the sympathetic stimulation of the heart from the cardioregulatory center causes the heart rate and stroke volume to increase. This increase is accompanied by a decrease in parasympathetic stimulation of the heart (see figure 21.39).

The carotid sinus and aortic arch baroreceptor reflexes are important in regulating blood pressure moment to moment. When a person rises rapidly from a sitting or lying position to a standing position, a dramatic drop in blood pressure in the neck and thoracic regions occurs because of the pull of gravity on the blood. This reduction can be so great that blood flow to the brain becomes sufficiently sluggish to cause dizziness or loss of consciousness. The falling blood pressure initiates the baroreceptor reflexes, however, which reestablish the normal blood pressure within a few seconds. A healthy person may experience only a temporary sensation of dizziness.

10	P R E D I C T

Explain how the baroreceptor reflex responds when a person does a headstand.

✔ *Answer in Appendix F*

The baroreceptor reflexes are short term and rapid acting. They do not change the average blood pressure in the long run. The baroreceptors adapt within 1–3 days to any new blood pressure to which they are exposed. If the blood pressure is elevated for more than a few days, the baroreceptors adapt to the elevated pressure and do not reduce blood pressure to its original value. This adaptation is common in people who have hypertension.

Clinical Note

Occasionally the application of pressure to the carotid arteries in the upper neck results in a dramatic decrease in blood pressure. This condition, called the **carotid sinus syndrome,** is most common in patients in whom arteriosclerosis of the carotid artery is advanced. In such patients a tight collar can apply enough pressure to the region of the carotid sinuses to stimulate the baroreceptors. The increased action potentials from the baroreceptors initiate reflexes that result in a decrease in vasomotor tone and an increase in parasympathetic action potentials to the heart. As a result of the decreased peripheral resistance and heart rate, blood pressure decreases dramatically. As a consequence, blood flow to the brain decreases to such a low level that the person becomes dizzy or may even faint. People suffering from this condition must avoid applying external pressure to the neck region. If the carotid sinus becomes too sensitive, a treatment for this condition is surgical destruction of the innervation to the carotid sinuses.

Adrenal Medullary Mechanism

The adrenal medullary mechanism is activated when stimuli that result in increased sympathetic stimulation of the heart and the blood vessels also cause increased stimulation of the adrenal medulla, which results in increased secretion of epinephrine and some norepinephrine from the adrenal medulla (figure 21.40). These hormones affect the cardiovascular system in a fashion similar to direct sympathetic stimulation, causing increased heart rate, increased stroke volume, vasoconstriction in blood vessels to the skin and viscera, and vasodilation in blood vessels to cardiac muscle. The adrenal medullary mechanism is short term and rapid-acting, whereas other hormonal mechanisms are long term and slow-acting (see below).

Chemoreceptor Reflexes

The **chemoreceptor** (kem′ō-rē-sep′tŏr) **reflexes** help maintain homeostasis when the oxygen tension in the blood decreases or when carbon dioxide and hydrogen ion concentrations increase (figure 21.41).

Carotid bodies, small organs approximately 1–2 mm in diameter, lie near the carotid sinuses, and several **aortic bodies** lie adjacent to the aorta. Chemoreceptors are located in the carotid and aortic bodies. Afferent nerve fibers pass to the medulla oblongata through the glossopharyngeal nerve (IX) from the carotid bodies and through the vagus nerve (X) from the aortic bodies.

The chemoreceptors receive an abundant blood supply. When oxygen availability decreases in the chemoreceptor cells, the frequency of action potentials increases and stimulates the vasomotor center, resulting in increased vasomotor tone. The chemoreceptors act under emergency conditions and do not regulate the cardiovascular system under resting conditions. They normally do not respond strongly unless oxygen tension in the blood decreases markedly. The chemoreceptor cells are also stimulated by increased carbon dioxide and hydrogen ion concentrations, to increase vasomotor tone. The increased vasomotor tone increases the mean arterial pressure. The increased mean arterial pressure increases blood flow through the lungs, which helps eliminate excess carbon dioxide and hydrogen ions from the body and increases oxygen uptake.

A decrease in blood pressure can indirectly activate the chemoreceptors because with decreased blood pressure the blood flow to the lungs and tissues can be inadequate. Blood oxygen levels can decrease sufficiently to stimulate the chemoreceptors.

Central Nervous System Ischemic Response

When blood flow to the vasomotor center decreases enough, the neurons within the medulla are strongly excited by a buildup of carbon dioxide and an increase in hydrogen ions. As a result, vasoconstriction is stimulated by the vasomotor center, and the systemic blood pressure rises dramatically. The elevation in blood pressure in response to a lack of blood flow to the medulla is called the **central nervous system (CNS) ischemic response.**

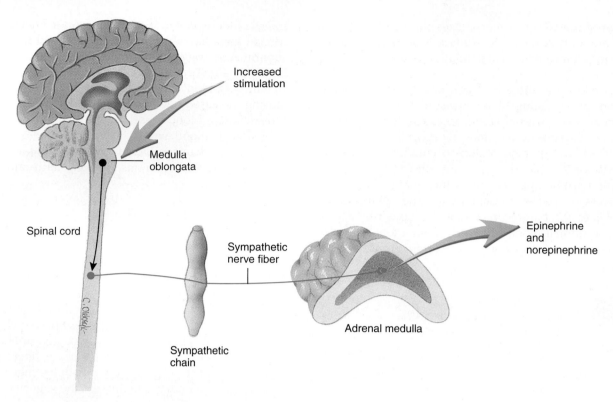

Figure 21.40   The Adrenal Medullary Mechanism

Stimuli that increase sympathetic stimulation of the heart and blood vessels also result in increased sympathetic stimulation of the adrenal medulla and result in epinephrine and some norepinephrine secretion.

1. Chemoreceptors in the carotid sinus and aortic bodies monitor blood O_2, CO_2, and pH.

2. Chemoreceptors in the medulla oblongata monitor blood CO_2 and pH.

3. Increased parasympathetic stimulation of the heart decreases the heart rate.

4. Increased sympathetic stimulation of the heart increases the heart rate and stroke volume.

5. Increased sympathetic stimulation of blood vessels increases vasoconstriction.

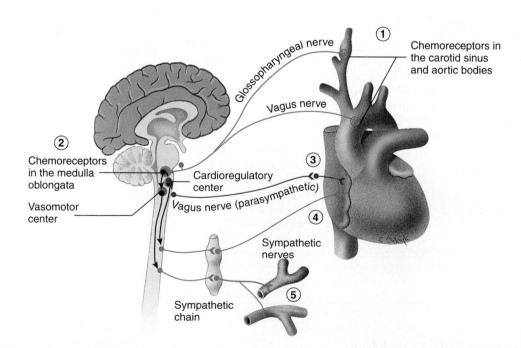

Figure 21.41   Chemoreceptor Reflex Control of Blood Pressure

An increase in blood CO_2 and a decrease in pH and blood O_2 result in an increased heart rate and vasoconstriction. A decrease in blood CO_2 and an increase blood pH result in a decreased heart rate and vasodilation.

The CNS ischemic response is important only if the blood pressure falls below 50 mm Hg. It therefore does not play an important role in regulating blood pressure under normal conditions and functions primarily in response to emergency situations in which blood flow to the brain is severely restricted. The increase in blood pressure that occurs in response to CNS ischemia increases blood flow to the CNS, provided the blood vessels are intact.

If severe ischemia lasts longer than a few minutes, metabolism in the brain fails because of the lack of oxygen. The vasomotor center becomes inactive, and extensive vasodilation occurs in the periphery as vasomotor tone decreases. Prolonged ischemia of the medulla oblongata leads to a massive decline in blood pressure and death.

The mechanisms responsible for the short-term regulation of the mean arterial blood pressure are summarized in figure 21.42 and figure 21.43.

Long-Term Regulation of Blood Pressure

In addition to the rapidly acting nervous mechanisms that regulate arterial pressure, the following hormonal mechanisms control it in the longer run: (1) the renin-angiotensin-aldosterone mechanism, (2) the vasopressin mechanism, (3) the atrial natriuretic mechanism, (4) fluid shift, and (5) the stress–relaxation response.

Renin-Angiotensin-Aldosterone Mechanism

The kidneys release an enzyme called **renin** (rē'nin) into the circulatory system (see chapter 26) from specialized structures called the **juxtaglomerular** (jŭks'tă-glō-mer'yū-lăr) **apparatuses.** Renin acts on plasma proteins called **angiotensinogen** (an-jē-ō-ten'sin'ō-jen) to split a fragment off one end. The fragment, called **angiotensin** (an-jē-ō-ten'sin) **I,** contains 10 amino acid molecules. Another enzyme, called **angiotensin-converting enzyme,** found primarily in small blood vessels of the lung, cleaves two additional amino acid molecules from angiotensin I to produce a fragment consisting of eight amino acids called **angiotensin II,** or **active angiotensin** (figure 21.44).

Angiotensin II causes vasoconstriction in arterioles and to some degree in veins. As a result, it increases peripheral resistance and venous return to the heart, both of which function to raise the blood pressure. Angiotensin II also stimulates aldosterone secretion from the adrenal cortex. **Aldosterone** (al-dos'ter-ōn) acts on the kidneys to increase the reabsorption of sodium and chloride ions and water. The results are to decrease the production of urine and to conserve water to prevent further reduction in blood volume caused by the formation of urine (see chapter 26). Angiotensin II also stimulates the sensation of thirst and increases salt appetite and antidiuretic hormone (ADH) secretion (see chapter 18).

Decreased blood pressure stimulates renin secretion. Elevated plasma concentration of potassium ions and reduced plasma concentration of sodium ions directly stimulate aldosterone secretion. Decreased blood pressure and elevated potassium ion concentration occur during plasma loss and dehydration and in response to tissue damage such as burns or crushing injuries.

The **renin-angiotensin-aldosterone mechanism** is important in maintaining blood pressure on a daily basis. It also reacts strongly under conditions of circulatory shock, in response to which it requires approximately 20 min to become maximally effective. Its onset of action is not as fast as nervous reflexes or the adrenal medullary response, but its duration of action is longer. Once renin is secreted, it remains active for approximately 1 h and the effect of aldosterone is much longer (many hours).

Vasopressin Mechanism

When the concentration of solutes in the plasma increases or when there is a decrease in blood pressure, hypothalamic neurons increase the frequency of impulses transmitted to the posterior pituitary and increase the secretion of **vasopressin** (vā-sō-pres'in), or ADH. Increases in the plasma concentration of solutes directly affect hypothalamic neurons that stimulate ADH secretion. Changes in blood pressure affect the frequency of afferent action potentials from baroreceptors, which influence the activity of the hypothalamic neurons (figure 21.45). Under normal conditions small changes in the concentration of solutes in the plasma play a greater role in regulating the rate of ADH secretion than do changes in blood pressure. Under conditions of circulatory shock, however, the large decrease in blood pressure results in greatly elevated ADH secretion.

ADH acts directly on blood vessels to cause vasoconstriction, although it is not as potent as other vasoconstrictor agents. Evidence indicates that within minutes after a rapid decline in blood pressure, ADH is released in sufficient quantities to affect the reestablishment of normal blood pressure. ADH also decreases the rate of urine production by the kidneys, helping to maintain blood volume and blood pressure.

Atrial Natriuretic Mechanism

A polypeptide called **atrial natriuretic** (ā'trē-ăl nā'trē-yū-ret'ik) **hormone** is released from cells in the atria of the heart. A major stimulus for its release is increased venous return, which stretches atrial cardiac muscle cells. Atrial natriuretic hormone acts on the kidneys to increase the rate of urine production and sodium loss in the urine. It also dilates arteries and veins. Loss of water and sodium ions in the urine causes the blood volume to decrease, which decreases venous return, and vasodilation results in a decrease in peripheral resistance. These effects function to cause a decrease in blood pressure.

The renin-angiotensin-aldosterone system and the atrial natriuretic hormone mechanism work simultaneously to regulate blood pressure by influencing kidney function.

When the blood pressure drops below 50 mm Hg, the volume of urine produced by the kidneys is close to zero. At

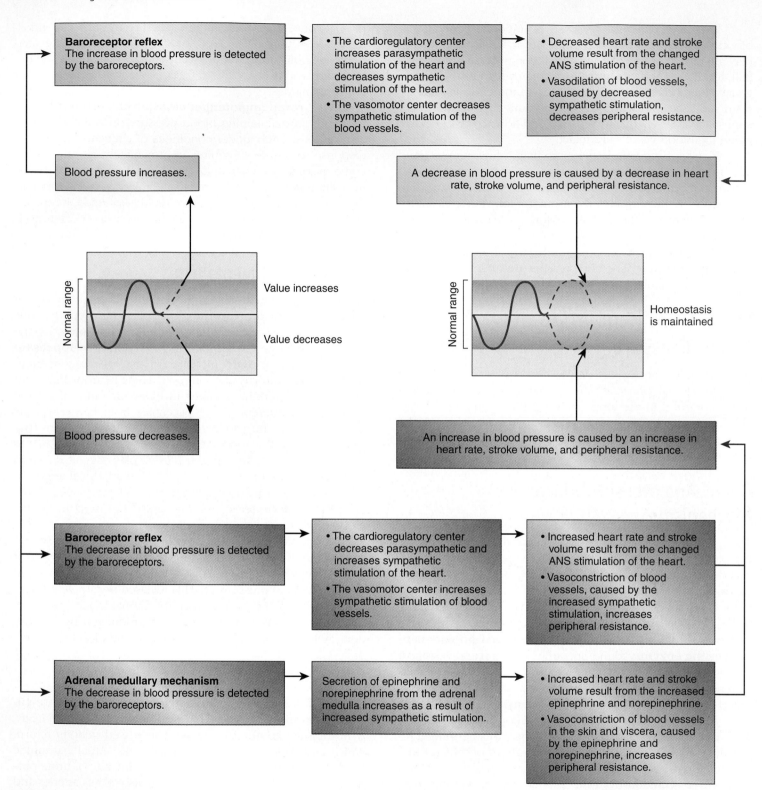

Figure 21.42   Homeostasis: Baroreceptor Effects on Blood Pressure

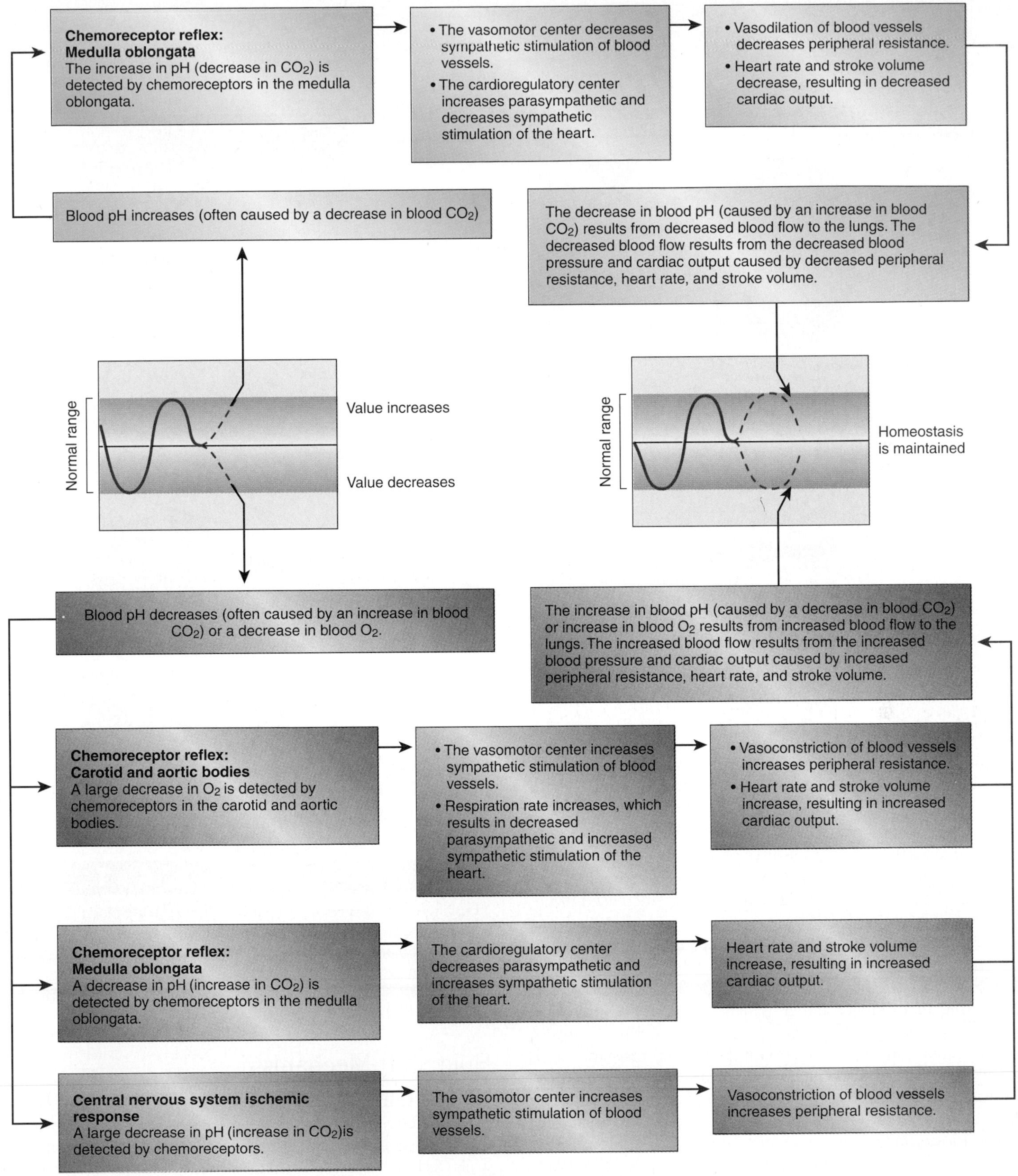

**Chemoreceptor reflex:
Medulla oblongata**
The increase in pH (decrease in CO_2) is detected by chemoreceptors in the medulla oblongata.

- The vasomotor center decreases sympathetic stimulation of blood vessels.
- The cardioregulatory center increases parasympathetic and decreases sympathetic stimulation of the heart.

- Vasodilation of blood vessels decreases peripheral resistance.
- Heart rate and stroke volume decrease, resulting in decreased cardiac output.

Blood pH increases (often caused by a decrease in blood CO_2)

The decrease in blood pH (caused by an increase in blood CO_2) results from decreased blood flow to the lungs. The decreased blood flow results from the decreased blood pressure and cardiac output caused by decreased peripheral resistance, heart rate, and stroke volume.

Normal range

Value increases

Value decreases

Normal range

Homeostasis is maintained

Blood pH decreases (often caused by an increase in blood CO_2) or a decrease in blood O_2.

The increase in blood pH (caused by a decrease in blood CO_2) or increase in blood O_2 results from increased blood flow to the lungs. The increased blood flow results from the increased blood pressure and cardiac output caused by increased peripheral resistance, heart rate, and stroke volume.

**Chemoreceptor reflex:
Carotid and aortic bodies**
A large decrease in O_2 is detected by chemoreceptors in the carotid and aortic bodies.

- The vasomotor center increases sympathetic stimulation of blood vessels.
- Respiration rate increases, which results in decreased parasympathetic and increased sympathetic stimulation of the heart.

- Vasoconstriction of blood vessels increases peripheral resistance.
- Heart rate and stroke volume increase, resulting in increased cardiac output.

**Chemoreceptor reflex:
Medulla oblongata**
A decrease in pH (increase in CO_2) is detected by chemoreceptors in the medulla oblongata.

The cardioregulatory center decreases parasympathetic and increases sympathetic stimulation of the heart.

Heart rate and stroke volume increase, resulting in increased cardiac output.

Central nervous system ischemic response
A large decrease in pH (increase in CO_2) is detected by chemoreceptors.

The vasomotor center increases sympathetic stimulation of blood vessels.

Vasoconstriction of blood vessels increases peripheral resistance.

Figure 21.43   Homeostasis: Effects of pH and Gases on Blood Pressure

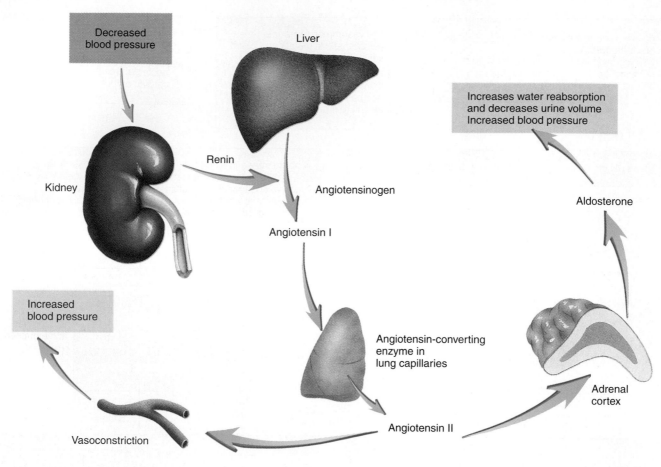

Figure 21.44   The Renin-Angiotensin-Aldosterone Mechanism

Decreased blood pressure is detected by the kidney, resulting in increased renin secretion. The result is vasoconstriction, increased water reabsorption, and decreased urine volume. These changes function to maintain blood pressure.

200 mm Hg the volume produced is approximately six to eight times greater than normal.

The kidney is also sensitive to small changes in blood volume. An acute increase of a few hundred milliliters of volume increases the blood pressure significantly. Within several hours the small increase in blood volume is eliminated as urine because the elevated pressure results in increased urine production (see chapter 26).

When the intake of water and salt increases, the blood volume and blood pressure increase, and the amount of renin secreted by the kidney decreases. The decreased renin secretion results in a reduced rate at which angiotensinogen is ultimately converted to angiotensin II. As a consequence, vasodilation occurs, causing a reduction in peripheral resistance and blood pressure but also causing increased urine formation by the kidneys. The decrease in renin secretion also causes a reduction in aldosterone secretion by the adrenal cortex. Because aldosterone promotes sodium and water retention by the kidney, its decrease results in a greater-than-normal loss of sodium and water through the kidneys and a reduction in blood volume, causing the blood pressure to return to normal.

An increase in total blood volume also leads to a small increase in blood volume within the atria of the heart, caus-

ing an elevated natriuretic hormone secretion. Natriuretic hormone causes increased urine production by acting on the kidneys and inhibiting ADH secretion.

11	**P R E D I C T**

Explain the differences in mechanisms that regulate blood pressure in response to hemorrhage that results in the rapid loss of a large volume of blood compared with hemorrhage that results in the loss of the same volume of blood but over a period of several hours.

✔ *Answer in Appendix F*

Fluid Shift Mechanism

The **fluid shift mechanism** begins to act within a few minutes but requires hours to achieve its full functional capacity. It occurs in response to changes in the pressures across the capillary walls. As blood pressure increases, some fluid is forced from the blood vessels into the interstitial spaces. The movement of fluid into the interstitial spaces helps prevent the development of very high blood pressures. As blood pressure falls, interstitial fluid moves into the capillaries, which resist a further decline in the

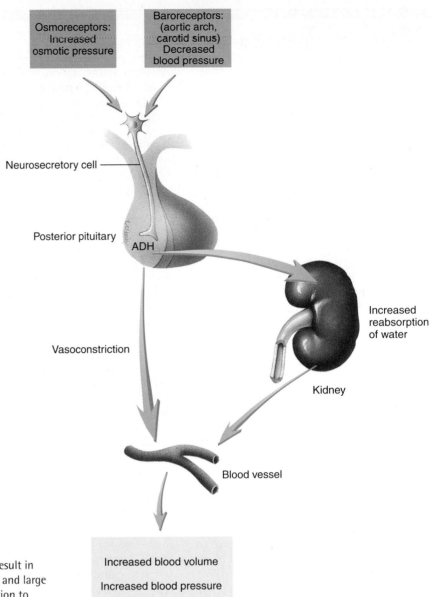

Figure 21.45   The Vasopressin (ADH) Mechanism

Increases in osmolality of blood or decreases in blood pressure result in ADH secretion. ADH increases water reabsorption by the kidney, and large amounts of ADH result in vasoconstriction. These changes function to maintain blood pressure.

blood pressure. The fluid shift mechanism is a powerful method through which blood pressure is maintained because the interstitial fluid volume acts as a reservoir, and it is in equilibrium with the large volume of intracellular fluid. The fluid shift mechanism plays a very important role in the maintenance of blood volume when dehydration develops over several hours.

Stress–Relaxation Response

A **stress–relaxation response** is characteristic of smooth muscle cells (see chapter 10). When blood volume suddenly declines, blood pressure also decreases, causing a reduction in the force applied to smooth muscle cells in the blood vessel walls. As a result, during the next few minutes to an hour the smooth muscle cells contract, reducing the volume of the blood vessels and thus resisting a further decline in blood pressure. Conversely, when the blood volume increases rapidly such as during a transfusion, blood pressure increases and the smooth muscle cells of the blood vessel walls relax, resulting in a more gradual increase in blood pressure. The stress–relaxation mechanism is most effective when changes in blood pressure occur over a period of many minutes.

The mechanisms that regulate blood pressure in the long term are summarized in figure 21.46.

Clinical Focus Shock

Circulatory shock is defined as an inadequate blood flow throughout the body. Failure of mechanisms that function to maintain blood pressure within a normal range of values result in dramatic decreases in blood pressure. As a consequence, tissues can suffer damage as a result of too little delivery of oxygen to cells. Severe circulatory shock can damage vital body tissues to the extent that the individual dies.

Depending on its severity, shock can be divided into three separate stages: (1) the nonprogressive or compensated stage, (2) the progressive stage, and (3) the irreversible stage. All types of circulatory shock exhibit one or more of these stages, regardless of their cause. There are several causes of shock, but hemorrhagic, or hypovolemic, shock is used to illustrate the characteristics of each stage.

In compensated shock, the blood pressure decreases only a moderate amount, and the mechanisms that regulate blood pressure function successfully to reestablish normal blood pressure and blood flow. The baroreceptor reflexes, chemoreceptor reflexes, and ischemia within the medulla oblongata initiate strong sympathetic responses that result in intense vasoconstriction and increased heart rate. As the blood volume decreases, the stress–relaxation response of blood vessels causes the blood vessels to contract and helps sustain blood pressure. In response to reduced blood flow through the kidneys, increased amounts of renin are released. The elevated renin release results in a greater rate of angiotensin II formation, causing vasoconstriction and increased aldosterone release from the adrenal cortex. The aldosterone, in turn, promotes water and salt retention by the kidneys, conserving water. In addition, ADH is released from the posterior pituitary gland and enhances the retention of water by the kidneys. Because of the fluid shift mechanism, water also moves from the interstitial spaces and the intestinal lumen to restore the normal blood volume. An intense sensation of thirst increases water intake, also helping to elevate normal blood volume.

In mild cases of compensated shock, the baroreceptor reflexes can be adequate to compensate for blood loss until the blood volume is restored, but in more severe cases all of the mechanisms described are required to compensate for the blood loss.

In progressive shock, the compensatory mechanisms are inadequate to compensate for the loss of blood volume. As a consequence, a positive-feedback cycle develops in which the blood pressure regulatory mechanisms are unable to compensate for circulatory shock. As circulatory shock worsens, regulatory mechanisms become even less able to compensate for the increasing severity of the circulatory shock. The cycle proceeds until the next stage of shock is reached or until medical treatment is applied that assists the regulatory mechanisms in reestablishing adequate blood flow to tissues.

During progressive shock, the blood pressure declines to a very low level that is inadequate to maintain blood flow to the cardiac muscle; thus the heart begins to deteriorate. Substances that are toxic to the heart are released from tissues that suffer from severe ischemia. When the blood pressure declines to a very low level, blood begins to clot in the small vessels. Eventually blood vessel dilation begins as a result of decreased sympathetic activity and because of the lack of oxygen in capillary beds. Capillary permeability increases under ischemic conditions, allowing fluid to leave the blood vessels and enter the interstitial spaces, and finally intense tissue deterioration begins in response to inadequate blood flow.

Without medical intervention, progressive shock leads to irreversible shock. Irreversible shock leads to death, regardless of the amount or type of medical treatment applied. In this stage of shock, the damage to tissues, including cardiac muscle, is so extensive that the patient is destined to die, even if adequate blood volume is reestablished and blood pressure is elevated to its normal value. Irreversible shock is characterized by decreasing heart function and progressive dilation of and increased permeability of peripheral blood vessels.

Patients suffering from shock are normally placed in a horizontal plane, usually with the head slightly lower than the feet, and oxygen is often supplied. Replacement therapy consists of transfusions of whole blood, plasma, artificial solutions called plasma substitutes, and physiologic saline solutions administered to increase blood volume. In some circumstances, drugs that enhance vasoconstriction are also administered. Occasionally, such as in patients in anaphylactic (an'ă-fĭ-lak'tik; allergic) shock, anti-inflammatory substances such as glucocorticoids and antihistamines are administered. The basic objective in treating shock is to reverse the condition so that progressive shock is arrested, to prevent it from progressing to the irreversible stage, and to reverse the condition so that normal blood flow through tissues is reestablished.

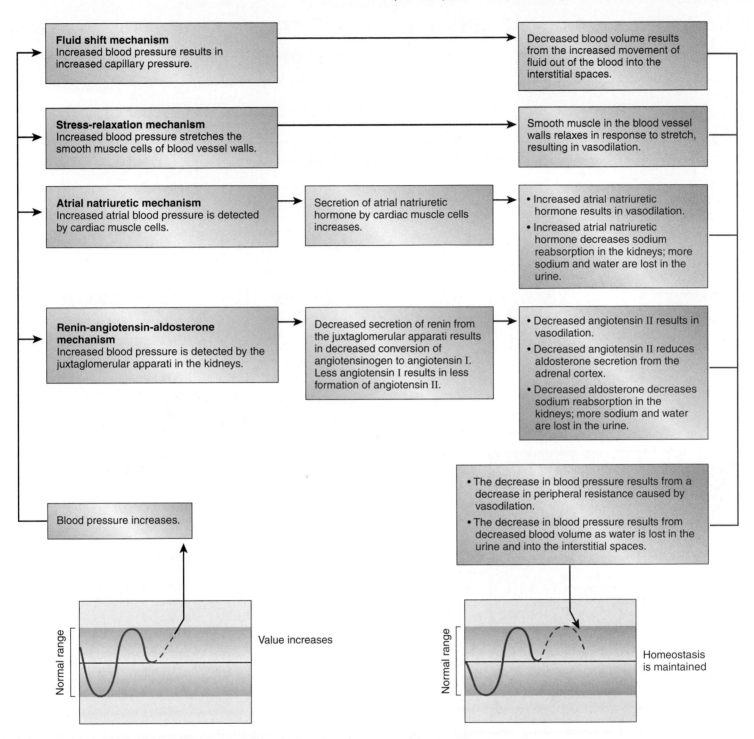

Figure 21.46 Homeostasis: Control of Blood Pressure Long-Term (Slow-Acting) Mechanisms

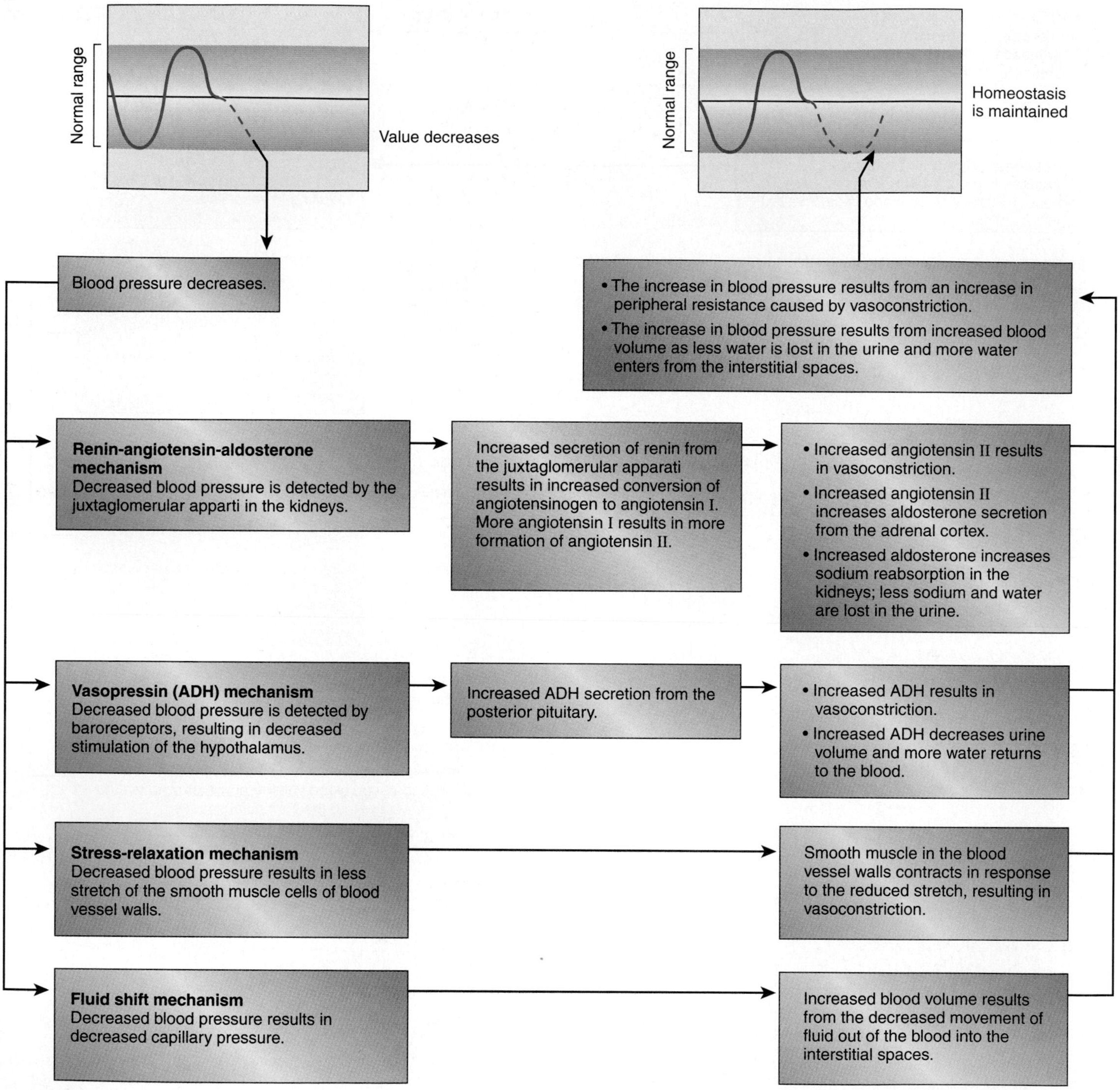

Normal range

Value decreases

Normal range

Homeostasis is maintained

Blood pressure decreases.

• The increase in blood pressure results from an increase in peripheral resistance caused by vasoconstriction.
• The increase in blood pressure results from increased blood volume as less water is lost in the urine and more water enters from the interstitial spaces.

Renin-angiotensin-aldosterone mechanism
Decreased blood pressure is detected by the juxtaglomerular apparti in the kidneys.

Increased secretion of renin from the juxtaglomerular apparati results in increased conversion of angiotensinogen to angiotensin I. More angiotensin I results in more formation of angiotensin II.

• Increased angiotensin II results in vasoconstriction.
• Increased angiotensin II increases aldosterone secretion from the adrenal cortex.
• Increased aldosterone increases sodium reabsorption in the kidneys; less sodium and water are lost in the urine.

Vasopressin (ADH) mechanism
Decreased blood pressure is detected by baroreceptors, resulting in decreased stimulation of the hypothalamus.

Increased ADH secretion from the posterior pituitary.

• Increased ADH results in vasoconstriction.
• Increased ADH decreases urine volume and more water returns to the blood.

Stress-relaxation mechanism
Decreased blood pressure results in less stretch of the smooth muscle cells of blood vessel walls.

Smooth muscle in the blood vessel walls contracts in response to the reduced stretch, resulting in vasoconstriction.

Fluid shift mechanism
Decreased blood pressure results in decreased capillary pressure.

Increased blood volume results from the decreased movement of fluid out of the blood into the interstitial spaces.

Figure 21.46   *(continued)*

Clinical Focus   Types of Shock

Several types of shock are classified here by cause:

- **Hemorrhagic shock**—reduced blood volume caused by either external or internal bleeding.
- **Plasma loss shock**—reduced blood volume results from a loss of plasma into the interstitial spaces and greatly increased blood viscosity.
- **Intestinal obstruction**—results in the movement of a large amount of plasma from the blood into the intestine.
- **Severe burns**—loss of large amounts of plasma from the burned surface.
- **Dehydration**—results from a severe and prolonged shortage of fluid intake.

- **Severe diarrhea or vomiting**—loss of plasma through the intestinal wall.
- **Neurogenic shock**—rapid loss of vasomotor tone that leads to vasodilation so extensive that a severe decrease in blood pressure results.
- **Anesthesia**—deep general anesthesia or spinal anesthesia that decreases the activity of the medullary vasomotor center or the sympathetic nerve fibers.
- **Brain damage**—leads to an ineffective medullary vasomotor function.
- **Emotional shock (vasovagal syncope)**—results from emotions that cause strong parasympathetic stimulation of the heart and results in vasodilation in skeletal muscles and in the viscera.

- **Anaphylactic shock**—results from an allergic response that causes the release of inflammatory substances that increase vasodilation and capillary permeability.
- **Septic shock, or "blood poisoning"**—results from peritoneal, systemic, and gangrenous infections that cause the release of toxic substances into the circulatory system, depressing the activity of the heart, leading to vasodilation, and increasing capillary permeability.
- **Cardiogenic shock**—occurs when the heart stops pumping in response to conditions such as heart attack or electrocution.

Summary

General Features of Blood Vessel Structure

1. Blood flows from the heart through elastic arteries, muscular arteries, and arterioles to the capillaries.
2. Blood returns to the heart from the capillaries through venules, small veins, and large veins.

Capillaries

1. The entire circulatory system is lined with simple squamous epithelium called endothelium. Capillaries consist only of endothelium.
2. Capillaries are surrounded by loose connective tissue, the adventitia, which contains pericapillary cells.
3. There are three types of capillaries.
   - Fenestrated capillaries have pores called fenestrae that extend completely through the cell.
   - Sinusoidal capillaries are large-diameter capillaries with large fenestrae.
   - Continuous capillaries do not have fenestrae.
4. Materials pass through the capillaries in several ways: between the endothelial cells, through the fenestrae, and through the cell membrane.
5. Blood flows from arterioles through metarterioles and then through the capillary network. Venules drain the capillary network.
   - Smooth muscle in the arterioles, metarterioles, and precapillary sphincters regulates blood flow into the capillaries.
   - Blood can pass rapidly through the thoroughfare channel.

Structure of Arteries and Veins

1. Except for capillaries and venules, blood vessels have three layers. The inner tunica intima consists of endothelium, basement membrane, and internal elastic lamina.
   - The tunica media, the middle layer, contains circular smooth muscle and elastic fibers.
   - The outer tunica adventitia is connective tissue.
2. The thickness and the composition of the layers vary with blood vessel type and diameter.
   - Large elastic arteries are thin-walled with large diameters. The tunica media has many elastic fibers and little smooth muscle.
   - Muscular arteries are thick-walled with small diameters. The tunica media has abundant smooth muscle and some elastic fibers.
   - Arterioles are the smallest arteries. The tunica media consists of smooth muscle cells and a few elastic fibers.
   - Venules are composed of endothelium surrounded by a few smooth muscle cells.
   - Small veins are venules covered with a layer of smooth muscle.
   - Medium-sized veins and large veins contain less smooth muscle and fewer elastic fibers than arteries of the same size.
3. Valves prevent the backflow of blood in the veins.
4. Vasa vasorum are blood vessels that supply the tunica adventitia and tunica media.
5. Arteriovenous anastomoses allow blood to flow from arteries to veins without passing through the capillaries. They function in temperature regulation.

Nerves

The smooth muscle of the tunica media is supplied by sympathetic nerve fibers.

Aging of the Arteries

Arteriosclerosis results from a loss of elasticity in the aorta, large arteries, and coronary arteries.

Pulmonary Circulation

The pulmonary circulation moves blood to and from the lungs. The pulmonary trunk arises from the right ventricle and divides to form the pulmonary arteries, which project to the lungs. From the lungs the pulmonary veins return to the left atrium.

Systemic Circulation: Arteries

Aorta

The aorta leaves the left ventricle to form the ascending aorta, aortic arch, and descending aorta (consisting of the thoracic and abdominal aortae).

Coronary Arteries

Coronary arteries supply the heart.

Arteries to the Head and the Neck

1. The brachiocephalic, left common carotid, and left subclavian arteries branch from the aortic arch to supply the head and the upper limbs. The brachiocephalic artery divides to form the right common carotid and the right subclavian arteries. The vertebral arteries branch from the subclavian arteries.
2. The common carotid arteries and the vertebral arteries supply the head.
   - The common carotid arteries divide to form the external carotids, which supply the face and mouth, and the internal carotids, which supply the brain.
   - The vertebral arteries join within the cranial vault to form the basilar artery, which supplies the brain.

Arteries of the Upper Limb

1. The subclavian artery continues (without branching) as the axillary artery and then as the brachial artery. The brachial artery divides into the radial and ulnar arteries.
2. The radial artery supplies the deep palmar arch, and the ulnar artery supplies the superficial palmar arch. Both arches give rise to the digital arteries.

Thoracic Aorta and Its Branches

The thoracic aorta has visceral branches that supply the thoracic organs and parietal branches that supply the thoracic wall.

Abdominal Aorta and Its Branches

1. The abdominal aorta has visceral branches that supply the abdominal organs and parietal branches that supply the abdominal wall.
2. The visceral branches are paired and unpaired. The paired arteries supply the kidneys, adrenal glands, and gonads. The unpaired arteries supply the stomach, spleen, and liver (celiac trunk); the small intestine and upper part of the large intestine (superior mesenteric); and the lower part of the large intestine (inferior mesenteric).

Arteries of the Pelvis

1. The common iliac arteries arise from the abdominal aorta, and the internal iliac arteries branch from the common iliac arteries.
2. The visceral branches of the internal iliac arteries supply pelvic organs, and the parietal branches supply the pelvic wall and floor and the external genitalia.

Arteries of the Lower Limb

1. The external iliac arteries branch from the common iliac arteries.
2. The external iliac artery continues (without branching) as the femoral artery and then as the popliteal artery. The popliteal artery divides to form the anterior and posterior tibial arteries.
3. The posterior tibial artery gives rise to the fibular (peroneal) and plantar arteries. The plantar arteries form the plantar arch from which the digital arteries arise.

Systemic Circulation: Veins

1. The three major veins returning blood to the heart are the superior vena cava (head, neck, thorax, and upper limbs), the inferior vena cava (abdomen, pelvis, and lower limbs), and the coronary sinus (heart).
2. Veins are of three types: superficial, deep, and sinuses.

Veins Draining the Heart

Coronary veins enter the coronary sinus or the right atrium.

Veins of the Head and Neck

1. The internal jugular veins drain the venous sinuses of the anterior head and neck.
2. The external jugular veins and the vertebral veins drain the posterior head and neck.

Veins of the Upper Limb

1. The deep veins are the small ulnar and radial veins of the forearm, which join the brachial veins of the arm. The brachial veins drain into the axillary vein.
2. The superficial veins are the basilic, cephalic, and median cubital. The basilic vein becomes the axillary vein, which then becomes the subclavian vein. The cephalic vein drains into the axillary vein.

Veins of the Thorax

The left and right brachiocephalic veins and the azygos veins return blood to the superior vena cava.

Veins of the Abdomen and Pelvis

1. Ascending lumbar veins from the abdomen join the azygos and hemiazygos veins.
2. Vessels from the kidneys, adrenal gland, and gonads directly enter the inferior vena cava.
3. Vessels from the stomach, intestines, spleen, and pancreas connect with the hepatic portal vein. The hepatic portal vein transports blood to the liver for processing. Hepatic veins from the liver join the inferior vena cava.

Veins of the Lower Limb

1. The deep veins are the fibular (peroneal), anterior and posterior tibials, popliteal, femoral, and external iliac.
2. The superficial veins are the small and great saphenous veins.

Lymphatic Vessels

1. Lymphatic vessels carry lymph away from tissues.
2. Lymphatic capillaries lack a basement membrane and have loosely overlapping epithelial cells. Fluids and other substances easily enter the lymph capillary.
3. Lymphatic vessels are formed by the joining of lymph capillaries.

- Lymphatic vessels have valves that ensure one-way flow of lymph.
- Skeletal muscle action, contraction of lymphatic vessel smooth muscle, and thoracic pressure changes move the lymph.

4. Lymph nodes are along the lymphatic vessels from the abdomen and lower limbs, the left thorax, the upper-left limb, and the left side of the head and the neck. The expanded end of the thoracic duct is the cisterna chyli. The thoracic duct empties into the left subclavian vein.

5. The right lymphatic duct receives lymphatic vessels from the right thorax, the upper-right limb, and the right side of the head and the neck. The right lymphatic duct empties into the right subclavian vein.

Physics of Circulation
Laminar and Turbulent Flow in Vessels

Blood flow through vessels normally is streamlined, or laminar. Turbulent flow is disruption of laminar flow.

Blood Pressure

1. Blood pressure is a measure of the force exerted by blood against the blood vessel wall. Blood moves through vessels because of blood pressure.
2. Blood pressure can be measured by listening for Korotkoff sounds produced by turbulent flow in arteries as pressure is released from a blood pressure cuff.

Rate of Blood Flow

Blood flow is the amount of blood that moves through a vessel in a given period. Blood flow is directly proportional to pressure differences and is inversely proportional to resistance.

Poiseuille's Law

Resistance is the sum of all the factors that inhibit blood flow. Resistance increases when viscosity increases and when blood vessels become smaller in diameter or longer.

Viscosity

1. Viscosity is the resistance of a liquid to flow. Most of the viscosity of blood results from erythrocytes.
2. The viscosity of blood increases when the hematocrit increases.

Critical Closing Pressure and Laplace's Law

1. As pressure in a vessel decreases, the force holding it open decreases, and the vessel tends to collapse. The critical closing pressure is the pressure at which a blood vessel closes.
2. Laplace's law states that the force acting on the wall of a blood vessel is proportional to the diameter of the vessel times the blood pressure.

Vascular Compliance

Vascular compliance is a measure of the change in volume of blood vessels produced by a change in pressure. The venous system has a large compliance and acts as a blood reservoir.

Physiology of Systemic Circulation

The greatest volume of blood is contained in the veins. The smallest volume is in the arterioles.

Cross-Sectional Area of Blood Vessels

As the diameter of vessels decreases, their total cross-sectional area increases, and the velocity of blood flow through them decreases.

Pressure and Resistance

Blood pressure averages 100 mm Hg in the aorta and drops to 0 mm Hg in the right atrium. The greatest drop occurs in the arterioles, which regulate blood flow through tissues.

Pulse Pressure

1. Pulse pressure is the difference between systolic and diastolic pressures. Pulse pressure increases when stroke volume increases or vascular compliance decreases.
2. Pulse pressure waves travel through the vascular system faster than the blood flows. Pulse pressure can be used to take the pulse.

Capillary Exchange

1. Blood pressure, capillary permeability, and osmosis affect movement of fluid from the capillaries.
2. There is a net movement of fluid from the blood into the tissues. The fluid gained by the tissues is removed by the lymphatic system.

Functional Characteristics of Veins

Venous return to the heart increases because of an increase in blood volume, venous tone, and arteriole dilation.

Blood Pressure and the Effect of Gravity

In a standing person hydrostatic pressure caused by gravity increases blood pressure below the heart and decreases pressure above the heart.

Control of Blood Flow in Tissues
Local Control of Blood Flow by the Tissues

1. Blood flow through a tissue is usually proportional to the metabolic needs of the tissue. Exceptions are tissues that perform functions that require additional blood.
2. Control of blood flow by the metarterioles and precapillary sphincters can be regulated by vasodilator substances or by lack of nutrients.
3. Only large changes in blood pressure have an effect on blood flow through tissues.
4. If the metabolic activity of a tissue increases, the number and the diameter of capillaries in the tissue increases over time.

Nervous And Hormonal Regulation of Local Circulation

1. The sympathetic nervous system (vasomotor center in the medulla) controls blood vessel diameter. Other brain areas can excite or inhibit the vasomotor center.
2. Vasomotor tone is a state of partial contraction of blood vessels.
3. The nervous system is responsible for routing the flow of blood and maintaining blood pressure.
4. Sympathetic action potentials stimulate epinephrine and norepinephrine release from the adrenal medulla, and these hormones cause vasoconstriction in most blood vessels.

Regulation of Mean Arterial Pressure

Mean blood pressure is proportional to cardiac output times the peripheral resistance.

Short-Term Regulation of Blood Pressure

1. Baroreceptors are sensory receptors sensitive to stretch.
   - Baroreceptors are located in the carotid sinuses and the aortic arch.
   - The baroreceptor reflex changes peripheral resistance, heart rate, and stroke volume in response to changes in blood pressure.
2. Chemoreceptors are sensory receptors sensitive to oxygen, carbon dioxide, and pH levels in the blood.
3. The CNS ischemic response results from high carbon dioxide or low pH levels in the medulla and increases peripheral resistance.
4. Epinephrine and norepinephrine are released from the adrenal medulla as a result of sympathetic stimulation. They increase heart rate, stroke volume, and vasoconstriction.
5. Renin is released by the kidneys in response to low blood pressure. Renin promotes the production of angiotensin II, which causes vasoconstriction and an increase in aldosterone secretion.
6. ADH released from the posterior pituitary causes vasoconstriction.
7. Atrial natriuretic hormone is released from the heart when atrial blood pressure increases. It stimulates an increase in urinary production, causing a decrease in blood volume and blood pressure.
8. Fluid shift is a movement of fluid from the interstitial spaces to maintain blood volume.
9. The stress–relaxation response is an adjustment of the smooth muscles of blood vessels in response to a change in blood volume.

Long-Term Regulation of Blood Pressure

1. The kidneys regulate blood pressure by controlling blood volume.
2. In response to an increase in blood volume, the kidneys produce more urine and decrease blood volume. Renin, angiotensin II, aldosterone, vasopressin, atrial natriuretic hormone, and sympathetic stimulation play a role in controlling urinary volume.
3. Fluid shift and stress–relaxation responses help control blood pressure.

Content Review

1. Name, in order, all the types of blood vessels, starting at the heart, going to the tissues, and returning to the heart.
2. Describe the three types of capillaries. Explain the ways that materials pass through the capillary wall.
3. Describe a capillary network. Where is the smooth muscle that regulates blood flow into and through the capillary network located? What is the function of the thoroughfare channel?
4. Name the three layers of a blood vessel. What kinds of tissue are in each layer?
5. For the different types of arterial and venous blood vessels, compare the amount of elastic fibers and smooth muscle in each.
6. What is the function of valves in blood vessels? In which blood vessels are they found?
7. Define the terms vasa vasorum and arteriovenous anastomoses, and give their function.
8. Name the different parts of the aorta. Name the major arteries that branch from the aorta to supply the heart, the head and upper limbs, and the lower limbs.
9. What areas of the body are supplied by the paired arteries that branch from the abdominal aorta? The unpaired arteries?
10. Name the three major vessels that return blood to the heart. What areas of the body do they drain?
11. Name the three major veins that return blood to the superior vena cava.
12. Explain the three ways that blood from the abdomen returns to the heart.
13. List the major deep and superficial veins of the upper and lower limbs.
14. Describe the structure of a lymph capillary. Explain why the structure makes it easy for fluid and other substances to enter the capillary.
15. What is the function of the valves in lymph vessels? What causes lymph to move through the lymph vessels?
16. What parts of the body are drained by the thoracic duct and the right lymphatic duct? What is the cisterna chyli?
17. Define the term viscosity, and state the effect of hematocrit on viscosity. Define the terms laminar flow and turbulent flow.
18. Define the terms blood pressure, blood flow, and resistance. How can each be determined?
19. State Poiseuille's law. What effect do viscosity, blood vessel diameter, and blood vessel length have on resistance? On blood flow?
20. State Laplace's law. How does it explain critical closing pressure and aneurysms?
21. Define the term vascular compliance. Do veins or arteries have the greater compliance?
22. Describe the distribution of blood volumes throughout the circulatory system.
23. What is the relationship between blood vessel diameter, total cross-sectional area, and blood flow velocity?
24. Describe the changes in blood pressure, starting in the aorta, moving through the vascular system, and returning to the right atrium.
25. What is pulse pressure? How do stroke volume and vascular compliance affect pulse pressure?
26. Describe the factors that influence the movement of fluid from capillaries into the tissues. What happens to the fluid in the tissues? What is edema?
27. How do blood volume and tone in large blood vessels and arterioles affect cardiac output?
28. What effect does standing have on blood pressure in the feet and the head? Explain why this effect occurs.
29. Explain how vasodilator substances and nutrients are involved with local control of blood flow. What is autoregulation of local blood flow? How is long-term regulation of blood flow through tissues accomplished?
30. Describe nervous control of blood flow. Define the term vasomotor tone.

31. Where are baroreceptors located? Describe the response of the baroreceptor reflex when blood pressure increases and decreases.
32. Where are the chemoreceptors for carbon dioxide, pH changes, and oxygen located? Describe what happens when oxygen levels in the blood decrease.
33. Describe the CNS ischemic response.
34. For each of the following hormones—epinephrine, norepinephrine, renin, angiotensin, aldosterone, antidiuretic hormone, and atrial natriuretic hormone—state where the hormone is produced and what effects it has on the circulatory system.
35. What is fluid shift, and what does it accomplish? Describe the stress–relaxation response of a blood vessel.
36. Discuss two ways that the kidneys are involved in the long-term regulation of blood pressure.

Develop Your Reasoning Skills

1. For each of the following destinations, name all the arteries that an erythrocyte would encounter if it started its journey in the left ventricle.
   a. Posterior interventricular groove of the heart
   b. Anterior neck to the brain (give two ways)
   c. Posterior neck to the brain (give two ways)
   d. External skull
   e. Tip of the fingers of the left hand (What other blood vessel would be encountered if the trip were through the right upper limb?)
   f. Anterior compartment of the leg
   g. Liver
   h. Small intestine
   i. Urinary bladder
2. For each of the following starting places, name all the veins that an erythrocyte would encounter on its way back to the right atrium.
   a. Anterior interventricular groove of the heart (give two ways)
   b. Venous sinus near the brain
   c. External posterior of skull
   d. Hand (return deep and superficial)
   e. Foot (return deep and superficial)
   f. Stomach
   g. Kidney
   h. Left inferior wall of the thorax
3. In a study of heart valve functions, it is necessary to inject a dye into the right atrium of the heart by inserting a catheter into a blood vessel and moving the catheter into the right atrium. What route would you suggest? If you wanted to do this procedure into the left atrium, what would you do differently?
4. In endurance-trained athletes, the hematocrit can be lower than normal because plasma volume increases more than erythrocyte numbers increase. Explain why this condition would be beneficial.
5. All the blood that passes through the aorta, except the blood that flows into the coronary vessels, returns to the heart through the venae cavae. (*Hint:* The diameter of the aorta is 26 mm, and the diameter of a vena cava is 32 mm.) Explain why the resistance to blood flow in the aorta is greater than the resistance to blood flow in the venae cavae. Because the resistances are different, explain why blood flow can be the same.
6. As blood vessels increase in diameter, the amount of smooth muscle decreases, and the amount of connective tissue increases. Explain why. (*Hint:* Remember Laplace's law.)
7. A patient is suffering from edema in the lower right limb. Explain why massage would help remove the excess fluid.
8. A very short nursing student was asked to measure the blood pressure of a very tall person. She decides to measure the blood pressure at the level of his foot while the tall person is standing. What artery does she use? After taking the blood pressure, she decides that the tall person is suffering from hypertension because the systolic pressure was 200 mm Hg. Is her diagnosis correct? Why or why not?
9. During hyperventilation, carbon dioxide is "blown off," and carbon dioxide levels in the blood decrease. What effect does this decrease have on blood pressure? Explain. What symptoms do you expect to see as a result?
10. Epinephrine causes vasodilation of blood vessels in cardiac muscle but vasoconstriction of blood vessels in the skin. Explain why this is a beneficial arrangement.
11. One cool evening Skinny Dip jumps into a hot Jacuzzi. Predict what happens to Skinny's heart rate.

Web Site Link

For a listing of the most current web sites related to this chapter, please visit the Seeley home page at:
http://www.mhhe.com/biosci/ap/seeleyap/

Chapter Twenty-Two

Lymphatic System and Immunity

Objectives

1. Describe the functions of the lymphatic system.

2. Describe the structure and functions of diffuse lymphatic tissue, lymph nodules, lymph nodes, tonsils, spleen, and thymus gland.

3. Define the term innate immunity, and describe the cells and chemicals involved.

4. Name the phagocytic cells of the immune system, and describe their location and activities.

5. List the events that occur during an inflammatory response, and explain their significance.

6. Define the term antigen.

7. Describe the origin, development, activation, and inhibition of lymphocytes.

8. Define the term antigenic determinant, and describe antigen receptors.

9. Explain the importance of the major histocompatibility complex antigens and costimulation.

10. Define the terms antibody–mediated immunity and cell–mediated immunity, and name the cells responsible for each.

11. Describe the structure of an antibody, list the different types of antibodies, and describe the effects produced by them.

12. Discuss the primary and secondary response to an antigen. Explain the basis for long-lasting immunity.

13. Describe the functions of T cells.

14. Explain how innate, antibody-mediated, and cell-mediated immunity can function together to eliminate an antigen.

15. Define and give examples of immunotherapy.

16. Explain four ways by which adaptive immunity can be acquired.

One of the basic themes of life is that many organisms consume or use other organisms to survive. Some microorganisms, such as certain bacteria or viruses, use humans as a source of nutrients and as a sheltered environment where they can survive and reproduce. As a result, they can damage the body, causing disease and sometimes death. Not surprisingly, our bodies have ways to resist or destroy harmful microorganisms. This chapter considers how the lymphatic system continually protects us against invading microorganisms.

Lymphatic System

The **lymphatic** (lim-fat'ik) **system** includes lymph, lymphocytes, lymphatic vessels, lymph nodules, lymph nodes, tonsils, the spleen, and the thymus gland (figure 22.1).

Functions of the Lymphatic System

The lymphatic system helps to maintain fluid balance in tissues and absorb fats from the digestive tract. It is also part of the body's defense system against microorganisms and other harmful substances.

1. *Fluid balance.* Approximately 30 L of fluid passes from the blood capillaries into the interstitial spaces each day, whereas only 27 L passes from the interstitial spaces back into the blood capillaries. If the extra 3 L of interstitial fluid was to remain in the interstitial spaces, edema would result, causing tissue damage and eventual death. Instead, this 3 L of fluid enters the lymphatic capillaries, where the fluid is called **lymph** (limf, meaning clear spring water), and passes through the lymphatic vessels back to the blood (see chapter 21). In addition to water, lymph contains solutes derived from two sources: (1) substances in plasma, such as ions, nutrients, gases, and some proteins, pass from blood capillaries into the interstitial spaces and become part of the lymph; and (2) substances derived from cells, such as hormones, enzymes, and waste products, are also found in the lymph.
2. *Fat absorption.* The lymphatic system absorbs fats and other substances from the digestive tract (see chapter 24). Special lymphatic vessels called **lacteals** (lak'tē-ălz) are located in the lining of the small intestine. Fats enter the lacteals and pass through the lymphatic vessels to the venous circulation. The lymph passing through these lymphatic vessels has a milky appearance because of its fat content and is called **chyle** (kīl).
3. *Defense.* Microorganisms and other foreign substances are filtered from lymph by lymph nodes and from blood by the spleen. In addition, lymphocytes and other cells are capable of destroying microorganisms and other foreign substances.

Lymphatic Organs

Lymphatic organs contain **lymphatic tissue,** which consists primarily of lymphocytes; but it also includes macrophages, dendritic cells, reticular cells, and other cell types. **Lymphocytes** are a type of white blood cell (see chapter 19). The lymphocytes originate from red bone marrow and are carried by the blood to lymphatic organs and other tissues, where they are part of the immune response that destroys microorganisms and foreign substances. When the body is exposed to microorganisms or foreign substances, the lymphocytes divide and increase in number. Lymphatic tissue also has very fine collagen fibers, called **reticular fibers,** which are produced by **reticular cells.** The lymphocytes and other cells attach to these fibers. When lymph or blood filters through lymphatic organs, the fiber network traps microorganisms and other particles in the fluid.

Diffuse Lymphatic Tissue and Lymph Nodules

Diffuse lymphatic tissue contains dispersed lymphocytes, macrophages, and other cells; has no clear boundary; and blends with surrounding tissues (figure 22.2). It is located deep to mucous membranes, around lymph nodules, and within the spleen.

Lymph nodules are denser arrangements of lymphoid tissue organized into compact, somewhat spherical structures, ranging in size from a few hundred microns to a few millimeters or more in diameter (see figure 22.2). Lymph nodules are numerous in the loose connective tissue of the digestive, respiratory, and urinary systems. In lymph nodes and the spleen, lymph nodules are usually referred to as lymph follicles. In some locations many lymph nodules join together to form larger structures. For example, Peyer's patches are aggregations of lymph nodules found in the distal half of the small intestine and the appendix.

Tonsils

Tonsils are unusually large groups of lymph nodules and diffuse lymphatic tissue located deep to the mucous membranes within the oral cavity and the nasopharynx (back of the throat) (figure 22.3; see figure 23.2). They form a protective ring of lymphatic tissue around the openings between the nasal and oral cavities and the pharynx (throat). The tonsils provide protection against bacteria and other potentially harmful material in the nose and mouth. In adults the tonsils decrease in size and eventually may disappear.

Of the three groups of tonsils, the **palatine tonsils** usually are referred to as "the tonsils." They are relatively large, oval lymphoid masses on each side of the junction between the oral cavity and the pharynx. The **pharyngeal** (fă-rin'jē-ăl) **tonsil,** or adenoid (ad'ĕ-noyd), is a collection of somewhat closely aggregated lymph nodules near the junction between

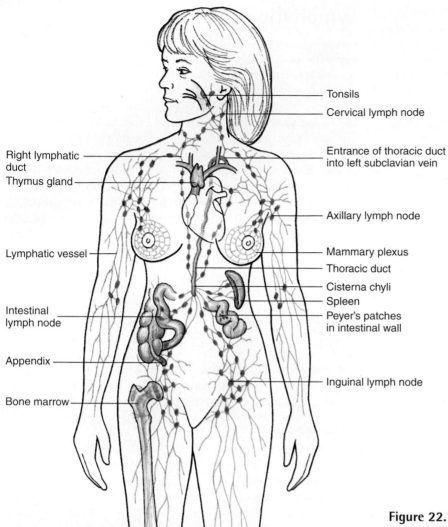

Tonsils

Cervical lymph node

Right lymphatic duct

Entrance of thoracic duct into left subclavian vein

Thymus gland

Axillary lymph node

Lymphatic vessel

Mammary plexus

Thoracic duct

Cisterna chyli

Spleen

Peyer's patches in intestinal wall

Intestinal lymph node

Appendix

Inguinal lymph node

Bone marrow

Figure 22.1   Lymphatic System
The major lymphatic organs and vessels are shown.

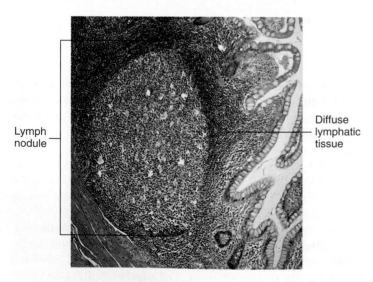

Lymph nodule

Diffuse lymphatic tissue

Figure 22.2   Lymph Nodule
Diffuse lymphatic tissue surrounding lymph nodules.

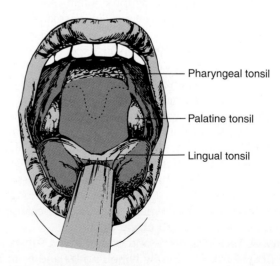

Pharyngeal tonsil

Palatine tonsil

Lingual tonsil

Figure 22.3   Location of the Tonsils
Anterior view through the oral cavity with part of the hard and soft palates removed (*dashed line*) to show the pharyngeal tonsil.

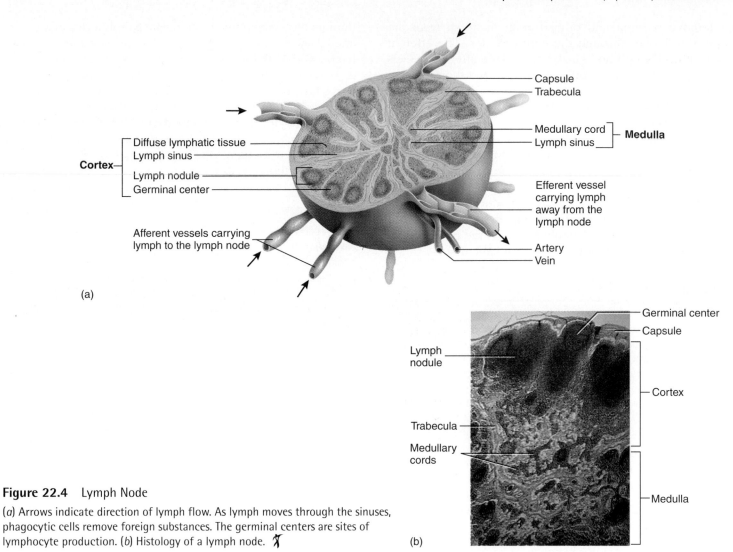

Cortex—
- Diffuse lymphatic tissue
- Lymph sinus
- Lymph nodule
- Germinal center

Afferent vessels carrying
lymph to the lymph node

Capsule
Trabecula

Medullary cord ⎫ **Medulla**
Lymph sinus ⎭

Efferent vessel
carrying lymph
away from the
lymph node

Artery
Vein

(a)

Germinal center
Capsule

Lymph
nodule

Cortex

Trabecula

Medullary
cords

Medulla

(b)

Figure 22.4   Lymph Node

(*a*) Arrows indicate direction of lymph flow. As lymph moves through the sinuses, phagocytic cells remove foreign substances. The germinal centers are sites of lymphocyte production. (*b*) Histology of a lymph node.

the nasal cavity and the pharynx. An enlarged pharyngeal tonsil can interfere with normal breathing. The **lingual tonsil** is a loosely associated collection of lymph nodules on the posterior surface of the tongue.

Sometimes the palatine or pharyngeal tonsils become chronically infected and must be removed. The lingual tonsil becomes infected less often than the other tonsils and is more difficult to remove.

Lymph Nodes

Lymph nodes are small, round, or bean-shaped structures, ranging in size from 1 to 25 mm long, and are distributed along the course of the lymphatic vessels (figure 22.4; see figure 22.1). They filter the lymph, removing bacteria and other materials. In addition, lymphocytes congregate, function, and proliferate within lymph nodes.

Lymph nodes are found throughout the body. Three superficial aggregations of lymph nodes exist on each side of the body: the inguinal nodes in the groin, the axillary nodes in the axillary (armpit) region, and the cervical nodes of the neck. Aggregations of lymph nodes are also in the connective tissue

around the intestines and along large blood vessels in the thorax and abdomen.

Lymph nodes are surrounded by a dense connective tissue **capsule.** Extensions of the capsule, called **trabeculae** (tră-bek′yū-lē), form a delicate internal skeleton in the lymph node. Reticular fibers extend from the capsule and trabeculae to form a fibrous network throughout the entire node. In some areas of the lymph node, lymphocytes and macrophages are packed around the reticular fibers to form lymphatic tissue, and in other areas the reticular fibers extend across open spaces called **lymph sinuses.** The lymphatic tissue and sinuses within the node are arranged into two somewhat indistinct layers. The outer **cortex** consists of lymph nodules separated by diffuse lymphatic tissue, trabeculae, and lymph sinuses. The inner **medulla** is organized into branching, irregular strands of diffuse lymphatic tissue, the **medullary cords,** separated by sinuses.

Lymph nodes are the only structures to filter lymph and to have both efferent and afferent lymphatic vessels. Lymph enters the lymph nodes through **afferent lymphatic vessels** and exits through **efferent lymphatic vessels.** The lymph sinuses of each node are lined with phagocytic cells that remove bacteria and other foreign material from the lymph as it moves through the lymph nodes. The efferent vessels of one lymph node may become the afferent vessels of another node or may converge toward the thoracic duct or right lymphatic duct (see chapter 21).

Cells of the lymph nodes consist primarily of lymphocytes, macrophages, and reticular cells. Microorganisms or other foreign substances in the lymph can stimulate lymphocytes throughout the lymph node to undergo cell division, with proliferation especially evident in the lymph nodules of the cortex. These areas of rapid lymphocyte division are called **germinal centers.** The newly produced lymphocytes are released into the lymph and eventually reach the bloodstream, where they circulate. Subsequently, the lymphocytes can leave the blood and enter other lymphatic tissues.

Clinical Note

Cancer cells can spread from a tumor site to the lymphatic system, where they are trapped in the lymph nodes. If the cancer cells escape from the lymph nodes, they may pass through the lymphatic system to the blood and eventually reach other parts of the body. During cancer surgery, malignant (cancerous) lymph nodes are often removed, and their vessels are tied off and cut to prevent the spread of the cancer.

Spleen

The **spleen,** which is roughly the size of a clenched fist, is located on the left side in the extreme superior, posterior part of the abdominal cavity (figure 22.5). It has a fibrous **capsule** with **trabeculae** extending from the capsule into the tissue of the spleen, and it has an internal network of reticular fibers. The spleen contains two types of lymphatic tissue: **red pulp** and **white pulp.** White pulp is associated with the arterial supply to the spleen, and red pulp is associated with the venous supply.

The splenic (splen′ik) arteries enter the spleen at the **hilum,** and their branches follow the various trabeculae into the spleen. Arterial branches leave the trabeculae and subdivide, eventually forming arterioles. Small arteries and arterioles are surrounded by the white pulp, which consists of diffuse lymphatic tissue and lymph nodules, which resemble those in the cortex of lymph nodes. The diffuse lymphatic tissue around the artery is called the **periarterial sheath.** The arterioles branch to form capillaries that supply the red pulp, which consists of the splenic cords and venous sinuses. The **splenic cords** are a network of reticular fibers filled with blood cells that have come from the capillaries, and the **venous sinuses** are enlarged blood vessels between the splenic cords. The venous sinuses unite to form veins that eventually return to the trabeculae and leave the spleen as splenic veins.

The spleen detects and responds to foreign substances in the blood, destroys worn-out erythrocytes, and acts as a blood reservoir. Foreign substances in the blood passing

through the white pulp can stimulate lymphocytes in the periarterial sheath or the lymph nodules in the same manner as in lymph nodes. Before blood leaves the spleen through veins, it passes into the red pulp. Macrophages in the red pulp remove foreign substances and worn-out erythrocytes through phagocytosis. In emergency situations such as hemorrhage, smooth muscle in splenic blood vessels and in the splenic capsule contract in response to sympathetic stimulation. The result is the movement of a small amount of blood from the spleen into the general circulation.

Clinical Note

Although the spleen is protected by the ribs, it can be ruptured in traumatic abdominal injuries. Injury to the spleen can cause severe bleeding, shock, and possibly death. A **splenectomy** (splē-nek′tō-mē), removal of the spleen, is performed to stop the bleeding. The liver and other lymphatic tissues can compensate for loss of the functions of the spleen.

Thymus Gland

The **thymus gland** is a roughly triangular bilobed gland (figure 22.6) located primarily in the superior mediastinum, deep to the manubrium of the sternum. The size of the thymus gland differs markedly, depending on the age of the individual. In a newborn the thymus gland can extend halfway down the length of the thorax. It continues to grow until puberty, although not as rapidly as other structures of the body. After puberty it gradually decreases in size, and in older adults the thymus gland can be so small that it is difficult to find during dissection.

Each lobe of the thymus gland is surrounded by a thin connective tissue **capsule. Trabeculae** extend from the capsule into the substance of the gland, dividing it into **lobules.** Lymphocytes are concentrated near the capsule or trabeculae of each lobule and constitute the cortex. The relatively lymphocyte-free core of each lobule is the medulla. The medulla contains rounded epithelial structures, called **thymic corpuscles** (Hassall's corpuscles), whose function is unknown.

Unlike other lymphatic tissues, the thymus gland has few reticular fibers. Instead, reticular cells have long, branching processes that join to form an interconnected network of cells. In the cortex the reticular cells surround capillaries to form a **blood–thymic barrier,** which prevents large molecules from leaving the blood and entering the cortex.

The thymus gland produces lymphocytes, which then move to other lymphatic tissues, where they can respond to foreign substances. Lymphocytes within the thymus gland do not respond to foreign substances because the blood–thymic barrier prevents the entry of foreign substances into the thymus. Large numbers of lymphocytes are produced in the thymus gland, but most degenerate. The lymphocytes that survive are capable of reacting to foreign substances, but they normally do not react to and destroy healthy body cells (see section on the

Figure 22.5   Spleen

(*a*) Inferior view of the spleen. (*b*) Section showing the arrangement of arteries, veins, white pulp, and red pulp. White pulp is associated with arteries, and red pulp is associated with veins. (*c*) Histology of spleen.

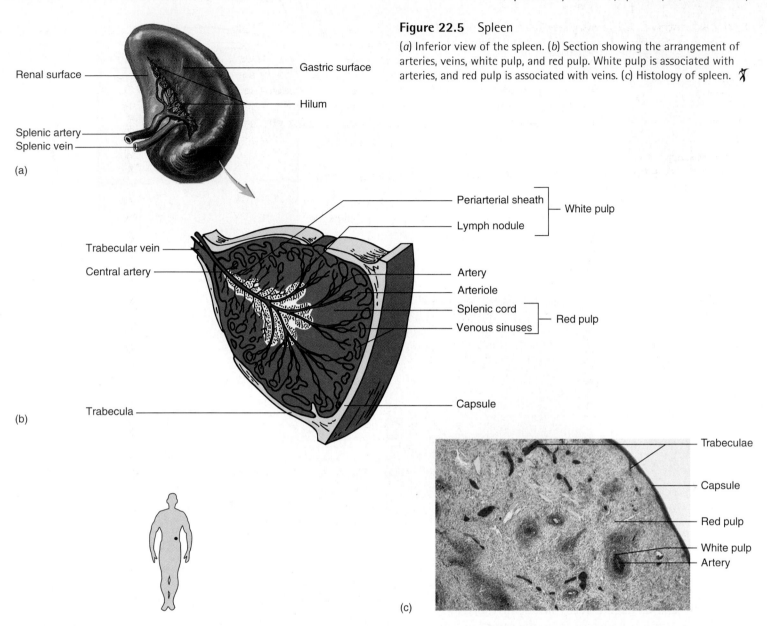

Origin and Development of Lymphocytes later in this chapter). These surviving thymic lymphocytes migrate to the medulla, enter the blood, and travel to other lymphatic tissues.

Immunity

Immunity is the ability to resist damage from foreign substances such as microorganisms and harmful chemicals such as toxins released by microorganisms. Immunity is categorized as **innate immunity** (also called nonspecific resistance) or **adaptive immunity** (also called specific immunity). In innate immunity, the body is born with the ability to recognize and destroy certain foreign substances, but the ability to destroy them does not improve each time the body is exposed to them. In adaptive immunity, the body's ability to recognize and destroy foreign substances improves each time the foreign substance is encountered.

The distinction between innate immunity and adaptive immunity involves the concepts of specificity and memory. **Specificity** is the ability of the immune system to recognize a particular substance. For example, innate immunity can act against bacteria in general, whereas adaptive immunity can distinguish among different kinds of bacteria. **Memory** is the ability of the immune system to remember previous encounters with a particular substance and, as a result, to respond to it more rapidly.

In innate immunity, each time the body is exposed to a substance, the response is the same because there is no specificity and memory of previous encounters. For example, each time a bacterial cell is introduced into the body, it is phagocytized with the same speed and efficiency. In adaptive immunity, the response during the second exposure is faster and stronger than the response to the first exposure because the immune system remembers the bacteria from the first exposure. For example, following initial exposure to the bacteria,

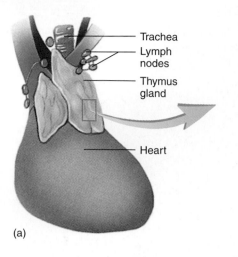

(a)

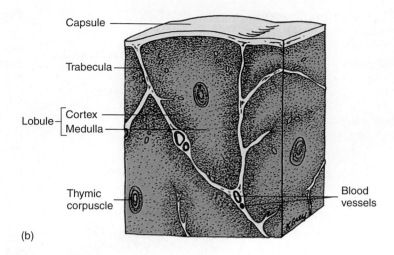

(b)

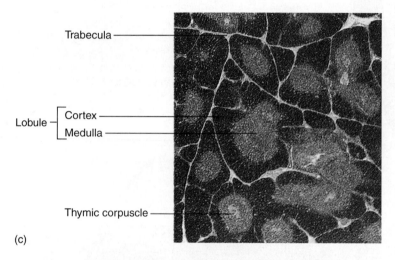

(c)

Figure 22.6   Thymus Gland

(*a*) Location and shape of the thymus gland. (*b*) Section showing a thymic lobule. (*c*) Histology of the thymus gland, showing outer cortex and inner medulla. ✗

the body can take many days to destroy them. During this time the bacteria damage tissues, producing the symptoms of disease. After the second exposure to the same bacteria, however, the response is very rapid and effective. Bacteria are destroyed before any symptoms develop, and the person is said to be **immune.** Adaptive immunity is possible because of the immune system's ability to recognize and remember a particular substance.

Innate Immunity

The main components of innate immunity include (1) mechanical mechanisms that prevent the entry of microbes into the body or that physically remove them from body surfaces; (2) chemical mediators that act directly against microorganisms or that activate other mechanisms, leading to the destruction of the microorganisms; (3) cells involved in phagocytosis and the production of chemicals that participate in the response of the immune system; and (4) inflammation that mobilizes the immune system and isolates microorganisms until they can be destroyed.

Mechanical Mechanisms

Mechanical mechanisms such as the skin and mucous membranes form barriers that prevent the entry of microorganisms and chemicals into the tissues of the body. They also remove microorganisms and other substances from the surface of the body in several ways. The substances are washed from the eyes by tears, from the mouth by saliva, and from the urinary tract by urine. In the respiratory tract, ciliated mucous membranes sweep microbes trapped in the mucus to the back of the throat, where they are swallowed. Coughing and sneezing also remove microorganisms from the respiratory tract. Microorganisms cannot cause disease if they cannot get into the body.

Chemical Mediators

Chemical mediators are molecules involved with the development of immunity (table 22.1). Some chemical mediators found on the surface of cells, such as lysozyme, sebum, and mucus, kill microorganisms or prevent their entry into the cells. Other chemical mediators, such as histamine, complement,

Clinical Focus   Disorders of the Lymphatic System

It is not surprising that many infectious diseases produce symptoms associated with the lymphatic system, because the lymphatic system is involved with the production of lymphocytes that fight infectious diseases, and the lymphatic system filters blood and lymph to remove microorganisms. **Lymphadenitis** (lim-fad′ĕ-nī′tis) is an inflammation of the lymph nodes, which causes them to become enlarged and tender. This inflammation is an indication that microorganisms are being trapped and destroyed within the lymph nodes. Sometimes the lymphatic vessels become inflamed to produce **lymphangitis** (lim-fan-jī′tis). This often results in visible red streaks in the skin that extend away from the site of infection. If the microorganisms pass through the lymphatic vessels and nodes to reach the blood, **septicemia** (sep-ti-sē′mē-ă), or blood poisoning, can result (see chapter 19).

Bubonic plague and elephantiasis are diseases of the lymphatic system. **Bubonic** (bū-bon′ik) **plague** is caused by bacteria (*Yersinia pestis*) that are transferred to humans from rats by the bite of the rat flea (*Xenopsylla*). The bacteria localize in the lymph nodes, causing them to enlarge. The term bubonic is derived from a Greek word referring to the groin because the disease often causes the inguinal lymph nodes of the groin to swell. Without treatment, the bacteria enter the blood, multiply, and infect tissues throughout the body, rapidly causing death in 70%–90% of those infected. In the sixth, fourteenth, and nineteenth centuries the bubonic plague killed large numbers of people. Because of improved sanitation and the advent of antibiotics there are fortunately relatively few cases today. **Elephantiasis** (el-ĕ-fan-tī′ă-sis) is caused by long, slender roundworms (*Wuchereria bancrofti*). The adult worms lodge in the lymphatic vessels and block lymph flow. The resulting accumulation of fluid in the interstitial spaces and lymphatic vessels can cause permanent swelling and enlargement of a limb. The affected limb supposedly resembles an elephant's leg, providing the basis for the name of the disease. The offspring of the adult worms pass through the lymphatic system into the blood, from which they can be transferred to another human by mosquitoes.

A **lymphoma** (lim-fō′mă) is a neoplasm (tumor) of lymphatic tissue. Lymphomas are usually divided into two groups: (1) Hodgkin's disease, and (2) all other lymphomas, which are called non-Hodgkin's lymphomas. Typically, lymphomas begin as an enlarged, painless mass of lymph nodes. The immune system is depressed, and the patient has an increased susceptibility to infections. Enlargement of the lymph nodes can also compress surrounding structures and produce complications. Fortunately, treatment with drugs and radiation is effective for many people who suffer from lymphoma.

Table 22.1   Chemicals of Innate Immunity and Their Functions

Chemical	Description	Chemical	Description
Surface chemicals	Lysozymes (in tears, saliva, nasal secretions, and sweat) lyse cells; acid secretions (sebum in the skin and hydrochloric acid in the stomach) prevent microbial growth or kill microorganisms; mucus on the mucous membranes traps microorganisms until they can be destroyed	Complement	A group of plasma proteins that increase vascular permeability, stimulate the release of histamine, activate kinins, lyse cells, promote phagocytosis, and attract neutrophils, monocytes, macrophages, and eosinophils
Histamine	An amine released from mast cells, basophils, and platelets; histamine causes vasodilation, increases vascular permeability, stimulates gland secretions (especially mucus and tear production), causes smooth muscle contraction of airway passages (bronchioles) in the lungs, and attracts eosinophils	Prostaglandins	A group of lipids (PGEs, PGFs, thromboxanes, and prostacyclins), produced by mast cells, that causes smooth muscle relaxation and vasodilation, increases vascular permeability, and stimulates pain receptors
Kinins	Polypeptides derived from plasma proteins; kinins cause vasodilation, increase vascular permeability, stimulate pain receptors, and attract neutrophils	Leukotrienes	A group of lipids, produced primarily by mast cells and basophils, that causes prolonged smooth muscle contraction (especially in the lung bronchioles), increases vascular permeability, and attracts neutrophils and eosinophils
Interferon	A protein, produced by most cells, that interferes with virus production and infection	Pyrogens	Chemicals, released by neutrophils, monocytes, and other cells, that stimulate fever production

Abbreviations: PGE = prostaglandin E; PGF = prostaglandin F.

prostaglandins, and leukotrienes, promote inflammation by causing vasodilation, increasing vascular permeability, attracting leukocytes, and stimulating phagocytosis. In addition, interferons protect cells against viral infections.

Complement

Complement is a group of approximately 20 proteins that makes up approximately 10% of the globulin part of serum. They include proteins named C1–C9 and factors B, D, and P (properdin). Normally, complement proteins circulate in the blood in an inactive, nonfunctional form. They become activated in the **complement cascade,** a series of reactions in which each component of the series activates the next component (figure 22.7). The complement cascade begins through either the alternative pathway or the classical pathway. The **alternative pathway** is part of innate immunity and is initiated when the complement protein C3 becomes spontaneously active. Activated C3 normally is quickly inactivated by proteins on the surface of the body's cells. If activated, however, C3 combines with some foreign substance such as part of a bacterial cell or virus, it can become stabilized and cause activation of the complement cascade. The **classical pathway** is part of the adaptive immune system discussed later in this chapter.

Activated complement proteins provide protection in several ways (see figure 22.7). Some attach to and form a hole in the membrane of bacterial cells, resulting in the lysis of the bacterial cell. Complement proteins attached to the surface of bacterial cells can also stimulate macrophages to phagocytize the bacteria. In addition, complement proteins attract immune system cells to sites of infection and promote inflammation.

Interferons

Interferons (in-ter-fēr′onz) are proteins that protect the body against viral infection and perhaps some forms of cancer. When a virus infects a cell, viral nucleic acids take control of directing the cell's activities. The cell produces new nucleic acids and proteins, which are assembled into new virus particles, and the new viruses are released from the infected cell to infect other cells. Because infected cells usually stop their normal functions or die during viral replication, viral infections are clearly harmful to the body. Fortunately, viruses and other substances can also stimulate infected cells to produce interferons. Interferons neither protect the cell that produces them nor act directly against viruses. Instead, they bind to the surface of neighboring cells, where they stimulate them to produce antiviral proteins. These antiviral proteins stop viral reproduction in the neighboring cells by preventing the production of new viral nucleic acids and proteins. Interferon viral resistance is innate rather than adaptive, and the same interferons act against many different viruses. Infection by one kind of virus actually can produce protection against infection by other kinds of viruses.

Clinical Note

Because some cancers are induced by viruses, interferons may play a role in controlling cancers. Interferons activate macrophages and natural killer cells (a type of lymphocyte) that attack tumor cells. Through genetic engineering, interferons currently are produced in sufficient quantities for clinical use and, along with other therapies, have been effective in treating certain viral infections and cancers. For example, interferons are used to treat hepatitis C, a viral disorder that can cause cirrhosis and cancer of the liver, and to treat genital warts, caused by the herpes virus. Interferons are also approved for the treatment of Kaposi's sarcoma, a cancer that can develop in AIDS patients.

Cells

Leukocytes and the cells derived from them (see table 19.2) are the most important cellular components of the immune system (table 22.2). Leukocytes are produced in red bone marrow and lymphatic tissue and are released into the blood, where they are transported throughout the body. To be effective, leukocytes must move into the tissues where they are needed. **Chemotactic** (kem-ō-tak′tik) **factors** are parts of microbes or chemicals released by tissue cells that act as chemical signals to attract leukocytes. Important chemotactic factors include complement, leukotrienes, kinins, and histamine. They diffuse from the area where they are released. Leukocytes can detect small differences in chemotactic factor concentration and move from areas of lower chemotactic factor concentration to areas of higher concentration. Thus they move toward the source of these substances, an ability called **chemotaxis.** Leukocytes can move by ameboid movement over the surface of cells, can squeeze between cells, and sometimes pass directly through other cells.

Phagocytosis (fag′ō-sī-tō′sis) is the endocytosis and destruction of particles by cells called **phagocytes** (see figure 3.25 *a* and *b*). The particles can be microorganisms or their parts, foreign substances, or dead cells from the individual's body. The most important phagocytic cells are neutrophils and macrophages.

Neutrophils

Neutrophils are small phagocytic cells produced in large numbers in red bone marrow that are released into the blood, where they circulate for a few hours. Approximately 126 billion neutrophils per day leave the blood and pass through the wall of the gastrointestinal tract, where they provide phagocytic protection. The neutrophils are then eliminated as part of the feces. Neutrophils are usually the first cells to enter infected tissues, and they often die after a single phagocytic event.

Neutrophils also release lysosomal enzymes that kill microorganisms and also cause tissue damage and inflammation.

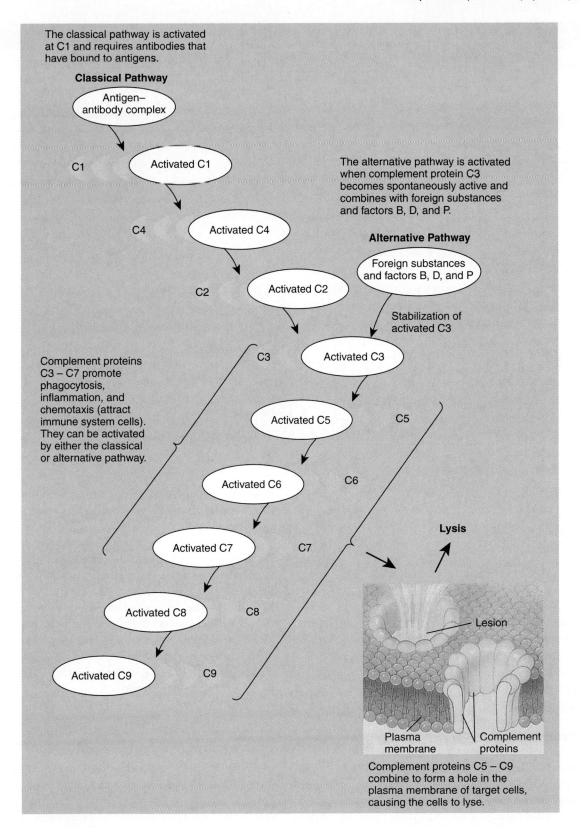

The classical pathway is activated at C1 and requires antibodies that have bound to antigens.

Classical Pathway

Antigen–antibody complex

C1   Activated C1

C4   Activated C4

The alternative pathway is activated when complement protein C3 becomes spontaneously active and combines with foreign substances and factors B, D, and P.

Alternative Pathway

Foreign substances and factors B, D, and P

C2   Activated C2

Stabilization of activated C3

Complement proteins C3 – C7 promote phagocytosis, inflammation, and chemotaxis (attract immune system cells). They can be activated by either the classical or alternative pathway.

C3   Activated C3

Activated C5   C5

Activated C6   C6

Lysis

Activated C7   C7

Activated C8   C8

Lesion

Activated C9   C9

Plasma membrane   Complement proteins

Complement proteins C5 – C9 combine to form a hole in the plasma membrane of target cells, causing the cells to lyse.

Figure 22.7   Complement Cascade

Inactive complement proteins (shown as shaded ovals) become active complement proteins (shown as white ovals) in a cascade reaction: each activated complement protein activates the next protein in the sequence.

Table 22.2   Summary of Immune System Cells and Their Primary Functions

Cell	Primary Function	Cell	Primary Function
Innate Immunity		**Adaptive Immunity**	
Neutrophil	Phagocytosis and inflammation; usually the first cell to leave the blood and enter infected tissues	B cell	After activation, differentiates to become plasma cell or memory B cell
Monocyte	Leaves the blood and enters tissues to become a macrophage	Plasma cell	Produces antibodies that are directly or indirectly responsible for the destruction of the antigen
Macrophage	Most effective phagocyte; important in later stages of infection and in tissue repair; located throughout the body to "intercept" foreign substances; processes antigens; involved in the activation of B and T cells	Memory B cell	Quick and effective response to an antigen against which the immune system has previously reacted; responsible for immunity
Basophil	Motile cell that leaves the blood, enters tissues, and releases chemicals that promote inflammation	Cytotoxic T cell	Responsible for the destruction of cells by lysis or by the production of cytokines
		Delayed hypersensitivity T cell	Produces cytokines that promote inflammation
Mast cell	Nonmotile cell in connective tissues that promotes inflammation through the release of chemicals	Helper T cell	Activates B and effector T cells
		Suppressor T cell	Inhibits B and effector T cells
Eosinophil	Enters tissues from the blood and releases chemicals that inhibit inflammation	Memory T cell	Quick and effective response to an antigen against which the immune system has previously reacted; responsible for adaptive immunity
Natural killer cell	Lyses tumor and virus-infected cells	Dendritic cell	Processes antigen and is involved in the activation of B and T cells

Pus is an accumulation of dead neutrophils, dead microorganisms, debris from dead tissue, and fluid.

Macrophages

Macrophages are monocytes that leave the blood, enter tissues, enlarge about fivefold, and increase their number of lysosomes and mitochondria. They are large phagocytic cells that outlive neutrophils, and they can ingest more and larger phagocytic particles than neutrophils. Macrophages usually appear in tissues after neutrophils and are responsible for most of the phagocytic activity in the late stages of an infection, including the cleanup of dead neutrophils and other cellular debris. In addition to their phagocytic role, macrophages produce a variety of chemicals, such as interferons, prostaglandins, and complement, that enhance the immune system response.

Macrophages are beneath the free surfaces of the body, such as the skin (dermis), hypodermis, mucous membranes, serous membranes, and around blood and lymphatic vessels. In these locations, macrophages provide protection by trapping and destroying microorganisms entering the tissues.

If microbes do gain entry to the blood or lymphatic system, macrophages are waiting within enlarged spaces, called sinuses, to phagocytize them. Blood vessels in the spleen, bone marrow, and liver have sinuses, as do lymph nodes. Within the sinuses reticular cells produce a fine network of reticular fibers that slows the flow of blood or lymph and provides a large surface area for the attachment of macrophages. In addition, macrophages are on the endothelial lining of the sinuses.

Because macrophages on the reticular fibers and endothelial lining of the sinuses were among the first macrophages studied, these cells were referred to as the **reticuloendothelial system.** It is now recognized that macrophages are derived from monocytes and are in locations other than the sinuses. Because monocytes and macrophages have a single, unlobed nucleus, they are now called the **mononuclear phagocytic system.** Sometimes macrophages are given specific names such as dust cells in the lungs, Kupffer cells in the liver, and microglia in the central nervous system.

Basophils, Mast Cells, and Eosinophils

Basophils, which are derived from red bone marrow, are motile white blood cells that can leave the blood and enter infected tissues. **Mast cells,** which are also derived from red bone marrow, are nonmotile cells in connective tissue, especially near capillaries. Like macrophages, mast cells are located at potential points of entry of microorganisms into the body such as the skin, lungs, gastrointestinal tract, and urogenital tract.

Basophils and mast cells can be activated through innate immunity (e.g., by complement) or through adaptive immunity (see section on Antibodies). When activated, they release chemicals such as histamine and leukotrienes that produce an inflammatory response or activate other mechanisms such as smooth muscle contraction in the lungs.

Eosinophils are produced in red bone marrow, enter the blood, and within a few minutes enter tissues. Enzymes released by eosinophils break down chemicals released by basophils and mast cells. Thus at the same time that inflammation is initiated, mechanisms are activated that contain and reduce the inflammatory response. This process is similar to the blood clotting system in which clot prevention and removal mechanisms are activated while the clot is being formed (see chapter 19). In patients with parasitic infections or allergic reactions with much inflammation, eosinophil numbers greatly increase. Eosinophils also secrete enzymes that effectively kill some parasites.

Natural Killer Cells

Natural killer (NK) cells are a type of lymphocyte produced in red bone marrow, and they account for up to 15% of lymphocytes. NK cells recognize classes of cells, such as tumor cells or virus-infected cells in general, rather than specific tumor cells or cells infected by a specific virus. For this reason and because NK cells do not exhibit a memory response, NK cells are classified as part of innate immunity. NK cells use a variety of methods to kill their target cells, including the release of chemicals that damage cell membranes, causing the cells to lyse.

Inflammatory Response

The **inflammatory response** is a complex sequence of events involving many of the chemicals and cells previously discussed. A bacterial infection is used here to illustrate an inflammatory response (figure 22.8). The bacteria, or damage to tissues, cause the release or activation of chemical mediators, such as histamine, prostaglandins, leukotrienes, complement, kinins, and others. The chemical mediators produce several effects: (1) vasodilation, which increases blood flow and brings phagocytes and other leukocytes to the area, (2) chemotactic attraction of phagocytes, which leave the blood and enter the tissue, and (3) increased vascular permeability, allowing fibrinogen and complement to enter the tissue from the blood. Fibrinogen is converted to fibrin, which prevents the spread of infection by walling off the infected area. Complement further enhances the inflammatory response and attracts additional phagocytes. The process of releasing chemical mediators and attracting phagocytes and other leukocytes continues until the bacteria are destroyed. Phagocytes (mainly macrophages) remove microorganisms and dead tissue, and the damaged tissues are repaired.

Inflammation can be localized or systemic. **Local inflammation** is an inflammatory response confined to a specific area of the body. Symptoms of local inflammation include redness, heat, swelling, pain, and loss of function. Redness, heat, and swelling result from increased blood flow and increased vascular permeability. Pain is caused by swelling and by chemicals acting on sensory or pain recep-

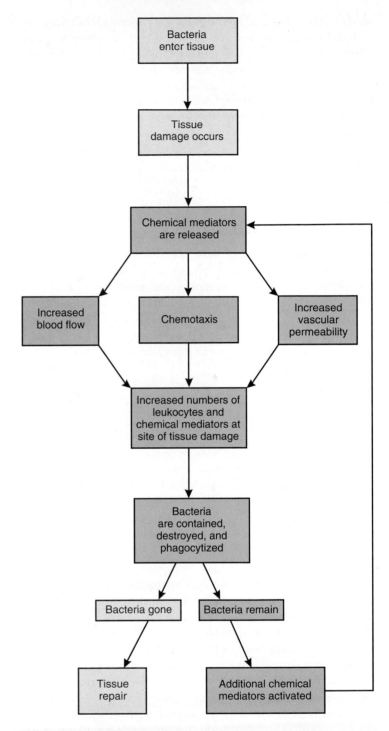

Figure 22.8   Inflammatory Response

Flow diagram of the inflammatory response. Bacteria cause tissue damage and release of chemical mediators that initiate inflammation, resulting in the destruction of the bacteria.

tors. Loss of function results from tissue destruction, swelling, and pain.

Systemic inflammation is an inflammatory response that occurs in many parts of the body. In addition to the local symptoms at the sites of inflammation, three additional features can be present. First, red bone marrow produces and releases large numbers of neutrophils that promote phagocytosis. Second, **pyrogens** (pī′rō-jenz), chemicals released by microorganisms, macrophages, neutrophils, and other cells, stimulate fever production. Pyrogens affect the body's temperature-regulating mechanism in the hypothalamus, heat is conserved, and body temperature increases. Fever promotes the activities of the immune system such as phagocytosis and inhibits the growth of some microorganisms. Third, in severe cases of systemic inflammation, vascular permeability is so widespread that large amounts of fluid are lost from the blood into the tissues. The decreased blood volume can cause shock and death.

Adaptive Immunity

Adaptive immunity involves the ability to recognize, respond to, and remember a particular substance. Substances that stimulate adaptive immunity are **antigens** (an′ti-jenz), which usually are large molecules with a molecular weight of 10,000 or more. **Haptens** (hap′tenz) are small molecules (low molecular weight) capable of combining with larger molecules such as blood proteins to stimulate an adaptive immune system response.

Clinical Note

Penicillin is an example of a hapten of clinical importance. It is a small molecule that does not evoke an immune system response. Penicillin can, however, break down and bind to serum proteins to form a combined molecule that can produce an allergic reaction. Most commonly, the reaction produces a rash and fever, but rarely a severe reaction can cause death.

Antigens can be divided into two groups: foreign antigens and self-antigens. **Foreign antigens** are not produced by the body but are introduced from outside it. Components of bacteria, viruses, and other microorganisms are examples of foreign antigens that cause disease. Pollen, animal dander (scaly, dried skin), feces of house dust mites, foods, and drugs are also foreign antigens and can trigger an overreaction of the immune system in some people, called an **allergic reaction.** Transplanted tissues and organs that contain foreign antigens result in the rejection of the transplant. **Self-antigens** are molecules produced by the body that stimulate an adaptive immune system response. The response to self-antigens can be beneficial or harmful. For example, the recognition of tumor antigens can result in tumor destruction, whereas **autoimmune disease** can result when self-antigens stimulate unwanted tissue destruction.

Adaptive immunity historically has been divided into two types: **humoral** (hyū′mōr-ăl) and **cell-mediated immunity.** Early investigators of the immune system found that, when plasma from an immune animal was injected into the blood of a nonimmune animal, the nonimmune animal became immune. Because this process involved body fluids (humors), it was called humoral immunity. It was also dis-

covered that blood cells transferred from an immune animal could be responsible for immunity, and this process was called cell-mediated immunity.

It is now known that immunity results from the activities of lymphocytes called B and T cells (see table 22.2). **B cells** give rise to cells that produce proteins called **antibodies,** which are found in the plasma. Because antibodies are responsible, humoral immunity is now called **antibody-mediated immunity.**

T cells are responsible for cell-mediated immunity. There are several subpopulations of T cells, each of which is responsible for a particular aspect of cell-mediated immunity. For example, **effector T cells,** such as **cytotoxic T cells** and **delayed hypersensitivity T cells,** are responsible for producing the effects of cell-mediated immunity; whereas **regulatory T cells,** such as **helper T cells** and **suppressor T cells,** can control the activities of both antibody-mediated immunity and cell-mediated immunity.

Table 22.3 summarizes and contrasts the main features of innate, antibody-mediated, and cell-mediated immunity.

Origin and Development of Lymphocytes

All blood cells, including lymphocytes, are derived from stem cells in the red bone marrow (see chapter 19). The process of blood cell formation begins during embryonic development and continues throughout life. Some stem cells give rise to pre-T cells that migrate through the blood to the thymus gland, where they divide and are processed into T cells. The thymus gland produces hormones such as thymosin, which stimulates T-cell maturation. Other stem cells produce pre-B cells, which are processed in the red bone marrow into B cells (figure 22.9). A **positive selection** process results in the survival of pre-B and pre-T cells that are capable of an immune response. Cells that are incapable of an immune response die.

The B and T cells that can respond to antigens are composed of small groups of identical lymphocytes called **clones.** Although each clone can respond only to a particular antigen, there is such a large number of clones that the immune system can react to most molecules. But the clones can also respond to self-antigens. A **negative selection** process eliminates or suppresses clones acting against self-antigens because this response could destroy self-cells. Although most of this process occurs during prenatal development, it continues throughout life (see section on Inhibition of Lymphocytes later in the chapter).

B cells are released from red bone marrow, T cells are released from the thymus gland, and both types of cells move through the blood to lymphatic tissue. Normally there are approximately five T cells for every B cell in the blood. These lymphocytes live for a few months to many years and continually circulate between the blood and the lymphatic tissues. Antigens can come into contact with and activate lymphocytes, resulting in cell divisions that increase the number of

Table 22.3   Comparison of Innate Immunity, Antibody-Mediated Immunity, and Cell-Mediated Immunity

Characteristics	Innate Immunity	Antibody-Mediated Immunity	Cell-Mediated Immunity
Primary cells	Neutrophils, eosinophils, basophils, mast cells, monocytes, and macrophages	B cells	T cells
Origin of cells	Red bone marrow	Red bone marrow	Red bone marrow
Site of maturation	Red bone marrow (neutrophils, eosinophils, basophils, monocytes) and tissues (mast cells and macrophages)	Red bone marrow	Thymus gland
Location of mature cells	Blood, connective tissue, and lymphatic tissue	Blood and lymphatic tissue	Blood and lymphatic tissue
Primary secretory products	Histamine, kinins, complement, prostaglandins, leukotrienes, and interferon	Antibodies	Cytokines
Primary actions	Inflammatory response and phagocytosis	Protection against extracellular antigens (bacteria, toxins, parasites, and viruses outside of cells)	Protection against intracellular antigens (viruses, intracellular bacteria, and intracellular fungi) and tumors: regulates antibody-mediated immunity and cell-mediated immunity responses (helper T and suppressor T cells)
Hypersensitivity reactions	None	Immediate hypersensitivity (atopy, anaphylaxis, cytotoxic reactions, and immune complex disease)	Delayed hypersensitivity (allergy of infection and contact hypersensitivity)

lymphocytes that can recognize the antigen. These lymphocytes can circulate in blood and lymph to reach antigens in tissues throughout the body.

Activation of Lymphocytes

Lymphocytes can be activated by antigens in different ways, depending on the type of lymphocyte and the type of antigen involved. Despite these differences, however, there are two general principles of lymphocyte activation: (1) lymphocytes must be able to recognize the antigen, and (2) after recognition the lymphocytes must increase in number to effectively destroy the antigen.

Antigenic Determinants and Antigen Receptors

If an adaptive immune system response is to occur, lymphocytes must recognize an antigen. Lymphocytes do not interact with an entire antigen, however. Instead, **antigenic determinants,** or **epitopes** (ep′i-tōpz), are specific regions of a given antigen that activate a lymphocyte, and each antigen has many different antigenic determinants (figure 22.10). All the lymphocytes of a given clone have on their surfaces identical proteins called **antigen receptors,** which combine with antigenic determinants. The immune system response to an antigen with a particular antigenic determinant is similar to the

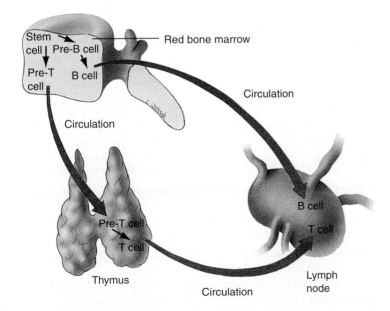

Figure 22.9   Origin and Processing of B and T Cells

Both B and T cells originate in red bone marrow. B cells are processed in the red marrow, whereas T cells are processed in the thymus gland. Both cell types circulate to other lymphatic tissues, where they can divide and increase in number in response to antigens.

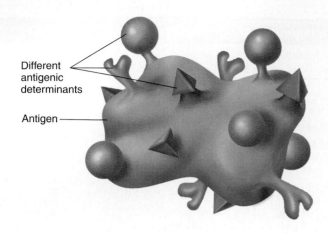

Figure 22.10   Antigenic Determinants

An antigen has many antigenic determinants to which lymphocytes can respond.

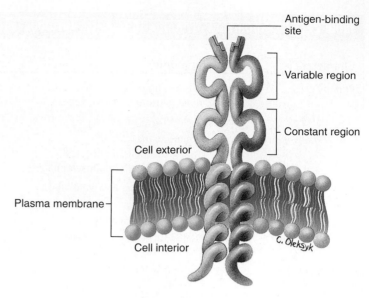

Figure 22.11   The T-Cell Receptor

The T-cell receptor consists of two polypeptide chains. The variable region of each type of T-cell receptor is specific for a given antigen. The constant region attaches the T-cell receptor to the plasma membrane.

lock-and-key model for enzymes (see chapter 3), and any given antigenic determinant can combine only with a specific antigen receptor on a lymphocyte. For example, the **T-cell receptor** consists of two polypeptide chains, which are subdivided into a variable and a constant region (figure 22.11). The variable region can bind to an antigen, and different T-cell receptors respond to different antigens because they have different variable regions. The **B-cell receptor** consists of four polypeptide chains with two identical variable regions. It is a type of antibody and is considered in greater detail later in this chapter.

Major Histocompatibility Complex Molecules

Although some antigens bind to their receptors and directly activate B cells and some T cells, most lymphocyte activation involves glycoproteins on the surfaces of cells called **major histocompatibility complex (MHC) molecules.** The MHC molecules are attached to the plasma membrane, and they have a variable region that can bind to foreign and self-antigens.

MHC class I molecules are found on nucleated cells and function to display antigens produced inside the cells on their surfaces (figure 22.12a). This is necessary because the immune system cannot directly respond to an antigen inside a cell. For example, viruses reproduce inside cells, forming viral proteins that are foreign antigens. Some of these viral proteins are broken down in the cytoplasm. The protein fragments enter the rough endoplasmic reticulum and combine with MHC class I molecules to form complexes that move through the Golgi apparatus to be distributed on the surface of the cell (see chapter 3).

On the surface of the cell, the MHC class I/antigen complex can bind to a T-cell receptor, activating the T cell. As will be described later in this chapter, the activated T cell can destroy the infected cell, which effectively stops viral replication. Thus the MHC class I/antigen complex functions as a signal,

or "red flag," that prompts the immune system to destroy the displaying cell. In essence the cell is displaying a sign that says "Kill me!" This process is said to be **MHC-restricted,** because both the antigen and the individual organism's own MHC molecule are required.

The same process that moves foreign protein fragments to the surface of cells can also inadvertently transport self-protein fragments (see figure 22.12a). As part of normal protein metabolism, cells continually break down old proteins and synthesize new ones. Some self-protein fragments that result from protein breakdown can combine with MHC class I molecules and be displayed on the surface of the cell, thus becoming self-antigens. Normally the immune system does not respond to self-antigens in combination with MHC molecules because the lymphocytes that could respond have been eliminated or inactivated (see section on Inhibition of Lymphocytes later in the chapter).

MHC class II molecules are found on **antigen-presenting cells,** which include B cells, macrophages, monocytes, and dendritic cells. **Dendritic** (den-drit′ik) **cells** are large, motile cells with long cytoplasmic extensions, and they are scattered throughout most tissues (except the brain), with their highest concentrations in lymphatic tissues and the skin. Dendritic cells in the skin are often called **Langerhans' cells.**

1. Foreign proteins or self-proteins within the cytosol are broken down into fragments that are antigens.
2. Antigens are transported into the rough endoplasmic reticulum.
3. Antigens combine with MHC class I molecules.
4. The MHC class I/antigen complex is transported to the Golgi apparatus, packaged into a vesicle, and transported to the plasma membrane.
5. Foreign antigens combined with MHC class I molecules stimulate cell destruction.
6. Self-antigens combined with MHC class I molecules do not stimulate cell destruction.

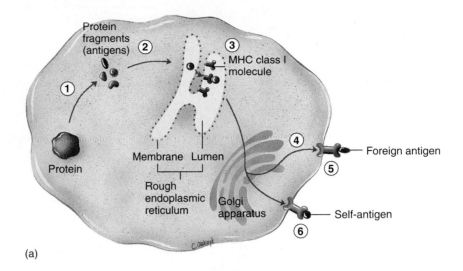

(a)

1. The unprocessed extracellular antigen is ingested by endocytosis and is within a vesicle.
2. The antigen is broken down into fragments to form processed antigens.
3. The vesicle containing the processed antigen fuses with vesicles produced by the Golgi apparatus that contain MHC class II molecules. The processed antigen and the MHC class II molecule combine.
4. The MHC class II/antigen complex is transported to the plasma membrane.
5. The displayed MHC class II/antigen complex can stimulate immune cells.

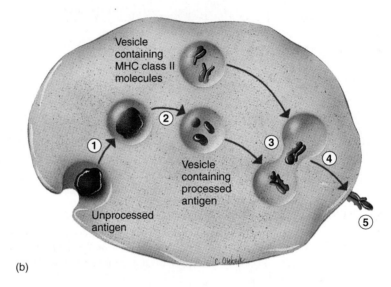

(b)

Figure 22.12 Antigen Processing

(a) Foreign proteins, such as viral proteins, or self-proteins in the cytosol, are processed and presented at the cell surface by MHC class I molecules. (b) Extracellular antigens are taken into an antigen-presenting cell, processed, and presented at the cell surface by MHC class II molecules. ✗

Antigen-presenting cells are specialized to take in foreign antigens, to process the antigens, and to use MHC class II molecules to display the foreign antigens to other immune system cells (figure 22.12b). For example, the MHC class II/antigen complex can bind with a T-cell receptor. Because both the antigen and the individual's own MHC class II molecule are required, this process is MHC-restricted. Unlike MHC class I molecules, however, this display does not result in the destruction of the antigen-presenting cell. Instead the MHC class II/antigen complex is a "rally around the flag" signal that stimulates other immune system cells to respond to the antigen. The displaying cell is like Paul Revere, who spread the alarm for the militia to arm and organize. The militia then went out and killed the enemy. For example, when the lymphocytes of the B-cell clone that can recognize the antigen come into contact with the MHC class II/antigen complex, they are stimulated to divide. The activities of these lymphocytes, such as the production of antibodies, then result in the destruction of the antigen.

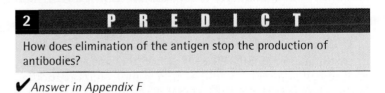

2   PREDICT

How does elimination of the antigen stop the production of antibodies?

✔ *Answer in Appendix F*

Costimulation

The combination of an MHC class II/antigen complex with an antigen receptor is usually only the first signal necessary to produce a response from a B or T cell. In many cases, **costimulation** by additional signals is also required. Costimulation can be achieved by **cytokines** (sī'tō-kīnz),

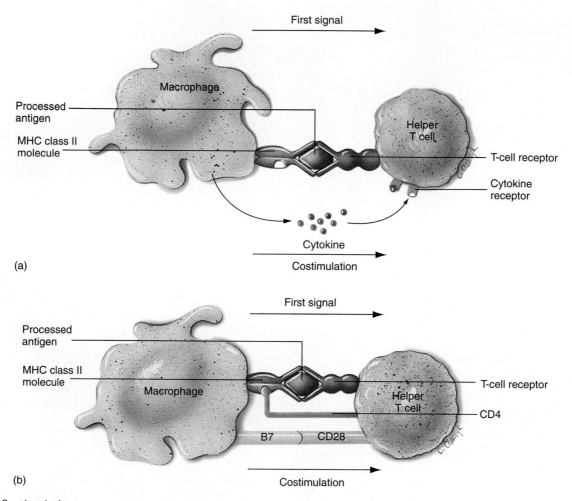

Figure 22.13   Costimulation

The first signal required for activation of a helper T cell is the binding of the MHC class II/antigen complex to the T-cell receptor. (*a*) One costimulatory signal is the release by the macrophage of a cytokine that binds to a receptor on the helper T cell. (*b*) Another costimulatory signal is the binding of a B7 molecule of the macrophage with a CD28 molecule of the helper T cell. The CD4 molecule of the helper T cell binds to the macrophage's MHC class II molecule and helps to hold the cells together. ✗

which are proteins or peptides secreted by one cell as a regulator of neighboring cells (figure 22.13*a*). Cytokines produced by lymphocytes are often called **lymphokines** (lim′fō-kīnz). Cytokines are involved in the regulation of immunity, inflammation, tissue repair, cell growth, and other processes. Table 22.4 lists some important cytokines and their functions.

Certain pairs of surface molecules can also be involved in costimulation (figure 22.13*b*). When the surface molecule on one cell combines with the surface molecule on another, the combination can act as a signal that stimulates a response from one of the cells, or the combination can hold the cells together. Typically, several different kinds of surface molecules are necessary to produce a response. For example, a molecule called B7 on macrophages must bind with a molecule called CD28 on helper T cells before the helper T cells can respond to the antigen presented by the macrophage. In addition, helper T cells have a glycoprotein called CD4, which helps to connect helper T cells to the macrophage by

binding to MHC class II molecules. For this reason, helper T cells are sometimes referred to as **CD4, or T4, cells.** In a similar fashion, cytotoxic T cells are sometimes called **CD8, or T8, cells** because they have a glycoprotein called CD8, which helps to connect cytotoxic T cells to body cells displaying MHC class I molecules. The CD designation stands for "cluster of differentiation," which is a system used to classify many surface molecules.

Lymphocyte Proliferation

Before exposure to an antigen, the number of lymphocytes in a clone is too small to produce an effective response against the antigen. After recognition of an antigen, the lymphocytes are stimulated to divide and increase in number. Lymphocyte proliferation begins with an increase in the number of helper T cells (figure 22.14). The increased number of helper T cells then stimulates an increase in the number of B cells or effector T cells (figure 22.15).

Table 22.4   Cytokines and Their Functions

Cytokine*	Description
Interferon alpha (IFNα)	Prevents viral replication and inhibits cell growth; secreted by virus-infected cells
Interferon beta (IFNβ)	Prevents viral replication, inhibits cell growth, and decreases the expression of major histocompatibility complex (MHC) class I and II molecules; secreted by virus-infected fibroblasts
Interferon gamma (IFNγ)	About 20 different proteins that activate macrophages and natural killer (NK) cells, stimulate adaptive immunity by increasing the expression of MHC class I and II molecules, and prevent viral replication; secreted by helper T, cytotoxic T, and NK cells
Interleukin-1 (IL-1)	Costimulation of B and T cells, promotes inflammation through prostaglandin production, and induces fever acting through the hypothalamus (pyrogen); secreted by macrophages, B cells, and fibroblasts
Interleukin-2 (IL-2)	Costimulation of B and T cells, activation of macrophages and NK cells; secreted by helper T cells
Interleukin-4 (IL-4)	Plays a role in allergic reactions by activation of B cells, resulting in the production of immunoglobulin E (IgE); secreted by helper T cells
Interleukin-5 (IL-5)	Part of the response against parasites by stimulating eosinophil production; secreted by helper T cells
Interleukin-8 (IL-8)	Chemotactic factor that promotes inflammation by attracting neutrophils and basophils; secreted by macrophages
Interleukin-10 (IL-10)	Inhibits the secretion of interferon gamma and interleukins; secreted by suppressor T cells
Lymphotoxin	Kills target cells; secreted by cytotoxic T cells
Perforin	Makes a hole in the membrane of target cells, resulting in lysis of the cell; secreted by cytotoxic T cells
Tumor necrosis factor (TNF)	Activates macrophages and promotes fever (pyrogen); secreted by macrophages

*Some cytokines were named according to the laboratory test first used to identify them; however, these names rarely are a good description of the actual function of the cytokine.

Inhibition of Lymphocytes

Tolerance is a state of unresponsiveness of lymphocytes to a specific antigen. Although foreign antigens can induce tolerance, the most important function of tolerance is to prevent the immune system from responding to self-antigens. The need to maintain tolerance and to avoid the development of autoimmune disease is obvious. Without tolerance a female would recognize the antigens of her developing fetus as foreign, and spontaneous abortion would result. Apparently tolerance can be induced in many different ways.

1. *Deletion of self-reactive lymphocytes.* During prenatal development and after birth, stem cells in red bone marrow and the thymus gland give rise to immature lymphocytes that develop into mature lymphocytes capable of an immune response. When immature lymphocytes are exposed to antigens, instead of responding in ways that result in the elimination of the antigen, they respond by dying. Because immature lymphocytes are exposed to self-antigens, this process eliminates self-reactive lymphocytes. In addition, immature lymphocytes that escape deletion during their development and become mature, self-reacting lymphocytes can still be deleted in ways that are not clearly understood.
2. *Preventing activation of lymphocytes.* For activation of lymphocytes to take place, two signals are usually required: (1) the MHC–antigen complex binding with an antigen receptor and (2) costimulation. Preventing either of these events stops lymphocyte activation. For

example, blocking, altering, or deleting an antigen receptor prevents activation. **Anergy** (an'er-jē), which means "without working," is a condition of inactivity in which a B or T cell does not respond to an antigen. Anergy develops when an MHC–antigen complex binds to an antigen receptor, and there is no costimulation. For example, if a T cell encounters a self-antigen on a cell that cannot provide costimulation, the T cell is turned off. It is likely that only antigen-presenting cells can provide costimulation.

> **Clinical Note**
>
> Decreasing the production or activity of cytokines can suppress the immune system. For example, cyclosporine, a drug used to prevent the rejection of transplanted organs, inhibits the production of interleukin-2. Conversely, genetically engineered interleukins can be used to stimulate the immune system. Administering interleukin-2 has promoted the destruction of cancer cells in some cases by increasing the activities of effector T cells.

3. *Activation of suppressor T cells.* **Suppressor T cells** are a poorly understood group of T cells that are defined by their ability to suppress immune responses. It is likely that suppressor T cells are subpopulations of helper T cells and cytotoxic T cells. The suppressor (helper) T cells release suppressive cytokines, or the suppressor (cytotoxic) T cells kill antigen-presenting cells.

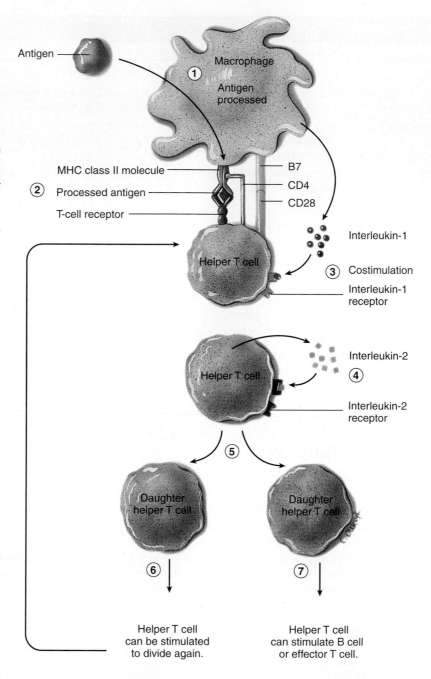

1. Antigen-presenting cells such as macrophages take in, process, and display antigens on the cell's surface.

2. The antigens are bound to MHC class II molecules, which function to present the processed antigen to the T-cell receptor of the helper T cell for recognition.

3. Costimulation occurs by the CD4 glycoprotein of the helper T cell and by cytokines. The macrophage secretes a cytokine called interleukin-1.

4. Interleukin-1 stimulates the helper T cell to secrete the cytokine interleukin-2 and to produce interleukin-2 receptors.

5. The helper T cell stimulates itself to divide when interleukin-2 binds to interleukin-2 receptors.

6. The "daughter" helper T cells resulting from this division can be stimulated to divide again if they are exposed to the same antigen that stimulated the "parent" helper T cell. This greatly increases the number of helper T cells.

7. The increased number of helper T cells can facilitate the activation of B cells or effector T cells.

Figure 22.14   Proliferation of Helper T Cells

Antigen-presenting cell (macrophage) stimulates helper T cells to divide.

Antibody-Mediated Immunity

Exposure of the body to an antigen can lead to activation of B cells and to production of antibodies, which are responsible for destruction of the antigen. Because antibodies occur in body fluids, **antibody-mediated immunity** is effective against extracellular antigens, such as bacteria, viruses when they are not inside cells, toxins, and parasites. Antibody-mediated immunity can also cause immediate hypersensitivity reactions (see Clinical Focus: Immune System Problems of Clinical Significance).

Antibodies

Antibodies are proteins produced in response to an antigen. Large amounts of antibodies occur in plasma, although plasma also contains other proteins. On the basis of protein type and associated lipids, the plasma proteins can be separated into albumin and alpha-(α), beta-(β), and gamma-(γ) globulin parts. As a group, antibodies are sometimes called **gamma globulins** because they are mostly found in the γ-globulin part of plasma. They are also called **immunoglobulins** (Ig) because they are globulin proteins involved in immunity.

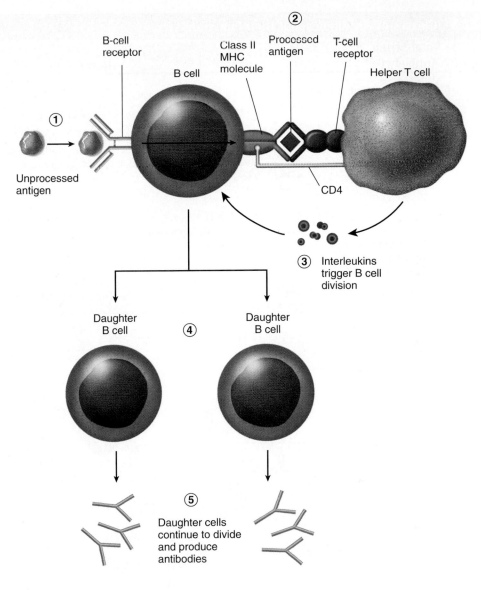

1. Before a B cell can be activated by a helper T cell, the B cell must process the same antigen that activated the helper T cell. The antigen binds to a B-cell receptor, and the receptor and antigen are both taken into the cell by endocytosis.

2. The B cell uses a MHC class II molecule to present the processed antigen to the helper T cell.

3. The helper T cell responds by releasing various interleukins that stimulate the B cell to divide.

4. The B cell divides, and the resulting daughter cells divide, and so on, eventually producing many cells (only two are shown here).

5. The increased number of cells produce antibodies, which are part of the antibody-mediated immune system response that eliminates the antigen.

Figure 22.15   Proliferation of B Cells

Helper T cell stimulates a B cell to divide.

The five general classes of immunoglobulins are denoted IgG, IgM, IgA, IgE, and IgD (table 22.5). All classes of antibodies have a similar structure, consisting of four polypeptide chains (figure 22.16): two identical heavy chains and two identical light chains. Each light chain is attached to a heavy chain, and the ends of the combined heavy and light chains form the **variable region** of the antibody, which is the part that combines with the antigenic determinant of the antigen. Different antibodies have different variable regions, and they are specific for different antigens. The rest of the antibody is the **constant region,** which is responsible for activities of antibodies such as the ability to activate complement or to attach the antibody to cells such as macrophages, basophils, mast cells, and eosinophils (see figure 22.16). All the antibodies of a particular class have nearly the same constant regions.

Effects of Antibodies

Antibodies can directly affect antigens in two ways (figure 22.17). The antibody can bind to the antigenic determinant of an antigen and interfere with the ability of the antigen to function. Alternatively, the antibody can combine with an antigenic determinant on two different antigens, rendering the antigens ineffective and making them more susceptible to phagocytosis. The ability of antibodies to join antigens together is the basis for many clinical tests such as blood typing because, when enough antigens are bound together, they become visible as a clump or a precipitate.

Although antibodies can directly affect antigens, most of the effectiveness of antibodies results from other mechanisms, such as opsonization, activation of complement, and stimulation of the inflammatory response (see figure 22.17).

Table 22.5   Classes of Antibodies and Their Functions

Antibody	Total Serum Antibody (%)	Structure	Description
IgG	80–85		Activates complement and functions as an opsonin to increase phagocytosis; can cross the placenta and provide immune protection to the fetus and newborn; responsible for Rh reactions such as hemolytic disease of the newborn
IgM	5–10		Activates complement and acts as an antigen-binding receptor on the surface of B cells; responsible for transfusion reactions in the ABO blood system; often the first antibody produced in response to an antigen
IgA	15		Secreted into saliva, tears, and onto mucous membranes to provide protection on body surfaces; found in colostrum and milk to provide immune protection to the newborn
IgE	0.002		Binds to mast cells and basophils and stimulates the inflammatory response
IgD	0.2		Functions as antigen-binding receptors on B cells

IgG   IgM   IgA

IgE   IgD

■ Light chain

■ Heavy chain

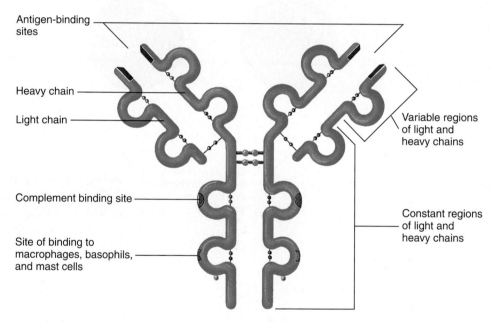

Figure 22.16   Structure of an Antibody

Antibodies consist of two heavy and two light polypeptide chains. The variable region of the antibody binds to the antigen. The constant region of the antibody can activate the classical pathway of the complement cascade. The constant region of the antibody can also attach the antibody to the cell membrane of cells such as macrophages, basophils, or mast cells.

Labels on figure: Antigen-binding sites; Heavy chain; Light chain; Complement binding site; Site of binding to macrophages, basophils, and mast cells; Variable regions of light and heavy chains; Constant regions of light and heavy chains

Opsonins (op'sŏ-ninz) are substances that make an antigen more susceptible to phagocytosis. An antibody (IgG) acts as an opsonin by connecting to an antigen through the variable region of the antibody and to a macrophage through the constant region of the antibody. The macrophage then phagocytizes the antigen and the antibody.

When an antibody (IgG or IgM) combines with an antigen through the variable region, the constant region can activate the complement cascade through the classical pathway

(see figure 22.7). Activated complement stimulates inflammation; attracts neutrophils, monocytes, macrophages, and eosinophils to sites of infection; and kills bacteria by lysis.

Antibodies (IgE) can initiate an inflammatory response. The antibodies attach to mast cells or basophils through their constant region. When antigens combine with the variable region of the antibodies, the mast cells or basophils release chemicals through exocytosis, and inflammation results.

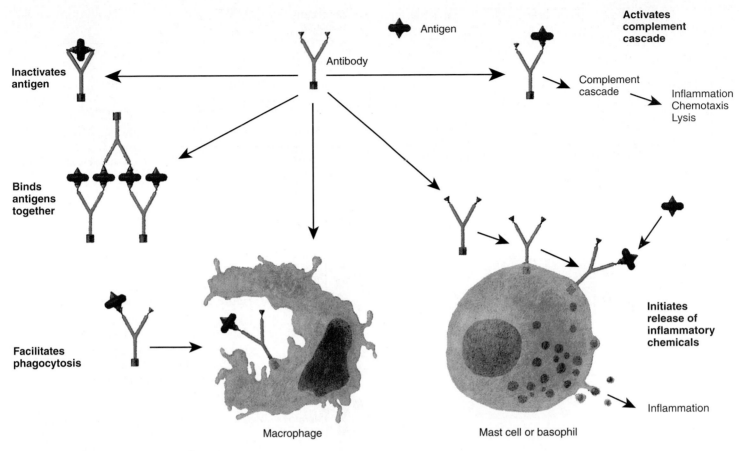

Figure 22.17   Actions of Antibodies

Antibodies can inactivate antigens, promote phagocytosis (binding antigens together or opsonization), and cause inflammation (release of chemicals from mast cells or basophils and activation of the complement cascade).

Each type of **monoclonal antibody** is a pure antibody preparation that is specific for only one antigen. When the antigen is injected into a laboratory animal, it activates a B-cell clone against the antigen. The B cells are removed from the animal and fused with tumor cells. The resulting hybridoma cells have two ideal characteristics: they divide to form large numbers of cells, and the cells of a given clone produce only one kind of antibody.

Monoclonal antibodies are used for determining pregnancy and for diagnosing diseases such as gonorrhea, syphilis, hepatitis, rabies, and cancer. These tests are specific and rapid because the monoclonal antibodies bind only to the antigen being tested. Monoclonal antibodies may someday be used to effectively treat cancer by delivering drugs to cancer cells (see Immunotherapy later in this chapter).

Antibody Production

The production of antibodies after the first exposure to an antigen is different from that after a second or subsequent exposure. The **primary response** results from the first exposure of a B cell to an antigen for which it is specific and includes a series of cell divisions, cell differentiation, and antibody production. The B-cell receptors on the surface of B cells are antibodies, usually IgM and IgD. The receptors have the same variable region as the antibodies that are eventually produced by the B cell. Before stimulation by an antigen, B cells are small lymphocytes. After activation the B cells undergo a series of divisions to produce large lymphocytes. Some of these enlarged cells become **plasma cells,** which produce antibodies, and others revert back to small lymphocytes and become **memory B cells** (figure 22.18). Usually IgM is the first antibody produced in response to an antigen, but later other classes of antibodies are produced as well. The primary response normally takes 3–14 days to produce enough antibodies to be effective against the antigen. In the meantime, the individual usually develops disease symptoms because the antigen has had time to cause tissue damage.

The **secondary,** or **memory, response** occurs when the immune system is exposed to an antigen against which it has already produced a primary response. The secondary response results from memory B cells, which rapidly divide to produce plasma cells and large amounts of antibody when exposed to the antigen. The secondary response provides better protection than the primary response for two reasons. First, the time required to start producing antibodies is less (hours to a few days); and second, the amount of antibody produced is much larger. As a consequence, the antigen is quickly destroyed, no disease symptoms develop, and the person is immune.

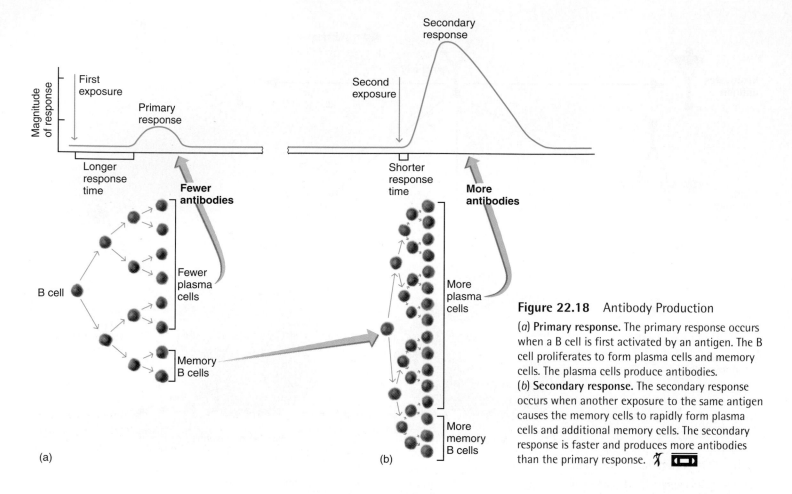

Figure 22.18   Antibody Production

(*a*) **Primary response.** The primary response occurs when a B cell is first activated by an antigen. The B cell proliferates to form plasma cells and memory cells. The plasma cells produce antibodies. (*b*) **Secondary response.** The secondary response occurs when another exposure to the same antigen causes the memory cells to rapidly form plasma cells and additional memory cells. The secondary response is faster and produces more antibodies than the primary response.

The memory response also includes the formation of new memory B cells, which provide protection against additional exposures to the antigen. Memory B cells are the basis for adaptive immunity. After destruction of the antigen, plasma cells die, the antibodies they released are degraded, and antibody levels decline to the point at which they can no longer provide adequate protection. Memory B cells may persist for many years and probably for life in some cases. If memory cell production is not stimulated, however, or if the memory B cells produced are short-lived, repeated infections of the same disease are possible. For example, the same cold virus can cause the common cold more than once in the same person.

---

3   P R E D I C T

One theory for long-lasting immunity assumes that humans are continually exposed to the disease-causing agent. Explain how this exposure could produce lifelong immunity.

✔ *Answer in Appendix F*

Cell-Mediated Immunity

Cell-mediated immunity is a function of T cells and is most effective against intracellular microorganisms such as viruses, fungi, intracellular bacteria, and parasites. Delayed hypersensi-tivity reactions and control of tumors also involve cell-mediated immunity (see Clinical Focus: Immune System Problems of Clinical Significance).

Activation of T cells to antigens is regulated by antigen-presenting cells and helper T cells. Once activated, T cells undergo a series of divisions and produce effector T cells and memory T cells (figure 22.19). Effector T cells such as cytotoxic T cells are responsible for the cell-mediated immunity response. Memory T cells can provide a secondary response and long-lasting immunity in the same fashion as memory B cells.

Cytotoxic T Cells

Cytotoxic T cells have two main effects: they lyse cells and they produce cytokines. Cytotoxic T cells can come into contact with other cells and cause them to lyse. Virus-infected cells have viral antigens, tumor cells have tumor antigens, and tissue transplants have foreign antigens on their surfaces that can stimulate cytotoxic T-cell activity. The cytotoxic T cell binds to the target cell and releases chemicals that cause the target cell to lyse. In one method of lysis the chemicals, like complement, form a pore in the membrane of the target cell. The cytotoxic T cell then moves on to destroy additional target cells.

In addition to lysing cells, cytotoxic T cells release cytokines that activate additional components of the immune

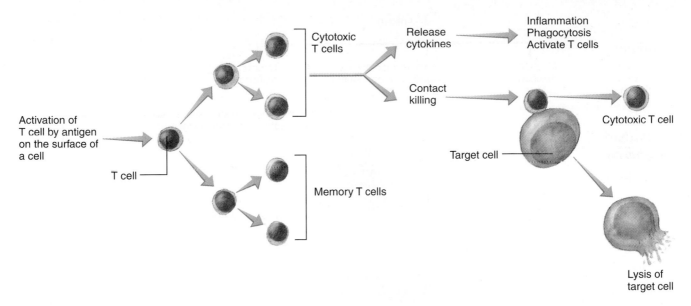

Figure 22.19   Stimulation and Effects of T Cells

When T cells are presented with a processed antigen, they can form memory T and cytotoxic T cells. Memory T cells are responsible for the secondary response, and cytotoxic T cells cause contact killing or release cytokines that promote the destruction of the antigen.

system. For example, one important function of cytokines is the recruitment of cells such as macrophages. These cells are then responsible for phagocytosis and inflammation.

> **4   P R E D I C T**
>
> In patients with acquired immune deficiency syndrome (AIDS), helper T cells are destroyed by a viral infection. The patients can die of pneumonia caused by an intracellular protozoan (*Pneumocystis carinii*) or from Kaposi's sarcoma, which consists of tumorous growths in the skin and lymph nodes. Explain what is happening.

✔ *Answer in Appendix F*

Delayed Hypersensitivity T Cells

Delayed hypersensitivity T cells respond to antigens by releasing cytokines. Consequently, they promote phagocytosis and inflammation, especially in allergic reactions (see Clinical Focus: Immune System Problems of Clinical Significance). For example, poison ivy antigens can be processed by Langerhans' cells in the skin, which present the antigen to delayed hypersensitivity T cells, resulting in an intense inflammatory response.

Immune Interactions

Although the immune system can be described in terms of innate, antibody-mediated, and cell-mediated immunity, there is really only one immune system. These categories are an artificial division that is used to emphasize particular aspects of immunity. Actually, immune system responses often involve components of more than one type of immunity (figure 22.20).

For example, although adaptive immunity can recognize and remember specific antigens, once recognition has occurred, many of the events that lead to the destruction of the antigen are innate immunity activities, such as inflammation and phagocytosis.

Immunotherapy

Knowledge of the basic ways that the immune system operates has produced two fundamental benefits: (1) an understanding of the cause and progression of many diseases, and (2) the development or proposed development of effective methods to prevent, stop, or even reverse diseases.

Immunotherapy treats disease by altering immune system function or by directly attacking harmful cells. Some approaches attempt to boost immune system function in general. For example, administering cytokines or other agents can promote inflammation and the activation of immune cells, which can help in the destruction of tumor cells. On the other hand, sometimes inhibiting the immune system is helpful. For example, multiple sclerosis is an autoimmune disease in which the immune system treats self-antigens as foreign antigens, destroying the myelin that covers axons. Interferon beta (IFNβ) blocks the expression of MHC molecules that display self-antigens and is now being used to treat multiple sclerosis.

Some immunotherapy takes a more specific approach. For example, vaccination can prevent many diseases (see section on Acquired Immunity later in this chapter). The ability to produce monoclonal antibodies may result in therapies that are effective for treating tumors. If an antigen unique to tumor cells can be found, then monoclonal antibodies could be used to deliver radioactive isotopes, drugs, toxins, enzymes, or cytokines that can kill the tumor cell or can activate the immune

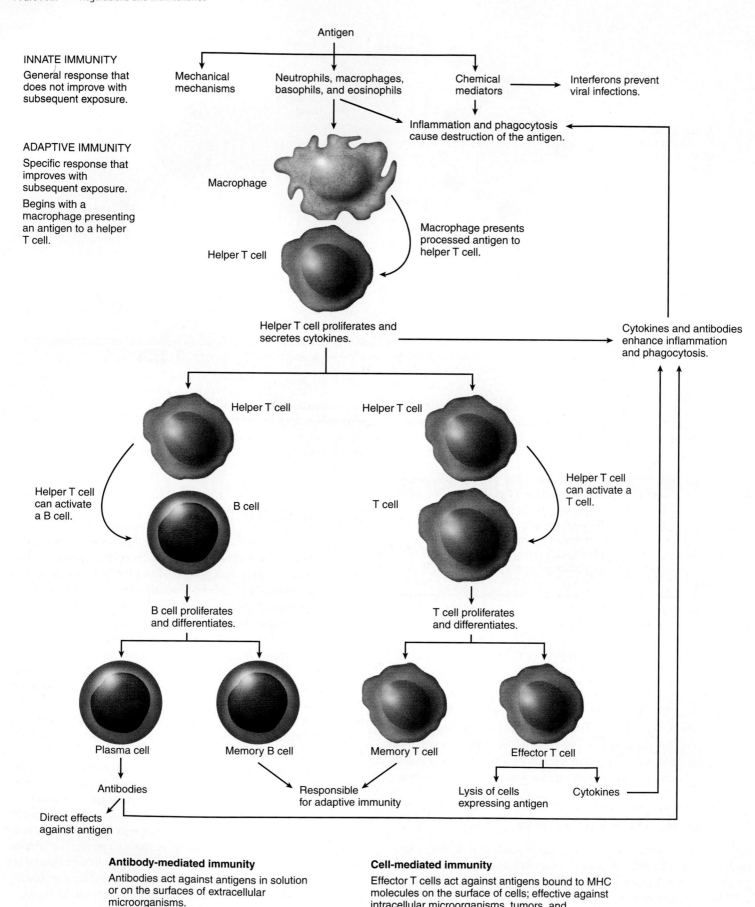

INNATE IMMUNITY
General response that does not improve with subsequent exposure.

ADAPTIVE IMMUNITY
Specific response that improves with subsequent exposure.

Begins with a macrophage presenting an antigen to a helper T cell.

Antigen

Mechanical mechanisms

Neutrophils, macrophages, basophils, and eosinophils

Chemical mediators

Interferons prevent viral infections.

Inflammation and phagocytosis cause destruction of the antigen.

Macrophage

Helper T cell

Macrophage presents processed antigen to helper T cell.

Helper T cell proliferates and secretes cytokines.

Cytokines and antibodies enhance inflammation and phagocytosis.

Helper T cell

Helper T cell

Helper T cell can activate a B cell.

B cell

T cell

Helper T cell can activate a T cell.

B cell proliferates and differentiates.

T cell proliferates and differentiates.

Plasma cell

Memory B cell

Memory T cell

Effector T cell

Antibodies

Responsible for adaptive immunity

Lysis of cells expressing antigen

Cytokines

Direct effects against antigen

Antibody-mediated immunity
Antibodies act against antigens in solution or on the surfaces of extracellular microorganisms.

Cell-mediated immunity
Effector T cells act against antigens bound to MHC molecules on the surface of cells; effective against intracellular microorganisms, tumors, and transplanted cells.

Figure 22.20 The Major Interactions and Responses of Innate and Adaptive Immunity to an Antigen

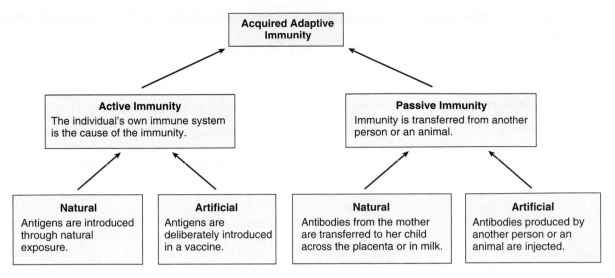

Figure 22.21   Ways to Acquire Adaptive Immunity

system to kill the cell. Unfortunately, no antigen on tumor cells has been found that is not also found on normal cells. Nonetheless, this approach may be useful if damage to normal cells is minimal. For example, tumor cells may have more surface antigens of a particular type than normal cells, resulting in greater treatment delivery. Tumor cells may also be more susceptible to damage, or normal cells may be better able to recover from the treatment.

One problem with monoclonal antibody delivery systems is that the immune system recognizes the monoclonal antibody as a foreign antigen. After the first exposure, a memory response quickly destroys the monoclonal antibodies, rendering the treatment ineffective. In a process called **humanization,** the monoclonal antibodies are modified to resemble human antibodies. This approach has allowed monoclonal antibodies to sneak past the immune system.

The use of monoclonal antibodies to treat tumors is mostly in the research stage of development, but a few clinical trials are now yielding promising results. For example, monoclonal antibodies with radioactive iodine (^{131}I) caused regression of B-cell lymphomas and produced few side effects.

Many other approaches for immunotherapy are being studied, and the development of treatments that use the immune system are certain to increase in the future. Your knowledge of the immune system will enable you to understand and appreciate these therapies.

Clinical Note

An intriguing possibility for reducing the severity of diseases or even curing diseases is to use neuroendocrine regulation of the immune system. The nervous system regulates the secretion of hormones such as cortisol, epinephrine, endorphins, and enkephalins, for which lymphocytes have receptors. For example, cortisol released during times of stress inhibits the immune system. In addition, most lymphatic tissues, including some individual lymphocytes, receive

sympathetic innervation. That a neuroendocrine connection exists with the immune system is clear. The question we need to answer is: Can we use this connection to control our own immunotherapy?

Acquired Immunity

There are four ways to acquire adaptive immunity: active natural, active artificial, passive natural, and passive artificial (figure 22.21). The terms natural and artificial refer to the method of exposure. Natural exposure implies that contact with an antigen or antibody occurs as part of everyday living and is not deliberate. Artificial exposure, also called **immunization,** is a deliberate introduction of an antigen or antibody into the body.

The terms active and passive indicate whether or not an individual's immune system is directly responding to the antigen. When an individual is naturally or artificially exposed to an antigen, an adaptive immune system response can occur that produces antibodies. This is called **active immunity** because the individual's own immune system is the cause of the immunity. **Passive immunity** occurs when another person or animal develops antibodies and the antibodies are transferred to a nonimmune individual. This is called passive immunity because the nonimmune individual did not produce the antibodies.

How long the immunity lasts differs for active and passive immunity. Active immunity can persist for a few weeks (common cold) to a lifetime (whooping cough and chickenpox). Immunity can be long lasting if enough B or T memory cells are produced and persist to respond to later antigen exposure. Passive immunity is not long lasting because the individual does not produce his or her own memory cells. Because active immunity can last longer than passive immunity, it is the preferred method. Passive immunity is preferred, however, in some situations when immediate protection is needed.

Clinical Focus   Acquired Immunodeficiency Syndrome

Acquired immune deficiency syndrome (AIDS) is a life-threatening disease caused by the **human immunodeficiency virus (HIV).** Two strains of HIV have been recognized: HIV-1 is responsible for most cases of AIDS, whereas HIV-2 is increasingly being found in West Africa. AIDS was first reported in 1981 in the United States. Since then over 650,000 cases have been reported in the United States to the Centers for Disease Control and Prevention (CDC). The World Health Organization (WHO) estimates that 40 million people have been infected by HIV worldwide, and almost 12 million have died. The course of HIV infection varies. After contracting HIV, some people die within a year, most, however, survive for 10 or 11 years, and some have survived beyond 20 years.

HIV is transmitted from an infected to a noninfected person in body fluids, such as blood, semen, or vaginal secretions. The major methods of transmission are unprotected intimate sexual contact, contaminated needles used by intravenous drug users, tainted blood products, and from a pregnant woman to her fetus. Present evidence indicates that household, school, or work contacts do not result in transmission.

In the United States, most cases of AIDS appear in homosexual or bisexual men and in intravenous drug users. A small percentage of cases have resulted from transfusions or contaminated clotting factors used by hemophiliacs. Children can be infected before birth, during delivery, or after birth from breast feeding. A few cases of AIDS have occurred in health-care workers accidentally exposed to HIV-infected blood or body fluids, and even fewer cases of health-care workers infecting patients have been documented. The most rapidly increasing group of AIDS patients in the United States is heterosexual males and females who have had sexual contact with infected people.

In other countries the pattern of AIDS cases can be different from that in the United States. For example, in Haiti and central Africa, heterosexual transmission is the major route of spread of HIV. Worldwide, approximately 40% of AIDS cases occur in females.

Preventing transmission of HIV is presently the only way to prevent AIDS. The risk of transmission can be reduced by educating the public about relatively safe sexual practices such as reducing the number of one's sexual partners, avoiding anal intercourse, and using a condom. Public education also includes warnings to intravenous drug users of the dangers of using contaminated needles. Ensuring the safety of the blood supply is another important preventive measure. In 1985, a test for HIV antibodies in blood became available. Heat treatment of clotting factors taken from blood has also been effective in preventing transmission of HIV to hemophiliacs.

HIV infection begins when a viral protein called gp120 binds to a cell surface molecule known as CD4. The CD4 molecule is found primarily on helper T cells, and it normally enables helper T cells to adhere to other lymphocytes, for example, during the process of antigen presentation. Certain monocytes, macrophages, neurons, and neuroglial cells also have CD4 molecules. Once attached to the CD4 molecules, the virus injects its genetic material (RNA) and enzymes into the cell. A viral **reverse transcriptase enzyme** uses the viral RNA to produce DNA, which directs the formation of new HIV RNA and proteins. A viral **protease** (prō'tē-ās) **enzyme** modifies and combines the proteins with the viral RNA to produce additional viruses that can infect other cells.

Following infection by HIV, within 3 weeks to 3 months many patients develop mononuculeosis-like symptoms such as fever, sweats, fatigue, muscle and joint aches, headache, sore throat, diarrhea, rash, and swollen lymph nodes. Within 1–3 weeks these symptoms disappear as the immune system responds to the virus by producing antibodies and activating cytotoxic T cells that kill HIV-infected cells. The immune system is not able to completely eliminate HIV, however, and by about 6 months a kind of "set point" is achieved in which the virus continues to replicate at a low, but steady rate. This chronic stage of infection lasts, on the average, for 8–10 years, and the infected person feels good and exhibits few, if any, symptoms.

Although helper T cells are infected and destroyed during the chronic stage of HIV infection, the body responds by producing large numbers of helper T cells. Nonetheless, over a period of years the HIV numbers gradually increase and helper T cell numbers decrease. Normally there are approximately 1200 helper T cells per cubic millimeter of blood. An HIV-infected person is considered to have AIDS when one or more of the following conditions appear: the helper T cell count falls below 200 cells/mm^3, an opportunistic infection occurs, or Kaposi's sarcoma develops.

Opportunistic infections involve organisms that normally do not cause disease, but can do so when the immune system is depressed. Without helper T cells, cytotoxic T- and B-cell activation is impaired, and adaptive resistance is suppressed. Examples of opportunistic infections include pneumocystis (nū-mō-sis'tis) pneumonia (caused by the intracellular protozoan *Pneumocystis carinii*); tuberculosis (caused by an intracellular bacterium, *Myocobacterium tuberculosis*); syphilis (caused by a sexually transmitted bacterium, *Treponema pallidum*); candidiasis (kan-di-dī'ǎ-sis; a yeast infection of the mouth or vagina caused by *Candida albicans*); and protozoans that cause severe, persistent diarrhea. Kaposi's sarcoma is a type of cancer that produces lesions in the skin, lymph nodes, and visceral organs. Also associated with AIDS are symptoms resulting from the effects of HIV on the nervous system, including motor retardation, behavioral changes, progressive dementia, and possibly psychosis.

There currently is no cure for AIDS. Management of AIDS can be divided into two categories: (1) management of secondary infections or malignancies associated with AIDS, and (2) treatment of the HIV infection itself. The first effective treatment of AIDS was the drug azidothymidine (AZT) (az'i-dō-thī'mi-dēn), also called zidovudine (zī-dō'vū-dēn). AZT is classified as a **nucleoside analog.** A nucleoside is a building block of nucleic acids (see chapter 2), and a nucleoside analog is a drug that resembles a nucleoside. When reverse transcriptase adds the nucleoside analog to a developing DNA strand of HIV, the analog stops further formation of the DNA, and thus prevents viral replication. AZT can delay the onset of AIDS but does not appear to increase the survival time of AIDS patients. The number of babies who contract AIDS from their HIV-infected mothers, however, can be dramatically reduced by giving AZT to the mothers during pregnancy and to the babies following birth. AZT can produce serious side effects such as anemia or even total bone marrow failure.

(*continued*)

It is now known that HIV undergoes mutations that make the virus resistant to a particular drug, such as AZT. Fortunately, other drugs have been developed. For example, other nucleoside analogs are effective against HIV that are resistant to AZT. Another strategy has been to develop drugs that prevent viral replication in different ways. For example, the **nonnucleoside reverse transcriptase inhibitors** bind to reverse transcriptase and prevent it from functioning. **Protease inhibitors,** which interfere with the protease enzymes necessary for viral protein production, are very effective.

The current treatment for suppressing HIV replication is to use a combination of three drugs, such as two nucleoside analogs and a protease inhibitor. It is less likely that HIV will develop resistance to all three drugs. This strategy has proven very effective, reducing the death rate from AIDS and partially restoring health in some individuals. Unfortunately, at a yearly cost of over $10,000 per patient, these drugs are not available in developing countries, which have over 90% of the HIV-infected people.

Another advance in AIDS treatment is a test for measuring **viral load,** which measures the number of viral RNA molecules in a milliliter of blood. The actual level of HIV is one-half the RNA count because each HIV has two RNA strands. It has been learned that viral load is a good predictor of how soon a person will develop AIDS. If viral load is high, the onset of AIDS is much sooner than if viral load is low. It is also possible to detect developing viral resistance by an increase in viral load. In response, a change in drug dose or type may contain the virus. Current treatment guidelines are to keep viral load below 500 RNA molecules per milliliter of blood.

An effective treatment for AIDS is not a cure. Even if viral load decreases to the point that the virus is undetected in the blood, the virus still remains in cells throughout the body. It is possible that the virus will eventually mutate and escape drug suppression. In addition, the long-term effects of these drug therapies are unknown.

The long-term goal for dealing with AIDS is to develop a vaccine that prevents HIV infection. Vaccines under development stimulate the production of antibodies against HIV, stimulate a cell-mediated response against HIV-infected cells, or both. In June, 1998, the first large-scale testing of a vaccine that stimulates antibody production against HIV gp120 protein began in the United States, Canada, and Thailand. The study will be completed in 3 years.

Active Natural Immunity

Natural exposure to an antigen such as a disease-causing microorganism can cause an individual's immune system to mount an adaptive immune system response against the antigen. Because the individual is not immune during the first exposure, he or she usually develops the symptoms of the disease. Interestingly, exposure to an antigen does not always produce symptoms. Many people, if exposed to the poliomyelitis virus at an early age, have an immune system response and produce poliomyelitis antibodies, yet they do not exhibit any disease symptoms.

Active Artificial Immunity

In active artificial immunity an antigen is deliberately introduced into an individual to stimulate his or her immune system. This process is **vaccination,** and the introduced antigen is a **vaccine.** Injection of the vaccine is the usual mode of administration (tetanus toxoid, diphtheria, and whooping cough), although ingestion (Sabin poliomyelitis vaccine) is sometimes used. The vaccine usually consists of some part of the microbe; a dead microbe; or a live, altered microbe. The antigen has been changed so that it stimulates the immune system but does not cause the symptoms of the disease. Active artificial immunity is the preferred method of acquiring adaptive immunity because it produces long-lasting immunity without disease symptoms.

The memory response is the basis for many immunization schemes. The first injection of the antigen stimulates a primary response, and the booster shot causes a memory response, which produces high levels of antibody, many memory cells, and long-lasting protection.

Passive Natural Immunity

Passive natural immunity results when antibodies are transferred from a mother to her child. During her life the mother has been exposed to many antigens, either naturally or artificially, and she therefore has antibodies against many of these antigens. These antibodies protect both the mother and the developing fetus against disease. The antibody IgG can cross the placenta and enter the fetal circulation. After birth the antibodies provide protection for the first few months of the infant's life. Eventually the antibodies are broken down, and the infant must rely on its own immune system. If the mother nurses her child, IgA in the mother's milk also provides some protection for the infant.

Passive Artificial Immunity

Achieving passive artificial immunity usually begins with vaccinating an animal such as a horse. After the animal's immune system responds to the antigen, antibodies (sometimes T cells) are removed from the animal and injected into the individual requiring immunity. In some cases a human who has developed immunity is used as a source. Passive artificial immunity provides immediate protection for the individual receiving the antibodies and is therefore preferred when there might not be time for the individual to develop his or her own immunity. This technique provides only temporary immunity, however, because the antibodies are used or eliminated by the recipient. Examples of passive artificial immunity include antibodies that protect against hepatitis, rabies, and tetanus.

Clinical Focus   Immune System Problems of Clinical Significance

Hypersensitivity Reactions

Immune and hypersensitivity (allergy) reactions involve the same mechanisms, but the differences between them are unclear. Both require exposure to an antigen and subsequent stimulation of antibody-mediated immunity or cell-mediated immunity (or both). If immunity to an antigen is established, later exposure to the antigen results in an immune system response that eliminates the antigen, and no symptoms appear. In **hypersensitivity reactions** the antigen is called an **allergen,** and later exposure to the allergen stimulates much the same process that occurs during the normal immune system response. The processes that eliminate the allergen, however, also produce undesirable side effects such as a very strong inflammatory reaction. This immune system response can be more harmful than beneficial and can produce many unpleasant symptoms. Hypersensitivity reactions are categorized as immediate or delayed.

Immediate Hypersensitivities

Immediate hypersensitivities are caused by antibodies interacting with the allergen, and symptoms appear within a few minutes of exposure to the allergen. Immediate hypersensitivity reactions include atopy, anaphylaxis, cytotoxic reactions, and immune complex disease.

Atopy (at'ō-pē) is a localized IgE-mediated hypersensitivity reaction. For example, plant pollens can be allergens that cause hay fever when they are inhaled and absorbed through the respiratory mucosa. The resulting localized inflammatory response produces swelling of the mucosa and excess mucus production. In asthma patients, allergens can stimulate the release of leukotrienes and histamine in the bronchioles of the lung, causing constriction of the smooth muscles of the bronchioles and difficulty in breathing. Hives (urticaria) is an allergic reaction that results in a skin rash or localized swellings and is usually caused by an ingested allergen.

Anaphylaxis (an'ă-fī-lak'sis) is a systemic IgE-mediated reaction and can be life-threatening. Introduction of allergens such as drugs (e.g., penicillin) and insect stings is the most common cause. The chemicals released from mast cells and basophils cause systemic vasodilation, a drop in blood pressure, and cardiac failure. Symptoms of hay fever, asthma, and hives may also be observed.

In **cytotoxic reactions,** IgG or IgM combines with the antigen on the surface of a cell, resulting in the activation of complement and subsequent lysis of the cell. Transfusion reactions caused by incompatible blood types, hemolytic disease of the newborn (see chapter 19), and some types of autoimmune disease are examples.

Immune complex disease occurs when too many immune complexes are formed. Immune complexes are combinations of soluble antigens and IgG or IgM. When there are too many immune complexes, too much complement is activated, and an acute inflammatory response develops. Complement attracts neutrophils to the area of inflammation and stimulates the release of lysosomal enzymes. This release causes tissue damage, especially in small blood vessels, where the immune complexes tend to lodge; and lack of blood supply causes tissue necrosis. Arthus reactions, serum sickness, some autoimmune diseases, and chronic graft rejection are examples of immune complex diseases.

An **Arthus reaction** is a localized immune complex reaction. For example, if an individual has been sensitized to antigens in the tetanus toxoid vaccine because of repeated vaccinations and if that individual were vaccinated again, at the injection site there would be large amounts of antigen with which the antibody could complex, causing a localized inflammatory response, neutrophil infiltration, and tissue necrosis.

Serum sickness is a systemic Arthus reaction in which the antibody–antigen complexes circulate and lodge in many different tissues. Serum sickness can develop from prolonged exposure to an antigen that provides enough time for an antibody response and the formation of many immune complexes. Examples of antigens include long-lasting drugs and passive artificial immunity. Symptoms include fever, swollen lymph nodes and spleen, and arthritis. Symptoms of anaphylaxis such as hives may also be present because IgE involvement is a part of serum sickness. If large numbers of the circulating antibody–antigen complexes are removed from the blood by the kidney, immune complex glomerulonephritis can develop, in which kidney blood vessels are destroyed and the kidneys fail to function.

Delayed Hypersensitivity

Delayed hypersensitivity is mediated by T cells, and symptoms usually take several hours or days to develop. Like immediate hypersensitivity, delayed hypersensitivity is an acute extension of the normal operation of the immune system. Exposure to the allergen causes activation of T cells and the production of cytokines. The cytokines attract basophils and monocytes, which differentiate into macrophages. The activities of these cells result in progressive tissue destruction, loss of function, and scarring.

Delayed hypersensitivity can develop as allergy of infection and contact hypersensitivity. **Allergy of infection** is a side effect of cell-mediated efforts to eliminate intracellular microorganisms, and the amount of tissue destroyed is determined by the persistence and distribution of the antigen. The minor rash of measles results from tissue damage as cell-mediated immunity destroys virus-infected cells.

In patients with chronic infections with long-term antigenic stimulation, the allergy-of-infection response can cause extensive tissue damage. The destruction of lung tissue in tuberculosis is an example.

Contact hypersensitivity is a delayed hypersensitivity reaction to allergens that contact the skin or mucous membranes. Poison ivy, poison oak, soaps, cosmetics, drugs, and a variety of chemicals can induce contact hypersensitivity, usually after prolonged exposure. The allergen is absorbed by epithelial cells, and T cells invade the affected area, causing inflammation and tissue destruction. Although itching can be intense, scratching is harmful because it damages tissues and causes additional inflammation.

Autoimmune Diseases

In **autoimmune disease** the immune system fails to differentiate between self-antigens and foreign antigens. Consequently, an immune system response is produced against some self-antigens, resulting in tissue destruction. In many instances, autoimmunity probably results from a breakdown of tolerance, which normally prevents an immune system response to self-antigens. In a situation called molecular mimicry, a foreign antigen that is very similar to a self-antigen

(continued)

stimulates an immune system response. After the foreign antigen is eliminated, the immune system continues to act against the self-antigen. It is hypothesized that type I diabetes (see chapter 18) develops in this fashion. In susceptible people, a foreign antigen can stimulate adaptive immunity, especially cell-mediated immunity, which destroys the insulin-producing beta cells of the pancreas. Other autoimmune diseases that involve antibodies are rheumatoid arthritis, rheumatic fever, Graves' disease, systemic lupus erythematosus, and myasthenia gravis.

Immunodeficiencies

Immunodeficiency is a failure of some part of the immune system to function properly. A deficient immune system is not uncommon because it can have many causes. Inadequate protein in the diet inhibits protein synthesis, allowing antibody levels to decrease. Stress can depress the immune system, and fighting an infection can deplete lymphocyte and granulocyte reserves, making a person more susceptible to further infection. Diseases that cause proliferation of lymphocytes, such as mononucleosis, leukemias, and myelomas, can result in an abundance of lymphocytes that do not function properly. Finally, the immune system can purposefully be suppressed by drugs to prevent graft rejection.

Congenital (present at birth) immunodeficiencies can involve inadequate B-cell formation, inadequate T-cell formation, or both. **Severe combined immunodeficiency disease (SCID)** in which both B and T cells fail to differentiate is probably the best known. Unless the person suffering from SCID is kept in a sterile environment or is provided with a compatible bone marrow transplant, death from infection results.

Tumor Control

Tumor cells have tumor antigens that distinguish them from normal cells. According to the concept of **immune surveillance,** the immune system detects tumor cells and destroys them before a tumor can form. T cells, NK cells, and macrophages are involved in the destruction of tumor cells. Immune surveillance may exist for some forms of cancer caused by viruses. The immune response appears to be directed more against the viruses, however, than against tumors in general. Only a few cancers are known to be caused by viruses in humans. For most tumors the response of the immune system may be ineffective and too late.

Transplantation

The genes that code for the production of the MHC molecules are generally called the major histocompatibility complex genes. Histocompatibility refers to the ability of tissues (Greek, *histo*) to get along (compatibility) when tissues are transplanted from one individual to another. In humans, the major histocompatibility complex genes are often referred to as **human leukocyte antigen (HLA) genes** because they were first identified in leukocytes. The HLA genes control the production of HLAs, also called MHC antigens, which are inserted onto the surface of cells. The immune system can distinguish between self- and foreign cells because they are both marked with HLAs. Rejection of a transplanted tissue is caused by a normal immune system response to the foreign HLAs. There are millions of different possible combinations of the HLA genes, and it is very rare for two individuals (except identical twins) to have the same set of HLA genes. Because they are genetically determined, however, the closer the relationship between two individuals, the greater the likelihood of sharing the same HLA genes.

Acute rejection of a graft occurs several weeks after transplantation and results from a delayed hypersensitivity reaction and cell lysis. Lymphocytes and macrophages infiltrate the area, a strong inflammatory response occurs, and the foreign tissue is destroyed. If acute rejection does not develop, **chronic rejection** may occur at a later time. In chronic rejection, immune complexes form in the arteries supplying the graft, blood supply fails, and the graft is rejected.

Graft rejection can occur in two different directions. In **host-versus-graft rejection,** the recipient's immune system recognizes the donor's tissue as foreign and rejects the transplant. In a **graft-versus-host rejection,** the donor tissue recognizes the recipient's tissue as foreign, and the transplant rejects the recipient, causing destruction of the recipient's tissues and death.

To reduce graft rejection, a tissue match is performed. Only tissues with HLAs similar to the recipient's have a chance of acceptance. Even when the match is close, immunosuppressive drugs must be administered throughout the person's life to prevent rejection. Unfortunately, the person then has a drug-produced immunodeficiency and is more susceptible to infections. An exact match is possible only for a graft from one part to another part of the same person's body or between identical twins.

HLAs are important in ways in addition to organ transplants. Because they are genetically determined, characterization of HLAs can help resolve paternity suits. In forensic medicine, the HLAs in blood, semen, and other tissues help identify the person from whom the tissue came.

Systems Pathology

systemic lupus erythematosus

SYSTEMIC LUPUS ERYTHEMATOSUS

Mrs. L is a 30-year-old divorced woman with two children. Despite the fact that she has to work to support herself and the children, she entered college, determined to become a nurse and provide a better life for her family. Mrs. L was an excellent student, but her class attendance and her performance on tests were somewhat erratic. Sometimes she seemed very energetic and earned high grades, but other times she seemed depressed and did not do as well. Toward the end of the course, she developed a rash on her face (figure A), a large red lesion on her arm, and was obviously not feeling well. Mrs. L went to the instructor to ask if she could take an incomplete grade and take the last exam at a later time. She explained that she has had lupus since she was 25 years old. Normally, medication helps to control her symptoms, but the stress of being a single parent combined with the challenges of school seemed to be making her condition worse. She further explained that the symptoms of lupus come and go and bed rest was often helpful. Mrs. L finished the course requirements later that summer. She went on to complete her education and now has a full-time job as a nurse at a local hospital.

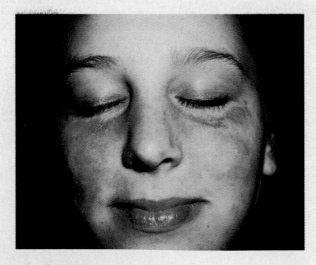

Figure A   Systemic Lupus Erythematosus

The butterfly rash resulting from inflammation in the skin caused by systemic lupus erythematosus.

BACKGROUND INFORMATION

Systemic lupus erythematosus (SLE) is a disease of unknown cause in which tissues and cells are damaged by the immune system. The name describes some of the characteristics of the disease. The term lupus literally means wolf and was originally used to refer to eroded (as if gnawed by a wolf) lesions of the skin. Erythematosus refers to a redness of the skin resulting from inflammation. Unfortunately, as the term systemic implies, the disorder is not confined to the skin but can affect tissues and cells throughout the body. Another systemic effect is the presence of low-grade fever in most cases of active SLE.

SLE is an autoimmune disorder in which a large variety of antibodies are produced that recognize self-antigens, such as nucleic acids, phospholipids, coagulation factors, erythrocytes, and platelets. The combination of the antibodies with self-antigens forms immune complexes that circulate throughout the body to be deposited in various tissues, in which they stimulate inflammation and tissue destruction. Thus, SLE is a disease that can affect many different systems of the body. For example, the most common antibodies act against DNA that is released from damaged cells. Normally the liver removes the DNA, but when DNA and antibodies form immune complexes, they tend to be deposited in the kidneys and other tissues. Approximately 40%–50% of individuals with SLE develop renal disease. In some cases, the antibodies can bind to antigens on cells, resulting in lysis of the cells. For example, the binding of antibodies to erythrocytes results in hemolysis and the development of anemia.

The cause of SLE is unknown. The most popular hypothesis is that a viral infection disrupts the function of suppressor T cells, resulting in loss of tolerance to self-antigens. The picture is probably more complicated, however, because there is evidence that not all SLE patients have reduced numbers of suppressor T cells. In addition, some SLE patients have decreased numbers of the helper T cells that normally stimulate suppressor T-cell activity.

Genetic factors probably contribute to the development of the disease. The likelihood of developing SLE is much higher if a family member also has SLE. In addition, family members of SLE patients who don't have SLE are much more likely to have DNA antibodies than does the general population.

Systemic Interactions

System	Interactions
Integumentary	Skin lesions frequently occur and are made worse by exposure to the sun. There are three forms: (1) an inflammatory redness that can take the form of the butterfly rash, which extends from the bridge of the nose to the cheeks; (2) small, localized pimplelike eruptions accompanied by scaling of the skin; (3) areas of atrophied, depigmented skin with borders of increased pigmentation. Diffuse thinning of the hair results from hair loss.
Skeletal	Arthritis, tendonitis, and death of bone tissue can occur.
Muscular	Destruction of muscle tissue and muscular weakness can occur.
Nervous	Memory loss, intellectual deterioration, disorientation, psychosis, reactive depression, headache, seizures, nausea, and loss of appetite can occur. Stroke is a major cause of dysfunction and death. Cranial nerve involvement results in facial muscle weakness, drooping of the eyelid, and double vision. Central nervous system lesion can cause paralysis.
Endocrine	Sex hormones may play a role in SLE because 90% of the cases occur in females and females with SLE have reduced levels of androgens.
Cardiovascular	Inflammation of the pericardium (pericarditis) with chest pain can develop. Damage to heart valves, inflammation of cardiac tissue, tachycardia, arrhythmias, angina, and myocardial infarction can also occur. Hemolytic anemia, and leukopenia can be present (see chapter 19). Antiphospholipid antibody syndrome, through an unknown mechanism, increases coagulation and thrombus formation, which increases the risk of stroke and heart attack.
Respiratory	Chest pain caused by inflammation of the pleural membranes; fever, shortness of breath, and hypoxemia caused by inflammation of the lungs; and alveolar hemorrhage can develop.
Digestive	Ulcers develop in the oral cavity and pharynx. Abdominal pain and vomiting are common, but no cause can be found. Inflammation of the pancreas and occasionally enlargement of the liver and minor abnormalities in liver function tests occur.
Urinary	Renal lesions and glomerulonephritis can result in progressive failure of kidney function. Excess proteins are lost in the urine, resulting in lower than normal blood proteins, which can produce edema.

Approximately 1 out of 2000 individuals in the United States has SLE. The first symptoms of SLE usually appear between 15 and 25 years of age, affecting women approximately nine times as often as men. The progress of the disease is unpredictable, with flare-ups of symptoms followed by periods of remission. The survival after diagnosis is greater than 90% after 10 years. The most frequent causes of death involve kidney failure, CNS dysfunction, infections, and cardiovascular disease.

There is no cure for SLE, nor is there one standard of treatment because the course of the disease is highly variable and there are many differences between patients with SLE. Treatment usually begins with mild medications and proceeds to more and more potent therapies as conditions warrant. Aspirin and nonsteroidal anti-inflammatory drugs are used to suppress inflammation. Antimalarial drugs are used to treat skin rash and arthritis in SLE, but the mechanism of action is unknown. Patients who do not respond to these drugs or patients with severe SLE are helped by steroids. Although steroids effectively suppress inflammation, they can produce undesirable side effects, including suppression of normal adrenal gland functions. In patients with life-threatening SLE, very high doses of steroids are used.

5 P R E D I C T

The red lesion Mrs. L developed on her arm is called purpura (pŭr′pū-ră), which is caused by bleeding into the skin. The lesions gradually change color and disappear in 2–3 weeks. Explain how SLE produces purpura.

✔ *Answer in Appendix F*

Summary

Lymphatic System

The lymphatic system consists of lymph, lymphatic vessels, lymphocytes, lymph nodules, lymph nodes, tonsils, the spleen, and the thymus gland.

Functions of the Lymphatic System

The lymphatic system maintains fluid balance in tissues, absorbs fats from the small intestine, and defends against microorganisms and foreign substances.

Lymphatic Organs

Lymphatic tissue is reticular connective tissue that contains lymphocytes and other cells.

Diffuse Lymphatic Tissue and Lymph Nodules

1. Diffuse lymphatic tissue consists of dispersed lymphocytes and has no clear boundaries.
2. Lymph nodules are small aggregates of lymphatic tissue (e.g., Peyer's patches in the small intestines).

Tonsils

1. Tonsils are large groups of lymph nodules in the oral cavity and nasopharynx.
2. The three groups of tonsils are the palatine, pharyngeal, and lingual tonsils.

Lymph Nodes

1. Lymphatic tissue in the node is organized into the cortex and the medulla. Lymph sinuses extend through the lymphatic tissue.
2. Substances in lymph are removed by phagocytosis, or they stimulate lymphocytes (or both).
3. Lymphocytes leave the lymph node and circulate to other tissues.

Spleen

1. The spleen is in the left superior side of the abdomen.
2. Foreign substances stimulate lymphocytes in the white pulp.
3. Foreign substances and defective erythrocytes are removed from the blood by phagocytes in the red pulp.

Thymus Gland

1. The thymus gland is in the superior mediastinum and is divided into a cortex and a medulla.
2. Lymphocytes in the cortex are separated from the blood by reticular cells.
3. Lymphocytes produced in the cortex migrate through the medulla, enter the blood, and travel to other lymphatic tissues where they can proliferate.

Immunity

Immunity is the ability to resist the harmful effects of microorganisms and other foreign substances.

Innate Immunity
Mechanical Mechanisms

Mechanical mechanisms prevent the entry of microbes (skin and mucous membranes) or remove them (tears, saliva, and mucus).

Chemical Mediators

1. Chemical mediators promote phagocytosis and inflammation.
2. Complement can be activated by either the alternative or the classical pathway. Complement lyses cells, increases phagocytosis, attracts immune system cells, and promotes inflammation.
3. Interferons prevent viral replication. Interferons are produced by virally infected cells and moves to other cells, which are then protected.

Cells

1. Chemotactic factors are parts of microorganisms or chemicals that are released by damaged tissues. Chemotaxis is the ability of leukocytes to move to tissues that release chemotactic factors.
2. Phagocytosis is the ingestion and destruction of materials.
3. Neutrophils are small phagocytic cells.
4. Macrophages are large phagocytic cells.
   - Macrophages can engulf more than neutrophils can.
   - Macrophages in connective tissue protect the body at locations where microbes are likely to enter, and macrophages clean blood and lymph.
5. Basophils and mast cells release chemicals that promote inflammation.
6. Eosinophils release enzymes that reduce inflammation.
7. NK cells lyse tumor cells and virus-infected cells.

Inflammatory Response

1. The inflammatory response can be initiated in many ways.
   - Chemical mediators cause vasodilation and increase vascular permeability, allowing the entry of other chemical mediators.
   - Chemical mediators attract phagocytes.
   - The amount of chemical mediators and phagocytes increases until the cause of the inflammation is destroyed. Then the tissue undergoes repair.
2. Local inflammation produces the symptoms of redness, heat, swelling, pain, and loss of function. Symptoms of systemic inflammation include an increase in neutrophil numbers, fever, and shock.

Adaptive Immunity

1. Antigens are large molecules that stimulate an adaptive immune system response. Haptens are small molecules that combine with large molecules to stimulate an adaptive immune system response.
2. B cells are responsible for humoral, or antibody-mediated, immunity. T cells are involved with cell-mediated immunity.

Origin and Development of Lymphocytes

1. B cells and T cells originate in red bone marrow. T cells are processed in the thymus gland, and B cells are processed in bone marrow.
2. Positive selection ensures the survival of lymphocytes that can react against antigens, and negative selection eliminates lymphocytes that react against self-antigens.
3. A clone is a group of identical lymphocytes that can respond to a specific antigen.
4. B cells and T cells move to lymphatic tissue from their processing sites. They continually circulate from one lymphatic tissue to another.

Activation of Lymphocytes

1. The antigenic determinant is the specific part of the antigen to which the lymphocyte responds. The antigen receptor (T-cell receptor or B-cell receptor) on the surface of lymphocytes combines with the antigenic determinant.
2. MHC class I molecules display antigens on the surface of nucleated cells, resulting in the destruction of the cells.
3. MHC class II molecules display antigens on the surface of antigen-presenting cells, resulting in the activation of immune cells.
4. MHC–antigen complex and costimulation are usually necessary to activate lymphocytes. Costimulation involves cytokines and certain surface molecules.
5. Antigen-presenting cells stimulate the proliferation of helper T cells, which stimulate the proliferation of B or T effector cells.

Inhibition of Lymphocytes

1. Tolerance is suppression of the immune system's response to an antigen.
2. Tolerance is produced by deletion of self-reactive cells, by preventing lymphocyte activation, and by suppressor T cells.

Antibody-Mediated Immunity

1. Antibodies are proteins.
   - The variable region of an antibody combines with the antigen. The constant region activates complement or binds to cells.
   - There are five classes of antibodies: IgG, IgM, IgA, IgE, and IgD.
2. Antibodies affect the antigen in many ways.
   - Antibodies bind to the antigen and interfere with antigen activity or bind the antigens together.
   - Antibodies act as opsonins (a substance that increases phagocytosis) by binding to the antigen and to macrophages.
   - Antibodies can activate complement through the classical pathway.

   - Antibodies attach to mast cells or basophils and cause the release of inflammatory chemicals when the antibody combines with the antigen.
3. The primary response results from the first exposure to an antigen. B cells form plasma cells, which produce antibodies and memory cells.
4. The secondary response results from exposure to an antigen after a primary response, and memory B cells quickly form plasma cells and additional memory cells.

Cell-Mediated Immunity

1. Antigen activates effector T cells and produces memory T cells.
2. Cytotoxic T cells lyse virus-infected cells, tumor cells, and tissue transplants.
3. Cytotoxic T cells produce cytokines, which promote phagocytosis and inflammation.

Immune Interactions

Innate immunity, antibody-mediated immunity, and cell-mediated immunity can function together to eliminate an antigen.

Immunotherapy

Immunotherapy stimulates or inhibits the immune system to treat diseases.

Acquired Immunity

1. Active natural immunity results from natural exposure to an antigen.
2. Active artificial immunity results from deliberate exposure to an antigen.
3. Passive natural immunity results from the transfer of antibodies from a mother to her fetus or baby.
4. Passive artificial immunity results from transfer of antibodies (or cells) from an immune animal to a nonimmune animal.

Content Review

1. List the parts of the lymphatic system, and describe the three main functions of the lymphatic system.
2. What are diffuse lymphatic tissues and lymph nodules, and where are they found?
3. Describe the structure, location, and function of the Peyer's patches and the tonsils.
4. Describe the structure of a lymph node. What happens to substances in the lymph as lymph passes through the lymph node? What effect can foreign materials in lymph have on lymphocytes?
5. Describe the structure and location of the spleen. What are its functions?
6. Where is the thymus gland located? Describe its structure and function.
7. What is the difference between innate immunity and adaptive immunity?
8. List some mechanical barriers and chemicals, and explain how they provide protection against microorganisms.
9. Define the terms chemotactic factor and chemotaxis.
10. What are the functions of neutrophils and macrophages?
11. What effects are produced by the chemicals released from basophils, mast cells, and eosinophils?
12. Describe the function of NK cells.
13. Describe the events that take place during an inflammatory response. What are the symptoms of local and systemic inflammations?
14. Define the terms antigen and hapten. Distinguish between a foreign antigen and a self-antigen.
15. Describe the origin and development of B and T cells. Distinguish between positive and negative lymphocyte selection.
16. Define the terms antigenic determinant and antigen-binding receptor. Describe a T-cell receptor.
17. Explain how antigens are processed by MHC class I and class II molecules, and describe what happens when these MHC molecules form a complex with an antigen and bind to an antigen receptor.
18. What is costimulation? State two ways in which it can happen.

19. Describe the role of antigen-presenting cells and helper T cells in the activation of B cells.
20. What is tolerance, and how is it accomplished?
21. What are the functions of the constant and variable regions of an antibody? List the five classes of antibodies, and state their functions.
22. Describe the different ways that antibodies participate in the destruction of antigens.
23. What are plasma cells and memory cells, and what are their functions?

24. What are the primary and secondary antibody responses?
25. What are the functions of cytotoxic T cells, delayed hypersensitivity T cells, and memory T cells?
26. Describe the interactions between innate, antibody-mediated, and cell-mediated immunity that can result in the destruction of an antigen.
27. What is immunotherapy?
28. State four general ways of acquiring adaptive immunity. Which two provide the longest lasting immunity?

Develop Your Reasoning Skills

1. A patient had many allergic reactions (see Clinical Focus: Immune System Problems of Clinical Significance). As part of the treatment scheme, it was decided to try to identify the allergen that stimulated the immune system's response. A series of solutions, each containing an allergen that commonly causes a reaction, was composed. Each solution was injected into the skin at different locations on the patient's back. The following results were obtained: (a) at one location the injection site became red and swollen within a few minutes; (b) at another injection site swelling and redness did not appear until 2 days later; and (c) no redness or swelling developed at the other sites. Explain what happened for each observation by describing what part of the immune system was involved and what caused the redness and swelling.

2. Suppose you want to test a woman's serum to see if she has antibodies against the measles virus. Assume that you (a) can take a sample of the woman's serum, (b) have a supply of the measles virus, and (c) have test cells that can be grown in a test tube. When the test cells are infected by the measles virus, they lyse, a reaction that is easily observed. Design a test procedure that can verify or refute the presence of the measles antibody in the woman's serum.

3. If the thymus gland of an experimental animal is removed immediately after its birth, the animal exhibits the following characteristics: (a) it is more susceptible to infections; (b) it has decreased numbers of lymphocytes in lymphatic tissue; and (c) its ability to reject grafts is greatly decreased. Explain these observations.

4. If the thymus gland of an adult experimental animal is removed, the following observations can be made: (a) there is no immediate effect; and (b) after 1 year the number of lymphocytes in the blood decreases, the ability to reject grafts decreases, and the ability to produce antibodies decreases. Explain these observations.

5. Adjuvants are substances that slow but do not stop the release of an antigen from an injection site into the blood. Suppose injection A of a given amount of antigen is given without an adjuvant and injection B of the same amount of antigen is given with an adjuvant that causes the release of antigen over a period of 2–3 weeks. Does injection A or B result in the greater amount of antibody production? Explain.

6. A researcher obtained two samples of blood from a heart attack victim. Sample A was taken the day after the heart attack, and sample B was taken several weeks later. The researcher tested the blood for the presence of antibodies that bind to mitochondria. She found no evidence for the antibodies in sample A but did find the antibodies in sample B. Explain these observations.

Web Site Link

For a listing of the most current web sites related to this chapter, please visit the Seeley home page at:
http://www.mhhe.com/biosci/ap/seeleyap/

Chapter Twenty-Three

Respiratory System

Objectives

1. Describe the anatomy and histology of the respiratory passages.

2. Distinguish between vestibular folds and vocal folds, and explain how sounds of different loudness and pitch are produced.

3. Describe the conducting and respiratory zones of the respiratory passages.

4. List the muscles of respiration and their functions.

5. Describe the lungs, their membranes, and the cavities in which they lie.

6. Describe the factors that affect the flow of air through a tube and the factors that determine the pressure of a gas.

7. Explain the movement of air into and out of the lungs.

8. Describe the factors that cause the alveoli to collapse and expand.

9. Define the term compliance, and give its significance.

10. List the pulmonary volumes and capacities, and define each of them.

11. Explain the significance of forced expiratory vital capacity, minute ventilation, and alveolar ventilation.

12. Define the terms partial pressure of a gas and water vapor pressure.

13. Explain why the partial pressures of oxygen and of carbon dioxide are different for dry atmospheric air, alveolar air, and expired air.

14. Describe the factors affecting movement of a gas into and through a liquid.

15. List the components of the respiratory membrane, and explain the factors that affect gas movement through it.

16. Explain the significance of the oxygen–hemoglobin dissociation curve, and illustrate how it is influenced by exercise.

17. Describe how carbon dioxide is transported in the blood, and discuss the chloride shift and how respiration can affect blood pH.

18. Describe the brainstem structures that regulate respiration. Explain how rhythmic ventilation is produced.

19. Describe the different ways by which rhythmic ventilation can be altered.

Part Four

tudying, sleeping, talking, eating, and exercising all have one common feature—they all involve breathing. From our first breath at birth, the rate and depth of our respiration is unconsciously matched to our activities. Although we can voluntarily stop breathing, within a few seconds we must breathe again. Breathing is so characteristic of life that, along with the pulse, it is one of the first things we check for to determine if an unconscious person is alive.

Breathing is necessary because all living cells of the body require oxygen and produce carbon dioxide. The respiratory system allows oxygen from the air to enter the blood in the lungs and carbon dioxide to leave the blood and enter the air. The cardiovascular system transports oxygen from the lungs to the cells of the body and carbon dioxide from the cells of the body to the lungs. Thus the respiratory and cardiovascular systems work together

to supply oxygen to cells and remove carbon dioxide from them. These systems also work together to regulate the pH of the body fluids. Without healthy respiratory and cardiovascular systems, the capacity to carry out normal activity is reduced, and without adequate respiratory and cardiovascular system functions, life is impossible.

Respiration includes the following processes: (1) ventilation, the movement of air into and out of the lungs; (2) gas exchange between the air in the lungs and the blood, sometimes called external respiration; (3) transport of oxygen and carbon dioxide in the blood; and (4) gas exchange between the blood and the tissues, sometimes called internal respiration. The term respiration is also used in reference to cell metabolism. In aerobic respiration, for example, cells use oxygen and produce carbon dioxide. Cellular respiration is considered in chapter 25.

nasal cavity into two halves (see figure 7.10a). The anterior part of this septum is cartilage, and the posterior part consists of the vomer bone and the perpendicular plate of the ethmoid bone. The external openings to the nasal cavity are the **external nares** (nā′rēs), or **nostrils,** and the posterior openings from the nasal cavity into the pharynx are the **internal nares,** or **choanae** (kō-ā′nē). The anterior part of the nasal cavity, just inside each external naris, is the **vestibule** (ves′ti-būl, meaning entry room). The vestibule is lined with stratified squamous epithelium that is continuous with the stratified squamous epithelium of the skin.

The floor of the nasal cavity is separated from the oral cavity by the **hard palate,** a bony plate covered by a mucous membrane (see figure 23.2). The lateral wall of the nasal cavity is modified by the presence of three bony ridges called **conchae** (kon′kē, meaning resembling a conch shell). Beneath each concha is a passageway called a **meatus** (mē-ā′tŭs, meaning a tunnel or passageway). Within the superior and middle meatus are openings from the various **paranasal sinuses** (see figure 7.11), and the opening of the **nasolacrimal** (nā-zō-lak′ri-măl) **duct** is within the inferior meatus (see figure 15.10).

The nasal cavity has several important functions:

1. The nasal cavity is a passageway for air that is open even when the mouth is full of food.
2. The nasal cavity cleans the air. The vestibule is lined with hairs that trap some of the large particles of dust in the air. The nasal septum and nasal conchae increase the surface area of the nasal cavity and make air flow within the cavity more turbulent, increasing the likelihood that air comes into contact with the mucous membrane lining the nasal cavity. This mucous membrane consists of pseudostratified ciliated columnar epithelium with goblet cells, which secrete a layer of mucus. The mucus traps debris in the air, and the cilia on the surface of the mucous membrane sweep the mucus posteriorly to the pharynx, where it is swallowed and eliminated by the digestive system.
3. The nasal cavity humidifies and warms the air. Moisture from the mucous epithelium and from excess tears that drain into the nasal cavity through the nasolacrimal duct is added to the air as it passes through the nasal cavity. Warm blood flowing through the mucous membrane warms the air within the nasal cavity before it passes into the pharynx, preventing damage from cold air to the rest of the respiratory passages.

Anatomy and Histology

The respiratory system consists of the nasal cavity, the pharynx, the larynx, the trachea, the bronchi, and the lungs (figure 23.1). The term **upper respiratory tract** refers to the nasal cavity, the pharynx, and associated structures; and the **lower respiratory tract** includes the larynx, trachea, bronchi, and lungs. Respiratory movements are accomplished by the diaphragm and the muscles of the thoracic and abdominal walls.

Nose and Nasal Cavity

The **nasus** (nā′sŭs), or **nose,** consists of the external nose and the nasal cavity. The **external nose** is the visible structure that forms a prominent feature of the face. The largest part of the external nose is composed of cartilage plates (see figure 7.10b). The bridge of the nose consists of the nasal bones plus extensions of the frontal and maxillary bones.

The **nasal cavity** is located inside the external nose and joins the pharynx (figure 23.2). The **nasal septum** divides the

1 **P R E D I C T**

Explain what happens to your throat when you sleep with your mouth open, especially when your nasal passages are plugged as a result of having a cold. Explain what may happen to your lungs when you run a long way in very cold weather while breathing rapidly through your mouth.

✔ *Answer in Appendix F*

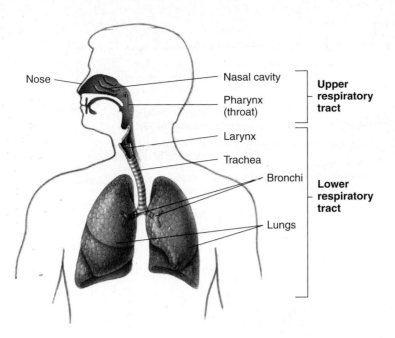

Figure 23.1 The Respiratory System

The upper respiratory tract consists of the nasal cavity and pharynx (throat). The lower respiratory tract consists of the larynx, trachea, bronchi, and lungs.

4. The olfactory epithelium, the sensory organ for smell, is located in the most superior part of the nasal cavity (see figure 15.4).

5. The nasal cavity and paranasal sinuses are resonating chambers for speech.

Pharynx

The **pharynx** (far'ingks, meaning throat) is the common opening of both the digestive and respiratory systems. It receives air from the nasal cavity and air, food, and water from the mouth. Inferiorly, the pharynx is connected to the respiratory system at the larynx and to the digestive system at the esophagus. The pharynx can be divided into three regions: the nasopharynx, the oropharynx, and the laryngopharynx (see figure 23.2).

The **nasopharynx** (nā'zō-far'ingks) is the superior part of the pharynx and extends from the internal nares to the level of the **uvula** (yū'vyū-lă, meaning a grape), a soft process that extends from the posterior edge of the soft palate. The soft palate and uvula prevent swallowed materials from entering the nasopharynx and nasal cavity. The nasopharynx is lined with a mucous membrane containing pseudostratified ciliated columnar epithelium with goblet cells. Debris-laden mucus from the nasal cavity is moved through the nasopharynx and swallowed. Two auditory tubes from the middle ears open into the nasopharynx (see figure 15.23), allowing air passage, which equalizes air pressure between the atmosphere and the middle ears. The posterior surface of the nasopharynx contains the pharyngeal

tonsil, or adenoid (ad'ĕ-noyd), which aids in defending the body against infection (see chapter 22). An enlarged pharyngeal tonsil can interfere with normal breathing.

The **oropharynx** (ōr'ō-far'ingks) extends from the uvula to the epiglottis. The oral cavity opens into the oropharynx through the **fauces** (faw'sēz). Thus food, drink, and air all pass through the oropharynx. The oropharynx is lined by moist stratified squamous epithelium, which provides protection against abrasion. Located near the fauces are two sets of tonsils called the palatine tonsils and the lingual tonsil.

The **laryngopharynx** (lă-ring'gō-far'ingks) passes posterior to the larynx and extends from the tip of the epiglottis to the esophagus. The laryngopharynx is lined with moist stratified squamous epithelium.

Larynx

The **larynx** (lar'ingks) consists of an outer casing of nine cartilages that are connected to one another by muscles and ligaments (figure 23.3). Six of the nine cartilages are paired, and three are unpaired. The largest of the cartilages is the unpaired **thyroid** (meaning shield; refers to the shape of the cartilage) **cartilage,** or Adam's apple.

The most inferior cartilage of the larynx is the unpaired **cricoid** (krī'koyd, meaning ring-shaped) **cartilage,** which forms the base of the larynx on which the other cartilages rest.

The third unpaired cartilage is the **epiglottis** (ep-i-glot'is, meaning on the glottis). It is attached to the thyroid cartilage and projects as a free flap toward the tongue. The epiglottis differs from the other cartilages in that it consists of elastic rather than hyaline cartilage. During swallowing the epiglottis covers the opening of the larynx and prevents materials from entering it.

The paired **arytenoid** (ar-i-tē'noyd, meaning ladle-shaped) **cartilages** articulate with the posterior, superior border of the cricoid cartilage, and the paired **corniculate** (kōr-nik'yū-lāt, meaning horn-shaped) **cartilages** are attached to the superior tips of the arytenoid cartilages (figure 23.3b). The paired **cuneiform** (kyū'nē-i-fōrm, meaning wedge-shaped) **cartilages** are contained in a mucous membrane anterior to the corniculate cartilages (figure 23.4).

Two pairs of ligaments extend from the anterior surface of the arytenoid cartilages to the posterior surface of the thyroid cartilage (see figures 23.3c and 23.4). The superior ligaments form the **vestibular folds,** or **false vocal cords.** When the vestibular folds come together, they prevent food and liquids from entering the larynx during swallowing and prevent air from leaving the lungs, as when a person holds his or her breath.

The inferior ligaments form the **vocal folds,** or **true vocal cords.** The true vocal cords and the opening between them are called the **glottis** (glot'is). The vestibular folds and the vocal cords are lined with stratified squamous epithelium. The remainder of the larynx is lined with pseudostratified ciliated columnar epithelium. An inflammation of the mucosal epithelium of the vocal cords is called **laryngitis** (lar-in-jī'tis).

The larynx performs three important functions.

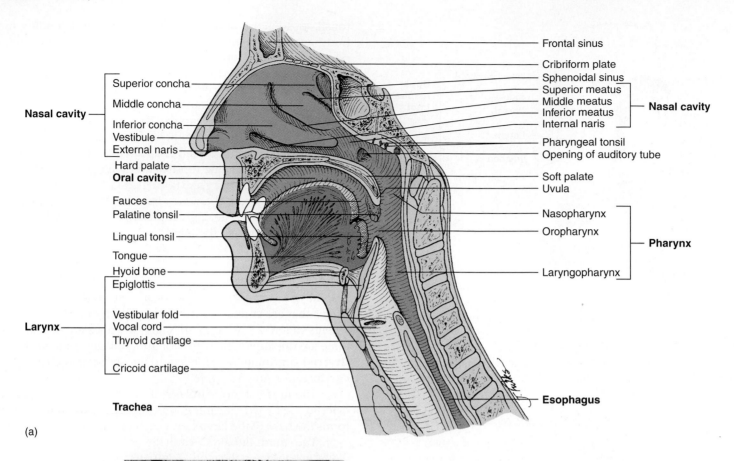

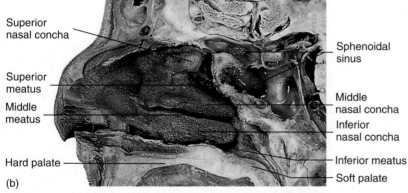

Figure 23.2   Nasal Cavity and Pharynx

(*a*) Sagittal section through the nasal cavity and pharynx viewed from the medial side. (*b*) Photograph of sagittal section of nasal cavity.

1. The thyroid and cricoid cartilages maintain an open passageway for air movement.
2. The movement of swallowed materials into the larynx is prevented by the epiglottis and the vestibular folds.
3. The vocal cords are involved with sound production. When air moves past the vocal cords causing them to vibrate, sound is produced. The greater the amplitude of the vibration, the louder is the sound. Pitch is controlled by the frequency of vibrations. Variations in the length of the vibrating segments of the vocal cords affect the frequency of the vibrations. The length of the vocal cords can be changed by the contraction of skeletal muscles that cause the arytenoid and other cartilages to move. Higher-pitched tones are produced when only the anterior parts of the cords vibrate, and progressively

lower tones result when longer sections of the cords vibrate. Because males usually have longer vocal cords than females, they usually have lower-pitched voices. The sound produced by the vibrating vocal cords is modified by the tongue, lips, teeth, and other structures to form words. A person whose larynx has been removed because of carcinoma of the larynx can produce sound by swallowing air and causing the esophagus to vibrate.

Trachea

The **trachea** (trā′kē-ă), or windpipe, is a membranous tube that consists of dense regular connective tissue and smooth muscle reinforced with 15–20 **C**-shaped pieces of cartilage.

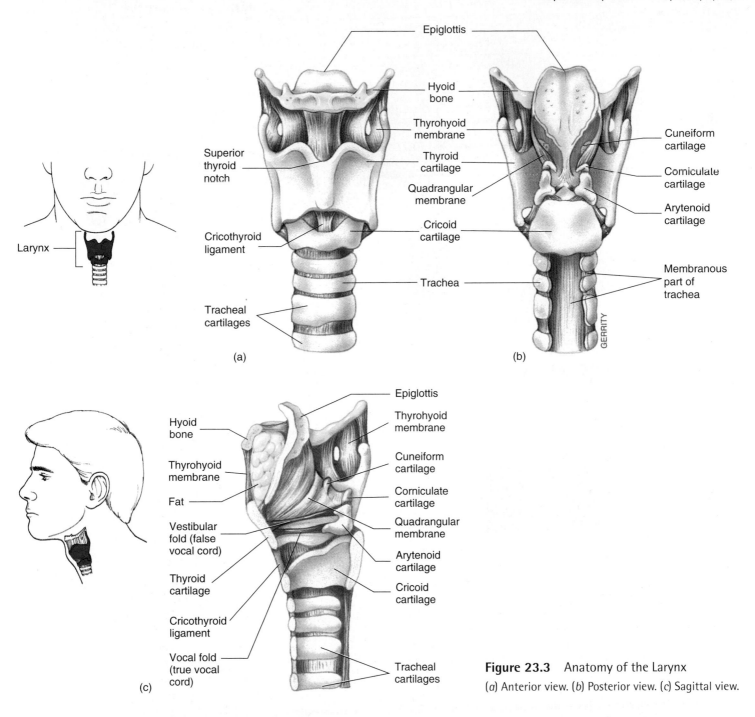

Figure 23.3   Anatomy of the Larynx
(*a*) Anterior view. (*b*) Posterior view. (*c*) Sagittal view.

The cartilages form the anterior and lateral sides of the trachea (figure 23.5*a*). They protect the trachea and maintain an open passageway for air. The posterior wall of the trachea is devoid of cartilage and consists of an elastic ligamentous membrane and bundles of smooth muscle called the **trachealis** (trā'kē-ā-lis) **muscle.** Contraction of the smooth muscle can narrow the diameter of the trachea. During coughing, this action causes air to move more rapidly through the trachea, which helps to expel mucus and foreign objects. The esophagus lies immediately posterior to the cartilage-free posterior wall of the trachea.

2   P R E D I C T

Explain what happens to the shape of the trachea when a person swallows a large mouthful of food. Why is this change of shape advantageous?

✔ *Answer in Appendix F*

The mucous membrane lining the trachea consists of pseudostratified ciliated columnar epithelium with numerous goblet cells (figure 23.5*b*). The cilia propel mucus and foreign particles toward the larynx, where they can enter the pharynx

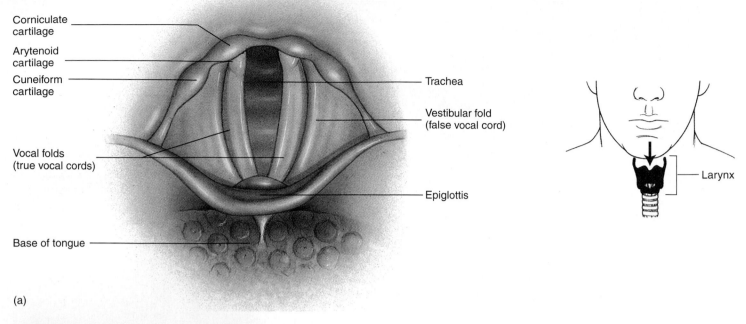

Corniculate cartilage

Arytenoid cartilage

Cuneiform cartilage

Vocal folds (true vocal cords)

Base of tongue

Trachea

Vestibular fold (false vocal cord)

Epiglottis

Larynx

(a)

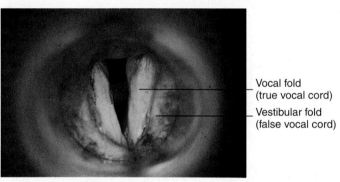

Vocal fold (true vocal cord)

Vestibular fold (false vocal cord)

(b)

Figure 23.4   Vocal Cords

Arrow shows the direction of viewing the vocal cords. (*a*) The relationship of the vocal cords to the paired cartilages of the larynx and the epiglottis. (*b*) Laryngoscopic view of the vocal cords.

and be swallowed. Constant irritation to the trachea, such as occurs in smokers, can cause the tracheal epithelium to become moist stratified squamous epithelium that lacks cilia and goblet cells. Consequently the normal function of the tracheal epithelium is lost.

Clinical Note

In cases of extreme emergency when the upper air passageway is blocked by a foreign object to the extent that the victim cannot breathe, quick reaction is required to save the person's life. The **Heimlich maneuver** is designed to force such an object out of the air passage by the sudden application of pressure to the abdomen, forcing air up the trachea to dislodge the obstruction. The person who performs the maneuver stands behind the victim with his arms under the victim's arms and his hands over the victim's abdomen between the navel and the rib cage. With one hand formed into a fist and the other hand over it, both hands are suddenly pulled toward the abdomen with an accompanying upward motion. This maneuver, if done properly, dislodges most foreign objects.

In rare cases, when the obstruction cannot be removed using the Heimlich maneuver, it may be necessary to form an artificial opening in the victim's air passageway, followed with insertion of a tube to facilitate the passage of air. The preferred point of entry in emergency cases is through the membrane between the cricoid and thyroid cartilages, a procedure referred to as a **cricothyrotomy** (krī′kō-thī-rot′ō-mē). A **tracheotomy** (trā′kē-ot′ō-mē) makes an opening in the trachea, usually between the second and third cartilage ring. Because arteries, nerves, and the thyroid gland overlie the anterior surface of the trachea, it is therefore not advisable to enter the air passageway through the trachea in emergency cases.

The trachea has an inside diameter of 12 mm and a length of 10–12 cm, descending from the larynx to the level of the fifth thoracic vertebra (figure 23.6). The trachea divides to form two smaller tubes called **primary bronchi** (brong′kī; sing. bronchus, brong′kŭs, meaning windpipe). The most inferior tracheal cartilage forms a ridge called the **carina** (kă-rī′nă), which separates the openings into the primary bronchi. The carina is an important radiologic landmark. In addition, the mucous membrane of the carina is very sensitive to mechanical stimulation, and foreign objects reaching the carina stimulate a powerful cough reflex. Once a foreign object passes the carina, coughing usually stops.

Tracheobronchial Tree

The trachea divides to form primary bronchi, which divide to form smaller bronchi, until eventually many microscopically small tubes and sacs are formed. Beginning with the trachea, all of these structures are called the **tracheobronchial** (trā′kē-ō-brong′kē-ăl) **tree** (see figure 23.6). Based on function, the tracheobronchial tree can be subdivided into the conducting zone and the respiratory zone.

Conducting Zone

The **conducting zone** extends from the trachea to small tubes called terminal bronchioles (see figure 23.6). There are approximately 16 generations of branching from the trachea to the terminal bronchioles. The conducting zone functions as a passageway for air movement and contains epithelial tissue that helps to remove debris from the air.

The trachea divides into the left and right primary bronchi, which extend to each lung (see figure 23.6). The right primary bronchus is shorter, has a wider diameter, and is more vertical than the left primary bronchus.

> **3   P R E D I C T**
>
> Into which lung would a foreign object that is small enough to pass into a primary bronchus most likely become lodged and block air movement?

✔ *Answer in Appendix F*

The primary bronchi divide into **secondary (lobar) bronchi** within each lung. There are two secondary bronchi in the left lung and three in the right lung. The secondary bronchi, in turn, give rise to **tertiary (segmental) bronchi.** The bronchi continue to branch, finally giving rise to **bronchioles** (brong′kē-ōlz), which are less than 1 mm in diameter. The bronchioles also subdivide several times to become even smaller **terminal bronchioles.**

As the air passageways of the lungs become smaller, the structure of their walls changes. Like the trachea, the primary bronchi are supported by **C**-shaped cartilage connected by smooth muscle. In the secondary bronchi, the **C**-shaped cartilages are replaced with cartilage plates, and smooth muscle forms a layer between the cartilage and the mucous membrane. As the bronchi become smaller, the cartilage becomes sparser and smooth muscle becomes more abundant. The terminal bronchioles have no cartilage, and the smooth muscle layer is prominent. Relaxation and contraction of the smooth muscle within the bronchi and bronchioles can change the diameter of

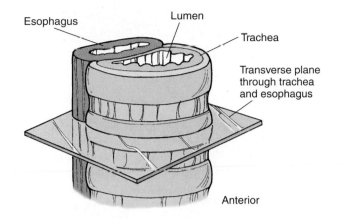

Esophagus   Lumen
Trachea
Transverse plane through trachea and esophagus
Anterior

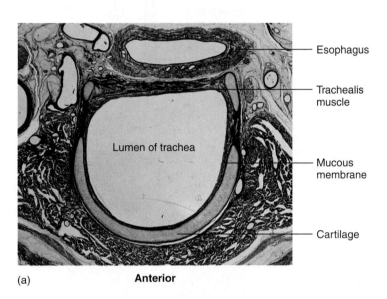

Esophagus
Trachealis muscle
Lumen of trachea
Mucous membrane
Cartilage

(a)   **Anterior**

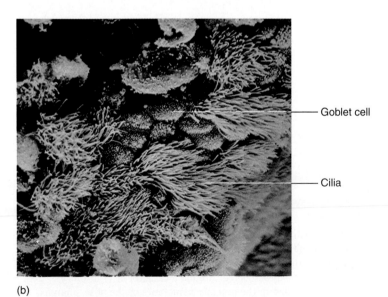

Goblet cell
Cilia

(b)

Figure 23.5   Trachea

(*a*) Photomicrograph of a transverse section of the trachea. The esophagus is next to the trachealis muscle, which connects the ends of the cartilage. (*b*) Scanning electron micrograph of the surface of the mucous membrane lining the trachea. Goblet cells with short microvilli are interspersed between ciliated cells. ⚹

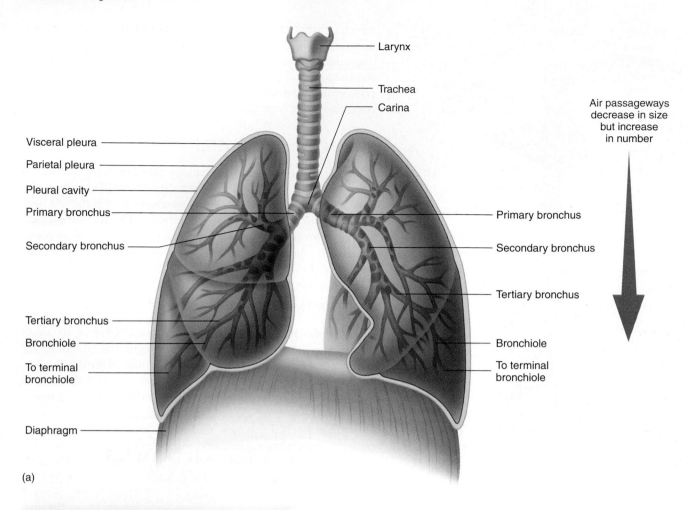

Larynx

Trachea

Carina

Air passageways decrease in size but increase in number

Visceral pleura

Parietal pleura

Pleural cavity

Primary bronchus

Secondary bronchus

Primary bronchus

Secondary bronchus

Tertiary bronchus

Tertiary bronchus

Bronchiole

Bronchiole

To terminal bronchiole

To terminal bronchiole

Diaphragm

(a)

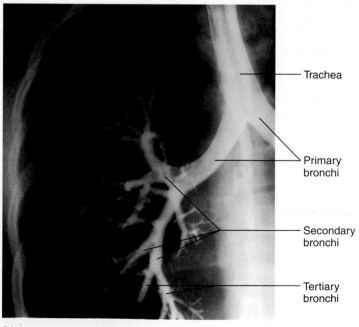

Trachea

Primary bronchi

Secondary bronchi

Tertiary bronchi

(b)

Figure 23.6   Tracheobronchial Tree

(*a*) The conducting zone of the tracheobronchial tree begins at the trachea and ends at the terminal bronchioles. (*b*) A bronchogram is a radiograph of the respiratory tree. A contrast medium, which makes the passageways visible, is injected through a catheter after a topical anesthetic is applied to the mucous membranes of the nose, pharynx, larynx, and trachea. 🏃

the air passageways. For example, during exercise the diameter can increase, which increases the volume of air moved. During an **asthma attack,** however, contraction of the smooth muscle in the terminal bronchioles, which have no cartilage in their walls, can result in greatly reduced air flow. In severe cases, air movement can be so restricted that death results.

The bronchi are lined with a pseudostratified ciliated columnar epithelium. The larger bronchioles are lined with ciliated simple columnar epithelium, which changes to ciliated simple cuboidal epithelium in the terminal bronchioles. The epithelium in the conducting part of the air passageways functions as a mucus–cilia escalator, which traps debris in the air and removes it from the respiratory system.

Respiratory Zone

The **respiratory zone** extends from the terminal bronchioles to small air-filled chambers called **alveoli** (al-vē'ō-lī, meaning hollow cavity), which are the sites of gas exchange between the air and blood. There are approximately seven generations of branching in the respiratory zone. The terminal bronchioles divide to form **respiratory bronchioles** (figure 23.7), which have a limited ability for gas exchange because of a few attached alveoli. As the respiratory bronchioles divide to form smaller respiratory bronchioles, the number of attached alveoli increases. The respiratory bronchioles give rise to **alveolar** (al-vē'ō-lăr) **ducts,** which are like long branching hallways with many open doorways. The doorways open into alveoli, which become so numerous that the alveolar duct wall is little more than a succession of alveoli. The alveolar ducts end as two or three **alveolar sacs,** which are chambers connected to two or more alveoli (see figure 23.7).

The epithelium in the respiratory bronchioles is a simple cuboidal epithelium, which changes to a squamous epithelium in the alveolar ducts and alveoli. Gas exchange between the air and the blood takes place across the thin walls of the squamous epithelium. Some of the epithelial cells of the alveoli are round or cuboidal-shaped secretory cells that produce surfactant (see section on Lung Recoil later in this chapter). Although the epithelium of the respiratory zone is not ciliated, debris from the air can be removed by macrophages that move over the surfaces of the cells. The macrophages do not accumulate in the respiratory zone because they either move into nearby lymphatic vessels or enter terminal bronchioles, becoming entrapped in mucus that is swept to the pharynx.

The tissue surrounding the alveoli contains elastic fibers, which allow the alveoli to expand during inspiration and recoil during expiration. The lungs are very elastic, and when inflated, they are capable of expelling the air and returning to their original, uninflated state. Even when not inflated, however, the lungs retain some air, which gives them a spongy quality.

Lungs

The **lungs** are the principal organs of respiration, and on a volume basis they are among the largest organs of the body. Each lung is conical in shape, with its base resting on the di-

aphragm and its apex extending superiorly to a point approximately 2.5 cm superior to the clavicle. The right lung is larger than the left and weighs an average of 620 g, whereas the left lung weighs 560 g.

The **hilum** (hī'lŭm) is a region on the medial surface of the lung where structures, such as the primary bronchi, blood vessels, nerves, and lymphatic vessels, enter or exit the lung. All the structures passing through the hilum are referred to as the **root of the lung.**

The right lung has three **lobes,** and the left lung has two (figure 23.8). The lobes are separated by deep, prominent **fissures** on the surface of the lung, and each lobe is supplied by a secondary bronchus. The lobes are subdivided into **bronchopulmonary segments,** which are supplied by the tertiary bronchi. There are 9 bronchopulmonary segments in the left lung and 10 bronchopulmonary segments in the right lung. The bronchopulmonary segments are separated from each other by connective tissue partitions, which are not visible as surface fissures. Because major blood vessels and bronchi do not cross the connective tissue partitions, individual diseased bronchopulmonary segments can be surgically removed, leaving the rest of the lung relatively intact (see figure 23.8). The bronchopulmonary segments are subdivided into **lobules** by incomplete connective tissue walls. The lobules are supplied by the bronchioles.

▉ Thoracic Wall and Muscles of Respiration

The **thoracic wall** consists of the thoracic vertebrae, the ribs, the costal cartilages, the sternum, and associated muscles (see chapters 7 and 11). The **thoracic cavity** is the space enclosed by the thoracic wall and the diaphragm. Changes in the shape of the diaphragm and the thoracic wall produce the changes in thoracic volume necessary for respiration.

Muscles associated with the thoracic cage are responsible for respiration (figure 23.9a). The **muscles of inspiration** include the diaphragm and muscles that elevate the ribs and sternum, such as the external intercostals, pectoralis minor, and scalenes. The **diaphragm** (dī'ă-fram, meaning partition) is a large dome of skeletal muscle that separates the thoracic cavity from the abdominal cavity (see figure 11.14). The **muscles of expiration** consist of muscles that depress the ribs and sternum, such as the internal intercostals and the abdominal muscles.

At the end of a normal, quiet expiration, the respiratory muscles are relaxed (see figure 23.9a). During quiet inspiration, contraction of the diaphragm causes the dome shape to flatten, which increases the superior–inferior dimension of the thoracic cavity (figure 23.9b). The largest change in thoracic volume results from movement of the diaphragm. Inferior movement of the diaphragm can occur because of relaxation of the abdominal muscles, which allows the abdominal organs to move out of the way of the diaphragm.

During quiet inspiration, contraction of the external intercostals results in expansion of the thorax and thoracic cavity (see figure 23.9b). As the ribs are elevated, the costal cartilages allow

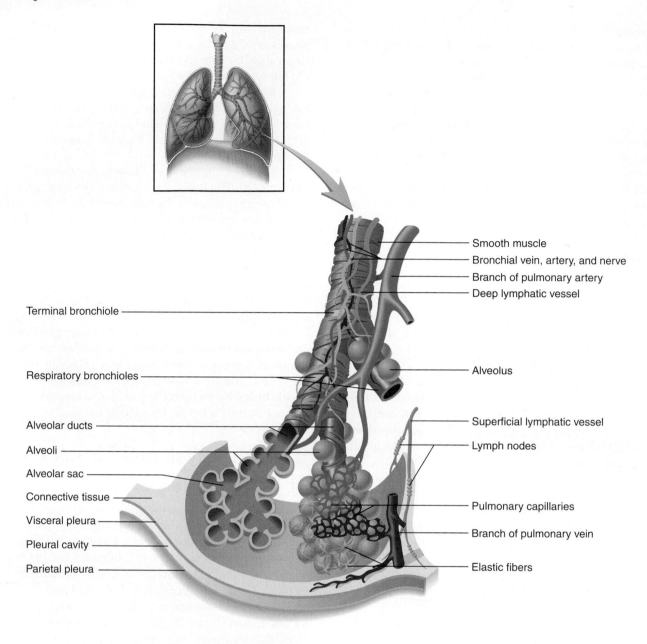

Smooth muscle

Bronchial vein, artery, and nerve

Branch of pulmonary artery

Deep lymphatic vessel

Terminal bronchiole

Respiratory bronchioles

Alveolus

Alveolar ducts

Alveoli

Superficial lymphatic vessel

Lymph nodes

Alveolar sac

Connective tissue

Visceral pleura

Pulmonary capillaries

Pleural cavity

Branch of pulmonary vein

Parietal pleura

Elastic fibers

(a)

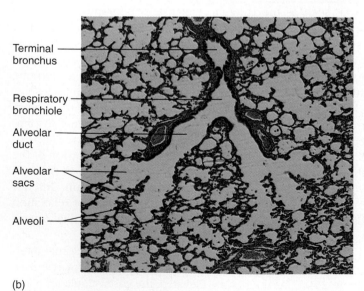

Terminal bronchus

Respiratory bronchiole

Alveolar duct

Alveolar sacs

Alveoli

(b)

Figure 23.7   Bronchioles and Alveoli

(*a*) Alveoli, the sites of gas exchange between air and blood, are connected to respiratory bronchioles, and alveolar ducts and are surrounded by capillaries. (*b*) Photomicrograph of lung tissue.

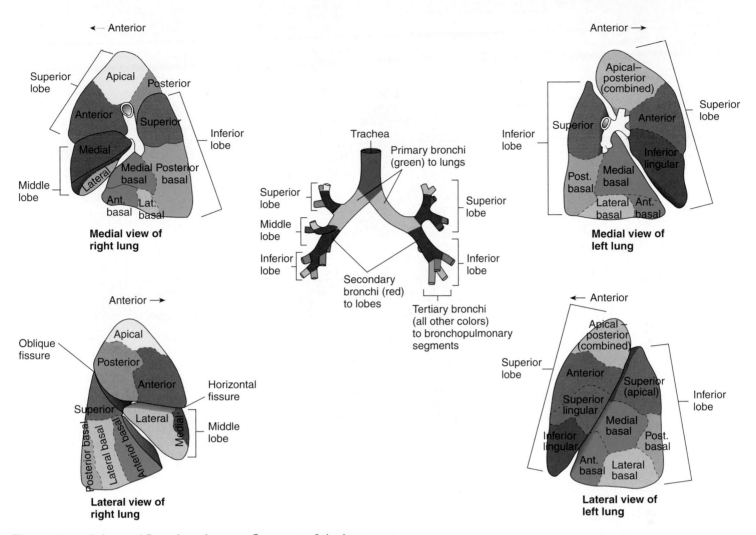

Figure 23.8 Lobes and Bronchopulmonary Segments of the Lungs

The trachea (*blue*), primary bronchi (*green*), secondary bronchi (*red*), and tertiary bronchi (*all other colors*) are in the center of the figure, surrounded by two views of each lung, showing the bronchopulmonary segments. In general, each bronchopulmonary segment is supplied by a tertiary bronchus (*color-coded to match the bronchopulmonary segment it supplies*).

lateral rib movement and lateral expansion of the thoracic cavity. The ribs slope inferiorly from the vertebrae to the sternum, and elevation of the ribs can also increase the anterior–posterior dimension of the thoracic cavity (figure 23.9*c*).

Expiration during quiet breathing occurs when the diaphragm and external intercostals relax and the elastic properties of the thorax and lungs cause a passive decrease in thoracic volume. In addition, contraction of the abdominal muscles helps to push abdominal organs and the diaphragm in a superior direction.

Clinical Note

The importance of the abdominal muscles in breathing can be observed in a person with a spinal cord injury that causes flaccid paralysis of the abdominal muscles. In the upright position, the abdominal organs and diaphragm are not pushed superiorly and passive recoil of the thorax and lungs is inadequate for normal expiration. An elastic binder around the abdomen can help such patients. When lying down, the weight of the abdominal organs can assist in expiration.

There are several differences between normal, quiet breathing and labored breathing. During labored breathing all of the inspiratory muscles are active and they contract more forcefully than during quiet breathing, causing a greater increase in thoracic volume (see figure 23.9*b*). During labored breathing, forceful contraction of the internal intercostals and the abdominal muscles produces a more rapid and greater decrease in thoracic volume than would be produced by the passive recoil of the thorax and lungs.

Pleura

The lungs are contained within the thoracic cavity, but each lung is surrounded by a separate **pleural** (plūr´ăl, meaning

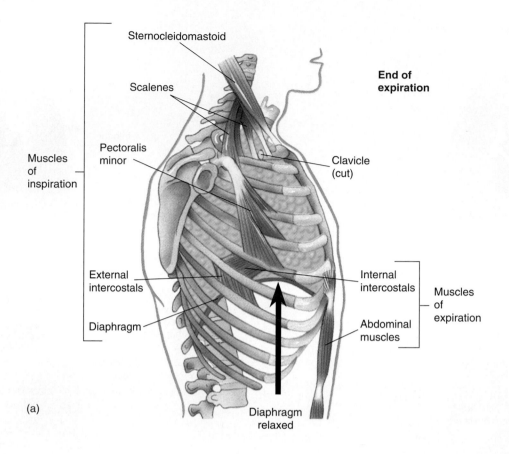

(a)

Sternocleidomastoid

Scalenes

Pectoralis
minor

Muscles
of
inspiration

Clavicle
(cut)

External
intercostals

Internal
intercostals

Muscles
of
expiration

Diaphragm

Abdominal
muscles

Diaphragm
relaxed

**End of
expiration**

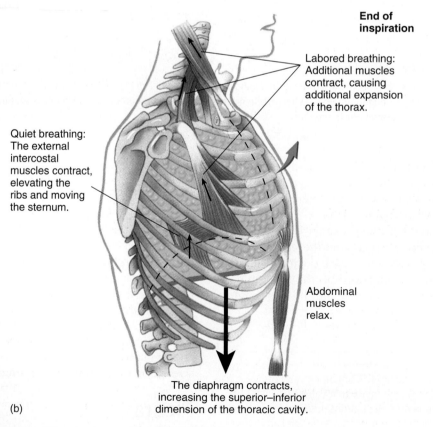

(b)

**End of
inspiration**

Labored breathing:
Additional muscles
contract, causing
additional expansion
of the thorax.

Quiet breathing:
The external
intercostal
muscles contract,
elevating the
ribs and moving
the sternum.

Abdominal
muscles
relax.

The diaphragm contracts,
increasing the superior–inferior
dimension of the thoracic cavity.

Figure 23.9   Effect of the Muscles of Respiration on Thoracic Volume

(a) Muscles of respiration at the end of expiration. (b) Muscles of respiration at the end of inspiration. 🏃

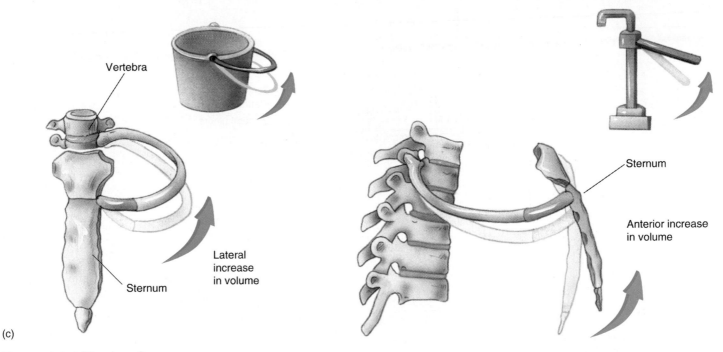

Vertebra

Sternum

Lateral increase in volume

Sternum

Anterior increase in volume

(c)

Figure 23.9 (*Continued*)

(c) Elevation of the rib in the "bucket-handle" movement laterally increases thoracic volume. As the rib is elevated, rotation of the rib in the "pump-handle" movement increases thoracic volume anteriorly.

relating to the ribs) **cavity** formed by the pleural serous membranes (figure 23.10). The pleural cavities are separated by the **mediastinum** (mē′dē-as-tī′nŭm), a midline partition formed by the heart, trachea, esophagus, and associated structures. The **parietal pleura** covers the inner thoracic wall, the superior surface of the diaphragm, and the mediastinum. At the hilum the parietal pleura is continuous with the **visceral pleura,** which covers the surface of the lung.

The pleural cavity is filled with pleural fluid, which is produced by the pleural membranes. The pleural fluid does two things: (1) it acts as a lubricant, allowing the parietal and visceral pleural membranes to slide past each other as the lungs and the thorax change shape during respiration, and (2) it helps hold the parietal and visceral pleural membranes together. Because the parietal pleura is attached to the diaphragm and inner thoracic wall, and the visceral pleura is attached to the lungs, when thoracic volume changes during respiration, lung volume is also changed. The pleural fluid is analogous to a thin film of water between two sheets of glass (the visceral and parietal pleurae); the glass sheets can easily slide over each other, but it is difficult to separate them.

Blood Supply

Blood that has passed through the lungs and picked up oxygen is called **oxygenated blood,** and blood that has passed through the tissues and released some of its oxygen is called **deoxygenated blood.** There are two important blood flow routes to the lungs. The major route brings deoxygenated

blood to the lungs, where it is oxygenated (see chapter 21 and figure 23.10b). The deoxygenated blood flows through the pulmonary arteries to the pulmonary capillaries, becomes oxygenated, and returns to the heart through the pulmonary veins. The other route brings oxygenated blood to the tissues of the bronchi down to the respiratory bronchioles. The oxygenated blood flows from the thoracic aorta through bronchial arteries to capillaries where oxygen is released. Deoxygenated blood from the proximal part of the bronchi returns to the heart through the bronchial veins and the azygos venous system (see chapter 21). More distally, the venous drainage from the bronchi enters the pulmonary veins. Thus the oxygenated blood returning from the alveoli in the pulmonary veins is mixed with a small amount of deoxygenated blood returning from the bronchi.

Lymphatic Supply

The lungs have two lymphatic supplies. The **superficial lymphatic vessels** are deep to the visceral pleura and function to drain lymph from the superficial lung tissue and the visceral pleura. The **deep lymphatic vessels** follow the bronchi and function to drain lymph from the bronchi and associated connective tissues. There are no lymphatic vessels located in the walls of the alveoli. Both the superficial and deep lymphatic vessels exit the lung at the hilum.

Phagocytic cells pick up carbon particles and other debris from inspired air and move it to the lymphatic vessels. In older people, the surface of the lungs can appear gray to black

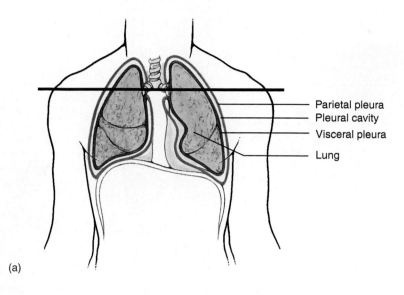

(a)

- Parietal pleura
- Pleural cavity
- Visceral pleura
- Lung

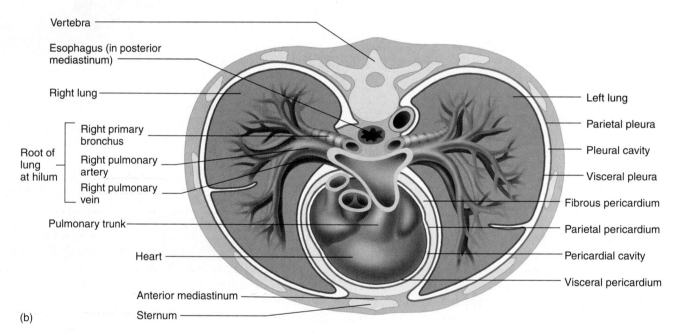

Vertebra

Esophagus (in posterior mediastinum)

Right lung

Root of lung at hilum
- Right primary bronchus
- Right pulmonary artery
- Right pulmonary vein

Pulmonary trunk

Heart

Anterior mediastinum

Sternum

(b)

Left lung

Parietal pleura

Pleural cavity

Visceral pleura

Fibrous pericardium

Parietal pericardium

Pericardial cavity

Visceral pericardium

Figure 23.10   Pleural Cavities and Membranes

(a) Each lung is surrounded by a pleural cavity. The parietal pleura lines the wall of each pleural cavity, and the visceral pleura covers the surface of the lungs. The space between the parietal and visceral pleurae is small and filled with pleural fluid. (b) Transverse section of the thorax, at the level indicated in part (a), showing the relationship of the pleural cavities to the thoracic organs. ⚕

Clinical Focus   Cough and Sneeze Reflexes

The function of both the cough reflex and the sneeze reflex is to dislodge foreign matter or irritating material from respiratory passages. The bronchi and trachea contain sensory receptors that are sensitive to foreign particles and irritating substances. The cough reflex is initiated when the sensory receptors detect such substances and initiate action potentials that pass along the vagus nerves to the medulla oblongata, where the cough reflex is triggered.

The movements resulting in a cough occur as follows: approximately 2.5 L of air is inspired, the epiglottis closes, and the vestibular folds and vocal cords close tightly to trap the inspired air in the lung; the abdominal muscles contract to force the abdominal contents up against the diaphragm; and the muscles of expiration contract forcefully. As a consequence, the pressure in the lungs increases to approximately 100 mm Hg. Then the vestibular folds, the vocal cords, and the epiglottis open suddenly, and the air rushes from the lungs at a high velocity, carrying foreign particles with it.

The sneeze reflex is similar to the cough reflex, but it differs in several ways. The source of irritation that initiates the sneeze reflex is in the nasal passages instead of in the trachea and bronchi, and the action potentials are conducted along the trigeminal nerves to the medulla, where the reflex is triggered. During the sneeze reflex the uvula and the soft palate are depressed so that air is directed primarily through the nasal passages, although a considerable amount passes through the oral cavity. The rapidly flowing air dislodges particulate matter from the nasal passages and can propel it a considerable distance from the nose.

---

because of the accumulation of these particles, especially if the person smoked or lived most of his or her life in a city with air pollution. Cancer cells from the lungs can also spread to other parts of the body through the lymphatic vessels.

◫ Ventilation

Pressure Differences and Air Flow

Ventilation is the process of moving air into and out of the lungs. The flow of air into the lungs requires a pressure gradient from the outside of the body to the alveoli, and air flow from the lungs requires a pressure gradient in the opposite direction. The physics of air flow in tubes such as the ones that make up the respiratory passages is the same as the flow of blood in blood vessels (see chapter 21). Thus the following relationships hold:

$$F = \frac{P_1 - P_2}{R}$$

where F = air flow (milliliters per minute) in a tube; P_1 = pressure at a point called P_1; P_2 = pressure at another point called P_2; and R = resistance to air flow.

Air moves through tubes because of a pressure difference. When P_1 is greater than P_2, gas flows from P_1 to P_2 at a rate that is proportional to the pressure difference. For example, during inspiration, air pressure outside the body is greater than air pressure in the alveoli, and air flows through the trachea and bronchi to the alveoli.

Clinical Note

The flow of air decreases when the resistance to air flow is increased by conditions that reduce the radius of the respiratory passageways. According to Poiseuille's law (see chapter 21), the resistance to air flow is proportional to the radius (r) of a tube raised to the fourth power (r^4). Thus, a small change in radius results in a large change in resistance, which greatly decreases air flow. For example, asthma results in the release of inflammatory chemicals such as leukotrienes that cause severe constriction of the bronchioles. Emphysema produces increased airway resistance because the bronchioles are obstructed as a result of inflammation and because damaged bronchioles collapse during expiration, trapping air within the alveoli. Cancer can also occlude respiratory passages as the tumor replaces lung tissue. Air flow can be maintained despite increased resistance by increasing the pressure difference between alveoli and the atmosphere. Within limits, this can be accomplished by increased contraction of the muscles of respiration.

Pressure and Volume

The pressure in a container, such as the thoracic cavity or an alveolus, can be described according to the **general gas law:**

$$P = \frac{nRT}{V}$$

where P = pressure; n = number of gram moles of gas (a measure of the number of molecules present); R = gas constant; T = absolute temperature; and V = volume.

The value of R is a constant, and the values of n and T (body temperature) are considered constants in humans. Thus the general gas law reveals that air pressure is inversely proportional to the volume. As volume increases, pressure decreases; and as volume decreases, pressure increases (table 23.1).

Air Flow into and out of Alveoli

Respiratory physiologists use three conventions to help simplify the numbers used to express pressures. First, **barometric air pressure (P_B),** which is atmospheric air pressure outside the

Table 23.1   Gas Law

Description	Importance
General Gas Law	
The pressure of a gas is inversely proportional to its volume (at a constant temperature, this is referred to as Boyle's law).	Air flow from areas of higher to lower pressure. When alveolar volume increases, causing pleural pressure to decrease below atmospheric pressure, air moves into the lungs. When alveolar volume decreases, causing pleural pressure to increase above atmospheric pressure, air moves out of the lungs.
Dalton's Law	
The partial pressure of a gas in a mixture of gases is the percentage of the gas in the mixture times the total pressure of the mixture of gases.	Gases move from areas of higher to areas of lower partial pressures. The greater the difference in partial pressure between two points, the greater the rate of gas movement. Maintaining partial pressure differences ensures gas movements.
Henry's Law	
The concentration of a gas dissolved in a liquid is equal to the partial pressure of the gas over the liquid times the solubility coefficient of the gas.	Only a small amount of the gases in air dissolves in the fluid lining the alveoli. Carbon dioxide, however, is 24 times more soluble than oxygen; therefore carbon dioxide passes out through the respiratory membrane more readily than oxygen enters.

1. *End of expiration* (figure 23.11*a*). At the end of expiration, barometric air pressure and alveolar pressure are equal. There is therefore no movement of air into or out of the lungs.
2. *During inspiration* (figure 23.11*b*). As inspiration begins, contraction of inspiratory muscles increases thoracic volume, which results in expansion of the lungs and an increase in alveolar volume (see following section on Changing Alveolar Volume). The increased alveolar volume causes an approximate 1-cm H_2O decrease in alveolar pressure below barometric air pressure. Air flows into the lungs because barometric air pressure is greater than alveolar pressure.
3. *End of inspiration* (figure 23.11*c*). At the end of inspiration the thorax stops expanding, the alveoli stop expanding, and alveolar pressure becomes equal to barometric air pressure. There is no movement of air, but the volume of the lungs is larger than at the end of expiration.
4. *During expiration* (figure 23.11*d*). During expiration the volume of the thorax decreases as the diaphragm relaxes, and the thorax and lungs recoil. The decreased thoracic volume results in a decrease in alveolar volume and an approximate 1-cm H_2O increase in alveolar pressure over barometric air pressure. Air flows out of the lungs because alveolar pressure is greater than barometric air pressure. As expiration ends, the decrease in thoracic volume stops, and the process repeats beginning at step 1.

Changing Alveolar Volume

It is important to understand how alveolar volume is changed because changes in alveolar volume cause the pressure differences that result in ventilation. In addition, many respiratory disorders affect how alveolar volume changes. Lung recoil and pleural pressure cause the alveoli to collapse and expand.

Lung Recoil

Lung recoil causes the alveoli to collapse, and it results from (1) elastic recoil caused by the elastic fibers in the alveolar walls and (2) surface tension of the film of fluid that lines the alveoli. Surface tension occurs at the boundary between water and air because the polar water molecules are attracted to one another more than they are attracted to the air molecules. Consequently, the water molecules are drawn together, tending to form a droplet. Because the water molecules of the alveolar fluid are also attracted to the surface of the alveoli, formation of a droplet causes the alveoli to collapse, producing fluid-filled alveoli with smaller volumes than air-filled alveoli.

Surfactant (ser-fak'tănt) is a mixture of lipoprotein molecules produced by the secretory cells of the alveolar epithelium. The surfactant molecules form a monomolecular layer over the surface of the fluid within the alveoli to reduce the surface tension. With surfactant the force produced by surface

body, is assigned a value of zero. Thus, whether at sea level, with a pressure of 760 mm Hg, or at 10,000 feet above sea level on a mountaintop, with a pressure of 523 mm Hg, P_B is always zero. Second, the small pressures in respiratory physiology are usually expressed in centimeters of water (cm H_2O). A pressure of 1 cm H_2O is equal to 0.74 mm Hg. Third, other pressures are measured in reference to barometric air pressure. For example, **alveolar pressure (P_{alv}),** is the pressure inside an alveolus. An alveolar pressure of 1 cm H_2O is 1 cm H_2O greater pressure than barometric air pressure, and an alveolar pressure of -1 cm H_2O is 1 cm H_2O less pressure than barometric air pressure.

Movement of air into and out of the lungs results from changes in thoracic volume, which cause changes in alveolar volume. The changes in alveolar volume produce changes in alveolar pressure. The pressure difference between barometric air pressure and alveolar pressure ($P_B - P_{alv}$) results in air movement. The details of this process during quiet breathing are described as follows:

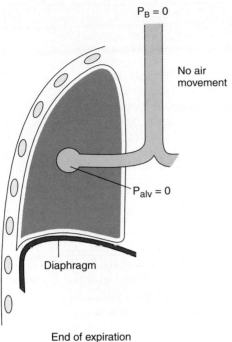

$P_B = 0$

No air movement

$P_{alv} = 0$

Diaphragm

End of expiration
$P_B = P_{alv}$

(a) Barometric air pressure (P_B) is equal to alveolar pressure (P_{alv}) and there is no air movement.

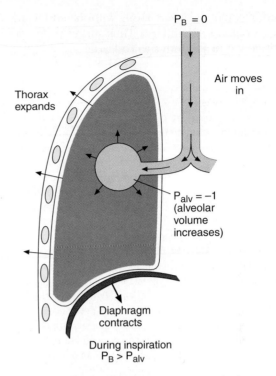

$P_B = 0$

Air moves in

Thorax expands

$P_{alv} = -1$ (alveolar volume increases)

Diaphragm contracts

During inspiration
$P_B > P_{alv}$

(b) Increased thoracic volume results in increased alveolar volume and decreased alveolar pressure. Barometric air presure is greater than alveolar pressure, and air moves into the lungs.

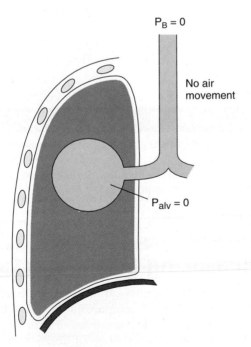

$P_B = 0$

No air movement

$P_{alv} = 0$

(c) End of inspiration.

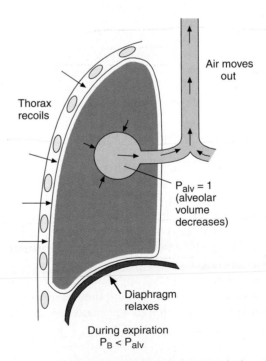

Air moves out

Thorax recoils

$P_{alv} = 1$ (alveolar volume decreases)

Diaphragm relaxes

During expiration
$P_B < P_{alv}$

(d) Decreased thoracic volume results in decreased alveolar volume and increased alveolar pressure. Alveolar pressure is greater than barometric air pressure, and air moves out of the lungs.

Figure 23.11   Alveolar Pressure Changes During Inspiration and Expiration

The combined space of all the alveoli is represented by a large "bubble." The alveoli are actually microscopic in size and cannot be seen in the illustration. ✗ ⬛

tension is approximately 4 cm H_2O; without surfactant the force can be as high as 40 cm H_2O. Thus surfactant greatly reduces the tendency of the lungs to collapse.

Pleural Pressure

Pleural pressure is the pressure in the pleural cavity. When pleural pressure is less than alveolar pressure, the alveoli tend to expand, and the greater the difference in pressure between the pleural cavity and the alveoli, the greater is the force of expansion. The expansion of the alveoli is opposed by lung recoil. If pleural pressure becomes sufficiently lower than alveolar pressure, lung recoil is overcome and the alveoli expand.

Normally the alveoli are expanded because of a negative pleural pressure that is lower than alveolar pressure. At the end of a normal expiration, pleural pressure is −5 cm H_2O, and alveolar pressure is 0 cm H_2O. The negative pleural pressure results from a "suction effect" caused by the tendency of the lungs to pull away from the thoracic wall. Normally the lungs do not pull away from the thoracic wall because pleural fluid holds the visceral and parietal pleurae together. The effect of lung recoil in producing the negative pressure in the pleural space can be appreciated by putting water on the palms of the hands and putting them together. A sensation of negative pressure is felt as the hands are gently pulled apart.

Alveolar Pressure Changes During Inspiration and Expiration

During inspiration, the lungs expand because they adhere to the thoracic wall and because a decrease in pleural pressure causes the alveoli to expand. At the end of a normal inspiration pleural pressure decreases to −8 cm H_2O. Consequently, the alveolar volume increases, alveolar pressure decreases, and air flows into the lungs (figure 23.12).

The decrease in pleural pressure occurs for two reasons. First, because of the effect of changing volume on pressure (general gas law), increasing the volume of the thoracic cavity results in a decrease in pleural pressure. Second, as the lungs expand, the tendency for the lungs to recoil increases, resulting in an increased suction effect and a lowering of pleural pressure. The tendency for the lungs to recoil increases as the lungs are stretched, similar to the increased force generated in a stretched rubber band.

During expiration, pleural pressure increases because of decreased thoracic volume and decreased lung recoil. As pleural pressure increases, alveolar volume decreases, alveolar pressure increases, and air flows out of the lungs (see figure 23.12).

4   P R E D I C T

How does the pleural pressure at the end of expiration in a newborn with respiratory distress syndrome compare with that of a healthy newborn? How does the pleural pressure compare during inspiration. Explain.

✔ *Answer in Appendix F*

Measuring Lung Function

A variety of measurements can be used to assess lung function. Each of these tests compares a subject's measurements to a normal range. These measurements can be used to diagnose diseases, track the progress of diseases, or track recovery from diseases.

Compliance of the Lungs and the Thorax

Compliance is a measure of the ease with which the lungs and the thorax expand. The compliance of the lungs and

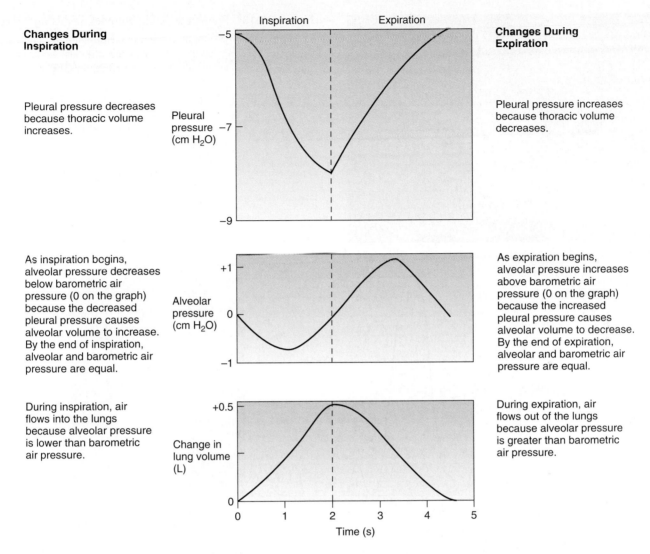

Changes During Inspiration

Pleural pressure decreases because thoracic volume increases.

As inspiration begins, alveolar pressure decreases below barometric air pressure (0 on the graph) because the decreased pleural pressure causes alveolar volume to increase. By the end of inspiration, alveolar and barometric air pressure are equal.

During inspiration, air flows into the lungs because alveolar pressure is lower than barometric air pressure.

Changes During Expiration

Pleural pressure increases because thoracic volume decreases.

As expiration begins, alveolar pressure increases above barometric air pressure (0 on the graph) because the increased pleural pressure causes alveolar volume to decrease. By the end of expiration, alveolar and barometric air pressure are equal.

During expiration, air flows out of the lungs because alveolar pressure is greater than barometric air pressure.

Figure 23.12   Dynamics of a Normal Breathing Cycle

thorax is the volume by which the lungs and the thorax increase for each unit of pressure change in the alveolar pressure. It is usually expressed in liters (volume of air) per centimeter of water (pressure), and for the normal person the compliance of the lungs and thorax is 0.13 L/cm H_2O. That is, for every 1-cm H_2O change in alveolar pressure, the volume changes by 0.13 L.

The greater the compliance, the easier it is for a change in pressure to cause expansion of the lungs and thorax. For example, one possible result of emphysema is the destruction of elastic lung tissue. This reduces the elastic recoil force of the lungs, making expansion of the lungs easier and resulting in a higher-than-normal compliance. A lower-than-normal compliance means that it is harder to expand the lungs and thorax. Conditions that decrease compliance include deposition of inelastic fibers in lung tissue (pulmonary fibrosis), collapse of the alveoli (respiratory distress syndrome and pulmonary edema), increased resistance to air flow caused by airway obstruction (asthma, bronchitis, and lung cancer), and deformities of the thoracic wall that reduce the ability of the thoracic volume to increase (kyphosis and scoliosis).

Clinical Note

Pulmonary diseases markedly affect the total amount of energy required for ventilation, as well as the percentage of the total amount of energy expended by the body. Diseases that decrease compliance can increase the energy required for breathing up to 30% of the total energy expended by the body.

Pulmonary Volumes and Capacities

Spirometry (spī-rom′ĕ-trē) is the process of measuring volumes of air that move into and out of the respiratory system, and a **spirometer** (spī-rom′ĕ-ter) is a device used to measure these pulmonary volumes (figure 23.13a). The four pulmonary volumes and representative values (figure 23.13b) for a young adult male follow:

1. **Tidal volume** is the volume of air inspired or expired during a normal inspiration or expiration (approximately 500 mL).

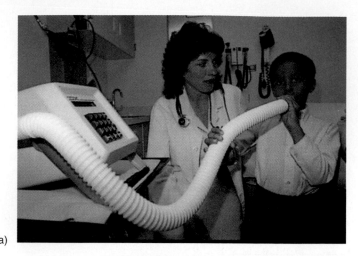

(a)

Figure 23.13 Spirometer, Lung Volumes, and Lung Capacities
(*a*) Photograph of a spirometer used to measure lung volumes and capacities. (*b*) Lung volumes and capacities. The tidal volume in the figure is the tidal volume during resting conditions.

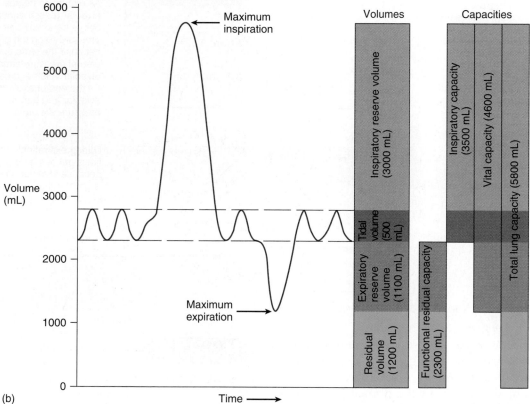

(b)

2. **Inspiratory reserve volume** is the amount of air that can be inspired forcefully after inspiration of the normal tidal volume (approximately 3000 mL).
3. **Expiratory reserve volume** is the amount of air that can be forcefully expired after expiration of the normal tidal volume (approximately 1100 mL).
4. **Residual volume** is the volume of air still remaining in the respiratory passages and lungs after the most forceful expiration (approximately 1200 mL).

   Pulmonary capacities are the sum of two or more pulmonary volumes (see figure 23.13*b*). Some pulmonary capacities follow:

1. **Inspiratory capacity** is the tidal volume plus the inspiratory reserve volume, which is the amount of air that a person can inspire maximally after a normal expiration (approximately 3500 mL).
2. **Functional residual capacity** is the expiratory reserve volume plus the residual volume, which is the amount of air remaining in the lungs at the end of a normal expiration (approximately 2300 mL).
3. **Vital capacity** is the sum of the inspiratory reserve volume, the tidal volume, and the expiratory reserve volume, which is the maximum volume of air that a person can expel from the respiratory tract after a maximum inspiration (approximately 4600 mL).
4. **Total lung capacity** is the sum of the inspiratory and expiratory reserve volumes plus the tidal volume and the residual volume (approximately 5800 mL).

Factors such as sex, age, body size, and physical conditioning cause variations in respiratory volumes and capacities from one individual to another. For example, the vital capacity of adult females is usually 20%–25% less than that of adult males. The vital capacity reaches its maximum amount in the young adult and gradually decreases in the elderly. Tall people usually have a greater vital capacity than short people, and thin people have a greater vital capacity than obese people. Well-trained athletes can have a vital capacity 30%–40% above normal. In patients whose respiratory muscles are paralyzed by spinal cord injury or diseases such as poliomyelitis or muscular dystrophy, the vital capacity can be reduced to values not consistent with survival (less than 500–1000 mL). Factors that reduce compliance also reduce the vital capacity.

The **forced expiratory vital capacity** is a simple and clinically important pulmonary test. The individual inspires maximally and then exhales maximally into a spirometer as rapidly as possible. The spirometer records the volume of air that enters it per second. In some conditions the volume to which the lungs are inflated may not be dramatically affected, but the rate at which air can be expired can be greatly decreased. For example, the expiratory flow rate is decreased by disorders that reduce the ability of the lungs or chest wall to deflate, such as pulmonary fibrosis, silicosis, kyphosis, and scoliosis. Airway obstruction caused by asthma, collapse of bronchi in emphysema, or by a tumor can also decrease expiratory flow rate.

Minute Ventilation and Alveolar Ventilation

The **minute ventilation** is the total amount of air moved into and out of the respiratory system each minute, and it is equal to the tidal volume times the respiratory rate. The **respiratory rate**, or **respiratory frequency**, is the number of breaths taken per minute. Because resting tidal volume is approximately 500 mL and respiratory rate is approximately 12 breaths per minute, the minute ventilation averages approximately 6 L/min.

Although the minute ventilation measures the amount of air moving into and out of the lungs per minute, it is not a measure of the amount of air available for gas exchange because gas exchange takes place mainly in the alveoli and to a lesser extent in the alveolar ducts and the respiratory bronchioles. The part of the respiratory system where gas exchange does not take place is called the dead space. A distinction can be made between the anatomic and physiologic dead space. The **anatomic dead space,** which measures 150 mL, is formed by the nasal cavity, pharynx, larynx, trachea, bronchi, bronchioles, and terminal bronchioles. The **physiologic dead space** is the anatomic dead space plus the volume of any alveoli in which gas exchange is less than normal. In a healthy person, the anatomic and physiologic dead spaces are nearly the same, meaning there are few nonfunctional alveoli.

Clinical Note

In patients with **emphysema,** alveolar walls degenerate, and small alveoli combine to form larger alveoli. The result is fewer alveoli, but alveoli with an increased volume and decreased surface area. Although the enlarged alveoli are still ventilated, there is inadequate surface area for complete gas exchange, and the physiologic dead space increases.

During inspiration much of the inspired air fills the dead space first before reaching the alveoli and thus is unavailable for gas exchange. The volume of air available for gas exchange per minute is called the **alveolar ventilation** ($\dot{V}_A$), and it is calculated as follows:

$$\dot{V}_A = f(V_T - V_D)$$

where $\dot{V}_A$ = alveolar ventilation (milliliters per minute); f = respiratory rate (frequency; breaths per minute); V_T = tidal volume (milliliters per respiration); and V_D = dead space (milliliters per respiration).

5   P R E D I C T

What is the alveolar ventilation of a resting person with a tidal volume of 500 mL, a dead space of 150 mL, and a respiratory rate of 12 breaths per minute? If the person exercises and tidal volume increases to 4000 mL, dead space increases to 300 mL as a result of dilation of the respiratory passageways, and respiratory rate increases to 24 breaths per minute, what is the alveolar ventilation? How is the change in alveolar ventilation beneficial for doing exercise?

✔ *Answer in Appendix F*

▌Physical Principles of Gas Exchange

Ventilation supplies atmospheric air to the alveoli. The next step in the process of respiration is the diffusion of gases between the alveoli and the blood in the pulmonary capillaries. The molecules of gas move randomly, and if a gas is in a higher concentration at one point than at another, random motion ensures that the net movement of gas is from the higher concentration toward the lower concentration until a homogeneous mixture of gases is achieved. One measurement of the concentration of gases is partial pressure.

Partial Pressure

At sea level the atmospheric pressure is approximately 760 mm Hg, which means that the mixture of gases that constitute atmospheric air exerts a total pressure of 760 mm Hg. The major components of dry air are nitrogen (approximately 79%) and oxygen (approximately 21%). According to **Dalton's law,**

Table 23.2 Partial Pressures of Gases at Sea Level

Gases	Dry Air mm Hg	Dry Air %	Humidified Air mm Hg	Humidified Air %	Alveolar Air mm Hg	Alveolar Air %	Expired Air mm Hg	Expired Air %
Nitrogen	600.2	78.98	563.4	74.09	569.0	74.9	566.0	74.5
Oxygen	159.5	20.98	149.3	19.67	104.0	13.6	120.0	15.7
Carbon dioxide	0.3	0.04	0.3	0.04	40.0	5.3	27.0	3.6
Water vapor	0.0	0.0	47.0	6.20	47.0	6.2	47.0	6.2

in a mixture of gases the part of the total pressure resulting from each type of gas is determined by the percentage of the total volume represented by each gas type (see table 23.1). The pressure exerted by each type of gas in a mixture is referred to as the **partial pressure** of that gas. Because nitrogen constitutes 79% of the volume of atmospheric air, the partial pressure resulting from nitrogen is 0.79 times 760 mm Hg, or 600.2 mm Hg. Because oxygen is approximately 21% of the volume of atmospheric air, the partial pressure resulting from oxygen is 0.21 times 760 mm Hg, or 159.5 mm Hg. It is traditional to designate the partial pressure of individual gases in a mixture with a capital P followed by the symbol for the gas. Thus the partial pressure of nitrogen is denoted P_{N_2}, oxygen is P_{O_2}, and carbon dioxide is P_{CO_2}.

When air comes into contact with water, some of the water turns into a gas and evaporates into the air. Water molecules in the gaseous form also exert a partial pressure. This partial pressure (P_{H_2O}) is sometimes referred to as the **water vapor pressure.** The composition of dry, humidified, alveolar, and expired air is presented in table 23.2. The composition of alveolar air and of expired air is not identical to the composition of dry atmospheric air for three reasons. First, air entering the respiratory system during inspiration is humidified; second, oxygen diffuses from the alveoli into the blood, and carbon dioxide diffuses from the pulmonary capillaries into the alveoli; and third, the air within the alveoli is only partially replaced with atmospheric air during each inspiration.

Diffusion of Gases Through Liquids

When a gas comes into contact with a liquid such as water, the gas tends to dissolve in the liquid. At equilibrium the concentration of a gas in the liquid is determined by its partial pressure in the gas and by its solubility in the liquid. This relationship is described by **Henry's law** (see table 23.1):

$$\text{Concentration of dissolved gas} = \text{Partial pressure of gas} \times \text{Solubility coefficient}$$

The solubility coefficient is a measure of how easily the gas dissolves in the liquid. In water the solubility coefficient for oxygen is 0.024, and for carbon dioxide it is 0.57. Thus carbon dioxide is approximately 24 times more soluble in water than oxygen.

Gases do not actually produce a partial pressure in a liquid as they do when in the gaseous state. If the concentration

of a gas in a liquid is used, however, it is possible to use the general gas law equation to mathematically determine the gas's partial pressures as if it were in a gaseous state. Because the calculated partial pressure of a gas in a liquid is a measure of concentration, it can be used to determine the direction of diffusion of the gas through the liquid: the gas moves from areas of higher to areas of lower partial pressure.

6   P R E D I C T

As a SCUBA diver descends, the pressure of the water on his body prevents normal expansion of the lungs. To compensate, the diver breathes pressurized air, which has a greater pressure than air at sea level. What effect does the increased pressure have on the amount of gas dissolved in the diver's body fluids? A SCUBA diver who suddenly ascends to the surface from a great depth can develop decompression sickness (the bends) in which bubbles of nitrogen gas form. The expanding bubbles damage tissues or block blood flow through small blood vessels. Explain the development of the bubbles.

✔ *Answer in Appendix F*

Diffusion of Gases Through the Respiratory Membrane

The **respiratory membrane** of the lungs is in the respiratory bronchioles, alveolar ducts, and alveoli. Approximately 300 million alveoli are in the two lungs. The average diameter of each alveolus is approximately 0.25 mm, and its walls are extremely thin. Surrounding each alveolus is a network of capillaries arranged so that air within the alveoli is separated by a thin respiratory membrane from the blood contained within the pulmonary capillaries (figure 23.14a).

The respiratory membrane (figure 23.14b) consists of

1. A thin layer of fluid lining the alveolus
2. The alveolar epithelium composed of simple squamous epithelium
3. The basement membrane of the alveolar epithelium
4. A thin interstitial space
5. The basement membrane of the capillary endothelium, and
6. The capillary endothelium composed of simple squamous epithelium

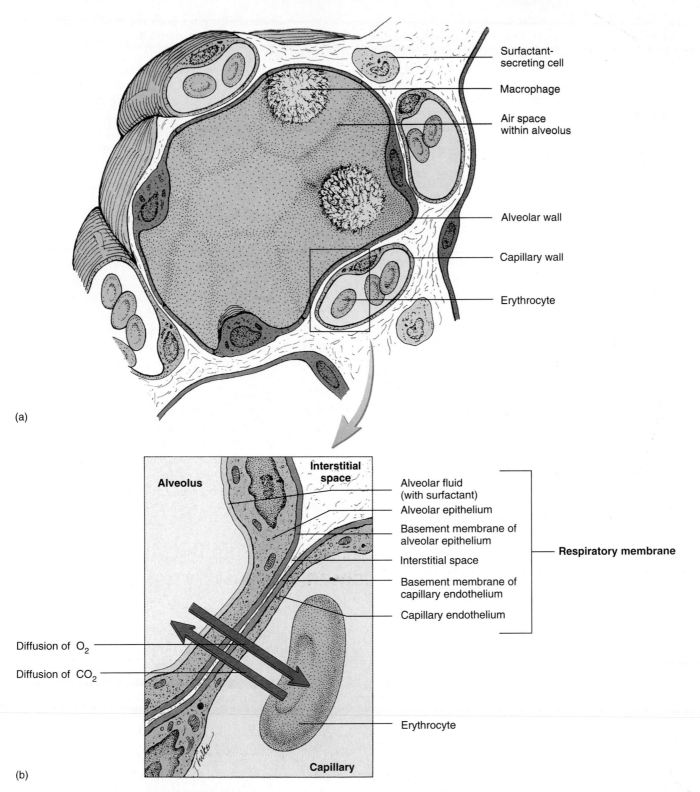

Surfactant-secreting cell

Macrophage

Air space within alveolus

Alveolar wall

Capillary wall

Erythrocyte

(a)

Interstitial space

Alveolus

Alveolar fluid (with surfactant)

Alveolar epithelium

Basement membrane of alveolar epithelium

Interstitial space

Basement membrane of capillary endothelium

Capillary endothelium

Respiratory membrane

Diffusion of O_2

Diffusion of CO_2

Erythrocyte

Capillary

(b)

Figure 23.14   Alveolus and the Respiratory Membrane

(*a*) Section of an alveolus showing the air-filled interior and thin walls composed of simple squamous epithelium. The alveolus is surrounded by elastic connective tissue and blood capillaries. (*b*) Diffusion of oxygen and carbon dioxide across the six thin layers of the respiratory membrane.

The factors that influence the rate of gas diffusion across the respiratory membrane include (1) the thickness of the membrane, (2) the diffusion coefficient of the gas in the substance of the membrane, which is approximately the same as the diffusion coefficient for the gas through water, (3) the surface area of the membrane, and (4) the difference of the partial pressures of the gas between the two sides of the membrane.

Respiratory Membrane Thickness

Increasing the thickness of the respiratory membrane decreases the rate of diffusion. The thickness of the respiratory membrane normally averages 0.5 mm, but the thickness can be increased by respiratory diseases. Pulmonary edema caused by failure of the left side of the heart is the most common cause of an increase in the thickness of the respiratory membrane. Increased venous pressure in the pulmonary capillaries results in the accumulation of fluid in the alveoli. Conditions that result in inflammation of the lung tissues, such as tuberculosis, pneumonia, or advanced silicosis, can cause fluid buildup within the alveoli. As a consequence, the efficiency of gas diffusion across the alveolar membrane decreases. If the thickness of the respiratory membrane increases two or three times, the rate of gas exchange markedly decreases.

Diffusion Coefficient

The **diffusion coefficient** is a measure of how easily a gas diffuses through a liquid or tissue, taking into account the solubility of the gas in the liquid and the size of the gas molecule (molecular weight). If the diffusion coefficient of oxygen is assigned a value of 1, then the relative diffusion coefficient of carbon dioxide is 20, which means carbon dioxide diffuses through the respiratory membrane about 20 times more readily than oxygen.

When the respiratory membrane becomes progressively damaged as a result of disease, its capacity for allowing the movement of oxygen into the blood is often impaired enough to cause death from oxygen deprivation before the diffusion of carbon dioxide is dramatically reduced. If life is being maintained by extensive oxygen therapy, which increases the concentration of oxygen in the lung alveoli, the reduced capacity for the diffusion of carbon dioxide across the respiratory membrane can result in substantial increases of carbon dioxide in the blood.

Surface Area

In a normal adult the total surface area of the respiratory membrane is approximately 70 m^2 (approximately the size of one-half of a single's tennis court). The surface area of the respiratory membrane is decreased by several respiratory diseases, including emphysema and lung cancer. Even small decreases in this surface area adversely affect the respiratory exchange of gases during strenuous exercise. When the total surface area of the respiratory membrane is decreased to one-third or one-fourth of normal, the exchange of gases is significantly restricted even under resting conditions.

A decreased surface area for gas exchange results from the surgical removal of lung tissue, the destruction of lung tissue by cancer, the degeneration of the alveolar walls by emphysema, or the replacement of lung tissue by connective tissue caused by tuberculosis. More acute conditions that cause the alveoli to fill with fluid also reduce the surface area for gas exchange. Examples include pneumonia, pulmonary edema resulting from failure of the left ventricle, and atelectasis (at-ĕ-lek'tă-sis, meaning collapse of the lung).

Partial Pressure Difference

The partial pressure difference of a gas across the respiratory membrane is the difference between the partial pressure of the gas in the alveoli and the partial pressure of the gas in the blood of the pulmonary capillaries. When the partial pressure of a gas is greater on one side of the respiratory membrane than on the other side, net diffusion occurs from the higher to the lower partial pressure (see figure 23.14b). Normally the partial pressure of oxygen (PO_2) is greater in the alveoli than in the blood of the pulmonary capillaries, and the partial pressure of carbon dioxide (PCO_2) is greater in the blood than in the alveolar air.

The partial pressure difference for oxygen and carbon dioxide can be increased by increasing the alveolar ventilation. The greater volume of atmospheric air exchanged with the residual volume raises alveolar PO_2, lowers alveolar PCO_2, and thus promotes gas exchange. Conversely, inadequate ventilation causes a lower-than-normal partial pressure difference for oxygen and carbon dioxide, resulting in inadequate gas exchange.

Relationship Between Ventilation and Pulmonary Capillary Blood Flow

Under normal conditions, the ventilation of the alveoli and the blood flow through pulmonary capillaries is such that effective gas exchange occurs between the air and the blood. During exercise, effective gas exchange is maintained because both ventilation and cardiac output increase.

The normal relationship between ventilation and pulmonary capillary blood flow can be disrupted in two different ways. One is when ventilation exceeds the ability of the blood to pick up oxygen, which can happen because of inadequate cardiac output after a heart attack. Another way is when ventilation is not great enough to provide the oxygen needed to oxygenate the blood flowing through the pulmonary capillaries. For example, constriction of the bronchioles in asthma can decrease air delivery to the alveoli.

Blood that is not completely oxygenated is called shunted blood. There are two sources of shunted blood in the lungs. An **anatomic shunt** results from deoxygenated blood from the bronchi and bronchioles mixing with blood in the pulmonary veins (see section on Blood Supply earlier in this chapter). The other source of shunted blood is blood that passes through pulmonary capillaries but does not become fully oxygenated. The **physiologic shunt** is the deoxygenated blood from the pulmonary capillaries plus the deoxygenated blood from the anatomic shunt. Normally, 1%–2% of cardiac output passes through the physiologic shunt.

When a person is standing, greater blood flow and ventilation occur in the base of the lung than in the top of the lung because of the effects of gravity. The arterial pressure at the base of the lung is 22 mm Hg greater than at the top of the lung because of hydrostatic pressure caused by gravity (see chapter 21). The greater pressure increases blood flow and distends blood vessels. The decreased pressure at the top of the lung results in less blood flow and vessels that are less distended, some of which are even collapsed during diastole.

During exercise, cardiac output and ventilation increase. The increased cardiac output increases pulmonary blood pressure throughout the lung, which increases blood flow. Blood flow increases the most at the top of the lung, however, because the increased pressure expands the less distended vessels and opens the collapsed vessels. Thus, the effectiveness of gas exchange at the top of the lung increases because of greater blood flow.

Although gravity is the major factor affecting regional blood flow in the lung, under certain circumstances alveolar P_{O_2} can have an effect. In most tissues, low P_{O_2} results in increased blood flow through the tissues (see chapter 21). In the lung, low P_{O_2} has the opposite effect, causing arterioles to constrict and reducing blood flow. This response reroutes blood away from areas of low oxygen toward parts of the lung that are better oxygenated. For example, if a bronchus becomes partially blocked, ventilation of alveoli past the blockage site decreases, which decreases gas exchange between the air and blood. The effect of this decreased gas exchange on overall gas exchange in the lungs is reduced by rerouting the blood to better-ventilated alveoli.

7   P R E D I C T

Even people in "good shape," can have trouble breathing at high altitudes. Explain how this can happen, even when ventilation of the lungs increases.

✔ *Answer in Appendix F*

Oxygen and Carbon Dioxide Transport in the Blood

Once oxygen diffuses across the respiratory membrane into the blood, most of the oxygen combines reversibly with hemoglobin, and a smaller amount dissolves in the plasma. Hemoglobin transports oxygen from the pulmonary capillaries through the blood vessels to the tissue capillaries, where some of the oxygen is released. The oxygen diffuses from the blood to tissue cells, where the oxygen is used in aerobic respiration.

Cells produce carbon dioxide during aerobic metabolism, and it diffuses from the cells into the tissue capillaries. Once carbon dioxide enters the blood, it is transported dissolved in the plasma, in combination with hemoglobin, and in the form of bicarbonate ions.

Oxygen Diffusion Gradients

The P_{O_2} within the alveoli averages approximately 104 mm Hg, and as blood flows into the pulmonary capillaries, it has a P_{O_2} of approximately 40 mm Hg (figure 23.15). Consequently, oxygen diffuses from the alveoli into the pulmonary capillary blood because the P_{O_2} is greater in the alveoli than in the capillary blood. By the time blood flows through the first third of the pulmonary capillary beds, an equilibrium is achieved, and the P_{O_2} in the blood is 104 mm Hg, which is equivalent to the P_{O_2} in the alveoli. Even with the greater velocity of blood flow associated with exercise, by the time blood reaches the venous ends of the pulmonary capillaries, the P_{O_2} in the capillaries has achieved the same value as that in the alveoli.

Blood leaving the pulmonary capillaries has a P_{O_2} of 104 mm Hg, but blood leaving the lungs in the pulmonary veins has a P_{O_2} of approximately 95 mm Hg. The decrease in the P_{O_2} occurs because the blood from the pulmonary capillaries mixes with deoxygenated (shunted) blood from the bronchial veins.

The blood that enters the arterial end of the tissue capillaries has a P_{O_2} of approximately 95 mm Hg. The P_{O_2} of the interstitial spaces, in contrast, is close to 40 mm Hg and is probably near 20 mm Hg in the individual cells. Oxygen diffuses from the tissue capillaries to the interstitial fluid and from the interstitial fluid into the cells of the body, where it is used in aerobic metabolism. Because oxygen is continuously used by cells, a constant diffusion gradient exists for oxygen from the tissue capillaries to the cells.

Carbon Dioxide Diffusion Gradients

Carbon dioxide is continually produced as a by-product of cellular respiration, and a diffusion gradient is established from tissue cells to the blood within the tissue capillaries. The intracellular P_{CO_2} is approximately 46 mm Hg, and the interstitial fluid P_{CO_2} is approximately 45 mm Hg. At the arteriolar end of the tissue capillaries the P_{CO_2} is close to 40 mm Hg. As blood flows through the tissue capillaries, carbon dioxide diffuses

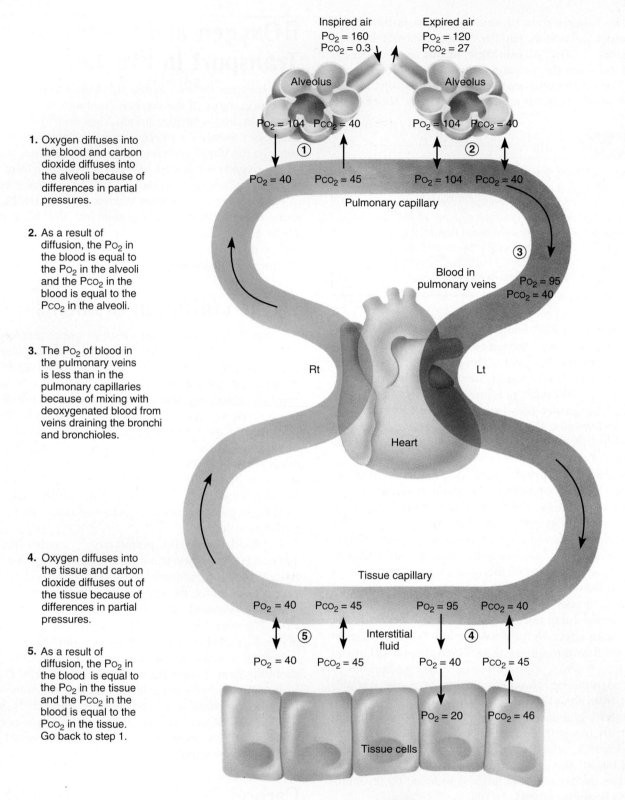

Inspired air
$Po_2 = 160$
$Pco_2 = 0.3$

Expired air
$Po_2 = 120$
$Pco_2 = 27$

Alveolus

Alveolus

$Po_2 = 104$ $Pco_2 = 40$

$Po_2 = 104$ $Pco_2 = 40$

1. Oxygen diffuses into the blood and carbon dioxide diffuses into the alveoli because of differences in partial pressures.

①

②

$Po_2 = 40$ $Pco_2 = 45$

$Po_2 = 104$ $Pco_2 = 40$

Pulmonary capillary

2. As a result of diffusion, the Po_2 in the blood is equal to the Po_2 in the alveoli and the Pco_2 in the blood is equal to the Pco_2 in the alveoli.

③

Blood in pulmonary veins

$Po_2 = 95$
$Pco_2 = 40$

3. The Po_2 of blood in the pulmonary veins is less than in the pulmonary capillaries because of mixing with deoxygenated blood from veins draining the bronchi and bronchioles.

Rt

Lt

Heart

4. Oxygen diffuses into the tissue and carbon dioxide diffuses out of the tissue because of differences in partial pressures.

Tissue capillary

$Po_2 = 40$ $Pco_2 = 45$

$Po_2 = 95$ $Pco_2 = 40$

⑤

Interstitial fluid

④

5. As a result of diffusion, the Po_2 in the blood is equal to the Po_2 in the tissue and the Pco_2 in the blood is equal to the Pco_2 in the tissue. Go back to step 1.

$Po_2 = 40$ $Pco_2 = 45$

$Po_2 = 40$ $Pco_2 = 45$

$Po_2 = 20$ $Pco_2 = 46$

Tissue cells

Figure 23.15   Changes in the Partial Pressures of Oxygen and Carbon Dioxide

from a higher Pco_2 to a lower Pco_2 until an equilibrium in Pco_2 is established. At the venous end of the capillaries, blood has a Pco_2 of 45 mm Hg (see figure 23.15).

After blood leaves the venous end of the capillaries, it is transported through the cardiovascular system to the lungs. At

the arteriolar end of the pulmonary capillaries the Pco_2 is 45 mm Hg. Because the Pco_2 is approximately 40 mm Hg in the alveoli, carbon dioxide diffuses from the pulmonary capillaries into the alveoli. At the venous end of the pulmonary capillaries the Pco_2 has again decreased to 40 mm Hg.

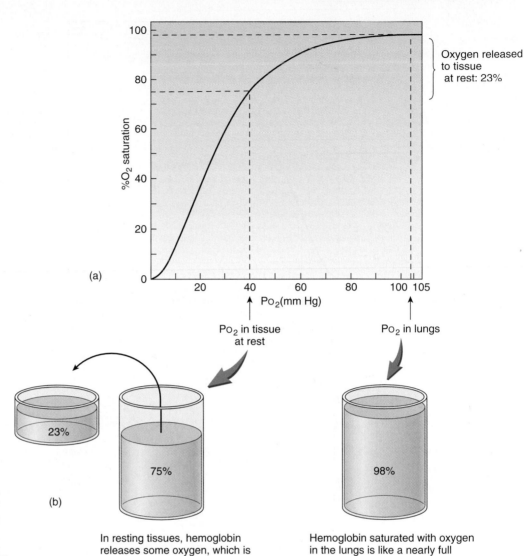

Figure 23.16   Oxygen–Hemoglobin Dissociation Curve at Rest

(*a*) At the Po$_2$ in the lungs, hemoglobin is 98% saturated. At the Po$_2$ of resting tissues, hemoglobin is 75% saturated. Consequently 23% of the oxygen picked up in the lungs is released to the tissues. (*b*) **Analogy:** The ability of hemoglobin to pick up and release oxygen is like a glass filling and emptying.

In resting tissues, hemoglobin releases some oxygen, which is like partially emptying the glass.

Hemoglobin saturated with oxygen in the lungs is like a nearly full glass.

Hemoglobin and Oxygen Transport

Approximately 98.5% of the oxygen transported in the blood from the lungs to the tissues is transported in combination with the hemoglobin in the erythrocytes, and the remaining 1.5% is dissolved in the water part of the plasma. The combination of oxygen with hemoglobin is reversible. In the pulmonary capillaries, oxygen binds to hemoglobin, and in the tissue spaces oxygen diffuses away from hemoglobin and enters the tissues.

Effect of Po$_2$

The **oxygen–hemoglobin dissociation curve** describes the percentage of hemoglobin saturated with oxygen at any given Po$_2$. Hemoglobin is saturated when an oxygen molecule is bound to each of its four heme groups (see chapter 19). At any Po$_2$ above 80 mm Hg, approximately 95% of the hemoglobin is saturated with oxygen (figure 23.16). Because the Po$_2$ in the pulmonary capillaries is normally 104 mm Hg, the hemoglobin is 98% saturated.

In a resting person the normal Po$_2$ of blood leaving the tissue capillaries of skeletal muscle is 40 mm Hg. At a Po$_2$ of 40 mm Hg, hemoglobin is approximately 75% saturated. Thus, approximately 23% of the oxygen bound to hemoglobin is released into the blood and can diffuse into the tissue spaces. During conditions of vigorous exercise the blood Po$_2$ can decline to levels as low as 15 mm Hg because the skeletal muscle cells are using the oxygen in aerobic respiration (see chapter 10). At a Po$_2$ of 15 mm Hg, approximately 25% of the hemoglobin is saturated with oxygen, and the hemoglobin releases 73% of the bound oxygen (figure 23.17). Thus, when the oxygen needs of tissues increase, blood Po$_2$ decreases, and more oxygen is released for use by the tissues.

Effect of pH, Pco$_2$, and Temperature

In addition to Po$_2$, other factors influence the degree to which oxygen binds to hemoglobin. As the pH of the blood declines, the amount of oxygen bound to hemoglobin at any given Po$_2$ also declines. This occurs because decreased pH results from an increase in hydrogen ions, and the

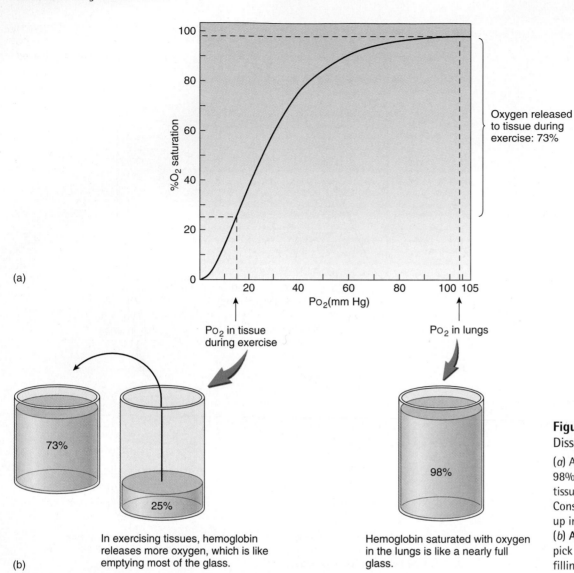

(a)

(b)

In exercising tissues, hemoglobin releases more oxygen, which is like emptying most of the glass.

Hemoglobin saturated with oxygen in the lungs is like a nearly full glass.

Figure 23.17   Oxygen–Hemoglobin Dissociation Curve During Exercise

(*a*) At the P_{O_2} in the lungs, hemoglobin is 98% saturated. At the P_{O_2} of exercising tissues, hemoglobin is 25% saturated. Consequently 73% of the oxygen picked up in the lungs is released to the tissues. (*b*) **Analogy:** The ability of hemoglobin to pick up and release oxygen is like a glass filling and emptying.

hydrogen ions combine with the protein part of the hemoglobin molecule and change its three-dimensional structure, causing a decrease in the ability of hemoglobin to bind oxygen. Conversely, an increase in blood pH results in an increased ability of hemoglobin to bind oxygen. The effect of pH (hydrogen ions) on the oxygen–hemoglobin dissociation curve is called the **Bohr effect** after its discoverer, Christian Bohr.

An increase in P_{CO_2} also decreases the ability of hemoglobin to bind oxygen because of the effect of carbon dioxide on pH. Within erythrocytes an enzyme called **carbonic anhydrase** catalyzes this reversible reaction.

$$\underset{\substack{\text{Carbon} \\ \text{dioxide}}}{CO_2} + \underset{\text{Water}}{H_2O} \underset{\substack{\text{Carbonic} \\ \text{anhydrase}}}{\rightleftarrows} \underset{\substack{\text{Carbonic} \\ \text{acid}}}{H_2CO_3} \rightleftarrows \underset{\substack{\text{Hydrogen} \\ \text{ion}}}{H^+} + \underset{\substack{\text{Bicarbonate} \\ \text{ion}}}{HCO_3^-}$$

As the carbon dioxide levels increase, more hydrogen ions are produced, and the pH declines. As the carbon dioxide levels decline, the reaction proceeds in the opposite direction, resulting in a decrease in hydrogen ion concentration and an increase in pH.

As blood passes through tissue capillaries, carbon dioxide enters the blood from the tissues. As a consequence, blood carbon dioxide levels increase, hemoglobin has less affinity for oxygen in the tissue capillaries, and a greater amount of oxygen is released in the tissue capillaries than would be released if carbon dioxide were not present. When blood is returned to the lungs and passes through the pulmonary capillaries, carbon dioxide leaves the capillaries and enters the alveoli. As a consequence, carbon dioxide levels in the pulmonary capillaries are reduced, and the affinity of hemoglobin for oxygen increases.

An increase in temperature also decreases the tendency for oxygen to remain bound to hemoglobin. Elevated temperatures resulting from increased metabolism therefore increase the amount of oxygen released into the tissues by hemoglobin. In less metabolically active tissues in which the temperature is lower, less oxygen is released from hemoglobin.

When the affinity of hemoglobin for oxygen decreases, the oxygen–hemoglobin dissociation curve is shifted to the

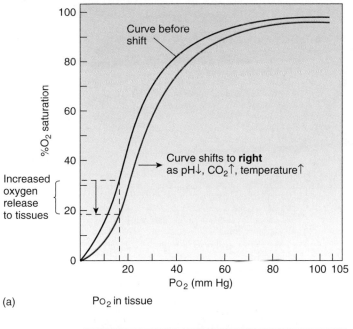

(a)   Po₂ in tissue

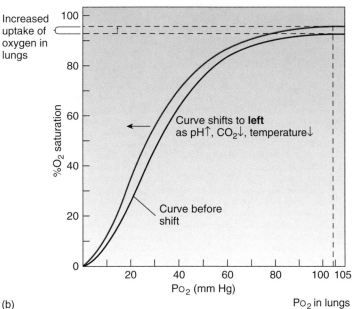

(b)

Figure 23.18   Effects of Shifting the Oxygen–Hemoblogin Dissociation Curve

(*a*) In the tissues, as pH decreases, Pco₂ increases, or temperature increases, the curve (*black*) shifts to the right (*red*), resulting in an increased release of oxygen. (*b*) In the lungs, as pH increases, Pco₂ decreases, or temperature decreases, the curve (*black*) shifts to the left (*red*), resulting in an increased ability of hemoglobin to pick up oxygen.

right, and hemoglobin releases more oxygen (figure 23.18*a*). During exercise when carbon dioxide and acidic substances such as lactic acid accumulate and the temperature increases in the tissue spaces, the oxygen–hemoglobin curve shifts to the right. Under these conditions as much as 75%–85% of the oxygen is released from the hemoglobin. In the lungs, however, the curve shifts to the left because of the lower carbon dioxide levels, lower temperature, and lower lactic acid levels. The affinity of hemoglobin for oxygen therefore increases, and it becomes easily saturated (figure 23.18*b*).

During resting conditions approximately 5 mL of oxygen is transported to the tissues in each 100 mL of blood, and the cardiac output is approximately 5000 mL/min. Consequently, 250 mL of oxygen is delivered to the tissues each minute. During conditions of exercise this value can increase up to 15 times. Oxygen transport can be increased threefold because of a greater degree of oxygen release from hemoglobin in the tissue spaces, and the rate of oxygen transport is increased another five times because of the increase in cardiac output. Consequently, the volume of oxygen delivered to the tissues can be as high as 3750 mL/min (15 × 250 mL/min). Highly trained athletes can increase this volume to as high as 5000 mL/min.

---

8   P R E D I C T

In carbon monoxide (CO) poisoning, CO binds to hemoglobin, preventing the uptake of oxygen by hemoglobin. In addition, when CO binds to hemoglobin, the oxygen–hemoglobin dissociation curve shifts to the left. What are the consequences of this shift on the ability of tissues to get oxygen? Explain.

✔ *Answer in Appendix F*

Effect of BPG

The substance **2,3-bisphosphoglycerate** (**BPG;** formerly called diphosphoglycerate) is produced by erythrocytes as they break down glucose for energy. BPG binds to hemoglobin and increases the ability of hemoglobin to release oxygen. Consequently, when BPG levels increase, hemoglobin releases more oxygen, and when BPG levels decrease, hemoglobin releases less oxygen. For example, people living at high altitudes have increased levels of BPG, which increases oxygen delivery to tissues by causing hemoglobin to release more oxygen. On the other hand, when blood is removed from the body and stored in a blood bank, the BPG levels in the stored blood gradually decrease. As BPG levels decrease, the blood becomes unsuitable for transfusion purposes because the hemoglobin releases less oxygen to the tissues.

---

9   P R E D I C T

If a person lacks the enzyme necessary for BPG synthesis, would she exhibit anemia (lower-than-normal number of erythrocytes) or polycythemia (higher-than-normal number of erythrocytes)? Explain.

✔ *Answer in Appendix F*

Transport of Carbon Dioxide

Carbon dioxide is transported in the blood in three major ways: approximately 7% is transported as carbon dioxide dissolved in the plasma, approximately 23% is transported in combination with blood proteins (mostly hemoglobin), and 70% is transported in the form of bicarbonate ions.

The most abundant protein to which carbon dioxide binds in the blood is hemoglobin. Carbon dioxide binds in a reversible fashion to the globin part of the hemoglobin molecule,

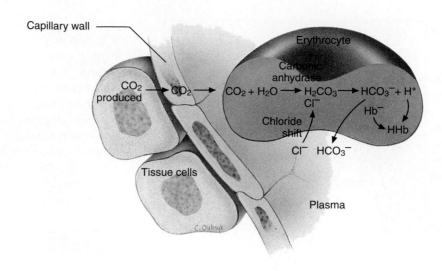

(a) In tissues, carbon dioxide enters erythrocytes and reacts with water to form carbonic acid, which dissociates to form bicarbonate and hydrogen ions. In the chloride shift, bicarbonate ions are exchanged for chloride ions. Hydrogen ions combine with hemoglobin. Lowering the concentration of bicarbonate and hydrogen ions inside erythrocytes promotes the conversion of carbon dioxide to bicarbonate ions.

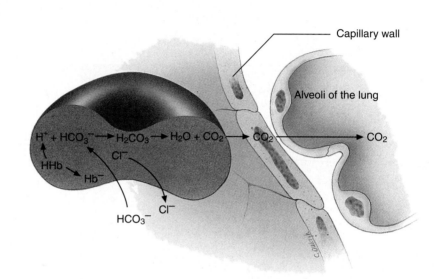

(b) In the lungs, carbon dioxide leaves erythroctyes, resulting in the formation of additional carbon dioxide from carbonic acid. Bicarbonate and hydrogen ions combine to replace the carbonic acid. The bicarbonate ions are exchanged for chloride ions, and the hydrogen ions are released from hemoglobin.

Figure 23.19   Carbon Dioxide Transport and Chloride Movement

and many carbon dioxide molecules can combine to a single hemoglobin molecule.

Hemoglobin that has released its oxygen binds more readily to carbon dioxide than hemoglobin that has oxygen bound to it. This is called the **Haldane effect.** In tissues, after hemoglobin has released oxygen, the hemoglobin has an increased ability to pick up carbon dioxide. In the lungs, as hemoglobin binds to oxygen, the hemoglobin more readily releases carbon dioxide.

Chloride Shift

Carbon dioxide from tissues diffuses into erythrocytes within the capillaries (figure 23.19a). Some of the carbon dioxide (CO_2) binds to hemoglobin, but most of it reacts with water (H_2O) inside the erythrocytes to form carbonic acid (H_2CO_3), a reaction catalyzed by carbonic anhydrase. The carbonic acid then dissociates to form bicarbonate (HCO_3^-) and hydrogen (H^+) ions. Thus, most of the carbon dioxide becomes part of a bicarbonate ion.

Lowering the amount of bicarbonate and hydrogen ions inside erythrocytes promotes carbon dioxide transport, because as these reaction products are removed and their ion concentrations decrease, more carbon dioxide combines with water to form additional bicarbonate and hydrogen ions (see section on Reversible Reactions in chapter 2). Bicarbonate ion concentration inside erythrocytes is decreased by exchanging bicarbonate ions for chloride ions in a process called the **chloride shift** (see figure 23.19a). As bicarbonate ions are produced, carrier molecules in erythrocyte membranes move bicarbonate ions out of the erythrocytes and chloride ions (Cl^-) into the erythrocytes. The exchange of negatively charged ions maintains electrical balance in erythrocytes and the plasma. The concentration of hydrogen ions inside the erythrocytes is decreased by hemoglobin, which binds hydrogen ions. Thus, hemoglobin functions as a buffer and resists an increase in pH within the erythrocytes.

The reverse of the previous events occurs in the lungs (figure 23.19b). Carbon dioxide diffuses from the erythrocytes into the alveoli. As carbon dioxide levels in the erythrocytes

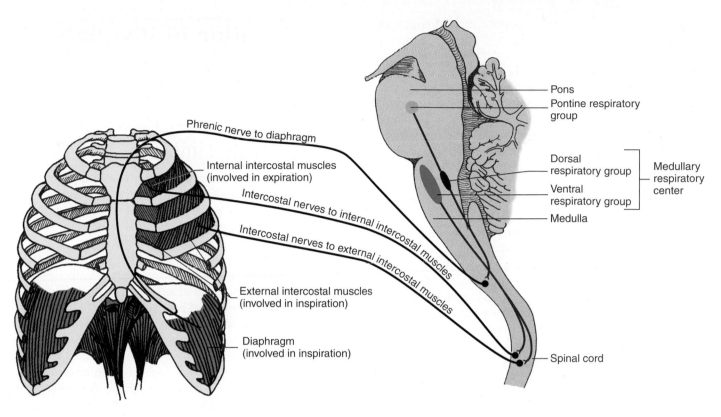

Figure 23.20   Respiratory Structures in the Brainstem

The relationship of respiratory structures to each other and to the nerves innervating the muscles of respiration.

decrease, carbonic acid is converted to carbon dioxide and water. In response, bicarbonate ions join with hydrogen ions to form carbonic acid. As the bicarbonate and hydrogen ions decrease because of this reaction, they are replaced. Bicarbonate ions enter the erythrocyte in exchange for chloride ions, and hydrogen ions are released from hemoglobin.

Carbon Dioxide and Blood pH

Blood pH refers to the pH in plasma, not inside erythrocytes. In plasma, carbon dioxide can combine with water to form carbonic acid, a reaction that is catalyzed by carbonic anhydrase on the surface of capillary endothelial cells. The carbonic acid then dissociates to form bicarbonate and hydrogen ions. Thus, as plasma carbon dioxide levels increase, hydrogen ion levels increase, and blood pH decreases. An important function of the respiratory system is to regulate blood pH by changing plasma carbon dioxide levels (see chapter 27). Hyperventilation decreases plasma carbon dioxide, and hypoventilation increases plasma carbon dioxide.

10   P R E D I C T

What effect does hyperventilation and holding one's breath have on blood pH? Explain.

✔*Answer in Appendix F*

Rhythmic Ventilation

The generation of the basic rhythm of ventilation is controlled by neurons within the medulla oblongata that stimulate the muscles of respiration. Recruitment of muscle fibers and increased frequency of stimulation of muscle fibers results in stronger contractions of the muscles and an increased depth of respiration. The rate of respiration is determined by how frequently the respiratory muscles are stimulated.

Respiratory Areas in the Brainstem

The classic view of respiratory areas held that distinct inspiratory and expiratory centers were located in the brainstem. This view is now known to be too simplistic. Although neurons involved with respiration are aggregated in certain parts of the brainstem, neurons that are active during inspiration are intermingled with neurons active during expiration.

The **medullary respiratory center** consists of **two dorsal respiratory groups,** each forming a longitudinal column of cells located bilaterally in the dorsal part of the medulla oblongata, and two **ventral respiratory groups,** each forming a longitudinal column of cells located bilaterally in the ventral part of the medulla oblongata (figure 23.20). Although the dorsal and ventral respiratory groups are bilaterally paired, cross communication exists between the pairs so that respiratory movements are symmetric. In addition, communication exists between the dorsal and ventral respiratory groups.

Each dorsal respiratory group is a collection of neurons that are most active during inspiration, but some are active during expiration. The dorsal respiratory groups are primarily responsible for stimulating contraction of the diaphragm. They receive input from other parts of the brain and peripheral receptors that allows modification of respiration.

Each ventral respiratory group is a collection of neurons that are active during inspiration and expiration. These neurons primarily stimulate the external intercostal, internal intercostal, and abdominal muscles.

The **pontine respiratory group** is a collection of neurons in the pons (see figure 23.20). Some of the neurons are only active during inspiration, some only during expiration, and some during both inspiration and expiration. The precise function of the pontine respiratory group is unknown, but it has connections with the medullary respiratory center and appears to play a role in switching between inspiration and expiration.

Generation of Rhythmic Ventilation

The exact locations of neurons in the medullary respiratory center responsible for rhythmic ventilation are unknown. Nor, is it well understood how they generate the basic pattern of spontaneous, rhythmic ventilation at rest. One explanation involves integration of stimuli that start and stop inspiration.

1. *Starting inspiration.* Neurons in the medullary respiratory center constantly receive stimulation from many sources, such as receptors that monitor blood gas levels, blood temperature, and movements of muscles and joints. In addition, stimulation from parts of the brain concerned with voluntary respiratory movements and emotions can occur. When the combined input from all these sources reaches a threshold level, neurons that stimulate respiratory muscles produce action potentials and inspiration starts.

2. *Increasing inspiration.* Once inspiration begins, more and more neurons are gradually activated. The result is progressively stronger stimulation of the respiratory muscles that lasts for approximately 2 s.

3. *Stopping inspiration.* The neurons stimulating the muscles of respiration also stimulate other neurons in the medullary respiratory center that are responsible for stopping inspiration. The neurons responsible for stopping inspiration also receive input from the pontine respiratory group, stretch receptors in the lungs, and probably other sources. When the input to these neurons exceeds a threshold level, they cause the neurons stimulating respiratory muscles to be inhibited. Relaxation of respiratory muscles results in expiration, which lasts approximately 3 s. For the next inspiration, go back to step 1.

Modification of Ventilation

Although the medullary neurons establish the basic rate and depth of breathing, their activities can be influenced by input from other parts of the brain and by input from peripherally located receptors.

Cerebral and Limbic System Control

Through the cerebral cortex it is possible to consciously or unconsciously increase or decrease the rate and depth of the respiratory movements (figure 23.21). For example, during talking or singing, air movement is controlled to produce sounds as well as to facilitate gas exchange.

A person may also stop breathing voluntarily. As the period of voluntary apnea increases, a greater and greater urge to breathe develops. That urge is primarily associated with increasing P_{CO_2} levels in the arterial blood. Finally the P_{CO_2} reaches levels that cause the respiratory center to override the conscious influence from the cerebrum. Occasionally people are able to hold their breath until the blood P_{O_2} declines to a level low enough that they lose consciousness. After consciousness is lost, the respiratory center resumes its normal function in automatically controlling respiration.

Voluntary hyperventilation can decrease the blood P_{CO_2} levels sufficiently to cause vasodilation of the peripheral blood vessels and a decrease in blood pressure (see chapter 21). Dizziness or a giddy feeling can result because of a decreased rate of blood flow to the brain after the blood pressure drops.

Emotions acting through the limbic system of the brain can also affect the respiratory center (see figure 23.21). For example, strong emotions can cause hyperventilation or produce the sobs and gasps of crying.

Chemical Control of Ventilation

The respiratory system maintains blood oxygen and carbon dioxide concentrations and blood pH within a normal range of values. A deviation by any of these parameters from their normal range has a marked influence on respiratory movements. The effect of changes in oxygen and carbon dioxide concentrations and in pH is superimposed on the neural mechanisms that establish rhythmic ventilation.

Chemoreceptors

Chemoreceptors are specialized neurons that respond to changes in chemicals in solution. The chemoreceptors involved with the regulation of respiration respond to changes in hydrogen ion concentrations or changes in P_{O_2} (or both) (see figure 23.21 and figure 23.22). **Central chemoreceptors** are located bilaterally and ventrally in the **chemosensitive area** of the medulla oblongata, and they are connected to the respiratory center. **Peripheral chemoreceptors** are found in the carotid and aortic bodies. These structures are small vas-

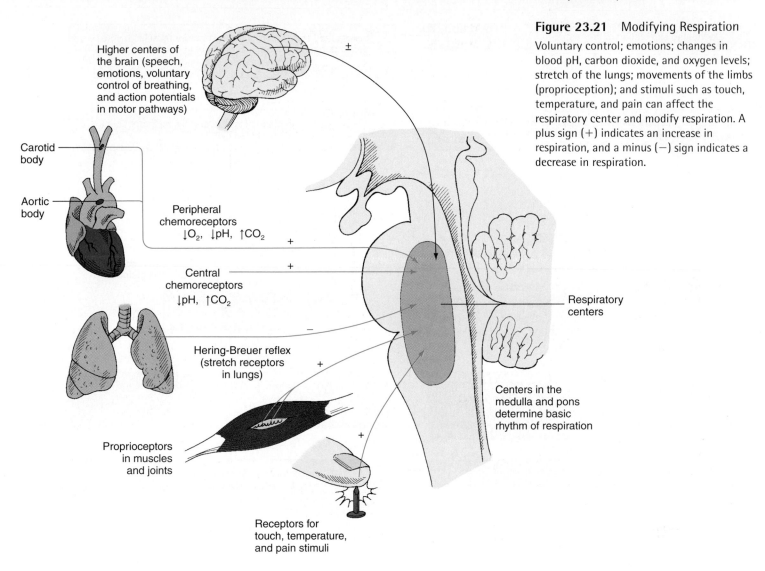

Figure 23.21   Modifying Respiration

Voluntary control; emotions; changes in blood pH, carbon dioxide, and oxygen levels; stretch of the lungs; movements of the limbs (proprioception); and stimuli such as touch, temperature, and pain can affect the respiratory center and modify respiration. A plus sign (+) indicates an increase in respiration, and a minus (−) sign indicates a decrease in respiration.

Higher centers of the brain (speech, emotions, voluntary control of breathing, and action potentials in motor pathways)

Carotid body

Aortic body

Peripheral chemoreceptors
$\downarrow O_2$, $\downarrow pH$, $\uparrow CO_2$

Central chemoreceptors
$\downarrow pH$, $\uparrow CO_2$

Respiratory centers

Hering-Breuer reflex (stretch receptors in lungs)

Centers in the medulla and pons determine basic rhythm of respiration

Proprioceptors in muscles and joints

Receptors for touch, temperature, and pain stimuli

cular sensory organs, which are encapsulated in connective tissue and located near the carotid sinuses and the aortic arch (see chapter 21). The respiratory center is connected to the carotid body chemoreceptors through the glossopharyngeal nerve (IX) and to the aortic body chemoreceptors by the vagus nerve (X).

Effect of pH

The chemosensitive area is bathed by cerebrospinal fluid and is sensitive to changes in pH of the fluid. Because the blood–brain barrier separates the chemosensitive area from the blood, the chemosensitive area does not directly detect changes in blood pH. Changes in blood pH can alter cerebrospinal fluid pH, however, so the chemosensitive area responds indirectly to changes in blood pH. In addition, the carotid and aortic bodies have a rich vascular supply and are directly sensitive to changes in blood pH.

Maintaining body pH levels within normal parameters is necessary for the proper functioning of cells. Because changes

in carbon dioxide levels can change pH, the respiratory system plays an important role in acid–base balance. For example, if pH decreases, the respiratory center is stimulated, resulting in elimination of carbon dioxide and an increase in pH back to normal levels. Conversely, if pH increases, the respiratory rate decreases, and carbon dioxide levels increase, causing pH to decrease back to normal levels. The role of the respiratory system in maintaining pH is considered in greater detail in chapter 27.

Effect of Carbon Dioxide

Carbon dioxide levels are a major regulator of respiration during resting conditions and conditions when the carbon dioxide levels are elevated such as during intense exercise. Even a small increase in carbon dioxide in the circulatory system triggers a large increase in the rate and depth of respiration. An increase in P_{CO_2} of 5 mm Hg, for example, causes an increase in ventilation of 100%. A greater-than-normal amount of carbon dioxide in the blood is called

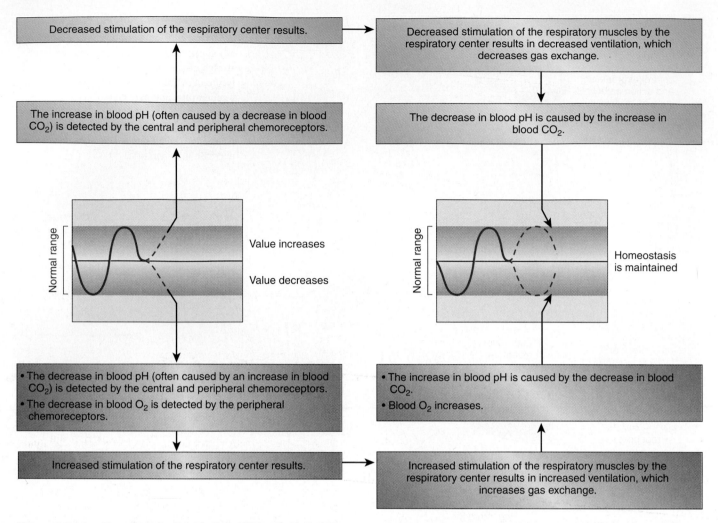

Figure 23.22   Homeostasis: Regulation of Blood pH and Gases

hypercapnia (hī-per-kap′nē-ă). Conversely, lower-than-normal carbon dioxide levels, a condition called **hypocapnia** (hī-pō-kap′nē-ă), result in periods in which respiratory movements do not occur.

Carbon dioxide apparently does not directly affect the chemosensitive area. Instead, carbon dioxide exerts its effect by changing hydrogen ion levels, which can affect the chemosensitive area (see figure 23.22). For example, if blood carbon dioxide levels increase, the carbon dioxide diffuses across the blood–brain barrier into the cerebrospinal fluid. The carbon dioxide combines with water to form carbonic acid, which dissociates into hydrogen ions and bicarbonate ions. The increased levels of hydrogen ions stimulate the chemosensitive area, which then stimulates the respiratory center, resulting in a greater rate and depth of breathing. Consequently, carbon dioxide levels decrease as carbon dioxide is eliminated from the body.

---

11   P R E D I C T

Explain why a person who breathes rapidly and deeply (hyperventilates) for several seconds experiences a short period during which respiration does not occur (apnea) before normal breathing resumes.

✔ *Answer in Appendix F*

The chemoreceptors in the carotid and aortic bodies also respond to changes in carbon dioxide because of the effects of carbon dioxide on blood pH. The carotid and aortic bodies, however, are responsible for, at most, 15%–20% of the total response to changes in P_{CO_2} or pH. The chemosensitive area in the medulla oblongata is far more important for the regulation of P_{CO_2} and pH than are the carotid and aortic bodies. During intense exercise, however, the carotid bodies respond more rapidly to changes in blood pH than does the chemosensitive area of the medulla.

Effect of Oxygen

Although P_{CO_2} levels detected by the chemosensitive area are responsible for most changes in respiration, changes in P_{O_2} can also affect respiration (see figure 23.22). A decrease in oxygen levels below normal values is called **hypoxia** (hī-pok'sē-ă). If P_{O_2} levels in the arterial blood are markedly reduced while the pH and P_{CO_2} are held constant, an increase in ventilation occurs. Within a normal range of P_{O_2} levels, however, the effect of oxygen on the regulation of respiration is small. Only after arterial P_{O_2} decreases to approximately 50% of its normal value does it begin to have a large stimulatory effect on respiratory movements.

At first it is somewhat surprising that small changes in P_{O_2} do not cause changes in respiratory frequencies. Consideration of the oxygen–hemoglobin dissociation curve, however, provides an explanation. Because of the **S**-shape of the curve, at any P_{O_2} above 80 mm Hg nearly all of the hemoglobin is saturated with oxygen. Consequently, until P_{O_2} levels change significantly, the oxygen-carrying capacity of the blood is unaffected.

The carotid and aortic body chemoreceptors respond to decreased P_{O_2} by increased stimulation of the respiratory center, which can keep it active, despite decreasing oxygen levels. If P_{O_2} decreases sufficiently, however, the respiratory center can fail to function, resulting in death.

Clinical Note

Carbon dioxide is much more important than oxygen as a regulator of normal alveolar ventilation, but under certain circumstances a reduced P_{O_2} in the arterial blood does play an important stimulatory role. During conditions of shock in which blood pressure is very low, the P_{O_2} in arterial blood can decrease to levels sufficiently low to strongly stimulate carotid and aortic body sensory receptors. At high altitudes where barometric air pressure is low, the P_{O_2} in arterial blood can also decrease to levels sufficiently low to stimulate carotid and aortic bodies. Although P_{O_2} levels in the blood are reduced, the ability of the respiratory system to eliminate carbon dioxide is not greatly affected by low barometric air pressure. Thus blood carbon dioxide levels become lower than normal because of the increased alveolar ventilation initiated in response to low P_{O_2}.

A similar situation exists in people who have emphysema. Because carbon dioxide diffuses across the respiratory membrane more readily than oxygen, the decreased surface area of the respiratory membrane caused by the disease results in low arterial P_{O_2} without elevated arterial P_{CO_2}. The elevated rate and depth of respiration are due, to a large degree, to the stimulatory effect of low arterial P_{O_2} levels on carotid and aortic bodies. More severe emphysema, in which the surface area of the respiratory membrane is reduced to a minimum, can also result in elevated P_{CO_2} levels in arterial blood.

Hering–Breuer Reflex

The **Hering–Breuer** (her'ing-broy'er) **reflex** limits the degree to which inspiration proceeds and prevents overinflation of the lungs (see figure 23.21). This reflex depends on stretch receptors in the walls of the bronchi and bronchioles of the lung. Action potentials are initiated in these stretch receptors when the lungs are inflated and are passed along afferent neurons within the vagus nerves to the medulla oblongata. The action potentials have an inhibitory influence on the respiratory center and result in expiration. As expiration proceeds, the stretch receptors are no longer stimulated, and the decreased inhibitory effect on the respiratory center allows inspiration to begin again.

In infants the Hering-Breuer reflex plays an important role in regulating the basic rhythm of breathing and in preventing overinflation of the lungs. In adults, however, the reflex is important only when the tidal volume is large, such as during exercise.

Effect of Exercise on Ventilation

The mechanisms by which ventilation is regulated during exercise is controversial, and no one factor can account for all of the observed responses. Ventilation during exercise can be divided into two phases.

1. *Ventilation increases abruptly.* At the onset of exercise, ventilation immediately increases. This initial increase can be as much as 50% of the total increase that occurs during exercise. The immediate increase in ventilation occurs too quickly to be explained by changes in metabolism or blood gases. As axons pass from the motor cortex of the cerebrum through the motor pathways, numerous collateral fibers project into the reticular formation of the brain. During exercise, action potentials in the motor pathways stimulate skeletal muscle contractions, and action potentials in the collateral fibers stimulate the respiratory center (see figure 23.21).

   Furthermore, during exercise body movements stimulate proprioceptors in the joints of the limbs. Action potentials from the proprioceptors pass along afferent nerve fibers to the spinal cord and along ascending nerve tracts (the dorsal-column/medial-lemniscal system) of the spinal cord to the brain. Collateral fibers project from these ascending pathways to the respiratory center in the medulla. Movement of the limbs has a strong stimulatory influence on the respiratory center (see figure 23.21).

   There may also be a learned component to the ventilation response during exercise. After a period of training the brain "learns" to match ventilation with the intensity of the exercise. Well-trained athletes match their respiratory movements more efficiently with their level of physical activity than do untrained individuals. Thus centers of the brain involved in learning have an indirect influence on the respiratory center, but the exact mechanism for this kind of regulation is unclear.

2. *Ventilation increases gradually.* After the immediate increase in ventilation, there is a gradual increase in ventilation that levels off within 4–6 min after the onset

of exercise. Factors responsible for the immediate increase in ventilation may play a role in the gradual increase as well.

Despite large changes in oxygen consumption and carbon dioxide production during exercise, the *average* arterial Po_2, Pco_2, and pH remain constant and close to resting levels as long as the exercise is aerobic (see chapter 10). This suggests that changes in blood gases and pH do not play an important role in regulating ventilation during aerobic exercise. During exercise, however, the values of arterial Po_2, Pco_2, and pH rise and fall more than at rest. Thus, even though their average values do not change, their oscillations may be a signal for helping to control ventilation.

The highest level of exercise that can be performed without causing a significant change in blood pH is called the **anaerobic threshold.** If the exercise intensity is high enough to exceed the anaerobic threshold, then skeletal muscles produce and release lactic acid into the blood. The resulting change in blood pH stimulates the carotid bodies, resulting in increased ventilation. In fact, ventilation can increase so much that arterial Pco_2 decreases below resting levels and arterial Po_2 increases above resting levels.

Other Modifications of Ventilation

The activation of touch, thermal, and pain receptors can also affect the respiratory center (see figure 23.21). For example, irritants in the nasal cavity can initiate a sneeze reflex, and irritants in the lungs can stimulate a cough reflex. An increase in body temperature can stimulate increased ventilation.

12	P R E D I C T

Describe the respiratory response when cold water is splashed onto a person. In the past, newborn babies were sometimes swatted on the buttocks. Explain the rationale for this procedure.

✔ *Answer in Appendix F*

Respiratory Adaptations to Exercise

In response to training, athletic performance increases because the cardiovascular and respiratory systems become more efficient at delivering oxygen and picking up carbon dioxide. Ventilation in most individuals does not limit performance because ventilation can increase to a greater extent than does cardiovascular function.

After training, vital capacity increases slightly and residual volume decreases slightly. Tidal volume at rest and during standardized submaximal exercise does not change. At maximal exercise, however, tidal volume increases. After training, respiratory rate at rest or during standardized submaximal exercise is slightly lower than in an untrained person, but at maximal exercise respiratory rate is generally increased.

Minute ventilation is affected by the changes in tidal volume and respiratory rate. After training, minute ventilation is essentially unchanged or slightly reduced at rest and is slightly reduced during standardized submaximal exercise. Minute ventilation is greatly increased at maximal exercise. For example, an untrained person with a minute ventilation of 120 L/min can increase to 150 L/min after training. Increases to 180 L/min are typical of highly trained athletes.

Gas exchange between the alveoli and blood increases at maximal exercise following training. The increased minute ventilation results in increased alveolar ventilation. In addition, increased cardiovascular efficiency results in greater blood flow through the lungs, especially in the superior parts of the lungs.

▐ Clinical Focus   Disorders of the Respiratory System

Bronchi and Lungs

Bronchitis (brong-kī′tis) is an inflammation of the bronchi caused by irritants, such as cigarette smoke, air pollution, or infections. The inflammation results in swelling of the mucous membrane lining the bronchi, increased mucus production, and decreased movement of mucus by cilia. Consequently, the diameter of the bronchi is decreased, and ventilation is impaired. Bronchitis can progress to emphysema.

Emphysema (em-fi-sē′mă) results in the destruction of the alveolar walls. Many smokers have both bronchitis and emphysema, which are often referred to as **chronic obstructive pulmonary disease (COPD)**. Chronic inflammation of the bronchioles, usually caused by cigarette smoke or air pollution, probably initiates emphysema. Narrowing of the bronchioles restricts air movement, and air tends to be retained in the lungs. Coughing to remove accumulated mucus increases pressure in the alveoli, resulting in rupture and destruction of alveolar walls. Loss of alveolar walls has two important consequences. The respiratory membrane has a decreased surface area, which decreases gas exchange, and loss of elastic fibers decreases the ability of the lungs to recoil and expel air. Symptoms of emphysema include shortness of breath and enlargement of the thoracic cavity. Treatment involves removing sources of irritants (e.g., stopping smoking), promoting the removal of bronchial secretions, bronchiodilators, retraining people to breathe so that expiration of air is maximized, and using antibiotics to prevent infections. The progress of emphysema can be slowed, but there is no cure.

Cystic fibrosis is an inherited disease that affects the secretory cells lining the lungs, pancreas, sweat glands, and salivary glands. The defect produces an abnormal chloride transport protein that does not reach the cell surface or does not function normally if it does reach the cell surface. The result is decreased chloride ion secretion out of cells. Normally, the diffusion of chloride and sodium ions out of the cells causes water to follow by osmosis. In the lungs, the water forms a thin fluid layer over which mucus is moved by ciliated cells. In cystic fibrosis, the decreased chloride ion diffusion results in dehydrated respiratory secretions.

The mucus is more viscous, resisting movement by cilia, and it accumulates in the lungs. For reasons not completely understood, the mucus accumulation increases the likelihood of infections. Chronic airflow obstruction causes difficulty in breathing, and coughing in an attempt to remove the mucus can result in pneumothorax and bleeding within the lungs. Once fatal during early childhood, many victims of cystic fibrosis are now surviving into young adulthood. Future treatments could include the development of drugs that correct or assist the normal ion transport mechanism. Alternatively, cystic fibrosis may someday be cured through genetic engineering by inserting a functional copy of the defective gene into a person with the disease. Research on this exciting possibility is currently underway.

Pulmonary fibrosis is the replacement of lung tissue with fibrous connective tissue, making the lungs less elastic and breathing more difficult. Exposure to asbestos, silica, or coal dust is the most common cause.

Lung cancer arises from the epithelium of the respiratory tract. Cancers arising from tissues other than respiratory epithelium are not called lung cancer, even though they occur in the lungs. Lung cancer is the most common cause of cancer death in males and females in the United States, and almost all cases occur in smokers. Because of the rich lymph and blood supply in the lungs, cancer in the lung can readily spread to other parts of the lung or body. In addition, the disease is often advanced before symptoms become severe enough for the victim to seek medical aid. Typical symptoms include coughing, sputum production, and blockage of the airways. Treatments include removal of part or all of the lung, chemotherapy, and radiation.

Nervous System

Sudden infant death syndrome (SIDS), or crib death, is the most frequent cause of death of infants between 2 weeks and 1 year of age. Death results when the infant stops breathing during sleep. Although the cause of SIDS remains controversial, there is evidence that damage to the respiratory center during development is a factor. There is no treatment, but at-risk babies can be placed on monitors that sound an alarm if the baby stops breathing.

Paralysis of the respiratory muscles can result from damage of the spinal cord in the cervical or thoracic regions. The damage interrupts nerve tracts that transmit action potentials to the muscles of respiration. Transection of the spinal cord can result from trauma such as automobile accidents or diving into water that is too shallow. Another cause of paralysis is poliomyelitis, a viral infection that damages neurons of the respiratory center or motor neurons that stimulate the muscles of respiration. Anesthetics or central nervous system depressants can also depress the function of the respiratory center if they are taken or administered in large enough doses.

Diseases of the Upper Respiratory Tract

Strep throat is caused by a streptococcal bacteria (*Streptococcus pyogenes*) and is characterized by inflammation of the pharynx and by fever. Frequently, inflammation of the tonsils and middle ear is involved. Without a throat analysis, the infection cannot be distinguished from viral causes of pharyngeal inflammation. Current techniques allow rapid diagnosis within minutes to hours, and antibiotics are an effective treatment.

Diphtheria (dif-thēr′ē-ă) was once a major cause of death among children. It is caused by a bacterium (*Corynebacterium diphtheriae*). A grayish membrane forms in the throat and can block the respiratory passages totally. A vaccine against diphtheria is part of the normal immunization program for children in the United States.

The **common cold** is the result of a viral infection. Symptoms include sneezing, excessive nasal secretions, and congestion. The infection easily can spread to sinus cavities, lower respiratory passages, and the middle ear. Laryngitis and middle ear infections are common complications. The common cold usually runs its course to recovery in about 1 week.

Diseases of the Lower Respiratory Tract

Laryngitis (lar-in-jī′tis) is an inflammation of the larynx, especially the vocal cords, and **bronchitis** is an inflammation of the bronchi. Bacterial or viral infections can move from the upper respiratory tract to cause laryngitis or bronchitis. Bronchitis is also often caused by continually breathing air containing harmful chemicals, such as those found in cigarette smoke.

Whooping cough (pertussis; per-tŭs'is) is a bacterial infection (*Bordetella pertusis*), which causes a loss of cilia of the respiratory epithelium. Mucus accumulates, and the infected person attempts to cough up the mucous accumulations. The coughing can be severe. A vaccine for whooping cough is part of the normal vaccination procedure for children in the United States.

Tuberculosis (tū-ber-kyū-lō'sis) is caused by a tuberculosis bacterium (*Mycobacterium tuberculosis*). In the lung, the bacteria forms lesions called tubercles. The small lumps contain degenerating macrophages and tuberculosis bacteria. An immune reaction is directed against the tubercles, which causes the formation of larger lesions and inflammation. The tubercles can rupture, releasing bacteria that infect other parts of the lung or body. Recently, a strain of the tuberculosis bacteria has developed that is resistant to treatment, and there is concern that tuberculosis will again become a widespread infectious disease.

Pneumonia (nū-mō'nē-ă) is a general term that refers to many infections of the lung. Most pneumonias are caused by bacteria, but some result from viral, fungal, or protozoan infections. Symptoms include fever, difficulty in breathing, and chest pain. Inflammation of the lungs results in the accumulation of fluid within alveoli (pulmonary edema) and poor inflation of the lungs with air. A protozoal infection (*Pneumocystis carinii*) that results in pneumocystosis pneumonia is rare, except in persons who have a compromised immune system. This type of pneumonia has become one of the infections commonly suffered by persons who have AIDS.

Flu (influenza) is a viral infection of the respiratory system and does not affect the digestive system as is commonly assumed. Flu is characterized by chills, fever, headache, and muscular aches, in addition to coldlike symptoms. There are several strains of flu viruses. The mortality rate from flu is approximately 1%, and most of those deaths occur among the very old and very young. During a flu epidemic the infection rate is so rapid and the disease so widespread that the total number of deaths is substantial, even though the percentage of deaths is relatively low. Flu vaccines can provide some protection against the flu.

A number of fungal diseases, such as **histoplasmosis** (his'tō-plaz-mō'sis) and **coccidioidomycosis** (kok-sid-ē-oy'dō-mī-kō'sis), affect the respiratory system. The fungal spores (*Histoplasma capsulatum; Coccidioides immitis*) usually enter the respiratory system through dust particles. Spores in soil and feces of certain animals make the rate of infection higher in farm workers and in gardeners. The infections usually result in minor respiratory infections, but in some cases they can cause infections throughout the body.

Systems Pathology

ASTHMA

Mr. W is an 18-year-old track athlete in seemingly good health. One day he came down with a common cold, resulting in the typical symptoms of nasal congestion and discomfort. After several days he began to cough and wheeze and he thought that his cold had progressed to his lungs. Determined not to get "out of shape" because of his cold, Mr. W took a few aspirin to relieve his discomfort and went to the track to do some jogging. After a few minutes of exercise he began to wheeze very forcefully and rapidly, and he felt that he could hardly get enough air. Even though he stopped jogging, his condition did not improve (figure A). Fortunately, a concerned friend who was also at the track, took him to the emergency room.

Although Mr. W had no previous history of asthma, careful evaluation by the emergency room doctor convinced her that he probably was having an asthma attack. Mr. W inhaled a bronchiodilator drug, which resulted in rapid improvement of his condition. He was released from the emergency room and referred to his personal physician for further treatment and education about asthma.

BACKGROUND INFORMATION

Asthma (az′mă) is a disease characterized by increased constriction of the trachea and bronchi in response to various stimuli, resulting in a narrowing of the air passageways and decreased ventilation efficiency. Symptoms include wheezing, coughing, and shortness of breath. In contrast to many other respiratory disorders, however, the symptoms of asthma typically reverse either spontaneously or with therapy.

It is estimated that the prevalence of asthma in the United States is from 3% to 6% of the general population. Approximately half the cases first appear before age 10, and twice as many boys as girls develop asthma. Anywhere from 25% to 50% of childhood asthmatics are symptom-free from adolescence onward.

The exact cause or causes of asthma is unknown, but asthma and allergies run strongly in some families. There is no definitive pathologic feature or diagnostic test for asthma, but three important features of the disease are chronic airway inflammation, airway hyperreactivity, and airflow obstruction. The inflammatory response results in tissue damage, edema, and mucous buildup, which can block air flow through the bronchi. Airway hyperreactivity is greatly increased contraction of the smooth muscle in the trachea and bronchi in response to a stimulus. As a result of airway hyperactivity, the diameter of the airway decreases, and resistance to air flow increases. The effects of inflammation and airway hyperreactivity combine to cause airflow obstruction.

Many cases of asthma appear to be associated with a chronic inflammatory response by the immune system. The number of immune cells in the bronchi increases, including mast cells, eosinophils, neutrophils, macrophages, and lymphocytes. These cells release chemical mediators, such as interleukins, leukotrienes, prostaglandins, platelet-activating factor, thromboxanes, and chemotactic factors. These chemical mediators promote inflammation, increase mucous secretion, and attract additional immune cells to the bronchi, resulting in chronic airway inflammation. Airway hyperreactivity and inflammation appear to be linked by some of the chemical mediators, which increase the sensitivity of the airway to stimulation and cause smooth muscle contraction.

The stimuli that prompt airflow obstruction varies from one individual to another. Some asthmatics have reactions to particular allergens, which are foreign substances that evoke an inappropriate immune system response (see chapter 22). Examples include inhaled pollen, animal dander, and dust mites. Many cases of asthma may be caused by an allergic reaction to substances in the droppings and carcasses of cockroaches, which may explain the higher rate of asthma in poor, urban areas.

On the other hand, inhaled substances, such as chemicals in the workplace or cigarette smoke, can provoke an asthma attack without stimulating an allergic reaction. Over 200 substances have been associated with occupational asthma. An asthma attack can also be stimulated by ingested substances such as aspirin, nonsteroidal anti-inflammatory compounds such as ibuprofen (i-bū′prō-fen), sulfites in food preservative, and tartrazine (tar′tra-zēn) in food colorings. Asthmatics can substitute acetaminophen (as-et-a-mē′nō-fen; Tylenol) for aspirin.

Other stimuli, such as strenuous exercise, especially in cold weather, can precipitate an asthma attack. Such episodes can often be avoided by using a bronchiodilator drug prior to exercise. Viral infections, emotional upset, stress, and even reflux of stomach acid into the esophagus are known to elicit an asthma attack.

Treatment of asthma involves avoiding the causative stimulus and administering drug therapy. Steroids and mast

Figure A Jogger with Asthma

cell-stabilizing agents, which prevent the release of chemical mediators from mast cells, are used to reduce airway inflammation. Theophylline (thē-of'i-lēn) and beta-adrenergic agents (see chapter 16) are commonly used to cause bronchiolar dilation. Although treatment is generally effective in controlling asthma, death by asphyxiation rarely can occur. Most of these deaths probably could have been prevented by earlier and more intensive therapy.

System Interactions	
System	**Interactions**
Integumentary	Cyanosis, a bluish skin color, results from a decreased blood oxygen content.
Muscular	Skeletal muscles are necessary for respiratory movements and the cough reflex. Increased muscular work during a severe asthma attack can cause metabolic acidosis because of anaerobic respiration and excessive lactic acid production.
Skeletal	Red bone marrow is the site of production of many of the immune cells responsible for the inflammatory response of asthma. The thoracic cage is necessary for respiration.
Nervous	Emotional upset or stress can evoke an asthma attack. Peripheral and central chemoreceptor reflexes affect ventilation. The cough reflex helps to remove mucus from respiratory passages. Pain, anxiety, and death from asphyxiation can result from the altered gas exchange caused by asthma. One theory of the cause of asthma is an imbalance of the autonomic nervous system (ANS) control of bronchiolar smooth muscle, and drugs that enhance sympathetic effects or block parasympathetic effects are used in asthma treatment.
Endocrine	Steroids from the adrenal gland play a role in regulating inflammation, and they are used in asthma therapy.
Cardiovascular	Increased vascular permeability of lung blood vessels results in edema. Blood carries ingested substances that provoke an asthma attack to the lungs. Blood carries immune cells from the red bone marrow to the lungs. Tachycardia commonly occurs, and the normal effects of respiration on venous return of blood to the heart are exaggerated, resulting in large fluctuations of blood pressure.
Lymphatic and Immune	Immune cells release chemical mediators that promote inflammation, increase mucous production, and cause bronchiolar constriction (believed to be a major factor in asthma). Ingested allergens, such as aspirin or sulfites in food, can evoke an asthma attack.
Digestive	Reflux of stomach acid into the esophagus can evoke an asthma attack.
Urinary	Modifying hydrogen ion secretion into the urine helps to compensate for acid–base imbalances caused by asthma.

Summary

Respiration includes the movement of air into and out of the lungs, the exchange of gases between the air and the blood, the transport of gases in the blood, and the exchange of gases between the blood and tissues.

Anatomy and Histology
Nose and Nasal Cavity

1. The bridge of the nose is bone, and most of the external nose is cartilage.
2. Openings of the nasal cavity
   - The external nares open to the outside, and the internal nares lead to the pharynx.
   - The paranasal sinuses and the nasolacrimal duct open into the nasal cavity.
3. Divisions of the nasal cavity
   - The nasal cavity is divided by the nasal septum.
   - The anterior vestibule contains hairs that trap debris.
   - The nasal cavity is lined with pseudostratified ciliated epithelium that traps debris and moves it to the pharynx.
   - The superior part of the nasal cavity contains the olfactory epithelium.

Pharynx

1. The nasopharynx joins the nasal cavity through the internal nares and contains the openings to the auditory tube and the pharyngeal tonsils.
2. The oropharynx joins the oral cavity and contains the palatine and lingual tonsils.
3. The laryngopharynx opens into the larynx and the esophagus.

Larynx

1. Cartilage
   - There are three unpaired cartilages. The thyroid cartilage and cricoid cartilage form most of the larynx. The epiglottis covers the opening of the larynx during swallowing.
   - There are six paired cartilages. The vocal cords attach to the arytenoid cartilages.
2. Sounds are produced as the vocal cords vibrate when air passes through the larynx. Tightening the cords produces sounds of different pitches by controlling the length of the cord that is allowed to vibrate.

Trachea

The trachea connects the larynx to the primary bronchi.

Tracheobronchial Tree

1. The conducting zone, from the trachea to the terminal bronchioles, is a passageway for air movement.
   - The area from the trachea to the terminal bronchioles is ciliated to facilitate removal of debris.
   - Cartilage helps to hold the tube system open (from the trachea to the bronchioles).
   - Smooth muscle controls the diameter of the tubes (terminal bronchioles).
2. The respiratory zone, from the respiratory bronchioles to the alveoli, is a site of gas exchange.

Lungs

1. There are two lungs.
2. The lungs are divided into lobes, bronchopulmonary segments, and lobules.

Thoracic Wall and Muscles of Respiration

1. The thoracic wall consists of vertebrae, ribs, sternum, and muscles that allow expansion of the thoracic cavity.
2. Contraction of the diaphragm increases thoracic volume.
3. Muscles can elevate the ribs and increase thoracic volume or can depress the ribs and decrease thoracic volume.

Pleura

The pleural membranes surround the lungs and provide protection against friction.

Blood Supply

1. Deoxygenated blood is transported to the lungs through the pulmonary arteries, and oxygenated blood leaves through the pulmonary veins.
2. Oxygenated blood is mixed with a small amount of deoxygenated blood from the bronchi.

Lymphatic Supply

The superficial and deep lymphatic vessels drain lymph from the lungs.

Ventilation
Pressure Differences and Air Flow

1. Ventilation is the movement of air into and out of the lungs.
2. Air moves from an area of higher pressure to an area of lower pressure.

Pressure and Volume

Pressure is inversely related to volume.

Air Flow into and out of Alveoli

1. Inspiration results when barometric air pressure is greater than alveolar pressure.
2. Expiration results when barometric air pressure is less than alveolar pressure.

Changing Alveolar Volume

1. Lung recoil causes alveoli to collapse.
   - Lung recoil results from elastic fibers and water surface tension.
   - Surfactant reduces water surface tension.
2. Pleural pressure is the pressure in the pleural cavity.
   - A negative pleural pressure can cause the alveoli to expand.
   - Pneumothorax is an opening between the pleural cavity and the air that causes a loss of pleural pressure.
3. Changes in thoracic volume cause changes in pleural pressure, resulting in changes in alveolar volume, alveolar pressure, and air flow.

Measuring Lung Function
Compliance of the Lungs and the Thorax

1. Compliance is a measure of lung expansion caused by alveolar pressure.
2. Reduced compliance means that it is more difficult than normal to expand the lungs.

Pulmonary Volumes and Capacities

1. There are four pulmonary volumes: tidal volume, inspiratory reserve, expiratory reserve, and residual volume.
2. Pulmonary capacities are the sum of two or more pulmonary volumes and include inspiratory capacity, functional residual capacity, vital capacity, and total lung capacity.
3. The forced expiratory vital capacity measures vital capacity as the individual exhales as rapidly as possible.

Minute Ventilation and Alveolar Ventilation

1. The minute ventilation is the total amount of air moved in and out of the respiratory system per minute.
2. Dead space is the part of the respiratory system in which gas exchange does not take place.
3. Alveolar ventilation is how much air per minute enters the parts of the respiratory system in which gas exchange takes place.

Physical Principles of Gas Exchange
Partial Pressure

1. Partial pressure is the contribution of a gas to the total pressure of a mixture of gases (Dalton's law).
2. Water vapor pressure is the partial pressure produced by water.
3. Atmospheric air, alveolar air, and expired air have different compositions.

Diffusion of Gases Through Liquids

The concentration of a gas in a liquid is determined by its partial pressure and by its solubility coefficient (Henry's law).

Diffusion of Gases Through the Respiratory Membrane

1. The respiratory membrane is thin and has a large surface area that facilitates gas exchange.
2. The components of the respiratory membrane include a film of water, the walls of the alveolus and the capillary, and an interstitial space.
3. The rate of diffusion depends on the thickness of the respiratory membrane, the diffusion coefficient of the gas, the surface area of the membrane, and the partial pressure of the gases in the alveoli and the blood.

Relationship Between Ventilation and Pulmonary Capillary Blood Flow

1. Increased ventilation or increased pulmonary capillary blood flow increases gas exchange.
2. The physiologic shunt is the deoxygenated blood returning from the lungs.

Oxygen and Carbon Dioxide Transport in the Blood
Oxygen Diffusion Gradients

1. Oxygen moves from the alveoli (P_{O_2} = 104 mm Hg) into the blood (P_{O_2} = 40 mm Hg). Blood is almost completely saturated with oxygen when it leaves the capillary.

2. The P_{O_2} in the blood decreases (P_{O_2} = 95 mm Hg) because of mixing with deoxygenated blood.
3. Oxygen moves from the tissue capillaries (P_{O_2} = 95 mm Hg) into the tissues (P_{O_2} = 40 mm Hg).

Carbon Dioxide Diffusion Gradients

1. Carbon dioxide moves from the tissues (P_{CO_2} = 45 mm Hg) into tissue capillaries (P_{CO_2} = 40 mm Hg).
2. Carbon dioxide moves from the pulmonary capillaries (P_{CO_2} = 45 mm Hg) into the alveoli (P_{CO_2} = 40 mm Hg).

Hemoglobin and Oxygen Transport

1. Oxygen is transported by hemoglobin (98.5%) and is dissolved in plasma (1.5%).
2. The oxygen–hemoglobin dissociation curve shows that hemoglobin is almost completely saturated when P_{O_2} is 80 mm Hg or above. At lower partial pressures the hemoglobin releases oxygen.
3. A shift of the oxygen–hemoglobin dissociation curve to the right because of a decrease in pH (Bohr effect), an increase in carbon dioxide, or an increase in temperature results in a decrease in the ability of hemoglobin to hold oxygen.
4. A shift of the oxygen–hemoglobin dissociation curve to the left because of an increase in pH (Bohr effect), a decrease in carbon dioxide, or a decrease in temperature results in an increase in the ability of hemoglobin to hold oxygen.
5. The substance 2,3-bisphosphoglycerate increases the ability of hemoglobin to release oxygen.

Transport of Carbon Dioxide

1. Carbon dioxide is transported as bicarbonate ions (70%), in combination with blood proteins (23%), and in solution in plasma (7%).
2. Hemoglobin that has released oxygen binds more readily to carbon dioxide than hemoglobin that has oxygen bound to it (Haldane effect).
3. In tissue capillaries, carbon dioxide combines with water inside the erythrocytes to form carbonic acid that dissociates to form bicarbonate ions and hydrogen ions.
4. The chloride shift is the movement of chloride ions into erythrocytes as bicarbonate ions move out.
5. In lung capillaries, bicarbonate ions and hydrogen ions move into erythrocytes, and chloride ions move out. Bicarbonate ions combine with hydrogen ions to form carbonic acid. The carbonic acid dissociates to form carbon dioxide that diffuses out of the erythrocyte.
6. Increased plasma carbon dioxide lowers blood pH. The respiratory system regulates blood pH by regulating plasma carbon dioxide levels.

Rhythmic Ventilation
Respiratory Areas in the Brainstem

1. The medullary respiratory center consists of the dorsal and ventral respiratory groups.
   - The dorsal respiratory groups stimulate the diaphragm.
   - The ventral respiratory groups stimulate the intercostal and abdominal muscles.
2. The pontine respiratory group is involved with switching between inspiration and expiration.

Generation of Rhythmic Ventilation

1. When stimuli from receptors or other parts of the brain exceed a threshold level, inspiration begins.
2. At the same time that respiratory muscles are stimulated, neurons that stop inspiration are stimulated. When the stimulation of these neurons exceeds a threshold level, inspiration is inhibited.

Modification of Ventilation

Cerebral and Limbic System Control

Respiration can be voluntarily controlled and can be modified by emotions.

Chemical Control of Ventilation

1. Carbon dioxide is the major regulator of respiration. An increase in carbon dioxide or a decrease in pH can stimulate the chemosensitive area, causing a greater rate and depth of respiration.
2. Oxygen levels in the blood affect respiration when there is a 50% or greater decrease from normal levels. Decreased oxygen is detected by receptors in the carotid and aortic bodies, which then stimulate the respiratory center.

Hering-Breuer Reflex

Stretch of the lungs during inspiration can inhibit the respiratory center, contributing to a cessation of inspiration.

Effect of Exercise on Ventilation

1. Collateral fibers from motor neurons and from proprioceptors stimulate the respiratory centers.
2. Chemosensitive mechanisms and learning fine-tune the effects produced through the motor neurons and proprioceptors.

Other Modifications of Ventilation

Touch, thermal, and pain sensations can modify ventilation.

Respiratory Adaptations to Exercise

Tidal volume, respiratory rate, minute ventilation, and gas exchange between the alveoli and blood remains unchanged or slightly lower at rest or during standardized submaximal exercise, but increase at maximal exercise.

Content Review

1. What are the functions of the respiratory system?
2. Define the term respiration.
3. Describe the structure of the respiratory passages.
4. Name the three parts of the pharynx. With what structures does each part communicate?
5. Name and describe the three unpaired cartilages of the larynx. How are sounds of different pitch produced by the vocal cords?
6. Name the parts of the conducting zone and respiratory zone of the tracheobronchial tree.
7. Describe the arrangement of cartilage, smooth muscle, and epithelium in the tracheobronchial tree. Explain why breathing becomes more difficult during an asthma attack.
8. Distinguish among the lungs, a lung lobe, a bronchopulmonary segment, and a lobule.
9. Explain how thoracic volume is changed during quiet inspiration and expiration. How does this change during labored breathing?
10. Name the pleurae of the lungs. What is their function?
11. What are the two major routes of blood flow to and from the lungs?
12. Describe the lymphatic supply of the lungs.
13. Define the term ventilation. Describe the factors that result in air flow through a tube, and explain the significance of the general gas law to respiration.
14. Explain how the alveoli change volume. Define the term pneumothorax, and explain how it causes the lungs to collapse.
15. Explain how changes in thoracic volume cause the changes in alveolar volume that result in air movement.
16. Define the term compliance. What is the effect on lung expansion when compliance is increased or decreased?
17. Define the terms tidal volume, inspiratory reserve, expiratory reserve, and residual volume. Define the terms inspiratory capacity, functional residual capacity, vital capacity, and total lung capacity.
18. Define the terms minute ventilation and alveolar ventilation.
19. What is dead space? What is the difference between anatomic and physiologic dead space? What conditions increase dead space?
20. What is the partial pressure of a gas? What is water vapor pressure?
21. Describe the factors that make inspired air, air in the alveoli, and expired air have different gas concentrations.
22. How do the partial pressure and the solubility of a gas affect the concentration of the gas in a liquid? How do they affect the rate of diffusion of the gas through the liquid?
23. List the components of the respiratory membrane. Describe the factors that affect the diffusion of gases across the respiratory membrane. Give some examples of diseases that decrease diffusion by altering these factors.
24. What effect do ventilation and pulmonary capillary blood flow have on gas exchange? What is the physiologic shunt?
25. Describe the partial pressures for oxygen and carbon dioxide in the alveoli, lung capillaries, tissue capillaries, and tissues. How do these partial pressures account for the movement of oxygen and carbon dioxide?
26. List two ways that oxygen is transported in the blood.
27. What is the oxygen–hemoglobin dissociation curve? How does it explain the release and uptake of oxygen? How does a shift of the curve to the left or right affect the ability of hemoglobin to bind to oxygen? Where do such shifts take place in the body?

.hree ways that carbon dioxide is transported in the blood. What is the chloride shift? Where and why does it take place?

29. How can changes in respiration affect blood pH?

30. Describe the structures in the brainstem that control ventilation. How is rhythmic ventilation generated?

31. How do carbon dioxide, pH, and oxygen influence the regulation of ventilation?

32. Describe the Hering-Breuer reflex and its function.

33. During exercise how is ventilation regulated?

34. In what ways does the respiratory system change following training?

Develop Your Reasoning Skills

1. What effect does rapid (respiratory rate equals 24 breaths per minute), shallow (tidal volume equals 250 mL per breath) breathing have on minute ventilation, alveolar ventilation, and alveolar P_{O_2} and P_{CO_2}?

2. A person's vital capacity is measured while he was standing and while he is lying down. What difference, if any, in the measurement do you predict and why?

3. Ima Diver wanted to do some underwater exploration. She did not want to buy expensive SCUBA equipment, however. Instead, she bought a long hose and an inner tube. She attached one end of the hose to the inner tube so that the end was always out of the water, and she inserted the other end of the hose in her mouth and went diving. What happened to her alveolar ventilation and why? How would she compensate for this change? How would diving affect lung compliance and the work of ventilation?

4. The bacteria that cause gangrene (*Clostridium perfringens*) are anaerobic microorganisms that do not thrive in the presence of oxygen. Hyperbaric oxygenation (HBO) treatment places a person in a chamber that contains oxygen at three to four times normal atmospheric pressure. Explain how HBO helps in the treatment of gangrene.

5. Cardiopulmonary resuscitation (CPR) has replaced older, less efficient methods of sustaining respiration. The back-pressure/arm-lift method is one such technique that is no longer used. This procedure is performed with the victim lying face down. The rescuer presses firmly on the base of the scapulae for several seconds, then grasps the arms and lifts them. The sequence is then repeated. Explain why this procedure results in ventilation of the lungs.

6. Another technique for artificial respiration is mouth-to-mouth resuscitation. The rescuer takes a deep breath, blows air into the victim's mouth, and then lets air flow out of the victim. The process is repeated. Explain the following: (1) Why do the victim's lungs expand? (2) Why does air move out of the victim's lungs? and (3) What effect do the P_{O_2} and the P_{CO_2} of the rescuer's air have on the victim?

7. During normal quiet respiration, when does the maximum rate of diffusion of oxygen in the pulmonary capillaries occur? The maximum rate of diffusion of carbon dioxide?

8. Is the oxygen–hemoglobin dissociation curve in humans who live at high altitudes to the left or to the right of a person who lives at low altitudes?

9. Predict what would happen to tidal volume if the vagus nerves were cut. The phrenic nerves? The intercostal nerves?

10. You and your physiology instructor are trapped in an overturned ship. To escape you must swim underwater a long distance. You tell your instructor it would be a good idea to hyperventilate before making the escape attempt. Your instructor calmly replies, "What good would that do, since your pulmonary capillaries are already 100% saturated with oxygen?" What would you do and why?

Web Site Link

For a listing of the most current web sites related to this chapter, please visit the Seeley home page at: http://www.mhhe.com/biosci/ap/seeleyap/

Chapter Twenty-Four

Digestive System

Objectives

1. Describe the general anatomic features of the digestive tract.

2. Outline the basic histologic characteristics of the digestive tract.

3. List and describe the major functions of the digestive tract.

4. List the major structures of the oral cavity, and describe them.

5. List the major types of teeth, and describe the structure of an individual tooth.

6. Describe the functional differences between the major salivary glands.

7. Describe the anatomy of the esophagus.

8. List the anatomic and physiologic characteristics of the stomach that are most important to its function.

9. List the anatomic and histologic characteristics of the small intestine that account for its large surface area.

10. Describe the structure of the liver, the gallbladder, and the pancreas.

11. Describe the anatomy of the large intestine.

12. Describe the peritoneum and the mesenteries.

13. Explain the functions of the oral cavity.

14. Describe mastication and deglutition.

15. Describe the stomach secretions and their functions, and explain how they are regulated.

16. Outline the three phases of stomach secretion regulation: cephalic, gastric, and intestinal.

17. Describe gastric movements, stomach emptying, and their regulation.

18. Explain the functions of the secretions of the small intestine, and describe their regulation.

19. List the major functions of the pancreas, the liver, and the gallbladder, and explain how they are regulated.

20. Describe the functions of the large intestine.

21. Describe the process of digestion for carbohydrates, lipids, and proteins, and list the products of digestion for each.

Part Four

Every cell of the body needs nourishment, yet most cells cannot leave their position in the body and travel to a food source, so the food must be delivered. The digestive system, with the help of the circulatory system, is like a gigantic "meal on wheels," serving over a hundred trillion customers the nutrients they need. It also has its own quality control and waste disposal system.

The digestive system provides the body with water, electrolytes, and other nutrients. To do this, the digestive system is specialized to ingest food; propel it through the digestive tract; digest it; and absorb water, electrolytes, and other nutrients from the lumen of the gastrointestinal tract. Once these useful substances are absorbed, they are transported through the circulatory system to cells, where they are used. Undigested matter from the food is moved through the digestive tract and eliminated through the anus.

This chapter presents a general overview of the digestive system, describes in detail the anatomy and histology of each section of the digestive tract, and describes the physiology of each section.

General Overview

The **digestive system** (figure 24.1) consists of the **digestive tract,** a tube extending from the mouth to the anus, and its associated **accessory organs,** primarily glands, which secrete fluids into the digestive tract. The digestive tract is also called the **alimentary tract,** or **alimentary canal.** The term **gastrointestinal (GI;** gas′trō-in-tes′ti-năl) **tract** technically only refers to the stomach and intestines but is often used as a synonym for the digestive tract.

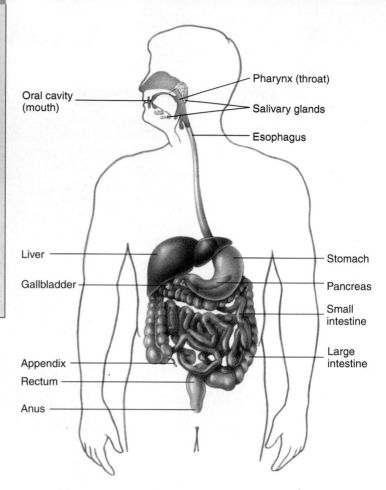

Figure 24.1   Digestive System Depicted in Place in the Body ✗

Anatomy Overview

The first section of the digestive tract is the mouth, or **oral cavity.** It is surrounded by the lips, cheeks, teeth, and palate; and it contains the tongue. The **salivary glands** and **tonsils** are accessory organs of the oral cavity.

The oral cavity opens posteriorly into the **pharynx,** which, in turn, continues inferiorly into the **esophagus** (ē-sof′ă-gŭs). The major accessory structures are small, simple, tubular mucous glands distributed the length of the pharynx and the esophagus.

The esophagus opens inferiorly into the stomach. The stomach wall contains many tubelike glands from which acid and enzymes are released into the stomach and mixed with ingested food.

The stomach opens inferiorly into the small intestine. The first segment of the small intestine is the duodenum (dū-ō-dē′nŭm). The major accessory structures in this segment of the digestive tract are the liver, the gallbladder, and the pancreas. The next segment of the small intestine is the jejunum (jĕ-jū′nŭm, meaning empty). Small glands exist along its length, and it is the major site of absorption. The last segment of the small intestine is the ileum (il′ē-ŭm, meaning twisted), which is similar to the jejunum, except that fewer digestive enzymes and more mucus are secreted and less absorption occurs.

The last section of the digestive tract is the large intestine. Its major accessory glands secrete mucus. It absorbs water and salts and concentrates undigested food into feces. The first segment is the cecum (sē′kŭm, meaning blind), with the attached appendix. The cecum is followed by the ascending, transverse, descending, and sigmoid colons (kō′lonz) and the rectum (rek′tŭm, meaning straight). The rectum joins the anal canal, which ends at the anus, the inferior termination of the digestive tract.

Histology Overview

Figure 24.2 depicts a generalized view of the digestive tract histology. The digestive tube consists of four major layers, or tunics: an internal mucosa and an external serosa with a submucosa and muscularis in between. These four tunics are present in all areas of the digestive tract from the esophagus to the anus. Three major types of glands are associated with the intestinal tract: (1) unicellular mucous glands in the mucosa, (2) multicellular glands in the mucosa and submucosa, and (3) multicellular glands (accessory glands) outside the digestive tract.

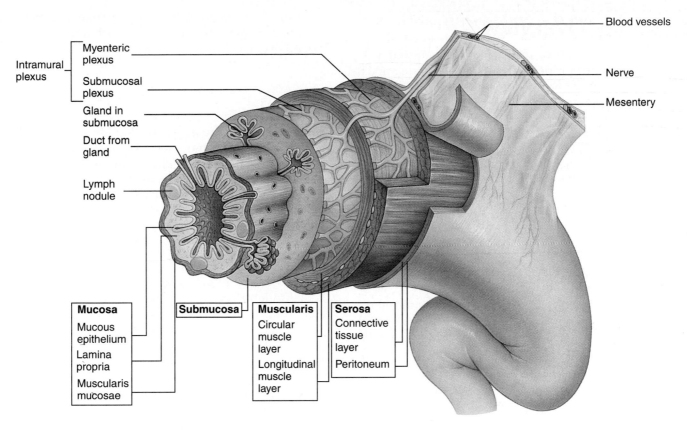

Figure 24.2   Digestive Tract Histology

The four tunics are the mucosa, submucosa, muscularis, and serosa or adventitia. Glands may exist along the digestive tract as part of the epithelium, within the submucosa, or as large glands that are outside the digestive tract.

Mucosa

The innermost tunic, the **mucosa** (myū-kō′să), consists of three layers: (1) the inner **mucous epithelium,** which is moist stratified squamous epithelium in the mouth, oropharynx, esophagus, and anal canal and simple columnar epithelium in the remainder of the digestive tract; (2) a loose connective tissue called the **lamina propria** (lam′i-nă prō′prē-ă); and (3) an outer thin smooth muscle layer, the **muscularis mucosae.**

Submucosa

The **submucosa** is a thick connective tissue layer containing nerves, blood vessels, and small glands that lies beneath the mucosa. The nerves of the submucosa form the **submucosal plexus** (plek′sŭs; Meissner's plexus), a parasympathetic ganglionic plexus.

Muscularis

The next tunic is the **muscularis,** which consists of an inner layer of circular smooth muscle and an outer layer of longitudinal smooth muscle. Two exceptions are the upper esophagus, where the muscles are striated, and the stomach, where there are three layers of smooth muscle. Another nerve plexus, the **myenteric plexus** (mī′en-ter′ik; Auerbach's plexus), which consists of nerve fibers and parasympathetic cell bodies, is between these two muscle layers (see figure 24.2). Together, the submucosal and myenteric plexuses constitute the **intramural** (in′tră-myū′răl, meaning within the walls) **plexus.** The intramural plexus is extremely important in the control of movement and secretion.

Serosa or Adventitia

The fourth layer of the digestive tract is a connective tissue layer called either the **serosa** or the **adventitia** (ad ven-tish′ă, meaning foreign or coming from outside), depending on the structure of the layer. Parts of the digestive tract that protrude into the peritoneal cavity have a serosa as the outermost layer. This serosa is called the visceral peritoneum. It consists of a thin layer of connective tissue and a simple squamous epithelium. When the outer layer of the digestive tract is derived from adjacent connective tissue, the tunic is called the adventitia and consists of a connective tissue covering that blends with the surrounding connective tissue. These areas include the esophagus and the retroperitoneal organs (discussed later in relation to the peritoneum).

Physiology Overview

The functions of the digestive system are to take in and masticate the food, and to propel it through the digestive tract, mixing it as it moves along. Secretions, which lubricate, liquefy, and digest the food, are added to the food by the digestive tract. Water, electrolytes, and other nutrients are absorbed from the digested food (table 24.1). Once these useful substances are absorbed, they are transported through the circulatory system to cells, where they are used. Undigested matter is moved out of the digestive tract and eliminated through the anus. The processes of propulsion, secretion, and absorption are regulated by elaborate nervous and hormonal mechanisms.

Ingestion

Ingestion (in-jes′chŭn, meaning a pouring in) is the introduction of solid or liquid food into the stomach. The normal route of ingestion is through the oral cavity, but food can be introduced directly into the stomach by a nasogastric, or stomach, tube.

Mastication

Mastication (mas-ti-kā′shŭn, meaning chewing) is the process by which food taken into the mouth is chewed by the teeth. Digestive enzymes cannot easily penetrate solid food particles and can only work effectively on the surfaces of the particles. It is therefore vital to normal digestive function that solid foods be mechanically broken down into small particles. Mastication breaks large food particles into many smaller particles, which have a much larger total surface area than do a few large particles.

Propulsion

Propulsion (prō-pŭl′shŭn) in the digestive tract is the movement of food from one end of the digestive tract to the other. The total time that it takes food to travel the length of the digestive tract is usually about 24–36 h (table 24.2). Each segment of the digestive tract is specialized to assist in moving its contents from the oral end to the anal end. **Deglutition** (dē′glŭ-tish′ŭn), or swallowing, moves mouthfuls of food and liquids, called a **bolus,** from the oral cavity into the esophagus. **Peristalsis** (per-i-stal′sis; figure 24.3a) is responsible for moving material through most of the digestive tract. Muscular contractions occur in peristaltic waves, consisting of a wave of relaxation of the circular muscles, which forms a leading wave of distention in front of the bolus, followed by a wave of strong contraction of the circular muscles behind the bolus, which forces the bolus along the digestive tube. Each peristaltic wave travels the length of the esophagus in about 10 s. Peristaltic waves in the small intestine usually only travel for short distances. In some parts of the large intestine, material is moved by **mass movements,** which are contractions that extend over much larger parts of the digestive tract than peristaltic movements.

Mixing

Some contractions do not propel food from one end of the digestive tract to the other but rather move the food back and forth within the digestive tract to **mix** it with digestive secretions and to help break it into smaller pieces. **Segmental contractions** (figure 24.3b) are mixing contractions that occur in the small intestine.

Secretion

As food moves through the digestive tract, **secretions** are added to lubricate, liquefy, and digest the food. **Mucus,** secreted along the entire digestive tract, lubricates the food and the lining of the digestive tract. The mucus coats and protects the epithelial cells of the digestive tract from mechanical abrasion, from the damaging effect of acid in the stomach, and from the digestive enzymes of the digestive tract. The secretions also contain large amounts of **water,** which liquefies the food, making it easier to digest and absorb. Water also moves into the intestine by osmosis. **Enzymes** secreted by the oral cavity, stomach, intestine, liver, and pancreas break food down into small molecules that can be absorbed by the intestinal wall.

Digestion

Digestion (di-jes′chŭn) is the breakdown of organic molecules into their component parts: carbohydrates into monosaccharides, proteins into amino acids, and triacylglycerols into fatty acids and glycerol. Digestion consists of **mechanical digestion,** which involves mastication and mixing of food, and **chemical digestion,** which is accomplished by digestive enzymes that are secreted along the digestive tract. Digestion of large molecules into their component parts must be accomplished before they can be absorbed by the digestive tract. Molecules such as vitamins, minerals, and water are not broken down before being absorbed. They are absorbed without digestion and lose their function if they are digested. This is especially important in the case of vitamins.

Absorption

Absorption is the movement of molecules out of the digestive tract and into the circulation or into the lymphatic system. The mechanism by which absorption occurs depends on the type of molecule involved. Molecules pass out of the digestive tract by simple diffusion, facilitated diffusion, active transport, or cotransport (see chapter 3).

Table 24.1   Functions of the Digestive Organs

Organ	Function	Secretion
Oral Cavity		
Teeth	Mastication (cutting and grinding of food); communication	None
Lips and cheeks	Manipulation of food; hold food in position between the teeth; communication	Saliva from buccal glands (mucus only)
Tongue	Manipulation of food; holds food in position between the teeth; cleaning teeth; taste; communication	Some mucus; small amounts of serous fluid
Salivary Glands		
Parotid gland	Secretion of saliva through ducts to superior and posterior portions of oral cavity	Serous saliva only, with amylase
Submandibular glands	Secretion of saliva in floor of oral cavity	Serous saliva, with amylase; mucous saliva
Sublingual glands	Secretion of saliva in floor of oral cavity	Mucous saliva only
Pharynx	Deglutition (movement of food from oral cavity to esophagus); breathing	Some mucus
Esophagus	Movement of food by peristalsis from pharynx to stomach	Mucus
Stomach	Mechanical mixing of food; enzymatic digestion; storage; absorption	
Mucous cells	Protection of stomach wall by mucus production	Mucus
Parietal cells	Decrease in stomach pH, vitamin B_{12} absorption	Hydrochloric acid, intrinsic factor
Chief cells	Protein digestion	Pepsin
Endocrine cells	Regulation of secretion and motility	Gastrin
Accessory Glands		
Liver	Secretion of bile into duodenum	Bile
Gallbladder	Bile storage; absorbs water and electrolytes to concentrate bile	No secretions of its own, stores and concentrates bile
Pancreas	Secretion of several digestive enzymes and bicarbonate ions into duodenum	Trypsin, chymotrypsin, carboxypeptidase, pancreatic amylase, pancreatic lipase, ribonuclease, deoxyribonuclease, cholesterol esterase, bicarbonate ions
Small Intestine		
Duodenal glands	Protection	Mucus
Goblet cells	Protection	Mucus
Absorptive cells	Secretion of digestive enzymes and absorption of digested materials	Enterokinase, amylase, peptidases, sucrase, maltase, isomaltase, lactase, lipase
Endocrine cells	Regulation of secretion and motility	Gastrin, secretin, cholecystokinin, gastric inhibitory peptide
Large Intestine	Absorption, storage, and food movement	
Goblet cells	Protection	Mucus

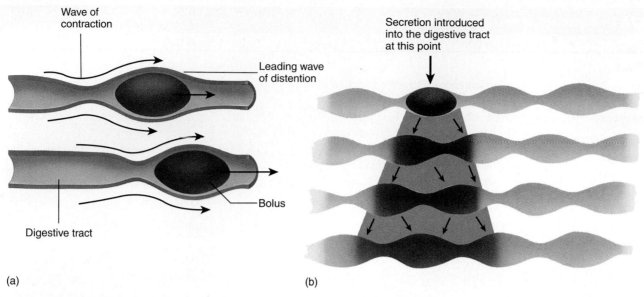

Figure 24.3   Movements in the Digestive Tract

(*a*) Peristalsis. A wave of relaxation of the circular muscles is followed by a wave of strong contraction of the circular muscles, which propels the bolus of food through the digestive tract. (*b*) Segmental contractions. Each section of the digestive tract involved in segmental contractions alternates between contraction and relaxation. The series of figures from top to bottom depicts a temporal sequence for one part of the small intestine. The arrows indicate the direction material in a given part of the intestine moves with each contraction. Material (*brown*) introduced at the beginning of the sequence (*top figure*) is spread out and becomes more diffuse (*lighter color*) through time. 🏃 ▭

Table 24.2	Time Food Spends in Each Part of the Digestive Tract
Region	**Time Spent**
Oral cavity	10–20 s
Pharynx	1–2 s
Esophagus	5–8 s
Stomach	
Liquid	1.3–2.5 h
Solid	3–4 h
Small intestine (pyloric valve to ileocecal valve; proximal end most rapid movement)	3–5 h
Large intestine	
Ileocecal valve to transverse colon	8–15 h
Entire length (ileocecal valve to anus)	18–24 h

Transportation

Transportation is the process by which absorbed molecules are distributed throughout the body. This distribution can occur either directly by way of the circulation or indirectly by first entering the lymphatic system and then passing to the circulatory system.

Elimination

Elimination is the process by which the waste products of digestion are removed from the body. During this process, occurring primarily in the large intestine, water and salts are absorbed, changing the material in the digestive tract from a liquefied state to a semisolid state. These semisolid waste products, called feces, are then eliminated from the digestive tract by the process of **defecation.**

Regulation

The processes of propulsion, secretion, absorption, and elimination are **regulated** by elaborate nervous and hormonal mechanisms. Some of the nervous control is local, occurring as the result of local reflexes within the intramural plexus, and some is more general, mediated largely by the vagus nerve.

▊Anatomy and Histology of the Digestive Tract
Oral Cavity

The **oral cavity,** or mouth, is that part of the digestive tract bounded by the lips anteriorly, the **fauces** (faw'sēz, meaning throat; opening into the pharynx) posteriorly, the cheeks lat-

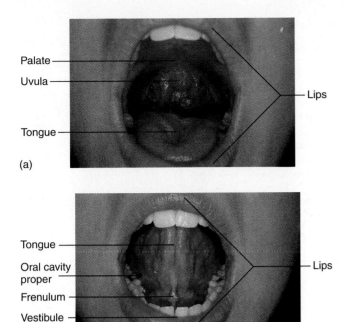

Palate

Uvula

Tongue

(a)

Lips

Tongue

Oral cavity
proper

Frenulum

Vestibule

(b)

Lips

Figure 24.4   Oral Cavity

(*a*) With the tongue depressed. (*b*) With the tongue elevated.

erally, the palate superiorly, and a muscular floor inferiorly. The oral cavity can be divided into two regions: (1) the **vestibule** (ves′ti-būl, meaning entry), which is the space between the lips or cheeks and the alveolar processes, which contain the teeth; and (2) the **oral cavity proper,** which lies medial to the alveolar processes. The oral cavity is lined with moist stratified squamous epithelium, which provides protection against abrasion.

Lips and Cheeks

The **lips,** or **labia** (lā′bē-ă) (figure 24.4), are muscular folds covered internally by mucosa and externally by stratified squamous epithelium. The epithelial covering of the lips is relatively thin and not as highly keratinized as the epithelium of the skin (see chapter 5); consequently, it is more transparent than the epithelium over the rest of the body surface. The color from the underlying blood vessels can be seen through the relatively transparent epithelium, giving the lips a reddish pink appearance, even if the epithelium is heavily pigmented.

The **cheeks** form the lateral walls of the oral cavity. They consist of an interior lining of moist stratified squamous epithelium and an exterior covering of skin. The substance of the cheek includes the **buccinator muscle** (see chapter 11), which flattens the cheek against the teeth, and the **buccal fat pad,** which rounds out the profile on the side of the face.

The lips and cheeks are important in the processes of mastication and speech. They help manipulate the food within the mouth and hold the food in place while the teeth crush or tear it. They also help form words during the speech

process. A large number of the muscles of facial expression are involved in movement of the lips. They are listed in chapter 11.

Tongue

The **tongue** is a large, muscular organ that occupies most of the oral cavity proper when the mouth is closed. Its major attachment in the oral cavity is through its posterior part. The anterior part of the tongue is relatively free and is attached to the floor of the mouth by a thin fold of tissue called the **frenulum** (fren′ū-lŭm, meaning bridle; see figure 24.4*b*). The muscles associated with the tongue are divided into two categories: **intrinsic muscles,** which are within the tongue itself; and **extrinsic muscles,** which are outside the tongue but attached to it. The intrinsic muscles are largely responsible for changing the shape of the tongue, such as flattening and elevating the tongue during drinking and swallowing. The extrinsic tongue muscles protrude and retract the tongue, move it from side to side, and change its shape (see chapter 11).

> **Clinical Note**
>
> A person is "tongue-tied" in a more literal sense if the frenulum extends too far toward the tip of the tongue, inhibiting normal movement of the tongue and interfering with normal speech. Surgically cutting the frenulum can correct the condition.

The tongue is divided into two parts by a groove called the **terminal sulcus.** The part anterior to the terminal sulcus accounts for about two-thirds of the surface area and is covered by papillae, some of which contain taste buds (see chapter 15). The posterior one-third of the tongue is devoid of papillae and has only a few scattered taste buds. It has, instead, a few small glands and a large amount of lymphoid tissue, the **lingual tonsil** (see chapter 22). The tongue is covered by moist stratified squamous epithelium.

> **Clinical Note**
>
> Certain drugs, which are lipid-soluble and can diffuse through the cell membranes of the oral cavity, can be quickly absorbed into the circulation. An example is nitroglycerin, which is a vasodilator used to treat cases of angina pectoris. The drug is placed under the tongue, where, in less than 1 min, it dissolves and passes through the very thin oral mucosa into the lingual veins.

The tongue moves food in the mouth and, in cooperation with the lips and gums, holds the food in place during mastication. It also plays a major role in the mechanism of swallowing (discussed later in this chapter). It is a major sensory organ for taste (see chapter 15) and one of the primary organs of speech.

> ### Clinical Note
>
> Patients who have undergone glossectomies (tongue removal) as a result of glossal carcinoma can compensate for loss of the tongue's function in speech, and they can learn to speak fairly well. These patients still have a major problem, which they cannot entirely overcome, with chewing and swallowing food.

Teeth

There are 32 **teeth** in the normal adult mouth, which are distributed in two **dental arches,** one maxillary and one mandibular. The teeth in the right and left halves of each dental arch are roughly mirror images of each other. As a result, the teeth can be divided into four quadrants: right upper, left upper, right lower, and left lower. The teeth in each quadrant include one central and one lateral **incisor;** one **canine;** first and second **premolars;** and first, second, and third **molars** (figure 24.5*a*). The third molars are referred to as **wisdom teeth** because they usually appear when the person is in his late teens or early twenties and has supposedly acquired a little wisdom.

> ### Clinical Note
>
> In some people with small mouths, the third molars may not have room to erupt into the oral cavity and remain embedded within the jaw. These embedded teeth are impacted, and their surgical removal is often necessary.

The teeth of the adult mouth are **permanent,** or **secondary, teeth.** Most of them are replacements for **primary,** or **deciduous, teeth** (dē-sid′yū-ŭs, meaning those that fall out; also called milk teeth) that are lost during childhood (figure 24.5*b*).

Each tooth consists of a **crown** with one or more **cusps** (points), a **neck,** and a **root** (figure 24.6). The center of the tooth is a **pulp cavity,** which is filled with blood vessels, nerves, and connective tissue called **pulp.** The pulp cavity within the root is called the **root canal.** The nerves and blood vessels of the tooth enter and exit the pulp through a hole at the point of each root called the **apical foramen.** The pulp cavity is surrounded by a living, cellular, and calcified tissue called **dentin.** The dentin of the tooth crown is covered by an extremely hard, nonliving, acellular substance called **enamel,** which protects the tooth against abrasion and acids produced by bacteria in the mouth. The surface of the dentin in the root is covered with a cellular, bonelike substance, called **cementum,** which helps anchor the tooth in the jaw.

The teeth are set in **alveoli** (al-vē′ō-lī, meaning sockets) along the alveolar ridges of the mandible and maxilla.

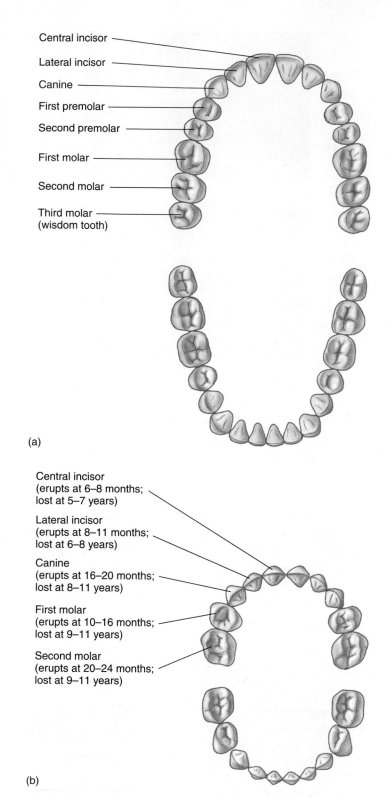

(a)

(b)

Figure 24.5   Teeth

(*a*) Permanent teeth. (*b*) Deciduous teeth.

The alveolar ridges are covered by dense fibrous connective tissue and stratified squamous epithelium, referred to as the **gingiva** (jin′ji-vă, meaning gums). The teeth are held in the alveoli by **periodontal** (per′ē-ō-don′tăl, meaning around the teeth) **ligaments.**

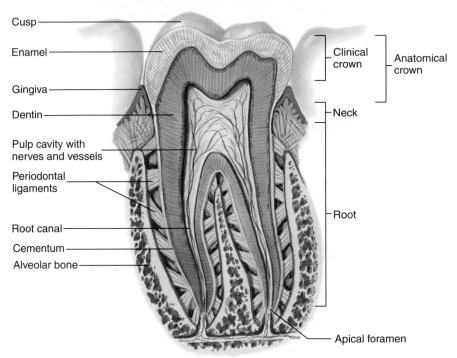

Cusp
Enamel
Gingiva
Dentin
Pulp cavity with nerves and vessels
Periodontal ligaments
Root canal
Cementum
Alveolar bone
Clinical crown
Anatomical crown
Neck
Root
Apical foramen

Figure 24.6   Molar Tooth in Place in the Alveolar Bone

The tooth consists of a crown and root. The clinical crown is that part exposed in the mouth. The anatomical crown is the entire enamel-covered part of the tooth. The root is covered with cementum, and the tooth is held in the socket by periodontal ligaments. Nerves and vessels enter and exit the tooth through the apical foramen.

The teeth play an important role in mastication and a role in speech.

Clinical Note

Dental caries, or tooth decay, is caused by a breakdown of enamel by acids produced by bacteria on the tooth surface. Because the enamel is nonliving and cannot repair itself, a dental filling is necessary to prevent further damage. If the decay reaches the pulp cavity with its rich supply of nerves, toothache pain may result. In some cases in which decay has reached the pulp cavity, it may be necessary to perform a dental procedure called a "root canal," which consists of removing the pulp from the tooth.

Periodontal disease is the inflammation and degradation of the periodontal ligaments, gingiva, and alveolar bone. This disease is the most common cause of tooth loss in adults. Gingivitis is an inflammation of the gingiva, often caused by food deposited in gingival crevices and not promptly removed by brushing and flossing. Gingivitis may eventually lead to periodontal disease. Pyorrhea (pī-ō-rē′ă) is a condition in which pus occurs with periodontal disease. Halitosis, or bad breath, often occurs with periodontal disease and pyorrhea.

Muscles of Mastication

Four pairs of muscles move the mandible during mastication: the **temporalis, masseter, medial pterygoid,** and **lateral pterygoid** (see chapter 11 and figure 11.8). The temporalis, masseter, and medial pterygoid muscles close the jaw; and the lateral pterygoid muscle opens it. Protrusion of the jaw and right and left excursion of the jaw are accomplished by the medial and lateral pterygoids and the masseter. Retraction of the jaw is accomplished by the temporalis. All these movements are involved in tearing, crushing, and grinding food.

Palate and Palatine Tonsils

The **palate** (see figure 24.4a) consists of two parts, an anterior bony part, the **hard palate** (see chapter 7), and a posterior, nonbony part, the **soft palate,** which consists of skeletal muscle and connective tissue. The **uvula** (yū′vyū-lă, meaning a grape) is the projection from the posterior edge of the soft palate. The palate is important in the swallowing process, preventing food from passing into the nasal cavity.

Palatine tonsils are located in the lateral wall of the fauces (see chapter 22).

Salivary Glands

A considerable number of **salivary glands** are scattered throughout the oral cavity. There are three pairs of large multicellular glands: the parotid, the submandibular and the sublingual glands (figure 24.7). In addition to these large consolidations of glandular tissue, numerous small, coiled tubular glands are in the tongue (lingual glands), palate (palatine glands), cheeks (buccal glands), and lips (labial glands). The secretions from all these glands help keep the oral cavity moist and begin the process of digestion.

All of the major large salivary glands are compound **alveolar glands,** which are branching glands with clusters of alveoli that resemble grapes (see chapter 4). They produce thin serous secretions or thicker mucous secretions. Thus saliva is a combination of serous fluids and mucus.

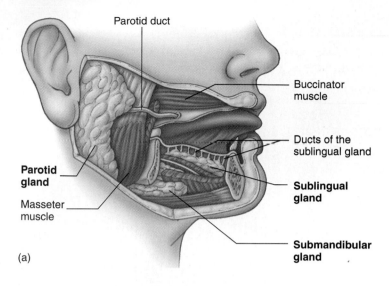

(a)

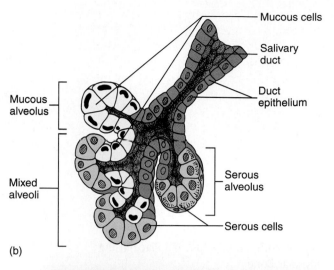

(b)

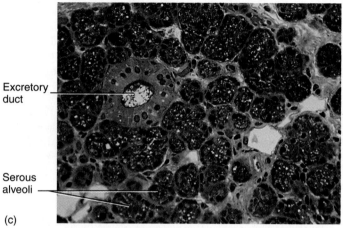

(c)

Figure 24.7   Salivary Glands

(*a*) The large salivary glands are the parotid glands, the submandibular glands, and the sublingual glands. The minor salivary glands include the buccal and labial glands. The parotid duct extends anteriorly from the parotid gland. (*b*) An idealized schematic drawing of the histology of the large salivary glands. The figure is representative of all the glands and does not depict any one salivary gland. (*c*) Photomicrograph of the parotid gland.

The largest salivary glands, the **parotid** (pa-rot′id, meaning beside the ear) **glands,** are serous glands, which produce mostly watery saliva, and are located just anterior to the ear on each side of the head. Each **parotid duct** exits the gland on its anterior margin, crosses the lateral surface of the masseter muscle, pierces the buccinator muscle, and enters the oral cavity adjacent to the second upper molar (see figure 24.7).

The **submandibular** (meaning below the mandible) **glands** are mixed glands with more serous than mucous alve-

oli. Each gland can be felt as a soft lump along the inferior border of the posterior half of the mandible. A submandibular duct exits each gland, passes anteriorly deep to the mucous membrane on the floor of the oral cavity, and opens into the oral cavity beside the frenulum of the tongue (see figure 24.4*b*). In certain people, if the mouth is opened and the tip of the tongue is elevated, saliva may squirt out of the mouth from the openings of these ducts.

The **sublingual** (meaning below the tongue) **glands,** the smallest of the three large, paired salivary glands, are mixed glands containing some serous alveoli but consisting primarily of mucous alveoli. They lie immediately below the mucous membrane in the floor of the mouth. These glands do not have single, well-defined ducts like those of the submandibular and parotid glands. Instead, each sublingual gland opens into the floor of the oral cavity through 10–12 small ducts.

Pharynx

The **pharynx** was described in detail in chapter 23; thus only a brief description is provided here. The pharynx consists of three parts: the nasopharynx, the oropharynx, and the laryngopharynx. Normally, only the oropharynx and laryngopharynx transmit food. The **oropharynx** communicates with the nasopharynx superiorly, the larynx and **laryngopharynx** inferiorly, and the mouth anteriorly. The laryngopharynx extends from the oropharynx to the esophagus and is posterior to the larynx. The posterior walls of the oropharynx and laryngopharynx consist of three muscles: the superior, middle, and inferior **pharyngeal constrictors,** which are arranged like three stacked flowerpots, one inside the other. The oropharynx and the laryngopharynx are lined with moist stratified squamous epithelium, and the nasopharynx is lined with ciliated pseudostratified columnar epithelium.

PREDICT

1

Explain the functional significance of the differences in epithelial types among the three pharyngeal regions.

✔ *Answer in Appendix F*

Esophagus

The **esophagus** is that part of the digestive tube that extends between the pharynx and the stomach. It is about 25 cm long and lies in the mediastinum, anterior to the vertebrae and posterior to the trachea. It passes through the esophageal hiatus (opening) of the diaphragm and ends at the stomach. The esophagus transports food from the pharynx to the stomach.

Clinical Note

A **hiatal hernia** is a widening of the esophageal hiatus, occurring most commonly in adults, which allows part of the stomach to extend through the opening into the thorax. The hernia can decrease the resting pressure in the lower esophageal sphincter, allowing gastroesophageal reflux and subsequent esophagitis to occur. Hiatal herniation can also compress the blood vessels in the stomach mucosa, which can lead to gastritis or ulcer formation. Esophagitis, gastritis, or ulceration are very painful.

The esophagus has thick walls consisting of the four tunics common to the digestive tract: mucosa, submucosa, muscularis, and adventitia. The muscular tunic has an outer longitudinal layer and an inner circular layer, as is true of most parts of the digestive tract, but it is different because it consists of skeletal muscle in the superior part of the esophagus and smooth muscle in the inferior part. An **upper esophageal sphincter** and a **lower esophageal sphincter,** at the upper and lower ends of the esophagus, respectively, regulate the movement of materials into and out of the esophagus. The mucosal lining of the esophagus is moist stratified squamous epithelium. Numerous mucous glands in the submucosal layer produce a thick, lubricating mucus that passes through ducts to the surface of the esophageal mucosa.

Stomach

The **stomach** is an enlarged segment of the digestive tract in the left superior part of the abdomen (see figure 24.1). Its shape and size vary from person to person; even within the same individual its size and shape change from time to time, depending on its food content and the posture of the body. Nonetheless, several general anatomic features can be described.

Stomach Anatomy

The opening from the esophagus into the stomach is the **gastroesophageal,** or **cardiac** (located near the heart), **opening,** and the region of the stomach around the cardiac opening is the **cardiac region** (figure 24.8). The lower esophageal sphincter, also called the **cardiac sphincter,** surrounds the cardiac opening. Although this is an important structure in the normal function of the stomach, it is a physiologic constrictor only and cannot be seen anatomically. A part of the stomach to the left of the cardiac region, the **fundus** (fŭn'dŭs, meaning the bottom of a round-bottomed leather bottle), is actually superior to the cardiac opening. The largest part of the stomach is the **body,** which turns to the right, thus creating a **greater curvature** and a **lesser curvature.** The body narrows to form the **pyloric** (pī-lōr'ik, meaning gatekeeper) **region,** which joins the small intestine. The opening between the stomach and the small intestine is the **pyloric opening,** which is surrounded by a relatively thick ring of smooth muscle called the **pyloric sphincter.**

Clinical Note

Hypertrophic pyloric stenosis is a common defect of the stomach in infants, occurring in 1 in 150 males and 1 in 750 females, in which the pylorus is greatly thickened, resulting in interference with normal stomach emptying. Infants with this defect exhibit projectile vomiting. Because the pylorus is blocked, little food enters the intestine, and the infant fails to gain weight. Constipation is also a frequent complication.

Stomach Histology

The **serosa,** or visceral peritoneum, is the outermost layer of the stomach. It consists of an inner layer of connective tissue and an outer layer of simple squamous epithelium. The **muscularis** of the stomach consists of three layers: an outer longitudinal layer, a middle circular layer, and an inner oblique layer (see figure 24.8*a*). In some areas of the stomach, such as in the fundus, the three layers blend with one another and cannot be separated. Deep to the muscular layer are the submucosa and the mucosa, which are thrown into large folds called **rugae** (rū'gē, meaning wrinkles) when the stomach is

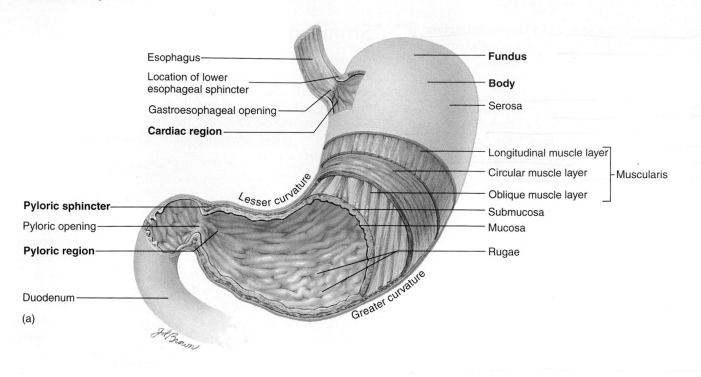

(a)

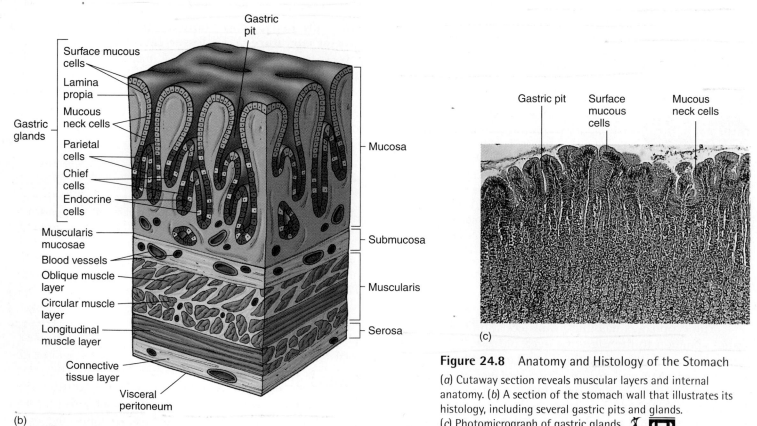

(b)

Gastric pit   Surface mucous cells   Mucous neck cells

(c)

Figure 24.8   Anatomy and Histology of the Stomach

(*a*) Cutaway section reveals muscular layers and internal anatomy. (*b*) A section of the stomach wall that illustrates its histology, including several gastric pits and glands. (*c*) Photomicrograph of gastric glands.

empty (see figure 24.8*a*). These folds allow the mucosa and submucosa to stretch, and the folds disappear as the stomach is filled.

The stomach is lined with simple columnar epithelium. The mucosal surface forms numerous tubelike **gastric pits,** which are the openings for the **gastric glands** (figure 24.8*b*). The epithelial cells of the stomach can be divided into five groups. The first group, **surface mucous cells,** which produce mucus, is on the surface and lines the gastric pit. The remaining four cell types are in the gastric glands. They are **mu-**

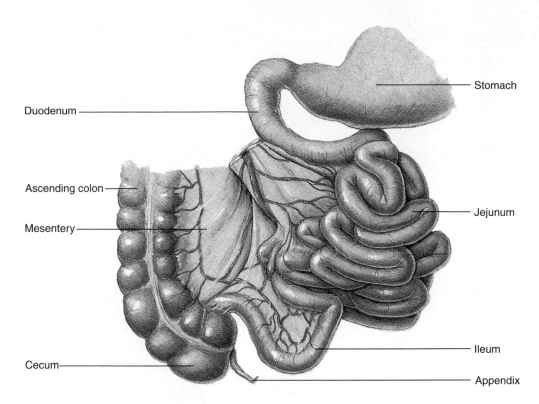

Figure 24.9 The Small Intestine 🏃

cous neck cells, which produce mucus; **parietal** (oxyntic) **cells,** which produce hydrochloric acid and intrinsic factor; **chief** (zymogenic) **cells,** which produce pepsinogen; and **endocrine cells,** which produce regulatory hormones. The mucous neck cells are located near the openings of the glands; whereas the parietal, chief, and endocrine cells are interspersed in the deeper parts of the glands.

Small Intestine

The **small intestine** consists of three parts: the duodenum, the jejunum, and the ileum (figure 24.9). The entire small intestine is about 6 m long (range: 4.6–9 m). The duodenum is about 25 cm long (the term duodenum means 12, suggesting that it is 12 inches long). The jejunum, constituting about two-fifths of the total length of the small intestine, is about 2.5 m long; and the ileum, constituting three-fifths of the small intestine, is about 3.5 m long. Two major accessory glands, the liver and the pancreas, are associated with the duodenum.

Duodenum

The **duodenum** nearly completes a 180-degree arc as it curves within the abdominal cavity (figure 24.10), and the head of the pancreas lies within this arc. The duodenum begins with a short superior part, which is where it exits the pylorus of the stomach, and ends in a sharp bend, which is where it joins the jejunum.

Two small mounds are within the duodenum about two-thirds of the way down the descending part: the **major duodenal papilla** and the **lesser duodenal papilla.** At the ma-

jor papilla the **common bile duct** and **pancreatic duct** join to form the **hepatopancreatic ampulla** (Vater's ampulla), which empties into the duodenum. The opening of the ampulla is usually kept closed by a smooth muscle sphincter, the **hepatopancreatic ampullar sphincter** (sphincter of Oddi). An accessory pancreatic duct, present in most people, opens at the tip of the lesser duodenal papilla.

The surface of the duodenum has several modifications that increase its surface area about 600-fold, allowing for more efficient digestion and absorption of food. The mucosa and submucosa form a series of folds called the **plicae** (plī'sē, meaning folds) **circulares,** or **circular folds** (figure 24.11*a*), which run perpendicular to the long axis of the digestive tract. Tiny fingerlike projections of the mucosa form numerous **villi** (vil'ī, meaning shaggy hair), which are 0.5–1.5 mm in length (figure 24.11*b*). Each villus is covered by simple columnar epithelium and contains a blood capillary network and a lymph capillary called a **lacteal** (lak'tē-ăl) (figure 24.11*c*). Most of the cells that make up the surface of the villi have numerous cytoplasmic extensions (about 1 μm long) called microvilli, which further increase the surface area (figure 24.11*d*). The combined microvilli on the entire epithelial surface form the **brush border.** These various modifications greatly increase the surface area of the small intestine and, as a result, greatly enhance absorption.

The mucosa of the duodenum is simple columnar epithelium with four major cell types: (1) **absorptive cells** are cells with microvilli, which produce digestive enzymes and absorb digested food; (2) **goblet cells,** which produce a protective mucus; (3) **granular cells** (Paneth's cells), which may help protect the intestinal epithelium from bacteria; and (4) **endocrine cells,** which produce regulatory hormones.

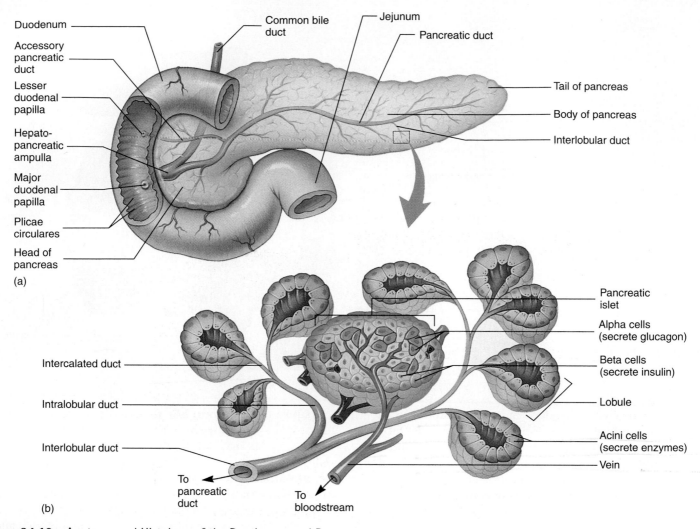

Figure 24.10   Anatomy and Histology of the Duodenum and Pancreas

(*a*) The head of the pancreas lies within the duodenal curvature, with the pancreatic duct emptying into the duodenum. (*b*) Histology of the pancreas showing both the acini and the pancreatic duct system.

The epithelial cells are produced within tubular invaginations of the mucosa, called **intestinal glands** (crypts of Lieberkühn), at the base of the villi. The absorptive and goblet cells migrate from the intestinal glands to cover the surface of the villi and eventually are shed from its tip. The granular and endocrine cells remain in the bottom of the glands. The submucosa of the duodenum contains coiled tubular mucous glands called **duodenal glands** (Brunner's glands), which open into the base of the intestinal glands.

Jejunum and Ileum

The **jejunum** and **ileum** are similar in structure to the duodenum (see figure 24.9), except that there is a gradual decrease in the diameter of the small intestine, the thickness of the intestinal wall, the number of circular folds, and the number of villi as one progresses through the small intestine. The duodenum and jejunum are the major sites of nutrient absorption, although some absorption occurs in the ileum.

Lymph nodules called **Peyer's patches** are numerous in the mucosa and submucosa of the ileum.

The junction between the ileum and the large intestine is the **ileocecal junction.** It has a ring of smooth muscle, the **ileocecal sphincter,** and a one-way **ileocecal valve** (see figure 24.14).

Liver

Liver Anatomy

The **liver** is the largest internal organ of the body, weighing about 1.36 kg (3 pounds), and it is in the right upper quadrant of the abdomen, tucked against the inferior surface of the diaphragm (see figure 24.1; figure 24.12). The liver consists of two major **lobes, left** and **right,** and two minor lobes, **caudate** and **quadrate.**

A **porta** (gate) is on the inferior surface of the liver, where the various vessels, ducts, and nerves enter and exit the

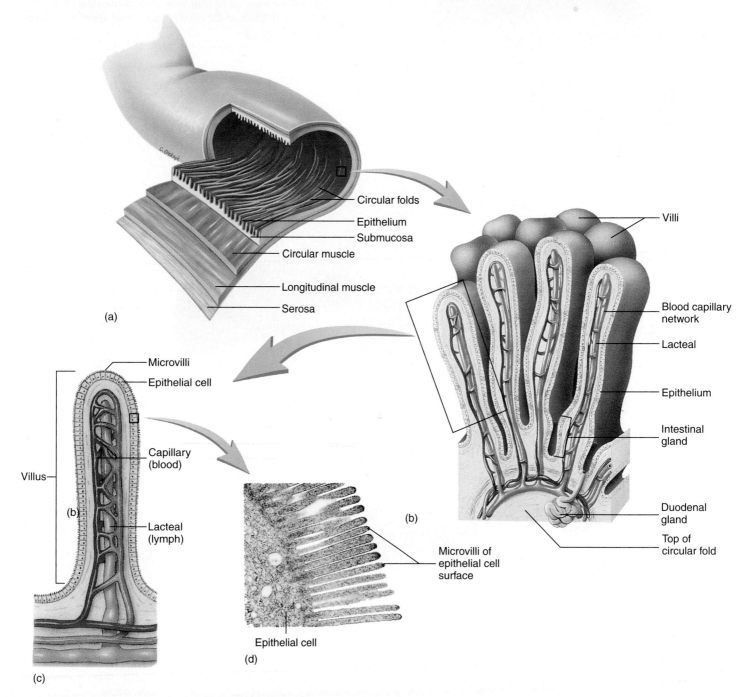

Circular folds
Epithelium
Submucosa
Circular muscle
Longitudinal muscle
Serosa
(a)

Villi
Blood capillary network
Lacteal
Epithelium
Intestinal gland
Duodenal gland
Top of circular fold
(b)

Microvilli
Epithelial cell
Capillary (blood)
Villus
(b)
Lacteal (lymph)
(c)

Microvilli of epithelial cell surface
Epithelial cell
(d)

Figure 24.11   Anatomy and Histology of the Duodenum

(a) Wall of the duodenum, showing the circular folds. (b) The villi on a circular fold. (c) A single villus, showing the lacteal and capillary network.
(d) Transmission electron micrograph of microvilli on the surface of a villus.

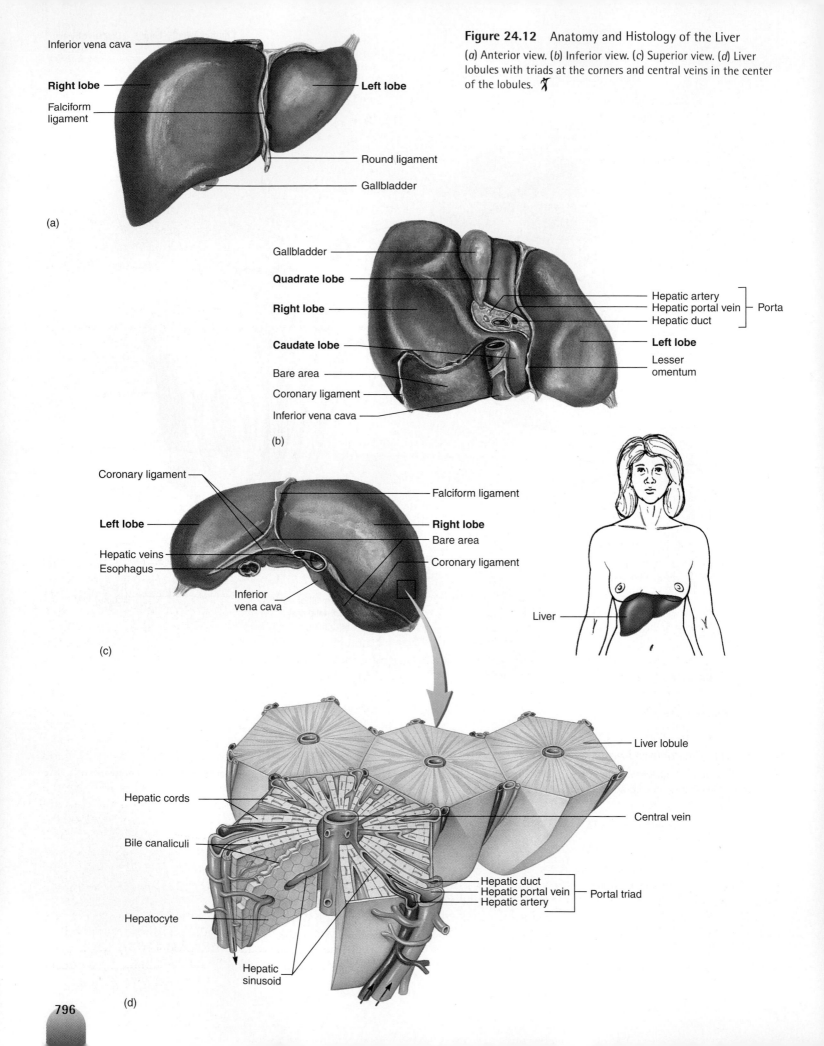

Figure 24.12 Anatomy and Histology of the Liver
(*a*) Anterior view. (*b*) Inferior view. (*c*) Superior view. (*d*) Liver lobules with triads at the corners and central veins in the center of the lobules. ⳨

(a)

Inferior vena cava

Right lobe

Falciform ligament

Left lobe

Round ligament

Gallbladder

(b)

Gallbladder

Quadrate lobe

Right lobe

Caudate lobe

Bare area

Coronary ligament

Inferior vena cava

Hepatic artery
Hepatic portal vein — Porta
Hepatic duct

Left lobe

Lesser omentum

(c)

Coronary ligament

Left lobe

Hepatic veins

Esophagus

Inferior vena cava

Falciform ligament

Right lobe

Bare area

Coronary ligament

Liver

(d)

Hepatic cords

Bile canaliculi

Hepatocyte

Hepatic sinusoid

Liver lobule

Central vein

Hepatic duct
Hepatic portal vein — Portal triad
Hepatic artery

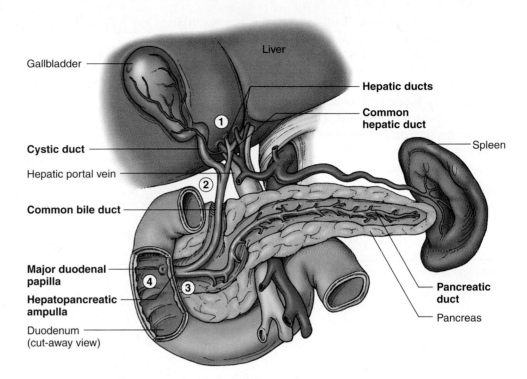

1. The hepatic ducts from the liver lobes combine to form the common hepatic duct.

2. The common hepatic duct combines with the cystic duct from the gallbladder to form the common bile duct.

3. The common bile duct and the pancreatic duct combine to form the hepatopancreatic ampulla.

4. The hepatopancreatic ampulla empties into the duodenum at the major duodenal papilla.

Figure 24.13   Duct System of the Major Abdominal Digestive Glands

---

liver. The **hepatic** (he-pat′ik, meaning associated with the liver) **portal vein,** the **hepatic artery,** and a small hepatic nerve plexus enter the liver through the porta. Lymphatic vessels and two hepatic ducts, one each from the right and left lobes, exit the liver at the porta. The hepatic ducts transport bile out of the liver. The right and left hepatic ducts unite to form a single common hepatic duct. The **common hepatic duct** is joined by the **cystic duct** from the gallbladder to form the **common bile duct,** which empties into the duodenum at the major duodenal papilla, in union with the pancreatic duct (figure 24.13; see figure 24.10a). The gallbladder is a small sac on the inferior surface of the liver that stores bile.

Liver Histology

The liver is covered by a connective tissue capsule and visceral peritoneum, except for the **bare area,** which is a small area on the diaphragmatic surface surrounded by the coronary ligament (see figure 24.12c). At the porta the connective tissue capsule sends a branching network of septa (walls) into the substance of the liver to provide its main support. Vessels, nerves, and ducts follow the connective tissue branches throughout the liver.

The connective tissue septa divide the liver into hexagon-shaped **lobules** with a **portal triad** at each corner. The triads are so named because three vessels—the hepatic portal vein, hepatic artery, and hepatic duct—are commonly located in them (see figure 24.12d). Hepatic nerves and lymph vessels, often too small to be easily seen in light micrographs, are also located in these areas. A **central vein** is in the center of each lob-

ule. Central veins unite to form **hepatic veins,** which exit the liver on its posterior and superior surfaces and empty into the inferior vena cava.

Hepatic cords radiate out from the central vein of each lobule like the spokes of a wheel. The hepatic cords are composed of **hepatocytes,** the functional cells of the liver. The spaces between the hepatic cords are blood channels called **hepatic sinusoids.** The sinusoids are lined with a very thin, irregular squamous endothelium consisting of two cell populations: (1) extremely thin, sparse **endothelial cells** and (2) **hepatic phagocytic cells** (Kupffer cells). A cleftlike lumen, the **bile canaliculus** (kan-ă-lik′yū-lŭs, meaning little canal), lies between the cells within each cord (see figure 24.12d).

Hepatocytes have four major functions (described in more detail later in this chapter): (1) synthesis of bile, (2) storage, (3) biotransformation, and (4) synthesis of blood components. Nutrient-rich, oxygen-poor blood from the viscera enters the hepatic sinusoids from branches of the hepatic portal vein and mixes with oxygen-rich, nutrient-depleted blood from the hepatic arteries. From the blood the hepatocytes can take up the oxygen and nutrients, which can be stored, detoxified, used for energy, or used to synthesize new molecules. Molecules produced by or modified in the hepatocytes are released into the hepatic sinusoids or into the bile canaliculi.

Mixed blood in the hepatic sinusoids flows to the central vein where it exits the lobule and then exits the liver through the hepatic veins. **Bile,** produced by the hepatocytes and consisting primarily of metabolic by-products, flows through the bile canaliculi toward the hepatic triad and exits the liver through the hepatic ducts. Blood therefore flows from the triad toward the center of each lobule, whereas bile flows away from the center of the lobule toward the triad.

In the fetus, blood is shunted through the liver by special vessels that bypass the sinusoids. The remnants of fetal blood vessels can be seen in the adult as the round ligament (ligamentum teres) and the ligamentum venosum (see chapter 29).

> **Clinical Note**
>
> The liver is easily ruptured because it is large, fixed in position, and fragile, or it can be lacerated by a broken rib. Liver rupture or laceration results in severe internal bleeding.
>
> The liver may become enlarged as a result of heart failure, hepatic cancer, cirrhosis, or Hodgkin's disease (a lymphatic cancer).

Gallbladder

The **gallbladder** is a saclike structure on the inferior surface of the liver that is about 8 cm long and 4 cm wide (see figure 24.13). Three tunics form the gallbladder wall: (1) an inner mucosa folded into rugae that allow the gallbladder to expand; (2) a muscularis, which is a layer of smooth muscle that allows the gallbladder to contract; and (3) an outer covering of serosa. The gallbladder is connected to the common bile duct by the cystic duct.

Pancreas

The **pancreas** is a complex organ composed of both endocrine and exocrine tissues that perform several functions. The pancreas consists of a **head,** located within the curvature of the duodenum (see figure 24.10a), a **body,** and a **tail,** which extends to the spleen.

The endocrine part of the pancreas consists of **pancreatic islets** (islets of Langerhans; see figure 24.10b). The islet cells produce insulin and glucagon, which are very important in controlling blood levels of nutrients such as glucose and amino acids, and somatostatin, which regulates insulin and glucagon secretion and may inhibit growth hormone secretion (see chapter 18).

The exocrine part of the pancreas consists of **acini** (as'i-nī, meaning grapes; see figure 24.10b), which produce digestive enzymes. Clusters of acini form lobules that are separated by thin septa. Lobules are connected by small **intercalated ducts** to **intralobular ducts,** which leave the lobules to join **interlobular ducts** between the lobules. The interlobular ducts attach to the main pancreatic duct, which joins the common bile duct at the hepatopancreatic ampulla (see figures 24.10a and 24.13). The ducts are lined with simple cuboidal epithelium, and the epithelial cells of the acini are pyramid-shaped.

Large Intestine
Cecum

The **cecum** (sē'kŭm, meaning blind), which is the proximal end of the **large intestine,** is where the large and small intestines meet. The cecum extends inferiorly about 6 cm past the ileocecal junction in the form of a blind sac (figure 24.14). Attached to the cecum is a small blind tube about 9 cm long called the **vermiform** (ver'mi-fōrm, meaning worm-shaped) **appendix.** The walls of the appendix contain many lymph nodules.

> **Clinical Note**
>
> Appendicitis is an inflammation of the vermiform appendix and usually occurs because of obstruction of the appendix. Secretions from the appendix cannot pass the obstruction and accumulate, causing enlargement and pain. Bacteria in the area can cause infection of the appendix. Symptoms include sudden abdominal pain, particularly in the right lower portion of the abdomen, along with a slight fever, loss of appetite, constipation or diarrhea, nausea, and vomiting. If the appendix bursts, the infection can spread throughout the peritoneal cavity, causing peritonitis, with life-threatening results. In the right lower quadrant of the abdomen, about midway along a line between the umbilicus and the right anterior superior iliac spine, is an area on the body's surface called McBurney's point. This area becomes very tender in patients with acute appendicitis because of pain referred from the inflamed appendix to the body's surface. Each year, 500,000 people in the United States suffer an appendicitis. An appendectomy is removal of the appendix.

Colon

The **colon** (kō'lon) is about 1.5–1.8 m long and consists of four parts: the ascending colon, transverse colon, descending colon, and sigmoid colon (see figure 24.14). The **ascending colon** extends superiorly from the cecum and ends at the right colic flexure (hepatic flexure) near the right inferior margin of the liver. The **transverse colon** extends from the right colic flexure to the left colic flexure (splenic flexure), and the **descending colon** extends from the left colic flexure to the superior opening of the true pelvis, where it becomes the sigmoid colon. The **sigmoid colon** forms an S-shaped tube that extends into the pelvis and ends at the rectum.

The circular muscle layer of the colon is complete, but the longitudinal muscle layer is incomplete. The longitudinal layer does not completely envelop the intestinal wall but forms three bands, called the **teniae coli** (tē'nē-ē kō'lī, meaning a band or tape along the colon), that run the length of the colon (figure 24.15; see figure 24.14). Contractions of the teniae coli cause pouches called **haustra** (haw'stră, meaning to draw up) to form along the length of the colon, giving it a puckered appearance. Small, fat-filled connective tissue pouches called **epiploic** (ep'i-plō'ik, meaning related to the

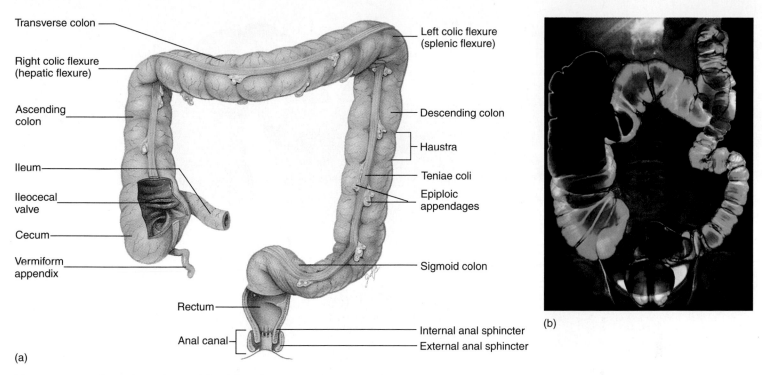

Figure 24.14   Large Intestine and Anal Canal

(*a*) Large intestine (that is, cecum, colon, and rectum) and anal canal. The teniae coli and epiploic appendages are along the length of the colon. (*b*) A radiograph of the large intestine following a barium enema. 🕺

omentum) **appendages** are attached to the outer surface of the colon along its length.

The mucosal lining of the large intestine consists of simple columnar epithelium. This epithelium is not formed into folds or villi like that of the small intestine but has numerous straight tubular glands called **crypts** (see figure 24.15). The crypts are somewhat similar to the intestinal glands of the small intestine, with three cell types that include absorptive, goblet, and granular. The major difference is that in the large intestine goblet cells predominate and the other three cell types are greatly reduced in number.

Rectum

The **rectum** is a straight, muscular tube that begins at the termination of the sigmoid colon and ends at the anal canal (see figure 24.14). The mucosal lining of the rectum is simple columnar epithelium, and the muscular tunic is relatively thick compared with the rest of the digestive tract.

Anal Canal

The last 2–3 cm of the digestive tract is the **anal canal** (see figure 24.14). It begins at the inferior end of the rectum and ends at the **anus** (external GI tract opening). The smooth muscle layer of the anal canal is even thicker than that of the rectum and forms the **internal anal sphincter** at the superior end of the anal canal. The **external anal sphincter** at

the inferior end of the canal is formed by skeletal muscle. The epithelium of the superior part of the anal canal is simple columnar, and that of the inferior part is stratified squamous.

Clinical Note

Hemorrhoids are the enlargement, or inflammation, of the hemorrhoidal veins, which supply the anal canal. The condition is also called varicose hemorrhoidal veins. Hemorrhoids cause pain, itching, and bleeding around the anus. Treatments include increasing the bulk (indigestible fiber) in the diet, taking sitz baths, and using hydrocortizone suppositories. Surgery may be necessary if the condition is extreme and does not respond to other treatments.

Peritoneum

The body walls and organs of the abdominal cavity are lined with **serous membranes.** These membranes are very smooth and secrete a serous fluid that provides a lubricating film between the layers of membranes. These membranes and fluid reduce the friction as organs move within the abdomen. The serous membrane that covers the organs is the **visceral peritoneum** (péri-tō-nē′ŭm, meaning to stretch over), and the one that covers the interior surface of the body wall is the **parietal peritoneum** (figure 24.16).

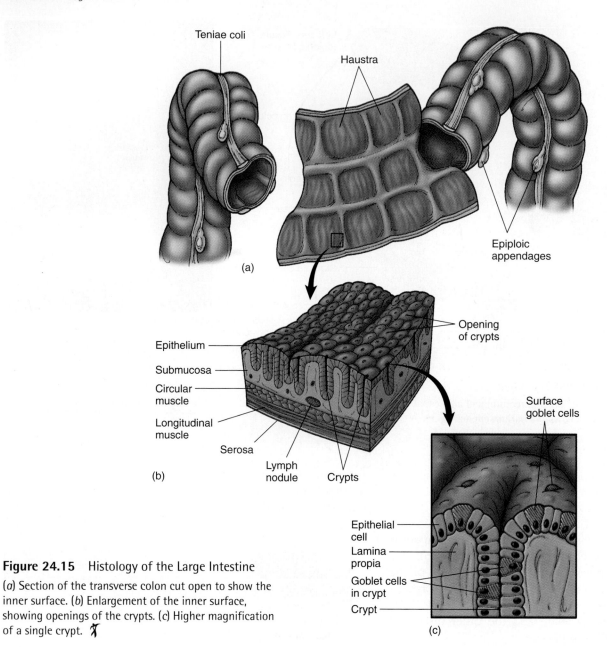

Teniae coli

Haustra

Epiploic
appendages

(a)

Opening
of crypts

Epithelium

Submucosa

Circular
muscle

Longitudinal
muscle

Serosa

Surface
goblet cells

(b)

Lymph
nodule   Crypts

Epithelial
cell

Lamina
propia

Goblet cells
in crypt

Crypt

(c)

Figure 24.15   Histology of the Large Intestine

(*a*) Section of the transverse colon cut open to show the
inner surface. (*b*) Enlargement of the inner surface,
showing openings of the crypts. (*c*) Higher magnification
of a single crypt.

Many of the organs of the abdominal cavity are held in
place by connective tissue sheets called **mesenteries** (mes′en-
ter′ēz, meaning middle intestine). The mesenteries consist of
two layers of serous membranes with a thin layer of loose
connective tissue between them. They provide a route by
which vessels and nerves can pass from the body wall to the
organs. Other abdominal organs lie against the abdominal
wall, have no mesenteries, and are referred to as **retroperitoneal** (re′trō-pe′ri-tō-nē′ăl, meaning behind the peritoneum;
see chapter 1). The retroperitoneal organs include the duodenum, the pancreas, the ascending colon, the descending
colon, the rectum, the kidneys, the adrenal glands, and the
urinary bladder.

Some mesenteries are given specific names. The mesentery connecting the lesser curvature of the stomach and the
proximal end of the duodenum to the liver and diaphragm is
called the **lesser omentum** (ō-men′tŭm, meaning membrane
of the bowels), and the mesentery extending as a fold from

Figure 24.16   Peritoneum and Mesenteries

Sagittal section through the trunk showing the peritoneum and mesenteries associated with some abdominal organs. ⚆

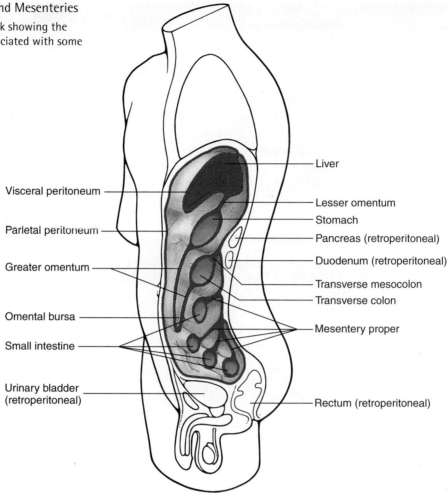

Visceral peritoneum

Parietal peritoneum

Greater omentum

Omental bursa

Small intestine

Urinary bladder (retroperitoneal)

Liver

Lesser omentum

Stomach

Pancreas (retroperitoneal)

Duodenum (retroperitoneal)

Transverse mesocolon

Transverse colon

Mesentery proper

Rectum (retroperitoneal)

---

the greater curvature and then to the transverse colon and posterior body wall is called the **greater omentum** (see figure 24.16). The greater omentum forms a long, double fold of mesentery that extends inferiorly from the stomach over the surface of the small intestine. Because of this folding, a cavity, or pocket, called the **omental bursa** (ber′să, meaning pocket) is formed between the two layers of mesentery. A large amount of fat accumulates in the greater omentum, and it is sometimes referred to as the fatty apron. The greater omentum has considerable mobility in the abdomen.

2	P R E D I C T

If you placed a pin through the greater omentum, through how many layers of simple squamous epithelium would the pin pass?

✔ *Answer in Appendix F*

The **coronary ligament** attaches the liver to the diaphragm. Unlike other mesenteries, the coronary ligament has a wide space in the center, the bare area of the liver, where there is no peritoneum. The **falciform ligament** attaches the liver to the anterior abdominal wall (see figure 24.12).

Although the term mesentery is a general term referring to the serous membranes attached to the abdominal organs, the term is also used specifically to refer to the mesentery associated with the small intestine, sometimes called the **mesentery proper.** The mesenteries of parts of the colon are the **transverse mesocolon,** which is actually a continuation of the posterior side of the greater omentum, and the **sigmoid mesocolon.** The vermiform appendix even has its own little mesentery called the **mesoappendix.**

Functions of the Digestive System

As food moves through the digestive tract, secretions are added to liquefy and digest it and to provide lubrication (table 24.3). Each segment of the digestive tract is specialized to assist in moving its contents from the oral end to the anal end. Parts of the digestive system are also specialized to transport molecules from the lumen of the digestive tract into the extracellular spaces. The processes of secretion, movement, and absorption are regulated by elaborate nervous and hormonal mechanisms (see chapters 12, 16–18).

Table 24.3 Functions of Various Digestive Secretions

Fluid or Enzyme	Function
Saliva	
Serous (watery)	Moistens food and mucous membrane; lysozyme kills bacteria
Salivary amylase	Starch digestion (conversion to maltose and isomaltose)
Mucus	Lubricates food; protects gastrointestinal tract from digestion by enzymes
Gastric Secretions	
Hydrochloric acid	Decreases stomach pH to activate pepsinogen
Pepsinogen	Active form (pepsin) digests protein into smaller peptide chains
Mucus	Protects stomach lining from digestion
Liver	

Bile
  Sodium glycocholate (bile salt)
  Sodium taurocholate (bile salt)
  Cholesterol
  Biliverdin
  Bilirubin
  Mucus
  Fat
  Lecithin
  Cells and cell debris
	Bile salts emulsify fats, making them available to intestinal lipases; help make end products soluble and available for absorption by the intestinal mucosa; aid peristalsis. Many of the other bile contents are waste products transported to the intestine for disposal

Pancreas	
Trypsin	Digests proteins (breaks polypeptide chains at arginine or lysine residues)
Chymotrypsin	Digests proteins (cleaves carboxyl links of hydrophobic amino acids)
Carboxypeptidase	Digests proteins (removes amino acids from carboxyl end of peptide chains)
Pancreatic amylase	Digests carbohydrates (hydrolyzes starches and glycogen to form maltose and isomaltose)
Pancreatic lipase	Digests fat (hydrolyzes fats—mostly triacylglycerols—into glycerol and fatty acids)
Ribonuclease	Digests ribonucleic acid
Deoxyribonuclease	Digests deoxyribonucleic acid (hydrolyzes phosphodiester bonds)
Cholesterol esterase	Hydrolyzes cholesterol esters to form cholesterol and free fatty acids
Bicarbonate ions	Provides appropriate pH for pancreatic enzymes
Small Intestine Secretions	
Mucus	Protects duodenum from stomach acid, gastric enzymes, and intestinal enzymes; provides adhesion for fecal matter; protects intestinal wall from bacterial action and acid produced in the feces
Aminopeptidase	Splits polypeptides into amino acids (from amino end of chain)
Peptidase	Splits amino acids from polypeptides
Enterokinase	Activates trypsin from trypsinogen
Amylase	Digests carbohydrates
Sucrase	Splits sucrose into glucose and fructose
Maltase	Splits maltose into two glucose molecules
Isomaltase	Splits isomaltose into two glucose molecules
Lactase	Splits lactose into glucose and galactose
Lipase	Splits fats into glycerol and fatty acids

The digestive tract also has well-developed **local reflexes** in the intramural plexus. Stimuli, such as distention of the digestive tract, activate receptors within the wall of the digestive tract, and action potentials are generated in the neurons of the intramural plexus. The action potentials travel up or down the intramural plexus and produce a response in an effector organ, such as the smooth muscle of a digestive organ, or a gland.

Functions of the Oral Cavity
Secretions of the Oral Cavity

Saliva is secreted at the rate of about 1–1.5 L/day. The serous part of saliva contains a digestive enzyme called **salivary amylase** (am′il-ās, meaning starch-splitting enzyme), which breaks the covalent bonds between glucose molecules in starch and other polysaccharides to produce the disaccharides maltose and isomaltose (see table 24.3). The release of maltose and isomaltose gives starches a sweet taste in the mouth. Only about 3%–5% of the total carbohydrates are digested in the mouth, however. Most of the starches are bound up with cellulose in plant tissues and are inaccessible to salivary amylase. Cooking and thorough chewing of food destroy the cellulose covering and increase the efficiency of the digestive process.

Saliva prevents bacterial infection in the mouth by washing the oral cavity. Saliva also contains substances, such as **lysozyme,** which has a weak antibacterial action, and immunoglobulin A, which helps prevent bacterial infection. Any lack of salivary gland secretion increases the chance of ulceration and infection of the oral mucosa and of caries in the teeth.

The mucous secretions of the submandibular and sublingual glands contain a large amount of **mucin** (myū′sin), a proteoglycan that gives a lubricating quality to the secretions of the salivary glands.

Salivary gland secretion is stimulated by the parasympathetic and sympathetic nervous systems, with the parasympathetic system being more important. Salivary nuclei in the brainstem increase salivary secretions by sending action potentials through parasympathetic fibers of the facial (VII) and glossopharyngeal (IX) cranial nerves in response to a variety of stimuli, such as tactile stimulation in the oral cavity or certain tastes, especially sour. Higher centers of the brain also affect the activity of the salivary glands. Odors that trigger thoughts of food or the sensation of hunger can increase salivary secretions.

Mastication

Food taken into the mouth is **chewed,** or **masticated,** by the teeth. The anterior teeth, the incisors, and the canines primarily cut and tear food, whereas the premolars and molars primarily crush and grind it. Mastication breaks large food particles into smaller ones, which have a much larger total surface area. Because digestive enzymes digest food molecules only at the surface of the particles, mastication increases the efficiency of digestion.

Chewing is controlled primarily by the **chewing,** or **mastication, reflex,** which is integrated in the medulla oblongata. The presence of food in the mouth stimulates sensory receptors, which activate a reflex that causes the muscles of mastication to relax. The muscles are stretched as the mandible is lowered, and the stretch of the muscles activates a reflex that causes contraction of the muscles of mastication. Once the mouth is closed, the food again stimulates the muscles of mastication to relax, and the cycle is repeated. The cerebrum can influence the activity of the mastication reflex so that chewing can be initiated or stopped consciously.

Deglutition

Deglutition (dē′glū-tish′ŭn), or **swallowing,** can be divided into three separate phases: voluntary, pharyngeal, and esophageal. During the **voluntary phase** (figure 24.17*a*) a bolus of food is formed in the mouth and pushed by the tongue against the hard palate, forcing the bolus toward the posterior part of the mouth and into the oropharynx.

The **pharyngeal phase** (figure 24.17*b–d*) of swallowing is a reflex that is initiated by stimulation of tactile receptors in the area of the oropharynx. Afferent action potentials travel through the trigeminal (V) and glossopharyngeal (IX) nerves to the **swallowing center** in the medulla oblongata. There they initiate action potentials in motor neurons, which pass through the trigeminal (V), glossopharyngeal (IX), vagus (X), and accessory (XI) nerves to the soft palate and pharynx. This phase of swallowing begins with the elevation of the soft palate, which closes the passage between the nasopharynx and oropharynx. The pharynx elevates to receive the bolus of food from the mouth and moves the bolus down the pharynx into the esophagus. The superior, middle, and inferior **pharyngeal constrictor muscles** contract in succession, forcing the food through the pharynx. At the same time, the upper esophageal sphincter relaxes, the elevated pharynx opens the esophagus, and food is pushed into the esophagus. This phase of swallowing is unconscious and is controlled automatically, even though the muscles involved are skeletal. The pharyngeal phase of swallowing lasts about 1–2 s.

---

3 ▌ **P R E D I C T**

Why is it important to close the opening between the nasopharynx and oropharynx during swallowing? What may happen if a person has an explosive burst of laughter while trying to swallow a liquid?

✔ *Answer in Appendix F*

During the pharyngeal phase, the vocal folds are moved medially, the **epiglottis** (ep-i-glot′is, meaning on the glottis) is tipped posteriorly so that the epiglottic cartilage covers the opening into the larynx and the larynx is elevated. These movements of the larynx prevent food from passing through the opening into the larynx.

---

4 ▌ **P R E D I C T**

What happens if you try to swallow and speak at the same time?

✔ *Answer in Appendix F*

The **esophageal phase** (figure 24.17*e*) of swallowing takes about 5–8 s and is responsible for moving food from the pharynx to the stomach. Muscular contractions in the wall of the esophagus occur in **peristaltic** (per-i-stal′tik) **waves.**

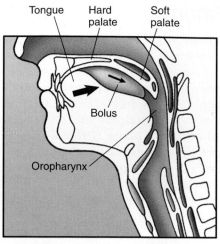

Tongue   Hard palate   Soft palate

Bolus

Oropharynx

(a) During the voluntary phase, a bolus of food (*yellow*) is pushed by the tongue against the hard palate and posteriorly toward the oropharynx (*red arrow* indicates tongue movement; *black arrow* indicates movement of the bolus).

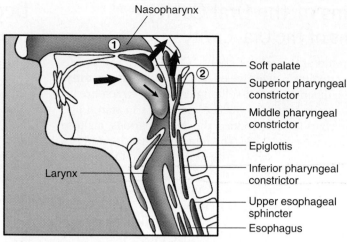

Nasopharynx
①
Soft palate
②
Superior pharyngeal constrictor
Middle pharyngeal constrictor
Epiglottis
Inferior pharyngeal constrictor
Larynx
Upper esophageal sphincter
Esophagus

(b) During the pharyngeal phase, ① the soft palate is elevated, closing off the nasopharynx. ② The pharynx is elevated (*red arrows* indicate muscle movement).

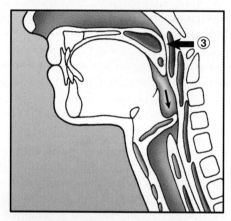

③

Epiglottis
Opening of larynx

(c) Successive constriction of the pharyngeal constrictors ③ from superior to inferior (*red arrows*) forces the bolus through the pharynx and into the esophagus. As this occurs, the epiglottis is bent down over the opening of the larynx largely by the force of the bolus pressing against it.

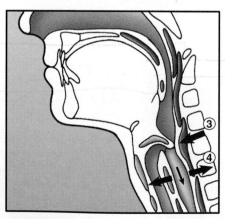

③
④

(d) As the inferior pharyngeal constrictor contracts ③, the upper esophageal sphincter relaxes ④ (*outwardly directed red arrows*), allowing the bolus to enter the esophagus.

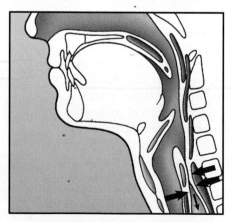

(e) During the esophageal phase, the bolus is moved by peristaltic contractions of the esophagus toward the stomach (*inwardly directed red arrows*).

Figure 24.17   Three Phases of Swallowing (Deglutition)

The peristaltic contractions associated with swallowing cause relaxation of the lower esophageal sphincter in the esophagus as the peristaltic waves, and bolus of food, approach the stomach. This sphincter is not anatomically distinct from the rest of the esophagus, but it can be identified physiologically because it remains tonically constricted to prevent the reflux of stomach contents into the lower part of the esophagus.

The presence of food in the esophagus stimulates the intramural plexus, which initiates the peristaltic waves. The presence of food in the esophagus also stimulates tactile receptors, which send afferent impulses to the medulla oblongata through the vagus nerves. Motor impulses, in turn, pass along the vagal efferent fibers to the striated and smooth muscles within the esophagus, stimulating their contractions, and reinforcing the peristaltic contractions.

Clinical Note

Gravity assists the movement of material through the esophagus, especially when liquids are swallowed. The peristaltic contractions that move material through the esophagus are sufficiently forceful, however, to allow a person to swallow, even while doing a headstand or floating in the zero-gravity environment of space.

Stomach Functions
Secretions of the Stomach

Ingested food and stomach secretions, mixed together, form a semifluid material called **chyme** (kīm, meaning juice). The stomach functions primarily as a storage and mixing chamber for the chyme. Although some digestion and absorption occur in the stomach, they are not its major functions.

Stomach secretions include mucus, hydrochloric acid, gastrin, intrinsic factor, and pepsinogen. Pepsinogen is the inactive form of the protein-digesting enzyme pepsin.

The surface mucous cells and mucous neck cells secrete a viscous and alkaline **mucus** that covers the surface of the epithelial cells and forms a layer 1–1.5 mm thick. The thick layer of mucus lubricates and protects the epithelial cells of the stomach wall from the damaging effect of the acidic chyme and pepsin. Irritation of the stomach mucosa results in stimulation of the secretion of a greater volume of mucus.

Parietal cells in the gastric glands of the pyloric region secrete intrinsic factor and a concentrated solution of hydrochloric acid. **Intrinsic factor** is a glycoprotein that binds with vitamin B_{12} and makes the vitamin more readily absorbed in the ileum. Vitamin B_{12} is important in deoxyribonucleic acid (DNA) synthesis.

Hydrochloric acid produces the low pH of the stomach, which is normally between 1 and 3. Although the hydrochloric acid secreted into the stomach has a minor digestive effect on ingested food, one of its main functions is to kill bacteria that are ingested with essentially everything humans put into their mouths. Some pathogenic bacteria taken in through the mouth may avoid digestion in the stomach, how-

ever, because they have an outer coat that resists stomach acids. The low pH of the stomach also stops carbohydrate digestion by inactivating salivary amylase. The low pH also denatures many proteins so that proteolytic enzymes can reach internal peptide bonds, and it provides the proper pH environment for the function of pepsin.

Hydrogen ions are derived from carbon dioxide and water, which enter the parietal cell from its serosal surface, which is the side opposite the lumen of the gastric pit (figure 24.18). Once inside the cell, carbonic anhydrase catalyzes the reaction between carbon dioxide and water to form carbonic acid. Some of the carbonic acid molecules then dissociate to form hydrogen ions and bicarbonate ions. The hydrogen ions are actively transported across the mucosal surface of the parietal cell into the lumen of the stomach, moving some potassium ions into the cell in exchange for the hydrogen ions. Although hydrogen ions are actively transported against a steep concentration gradient, chloride ions diffuse with the hydrogen ions from the cell through the cell membrane. Diffusion of chloride ions with the positively charged hydrogen ions reduces the amount of energy needed to transport the hydrogen ions against both a concentration gradient and an electrical gradient. Bicarbonate ions move down their concentration gradient from the parietal cell into the extracellular fluid. During this process, bicarbonate ions are exchanged for chloride ions through an anion exchange molecule, which is located in the cell membrane, and the chloride ions subsequently move into the cell.

5 PREDICT

Explain why a slight increase in the blood pH may occur following a heavy meal. The elevated pH of blood, especially in the veins that carry blood away from the stomach, is called "the postenteric alkaline tide."

✔ *Answer in Appendix F*

Chief cells within the gastric glands secrete **pepsinogen** (pep-sin′ō-jen). Pepsinogen is packaged in **zymogen** (zī′mō-jen, meaning related to enzymes) **granules,** which are released by exocytosis when pepsinogen secretion is stimulated. Once pepsinogen enters the lumen of the stomach, it is converted to **pepsin** by hydrochloric acid and previously formed pepsin molecules. Pepsin exhibits optimum enzymatic activity at a pH of 3 or less. Pepsin catalyzes the cleavage of some covalent bonds in proteins, breaking them into smaller peptide chains.

Clinical Note

Heartburn, or pyrosis, is a painful or burning sensation in the chest usually associated with backflush of acidic chyme into the esophagus. The pain is usually short-lived but may be confused with the pain of an ulcer or a heart attack. Overeating, eating fatty foods, lying down immediately after a meal, consuming too much alcohol or caffeine, smoking, or wearing extremely tight clothing can all cause heartburn. A hiatal hernia can also cause heartburn, especially in older people.

1. Carbon dioxide (CO_2) is taken into the cell.

2. CO_2 is combined with water (H_2O) in an enzymatic reaction that is catalyzed by carbon anhydrase (CA) to form carbonic acid.

3. Carbonic acid (H_2CO_3) dissociates into a bicarbonate ion (HCO_3^-) and a hydrogen ion (H^+).

4. The bicarbonate ion (HCO_3^-) moves back into the bloodstream.

5. An anion exchange molecule in the cell membrane changes HCO_3^- for a chloride ion (Cl^-).

6. The hydrogen ion (H^+) is actively transported into the stomach.

7. Chloride ions (Cl^-) diffuse with the charged hydrogen ions.

8. Some potassium ions (K^+) are transported into the cell in exchange for the hydrogen ions.

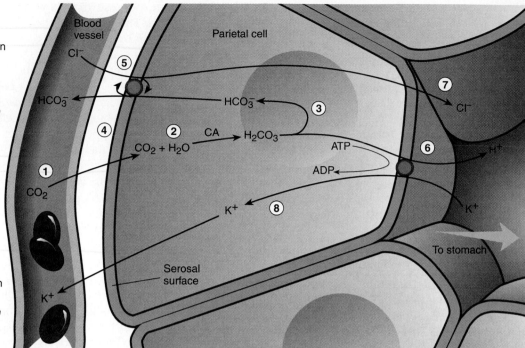

Figure 24.18   Hydrochloric Acid Production by Parietal Cells in the Gastric Glands of the Stomach

Regulation of Stomach Secretion

Approximately 2–3 L of gastric secretions (gastric juice) are produced each day. The amount and type of food entering the stomach dramatically affects the secretion amount, but up to 700 mL is secreted as a result of a typical meal. Both nervous and hormonal mechanisms regulate gastric secretions. The neural mechanisms involve reflexes integrated within the medulla oblongata and local reflexes integrated within the intramural plexuses of the GI tract. In addition, higher brain centers influence the reflexes. Hormones that regulate stomach secretions include gastrin, secretin, gastric-inhibitory polypeptide, and cholecystokinin (table 24.4).

Regulation of stomach secretion can be divided into three phases: the cephalic, gastric, and intestinal.

1. **Cephalic phase.** In the cephalic phase of gastric regulation the sensations of the taste and smell of food, stimulation of tactile receptors during the process of chewing and swallowing, and pleasant thoughts of food stimulate centers within the medulla that influence gastric secretions (figure 24.19*a*). Action potentials are sent from the medulla along parasympathetic neurons within the vagus (X) nerves to the stomach. Within the stomach wall the preganglionic neurons stimulate postganglionic neurons in the intramural plexuses. The postganglionic neurons, which are primarily cholinergic, stimulate secretory activity in the cells of the stomach mucosa.

Neuronal stimulation of the stomach mucosa results in the secretion of acetylcholine, which stimulates the secretory activity of both the parietal and chief cells and stimulates the secretion of **gastrin** (gas′trin) from

endocrine cells. Gastrin is released into the circulation and travels to the parietal cells, where it stimulates additional hydrochloric acid and pepsinogen secretion. In addition, histamine also stimulates parietal cells to produce hydrochloric acid. The histamine receptors on the parietal cells are different from those involved in allergic reactions, and drugs that block allergic reactions do not affect histamine-mediated stomach acid secretion. Acetylcholine, histamine, and gastrin have a synergistic effect, which is greater than the sum of the individual responses.

Clinical Note

Cimetidine (Tagamet) and ranitidine (Zantac) are synthetic analogs of histamine that can bind to histamine receptors on parietal cells, blocking histamine binding, without stimulating the cell. These chemicals are used as histamine blockers that are extremely effective inhibitors of gastric acid secretion. Cimetidine, one of the most commonly prescribed drugs, is used to treat cases of gastric acid hypersecretion associated with gastritis and gastric ulcers.

2. **Gastric phase.** The greatest volume of gastric secretions is produced during the gastric phase of gastric regulation, which is initiated by the presence of food in the stomach (figure 24.19*b*). The primary stimuli are distention of the stomach and the presence of amino acids and peptides in the stomach.

Distention of the stomach wall, especially in the body or fundus, results in the stimulation of mechanoreceptors. Action potentials generated by these receptors initiate

Table 24.4   Functions of the Gastrointestinal Hormones

Site of Production	Method of Stimulation	Secretory Effects	Motility Effects
Gastrin			
Stomach and duodenum	Distention; partially digested proteins, autonomic stimulation, ingestion of alcohol or caffeine	Increases gastric secretion	Increases gastric emptying by increasing stomach motility and relaxing the pyloric sphincter
Secretin			
Duodenum	Acidity of chyme	Inhibits gastric secretion; stimulates pancreatic secretions high in bicarbonate ions; increases the rate of bile secretion; and increases intestinal mucus secretion	Decreases gastric motility
Cholecystokinin			
Intestine	Fatty acids and other lipids	Slightly inhibits gastric secretion; stimulates pancreatic secretions high in digestive enzymes; causes contraction of the gallbladder and relaxation of the hepatopancreatic ampullar sphincter	Decreases gastric motility
Gastric Inhibitory Peptide			
Duodenum and proximal jejunum	Fatty acids and other lipids	Inhibits gastric secretions	Decreases gastric motility

reflexes that involve both the central nervous system and the local intramural reflexes, resulting in secretion of mucus, hydrochloric acid, pepsinogen, intrinsic factor, and gastrin. Gastrin secretion is also stimulated by the presence of partially digested proteins or moderate amounts of alcohol or caffeine in the stomach.

When the pH of the stomach contents falls below 2, the increased gastric secretion produced by distention of the stomach is blocked, an inhibitory effect that constitutes a negative-feedback mechanism that limits the secretion of gastric juice.

Amino acids and peptides that result from the digestive action of pepsin on proteins directly stimulate parietal cells of the stomach to secrete hydrochloric acid. The mechanism by which this response is mediated is not clearly understood. It does not involve any known neurotransmitters, and, when the pH drops below 2, this response is inhibited. Histamine also stimulates the secretory activity of parietal cells.

3. **Intestinal phase.** The intestinal phase of gastric regulation is controlled by the entrance of acidic stomach contents into the duodenum of the small intestine (figure 24.19c). The presence of chyme in the duodenum initiates both neural and hormonal mechanisms, which either stimulate or inhibit gastric secretions. When the pH of the chyme in the duodenum

is 3.0 or above, the stimulatory response predominates; but when the pH of the chyme in the duodenum is 2.0 or below, the inhibitory influence predominates.

Increased gastric secretion during the intestinal phase is due to the release of gastrin from the upper part of the duodenum. The gastrin is carried in the blood to the stomach, where it stimulates gastric secretions.

Inhibition of gastric secretions is controlled by several hormonal mechanisms that are initiated within the duodenum and jejunum of the small intestine during the intestinal phase of gastric regulation. Acidic solutions in the duodenum cause the release of the hormone **secretin** (se-krē′tin) into the circulatory system. Secretin inhibits gastric secretion by inhibiting both parietal and chief cells. Acidic solutions also initiate a local intramural reflex, which inhibits gastric secretions.

Fatty acids and certain other lipids in the duodenum and the proximal jejunum initiate the release of two hormones: gastric inhibitory peptide and **cholecystokinin** (kō′lē-sis-tō-kī′nin). **Gastric inhibitory peptide** strongly inhibits gastric secretion, and cholecystokinin inhibits gastric secretions to a lesser degree. Hypertonic solutions in the duodenum and jejunum also inhibit gastric secretions. The mechanism appears to involve the secretion of a hormone referred to as **enterogastrone** (en′ter-ō-gas′trōn), but the actual existence of this hormone has never been established.

Figure 24.19 Three Phases of Gastric Secretion

(a) Cephalic phase. (b) Gastric phase. (c) Intestinal phase.

1. The taste or smell of food, tactile sensations of food in the mouth, or even thoughts of food stimulate the medulla oblongata *(blue arrow)*.
2. Parasympathetic action potentials are carried by the vagus nerves to the stomach *(red arrow)*.
3. Preganglionic parasympathetic vagus nerve fibers synapse with postganglionic neurons in the intramural plexus of the stomach.
4. Postganglionic neurons directly stimulate secretion by parietal and chief cells and stimulate gastrin secretion by endocrine cells.
5. Gastrin is carried through the circulation back to other parts of the stomach *(lavender arrow)*, where it also stimulates secretion by parietal and chief cells.

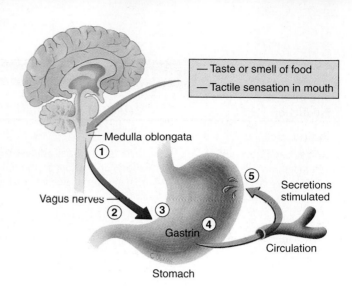

(a)

1. Distention of the stomach activates a parasympathetic reflex. Action potentials are carried by the vagus nerves to the medulla oblongata *(green arrow)*.

2. The medulla oblongata stimulates stomach secretions *(red arrow)*.

3. Distention of the stomach also activates local reflexes that increase stomach secretions *(pink arrow)*.

(b)

1. The presence in the duodenum of chyme with a pH greater than 3 *(blue arrow)* or containing amino acids and peptides stimulates gastric secretions through vagus nerve pathways *(blue and red arrows)* and through secretion of gastrin *(pink arrows to the right)*.
2. The presence in the duodenum of hypotonic chyme with a pH less than 2 or chyme containing fat digestion products inhibits gastric secretions by three mechanisms (3–5).
3. Afferent vagal action potentials *(green arrow)* inhibit efferent action potentials from the vagal nuclei of the medulla oblongata *(purple arrow)*.
4. Secretin inhibits gastrin secretion in the duodenum *(orange arrow)*.
5. Secretin, gastric inhibitory polypeptide, and cholecystokinin produced by the duodenum *(brown arrows)* inhibit gastric secretions in the stomach.

(c)

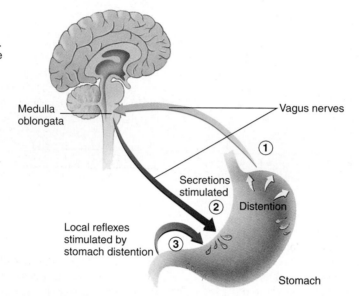

▣ Clinical Focus   Peptic Ulcer

Approximately 5%–12% of the population is affected by peptic ulcers. Most cases of peptic ulcer are apparently due to the infection of a specific bacterium, *Helicobacter pylori*. It is also thought that the bacterium is involved in many cases of gastritis and gastric cancer. Conventional wisdom has focused for years on the notion that stress, diet, smoking, or alcohol cause excess acid secretion in the stomach, resulting in ulcers. Even today, antacids are used to treat 90% of all ulcers, with $4.4 billion spent on antacids in the United States during 1992. Antacid therapy does relieve the ulcer in most cases. There is a 50% incidence of relapse within 6 months with antacid treatment, and a 95% incidence of relapse after 2 years. On the other hand, studies using antibiotic therapy in addition to bismuth and ranitidine have demonstrated a 95% eradication of gastric ulcers and 74% healing of duodenal ulcers within 2 months. Dramatically reduced relapse rates have also been obtained. One such study reported a recurrence rate of 8% following antibiotic therapy, compared with a recurrence rate of 86% in controls.

The infection rate from *H. pylori* in the United States population is about 1% per year of age: 30% of people that are 30 years old have the bacterium, and 80% of those age 80 are infected. In third world countries, as many as 100% of people age 25 or older are infected. This may relate to the high rates of stomach cancer in some of those countries. We still have much to learn in understanding this bacterium. Very little

is known concerning how people become infected. Also, with such high rates of infection it is not known why only a small fraction of those infected actually develop ulcers. It may be that factors of diet and stress and so forth predispose a person who is infected by the bacterium to actually develop an ulcer.

Peptic ulcer is classically viewed as a condition in which the stomach acids digest the mucosal lining of the GI tract itself. The most common site of a peptic ulcer is near the pylorus, usually on the duodenal side (i.e., a duodenal ulcer; 80% of peptic ulcers are duodenal). Ulcers occur less frequently along the lesser curvature of the stomach or at the point at which the esophagus enters the stomach. The most common presumed cause of peptic ulcers is the oversecretion of gastric juice relative to the degree of mucous and alkaline protection of the small intestine. One reason that bacterial involvement in ulcers has been dismissed for such a long time is that is was assumed that the extreme acid environment killed all bacteria. Apparently not only can *H. pylori* survive in such an environment, but it may even thrive there.

People experiencing severe anxiety for a long time are the most prone to develop duodenal ulcers. They often have a high rate of gastric secretion (as much as 15 times the normal amount) between meals. This secretion results in highly acidic chyme entering the duodenum. The duodenum is usually protected by sodium bicarbonate (secreted mainly by the pancreas), which neutralizes

the chyme. When large amounts of acid enter the duodenum, however, the sodium bicarbonate is not adequate to neutralize it. The acid tends to reduce the mucous protection of the duodenum, perhaps leaving that part of the digestive tract open to action of *H. pylori*, which may further destroy the mucous lining.

In one study, it was determined that ulcer patients prefer their hot drinks extra hot, 62°C compared with 56°C for a control group without ulcers. The high temperatures of the drinks may cause thinning of the mucous lining of the stomach, making those people more susceptible to ulcers, again perhaps by increasing their sensitivity to *H. pylori* invasion.

In some patients with gastric ulcers, there are often normal or even low levels of gastric hydrochloric acid secretion. The stomach has a reduced resistance to its own acid, however. Such inhibited resistance can result from excessive ingestion of alcohol or aspirin.

Reflux of duodenal contents into the pylorus can also cause gastric ulcers. In this case, bile, which is present in the reflux, has a detergent effect that reduces gastric mucosal resistance to acid and bacteria.

An ulcer may become perforated (a hole in the stomach or duodenum), causing peritonitis. The perforation must be corrected surgically. Selective vagotomy, cutting branches of the vagus (X) nerve going to the stomach, is sometimes performed at the time of surgery to reduce acid production in the stomach.

Inhibition of gastric secretions is also under nervous control. Distention of the duodenal wall, the presence of irritating substances in the duodenum, reduced pH, and hypertonic or hypotonic solutions in the duodenum can activate the enterogastric reflex. The enterogastric reflex is composed of a local reflex and a reflex integrated within the medulla oblongata, and it reduces gastric secretions.

Stomach Filling

As food enters the stomach, the rugae flatten, and the stomach volume increases. Despite the increase in volume, the pressure within the stomach does not increase until the volume nears maximum capacity because smooth muscle can stretch without an increase in tension (see chapter 10) and because of a reflex integrated within the medulla oblon-

gata. This reflex inhibits muscle tone in the body of the stomach.

Mixing of Stomach Contents

Ingested food is thoroughly mixed with the secretions of the stomach glands to form chyme. This mixing is accomplished by gentle **mixing waves,** which are peristaltic-like contractions that occur about every 20 s and proceed from the body toward the pyloric sphincter to mix the ingested material with the secretions of the stomach. **Peristaltic waves** occur less frequently, are significantly more powerful than mixing waves, and force the chyme near the periphery of the stomach toward the pyloric sphincter. The more solid material near the center of the stomach is pushed superiorly toward the cardiac region for further digestion (figure 24.20). Roughly 80% of the contractions are mixing waves, and 20% are peristaltic waves.

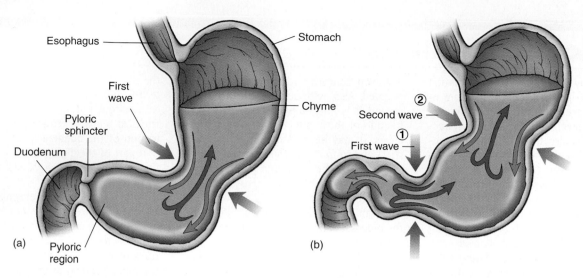

Figure 24.20   Movements in the Stomach

(*a*) Mixing waves that are initiated in the body of the stomach progress toward the pyloric region (*pink arrows directed inward*). The more fluid part of the chyme is pushed toward the pyloric region (*blue arrows*), whereas the more solid center of the chyme squeezes past the peristaltic constriction back toward the body (*orange arrow*). (*b*) Additional mixing waves (2, *purple arrows*) move in the same direction and in the same way as described in part (*a*) for earlier waves (1) that reach the pyloric region. Some of the most fluid chyme is squeezed through the pyloric opening into the duodenum (*small blue arrows*), whereas most of the chyme is forced back toward the body for further mixing (*orange arrows*). ⚕ ▭

Stomach Emptying

The amount of time food remains in the stomach depends on a number of factors, including the type and volume of food. Liquids exit the stomach within 1½–2½ h after ingestion. After a typical meal the stomach is usually empty within 3–4 h. The pyloric sphincter usually remains partially closed because of mild tonic contraction. Each peristaltic contraction is sufficiently strong to force a small amount of chyme through the pyloric opening and into the duodenum. The peristaltic contractions responsible for movement of chyme through the partially closed pyloric opening are called the **pyloric pump.**

Regulation of Stomach Movements

If the stomach empties too fast, the efficiency of digestion and absorption is reduced; and, if the rate of emptying is too slow, the highly acidic contents of the stomach may damage the stomach wall and reduce the rate at which nutrients are digested and absorbed. Stomach emptying is regulated to prevent these two extremes. Many of the hormonal and neural mechanisms that stimulate stomach secretions also are involved with increasing stomach motility. For example, during the gastric phase of stomach secretion, distention of the stomach stimulates local reflexes, central nervous system reflexes, and the release of gastrin, all of which increase stomach motility and cause relaxation of the pyloric sphincter. The result is an increase in stomach emptying. Conversely, the hormonal and neural mechanisms that decrease gastric secretions also inhibit gastric motility, increase constriction of the pyloric sphincter, and decrease the rate of stomach emptying.

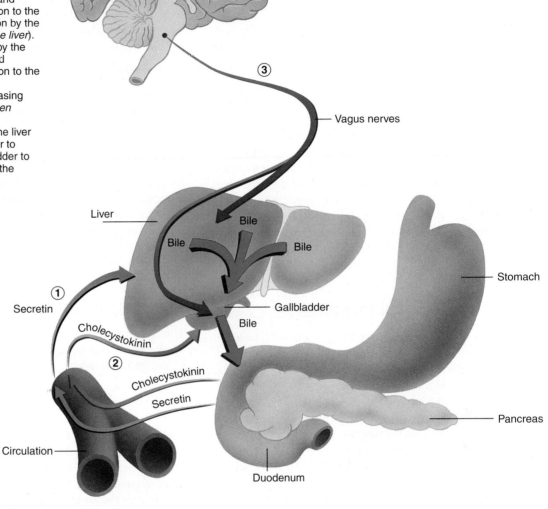

1. Secretin, produced by the duodenum (*purple arrows*) and carried through the circulation to the liver, stimulates bile secretion by the liver (*green arrows inside the liver*).
2. Cholecystokinin, produced by the duodenum (*pink arrows*) and carried through the circulation to the gallbladder, stimulates the gallbladder to contract, releasing bile into the duodenum (*green arrow outside the liver*).
3. Vagal nerve stimulation of the liver (*red arrows*) causes the liver to secrete bile and the gallbladder to contract, releasing bile into the duodenum.

Figure 24.21   Control of Bile Secretion

Functions of the Small Intestine

The small intestine is the site at which the greatest amount of digestion and absorption occurs.

Secretions of the Small Intestine

The mucosa of the small intestine produces secretions that primarily contain mucus, electrolytes, and water. Intestinal secretions lubricate and protect the intestinal wall from the acidic chyme and the action of digestive enzymes. They also keep the chyme in the small intestine in a liquid form to facilitate the digestive process (see table 24.3). Most of the secretions that enter the small intestine are produced by the intestinal mucosa, but the secretions of the liver and the pancreas also enter the small intestine and play an important role in the process of digestion. Most of the digestive enzymes that enter the small intestine come from the pancreas. The intestinal mucosa also produces enzymes that remain associated with the intestinal epithelial surface.

Mucus is secreted in large amounts by duodenal glands, intestinal glands, and goblet cells. The mucus provides the wall of the intestine with protection against the irritating effects of acidic chyme and against the digestive enzymes that enter the duodenum from the pancreas. Secretin and cholecystokinin are released from the intestinal mucosa and stimulate hepatic and pancreatic secretions (figure 24.21; see figure 24.23).

Secretion by duodenal glands is stimulated by the vagus nerve, secretin, and chemical or tactile irritation of the duodenal mucosa. Goblet cells produce mucus in response to the tactile and chemical stimulation of the mucosa.

Enzymes are bound to the membranes of the absorptive cell microvilli. These surface-bound enzymes include **disaccharidases,** which break disaccharides down to monosaccharides; **peptidases,** which hydrolyze the peptide bonds between small amino acid chains; and **nucleases,** which break down nucleic acids (see table 24.3). Although these enzymes are not secreted into the intestine, they influence the digestive process significantly, and the large surface area of the intestinal epithelium brings these enzymes into contact with the intestinal contents. Small molecules, which are breakdown products of digestion, are absorbed through the microvilli and enter the circulatory or lymphatic systems.

Movement in the Small Intestine

Mixing and propulsion of chyme are the primary mechanical events that occur in the small intestine. These functions are the result of segmental or peristalic contractions, which are accomplished by the smooth muscle in the wall of the small intestine and which are only propagated for short distances. **Segmental contractions** (see figure 24.3*b*) mix the intestinal contents, and **peristaltic contractions** propel the intestinal contents along the digestive tract. A few peristaltic contractions may proceed the entire length of the intestine. Frequently, intestinal peristaltic contractions are continuations of peristaltic contractions that begin in the stomach. These contractions both mix and propel substances through the small intestine as the wave of contraction proceeds. The contractions move at a rate of about 1 cm/min. The movements are slightly faster at the proximal end of the small intestine and slightly slower at the distal end. It usually takes 3–5 h for chyme to move from the pylorus to the ileocecal junction.

Local mechanical and chemical stimuli are especially important in regulating the motility of the small intestine. Smooth muscle contraction increases in response to distention of the intestinal wall. Solutions that are either hypertonic or hypotonic, solutions with a low pH, and certain products of digestion such as amino acids and peptides also stimulate contractions of the small intestine. Local reflexes, which are integrated within the intramural nervous plexuses of the small intestine, mediate the response of the small intestine to these mechanical and chemical stimuli. Stimulation through parasympathetic nerve fibers may also increase the motility of the small intestine, but the parasympathetic influences in the small intestine are not as important as those in the stomach.

The ileocecal sphincter at the juncture between the ileum and the large intestine remains mildly contracted most of the time, but peristaltic contractions reaching it from the small intestine cause it to relax and allow movement of chyme from the small intestine into the cecum. Cecal distention, however, initiates a local reflex that causes more intense constriction of the ileocecal sphincter. Closure of the sphincter facilitates digestion and absorption in the small intestine by slowing the rate of chyme movement from the small intestine into the large intestine and prevents material from returning to the ileum from the cecum.

Absorption from the Small Intestine

Each day, about 9 L of water enters the digestive system. It comes from water that is ingested and from fluid secretions produced by glands along the length of the digestive tract. Most of the water, 8–8.5 L, moves by osmosis, with the absorbed solutes, out of the small intestine. A small part, 0.5–1 L, enters the colon.

Liver Functions

The liver performs important digestive and excretory functions, stores and processes nutrients, synthesizes new molecules, and detoxifies harmful chemicals.

Bile Production

The liver produces and secretes about 600–1000 mL of **bile** each day (see table 24.3). Although bile contains no digestive enzymes, it plays a role in digestion by diluting and neutralizing stomach acid, and by emulsifying fats. The pH of chyme as it leaves the stomach is too low for the normal function of pancreatic enzymes. Bile helps to neutralize the acidic chyme and to bring the pH up to a level at which pancreatic enzymes can function. Bile salts **emulsify** fats (described in more detail later in the chapter). Bile also contains excretory products such as bile pigments. Bilirubin is a bile pigment that results from the breakdown of hemoglobin. Bile also contains cholesterol, fats, fat-soluble hormones, and lecithin.

The rate of bile secretion is controlled in a variety of ways (see figure 24.21). Secretin stimulates bile secretion, primarily by increasing the water and bicarbonate ion content of bile. Bile secretion is increased by parasympathetic stimulation through the vagus nerve and by increased blood flow through the liver. Bile salts also increase bile secretion through a positive-feedback system. Most bile salts are reabsorbed in the ileum and carried in the blood back to the liver, where they stimulate further bile secretion. The loss of bile salts in the feces is reduced by this recycling process. Bile secretion into the duodenum continues until the duodenum empties.

Storage

Hepatocytes can remove sugar from the blood and store it in the form of **glycogen.** They can also store fat, vitamins (A, B_{12}, D, E, and K), copper, and iron. This storage function is usually short term, and the amount of stored material in the hepatocytes and thus the cell size fluctuate during a given day.

If a large amount of sugar were dumped into the general circulation after a meal, it would increase the osmolality of the blood and produce hyperglycemia. These results are prevented because the blood from the intestine passes through the hepatic portal vein to the liver, where glucose and other substances are removed from the blood by hepatocytes, stored, and secreted back into the circulation when needed. By this means the hepatocytes can help control blood sugar levels within very narrow limits.

Nutrient Interconversion

Another function that the liver performs is the interconversion of nutrients. Ingested foods are not always in the proportion needed by the tissues. If this is the case, the liver can convert some nutrients into others. If, for example, a person is on a diet that is excessively high in protein, an oversupply of protein and an undersupply of lipids and carbohydrates may be delivered to the liver. The hepatocytes break down the amino acids and cycle many of them through metabolic pathways so they can be used to produce adenosine triphosphate, lipids, and glucose (see chapter 25).

Hepatocytes also transform substances that cannot be used by the cells into more readily usable substances. For example, ingested fats are combined with choline and phosphorus in the liver to produce phospholipids, which are essential components of cell membranes. Vitamin D is hydroxylated in the liver. The hydroxylated form of vitamin D is the major circulating form of vitamin D, which is transported through the circulation to the kidney, where it is again hydroxylated. The double-hydroxylated vitamin D is the active form of the vitamin, which functions in calcium maintenance.

Detoxification

Many ingested substances are harmful to the cells of the body. In addition, the body itself produces many by-products of metabolism that, if accumulated, are toxic. The liver is one line of defense against many of these harmful substances. It detoxifies many substances by altering their structure, making them less toxic or making their elimination easier. Ammonia, for example, a by-product of amino acid metabolism, is toxic and is not readily removed from the circulation by the kidneys. Hepatocytes remove ammonia from the circulation and convert it to urea, which is secreted into the circulation and then eliminated by the kidneys in the urine. Other substances are removed from the circulation and excreted by the hepatocytes into the bile.

Phagocytosis

Hepatic phagocytic cells (Kupffer cells), which lie along the sinusoid walls of the liver, phagocytize worn out and dying red and white blood cells, some bacteria, and other debris that enters the liver through the circulation.

Synthesis

The liver can also produce its own unique new compounds. Many of the blood proteins, such as albumins, fibrinogen, globulins, heparin, and clotting factors, are produced by the liver and released into the circulation (see chapter 19).

Clinical Note

Strictly defined, **hepatitis** is an inflammation of the liver and does not necessarily result from an infection. Hepatitis can be caused by alcohol consumption or infection. Infectious hepatitis is caused by viral infections. Hepatitis A, also called infectious hepatitis, is the most common form of viral hepatitis. Hepatitis B, also called serum hepatitis, is a more chronic infection responsible for 60%–90% of all liver carcinomas. If hepatitis is not treated, liver cells can die and be replaced by scar tissue, resulting in loss of liver function. Death caused by liver failure can occur.

Cirrhosis of the liver involves the death of hepatocytes and their replacement by fibrous connective tissue. The liver becomes pale in color (the term cirrhosis means a tawny or orange condition) because of the presence of excess white connective tissue. It also becomes firmer, and the surface becomes nodular. The loss of hepatocytes eliminates the function of the liver, often resulting in jaundice, and the buildup of connective tissue can impede blood flow through the liver. Cirrhosis frequently develops in alcoholics and may develop as a result of biliary obstruction, hepatitis, or nutritional deficiencies.

Under most conditions, mature hepatocytes can proliferate and replace lost parts of the liver. If the liver is severely damaged, however, the hepatocytes may not have enough regenerative power to replace the lost parts. In this case a liver transplant may be necessary. Recent evidence also suggests that the liver also maintains an undifferentiated stem cell population, called "oval" cells, which gives rise to two cell lines, one forming bile duct epithelium and the other producing hepatocytes. It is hoped that these stem cells can be used to reconstitute a severely damaged liver. It may even be possible at some time in the future to remove stem cells from a person with hemophilia, genetically engineer the cells to produce the missing clotting factors, and then reintroduce the altered stem cells back into the person's liver.

Functions of the Gallbladder

Bile is continually secreted by the liver and stored in the **gallbladder,** which can store 40–70 mL of bile. While the bile is in the gallbladder, water and electrolytes are absorbed, and bile salts and pigments become as much as 5 to 10 times more concentrated than they were when secreted by the liver.

Shortly after a meal, the gallbladder contracts in response to stimulation by cholecystokinin and, to a lesser degree, in response to vagal stimulation, dumping large amounts of concentrated bile into the small intestine (see figure 24.21).

Clinical Note

Cholesterol, secreted by the liver, may precipitate in the gallbladder to produce **gallstones** (figure A). Occasionally a gallstone can pass out of the gallbladder and enter the cystic duct, blocking release of bile. Such a condition interferes with normal digestion, and the gallstone often must be removed surgically. If the gallstone moves far enough down the duct, it can also block the pancreatic duct, resulting in pancreatitis.

Figure A   Gallstones

Drastic dieting with rapid weight loss may lead to gallstone production. In one study, 25% of obese people participating in an 8-week, quick-weight-loss program developed gallstones. Six percent required surgical removal of the stones. No gallstones developed in nondieting obese controls.

Functions of the Pancreas

The exocrine secretions of the pancreas are called **pancreatic juice** and have two major components: an aqueous component and an enzymatic component. Pancreatic juice is produced in the pancreas and is then delivered through the pancreatic ducts to the small intestine, where it functions in digestion. The **aqueous component** is produced principally by columnar epithelial cells that line the smaller ducts of the pancreas. It contains sodium and potassium ions in about the same concentration found in extracellular fluid. Bicarbonate ions are a major part of the aqueous component, and they neutralize the acidic chyme that enters the small intestine from the stomach. The increased pH caused by pancreatic secretions in the duodenum stops pepsin digestion but provides the proper environment for the function of pancreatic enzymes. Bicarbonate ions are actively secreted by the duct epithelium, and water follows passively to make the pancreatic juice isotonic. The cellular mechanism that is responsible for the secretion of bicarbonate ions is diagrammed in figure 24.22.

1. Water (H_2O) and carbon dioxide (CO_2) combine under the influence of carbon anhydrase (CA) to form carbonic acid.
2. Carbonic acid (H_2CO_3) dissociates to form hydrogen ions (H^+) and bicarbonate ions (HCO_3^-).
3. The hydrogen ions are removed in the bloodstream.
4. The bicarbonate ions are actively transported into the pancreatic ducts. Sodium, potassium, and water follow the bicarbonate ions into the ducts.

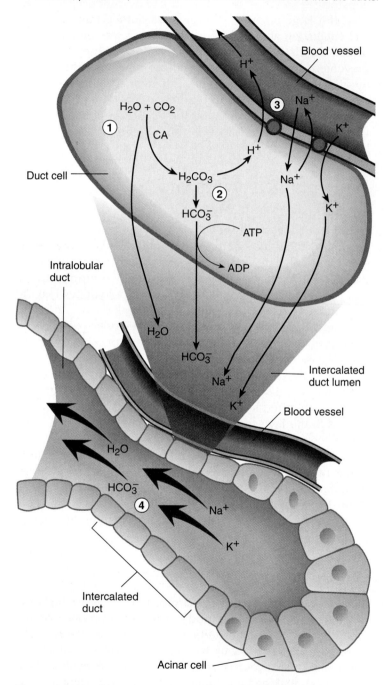

Figure 24.22   Bicarbonate Ion Production in the Pancreas

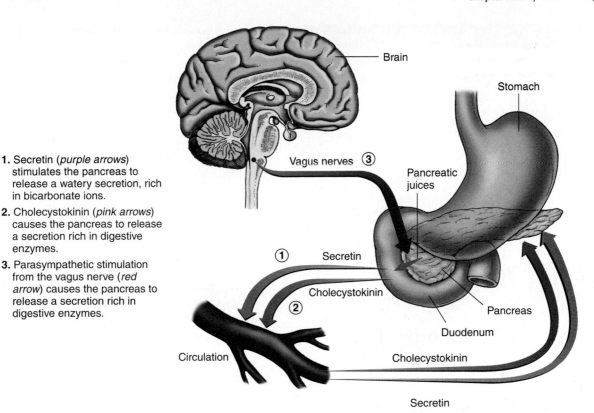

1. Secretin (*purple arrows*) stimulates the pancreas to release a watery secretion, rich in bicarbonate ions.
2. Cholecystokinin (*pink arrows*) causes the pancreas to release a secretion rich in digestive enzymes.
3. Parasympathetic stimulation from the vagus nerve (*red arrow*) causes the pancreas to release a secretion rich in digestive enzymes.

Figure 24.23   Control of Pancreatic Secretion

Pancreatic Enzymes

The **enzymatic component** of the pancreatic juice is produced by the acinar cells of the pancreas and is important for the digestion of all major classes of food. Without the enzymes produced by the pancreas, lipids, proteins, and carbohydrates are not adequately digested (see tables 24.1 and 24.3).

The proteolytic pancreatic enzymes, which digest proteins, are secreted in inactive forms, whereas many of the other enzymes are secreted in active form. The major proteolytic enzymes are **trypsin, chymotrypsin,** and **carboxypeptidase.** They are secreted in their inactive forms as trypsinogen, chymotrypsinogen, and procarboxypeptidase and are activated by the removal of certain peptides from the larger precursor proteins. If these were produced in their active forms, they would digest the tissues producing them. Trypsinogen is activated by the proteolytic enzyme **enterokinase** (en'tēr-ō-kī'nās, meaning intestinal enzyme), which is an enzyme attached to the brush border of the small intestine. Trypsin then activates more trypsinogen, as well as chymotrypsinogen and procarboxypeptidase.

Pancreatic juice also contains amylase, which continues the polysaccharide digestion that was initiated in the oral cavity.

In addition, pancreatic juice contains a group of lipid-digesting enzymes called pancreatic lipases, which break down lipids into free fatty acids, glycerides, cholesterol, and other components.

Enzymes that reduce DNA and ribonucleic acid to their component nucleotides, **deoxyribonucleases** and **ribonucleases,** respectively, are also present in pancreatic juice.

Control of Pancreatic Secretion

The exocrine secretions of the pancreas are controlled by both hormonal and neural mechanisms (figure 24.23). Secretin stimulates the secretion of a watery solution that contains a large amount of bicarbonate ions from the pancreas. The primary stimulus for secretin release is the presence of acidic chyme in the duodenum.

Cholecystokinin stimulates the secretion of bile from the liver and the secretion of pancreatic juice rich in digestive enzymes. The major stimulus for the release of cholecystokinin is the presence of fatty acids and amino acids in the intestine.

Parasympathetic stimulation through the vagus (X) nerves also stimulates the secretion of pancreatic juices rich in pancreatic enzymes, and sympathetic impulses inhibit secretion. The effect of vagal stimulation on pancreatic juice secretion is greatest during the cephalic and gastric phases of stomach secretion.

Functions of the Large Intestine

Normally 18–24 h is required for material to pass through the large intestine, in contrast to the 3–5 h required for movement of chyme through the small intestine. Thus the movements of the colon are more sluggish than those of the small intestine. While in the colon, chyme is converted to feces. Absorption of water and salts, the secretion of mucus, and extensive action of microorganisms are involved in the formation of feces, which the colon stores until the feces are eliminated by the process of defecation. About 1500 mL of chyme enters the cecum each day, but more than 90% of the volume is reabsorbed so that only 80–150 mL of feces is normally eliminated by defecation.

Secretions of the Large Intestine

The mucosa of the colon has numerous goblet cells that are scattered along its length and numerous crypts that are lined almost entirely with goblet cells. Little enzymatic activity is associated with secretions of the colon when mucus is the major secretory product (see table 24.3). Mucus lubricates the wall of the colon and helps the fecal matter stick together. Tactile stimuli and irritation of the wall of the colon trigger local intramural reflexes that increase mucous secretion. Parasympathetic stimulation also increases the secretory rate of the goblet cells.

Clinical Note

When the large intestine is irritated and inflamed, such as in patients with bacterial enteritis (inflamed intestine resulting from bacterial infection of the bowel), the intestinal mucosa secretes large amounts of mucus and electrolytes, and water moves by osmosis into the colon. An abnormally frequent discharge of fluid feces is called diarrhea. Although such discharge increases fluid and electrolyte loss, it also moves the infected feces out of the intestine more rapidly and speeds recovery from the disease.

Bicarbonate ions are exchanged by an exchange pump for chloride ions in epithelial cells of the colon in response to the acid produced by colic bacteria. Sodium ions are exchanged by another exchange pump for hydrogen ions. Water crosses the wall of the colon through osmosis with the sodium chloride gradient.

The feces that leave the digestive tract consist of water, solid substances (e.g., undigested food), microorganisms, and sloughed-off epithelial cells.

Numerous microorganisms inhabit the colon. They reproduce rapidly and ultimately constitute about 30% of the dry weight of the feces. Some bacteria in the intestine synthesize vitamin K, which is passively absorbed in the colon, and break down a small amount of cellulose to glucose.

Gases called **flatus** (flā′tŭs, meaning blowing) are produced by bacterial actions in the colon. The amount of flatus depends partly on the bacterial population present in the colon and partly on the type of food consumed. For example, beans, which contain certain complex carbohydrates, are well known for their flatus-producing effect.

Movement in the Large Intestine

Segmental mixing movements occur in the colon much less often than in the small intestine. Peristaltic waves are largely responsible for moving chyme along the ascending colon. At widely spaced intervals (normally three or four times each day), large parts of the transverse and descending colon undergo several strong peristaltic contractions, called **mass movements.** Each mass movement contraction extends over a much longer part of the digestive tract (≥ 20 cm) than does a peristaltic contraction and propels the colon contents a considerable distance toward the anus (figure 24.24). Mass movements are very common after meals because the presence of food in the stomach initiates strong peristaltic contractions in the colon. Mass movements are most common about 15 min after breakfast. They usually persist for 10–30 min and then stop for perhaps half a day. Mass movements are integrated by local reflexes in the intramural plexus, which are called **gastrocolic reflexes** if initiated by the stomach or **duodenocolic reflexes** if initiated by the duodenum (see figure 24.24).

Distention of the rectal wall by feces acts as a stimulus that initiates the **defecation reflex.** Local reflexes cause weak contractions of the rectum and relaxation of the internal anal sphincter. Parasympathetic reflexes cause strong contractions of the rectum and are normally responsible for most of the defecation reflex. Action potentials produced in response to the distention travel along afferent nerve fibers to the sacral region of the spinal cord, where efferent action potentials are initiated that reinforce peristaltic contractions in the lower colon and rectum. The defecation reflex reduces action potentials to the internal anal sphincter, causing it to relax. The external anal sphincter, which is composed of skeletal muscle and is under conscious cerebral control, prevents the movement of feces out of the rectum and through the anal opening. If this sphincter is relaxed voluntarily, feces are expelled.

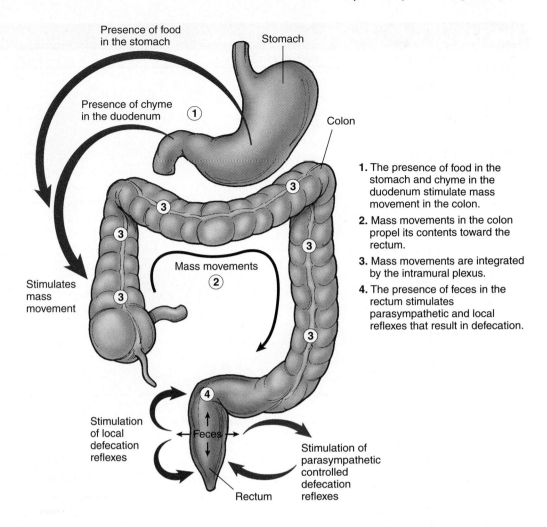

Figure 24.24 Reflexes in the Colon and Rectum

Presence of food in the stomach

Presence of chyme in the duodenum ①

Stomach

Colon

Stimulates mass movement

Mass movements ②

Stimulation of local defecation reflexes

Feces ④

Stimulation of parasympathetic controlled defecation reflexes

Rectum

1. The presence of food in the stomach and chyme in the duodenum stimulate mass movement in the colon.

2. Mass movements in the colon propel its contents toward the rectum.

3. Mass movements are integrated by the intramural plexus.

4. The presence of feces in the rectum stimulates parasympathetic and local reflexes that result in defecation.

The defecation reflex persists for only a few minutes and quickly dies. Generally the reflex is reinitiated after a period that may be as long as several hours. Mass movements in the colon are usually the reason for the reinitiation of the defecation reflex.

Defecation is usually accompanied by voluntary movements that support the expulsion of feces. These voluntary movements include a large inspiration of air followed by closure of the larynx and forceful contraction of the abdominal muscles. As a consequence, the pressure in the abdominal cavity increases, helping force the contents of the colon through the anal canal and out of the anus.

Clinical Note

The importance of regularity of defecation has been greatly overestimated. Many people have the misleading notion that a daily bowel movement is critical for good health. As with many other body functions, what is "normal" differs from person to person. Whereas many people defecate one or more times per day, some normal, healthy adults defecate on the average only every other day. A defecation rate of only twice per week, however, is usually described as constipation. Habitually postponing defecation when the

defecation reflex occurs can lead to constipation and may eventually result in desensitization of the rectum so that the defecation reflex is greatly diminished.

Digestion, Absorption, and Transport

Digestion is the breakdown of organic molecules into their component parts: carbohydrates into monosaccharides, proteins into amino acids, and fats into fatty acids and glycerol. Absorption and transport are the means by which molecules are moved out of the digestive tract and into the circulation for distribution throughout the body. Not all molecules (e.g., vitamins, minerals, and water) are broken down before being absorbed. Digestion begins in the oral cavity and continues in the stomach, but most digestion occurs in the proximal end of the small intestine, especially in the duodenum.

Absorption of certain molecules can occur all along the digestive tract. A few chemicals, such as nitroglycerin, can be absorbed through the thin mucosa of the oral cavity below the

Table 24.5   Digestion of the Three Major Food Types

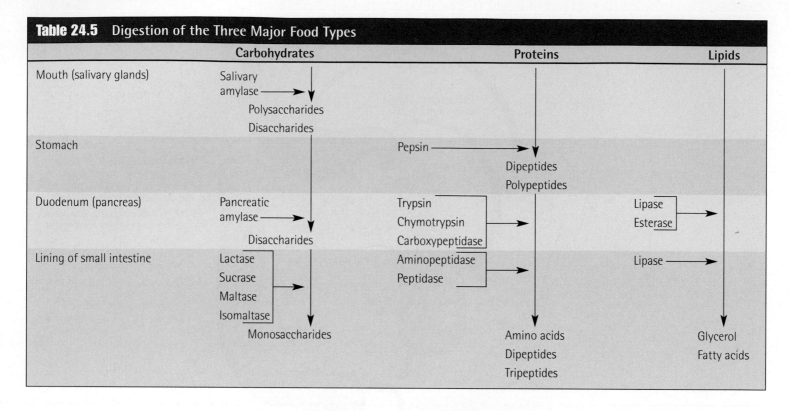

	Carbohydrates	Proteins	Lipids
Mouth (salivary glands)	Salivary amylase → Polysaccharides Disaccharides		
Stomach		Pepsin → Dipeptides Polypeptides	
Duodenum (pancreas)	Pancreatic amylase → Disaccharides	Trypsin Chymotrypsin Carboxypeptidase →	Lipase Esterase →
Lining of small intestine	Lactase Sucrase Maltase Isomaltase → Monosaccharides	Aminopeptidase Peptidase → Amino acids Dipeptides Tripeptides	Lipase → Glycerol Fatty acids

tongue. Some small molecules (e.g., alcohol and aspirin) can pass through the stomach epithelium into the circulation. Most absorption, however, occurs in the duodenum and jejunum, although some absorption occurs in the ileum.

Once the digestive products have been absorbed, they are transported to other parts of the body by two different routes. Water, ions, and water-soluble digestion products such as glucose and amino acids enter the hepatic portal system (see chapter 21) and are transported to the liver. The products of lipid metabolism are coated with proteins and transported into lacteals. The lacteals are connected by lymph vessels to the thoracic duct (see chapter 21), which empties into the left subclavian vein. The protein-coated lipid products then travel in the circulation to adipose tissue or to the liver.

Carbohydrates

Ingested **carbohydrates** consist primarily of polysaccharides such as starches and glycogen, disaccharides such as sucrose (table sugar) and lactose (milk sugar), and monosaccharides such as glucose and fructose (found in many fruits). During the digestion process polysaccharides are broken down into smaller chains and finally into disaccharides and monosaccharides. Disaccharides are broken down into monosaccharides. Carbohydrate digestion begins in the oral cavity with the partial digestion of starches by **salivary amylase** (am′il-ās) and is completed in the intestine by **pancreatic amylase** (table 24.5). The digestion of disaccharides into monosaccharides is accomplished by a series of **disaccharidases** that are bound to the microvilli of the intestinal epithelium.

> **Clinical Note**
>
> **Lactase deficiency** results in **lactose intolerance,** which is an inability to digest milk products. This disorder is primarily hereditary, affecting 5%–15% of Europeans and 80%–90% of Africans and Asians. Symptoms include cramps, bloating, and diarrhea.

Monosaccharides such as glucose and galactose are taken up into intestinal epithelial cells by cotransport, powered by a sodium ion gradient (figure 24.25). Monosaccharides such as fructose are taken up by facilitated diffusion. The monosaccharides are transferred by facilitated diffusion to the capillaries of the intestinal villi and are carried by the hepatic portal system to the liver, where the nonglucose sugars are converted to glucose. Glucose enters the cells through facilitated diffusion. The rate of glucose transport into most types of cells is greatly influenced by **insulin** and may increase 10-fold in its presence.

> **Clinical Note**
>
> In patients with type I diabetes mellitus, insulin is lacking, and insufficient glucose is transported into the cells of the body. As a result, the cells do not have enough energy for normal function, blood glucose levels become significantly elevated, and abnormal amounts of glucose are released into the urine. This condition is discussed more fully in chapter 18.

1. Monosaccharides are absorbed by secondary active transport into intestinal epithelial cells.

2. Monosaccharides move out of intestinal epithelial cells by facilitated diffusion.

3. They enter the capillaries of the intestinal villi and are carried through the hepatic portal vein to the liver.

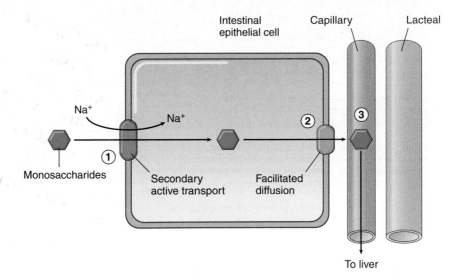

Figure 24.25   Glucose and Galactose Absorption

Lipids

Lipids are molecules that are insoluble or only slightly soluble in water. They include triacylglycerol, phospholipids, cholesterol, steroids, and fat-soluble vitamins. **Triacylglycerol** (trī-as'il-glis'er-ol; this term is a more contemporary replacement for the term triglycerides) consists of three fatty acids and one glycerol molecule covalently bound together. The first step in lipid digestion is **emulsification** (ē-mŭl'si-fi-kā'shŭn), which is the transformation of large lipid droplets into much smaller droplets. The enzymes that digest lipids are water-soluble and can digest the lipids only by acting at the surface of the droplets. The emulsification process increases the surface area of the lipid exposed to the digestive enzymes by decreasing the droplet size. Emulsification is accomplished by **bile salts** secreted by the liver and stored in the gallbladder.

    Lipase (lī'pās) secreted by the pancreas digests lipid molecules (see table 24.5). The primary products of this digestive process are free fatty acids and glycerol. Cholesterol and phospholipids also constitute part of the lipid digestion products.

Cystic fibrosis is a hereditary disorder that occurs in 1 of every 2000 births and affects 33,000 people in the United States; it is the most common lethal genetic disorder among whites. The most critical effects of the disease, accounting for 90% of the deaths, are on the respiratory system. Several other problems occur, however, in affected people. Because the disease is a disorder in chloride ion transport channel proteins, which affects chloride transport and, as a result, movement of water, all exocrine glands are affected. The buildup of thick mucus in the pancreatic and hepatic ducts causes blockage of the ducts so that bile salts and pancreatic digestive enzymes are prevented from reaching the duodenum. As a result, fats and fat-soluble vitamins, which require bile salts to form micelles and which cannot be adequately digested without pancreatic enzymes, are not well digested and absorbed. The patient suffers from vitamin A, D, E, and K deficiencies, which result in conditions such as night blindness, skin disorders, rickets, and excessive bleeding. Therapy includes administering the missing vitamins to the patient and reducing dietary fat intake.

Once lipids are digested in the intestine, bile salts aggregate around the small droplets to form **micelles** (mī-selz', meaning a small morsel; figure 24.26). The hydrophobic ends of the bile salts are directed toward the free fatty acids, cholesterol, and glycerides at the center of the micelle; and the hydrophilic ends are directed outward toward the water environment. When a micelle comes into contact with the epithelial cells of the small intestine, the contents of the micelle pass by means of simple diffusion through the lipid cell membrane of the epithelial cells.

Lipid Transport

Within the smooth endoplasmic reticulum of the intestinal epithelial cells, free fatty acids are combined with glycerol molecules to form triacylglycerol. Proteins synthesized in the epithelial cells attach to droplets of triacylglycerol, phospholipids, and cholesterol to form **chylomicrons** (kī-lō-mī'kronz, meaning small particles in the chyle, or fat-filled lymph). The chylomicrons leave the epithelial cells and enter the lacteals of the lymphatic system within the villi. Chylomicrons enter the lymph capillaries rather than the blood capillaries because the lymph capillaries lack a basement membrane and are more permeable to large particles such as chylomicrons (about 0.3 mm in diameter). Chylomicrons are

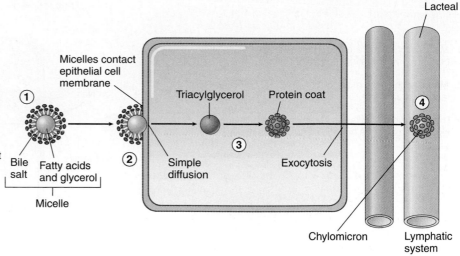

1. Bile salts surround fatty acids and glycerol to form micelles.

2. Micelles attach to the cell membranes of intestinal epithelial cells, and the fatty acids and glycerol pass by simple diffusion into the intestinal epithelial cells.

3. Within the intestinal epithelial cell, the fatty acids and glycerol are converted to triacylglycerol; proteins coat the triacylglycerol to form chylomicrons, which move out of the intestinal epithelial cells by exocytosis.

4. The chylomicrons enter the lacteals of the intestinal villi and is carried through the lymphatic system to the general circulation.

Figure 24.26   Lipid Absorption

---

about 90% triacylglycerol, 5% cholesterol, 4% phospholipid, and 1% protein (figure 24.27). They are carried through the lymphatic system to the bloodstream and then by the blood to adipose tissue (figure 24.28). Before entering the adipose cells, triacylglycerol is broken back down into fatty acids and glycerol, which enter the fat cells and are once more converted back to triacylglycerol. Triacylglycerol is stored in adipose tissue until an energy source is needed elsewhere in the body. In the liver the chylomicron lipids are stored, converted into other molecules, or used as energy. The chylomicron remnant, minus the triacylglycerol, is conveyed through the circulation to the liver, where it is broken up.

Because lipids are either insoluble or only slightly soluble in water, they are transported through the blood in combination with proteins, which are water-soluble. Lipids combined with proteins are called **lipoproteins.** Chylomicrons are one type of lipoprotein. Other lipoproteins are referred to as high- or low-density lipoproteins. Density describes the compactness of a substance and is the ratio of mass to volume. Lipids are less dense than water and tend to float in water. Proteins, which are denser than water, tend to sink in water. A lipoprotein with a high lipid content has a very low density, whereas a lipoprotein with a high protein content has a relatively high density. Chylomicrons, which are made up of 99% lipid and only 1% protein, have an extremely low density. The other major transport lipoproteins are **very low density lipoprotein (VLDL),** which is 92% lipid and 8% protein, **low-density lipoprotein (LDL),** which is 75% lipid and 25% protein, and **high-density lipoprotein (HDL),** which is 55% lipid and 45% protein (see figure 24.27).

About 15% of the cholesterol in the body is ingested in the food we eat, the remaining 85% is manufactured in the cells of the body, mostly in the liver and intestinal mucosa. Most of the lipid taken into or manufactured in the liver leaves the liver in the form of VLDL. Most of the triacylglycerol is re-

moved from the VLDL to be stored in adipose tissue and, as a result, VLDL becomes LDL (see figure 24.28).

The cholesterol in LDL is critical for the production of steroid hormones in the adrenal cortex and the production of bile acids in the liver. It is also an important component of cell membranes. LDL is delivered to cells of various tissues through the circulation. Cells have **LDL receptors** in "pits" on their surfaces, which bind the LDL. Once LDL is bound to the receptors, the pits on the cell surface become endocytotic vesicles, and the LDL is taken into the cell by receptor-mediated endocytosis (figure 24.29). Each fibroblast, as an example of a tissue cell, has 20,000–50,000 LDL receptors on the surface. Those receptors are confined to cell surface pits, however, which occupy only 2% of the cell surface. Once inside the cell, the endocytotic vesicle combines with a lysosome, and the LDL components are separated for use in the cell.

Cells not only take in cholesterol and other lipids from LDLs, but they also make their own cholesterol. When the combined intake and manufacture of cholesterol exceeds a cell's needs, a negative-feedback system functions, which reduces the amount of LDL receptors and cholesterol manufactured by the cell. Excess lipids are also packaged into HDLs by the cells. These are transported back to the liver for recycling or disposal (see figure 24.28).

Clinical Note

Cholesterol is a major component of atherosclerotic plaques. The level of plasma cholesterol is positively linked to coronary heart disease (CHD). Cholesterol levels of over 200 mg/100 mL increase the risk of CHD. Other risk factors, which are additive to high cholesterol levels, are hypertension, diabetes mellitus, cigarette smoking, and low plasma HDL levels. Low HDL levels are linked to obesity, and weight reduction increases HDL levels. Aerobic exercise can decrease LDL levels and increase HDL levels. Ingestion of saturated fatty acids raises plasma cholesterol levels by stimulating LDL production and inhibiting LDL receptor production, which would enhance HDL production and cholesterol clearance. Ingestion of unsaturated fatty acids lowers plasma cholesterol levels. Replacing fats by carbohydrates in the diet can also reduce blood cholesterol levels. The American Heart Association recommends that no more than 30% of an adult's total caloric intake should be from fats and that only 10% be from saturated fats. Our total cholesterol intake should be no more than 300 mg/day. People should eat no more than 7 ounces of meat per day, and that should be chicken, fish, or lean meat. We should eat only two eggs per week and drink milk with 1% or less butter fat. Young children, however, require more fats in their diet to stimulate normal brain development, and whole milk is recommended in their diets. There is also some evidence that severely reducing plasma cholesterol levels, below about 180 mg/100 mL may be harmful in adults. Cholesterol is required for normal membrane structure in cells. Abnormally low cholesterol levels may lead to weakened blood vessel walls and an increased risk for cerebral hemorrhage.

A small number of people have a genetic disorder in the production or function of LDL receptors, resulting in poor LDL clearing, and, as a result, have what is called **familial hypercholesterolemia.** These people commonly develop premature atherosclerosis and are prone to die at an early age of a heart attack. In some of these disorders, the LDL receptor is not produced. In other cases, the receptor is produced but it has a lower-than-normal affinity for LDL. In yet other cases, the receptor binds to LDL but the receptor–LDL complex is not taken into the cell by endocytosis.

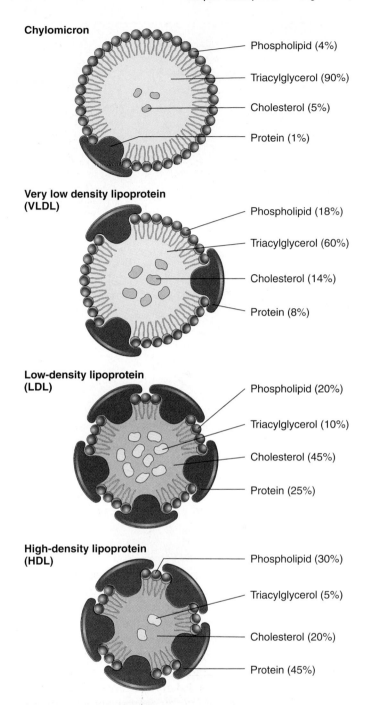

Figure 24.27   Lipoproteins

Proteins

Proteins are taken into the body from a number of dietary sources. **Pepsin** secreted by the stomach (see table 24.5) catalyzes the cleavage of covalent bonds in proteins, producing smaller polypeptide chains. As much as 10%–20% of the total ingested protein is digested by gastric pepsin. Once the proteins and polypeptide chains leave the stomach, proteolytic enzymes produced in the pancreas continue the digestive process, producing small peptide chains. These are broken down into dipeptides, tripeptides, and amino acids by **peptidases** bound to the microvilli of the small intestine. Each peptidase is specific for a certain peptide chain length or for a certain peptide bond.

Dipeptides and tripeptides enter intestinal epithelial cells through a group of related carrier molecules, by a cotransport mechanism, powered by a sodium ion concentration gradient similar to that described for glucose. Transport of amino acids into the epithelial cells is accomplished by separate transport molecules for basic, acidic, and neutral amino acids. Acidic and most neutral amino acids are cotransported with a sodium ion gradient, whereas basic amino acids enter the epithelial cells by facilitated diffusion. The total amount of each amino acid that enters the intestinal epithelial cells as dipeptides or tripeptides is considerably more than the amount that enters as single amino acids. Once inside the cells, dipeptidases and tripeptidases split the dipeptides and tripeptides into their component amino acids. Individual amino acids then leave the epithelial cells and enter the hepatic portal system, which transports them to the liver (figure 24.30). The amino acids may be modified in the liver or released into the bloodstream and distributed throughout the body.

Amino acids are actively transported into the various cells of the body. This transport is stimulated by growth hormone and insulin. Most amino acids are used as building blocks to form new proteins (see chapter 2), but some amino acids may be used for energy.

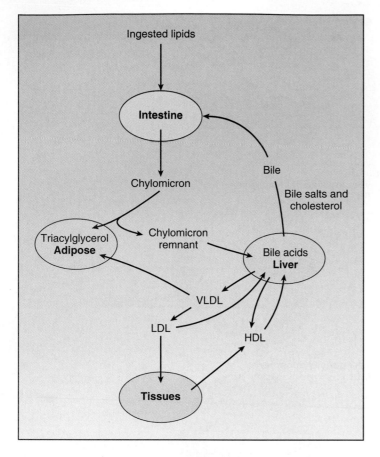

Figure 24.28   Transport of Lipoproteins

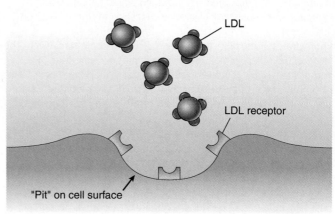

Cells have pits on the surface, which contain LDL receptors.

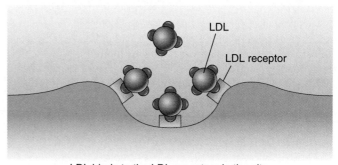

LDL binds to the LDL receptors in the pits.

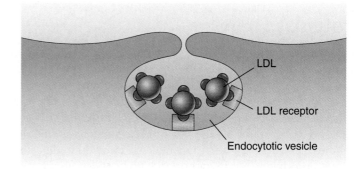

The LDL, bound to LDL receptors, is taken into the cell by endocytosis.

Figure 24.29   Transport of LDL into Cells

Water

About 9 L of **water** enters the digestive tract each day, of which about 92% is absorbed in the small intestine, and another 6%–7% is absorbed in the large intestine (figure 24.31). Water can move in either direction across the wall of the small intestine. The direction of its diffusion is determined by osmotic gradients across the epithelium. When the chyme is dilute, water is absorbed by osmosis across the intestinal wall into the blood. When the chyme is very concentrated and contains very little water, water moves by osmosis into the lumen of the small intestine. As nutrients are absorbed in the small intestine, its osmotic pressure decreases; as a consequence, water moves from the intestine into the surrounding extracellular fluid. Water in the extracellular fluid can then enter the circulation. Because of the osmotic gradient produced as nutrients are absorbed in the small intestine, nearly 90% of the water that enters the small intestine by way of the stomach or intestinal secretions is reabsorbed.

Ions

Active transport mechanisms for **sodium** ions are present within the epithelial cells of the small intestine. **Potassium, calcium, magnesium,** and **phosphate** are also actively transported. **Chloride** ions move passively through the intestinal wall of the duodenum and the jejunum following the positively charged sodium ions, but chloride ions are actively transported from the ileum. Although calcium ions are actively transported along the entire length of the small intestine, vitamin D is required for that transport process. The absorption of calcium is under hormonal control, as is its excretion and storage. Parathyroid hormones, calcitonin, and vitamin D all play a role in regulating blood levels of calcium in the circulatory system (see chapters 6, 18, and 27).

1. Amino acids are absorbed by secondary active transport into intestinal epithelial cells.

2. Amino acids move out of intestinal epithelial cells by active transport.

3. They enter the capillaries of the intestinal villi, and are carried through the hepatic portal vein to the liver.

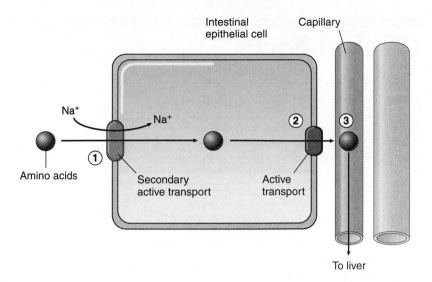

Figure 24.30   Mechanisms of Amino Acid Absorption

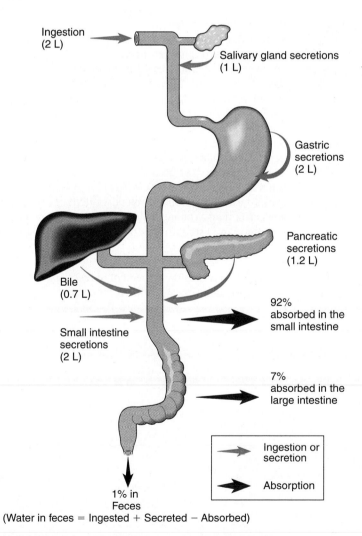

Figure 24.31   Fluid Volumes in the Digestive Tract

Clinical Focus   Intestinal Disorders

Malabsorption Syndrome

Malabsorption syndrome (sprue) is a spectrum of disorders of the small intestine that result in abnormal nutrient absorption. One type of malabsorption results from an immune response to gluten, which is present in certain types of grains and involves the destruction of newly formed epithelial cells in the intestinal glands. These cells fail to migrate to the villi surface, the villi become blunted, and the surface area decreases. As a result, the intestinal epithelium is less capable of absorbing nutrients. Another type of malabsorption (called tropical malabsorption) is apparently caused by bacteria, although no specific bacterium has been identified.

Enteritis

Enteritis is any inflammation of the intestines that can result in diarrhea, dehydration, fatigue, and weight loss. It may result from an infection, chemical irritation, or from some unknown cause. **Regional enteritis,** or Crohn's disease, is a local enteritis of unknown cause characterized by patchy, deep ulcers developing in the intestinal wall, usually in the distal end of the ileum. The disease results in overproliferation of connective tissue and invasion of lymphatic tissue into the involved area, with a subsequent thickening of the intestinal wall and narrowing of the lumen.

Colitis is an inflammation of the colon.

Colon Cancer

Colon cancer is the second leading cause of cancer-related deaths in the United States, accounting for 55,000 deaths a year. Susceptibility to colon cancer can be familial; however, there is a correlation between colon cancer and diets low in fiber and high in fat. People who eat beef, pork, or lamb daily have 2.5 times the risk of developing colon cancer compared with people who eat these meats less than once per month. Eating processed meats increases the risk by an additional 50%–100%. Ingesting calcium in the form of calcium carbonate antacid tablets at twice the recommended daily allowances may prevent 75% of colon cancers. Greatly increased calcium levels may also cause constipation.

A gene for colon cancer may be present in as many as 1 in 200 people, making colon cancer one of the most common inherited diseases. Nine different genes have been found to be associated with colon cancer. Most of those genes are involved in cell regulation, that is, keeping cell growth in check, but one gene mutation results in a high degree of genetic instability. As a result of this mutation, the DNA is not copied accurately during cell division of the colon cancer cells, causing wholesale errors and mutations throughout the genome (all the genes). Such genetic instability has been identified in 13% of sporadic (not occurring in families) colon cancer. Screening for colon cancer includes testing the stool for blood content and performing a colonoscopy, which allows the physician to see into the colon.

Constipation

Constipation is the slow movement of feces through the large intestine. The feces often become dry and hard because of increased fluid absorption during the extended time they are retained in the large intestine. In the United States, there are 2.5 million doctor visits each year from people complaining of constipation, and $400 million dollars is spent each year on laxatives.

Constipation often results after a prolonged time of inhibiting normal defecation reflexes. A change in habits, such as travel, dehydration, depression, disease, metabolic disturbances, certain medications, pregnancy, or dependency on laxatives can all cause constipation. Irritable bowel syndrome, also called spastic colon, which is of unknown cause but is stress-related, can also cause constipation. Constipation can also occur with diabetes, kidney failure, colon nerve damage, or spinal cord injuries or as the result of an obstructed bowel; of greatest concern, the obstruction could be caused by colon cancer. Chronic constipation can result from the slow movement of feces through the entire colon, in just the distal part (descending colon and rectum), or in just the rectum. Interestingly, in one large study of people who claimed to be suffering from chronic constipation, one-third were found to have normal movement of feces through the large intestine. Defecation frequency was often normal. Many of those people were suffering from psychologic distress, anxiety, or depression and just thought they had abnormal defecation frequencies.

Systems Pathology

diarrhea

DIARRHEA

While on vacation in Mexico, Mr. T was shopping with his wife, when he started to experience sharp pains in his abdominal region (figure B). He also began to feel hot and sweaty, and felt an extreme urge to defecate. His wife quickly looked up the word toilet in their handy Spanish–English pocket travel dictionary, and Mr. T anxiously inquired of a local resident where the nearest facility could be found. Once the immediate need was taken care of, Mr. and Mrs. T went back to their hotel room, where they remained while Mr. T recovered. During the next 2 days his stools were frequent and watery. He also vomited a couple of times during the next day. As they were in a foreign country, Mr. T did not consult a physician. Mr. T rested, took plenty of fluids, and was feeling much better, although a little weak, in a couple of days.

BACKGROUND

Diarrhea is one of the most common complaints in clinical medicine and affects more than half of the tourists in developing countries. **Diarrhea** is defined as any change in bowel habits in which stool frequency or volume is increased or in which stool fluidity is increased. Diarrhea is not itself a disease, but is a symptom of a wide variety of disorders. Normally, about 600 mL of fluid enters the colon each day and all but 150 mL is reabsorbed. The loss of more than 200 mL of stool per day is considered abnormal.

Mucus secretion by the colon increases dramatically in response to diarrhea. This mucus contains large quantities of bicarbonate ions, which comes from the dissociation of carbonic acid into bicarbonate ions (HCO_3^-) and hydrogen (H^+) ions within the blood supply to the colon. The HCO_3^- ions enter the mucus secreted by the colon, whereas the H^+ ions remain in the circulation and, as a result, the blood pH decreases. Thus, a condition called metabolic acidosis can develop (see chapter 27).

Diarrhea in tourists usually results from the ingestion of food or water contaminated with bacteria or bacterial toxins. Acute diarrhea is defined as lasting less than 2–3 weeks, and diarrhea lasting longer than that is considered chronic. Acute diarrhea is usually self-limiting, but some forms of diarrhea can be fatal if not treated. Diarrhea results from either a decrease in fluid absorption in the gut or an increase in fluid secretion. Some bacterial toxins and other chemicals can also cause an increase in bowel motor activity. As a result, chyme is moved more rapidly through the digestive tract, fewer nutrients and water are absorbed out of the small intestine, and more water enters the colon. Symptoms can occur in as little as 1–2 h after bacterial toxins are ingested to as long as 24 h or more for some strains of bacteria.

In cases of short-term acute diarrhea, the infectious agent is seldom identified. Nearly any bacterial species is capable of causing diarrhea. Some types of bacterial diarrhea include severe vomiting, whereas others do not. Some bacterial toxins also induce fever. Some viruses and amebic parasites can also cause diarrhea. In most cases, laboratory analysis of food or stool is necessary to identify the causal organism. In cases of mild diarrhea away from home, laboratory evaluation is not practical, and empiric therapy is usually applied. Fluids and electrolytes must be replaced, and consumption of fluids with electrolytes is important. The diet should be limited to clear fluids during at least the first day or so. Bismuth subsalicylate (Pepto-Bismol) or loperamide (Imodium; except in cases of fever) may also be used to help combat secretory diarrhea. Milk and milk products should be avoided. Breads, toast, rice, and baked fish or chicken can be added to the diet with improvement. A normal diet can be resumed after 2–3 days.

7   P R E D I C T

Predict the effects of prolonged diarrhea.

✔ *Answer in Appendix F*

Figure B Some tourists develop diarrhea by ingesting contaminated food or water.

System Interactions

System	Interactions with Digestive System
Integumentary	Pallor occurs due to vasoconstriction of blood vessels in the skin, resulting from a decrease in blood fluid levels. Pallor and sweating increase in response to abdominal pain and anxiety.
Muscular	Muscular weakness may result due to electrolyte loss, metabolic acidosis, fever, and general malaise. The involuntary stimulus to defecate may become so strong as to overcome the voluntary control mechanisms.
Nervous	Local reflexes in the colon respond to increased colon fluid volume by stimulating mass movements and the defecation reflex. Abdominal pain, much of which is felt as referred pain, can occur as the result of inflammation and distention of the colon. Decreased function is due to electrolyte loss. Reduced blood fluid levels stimulate a sensation of thirst in the CNS.
Endocrine	A decrease in extracellular fluid volume, due to the loss of fluid in the feces, stimulates the release of hormones (antidiuretic hormone from the posterior pituitary and aldosterone from the adrenal cortex) that increase water retention and electrolyte reabsorption in the kidney. In addition, decreased extracellular fluid volume and anxiety result in increased release of epinephrine and norepinephrine from the adrenal medulla.
Cardiovascular	Movement of extracellular fluid into the colon results in a decreased blood volume. The reduced blood volume activates the baroreceptor reflex, antidiuretic hormone release, the renin-angiotensin-aldosterone mechanism, and the fluid shift mechanism, which all function to elevate blood volume or increase blood pressure.
Lymphatic and Immune	White blood cells migrate to the colon in response to infection and inflammation. In the case of bacterial diarrhea, the immune response is initiated to begin production of antibodies against bacteria and bacterial toxins.
Respiratory	As the result of reduced blood pH, the rate of respiration increases to eliminate carbon dioxide, which helps eliminate excess H^+ ions.
Urinary	A decrease in urine volume and an increase in urine concentration results from activation of the baroreceptor reflex, which decreases blood flow to the kidney; antidiuretic hormone secretion, which increases water reabsorption in the kidney; and aldosterone secretion, which increases electrolyte and water reabsorption in the kidney. After a period of approximately 24 h the kidney is activated to compensate for metabolic acidosis by increasing hydrogen ion secretion and bicarbonate ion reabsorption.

Summary

The digestive system provides the body with water, electrolytes, and other nutrients.

General Overview

The digestive system consists of a digestive tube and its associated accessory organs.

Anatomy Overview

1. The digestive system consists of the oral cavity, pharynx, esophagus, stomach, small intestine, large intestine, and anus.
2. Accessory organs such as the salivary glands, liver, gallbladder, and pancreas are located along the digestive tract.

Histology Overview

The digestive tract is composed of four tunics: mucosa, submucosa, muscularis, and serosa or adventitia.

Mucosa

The mucosa consists of a mucous epithelium, a lamina propria, and a muscularis mucosae.

Submucosa

The submucosa is a connective tissue layer containing nerves, blood vessels, and small glands.

Muscularis

1. The muscularis consists of an inner layer of circular smooth muscle and an outer layer of longitudinal smooth muscle.
2. The myenteric plexus is between the two muscle layers.

Serosa or Adventitia

The serosa or adventitia forms the outermost layer of the digestive tract.

Physiology Overview

1. The functions of the digestive system are ingestion, mastication, propulsion, mixing, secretion, digestion, absorption, transportation, elimination, and regulation.
2. The functions of the digestive system are regulated by elaborate nervous and hormonal mechanisms.

Ingestion

Ingestion is the introduction of food into the stomach.

Mastication

1. Mastication is the chewing of food and its mechanical digestion.
2. Additional mechanical digestion occurs as food is mixed in the stomach and intestines.

Propulsion

Propulsion is the movement of food from one end of the digestive tract to the other.

Mixing

Some peristaltic contractions mix food with digestive secretions and break it down.

Secretion

Secretions are added to lubricate, liquefy, and digest the food as it moves through the digestive tract.

Digestion

1. Digestion is the breakdown of organic molecules into their component parts.
2. Digestion consists of mechanical digestion and chemical digestion.

Absorption

Absorption is the means by which molecules are moved out of the digestive tract and into the circulation or extracellular spaces.

Transportation

Transportation is the means by which molecules are distributed throughout the body.

Elimination

Elimination is the means by which the waste products of digestion are removed from the body by defecation.

Regulation

1. The processes of propulsion, secretion, absorption, and elimination are regulated by nervous and hormonal mechanisms.
2. Nervous control occurs through local reflexes and through the vagus nerve.

Anatomy and Histology of the Digestive Tract
Oral Cavity

1. The lips and checks are involved in facial expression, mastication, and speech.
2. The tongue is involved in speech, taste, mastication, and swallowing.
   - The intrinsic tongue muscles change the shape of the tongue, and the extrinsic tongue muscles move the tongue.
   - The anterior two-thirds of the tongue is covered with papillae, the posterior one-third is devoid of papillae.
3. There are 20 deciduous teeth that are replaced by 32 permanent teeth.
   - The types of teeth are incisors, canines, premolars, and molars.

- A tooth consists of a crown, a neck, and a root.
- The root is composed of dentin. Within the dentin of the root is the pulp cavity, which is filled with pulp, blood vessels, and nerves. The crown is dentin covered by enamel.
- Teeth are held in the alveoli by the periodontal ligaments.

4. The muscles of mastication are the masseter, the temporalis, the medial pterygoid, and the lateral pterygoid.
5. The roof of the oral cavity is divided into the hard and soft palates.
6. Salivary glands produce serous and mucous secretions. The three pairs of large salivary glands are the parotid, submandibular, and sublingual.

Pharynx

The pharynx consists of the nasopharynx, oropharynx, and laryngopharynx.

Esophagus

1. The esophagus connects the pharynx to the stomach. The upper and lower esophageal sphincters regulate movement.
2. The esophagus consists of an outer adventitia, a muscular layer (longitudinal and circular), a submucosal layer (with mucous glands), and a stratified squamous epithelium.

Stomach

1. Structures of the stomach.
   - The openings of the stomach are the gastroesophageal (to the esophagus) and the pyloric (to the duodenum).
   - The wall of the stomach consists of an external serosa, a muscle layer (longitudinal, circular, and oblique), a submucosa, and simple columnar epithelium (surface mucous cells).
   - Rugae are the folds in the stomach when it is empty.
2. Gastric pits are the openings to the gastric glands that contain mucous neck cells, parietal cells, chief cells, and endocrine cells.

Small Intestine

1. The small intestine is divided into the duodenum, jejunum, and ileum.
2. The wall of the small intestine consists of an external serosa, muscles (longitudinal and circular), submucosa, and simple columnar epithelium.
3. Circular folds, villi, and microvilli greatly increase the surface area of the intestinal lining.
4. Absorptive, goblet, and endocrine cells are in intestinal glands. Duodenal glands produce mucus.

Liver

1. The liver has four lobes: right, left, caudate, and quadrate.
2. The liver is divided into lobules.
   - The hepatic cords are composed of columns of hepatocytes that are separated by the bile canaliculi.
   - The sinusoids are enlarged spaces filled with blood and lined with endothelium and hepatic phagocytic cells.
3. The portal triads supply the lobules.
   - The hepatic arteries and the hepatic portal veins bring blood to the lobules and empty into the sinusoids.
   - The sinusoids empty into central veins, which join to form the hepatic veins, which leave the liver.
   - Bile canaliculi converge to form hepatic ducts, which leave the liver.

4. Bile leaves the liver through the hepatic duct system.
   - The hepatic ducts receive bile from the lobules.
   - The cystic duct from the gallbladder joins the hepatic duct to form the common bile duct.
   - The common bile duct joins the pancreatic duct at the point at which it empties into the duodenum.

Gallbladder

The gallbladder is a small sac on the inferior surface of the liver.

Pancreas

1. The pancreas is an endocrine and an exocrine gland. Its exocrine function is the production of digestive enzymes.
2. The pancreas is divided into lobules that contain acini. The acini connect to a duct system that eventually forms the pancreatic duct, which empties into the duodenum.

Large Intestine

1. The cecum forms a blind sac at the junction of the small and large intestines. The vermiform appendix is a blind tube off the cecum.
2. The ascending colon extends from the cecum superiorly to the right colic flexure. The transverse colon extends from the right to the left colic flexure. The descending colon extends inferiorly to join the sigmoid colon.
3. The sigmoid colon is an S-shaped tube that ends at the rectum.
4. Longitudinal smooth muscles of the large intestine wall are arranged into bands called teniae coli that contract to produce pouches called haustra.
5. The mucosal lining of the large intestine is simple columnar epithelium with mucus-producing crypts.
6. The rectum is a straight tube that ends at the anus.
7. The anal canal is surrounded by an internal anal sphincter (smooth muscle) and an external anal sphincter (skeletal muscle).

Peritoneum

1. The peritoneum is a serous membrane that lines the abdominal cavity and organs.
2. Mesenteries are peritoneum that extends from the body wall to many of the abdominal organs.
3. Retroperitoneal organs are located behind the peritoneum.

Functions of the Digestive System

The digestive system is regulated by neural and hormonal mechanisms. The intramural plexus is responsible for local reflexes.

Functions of the Oral Cavity

1. Amylase in saliva starts starch digestion. Mucin provides lubrication.
2. Chewing is primarily a reflex activity. The teeth cut, tear, and crush the food.

Deglutition

1. During the voluntary phase of deglutition, a bolus of food is moved by the tongue from the oral cavity to the pharynx.
2. The pharyngeal phase is a reflex caused by stimulation of stretch receptors in the pharynx.
   - The soft palate closes the nasopharynx, and the epiglottis closes the opening into the larynx.
   - Pharyngeal muscles move the bolus to the esophagus.

3. The esophageal phase is a reflex initiated by the stimulation of stretch receptors in the esophagus. A wave of contraction (peristalsis) moves the food to the stomach.

Stomach Functions

1. Stomach secretions
   - Mucus protects the stomach lining.
   - Pepsinogen is converted to pepsin, which digests proteins.
   - Hydrochloric acid promotes pepsin activity and kills microorganisms.
   - Intrinsic factor is necessary for vitamin B_{12} absorption.
2. Regulation of stomach secretions.
   - The cephalic phase is initiated by the sight, smell, taste, or thought of food. Nerve impulses from the medulla stimulate hydrochloric acid, pepsinogen, and gastrin secretion.
   - The gastric phase is initiated by distention of the stomach, which stimulates gastrin secretion and activates central nervous system and local reflexes that promote secretion.
   - The intestinal phase is initiated by acidic chyme, which enters the duodenum and stimulates neuronal reflexes and the secretion of hormones that induce and then inhibit gastric secretions.
3. Movement in the stomach
   - The stomach stretches and relaxes to increase volume.
   - Mixing waves mix the stomach contents with stomach secretions to form chyme.
   - Peristaltic waves move the chyme into the duodenum.
4. Regulation of stomach emptying
   - Gastrin and stretching of the stomach stimulate stomach emptying.
   - Chyme entering the duodenum inhibits movement through neuronal reflexes and the release of hormones.

Functions of the Small Intestine

1. Secretions of the small intestine
   - Mucus protects against digestive enzymes and stomach acids.
   - Digestive enzymes (disaccharidases and peptidases) are bound to the intestinal wall.
   - Chemical or tactile irritation, vagal stimulation, and secretin stimulate intestinal secretion.
2. Movement in the small intestine
   - Segmental contractions mix intestinal contents. Peristaltic contractions move materials distally.
   - Stretch of smooth muscles, local reflexes, and the parasympathetic nervous system stimulate contractions. Distention of the cecum initiates a reflex that inhibits peristalsis.

Liver Functions

1. The liver produces bile, which contains bile salts that emulsify fats. Secretin and parasympathetic stimulation increase bile production.
2. The liver stores and processes nutrients, produces new molecules, and detoxifies molecules.
3. The liver produces blood components.

Functions of the Gallbladder

1. The gallbladder stores and concentrates bile.
2. Cholecystokinin stimulates gallbladder contraction.

Functions of the Pancreas

1. Secretin stimulates the release of a watery bicarbonate solution that neutralizes acidic chyme.
2. Cholecystokinin stimulates the release of digestive enzymes.

Functions of the Large Intestine

1. Secretion and absorption.
   - Mucus provides protection to the intestinal lining.
   - Bicarbonate ions are secreted by epithelial cells. Sodium is absorbed by active transport, and water is absorbed by osmosis.
2. Microorganisms are responsible for vitamin K production, gas production, and much of the bulk of feces.
3. Movement in the large intestine
   - Segmental movements mix the colon's contents.
   - Mass movements are strong peristaltic contractions that occur three to four times a day.
   - Defecation is the elimination of feces. Reflex activity moves feces through the internal anal sphincter. Voluntary activity regulates movement through the external anal sphincter.

Digestion, Absorption, and Transport

1. Digestion is the breakdown of organic molecules into their component parts.
2. Absorption and transport are the means by which molecules are moved out of the digestive tract and are distributed throughout the body.
3. Transportation occurs by two different routes.
   - Water, ions, and water-soluble products of digestion are transported to the liver through the hepatic portal system.
   - The products of lipid metabolism are transported through the lymphatic system to the circulatory system.

Carbohydrates

1. Carbohydrates consist of starches, glycogen, sucrose, lactose, glucose, and fructose.
2. Polysaccharides are broken down into monosaccharides by a number of different enzymes.
3. Monosaccharides are taken up by intestinal epithelial cells by active transport or by facilitated diffusion.
4. The monosaccharides are carried to the liver where the nonglucose sugars are converted to glucose.
5. Glucose is transported to the cells that require energy.
6. Glucose enters the cells through facilitated diffusion.
7. The rate of transport is influenced by insulin.

Lipids

1. Lipids include triacylglycerol, phospholipids, steroids, and fat-soluble vitamins.
2. Emulsification is the transformation of large lipid droplets into smaller droplets and is accomplished by bile salts.
3. Lipase digests lipid molecules to form free fatty acids and glycerol.
4. Micelles form around lipid digestion products and move to epithelial cells of the small intestine where the products pass into the cells by simple diffusion.
5. Within the epithelial cells, free fatty acids are combined with glycerol to form triacylglycerol.
6. Proteins coat triacylglycerol, phospholipids, and cholesterol to form chylomicrons.
7. Chylomicrons enter lacteals within intestinal villi and are carried through the lymphatic system to the bloodstream.
8. Triacylglycerol is stored in adipose tissue, converted into other molecules, or used as energy.
9. Lipoproteins include chylomicrons, VLDL, LDL, and HDL.
10. Cholesterol is transported to cells by LDL and from cells to the liver by HDL.
11. LDLs are taken into cells by receptor-mediated endocytosis, which is controlled by a negative-feedback mechanism.

Proteins

1. Pepsin in the stomach breaks proteins into smaller polypeptide chains.
2. Proteolytic enzymes from the pancreas produce small peptide chains.
3. Peptides are broken down by peptidases bound to the microvilli of the small intestine.
4. Amino acids are absorbed by cotransport, which requires transport of sodium.
5. Amino acids are transported to the liver, where the amino acids can be modified or released into the bloodstream.
6. Amino acids are actively transported into cells under the stimulation of growth hormone and insulin.
7. Amino acids are used as building blocks or for energy.

Water

Water can move in either direction across the wall of the small intestine, depending on the osmotic gradients across the epithelium.

Ions

1. Sodium, potassium, calcium, magnesium, and phosphate are actively transported.
2. Chloride ions move passively through the wall of the duodenum and jejunum but are actively transported from the ileum.
3. Calcium ions are actively transported, but vitamin D is required for transport, and the transport is under hormonal control.

Content Review

1. List the major digestive organs.
2. What are the major layers of the digestive tract? How do the serosa and the adventitia differ?
3. What are the general functions of the digestive system?
4. What are the functions of the lips and cheeks?
5. List the functions of the tongue. Distinguish between intrinsic and extrinsic tongue muscles.
6. What are deciduous and permanent teeth? Name the different kinds of teeth.
7. Describe the parts of a tooth. What are dentin, enamel, cementum, and pulp?
8. List the muscles of mastication and the actions they produce.
9. What are the hard and the soft palates?

10. Name and give the location of the three largest salivary glands. Name the other kinds of salivary glands. What is the difference between serous and mucous saliva?

11. Name the three parts of the pharynx.

12. Where is the esophagus located? Describe the layers of the esophageal wall and the esophageal sphincters.

13. Describe the parts of the stomach. List the layers of the stomach wall. How is the stomach different from the esophagus?

14. What are gastric pits and gastric glands? Name the different cell types in the stomach and the secretions they produce.

15. Name and describe the three parts of the small intestine. What are the greater and lesser duodenal papilla?

16. What are circular folds, villi, and microvilli in the small intestine? What are their functions?

17. What is the function of the ileocecal sphincter?

18. What are the hepatic cords and the sinusoids?

19. Describe the flow of blood to and through the liver. Describe the flow of bile away from the liver.

20. What kind of gland is the pancreas? Describe the acini and the duct system of the pancreas.

21. Describe the parts of the large intestine. What are teniae coli, haustra, and crypts?

22. What are the peritoneum, the mesenteries, and the retroperitoneal organs?

23. What are the functions of saliva? How is salivary secretion regulated?

24. Describe the mastication reflex and the three stages of swallowing.

25. List five stomach secretions, and give their functions.

26. Name the three stages of stimulation of gastric secretions, and discuss the cause and result of each stage.

27. How are gastric secretions inhibited? Why is this inhibition necessary?

28. Why does pressure in the stomach not greatly increase as the stomach fills?

29. What are the two kinds of stomach movements? How are stomach movements regulated by hormones and nervous control?

30. List the enzymes of the small intestine wall, and state their functions.

31. What are the two kinds of movements of the small intestine? How are they regulated?

32. Describe the functions of the liver and the gallbladder. What stimulates the release of bile from the liver and the gallbladder?

33. Name the two kinds of exocrine secretions that are produced by the pancreas. Where are they produced, what stimulates their production, and what is their function?

34. What kinds of movements occur in the colon? Describe the defecation reflex.

35. Name the substances secreted and absorbed by the colon. What is the role of microorganisms in the colon?

36. Describe the mechanism of absorption and the route of transport for water-soluble and lipid-soluble molecules.

37. Describe the enzymatic digestion of carbohydrates, lipids, and proteins, and list the breakdown products of each.

38. Explain the function of LDLs and HDLs.

39. Describe the movement of water through the intestinal wall.

40. Where and how are the various ions absorbed?

Develop Your Reasoning Skills

1. While anesthetized, patients sometimes vomit. Given that the anesthetic eliminates the swallowing reflex, explain why it is dangerous for an anesthetized patient to vomit.

2. Achlorhydria is a condition in which the stomach stops producing hydrochloric acid and other secretions. What effect would achlorhydria have on the digestive process? On red blood cell count?

3. Victor Worrystudent experienced the pain of a duodenal ulcer during final examination week. Describe the possible reasons. Explain what habits could have caused the ulcer, and recommend a reasonable remedy.

4. Gallstones sometimes obstruct the common bile duct. What is the consequences of such a blockage?

5. A patient has a spinal cord injury at level L2 of the spinal cord. How will this injury affect his ability to defecate? What components of his defecation response are still present, and which are lost?

6. The bowel (colon) occasionally can become impacted. Given what you know about the functions of the colon and the factors that determine the movement of substances across the colon wall, predict the effect of the impaction on the contents of the colon above the point of impaction.

Web Site Link

For a listing of the most current web sites related to this chapter, please visit the Seeley home page at:
http://www.mhhe.com/biosci/ap/seeleyap/

Chapter Twenty-Five

Nutrition, Metabolism, and Temperature Regulation

Objectives

1. Define the term nutrition.

2. Define the term kilocalorie, and list the kilocalories supplied by a gram each of carbohydrate, lipid, and protein.

3. Describe for carbohydrates, lipids, and proteins their dietary sources, their uses in the body, and the daily recommended amounts of each in the diet.

4. List the vitamins, and indicate the function of each.

5. List the most common minerals, and indicate the major function of each.

6. Describe the basic steps in glycolysis, and indicate its major products.

7. Describe the citric acid cycle and its major products.

8. Describe the electron-transport chain and how ATP is produced in the process.

9. Explain how 2 ATP molecules are produced in anaerobic respiration and 38 ATP molecules are produced in aerobic respiration from one molecule of glucose.

10. Describe the basic steps involved in using lipids as an energy source.

11. Explain how amino acids can be used for energy.

12. Describe the conversion of lipids and protein into glucose and the conversion of glucose into glycogen.

13. Differentiate between the absorptive and postabsorptive metabolic states.

14. Define the term metabolic rate.

15. Describe heat production and regulation in the body.

Part Four

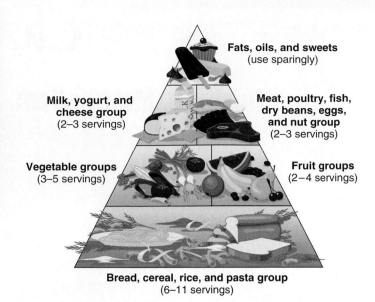

Figure 25.1 Food Guide Pyramid

The pyramid suggests three approaches to a healthy diet: eat different amounts of foods from each basic food group, use fats and sugars sparingly, and choose variety by eating the indicated number of servings per day of the different foods from each major food group.

Fats, oils, and sweets (use sparingly)

Milk, yogurt, and cheese group (2–3 servings)

Meat, poultry, fish, dry beans, eggs, and nut group (2–3 servings)

Vegetable groups (3–5 servings)

Fruit groups (2–4 servings)

Bread, cereal, rice, and pasta group (6–11 servings)

"What's for dinner?" When we ask this question we are usually more concerned with the taste of the food than with the nutritional value of the meal. But the food we love to eat provides us with energy and the building blocks necessary to synthesize new molecules. Because we literally are what we eat, we need to know about nutrition. What happens if we don't obtain enough vitamins, or if we eat too much sugar and fats? Health claims about foods and food supplements bombard us every day. Which ones are ridiculous, and which ones have merit? A basic understanding of nutrition can help us to answer these and other questions so that we can develop a healthy diet.

Nutrition

Nutrition is the process by which certain components of food are obtained and used by the body. The process includes digestion, absorption, transportation, and cell metabolism. Nutrition can also be defined as the evaluation of food and drink requirements for normal body function.

Nutrients

Nutrients are the chemicals taken into the body that are used to produce energy, provide building blocks for new molecules, or function in other chemical reactions. Some substances in food, such as nondigestible plant fibers, are not nutrients. Nutrients can be divided into six major classes: carbohydrates, proteins, lipids, vitamins, minerals, and water. Carbohydrates, proteins, and lipids are the major organic nutrients and are broken down by enzymes into their individual components during digestion. Many of these subunits are broken down further to supply energy, whereas others are used as building blocks for other macromolecules. Carbohydrates, proteins, lipids, and water are required in fairly substantial quantities, whereas vitamins and minerals are required in only small amounts. Vitamins, minerals, and water are taken into the body without being digested.

Essential nutrients are nutrients that must be ingested because the body cannot manufacture them or is unable to manufacture adequate amounts of them. The essential nutrients include certain amino acids, certain fatty acids, most vitamins, minerals, water, and a minimum amount of carbohydrates. The term essential does not mean, however, that only the essential nutrients are required by the body. Other nutrients are necessary, but, if they are not part of the diet, they can be synthesized from other ingested nutrients. Most of this synthesis takes place in the liver, which has a remarkable ability to transform and manufacture molecules.

The U.S. Department of Agriculture provides directions for obtaining the proper amounts of carbohydrates, lipids, proteins, vitamins, minerals, and fiber in the form of a "food guide pyramid" (figure 25.1). The six food groups shown in the pyramid are (1) grains; (2) vegetables; (3) fruits; (4) dairy products; (5) meat, poultry, fish, dry beans, eggs, and nuts; and (6) fats, oils, and sweets. The shape of the pyramid suggests that grains, vegetables, and fruits should be the main part of the diet. Fats, oils, and sweets can be used in moderation to improve the flavor of foods. A balanced diet includes a variety of foods from each of the major food groups. Variety is necessary because no one food contains all the nutrients necessary for good health.

Kilocalories

The energy stored within the chemical bonds of certain nutrients can be used by the body. A **calorie** (kal´ō-rē) **(cal)** is the amount of energy (heat) necessary to raise the temperature of 1 g of water 1°C. A **kilocalorie** (kil´ō-kal-ō-rē) **(kcal)** is 1000 calories and is used to express the larger amounts of energy supplied by foods and released through metabolism.

> **Clinical Note**
>
> A kilocalorie is often called a Calorie (with a capital "C"). Unfortunately, this usage has resulted in confusion between the term calorie (with a lowercase "C") and Calorie (with a capital "C"). It is common practice on food labels and in nutrition books to use calorie when Calorie (kilocalorie) is the proper term.

Almost all of the kilocalories supplied by food come from carbohydrates, proteins, or fats. For each gram of carbohydrate or protein metabolized by the body, about 4 kcal of energy is

Table 25.1   Food Composition

Food	Quantity	Food Energy (kcal)	Carbohydrate (g)	Fat (g)	Protein (g)
Dairy Products					
Whole milk (3.3% fat)	1 cup	150	11	8	8
Low fat milk (2% fat)	1 cup	120	12	5	8
Butter	1 T	100	—	12	—
Grain					
Bread, white enriched	1 slice	75	24	1	2
Bread, whole wheat	1 slice	65	14	1	3
Fruit					
Apple	1	80	20	1	—
Banana	1	100	26	—	1
Orange	1	65	16	—	1
Vegetables					
Corn, canned	1 cup	140	33	1	4
Peas, canned	1 cup	150	29	1	8
Lettuce	1 cup	5	2	—	—
Celery	1 cup	20	5	—	1
Potato, baked	1 large	145	33	—	4
Meat, Fish, and Poultry					
Lean ground beef (10% fat)	3 oz	185	—	10	23
Shrimp, french fried	3 oz	190	9	9	17
Tuna, canned	3 oz	170	—	7	24
Chicken breast, fried	3 oz	160	1	5	26
Bacon	2 slices	85	—	8	4
Hot dog	1	170	1	15	7

continued next page

released. Fats contain more energy per unit of weight than carbohydrates and proteins and yield about 9 kcal/g. Table 25.1 lists the kilocalories supplied by some typical foods. A typical American diet consists of 50%–60% carbohydrates, 35%–45% fats, and 10%–15% protein. Table 25.1 also lists the carbohydrate, fat, and protein composition of some foods.

Carbohydrates
Sources in the Diet

Carbohydrates include monosaccharides, disaccharides, and polysaccharides (see chapter 2). Most of the carbohydrates humans ingest come from plants. An exception is lactose (milk sugar), which is found in animal and human milk.

The most common monosaccharides in the diet are glucose and fructose. Plants capture the energy in sunlight and use the energy to produce glucose, which can be found in vegetables. Fructose (fruit sugar), an isomer of glucose (see figure 2.13), is most often derived from fruits and berries.

The disaccharide sucrose (table sugar) is what most people think of when they use the term sugar. Sucrose is a glucose and a fructose molecule joined together, and its principal sources are sugarcane, sugar beets, maple sugar, and honey. Maltose (malt sugar), derived from germinating cereals, is a combination of two glucose molecules, and lactose (in milk) consists of a glucose and a galactose molecule (see figure 2.13).

The **complex carbohydrates** are the polysaccharides: starch, glycogen, and cellulose. These polysaccharides consist of many glucose molecules bound together to form long chains. Starch is an energy storage molecule in plants and is found primarily in vegetables, fruits, and grains. Glycogen is an energy storage molecule in animals and is located in muscle and in the liver. By the time meats are processed and cooked, they contain little, if any, glycogen. Cellulose forms cell walls, which surround plant cells.

Table 25.1   Food Composition—cont'd

Food	Quantity	Food Energy (kcal)	Carbohydrate (g)	Fat (g)	Protein (g)
Fast Foods					
McDonald's Egg McMuffin	1	327	31	15	19
McDonald's Big Mac	1	563	41	33	26
Taco Bell's beef burrito	1	466	37	21	30
Arby's roast beef	1	350	32	15	22
Pizza Hut Super Supreme	1 slice	260	23	13	15
Long John Silver's fish	2 pieces	366	21	22	22
McDonald's fish fillet	1	432	37	25	14
Dairy Queen malt, large	1	840	125	28	22
Desserts					
Cupcake with icing	1	130	21	5	2
Chocolate chip cookie	4	200	29	9	2
Apple pie	1 piece	345	51	15	3
Dairy Queen cone, large	1	340	52	10	10
Beverage					
Cola soft drink	12 oz	145	37	–	–
Beer	12 oz	144	13	–	1
Wine	3½ oz	73	2	–	–
Hard liquor (86 proof)	1½ oz	105	–	–	–
Miscellaneous					
Egg	1	80	1	6	6
Mayonnaise	1 T	100	–	11	–
Sugar	1 T	45	12	–	–

Uses in the Body

During digestion, polysaccharides and disaccharides are split into monosaccharides, which are absorbed into the blood (see chapter 24). Humans can break the bonds between the glucose molecules of starch and glycogen, but cellulose is indigestible. Instead, it provides fiber, or "roughage," increasing the bulk of feces and promoting defecation.

Fructose, galactose, and other monosaccharides absorbed into the blood are converted into glucose by the liver. Glucose, whether absorbed from the digestive tract or produced by the liver, is a primary energy source for most cells, which use it to produce **adenosine triphosphate (ATP)** molecules (see sections on Anaerobic Respiration and Aerobic Respiration later in the chapter). Because the brain relies almost entirely on glucose for its energy, blood glucose levels are carefully regulated (see chapter 18).

If excess amounts of glucose are present, the glucose is converted into glycogen that is stored in muscle and in the liver. The glycogen can be rapidly converted back to glucose when energy is needed. Because cells can store only a limited amount of glycogen, any additional glucose is converted into fat that is stored in adipose tissue.

In addition to being used as a source of energy, sugars have other functions. They form part of deoxyribonucleic acid (DNA), ribonucleic acid (RNA), and ATP molecules (see chapter 2); and they combine with proteins to form glycoprotein receptor molecules on the outer surface of the plasma membrane (see chapter 3).

Recommended Amounts

It is recommended that 50–100 g of carbohydrates be ingested every day. Although a minimum acceptable level of carbohydrate ingestion is unknown, consumption of too few carbohydrates per day results in overuse of proteins and fats for energy sources. Because muscles are primarily protein, the use of proteins for energy can result in the breakdown of muscle tissue, and the use of fats can result in acidosis (see chapter 27).

Complex carbohydrates are recommended because starchy foods often contain other valuable nutrients such as vitamins and minerals. Although foods such as soft drinks and candy are rich in carbohydrates, they are mostly sugar and they may have little other nutritive value. For example, a typical soft drink contains 9 teaspoons of sugar. In excess, the consumption of these kinds of foods can result in obesity and tooth decay.

Lipids
Sources in the Diet

About 95% of the lipids in the human diet are **triacylglycerols** (trī-as′il-glis′er-olz). Triacylglycerols, which are sometimes called triglycerides (trī-glis′er-īdz), consist of three fatty acids attached to a glycerol molecule (see chapter 2). Triacylglycerols are often referred to as fats, which can be divided into saturated and unsaturated fats. Fats are saturated if their fatty acids have only single covalent bonds between their carbon atoms, and they are unsaturated if they have one (monounsaturated) or more (polyunsaturated) double covalent bonds between their carbon atoms (see figure 2.16). Saturated fats are found in the fats of meats (e.g., beef, pork), dairy products (e.g., whole milk, cheese, butter), eggs, coconut oil, and palm oil. Monounsaturated fats include olive and peanut oils; and polyunsaturated fats occur in fish, safflower, sunflower, and corn oils.

The remaining 5% of lipids include cholesterol and phospholipids such as **lecithin** (les′i-thin). Cholesterol is a

Clinical Note

Solid fats, such as shortening and margarine, work better than liquid oils for preparing some foods such as pastries. Polyunsaturated vegetable oils can be changed from a liquid to a solid by making them more saturated, that is, by decreasing the number of double covalent bonds in their polyunsaturated fatty acids. Hydrogen gas is bubbled through the oil. As hydrogen binds to the fatty acids, double covalent bonds are converted to single covalent bonds, producing a change in molecular shape that solidifies the oil. The more saturated the product, the harder it becomes at room temperature.

steroid (see chapter 2) found in high concentrations in the brain, the liver, and egg yolks; but it is also present in whole milk, cheese, butter, and meats. Cholesterol is not found in plants. Phospholipids are major components of cell membranes, and they are found in a variety of foods. A good source of lecithin is egg yolks.

Uses in the Body

Triacylglycerols are important sources of energy that can be used to produce ATP molecules. A gram of triacylglycerol delivers more than twice as many kilocalories as a gram of carbohydrate. Some cells such as skeletal muscle cells derive most of their energy from triacylglycerols.

After a meal, excess triacylglycerols that are not immediately used are stored in adipose tissue or the liver. Later, when energy is required, the triacylglycerols are broken down, and their fatty acids are released into the blood, where they can be taken up and used by various tissues. In addition to storing energy, adipose tissue surrounds and pads organs, and under the skin adipose tissue is an insulator, which prevents heat loss.

Cholesterol is an important molecule with many functions in the body. It can either be obtained in food or manufactured by the liver and most other tissues. Cholesterol is a component of the plasma membrane, and it can be modified to form other useful molecules such as bile salts and steroid hormones. Bile salts are necessary for fat digestion and absorption. Steroid hormones include the sex hormones estrogen, progesterone, and testosterone, which regulate the reproductive system.

Prostaglandins, which are derived from fatty acids, are involved in activities such as inflammation, blood clotting, tissue repair, and smooth muscle contraction. Phospholipids such as lecithin are part of the plasma membrane and are used to construct the myelin sheath around the axons of nerve cells.

Recommended Amounts

The American Heart Association recommends that fats account for 30% or less of the total kilocaloric intake. Furthermore, saturated fats should contribute no more than 10% of total fat intake, and cholesterol should be limited to 300 mg (the amount in an egg yolk) or less per day. These guidelines reflect the belief that excess amounts of fats, especially saturated fats and cholesterol, contribute to cardiovascular disease. Evidence also suggests that high fat intake is associated with colon cancer. The typical American diet derives 35%–45% of its kilocalories from fats, indicating that most Americans

Clinical Note

The essential fatty acids can be used to synthesize prostaglandins that affect blood clotting. Linoleic acid can be converted to **arachidonic** (ă-rak-i-don′ik) **acid,** which is used to produce prostaglandins that increase blood clotting. Alpha-linolenic acid can be converted to **eicosapentaenoic** (ī-kō′să-pen-tă-nō′ik) **acid (EPA),** which is used to produce prostaglandins that decrease blood clotting. Normally, most prostaglandins are synthesized from linoleic acid because it is more plentiful in the body. Individuals who consume foods rich in EPA, however, such as herring, salmon, tuna, and sardines, increase the synthesis of prostaglandins from EPA. Individuals who eat these fish twice or more times per week have a lower risk of heart attack than those who don't, probably because of reduced blood clotting. Although EPA can be obtained using fish oil supplements, this is not currently recommended because fish oil supplements contain high amounts of cholesterol, vitamins A and D, and uncommon fatty acids, all of which can cause health problems when taken in large amounts.

need to reduce fat consumption. On the other hand, fat intake can account for as little as 10% of the kilocalories in a healthy person's diet.

Most of the lecithin consumed in the diet is broken down in the digestive tract. The liver has the ability to manufacture all of the lecithin necessary to meet the body's needs, and it is not necessary to consume lecithin supplements.

Linoleic (lin-ō-lē′ik) **acid** and **α-linolenic** (lin-ō-len′ik) **acid** are **essential fatty acids** because the body cannot synthesize them and they must be ingested. They are found in plant oils, such as canola or soybean oils.

Proteins
Sources in the Diet

Proteins are chains of amino acids (see chapter 2). Proteins in the human body are constructed of 20 different kinds of amino acids, which can be divided into two groups. **Essential amino acids** cannot be synthesized by the body and must be obtained in the diet. The nine essential amino acids are histidine, isoleucine, leucine, lysine, methionine, phenylalanine, threonine, tryptophan, and valine. **Nonessential amino acids** can be produced by the body from other molecules. If adequate amounts of the essential amino acids are ingested, they can be used to manufacture the nonessential amino acids, which are also necessary for good health.

A **complete protein** food contains adequate amounts of all nine essential amino acids, whereas an **incomplete protein** food does not. Examples of complete proteins are meat, fish, poultry, milk, cheese, and eggs; and examples of incomplete proteins are leafy green vegetables, grains, and legumes (peas and beans).

Uses in the Body

Proteins perform numerous functions in the human body as the following examples illustrate. Collagen provides structural strength in connective tissue as does keratin in the skin, and the combination of actin and myosin makes muscle contraction possible. Enzymes are responsible for regulating the rate of chemical reactions, and protein hormones regulate many physiologic processes (see chapter 18). Proteins in the blood act as buffers to prevent changes in pH, and hemoglobin transports oxygen and carbon dioxide in the blood. Proteins also function as carrier molecules to move materials across plasma membranes, and other proteins in the plasma membrane function as receptor molecules and ion channels. Antibodies, lymphokines, and complement are part of the immune system response that protects against microorganisms and other foreign substances.

Proteins can also be used as a source of energy, yielding the same amount of energy as carbohydrates. If excess proteins are ingested, the energy in the proteins can be stored by converting their amino acids into glycogen or fats.

Recommended Amounts

The recommended daily consumption of protein for a healthy adult is 0.8 g/kg of body weight, or about 12% of total kilocalories. For a 58-kg (128 pounds) female this is 46 g/day, and for a 70-kg male (154 pounds) it is 56 g/day. A cup of skim milk contains 8 g protein, 1 ounce of meat contains 7 g protein, and a slice of bread provides 2 g protein. If two incomplete proteins, such as rice and beans are ingested, each can provide amino acids lacking in the other. Thus a correctly balanced vegetarian diet can provide all of the essential amino acids.

When protein intake is adequate, the synthesis and breakdown of proteins in a healthy adult occurs at the same rate. The amino acids of proteins contain nitrogen; so saying that a person is in **nitrogen balance** means that the nitrogen content of ingested protein is equal to the nitrogen excreted in urine and feces. A starving person is in negative nitrogen balance because the nitrogen gained in the diet is less than that lost by excretion. In other words, when proteins are broken down for energy, more nitrogen is lost than is replaced in the diet. A growing child or a healthy pregnant woman, on the other hand, is in positive nitrogen balance because more nitrogen is going into the body to produce new tissues than is lost by excretion.

Vitamins

Vitamins (vīt′ă-minz) exist in minute quantities in food and are essential to normal metabolism (table 25.2). Most vitamins cannot be produced by the body and must be obtained through the diet. The absence of a specific vitamin in the diet can result in a specific deficiency disease. A few vitamins such as vitamin K are produced by intestinal bacteria, and a few can be formed by the body from substances called provitamins. A **provitamin** is a part of a vitamin that can be assembled or modified by the body into a functional vitamin. **Carotenoids** (ka-rot′e-noydz) are examples of provitamins that can be modified by the body to form vitamin A. Of the approximately 50 different carotenoids that can function as provitamins, the most important is beta-carotene. The other provitamins are 7-dehydrocholesterol, which can be converted to vitamin D, and tryptophan (trip′tō-fan), which can be converted to niacin.

Vitamins are not broken down by catabolism but are used by the body in their original or slightly modified forms. Once the chemical structure of a vitamin is destroyed, its function is usually lost. The chemical structure of many vitamins is destroyed by heat (e.g., when food is overcooked). Many vitamins function as coenzymes, which combine with enzymes to make the enzymes functional (see chapter 2). Vitamins such as riboflavin, pantothenic acid, niacin, and biotin are critical to the production of energy, whereas folate and vitamin B_{12} are involved in nucleic acid synthesis. Retinol, thiamine, and vitamins C, D, and E are necessary for general growth. Vitamin K is necessary for the synthesis of blood clotting proteins.

Table 25.2 The Principal Vitamins

Vitamin	Fat- (F) or Water- (W) Soluble	Source	Function	Symptoms of Deficiency	Reference Daily Intake*
A (retinoids, carotenoids)	F	From provitamin carotene found in yellow and green vegetables: preformed in liver, egg yolk, butter, and milk	Necessary for rhodopsin synthesis, normal health of epithelial cells, and bone and tooth growth	Rhodopsin deficiency, night blindness, retarded growth, skin disorders, and increased infection risk	875 μg
B_1 (thiamine)	W	Yeast, grains, and milk	Involved in carbohydrate and amino acid metabolism; necessary for growth	Beriberi—muscle weakness (including cardiac muscle), neuritis, and paralysis	1.2 mg
B_2 (riboflavin)	W	Green vegetables, liver, wheat germ, milk, and eggs	Component of flavin adenine dinucleotide (FAD); involved in citric acid cycle	Eye disorders and skin cracking, especially at corners of the mouth	1.4 mg
B_3 (niacin)	W	Fish, liver, red meat, yeast, grains, peas, beans, and nuts	Component of nicotinamide adenine dinucleotide; involved in glycolysis and citric acid cycle	Pellagra—diarrhea, dermatitis, and mental disturbance	16 mg
Pantothenic acid	W	Liver, yeast, green vegetables, grains and intestinal bacteria	Constituent of coenzyme A, glucose production from lipids and amino acids, and steroid hormone synthesis	Neuromuscular dysfunction and fatigue	5.5 mg
Biotin	W	Liver, yeast, eggs, and intestinal bacteria	Fatty acid and purine synthesis; movement of pyruvic acid into citric acid cycle	Mental and muscle dysfunction, fatigue, and nausea	60 μg
B_6 (pyridoxine)	W	Fish, liver, yeast, tomatoes, and intestinal bacteria	Involved in amino acid metabolism	Dermatitis, retarded growth, and nausea	1.5 mg
Folate (folic acid)	W	Liver, green leafy vegetables, and intestinal bacteria	Nucleic acid synthesis; hematopoiesis	Macrocytic anemia (enlarged red blood cells) and spina bifida	180 μg
B_{12} (cobalamins)	W	Liver, red meat, milk and eggs	Necessary for erythrocyte production; some nucleic acid and amino acid metabolism	Pernicious anemia and nervous system disorders	2 μg

continued next page

*For adults and children 4 or more years of age. The Reference Daily Intakes (RDIs) are the average values of the Recommended Daily Allowances (RDAs) over the age range indicated. The RDA is an estimate of the quantity of a nutrient recommended to meet the needs of most members of the population on the basis of their age and sex.

Table 25.2   The Principal Vitamins—cont'd

Vitamin	Fat- (F) or Water- (W) Soluble	Source	Function	Symptoms of Deficiency	Reference Daily Intake*
C (ascorbic acid)	W	Citrus fruit, tomatoes, and green vegetables	Collagen synthesis; general protein metabolism	Scurvy—defective collagen formation and poor wound healing	60 mg
D (cholecalciferol, ergocalciferol)	F	Fish liver oil, enriched milk, and eggs; provitamin D converted by sunlight to cholecalciferol in the skin	Promotes calcium and phosphorus use; normal growth and bone and teeth formation	Rickets—poorly developed, weak bones; osteomalacia; bone reabsorption	6.5 μg
E (tocopherols, tocotrienols)	F	Wheat germ; cottonseed, palm, and rice oils; grain, liver, and lettuce	Prevents catabolism of certain fatty acids; may prevent miscarriage	Hemolysis of erythrocytes and nerve destruction	9 mg
K (phylloquinone)	F	Alfalfa, liver, spinach, vegetable oils, cabbage, and intestinal bacteria	Required for synthesis of a number of clotting factors	Excessive bleeding because of retarded blood clotting	65 μg

Clinical Note

As part of normal metabolism cells produce molecules called **free radicals,** which are missing an electron. Free radicals can replace the missing electron by taking an electron from cell components, such as fats, proteins, or DNA, resulting in damage to the cell. The loss of an electron from a molecule is called oxidation (see chapter 2). **Antioxidants** are substances that prevent oxidation of cell components by donating an electron to free radicals. Examples of antioxidants include beta-carotene (provitamin A), vitamin C, and vitamin E.

There is speculation that damage resulting from free radicals contributes to aging and certain diseases, such as atherosclerosis, cancer, and cataracts. Antioxidants may provide protection against these processes. There is evidence for and against the antioxidant hypothesis, and it may be that antioxidants are effective in preventing some disorders, but not others. For example, vitamins C and E help to prevent cataracts, but in recent studies, the effectiveness of vitamin E and beta-carotene supplements in preventing heart disease and cancer have been disappointing. Nonetheless, a balanced diet that includes fruits and vegetables rich in antioxidants is recommended. It may be that popping pills is not a substitute for the complex mix of chemicals in foods.

There are two major classes of vitamins: **fat-soluble** and **water-soluble.** Fat-soluble vitamins such as vitamins A, D, E, and K are absorbed from the intestine along with lipids, and some of them can be stored in the body for long periods. Because they can be stored, it is possible to accumulate these vitamins in the body to the point of toxicity, a condition called **hypervitaminosis** (hī′per-vī′tă-mi-nō′sis). Water-soluble vitamins, such as the B vitamins and vitamin C, are absorbed with water from the intestinal tract and remain in the body only a short time before being excreted. The B vitamins are thiamine, riboflavin, pantothenic acid, biotin, vitamin B_6, folate, and vitamin B_{12}.

1   P R E D I C T

Predict what would happen if vitamins were broken down during the process of digestion rather than being absorbed intact into the circulation.

✔ *Answer in Appendix F*

Minerals

A number of inorganic nutrients, called **minerals,** are also necessary for normal metabolic functions. They constitute about 4%–5% of the total body weight and are involved in a number of important functions. Minerals are components of enzymes, a few vitamins, hemoglobin, and other organic molecules. They add mechanical strength to bone, are necessary for nerve and muscle activity, function as buffers, and are involved with energy transfer processes (ATP) and osmosis. Some of the important minerals and their functions are listed in table 25.3.

Minerals are taken into the body by themselves or in combination with organic molecules. Minerals can be obtained from animal and plant sources. Mineral absorption from plants, however, can be limited because the minerals tend to bind to plant fibers. Refined cereals and breads, fats, and sugar have hardly any minerals. A balanced diet can provide all the necessary minerals, with a few possible excep-

Table 25.3   Important Minerals

Mineral	Function	Symptoms of Deficiency	Reference Daily Intake*
Calcium	Bone and teeth formation, blood clotting, muscle activity, and nerve function	Spontaneous nerve discharge and tetany	900 mg
Chlorine	Blood acid–base balance; hydrochloric acid production in stomach	Acid–base imbalance	3.2 g
Chromium	Associated with enzymes in glucose metabolism	Unknown	120 μg
Cobalt	Component of vitamin B_{12}; erythrocyte production	Anemia	Unknown
Copper	Hemoglobin and melanin production; electron-transport system	Anemia and loss of energy	2.0 mg
Fluorine	Provides extra strength in teeth; prevents dental caries	No real pathology	2.5 mg
Iodine	Thyroid hormone production; maintenance of normal metabolic rate	Decrease of normal metabolism	150 μg
Iron	Component of hemoglobin; ATP production in electron-transport system	Anemia, decreased oxygen transport, and energy loss	12 mg
Magnesium	Coenzyme constituent; bone formation; muscle and nerve function	Increased nervous system irritability, vasodilation, and arrhythmias	3.5 mg
Manganese	Hemoglobin synthesis; growth; activation of several enzymes	Tremors and convulsions	3.5 mg
Molybdenum	Enzyme component	Unknown	150 μg
Phosphorus	Bone and teeth formation; important in energy transfer (ATP); component of nucleic acids	Loss of energy and cellular function	900 mg
Potassium	Muscle and nerve function	Muscle weakness, abnormal electrocardiogram, and alkaline urine	2 g
Selenium	Component of many enzymes	Unknown	55 μg
Sodium	Osmotic pressure regulation; nerve and muscle function	Nausea, vomiting, exhaustion, and dizziness	500 mg
Sulfur	Component of hormones, several vitamins, and proteins	Unknown	Unknown
Zinc	Component of several enzymes; carbon dioxide transport and metabolism; necessary for protein metabolism	Deficient carbon dioxide transport and deficient protein metabolism	13 mg

*For adults and children 4 or more years of age.

tions. For example, women who suffer from excessive menstrual bleeding may need an iron supplement.

Clinical Note

Studies in mice, rats, and other animals indicate that life span can be increased by approximately one-third by decreasing normal caloric intake 30%–50%, provided that the diet includes enough protein, fat, vitamins, and minerals. Why life span increases is not understood, but one proposed explanation for this phenomenon is that decreased caloric intake in some way reduces free radical damage to mitochondria. It has been suggested that humans might derive a similar benefit by reducing caloric input, starting at age 20. Unlike laboratory animals, however, humans would have to voluntarily restrict their caloric intake by 30%–50%, which is an unlikely behavioral change for most humans. Restriction of calorie intake to increase longevity is an interesting idea, but much more needs to be learned before it will be known if this approach is beneficial to humans.

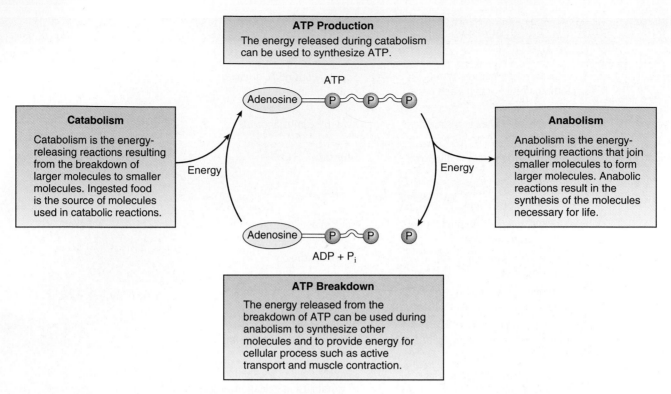

Figure 25.2   ATP Coupling of Catabolic and Anabolic Reactions

Energy released by catabolic reactions is used to form ATP, which releases the energy for use in anabolic reactions.

Metabolism

Metabolism (mě-tab′ō-lizm, meaning change) is the total of all the chemical changes that occur in the body. It consists of **anabolism** (ă-nab′ō-lizm), the energy-requiring process by which small molecules are joined to form larger molecules, and **catabolism** (kă-tab′ō-lizm), the energy-releasing process by which large molecules are broken down into smaller molecules. Anabolism occurs in all cells of the body as they divide to form new cells, maintain their own intracellular structure, and produce molecules such as hormones, neurotransmitters, or extracellular matrix molecules for export. Catabolism begins during the process of digestion and is concluded within individual cells. The energy derived from catabolism is used to drive anabolic reactions.

The cellular metabolic processes are often referred to as cellular metabolism or cellular respiration. The digestive products of carbohydrates, proteins, and lipids taken into body cells are catabolized, and the released energy is used to combine adenosine diphosphate (ADP) and an inorganic phosphate group (P_i) to form ATP (figure 25.2).

$$ADP + P_i + Energy \rightarrow ATP$$

ATP is often called the energy currency of the cell, and it is used to drive such cell activities as active transport and muscle contraction.

The chemical reactions responsible for the transfer of energy from the chemical bonds of nutrient molecules to ATP

molecules involve oxidation–reduction reactions (see chapter 2). A molecule is reduced when it gains electrons and is oxidized when it loses electrons. A nutrient molecule has many hydrogen atoms covalently bonded to the carbon atoms that form the "backbone" of the molecule. Because a hydrogen atom is a hydrogen ion (proton) and an electron, the nutrient molecule has many electrons and is therefore highly reduced. When a hydrogen ion and an associated electron are lost from the nutrient molecule, the molecule loses energy and becomes oxidized. The energy in the electron is used to synthesize ATP. The major events of cellular metabolism are summarized in figure 25.3.

Carbohydrate Metabolism
Glycolysis

Carbohydrate metabolism begins with **glycolysis** (glī-kol′i-sis), which is a series of chemical reactions in the cytosol that results in the breakdown of glucose to two **pyruvic** (pī-rū′vik) **acid** molecules (figure 25.4).

Glycolysis can be divided into four phases.

1. *Input of ATP.* The first steps in glycolysis require the input of energy in the form of two ATP molecules. A phosphate group is transferred from ATP to the glucose molecule, a process called **phosphorylation** (fos′fōr-i-lā′shŭn) to form glucose-6-phosphate. The glucose-6-

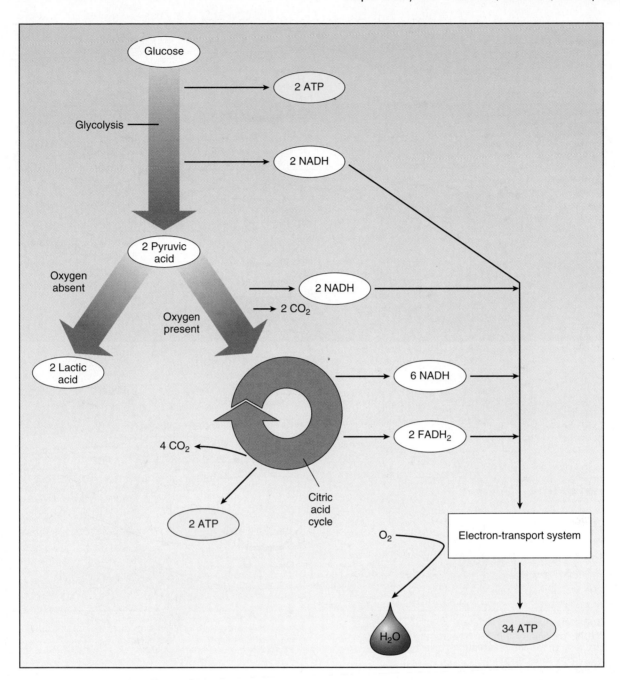

Figure 25.3   Cellular Metabolism

Overview of cellular metabolism, including glycolysis, citric acid cycle, and electron-transport system.

phosphate atoms are rearranged to form fructose-6-phosphate, which is then converted to fructose-1,6-bisphosphate by the addition of another phosphate group from another ATP.

2. *Sugar cleavage.* Fructose-1,6-bisphosphate is cleaved into two three-carbon molecules, glyceraldehyde (glis-er-al′dĕ-hīd)-3-phosphate and dihydroxyacetone (dī′hī-drok-sē-as′e-tōn) phosphate. Dihydroxyacetone phosphate is rearranged to form glyceraldehyde-3-phosphate; consequently, two molecules of glyceraldehyde-3-phosphate result.

3. *NADH production.* Each glyceraldehyde-3-phosphate molecule is oxidized (loses two electrons) to form 1,3-bisphosphoglyceric (biz′phos-fo-gli′sēr′ik) acid, and **nicotinamide adenine dinucleotide (NAD$^+$)** (nik-ō-tin′ă-mīd ad′ĕ-nēn) is reduced (gains two electrons) to **NADH.** Glyceraldehyde-3-phosphate also loses two hydrogen ions, one of which binds to NAD$^+$.

$$NAD^+ + 2\,e^- + 2\,H^+ \rightarrow NADH + H^+$$

NAD$^+$ is the oxidized form of nicotinamide adenine dinucleotide, and NADH is the reduced form. NADH is

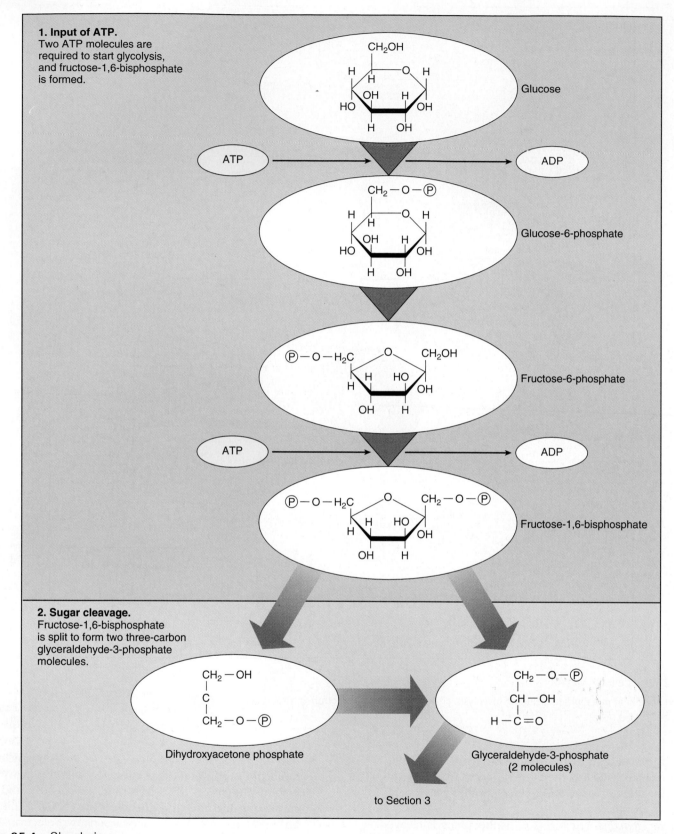

1. Input of ATP.
Two ATP molecules are required to start glycolysis, and fructose-1,6-bisphosphate is formed.

CH₂OH

Glucose

ATP → ADP

Glucose-6-phosphate

Fructose-6-phosphate

ATP → ADP

Fructose-1,6-bisphosphate

2. Sugar cleavage.
Fructose-1,6-bisphosphate is split to form two three-carbon glyceraldehyde-3-phosphate molecules.

Dihydroxyacetone phosphate

Glyceraldehyde-3-phosphate
(2 molecules)

to Section 3

Figure 25.4   Glycolysis

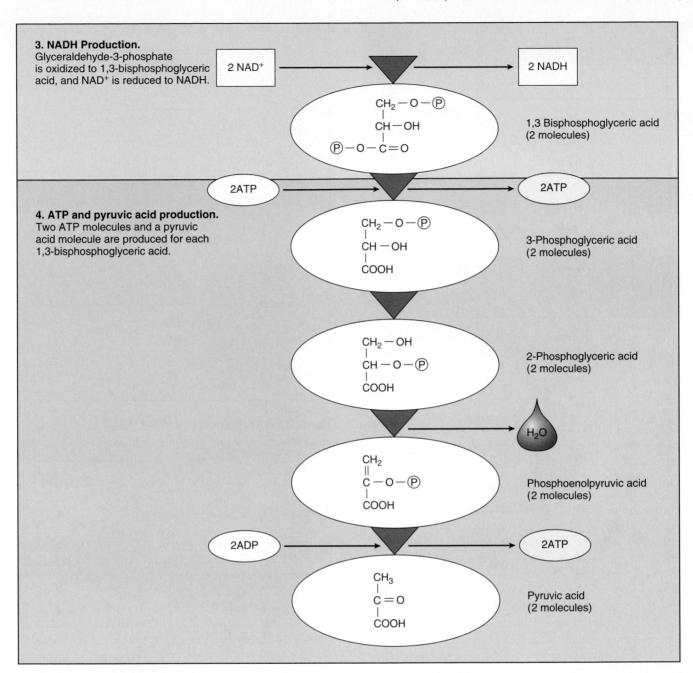

3. NADH Production.
Glyceraldehyde-3-phosphate is oxidized to 1,3-bisphosphoglyceric acid, and NAD⁺ is reduced to NADH.

2 NAD⁺ → → 2 NADH

CH₂—O—Ⓟ
|
CH—OH
|
Ⓟ—O—C=O

1,3 Bisphosphoglyceric acid (2 molecules)

2ATP → → 2ATP

4. ATP and pyruvic acid production.
Two ATP molecules and a pyruvic acid molecule are produced for each 1,3-bisphosphoglyceric acid.

CH₂—O—Ⓟ
|
CH—OH
|
COOH

3-Phosphoglyceric acid (2 molecules)

CH₂—OH
|
CH—O—Ⓟ
|
COOH

2-Phosphoglyceric acid (2 molecules)

H₂O

CH₂
‖
C—O—Ⓟ
|
COOH

Phosphoenolpyruvic acid (2 molecules)

2ADP → → 2ATP

CH₃
|
C=O
|
COOH

Pyruvic acid (2 molecules)

Figure 25.4 (continued)

a carrier molecule with two high-energy electrons (e⁻) that can be used to produce ATP molecules through the electron-transport chain (described later in this chapter).

4. *ATP and pyruvic acid production.* The last four steps of glycolysis produce two ATP molecules and one pyruvic acid molecule from each 1,3-bisphosphoglyceric acid molecule.

The events of glycolysis are summarized in table 25.4. Each glucose molecule that enters glycolysis forms two glyceraldehyde-3-phosphate molecules at the sugar cleavage phase. Each glyceraldehyde-3-phosphate molecule produces two ATP molecules, one NADH molecule, and one pyruvic acid molecule. Each glucose molecule therefore forms four

ATP, two NADH, and two pyruvic acid molecules. Because the start of glycolysis requires the input of two ATP molecules, however, the final yield of each glucose molecule is two ATP, two NADH, and two pyruvic acid molecules (see figure 25.3).

If the cell has adequate amounts of oxygen, the NADH and pyruvic acid molecules are used in aerobic respiration to produce ATP. In the absence of sufficient oxygen they are used in anaerobic respiration.

Anaerobic Respiration

Anaerobic (an-ār-ō'bik) **respiration** is the breakdown of glucose in the absence of oxygen to produce two molecules

Table 25.4   ATP Production from one Glucose Molecule

Process	Product	Total ATP Produced*
Glycolysis	4 ATP	2 ATP (4 ATP produced minus 2 ATP to start)
	2 NADH	6 ATP (or 4 ATP; see text)
Acetyl-CoA production	2 NADH	6 ATP
Citric acid cycle	2 ATP	2 ATP
	6 NADH	18 ATP
	2 FADH$_2$	4 ATP
Total		38 ATP (or 36 ATP; see text)

*NADH and FADH$_2$ are used in the production of ATP in the electron-transport chain.
Abbreviations: ATP = adenosine triphosphate, NADH = reduced nicotinamide adenine dinucleotide, FADH$_2$ = reduced flavin adenine diphosphate, acetyl-CoA = acetyl coenzyme A.

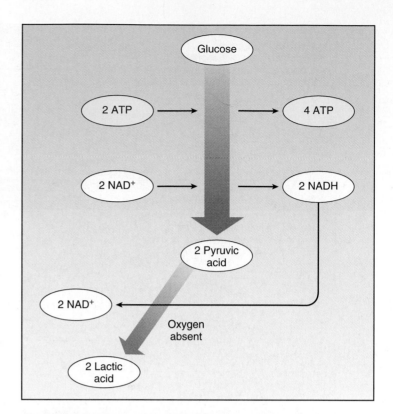

Figure 25.5   Anaerobic Respiration
In the absence of oxygen, the pyruvic acid produced in glycolysis is converted to lactic acid. The NADH produced in glycolysis is converted back to NAD$^+$.

of lactic acid and two ATP molecules (figure 25.5). The ATP molecules are a source of energy during activities such as intense exercise when insufficient oxygen is delivered to tissues. The first phase of anaerobic respiration is glycolysis, and the second phase is the reduction of pyruvic acid to lactic acid. In this reaction the pyruvic acid gains electrons and hydrogen ions through the oxidation of the NADH molecules produced by glycolysis. Because there is a net gain of two ATP molecules during glycolysis, anaerobic respiration produces two ATP molecules for each molecule of glucose converted into two lactic acid molecules.

Lactic acid is released from the cells that produce it and is transported by the blood to the liver. If oxygen becomes available, the lactic acid in the liver can be converted through a series of chemical reactions into glucose. The glucose is released from the liver and transported in the blood to cells that use the glucose as an energy source. This process of converting lactic acid to glucose is called the **Cori cycle.** Some of the reactions involved in converting lactic acid into glucose require the input of ATP (energy) produced by aerobic respiration. The oxygen necessary for the synthesis of the ATP is part of the **oxygen debt** (see chapter 10).

Aerobic Respiration

Aerobic (ār-ō′bik) **respiration** is the breakdown of glucose in the presence of oxygen to produce carbon dioxide, water, and 38 ATP molecules. Most of the ATP molecules required to sustain life are produced through aerobic respiration, which can be considered in four phases: glycolysis, acetyl-CoA formation, the citric acid cycle, and the electron-transport chain. The first phase of aerobic respiration, as in anaerobic respiration, is glycolysis.

Acetyl-CoA Formation

In the second phase of aerobic respiration pyruvic acid moves from the cytosol into a mitochondrion, which is separated into an inner and outer compartment by the inner mitochondrial membrane. Within the inner compartment, enzymes remove a carbon atom from the three-carbon pyruvic acid molecule to form carbon dioxide and a two-carbon acetyl (as′e-til) group (figure 25.6). Energy is released in the reaction and is used to reduce NAD$^+$ to NADH. The acetyl group combines with coenzyme A (CoA) to form acetyl-CoA. For each two pyruvic acid molecules from glycolysis, two NADH and two carbon dioxide molecules are formed (see figure 25.3).

Citric Acid Cycle

The third phase of aerobic respiration is the **citric acid cycle,** which is named after the six-carbon citric acid molecule formed in the first step of the cycle (see figure 25.6). It is also called the Krebs cycle after its discoverer, the British biochemist Sir Hans Krebs. The citric acid cycle begins with the production of citric acid from the combination of acetyl-CoA and a four-carbon molecule called oxaloacetic (ok′să-lō-ă-sē′tik) acid. A series of reactions occurs, resulting in the formation of another oxaloacetic acid, which can start the cycle again by combining with another acetyl-CoA. During the reactions of the citric acid cycle, three important events occur.

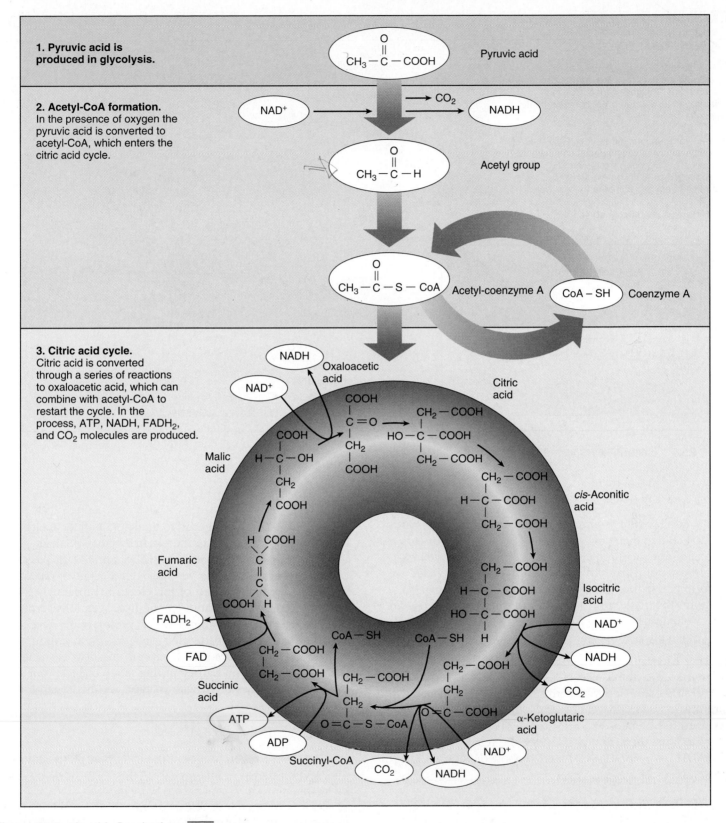

1. Pyruvic acid is produced in glycolysis.

Pyruvic acid

2. Acetyl-CoA formation. In the presence of oxygen the pyruvic acid is converted to acetyl-CoA, which enters the citric acid cycle.

Acetyl group

Acetyl-coenzyme A

Coenzyme A

3. Citric acid cycle. Citric acid is converted through a series of reactions to oxaloacetic acid, which can combine with acetyl-CoA to restart the cycle. In the process, ATP, NADH, $FADH_2$, and CO_2 molecules are produced.

Oxaloacetic acid

Citric acid

Malic acid

cis-Aconitic acid

Fumaric acid

Isocitric acid

Succinic acid

α-Ketoglutaric acid

Succinyl-CoA

Figure 25.6   Aerobic Respiration

1. *ATP production.* For each citric acid molecule one ATP is formed.
2. *NADH and FADH₂ production.* For each citric acid molecule three NAD⁺ molecules are converted to NADH molecules, and one flavin (flā′vin) adenine dinucleotide (FAD) molecule is converted to FADH₂. The NADH and FADH₂ molecules are electron carriers that enter the electron-transport chain and are used to produce ATP.

1. NADH and FADH$_2$ transfer their electrons to electron carriers located on the inner mitochondrial membrane.

2. The electron's energy is used to pump hydrogen ions across the inner mitochondrial membrane. A higher concentration of hydrogen ions in the outer compartment results.

3. The hydrogen ions diffuse back into the inner compartment through special channels that couple the hydrogen ion movement with the production of ATP. The electrons, hydrogen ions, and oxygen combine to form water.

4. ATP leaves the mitochondria.

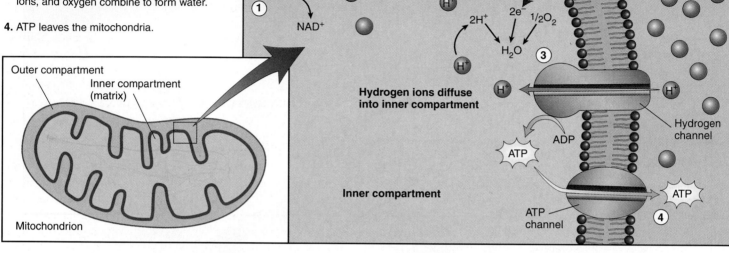

Figure 25.7   Electron-Transport Chain

3. *Carbon dioxide production.* Each six-carbon citric acid molecule at the start of the cycle becomes a four-carbon oxaloacetic acid molecule at the end of the cycle. Two carbon atoms from the citric acid molecule are used to form two carbon dioxide molecules. Thus the carbon atoms that make up food molecules such as glucose are eventually eliminated from the body as carbon dioxide. We literally breathe out part of the food we eat!

For each glucose molecule that begins aerobic respiration, two pyruvic acid molecules are produced in glycolysis, and they are converted into two acetyl-CoA molecules that enter the citric acid cycle. To determine the number of molecules produced from glucose by the citric acid cycle, two "turns" of the cycle must therefore be counted; the results are two ATP, six NADH, two FADH$_2$, and four carbon dioxide molecules (see figure 25.3).

Electron-Transport Chain

The fourth phase of aerobic respiration involves the **electron-transport chain** (figure 25.7), which is a series of electron carriers in the inner mitochondrial membrane. Electrons are transferred from NADH and FADH$_2$ to the electron-transport carriers, and hydrogen ions are released from NADH and FADH$_2$. After the loss of the electrons and the hydrogen ions, the oxidized

NAD$^+$ and FAD can be reused to transport additional electrons from the citric acid cycle to the electron-transport chain.

The electrons released from NADH and FADH$_2$ pass from one electron carrier to the next through a series of oxidation–reduction reactions. Three of the electron carriers also function as proton pumps that move the hydrogen ions across the inner mitochondrial membrane. Each proton pump accepts an electron, uses some of the electron's energy to export a hydrogen ion, and passes the electron to the next electron carrier. The last electron carrier in the series collects four electrons and combines them with oxygen and hydrogen ions to form water:

$$\tfrac{1}{2}\,O_2 + 2\,H^+ + 2\,e^- \rightarrow H_2O$$

Without oxygen to accept the electrons, the reactions of the electron-transport chain cease, effectively stopping aerobic respiration.

The hydrogen ions released from NADH and FADH$_2$ are moved from the inner mitochondrial compartment to the outer mitochondrial compartment by the proton pumps. As a result, the concentration of hydrogen ions in the outer compartment exceeds that of the inner compartment, and hydrogen ions diffuse back into the inner compartment. The hydrogen ions pass through special channels in the inner mitochondrial membrane that couple the movement of the hydrogen ions to ATP production. This process is called the **chemiosmotic** (kem-ē-os-mot′ik) **model** because the chemical formation of ATP is coupled to a diffusion force similar to osmosis.

---

2   P R E D I C T

Many poisons function by blocking certain steps in the metabolic pathways. For example, cyanide blocks the last step in the electron-transport chain. Explain why this blockage would cause death.

✔ *Answer in Appendix F*

Summary of ATP Production

For each glucose molecule, aerobic respiration produces a net gain of 38 ATP molecules: 2 from glycolysis, 2 from the citric acid cycle, and 34 from the NADH molecules and FADH$_2$ molecules that pass through the electron-transport chain (see table 25.4). For each NADH molecule formed, three ATP molecules are produced by the electron-transport chain, and for each FADH$_2$ molecule, two ATP molecules are produced.

The number of ATP molecules produced can also be reported as 36 ATP molecules. The two NADH molecules produced by glycolysis in the cytosol cannot cross the inner mitochondrial membrane; thus their electrons are donated to a shuttle molecule that carries the electrons to the electron-transport chain. Depending on the shuttle molecule, each glycolytic NADH molecule can produce 2 or 3 ATP molecules. In skeletal muscle and the brain 2 ATP molecules are produced for each NADH molecule, resulting in a total number of 36 ATP molecules; but in the liver, kidneys, and heart 3 ATP molecules are produced for each NADH molecule, and the total number of ATP molecules formed is 38.

Six carbon dioxide molecules and six molecules of water are also produced in aerobic respiration. Thus aerobic respiration can be summarized as follows:

$$C_6H_{12}O_6 + 6\ O_2 + 38\ ADP + 38\ P_i \rightarrow$$
$$6\ CO_2 + 6\ H_2O + 38\ ATP$$

Clinical Note

The number of ATP molecules produced per glucose molecule is a theoretical number that assumes two hydrogen ions are necessary for the formation of each ATP. If the number required is more than two, the efficiency of aerobic respiration decreases. In addition, it is now understood that it costs energy to get ADP and phosphates into the mitochondria and to get ATP out. Considering all these factors, it is currently estimated that each glucose molecule yields about 25 ATP molecules instead of 38 ATP molecules.

Lipid Metabolism

Lipids are the body's main energy storage molecules. In a healthy person, lipids are responsible for about 99% of the body's energy storage, and glycogen accounts for about 1%. Although proteins can be used as an energy source, they are not considered storage molecules because the breakdown of proteins normally involves the loss of necessary tissue.

Lipids are stored primarily as triacylglycerols in adipose tissue. There is a constant synthesis and breakdown of triacylglycerols; thus the fat present in adipose tissue today is not the same fat that was there a few weeks ago. Between meals when triacylglycerols are broken down in adipose tissue, some of the fatty acids produced are released into the blood, where they are called **free fatty acids.** Other tissues, especially skeletal muscle and the liver, use the free fatty acids as a source of energy.

The metabolism of fatty acids occurs by **beta-oxidation,** a series of reactions in which two carbon atoms are removed from the end of a fatty acid chain to form acetyl-CoA. The process of beta-oxidation continues to remove two carbon atoms at a time until the entire fatty acid chain is converted into acetyl-CoA molecules. Acetyl-CoA can enter the citric acid cycle and be used to generate ATP (figure 25.8).

Acetyl-CoA can also be used in **ketogenesis** (kē-tō-jen′ě-sis), the formation of ketone bodies. In the liver when large amounts of acetyl-CoA are produced, not all of the acetyl-CoA enters the citric acid cycle. Instead, two acetyl-CoA molecules combine to form a molecule of acetoacetic (as′e-tō-a-sē′tik) acid, which is converted mainly into β-hydroxybutyric (hī-drōk′sē-byū-tir′ik) acid and a smaller amount of acetone (as′e-tōn). Acetoacetic acid, β-hydroxybutyric acid, and acetone are called **ketone** (kē′tōn) **bodies** and are released into the blood, where they travel to other tissues, especially skeletal muscle. In these tissues the ketone bodies are converted back into acetyl-CoA that enters the citric acid cycle to produce ATP.

Clinical Note

Normally the blood contains only small amounts of ketone bodies. During starvation (see Clinical Focus: Starvation), however, or in patients with diabetes mellitus, the quantity of ketone bodies can increase to produce the condition called **ketosis.** The increased number of ketone bodies can exceed the capacity of the body's buffering system, resulting in acidosis, a decrease in blood pH (see chapter 27).

Protein Metabolism

Once absorbed into the body, amino acids, the products of protein digestion, are quickly taken up by cells, especially in the liver. Amino acids can be used to synthesize needed proteins (see chapter 3) or as a source of energy (figure 25.9). Unlike glycogen and triacylglycerols, amino acids are not stored in the body.

The synthesis of nonessential amino acids usually begins with keto acids (figure 25.10). A keto acid can be converted

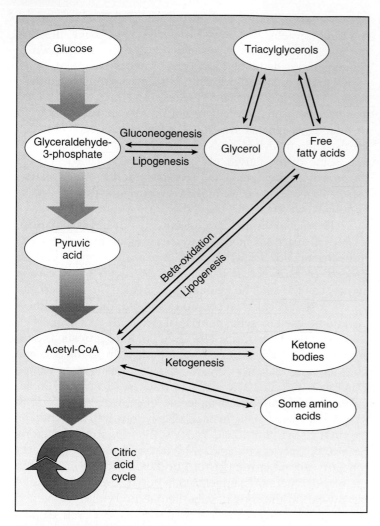

Figure 25.8 Lipid Metabolism

Triacylglycerol is broken down into glycerol and fatty acids. Glycerol enters glycolysis to produce ATP. The fatty acids are broken down by beta-oxidation into acetyl-CoA, which enters the citric acid cycle to produce ATP. Acetyl-CoA can also be used to produce ketone bodies (ketogenesis). Lipogenesis is the production of lipids. Glucose is converted to glycerol, and amino acids are converted to acetyl-CoA molecules. Acetyl-CoA molecules can combine to form fatty acids. Glycerol and fatty acids join to form triacylglycerols.

into an amino acid by replacing its oxygen with an amine group. Usually this conversion is accomplished by transferring an amine group from an amino acid to the keto acid, a reaction called **transamination** (trans-am′i-nā′shŭn). For example, α-Ketoglutaric acid (a keto acid) can react with an amino acid to form glutamic acid (an amino acid; figure 25.11a). Most amino acids can undergo transamination to produce glutamic acid. The glutamic acid can be used as a source of an amine group to construct most of the nonessential amino acids. A few nonessential amino acids are formed in other ways from the essential amino acids.

Amino acids can be used as a source of energy. In **oxidative deamination** (dē-am-i-nā′shŭn) an amine group is removed from an amino acid (usually glutamic acid), leaving

ammonia and a keto acid (figure 25.11b). In the process, NAD^+ is reduced to NADH, which can enter the electron-transport chain to produce ATP. Ammonia is toxic to cells and is converted by the liver into urea, which is carried by the blood to the kidneys, where the urea is eliminated (figure 25.11c; see chapter 26).

Amino acids can also be used as a source of energy by converting them into the intermediate molecules of carbohydrate metabolism (see figure 25.9). These molecules then are metabolized to yield ATP. The conversion of an amino acid often begins with a transamination or oxidative deamination reaction in which the amino acid is converted into a keto acid (see figure 25.11). The keto acid can enter the citric acid cycle or be converted into pyruvic acid or acetyl-CoA.

Interconversion of Nutrient Molecules

Blood glucose enters most cells by facilitated diffusion and is immediately converted to glucose-6-phosphate, which cannot recross the cell membrane (figure 25.12a). Glucose-6-phosphate can then continue through glycolysis to produce ATP. If, however, there is excess glucose (e.g., after a meal), it can be used to form glycogen through a process called **glycogenesis** (glī-kō-jen′ĕ-sis). Most of the body's glycogen is contained in skeletal muscle and the liver.

Once glycogen stores, which are quite limited, are filled, glucose and amino acids are used to synthesize lipids, a process called **lipogenesis** (lip-ō-jen′ĕ-sis; see figure 25.8). Glucose molecules can be used to form glyceraldehyde-3-phosphate and acetyl-CoA. Amino acids can also be converted to acetyl-CoA. Glyceraldehyde-3-phosphate can be converted to glycerol, and the two-carbon acetyl-CoA molecules can be joined together to form fatty acid chains. Glycerol and three fatty acids then combine to form triacylglycerols.

Clinical Note

Enzymes in the liver convert ethanol (beverage alcohol) into acetyl-CoA, and in the process two NADH molecules are produced. The NADH molecules enter the electron-transport chain and are used to produce ATP molecules. Each gram of ethanol provides 7 kcal of energy. Because of the high level of NADH in the cell that results from the metabolism of ethanol, the production of NADH by glycolysis and by the citric acid cycle is inhibited. Consequently, sugars and amino acids are not broken down but are converted into fats that accumulate in the liver. Chronic alcohol abuse can therefore result in **cirrhosis** (sir-rō′sis) **of the liver,** which involves fat deposition, cell death, inflammation, and scar tissue formation. Death can occur because the liver is unable to carry out its normal functions.

When glucose is needed, glycogen can be broken down into glucose-6-phosphate through a set of reactions

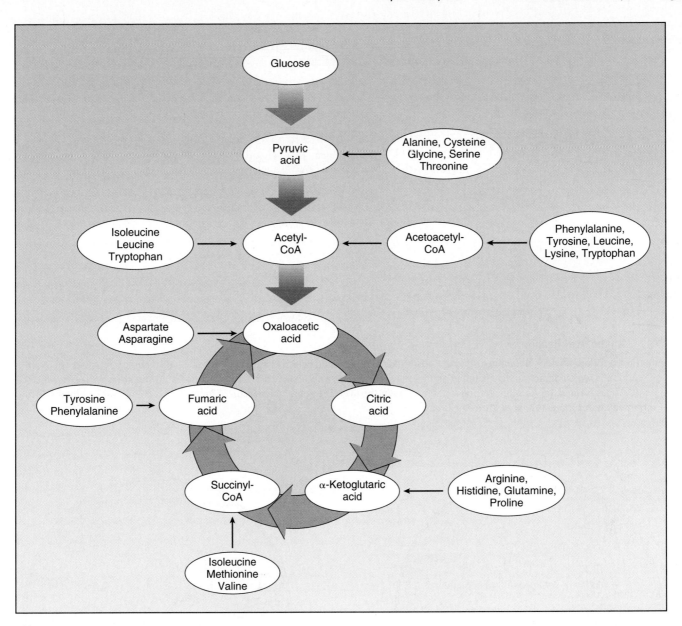

Figure 25.9   Amino Acid Metabolism

Various entry points for amino acids into carbohydrate metabolism.

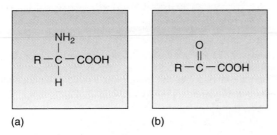

(a)   (b)

Figure 25.10   General Formulas of an Amino Acid and a Keto Acid

(a) Amino acid with a carboxyl group (—COOH), an amine group (NH₂), a hydrogen atom (H), and a group called "R" that represents the rest of the molecule. (b) Keto acid with a double-bonded oxygen replacing the amine group and the hydrogen atom of the amino acid.

called **glycogenolysis** (glī′kō-jě-nol′i-sis; figure 25.12b). In skeletal muscle, glucose-6-phosphate continues through glycolysis to produce ATP. The liver can use glucose-6-phosphate for energy or can convert it to glucose, which diffuses into the blood. The liver can release glucose, but skeletal muscle cannot because it lacks the necessary enzymes to convert glucose-6-phosphate into glucose.

Release of glucose from the liver is necessary to maintain blood glucose levels between meals. Maintaining these levels is especially important to the brain, which normally uses only glucose for an energy source and consumes about two-thirds of the total glucose used each day. When liver glycogen levels are inadequate to supply glucose, amino acids from proteins and glycerol from triacylglycerols are used to produce glucose in a process called **gluconeogenesis**

(a)

(b)

(c)

Figure 25.11   Amino Acid Reactions

(*a*) Transamination reaction in which an amine group is transferred from an amino acid to a keto acid to form a different amino acid. (*b*) Oxidative deamination reaction in which an amino acid loses an amine group to become a keto acid and to form ammonia. In the process, NADH, which can be used to generate ATP, is formed. (*c*) Ammonia is converted to urea in the liver. The actual conversion of ammonia to urea is more complex, involving a number of intermediate reactions that constitute the urea cycle.

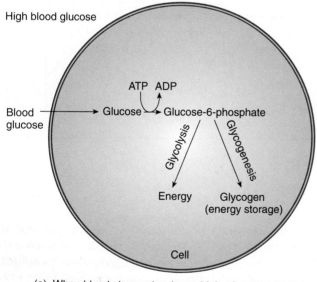

(a) When blood glucose levels are high, glucose enters the cell and is phosphorylated to form glucose-6-phosphate, which can enter glycolysis or glycogenesis.

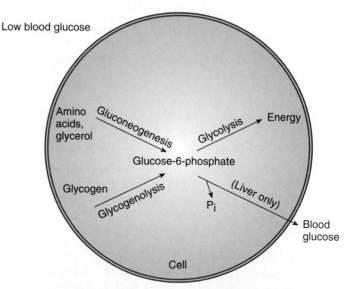

(b) When blood glucose levels drop, glucose-6-phosphate can be produced through glycogenolysis or gluconeogenesis. Glucose-6-phosphate can enter glycolysis, or the phosphate group can be removed in liver tissue, and glucose released into the blood.

Figure 25.12   Interconversion of Nutrient Molecules

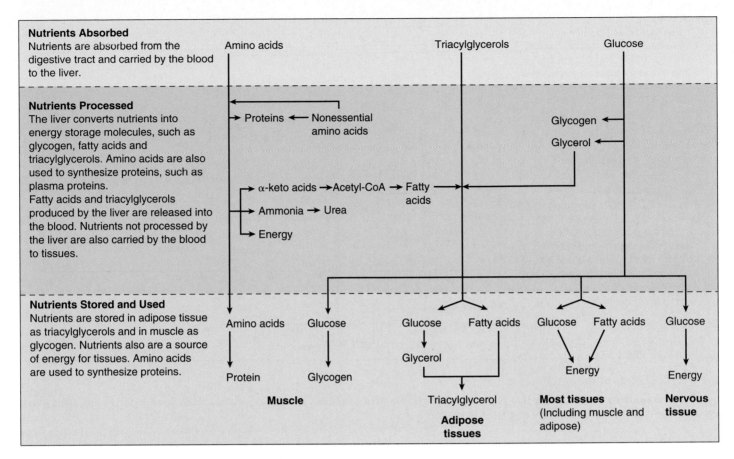

Nutrients Absorbed
Nutrients are absorbed from the digestive tract and carried by the blood to the liver.

Nutrients Processed
The liver converts nutrients into energy storage molecules, such as glycogen, fatty acids and triacylglycerols. Amino acids are also used to synthesize proteins, such as plasma proteins.
Fatty acids and triacylglycerols produced by the liver are released into the blood. Nutrients not processed by the liver are also carried by the blood to tissues.

Nutrients Stored and Used
Nutrients are stored in adipose tissue as triacylglycerols and in muscle as glycogen. Nutrients also are a source of energy for tissues. Amino acids are used to synthesize proteins.

Figure 25.13   Events of the Absorptive State

Absorbed molecules, especially glucose, are used as sources of energy. Molecules not immediately needed for energy are stored: glucose is converted to glycogen or triacylglycerols, triacylglycerols are deposited in adipose tissue, and amino acids are converted to triacylglycerols or carbohydrates.

(glū′kō-nē-ō-jen′ě-sis). Most amino acids can be converted into citric acid cycle molecules, acetyl-CoA, or pyruvic acid (see figure 25.9). Through a series of chemical reactions, these molecules can be converted into glucose. Glycerol can enter glycolysis by becoming glyceraldehyde-3-phosphate.

Metabolic States

There are two major metabolic states in the body. The first is the **absorptive state,** the period immediately after a meal when nutrients are being absorbed through the intestinal wall into the circulatory and lymphatic systems (figure 25.13). The absorptive state usually lasts about 4 h after each meal, and most of the glucose that enters the circulation is used by cells to provide the energy they require. The remainder of the glucose is converted into glycogen or fats. Most of the absorbed fats are deposited in adipose tissue. Many of the absorbed amino acids are used by cells in protein synthesis, some are used for energy, and others enter the liver and are converted to fats or carbohydrates.

The second state, the **postabsorptive state,** occurs late in the morning, late in the afternoon, or during the night after

each absorptive state is concluded (figure 25.14). Normal blood glucose levels range between 70 and 110 mg/100 mL, and it is vital to the body's homeostasis that this range be maintained. During the postabsorptive state blood glucose levels are maintained by the conversion of other molecules to glucose. The first source of blood glucose during the postabsorptive state is the glycogen stored in the liver. This glycogen supply, however, can provide glucose for only about 4 h. The glycogen stored in skeletal muscles can also be used during times of vigorous exercise. As the glycogen stores are depleted, fats are used as an energy source. The glycerol from triacylglycerols can be converted to glucose. The fatty acids from fat can be converted to acetyl-CoA, moved into the citric acid cycle, and used as a source of energy to produce ATP. In the liver, acetyl-CoA can be used to produce ketone bodies that other tissues can use for energy. The use of fatty acids as an energy source can partly eliminate the need to use glucose for energy, resulting in reduced glucose removal from the blood and maintaining blood glucose levels at homeostatic levels. Proteins can also be used as a source of glucose or can be directly used for energy production, again sparing blood glucose.

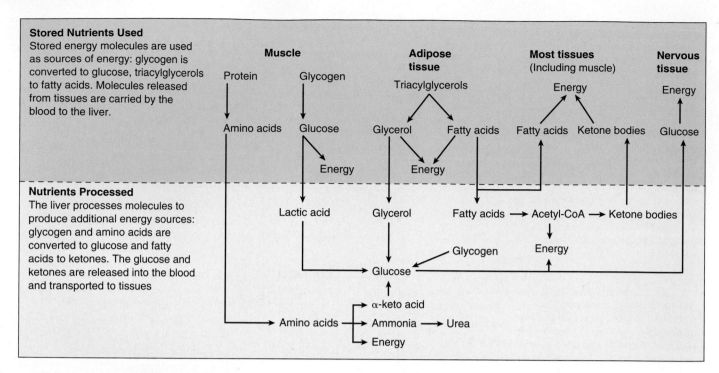

Stored Nutrients Used
Stored energy molecules are used as sources of energy: glycogen is converted to glucose, triacylglycerols to fatty acids. Molecules released from tissues are carried by the blood to the liver.

Nutrients Processed
The liver processes molecules to produce additional energy sources: glycogen and amino acids are converted to glucose and fatty acids to ketones. The glucose and ketones are released into the blood and transported to tissues

Figure 25.14 Events of the Postabsorptive State

Stored energy molecules are used as sources of energy: glycogen is converted to glucose, triacylglycerols are broken down to fatty acids, some of which are converted to ketones, and proteins are converted to glucose.

Clinical Focus Starvation

Starvation results from the inadequate intake of nutrients or the inability to metabolize or absorb nutrients. It can have a number of causes such as prolonged fasting, anorexia, deprivation, or disease. No matter what the cause, starvation takes about the same course and consists of three phases. The events of the first two phases occur even during relatively short periods of fasting or dieting, but the third phase occurs only in prolonged starvation and can end in death.

During the first phase of starvation, blood glucose levels are maintained through the production of glucose from glycogen, proteins, and fats. At first glycogen is broken down into glucose; however, only enough glycogen is stored in the liver to last a few hours. Thereafter, blood glucose levels are maintained by the breakdown of proteins and fats. Fats are decomposed into fatty acids and glycerol. Fatty acids can be used as a source of energy, especially by skeletal muscle, thus decreasing the use of glucose by tissues other than the

brain. Glycerol can be used to make a small amount of glucose, but most of the glucose is formed from the amino acids of proteins. In addition, some amino acids can be used directly for energy.

In the second stage, which can last for several weeks, fats are the primary energy source. The liver metabolizes fatty acids into ketone bodies that can be used as a source of energy. After about a week of fasting, the brain begins to use ketone bodies, as well as glucose, for energy. This usage decreases the demand for glucose, and the rate of protein breakdown diminishes but does not stop. In addition, the proteins not essential for survival are used first.

The third stage of starvation begins when the fat reserves are depleted and there is a switch to proteins as the major energy source. Muscles, the largest source of protein in the body, are rapidly depleted. At the end of this stage, proteins essential for cellular functions are broken down, and cell function degenerates.

In addition to weight loss, symptoms of starvation include apathy, listlessness, withdrawal, and increased susceptibility to infectious disease. Few people die directly from starvation because they usually die of some infectious disease first. Other signs of starvation can include changes in hair color, flaky skin, and massive edema in the abdomen and lower limbs, causing the abdomen to appear bloated.

During the process of starvation, the ability of the body to consume normal volumes of food also decreases. Foods high in bulk but low in protein content often cannot reverse the process of starvation. Intervention involves feeding the starving person low-bulk food that provides ample proteins and kilocalories and is fortified with vitamins and minerals. The process of starvation also results in dehydration, and rehydration is an important part of intervention. Even with intervention, a victim may be so affected by disease or weakness that he or she cannot recover.

Metabolic Rate

The **metabolic rate** is the total amount of energy produced and used by the body per unit of time. A molecule of ATP exists for less than 1 min before it is degraded back to ADP and inorganic phosphate. For this reason, ATP is produced in cells at about the same rate as it is used. Thus, in examining metabolic rate, ATP production and use can be roughly equated. Metabolic rate is usually estimated by measuring the amount of oxygen used per minute because most ATP production involves the use of oxygen. One liter of oxygen consumed by the body is assumed to produce 4.825 kcal of energy.

The daily input of energy should equal the metabolic expenditure of energy; otherwise, a person will gain or lose weight. For a typical 23-year-old, 70-kg (154 pounds) male to maintain his weight, the daily input should be 2700 kcal/day; for a typical 58-kg (128 pounds) female of the same age 2000 kcal/day is necessary. A pound of body fat provides about 3500 kcal. Reducing kilocaloric intake by 500 kcal/day can result in the loss of 1 pound of fat per week. Clearly, adjusting kilocaloric input is an important way to control body weight.

> ### Clinical Note
>
> Not only the number of kilocalories ingested but the proportion of fat in the diet has an effect on body weight. To convert dietary fat into body fat, 3% of the energy in the dietary fat is used, leaving 97% for storage as fat deposits. On the other hand, the conversion of dietary carbohydrate to fat requires 23% of the energy in the carbohydrate, leaving just 77% as body fat. If two people have the same kilocaloric intake, the one with the higher proportion of fat in his diet is more likely to gain weight because fewer kilocalories are used to convert the dietary fat into body fat.

Metabolic energy can be used in three ways: for basal metabolism, for the thermic effect of food, and for muscular activity.

Basal Metabolic Rate

The **basal metabolic rate (BMR)** is the metabolic rate calculated in expended kilocalories per square meter of body surface area per hour. It is determined by measuring the oxygen consumption of a person who is awake but restful and has not eaten for 12 h. The liters of oxygen consumed are then multiplied by 4.825 because each liter of oxygen used results in the production of 4.825 kcal of energy. A typical BMR for a 70-kg (154 pounds) male is 38 kcal/m²/h.

BMR is the energy needed to keep the resting body functional. In the average person, basal metabolism accounts for about 60% of energy expenditure. Active transport mechanisms, muscle tone, maintenance of body temperature, beating of the heart, and other activities are supported by basal

metabolism. A number of factors can affect the BMR. Muscle tissue is metabolically more active than adipose tissue, even at rest. Younger people have a higher BMR than older people because of increased cell activity, especially during growth. Fever can increase BMR 7% for each degree Fahrenheit increase in body temperature. During dieting or fasting, greatly reduced kilocaloric input can depress BMR, which apparently is a protective mechanism to prevent weight loss. BMR can be increased on a long-term basis by thyroid hormones and on a short-term basis by epinephrine (see chapter 18). Males have a greater BMR than females because men have proportionately more muscle tissue and less adipose tissue than women. During pregnancy a woman's BMR can increase 20% because of the metabolic activity of the fetus.

Thermic Effect of Food

The second component of metabolic energy concerns the assimilation of food. When food is ingested, the accessory digestive organs and the intestinal lining produce secretions, the motility of the digestive tract increases, active transport increases, and the liver is involved in the synthesis of new molecules. The energy cost of these events is called the **thermic effect of food** and accounts for about 10% of the body's energy expenditure.

Muscular Activity

Muscular activity consumes about 30% of the body's energy. Physical activity resulting from skeletal muscle movement requires the expenditure of energy. In addition, energy must be provided for increased contraction of the heart and of the muscles of respiration. The number of kilocalories used in an activity depends almost entirely on the amount of muscular work performed and on the duration of the activity. Despite the fact that studying can make a person feel tired, intense mental concentration produces little change in the BMR.

Energy loss through muscular activity is the only component of energy expenditure that a person can reasonably control. A comparison of the number of kilocalories gained from food versus the number of kilocalories lost in exercise reveals why losing weight can be difficult. For example, walking (3 mph) for 20 min burns the kilocalories supplied by one slice of bread, whereas jogging (5 mph) for the same time eliminates the kilocalories obtained from a soft drink or a beer (see table 25.1). Nonetheless, weight loss through exercise and dieting is possible.

Body Temperature Regulation

Humans are **homeotherms** (hō′mē-ō-thermz; meaning, uniform warming), or **warm-blooded animals,** and can regulate body temperature rather than have it adjusted by the external environment. Maintenance of a constant body temperature is very important to homeostasis. Most enzymes

Clinical Focus Obesity

Obesity is the storage of excess fat, and it can be classified according to the number and size of fat cells. The greater the amount of lipids stored in the fat cells, the larger their size. In **hyperplastic obesity,** a greater-than-normal number of fat cells occur that are also larger than normal. This type of obesity is associated with massive obesity and begins at an early age. In nonobese children, the number of fat cells triples or quadruples between birth and 2 years of age and then remains relatively stable until puberty, when a further increase in number occurs. In obese children, however, between 2 years of age and puberty, there is also an increase in the number of fat cells. **Hypertrophic obesity** results from a normal number of fat cells that have increased in size. This type of obesity is more common, is associated with moderate obesity or being "overweight," and typically develops in adults. People who were thin or of average weight and quite active when they were young become less active as they become older. They begin to gain weight between age 20 and 40, and, although they no longer use as many kilocalories, they still take in the same amount of food as when they were younger. The unused kilocalories are turned into fat, causing fat cells to increase in size. At one time it was believed that the number of fat cells did not increase after adulthood. It is now known that the number of fat cells can increase in adults. Apparently if all the existing fat cells are filled to capacity with lipids, new fat cells are formed to store the excess lipids. Once fat cells are formed, however, dieting and weight loss do not result in a decrease in the number of fat cells—instead, they become smaller in size as their lipid content decreases.

The distribution of fat in obese individuals can vary. Fat can be found mainly in the upper body, such as in the abdominal region, or it can be associated with the hips and buttocks. These distribution differences can be clinically significant because upper body obesity is associated with an increased likelihood of diabetes mellitus, cardiovascular disease, stroke, and death.

In some cases, a specific cause of obesity can be identified. For example, a tumor in the hypothalamus can stimulate overeating. In most cases, however, no specific cause is apparent. In fact, obesity can occur for many reasons, and obesity in an individual can have more than one cause. There seems to be a genetic component to obesity, and, if one or both parents are obese, their children are more likely to also be obese. Environmental factors such as eating habits, however, can also play an important role. For example, adopted children can exhibit similarities in obesity to their adoptive parents. In addition, psychologic factors such as overeating as a means for dealing with stress can contribute to obesity.

Regulation of body weight is actually a matter of regulating body fat because most changes in body weight reflect changes in the amount of fat in the body. According to the "set point" theory of weight control, the body maintains a certain amount of body fat. If the amount of body fat decreases below or increases above this level, mechanisms are activated to return the amount of body fat to its normal value.

The two factors that affect the amount of adipose tissue in the body are energy intake and energy expenditure. The regulation of energy intake is poorly understood. Apparently appetite and food-seeking behaviors are continually and spontaneously stimulated by neurons originating in or passing through the hypothalamus. After food is consumed, several mechanisms can be responsible for decreasing further food intake. Neural mechanisms such as distension of the stomach are known to inhibit feeding, and a number of hormones released from the gastrointestinal tract or pancreas also decrease appetite. For example, somatostatin, cholecystokinin, glucagon, insulin, and other hormones have been shown to reduce food intake. The level of fatty acids, glucose, or amino acids in the blood can also provide the brain with information necessary to adjust appetite. Low levels of fatty acids, glucose, and amino acids stimulate appetite, whereas high levels of these substances inhibit appetite.

Some scientists believe that the number of fat cells in the body can also affect appetite. According to this line of reasoning, fat cells maintain their size, and, once a "fat plateau" is attained, the body stays at that plateau. Fat cells can accomplish this by effectively taking up triacylglycerols and converting them to fat. Consequently, less energy is available for muscle and body organs, and, to compensate, appetite increases to provide needed energy. In support of this hypothesis, it is known that obese in-

dividuals have an increased amount of the enzyme lipoprotein lipase, which is responsible for the uptake and storage of triacylglycerols in fat cells. Furthermore, in obese individuals who have lost weight, the levels of lipoprotein lipase increase even more.

It is a common belief that the main cause of obesity is overeating. Certainly for obesity to occur, at some time energy intake must have exceeded energy expenditure. A comparison of the kilocaloric intake of obese and lean individuals at their usual weights, however, reveals that on a per kilogram basis, obese people consume fewer kilocalories than lean people.

When people lose a large amount of weight their feeding behavior changes. They become hyperresponsive to external food cues, think of food often, and cannot get enough to eat without gaining weight. It is now understood that this behavior is typical of both lean and obese individuals who are below their relative set point for weight. Other changes such as a decrease in basal metabolic rate take place in a person who has lost a large amount of weight. Most of this decrease in BMR probably results from a decrease in muscle mass associated with weight loss. In addition, there is some evidence that energy lost through exercise and the thermic effect of food are also reduced.

Thus a person who has lost a large amount of weight is a person with an increased appetite and a decreased ability to expend energy. It is no surprise that only a small percentage of obese people maintain weight loss over the long term. Instead, the typical pattern is one of repeated cycles of weight loss followed by a rapid regain of the lost weight.

Current research is attempting to find ways to help manage obesity. Unfortunately, most appetite suppressants can only be used for a short time. Dexfenfluramine (deks-fen-flū′ră-mēn), which had been approved by the FDA for long-term use, has been recalled because of harmful side effects.

There is a gene in mice that can cause obesity if it is absent or defective. The gene is called the obese (*Ob*) gene, and it causes adipose cells to produce a protein called leptin (lep′tin). Leptin suppresses appetite in mice, and even normal mice lose weight when given leptin. Humans have an *Ob* gene that is normal and humans also produce leptin. In fact, some obese people have 20–30 times higher-than-normal levels of leptin. It has been hypothesized that some obese people may have defective receptors for leptin or in some other way do not respond appropriately to leptin. This is analogous to people with noninsulin-dependent diabetes mellitus (see chapter 18), who have increased levels of insulin but do not respond to it.

The message emerging from current research is that body weight results from many complicated genetic and metabolic factors that can go awry in many different ways. Obesity is being regarded as a chronic condition that may someday respond to medication in much the same way that diabetes does. Nonetheless, medication will only be part of the story. Drugs can help, but eating less and exercising more will still be necessary for optimal health.

---

are very temperature sensitive and function only in narrow temperature ranges. Environmental temperatures are too low for normal enzyme function, and the heat produced by metabolism helps maintain the body temperature at a steady, elevated level that is high enough for normal enzyme function.

Free energy is the total amount of energy that can be liberated by the complete catabolism of food. It is usually expressed in terms of kilocalories (kcal) per mole of food consumed. For example, the complete catabolism of 1 mol of glucose (168 g; see chapter 2) releases 686 kcal of free energy. About 43% of the total energy released by catabolism is used to produce ATP and to accomplish biologic work such as anabolism, muscular contraction, and other cellular activities. The remaining energy is lost as **heat.**

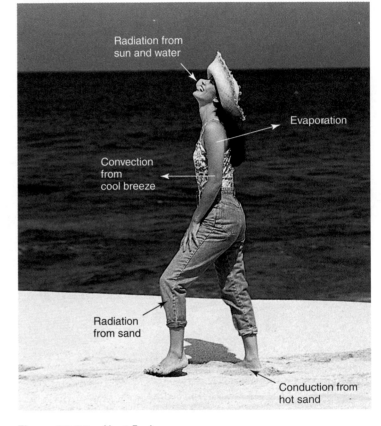

Figure 25.15   Heat Exchange

Heat exchange between a person and the environment occurs by convection, radiation, evaporation, and conduction.

3	P R E D I C T

Explain why we become warm during exercise and why we shiver when it is cold.

✔ *Answer in Appendix F*

Normal body temperature is a range like any other homeostatically controlled condition in the body. The average normal temperature usually is considered 37°C (98.6°F) when it is measured orally and 37.6°C (99.7°F) when it is measured rectally. Rectal temperature comes closer to the true core body temperature, but an oral temperature is more easily obtained in older children and adults and therefore is the preferred measure.

Body temperature is maintained by balancing heat input with heat loss. Heat can be exchanged with the environment in a number of ways (figure 25.15). **Radiation** is the loss of heat as infrared radiation, a type of electromagnetic radiation. For example, the coals in a fire give off radiant heat that can be felt some distance away from the fire. **Conduction** is the exchange of heat between objects in direct contact with each other such as the bottom of the feet and the floor. **Convection** is a transfer of heat between the body and the air. A cool breeze results in movement of air over the body and loss of heat from the body. **Evaporation** is the conversion of water from a liquid to a gaseous form, a process that requires heat. The evaporation of 1 g of water from the body's surface results in the loss of 580 cal of heat.

The amount of heat exchanged between the environment and the body is determined by the difference in temperature between the body and the environment. The greater the temperature difference, the greater the rate of heat exchange. Control of the temperature difference can be used to regulate body temperature. For example, if environmental temperature is very cold such as on a cold winter day, there is a large temperature difference between the body and the environment, and there is a large loss of heat. The loss of heat can be decreased by behaviorally selecting a warmer environment, for example, by going inside a heated house. Heat loss can also be decreased by insulating the exchange surface, such as by

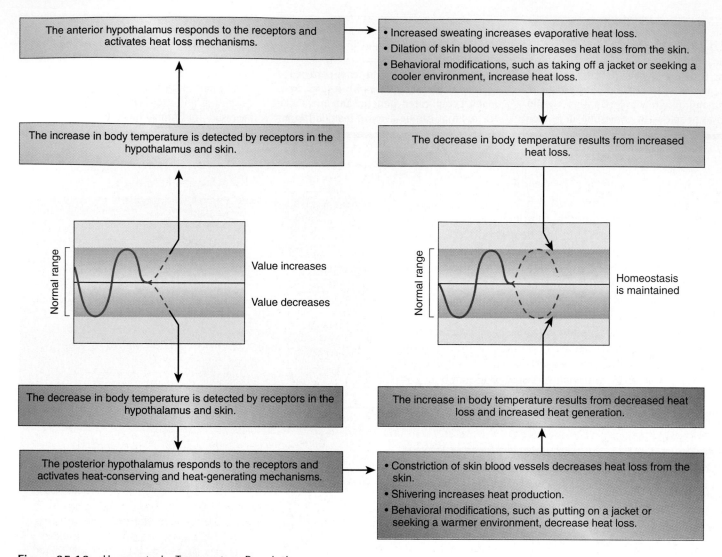

The anterior hypothalamus responds to the receptors and activates heat loss mechanisms.

- Increased sweating increases evaporative heat loss.
- Dilation of skin blood vessels increases heat loss from the skin.
- Behavioral modifications, such as taking off a jacket or seeking a cooler environment, increase heat loss.

The increase in body temperature is detected by receptors in the hypothalamus and skin.

The decrease in body temperature results from increased heat loss.

Normal range

Value increases

Value decreases

Normal range

Homeostasis is maintained

The decrease in body temperature is detected by receptors in the hypothalamus and skin.

The increase in body temperature results from decreased heat loss and increased heat generation.

The posterior hypothalamus responds to the receptors and activates heat-conserving and heat-generating mechanisms.

- Constriction of skin blood vessels decreases heat loss from the skin.
- Shivering increases heat production.
- Behavioral modifications, such as putting on a jacket or seeking a warmer environment, decrease heat loss.

Figure 25.16   Homeostasis: Temperature Regulation

putting on extra clothes. Physiologically, temperature difference can be controlled through dilation and constriction of blood vessels in the skin. When these blood vessels dilate, they bring warm blood to the surface of the body, raising skin temperature; conversely, vasoconstriction decreases blood flow and lowers skin temperature.

4   P R E D I C T

Explain why vasoconstriction of the skin's blood vessels on a cool day is beneficial.

✔ *Answer in Appendix F*

When environmental temperature is greater than body temperature, vasodilation brings warm blood to the skin, causing an increase in skin temperature that decreases heat gain from the environment. At the same time, evaporation carries away excess heat to prevent heat gain and overheating.

Body temperature regulation is an example of a negative-feedback system that is controlled by a "set point." A small area in the anterior part of the hypothalamus can detect slight increases in body temperature through changes in blood temperature (figure 25.16). As a result, mechanisms are activated that cause heat loss, such as vasodilation and sweating, and body temperature decreases. A small area in the posterior hypothalamus can detect slight decreases in body temperature and can initiate heat gain by increasing muscular activity (shivering) and vasoconstriction.

Under some conditions the set point of the hypothalamus is actually changed. For example, during a fever the set point is raised, heat-conserving and heat-producing mechanisms are stimulated, and body temperature increases. In recovery from a fever the set point is lowered to normal, heat loss mechanisms are initiated, and body temperature decreases.

Clinical Focus   Hyperthermia and Hypothermia

Hyperthermia

If heat gain exceeds the ability of the body to lose heat, then body temperature increases above normal levels, a condition called **hyperthermia.** Hyperthermia can result from exposure to hot environments, exercise, fever, and anesthesia.

Exposure to a hot environment normally results in the activation of heat loss mechanisms, and body temperature is maintained at normal levels. This is an excellent example of a negative-feedback mechanism. Prolonged exposure to a hot environment, however, can result in **heat exhaustion.** The normal negative-feedback mechanisms for controlling body temperature are operating, but they are unable to maintain a normal body temperature. Heavy sweating results in dehydration, decreased blood volume, decreased blood pressure, and increased heart rate. Individuals suffering from heat exhaustion have a wet, cool skin because of the heavy sweating. They usually feel weak, dizzy, and nauseated. Treatment includes reducing heat gain by moving to a cooler environment, reducing heat production by muscles by ceasing activity, and restoring blood volume by drinking fluids.

Heat stroke is a breakdown of the normal negative-feedback mechanisms of temperature regulation. If the temperature of the hypothalamus becomes too high, it no longer functions appropriately. Sweating stops, and the skin becomes dry and flushed. The person becomes confused, irritable, or even comatose. In addition to the treatment for heat exhaustion, heat loss from the skin should be increased. This can be accomplished by increasing evaporation from the skin by applying wet cloths or by increasing conductive heat loss by immersing the person in a cool bath.

Exercise increases body temperature because of the heat produced as a by-product of muscle activity (see chapter 10). Normally vasodilation and increased sweating prevent body temperature increases that are harmful. In a hot, humid environment the evaporation of sweat is decreased, and exercise levels have to be reduced to prevent overheating.

Fever is the development of a higher-than-normal body temperature following the invasion of the body by microorganisms or foreign substances. Lymphocytes, neutrophils, and macrophages release chemicals called **pyrogens** (pī′rō-jenz) that raise the temperature set point of the hypothalamus. Consequently body temperature and metabolic rate increase. Fever is believed to be beneficial because it speeds up the chemical reactions of the immune system (see chapter 22) and inhibits the growth of some microorganisms. Although beneficial, body temperatures greater than 41°C (106°F) can be harmful. Aspirin lowers body temperature by affecting the hypothalamus, resulting in dilation of skin blood vessels and sweating.

Malignant hyperthermia is an inherited muscle disorder. Drugs used to induce general anesthesia for surgery cause sustained, uncoordinated muscle contractions in some individuals. Consequently body temperature increases.

Therapeutic hyperthermia is an induced local or general body increase in temperature. It is a treatment sometimes used on tumors and infections.

Hypothermia

If heat loss exceeds the ability of the body to produce heat, body temperature decreases below normal levels. **Hypothermia** is a decrease in body temperature to 35°C (95°F) or below. Hypothermia usually results from prolonged exposure to cold environments. At first, normal negative-feedback mechanisms maintain body temperature. Heat loss is decreased by constricting blood vessels in the skin, and heat production is increased by shivering. If body temperature decreases despite these mechanisms, hypothermia develops. The individual's thinking becomes sluggish, and movements are uncoordinated. Heart, respiratory, and metabolic rates decline, and death results unless body temperature is restored to normal. Rewarming should occur at a rate of a few degrees per hour.

Frostbite is damage to the skin and deeper tissues resulting from prolonged exposure to the cold. Damage results from direct cold injury to cells, injury from ice crystal formation, and reduced blood flow to affected tissues. The fingers, toes, ears, nose, and cheeks are most commonly affected. Damage from frostbite can range from redness and discomfort to loss of the affected part. The best treatment is immersion in a warm water bath. Rubbing the affected area and local, dry heat should be avoided.

Therapeutic hypothermia is sometimes used to slow metabolic rate during surgical procedures such as heart surgery. Because metabolic rate is decreased, tissues do not require as much oxygen as normal and are less likely to be damaged.

Summary

Nutrition

Nutrition is the taking in and use of food.

Nutrients

1. Nutrients are the chemicals used by the body and consist of carbohydrates, lipids, proteins, vitamins, minerals, and water.
2. Essential nutrients are nutrients that must be ingested because the body cannot manufacture them or is unable to manufacture adequate amounts of them.

Kilocalories

1. A kilocalorie (kcal) is 1000 calories. A calorie is the heat (energy) necessary to raise the temperature of 1 g of water 1°C.
2. A gram of carbohydrate or protein yields 4 kcal, and a gram of fat yields 9 kcal.

Carbohydrates

1. Carbohydrates are ingested as monosaccharides (glucose, fructose), disaccharides (sucrose, maltose, lactose), and polysaccharides (starch, glycogen, cellulose).

2. Polysaccharides and disaccharides are converted to glucose. Glucose can be used for energy or stored as glycogen or fats.
3. About 125–175 g of carbohydrates should be ingested each day.

Lipids

1. Lipids are ingested as triacylglycerols (95%) or cholesterol and phospholipids (5%).
2. Triacylglycerols are used for energy or are stored in adipose tissue. Cholesterol forms other molecules such as steroid hormones. Cholesterol and phospholipids are part of the plasma membrane.
3. The daily diet should derive no more than 30% of its kilocalories from lipids, and no more than 300 mg should be in the form of cholesterol.

Proteins

1. Proteins are ingested and broken down into amino acids.
2. Proteins perform many functions: protection (antibodies), regulation (enzymes, hormones), structure (collagen), muscle contraction (actin and myosin), and transportation (hemoglobin, carrier molecules, ion channels).
3. An adult should consume 0.8 g of protein per kilogram of body weight each day.

Vitamins

1. Many vitamins function as coenzymes or as parts of coenzymes.
2. Most vitamins are not produced by the body and must be obtained in the diet. Some vitamins can be formed from provitamins.
3. Vitamins are classified as either fat-soluble or water-soluble.

Minerals

Minerals are necessary for normal metabolism, add mechanical strength to bones and teeth, function as buffers, and are involved in osmotic balance.

Metabolism

1. Metabolism consists of anabolism and catabolism. Anabolism is the building up of molecules and requires energy. Catabolism is the breaking down of molecules and gives off energy.
2. The energy in carbohydrates, lipids, and proteins is used to produce ATP through oxidation–reduction reactions.

Carbohydrate Metabolism

1. Glycolysis is the breakdown of glucose to two pyruvic acid molecules. Also produced are two NADH molecules and two ATP molecules.
2. Anaerobic respiration is the breakdown of glucose in the absence of oxygen to two lactic acid and two ATP molecules.
3. Lactic acid can be converted to glucose (Cori cycle) using aerobically produced ATP (oxygen debt).
4. Aerobic respiration is the breakdown of glucose in the presence of oxygen to produce carbon dioxide, water, and 38 (or 36) ATP molecules.
   - The first phase is glycolysis, which produces two ATP, two NADH, and two pyruvic acid molecules.
   - The second phase is the conversion of the two pyruvic acid molecules to two molecules of acetyl-CoA. These reactions also produce two NADH and two carbon dioxide molecules.
   - The third phase is the citric acid cycle, which produces two ATP, six NADH, two $FADH_2$, and four carbon dioxide molecules.

- The fourth phase is the electron-transport chain. The high-energy electrons in NADH and $FADH_2$ enter the electron-transport chain and are used in the synthesis of ATP and water.

Lipid Metabolism

1. Adipose triacylglycerols are broken down and released as free fatty acids.
2. Free fatty acids are taken up by cells and broken down by beta-oxidation into acetyl-CoA.
   - Acetyl-CoA can enter the citric acid cycle.
   - Acetyl-CoA can be converted into ketone bodies.

Protein Metabolism

1. New amino acids are formed by transamination, the transfer of an amine group to a keto acid.
2. Amino acids are used to synthesize proteins. If used for energy, ammonia is produced as a by-product of oxidative deamination. Ammonia is converted to urea and is excreted.

Interconversion of Nutrient Molecules

1. Glycogenesis is the formation of glycogen from glucose.
2. Lipogenesis is the formation of lipids from glucose and amino acids.
3. Glycogenolysis is the breakdown of glycogen to glucose.
4. Gluconeogenesis is the formation of glucose from amino acids and glycerol.

Metabolic States

1. In the absorptive state nutrients are used as energy or stored.
2. In the postabsorptive state stored nutrients are used for energy.

Metabolic Rate

Metabolic rate is the total energy expenditure per unit of time, and it has three components.

Basal Metabolic Rate

Basal metabolic rate is the energy used at rest. It is 60% of the metabolic rate.

Thermic Effect of Food

The thermic effect of food is the energy used to digest and absorb food. It is 10% of the metabolic rate.

Muscular Activity

Muscular energy is used for muscle contraction. It is 30% of the metabolic rate.

Body Temperature Regulation

1. Body temperature is a balance between heat gain and heat loss.
   - Heat is produced through metabolism.
   - Heat is exchanged through radiation, conduction, convection, and evaporation.
2. The greater the temperature difference between the body and the environment, the greater the rate of heat exchange.
3. Body temperature is regulated by a "set point" in the hypothalamus.

Content Review

1. Define the term nutrient, and list the six major classes of nutrients.
2. Define the term kilocalorie, and state the number of kilocalories supplied by a gram of carbohydrate, lipid, and protein.
3. List the dietary sources of carbohydrates. After they are converted to glucose, what happens to the glucose? What quantities of carbohydrate should be ingested daily?
4. List the dietary sources of lipids; explain how triacylglycerols, cholesterol, and phospholipids are used in the body, and describe the recommended dietary intake of lipids.
5. List the dietary sources of complete and incomplete protein foods. Describe some of the functions performed by proteins in the body. What is the recommended daily consumption of proteins?
6. What are vitamins and provitamins? Name the water-soluble vitamins and the fat-soluble vitamins. List some of the functions of vitamins.
7. List some of the minerals, and give their functions.
8. Define the terms metabolism, anabolism, and catabolism.
9. How does the removal of hydrogen atoms from nutrient molecules result in a loss of energy from the nutrient molecule?
10. Describe glycolysis. Although four ATP molecules are produced in glycolysis, explain why there is a net gain of only two ATP molecules.
11. What determines whether the pyruvic acid produced in glycolysis becomes lactic acid or acetyl-CoA?
12. Describe the two phases of anaerobic respiration. How many ATP molecules are produced? What happens to the lactic acid produced when oxygen becomes available?
13. Define the term aerobic respiration, and list the products produced by it. Name the four phases of aerobic respiration.
14. Why is the citric acid cycle a cycle? What molecules are produced as a result of the citric acid cycle?
15. What is the function of the electron-transport chain? Describe the chemiosmotic model of ATP production.
16. Define the term beta-oxidation, and explain how it results in ATP production.
17. What are ketone bodies, how are they produced, and for what are they used?
18. Distinguish between an essential and a nonessential amino acid.
19. Define the terms transamination and oxidative deamination. How are proteins used to produce energy?
20. Define the terms glycogenesis, lipogenesis, glycogenolysis, and gluconeogenesis.
21. Describe the events of the absorptive and the postabsorptive states. Why is it important to maintain blood glucose levels?
22. Define the term metabolic rate, and describe its three components.
23. How are kilocaloric input and output adjusted to maintain body weight?
24. Explain the ways that heat is produced and lost by the body. How does the hypothalamus regulate body temperature?

Develop Your Reasoning Skills

1. Why does a vegetarian usually have to be more careful about his or her diet than a person who includes meat in the diet?
2. Explain why a person suffering from copper deficiency feels tired all the time.
3. Some people claim that fasting occasionally for short periods can be beneficial. How can fasts be damaging?
4. Why can some people lose weight on a 1200-kcal/day diet and others cannot?
5. Lotta Bulk, a muscle builder, wanted to increase her muscle mass. Knowing that proteins are the main components of muscle, she began a high-protein diet in which most of her daily kilocalories were supplied by proteins. She also exercised regularly with heavy weights. After 3 months of this diet and exercise program, Lotta increased her muscle mass, but not any more than her friend, who did the same exercises but did not have a high-protein diet. Explain what happened. Was Lotta in positive or negative nitrogen balance?
6. On learning that sweat evaporation results in the loss of calories, an anatomy and physiology student enters a sauna in an attempt to lose weight. He reasons that a liter (about a quart) of water weighs 1000 g, which is equivalent to 580,000 cal or 580 kcal of heat when lost as sweat. Instead of reducing his diet by 580 kcal/day, if he loses a liter of sweat every day in the sauna, he believes he will lose about a pound of fat a week. Will this approach work? Explain.

Web Site Link

For a listing of the most current web sites related to this chapter, please visit the Seeley home page at:
http://www.mhhe.com/biosci/ap/seeleyap/

Urinary System

Objectives

1. List the components of the urinary system, and describe its overall functions.

2. Describe the location, size, shape, and internal anatomy of the kidneys.

3. Describe the structure of the nephron and the orientation of its parts within the kidney.

4. Describe the course of blood flow through the kidney, and identify the blood volume that flows through the kidney.

5. List the components of the filtration barrier, and describe its structure and the composition of the filtrate.

6. List factors that influence filtration pressure and the rate of filtrate formation.

7. Explain how tubular reabsorption in the proximal tubule is accomplished and how it influences filtrate composition.

8. Describe the permeability characteristics of the descending limb of the loop of Henle, and discuss how the movement of substances across its wall influences the composition of the filtrate.

9. Describe permeability and transport characteristics of the ascending limb of the loop of Henle, and explain how they influence filtrate composition.

10. Describe the permeability and transport characteristics of the distal tubule and the collecting duct, and explain how they influence filtrate composition.

11. Explain the function of the vasa recta.

12. Describe the role of the loops of Henle and the collecting ducts in producing concentrated urine.

13. Define autoregulation, and explain how it influences renal function.

14. Explain the effect that sympathetic stimulation has on the kidney during rest, exercise, and shock.

15. Define and explain tubular maximum and plasma clearance.

16. Describe the micturition reflex.

The kidneys can suffer extensive damage and still maintain their extremely important role in the maintenance of homeostasis. As long as about one-third of one kidney remains functional, survival is possible. If the kidneys stop functioning completely, however, death results without special medical treatment. The role the kidneys play in controlling the composition and volume of body fluids is essential.

The kidneys are the major excretory organs of the body. They remove most waste products, many of which are toxic, from the blood, and they play a major role in controlling blood volume, the concentration of ions in the blood, the pH of the blood, red blood cell production, and vitamin D metabolism. The skin, liver, lungs, and intestines eliminate some waste products, but if the kidneys fail to function, other excretory organs cannot adequately compensate.

The **urinary system** consists of the following organs: (1) two kidneys; (2) a single, midline urinary bladder; (3) two ureters, which carry urine from the kidneys to the urinary bladder (figure 26.1a); and (4) a single urethra, which carries urine from the bladder to the outside of the body.

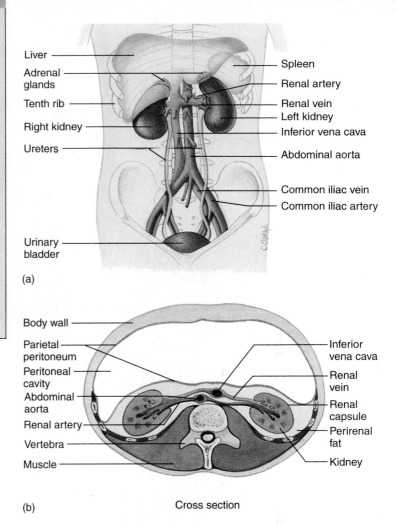

(a)

(b)   Cross section

Figure 26.1   Anatomy of the Urinary System

The urinary system consists of two kidneys, two ureters, a single urinary bladder, and a single urethra. (a) The kidneys are located in the abdominal cavity, with the right kidney just below the liver and the left kidney below the spleen. The ureters extend from the kidneys to the urinary bladder within the pelvic cavity. An adrenal gland is located at the superior pole of each kidney. (b) The kidneys are located behind the parietal peritoneum. Surrounding each kidney is the renal fat pad. The renal arteries extend from the abdominal aorta to each kidney, and the renal veins extend from the kidneys to the inferior vena cava.

Urinary System

Kidneys

The **kidneys** are bean-shaped organs, each about the size of a tightly clenched fist. They lie on the posterior abdominal wall behind the peritoneum and on either side of the vertebral column near the lateral borders of the psoas muscles (see figure 26.1a and b). The superior pole of each kidney is protected by the rib cage, and the right kidney is slightly lower than the left because of the presence of the liver superior to it. Each kidney measures about 11 cm long, 5 cm wide, and 3 cm thick and weighs about 130 g. A fibrous connective tissue layer, called the **renal capsule,** encloses each kidney, and around the renal capsule is a dense deposit of adipose tissue, the **perirenal fat,** which protects the kidney from mechanical shock. The kidneys and surrounding adipose tissue are anchored to the abdominal wall by a thin layer of loose connective tissue, the **renal fascia.**

On the medial side of each kidney is a small area called the **hilum** (hī′lŭm, meaning a small amount), where the renal artery and nerves enter and the renal vein and the ureter exit. The hilum opens into a cavity called the **renal sinus,** which contains fat and connective tissue (figure 26.2).

The kidney is divided into an outer **cortex** and an inner **medulla** that surrounds the renal sinus. The medulla consists of a number of cone-shaped **renal pyramids,** which appear triangular when seen in a longitudinal section of the kidney. Extensions of the pyramids, called **medullary rays,** project from the pyramids into the cortex, and extensions of the cortex, called **renal columns,** project between the pyramids. The base of each pyramid is located at the boundary between the cortex and the medulla, and the tips of the pyramids, the **renal papillae,** are pointed toward the renal sinus. Funnel-shaped structures called **minor calyces** (kal′-i-sēz, meaning cup of a flower) sur-

round the renal papillae. The minor calyces from several pyramids join together to form larger funnels called **major calyces.** There are 8–20 minor calyces and 2 or 3 major calyces per kidney. The major calyces converge to form an enlarged channel called the **renal pelvis,** which is located in the renal sinus. The renal pelvis then narrows to form a small-diameter tube, the **ureter** (yūr-rē′-ter), which exits the kidney and connects to the urinary bladder. Urine formed within the kidneys passes from the renal papillae into the minor calyces. From the minor calyces, urine moves into the major calyces, collects in the renal pelvis, and exits the kidney through the ureter.

The basic histologic and functional unit of the kidney is the **nephron** (nef′ron; figure 26.3), which consists of an enlarged terminal end called the renal corpuscle, a proximal tubule, a loop of Henle (nephric loop), and a distal tubule.

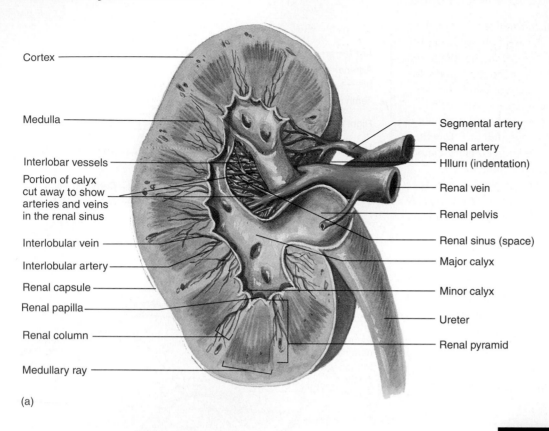

Cortex

Medulla

Interlobar vessels

Portion of calyx cut away to show arteries and veins in the renal sinus

Interlobular vein

Interlobular artery

Renal capsule

Renal papilla

Renal column

Medullary ray

(a)

Segmental artery

Renal artery

Hilum (indentation)

Renal vein

Renal pelvis

Renal sinus (space)

Major calyx

Minor calyx

Ureter

Renal pyramid

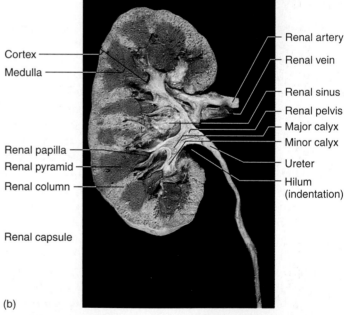

Cortex

Medulla

Renal papilla

Renal pyramid

Renal column

Renal capsule

(b)

Renal artery

Renal vein

Renal sinus

Renal pelvis

Major calyx

Minor calyx

Ureter

Hilum (indentation)

Figure 26.2   Longitudinal Section of the Kidney and Ureter

(a) The cortex forms the outer part of the kidney, and the medulla forms the inner part. A central cavity called the renal sinus contains the renal pelvis. The renal columns of the kidney project from the cortex into the medulla and separate the pyramids. (b) Photograph of a longitudinal section of a human kidney and ureter.

The distal tubule empties into a collecting duct, which carries the urine from the cortex of the kidney to a minor calyx. The renal corpuscles, proximal tubules, and distal tubules are in the renal cortex. The collecting tubules and parts of the loops of Henle enter the renal medulla.

There are approximately 1.3 million nephrons in each kidney, and one-third of them must be functional to ensure survival. Most nephrons measure 50–55 mm in length, although the nephrons with renal corpuscles located within the cortex near the medulla are longer than the those in the cortex nearer to the exterior of the kidney. Nephrons whose renal corpuscles lie near the medulla are called **juxtamedullary** (jŭks′tă-med′ū-lăr-ē; *juxta* is Latin meaning next to) **nephrons** and make up about 15% of all the nephrons. The juxtamedullary nephrons have longer loops of Henle, which extend farther into the medulla than the loops of Henle of nephrons whose renal corpuscles originate in the superficial cortex (see figure 26.3).

Each **renal corpuscle** consists of the enlarged end of a nephron called **Bowman's capsule** and a network of capil-

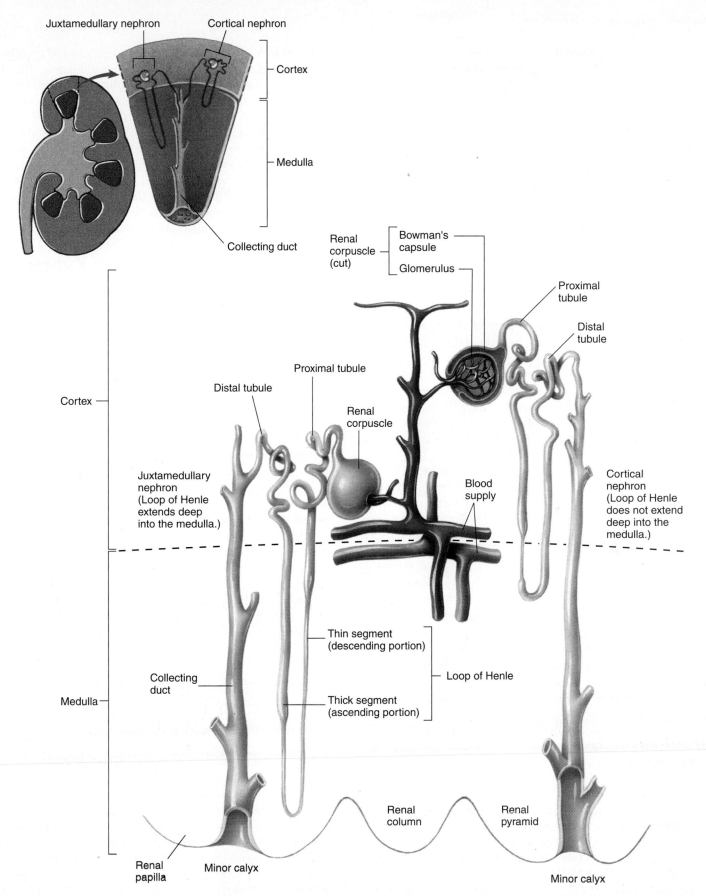

Figure 26.3 Functional Unit of the Kidney—the Nephron

A nephron consists of a renal corpuscle, proximal tubule, loop of Henle, and distal tubule. The distal tubule empties into a collecting duct. Juxtamedullary nephrons (those near the medulla of the kidney) have loops of Henle that extend deep into the medulla of the kidney, whereas other nephrons do not.

laries called the **glomerulus** (glō-măr′yū-lŭs). The wall of Bowman's capsule is indented to form a double-walled chamber, and the indentation is occupied by the glomerulus, which resembles a ball of yarn (figure 26.4*a* and *b*). Fluid passes from the glomerulus into Bowman's capsule.

The cavity of Bowman's capsule opens into the proximal tubule, which carries fluid away from Bowman's capsule (see figure 26.4*b*). The outer **parietal layer** of Bowman's capsule is composed of simple squamous epithelium, which becomes cuboidal at the beginning of the proximal tubule. Surrounding the glomerulus is the inner layer of Bowman's capsule, called the **visceral layer.** It consists of specialized cells called **podocytes** (pŏd′o-sītz).

The walls of the glomerular capillaries are lined with endothelial cells that have openings called **fenestrae** (fe-nes′trē, meaning windows). The visceral layer of Bowman's capsule surrounds the glomerular capillaries with gaps, called **filtration slits,** between the podocyte processes surrounding the capillaries. There is a basement membrane between the endothelial cells of the glomerular capillaries and the podocytes of Bowman's capsule. The capillary endothelium, the basement membrane, and the podocytes of Bowman's capsule make up the **filtration membrane** (figure 26.4*c* and *d*). In the first step of urine formation, fluid passes from the glomerular capillaries into the lumen of Bowman's capsule across the filtration membrane.

The glomerulus is supplied by an **afferent** (af′er-ent) **arteriole** and is drained by an **efferent** (ef′er-ent) **arteriole.** The afferent and efferent arterioles both have a layer of smooth muscle. At the point where the afferent arteriole enters the renal corpuscle, the smooth muscle cells are modified to form a cuff around the arteriole. These modified cells are called **juxtaglomerular cells.** A part of the distal tubule of the nephron lies adjacent to the renal corpuscle between the afferent and efferent arterioles. The specialized tubule cells in that area are collectively called the **macula** (mak′yū-lă) **densa.** The juxtaglomerular cells of the afferent arteriole and the macula densa cells are in intimate contact with one another, and together they are called the **juxtaglomerular apparatus** (see figure 26.4*b*).

The **proximal tubule,** also called the **proximal convoluted tubule,** is approximately 14 mm long and 60 μm in diameter, and its wall is composed of simple cuboidal epithelium. The cells are broader at their base, which lies away from the lumen, than they are at the surface of the lumen, and they have microvilli at their luminal surface (figure 26.5*a* and *b*).

The **loops of Henle** (nephric loops) are continuations of the proximal tubules. Each loop has a **descending limb** and an **ascending limb.** The first part of the descending limb is similar in structure to the proximal tubules. The loops of Henle that extend into the medulla become very thin near the end of the loop (figure 26.5*a* and *c*). In the thin part, the lumen becomes narrow, and there is an abrupt transition from simple cuboidal epithelium to simple squamous epithelium. The first part of the ascending limb is also very thin and consists of simple squamous epithelium, but it soon becomes thicker and is again composed of simple cuboidal epithelium. The thick part

of the loop returns toward the renal corpuscle and ends by giving rise to the distal tubule near the macula densa. The **distal tubules,** also called the **distal convoluted tubules**, are not as long as the proximal tubules. The epithelium is simple cuboidal, but the cells are smaller than the epithelial cells in the proximal tubules and do not possess a large number of microvilli (figure 26.5*a* and *d*). The **collecting ducts** are composed of simple cuboidal epithelium, are joined by the distal tubules of many nephrons, and are larger in diameter than other segments of the nephron (figure 26.5*a* and *e*). The collecting ducts form much of the medullary rays, and they extend through the medulla to the tips of the renal pyramids.

Arteries and Veins of the Kidneys

A **renal artery** branches off the abdominal aorta and enters the renal sinus of each kidney (figure 26.6*a*). **Segmental arteries** diverge from the renal artery to form **interlobar arteries,** which ascend within the renal columns toward the renal cortex. Branches from the interlobar arteries diverge near the base of each pyramid and arch over the base of the pyramids to form the **arcuate** (ar′kū-āt) **arteries. Interlobular arteries** project from the arcuate arteries into the cortex, and the afferent arterioles are derived from the interlobular arteries or their branches. The afferent arterioles supply blood to the glomerular capillaries of the renal corpuscles. Efferent arterioles arise from the glomerular capillaries and carry blood away from the glomeruli. After each efferent arteriole exits the glomerulus, it gives rise to a plexus of capillaries called the **peritubular capillaries** around the proximal and distal tubules. Specialized parts of the peritubular capillaries, called **vasa recta** (vāsă rek′tă), course into the medulla along with the loops of Henle (figure 26.6*b*) and then back toward the cortex. The peritubular capillaries drain into **interlobular veins,** which in turn drain into the **arcuate veins.** The arcuate veins empty into the **interlobar veins,** which drain into the **renal vein.** The renal vein exits the kidney and connects to the inferior vena cava.

Ureters and Urinary Bladder

The **ureters** extend inferiorly and medially from the renal pelvis at the renal hilum to reach the urinary bladder (see figure 26.1; figure 26.7), which stores urine. The **urinary bladder** is a hollow muscular container that lies in the pelvic cavity just posterior to the symphysis pubis. In the male it is just anterior to the rectum, and in the female it is just anterior to the vagina and inferior and anterior to the uterus. The size of the bladder depends on the presence or absence of urine. The ureters enter the bladder inferiorly on its posterolateral surface, and the urethra exits the bladder inferiorly and anteriorly (see figure 26.7*a*). The triangular area of the bladder wall between the two ureters posteriorly and the urethra anteriorly is called the **trigone** (trī′gōn). This region differs histologically from the rest of the bladder wall and expands minimally during bladder filling.

The ureters and urinary bladder are lined with transitional epithelium, which is surrounded by a lamina propria, a

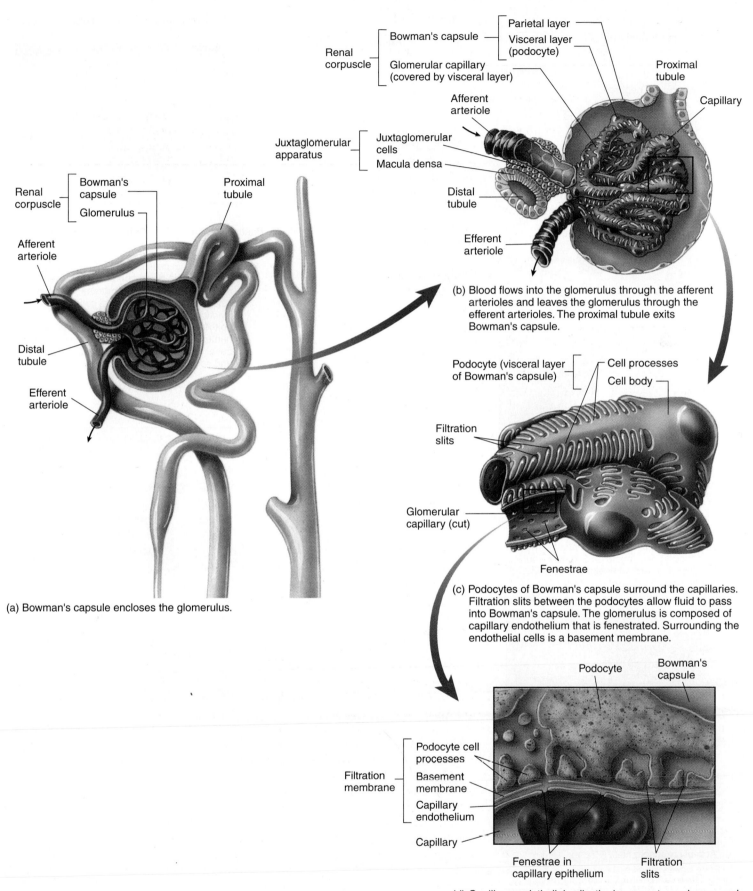

Parietal layer
Bowman's capsule
Visceral layer
(podocyte)
Renal
corpuscle
Glomerular capillary
(covered by visceral layer)

Proximal
tubule

Capillary

Afferent
arteriole

Juxtaglomerular
apparatus
Juxtaglomerular
cells
Macula densa

Distal
tubule

Efferent
arteriole

(b) Blood flows into the glomerulus through the afferent
arterioles and leaves the glomerulus through the
efferent arterioles. The proximal tubule exits
Bowman's capsule.

Renal
corpuscle
Bowman's
capsule
Glomerulus

Proximal
tubule

Afferent
arteriole

Distal
tubule

Efferent
arteriole

(a) Bowman's capsule encloses the glomerulus.

Podocyte (visceral layer
of Bowman's capsule)
Cell processes
Cell body

Filtration
slits

Glomerular
capillary (cut)

Fenestrae

(c) Podocytes of Bowman's capsule surround the capillaries.
Filtration slits between the podocytes allow fluid to pass
into Bowman's capsule. The glomerulus is composed of
capillary endothelium that is fenestrated. Surrounding the
endothelial cells is a basement membrane.

Podocyte

Bowman's
capsule

Filtration
membrane
Podocyte cell
processes
Basement
membrane
Capillary
endothelium

Capillary

Fenestrae in
capillary epithelium
Filtration
slits

(d) Capillary endothelial cells, the basement membrane, and
podocytes make up the filtration membrane of the kidney.

Figure 26.4   Renal Corpuscle

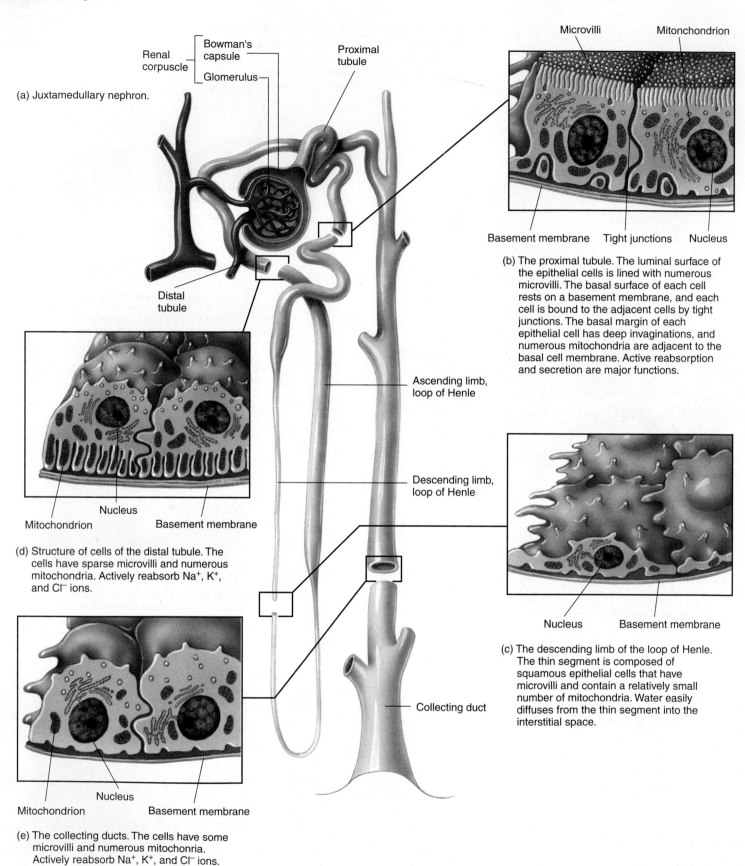

(a) Juxtamedullary nephron.

Renal corpuscle
Bowman's capsule
Glomerulus
Proximal tubule

Microvilli   Mitonchondrion

Basement membrane   Tight junctions   Nucleus

(b) The proximal tubule. The luminal surface of the epithelial cells is lined with numerous microvilli. The basal surface of each cell rests on a basement membrane, and each cell is bound to the adjacent cells by tight junctions. The basal margin of each epithelial cell has deep invaginations, and numerous mitochondria are adjacent to the basal cell membrane. Active reabsorption and secretion are major functions.

Distal tubule

Ascending limb, loop of Henle

Descending limb, loop of Henle

Nucleus
Mitochondrion   Basement membrane

(d) Structure of cells of the distal tubule. The cells have sparse microvilli and numerous mitochondria. Actively reabsorb Na^+, K^+, and Cl^- ions.

Nucleus   Basement membrane

(c) The descending limb of the loop of Henle. The thin segment is composed of squamous epithelial cells that have microvilli and contain a relatively small number of mitochondria. Water easily diffuses from the thin segment into the interstitial space.

Collecting duct

Nucleus
Mitochondrion   Basement membrane

(e) The collecting ducts. The cells have some microvilli and numerous mitochonria. Actively reabsorb Na^+, K^+, and Cl^- ions.

Figure 26.5   Histology of the Nephron

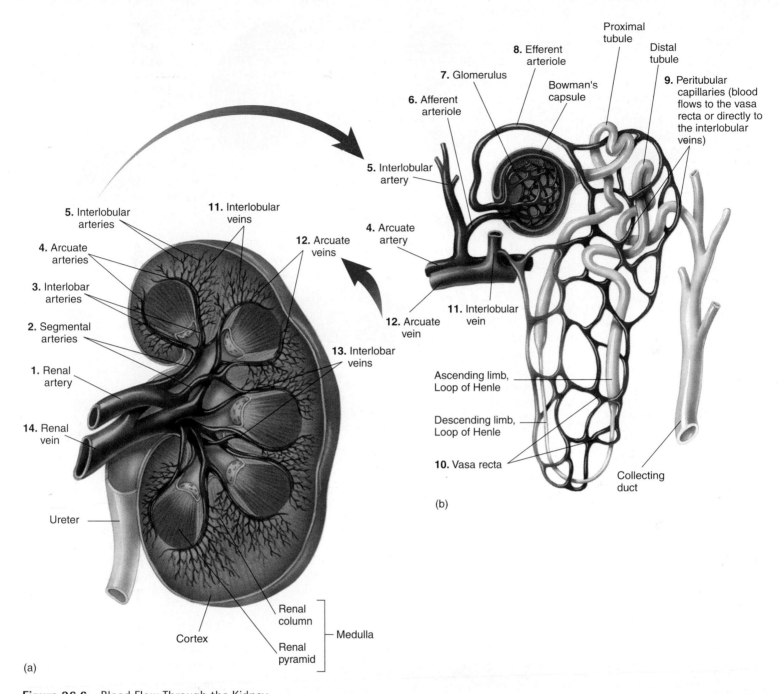

Figure 26.6 Blood Flow Through the Kidney

(*a*) Blood flow through the larger arteries and veins of the kidney is illustrated. (*b*) Blood flow through the arteries, capillaries, and veins that provide circulation to the nephrons is illustrated. 𝕏

muscular coat, and a fibrous adventitia (see figure 26.7*b* and *c*). The wall of the bladder is much thicker than the wall of a ureter. This thickness is caused by the layers, composed primarily of smooth muscle, that are external to the epithelium. The epithelium itself ranges from four or five cells thick in the empty urinary bladder to two or three cells thick when the bladder is distended. Transitional epithelium is specialized so that the cells slide past one another, and the number of cell layers decreases as the volume of the urinary bladder increases. The epithelium of the urethra is stratified or pseudostratified columnar epithelium.

At the junction of the urethra with the urinary bladder, smooth muscle of the bladder forms the **internal urinary sphincter.** The **external urinary sphincter** is skeletal muscle that surrounds the urethra as the urethra extends through the pelvic floor. The sphincters control the flow of urine through the urethra.

In the male the urethra extends to the end of the penis, where it opens to the outside (see chapter 28). The urethra is much shorter in the female than in the male and it opens into the vestibule anterior to the vaginal opening.

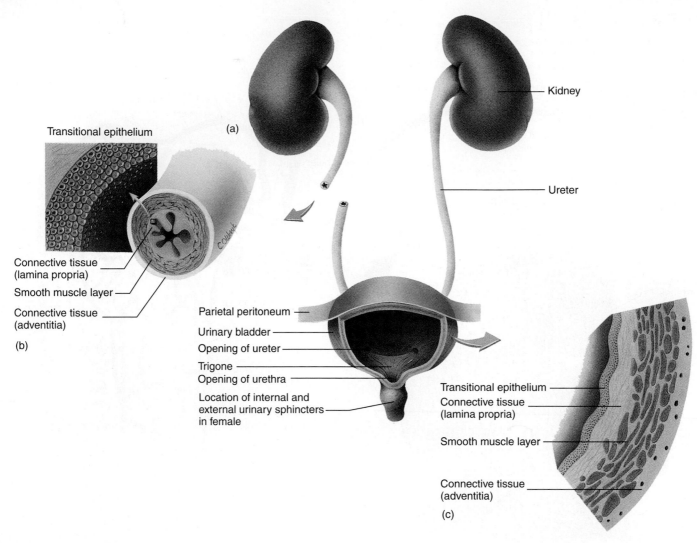

Figure 26.7   Ureters and Urinary Bladder

(*a*) Ureters extend from the pelvis of the kidney to the urinary bladder. (*b*) The walls of the ureters and the urinary bladder are lined with transitional epithelium, which is surrounded by a connective tissue layer (lamina propria), smooth muscle layers, and a fibrous adventitia. (*c*) Section through the wall of the urinary bladder.

---

1   **P R E D I C T**

Cystitis (sis-tī'tis) is inflammation of the urinary bladder that typically results from infections, and urinary bladder infections often occur when bacteria from outside the body enter the bladder. Are males or females more prone to urinary bladder infections? Explain.

✔ *Answer in Appendix F*

Urine Production

Because nephrons are the smallest structural components capable of producing urine, they are called the **functional units of the kidney.** Filtration, reabsorption, and secretion are the three major processes critical to the formation of urine

(figure 26.8). **Filtration** is movement of fluid across the filtration membrane as a result of a pressure difference. The fluid entering the nephron becomes the **filtrate. Reabsorption** is the movement of substances from the filtrate back into the blood. In general, most of the water and useful solutes are reabsorbed, while waste products, excess solutes, and a small amount of water are not (table 26.1). **Secretion** is the active transport of solutes into the nephron. Urine produced by the nephrons consists of solutes and water filtered and solutes secreted into the nephron minus the solutes and water that are reabsorbed.

Filtration

The part of the total cardiac output that passes through the kidneys is called the **renal fraction.** Table 26.2 presents the calculation of renal blood flow rate and other kidney flow

Urine formation results from the following three processes:

1. Filtration   Filtration (*blue arrow*) is the movement of materials across the filtration membrane into the lumen of Bowman's capsule to form filtrate.

2. Reabsorption   Solutes are reabsorbed (*purple arrow*) across the wall of the nephron by transport processes, such as active transport and cotransport.

Water is reabsorbed (*green arrow*) across the wall of the nephron by osmosis.

3. Secretion   Solutes are secreted (*orange arrow*) across the wall of the nephron into the filtrate.

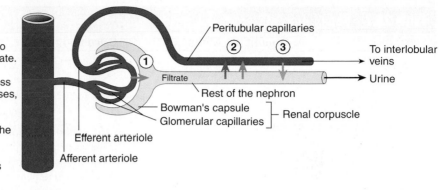

Figure 26.8   Urine Formation

Table 26.1   Concentrations of Major Solutes

Substance	Plasma	Filtrate	Net Movement of Solute*	Urine	Concentration Urine / Concentration Plasma
Water (L)	180	180	178.6	1.4	—
Organic molecules (mg/100 mL)					
Protein	3900–5000	6–11	−100.0	0†	0
Glucose	100	100	−100.0	0	0
Urea	26	26	−11.4	1820	70
Uric acid	3	3	−2.7	42	14
Creatinine	1.1	1.1	0.5	196	140
Ions (mEq/L)					
Na^+	142	142	−141.0	128	0.9
K^+	5	5	−4.5	60	12.0
Cl^-	103	103	−101.9	134	1.3
HCO_3^-	28	28	−27.9	14	0.5

*In many cases solute moves into and out of the nephron. Figures indicate net movement. Negative numbers are net movement out of the filtrate, and positive numbers are net movement into the filtrate.

†Trace amounts of protein can be found in the urine. A value of zero is assumed here.

rates. Although the renal fraction varies from 12% to 30% of the cardiac output in healthy resting adults, it averages 21%, to produce a **renal blood flow rate** of 1176 mL/min. The plasma that flows through the kidneys each minute, called the **renal plasma flow rate,** is equal to the renal blood flow rate multiplied by the portion of the blood that is made up of plasma, which is approximately 55% (1176 mL/min × 0.55 = 646.8 mL plasma/min, or approximately 650 mL plasma/min).

The part of the plasma flowing through the kidney that is filtered through the filtration membranes into the lumen of Bowman's capsules to become filtrate is called the **filtration fraction.** The filtration fraction averages 19% of the plasma flowing through the kidney (650 mL plasma/min × 0.19 = 123.5 mL plasma/min). Thus, approximately 125 mL of filtrate is produced each minute. The amount of filtrate produced each minute is called the **glomerular filtration rate (GFR),** which is equivalent to approximately 180 L of filtrate produced daily. Because only 1–2 L of urine is produced

each day by a healthy person, it is obvious that not all of the filtrate becomes urine. Approximately 99% of the filtrate volume is reabsorbed in the nephron, and less than 1% becomes urine.

2   P R E D I C T

If the filtration fraction increases from 19% to 22% and if 99.2% of the filtrate is reabsorbed, how much urine is produced in a normal person with a cardiac output of 5600 mL/min (*Hint:* See table 26.2)?

✔ *Answer in Appendix F*

Filtration Barrier

The filtration membrane is a **filtration barrier** that prevents the entry of blood cells and proteins into the lumen

Table 26.2   Calculation of Renal Flow Rates

Substance	Amount per Minute (mL)	Calculation
Renal blood flow	1176	Amount of blood flowing through the kidneys per minute; equals cardiac output (5600 mL blood/min) times the percent (21%; renal fraction) of cardiac output that enters the kidneys. 5600 mL blood/min × 0.21 = 1176 blood/min
Renal plasma flow	650	Amount of plasma flowing through the kidneys per minute; equals renal blood flow times percent of the blood that is plasma. Because the hematocrit is the percent of the blood that consists of formed elements, the percent of the blood that is plasma is 100 minus the hematocrit. Assuming a hematocrit of 45, the percent of the blood that is plasma is 55% (100 − 45). Renal plasma flow is therefore 55% of renal blood flow. 1176 mL blood/min × 0.55 ≈ 650 mL plasma/min
Glomerular filtration rate	125	Amount of plasma (filtrate) that enters Bowman's capsule per minute; equals renal plasma flow times the percent (19%; filtration fraction) of the plasma that enters the renal capsule. 650 mL plasma/min × 0.19 ≈ 125 mL filtrate/min
Urine	1	Nonreabsorbed filtrate that leaves the kidneys per minute; equals glomerular filtration rate times the percent (0.8%) of the filtrate that is not reabsorbed into the blood. 125 mL filtrate/min × 0.008 = 1 mL urine/min Milliliters of urine per minute can be converted to liters of urine per day by multiplying by 1.44. 1 mL urine/min × 1.44 = 1.4 L/day

of Bowman's capsule but allows other blood components to enter. The filtration membrane is many times more permeable than a typical capillary. Water and solutes of a small molecular diameter readily pass from the glomerular capillaries through the filtration membrane into Bowman's capsule, but larger molecules do not. The fenestrae of the glomerular capillary, the basement membrane, and the podocyte cells (see figure 26.4d) prevent molecules larger than 7 nm in diameter or with a molecular mass of 40,000 daltons from passing through. Most plasma proteins are slightly larger than 7 nm in diameter; thus they are retained in the glomerular capillaries. Albumin, which has a diameter just slightly less than 7 nm, enters the filtrate in small amounts so that the filtrate contains about 0.03% protein. Protein hormones are also small enough to pass through the filtration barrier. Proteins that do pass through the filtration membrane are actively reabsorbed by endocytosis and metabolized by the cells in the proximal tubule. Consequently, little protein is found in the urine of healthy people.

3   P R E D I C T

Hemoglobin has a smaller diameter than albumin, but very little hemoglobin passes from the blood into the filtrate. Explain why. Under what circumstances would large amounts of hemoglobin enter the filtrate?

✔ *Answer in Appendix F*

Clinical Note

Hematuria (hē-mă-tū′rē-ă) occurs when erythrocytes are found in the urine. The source of erythrocytes in the urine can be outside of the kidney or from the kidney. Conditions outside the kidney that result in hematuria include kidney stones or tumors in the renal pelvis, ureter, urinary bladder, prostate, or urethra. Infections of the urinary tract, such as cystitis, prostatitis (pros-tă-tī′tis; prostate gland), and urethritis (yū-rē-thrī′tis; urethra) can also cause hematuria. Conditions inside the kidney include those that affect the filtration membrane or other areas of the kidney. Inflammation of the glomeruli called **glomerulonephritis** (glō-măr′yū-lō-nef-rī′tis) can increase the permeability of the filtration membrane and allow blood cells to cross. Other areas of the kidney can be a source of blood in response to inflammation of the nephrons due to infections, tumors in the kidney tissue, and infarcted areas of the kidney, where an artery is blocked, resulting in necrosis of part of the kidney tissue.

Filtration Pressure

The formation of filtrate depends on a pressure gradient, called the **filtration pressure,** which forces fluid from the glomerular capillary across through the filtration membrane into the lumen of Bowman's capsule. The filtration pressure results from the sum of the forces that move fluid out of the glomerular capillary into the lumen of Bowman's capsule and those that move fluid out of the lumen of Bowman's capsule into the glomerular capillary (figure 26.9). The **glomerular**

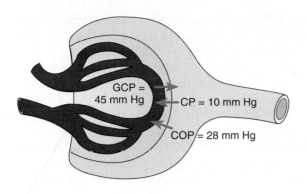

Filtration pressure =
  45 mm Hg GCP (glomerular capillary pressure)
  −28 mm Hg COP (colloid osmotic pressure)
  −10 mm Hg CP (capsule pressure)

  7 mm Hg filtration pressure

Figure 26.9   Filtration Pressure

Filtration pressure across the filtration membrane is equal to the glomerular capillary pressure (GCP) minus the colloid osmotic pressure (COP) in the glomerular capillary and minus the pressure in Bowman's capsule (CP).

capillary pressure (GCP), the blood pressure inside the capillary, moves fluid out of the capillary into Bowman's capsule. The glomerular capillary pressure averages approximately 45 mm Hg, which is much higher than that in most capillaries. Opposing the movement of fluid into the lumen of Bowman's capsule is the **capsule pressure (CP),** which is approximately 10 mm Hg, caused by the pressure of filtrate already inside Bowman's capsule. The **colloid osmotic pressure (COP),** resulting from the presence of unfiltered plasma proteins remaining within the glomerular capillary, produces an osmotic force of about 28 mm Hg that also moves fluid to the glomerular capillary from Bowman's capsule. The filtration pressure is therefore approximately 7 mm Hg.

Filtration capillary pressure (7 mm Hg)	=	Glomerular pressure (45 mm Hg)	−	Capsule osmotic pressure (10 mm Hg)	−	Colloid osmotic pressure (28 mm Hg)

The high glomerular capillary pressure results from a low resistance to blood flow in the afferent arterioles and glomerular capillaries and from a higher resistance to blood flow in the efferent arterioles. As the diameter of a vessel decreases, resistance to flow through the vessel increases (see chapter 21), and pressure upstream from the point of decreased vessel diameter is higher than the pressure downstream from the point of decreased diameter. For example, in the extreme case of a completely closed vessel, pressure is higher upstream from the constriction and it falls to zero downstream from the constriction. The efferent arteriole has a small diameter, and the blood pressure is higher within the glomerular capillaries because of the low resistance to blood flow in the afferent arterioles and glomerular capillaries and

because of the higher resistance to blood flow in the efferent arteriole. Also the pressure is lower in the peritubular capillaries which are downstream from the efferent arterioles. Consequently, filtrate is forced across the filtration membrane into the lumen of Bowman's capsule. The low pressure in the peritubular capillaries allows fluid to move into them from the interstitial fluid.

The smooth muscle cells in the walls of the afferent and efferent arterioles can alter the vessel diameter and the glomerular filtration pressure. For example, dilation of the afferent arterioles or constriction of the efferent arterioles increases glomerular capillary pressure, increasing filtration pressure and glomerular filtration.

4	**P R E D I C T**

What effect does constriction of the afferent arteriole have on the filtration pressure? What effect does a decrease in the concentration of plasma proteins have on filtration pressure?

✔ *Answer in Appendix F*

Tubular Reabsorption

The filtrate leaves the lumen of Bowman's capsule and flows through the proximal tubule, the loop of Henle, and the distal tubule, and then into the collecting ducts. As it passes through these structures, many of the substances in the filtrate undergo **tubular reabsorption.** Tubular reabsorption results from processes such as diffusion, facilitated diffusion, active transport, cotransport, and osmosis. Inorganic salts, organic molecules, and about 99% of the filtrate volume leave the nephron and enter the interstitial fluid. These substances then enter the low-pressure peritubular capillaries and flow through the renal veins to enter the general circulation (see figure 26.8).

Solutes reabsorbed from the lumen of the nephron to the interstitial fluid include amino acids, glucose, and fructose, as well as sodium, potassium, calcium, bicarbonate, and chloride ions. A more complete list is provided in table 26.3 for each part of the nephron.

Water follows the solutes that are reabsorbed across the wall of the nephron. As the solutes in the nephron are reabsorbed, water follows the solutes by the process of osmosis. The transport processes and the permeability characteristics of each portion of the nephron allow up to 99% of the volume of the filtrate to be reabsorbed. The small volume of the filtrate that forms urine contains a relatively high concentration of urea, uric acid, creatinine, potassium ions, and other substances that are toxic in high concentrations. Regulation of solute reabsorption and the permeability characteristics of portions of the nephron allow for the production of a small volume of very concentrated urine or a large volume of very dilute urine. The mechanisms in the wall of the nephron responsible for reabsorption are described in the following section. Mechanisms that regulate urine concentration are described in a later section of this chapter.

Table 26.3 Reabsorption of Major Solutes from the Nephron

Apical Membrane	Basal Membrane
Proximal Nephron	
Substances cotransported with Na^+	Active transport Na^+ (exchanged for K^+)
K^+	Facilitated diffusion
Cl^-	K^+
Ca^{2+}	Cl^-
Mg^{2+}	Ca_2^+
HCO_3^-	Ca_2^+
PO_4^{3-}	HCO_3^-
Amino acids	PO_4^{3-}
Glucose	Amino acids
Fructose	Glucose
Galactose	Fructose
Lactate	Galactose
Succinate	Lactate
Citrate	Succinate
Diffusion between nephron cells	Citrate
K^+	
Ca^{2+}	
Mg^{2+}	
Thick Ascending Limb of Loop of Henle	
Substances cotransported with Na^+	Active transport Na^+ (exchanged for K^+)
K^+	Facilitated diffusion
Cl^-	K^+
K^+	Cl^-
Diffusion between nephron cells	
K^+	
Ca^{2+}	
Mg^{2+}	
Distal Nephron and Collecting Duct	
Substances cotransported with Na^+	Active Transport Na^+ (exchanged for K^+)
Cl^-	Facilitated diffusion
K^+	K^+
	Cl^-

Reabsorption in the Proximal Tubule

Reabsorption of most solute molecules from the proximal tubule is linked to the primary active transport of sodium ions across the **basal membrane** of the nephron epithelial cells from the cytoplasm into the interstitial fluid creating a low concentration of sodium ions inside the cells (figure 26.10). At the basal cell membrane, ATP provides the energy required to move sodium ions out of the cell in exchange for potassium ions by countertransport. Because the concentration of sodium ions in the lumen of the tubule is high, there is a large concentration gradient from the lumen of the nephron to the intracellular fluid of the cells lining the nephron. This con-

centration gradient for sodium ions is the source of energy for the cotransport of many solute molecules from the lumen of the nephron into the nephron cells.

Within the **apical membrane,** which separates the lumen of the nephron from the cytoplasm of the nephron cells, there are carrier molecules for amino acids, glucose, and other solutes. Each of these carrier molecules binds specifically to one of the substances to be transported and to sodium ions. The concentration gradient for sodium ions provides the source of energy that moves both the sodium ions and the other molecules or ions bound to the carrier molecule from the lumen into the cell of the nephron (see figure 26.10). Once the cotransported molecules are inside the cell, they cross the basal membrane of the cell by facilitated diffusion.

Some solutes also diffuse between the cells from the lumen of the nephron into the interstitial fluid. The concentration of these solutes increases as other solutes are cotransported and water moves by osmosis from the lumen into interstitial fluid. As the concentration gradient for these solutes increases above the concentration in the interstitial fluid, they diffuse between the epithelial cells. Some potassium, calcium, and magnesium ions diffuse between the cells of the proximal tubule wall from the lumen of the tubule to the interstitial fluid. Reabsorption of these solutes by diffusion occurs even though these same ions are also reabsorbed by cotransport processes.

In the proximal tubule, reabsorption of solutes is extensive, and the proximal tubule is permeable to water. As solute molecules are transported from the nephron to the interstitial fluid, water moves by osmosis in the same direction. By the time the filtrate has reached the end of the proximal tubule, its volume has been reduced by approximately 65%. Because the wall of the nephron is permeable to water, the concentration of the filtrate in the proximal tubule remains about the same as that of the interstitial fluid (300 mOsm/kg).

Reabsorption in the Loop of Henle

The loop of Henle descends into the medulla of the kidney, where the concentration of solutes in the interstitial fluid is very high. The thin segment of the loop of Henle (figure 26.11) is highly permeable to water and moderately permeable to urea, sodium, and most other ions. It is adapted to allow passive movement of solutes through its wall, although water passes through much more rapidly than solutes. As the filtrate passes through the thin segment of the loop of Henle, water moves out of the nephron by osmosis. Some solutes move into the nephron. By the time the filtrate has reached the end of the thin segment of the loop of Henle, the volume of the filtrate has been reduced by another 15%, and the concentration of the filtrate is equal to the high concentration of the interstitial fluid (1200 mOsm/L).

The ascending limb of the loop of Henle is impermeable to water. Therefore, no additional water diffuses from the nephron as it passes through the ascending limb of the loop of Henle. Solute molecules such as sodium, potassium, and chloride ions, however, are transported from the nephron into

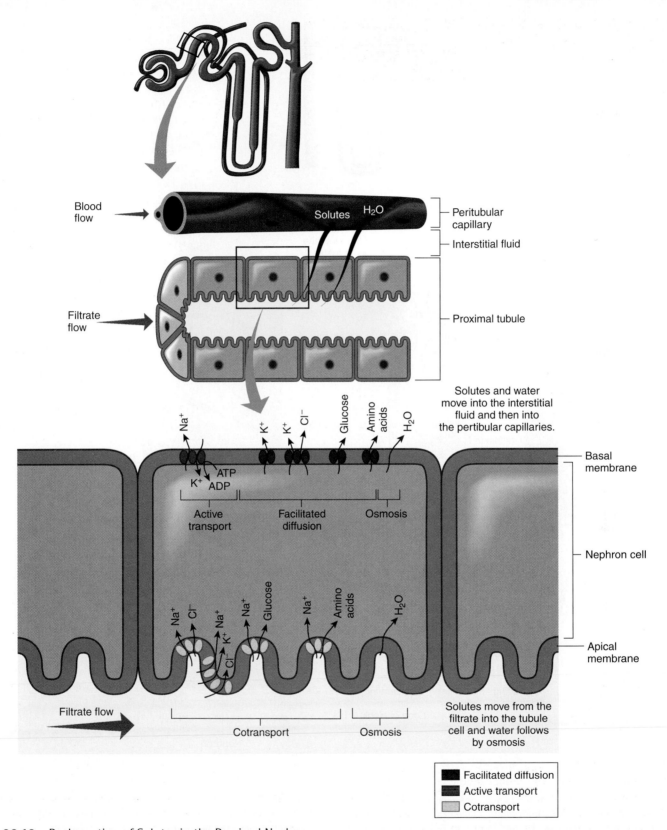

Figure 26.10   Reabsorption of Solutes in the Proximal Nephron

Cotransport of molecules and ions across the epithelial lining of the nephron depends on the active transport of sodium ions, in exchange for potassium ions across the basal membrane. Cotransport is the process by which carrier proteins move molecules or ions with sodium ions across the apical membrane. The sodium ion concentration gradient provides the energy for cotransport. Amino acids, glucose, potassium ions, chloride ions, and most other solutes are transported into the cells of the nephron with sodium ions. Water enters and leaves the cell by osmosis. Glucose, amino acids, sodium ions, chloride ions, and many other solutes leave the cells across the apical membrane by facilitated diffusion.

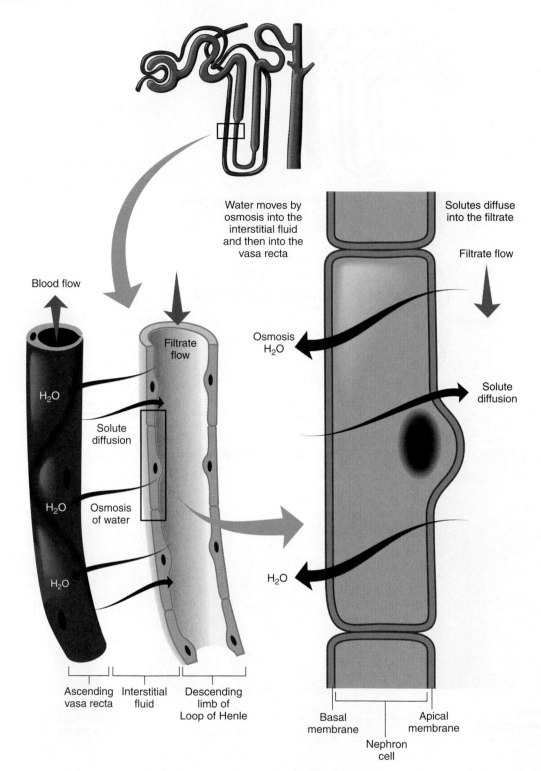

Figure 26.11   Reabsorption in the Descending Limb of the Loop of Henle

The wall of the descending limb of the loop of Henle is permeable to water and, to a lesser extent, to solutes. The interstitial fluid and the vasa recta in the medulla of the kidney have a high solute concentration (high osmolality). Water therefore diffuses from the nephron into the interstitial fluid and into the vasa recta. To a lesser extent, solutes diffuse from the vasa recta and interstitial fluid into the nephron.

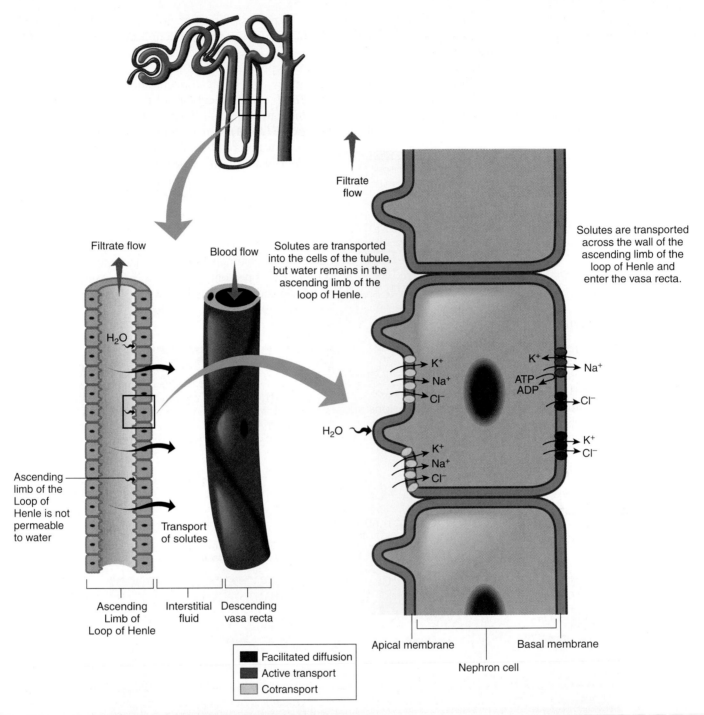

Figure 26.12   Reabsorption in the Ascending Limb of the Loop of Henle

The wall of the ascending limb of the loop of Henle is impermeable to water. Sodium and chloride ions move across the wall of the ascending limb of Henle's loop. Potassium and chloride ions are cotransported with sodium ions across the apical membrane. The concentration gradient for sodium ions is created by active transport of sodium ions across the basal cell membrane. Ions pass by facilitated diffusion across the basal cell membrane of the nephron cells.

the interstitial fluid. Cotransport is responsible for the movement of potassium and chloride ions with sodium ions across the apical membrane of the ascending limb of the loop of Henle (figure 26.12).

Once inside the cells of the ascending limb, chloride and potassium ions cross the basal cell membrane into the in-

terstitial fluid from a higher concentration inside the cells to a lower concentration outside of the cells by facilitated diffusion. The concentration gradient for sodium ions is created by the active transport of sodium ions out of the cell in exchange for potassium ions across the basal membrane (see figure 26.12).

Because the ascending limb of the loop of Henle is impermeable to water and because ions are transported out of the nephron, the concentration of solutes in the tubule is reduced to about 100 mOsm/kg by the time the fluid reaches the distal tubule. In contrast, the concentration of the interstitial fluid in the cortex is about 300 mOsm/kg. Thus the filtrate entering the distal tubule is much more dilute than the interstitial fluid surrounding it.

An **osmole** is a measure of the number of particles in solution. One osmole is the molecular weight, in grams, of a solute divided by the number of ions or particles into which it dissociates in solution. A milliosmole (mOsm) is 1/1000 of an osmole. The osmolality of a solution is the number of osmoles in a kilogram of solution. Water moves by osmosis from a solution with a lower osmolality to a solution with a higher osmolality. Thus water moves by osmosis from a solution of 100 mOsm/kg toward a solution of 300 mOsm/kg (see appendix C).

Reabsorption in the Distal Tubule and Collecting Duct

Sodium and chloride ions are transported across the wall of the distal tubules and collecting ducts. Chloride ions are cotransported across the apical membrane with sodium ions. The concentration gradient for sodium ions is a result of the active transport of sodium ions across the basal cell membrane. In addition, the collecting ducts extend from the cortex of the kidney, where the concentration of the interstitial fluid is approximately 300 mOsm/kg, through the medulla of the kidney, where the concentration of the interstitial fluid is very high. The permeability of the distal tubules and collecting ducts to water is under hormonal control. **Antidiuretic hormone (ADH)** increases the permeability of the cell membranes to water, but the cell membranes are relatively impermeable to water in the absence of ADH.

ADH binds to a membrane-bound receptor, activating a G-protein mechanism that increases cAMP systhesis in the cells of the distal tubules and collecting ducts. Cyclic AMP increases the permeability of the cell membranes of the distal tubules and collecting ducts to water by increasing the number of water channels in the plasma membrane (figure 26.13). When ADH is present, water moves by osmosis out of the distal tubule and collecting duct, whereas in the absence of ADH, water remains within the nephron.

5   **P R E D I C T**

What effect does a lack of ADH secretion have on the volume and concentration of urine produced by the kidney?

✔ *Answer in Appendix F*

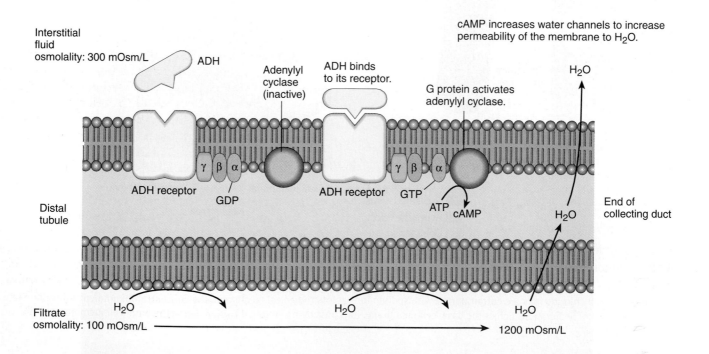

Figure 26.13   The Effect of Antidiuretic Hormone (ADH) on the Nephron

ADH binds to ADH receptors on the plasma membrane of the distal nephron and the collecting duct. When ADH is bound to its receptor, G proteins are activated, which, in turn, activate adenylyl cyclase. The increase in the rate of cyclic adenosine monophosphate (cAMP) synthesis increases the permeability of the epithelial cells to water by increasing the number of water channels in their plasma membranes. Water then moves out of the tubule into the interstitital spaces by osmosis, both decreasing the volume of the filtrate and increasing its concentration (osmolality).

In the proximal tubules, 65% of the filtrate volume is reabsorbed, and 15% of the filtrate volume is reabsorbed in the thin segment of the loops of Henle. About 80% of the volume of the filtrate is therefore reabsorbed in these structures. When ADH is present, another 19% is reabsorbed in the distal tubules and collecting ducts.

Changes in the Concentration of Solutes in the Nephron

Urea enters the glomerular filtrate and is present in the same concentration as it is in the plasma. As the volume of the filtrate decreases in the nephron, the concentration of urea increases because renal tubules are not as permeable to urea as they are to water. Only 40%–60% of the urea is passively reabsorbed in the nephron, although about 99% of the water is reabsorbed. In addition to urea, urate ions, creatinine, sulfates, phosphates, and nitrates are reabsorbed but not to the same extent as water. They therefore become more concentrated in the filtrate as the volume of the filtrate becomes smaller. These substances are toxic if they accumulate in the body, so their accumulation in the filtrate and elimination in urine help maintain homeostasis.

Clinical Note

Some drugs, environmental pollutants, and other foreign substances that gain access to the circulatory system are reabsorbed from the nephron. These substances are usually lipid-soluble, nonpolar compounds. They enter the glomerular filtrate and are reabsorbed passively by a process similar to that by which urea is reabsorbed. Because these substances are passively reabsorbed within the nephron, they are not rapidly excreted. The liver cells attach other molecules to them by a process called **conjugation** (kon-jū-gā′shŭn), which converts them to more water-soluble molecules. These more water-soluble substances enter the filtrate , but do not pass as readily through the wall of the nephron, are not reabsorbed from the renal tubules, and consequently are more rapidly excreted in the urine. One of the important functions of the liver is to convert nonpolar toxic substances to more water-soluble forms, thus increasing the rate at which they are excreted in the urine.

Tubular Secretion

Some substances, including by-products of metabolism that become toxic in high concentrations and drugs or molecules not normally produced by the body, are moved into the nephron (table 26.4) by **tubular secretion.** As with tubular reabsorption, tubular secretion can be either active or passive. Ammonia is synthesized in the epithelial cells of the nephron and diffuses into the lumen of the nephron. Substances that are actively secreted by either active transport or counter-transport processes into the nephron include hydrogen ions, potassium ions, penicillin, and ***para*-aminohippuric acid** (par-ă-ă-mē′nō-hi-pyūr′ik; *p*-aminohippuric acid; **PAH**).

Table 26.4	Secretion of Substances into the Nephron
Transport Process	**Substance Transported**
Proximal Convoluted Tubule	
Active transport	Hydrogen ions
	Hydroxybenzoates
	para-Aminohippuric acid
	Neurotransmitters Dopamine Acetylcholine Epinephrine
	Bile pigments
	Uric acid
	Drugs and toxins Penicillin Atropine Morphine Saccharin
Passive transport	Ammonia
Distal Convoluted Tubule	
Active transport	Potassium ions
Passive transport	Potassium ions Hydrogen ions

For example, hydrogen ions are transported from the cells of the nephron into the lumen of the nephron by a countertransport process. Hydrogen ions bind to a carrier molecule on the inside of the cell, and sodium ions bind to the carrier molecule on the outside of the cell membrane. As the sodium ion is moved into the cell, the hydrogen ion is moved to the outside of the cell (figure 26.14). The hydrogen ions that are secreted are produced as a result of carbon dioxide and water reacting to form hydrogen ions and bicarbonate ions. The countertransport molecules secrete hydrogen ions into the nephron lumen, and sodium ions enter the nephron cell. Sodium ions and bicarbonate ions are cotransported across the basal membrane of the cell and enter the peritubular capillaries. Hydrogen ions are secreted into the proximal and distal tubules, and potassium ions are actively secreted in the distal tubule (see chapter 27). Penicillin and *p*-aminohippuric acid are examples of substances not normally produced by the body that are actively secreted into the proximal tubules.

Urine Concentration Mechanism

When a large volume of water is consumed, it is necessary to eliminate the excess without losing excessive electrolytes or other substances essential for the maintenance of homeostasis. The response of the kidneys is to produce a large volume of dilute urine. On the other hand, when drinking water is not

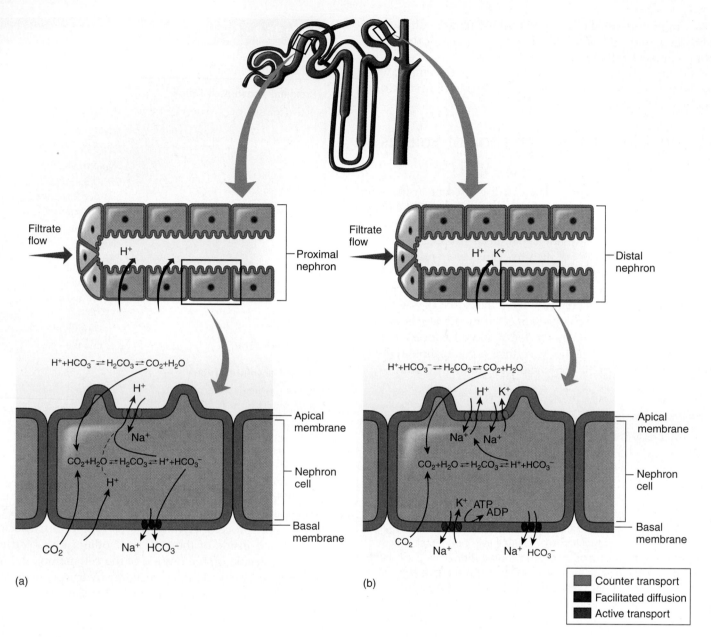

Figure 26.14   Secretion of Hydrogen Ions and Potassium Ions into the Nephron

(a) Hydrogen ions are secreted by a countertransport mechanism in the proximal nephron in which hydrogen ions are exchanged for sodium ions. The hydrogen ions secreted either diffuse into the nephron cell from the peritubular capillaries into the interstitial fluid and then into the nephron, or they are derived from the reaction between carbon dioxide and water in the cell. Sodium ions and bicarbonate ions are cotransported across the basal membrane into the interstitial fluid and then diffuse into the peritubular capillaries. (b) Hydrogen and potassium ions are secreted by a countertransport mechanism in the distal nephron tubule. Sodium and potassium ions are moved by active transport across the basal membrane of the nephron cell. The sodium ions and the bicarbonate ions then diffuse into the peritubular capillaries.

available, producing a large volume of dilute urine would lead to rapid dehydration. When water intake is restricted, the kidneys produce a small volume of concentrated urine that contains sufficient waste products to prevent their accumulation in the circulatory system. The kidneys are able to produce urine with concentrations that vary between 65 and 1200 mOsm/kg while maintaining an extracellular fluid osmolality very close to 300 mOsm/kg.

Medullary Concentration Gradient

The ability of the kidney to concentrate urine depends on maintaining a high medullary concentration gradient. The interstitial fluid concentration is about 300 mOsm/kg in the cortical region of the kidney and becomes progressively higher in the medulla. The interstitial osmolality reaches about 1200 mOsm/kg near the tips of the renal pyramids

Clinical Focus   Kidney Dialysis

The artificial kidney (renal dialysis machine) is a machine used to treat patients suffering from renal failure. The use of this machine often allows people with severe acute renal failure to recover without developing the side effects of renal failure, and the machine can substitute for the kidneys for long periods in people suffering from chronic renal failure.

Renal dialysis is based on blood flow though tubes made of a selectively permeable membrane. On the outside of the dialysis tubes is a fluid that contains the same concentration of solutes as the plasma, except for the metabolic waste products. As a consequence, a diffusion gradient exists for the metabolic waste products from the blood to the dialysis fluid. The dialysis membrane has pores that are too small to allow the plasma proteins to pass through them. For smaller solutes the dialysis fluid contains the same beneficial solutes as the plasma, so the net movement of these substances is zero. In contrast, the metabolic waste products diffuse rapidly from the blood into the dialysis fluid.

Blood usually is taken from an artery, passed through tubes of the dialysis machine, and then returned to a vein. The rate of blood flow is normally several hundred milliliters per minute, and the total surface area of exchange in the machine is close to $10,000–20,000 \text{ cm}^2$ (figure A).

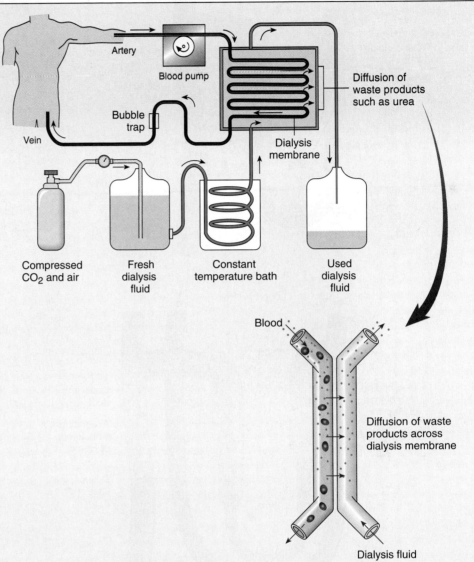

Figure A   Kidney Dialysis

During kidney dialysis blood flows through a system of tubes composed of a selectively permeable membrane. Dialysis fluid, the composition of which is similar to that of blood, except that the concentration of waste products is very low, flows in the opposite direction on the outside of the dialysis tubes. Consequently, waste products such as urea diffuse from the blood into the dialysis fluid. Other substances such as sodium, potassium, and glucose do not rapidly diffuse from the blood into the dialysis fluid because there is no concentration gradient for these substances between the blood and the dialysis fluid.

(figure 26.15). Maintenance and production of the high solute concentration in the kidney medulla depend on the loops of Henle, the vasa recta, and the distribution of urea. Water and solutes move from the loop of Henle into the medullary interstitial fluid. The vasa recta carry the excess water and solutes from the medullary interstitial fluid. The high concentration of interstitial solutes is not diminished because the walls of the vasa recta are more permeable to water than to solutes. Urea circulates repeatedly through the medullary interstitial fluid. Some urea is lost in the urine, but much of it is circulated many times through the interstitial fluid of the medulla. A more detailed description of the

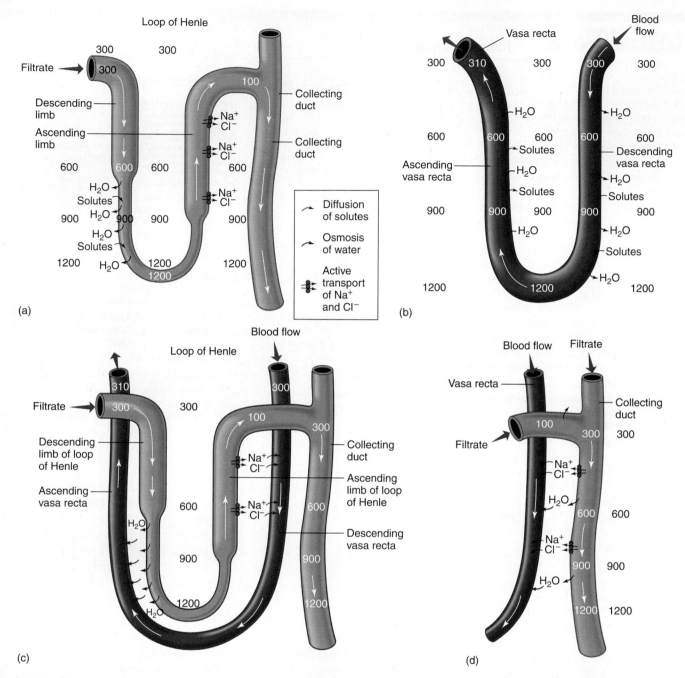

Figure 26.15 Filtrate Concentration and the Medullary Concentration Gradient

The loop of Henle and the vasa recta function together to maintain a high concentration of solutes in the medulla of the kidney. See the text for a description of water and solute movements for each part of this figure. (*a*) The movement of water and solutes across the wall of the loop of Henle. (*b*) The movement of water and solutes across the wall of the vasa recta. (*c*) The loop of Henle and the vasa recta function together. Water moves out of the descending limb of the loop of Henle and enters the vasa recta. Solutes transported out of the ascending limb of the loop of Henle enter the vasa recta. Excess water and solutes are carried away from the medulla without reducing the high concentration of solutes. The concentration of the filtrate is reduced to 100 mOsm/kg by the time it reaches the distal nephron. (*d*) Water and solutes move out of the collecting duct into the vasa recta.

function of the loops of Henle and vasa recta, as well as the circulation of urea follows.

1. *Loops of Henle.* The walls of the descending limbs of the loops of Henle are permeable to water. As filtrate flows into the medulla of the kidney through the descending

limbs of the loops of Henle, water diffuses out of the nephrons into the more concentrated interstitial fluid. The walls of the ascending limbs of the loops of Henle are impermeable to water, and active transport mechanisms transport solutes such as sodium and chloride ions out of the filtrate and into the interstitial fluid. Thus, both water and solutes enter the medullary interstitial fluid from the loops of Henle (figure 26.15a).

2. *The vasa recta.* The vasa recta function to remove excess water and solutes from the medulla of the kidney without changing the high concentration of solutes in the medullary interstitial fluid. The vasa recta also supply blood to the medulla of the kidney. The vasa recta are countercurrent systems. **Countercurrent** (kown′ter-ker′ent) **systems** consist of parallel tubes in which fluid flows, but in opposite directions, and heat or substances such as water or solutes diffuse from tubes carrying fluid in one direction to tubes carrying fluid in the opposite direction so that the fluid in both sets of tubes have nearly the same composition. The vasa recta make up a countercurrent system because blood flows through them to the kidney medulla and, after the vessels turn near the tip of the renal pyramid, the blood is carried in the opposite direction. The walls of the vasa recta are permeable to water and to solutes. As blood flows toward the medulla, water moves out of the vasa recta, and some solutes diffuse into them. As blood flows back toward the cortex, water moves into the vasa recta, and some solutes diffuse out of them (see figure 26.15b). The rates of diffusion are such that slightly more water and slightly more solute are carried from the medulla by the vasa recta than enters it. Thus, the composition of the blood at both ends of the vasa recta is nearly the same, with the volume and osmolality being slightly greater as the blood once again reaches the cortex. If excess solutes and water were not carried from the medulla by the vasa recta, the volume of the medulla would have to increase continually.

The loops of Henle and the vasa recta are in parallel with one another, and their functions are closely related. The water and solutes that leave the loops of Henle enter the vasa recta. The vasa recta carry away the water and solutes without diminishing the high concentration of solutes in the medulla of the kidney (figure 26.15c). Water and solutes also leave the collecting ducts and enter the medulla of the kidney (figure 26.15d). Like the water and solutes that leave the loops of Henle, the water and solutes that leave the collecting ducts enter the vasa recta and are carried from the medulla.

The loops of Henle and vasa recta are sometimes described as a countercurrent multiplier system. A **countercurrent multiplier system** is a countercurrent system that is assisted by active transport mechanisms. It is the active transport of solutes from the ascending limbs of the loops of Henle that make the loops of Henle and vasa recta a countercurrent multiplier system. The loops of Henle and vasa recta function together to maintain a concentration of solutes in the interstitial fluids of the medulla and to carry away the water and solutes that enter the medulla from the loops of Henle and collecting ducts.

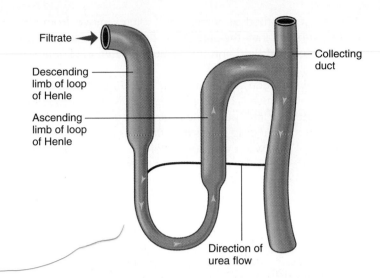

Figure 26.16   The Medullary Concentration Gradient and Urea Cycling

The concentration of urea in the medulla of the kidney is high and contributes to the overall high concentration of solutes in the medulla of the kidney. The wall of the collecting duct is permeable to urea. Urea diffuses out of the collecting duct into the interstitial fluid of the medulla. The wall of the descending limb of the loop of Henle is also permeable to urea. Urea diffuses from the interstitial fluid into the descending limb of the loop of Henle. Thus a cycle is produced in which urea flows into the descending limb of the loop of Henle, through the ascending limb, through the distal nephron, through the collecting duct, out of the collecting duct, and back into the descending limb of the loop of Henle.

3. *Urea.* **Urea** (yū-rē′ă) molecules are responsible for a substantial part of the high osmolality in the medulla of the kidney (figure 26.16). The walls of the descending limbs of the loops of Henle are permeable to urea, and urea diffuses into the descending limbs from the interstitial fluid. The ascending limbs of the loops of Henle and the distal tubules are impermeable to urea. The collecting ducts are permeable to urea, however, and some urea diffuses out of the collecting ducts into the interstitial fluid of the medulla. Thus, urea flows in a cycle from the interstitial fluid into the descending limbs of Henle's loops, through the ascending limbs, the distal tubules, and into the collecting ducts. Urea then diffuses from the collecting ducts back into the interstitial fluid of the medulla. Consequently, a high urea concentration is maintained in the medulla of the kidney.

Summary of Changes in Filtrate Volume and Concentration

In the average person, about 180 L of filtrate enters the proximal tubules daily. Substances such as glucose, amino acids, sodium ions, calcium ions, potassium ions, and chloride ions (see table 26.3) are actively transported, and water moves by osmosis from the lumens of the proximal tubules into the interstitial fluid. The excess solutes and water then enter the peritubular capillaries. Consequently, approximately 65% of the filtrate volume is reabsorbed as water, and solutes move from the proximal tubules into the interstitial fluid. The osmolality of both the interstitial fluid and the filtrate is maintained at about 300 mOsm/L.

The filtrate then passes into the descending limb of the loops of Henle, which is highly permeable to water and solutes. As the descending limb penetrates deep into the medulla of the kidney, the surrounding interstitial fluid has a progressively greater osmolality. Water diffuses out of the nephron as solutes slowly diffuse into it. By the time the filtrate reaches the deepest part of the loops of Henle, its volume has been reduced by an additional 15% of the original volume and its osmolality has increased to about 1200 mOsm/kg (figure 26.17). By the time the filtrate has reached the tip of the loops of Henle, at least 80% of the filtrate volume has been reabsorbed.

After passing through the descending limbs of the loops of Henle, the filtrate enters the ascending limbs, or the thick segments. The thick segments are impermeable to water; but sodium, chloride, and potassium ions are transported from the filtrate into the interstitial fluid (see figure 26.17). The movement of ions, but not water, across the wall of the ascending limbs, causes the osmolality of the filtrate to decrease from 1200 to about 100 mOsm/kg by the time the filtrate again reaches the cortex of the kidney. As a result, the filtrate in the nephron is dilute compared with the concentration of the surrounding interstitial fluid, which has an osmolality of about 300 mOsm/kg.

The changes just described are obligatory; that is, they occur regardless of the concentration and the volume of the urine that is finally produced by the kidney. The mechanisms by which concentrated and dilute urine are formed by the kidney are described in the following sections.

Formation of Concentrated Urine

The filtrate enters the distal tubules after passing through the loops of Henle. Near the end of the distal tubules, the wall of the tubules can be permeable to water, providing ADH is present. Water diffuses from the lumen of the nephron to the interstitial spaces.

The filtrate then flows into the collecting ducts which pass through the medulla of the kidney with its high concentration of solutes. If ADH is present, water diffuses from the collecting ducts into the interstitial fluid. By the time the filtrate passes through the collecting ducts, another 19% of the filtrate is reabsorbed. Thus, 1% of the filtrate remains as urine, and 99% of the filtrate is reabsorbed. The osmolality of the filtrate at the end of the collecting ducts is approximately 1200 mOsm/kg (see figure 26.17).

In addition to the dramatic decrease in filtrate volume and increase in filtrate osmolality, there is a marked alteration in the filtrate composition. Waste products, such as creatinine and urea as well as potassium, hydrogen, phosphate, and sulfate ions, are present at a much higher concentration in urine than in the original filtrate because of the removal of water from the filtrate. Many substances are selectively reabsorbed from the nephron, and others are secreted into the nephron

so that beneficial substances are retained in the body and toxic substances are eliminated.

Formation of Dilute Urine

If ADH is not present or if its concentration is reduced, the distal tubules and collecting ducts have a lower permeability to water. The amount of water moving by osmosis from the distal nephron and collecting duct, therefore, is reduced. The concentration of the urine produced is less than 1200 mOsm/kg, and the volume is increased. The value can be much larger than 1% of the filtrate formed each day. If no ADH is secreted, the osmolality of the urine may be close to the osmolality of the filtrate in the distal nephron, and the volume of urine may approach 20–30 L/day.

In a healthy person, even when the kidney produces dilute urine, the concentration of waste products in the urine is large enough to maintain homeostasis in the body. Consequently, beneficial substances are retained, and both toxic substances and excess water are eliminated in the urine.

Clinical Note

Only the juxtamedullary nephrons have loops of Henle that descend deep into the medulla, but there are enough of them to maintain a high interstitial concentration of solutes in the interstitial fluid of the medulla. Not all of the nephrons need to have loops of Henle that descend into the medulla in order to effectively concentrate urine. The cortical nephrons function like the juxtamedullary nephrons, with the exception that their loops of Henle are not as efficient at concentrating urine. Because the filtrate from the cortical nephrons passes through the collecting ducts, however, water can diffuse out of the collecting ducts into the interstitial fluid. Thus the filtrate becomes concentrated. Animals that concentrate urine more effectively than humans have a greater percentage of nephrons that descend into the medulla of the kidney. For example, mammals that live in deserts have many nephrons that descend into the medulla of the kidney, and the renal pyramids are longer than in humans and most other mammals.

Regulation of Urine Concentration and Volume

Urine can be dilute or very concentrated, and it can be produced in large or small amounts. Urine concentration and volume are regulated by mechanisms that maintain the extracellular fluid osmolality and volume within narrow limits.

Filtrate reabsorption in the proximal tubules and the descending limbs of the loops of Henle is obligatory and therefore remains relatively constant. Filtrate reabsorption in the distal tubules and collecting ducts is regulated, however, and can change dramatically, depending on the conditions to which the body is exposed. If homeostasis requires the elimination of a large volume of dilute urine, a large volume of filtrate is produced, and the dilute filtrate in the distal tubules and collecting ducts can pass through them with little change

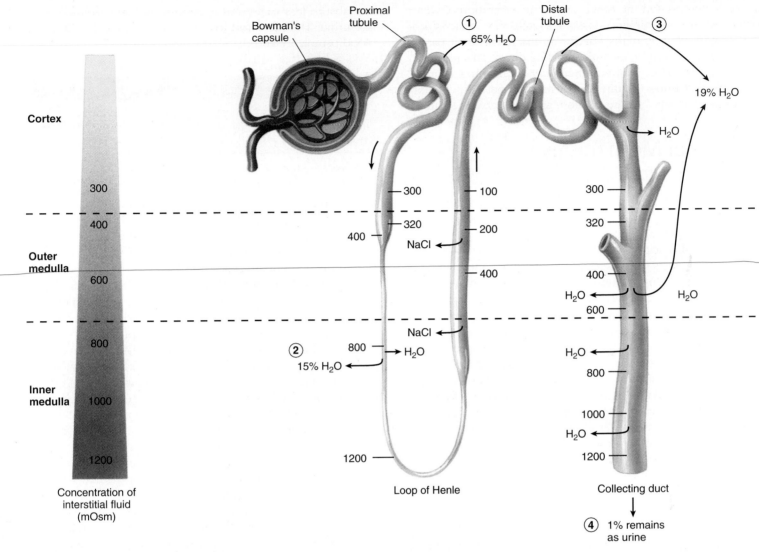

Cortex

Outer medulla

Inner medulla

Concentration of interstitial fluid (mOsm)

Loop of Henle

Collecting duct

4. 1% remains as urine

Figure 26.17   Urine Concentrating Mechanism

Summary of the mechanisms responsible for the concentration of urine.

1. Approximately 180 L of filtrate enters the nephrons each day; however, 65% of that volume is reabsorbed in the proximal tubule. In the proximal tubule, solute molecules are transported from the lumen of the tubule into the interstitial fluid. Water follows the reabsorbed solutes because the cells of the tubule wall are permeable to water.

2. Approximatley 15% of the filtrate volume is reabsorbed in the descending limb of Henle's loop. The descending limb of Henle's loop passes through the concentrated fluid of the medulla. Because the wall of the descending limb of Henle's loop is permeable to water, water diffuses from the tubule into the more concentrated interstitial fluid. By the time the filtrate reaches the apex of the medulla of the kidney, the concentration of the filtrate is equal to the concentration of the the interstitial fluid.

The ascending limb of Henle's loop is not permeable to water, and Na^+, K^+, and Cl^- ions are actively transported from the filtrate into the interstitial fluid. Consequently, the volume of the filtrate does not change as it passes through the interstitial fluid, but the concentration is greatly reduced. By the time the filtrate reaches the cortex of the kidney, the concentration is approximately 100 mOsm/L, which is less concentrated than the interstitial fluid of the cortex (300 mOsm/L).

3. The distal nephron and collecting duct are permeable to water if ADH is present. If ADH is present, water diffuses from the more concentrated filtrate into the interstitial fluid. By the time the filtrate has reached the apex of the medulla an additional 19% of the filtrate has been reabsorbed, and

4. 1% or less remains as urine.

in concentration. On the other hand, if conservation of water is required to maintain homeostasis, slightly less filtrate is produced, and water is reabsorbed from the filtrate as it passes through the distal tubules and collecting ducts. This results in the production of a small volume of very concentrated urine. Regulation of urine volume and concentration involves hormonal mechanisms, autoregulation, and the nervous system.

Hormonal Mechanisms
Antidiuretic Hormone

The distal tubules and the collecting ducts remain relatively impermeable to water in the absence of ADH (see figure 26.13). When little ADH is secreted, a large part of the 19% of the filtrate that is normally reabsorbed in the distal tubules and the collecting ducts becomes part of the urine. People who do not secrete sufficient ADH often produce 10–20 L of urine per day and develop major problems such as dehydration and ion imbalances. Insufficient ADH secretion results in a condition called **diabetes insipidus** (dī-ă-bē′tēz in-sip′i-dŭs); the word diabetes implies the production of a large volume of urine, and the word insipidus implies the production of a clear, tasteless, dilute urine. This condition is in contrast to **diabetes mellitus** (me-lī′tŭs), which implies the production of a large volume of urine that contains a high concentration of glucose. The word mellitus means "honeyed," or "sweet."

ADH is secreted from the posterior pituitary, or neurohypophysis. Neurons with cell bodies primarily in the supraoptic nuclei of the hypothalamus have axons that course to the posterior pituitary gland. From these neuron terminals, ADH is released into the circulatory system. Cells of the supraoptic nuclei are sensitive to slight changes in the osmolality of the interstitial fluid. If the osmolality of the blood and interstitial fluid increases, these cells stimulate the ADH-secreting neurons. Action potentials are then propagated along the axons of the ADH-secreting neurons to the posterior pituitary gland, where ADH is released from the ends of the axons. Reduced osmolality of the interstitial fluid within the supraoptic nuclei inhibits ADH secretion from the posterior pituitary gland (see figure 18.5).

Pressure receptors that monitor blood pressure in the atria of the heart, large veins, carotid sinuses, and aortic arch also influence ADH secretion when the blood pressure increases or decreases in excess of 5%–10%. Decreases in blood pressure are detected by the pressure receptors, which consequently decrease the frequency of action potentials sent along the afferent pathways that ultimately extend to the supraoptic region of the hypothalamus. The result is an increase in ADH secretion.

When blood osmolality increases or when blood pressure declines significantly, ADH secretion increases and acts on the kidneys to increase the reabsorption of water. The reabsorption of water by the kidneys decreases blood osmolality and increases blood pressure. Conversely, when blood osmolality decreases or when blood pressure increases, ADH secretion declines. The reduced ADH levels cause the kidneys

to reabsorb less water and to produce a larger volume of dilute urine. The increased loss of water in the urine increases blood osmolality and decreases blood pressure.

Renin-Angiotensin-Aldosterone

Renin is an enzyme secreted by cells of the juxtaglomerular apparatus. The rate of renin secretion increases if blood pressure in the afferent arteriole decreases, or if the Na^+ ion concentration of the filtrate passing by the macula densa cells of the juxtaglomerular apparatuses decreases. Renin enters the general circulation, acting on **angiotensinogen** and converting it to **angiotensin I.** Subsequently, a proteolytic enzyme called **angiotensin-converting enzyme (ACE)** converts angiotensin I to **angiotensin II** (see figure 21.44). Angiotensin II is a potent vasoconstrictor substance that increases the peripheral resistance, causing blood pressure to increase. Angiotensin II also increases the rate of aldosterone secretion, the sensation of thirst, salt appetite, and ADH secretion.

The rate of renin secretion decreases if blood pressure in the afferent arteriole increases, or if the Na^+ ion concentration of the filtrate increases as it passes by the macula densa of the juxtaglomerular apparatuses.

A large decrease in the concentration of Na^+ ions in the interstitial fluids acts directly on the aldosterone-secreting cells of the adrenal cortex to increase the rate of aldosterone secretion. Angiotensin II is much more important than the blood level of Na^+ ions, however, in regulating aldosterone secretion.

Aldosterone, a steroid hormone secreted by the cortical cells of the adrenal glands (see chapter 18), passes through the circulatory system from the adrenal glands to the cells in the distal tubules and the collecting ducts. Aldosterone molecules diffuse through the plasma membranes and bind to receptor molecules within the cells. The combination of aldosterone molecules with their receptor molecules increases synthesis of the protein molecules that are responsible for the active transport of Na^+ ions across the epithelial cells of the nephron. As a result, the rate of Na^+ ion transport out of the filtrate back into the blood increases (figure 26.18).

Decreased secretion of aldosterone decreases the rate of Na^+ ion transport. As a consequence, the concentration of Na^+ ions in the distal tubules and the collecting ducts remains high. Because the concentration of filtrate passing through the distal tubules and the collecting ducts has a greater-than-normal concentration of solutes, the capacity for water to move by osmosis from the distal tubules and the collecting ducts is diminished, urine volume increases, and the urine has a greater-than-normal concentration of Na^+ ions.

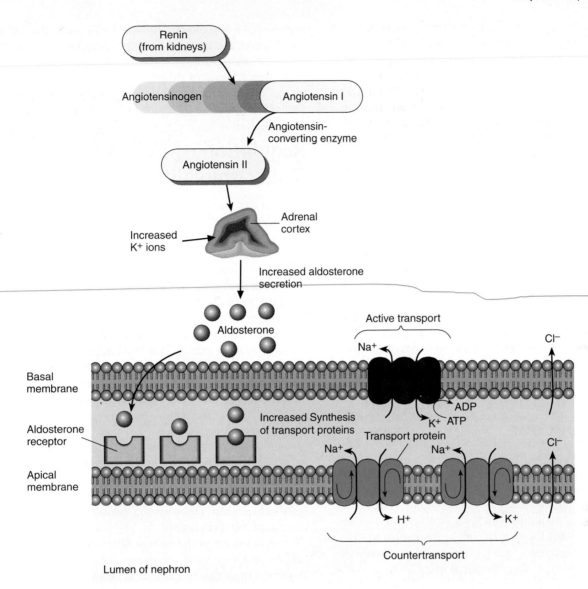

Figure 26.18   Effect of Aldosterone on the Distal Tubule

Aldosterone increases the rate at which sodium ions are absorbed and potassium and hydrogen ions are secreted. Chloride ions move with the sodium ions because they are attracted to the positive charge of the sodium ions.

---

7   **P R E D I C T**

Drugs that increase the urine volume are called diuretics. Some diuretics inhibit the active transport of sodium and ions in the nephron. Explain how these diuretic drugs could cause increased urine volume.

✔ *Answer in Appendix F*

Other Hormones

A polypeptide hormone called **atrial natriuretic** (nā′trē-yū-ret′ik) **hormone** is secreted from cardiac muscle cells in the right atrium of the heart when blood volume in the right atrium increases and stretches the cardiac muscle cells (see

chapter 21). Atrial natriuretic hormone inhibits ADH secretion from the posterior pituitary gland and Na^+ ion reabsorption in the kidney, which leads to the production of a large volume of dilute urine. The resulting decrease in blood volume decreases blood pressure. Atrial natriuretic hormone also dilates arteries and veins, which reduces peripheral resistance and lowers blood pressure. Thus, there is a decrease in venous return and blood volume in the right atrium.

Two other substances, prostaglandins and kinins, are formed in the kidneys and affect kidney function. Their roles are unclear, but both substances influence the rate of filtrate formation and Na^+ ion reabsorption. Prostaglandins probably moderate the sensitivity of the renal blood vessels to neural stimuli and to angiotensin II.

Autoregulation

Autoregulation is the maintenance, within the kidneys, of a relatively stable GFR over a wide range of systemic blood pressures. For example, the GFR is relatively constant as the systemic blood pressure changes between 90 and 180 mm Hg.

Autoregulation involves changes in the degree of constriction in the afferent arterioles. The precise mechanism by which autoregulation is achieved is unclear, but, as systemic blood pressure increases, the afferent arterioles constrict and prevent an increase in renal blood flow and filtration pressure across the filtration membrane of the renal corpuscle. Conversely, a decrease in systemic blood pressure results in dilation of the afferent arterioles, thus preventing a decrease in the renal blood flow and filtration pressure across the filtration membrane of the renal corpuscle.

Autoregulation is also influenced by the rate of flow of filtrate by cells of the macula densa. An increased flow rate is detected by the macula densa, which sends a signal to the juxtaglomerular apparatus to constrict the afferent arteriole. The result is a decrease in the filtration pressure across the filtration membrane of the renal corpuscle.

Effect of Sympathetic Stimulation on Kidney Function

Sympathetic neurons with norepinephrine as their neurotransmitter innervate the blood vessels of the kidneys. Sympathetic stimulation constricts the small arteries and afferent arterioles, decreasing renal blood flow and filtrate formation. Intense sympathetic stimulation, such as during shock or intense exercise, decreases the rate of filtrate formation to only a few milliliters per minute. Small changes in sympathetic stimulation have a minimal effect on renal blood flow and filtrate formation. Autoregulation maintains renal blood flow and filtrate formation at a relatively constant rate unless sympathetic stimulation is intense.

In response to severe stress or circulatory shock, renal blood flow can decrease to such low levels that the blood supply to the kidney is inadequate to maintain normal kidney metabolism. As a consequence, kidney tissues can be damaged and thus be unable to perform their normal functions. This is one of the reasons why shock should be treated quickly.

Regulation of Body Fluid Concentration and Volume

Water intake varies from person to person and is strongly influenced by habit. Most water enters the body through ingested liquids and solid foods, but approximately 10% of body water is produced by cellular metabolism (see chapter 24). Water output occurs by several routes and is equal to water intake. Approximately 40% of water loss occurs through evaporation from the lungs, diffusion through the skin, secretion of glands, perspiration, and in the feces; whereas approxi-

mately 60% is excreted by the kidneys in the urine. Major functions of the kidneys are the production of a small volume of concentrated urine, when it is necessary to conserve water, or a large volume of dilute urine, when it is necessary to lose water. Thus, the kidneys help maintain body fluid osmolality and volume within a narrow range of values.

Regulation of Extracellular Fluid Osmolality

Given a solution in a container, such as a pan on a stove, it is possible to decrease the osmolality of the solution by adding water to it. It is also possible to increase the osmolality of the solution by boiling the water in the pan, thus removing water from the solution by evaporation. Similarly, the kidneys function to maintain the concentration of the body fluids between 285 and 300 mOsm/kg by increasing water reabsorption from the filtrate when extracellular fluid osmolality increases and by reducing water reabsorption from the filtrate when the osmolality of the extracellular fluid decreases.

An increase in the osmolality of the extracellular fluid triggers thirst and ADH secretion. Water that is consumed is absorbed from the intestine and enters the extracellular fluid. ADH acts on the distal tubules and collecting ducts to increase the reabsorption of water. The increase in the amount of water entering the extracellular fluid causes a decrease in osmolality (figure 26.19), and a small volume of concentrated urine is produced. The ADH and thirst mechanisms are sensitive to small changes in extracellular fluid osmolality to which they respond quickly. Dehydration results in increased thirst and increased ADH secretion.

A decrease in the osmolality of the extracellular fluid inhibits thirst and ADH secretion. Less water is consumed and absorbed from the intestine, and less water is reabsorbed from the distal tubules and collecting ducts of the kidneys. Consequently, more water is lost in the form of a large volume of dilute urine and there is an increase in the osmolality of the extracellular fluid (see figure 26.19). For example, consumption of a large volume of water in a beverage results in less reabsorption of water in the kidneys and the production of a large volume of dilute urine, while the osmolality of the extracellular fluid is maintained within a normal range of values.

Regulation of Extracellular Fluid Volume

It is possible for the volume of extracellular fluid to increase or decrease, even if the osmolality of the extracellular fluid is maintained within a narrow range of values. Mechanisms exist to regulate the extracellular fluid volume.

Cells that are sensitive to changes in blood pressure are important in the regulation of extracellular fluid volume. Carotid sinus and aortic arch baroreceptors monitor blood pressure in large arteries, and cells of the juxtaglomerular apparatuses are sensitive to pressure changes within the afferent

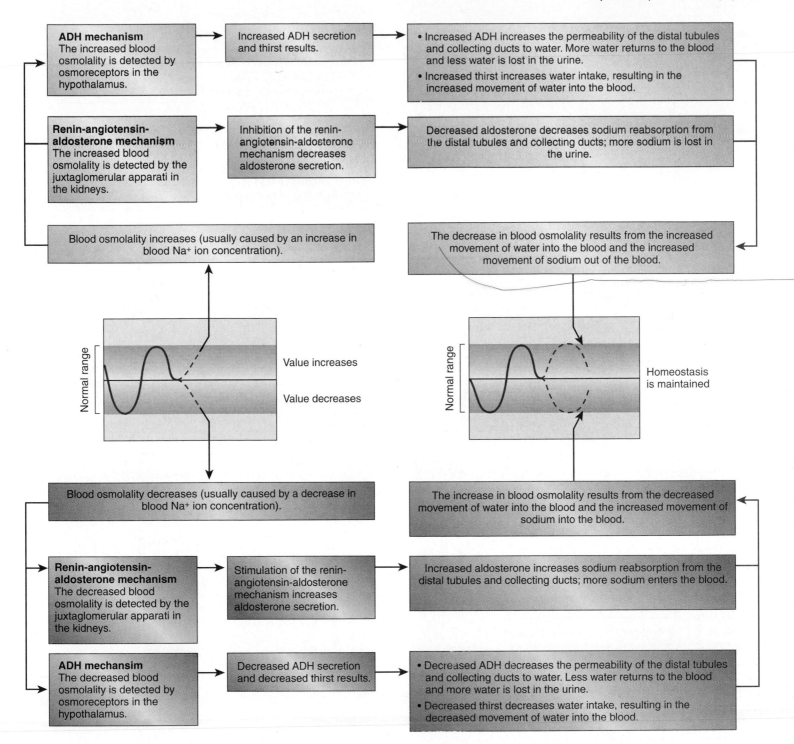

ADH mechanism
The increased blood osmolality is detected by osmoreceptors in the hypothalamus.

Increased ADH secretion and thirst results.

• Increased ADH increases the permeability of the distal tubules and collecting ducts to water. More water returns to the blood and less water is lost in the urine.
• Increased thirst increases water intake, resulting in the increased movement of water into the blood.

Renin-angiotensin-aldosterone mechanism
The increased blood osmolality is detected by the juxtaglomerular apparati in the kidneys.

Inhibition of the renin-angiotensin-aldosterone mechanism decreases aldosterone secretion.

Decreased aldosterone decreases sodium reabsorption from the distal tubules and collecting ducts; more sodium is lost in the urine.

Blood osmolality increases (usually caused by an increase in blood Na+ ion concentration).

The decrease in blood osmolality results from the increased movement of water into the blood and the increased movement of sodium out of the blood.

Normal range

Value increases

Value decreases

Normal range

Homeostasis is maintained

Blood osmolality decreases (usually caused by a decrease in blood Na+ ion concentration).

The increase in blood osmolality results from the decreased movement of water into the blood and the increased movement of sodium into the blood.

Renin-angiotensin-aldosterone mechanism
The decreased blood osmolality is detected by the juxtaglomerular apparati in the kidneys.

Stimulation of the renin-angiotensin-aldosterone mechanism increases aldosterone secretion.

Increased aldosterone increases sodium reabsorption from the distal tubules and collecting ducts; more sodium enters the blood.

ADH mechansim
The decreased blood osmolality is detected by osmoreceptors in the hypothalamus.

Decreased ADH secretion and decreased thirst results.

• Decreased ADH decreases the permeability of the distal tubules and collecting ducts to water. Less water returns to the blood and more water is lost in the urine.
• Decreased thirst decreases water intake, resulting in the decreased movement of water into the blood.

Figure 26.19   Homeostasis: Hormonal Regulation of Blood Osmolality

arterioles. These pressure receptors respond to changes in arterial blood pressure, including pressure changes resulting from increases or decreases in blood volume. Receptors are also located in the walls of the atria and large veins that are sensitive to forces that stretch their walls. The small changes in pressure that occur in response to increases or decreases in the volume of venous blood are examples. In addition, cells of the macula densa are sensitive to the Na+ ion concentra-

tion in the filtrate. Together these receptors play important roles in the regulation of the extracellular fluid volume.

An increase or decrease in the extracellular fluid volume increases or decreases the pressure of arterial and venous blood. The pressure receptors of the aortic arch, carotid sinus, atria, large veins, and juxtaglomerular apparatuses detect the pressure changes and activate neural mechanisms and three major hormonal mechanisms (figure 26.20).

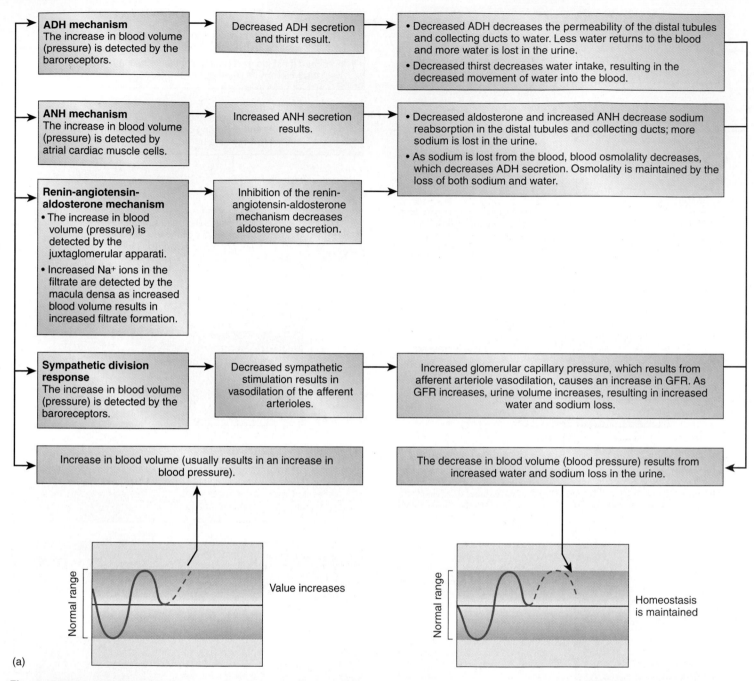

ADH mechanism
The increase in blood volume (pressure) is detected by the baroreceptors.

→ Decreased ADH secretion and thirst result.

→ • Decreased ADH decreases the permeability of the distal tubules and collecting ducts to water. Less water returns to the blood and more water is lost in the urine.
• Decreased thirst decreases water intake, resulting in the decreased movement of water into the blood.

ANH mechanism
The increase in blood volume (pressure) is detected by atrial cardiac muscle cells.

→ Increased ANH secretion results.

→ • Decreased aldosterone and increased ANH decrease sodium reabsorption in the distal tubules and collecting ducts; more sodium is lost in the urine.
• As sodium is lost from the blood, blood osmolality decreases, which decreases ADH secretion. Osmolality is maintained by the loss of both sodium and water.

Renin-angiotensin-aldosterone mechanism
• The increase in blood volume (pressure) is detected by the juxtaglomerular apparati.
• Increased Na+ ions in the filtrate are detected by the macula densa as increased blood volume results in increased filtrate formation.

→ Inhibition of the renin-angiotensin-aldosterone mechanism decreases aldosterone secretion.

Sympathetic division response
The increase in blood volume (pressure) is detected by the baroreceptors.

→ Decreased sympathetic stimulation results in vasodilation of the afferent arterioles.

→ Increased glomerular capillary pressure, which results from afferent arteriole vasodilation, causes an increase in GFR. As GFR increases, urine volume increases, resulting in increased water and sodium loss.

Increase in blood volume (usually results in an increase in blood pressure).

The decrease in blood volume (blood pressure) results from increased water and sodium loss in the urine.

Normal range — Value increases

Normal range — Homeostasis is maintained

(a)

Figure 26.20   Homeostasis: Hormonal Regulation of Blood Volume
(*a*) Regulation of blood volume (responses to increased blood volume) (continued on next page).

1. *Neural mechanisms.* Neural mechanisms change the frequency of action potentials carried by sympathetic neurons to the afferent arterioles of the kidney in response to increases or decreases in blood volume. When the pressure receptors detect an increase in arterial and venous blood pressure, the frequency of action potentials carried by sympathetic neurons to the afferent arterioles decreases. Consequently, the afferent arterioles dilate. This increases the glomerular capillary pressure, which increases the filtration pressure, resulting in an increase in the GFR and an increase in

the filtrate volume and urine volume. Because of autoregulation, a large increase in blood pressure is required to substantially increase the filtration pressure.

When the pressure receptors detect a decrease in arterial and venous blood pressure, there is an increase in the frequency of action potentials carried by sympathetic neurons to the afferent arterioles. Consequently, the afferent arterioles constrict. This decreases the glomerular capillary pressure, which decreases the filtration pressure, resulting in a decrease in the GFR and a decrease in the filtrate volume and

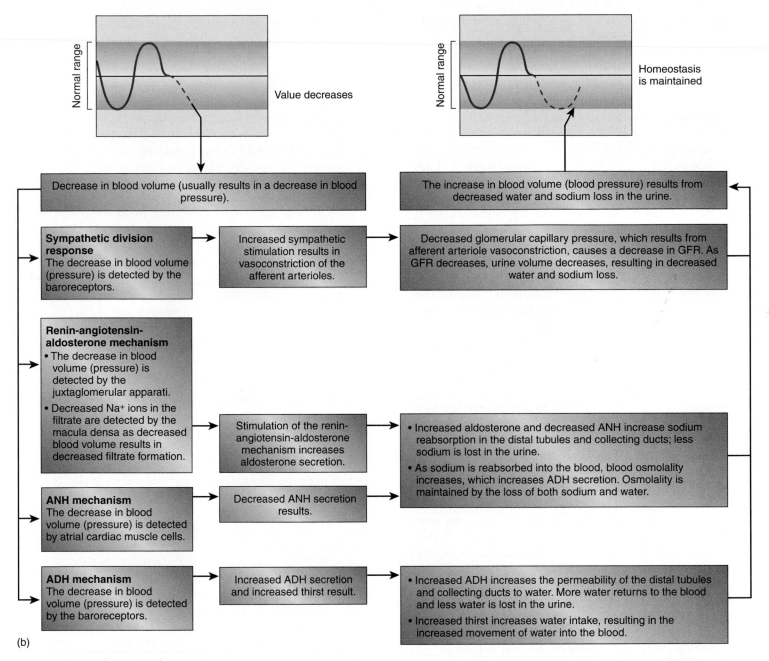

Figure 26.20 (*continued*)
(*b*) Regulation of blood volume (responses to decreased blood volume).

urine volume. Because of autoregulation, a large decrease in blood pressure is required to substantially decrease the filtration pressure.

2. *Renin-angiotensin-aldosterone mechanism.* The renin-angiotensin-aldosterone mechanism responds to (1) small changes in blood volume and (2) changes in the Na$^+$ ion concentration in the filtrate. An increase in blood volume can cause an increased blood pressure in the afferent arterioles, which results in a decreased rate of renin secretion by the juxtaglomerular cells. The concentration of Na$^+$ ions in the filtrate passing the cells

of the macula densa increases when the volume of filtrate flowing through the nephron increases. The macula densa cells respond to the increased Na$^+$ ion concentration by decreasing renin secretion from the juxtaglomerular cells (see figure 26.20*a*).

The decrease in renin secretion results in a decreased conversion of angiotensinogen to angiotensin I, which results in a decrease in the conversion of angiotensin I to angiotensin II. The reduced angiotensin II causes a decrease in the rate of aldosterone secretion from the adrenal cortex. The decreased aldosterone reduces the rate of Na$^+$ ion reabsorption, primarily from the distal tubules and collecting ducts. Consequently, more Na$^+$ ions remain in the filtrate and fewer Na$^+$ ions are reabsorbed. The effect is to increase the osmolality of the filtrate and to reduce the osmolality of the extracellular fluid. Because the mechanisms that regulate extracellular fluid osmolality function simultaneously, ADH secretion decreases in response to the reduced osmolality of the extracellular fluid and, consequently, less water is reabsorbed in the distal tubules and collecting ducts of the kidney. The water remains, with the excess Na$^+$ ions, in the filtrate. Thus the volume of urine produced by the kidney increases and the extracellular fluid volume decreases (see figure 26.20a).

A decrease in blood volume can cause a decrease in blood pressure in the afferent arterioles, which results in an increased rate of renin secretion by the juxtaglomerular cells. The concentration of Na$^+$ ions in the filtrate passing the cells of the macula densa decrease when the volume of filtrate flowing through the nephron decreases. The macula densa cells respond to the decreased Na$^+$ ion concentration by increasing renin secretion from the juxtaglomerular cells (see figure 26.20b).

The increase in renin secretion results in an increased conversion of angiotensinogen to angiotensin I, which results in an increase in the conversion of angiotensin I to angiotensin II. The increased angiotensin II causes an increase in the rate of aldosterone secretion from the adrenal cortex. The increased aldosterone increases the rate of Na$^+$ ion reabsorption, primarily from the distal tubules and collecting ducts. Consequently, fewer Na$^+$ ions remain in the filtrate and more Na$^+$ ions are reabsorbed. The effect is to decrease the osmolality of the filtrate and to increase the osmolality of the extracellular fluid. Because the mechanisms that regulate extracellular fluid osmolality function simultaneously, there is an increase in ADH secretion in response to the increased osmolality of the extracellular fluid and, consequently, more water is reabsorbed in the distal tubules and collecting ducts of the kidney. Thus the volume of urine produced by the kidney decreases and the extracellular fluid volume increases (see figure 26.20b).

3. *Atrial natriuretic hormone mechanism.* This mechanism is important in the regulation of extracellular fluid volume, especially in response to increases in extracellular fluid volume. An increase in pressure in the atria of the heart, which usually results from an increase in blood volume, stimulates the secretion of atrial natriuretic hormone. Atrial natriuretic hormone decreases Na$^+$ ion reabsorption in the distal tubules and collecting ducts and, therefore, increases the rate at which Na$^+$ ions and water are lost in the urine. Thus, increased atrial natriuretic hormone decreases the extracellular fluid volume (see figure 26.20a).

A decrease in pressure in the atria of the heart inhibits the secretion of atrial natriuretic hormone. The decreased atrial natriuretic hormone decreases the

inhibition on Na$^+$ ion reabsorption in the distal tubules and collecting ducts and, therefore, the rate at which Na$^+$ ions are reabsorbed increases. As Na$^+$ ion reabsorption increases, water reabsorption also increases. Thus, decreased atrial natriuretic hormone tends to result in decreased urine volume and an increase in the extracellular fluid volume (see figure 26.20b).

4. *Antidiuretic hormone mechanism.* The ADH mechanism plays an important role in regulating extracellular fluid volume in response to large changes in blood pressure (of 5%–10%). An increase in blood pressure results in a decrease in ADH secretion. As a result the reabsorption of water from the lumen of the distal tubules and collecting ducts decreases, resulting in a larger volume of dilute urine. This response helps decrease the extracellular fluid volume and blood pressure (see figure 26.20a).

A decrease in blood pressure results in an increase in ADH secretion. Consequently, the reabsorption of water from the lumen of the distal tubules and collecting ducts increases, resulting in a smaller volume of concentrated urine. This response helps increase the extracellular fluid volume and blood pressure (see figure 26.20b).

The mechanisms that maintain extracellular fluid concentration and volume function together. When mechanisms that maintain fluid volume do not function normally, it is possible to have an increased extracellular fluid volume even though the extracellular concentration of fluids is maintained within a normal range of values. For example, if aldosterone secretion by the adrenal cortex increases abnormally, Na$^+$ ion reabsorption by the kidney increases, and the total volume of extracellular fluid increases because the mechanisms that keep the concentration of the body fluids constant, such as the regulation of ADH secretion, still operate. The blood pressure can be elevated and edema can result, but the osmolality of the extracellular fluid is maintained between 185 and 300 mOsm/kg. In people suffering from heart failure, the resulting reduced blood pressure activates mechanisms that function to increase the blood pressure to their normal range of values. Those mechanisms include the release of renin from the kidneys. Consequently, the renin-angiotensin-aldosterone mechanism functions to increase Na$^+$ ion reabsorption. Water reabsorption also increases, and the osmolality of the extracellular fluid is maintained between 185 and 300 mOsm/kg. The consequence is an increase in the extracellular fluid volume, which results in edema in the periphery and possibly in the lungs (congestive heart failure).

Clearance and Tubular Maximum

Plasma clearance is a calculated value representing the volume of plasma that is cleared of a specific substance each minute. For example, if the clearance value is 100 mL/min for a substance, the substance is completely removed from 100 mL of plasma each minute.

Clinical Focus   Diuretics

Diuretics (dī-yū-ret′iks) are agents that increase the rate of urine formation. Although the definition is simple, a number of different physiologic mechanisms are involved.

Diuretics are used to treat disorders such as hypertension and several types of edema that are caused by conditions such as congestive heart failure and cirrhosis of the liver. Use of diuretics can lead to complications, however, including dehydration and electrolyte imbalances.

The action of **carbonic anhydrase** (kar-bon′ik an-hī′drās) **inhibitors** reduces the rate of hydrogen ion secretion and the reabsorption of bicarbonate ions. The bicarbonate ions increase tubular osmotic pressure, causing osmotic diuresis. With long-term use, the diuretic effect of carbonic anhydrase inhibitors tends to be lost. The diuretic effect of carbonic anhydrase inhibitors is useful in treating conditions such as glaucoma and altitude sickness.

Inhibitors of sodium ion reabsorption include thiazide-type diuretics. They promote the loss of Na^+ ions, chloride ions, and water in urine. These diuretics are given to some people who have hypertension. Inhibitors of Na^+ ion reabsorption, such as bumetanide, furosemide, and ethacrynic acid, specifically inhibit transport in the ascending limb of the loop of Henle. These diuretics are frequently used to treat congestive heart failure, cirrhosis of the liver, and renal disease.

Potassium-sparing diuretics are antagonists to aldosterone or directly prevent Na^+ ion reabsorption in the distal tubules and collecting ducts. Thus they promote Na^+ ion and water loss in the urine. These diuretics are used to reduce the loss of potassium ions in the urine and therefore preserve, or "spare," potassium ions. They are often used in combination with inhibitors of Na^+ ion reabsorption and are effective in preventing excess potassium loss in the urine.

Osmotic diuretics freely pass by filtration into the filtrate, and they undergo limited reabsorption by the nephron. These diuretics increase urine volume by elevating the osmotic concentration of the filtrate, thus reducing the amount of water moving by osmosis out of the nephron. Urea, mannitol, and glycerine have been used as osmotic diuretics. Although they are not commonly used, they are effective in treating people who are suffering from cerebral edema and edema in acute renal failure.

Xanthines (zan′thēnz), including caffeine and related substances, act as diuretics, partly because they increase renal blood flow and the rate of glomerular filtrate formation. They also influence the nephron by decreasing Na^+ and chloride reabsorption.

Alcohol acts as a diuretic, although it is not used clinically for that purpose. It inhibits ADH secretion from the posterior pituitary and results in increased urine volume.

Clinical Note

The plasma clearance can be calculated for any substance that enters the circulatory system according to the following formula:

$$\text{Plasma clearance (mL/min)} = \text{Quantity of urine (mL/min)} \times \frac{\text{Concentration of substance in urine}}{\text{Concentration of substance in plasma}}$$

8   P R E D I C T

During surgery, a patient's blood pressure drops to very low levels, and ischemia of the kidney develops. Within 1 day following the surgery, the GFR and the urine volume decrease to very low levels. Given that the structure of the glomeruli did not dramatically change, but that the epithelium of nephrons suffered from ischemia and sloughed into the nephron to form casts of epithelial cells in the nephron, explain why the GFR was reduced.

✔ *Answer in Appendix F*

Plasma clearance can be used to estimate GFR if the appropriate substance is monitored (see table 26.2). Such a substance must have the following characteristics: (1) it must pass through the filtration membrane of the renal corpuscle as freely as water or other small molecules, (2) it must not be reabsorbed, (3) it must not be secreted into the nephron, and (4) it must not be either metabolized or produced in the kidney. **Inulin** (in′yū-lin) is a polysaccharide that has these characteristics. As filtrate is formed, it has the same concentration of inulin as plasma; but, as the filtrate flows through the nephron, all of the inulin remains in the nephron to enter the urine. As a consequence, all of the volume of plasma that becomes filtrate is cleared of inulin, and the plasma clearance for inulin is equal to the rate of glomerular filtrate formation. The GFR is reduced when the kidney fails. Measurement of the GFR indicates the degree to which kidney damage has occurred.

Plasma clearance can also be used to calculate renal plasma flow (see table 26.2). Substances with the following characteristics, however, must be used: (1) the substance must pass through the filtration membrane of the renal corpuscle, and (2) it must be secreted into the nephron at a sufficient rate so that very little of it remains in the blood as the blood leaves the kidney. PAH meets these requirements. As blood flows through the kidney, essentially all of the PAH is either filtered or secreted into the nephron. The clearance calculation for PAH is therefore a good estimate of the volume of plasma flowing through the kidney each minute. Also, if the hematocrit is known, the total volume of blood flowing through the kidney each minute can be easily calculated.

The concept of plasma clearance can be used to make the measurements described previously, or it can be used to determine the means by which drugs or other substances are excreted by the kidney. A plasma clearance value greater than the inulin clearance value suggests that the substance is secreted by the nephron into the filtrate.

---

9 **P R E D I C T**

A person is suspected of suffering from chronic renal failure. To assess kidney function, urea clearance is measured and found to be very low. Explain what a very low urea clearance indicates in a person suffering from chronic renal failure.

✔ *Answer in Appendix F*

The **tubular load** of a substance is the total amount of the substance that passes through the filtration membrane into the nephrons each minute. Normally, glucose is almost completely reabsorbed from the nephron by the process of active transport. The capacity of the nephron to actively transport glucose across the epithelium of the nephron is limited, however. If the tubular load is greater than the capacity of the nephron to reabsorb it, the excess glucose remains in the urine.

The maximum rate at which a substance can be actively reabsorbed is called the **tubular maximum** (figure 26.21). Each substance that is reabsorbed has its own tubular maximum, determined by the number of active transport carrier molecules and the rate at which they are able to transport molecules of the substance. For example, in people suffering from diabetes mellitus the tubular load for glucose can exceed the tubular maximum by a substantial amount, allowing glucose to appear in the urine. Urine volume is also greater than normal because the glucose molecules in the filtrate reduce the effectiveness of water reabsorption by osmosis.

Urine Movement
Urine Flow Through the Nephron and the Ureters

Hydrostatic pressure averages 10 mm Hg in Bowman's capsule and nearly 0 mm Hg in the renal pelvis. This pressure gradient forces the filtrate to flow from Bowman's capsule through the nephron into the renal pelvis. Because the pressure is 0 mm Hg in the renal pelvis, there is no pressure gradient to force urine to flow to the urinary bladder through the ureters. The circular smooth muscle in the walls of the ureters exhibits peristaltic contractions that force the urine to flow through the ureters. The peristaltic waves progress from the region of the renal pelvis to the urinary bladder. They occur from once every few seconds to once every 2–3 min. Parasympathetic stimulation increases their frequency, and sympathetic stimulation decreases it.

The peristaltic contractions of each ureter proceed at a velocity of approximately 3 cm/s and can generate pressures in excess of 50 mm Hg. At the point where the ureters penetrate the bladder, they course obliquely through the trigone. Pressure inside the bladder compresses that part of the ureter, preventing backflow of urine.

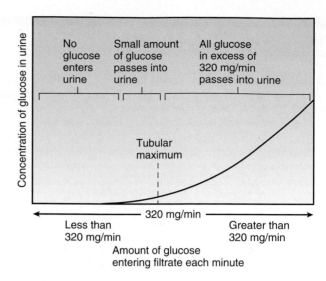

Figure 26.21   Tubular Maximum for Glucose

As the concentration of glucose increases in the filtrate, it reaches a point that exceeds the ability of the nephron to actively reabsorb it. That concentration is called the tubular maximum. Beyond that concentration, the excess glucose enters the urine.

When no urine is present in the urinary bladder, internal pressure is about 0 mm Hg. When the volume is 100 mL of urine, pressure is elevated to only 10 mm Hg. Pressure in the urinary bladder increases slowly as its volume increases to 400–500 mL, but above bladder volumes of 500 mL the pressure rises rapidly.

Clinical Note

Kidney stones are hard objects usually found in the pelvis of the kidney. They are normally 2–3 mm in diameter with either a smooth or jagged surface, but occasionally a large branching kidney stone called a **staghorn stone** forms in the renal pelvis. About 1% of all autopsies reveal kidney stones, and many of the stones occur without causing symptoms. The symptoms associated with kidney stones occur when a stone passes into the ureter, resulting in referred pain down the back, side, and groin area. The ureter contracts around the stone, causing the stone to irritate the epithelium and produce bleeding, which appears as blood in the urine, a condition called **hematuria.** In addition to causing intense pain, kidney stones can block the ureter, cause ulceration in the ureter, and increase the probability of bacterial infections.

About 65% of all kidney stones are composed of calcium oxalate mixed with calcium phosphate, 15% are magnesium ammonium phosphate, and 10% are uric acid or cystine; approximately 2.5% of each kidney stone is composed of mucoprotein.

The cause of kidney stones is usually obscure. Predisposing conditions include a concentrated urine and an abnormally high calcium concentration in the urine, although the cause of the high calcium concentration is usually unknown. Magnesium ammonium phosphate stones are often found in people with recurrent kidney infections, and uric acid stones often occur in people suffering from gout. Severe kidney stones must be removed surgically. Instruments that pulverize kidney stones with ultrasound or lasers, however, have replaced most traditional surgical procedures.

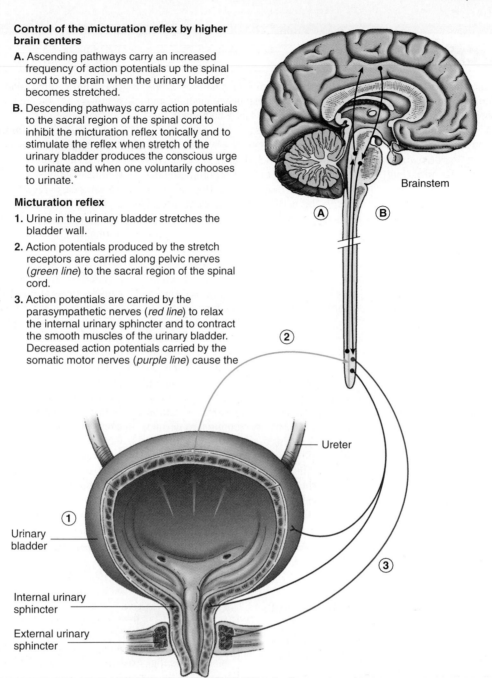

Control of the micturation reflex by higher brain centers

A. Ascending pathways carry an increased frequency of action potentials up the spinal cord to the brain when the urinary bladder becomes stretched.

B. Descending pathways carry action potentials to the sacral region of the spinal cord to inhibit the micturation reflex tonically and to stimulate the reflex when stretch of the urinary bladder produces the conscious urge to urinate and when one voluntarily chooses to urinate.°

Micturation reflex

1. Urine in the urinary bladder stretches the bladder wall.

2. Action potentials produced by the stretch receptors are carried along pelvic nerves (*green line*) to the sacral region of the spinal cord.

3. Action potentials are carried by the parasympathetic nerves (*red line*) to relax the internal urinary sphincter and to contract the smooth muscles of the urinary bladder. Decreased action potentials carried by the somatic motor nerves (*purple line*) cause the

Brainstem

Ureter

Urinary bladder

Internal urinary sphincter

External urinary sphincter

Figure 26.22   Micturition Reflex

Micturition Reflex

The **micturition** (mik-chū-rish′ŭn) **reflex** is initiated when the bladder wall stretches and results in **micturition,** which is the elimination of urine from the bladder. Integration of the micturition reflex occurs both in the sacral region of the spinal cord and in the brainstem.

As the bladder fills with urine, stretch receptors are stimulated. Afferent action potentials are conducted to the sacral segments of the spinal cord through the pelvic nerves. In response, efferent action potentials are sent to the urinary bladder through parasympathetic fibers in the pelvic nerves (fig-

ure 26.22). The parasympathetic action potentials cause the bladder to contract and the internal (smooth muscle) urinary sphincter to relax. Decreased somatic motor action potentials cause the external (skeletal muscle) urinary sphincter to relax. The micturition reflex normally produces a series of contractions of the urinary bladder.

Afferent action potentials from stretch receptors also initiate action potentials that ascend the spinal cord to a micturition center in the pons. In response, descending action potentials are sent to the sacral region of the spinal cord, where they initiate parasympathetic action potentials that cause the bladder to contract and the internal urinary sphincter to relax.

Clinical Focus   Renal Pathologies

Glomerular nephritis (glō-măr′yū-lăr ne-frī′tis) results from inflammation of the filtration membrane within the renal corpuscle. It is characterized by an increased permeability of the filtration membrane and the accumulation of numerous white blood cells in the area of the filtration membrane. As a consequence, a high concentration of plasma proteins enters the filtrate along with numerous white blood cells. A greater-than-normal urine volume accompanies the increase in plasma proteins in the urine.

Acute glomerular nephritis often occurs 1–3 weeks after a severe bacterial infection such as streptococcal sore throat or scarlet fever. Antigen–antibody complexes associated with the disease become deposited in the filtration membrane and cause its inflammation. This acute inflammation normally subsides after several days.

Chronic glomerular nephritis is long term and usually progressive. The filtration membrane thickens and eventually is replaced by connective tissue. Although in the early stages chronic glomerular nephritis resembles the acute form, in the advanced stages many of the renal corpuscles are replaced by fibrous connective tissue, and the kidney eventually ceases to function.

Pyelonephritis (pī′ĕ-lō-ne-frī′tis) is inflammation of the renal pelvis, medulla, and cortex. It often begins as a bacterial infection of the renal pelvis and then extends into the kidney itself. It can result from several types of bacteria, including *Escherichia coli*. Pyelonephritis can destroy nephrons and renal corpuscles, but, because the infection starts in the pelvis of the kidney, it affects the medulla more than the cortex. As a consequence, the ability of the kidney to concentrate urine is dramatically affected.

Renal failure can result from any condition that interferes with kidney function. **Acute renal failure** occurs when kidney damage is extensive and leads to the accumulation of urea in the blood and to acidosis (see chapter 27). In complete renal failure death can occur in 1–2 weeks. Acute renal failure can result from acute glomerular nephritis, or it can be caused by damage to or blockage of the renal tubules. Some poisons, such as mercuric ions or carbon tetrachloride, that are common to certain industrial processes cause necrosis of the nephron epithelium. If the damage does not interrupt the basement membrane surrounding the nephrons, extensive regeneration can occur within 2–3 weeks. Severe ischemia associated with circulatory shock resulting from sympathetic vasoconstric-

tion of the renal blood vessels can cause necrosis of the epithelial cells of the nephron.

Chronic renal failure results when so many nephrons are permanently damaged that those nephrons remaining functional cannot adequately compensate. Chronic renal failure can result from chronic glomerular nephritis, trauma to the kidneys, absence of kidney tissue caused by congenital abnormalities, or tumors. Urinary tract obstruction by kidney stones, damage resulting from pyelonephritis, and severe arteriosclerosis of the renal arteries also cause degeneration of the kidney.

In chronic renal failure the GFR is dramatically reduced, and the kidney is unable to excrete excess excretory products, including electrolytes and metabolic waste products. The accumulation of solutes in the body fluids causes water retention and edema. Potassium levels in the extracellular fluid are elevated, and acidosis occurs because the distal convoluted tubules and collecting ducts cannot excrete sufficient quantities of potassium and hydrogen ions. Acidosis, elevated potassium levels in the body fluids, and the toxic effects of metabolic waste products cause mental confusion, coma, and finally death when chronic renal failure is severe.

Decreased somatic motor action potentials cause the external urinary sphincter to relax.

The micturition reflex integrated in the spinal cord predominates in infants, but the micturition reflex integrated in the brainstem predominates in adults.

The micturition reflex is automatic, but it can be either inhibited or stimulated by higher centers in the brain. The higher brain centers prevent micturition by sending action potentials that inhibit the parasympathetic stimulation of the urinary bladder to contract and by stimulating the somatic motor neurons that keep the external urinary sphincter tonically contracted. The ability to voluntarily inhibit micturition develops at the age of 2–3 years.

When the desire to urinate exists, the higher brain centers send action potentials to facilitate the micturition reflex and to voluntarily relax the external urinary sphincter. The desire to urinate is initiated because stretch of the urinary bladder stimulates ascending fibers, which send action potentials to the higher centers of the brain. Irritation of the urinary bladder or the urethra by bacterial infections or other conditions can also initiate the urge to urinate, even though the bladder may be empty.

Clinical Note

If the spinal cord is damaged above the sacral region, no micturition reflex exists for a time; but if the bladder is emptied frequently, the micturition reflex eventually becomes adequate to cause the bladder to empty. Time is required for the micturition reflex integrated within the spinal cord to begin to operate. A typical micturition reflex can exist, but there is no conscious control over the onset or duration of it. This condition is called the **automatic bladder.**

A bladder that does not contract can result from damage to the sacral region of the spinal cord or to the nerves that carry action potentials between the spinal cord and the urinary bladder. As a result, the micturition reflex cannot occur. The bladder fills to capacity, and urine is forced in a slow dribble through the urinary sphincters. In elderly people or in patients with damage to the brainstem or spinal cord, there can be a loss of inhibitory action potentials to the sacral region of the spinal cord. Without inhibition, the sacral centers are hyperexcitable, and even a small amount of urine in the bladder can elicit an uncontrollable micturition reflex.

Systems Pathology
acute renal failure

ACUTE RENAL FAILURE

A large piece of machinery overturned at the construction site where Mr. H worked, trapping him beneath it. His legs were severely crushed, although they healed after several months. Mr. H nearly lost his life, however, because of the acute renal failure that developed because of his injury. Mr. H was trapped for several hours in a very difficult place to reach. During that time his blood pressure decreased to very low levels because of the blood loss, the edema in the inflamed tissues, and emotional shock. After he was rescued, fluid replacement in the form of both intravenous saline solutions and blood transfusions were given, and his blood pressure was successfully returned to its normal range. Twenty-four hours after the accident, however, his urine volume began to decrease. His urinary Na^+ ion concentration increased, his urine osmolality decreased, his urine specific gravity decreased, and casts and cellular debris were evident in his urine.

For approximately 7 days Mr. H exhibited oliguria (reduced urine production). During this period, renal dialysis was required to maintain Mr. H's blood volume and electrolyte concentrations within normal ranges. After approximately 7 days, his kidneys gradually began to produce large quantities of urine (a diuretic phase). Careful observation was required to keep his blood pressure and electrolyte concentrations within normal ranges. Substantial water, Na^+ ions, and K^+ ions had to be administered to him. After about 3 weeks, the function of his kidneys slowly began to improve, although many months passed before his kidney function had returned to normal.

BACKGROUND INFORMATION

The events after 24 h are consistent with acute renal failure caused by prolonged hypotension and ischemia of the kidney. While Mr. H was suffering from hypotension, blood flow to his kidneys was very low. The reduced blood flow to the kidneys was severe enough to result in damage to the epithelial lining of the kidney tubules (tubular necrosis). The period of reduced urine volume resulted from tubular damage. Renal ischemia results in necrosis of tubular cells, which then slough off into the tubules and block them so that filtrate cannot flow through the tubules. In addition, the filtrate leaks from the blocked or partially blocked tubules back into the interstitial spaces and, therefore, back into the circulatory system. As a result the amount of filtrate that becomes urine is markedly reduced. Blood levels of urea and of creatine increase because of the reduction in filtrate formation and reduced function of the tubular epithelium. The kidneys' ability to eliminate metabolic waste

products is therefore reduced. The small amount of urine produced has a high Na^+ ion concentration but an osmolality that is close to the concentration of the body fluids because the kidneys are not able to reabsorb Na^+ ions and because the urine-concentrating ability of the kidneys is severely damaged.

During the diuretic phase, the large quantities of urine were produced because the nephrons were partially healed and could produce urine, but the ability of the nephrons to concentrate urine was not yet normal. Large volumes of urine that contained significant amounts of Na^+ and K^+ ions were therefore produced. The kidneys were able to produce urine that was more concentrated than the body fluids, but the concentrating ability of the kidneys was still below normal. As time passed, the concentrating ability of the kidneys improved and eventually became normal once again.

---

10 **P R E D I C T**

Nine days after the accident, Mr. H began to appear pale, he became dizzy and lethargic. His hematocrit was elevated and his heart was arrythmic. He was very weak. Explain these manifestations.

✔ *Answer in Appendix F*

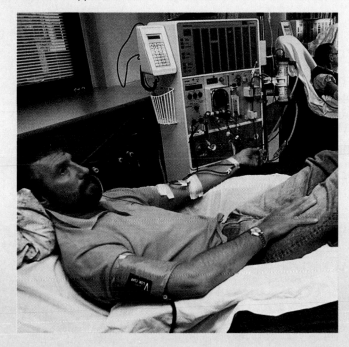

Figure B Dialysis
A patient undergoing dialysis.

System Interactions

System	Interactions
Integumentary	Pallor results from anemia, and bruising results from reduced clotting proteins in the blood because they are lost in the urine. A waxy yellow caste develops to the skin of light-skinned people, an ashen gray caste in black-skinned people, or a yellowish brown caste in brown-skinned people due to accumulation of urinary pigments. When the urea concentration in the blood is very high, white crystals of urea, called uremic frost, may appear on areas of the skin where there is heavy perspiration.
Skeletal	Changes in the skeletal system are not marked unless kidney damage results in chronic kidney failure. Bone resorption may result during a prolonged diuretic phase because of excessive loss of Ca^{2+} ions in the urine. Also, vitamin D levels may be reduced during both the oliguric and diuretic phases.
Muscular	Neuromuscular irritability results from the toxic effect of metabolic wastes on the central nervous system and ionic imbalances such as hyperkalemia. Involuntary jerking and twitching may occur as neuromuscular irritability develops. Tremors of the hands are an indication of the toxic effects of metabolic wastes on the cerebrum.
Nervous	Elevated blood K^+ ion levels and the toxic effects of metabolic wastes result in depolarization of neurons. Slowing of nerve conduction, burning sensations, pain, numbness, or tingling results. Also, decreased mental acuity, reduced ability to concentrate, apathy, and lethargy result. Periods of lethargy may alternate with restlessness and insomnia. In severe cases the patient may become confused and comatose.
Endocrine	Major predictable hormone deficiencies include vitamin D deficiency. In addition, there is a decrease in the secretion of reproductive hormones due to the effects of metabolic wastes and ionic imbalances on the hypothalamus.
Cardiovascular	Water and Na^+ ion retention may result in edema in peripheral tissues and in the lung. Also, hypertension and congestive heart failure may result. Hyperkalemia results in dysrhythmias and may cause cardiac arrest. Anemia due to decreased erythropoietin production by the damaged kidney and decreased half-life of red blood cells may result. Anemia is more likely because of the blood lost as a result of the crushing injury. Nosebleeds and bruising occur due to reduced concentration of clotting factors because they are lost in the urine.
Lymphatic	There are no major direct effects on the lymphatic system with the exception that increased lymph flow occurs as a result of edema.
Respiratory	Early during acute renal failure the depth of breathing increases, becoming labored as acidosis develops because the kidney is not able to secrete H^+ ions. Pulmonary edema often develops because of water and Na^+ ion retention. The likelihood of pulmonary infection increases secondary to pulmonary edema.
Digestive	Anorexia, nausea, and vomiting result from altered gastrointestinal functions due to the effects of ionic imbalances on the nervous system. There may be an odor of ammonia on the breath and metallic taste in the mouth. These effects are the result of the accumulation of metabolic waste products in the gastrointestinal tract, and the action of the normal gastrointestinal microorganisms on the waste products, which convert urea to ammonia. The ammonia and other metabolic waste products predispose the mouth to inflammation and infection.

Summary

The urinary system eliminates wastes; regulates blood volume, ion concentration, and pH; it is also involved with red blood cell and vitamin D production. The excretory system consists of the kidneys, the ureters, the urinary bladder, and the urethra.

Urinary System
Kidneys

1. The kidney is surrounded by a renal capsule and a renal fat pad and is held in place by the renal fascia.
2. The two layers of the kidney are the cortex and the medulla.
   - The renal columns extend toward the medulla between the renal pyramids.
   - The renal pyramids of the medulla project to the minor calyces.

3. The minor calyces open into the major calyces, which open into the renal pelvis. The renal pelvis leads to the ureter.
4. The functional unit of the kidney is the nephron. The parts of a nephron are the renal corpuscle, the proximal tubule, the loop of Henle, and the distal tubule.
   - The renal corpuscle is Bowman's capsule and the glomerulus. Materials leave the blood in the glomerulus and enter Bowman's capsule through the filtration membrane.
   - The nephron empties through the distal tubule into a collecting duct.
5. The juxtaglomerular apparatus consists of the macula densa (part of the distal tubule) and the juxtaglomerular cells of the afferent arteriole.

Arteries and Veins of the Kidney

1. Arteries branch as follows: renal artery to segmental artery to interlobar artery to arcuate artery to interlobular artery to afferent arteriole.
2. Afferent arterioles supply the glomeruli.
3. Efferent arteries from the glomeruli supply the peritubular capillaries and vasa recta.
4. Veins form from the peritubular capillaries as follows: interlobular vein to arcuate vein to interlobar vein to renal vein.

Ureters and Urinary Bladder

1. Structure
   - The walls of the ureter and urinary bladder consist of the epithelium, the lamina propria, a muscular coat, and a fibrous adventitia.
   - The transitional epithelium permits changes in size.
2. Function
   - The ureters transport urine from the kidney to the urinary bladder.
   - The urinary bladder stores urine.

Urine Production

Urine is produced by the processes of filtration, absorption, and secretion.

Filtration

1. The renal filtrate is plasma minus blood cells and blood proteins. Most (99%) of the filtrate is reabsorbed.
2. The filtration membrane is fenestrated endothelium, basement membrane, and the slitlike pores formed by podocytes.
3. Filtration pressure is responsible for filtrate formation.
   - Filtration pressure is glomerular capillary pressure minus capsule pressure minus colloid osmotic pressure.
   - Filtration pressure changes are primarily caused by changes in glomerular capillary pressure.

Tubular Reabsorption

1. Filtrate is reabsorbed by passive transport, including simple diffusion, facilitated diffusion, active transport, and cotransport from the nephron into the peritubular capillaries.
2. Specialization of tubule segments
   - The thin segment of the loop of Henle is specialized for passive transport.
   - The rest of the nephron and collecting tubules perform active transport, cotransport, and passive transport.
3. Substances transported
   - Active transport moves mainly Na^+ sodium ions across the wall of the nephron. Other ions and molecules are moved primarily by cotransport.
   - Passive transport moves water; urea; and lipid-soluble, nonpolar compounds.

Tubular Secretion

1. Substances enter the proximal or distal tubules and the collecting ducts.
2. Hydrogen ions, potassium ions, and some substances not produced in the body are secreted by countertransport mechanisms.

Urine Concentration Mechanism

1. Countercurrent systems (e.g., vasa recta and loop of Henle) and the distribution of urea are responsible for the concentration gradient in the medulla. The concentration gradient is necessary for the production of concentrated urine.
2. Production of urine
   - In the proximal tubule Na^+ ions and other substances are removed by active transport. Water follows passively, filtrate volume is reduced 65%, and the filtrate concentration is 300 mOsm/L.
   - In the descending limb of the loop of Henle water exits passively, and solute enters. The filtrate volume is reduced 15%, and the filtrate concentration is 1200 mOsm/L.
   - In the ascending limb of the loop of Henle, Na^+, Cl^-, and K^+ ions are transported out of the filtrate, but water remains because this segment of the nephron is impermeable to water. The filtrate concentration is 100 mOsm/L.
   - In the distal tubules and collecting ducts, water movement out of them is regulated by ADH. If ADH is absent, water is not reabsorbed, and a dilute urine is produced. If ADH is present, water moves out, and a concentrated urine is produced.

Regulation of Urine Concentration and Volume

Hormonal Mechanisms

1. ADH is secreted by the posterior pituitary and increases water permeability in the distal tubules and the collecting ducts.
   - ADH decreases urine volume, increases blood volume, and thus increases blood pressure.
   - ADH release is stimulated by increased blood osmolality or a decrease in blood pressure.
2. Aldosterone is produced in the adrenal cortex and affects Na^+ and Cl^- ion transport in the nephron and the collecting ducts.
   - A decrease in aldosterone results in less Na^+ ion reabsorption and an increase in urine concentration and volume. An increase in aldosterone results in greater Na^+ ion reabsorption and a decrease in urine concentration and volume.
   - Aldosterone production is stimulated by angiotensin II, increased blood K^+ ion concentration, and decreased blood Na^+ ion concentration.
3. Renin, produced by the kidneys, causes the production of angiotensin II.
   - Angiotensin II acts as a vasoconstrictor and stimulates aldosterone secretion, causing a decrease in urine production and an increase in blood volume.
   - Decreased blood pressure or decreased Na^+ ion concentration stimulates renin production.
4. Atrial natriuretic hormone, produced by the heart when blood pressure increases, inhibits ADH production and reduces the ability of the kidney to concentrate urine.

Autoregulation

Autoregulation dampens systemic blood pressure changes by altering afferent arteriole diameter.

Effects of Sympathetic Stimulation on Kidney Function

Sympathetic stimulation decreases afferent arteriole diameter.

Regulation of Body Fluid Concentration and Volume

Regulation of Extracellular Fluid Osmolality

1. Increased water consumption and ADH secretion occur in response to increases in extracellular fluid osmolality.
   - Decreased water consumption and ADH secretion occur in response to decreases in extracellular fluid osmolality.

2. Increased water consumption and ADH decrease extracellular fluid osmolality by increasing water absorption from the intestine and water reabsorption from the nephrons.
   • Decreased water consumption and ADH increase extracellular fluid osmolality by decreasing absorption from the intestine and water reabsorption from the nephrons.

Regulation of Extracellular Fluid Volume

1. Increased extracellular fluid volume results in decreased aldosterone secretion, increased atrial natriuretic hormone secretion, decreased ADH secretion, and decreased sympathetic stimulation of afferent arterioles.
   • The effect of these changes is to decrease Na^+ ion reabsorption and to increase urine volume to decrease extracellular fluid volume.
2. Decreased extracellular fluid volume results in increased aldosterone secretion, decreased atrial natriuretic hormone secretion, increased ADH secretion, and increased sympathetic stimulation of the afferent arterioles.
   • The effect of these changes is to increase Na^+ ion reabsorption and to decrease urine volume so as to increase extracellular fluid volume.

Clearance and Tubular Maximum

1. Plasma clearance is the volume of plasma that is cleared of a specific substance each minute.
2. The tubular load is the total amount of substance that enters the nephron each minute.
3. Tubular maximum is the fastest rate at which a substance is reabsorbed from the nephron.

Urine Movement
Urine Flow Through the Nephron and the Ureters

1. Hydrostatic pressure forces urine through the nephron.
2. Peristalsis moves urine through the ureters.

Micturition Reflex

1. Stretch of the urinary bladder stimulates a reflex that causes the bladder to contract and inhibits the urinary sphincters.
2. Higher brain centers can stimulate or inhibit the micturition reflex.

Content Review

1. What functions are performed by the urinary system? Name the structures that make up the urinary system.
2. What structures surround the kidney?
3. Name the two layers of the kidney. What are the renal columns and the renal pyramids?
4. Describe the relationship between the calyces, the renal pelvis, and the ureter.
5. What is the functional unit of the kidney? Name its parts.
6. What is the juxtaglomerular complex?
7. Describe the type of epithelium found in the nephron and the collecting duct.
8. Describe the blood supply for the kidney.
9. What are the functions of the ureters and the urinary bladder? Describe their structure.
10. Name the three general processes involved in the production of urine.
11. Define the terms renal blood flow, renal plasma flow, and GFR. How do they affect urine production?
12. Describe the filtration barrier. What substances do not pass through it?
13. What is filtration pressure? How does glomerular capillary pressure affect filtration pressure and the amount of urine produced?
14. How do systemic blood pressure and afferent arteriole diameter affect glomerular capillary pressure?
15. What happens to most of the filtrate that enters the nephron?
16. On what side of the nephron tubule cell does active transport take place during reabsorption and secretion of materials?
17. Describe how cotransport works in the nephron.
18. Name the substances that are moved by active and passive transport. In what part of the nephron does this movement take place?

19. Where does tubular secretion take place? What substances are secreted? Are these substances secreted by active or passive transport?
20. What is a countercurrent system? How does the vasa recta help maintain the concentration gradient in the medulla?
21. What is a countercurrent multiplier system? How do the loop of Henle and urea contribute to the production of a medullary concentration gradient?
22. Describe the net movement of solutes and water in the nephron and the collecting duct.
23. What are the effects of aldosterone on Na^+ and Cl^- ion transport? How does aldosterone affect urinary concentration, urinary volume, and blood pressure?
24. Where is aldosterone produced? What factors stimulate aldosterone secretion?
25. How is angiotensin II activated? What effects does it produce?
26. What factors cause an increase in renin production?
27. Where is ADH produced? What factors stimulate an increase in ADH secretion?
28. What effect does ADH have on urine volume and concentration?
29. Where is atrial natriuretic hormone produced, and what effect does it have on urine production?
30. Describe autoregulation.
31. How does sympathetic stimulation affect filtrate production?
32. Define the terms plasma clearance, tubular load, and tubular maximum.
33. What is responsible for the movement of urine through the nephron and the ureter?
34. Describe the micturition reflex. How is voluntary control of micturition accomplished?

Develop Your Reasoning Skills

1. To relax after an anatomy and physiology examination, Mucho Gusto goes to a local bistro and drinks 2 quarts of low-sodium beer. What effect does this beer have on urine concentration and volume? Explain the mechanisms involved.

2. A male eats a full bag of salty potato chips. What effect does this have on urine concentration and volume? Explain the mechanisms involved.

3. During severe exertion in a hot environment a person can lose up to 4 L of hypoosmotic (less concentrated than plasma) sweat per hour. What effect does this loss have on urine concentration and volume? Explain the mechanisms involved.

4. Harry Macho is doing yard work one hot summer day and refuses to drink anything until he is finished. He then drinks glass after glass of plain water. Assume that he drinks enough water to replace all the water he lost as sweat. How does this much water affect urine concentration and volume? Explain the mechanisms involved.

5. Which of the following symptoms are consistent with hyposecretion of aldosterone: polyuria (excessive urine production), low blood pressure, high plasma sodium levels, low plasma potassium levels, and muscle weakness?

6. A patient has the following symptoms: slight increase in extracellular fluid volume, large decrease in plasma sodium concentration, very concentrated urine, and cardiac fibrillation. An imbalance of what hormone is responsible for these symptoms? Are the symptoms caused by oversecretion or undersecretion of the hormone?

7. Propose as many ways as you can to decrease the GFR.

8. Design a kidney that can produce hypoosmotic urine, which is less concentrated than plasma, or hyperosmotic urine, which is more concentrated than plasma, by the active transport of water instead of Na^+ ions. Assume that the anatomic structure of the kidney is the same as that in humans. Feel free to change anything else you choose.

9. If only a very small amount of urea was present in the interstitial fluid of the kidney instead of its normal concentration, how would it affect the kidney's ability to concentrate urine?

Web Site Link

For a listing of the most current web sites related to this chapter, please visit the Seeley home page at:
http://www.mhhe.com/biosci/abio/ap/seeleyap/

Chapter Twenty-Seven

Water, Electrolytes and Acid–Base Balance Headings

Physiology

Human Anatomy

Objectives

1. List the major body fluid compartments and the approximate percent of body weight contributed by the fluid within each compartment, and describe how the compartments are influenced by age and body fat.

2. Compare the composition of intracellular and extracellular fluids.

3. Diagram the mechanisms by which sodium, potassium, calcium, and chloride ions are regulated in the extracellular fluid.

4. Describe how the regulatory mechanisms respond to an increase or a decrease in extracellular sodium, potassium, chloride, calcium, or phosphate ion concentration.

5. Diagram the mechanisms by which the water content of the body fluids is regulated.

6. Describe how the regulatory mechanisms respond to either an increase or a decrease in the water content of body fluids.

7. Define the terms acid, base, acidosis, alkalosis, and buffer.

8. Explain how buffers regulate the body fluid pH, and list the major buffers that exist in the body fluids.

9. Diagram the mechanisms that regulate the body fluid pH, and describe how they respond to either acidosis or alkalosis.

10. Describe how acidosis and alkalosis are classified, and provide specific examples.

Body Fluids

The proportion of body weight composed of water decreases from birth to old age, with the greatest decrease occurring during the first 10 years of life (table 27.1). Because the water content of fat is relatively low, the fraction of the body's weight composed of water decreases as fat content increases. The relatively lower water content of adult females when compared to adult males reflects the greater development of subcutaneous adipose tissue characteristic of women.

For people of all ages and body compositions, the two major fluid compartments are the intracellular and extracellular fluid compartments. Each of the several trillion cells contains a small volume of intracellular fluid, which accounts for approximately 40% of total body weight and makes up the **intracellular** (in-trǎ-sel'yū-lǎr) **fluid compartment.**

The **extracellular** (eks-trǎ-sel'yū-lǎr) **fluid compartment** includes all of the fluid outside the cells and constitutes nearly 20% of total body weight. The extracellular fluid can be divided into several subcompartments. The major ones are the interstitial fluid and plasma; others include lymph, cerebrospinal fluid, and synovial fluid. **Interstitial** (in-ter-stish'ǎl) **fluid** occupies the extracellular spaces outside the blood vessels, and **plasma** (plaz'mǎ) occupies the extracellular space within blood vessels. All of the other subcompartments of the extracellular compartment constitute relatively small volumes.

Although the fluid contained in each subcompartment differs somewhat in composition from that in the others, continuous and extensive exchange occurs between the subcompartments. Water diffuses from one subcompartment to another, and small molecules and ions are either transported or diffuse freely between them. Large molecules such as proteins are much more restricted in their movement because of the permeability characteristics of the membranes that separate the fluid subcompartments (table 27.2).

The osmotic concentration of most fluid compartments is approximately equal. For example, the osmotic concentration of hyaluronic acid in synovial joints roughly equals that of the proteins in intraocular fluid.

Regulation of Intracellular Fluid Composition

The composition of intracellular fluid is substantially different from that of extracellular fluid. Cell membranes, which separate the two compartments, are selectively permeable, being relatively impermeable to proteins and other large molecules and having limited permeability to smaller molecules and ions. Consequently, most large molecules synthesized within cells, such as proteins, remain within the intracellular fluid. Some substances such as electrolytes are actively transported across the cell membrane, and their concentrations in the intracellular fluid are determined by the transport processes and by the electric charge difference across the cell membrane (figure 27.1).

Water movement across the cell membrane is controlled by osmosis. Thus the net movement of water is affected by changes in the concentration of solutes in the extracellular and intracellular fluids. For example, as dehydration develops, the concentration of solutes in the extracellular fluid increases, resulting in the movement of water by osmosis from the intracellular space into the extracellular space. If dehydration is severe, enough water moves from the intracellular space to cause the cells to function abnormally. If water intake increases after a period of dehydration, the concentration of solutes in the extracellular fluids decreases, which results in the movement of water back into the cells.

Regulation of Extracellular Fluid Composition

Maintenance of homeostasis requires that the intake of substances must equal their elimination. Thus the intake of water and **electrolytes** (ē-lek'trō-lītz), which are molecules or ions with an electric charge, must equal their elimination. Ingestion of water and electrolytes adds them to the body, whereas organs such as the kidneys and, to a lesser degree, the liver, skin, and lung remove them from the body. Over a long period, the total amount of water and electrolytes in the body does not change unless the individual is growing, gaining weight, or losing weight. Regulation of water and electrolytes involves the coordinated participation of several organ systems.

Regulation of Ion Concentrations
Sodium Ions

Sodium (Na^+) **ions** are the dominant extracellular cations. Because of their abundance in the extracellular fluids, they exert substantial osmotic pressure. Approximately 90%–95% of the osmotic pressure of the extracellular fluid is caused by Na^+ ions and the negative ions associated with them.

Table 27.1   Approximate Volumes of Body Fluid Compartments*

Age of Person	Total Body Water	Intracellular Fluid	Extracellular Fluid		
			Plasma	Interstitial	Total
Infants	75	45	4	26	30
Adult males	60	40	5	15	20
Adult females	50	35	5	10	15

*Expressed as percentage of body weight.

Table 27.2   Approximate Concentration of Major Solutes in Body Fluid Compartments*

Solute	Plasma	Interstitial Fluid	Intracellular Fluid[†]
Cations			
Sodium (Na^+)	153.2	145.1	12.0
Potassium (K^+)	4.3	4.1	150.0
Calcium (Ca^{2+})	3.8	3.4	4.0
Magnesium (Mg^{2+})	1.4	1.3	34.0
TOTAL	162.7	153.9	200.0
Anions			
Chloride (Cl^-)	111.5	118.0	4.0
Bicarbonate (HCO_3^-)	25.7	27.0	12.0
Phosphate (HPO_4^{2-} plus HPO_4^-)	2.2	2.3	40.0
Protein	17.0	0.0	54.0
Other	6.3	6.6	90.0
TOTAL	162.7	153.9	200.0

*Expressed as milliequivalents per liter (mEq/L).
[†]Data are from skeletal muscle.

1. Large organic molecules such as proteins, which cannot cross the plasma membrane, are synthesized inside cells and influence the concentration of solutes inside the cells.

2. The transport of solutes across the plasma membrane, such as Na^+, K^+, and Ca^{2+} ions, influences the concentration of solutes inside and outside the cell.

3. The electric charge across the plasma membrane influences the distribution of ions inside and outside the cell.

4. The distribution of water inside and outside the cell is determined by osmosis.

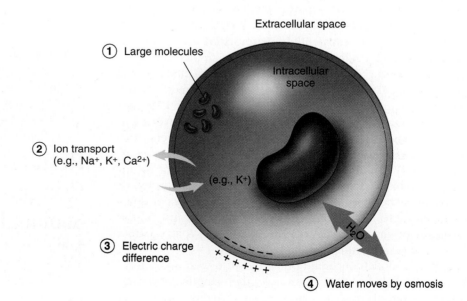

Figure 27.1   Regulation of Intracellular and Extracellular Distribution of Water and Solutes

Clinical Note

In the United States the quantity of Na^+ ions ingested each day is 20–30 times the amount needed. Less than 0.5 g is required to maintain homeostasis, but approximately 10–15 g of sodium chloride is ingested daily by the average individual. Regulation of the Na^+ ion content in the body therefore depends primarily on the excretion of excess quantities of Na^+ ions. The mechanisms for conserving Na^+ ions in the body are effective, however, when the Na^+ ion intake is very low.

The kidneys are the major route by which Na^+ ions are excreted. Na^+ ions readily pass from the glomerulus into the lumen of Bowman's capsule and are present in the same concentration in the filtrate as in the plasma. The concentration of Na^+ ions excreted in the urine is determined by the amount of Na^+ ions and water reabsorbed from filtrate in the nephron. If fewer Na^+ ions are reabsorbed from the nephron, a large amount is lost in the urine. If Na^+ ions are intensively reabsorbed by the nephron, however, few Na^+ ions are lost in the urine.

The rate of Na^+ ion transport in the proximal tubule is relatively constant, but the Na^+ ion transport mechanisms of the distal tubule and the collecting duct are under hormonal control. When **aldosterone** is present, the reabsorption of Na^+ ions from the distal tubule and the collecting duct is very efficient. As little as 0.1 g of sodium is excreted in the urine each day in the presence of high blood levels of aldosterone. When aldosterone is absent, Na^+ ion reabsorption in the nephron is greatly reduced, and as much as 30–40 g of sodium is lost in the urine daily.

Na^+ ions are also excreted from the body in **sweat.** Normally only a small quantity of Na^+ ions is lost each day in the form of sweat, but the amount increases during conditions of heavy exercise in a warm environment. The quantity of Na^+ ions excreted through the skin is controlled by the mechanisms that regulate sweating. As the body temperature increases, the thermoreceptor neurons within the hypothalamus increase the rate of sweat production. As the rate of sweat production increases, the quantity of Na^+ ions lost in the urine decreases to keep the extracellular concentration of Na^+ ions constant. The loss of Na^+ ions in sweat is rarely physiologically significant.

The primary mechanisms that regulate the Na^+ ion concentration in the extracellular fluid do not directly monitor Na^+ ion levels but are sensitive to changes in extracellular fluid osmolality or blood pressure (figure 27.2 and table 27.3). The quantity of Na^+ ions in the body has a dramatic effect on the extracellular osmotic pressure and the extracellular fluid volume. For example, if the Na^+ ion content increases, either the osmolality or the volume of the extracellular fluid increases. An increase in the osmolality of the extracellular fluids causes a small volume of concentrated urine to be produced and increases the sensation of thirst. An increase in extracellular fluid volume, or a decreased osmolality of the extracellular fluids causes a large volume of dilute urine to be produced and decreases the sensation of thirst (see chapter

26). By regulating the extracellular fluid osmolality and the extracellular fluid volume, the concentration of Na^+ ions in the body fluids is kept with a narrow range of values.

Elevated blood pressure under resting conditions increases Na^+ ion and water excretion because it inhibits renin secretion from the juxtaglomerular apparatuses in the kidney. Renin causes the conversion of angiotensinogen to angiotensin I. Angiotensin-converting enzyme present in the plasma and especially in the lungs converts angiotensin I to angiotensin II. Angiotensin II also stimulates aldosterone secretion, and it causes vasoconstriction. A reduced rate of renin secretion leads to a reduced rate of angiotensin II formation. In response to the lower levels of angiotensin II, aldosterone secretion declines, reducing the rate of Na^+ ion reabsorption from nephrons in the kidneys and allowing excretion of large quantities of Na^+ ions in the urine. Elevated blood pressure also stimulates baroreceptors, which in turn send signals to the hypothalamus of the brain to reduce **antidiuretic hormone (ADH)** secretion, providing ADH secretion is elevated. Large increases in blood pressure are required to influence ADH secretion. The combined effect of these mechanisms is regulation of Na^+ ion concentration and water content of the extracellular fluid.

If blood pressure is low, the total Na^+ ion content of the body is usually also low. In response to low blood pressure, the juxtaglomerular apparatuses of the kidney increase their rate of renin secretion. At the same time, if the decrease in blood pressure is substantial, baroreceptors send fewer nerve action potentials to the hypothalamus, resulting in an increased rate of ADH secretion.

1	P R E D I C T

In response to hemorrhagic shock, the kidneys produce a small volume of very concentrated urine. Explain how the rate of filtrate formation changes and how Na^+ ion transport changes in the distal part of the nephron in response to hemorrhagic shock.

✔ *Answer in Appendix F*

Atrial natriuretic hormone is synthesized by cells in the walls of the atria and secreted in response to an elevation of blood pressure within the right atrium. Atrial natriuretic hormone acts on the kidneys to increase urine production by inhibiting the reabsorption of Na^+ ions, inhibiting the effect of ADH on the distal tubules and collecting ducts, and inhibiting ADH secretion (see chapter 26, figure 27.2 and table 27.3).

Deviations from the normal concentration range for Na^+ ions in body fluids result in significant symptoms. Some major causes of elevated Na^+ ion concentrations **(hypernatremia)** and reduced Na^+ ion concentrations **(hyponatremia)** and the major symptoms of each are listed in table 27.4.

2	P R E D I C T

If a person consumes an excess amount of sodium and water, predict the effect on (a) blood pressure, (b) urine volume, and (c) urine concentration.

✔ *Answer in Appendix F*

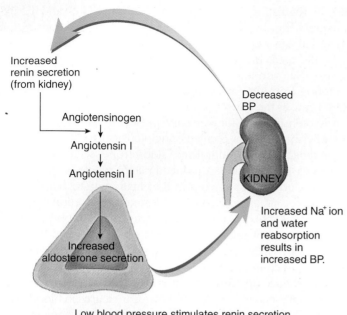

Low blood pressure stimulates renin secretion from the kidney. Renin stimulates the production of angiotensin I, which is converted to angiotensin II, which in turn stimulates aldosterone secretion from the adrenal cortex. Aldosterone stimulates Na^+ ion and water reabsorption in the kidney.

(a)

Increased blood pressure (BP) in the right atrium of the heart causes increased secretion of atrial natriuretic hormone, which increases sodium ion excretion and water loss in the form of urine.

(b)

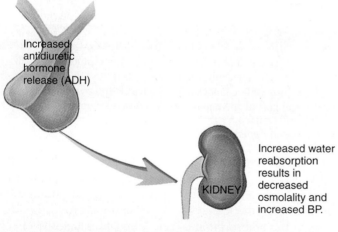

Increased blood osmolality affects hypothalamic neurons, and decreased blood pressure affects baroreceptors in the aortic arch, carotid sinuses, and atrium. As a result of these stimuli, an increased rate of ADH secretion from the posterior pituitary results, which increases water reabsorption.

(c)

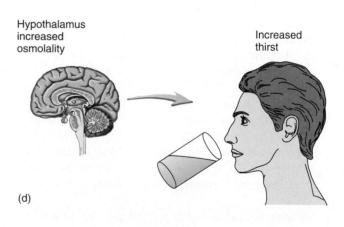

(d)

Figure 27.2   Major Mechanisms Regulating Sodium Ion Levels in the Extracellular Fluids

Chloride Ions

The predominant anions in the extracellular fluid are **chloride (Cl^-) ions.** The electrical attraction of anions and cations makes it expensive in terms of energy to separate these charged particles. Consequently, the regulatory mechanisms that influence the concentration of cations in the extracellular fluid also influence the concentration of anions. The mechanisms that regulate Na^+, K^+, and Ca^{2+} ion levels in the body are important in influencing Cl^- ion levels.

Table 27.3   Homeostasis: Mechanisms Regulating Blood Sodium

Mechanism	Stimulus	Response to Stimulus	Effect of Response	Result
Response to Changes in Blood Osmolality				
Antidiuretic hormone (ADH); the most important regulator of blood osmolality	Increased blood osmolality (e.g., increased sodium ion concentration)	Increased ADH secretion from the posterior pituitary; mediated through cells in the hypothalamus	Increased water reabsorption in the kidney; production of a small volume of concentrated urine	Decreased blood osmolality as reabsorbed water dilutes the blood
	Decreased blood osmolality (e.g., decreased sodium ion concentration)	Decreased ADH secretion from the posterior pituitary; mediated through cells in the hypothalamus	Decreased water reabsorption in the kidney; production of a large volume of dilute urine	Increased blood osmolality as water is lost from the blood into the urine
Response to Changes in Blood Pressure				
Renin-angiotensin-aldosterone	Decreased blood pressure in the kidney's afferent arterioles	Increased renin release from the juxtaglomerular glomerular apparatuses; renin initiates the conversion of angiotensinogen to angiotensin; angiotensin I is converted to angiotensin II, which increases aldosterone secretion from the adrenal cortex	Increased Na^+ ion reabsorption in the kidney (because of increased aldosterone); increased water reabsorption as water follows the Na^+ ions; decreased urine volume	Increased blood pressure as blood volume increases because of increased water reabsorption; blood osmolality is maintained because both Na^+ ions and water are reabsorbed*
	Increased blood pressure in the kidney's afferent arterioles	Decreased renin release from the juxtaglomerular apparatuses, resulting in reduced formation of angiotensin I; reduced angiotensin I leads to reduced angiotensin II, which causes a decrease in aldosterone secretion from the adrenal cortex	Decreased Na^+ ion reabsorption in the kidney (because of decreased aldosterone); decreased water reabsorption as fewer Na^+ ions are reabsorbed; increased urine volume	Decreased blood pressure as blood volume decreases because water is lost in the urine; blood osmolality is maintained because both Na^+ ions and water are lost in the urine*
Atrial natriuretic hormone	Decreased blood pressure in the atria of the heart	Decreased atrial natriuretic hormone released from the atria	Increased Na^+ ion reabsorption in the kidney; increased water reabsorption as water follows the Na^+ ions; decreased urinary volume	Increased blood pressure as blood volume increases because of increased water reabsorption; blood osmolality is maintained because both Na^+ ions and water are reabsorbed*

continued next page

Potassium Ions

The extracellular concentration of **potassium (K^+) ions** must be maintained within a narrow range. The concentration gradient of K^+ ions across the cell membrane has a major influence on the resting membrane potential, and cells that are electrically excitable are highly sensitive to slight changes in that concentration gradient. An increase in extracellular K^+ ion concentration leads to depolarization, and a decrease in extracellular K^+ ion concentration leads to hyperpolarization

Table 27.3   Homeostasis: Mechanisms Regulating Blood Sodium—cont'd

Mechanism	Stimulus	Response to Stimulus	Effect of Response	Result
Response to Changes in Blood Pressure—cont'd				
	Increased blood pressure in the atria of the heart	Increased atrial natriuretic hormone released from the atria	Decreased sodium ion reabsorption in the kidney; decreased water reabsorption as water is lost with Na^+ ions in the urine; increased urinary volume	Decreased blood osmolality as blood volume decreases because water is lost in the urine; blood osmolality is maintained because both Na^+ ions and water are lost in the urine*
ADH—activated by significant decreases in blood pressure; normally regulates blood osmolality (see above)	Decreased arterial blood pressure	Increased ADH secretion from the posterior pituitary; mediated through baroreceptors	Increased water reabsorption in the kidney; production of a small volume of concentrated urine	Increased blood pressure resulting from increased blood volume; decreased blood osmolality
	Increased arterial blood pressure	Decreased ADH secretion from the posterior pituitary; mediated through baroreceptors	Decreased water reabsorption in the kidney; production of a large volume of dilute urine	Decreased blood pressure resulting from decreased blood volume; increased blood osmolality

Abbreviations: ADH = antidiuretic hormone.
*Assumes normal levels of ADH.

Table 27.4   Consequences of Abnormal Plasma Levels of Sodium Ions

Major Causes	Symptoms
Hypernatremia	
High dietary sodium rarely causes symptoms	Thirst, fever, dry mucous membranes, restlessness; most serious symptoms—convulsions and pulmonary edema
Administration of hypertonic saline solutions (e.g., sodium bicarbonate treatment for acidosis)	
Oversecretion of aldosterone (i.e., aldosteronism)	When occurring with an increased water volume—weight gain, edema, elevated blood pressure, and bounding pulse
Water loss (e.g., because of fever, respiratory infections, diabetes insipidus, diabetes mellitus, diarrhea)	
Hyponatremia	
Inadequate dietary intake of sodium rarely causes symptoms—can occur in those on low-sodium diets and those taking diuretics	Lethargy, confusion, apprehension, seizures, and coma
Extrarenal losses—vomiting, prolonged diarrhea, gastrointestinal suctioning, burns	When accompanied by reduced blood volume—reduced blood pressure, tachycardia, and decreased urine output
Dilution—intake of large water volume after excessive sweating	When accompanied by increased blood volume—weight gain, edema, and distension of veins
Hyperglycemia, which attracts water into the circulatory system but reduces the concentration of sodium ions	

Table 27.5 Consequences of Abnormal Concentrations of Potassium Ions

Major Causes	Symptoms
Hyperkalemia	
Movement of K$^+$ ions from intracellular to extracellular fluid resulting from cell trauma (e.g., burns or crushing injuries) and alterations in cell membrane permeability (e.g., acidosis, insulin deficiency, and cell hypoxia)	Mild hyperkalemia (caused mainly by partial depolarization of cell membranes): Increased neuromuscular irritability, restlessness Intestinal cramping and diarrhea Electrocardiogram—alterations, including rapid repolarization with narrower and taller T waves and shortened QT intervals
Decreased renal excretion of K$^+$ ions (e.g., from decreased secretion of aldosterone in persons with Addison's disease)	
	Severe hyperkalemia (caused mainly by partial depolarization of cell membranes severe enough to hamper action potential conduction): Muscle weakness, loss of muscle tone, and paralysis Electrocardiogram—alterations, including changes caused by reduced rate of action potential conduction (e.g., depressed ST segment, prolonged PR interval, wide QRS complex, arrhythmias, and cardiac arrest)
Hypokalemia	
Alkalosis (K$^+$ ions shift into cell in exchange for H$^+$ ions)	Symptoms are mainly due to hyperpolarization of membranes
Insulin administration (promotes cellular uptake of K$^+$ ions)	Decreased neuromuscular excitability—skeletal muscle weakness
Reduced K$^+$ ion intake (especially with anorexia nervosa and alcoholism)	Decreased tone in smooth muscle
Increased renal loss (excessive aldosterone secretion, improper use of diuretics, kidney diseases that result in reduced ability to reabsorb Na$^+$ ions)	Cardiac muscle—delayed ventricular repolarization, bradycardia, and atrioventricular block

of the resting membrane potential. **Hyperkalemia** (hī′per-ka-lē′mē-ă) is an abnormally high level of potassium ions in the extracellular fluid, and **hypokalemia** (hī-pō-ka-lē′mē-ă) is an abnormally low level of K$^+$ ions in the extracellular fluid. Some major causes of hyperkalemia and hypokalemia and their symptoms are listed in table 27.5.

K$^+$ ions pass freely through the filtration membrane of the renal corpuscle. They are actively reabsorbed in the proximal tubules and actively secreted in the distal tubules and collecting ducts. K$^+$ ion secretion into the distal tubule and collecting duct is highly regulated and primarily responsible for controlling the extracellular concentration of K$^+$ ions.

Aldosterone plays a major role in regulating the concentration of K$^+$ ions in the extracellular fluid by increasing the rate of K$^+$ ion secretion in the distal tubule and collecting duct. Aldosterone secretion from the adrenal cortex is stimulated by elevated K$^+$ blood levels and by angiotensin II (figure 27.3; see chapter 26). The elevated aldosterone concentrations in the circulatory system increase potassium ion secretion into the nephron, lowering the blood level of K$^+$ ions.

Circulatory system shock can result from plasma loss, dehydration, and tissue damage, such as occurs in burn patients. This shock causes the extracellular K$^+$ ions to be more concentrated than normal, which stimulates aldosterone secretion from the adrenal cortex. Aldosterone secretion also

occurs in response to decreased blood pressure, which stimulates the renin-angiotensin-aldosterone mechanism (see chapter 26). Homeostasis is reestablished as K$^+$ ion excretion increases. Also, increased Na$^+$ and water reabsorption stimulated by aldosterone results in an increase in extracellular fluid volume which dilutes the K$^+$ ions in the body fluids. Blood pressure increases toward normal as water reabsorption increases, and when vasoconstriction is stimulated by angiotensin II.

Calcium Ions

The extracellular concentration of **calcium** (Ca^{2+}) **ions,** like that of K$^+$ ions, is regulated within a narrow range. The normal concentration of Ca^{2+} ions in plasma is 9.4 mg/100 mL. **Hypocalcemia** (hī-pō-kal-sē′mē-ă) is a below-normal level of Ca^{2+} ions in the extracellular fluid, and **hypercalcemia** (hī-per-kal-sē′mē-ă) is an above-normal level of Ca^{2+} ions in the extracellular fluid. Major symptoms develop when the extracellular concentration of Ca^{2+} ions declines to 6 mg/100 mL, and major symptoms develop when concentrations reach 12 mg/100 mL. Increases and decreases in the extracellular concentration of Ca^{2+} ions markedly affect the electrical properties of excitable tissues. Hypercalcemia decreases the permeability of the cell membrane to Na$^+$ ions, thus preventing

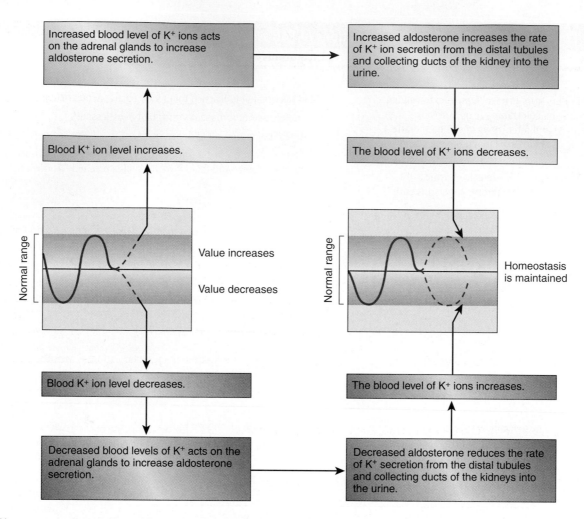

Figure 27.3   Homeostasis: Regulation of Potassium Ions in the Extracellular Fluid.
Aldosterone acts on the kidney to regulate the extracellular concentration of K$^+$ ions.

normal depolarization of nerve and muscle cells. High extra-cellular Ca^{2+} ion levels cause the deposition of calcium carbonate salts in soft tissues, resulting in irritation and inflammation of those tissues. Hypocalcemia increases the permeability of cell membranes to Na$^+$ ions. As a result, nerve and muscle tissues undergo spontaneous action potential generation. Table 27.6 lists some of the major causes and symptoms of hypercalcemia and hypocalcemia.

The kidneys, intestinal tract, and bones are important in maintaining extracellular Ca^{2+} ion levels (figure 27.4). Almost 99% of total body calcium is contained in bone. Part of the extracellular Ca^{2+} ion regulation involves the regulation of Ca^{2+} ion deposition into and resorption from bone (see chapter 6). Long-term regulation of Ca^{2+} levels, however, depends on maintaining a balance between Ca^{2+} ion absorption across the wall of the intestinal tract and Ca^{2+} ion excretion by the kidneys.

Parathyroid (par'ă-thī'royd) **hormone,** secreted by the parathyroid glands, increases extracellular Ca^{2+} ion levels and reduces extracellular phosphate levels (see figure 27.4). The rate of parathyroid hormone secretion is regulated by the extracellular Ca^{2+} ion levels. Elevated Ca^{2+} ion levels inhibit

and reduced levels stimulate its secretion. Parathyroid hormone causes increased osteoclast activity, which results in the degradation of bone and the release of Ca^{2+} and phosphate ions into the body fluids. Parathyroid hormone increases the rate of calcium ion reabsorption from nephrons in the kidneys and increases the concentration of phosphate ions in the urine. It also increases the rate at which vitamin D is converted to 1,25-dihydroxycholecalciferol, or **active vitamin D.** Active vitamin D acts on the intestinal tract to increase Ca^{2+} ion absorption across the intestinal mucosa.

A lack of parathyroid hormone secretion results in a rapid decline in extracellular Ca^{2+} ion concentration. This decline is caused by a reduction in the rate of absorption of Ca^{2+} ions from the intestinal tract, increased Ca^{2+} ion excretion by the kidneys, and reduced bone resorption. A lack of parathyroid hormone secretion can result in death because of tetany of the respiratory muscles caused by hypocalcemia.

Vitamin D can be obtained from food or from vitamin D biosynthesis. Normally vitamin D biosynthesis is adequate, but prolonged lack of exposure to sunlight reduces the biosynthesis because ultraviolet light is required for one step

Table 27.6 Consequences of Abnormal Concentrations of Calcium

Major Causes	Symptoms
Hypercalcemia	
Excessive parathyroid hormone secretion	Symptoms are mainly due to decreased permeability of cell membranes to Na^+ ions
Excess vitamin D	Loss of membrane excitability—fatigue, weakness, lethargy, anorexia, nausea, and constipation
	Electrocardiogram—shortened QT segment and depressed T waves
	Kidney stones
Hypocalcemia	
Nutritional deficiencies	Symptoms are mainly due to increased permeability of cell membranes to Na^+ ions.
Vitamin D deficiency	Increase in neuromuscular excitability—confusion, muscle spasms, hyperreflexia, and intestinal cramping
Decreased parathyroid hormone secretion	
Malabsorption of fats (reduced vitamin D absorption)	Severe neuromuscular excitability—convulsions, tetany, inadequate respiratory movements
Bone tumors that increase Ca^{2+} ion deposition	Electrocardiogram—prolonged QT interval (prolonged ventricular depolarization)

in the process (see chapter 5). The consumption of dietary vitamin D can involve the ingestion of active vitamin D or one of its precursors.

Without vitamin D, the transport of Ca^{2+} ions across the wall of the intestinal tract is negligible. This leads to inadequate Ca^{2+} ion intake, even though large amounts of Ca^{2+} ions may be present in the diet. Thus Ca^{2+} ion absorption depends on both the consumption of an adequate amount of calcium in food and the presence of an adequate amount of vitamin D.

Calcitonin (kal-si-tō'nin), which is secreted by the parafollicular cells of the thyroid gland, reduces extracellular Ca^{2+} ion levels. It is most effective when Ca^{2+} levels are elevated, although greater-than-normal calcitonin levels in the blood are not consistently effective in causing blood levels of Ca^{2+} ions to decline below normal values. The major effect of calcitonin is on bone. It inhibits osteoclast activity and prolongs the activity of osteoblasts. Thus it decreases bone demineralization and increases bone mineralization.

Elevated Ca^{2+} ion levels stimulate calcitonin secretion, whereas reduced Ca^{2+} ion levels inhibit it. Increased calcitonin secretion reduces blood levels of Ca^{2+} ions, but large doses of calcitonin do not consistently reduce blood levels of Ca^{2+} ions below normal levels. Although calcitonin reduces the blood levels of Ca^{2+} ions when they are elevated, it is not as important as parathyroid hormone in the regulation of blood Ca^{2+} ion levels (see figure 27.4).

Phosphate Ions

The nephrons of the kidney have a maximum rate at which phosphate ions are reabsorbed from the filtrate. Under normal conditions this transport mechanism functions near its maximum rate. If the concentration of **phosphate ions** increases in the plasma, phosphate ions are lost in the urine. If the phosphate ions entering the filtrate exceed the ability of the nephron to reabsorb phosphate, the excess phosphate is lost in the urine. Consequently, the major mechanism that controls the plasma levels of phosphate ions is the transport mechanism for phosphate ions in the nephrons.

Elevated blood levels of phosphate may occur with acute or chronic renal failure as a result of a very reduced rate of filtrate formation by the kidney. The rate of phosphate excretion is consequently reduced. Also, the chronic use of laxatives containing phosphates may result in elevated blood levels of phosphate. Symptoms of elevated phosphate levels are related to reduced blood Ca^{2+} ion levels because phosphate ions and Ca^{2+} ions precipitate out of solution and are deposited in soft tissues of the body. Prolonged elevation of blood levels of phosphate can result in calcium phosphate deposits in the lungs, kidneys, joints, and other soft tissues.

Regulation of Water Content

The body's water content is regulated so that the total volume of water in the body remains constant (figure 27.5). Thus the volume of water taken into the body is equal to the volume lost each day. Changes in the volume of water in body fluids alter the osmolality of body fluids, the blood pressure, and the interstitial fluid pressure. The total volume of water entering the body each day is 1500–3000 mL. Most of that volume comes from ingested fluids, some comes from food, and a smaller amount is derived from the water produced during cellular metabolism (table 27.7).

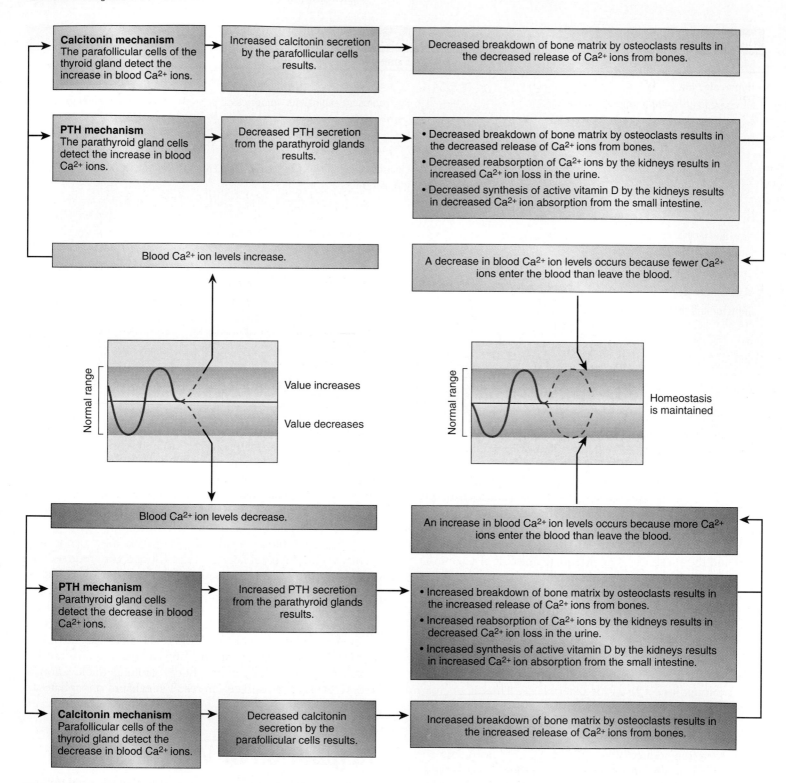

Figure 27.4   Homeostasis: Regulation of Calcium Ions in the Extracellular Fluid

Parathyroid hormone and calcitonin play major roles in regulating the extracellular concentration of Ca^{2+} ions.

The movement of water across the wall of the gastrointestinal tract depends on osmosis, and the volume of water entering the body depends, to a large degree, on the volume of water consumed. If a large volume of dilute liquid is consumed, the rate at which water enters the body fluids increases. If a small volume of concentrated liquid is consumed, the rate decreases.

Although fluid consumption is heavily influenced by habit and by social phenomena, water ingestion does depend, at least in part, on regulatory mechanisms. The sensation of

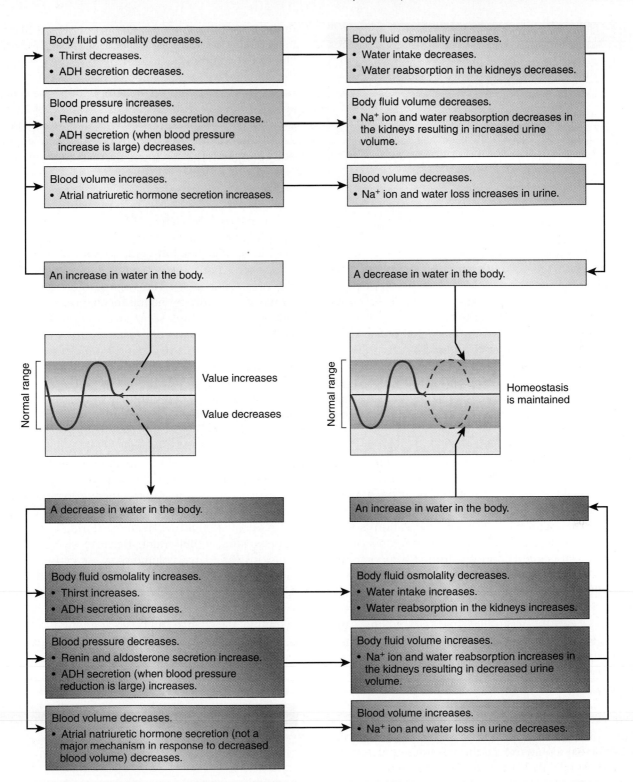

Figure 27.5   Homeostasis: Regulation of Water Content in the Body.

Changes in blood volume, blood pressure, and body fluid osmolality play major roles in regulating water content in the body.

thirst results from an increase in the osmolality of the extracellular fluids and from a reduction in plasma volume. Cells of the supraoptic nucleus can, within the hypothalamus, detect an increased extracellular fluid osmolality and initiate

activity in neural circuits that results in a conscious sensation of thirst.

Baroreceptors can also influence the sensation of thirst. When they detect a decrease in blood pressure, action

Table 27.7	Summary of Water Intake and Loss
Sources of Water	**Routes by Which Water Is Lost**
Ingestion	Evaporation
Metabolism (small volume)	Perspiration
	Insensible
	Sensible
	Respiratory passages
	Urine
	Feces

potentials are conducted to the brain along afferent pathways to influence the sensation of thirst. Low blood pressure associated with hemorrhagic shock, for example, is correlated with an intense sensation of thirst.

When renin is released from the juxtaglomerular apparatuses, it increases the formation of angiotensin II in the circulatory system (see chapters 21 and 26). Angiotensin II stimulates the sensation of thirst by acting on the brain. It reverses any decreases in blood pressure by stimulating the sensation of thirst and increasing aldosterone secretion and vasoconstriction.

When people who are dehydrated are allowed to drink water, they eventually consume a quantity sufficient to reduce the osmolality of the extracellular fluid to its normal value. They do not normally consume the water all at once, however, but drink intermittently until the proper osmolality of the extracellular fluid is established. The thirst sensation is temporarily reduced after the ingestion of small amounts of liquid. At least two factors are responsible for this temporary interruption of the thirst sensation. First, when the oral mucosa becomes wet after it has been dry, inhibitory action potentials are sent to the thirst center of the hypothalamus. Second, consumed fluid increases the gastrointestinal tract volume, and stretch of the gastrointestinal wall initiates afferent action potentials in stretch receptors. The action potentials are transmitted to the brain, where they temporarily suppress the sensation of thirst.

Because the absorption of water from the gastrointestinal tract requires time, mechanisms that temporarily suppress the sensation of thirst prevent the consumption of extreme volumes of fluid that would exceed the amount required to reduce the blood osmolality. Long-term suppression of the thirst sensation, however, requires that extracellular fluid osmolality and blood pressure come within their normal ranges.

Learned behavior can be very important in avoiding periodic dehydration through the consumption of fluids either with or without food, even though blood osmolality is not reduced. The volume of fluid ingested by a healthy person usually exceeds the minimum volume required to maintain homeostasis, and the kidneys eliminate the excess water in urine.

Water loss from the body occurs through three major routes (see table 27.7). The greater amount of water is lost

through the urine and through evaporation, and a smaller volume is lost in the feces. Water lost through evaporation includes the volume lost from the respiratory passages and from the skin surface. The volume of water lost through the respiratory system depends on the temperature and humidity of the air, the body temperature, and the volume of air expired.

Water lost through simple evaporation from the skin is called **insensible perspiration.** For each degree that the body temperature rises above normal, an increased volume of 100–150 mL of water is lost each day in the form of insensible perspiration.

Sweat, or **sensible perspiration,** is secreted by the sweat glands (see chapter 5), and, in contrast to insensible perspiration, it contains solutes. Sweat resembles extracellular fluid in its composition, with sodium chloride as the major component, but it also contains some potassium, ammonia, and urea (table 27.8). The volume of sweat produced is determined primarily by neural mechanisms that regulate body temperature, although some sweat is produced as a result of sympathetic stimulation in response to stress. The volume of fluid lost as sweat is negligible for a person at rest in a cool environment. Under conditions of exercise, elevated environmental temperature, or fever, the volume increases substantially. Sweat losses of 8–10 L/day have been measured in outdoor workers in the summertime.

Evaporation of water is a major mechanism by which body temperature is regulated. Approximately one-fourth of the total heat produced by metabolism is lost from the body through the evaporation of water. The rate of evaporation for a person exercising in a hot, dry environment is much greater than for a person at rest in a cool, humid environment.

Adequate fluid replacement during conditions of extensive sweating is important. Because sweat is usually hypotonic, loss of a large volume of sweat causes hypertonicity in the body fluids. Fluid volume is lost primarily from the extracellular space, which leads to a reduction in plasma volume and an increase in hematocrit. During conditions of severe dehydration, the change can be great enough to cause blood viscosity to increase substantially. The increased workload created for the heart by that increase in viscosity can result in heart failure.

3   P R E D I C T

Mary Thon runs several miles each day. List the mechanisms through which water loss changes during her run.

✔ *Answer in Appendix F*

Relatively little water is lost by way of the digestive tract. Although the total volume of fluid secreted into the gastrointestinal tract is large, nearly all of the fluid is reabsorbed under normal conditions (see chapter 24). Severe vomiting or diarrhea, however, can result in a large volume of fluid loss.

The kidneys are the primary organs that regulate the composition and volume of body fluids by controlling the vol-

Table 27.8	Composition of Sweat
Solute	Concentration (mM)
Sodium	9.8–77.2
Potassium	3.9–9.2
Chloride	5.2–65.1
Ammonia	1.7–5.6
Urea	6.2–12.1

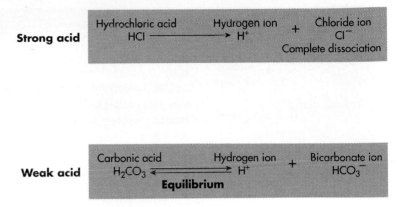

Figure 27.6   Comparison of Strong and Weak Acids

Strong acids completely dissociate when dissolved in water, whereas weak acids do not. Weak acids partially dissociate so that an equilibrium is established between the acid and the ions that are formed when the dissociation occurs.

ume and concentration of water excreted in the form of urine. Urine production varies greatly in response to mechanisms that regulate the body's water content. Reduced extracellular fluid osmolality and large increases in elevated blood pressure inhibit ADH secretion from the posterior pituitary, and elevated blood pressure also reduces renin secretion. These hormonal changes produce a large volume of dilute urine. If the blood osmolality increases, ADH levels also increase, producing a smaller volume of more concentrated urine. Reduced blood pressure also results in elevated ADH and renin secretion (see figure 27.5).

Regulation of Acid–Base Balance

Hydrogen ions affect the activity of enzymes and interact with many electrically charged molecules. Consequently, most chemical reactions within the body are highly sensitive to the hydrogen ion concentration of the fluid in which they occur, and maintenance of hydrogen ion concentration within a narrow range of values is essential for normal metabolic reactions. The two major components of the mechanism of pH regulation are the buffer systems and the regulatory mechanisms.

Acids and Bases

The acidity of a solution depends on its hydrogen ion concentration. The greater the hydrogen ion concentration, the greater its acidity. Normally, the acidity of a solution is measured on the **pH scale** (see chapter 2 and appendix D). As the acidity of the solution becomes greater, the pH value becomes smaller, and as the solution becomes more basic, the pH value becomes larger. A solution is considered **neutral** at a pH of 7. Below 7 the pH is **acidic,** and above 7 it is **basic.**

For most purposes **acids** can be defined as substances that release hydrogen ions (H^+) into a solution; **bases** bind to hydrogen ions and remove them from solution. Acids can be grouped as either strong or weak. Strong acids completely dissociate in solution so that all the H^+ ions are released into the solution (figure 27.6). Weak acids release H^+ ions into the solution, but fewer of the acid molecules dissociate. Some of the acid molecules remain intact and do not

release H^+ ions into the solution. The proportion of weak acid molecules that do release H^+ ions into solution is very predictable and is influenced by the pH of the solution into which the weak acid is placed. Weak acids are common in living systems and play an important role in preventing large changes in body fluid pH.

Buffer Systems

Buffers (bŭf′erz) resist changes in the pH of a solution. Buffers within body fluids stabilize pH by chemically binding to excess hydrogen ions when they are added to a solution or by releasing H^+ ions when their concentration in a solution begins to fall.

A solution of carbonic acid (H_2CO_3) and sodium bicarbonate ($NaHCO_3$) is an example of a buffer system. Carbonic acid is a weak acid, and sodium bicarbonate is the salt of this weak acid. When these substances are dissolved in solution, the following equilibrium is established:

$$H_2CO_3 \rightleftarrows HCO_3^- + H^+$$

Many carbonic acid molecules and many bicarbonate ions are in such a solution. When H^+ ions are added, a large proportion of them bind to bicarbonate ions to form carbonic acid, and only a small percentage remain as free hydrogen ions. Thus a large decrease in pH is prevented when H^+ ions are added to a solution. If a large number of H^+ ions are removed from solution, many carbonic acid molecules will dissociate to form bicarbonate and hydrogen ions; thus a large increase in pH is resisted by releasing H^+ ions into solution.

Several important buffer systems in the body work together to resist changes in the pH of body fluids (table 27.9). The carbonic acid/bicarbonate buffer system, protein molecules such as hemoglobin and plasma protein, and phosphate compounds all act as buffers.

Table 27.9 Buffer Systems	
Protein buffer system	Intracellular proteins and plasma proteins form a large pool of protein molecules that can act as buffer molecules. Because of their high concentration, they provide approximately three-fourths of the buffer capacity of the body. Hemoglobin in red blood cells is an important intracellular protein. Other intracellular molecules such as histone proteins and nucleic acids also act as buffers.
Bicarbonate buffer system	Components of the bicarbonate buffer system are not present in high enough concentrations in the extracellular fluid to constitute a powerful buffer system. Because the concentrations of the components of the buffer system are regulated, however, it plays an exceptionally important role in controlling the pH of extracellular fluid.
Phosphate buffer system	Concentration of the phosphate buffer components is low in the extracellular fluids compared with the other buffer systems, but it is an important intracellular buffer system.

Mechanisms of Acid–Base Balance Regulation

Buffers, the respiratory system, and the kidneys play essential roles in the regulation of acid–base balance. Buffers resist changes in the pH of body fluids (figure 27.7). The respiratory system responds rapidly to a change in pH to bring the pH of body fluids back toward its normal range. The respiratory system's capacity to regulate pH, however, is not as great as that of the kidneys, nor does the respiratory system have the same ability to return the pH to its precise range of normal values. In contrast, the kidneys respond more slowly than the respiratory system to a change in body fluid pH.

Respiratory Regulation of Acid–Base Balance

The ability of the respiratory system to regulate acid–base balance depends on the **carbonic acid/bicarbonate buffer system.** Carbon dioxide (CO_2) reacts with water (H_2O) to form carbonic acid (H_2CO_3), which in turn dissociates to form hydrogen ions (H^+) and bicarbonate ions (HCO_3^-) as follows:

$$H_2O + CO_2 \rightleftarrows H_2CO_3 \rightleftarrows H^+ + HCO_3^-$$

This reaction is in equilibrium. The higher the concentration of carbon dioxide, the greater the amount of carbonic acid and the greater the number of H^+ and bicarbonate ions. If carbon dioxide levels decline, however, equilibrium shifts in the opposite direction because H^+ and bicarbonate ions combine to form carbonic acid, which then dissociates to form carbon dioxide and water.

The reaction between carbon dioxide and water is catalyzed by an enzyme, **carbonic anhydrase,** which is found in relatively high concentration in red blood cells and on the surface of capillary epithelial cells (figure 27.8). This enzyme does not influence equilibrium but accelerates the rate at which the reaction proceeds in either direction.

Increasing carbon dioxide levels and decreasing body fluid pH stimulate neurons in the medullary respiratory center of the brain and cause the rate and depth of ventilation to increase. In response to the increased rate and depth of ventilation, carbon dioxide is eliminated from the body through the lungs at a greater rate, and the concentration of carbon dioxide in the body fluids decreases. As carbon dioxide levels decline, the concentration of H^+ ions and therefore the pH are returned to their normal range (see figure 27.7). The responses of the respiratory system increases changes in blood levels of carbon dioxide and decreases in pH are listed in figure 27.8.

If carbon dioxide levels become too low or the pH of the body fluids is elevated, the rate and the depth of respiration decline. As a consequence, the rate at which carbon dioxide is eliminated from the body is reduced. Carbon dioxide then accumulates in the body fluids because it is continually produced as a by-product of metabolism. As carbon dioxide increases in the body fluids, it reacts with water to form carbonic acid, which dissociates to form H^+ and bicarbonate ions, and the pH is increased toward its normal value (see figure 27.7).

Renal Regulation of Acid–Base Balance

The nephrons directly regulate acid–base balance by secreting hydrogen ions into the filtrate (see figure 27.7). Cells in the walls of the distal tubule and collecting duct of the nephron are primarily responsible for the secretion of H^+ ions. Carbonic anhydrase is present within these cells and catalyzes the formation of carbonic acid from carbon dioxide and water. The carbonic acid molecules dissociate to form H^+ and bicarbonate ions, and the H^+ ions are then transported into the lumen of the nephron by a countertransport mechanism that exchanges sodium ions for the H^+ ions. The bicarbonate ions and sodium ions then pass from the nephron cells into the extracellular fluid. As a result, H^+ ions are secreted into the lumen of the nephron, and bicarbonate ions pass into the extracellular fluid, where they combine with excess H^+ ions. Both secretion of H^+ ions into the nephron and the movement of bicarbonate ions into the extracellular fluid raise the body fluid pH. The rate of this process increases as the pH of the body fluids decreases, and it slows as the pH of the body fluids increases (figure 27.9).

Some of the H^+ ions secreted into the filtrate combine with bicarbonate ions, which enter the filtrate through the renal corpuscle, to form carbonic acid. The carbonic acid molecules dissociate to form carbon dioxide and water. Carbon

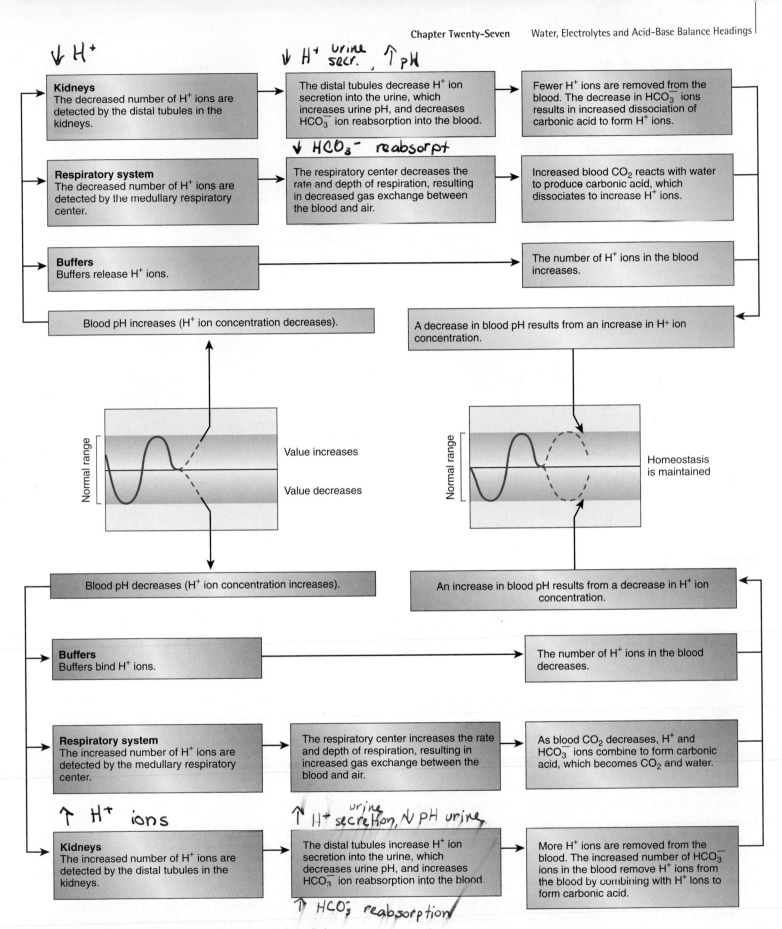

Figure 27.7 Homeostasis: Regulation of Acid–Base Balance

Important mechanism through which the pH of the body fluids is regulated by the lungs and the kidneys.

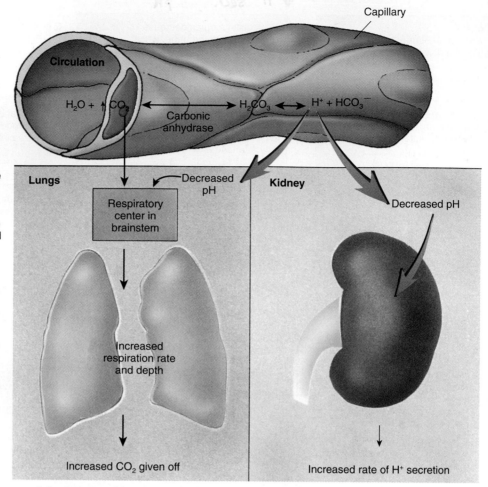

An equilibrium exists between carbon dioxide, carbonic acid, and bicarbonate and hydrogen ions in the blood. Carbon dioxide reacts with water to form carbonic acid. An enzyme, carbonic anhydrase, found in erythrocytes and on the surface of blood vessel epithelium, catalyzes the reaction. Carbonic acid then dissociates to form hydrogen ions and bicarbonate ions. An equilibrium exists so that an increase in carbon dioxide causes more carbonic acid formation. A decrease in carbon dioxide causes some of the carbonic acid to dissociate to form carbon dioxide and water.

A decreased pH and increased PCO_2 stimulate the respiratory center and increase the rate of respiration resulting in an increased rate of carbon dioxide elimination.

A decrease in pH increases the rate of hydrogen ion secretion into the filtrate in the kidney.

Figure 27.8   Homeostasis: Respiratory Regulation of Acid–Base Balance

dioxide diffuses from the nephron into the tubular cells, where it reacts with water to form carbonic acid, which subsequently dissociates to form H^+ ions and bicarbonate ions. The H^+ ions are actively transported into the lumen of the nephron, whereas bicarbonate ions reenter the extracellular fluid. As a result, many of the bicarbonate ions that enter the filtrate reenter the extracellular fluid. Normally the H^+ ions secreted into the nephron exceed the amount of bicarbonate that is filtered so that almost all of the bicarbonate is reabsorbed. Few bicarbonate ions are lost in the urine unless the pH of the body fluids is elevated above its normal value. Excess H^+ ions remain in the urine, and the pH of the urine decreases.

If the pH of the body fluids increases, the rate of H^+ ion secretion into the nephron decreases. As a result, the amount of bicarbonate filtered into the nephron exceeds the amount of secreted hydrogen ion, and the excess bicarbonate ions pass into the urine. Excretion of excess bicarbonate ions in the urine diminishes the amount of bicarbonate ions in the extracellular fluid, and, as a consequence, the pH of the body fluids decreases.

Figure 27.7 summarizes the response of buffers, the respiratory system, and the kidneys to changes in body fluid pH.

Clinical Note

Aldosterone increases the rate of Na^+ ion reabsorption and K^+ ion secretion by the kidneys, but in high concentrations aldosterone also stimulates H^+ ion secretion. Elevated aldosterone levels, such as occur in patients with Cushing's syndrome, can therefore elevate body fluid pH above normal (alkalosis). The major factor that influences the rate of H^+ ion secretion, however, is the pH of the body fluids.

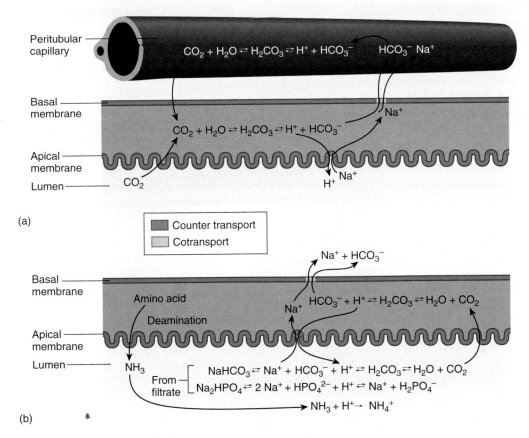

Figure 27.9   Kidney Regulation of Body Fluid pH

(a) If blood pH increases, hydrogen ions combine with bicarbonate ions to produce carbonic acid that becomes carbon dioxide and water. The carbon dioxide diffuses into the nephron cell and combines with water to form carbonic acid that dissociates to form hydrogen and bicarbonate ions. A countertransport mechanism exchanges the hydrogen ions for sodium ions, resulting in the secretion of the hydrogen ions into the lumen of the nephron. The sodium and bicarbonate ions are cotransported into the blood. (b) Buffering hydrogen ions in the filtrate increases the ability of the nephron to secrete additional hydrogen ions. Bicarbonate and phosphate compounds in the filtrate, such as $NaHCO_3$ and Na_2HPO_4, can dissociate and combine with hydrogen ions. Ammonia, produced as part of protein metabolism in the nephron cell, can diffuse into the lumen of the nephron and combine with hydrogen ions to form ammonium ions (NH_4^+).

4   P R E D I C T

Predict the effect of aldosterone hyposecretion on body fluid pH.

✔ *Answer in Appendix F*

The mechanisms that cause increased H^+ ion secretion into the urine can achieve a urine pH of 4.5. Urine pH lower than 4.5 inhibits the secretion of additional H^+ ions. The total amount of H^+ ions that pass into the urine is greater than the quantity required to lower the pH of an unbuffered solution below 4.5. Buffers found in the filtrate combine with the excess H^+ ions. Bicarbonate ions, phosphate ions (HPO_4^-), and ammonia (NH_3) act as buffers in the urine. Bicarbonate ions and phosphate ions enter the nephron through the filtration membrane along with the rest of the filtrate, and ammonia passes across the wall of the nephron. These ions combine with hydrogen ions secreted by the nephron (see figure 27.9).

Ammonia is produced in the cells of the nephron when amino acids such as glycine are deaminated. The ammonia molecules then diffuse into the filtrate. Subsequently, they combine with free H^+ ions in the filtrate to form ammonium ions (NH_4^+), which are secreted in the urine in combination with chloride ions. The rate of ammonia production increases when the pH of the body fluids has been depressed for 2–3 days, such as during prolonged respiratory or metabolic acidosis. The elevated ammonia production increases the buffering capacity of the filtrate, allowing secretion of additional H^+ ions into the urine.

Although bicarbonate, phosphate, and ammonia constitute major buffers within the filtrate, other weak acids such as lactic acid also combine with H^+ ions in the nephron and increase the amount of H^+ ions that can be secreted.

Clinical Focus   Acidosis and Alkalosis

The normal pH value for the body fluids is between 7.35 and 7.45. When the pH value of body fluids is below 7.35, the condition is called **acidosis** (as-i-dō'sis), and when the pH is above 7.45, it is called **alkalosis** (al'kă-lō'sis).

Metabolism produces acidic products that lower the pH of the body fluids. For example, carbon dioxide is a by-product of metabolism, and carbon dioxide combines with water to form carbonic acid. Also, lactic acid is a product of anaerobic metabolism, protein metabolism produces phosphoric and sulfuric acids, and lipid metabolism produces fatty acids. These acidic substances must continuously be eliminated from the body to maintain pH homeostasis. Rapid elimination of acidic products of metabolism may result in alkalosis, however, and the failure to eliminate acidic products of metabolism results in acidosis.

The major effect of acidosis is depression of the central nervous system. When the pH of the blood falls below 7.35, the central nervous system malfunctions, and the individual becomes disoriented and possibly comatose as the condition worsens.

A major effect of alkalosis is hyperexcitability of the nervous system. Peripheral nerves are affected first, resulting in spontaneous nervous stimulation of muscles. Spasms and tetanic contractions and possibly extreme nervousness or convulsions result. Severe alkalosis can cause death as a result of tetany of the respiratory muscles.

Although buffers in the body fluids help resist changes in the pH of body fluids, the respiratory system and the kidneys regulate the pH of the body fluids. Malfunctions of either the respiratory system or the kidneys can result in acidosis or alkalosis.

Acidosis and alkalosis are categorized by the cause of the condition. **Respiratory acidosis** or **respiratory alkalosis** results from abnormalities of the respiratory system. **Metabolic acidosis** or **metabolic alkalosis** results from all causes other than abnormal respiratory functions.

Inadequate ventilation of the lungs causes respiratory acidosis (table A). The rate at which carbon dioxide is eliminated from the body fluids through the lungs falls. This increases the concentration of carbon dioxide in the body fluids. As carbon dioxide levels increase, excess carbon dioxide reacts with water to form carbonic acid. The carbonic acid dissociates to form hydrogen ions and bicarbonate ions. The increase in hydrogen ion concentration causes the pH of the body fluids to decrease. If the pH of the body fluids falls below 7.35, symptoms of respiratory acidosis become apparent.

Buffers help resist a decrease in pH, and the kidneys help compensate for failure of the lungs to prevent respiratory acidosis by increasing the rate at which they secrete hydrogen ions into the filtrate and reabsorb bicarbonate ions. The capacity of buffers to resist changes in pH can be exceeded, however, and a period of 1–2 days is required for the kidney to become maximally functional. Thus the kidneys are not effective if respiratory acidosis develops quickly, but they are very effective if respiratory acidosis develops slowly or if it lasts long enough for the

Table A   Acidosis and Alkalosis

Acidosis

Respiratory Acidosis

Reduced elimination of carbon dioxide from the body fluids

Asphyxia   Hypoventilation (e.g., impaired respiratory center function due to trauma, tumor, shock, or renal failure)

Advanced asthma

Severe emphysema

Metabolic Acidosis

Elimination of large amounts of bicarbonate ions resulting from mucous secretion (e.g., severe diarrhea and vomiting of lower intestinal contents)

Direct reduction of the body fluid pH as acid is absorbed (e.g., ingestion of acidic drugs such as aspirin)

Production of large amounts of fatty acids and other acidic metabolites such as ketone bodies (e.g., untreated diabetes mellitus)

Inadequate oxygen delivery to tissue resulting in anaerobic respiration and lactic acid buildup (e.g., exercise, heart failure, or shock)

Alkalosis

Respiratory Alkalosis

Reduced carbon dioxide levels in the extracellular fluid (e.g., hyperventilation due to emotions)

Decreased atmospheric pressure reduces oxygen levels, which stimulates the chemoreceptor reflex, causing hyperventilation (e.g., high altitudes)

Metabolic Alkalosis

Elimination of hydrogen and reabsorption of bicarbonate ions in the stomach or kidney (e.g., severe vomiting or formation of acidic urine in response to excess aldosterone)

Ingestion of alkaline substances (e.g., large amounts of sodium bicarbonate)

kidneys to respond. For example, the kidneys cannot compensate for respiratory acidosis occurring in response to a severe asthma attack that begins quickly and subsides within hours. If, however, respiratory acidosis results from emphysema, which develops over a long time, the kidneys play a significant role in helping to compensate.

Respiratory alkalosis results from hyperventilation of the lungs (see table A). This increases the rate at which carbon dioxide is eliminated from the body fluids and results in a decrease in the concentration of carbon dioxide in the body fluids. As carbon dioxide levels decrease, hydrogen ions react with bicarbonate ions to form carbonic acid. The carbonic acid dissociates to form water and carbon dioxide. The resulting decrease in the concentration of hydrogen ions causes the pH of the body fluids to increase. If the pH of body fluids increases above 7.45, symptoms of respiratory alkalosis become apparent.

The kidneys help to compensate for respiratory alkalosis by decreasing the rate of H^+ ion secretion into the urine and the rate of bicarbonate ion reabsorption. If an increase in pH occurs, the kidneys need 1–2 days to compensate. Thus the kidneys are not effective if respiratory alkalosis develops quickly. They are very effective, however, if respiratory alkalosis develops slowly. For example, the kidneys are not effective in compensating for respiratory alkalosis that occurs in response to hyperventilation triggered by emotions, which usually begins quickly and subsides within minutes or hours. If alkalosis results, however, from staying at a high altitude over a 2- or 3-day period, the kidneys play a significant role in helping to compensate.

Metabolic acidosis results from all conditions that decrease the pH of the body fluids below 7.35, with the exception of conditions resulting from altered function of the respiratory system (see table A). As hydrogen ions accumulate in the body fluids, buffers first resist a decline in pH. If the buffers cannot compensate for the increase in hydrogen ions, the respiratory center helps regulate the body fluid pH. The reduced pH stimulates the respiratory center, which causes hyperventilation. During hyperventilation, carbon dioxide is eliminated at a greater rate. The elimination of carbon dioxide also eliminates excess hydrogen ions and helps maintain the pH of the body fluids within a normal range.

If metabolic acidosis persists for many hours and if the kidneys are functional, the kidneys can also help compensate for metabolic acidosis by secreting H^+ ions at a greater rate and increasing the rate of bicarbonate ion reabsorption. Symptoms of metabolic acidosis appear if the respiratory and renal systems are not able to maintain the pH of the body fluids within its normal range.

Metabolic alkalosis results from all conditions that increase the pH of the body fluids above 7.45, with the exception of conditions resulting from altered function of the respiratory system. As hydrogen ions decrease in the body fluids, buffers first resist an increase in pH. If the buffers cannot compensate for the decrease in H^+ ions, the respiratory center helps regulate the body fluid pH. The increased pH inhibits respiration. Reduced respiration allows carbon dioxide to accumulate in the body fluids. Carbon dioxide reacts with water to produce carbonic acid. If metabolic alkalosis persists for several hours, and if the kidneys are functional, the kidneys reduce the rate of H^+ ion secretion to help reverse alkalosis (see table A).

Summary

Water, acid, base, and electrolyte levels are maintained within a narrow range of concentrations. The urinary, respiratory, gastrointestinal, integumentary, nervous, and endocrine systems play a role in maintaining fluid, electrolyte, and pH balance.

Body Fluids

1. Intracellular fluid is inside cells.
2. Extracellular fluid is outside cells and includes interstitial fluid and plasma.

Regulation of Intracellular Fluid Composition

1. Intracellular fluid composition is determined by substances used or produced inside the cell and substances exchanged with the extracellular fluid.
2. Intracellular fluid is different from extracellular fluid because the cell membrane regulates the movement of materials.
3. Water movement is determined by the difference between intracellular and extracellular fluid concentrations.

Regulation of Extracellular Fluid Composition

Extracellular fluid composition is determined by the intake and elimination of substances from the body and the exchange of substances between the extracellular and intracellular fluids.

Regulation of Ion Concentrations

Sodium Ions

1. Sodium is responsible for 90%–95% of extracellular osmotic pressure.
2. The amount of Na^+ ions excreted in the kidneys is the difference between the amount of Na^+ ions that enters the nephron and the amount that is reabsorbed from the nephron.
   - Glomerular filtration rate determines the amount of Na^+ ions entering the nephron.
   - Aldosterone determines the amount of Na^+ ions reabsorbed.
3. Small quantities of Na^+ ions are lost in sweat.
4. Increased blood osmolality leads to the production of a small volume of concentrated urine and to thirst. Decreased blood osmolality leads to the production of a large volume of dilute urine and to decreased thirst.
5. Increased blood pressure increases water and salt loss.
   - Baroreceptor reflexes reduce ADH secretion.
   - Renin secretion is inhibited, leading to reduced aldosterone production.

Chloride Ions

Chloride ions are the dominant negatively charged ions in extracellular fluid.

Potassium Ions

1. The extracellular concentration of K^+ ions affects resting membrane potentials.
2. The amount of K^+ ions excreted depends on the amount that enters with the glomerular filtrate, the amount actively reabsorbed by the nephron, and the amount secreted into the distal convoluted tubule.
3. Aldosterone increases the amount of K^+ ions secreted.

Calcium Ions

1. Elevated extracellular calcium levels prevent membrane depolarization. Decreased levels lead to spontaneous action potential generation.
2. Parathyroid hormone increases extracellular Ca^{2+} ion levels and decreases extracellular phosphate levels. It stimulates osteoclast activity, increases calcium reabsorption from the kidneys, and stimulates active vitamin D production.
3. Vitamin D stimulates Ca^{2+} ion uptake in the intestines.
4. Calcitonin decreases extracellular Ca^{2+} ion levels.

Phosphate Ions

1. Under normal conditions reabsorption of phosphate occurs at a maximum rate in the nephron.
2. An increase in plasma phosphate increases the amount of phosphate in the nephron beyond that which can be reabsorbed, and the excess is lost in the urine.

Regulation of Water Content

1. Water crosses the gastrointestinal tract through osmosis.
2. The sense of thirst is stimulated by an increase in extracellular osmolality or by a decrease in blood pressure.
3. Thirst is inhibited by wetting the oral mucosa or by stretch of the gastrointestinal tract.
4. Learned behavior plays a role in the amount of fluid ingested.
5. Routes of water loss.
   - Water is lost through evaporation from the respiratory system and the skin (insensible perspiration and sweat).
   - Water loss into the gastrointestinal tract normally is small. Vomiting or diarrhea can significantly increase this loss.
   - The kidneys are the primary regulator of water excretion. Urine output can vary from a small amount of concentrated urine to a large amount of dilute urine.

Regulation of Acid–Base Balance

Acids and Bases

Acids release H^+ ions into solution, and bases remove them.

Buffer Systems

1. A buffer resists changes in pH.
   - When H^+ ions are added to a solution, the buffer removes them.
   - When H^+ ions are removed from a solution, the buffer replaces them.
2. Proteins, carbonic acid/bicarbonate, and phosphate compounds are important buffers.

Mechanisms of Acid–Base Balance Regulation

1. Respiratory regulation of pH is achieved through the carbonic acid/bicarbonate buffer system.
   - As carbon dioxide levels increase, pH decreases.
   - As carbon dioxide levels decrease, pH increases.
   - Carbon dioxide levels and pH affect the respiratory centers. Hypoventilation increases blood carbon dioxide levels, and hyperventilation decreases blood carbon dioxide levels.
2. The loss of H^+ ions into urine and the gain of bicarbonate ions into blood cause extracellular pH to increase.
   - Carbonic acid dissociates to form H^+ ions and bicarbonate ions in nephron cells.
   - Active transport pumps H^+ ions into the nephron lumen and Na^+ ions into the nephron cell.
   - Na^+ ions and bicarbonate diffuse into the extracellular fluid.
3. Bicarbonate ions in the filtrate are reabsorbed.
   - Bicarbonate ions combine with H^+ ions to form carbonic acid that dissociates to form carbon dioxide and water.
   - Carbon dioxide diffuses into nephron cells and forms carbonic acid, which dissociates to form bicarbonate ions and H^+ ions.
   - Bicarbonate ions diffuse into the extracellular fluid, and H^+ ions are pumped into the nephron lumen.
4. The rate of H^+ ion secretion increases as body fluid pH decreases or as aldosterone levels increase.
5. Secretion of H^+ ions is inhibited when urine pH falls below 4.5.
   - Ammonia and phosphate buffers in the urine resist a drop in pH.
   - As the buffers absorb H^+ ions, more H^+ ions are pumped into the urine.

Content Review

1. What systems are involved with the regulation of fluid, electrolyte, and pH balance?
2. Define the terms intracellular fluid, extracellular fluid, interstitial fluid, and plasma.
3. What factors determine the composition of intracellular fluid and extracellular fluid?
4. Name the substance that is responsible for most of the osmotic pressure of extracellular fluid.
5. How do the glomerular filtration rate and aldosterone affect the amount of sodium in the urine?
6. What role does sweating play in Na^+ ion balance?
7. How does increased blood pressure lead to an increased loss of water and salt? What happens when blood pressure decreases?
8. What effect does atrial natriuretic hormone have on Na^+ ion and water loss in urine?
9. How are chloride ion concentrations regulated?
10. What effect does an increase or a decrease in extracellular K^+ ion concentration have on resting membrane potentials?
11. Where are K^+ ions secreted in the nephron? How is its secretion regulated?
12. What effects are produced by an increase or a decrease in extracellular calcium concentration?

13. What effects on extracellular Ca^{2+} ion concentrations do an increase or a decrease in parathyroid hormone have? What causes these effects?

14. What effect does calcitonin have on extracellular Ca^{2+} ion levels?

15. How does an increase in extracellular osmolality or a decrease in blood pressure affect the sensation of thirst? Name two things that inhibit the sense of thirst.

16. Explain how the kidneys control plasma levels of phosphate ions.

17. Describe three routes for the loss of water from the body.

18. Define the terms acid and base. What is normal blood pH? Define the terms acidosis and alkalosis.

19. Define the term buffer. Describe how a buffer works when H^+ ions are added to a solution or when they are removed from a solution. Name the three buffer systems of the body.

20. What happens to blood pH when blood carbon dioxide levels go up or down? What causes this change?

21. What effect do increased blood carbon dioxide levels or decreased pH have on respiration? How does this change in respiration affect blood pH?

22. Describe the process by which nephron cells move H^+ ions into the nephron lumen and bicarbonate ions into the extracellular fluid.

23. Describe the process by which bicarbonate ions are reabsorbed from the nephron lumen.

24. Name the factors that can cause an increase and a decrease in H^+ ion secretion.

25. What is the purpose of buffers in the urine? Describe how the ammonia buffer system operates.

Develop Your Reasoning Skills

1. In patients with diabetes mellitus, not enough insulin is produced; as a consequence, blood glucose levels increase. If blood glucose levels rise high enough, the kidneys are unable to absorb the glucose from the glomerular filtrate, and glucose "spills over" into the urine. What effect does this glucose have on urine concentration and volume? How does the body adjust to the excess glucose in the urine?

2. A patient suffering from a tumor in the hypothalamus produces excessive amounts of ADH, a condition called syndrome of inappropriate ADH (SIADH) production. For this patient the excessive ADH production is chronic and has persisted for many months. A student nurse keeps a fluid intake–output record on the patient. She is surprised to find that fluid intake and urinary output are normal. What effect was she expecting? Can you explain why urinary output is normal?

3. A patient exhibits the following symptoms: elevated urine ammonia and increased rate of respiration. Does the patient have metabolic acidosis or metabolic alkalosis?

4. Swifty Trotts has an enteropathogenic *Escherichia coli* infection that produces severe diarrhea. What does this diarrhea do to his blood pH, urine pH, and respiratory rate?

5. Acetazolamide is a diuretic that blocks the activity of the enzyme carbonic anhydrase inside kidney tubule cells. This blockage prevents the formation of carbonic acid from carbon dioxide and water. Normally carbonic acid dissociates to form H^+ ions and bicarbonate ions, and the H^+ ions are exchanged for Na^+ ions from the urine. Blocking the formation of H^+ ions in the cells of the nephron tubule blocks sodium reabsorption, inhibiting water reabsorption and producing the diuretic effect. With this information in mind, what effect does acetazolamide have on blood pH, urine pH, and respiratory rate?

6. As part of a physiology experiment, Hardy Breath, an anatomy and physiology student, is asked to breathe through a 3-foot long glass tube. What effect does this action have on his blood pH, urine pH, and respiratory rate?

7. A young boy is suspected of having epilepsy and therefore is prone to having convulsions. On the basis of your knowledge of acid–base balance and respiration, propose a hypothetical experiment which might suggest that the boy is susceptible to convulsions.

8. Hardy Explorer climbed to the top of a very high mountain. To celebrate, he drank a glass of whiskey. Alcohol stimulates hydrochloric acid secretion in the stomach. What do you expect to happen to Hardy's respiratory rate and the pH of his urine?

Web Site Link

For a listing of the most current web sites related to this chapter, please visit the Seeley home page at:
http://www.mhhe.com/biosci/abio/ap/seeleyap/

Chapter Twenty-Eight

Reputational System

Reproductive System

Objectives

1. Describe the scrotum, and explain the role of the dartos and cremaster muscles in temperature regulation of the testes.

2. Describe the structure of the testes.

3. Describe the process of sperm cell formation.

4. Describe the route sperm cells follow from the site of their production to the outside of the body.

5. Name the parts of the spermatic cord.

6. Describe the parts of the penis.

7. Name the male reproductive glands, state where they empty into the duct system, and describe their secretions.

8. List the hormones that influence the male reproductive system, and explain how reproductive hormone secretions are regulated.

9. Explain the role of psychic stimulation, tactile stimulation, and the parasympathetic and sympathetic nervous systems in the male sex act.

10. Describe the anatomy and histology of the ovaries.

11. Discuss the development of the follicle and the oocyte, the process of ovulation, and fertilization.

12. Name and describe the parts of the uterine tube, uterus, vagina, external genitalia, and mammae.

13. Describe the phases of the ovarian and uterine cycles.

14. List the hormones of the female reproductive system, and explain how reproductive hormone secretions are regulated.

15. Discuss the effects of the ovarian hormones on the uterus.

16. Explain what happens to the ovaries and the uterus if fertilization occurs and if fertilization does not occur.

17. Describe the role of the nervous system in the female sex act.

18. Define the term menopause, and describe the changes that occur because of it.

Part Five

Although not essential for survival of the individual human being, the reproductive system does affect the structural and functional characteristics of adults. Structural differences between males and females and the important role these differences play in human behavior reflect the significance of the reproductive system. Also, the role of reproductive systems in the production of offspring, by some individuals, is required for the survival of the species.

Most organ systems of the body show little difference between males and females. This is not the case with the reproductive systems. The male reproductive system produces sperm cells and can transfer them to the female. The female reproductive system produces oocytes and can receive sperm cells, one of which may unite with an oocyte. The female reproductive system is then intimately involved with nurturing the development of a new individual until birth and usually for some considerable time after birth.

Although the male and female reproductive systems show such striking differences, they also share a number of similarities. Many reproductive organs of males and females are derived from the same embryologic structures (see chapter 29). In addition, some hormones are the same in males and females, even though they act in very different ways (table 28.1).

Male Reproductive System

The male reproductive system consists of the testes (sing., testis), epididymides (sing., epididymis), ductus deferentia (sing., deferens, also vas deferens), urethra, seminal vesicles, prostate gland, bulbourethral glands, scrotum, and penis (figure 28.1a). Sperm cells are very temperature-sensitive and do not develop normally at usual body temperatures. The testes and epididymides, in which the sperm cells develop, are located outside the body cavity in the scrotum, where the temperature is lower. The ductus deferentia lead from the testes into the pelvis, where they join the ducts of the seminal vesicles to form the ampullae. Extensions of the ampullae, called the ejaculatory ducts, pass through the prostate and empty into the urethra within the prostate. The urethra, in turn, exits from the pelvis and passes through the penis to the outside of the body.

Scrotum

The **scrotum** (skrō′tum) contains the testes and is divided into two internal compartments by a connective tissue septum. Externally the scrotum is marked in the midline by an irregular ridge, the **raphe** (rā′fē, meaning a seam), which continues posteriorly to the anus and anteriorly onto the inferior surface of the penis. The outer layer of the scrotum includes the skin, a layer of superficial fascia consisting of loose connective tissue, and a layer of smooth muscle called the **dartos** (dar′tōs, meaning to skin) **muscle.**

When the scrotum is exposed to cool temperatures, the dartos muscle contracts, causing the skin of the scrotum to become firm and wrinkled and reducing its overall size. At the same time the **cremaster** (krē-mas′ter) **muscles** (see figure 28.5), which are extensions of abdominal muscles into the scrotum, contract and help pull the testes nearer the body. When the scrotum is exposed to warm temperatures or becomes warm because of exercise, the dartos and cremaster muscles relax, and the skin of the scrotum becomes loose and thin, allowing the testes to descend away from the body. The response of the dartos and cremaster muscles is important in the regulation of temperature in the testes. If the testes become too warm or too cold, normal sperm cell formation does not occur.

Perineum

The area between the thighs, which is bounded by the pubis anteriorly, the coccyx posteriorly, and the ischial tuberosities laterally, is called the **perineum** (per′i-nē′ŭm). The perineum is divided into two triangles by a set of muscles, the superficial transverse and deep transverse perineal muscles, which run transversely between the two ischial tuberosities. The anterior, or **urogenital** (yū′rō-jen′i-tăl), **triangle,** contains the base of the penis and the scrotum. The smaller, posterior, or **anal, triangle,** contains the anal opening (figure 28.1b).

Testes

Testicular Histology

The **testes** (test′tēs) are small ovoid organs, each about 4–5 cm long, within the scrotum (see figure 28.1). They are both exocrine and endocrine glands. Sperm cells form a major part of the exocrine secretions of the testes, and testosterone is the major endocrine secretion of the testes.

The outer part of each testis is a thick, white capsule called the **tunica albuginea** (al-byū-jin′ē-ă, meaning white). Connective tissue of the tunica albuginea enters the inferior part of the testis as incomplete **septa** (sep′tă) (figure 28.2a). The septa divide each testis into about 300–400 cone-shaped **lobules.** The substance of the testis between the septa includes two types of tissue: **seminiferous** (sem′i-nif′er-ŭs, meaning seed carriers) **tubules** in which sperm cells develop and a loose connective tissue stroma that surrounds the tubules and contains clusters of endocrine cells called **interstitial cells,** or **Leydig cells,** which secrete testosterone.

The combined length of the seminiferous tubules in both testes is nearly half a mile. The seminiferous tubules empty into a set of short, straight tubules, which in turn empty into a tubular network called the **rete** (rē′tē, meaning net) **testis.** The rete testis empties into 15–20 tubules called **efferent ductules** (dŭk′tūls). They have a ciliated pseudostratified columnar epithelium that helps move sperm cells out of the testis. The efferent ductules pierce the tunica albuginea to exit the testis.

Table 28.1   Major Reproductive Hormones

Hormone	Source	Target Tissue	Response
Males			
Gonadotropin-releasing hormone (GnRH)	Hypothalamus	Anterior pituitary	Stimulates secretion of LH and FSH
Luteinizing hormone (LH) (also called interstitial cell-stimulating hormone [ICSH] in males)	Anterior pituitary	Leydig cells in the testes	Stimulates synthesis and secretion of testosterone
Follicle-stimulating hormone (FSH)	Anterior pituitary	Seminiferous tubules (Sertoli's cells)	Supports spermatogenesis
Testosterone	Leydig cells in the testes	Testes and body tissues	Supports spermatogenesis, development and maintenance of reproductive organs and secondary sexual characteristics
		Anterior pituitary and hypothalamus	Inhibits GnRH, LH, and FSH secretion through negative feedback
Females			
Gonadotropin-releasing hormone (GnRH)	Hypothalamus	Anterior pituitary	Stimulates secretion of LH and FSH
Luteinizing hormone (LH)	Anterior pituitary	Ovaries	Causes follicles to complete maturation and undergo ovulation; causes the ovulated follicle to become the corpus luteum
Follicle-stimulating hormone (FSH)	Anterior pituitary	Ovaries	Causes follicles to begin development
Prolactin	Anterior pituitary	Mammary glands	Stimulates milk secretion following parturition
Estrogens	Follicles of ovaries	Uterus	Proliferation of endometrial cells
		Mammary glands	Development of the mammary glands (especially duct systems)
		Anterior pituitary and hypothalamus	Positive feedback before ovulation, resulting in increased LH and FSH secretion; negative feedback with progesterone on the hypothalamus and anterior pituitary after ovulation, resulting in decreased LH and FSH secretion
		Other tissues	Secondary sexual characteristics
Progesterone	Corpus luteum of ovaries	Uterus	Hypertrophy of endometrial cells and secretion of fluid from uterine glands
		Mammary glands	Development of the mammary glands (especially alveoli)
		Anterior pituitary	Negative feedback with estrogens on the hypothalamus and anterior pituitary after ovulation, resulting in decreased LH and FSH secretion
		Other tissues	Secondary sexual characteristics
Oxytocin*	Posterior pituitary	Uterus and mammary glands	Contraction of uterine smooth muscle during intercourse and childbirth; contraction of myoepithelial cells in the breast resulting in milk letdown in lactating women
Human chorionic gonadotropin (HCG)	Placenta	Corpus luteum of ovaries	Maintains the corpus luteum and increases its rate of progesterone secretion during the first one-third (first trimester) of pregnancy

*Covered in chapter 29.

924

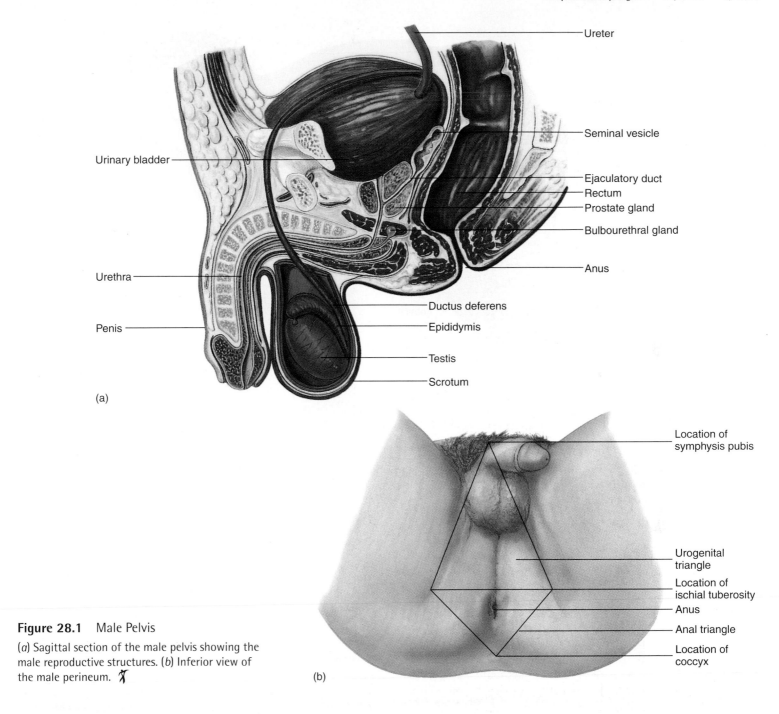

Ureter

Seminal vesicle

Urinary bladder

Ejaculatory duct

Rectum

Prostate gland

Bulbourethral gland

Urethra

Anus

Ductus deferens

Epididymis

Penis

Testis

Scrotum

(a)

Location of symphysis pubis

Urogenital triangle

Location of ischial tuberosity

Anus

Anal triangle

Location of coccyx

Figure 28.1   Male Pelvis

(*a*) Sagittal section of the male pelvis showing the male reproductive structures. (*b*) Inferior view of the male perineum.

(b)

Descent of the Testes

The testes develop as retroperitoneal organs in the abdominopelvic cavity and are connected to the scrotum by the **gubernaculum** (gū′ber-nak′yū-lŭm), a fibromuscular cord (figure 28.3*a*; see chapter 29). The testes move from the abdominal cavity through the **inguinal** (ing′gwi-năl) **canal** (figure 28.3*b*) to the scrotum (figure 28.3*c*). As they move into the scrotum, the testes are preceded by outpocketings of the peritoneum called the **process vaginalis** (vaj′i-nă-lis). The superior part of the process vaginalis usually becomes obliterated, and the inferior part remains as a small, closed sac, the **tunica** (tū′ni-kă) **vaginalis,** which covers most of the testes.

Clinical Note

Normally the inguinal canal is closed, but it does represent a weak spot in the abdominal wall. If the inguinal canal weakens or ruptures, an **inguinal hernia** (ing′gwi-năl her′nē-ă) can result, and a loop of intestine can protrude into or even pass through the inguinal canal. This herniation can be quite painful and even very dangerous, especially if the inguinal canal compresses the intestine and cuts off its blood supply. Fortunately, inguinal hernias can be repaired surgically. Males are much more prone to inguinal hernias than females, apparently because a male's inguinal canal is weakened as the testis passes through it on its way into the scrotum.

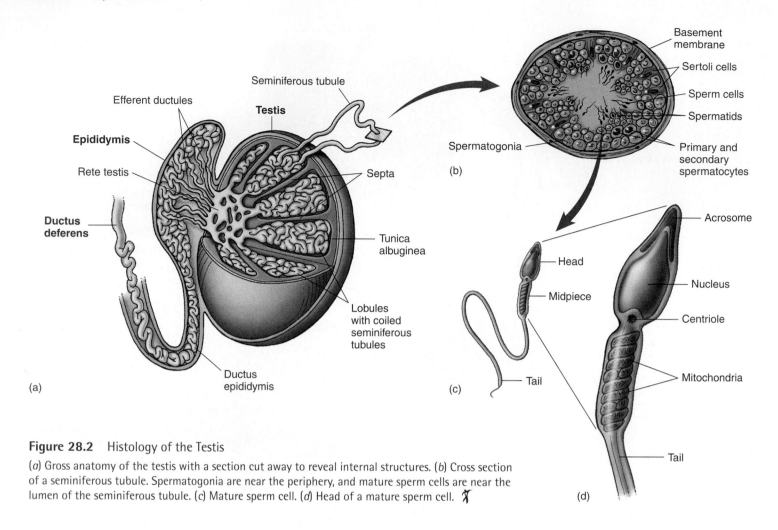

Figure 28.2   Histology of the Testis

(*a*) Gross anatomy of the testis with a section cut away to reveal internal structures. (*b*) Cross section of a seminiferous tubule. Spermatogonia are near the periphery, and mature sperm cells are near the lumen of the seminiferous tubule. (*c*) Mature sperm cell. (*d*) Head of a mature sperm cell. ✗

Sperm Cell Development

Before puberty the testes remain relatively simple and unchanged from the time of their initial development. The interstitial cells are not particularly prominent during this period, and the seminiferous tubules lack a lumen and are not yet functional. At 12–14 years of age, the interstitial cells increase in number and size, a lumen develops in each seminiferous tubule, and sperm cell production begins.

A cross section of a mature seminiferous tubule reveals the various stages of sperm cell development, a process called **spermatogenesis** (sper′mă-tō-jen′ĕ-sis; see figure 28.2*b*; figure 28.4). The seminiferous tubules contain two types of cells, **germ cells** and **Sertoli** (sēr-tō′lē; named for an Italian histologist) **cells.** Sertoli cells are also sometimes referred to as **sustentacular** (sŭs′ten-tak′yū-lăr), or **nurse, cells.**

Sertoli cells are large cells that extend from the periphery to the lumen of the seminiferous tubule. They nourish the germ cells and probably produce, together with the Leydig cells, a number of hormones, such as androgens, estrogens, and inhibins. In addition, tight junctions between the Sertoli cells form a **blood–testes barrier,** which isolates the sperm cells from the immune system (see figure 28.4). This barrier is

necessary because, as the sperm cells develop, they form surface antigens that could stimulate an immune response, resulting in their destruction.

Testosterone, produced by the Leydig cells, passes into the Sertoli cells and binds to receptors. The combination of testosterone with the receptors is required for the Sertoli cells to function normally. In addition, testosterone is converted to two other steroids in the Sertoli cells: **dihydrotestosterone** (dī-hī′drō-tes-tos′ter-ōn) and estradiol. The Sertoli cells also secrete a protein called **androgen-binding** (an′drō-jen) **protein** into the seminiferous tubule. Testosterone and dihydrotestosterone bind to the androgen-binding protein and are carried along with other secretions of the seminiferous tubule to the epididymis. The estradiol and dihydrotestosterone may be the active hormones that promote sperm cell formation.

Scattered between the Sertoli cells are smaller germ cells from which sperm cells are derived. The germ cells are arranged according to maturity from the periphery to the lumen of the seminiferous tubules. The most peripheral cells, those adjacent to the basement membrane of the seminiferous tubules, are **spermatogonia** (sper′ma-tō-gō′nē-ă), which divide by mitosis (see figure 28.4). Some of the daughter cells produced from these mitotic divisions remain spermatogonia

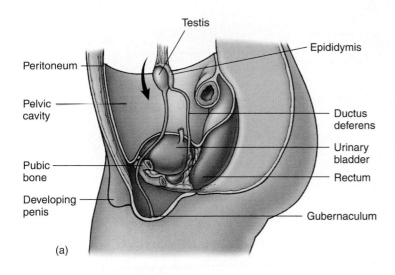

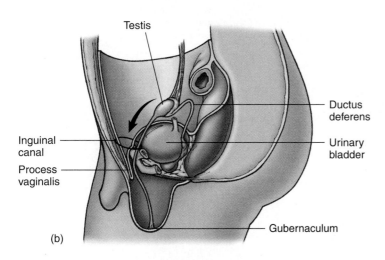

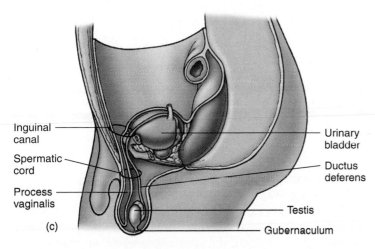

Figure 28.3   Descent of the Testes

(a) Testes form as retroperitoneal structures near the level of each kidney. (b) The testis descends through the inguinal canal. (c) The testis descends into the scrotum.

and continue to produce additional spermatogonia. The others divide through mitosis and differentiate to form **primary spermatocytes** (sper′mă-tō-sītz).

Meiosis (see the Clinical Focus in this chapter or in chapter 3) begins when the primary spermatocytes divide. Primary spermatocytes pass through the first meiotic division to become two **secondary spermatocytes.** Each secondary spermatocyte undergoes a second meiotic division to produce two even smaller cells called **spermatids** (sper′mă-tidz). Each spermatid undergoes the last phase of spermatogenesis to form a sperm cell, or **spermatozoon** (sper′mă-tō-zō′on; pl., spermatozoa, sper′mă-tō-zō′ă) (see figures 28.2c and d, and 28.4). Each spermatid develops a head and a tail. The head contains the chromosomes, and at the leading end it has a cap, the **acrosome** (ak′rō-sōm), which contains enzymes necessary for the sperm cell to penetrate the female sex cell. The flagellum is composed of a middle piece and a tail. The flagellum is similar to a cilium (see chapter 3), and movement of microtubules past one another within the tail causes the tail to move and propel the sperm cell forward. The middle piece has large numbers of mitochondria, which produce the adenosine triphosphate necessary for microtubule movement.

At the end of spermatogenesis, the developing sperm cells gather around the lumen of the seminiferous tubules with their heads directed toward the surrounding Sertoli cells and their tails directed toward the center of the lumen (see figures 28.2b and 28.4). Finally, sperm cells are released into the lumen of the seminiferous tubules.

Ducts

After their release into the seminiferous tubules, the sperm cells leave the testes through the efferent ductules and pass through a series of ducts to reach the exterior of the body.

Epididymis

The **efferent ductules** from each testis become extremely convoluted and form a comma-shaped structure on the posterior side of the testis called the **epididymis** (ep-i-did′i-mis; pl., epididymides, ep-i-di-dim′i-dēz, meaning on the twin; "twin" refers to the paired, or twin, testes). The final maturation of the sperm cells occurs within the ductules of the epididymides. Sperm cells taken directly from the testes in experimental animals are not capable of fertilization, but, after spending 1 to several days in the epididymides, they develop the capacity to fertilize the female sex cell.

Each epididymis consists of a head, a body, and a long tail (see figure 28.2a; figure 28.5). The head contains the convoluted efferent ductules, which empty into a single convoluted ductule, the **ductus epididymis,** located primarily within the body of the epididymis (see figure 28.2a). This ductile alone, if unraveled, would extend for several meters. The epididymis has a pseudostratified columnar epithelium with microvilli called stereocilia (ster′ē-ō-sil′ē-ă). The stereocilia function to increase the surface area of epithelial cells

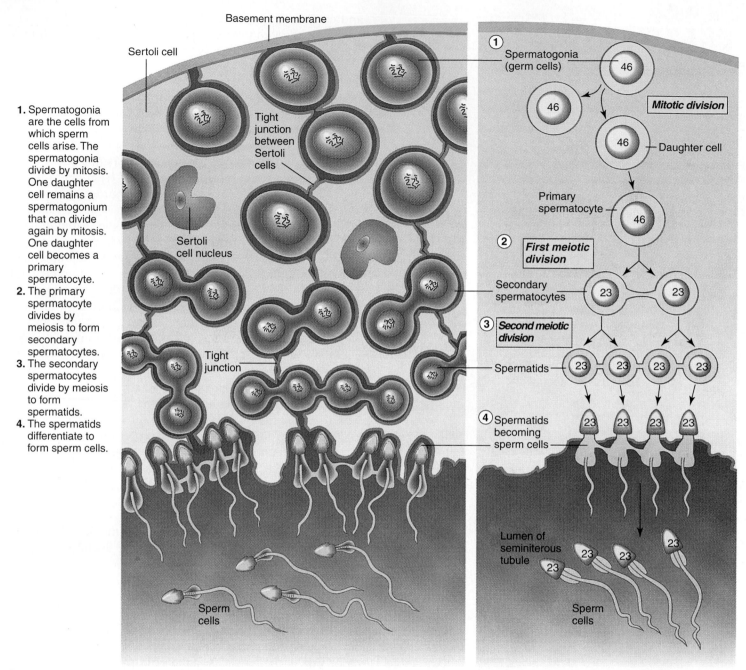

1. Spermatogonia are the cells from which sperm cells arise. The spermatogonia divide by mitosis. One daughter cell remains a spermatogonium that can divide again by mitosis. One daughter cell becomes a primary spermatocyte.
2. The primary spermatocyte divides by meiosis to form secondary spermatocytes.
3. The secondary spermatocytes divide by meiosis to form spermatids.
4. The spermatids differentiate to form sperm cells.

Figure 28.4   A Seminiferous Tubule

The process of meiosis and sperm cell formation. A section of the seminiferous tubule illustrating the process of meiosis and sperm cell formation. The tight junctions that form between adjacent Sertoli cells form the blood–testis barrier. Spermatogonia are peripheral to the blood–testis barrier, and spermatocytes are central to it.

that absorb fluid from the ductus epididymis. The ductus epididymis ends at the tail of the epididymis, which is located at the inferior border of the testis.

Ductus Deferens

The **ductus deferens,** or **vas deferens,** emerges (see figures 28.1, 28.2, and 28.5) from the tail of the epididymis and ascends along the posterior side of the testis medial to the epididymis and becomes associated with the blood vessels and nerves that

supply the testis. These structures and their coverings constitute the **spermatic cord.** The spermatic cord consists of (1) the ductus deferens, (2) the testicular artery and venous plexus, (3) lymph vessels, (4) nerves, (5) fibrous remnants of the process vaginalis, and (6) three coats, which include the **external spermatic fascia** (fash′ē-ă); the cremaster muscle, an extension of the muscle fibers of the internal oblique muscle of the abdomen; and the **internal spermatic fascia** (see figure 28.5).

The spermatic cord passes obliquely through the inferior abdominal wall by way of the inguinal canal. The superficial

Sperm cell development and oocyte development involve meiosis (see chapter 3). This kind of cell division occurs only in the gonads. It consists of two consecutive nuclear divisions without a second replication of the genetic material between the divisions. Four daughter cells are produced, and each has half as many chromosomes as the parent cell (figure A).

First Division (Meiosis I)

Second Division (Meiosis II)

Early prophase I
The duplicated chromosomes become visible chromatids (shown separated for emphasis, they actually are so close together that they appear as a single strand)

Chromosomes

Nucleus

Chromatids

Centrioles

Prophase II
Each chromosome consists of two chromatids

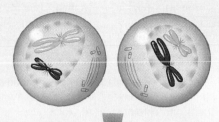

Middle prophase I
Homologous chromosomes synapse to form tetrads

Crossing over may occur at this stage

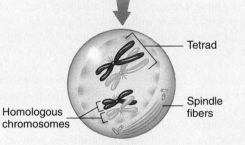

Tetrad

Homologous chromosomes

Spindle fibers

Metaphase II
Chromosomes align at the equatorial plane

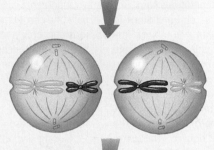

Metaphase I
Tetrad align at the equatorial plane

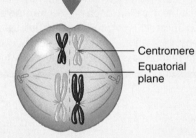

Centromere

Equatorial plane

Anaphase II
Chromatids separate and each is now called a chromosome

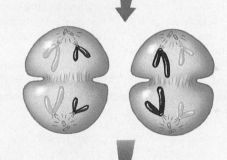

Anaphase I
Homologous chromosomes move apart to opposite sides of the cell

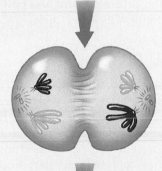

Telophase II
New nuclei form around the chromosomes

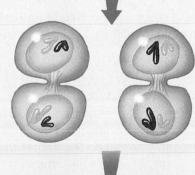

Telophase I
New nuclei form, and the cell divides; during interkinesis (not shown) there is no duplication of chromosomes

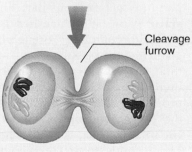

Cleavage furrow

Haploid cells
The chromosomes are about to unravel and become less distinct chromatin

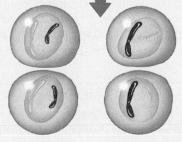

Figure A Meiosis

(continued)

The normal chromosome number in human cells is 46. This number is called a **diploid** (dip'loyd), or a 2*n* number of chromosomes. The chromosomes consist of 23 pairs. Each pair of chromosomes is called a **homologous** (hŏ-mol'ō-gŭs) **pair.** One chromosome of each homologous pair is from the male parent, and the other is from the female parent. The chromosomes of each homologous pair look alike, and they contain genes for the same traits.

In sperm cells and oocytes the number of chromosomes is 23. This number is called a **haploid** (hap'loyd), or *n* number of chromosomes. Each gamete contains one chromosome from each of the homologous pairs. Reduction of the number of chromosomes in sperm cells or oocytes to an *n* number is important. When a sperm cell and an oocyte fuse to form a fertilized egg, each provides an *n* number of chromosomes, which reestablishes a 2*n* number of chromosomes. If meiosis did not occur, each time fertilization occurred the number of chromosomes in the fertilized oocyte would double. The extra chromosomal material would be lethal to the developing offspring.

The two divisions of meiosis are called **meiosis** (mī-ō'sis) **I** and **meiosis II.** The stages of meiosis have the same names as these stages in mitosis, that is, prophase, metaphase, anaphase, and telophase; but there are distinct differences between mitosis and meiosis.

Before meiosis begins, all the deoxyribonucleic acid in the chromosomes is duplicated. At the beginning of meiosis each of the 46 chromosomes consists of two sister **chromatids** (krō'mă-tid) connected by a **centromere** (sen'trō-mēr) (see figure A). In prophase of meiosis I the chromosomes align with their homologous pairs near the middle of the cell. This process is called **synapsis** (si-nap'sis). Because each chromosome consists of two chromatids, the pairing of the homologous chromosomes brings two chromatids of each chromosome close together, an arrangement called a **tetrad.** Occasionally part of a chromatid of one homologous chromosome breaks off and is exchanged with part of another chromatid from the other homologous chromosome of the tetrad. This exchange of genetic material is called **crossing over.** Crossing over allows the exchange of genetic material between maternal and paternal chromosomes (see figure A).

During synapsis homologous pairs of chromosomes line up near the center of the cell undergoing meiosis. For each pair of homologous chromosomes, however, the side of the cell on which the maternal or paternal chromosome is located is random. The way the chromosomes align during synapsis results in the random assortment of maternal and paternal chromosomes in the daughter cells during meiosis. Crossing over and the random assortment of maternal and paternal chromosomes are responsible for the large degree of diversity in the genetic composition of sperm cells and oocytes produced by each individual.

During anaphase I the homologous pairs are separated to each side of the cell. As a consequence, when meiosis I is complete, each daughter cell has one chromosome from each of the homologous pairs. Each of the 23 chromosomes in each daughter cell consists of two chromatids joined by a centromere.

It is during the first meiotic division that the chromosome number is reduced from a 2*n* number (46 chromosomes, or 23 pairs) to an *n* number (23 chromosomes, or one from each homologous pair). The first meiotic division is therefore called a **reduction division.**

The second meiotic division is similar to mitosis. The chromosomes, each consisting of two chromatids, line up near the middle of the cell (see figure A). Then the chromatids separate at the centromere, and each daughter cell receives one of the chromatids from each chromosome. When the centromere separates, each of the chromatids is called a chromosome. Consequently, each of the four daughter cells produced by meiosis contains 23 chromosomes.

---

opening of the inguinal canal, called the **superficial inguinal ring,** is medial, whereas the deep opening, the **deep inguinal ring,** is lateral.

The ductus deferens and the rest of the spermatic cord structures ascend and pass through the inguinal canal to enter the pelvic cavity (see figures 28.1 and 28.5). The ductus deferens crosses the lateral wall of the pelvic cavity, travels over the ureter, and loops over the posterior surface of the urinary bladder to approach the prostate gland. The end of the ductus deferens enlarges to form the **ampulla** (am-pul'lă). The ductus deferens has a pseudostratified columnar epithelium and is surrounded by smooth muscle. Peristaltic contractions of this smooth muscle tissue help propel the sperm cells through the ductus deferens.

Ejaculatory Duct

Adjacent to the ampulla of each ductus deferens is a sac-shaped gland that is called the seminal vesicle. A short duct from the seminal vesicle joins the ductus deferens to form the **ejaculatory** (ēj-jak'yū-lă-tōr-ē) **duct.** The ejaculatory ducts are approximately 2.5 cm long. These ducts project into the prostate gland and end by opening into the urethra (see figures 28.1 and 28.5).

Urethra

The **male urethra** (yū-rē'thra) is about 20 cm long and extends from the urinary bladder to the distal end of the penis (see figures 28.1 and 28.5; figure 28.6). The urethra is a passageway for both urine and male reproductive fluids. The urethra can be divided into three parts: the prostatic part, the membranous part, and the spongy part. The **prostatic** (pros-tat'ik) **urethra** is closest to the bladder and passes through the prostate gland. Ducts from the prostate gland and the ejaculatory ducts empty into the prostatic urethra. The **membranous urethra** is the shortest part of the urethra and extends from the prostatic urethra through the urogenital diaphragm, which is part of the muscular floor of the pelvis. The **spongy urethra,** also called the **penile** (pē'nīl) **urethra,** is by far the longest part of the urethra, and extends from the membranous urethra through the length of the penis. Most of the urethra is lined by stratified columnar epithelium, but transitional epithelium is in the

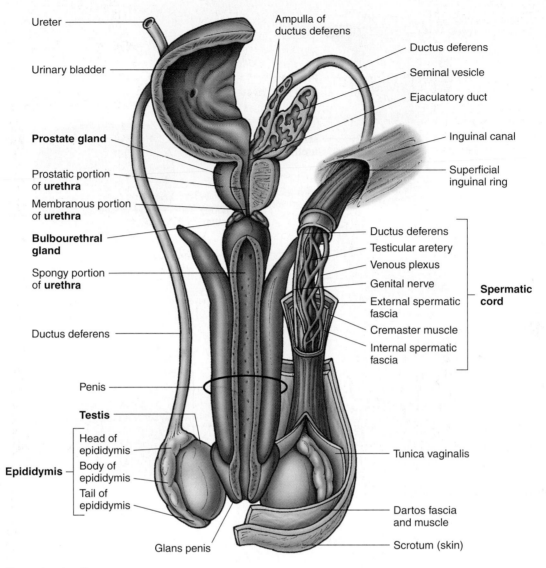

Figure 28.5   Male Reproductive Structures

Frontal view of the testes, epididymis, ductus deferens, and glands of the male reproductive system. The urethra has been cut open along its dorsal side.

prostatic urethra near the bladder, and stratified squamous epithelium is near the opening of the spongy urethra. There are several minute mucus-secreting **urethral glands** that empty into the urethra.

Penis

The **penis** consists of three columns of erectile tissue (see figure 28.6), and engorgement of this erectile tissue with blood causes the penis to enlarge and become firm, a process called erection. The penis is the male organ of copulation through which sperm cells are transferred from the male to the female. Two of the erectile columns form the dorsum and sides of the penis and are called the **corpora cavernosa** (kōr′pōr-ă kav-er-nos′ă). The third column, the **corpus spongiosum**

(kōr′pŭs spŭn′jē-ō′sŭm), expands to form a cap, the **glans penis,** over the distal end of the penis. The spongy urethra passes through the corpus spongiosum, penetrates the glans penis, and opens as the **external urethral orifice.** At the base of the penis the corpus spongiosum expands to form the **bulb of the penis,** and each corpus cavernosum expands to form a **crus** (krŭs) **of the penis.** Together these structures constitute the **root of the penis** and attach the penis to the coxae.

The shaft of the penis is covered by skin that is loosely attached to the connective tissue surrounding the penis. The skin is firmly attached at the base of the glans penis, and a thinner layer of skin tightly covers the glans penis. The skin of the penis, especially the glans penis, is well supplied with sensory receptors. A loose fold of skin called the **prepuce** (prē′pūs), or **foreskin,** covers the glans penis.

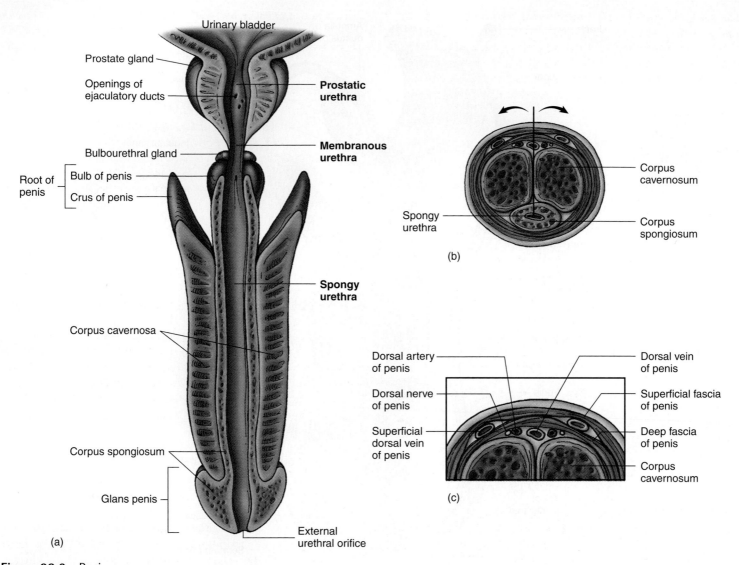

Figure 28.6 Penis

(*a*) Sagittal section through the spongy or penile urethra laid open and viewed from above. The prostate is also cut open to show the prostatic urethra. (*b*) Cross section of the penis. The line and arrows depict the manner in which (*a*) is cut and laid open. (*c*) Principal nerves, arteries, and veins along the dorsum of the penis. ⚡

The primary nerves, arteries, and veins of the penis pass along its dorsal surface (see figure 28.6). A single, midline dorsal vein is flanked on each side by dorsal arteries, with dorsal nerves lateral to them. Additional deep arteries lie within the corpora cavernosa.

Accessory Glands
Seminal Vesicles

The **seminal vesicles** (sem′i-năl ves′i-klz) are sac-shaped glands located next to the ampullae of the ductus deferentia (see figure 28.5). Each gland is about 5 cm long and tapers into a short duct that joins the ductus deferens to form the ejaculatory duct.

Prostate Gland

The **prostate** (pros′tāt, meaning one standing before) gland consists of both glandular and muscular tissue and is about the size and shape of a walnut; that is, about 4 cm long and 2 cm wide. The prostate gland is dorsal to the symphysis pubis at the base of the bladder, where it surrounds the prostatic urethra and the two ejaculatory ducts (see figure 28.1).

The gland is composed of an indistinct smooth muscle capsule and numerous smooth muscle partitions that radiate inward toward the urethra. Covering these muscular partitions is a layer of columnar epithelial cells that form saccular dilations into which the cells secrete prostatic fluid. Twenty to 30 small prostatic ducts transport these secretions into the prostatic urethra.

1   P R E D I C T

The prostate gland can enlarge for several reasons, including infections, tumor, and old age. The detection of enlargement or changes in the prostate is an important way to detect prostatic cancer. Suggest a way other than surgery that the prostate gland can be examined by palpation for any abnormal changes.

✔ *Answer in Appendix F*

Bulbourethral Glands

The **bulbourethral** (bŭl′bō-yū-rē′thrăl) **glands** are a pair of small glands located near the membranous part of the urethra (see figures 28.1 and 28.5). In young males each is about the size of a pea, but they decrease in size with age and are almost impossible to see in old men. Each bulbourethral gland is a compound mucous gland (see chapter 4). The small ducts of each gland unite to form a single duct. The single duct from each bulbourethral gland then enters the spongy urethra at the base of the penis.

Secretions

Semen (sē′men) is a composite of sperm cells and secretions from the male reproductive glands. The seminal vesicles produce about 60% of the fluid, the prostate gland contributes about 30%, the testes contribute 5%, and the bulbourethral glands contribute 5%. Emission is the discharge of semen into the prostatic urethra. Ejaculation is the forceful expulsion of semen from the urethra caused by the contraction of the urethra, the skeletal muscles in the floor of the pelvis, and the muscles at the base of the penis.

The bulbourethral glands and urethral mucous glands produce a mucous secretion just before ejaculation. This mucus lubricates the urethra, neutralizes the contents of the normally acidic spongy urethra, provides a small amount of lubrication during intercourse, and helps reduce vaginal acidity.

Testicular secretions include sperm cells, a small amount of fluid, and metabolic by-products. The thick, mucoid secretions of the seminal vesicles contain large amounts of fructose and other nutrients that nourish the sperm cells. The seminal vesicle secretions also contain fibrinogen, which is involved in a weak coagulation reaction of the semen after ejaculation, and prostaglandins, which can cause uterine contractions.

The thin, milky secretions of the prostate have a rather high pH and, with secretions of the seminal vesicles, help to neutralize the acidic urethra, the acidic secretions of the testes, and those of the vagina. The prostatic secretions are also important in the transient coagulation of semen because they contain clotting factors that convert fibrinogen from the seminal vesicles to fibrin, resulting in coagulation. The coagulated material keeps the semen as a single, sticky mass for a few minutes after ejaculation, and then fibrinolysin from the prostate causes the coagulum to dissolve, releasing the sperm cells to make their way up the female reproductive tract as somewhat free, motile cells.

Before ejaculation, the ductus deferens begins to contract rhythmically, propelling sperm cells and testicular fluid from the tail of the epididymis to the ampulla of the ductus deferens. Contractions of the ampullae, seminal vesicles, and ejaculatory ducts cause the sperm cells, testicular secretions, and seminal fluid to move into the prostatic urethra, where they mix with prostatic secretions released as a result of contractions of the prostate gland.

2   P R E D I C T

Explain a possible reason for having the coagulation reaction.

✔ *Answer in Appendix F*

Normal sperm cell counts in the semen range from 75 to 400 million sperm cells per milliliter of semen, and a normal ejaculation usually consists of about 2–5 mL of semen. Most of the sperm cells (millions) are expended in moving the general group of sperm cells through the female reproductive system. Enzymes carried in the acrosomal cap of each sperm cell help to digest a path through the mucoid fluids of the female reproductive tract and through materials surrounding the oocyte. Once the acrosomal fluid is depleted from a sperm cell, the sperm cell is no longer capable of fertilization.

Physiology of Male Reproduction

Normal function of the male reproductive system depends on both hormonal and neural mechanisms. Hormones are primarily responsible for the development of reproductive structures and maintenance of their functional capacities, development of secondary sexual characteristics, control of sperm cell formation, and influencing sexual behavior. Neural mechanisms are primarily involved in sexual behavior and controlling the sexual act.

Regulation of Sex Hormone Secretion

Hormonal mechanisms that influence the male reproductive system involve the hypothalamus, the pituitary gland, and the testes (figure 28.7). A small peptide hormone called **gonadotropin-releasing hormone (GnRH)** or **luteinizing hormone-releasing hormone (LH/RH)** is released from neurons in the hypothalamus. GnRH passes through the hypothalamohypophyseal portal system to the anterior pituitary gland (see chapter 18). In response to GnRH, cells within the anterior pituitary gland secrete two hormones referred to as **gonadotropins** (gō′nad-ō-trō′pinz) because they influence the function of the **gonads** (gō′nadz; the testes or ovaries).

The two gonadotropins are **luteinizing hormone (LH)** and **follicle-stimulating hormone (FSH).** They are named for their functions in females, but they also perform important functions in males. LH in males is sometimes called **interstitial cell-stimulating hormone (ICSH).** LH binds to the Leydig cells in the testes and causes them to increase their rate of testosterone synthesis and secretion. FSH binds primarily to Sertoli cells in the seminiferous tubules and promotes sperm cell development. Both gonadotropins bind to specific receptor molecules on the membranes of the cells that they influence, and cyclic adenosine monophosphate is an important intracellular mediator in those cells.

For GnRH to stimulate the secretion of large quantities of LH and FSH, the anterior pituitary must be exposed to a series of brief increases and decreases in GnRH. Chronically elevated GnRH levels in the blood cause the anterior pituitary cells to become insensitive to stimulation by GnRH molecules, and little LH or FSH is secreted.

Clinical Note

GnRH can be produced synthetically and is useful in treating males who are infertile if it is administered in small amounts in frequent pulses or surges. GnRH can also inhibit reproduction because chronic administration of GnRH can sufficiently reduce LH and FSH levels to prevent sperm cell production in males or ovulation in females.

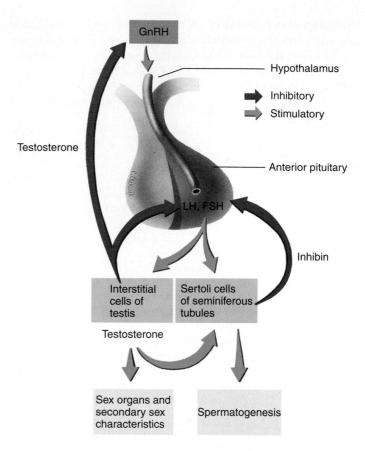

Figure 28.7   Regulation of Reproductive Hormone Secretion in Males

GnRH from the hypothalamus stimulates the secretion of LH and FSH from the anterior pituitary. LH and FSH stimulate spermatogenesis, secretion of testosterone, and secretion of inhibin in the testes. Testosterone has a negative-feedback effect on the hypothalamus and pituitary to reduce LH and FSH secretion, whereas inhibin specifically inhibits FSH secretion. Testosterone has a stimulatory effect on the sex organs, secondary sex characteristics, and the Sertoli cells.

Testosterone is the major male hormone secreted by the testes. It is classified as an **androgen** (*andros* is Greek for male human being) because it stimulates the development of reproductive structures (see chapter 29) and male secondary sexual characteristics. Other androgens are secreted by the testes, but they are produced in smaller concentrations and are less potent than testosterone. In addition, small amounts of estrogen and progesterone are secreted by the testes.

Testosterone has a major influence on many tissues. It plays an essential role in the embryonic development of reproductive structures, their further development during puberty, the development of secondary sexual characteristics during puberty, the maintenance of sperm cell production, and the regulation of gonadotropin secretion. It also influences behavior.

Inhibin (in-hib′in) is a polypeptide hormone secreted by the Sertoli cells of the testis. Inhibin inhibits FSH secretion from the anterior pituitary.

Clinical Focus   Male Infertility

Infertility (in-fer-til'i-tē) is reduced or diminished fertility. The most common cause of infertility in males is a low sperm cell count. If the sperm cell count drops to below 20 million sperm cells per milliliter, the male is usually sterile.

A decreased sperm cell count can occur because of damage to the testes as a result of trauma, radiation, cryptorchidism, or infections such as mumps. Reduced sperm cell counts can result from inadequate secretion of luteinizing hormone and follicle-stimulating hormone, which can be caused by hypothyroidism, trauma to the hypothalamus, infarctions of the hypothalamus or anterior pituitary gland, and tumors. Decreased testosterone secretion also reduces the sperm cell count.

Fertility is reduced if the sperm cell count is normal but sperm cell structure is abnormal, such as from chromosomal abnormalities caused by genetic factors. Reduced sperm cell motility also results in infertility. A major cause of reduced sperm cell motility is anti-sperm antibodies produced by the immune system, which bind to sperm cells. Some reports suggest that the average sperm count has decreased substantially since the end of World War II (1945). It is speculated that certain synthetic chemicals are responsible.

Fertility can sometimes be achieved by collecting several ejaculations, concentrating the sperm cells, and inserting the sperm cells into the female's reproductive tract, a process called **artificial insemination** (in-sem-i-nā'shŭn).

Puberty

A gonadotropin-like hormone called **human chorionic** (kō-rē-on'ik) **gonadotropin (HCG),** which is secreted by the maternal placenta, stimulates the synthesis and secretion of testosterone by the fetal testes before birth. After birth, however, no source of stimulation is present, and the testes of the newborn baby atrophy slightly and secrete only small amounts of testosterone until puberty, which normally begins when a boy is 12–14 years old.

Puberty (pyū'ber-tē) is the age at which individuals become capable of sexual reproduction. Before puberty small amounts of testosterone and other androgens in males inhibit GnRH release from the hypothalamus. At puberty the hypothalamus becomes much less sensitive to the inhibitory effect of androgens, and the rate of GnRH secretion increases, leading to increased LH and FSH release. Elevated FSH levels promote sperm cell formation, and elevated LH levels cause the interstitial Leydig cells to secrete larger amounts of testosterone. Testosterone still has a negative-feedback effect on GnRH secretion after puberty but is not capable of completely suppressing it.

Clinical Note

Some men have a genetic tendency called **male pattern baldness,** which develops in response to testosterone and other androgens. When testosterone levels increase at puberty, the density of hair on the top of the head begins to decrease. Baldness usually reaches its maximum rate of development when the individual is in the third or fourth decade of life.

Testosterone also causes the texture of the skin and hair to become rougher or coarser. The quantity of melanin in the skin also increases, making the skin darker. Testosterone increases the rate of secretion from the sebaceous glands, especially in the region of the face, frequently resulting near the time of puberty in the development of acne. Beginning near the time of puberty, testosterone also causes hypertrophy of the larynx. The structural changes can first result in a voice that is difficult to control, but ultimately the voice reaches its normal masculine quality.

Testosterone has a general stimulatory effect on metabolism so that males have a slightly higher metabolic rate than females. The red blood cell count is increased by nearly 20% as a result of the effects of testosterone on erythropoietin production. Testosterone also has a minor mineralocorticoidlike effect, causing the retention of sodium in the body and, consequently, an increase in the volume of body fluids. Testosterone promotes protein synthesis in most tissues of the body; as a result, skeletal muscle mass increases at puberty. The average percentage of the body weight composed of skeletal muscle is greater for males than for females because of the effect of androgens.

Effects of Testosterone

Testosterone is by far the major androgen in males. Nearly all of the androgens, including testosterone, are produced by the Leydig cells, with small amounts produced by the adrenal cortex and possibly by the Sertoli cells. Testosterone causes the enlargement and differentiation of the male genitals and reproductive duct system, is necessary for sperm cell formation, and is required for the descent of the testes near the end of fetal development. Testosterone stimulates hair growth in the following regions: (1) the pubic area and extending up the linea alba, (2) the legs, (3) the chest, (4) the axillary region, (5) the face, and (6) occasionally, the back.

Clinical Note

Some athletes, especially weight lifters, ingest synthetic androgens in an attempt to increase muscle mass. The side effects of the large doses of androgens are often substantial and include testicular atrophy, kidney damage, liver damage, heart attacks, and strokes. Administration of synthetic androgens is highly discouraged by the medical profession and is a violation of the rules of most athletic organizations.

Testosterone causes rapid bone growth and increases the deposition of calcium in bone, resulting in an increase in height. The growth in height is limited, however, because testosterone also causes early closure of the epiphyseal plates of long bones (see chapter 6). Males who mature sexually at an earlier age grow rapidly but reach their maximum height earlier. Males who mature sexually at a later age do not exhibit a rapid period of growth, but they grow for a longer period and can become taller than those who mature sexually at an earlier age.

Male Sexual Behavior and the Male Sex Act

Testosterone is required to initiate and maintain male sexual behavior. Testosterone enters cells within the hypothalamus and the surrounding areas of the brain and influences their function, resulting in sexual behavior. Male sexual behavior may depend, in part, however, on the conversion of testosterone to other substances in the cells of the brain.

The blood levels of testosterone remain relatively constant throughout the lifetime of a male from puberty until about 40 years of age. Thereafter, the levels slowly decline to about 20% of this value by 80 years of age, causing a slow decrease in sex drive and fertility.

The male sex act is a complex series of reflexes that result in erection of the penis, secretion of mucus into the urethra, emission, and ejaculation. Sensations that are normally interpreted as pleasurable occur during the male sexual act and result in a climactic sensation, **orgasm** (or′gazm), associated with ejaculation. After ejaculation a phase called **resolution** occurs, in which the penis becomes flaccid, an overall feeling of satisfaction exists, and the male is unable to achieve erection and a second ejaculation for many minutes to many hours.

Afferent Impulses and Integration

Afferent impulses from the genitals are propagated through the pudendal nerve to the sacral region of the spinal cord, where reflexes that result in the male sexual act are integrated. Impulses travel from the spinal cord to the cerebrum to produce the conscious sexual sensations.

Rhythmic massage of the penis, especially the glans, provides an extremely important source of afferent impulses that initiate erection and ejaculation. Sensory impulses produced in surrounding tissues such as the scrotum and the anal, perineal, and pubic regions reinforce sexual sensations. Engorgement of the prostate and seminal vesicles with secretions and irritation of the urethra, urinary bladder, ductus deferens, and testes can also cause sexual sensations.

Psychic stimuli, such as sight, sound, odor, or thoughts, have a major effect on sexual reflexes. Thinking sexual thoughts or dreaming about erotic events tends to reinforce stimuli that trigger sexual reflexes such as erection and ejaculation. Ejaculation while sleeping is a relatively common event in young males and is thought to be triggered by psychic stimuli associated with dreaming. Psychic stimuli can also inhibit the sexual act, and thoughts that are not sexual in nature tend to decrease the effectiveness of the male sexual act. The inability to concentrate on sexual sensations is one of the causes of **impotence** (im′pŏ-tens), the inability to achieve or maintain an erection and to accomplish the male sexual act. Impotence can also be caused by physical factors such as inability of the erectile tissue to fill with blood.

Impulses from the cerebrum that reinforce the sacral reflexes are not absolutely required for the culmination of the male sexual act, and the male sexual act can occasionally be accomplished by males who have suffered spinal cord injuries superior to the sacral region.

Erection, Emission, and Ejaculation

When **erection** (ē-rek′shŭn) occurs, the penis becomes enlarged and rigid. Erection is the first major component of the male sexual act. Nerve impulses from the spinal cord cause the arteries that supply blood to the erectile tissues to dilate. As a consequence, blood fills the sinusoids of the erectile tissue and compresses the veins. Because venous outflow is partially occluded, the blood pressure in the sinusoids causes inflation and rigidity of the erectile tissue. Nerve impulses that result in erection can come from parasympathetic centers (S2–S4) or sympathetic centers (T2–L1) in the spinal cord. Normally the parasympathetic centers are more important, but in cases of damage to the sacral region of the spinal cord, it is possible for erection to occur through the sympathetic system.

Parasympathetic impulses also cause the mucous glands within the penile urethra and the bulbourethral glands at the base of the penis to secrete mucus.

> ### Clinical Note
>
> Failure to achieve erections can be a major source of frustration for some men and can contribute to disharmony in relationships. The inability to achieve erections can be due to reduced testosterone secretion that can result from hypothalamic, pituitary, or testicular complications. In other cases the inability to achieve erections can be due to defective stimulation of the erectile tissue by nerve fibers or reduced response of the blood vessels to neural stimulation. Erection can be achieved in some people by oral medication or by the injection of specific drugs into the base of the penis, which functions to increase blood flow into the sinusoids of the erectile tissue of the penis, resulting in erection for many minutes.

Emission (ē-mish′ŭn) is the accumulation of sperm cells and secretions of the prostate gland and seminal vesicles in the urethra. Emission is controlled by sympathetic centers (T12–L1) in the spinal cord, which are stimulated as the level of sexual tension increases. Efferent sympathetic impulses cause peristaltic contractions of the reproductive ducts and stimulate the seminal vesicles and the prostate gland to release their secretions. Consequently, semen accumulates in the prostatic urethra, producing afferent impulses that pass through the pudendal nerves to the spinal cord. Integration

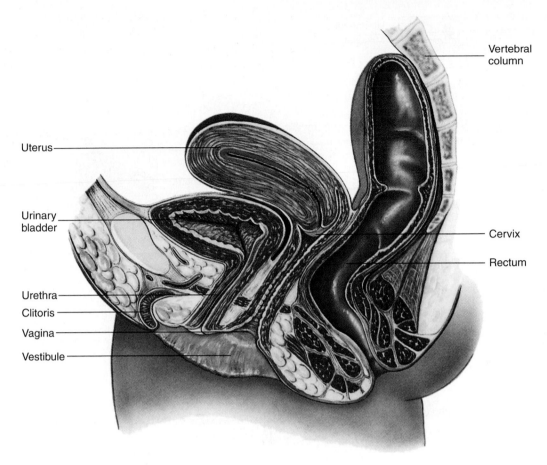

Figure 28.8 Sagittal Section of the Female Pelvis. ✗

Uterus

Urinary bladder

Urethra

Clitoris

Vagina

Vestibule

Vertebral column

Cervix

Rectum

of these impulses results in both sympathetic and somatic motor output. Sympathetic impulses cause constriction of the internal sphincter of the urinary bladder so that semen and urine are not mixed. Somatic motor impulses are sent to the skeletal muscles of the urogenital diaphragm and the base of the penis, causing several rhythmic contractions that force the semen out of the urethra. The movement of semen out of the urethra is called **ejaculation** (ē-jak-yū-lā′shŭn). In addition, there is an increase in muscle tension throughout the body.

Female Reproductive System

The female reproductive organs consist of the ovaries, uterine tubes, uterus, vagina, external genital organs, and mammary glands. The internal reproductive organs of the female (figures 28.8 and 28.9) are within the pelvis between the urinary bladder and the rectum. The uterus and the vagina are in the midline, with the ovaries to each side of the uterus. The internal reproductive organs are held in place within the pelvis by a group of ligaments. The most conspicuous is the **broad ligament,** an extension of the peritoneum that spreads out on both sides of the uterus and to which the ovaries and uterine tubes are attached.

Ovaries

The two **ovaries** (ō-var′ēz) are small organs about 2–3.5 cm long and 1–1.5 cm wide (see figure 28.9). Each is attached to the posterior surface of the broad ligament by a peritoneal fold called the **mesovarium** (mez′ō-vā′rē-ŭm, meaning mesentery of the ovary). Two other ligaments are associated with the ovary: the **suspensory ligament,** which extends from the mesovarium to the body wall, and the **ovarian ligament,** which attaches the ovary to the superior margin of the uterus. The ovarian arteries, veins, and nerves traverse the suspensory ligament and enter the ovary through the mesovarium.

Ovarian Histology

The peritoneum covering the surface of the ovary is called the **ovarian,** or **germinal, epithelium** because it was once thought to produce oocytes. Immediately below the epithelium a layer of dense, fibrous connective tissue, the **tunica albuginea** (al-byū-jin′e-ă), surrounds the ovary. The ovary itself consists of a dense outer part called the **cortex** and a looser inner part called the **medulla** (figure 28.10). Blood vessels, lymph vessels, and nerves from the mesovarium enter the medulla. Numerous small vesicles called ovarian follicles, each of which contains an **oocyte** (ō′ō-sīt), are distributed throughout the cortex.

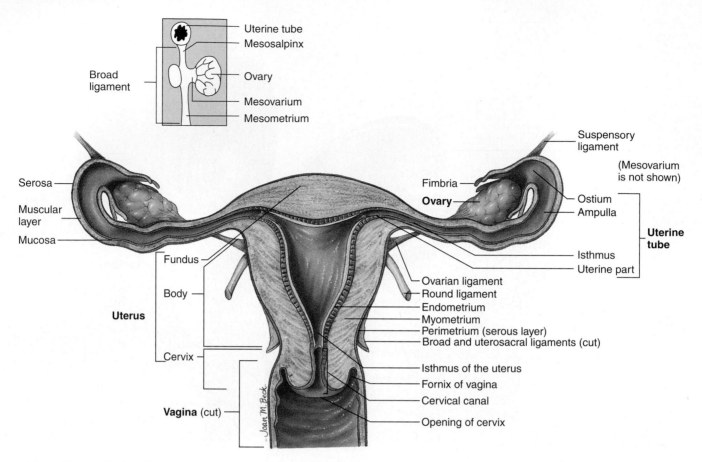

Figure 28.9   Uterus, Vagina, Uterine Tubes, Ovaries, and Supporting Ligaments

The uterus and uterine tubes are cut in section, and the vagina is cut to show the internal anatomy.

Follicle and Oocyte Development

Oogenesis (ō-ō-jen'ĕ-sis) is the production of a secondary oocyte within the ovaries. By the fourth month of prenatal life, the ovaries can contain 5 million **oogonia** (ō-ō-gō'nē-ă), the cells from which oocytes develop. By the time of birth many of the oogonia have degenerated, and those remaining have begun meiosis. Meiosis stops, however, during the first meiotic division at a stage called prophase I (figure 28.11). The cell at this stage is called a **primary oocyte,** and at birth there are about 2 million of them. The primary oocyte is surrounded by a single layer of flat cells called **granulosa** (gran-yū-lō'să) **cells,** and the structure is called a **primordial follicle.** From birth to puberty the number of primordial follicles declines to around 300,000–400,000; of these only about 400 continue oogenesis and are released from the ovary. At puberty, the cyclical secretion of FSH stimulates the further development of a small number of primordial follicles. The primordial follicle is converted to a **primary follicle** when the oocyte enlarges and the single layer of granulosa cells first become enlarged and cuboidal. Subsequently several layers of granulosa cells form, and a layer of clear material is deposited around the primary oocyte called the **zona pellucida** (zō'nă pe-lū'si-dă).

Some of the primary follicles continue development and become **secondary follicles.** The granulosa cells multiply and form an increasing number of layers around the oocyte.

Irregular small spaces called **vesicles,** which are fluid-filled, form among the granulosa cells. The vesicles ultimately fuse to form a single chamber called the **antrum** (an'trŭm). As the secondary follicle enlarges, surrounding cells are molded around it to form the **theca** (thē'kă), or **capsule.** Two layers of thecae can be recognized around the secondary follicle: the vascular **theca interna** and the fibrous **theca externa** (see figure 28.10).

The secondary follicle continues to enlarge, and, when the fluid-filled spaces fuse to form a single fluid-filled antrum, the follicle is called the **mature,** or **graafian** (graf'ē-ăn), **follicle.** The antrum progressively increases in size and fills with additional fluid, and the follicle forms a lump on the surface of the ovary after reaching its maximum size (see figure 28.10).

As the antrum forms, it is filled with fluid produced by the granulosa cells. The oocyte is pushed off to one side of the follicle and lies in a mass of follicular cells called the **cumulus mass,** or **cumulus oophorus** (kyū'myū-lŭs ō-of'ōr-ŭs) (see figure 28.10). The innermost cells of this mass resemble a crown radiating from the oocyte and are thus called the **corona radiata.**

Usually only one graafian follicle reaches the most advanced stages of development and is ovulated. The other follicles degenerate. In a graafian follicle, just before ovulation, the primary oocyte completes the first meiotic division to produce a **secondary oocyte** and a **polar body** (figure 28.12).

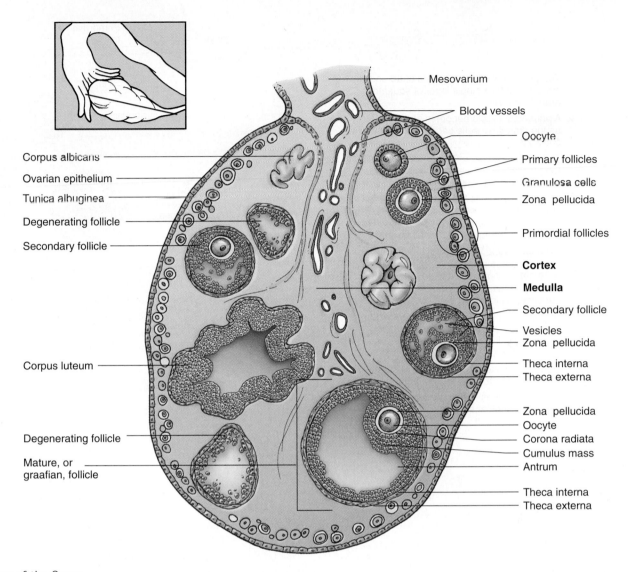

Corpus albicans
Ovarian epithelium
Tunica albuginea
Degenerating follicle
Secondary follicle

Mesovarium
Blood vessels
Oocyte
Primary follicles
Granulosa cells
Zona pellucida

Primordial follicles

Cortex
Medulla
Secondary follicle
Vesicles
Zona pellucida
Theca interna
Theca externa

Corpus luteum

Zona pellucida
Oocyte
Corona radiata
Cumulus mass
Antrum

Degenerating follicle

Mature, or
graafian, follicle

Theca interna
Theca externa

Figure 28.10   Histology of the Ovary

The ovary is sectioned to illustrate its internal structure (inset shows plane of section). Ovarian follicles from each major stage of development are present.

Division of the cytoplasm is unequal, and most of it is given to the secondary oocyte, whereas the polar body receives very little. The secondary oocyte begins the second meiotic division, which stops in metaphase II.

Ovulation

As the mature follicle continues to swell, it can be seen on the surface of the ovary as a tight, translucent blister. The follicular cells secrete a thinner fluid than previously and at an increased rate so that the follicle swells more rapidly than can be accommodated by follicular growth. As a result, the granulosa cells and theca become very thin over the area exposed to the ovarian surface.

The mature follicle expands and ruptures, forcing a small amount of blood and follicular fluid out of the vesicle. Shortly after this initial burst of fluid, the secondary oocyte, surrounded by the cumulus mass and the zona pellucida, escapes from the follicle. The release of the secondary oocyte is called **ovulation** (ov′yū-lā′shun).

During ovulation, development of the secondary oocyte has stopped at metaphase II. If sperm cell penetration does not occur, the secondary oocyte never completes this second division and simply degenerates and passes out of the system. Continuation of the second meiotic division is triggered by **fertilization,** the entry of a sperm cell into the secondary oocyte. Once the sperm cell penetrates the secondary oocyte, the second meiotic division is completed, and a second polar body is formed. The fertilized oocyte is now called a **zygote** (zī′gōt; see figure 28.12).

Fate of the Follicle

After ovulation, the follicle still has an important function. It becomes transformed into a glandular structure called the **corpus luteum** (kōr′pus lū′tē-ŭm, meaning yellow body), which has a convoluted appearance as a result of its collapse after ovulation (see figure 28.11). The granulosa cells and the theca interna, now called **luteal cells,** enlarge and begin to secrete hormones—progesterone and smaller amounts of estrogen.

1. The primordial follicle consists of an oocyte surrounded by a single layer of squamous granulosa cells.

2. A primordial follicle becomes a primary follicle as the granulosa cells become enlarged and cuboidal.

3. The primary follicle enlarges. Granulosa cells form more than one layer of cells. The zona pellucida forms around the oocyte.

4. A secondary follicle forms when fluid-filled vesicles (spaces) develop among the granulosa cells and a well developed theca becomes apparent around the granulosa cells.

5. A mature follicle forms when the fluid-filled vesicles form a single antrum. When a follicle becomes fully mature, it is enlarged to its maximum size, a large antrum is present, and the oocyte is located in the cumulus mass.

6. During ovulation the oocyte is released from the follicle, along with some surrounding granulosa cells of the cumulus mass called the corona radiata.

7. Following ovulation, the granulosa cells divide rapidly and enlarge to form the corpus luteum.

8. When the corpus luteum degenerates, it forms the corpus albicans.

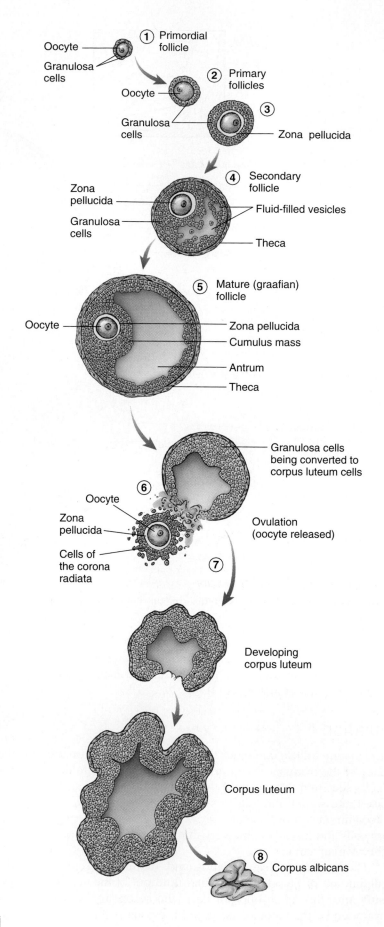

Figure 28.11   Maturation of the Follicle and Oocyte

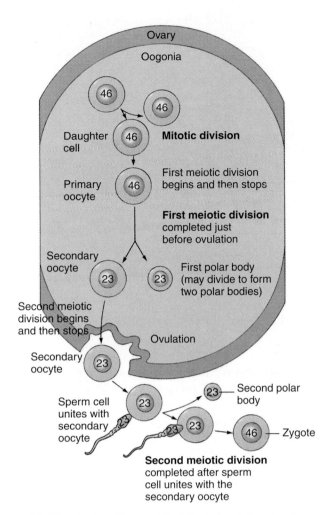

Figure 28.12 Maturation and Fertilization of the Oocyte

The primary oocyte undergoes meiosis and gives off the first polar body to become a secondary oocyte just before ovulation. Sperm cell penetration initiates the completion of the second meiotic division and the expulsion of a second polar body. The nuclei of the oocyte and the sperm cell unite. Fertilization results in the formation of a zygote. 🏃 ▭

If pregnancy occurs, the corpus luteum enlarges and remains throughout pregnancy as the **corpus luteum of pregnancy.** If pregnancy does not occur, the corpus luteum lasts for about 10–12 days and then begins to degenerate. The connective tissue cells become enlarged and clear, giving the whole structure a whitish color; it is therefore called the **corpus albicans** (al'bĭ-kanz, meaning white body). The corpus albicans continues to shrink and eventually disappears after several months or even years.

Uterine Tubes

There are two **uterine tubes,** also called **fallopian** (fa-lō'pē-an) **tubes,** or **oviducts** (ō'vi-dŭkts), one on each side of the uterus and each associated with one ovary (see figure 28.9). Each tube is located along the superior margin of the broad ligament. The part of the broad ligament most directly associated

with the tube is called the **mesosalpinx** (mez'ō-sal'pinks, meaning mesothelium of the trumpet-shaped uterine tube).

The uterine tube opens directly into the peritoneal cavity to receive the oocyte and expands to form the **infundibulum** (in-fŭn-dib'yū-lŭm, meaning funnel). The opening of the infundibulum, the **ostium** (os'tē-ŭm), is surrounded by long, thin processes called **fimbriae** (fim'brē-ē, meaning fringe). The inner surfaces of the fimbriae consist of a ciliated mucous membrane.

The part of the uterine tube that is nearest to the infundibulum is called the **ampulla.** It is the widest and longest part of the tube and accounts for about 7.5–8 cm of the total 10 cm length of the tube. The part of the tube nearest the uterus, the **isthmus,** is much narrower and has thinner walls than does the ampulla. The **uterine,** or **intramural, part** of the tube traverses the uterine wall and ends in a very small uterine opening.

The wall of each uterine tube consists of three layers. The outer **serosa** is formed by the peritoneum, the middle **muscular layer** consists of longitudinal and circular smooth muscle fibers, and the inner **mucosa** consists of a mucous membrane of simple ciliated columnar epithelium (see figure 28.9). The mucosa is arranged into numerous longitudinal folds.

The mucosa of the uterine tubes provides nutrients for the oocyte, or developing embryonic mass (see chapter 29) if fertilization has occurred, as long as it is traversing the uterine tubes. The ciliated epithelium helps move the small amount of fluid and the oocyte through the uterine tubes.

Uterus

The **uterus** (yū'ter-ŭs) is the size and shape of a medium-sized pear and is about 7.5 cm long and 5 cm wide (see figures 28.8 and 28.9). It is slightly flattened anteroposteriorly and is oriented in the pelvic cavity with the larger, rounded part, the **fundus** (fŭn'dŭs, meaning bottom of a rounded flask), directed superiorly and the narrower part, the **cervix** (ser'viks, meaning neck), directed inferiorly. The main part of the uterus, the **body,** is between the fundus and the cervix. A slight constriction called the **isthmus** marks the junction of the cervix and the body. Internally, the uterine cavity continues as the **cervical canal,** which opens through the **ostium** into the vagina.

> **Clinical Note**
>
> **Cancer of the cervix** is common in females and fortunately can be detected and treated. Early in the development of cervical cancer, the cells of the cervix change in a characteristic way. This change can be observed by taking a cell sample and examining the cells microscopically. The most common technique is to obtain a **Papanicolaou (Pap) smear,** which has a reliability of 90% for detecting cervical cancer.

The major ligaments holding the uterus in place are the **broad ligament,** the **round ligaments,** and the **uterosacral ligaments** (see figure 28.9). The round ligaments extend from the uterus through the inguinal canals to the labia majora of the external genitalia, and the uterosacral ligaments attach the uterus to the sacrum. Normally the uterus is anteverted, meaning that the body of the uterus is tipped slightly anteriorly. In some women the uterus can be retroverted, or tipped posteriorly. In addition to the ligaments, much support is provided inferiorly to the uterus by the skeletal muscles of the pelvic floor. If these muscles are weakened (e.g., in childbirth), the uterus can extend inferiorly into the vagina, a condition called a **prolapsed uterus.**

The uterine wall is composed of three layers: serous, muscular, and mucous (see figure 28.9). The **perimetrium** (pĕr′i-mē′trē-ŭm), or **serous layer,** of the uterus is the peritoneum that covers the uterus. The next layer, just deep to the perimetrium, is the **myometrium** (mī′ō-mē′trē-ŭm), or **muscular coat,** which consists of a thick layer of smooth muscle. The myometrium accounts for the bulk of the uterine wall and is the thickest layer of smooth muscle in the body. In the cervix the muscular layer contains less muscle and more dense connective tissue. The cervix is therefore more rigid and less contractile than the rest of the uterus. The innermost layer of the uterus is the **endometrium** (en′dō-mē′trē-ŭm), or **mucous membrane.** The endometrium consists of a simple columnar epithelial lining and a connective tissue, the lamina propria. Simple tubular glands are scattered about the lamina propria and open through the epithelium into the uterine cavity. The endometrium consists of two layers: a thin, deep **basal layer,** which is the deepest part of the lamina propria and is continuous with the myometrium; and a thicker, superficial **functional layer,** which consists of most of the lamina propria and the endothelium and lines the cavity itself. The functional layer is so named because it undergoes menstrual changes and sloughing during the female sex cycle.

The cervical canal is lined by columnar epithelial cells and contains **cervical mucous glands.** The mucus fills the cervical canal and acts as a barrier to substances that could pass from the vagina into the uterus. Near the time of ovulation the consistency of the mucus changes, making the passage of sperm cells from the vagina into the uterus easier.

Vagina

The **vagina** (vă-jī′nă) is a tube about 10 cm long that extends from the uterus to the outside of the body (see figure 28.9). The vagina is the female organ of copulation, functioning to receive the penis during intercourse, and it allows menstrual flow and childbirth. Longitudinal ridges called **columns** extend the length of the anterior and posterior vaginal walls, and several transverse ridges called **rugae** (rū′gē) extend between the anterior and posterior columns. The superior, domed part of the vagina, the **fornix** (fōr′niks, meaning domed), is attached to the sides of the cervix so that a part of the cervix extends into the vagina.

The wall of the vagina consists of an outer muscular layer and an inner mucous membrane. The muscular layer is smooth muscle that allows the vagina to increase in size to accommodate the penis during intercourse and to stretch greatly during delivery. The mucous membrane is moist stratified squamous

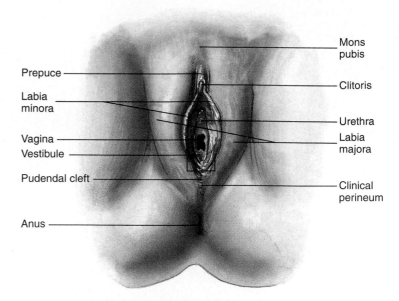

Figure 28.13   Female External Genitalia

epithelium that forms a protective surface layer. Most of the lubricating secretions produced by the female during intercourse are released by the vaginal mucous membrane.

The **vaginal opening,** or **orifice,** is covered by a thin mucous membrane called the **hymen** (hī′men). Sometimes the hymen completely closes the vaginal opening (a condition called **imperforate hymen**), and it must be removed to allow menstrual flow. More commonly, the hymen is perforated by one or several holes. The openings in the hymen are usually greatly enlarged during the first sexual intercourse. In addition, the hymen can be perforated or torn at some earlier time in a young woman's life, such as during strenuous physical exercise. Thus the absence of an intact hymen does not necessarily indicate that a woman has had sexual intercourse, as was once thought.

External Genitalia

The external female genitalia, also referred to as the **vulva** (vŭl′vă) or **pudendum** (pyū-den′dum), consist of the vestibule and its surrounding structures (figure 28.13). The **vestibule** (ves′ti-būl) is the space into which the vagina opens posteriorly and the urethra opens anteriorly. It is bordered by a pair of thin, longitudinal skin folds called the **labia** (lā′bē-ă, meaning lips) **minora** (sing., labium minus). A small erectile structure called the **clitoris** (klit′ō-ris) is located in the anterior margin of the vestibule. Anteriorly, the two labia minora unite over the clitoris to form a fold of skin called the **prepuce.**

The clitoris is usually less than 2 cm in length and consists of a shaft and a distal glans. It is well supplied with sensory receptors and functions to initiate and intensify levels of sexual tension. The clitoris contains two erectile structures, the **corpora cavernosa,** each of which expands at the base end of the clitoris to form the **crus of the clitoris** and attaches the clitoris to the coxae. The corpora cavernosa of the clitoris are

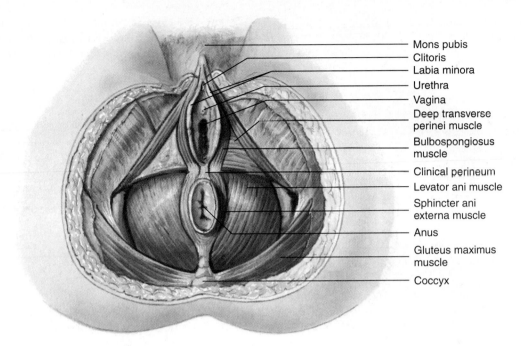

Mons pubis
Clitoris
Labia minora
Urethra
Vagina
Deep transverse
perinei muscle
Bulbospongiosus
muscle
Clinical perineum
Levator ani muscle
Sphincter ani
externa muscle
Anus
Gluteus maximus
muscle
Coccyx

Figure 28.14   Inferior
View of the Female Perineum

comparable to the corpora cavernosa of the penis, and they become engorged with blood as a result of sexual excitement. In most women this engorgement results in an increase in the diameter, but not the length, of the clitoris. With increased diameter, the clitoris makes better contact with the prepuce and surrounding tissues and is more easily stimulated.

Erectile tissue that corresponds to the corpus spongiosum of the male lies deep to and on the lateral margins of the vestibular floor on either side of the vaginal orifice. Each erectile body is called a **bulb of the vestibule.** Like other erectile tissue, it becomes engorged with blood and is more sensitive during sexual arousal. Expansion of the bulbs causes narrowing of the vaginal orifice, producing better contact of the vagina with the penis during intercourse.

On each side of the vestibule, between the vaginal opening and the labia minora, is an opening of the duct of the **greater vestibular gland.** Additional small mucous glands, the **lesser vestibular glands,** or **paraurethral glands,** are located near the clitoris and urethral opening. They produce a lubricating fluid that helps to maintain the moistness of the vestibule.

Lateral to the labia minora are two prominent, rounded folds of skin called the **labia majora** (lā′bē-ă; sing., labium majus). The prominence of the labia majora is primarily caused by the presence of subcutaneous fat within the labia. The two labia majora unite anteriorly in an elevation over the symphysis pubis called the **mons pubis** (monz pyū′bis). The lateral surfaces of the labia majora and the surface of the mons pubis are covered with coarse hair. The medial surfaces are covered with numerous sebaceous and sweat glands. The space between the labia majora is called the **pudendal** (pyū-den′dal) **cleft.** Most of the time, the labia majora are in contact with each other across the midline, closing the pudendal cleft and concealing the deeper structures within the vestibule.

Perineum

The **perineum** (figure 28.14), as in the male, is divided into two triangles by the superficial and deep transverse perineal muscles. The anterior, urogenital triangle contains the external genitalia, and the posterior, anal triangle contains the anal opening. The region between the vagina and the anus is the **clinical perineum.** The skin and muscle of this region can tear during childbirth. To prevent such tearing, an incision called an **episiotomy** (e-piz′ē-ot′ō-mē) is sometimes made in the clinical perineum. This clean, straight incision is easier to repair than a tear would be. Alternatively, allowing the perineum to stretch slowly during the delivery may prevent tearing, making an episiotomy unnecessary.

Mammary Glands

The **mammary glands** are the organs of milk production and are located within the **mammae** (mam′ē), or **breasts** (figure 28.15). The mammary glands are modified sweat glands. Externally, the breasts of both males and females have a raised **nipple** surrounded by a circular, pigmented **areola** (ă-rē′ō-lă). The areolae normally have a slightly bumpy surface caused by the presence of rudimentary mammary glands, called **areolar glands,** just below the surface. Secretions from these glands protect the nipple and the areola from chafing during nursing.

In prepubescent children the general structure of the breasts is similar, and both males and females possess a rudimentary glandular system, which consists mainly of ducts with sparse alveoli. The female breasts begin to enlarge during puberty, primarily under the influence of estrogen and progesterone. This enlargement is often accompanied by increased sensitivity or pain in the breasts. Males often experience these same sensations during early puberty, and their breasts can

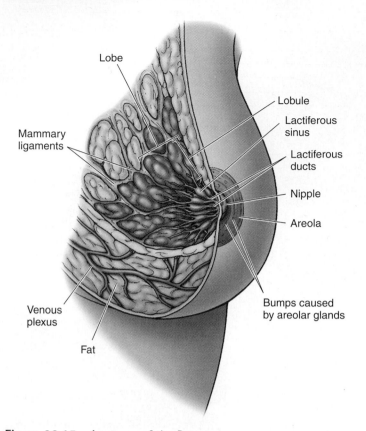

Lobe

Lobule

Lactiferous sinus

Lactiferous ducts

Mammary ligaments

Nipple

Areola

Venous plexus

Bumps caused by areolar glands

Fat

Figure 28.15   Anatomy of the Breast

The section illustrates the blood supply, the mammary glands, and the duct system.

even develop slight swellings; however, these symptoms usually disappear fairly quickly. On rare occasions the breasts of a male become enlarged, a condition called **gynecomastia** (gī′nĕ-kō-mas′tē-ă).

Each adult female mammary gland usually consists of 15–20 glandular **lobes** covered by a considerable amount of adipose tissue. It is primarily this superficial fat that gives the breast its form. The lobes of each mammary gland form a conical mass, with the nipple located at the apex. Each lobe possesses a single **lactiferous** (lak-tif′er-ŭs, meaning milk-producing) **duct,** which opens independently of other lactiferous ducts on the surface of the nipple. Just deep to the surface, each lactiferous duct enlarges to form a small, spindle-shaped **lactiferous sinus,** which accumulates milk during milk production. The lactiferous duct supplying a lobe subdivides to form smaller ducts, each of which supplies a **lobule.** Within a lobule the ducts branch and become even smaller. In the milk-producing breast the ends of these small ducts expand to form secretory sacs called **alveoli.**

The mammae are supported and held in place by a group of **mammary,** or **Cooper's, ligaments.** These ligaments extend from the fascia over the pectoralis major muscles to the skin over the mammary glands and prevent the mammary glands from excessive sagging. In older adults, however, these ligaments weaken and elongate, allowing the breasts to sag to a greater extent than when the person was younger.

The nipples are very sensitive to tactile stimulation and contain smooth muscle that can contract, causing the nipple to become erect in response to stimulation. These smooth muscle fibers respond, like other erectile tissues, during sexual arousal.

Clinical Note

Cancer of the breast is a serious, often fatal disease in women. The use of mammography and regular self-examination of the breast can help in early detection of breast cancer and effective treatment.

Physiology of Female Reproduction

As in the male, female reproduction is under the control of hormonal and nervous regulation. Development of the female reproductive organs and normal function depend on the relative levels of a number of hormones in the body.

Puberty

Puberty in females is marked by the first episode of menstrual bleeding, which is called **menarche** (me-nar′kē). During puberty the vagina, uterus, uterine tubes, and external genitalia begin to enlarge. Fat is deposited in the breasts and around the hips, causing them to enlarge and assume an adult form. The ducts of the breasts develop, pubic and axillary hair grows, and the voice changes, although this is a more subtle change than in males. Development of the sexual drive is also associated with puberty.

The changes associated with puberty are caused primarily by the elevated rate of estrogen and progesterone secretion by the ovaries. Before puberty estrogen and progesterone are secreted in very small amounts. LH and FSH levels also remain very low. The low secretory rates are due to a lack of GnRH released from the hypothalamus. At puberty not only are GnRH, LH, and FSH secreted in greater quantities than before puberty, but the adult pattern is established in which a cyclic pattern of FSH and LH secretion occurs. The cyclic secretion of LH and FSH, ovulation, the monthly changes in secretion of estrogen and progesterone, and the resultant changes in the uterus characterize the menstrual cycle.

Menstrual Cycle

The term **menstrual** (men′strū-ăl) **cycle** technically refers to the cyclic changes that occur in sexually mature, nonpregnant females and culminate in menses. Typically the menstrual cycle is about 28 days long, although it can be as short as 18 days in some women and as long as 40 days in others (figure 28.16 and table 28.2). **Menses** (men′sēz) is derived from a Latin word meaning month and is a period of mild hemorrhage during which the uterine epithelium is sloughed

Table 28.2 The Menstrual Cycle

Menses	Proliferative Phase	Ovulation	Secretory Phase
Pituitary Hormones			
LH levels are low and remain low; FSH increases somewhat.	LH and FSH levels begin to increase rapidly in response to increases in estrogen near the end of the proliferative phase.	Ovulation is triggered by increasing LH levels. Ovulation generally occurs after LH levels have reached their peak. FSH reaches a peak about the time of ovulation and initiates development of follicles that may complete maturation during a later cycle.	LH and FSH levels decline to low levels following ovulation and remain at low levels during the secretory phase in response to increases in estrogen and progesterone.
Developing Follicles			
FSH secreted during menses causes several follicles to begin to enlarge.	Several follicles continue to enlarge. As they enlarge they begin to secrete estrogen. In addition, many of them degenerate. Only one of the follicles becomes a mature follicle that is capable of ovulating by the end of the proliferative phase.	Normally a single follicle reaches maturity and ovulates in response to LH. The ovum and some cumulus cells are released during ovulation. The remaining granulosa cells become luteal cells and form the corpus luteum.	Following ovulation, the granulosa cells of the ovulated follicle change to luteal cells and begin secreting large amounts of progesterone and some estrogen.
Estrogen			
Very little estrogen is secreted by the ovarian follicles.	Near the end of the proliferative phase the enlarging follicles begin to secrete increasing amounts of estrogen. The estrogen causes the pituitary gland to secrete increasing quantities of LH and smaller quantities of FSH. The positive-feedback relationship between estrogen and LH results in rapidly increasing LH and estrogen levels several days prior to ovulation. The rapid increase in LH triggers ovulation.	Estrogen, secreted by developing follicles, reaches a peak at ovulation and falls as the cells of the follicle are disrupted during ovulation.	Following ovulation, estrogen levels decline. After the luteal cells have been established, smaller amounts of estrogen are secreted by the corpus luteum.
Progesterone			
Very little progesterone is secreted by the ovarian follicles.	Progesterone levels are low during the proliferative phase.	Progesterone levels are low and do not increase until granulosa cells become luteinized after ovulation.	Following ovulation, progesterone levels increase due to the secretion of progesterone by the corpus luteum. The progesterone levels remain high throughout the secretory phase and fall rapidly just before menses unless pregnancy occurs.
Uterine Endometrium			
The endometrium of the uterus undergoes necrosis and is eliminated in the menstrual fluid during menses. The necrosis is a result of decreasing progesterone concentrations near the end of the proliferative phase.	In response to estrogen, endometrial cells of the uterus undergo rapid cell division and proliferate rapidly. In addition, the number of progesterone receptors in the endometrial cells increases in response to estrogen.	Ovulation occurs over a short time, and it signals the end of the proliferative phase as estrogen levels decline and the onset of the secretory phase as the progesterone levels begin to increase.	Progesterone causes the endometrial cells to enlarge and secrete a small amount of fluid. The endometrium continues to thicken throughout the secretory phase. Near the end of the secretory phase declining progesterone levels allow the spiral arteries of the endometrium to constrict, causing ischemia, and the endometrium becomes necrotic unless pregnancy occurs.

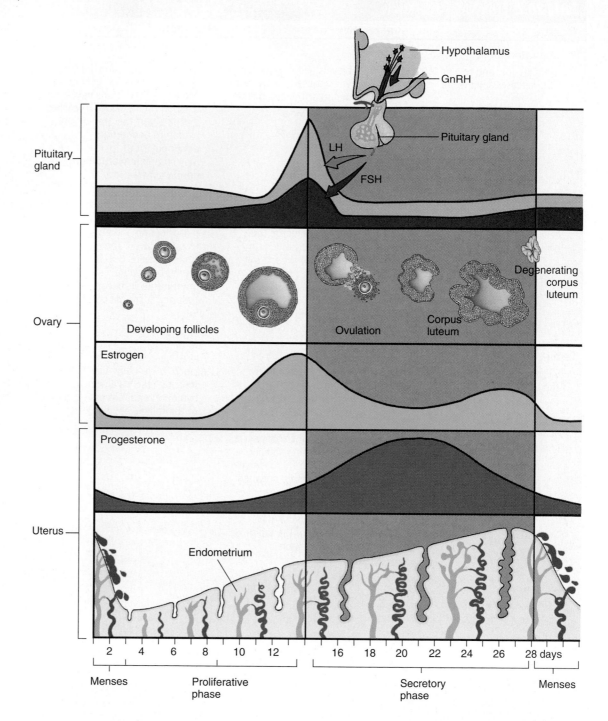

Figure 28.16 The Menstrual Cycle

The various lines depict the changes in blood hormone levels, the development of the follicles, and the changes in the endometrium during the cycle. (Figure continued on page 947.) (Figure continued on page 947.)

and expelled from the uterus. Although the term menstrual cycle refers specifically to changes that occur in the uterus, several other cyclic changes are associated with it, and the term is often used to refer to all of the cyclic events that occur in the female reproductive system. These changes include cyclic changes in hormone secretion, in the ovary, and in the uterus.

The first day of menses is day 1 of the menstrual cycle, and menses typically lasts 4–5 days. Ovulation occurs on about day 14 of a 28-day menstrual cycle, although the timing of ovulation varies from individual to individual and varies within a single individual from one menstrual cycle to the next.

The time between ovulation and the next menses is typically 14 days, and the time between the first day of menses and the next ovulation is more variable. The time between the ending of menses and ovulation is called the **follicular phase** because of the rapid development of ovarian follicles, or the **proliferative phase** because of the rapid proliferation of the uterine mucosa. The period after ovulation and before the next menses is called the **luteal phase,** because of the existence of the corpus luteum, or the **secretory phase,** because of maturation of and secretion by uterine glands. About 28 days after another period of menses begins, a new menstrual cycle is initiated.

Although several follicles begin to mature during each cycle, normally only one is ovulated. The remaining follicles degenerate. Larger and more mature follicles appear to secrete estrogen and other substances that have an inhibitory effect on other less mature follicles.

Early in the menstrual cycle, the release of GnRH from the hypothalamus increases, and the sensitivity of anterior pituitary to GnRH increases. These changes stimulate the production and release of a small amount of FSH and LH by the anterior pituitary.

FSH and LH stimulate follicular growth and maturation and an increase in estradiol secretion by the developing follicles. FSH exerts its main effect on the granulosa cells, and LH exerts its initial effect on the cells of the theca interna and later on the granulosa cells.

LH stimulates the theca interna cells to produce androgens, which diffuse from the theca interna cells to the granulosa cells. FSH stimulates the granulosa cells to convert the androgens to estrogen. In addition, FSH gradually increases LH receptors in the granulosa cells, and estrogen produced by the granulosa cells increases LH receptors in the theca interna cells. After LH receptors in the granulosa cells have increased, LH stimulates the theca interna cells to produce some progesterone, which diffuses from the granulosa cells to the theca interna cells, where it is converted to androgens. Thus, the production of androgens by the theca interna cells increases, and the conversion of androgens to estrogen by the granulosa cells is responsible for a gradual increase in estrogen secretion by the granulosa cells throughout the follicular phase, even though there is only a small increase in LH secretion. FSH levels actually decrease during the follicular phase because inhibin is produced by developing follicles, and the inhibin has a negative-feedback effect on FSH secretion.

As estrogen levels begin to increase in the follicular phase, they have a negative-feedback effect on the secretion of LH and FSH by the anterior pituitary. Late in the follicular phase, the higher estrogen levels begin to have a positive-feedback effect on LH and FSH release from the anterior pituitary. In response to the positive-feedback effect of estrogen on the anterior pituitary, LH and FSH secretion increase rapidly and in large amounts just before ovulation (figure 28.17). The increase in blood levels of both LH and FSH is called the **LH surge,** and the increase in FSH is called the **FSH surge.** The LH surge occurs several hours earlier and to a greater degree than the FSH surge, and the LH surge can last up to 24 h.

The LH surge initiates ovulation and causes the ovulated follicle to become the corpus luteum. FSH can make the follicle more sensitive to the influence of LH by stimulating the synthesis of additional LH receptors in the follicles and by stimulating the development of follicles that may ovulate during later ovarian cycles.

The LH surge causes the primary oocyte to complete the first meiotic division just before or during the process of ovulation. Also, the LH surge triggers several events that are very much like inflammation in the mature follicle and that result in ovulation. The follicle becomes edematous, proteolytic

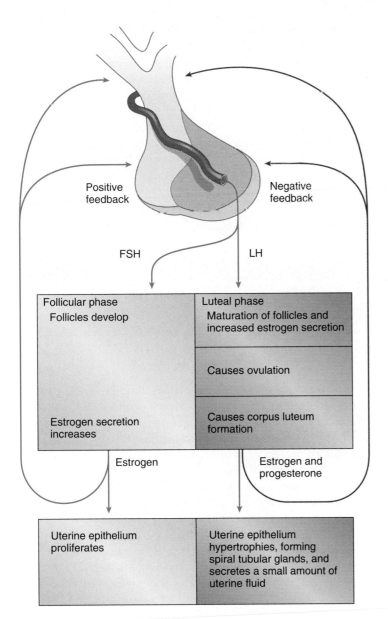

Figure 28.17   Regulation of Hormone Secretion during the Menstrual Cycle

The regulation of hormone secretion from the anterior pituitary and the ovary during the menstrual cycle before and after ovulation is depicted.

Ovarian Cycle

The **ovarian cycle** specifically refers to the events that occur in a regular fashion in the ovaries of sexually mature, non-pregnant women during the menstrual cycle. These events are controlled by hormones released from the hypothalamus and anterior pituitary. FSH from the anterior pituitary is primarily responsible for initiating the development of primary follicles, and as many as 25 begin to mature during each menstrual cycle. The follicles that start to develop in response to FSH may not ovulate during the same menstrual cycle in which they begin to mature, but they are ovulated one or two cycles later.

enzymes cause the degeneration of the ovarian tissue around the follicle, the follicle ruptures, and the oocyte and some surrounding cells are slowly extruded from the ovary.

Shortly after ovulation, production of estrogen by the follicle decreases, and production of progesterone increases as granulosa cells are converted to corpus luteum cells. After the corpus luteum forms, both estrogen and progesterone levels become much higher than before ovulation. The increased estrogen and progesterone have a negative-feedback effect on GnRH release from the hypothalamus. As a result, LH and FSH release from the anterior pituitary decreases. Estrogen and progesterone cause down-regulation of GnRH receptors in the anterior pituitary, and the anterior pituitary cells become less sensitive to GnRH. Because of the decreased secretion of GnRH and decreased sensitivity of the anterior pituitary to GnRH, the rate of LH and FSH secretion declines to very low levels after ovulation (see figure 28.16 and 28.17).

If fertilization of the ovulated oocyte does take place, the developing embryonic mass begins to secrete the LH-like substance, human chorionic gonadotropin (HCG), which keeps the corpus luteum from degenerating. As a result, blood levels of estrogen and progesterone do not decrease, and menses does not occur. If fertilization does not occur, HCG is not produced. The cells of the corpus luteum begin to atrophy after day 25 or 26, and the blood levels of estrogen and progesterone decrease rapidly, which results in menses.

<table>
<tr><td>3</td><td>P R E D I C T</td></tr>
</table>

Predict the effect on the ovarian cycle of administering a relatively large amount of estrogen and progesterone just before the preovulatory LH surge. Also predict the consequences of continually administering high concentrations of GnRH.

✔ *Answer in Appendix F*

Uterine Cycle

The term **uterine cycle** refers to changes that occur primarily in the endometrium of the uterus during the menstrual cycle (see figure 28.16). Other, more subtle changes also occur in the vagina and other structures during the menstrual cycle. These changes are caused primarily by cyclic secretions of estrogen and progesterone.

The endometrium of the uterus begins to proliferate after menses. The epithelial cells of the basal layer rapidly divide and replace the cells of the functional layer that was sloughed during the last menses. A relatively uniform layer of low cuboidal endometrial cells is produced. It later becomes columnar and is thrown into folds to form tubular **spiral glands.** Blood vessels called **spiral arteries** project through the delicate connective tissue that separates the individual glands to supply nutrients to the endometrial cells. After ovulation the endometrium becomes thicker, and the spiral glands develop to a greater extent and begin to secrete small amounts of a fluid rich in glycogen. Approximately 7 days after ovulation, or about day 21 of the menstrual cycle, the endometrium

is prepared to receive the developing embryonic mass, if fertilization has occurred. If the developing embryonic mass arrives in the uterus too early or too late, the endometrium does not provide a suitable environment for implantation.

Estrogen causes the endometrial cells and, to a lesser degree, the myometrial cells to proliferate. It also makes the uterine tissue more sensitive to progesterone by stimulating the synthesis of progesterone receptor molecules within the uterine cells. After ovulation, progesterone from the corpus luteum binds to the progesterone receptors, resulting in cellular hypertrophy in the endometrium and myometrium and causing the endometrial cells to become secretory. Progesterone also inhibits smooth muscle contractions.

<table>
<tr><td>4</td><td>P R E D I C T</td></tr>
</table>

Predict the effect on the endometrium of elevated progesterone levels in the circulatory system before the estrogen surge that occurs after menstruation.

✔ *Answer in Appendix F*

If pregnancy does not occur by day 24 or 25, progesterone and estrogen levels decline to low levels as the corpus luteum degenerates. As a consequence, the uterine lining also begins to degenerate. The spiral arteries constrict in a rhythmic pattern for longer and longer periods as progesterone levels fall. As a result, all but the basal parts of the spiral glands become ischemic and then necrotic. As the cells become necrotic, they slough into the uterine lumen. The necrotic endometrium, mucous secretions, and a small amount of blood released from the spiral arteries make up the menstrual fluid. Decreases in progesterone levels and increases in inflammatory substances that stimulate myometrial smooth muscle cells cause uterine contractions that expel the menstrual fluid from the uterus through the cervix and into the vagina.

Female Sexual Behavior and the Female Sex Act

Sexual drive in females, like sexual drive in males, depends on hormones. Androgens are produced in the adrenal gland and other tissues such as the liver by the conversion of other steroids, such as progesterone, to androgens. Androgens and possibly estrogens affect cells in the brain, especially in the hypothalamus, and influence sexual behavior. Sexual drive is not controlled by androgens and estrogen alone, however. For example, sexual drive cannot be predictably increased simply by injecting these hormones into normal women or men. Psychologic factors also affect sexual behavior. For example, after removal of the ovaries or menopause, many women report having an increased sex drive because they no longer fear pregnancy.

The neural pathways, both afferent and efferent, involved in controlling sexual responses are the same for males and females. Afferent impulses are transported from the genitals to the sacral region of the spinal cord, where reflexes that govern sexual responses are integrated. Ascending pathways, primarily the spinothalamic tracts (see chapter 13), transport sensory information through the spinal cord to the brain, and descending pathways transport impulses back to the sacrum. As a result, the sacral reflexes are modulated by cerebral influences. Motor impulses are transported from the spinal cord to the reproductive organs by both parasympathetic and sympathetic fibers and to skeletal muscles by the somatic motor system.

During sexual excitement erectile tissue within the clitoris and around the vaginal opening becomes engorged with blood as a result of parasympathetic stimulation. The nipples of the breast often become erect as well. The mucous glands within the vestibule, especially the vestibular glands, secrete small amounts of mucus. Large amounts of mucuslike fluid are also extruded into the vagina through its wall, although no well-developed mucous glands are within the vaginal wall. These secretions provide lubrication that allows for easy entry of the penis into the vagina and easy movement of the penis during intercourse. The tactile stimulation of the female's genitals that occurs during sexual intercourse along with psychologic stimuli normally trigger an orgasm. The vaginal, uterine, and perineal muscles contract rhythmically, and muscle tension increases throughout much of the body. After the sexual act there is a period of resolution characterized by an overall sense of satisfaction and relaxation. The female can be receptive to further stimulation and can experience successive orgasms. Although orgasm is a pleasurable component of sexual intercourse, it is not necessary for females to experience an orgasm for fertilization to occur.

Clinical Note

Menstrual cramps are the result of strong myometrial contractions that occur before and during menstruation. The cramps can result from excessive prostaglandin secretion. Sloughing of the endometrium of the uterus results in an inflammation in the endometrial layer of the uterus, and prostaglandins are produced as part of the inflammatory process. Sloughing of the endometrium is inhibited by progesterone but stimulated by estrogen. In some women, menstrual cramps are extremely uncomfortable. Many women can alleviate painful menstruation by taking anti-inflammatory drugs, such as ibuprofen or other aspirinlike drugs, that inhibit prostaglandin biosynthesis just before the onset of menstruation. These treatments, however, are not effective in treating all painful menstruation, especially when the causes of pain are more complicated than inflammation associated with normal menstruation such as from tumors of the myometrium or obstruction of the cervical canal.

A topic of current research emphasis concerns a phenomenon called the **premenstrual syndrome (PMS).** Some women suffer from severe changes in mood that often result in aggression and other socially unacceptable behaviors just before menses. It has been hypothesized that hormonal changes associated with the menstrual cycle trigger these mood changes. Although some women appear to have been successfully treated with steroid hormones, this treatment does not appear to be effective for everyone with the condition. Similarly,

reducing caffeine, alcohol, sugar, and animal fat consumption helps some people. It is unclear how many women are affected by PMS, and it is a controversial condition. The definition of the premenstrual period is not well established, the symptoms of the condition are not easily monitored, and its precise cause and physiologic mechanisms are unknown. In addition, it is unclear whether all women diagnosed as having PMS are suffering from the same condition. Additional research is needed to resolve these uncertainties.

The absence of a menstrual cycle is called **amenorrhea** (ă-men-ō-rē′ă). If the pituitary gland does not function properly because of abnormal development, the woman will not begin to menstruate at puberty. This condition is called **primary amenorrhea.** In contrast, if a woman has had normal menstrual cycles and later stops menstruating, the condition is called **secondary amenorrhea.** One cause of secondary amenorrhea is anorexia, a condition in which lack of food causes the hypothalamus of the brain to decrease GnRH secretion to levels so low that the menstrual cycle cannot occur. Female athletes or ballet dancers who have rigorous training schedules have a high frequency of secondary amenorrhea. The physical stress that can be coupled with an inadequate food intake also results in very low GnRH secretion. Increased food intake for anorexic women and reduced training generally restores normal hormone secretion and normal menstrual cycles.

Secondary amenorrhea can also be the result of pituitary tumors, which decrease FSH and LH secretion, or from a lack of GnRH secretion from the hypothalamus. Head trauma and tumors that affect the hypothalamus can result in lack of GnRH secretion.

Secondary amenorrhea can result from a lack of normal hormone secretion from the ovaries, which can be caused by autoimmune diseases that attack the ovary or by polycystic ovarian disease, in which the cysts in the ovary produce large amounts of androgen that are converted to estrogens by other tissues in the body. The increased estrogen prevents the normal cycle of FSH and LH secretion required for ovulation to occur. Other hormone-secreting tumors of the ovary can also disrupt the normal menstrual cycle and result in amenorrhea.

Female Fertility and Pregnancy

After the sperm cells are ejaculated into the vagina during sexual intercourse, they are transported through the cervix, the body of the uterus, and the uterine tubes to the ampulla (figure 28.18). The forces responsible for the movement of sperm cells through the female reproductive tract involve the swimming ability of the sperm cells and, possibly, the muscular contraction of the uterus and the uterine tubes. During sexual intercourse, oxytocin is released from the posterior pituitary of the female, and the semen introduced into the vagina contains prostaglandins. Both of these hormones stimulate smooth muscle contractions in the uterus and the uterine tubes.

While passing through the vagina, uterus, and the uterine tubes, the sperm cells undergo **capacitation** (kă-pas′i-tā′shŭn), a process that enables them to release acrosomal enzymes that allow penetration of the cervical mucus, cumulus mass cells, and the oocyte cell membrane.

The oocyte can be fertilized for up to 24 h after ovulation, and some sperm cells remain viable in the female reproductive tract for up to 72 h, although most of them have degenerated

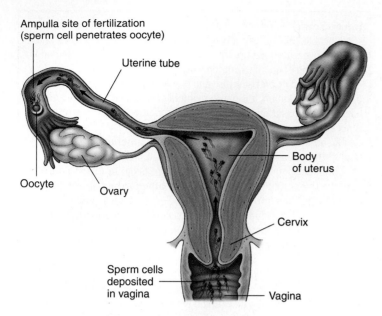

Ampulla site of fertilization
(sperm cell penetrates oocyte)

Uterine tube

Oocyte

Ovary

Body
of uterus

Cervix

Sperm cells
deposited
in vagina

Vagina

Figure 28.18   Sperm Cell Movement

Sperm cells are deposited in the vagina as part of the semen when the
male ejaculates. Sperm cells pass through the cervix, the body of the
uterus, and the uterine tube. Fertilization normally occurs when the
oocyte is in the upper third of the uterine tube (the ampulla).

after 24 h. For fertilization to occur successfully, sexual intercourse must therefore occur approximately between 3 days before and 1 day after ovulation.

One sperm cell enters the secondary oocyte, and fertilization occurs (see chapter 29). For the next several days a sequence of cell divisions occurs while the developing cells pass
through the uterine tube to the uterus. By 7 or 8 days after
ovulation, which is day 21 or 22 of the average menstrual cycle, the endometrium of the uterus is prepared for implantation. Estrogen and progesterone have caused it to reach its
maximum thickness and secretory activity, and the developing embryonic mass begins to implant. The outer layer of the
developing embryonic mass, the **trophoblast** (trof'ō-blast),
secretes proteolytic enzymes that digest the cells of the thickened endometrium (see chapter 29), and the mass digests its
way into the endometrium.

> ### Clinical Note
>
> An ectopic pregnancy results if implantation occurs anywhere other
> than in the uterine cavity. The most common site of ectopic
> pregnancy is the uterine tube. Implantation in the uterine tube
> eventually is fatal to the fetus and can cause the tube to rupture. In
> some cases implantation can occur in the mesenteries of the
> abdominal cavity, and the fetus can develop normally but must be
> delivered by caesarean section.

The trophoblast secretes HCG, which is transported in
the blood to the ovary and causes the corpus luteum to remain functional. As a consequence, both estrogen and progesterone levels continue to increase rather than decrease.
The secretion of HCG increases rapidly and reaches a peak
about 8–9 weeks after fertilization. Subsequently, HCG levels
in the circulatory system decline to a lower level by 16 weeks
and remain at a relatively constant level throughout the remainder of pregnancy. Detection of HCG excreted in the urine
is the basis for some pregnancy tests.

The estrogen and progesterone secreted by the corpus
luteum are essential for the maintenance of pregnancy. After
the **placenta** (plă-sen'tă) forms from the trophoblast and
uterine tissue, however, it also begins to secrete estrogen
and progesterone. By the time the first 3 months of pregnancy are complete, the corpus luteum is no longer needed
to maintain pregnancy; the placenta has become an endocrine gland that secretes sufficient quantities of estrogen
and progesterone to maintain pregnancy. Estrogen and progesterone levels increase in the woman's blood throughout
pregnancy (figure 28.19).

Menopause

When a female is 40–50 years old, menstrual cycles become
less regular, and ovulation does not occur. Eventually menstrual cycles stop completely. The cessation of menstrual cycles is called **menopause** (men'ō-pawz), and the time from
the onset of irregular cycles to their complete cessation is
called the **female climacteric** (klī-mak'ter-ik).

Menopause is associated with changes in the ovary. The
number of follicles remaining in the ovaries of menopausal
women is small. In addition, the follicles that remain become
less sensitive to stimulation by LH and FSH, even though LH
and FSH levels are elevated. As the ovaries become less responsive to stimulation by FSH and LH, fewer mature follicles
and corpora lutea are produced. Gradual morphologic
changes occur in the female in response to the reduced
amount of estrogen and progesterone produced by the
ovaries (table 28.3).

A variety of symptoms occur in some females during the
climacteric, including "hot flashes," irritability, fatigue, anxiety, and occasionally severe emotional disturbances. Many of
these symptoms can be treated successfully by administering
small amounts of estrogen and then gradually decreasing the
treatment over time or by providing psychologic counseling.
It appears that administering estrogen following menopause
also helps to prevent osteoporosis and heart disease. Although estrogen therapy has been successful, it prolongs
symptoms associated with menopause in many cases. Some
potential side effects of estrogen therapy are of concern, such
as a small increase in the possibility for the development of
breast and uterine cancer.

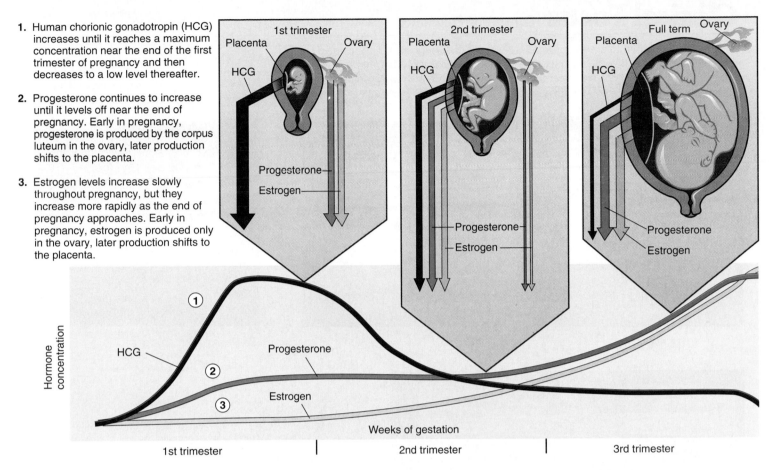

1. Human chorionic gonadotropin (HCG) increases until it reaches a maximum concentration near the end of the first trimester of pregnancy and then decreases to a low level thereafter.

2. Progesterone continues to increase until it levels off near the end of pregnancy. Early in pregnancy, progesterone is produced by the corpus luteum in the ovary, later production shifts to the placenta.

3. Estrogen levels increase slowly throughout pregnancy, but they increase more rapidly as the end of pregnancy approaches. Early in pregnancy, estrogen is produced only in the ovary, later production shifts to the placenta.

Figure 28.19 Changes in Hormone Concentration During Pregnancy

HCG, progesterone, and estrogens are secreted from the placenta during pregnancy. Early in pregnancy estrogen and progesterone are secreted by the ovary. During midpregnancy there is a shift toward estrogen and progesterone secretion by the placenta. Late in pregnancy these two hormones are secreted by the placenta.

Table 28.3 Possible Changes Caused by Decreased Ovarian Hormone Secretion in Postmenopausal Women

Affected Structures and Functions	Changes
Menstrual cycle	Five to 7 years before menopause the cycle becomes more irregular; finally the number of cycles in which ovulation does not occur increases, and corpora lutea do not develop
Oviduct	Little change
Uterus	Irregular menstruation gradually is followed by no menstruation; chance of cystic glandular hypertrophy of the endometrium increases; the endometrium finally atrophies, and the uterus becomes smaller.
Vagina and external genitalia	Dermis and epithelial lining become thinner; vulva becomes thinner and less elastic; labia majora become smaller; pubic hair decreases; vaginal epithelium produces less glycogen; vaginal pH increases; reduced secretion leads to dryness; the vagina is more easily inflamed and infected
Skin	Epidermis becomes thinner; melanin synthesis increases
Cardiovascular system	Hypertension and atherosclerosis occur more frequently
Vasomotor instability	Hot flashes and increased sweating are correlated with vasodilation of cutaneous blood vessels; hot flashes are not caused by abnormal FSH and LH secretion but are related to decreased estrogen levels
Libido	Temporary changes, usually a decrease, or a brief increase for some, in libido are associated with the onset of menopause
Fertility	Fertility begins to decline approximately 10 years before the onset of menopause; by age 50 almost all germ cells and follicles are lost; loss is gradual, and no increased follicular degeneration is associated with the onset of menopause

Many methods are used to prevent or terminate pregnancy (figure B), including methods that prevent fertilization (contraception), prevent implantation of the developing embryo (IUDs), or remove the implanted embryo or fetus (abortion). Many of these techniques are quite effective when done properly and used consistently (table A).

Behaviorial Methods

Abstinence, or refraining from sexual intercourse, is a sure way to prevent pregnancy when practiced consistently. It is

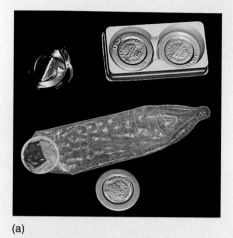

(a)

(b)

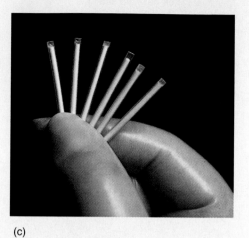

(c)

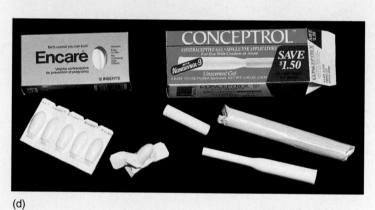

(d)

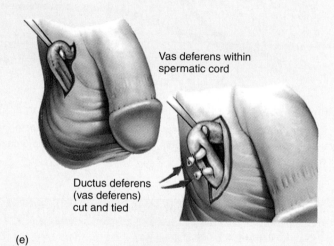

Vas deferens within spermatic cord

Ductus deferens (vas deferens) cut and tied

(e)

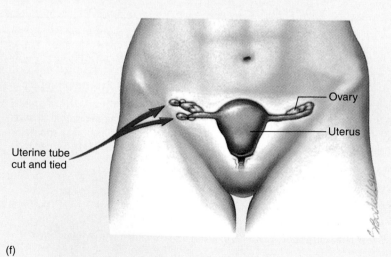

Ovary

Uterus

Uterine tube cut and tied

(f)

(g)

Figure B Contraceptive Devices and Techniques

(*a*) Condom. (*b*) Diaphragm used with spermicidal jelly. (*c*) Norplant system. (*d*) Spermicidal foam. (*e*) Vasectomy. (*f*) Tubal ligation. (*g*) Oral contraceptives. ✗

Table A   Effectiveness of Various Methods for Preventing Pregnancy

Technique	Effectiveness When Used Properly (%)	Actual Effectiveness (%)
Abortion	100	Unknown
Sterilization	100	99.9
Combination (estrogens and progesterones) pill	99.9	98
Intrauterine device	98	98
Mini pill (low dose of estrogens and progesterones)	99	97
Condom plus spermicide	99	96
Condom alone	97	90
Diaphragm plus spermicide	97	85
Foam	97	80
Rhythm	97	70

not an effective method when used only occasionally.

Coitus (kō'i-tŭs) **interruptus** is removal of the penis from the vagina just before ejaculation. This is a very unreliable method of preventing pregnancy, since it requires perfect awareness, and willingness to withdraw the penis at the correct time. It also ignores the fact that some sperm cells are found in pre-ejaculatory emissions.

Periodic abstinence, the **natural family planning method,** or the **rhythm method** requires abstaining from sexual intercourse near the time of ovulation. A major factor in the success of this method is the ability to predict accurately the time of ovulation. Although the rhythm method provides some protection against becoming pregnant, it has a relatively high rate of failure, resulting from both the inability to predict the time of ovulation and the failure to abstain around the time of ovulation.

Barrier Methods

A **condom** (kon'dom) is a sheath of animal membrane, rubber, or latex. Placed over the erect penis, the condom is a barrier device, because the semen is collected within the condom instead of within the vagina. Condoms also provide protection against sexually transmitted diseases.

A **vaginal condom** also acts as a barrier device. The vaginal condom can be placed into the vagina by the woman before sexual intercourse.

Methods to prevent sperm cells from reaching the oocyte once they are in the vagina include use of a diaphragm, cervical cap, spermicidal agents, and a vaginal sponge. The diaphragm and cervical cap are flexible plastic or rubber domes that are placed over the cervix within the vagina, where they prevent passage of sperm cells from the vagina through the cervical canal of the uterus. The diaphragm is larger than the cervical cap. The most commonly used **spermicidal agents** are foams or creams that kill the sperm cells. They are inserted into the vagina before sexual intercourse. A **sponge,** either natural or synthetic, is permeated with spermicidal agents and placed over the cervix, where it acts as a barrier and kills the sperm cells. When used in combination, a condom and foam, cream, or sponge are much more effective than when they are used alone. A **spermicidal douche** (dūsh) is a stream of fluid containing a chemical toxic to sperm cells that is injected into the vagina. The stream of fluid removes and kills sperm cells. Spermicidal douches used alone are not very effective.

Lactation

Lactation (lak-tā'shŭn) prevents the menstrual cycle for a few months after childbirth. Action potentials sent to the hypothalamus in response to suckling that cause the release of oxytocin and prolactin also inhibit FSH and LH release from the anterior pituitary. Lactation therefore prevents the development of ovarian follicles and ovulation. Despite continual lactation, the ovarian and uterine cycles eventually resume. Because ovulation occurs before menstruation, relying on lactation to prevent pregnancy is not consistently effective.

Chemical Methods

Synthetic estrogen and progesterone in **oral contraceptives** (birth control pills) effectively suppress fertility in females. These substances may have more than one action, but they reduce LH and FSH release from the anterior pituitary. Estrogen and progesterone are present in high enough concentrations to have a negative-feedback effect on the pituitary, which prevents the large increase in LH and FSH secretion that triggers ovulation. Over the years, the dose of estrogen and progesterone in birth control pills has been reduced. The current lower-dose birth control pills have fewer side effects than earlier dosages. There is an increased risk of heart attack or stroke in women using oral contraceptives who smoke or have a history of hypertension or coagulation disorders. For most women, the pill is effective and has a minimum frequency of complications, until at least age 35.

Progesteronelike chemicals, such as medroxy progesterone (Depo-Provera), which are injected intramuscularly and slowly released into the circulatory system, can act as effective contraceptives. Injected progesteronelike chemicals can provide protection from pregnancy for approximately a month, depending on the amount injected. A thin silastic tube containing these chemicals, such as the Norplant system, can be implanted beneath the skin, usually in the upper arm, from which they are slowly released into the circulatory system. The implants can be effective for up to 5 years.

(continued)

Advantages of the injected and implanted progesteronelike contraceptives over other chemical methods of birth control are that they do not require taking pills on a daily basis. The long-term effects of the injected and implanted progesterone-like chemicals have not been as thoroughly studied as birth control pills, and they are still being evaluated.

Mifepristone (RU486), blocks the action of progesterone, causing the endometrium of the uterus to slough off as it does at the time of menstruation. It is therefore used to induce menstruation and reduce the possibility of implantation when sexual intercourse has occurred near the time of ovulation. It can also be used to terminate pregnancies. **Morning–after pills,** similar in composition to birth control pills, are available. This technique can be used after intercourse but is only about 75% effective.

Surgical Methods

Vasectomy (va-sek′tō-mē) is a common method used to render males permanently incapable of fertilization without affecting the performance of the sex act. Vasectomy is a surgical procedure used to cut and tie the ductus deferens from each testis within the scrotal sac. This procedure prevents sperm cells from passing through the ductus deferens and becoming part of the ejaculate. Because such a small volume of ejaculate comes from the testis and epididymis, vasectomy has little effect on the volume of the ejaculated semen. The sperm cells are reabsorbed in the epididymis.

A common method of permanent birth control in females is **tubal ligation** (lī-gā′shŭn), a procedure in which the uterine tubes are tied and cut or clamped through an incision made through the wall of the abdomen. This procedure closes off the pathway between the sperm cells and the oocyte. **Laparoscopy** (lap-ă-ros′kŏ-pē), a procedure in which a special instrument is inserted into the abdomen through a small incision, is commonly used so that only small openings are required to perform the operation.

In some cases, pregnancies are terminated by surgical procedures called **abortions.** The most common method for performing abortions is the insertion of an instrument through the cervix into the uterus. The instrument scrapes the endometrial surface while a strong suction is applied. The endometrium and the embedded embryo are disrupted and sucked out of the uterus. This technique is normally used only in pregnancies that have progressed less than 3 months.

Prevention of Implantation

Intrauterine devices (IUDs) are inserted into the uterus through the cervix to prevent normal implantation of the developing embryonic mass within the endometrium. Some early IUD designs produced serious side effects such as perforation of the uterus, and, as a result, many IUDs have been removed from the market. Data indicate, however, that IUDs are effective in preventing pregnancy.

Clinical Focus   Causes of Female Infertility

Causes of infertility in females include malfunctions of the uterine tubes, reduced hormone secretion from the pituitary or ovary, and interruption of implantation.

Adhesions from pelvic inflammatory conditions caused by a variety of infections can cause blockage of one or both uterine tubes and is a relatively common cause of infertility in women.

Reduced ovulation can result from inadequate secretion of LH and FSH, which can be caused by hypothyroidism, trauma to the hypothalamus, infarctions of the hypothalamus or anterior pituitary gland, and tumors.

Interruption of implantation may result from uterine tumors or conditions causing abnormal ovarian hormone secretion. For example, premature degeneration of the corpus luteum causes progesterone levels to decline and menses to occur. If the corpus luteum degenerates before the placenta begins to secrete progesterone, the endometrium and the developing embryonic mass degenerate and are eliminated from the uterus. The conditions that result in secondary amenorrhea also reduce fertility (see Clinical Note on menstrual problems earlier in the chapter).

Endometriosis (en′dō-mē-trē-ō′sis), a condition in which endometrial tissue is found in abnormal locations, reduces fertility. Generally endometriosis is thought to result from some endometrial cells passing from the uterus through the uterine tubes into the pelvic cavity. The endometrial cells invade the peritoneum of the pelvic cavity. Because the endometrium is sensitive to estrogen and progesterone, periodic inflammation of the areas where the endometrial cells implant occurs. Endometriosis is a cause of painful menstruation and can reduce fertility.

Clinical Focus Infectious Diseases

Sexually Transmitted Diseases

Sexually transmitted diseases (STDs) are a class of infectious diseases spread by intimate sexual contact between individuals. These diseases include the major venereal diseases such as nongonococcal urethritis, trichomoniasis, gonorrhea, genital herpes, genital warts, syphilis, and acquired immunodeficiency syndrome.

Nongonococcal urethritis refers to any inflammation of the urethra that is not caused by gonorrhea. Factors such as trauma or passage of a nonsterile catheter through the urethra can cause this condition, but many cases are acquired through sexual contact. In most cases the bacterium *Chlamydia trachomatis* is responsible, but other bacteria may be involved. *Chlamydia trachomatis* infection is one of the most common sexually transmitted diseases. It often is unrecognized in people who have it and is responsible for many cases of pelvic inflammatory disease. Left untreated, it can also result in sterility, but antibiotics are usually effective in curing the condition.

Trichomonas vaginalis is a protozoan commonly found in the vagina of females and the urethra of males. If the normal acidity of the vagina is disturbed, *Trichomonas* can grow rapidly. *Trichomonas* infection is more common in females than in males because the vagina provides a suitable environment in which these organisms can survive. The rapid growth of these organisms results in inflammation and a greenish yellow discharge characterized by a foul odor.

Gonorrhea (gon-ō-rē′ă) is caused by *Neisseria gonorrhoeae.* The organisms attach to the epithelial cells of the vagina or to the male urethra. The invasion of bacteria establishes an inflammatory response in which pus is formed. Males become aware of a gonorrheal infection by painful urination and the discharge of pus-containing material from the urethra. Symptoms appear within a few days to a week. Recovery may eventually occur without complication, but, when complications do occur, they can be serious. The urethra can become partially blocked, or sterility can result from blockage of reproductive ducts with scar tissue. In some cases, other organ systems such as the heart, meninges of the brain, or joints may become infected. In females the early stages of infection may pass unnoticed, but the infection can lead to pelvic inflammatory disease. Gonorrheal eye infections may occur in newborn children of females with gonorrheal infections. Antibiotics are usually effective in treating gonorrheal infections, and the immune system often successfully combats gonorrheal infections in untreated individuals.

Genital herpes (her′pēz) is a viral infection by herpes simplex type 2. Lesions appear after an incubation period of about 1 week and cause a burning sensation. After this, blisterlike areas of inflammation appear. In males and females, urination can be painful, and walking or sitting can be unpleasant, depending on the location of the lesions. The blisterlike areas heal in about 2 weeks. The lesions may reoccur. The viruses exist in a latent condition in the infected tissues and may produce inflamed lesions on the genitals in response to factors such as menstruation, emotional stress, or illness. If active lesions are present in the mother's vagina or external genitalia, a caesarean delivery is performed to prevent newborns from becoming infected with the herpesvirus. Because genital herpes is caused by a virus, there is no effective antibiotic cure for it.

Genital warts also result from a viral infection (human papillomavirus) and are quite contagious. Genital warts are common, and their frequency is increasing. They can also be transmitted from infected mothers to their infants. Genital warts vary from separate, small, warty growths to large cauliflowerlike clusters. The lesions are usually not painful, but they can cause painful intercourse, and they bleed easily. For women who have genital warts, there is an increased risk of developing cervical cancer. Treatments for genital warts include topical agents, cryosurgery, or other surgical methods.

Syphilis (sif′i-lis) is caused by the bacterium *Treponema pallidum,* which can be spread by sexual contact of all kinds. Syphilis exhibits an incubation period from 2 weeks to several months. The disease progresses through several recognized stages. In the primary stage the initial symptom is a small, hard-based **chancre** (shang′ker), or sore, which usually appears at the site of infection. Several weeks after the primary stage, the disease enters the secondary stage, characterized mainly by skin rashes and mild fever. The symptoms of secondary syphilis usually subside after a few weeks, and the disease enters a latent period, in which no symptoms are present. In less than half the cases, a tertiary stage develops after many years. In the tertiary stage, many lesions develop that can cause extensive tissue damage and can lead to paralysis, insanity, and even death. Syphilis can be passed on to newborns through an infected mother. Damage to mental development and other neurologic symptoms are among the more serious consequences. Females who have syphilis in the latent phase are most likely to have babies who are infected. Antibiotics are used to treat syphilis, although some strains are very resistant to certain antibiotics.

Acquired immunodeficiency syndrome (AIDS) is caused by infection with the human immunodeficiency virus (HIV), which appears to ultimately result in destruction of the immune system (see chapter 22). The most common mechanisms of transmission of the virus are through sexual contact with a person infected with HIV and through sharing needles with an infected person during the administration of illicit drugs. Although transmission did occur during the early 1980s through tainted blood products, screening techniques now make the transmission of HIV through blood transfusions very rare. Some rare cases of transmission of HIV through accidental needlesticks in hospitals and other health care facilities have been documented. There is no evidence that casual contact with a person who has AIDS or who is infected with HIV results in transmission of the disease. Transmission appears to require exposure to body fluids of an infected person in a way that allows HIV into the interior of another person. Normal casual contact, including touching an HIV-infected person, does not increase the risk of infection.

Other Infectious Diseases

Pelvic inflammatory disease (PID) is a bacterial infection of the female pelvic organs. It usually involves the uterus, uterine tubes, or ovaries. A vaginal or uterine infection may spread throughout the pelvis. PID is commonly caused by gonorrhea or chlamydia, but other bacteria can be involved. Early symptoms of PID include increased vaginal discharge and pelvic pain. Early treatment with antibiotics can stop the spread of PID, but lack of treatment results in a life-threatening infection. PID can also lead to sterility.

BENIGN UTERINE TUMORS

Mrs. M had four children and was 43 years old. She noticed that menstruation was becoming gradually more severe and lasting up to several days longer each time menstruation started. After she menstruated almost continuously for 2 months she made an appointment with her physician. She performed a pelvic examination of Mrs. M, including tests for conditions such as cervical cancer and uterine cancer. Palpation of the uterus indicated the presence of enlarged masses in Mrs. M's uterus. The results of a dilation and curettage (D&C—dilation of the cervix and scraping [cutterage] of the endometrium to remove growths or other abnormal tissues) indicated that Mrs. M suffered from leiomyomas, or fibroid tumors of the uterus.

BACKGROUND INFORMATION

Uterine **leiomyomas** (lī'ō-mī-ō'mǎs), also called uterine fibroids, are one of the most common disorders of the uterus, and the most frequent tumor in women, affecting one of every four. Three-fourths of the women with this condition, however, experience no symptoms. The enlarged mass compresses the uterine lining (endometrium) resulting in ischemia and inflammation and results in frequent and severe menstruations. Abdominal cramping because of strong uterine contractions can be present. Constant menstruation is a frequent manifestation of these tumors, and it is one of the most common reasons why women elect to have the uterus removed, a procedure called a **hysterectomy** (his-ter-ek'tō-mē).

5   P R E D I C T

When discussing her condition with her mother, Mrs. M. discovered that her mother recalled frequent menstruations that were irregular and prolonged when she was in her late forties. Her mother did not have a hysterectomy, and in a few years the frequency of menstruation began to gradually subside. Explain.

✔ *Answer in Appendix F*

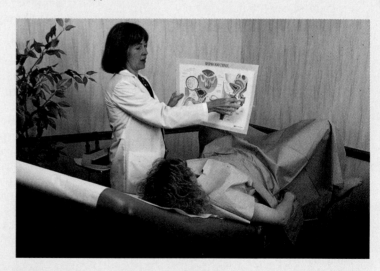

Figure C   Physician examining a patient.

System Interactions

System	Interactions
Integumentary	If anemia does not develop, skin appearance is normal, but if anemia does develop, the skin can appear pale because of the reduced hemoglobin in erythrocytes. The continual loss of blood often results in iron-deficiency anemia. The hemoglobin concentration of blood is therefore reduced as well as the hematocrit.
Muscular	If anemia develops and is severe, muscle weakness may result because of the reduced ability of the cardiovascular system to deliver adequate oxygen to muscles.
Skeletal	The rate of erythrocyte synthesis in the red bone marrow increases.
Digestive	An enlarged tumor can put pressure on the rectum or sigmoid colon, resulting in constipation.
Cardiovascular	A chronic loss of blood as in prolonged menstruation over many months to years frequently results in iron-deficiency anemia. Manifestations of anemia include reduced hematocrit, reduced hemoglobin concentration, smaller-than-normal erythrocytes (microcytic anemia), and increased heart rate.
Respiratory	Because of anemia, the oxygen-carrying capacity of the blood is reduced. Increased respiration during physical exertion and rapid fatigue are likely to occur if anemia develops.
Urinary	The kidneys increase erythropoietin secretion in response to the loss of erythrocytes. The erythropoietin increases erythrocyte synthesis in red bone marrow. An enlarged tumor can put pressure on the urinary bladder, resulting in increased frequency of and painful urination.

Summary

The male reproductive system produces sperm cells and transfers them to the female. The female reproductive system produces the oocyte and nurtures the developing child.

Male Reproductive System

Scrotum

1. The scrotum is a two-chambered sac that contains the testes.
2. The dartos and cremaster muscles help to regulate testicular temperature.

Perineum

The perineum is the diamond-shaped area between the thighs and consists of a urogenital triangle and an anal triangle.

Testes

1. The tunica albuginea is the outer connective tissue capsule of the testes.
2. The testes are divided by septa into compartments that contain the seminiferous tubules and the Leydig cells.
3. The seminiferous tubules empty into short ducts that lead to the rete testis. The rete testis opens into the efferent ductules of the epididymis.
4. During development the testes pass from the abdominal cavity through the inguinal canal to the scrotum.
5. Sperm cells (spermatozoa) are produced in the seminiferous tubules.
   - Spermatogonia divide (mitosis) to form primary spermatocytes.
   - Primary spermatocytes divide (first division of meiosis) to form secondary spermatocytes that divide (second division of meiosis) to form spermatids.
   - Spermatids develop an acrosome and a flagellum to become sperm cells.
   - Sertoli cells nourish the sperm cells, form a blood–testes barrier, and produce hormones.

Ducts

1. The epididymis is a coiled tube system located on the testis that is the site of sperm cell maturation.
2. The ductus deferens passes from the epididymis into the abdominal cavity.
3. The ejaculatory duct is formed by the joining of the ductus deferens and the duct from the seminal vesicle.
4. The prostatic urethra extends from the urinary bladder to join with the ejaculatory ducts to form the membranous urethra.
5. The membranous urethra extends through the urogenital diaphragm and becomes the spongy urethra, which continues through the penis.
6. The spermatic cord consists of the ductus deferens, blood and lymph vessels, nerves, remnants of the process vaginalis, cremaster muscle, and fascia.
7. The spermatic cord passes through the inguinal canal into the abdominal cavity.

Penis

1. The penis consists of erectile tissue.
   - The two corpora cavernosa form the dorsum and the sides of the penis.
   - The corpus spongiosum forms the ventral part and the glans penis.
2. The root of the penis attaches to the coxae.
3. The prepuce covers the glans penis.

Accessory Glands

1. The seminal vesicles empty into the ejaculatory ducts.
2. The prostate gland consists of glandular and muscular tissue and empties into the prostatic urethra.
3. The bulbourethral glands are compound mucous glands that empty into the spongy urethra.
4. Secretions
   - Semen is a mixture of gland secretions and sperm cells.
   - The bulbourethral glands and the urethral mucous glands produce mucus that neutralizes the acidic pH of the urethra.
   - The testicular secretions contain sperm cells.
   - The seminal vesicle fluid contains fructose and fibrinogen.
   - The prostate secretions make the seminal fluid more pH-neutral. Clotting factors activate fibrinogen, and fibrinolysin breaks down fibrin.

Physiology of Male Reproduction

Regulation of Sex Hormone Secretion

1. GnRH is produced in the hypothalamus and released in surges.
2. GnRH stimulates LH and FSH release from the anterior pituitary.
   - LH stimulates the Leydig cells to produce testosterone.
   - FSH stimulates sperm cell formation.
3. Inhibin, produced by Sertoli cells, inhibits FSH secretion.

Puberty

1. Before puberty small amounts of testosterone inhibit GnRH release.
2. During puberty testosterone does not completely suppress GnRH release, resulting in increased production of FSH, LH, and testosterone.

Effects of Testosterone

1. Testosterone is produced by the Leydig cells, the adrenal cortex, and possibly the Sertoli cells.
2. Testosterone causes the development of male sex organs in the embryo and stimulates the descent of the testes.
3. Testosterone causes enlargement of the genitals and is necessary for sperm cell formation.
4. Other effects of testosterone
   - Hair growth stimulation (pubic area, axilla, and beard) and inhibition (male pattern baldness)
   - Enlargement of the larynx and deepening of the voice
   - Increased skin thickness and melanin and sebum production
   - Increased protein synthesis (muscle), bone growth, blood cell synthesis, and blood volume
   - Increased metabolic rate

Male Sexual Behavior and the Male Sex Act

1. Testosterone is required for normal sex drive.
2. Stimulation of the sexual act can be tactile or psychologic.

3. Afferent impulses pass through the pudendal nerve to the sacral region of the spinal cord.
4. Parasympathetic stimulation
   - Erection is due to vasodilation of the blood vessels that supply the erectile tissue.
   - Mucus is produced by the glands of the urethra and the bulbourethral glands.
5. Sympathetic stimulation causes erection, emission, and ejaculation.

Female Reproductive System
Ovaries

1. The ovaries are held in place by the broad ligament, the mesovarium, the suspensory ligaments, and the ovarian ligaments.
2. The ovaries are covered by the peritoneum (ovarian epithelium) and the tunica albuginea.
3. The ovary is divided into a cortex (contains follicles) and a medulla (receives blood and lymph vessels and nerves).
4. Follicular development
   - Oogonia proliferate and become primary oocytes that are in prophase I of meiosis.
   - Primary follicles are primary oocytes surrounded by granulosa cells.
   - During puberty primary follicles become secondary follicles.
   - The primary oocytes continue meiosis to metaphase II and become secondary oocytes surrounded by the zona pellucida. The center of the follicle fills with fluid to form the antrum, the granulosa cells increase in number, and theca cells form around the secondary follicle.
   - Graafian follicles are enlarged secondary follicles at the surface of the ovary.
5. Ovulation
   - The follicle swells and ruptures, and the secondary oocyte is released from the ovary.
   - The second meiotic division is completed when the secondary oocyte unites with a sperm cell to form a zygote.
6. Fate of the follicle
   - The graafian follicle becomes the corpus luteum.
   - If fertilization occurs, the corpus luteum persists. If there is no fertilization, it becomes the corpus albicans.

Uterine Tubes

1. The mesosalpinx holds the uterine tubes.
2. The uterine tubes transport the oocyte or zygote from the ovary to the uterus.
3. Structures
   - The ovarian end of the uterine tube is expanded as the infundibulum. The opening of the infundibulum is the ostium, which is surrounded by fimbriae.
   - The infundibulum connects to the ampulla that narrows to become the isthmus. The isthmus becomes the uterine part of the uterine tube and passes through the uterus.
4. The uterine tube consists of an outer serosa, a middle muscular layer, and an inner mucosa with simple ciliated columnar epithelium.
5. Movement of the oocyte
   - Cilia move the oocyte over the fimbriae surface into the infundibulum.
   - Peristaltic contractions and cilia move the oocyte within the uterine tube.

- Fertilization occurs in the ampulla, where the zygote remains for several days.

Uterus

1. The uterus consists of the body, the isthmus, and the cervix. The uterine cavity and the cervical canal are the spaces formed by the uterus.
2. The uterus is held in place by the broad, round, and uterosacral ligaments.
3. The wall of the uterus consists of the perimetrium (serous membrane), the myometrium (smooth muscle), and the endometrium (mucous membrane).

Vagina

1. The vagina connects the uterus (cervix) to the vestibule.
2. The vagina consists of a layer of smooth muscle and an inner lining of moist stratified squamous epithelium.
3. The vagina is folded into rugae and longitudinal folds.
4. The hymen covers the vestibular opening of the vagina.

External Genitalia

1. The vulva, or pudendum, comprises the external genitalia.
2. The vestibule is the space into which the vagina and the urethra open.
3. Erectile tissue
   - The clitoris is formed by the two corpora cavernosa.
   - The bulbs of the vestibule are formed by the corpora spongiosa.
4. The labia minora are folds that cover the vestibule and form the prepuce.
5. The greater and lesser vestibular glands produce a mucous fluid.
6. When closed the labia majora cover the labia minora.
   - The pudendal cleft is a space between the labia majora.
   - The mons pubis is an elevated fat deposit superior to the labia majora.

Perineum

The clinical perineum is the region between the vagina and the anus.

Mammary Glands

1. The mammary glands are modified sweat glands.
   - The mammary glands consist of glandular lobes and adipose tissue.
   - The lobes consist of lobules that are divided into alveoli.
   - The lobes connect to the nipple through the lactiferous ducts.
   - The nipple is surrounded by the areola.
2. The breast is supported by Cooper's ligaments.

Physiology of Female Reproduction
Puberty

1. Puberty begins with the first menstrual bleeding (menarche).
2. Puberty begins when GnRH levels increase.

Menstrual Cycle

1. Ovarian cycle
   - FSH initiates development of the primary follicles.

- The follicles secrete a substance that inhibits the development of other follicles.
- LH stimulates ovulation and completion of the first meiotic division by the primary oocyte.
- The LH surge stimulates the formation of the corpus luteum. If fertilization occurs, HCG stimulates the corpus luteum to persist. If fertilization does not occur, the corpus luteum becomes the corpus albicans.

2. A positive-feedback mechanism causes FSH and LH levels to increase near the time of ovulation.
   - Estrogen produced by the theca cells of the follicle stimulates GnRH secretion.
   - GnRH stimulates FSH and LH, which stimulate more estrogen secretion, and so on.
   - Inhibition of GnRH levels causes FSH and LH levels to decrease after ovulation. Inhibition is due to the high levels of estrogen and progesterone produced by the corpus luteum.

3. Uterine cycle
   - Menses (day 1 to days 4 or 5). The spiral arteries constrict, and endometrial cells die. The menstrual fluid is composed of sloughed cells, secretions, and blood.
   - Proliferation phase (day 5 to day 14). Epithelial cells multiply and form glands, and the spiral arteries supply the glands.
   - Secretory phase (day 15 to day 28). The endometrium becomes thicker, and the endometrial glands secrete.
   - Estrogen stimulates proliferation of the endometrium and synthesis of progesterone receptors.
   - Increased progesterone levels cause hypertrophy of the endometrium, stimulate gland secretion, and inhibit uterine

contractions. Decreased progesterone levels cause the spiral arteries to constrict and start menses.

Female Sexual Behavior and the Female Sex Act

1. Female sex drive is partially influenced by androgens (produced by the adrenal gland) and steroids (produced by the ovaries).
2. Parasympathetic effects
   - The erectile tissue of the clitoris and the bulbs of the vestibule become filled with blood.
   - The vestibular glands secrete mucus, and the vagina extrudes a mucuslike substance.

Female Fertility and Pregnancy

1. Intercourse must take place 3 days before to 1 day after ovulation if fertilization is to occur.
2. Sperm cell transport to the ampulla depends on the ability of the sperm cells to swim and possibly on contractions of the uterus and the uterine tubes.
3. Implantation of the developing embryonic mass into the uterine wall occurs when the uterus is most receptive.
4. Estrogen and progesterone secreted first by the corpus luteum and later by the placenta are essential for the maintenance of pregnancy.

Menopause

The female climacteric begins with irregular menstrual cycles and ends with menopause, the cessation of the menstrual cycle.

Content Review

1. What is the scrotum? Explain the function of the dartos and cremaster muscles.
2. Describe the covering and the structure of a testis.
3. When and how do the testes descend into the scrotum?
4. Where, specifically, are sperm cells produced in the testes? Describe the process of sperm cell formation.
5. Name all the ducts the sperm cells traverse to go from their site of production to the outside.
6. Where do sperm cells undergo maturation?
7. Distinguish between the prostatic, membranous, and spongy parts of the male urethra.
8. Name the parts of the spermatic cord.
9. Describe the erectile tissue of the penis. Define the terms glans penis, crus, bulb, and prepuce.
10. State where the seminal vesicles, prostate gland, and bulbourethral glands empty into the male reproductive duct system.
11. Define the terms emission and ejaculation.
12. Define the term semen. Describe the contribution to semen of the accessory sex glands. What is the function of each secretion?
13. Where are GnRH, FSH, LH, and inhibin produced? What effects do they produce?
14. What changes in hormone production occur at puberty?
15. Where is testosterone produced? Describe the effects of testosterone on the embryo, during puberty, and on the adult male.

16. What effects do psychologic, parasympathetic, and sympathetic stimulation have on the male sex act?
17. Name and describe the ligaments that hold the uterus, uterine tubes, and ovaries in place.
18. Describe the coverings and structure of the ovary.
19. Starting with the oogonia, describe the development and production of a graafian follicle that contains a secondary oocyte.
20. Describe the process of ovulation.
21. What is the corpus luteum? What happens to the corpus luteum if fertilization occurs? If fertilization does not occur?
22. Describe the structures of the uterine tube. How are they involved in moving the oocyte or the zygote?
23. Where does fertilization usually take place?
24. Name the parts of the uterus. Describe the layers of the uterine wall.
25. Where is the vagina located? Describe the layers of the vaginal wall. What are rugae and longitudinal folds?
26. What are the hymen, vulva, pudendum, and vestibule?
27. What erectile tissue is in the clitoris and the bulb of the vestibule? What is the function of the clitoris and the bulb of the vestibule?
28. Describe the labia minora, the prepuce, the labia majora, the pudendal cleft, and the mons pubis.
29. Where are the greater and lesser vestibular glands located? What is their function?
30. Define the term perineum. What is the anterior clinical perineum? Define and give the purpose of an episiotomy.

31. Describe the route taken by a drop of milk from its site of production to the outside of the body. What are Cooper's ligaments?

32. Describe the events of the ovarian cycle. What role do FSH and LH play in the ovarian cycle? Where is HCG produced, and what effect does it have on the ovary?

33. Describe how the cyclic increase and decrease in FSH and LH is produced.

34. Name the stages of the uterine cycle, and describe the events that take place in each stage. What are the effects of estrogen and progesterone on the uterus?

35. When must intercourse take place for fertilization to occur?

36. Define the terms menopause and female climacteric. What causes these changes, and what symptoms commonly occur?

Develop Your Reasoning Skills

1. If an adult male were castrated, what would happen to the levels of GnRH, FSH, LH, and testosterone in his blood? What effect would these hormonal changes have on sexual characteristics and behavior?

2. If a 9-year-old boy were castrated, what would happen to the levels of GnRH, FSH, LH, and testosterone in his blood? What effect would these hormonal changes have on sexual characteristics and behavior?

3. Suppose you want to produce a birth control pill for men. On the basis of what you know about the male hormone system, what do you want the pill to do? Discuss any possible side effects that could be produced by your pill.

4. If the ovaries are removed from a postmenopausal woman, what happens to the levels of GnRH, FSH, LH, estrogen, and progesterone in her blood? What symptoms do you expect to observe?

5. During the secretory phase of the menstrual cycle, you normally expect
   a. The highest levels of progesterone that occur during the menstrual cycle
   b. A follicle present in the ovary that is ready to undergo ovulation
   c. That the endometrium reaches its greatest degree of development
   d. a and b
   e. a and c

6. If the ovaries are removed from a 20-year-old female, what happens to the levels of GnRH, FSH, LH, estrogen, and progesterone in her blood? What side effects do these hormonal changes have on her sexual characteristics and behavior?

7. A study divides normal adult females into two groups (A and B). Both groups are composed of females who have been married for at least 2 years and are not pregnant at the beginning of the experiment. The subjects weigh about the same amount, and none smoke cigarettes, although some do drink alcohol occasionally. Group A women receive a placebo in the form of a sugar pill each morning during their menstrual cycles. Group B women receive a pill containing estrogen and progesterone each morning of their menstrual cycles. Then plasma LH levels are measured before, during, and after ovulation. The results are as follows:

Group	4 Days Before Ovulation	The Day of Ovulation	4 Days After Ovulation
A	18 mg/100 mL	300 mg/100 mL	17 mg/100 mL
B	21 mg/100 mL	157 mg/100 mL	15 mg/100 mL

The number of pregnancies in group A was 37/100 females/year. The number of pregnancies in group B was 1.5/100 females/year. What conclusion can you reach on the basis of these data? Explain the mechanism involved.

8. A woman who is taking birth control pills that consist of only progesterone experiences the hot flash symptoms of menopause. Explain why.

9. GnRH can be used to treat some women who want to have children but have not been able to get pregnant. Explain why it is critical to administer the correct concentration of GnRH at the right time during the menstrual cycle.

Web Site Link

For a listing of the most current web sites related to this chapter, please visit the Seeley home page at:
http://www.mhhe.com/biosci/ap/seeleyap/

Chapter Twenty-Nine

Development, Growth, Aging, and Genetics

Objectives

1. List the three prenatal periods, and state major events associated with each.

2. Describe the events of fertilization.

3. Define the term pluripotent, and explain what it means with regard to development.

4. Describe the morula and the blastocyst.

5. State the derivatives of the inner cell mass and the trophoblast.

6. Describe the process of implantation and placental formation.

7. List the three germ layers, describe their formation, and list their adult derivatives.

8. Describe the formation of the neural tube and the somites.

9. Describe the formation of the gastrointestinal tract and the body cavities.

10. Describe the formation of the limbs, face, and palate.

11. Briefly describe the formation of the following major organ systems: integumentary, skeletal, muscular, nervous, endocrine, circulatory, respiratory, digestive, urinary, and reproductive.

12. Explain the process by which a one-chambered heart becomes a four-chambered heart.

13. Describe the effects of hormones on the development of the male and the female reproductive systems.

14. Explain the hormonal and nervous system factors responsible for parturition.

15. Discuss circulatory, digestive, and other changes that occur at the time of birth.

16. Explain the hormonal and nervous system factors responsible for lactation.

17. List the stages of life and describe the major events associated with each stage.

18. Describe the major changes associated with aging and death.

19. Explain the major inheritance patterns.

20. Define the term genetics, and explain how chromosomes are related to genes.

21. Define the term gene, and explain how genes control cell functions.

22. Describe the different ways in which genes are expressed.

23. Give examples of different genetic disorders.

Part Five

The identification and discussion of life stages has been a popular topic during the past few years. We tend to view life stages very differently today from how we did just a few years ago. For example, the percentage of high school graduates attending college has increased dramatically in the past 30 years, especially among women. In 1960, about 20% of males and 12% of females graduating from high school attended college. Today, just over half of all people over age 25 have attended some college. In addition, there are many more nontraditional college students than there were just a few years ago. In 1900, only 5% of the U.S. population was over age 65. Today, about 16% of the population is over age 65, and by 2030 more than 20% will be older than 65. The average life expectancy in 1900 was about 47 years, in 1940 it was about 63 years, and today it is about 78 years. In 1900, nearly 70% of all males over age 65 were still working; today only about 20% are still working past age 65. Older people are healthier and more active than they have ever been. Many more older adults are actively pursuing additional education, new vocations, avocations, and leisure activities.

The life span is usually considered to be the period between birth and death; however, the 9 months before birth are a critical part of a person's existence. What happens in these 9 months profoundly affects the rest of a person's life. This chapter describes the major events that occur during prenatal development. Although most people develop normally and are born without defects, approximately 10 out of every 100 people are born with some type of birth defect, most of which are minor. Three out of every 100 people, however, are born with a birth defect so severe that it requires medical attention during the first year of life. Later in life many more people discover previously unknown problems such as the tendency to develop asthma, certain brain disorders, or cancer. Genetics is also discussed in this chapter, as well as the life stages that occur after birth.

Prenatal Development

The prenatal period is the period from conception until birth, and it can be divided into three parts: (1) the **germinal period**—approximately the first 2 weeks of development during which the primitive germ layers are formed; (2) the **embryonic period**—from about the second to the end of the eighth week of development, during which the major organ systems come into existence; and (3) the **fetal period**—the last 30 weeks of the prenatal period, during which the organ systems grow and become more mature.

The medical community in general uses the mother's **last menstrual period (LMP)** to calculate the **clinical age** of the unborn child. Most embryologists, on the other hand, use **postovulatory age** to describe the timing of developmental events. Postovulatory age is used in this book. Because ovulation occurs about 14 days after LMP and fertilization occurs near the time of ovulation, it is assumed that postovulatory age is 14 days less than clinical age.

Fertilization

Dozens of sperm cells reach the oocyte and help digest a path through the cumulus mass cells and zona pellucida, but several mechanisms provide that normally only one sperm cell penetrates the oocyte cell membrane and enters the cytoplasm in the process of **fertilization.** Entrance of a sperm cell into the oocyte stimulates the female nucleus to undergo the second meiotic division, and the second polar body is formed. The nucleus that remains after the second meiotic division, called the **female pronucleus,** moves to the center of the oocyte, where it meets the enlarged head of the sperm cell, the **male pronucleus.** Both the male and female pronuclei are haploid, each having one-half of each chromosome pair (see chapter 3). Fusion of the pronuclei completes the process of fertilization and restores the diploid number of chromosomes. The product of fertilization is the **zygote** (zī′gōt) (figure 29.1a).

Early Cell Division

About 18–39 h after fertilization the zygote divides to form two cells. Those two cells divide to form four cells, which divide to form eight cells, and so on (figure 29.1b–d). The cells of this dividing embryonic mass are referred to as **pluripotent** (plū-rip′ō-tent, meaning multiple-powered), which means that any cell of the mass has the ability to develop into a wide range of tissues. As a result, the total number of embryonic cells can be decreased, increased, or reorganized without affecting the normal development of the embryo.

> **Clinical Note**
>
> In rare cases, following early cell divisions, the cells may separate and develop to form two individuals, called "identical," or **monozygotic, twins.** Identical twins have identical genetic information in their cells. Identical twins can also occur by other mechanisms, which occur a little later in development. Occasionally a woman may ovulate two or more secondary oocytes at the same time. Fertilization of two oocytes by different sperm cells results in "fraternal," or **dizygotic, twins.** Multiple ovulations can occur naturally or be stimulated by injection of drugs that stimulate gonadotropin release. These drugs are sometimes used to treat certain forms of infertility.

Morula and Blastocyst

Once the dividing embryonic mass is a solid ball of 12 or more cells, it is a sphere composed of numerous smaller spheres and is therefore called a **morula** (mōr′ū-lă, meaning mulberry). Three or 4 days after ovulation, the morula consists of about 32 cells. Near this time a fluid-filled cavity called the **blastocele** (blas′tō-sēl) begins to appear approximately in the center of the cellular mass. The hollow sphere that results is

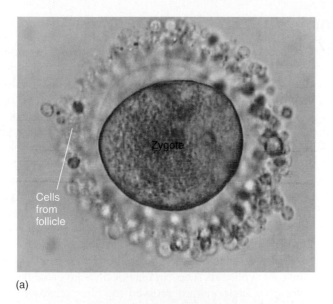

(a)

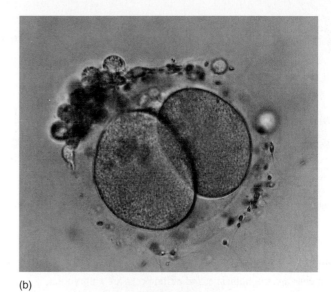

(b)

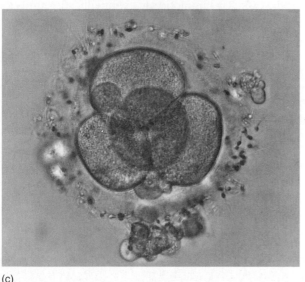

(c)

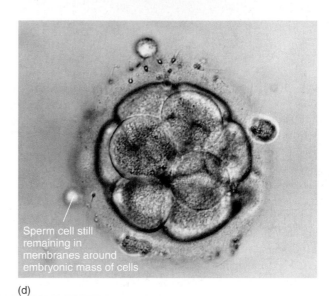

(d)

Figure 29.1   Early Stages of Human Development

(*a*) Zygote (120 μm in diameter). (*b*)–(*d*) During the early cell divisions, the zygote divides into more and more cells, but the total mass remains relatively constant. (*b*) Two cells, about 18–36 h after fertilization. (*c*) Four cells, about 36–48 h after fertilization. (*d*) 16 cells, about 48–72 h after fertilization.

called a **blastocyst** (blas′tō-sist) (figure 29.2). A single layer of cells, the **trophoblast** (trō′fō-blast, meaning feeding layer), surrounds most of the blastocele, but at one end of the blastocyst the cells are several layers thick. The thickened area is the **inner cell mass** and is the tissue from which the embryo develops. The trophoblast forms the placenta and the membranes (chorion and amnion) surrounding the embryo.

Implantation of the Blastocyst and Development of the Placenta

All of the events of the early germinal phase, including the first cell division through formation of the blastocele and the inner cell mass, occur as the embryonic mass moves from the

site of fertilization in the ampulla of the uterine tube to the site of implantation in the uterus. About 7 days after fertilization, the blastocyst attaches itself to the uterine wall, usually in the area of the uterine fundus, and begins the process of **implantation,** which is the burrowing of the blastocyst into the uterine wall.

As the blastocyst invades the uterine wall, two populations of trophoblast cells develop and form the **placenta** (figure 29.3), the organ of nutrient and waste product exchange between the fetus and the mother. The first is a proliferating population of individual trophoblast cells called the **cytotrophoblast** (sī-tō-trō′fō-blast). The other is a nondividing syncytium, or multinucleated cell, called the **syncytiotrophoblast** (sin-sish′ē-ō-trō′fō-blast). The cytotrophoblast remains nearer the other embryonic tissues,

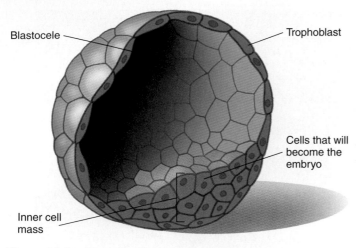

Blastocele

Trophoblast

Cells that will become the embryo

Inner cell mass

Figure 29.2   Blastocyst

Green cells are trophoblastic, and orange cells are embryonic.

and the syncytiotrophoblast invades the endometrium of the uterus. The syncytiotrophoblast is nonantigenic, which means that as it invades the maternal tissue, no immune reaction is triggered.

As the syncytiotrophoblast encounters maternal blood vessels, it surrounds them and digests the vessel wall, forming pools of maternal blood within cavities called **lacunae** (la-kū′nē; see figure 29.3*b*). The lacunae are still connected to intact maternal vessels so that blood circulates from the maternal vessels through the lacunae. Cords of cytotrophoblast surround the syncytiotrophoblast and lacunae (figure 29.3*c*). Branches sprout from these cords and protrude into the lacunaelike fingers called **chorionic** (kō-rē-on′ik) **villi,** and the entire embryonic structure facing the maternal tissues is called the **chorion** (kō′rē-on). Embryonic blood vessels follow the cords into the lacunae. In the mature placenta (figure 29.4) the cytotrophoblast disappears, so that the embryonic blood supply is separated from the maternal blood supply by only the embryonic capillary wall, a basement membrane, and a thin layer of syncytiotrophoblast.

Clinical Note

If the embryo implants near the cervix, a condition called **placenta previa** (prē′vē-ă) can occur. In this condition, as the placenta grows, it may extend partially or completely across the internal cervical opening. As the fetus and placenta continue to grow and the uterus stretches, the region of the placenta over the cervical opening may be torn, and hemorrhaging may occur. **Abruptio** (ab-rŭp′shē-ō) **placentae** is a tearing away of a normally positioned placenta from the uterine wall accompanied by hemorrhaging. Both of these conditions can result in miscarriage and can also be life-threatening to the mother.

Formation of the Germ Layers

After implantation, a new cavity called the **amniotic** (am-nē-ot′ik) **cavity** forms inside the inner cell mass and is surrounded by a layer of cells called the **amnion** (am′nē-on), or **amniotic sac.** Formation of the amniotic cavity causes part of the inner cell mass nearest the blastocele to separate as a flat disk of tissue called the **embryonic disk** (figure 29.5). This embryonic disk is composed of two layers of cells: an **ectoderm** (ek′tō-derm, meaning outside layer) adjacent to the amniotic cavity and an **endoderm** (en′dō-derm, meaning inside layer) on the side of the disk opposite the amnion. A third cavity, the **yolk sac,** forms inside the blastocele from the endoderm. The amniotic sac, yolk sac, and intervening double-layered embryonic disk can be thought of as resembling two balloons pushed together. One balloon represents the amniotic sac, and the other represents the yolk sac. The circular double layer of balloon where the two balloons are pressed together represents the embryonic disk. The amniotic sac eventually surrounds the developing embryo, providing it with a protective fluid environment, the "bag of waters," where the embryo can form.

About 13 or 14 days after fertilization, the embryonic disk becomes a slightly elongated oval structure. Proliferating cells of the ectoderm migrate toward the center and the caudal end of the disk, forming a thickened line called the **primitive streak.** Some ectoderm cells leave the ectoderm, migrate through the primitive streak, and emerge between the ectoderm and endoderm as a new germ layer, the **mesoderm** (mez′o-derm, meaning middle layer; figure 29.6). These three germ layers, the ectoderm, mesoderm, and endoderm, are the beginning of the **embryo.** All tissues of the adult can be traced to them (table 29.1). A cordlike structure called the **notochord** extends from the cephalic end of the primitive streak.

1   P R E D I C T

Predict the results of two primitive streaks forming in one embryonic disk. What if the two primitive streaks are touching each other?

✔ *Answer in Appendix F*

Clinical Note

During the first 2 weeks of development the embryo is quite resistant to outside influences that may cause malformations. Factors that adversely affect the embryo at this age are more likely to kill it. Between 2 weeks and the next 4–7 weeks (depending on the structure considered) the embryo is more sensitive to outside influences that cause malformations than at any other time.

Neural Tube and Neural Crest Formation

The ectoderm near the cephalic end of the primitive streak is stimulated about 18 days after fertilization to form a thickened

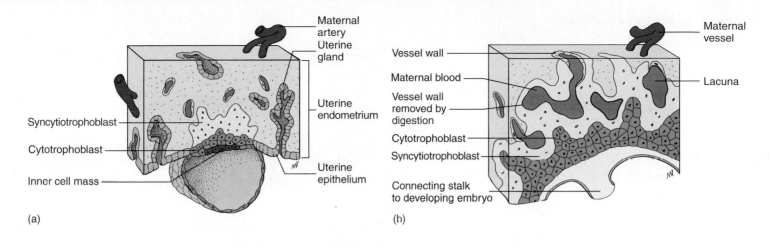

(a)

(b)

Figure 29.3 Formation of the Placenta

Implantation of the blastocyst and invasion of the trophoblast to form the placenta. (*a*) Implantation of the blastocyst with syncytiotrophoblast columns beginning to invade the uterine wall (at about 8–12 days). (*b*) Intermediate stage of placental formation (at about 14–20 days). As maternal blood vessels are encountered by the syncytiotrophoblast, lacunae are formed and filled with maternal blood. (*c*) Cytotrophoblast cords surround the syncytiotrophoblast and lacunae, and embryonic mesoderm enters the cord (at about 1 month).

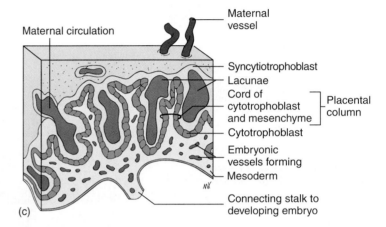

(c)

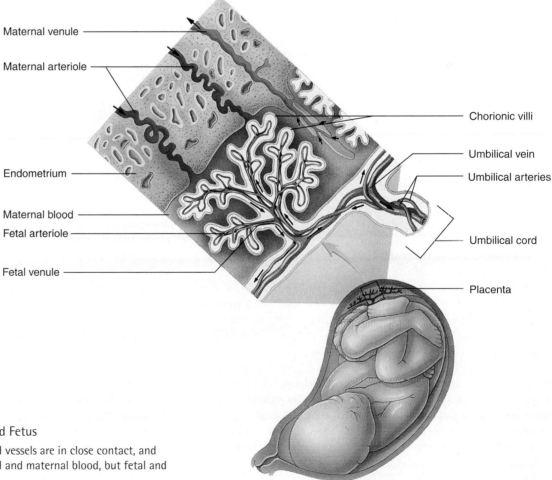

Figure 29.4 Mature Placenta and Fetus

Fetal blood vessels and maternal blood vessels are in close contact, and nutrients are exchanged between fetal and maternal blood, but fetal and maternal blood do not mix.

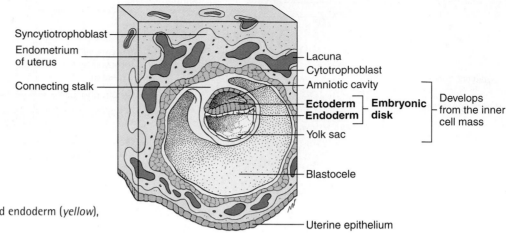

Figure 29.5 Embryonic Disk

Embryonic disk consisting of ectoderm (*blue*) and endoderm (*yellow*), with the amniotic cavity and yolk sac. ◼▬▭

1. Cells in the surface ectoderm move toward the primitive streak and fold into the streak (*blue arrow tails*).

2. Cells that enter the primitive streak come out the other side of the streak as mesodermal cells (*red arrows*).

3. The mesoderm (*red*) lies between the ectoderm (*blue*) and endoderm (*yellow*).

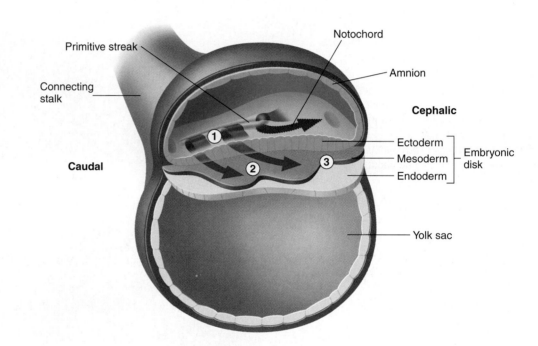

Figure 29.6 Primitive Streak

Embryonic disk with a primitive streak. The head of the embryo will develop over the notochord. ◼▬▭

neural plate. The lateral edges of the plate begin to rise like two ocean waves coming together. These edges are called the **neural folds,** and a **neural groove** lies between them (figure 29.7). The folding of the neural plate at the neural groove is stimulated by the underlying notochord. The crests of the neural folds begin to meet in the midline and fuse into a **neural tube,** which is completely closed by 26 days. The neural tube becomes the brain and the spinal cord, and the cells of the neural tube are called **neuroectoderm** (see table 29.1).

As the neural folds come together and fuse, a population of cells breaks away from the neuroectoderm all along the crests of the folds. These **neural crest cells** migrate down along the side of the developing neural tube to become part of the peripheral nervous system and the adrenal medulla and migrate laterally to just below the ectoderm, where they become melanocytes of the skin. In the head, neural crest cells perform additional functions; they contribute to the skull, the dentin of teeth, blood vessels, a few small muscles, and general connective tissue. Because neural crest cells in the head give rise to many of the same tissues as the mesoderm in the head and trunk, the general term **mesenchyme** (mez′en-kīm) is sometimes applied to cells of either neural crest or mesoderm origin.

Somite Formation

As the neural tube forms, the mesoderm immediately adjacent to the tube forms distinct segments called **somites** (so′mītz). In the head the first few somites never become clearly divided but develop into indistinct segmented structures called **somitomeres.** The somites and somitomeres eventually give rise to a part of the skull, the vertebral column, and skeletal muscle. Most of the head muscles are derived from the somitomeres.

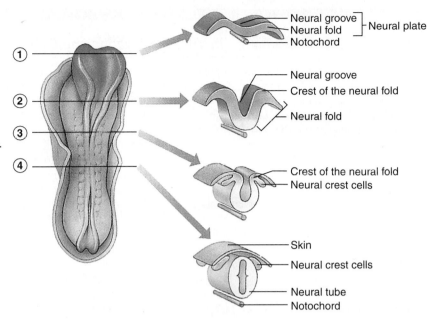

1. The neural plate is formed from ectoderm.

2. Neural folds form as parallel ridges along the embryo.

3. Neural crest cells break away from the crest of the neural folds.

4. The neural folds meet at the midline to form the neural tube.

Figure 29.7 Formation of the Neural Tube

The neural folds come together in the midline and fuse to form a neural tube. This fusion begins in the center and moves both cranially and caudally. The embryo shown is about 21 days after fertilization. The insets to the right show progressive closure of the neural tube.

Table 29.1 Germ Layer Derivatives

Ectoderm	Mesoderm
Epidermis of skin	Dermis of skin
Tooth enamel	Circulatory system
Lens and cornea of eye	Parenchyma of glands
Outer ear	Muscle
Nasal cavity	Bones (except facial)
Anterior pituitary	Microglia
Neuroectoderm	
Brain and spinal cord	**Endoderm**
Somatic motor neurons	Lining of gastrointestinal tract
Preganglionic autonomic neurons	Lining of lungs
Neuroglia cells (except microglia)	Lining of hepatic, pancreatic, and other exocrine ducts
Neural crest cells	Urinary bladder
Melanocytes	Thymus
Sensory neurons	Thyroid
Postganglionic autonomic neurons	Parathyroid
Adrenal medulla	Tonsils
Facial bones	
Teeth (dentin, pulp, and cementum) and gingiva	
A few skeletal muscles in head	

Formation of the Gut and Body Cavities

At the same time the neural tube is forming, the embryo itself is becoming a tube along the upper part of the yolk sac. The **foregut** and **hindgut** develop as the cephalic and caudal ends of the yolk sac are separated from the main yolk sac. This is the beginning of the digestive tract (figure 29.8*a*). The developing digestive tract pinches off from the yolk sac as a tube, remaining attached in the center to the yolk sac by a yolk stalk.

The foregut and hindgut (figure 29.8*b*) are in close relationship to the overlying ectoderm and form membranes

Clinical Focus   In Vitro Fertilization and Embryo Transfer

In a small number of women, normal pregnancy is not possible because of some anatomic or physiologic condition. In 87% of these cases the uterine tubes are incapable of transporting the zygote to the uterus or of allowing sperm cells to reach the oocyte. In vitro fertilization and embryo transfer have made pregnancy possible in hundreds of such women since 1978. **In vitro fertilization (IVF)** involves removal of secondary oocytes from a woman, placing the oocytes into a petri dish, and adding sperm cells to the dish, allowing fertilization and early development to occur in vitro, which means "in glass." **Embryo transfer** involves the removal of the developing embryonic cellular mass (not yet technically an embryo) from the petri dish and introduction of the mass into the uterus of a recipient female.

For IVF and embryo transfer to be accomplished, a woman is first injected with an LH-like substance, which causes more than one follicle to ovulate at one time. Just before the follicles rupture, the secondary oocytes are surgically removed from the ovary. The oocytes are then incubated in a dish and maintained at body temperature for 6 h. Then sperm cells are added to the dish.

After 24–48 h, when the zygotes have divided to form two- to eight-cell masses, several of the embryonic masses are transferred to the uterus. Several cell masses are transferred, because only a small percentage of them survives. Implantation and subsequent development then proceed in the uterus as they would for natural implantation; however, the woman is usually required to lie perfectly still for several hours after the cell masses have been introduced into the uterus to prevent possible expulsion before implantation can occur, which happens within 2–3 days after transfer. It is not fully understood why such expulsion does not occur in natural fertilization and implantation.

The implantation rate of embryo transfer is about 30%. The success rate varies with the number of embryonic masses implanted per transfer. Typically, three embryonic masses are transferred at a time. The rate of complications, such as multiple pregnancies, miscarriage, and prematurity, however, also increases with increased numbers of embryonic masses per transfer. About one-third of transfers of three embryonic masses end in multiple pregnancies. Of triplets born as a result of IVF, 64% required intensive care after birth, and 75% of quadruplets required intensive care, often for several weeks. Prematurity from IVF pregnancies in the United Kingdom resulted in newborn mortality in 2.7% of cases, a rate three times that of natural pregnancies. As a result of these complications, no more than two to three embryonic masses are now transferred per IVF in the United Kingdom.

The success rate has dramatically increased through time. The success rate at the best U.S. clinics was 20% in 1982, 30% in 1995, and 50% in 1997. This success rate may be approaching the natural limits, because only 50% or less of natural fertilizations result in a successful delivery.

---

called the oropharyngeal membrane and the cloacal membrane, respectively. The **oropharyngeal membrane** opens to form the mouth, and the **cloacal membrane** opens to form the urethra and anus. Thus the digestive tract becomes a tube that is open to the outside at both ends.

A considerable number of **evaginations** (ē-vaj-i-nā′shŭnz, meaning outpocketings) occur along the early digestive tract (figure 29.8c). They develop into structures such as the anterior pituitary, the thyroid gland, the lungs, the liver, the pancreas, and the urinary bladder. At the same time solid bars of tissue known as **branchial arches** (figures 29.8c and 29.9) form along the lateral sides of the head, and the sides of the foregut expand as pockets between the branchial arches. The central expanded foregut is called the **pharynx,** and the pockets along both sides of the pharynx are called **pharyngeal pouches.** Adult derivatives of the pharyngeal pouches include the auditory tube, tonsils, thymus, and parathyroids.

At about the same time, a series of isolated cavities starts to form within the embryo, thus beginning development of the **celom** (sē′lom; see figure 29.8), or body cavities. The most cranial group of cavities enlarges and fuses to form the **pericardial cavity.** Shortly thereafter the celomic cavity extends toward the caudal end of the embryo as the **pleural** and **peritoneal cavities.** Initially all three of these cavities are continuous, but they eventually separate into three distinct adult cavities (see chapter 1).

Limb Bud Development

Arms and legs first appear as limb buds (see figure 29.9). The **apical ectodermal ridge,** a specialized thickening of the ectoderm, develops on the lateral margin of each limb bud and stimulates its outgrowth. As the buds elongate, limb tissues are laid down in a proximal-to-distal sequence. For example, in the upper limb the arm is formed before the forearm, which is formed before the hand.

Development of the Face

The face develops by fusion of five embryonic structures: the **frontonasal process,** which forms the forehead, nose, and midportion of the upper jaw and lip; two **maxillary processes,** which form the lateral parts of the upper jaw and lip; and two **mandibular processes,** which form the lower jaw and lip (figure 29.10a). **Nasal placodes** (plak′ōdz), which develop at the lateral margins of the frontonasal

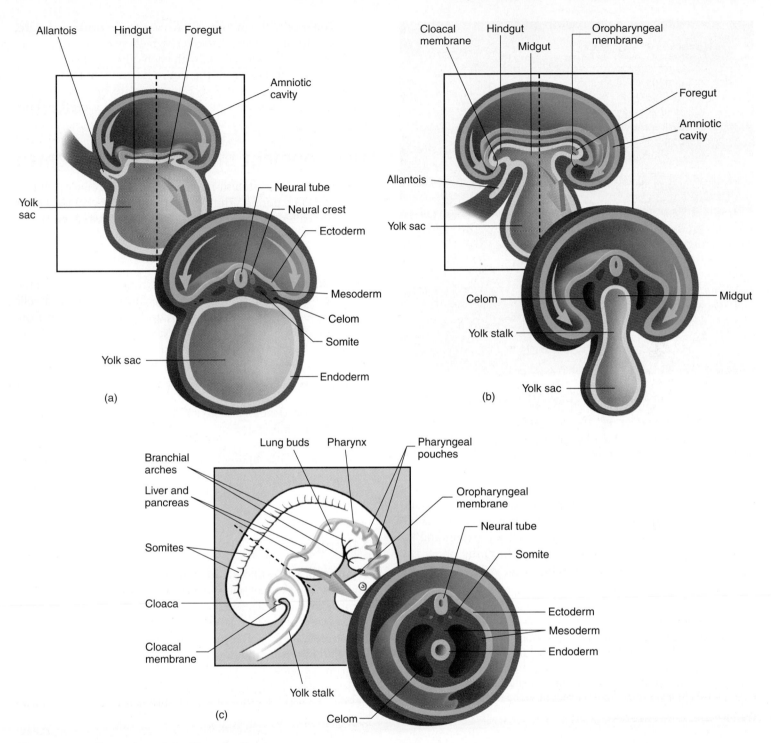

Figure 29.8 Formation of the Digestive Tract

Blue arrows show the folding of the digestive tract into a tube. Dotted lines show the plane of the section from which insets were taken. (*a*) 20 days after fertilization. (*b*) 25 days after fertilization. (*c*) 30 days after fertilization. Evaginations are identified along the pharynx and digestive tract in part (*c*).

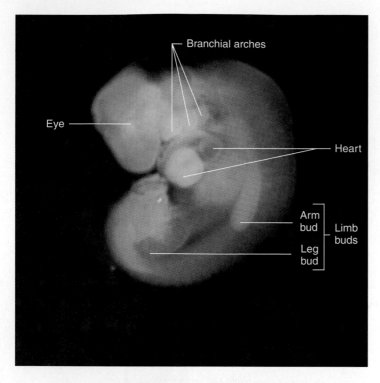

Figure 29.9   Human Embryo 35 Days After Fertilization

process, develop into the nose and the center of the upper jaw and lip (figure 29.10*b*).

As the brain enlarges and the face matures, the nasal placodes approach each other in the midline. The medial edges of the placodes fuse to form the midportion of the upper jaw and lip (figure 29.10*c* and *d*). This part of the frontal process is between the two maxillary processes, which are expanding toward the midline, and fuses with them to form the upper jaw and lip, known as the **primary palate.**

Clinical Note

A **cleft lip** results from failure of the frontonasal and two maxillary processes to fuse (see figure 29.10). Because three structures—one midline and two lateral—are involved in formation of the primary palate, cleft lips usually do not occur in the midline but to one side (or both sides) and extend from the mouth to the naris (nostril).

At about the same time the primary palate is forming, the lateral edges of the nasal placodes fuse with the maxillary processes to close off the groove extending from the mouth to the eye (figure 29.10*d* and *e*). On rare occasions these structures fail to meet, resulting in a facial cleft extending from the mouth to the eye.

The inferior margins of the maxillary processes fuse with the superior margins of the mandibular processes to decrease the size of the mouth.

All of the previously described fusions and the growth of the brain give the face a decidedly "human" appearance by about 50 days.

The roof of the mouth, known as the **secondary palate,** begins to form as vertical shelves, which swing to a horizontal position and begin to fuse with each other at about 56 days of development. Fusion of the entire palate is not completed until about 90 days. If the secondary palate does not fuse, a midline cleft in the roof of the mouth, called a **cleft palate,** results.

Development of the Organ Systems

The major organ systems appear and begin to develop during the embryonic period. The period between 14 and 60 days is therefore called the period of organogenesis (table 29.2).

Skin

The **epidermis** of the skin is derived from ectoderm, and the **dermis** is derived from mesoderm, or from neural crest cells in the case of the face. Nails, hair, and glands develop from the epidermis (see chapter 5). Melanocytes and sensory receptors in the skin are derived from neural crest cells.

Skeleton

The skeleton develops from either mesoderm or the neural crest cells by intramembranous or endochondral bone formation (see chapter 6). The bones of the face develop from neural crest cells, whereas the rest of the skull, the vertebral column, and ribs develop from somite- or somitomere-derived mesoderm. The appendicular skeleton develops from limb bud mesoderm.

Muscle

Myoblasts (mī′ō-blastz) are the early, embryonic cells that give rise to skeletal muscle fibers. Myoblasts migrate from somites or somitomeres to sites of future muscle development, where they begin to fuse and form multinuclear cells called **myotubes.** Shortly after myotubes form, nerves grow into the area and innervate the developing muscle fibers. After the basic form of each muscle is established, continued growth of the muscle occurs by an increase in the number of muscle fibers. The total number of muscle fibers is established before birth and remains relatively constant thereafter. Muscle enlargement after birth results from an increase in the size of individual fibers.

Nervous System

The nervous system is derived from the neural tube and neural crest cells. Neural tube closure begins in the upper cervical region and proceeds into the head and down the spinal cord. Soon after the neural tube has closed, the part of the neural tube that will become the brain begins to expand and develops a series of pouches (see figure 13.3). The central cavity of the neural tube becomes the ventricles of the brain and the central canal of the spinal cord.

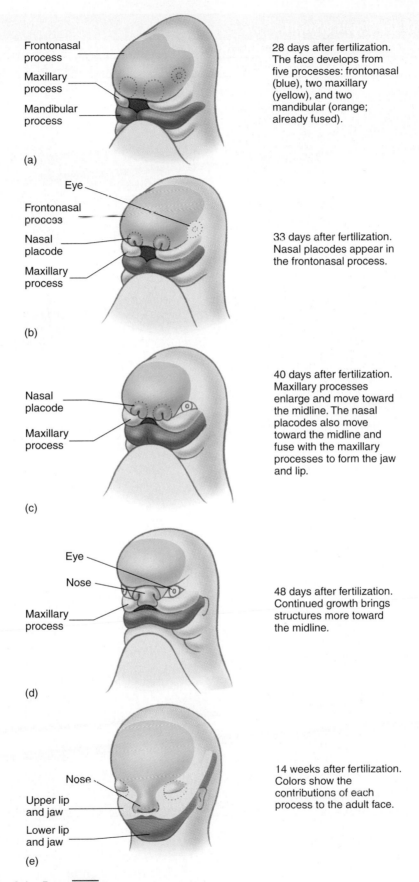

Frontonasal process
Maxillary process
Mandibular process

(a)

28 days after fertilization. The face develops from five processes: frontonasal (blue), two maxillary (yellow), and two mandibular (orange; already fused).

Eye
Frontonasal process
Nasal placode
Maxillary process

(b)

33 days after fertilization. Nasal placodes appear in the frontonasal process.

Nasal placode
Maxillary process

(c)

40 days after fertilization. Maxillary processes enlarge and move toward the midline. The nasal placodes also move toward the midline and fuse with the maxillary processes to form the jaw and lip.

Eye
Nose
Maxillary process

(d)

48 days after fertilization. Continued growth brings structures more toward the midline.

Nose
Upper lip and jaw
Lower lip and jaw

(e)

14 weeks after fertilization. Colors show the contributions of each process to the adult face.

Figure 29.10   Development of the Face

Table 29.2   Development of the Organ Systems

	Age (days since fertilization)					
	1–5	**6–10**	**11–15**	**16–20**	**21–25**	**26–30**
General features	Fertilization Morula Blastocyst	Blastocyst implants	Primitive streak Three germ layers	Neural plate	Neural tube closed	Limb buds and other "buds" appear
Integumentary system			Ectoderm Mesoderm			Melanocytes from neural crest
Skeletal system			Mesoderm		Neural crest (will form facial bones)	Limb buds
Muscular system			Mesoderm	Somites begin to form		Somites all present
Nervous system			Ectoderm	Neural plate	Neural tube complete Neural crest Eyes and ears begin	Lens begins to form
Endocrine system			Ectoderm Mesoderm Endoderm	Thyroid begins to develop		Parathyroids appear
Cardiovascular system			Mesoderm	Blood islands form Two heart tubes	Single-tubed heart begins to beat	Interatrial septum begins to form
Lymphatic system			Mesoderm			Thymus appears
Respiratory system			Mesoderm Endoderm		Diaphragm begins to form	Trachea forms as single bud Lung buds (primary bronchi)
Digestive system			Mesoderm Endoderm		Neural crest (will form tooth dentin) Foregut and hindgut form	Liver and pancreas appear as buds Tongue bud appears
Urinary system			Mesoderm Endoderm		Pronephros develops Allantois appears	Mesonephros appears
Reproductive system			Mesoderm Endoderm		Primordial germ cells on yolk sac	Mesonephros appears Genital tubercle forms

Table 29.2—cont'd

Age (days since fertilization)					
31–35	**36–40**	**41–45**	**46–50**	**51–55**	**56–60**
Hand and foot plates on limbs	Fingers and toes appear Lips formed Embryo 15 mm	External ear forming Embryo 20 mm	Embryo 25 mm	Limbs elongate to a more adult relationship Embryo 35 mm	Face is distinctly human in appearance
Sensory receptors appear in skin		Collagen fibers clearly present in skin		Extensive sensory endings in skin	
Mesoderm condensation in areas of future bone	Cartilage in site of future humerus	Cartilage in site of future ulna and radius	Cartilage in site of hand and fingers		Ossification begins in clavicle and then in other bones
Muscle precursor cells enter limb buds			Functional muscle		Nearly all muscles appear in adult form
Nerve processes enter limb buds		External ear forming Olfactory nerve begins to form		Semicircular canals in inner ear complete	Eyelids form Cochlea in inner ear complete
Pituitary appears as evaginations from brain and mouth	Gonadal ridges form Adrenal glands forming		Pineal body appears	Thyroid gland in adult position and attachment to tongue lost	Anterior pituitary loses its connection to the mouth
Interventricular septum begins to form		Interventricular septum complete	Interatrial septum complete but still has opening until birth		
Large lymphatic vessels form in neck	Spleen appears			Adult lymph pattern form	
Secondary bronchi to lobes form	Tertiary bronchi to bronchopulmonary segments form		Tracheal cartilage begins to form		
Oropharyngeal membrane ruptures		Secondary palate begins to form Tooth buds begin to form			Secondary palate begins to fuse (fusion complete by 90 days)
Metanephros begins to develop				Mesonephros degenerates	Anal portion of cloacal membrane ruptures
	Gonadal ridges form	Primordial germ cells enter gonadal ridges	Paramesonephric ducts appear		Uterus forming Beginning of differentiation of external genitalia in male and female

Anencephaly (an-en-sef′a-lē, meaning no brain) is a birth defect in which much of the brain fails to form because the neural tube fails to close in the region of the head. A baby born with anencephaly cannot survive. **Spina bifida** (spī′nă bi′fi-dă, meaning split spine) is a general term describing defects of the spinal cord or vertebral column (or both). Spina bifida can range from a simple defect with no clinical manifestations and with one or more vertebral spinous processes split or missing to a more severe defect that can result in paralysis of the limbs or the bowels and bladder, depending on where the defect occurs.

It has now been well documented that the inclusion of **folic acid** in the diet of a woman during the early stages of her pregnancy can significantly reduce the risk of neural tube defects in her developing child.

The neuron cell bodies of somatic motor neurons and preganglionic neurons of the autonomic nervous system, which provide axons to the peripheral nervous system, are located within the neural tube. Sensory nerves and postganglionic neurons of the autonomic nervous system are derived from neural crest cells.

A number of drugs and other chemicals are known to affect the embryo and fetus during development. The two most common are alcohol and cigarette smoke. Alcoholism or binge drinking can result in **fetal alcohol syndrome,** which includes decreased mental function. Exposure of the fetus to **cigarette smoke** throughout pregnancy can stunt the physical growth and mental development of the fetus.

Special Senses

The **olfactory bulb** and **nerve** develop as an evagination from the telencephalon (see figure 13.3). The eyes develop as evaginations from the diencephalon. Each evagination elongates to form an **optic stalk,** and a bulb called the **optic vesicle** develops at its terminal end. The optic vesicle reaches the side of the head and stimulates the overlying ectoderm to thicken into a **lens.** The sensory part of the ear appears as an ectodermal thickening or placode that invaginates and pinches off from the overlying ectoderm.

Endocrine System

The **posterior pituitary gland** is formed by an evagination from the floor of the diencephalon. The **anterior pituitary gland** develops from an evagination of ectoderm in the roof of the embryonic oral cavity and grows toward the floor of the brain. It eventually loses its connection with the oral cavity and becomes attached to the posterior pituitary gland (see chapter 18).

The **thyroid gland** originates as an evagination from the floor of the pharynx in the region of the developing tongue and moves into the lower neck, eventually losing its connection with the pharynx. The **parathyroid glands,** which are derived from the third and fourth pharyngeal pouches, migrate inferiorly and become associated with the thyroid gland.

The **adrenal medulla** arises from neural crest cells and consists of specialized postganglionic neurons of the sympathetic division of the autonomic nervous system (see chapter 16). The **adrenal cortex** is derived from mesoderm.

The **pancreas** originates as two evaginations from the duodenum, which come together to form a single gland (see figure 29.8c).

Circulatory System

The heart develops from two endothelial tubes (figure 29.11a), which fuse into a single, midline heart tube (figure 29.11b). Blood vessels form from blood islands on the surface of the yolk sac and inside the embryo. **Blood islands** are small masses of mesoderm that become blood vessels on the outside and blood cells on the inside. These islands expand and fuse to form the circulatory system. A series of dilations appears along the length of the primitive heart tube, and four major regions can be identified: the **sinus venosus,** the site where blood enters the heart; a single **atrium;** a single **ventricle;** and the **bulbus cordis,** where blood exits the heart (see figure 29.11b).

The elongating heart, confined within the pericardium, becomes bent into a loop, the apex of which is the ventricle (see figure 29.11b). The major chambers of the heart, the atrium and the ventricle, expand rapidly. The right part of the sinus venosus becomes absorbed into the atrium, and the bulbus cordis is absorbed into the ventricle. The embryonic sinus venosus initiates contraction at one end of the tubular heart. Later in development, part of the sinus venosus becomes the sinoatrial node, which is the adult pacemaker.

2   P R E D I C T

What would happen if the sinus venosus did not contract before other areas of the primitive heart?

✔ *Answer in Appendix F*

The single ventricle is divided into two chambers by the development of an **interventricular septum** (figure 29.11c–e). The **interatrial septum** (see figure 29.11c–e), which separates the two atria in the adult heart, is formed from two parts: the **septum primum** (primary septum) and the **septum secundum** (secondary septum). An opening in the interatrial septum called the **foramen ovale** (ō-val′ē) connects the two atria and allows blood to flow from the right to the left atrium in the embryo and fetus.

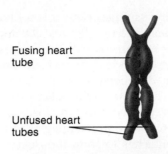

Fusing heart tube

Unfused heart tubes

(a) 20 days after fertilization. At this age, the heart consists of two parallel tubes.

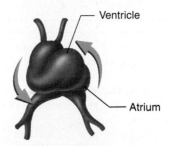

Ventricle

Atrium

(b) 22 days after fertilization. Fused, bent heart tube (*blue arrows suggest the direction of bending*) results from the elongation of the heart within the confined space of the pericardium.

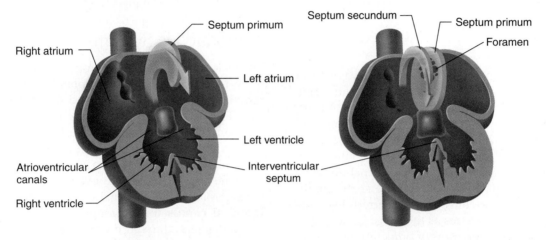

Septum primum

Right atrium

Left atrium

Left ventricle

Atrioventricular canals

Interventricular septum

Right ventricle

(c) 31 days after fertilization. The septum primum of the interatrial septum and the interventricular septum grow toward the center of the heart.

Septum secundum

Septum primum

Foramen

(d) 35 days after fertilization. The septum primum is complete and a foramen opens in the septum. The interventricular septum is nearly complete.

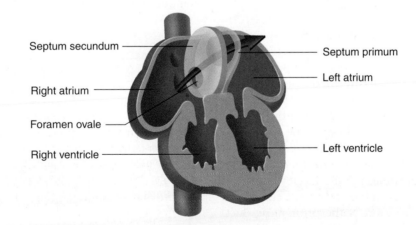

Septum secundum

Right atrium

Foramen ovale

Right ventricle

Septum primum

Left atrium

Left ventricle

(e) The final embryonic condition of the interatrial septum. Blood from the right atrium can flow through the foramen ovale into the left atrium. As blood begins to flow in the other direction, the septum primum is forced against the septum secundum, closing the foramen ovale.

Figure 29.11   Development of the Heart

Respiratory System

The lungs begin to develop as a single midline evagination from the foregut in the region of the future esophagus. This evagination branches to form two **lung buds** (figure 29.12a). The lung buds elongate and branch, first forming the bronchi that project to the lobes of the lungs (figure 29.12b) and then the bronchi that project to the bronchopulmonary segments of the lungs (figure 29.12c). This branching continues (figure 29.12d) until, by the end of the sixth month, about 17 generations of branching have occurred. Even after birth some branching continues as the lungs grow larger, and in the adult about 24 generations of branches have been established.

Urinary System

The kidneys develop from mesoderm located between the somites and the lateral part of the embryo. About 21 days after fertilization the mesoderm in the cervical region differentiates into a structure called the **pronephros** (meaning the most forward or earliest kidney) (figure 29.13a), which consists of a duct and simple tubules connecting the duct to the open celomic cavity. This type of kidney is the functional adult kidney in some lower chordates, but it is probably not functional in the human embryo and soon disappears.

The **mesonephros** (meaning middle kidney) (see figure 29.13a) is a functional organ in the embryo. It consists of a duct, which is a caudal extension of the pronephric duct, and a number of minute tubules, which are smaller and more complex than those of the pronephros. One end of each tubule opens into the mesonephric duct, and the other end forms a glomerulus (see chapter 26).

As the mesonephros is developing, the caudal end of the hindgut begins to enlarge to form the **cloaca** (klō-ā′kǎ, meaning sewer), the common junction of the digestive, urinary, and genital systems (figure 29.13b). The cloaca becomes divided by a **urorectal septum** into two parts: a digestive part called the **rectum** and a urogenital part called the **urethra** (figure 29.13c). The cloaca has two tubes associated with it: the hindgut and the **allantois** (ā-lan′tō-is,

meaning sausage), which is a blind tube extending into the umbilical cord (see figures 29.8 and 29.13). The part of the allantois nearest the cloaca enlarges to form the urinary bladder, and the remainder, which is from the bladder to the umbilicus, degenerates.

The mesonephric duct extends caudally as it develops and eventually joins the cloaca. At the point of junction another tube, the **ureter,** begins to form. Its distal end enlarges and branches to form the duct system of the adult kidney, called the **metanephros** (meaning last kidney), which takes over the function of the degenerating mesonephros. The mesonephric duct and a few tubules remain in the male as part of the reproductive system but almost completely disappear in the female (figure 29.13d).

Reproductive System

The male and female gonads appear as **gonadal ridges** along the ventral border of each mesonephros (figure 29.14a). **Primordial germ cells,** destined to become oocytes or sperm cells, form on the surface of the yolk sac, migrate into the embryo, and enter the gonadal ridge.

In the female the ovaries descend from their original position high in the abdomen to a position within the pelvis. In the male the testes descend even farther. As the testes reach the anteroinferior abdominal wall, a pair of tunnels called the **inguinal canals** form through the abdominal musculature. The testes pass through these canals, leaving the abdominal cavity and coming to lie within the **scrotum** (see figure 28.3). Descent of the testes through the canals begins about 7 months after conception, and the testes enter the scrotum about 1 month before the infant is born.

Paramesonephric ducts begin to develop just lateral to the mesonephric ducts and grow inferiorly to meet one another, where they enter the cloaca as a single, midline tube.

Testosterone, secreted by the testes, causes the mesonephric duct system to enlarge and differentiate to form the ductus deferens, the seminal vesicles, and the prostate gland (figure 29.14b). Müllerian-inhibiting hormone, also secreted by the testes, causes the paramesonephric ducts to degenerate. The paramesonephric ducts are also called the Müllerian ducts, and they give rise to the uterine tubes, the uterus, and part of the

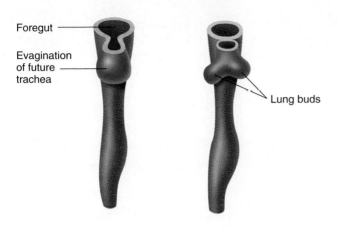

(a) 28 days after fertilization. A single bud forms and divides into two buds, which will become the lungs and primary bronchi.

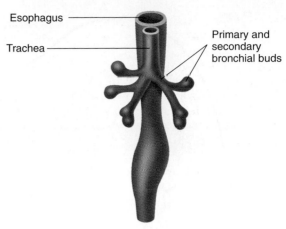

(b) 32 days after fertilization. Primary bronchi branch to form secondary bronchi, which supply the lobes.

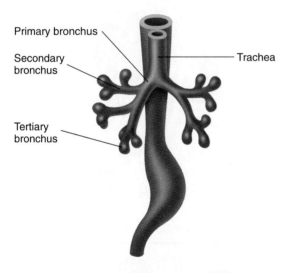

(c) 35 days after fertilization. Secondary bronchi branch to form tertiary bronchi, which supply the bronchopulmonary segments.

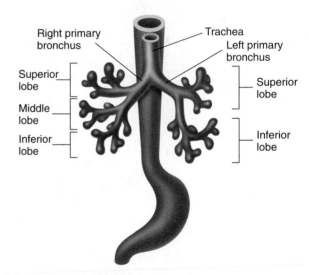

(d) 50 days after fertilization. Continued branching.

Figure 29.12   Development of the Lung

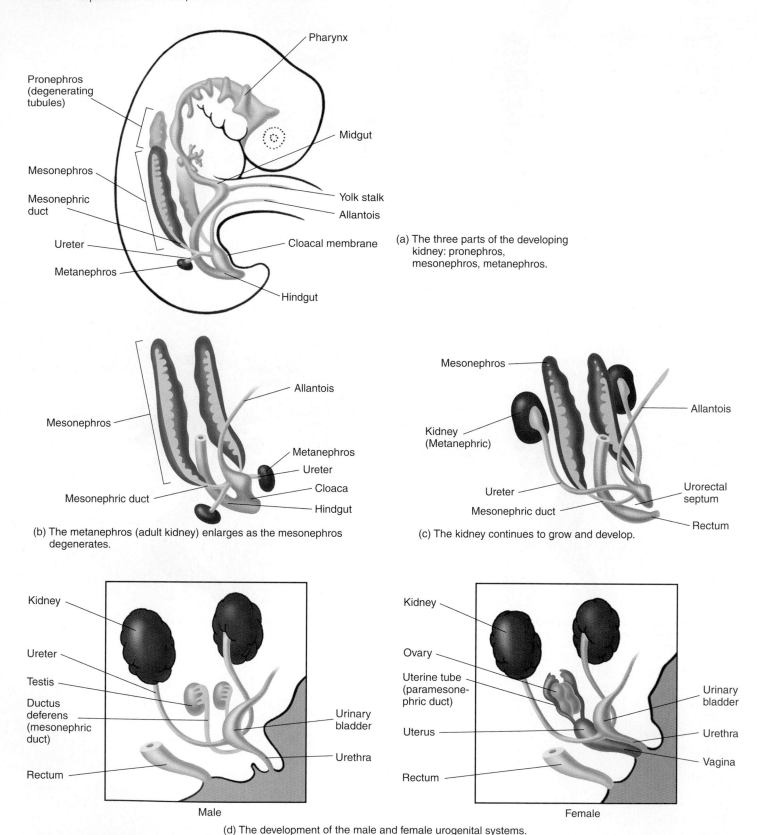

(a) The three parts of the developing kidney: pronephros, mesonephros, metanephros.

(b) The metanephros (adult kidney) enlarges as the mesonephros degenerates.

(c) The kidney continues to grow and develop.

(d) The development of the male and female urogenital systems.

Figure 29.13   Development of the Kidney and Urinary Bladder

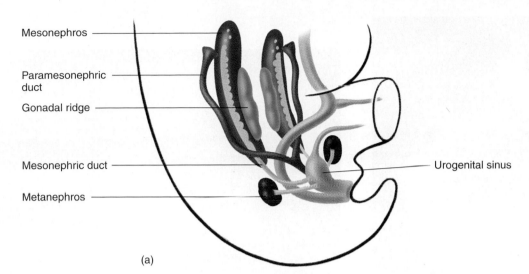

Mesonephros

Paramesonephric duct

Gonadal ridge

Mesonephric duct

Metanephros

Urogenital sinus

(a)

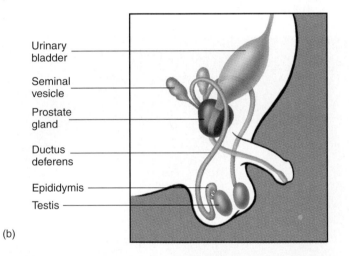

Urinary bladder

Seminal vesicle

Prostate gland

Ductus deferens

Epididymis

Testis

(b)

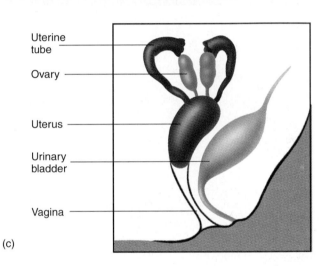

Uterine tube

Ovary

Uterus

Urinary bladder

Vagina

(c)

Figure 29.14 Development of the Reproductive System

(*a*) Indifferent stage. (*b*) The male, under the influence of male hormones, develops a ductus deferens from the mesonephric duct, and the paramesonephric duct degenerates. (*c*) The female, without male hormones, develops a uterus and uterine tubes from the paramesonephric duct, and the mesonephros disappears.

vagina in females (figure 29.14*c*). If neither testosterone nor Müllerian-inhibiting hormone is secreted, the mesonephric duct system atrophies, and the paramesonephric duct system develops to form the internal female reproductive structures.

Like the other sexual organs, the external genitalia begin as the same structures in the male and female and then diverge. An enlargement called the **genital tubercle** develops in the groin of the embryo. **Urogenital folds** develop on each side of the urogenital opening, and **labioscrotal swellings**

develop lateral to the folds. A **urethral groove** develops along the ventral surface of the genital tubercle.

In the male, under the influence of testosterone the genital tubercle and the urogenital folds close over the urogenital opening and the urethral groove to form the penis. If this closure does not proceed all the way to the end of the penis, a defect known as **hypospadias** (hī′pō-spā′dē-ăs) results. The testes move into the labioscrotal swellings, which become the scrotum of the male.

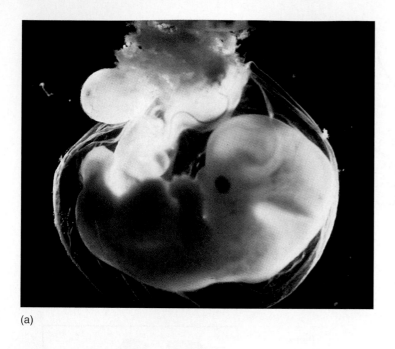

(a)

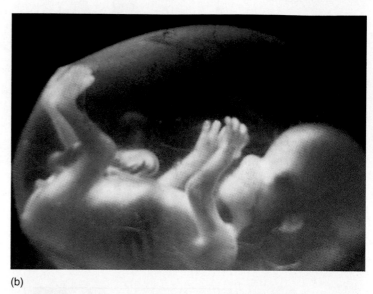

(b)

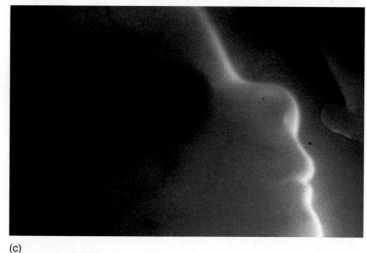

(c)

Figure 29.15   Embryos and Fetuses at Different Ages
(a) 50 days after fertilization. (b) 3 months after fertilization. (c) 4 months after fertilization.

In the female, in the absence of testosterone the genital tubercle becomes the clitoris. The urethral groove disappears, urogenital folds do not fuse. As a result, the urethra opens somewhat posterior to the clitoris but anterior to the vaginal opening. The unfused urogenital folds become the labia minora, and the labioscrotal folds become the labia majora.

Growth of the Fetus

The embryo becomes a **fetus** approximately 60 days after fertilization (a 50-day-old embryo is shown in figure 29.15a). The major difference between the embryo and the fetus is that in the embryo most of the organ systems are developing, whereas in the fetus the organs are present. Most morphologic changes occur in the embryonic phase of development, whereas the fetal period is primarily a "growing phase."

The fetus grows from about 3 cm and 2.5 g at 60 days to 50 cm and 3300 g at term—more than a 15-fold increase in length and a 1300-fold increase in weight (figure 29.16). Although growth is certainly a major feature of the fetal period, it is not the only feature. The major organ systems still continue to develop during the fetal period.

Fine, soft hair called **lanugo** (la-nū′gō) covers the fetus, and a waxy coat of sloughed epithelial cells called **vernix caseosa** (ver′niks kā-se-ō′sǎ) protects the fetus from the somewhat toxic nature of the amniotic fluid formed by the accumulation of waste products from the fetus.

Subcutaneous fat that accumulates in the older fetus and newborn provides a nutrient reserve, helps insulate the baby, and aids the baby in sucking by strengthening and supporting the cheeks so that negative pressure can be developed in the oral cavity.

Peak body growth occurs late in gestation, but, as placental size and blood supply limits are approached, the growth rate slows. Growth of the placenta essentially stops at about 35 weeks, restricting further intrauterine growth.

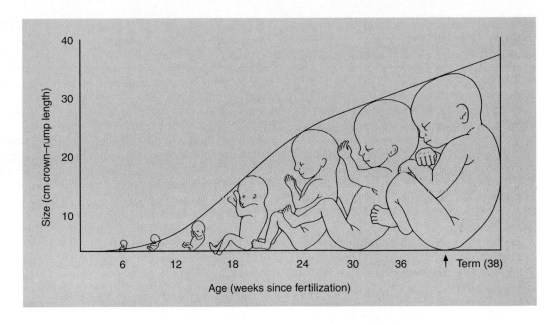

Figure 29.16   Growth of the Fetus

Fetal surgery performed while the fetus is still in the uterus was first done in the United States in 1979 to drain the excess fluid associated with hydrocephalus. These surgeries did not usually solve the underlying neurologic problems and have been discontinued. Since 1981, in utero surgeries have successfully removed excess fluid from enlarged urinary bladders of male fetuses. The fluid buildup occurs in 1 in every 2000 male fetuses when a flap of tissue grows over the internal opening of the urethra. Without treatment the amount of amniotic fluid is greatly reduced, and most of those babies die shortly after birth. Since 1989, more advanced surgeries have repaired diaphragmatic hernia, in which part of the abdominal organs push up through a hole in the left side of the diaphragm into the left pleural space, so that the left lung fails to develop fully. The defect occurs in 1 of every 2000 babies, and without surgery babies with this defect run a 75% risk of dying before or soon after birth. During surgery, the uterus is cut open, and the fetus is pulled far enough out of the opening so that a small incision can be made in its side. The abdominal organs are moved back into the abdomen, the hole in the diaphragm is covered with a surgical patching material called Gore-Tex, the incision in the fetus is closed, and the fetus is tucked back into the uterus. The amniotic fluid, which was removed and saved earlier in the surgery, is replaced, and the incision in the uterus and mother's skin is repaired.

At about 38 weeks of development the fetus has progressed to the point at which it can survive outside the mother. The average weight at this point is 3250 g for a female fetus and 3300 g for a male fetus.

Parturition

Parturition (par-tūr-ish′ŭn) refers to the process by which the baby is born. Physicians usually calculate the gestation period, or length of the pregnancy, as 280 days (40 weeks or 10 lunar months) from the last menstrual period (LMP) to the date of confinement, which is the date of delivery of the infant.

| 3 | P R E D I C T |

How many days (postovulatory age) does it take an infant to develop from fertilization to parturition?

✔ *Answer in Appendix F*

Occasionally the fetus is delivered before it has sufficiently matured. It is then considered to be **premature.** Prematurity is one of the most significant problems in pediatrics, the branch of medical science dealing with children, because of all the complications associated with prematurity. The most significant of these complications is **respiratory distress syndrome,** which occurs because very young premature infants cannot produce **surfactant,** a mixture of phospholipids and protein that lines the inner surface of the lungs, allowing the lungs to expand as we breathe. Each year, 65,000 premature infants suffer from respiratory distress syndrome in the United States. Until recently, 10% of those infants died. Now surfactant substitutes are being developed, and glucocorticoid administration can stimulate surfactant production. These therapies have cut the death rate in half, and more effective replacements are being investigated.

Near the end of pregnancy the uterus becomes progressively more irritable and usually exhibits occasional contractions that become stronger and more frequent until parturition is initiated. The cervix gradually dilates, and strong uterine contractions help expel the fetus from the uterus through the vagina (figure 29.17). Before expulsion of the fetus from the uterus, the amniotic sac ruptures, and amniotic fluid flows through the vagina to the exterior of the woman's body.

Clinical Focus   Fetal Monitoring

Amniocentesis (am'nē-ō-sen-tē'sis) is the removal of amniotic fluid from the amniotic cavity (figure A). As the fetus develops, molecules of various types, as well as living cells, are expelled into the amniotic fluid. These molecules and cells can be collected and analyzed. A number of normal conditions can be evaluated, and a number of metabolic disorders can be detected by analysis of the types of molecules expelled by the fetus. The cells collected by amniocentesis can be grown in culture, and additional metabolic disorders can be evaluated. Chromosome analysis, called a **karyotype,** can also be performed on the cultured cells. Amniocentesis has been done as early as 10 weeks after fertilization, but the success rate at that time is quite low. It is most commonly performed at 13–14 weeks after fertilization.

Fetal tissue samples may also be obtained by **chorionic villus sampling,** in which a probe is introduced into the uterine cavity through the cervix and a small piece of chorion is removed. This technique has an advantage over amniocentesis in that it can be used earlier in development, as early as the seventh to ninth week after fertilization.

One of the molecules normally produced by the fetus and released into the amniotic fluid is **α (alpha)-fetoprotein.** If the fetus has tissues exposed to the amniotic fluid that are normally covered by skin, such as nervous tissue, resulting from failure of the neural tube to close, or abdominal tissues, resulting from failure of the abdominal wall to fully form, an excessive amount of α-fetoprotein is lost into the amniotic fluid.

Some of the metabolic by-products from the fetus, such as α-fetoprotein and estriol, a weak form of estrogen produced in the placenta after 20 weeks of gestation, can enter the maternal blood. In some cases the by-products can be processed and passed to the maternal urine. The levels of these fetal products can then be measured in the mother's blood or urine.

The fetus can be seen within the uterus by **ultrasound,** which uses sound waves that are bounced off the fetus like sonar and then analyzed and enhanced by computer; or by **fetoscopy,** in which a fiberoptic probe is introduced into the amniotic cavity. Because of the constantly increasing resolution in ultrasound and because it is noninvasive compared with fetoscopy, the latter technique is not commonly used at present. Ultrasound has not been found to pose any risk to the fetus or mother. It is accomplished by placing a transducer on the abdominal wall (transabdominal) or by inserting the transducer into the woman's vagina (transvaginal). The latter technique produces much higher resolution because there are fewer layers of tissue between the transducer and the uterine cavity. Transvaginal ultrasound can be used to identify the yolk sac of a developing embryo as early as 17 days after fertilization, and the embryo can be visualized by 25 days. Transabdominal ultrasound allows for fetal monitoring by 6–8 weeks after fertilization.

Fetal heart rate can be detected with an ultrasound stethoscope by the 10th week after fertilization and with a conventional stethoscope by 20 weeks. The normal fetal heart rate is 140 bpm (normal range 110–160).

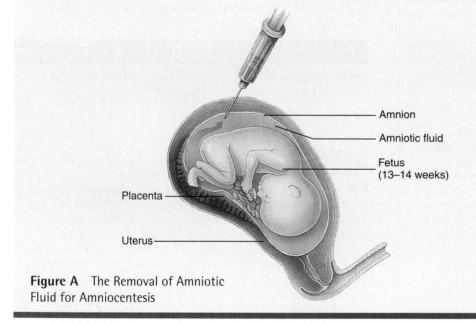

Figure A   The Removal of Amniotic Fluid for Amniocentesis

Labels on figure:
- Amnion
- Amniotic fluid
- Fetus (13–14 weeks)
- Placenta
- Uterus

Labor is the period during which the contractions occur that result in expulsion of the fetus from the uterus. It occurs as three stages.

1. *First stage.* The first stage begins with the onset of regular uterine contractions and extends until the cervix dilates to a diameter about the size of the fetus' head. This stage of labor commonly lasts from 8–24 h, but it may be as short as a few minutes in some women who have had more than one child. Normally (95% of the time) the head of the fetus is in an inferior position within the woman's pelvis during labor. The head acts as a wedge, forcing the cervix and vagina to open as the uterine contractions push against the fetus.

Clinical Note

The **central tendon of the perineum** (see figure 11.18*b*) is very important in supporting the uterus and vagina. Tearing or stretching of the tendon during childbirth may weaken the inferior support of these organs, and prolapse of the uterus may occur. **Prolapse** is a "sinking" of the uterus so that the uterine cervix moves down into the vagina (first degree), moves down near the vaginal orifice (second degree), or may protrude through the vaginal orifice (third degree). During childbirth an **episiotomy,** a cut through the perineal central tendon, prevents tearing of the perineum. The cut relieves the pressure and heals better than would a tear; however, there appears to be evidence that many episiotomies may be unnecessary.

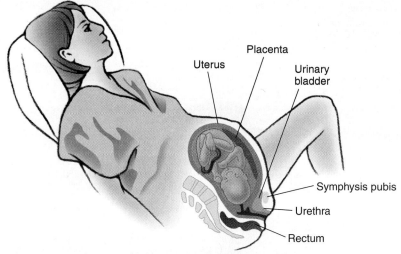

(a) The position of the fetus before parturition.

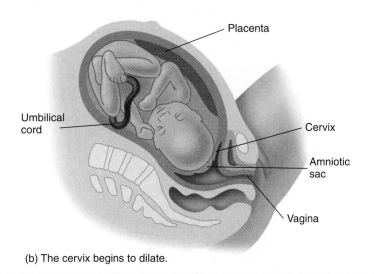

(b) The cervix begins to dilate.

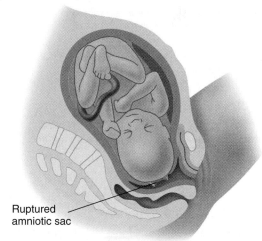

(c) Further dilation of the cervix and rupture of the amniotic sac occur.

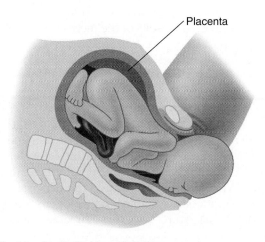

(d) The fetus is expelled from the uterus.

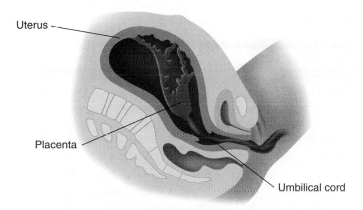

(e) The placenta is then expelled.

Figure 29.17   Process of Parturition

1. The fetal hypothalamus secretes CRH, which stimulates ACTH secretion from the pituitary. The fetal pituitary secretes ACTH in greater amounts near parturition.

2. ACTH causes the fetal adrenal gland to secrete greater quantities of adrenal glucocorticoids.

3. Glucocorticoids travel in the umbilical blood to the placenta.

4. In the placenta the adrenal glucocorticoids cause progesterone synthesis to level off and estrogen and prostaglandin synthesis to increase, making the uterus more irritable.

5. The stretching of the uterus produces action potentials that are transmitted to the brain through ascending pathways.

6. Action potentials stimulate the secretion of oxytocin by the posterior pituitary.

7. Oxytocin causes the uterine smooth muscle to contract.

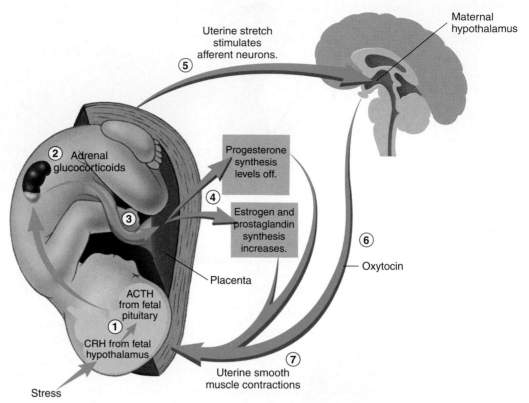

Figure 29.18 Factors That Influence the Process of Parturition

Although the precise control of parturition in humans is unknown, these changes appear to play a role. *Abbreviation:* CRH = Corticotropin-releasing hormone.

2. *Second stage.* The second stage of labor lasts from the time of maximum cervical dilation until the baby exits the vagina. This stage may last from a minute to up to an hour. During this stage contractions of the abdominal muscles assist the uterine contractions. The contractions generate enough pressure to compress blood vessels in the placenta so that blood flow to the fetus is stopped. During periods of relaxation blood flow to the placenta resumes.

Clinical Note

Occasionally drugs such as oxytocin are administered to women during labor to increase the force of the uterine contractions. Caution must be exercised in the use of this drug, however, so that tetaniclike contractions, which would drastically reduce the blood flow through the placenta, do not occur.

3. *Third stage.* The third stage of labor involves the expulsion of the placenta from the uterus. Contractions of the uterus cause the placenta to tear away from the wall of the uterus. Some bleeding occurs because of the intimate contact between the placenta and the uterus; however, bleeding normally is restricted because uterine smooth muscle contractions compress the blood vessels to the placenta.

Blood levels of estrogen and progesterone fall dramatically after parturition. Once the placenta has been dislodged from the uterus, the source of these hormones is gone. In addition, during the 4 or 5 weeks after parturition the uterus becomes much smaller, but it remains somewhat larger than it was before pregnancy. The cells of the uterus become smaller, and many of them degenerate. A vaginal discharge composed of small amounts of blood and degenerating endometrium persists for 1 week or more after parturition.

The precise signal that triggers parturition is unknown, but many of the factors that support parturition have been identified (figure 29.18). Before parturition the progesterone concentration in the maternal circulation is at its highest level (see figure 28.18). Progesterone has an inhibitory effect on uterine smooth muscle cells. Near the end of pregnancy, however, estrogen levels rapidly increase in the maternal circulation, and the excitatory influence of estrogens on uterine smooth muscle cells overcomes the inhibitory influence of progesterone.

The adrenal gland of the fetus is greatly enlarged before parturition. The stress of the confined space of the uterus and the limited oxygen supply resulting from a more rapid increase in the size of the fetus than in the size of the placenta increase the rate of adrenocorticotropic hormone (ACTH) secretion by the fetus' anterior pituitary gland. ACTH causes the fetal adrenal cortex to produce glucocorticoids, which travel to the placenta, where they decrease the rate of progesterone secretion and increase the rate of estrogen synthesis. In addition, prostaglandin synthesis is initiated. Prostaglandins strongly stimulate uterine contractions.

During parturition, nervous reflexes initiated by stretch of the uterine cervix cause the release of oxytocin from the woman's posterior pituitary gland. Oxytocin stimulates uterine contractions, which move the fetus farther into the cervix, causing further stretch.

Thus a positive-feedback mechanism is established in which stretch stimulates oxytocin release and oxytocin causes further stretch. This positive-feedback system stops after delivery, when the cervix is no longer stretched.

Progesterone inhibits oxytocin release; thus decreased progesterone levels in the maternal circulation can support the increased secretion rate of oxytocin. In addition, estrogens make the uterus more sensitive to oxytocin stimulation by increasing the synthesis of receptor sites for oxytocin. Some evidence suggests that oxytocin also stimulates prostaglandin synthesis in the uterus. All these events support the development of strong uterine contractions.

4	P R E D I C T

A woman is having an extremely prolonged labor. From her anatomy and physiology course she remembers the role of calcium in muscle contraction and asks the doctor to give her a calcium injection to speed the delivery. Explain why the doctor would or would not do as she requested.

✔ *Answer in Appendix F*

The Newborn

The newborn baby, or **neonate,** experiences several dramatic changes at the time of birth. The major and earliest changes in the infant are separation from the maternal circulation and transfer from a fluid to a gaseous environment. The large, forced gasps of air that occur when the infant cries at the time of delivery help inflate the lungs.

Circulatory Changes

The initial inflation of the lungs causes important changes in the circulatory system (figure 29.19). Expansion of the lungs reduces the resistance to blood flow through the lungs, resulting in increased blood flow through the pulmonary arteries. Consequently, more blood flows from the right atrium to the right ventricle and into the pulmonary arteries, and less blood flows from the right atrium through the foramen ovale to the left atrium. In addition, an increased volume of blood returns from the lungs through the pulmonary veins to the left atrium, which increases the pressure in the left atrium. The increased left atrial pressure and decreased right atrial pressure, resulting from decreased pulmonary resistance, forces blood against the septum primum, causing the foramen ovale to close. This action functionally completes the separation of the heart into two pumps: the right side of the heart and the left side of the heart. The closed foramen ovale becomes the **fossa ovalis.**

The **ductus arteriosus,** which connects the pulmonary trunk to the aorta and allows blood to flow from the pulmonary trunk to the systemic circulation, closes off within 1 or 2 days after birth. This closure occurs because of the sphincterlike constriction of the artery and is probably stimulated by local changes in blood pressure and blood oxygen content. Once closed, the ductus arteriosus is replaced by connective tissue and is known as the **ligamentum arteriosum.**

> ### Clinical Note
>
> If the ductus arteriosus does not close completely, it is said to be **patent.** This is a serious birth defect, resulting in marked elevation in pulmonary blood pressure because blood flows from the left ventricle to the aorta, through the ductus arteriosus to the pulmonary arteries. If not corrected, it can lead to irreversible degenerative changes in the heart and lungs.

The fetal blood supply passes to the placenta through umbilical arteries from the internal iliac arteries and returns through an umbilical vein. The blood passes through the liver via the ductus venosus, which joins the inferior vena cava. When the umbilical cord is tied and cut, no more blood flows through the umbilical vein and arteries, and they degenerate. The remnant of the umbilical vein becomes the **ligamentum teres,** or **round ligament,** of the liver, and the ductus venosus becomes the **ligamentum venosum.**

Digestive Changes

When a baby is born, it is suddenly separated from its source of nutrients provided by the maternal circulation. Because of this separation and the shock of birth and new life, the neonate usually loses 5%–10% of its total body weight during the first few days of life. Although the digestive system of the fetus becomes somewhat functional late in development, it is still very immature in comparison with that of the adult and can digest only a limited number of food types.

Late in gestation, the fetus swallows amniotic fluid from time to time. Shortly after birth this swallowed fluid plus cells sloughed from the mucosal lining, mucus produced by intestinal mucous glands, and bile from the liver pass from the digestive tract as a greenish anal discharge called **meconium** (mē-kō′nē-ŭm).

The pH of the stomach at birth is nearly neutral because of the presence of swallowed alkaline amniotic fluid. Within the first 8 hours of life, a striking increase in gastric acid secretion occurs, causing the stomach pH to decrease. Maximum acidity is reached at 4–10 days, and the pH gradually increases for the next 10–30 days.

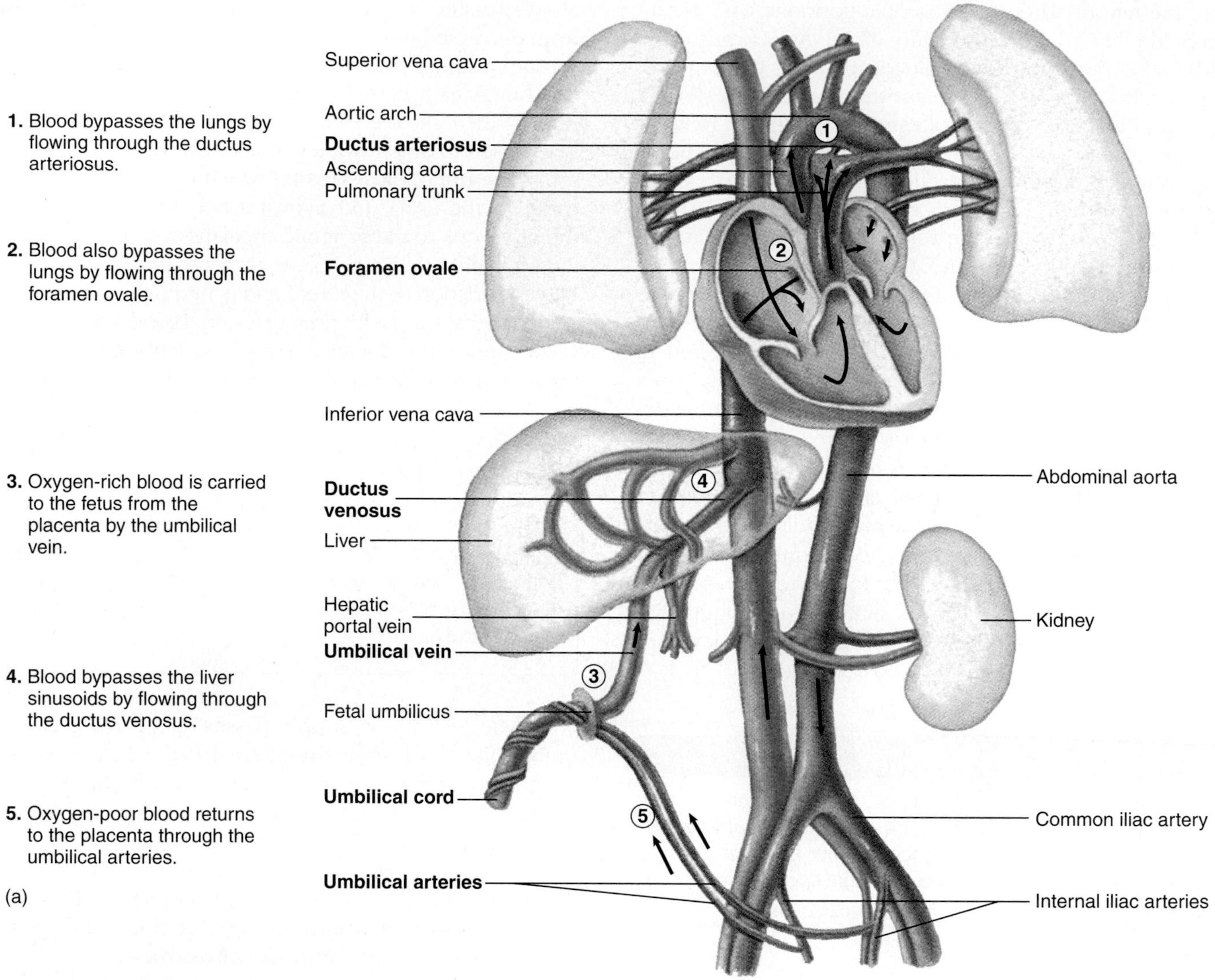

1. Blood bypasses the lungs by flowing through the ductus arteriosus.

2. Blood also bypasses the lungs by flowing through the foramen ovale.

3. Oxygen-rich blood is carried to the fetus from the placenta by the umbilical vein.

4. Blood bypasses the liver sinusoids by flowing through the ductus venosus.

5. Oxygen-poor blood returns to the placenta through the umbilical arteries.

(a)

Superior vena cava
Aortic arch
Ductus arteriosus
Ascending aorta
Pulmonary trunk
Foramen ovale
Inferior vena cava
Ductus venosus
Liver
Hepatic portal vein
Umbilical vein
Fetal umbilicus
Umbilical cord
Umbilical arteries

Abdominal aorta
Kidney
Common iliac artery
Internal iliac arteries

Figure 29.19   Circulatory Changes at Birth

(a) Circulatory conditions in the fetus (*continued on next page*)

The neonatal liver is also functionally immature. It lacks adequate amounts of the enzyme required in the production of bilirubin. This enzyme system usually develops within 2 weeks after birth in a healthy neonate, but because this enzyme system is not fully developed at birth, some full-term babies may temporarily develop jaundice. Jaundice often occurs in premature babies.

The newborn digestive system is capable of digesting lactose (milk sugar) from the time of birth. The pancreatic secretions are sufficiently mature for a milk diet, but the digestive system only gradually develops the ability to digest more solid foods over the first year or two; therefore, new foods should be introduced gradually during the first 2 years. It is also advised that only one new food be introduced at a time into the infant's diet so that, if an allergic reaction occurs, the cause is more easily determined.

Amylase secretion by the salivary glands and the pancreas remains low until after the first year. Lactase activity in the small intestine is high at birth but declines during infancy,

although the levels still exceed those in adults. Lactase activity is lost in many adults (see chapter 24).

Apgar Scores

The newborn baby may be evaluated soon after birth to assess its physiologic condition. This assessment is referred to as the **Apgar score.** Apgar, which was named for Virginia Apgar, the physician who developed it, also stands for **a**ppearance, **p**ulse, **g**rimace, **a**ctivity, and **r**espiratory effort. Each of these characteristics is rated on a scale of 0 to 2: 2 denotes normal function, 1 denotes reduced function, and 0 denotes seriously impaired function. The total Apgar score is the sum of the scores from the five characteristics, ranging therefore from 0 to 10 (table 29.3). A total Apgar score of 8–10 at 1–5 minutes after birth is considered normal. Other scoring systems to estimate normal growth and development, including general external appearance and neurologic development, may also be applied to the neonate.

1. When air enters the lungs, the ductus arteriosus closes.

2. The foramen ovale closes.

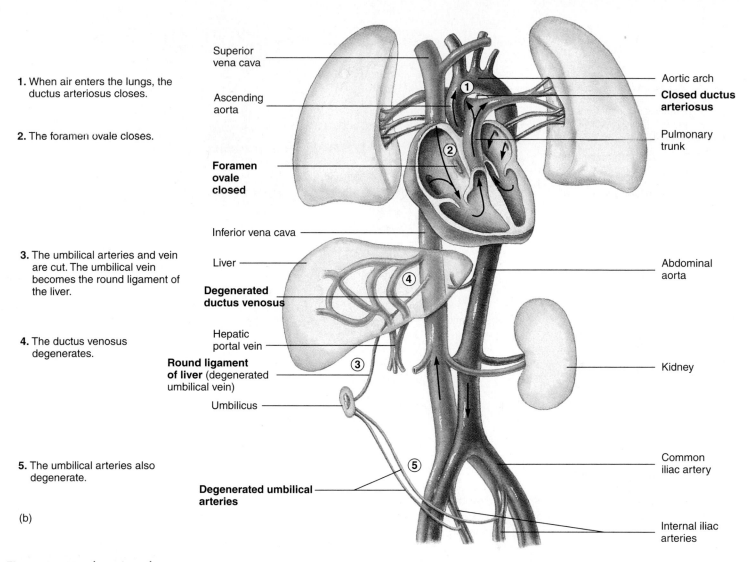

3. The umbilical arteries and vein are cut. The umbilical vein becomes the round ligament of the liver.

4. The ductus venosus degenerates.

5. The umbilical arteries also degenerate.

(b)

Figure 29.19   (*continued*)

(*b*) Circulatory changes that occur at birth.

Table 29.3 Examples of Apgar Rating Scales	0	1	2
Appearance (skin color)	White or blue	Limbs blue, body pink	Pink
Pulse (rate)	No pulse	100 bpm	>100 bpm
Grimace (reflexive grimace initiated by stimulating the plantar surface of the foot)	No response	Facial grimaces, slight body movement	Facial grimaces, extensive body movement
Activity (muscle tone)	No movement, muscles flaccid	Limbs partially flexed, little movement, poor muscle tone	Active movement, good muscle tone
Respiratory effort (amount of respiratory activity)	No respiration	Slow, irregular respiration	Good, regular respiration, strong cry

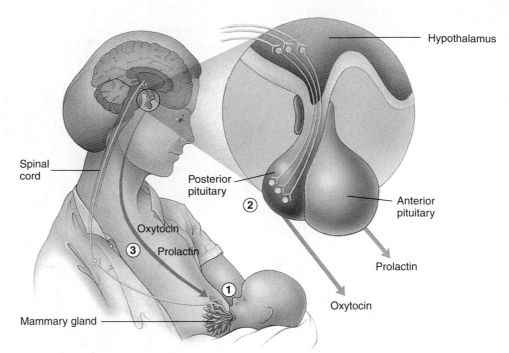

1. Stimulation of the nipple by the baby's suckling initiates action potentials in the afferent neurons that connect with the hypothalamus.

2. The hypothalamus stimulates the posterior pituitary to release oxytocin and the anterior pituitary to release prolactin.

3. Oxytocin stimulates milk release from the breast. Prolactin stimulates additional milk production.

Figure 29.20   Hormonal Control of Lactation

Lactation

Lactation is the production of milk by the mother's breasts (mammary glands; figure 29.20). It normally occurs in females after parturition and may continue for 2 or 3 years, provided suckling occurs often and regularly.

During pregnancy the high concentration and continuous presence of estrogens and progesterone cause expansion of the duct system and secretory units of the breasts. The ducts grow and branch repeatedly to form an extensive network. Additional adipose tissue is deposited also; thus the size of the breasts increases substantially throughout pregnancy. Estrogen is primarily responsible for breast growth during pregnancy, but normal development of the breast does not occur without the presence of several other hormones. Progesterone causes development of the breasts' secretory alveoli, which enlarge but do not secrete milk during pregnancy. The other hormones include growth hormone, prolactin, thyroid hormones, glucocorticoids, and insulin. A growth hormonelike substance (human somatotropin) and a prolactinlike substance (human placental lactogen) are secreted by the placenta, and these substances also help support the development of the breasts.

Prolactin, which is produced by the anterior pituitary gland, is the hormone responsible for milk production. Before parturition, high levels of estrogen stimulate an increase in prolactin production. Milk production is inhibited during pregnancy, however, because high levels of estrogen and progesterone inhibit the effect of prolactin on the mammary gland.

After parturition, estrogen, progesterone, and prolactin levels decrease; and, with lower estrogen and progesterone levels, prolactin can stimulate milk production. Despite a decrease in the basal levels of prolactin, a reflex response produces surges of prolactin release. During suckling, mechanical stimulation of the breasts initiates nerve impulses that reach the hypothalamus, causing the secretion of **prolactin-releasing factor (PRF)** and inhibiting the release of **prolactin-inhibiting factor (PIF).** Consequently, prolactin levels temporarily increase and stimulate milk production.

For the first few days after birth the mammary glands secrete **colostrum** (kō-los′trŭm), which contains little fat and less lactose than milk. Eventually more nutritious milk is produced. Colostrum and milk not only provide nutrition, but they also contain antibodies (see chapter 22) that help protect the nursing baby from infections.

Clinical Note

The human immunodeficiency virus (HIV) can be transmitted from a mother to her child in utero, during parturition, or during breast-feeding. HIV has been isolated from human breast milk and colostrum.

In a study of 212 mothers in Africa who were seronegative (no HIV antibodies found in the serum) at the time of delivery, 16 seroconverted (developed HIV antibodies) during the time that they were breast feeding. Nine of the nursing babies also seroconverted. At least four of the babies were confirmed to be seronegative at the time of birth.

Repeated stimulation of prolactin release makes nursing possible for several years. If nursing stops, however, within a few days the ability to produce prolactin ceases, and milk production stops.

Because it takes time to produce milk, an increase in prolactin results in the production of milk to be used in the next nursing period. At the time of nursing, stored milk is released as a result of a reflex response. Mechanical stimulation of the breasts produces nerve impulses that cause the release of **oxytocin** from the posterior pituitary, which stimulates cells surrounding the alveoli to contract. Milk is then released from the breasts, a process that is called **milk letdown.** In addition, higher brain centers can stimulate oxytocin release, and such things as hearing an infant cry can result in milk letdown.

First Year After Birth

Many changes occur in the life of the newborn from birth until 1 year of age. The time of these changes may vary considerably from child to child, and the dates given are only rough estimates. The brain is still developing, and much of what the neonate can do depends on how much brain development has occurred. It is estimated that the total adult number of neurons is present in the central nervous system at birth, but subsequent growth and maturation of the brain involve the addition of new neuroglial cells, some of which form new myelin sheaths, and the addition of new connections between neurons, which may continue throughout life.

By 6 weeks, the infant usually can hold up its head when placed in a prone position and begins to smile in response to people or objects. At 3 months of age, the infant's limbs move apparently aimlessly. The infant has enough control of the arms and hands, however, that voluntary thumb sucking can occur. The infant can follow a moving person with its eyes. At 4 months the infant begins to raise itself by its arms. It can begin to grasp objects placed in its hand, coo and gurgle, roll from its back to its side, listen quietly when hearing a person's voice or music, hold its head erect, and play with its hands. At 5 months the infant can usually laugh, reach for objects, turn its head to follow an object, lift its head and shoulders, sit with support, and roll over. At 8 months the infant recognizes familiar people, sits up without support, and reaches for specific objects. At 12 months the infant may pull itself to a standing position and may be able to walk without support. It can pick up objects in its hands and examine them carefully. It can understand much of what is said to it and may say several words of its own.

Figure 29.21   Active Older Adults
A conspicuous feature of the population of older adults is the range of variability. In some adults over 70 years, many systems are beginning to fail. Others can look forward to at least 10 more years of healthy living.

Life Stages

The prenatal and neonatal periods of life previously described are only a small part of the total life span. The life stages from fertilization to death are as follows: (1) the germinal period: fertilization to 14 days; (2) embryo: 14–56 days after fertilization; (3) fetus: 56 days after fertilization to birth; (4) neonate: birth to 1 month after birth; (5) infant: 1 month to 1 or 2 years (the end of infancy is sometimes set at the time that the child begins to walk); (6) child: 1 or 2 years to puberty; (7) adolescent: puberty (age 11–14) to 20 years; and (8) adult: age 20 to death. Adulthood is sometimes divided into three periods: young adult, age 20–40; middle age, age 40–65; and older adult or senior citizen, age 65 to death (figure 29.21). Much of this designation is associated more with social norms than with physiology.

During childhood the individual develops considerably. Many of the emotional characteristics that a person possesses throughout life are formed during early childhood.

Major physical and physiologic changes occur during adolescence that also affect the emotions and behavior of the individual. Other emotional changes occur as the adolescent attempts to fit into an adult world. Puberty usually occurs somewhat earlier in females (at about 11–13 years) than in males (at about 12–14 years). The onset of puberty is usually accompanied by a growth spurt, followed by a period of slower growth. Full adult stature is usually achieved by age 17 or 18 in females and 19 or 20 years in males.

Aging

Development of a new human being begins at fertilization, as does the process of aging. Cells proliferate at an extremely rapid rate during early development and then the process begins to slow as various cells become committed to specific functions within the body.

Some cells of the body, such as liver and skin cells, continue to proliferate throughout life, replacing dead or damaged tissue. Many other cells, however, such as the neurons in the central nervous system, cease to proliferate once they have reached a certain number and dead cells are not replaced. After the number of neurons reaches a peak, at about the time of birth, their numbers begin to decline. Neuronal loss is most rapid early in life and later decreases to a slower, steadier rate.

There is a natural, but as yet unexplained, decline in mitochondrial deoxyribonucleic acid (DNA) function with age. If this decline reaches a threshold, which apparently differs from tissue to tissue, the normal function of the mitochondrion is lost, and the tissue or organ may exhibit disease symptoms. In a small number of people, this mitochondrial degeneration occurs very early in life, resulting in premature aging.

The physical plasticity (i.e., the state of being soft and pliable) of young embryonic tissues results largely from the presence of large amounts of hyaluronate and relatively small amounts of collagen; furthermore, the collagen and other, related proteins that are present are not highly cross-linked; thus the tissues are very flexible and elastic. Many of these proteins produced during development are permanent components of the individual, however; and, as the individual ages, more and more cross-links form between these protein molecules, rendering the tissues more rigid and less elastic.

The tissues with the highest content of collagen and other, related proteins are those most severely affected by the collagen cross-linking and tissue rigidity associated with aging. The lens of the eye is one of the first structures to exhibit pathologic changes as a result of this increased rigidity. Vision of close objects becomes more difficult with advancing age until most middle-aged people require reading glasses (see chapter 15). Loss of elasticity also affects other tissues, including the joints, blood vessels, kidneys, lungs, and heart, and greatly reduces the functional ability of these organs.

Like nervous tissue, mature muscle cells do not normally proliferate after terminal differentiation occurs before birth. As a result, the total number of skeletal and cardiac muscle fibers declines with age. The strength of skeletal muscle reaches a peak between 20 and 30 years of life and declines steadily thereafter. Furthermore, like the collagen of connective tissue, the macromolecules of muscle undergo biochemical changes during aging, rendering the muscle tissue less functional. A good exercise program, however, can slow or even reverse this process.

The decline in muscular function also contributes to the decline in cardiac function with advancing age. The heart loses elastic recoil ability and muscular contractility. As a result, total cardiac output declines, and less oxygen and fewer nutrients reach cells such as neurons of the brain and cartilage cells of the joints, contributing to the decline in these tissues. Reduced cardiac function also may result in decreased blood flow to the kidneys, contributing to decreases in their filtration ability. Degeneration of the connective tissues as a result of collagen cross-linking and other factors also decreases the filtration efficiency of the glomerular basement membrane.

Atherosclerosis (ath′er-ō-skler-ō′sis) is the deposit and subsequent hardening of a soft, gruellike material in lesions of the intima of large- and medium-sized arteries. These deposits then become fibrotic and calcified, resulting in **arteriosclerosis** (ar-tēr′ē-ō-skler-ō′sis, meaning hardening of the arteries). Arteriosclerosis interferes with normal blood flow and can result in a **thrombus,** which is a clot or plaque formed inside a vessel. A piece of the plaque, called an **embolus** (em′bō-lŭs) can break loose, float through the circulation, and lodge in smaller arteries to cause myocardial infarctions or strokes. Although atherosclerosis occurs to some extent in all middle-aged and elderly people and even may occur in certain young people, some people appear more at risk because of high blood cholesterol levels. In addition to dietary influences, this condition seems to have a heritable component, and blood tests are available to screen people for high blood cholesterol levels.

Many other organs, such as the liver, pancreas, stomach, and colon, undergo degenerative changes with age. The ingestion of harmful agents may accelerate such changes. Examples include the degenerative changes induced in the lungs, aside from lung cancer, by cigarette smoke and sclerotic changes in the liver as a result of alcohol consumption.

In addition to the previously described changes associated with aging, cellular wear and tear, or cytologic aging, is another factor that contributes to aging. Progressive damage from many sources such as radiation and toxic substances may result in irreversible cellular insults and may be one of the major factors leading to aging. It has been speculated that ingestion of the antioxidant vitamins C and E in combination may help slow this part of aging by stimulating cell repair. Vitamin C also stimulates collagen production and may slow the loss of tissue plasticity associated with aging collagen.

As a result of poor diet, many people over age 50 do not get the minimum daily allotment of several vitamins and minerals. Feeling "bad" is not necessarily a part of aging but is mostly a result of poor nutrition and lack of exercise. Moderate exercise and avoiding overeating can prolong life. Moderate exercise can reduce the risk of heart attack by as much as 20%. It can also reduce the risk of stroke, high blood pressure, and some forms of cancer. Exercise can also increase a person's ability to reason and remember. Walking 30 min/day is recommended.

Immune changes may also be a major factor contributing to aging. The aging immune system loses its ability to respond to outside antigens but becomes more sensitive to the body's own antigens. These autoimmune changes add to the degeneration of the tissues already described and may be responsible for such things as arthritic joint disorders, chronic glomerular nephritis, and hyperthyroidism. In addition, T lymphocytes tend to lose their functional capacity with aging and cannot destroy abnormal cells as efficiently. This change may be one reason that certain types of cancer occur more readily in older people.

Many changes associated with aging may be caused by genetic traits. As a general rule, animals with a very high metabolic rate have a shorter life span than those with a lower metabolic rate. In humans, a very small number of exceptional people have a slightly reduced normal body temperature, suggesting a lower metabolic rate. These same people often have an unusually long life span. This tendency appears to run in families and probably has some genetic basis. Studies of the general population suggest that if your parents and grandparents have lived long, so will you. Conversely, if your parents and grandparents died young, you may expect the same.

Another piece of evidence suggesting that there is a strong genetic component to aging comes from a disorder called **progeria** (prō-jēr'ē-ă, meaning premature aging). This apparent genetic trait causes the degenerative changes of aging to occur shortly after the first year of life, and the child may look like a very old person by age 7.

One of the greatest disadvantages of aging is the increasing lack of ability to adjust to stress. Older people have a far more precarious homeostatic balance than younger people, and eventually some stress is encountered that is so great that the body's ability to recover is surpassed, and death results.

Death

Death is usually not attributed to old age. Some other problem such as heart failure, renal failure, or stroke is usually listed as the cause of death.

Death was once defined as the loss of heartbeat and respiration. In recent years, however, more precise definitions of death have been developed because both the heart and the lungs can be kept working artificially, and the heart can even be replaced by an artificial device. Modern definitions of death are based on the permanent cessation of life functions and the cessation of integrated tissue and organ function. The most widely accepted indication of death in humans is **brain death,** which is defined as irreparable brain damage manifested clinically by the absence of response to stimulation, the absence of spontaneous respiration and heart beat, and an isoelectric ("flat") electroencephalogram for at least 30 min, in the absence of known central nervous system poisoning or hypothermia.

Genetics

Genetics is the study of heredity, that is, those characteristics inherited by children from their parents. Although the environment can influence gene expression, a person's physical characteristics and abilities are largely determined by their genetic makeup. Many of a person's abilities, susceptibility to disease, and even their life span are influenced by heredity. Because many of the diseases caused by microorganisms now are preventable or treatable, diseases that have a genetic basis are receiving more attention.

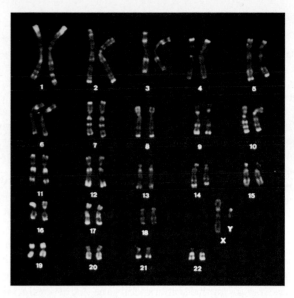

Figure 29.22   Human Karyotype

The 23 pairs of chromosomes in humans consists of 22 pairs of autosomal chromosomes (numbered 1–22) and 1 pair of sex chromosomes. This karyotype is of a male and has an X and a Y sex chromosome. A female karyotype would have two X chromosomes.

Chromosomes

Deoxyribonucleic (dē-oks'ē-rī'bō-nū-klē'ik) **acid (DNA)** is the hereditary material of cells and is responsible for controlling cell activities. DNA molecules and their associated proteins become visible as densely stained bodies, called **chromosomes** (krō'mō-sōmz, meaning colored bodies), during cell division (see figure 2.26). Somatic cells contain 23 pairs of chromosomes, or 46 total chromosomes, and gametes contain 23 chromosomes. **Somatic** (sō-mat'ik) **cells** are all the cells of the body except for the **gametes** (gam'ētz), or sex cells. Examples of somatic cells are epithelial cells, muscle cells, neurons, fibroblasts, lymphocytes, and macrophages. In the male, the gametes are sperm cells, and in the female, the gametes are oocytes (see chapter 28).

A **karyotype** (kar'ē-ō-tīp), or display of the chromosomes in a somatic cell, can be produced by photographing the chromosomes through a microscope, cutting the pictures of the chromosomes out of the photograph, and arranging the chromosomes in pairs (figure 29.22). The 23 pairs of chromosomes are divided into two groups. There are 22 pairs of **autosomal** (aw-tō-sō'măl) **chromosomes,** all the chromosomes but the sex chromosomes, and one pair of **sex chromosomes,** which determine the sex of the individual. For convenience, the autosomes are numbered in pairs from 1 through 22, and sex chromosomes are denoted as **X** or **Y chromosomes.** A normal female has two X chromosomes (XX) in each somatic cell, whereas a normal male has one X and one Y chromosome (XY) in each somatic cell.

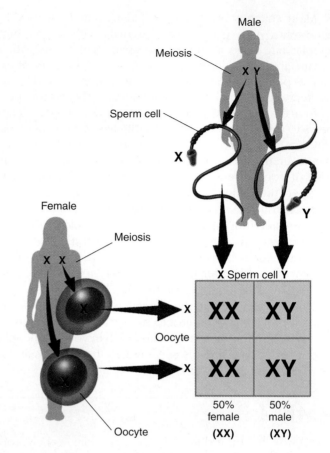

Figure 29.23   Inheritance of Sex

The female produces oocytes containing one X chromosome, whereas the male produces sperm cells with either an X or a Y chromosome. There are four possible combinations of an oocyte with a sperm cell, half of which produce females and half of which produce males.

Gametes are derived from somatic cells by **meiosis** (mī-ō′sis). In this process the somatic cells divide twice, and the chromosomes from the somatic cells are distributed to the gametes. Meiosis is called a reduction division because the number of chromosomes in the gametes is half the number in the somatic cells. When a sperm cell and an oocyte fuse during fertilization, each contributes one-half of the chromosomes necessary to produce new somatic cells; therefore half of an individual's genetic makeup comes from the father and half from the mother.

During meiosis, the chromosomes are distributed in such a way that each gamete receives only one chromosome from each **homologous** (hŏ-mol′ŏ-gŭs) pair of chromosomes. Homologous chromosomes contain the same compliment of genetic information. The inheritance of sex illustrates, in part, how chromosomes are distributed during gamete formation and fertilization. During meiosis and gamete formation, the pair of sex chromosomes separates so that each oocyte receives one of a homologous pair of X chromosomes, whereas each sperm cell receives either an X chromosome or a Y chromosome (figure 29.23). When a sperm cell fertilizes an oocyte to form a single cell, the sex of the individual is determined randomly. If the oocyte is fertilized by a sperm cell with a Y chromosome, a male results, but if the oocyte is fertilized by a sperm cell with an X chromosome, a female results. Estimating the probability of any given zygote being male or female is much like flipping a coin. When all the possible combinations of sperm cells with oocytes are considered, about half the individuals should be female and the other half should be male.

Genes

The functional unit of heredity is the gene. Each **gene** is a certain portion of a DNA molecule but not necessarily a continuous stretch of DNA (see figure 3.30). Each chromosome contains thousands of genes, and each gene occupies a specific **locus** on a chromosome. Both chromosomes of a given pair contain similar but not necessarily identical genes. The genes occupying the same locus on homologous chromosomes are called **alleles** (ă-lēlz′). If the two allelic genes are identical, the person is **homozygous** (hō-mō-zī′gŭs) for the trait specified by that gene locus. If the two alleles are slightly different, the person is **heterozygous** (het′er-ō-zī′gŭs) for the trait.

There are two major types of genes: structural and regulatory. **Structural genes** are those DNA sequences that serve as a template for mRNA and code for specific amino acid sequences in proteins such as enzymes, hormones, or structural proteins like collagen. **Regulatory genes** are segments of DNA involved in controlling which structural genes are transcribed in a given tissue. By determining the structure of proteins and which proteins are produced by which cells, genes are responsible for the characteristics of cells and, therefore, the characteristics of the entire organism. All the genes in one homologous set of 23 chromosomes in one individual, taken together is called the **genome** (jē′nōm). The two combined genomes of a person are responsible for all of that person's genetic traits.

The importance of genes is dramatically illustrated by situations in which the alteration of a single gene results in a genetic disorder. For example, in **phenylketonuria** (fen'il-kē'tō-nū'rē-ă) (**PKU**) the gene responsible for producing an enzyme that converts the amino acid phenylalanine to the amino acid tyrosine is defective. Phenylalanine therefore accumulates in the blood and is eventually converted to harmful substances that can cause mental retardation.

Through the processes of meiosis, gamete formation, and fertilization there is essentially a random distribution of genes received from each parent, a process called **independent assortment** of the genes. This random distribution is influenced by several factors, however. For example, all of the genes on a given chromosome are **linked,** that is, they tend to be inherited as a set rather than as individual genes because chromosomes, not individual genes segregate during meiosis. Sets of linked genes can be broken up, however. When tetrads are formed during meiosis (see chapter 3), homologous chromosomes may exchange genetic information by **crossing over** (see figure 3.36).

Furthermore, segregation errors may occur during meiosis. As the chromosomes separate during meiosis, the two members of a homologous pair may become "sticky" and not segregate as they normally do. As a result, one of the daughter cells receives both chromosome pairs and the other daughter cell receives none. This event is called **nondisjunction.** When the gametes are fertilized, the resulting zygote has either 47 chromosomes or 45 chromosomes rather than the normal 46, a condition called **aneuploidy** (an'yū-ploy-dē). This condition is usually lethal and is one reason for a high rate of early embryo loss. Some types of aneuploidy are not lethal, however. The sex chromosome abnormalities described in the section dealing with chromosomes are examples. Another example is **Down's syndrome,** or **trisomy 21,** a type of aneuploidy in which there are three chromosomes 21.

Dominant and Recessive Genes

Most human genetic traits that we are aware of are recognized because defective alleles for those traits exist in the population. For example, on chromosome 11 there is a gene that produces an enzyme necessary for the synthesis of melanin, the pigment responsible for skin, hair, and eye color (see chapter 5). The normal allele for the melanin gene produces a normal, functional enzyme. Another, abnormal allele, however, produces a defective enzyme not capable of catalyzing one of the normal steps in melanin synthesis. If a given person inherits two defective alleles at that melanin-producing enzyme locus, a homozygous condition, the person is unable to produce melanin and therefore lacks normal pigment in the skin, hair, and eyes. This condition is referred to as **albinism.** Instead of the normal coloration, the coloration of a person with albinism consists of shades of pink,

blue, and yellow. The pink and blue colors result from blood seen through the skin (see chapter 5), and the yellow color is from the natural accumulation of ingested yellow plant pigments in the skin.

For many genetic traits, the effects of one allele for that trait can mask the effect of another allele for that same trait. For example, a person who is heterozygous for the melanin-producing enzyme gene on chromosome 11 has a normal gene for melanin production on one chromosome 11 and the defective gene for melanin production at the same locus on the other chromosome 11. In the case of this melanin-producing enzyme, one copy of the gene and its resulting enzymes are enough to make normal melanin. As a result, the person who is heterozygous produces melanin and appears normal. In this case, the allele that produces the normal enzyme and is responsible for normal appearance is said to be **dominant,** whereas the allele producing the abnormal enzyme is **recessive.** The lost function of the defective enzyme is masked by the dominant, normal allele. Thus normal pigmentation is a dominant trait and albinism is a recessive trait. By convention, dominant traits are indicated by uppercase letters, and recessive traits are indicated by lowercase letters. In this example, the letter "A" designates the dominant normal, pigmented condition and the letter "a" the recessive albino condition. It is important to note that not all dominant traits are the normal condition and that not all recessive traits are abnormal. There are many examples in which the dominant trait is abnormal.

The possible combinations of dominant and recessive alleles for normal melanin production versus albinism are *AA* (homozygous dominant), *Aa* (heterozygous), and *aa* (homozygous recessive). The actual set of alleles that a person has for a given trait is called the **genotype** (jen'ō-tīp). The person's appearance is called the **phenotype** (fē'nō-tīp). A person with the genotype *AA* or *Aa* has the phenotype of normal pigmentation, whereas a person with the genotype *aa* has the phenotype of albinism. Note that the recessive trait is expressed when it is not masked by the dominant trait.

6   P R E D I C T

Polydactyly (pol-ē-dak'ti-lē) is a condition in which a person has extra fingers or toes. Given that polydactyly is a dominant trait, list all the possible genotypes and phenotypes for polydactyly. Use the letters "*D*" and "*d*" for the genotypes.

✔ *Answer in Appendix F*

The inheritance of dominant and recessive traits can be determined if the genotypes of the parents are known. For example, if an albino person (*aa*) mates with a heterozygous normal person (*Aa*), the probability is that half of the children will be albino (*aa*), and half will be normal heterozygous carriers (*Aa*). If two carriers (*Aa*) mate, the probability is that 1 in 4 will be homozygous dominant (*AA*), 1 in 4 will

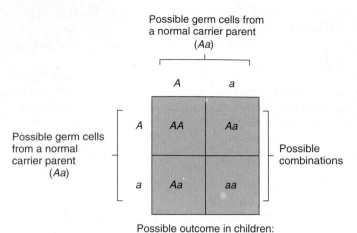

Possible germ cells from
a normal carrier parent
(*Aa*)

Possible germ cells
from a normal
carrier parent
(*Aa*)

Possible
combinations

Possible outcome in children:
$\frac{1}{4}$ *AA* (normal) : $\frac{1}{2}$ *Aa* (normal carrier) : $\frac{1}{4}$ *aa* (albino)

Figure 29.24   Inheritance of a Recessive Trait: Albinism
A represents the normal pigmented condition, and *a* represents the recessive unpigmented condition. The figure represents a mating between two normal carriers.

be homozygous recessive (*aa*), and 1 in 2 will be heterozygous (*Aa*). Such a probability can be easily determined by the use of a table called a **Punnett square** (figure 29.24). **Carriers** are heterozygous persons with an abnormal recessive gene but with a normal phenotype because they also have a normal dominant allele for that gene.

7 . P R E D I C T

If a carrier for albinism mates with a homozygous normal person, what is the likelihood that any of their children will be albinos? Explain.

✔ *Answer in Appendix F*

Sex-Linked Traits

Traits affected by genes on the sex chromosomes are called **sex-linked traits.** Most sex-linked traits are **X-linked,** that is, they are on the X chromosome, whereas, there only a few **Y-linked** traits, largely because the Y chromosome is very small. An example of an X-linked trait is **hemophilia A** (classic hemophilia) in which there is an inability to produce one of the clotting factors (see chapter 19). Consequently, clotting is impaired and persistent bleeding can occur either spontaneously or as a result of an injury. Hemophilia A is a recessive trait located on the X chromosome. The possible genotypes and phenotypes are therefore $X^H X^H$ (normal homozygous female), $X^H X^b$ (normal heterozygous female), $X^b X^b$ (hemophiliac homozygous female), $X^H Y$ (normal male), and $X^b Y$ (hemophiliac male). Note that a female must have both recessive genes to exhibit hemophilia, whereas a male, because he has only one X chromosome, has hemophilia if he has only one of the recessive genes. An example of the inheritance of hemophilia is il-

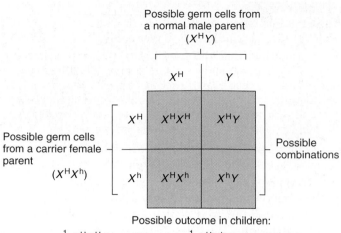

Possible germ cells from
a normal male parent
($X^H Y$)

Possible germ cells
from a carrier female
parent
($X^H X^h$)

Possible
combinations

Possible outcome in children:
$\frac{1}{4}$ $X^H X^H$ (normal female) : $\frac{1}{4}$ $X^H X^h$ (carrier female) :
$\frac{1}{4}$ $X^H Y$ (normal male) : $\frac{1}{4}$ $X^h Y$ (male with hemophilia)

Figure 29.25   Inheritance of an X-Linked Trait: Hemophilia
X^H represents the normal X chromosome condition with all clotting factors, and X^h represents the X chromosome lacking a gene for one clotting factor. The figure represents a mating between a normal male and a normal carrier female.

lustrated in figure 29.25. If a woman who is a carrier for hemophilia mates with a man who does not have hemophilia, none of their daughters but half of their sons will have hemophilia.

Other Types of Gene Expression

The expression of a dominant over a recessive gene is the simplest manner in which genes determine a person's phenotype. There are many other ways in which genes influence the expression of a trait. In some cases the dominant gene does not completely mask the effects of the recessive gene, a phenomenon called **incomplete dominance.** An example of incomplete dominance is **sickle cell anemia,** in which a gene responsible for producing hemoglobin in red blood cells is abnormal. Consequently, the hemoglobin produced by the gene is abnormal. The result is red blood cells that are stretched into an elongated sickle shape. These red blood cells tend to stick in capillaries, blocking blood flow to tissues. In addition, the sickle-shaped cells tend to rupture more easily than normal red blood cells. The normal allele (*S*) for producing normal hemoglobin is dominant over the sickle cell allele (*s*) responsible for producing the abnormal hemoglobin. A normal person has the genotype *SS* and has normal hemoglobin. A person with sickle cell anemia has genotype *ss* and has abnormal hemoglobin. A person who is heterozygous has the genotype *Ss,* has half normal hemoglobin and half abnormal hemoglobin, and usually has only a few sickle-shaped red blood cells. This condition is called **sickle cell trait.** Usually a person with sickle cell trait exhibits no adverse symptoms. For this set of alleles, expressing incomplete dominance, however, each genotype presents a unique, recognizable phenotype.

In another type of gene expression, called **codominance,** two alleles can combine to produce an effect without either of them being dominant or recessive. For example, a person with type AB blood has A antigens and B antigens on the surface of the red blood cells (see chapter 19). The antigens result from a gene that causes the production of the A antigen and a different gene that causes the production of the B antigen. In this case A and B are neither dominant or recessive in relation to each other.

Many traits, called **polygenic traits,** are determined by the expression of multiple genes on different chromosomes. Examples are a person's height, intelligence, eye color, and skin color. Polygenic traits typically are characterized by having a great amount of variability. For example, there are many different shades of eye color and skin color (figure 29.26). Because of the many genes involved, it is difficult to predict how a polygenic trait will be passed from one generation to the next. Notice, however, that even though skin color is determined by a complex combination of genes, one single defective gene can eliminate skin color completely, resulting in albinism, which is inherited as a simple dominant trait.

Genetic Disorders

Genetic disorders are caused by abnormalities in a person's genetic makeup, that is, in his or her DNA. They may involve a single gene or an entire chromosome, as in the case of aneuploidy (table 29.4). Genetic disorders are often confused with **congenital disorders.** Congenital means "present at birth" and is commonly referred to as a birth defect, but all congenital disorders are not necessarily genetic. Approximately 15% of all congenital disorders have a known genetic cause, and approximately 70% of all birth defects are of unknown cause. The remaining 15% are the result of environmental causes or a combination of environmental and genetic causes. In the case of environmental causes, the birth defect results from damage to the fetus during development. Agents that can cause birth defects are called **teratogens** (ter-at′ō-jenz). For example, fetal alcohol syndrome results when a pregnant woman drinks alcohol, which crosses the placenta and damages the fetus. The baby is born with a smaller than normal head and mental retardation, and may exhibit other birth defects.

One cause of genetic disorders is a **mutation,** a change in a gene that usually involves a change in the number or kinds of nucleotides composing the DNA (see chapter 2). Mutations are known to occur by chance (randomly without known cause) or may be caused by chemicals, radiation, or viruses. Agents that cause mutations are called **mutagens** (mū′ta-jenz). In most cases, a specific cause of a mutation cannot be determined. Once a mutation has occurred, however, the abnormal trait can be passed from one generation to the next.

Cancer is a tumor resulting from uncontrolled cell divisions. **Oncogenes** (ong′kō-jēnz) are genes associated with cancer. Many oncogenes are actually control genes involved in regulating cell proliferation and differentiation in the embryo and fetus. A change in an oncogene or in the regula-

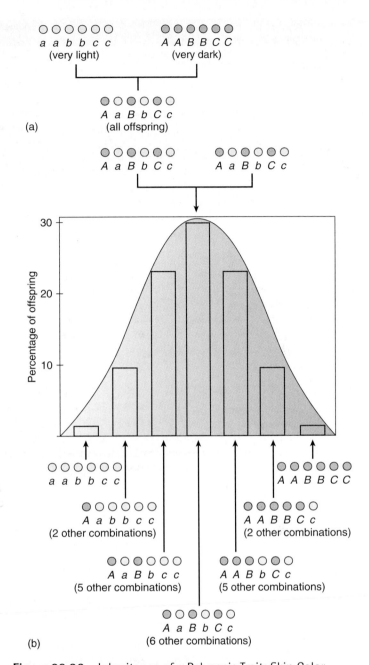

Figure 29.26   Inheritance of a Polygenic Trait: Skin Color

In this example, three genes for skin color are shown. The dominant alleles (*A, B, C*), each of which contributes one "unit of dark color" to the offspring (indicated by a dark dot), are incompletely dominant over the recessive alleles (*a, b, c*), each of which contributes one "unit of light color" to the offspring (indicated by a light dot). (*a*) A mating between a very light-skinned person (*aabbcc*) and a very dark-skinned person (*AABBCC*) is shown. All the offspring are of intermediate color. (*b*) A mating between two people of intermediate skin color (*AaBbCc*). The possible offspring skin color falls within a normal distribution in which a very low percentage (<2%) are either very light or very dark, and most of the offspring are of intermediate color.

tion of an oncogene can result in uncontrolled cell proliferation and the development of cancerous tumors. It is believed that certain chemicals called **carcinogens** (kar-sin′ō-jenz) can induce such changes and thereby initiate the

Table 29.4   Genetic Disorders

Disorder	Description
Dominant Traits	
Achondroplasia	Dwarfism characterized by shortening of the upper and lower limbs
Huntington's disease	Severe degeneration of the basal nuclei and frontal cerebral cortex; characterized by purposeless movements and mental deterioration; onset is usually between 40 and 50 years of age
Hypercholesterolemia	Elevated blood cholesterol levels that contribute to atherosclerosis and cardiovascular disease
Marfan's syndrome	Abnormal connective tissue results in increased height, elongated digits, and weakness in the aortic wall
Neurofibromatosis	Small pigmented lesions (café-au-lait spots) in the skin and disfiguring tumors (noncancerous) caused by proliferation of Schwann cells along nerves
Osteogenesis imperfecta	Abnormal phosphate metabolism results in brittle bones that repeatedly break
Recessive Traits	
Albinism	Lack of the enzyme necessary to produce the pigment melanin; characterized by lack of skin, hair, and eye coloration
Cystic fibrosis	Impaired transport of chloride ions across cell membranes; results in excessive production of thick mucus that blocks the respiratory and gastrointestinal tract; the most common fatal genetic disorder
Phenylketonuria	Lack of the enzyme necessary to convert the amino acid phenylalanine to the amino acid tyrosine; an accumulation of phenylalanine leads to mental retardation
Severe combined immune deficiency	Inability to form the white blood cells (B cells, T cells, and phagocytes) necessary for an immune system response
Sickle cell anemia	Inability to produce normal hemoglobin; results in abnormally shaped red blood cells that clog capillaries or rupture
Tay-Sachs disease	Lack of the enzyme necessary to break down certain fatty substances; an accumulation of fatty substances impairs action potential propagation, resulting in deterioration of mental and physical functions and death by 3–4 years of age
Thalassemia	Decreased rate of hemoglobin synthesis; results in anemia, enlargement of the spleen, increased cell numbers in red bone marrow, and congestive heart failure
Sex-Linked Traits	
Hemophilia	Most commonly, a recessive gene causes a failure to produce blood clotting factors; resulting in prolonged bleeding
Red-green color blindness	Most commonly, a recessive gene causes a deficiency of functional green-sensitive cones; inability to distinguish between red and green colors
Chromosomal Disorders	
Down's syndrome	Caused by having three chromosomes 21; results in mental retardation, short stature, and poor muscle tone
Duchenne's muscular dystrophy	Caused by deletion or alteration of part of the X chromosome; results in progressive weakness and wasting of muscles
Klinefelter's syndrome	Caused by two or more X chromosomes in a male (XXY); results in small testes, sterility, and development of female-like breasts
Turner's syndrome	Caused by having only one X chromosome; results in immature uterus, lack of ovaries, and short stature

development of cancer. For example, chemicals in cigarette smoke are known to cause lung cancer. Thus, even though everyone has oncogenes, an outside agent may be necessary for a cancer to begin.

A change in somatic cells that results in cancer is not usually inheritable; nonetheless, there may be a genetic basis that allows cancer development, especially under the right environmental conditions. In this sense, the inheritance of cancer and other abnormalities has been described as **genetic susceptibility,** or **genetic predisposition.** For example, if a woman's close relatives, such as her mother or sister, have breast cancer, she has a greater-than-average risk of de-

veloping breast cancer. Similar genetic susceptibilities have been found for diabetes mellitus, schizophrenia, and other disorders.

Treatment of genetic disorders usually involves treatment of symptoms, but does not result in a cure because the basic genetic material is unchanged by the treatment. For example, treating the problems associated with mucus buildup in cystic fibrosis does not cure the disorder. Currently, research is underway to actually alter a person's genetic makeup. Ultimately it may be possible to insert a normal gene into cells to replace an abnormal or missing gene. Then, if the normal gene functions, the disorder will be cured.

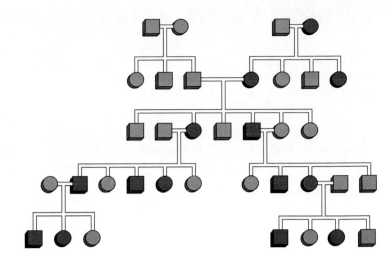

Figure 29.27   Pedigree of a Simple Dominant Trait

Males are indicated by squares, females by circles. Affected people are indicated by the purple symbols. The horizontal line between symbols represents a mating. The symbols connected to the mating line by vertical and horizontal lines represent the children resulting from the mating in order of birth from left to right. Matings not related to the pedigree are not shown.

Genetic Counseling

Genetic counseling includes predicting the possible results of matings involving carriers of harmful genes and talking to parents or prospective parents about the possible outcomes and treatments of a genetic disorder. With this knowledge, prospective parents can make informed decisions about having children.

A first step in genetic counseling is to attempt to determine the genotype of the individuals involved. For example, a couple may suspect that they are carriers for a genetic disorder. A family tree, or **pedigree,** provides historical information about family members (figure 29.27). Sometimes by knowing the phenotypes of relatives it is possible to determine a person's genotype. Direct means of obtaining genetic information are also available. A karyotype can be taken from white blood cells or the epithelial cells lining the inside of the cheek. Alternatively, the amount of a given substance, such as an enzyme, produced by a carrier can be tested. Sometimes a

carrier produces slightly more or less of a given substance because they are heterozygous and have only one dominant gene for the normal or abnormal trait. For example, carriers for cystic fibrosis produce more salt in their sweat than is normal.

Sometimes it is suspected that a fetus may have a genetic abnormality. Fetal cells can be tested by amniocentesis, which takes cells floating in the amniotic fluid (see figure A), or chorionic villus sampling, which takes cells from the fetal side of the placenta.

---

8   P R E D I C T

It is estimated that the Human Genome Project will cost approximately $3 billion. Explain why you believe this will be money well spent or why you believe the project should be stopped?

✔ *Answer in Appendix F*

Clinical Focus The Human Genome Project

The human genome is all of the genes found in one homologous set of human chromosomes. It is estimated that humans have 60,000–80,000 genes. A **genomic map** is a description of the DNA nucleotide sequences of the genes and their locations on the chromosomes (figure B). To date, approximately 7000 genes have been mapped at least to their location on chromosomes. The goal of the **Human Genome Project** is to have a complete genomic map of all the genes by the year 2005.

Armed with a knowledge of the human genome and what effects the genome has on a person's physical, mental, and behavioral abilities, medicine and society will be transformed in many ways. Medicine, for

example, will shift emphasis from the curative to the preventative. The potential disorders or diseases a person is likely to develop can be prevented or their severity lessened. When prevention is not possible, knowledge of the enzymes or other molecules involved in a disorder may result in new drugs and techniques that can compensate for the genetic disorder. Knowledge of the genes involved in a disorder may result in **gene therapy**, or **genetic engineering**, that repairs or replaces defective genes, resulting in cures of genetic disorders.

Despite the great promise of benefits from the Human Genome Project, the knowledge that will be produced has raised a number of ethical and legal questions for

society. Should a person's genomic information be public knowledge? Should persons with a genome that predisposes them to cancer or behavioral disorders be barred from certain types of employment or be refused medical insurance because they are a high risk? Can a person demand to know a prospective mate's genome? Should parents know the genome of their fetus and be allowed to make decisions regarding abortion based on this knowledge? Should the same genetic-engineering techniques that provide alteration of the genome to cure genetic disorders be used to create genomes that are deemed to be superior? Such questions raise the specter of genetic discrimination.

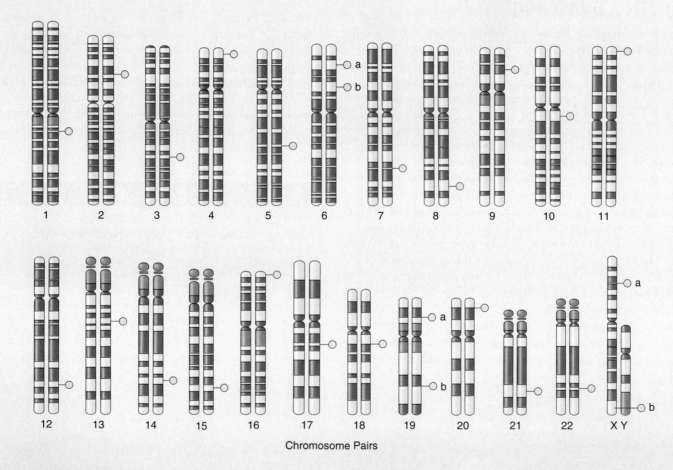

Chromosome Pairs

1. Gaucher disease
2. Familial colon cancer*
3. Retinitis pigmentosa*
4. Huntington disease
5. Familial polyposis of the colon
6a. Spinocerebellar ataxia
6b. Hemochromatosis

7. Cystic fibrosis
8. Multiple exostoses*
9. Malignant melanoma
10. Multiple endocrine neoplasia, type 2
11. Sickle cell disease
12. PKU (phenylketonuria)

13. Retinoblastoma
14. Alzheimer's disease*
15. Tay-Sachs disease
16. Polycystic kidney disease
17. Breast cancer*
18. Amyloidosis
19a. Familial hypercholesterolemia
19b. Myotonic dystrophy

20. ADA deficiency
21. Amyotrophic lateral sclerosis*
22. Neurofibromatosis, type 2
Xa. Muscular dystrophy
Xb. Factor VIII deficiency (hemophilia A)

*Gene responsible for only some cases.

Figure B   The Human Genomic Map

Representative genetic defects mapped to date. The blue balls indicate the location of the genes listed for each chromosome.

Summary

Prenatal development is an important part of an individual's life.

Prenatal Development

1. Prenatal development is divided into the germinal, embryonic, and fetal periods.
2. Postovulatory age is 14 days less than clinical age.
3. Fertilization, the union of the oocyte and sperm, results in a zygote.
4. The product of fertilization undergoes divisions until it becomes a mass, called a morula, and then a hollow ball of cells, called a blastocyst.
5. The cells of the morula are pluripotent (capable of making any cell of the body).
6. The blastocyst implants into the uterus about 7 days after fertilization. The placenta is derived from the trophoblast of the blastocyst.
7. All tissues of the body are derived from three primary germ layers: ectoderm, mesoderm, and endoderm.
8. The nervous system develops from a neural tube that forms in the ectodermal surface of the embryo and from neural crest cells derived from the developing neural tube.
9. Segments called somites that develop along the neural tube give rise to the musculature, vertebral column, and ribs.
10. The gastrointestinal tract develops as the developing embryo closes off part of the yolk sac.
11. The celom develops from small cavities that fuse within the embryo.
12. The limbs develop from proximal to distal as outgrowths called limb buds.
13. The face develops by the fusion of five major tissue processes.

Development of the Organ Systems

1. The skin develops from ectoderm (epithelium), mesoderm and neural crest (dermis), and the neural crest (melanocytes).
2. The skeletal system develops from mesoderm or neural crest cells.
3. Muscle develops from myoblasts, which migrate from somites.
4. The brain and spinal cord develop from the neural tube, and the peripheral nervous system develops from the neural tube and the neural crest cells.
5. The special senses develop mainly as neural tube or neural crest cell derivatives.
6. Many endocrine organs develop mainly as outpocketings of the brain or digestive tract.
7. The heart develops as two tubes fuse into a single tube that bends and develops septa to form four chambers.
8. The peripheral circulation develops from mesoderm as blood islands become hollow and fuse to form a network.
9. The lungs form as evaginations of the digestive tract. These evaginations undergo repeated branching.
10. The urinary system develops in three stages—pronephros, mesonephros, and metanephros—from the head to the tail of the embryo. The ducts join the allantois, part of which becomes the urinary bladder.
11. The reproductive system develops in conjunction with the urinary system. The presence or absence of certain hormones are very important to sexual development.

Growth of the Fetus

1. The embryo becomes a fetus at 60 days.
2. The fetal period is from day 60 to birth. It is a time of rapid growth.

Parturition

1. The total length of gestation is 280 days (clinical age).
2. Uterine contractions force the baby out of the uterus during labor.
3. Increased estrogen levels and decreased progesterone levels help initiate parturition.
4. Fetal glucocorticoids act on the placenta to decrease progesterone synthesis and to increase estrogen and prostaglandin synthesis.
5. Stretching of the uterus and decreased progesterone levels stimulate oxytocin secretion, which stimulates uterine contraction.

The Newborn
Circulatory Changes

1. The foramen ovale closes, separating the two atria.
2. The ductus arteriosus closes, and blood no longer flows between the pulmonary trunk and the aorta.
3. The umbilical vein and arteries degenerate.

Digestive Changes

1. Meconium is a mixture of cells from the digestive tract, amniotic fluid, bile, and mucus excreted by the newborn.
2. The stomach begins to secrete acid.
3. The liver does not form adult bilirubin for the first 2 weeks.
4. Lactose can be digested, but other foods must be gradually introduced.

Apgar Scores

1. Apgar represents appearance, pulse, grimace, activity, and respiratory effort.
2. Apgar and other methods are used to assess the physiologic condition of the newborn.

Lactation

1. Estrogen, progesterone, and other hormones stimulate the growth of the breasts during pregnancy.
2. Suckling stimulates prolactin and oxytocin synthesis. Prolactin stimulates milk production, and oxytocin stimulates milk letdown.

First Year After Birth

1. The number of neuron connections and glial cells increases.
2. Motor skills gradually develop, especially head, eye, and hand movements.

Life Stages

The life stages include the following: germinal, embryo, fetus, neonate, infant, child, adolescent, and adult.

Aging

1. Loss of cells that are not replaced contributes to aging.
   - There is a loss of neurons.
   - Loss of muscle cells can affect skeletal and cardiac muscle function.
2. Loss of tissue plasticity results from cross-link formation between collagen molecules. The lens of the eye loses the ability to accommodate. Other organs, such as the joints, kidneys, lungs, and heart, also have reduced efficiency with advancing age.
3. The immune system loses the ability to act against foreign antigens and may attack self-antigens.
4. Many aging changes are probably genetic.

Death

Death is the loss of brain functions.

Genetics

Chromosomes

1. Humans have 46 chromosomes in 23 pairs; 22 pairs of autosomes and 1 pair of sex chromosomes.
2. Males have the sex chromosomes XY and females XX.
3. During gamete formation, the chromosomes of each pair of chromosomes separate; therefore, half of a person's genetic makeup comes from the father and half from the mother.

Genes

1. A gene is a portion of a DNA molecule. Genes determine the proteins in a cell.
2. Genes are paired (located on the paired chromosomes).
3. Dominant genes mask the effects of recessive genes.
4. Sex-linked traits result from genes on the sex chromosomes.
5. In incomplete dominance, the heterozygote expresses a trait that is intermediate between the two homozygous traits.
6. In codominance, neither gene is dominant or recessive, but both are fully expressed.
7. Polygenic traits result from the expression of multiple genes.

Genetic Disorders

1. A mutation is a change in the number or kinds of nucleotides in DNA.
2. Some genetic disorders result from an abnormal distribution of chromosomes during gamete formation.
3. Oncogenes are genes associated with cancer.
4. Genetic predisposition makes it more likely a person will develop a disorder.

Genetic Counseling

1. A pedigree (family history) can be used to determine the risk of having children with a genetic disorder.
2. Specific chemical testing or examination of a person's karyotype can be used to determine a person's genotype.

Content Review

1. Define clinical age and postovulatory age, and distinguish between the two.
2. What are the events during the first week after fertilization? Define the terms zygote, morula, and blastocyst.
3. What is meant by the term pluripotent?
4. How does the placenta develop?
5. Describe the formation of the germ layers and the role of the primitive streak.
6. How are the neural tube and the neural crest formed? What do they become?
7. What is a somite?
8. Describe the formation of the gut and the body cavities.
9. What does proximal-to-distal growth mean in relation to the limbs?
10. Describe the processes involved in formation of the face. What clefts are closed by fusion of these processes?
11. Describe the formation of these major organs: skin, bones, skeletal muscles, eyes, pituitary gland, thyroid gland, pancreas, heart, lungs, kidneys, and gonads.
12. What major events distinguish embryonic and fetal development?
13. Describe the hormonal changes that take place before and during delivery. How is stretch of the cervix involved in delivery?
14. What changes take place in the newborn's circulatory and digestive systems shortly after birth?
15. Define the term meconium. Why does jaundice often develop after birth?
16. What is the Apgar score?
17. What hormones are involved in preparing the breast for lactation? Describe the events involved in milk production and milk letdown.
18. List the major changes that take place during the first year of life.
19. Define the different life stages, starting with the germinal stage and ending with the adult.
20. How does the loss of cells that are not replaced affect the aging process? Give examples.
21. How does loss of tissue plasticity affect the aging process? Give examples.
22. How does aging affect the immune system?
23. What role does genetics play in aging?
24. Define death.
25. What is the number and type of chromosomes in the karyotype of a human somatic cell. How do the chromosomes of a male and female differ from each other?
26. How do the chromosomes in somatic cells and gametes differ from each other?
27. What is a gene, and how are genes responsible for the structure and function of cells?
28. Define the terms homozygous dominant, heterozygous, and homozygous recessive.
29. What is the difference between genotype and phenotype?

30. What is a sex-linked trait? Give an example.
31. How does sickle cell anemia, type AB blood, and a person's height result from the expression of genes?
32. What is a mutation?
33. What is the cause of the genetic disorder called Down's syndrome?

34. What are oncogenes and carcinogens?
35. What is genetic susceptibility?
36. How are pedigrees, karyotypes, chemical tests, amniocentesis, and chorionic villus sampling used in genetic counseling?

Develop Your Reasoning Skills

1. A woman is told by her physician that she is pregnant and that she is 44 days past her LMP. How many days has the embryo been developing, and what developmental events are occurring?
2. A high fever can prevent neural tube closure. If a woman has a high fever approximately 35–45 days after her LMP, what kinds of birth defects may be seen in the developing embryo?
3. If the apical ectodermal ridge is damaged during embryonic development when the limb bud is about one-half grown, what kinds of birth defects might be expected? Describe the anatomy of the affected structure.
4. What are the results of exposing a female embryo to high levels of testosterone while she is developing?
5. Three minutes after birth, a newborn has an Apgar score of 5 as follows: A, 0; P, 1; G, 1; A, 1; and R, 2. What are some of the possible causes for this low score? What might be done for this neonate?

6. When a woman nurses, it is possible for milk letdown to occur in the breast that is not being suckled. Explain how this response happens.
7. Dimpled cheeks are inherited as a dominant trait. Is it possible for two parents, each of whom have dimpled cheeks, to have a child without dimpled cheeks? Explain.
8. The ability to roll the tongue to form a "tube" results from a dominant gene. Suppose that a woman and her son can both roll their tongues, but her husband cannot. Is it possible to determine if the husband is the father of her son?
9. A woman who does not have hemophilia marries a man who has the disorder. Determine the genotype of both parents if half of their children have hemophilia.

Web Site Link

For a listing of the most current web sites related to this chapter, please visit the Seeley home page at:
http://www.mhhe.com/biosci/ap/seeleyap/

Appendix A

Table of Measurements

Appendix A	Table of Measurements		
Unit	**Metric Equivalent**	**Symbol**	**U.S. Equivalent**
Measures of Length			
1 kilometer	= 1000 meters	km	0.62137 mile
1 meter	= 10 decimeters or 100 centimeters	m	39.37 inches
1 decimeter	= 10 centimeters	dm	3.937 inches
1 centimeter	= 10 millimeters	cm	0.3937 inch
1 millimeter	= 1000 micrometers	mm	
1 micrometer	= 1/1000 millimeter or 1000 nanometers	μm	
1 nanometer	= 10 angstroms or 1000 picometers	nm	
1 angstrom	= 1/10,000,000 millimeter	Å	
1 picometer	= 1/1,000,000,000 millimeter	pm	
Measures of Volume			
1 cubic meter	= 1000 cubic decimeters	m^3	1.308 cubic yards
1 cubic decimeter	= 1000 cubic centimeters	dm^3	0.03531 cubic foot
1 cubic centimeter	= 1000 cubic millimeters or 1 milliliter	cm^3 (cc)	0.06102 cubic inch
Measures of Capacity			
1 kiloliter	= 1000 liters	kL	264.18 gallons
1 liter	= 10 deciliters	L	1.0567 quarts
1 deciliter	= 100 milliliters	dL	0.4227 cup
1 milliliter	= volume of 1 gram of water at standard temperature and pressure	mL	0.3381 ounce
Measures of Mass			
1 kilogram	= 1000 grams	kg	2.2046 pounds
1 gram	= 100 centigrams or 1000 milligrams	g	0.0353 ounce
1 centigram	= 10 milligrams	cg	0.1543 grain
1 milligram	= 1/1000 gram	mg	

Note that a micrometer was formerly called a micron (μ), and a nanometer was formerly called a millimicron (mμ).

Appendix B

Scientific Notation

Very large numbers with many zeros such as 1,000,000,000,000,000 or very small numbers such as 0.0000000000000001 are very cumbersome to work with. Consequently, the numbers are expressed in a kind of mathematical shorthand known as scientific notation. Scientific notation has the following form:

$$M \times 10^n$$

where n specifies how many times the number M is raised to the power of 10. The exponent n has two meanings, depending on its sign. If n is positive, M is multiplied by 10 n times. For example, if $n = 2$ and $M = 1.2$, then

$$1.2 \times 10^2 = 1.2 \times 10 \times 10 = 120$$

In other words, if n is positive, the decimal point of M is moved to the right n times. In this case the decimal point of 1.2 is moved two places to the right.

1.20.

If n is negative, M is divided by 10 n times.

$$1.2 \times 10^{-2} = \frac{1.2}{(10 \times 10)} = \frac{1.2}{100} = 0.012$$

In other words, if n is negative, the decimal point of M is moved to the left n times. In this case the decimal point of 1.2 is moved two places to the left.

0.01.2

If M is the number 1.0, it often is not expressed in scientific notation. For example, 1.0×10^2 is the same thing as 10^2, and 1.0×10^{-2} is the same thing as 10^{-2}.

Two common examples of the use of scientific notation in chemistry are Avogadro's number and pH. Avogadro's number, 6.023×10^{23}, is the number of atoms in 1 molar mass of an element. Thus

$$6.023 \times 10^{23} =$$
$$602,300,000,000,000,000,000,000$$

which is a very large number of atoms.

The pH scale is a measure of the concentration of hydrogen ions in a solution. A neutral solution has 10^{-7} moles of hydrogen ions per liter. In other words

$$10^{-7} = 0.0000001$$

which is a very small amount (1 ten-millionth of a gram) of hydrogen ions.

Appendix C

Solution Concentrations

Physiologists often express solution concentration in terms of percent, molarity, molality, and equivalents.

Percent

The weight-volume method of expressing percent concentrations states the weight of a solute in a given volume of solvent. For example, to prepare a 10% solution of sodium chloride, 10 g of sodium chloride is dissolved in a small amount of water (solvent) to form a salt solution. Then additional water is added to the salt solution to form 100 mL of salt solution. Note that the sodium chloride was dissolved in water and then diluted to the required volume. The sodium chloride was not dissolved directly in 100 mL of water.

Molarity

Molarity determines the number of moles of solute dissolved in a given volume of solvent. A 1 molar (1 M) solution is made by dissolving 1 mole (mol) of a substance in enough water to make 1 L of solution. For example, 1 mol of sodium chloride solution is made by dissolving 58.44 g of sodium chloride in enough water to make 1 L of solution. One mol of glucose solution is made by dissolving 180.2 g of glucose in enough water to make 1 L of solution. Both solutions have the same number (Avogadro's number) of formula units (NaCl) and molecules (glucose) in solution.

Molality

Although 1 M solutions have the same number of solute molecules, they do not have the same number of solvent (water) molecules. Because 58.5 g of sodium chloride occupies less volume than 180 g of glucose, the sodium chloride solution has more water molecules. **Molality** is a method of calculating concentrations that takes into account the number of solute and solvent molecules. A 1 molal solution (1 m) is 1 mol of a substance dissolved in 1 kg of water. Thus all 1-molal solutions have the same number of solvent molecules.

When sodium chloride, which is an ionic compound, is dissolved in water it dissociates to form two ions, a sodium cation (Na^+) and a chloride anion (Cl^-). Glucose does not dissociate when dissolved in water, however, because it is a molecule. Thus, the sodium chloride solution contains twice as many particles as the glucose solution (one Na^+ ion and one Cl^- ion for each glucose molecule). To report the concentration of these substances in a way that reflects the number of particles in a given mass of solvent the concept of **osmolality** is used. The osmolality of a solution is the molality of the solution times the number of particles into which the solute dissociates in 1 kg of solvent. Thus 1 mol of sodium chloride in 1 kg of water is a 2 osmolal solution because sodium chloride dissociates to form two ions.

The osmolality of a solution is a reflection of the number, not the type, of particles in a solution. Thus a 1 osmolal (1 Osm) solution contains 1 Osm of particles per kilogram of solvent, but the particles may be all one type or a complex mixture of different types.

The concentration of particles in body fluids is so low that the measurement milliosmole (mOsm), 1/1000 of an osmole, is used. Most body fluids have an osmotic concentration of approximately 300 mOsm and consist of many different ions and molecules. The osmotic concentration of body fluids is important because it influences the movement of water into or out of cells (see chapter 3).

Equivalents

Equivalents are a measure of the concentrations of ionized substances. One equivalent (Eq) is 1 mol of an ionized substance multiplied by the absolute value of its charge. For example, 1 mol of NaCl dissociates into 1 mol of Na^+ and 1 mol of Cl^-. Thus there is 1 Eq of Na^+ (1 mol × 1) and 1 Eq of Cl^- (1 mol × 1). One mole of $CaCl_2$ dissociates into 1 mol of Ca^{2+} and 2 mol of Cl^-. Thus there are 2 Eq of Ca^{2+} (1 mol × 2) and 2 Eq of Cl^- (2 mol × 1). In an electrically neutral solution the equivalent concentration of positively charged ions is equal to the equivalent concentration of the negatively charged ions. One milliequivalent (mEq) is 1/1000 of an equivalent.

Appendix D

pH

Pure water weakly dissociates to form small numbers of hydrogen and hydroxide ions:

$$H_2O \longleftrightarrow H^+ + OH^-$$

At 25°C the concentration of both hydrogen ions and hydroxide ions is 10^{-7} mol/L. Any solution that has equal concentrations of hydrogen and hydroxide ions is considered **neutral.** A solution is an **acid** if it has a higher concentration of hydrogen ions than hydroxide ions, and a solution is a **base** if it has a lower concentration of hydrogen ions than hydroxide ions. In any aqueous solution (at 25°C) the hydrogen ion concentration $[H^+]$ times the hydroxide ion concentration $[OH^-]$ is a constant that is equal to 10^{-14}.

$$[H^+] \times [OH^-] = 10^{-14}$$

Consequently, as the hydrogen ion concentration decreases, the hydroxide ion concentration increases, and vice versa. For example:

	$[H^+]$	$[OH^-]$
Acidic solution	10^{-3}	10^{-11}
Neutral solution	10^{-7}	10^{-7}
Basic solution	10^{-12}	10^{-2}

Although the acidity or basicity of a solution could be expressed in terms of either hydrogen or hydroxide ion concentration, it is customary to use hydrogen ion concentration. The pH of a solution is defined as

$$pH = -\log_{10}(H^+)$$

Thus a neutral solution with 10^{-7} mol of hydrogen ions per liter has a pH of 7

$$pH = -\log_{10}(H^+)$$
$$= -\log_{10}(10^{-7})$$
$$= -(-7)$$
$$= 7$$

In simple terms, to convert the hydrogen ion concentration to the pH scale, the exponent of the concentration (e.g., -7) is used, and it is changed from a negative to a positive number. Thus an acidic solution with 10^{-3} mol of hydrogen ions/L has a pH of 3, whereas a basic solution with 10^{-12} hydrogen ions/L has a pH of 12.

Appendix E

Some Reference Laboratory Values

Table E-1 Blood, Plasma, or Serum Values

Test	Normal Values	Clinical Significance
Acetoacetate plus acetone	0.32–2 mg/100 mL	Values increase in diabetic acidosis, fasting, high-fat diet, and toxemia of pregnancy.
Ammonia	80–110 μg/100 mL	Values decrease with proteinuria and as a result of severe burns and increase in multiple myeloma
Amylase	4–25 U/mL*	Values increase in acute pancreatitis, intestinal obstruction, and mumps; values decrease in cirrhosis of the liver, toxemia of pregnancy, and chronic pancreatitis
Barbiturate	0	Coma level: phenobarbital, approximately 10 mg/100 mL; most other drugs, 1–3 mg/100 mL
Bilirubin	0.4 mg/100 mL	Values increase in conditions causing red blood cell destruction of biliary obstruction or liver inflammation
Blood volume	8.5%–9% of body weight in kilograms	
Calcium	8.5–10.5 mg/mL	Values increase in hyperparathyroidism, vitamin D hypervitaminosis; values decrease in hypoparathyroidism, malnutrition, and severe diarrhea
Carbon dioxide content	24–30 mEq/L 20–26 mEq/L in infants (as HCO_3^-)	Values increase in respiratory diseases, vomiting, and intestinal obstruction; they decrease in acidosis, nephritis, and diarrhea
Carbon monoxide	0	Symptoms with over 20% saturation
Chloride	100–106 mEq/L	Values increase in Cushing's syndrome, nephritis, and hyperventilation; they decrease in diabetic acidosis, Addison's disease, and diarrhea and after severe burns
Creatine phosphokinase (CPK)	Female 5–35 mU/mL Male 5–55 mU/mL	Values increase in myocardial infarction and skeletal muscle diseases such as muscular dystrophy
Creatinine	0.6–1.5 mg/100 mL	Values increase in certain kidney diseases
Ethanol	0	0.3%–0.4%, marked intoxication 0.4%–0.5%, alcoholic stupor 0.5% or over, alcoholic coma
Glucose	Fasting 70–110 mg/100 mL	Values increase in diabetes mellitus, liver diseases, nephritis, hyperthyroidism, and pregnancy; they decrease in hyperinsulinism, hypothyroidism, and Addison's disease
Iron	50–150 μg/100 mL	Values increase in various anemias and liver disease; they decrease in iron deficiency anemia
Lactic acid	0.6–1.8 mEq/L	Values increase with muscular activity and in congestive heart failure, severe hemorrhage, shock, and anaerobic exercise
Lactic dehydrogenase	60–120 U/mL	Values increase in pernicious anemia, myocardial infarction, liver diseases, acute leukemia, and widespread carcinoma

*A unit (U) is the quantity of a substance that has a physiologic effect.

Table E-1 Blood, Plasma, or Serum Values—cont'd

Test	Normal Values	Clinical Significance
Lipids	Cholesterol 120–220 mg/100 mL Cholesterol esters 60%–75% of cholesterol Phospholipids 9–16 mg/100 mL as lipid phosphorus Total fatty acids 190–420 mg/100 mL Total lipids 450–1000 mg/100 mL Triglycerides 40–150 mg/100 mL	Increased values for cholesterol and triglycerides are connected with increased risk of cardiovascular disease, such as heart attack and stroke
Lithium	Toxic levels 2 mEq/L	
Osmolality	285–295 mOsm/kg water	
Oxygen saturation (arterial) see Po_2	96%–100%	
Pco_2	35–43 mm Hg	Values decrease in acidosis, nephritis, and diarrhea; they increase in respiratory diseases, intestinal obstruction, and vomiting
pH	7.35–7.45	Values decrease as a result of hypoventilation, severe diarrhea, Addison's disease, and diabetic acidosis; values increase due to hyperventilation, Cushing's syndrome, and vomiting
Po_2	75–100 mm Hg (breathing room air)	Values increase in polycythemia and decrease in anemia and obstructive pulmonary diseases
Phosphatase (acid)	Male: total 0.13–0.63 U/mL Female: total 0.01–0.56 U/mL	Values increase in cancer of the prostate gland, hyperparathyroidism, some liver diseases, myocardial infarction, and pulmonary embolism
Phosphatase (alkaline)	13–39 IU/L* (infants and adolescents up to 104 IU/L)	Values increase in hyperparathyroidism, some liver diseases, and pregnancy
Phosphorus (inorganic)	3–4.5 mg/100 mL (infants in first year up to 6 mg/100 mL)	Values increase in hypoparathyroidism, acromegaly, vitamin D hypervitaminosis, and kidney diseases; they decrease in hyperparathyroidism
Potassium	3.5–5 mEq/100 mL	
Protein	Total 6–8.4 g/100 mL Albumin 3.5–5 g/100 mL Globulin 2.3–3.5 g/100 mL	Total protein values increase in severe dehydration and shock; they decrease in severe malnutrition and hemorrhage
Salicylate Therapeutic Toxic	0	20–25 mg/100 mL Over 30 mg/100 mL Over 20 mg/100 mL after age 60
Sodium	135–145 mEq/L	Values increase in nephritis and severe dehydration; they decrease in Addison's disease, myxedema, kidney disease, and diarrhea
Sulfonamide Therapeutic	0	5–15 mg/100 mL
Urea nitrogen	8–25 mg/100 mL	Values increase in response to increased dietary protein intake; values decrease in impaired renal function
Uric acid	3–7 mg/100 mL	Values increase in gout and toxemia of pregnancy and as a result of tissue damage

Table E-2 Blood Count Values

Test	Normal Values	Clinical Significance
Clotting (coagulation) time	5–10 min	Values increase in afibrinogenemia and hyperheparinemia, severe liver damage
Fetal hemoglobin	Newborns: 60%–90% Before age 2: 0%–4% Adults: 0%–2%	Values increase in thalassemia, sickle cell anemia, and leakage of fetal blood into maternal bloodstream during pregnancy
Hemoglobin	Male: 14–16.5 g/100 mL Female: 12–15 g/100 mL Newborn: 14–20 g/100 mL	Values decrease in anemia, hyperthyroidism, cirrhosis of the liver, and severe hemorrhage; values increase in polycythemia, congestive heart failure, obstructive pulmonary disease, high altitudes
Hematocrit	Male: 40%–54% Female: 38%–47%	Values increase in polycythemia, severe dehydration, and shock; values decrease in anemia, leukemia, cirrhosis, and hyperthyroidism
Ketone bodies	0.3–2 mg/100 mL Toxic level: 20 mg/100 mL	Values increase in ketoacidosis, fever, anorexia, fasting, starvation, high fat diet
Platelet count	250,000–400,000/mm^3	Values decrease in anemias and allergic conditions and during cancer chemotherapy; values increase in cancer, trauma, heart disease and cirrhosis
Prothrombin time	11–15 s	Values increase in prothrombin and vitamin deficiency, liver disease, and hypervitaminosis A
Red blood cell count	Males: 4.6–6.2 million/mm^3 Females: 4.2–5.4 million/mm^3	Values decrease in systemic lupus erythematosus, anemias, and Addison's disease; values increase in polycythemia and dehydration and following hemorrhage
Reticulocyte count	1%–3%	Values decrease in iron-deficiency and pernicious anemia and radiation therapy; values increase in hemolytic anemia, leukemia, and metastatic carcinoma
White blood cell count, differential	Neutrophils 60%–70% Eosinophils 2%–4% Basophils 0.5%–1% Lymphocytes 20%–25% Monocytes 3%–8%	Neutrophils increase in acute infections; eosinophils and basophils increase in allergic reactions; monocytes increase in chronic infections; lymphocytes increase during antigen–antibody reactions
White blood cell count, total	5000–9000/mm^3	Values decrease in diabetes mellitus, anemias, and following cancer chemotherapy; values increase in acute infections, trauma, some malignant diseases, and some cardiovascular diseases

Table E-3 Urine Values

Test	Normal Values	Clinical Significance
Acetone and acetoacetate	0	Values increase in diabetic acidosis and during fasting
Albumin	0 to trace	Values increase in glomerular nephritis and hypertension
Ammonia	20–70 mEq/L	Values increase in diabetes mellitus and liver disease
Bacterial count	Under 10,000/mL	Values increase in urinary tract infection
Bile and bilirubin	0	Values increase in biliary tract obstruction
Calcium	Under 250 mg/24 h	Values increase in hyperparathyroidism and decrease in hypoparathyroidism
Chloride	110–254 mEq/24 h	Values decrease in pyloric obstruction, diarrhea; values increase in Addison's disease and dehydration
Potassium	25–100 mEq/L	Values decrease in diarrhea, malabsorption syndrome, and adrenal cortical insufficiency; values increase in chronic renal failure, dehydration, and Cushing's syndrome
Sodium	75–200 mg/24 h	Values decrease in diarrhea, acute renal failure, and Cushing's syndrome; values increase in dehydration, starvation, and diabetic acidosis
Creatinine clearance	100–140 mL/min	Values increase in renal diseases
Creatinine	1–2 g/24 h	Values increase in infections and decrease in muscular atrophy, anemia, and certain kidney diseases
Glucose	0	Values increase in diabetes mellitus and certain pituitary gland disorders
Urea clearance	Over 40 mL of blood cleared of urea per minute	Values increase in certain kidney diseases
Urea	25–35 g/24 h	Values decrease in complete biliary obstruction and severe diarrhea; values increase in liver diseases and hemolytic anemia
Uric acid	0.6– 1 g/24 h	Values increase in gout and decrease in certain kidney diseases
Casts		
Epithelial	Occasional	Increase in nephrosis and heavy-metal poisoning
Granular	Occasional	Increase in nephritis and pyelonephritis
Hyaline	Occasional	Increase in glomerular membrane damage and fever
Red blood cell	Occasional	Values increase in pyelonephritis; blood cells appear in urine in response to kidney stones and cystitis
White blood cell	Occasional	Values increase in kidney infections
Color	Amber, straw, transparent yellow	Varies with hydration, diet, and disease states
Odor	Aromatic	Becomes acetonelike in diabetic ketosis
Osmolality	500–800 mOsm/kg water	Values decrease in aldosteronism and diabetes insipidus; values increase in high protein diets, heart failure, and dehydration
pH	4.6–8	Values decrease in acidosis, emphysema, starvation, and dehydration; values increase in urinary tract infections and severe alkalosis

Table E-4 Hormone Levels

Test	Normal Values
Steroid hormones	
Aldosterone	Excretion: 5–19 μg/24 h*
Fasting at rest, 210 mEq sodium diet	Supine: 48 ± 29 pg/mL†
	Upright: 65 ± 23 pg/mL
Fasting at rest, 10 mEq sodium diet	Supine: 175 ± 75 pg/m†
	Upright: 532 ± 228 pg/mL
Cortisol	
Fasting	8 AM: 5–25 μg/100 mL
At rest	8 PM: Below 10 μg/100 mL
Testosterone	Adult male: 300–1100 ng/100 mL†
	Adolescent male: over 100 ng/100 mL
	Female: 25–90 ng/100 mL
Peptide hormones	
Adrenocorticotropin (ACTH)	15–170 pg/mL
Calcitonin	Undetectable in normals
Growth hormone (GH)	
Fasting, at rest	Below 5 ng/mL
After exercise	Children: over 10 ng/mL
	Male: below 5 ng/mL
	Female: up to 30 ng/mL
Insulin	
Fasting	6–26 μU/mL
During hypoglycemia	Below 20 μU/mL
After glucose	Up t 150 μU/mL
Luteinizing hormone (LH)	Male: 6–18 mU/mL
	Preovulatory or postovulatory female: 5–22 mU/mL
	Midcycle peak 30–250 mU/mL
Parathyroid hormone	Less than 10 microl equiv/L
Prolactin	2–15 ng/mL
Renin activity	
Normal diet	
Supine	1.1 ± 0.8 ng/mL/h
Upright	1.9 ± 1.7 ng/mL/h
Low-sodium diet	
Supine	2.7 ± 1.8 ng/mL/h
Upright	6.6 ± 2.5 ng/mL/h
Thyroid-stimulating hormone (TSH)	0.5–3.5 μU/mL
Thyroxine-binding globulin	15.25 μg T_4/100 mL
Total thyroxine	4–12 μg/100 mL

*1 microgram (1 μg) is equal to 10^{-6} g.
†1 picogram (1 pg) is equal to 10^{-12} g.
†1 nanogram (1 ng) is equal to 10^{-9} g.

Appendix F

Answers to Predict Questions

Chapter 1

1. The chemical level is the level at which correction is currently being accomplished. Insulin can be purchased and injected into the circulation to replace the insulin normally produced by the pancreas. Another approach is drugs that stimulate pancreatic cells to produce insulin. Current research is directed at transplanting cells that can produce insulin. Another possibility is a partial transplant of tissue or a complete organ transplant.

2. Negative-feedback mechanisms work to control respiratory rates so that body cells have adequate oxygen and are able to eliminate carbon dioxide. The greater the respiratory rate, the greater the exchange of gases between the body and the air. When a person is at rest, there is less of a demand for oxygen, and less carbon dioxide is produced than during exercise. At rest, homeostasis can be maintained with a low respiration rate. During exercise there is a greater demand for oxygen, and more carbon dioxide must be eliminated. Consequently, to maintain homeostasis during exercise, the respiratory rate increases.

3. The sensation of thirst is involved in a negative-feedback mechanism that maintains body fluids. The sensation of thirst increases with a decrease in body fluids. The thirst mechanism causes a person to drink fluids, which returns body fluid levels to normal, maintaining homeostasis.

4. In the cat, cephalic and anterior are toward the head; dorsal and superior are toward the back. In humans, cephalic and superior are toward the head; dorsal and posterior are toward the back.

5. Your kneecap is both proximal and superior to the heel. It is also anterior to the heel because it is on the anterior side of the lower limb, whereas the heel is on the posterior side.

6. The spleen is in the left upper quadrant, the gallbladder is in the right upper quadrant, the left kidney is in the left upper quadrant, the right kidney is in the right upper quadrant, the stomach is mostly in the left upper quadrant, and the liver is mostly in the right upper quadrant.

7. There are two ways in which an organ can be located within the abdominopelvic cavity but not be within the peritoneal cavity. First, the visceral peritoneum wraps around organs. Thus the peritoneal cavity surrounds the organ, but the organ is not inside the peritoneal cavity. The peritoneal cavity contains only peritoneal fluid. Second, retroperitoneal organs are in the abdominopelvic cavity, but they are between the wall of the abdominopelvic cavity and the parietal peritoneal membrane.

Chapter 2

1. The mass (amount of matter) of the astronaut on the surface of the earth and in outer space does not change. In outer space where the force of gravity from the earth is very small, the astronaut is "weightless" compared with his weight on the earth's surface.

2. Potassium has 19 protons (the atomic number), 20 neutrons (the mass number minus the atomic number), and 19 electrons (because the number of electrons equals the number of protons).

3. There is Avogadro's number of atoms in 12.01 g of carbon and in 24.305 g of magnesium. In 12.01 g of magnesium, which is about half a mole of magnesium, there is about one-half of Avogadro's number of atoms.

4. The molecular formula for glucose is $C_6H_{12}O_6$. The atomic mass of carbon is 12.01, hydrogen is 1.008, and oxygen is 16.00. The molecular mass of glucose is therefore $(6 \times 12.01) + (12 \times 1.008) + (6 \times 16.00)$, or 180.2.

5. When two hydrogen atoms combine with an oxygen atom to form water, a polar covalent bond forms between each hydrogen atom and the oxygen atom. There is unequal sharing of electrons, and the electrons are associated with the oxygen atom more than with the hydrogen atoms. In this sense, the hydrogen atoms lose their electrons, and the oxygen atom gains electrons. The hydrogen atoms are therefore oxidized, and the oxygen atom is reduced.

6. A decrease in blood carbon dioxide decreases the amount of carbonic acid and therefore the blood hydrogen ion level. Because carbon dioxide and water are in equilibrium with hydrogen ions and bicarbonate ions, with carbonic acid as an intermediate, a decrease in carbon dioxide causes some hydrogen ions and bicarbonate ions to join together to form carbonic acid, which then forms carbon dioxide and water. Consequently, the hydrogen ion concentration decreases.

7. During exercise, muscle contractions increase, which requires energy. This energy is obtained from the energy in the chemical bonds of ATP. As ATP is broken down, energy is released. Some of the energy is used to drive muscle contractions, and some becomes heat. Because the rate of these reactions increases during exercise, more heat is produced than when at rest, and body temperature increases.

8. Monohydrogen phosphate ion (HPO_4^{2-}) is the conjugate base formed when the conjugate acid, dihydrogen phosphate ion ($H_2PO_4^-$)

loses a hydrogen ion. If hydrogen ions are added to the solution, they combine with the conjugate base, monohydrogen phosphate ions, to form dihydrogen phosphate ions, which helps to prevent an increase in hydrogen ion concentration. If hydroxide ions are added to the solution, they combine with hydrogen ions to form water. Then the conjugate acid, dihydrogen phosphate ions, dissociate to replace the hydrogen ions, which helps to prevent a decrease in hydrogen ion concentration.

9. Changing one amino acid in a protein chain can alter the three-dimensional structure of the protein chain. If the three-dimensional structure of an enzyme is changed, the function of the enzyme can decrease.

Chapter 3

1. (a) Cells highly specialized to synthesize and secrete proteins have large amounts of rough endoplasmic reticulum (ribosomes attached to endoplasmic reticulum) because these organelles are important for protein synthesis. Golgi apparatuses are well developed because they package materials for release in secretory vesicles. Also, there are numerous secretory vesicles in the cytoplasm.

    (b) Cells highly specialized to actively transport substances into the cell have a large surface area exposed to the fluid from which substances are actively transported, and numerous mitochondria are present near the membrane across which active transport occurs.

    (c) Cells highly specialized to synthesize lipids have large amounts of smooth endoplasmic reticulum. Depending on the kind of lipid produced, lipid droplets may accumulate in the cytoplasm.

    (d) Cells highly specialized to phagocytize foreign substances have numerous lysosomes in their cytoplasm and evidence of phagocytic vesicles.

2. Urea is continually produced by metabolizing cells and diffuses from the cells into the interstitial spaces and from the interstitial spaces into the blood. If the kidneys stop eliminating urea, it begins to accumulate in the blood. Because the concentration of urea increases in the blood, urea cannot diffuse from the interstitial spaces. As urea accumulates in the interstitial spaces, the rate of diffusion from cells into the interstitial spaces slows because the urea must pass from a higher to a lower concentration by the process of diffusion. The urea finally reaches concentrations high enough to be toxic to cells, causing cell damage followed by cell death.

3. If the membrane is freely permeable, the solutes in the tube diffuse from the tube (higher concentration of solutes) into the beaker (lower concentration of solutes) until there are equal amounts of solutes inside the tube and beaker (i.e., equilibrium). In a similar fashion, water in the beaker diffuses from the beaker (higher concentration of water) into the tube (lower concentration of water) until equal amounts of water are inside the tube and beaker. Consequently, the solution concentrations inside the tube and beaker are the same because they both contain the same amounts of solutes and water. Under these conditions, there is no net movement of water into the tube. This simple experiment demonstrates that osmosis and osmotic pressure require a membrane that is selectively permeable.

4. Glucose transported by facilitated diffusion across the plasma membrane moves from a higher to a lower concentration. If glucose molecules are quickly converted to some other molecule as they enter the cell, a steep concentration gradient is maintained. The rate of glucose transport into the cell is directly proportional to the magnitude of the concentration gradient.

5. Digitalis should increase the force of heart contraction. By interfering with Na^+ ion transport, digitalis decreases the concentration gradient for Na^+ ions because fewer Na^+ ions are pumped out of cells by active transport. Consequently, fewer Na^+ ions diffuse into cells, and fewer Ca^{2+} ions move out of the cells by countertransport. The higher intracellular levels of Ca^{2+} promote more forceful contractions.

6. By changing a single nucleotide within a DNA molecule, a change in the nucleotide of messenger RNA produced from that segment of DNA also occurs, and a different amino acid is placed in the amino acid chain for which the messenger RNA provides direction. Because a change in the amino acid sequence of a protein could change its structure, one substitution of a nucleotide in a DNA chain could result in altered protein structure and function.

7. Because adenine pairs with thymine (there is no uracil in DNA) and cytosine pairs with guanine, the sequence of DNA replicated from strand one is TACGAT. This sequence is also the sequence of DNA in the original strand two. A replicate of strand two is therefore ATGCTA, which is the same as the original strand one.

Chapter 4

1. (a) Secretion of mucus and digestive enzymes and the absorption of nutrients normally occurs in the digestive tract. Simple columnar epithelial cells contain organelles that are specialized to carry out nutrient absorption and secretion of mucus and digestive enzymes. Stratified squamous epithelium is not specialized to either absorb or secrete, and the layers of epithelial cells reduce the ability of nutrient molecules to be absorbed and the ability of digestive enzymes to be secreted.

    (b) Keratinized stratified epithelium forms a tough layer that is a barrier to the movement of water. Replacing the epithelium of skin with moist stratified squamous epithelium increases the loss of water across the skin because water can diffuse through moist stratified squamous epithelium, and it is more delicate and provides less protection than keratinized stratified squamous epithelium.

    (c) The stratified squamous epithelium that lines the mouth provides protection. Replacement of it with simple columnar epithelium makes the lining of the mouth much more susceptible to damage because the single layer of epithelial cells is easier to damage.

2. Collagen synthesis is required for scar formation. If collagen synthesis does

not occur because of a lack of vitamin C or if collagen synthesis is slowed, wound healing does not occur or is slower than normal. One might expect that the density of collagen fibers in a scar is reduced and the scar is not as durable as a normal scar.

3. Elastic ligaments attached to the vertebrae help the vertebral column return to its normal upright position after it is flexed. The elastic ligaments act much like elastic bands. Tendons attach muscles to bones. When muscles contract, they pull on the tendons, which in turn pull on bones. Because they are not elastic, when the muscle pulls on the tendon, all of the force is applied to the bone, causing it to move. If tendons were elastic, when the muscle contracted, the tendon would stretch, and not all of the tension would be applied to the bone.

4. Hyaline cartilage provides a smooth surface so that bones in joints can move easily. When the smooth surface provided by hyaline cartilage is replaced by dense fibrous connective tissue, the smooth surface is replaced by a less smooth surface, and the movement of bones in joints is much more difficult. The increased friction helps to increase inflammation that occurs in the joints of people who have rheumatoid arthritis.

5. In severely damaged tissues in which cells are killed and blood vessels are destroyed, the usual symptoms of inflammation cannot occur. Surrounding these areas of severe tissue damage, however, where blood vessels are still intact and cells are still living, the classic signs of inflammation do develop. The signs of inflammation therefore appear around the periphery of severely injured tissues.

Chapter 5

1. Because the permeability barrier is mainly composed of lipids surrounding the epidermal cells, substances that are lipid-soluble easily pass through, whereas water-soluble substances have difficulty.

2. (a) The lips are pinker or redder than the palms of the hand. Several explanations for this are possible. There could be more blood vessels in the lips, there could be increased blood flow in the lips, or the blood vessels could be easier to see through the epidermis of the lips. The last possibility explains most of the difference in color between the lips and the palms. The epidermis of the lips is thinner and not as heavily keratinized as that of the palms. In addition, the papillae containing the blood vessels in the lips are "high" and closer to the surface.

(b) A person who does manual labor has a thicker stratum corneum on the palms (and possibly calluses) than a person who does not perform manual labor. The thicker epidermis masks the underlying blood vessels, and the palms do not appear as pink. In addition, carotene accumulating in the lipids of the stratum corneum might impart a yellowish cast to the palms.

(c) The posterior surface of the forearm appears darker because of the tanning effect of ultraviolet light from the sun.

(d) The genitals normally have more melanin and appear darker than the soles of the feet.

3. The story is not true. Hair color results from melanin that is added to the hair in the hair matrix as the hair grows. The hair itself is dead. To turn white, the hair must grow out without the addition of melanin, a process that takes weeks.

4. On cold days, skin blood vessels of the ears and nose can dilate, bringing warm blood to the ears and nose and thus preventing tissue damage from the cold. The increased blood flow makes the ears and nose appear red.

5. Reducing water loss is one of the normal functions of the skin. Loss of skin, or damage to the skin, can greatly increase water loss. In addition, burning large areas of the skin results in increased capillary permeability and additional loss of fluid from the burn and into tissue spaces. The loss of fluid reduces blood volume, which results in reduced blood flow to the kidneys. Consequently, urine output by the kidneys decreases, which reduces fluid loss and thereby helps to compensate for the fluid loss caused by the burn. The reduced blood flow to the kidneys can cause tissue damage, however. To counteract this effect, during the first 24 h following the injury, part of the treatment for burn victims is the administration of large volumes of fluid. But, how much fluid should be given? The amount of fluid given should be sufficient to match that lost plus enough to prevent kidney damage and allow the kidneys to function. Urine output is therefore monitored. If it is too low, more fluid is administered, and if it is too high, less fluid is given. An adult receiving intravenous fluids should produce 30–50 mL of urine/h, and children should produce 1 mL/kg of body weight/h.

Chapter 6

1. In the absence of a good blood supply, nutrients, chemicals, and cells involved in tissue repair enter cartilage tissue very slowly. As a result, the ability of cartilage to undergo repair is poor. Within a joint, the articular cartilage of one bone presses against and moves against the articular cartilage of another bone. If the articular cartilages were covered by perichondrium, or contained blood vessels and nerves, the resulting pressure and friction could damage these structures.

2. In the elderly, the bone matrix contains proportionately less collagen than hydroxyapatite compared with the bones of younger people. Collagen provides bone with flexible strength, and a reduction in collagen results in brittle bones. In addition, the elderly have less dense bones with less matrix. The combination of reduced matrix that is more brittle results in a greater likelihood of bones breaking.

3. Cancellous bone consists of trabeculae with spaces between the trabeculae. Blood vessels can pass through these spaces. In compact bone, the blood vessels pass through the perforating and central canals. The trabeculae in cancellous bone are thin enough that nutrients and gases can diffuse from blood vessels around the trabeculae to the osteocytes through the canaliculi.

4. Chondroblasts are surrounded by cartilage matrix and receive oxygen and nutrients by diffusion through the matrix. When the matrix becomes calcified, diffusion is reduced to the point the cells die. When osteoblasts form bone matrix, they connect to

one another by their cell processes. Thus, when the matrix is laid down, canaliculi are formed. Even though the ossified bone matrix is dense and prevents significant diffusion, it is possible for the osteocytes to receive gases and nutrients through the canaliculi or by movement from one osteocyte to another.

5. Interstitial growth of cartilage results from the division of chondrocytes within the cartilage followed by the production of new cartilage matrix. The resulting expansion of the cartilage matrix results in cartilage growth. Bones cannot undergo interstitial growth because bone matrix is rigid and cannot expand from within. New bone must therefore be added to the surface by apposition.

6. Damage to the epiphyseal plate interferes with bone elongation, and as a result the bone, and therefore the thigh, will be shorter than normal. Recovery is difficult because cartilage repairs very slowly.

7. Growth of articular cartilage results in an increase in the size of epiphyses. This is only one of the functions of articular cartilage, however; it also forms a smooth, resilient covering over the ends of the epiphyses within joints. Ossified articular cartilage could not perform that function.

8. Her growth for the next few months increases, and she may be taller than a typical 12-year-old female. Because the epiphyseal plates ossify earlier than normal, however, her height at age 18 will be less than otherwise expected.

9. Taking in adequate calcium and vitamin D through the digestive system during adulthood increases calcium absorption from the small intestine. The increased calcium is used to increase bone mass. The greater the bone mass before the onset of osteoporosis, the greater the tolerance for bone loss later in life. For this reason it is important for adults, especially women in their twenties and thirties, to ingest adequate amounts of calcium. Exercising the muscular system places stress on bone, which also increases bone density. The granddaughter should not smoke because this reduces estrogen levels. Following menopause, estrogen replacement therapy can reduce bone loss.

Chapter 7

1. The sagittal suture is so named because it is in line with the midsagittal plane of the head. The coronal suture is so named because it is in line with the coronal plane (see chapter 1).

2. The bones most often broken in a "broken nose" are the nasals, ethmoid, vomer, and maxillae.

3. The lumbar vertebrae support a greater weight than the other vertebrae. The arrangement of the lumbar articular processes, with the inferior facets facing laterally and the superior facets facing medially, limits rotational movement and provides greater stability and strength. The vertebrae are more massive because of the greater weight they support.

4. The anterior support of the scapula is lost with a broken clavicle, and the shoulder is located more inferiorly and anteriorly than normal. In addition, since the clavicle normally holds the upper limb away from the body, the upper limb moves medially and rests against the side of the body.

5. The olecranon process moves into the olecranon fossa as the elbow is straightened. The coronoid process moves into the coronoid fossa as the elbow is bent.

6. The dried skeleton seems to have longer "fingers" than the hand with soft tissue intact because the soft tissue fills in the space between the metacarpals. With the soft tissue gone, the metacarpals seem to be an extension of the fingers, which appear to extend from the most distal phalanx to the carpals.

7. The depth of the hip socket is deeper, the bone is more massive, and the tubercles are larger than similar structures in the upper limb. All of this correlates with the weight-bearing nature of the lower limb and the more massive muscles necessary for moving the lower limb compared with the upper limb.

8. The top of modern ski boots is placed high up the leg in order to protect the weakest point of the fibula and make it less susceptible to great strain during a fall. Modern ski boots are also designed to reduce ankle mobility, which increases comfort and performance.

Chapter 8

1. The joint between the metacarpals and the phalanges is the metacarpophalangeal joint.

2. Premature sutural synostosis can result in abnormal skull shape, can interfere with normal brain growth, and can result in brain damage if not corrected. Such an abnormality is usually corrected surgically by removing some of the bone around the suture and creating an artificial fontanel, which then undergoes normal synostosis.

3. The synovial membrane is very thin and delicate. There is a considerable amount of pressure on the articular cartilages within a joint, and the articular cartilage is very tough, yet flexible, to withstand the pressure. If the synovial membrane covered the articular cartilage, it would be easily damaged during movement.

4. The movements required are abduction of the arm and flexion of the forearm, or flexion of the arm and forearm and pronation of the hand.

5. A shoulder separation involves stretching or tearing of the acromioclavicular ligament and may involve tearing of the coracoclavicular ligament as well. Because the only bony attachment of the upper limb to the body is from the scapula through the clavicle to the sternum, separation of the acromioclavicular joint greatly reduces the stability of the shoulder. The scapula and humerus tend to be displaced inferiorly, and the proximal pivot point for the upper limb is destabilized.

Chapter 9

1. A drug called theophylline is used to treat asthma. A possible effect of this drug is to bind to the phosphodiesterase enzyme that breaks down cAMP to AMP and inhibit it. Once the phosphodiesterase enzymes are inhibited, cAMP levels increase in the cell because they are being produced by the adenylyl cyclase enzymes. It might also be possible to directly stimulate the adenylyl cyclase enzymes with a drug without affecting the receptors or G proteins.

2. Drugs that bind the membrane-bound receptors generally produce rapid responses. The drugs bind to receptors and either alter membrane permeability or increase intracellular mediator production. The changed membrane permeability produces a rapid response, and the intracellular mediators function to alter the activity

of enzymes that are already present in the cytoplasm of the cell. As soon as the drug is removed, the permeability of the plasma membrane and the concentration of the intracellular mediator inside of the cell returns relatively rapidly to their resting levels. The response of the cell therefore disappears fairly rapidly after the drug disappears. In contrast, when a drug binds to an intracellular receptor and causes an increase in protein synthesis, the production of mRNA and the synthesis of proteins takes time, up to many hours. After the proteins are produced, they remain in the cell for a substantial length of time. The onset of the response to drugs that affect intracellular receptors is therefore slow, requiring many hours, and the response may persist for a relatively long period after the drug leaves the system.

3. If the intracellular concentration of K^+ ions is increased, the concentration gradient from the inside to the outside of the plasma membrane increases. This situation is similar to decreasing the extracellular concentration of potassium ions. The greater concentration gradient for K^+ ions increases their tendency to diffuse out of the cell across the plasma membrane. A greater negative charge then develops inside the cell (hyperpolarization). At equilibrium the greater negative charge is just enough to prevent the diffusion of additional potassium ions from the cell.

4. A decrease in the plasma membrane's permeability to K^+ ions results in depolarization of the plasma membrane. When the permeability of the plasma membrane to K^+ ions decreases, the tendency for K^+ ions to diffuse out of the cell decreases; fewer K^+ ions line up on the outside of the plasma membrane, and a smaller negative charge is required inside the cell to prevent K^+ ions from leaving it. Thus a new equilibrium is established in which the membrane potential is less polar (is depolarized relative to the resting membrane potential).

5. If the extracellular concentration of Ca^{2+} ions decreases, the resting membrane potential becomes depolarized. When extracellular Ca^{2+} ion concentrations decline, voltage-

gated Na^+ ion channels open. The open Na^+ ion channels allow Na^+ ions to diffuse into the cell and cause depolarization of the plasma membrane. Ca^{2+} ions bind to gating proteins that regulate the voltage-gated Na^+ ion channels. Low concentrations of Ca^{2+} ions cause the voltage-gated Na^+ ion channels to open, and high concentrations of Ca^{2+} ions cause the voltage-gated Na^+ ion channels to close.

6. If a cell is stimulated, an increase in the permeability of the plasma membrane to Na^+ ions usually results, with the degree of permeability depending on the strength and frequency of the stimulus. The greater the stimulus strength, the greater the permeability of the membrane to Na^+ ions. Na^+ ions diffuse into the cell down their concentration gradient and cause depolarization of the plasma membrane. If the concentration gradient for Na^+ ions is reduced, the tendency for Na^+ ions to diffuse into the cell decreases in comparison with the normal condition. Thus two stimuli of the same strength result in local potentials of differing magnitudes. In the cell with the reduced Na^+ ion concentration gradient, the local depolarization is of a smaller magnitude because fewer Na^+ ions are able to diffuse into the cell in response to the stimulus, even though the increase in the permeability of the plasma membrane to Na^+ ions increases to the same value in both situations.

7. If the extracellular concentration of Na^+ ions decreases, the magnitude of the action potential is reduced. For example, if depolarization occurs from a resting membrane potential of -80 to $+20$ mV during an action potential, the depolarization may be only from -80 to $+5$ mV or 10 mV if the extracellular concentration of Na^+ ions is reduced. The smaller concentration of Na^+ ions reduces the tendency for Na^+ ions to diffuse into the cell when the Na^+ ion channels are open during an action potential. Consequently, the inside of the plasma membrane does not become as positive as it does in cells with a high extracellular concentration of Na^+ ions. An increase in the extracellular concentration of Na^+ ions increases the magnitude of the action potential. The larger concentration of Na^+ ions

increases the tendency for Na^+ ions to diffuse into the cell when the Na^+ ion channels are open during an action potential. Consequently, the inside of the plasma membrane becomes more positive.

8. A prolonged stronger-than-threshold stimulus produces more action potentials than a prolonged threshold stimulus of the same duration. A prolonged stronger-than-threshold stimulus can stimulate more action potentials because the permeability of the membrane to Na^+ ions is increased. A very strong stimulus can even stimulate action potentials during the relative refractory period, whereas a prolonged threshold stimulus stimulates a low frequency of action potentials. Thus, when a prolonged stronger-than-threshold stimulus is applied, less time elapses between the production of one action potential and the next, resulting in the production of a greater number of action potentials.

9. Recording 1 had an action potential frequency of 200/s for 4 s, recording 2 had an action potential frequency of 400/s for 2 s, and recording 3 produced an action potential frequency of 600/s for 1 s. The recording with the smallest frequency is in response to the weakest stimulus, which is recording 1. The recording with the greatest frequency is in response to the strongest stimulus, which is recording 3. Recording 3 was applied for the shortest time and recording 1 was applied for the longest time.

10. The data indicate that the neuron exhibited a marked decrease in action potential frequency, even though a stimulus of constant strength was applied to the neuron for a long time. The data are consistent with accommodation. In neurons that exhibit accommodation, the local potential declines in magnitude, even though the stimulus is applied for a long time. As the local potential declines in magnitude, the frequency of the action potential declines also. When the local potential declines below threshold, no more action potentials are produced.

Chapter 10

1. When a muscle changes length, the I bands and the H zones change in width, but the A band does not.

When a muscle is stretched, the I bands and the H zones increase in width as the length of the sarcomere increases. When a muscle contracts, cross-bridges form and cause the actin myofilaments to slide over the myosin myofilaments. The result is that the I bands and H zones decrease in width as the sarcomeres shorten. When a muscle relaxes, cross-bridges release, and actin myofilaments slide past myosin myofilaments as the sarcomeres lengthen. The I bands and H zones increase in width.

2. If insufficient acetylcholine is released from the presynaptic terminal of an axon, an action potential is not produced in the muscle fiber and the muscle cannot contract. An action potential must be produced in the muscle fiber for contraction to occur. If inadequate acetylcholine is released from the presynaptic terminal of an axon, several action potentials in the axons would have to occur to cause the presynaptic terminal neurons to release enough acetylcholine to produce an action potential in the muscle fibers. Each action potential would release some acetylcholine, and, in response, a local potential may be produced in the postsynaptic membrane. If the local potentials were produced over a short period, they could summate (see chapter 9) and reach threshold. If threshold is reached, an action potential is produced.

3. (a) Organophosphate poisons inhibit the activity of acetylcholinesterase, which breaks down acetylcholine at the neuromuscular junction and limits the length of time the acetylcholine stimulates the postsynaptic terminal of the muscle fiber. Consequently, acetylcholine accumulates in the synaptic cleft and continuously stimulates the muscle fiber. As a result, the muscle remains contracted until it fatigues. Death is caused by the inability of the victim to breathe. Either the respiratory muscles are in spastic paralysis, or they are so depleted of ATP that they cannot contract at all.

   (b) Curare binds to acetylcholine receptors and thus prevents acetylcholine from binding to them. Because curare does not

activate the receptors, the muscles do not respond to nervous stimulation. The person suffers from flaccid paralysis and dies from suffocation because the respiratory muscles are not able to contract.

4. (a) If Na^+ ions cannot enter the muscle fiber, no action potentials are produced in the muscle fiber because the influx of Na^+ ions causes the depolarization phase of the action potential. Without action potentials, the muscle fiber cannot contract at all. The result is flaccid paralysis.

   (b) If ATP levels are low in a muscle fiber before stimulation, the following events occur. Energy from the breakdown of ATP already is stored in the heads of the myosin molecules. After stimulation, cross-bridges form. If not enough additional ATP molecules are in the muscle cells to bind to the myosin molecules to allow for cross-bridge release, however, the muscle becomes stiff without contracting or relaxing.

   (c) If ATP levels in the muscle fiber are adequate but the action potential frequency is so high that Ca^{2+} ions accumulate around the myofilaments, the muscle contracts continuously without relaxing. As long as Ca^{2+} ions are numerous within the sarcoplasm in the area of the myofilaments, cross-bridge formation is possible. If ATP levels are adequate, cross-bridge formation, release, and formation can proceed again, resulting in a continuously contracting muscle.

5. There is a decrease in muscle control when reinnervation of muscle fibers occurs after poliomyelitis because the number of motor units in the muscle is decreased. Reinnervation results in a greater number of muscle fibers per motor unit. Control is reduced because the number of motor units that can be recruited is decreased. The greater the number of motor units in a muscle, the greater is the ability to have fine gradations of muscle contraction as motor units are recruited. A smaller number of motor units means that gradations of muscle contraction are not as fine.

6. As a weight is lifted, the muscle contractions are concentric contractions. When a weight lifter lifts a heavy weight above his head, most of the muscle groups contract with a force while the muscle is shortening. Concentric contractions are a category of isotonic contractions in which tension in the muscle increases or remains about the same while the muscle shortens. While the weight is held above his head, the contractions are isometric contractions, because the length of the muscles does not change. While the weight is lowered, unless the weight lifter simply drops the weight, the length of the muscles increases as the weight is lowered for most of the muscle groups. Eccentric contractions are contractions in which tension is maintained in a muscle while the muscle increases in length. The major muscle groups are therefore contracting eccentrically while the weight is lowered.

7. During a 10-km run, aerobic metabolism is the primary source of ATP production for muscle contraction. Anaerobic metabolism provides enough ATP for up to 3 min during vigorous anaerobic exercise, but running a 10-km race takes much longer. If the runner sprints at the end of the 10-km run, however, anaerobic metabolism accounts for some of the energy production. After the run, aerobic metabolism is elevated for a time to pay back the oxygen debt. Anaerobic metabolism near the end of the run produced lactic acid, which is converted back to glucose after the run, a process that requires ATP. ATP is also required to restore the normal creatine phosphate levels in the muscle fibers and is produced through aerobic metabolism, which uses oxygen. The amount of oxygen used to produce the necessary ATP is the oxygen debt.

8. Long-distance runners should concentrate on running long distances. Slow-twitch muscles function very well for long-distance running. In addition, exercise that causes the muscles to perform aerobic metabolism improves the ability to carry on aerobic exercise and is more effective than exercise done under anaerobic conditions. Aerobic exercise combined with a large percent of slow-twitch muscle fibers is the best combination for long-distance

runners. Aerobic exercise, however, increases the ability of even fast-twitch muscle to resist fatigue.

9. A ligand that binds to its receptor and results in a sustained increase in the permeability of the plasma membrane to Ca^{2+} ions results in a sustained contraction without a large increase in ATP breakdown. The increased intracellular concentration of Ca^{2+} ions increases the number of phosphate groups removed from the myosin molecules while cross-bridges are attached. Because these cross-bridges release slowly, the result is a sustained contraction.

10. Muscular dystrophy affects the muscles of respiration and causes deformity of the thoracic cavity. The reduced capacity of muscle tissue to contract is one factor that reduces the ability to breath deeply or cough effectively. In addition, the thoracic cavity can become severely deformed because of the replacement of skeletal muscle with connective tissue. The deformity can result in severe kyphoscoliosis, which also reduces the ability to breath deeply. In addition, muscular dystrophy can affect the muscle of the heart and cause heart failure. The persistent edema in the lung increases the chance of bacteria multiplying in the lung tissue.

Chapter 11

1. Shortening the right sternocleidomastoid muscle rotates the head to the left. It also slightly elevates the chin.

2. Raising eyebrows—occipitofrontalis; winking—orbicularis oculi and then levator palpebrae superioris; whistling—orbicularis oris and buccinator; smiling—levator anguli oris, risorius, zygomaticus major, and zygomaticus minor; frowning—corrugator supercilii and procerus; flaring nostrils—levator labii superioris alaeque nasi and nasalis.

3. Weakness of the lateral rectus allows the eye to deviate medially.

4. Pain in one of the four rotator cuff muscles associated with abduction involves the supraspinatus.

5. Two arm muscles are involved in flexion of the forearm: the brachialis and the biceps brachii. The brachialis only flexes, whereas the biceps brachii both flexes and supinates the forearm. With the forearm supinated,

both muscles can flex the forearm optimally; when pronated, the biceps brachii does less to flex the forearm. Chin-ups with the forearm supinated are therefore easier because both muscles flex the forearm optimally in this position. Bodybuilders who wish to build up the brachialis muscle perform chin-ups with the forearms pronated.

Chapter 12

1. When the axon of a neuron is severed, the proximal portion of the axon remains attached to the neuron cell body. The distal portion of the axon is detached, however, and has no way to replenish the enzymes and other proteins essential to its survival. Because the DNA in the nucleus provides the information that determines the structure of proteins by directing mRNA synthesis, the distal portion of the axon has no source of new proteins. Consequently, it degenerates and dies. On the other hand, the proximal portion of the axon is still attached to the nucleus and therefore has a source of new proteins. It remains alive and, in many cases, grows to replace the severed distal axon.

2. Myelinated axons conduct action potentials much more rapidly than unmyelinated axons. In fact, the smallest diameter myelinated axons conduct action potentials more rapidly than the largest diameter unmyelinated axons. Small-diameter myelinated axons have the advantage of conducting action potentials more rapidly (myelinated) while taking up less space (small diameter). For unmyelinated axons to conduct as rapidly, their diameters would have to be so large that nerves and nerve tracts would have to be much larger than they are. The spinal cord, for example, would have to be many times larger in diameter than it is. Animals with unmyelinated axons would therefore have to have much larger nervous systems. Otherwise, they would move much more slowly, and their responses to stimuli would be much slower.

3. Temporal summation resulting from stimulation by neuron B produces more action potentials in the postsynaptic neuron than temporal summation resulting from stimulation by neuron A. The neuromodulator

from neuron B produces EPSPs, which depolarize the membrane potential of neuron C, bringing the membrane potential closer to threshold. A smaller amount of neurotransmitter is therefore required to produce an action potential. Although neuron A and B release the same amount and type of neurotransmitter, the neuromodulator makes the neurotransmitter from neuron B more effective, resulting in more action potentials.

4. Parallel processing is possible because of divergence and convergence. Sensory input to the CNS can go to many different parts of the brain because of divergence. After integrating the sensory information, the different parts of the brain can interact to produce a response through convergence.

Chapter 13

1. The reticular activating system receives afferent stimulation primarily from the face, which plays an important role in arousing consciousness and stimulating the wakeful part of the sleep–wake cycle. An effective technique for arousing a sleeping person is therefore to stimulate afferent nerve fibers in the face. This can be accomplished by lightly slapping or tickling the person's face, or splashing water in the person's face.

2. In the visual cortex the brain "sees" an object. Without a functional visual cortex, a person is blind. The visual association areas allow one to relate objects seen to previous experiences and to interpret what has been seen. Similarly, other association areas allow one to relate the sensory information integrated in the primary sensory areas with previous experiences and to make judgments about the information. The association cortices therefore function to interpret and recognize information coming into the primary sensory areas of the brain.

3. If a person holds an object in her right hand, tactile sensations of various types travel up the spinal cord to the brain, where they reach the somatic sensory cortex of the left hemisphere. The information is passed in the form of action potentials to the somatic sensory association cortex, where the object is recognized. Action potentials then

travel to Wernicke's area (probably on both sides of the cerebrum), where the object is given a name. From there action potentials travel to Broca's area, where the spoken word is initiated. Action potentials from Broca's area travel to the premotor area and primary motor cortex, where action potentials are initiated that stimulate the muscles necessary to form the word.

4. The cord is enlarged in the inferior cervical and superior lumbar regions because of the large numbers of nerve fibers exiting from the cord to the limbs and entering the cord from the limbs. Also, more neuron cell bodies in the cord regions are associated with the increased numbers of afferent and efferent fibers.

5. Dorsal root ganglia contain neuron cell bodies, which are larger in diameter than the axons of the dorsal roots.

6. Reciprocal innervation in an extended leg causes the flexor muscles to relax.

7. Collateral branches in the anterior spinothalamic tracts result in increased light touch sensitivity because collaterals from a number of sensory nerve endings can converge onto one ascending neuron and enhance its afferent conduction. As a result, light touch requires less peripheral stimulation to produce action potentials in the ascending pathway. Collateral, converging pathways, however, result in less discriminative information because sensory receptors from more than one point of the skin have input onto the same ascending neuron, and the neuron cannot distinguish one small area of skin from another within the zone where its afferent receptors are located.

8. Constipation, with painful distention and cramping of the colon, results in the sensation of diffuse pain. Deep, visceral pain is not highly localized because there are few mechanoreceptors in deeper structures such as the colon. The pain is perceived as occurring in the skin over the lower central portion of the abdomen (in the hypogastric region) because it is referred to that location because of converging CNS pathways.

9. The damage to Bill's spinal cord would be on the left side. The fasciculus gracilis conveys sensations of proprioception, fine touch, and vibration through the spinal cord on the same side of the body as the sensory nerve endings. The damage to Mary's brainstem would be on the right side if the damage occurred above the medulla oblongata or on the left if it occurred in the caudal part of the medulla oblongata. The secondary neurons in the nucleus gracilis cross over in the medulla through the decussations of the medial lemniscus, and once crossed, are on the opposite side of the body from the nerve endings where the sensations would be initiated.

10. Most proprioception from the lower limbs is unconscious, whereas that from the upper limbs is mostly conscious. This difference is valuable because walking and standing (balance) are not activities on which we want to focus our attention, whereas proprioceptive activities of the arms and hands are essential for gaining information about the environment.

11. The stroke was on the left side of the brainstem. Both the motor and sensory neurons to the right side of the body are located in the left cerebral cortex. At the level of the upper medulla oblongata, neither the motor nor sensory pathways to the limbs have yet crossed over to the left side of the CNS. Most of the motor fibers cross at the inferior end of the medulla oblongata, whereas sensory pain and temperature fibers cross over at the level where they enter the CNS. Loss of pain and temperature to the left side of the face indicates that the lesion occurred at a level where the nerve fibers from the face had entered the CNS but had not yet crossed (in the brainstem).

Chapter 14

1. The oculomotor nerve innervates four eye muscles and the levator palpebrae superioris muscle. One cause of ptosis, a drooping upper eyelid, can be oculomotor nerve damage and subsequent paralysis of the levator palpebrae superioris muscle. The four eye muscles innervated by the oculomotor nerve move the eyeball so that the gaze is directed superiorly, inferiorly, medially, or superolaterally. Damage to this nerve can be tested by having the patient look in these directions. The abducens nerve directs the gaze laterally, and the trochlear nerve directs the gaze inferolaterally. If the patient can move his eyes in these directions, the associated nerves are intact.

2. The sternocleidomastoid muscle pulls the mastoid process (located behind the ear) toward the sternum, thus turning the face to the opposite side. If the innervation to one sternocleidomastoid muscle is eliminated (accessory nerve injury), the opposite muscle is unopposed and turns the face toward the side of injury. A person with wry neck whose head is turned to the left, most likely has an injured left accessory nerve.

3. The tongue is protruded by contraction of the geniohyoid muscle, which pulls the back of the tongue forward, pushing the muscle mass of the tongue forward. With one side pushed forward and unopposed by muscles of the opposite side, the tongue deviates toward the nonfunctional side. In the example, therefore, the right hypoglossal nerve is damaged.

4. Nerves C5–T1, which innervate the left arm, forearm, and hand, were damaged.

5. Damage to the right phrenic nerve results in the absence of muscular contraction in the right half of the diaphragm. Because the phrenic nerves originate from C3 to C5, damage to the upper cervical region of the spinal cord eliminates their function; damage in the lower cord below the point where the spinal nerves originate does not affect the nerves to the diaphragm. Breathing is affected, however, because the intercostal nerves to the intercostal muscles, which move the ribs, are paralyzed.

6. The radial nerve lies along the shaft of the humerus about midway along its length. If the humerus is fractured, the radial nerve may be lacerated by bone fragments or, more commonly, pinched between two fragments of bone, decreasing or eliminating the function of the nerve.

Chapter 15

1. Directly stimulating a neuron in the "finger region" of the somatic sensory cortex gives the same result as stimulating the receptor or the nerve. The sensation of touch is projected to the finger that is represented in the portion of the cortex stimulated.

2. Because hot and cold objects may not be perceived any differently for temperatures of 0°–12°C or above 47°C (both temperature ranges stimulate pain fibers), the nervous system may not be able to discriminate between the two temperatures. At low temperatures, both cold and pain receptors are stimulated; thus, after the object has been in the hand for a very short time, it is possible to discriminate between cold and pain. If, however, the CNS has been preprogrammed to think that the object to be placed in the hand is hot, a cold object can elicit a rapid withdrawal reflex.

3. Inhaling slowly and deeply allows a large amount of air to be drawn into the olfactory recess, whereas not as much air enters during normal breaths. Sniffing (rapid, repeated air intake) is effective for the same reason.

4. Adaptation can occur at several levels in the olfactory system. First, adaptation can occur at the receptor cell membrane, where receptor sites are filled or become less sensitive to a specific odor. Second, association neurons within the olfactory bulb can modify sensitivity to an odor by inhibiting mitral cells or tufted cells. Third, neurons from the intermediate olfactory area of the cerebrum can send action potentials to the association neurons in the olfactory bulb to inhibit further afferent action potentials.

5. Eyedrops placed into the eye tend to drain through the nasolacrimal duct into the nasal cavity. Recall that much of what is considered "taste" is actually smell. The medication is detected by the olfactory neurons and is interpreted by the brain as taste sensation. Crying produces extra tears, which are conducted to the nasal cavity, causing a "runny" nose.

6. Inflammation of the cornea involves edema, the accumulation of fluid. Fluid accumulation in the cornea increases its water content, and because water causes the proteoglycans to expand, the transparency of the lens decreases, interfering with normal vision.

7. Eye strain, or eye fatigue, occurs primarily in the ciliary muscles. It occurs because close vision requires accommodation. Accommodation occurs as the ciliary muscles contract, releasing the tension of the suspensory ligaments, and allowing the lens to become more rounded. Continued close vision requires maintenance of accommodation, which requires that the ciliary muscles remain contracted for a long time, resulting in their fatigue.

8. Rhodopsin breakdown is associated with adaptation to bright light and occurs rapidly, whereas rhodopsin production occurs slowly and is associated with adaptation to conditions of little light. Eyes adapt rather quickly to bright light but quite slowly to very dim light.

9. Rod cells distributed over most of the retina are involved in both peripheral vision (out of the corner of the eye) and vision under conditions of very dim light. When attempting to focus directly on an object, however, a person relies on the cones within the macula lutea; although the cones are involved in visual acuity, they do not function well in dim light; thus the object may not be seen at all.

10. A lesion in the right optic nerve at (B) results in loss of vision in the right visual field (see illustration).

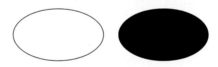

11. The stapedius muscle, attached to the stapes, is innervated by the facial nerve (VII). Loss of facial nerve function eliminates part of the sound attenuation reflex, although not all of it, because the tensor tympani muscle, innervated by the trigeminal nerve, is still functional. A reduction in the sound attenuation reflex results in sounds being excessively loud in the affected ear. A reduced reflex can also leave the ear more susceptible to damage by prolonged loud sounds.

12. "Perfect pitch" is the ability to precisely reproduce a pitch just by being told its name or reading it on a sheet of music, with no other musical support, such as from piano accompaniment. This remarkable talent as well as conditions such as tone deafness (the complete inability to recognize or reproduce musical pitches) or a decreased ability to perceive tone differences could occur at a number of locations. The structure of the basilar membrane may be such that tones are not adequately spaced along the cochlear duct in some people to facilitate clear separation of tones. The reflex from the superior olive to the spiral organ may have a very narrow "window of function" for people with perfect pitch but may not be functioning in some other people. The auditory cortex may not be able to translate as accurately in some people to distinguish differences in tones.

13. It is much easier to perceive subtle musical tones when music is played somewhat softly as opposed to very loudly because loud sounds have sound waves with a greater amplitude, which causes the basilar membrane to vibrate more violently over a wider range. The spreading of the wave in the basilar membrane to some extent counteracts the reflex from the superior olive that is responsible for enabling a person to hear subtle tone differences.

Chapter 16

1. Nicotinic receptors are located within the autonomic ganglia as components of the membranes of the postganglionic neurons of the sympathetic and parasympathetic divisions. Nicotine binds to the nicotinic receptors of the postganglionic neurons, resulting in action potentials. Consequently, the postganglionic neurons stimulate their effector organs. After consumption of nicotine, structures innervated by both the sympathetic and parasympathetic divisions are stimulated.

   After the consumption of muscarine, only the effector organs that respond to acetylcholine are affected. This includes all the effector organs innervated by the parasympathetic division, and the sweat glands, which are innervated by the sympathetic division.

2. The frequency of action potentials in sympathetic neurons to the sweat glands increases as the body temperature increases. The increasing body temperature is detected by the hypothalamus, which activates the sympathetic neurons. Sweating cools the body by evaporation. As the body temperature declines, the frequency of action potentials in sympathetic neurons to the sweat glands decreases. A lack of sweating helps prevent heat loss from the body.

3. In response to an increase in blood pressure, information is transmitted in the form of action potentials along afferent neurons to the medulla oblongata. Within the medulla oblongata the frequency of action potentials delivered along sympathetic nerve fibers to blood vessels decreases. As a result, blood vessels dilate, causing the blood pressure to decrease.

4. (a) Responses in a person who is extremely angry are primarily controlled by the sympathetic division of the autonomic nervous system. These responses include increased heart rate and blood pressure, decreased blood flow to the internal organs, increased blood flow to skeletal muscles, decreased contractions of the intestinal smooth muscle, flushed skin in the face and neck region, and dilation of the pupils of the eyes.

   (b) For a person who has just finished eating and is now relaxing, the parasympathetic reflexes are more important than sympathetic reflexes. The blood pressure and heart rate are a normal resting level, the blood flow to the internal organs is greater, contractions of smooth muscle in the intestines is greater, and secretions that achieve digestion are more active. If the urinary bladder or the colon becomes distended, autonomic reflexes that result in urination or defecation can result. Blood flow to the skeletal muscles is reduced.

Chapter 17

1. Because the abnormal substance acts like TSH, it acts on the thyroid gland to increase the rate of secretion of the thyroid hormones that increase in concentration in the circulatory system. The thyroid hormones have a negative-feedback effect on the secretion of TSH, thereby decreasing the concentration of TSH in the circulatory system. Because the abnormal substance is not regulated it may cause thyroid hormone levels to become very elevated.

2. A major function of plasma proteins to which hormones bind is to increase the half-life of the hormone. If the concentration of the plasma protein decreases, the half-life and,

consequently, the concentration of the hormone in the circulatory system decrease also. Because the half-life of the hormone is decreased, the rate at which the hormone is removed from the circulatory system increases; and, if the secretion rate for the hormone does not increase, its concentration in the blood declines.

3. If too little estrogen is secreted, the up-regulation of receptors in the uterus for progesterone cannot occur. As a result, the uterus is not prepared for the embryo to attach to its wall following ovulation, and pregnancy cannot occur. Because of the lack of up-regulation, the uterus probably will not respond to progesterone, regardless of how much progesterone is secreted. If some progesterone receptors are present, however, the uterus will require a much larger amount of progesterone to produce the normal response.

4. A drug could increase the cAMP concentration in a cell by stimulating its synthesis or by inhibiting its breakdown. Drugs which increase adenylyl cyclase activity by binding to a receptor which stimulates its activity will increase cAMP synthesis. Because phosphodiesterase normally causes the breakdown of cAMP, an inhibitor of phosphodiesterase causes cAMP to increase in the smooth muscle cells of the airway and produces relaxation.

5. Intracellular receptor mechanisms result in the synthesis of new proteins that exist within the cell for a considerable amount of time. Intracellular receptors are therefore better adapted for mediating responses that last a relatively long time (i.e., for many minutes, hours, or longer). On the other hand, membrane-bound receptors which increase the synthesis of intracellular mediators such as cAMP normally activate enzymes already existing in the cytoplasm of the cell for shorter periods. The synthesis of cAMP occurs quickly, but the duration is short because cAMP is broken down quickly and the activated enzymes are then deactivated. Membrane-bound receptor mechanisms are therefore better adapted to short-term and rapid responses.

Chapter 18

1. The cell bodies of the neurosecretory cells that produce ADH are in the hypothalamus, and their axons extend

into the posterior pituitary, where ADH is stored and secreted. Removing the posterior pituitary severs the axons, resulting in a temporary reduction in secretion. The cell bodies still produce ADH, however, and as the ADH accumulates at the ends of severed axons, ADH secretion resumes.

2. If GH is administered to young people before growth of their long bones is complete, it causes their long bones to grow and they will grow taller. To accomplish this, however, GH would have to be administered over a considerable length of time. It is likely that some symptoms of acromegaly would develop. In addition to undesirable changes in the skeleton, nerves frequently are compressed as a result of the proliferation of connective tissue. Because GH spares glucose usage, chronic hyperglycemia results, frequently leading to diabetes mellitus and the development of severe atherosclerosis. Mr. Hoops' doctor would therefore not prescribe GH.

3. The thyroid gland enlarges in response to iodine deficiency because without iodine, thyroid hormones cannot be synthesized. Consequently, TSH levels in the circulatory system increase because of the lower-than-normal levels of thyroid hormones in the blood. Increased TSH levels cause the thyroid gland to enlarge because it continues to stimulate thyroglobulin synthesis in large amounts. The thyroid follicles enlarge, even though thyroid hormones cannot be produced.

4. In response to a reduced dietary intake of calcium, the blood levels of calcium begin to decline. In response to the decline in blood levels of calcium, there is an increase of PTH from the parathyroid glands. The PTH functions to increase calcium resorption from bone. Consequently, blood levels of calcium are maintained within the normal range but at the same time bones are being decalcified. Severe dietary calcium deficiency results in bones that become soft and eaten away because of the decrease in calcium content.

5. Removal of the thyroid gland means that the tissue responsible for thyroid hormone production (follicles), calcitonin (parafollicular cells), and PTH (parathyroid glands are

embedded in the thyroid gland) are also removed. Thyroid hormones, calcitonin, and PTH are therefore no longer found in the blood. Without the negative-feedback effect of thyroid hormones, TRH and TSH levels in the blood increase. Without PTH, blood levels of calcium fall. When the blood levels of calcium fall below normal, the permeability of nerve and muscle cells to Na^+ ions increases. As a consequence, spontaneous action potentials are produced that cause tetany of muscles. Death can result from tetany of respiratory muscles.

6. High aldosterone levels in the blood lead to elevated Na^+ ion levels in the circulatory system and low blood levels of K^+ ions. The effect of low blood levels of K^+ ions is hyperpolarization of muscle and neurons. The hyperpolarization results from the lower levels of K^+ ions in the extracellular fluid and a greater tendency for K^+ ions to diffuse from the cell. As a result, a greater-than-normal stimulus is required to cause the cells to depolarize to threshold and generate an action potential. Symptoms of low serum K^+ ion levels therefore include lethargy and muscle weakness. Elevated Na^+ ion concentrations result in a greater-than-normal amount of water retention in the circulatory system, which can result in elevated blood pressure. The major effect of a low rate of aldosterone secretion is elevated blood K^+ ion levels. As a result, nerve and muscle cells partially depolarize. Because of their partial depolarization, they produce action potentials spontaneously or in response to very small stimuli. The result is muscle spasms, or tetany.

7. Large doses of cortisone can damage the adrenal cortex because cortisone inhibits ACTH secretion from the anterior pituitary. ACTH is required to keep the adrenal cortex from undergoing atrophy. Prolonged use of large doses of cortisone can cause the adrenal gland to atrophy to the point at which it cannot recover if ACTH levels do increase again.

8. An increase in insulin secretion in response to parasympathetic stimulation and gastrointestinal hormones is consistent with the maintenance of homeostasis because parasympathetic stimulation and increased gastrointestinal hormones result from conditions such as eating a meal. Insulin levels therefore increase just before large amounts of glucose and amino acids enter the circulatory system. The elevated insulin levels prevent a large increase in blood glucose and the loss of glucose in the urine.

9. In response to a meal high in carbohydrates, insulin secretion is increased, and glucagon secretion is reduced. The stimulus for the insulin secretion comes from parasympathetic stimulation and, more importantly, from elevated blood levels of glucose. In response to a meal high in protein but low in carbohydrates, insulin secretion is increased slightly, and glucagon secretion is also increased. Insulin secretion is stimulated by the parasympathetic system and an increase in blood amino acid levels. Glucagon secretion is stimulated by low blood glucose levels and by some amino acids. During periods of exercise, sympathetic stimulation inhibits insulin secretion. As blood glucose levels decline, there is an increase of glucagon secretion.

10. Sympathetic stimulation during exercise inhibits insulin secretion. Blood glucose levels are not high because skeletal muscle tissue continues to take up some glucose and metabolizes it. Muscle contraction depends on glucose stored in the form of glycogen in muscles and fatty acid metabolism. During a long run glycogen levels are depleted. The "kick" at the end of the race results from increased energy production through anaerobic respiration, which uses glucose or glycogen as an energy source. Because blood glucose levels and glycogen levels are low, there is an insufficient source of energy for greatly increased muscle activity.

11. Increased sugar intake will result in elevated blood glucose levels. The elevated blood glucose levels can lead to polyurea and to increased osmolality of the body fluids. That results in dehydration of neurons. As a result some of the neural symptoms of untreated diabetes, such as irritability and a general sensation of not feeling well, occur. Billy may also experience a sudden increase in weight gain because of increased sugar intake and insulin administration. In addition, he may have an increased chance of infections such as urinary tract infections. Many of the long-term consequences of diabetes such as nephropathies, neuropathies, atherosclerosis, and others, develop much more rapidly.

Chapter 19

1. The reason fetal hemoglobin must be more effective at binding oxygen than adult hemoglobin is so that the fetal circulation can draw the needed oxygen away from the maternal circulation. If maternal blood had an equal or greater oxygen affinity, the fetal blood would not be able to draw away the required oxygen, and the fetus would die.

2. An elevated reticulocyte count indicates that erythropoiesis and the demand for erythrocytes are increased and that immature erythrocytes (reticulocytes) are entering the circulation in large numbers. An elevated reticulocyte count can occur for a number of reasons, including loss of blood; therefore, after a person donates a unit of blood, his reticulocyte count increases.

3. Carbon monoxide binds to the iron of hemoglobin and prevents the transport of oxygen. The decreased oxygen stimulates the release of erythropoietin, which increases erythrocyte production in red bone marrow, causing the number of erythrocytes in the blood to increase.

4. The leukocytes shown in figure 19.8 are (a) Lymphocyte; (b) Basophil; (c) Monocyte; (d) Neutrophil; (e) Eosinophil.

5. Platelets become activated at sites of tissue damage, which is the location where it is advantageous to form a clot in order to stop bleeding.

6. People with type AB blood were called universal recipients because they could receive type A, B, AB, or O blood with little likelihood of a transfusion reaction. Type AB blood does not have antibodies against type A or B antigens; therefore, transfusion of these antigens in type A, B, or AB blood does not cause a transfusion reaction in a person with type AB blood. The term is misleading, however, for two reasons. First, other blood groups can cause a transfusion reaction. Second, antibodies in the donor's blood can cause a transfusion reaction. For example, type O blood

contains A and B antibodies that can react against the A and B antigens in type AB blood.

7. A white blood cell count (WBC) should be done. An elevated WBC, leukocytosis, can be an indication of bacterial infections. A white blood cell differential count should also be done. An increase in the number of neutrophils supports the diagnosis of a bacterial infection. Coupled with other symptoms, this could mean appendicitis. If these tests are normal, appendicitis is still a possibility and the physician must rely on other clinical signs. Diagnostic accuracy for appendicitis is approximately 75%–85% for experienced physicians.

Chapter 20

1. The heart tissues supplied by the artery lose their oxygen and nutrient supply and die. This part of the heart (and possibly the entire heart) stops functioning. If this condition develops rapidly, it is called a heart attack, or myocardial infarction.

2. The heart must continue to function under all conditions and requires energy in the form of ATP. During heavy exercise lactic acid is produced in skeletal muscle as a by-product of anaerobic metabolism. The ability to use lactic acid provides the heart with an additional energy source.

3. Contraction of the ventricles, beginning at the apex and moving toward the base of the heart, forces blood out of the ventricles and toward their outflow vessels—the aorta and pulmonary trunks.

4. Ectopic foci cause various regions of the heart to contract at different times. As a result, pumping effectiveness is reduced. Cardiac muscle contraction is not coordinated, which interrupts the cyclic filling and emptying of the ventricles.

5. If cardiac muscle could undergo tetanic contraction, it would contract for a long time without relaxing. Its pumping action then would stop because that action requires alternating contraction and relaxation.

6. In artificial heart implants it is most important to replace the ventricles because they are the major pumps of the heart. The heart can function fairly well without the pumping action of the atria.

7. It is important for each ventricle to pump the same amount of blood because, with two connected circulation loops, the blood flowing into one must equal the blood flowing into the other so that one does not become overfilled with blood at the expense of the other. For example, if the right ventricle pumps less blood than the left ventricle, blood must accumulate in the systemic blood vessels. If the left ventricle pumps less blood than the right ventricle, blood accumulates in the pulmonary blood vessels.

8. Sympathetic stimulation increases heart rate. If venous return remains constant, stroke volume decreases as the number of beats per minute increases. Dilation of the coronary arteries is important because, as the heart does more work, the cardiac tissue requires more energy and therefore a greater blood supply to carry more oxygen.

9. Rupture of the left ventricle, as experienced by Mr. P, is more likely several days after a myocardial infarction. As the necrotic tissues are removed by macrophages, the wall of the ventricle becomes thinner and may bulge during systole. If the wall of the ventricle becomes very thin before new connective tissue is deposited, it may rupture. If the left ventricle ruptures, blood flows from the left ventricle into the pericardial sac. As blood fills the pericardial sac, it compresses the ventricle from the outside. Thus the ventricle is not able to fill with blood and its pumping ability is eliminated. Death occurs quickly in response to a ruptured wall of the left ventricle.

Chapter 21

1. Arteriosclerosis slowly reduces blood flow through the carotid arteries and therefore the amount of blood that flows to the brain. As the resistance to flow increases in the carotid arteries during the late stages of arteriosclerosis, the blood flow to the brain is compromised resulting in reduced oxygen delivery. Confusion, loss of memory, and loss of the ability to perform other normal brain functions occur.

2. Cutting and tying off the lymph vessels prevents the movement of interstitial fluid from the interstitial spaces. The small amount of fluid that fails to reenter the venous end of the capillaries after it leaves the arteriolar end of the capillaries is normally carried by the lymphatic vessels away from the tissue spaces and back to the general circulation. If the lymphatic vessels are tied off, the fluid accumulates in the interstitial spaces resulting in edema.

3. (a) Vasoconstriction of blood vessels in the skin in response to exposure to cold results in a decreased flow of blood through the skin and in a dramatic increase in resistance (see Poiseuille's law). Vasoconstriction makes the skin appear pale.

   (b) Vasodilation of blood vessels in the skin results in increased blood flow through the skin. Vasodilation makes the skin appear flushed or red in color.

   (c) In a patient with polycythemia vera, the hematocrit increases dramatically. As a result, the viscosity of the blood increases, which increases resistance to flow. Consequently, flow decreases or a greater pressure is needed to maintain the same flow.

4. An aneurysm in the aorta is a major problem because the tension applied to the aneurysm becomes greater as its size increases (see Laplace's law). The aneurysm usually develops because of a weakness in the wall of the aorta. Arteriosclerosis complicates the matter by making the wall of the artery less elastic and by increasing the systolic blood pressure. The decreased elasticity and the increased blood pressure increase the probability that the aneurysm will rupture.

5. Premature beats of the heart and ectopic beats result in contraction of the heart muscle before the heart has had time to fill to its normal capacity. Consequently, a reduced stroke volume occurs, which results in a weak pulse. Strong bounding pulses in a person who received too much saline solution in an intravenous transfusion result from an increase in venous return to the heart because of the increased volume of fluid in the circulatory system. Because of the increased preload, the heart contracts with a greater force, producing a larger stroke volume. The strong bounding pulse results from the increased stroke volume. Weak pulses occur in response to hemorrhagic shock because of a decreased venous

return. The heart does not fill with blood between contractions (decreased preload); the stroke volume is therefore reduced, and the pulse is weak.

6. (a) Decreased plasma protein concentration reduces the colloid osmotic pressure of the blood. Edema results because less fluid reenters the venous end of the capillary and more fluid remains in the interstitial spaces.

(b) The increased permeability allows plasma protein to leak into the interstitial spaces, causing an increase in the colloid osmotic pressure in the interstitial spaces. Consequently, the inwardly directed osmotic force that moves fluid from the interstitial spaces into the capillaries is reduced. More fluid leaves the arterial end of the capillary, and less fluid enters the venous end of the capillary, causing a buildup of interstitial fluid (i.e., edema).

(c) Increased blood pressure within the capillary increases the amount of fluid that is forced from the arteriolar end of the capillary and reduces the amount of the fluid reentering the capillary at its venous end.

7. Keeping the legs elevated reduces the blood pressure in the venous ends of the capillaries in the legs because of the effect of gravity on blood flow. The force that moves fluid out of the capillary at its venous end is therefore decreased. As a result, the net movement of interstitial fluid into the venous ends of the capillaries increases, and the excess interstitial fluid is carried away from the legs. In addition, the effect of gravity increases lymph flow through the lymphatic vessels, which also increases the rate that interstitial fluid is drained from the legs.

8. Reactive hyperemia can be explained on the basis of any of the theories for the local control of blood pressure. When a blood vessel is occluded, nutrients are depleted, and waste products accumulate in tissue that is suffering from a lack of adequate blood supply. Both of these effects cause vasodilation and a greatly increased blood flow through the area after the occlusion has been removed.

9. While she was relaxing, the sympathetic stimulation of arteries in her skeletal muscles decreased, of arteries in her digestive system decreased, and of large veins decreased. As a result, vasoconstriction increased in the arteries of her muscles, and vasodilation occurred in blood vessels of her digestive system and in the large veins. Blood flow decreased to her skeletal muscles, and blood flow increased to her digestive system. In addition, more blood accumulated in the large veins. Consequently, venous return to the heart decreased, which is consistent with the reduced cardiac output.

10. During a headstand, gravity acting on the blood causes the blood pressure in the area of the aortic arch and carotid sinus baroreceptors to increase. The increased pressure activates the baroreceptor reflexes, increasing parasympathetic stimulation of the heart and decreasing sympathetic stimulation. Thus the heart rate decreases. Because standing on one's head also causes blood from the periphery to run downhill to the heart, the venous return increases, causing the stroke volume to increase because of Starling's law of the heart. Some peripheral vasodilation also can occur because the elevated baroreceptor pressure causes a decrease in vasomotor tone.

11. The baroreceptor reflex, ADH, and renin-angiotensin-aldosterone mechanisms function similarly in both cases. The fluid shift mechanism, however, is important when the loss of blood occurs over several hours, but it does not operate within a short period. The fluid shift mechanism plays a very important role in the maintenance of blood volume when blood loss or dehydration develops over several hours. When the blood pressure decreases, interstitial fluids pass into the capillaries, which prevents a further decline in the blood pressure. The fluid shift mechanism is a powerful method through which blood pressure is maintained because the interstitial fluid acts as a fluid reservoir.

Chapter 22

1. The T cells transferred to mouse B do not respond to the antigen. The T cells are MHC-restricted and must have the MHC proteins of mouse A as well as antigen X in order to respond.

2. When the antigen is eliminated, it is no longer available for processing and combining with MHC class II molecules. Consequently there is no signal to cause lymphocytes to proliferate and produce antibodies.

3. The first exposure to the disease-causing agent (antigen) evokes a primary immune system response. Gradually, however, antibodies degrade, and memory cells die. If, before all the memory cells are eliminated, a second exposure to the antigen occurs, a secondary immune system response results. The memory cells produced then could provide immunity until the next exposure to the antigen.

4. With depression of helper T-cell activity, the ability of antigens to activate effector T cells is greatly decreased. Depression of cell-mediated immunity results in an inability to resist intracellular microorganisms and cancer.

5. SLE is an autoimmune disorder in which self-antigens activate immune responses. Often, this results in the formation of immune complexes and inflammation. But sometimes antibodies bind to antigens on cells, resulting in the lysis of the cells. Purpura results from bleeding into the skin, which means that platelet plug formation, the normal mechanism for repairing small breaks in blood vessels, is not working. In this case of SLE, antibodies are causing the destruction of platelets, and the decreased number of platelets results in decreased platelet plug formation and coagulation (see chapter 19). The condition is called thrombocytopenia.

Chapter 23

1. Air moving through the mouth is not as efficiently warmed and moistened as air moving through the nasal cavity, and the throat or lung tissue can become dehydrated or damaged by the cold air.

2. When food moves down the esophagus, the normally collapsed esophagus expands. If the cartilage rings were solid, expansion of the esophagus, and therefore swallowing, would be more difficult.

3. A foreign object is more likely to become lodged in the right primary bronchus because it has a larger diameter and is more directly in line with the trachea.

4. Respiratory distress syndrome results from inadequate surfactant, which results in increased water surface tension. Consequently, lung recoil is increased. At the end of expiration pleural pressure is lower than normal because of the increased lung recoil. Although the decreased pleural pressure increases the tendency for the alveoli to expand, the alveoli do not expand because the increased force of expansion is only counteracting the increased lung recoil. The alveoli collapse if the lung recoil becomes larger than the force of expansion caused by the difference between alveolar and pleural pressure.

   During inspiration, pleural pressure has to be lower than normal to overcome the effect of the larger-than-normal lung recoil. A larger-than-normal increase in thoracic volume can cause a greater-than-normal decrease in pleural pressure. The effort of overcoming the increased lung recoil, however, can cause muscular fatigue and death.

5. The alveolar ventilation is 4200 mL/min ($12 \times [500 - 150]$). During exercise the alveolar ventilation is 88,800 mL/min ($24 \times [4000 - 300]$), a 21-fold increase. The increased air exchange increases P_{O_2} and decreases P_{CO_2} in the alveoli, thus increasing gas exchange between the alveoli and the blood.

6. The air the diver is breathing has a greater total pressure than atmospheric pressure at sea level. Consequently, the partial pressure of each gas in the air increases. According to Henry's law, as the partial pressure of a gas increases, the amount (concentration) of gas dissolved in the liquid (e.g., body fluids) with which the gas is in contact increases. When the diver suddenly ascends, the partial pressure of gases in the body returns toward sea level barometric pressure. As a result, the amount (concentration) of gas that can be dissolved in body fluids suddenly decreases. When the fluids can no longer hold all the gas, gas bubbles form.

7. At high altitudes the atmospheric P_{O_2} decreases because of a decrease in atmospheric pressure. The decreased atmospheric P_{O_2} results in a decrease in alveolar P_{O_2} and less oxygen diffusion into lung tissue. If the person's arterioles are especially sensitive to the decreased oxygen levels, constriction of the arterioles reduces blood flow through the lungs, and the ability to oxygenate blood decreases. Such generalized hypoxemia can also be caused by certain respiratory diseases, such as emphysema and cystic fibrosis.

8. Remember that the oxygen–hemoglobin dissociation curve normally shifts to the right in tissues. The shift of the oxygen–hemoglobin dissociation curve to the left caused by CO reduces the ability of hemoglobin to release oxygen to tissues, which contributes to the detrimental effects of CO poisoning. In the lungs, the shift to the left could slightly increase the ability of hemoglobin to pick up oxygen, but this effect is offset by the decreased ability of hemoglobin to release oxygen to tissues.

9. A person who cannot synthesize BPG has mild polycythemia. Her hemoglobin releases less oxygen to tissues. Consequently, one would expect increased erythropoietin release from the kidneys and increased erythrocyte production in red bone marrow.

10. Hyperventilation decreases blood carbon dioxide levels, causing an increase in blood pH. Holding one's breath increases blood carbon dioxide levels and decreases blood pH.

11. When a person hyperventilates, P_{CO_2} in the blood decreases. Consequently, carbon dioxide moves out of cerebrospinal fluid into the blood. As carbon dioxide levels in cerebrospinal fluid decrease, hydrogen ions and bicarbonate ions combine to form carbonic acid that forms carbon dioxide. The result is a decrease in hydrogen ion concentration in cerebrospinal fluid and decreased stimulation of the respiratory center by the chemosensitive area. Until blood P_{CO_2} levels increase, the chemosensitive area is not stimulated, and apnea results.

12. Through touch, thermal, or pain receptors the respiratory center can be stimulated to cause a sudden inspiration of air.

13. A P_{O_2} of 60 mm Hg and a P_{CO_2} of 30 mm Hg are both below normal. The movement of air into and out the lungs is restricted because of the asthma and there is a mismatch between ventilation of the alveoli and blood flow to the alveoli. Consequently, because of the ineffective ventilation, blood oxygen levels decrease. Mr. W hyperventilates, which helps to maintain blood oxygen levels, but also results in lower than normal blood carbon dioxide levels. (If there was no hyperventilation, one would expect decreased blood oxygen but increased blood carbon dioxide levels.)

Chapter 24

1. The moist, stratified squamous epithelium of the oropharynx and the laryngopharynx protects these regions from abrasive food when it is first swallowed. The ciliated pseudostratified epithelium of the nasopharynx helps move mucus produced in the nasal cavity and the nasopharynx into the oropharynx and esophagus. It is not as necessary to protect the nasopharynx from abrasion because food does not normally pass through this cavity.

2. A pin placed through the greater omentum passes through four layers of simple squamous epithelium. The greater omentum is actually a folded mesentery, with each part consisting of two layers of serous squamous epithelium.

3. It is important for the nasopharynx to be closed during swallowing so that food will not reflux into it or the nasal cavity. An explosive burst of laughter can relax the soft palate, open the nasopharynx, and cause the liquid to enter the nasal cavity.

4. Usually if a person tries to swallow and speak at the same time, the epiglottis is elevated, the laryngeal muscles closing the opening to the larynx are mostly relaxed, and food or liquid could enter the larynx, causing the person to choke.

5. After a heavy meal, blood pH may increase because, as bicarbonate ions pass from the cells of the stomach into the extracellular fluid, the pH of the extracellular fluid increases. As the extracellular fluid exchanges ions with the blood, the blood pH also increases.

6. Secretin production and its stimulation of bicarbonate ion secretion constitute a negative-feedback mechanism because, as the pH of the chyme in the duodenum decreases as a result of the presence of acid, secretin causes

an increase in bicarbonate ion secretion that increases the pH, restoring the proper pH balance in the duodenum.

7. The major effect of prolonged diarrhea is on the cardiovascular system and is much like massive blood loss. Hypovolemia continues to increase. Blood pressure declines in a positive-feedback cycle, and without intervention can lead to heart failure.

Chapter 25

1. If vitamins were broken down during the process of digestion, their structures would be destroyed, and, as a result, their ability to function would be lost.

2. If the electron of the electron-transport chain cannot be donated to oxygen, the entire electron-transport chain stops, no ATP can be produced aerobically, and the patient dies because too little energy is available for the body to perform vital functions. Anaerobic respiration is not adequate to provide all the energy needed to maintain human life, except for a short time.

3. When muscles contract, they use ATP. As a result of the chemical reactions necessary to synthesize ATP, heat is also produced. During exercise the large amounts of heat can raise body temperature, and we feel warm. Shivering consists of small, rapid muscle contractions that produce heat in an effort to prevent a decrease in body temperature in the cold.

4. Vasoconstriction reduces blood flow to the skin, which reduces skin temperature because less warm blood from the deeper parts of the body reach the skin. As the difference in temperature between the skin and the environment decreases, there is less loss of heat.

Chapter 26

1. The urethra of females is much shorter than the urethra of males. In addition, the opening of the urethra in females is closer to the anus, which is a potential source of bacteria. The female urinary bladder is therefore more accessible to bacteria from the exterior. This accessibility is a major reason that urinary bladder infections are more common in females than in males.

2. If the cardiac output is 5600 mL of blood per minute, and the hematocrit is 45, renal plasma flow is 650 mL of plasma per minute (see table 26.2). If the filtration fraction increased from 19% to 22%, the GFR would be 143 mL of filtrate per minute (650 mL of plasma × 0.22). If 99.2% of the filtrate is reabsorbed, 0.8% becomes urine. Thus the urine produced is 1.14 mL of urine per minute (143 mL of filtrate × 0.008). Compared to the rate of urine production when the filtrate fraction was 19% (that is, 1 mL per minute), the 3% increase in filtration fraction has caused a 14% increase in urine production. Converting 1.14 mL of urine per minute to liters of urine produced per day yields 1.64 L/day (1.14 ml/min × 1 L/1000 mL × 1440 min/day).

3. Even though hemoglobin is a smaller molecule than albumin, it does not normally enter the filtrate because hemoglobin is contained within erythrocytes and red blood cells cannot pass through the filtration membrane. If erythrocytes rupture, however, a process called hemolysis, the hemoglobin is released into the plasma, and large amounts of hemoglobin enter the filtrate. Conditions that cause erythrocytes to rupture in the circulatory system result in large amounts of hemoglobin entering the urine.

4. Constriction of the afferent arteriole decreases the blood pressure in the glomerulus. As a consequence, the total filtration pressure decreases. A decrease in the concentration of plasma proteins reduces the colloid osmotic pressure within the glomerular capillary. Because the total filtration pressure is determined by the glomerular blood pressure minus the colloid osmotic pressure minus the glomerular capsule pressure, a decrease in the colloid osmotic pressure increases the total filtration pressure. As a result, the total volume of filtrate produced per minute increases.

5. Without ADH, the distal convoluted tubule and the collecting duct are impermeable to water. Consequently, water cannot move by osmosis from the nephron into the interstitial spaces and therefore remains in the nephron to become urine. Because about 19% of the filtrate volume leaves the nephron in the distal convoluted tubule and the collecting duct, much of that volume appears as urine. As a result, urine volume increases, but its concentration decreases dramatically.

6. Inhibition of ADH secretion is one of the numerous effects alcohol has on the body. The lack of ADH secretion causes the distal tubules and the collecting ducts to be relatively impermeable to water. The water cannot therefore move by osmosis from the distal nephrons and collecting ducts and remains in the nephrons to become urine. In addition, because other fluids are normally consumed with the alcohol, the increased water intake also results in an increase in dilute urine production.

7. Without the normal active transport of sodium ions, the concentration of sodium ions and ions cotransported with them remains elevated in the nephron. Movement of water by osmosis out of the nephron into the interstitial spaces is decreased, resulting in an increased volume of urine.

8. Anything that reduces the formation of filtrate reduces the glomerular filtration rate. If the epithelium of the nephrons sloughs off and forms casts in the nephrons, normal flow of filtrate through them is blocked. Consequently, the blocked flow of filtrate in the nephron causes the pressure in Bowman's capsule to increase enough so that the pressure inside of Bowman's capsule is close to the pressure in the glomerulus. Unless the pressure in the glomerulus is higher than the pressure in Bowman's capsule, no filtrate forms, and the GFR is very low. If very little filtrate forms, the volume of urine produced is reduced.

9. Low urea clearance indicates that the amount of blood cleared of urea, a metabolic waste product, per minute is lower than normal. It is consistent with the reduction of the number of functional nephrons that occurs in advanced cases of renal failure. In addition, a low urea clearance is an indication that the GFR is reduced and that the blood levels of urea are increasing.

10. After 7 days Mr. H's kidney's began to produce a large volume of urine with larger-than-normal Na^+ and K^+ ion concentrations. The observations are consistent with Mr. H becoming dehydrated by day 9. Dehydration results in reduced blood volume. The pale skin was the result of vasoconstriction, which was triggered

by the reduced blood pressure. Dizziness resulted from reduced blood flow to the brain when Mr. H tried to stand and walk. He was lethargic in part because of reduced blood volume but also because low blood levels of K^+ and Na^+ ions. The arrythmia of his heart was due to low blood levels of K^+ ions and increased sympathetic stimulation, which also was triggered by low blood pressure.

Chapter 27

1. During hemorrhagic shock, blood pressure decreases, and visceral blood vessels constrict (see chapter 21). As a consequence, blood flow to the kidneys and the blood pressure in the glomeruli decrease dramatically. The total filtration pressure decreases, and the amount of filtrate formed each minute decreases. The rate at which Na^+ ions enter the nephron therefore decreases. In addition, renin is secreted from the kidneys in large amounts. Renin causes the formation of angiotensin I from angiotensinogen. Angiotensin I converts to angiotensin II, which stimulates aldosterone secretion. Aldosterone increases the rate at which Na^+ ions are reabsorbed from the filtrate in the distal tubule and collecting ducts.

2. (a) If the amount of Na^+ ions and water ingested in food exceeds that needed to maintain a constant extracellular fluid composition, it increases the total blood volume and also increases the blood pressure.

   (b) Excessive Na^+ ion and water intake causes an increase in total blood volume and blood pressure. The elevated blood pressure causes a reflex response that results in decreased ADH secretion. The elevated pressure also causes reduced renin secretion from the kidneys, resulting in a reduction in the rate at which angiotensin II is formed. The reduction in angiotensin II reduces the rate of aldosterone secretion. Together these changes cause increased loss of Na^+ ions in the urine and an increase in the volume of urine produced. Increased Na^+ ions and increased blood pressure also cause the secretion of atrial natriuretic hormone, which inhibits ADH

secretion and Na^+ ion reabsorption in the nephron.

   (c) If the amount of water ingested is large, the urine concentration is reduced, the urine volume is increased, and the concentration of Na^+ ions in the urine is low. If the amount of salt ingested is great, the concentration of the salt in the urine can be high, and the urine volume is larger and contains a substantial concentration of salt.

3. During conditions of exercise the amount of water lost is increased because of increased evaporation from the respiratory system, increased insensible perspiration, and increased sweat. The amount of water lost in the form of sweat can increase substantially. The amount of urine formed decreases during conditions of exercise.

4. Aldosterone hyposecretion results in acidosis. Aldosterone increases the rate at which Na^+ ions are reabsorbed from nephrons, but it also increases the rate at which K^+ and H^+ ions are secreted. Hyposecretion of aldosterone decreases the rate at which H^+ ions are secreted by the nephrons and therefore can result in acidosis.

Chapter 28

1. The prostate gland is anterior to the wall of the rectum. A finger inserted into the rectum can palpate the prostate gland through the rectal wall.

2. Coagulation may help keep the sperm cells within the female reproductive tract, increasing the likelihood of fertilization.

3. If administered before the preovulatory LH surge, estrogen stimulates the hypothalamus to secrete GnRH. Estrogen and progesterone, in large amounts, inhibit GnRH and LH releases. A large amount of estrogen and progesterone administered at this time should therefore reduce the surge of LH. Continual administration of high levels of GnRH causes anterior pituitary cells to become insensitive to GnRH. Thus LH and FSH levels remain low, and the ovarian cycle stops.

4. High progesterone levels after menses inhibit GnRH secretion from the hypothalamus and therefore FSH and LH secretion from the anterior pituitary. Without FSH and LH, the

events of the ovarian cycle, including estrogen production, are inhibited. Because estrogen causes proliferation of the endometrium, thickening of the endometrium is not expected. Also, estrogen increases the synthesis of uterine progesterone receptors, and without estrogen the secretory response of the endometrium to the elevated progesterone is inhibited.

5. Mrs. M's mother could have had leiomyomas also, although, without direct data from medical examinations, one cannot be certain. If that was the cause of her irregular menstruations, they may have become less frequent as Mrs. M's mother experienced menopause. During menopause, the uterus gradually becomes smaller, and eventually the cyclic changes in the endometrial lining cease. If the leiomyomas were relatively mild, the onset of menopause could explain the gradual disappearance of the irregular and prolonged menstruations (*Note:* If the tumors are large, constant and severe menstruations are likely even if regular menstrual cycles stop due to menopause).

Chapter 29

1. Two primitive streaks on one embryonic disk result in the development of two embryos. If the two primitive streaks are touching or are very close to each other, the embryos may be joined. This condition is called conjoined (Siamese) twins.

2. Because the early embryonic heart is a simple tube, blood must be forced through the heart in almost a peristaltic fashion, and the contraction begins in the sinus venosus. If the sinus venosus did not contract first, blood could flow in the opposite direction.

3. Postovulatory age is 14 days less than clinical age, which is 280 days to parturition. Parturition is therefore 266 days after ovulation (280 days minus 14 days).

4. Elevation of calcium levels might cause the uterine muscles to contract tetanically. This tetanic contraction could compress blood vessels and cut the blood supply to the fetus. Hypercalcemia can also result in arrythmias and muscle weakness (see chapter 27). The doctor would therefore not administer calcium to

the woman in labor but may give oxytocin, which strengthens contractions but is less likely to produce tetany.

5. Nursing stimulates the release of oxytocin from the mother's posterior pituitary gland, which is responsible for milk letdown. Oxytocin can also cause uterine contractions and cramps.

6. Genotype *DD* (homozygous dominant) would have the polydactyly phenotype, genotype *Dd* (heterozygous) would have the polydactyly phenotype, and genotype *dd* (homozygous recessive) would have the normal phenotype.

7. None. One in two will be homozygous normal and one in two will be normal heterozygous carriers.

	A	*a*
A	*AA*	*Aa*
A	*AA*	*Aa*

8. The answer to this question depends on your values and a weighing of the possible benefits against the possible harm that could result from a complete knowledge of the human genome. You might also consider whether or not this is a function of the government or if the money could be better spent on other projects. Discuss the issue with others.

Glossary

Many of the words in this glossary and text are followed by a simplified phonetic spelling showing pronunciation. The pronunciation key reflects standard clinical usage as presented in *Stedman's Medical Dictionary* (26th edition), a leading reference volume in the health sciences.

Generally, vowels are unmarked. Page numbers indicate where entries are in the text.

ā as in day, ate, way
a as in mat, hat, act
ă as in alone, abortion, media
ah as in father
ar as in far
aw as in fall
ē as in be, bee, meet
ĕ as in taken, genesis
er as in term, earn, learn
ī as in pie, pine, side
i as in pit, tip, fit
ĭ as in pencil
ō as in no, note, toe
o as in not, box, cot
ŏ as in occult, lemon, son
ow as in cow, brow, plow, now
oy as in boy, toy, oil
ū as in food, to, tool
u as in wood, foot, took
ŭ as in but, sun, bud, cup, up
yū as in pure, unit, union, future

A

A band Length of the myosin myofilament in a sarcomere. (278)

abdomen (ab-do'men, ab'dō-men) Belly, between the thorax and the pelvis. (15)

abduction (ab-dŭk'shun) [L., *abductio*, take away] Movement away from the midline. (232)

absolute refractory period Portion of the action potential during which the membrane is insensitive to all stimuli, regardless of their strength. (267)

absorptive cell Cell on the surface of villi of the small intestines and the luminal surface of the large intestine that is characterized by having microvilli; secretes digestive enzymes and absorbs digested materials on its free surface. (793)

absorptive state Immediately after a meal when nutrients are being absorbed from the intestine into the circulatory system. (851)

accommodation [L., *ac + commodo*, to adapt] Ability of electrically excitable tissues, such as nerve or muscle cells, to adjust to a constant stimulus so that the magnitude of the local potential decreases through time. (270)

acetabulum (as-ĕ-tab'yū-lum) [L., shallow vinegar vessel or cup] Cup-shaped depression on the external surface of the coxa. (214)

acetylcholine (as-e-til-kō'lēn) Neurotransmitter substance released from motor neurons, all preganglionic neurons of the parasympathetic and sympathetic divisions, all postganglionic neurons of the parasympathetic division, some postganglionic neurons of the sympathetic division, and some central nervous system neurons. (283)

acetylcholinesterase (as'ē-til-kō-lin-es'ter-ās) Enzyme found in the synaptic cleft that causes the breakdown of acetylcholine to acetic acid and choline, thus limiting the stimulatory effect of acetylcholine. (283)

Achilles tendon See calcaneal tendon.

acid Molecule that is a proton donor; any substance that releases hydrogen ions (H^+). (38)

acidic Solution containing more than 10^{-7} mol of hydrogen ions per liter; has a pH less than 7. (39)

acinus, pl. acini (as'i-nŭs, as'ĭ-nī) [L., berry, grape] Grape-shaped secretory portion of a gland. The terms acinus and alveolus are sometimes used interchangeably. Some authorities differentiate the terms: acini have a constricted opening into the excretory duct, whereas alveoli have an enlarged opening. (111)

acromion (ă-krō'mē-on) [Gr. *akron*, extremity + *omos*, shoulder] Bone comprising the tip of the shoulder. (210)

acrosome (ak'rō-sōm) [Gr. *akron*, extremity + *soma*, body] Cap on the head of the spermatozoon, with hydrolytic enzymes that help the spermatozoon to penetrate the ovum. (927)

actin myofilament (ak'tin) Thin myofilament within the sarcomere; composed of two F actin molecules, tropomyosin, and troponin molecules. (278)

action potential [L. *potentia*, power, potency] Change in membrane potential in an excitable tissue that acts as an electric signal and is propagated in an all-or-none fashion. (256)

activation energy Energy that must be added to molecules to initiate a reaction. (48)

active site Portion of an enzyme in which reactants are brought into close proximity and that plays a role in reducing activation energy of the reaction. (48)

active tension Tension produced by the contraction of a muscle. (293)

active transport Carrier-mediated process that requires ATP and can move substances against a concentration gradient. (79)

adaptive immunity Immune status in which there is an ability to recognize, remember, and destroy a specific antigen. (707)

adenohypophysis (ad'ĕ-nō-hī-pof'ĭ-sis) Portion of the hypophysis derived from the oral ectoderm; commonly called the anterior pituitary. (547)

adenosine diphosphate (ADP) (ă-den'ō-sēn) Adenosine, an organic base, with two phosphate groups attached to it. Adenosine diphosphate combines with a phosphate group to form adenosine triphosphate. (33)

adenosine triphosphate (ATP) Adenosine, an organic base, with three phosphate groups attached to it. Energy stored in

ATP is used in nearly all of the endergonic reactions in cells. (834)

adipocyte (ad′i-pō-sīt) Fat cell. (122)

adipose (ad′i-pōs) [L. *adeps,* fat] Fat. (122)

ADP See adenosine diphosphate.

adrenal gland (ă-drē′năl) [L. *ad,* to + *ren,* kidney] Also called the suprarenal gland. Located near the superior pole of each kidney, it is composed of a cortex and a medulla. The adrenal medulla is a highly modified sympathetic ganglion that secretes the hormones epinephrine and norepinephrine; the cortex secretes aldosterone and cortisol as its major secretory products. (561)

adrenaline (ă-dren′ă-lin) See epinephrine.

adrenergic receptor (ad-rĕ-ner′jik) Receptor molecule that binds to adrenergic agents such as epinephrine and norepinephrine. (514)

adrenocorticotropic hormone (ACTH) (ă-drē′no-kōr′ti-kō-trō′pik) Hormone of the adenohypophysis that governs the nutrition and growth of the adrenal cortex, stimulates it to functional activity, and causes it to secrete cortisol. (554)

adventitia (ad-ven-tish′ă) [L. *adventicius,* coming from abroad, foreign] Outermost covering of any organ or structure that is properly derived from outside the organ and does not form an integral part of the organ. (783)

aerobic respiration (ār-ō′bik), also (ă-ro′bik) Breakdown of glucose in the presence of oxygen to produce carbon dioxide, water, and approximately 38 ATPs; includes glycolysis, the citric acid cycle, and the electron transport chain. (82)

afferent arteriole (af′er-ent) Branch of an interlobular artery of the kidney that conveys blood to the glomerulus. (864)

afferent division Nerve fibers that send impulses from the periphery to the central nervous system. (360)

agglutination (ă-glū′ti-nā′shun) [L. *ad,* to + *gluten,* glue] Process by which blood cells, bacteria, or other particles are caused to adhere to one another and form clumps. (599)

agglutinin (ă-glū′ti-nin) Antibody that binds to an antigen and causes agglutination. (599)

agglutinogen (ă-glū-tin′ō-jen) Antigen on surface of erythrocytes that can stimulate the production of antibodies (agglutinins) that combine with the antigen and cause agglutination. (599)

agranulocyte (ă-gran′yū-lō-sīt) Nongranular leukocyte (monocyte or lymphocyte). (584)

ala, pl. alae (ā′lă, ā′lē) [L., a wing] Wing-shaped structure. (208)

aldosterone (al-dos′ter-ōn) Steroid hormone produced by the zona glomerulosa of the adrenal cortex that facilitates potassium exchange for sodium in the distal renal tubule, causing sodium reabsorption and potassium and hydrogen secretion. (689)

alkaline (al′kă-līn) Solution containing less than 10^{-7} mol of hydrogen ions per liter; has a pH greater than 7.0. (39)

alkalosis (al-kă-lō′sis) Condition characterized by blood pH of 7.45 or above. (918)

allantois (ă-lan′tō-is) Tube extending from the embryonic hindgut into the umbilical cord; forms the urinary bladder. (976)

allele (ă-lēl′) [Gr. *allelon,* reciprocally] Any one of a series of two or more different genes that may occupy the same position or locus on a specific chromosome. (992)

all–or–none When a stimulus is applied to a cell, an action potential is either produced or not. In muscle cells the cell either contracts to the maximum extent possible (for a given condition) or does not contract. (264)

alternative pathway Part of the nonspecific immune system for activation of complement. (710)

alveolar duct (al-vē′ō-lăr) Part of the respiratory passages beyond a respiratory bronchiole; from it arise alveolar sacs and alveoli. (745)

alveolar gland One in which the secretory unit has a saclike form and an obvious lumen. (789)

alveolar sac Two or more alveoli that share a common opening. (745)

alveolus, pl. alveoli (al-vē′ō-lus, al-ve-ō-lī) Cavity. Examples include the sockets into which teeth fit, the endings of the respiratory system, and the terminal endings of secretory glands. (111)

amino acid (ă-mēn′ō) Class of organic acids that constitute the building blocks for proteins. (46)

amplitude–modulated signal (am′pli-tūd) Signal that varies in magnitude or intensity such as with large versus small concentrations of hormones. (527)

ampulla (am-pul′lă, -ē) [L., two-handled bottle] Saclike dilatation of a semicircular canal; contains the crista ampullaris. Wide portion of the uterine tube between the infundibulum and the isthmus. (500)

amylase (am′il-ās) One of a group of starch-splitting enzymes that cleave starch, glycogen, and related polysaccharides. (803)

anabolism (ă-nab′ō-lizm) [Gr. *anabole,* a raising up] All of the synthesis reactions that occur within the body; requires energy. (32)

anaerobic respiration (an-ār-ō′bik) Breakdown of glucose in the absence of oxygen to produce lactic acid and two ATPs; consists of glycolysis and the reduction of pyruvic acid to lactic acid. (84)

anal canal (ā′năl) Terminal portion of the digestive tract. (799)

anal triangle Posterior portion of the perineal region through which the anal canal opens. (923)

anaphase (an′ă-fāz) Time during cell division when chromatids divide (or in the case of first meiosis, when the chromosome pairs divide). (92)

anastomoses (ă-nas′tō-mō′sez) A natural communication, direct or indirect, between two blood vessels or other tubular structures. An opening created by surgery, trauma, or disease between two or more normally separate spaces or organs. (651)

anatomic dead air space Volume of the conducting airways from the external environment down to the terminal bronchioles. (757)

androstenedione (an-drō-stēn′dī-ōn) Androgenic steroid of weaker potency than testosterone; secreted by the testis, ovary, and adrenal cortex. (566)

anencephaly (an′en-sef-ă-lē) [Gr. *an* + *enkephalos,* no brain] Defective development of the brain and absence of the bones of the cranial vault. Only a rudimentary brainstem and some trace of basal ganglia are present. (974)

aneurysm (an′yū-rizm) [Gr. *eurys,* wide] Dilated portion of an artery. (652)

angiotensin I (an-jē-ō-ten′sin) Peptide derived when renin acts on angiotensinogen. (689)

angiotensin II Peptide derived from angiotensin I; stimulates vasoconstriction and aldosterone secretion. (689)

anion (an′ī-on) Ion carrying a negative charge. (28)

antagonist (an-tag′ō-nist) Muscle that works in opposition to another muscle. (309)

anterior chamber of eye Chamber of the eye between the cornea and the iris. (477)

anterior interventricular sulcus Groove on the anterior surface of the heart, marking the location of the septum between the two ventricles. (611)

anterior pituitary   See adenohypophysis.

antibody   (an'tē-bod-ē) Protein found in the plasma that is responsible for humoral immunity; binds specifically to antigen. (593)

antibody–mediated immunity   Immunity due to B cells and the production of antibodies. (714)

anticoagulant   (an'tē-kō-ag'yū-lant) Agent that prevents coagulation. (598)

antidiuretic hormone (ADH)   (an'tē-dī-yū-ret'ik) Hormone secreted from the neurohypophysis that acts on the kidney to reduce the output of urine; also called vasopressin because it causes vasoconstriction. (549)

antigen   (an'ti-jen) [anti(body) + Gr. *-gen,* producing] Any substance that induces a state of sensitivity or resistance to infection or toxic substances after a latent period; substance that stimulates the specific immune system. (599)

antigenic determinant   (an-ti-jen'ik) The specific part of an antigen that stimulates an immune system response by binding to receptors on the surface of lymphocytes. (715)

antithrombin   (an-tē-throm'bin) Any substance that inhibits or prevents the effects of thrombin so that blood does not coagulate. (598)

antrum   (an'trūm) [Gr. *antron,* a cave] Cavity of an ovarian follicle filled with fluid containing estrogen. (938)

anulus fibrosus   (an'yu-lŭs fī-brō'sus) [L., fibrous ring] Fibrous material forming the outer portion of an intervertebral disk. (204)

anus   (ā'nŭs) Lower opening of the digestive tract through which fecal matter is extruded. (799)

aorta   (ā-ōr'tă) [Gr. *aorte* from *aeiro,* to lift up] Large elastic artery that is the main trunk of the systemic arterial system; carries blood from the left ventricle of the heart and passes through the thorax and abdomen. (611)

aortic arch   (ā-ōr'tik) [L., bow] Curve between the ascending and descending portions of the aorta. (652)

aortic body   One of the smallest bilateral structures, similar to the carotid bodies, attached to a small branch of the aorta near its arch; contains chemoreceptors that respond primarily to decreases in blood oxygen; less sensitive to decreases in blood pH or increases in carbon dioxide. (687)

apex   (ā'peks) [L, summit or tip] Extremity of a conical or pyramidal structure. The apex of the heart is the rounded tip directed anteriorly and slightly inferiorly. (608)

Apgar score   Named for the U.S. anesthesiologist, Virginia Apgar (1909–1974). Evaluation of a newborn infant's physical status by assigning numerical values to each of five criteria; appearance (skin color), pulse (heart rate), grimace (response to stimulation), activity (muscle tone), and respiratory effort; a score of 10 indicates the best possible condition. (986)

apical ectodermal ridge   Layer of surface ectodermal cells at the lateral margin of the embryonic limb bud; they stimulate growth of the limb. (968)

apical foramen (tooth)   [L., aperture] Opening at the apex of the root of a tooth that gives passage to the nerve and blood vessels. (788)

apocrine   (ap'ō-krin) [Gr. *apo,* away from + *krino,* to separate] Gland whose cells contribute cytoplasm to its secretion (e.g., mammary glands). Sweat glands that produce organic secretions traditionally are called apocrine. These sweat glands, however, are actually merocrine glands; see also merocrine and holocrine. (111)

appendicular skeleton   (ap'en-dik'yū-lăr) The portion of the skeleton consisting of the upper limbs and the lower limbs and their girdles. (210)

appositional growth   (ap-ō-zish'ŭn-al) [L. *ap + pono,* to put or place] To place one layer of bone, cartilage, or other connective tissue against an existing layer. (157)

aqueous humor   (ak'wē-ŭs or ā'kwē-ŭs) Watery, clear solution that fills the anterior and posterior chambers of the eye. (477)

arachnoid   (ă-rak'noyd) [Gr. *arachne,* spider, cobweb] Thin, cobweb-appearing meningeal layer surrounding the brain; the middle of the three layers. (422)

arcuate artery   (ar'kyū-āt) Originates from the interlobar arteries of the kidney and forms an arch between the cortex and medulla of the kidney. (864)

areola   (ă-rē'ō-lă, -lē) [L., area] Circular pigmented area surrounding the nipple; its surface is dotted with little projections caused by the presence of the areolar glands beneath. (943)

areolar gland   (ă-rē'ō-lăr) Gland forming small, rounded projections from the surface of the areola of the mamma. (943)

arrectores pilorum, pl. arrector pili   (ă-rek'tō-rēz pī-lōr'um, ă-rek'tōr pī'lī) [L., that which raises; hair] Smooth muscle attached to the hair follicle and dermis that raises the hair when it contracts. (146)

arterial capillary   (ar-tē'rē-ăl) Capillary opening from an arteriole or metarteriole. (648)

arteriole   (ar-tēr'ē-ōl) Minute artery with all three tunics that transports blood to a capillary. (649)

arteriosclerosis   (ar-tēr'ē-ō-skler-ō'sis) [L. *arterio* + Gr. *sklerosis,* hardness] Hardening of the arteries. (651)

arteriovenous anastomosis   (ar-tēr'ē-ō-vē'nŭs ă-nas'tō-mō'sis) Vessel through which blood is shunted from an arteriole to a venule without passing through the capillaries. (651)

artery   Blood vessel that carries blood away from the heart. (652)

articular cartilage   Hyaline cartilage covering the ends of bones within a synovial joint. (157)

articulation   Place where two bones come together—a joint. (225)

arytenoid cartilages   (ar-i-tē'noyd) Small pyramidal laryngeal cartilages that articulate with the cricoid cartilage. (739)

ascending aorta   Part of the aorta from which the coronary arteries arise. (652)

ascending colon   (kō'lon) Portion of the colon between the small intestine and the right colic flexure. (798)

asthma   (az'mă) Condition of the lungs in which widespread narrowing of airways occurs caused by contraction of smooth muscle, edema of the mucosa, and mucus in the lumen of the bronchi and bronchioles. (775)

astrocyte   (as'trō-sīt) [Gr. *astron,* star + *kytos,* a hollow, a cell] Star-shaped neuroglia cell involved with forming the blood–brain barrier. (364)

atherosclerosis   (ath'er-ō-sker-ō'sis) Arteriosclerosis characterized by irregularly distributed lipid deposits in the intima of large and medium-sized arteries. (651)

atomic number   Number of protons in each type of atom. (24)

ATP   See adenosine triphosphate.

atrial diastole   Dilation of the heart's atria. (625)

atrial natriuretic hormone   (ā'trē-ăl nă'trē-yū-ret'ik) Peptide released from the atria when atrial blood pressure is increased; acts to lower blood pressure by increasing the rate of urinary production, thus reducing blood volume. (689)

atrial systole   (ā'trē-ăl sis'tō-lē) Contraction of the atria. (625)

atrioventricular (AV) node (ā′trē-ō-ven-trik′yū-lar) Small node of specialized cardiac muscle fibers that gives rise to the atrioventricular bundle of the conduction system of the heart. (619)

atrioventricular bundle Bundle of modified cardiac muscle fibers that projects from the AV node through the interventricular septum. (619)

atrioventricular valve One of two valves closing the openings between the atria and ventricles. (615)

atrium, pl. atria (ā′trē-ŭm, ā′trē-ă) [L., entrance hall] One of two chambers of the heart into which veins carry blood. (611)

auditory cortex Portion of the cerebral cortex that is responsible for the conscious sensation of sound; in the dorsal portion of the temporal lobe within the lateral fissure and on the superolateral surface of the temporal lobe. (499)

auditory ossicle Bone of the middle ear: includes the malleus, incus, and stapes. (486)

auricle (aw′rĭ-kl) [L. auris, ear] Part of the external ear that protrudes from the side of the head. Small pouch projecting from the superior, anterior portion of each atrium of the heart. (486)

auscultatory (aws-kŭl′tăh-tō′rē) Relating to auscultation, listening to the sounds made by the various body structures as a diagnostic method. (674)

autoimmune disease Disease resulting from a specific immune system reaction against self-antigens. (714)

autonomic ganglia Ganglia containing the nerve cell bodies of the autoimmune division of the nervous system. (509)

autonomic nervous system Composed of nerve fibers that send impulses from the central nervous system to smooth muscle, cardiac muscle, and glands. (360)

autophagia (aw-tō-fā′jē-ă) [Gr. auto, self + phagein, to eat] Segregation and disposal of organelles within a cell. (68)

autoregulation Maintenance of a relatively constant blood flow through a tissue despite relatively large changes in blood pressure; maintenance of a relatively constant glomerular filtration rate despite relatively large changes in blood pressure. (682)

autorhythmic Spontaneous and periodic; for example, in smooth muscle it implies spontaneous (without nervous or hormonal stimulation) and periodic contractions. (621)

autosome [Gr. auto, self + soma, body] Any chromosome other than a sex chromosome; normally occur in pairs in somatic cells and singly in gametes. (92)

axial skeleton Skull, vertebral column, and rib cage. (184)

axillary (ak′si-lār ē) Relating to the axilla. The space below the shoulder joint, bounded by the pectoralis major anteriorly, the latissimus dorsi posteriorly, the serratus anterior medially, and the humerus laterally. (446)

axolemma [Gr. axo + lemma, husk] Cell membrane of the axon. (363)

axon (ak′son) [Gr., axis] Main central process of a neuron that normally conducts action potentials away from the neuron cell body. (126)

axon hillock Area of origin of the axon from the nerve cell body. (363)

axoplasm (ak′sō-plazm) Neuroplasm or cytoplasm of the axon. (363)

B

B cell Type of lymphocyte responsible for antibody-mediated immunity. (593)

baroreceptor (bar′ō-rē-sep′ter, -tor) (pressoreceptor) Sensory nerve ending in the walls of the atria of the heart, venae cavae, aortic arch, and carotid sinuses; sensitive to stretching of the wall caused by increased blood pressure. (518)

baroreceptor reflex Detects changes in blood pressure and produces changes in heart rate, heart force of contraction, and blood vessel diameter that return blood pressure to homeostatic levels. (635)

basal ganglia Nuclei at the base of the cerebrum involved in controlling motor functions. (398)

base Molecule that is a proton acceptor; any substance that binds to hydrogen ions. (39)

base [L. and Gr. basis] Lower part or bottom of a structure. The base of the heart is the flat portion directed posteriorly and superiorly. Veins and arteries project into and out of the base, respectively. (608)

basement membrane Specialized extracellular material located at the base of epithelial cells and separating them from the underlying connective tissues. (103)

basic See alkaline.

basilar membrane Wall of the membranous labyrinth bordering the scala tympani; supports the organ of Corti. (493)

basophil (bā′sō-fil) [Gr. basis, baso + phileo, to love] White blood cell with granules that stain specifically with basic dyes; promotes inflammation. (593)

belly Largest portion of muscle between the origin and insertion. (309)

beta-oxidation Metabolism of fatty acids by removing a series of two-carbon units to form acetyl-CoA. (847)

bicarbonate ion (bī-kar′bon-āt) Anion (HCO_3^-) remaining after the dissociation of carbonic acid. (588)

bicuspid [mitral] valve (bī-kŭs′pid) Valve closing the orifice between left atrium and left ventricle of the heart. (615)

bile (bīl) Fluid secreted from the liver into the duodenum; consists of bile salts, bile pigments, bicarbonate ions, cholesterol, fats, fat-soluble hormones, and lecithin. (797)

bile canaliculus (bīl kan′ă-lik′yu-lŭs) One of the intercellular channels approximately 1 μm or less in diameter that occurs between liver cells into which bile is secreted; empties into the hepatic ducts. (797)

bile salt Organic salt secreted by the liver that functions as an emulsifying agent. (819)

bilirubin (bil-i-rū′bin) [L. bili + ruber, red] Bile pigment derived from hemoglobin during destruction of erythrocytes. (592)

biliverdin (bil-i-ver′din) Green bile pigment formed from the oxidation of bilirubin. (592)

binocular vision (bin-ok′yū-lar) [L. bini, paired + oculus, eye] Vision using two eyes at the same time; responsible for depth perception when the visual fields of each eye overlaps. (486)

bipolar neuron (bī-pō′ler) One of the three categories of neurons consisting of a neuron with two processes—one dendrite and one axon—arising from opposite poles of the cell body. (126)

blastocele (blas′tō-sēl) [Gr. blastos, germ + koilos, hollow] Cavity in the blastocyst. (962)

blastocyst (blas′tō-sist) [Gr. blastos, germ + kystis, bladder] Stage of mammalian embryos that consists of the inner cell mass and a thin trophoblast layer enclosing the blastocele. (963)

bleaching In response to light, retinal separates from opsin. (480)

blind spot Point in the retina where the optic nerve penetrates the fibrous tunic; contains no rods or cones and therefore does not respond to light. (476)

blood–brain barrier Permeability barrier controlling the passage of most large-molecular compounds from the blood to the cerebrospinal fluid and brain tissue; consists of capillary endothelium and may include the astrocytes. (364)

blood clot Coagulated phase of blood. (595)

blood groups Classification of blood based on the type of antigen found on the surface of erythrocytes. (599)

blood island Aggregation of mesodermal cells in the embryonic yolk sac that forms vascular endothelium and primitive blood cells. (974)

blood pressure [L. *pressus,* to press] Tension of the blood within the blood vessels; commonly expressed in units of millimeters of mercury (mm Hg). (674)

blood–thymic barrier Layer of reticular cells that separates capillaries from thymic tissue in the cortex of the thymus gland; prevents large molecules from leaving the blood and entering the cortex. (706)

Bohr effect Named for the Danish physiologist, Christian Bohr (1855–1911). Shift of the oxygen–hemoglobin dissociation curve to the right or left because of changes in blood pH. The definition sometimes is extended to include shifts caused by changes in blood carbon dioxide levels. (764)

bony labyrinth (lab′i-rinth) Part of the inner ear; contains the membranous labyrinth that forms the cochlea, vestibule, and semicircular canals. (491)

brachial (brā′kē-ăl) [L. *brachium,* arm] Relating to the arm. (656)

branchial arch Typically, six arches in vertebrates; in the lower vertebrates they bear gills, but they appear transiently in the higher vertebrates and give rise to structures in the head and neck. (968)

broad ligament Peritoneal fold passing from the lateral margin of the uterus to the wall of the pelvis on either side. (937)

bronchiole (brong′kē-ōl) One of the finer subdivisions of the bronchial tubes, less than 1 mm in diameter; has no cartilage in its wall but does have relatively more smooth muscle and elastic fibers. (743)

brush border Epithelial surface consisting of microvilli. (793)

buffer Mixture of an acid and base that reduces any changes in pH that would otherwise occur in a solution when acid or base is added to the solution. (40)

bulb of the penis Expanded posterior part of the corpus spongiosum of the penis. (931)

bulb of the vestibule Mass of erectile tissue on either side of the vagina. (943)

bulbar conjunctiva (bŭl′bar kon-jŭnk-tī′-vă) Conjunctiva that covers the surface of the eyeball. (471)

bulbourethral gland (bŭl′bō-yū-rē′thrăl) One of two small compound glands that produce a mucoid secretion; it discharges through a small duct into the spongy urethra. (933)

bulbus cordis (bŭl′bŭs) [L., plant bulb] End of the embryonic cardiac tube where blood leaves the heart; becomes part of the ventricle. (974)

bursa, pl. bursae (ber′să, ber′sē) [L., purse] Closed sac or pocket containing synovial fluid, usually found in areas where friction occurs. (230)

bursitis (ber-sī′tis) [L. *purse* + Gr. *ites,* inflammation] Inflammation of a bursa. (230)

C

calcaneal tendon (kal-kā′nē-al) Common tendon of the gastrocnemius, soleus, and plantaris muscle that attaches to the calcaneus. (344)

calcitonin (kal-si-tō′nin) Hormone released from parafollicular cells that acts on tissues to cause a decrease in blood levels of calcium ions. (175)

calmodulin (kal-mod′yū-lin) [calcium + modulate] Protein receptor for Ca^{2+} ions that plays a role in many Ca^{2+}-regulated processes such as smooth muscle contraction. (540)

calorie (kal′ō-rē) [L. *calor,* heat] Unit of heat content or energy. The quantity of energy required to raise the temperature of 1 g of water 1°C. (832)

calpain (kal′pān) Enzyme involved in changing the shape of dendrites; involved with long-term memory. (397)

calyx, pl. calyces (kā′liks, kal′i-sēz) [Gr., cup of a flower] Flower-shaped or funnel-shaped structure; specifically, one of the branches or recesses of a renal pelvis into which the tips of the renal pyramids project. (861)

canal of Schlemm Named for the German anatomist Friedrich Schlemm (1795–1858). Series of veins at the base of the cornea that drain excess aqueous humor from the eye. (477)

cancellous bone (kan-sĕl′ŭs, -lī) [L., grating or lattice] Bone with a latticelike appearance; spongy bone. (158)

cancer (kan′ser) General term frequently used to indicate any of various types of malignant neoplasms, most of which invade surrounding tissues, may metastasize to several sites, and are likely to recur after attempted removal and to cause death of the patient unless adequately treated. (995)

canine (kā′nīn) Referring to the cuspid tooth. (788)

cannula (kan′yū-lă) [L. *canna,* reed] Tube; often inserted into an artery or vein. (674)

capacitation (kă-pas′i-tā′shun) [L. *capax,* capable of] Process whereby spermatozoa acquire the ability to fertilize ova. This process occurs in the female genital tract. (949)

capitulum (kă-pit′yū-lŭm, -lă) [L. *caput,* head] Head-shaped structure. (212)

carbaminohemoglobin (kar-bam′i-nō-hē-mō-glō′bin) Carbon dioxide bound to hemoglobin by means of a reactive amino group on the hemoglobin. (588)

carbohydrate (kar-bō-hī′drăt) Monosaccharide (simple sugar) or the organic molecules composed of monosaccharides bound together by chemical bonds, e.g., glycogen. For each carbon atom in the molecule there are typically one oxygen molecule and two hydrogen molecules. (41)

carbonic acid/bicarbonate buffer system (kar-bon′ik) One of the major buffer systems in the body; major components are carbonic acid and bicarbonate ions. (914)

carbonic anhydrase Enzyme that catalyzes the reaction between carbon dioxide and water to form carbonic acid. (588)

carcinoma (kar-si-nō′mă) A malignant neoplasm derived from epithelial tissue. (151)

cardiac (kar′dē-ak) [Gr. *kardia,* heart] Related to the heart. (79)

cardiac cycle [Gr. *kyklos,* circle] Complete round of cardiac systole and diastole. (625)

cardiac nerve   Nerve that extends from the sympathetic chain ganglia to the heart. (634)

cardiac output (minute volume)   Volume of blood pumped by the heart per minute. (631)

cardiac region   Region of the stomach near the opening of the esophagus. (791)

cardiac reserve   [L. *re* + *servo,* to keep back, reserve] Work that the heart is able to perform beyond that required during ordinary circumstances of daily life. (632)

carotid body   (kar-rot′id) One of the small organs near the carotid sinuses; contains chemoreceptors that respond primarily to decreases in blood oxygen; less sensitive to decreases in blood pH or increases in carbon dioxide. (687)

carotid sinus   Enlargement of the internal carotid artery near the point where the internal carotid artery branches from the common carotid artery; contains baroreceptors. (655)

carpal   (kar′păl) [Gr. *karpos,* wrist] Bone of the wrist. (213)

carrier   Person in apparent health whose chromosomes contain a pathologic mutant gene that may be transmitted to his or her children. (994)

cartilage   (kar′ti-lij) [L. *cartilage,* gristle] Firm, smooth, resilient, nonvascular connective tissue. (122)

cartilaginous joint   (kar-ti-laj′i-nus) Bones connected by cartilage; includes synchondroses and symphyses. (226)

catabolism   (kă-tab′ō-lizm) [Gr. *katabole,* a casting down] All of the decomposition reactions that occur in the body; releases energy. (33)

catalyst   (kat′ă-list) Substance that increases the rate at which a chemical reaction proceeds without being changed permanently. (35)

cations   (kat′ī-on) [Gr. *kation,* going down] Ions carrying a positive charge. (28)

caveola, pl. caveolae   (kav-ē-ō′la, -lē) [L., a small pocket] Shallow invagination in the membranes of smooth muscle cells that may perform a function similar to both the T tubules and sarcoplasmic reticulum of skeletal muscle. (298)

cecum   (sē′kum, sē′kă) [L. *caecus,* blind] Cul-de-sac forming the first part of the large intestine. (798)

cell–mediated immunity   Immunity due to the actions of T cells and null cells. (714)

celom   (sē′lōm, sē′lōm′mă) [Gr. *koilo* + *amma,* a hollow] Principal cavities of the trunk, e.g., the pericardial, pleural, and peritoneal cavities. Separate in the adult, they are continuous in the embryo. (968)

cementum   (se-men′tŭm) [L. *caementum,* rough quarry stone] Layer of modified bone covering the dentin of the root and neck of a tooth; blends with the fibers of the periodontal membrane. (788)

central nervous system (CNS)   Major subdivision of the nervous system consisting of the brain and spinal cord. (360)

central vein   Terminal branches of the hepatic veins that lie centrally in the hepatic lobules and receive blood from the liver sinusoids. (670)

centrosome   (sen′trō-sōm) Specialized zone of cytoplasm close to the nucleus and containing two centrioles. (70)

cerebellum   (ser-e-bel′ŭm, bel′a) [L., little brain] Separate portion of the brain attached to the brainstem at the pons; important in maintaining muscle tone, balance, and coordination of movement. (400)

cerebrospinal fluid   (ser-ĕ′brō-spī-năl, sĕ-rē′brō-spī-năl) Fluid filling the ventricles and surrounding the brain and spinal cord. (423)

ceruminous glands   (sĕ-rū′mi-nŭs) Modified sebaceous glands in the external auditory meatus that produce cerumen (earwax). (486)

cervical canal   (ser′vĭ-kal) Canal extending from the isthmus of the uterus to the opening of the uterus into the vagina. (941)

cervix, pl. cervices   (ser′viks, ser′vi-sēz) [L., neck] Lower part of the uterus extending from the isthmus of the uterus into the vagina. (941)

chalazion   (ka-lā′zē-on) A chronic inflammation of a meibomian gland. (471)

cheek   (chēk) Side of the face forming the lateral wall of the mouth. (787)

chemoreceptor   (kem′ō-rē-sep′tŏr, kē′mō-) Sensory cell that is stimulated by a change in the concentration of chemicals to produce action potentials. Examples include taste receptors, olfactory receptors, and carotid bodies. (462)

chemoreceptor reflex   Chemoreceptors detect decrease in blood oxygen, increase in carbon dioxide, or decrease in pH and produce an increased rate and depth of respiration, and by means of the vasomotor center, vasoconstriction. (636)

chemosensitive area   (kem-ō-sen′si-tiv, kē-mō-) Chemosensitive neurons in the medulla oblongata detect changes in blood, carbon dioxide, and pH. (768)

chemotactic factor   (kem-ō-tak′tik, kē-mo-) Part of a microorganism or chemical released by tissues and cells that act as chemical signals to attract leukocytes. (710)

chemotaxis   (kem-ō-tak′sis, kē-mō-) [Gr. *chemo* + *taxis,* orderly arrangement] Attraction of living protoplasm (cells) to chemical stimuli. (593)

chief cell   Cell of the parathyroid gland that secretes parathyroid hormone. (560) Cell of a gastric gland that secretes pepsinogen. (793)

chloride   (klōr′īd) Compound containing chlorine, e.g., salts of hydrochloric acid. (822)

chloride shift   Diffusion of chloride ions into red blood cells as bicarbonate ions diffuse out; maintains electrical neutrality inside and outside the red blood cells. (766)

choana, pl. choanae   (kō′an-ă, kō-ā′nē) See internal naris.

cholecystokinin   (kō′lē-sis-tō-kī′nin) Hormone liberated by the upper intestinal mucosa on contact with gastric contents; stimulates the contraction of the gallbladder and the secretion of pancreatic juice high in digestive enzymes. (807)

cholinergic neuron   (kol-in-er′jik) Refers to nerve fibers that secrete acetylcholine as a neurotransmitter substance. (514)

chondroblast   (kon′drō-blast) [Gr. *chondros,* gristle, cartilage + *blastos,* germ] Cartilage-producing cell. (157)

chondrocyte   (kon′drō-sīt) [Gr. *chondros,* gristle, cartilage + *kytos,* a cell] Mature cartilage cell. (122)

chorda tympani, pl. chordae   (kōr′da tym′pan-ē, kōr′dē) Branch of the facial nerve that conveys taste sensation from the front two-thirds of the tongue. (470)

chordae tendineae   (kōr′dă ten′di-nē-ē) [L., cord] Tendinous strands running from the papillary muscles to the atrioventricular valves. (615)

choroid   (ko′royd) Portion of the vascular tunic associated with the sclera of the eye. (365)

choroid plexus   [Gr. *chorioeides,* membranelike] Specialized plexus located within the ventricles of the brain that secretes cerebrospinal fluid. (365)

chromatid   (krō′mă-tid) One half of a chromosome; separates from its partner during cell division. (94)

chromatin (krō′ma-tin) Colored material; the genetic material in the nucleus. (52)

chromosome (krō′mō-sōm) Colored body in the nucleus, composed of DNA and proteins and containing the primary genetic information of the cell; 23 pairs in humans. (991)

chylomicron (kī-lō-mī′kron) [Gr. *chylos,* juice + *micros,* small] Microscopic particle of lipid surrounded by protein, occurring in chyle and in blood. (819)

chymotrypsin (kī-mō-trip′sin) Proteolytic enzyme formed in the small intestine from the pancreatic precursor chymotrypsinogen. (815)

ciliary body (sil′ē-ar-ē) Structure continuous with the choroid layer at its anterior margin that contains smooth muscle cells and functions in accommodation. (475)

ciliary gland Modified sweat gland that opens into the follicle of an eyelash, keeping it lubricated. (470)

ciliary muscle Smooth muscle in the ciliary body of the eye. (475)

ciliary process Portion of the ciliary body of the eye that attaches by suspensory ligaments to the lens. (475)

ciliary ring Portion of the ciliary body of the eye that contains smooth muscle cells. (475)

circumduction (ser-kŭm-dŭk′shŭn) [L., around + *ductus,* to draw] Movement in a circular motion. (232)

circumferential lamellae (ser-kŭm-fer-en′shē-al lă-mel′ē) Lamellae covering the surface of and extending around compact bone inside the periosteum. (163)

circumvallate papilla (ser-kŭm-val′āt pă-pil′ă) Type of papilla on the surface of the tongue surrounded by a groove. (468)

cisterna, pl. cisternae (sis-ter′nă, sis-ter′nē) Interior space of the endoplasmic reticulum. (66)

cisterna chyli (kī′lē) [L., tank + Gr. *chylos,* juice] Enlarged inferior end of the thoracic duct that receives chyle from the intestine. (672)

citric acid cycle (sit′rik) Series of chemical reactions in which citric acid is converted into oxaloacetic acid, carbon dioxide is formed, and energy is released. The oxaloacetic acid can combine with acetyl-CoA to form citric acid and restart the cycle. The energy released is used to form NADH, FADH, and ATP. (844)

classical pathway Part of the specific immune system for activation of complement. (710)

clavicle (klav′i-kl) The collar bone, between the sternum and scapula. (210)

cleavage furrow (klēv′ij) Inward pinching of the plasma membrane that divides a cell into two halves, which separate from each other to form two new cells. (92)

cleft palate (kleft) Failure of the embryonic palate to fuse along the midline, resulting in an opening through the roof of the mouth. (970)

clinical age (klin′i-kl) Age of the developing fetus from the time of the mother's last menstrual period before pregnancy. (962)

clinical perineum (klin′i-kl per′i-nē′ŭm) Portion of the perineum between the vaginal and anal openings. (943)

clitoris (klit′ō-ris) Small cylindrical, erectile body, rarely exceeding 2 cm in length, situated at the most anterior portion of the vulva and projecting beneath the prepuce. (942)

cloaca (klō-ā′kă) [L., sewer] In early embryos the endodermally lined chamber into which the hindgut and allantois empty. (976)

clot retraction Condensation of the clot into a denser, compact structure; caused by the elastic nature of fibrin. (598)

coagulation (kō-ag-yū-lā′shŭn) Process of changing from liquid to solid, especially of blood; formation of a blood clot. (595)

cochlear duct (kok′lē-ăr) Interior of the membranous labyrinth of the cochlea; cochlear canal. (493)

cochlear nerve Nerve that carries sensory impulses from the organ of Corti to the vestibulocochlear nerve. (493)

cochlear nucleus Neurons from the cochlear nerve synapse within the dorsal or ventral cochlear nucleus in the superior medulla oblongata. (498)

codon (kō′don) Sequence of three nucleotides in mRNA or DNA that codes for a specific amino acid in a protein. (87)

cofactor (kō′fak′ter, tōr) Nonprotein component of an enzyme such as coenzymes and inorganic ions essential for enzyme action. (49)

collagen (kol′lă-jen) [Gr. *koila,* glue + *gen,* producing] Ropelike protein of the extracellular matrix. (113)

collateral ganglia (ko-lat′er-ăl gang′glē-ă) Sympathetic ganglia that are found at the origin of large abdominal arteries; include the celiac, superior, and inferior mesenteric arteries. (513)

collecting duct Straight tubule that extends from the cortex of the kidney to the tip of the renal pyramid. Filtrate from the distal convoluted tubes enters the collecting duct and is carried to the calyces. (864)

colloid (kol′loyd) [Gr. *kolla,* glue + *eidos,* appearance] Atoms or molecules dispersed in a gaseous, liquid, or solid medium that resist separation from the liquid or gas. (38)

colloid osmotic pressure Osmotic pressure due to the concentration difference of proteins across a membrane that does not allow passage of the proteins. (871)

colloidal solution (ko-loy′del) Fine particles suspended in a liquid; particles are resistant to sedimentation or filtration. (583)

colon (kō′lon) Division of the large intestine that extends from the cecum to the rectum. (798)

colostrum (kō-los′trŭm) Thin, white fluid; the first milk secreted by the breast at the termination of pregnancy; contains less fat and lactose than the milk secreted later. (988)

columnar Shaped like a column. (104)

commissure (kom′i-syūr) [L. *commissura,* a joining together] Connection of nerve fibers between the cerebral hemispheres or from one side of the spinal cord to the other. (397)

common bile duct Duct formed by the union of the common hepatic and cystic ducts; it empties into the small intestine. (793)

common hepatic duct Part of the biliary duct system that is formed by the joining of the right and left hepatic ducts. (797)

compact bone Bone that is more dense and has fewer spaces than cancellous bone. (158)

competition Similar molecules binding to the same carrier molecule or receptor site. (78)

complement Group of serum proteins that stimulates phagocytosis and inflammation. (710)

complement cascade Series of reactions in which each component activates the next component, resulting in activation of complement proteins. (710)

compliance (kom-plī′ans) Change in volume (e.g., in lungs or blood vessels) caused by a given change in pressure. (676)

compound A substance composed of two or more different types of atoms that are chemically combined. (30)

concha, pl. conchae (kon′kă, kon′kē) [L., shell] Structure comparable to a shell in shape; the three bony ridges on the lateral wall of the nasal cavity. (738)

conduction (kon-dŭk′shŭn) [L. *con* + *ductus*, to lead, conduct] Transfer of energy such as heat from one point to another without evident movement in the conducting body. (855)

cone Photoreceptor in the retina of the eye; responsible for color vision. (476)

congenital (kon-jen′i-tăl) [L. *congenitus*, born with] Occurring at birth; may be genetic or due to some influence (e.g., drugs) during development. (995)

conjunctiva (kon-jŭnk-tī′vă, -vē) [L. *conjungo*, to bind together] Mucous membrane covering the anterior surface of the eyeball and lining the lids. (471)

conjunctival fornix (kon-jŭnk-tī′văl fōr′niks) Area in which the palpebral and bulbar conjunctiva meet. (471)

constant region Portion of the antibody that does not combine with the antigen and is the same in different antibodies. (721)

continuous capillary [L. *capillaris*, relating to hair] Capillary in which pores are absent; less permeable to large molecules than other types of capillaries. (647)

contraction phase One of the three phases of muscle contraction; the time during which tension is produced by the contraction of muscle. (288)

convection (kon-vek′shŭn) [L. *con* + *vectus*, to carry or bring together] Transfer of heat in liquids or gases by movement of the heated particles. (855)

coracoid (kōr′ă-koyd) [Gr. *korakodes*, crow's beak] Resembling a crow's beak, for example, a process on the scapula. (210)

Cori cycle Named for the Czech-U.S. biochemist and Nobel laureate, Carl F. Cori (1896–1984). Lactic acid, produced by skeletal muscle, is carried in the blood to the liver, where it is aerobically converted into glucose. The glucose may return through the blood to skeletal muscle or may be stored as glycogen in the liver. (844)

cornea (kor′nē-ă) Transparent portion of the fibrous tunic that makes up the outer wall of the anterior portion of the eye. (473)

corniculate cartilage (kōr-nik′yū-lăt) Conical nodule of elastic cartilage surmounting the apex of each arytenoid cartilage. (739)

coronary (kōr′o-nār-ē) [L. *coronarius*, a crown] Resembling a crown; encircling. (611)

coronary artery One of two arteries that arise from the base of the aorta and carry blood to the muscle of the heart. (611)

coronary ligament Peritoneal reflection from the liver to the diaphragm at the margins of the bare area of the liver. (801)

coronary sinus Short trunk that receives most of the veins of the heart and empties into the right atrium. (614)

coronoid (kōr′ŏ-noyd) [Gr. *korone*, a crow] Shaped like a crow's beak, for example, a process on the mandible. (189)

corpus, pl. corpora (kōr′pus, -pōr-ă) [L. *body*] Any body or mass; the main part of an organ. (386)

corpus albicans (al′bĭ-kanz) Atrophied corpus luteum leaving a connective tissue scar in the ovary. (941)

corpus callosum (kăl-lō′sŭm) [L., *body* + callous] Largest commissure of the brain, connecting the cerebral hemispheres. (397)

corpus cavernosum, pl. corpora cavernosa One of two parallel columns of erectile tissue forming the dorsal part of the body of the penis or the body of the clitoris. (931)

corpus luteum (lū′tē-ŭm) Yellow endocrine body formed in the ovary in the site of a ruptured vesicular follicle immediately after ovulation; secretes progesterone and estrogen. (939)

corpus luteum of pregnancy Large corpus luteum in the ovary of a pregnant female; secretes large amounts of progesterone and estrogen. (941)

corpus spongiosum (spŭn′jē-ō′sŭm) Median column of erectile tissue located between and ventral to the two corpora cavernosa in the penis; posteriorly it forms the bulb of the penis, and anteriorly it terminates as the glans penis; it is traversed by the urethra. In the female it forms the bulb of the vestibule. (931)

corpus striatum (strī-ā′tŭm) [L. *corpus*, body + *striatus*, striated or furrowed] Collective term for the caudate nucleus, putamen, and globus pallidus; so named because of the striations caused by intermixing of gray and white matter that results from the number of tracts crossing the anterior portion of the corpus striatum. (398)

cortex, pl. cortices (kōr′teks, kōr′ti-sēz) [L., bark] Outer portion of an organ (e.g., adrenal cortex or cortex of the kidney). (368)

corticotropin–releasing hormone (kōr′ti-kō-trō′pin) Hormone from the hypothalamus that stimulates the anterior pituitary gland to release adrenocorticotropic hormone. (566)

cortisol (kōr′ti-sol) Steroid hormone released by the zona fasciculata of the adrenal cortex; increases blood glucose and inhibits inflammation. (564)

cotransport Carrier-mediated simultaneous movement of two substances across a membrane in the same direction. (79)

covalent bond (kō-vāl′ent) Chemical bond characterized by the sharing of electrons. (28)

coxa (kok′să, -sē) Hip bone. (214)

cranial nerve (krā′nē-ăl) Nerve that originates from a nucleus within the brain; there are 12 pairs of cranial nerves. (434)

cranial vault Eight skull bones that surround and protect the brain; brain case. (196)

craniosacral division (krā′nē-ō-sā′krăl) Synonym for the parasympathetic division of the autonomic nervous system. (513)

cranium (krā′nē-ŭm) [Gr. *kranion*, skull] Skull; in a more limited sense, the brain case. (196)

cremaster muscle (krē-mas′ter) Extension of abdominal muscles originating from the internal oblique muscles; in the male, raises the testicles; in the female, envelops the round ligament of the uterus. (923)

crenation (krē-nā′shŭn) [L. *crena*, notched] Denoting the outline of a shrunken cell. (77)

cricoid cartilage (krī′koyd) Most inferior laryngeal cartilage. (739)

cricothyrotomy (krī′kō-thī-rot′ō-mī) Incision through the skin and cricothyroid membrane for relief of respiratory obstruction. (742)

crista ampullaris (kris′tă am-pul′ăr-ĭs) [L., crest] Elevation on the inner surface of the ampulla of each semicircular duct for dynamic or kinetic equilibrium. (500)

cristae (kris′tă, kris′tē) [L., crest] Shelflike infoldings of the inner membrane of a mitochondrion. (69)

critical closing pressure Pressure in a blood vessel below which the vessel collapses, occluding the lumen and preventing blood flow. (676)

crown (tooth)   That part of a tooth that is covered with enamel. (788)

cruciate   (krū'shē-āt) [L. *cruciatus,* cross] Resembling or shaped like a cross. (242)

crus of the penis   (krŭs) Posterior portion of the corpus cavernosum of the penis attached to the ischiopubic ramus. (931)

cryptorchidism   (krip-tōr'ki-dizm) Failure of the testis to descend. (976)

crystalline   Protein that fills the epithelial cells of the lens in the eye. (477)

cuboidal   Something that resembles a cube. (104)

cumulus mass   (kyū'myū-lŭs) See cumulus oophorus.

cumulus oophorus   (ō-of'ōr-ŭs) [L., a heap] Mass of epithelial cells surrounding the oocyte; also called the cumulus mass. (938)

cuneiform cartilage   (kyū'nē-i-fōrm) Small rod of elastic cartilage above each corniculate cartilage in the larynx. (739)

cupula, pl. cupulae   (kū'pū-lă, -lē) [L. *cupa,* a tub] Gelatinous mass that overlies the hair cells of the cristae ampullares of the semicircular ducts. (500)

cuticle   (kyū'ti-kl) [L. *cutis,* skin] Outer thin layer, usually horny, e.g., the outer covering of hair or the growth of the stratum corneum onto the nail. (145)

cystic duct   (sis'tik) Duct leading from the gallbladder; joins the common hepatic duct to form the common bile duct. (797)

cytokine   (sī'tō-kīn) A protein or peptide secreted by a cell that functions to regulate the activity of neighboring cells. (717)

cytokinesis   (sī'tō-ki-nē'sis) [Gr. *cyto,* cell + *kinsis,* movement] Division of the cytoplasm during cell division. (92)

cytology   (sī-tol'ō-jē) [Gr. *kytos,* a hollow (cell) + *logos,* study] Study of anatomy, physiology, pathology, and chemistry of the cell. (2)

cytoplasm   (sī'tō-plazm) Protoplasm of the cell surrounding the nucleus. (63)

cytoplasmic inclusion   (sī'tō-plaz'mik) Any foreign or other substance contained in the cytoplasm of a cell. (63)

cytotoxic reaction   (sī'tō-tok'sik) [Gr. *cyto,* cell + L, *toxic,* poison] Antibodies (IgG or IgM) combine with cells and activate complement, and cell lysis occurs. (730)

cytotrophoblast   (sī'tō-trof'ō-blast) Inner layer of the trophoblast composed of individual cells. (963)

D

Dalton's law   Named for the English chemist, John Dalton (1766–1844). In a mixture of gases the portion of the total pressure resulting from each type of gas is determined by the percentage of the total volume represented by each gas type. (757)

dartos muscle   (dar'tōs) Layer of smooth muscle in the skin of the scrotum; contracts in response to lower temperature and relaxes in response to higher temperature; raises and lowers testes in the scrotum. (923)

deciduous tooth   (dē-sid'yū-ŭs) Tooth of the first set of teeth; primary tooth. (788)

decussate   (dē'kŭ-sāt, dē'kŭs-āt) [L. *decusso,* X-shaped, from *decussis,* ten(X)] To cross. (386)

deep inguinal ring   (ing'gwi-năl) Opening in the transverse fascia through which the spermatic cord (or round ligament in the female) enters the inguinal canal. (930)

defecation   (def-ē-kā'shŭn) [L. *defaeco,* to remove the dregs, purify] Discharge of feces from the rectum. (786)

defecation reflex   Combination of local and central nervous system reflexes initiated by distention of the rectum and resulting in movement of feces out of the lower colon. (816)

deglutition   (dē-glū-tish'ŭn) [L. *de + glutio,* to swallow] Act of swallowing. (784)

dendrite   (den'drīt) [Gr. *dendrites,* tree] Branching processes of a neuron that receives stimuli and conducts potentials toward the cell body. (126)

dendritic cell   (den-drit'ik) Large cells with long cytoplasmic extensions that are capable of taking up and concentrating antigens leading to activation of B or T lymphocytes. (716)

dendritic spine   Extension of nerve cell dendrites where axons form synapses with the dendrites; also called gemmule. (362)

dental arch   (den'tăl) [L. *arcus,* bow] Curved maxillary or mandibular arch in which the teeth are located. (788)

dentin   (den'tin) Bony material forming the mass of the tooth. (788)

deoxyhemoglobin   (dē-oks'ē-hē-mō-glō'bin) Hemoglobin without oxygen bound to it. (588)

deoxyribonuclease   Enzyme that splits DNA into its component nucleotides. (815)

deoxyribonucleic acid (DNA)   (dē-oks'ē-rī'bō-nū-klē'ik) Type of nucleic acid containing deoxyribose as the sugar component, found principally in the nuclei of cells; constitutes the genetic material of cells. (50)

depolarization   (dē-pō'lăr-i-zā'shŭn) Change in the electric charge difference across the cell membrane that causes the difference to be smaller or closer to 0 mV; phase of the action potential in which the membrane potential moves toward zero, or becomes positive. (261)

depression   (dē-presh'ŭn) Movement of a structure in an inferior direction. (232)

depth perception   (per-sep'shun) Ability to distinguish between near and far objects and to judge their distance. (486)

dermatome   (der'mă-tōm) Area of skin supplied by a spinal nerve. (445)

dermis   (der'mis) [Gr. *derma,* skin] Dense, irregular connective tissue that forms the deep layer of the skin. (137)

descending aorta   Part of the aorta, further divided into the thoracic aorta and abdominal aorta. (652)

descending colon   Part of the colon extending from the left colonic flexure to the sigmoid colon. (798)

desmosome   (dez'mō-sōm) [Gr. *desmos,* a band + *soma,* body] Point of adhesion between cells. Each contains a dense plate at the point of adhesion and a cementing extracellular material between the cells. (110)

desquamate   (des'kwă-māt) [L. *desquamo,* to scale off] Peeling or scaling off of the superficial cells of the stratum corneum. (138)

diabetes insipidus   (dī-ă-bē'tēz in-sip'ĭ-dŭs) Chronic excretion of large amounts of urine of low specific gravity accompanied by extreme thirst; results from inadequate output of antidiuretic hormone. (884)

diabetes mellitus   (me-lī'tŭs) Metabolic disease in which carbohydrate use is reduced and that of lipid and protein enhanced; caused by deficiency of insulin or an inability to respond to insulin and is characterized, in more severe cases, by hyperglycemia, glycosuria, water and electrolyte loss, ketoacidosis, and coma. (884)

diapedesis   (dī'ă-pĕ-dē'sis) [Gr. *dia,* through + *pedesis,* a leaping] Passage of blood or any of its formed elements through the intact walls of blood vessels. (593)

diaphragm (dī′ă-fram) Musculomembranous partition between the abdominal and thoracic cavities. (328)

diaphysis (dī-af′i-sis, -sēz) [Gr., growing between] Shaft of a long bone. (158)

diastole (dī-as′tō-lē) [Gr. *diastole,* dilation] Relaxation of the heart chambers during which they fill with blood; usually refers to ventricular relaxation. (625)

diencephalon (dī-en-sef′ă-lon) [Gr. *dia,* through + *enkephalos,* brain] Second portion of the embryonic brain; in the inferior core of the adult cerebrum. (384)

diffuse lymphatic tissue Dispersed lymphocytes and other cells with no clear boundary; found beneath mucous membranes, around lymph nodules, and within lymph nodes and spleen. (703)

diffusion (di-fyū′zhŭn) [L. *diffundo,* to pour in different directions] Tendency for solute molecules to move from an area of high concentration to an area of low concentration in solution; the product of the constant random motion of all atoms, molecules, or ions in a solution. (74)

diffusion coefficient Measure of how easily a gas diffuses through a liquid or tissue. (760)

digestive tract (di-jes′tiv, dī-) Mouth, oropharynx, esophagus, stomach, small intestine, and large intestine. (782)

digit (dij′it) Finger, thumb, or toe. (214)

dilator pupillae (dī′lă-tĕr pyū-pil′ē) Radial smooth muscle cells of the iris diaphragm that cause the pupil of the eye to dilate. (476)

diploid (dip′loyd) Normal number of chromosomes (in humans, 46 chromosomes) in somatic cells. (92)

disaccharide (dī-sak′ă-rīd) Condensation product of two monosaccharides by elimination of water. (41)

dissociate (di-sō-sē-āt′) [L. *dis* + *socio,* to disjoin, separate] Ionization in which ions are dissolved in water and the cations and anions are surrounded by water molecules. (31)

distal tubule Convoluted tubule of the nephron that extends from the ascending limb of the loop of Henle and ends in a collecting duct. (864)

distributing artery Medium-sized artery with a tunica media composed principally of smooth muscle; regulates blood flow to different regions of the body. (649)

DNA See deoxyribonucleic acid.

dominant [L. *dominus,* a master] In genetics a gene that is expressed phenotypically to the exclusion of a contrasting recessive gene. (993)

dorsal root Sensory (afferent) root of a spinal nerve. (403)

dorsal root ganglion (gang′glē-on) Collection of sensory neuron cell bodies within the dorsal root of a spinal nerve. (403)

ductus arteriosus (dŭk′tŭs ar-tēr′ē-ō-sŭs) Fetal vessel connecting the left pulmonary artery with the descending aorta. (985)

ductus deferens (def′er-enz) Duct of the testicle, running from the epididymis to the ejaculatory duct; also called the vas deferens. (928)

duodenal gland (dū′ō-dē′năl, dū-od′ĕ-năl) Small gland that opens into the base of intestinal glands; secretes a mucoid alkaline substance. (794)

duodenocolic reflex (dū-ō-dē′nō-kō-lik) Local reflex resulting in a mass movement of the contents of the colon; produced by stimuli in the duodenum. (816)

duodenum (dū-ō-dē′nŭm, dū-od′ĕ-nŭm) [L. *duodeni,* twelve] First division of the small intestine; connects to the stomach. (793)

dura mater (dū′ră mā′ter) [L., hard mother] Tough, fibrous membrane forming the outer covering of the brain and spinal cord. (420)

E

eardrum (ēr′drŭm) Tympanic membrane; cellular membrane that separates the external from the middle ear; vibrates in response to sound waves. (486)

ectoderm (ek′tō-derm) Outermost of the three germ layers of an embryo. (127)

ectopic focus, pl. foci (ek-top′ik fō′kŭs; fō′sī) Any pacemaker other than the sinus node of the heart; abnormal pacemaker; an ectopic pacemaker. (622)

edema (e-dē′-mă) [Gr. *oidema,* a swelling] Excessive accumulation of fluid, usually causing swelling. (128)

effector T cell Subset of T lymphocytes that is responsible for cell-mediated immunity. (714)

efferent arteriole (ef′er-ent ar-tēr′ē-ōl) Vessel that carries blood from the glomerulus to the peritubular capillaries. (864)

efferent division Nerve fibers that send impulses from the central nervous system to the periphery. (360)

efferent ductule (ef′er-ent dŭk′tīl) [L. *ductus,* duct] One of a number of small ducts leading from the testis to the head of the epididymis. (923)

ejaculation (ē-jak-yū-lā′shŭn) Reflexive expulsion of semen from the penis. (937)

ejaculatory duct (ē-jak′yū-lă-tōr-ē) Duct formed by the union of the ductus deferens and the excretory duct of the seminal vesicle; opens into the prostatic urethra. (930)

ejection period (ē-jek′shun) Time in the cardiac cycle when the semilunar valves are open and blood is being ejected from the ventricles into the arterial system. (625)

elastin (ē-las′tin) A yellow elastic fibrous mucoprotein that is the major connective tissue protein of elastic structures (e.g., large blood vessels and elastic ligaments, etc.) (113)

electrocardiogram (ECG) (ē-lek-trō-kar′dē-ō-gram) [Gr. *elektron,* amber + *kardia,* heart + *gramma,* a drawing] Graphic record of the heart's electric currents obtained with the electrocardiograph. (608)

electrolyte (ē-lek′trō-līt) [Gr. *electro* + *lytos,* soluble] Cation or anion in solution that conducts an electric current. (31)

electron (ē-lek′tron) Negatively charged subatomic particle in an atom. (24)

electron–transport chain Series of electron carriers in the inner mitochondrial membrane; they receive electrons from NADH and $FADH_2$, using the electrons in the formation of ATP and water. (846)

element (e′lĕ-ment) [L. *elementum,* a rudiment, beginning] Substance composed of atoms of only one kind. (24)

elevation (el-ĕ-vā′shŭn) Movement of a structure in a superior direction. (232)

embolism (em′bō-lizm) [Gr. *embolisma,* a piece of patch, literally something thrust in] Obstruction or occlusion of a vessel by a transported clot, a mass of bacteria, or other foreign material. (656)

embolus, pl. emboli (em′bō-lŭs, -lī) [Gr. *embolos,* plug, wedge, or stopper] Plug, composed of a detached clot, mass of bacteria, or other foreign body, occluding a blood vessel. (990)

embryo (em′brē-ō) Developing human from the second to the eighth week of development. (964)

embryonic disk (em-brē-on′ik) Point in the inner cell mass at which the embryo begins to be formed. (964)

embryonic period From approximately the second to the eighth week of development, during which the major organ systems are organized. (962)

emission (ē-mish′ŭn) [L. *emissio*, to send out] Discharge; accumulation of semen in the urethra prior to ejaculation. A nocturnal emission refers to a discharge of semen while asleep. (936)

emmetropia (em-ĕ-trō′pē-ă) [Gr. *emmetros*, according to measure + *ops*, eye] In the eye the state of refraction in which parallel rays are focused exactly on the retina; no accommodation is necessary. (478)

emulsify (ē-mŭl′si-fī) To form an emulsion. (812)

enamel (ē-nam′ĕl) Hard substance covering the exposed portion of the tooth. (788)

endocardium, pl. endocardia (en′dō-kar′dē-ŭm, -ē-ă) Innermost layer of the heart, including endothelium and connective tissue. (611)

endocrine gland (en′dō-krin) [Gr. *endon*, inside + *krino*, to separate] Ductless gland that secretes a hormone internally, usually into the circulation. (111)

endocytosis (en′dō-sī-tō′sis) Bulk uptake of material through the cell membrane. (81)

endoderm (en′dō-derm) Innermost of the three germ layers of an embryo. (126)

endolymph (en′dō-limf) [Gr. *endo* + L. *lympha*, clear fluid] Fluid found within the membranous labyrinth of the inner ear. (492)

endometrium, pl. endometria (en′dō-mē′trē-ŭm, -trē-ă) Mucous membrane composing the inner layer of the uterine wall; consists of a simple columnar epithelium and a lamina propria that contains simple tubular uterine glands. (942)

endomysium (en′dō-miz′ē-ŭm, -mis′ē-ŭm) [Gr. *endo*, within + *mys*, muscle] Fine connective tissue sheath surrounding a muscle fiber. (277)

endoneurium (en′dō-nū′rē-ŭm) [Gr. *endo*, within + *neuron*, nerve] Delicate connective tissue surrounding individual nerve fibers within a peripheral nerve. (368)

endoplasmic reticulum, pl. reticula (en′dō-plas′mik re-tik′-yŭ-lŭm, -lă) Double-walled membranous network inside the cytoplasm; rough has ribosomes attached to the surface; smooth does not have ribosomes attached. (66)

endorphin (en′-dōr-phin) Opiate-like polypeptide found in the brain and other parts of the body; binds in the brain to the same receptors that bind exogenous opiates. (575)

endosteum (en-dos′tē-ŭm) [Gr. *endo*, within + *osteon*, bone] Membranous lining of the medullary cavity and the cavities of spongy bone. (160)

endothelium, pl. endothelia (en-dō-thē′lē-ŭm, -lē-ă) [Gr. *endo*, within + *thele*, nipple] Layer of flat cells lining blood and lymphatic vessels and the chambers of the heart. (647)

enkephalin (en-kef′ă-lin) Pentapeptide found in the brain; binds to specific receptor sites, some of which may be pain-related opiate receptors. (575)

enterokinase (en′tēr-ō-kī′nas) Intestinal proteolytic enzyme that converts trypsinogen into trypsin. (815)

enzyme (en′zīm) [Gr. *en*, in + *zyme*, leaven] Protein that acts as a catalyst. (35)

eosinophil (ē-ō-sin′ō-fil) [Gr. *eos*, dawn + *philos*, fond] White blood cell that stains with acidic dyes; inhibits inflammation. (593)

epicardium (epi-kar′dē-ŭm) [Gr. *epi*, on + *kardia*, heart] Serous membrane covering the surface of the heart. Also called the visceral pericardium. (610)

epidermis (ep-i-derm′is) [Gr. *epi*, on + *derma*, skin] Outer portion of the skin formed of epithelial tissue that rests on or covers the dermis. (137)

epididymis, pl. epididymides (ep-i-did′i-mis, -di-dim′i-dēz) [Gr. *epi*, on + *didymos*, twin] Elongated structure connected to the posterior surface of the testis, which consists of the head, body, and tail; site of storage and maturation of the spermatozoa. (927)

epiglottis (ep-i-glot′is) [Gr. *epi*, on + *glottis*, the mouth of the windpipe] Plate of elastic cartilage covered with mucous membrane; serves as a valve over the glottis of the larynx during swallowing. (739)

epimysium (ep-i-mis′ē-ŭm) [Gr. *epi*, on + *mys*, muscle] Fibrous envelope surrounding a skeletal muscle. (277)

epinephrine (ep′i-nef′rin) Hormone (amino acid derivative) similar in structure to the neurotransmitter norepinephrine; major hormone released from the adrenal medulla; increases cardiac output and blood glucose levels. (513)

epineurium (ep-i-nū′rē-ŭm) [Gr. *epi*, upon + *neuron*, nerve] Connective tissue sheath surrounding a nerve. (368)

epiphyseal line (ep-i-fiz′ē-ăl) Dense plate of bone in a bone that is no longer growing, indicating the former site of the epiphyseal plate. (158)

epiphyseal plate Site at which bone growth in length occurs; located between the epiphysis and diaphysis of a long bone; area of hyaline cartilage where cartilage growth is followed by endochondral ossification; also called the metaphysis or growth plate. (158)

epiphysis, pl. epiphyses (e-pif′i-sis, e-pif′i-sēz) [Gr. *epi*, on + *physis*, growth] Portion of a bone developed from a secondary ossification center and separated from the remainder of the bone by the epiphyseal plate. (158)

epiploic appendage (ep′i-plō′ik) One of a number of little processes of peritoneum projecting from the serous coat of the large intestine except the rectum; they are generally distended with fat. (798)

epithelium, pl. epithelia (ep-i-thē′lē-ŭm, -ă) [Gr. *epi*, on + *thele*, nipple] One of the four primary tissue types. "Nipples" refers to the tiny capillary-containing connective tissue in the lips, which is where the term was first used. The use of the term was later expanded to include all covering and lining surfaces of the body. (103)

epitope (ep′i-tōp) [Gr. *epi*, upon + *top*, place] See antigenic determinant.

eponychium (ep-ō-n-nik′ē-ŭm) [Gr. *epi*, on + *onyx*, nail] Outgrowth of the skin that covers the proximal and lateral borders of the nail. Cuticle. (147)

erection (ē-rek′shŭn) [L. *erectio*, to set up] Condition of erectile tissue when filled with blood; becomes hard and unyielding; especially refers to this state of the penis. (936)

erythroblastosis fetalis (ĕ-rith′rō-blas-tō′sis fē-tă′lis) [erythroblast + *osis*, condition] Destruction of erythrocytes in the fetus or newborn caused by antibodies produced in the Rh-negative mother acting on Rh-positive blood of the fetus or newborn. (601)

erythrocyte (ĕ-rith′rō-sīt) [Gr. *erythros,* red + *kytos,* cell] Red blood cell; biconcave disk containing hemoglobin. (586)

erythropoiesis (ĕ-rith′rō-poy-ē′sis) [erythrocyte + Gr. *poiesis,* a making] Production of erythrocytes. (589)

erythropoietin (ĕ-rith′rō-poy′ĕ-tin) Protein that enhances erythropoiesis by stimulating formation of proerythroblasts and release of reticulocytes from bone marrow. (591)

esophagus, pl. esophagi (ē-sof′ă-gŭs, -gī, -jī) [Gr. *oisophagos,* gullet] Portion of the digestive tract between the pharynx and stomach. (782)

essential amino acid Amino acid required by animals that must be supplied in the diet. (836)

eustachian tube (yū-stā′shŭn, yū-stā′kē-ăn) Named for the Italian anatomist, Bartolommeo Eustachio (1524–1574). Auditory canal; extends from the middle ear to the nasopharynx. (490)

evagination (ē′vaj-i-nā′shŭn) [L. *e,* out + *vagina,* sheath] Protrusion of some part or organ from its normal position. (968)

evaporation (i-vap-ə-′rā-shən) [L. *e,* out + *vaporare,* to emit vapor] Change from liquid to vapor form. (855)

eversion (ē-ver′zhun) [L. *everto,* to overturn] Turning outward. (236)

excitation–contraction coupling (ek-sī-tā′shŭn kon-trak′shŭn kŭp′ling) Stimulation of a muscle fiber produces an action potential that results in contraction of the muscle fiber. (283)

excitatory postsynaptic potential (EPSP) (ek-sī′tă-tō-rē pōst-si-nap′tic pō-ten′shal) Depolarization in the postsynaptic membrane that brings the membrane potential close to threshold. (270)

exocrine gland (ek′sō-krin) [Gr. *exo,* outside + *krino,* to separate] Gland that secretes to a surface or outward through a duct. (111)

exocytosis (ek′sō-sī-to′sis) Elimination of material from a cell through the formation of vacuoles. (82)

expiratory reserve volume Maximum volume of air that can be expelled from the lungs after a normal expiration. (756)

extension [L. *extensio,* to stretch out] To stretch out. (232)

external anal sphincter Ring of striated muscular fibers surrounding the anus. (799)

external auditory meatus (mē-ā′tŭs) Short canal that opens to the exterior environment and terminates at the eardrum; part of the external ear. (486)

external ear Portion of the ear that includes the auricle and external auditory meatus; terminates at the eardrum. (486)

external naris, pl. nares (nā′ris, -res) Nostril; anterior or external opening of the nasal cavity. (738)

external spermatic fascia Outer fascial covering of the spermatic cord. (928)

external urethral orifice Slitlike opening of the urethra in the glans penis. (931)

external urinary sphincter Sphincter skeletal muscle around the base of the urethra external to the internal urinary sphincter. (867)

exteroceptor (eks′ter-ō-sep′ter, -tor) [L. *exterus,* external + *receptor,* receiver] Sensory receptor in the skin or mucous membranes, which responds to stimulation by external agents or forces. (463)

extracellular (eks-tră-sel′yū-lăr) Outside the cell. (58)

extracellular matrix, pl. matrices (eks-tră-sel′yū-lăr mā′triks, mā′tri-sēz) Nonliving chemical substances located between connective tissue cells. (103)

extrinsic clotting pathway (eks-trin′sik) Series of chemical reactions resulting in clot formation; begins with chemicals (e.g., tissue thromboplastin) found outside the blood. (596)

extrinsic muscle Muscle located outside the structure being moved. (319)

eyebrow Short hairs on the bony ridge above the eyes. (470)

eyelash Hair at the margins of the eyelids. (470)

eyelid Palpebra; Movable fold of skin in front of the eyeball. (470)

F

F actin (F ak′tin) Fibrous actin molecule that is composed of a series of globular actin molecules (G actin). (278)

facilitated diffusion Carrier-mediated process that does not require ATP and moves substances into or out of cells from a high to a low concentration. (79)

falciform ligament (fal′si form lig′ă-ment) Fold of peritoneum extending to the surface of the liver from the diaphragm and anterior abdominal wall. (801)

fallopian tube (fa-lō′pē-an) See uterine tube.

false pelvis Portion of the pelvis superior to the pelvic brim; composed of the bone on the posterior and lateral sides and by muscle on the anterior side. (216)

falx cerebelli (falks ser-e-bel′i′) Dural fold between the two cerebellar hemispheres. (420)

falx cerebri (falx ser-ĕ-brī′) Dural fold between the two cerebral hemispheres. (420)

far point of vision Distance from the eye where accommodation is not needed to have the image focused on the retina. (479)

fascia, pl. fasciae (fash′ē-ă, -ē-ē) [L., band or fillet] Loose areolar connective tissue found beneath the skin (hypodermis) or dense connective tissue that encloses and separates muscles. (278)

fasciculus (fă-sik′yū-lus) [L. *fascis,* bundle] Band or bundle of nerve or muscle fibers bound together by connective tissue. (277)

fat [A.S. *faet*] Greasy, soft-solid material found in animal tissues and many plants; composed of two types of molecules: glycerol and fatty acids. (43)

fatigue (fă-tēg′) [L. *fatigo,* to tire] Period characterized by a reduced capacity to do work. (293)

fat-soluble vitamin Vitamin such as A, D, E, and K that is soluble in lipids and absorbed from the intestine along with lipids. (45)

fauces (faw′sēz) [L., throat] Space between the cavity of the mouth and the pharynx. (739)

female climacteric Period of life occurring in women, encompassing termination of the reproductive period and characterized by endocrine, somatic, and transitory psychologic changes and ultimately menopause. (950)

female pronucleus Nuclear material of the ovum after the ovum has been penetrated by the spermatozoon. Each pronucleus carries the haploid number of chromosomes. (962)

fertilization Process that begins with the penetration of the secondary oocyte by the spermatozoon and is completed with the fusion of the male and female pronuclei. (939)

fetal period The last 7 months of development, during which the organ systems grow and become functionally mature. (962)

fibrin (fī′brin) An elastic filamentous protein derived from fibrinogen by the action of thrombin, which releases peptides from fibrinogen in coagulation of the blood. (595)

fibroblast (fī′brō-blast) [L. *fibra*, fiber + Gr. *blastos*, germ] Spindle-shaped or stellate cells that form connective tissue. (114)

fibrocyte (fī′brō-sīt) Mature cell of fibrous connective tissue. (114)

fibrous joint (fī′brŭs) Bones connected by fibrous tissue with no joint cavity; includes sutures, syndesmoses, and gomphoses. (225)

fibrous tunic Outer layer of the eye; composed of the sclera and the cornea. (473)

filiform papilla (fil′i-fōrm) Filament-shaped papilla on the surface of the tongue. (468)

filtrate (fil′trāt) Liquid that has passed through a filter; for example, fluid that enters the nephron through the filtration membrane of the glomerulus. (868)

filtration (fil-trā′shŭn) Movement, due to a pressure difference, of a liquid through a filter that prevents some or all of the substances in the liquid from passing through. (868)

filtration fraction Fraction of the plasma entering the kidney that filters into Bowman's capsule. Normally it is around 19%. (869)

filtration membrane Membrane formed by the glomerular capillary endothelium, the basement membrane, and the podocytes of Bowman's capsule. (864)

filtration pressure Pressure gradient that forces fluid from the glomerular capillary through the filtration membrane into Bowman's capsule; glomerular capillary pressure minus glomerular capsule pressure minus colloid osmotic pressure. (870)

fimbria, pl. fimbriae (fim′brē-ă, -brē-ē) [L., fringe] Fringelike structure located at the ostium of the uterine tube. (941)

first messenger See intercellular chemical signal.

fixator (fik-sā′ter) Muscle that stabilizes the origin of a prime mover. (309)

flagellum, pl. flagella (flă-jel′ŭm, -ă) [L., whip] Whiplike locomotory organelle of constant structural arrangement consisting of double peripheral microtubules and two single central microtubules. (71)

flatus (flā′tŭs) [L., a blowing] Gas or air in the gastrointestinal tract that may be expelled through the anus. (816)

flexion (flek′shŭn) [L. *flectus*] To bend. (232)

focal point Point at which light rays cross after passing through a concave lens such as the lens of the eye. (478)

foliate papilla (fō′lē-āt) Leaf-shaped papilla on the lateral surface of the tongue. (468)

follicle-stimulating hormone (FSH) (fol′i-kl) Hormone of the adenohypophysis that, in females, stimulates the graafian follicles of the ovary and assists in follicular maturation and the secretion of estrogen; in males, FSH stimulates the epithelium of the seminiferous tubules and is partially responsible for inducing spermatogenesis. (554)

follicular phase (fŏ-lik′yū-lăr) Time between the end of menses and ovulation characterized by rapid division of endometrial cells and development of follicles in the ovary; the proliferative phase. (946)

foramen, pl. foramina (fō-rā′men, fō-ram′i-nă) A hole. (184)

foramen ovale (o-val′ē) In the fetal heart the oval opening in the septum secundum; the persistent part of septum primum acts as a valve for this interatrial communication during fetal life; postnatally the septum primum becomes fused to the septum secundum to close the foramen ovale, forming the fossa ovale. (614)

force That which produces a motion in the body; pull. (311)

foregut Cephalic portion of the primitive digestive tube in the embryo. (967)

foreskin See prepuce.

formed elements Cells (i.e., erythrocytes and leukocytes) and cell fragments (i.e., platelets) of blood. (584)

formula unit A description of the relative number of cations and ions in an ionic compound. (30)

fornix (fōr′niks) [L. arch, vault] Recess at the cervical end of the vagina; also the recess deep to each eyelid where the palpebral and bulbar conjunctivae meet. (399)

fovea centralis (fō′vē-ă) Depression in the middle of the macula where there are only cones and no blood vessels. (476)

free energy Total amount of energy that can be liberated by the complete catabolism of food. (855)

frenulum (fren′yū-lŭm) [L. *frenum*, bridle] Fold extending from the floor of the mouth to the midline of the undersurface of the tongue. (787)

frequency-modulated signals Signals, all of which are identical in amplitude, that differ in their frequency; for example, strong stimuli may initiate a high frequency of action potentials and weak stimuli may initiate a low frequency of action potentials. (527)

FSH See follicle-stimulating hormone.

FSH surge Increase in plasma follicle-stimulating hormone (FSH) levels before ovulation. (947)

fulcrum Pivot point. (311)

fundus (fŭn′dŭs) [L., bottom] "Bottom," or rounded end, of a hollow organ, for example, the fundus of the stomach or uterus. (791)

fungiform papilla (fŭn′ji-fōrm) Mushroom-shaped papilla on the surface of the tongue. (468)

G

G actin (jē ak′tin) Globular protein molecules that, when bound together, form fibrous actin (F actin). (278)

gallbladder Pear-shaped receptacle on the inferior surface of the liver; serves as a storage reservoir for bile. (798)

gamete (gam′ēt) Ovum or spermatozoon. (92)

gamma (γ) globulin (gam′ă glob′yū-lin) [L. *globulus*, globule] Plasma proteins that include the antibodies. (720)

ganglion, pl. ganglia (gang′glē-on, -glē-ă) [Gr., swelling, or knot] Any group of nerve cell bodies in the peripheral nervous system. (360)

gap junction Small channel between cells that allows the passage of ions and small molecules between cells; provides means of intercellular communication. (110)

gastric gland Gland located in the mucosa of the fundus and body of the stomach. (792)

gastric inhibitory polypeptide Hormone secreted by the duodenum that inhibits gastric acid secretion. (807)

gastric pit Small pit in the mucous membrane of the stomach at the bottom of which are the mouths of the gastric glands that secrete mucus, hydrochloric acid, intrinsic factor, pepsinogen, and hormones. (792)

gastrin (gas′-trin) Hormone secreted in the mucosa of the stomach and duodenum that stimulates secretion of hydrochloric acid by the parietal cells of the gastric glands. (806)

gastrocolic reflex (gas′trō-kol′ik) Local reflex resulting in mass movement of the contents of the colon that occurs after the entrance of food into the stomach. (816)

gastroesophageal (cardiac) opening (gas'trō-ē-sof'ă-jē'ăl) Opening of the esophagus into the stomach. (791)

gene [Gr. *genos,* birth] Functional unit of heredity. Each gene occupies a specific place, or locus, on a chromosome, is capable of reproducing itself exactly at each cell division, and often is capable of directing the formation of an enzyme or other protein. (86)

general gas law The pressure of a gas is equal to the number of gram moles of the gas times the gas constant times the absolute temperature divided by the volume of the gas. Assuming a constant temperature, the pressure of a given amount of gas is inversely proportional to its volume. This relationship also is called Boyle's law. (751)

genetics (jĕ-net'iks) [Gr. *genesis,* origin or production] Branch of science that deals with heredity. (991)

genital tubercle (jen'i-tăl) Median elevation just cephalic to the urogenital orifice of an embryo; gives rise to the penis of the male or the clitoris of the female. (979)

genotype (jen'ō-tīp) [Gr. *genos,* birth, descent + *typos,* type] Genetic makeup of an individual. (993)

germ cell Spermatozoon or ovum. (926)

germ layer One of three layers in the embryo (ectoderm, endoderm, or mesoderm) from which the four primary tissue types arise. (964)

germinal center Lighter-staining center of a lymphatic nodule; area of rapid lymphocyte division. (706)

germinal period Approximately the first 2 weeks of development. (962)

gingiva (jin'ji-vă) Dense fibrous tissue, covered by mucous membrane, that covers the alveolar processes of the upper and lower jaws and surrounds the necks of the teeth. (788)

girdle Belt or zone; the bony region where the limbs attach to the body. (210)

gland [L. *glans,* acorn] Secretory organ from which secretions may be released into the blood, a cavity, or onto a surface. (111)

glans penis [L., acorn] Conical expansion of the corpus spongiosum that forms the head of the penis. (931)

globin Protein portion of hemoglobin. (588)

glomerular capillary pressure (glō-măr'yū-lăr) Blood pressure within the glomerulus. (870)

glomerular filtration rate (GFR) Amount of plasma (filtrate) that filters into Bowman's capsules per minute. (869)

glomerulus (glō-măr'yū-lŭs) [L. *glomus,* ball of yarn] Mass of capillary loops at the beginning of each nephron, nearly surrounded by Bowman's capsule. (864)

glottis (glot'is) [Gr., aperture of the larynx] Vocal apparatus; includes vocal folds and the cleft between them. (739)

glucocorticoid (glū-ko-kōr'ti-koyd) Steroid hormone (e.g., cortisol) released by zonula fasciculata of the adrenal cortex; increases blood glucose and inhibits inflammation. (563)

gluconeogenesis (glū'kō-nē-ō-jen'ĕ-sis) [Gr. *glykys,* sweet + *neos,* new + *genesis,* production] Formation of glucose from noncarbohydrates such as proteins (amino acids) or lipids (glycerol). (564)

glycogenesis (glī'kō-jen'ĕ-sis) Formation of glycogen from glucose molecules. (848)

glycolysis (glī-kol'i-sis) [Gr. *glykys,* sweet + *lysis,* a loosening] Anaerobic process during which glucose is converted to pyruvic acid; net of two ATP molecules is produced during glycolysis. (82)

goblet cell Mucous-producing epithelial cell that has its apical end distended with mucin. (109)

Golgi apparatus (gol'jē) Named for Camillo Golgi, Italian histologist and Nobel laureate, 1843–1926. Specialized endoplasmic reticulum that concentrates and packages materials for secretion from the cell. (66)

Golgi tendon organ Proprioceptive nerve ending in a tendon. (465)

gomphosis (gom-fō'sis) [Gr. *gomphos,* bolt, nail + *osis,* condition] Fibrous joint in which a peglike process fits into a hole. (226)

gonad (gō'nad) [Gr. *gone,* seed] Organ that produces sex cells; testis of a male or ovary of a female. (554)

gonadal ridge (gō-nad'ăl) Elevation on the embryonic mesonephros; primordial germ cells become embedded in it, establishing it as the testis or ovary. (976)

gonadotropin (gō'nad-ō-trō'pin, gon'ă-dō-) Hormone capable of promoting gonadal growth and function. Two major gonadotropins are luteinizing hormone (LH) and follicle-stimulating hormone (FSH). (554)

gonadotropin–releasing hormone (GnRH) Hypothalamic-releasing hormone that stimulates the secretion of gonadotropins (LH and FSH) from the adenohypophysis. (554)

graaffian follicle See mature follicle.

granulocyte (gran'yū-lō-sīt) Mature granular leukocyte (neutrophil, basophil, or eosinophil). (584)

granulosa cell (gran-yū-lō'să) Cell in the layer surrounding the primary follicle. (938)

gray matter Collections of nerve cell bodies, their dendritic processes, and associated neuroglial cells within the central nervous system. (368)

greater omentum Peritoneal fold passing from the greater curvature of the stomach to the transverse colon, hanging like an apron in front of the intestines. (801)

greater vestibular gland One of two mucus-secreting glands on either side of the lower part of the vagina. The equivalent of the bulbourethral glands in the male. (943)

growth hormone Somatotropin; stimulates general growth of the individual; stimulates cellular amino acid uptake and protein synthesis. (553)

gubernaculum (gū'ber-nak'yū-lŭm) [L., helm] Column of tissue that connects the fetal testis to the developing scrotum; involved in testicular descent. (925)

gustatory (gŭs'tă-tōr-ē) Associated with the sense of taste. (468)

gustatory hair Microvillus of gustatory cell in a taste bud. (468)

gynecomastia (gī'nĕ-kō-mas'tē-ă) [Gr. *gyne,* woman + *mastos,* breast] Excessive development of the male mammary glands, which sometimes secrete milk. (944)

H

H zone Area in the center of the A band in which there are no actin myofilaments; contains only myosin. (278)

hair [A.S., hear] Columns of dead keratinized epithelial cells. (144)

hair follicle Invagination of the epidermis into the dermis; contains the root of the hair and receives the ducts of sebaceous and apocrine glands. (146)

Haldane effect Named for the Scottish physiologist, John S. Haldane (1860–1936). Hemoglobin that is not bound to carbon dioxide binds more readily to oxygen than hemoglobin that is bound to carbon dioxide. (766)

half-life The time it takes for one half of an administered substance to be lost through biologic processes. (534)

haploid   (hap'loyd) Having only one set of chromosomes, in contrast to diploid; characteristic of gametes. (92)

hapten   (hap'ten) [Gr. *hapto*, to fasten] Small molecule that binds to a large molecule; together they stimulate the specific immune system. (714)

hard palate   Floor of the nasal cavity that separates the nasal cavity from the oral cavity; composed of the palatine processes of the maxillary bones and the horizontal plates of the palatine bones. (194)

haustra   (haw'strǎ) [L., machine for drawing water] Sacs of the colon, caused by contraction of the taeniae coli, which are slightly shorter than the gut, so that the latter is thrown into pouches. (798)

haversian canal   (ha-ver'shan) Named for seventeenth century English anatomist, Clopton Havers (1650–1702). Canal containing blood vessels, nerves, and loose connective tissue and running parallel to the long axis of the bone. (163)

haversian system   See osteon.

heart skeleton   Fibrous connective tissue that provides a point of attachment for cardiac muscle cells, electrically insulates the atria from the ventricles, and forms the fibrous rings around the valves. (618)

heat energy   Energy that results from the random movement of atoms, ions, or molecules; the greater the amount of heat energy in an object, the higher is the object's temperature. (37)

helicotrema   (hel'i-kō-trē'mǎ) [Gr. *helix*, spiral + *traema*, hole] Opening at the apex of the cochlea through which the scala vestibuli and the scala tympani of the cochlea connect. (493)

helper T cell   Subset of T lymphocytes that increases the activity of B cells and T cells. (714)

hematocrit   (hē'mǎ'tō-krit, hem'ǎ-) [Gr. *hemato*, blood + *krin*, to separate] Percentage of blood volume occupied by erythrocytes. (602)

hematopoiesis   (hē'mǎ-tō-poy-ē'sis) [Gr. *haima*, blood + *poiesis*, a making] Production of blood cells. (584)

heme   (hēm) Oxygen-carrying, color-furnishing part of hemoglobin. (588)

hemidesmosome   (hem-ē-des'mō-sōm) Similar to half a desmosome, attaching epithelial cells to the basement membrane. (110)

hemoglobin   (hē'mō-glō-bin) Red, respiratory protein of erythrocytes; consists of 6% heme and 94% globin;

transports oxygen and carbon dioxide. (586)

hemolysis   (hē-mol'i-sis) [Gr. *haima* + *lysis*, destruction] Destruction of red blood cells in such a manner that hemoglobin is released. (586)

hemopoiesis   (hē'mō-poy-ē'sis) [Gr. *haima*, blood + *poiesis*, a making] Formation of the formed elements of blood, that is, erythrocytes, leukocytes, and thrombocytes. (584)

hemopoietic tissue   (hē'mō-poy-et'ik) [Gr. *haima*, blood + *poiesis*, to make] Blood-forming tissue. (122)

hemostasis   (hē'mō-stā-sis) Arrest of bleeding. (594)

Henry's law   Named for the English chemist, William Henry (1775–1837). The concentration of a gas dissolved in a liquid is equal to the partial pressure of the gas over the liquid times the solubility coefficient of the gas. (758)

heparin   (hep'ǎ-rin) Anticoagulant that prevents platelet agglutination and thus prevents thrombus formation. (593)

hepatic artery   (he-pa'tik) Branch of the aorta that delivers blood to the liver. (797)

hepatic cord   Plate of liver cells that radiates away from the central vein of a liver lobule. (797)

hepatic portal system   System of portal veins that carry blood from the intestines, stomach, spleen, and pancreas to the liver. (667)

hepatic portal vein   Portal vein formed by the superior mesenteric and splenic veins and entering the liver. (667)

hepatic sinusoid   (si'nŭ-soyd) Terminal blood vessel having an irregular and larger caliber than an ordinary capillary within the liver lobule. (797)

hepatic vein   Vein that drains the liver into the inferior vena cava. (670)

hepatocyte   (hep'ǎ-tō-sīt) Liver cell. (797)

hepatopancreatic ampulla   Dilation within the major duodenal papilla that normally receives both the common bile duct and the main pancreatic duct. (793)

hepatopancreatic ampullar sphincter   Smooth muscle sphincter of the hepatopancreatic ampulla; sphincter of Oddi. (793)

Hering–Breuer reflex   (her'ing broy'er) Named for the German physiologist, Heinrich Ewald Hering (1866–1948), and the Austrian internist, Josef Breuer (1842–1925). Afferent impulses from stretch receptors in the lungs arrest inspiration; expiration then occurs. (771)

heterozygous   (het'er-ō-zī'gus) [Gr. *heteros*, other + *zygon*, yoke] State of

having different allelic genes at one or more paired loci in homologous chromosomes. (992)

hiatus   (hī-ā'tus) [L., aperture, to yawn] Opening. (208)

hilum   (hī'lŭm) [L., small bit or trifle] Indented surface on many organs, serving as a point where nerves and vessels enter or leave. (706)

hindgut   Caudal or terminal part of the embryonic gut. (967)

histamine   (his'tǎ-mēn) Amine released by mast cells and basophils that promotes inflammation. (593)

histology   (his-tol'ō-jē) [Gr. *histo*, web (tissue) + *logos*, study] The science that deals with the microscopic structure of cells, tissues, and organs in relation to their function. (2)

holocrine gland   (hol'ō-krin) [Gr. *holos*, complete + *krino*, to separate] Gland whose secretion is formed by the disintegration of entire cells, (e.g., sebaceous gland; see also apocrine and merocrine). (111)

homeostasis   (hō'mē-ō-stā'sis) [Gr. *homoio*, like + *stasis*, a standing] State of equilibrium in the body with respect to functions, composition of fluids and tissues. (9)

homeotherm   (hō'mē-ō-therm) (warm-blooded animals) [Gr. *homoiois*, like + *thermos*, warm] Any animal, including mammals and birds, that tends to maintain a constant body temperature. (853)

homologous   (hō-mol'ō-gŭs) [Gr., ratio or relation] Alike in structure or origin. (92)

homozygous   (hō-mō-zī'gŭs) [Gr. *homos*, the same + *zygon*, yoke] State of having identical allelic genes at one or more paired loci in homologous chromosomes. (992)

hormone   (hōr'mōn) [Gr. *hormon*, to set into motion] Substance secreted by endocrine tissues into the blood that acts on a target tissue to produce a specific response. (527)

hormone receptor   Protein or glycoprotein molecule of cells that specifically binds to hormones and produces a response. (536)

horn   Subdivision of gray matter in the spinal cord. The axons of sensory neurons synapse with neurons in the posterior horn, the cell bodies of motor neurons are in the anterior horn, and the cell bodies of autonomic neurons are in the lateral horn. (403)

human chorionic gonadotropin (HCG)   Hormone produced by the placenta; stimulates secretion of testosterone by the fetus; during the first trimester stimulates ovarian secretion from the corpus luteum of the estrogen and progesterone required for the maintenance of the placenta. In a male fetus, stimulates secretion of testosterone by the fetal testis. (935)

humoral immunity   [L. *humor,* a fluid] Immunity due to antibodies in serum. (714)

hyaline cartilage   (hī′ă-lin) [Gr. *hyalos,* glass] Gelatinous, glossy cartilage tissue consisting of cartilage cells and their matrix; contains collagen, proteoglycans, and water. (123)

hyaluronic acid   (hī′ă-lū-ron′ik; glassy appearance) A mucopolysaccharide made up of alternating β-(1,4)-linked residues of hyalobiuronic acid, forming a gelatinous material in the tissue spaces and acting as a lubricant and shock absorbant generally throughout the body. (113)

hydrochloric acid (HCl)   Acid of gastric juice. (805)

hydrogen bond   Hydrogen atoms bound covalently to either N or O atoms have a small positive charge that is weakly attracted to the small negative charge of other atoms such as O or N; can occur within a molecule or between different molecules. (30)

hydroxyapatite   (hī-drok′sē-ap′ă-tīt) Mineral with the empiric formula 3 $Ca_3(PO_4)_2 \cdot Ca(OH)_2$; the main mineral of bone and teeth. (123)

hymen   (hī′men) [Gr., membrane] Thin, membranous fold partly occluding the vaginal external orifice; normally disrupted by sexual intercourse or other mechanical phenomena. (942)

hyoid   (hī′oyd) [Gr. *hyoeides,* shaped like the Greek letter epsilon, ε] U-shaped bone between the mandible and larynx. (196)

hypercalcemia   (hī′pcr-kal-sē′mē-ă) Abnormally high levels of calcium in the blood. (907)

hypercapnia   (hī′per-kap′nē-ă) Higher-than-normal levels of carbon dioxide in the blood or tissues. (770)

hyperkalemia   (hī′per-kă-lē′mē-ă) A greater than normal concentration of potassium ions in the circulating blood. (907)

hypernatremia   (hī′per-nă-trē′mē-ă) An abnormally high plasma concentration of sodium ions. (903)

hyperosmotic   (hī′per-oz-mot′ik) [Gr. *hyper,* above + *osmos,* an impulsion] Having a greater osmotic concentration or pressure than a reference solution. (77)

hyperpolarization   (hī′per-pō′lăr-i-zā′shŭn) Increase in the charge difference across the cell membrane; causes the charge difference to move away from 0 mV. (262)

hypertonic   (hī-per-ton′ik) [Gr. *hyper,* above + *tonos,* tension] Solution that causes cells to shrink. (77)

hypertrophy   (hī-per′tro-fe) [Gr. *hyper,* above + *trophe,* nourishment] Increase in bulk or size; not due to an increase in number of individual elements. (166)

hypocalcemia   (hī-pō-kal-sē′mē-ă) Abnormally low levels of calcium in the blood. (907)

hypocapnia   (hī′pō-kap′nē-ă) Lower-than-normal levels of carbon dioxide in the blood or tissues. (770)

hypodermis   (hī′pō-der′mis) [Gr. *hypo,* under + *dermis,* skin] Loose areolar connective tissue found deep to the dermis that connects the skin to muscle or bone. (137)

hypokalemia   (hī′pō-ka-lē′mē-ă) Abnormally small concentration of potassium ions in the blood. (907)

hyponatremia   (hī′pō-nă-trē′mē-ă) An abnormally low plasma concentration of sodium ions. (903)

hyponychium   (hī′pō-nik′ē-ŭm) [Gr. *hypo,* under + *onyx,* nail] Thickened portion of the stratum corneum under the free edge of the nail. (147)

hypophysis   (hī-pof′i-sis) [Gr., an undergrowth] Endocrine gland attached to the hypothalamus by the infundibulum. Also called the pituitary gland. (547)

hypopolarization   Change in the electric charge difference across the cell membrane that causes the charge difference to be smaller or move closer to 0 mV. (261)

hyposmotic   (hī′pos-mot′ik) [Gr. *hypo,* under + *osmos,* an impulsion] Having a lower osmotic concentration or pressure than a reference solution. (77)

hypospadias   (hī-pō-spā′dē-ăs) [Gr., one having the orifice of the penis too low; *hypospao,* to draw away from under] Developmental anomaly in the wall of the urethra so that the canal is open for a greater or lesser distance on the undersurface of the penis; also a similar defect in the female in which the urethra opens into the vagina. (979)

hypothalamohypophyseal portal system   (hī′pō-thal′ă-mō-hī′pō-fiz′ē-ăl) Series of blood vessels that carry blood from the area of the hypothalamus to the anterior pituitary gland; originate from capillary beds in the hypothalamus and terminate as a capillary bed in the anterior pituitary gland. (548)

hypothalamohypophyseal tract   Nerve tract, consisting of the axons of neurosecretory cells, extending from the hypothalamus into the posterior pituitary gland. Hormones produced in the neurosecretory cell bodies in the hypothalamus are transported through the hypothalamohypophyseal tract to the posterior pituitary gland where they are stored for later release. (548)

hypothalamus   (hī-pō-thal′ă-mŭs) [Gr. *hypo,* under + *thalamus,* bedroom] Important autonomic and neuroendocrine control center beneath the thalamus. (390)

hypothenar   (hī-pō-thē′nar) [Gr. *hypo,* under + *thenar,* palm of the hand] Fleshy mass of tissue on the medial side of the palm; contains muscles responsible for moving the little finger. (341)

hypotonic   (hī-pō-ton′ik) [Gr. *hypo,* under + *tonos,* tension] Solution that causes cells to swell. (77)

I

I band   Area between the ends of two adjacent myosin myofilaments within a myofibril; Z disk divides the I band into two equal parts. (278)

ileocecal sphincter   (il′ē-ō-sē′kăl) Thickening of circular smooth muscle between the ileum and the cecum forming the ileocecal valve. (794)

ileocecal valve   Valve formed by the ileocecal sphincter between the ileum and the cecum. (794)

ileum   (il′ē-ŭm) [Gr. *eileo,* to roll up, twist] Third portion of the small intestine, extending from the jejunum to the ileocecal opening into the large intestine; the posterior inferior bone of the coxa. (794)

immunity   (i-myū′ni-tē) [L. *immunis,* free from service] Resistance to infectious disease and harmful substances. (707)

immunization   Process by which a subject is rendered immune by deliberately introducing an antigen or antibody into the subject. (727)

immunoglobulin   (im′yū-nō-glob′yū-lin) Antibody found in the γ-globulin portion of plasma. (720)

implantation   (im-plan-tā′shŭn) Attachment of the blastocyst to the endometrium of the uterus; occurring 6 or 7 days after fertilization of the ovum. (963)

impotence   (im′pŏ-tens) Inability to accomplish the male sexual act; caused by psychologic or physical factors. (936)

incisor   (in-sī′zŏr) [L. *incido,* to cut into] One of the anterior, cutting teeth. (788)

incisura   (in′sī-sū′ră) [L., a cutting into] Notch or indentation at the edge of any structure. (631)

incus   (ing′kus) [L., anvil] Middle of the three ossicles in the middle ear. (491)

inferior colliculus   (kol-lik′yū-lŭs) [L. *collis,* hill] One of two rounded eminences of the midbrain; involved with hearing. (484)

inferior vena cava   Vein that returns blood from the lower limbs and the greater part of the pelvic and abdominal organs to the right atrium. (611)

inflammatory response   Complex sequence of events involving chemicals and immune cells that results in the isolation and destruction of antigens and tissues near the antigens. See also local and systemic inflammation. (713)

infundibulum   (in-fŭn-dib′yū-lŭm) [L., funnel] Funnel-shaped structure or passage, for example, the infundibulum that attaches the hypophysis to the hypothalamus or the funnellike expansion of the uterine tube near the ovary. (390)

inguinal canal   (ing′gwi-năl) Passage through the lower abdominal wall that transmits the spermatic cord in the male and the round ligament in the female. (925)

inhibin   (in-hib′in) Polypeptide secreted from the testes that inhibits FSH secretion. (934)

inhibitory neuron   Neuron that produces IPSPs and has an inhibitory influence. (372)

inhibitory postsynaptic potential (IPSP)   Hyperpolarization in the postsynaptic membrane that causes the membrane potential to move away from threshold. (270)

innate immunity   Immune system response that is the same with each exposure to an antigen; there is no ability for the system to remember a previous exposure to the antigen. (707)

inner cell mass   Group of cells at one end of the blastocyst, part of which forms the body of the embryo. (963)

inner ear   Contains the sensory organs for hearing and balance; contains the bony and membranous labyrinth. (486)

insensible perspiration   [L. *per,* through + *spiro,* to breathe everywhere] Perspiration that evaporates before it is perceived as moisture on the skin; the term sometimes includes evaporation from the lungs. (912)

insertion   More movable attachment point of a muscle; usually the lateral or distal end of a muscle associated with the limbs. (309)

inspiratory capacity   (in-spī′ră-tō-rē) Volume of air that can be inspired after a normal expiration; the sum of the tidal volume and the inspiratory reserve volume. (756)

inspiratory reserve volume   Maximum volume of air that can be inspired after a normal inspiration. (756)

insulin   (in′sŭ-lin) Protein hormone secreted from the pancreas that increases the uptake of glucose and amino acids by most tissues. (818)

interatrial septum   (in-ter-ā′trē-ăl) [L. *saeptum,* a partition] Wall between the atria of the heart. (614)

intercalated disk   (in-ter′kă-lā-ted) Cell-to-cell attachment with gap junctions between cardiac muscle cells. (110)

intercalated duct   Minute duct of glands such as the salivary gland and the pancreas; leads from the acini to the interlobular ducts. (798)

intercellular   Between cells. (58)

intercellular chemical signal   Chemical that is released from cells and passes to other cells; acts as signal that allows cells to communicate with each other. (528)

interferon   (in-ter-fēr′on) Protein that prevents viral replication. (710)

interlobar artery   (in-ter-lō′bar) Branch of the segmental arteries of the kidney; runs between the renal pyramids and gives rise to the arcuate arteries. (864)

interlobular artery   (in-ter-lob′yū-lăr) Artery that passes between lobules of an organ; branches of the interlobar arteries of the kidney pass outward through the cortex from the arcuate arteries and supply the afferent arterioles. (864)

interlobular duct   Any duct leading from a lobule of a gland and formed by the junction of the intercalated ducts draining the acini. (798)

interlobular vein   Parallels the interlobular arteries; in the kidney drains the peritubular capillary plexus, emptying into arcuate veins. (864)

intermediate olfactory area   Part of the olfactory cortex responsible for modulation of olfactory sensations. (468)

internal anal sphincter   [Gr. *sphinkter,* band or lace] Smooth muscle ring at the upper end of the anal canal. (799)

internal naris, pl. nares   (nā′ris, -res) Opening from the nasal cavity into the nasopharynx. (738)

internal spermatic fascia   Inner connective tissue covering of the spermatic cord. (928)

internal urinary sphincter   Traditionally recognized as a sphincter composed of a thickening of the middle smooth muscle layer of the bladder around the urethral opening. (867)

interphase   (inter-fāz) Period between active cell divisions when DNA replication occurs. (90)

interstitial   [L. *inter,* between + *sisto,* to stand] Space within tissue. Interstitial growth means growth from within. (157)

interstitial cell   (Leydig cell) (in-ter-stish′al) Cell between the seminiferous tubules of the testes; secretes testosterone. (923)

interventricular septum   (in-ter-ven-trik′yū-lăr) Wall between the ventricles of the heart. (614)

intestinal gland   (in-tes′ti-năl) Tubular glands in the mucous membrane of the small and large intestines. (794)

intracellular   (in-tră-sel′yū-lăr) Inside a cell. (58)

intracellular mediator   Molecule that is produced in a cell in which an intercellular mediator interacts with a membrane-bound receptor molecule; the intercellular mediator then acts as a signal and carries information to a site within the cell; e.g., cyclic-AMP. (540)

intramural plexus   (in′tră-mu′ral plek′sus) Combined submucosal and myenteric plexuses. (783)

intrinsic clotting pathway   (in-trin′sik) Series of chemical reactions resulting in clot formation that begins with chemicals (e.g., plasma factor XII) found within the blood. (596)

intrinsic factor   Factor secreted by the parietal cells of gastric glands and required for adequate absorption of vitamin B_{12}. (805)

intrinsic muscles   Muscles located within the structure being moved. (319)

inversion   (in-ver′zhŭn) [L. *inverto,* to turn about] Turning inward. (236)

ion   (ī′on) [Gr. *ion,* going] Atom or group of atoms carrying a charge of electricity by virtue of having gained or lost one or more electrons. (26)

ion channel   Pore in the cell membrane through which ions, such as sodium and potassium, move. (258)

ionic bond (ī-on′ik) Chemical bond that is formed when one atom loses an electron and another accepts that electron. (26)

iris (ī′ris) Specialized portion of the vascular tunic; the "colored" portion of the eye that can be seen through the cornea. (475)

ischemia (is-kē′mē-ă) [Gr. *ischo,* to keep back + *haima,* blood] Reduced blood supply to some area of the body. (151)

ischium (is′kē-ŭm) Superior bone of the coxa. (214)

isomer (ī′sō-mer) [Gr. *isos,* equal + *meros,* part] Molecules having the same number and types of atoms but differing in their three-dimensional arrangement. (41)

isometric contraction (ī-sō-met′rik) [Gr. *isos,* equal + *metron,* measure] Muscle contraction in which the length of the muscle does not change but the tension produced increases. (292)

isosmotic (ī′sos-mot′ik) [Gr. *isos,* equal + *osmos,* an impulsion] Having the same osmotic concentration or pressure as a reference solution. (77)

isotonic solution (ī′sō-ton′ik) [Gr. *isos,* equal + *tonos,* tension] Solution that causes cells to neither shrink nor swell. (77)

isotope (īsō-tōp) [Gr. *isos,* equal + *topos,* part, place] Either of two or more atoms that have the same atomic number but a different number of neutrons. (25)

isthmus (is′mŭs) Constriction connecting two larger parts of an organ, such as the constriction between the body and the cervix of the uterus, or the portion of the uterine tube between the ampulla and the uterus. (555)

J

jaundice (jawn′dis) [Fr. *jaune,* yellow] Yellowish staining of the integument, sclerae, and the other tissues with bile pigments. (150)

jejunum (jĕ-jū′nŭm) [L. *jejunus,* empty] Second portion of the small intestine; located between the duodenum and the ileum. (794)

juxtaglomerular apparatus (jŭks′tă-glŏ-mer′yū-lăr) Complex consisting of juxtaglomerular cells of the afferent arteriole and macular densa cells of the distal convoluted tubule near the renal corpuscle; secretes renin. (689)

juxtaglomerular cell Modified smooth muscle cell of the afferent arteriole located at the renal corpuscle; a component of the juxtaglomerular apparatus. (864)

juxtamedullary nephron (jŭks′tă-med′ŭ-lăr-ē) Nephron located near the junction of the renal cortex and medulla. (862)

K

karyotype (kār′ē-o-tīp) A display of chromosomes arranged by pairs. (991)

keratinization (ker′ă-tin-i-zā′shŭn) Production of keratin and changes in the chemical and structural character of epithelial cells as they move to the skin surface. (138)

keratinized (ker′ă-ti-nizd) [Gr. *keras,* horn] Word means turned into a horn. In modern usage the term means to become a structure that contains keratin, a protein found in skin, hair, nails, and horns. (139)

keratinocyte (ke-rat′i-nō-sīt) [Gr. *keras,* horn + *kytos,* cell] Epidermal cell that produces keratin. (137)

keratohyalin (ker′ă-tō-hī′ă-lin) Nonmembrane-bound protein granules in the cytoplasm of stratum granulosum cells of the epidermis. (140)

ketogenesis (kē-tō-jen′ĕ-sis) Production of ketone bodies, such as from acetyl-CoA. (847)

ketone body (kē′tōn) One of a group of ketones, including acetoacetic acid, β-hydrobutyric acid, and acetone. (847)

kidney (kid′nē) [A.S. *cwith,* womb, belly + *neere,* kidney] One of the two organs that excrete urine. The kidneys are bean-shaped organs approximately 11 cm long, 5 cm wide, and 3 cm thick lying on either side of the spinal column, posterior to the peritoneum, approximately opposite the twelfth thoracic and first three lumbar vertebrae. (861)

kilocalorie (kil′o-kal-ō-rē) Quantity of energy required to raise the temperature of 1 kg of water 1°C; 1000 calories. Equal to one dietary calorie. (832)

kinetic energy (ki-net′ik) Motion energy or energy that can do work. (35)

kinetic labyrinth (lab′i-rinth) Part of the membranous labyrinth composed of the semicircular canals; detects dynamic or kinetic equilibrium, such as movement of the head. (499)

Korotkoff sounds (kō-rot′kof) Named for Russian physician Nikolai S. Korotkoff (1874–1920). Sounds heard over an artery when blood pressure is determined by the auscultatory method; caused by turbulent flow of blood. (674)

L

labium majus, pl. labia majora (lā′bē-ŭm, -bē-ă) One of two rounded folds of skin surrounding the labia minora and vestibule; homolog of the scrotum in males. (943)

labium minus, pl. labia minora One of two narrow longitudinal folds of mucous membrane enclosed by the labia majora and bounding the vestibule; anteriorly they unite to form the prepuce. (942)

lacrimal apparatus (lak′ri-măl) Lacrimal, or tear, gland in the superolateral corner of the orbit of the eye and a duct system that extends from the eye to the nasal cavity. (472)

lacrimal canaliculus Canal that carries excess tears away from the eye; located in the medial canthus and opening on a small lump called the lacrimal papilla. (472)

lacrimal gland Tear gland located in the superolateral corner of the orbit. (472)

lacrimal papilla Small lump of tissue in the medial canthus or corner of the eye; the lacrimal canal opens within the lacrimal papilla. (472)

lacrimal sac Enlargement in the lacrimal canal that leads into the nasolacrimal duct. (472)

lactation (lak-tā′shŭn) [L. *lactatio,* suckle] Period after childbirth during which milk is formed in the breasts. (988)

lacteal (lak′tē-ăl) Lymphatic vessel in the wall of the small intestine that carries chyle from the intestine and absorbs fat. (703)

lactiferous duct (lak-tif′er-ŭs) One of 15–20 ducts that drain the lobes of the mammary gland and open onto the surface of the nipple. (944)

lactiferous sinus Dilation of the lactiferous duct just before it enters the nipple. (944)

lacuna, pl. lacunae (lă-kū′nă, lă-kū′nē) [L. *lacus,* a hollow, a lake] Small space or cavity; potential space within the matrix of bone or cartilage normally occupied by a cell that can only be visualized when the cell shrinks away from the matrix during fixation; space containing maternal blood within the placenta. (122)

lag phase One of the three phases of muscle contraction; time between the application of the stimulus and the beginning of muscular contraction. Also called the latent phase. (288)

lamella, pl. lamellae (lă-mel′ă, lă-mel′ē) Thin sheet or layer of bone. (123)

lamellated corpuscle (lam′ĕ-lāt-ed) Pacinian corpuscle. Oval receptor found in the deep dermis or hypodermis (responsible for deep cutaneous pressure and vibration) and in tendons (responsible for proprioception). (465)

lamina, pl. laminae (lam′i-nă, lam′i-nē) [L. *lamina,* plate, leaf] Thin plate, for example, the thinner portion of the vertebral arch. (206)

lamina propria (prō′prē-ă) Layer of connective tissue underlying the epithelium of a mucous membrane. (127)

laminar flow (lam′i-nar) Relative motion of layers of a fluid along smooth concentric parallel paths. (674)

Langerhans cell Dendritic cell named after the German anatomist, Paul Langerhans (1847–1888); found in the skin. (138)

lanugo (lă-nū′gō) [L. *lana,* wool] Fine, soft, unpigmented fetal hair. (144)

Laplace's law Named for the French mathematician, Pierre S. de Laplace (1749–1827). Force that stretches the wall of a blood vessel is proportional to the radius of the vessel times the blood pressure. (676)

large intestine Portion of the digestive tract extending from the small intestine to the anus. (798)

laryngitis (lar-in-jī′tis) Inflammation of the mucous membrane of the larynx. (739)

laryngopharynx (lă-ring′gō-far-ingks) Part of the pharynx lying posterior to the larynx. (739)

larynx, pl. larynges (lar′ingks, lă-rin′jēz) Organ of voice production located between the pharynx and the trachea; it consists of a framework of cartilages and elastic membranes housing the vocal folds and the muscles that control the position and tension of these elements. (739)

last menstrual period (LMP) Beginning of the last menstruation before pregnancy; used clinically to time events during pregnancy. (962)

latent phase See lag phase.

lateral geniculate nucleus (je-nik′yū-lāt) Nucleus of the thalamus where fibers from the optic tract terminate. (390)

lateral olfactory area (ol-fak′tō-rē) Part of the olfactory cortex involved in the conscious perception of olfactory stimuli. (468)

lens Transparent biconvex structure lying between the iris and the vitreous humor. (477)

lens fiber Epithelial cell that makes up the lens of the eye. (477)

lesser duodenal papilla Site of the opening of the accessory pancreatic duct into the duodenum. (793)

lesser omentum (ō-men′tŭm) [L., membrane that encloses the bowels] Peritoneal fold passing from the liver to the lesser curvature of the stomach and to the upper border of the duodenum for a distance of approximately 2 cm beyond the pylorus. (800)

lesser vestibular gland (ves-tib′yū-lăr) Paraurethral gland. Number of minute mucous glands opening on the surface of the vestibule between the openings of the vagina and urethra. (943)

leukocyte (lū′kō-sīt) White blood cell. (593)

leukocytosis (lu-ko-si-tō′sis) Abnormally large number of leukocytes in the blood. (603)

leukopenia (lū-kō-pē′nē-ă) Lower-than-normal number of leukocytes in the blood. (603)

leukotriene (lū-kō-trī′ēn) Specific class of physiologically active fatty acid derivatives present in many tissues. (575)

lever Rigid shaft capable of turning about a fulcrum or pivot point. (311)

LH surge Increase in plasma luteinizing hormone (LH) levels before ovulation and responsible for initiating it. (947)

ligamentum arteriosum (lig′ă-men′tŭm) Remains of the ductus arteriosus. (985)

ligamentum venosum Remnant of the ductus venosus. (955)

limbic system (lim′bik) [L. *limbus,* border] Parts of the brain involved with emotions and olfaction; includes the cingulate gyrus, hippocampus, habenular nuclei, parts of the basal ganglia, the hypothalamus (especially the mammillary bodies, the olfactory cortex, and various nerve tracts (e.g., fornix). (399)

lingual tonsil (ling′gwăl) Collection of lymphoid tissue on the posterior portion of the dorsum of the tongue. (705)

lipase (lip′ās) In general, any fat-splitting enzyme. (49)

lipid [Gr. *lipos,* fat] Substance composed principally of carbon, oxygen, and hydrogen; contains a lower ratio of oxygen to carbon and is less polar than carbohydrates; generally soluble in nonpolar solvents. (43)

lipid bilayer Double layer of lipid molecules forming the plasma membrane and other cellular membranes. (58)

lipochrome (lip′ō-krōm) Lipid-containing pigment that is metabolically inert. (64)

lipotropin (li-pō-trō′pin) One of the peptide hormones released from the adenohypophysis; increases lipolysis in fat cells. (554)

liver Largest gland of the body, lying in the upper right quadrant of the abdomen just inferior to the diaphragm; secretes bile and is of great importance in carbohydrate and protein metabolism and in detoxifying chemicals. (794)

lobe (lōb) Rounded projecting part, such as the lobe of a lung, the liver, or a gland. (745)

lobule (lob′yūl) Small lobe or a subdivision of a lobe, such as a lobule of the lung or a gland. (706)

local inflammation Inflammation confined to a specific area of the body. Symptoms include redness, heat, swelling, pain, and loss of function. (713)

local potential Depolarization that is not propagated and that is graded or proportional to the strength of the stimulus. (263)

local reflex Reflex of the intramural plexus of the digestive tract that does not involve the brain or spinal cord. (802)

locus, pl. loci (lō′kŭs, lō′sī) Place; usually a specific site. (992)

loop of Henle Named for the German anatomist Friedrich G. J. Henle (1809–1885). U-shaped part of the nephron extending from the proximal to the distal convoluted tubule and consisting of descending and ascending limbs. Some of the loops of Henle extend into the renal pyramids. (864)

lower respiratory tract The larynx, trachea, and lungs. (738)

lunula, pl. lunulae (lū′nū-lă, -lē) [L. *luna,* moon] White, crescent-shaped portion of the nail matrix visible through the proximal end of the nail. (147)

luteal phase (lū′tē-ăl) That portion of the menstrual cycle extending from the time of formation of the corpus luteum after ovulation to the time when menstrual flow begins; usually 14 days in length; the secretory phase. (946)

luteinizing hormone (LH) (lū′tē-ĭ-nīz-ing) In females, hormone stimulating the final maturation of the follicles and the secretion of progesterone by them, with their rupture releasing the ovum, and the conversion of the ruptured follicle into the corpus luteum; in males, stimulates the secretion of testosterone in the testes. (554)

luteinizing hormone–releasing hormone (LHRH) See gonadotropin-releasing hormone.

lymph (limf) [L. *lympha,* clear spring water] Clear or yellowish fluid derived from interstitial fluid and found in lymph vessels. (671)

lymph capillary Beginning of the lymphatic system of vessels; lined with flattened endothelium lacking a basement membrane. (671)

lymph node Encapsulated mass of lymph tissue found among lymph vessels. (671)

lymph nodule Small accumulation of lymph tissue lacking a distinct boundary. (703)

lymph sinus Channels in a lymph node crossed by a reticulum of cells and fibers. (705)

lymph vessel One of the system of vessels carrying lymph from the lymph capillaries to the veins. (671)

lymphoblast (lim′fō-blast) Cell that matures into a lymphocyte. (585)

lymphocyte (lim′fō-sīt) Nongranulocytic white blood cell formed in lymphoid tissue. (593)

lymphokine (lim′fō-kīn) Chemical produced by lymphocytes that activates macrophages, attracts neutrophils, and promotes inflammation. (718)

lysis (lī′sis) [Gr. *lysis,* a loosening] Process by which a cell swells and ruptures. (77)

lysosome (lī′sō-sōm) [Gr. *lysis,* loosening + *soma,* body] Membrane-bound vesicle containing hydrolytic enzymes that function as intracellular digestive enzymes. (68)

lysozyme (lī′sō-zīm) Enzyme that is destructive to the cell walls of certain bacteria; present in tears and some other fluids of the body. (593)

M

M line Line in the center of the H zone made of delicate filaments that holds the myosin myofilaments in place in the sarcomere of muscle fibers. (278)

macrophage (mak′rō-fāj) [Gr. *makros,* large + *phagein,* to eat] Any large mononuclear phagocytic cell. (592)

macula, pl. maculae (mak′yū-lă, -yū-lē) [L., a spot] Sensory structures in the utricle and saccule, consisting of hair cells and a gelatinous mass embedded with otoliths. (500)

macula densa Cells of the distal convoluted tubule located at the renal corpuscle and forming part of the juxtaglomerular apparatus. (864)

macula lutea (mak′u-lah lu′te-ah) [L. *macula,* a spot + *luteus,* yellow]

Small spot different in color from surrounding tissue; spot in the retina directly behind the lens in which densely packed cones are located. (476)

major duodenal papilla Point of opening of the common bile duct and pancreatic duct into the duodenum. (793)

major histocompatibility complex Group of genes that control the production of major histocompatibility complex proteins, which are glycoproteins found on the surfaces of cells. The major histocompatibility proteins serve as self-markers for the immune system and are used by antigen-presenting cells to present antigens to lymphocytes. (716)

male pronucleus Nuclear material of the sperm cell after the ovum has been penetrated by the sperm cell. (962)

malignant (mă-lig′nănt) Resistant to treatment; occurring in severe form, and frequently fatal; having the property of locally invasive and destructive growth and metastasis. (132)

malleus, pl. mallei (mal′ē-ŭs, mal′ē-ī) [L., hammer] Largest of the three auditory ossicles; attached to the tympanic membrane. (491)

mamillary bodies (mam′i-lār-ē) [L., breast- or nipple-shaped] Nipple-shaped structures at the base of the hypothalamus. (390)

mamma, pl. mammae (mam′ă, mam′ē) Breast. The organ of milk secretion; one of two hemispheric projections of variable size situated in the subcutaneous layer over the pectoralis major muscle on either side of the chest; it is rudimentary in the male. (943)

mammary ligaments (mam′ă-rē) Cooper's ligaments. Well-developed ligaments that extend from the overlying skin to the fibrous stroma of mammary gland. (944)

manubrium, pl. manubria (mă-nū′brē-ŭm, -ă) [L., handle] Part of a bone representing the handle, such as the manubrium of the sternum representing the handle of a sword. (210)

marrow (mar′ō) A highly cellular hematopoietic connective tissue filling the medullary cavities and spongy epiphyses of bones that becomes predominantly fatty with age, particularly in the long bones of the limbs. (159)

mass movement Forcible peristaltic movement of short duration, occurring only three or four times a day, which moves the contents of the large intestine. (784)

mass number Equal to the number of protons plus the number of neutrons in each atom. (25)

mastication (mas′ti-kă′shŭn) [L. *mastico,* to chew] Process of chewing. (319)

mastication reflex Repetitive cycle of relaxation and contraction of the muscles of mastication that results in chewing of food. (803)

mastoid (mas′toyd) [Gr. *mastos,* breast] Resembling a breast. (186)

mastoid air cells Spaces within the mastoid process of the temporal bone connected to the middle ear by ducts. (490)

mature follicle An ovarian follicle in which the oocyte attains its full size. The follicle contains a fluid-filled antrum and is surrounded by the theca interna and externa. (938)

maximal stimulus Stimulus resulting in a local potential just large enough to produce the maximum frequency of action potentials. (270)

meatus (mē-ā′tŭs) [L., to go, pass] Passageway or tunnel. (738)

mechanoreceptor (mek′ă-nō-rē-sep′tŏr) A sensory receptor that has the role of responding to mechanical pressures. Examples are pressure receptors in the carotid sinus or touch receptors in the skin. (462)

meconium (mē-kō′nē-ŭm) [Gr. *mekon,* poppy] First intestinal discharges of the newborn infant, greenish in color and consisting of epithelial cells, mucus, and bile. (985)

medial olfactory area Part of the olfactory cortex responsible for the visceral and emotional reactions to odors. (468)

medulla oblongata (me-dūl′ă ob-long-gah′tă) Inferior portion of the brainstem that connects the spinal cord to the brain and contains autonomic centers controlling such functions as heart rate, respiration, and swallowing. (385)

medullary cavity (med′ū-lār-ē, mĕ-dul′er-ē, med′yū-lār-ē) Large, marrow-filled cavity in the diaphysis of a long bone. (158)

medullary ray Extension of the kidney medulla into the cortex, consisting of collecting ducts and loops of Henle. (861)

megakaryoblast (meg-ă-kar′ē-ō-blast) [Gr. *mega* + *karyon,* nut (nucleus) + *blastos,* germ)] Cell that gives rise to platelets or thrombocytes. (585)

meibomian cyst (mī-bō′mē-an) Named for German anatomist, Hendrik Meibom (1638–1700). A chronic inflammation of a meibomian gland; see also chalazion. (471)

meibomian gland   Sebaceous gland near the inner margins of the eyelid; secretes sebum that lubricates the eyelid and retains tears. (470)

meiosis   (mī-ō′sis) [Gr., a lessening] Process of cell division that results in the formation of gametes. Consists of two divisions that result in one (female) or four (male) gametes, each of which contains one half the number of chromosomes in the parent cell. (92)

Meissner's corpuscle   (mīs′nerz kōr′pŭs-l) Named for Georg Meissner, German histologist, (1829–1905). See tactile corpuscle.

melanin   (mel′ă-nin) [Gr. melas, black] A group of related molecules responsible for skin, hair, and eye color. Most melanins are brown to black pigments, some are yellowish or reddish. (141)

melanocyte   (mel′ă-nō-sīt) [Gr. melas, black + kytos, cell] Cell found mainly in the stratum basale that produces the brown or black pigment melanin. (138)

melanocyte–stimulating hormone (MSH)   Peptide hormone secreted by the anterior pituitary; increases melanin production by melanocytes, making the skin darker in color. (554)

melanosome   (mel′ă-nō-sōm) [Gr. melas, black + soma, body] Membranous organelle containing the pigment melanin. (141)

melatonin   (mel-ă-tōn′in) Hormone (amino acid derivative) secreted by the pineal body; inhibits secretion of gonadotropin-releasing hormone from the hypothalamus. (572)

membrane–bound receptor   Receptor molecule such as a hormone receptor that is bound to the cell membrane of the target cell. (537)

membranous labyrinth   (mem′bră-nŭs lab′i-rinth) Membranous structure within the inner ear consisting of the cochlea, vestibule, and semicircular canals. (492)

membranous urethra   (yū-rē′thră) Portion of the male urethra, approximately 1 cm in length, extending from the prostate gland to the beginning of the penile urethra. (930)

memory cell   Small lymphocytes that are derived from B cells or T cells and that rapidly respond to a subsequent exposure to the same antigen. (723)

menarche   (me-nar′kē) [Gr. mensis, month + arche, beginning] Establishment of menstrual function;

the time of the first menstrual period or flow. (944)

meninx, pl. meninges   (mē′ninks, mĕ-nin′jes) [Gr., membrane] Connective tissue membranes surrounding the brain. (420)

menopause   (men′ō-pawz) [Gr. mensis, month + pausis, cessation] Permanent cessation of the menstrual cycle. (950)

menses   (men′sēz) [L. mensis, month] Periodic hemorrhage from the uterine mucous membrane, occurring at approximately 28-day intervals. (944)

menstrual cycle   (men′strū-ăl) Series of changes that occur in sexually mature, nonpregnant women and result in menses. Specifically refers to the uterine cycle but is often used to include both the uterine and ovarian cycles. (944)

Merkel's disk   (mer′kelz) Named for Friedrich Merkel, German anatomist (1845–1919). See tactile disk.

merocrine   (mer′ō-krin) [Gr. meros, part + krino, to separate] Gland that secretes products with no loss of cellular material; an example is water-producing sweat glands; see also apocrine and holocrine. (111)

mesencephalon   (mez-en-sef′ă-lon) [Gr. mesos, middle + enkephalos, brain] Midbrain in both the embryo and adult; consists of the cerebral peduncle and the corpora quadrigemini. (384)

mesentery   (mes′en-ter′ē) [Gr. mesos, middle + enteron, intestine] Double layer of peritoneum extending from the abdominal wall to the abdominal viscera, conveying to it its vessels and nerves. (800)

mesoderm   (mez′ō-derm) Middle of the three germ layers of an embryo. (126)

mesonephros   (mez′ō-nef′ros) One of three excretory organs appearing during embryonic development; forms caudal to the pronephros as the pronephros disappears. It is well developed and is functional for a time before the establishment of the metanephros, which gives rise to the kidney; undergoes regression as an excretory organ, but its duct system is retained in the male as the efferent ductule and epididymis. (976)

mesosalpinx   (mez′ō-sal′pinks) [Gr. mesos, middle + salpinx, trumpet] Part of the broad ligament supporting the uterine tube. (941)

mesothelium   (mez-ō-thē′lē-ŭm) A single layer of flattened cells forming an epithelium that lines serous cavities,

such as peritoneum, pleura, pericardium. (127)

mesovarium   (mez′ō-vā′rē-ŭm) Short peritoneal fold connecting the ovary with the broad ligament of the uterus. (937)

messenger RNA (mRNA)   Type of RNA that moves out of the nucleus and into the cytoplasm where it is used as a template to determine the structure of proteins. (87)

metabolism   (mĕ-tab′ō-lizm) [Gr. metabole, change] Sum of all the chemical reactions that take place in the body, consisting of anabolism and catabolism. Cellular metabolism refers specifically to the chemical reactions within cells. (33)

metacarpal   (met′ă-kar′păl) Relating to the fine bones of the hand between the carpus (wrist) and the phalanges. (214)

metanephros   (met-ă-nef′ros) Most caudally located of the three excretory organs appearing during embryonic development; becomes the permanent kidney of mammals. In mammalian embryos it is formed caudal to the mesonephros and develops later as the mesonephros undergoes regression. (976)

metaphase   (met′ă-fās) Time during cell division when the chromosomes line up along the equator of the cell. (92)

metarteriole   (met′ar-tēr′ē-ōl) One of the small peripheral blood vessels that contain scattered groups of smooth muscle fibers in their walls; located between the arterioles and the true capillaries. (648)

metastasis   (mĕ-tas′tă-sis) The shifting of a disease or its local manifestations, or the spread of a disease from one part of the body to another as in a malignant neoplasm. (132)

metatarsal   (met′ă-tar′sal) [Gr. meta, after + tarsos, sole of the foot] Distal bone of the foot. (220)

metencephalon   (met′en-sef′ă-lon) [Gr. meta, after + enkephalos, brain] Second-most posterior division of the embryonic brain; becomes the pons and cerebellum in the adult. (384)

micelle   (mi-sel′, mī-sel′) [L. micella, small morsel] Droplets of lipid surrounded by bile salts in the small intestine. (819)

microfilament   (mī-krō-fil′ă-ment) Small fibril forming bundles, sheets, or networks in the cytoplasm of cells; provides structure to the cytoplasm and mechanical support for microvilli and stereocilia. (63)

microglia   (mī-krog′lē-ă) [Gr. micro + glia, glue] Small neuroglial cells that become phagocytic and mobile in response to inflammation; considered to be macrophages within the central nervous system. (365)

microtubule   (mī-krō-tū′byūl) Hollow tube composed of tubulin, measuring approximately 25 nm in diameter and usually several micrometers long. Helps provide support to the cytoplasm of the cell and is a component of certain cell organelles such as centrioles, spindle fibers, cilia, and flagella. (63)

microvillus, pl. microvilli   (mī′krō-vil′ŭs, -vil′ī) Minute projection of the cell membrane that greatly increases the surface area. (71)

micturition reflex   (mik-chū-rish′ŭn) Contraction of the urinary bladder stimulated by stretching of the bladder wall; results in emptying of the bladder. (893)

middle ear   Air-filled space within the temporal bone; contains auditory ossicles; between the external and internal ear. (486)

milk letdown   Expulsion of milk from the alveoli of the mammary glands; stimulated by oxytocin. (989)

mineral   Inorganic nutrient necessary for normal metabolic functions. (838)

mineralocorticoid   (min′er-al-ō-kōr′ti-koyd) Steroid hormone (e.g., aldosterone) produced by the zona glomerulosa of the adrenal cortex; facilitates exchange of potassium for sodium in the distal renal tubule, causing sodium reabsorption and potassium and hydrogen ion secretion. (563)

minute ventilation   Product of tidal volume times the respiratory rate. (757)

minute volume   Amount of blood pumped by either the left or right ventricle each minute. (631)

mitochondrion, pl. mitochondria   (mī-tō-kon′drē-on, -kon′drē-ă) [Gr. *mitos,* thread + *chandros,* granule] Small, spherical, rod-shaped or thin filamentous structure in the cytoplasm of cells that is a site of ATP production. (69)

mitosis   (mī-tō′sis) [Gr., thread] Cell division resulting in two daughter cells with exactly the same number and type of chromosomes as the mother cell. (92)

modiolus   (mō-dī′ō′lŭs) [L., nave of a wheel] Central core of spongy bone about which turns the spiral canal of the cochlea. (493)

molar   (mō′lăr) Tricuspid tooth; the three posterior teeth of each dental arch. (788)

molecule   A substance composed of two or more atoms chemically combined to form a structure that behaves as an independent unit. (29)

monoblast   (mon′ō-blast) Cell that matures into a monocyte. (585)

mononuclear phagocytic system   (mon-ō-nū′klē-ăr fag-ō-sit′ik) Phagocytic cells, each with a single nucleus; derived from monocytes. (712)

monosaccharide   Simple sugar carbohydrate that cannot form any simpler sugar by hydrolysis. (41)

mons pubis   (monz pyu′bis) [L., mountain] Prominence caused by a pad of fatty tissue over the symphysis pubis in the female. (943)

morula   (mōr′ū-la, mōr′yū-la) [L. *morus,* mulberry] Mass of 12 or more cells resulting from the early cleavage divisions of the zygote. (962)

motor neuron   Neuron that innervates skeletal, smooth, or cardiac muscle fibers. (282)

motor unit   Single neuron and the muscle fibers it innervates. (288)

mucosa   (myū-kō′să) [L. *mucosus,* mucous] Mucous membrane consisting of epithelium and lamina propria. In the digestive tract there is also a layer of smooth muscle. (783)

mucous membrane   (myū′kŭs) Thin sheet consisting of epithelium and connective tissue (lamina propria) that lines cavities that open to the outside of the body; many contain mucous glands that secrete mucus. (127)

mucous neck cell   One of the mucous-secreting cells in the neck of a gastric gland. (793)

mucus   (myū′kŭs) Viscous secretion produced by and covering mucous membranes; lubricates mucous membranes and traps foreign substances. (127)

multiple motor unit summation   Increased force of contraction of a muscle due to recruitment of motor units. (289)

multiple wave summation   Increased force of contraction of a muscle due to increased frequency of stimulation. (290)

multipolar neuron   One of three categories of neurons consisting of a neuron cell body, an axon, and two or more dendrites. (126)

muscarinic receptor   (mŭs′kă-rin′ik) Class of cholinergic receptor that is specifically activated by muscarine in addition to acetylcholine. (514)

muscle fiber   Muscle cell. (277)

muscle spindle   Three to 10 specialized muscle fibers supplied by gamma motor neurons and wrapped in sensory nerve endings; detects stretch of the muscle and is involved in maintaining muscle tone. (404)

muscle tone   Relatively constant tension produced by a muscle for long periods as a result of asynchronous contraction of motor units. (293)

muscle twitch   Contraction of a whole muscle in response to a stimulus that causes an action potential in one or more muscle fibers. (288)

muscular fatigue   Fatigue due to a depletion of ATP within the muscle fibers. (294)

muscularis   (mŭs-kyū-lā′ris) [Modern L., muscular] Muscular coat of a hollow organ or tubular structure. (783)

muscularis mucosa   Thin layer of smooth muscle found in most parts of the digestive tube; located outside the lamina propria and adjacent to the submucosa. (783)

musculi pectinati   (pek′tĭ-nă′tē) Prominent ridges of atrial myocardium located on the inner surface of much of the right atrium and both auricles. (611)

mutation   A change in the number or kinds of nucleotides in the DNA of a gene. (995)

myelencephalon   (mī′el-en-sef′ă-lon) [Gr. *myelos,* medulla, marrow + *enkephalos,* brain] Most caudal portion of the embryonic brain; medulla oblongata. (384)

myelin sheath   (mī′ĕ-lin) Envelope surrounding most axons; formed by Schwann cell membranes being wrapped around the axon. (366)

myelinated axon   (mī′ĕ-li-nāt-ed ak′son) Nerve fiber having a myelin sheath. (366)

myeloblast   (mī′ĕ-lō-blast) Immature cell from which the different granulocytes develop. (585)

myenteric plexus   (mī′en-ter′ik) Plexus of unmelinated fibers and postganglionic autonomic cell bodies lying in the muscular coat of the esophagus, stomach, and intestines; communicates with the submucosal plexuses. (783)

myoblast   (mī′ō-blast) [Gr. *mys,* muscle + *blastos,* germ] Primitive multinucleated cell with the potential of developing into a muscle fiber. (277)

myofilament   (mī-ō-fil′ă-ment) Extremely fine molecular thread helping to form the myofibrils of muscle; thick myofilaments are formed of myosin, and thin myofilaments are formed of actin. (278)

myometrium   (mī′ō-mē′trē-ŭm) Muscular wall of the uterus; composed of smooth muscle. (942)

myosin myofilament (mī'ō-sin mī-ō-fil'ă-ment) Thick myofilament of muscle fibrils; composed of myosin molecules. (278)

N

nail (nāl) [A.S., naegel] Several layers of dead epithelial cells containing hard keratin on the ends of the digits. (147)

nail matrix Portion of the nail bed from which the nail is formed. (147)

nasal cavity (nā'zăl) Cavity between the external nares and the pharynx. It is divided into two chambers by the nasal septum and is bounded inferiorly by the hard and soft palates. (190)

nasal septum Bony partition that separates the nasal cavity into left and right parts; composed of the vomer, the perpendicular plate of the ethmoid, and hyaline cartilage. (190)

nasolacrimal duct (nā-zō-lak'ri-măl) Duct that leads from the lacrimal sac to the nasal cavity. (472)

nasopharynx (nā-zō-far'ingks) Part of the pharynx that lies above the soft palate; anteriorly it opens into the nasal cavity. (739)

near point of vision Closest point from the eye at which an object can be held without appearing blurred. (479)

neck (tooth) Slightly constricted part of a tooth, between the crown and the root. (788)

neoplasm (nē'ō-plazm) An abnormal tissue that grows by cellular proliferation more rapidly than normal and continues to grow after the stimuli that initiated the new growth ceases. (132)

nephron (nef'ron) [Gr. nephros, kidney] Functional unit of the kidney, consisting of the renal corpuscle, the proximal convoluted tubule, the loop of Henle, and the distal convoluted tubule. (861)

nerve tract Bundles of parallel axons with their associated sheaths in the central nervous system. (368)

neural crest (nūr'ăl) Edge of the neural plate as it rises to meet at the midline to form the neural tube. (383)

neural crest cells Cells derived from the crests of the forming neural tube in the embryo; together with the mesoderm, form the mesenchyme of the embryo; give rise to part of the skull, the teeth, melanocytes, sensory neurons, and autonomic neurons. (127)

neural plate Region of the dorsal surface of the embryo that is transformed into the neural tube and neural crest. (383)

neural tube Tube formed from the neuroectoderm by the closure of the neural groove. The neural tube develops into the spinal cord and brain. (383)

neuroectoderm (nūr-ō-ek'tō-derm) That part of the ectoderm of an embryo giving rise to the brain and spinal cord. (127)

neuroglia (nū-rog'lē-ă) [Gr. neuro, nerve + glia, glue] Cells in the nervous system other than the neurons; includes astrocytes, ependymal cells, microglia, oligodendrocytes, satellite cells, and Schwann cells. (126)

neurohormone (nūr-ō-hōr'mōn) Hormone secreted by a neuron. (528)

neurohypophysis (nūr'ō-hī-pof'i-sis) Portion of the hypophysis derived from the brain; commonly called the posterior pituitary. Major secretions include antidiuretic hormone and oxytocin. (390)

neuromodulator Substance that influences the sensitivity of neurons to neurotransmitters but neither strongly stimulates nor strongly inhibits neurons by itself. (370)

neuromuscular junction (nūr-ō-mŭs'kyū-lăr) Specialized synapse between a motor neuron and a muscle fiber. (282)

neuron (nūr'on) [Gr., nerve] Morphologic and functional unit of the nervous system, consisting of the nerve cell body, the dendrites, and the axon. (126)

neuron cell body Enlarged portion of the neuron containing the nucleus and other organelles; also called nerve cell body. (362)

neurotransmitter (nūr'ō-trans-mit'er) [Gr. neuro, nerve + L. transmitto, to send across] Any specific chemical agent released by a presynaptic cell on excitation that crosses the synaptic cleft and stimulates or inhibits the postsynaptic cell. (283)

neutral solution Solution such as pure water that has 10^{-7} mol of hydrogen ions per liter and an equal concentration of hydroxide ions; has a pH of 7. (39)

neutron (nū'tron) [L. neuter, neither] Electrically neutral particle in the nuclei of atoms (except hydrogen). (24)

neutrophil (nū'trō-fil) [L. neuter, neither + Gr. philos, fond] Type of white blood cell; small phagocytic white blood cell with a lobed nucleus and small granules in the cytoplasm. (593)

nicotinic receptor (nik'ō-tin'ik) Class of cholinergic receptor molecule that is specifically activated by nicotine and by acetylcholine. (514)

nipple (nip'l) Projection at the apex of the mamma, on the surface of which the lactiferous ducts open; surrounded by a circular pigmented area, the areola. (943)

Nissl bodies (nis'l) Named after the German neurologist, Franz Nissl (1860–1919). Areas in the neuron cell body containing rough endoplasmic reticulum. (362)

nociceptor (nō'si-sep'ter, -tor) [L. noceo, to injure + capio, to take] A sensory receptor that detects painful or injurious stimuli. (462)

nonelectrolyte [Gr. electro + lytos, soluble] Molecules that do not dissociate and do not conduct electricity. (31)

norepinephrine (nōr'ep-i-nef'rin) Neurotransmitter substance released from most of the postganglionic neurons of the sympathetic division; hormone released from the adrenal cortex that increases cardiac output and blood glucose levels. (513)

nose, or nasus (nōz or nā'sŭs) Visible structure that forms a prominent feature of the face; can also refer to the nasal cavities. (738)

notochord (nō'tō-kōrd) [Gr. notor, back + chords, cord] Small rod of tissue lying ventral to the neural tube. A characteristic of all vertebrates, in humans it becomes the nucleus pulposus of the intervertebral disks. (383)

nuchal (nū'kăl) The back of the neck. (122)

nuclear envelope (nū'klē-er) Double membrane structure surrounding and enclosing the nucleus. (61)

nuclear pores Porelike openings in the nuclear envelope where the inner and outer membranes fuse. (61)

nucleic acid (nū-klē'ik) Polymer of nucleotides, consisting of DNA and RNA, forms a family of substances that comprise the genetic material of cells and control protein synthesis. (50)

nucleolus, pl. nucleoli (nū-klē'ō-lŭs, -lī) Somewhat rounded, dense, well-defined nuclear body with no surrounding membrane; contains ribosomal RNA and protein. (62)

nucleotide (nū'klē-ō-tīd) Basic building block of nucleic acids consisting of a sugar (either ribose or deoxyribose) and one of several types of organic bases. (50)

nucleus, pl. nuclei (nū'klē-ŭs, -ī) [L., inside of a thing] Cell organelle containing most of the genetic material of the cell; collection of nerve cell bodies within the central nervous system; center of an atom consisting of protons and neutrons. (61)

nucleus pulposus (pŭl-pō′sŭs) [L., central pulp] Soft central portion of the intervertebral disk. (204)

nutrient (nū′trē-ent) [L. *nutriens,* to nourish] Chemicals taken into the body that are used to produce energy, provide building blocks for new molecules, or function in other chemical reactions. (832)

O

olecranon (ō-lek′ră-non, ō-lē-krā′non) Process on the distal end of the ulna, forming the point of the elbow. (212)

olfaction (ol-fak′shŭn) [L. *olfactus,* smell] Sense of smell. (466)

olfactory bulb (ol-fak′tō-rē) Ganglionlike enlargement at the rostral end of the olfactory tract that lies over the cribriform plate; receives the olfactory nerves from the nasal cavity. (466)

olfactory cortex Termination of the olfactory tract in the cerebral cortex within the lateral fissure of the cerebrum. (391)

olfactory epithelium Epithelium of the olfactory recess containing olfactory receptors. (466)

olfactory recess Extreme superior region of the nasal cavity. (466)

olfactory tract Nerve tract that projects from the olfactory bulb to the olfactory cortex. (466)

oligodendrocyte (ol′i-gō-den′drō-sīt) Neuroglial cell that has cytoplasmic extensions that form myelin sheaths around axons in the central nervous system. (365)

oncogene (ong′kō-jēn) A gene that can change or be activated to cause cancer. (995)

oncology (ong-kol′ō-jē) The study of neoplasms. (132)

oocyte (ō′ō-sīt) [Gr. *oon,* egg + *kytos,* a hollow (cell)] Immature ovum. (92)

oogenesis (ō-ō-jen′ĕ-sis) Formation and development of a secondary oocyte or ovum. (938)

oogonium (ō-ō-gō′nē-ŭm, -ă) [Gr. *oon,* egg + *gone,* generation] Primitive cell from which oocytes are derived by meiosis. (938)

opposition Movement of the thumb and little finger toward each other; movement of the thumb toward any of the fingers. (236)

opsin (op′sin) Protein portion of the rhodopsin molecule. A class of proteins that bind to retinal to form the visual pigments of the rods and cones of the eye. (480)

opsonin (op′sŏ-nin) [Gr. *opsonein,* to prepare food] Substance such as antibody or complement that enhances phagocytosis. (722)

optic chiasma (op′tik kī′az-mă) [Gr., two crossing lines; *chi,* the letter ӿ] Point of crossing of the optic tracts. (484)

optic disc Point at which axons of ganglion cells of the retina converge to form the optic nerve, which then penetrates through the fibrous tunic of the eye. (476)

optic nerve Nerve carrying visual signals from the eye to the optic chiasm. (484)

optic stalk Constricted proximal portion of the optic vesicle in the embryo; develops into the optic nerve. (974)

optic tract Tract that extends from the optic chiasma to the lateral geniculate nucleus of the thalamus. (484)

optic vesicle One of the paired evaginations from the walls of the embryonic forebrain from which the retina develops. (974)

oral cavity (ōr′ăl) The mouth; consists of the space surrounded by the lips, cheeks, teeth, and palate; limited posteriorly by the fauces. (782)

orbit (ōr′bit) Eye socket; formed by seven skull bones that surround and protect the eye. (190)

organ of Corti Named for the Italian anatomist, Marquis Alfonso Corti (1822–1888). Spiral organ; rests on the basilar membrane and supports the hair cells that detect sounds. (493)

organelle (ōr′gă-nel) [Gr. *organon,* tool] Specialized part of a cell serving one or more specific individual functions. (3)

orgasm (ōr′gazm) [Gr. *orgao,* to swell, be excited] Climax of the sexual act, associated with a pleasurable sensation. (936)

origin Less movable attachment point of a muscle; usually the medial or proximal end of a muscle associated with the limbs. (309)

oropharynx (ōr′ō-far′ingks) Portion of the pharynx that lies posterior to the oral cavity; it is continuous above with the nasopharynx and below with the laryngopharynx. (739)

oscillating circuit Neuronal circuit arranged in a circular fashion that allows action potentials produced in the circuit to keep stimulating the neurons of the circuit. (377)

osmolality (os-mō-lal′i-tē) Osmotic concentration of a solution; the number of moles of solute in 1 kg of water

times the number of particles into which the solute dissociates. (38)

osmoreceptor cell (os′mō-rē-sep′ter, -tōr) [Gr. *osmos,* impulsion] Receptor in the central nervous system that responds to changes in the osmotic pressure of the blood. (544)

osmosis (os-mō′sis) [Gr. *osmos,* thrusting or an impulsion] Diffusion of solvent (water) through a membrane from a less concentrated solution to a more concentrated solution. (75)

osmotic pressure (os-mot′ik) Force required to prevent the movement of water across a selectively permeable membrane. (75)

ossification (os′i-fi-kā′shŭn) [L. *os,* bone + *facio,* to make] Bone formation. (162)

osteoblast (os′tē-ō-blast) [Gr. *osteon,* bone + *blastos,* germ] Bone-forming cell. (161)

osteoclast (os′tē-ō-klast) [Gr. *osteon,* bone + *klastos,* broken] Large multinucleated cell that absorbs bone. (162)

osteocyte (os′tē-ō-sīt) [Gr. *osteon,* bone + *kytos,* cell] Mature bone cell surrounded by bone matrix. (123)

osteomalacia (os′tē-ō-mă-lā′shē-ă) Softening of bones due to calcium depletion. Adult rickets. (171)

osteon (os′tē-on) A central canal containing blood capillaries and the concentric lamellae around it; occurs in compact bone. (163)

osteoporosis (os′tē-ō-pō-rō′sis) [Gr. *osteon,* bone + *poros,* pore + *osis,* condition] Reduction in quantity of bone, resulting in porous bone. (178)

ostium (os′tē-ŭm) [L., door, entrance, mouth] Small opening, for example, the opening of the uterine tube near the ovary or the opening of the uterus into the vagina. (941)

otolith (ō′tō-lith) Crystalline particles of calcium carbonate and protein embedded in the maculae. (500)

oval window Membranous structure to which the stapes attaches; transmits vibrations to the inner ear. (491)

ovarian cycle (ō-var′ē-an) Series of events that occur in a regular fashion in the ovaries of sexually mature, nonpregnant females; results in ovulation and the production of the hormones estrogen and progesterone. (947)

ovarian epithelium (germinal epithelium) Peritoneal covering of the ovary. (937)

ovarian ligament Bundle of fibers passing to the uterus from the ovary. (937)

ovary (ō′vă-rē) One of two female reproductive glands located in the pelvic cavity; produces the secondary oocyte, estrogen, and progesterone. (937)

oviduct   (ō′vi-dŭkt) See uterine tube.

ovulation   (ov′yū-lā′shun) Release of an ovum, or secondary oocyte, from the vesicular follicle. (939)

oxidation   (ox-si-dā′shŭn) Loss of one or more electrons from a molecule. (33)

oxidation–reduction reaction   Reaction in which one molecule is oxidized and another is reduced. (34)

oxidative deamination   Removal of the amine group of an amino acid to form a keto acid, ammonia, and NADH. (848)

oxygen debt   Oxygen necessary for the synthesis of the ATP required to remove lactic acid produced by anaerobic respiration. (296)

oxygen–hemoglobin dissociation curve   Graph describing the relationship between the percentage of hemoglobin saturated with oxygen and a range of oxygen partial pressures. (763)

oxyhemoglobin   (ox′sē-hē-mō-glō′bin) Oxygenated hemoglobin. (588)

P

P wave   First complex of the electrocardiogram representing depolarization of the atria. (623)

PQ interval   Time elapsing between the beginning of the P wave and the beginning of the QRS complex in the electrocardiogram; also called PR interval. (623)

PR interval   See PQ interval.

pacinian corpuscle   (pa-sin′ē-an, pa-chin′) Named for Filippo Pacini, Italian anatomist, (1812–1883). See lamellated corpuscle. (465)

palate   (pal′ăt) [L. palatum, palate] Roof of the mouth. (789)

palatine tonsil   (pal′ă-tīn) One of two large oval masses of lymphoid tissue embedded in the lateral wall of the oral pharynx. (703)

palpebra, pl. palpebrae   (pal-pē′bră, -pē′brē) [L., eyelid] An eyelid. (470)

palpebral conjunctiva   (pal-pē′brăl kon-jŭnk-tī′vă) Conjunctiva that covers the inner surface of the eyelids. (471)

palpebral fissure   Space between the upper and lower eyelids. (470)

pancreas   (pan′krē-as) [Gr. pankreas, the sweetbread] Abdominal gland that secretes pancreatic juice into the intestine and insulin and glucagon from the pancreatic islets into the bloodstream. (566)

pancreatic duct   (pan-krē-at′ik) Excretory duct of the pancreas that extends through the gland from tail to head where it empties into the duodenum at the greater duodenal papilla. (793)

pancreatic islet   Islets of Langerhans; cellular mass varying from a few to hundreds of cells lying in the interstitial tissue of the pancreas; composed of different cell types that make up the endocrine portion of the pancreas and are the source of insulin and glucagon. (566)

pancreatic juice   [L. jus, broth] External secretion of the pancreas; clear, alkaline fluid containing several enzymes. (814)

papilla   (pă-pil′ă) [L., nipple] A small nipplelike process. Projection of the dermis, containing blood vessels and nerves, into the epidermis. Projections on the surface of the tongue. (137)

papillary muscle   (pap′i-lăr′ē) Nipple-like conical projection of myocardium within the ventricle; the chordae tendineae are attached to the apex of the papillary muscle. (615)

parafollicular cell   (par-ă-fo-lik′yū-lar) Endocrine cell scattered throughout the thyroid gland; secretes the hormone calcitonin. (556)

paramesonephric duct   (par-ă-mes-ō-nef′rik) One of two embryonic tubes extending along the mesonephros and emptying into the cloaca; in the female the duct forms the uterine tube, the uterus, and part of the vagina; in the male it degenerates. (976)

paranasal sinus   (par-ă-nā′săl) Air-filled cavities within certain skull bones that connect to the nasal cavity; located in the frontal, maxillary, sphenoid, and ethmoid bones. (738)

parasympathetic   (par-ă-sim-pa-thet′ik) Subdivision of the autonomic nervous system; characterized by having the cell bodies of its preganglionic neurons located in the brainstem and the sacral region of the spinal cord (craniosacral division); usually involved in activating vegetative functions such as digestion, defecation, and urination. (434)

parathyroid gland   (par-ă-thī′royd) One of four glandular masses imbedded in the posterior surface of the thyroid gland; secretes parathyroid hormone. (560)

parathyroid hormone   Peptide hormone produced by the parathyroid gland; increases bone breakdown and blood calcium levels. (560)

parietal   (pă-rī′ĕ-tăl) [L. paries, wall] Relating to the wall of any cavity. (658)

parietal cell   Gastric gland cell that secretes hydrochloric acid. (793)

parietal pericardium   Serous membrane lining the fibrous portion of the pericardial sac. (610)

parietal peritoneum   Layer of peritoneum lining the abdominal walls. (799)

parietal pleura   Serous membrane that lines the different parts of the wall of the pleural cavity. (749)

parotid gland   (pă-rot′id) Largest of the salivary glands; situated anterior to each ear. (790)

partial pressure   Pressure exerted by a single gas in a mixture of gases. (758)

passive tension   Tension applied to a load by a muscle without contracting; produced when an external force stretches the muscle. (293)

patella   (pa-tel′ă) [L. patina, shallow disk] Kneecap. (213)

pectoral girdle   (pek′to-răl) Site of attachment of the upper limb to the trunk; consists of the scapula and the clavicle. (210)

pedicle   (ped′ĭ-kl) [L. pes, feet] Stalk or base of a structure, such as the pedicle of the vertebral arch. (206)

pelvic brim   (pel′vik) Imaginary plane passing from the sacral promontory to the pubic crest. (216)

pelvic girdle   Site of attachment of the lower limb to the trunk; ring of bone formed by the sacrum and the coxae. (214)

pelvic inlet   Superior opening of the true pelvis. (216)

pelvic outlet   Inferior opening of the true pelvis. (216)

pelvis, pl. pelves   (pel′vis, pel′vēz) [L., basin] Any basin-shaped structure; cup-shaped ring of bone at the lower end of the trunk, formed from the ossa coxae, sacrum, and coccyx. (15)

pennate   (pen′āt) [L. penna, feather] Muscles with fasciculi arranged like the barbs of a feather along a common tendon. (309)

pepsin   (pep′sin) [Gr. pepsis, digestion] Principal digestive enzyme of the gastric juice, formed from pepsinogen; digests proteins into smaller peptide chains. (805)

pepsinogen   (pep-sin′ō-jen) [pepsin + Gr. gen, producing] Proenzyme formed and secreted by the chief cells of the gastric mucosa; the acidity of the gastric juice and pepsin itself converts pepsinogen into pepsin. (805)

peptidase   (pep′ti-dās) An enzyme capable of hydrolyzing one of the peptide links of a peptide. (812)

peptide bond   (pep′tīd) Chemical bond between amino acids. (46)

perforating canal   Canal containing blood vessels and nerves and running through bone perpendicular to the haversian canals. (164)

periarterial sheath   (per′ē-ar-tē′rē-ăl) Dense accumulations of lymphocytes (white pulp) surrounding arteries within the spleen. (706)

pericapillary cell   One of the slender connective tissue cells in close relationship to the outside of the capillary wall; relatively undifferentiated and may become a fibroblast, macrophage, or smooth muscle cell. (647)

pericardial cavity   (per-i-kar′dē-ăl) Space within the mediastinum in which the heart is located. (610)

pericardial fluid   Viscous fluid contained within the pericardial cavity between the visceral and parietal pericardium; functions as a lubricant. (610)

pericardium   (per-i-kar′dē-ŭm) [Gr. *pericardion,* the membrane around the heart] Membrane covering the heart. (610)

perichondrium   (per-i-kon′drē-ŭm) [Gr. *peri,* around + *chondros,* cartilage] Double-layered connective tissue sheath surrounding cartilage. (123)

perilymph   (per′i-limf) [Gr. *peri,* around + L. *lympha,* a clear fluid (lymph)] Fluid contained within the bony labyrinth of the inner ear. (492)

perimetrium   (per-i-mē′trē-ŭm) Outer serous coat of the uterus. (942)

perimysium   (per-i-mis′ē-ŭm, -miz′ē-ŭm) [Gr. *peri,* around + *mys,* muscle] Fibrous sheath enveloping a bundle of skeletal muscle fibers (muscle fascicle). (277)

perineum   (per′i-nē′ŭm) Area inferior to the pelvic diaphragm between the thighs; extends from the coccyx to the pubis. (330)

perineurium   (per-i-nū′rē-ŭm) [L. *peri,* around + Gr. *neuron,* nerve] Connective tissue sheath surrounding a nerve fascicle. (368)

periodontal ligament   (per′ē-ō-don′tăl) Connective tissue that surrounds the tooth root and attaches it to its bony socket. (226)

periosteum   (per-ē-os′tē-ŭm) [Gr. *peri,* around + *osteon,* bone] Thick, double-layered connective tissue sheath covering the entire surface of a bone except the articular surface, which is covered with cartilage. (159)

peripheral nervous system (PNS)   (pĕ-rif′ĕ-răl) Major subdivision of the nervous system consisting of nerves and ganglia. (360)

peripheral resistance   Resistance to blood flow in all the blood vessels. (631)

peristaltic wave   (per-i-stal′tik) Contraction in a tube such as the intestine characterized by a wave of contraction in smooth muscle preceded by a wave of relaxation that moves along the tube. (803)

peritubular capillary   The capillary network located in the cortex of the kidney; associated with the distal and proximal convoluted tubules. (864)

permanent tooth   One of the 32 teeth belonging to the second, or permanent, dentition. (788)

peroneal   (per-ō-nē′ăl) [Gr. *perone,* fibula] Associated with the fibula. (660)

peroxisome   (per-ok′si-sōm) Membrane-bound body similar to a lysosome in appearance but often smaller and irregular in shape; contains enzymes that either decompose or synthesize hydrogen peroxide. (69)

Peyer's patch   Named for the Swiss anatomist, Johann K. Peyer (1653–1712). Lymph nodule found in the lower half of the small intestine and the appendix. (794)

phagocyte   (fag′ō-sīt) Cell possessing the property of ingesting bacteria, foreign particles, and other cells. (710)

phagocytosis   (fag′ō-sī-tō′sis) [Gr. *phagein,* to eat + *kytos,* cell + *osis,* condition] Process of ingestion by cells of solid substances, such as other cells, bacteria, bits of necrosed tissue, and foreign particles. (81)

phalange, pl. phalanges   (fă-lanj′, fă-lan′jēz) [Gr. *phalanx,* line of soldiers] Bone of the fingers or toes. (214)

pharyngeal pouch   (fă-rin′jē-ăl) Paired evagination of embryonic pharyngeal endoderm between the brachial arches that gives rise to the thymus, thyroid gland, tonsils, and parathyroid glands. (968)

pharyngeal tonsil   One of two collections of aggregated lymphoid nodules on the posterior wall of the nasopharynx. (703)

pharynx   (far′ingks) [Gr. *pharynx,* throat, the joint opening of the gullet and windpipe] Upper expanded portion of the digestive tube between the esophagus below and the oral and nasal cavities above and in front. (739)

phenotype   (fē′nō-tīp) [Gr. *phaino,* to display, show forth + *typos,* model] Characteristic observed in an individual due to expression of his genotype. (993)

phosphodiesterase   (fos′fō-dī-es′ter-ās) Enzymes that split phosphodiester bonds, that is, that break down cyclic-AMP to AMP. (539)

phospholipid   (fos-fō-lip′id) Lipid with phosphorus, resulting in a molecule with a polar end and a nonpolar end; main component of the lipid bilayer. (44)

phosphorylation   (fos′fōr-i-lā′shŭn) Addition of phosphate to an organic compound. (840)

photoreceptor   (fō′tō-rē-sep′ter, -tōr) [L. *photo,* light + L. *ceptus,* to receive] A sensory receptor that is sensitive to light. Examples are rods and cones of the retina. (462)

phrenic nerve   (fren′ik) Nerve derived from spinal nerves C3 to C5; supplies the diaphragm. (445)

physiologic contracture   (fiz-ē-ō-loj′-ik kon-trak′chūr) Temporary inability of a muscle to either contract or relax because of a depletion of ATP so that active transport of calcium ions into the sarcoplasmic reticulum cannot occur. (295)

physiologic dead space   Sum of anatomic dead air space plus the volume of any nonfunctional alveoli. (757)

physiologic shunt   Deoxygenated blood from the alveoli plus deoxygenated blood from the bronchi and bronchioles. (761)

pia mater   (pī′ă mā′ter, pē′a mah′ter) [L., tender mother] Delicate membrane forming the inner covering of the brain and spinal cord. (422)

pigmented retina   Pigmented portion of the retina. (476)

pineal body   (pin′ē-ăl) [L. *pineus,* relating to pine trees] A small pine cone-shaped structure that projects from the epiphysis of the diencephalon; produces melatonin. (390)

pinna   (pin′ă) [L. *pinna* or *penna,* feather, in plural wing] See auricle. (486)

pinocytosis   (pin′ō-sī-tō′sis, pī′no) [Gr. *pineo,* to drink + *kytos,* cell + *osis,* condition] Cell drinking; uptake of liquid by a cell. (81)

pituitary gland   (pi-tū′i-tār-rē) See hypophysis.

plane   (plān) [L. *planus,* flat] A flat surface. An imaginary surface formed by extension through any axis or two points. Examples include a midsagittal plane, a coronal plane, and a transverse plane. (14)

plasma   (plaz′mă) [Gr., something formed] Fluid portion of blood. (901)

plasma cell   Cell derived from B cells; produces antibodies. (723)

plasma clearance   Volume of plasma per minute from which a substance can be completely removed by the kidneys. (890)

plasmin (plaz'min) Enzyme derived from plasminogen; dissolves clots by converting fibrin into soluble products. (598)

plateau phase of action potential Prolongation of the depolarization phase of a cardiac muscle cell membrane; results in a prolonged refractory period. (620)

platelet (plāt'let) Irregularly shaped disk found in blood; contains granules in the central part and clear protoplasm peripherally but has no definite nucleus. (594)

platelet plug Accumulation of platelets that stick to each other and to connective tissue; functions to prevent blood loss from damaged blood vessels. (594)

pleural cavity (plūr'ăl) Potential space between the parietal and visceral layers of the pleura. (747)

plexus, pl. plexuses (plek'sŭs, -sŭs-ez) [L., a braid] Intertwining of nerves or blood vessels. (445)

plicae circulares (plī'kă, plī'sē) (circular folds) Numerous folds of the mucous membrane of the small intestine. (793)

pluripotent (plū-rip'ō-tent) [L. *pluris,* more + *potentia,* power] In development the term refers to a cell or group of cells that have not yet become fixed or determined as to what specific tissues they are going to become. (962)

podocyte (pod'ō-sīt) [Gr. *pous, podos,* foot + *kytos,* a hollow (cell)] Epithelial cell of Bowman's capsule attached to the outer surface of the glomerular capillary basement membrane by cytoplasmic foot processes. (864)

Poiseuille's law (pwah-zuh'yez) Named for the French physiologist and physicist, Jean Léonard Marie Poiseuille (1797–1869). The volume of a fluid passing per unit of time through a tube is directly proportional to the pressure difference between its ends and to the fourth power of the internal radius of the tube and inversely proportional to the tube's length and the viscosity of the fluid. (675)

polar body One of the two small cells formed during oogenesis because of unequal division of the cytoplasm. (938)

polar covalent bond Covalent bond in which atoms do not share their electrons equally. (28)

polycythemia (pol'ē-sī-thē'mē-ă) Increase in red blood cell number above the normal. (601)

polygenic (pol-ē-jen'ik) Relating to a hereditary disease or normal characteristic controlled by interaction of genes at more than one locus. (995)

polysaccharide (pol-ē-sak'ă-rīd) Carbohydrate containing a large number of monosaccharide molecules. (42)

polyunsaturated Fatty acid that contains two or more double covalent bonds between its carbon atoms. (44)

pons (ponz) [L., bridge] That portion of the brainstem between the medulla and midbrain. (386)

popliteal (pop-lit'ē-ăl, pop-li-tē'ăl) [L., ham] Posterior region of the knee. (660)

porta (pōr'tă) [L., gate] Fissure on the inferior surface of the liver where the portal vein, hepatic artery, hepatic nerve plexus, hepatic ducts, and lymphatic vessels enter or exit the liver. (794)

portal system (pōr'tal) System of vessels in which blood, after passing through one capillary bed, is conveyed through a second capillary network. (667)

portal triad Branches of the portal vein, hepatic artery, and hepatic duct bound together in the connective tissue that divides the liver into lobules. (797)

postabsorptive state Following the absorptive state; blood glucose levels are maintained because of conversion of other molecules to glucose. (851)

posterior chamber of the eye Chamber of the eye between the iris and the lens. (477)

posterior interventricular sulcus Groove on the diaphragmatic surface of the heart, marking the location of the septum between the two ventricles. (611)

posterior pituitary See neurohypophysis.

postganglionic neuron (pōst'gang-glē-on'ik) Autonomic neuron that has its cell body located within an autonomic ganglion and sends its axon to an effector organ. (509)

postovulatory age Age of the developing fetus based on the assumption that fertilization occurs 14 days after the last menstrual period before the pregnancy. (962)

postsynaptic (pōst-si-nap'tik) Refers to the membrane of a nerve, muscle, or gland that is in close association with a presynaptic terminal. The postsynaptic membrane has receptor molecules within it that bind to neurotransmitter molecules. (283)

potential difference (pō-ten'shăl) Difference in electrical potential, measured as the charge difference across the cell membrane. (259)

potential energy [Gr. *en,* in + *ergon,* work] Energy in a chemical bond that is not being exerted or used to do work. (35)

precapillary sphincter Smooth muscle sphincter that regulates blood flow through a capillary. (648)

preganglionic neuron (prē'gang-glē-on'ik) Autonomic neuron that has its cell body located within the central nervous system and sends its axon through a nerve to an autonomic ganglion where it synapses with postganglionic neurons. (509)

premolar (prē-mō'lăr) Bicuspid tooth. (788)

prepuce (prē'pūs) In males, the free fold of skin that more-or-less completely covers the glans penis; the foreskin. In females, the external fold of the labia minora that covers the clitoris. (931)

pressoreceptor (pres'ō-rē-sep'ter, -tōr) See baroreceptor.

presynaptic terminal (prē'si-nap'tik) Enlarged axon terminal or terminal bouton. (283)

primary bronchus, pl. bronchi (brong'kŭs, -kī) One of two tubes arising at the inferior end of the trachea; each primary bronchus extends into one of the lungs. (742)

primary palate In the early embryo, gives rise to the upper jaw and lips. (970)

primary response Immune response that occurs as a result of the first exposure to an antigen. (723)

primary spermatocyte (sper'mă-tō-sīt) Spermatocyte arising by a growth phase from a spermatogonium; gives rise to secondary spermatocytes after the first meiotic division. (927)

prime mover Muscle that plays a major role in accomplishing a movement. (309)

primitive streak Ectodermal ridge in the midline of the embryonic disk from which arises the mesoderm by inward and then lateral migration of cells. (964)

primordial germ cell (prī-mōr'dē-ăl) Most primitive undifferentiated sex cell, found initially outside the gonad on the surface of the yolk sac. (976)

process Projection on a bone. (184)

processus vaginalis (prō-ses'ŭs vaj'i-năl-ŭs) Peritoneal outpocketing in the embryonic lower anterior abdominal wall that traverses the inguinal canal; in the male it forms the tunica vaginalis testis and normally loses its connection with the peritoneal cavity. (925)

progeria (prō-jēr'ē-ă) [Gr. *pro,* before + *ge* + *amras,* old age] Severe retardation of growth after the first year accompanied by a senile appearance and death at an early age. (991)

prolactin　(prō-lak′tin) Hormone of the adenohypophysis that stimulates the production of milk. (988)

prolactin–inhibiting hormone (PIH)　Neurohormone released from the hypothalamus that inhibits prolactin release from the adenohypophysis. (555)

prolactin–releasing hormone (PRH)　Neurohormone released from the hypothalamus that stimulates prolactin release from the adenohypophysis. (555)

proliferative phase　(prō-lif′er-ă-tiv) See follicular phase.

pronation　(prō-nā′shŭn) [L. *pronare,* to bend forward] Rotation of the forearm so that the anterior surface is down (prone). (232)

pronephros　(prō-nef′ros) In the embryos of higher vertebrates, a series of tubules emptying into the celomic cavity. It is a temporary structure in the human embryo, followed by the mesonephros and still later by the metanephros, which gives rise to the kidney. (976)

prophase　(prō′fāz) First stage in cell division when chromatin strands condense to form chromosomes. (92)

proprioception　(prō-prē-ō-sep′shun) [L. *proprius,* one's own + *capio,* to take] Information about the position of the body and its various parts. (413)

proprioceptor　(prō′prē-ō-sep′ter) Sensory receptor associated with joints and tendons. (463)

prostaglandin　(pros′tă-glan′din) Class of physiologically active substances present in many tissues; among effects are those of vasodilation, stimulation and contraction of uterine smooth muscle, and promotion of inflammation and pain. (575)

prostate gland　(pros′tāt) [Gr. *prostates,* one standing before] Gland that surrounds the beginning of the urethra in the male. The secretion of the gland is a milky fluid that is discharged by 20–30 excretory ducts into the prostatic urethra as part of the semen. (932)

prostatic urethra　(pros-tat′ik) Part of the male urethra, approximately 2.5 cm in length, that passes through the prostate gland. (930)

protease　(prō′te-ās) Enzyme that breaks down proteins. (49)

protein　(prō′tēn, prō′tē-ĭn) [Gr. *proteios,* primary] Macromolecule consisting of long sequences of amino acids linked together by peptide bonds. (46)

proteoglycan　(prō′tē-ō-glī′kan) Macromolecule consisting of numerous polysaccharides attached to a common protein core. (113)

prothrombin　(prō-throm′bin) Glycoprotein present in blood that, in the presence of prothrombin activator, is converted to thrombin. (596)

proton　(prō′ton) [Gr. *protos,* first] Positively charged particle in the nuclei of atoms. (24)

protraction　(prō-trak′shŭn) [L. *protractus,* to draw forth] Movement forward or in the anterior direction. (232)

provitamin　(prō-vī′tă-min) Substance that may be converted into a vitamin. (836)

proximal tubule　Part of the nephron that extends from the glomerulus to the descending limb of the loop of Henle. (864)

pseudostratified epithelium　Epithelium consisting of a single layer of cells but having the appearance of multiple layers. (104)

psychologic fatigue　(sī-kō-loj′-ik) Fatigue caused by the central nervous system. (293)

ptosis　(tō′sis) [G. *ptosis,* a falling] Falling down of an organ, for example, drooping of the upper eyelid. (312)

puberty　(pyū′ber-tē) [L. *pubertas,* grown up] Series of events that transform a child into a sexually mature adult; involves an increase in the secretion of GnRH. (935)

pubis　(pyū′bis) Anterior inferior bone of the coxa. (214)

pudendal cleft　(pyū-den′dal) Cleft between the labia majora. (943)

pudendum　(pyū-den′dum) See vulva.

pulmonary artery　(pŭl′mō-năr-ē) One of the arteries that extend from the pulmonary trunk to the right or left lungs. (618)

pulmonary capacity　Sum of two or more pulmonary volumes. (756)

pulmonary trunk　Large elastic artery that carries blood from the right ventricle of the heart to the right and left pulmonary arteries. (611)

pulmonary vein　One of the veins that carry blood from the lungs to the left atrium of the heart. (611)

pulp　(tooth) (pŭlp) [L. *pulpa,* flesh] The soft tissue within the pulp cavity, consisting of connective tissue containing blood vessels, nerves, and lymphatics. (788)

pulse pressure　(pŭls) Difference between systolic and diastolic pressure. (677)

pupil　(pyŭ′pĭl) Circular opening in the iris through which light enters the eye. (476)

Purkinje fiber　(pŭr-kĭn′jē) Named for the Bohemian anatomist, Johannes E. von Purkinje (1787–1869). Modified cardiac muscle cells found beneath the endocardium of the ventricles. Specialized to conduct action potentials. (620)

pus　(pŭs) Fluid product of inflammation; contains leukocytes, the debris of dead cells, and tissue elements liquefied by enzymes. (130)

pyloric opening　(pī-lōr′ik) Opening between the stomach and the superior part of the duodenum. (791)

pyloric sphincter　Thickening of the circular layer of the gastric musculature encircling the junction between the stomach and duodenum. (791)

pyrogen　(pī′rō-jen) Chemical released by microorganisms, neutrophils, monocytes, and other cells that stimulates fever production by acting on the hypothalamus. (714)

Q

QRS complex　Principle deflection in the electrocardiogram, representing ventricular depolarization. (623)

QT interval　Time elapsing from the beginning of the QRS complex to the end of the T wave, representing the total duration of electrical activity of the ventricles. (623)

R

radial pulse　Pulse detected in the radial artery. (678)

radiation　(rā′dē-ā′shŭn) [L. *radius,* ray, beam] The sending forth of light, short radiowaves, ultraviolet or x-rays, or any other rays for treatment or diagnosis or for other reasons; radiant heat. (855)

radioactive isotope　(rā′dē-ō-ak′tiv) Isotope with a nuclear composition that is unstable from which subatomic particles and electromagnetic waves are emitted. (33)

ramus, pl. rami　(rā′mŭs, rā′mī) [L., branch] One of the primary subdivisions of a nerve or blood vessel. The part of a bone that forms an angle with the main body of the bone. (189)

raphe　(rā′fē) [Gr., *rhaphe,* suture, seam] Central line running over the scrotum from the anus to the root of the penis. (923)

recessive (rē-ses′iv) In genetics, a gene that may not be expressed because of suppression by a contrasting dominant gene. (993)

rectum (rek′tŭm) [L. *rectus,* straight] Portion of the digestive tract that extends from the sigmoid colon to the anal canal. (799)

red pulp [L. *pulpa,* flesh] Reddish brown substance of the spleen consisting of venous sinuses and the tissues intervening between them called pulp cords. (706)

reduction Gain of one or more electrons by a molecule. (33)

refraction (rē-frak-shŭn) Bending of a light ray when it passes from one medium into another of different density. (478)

refractory period (rē-frak′tōr-ē) [Gr. *periodos,* a way around, a cycle] Period following effective stimulation during which excitable tissue such as heart muscle fails to respond to a stimulus of threshold intensity. (267)

regeneration (rē′jen-er-ā′shŭn) Reproduction or reconstruction of a lost or injured part. (129)

regulatory gene Gene involved with controlling the activity of structural genes. (992)

relative refractory period Portion of the action potential following the absolute refractory period during which another action potential can be produced with a greater-than-threshold stimulus strength. (267)

relaxation phase Phase of muscle contraction following the contraction phase; the time from maximal tension production until tension decreases to its resting level. (288)

renal artery (rē′năl) Originates from the aorta and delivers blood to the kidney. (864)

renal blood flow rate Volume at which blood flows through the kidneys per minute; an average of approximately 1200 mL/min. (869)

renal column Cortical substance separating the renal pyramids. (861)

renal corpuscle Glomerulus and Bowman's capsule that encloses it. (862)

renal fascia Connective tissue surrounding the kidney that forms a sheath or capsule for the organ. (861)

renal fat pad Fat layer that surrounds the kidney and functions as a shock-absorbing material. (861)

renal fraction Portion of the cardiac output that flows through the kidneys; averages 21%. (868)

renal pelvis Funnel-shaped expansion of the upper end of the ureter that receives the calyces. (861)

renal pyramid One of a number of pyramidal masses seen on longitudinal section of the kidney; they contain part of the loops of Henle and the collecting tubules. (861)

renin (rē′nin) Enzyme secreted by the juxtaglomerular apparatus that converts angiotensinogen to angiotensin I. (689)

renin–angiotensin–aldosterone mechanism Renin, released from the kidneys in response to low blood pressure, converts angiotensinogen to angiotensin I. Angiotensin I is converted by angiotensin-converting enzyme to angiotensin II, which causes vasoconstriction, resulting in increased blood pressure. Angiotensin II also increases aldosterone secretion, which increases blood pressure by increasing blood volume. (689)

repolarization (re′pō-lăr-i-zā′shŭn) Phase of the action potential in which the membrane potential moves from its maximum degree of depolarization toward the value of the resting membrane potential. (264)

reposition Return of a structure to its original position. (236)

residual volume (rē-zid′yū-ăl) Volume of air remaining in the lungs after a maximum expiratory effort. (756)

resolution (rez-ō-lū′shun) [L. *resolutio,* a slackening] Phase of the male sexual act after ejaculation during which the penis becomes flaccid; feeling of satisfaction; inability to achieve erection and second ejaculation. Last phase of the female sexual act, characterized by an overall sense of satisfaction and relaxation. (936)

respiration (res-pi-rā′shŭn) [L. *respiratio,* to exhale, breathe] Process of life in which oxygen is used to oxidize organic fuel molecules, providing a source of energy, carbon dioxide, and water. (82) Movement of air into and out of the lungs, the exchange of gases with blood, the transportation of gases in the blood, and gas exchange between the blood and the tissues. (738)

respiratory bronchiole (res′pi-ră-tōr-ē, rĕ-spīr′ă-tōr-ē) Smallest bronchiole (0.5 mm in diameter) that connects the terminal bronchiole to the alveolar duct. (745)

respiratory membrane Membrane in the lungs across which gas exchange occurs with blood. (758)

resting membrane potential Electric charge difference inside a cell membrane, measured relative to just outside the cell membrane. (259)

reticular (rē-tik′yū-lăr) [L. *rete,* net] Relating to a fine network of cells or collagen fibers. (113)

reticular cell Cell with processes making contact with those of other similar cells to form a cellular network; along with the network of reticular fibers, the reticular cells form the framework of bone marrow and lymphatic tissues. (122)

reticulocyte (rē-tik′yū-lō-sīt) Young red blood cell with a network of basophilic endoplasmic reticulum occurring in larger numbers during the process of active red blood cell synthesis. (589)

reticuloendothelial system (re-tik′yū-lō-en-dō-thē′lē-ăl) See mononuclear phagocytic system. (712)

retina (ret′i-nă) Nervous tunic of the eyeball. (476)

retinaculum (ret-i-nak′yū-lŭm) [L., band, halter, to hold back] Dense, regular connective tissue sheath holding down the tendons at the wrist, ankle, or other sites. (338)

retraction (rē-trak′shŭn) [L., *retractio,* a drawing back] Movement in the posterior direction. (232)

retroperitoneal (re′trō-per′i-tō-nē′ăl) Behind the peritoneum. (17)

rhodopsin (rō-dop′sin) Light-sensitive substance found in the rods of the retina; composed of opsin loosely bound to retinal. (480)

ribonuclease (rī-bō-nū′klē-ās) Enzyme that splits RNA into its component nucleotides. (815)

ribonucleic acid (rī′bō-nū′klē′ik) Nucleic acid containing ribose as the sugar component; found in all cells in both nuclei and cytoplasm; helps direct protein synthesis. (50)

ribosomal RNA (rRNA) (rī′bō-sōm-ăl) RNA that is associated with certain proteins to form ribosomes. (64)

ribosome (ri′bō-sōm) Small, spherical, cytoplasmic organelle where protein synthesis occurs. (64)

right lymphatic duct Lymphatic duct that empties into the right subclavian vein; drains the right side of the head and neck, the right upper thorax, and the right upper limb. (672)

rigor mortis (rig′er mōr′tĭs) Increased rigidity of muscle after death due to cross-bridge formation between actin and myosin as calcium ions leak from the sarcoplasmic reticulum. (295)

rod Photoreceptor in the retina of the eye; responsible for noncolor vision in low-intensity light. (476)

root of the penis Proximal attached part of the penis, including the two crura and the bulb. (931)

root of the tooth That part below the neck of a tooth covered by cementum rather than enamel and attached by the periodontal ligament to the alveolar bone. (788)

rotation (rō-tā′shun) Movement of a structure about its axis. (232)

rotator cuff muscle (rō-tā′ter, -tor) One of four deep muscles that attach the humerus to the scapula. (336)

round ligament Fibromuscular band that is attached to the uterus on either side in front of and below the opening of the uterine tube; it passes through the inguinal canal to the labium majus. (941)

round ligament of the liver Remains of the umbilical vein. (985)

round window Membranous structure separating the scala tympani of the inner ear from the middle ear. (493)

Ruffini's end-organ (roo-fē′nēz) Named for Angelo Ruffini, Italian histologist (1864–1929); receptor located deep in the dermis and responding to continuous touch or pressure. (465)

ruga, pl. rugae (rū′gă, rū′gē) [L., a wrinkle] Fold or ridge; fold of the mucous membrane of the stomach when the organ is contracted; transverse ridge in the mucous membrane of the vagina. (942)

S

saccule (sak′yūl) Part of the membranous labyrinth; contains sensory structure, the macula, that detects static equilibrium. (499)

salivary amylase (sal′i-vār-ē am′il-ās) Enzyme secreted in the saliva that breaks down starch to maltose and isomaltose. (803)

salivary gland Gland that produces and secretes saliva into the oral cavity. The three major pairs of salivary glands are the parotid, submandibular, and sublingual glands. (782)

salt Molecule consisting of a cation other than hydrogen and an anion other than hydroxide. (39)

sarcolemma (sar′kō-lem′ă) [Gr. sarco, muscle + lemma, husk] Plasma membrane of a muscle fiber. (277)

sarcoma (sar-kō′mă) A malignant neoplasm derived from connective tissue. (132)

sarcomere (sar′kō-mēr) [Gr. sarco, muscle + meros, part] Part of a myofibril between adjacent Z disks. (278)

sarcoplasm (sar′kō-plazm) [Gr. sarco, muscle + plasma, a thing formed] Cytoplasm of a muscle fiber, excluding the myofilaments. (278)

sarcoplasmic reticulum (sar′kō-plaz′mik) [Gr. sarco, muscle + plasma, a thing formed + reticulum, net] Endoplasmic reticulum of muscle. (280)

satellite cell (sat′ĕ-līt) Specialized cell that surrounds the cell bodies of neurons within ganglia. (366)

saturated (satch′ŭ-rāt-ĕd) Fatty acid in which the carbon chain contains only single bonds between carbon atoms. (44)

saturation (satch-ŭ-rā′shun) Point when all carrier molecules or enzymes are attached to substrate molecules and no more molecules can be transported or reacted. (79)

scala tympani (skā′lă tim′pă-nī) [L., stairway] Division of the spiral canal of the cochlea lying below the spiral lamina and basilar membrane. (493)

scala vestibuli (skā′lă ves-tib′yū-lī) Division of the cochlea lying above the spiral lamina and vestibular membrane. (493)

scapula (skap′yū-lă) Bone forming the shoulder blade. (210)

scar (skar) [Gr. eschara, scab] Fibrous tissue replacing normal tissue; cicatrix. (131)

sciatic nerve (sī-at′ik) Tibial and common peroneal nerves bound together. (452)

sclera (sklēr′ă) White of the eye; white, opaque portion of the fibrous tunic of the eye. (473)

scrotum, pl. scrota, scrotums (skrō′tum, -tă, -tŭmz) Musculocutaneous sac containing the testes. (923)

sebaceous gland (se-ba′shus) [L. sebum, tallow] Gland of the skin, usually associated with a hair follicle, that produces sebum. (146)

second messenger See intracellular mediator.

secondary bronchus (brong′kus) Branch from a primary bronchus that conducts air to each lobe of the lungs. There are two branches in the left lung and three branches from the primary bronchus in the right lung. (743)

secondary follicle Follicle in which the secondary oocyte is surrounded by granulosa cells at the periphery; contains fluid-filled antral spaces. (938)

secondary (memory) response Immune response that occurs when the immune system is exposed to an antigen against which it has already produced a primary response. (723)

secondary oocyte (ō′ō-sīt) Ocyte in which the second meiotic division stops at metaphase II unless fertilization occurs. (938)

secondary palate Roof of the mouth in the early embryo that gives rise to the hard and the soft palates. (970)

secondary spermatocyte (sper′mă-tō-sīt) Spermatocyte derived from a primary spermatocyte by the first meiotic division; each secondary spermatocyte gives rise by the second meiotic division to two spermatids. (927)

secretin (se-krē′tin) Hormone formed by the epithelial cells of the duodenum; stimulates secretion of pancreatic juice high in bicarbonate ions. (807)

secretion (se-krē′shun) General term for a substance produced inside a cell and released from the cell. (784)

secretory phase (se-krēt′ĕ-rē, sē′krĕ-tōr-ē) See luteal phase.

segmental artery One of five branches of the renal artery, each supplying a segment of the kidney. (864)

self-antigen Antigen produced by the body that is capable of initiating an immune response against the body. (714)

semen (sē′men) [L., seed (of plants, men, animals)] Penile ejaculate; thick, yellowish white, viscous fluid containing spermatozoa and secretions of the testes, seminal vesicles, prostate, and bulbourethral glands. (933)

semicircular canal (sem′ē-sir′kyū-lăr) Canal in the petrous portion of the temporal bone that contains sensory organs that detect kinetic or dynamic equilibrium. Three semicircular canals are within each inner ear. (492)

seminal vesicle (sem′i-năl) One of two glandular structures that empty into the ejaculatory ducts; its secretion is one of the components of semen. (932)

seminiferous tubule (sem′i-nif′er-ŭs) Tubule in the testis in which spermatozoa develop. (923)

sensible perspiration (sen′si-bl pers-pi-rā′shŭn) Perspiration excreted by the sweat glands that appears as moisture on the skin; produced in large quantity when there is much humidity in the atmosphere. (912)

sensory retina (sen′sŏ-rē) Portion of the retina containing rods and cones. (476)

septum primum (sep′tŭm prī′mŭm) First septum in the embryonic heart that arises on the wall of the originally single atrium of the heart and separates it into right and left chambers. (974)

septum secundum (sek′ŭn-dŭm) Second of two major septal structures involved in the partitioning of the atrium, arising later than the septum primum and located to the right of it; it remains an incomplete partition until after birth, with its unclosed area constituting the foramen ovale. (974)

serosa   (se-rō′să) [L. *serosus,* serous] Outermost covering of an organ or structure that lies in a body cavity; see also adventitia. (783)

serous fluid   (ser′ŭs) Fluid similar to lymph that is produced by and covers serous membrane; lubricates the serous membrane. (128)

serous membrane   Thin sheet composed of epithelial and connective tissues; lines cavities that do not open to the outside of the body or contain glands but do secrete serous fluid. (17)

serous pericardium   Lining of the pericardial sac composed of a serous membrane. (610)

Sertoli cell   (ser-tō′lē) Named for the Italian histologist, Enrico Sertoli (1842–1910). Elongated cell in the wall of the seminiferous tubules to which spermatids are attached during spermatogenesis. (926)

serum   (sēr′ŭm) [L., whey] Fluid portion of blood after the removal of fibrin and blood cells. (598)

sesamoid bone   (ses′ă-moyd) [Gr. *sesamoceies,* like a sesame seed] Bone found within a tendon; such as the patella. (214)

sex chromosomes   Pair of chromosomes responsible for sex determination; XX in female and XY in male. (991)

sex-linked trait   Characteristic resulting from the expression of a gene on a sex chromosome. (994)

sigmoid colon   (sig′moyd) Part of the colon between the descending colon and the rectum. (798)

sigmoid mesocolon   Fold of peritoneum attaching the sigmoid colon to the posterior abdominal wall. (801)

simple epithelium   Epithelium consisting of a single layer of cells. (103)

sinoatrial (SA) node   (si′nō-ā′trē-ăl) Mass of specialized cardiac muscle fibers; acts as the "pacemaker" of the cardiac conduction system. (619)

sinus   (sī′nŭs) [L., cavity] Hollow in a bone or other tissue; enlarged channel for blood or lymph. (160)

sinus venosus   End of the embryonic cardiac tube where blood enters the heart; becomes a portion of the right atrium, including the SA node. (974)

sinusoid   (si′nŭ-soyd) [L. sinus + Gr. *eidos,* resemblance] Terminal blood

vessel having a larger diameter than an ordinary capillary. (648)

sinusoidal capillary   (si-nŭ-soy′dăl) Capillary with caliber of from 10 to 20 μm or more; lined with a fenestrated type of endothelium. (647)

small intestine   [L. *intestinus,* the entrails] Portion of the digestive tube between the stomach and the cecum; consists of the duodenum, jejunum, and ileum. (793)

sodium–potassium exchange pump   Biochemical mechanism that uses energy derived from ATP to achieve the active transport of potassium ions opposite to that of sodium ions. (257)

soft palate   Posterior muscular portion of the palate, forming an incomplete septum between the mouth and the oropharynx and between the oropharynx and the nasopharynx. (789)

solute   (sol′yūt, sō′lūt) [L. *solutus,* dissolved] Dissolved substance in a solution. (38)

solution   (sō-lū′shun) [L. *solutio*] Homogenous mixture formed when a solute is dissolved in a solvent. (38)

solvent   (sol′vent) [L. *solvens,* to dissolve] Liquid that holds another substance in solution. (38)

soma   (sō′mă) [Gr., body] Neuron cell body or the enlarged portion of the neuron containing the nucleus and other organelles. (362)

somatic   (sō-mat′ik) [Gr. *somatikos,* bodily] Relating to the body; the cells of the body except the reproductive cells. (92)

somatic nervous system   Composed of nerve fibers that send impulses from the central nervous system to skeletal muscle. (360)

somatomedin   (sō′mă-tō-mē′din) Peptide synthesized in the liver capable of stimulating certain anabolic processes in bone and cartilage such as synthesis of DNA, RNA, and protein. (553)

somatotropin   (sō′mă-tō-trō′pin) Protein hormone of the anterior pituitary gland; it promotes body growth, fat mobilization, and inhibition of glucose utilization. (553)

somite   (sō′mīt) [Gr. *soma,* body + *ite*] One of the paired segments consisting of cell masses formed in the early embryonic mesoderm on either side of the neural tube. (966)

somitomere   (sō′mīt-ō-mēr) An indistinct somite in the head region of the embryo. (966)

spatial summation   Summation of the local potentials in which two or more

action potentials arrive simultaneously at two or more presynaptic terminals that synapse with a single neuron. (373)

specific heat   Heat required to raise the temperature of any substance 1°C compared with the heat required to raise the same volume of water 1°C. (38)

speech   Use of the voice in conveying ideas. (788)

spermatic cord   (sper-mat′ik) Cord formed by the ductus deferens and its associated structures; extends through the inguinal canal into the scrotum. (928)

spermatid   (sper′mă-tid) [Gr. *sperma,* seed + *id*] Cell derived from the secondary spermatocyte; gives rise to a spermatozoon. (927)

spermatogenesis   (sper′mă-tō-jen′ĕ-sis) Formation and development of the spermatozoon. (926)

spermatogonium   (sper′mă-tō-gō′nē-um) [Gr. *sperma,* seed + *gone,* generation] Cell that divides by mitosis to form primary spermatocytes. (926)

spermatozoon, pl. spermatozoa   (sper′mă-tō-zō′on, sper′mă-to-zō′ă) [Gr. *sperma,* seed + *zoon,* animal] Sperm cell. Male gamete or sex cell, composed of a head and a tail. The spermatozoon contains the genetic information transmitted by the male. (927)

sphenoid   (sfē′noyd) [Gr. *shen,* wedge] Wedge-shaped. (189)

sphincter pupillae   (sfingk′ter pyū-pil′ē) Circular smooth muscle fibers of the iris' diaphragm that constrict the pupil of the eye. (476)

sphygmomanometer   (sfig′mō-mă-nom′ĕ-ter) [Gr. *sphygmos,* pulse + *manos,* thin, scanty + *metron,* measure] Instrument for measuring blood pressure. (674)

spinal nerve   (spi′năl) One of 31 pairs of nerves formed by the joining of the dorsal and ventral roots that arise from the spinal cord. (442)

spindle fiber   (spin′dl) Specialized microtubule that develops from each centrosome and extends toward the chromosomes during cell division. (71)

spiral artery   (spi′răl) One of the corkscrewlike arteries in premenstrual endometrium; most obvious during the secretory phase of the uterine cycle. (948)

spiral ganglion   Cell bodies of sensory neurons that innervate hair cells of the organ of Corti are located in the spiral ganglion. (493)

spiral lamina   Attached to the modiolus and supports the basilar and vestibular membranes. (493)

spiral ligament   Attachment of the basilar membrane to the lateral wall of the bony labyrinth. (493)

spiral organ   Organ of Corti; rests on the basilar membrane and consists of the hair cells that detect sound. (493)

spiral tubular gland   Well-developed simple or compound tubular glands that are spiral in shape within the endometrium of the uterus; prevalent in the secretory phase of the uterine cycle. (948)

spirometer   (spī-rom′ĕ-ter) [L. *spiro*, to breathe + Gr. *metron*, measure] Gasometer used for measuring the volume of respiratory gases; usually understood to consist of a counterbalanced cylindrical bell sealed by dipping into a circular trough of water. (755)

spirometry   (spī-rom′ĕ-trē) Making pulmonary measurements with a spirometer. (755)

spleen   (splēn) Large lymphatic organ in the upper part of the abdominal cavity on the left side between the stomach and diaphragm, composed of white and red pulp. It responds to foreign substances in the blood, destroys worn out erythrocytes, and is a storage site for blood cells. (706)

spongy urethra   Portion of the male urethra, approximately 15 cm in length, that traverses the corpus spongiosum of the penis. (930)

squamous   (skwā′mŭs) [L. *squama*, a scale] Scalelike, flat. (104)

stapedius   (stă-pē′dē-ŭs) Small skeletal muscles attached to the stapes. (495)

stapes   (stā′pēz) [L., stirrup] Smallest of the three auditory ossicles; attached to the oval window. (491)

Starling's law of the heart   Named for the English physiologist, Ernest H. Starling (1866–1927). Force of contraction of cardiac muscle is a function of the length of its muscle fibers at the end of diastole; the greater the ventricular filling, the greater the stroke volume produced by the heart. (633)

sternum   (ster′nŭm) [L. *sternon*, chest] Breast bone. (210)

steroid   (stēr′oyd, ster′oyd) Large family of lipids, including some reproductive hormones, vitamins, and cholesterol. (45)

stomach   (stŭm′ŭk) Large sac between the esophagus and the small intestine, lying just beneath the diaphragm. (791)

stratified epithelium   (strat′i-fīd ep-i-thē′lē-ŭm) Epithelium consisting of more than one layer of cells. (103)

stratum basale   (strat-ŭm băh-săl′ē) [L., layer; basal] Basal or deepest layer of the epidermis. (139)

stratum corneum   (kōr′nē-ŭm) [L., layer; *corneus*, horny] Most superficial layer of the epidermis consisting of flat, keratinized, dead cells. (140)

stratum granulosum   (gran′yū-lō′sŭm) [L., layer; granulum] Layer of cells in the epidermis filled with granules of keratohyalin. (139)

stratum lucidum   (lū′sid-ŭm) [L., layer; *lucidus*, clear] Clear layer of the epidermis found in thick skin between the stratum granulosum and the stratum corneum. (140)

stratum spinosum   (spī′nōs-ŭm) [L., layer; *spina*, spine] Layer of many-sided cells in the epidermis with intercellular connections (desmosomes) that give the cells a spiny appearance. (139)

stria, pl. striae   (strī′ă, strī′ē) [L., channel] Line or streak in the skin that is a different texture or color from the surrounding skin. Stretch mark. (137)

striated   (strī′āt-ĕd) [L. *striatus*, furrowed] Striped; marked by stripes or bands. (277)

stroke volume   [L. *volumen*, something rolled up, scroll, from *volvo*, to roll] Volume of blood pumped out of one ventricle of the heart in a single beat. (631)

structural gene   Gene with the function of determining the structure of a specific protein or peptide. (992)

sty   (stī) Inflamed ciliary gland of the eye. (470)

subcutaneous   (sŭb′kyū-tā′nē-ŭs) [L. *sub*, under + *cutis*, skin] Under the skin; same tissue as the hypodermis. (137)

sublingual gland   (sŭb-ling′gwăl) One of two salivary glands in the floor of the mouth beneath the tongue. (790)

submandibular gland   (sŭb-man-dib′yū-lăr) One of two salivary glands in the neck, located in the space bounded by the two bellies of the digastric muscle and the angle of the mandible. (790)

submucosa   (sŭb-mū-kō′să) Layer of tissue beneath a mucous membrane. (783)

submucosal plexus   (sŭb-mū-kō′săl) [L., a braid] Gangliated plexus of unmyelinated nerve fibers in the intestinal submucosa. (783)

substantia nigra   (sŭb-stan′shē-ă nī′gră) [L., substance; black] Black nuclear mass in the midbrain; involved in coordinating movement and maintaining muscle tone. (388)

subthreshold stimulus   Stimulus resulting in a local potential so small that it does not reach threshold and produce an action potential. (270)

sucrose   (sū′krōs) Disaccharide composed of glucose and fructose; table sugar. (42)

sulcus, pl. sulci   (sūl′kŭs, sul′si) [L., furrow or ditch] Furrow or groove on the surface of the brain between the gyri; may also refer to a fissure. (391)

superficial inguinal ring   (ing′gwi-năl) Slitlike opening in the aponeurosis of the external oblique muscle of the abdominal wall through which the spermatic cord (round ligament in the female) emerges from the inguinal canal. (930)

superior colliculus   (ko-lik′yū-lŭs) [L. *collis*, hill] One of two rounded eminences of the midbrain; aids in coordination of eye movements. (484)

superior vena cava   (vē′nă că′vă) Vein that returns blood from the head and neck, upper limbs, and thorax to the right atrium. (611)

supination   (sū′pi-nā′shun) [L. *supino*, to bend backward, place on back] Rotation of the forearm (when the forearm is parallel to the ground) so that the anterior surface is up (supine). (232)

supramaximal stimulus   Stimulus of greater magnitude than a maximal stimulus; however, the frequency of action potentials is not increased above that produced by a maximal stimulus. (270)

suppressor T cell   Subset of T lymphocytes that decreases the activity of B cells and T cells. (714)

surfactant   (ser-fak′tănt) Lipoproteins forming a monomolecular layer over pulmonary alveolar surfaces; stabilizes alveolar volume by reducing surface tension and the tendency for the alveoli to collapse. (752)

suspension   (sŭs-pen′shŭn) Liquid through which a solid is dispersed and from which the solid separates unless the liquid is kept in motion. (38)

suspensory ligament   (sŭs-pen′sŏ-rē) Band of peritoneum that extends from the ovary to the body wall; contains the ovarian vessels and nerves. Small ligament attached to the margin of the lens in the eye and the ciliary body to hold the lens in place. (475)

suture   (sū′chūr) [L. *sutura*, a seam] Junction between flat bones of the skull. (184)

sweat   (swet) [A.S. *swat*] Perspiration; secretions produced by the sweat glands of the skin. See also sensible and insensible perspiration. (903)

sweat gland   Usually means structures that produce a watery secretion called sweat. Some sweat glands, however, produce viscous organic secretions. (146)

sympathetic chain ganglion   (sim-pă-thet′ik) Collection of sympathetic postganglionic neurons that are connected to each other to form a chain along both sides of the spinal cord. (509)

sympathetic division   Subdivision of the autonomic division of the nervous system characterized by having the cell bodies of its preganglionic neurons located in the thoracic and upper lumbar regions of the spinal cord (thoracolumbar division); usually involved in preparing the body for physical activity. (509)

symphysis, pl. symphyses   (sim′fi-sis, sim′fă-sēz) [Gr., a growing together] Fibrocartilage joint between two bones. (228)

synapse, pl. synapses   (sin′aps, sǐ-naps′, sǐ-nap′sēz) [Gr. syn, together + haptein, to clasp] Functional membrane-to-membrane contact of a nerve cell with another nerve cell, muscle cell, gland cell, or sensory receptor; functions in the transmission of action potentials from one cell to another. (268)

synaptic cleft   (si-nap′tik) Space between the presynaptic and the postsynaptic membranes. (283)

synaptic fatigue   Fatigue due to depletion of neurotransmitter vesicles in the presynaptic terminals. (294)

synaptic vesicle   Secretory vesicle in the presynaptic terminal containing neurotransmitter substances. (283)

synchondrosis, pl. synchondroses   (sin′kon-drō′sis, -sēz) [Gr. syn, together + chondros, cartilage + osis, condition] Union between two bones formed by hyaline cartilage. (226)

syncytiotrophoblast   (sin-sish′ē-ō-trō′fō-blast) Outer layer of the trophoblast composed of multinucleated cells. (963)

syndesmosis, pl. syndesmoses   (sin′dez-mō′sis, -sēz) [Gr. syndeo, to bind + osis, condition] Form of fibrous joint in which opposing surfaces that are some distance apart are united by ligaments. (226)

synergist   (sin′er-jist) Muscle that works with other muscles to cause a movement. (309)

synovial   (si-nō′vē-ăl) [Gr. syn, together + oon, egg] Relating to or containing synovia (a substance that serves as a lubricant in a joint, tendon sheath, or bursa). (229)

synovial fluid   Slippery fluid found inside synovial joints and bursae; produced by the synovial membranes. (229)

systemic inflammation   (sis-tem′ik) Inflammation that occurs in many areas of the body. In addition to symptoms of local inflammation, increased neutrophil numbers in the blood, fever, and shock can occur. (713)

systole   (sis′tō-lē) [Gr. systole, a contracting] Contraction of the heart chambers during which blood leaves the chambers; usually refers to ventricular contraction. (625)

T

T cell   Thymus-derived lymphocyte of immunologic importance; it is of long life and is responsible for cell-mediated immunity. (714)

T tubule   (tū′byul) Tubelike invagination of the sarcolemma that conducts action potentials toward the center of the cylindrical muscle fibers. (278)

T wave   Deflection in the electrocardiogram following the QRS complex, representing ventricular repolarization. (623)

tactile corpuscle   (tak′til) Oval receptor found in the papillae of the dermis; responsible for fine, discriminative touch; Meissner's corpuscle. (465)

tactile disk   Cuplike receptor found in the epidermis; responsible for light touch and superficial pressure; Merkel's disk. (465)

talus   (tā′lŭs) [L., ankle bone, heel] Tarsal bone contributing to the ankle. (220)

target tissue   Tissue on which a hormone acts. (527)

tarsal   (tar′sal) [Gr. tarsos, sole of foot] One of seven ankle bones. (220)

tarsal plate   (tar′săl) Crescent-shaped layer of connective tissue that helps maintain the shape of the eyelid. (470)

taste   (tāst) Sensations created when a chemical stimulus is applied to the taste receptors in the tongue. (468)

taste bud   Sensory structure, mostly on the tongue, that functions as a taste receptor. (468)

tectum   (tek′tŭm) Roof of the midbrain. (386)

tegmentum   (teg-men′tŭm) Floor of the midbrain. (388)

telencephalon   (tel-en-sef′ă-lon) [Gr. telos, end + enkephalos, brain] Anterior division of the embryonic brain from which the cerebral hemispheres develop. (384)

telophase   (tel′ō-fāz) Time during cell division when the chromosomes are pulled by spindle fibers away from the cell equator and into the two halves of the dividing cell. (92)

temporal summation   Summation of the local potential that results when two or more action potentials arrive at a single synapse in rapid succession. (263)

tendon   Band or cord of dense connective tissue that connects a muscle to a bone or other structure. (309)

tensor tympani   (ten′sŏr tim′pa-nī) Small skeletal muscle attached to the malleus. (495)

tentorium cerebelli   (ten-tō′rē-ŭm ser′ĕ-bel′ī) Dural folds between the cerebrum and the cerebellum. (420)

terminal bouton   (bū-ton′) [Fr., button] Enlarged axon terminal or presynaptic terminal. (363)

terminal cisterna   (sis-ter′nă) [L. terminus, limit + cista, box] Enlarged end of the sarcoplasmic reticulum in the area of the T tubules. (280)

terminal hair   [L. terminus, a boundary, limit] Long, coarse, usually pigmented hair found in the scalp, eyebrows, and eyelids and replacing vellus hair. (144)

terminal sulcus   (sŭl′kŭs) [L., furrow or ditch] V-shaped groove on the surface of the tongue at the posterior margin. (787)

tertiary bronchus   (brong′kŭs) Extends from the secondary bronchus and conducts air to each lobule of the lungs. (743)

testis, pl. testes   (tes′tis, tes′tēz) One of two male reproductive glands located in the scrotum; produces spermatozoa, testosterone, and inhibin. (923)

testosterone   (tes-tos′tĕ-rōn) Steroid hormone secreted primarily by the testes; aids in spermatogenesis, maintenance and development of male reproductive organs, secondary sexual characteristics, and sexual behavior. (934)

tetraiodothyronine (T₄)   (tet′ră-ī-ō′dō-thī′rō-nēn) One of the iodine-containing thyroid hormones; also called thyroxine. (556)

thalamus   (thal′ă-mŭs) [Gr. thalamos, a bed, a bedroom] Large mass of gray matter that forms the larger dorsal subdivision of the diencephalon. (390)

theca   (thē′kă) [Gr. theke, a box] Sheath or capsule. (938)

theca externa   External fibrous layer of the theca or capsule of a vesicular follicle. (938)

theca interna Inner vascular layer of the theca or capsule of the secondary and mature follicle; produces estrogen and contributes to the formation of the corpus luteum after ovulation. (938)

thenar (thē′nar) [Gr., palm of the hand] Fleshy mass of tissue at the base of the thumb; contains muscles responsible for thumb movements. (339)

thick skin Found in the palms, soles, and tips of the digits and has all five epidermal strata. (140)

thin skin Found over most of the body, usually without a stratum lucidum, and has fewer layers of cells than thick skin. (140)

thoracic cavity (thō-ras′ik) Space within the thoracic walls, bounded below by the diaphragm and above by the neck. (745)

thoracic duct Largest lymph vessel in the body, beginning at the cisterna chyli and emptying into the left subclavian vein; drains the left side of the head and neck, the left upper thorax, the left upper limb, and the inferior half of the body. (672)

thoracolumbar division (thōr′ă-kō-lŭm′bar) Synonym for the sympathetic division of the autonomic nervous system. (509)

thoroughfare channel Channel for blood through a capillary bed from an arteriole to a venule. (648)

threshold potential Value of the membrane potential at which an action potential is produced as a result of depolarization in response to a stimulus. (264)

threshold stimulus Stimulus resulting in a local potential just large enough to reach threshold and produce an action potential. (270)

thrombocyte (throm′bō-sīt) Platelet. (594)

thrombocytopenia (throm′bō-sī′-tō-pē′nē-ă) [*thrombocyte* + Gr. *penia,* poverty] Condition in which there is an abnormally small number of platelets in the blood. (603)

thromboxane (throm′bok-zān) Specific class of physiologically active fatty acid derivatives present in many tissues. (575)

thrombus, pl. thrombi (throm′bŭs, -bī) [Gr. *thrombos,* a clot] Clot in the cardiovascular system formed from constituents of blood; may be occlusive or attached to the vessel or heart wall without obstructing the lumen. (990)

thymus gland (thī′mŭs) [Gr. *thymos,* sweetbread] Bilobed lymph organ located in the inferior neck and superior mediastinum; secretes the hormone thymosin. (574)

thyroid cartilage (thī′royd) Largest laryngeal cartilage. It forms the laryngeal prominence, or Adam's apple. (739)

thyroid gland [Gr. *thyreoeides,* shield] Endocrine gland located inferior to the larynx and consisting of two lobes connected by the isthmus; secretes the thyroid hormones triiodothyronine (T_3) and tetraiodothyronine (T_4). (555)

thyroid-stimulating hormone (TSH) Glycoprotein hormone released from the hypothalamus; stimulates thyroid hormone secretion from the thyroid gland. (553)

thyrotropin (thī-rot′rō-pin, thī-rō-trō′pin) See thyroid-stimulating hormone.

thyroxine (thi-rok′sēn, -sin) See tetraiodothyronine.

tidal volume Volume of air that is inspired or expired in a single breath during regular, quiet breathing. (755)

tissue repair Substitution of viable cells for damaged or dead cells by regeneration or replacement. (129)

tolerance Failure of the specific immune system to respond to an antigen. (719)

tongue (tŭng) Muscular organ occupying most of the oral cavity when the mouth is closed; major attachment is through its posterior portion. (787)

tonsil, pl. tonsils (ton′sil, ton′silz) [L. *tonsilla,* stake] Any collection of lymphoid tissue; usually refers to large collections of lymphatic tissue beneath the mucous membrane of the oral cavity and pharynx; lingual, pharyngeal, and palatine tonsils. (703)

total lung capacity Volume of air contained in the lungs at the end of a maximum inspiration; equals vital capacity plus residual volume. (756)

total tension Sum of active and passive tension. (293)

trabecula, pl. trabeculae (tră-bek′yū-lă, -lē) [L. *trabs,* beam] One of the supporting bundles of fibers traversing the substance of a structure, usually derived from the capsule or one of the fibrous septa, such as trabeculae of lymph nodes, testes; a beam or plate of cancellous bone. (123)

trachea (tră′kē-ă) [Gr. *tracheia arteria,* rough artery] Air tube extending from the larynx into the thorax where it divides to form the bronchi; composed of 16–20 rings of hyaline cartilage. (740)

transcription (tran-skrip′shŭn) Process of forming RNA from a DNA template. (86)

transfer RNA (tRNA) RNA that attaches to individual amino acids and transports them to the ribosomes where they are connected to form a protein polypeptide chain. (87)

transfusion (tran-fyū′zhun) [L. *trans,* across + *fundo,* to pour from one vessel to another] Transfer of blood from one person to another. (598)

transitional epithelium Stratified epithelium that may be either cuboidal or squamouslike, depending on the presence or absence of fluid in the organ (as in the urinary bladder). (108)

translation (trans-lā′shŭn) Synthesis of polypeptide chains at the ribosome in response to information contained in mRNA molecules. (87)

transverse colon (trans-vers′ kō′lon) Part of the colon between the right and left colic flexures. (798)

transverse mesocolon (mez′ō-kō′lon) Fold of peritoneum attaching the transverse colon to the posterior abdominal wall. (801)

transverse tubule [L. *tubus,* tube] Tubule that extends from the sarcolemma to a myofibril of striated muscles. (278)

treppe (trep′eh) [Ger., staircase] Series of successively stronger contractions that occur when a rested muscle fiber receives closely spaced stimuli of the same strength but with a sufficient stimulus interval to allow complete relaxation of the fiber between stimuli. (291)

triad (trī′ad) Two terminal cisternae and a T tubule between them. (280)

tricuspid valve (trī-kŭs′pid) Valve closing the orifice between the right atrium and the right ventricle of the heart. (615)

trigone (trī′gōn) [Gr. *trigonon,* triangle] Triangular smooth area at the base of the bladder between the openings of the two ureters and that of the urethra. (864)

triiodothyronine (trī-ī′ō-dō-thī′rō-nēn) One of the iodine-containing thyroid hormones. (556)

trochlea (trok′lē-ă) [L., pulley] Structure shaped like or serving as a pulley or spool. (212)

trochlear nerve (trok′lē-ar) [L. *trochlea,* pulley] Cranial nerve IV, to the muscle (superior oblique) turning around a pulley. (440)

trophoblast (trof′ō-blast) [Gr. *trophe,* nourishment + *blastos,* germ] Cell layer forming the outer layer of the blastocyst, which erodes the uterine mucosa during implantation; the trophoblast does not become part of the embryo but contributes to the formation of the placenta. (950)

tropomyosin (trō′pō-mī′ō-sin) Fibrous protein found as a component of the actin myofilament. (278)

troponin (trō′pō-nin) Globular protein component of the actin myofilament. (278)

true pelvis Portion of the pelvis inferior to the pelvic brim. (216)

true or **vertebrosternal rib** (ver-tě′brō-ster′năl) Rib that attaches by an independent costal cartilage directly to the sternum. (208)

trypsin (trip′sin) Proteolytic enzyme formed in the small intestine from the inactive pancreatic precursor trypsinogen. (815)

tubercle (tū′ber-kl) Lump on a bone. (184)

tubular load Amount of a substance per minute that crosses the filtration membrane into Bowman's capsule. (892)

tubular maximum Maximum rate of secretion or reabsorption of a substance by the renal tubules. (892)

tubular reabsorption Movement of materials, by means of diffusion, active transport, or cotransport, from the filtrate within a nephron to the blood. (871)

tubular secretion Movement of materials, by means of active transport, from the blood into the filtrate of a nephron. (877)

tumor (tū′mŏr) Any swelling or growth; a neoplasm. (132)

tunic (tū′nik) [L., coat] One of the enveloping layers of a part; one of the coats of a blood vessel; one of the coats of the eye; one of the coats of the digestive tract. (648)

tunica adventitia (tū′ni-kă ad-ven-tish′ă) Outermost fibrous coat of a vessel or an organ that is derived from the surrounding connective tissue. (648)

tunica albuginea (al-byū-jin′ē-ă) Dense, white, collagenous tunic surrounding a structure; such as the capsule around the testis. (923)

tunica intima (in′ti-mă) Innermost coat of a blood vessel; consists of endothelium, a lamina propria, and an inner elastic membrane. (648)

tunica media Middle, usually muscular, coat of an artery or other tubular structure. (648)

turbulent flow Flow characterized by eddy currents exhibiting nonparallel blood flow. (674)

tympanic membrane (tim-pan′ik) Eardrum; cellular membrane that separates the external from the middle ear; vibrates in response to sound waves. (486)

U

unipolar neuron (yū-ni-pō′lăr) One of the three categories of neurons consisting of a nerve cell body with a single axon projecting from it; also called a pseudounipolar neuron. (126)

unmyelinated axon (ŭn-mī′ě-li-nāt-ted) Nerve fibers lacking a myelin sheath. (366)

unsaturated (ŭn-satch′ŭ-rāt-ed) Carbon chain of a fatty acid that possesses one or more double or triple bonds. (44)

upper respiratory tract The nasal cavity, pharynx, and associated structures. (738)

up–regulation An increase in the concentration of receptors in response to a signal. (536)

ureter (yū-rē′ter, yū′rē′ter) [Gr. *oureter,* urinary canal] Tube conducting urine from the kidney to the urinary bladder. (861)

urethral gland (yū-rē′thrăl) One of numerous mucous glands in the wall of the spongy urethra in the male. (931)

urogenital fold (yū′rō-jen′i-tăl) Paired longitudinal ridges developing in the embryo on either side of the urogenital orifice. In the male they form part of the penis; in the female they form the labia minora. (979)

urogenital triangle Anterior portion of the perineal region containing the openings of the urethra and vagina in the female and the urethra and root structures of the penis in the male. (923)

uterine cycle (yū′ter-in, yū′ter-īn) Series of events that occur in a regular fashion in the uterus of sexually mature, nonpregnant females; prepares the uterine lining for implantation of the embryo. (948)

uterine part Portion of the uterine tube that passes through the wall of the uterus. (941)

uterine tube One of the tubes leading on either side from the uterus to the ovary; consists of the infundibulum, ampulla, isthmus, and uterine parts; also called the fallopian tube or oviduct. (941)

uterus (yū′ter-ŭs) Hollow muscular organ in which the fertilized ovum develops into a fetus. (941)

utricle (yū′tri-kl) Part of the membranous labyrinth; contains sensory structure, the macula, that detects static equilibrium. (499)

uvula (yū′vyū-lă) [L. *uva,* grape] Small grapelike appendage at posterior margin of soft palate. (739)

V

vaccination (vak′si-nā′shŭn) Deliberate introduction of an antigen into a subject to stimulate the immune system and produce immunity to the antigen. (729)

vaccine (vak′sēn, vak-sēn′) [L. *vaccinus,* relating to a cow] Preparation of killed microbes, altered microbes, or derivatives of microbes or microbial products intended to produce immunity. The method of administration is usually inoculation, but ingestion is preferred in some instances, and nasal spray is used occasionally. (729)

vagina (vă-jī′nă) [L., sheath] Genital canal in the female, extending from the uterus to the vulva. (942)

vapor pressure Partial pressure exerted by water vapor. (758)

variable region Part of the antibody that combines with the antigen. (721)

vas deferens (vas def′er-enz) See ductus deferens.

vasa recta (vā′să rek′tă) Specialized capillary that extends from the cortex of the kidney into the medulla and then back to the cortex. (864)

vasa vasorum (vā′sor-ŭm) [L., vessel, dish] Small vessels distributed to the outer and middle coats of larger blood vessels. (651)

vascular tunic (vas′kyū-lăr) Middle layer of the eye; contains many blood vessels. (473)

vasoconstriction (vā′sō-kon-strik′shun) Decreased diameter of blood vessels. (648)

vasodilation (vā′sō-dī-lā′shun) Increased diameter of blood vessels. (648)

vasomotion Periodic contraction and relaxation of the precapillary sphincter, resulting in cyclic blood flow through capillaries. (682)

vasomotor center (vā-sō-mō′ter, vas-ō-) Area within the medulla oblongata that regulates the diameter of blood vessels by way of the sympathetic nervous system. (684)

vasomotor tone Relatively constant frequency of sympathetic impulses that keep blood vessels partially constricted in the periphery. (684)

vasopressin (vā-sō-pres′in, vas-ō-) Hormone secreted from the neurohypophysis that causes vasoconstriction and acts on the kidney to reduce urinary volume; also called antidiuretic hormone. (549)

vellus (vel′ŭs) [L., fleece] Short, fine, usually unpigmented hair that covers the body except for the scalp, eyebrows, and eyelids. Much of the vellus is replaced at puberty by terminal hairs. (144)

venous capillary (vē′nŭs) Capillary opening into a venule. (648)

venous return Volume of blood returning to the heart. (632)

venous sinus Endothelium-lined venous channel in the dura mater that receives cerebrospinal fluid from the arachnoid granulations. (648)

ventilation (ven-ti-lā′shŭn) [L. *ventus,* the wind] Movement of gases into and out of the lungs. (751)

ventral root (ven′trăl) Motor (efferent) root of a spinal nerve. (403)

ventricle (ven′tri-kl) [L. *venter,* belly] Chamber of the heart that pumps blood into arteries (i.e., the left and right ventricles). In the brain, a fluid-filled cavity. (384)

ventricular diastole (ven-trik′yū-lăr) Dilation of the heart ventricles. (625)

ventricular systole Contraction of the ventricles. (625)

venule (ven′yūl, vē′nūl) Minute vein, consisting of endothelium and a few scattered smooth muscles, that carries blood away from capillaries. (649)

vermiform appendix (ver′mi-fōrm) [L. *vermis,* worm + *forma,* form; appendage] Wormlike sac extending from the blind end of the cecum. (798)

vesicle (ves′i-kl) [L. *vesica,* bladder] Small sac containing a liquid or gas, such as a blister in the skin or an intracellular, membrane-bound sac. (66)

vestibular fold (ves-tib′yū-lăr) (false vocal cord) One of two folds of mucous membrane stretching across the laryngeal cavity from the angle of the thyroid cartilage to the arytenoid cartilage superior to the vocal cords; helps close the glottis; false vocal cord. (739)

vestibular membrane Membrane separating the cochlear duct and the scala vestibuli. (493)

vestibule (ves′ti-būl) [L., antechamber, entrance court] Anterior part of the nasal cavity just inside the external nares that is enclosed by cartilage; space between the lips and the alveolar processes and teeth; middle region of the inner ear containing the utricle and saccule; space behind the labia minora containing the openings of the vagina, urethra, and vestibular glands. (787)

vestibulocochlear nerve (ves-tib′yū-lō-kok′lē-ăr) Formed by the cochlear and vestibular nerves and extends to the brain. (442)

villus, pl. villi (vil′ŭs, vil′ī) [L., shaggy hair (of beasts)] Projections of the mucous membrane of the intestine; they are leaf-shaped in the duodenum and become shorter, more finger-shaped, and sparser in the ileum. (793)

visceral (vis′er-ăl) Relating to the internal organs. (299)

visceral pericardium (per′i-kar′dē-ŭm) Serous membrane covering the surface of the heart. Also called the epicardium. (610)

visceral peritoneum (per′i-tō-nē′ŭm) [Gr. *periteino,* to stretch over] Layer of peritoneum covering the abdominal organs. (799)

visceral pleura (vis′er-ăl plūr′ă) Serous membrane investing the lungs and dipping into the fissures between the several lobes. (749)

visceroreceptor (vis′er-ō-rē-sep′tŏr) Sensory receptor associated with the organs. (463)

viscosity (vis-kos′i-tē) [L. *viscosus,* viscous] In general, the resistance to flow or alteration of shape by any substance as a result of molecular cohesion. (74)

visual cortex (vizh′ū-ăl) Area in the occipital lobe of the cerebral cortex that integrates visual information and produces the sensation of vision. (391)

visual field Area of vision for each eye. (484)

vital capacity (vīt′ăl) Greatest volume of air that can be exhaled from the lungs after a maximum inspiration. (756)

vitamin (vīt′ă-min) [L. *vita,* life + amine] One of a group of organic substances present in minute amounts in natural foodstuffs that are essential to normal metabolism; insufficient amounts in the diet may cause deficiency diseases. (836)

vitamin D Fat-soluble vitamin produced from precursor molecules in skin exposed to ultraviolet light; increases calcium and phosphate uptake from the intestines. (149)

vitreous humor (vit′rē-ŭs) Transparent jellylike material that fills the space between the lens and the retina. (477)

Volkmann's canal Named for the German surgeon, Richard Volkmann (1830–1889). Canal in bone containing blood vessels; not surrounded by lamellae; runs perpendicular to the long axis of the bone and the haversian canals, interconnecting the latter with each other and the exterior circulation. (164)

vulva (vŭl′vă) [L., wrapper or covering, seed covering, womb] External genitalia of the female composed of the mons pubis, the labia majora and minora, the clitoris, the vestibule of the vagina and its glands, and the opening of the urethra and of the vagina; the pudendum. (942)

W

water–soluble vitamin Vitamin such as B complex and C that is absorbed with water from the intestinal tract. (838)

white matter Bundles of parallel axons with their associated sheath in the central nervous system. (368)

white pulp That part of the spleen consisting of lymphatic nodules and diffuse lymphatic tissue; associated with arteries. (706)

white ramus communicans, pl. rami communicantes (rā′mŭs kŏ-myū′-nĭ-kans, rā′mī kŏ-myū′nĭ-kan′-tēz) Connection between spinal ganglia through which myelinated preganglionic axons project. (510)

wisdom tooth Third molar tooth on each side in each jaw. (788)

X

xiphoid (zif′oyd) [Gr. *xiphos,* sword] Sword-shaped, with special reference to the sword tip; the inferior part of the sternum. (210)

X-linked Gene located on an X chromosome. (994)

Y

Y-linked Gene located on a Y chromosome. (994)

yolk sac Highly vascular layer surrounding the yolk of an embryo. (964)

Z

Z disk Delicate membranelike structure found at either end of a sarcomere to which the actin myofilaments attach. (278)

zona fasciculata (zō′nă fa-sik′yū lă′tă) [L. *zone,* a girdle, one of the zones of the sphere] Middle layer of the adrenal cortex that secretes cortisol. (562)

zona glomerulosa (glō-mār-yū-lōs-ă) Outer layer of the adrenal cortex that secretes aldosterone. (562)

zona pellucida (pe-lū′sĭ-dă) Layer of viscous fluid surrounding the oocyte. (938)

zona reticularis (rē-tik′ū-lar′is) Inner layer of the adrenal cortex that secretes androgens and estrogens. (562)

zonula adherens (zō′nyū-lă ad-her′enz) [L., a small zone; adhering] Small zone holding or adhering cells together. (110)

zonula occludens (ō-klūd′enz) [L., occluding] Junction between cells in which the cell membranes may be fused; occludes or blocks off the space between the cells. (110)

zygomatic (zī-gō-mat′ik) [Gr. *zygon*, yoke] To yoke or join. Bony arch created by the junction of the zygomatic and temporal bones. (189)

zygote (zī′gōt) [Gr. *zygotos*, yoked] Diploid cell resulting from the union of a sperm cell and an oocyte. (939)

Credits

Photographs

Chapter: 1
1.1: © Pat Watson, photographer/McGraw-Hill Higher Education; **p.4A:** © SIU/ Photo Researchers, Inc.; **p.4B:** © St. Bartholomew's Hospital/ SPL/Photo Researchers, Inc.; **p.4C, D, E, G:** © Howard Sochurek/Woodfin Camp; **p.4F:** © Phillipe Plailly/Photo Researchers, Inc.; **1.2, (all); 1.9 (both):** © Terry Cockerham/ Cynthia Alexander/Synapse Media Production; **1.10a:** © Terry Cockerham/ Cynthia Alexander/Synapse Media Production; **1.10b-d:** © R.T. Hutchings; **1.12:** © Terry Cockerham/Cynthia Alexander/ Synapse Media Production

Chapter: 2
2.4c: © Pat Watson, photographer/McGraw-Hill Higher Education; **2.14:** © Barry King/Tom Stack & Associates

Chapter: 3
3.2b: © J. David Robertson, from Charles Flickinger, Medical Cell Biology, Philadelphia; **3.6b:** © Don Fawcett/Science Source/Photo Researchers, Inc.; **3.6c:** © Bernard Gilula/Photo Researchers, Inc.; **3.7b:** © Don Fawcett/Science Source/Photo Researchers, Inc.; **3.9b:** © J. David Robertson, from Charles Flickinger, Medical Cell Biology, Philadelphia; **3.10b:** © Robert Bollender/Don Fawcett/Visuals Unlimited;

3.13b: © Don Fawcett/Visuals Unlimited; **3.14b:** © Biology Media/Science Source/Photo Researchers, Inc.; **3.15 (both):** © Dr. Gopal Murti/SPL/Photo Researchers, Inc.; **3.16b:** © Courtesy Susumu Ito, Ph.D.; **3.25b:** © Biology Media/Photo Researchers; **3.25d:** © F. Hossler/Visuals Unlimited; **3.26c:** © Courtesy of Dr. Birgit H. Satir; **3.34a-f:** © Ed Reschke

Chapter: 4
4.2a, b: © Ed Reschke; **4.2c:** © Trent Stephens; **4.2d:** © Ed Reschke; **4.2e, f, g:** © Victor Eroschenko; **4.2h:** © R. Calentine/Visuals Unlimited; **4.6a:** © Ed Reschke; **4.6b, c:** © Trent Stephens; **4.6d:** © Victor Eroschenko; **4.6e:** © Ed Reschke; **4.6f:** © Carolina Biological/Phototake; **4.6g, h:** © Ed Reschke; **4.6I:** © Carolina Biological Supply/Phototake; **4.6j, k:** © Victor Eroschenko; **4.6l, m:** © Trent Stephens; **4.6n, 4.7a, b, c:** © Ed Reschke; **4.8a, b, 4.9:** © Trent Stephens

Chapter: 5
5.2a: © Alan Stevens, University of Nottingham, Med School, UK; **5.2b:** © The Bergman Collection; **5.9:** © John Cunningham/Visuals Unlimited; **p.151a-c:** © Thomas B. Habif; **p.153a:** © James Stevenson/ SPL/Photo Researchers, Inc.; **p.153b:** © Stan Levy/Photo Researchers, Inc.

Chapter: 6
6.1: © Ed Reschke; **6.6a, b, c:** © Trent Stephens; **6.7c:** © Bio-Photo Associates, Photo Researchers, Inc.; **6.10b:** © Trent Stephens; **6.11a, b:** © Victor

Eroshenko; **6.12:** © Visuals Unlimited; **6.14a:** © SPL/Photo Researchers, Inc.; **6.14b:** © Bio-Photo Associates, Photo Researchers, Inc.; **6.16:** © J.M. Booher; **p.177a:** © Ewing Galloway, Inc. All Rights Reserved; **p.177b:** © Hilt and Cogburn, Manual of Orthopedics; **p.177c:** © Brashear and Beverly, Handbook of Orthopedic Surgery; **p.177d:** © Hilt and Cogburn, Manual of Orthopedics; **p.178a:** © Princess Margaret Rose Orthopaedic Hospital/SPL/Photo Researchers, Inc.; **p.178b:** © Dr. Michael Klein/Peter Arnold

Chapter: 7
7.5, 7.7: © Terry Cockerham/ Cynthia Alexander/ Synapse Media Production; **7.10c:** © CNRI/SPL/Photo Researchers, Inc.; **7.16:** © Brashear and Beverly, Handbook of Orthopedic Surgery; **7.17c:** © Trent Stephens; **7.20:** © Terry Cockerham/Cynthia Alexander/Synapse Media Production; **7.22c:** © Trent Stephens; **7.23:** © Terry Cockerham/Cynthia Alexander/ Synapse Media Production; **7.24c:** © Trent Stephens; **7.27:** © R.T. Hutchings; **7.31, 7.36:** © Terry Cockerham/ Cynthia Alexander/Synapse Media Production

Chapter: 8
8.7 (all): © Terry Cockerham/ Cynthia Alexander/Synapse Media Production; **8.11e:** © R.T. Hutchings; **p.244a:** © James Stevenson/SPL/Photo Researchers, Inc.; **p.244b:** © CNRI/SPL/Photo Researchers, Inc.

Chapter: 10
10.1: © Ed Reschke; **10.4:** © Richard Rodewald; **10.10b:** © Fred Hossler/Visuals Unlimited; **10.15b:** © Don Fawcett/Science Source/Photo Researchers, Inc.; **10.20a:** © Ed Reschke; **p.303:** © Phillippe Plailly/SPL/Photo Researchers, Inc.

Chapter: 11
11.5a, b, 11.7 (all): © Terry Cockerham/Cynthia Alexander/ Synapse Media Production; **11.17, 11.21a:** © Terry Cockerham/Cynthia Alexander/ Synapse Media Production; **11.21b:** © R.T. Hutchings; **11.21c:** © Terry Cockerham/ Cynthia Alexander/Synapse Media Production; **11.21d, 11.23e:** © R.T. Hutchings; **11.25a, b, 11.30a, b:** © Terry Cockerham/ Cynthia Alexander/Synapse Media Production; **p.356:** © Sylvan LeGrand/Jean M Barnery/Photo Researchers, Inc.

Chapter: 13
13.1: © Howard Sochurek/ Woodfin Camp; **13.4:** © Courtesy Branislav Vidic; **13.8a, b:** © R.T. Hutchings; **13.9b:** © Trent Stephens; **13.12 (all):** © Marcus Raichle, MD., Washington University School of Medicine/ Peter Arnold, Inc.; **13.13:** © Deep Light Productions/SPL/ Photo Researchers, Inc.; **13.14b, 13.16a-c:** © Trent Stephens; **13.19b:** © John Cunningham/ Visuals Unlimited; **13.24a-c:** © Terry Cockerham/Cynthia Alexander/Synapse Media Production; **p.428:** © M. Bernsau/The Image Works

Chapter: 14
14.2b: © R.T. Hutchings

Chapter: 15
15.5g: © OMIKRAN/Science Source/Photo Researchers, Inc.; **15.8:** © Terry Cockerham/ Cynthia Alexander/Synapse Media Production; **15.12:** © R.T. Hutchings; **15.15:** © A.L. Blum/Visuals Unlimited; **15.22d:** © Trent Stephens; **15.24:** © Terry Cockerham/ Cynthia Alexander/Synapse Media Production; **15.27:** © Fred Hossler/Visuals Unlimited; **15.34a, b:** © Erika Stone/Peter Arnold, Inc.; **15.36:** © Jerry Wachter/ Photo Researchers, Inc.; **p.489 (all):** © National Eye Institute, Bethesda MD

Chapter: 18
18.7c: © Biophoto Associates/ Photo Researchers, Inc.; **18.10b, 18.12b:** © Victor Eroschenko; **p.565:** © Biophoto Associates/Photo Researchers, Inc.; **18.15:** © Bio-Photo

Associates/Photo Researchers, Inc.; **p.576 (child):** © Mark Clark/SPL/Photo Researchers, Inc.; **p.576 (insulin):** © SIU/ Visuals Unlimited

Chapter: 19
19.3: © National Cancer Institute/SPL/Photo Researchers, Inc.; **19.7:** © Ed Reschke; **19.8 (all):** © Victor Eroschenko

Chapter: 20
20.5b: © R.T. Hutchings; **20.8a, b:** © McMinn & Hutchings, *Color Atlas of Human Anatomy*/Mosby; **20.12b:** © Ed Reschke; **p.641:** © Hank Morgan/ Science Source/Photo Researchers, Inc.

Chapter: 22
22.2: © George V Wilder/ Visuals Unlimited; **22.4b:** © Trent Stephens; **22.5c:** © Victor Eroschenko; **22.6c:** © Trent Stephens; **p.732:** © Kenneth Greer/Visuals Unlimited

Chapter: 23
23.2b: © R.T. Hutchings; **23.4b:** © Custom Medical Stock Photos; **23.5a:** © John Cunningham/ Visuals Unlimited; **23.5b:** © Manfred Kage/Peter Arnold, Inc.; **23.6b:** © Walter Reiter/ Phototake; **23.7b:** © Carolina Biological Supply/ Phototake; **23.13a:** © Custom Medical Stock; **p.776:** © Deni McIntyre/ Photo Researchers, Inc.

Chapter: 24
24.4a, b © Trent Stephens; **24.7c:** © Ed Reschke; **24.8c:** © S. Eleins/Visuals Unlimited; **24.11d:** © David M Phillips/ Visuals Unlimited; **24.14b:** © Scott Camazine/Brian Camazine/Photo Researchers, Inc.; **p.814:** © SIU Biomedical/ Science Source/Photo Researchers, Inc. **p.825:** © Frederica Georgia/Photo Researchers, Inc.

Chapter: 25
25.15: © Rothenborg Pacific/ Pacific Stock

Chapter: 26
26.2b: © Trent Stephens; **p.895:** © Hank Morgan/Science Source/Photo Researchers, Inc.

Chapter: 28
p.952a: © M. Long/Visuals Unlimited; **p.952b:** © Ray Ellis/Photo Researchers, Inc.; **p.952c:** © Hank Morgan/ Science Source/Photo Researchers, Inc.; **p.952d:** © McGraw-Hill Higher Education; **p.952g:** © McGraw-Hill Higher Education Group, Inc./Bob Coyle photographer; **p.956:** © Blair Seitz/Photo Researchers, Inc.

Chapter: 29
29.1a-d: © Lucinda L. Veeck, Jones Institute for Reproductive Medicine, Norfold, VA; **29.9:** © Lennart Nilsson; **29.15a-c:** © Petit Format/Nestle/Photo Researchers, Inc.; **29.21:** © Stephen Marks/The Image Bank; **29.22:** © Phototake/ CNRI/Photo Researchers, Inc.

Index

Prefixes, Suffixes, And Combining Forms

Continued from inside front cover.

Term	Meaning	Example
intra-	Within	Intraocular (within the eye)
-ism	Condition, state of	Dimorphism (condition of two forms)
iso-	Equal	Isotonic (same tension)
-itis	Inflammation	Gastritis (inflammation of the stomach)
-ity	Expressing condition	Acidity (condition of acid)
kerato-	Cornea or horny tissue	Keratinization (formation of a hard tissue)
-kin-	Move	Kinesiology (study of movement)
leuko-	White	Leukocyte (white blood cell)
-liga-	Bind	Ligament (structure that binds bone to bone)
lip-	Fat	Lipolysis (breakdown of fats)
-logy	Study	Histology (study of tissue)
-lysis	Breaking up, dissolving	Glycolysis (breakdown of sugar)
macro-	Large	Macrophage (large phagocytic cell)
mal-	Bad	Malnutrition (bad nutrition)
malaco-	Soft	Osteomalacia (soft bone)
mast-	Breast	Mastectomy (excision of the breast)
mega-	Great	Megacolon (large colon)
melano-	Black	Melanocyte (black pigment producing skin cell)
meso-	Middle, mid	Mesoderm (middle skin)
meta-	Beyond, after, change	Metastasis (beyond original position)
micro-	Small	Microorganism (small organism)
mito-	Thread, filament	Mitosis (referring to threadlike chromosomes during cell division)
mono-	One, single	Monosaccharide (one sugar)
-morph-	Form	Morphology (study of form)
multi-	Many, much	Multinucleated (two or more nuclei)
myelo-	Marrow, spinal cord	Myeloid (derived from bone marrow)
myo-	Muscle	Myocardium (heart muscle)
narco-	Numbness	Narcotic (drug producing stupor or weakness)
neo-	New	Neonatal (first four weeks of life)
nephro-	Kidney	Nephrectomy (removal of a kidney)
neuro-	Nerve	Neuritis (inflammation of a nerve)
oculo-	Eye	Oculomotor (movement of the eye)
odonto-	Tooth or teeth	Odontomy (cutting a tooth)
-oid	Expressing resemblance	Epidermoid (resembling epidermis)
oligo-	Few, scanty, little	Oliguria (little urine)
-oma	Tumor	Carcinoma (cancerous tumor)
-op-	See	Myopia (nearsighted)
ophthalm-	Eye	Ophthalmology (study of the eye)
ortho-	Straight, normal	Orthodontics (discipline dealing with the straightening of teeth)
-ory	Referring to	Olfactory (relating to the sense of smell)
-ose	Full of	Adipose (full of fat)
-osis	A condition of	Osteoporosis (porous condition of bone)
osteo-	Bone	Osteocyte (bone cell)
oto-	Ear	Otolith (ear stone)
-ous	Expressing material	Serous (composed of serum)
para-	Beside, beyond, near to	Paranasal (near the nose)